WHEN a wea... ufacturer
can give you THE

FIELD POWER...

a world leader in small arms is at work

NOW SEE ON PAGES
5, 135, 295 and 576

Our tradition is innovation

FN HERSTAL S.A.

B-4400 HERSTAL (Belgium) - Tel: 32/41/40 81 11
Telex: 41 223 fabna - Telefax: 32/41/64 54 52

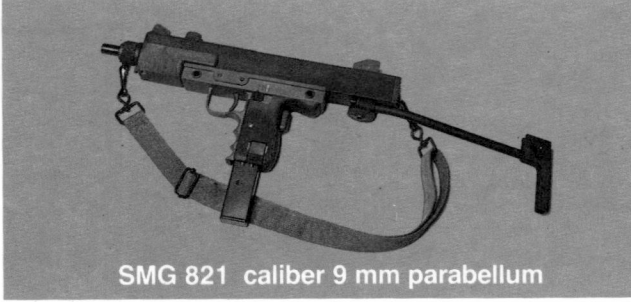

[iv]

JANE'S INFANTRY WEAPONS

SIXTEENTH EDITION

EDITED BY
IAN V HOGG

1990-91

ISBN 0 7106 0906 X
JANE'S DEFENCE DATA
''Jane's'' is a registered trademark

In the USA and its dependencies
Jane's Information Group Inc, 1340 Braddock Place, Suite 300, Alexandria, VA 22314-1651, USA

Printed in the United Kingdom

Contents

Alphabetical list of advertisers

Do not disturb

MILAN 2

HOT 2

ROLAND 3

[7]

Classified List of Advertisers

The companies advertising in this publication have informed us that they are involved in the fields of manufacture indicated below:-

Ammunition, anti-tank
BNJ Industries
Defex
Dince Hill (Holdings)
Eurometaal
FFV Ordnance
Luchaire Defense
Matra Manurhin
Raufoss

Ammunition, mortar
BNJ Industries
Defex
Dince Hill (Holdings)
Eurometaal
FFV Ordnance
Luchaire Defense
Raufoss

Ammunition, small arms
BNJ Industries
Defex
Dince Hill (Holdings)
Eurometaal
FFV Ordnance

FN Herstal
Lapua Cartridge Factory
Manroy Engineering
Raufoss
Santa Barbara
SFM
SOCIMI

Ammunition, sniper purposes
BNJ Industries
Lapua Cartridge Factory
Luchaire Defense
Manroy Engineering
SFM

Anti-tank grenades
BNJ Industries
Eurometaal
FFV Ordnance
Luchaire Defense

Anti-tank launchers
Defex
Eurometaal
FFV Ordnance
Luchaire Defense

Anti-tank missiles
Euromissile
Matra Manurhin Defense

Anti-tank systems
Euromissile

Armament systems
Defex
Luchaire Defense
Santa Barbara

Automatic pistols
Beretta Pietro
BNJ Industries
Defex
Dince Hill (Holdings)
FN Herstal
Manroy Engineering
SIG Swiss Industrial Company

Automatic rifles
Beretta Pietro
BNJ Industries
Defex
Dince Hill (Holdings)
Eurometaal
FN Herstal

Manroy Engineering
Santa Barbara
SIG Swiss Industrial Company

Bayonets
Beretta Pietro
Defex
Dince Hill (Holdings)
FN Herstal
Manroy Engineering
Santa Barbara

Blank firing attachments
Beretta Pietro
Defex
Dince Hill (Holdings)
FN Herstal
Manroy Engineering
SFM
SIG Swiss Industrial Company

Body armour
BNJ Industries
FFV Ordnance
Second Chance Body Armour

Cannons, 20-30mm calibre
BNJ Industries
Defex

Carbines
Beretta Pietro
FN Herstal
Manroy Engineering
SIG Swiss Industrial Company
SOCIMI

Clothing combat
Defex
Second Chance Body Armour

Components for electro-optic systems
Defex
Luchaire Defense
Oldelft
Varian Electro Optical Sensors Division
Varo Electron Devices

Components for infra-red equipment
Oldelft

Components for night vision systems
Defex
Oldelft
Varian Electro Optical Sensors Division
Varo Electron Devices

Electro-optical sights
Oldelft
Simrad Optronics
Varo Electron Devices

Electro-optical surveillance devices
Oldelft
Simrad Optronics
Varian Electro Optical Sensors Division
Varo Electron Devices

Explosives
BNJ Industries
Defex
Santa Barbara

Fire-control equipment, anti-armour
Defex
Dince Hill (Holdings)
Oldelft
Simrad Optronics

Fire control equipment, mortar
Defex
Dince Hill (Holdings)
Manroy Engineering
Oldelft
Simrad Optronics

Fuses, anti-tank systems
BNJ Industries
Defex
EMS Patvag
Junghans Feinwerktechnik
Luchaire Defense
Raufoss

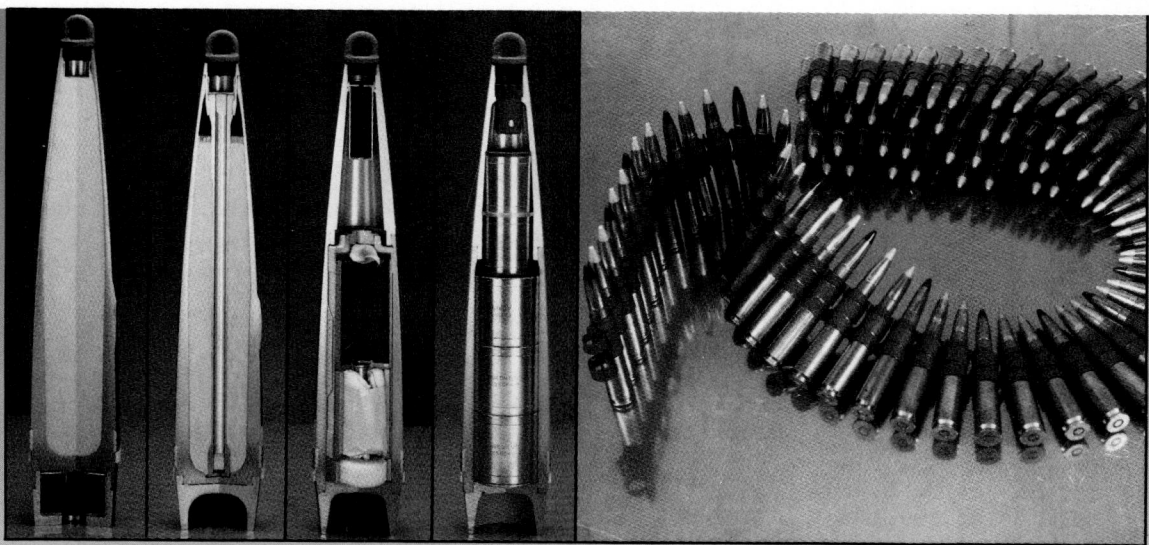

Fuses, grenade and mortar
BNJ Industries
Commerce International Spain
Defex
Junghans Feinwerktechnik
Dince Hill (Holdings)
Luchaire Defense

Grenades, hand and rifle
BNJ Industries
Defex
Dince Hill (Holdings)
Eurometaal
FN Herstal
Lochaire Defense

Grenade launchers
Beretta Pietro
BNJ Industries
Dince Hill (Holdings)
Santa Barbara
SOCIMI

Gun cleaning brushes/equipment
BNJ Industries
Defex
Dince Hill (Holdings)
FN Herstal
Lapua Cartridge Factory
Manroy Engineering
Steyr Mannlicher

Helmets
FFV Ordnance

Image intensifying sights
Defex
Oldelft
Varian Electro Optical Sensors Division
Varo Electron Devices

IR thermal imaging systems
Oldelft

Light anti-armour weapons
BNJ Industries
Defex
Dince Hill (Holdings)
FFV Ordnance
Manroy Engineering

Linking machines
Dince Hill (Holdings)
FN Herstal
Manroy Engineering

Machine guns
Beretta Pietro
BNJ Industries
Defex
Dince Hill (Holdings)
FN Herstal
Manroy Engineering
Santa Barbara
SOCIMI

Machine gun mounts and tripods
BNJ Industries
Defex
Dince Hill (Holdings)
FN Herstal
Manroy Engineering
Santa Barbara

Magazines
Accuracy International
Beretta Pietro
BNJ Industries
Defex
Dince Hill (Holdings)
FN Herstal
Manroy Engineering
Santa Barbara
SIG Swiss Industrial Company

Metal blank cartridges
BNJ Industries
Lapua Cartridge Factory
SFM

Military spares for small arms
Beretta Pietro
BNJ Industries
Dince Hill (Holdings)
Eurometaal
FN Herstal
Manroy Engineering

Mortars
BNJ Industries
Defex
Dince Hill (Holdings)
Eurometaal
Manroy Engineering

Obsolete or surplus small arms
BNJ Industries
Dince Hill (Holdings)
Manroy Engineering

Optical night vision systems
BNJ Industries
Defex
Oldelft
Simrad Optronics
Varian Electro Optical Sensors Division
Varo Electron Devices

Optical surveillance equipment
Defex
Kern
Oldelft
Simrad Optronics
Varian Electro Optical Sensors Division
Varo Electron Devices

Plastic blank cartridge
BNJ Industries
Defex
Dince Hill (Holdings)

Plastic blank cartridge and metal blank cartridge
SFM

Practice ammunition
BNJ Industries
Defex
Dince Hill (Holdings)
Eurometaal
FN Herstal
Luchaire Defense
Manroy Engineering
SFM

Propellants
BNJ Industries
Defex
FFV Ordnance
Luchaire Defense
Raufoss

Pyrotechnic devices
BNJ Industries
Defex
EMS Patvag
Eutometaal
Raufoss
Valsella Meccanotecnica

Revolvers
BNJ Industries
Dince Hill (Holdings)
FN Herstal
Manroy Engineering
SOCIMI

S & A units
BNJ Industries
Dince Hill (Holdings)
EMS Patvag
Junghans Feinwerktechnik

Semi-automatic rifles
Beretta Pietro
BNJ Industries
FN Herstal
Manroy Engineering
SIG Swiss Industrial Company

Shotguns, combat
Beretta Pietro
BNJ Industries
Defex
Dince Hill (Holdings)
Manroy Engineering
O.F. Mossberg & Sons
SOCIMI

Shotguns, over and under
Beretta Pietro

Sights, reflex sight for guns, zoom sight
BNJ Industries
Kern
Oldelft

Sights, weapon
BNJ Industries
Defex
Dince Hill (Holdings)
FN Herstal
Kern
Manroy Engineering
Oldelft
Varo Electron Devices

Silenced weapons
BNJ Industries
Dince Hill (Holdings)

Silencers/sound moderators
BNJ Industries

Smoke munitions
BNJ Industries
Defex

Eurometaal
FN Herstal
Luchaire Defense
Raufoss
Santa Barbara
Valsella Meccanotecnica

Sniping rifles
Beretta Pietro
BNJ Industries
Dince Hill (Holdings)
Manroy Engineering
SIG Swiss Industrial Company

Sub-calibre conversions
SIG Swiss Industrial Company

Sub-machine guns
Beretta Pietro
BNJ Industries
Defex
Dince Hill (Holdings)
FN Herstal

Manroy Engineering
Santa Barbara
SIG Swiss Industrial Company
SOCIMI

Thermal imaging sights
Oldelft

Thermal imaging surveillance equipment
Defex
Oldelft

Training aids and simulators
Dince Hill (Holdings)
FN Herstal

Zeroing devices, small arms
Dince Hill (Holdings)
Manroy Engineering
Varo Electron Devices

Mortar Fuzes

For smooth bore mortar rounds 51 mm through 120 mm

- DM 111 A5   PD   Fuze for HE Ammunition
- DM 111 A4   PD   Fuze for HE Ammunition
- DM 111 AZ-W   PD   Fuze for HE Ammunition
- DM 93/M 776   MTSQ Fuze for Smoke and Illumination
- DM 123   MTSQ Fuze for HE Ammunition
- M 772   MTSQ Fuze for Smoke and Illumination

JUNGHANS FEINWERKTECHNIK
Branch Company of Diehl GmbH & Co.
D-7230 Schramberg, West-Germany
Phone: 7422/18522
Fax: 7422/18400, Tlx: 760844 ju d

✵JUNGHANS   FEINWERKTECHNIK

[14]

Technically superior.

Steyr Pandur 6 x 6.
A highly versatile wheeled armoured vehicle. Light armour protection and low silhouette. One- or two-man turrets. Choice of armaments.

Steyr Military Trucks.
Suitable for the whole military sector. Excellent off-road performance with state-of-the-art all-wheel drive and single tyres. Supply of spare parts guaranteed through design compatibility with civilian models.

Steyr-Puch Pinzgauer.
Its unequaled off-road capability makes it the most mobile wheeled military vehicle there is. Factors such as a 100% climb capability. 70 cm fording depth, 30 cm ground clearance and 100 km/h maximum speed make the Pinzgauer unbeatable in every terrain.

Steyr AUG.
Components interchangeable in the field – fulfills requirements for an extremely wide variety of missions. Great versatility assured by a complete line of accessories.

STEYR DEFENCE PRODUCTS

For further information please contact: **Steyr-Daimler-Puch AG.** P.O.Box 100, A-1111 Vienna, Austria

[15]

SIG SAUER Pistols:
Superior technology for superb accuracy

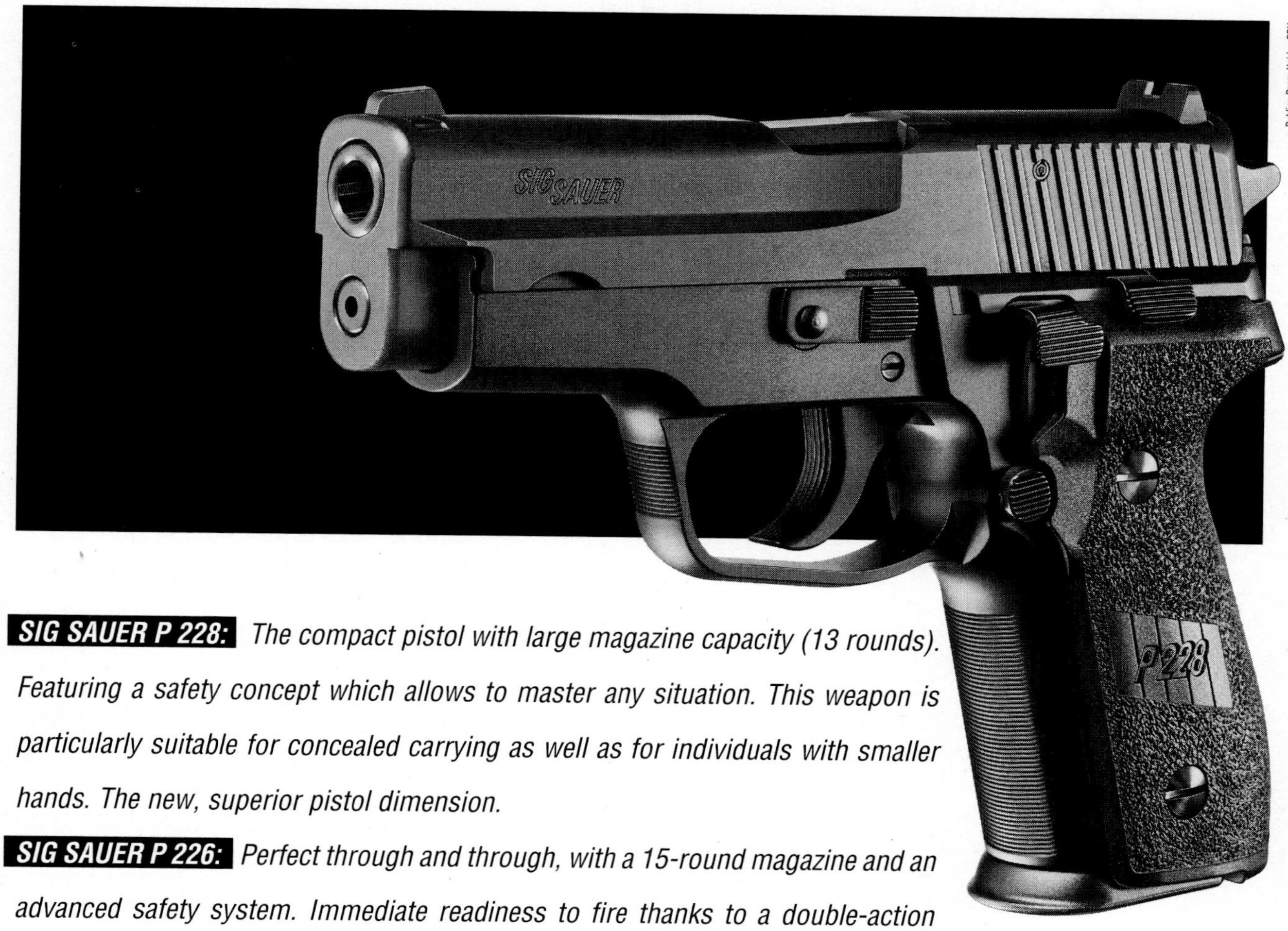

SIG SAUER P 228: The compact pistol with large magazine capacity (13 rounds). Featuring a safety concept which allows to master any situation. This weapon is particularly suitable for concealed carrying as well as for individuals with smaller hands. The new, superior pistol dimension.

SIG SAUER P 226: Perfect through and through, with a 15-round magazine and an advanced safety system. Immediate readiness to fire thanks to a double-action trigger with automatic firing pin lock. **SIG SAUER P 225:** A rugged and highly efficient defensive weapon with compact dimensions, specifically designed for the police. **SIG SAUER P 220:** The successful combat pistol with an unmatched service record. Over 100'000 currently in service. **SIG SAUER P 230:** The second weapon with superior fire power, extremely light and handy, particularly suited for concealed carrying.

P220

P230

P228   P226   P225

SIG Swiss Industrial Company   Small Arms Division
CH-8212 Neuhausen - Rhine Falls (Switzerland)
Telephone 053 21 61 11   Telefax 053 21 66 01
Telex 896 021 sig ch

Publica Press Heiden PPH

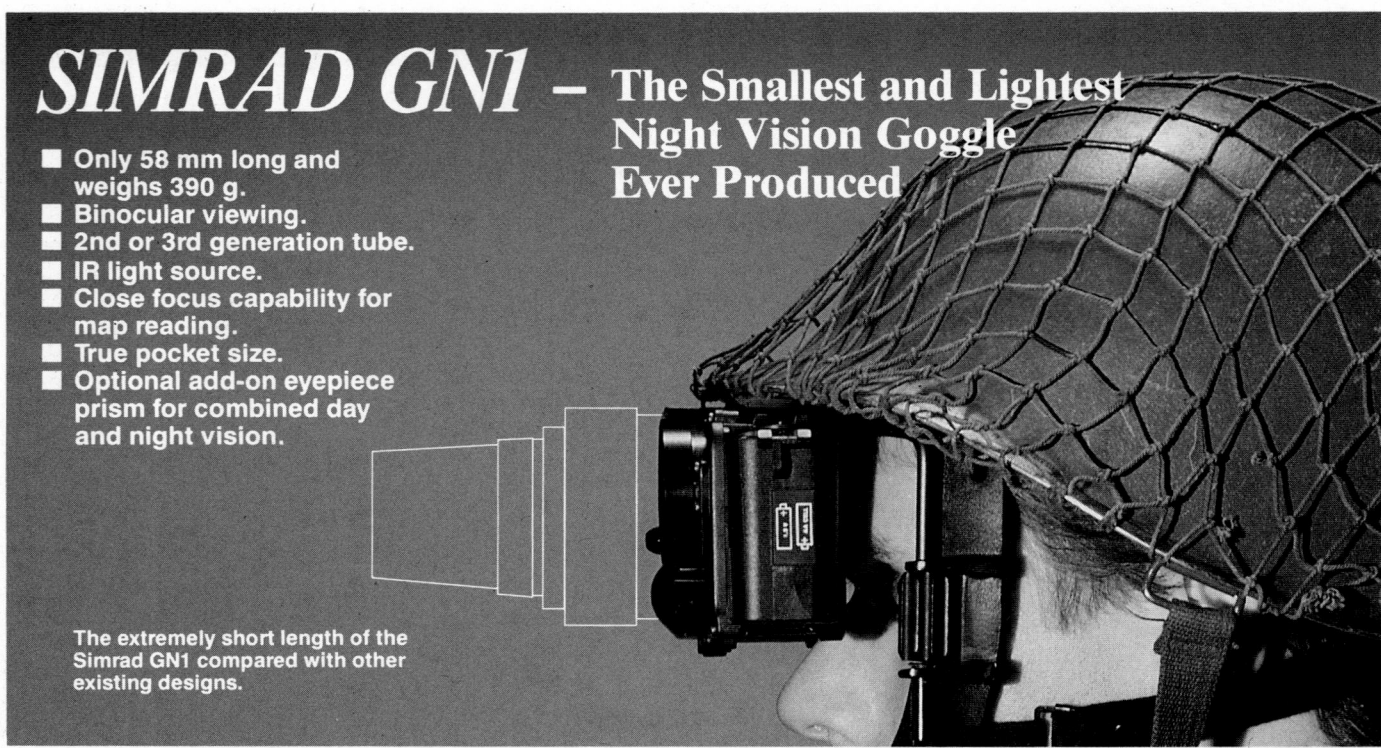

Key Functions

The M72 ILAW configuration provides the combat soldier with a significantly increased firepower potential which enables him to defeat a wide spectrum of targets. These include tanks, armoured vehicles, air defence systems, self-propelled artillery, gun emplacements, and other hard point fortifications.

This improved M72 Lightweight Multi-Purpose Assault Weapon also provides the combat soldier with an updated system which possesses a longer effective range, greater penetration, higher velocity and increased hit probability. The motor case and head closure safety margins are also increased.

This portable weapon facilitates personal ammunition issue and the launcher is expendable.

Advantages

■ Compact configuration, low system weight (3.3 kg) and improved lethality ensuring high destructive capability against armoured vehicles and fixed installations – combined with maximum portability and proliferation.

■ Suitable for airborne operations, the ILAW can be carried by the individual parachutist or air dropped in its outer pack.

■ One-man portable weapon, extremely simple to operate – thus requiring minimum training – and capable of being fired in any normal rifle-firing position.

■ Shoulder-fired, single shot, prepackaged at the factory in a disposable launcher which also serves as a tactical storage container.

M72 ILAW
Improved Lightweight Assault Weapon

Characteristics:

System Weight	7.6 lbs
Carry Length	30.6 in.
Firing Length	38.7 in.
Calibre	66 mm
Penetration	up to 355 mm
Effective Range	250 meters
Operational Range	350 meters
Dispersion	1.78 mil

Raufoss A/s
N-2831 Raufoss, Norway
Defence Products Division
Tel.: +47 61 52 000
Telex: 71144 ra n,
Telefax: +47 61 52 754

[19]

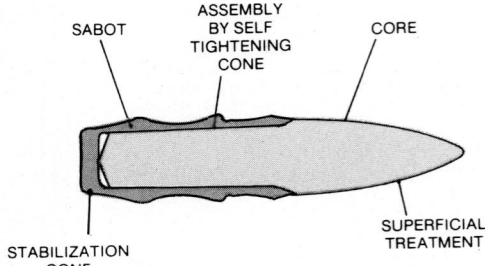

AR 70/90

92FS

M3 P

Studio Più

ON THE ALERT

Beretta defence programme: a complete range of hi-tech portable weapons, unique for their reliability, precision and safety in all environments and logistical situations. Pistols, sub-machine guns, assault rifles and anti-riot shotguns, Beretta has had your defence interests in mind since 1526.

PM 12S

P. BERETTA

Beretta
defence division

[21]

GEN III FROM VARIAN

Varian…The Technology Leader and primary supplier of GEN III image intensifiers to the European Community.

Ready today to deliver:
- Performance
- Reliability
- Quality

varian ⊕

electro optical sensors division

601 California Ave.
Palo Alto, CA 94304
415 493-1800

VARIAN Int'l AG
Steinhauserstrasse
Postfach
CH-6300 Zug, Switzerland
(41) 42 44 88 44

TAKE ANOTHER LOOK AT DARKNESS

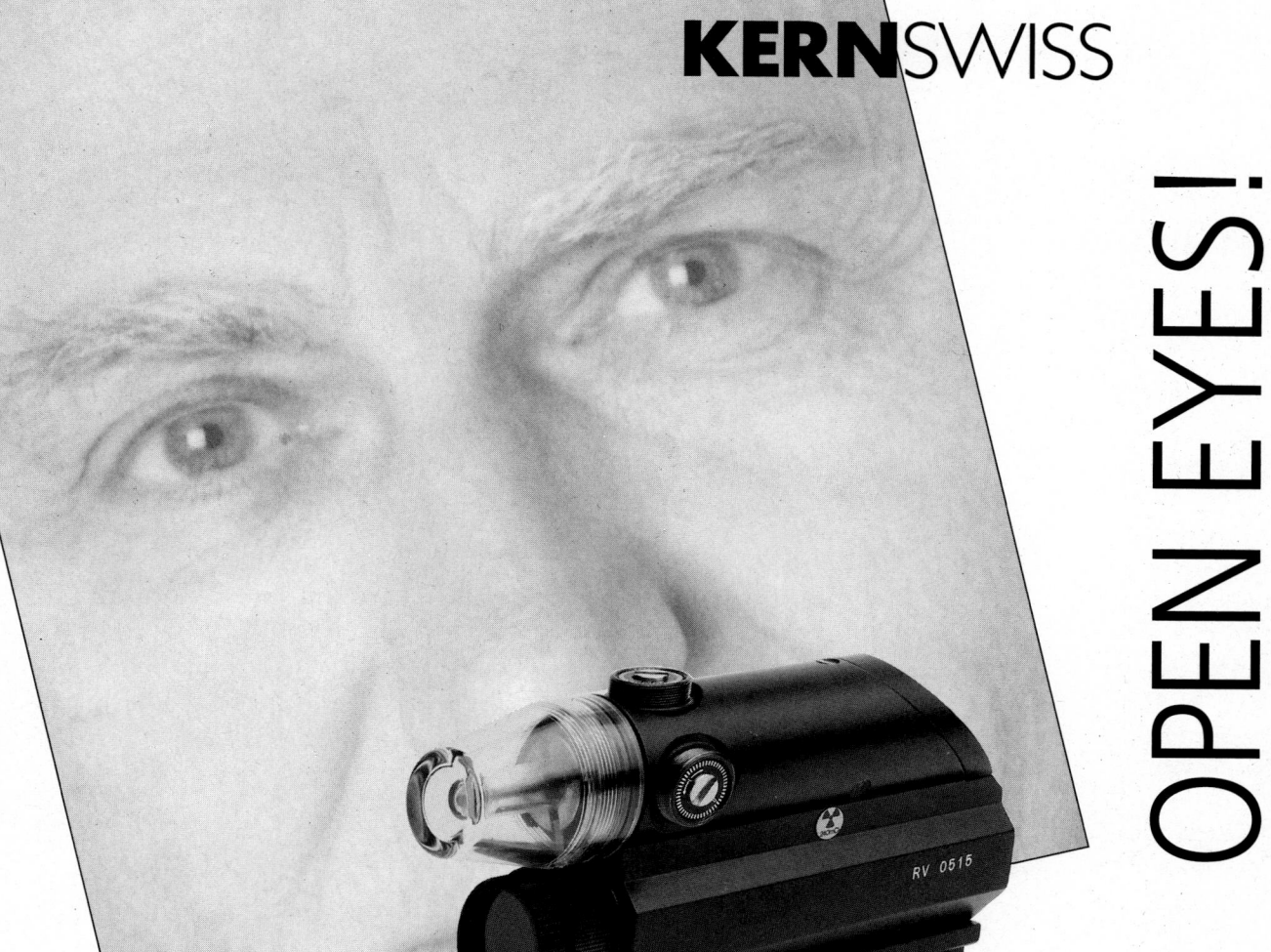

KERNSWISS

OPEN EYES!

REFLEX SIGHT RV-

Independent of the type of weapon to wich the REFLEX SIGHT RV is attached, new dimensions are opened to be on target in the twinkling of an eye and to raise the accuracy of shooting.

- Fatigue-free aiming (both eyes can be used!)
- Insensitive to parallax errors
- Reticle illumination by self-activated light source (no battery is needed!)
- Rapid target acquisition
- Weapon compatible mounting
- Different reticles available
- **Reliable - simple - comfortable**

KERNSWISS

Kern & Co. Ltd.
Special Products Division
CH-5001 Aarau Switzerland
Tel 0041-64-26 44 44
Fax 0041-64-26 41 03
Telex 981 106

[23]

One advantage of 5.56mm machine guns is that they can be used as assault rifles: the FN Minimi
in the shoulder-fired role

JANE'S INFANTRY WEAPONS

1990-91

Jane's Information Group, 163 Brighton Road, Coulsdon, Surrey CR5 2NH, UK
Jane's Information Group Inc, 1340 Braddock Place, Suite 300, Alexandria, VA 22314-1651, USA

TO BE A WINNER
takes skill, professionalism, determination and flair.

Strix, the new 120 mm mortar-launched anti-armor projectile, will be one of the most important anti-armor weapons in the 1990s.

Strix features fire-and-forget capability. A passive IR-seeker, terminal guidance and hight resistance to countermeasures provide very high hit probability. Strix is deployed with ordinary mortar platoons with normal fire control support.

A range exceeding 8000 m (5 miles) — considerably longer than for anti-armor guns and missiles — and top attack using an HEAT warhead producing behind armor effect, enable Strix to defeat all present and future combat vehicles.

FFV Ordnance

S-631 87 Eskilstuna, Sweden

Always in the front line – technical and tactical

Foreword

This year we have abstained from any new departures or additions; our principal task is to try and condense material in order to keep the size of the book within reasonable and economic bounds. Your attention is drawn to the Addenda on page 860, wherein the most recent information on weapons will be found. As usual, I will be pleased to hear from any correspondent who cares to comment on, add to or correct any of the information in the Yearbook. Letters addressed to me at Sentinel House, 163 Brighton Road, Coulsdon, Surrey CR5 2NH, England, will be carefully studied and acknowledged.

Of course, the subject which is pressing upon all military commentators today is the future shape of European defence in the aftermath of the upheavals which have taken place in Central Europe. For my part, I keep Josh Billings' famous admonition firmly in mind: "Never prophesy; for if you prophesy right nobody remembers, and if you prophesy wrong nobody forgets". However, I doubt that there will be much effect upon the basic tools of the soldier's trade which occupy us in this Yearbook. Those looking to make sweeping economies in defence expenditure will be aiming at the high-tech missile systems, those seeking to make economies in military forces will be swinging the axe against armoured divisions. Nobody is likely to be discussing the pros and cons of this rifle over that one or whether the sub-machine gun should be totally replaced by an assault carbine. There is even the faint possibility that if smaller armies become the norm, then they might even get waterproof boots, a general issue of fragmentation jackets and a reasonable amount of ammunition for practice. Every cloud has a silver lining.

To the Fourth Generation?

It has been suggested to me that we are about to enter the Fourth Generation of small arms, and I must say that the idea has some validity. In the beginning there was the hand-made weapon, constructed with loving care by a gunsmith and as individual as the man who made it. That we can consider as the First Generation. Then came the age of machine manufacture, giving us interchangeability of component parts and mass production; the Second Generation. However, the machines had to be set up and operated by skilled men, who became scarce and expensive in wartime, so next came the era of pressed sheet metal and welding, in which we live now; the Third Generation.

Weapons are still being made that way, and represented as being state-of-the-art, but technology which is the better part of 50 years old is still being used. What is worse, some of today's manufacturers have not yet grasped it, 50 years old though it may be.

The past 10 or 15 years have seen a revolution in metal-working, due to the adoption of computer-controlled machine tools and electronic measurement technology, and it is time this became more evident in firearms manufacture. True, many manufacturers are adopting this technology but too many of them are still applying it to the Third Generation when they should really be throwing previous notions out of the window and looking to the Fourth Generation.

It is now possible to revert to the Second Generation philosophies of manufacture but drive them with Fourth Generation technology. Many years ago a pundit at the Royal Small Arms Factory was asked how a rifle was made. "Very simple," he replied, "you take a block of steel and cut away anything that doesn't look like a rifle." It was a good joke, but it was literally true; what was less funny was that it required a highly skilled man to do the cutting so that what was left actually performed like a rifle. Now we can sit down at a computer and draw up a set of instructions to be fed to a machine which will do the cutting, and do it a great deal more quickly and accurately than the skilled man ever could. True, you still need a skilled man, but you only need him once - to write the computer instructions - after which the machine will produce a hundred, or a million, pieces to order.

This means that we can get rid of these pressed-steel-and-wire-spring wonders and go back to making small arms out of decent chunks of metal so that they will withstand what the soldier hands out to them and still look good after 20 years. 'But the economic life of a rifle is only 25 years,' I hear someone say. Re-phrase that; the economic life of a Third Generation rifle is 25 years. Given that the designers choose the right calibre, operating systems and dimensions, the economic life of a Fourth Generation rifle will get back to that of the Second Generation. In 1942 I was learning arms drill on a Vetterli-Vitali which had been made in Terni in 1888; goodness knows how the British Army acquired them - I suspect via the Italians in Somaliland - but they were still as strong as the day they left Terni. When it came to actual shooting, I

graduated to a 1913 Lee-Enfield and managed to qualify as a marksman quite comfortably with it. Given the ability to re-barrel if necessary there is no reason why the life of a small arm should be limited to 25 years, and considering the enormous expense generated by developing, manufacturing, withdrawing, modifying, re-issuing and finally mass producing a weapon, doubling the weapon's service life will show some substantial economic benefits.

I have nothing against some of today's pressed steel and plastic products; they perform exceptionally well and some of them are remarkably robust, but if the basic designs were translated into solid steel and machined by CNC machines, I suspect they would be even better and longer-lived into the bargain. Who will start the ball rolling and have a Fourth Generation assault rifle in service for the coming century?

Correction

I was severely taken to task by one or two manufacturers for my comments in last year's Foreword regarding the life of 5.56 mm machine gun barrels; each was under the impression that I was directing my remarks at them. Far from it; the remarks were not directed at anybody in particular, they were merely repeats of comments made to me by soldiers about weapons of several different types. (The fact that I was grossly misquoted in the popular press was no help either.)

Soldiers are not happy about forecast barrel life and that's all there is to it. Perhaps the most amazing remark in rebuttal came from an unidentified US Government spokesman who maintained that whilst a heavy rate of fire was known during the Second World War and the Korean War, such a thing would not happen in the next war. I look forward to his information about when and where the next war will take place, if he is so sure of how it will be fought. I have distrusted people who tell me about how the next war will be fought ever since June 1940.

Second correction

Some five or six years ago a correspondent in the Middle East sent us some varied information, amongst which were details of short-timed fuze assemblies in Soviet grenades. This sounded reasonable, since I had met the same thing in wartime German grenades, and the information was duly placed in the appropriate section. Further enquiry of the correspondent elicited no reply.

Last year I was informed that this information had originally been published in the American magazine 'Soldier of Fortune' and, apparently, my informant had merely repeated it. I was unaware of the original article, as I was also unaware that 'Soldier of Fortune' had found the information to be erroneous. My apologies, therefore, to 'Soldier of Fortune' for inadvertently stealing their thunder. And my thanks to them for recently providing me with some additional information on Chinese weapons, which has been incorporated into the relevant entries.

Current developments

After the fanfare which greeted the Advanced Combat Rifle technology demonstrators, this scene has become remarkably quiet, with virtually no information about how the trials are progressing. Perhaps the most interesting repercussion has been the general headshaking over the revival of the flechette as a small arms projectile. Colonel Martin Fackler, the well-known authority on wound ballistics, has pointed out with some force that the gradual reduction in rifle calibres throughout this century has seen a corresponding reduction in wound severity, and that taking this to its latest point, the 1.5 mm flechette, is really asking too much. I have recollections of some hair-raising film of flechettes fired against animals in the early 1960s, but one is inclined to wonder if, perhaps, we were being shown the 'best case' results, the failures being unrecorded. Many people will recall the fearsome tales which accompanied the 5.56 mm bullet when it first began to make its mark in the world. Knowing nothing of physiology I have to admit that I took much of the information being given out as gospel, as did most people; after all, the experts always know best. But over the years I have begun to have my doubts, and my aversion to the 5.56 mm bullet as the primary combat round is long-held. Colonel Fackler and his companions have, over the past three or four years, produced some very persuasive papers and quoted apparently irrefutable examples which reinforce what many practical soldiers have always feared: that the 5.56 mm bullet simply does not have the wounding ability of heavier calibres. I still think that the 7 mm

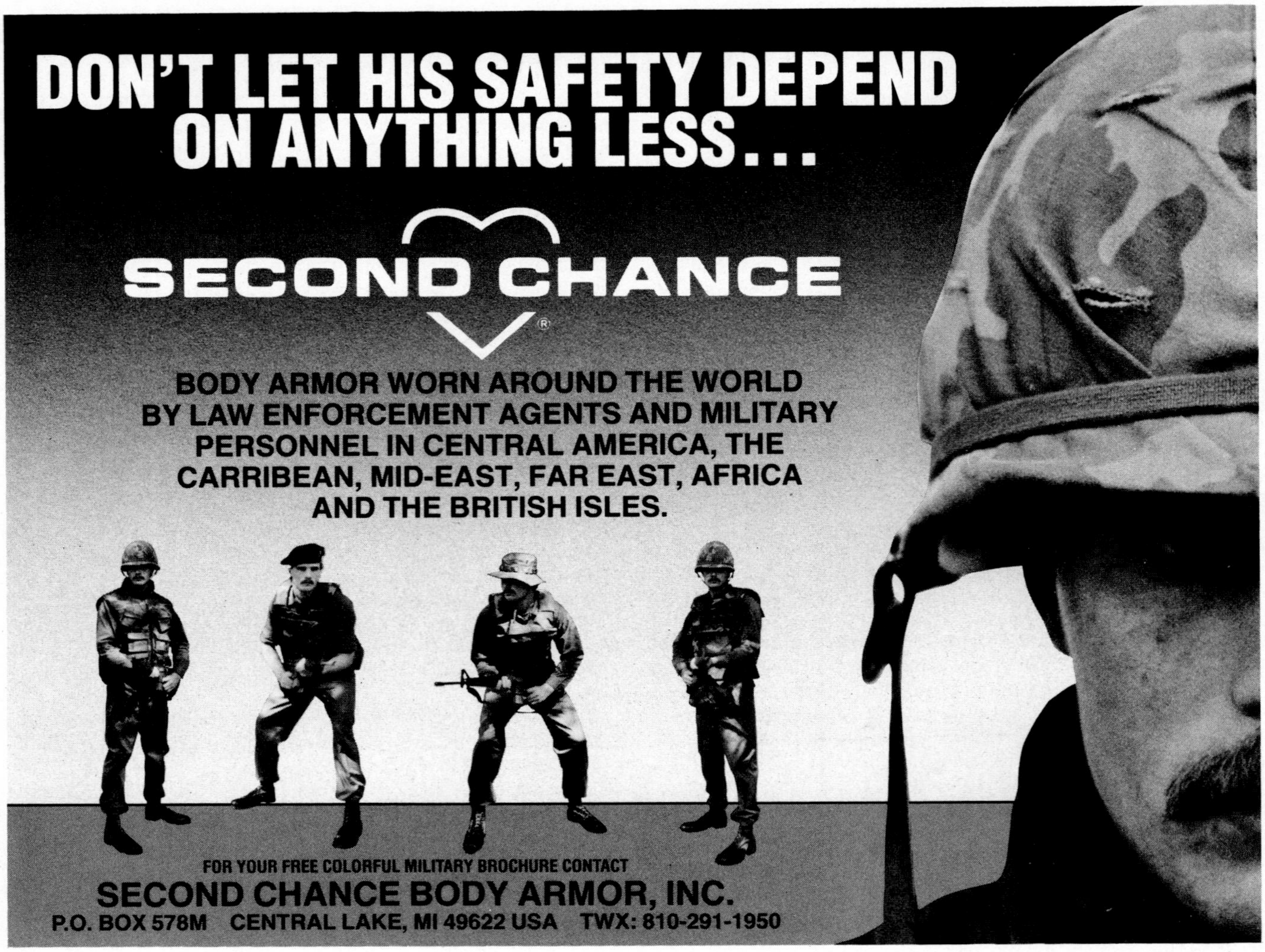
DON'T LET HIS SAFETY DEPEND ON ANYTHING LESS...
SECOND CHANCE
BODY ARMOR WORN AROUND THE WORLD BY LAW ENFORCEMENT AGENTS AND MILITARY PERSONNEL IN CENTRAL AMERICA, THE CARRIBEAN, MID-EAST, FAR EAST, AFRICA AND THE BRITISH ISLES.
FOR YOUR FREE COLORFUL MILITARY BROCHURE CONTACT
SECOND CHANCE BODY ARMOR, INC.
P.O. BOX 578M CENTRAL LAKE, MI 49622 USA TWX: 810-291-1950

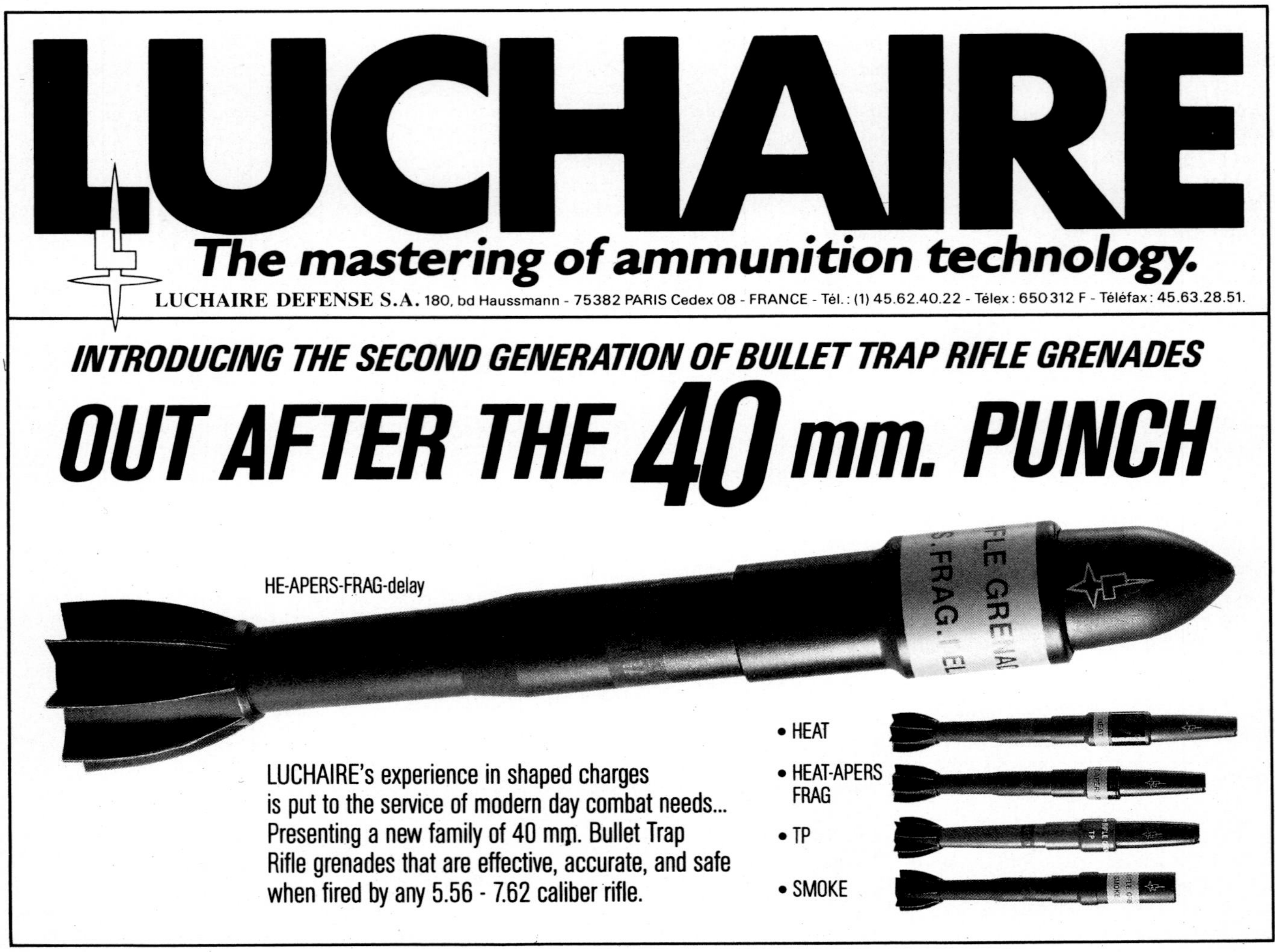
LUCHAIRE
The mastering of ammunition technology.
LUCHAIRE DEFENSE S.A. 180, bd Haussmann - 75382 PARIS Cedex 08 - FRANCE - Tél.: (1) 45.62.40.22 - Télex: 650312 F - Téléfax: 45.63.28.51.
INTRODUCING THE SECOND GENERATION OF BULLET TRAP RIFLE GRENADES
OUT AFTER THE 40 mm. PUNCH
HE-APERS-FRAG-delay
LUCHAIRE's experience in shaped charges is put to the service of modern day combat needs... Presenting a new family of 40 mm. Bullet Trap Rifle grenades that are effective, accurate, and safe when fired by any 5.56 - 7.62 caliber rifle.
• HEAT
• HEAT-APERS FRAG
• TP
• SMOKE

cartridge developed in Britain in the late 1940s would have been the ideal round, but I fancy we have some way to go down the diminishing calibre path before sanity returns. Another target for the end of the century perhaps?

It was also interesting to see the revival of the Duplex cartridge, an idea which comes round every 20 years or so with the regularity of a minor comet. Like the flechette, every time it comes around technology has advanced in the interim and the latest design is far better than the last. My only objection to it is that it is an attempt to use material as a substitute for training, a modern tendency which is not confined to small arms.

There does not appear to have been a rush to follow FN Herstal, GIAT and Steyr into the 'personal weapon' field; perhaps other manufacturers are waiting to see what sort of a reception these pioneers will get. The only people to have dipped a toe into this particular pool are an entirely new company who have developed the 'Bushman', which they term an Individual Defence Weapon, which, one supposes, adds up to the same thing. The interesting part about this design is that they have employed some original thinking and have allied a novel rate reducer to the natural dynamics of the weapon to produce a small automatic which really does keep its rounds in the target area and does not waste ammunition in the sky. Other novelties in the sub-machine gun field are few; the Yugoslavs have revived the 'American 180' .22 weapon and they have also developed a short 5.56 mm weapon which appears to have been inspired by the Soviet AKSU.

The pistol world has also been quiet during 1989. Ruger has just informed me that its P85 is now available with a de-cocking lever, and FN Herstal has re-launched its double-action version of the High-Power. The only new design of note is the 'Spitfire', developed by a small English gunsmith originally for practical target shooting, but which has developed into something which most military forces would find attractive, a no-frills 9 mm double-action pistol built to a high quality and, like the 'Bushman' weapon mentioned above, an example of 'Fourth Generation' manufacture at a competitive price.

As always, the thing to beat is the main battle tank, although the number of potential solutions displayed during the past year have been relatively few in comparison with former years. Purveyors of shoulder-fired devices have turned to improvement of warheads as their only salvation, but there is no doubt a limit to what any specified calibre of shaped charge can do, even with the most arcane wave-shapers and tandem charges, precursors and high-brisance fillings. The advocates of top attack have been confounded by the application of explosive reactive armour to turret tops, though there is room for doubt about the overall efficiency of this system of protection since it has never been tried in war. Moreover, the duplication of layers of reactive armour tends to suggest that the occupants of the tank might have an uncomfortable time when these multiple charges are exploded. A recent paper on the attack of tanks suggested that the one sure and certain Achilles Heel is the rear of the turret ring, so perhaps the next stage will be a missile which overflies the tank and then fires a shaped charge in the reverse direction to strike beneath the bustle; or perhaps one programmed to fly past, execute a 180° turn and strike the engine compartment. This, of course, could be hazardous if it missed.

An interesting byproduct of the anti-tank problem is the appearance of projectiles to be fired from shoulder launchers for purposes other than the attack of armour. Some of the smaller and older launchers are now quite redundant in the anti-armour role except against APCs and similar vehicles, but they have a useful lightweight ability to project quite sizeable anti-personnel and anti-materiel projectiles against field defences. The Falklands campaign introduced the use of such

expensive devices as MILAN for taking out machine gun posts, and adapting much cheaper one-shot systems for this task makes a good deal of sense. They may not have much range, but they certainly have the destructive power needed.

The other method of attacking materiel to have become attractive in the past few years is the heavy-calibre long-range rifle. These appeared in the USA some five or six years ago, invariably chambered for the .50 Browning cartridge. I commented at the time upon the inherent lack of accuracy in this cartridge for so-called sniping purposes, to which the usual reply was 'hand-loading', which is all very well for fun shooters but not very practical for soldiers. After that the accent appears to have moved away from 'sniping' to 'anti-materiel'; it may well be that this is what the inventors had in mind all the time but they chose to use the word 'sniping' which, in everyone's mind, equates with anti-personnel shooting. Certainly a jet fighter or a radar set presents a bigger target than a man, and at 1000 m or more there is far more chance of hitting such a target than there is of hitting an individual.

Steyr-Mannlicher has been careful to call its new weapon an Anti-Materiel Rifle, but at the same time the ballistic solution the company has chosen gives the weapon a degree of long-range accuracy which no .50 is ever going to match. The high-velocity flechette has formidable armour-piercing properties, though what it is likely to do to softer targets is less predictable; we have all heard of the shaped charge which failed to stop the tank because the jet passed entirely through empty space inside the hull, and one can see the possibility of a thin flechette doing something similar and penetrating both sides of a soft target quite cleanly without striking anything of significance inside it, though I'm sure Steyr-Mannlicher has thought of that one and has a suitable response up its sleeve.

But bearing in mind the number of heavy-calibre cartridges now available it seems probable that this type of weapon is going to receive more attention in the near future. I would be interested to see a heavy rifle firing FN Herstal's new 15.5 mm round, for example, and I wonder if anyone has contemplated reviving the wartime 14.5 mm PTRD anti-tank rifle for this role? And who will be the first to revive a 20 mm manportable rifle and then uprate it to 25 mm to take advantage of modern ammunition designs? The possibilities here are endless; but what is needed is a sure analysis of the tactical requirement before vanishing into the workshop to start building. To sell a weapon to an army you have to do more than show them how it works and what it does; you have to show them where it can be profitably used.

Acknowledgements

It remains now to say thank you to a host of people around the world who are kind enough to provide me with the information necessary to overhaul this Yearbook annually. Many prefer not to be identified, but 'thank you' to Jack Krcma, Waldemar Hoff, Kelley Barber, Pedro del Fierro . . . and several others. Special thanks to the editorial staff of the International Defense Review for providing a number of photographs first seen in their journal. Mention should also be made of the multitude of company representatives who check the information, tell us what is new and what is no longer new and generally provide basic facts which go to form the bulk of the text. Thanks also to the in-house staff of Jane's Defence Data who assist in the production: to Sally Znidericz, Kevin Borras, Ruth Simmance and Sarah Erskine who catch the mistakes; to the staff of Method Ltd who convert computer discs and piles of photographs into orderly pages: and to the printers and binders who turn everything into the volume you now hold.

May 1990

Ian V. Hogg

Data Tables

Pistols

Country	TITLE	Action	Locking System	Feed System	Weight Empty (g)	Length Overall (mm)	Barrel Length (mm)	Rifling	Muzzle Velocity (m/s)	Remarks
(1) Metric Calibres										
5.45 × 18 mm PSM										
USSR	PSM	BB; DA	—	8 box	460	160	85	6 rh	315	Security forces only
7.62 × 17 mm										
China	Type 64	BB or SS	Man RB	9 box	1810	222	95	—	205	Silenced; locked bolt option
China	Type 67	BB or SS	Man RB	9 box	1020	225	89	—	181	Silenced; locked bolt option
China	Type 77	BB	—	7 box	500	148	—	—	318	One-hand cocking
7.62 × 25										
China	Type 54	Recoil	Browning	8 box	890	195	115	4 rh	420	Tokarev copy
China	Type 80	Recoil	Bolt lug	10/20 box	1100	300	—	—	470	Mauser copy; selective-fire
CZ	Model 52	Recoil	Rollers	8 box	960	209	120	4 rh	396	—
N Korea	Type 68	Recoil	Browning	8 box	795	185	108	4 rh 1/40	395	Tokarev copy
USSR	Tokarev TT-33	Recoil	Browning	8 box	850	196	116	4 rh	420	—
Yugo	M57	Recoil	Browning	8 box	900	200	116	4 rh	450	Tokarev copy
7.65 × 17SR (7.65 Browning, .32 ACP)										
B	Browning 1910	BB	—	7 box	582	152	89	6 rh	295	—
B	Browning 140DA	BB; DA	—	12 box	650	170	96	6 rh 1/32	300	Beretta 81 basis
CZ	Model 50	BB; DA	—	8 box	681	173	97	6 rh	280	—
CZ	Model 70	BB; DA	—	8 box	681	173	97	6 rh	280	—
CZ	Model 83	BB; DA	—	15 box	650	173	96	6 rh 1/32	320	Improved Model 50
F	MAB Modele D	BB	—	9 box	725	176	103	6 rh	—	—
D	Mauser HsC	BB; DA	—	8 box	660	160	85	6 rh	290	—
D	Walther PP	BB; DA	—	8 box	660	170	98	6 rh	320	—
D	Walther PPK	BB; DA	—	7 box	590	155	83	6 rh	308	—
D	H&K HK4	BB; DA	—	8 box	520	157	85	6 rh 1/32	302	—
Hung	FEG Model R	BB; DA	—	6 box	450	140	72	—	300	—
I	Beretta 81	BB; DA	—	12 box	670	172	97	6 rh 1/32	300	—
I	Beretta 81BB	BB; DA	—	12 box	670	172	97	6 rh 1/32	300	Improved safety; also 81F with hammer-drop
I	Beretta 82BB	BB; DA	—	9 box	630	172	97	6 rh 1/32	300	Also 82F with hammer-drop safety
J	New Nambu 57B	BB	—	8 box	600	150	90	—	300	Prototypes only
N Korea	Type 64	BB	—	7 box	640	170	102	6 rh	300	Browning 1900 copy
CH	SIG P-230	BB; DA	—	8 box	465	168	92	6 rh 1/32	290	—
Turk	MKE	BB; DA	—	7 box	680	170	98	6 rh	280	—
Yugo	M70	BB	—	8 box	740	200	94	6 rh 1/31	300	—
7.65 × 21 (7.65 Parabellum)										
I	Beretta 98	Recoil; DA	Lock plate	13 box	900	197	109	6 rh	365	—
I	Beretta 99	Recoil; DA	Lock plate	8 box	900	197	109	6 rh	365	—
I	Tanfoglio GT30	Recoil; DA	Browning	15 box	1015	202	120	6 rh 1/32	385	—
CH	SIG P-210	Recoil	Browning	8 box	900	215	120	4 rh 1/32	385	—
CH	SIG P-220	Recoil; DA	Browning	9 box	765	198	112	4 rh 1/32	365	—
9 × 17 mm (9 mm Short; .380 Auto)										
B	Browning 1910	BB	—	6 box	582	152	89	6 rh	270	—
B	Browning DA140	BB; DA	—	13 box	650	170	96	6 rh 1/28	280	—
CZ	Model 83	BB; DA	—	13 box	650	173	96	6 rh 1/28	340	—
D	Mauser HsC	BB; DA	—	7 box	660	160	85	6 rh	280	—
D	Walther PP	BB; DA	—	8 box	665	170	98	6 rh	256	—
D	Walther PPK	BB; DA	—	6 box	590	155	83	—	244	—
D	H&K HK4	BB; DA	—	7 box	520	157	85	6 rh 1/28	299	Convertible to .22, 6.35, 7.65 ACP
D	H&K P7K3	BB; SC	—	8 box	750	160	96.5	Poly, rh	280	Convertible to .22, 7.65 ACP
I	Beretta 84	BB; DA	—	13 box	660	172	97	6 rh 1/28	280	—
I	Beretta 84BB	BB; DA	—	13 box	660	172	97	6 rh 1/28	280	Improved safety; also 84F with hammer-drop
I	Beretta 85BB	BB; DA	—	8 box	620	172	97	6 rh 1/28	280	Also 85F with hammer-drop safety
I	Beretta 86	BB; DA	—	8 box	660	185	111	6 rh	310	Tip-up barrel
I	Bernadelli P-060	BB	—	7 box	690	164	90	6 rh	275	—
E	Star DKL	Recoil	Browning	6 box	420	145	80.5	6 rh	270	—
E	Astra A-50	BB	—	7 box	650	168	89	—	275	—
E	Astra A-60	BB; DA	—	13 box	700	168	89	—	275	—
E	Astra Falcon	BB	—	7 box	646	164	98.5	—	280	—
CH	SIG P230	BB; DA	—	7 box	460	168	92	6 rh 1/28	275	—
Turk	MKE	BB; DA	—	7 box	680	170	98	6 rh	280	—
Yugo	M70(k)	BB	—	8 box	720	200	94	6 rh 1/27	260	—
9 × 18 mm Soviet (9 mm Makarov; Chinese Type 59)										
China	Type 59	BB; DA	—	8 box	730	162	93.5	—	314	—
E Ger	Pistole M	BB; DA	—	8 box	663	160	91	4 rh	315	—
Hung	PA-63	BB; DA	—	7 box	595	175	100	6 rh	310	—
Pol	P-64	BB; DA	—	6 box	636	155	84	—	310	—
USSR	Makarov PM	BB; DA	—	8 box	663	160	91	4 rh	315	—
USSR	Stechkin APS	BB; DA	—	20 box	1020	222	140	4 rh	340	Selective-fire option, 750 rds/min
9 × 18 mm Police										
I	Tanfoglio TA18	Recoil; DA	Browning	15 box	1015	204	120	6 rh 1/28	340	—

Country	TITLE	Action	Locking System	Feed System	Weight Empty (g)	Length Overall (mm)	Barrel Length (mm)	Rifling	Muzzle Velocity (m/s)	Remarks
9 × 19 mm Parabellum										
A	Steyr GB	Del BB; DA	Gas delay	18 box	845	216	136	—	400	Production ceased 1989
A	Glock 17	Recoil; SC	Browning	17 box	661	188	114	6 rh	350	Hexagonal rifling
A	Glock 18	Recoil; SC	Browning	17 box	636	223	114	6 rh	350	Selective fire
A	Glock 19	Recoil; SC	Browning	15 box	595	178	101	6 rh	350	Hexagonal rifling
B	Browning Hi-Power	Recoil	Browning	13 box	882	200	118	6 rh 1/28	350	—
B	Browning Mk 3	Recoil	Browning	—	—	—	—	—	—	—
B	Browning BDA	Recoil; DA	Browning	14 box	905	200	118	6 rh 1/28	350	—
CZ	Model 75	Recoil; DA	Browning	15 box	1000	203	120	6 rh 1/28	350	—
CZ	Model 85	Recoil; DA	Browning	15 box	1000	203	120	6 rh 1/28	350	Ambidextrous
SF	Lahti 35/40	Recoil	Bolt lugs	8 box	1220	246	107	6 rh 1/27	350	—
F	MAS Mle 50	Recoil	Browning	9 box	860	195	112	4 lh 1/27	354	—
F	MAB PA15	Recoil	Rot Brl	15 box	1090	203	114	6 rh	350	—
D	Pistole '08	Recoil	Toggle	8 box	835	220	100	6 rh 1/27.5	350	—
D	Walther P1	Recoil; DA	Drop lug	8 box	800	216	125	6 rh 1/28	356	—
D	Walther P1A1	Recoil; DA	Drop lug	8 box	808	179	90	6 rh 1/28	350	—
D	Walther P5	Recoil; DA	Drop lug	8 box	795	180	90	6 rh 1/28	350	—
D	Walther P5C	Recoil; DA	Drop lug	8 box	780	169	79	6 rh	350	—
D	Walther P88	Recoil; DA	Browning	15 box	900	187	102	—	350	—
D	H&K P9S	Del BB; DA	Rollers	9 box	880	192	102	Poly, rh	351	—
D	H&K VP70	BB; SC	—	18 box	823	204	116	6 rh	360	Selective-fire option
D	H&K P7M13	Del BB; SC	Gas delay	13 box	975	169	105	Poly, rh	351	P7M8 with 8-rd magazine
Hung	FEG P9R	Recoil; DA	Browning	14 box	1000	203	118.5	6 rh	380	Steel frame
Hung	FEG P9RA	Recoil; DA	Browning	14 box	820	203	118.5	6 rh	380	Alloy frame
Hung	Tokagypt	Recoil	Browning	7 box	910	194	114	6 rh	350	Tokarev copy
Indon	Pindad P1A	Recoil	Browning	13 box	880	196	112	6 rh	354	Browning Hi-Power copy
IL	Uzi	BB	—	20 box	1650	240	115	4 rh 1/28	350	—
I	Beretta 951	Recoil	Drop lug	8 Box	870	203	114	6 rh 1/28	350	—
I	Beretta 951R	Recoil	Drop lug	10 box	1350	170	125	6 rh 1/28	390	Selective-fire 750 rds/min
I	Beretta 92	Recoil; DA	Drop lug	15 box	950	217	125	6 rh 1/28	390	—
I	Beretta 92S	Recoil; DA	Drop lug	15 box	980	217	125	6 rh 1/28	390	Modified safety
I	Beretta 92F, SB	Recoil; DA	Drop lug	15 box	980	217	125	6 rh 1/28	390	Ambidextrous
I	Beretta 92 SB-C	Recoil; DA	Drop lug	13 box	—	197	109	6 rh 1/28	360	F compact similar, ambidextrous
I	Beretta 93R	Recoil; DA	Drop lug	15, 20 box	1120	240	156	6 rh 1/28	375	3-round burst fire
I	Bernadelli PO18	Recoil; DA	Browning	15 box	998	213	122	6 rh	355	—
I	Bernadelli PO18 Com.	Recoil; DA	Browning	14 box	950	190	102	6 rh	350	Compact model
I	Tanfoglio TA90	Recoil; DA	Browning	15 box	1015	202	120	6 rh 1/28	350	—
I	Tanfoglio Baby	Recoil; DA	Browning	12 box	850	175	90	6 rh 1/28	350	—
I	Benelli B76	Del BB; DA	Toggle	8 box	970	205	108	6 rh 1/28	355	—
I	Benelli MP3S	Del BB	Toggle	6 box	1175	237	140	6 rh 1/28	360	Target version of B76
J	New Nambu 57A	Recoil	Browning	8 box	890	198	118	—	350	Prototypes only
S. Africa	Z-88	Recoil; DA	Drop lug	15 box	950	217	125	6 rh 1/28	390	Provisional data
E	Astra A-70	Recoil; DA	Browning	7 box	840	166	89	—	—	—
E	Astra A-80	Recoil; DA	Browning	15 box	985	180	96.5	—	350	A-90, improved model, similar data
E	Llama M-82	Recoil; DA	Drop lug	15 box	1105	210	114	6 rh	360	—
E	Star Firestar	Recoil	Browning	7 box	798	163	86	—	380	—
E	Star Super B	Recoil	Browning	8 box	1020	222	130	4 rh	400	—
E	Star 30M	Recoil; DA	Browning	15 box	1140	205	110	—	380	—
E	Star 30PK	Recoil; DA	Browning	15 box	860	193	98	—	375	Alloy frame
E	Star BM	Recoil	Browning	8 box	965	182	99	6 rh 1/34.5	390	Also BKM with alloy frame
CH	SIG P-210	Recoil	Browning	8 box	900	215	120	6 rh 1/28	335	—
CH	SIG P220	Recoil; DA	Browning	9 box	830	198	112	6 rh 1/28	345	—
CH	SIG P225	Recoil; DA	Browning	8 box	740	180	131	6 rh 1/28	345	—
CH	SIG P226	Recoil; DA	Browning	15, 20 box	750	196	112	6 rh 1/28	350	—
CH	SIG P228	Recoil; DA	Browning	13 box	830	180	98	6 rh 1/28	350	—
CH	ITM AT-88S	Recoil; DA	Browning	15 box	1000	206	120	6 rh 1/28	352	—
CH	ITM AT-88P	Recoil; DA	Browning	13 box	910	184	93	6 rh 1/28	350	—
CH	ITM AT88H	Recoil; DA	Browning	10 box	740	172	87	6 rh 1/28	350	—
USA	ASP	Recoil	Browning	7 box	570	188	57	6 rh	—	Production ceased 1988
USA	Goncz	BB	—	18, 36 box	1410	384	241	6 rh	375	—
USA	Intratec TEC-9A	BB	—	20 box	1417	318	127	6 rh	355	—
USA	Ruger P85	Recoil; DA	Browning	15 box	907	200	114	6 rh	355	—
USA	S&W Model 469	Recoil	Browning	12 box	737	175	89	6 rh	360	—
USA	S&W Model 5900	Recoil; DA	Browning	14 box	978	191	102	6 rh	375	Also alloy frame version
USA	S&W Model 3900	Recoil; DA	Browning	8 box	964	191	102	6 rh	375	Also alloy frame version
USA	S&W Model 6900	Recoil; DA	Browning	12 box	666	175	89	6 rh	360	Also alloy frame version
USA	Springfield P9	Recoil; DA	Browning	16 box	1000	206	120	4 rh 1/28	365	CZ-75 based
USA	Springfield P9C	Recoil; DA	Browning	10 box	910	184	93	4 rh 1/28	350	—
USA	Calico M950	Del BB	2-pt bolt	50 helix	1000	356	152	6 rh	393	—
USA	Knight KAC	Recoil; SC	Rot.brl	13 box	822	171	94	6 rh	—	—
Yugo	M70	Recoil	Browning	9 box	900	200	116	6 rh	330	Tokarev copy. M70A has improved safety
9 × 20SR (9 mm Browning Long)										
S	Pistol m/07	BB	—	7 box	907	203	127	6 rh 1/50	320	Browning M1903 pistol
(2) Inch Calibres										
China	Combat Knife	SS	—	4 bbl	464	262	86	4 rh 1/40	—	Plus 140 mm knife blade
I	Beretta 87BB	BB	—	8 box	570	—	—	—	—	Single action only
I	Beretta 87BB/LB	BB	—	8 box	660	242	150	—	—	Single action only
I	Beretta M89	BB	—	8 box	1160	240	152	6 rh 1/62	345	—
USA	Ruger Gov't Target	BB	—	10 box	1247	283	175	6 rh 1/68	350	—
.357 Magnum										
B	Barracuda	Revolver	DA	6 cyl	1050	—	76.2	6 rh	—	Production ceased; also in other calibres
E	Astra Police	Revolver	DA	6 cyl	1040	212	77	—	430	Also .38 Spl, 9 mm Parabellum versions
F	Manurhin MR73	Revolver	DA	6 cyl	890	205	76.2	6 rh	—	Other barrel lengths, calibres
IL	Desert Eagle	Gas, s/auto	Rot Bolt	9 box	1760	260	152	6 rh 1/39	450	Other calibres; also alloy-frame version
USA	Ruger GP-100	Revolver	DA	6 cyl	1247	238	102	5 rh 1/52	425	Also .38 Spl, other barrel lengths
Yugo	M1983	Revolver	DA	6 cyl	900	188	64	—	—	Other barrel lengths

Country	TITLE	Action	Locking System	Feed System	Weight Empty (g)	Length Overall (mm)	Barrel Length (mm)	Rifling	Muzzle Velocity (m/s)	Remarks
.38 Special										
I	Franchi RF83	Revolver	DA	6 cyl	800	—	101	—	—	Four variant models
J	New Nambu Model 60	Revolver	DA	5 cyl	680	197	77	—	220	—
E	Astra 960	Revolver	DA	6 cyl	1150	241	102	—	265	—
CH	FAMAE	Revolver	DA	5 cyl	630	225	82	—	—	Other barrel lengths
USA	Ruger SP101	Revolver	DA	5 cyl	765	195	78	—	260	—
USA	S&W Model 64 M&P	Revolver	DA	6 cyl	865	235	102	—	265	—
.44 Magnum										
E	Astra 44	Revolver	DA	6 cyl	1280	193	152	—	450	Also 216 mm barrel
USA	S&W Model 29	Revolver	DA	6 cyl	1332	302	165	—	455	Other barrel lengths
.45 ACP										
E	Star PD	Recoil		Browning	6 box	710	180	100	6 rh 1/27	220	—
CH	SIG P-220-1	Recoil		Browning	7 box	800	198	112	6 rh 1/35	245	—
USA	Arminex Trifire	Recoil		Browning	8 box	1078	219	127	—	255	Also 9 Para and .38 Super
USA	Colt M1911A1	Recoil		Browning	8 box	1130	219	127	6 rh 1/35.5	253	—
USA	Detonics Mk VI	Recoil		Browning	6 box	820	171	89	—	—	—
USA	M15	Recoil		Browning	7 box	1020	200	106	6 rh	245	General Officers
USA	S&W 4506	Recoil; DA		Browning	8 box	1020	216	127	—	—	—
USA	Springfield M1911A1	Recoil		Browning	7 box	1010	218	128	6 rh 1/35.5	253	Also in 9 mm Parabellum
.45 Colt										
E	Astra 45	Revolver	DA	6 cyl	1240	293	152	—	270	Also 216 mm barrel

Abbreviations:
BB – Blowback; cyl – cylinder; DA – Double-action; Del BB – delayed blowback; lh – Left-hand twist; poly – Polygonal rifling; rh – Right-hand twist; SC – self-cocking; SS –Single-shot. Rifling twist is expressed as one turn in x calibres.

Sub-machine Guns

Country	TITLE	Operating System	Locking System	Feed System	Weight Empty (kg)	Length Stock Open (mm)	Length Stock Folded (mm)	Length Barrel (mm)	Rifling	Muzzle Velocity (m/s)	Rate of Fire (rds/min)	Remarks
(1) Metric Calibres												
5.45 × 39.5 mm												
USSR	AKSU-74	Gas	Rot bolt	30 box	—	675	420	200	—	800	800	Shortened Kalashnikov
5.56 × 45 mm												
D	H&K HK53	Del BB	Rollers	25 box	3.05	755	563	211	6 rh	750	700	—
USA	Ruger AC556F	Gas	Rot bolt	20,30 box	3.27	825	584	292	—	—	—	—
YU	Zastava M85	Gas	Rot. bolt	20,30 box	3.20	790	570	315	—	790	700	Resembles AKSU-74
5.7 × 28 mm												
B	FN P-90	BB	—	50 box	2.80	400	n/a	—	—	850	1000	Prototypes
7.62 × 25 mm												
China	Type 64	BB	—	30 box	3.40	843	635	244	4 rh	513	1300	Silenced
China	Type 79	BB	—	20 box	1.90	740	470	—	—	500	650	—
China	Type 85	BB	—	30 box	1.90	628	444	—	—	500	780	—
China	Type 85 (S)	BB	—	30 box	2.50	869	631	—	—	300	800	Silenced
CZ	Model 26	BB	—	32 box	3.41	686	445	284	4 rh	550	650	Mod 24 same but wood stock
Hung	M48M	BB	—	71 drum	3.56	840	n/a	269	4 rh	488	900	Copy of Soviet PPSh 41
Pol	M43/52	BB	—	35 box	3.63	836	n/a	240	4 rh	490	600	Based on Soviet PPS-42
USSR	PPD-40	BB	—	71 drum	3.63	787	n/a	267	4 rh	488	800	—
USSR	PPSh-41	BB	—	71 drum	3.56	840	n/a	269	4 rh	488	900	—
USSR	PPS-42	BB	—	35 box	2.95	907	641	273	4 rh	488	700	—
USSR	PPS-43	BB	—	35 box	3.35	819	622	254	4 rh	488	700	Improved PPS-42
Viet	K-50	BB	—	35 box	3.40	756	571	269	4 rh	488	700	Modified PPSh-41
Yugo	M49	BB	—	35 box	3.95	870	n/a	273	4 rh	500	700	Based on PPSh-41
Yugo	M57	BB	—	35 box	3.00	870	591	250	4 rh	500	600	—
7.65 × 17SR (7.65 mm ACP)												
CZ	Model 61 Skorpion	BB	—	10,20 box	1.59	513	269	112	6 rh 1/40	317	840	Also other calibres, but rarely
7.65 × 19.5 mm (7.65 mm Longue)												
F	MAS Modele 38	BB	—	32 box	2.87	734	n/a	224	4 rh	351	600	—
9 × 17 mm (9 mm Short/.380 Auto)												
USA	Ingram M11	BB	—	16,32 box	1.59	460	248	129	6 rh 1/34	293	1200	—
9 × 19 mm (9 mm Parabellum)												
Argent	FMK-3 Mod 2	BB	—	40 box	3.76	690	520	290	6 rh 1/28	400	600	—
Austr	Owen	BB	—	33 box	4.23	813	n/a	250	7 rh	366	700	Top-mounted magazine
Austr	F1	BB	—	34 box	3.27	714	n/a	213	—	366	640	Sterling-based; top-mounted magazine
A	Steyr MPi69	BB	—	25,32 box	3.13	670	465	260	6 rh 1/28	381	550	MPi81 similar, 700 rds/min
A	Steyr AUG 9 Para	BB	—	25,32 box	3.30	665	n/a	420	6 rh 1/28	400	700	AUG rifles can be converted
A	Steyr TMP	Recoil	Rot. brl	15,25 box	1.30	270	—	150	—	350	—	s/auto only
B	Vigneron M2	BB	—	32 box	3.29	886	706	305	6 rh	381	620	—
BR	M9-M1-CEV	BB	—	30 box	3.00	—	—	228	—	400	600	—
BR	Imbel MD-1	BB	—	30 box	—	—	—	211	4 rh	—	550	—
BR	Imbel MD-2	BB	—	15,30 box	3.60	770	n/a	221	4 rh 1/28	400	700	—
BR	Imbel MD-2A1	BB	—	15,30 box	3.18	680	430	160	4 rh 1/28	360	685	—
BR	Uru Mekanika	BB	—	30 box	3.01	700	470	175	6 rh	389	750	—

[33]

Country	TITLE	Operating System	Locking System	Feed System	Weight Empty (kg)	Length Stock Open (mm)	Length Stock Folded (mm)	Length Barrel (mm)	Rifling	Muzzle Velocity (m/s)	Rate of Fire (rds/min)	Remarks
BR	LAPA SM	BB	—	30,32 box	2.80	623	n/a	202	4 rh	405	485	—
BR	MSM	BB	—	32 box	2.70	495	290	150	4 rh	365	600	—
BR	CEL Madsen	BB	—	30 box	3.74	—	—	—	4 rh	392	600	—
Can	C1	BB	—	30 box	2.95	686	493	198	6 rh	366	550	Sterling-based
CZ	Model 25	BB	—	24,40 box	3.50	686	445	284	6 rh	381	650	Also Model 23 with wood butt
DK	Hovea 49	BB	—	36 box	3.40	810	550	215	—	380	600	Swedish design
DK	Madsen 1950	BB	—	32 box	3.20	794	528	198	4 rh	390	550	—
SF	Suomi M1931	BB	—	50 box	4.68	870	n/a	318	4 rh	399	900	Also 71-round drum magazine
SF	Model 1944	BB	—	36 box	2.90	831	622	249	4 rh	399	650	Modified Soviet PPS-43
SF	Jati-Matic	BB	—	20,40 box	1.65	375	n/a	203	6 rh 1/28	400	650	—
F	MAT-49	BB	—	20,32 box	3.50	720	460	228	4 1h	390	600	—
D	H&K MP5	Del BB	Rollers	15,30 box	2.55	660	490	225	6 rh	400	800	Also fixed butt model 680 mm long
D	H&K MP5SD3	Del BB	Rollers	15,30 box	2.90	780	610	146	6 rh	285	800	Silenced; six variant models
D	H&K MP5K	Del BB	Rollers	15,30 box	2.10	325	n/a	115	6 rh	375	900	—
D	H&K MP2000	BB	—	30 box	2.78	565	387	—	—	375	880	And silenced version
D	Walther MP-K	BB	—	32 box	2.82	653	368	171	6 rh	356	550	—
D	Walther MP-L	BB	—	32 box	3.00	737	455	257	6 rh	396	550	—
Indon	PM Model VII	BB	—	33 box	3.29	840	540	274	—	381	600	—
IL	Uzi	BB	—	25,32 box	3.70	650	470	260	4 rh 1/28	400	600	Also wood butt 3.80 kg same length
IL	Mini-Uzi	BB	—	20,25,32	2.70	600	360	197	4 rh 1/28	352	950	—
IL	Micro-Uzi	BB	—	20 box	1.95	460	250	117	4 rh 1/28	350	1250	Also in .45 ACP
I	Beretta 38/42	BB	—	20,40 box	3.27	800	n/a	213	6 rh	381	550	—
I	Beretta 12	BB	—	20,32,40	3.00	645	418	200	6 rh	381	550	Also wood butt model 660 mm, 3.4 kg
I	Beretta 12S	BB	—	20,32,40	3.20	660	418	200	6 rh 1/28	430	550	Also wood butt model 660 mm, 3.6 kg
I	Franchi LF57	BB	—	10,20,30	3.30	680	420	205	6 rh 1/28	400	450	—
I	Socimi 821	BB	—	32 box	2.45	600	400	200	6 rh 1/28	380	600	—
I	Spectre M4	BB	—	30,50 box	2.90	580	350	130	—	400	850	Double-action trigger
I	AGM-1	BB	—	13,20 box	3.00	670	n/a	410	—	385	—	Semi-auto only; bull-pup
Mex	Mendoza HM-3	BB	—	32 box	2.69	635	400	255	—	360	600	—
Peru	MGP-79A	BB	—	20,32 box	3.09	809	544	237	12 rh	410	850	—
Peru	MGP-87	BB	—	20,32 box	2.89	702	500	194	12 rh	410	850	—
Peru	MGP-84	BB	—	20,32 box	2.31	503	284	166	12 rh	410	700	—
Port	FBP M948	BB	—	32 box	3.77	807	645	249	6 rh	390	500	—
Port	Indep Lusa	BB	—	30 box	2.50	600	445	160	6 rh 1/28	390	900	—
Rom	Orita M1941	BB	—	25,32 box	3.46	894	n/a	287	6 rh	381	600	—
S Africa	BXP	BB	—	22,32 box	2.50	560	350	208	6 rh	380	800	—
E	Star Z-62, Z-70/B	BB	—	20,30,40	2.87	701	480	201	—	380	550	Also in 9 × 23 mm calibre
E	Star Z-84	BB	—	20,30 box	3.10	615	410	215	6 rh 1/35	400	600	Also 270 mm barrel available
E	C-2	BB	—	32 box	2.65	720	500	212	4 rh	325	600	Also in 9 × 23 mm calibre
S	Carl Gustav M45	BB	—	36 box	3.90	808	552	213	6 rh	410	600	—
CH	FAMAE	BB	—	43 box	2.44	—	—	175	4 rh 1/27	400	875	—
CH	SIG MP310	BB	—	40 box	3.15	735	610	200	—	365	900	—
Turk	M1968	BB	—	32 box	4.69	880	n/a	350	—	400	600	Modified Rexim-Favor
UK	Sten Mk II	BB	—	32 box	2.80	762	n/a	197	2rh 1/28	366	550	Other, less common, variants
UK	Sten Mk IIS	BB	—	32 box	3.50	857	n/a	91.4	6rh 1/28	305	—	Automatic fire not recommended
UK	Sten Mk V	BB	—	32 box	3.90	762	n/a	198	6rh 1/28	366	550	Silenced version Mk VI
UK	Sterling L2A3	BB	—	34 box	2.72	690	483	198	6rh 1/28	390	550	—
UK	Sterling L34A1	BB	—	34 box	3.60	864	660	198	6rh 1/28	300	550	Silenced
UK	Sterling Mk 6	BB	—	34 box	3.40	889	685	410	6rh 1/28	400	—	Semi-automatic only
UK	Sterling Mk7 Para	BB	—	10,15,34	2.20	335	n/a	89	6rh 1/28	355	500	Four variant models
USA	Ares FMG	BB	—	20,32 box	2.38	503	262	—	—	360	650	Folding weapon
USA	Calico	Del BB	—	50 helical	1.68	884	724	406	6 rh	—	—	s/auto only
USA	Saco 683	BB	—	25,32 box	3.31	699	520	203	—	396	650	—
USA	Colt	BB	—	20,32 box	2.59	730	650	260	—	397	900	Based on M16 rifle
USA	Ingram M10	BB	—	32 box	2.84	548	269	146	6rh 1/34	366	1090	—
USA	Weaver PKS-9	BB	—	25,30,42	2.77	—	416	181	—	396	1000	—
USA	KF-9-AMP	BB	—	20,36,60	1.13	603	273	76.2	6rh 1/44	350	800	Other calibres
USA	Viking	BB	—	20,36 box	3.52	600	400	220	—	400	800	—

9 × 23 mm (9 mm Bergmann-Bayard; 9 mm Largo)

Country	TITLE	Operating System	Locking System	Feed System	Weight Empty (kg)	Length Stock Open (mm)	Length Stock Folded (mm)	Length Barrel (mm)	Rifling	Muzzle Velocity (m/s)	Rate of Fire (rds/min)	Remarks
E	Star Z-45	BB	—	30 box	3.86	838	579	198	—	381	450	Based on MP40

(2) Inch Calibres

.45 ACP

Country	TITLE	Operating System	Locking System	Feed System	Weight Empty (kg)	Length Stock Open (mm)	Length Stock Folded (mm)	Length Barrel (mm)	Rifling	Muzzle Velocity (m/s)	Rate of Fire (rds/min)	Remarks
BR	INA MB50	BB	—	30 box	3.40	794	546	213	4 rh	280	650	Based on Madsen M1946
USA	Thompson M1928A1	Del BB	Blish	50 drum	—	857	n/a	267	6 rh	280	650	Other box & drum magazines
USA	Thompson M1	BB	—	20,30 box	4.74	813	n/a	267	6 rh	280	700	Simplified version
USA	M3A1	BB	—	30 box	3.69	757	579	203	4 rh	280	450	—
USA	Ingram M10	BB	—	30 box	2.84	548	269	146	6rh 1/44	280	1145	—

Abbreviations:
BB – blowback: Blish – Blish delay system using slipping inclined faces; Del BB – delayed blowback: lh – Left-hand twist: n/a – not applicable, usually meaning that the weapon uses a fixed wooden butt: rh – Right-hand twist: Rot Bolt –rotating bolt. Rifling is expressed as one turn in x calibres.

Rifles

Country	TITLE	Action	Lock	Feed System	Selective Fire?	Weight Empty (kg)	Length Butt ext (mm)	Length Butt fold (mm)	Length Barrel (mm)	Rifling	Rate of fire (rds/min)	Muzzle Velocity (m/s)	Remarks
4.7 × 33 mm Caseless													
D	H&K G11	Recoil	—	50 box	Y+3	3.60	750	—	540	Poly 1/33	450	930	—
5.45 × 39.5 mm Soviet													
USSR	AK74	Gas	RB	30 box	Y	3.60	930	691	400	4R 1/36	650	900	Also fixed butt model

Country	TITLE	Action	Lock	Feed System	Selective Fire?	Weight Empty (kg)	Length Butt ext (mm)	Length Butt fold (mm)	Length Barrel (mm)	Rifling	Rate of fire (rds/min)	Muzzle Velocity (m/s)	Remarks
5.56 × 45 mm													
A	Steyr AUG	Gas	RB	30 box	Y	3.60	790	—	508	6R 1/41	650	970	Four barrel lengths available
Argent	FARA 83	Gas	RB	30 box	Y	3.95	1000	745	452	6R 1/41	800	1005	
B	FN FNC	Gas	RB	30 box	Y	3.80	997	766	449	6R 1/32	750	915	Two barrel lengths available
B	FN Mod 7030	Gas	RB	30 box	No	3.80	997	766	449	6R 1/32	—	915	Semi-auto for Police use
BR	Imbel MD1	Gas	RB	20/30 box	Y	—	—	—	440	6R	750	—	
BR	LAPA 03	Gas	RB	20/30 box	Y	3.16	738	—	489	6R 1/55	700	1000	Development
China	CQ	Gas	RB	20 box	Y	3.20	987	—	505	1/55	—	990	M16 clone
SF	Valmet M76	Gas	RB	30 box	Y	3.60	914	—	420	4R 1/55	650	960	Kalashnikov-based
F	FA-MAS	DBB	2-pc bolt	25 box	Y+3	3.61	757	—	488	3R 1/55	960	—	
D	H&K HK33E	DBB	Rollers	25 box	Y	3.65	940	735	390	6R 1/55	750	920	
D	H&K G41	DBB	Rollers	30 box	Y+3	4.10	997	—	450	6R 1/32	850	960	
Hung	NGM	Gas	RB	30 box	Y	3.18	935	—	412	6R 1/36	600	900	—
IL	Galil ARM	Gas	RB	35 box	Y	3.95	979	742	460	6R 1/55	650	950	—
IL	Galil SAR	Gas	RB	35 box	Y	3.75	840	614	332	6R 1/55	650	900	—
India	—	Gas	RB	30 box	No+3	3.20	990	—	464	—	—	885	—
I	Beretta AR 70/223	Gas	RB	30 box	Y	3.80	955	—	450	4R 1/55	650	950	—
I	Beretta AR 70/90	Gas	RB	30 box	Y	3.99	998	—	450	6R 1/32	—	—	—
I	Socimi AR 871	Gas	RB	30 box	Y	3.60	970	—	450	6R 1/32	750	970	—
S Korea	K1A1	Gas	RB	20,30 box	Y	2.88	830	645	263	6R	800	820	—
S Korea	K2	Gas	RB	30 box	Y+3	3.26	900	730	465	6R	—	960	—
SP	SAR 80	Gas	RB	30 box	Y	3.70	970	—	459	—	700	970	—
SP	SR88	Gas	RB	20/30 box	Y	3.60	970	—	459	6R 1/32	750	970	—
S.Africa	R-4	Gas	RB	35 box	Y	4.30	1005	740	460	6R 1/55	650	980	—
S.Africa	R-5	Gas	RB	35 box	Y	3.70	877	615	332	6R 1/55	650	920	—
E	CETME L	DBB	Rollers	30 box	Y	3.40	925	—	400	6R 1/32	700	875	—
E	CETME LC	DBB	Rollers	30 box	Y	3.40	860	665	320	6R 1/32	750	832	—
S	AK-5	Gas	RB	30 box	Y	3.90	1008	753	450	6R 1/32	650	930	FN-FNC based
CH	SIG 540	Gas	RB	20 box	Y	3.26	950	—	460	6R 1/55	700	980	—
CH	SIG SG550	Gas	RB	30 box	Y	4.10	998	772	528	6R 1/32	700	—	—
CH	SIG SG551	Gas	RB	30 box	Y	3.50	827	601	362	6R 1/32	700	—	—
TW	Type 65	Gas	RB	30 box	Y+3	3.17	990	—	508	4R 1/32	750	990	—
UK	L85A1	Gas	RB	30 box	Y	3.80	785	—	518	6R 1/32	750	940	—
UK	Sterling SAR	Gas	RB	30 box	Y	3.08	865	—	457	6R 1/41	—	—	Future in doubt
USA	M16A2	Gas	RB	30 box	No+3	3.40	990	—	508	6R 1/32	800	1000	—
USA	M16A2 Carbine	Gas	RB	30 box	Y	2.70	838	757	368	6R 1/32	850	906	—
USA	Colt Model 733	Gas	RB	30 box	Y	2.59	762	680	254	6R 1/32	800	795	—
USA	Ruger Mini-14	Gas	RB	30 box	No	2.90	946	—	470	6R 1/32	—	1005	—
USA	Ruger AC-556	Gas	RB	30 box	Y	2.89	984	—	470	6R 1/55	750	1058	—
USA	La France	Gas	RB	20,30 box	Y	2.50	686	610	213	6R 1/32	600	762	—
YU	M80	Gas	RB	30 box	Y	3.50	990	—	460	6R 1/32	—	970	—
7.5 × 54 mm French													
F	M1949/56	Gas	Block	10 box	No	3.90	1010	—	521	4R 1/40	—	817	—
F	FR-F1	Bolt	—	10 box	No	5.20	1138	—	552	4R 1/40	—	852	Sniping rifle
7.62 × 33 mm (M1 Carbine)													
DR	Cristobal M2	DBB	2-pc bolt	30 box	Y	3.52	945	—	409	—	580	572	—
USA	Carbine M1	Gas	RB	15 box	No	2.36	904	—	458	4R 1/66	—	607	—
USA	Carbine M2	Gas	RB	30 box	Y	2.36	904	—	458	4R 1/66	750	607	—
7.62 × 39 mm Soviet													
China	Type 56 Carbine	Gas	Block	10 box	No	3.86	1021	—	521	—	—	735	SKS clone
China	Type 56 Rifle	Gas	RB	30 box	Y	3.80	874	645	414	—	600	710	Kalashnikov clone
China	Type 68 Rifle	Gas	RB	15 box	Y	3.49	1029	—	521	4R	750	730	—
CZ	Model 58	Gas	Block	30 box	Y	3.14	820	635	401	4R 1/32	800	710	—
SF	Valmet M76	Gas	RB	30 box	Y	3.60	914	—	420	4R	650	719	Kalashnikov-based
Hung	AMD-65	Gas	RB	30 box	Y	3.27	851	648	378	4R	600	700	Kalashnikov-based
USSR	Simonov SKS	Gas	Block	10 integ	No	3.85	1021	—	520	4R 1/33	—	735	—
USSR	Kalashnikov AK47	Gas	RB	30 box	Y	4.30	869	699	414	4R 1/31	600	710	—
USSR	Kalashnikov AKM	Gas	RB	30 box	Y	3.15	876	—	414	4R 1/31	600	715	—
USA	Ruger Mini-30	Gas	RB	5 integ	No	3.26	948	—	470	6R 1/33	—	713	—
YU	M59/66A1	Gas	Block	10 integ	No	4.10	1120	—	530	—	—	735	Simonov-based
YU	M70B1	Gas	RB	30 box	Y	3.70	900	—	—	4R	650	720	Kalashnikov-based
7.62 × 51 mm NATO													
A	Steyr SSG-69	Bolt	—	5 rotary	No	3.90	1140	—	650	4R	—	860	Sniping rifle
A	Steyr Police	Bolt	—	10 box	No	4.20	1135	—	650	—	—	860	Sniping rifle
B	FN FAL	Gas	Block	20 box	Y	4.25	1090	—	533	4R 1/40	700	840	—
B	FN Mod 30-11	Bolt	—	10 box	No	4.85	1117	—	502	—	—	850	Sniping rifle
Can	C3A1	Bolt	—	6 box	No	6.30	1140	—	660	4R 1/40	—	830	Sniping rifle
SF	Valmet M86	Bolt	—	9 integ	No	5.70	1210	—	720	4R 1/35	—	760	
SF	TRG21	Bolt	—	10 box	No	5.30	1200	—	660	4R 1/35	—	830	Sniping rifle
F	FR-F2	Bolt	—	10 box	No	5.20	1138	—	552	4R 1/40	—	850	Sniping rifle
D	Mauser SP-66	Bolt	—	3 integ	No	—	—	—	650	—	—	—	Sniping rifle
D	Mauser 86-SR	Bolt	—	9 box	No	4.90	1210	—	730	4R 1/40	—	850	Sniping rifle
D	HK G3	DBB	Rollers	20 box	Y	4.40	1025	—	450	4R 1/40	550	800	—
D	H&K PSG-1	DBB	Rollers	S.S.	No	8.10	1208	—	650	—	—	—	Sniping rifle
D	H&K MSG-90	DBB	Rollers	20 box	No	6.40	1165	—	600	—	—	—	Sniping rifle
D	H&K G8	DBB	Rollers	20 box	Y+3	8.15	1030	—	450	—	800	800	—
IL	Galil AR	Gas	RB	25 box	Y	3.95	1050	810	535	4R 1/40	650	850	—
IL	Galil SAR	Gas	RB	25 box	Y	3.75	915	675	400	4R 1/40	750	800	—
IL	Galil Sniper	Gas	RB	20 box	No	6.40	1115	840	508	4R 1/40	—	815	—
I	Beretta BM59	Gas	RB	20 box	Y	4.60	1095	—	490	4R	750	823	Garand-based; several variants
I	Beretta Sniper	Bolt	—	5 box	No	5.55	1165	—	586	4R 1/40	—	855	—
I	Socimi AR832	Gas	RB	30 box	Y	4.30	1030	810	450	4R 1/40	850	820	—
J	Type 64	Gas	Block	20 box	Y	4.40	990	—	450	4R	500	700	—
NO	NM149S	Bolt	—	5 box	No	5.60	1120	—	600	4R 1/40	—	855	Sniping rifle
E	CETME C	DBB	Rollers	20 box	Y	4.20	1015	—	450	4R 1/40	600	780	—
CH	SIG 510	DBB	Rollers	20 box	Y	4.25	1016	—	505	4R 1/40	600	790	—
CH	SIG SG542	Gas	RB	20 box	Y	3.55	1000	—	465	4R 1/40	700	820	—
UK	L1A1	Gas	Block	20 box	No	4.30	43	—	554	6R 1/40	—	838	FN-FAL

[35]

Country	TITLE	Action	Lock	Feed System	Selective Fire?	Weight Empty (kg)	Length Butt ext (mm)	Length Butt fold (mm)	Length Barrel (mm)	Rifling	Rate of fire (rds/min)	Muzzle Velocity (m/s)	Remarks
UK	L39A1	Bolt	—	10 box	No	4.42	1180	—	700	4R 1/40	—	841	Lee-Enfield sniping rifle
UK	L42A1	Bolt	—	10 box	No	4.43	1181	—	699	4R 1/40	—	838	Lee-Enfield sniping rifle
UK	Parker-Hale 82	Bolt	—	5 integ	No	4.80	1162	—	660	4R	—	830	Sniping rifle
UK	Parker-Hale 83	Bolt	—	S.S.	No	4.98	1187	—	660	4R 1/14	—	—	Sniping rifle
UK	Parker-Hale 85	Bolt	—	10 box	No	13.00	1150	—	700	—	—	—	Sniping rifle
UK	L96A1	Bolt	—	10 box	No	6.50	1194	—	655	—	—	—	Sniping rifle
UK	BRG Sniper	Bolt	—	10 box	No	6.60	1200	—	700	4R 1/37	—	—	
USA	M14/M21	Gas	RB	20 box	No	5.10	1120	—	559	4R 1/40	—	853	M21 is sniping rifle
USA	M40A1	Bolt	—	5 integ	No	6.57	1117	—	610	4R 1/40	—	777	Sniping rifle
USA	Grendel SRT	Bolt	—	9 box	No	3.00	1035	760	508	4R 1/33	—	—	Sniping rifle
USA	RAI	Bolt	—	5 box	No	5.67	—	—	610	4R 1/40	—	800	Sniping rifle
USA	Ruger M77V	Bolt	—	5 integ	No	4.08	1118	—	610	—	—	—	Sniping rifle
USA	La France	Gas	RB	20 box	Y	3.75	902	—	338	4R 1/40	650	765	—
YU	M77B1	Gas	RB	20 box	Y	4.80	990	—	500	6R 1/31	600	840	Kalashnikov-based

7.62 × 54R Soviet

Country	TITLE	Action	Lock	Feed System	Selective Fire?	Weight Empty (kg)	Length Butt ext (mm)	Length Butt fold (mm)	Length Barrel (mm)	Rifling	Rate of fire (rds/min)	Muzzle Velocity (m/s)	Remarks
USSR	Dragunov SVD	Gas	RB	10 box	No	4.30	1225	—	622	4R 1/33	—	830	—

7.62 × 63 mm (.30-06)

Country	TITLE	Action	Lock	Feed System	Selective Fire?	Weight Empty (kg)	Length Butt ext (mm)	Length Butt fold (mm)	Length Barrel (mm)	Rifling	Rate of fire (rds/min)	Muzzle Velocity (m/s)	Remarks
USA	M1 Garand	Gas	RB	8 clip	No	4.30	1106	—	610	4R 1/33	—	865	—

7.92 × 57 mm Mauser

Country	TITLE	Action	Lock	Feed System	Selective Fire?	Weight Empty (kg)	Length Butt ext (mm)	Length Butt fold (mm)	Length Barrel (mm)	Rifling	Rate of fire (rds/min)	Muzzle Velocity (m/s)	Remarks
B	M49	Gas	Block	10 integ	No	4.30	1110	—	589	—	—	725	—
YU	M76	Gas	RB	10 box	No	4.20	1135	—	550	4R 1/33	—	720	Kalashnikov-based sniping rifle

12.7 × 99 (.50 Browning)

Country	TITLE	Action	Lock	Feed System	Selective Fire?	Weight Empty (kg)	Length Butt ext (mm)	Length Butt fold (mm)	Length Barrel (mm)	Rifling	Rate of fire (rds/min)	Muzzle Velocity (m/s)	Remarks
USA	RAI Model 500	Bolt	—	S.S.	No	13.60		—	840	—	—	888	Long-range sniping rifle
USA	Barrett Light 50	Recoil	RB	11 box	No	14.70	1549	—	838	—	—	853	Long-range sniping rifle
USA	Peregrine	Gas	RB	10 box	No	15.20	1410	—	737	—	—	—	—

15 mm Special

Country	TITLE	Action	Lock	Feed System	Selective Fire?	Weight Empty (kg)	Length Butt ext (mm)	Length Butt fold (mm)	Length Barrel (mm)	Rifling	Rate of fire (rds/min)	Muzzle Velocity (m/s)	Remarks
A	AMR5075	Recoil	RB	5 box	No	20.00	—	—	1200	SB	—	1500	Anti-material rifle

Abbreviations:
rh – Right-hand twist; Inc – Increasing (Progressive) twist; RB – rotating bolt; DBB – delayed blowback.

Machine Guns

Country	TITLE	Action	Breech Lock	Feed System	Gun Weight (kg)	Gun Length (mm)	Barrel Length (mm)	Rifling	QCB?	Rate of Fire (rds/min)	Muzzle Velocity (m/s)	Remarks

5.45 × 39.5 mm

Country	TITLE	Action	Breech Lock	Feed System	Gun Weight (kg)	Gun Length (mm)	Barrel Length (mm)	Rifling	QCB?	Rate of Fire (rds/min)	Muzzle Velocity (m/s)	Remarks
USSR	RPK-74	Gas	Rot bolt	40 box	—	1039	—	—	No	—	—	—

5.56 × 45 mm

Country	TITLE	Action	Breech Lock	Feed System	Gun Weight (kg)	Gun Length (mm)	Barrel Length (mm)	Rifling	QCB?	Rate of Fire (rds/min)	Muzzle Velocity (m/s)	Remarks
A	Steyr LSW	Gas	Rot bolt	30/42 box	4.9	900	621	rh 1/32	Y	680	1000	—
B	FN Minimi	Gas	Rot bolt	belt/box	6.88	1040	466	6rh 1/32	Y	750/1000	965	—
Can	C7	Gas	Rot bolt	30 box	5.80	990	508	6rh 1/32	No	625	1000	—
D	H&K HK23E	DBB	Rollers	belt	9.30	1140	560	rh 1/40	Y	750	950	—
D	H&K HK13	DBB	Rollers	25 box	6.00	980	450	4rh 1/55	Y	750	950	—
D	H&K HK13E	DBB	Rollers	20/30 box	8.00	1030	450	4rh 1/32	Y	750	950	—
IL	Galil ARM	Gas	Rot bolt	50 box	4.00	1050	535	4rh 1/55	Y	650	850	—
IL	Negev	Gas	Rot bolt	box/drum	7.20	1020	460	6rh 1/50	Y	950	970	also short barrel
I	Beretta AR 70/84	Gas	Rot bolt	30 box	5.30	955	450	4rh 1/55	No	670	970	—
I	Beretta AS 70/70	Gas	Rot bolt	30 box	5.30	1000	465	6rh 1/32	No	800	—	—
S.Korea	Daewoo K3	Gas	Rot bolt	belt	6.10	1000	—	6rh	Y	800	1000	—
SP	Ultimax 100	Gas	Rot bolt	box/drum	4.90	1024	508	6rh 1/32	Y	540	970	—
E	Ameli	DBB	Rollers	belt/box	6.35	970	406	6rh 1/32	Y	1200	875	—
UK	L86A1	Gas	Rot bolt	30 box	5.40	900	646	6rh 1/32	No	850	970	—
USA	Ares light	Gas	Rot bolt	belt	4.90	1073	550	—	No	600	945	—
USA	Colt XM214	Mech	Gatling	belt	12.20	1041	686	—	No	4000	991	—
YU	M82	Gas	Rot bolt	30 box	4.00	1020	542	6rh 1/32	No	—	1000	—

7.5 × 54 mm

Country	TITLE	Action	Breech Lock	Feed System	Gun Weight (kg)	Gun Length (mm)	Barrel Length (mm)	Rifling	QCB?	Rate of Fire (rds/min)	Muzzle Velocity (m/s)	Remarks
F	Mle 24/29	Gas	Tilt block	26 box	9.20	1082	500	4lh	No	500	850	—
F	AA52	DBB	2-part bolt	belt	9.95	1145	500	4lh	Y	700	840	—

7.5 × 55 mm

Country	TITLE	Action	Breech Lock	Feed System	Gun Weight (kg)	Gun Length (mm)	Barrel Length (mm)	Rifling	QCB?	Rate of Fire (rds/min)	Muzzle Velocity (m/s)	Remarks
CH	M25	Recoil	Toggle	30 box	10.80	1163	583	4rh	No	450	746	—
CH	M51	Gas/recoil	Flaps	belt	16.00	1270	564	4rh	Yes	1000	792	—

7.62 × 39 mm

Country	TITLE	Action	Breech Lock	Feed System	Gun Weight (kg)	Gun Length (mm)	Barrel Length (mm)	Rifling	QCB?	Rate of Fire (rds/min)	Muzzle Velocity (m/s)	Remarks
SF	Valmet M62	Gas	Tilt block	belt	8.30	1085	470	4rh	Y	1000	730	—
USSR	RPD	Gas	Lug	belt	7.10	1036	521	4rh	No	700	700	—
USSR	RPK	Gas	Rot bolt	box/drum	5.00	1035	591	4rh	No	660	732	—
YU	M72B1	Gas	Rot bolt	box/drum	5.00	1025	540	4rh	No	600	745	—

7.62 × 45 mm

Country	TITLE	Action	Breech Lock	Feed System	Gun Weight (kg)	Gun Length (mm)	Barrel Length (mm)	Rifling	QCB?	Rate of Fire (rds/min)	Muzzle Velocity (m/s)	Remarks
CZ	Model 59	Gas	Swing block	belt	8.67	1116	593	4rh 1/32	Y	800	810	—

7.62 × 51 mm

Country	TITLE	Action	Breech Lock	Feed System	Gun Weight (kg)	Gun Length (mm)	Barrel Length (mm)	Rifling	QCB?	Rate of Fire (rds/min)	Muzzle Velocity (m/s)	Remarks
B	FN MAG	Gas	Drop lever	belt	10.85	1260	545	4rh 1/40	Y	600	840	—
BR	Uirapuru	Gas	Drop lever	belt	13.00	1300	600	6rh 1/40	Y	700	850	—

Country	TITLE	Action	Breech Lock	Feed System	Gun Weight (kg)	Gun Length (mm)	Barrel Length (mm)	Rifling	QCB?	Rate of Fire (rds/min)	Muzzle Velocity (m/s)	Remarks
F	AA52	DBB	2-part bolt	belt	9.95	1145	500	4lh	Y	700	840	—
D	H&K HK21	DBB	Rollers	belt	7.92	1021	450	4rh 1/40	Y	900	800	—
D	H&K HK21A1	DBB	Rollers	belt	8.30	1030	450	4rh 1/40	Y	900	800	—
D	H&K HK21E	DBB	Rollers	belt	9.30	1140	560	4rh 1/40	Y	800	840	—
D	H&K HK11A1	DBB	Rollers	20 box	7.70	1030	450	4rh 1/40	Y	800	800	—
D	H&K HK11E	DBB	Rollers	box/drum	8.15	1030	450	4rh 1/40	Y	800	800	—
D	MG3	Recoil	Rollers	belt	11.05	1225	565	4rh 1/40	Y	700	820	—
I	MG42/59	Recoil	Rollers	belt	12.00	1220	567	4rh 1/40	Y	800	820	—
J	Model 62	Gas	Tilt block	belt	10.70	1200	524	6rh	Y	600	855	—
J	Model 74	Gas	Tilt block	belt	20.40	1085	625	6rh	No	700	855	—
S.Africa	SS-77	Gas	Transverse	belt	9.60	1155	550	4rh 1/55	Y	900	840	—
CH	SIG 710-3	DBB	Rollers	belt	9.25	1143	559	4rh	Y	950	790	—
UK	L4A4	Gas	Tilt block	30 box	9.53	1133	536	6rh	Y	500	823	—
UK	L7A2	Gas	Tilt lever	belt	10.90	1232	679	6rh	Y	1000	838	—
USA	M60	Gas	Rot bolt	belt	10.50	1105	560	4rh	Y	550	855	—
USA	M60E3	Gas	Rot bolt	belt	8.60	1067	560	4rh	Y	550	855	—
USA	Ares EPG	Mech	Cam dwell	belt	12.50	808	597	—	No	650	—	—
USA	M134	Mech	Gatling	belt	15.90	800	559	—	No	6000	869	—
USA	EX-34	Mech	Chain	belt	13.70	940	559	—	No	570	856	—
YU	M77B1	Gas	Rot bolt	20 box	5.10	1025	535	6rh 1/31.5	No	600	840	—

7.62 × 54 mm

Country	TITLE	Action	Breech Lock	Feed System	Gun Weight (kg)	Gun Length (mm)	Barrel Length (mm)	Rifling	QCB?	Rate of Fire (rds/min)	Muzzle Velocity (m/s)	Remarks
USSR	DP	Gas	Lug	47 drum	9.10	1270	605	4rh	No	600	840	—
USSR	PKM	Gas	Rot bolt	belt	8.39	1194	603	4rh 1/36	Y	700	835	—
USSR	RP-46	Gas	Lug	belt	13.00	1283	607	4rh	Y	600	840	—
USSR	SG43/SGM	Gas	Tilt block	belt	13.60	1120	719	4rh	Y	650	800	—
YU	M84	Gas	Rot bolt	belt	10.00	1175	658	4rh 1/31.5	Y	700	825	—

7.62 × 63 mm

Country	TITLE	Action	Breech Lock	Feed System	Gun Weight (kg)	Gun Length (mm)	Barrel Length (mm)	Rifling	QCB?	Rate of Fire (rds/min)	Muzzle Velocity (m/s)	Remarks
Mex	Mendoza RM2	Gas	Rot bolt	20/32 box	6.40	1100	610	4rh	No	650	837	—
USA	M1919A4	Recoil	Lug	belt	14.06	1044	610	4rh	No	500	860	—

7.92 × 57 mm

Country	TITLE	Action	Breech Lock	Feed System	Gun Weight (kg)	Gun Length (mm)	Barrel Length (mm)	Rifling	QCB?	Rate of Fire (rds/min)	Muzzle Velocity (m/s)	Remarks
CZ	Model 37	Gas	Tilt block	belt	18.80	1105	678	4rh	Y	700	793	—
YU	M53	Recoil	Rollers	belt	11.50	1210	560	4rh	Y	1050	715	—

12.7 × 99 mm (.50)

Country	TITLE	Action	Breech Lock	Feed System	Gun Weight (kg)	Gun Length (mm)	Barrel Length (mm)	Rifling	QCB?	Rate of Fire (rds/min)	Muzzle Velocity (m/s)	Remarks
B	FN M2HB	Recoil	Lug	belt	38.15	1650	1143	—	No	—	930	—
B	FN M2HB/QCB	Recoil	Lug	belt	38.15	1650	1143	—	Y	—	930	—
SP	CIS-50	Gas	Rot bolt	dual belt	30.00	1778	1143	—	Y	600	893	—
USA	Browning M2HB	Recoil	Lug	belt	39.10	1653	1143	8rh 1/30	No	600	810	—
USA	Saco Fifty/.50	Recoil	Lug	belt	25.00	1560	—	—	Y	725	850	—
USA	GECAL	Mech	Gatling	belt	43.60	1181	914	—	No	8000	884	—

12.7 × 107 mm

Country	TITLE	Action	Breech Lock	Feed System	Gun Weight (kg)	Gun Length (mm)	Barrel Length (mm)	Rifling	QCB?	Rate of Fire (rds/min)	Muzzle Velocity (m/s)	Remarks
China	W-85	Gas	—	belt	18.50	1995	—	—	—	—	800	—
USSR	DShK 38	Gas	Lugs	belt	35.70	1588	1070	—	No	575	860	—
USSR	DShK 38/46	Gas	Lugs	belt	36.00	1588	1070	—	Y	575	860	—
USSR	NSV	Gas	Rot bolt	belt	25.00	1560	1000	—	Y	800	845	—

12.7 mm Special

Country	TITLE	Action	Breech Lock	Feed System	Gun Weight (kg)	Gun Length (mm)	Barrel Length (mm)	Rifling	QCB?	Rate of Fire (rds/min)	Muzzle Velocity (m/s)	Remarks
USA	Ares TARG	Gas	Rev	linkless	20.40	1128	927	—	No	1400+	900	—

14.5 × 114 mm

Country	TITLE	Action	Breech Lock	Feed System	Gun Weight (kg)	Gun Length (mm)	Barrel Length (mm)	Rifling	QCB?	Rate of Fire (rds/min)	Muzzle Velocity (m/s)	Remarks
China	75-1	Gas	—	belt	—	2390	—	—	—	550	995	—
USSR	KPV	Recoil	Rot bolt	belt	49.10	2006	1346	—	Y	600	1000	—

15 × 115 mm

Country	TITLE	Action	Breech Lock	Feed System	Gun Weight (kg)	Gun Length (mm)	Barrel Length (mm)	Rifling	QCB?	Rate of Fire (rds/min)	Muzzle Velocity (m/s)	Remarks
B	FN BRG	Gas	Rot bolt	dual belt	60.00	2000	—	—	Y	—	1035	—

Abbreviations:
Rot bolt – Rotating bolt; rh – Right-hand twist; lh –left-hand twist; Mech – mechanical.

Cannon

Country	TITLE	Action	Feed System	Gun Weight (kg)	Gun Length (mm)	Barrel Length (mm)	Rifling	Rate of fire (rds/min)	Muzzle Velocity (m/s)	Remarks
20 × 82										
S. Africa	GA1	Recoil	belt	39.00	1760	—	8 rh	700	720	Improved MG151
20 × 99R										
USSR	ShVaK	Gas	belt	67.88	2121	1648	8 rh 1/25.4	750	808	—
20 × 102 mm										
F	M621	DBB	belt	45.50	2207	1550	—	740	1026	—
USA	M39A3	Gas, revolver	belt	78.90	1836	1346	—	1800	—	—
USA	Mk 22 Mod 2	Rec/blowback	—	49.89	2000	1391	—	800	—	—
USA	M61A1	Gatling	linkless	115.70	1844	1524	—	7200	1110	—
USA	M168	Gatling	linked	136.00	1828	1524	—	3000	1030	—
USA	EX29	Recoil	belt	57.30	1930	1219	—	500	1006	—
USA	M197	Gatling	belt	66.50	1829	1524	—	1500	1030	—
20 × 110 mm										
CH	Hispano 820	Gas	box, belt	61.00	2565	1906	15 rh 1/26	1000	1050	—
USA	M3/M24	Gas/blowback	belt	53.50	1981	1334	—	850	—	—
YU	M1955	Blowback	60-rd drum	—	—	1400	—	800	835	—

Country	TITLE	Action	Feed System	Gun Weight (kg)	Gun Length (mm)	Barrel Length (mm)	Rifling	Rate of fire (rds/min)	Muzzle Velocity (m/s)	Remarks
20 × 110USN										
USA	Mk 11 Mod 5	Recoil/gas	belt	108.90	1994	1435	—	4200	—	—
USA	Mk 12 Mod 0	Gas/blowback	belt	52.16	1908	1219	—	1000	—	—
20 × 128 mm										
CH	Oerlikon KAA	Gas	belt	87.00	2627	1856	12 rh Inc	1000	1150	—
CH	Oerlikon KAB	Gas	20, 50 drum	109.00	3388	2400	12 rh Inc	1000	150	—
CH	Oerlikon 5TG	Gas	20, 50 drum	109.00	—	2400	12 rh Inc	1000	1200	—
20 × 139 mm										
F	M693	DBB	dual belt	70.50	2695	2065	—	900	1050	—
D	Rh202	Gas	dual belt	75.00	2612	1839	—	800	1050	—
D	Mauser B	Gas/blowback	belt	62.14	2454	1905	—	1050	—	—
S. Africa	G12	Gas/blowback	dual belt	73.50	2695	2065	15 rh	740	1050	—
CH	Hispano 804	Gas	60 rd drum	—	—	1510	—	800	825	—
CH	Oerlikon KAD	—	50 rd drum	61.00	2977	2316	15 rh 1/36	1000	1040	Formerly Hispano 820
USA	M139	Gas/blowback	belt	73.00	2565	1905	—	1050	—	—
20 × 145R										
S	Bofors M40	Recoil	magazine	—	—	1400	—	360	815	—
23 × 115 mm										
CH	Type 2H	Gas	belt	47.00	—	—	—	1150	705	—
USSR	AM-23	—	—	43.00	—	—	—	1000	690	—
USSR	GSh-23	—	—	60.00	—	—	—	3000	690	—
USSR	NR-23	Gas	—	79.00	2611	—	—	850	690	—
USSR	NS-23KM	Recoil	belt	37.00	1989	1454	10 rh 1/34.5	550	690	—
23 × 152B										
USSR	VYa	Gas	belt	67.10	2146	1651	10 rh Inc	720	970	—
USSR	ZU-23	Gas	belt	75.00	2555	1880	10 rh Inc	1000	970	—
25 × 137 mm										
F	25M811	Mech. cam	dual belt	120.00	2680	—	—	650	—	—
D	Mauser E	Gas	dual belt	109.00	2850	2100	—	1000	1100	—
CH	Oerlikon KBA	Gas	dual belt	108.00	2806	2173	18 rh Inc	600	1100	—
UK	Aden 25	Gas revolver	belt	—	2285	1700	16 rh Inc	1850	1050	—
USA	M242	Mech. chain	belt	110.50	2760	2032	—	500	—	—
USA	GE 2-barrel	Gas/ext pwr	belt	86.00	2210	—	—	2000	1097	—
USA	GAU-12/U	Gatling	belt	127.00	2113	—	—	4200	1097	—
25 × 184 mm										
CH	Oerlikon KBB	Gas	dual belt	146.00	3190	2300	18 rh Inc	800	1160	—
27 × 145B										
D	Mauser MRCA	Gas revolver	belt	92.80	2310	1700	—	1700	1050	—
30 × 86B										
UK	Aden Mk IV	Gas revolver	belt	87.00	1590	1080	16 rh Inc	1300	790	—
30 × 113B										
F	DEFA 553	Gas revolver	belt	85.00	1956	1516	—	1350	815	—
F	DEFA 554	Gas revolver	belt	85.00	2010	—	—	1800	820	—
UK	Aden Mk 5	Gas revolver	belt	87.50	1638	1080	—	1400	790	—
USA	M230	Mech. chain	belt	55.90	1638	1067	—	625	808	—
USA	ASP-30	Gas	belt	47.60	2026	1320	—	450	—	—
USA	SWC	Recoil	belt/mag	51.00	1700	—	—	450	—	—
USA	XM188E1	—	—	—	—	—	—	—	—	—
30 × 150B										
F	M781	Mech. cam	belt	65.00	1870	—	—	750	—	—
30 × 170mm										
CH	Oerlikon KCB	Gas	belt	135.50	3524	2555	18 rh 1/30	650	1080	—
UK	Rarden	Long recoil	3-shot clip	113.00	2959	2440	18 rh 1/30	single	1075	—
30 × 173 mm										
D	Mauser F	Gas	dual belt	146.50	3350	2438	—	800	1100	—
CH	Oerlikon KCA	Revolver	belt	136.00	2691	1976	21 rh Inc	1350	1030	—
USA	GAU-8	Gatling	linkless	286.00	2858	2160	—	4000	1036	—
30 × 210B										
YU	CZ30	Gas	belt	150.00	3065	2100	12 rh Inc	750	1100	—
30 × 220										
CZ	M53	Recoil	10-rd clip	—	—	2430	—	450	1000	—

Abbreviations:
rh – Right-hand twist; Inc – Increasing (Progressive) twist; Mech – mechanical; ext pwr – external power source.

Personal Weapons

Pistols
Sub-machine guns
Rifles
Light Support Weapons

Pistols

ARGENTINA

9 mm high-power FN pistol

This is now the standard pistol of the Argentinian armed services and is produced under licence from FN Herstal SA at the Argentinian Government Factory at Rosario. The pistol is very similar to the original Belgian design, but there are dimensional differences as can be seen from the data table below.

Argentine production covers three distinct variations; the 'Militar' model is the standard fixed sight, blued-finish pattern with plastic grip surfaces; the 'Comando' model is of blued external finish but with polished barrel, ramp foresight, fully adjustable rearsight, walnut grip surfaces , spur-type hammer, and no lanyard ring while the 'Trofeo' has silver-plated and engraved exterior, ramp foresight, adjustable rearsight, ring hammer and no lanyard ring.

DATA
Cartridge: 9 × 19 mm Parabellum
Operation: recoil, semi-automatic
Method of locking: projecting lugs
Feed: 13-round box magazine
Weight, empty: 910 g
Length: 197 mm
Barrel: 118 mm
Rifling: 6 grooves, rh, one turn in 27.7 calibres (250 mm)
Sights: (fore) adjustable blade; (rear) laterally adjustable notch (and see text)

Manufacturer
Fabrica Militar de Armas Portatiles, 'Domingo Matheu', Avenida Ovidio Lagos 5250, 2000-Rosario.

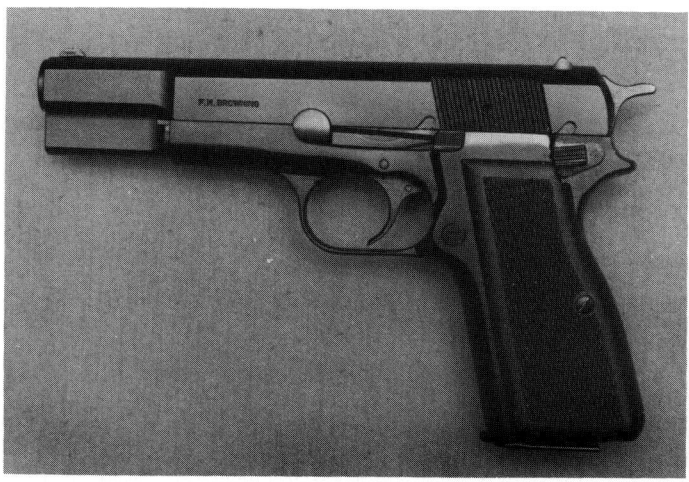
Argentine manufactured Browning GP35 'Militar'

Status
Production.
Service
Argentinian armed forces.

.45 automatic Model 1927 pistol

This pistol was a copy of the Colt Model 1911A1 (which see). It was manufactured at Fabrica Militar de Armas Portatiles, Rosario and is still in service with some units of the Argentinian Army. The total quantity supplied, between 1927 and 1942, is believed to have been 38 000.

.45 Ballester Molina pistol

This pistol was an unauthorised copy of the Colt M1911A1 manufactured in the late 1930s. It had some slight constructional differences from the Colt, but operated in the same basic manner. The Ballester Molina is now obsolete and is unlikely to be encountered, but a full description can be found in earlier editions of *Jane's Infantry Weapons*.

AUSTRIA

9 mm Steyr GB pistol

This is a delayed-blowback weapon in which the delay is achieved by gas action. The barrel exterior is shaped about halfway along its length into a piston-head which fits tightly into a cylinder formed in the rear of the muzzle-locking cap. In front of this piston-head is a small port connecting the interior of the barrel with the annular space between the barrel and the locking-cap cylinder. When the pistol is fired gas passes through this port and fills the space at high pressure; the normal recoil action tries to blow the slide back, pulling the muzzle cap and its cylinder over the barrel which is fixed to the frame. The pressure in the annular space resists this movement, so that the recoil of the slide is delayed until the bullet has left the muzzle and the chamber pressure has dropped to safe limits.

The lockwork is double-action; on loading, the hammer is cocked and can be safely lowered by operating the de-cocking lever which lies on the left side of the slide. This locks the firing pin to the rear, interposes a safety block and then drops the hammer. Thereafter the pistol can be fired by either thumb-cocking the hammer or by pulling the trigger. The magazine holds 18 rounds and an additional cartridge can be manually loaded into the chamber to give a capacity of 19 shots. The barrel is chrome-plated inside and out for maximum resistance to wear.

The early models were all-steel, including the butt-grip plates, but production models now have grip plates made of moulded plastic material. The sights are a modified form of the Stavenhagen pattern, using a blade front sight with a luminous dot inset in its rear surface and a square-notch rearsight with luminous dots at each slide of the notch.

To strip the GB pistol:
Remove magazine, operate slide, check chamber is empty.
Turn the dismantling lever (right side of frame) downward.
Turn the muzzle lock cap anti-clockwise (viewed from the front) until it unlocks and remove it.
Withdraw the return spring guide tube from front of frame.

Pull back the slide until it stops, lift rear end, slide it forward and remove it from the frame.
No further stripping should normally be necessary.
To re-assemble:
Place the side over the barrel, pull it to the rear until it can be pressed down into engagement with the frame, slide it forward.
Insert spring guide tube.

9 mm Steyr GB pistol

Press slide fully forward, insert muzzle lock cap and turn clockwise (seen from the front) until it locks.

Turn dismounting lever upward. Press de-cocking lever to allow the hammer to fall. Insert magazine.

DATA
Cartridge: 9 mm Parabellum
Operation: gas-delayed blowback
Method of locking: none
Feed: detachable 18-round magazine
Weight: (empty) 845 g; (full magazine) 340 g
Length: 216 mm

Barrel: 136 mm
Sights: (foresight) blade with luminous point; (rearsight) rectangular notch with 2 luminous points
Muzzle velocity: 360-420 m/s depending on cartridge
Trigger pull: 2.2 and 6.5 kg

Manufacturer
Steyr-Mannlicher GmbH, a member of the Steyr-Daimler-Puch AG group, Werke Steyr, Postfach 1000, A-4400 Steyr.
Status
Current. No longer manufactured.

9 mm Glock 17 and 17L pistols

This pistol has been adopted by the Austrian Army and Police, by Norway and by several international police and security forces. It is of light weight and simple design, with only 33 parts including the magazine, and can be completely stripped in under one minute using only a pin or nail. The complete receiver is made of high-resistance polymer material, in monocoque design, providing a low and smooth recoil. Heat resistance is effective up to 200°C.

The Glock 17 is a locked-breech pistol using the Colt-Browning system of cam-controlled dropping barrel; connection with the slide is by a squared section around the barrel breech which engages with the slide and ejection port. Firing is by striker, controlled by the operation of the trigger; the first pressure on the trigger disengages the trigger safety, then (5 mm of travel) cocks the striker (which is in the half-cocked and secure position at all times when the weapon is ready to fire) and releases two internal safety devices, the integral firing pin lock and the safety ramp; the second pressure (2.5 mm travel) releases the striker to fire the pistol. The pressure required to operate the trigger can be adjusted between 2 and 4 kg. There is no conventional manual safety catch, since the trigger safety, automatic firing pin lock and automatic trigger bar safety will not permit the firing of a cartridge unless the trigger is correctly pulled. After each shot the gun returns to a safe condition until the trigger is pulled again.

Model 17L
The Model 17L is to the same design as the Model 17 except that the barrel and slide are longer; with the exception of these two items, all other parts are completely interchangeable with the Model 17.

DATA
DATA FOR 17; WHERE 17L DIFFERS, SHOWN IN PARENTHESIS:
Calibre: 9 × 19 mm
Operation: short recoil, self-loading
Method of locking: cam-dropped barrel
Feed: 17-round box magazine
Weights
  Pistol: 620 g (661 g) empty, without magazine
  Empty magazine: 59 g
  Full magazine: 270 g
Lengths
  Overall: 188 mm (228 mm)
  Barrel: 114 mm (153 mm)
  Sight radius: 165 mm (204 mm)
Mechanical features
Rifling: hexagonal profile, rh twist, one turn in 250 mm
Sights: fixed or click-adjustable; front blade, rear notch
Performance
  Muzzle velocity: ca 360 m/s
  Muzzle energy: ca 500J

Manufacturer
Glock GmbH, Hausfeldstrasse 17, A2232 Deutsch-Wagram.

Glock 17 9 mm military pistol

(Top to bottom): 9 mm Glock 19, 17 and 17L pistols compared

Status
Current. Production.
Service
ONORM, DIN, CIP and SAAMI standards. NATO Stock Number 1005-25-133-6775. Introduced into NATO countries, armed forces, police departments and special forces units in more than 42 countries. Approximately 100,000 pistols are now in daily use by about 2000 US police departments and other government, federal and local agencies.

9 mm Glock 18 pistol

The Glock 18 has been developed from the Model 17 by adding a fire selector assembly and giving the weapon a larger magazine capacity, so turning it into a machine pistol capable of automatic fire. For security reasons the main components of the Glock 18 are not interchangeable with the Glock 17.

The Glock 18 provides for special forces, SWAT teams and military personnel the option of firing trigger-controlled bursts, if required, from a fully concealable lightweight pistol, without the need for a shoulder stock. The original 17-round magazine, plus the new 19-round magazine, of which two are considered to be carried in a separate double-magazine pouch, provide the officer with an immediate firepower of 56 rounds (one being in the chamber). Mass production of an optional 33-round magazine is now in progress.

The manufacturer recommends that only special law enforcement, SWAT and specially-trained military personnel should, as a general rule, be allowed to use the Glock 18, since the effective and safe operation of a pistol in full automatic mode requires special training and high personal discipline.

DATA
Cartridge: 9 × 19 mm Parabellum
Operation: short recoil, selective fire
Locking system: tilting barrel
Feed: 17, 19 or 33-round box magazines
Weight: 636 g empty, with empty magazine
Length: 223 mm
Barrel length: 114 mm
Rifling: hexagonal, rh twist
Rate of fire: ca 1300 rds/min

9 mm Glock 18; note fire selector at rear end of slide

Manufacturer
Glock GmbH, Hausfeldstrasse 17, A2232 Deutsch-Wagram GmbH.
Status
Current. Production.

9 mm Glock 19 pistol

The Glock 19 is generally to the same design as the Glock 17 but is smaller and more compact; as a result, there are several major components which are not interchangeable with those of the Glock 17 even though they are of the same basic design.

DATA
Cartridge: 9 × 19 mm Parabellum
Operation: short recoil, semi-automatic
Method of locking: cam-dropped barrel
Feed: 15-round box magazine (17-round optional)
Weights
  Pistol: 594 g empty, without magazine
  Empty magazine: 56 g
  Full magazine: 242 g
Lengths
  Overall: 177 mm
  Barrel: 102 mm
  Sight radius: 152 mm
Mechanical features
Rifling: hexagonal profile, rh twist, one turn in 250 mm
Sights: fixed or click-adjustable; front blade, rear notch
Performance
  Muzzle velocity: ca 360 m/s
  Muzzle energy: ca 500J

9 mm Glock 19 compact pistol

Manufacturer
Glock GmbH, Hausfeldstrasse 17, A2232 Deutsch-Wagram.

Status
Current. Production.

BELGIUM

Browning pistols

The most famous name in Belgian pistol design is that of John Moses Browning who left his native USA and went to Herstal to work with Fabrique Nationale d'Armes de Guerre.

This partnership produced a number of designs which achieved wide use and were manufactured on a large scale. Chief among these designs were the following:
Model 1900: blowback design, in 7.65 mm (.32 Automatic Colt Pistol(ACP)) calibre;
Model 1903: blowback design, in 9 mm Browning long calibre;
Model 1910: blowback design, in 7.65 mm (.32ACP) and 9 mm Short(.380ACP) calibre;
Model 1922: blowback design, in 7.65 mm (.32ACP) and 9 mm Short(.380ACP) calibre.

Of these self-loading pistols, only a few Model 1900 weapons, manufactured in North Korea, a number of Model 1903 in Swedish service,

9 mm Browning Model 1900 pistol

and possibly some Model 1903 pistols in Turkey, survive in official use. The Model 1922 was used in the Netherlands, Yugoslavia and by the Germans in the Second World War under the title of Pistole 626 (B).

9 mm Browning Model 1903 pistol

7.65 mm Model 1910 Browning pistol

This pistol was manufactured by FN Herstal SA using the 7.65 mm (.32 Automatic Colt Pistol (ACP)) cartridge and also the 9 mm Short (.380 ACP) cartridge. It was a striker fired, blowback-operated pistol with the return spring around the barrel. It had a grip safety, a manual safety and the Browning disconnector (see description of the High-power pistol).

DATA
Cartridge: 7.65 mm (.32ACP), 9 mm Short (.380ACP)
Operation: blowback, semi-automatic
Method of locking: nil
Feed: 7-round detachable magazine (7.65 mm)
Weight: 582 g
Length: 152 mm
Barrel: 89 mm
Rifling: 6 grooves rh
Sights: (foresight) blade; (rearsight) notch
Muzzle velocity: 299 m/s (.32ACP)
Rate of fire: 35 rds/min
Effective range: 40 m

Manufacturer
FN Herstal SA, Branche Défense et Sécurité, B-4400 Herstal.
Status
Obsolete. No longer in production.

7.65 mm Model 1910 Browning pistol

Service
Still in occasional use in some countries.

7.65 mm 140 double-action FN pistol

This pistol is intended as a general-purpose defensive pistol for military and police forces. It is a blowback-operated self-loading pistol with some new features and a high standard of reliability. The frame is light alloy, with a steel slide and the overall weight is kept as low as possible. The magazine capacity is unusually large for a pistol of this size, reflecting the experience gained over the years with the High-power pistol. The trigger-guard is generously large, to accommodate the largest finger without presenting any difficulties for a fast draw should this be needed. Another assistance in drawing the pistol from a pocket or holster is the fact that the corners are rounded as much as possible so that nothing will catch in clothing.

The trigger is double-action and the pistol can be carried with perfect safety when set on double-action. However, there is another safety as well and setting the manual safety lever will disconnect the trigger from the hammer. To fire all that is necessary is to move the lever with the thumb and pull the trigger. A particular feature of this safety lever is that it projects on both sides of the slide and so can be used by a left- or right-handed shooter without any alteration to the pistol.

The pistol is made in both 7.65 mm and 9 mm Short calibres.

DATA
Cartridge: 9 mm Short or 7.65 mm
Operation: blowback
Feed: (9 mm Short) 13-round box magazine; (7.65 mm) 12-round box magazine
Weight, unloaded: 640 g
Length: 173 mm
Rifling: 6 grooves rh
Muzzle velocity: (9 mm Short) 270—290 m/s; (7.65 mm) 290—300 m/s

Manufacturer
FN Herstal SA, Branche Défense et Sécurité, B-4400 Herstal.
Status
Current. Production.
Service
Belgian and some foreign police forces.

7.65 mm 140 double-action FN pistol

9 mm High-power FN Browning pistol

The Browning High-power pistol was designed in 1925 by John M Browning and granted a US Patent in February 1927, some three months after his death. It was produced in 1935 and in the USA it is still often referred to as the 1935 Model. When first introduced in that year, two models were available: the 'ordinary model' had fixed sights and the alternative, which had a tangent rearsight graduated to 500 m and a dovetail slot in the rear of the pistol grip to take a wooden shoulder stock attached to a leather holster, was known as the 'adjustable rearsight model'. In Belgium the pistol is known as the Grande Puissance. The High-power has been manufactured in Belgium by FN Herstal SA, near Liège, for sale world-wide. It was in service in Belgium, Denmark, the Netherlands, Lithuania and Romania. During the Second World War the pistol was made in Liège for the use of the German SS troops and by the John Inglis Company in Toronto for use by Australian, British, Canadian and Chinese troops.

Construction

The pistol consists of the frame, the slide and the barrel. The barrel fits into the slide and on its underside, beneath the chamber where the barrel is strongly reinforced, is a lug into which is cut a forward and upward sloping camway. A pin is riveted to the frame in such a way that when the barrel moves backwards the pin enters the upward sloping path and pulls the rear of the barrel down. The receiver also has a stop to arrest the slide, the firing mechanism and the magazine well with the magazine-retaining catch. The trigger mechanism consists of a trigger pivoted at its top front edge so that the rear end of the trigger assembly rises when the trigger is pulled. This forces a short straight trigger lever up under the front end of a long horizontal, centrally-pivoted sear lever mounted in the slide. The lowering of the rear end of this sear lever pivots the sear out of engagement with the bent of the hammer. The safety catch, mounted on the left of the frame behind the pistol grip, is pivoted to lock the slide and a projection on its lower side fits into the rear of the sear to prevent the release of the hammer. There is no grip safety fitted.

The slide contains the integral breech-block at the rear. The front end has a single recess for the muzzle and a solid face below in which the return spring is seated. The slide travels on the frame and has an ejection opening on the right side. Within the top of the slide are the recesses in which the barrel locking lugs engage.

9 mm High-power FN Browning pistol

Operation

On firing, the gas pressure forces the slide rearwards and the barrel is pulled back since its locking lugs are engaged in the recesses in the slide. After 5 mm of free travel the cam path below the barrel rides over the pin and the rear end of the barrel is pulled down to unlock it from the slide, and the barrel movement is terminated. The slide continues back to ride over the hammer and the extractor on the breech-block pulls out the empty case which hits the ejector and is thrown out of the ejection port on the right of the slide. As the slide recoils the return spring is compressed around its guide rod. Rearward movement ceases when the slide hits the stop in the frame. The slide is then driven forward by the compressed return spring and the breech-block feeds a round from the magazine into the chamber. The breech-block forces the barrel forward and the cam forces the breech up and holds it up when the locking lugs on the barrel enter the recesses in the slide ceiling.

When the firer pulls the trigger again, the trigger lever is raised, the tail of the sear lever is forced up and the nose is forced down to contact the sear in the

Trigger pressure maintained

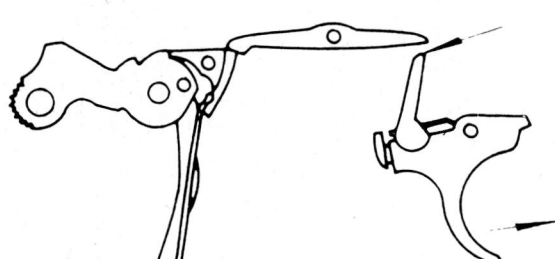

Trigger released

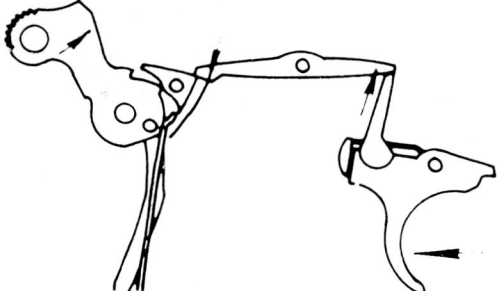

Trigger pressed

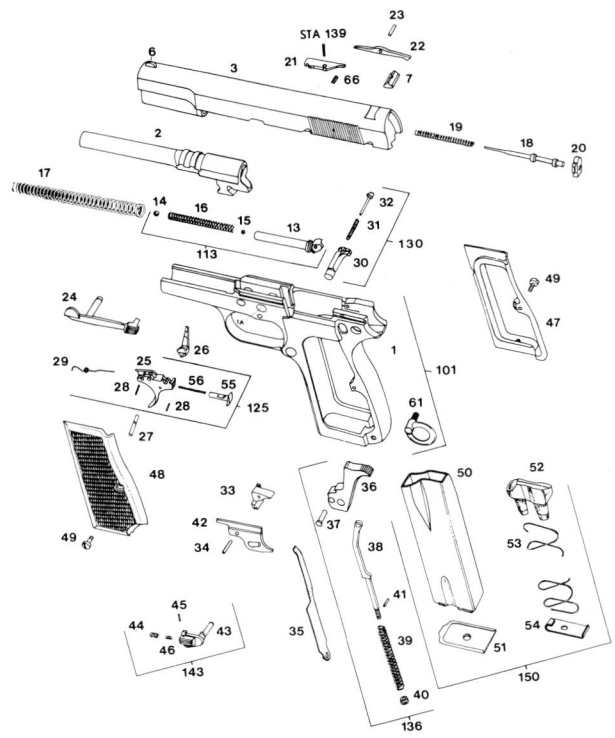

Component parts of High-power FN Browning pistol

(1) *frame with ring for lanyard* **(2)** *barrel* **(3)** *slide with foresight and cap* **(7)** *rearsight* **(13)** *return spring guide* **(14)** *return spring guide cap* **(15)** *ball* **(16)** *spring of return spring guide* **(17)** *return spring* **(18)** *firing pin* **(19)** *firing pin spring* **(20)** *firing pin retaining plate* **(21)** *pivoting extractor* **(22)** *sear lever* **(23)** *sear lever axis pin* **(24)** *slide stop* **(25)** *trigger* **(26)** *trigger lever* **(28)** *trigger and magazine safety pin* **(29)** *trigger spring* **(30)** *magazine stop* **(31)** *magazine stop spring* **(32)** *magazine stop spring guide* **(33)** *sear* **(34)** *sear pin* **(35)** *sear spring* **(36)** *hammer* **(37)** *hammer pin* **(38)** *hammer strut* **(39)** *hammer spring* **(40)** *hammer spring support* **(41)** *hammer strut pin* **(42)** *ejector* **(43)** *safety* **(50)** *magazine body* **(51)** *magazine base* **(52)** *magazine platform* **(53)** *magazine spring* **(54)** *magazine bottom plate catch* **(55)** *magazine safety* **(56)** *magazine safety spring*

LAST ROUND FIRED

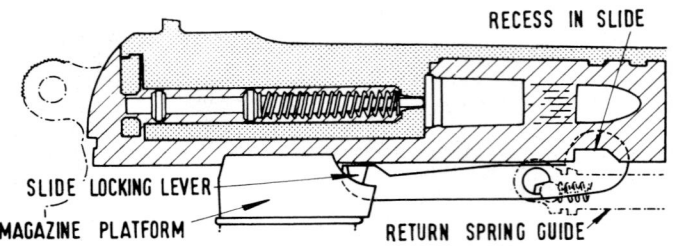

SLIDE HELD TO REAR

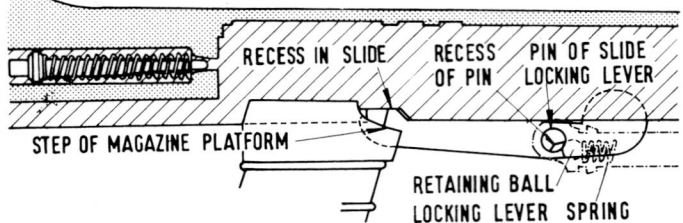

Holding-open device

Grasp slide, press catch out of engagement and remove slide and barrel assembly from the front.
Push recoil-spring guide towards muzzle to release the head of guide from the barrel and remove it.
Remove barrel from breech end.
Remove stocks.

DATA
Cartridge: 9 mm Parabellum
Operation: recoil, semi-automatic
Method of locking: projecting lug
Feed: 13-round box magazine
Weight, empty: 882 g
Length: 200 mm
Barrel: 118 mm
Rifling: 6 grooves, rh, one turn in 250 mm
Sights: (fore) adjustable barleycorn; (rear) square notch
Sight radius: 159 mm
Muzzle velocity: 350 m/s
Chamber pressure: 212 MP

Manufacturer
FN Herstal SA, Branche Défense et Sécurité, B-4400 Herstal. Also under licence by Fabrica Militar de Armas Portatiles, Domingo Matheu, Rosario, Argentina.
Status
Current. Production.
Service
Belgian forces and most Western nations. In use in more than 55 countries.

receiver and rotate it out of engagement with the notch of the hammer. The mainspring, lying in the rear of the pistol grip, drives the hammer forward to strike the firing pin driving it forward to hit the cap and fire the cartridge.

When the ammunition is expended, the magazine follower forces up the slide stop and the slide is held to the rear. When a fresh magazine has been inserted into the pistol grip, the slide stop can be pressed down to allow the return spring to force the slide forward. When the magazine is taken out a spring-loaded safety lever is forced out into the magazine well: this lever is linked to the trigger lever and forces it forward from beneath the tail of the sear lever. Thus, when the magazine is removed and a cartridge is left in the chamber, the pistol cannot be inadvertently discharged and the magazine must be replaced before firing is possible.

Stripping
Press magazine-release catch on receiver behind trigger and drop magazine out of pistol grip.
Operate slide and check chamber is empty.
Retract slide and push safety catch up into second notch in the slide.
Press up slide stop and withdraw it on the left of the receiver.

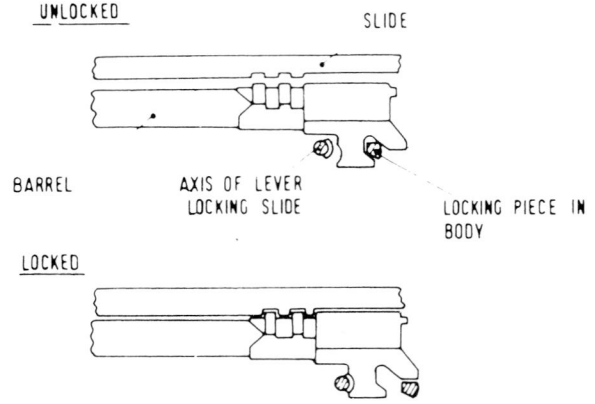

Breech locking and unlocking

9 mm FN High-power Mark 2 pistol

The High-power Mark 2 pistol is a variant of the standard 9 mm High-power described above. Its technical characteristics are the same. It is fitted with anatomical grip plates, an oblong ambidextrous safety catch, and wider sights and has an external phosphated anti-glare finish.

DATA
As for High-power Mark 1 (above)

Manufacturer
FN Herstal SA, B-4400 Herstal.
Status
Obsolete. Production discontinued.
Service
Belgian police forces.

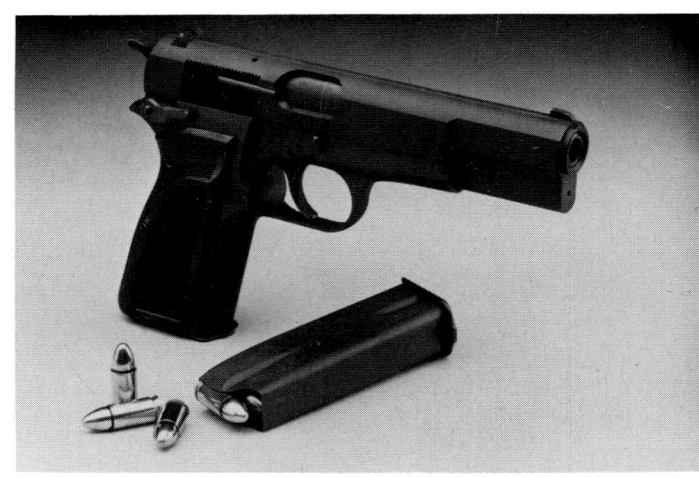

9 mm FN High-power Mark 2 pistol

9 mm FN High-Power Mark 3 pistol

This model is more or less the Mark 2 with improvements. These pistols are now made on new machinery, which has improved the quality, and the frame and slide have been slightly re-dimensioned to make them stronger and obviate the danger of cracking. The ejection port has been enlarged and re-contoured, the grips re-designed, and the sights are now mounted in

dovetails for easier adjustment and so that the rear sight can be removed and replaced with target sights should the user so wish.

A Mark 3S version is manufactured for police use; this is exactly the same but also incorporates an automatic firing pin safety system. This model is marketed by Browning SA.

9 mm Browning Mark 3 pistol

9 mm Browning Mark 3 dismantled

The dimensions of the Marks 3 and 3S are the same as those of the Mark 1 High-Power.

Manufacturer
FN Herstal SA, B-4400 Herstal.

Status
Current. Production.

9 mm FN BDA 9 pistol

The BDA 9 is derived from its predecessor, the FN High-Power and functions in the same way, using short recoil and a cam beneath the breech to disengage the barrel from the slide.

It differs in having a double-action trigger and a hammer decocking lever in place of the safety catch; this decocking lever is duplicated on both sides of the frame and can thus be used with either hand. The magazine release is normally fitted for right-handed users, but it can easily be removed and re-assembled for left-handed use. The trigger guard is shaped to facilitate a two-handed grip.

The pistol is loaded in the usual way, by inserting a magazine, pulling back the slide and releasing it to chamber the first round, leaving the hammer cocked. The pistol can then be fired, or, by pressing up the decocking lever, the hammer can be safely lowered. Pushing up the decocking lever causes a decocking safety to interpose itself between the hammer and the firing pin, and a braking lever slows down the descent of the hammer so as to place less strain upon it. There is an automatic firing pin safety system which locks the firing pin except during the last movement of pressing the trigger to fire the pistol. Once the hammer has been lowered, the pistol can be safely carried, and it can be instantly fired by simply pulling through on the trigger for

9 mm FN BDA 9 double-action pistol

double-action firing. Once a shot has been fired, subsequent shots are in the single-action mode.

Dismantling is carried out in exactly the same way as for the standard High-power pistol.

DATA
Cartridge: 9 × 19 mm Parabellum
Operation: short recoil
Locking: dropping barrel, cam actuated
Feed: 14-round box magazine
Weight: 905 g empty, with empty magazine
Length: 200 mm
Barrel: 118 mm
Rifling: 6 grooves, right hand, one turn in 250 mm
Sights: (fore) fixed blade; (rear) fixed notch, adjustable laterally by an armourer
Muzzle velocity: 350 m/s

Manufacturer
FN Herstal SA, B-4400 Herstal.
Status
Current. Production.

9 mm FN BDA 9 pistol dismantled

BRAZIL

9 mm and .45 IMBEL pistols

This is a re-designed version of the .45 M1911A1. The pistol is made in both .45 ACP and 9 mm Parabellum calibres, the former for competition and export, the latter for use by the Brazilian Armed Forces. As can be seen from the photographs, the standard pistol is in all respects similar to the original Colt; the competition model uses a special barrel bushing and barrel contour, available only to special order, to improve accuracy.

In 1986 IMBEL developed a .45 ACP pocket pistol, based upon the M973 design. The new pistol, known as the .45 ACP IMBEL—MD1, has a smaller slide and frame than standard and entered series production in 1989.

DATA
Cartridge: .45 ACP or 9 × 19 mm Parabellum
Operation: Recoil, semi-automatic, single action
Feed: 8-round (9 mm) or 7-round (.45) box magazine
Length: 218 mm (9 mm) or 216 mm (.45)
Barrel: 4 grooves rh (9 mm); 6 grooves lh (.45)
Muzzle velocity: 349 m/s (9 mm); 250 m/s (.45)

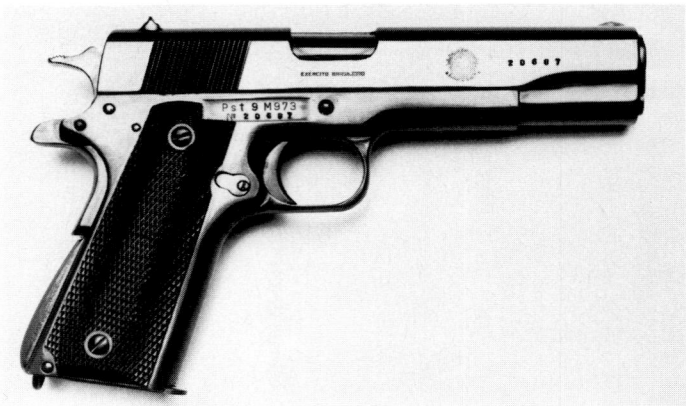

9 mm Imbel M973 pistol

Manufacturer
Indústria de Material Bélico do Brasil – IMBEL, Avenida dos Nações Unidas 13.797, Bloco III, 1° Andar, CEP 04794 São Paulo SP.
Status
Current. Production.
Service
Brazilian Armed Forces and export (see text).

9 mm Imbel M973 pistol partly dismantled

CANADA

9 mm High-power FN Browning pistol

This pistol was made under licence by John Inglis and Co of Toronto during the Second World War. Production was originally set up to supply the Chinese Nationalist Army, but the weapon was issued also to airborne forces and commandos of the British Army. Although Canadian manufacture was based strictly on the pattern of the original FN pistol there were several early variations, but the only version now issued to Canadian Forces is the No 2 Mark 1*. This has a non-adjustable rearsight with a U-notch, and it has an improved ejector and extractor which is not interchangeable with earlier patterns of pistol.

9 mm Browning No 2 Mark 1 pistol*

CHILE

.38 Special FAMAE revolver

This is a conventional design of solid-frame, double-action five shot revolver. The cylinder is released by a catch on the left side of the frame and swings out for extraction and reloading. A slightly smaller version in .32 Long Colt chambering is also manufactured.

DATA
Cartridge: .38 Special
Operation: double-action revolver

Feed: 5-shot cylinder
Weight, empty: 630 g
Length: 225 mm
Barrel: 82 mm
Sights: (fore) blade, (rear) frame notch

Manufacturer
FAMAE Fabricaciones Militares, Pedro Montt 1606, Santiago.
Status
Current. Production.

FAMAE .38 Special revolver

CHINA, PEOPLE'S REPUBLIC

7.62 mm Type 54 pistol

This is a direct copy of the Soviet 7.62 mm Tokarev TT33 pistol and is the standard pistol in the Chinese Army. The Chinese version may be distinguished from the Soviet or Polish pistols by the serrations on the slide. The Soviet and Polish pistols have a series of alternate wide and narrow vertical cuts and the Chinese pistols have uniform narrow vertical serrations. The Chinese pistol can be further distinguished from the Hungarian Model 48 and Yugoslav Model M57 (which also have uniform narrow slots) by the Chinese markings on the receiver or top of the slide. The Yugoslav pistol carries '7.62 mm M57' on the left side of the slide, and the Hungarian Model 48 has an emblem consisting of a star, wheatsheaf and hammer, in a wreath, on the grip.

DATA
Cartridge: 7.62 × 25 mm 'P' cartridge
Operation: recoil, semi-automatic
Method of locking: projecting lugs
Feed: 8-round box magazine
Weight, empty: 890 g
Length: 195 mm
Barrel: 115 mm
Rifling: 4 grooves rh
Sights: (foresight) blade; (rearsight) notch
Muzzle velocity: 420 m/s

Manufacturer
State arsenals.

7.62 mm Type 54 pistol

Status
Current. Production.
Service
Chinese armed forces.

7.65 mm Type 64 and Type 67 silenced pistols

The Type 64 is a pistol produced solely in silenced form. It may be used either as a manually operated single-shot arm or as a self-loader.

When the maximum silencing effect is required the selector bar is pushed to the left and the lugs of the rotating bolt in the slide engage in recesses in the receiver and the weapon fires from a locked-breech. After the round has been fired the slide is hand-operated to unlock the bolt, retract the slide and extract the fired case. When the selector bar is pushed to the right, the locking lugs do not engage in the recesses in the receiver and the pistol functions as a blowback-operated semi-automatic. This results in a more noisy method of operation as the slide reciprocates and the empty case is ejected. The cartridge is 7.65 mm × 17. It is rimless and unique. No other round can be used in this pistol.

The silencing effect is obtained by placing a large bulbous attachment on the front of the receiver extending well forward of the muzzle. The gases leave the muzzle and expand into a wire-mesh cylinder surrounded by an expanded metal sleeve. The bullet passes through a series of rubber discs which trap the gases. Used as a single-shot manually operated pistol it is extremely quiet but its reduced muzzle velocity greatly affects its powers of penetration.

7.65 mm Type 64 pistol

7.65 mm Type 64 pistol stripped

7.65 mm Type 67 pistol

7.65 mm Type 64 pistol: baffles, gauze and mesh make up interior of silencer

The Type 67 is an improved model; in essentials it is the same as the Type 64 above but the silencer has been re-shaped into a plain cylindrical unit which makes the weapon easier to carry in a holster and gives it better balance. There are some small changes in the assembly of the silencer but the principle of operation remains the same.

DATA
Type 64 (Type 67)
Cartridge: 7.65 mm × 17 rimless
Operation: (a) manual, single-shot or (b) blowback, semi-automatic
Method of locking: (a) rotating bolt or (b) nil
Feed: 9-round box magazine
Weight: 1.81 kg (1.02 kg)
Length: 222 mm (225 mm)
Barrel: 95 mm (89 mm)
Sights: (foresight) blade; (rearsight) notch
Muzzle velocity: 205 m/s (181 m/s)

Manufacturer
State arsenals.
Status
Current. Production.
Service
Chinese armed forces.

7.62 mm pistol Type 77

The Type 77 is a light pistol intended for equipping senior officers, military attachés and police. It fires the rimless 7.62 mm Type 64 cartridge, the round used with the Type 64 silenced pistol (above).

The design is rather unusual, reviving a long-defunct German system of operation. The pistol is a simple blowback weapon, and it can be carried with a loaded magazine and empty chamber. When required for use, the trigger finger is hooked around the front edge of the trigger-guard and the guard pulled to the rear. This moves the slide back against its return spring and if the finger is then released the slide will run back and chamber a cartridge. By simply transferring the finger to the trigger the weapon is thus ready to fire almost instantly. The trigger-guard is not permanently linked to the slide and thus does not move during normal firing. This method of operation also has the advantage that in the event of a misfire the cartridge can be ejected single-handed.

The chamber is fluted in an endeavour to reduce the slide recoil velocity by reducing the gas pressure acting on the base of the cartridge case.

There have been several 'single-handed cocking' pistol designs. This version, one of the few to have any commercial success, was originated in 1913-15 and marketed by Bergmann and later by Lignose in Germany in the early 1920s.

7.62 mm pistol Type 77

DATA
Cartridge: 7.62 mm × 17 mm
Operation: blowback
Feed: 7-round magazine
Weight: 500 g

Length: 148 mm
Muzzle velocity: 318 m/s

Manufacturer
China North Industries Corp, PO Box 2137, Beijing.

7.62 mm machine pistol Type 80

This pistol is based on the long-obsolete Mauser Model 712 (System Westinger) machine pistol. This was manufactured from 1932 to 1936 and an estimated 70 000 were supplied to China. Numbers were also locally manufactured in the late 1930s, but the design then fell into disuse.

The Type 80 uses the same basic mechanism, an internal reciprocating bolt locked by a plate beneath it, with an external hammer, but has a canted

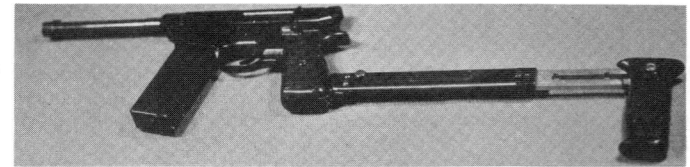

7.62 mm Type 80 pistol with folding stock attached

magazine (probably for more reliable feed) and a lengthened barrel. There are also a few minor differences in the contours of the frame and the grip angle (long a point of dispute with Mauser owners) has also been improved. A clip-on folding stock is available, as is a bayonet, and with the stock attached the weapon can be used at ranges up to 150 m with good effect.

DATA
Calibre: 7.62 Tokarev (7.63 × 25 mm Mauser)
Operation: locked breech, recoil operated, selective fire
Feed: 10- or 20-round detachable magazines
Weight: 1.10 kg
Length: 300 mm
Rate of fire: 60 rds/min (effective)
Muzzle velocity: 470 m/s

Manufacturer
China North Industries Corp, PO Box 2137, Beijing.
Status
Current. Production.

7.62 mm machine pistol Type 80

9 mm pistol Type 59

In spite of a slight difference in weight, this appears to be no more than a locally manufactured copy of the Soviet PM Makarov pistol. It is a blowback weapon with double-action trigger. For further information on the design, refer to the Soviet Makarov entry.

DATA
Cartridge: 9 × 18 Soviet (known as Type 59 in China)
Operation: blowback, double-action
Feed: 8-round magazine
Weight: 730 g
Length: 162 mm
Barrel: 93.5 mm
Sights: (fore) blade; (rear) U-notch
Sight radius: 130 mm
Muzzle velocity: 314 m/s

Manufacturer
China North Industries Corp, PO Box 2137, Beijing.

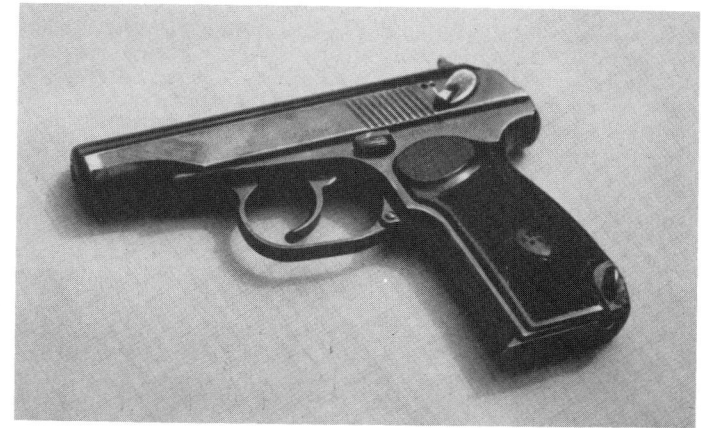

9 mm pistol Type 59

.22LR Firing Combat Knife

This unusual weapon appears to be a Chinese version of a Spetsnaz pistol/knife, the existence of which is known though not the details and which is said to be in use by Czech paratroops. The Chinese Firing Combat Knife contains four barrels in the grip, chambered for the standard .22LR rimfire cartridge. These barrels are aligned so that two lie each side of the blade, with the muzzles at the base of the blade. By pressing in a spring-loaded catch at the rear of the handle, the rear cap is removed, exposing the four chambers. After loading, the cap is replaced and locked in place. There is a serrated rotary catch at the front end of the handle which acts as a safety catch. Rotated clockwise, this locks the trigger (which forms half the handguard of the knife) and makes the pistol inactive, so that the knife can be used. When rotated anti-clockwise, the safety catch uncovers a red warning mark alongside the foresight and allows the trigger to function. Pulling the trigger then fires one barrel at a time. To reload, the rear cap is removed and the trigger pulled, whereupon the empty cases are ejected from two barrels; a further pull ejects the cases from the other two barrels. The safety catch is applied and the chambers can then be reloaded.

A notch rearsight is built into the rear end of the handle and a foresight forms part of the handguard. To fire, the pistol is held at arms length so that aim can be taken.

Removing cap of Firing Knife exposes loaded chambers. Note also sight and trigger which form handguard

DATA
Cartridge: .22LR rimfire
Operation: Single-shot
Feed: 4 pre-loaded barrels
Weight, empty: 464 g
Length overall: 262 mm
Barrels: 86.3 mm
Rifling: 4 groove, rh, one turn in 224 mm
Blade length: 140 mm
Effective range: 30 m

Manufacturer
China North Industries Co, PO Box 2137, Beijing.
Status
Development.

Firing Combat Knife in firing position

CZECHOSLOVAKIA

7.62 mm Model 52 pistol

Model 52 was designed to fire the Model 48 cartridge which has the same physical dimensions as the Mauser 7.63 × 25 mm round, and the Soviet Type P pistol cartridge. The Czechoslovak round, however, has a larger propellant load than either the Soviet military round or most commercial cartridges of this type. This is probably the reason for the use of the very unusual, but very rigid, roller locking system employed.

Operation
The roller locking system which is used is unique to the Model 52 pistol. There is a rectangular barrel lug over the chamber, and carried in the lower surface of this is a locking block and two rollers. Two slots in the barrel lug permit the rollers to move outwards. At the rear end of the locking block a tongue is formed and a projection, the unlocking lug, reaches down and rests against a projection in the frame. When barrel and slide are fully forward the locking block forces the two rollers outwards and they enter recesses in the slide, thus locking barrel and slide together. On firing, the barrel and slide begin moving rearwards. The unlocking lug remains in contact with the projection in the frame, and so the locking block remains still. After 5 mm of recoil the rollers are carried over the narrow tongue of the locking block and are forced into their housings in the barrel lug by the edges of the recesses in the slide, thus unlocking the barrel from the slide.

The slide continues to the rear, cocking the hammer, and the barrel comes up against the barrel stop in the frame and is halted. One advantage of this system is that the barrel movement is purely one of reciprocation and there is no tilting of the barrel to unlock. Thus there is no need for a loose fit at the front barrel bearing and there is very little wear apparent, even after a large number of rounds have been fired. This results in a long accuracy life for this weapon.

The backward movement of the slide terminates when the underside of the front barrel bearing in the slide comes up against the receiver and the return spring then forces the slide forward. The feed horn below the breech face picks up a round from the magazine and the breech face pushes it against the bullet guide and into the chamber.

When the recesses in the slide come up to the rollers the latter are forced outwards by the barrel lug, along the tongue of the locking block and in the final 5 mm of forward movement of both barrel and slide, the full width of the locking block holds the rollers so that they are engaging both the slide and the barrel lug. This provides a firm and rigid locking system.

7.62 mm Model 52 pistol

DATA
Cartridge: 7.62 mm bottleneck M48, 7.63 mm Mauser, 7.62 mm Type P
Operation: recoil, self-loading, single-action
Method of locking: rollers
Feed: 8-round box magazine
Weight, empty: 960 g
Length: 209 mm
Barrel: 120 mm
Rifling: 4 grooves rh
Sights: (fore) blade; (rear) square notch
Muzzle velocity: 396 m/s

Manufacturer
Ceská Zbrojovka, Strakonice.
Status
Obsolescent. Production ended mid-1950s.

7.62 mm Model 52 pistol's roller locking system

7.65 mm Model 50 pistol

This blowback pistol is generally similar to the West German Walther PP and PPK. The safety catch has been located at the top of the left grip instead of the slide and the trigger-guard is integral with the frame, so that dismantling is controlled by a stripping catch on the right side of the frame, ahead of the trigger guard. Together with the later Model 70 it is the main handgun of the Czechoslovak Security Forces, but is not issued to the Army. Production of the Model 50 ceased in 1969 when it was replaced by the Model 70 (below).

7.65 mm Model 70 pistol
The Model 70 was a somewhat improved version of the Model 50, which, in its early days, had a poor reputation for reliability. The Model 70 addressed these problems and was a more serviceable weapon. The changes are detail items and the appearance of the pistol is almost unchanged, except for the markings and the patterning on the butt-grips. Production of the Model 70 ceased in 1983.

DATA
Cartridge: 7.65 mm .32 Automatic Colt Pistol (ACP)
Operation: blowback, semi-automatic, double-action only
Method of locking: nil
Feed: 8-round detachable box magazine
Weight: 681 g
Length: 173 mm

7.65 mm Model 50 pistol

Barrel: 97 mm
Rifling: 6 grooves rh
Sights: (foresight) blade; (rearsight) notch
Muzzle velocity: 280 m/s

Manufacturer
Agrozet kp, Uherský Brod.
Status
Obsolescent. Production ceased (see text).

9 mm Model 75 pistol

This pistol is probably the best that has appeared in Czechoslovakia since the
end of the Second World War. It is not a development of the previous models
and it represents a completely new design in which the best features of several
other pistols, not necessarily Czechoslovak, have been incorporated. The
standards of workmanship are excellent and the 'balance' of the design is
exactly right. The designers are the brothers Josef and František Koucký and
they have not only aimed for a good product, but have also looked to the
manufacturing processes and the final cost.

The system of operation is by short recoil, the barrel being locked to the
slide for sufficient time to allow the bullet to clear the muzzle and the chamber
pressure to drop to a safe level. The slide and frame are die-castings, the barrel
is forged steel and the grips can be either walnut or plastic. The angle and size
of the grip is comfortable and convenient for shooting. The slide catch, safety
lever and magazine catch are all placed on the left side of the frame. Sights are
available with either a fixed back sight or one on a dovetail which allows for
lateral adjustment.

A particular feature of the pistol is the large double-row magazine holding
15 rounds. Another round can be carried in the chamber so that the total load
is 16. The firing pin is an inertia type and it is out of contact with the base of
the cartridge until fired. Drop tests have shown that repeated falls of 2 m onto
concrete, muzzle first, do not fire the pistol. The double-action fires the first
round without needing the slide to be cocked, hence the pistol can be carried
perfectly safely with a round in the chamber and, on drawing out of the holster
to shoot, the only action required is to trip the safety lever and pull the trigger.

The pistol is in full production but it has not been adopted by the
Czechoslovak forces, probably because they hold sufficient stocks of the
previous models. It is, however, sold in the West as a police or military arm.

DATA
Calibre: 9 mm Parabellum
Operation: short recoil
Method of locking: projecting lugs (Browning)
Feed: 15-round box magazine
Weight: (empty) 0.98 kg; (with empty magazine) 1 kg
Length: 203 mm
Height: 131 mm
Width: 33 mm
Barrel length: 120 mm
Rifling: 6 grooves rh. 250 mm pitch
Sight radius: 160 mm

Manufacturer
Česká Zbrojovka, Uherský Brod.

9 mm Model 75 pistol

9 mm Model 75 pistol showing thumb-operated levers

Status
Current. Production.
Service
Some units of Czechoslovak Police Force. Commercial sales and police use in
various countries.

9 mm CZ Model 85 pistol

The Model 85 is an updated version of the CZ75; in size, shape and general
design it retains the best features of the CZ75 but the manual safety and slide
stop have been remodelled to permit ambidextrous operation. The top surface
of the slide is ribbed, to reduce reflection, and minor internal mechanical
changes have been made to improve the action and reliability. The pistol can
be fitted with an adjustable rear sight to suit the user's requirements.

Dimensions and weights of the CZ85 are similar to those of the CZ75
(above).

Manufacturer
Česka Zbrojovká, Uherský Brod.
Status
Current. Production.

*9MM CZ Model 85 pistol, right side showing ambidextrous safety and
slide release*

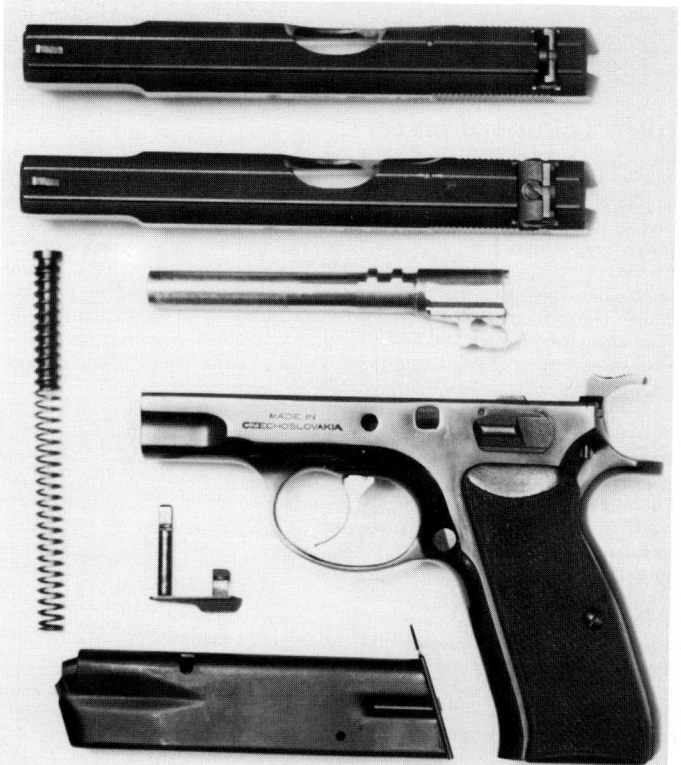

CZ Model 85 pistol dismantled

7.65 mm CZ Model 83 pistol

This pistol is a fixed-barrel blowback of conventional pattern. Of all-steel construction, it has an ambidextrous safety catch and a magazine catch behind the trigger-guard which is operable from both sides of the pistol. The double-action lockwork includes an automatic safety feature which blocks the hammer until the trigger has been fully squeezed. The trigger-guard is large enough to accommodate a gloved finger, and it also controls dismantling; a sprung detent holds the guard either open or closed, and it is interlinked so that the weapon cannot be stripped if the magazine is in place, and the magazine cannot be inserted unless the trigger-guard is closed into the frame. The sides of the slide are polished but the top is left matt to prevent sight glare.

DATA
Cartridge: 7.65 mm ACP; 9 mm Short/.380 ACP; 9 mm Makarov
Operation: blowback
Weight, empty: 650 g
Length overall: 173 mm
Barrel: 96 mm, 6 grooves, rh, one turn in 250 mm, except in 9 mm Makarov calibre when polygonal rifling is used.
Sight radius: 130 mm
Sights: (foresight) blade with white insert; (rearsight) square notch with 2 white spots
Magazine capacity: 7.65 mm–15 shots; 9 mm–13 shots
Muzzle velocity: 320 m/s (7.65 mm); 340 m/s (9 mm)

Manufacturer
Česká Zbrojovka, Uherský Brod.

7.65 mm Model 83 pistol

Status
Production.

EGYPT

9 mm Helwan pistol

The Helwan pistol is a licence-built copy of the Beretta Model 951 manufactured in Egypt during the mid-1960s. It is similar in every respect to the Model 951, to which further reference should be made.

9 mm Helwan pistol

9 mm Tokagypt pistol

This is a slightly improved copy of the Soviet TT33 Tokarev pistol, chambered for 9 mm Parabellum and manufactured for the Egyptian Government by Femaru es Szerszamgepgyar, Hungary. The improvements on the basic design are mainly cosmetic; the grip is a better shape being a one-piece plastic moulding, a safety catch has been added and the general quality and finish is rather higher than that normally found on Tokarev pistols. The principal, and most sensible, change is to 9 mm Parabellum chambering, which made the Tokagypt a very practical weapon.

For details of mechanism and functioning, refer to the entry on Tokarev under USSR.

DATA
Cartridge: 9 mm Parabellum
Operation: short recoil
Method of locking: Browning swinging link
Feed: 7-round detachable box magazine
Weight, empty: 0.91 kg
Length: 194 mm
Barrel: 114 mm
Rifling: 6 grooves rh
Sights: (foresight) blade; (rearsight) notch
Muzzle velocity: 350 m/s

Manufacturer
Femaru es Szerszamgepgyar NV, Budapest, Hungary.
Status
Obsolescent. No longer manufactured.

9 mm Tokagypt pistol

Service
Egyptian Army and police.

FINLAND

9 mm Model 35 and Model 40 pistols

In 1935 the Lahti Model 35 replaced the 9 mm Luger which had been the standard pistol in the Finnish Army since its adoption in 1923. The Model 35 was designed by Aimo Lahti and the weapon was manufactured by Valtion Kiväärithedas (VKT) Jyväskylä. It is a mixture of the Luger and the Bergmann-Bayard with a feature from Browning's designs. In appearance it resembles the Luger but its action is different. The barrel is screwed to the slide and the rear end of the slide is enlarged to hold the breech-block. The breech-block travels in the slide and cocks the hammer as it recoils. It has thumb pieces by which it can be pulled back to the rear to cock the action. When the breech-block is fully forward the locking piece is engaged in mortices on each side of the slide and its inner surfaces are engaged in slots in the breech-block. Cam faces on the sides of the locking piece raise and lower it during rearward and forward travel. One very unusual device is the accelerator which, although a common feature on a short recoil operated machine gun, is unique in a pistol. It is similar to that on the Browning .30 calibre guns and when the barrel ceases its rearward travel the accelerator is rotated and the breech-block thrown rearwards with increased velocity. The magazine, like the Luger, has a button which relieves the weight of the spring while loading is carried out. The button operates a pivoted lever so the breech-block remains open when the ammunition is expended.

The Lahti Model 35 is extremely well sealed to keep out mud, snow and sand and is very difficult to strip without workshop facilities. It is also very reliable and this probably accounts for the retention in service of a relatively heavy pistol of rather old design.

It was also used, in a slightly modified form, in Sweden where it was known as the Model 40. Some of the Swedish weapons, in turn, were taken to Denmark from Sweden by Danish troops after the Second World War and were there taken into service as the Model 40(S).

DATA
Cartridge: 9 mm Parabellum
Operation: short recoil, semi-automatic
Method of locking: dropping-block
Feed: 8-round detachable magazine

Weight: 1.22 kg
Length: 246 mm
Barrel: 107 mm
Rifling: 6 grooves rh
Sights: (foresight) blade; (rearsight) notch
Muzzle velocity: 350 m/s

Manufacturer
Valtion Kiväärithedas Jyväskylä. Formerly made by Husqvarna in Sweden as the Model 40.
Status
Obsolescent. No longer manufactured.
Service
Finland.

9 mm Model 35 pistol

FRANCE

Model D MAB pistol

Designed primarily for police use, but suitable also as a close-in defence weapon for military use, the Model D MAB is a robust hammerless self-loading pistol which is available either in 7.65 mm Short calibre or, by a simple change of the barrel, in 9 mm Short. Good accuracy at 25 m is claimed for the weapon by the manufacturers.

DATA
Calibre: 7.65 mm or 9 mm Short
Operation: self-loading hammerless
Feed: 9-round detachable box magazine
Weight, empty: 0.725 kg
Length: 176 mm
Barrel length: 103 mm
Rifling: 6 grooves rh
Sights: (fore) blade; (rear) notch

Manufacturer
Manufacture d'Armes Automatiques (MAB), Lotissement industriel des Pontots, 64100 Bayonne.
Status
Current. No longer manufactured.
Service
Police forces and some government agencies.

7.65 mm version of Model D MAB pistol

9 mm Model 1950 MAS self-loading pistol

This pistol was designed at the French Arsenal at St Étienne and manufactured there and at the Chatellerault factory.

Operation
The pistol is loaded by removing the magazine, inserting nine 9 mm Parabellum cartridges, and replacing the magazine in the butt.

The magazine release catch is on the left side of the receiver, behind the trigger. Pulling back on the rear end of the slide and then releasing it positions the cartridge in the chamber. There is an indicator to the rear of the ejection slot which projects above the slide to indicate that the chamber is loaded. The safety catch is on the left rear of the slide. When the lever is horizontal the pistol is safe. When it is below the horizontal a red dot is exposed to indicate

that this is the 'fire' position. The hammer can be lowered by pulling the trigger if the pistol is set to 'safe' because the safety blocks the hammer from hitting the firing pin.

When the pistol is fired, the gas pressure forces the slide rearwards. The locking ribs on top of the barrel are fitted into the grooves on the inside of the slide and so the barrel goes back with it. The lower rear end of the barrel is attached to the receiver by a swinging link. The barrel and slide move back together for a short distance and then, since the lower portion of the link is attached to the non-recoiling frame, the link pulls the rear end of the barrel down. This separates the locking ribs of the barrel from the recesses inside the slide and when the locking system is disconnected the barrel is at rest and the slide continues to the rear under its own momentum. The empty case comes

9 mm Model 1950 MAS self-loading pistol

9 mm Model 1950 MAS self-loading pistol, field-stripped

Stripping

Remove the magazine and pull the slide rearwards.
Hold it back with the slide stop above the trigger, on the left of the receiver. Hold the slide and with the stop still rotated up into the slide notch, pull the slide stop out to the left. Ease the slide forward off the receiver.
Invert the slide and take out the return spring and guide. Lift the rear end of the barrel up and pull it out of the slide.
The hammer mechanism pulls out of the receiver as an entire unit.

DATA
Cartridge: 9 mm Parabellum
Operation: recoil, self-loading, single-action
Method of locking: projecting lug
Feed: 9-round box magazine
Weight, empty: 860 g
Length: 195 mm
Barrel: 112 mm
Rifling: 4 grooves, lh, one turn in 254 mm
Sights: (fore) blade; (rear) square notch
Muzzle velocity: 354 m/s

Manufacturer
Manufacture Nationale d'Armes de Chatellerault; Manufacture Nationale d'Armes de St Étienne.
Status
Current. No longer manufactured.
Service
French and ex-French colonial forces.

out on the breech face and stays there until it strikes the fixed ejector. It is then rotated about the extractor and flung out to the right. The hammer is cocked and the return spring compressed.

The return spring forces the slide forward and the face of the breech-block carries the top cartridge in the magazine forward into the chamber. The breech-block contacts the barrel and pushes it forward. As the barrel goes forward the link raises the breech, and the ribs on the top of the barrel enter the grooves on the slide and lock the two together. Forward movement ceases when the lug on the bottom of the barrel contacts the slide stop pin.

9 mm PA15 MAB pistol

Manufacture d'Armes Automatiques (MAB) manufactured a number of commercial and target pistols. Among these was the 9 mm PA15 MAB which is used by the French Army. The pistol has the bulky grip required to accommodate 15 9 mm Parabellum cartridges. It has a very prominent spur at the rear of the receiver and a ring-type hammer.

Operation
PA15 has a delayed blowback action relying upon a rotating barrel. The barrel carries two lugs, above and below the chamber. The lower lug engages in a slot cut in the return spring guide, which is pinned to the frame, so that the lug can rotate but cannot move back or forward. The upper lug engages in a track cut in the inner surface of the slide; this track is shaped so that the initial opening movement of the slide will rotate the barrel through about 35°, after which the track is straight so that the slide can recoil while the barrel remains still. Rotation of the barrel is initially resisted by a combination of its inertia and the torque effect of the bullet passing through the rifling. When the chamber pressure has reached a safe level the slide is blown back, compressing the return spring, rotating the hammer and depressing the trigger bar. The empty case is extracted and ejected to the right of the gun.

The return spring forces the slide forward. The top round in the magazine is pushed forward up the bullet guide and enters the chamber. The cam groove on the slide then causes the barrel to rotate and the extractor grips the rim of the case. The trigger bar rises into its recess when the slide is fully forward and the pistol is ready to fire another round.

There is a magazine safety that prevents the hammer going forward when the magazine is out of the gun. There is also a holding-open device.

Stripping
The magazine catch is on the left of the grip immediately behind the trigger. When the magazine is out the slide should be retracted and the chamber and feedway checked.

With the slide held back about 6 mm, the slide stop pin can be pushed to the left and then removed.
The slide can be pulled forward off the receiver. With the slide upside down the return spring guide and barrel seat can be pressed towards the muzzle. They can then be disengaged and lifted out. The barrel will lift out of the slide if it is pushed forward about 8 mm.

9 mm PA15 MAB pistol

DATA
Cartridge: 9 mm Parabellum
Operation: delayed blowback, semi-automatic
Method of delay: barrel lug rotates in cam path in slide
Feed: 15-round box magazine
Weight: 1.09kg
Length: 203 mm
Barrel: 114 mm
Rifling: 6 grooves rh

Sights: (foresight) blade; (rearsight) notch
Muzzle velocity: 350 m/s

Manufacturer
Manufacture d'Armes Automatiques (MAB), Lotissement industriel des Pontots, 64100 Bayonne.
Status
No longer in production.
Service
French armed forces.

.357 in MR 73 Manurhin revolver

This compact double-action revolver is available in three different barrel lengths and can fire .38 Special ammunition as well as the .357 Magnum round. It has a six-shot cylinder which swings sideways for reloading. For competition shooting a series of target models are also made with adjustable rearsight and a choice of four barrel lengths. A replacement cylinder, enabling the weapon to be used with the 9 mm Parabellum round, is available.

Operation
Referring to the numbered components in the accompanying diagram, in the rest position the hammer (126) is under load by the spring (134), the tension of which is adjustable by the strain screw (135). The hammer lies on the rebound slide (114). The hammer nose (130) which is riveted to the hammer is thus held back from the cartridge.

The rebound slide (114) glides on four rollers (115) along the frame base (011) with negligible friction. It is connected to the trigger (111) by the trigger bar (112). The internal automatic safety (136) is connected to the rebound slide and is located between the hammer and the frame front, thus preventing the hammer from striking if the trigger is not pulled.

The flat trigger spring (118) is pivoted on a pin (119) and its tension can be adjusted by the strain screw (120). This spring bears on the rebound slide roller (116); and its design is such that as the trigger is pulled, carrying the slide (114) to compress the trigger spring, the point of contact between the spring and the roller rises. As a result, although the total force exerted by the trigger spring increases as the trigger is pulled, the force component directly opposed to the trigger pull remains substantially constant. The remainder of the action is conventional.

The .357 MR 73 Gendarmerie Manurhin revolver is a variant of the standard MR 73 model which differs in having an adjustable rearsight with rounded edges and a more prominent ramped foresight.

DATA
Calibre: .357 Magnum (will fire .38 Special) or 9 mm Parabellum (see text)
Operation: double- or single-action
Cylinder: side-loading 6-chamber
Weight: 2.5 in (63.5 mm) barrel: 880 g; 3 in (76.2 mm) barrel: 890 g; 4 in (101.6 mm) barrel: 950 g (combat versions)
Length: 195 mm, 205 mm or 233 mm overall (combat versions)
Barrel: combat versions 2.5 in (63.5 mm), 3 in (76.2 mm) or 4 in (101.6 mm). Competition versions 4 in (101.6 mm), 5.25 in (133.4 mm), 6 in (152.4 mm) and 8 in (203.2 mm)
Rifling: 6 grooves rh
Sights: (foresight) blade; (rearsight) notch for combat models. Adjustable sight for competition versions

Manufacturer
Matra Manurhin Defense, 21 ave Louis Bréguet, 78145 Velizy.
Status
Current. Production.
Service
French forces and government agencies.

.357 in MR 73 Manurhin revolver

.357 in MR 73 Manurhin revolver, Type Gendarmerie

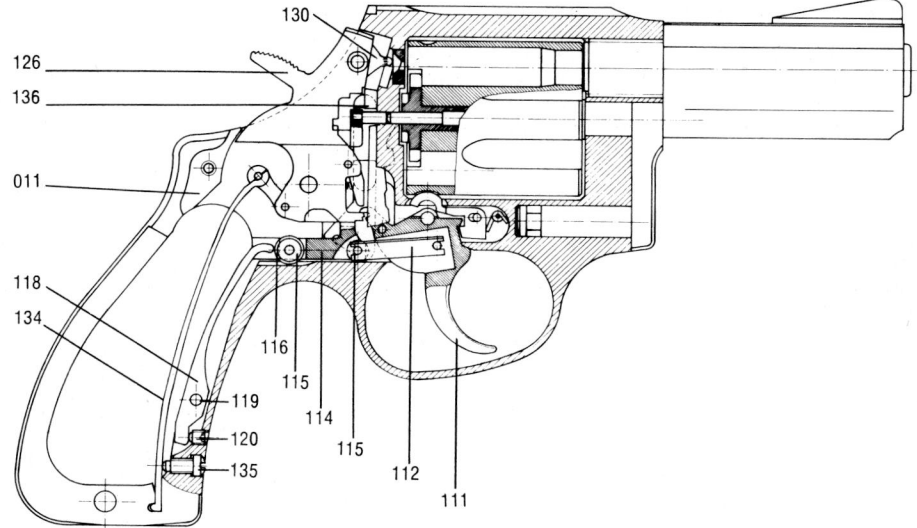

.357 in MR 73 Manurhin revolver mechanism
see text for key

GERMANY, EAST

Pistole M

The East German Army is equipped with the Pistole M. This is manufactured in the country and is a copy of the Soviet Makarov pistol. It is an eight-shot, blowback-operated, self-loading pistol using the 9 mm × 18 cartridge which was developed for the weapon. The trigger mechanism enables the pistol to be used as a single-action model by cocking the hammer manually: alternatively a single long pull can be used to cock and then fire.

In spite of superficial resemblance, there are some differences between the Pistole M and the Walther PP. The Pistole M does not have a loaded chamber indicator. The East German pistol has a leaf spring to rotate the hammer instead of a coil spring and there are differences in the trigger mechanism. The slide stop of the Walther is completely enclosed: that of the Pistole M is externally operated. Lastly, the safety catch of the East German pistol goes up for 'safe' and that of the Walther goes down.

DATA
Cartridge: 9 mm × 18
Operation: blowback, self-loading
Method of locking: nil

East German copy of Soviet 9 mm × 18 Makarov pistol known as Pistole M

Feed: 8-round detachable box magazine
Weight: 663 g
Length: 160 mm
Barrel: 91 mm
Rifling: 4 grooves rh
Sights: (foresight) blade; (rearsight) notch
Muzzle velocity: 315 m/s
Muzzle energy: 294 J

Manufacturer
State factory.
Status
Current.
Service
East German armed forces.

East German copy of 7.65 mm Walther PP known as Modele 1001-0

GERMANY, WEST

7.65 mm (or 9 mm) HSc Mauser pistol

This pistol was introduced in 1940 and is a double-action blowback weapon of attractive shape and advanced design. The hammer is almost entirely concealed and just sufficient is exposed to allow the firer to use his thumb to lower it on a loaded chamber. A few pistols were made with totally enclosed hammers. The safety device on this pistol is a lever which turns to lift the head of the firing pin into a recess and out of alignment with the hammer.

The action is straightforward: when the slide is retracted the hammer is cocked and the return spring, wrapped round the barrel, is compressed. When the slide is released the top round is fed into the chamber. The hammer can be lowered on the firing pin and fired with a long pull. When the ammunition is expended the slide is held open by the magazine follower. When the magazine is removed the slide will go forward as soon as a magazine, loaded or empty, is inserted. If the magazine is loaded the top round is fed into the chamber. When the magazine is out of the pistol, the weapon cannot be fired.

The HSc was used during the Second World War by the German Air Force and the Navy. After the war ended commercial production was resumed.

DATA
Cartridge: 7.65 mm .32 Automatic Colt Pistol (ACP) or 9 mm Short (.380 ACP)
Operation: blowback, semi-automatic
Method of locking: nil
Feed: 8 (7.65 mm) or 7 (9 mm) round detachable box magazine
Weight: Approximately 660 g with magazine
Length: 160 mm
Barrel: 85 mm
Rifling: 6 grooves rh
Sights: (foresight) blade; (rearsight) notch
Muzzle velocity: 290 m/s

7.65 mm HSc Mauser pistol.

Manufacturer
Mauser-Werke Oberndorf GmbH, Geschäftsbereich Waffensysteme, 7238 Oberndorf-Neckar.
Status
No longer in production by Mauser; currently manufactured under license by Societa Armi Bresciana, Gardone VT, Italy.
Service
No regular military service.

Mauser Luger Pistole '08

At the beginning of this century the Parabellum, or Luger, pistol – introduced into the German Army in 1908 and hence known as the Pistole '08 – was designed in various versions to meet the individual requirements of different services. It became world-famous and although manufacture ceased after 1945 and it is no longer in regular service, numbers still exist in police use and the pistol has become a focus for collectors. Mauser has resumed manufacture of this pistol either in standard finish, customised with engraving, or as commemorative models in limited series of 250 pieces. A full description of the mechanism can be found in earlier editions of *Jane's Infantry Weapons*.

DATA
Cartridge: 9 mm or 7.65 mm Parabellum
Operation: recoil; semi-automatic
Method of locking: toggle
Feed: 8-round detachable box magazine
Weight: 835 g
Length: 220 – 270 mm depending upon barrel
Barrel: 100 – 150 mm
Rifling: 6 grooves (9 mm), 4 grooves (7.65 mm), right hand, one turn in 250 mm
Sights: front blade, rear notch

Manufacturer
Mauser-Werke Oberndorf GmbH, 7238 Oberndorf-Neckar.

Mauser Luger Pistole '08

Status
Production.

7.65 mm Models PP and PPK Walther pistols

This was produced as the 'Police Pistol' in 1929 for carrying on the uniform belt and was widely adopted by police forces. Originally the pistol was made in 7.65 mm calibre but later it was constructed in .22 long rifle (lr) and 9 × 17 mm Short. A small number of 6.35 mm models were also made. Pistols in .22 and 6.35 mm calibres have not been made for several years.

In 1931 a smaller pistol called the PPK appeared in the same calibres. This was intended for police use as a concealed weapon and the initials stood for 'Polizei Pistole Kriminal'. This differed in size and also in the construction of the handgrip. In the PP the butt was forged to shape and had simple plastic side plates but in the PPK it was rectangular with a plastic butt providing the contour. Both pistols were blowback-operated and well constructed and finished. Both used a pin, with centre fire cartridges, to indicate a loaded chamber. This protruded from the rear of the slide. The pistols made before the Second World War were offered with some extras. Dural frames and slides were available to reduce weight; plastic magazine floor plates with spurs to increase the length of the grip were available for policemen with big hands. Luminous night sights could be added. The weapons were used extensively during the Second World War and often the loaded chamber indicator was omitted in wartime production. The finish in the war was often poor. After the war the pistols were copied and used by several countries including Turkey and Hungary. The French firm of Manurhin produced the pistols under licence in .22lr (PP), 7.65 mm (PPK) and 9 mm Short (both).

The PP and PPK are straight blowback pistols with external hammers, double-action triggers and very adequate safety arrangements. The hammer is prevented from reaching the firing pin until the sear movement, when the trigger is pulled, moves the block clear. The disconnector works into a recess in the slide and until the disconnector can rise, the sear cannot rotate. This only occurs when the slide is fully forward.

DATA
Cartridge: 7.65 mm (.32ACP), 9 mm Short (.38ACP)
Operation: blowback

7.65 mm Model PP Walther pistol

7.65 mm Model PP Walther pistol sectioned

7.65 mm Model PPK Walther pistol

Feed: detachable box magazine. (PP) 8 rounds; (PPK) 7 rounds
Weight: (PP) 682 g; (PPK) 568 g
Length: (PP) 173 mm; (PPK) 155 mm
Barrel: (PP) 99 mm; (PPK) 86 mm
Rifling: 6 grooves rh
Sights: (fore) blade; (rear) notch
Muzzle velocity (7.65 mm): (PP) 290 m/s; (PPK) 280 m/s

Manufacturer
Carl Walther Waffenfabrik, PO Box 4325, D-7900 Ulm.

Status
Current. Production in 9 mm and 7.65 mm calibres.

Service
Widespread use by armed forces, police and government agencies in West Germany and other European countries.

9 mm Model P1 Walther pistol

The P1 is the current version of the Walther P38 used by the German Army in the Second World War. After the war the P38 was produced in a lightweight frame and this has been continued in the P1. The difference in markings on the two weapons is small. Both have 'Walther' and then 'Carl Walther Waffenfabrik Ulm/Do'. The Army Pistol has 'P1 Cal 9 mm' and the commercial weapons 'P38 Cal 9 mm'.

The main parts of the pistol are the barrel, receiver, slide and the locking lug. There is a double-action trigger mechanism that enables the hammer to be cocked and released by a single trigger pull. The hammer can also be cocked manually to produce a single-action with a reduced trigger pull. If the weapon is set to 'safe' with the hammer cocked, the hammer will go forward but the spindle of the change lever locks the firing pin which cannot go forward.

In addition to the standard 9 mm Parabellum calibre, the P1 is also available in .22 Long Rifle and 7.65 mm Parabellum calibres.

Stripping
Remove the magazine and check that the chamber is empty.
The slide is retracted and held to the rear with the slide stop, and the take-down lever on the left side of the receiver forward of the trigger is rotated down.
The slide is eased back to release the slide stop then moved forward off the receiver.
The unlocking plunger behind the slide releases the barrel from the slide and the barrel can be pulled forward and clear.

DATA
Cartridge: 9 mm × 19 Parabellum
Operation: short recoil, self-loading, double-action
Method of locking: hinged locking piece
Feed: 8-round box magazine
Weight, empty: 772 g
Length: 218 mm
Barrel: 124 mm
Rifling: 6 grooves, rh, one turn in 254 mm
Sights: (fore) blade; (rear) U-notch
Muzzle velocity: 350 m/s

Manufacturer
Carl Walther Waffenfabrik, PO Box 4325, D-7900 Ulm.
Status
Current. Production.
Service
Chilean, Norwegian, Portuguese, West German and other armed forces and government agencies.

9 mm Model P1 Walther pistol

Sectional view of P38 Walther pistol

9 mm Walther P1A1 pistol

This pistol, announced in June 1989, can best be described as an improved version of the P5, details of which will be found overleaf. The principal difference is the adoption of a cross-bolt safety catch on the slide; when pushed so that the bolt protrudes on the left of the slide the pistol is safe; pushed to the right, the pistol is ready to fire. The operation of this cross-bolt safety is upon the firing pin; when pushed to the safe position the bolt forces the firing pin down, aligning it with the cutout on the face of the hammer and locking it against a lug in the slide. Thus, should the hammer fall the end of the firing pin will be enclosed in the cutout, and the hammer will strike the slide without placing any force on the firing pin. Moreover any pressure on the pin will be resisted by the slide lug, preventing any forward movement. Moving the cross-bolt to the fire position allows the firing pin to assume its correct position so that it will be struck effectively by the hammer.

DATA
Cartridge: 9 × 19 mm Parabellum
Operation: recoil, semi-automatic
Locking system: wedge
Feed: 8-round box magazine
Weight empty: 808 g
Length: 179 mm
Barrel: 90 mm
Rifling: 6 grooves RH, one turn in 27.7 calibres
Sights: (fore) blade; (rear) notch, adjustable for windage
Muzzle velocity: Approx 350 m/sec

Walther 9 mm P1A1 pistol

9 mm Model P5 Walther pistol

The Walther P5 was developed in order to provide a reliable and safe pistol for police and military use. The basic specification was provided by the West German police forces, who demanded a very high standard of safety in handling, together with the ability to fire double-action without having to release cumbersome safety devices before operating the trigger.

The P5 is a locked-breech recoil-operated pistol of conventional appearance, but with several unusual features incorporated in the lockwork. There are four built-in safety operations:

(a) the firing pin is held out of line with the hammer nose until the hammer is released by the trigger

(b) until the moment of firing the firing pin is held opposite a recess in the face of the hammer. Should the hammer be released by any means other than the trigger, it will not move the firing pin even though it strikes with full force

(c) the hammer has a safety notch

(d) the trigger bar is disconnected unless the slide is fully closed and the barrel locked to it.

The effect of these various safeties is that the hammer can strike the firing pin only when the slide is fully forward and locked to the barrel and when the trigger is pulled through to the limit of its movement. At any other time the pistol is completely safe from such actions as dropping, sharp blows, or inadvertent tripping of the hammer when cocking by hand.

A particular feature of the pistol is the de-cocking lever on the left side of the body, just behind the trigger. The function of this lever, as its name implies, is to take the spring tension off the hammer and moving parts after loading. The pistol is loaded with a magazine in the normal way and the slide allowed to go forward and chamber the first round. The de-cocking lever is now pushed down with the thumb and allowed to come up. This action operates all the safeties and releases the hammer which moves forward to rest against the rear of the slide with its recess enclosing the projecting end of the firing pin. The firing pin is then completely protected from any external blows and is also held in place by a notch in the pin engaging a lug in the slide. The hammer is held by the safety sear.

When the trigger is pulled through, its first action is to release the safety sear and then to cock the hammer. At full cock a trip is released which moves up to engage with the firing pin at the moment that the hammer moves. As the hammer arrives the pin is waiting for it, ready to be driven forward to fire the cartridge. For single-action shooting the hammer has to be fully cocked by the thumb. This will allow the safety sear to be pushed clear of the hammer breast and free it to move forward. If the hammer is not fully cocked, should the thumb slip off it for instance, the safety sear will catch it and will prevent the firing pin trip from working.

The locking system works by a loose locking piece held between two lugs beneath the barrel. A projection on each side of this locking piece engages with a slot in the slide on each side. When the locking piece is pushed upwards, by a ramp on the body when the slide is fully forward, the barrel and slide are locked together by the projections. Barrel and slide remain locked during the initial recoil movement, but after a short distance a sliding pin in the rear barrel lug meets a vertical face on the body. This pin then pushes against the locking piece, camming it down out of engagement.

In addition to the standard 9 mm Parabellum chambering, the P5 is available in 7.65 mm Parabellum and 9 × 21 mm chambering.

Dismantling

Remove the magazine and check the chamber is clear.
Press the muzzle down on to a soft surface until it reaches the stop.
Turn the barrel catch upwards as far as it will go.
Lift the barrel and slide assembly off the body.
The barrel is disconnected from the slide by pressing the locking pin fully forward. No further stripping should be undertaken, and the recoil springs should not be taken out.

Assembly

Reverse the above instructions, taking care to lock the barrel to the slide with the locking piece before replacing the assembly on the body.

DATA

Cartridge: 9 mm × 19 Parabellum; 7.65 mm Parabellum; 9 × 21 mm
Operation: recoil, locked breech
Method of locking: loose locking piece
Feed: 8-round detachable box magazine
Weight: (empty) 795 g; (loaded) 885 g

9 mm Model P5 Walther pistol

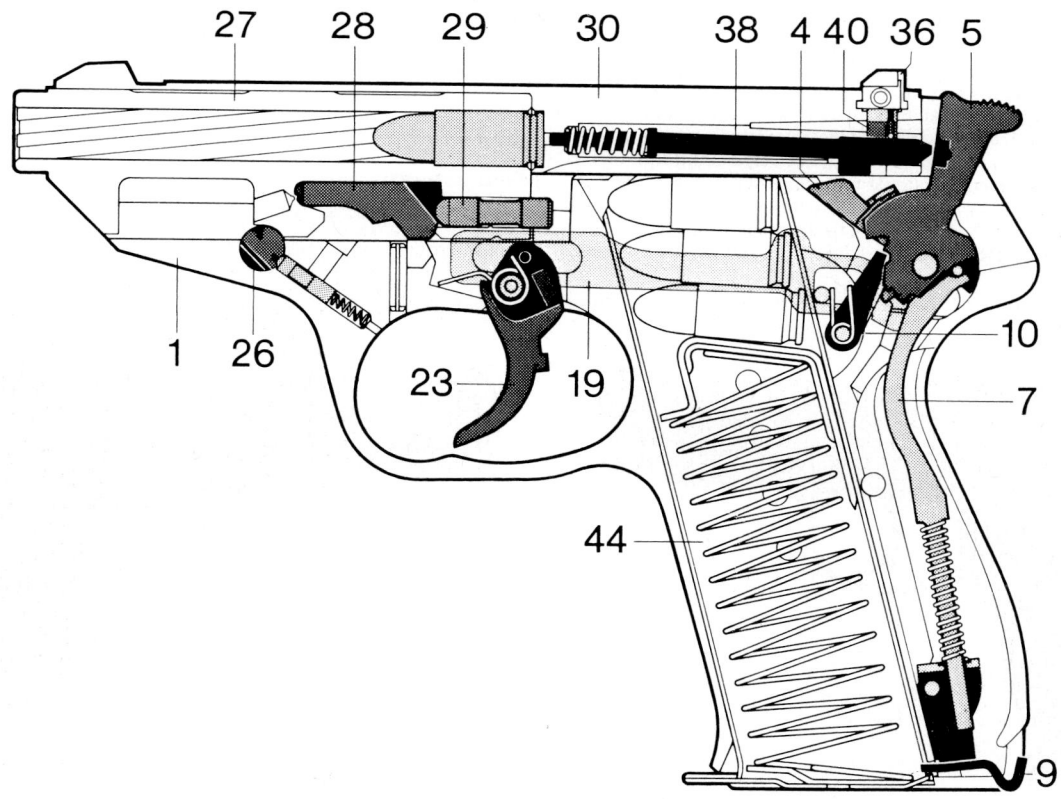

Cutaway view of 9 mm Model P5 Walther pistol, loaded and decocked
(1) *frame* **(4)** *trip lever* **(5)** *hammer* **(7)** *hammer strut* **(9)** *magazine catch* **(10)** *sear* **(19)** *trigger bar* **(23)** *trigger* **(26)** *barrel catch* **(27)** *barrel* **(28)** *locking piece* **(29)** *locking pin* **(30)** *slide assembly* **(35)** *detent* **(36)** *rearsight* **(37)** *sight adjustment screw* **(38)** *firing pin* **(40)** *insert* **(44)** *magazine*

Length: 180 mm
Barrel: 90 mm
Rifling: 6 grooves rh
Sights: (fore) blade; (rear) square notch with white contrast markings. Adjustable for elevation and line
Sight radius: 134 mm
Muzzle velocity: 350 m/s

Manufacturer
Carl Walther Waffenfabrik, PO Box 4325, D-7900 Ulm.
Status
Current. Production.
Service
Adopted by the Netherlands Police and the police forces of Baden-Württemberg and Rheinland-Pfalz. Notable export orders include Nigeria, Portugal, the United States and various South American countries.

9 mm Model P5 Compact Walther pistol

The Walther P5 Compact pistol is essentially the same as the P5 described above but is shorter and lighter, making it more convenient for concealed carrying. The size and shape make it well suited to the smaller hand. There is a lateral magazine release which falls conveniently under the thumb when required. The pistol has a light alloy frame, a polished finish and wooden grips, and the hammer has been rounded so as not to catch in clothing when drawn.

DATA
Cartridge: 9 × 19 mm Parabellum
Operation: recoil, locked breech
Method of locking: loose locking piece
Feed: 8-round detachable box magazine
Weight: (empty) 780 g; (loaded) 870 g
Length: 168.5 mm
Barrel: 79 mm
Rifling: 6 grooves rh
Sights: (fore) blade; (rear) square notch with white contrast markings. Adjustable for elevation and line
Sight radius: 129 mm
Muzzle velocity: 350 m/s

Manufacturer
Carl Walther Waffenfabrik, PO Box 4325, D-7900 Ulm.
Status
Current. Production.

9 mm Walther P9 Compact pistol

9 mm Model P-88 Walther pistol

This new pistol is a departure from previous Walther designs insofar as it employs a modified Colt-Browning method of breech locking; the squared-off chamber section of the barrel locks into the ejection recess and is released by a cam beneath the barrel striking an actuating lug. In consequence the exterior appearance lacks the exposed barrel common to the P-38 and its derivations and takes on a more conventional rectangular appearance. Normally supplied in 9 mm Parabellum calibre, it is also available chambered for the 9 × 21 mm cartridge.

The P-88 is a double-action, hammer-fired semi-automatic, with an ambidextrous de-cocking lever which also functions as the slide release. There is also an ambidextrous magazine catch located in the front edge of the butt, just below the trigger-guard. Safety is ensured by a complex firing pin system; the firing pin normally rests at an angle and the corresponding face of the hammer is recessed, so that even if it should fall no pressure will be placed on the firing pin. When the trigger is operated the rear end of the firing pin is moved up so as to align it with the solid face of the hammer, which then strikes the pin as it falls. Releasing pressure on the trigger then returns the firing pin

to its safe rest position until the next shot is required. There is no way in which the firing pin can be propelled forward by impact or accidental blows.

DATA
Cartridge: 9 × 19 mm Parabellum; 9 × 21 mm
Operation: locked-breech semi-automatic
Locking system: dropping barrel
Feed system: 15-round box magazine
Length: 187 mm
Weight, empty: 900 g
Barrel: 102 mm
Sights: (front) blade; (rear) adjustable square notch

Manufacturer
Carl Walther Waffenfabrik, PO Box 4325, D-7900 Ulm.
Status
Current. Production.

9 mm Walther Model P-88 pistol

HK4 self-loading Heckler and Koch pistol

The HK4 Heckler and Koch pistol is a self-loading, double-action pocket pistol designed for easy conversion from centre fire to rim fire, or vice versa, through a variety of calibres: 9 mm Short (.380), 7.65 mm (.32), 6.35 mm (.25) and .22 long rifle (lr). The safety catch on the left side of the receiver, is down for 'safe' and up for 'fire'. A white spot indicates 'safe', a red spot 'fire'. The magazine is slotted to show the number of rounds loaded. To chamber a round, the slide must be retracted and then released. When a round is in the chamber, the extractor on the right of the slide is proud and this can be seen by day and felt with the forefinger at night. When the pistol is set to 'safe' the hammer can be uncocked by pulling the trigger. The adoption of the 'safe' position blocks the firing pin from the hammer. When the magazine is emptied the slide remains to the rear. Putting in a loaded magazine releases the slide stop and the slide goes forward to chamber a round.

Change of calibre

To change from one centre fire barrel to another of a different calibre is merely a matter of changing barrels, springs and magazines. To change from centre to rim fire the firing pin must also be re-aligned. This is done by removing the barrel, holding the extractor clear of the breech face plate with a pin or a nail and removing the breech face plate. When using .22 long-rifle rim fire the face plate is turned so that the side marked 'R' is showing. The firing pin protrudes from the top hole. When converting back to centre fire the plate marking 'Z' is to the front and the firing pin comes through the lower holes.

DATA
COMMON TO ALL CALIBRES
Operation: blowback, self-loading, double-action
Method of locking: nil
Feed: box magazine
Weight, empty: 520g
Length: 157 mm
Barrel: 85 mm
Rifling: 6 grooves, rh, one turn in 254 mm
Sights: (fore) blade; (rear) U-notch

DATA

Calibre	Magazine capacity	Muzzle velocity
.22lr (.22lfB)	8	300 m/s
.25 Automatic Colt Pistol (ACP) (6.35 mm)	8	257 m/s
.32 ACP (7.65 mm)	8	302 m/s
.380 ACP (9 mm)	7	299 m/s

Manufacturer
Heckler and Koch GmbH, D-7238 Oberndorf-Neckar.
Status
No longer in production.
Service
Wide commercial sales and some military use.

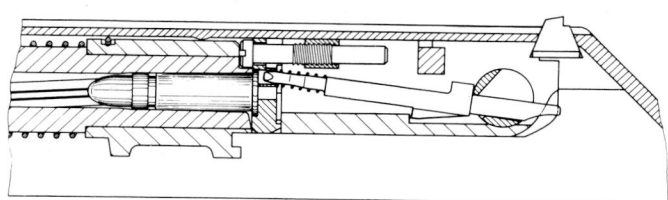

Sectioned view showing firing pin aligned for rim-fire cartridge

Alternative barrels for HK4 self-loading Heckler and Koch pistols

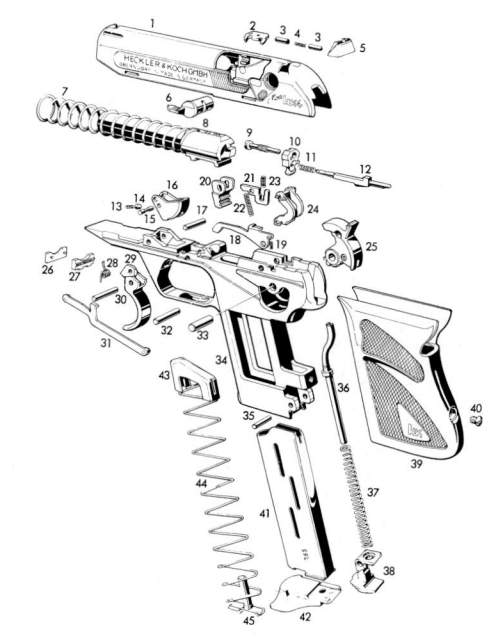

Exploded view of HK4 self-loading Heckler and Koch pistol

HK4 self-loading Heckler and Koch pistol

Cutaway view of HK4 self-loading Heckler and Koch pistol

9 mm P9S Heckler and Koch pistol

The 9 mm P9S self-loading pistol is a modern double-action weapon, though with only moderate magazine capacity. The barrel is polygonally rifled, which permits a somewhat higher muzzle velocity than obtained with normal rifling,

reduces barrel wear and bullet deformation and also reduces the accumulation of fouling.

The safety catch on the rear of the left side of the slide is pressed down for 'safe'. This uncovers a white spot. When rotated clockwise up to the horizontal position the weapon is set to 'fire' and a red spot is revealed.

To remove the magazine for loading, the magazine catch, below the pistol grip, is pushed to the rear. The magazine will then be eased out and can be withdrawn. The magazine is inserted into the pistol grip until the catch engages. To place a cartridge in the chamber the slide is pulled fully back and then released. This also cocks the hammer. When the hammer is cocked, a pin protrudes from the rear of the slide when the latter is fully forward. When a cartridge is chambered the extractor stands proud. Both of these indicators can be felt at night and seen by day. When the ammunition is expended the pistol will cease firing with the slide to the rear. As soon as a loaded magazine is inserted the slide may be released by pressing down on the cocking lever on the left side of the receiver behind the trigger or by pulling the slide back and letting the return spring drive it forward. As the slide goes forward a round is chambered and the hammer is left cocked. The cocked hammer may be released by first setting the safety, pressing down the cocking lever, pulling the trigger and holding it back while the cocking lever is allowed to rise again and releasing the trigger. After disengaging the safety, the model P9S can be fired from this position by a long trigger pull.

Although the main production is in 9 mm Parabellum calibre, an export version for the USA is manufactured in .45 Automatic Colt Pistol (ACP). For a short time some were made in 7.65 mm Parabellum, but this has now been discontinued.

Operation

The pistol operates by delayed blowback. The method of connecting the slide to the barrel is to use a two-part breech-block consisting of a bolt-head containing the two rollers and a heavy bolt-body which, by means of angled faces, forces the rollers out into barrel extensions. This method is derived from Heckler and Koch's G3 rifle and a more detailed account of the action will be found there. When the pistol fires, the gas pressure forces back the breech face but movement is severely limited because the projecting rollers are engaged in recesses in the barrel extension. The rollers must be free of these seatings before the breech face can move back significantly. The reaction of the recesses drives the rollers inwards but their inward movement is resisted by the angled faces of the bolt-body and the strength of the return spring. The velocity ratio obtained by the angles of the recesses in the barrel extension and the angle of the faces of the bolt-body results in the heavy bolt-body having a rearward movement four times as great as the bolt-face. Eventually the rollers are forced fully in, the bolt is then blown back by the residual pressure and the empty case comes out, held to the breech face by the extractor. The hammer is rotated back by the bolt-body, the empty case is thrown out by the ejector and the rearward movement of the bolt ceases on contact with the plastic buffer. The return spring around the barrel expands and the slide goes forward, feeding a round into the chamber. To fire another round the trigger must be released to allow the disconnector to bear on the sear which is engaged in the hammer notch.

Stripping

Engage the safety, remove the magazine, and inspect the chamber and feedway to ensure no cartridge remains in the weapon.
Return the slide to the forward position.
Press the barrel catch in the rear surface of the trigger-guard in front of the trigger.
Push the slide as far forward as it will go and lift it off.
The barrel is removed by pushing it forward against the return spring until it can be removed from the slide.
The bolt-head can be removed by using one side of the barrel extension to press between the bolt-head and the slide against the locking lever until the bolt-head springs forward and the bolt then can be removed.

DATA
Cartridge: 9 mm × 19 Parabellum and .45ACP
Operation: delayed blowback, self-loading, double-action
Method of delay: rollers
Feed: 9-round box magazine (.45ACP, 7 rounds)
Weight, empty: 880 g (750 g in .45 calibre)
Length: 192 mm
Barrel: polygonal rifling, rh
Sights: (fore) blade; (rear) square notch
Sight radius: 147 mm
Muzzle velocity: 351 m/s (.45ACP, 260 m/s)

Manufacturer
Heckler and Koch GmbH, D-7238 Oberndorf-Neckar.
Status
Current. Production.
Service
West German police forces. Military and police forces in many other countries. Wide commercial sales.

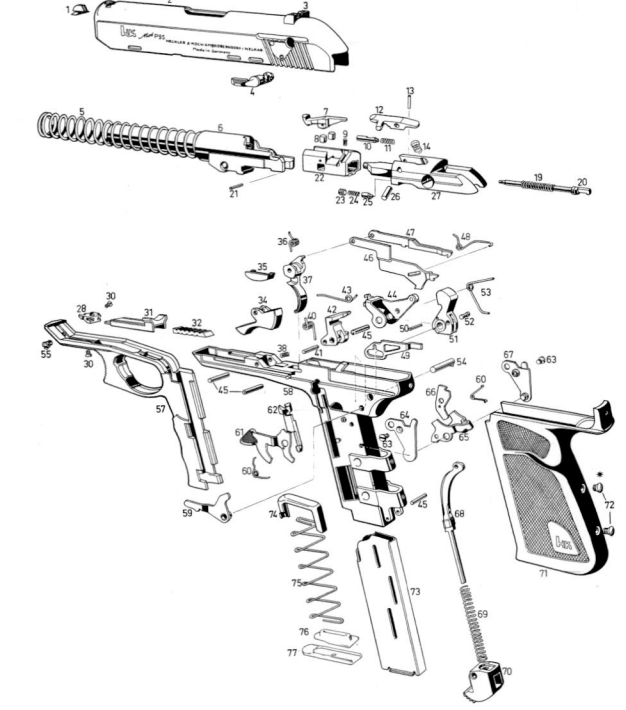

Exploded view of 9 mm P9S Heckler and Koch pistol

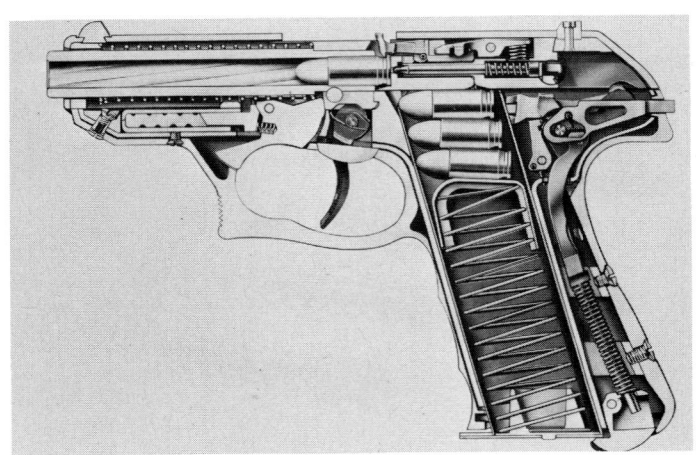

Cutaway view of 9 mm P9S Heckler and Koch pistol

9 mm P9S Heckler and Koch pistol

9 mm VP70M and VP70Z Heckler and Koch pistols

The VP70 is an automatic pistol of original and unusual design, making considerable use of plastics. As a hand-held weapon it fires only in the self-loading mode but in the case of the VP70M (M denoting military), as soon as the holster stock is attached it can also fire three-round bursts at a cyclic rate of 2200 rounds a minute. To make use of this burst fire ability a large capacity magazine is needed and the VP70 magazine takes 18 rounds.

The number of moving parts has been kept to a minimum by careful design; only four operating parts are necessary and the makers claim a life of 30 000 rounds. The receiver is plastic with support for the fixed barrel moulded in. The VP70Z (Zivil; meaning civilian), which does not accept a holster stock, is a variant of the VP70 and fires semi-automatically only.

Operation

To remove the magazine from the pistol for loading, the catch at the heel of the magazine is pressed back and the magazine is then ejected. The loaded magazine is pressed home in the grip until the catch engages and retains it. To cock the weapon the slide must be pulled back as far as possible and then released. The top round is picked up and chambered and the extractor snaps into the groove. When this is done the firer is aware that the chamber is loaded because the extractor is proud. The weapon can be carried in this condition with safety because it is not cocked.

The trigger is of unusual design. It is a double-action mechanism and when pulled straight back it has a first pressure that can be plainly felt. Further pressure causes the trigger bar to slip off the spring-loaded firing pin which goes forward to fire the cap.

The system of operation is pure blowback and the gas pressure generated inside the cartridge case pushes the bullet forward and the slide rearward. The empty case pushes the slide back and is then held to the breech face by the extractor until the ejector projects it out of the feedway, through the ejection port, to the right. The slide comes to rest, the compressed return spring re-asserts itself and as the slide goes forward the top round is fed into the chamber. The trigger is still held back and is disengaged from the firing pin. When it is released it moves forward and in front of the firing pin lug. Another trigger pull again pulls the firing pin back and the trigger bar slips off to allow the firing spring to drive the pin forward into the cap.

This type of trigger mechanism is very simple and, as in a revolver, no additional safety catch is fitted. When the ammunition is expended, the slide stops in the forward position and to renew firing a fresh magazine must be inserted and the slide withdrawn to chamber a round.

Stripping

To strip the weapon for maintenance and cleaning, the magazine is removed, the slide pulled back and the chamber and feedway visually inspected. The slide is released. Above the trigger-guard is a slide-retaining catch and when this is pulled down the slide can be pulled right back, lifted up and backwards and then allowed to slide forward. The firing pin is removed by rotating the end cap in the breech through 90° to the right and removing it. No further stripping is required. The barrel is permanently attached to the receiver and is cleaned *in situ*.

Sub-machine gun conversion

The holster plays a very important part in the concept of the VP70M. It provides a convenient transit holder enabling the firer to carry the pistol in almost any conceivable position. This is done by having a carrying plate to which the holster is attached and a harness that enables it to be worn on a waist belt, on either thigh, under either armpit with the butt facing forward or rearward at will. The holster can also be attached to the pistol to convert it into a sub-machine gun. The holster carries the selector lever in this mode, allowing single shots or bursts of three rounds for each trigger operation. The lever is on the left side of the holster/stock well forward and rotates down for single and up for multi-shots. To attach the holster to the pistol the fire selector lever is set to 1, that is single-shot fire, and the catch at the bottom of the holster is pressed back. There are two grooves at the back of the pistol and the stock is inserted into the grooves and pushed upwards until the latch engages.

9 mm VP70Z automatic Heckler and Koch pistol

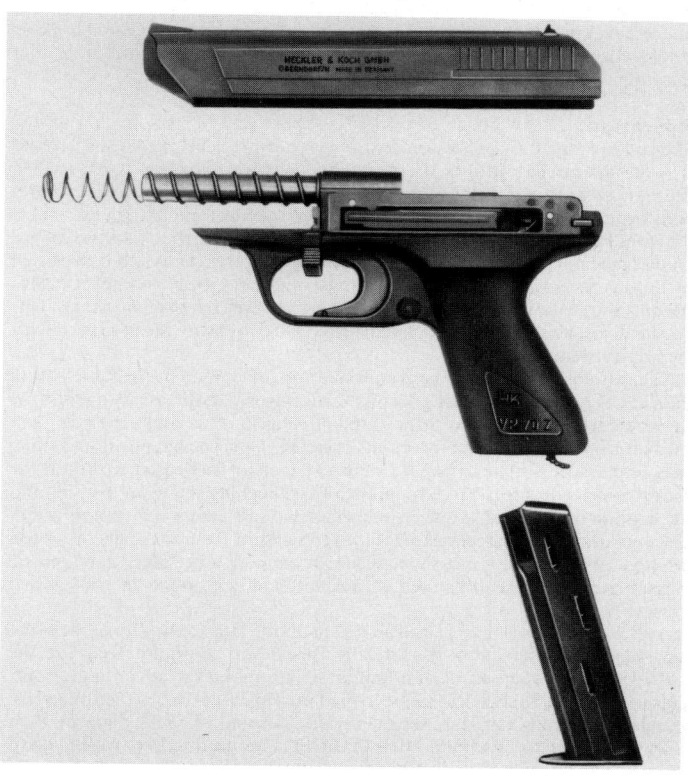
9 mm VP70M automatic Heckler and Koch pistol field-stripped

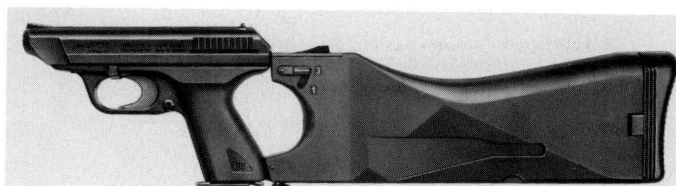
9 mm VP70M automatic Heckler and Koch pistol with holster stock attached to convert it to sub-machine gun

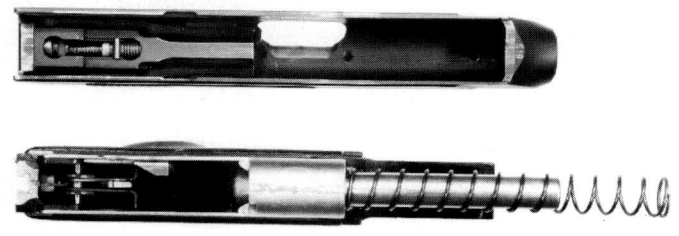
Slide removed from 9 mm VP70 automatic Heckler and Koch pistol to show trigger and spring-loaded firing pin

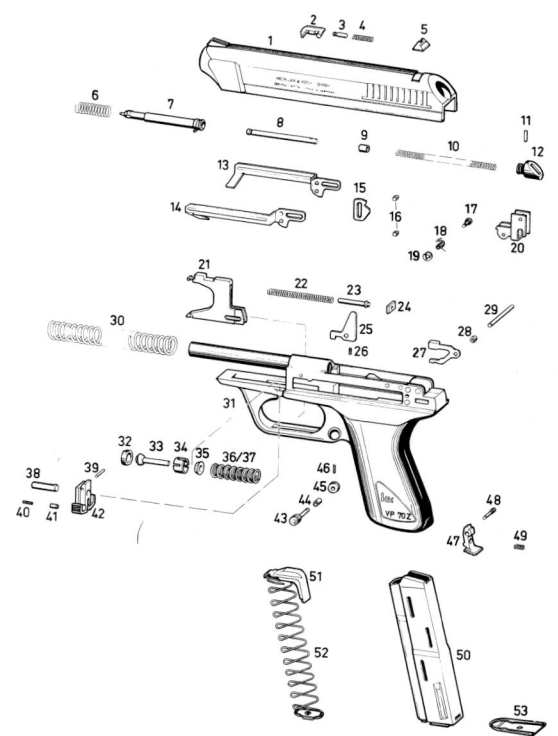
Exploded view of 9 mm VP70Z automatic Heckler and Koch pistol

DATA
PISTOL ONLY
Cartridge: 9 mm × 19
Operation: blowback, self-loading double-action
Method of locking: nil
Feed: 18-round box magazine
Weight, empty: 823 g
Length: 204 mm
Barrel: 116 mm
Rifling: 6 grooves rh
Sights: (fore) blade; (rear) square notch
Sight radius: 175 mm
Muzzle velocity: 360 m/s

DATA
PISTOL AND HOLSTER STOCK
Method of fire: single-shot or 3-round bursts
Weight, loaded: 1.59 kg
Length, with stock: 545 mm
Rate of fire: 2200 rds/min cyclic; 100rds/min bursts
Effective range: 150 m

Manufacturer
Heckler and Koch GmbH, D-7238 Oberndorf-Neckar.
Status
No longer in production.
Service
Commercial sales in Africa and Asia and in use by military and police forces in several countries.

9 mm P7 Heckler and Koch pistol

This weapon was developed by Heckler & Koch with the requirements of police forces primarily in mind. It is blowback operated, with a recoil braking system which delays breech opening, reduces the felt recoil and aids steadier shooting.

Operation
The action of the P7 is self-locked by the gas pressure developed when a round is fired. When the pistol is fired, part of the propellant gas is channelled through a small vent in the barrel ahead of the chamber and into a cylinder lying beneath the barrel. A piston, attached to the front end of the slide, enters the front end of this cylinder, and thus when the slide begins to move rearward under the recoil pressure, the movement of the piston in the cylinder is resisted by the gas pressure. This delays the movement of the slide, and hence delays the opening of the breech; it also tends to absorb some of the recoil shock. This system gives the advantage of a fixed barrel and does away with the need for a locking mechanism.

The pistol is loaded in the conventional way, by pulling back and releasing the slide. When the firer grasps the pistol, his fingers automatically depress the squeeze-cocking grip at the front of the pistol grip. This cocks the firing pin ready for the first shot, and it remains engaged in the cocked position so long as light pressure is maintained. As soon as the pistol is released, however, the cocking grip snaps forward, automatically de-cocking the firing pin. Should the weapon be dropped, it will be uncocked and safe before it hits the ground.' The cocking grip also releases the slide stop after a fresh magazine has been inserted. Since there is no slide release lever, nor any safety catch to be manipulated, the P7 can be used with equal facility by right- or left-handed firers.

Although the pistol grip is well angled at about 110° to the axis of the bore, the magazine enters almost vertically, providing optimum feed for the cartridges, even when using ammunition with unusual bullet configurations. This makes the P7 much less liable to feed malfunctions than other weapons. Should a misfire occur, the firer merely releases and re-cocks the grip, then pulls the trigger for a second attempt; there is no need to use two hands to re-cock. The pistol can be silently de-cocked by simply pulling back the slide about 10 mm, releasing the cocking grip, and then allowing the slide to go forward.

9 mm P7 M13 Heckler & Koch pistol

9 mm P7 M13 Heckler and Koch pistol cutaway

9 mm P7 M8 Heckler and Koch pistol

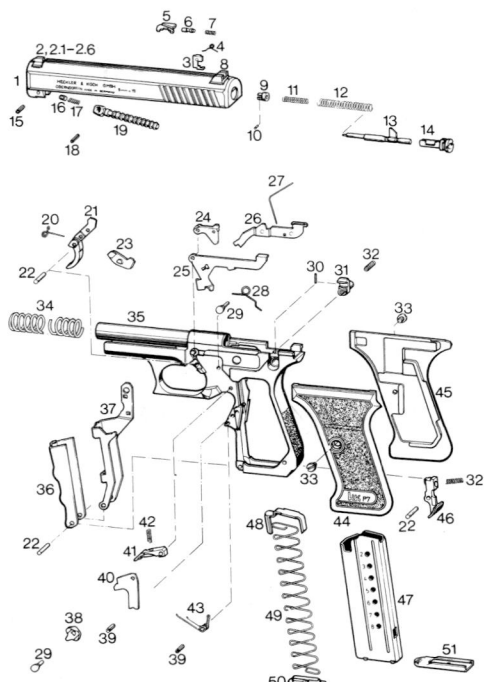

Exploded view of 9mm P7 Heckler and Koch pistol

Sights
The rearsight is adjustable for windage; adjustment for elevation is done by interchanging the foresight blade. These blades are available in different heights, and have their height engraved on the base. Both rear notch and front

blade are fitted with durable contrasting white markers for shooting in poor light, and 'Betalight' luminous markers are available if required.

There are two distinct models of the P7; the P7 M8 which takes an 8-round magazine, and the P7 M13 which takes a 13-round magazine. This obviously leads to some minor dimensional differences, which are noted in the following table in parenthesis for the P7 M13.

DATA
Cartridge: 9 × 19 mm Parabellum
Operation: delayed blowback, self-loading
Feed: 8 (13) round box magazine
Weight, loaded: 950 g (1135 g)
Length: 171 mm (175 mm)

Barrel: 105 mm
Sights: (fore) blade; (rear) notch (see text)
Sight radius: 148 mm
Muzzle velocity: ca. 351 m/s

Manufacturer
Heckler and Koch GmbH, D-7238 Oberndorf-Neckar.
Status
Current. Production.
Service
West German Police, Special Forces and Army; US Police forces; Military and police forces in many other countries, and wide commercial sales.

P7 K3 Heckler and Koch pistol

This is the most recent model in the P7 series, and its design and operation are broadly similar to the earlier P7 M8 and P7 M13 models. The principal differences are that the P7K3 is a blowback weapon without the gas delay system, using the less powerful 9 mm Short (.380 ACP) cartridge, and conversion kits are available to allow firing .22 Long Rifle or 7.65mm (.32 ACP) cartridges.

Operation
The P7 K3 is simple to operate. The filled magazine is pushed into the closed pistol until the magazine catch engages. The slide is fully retracted and then allowed to snap forward, chambering the top round from the magazine.

Instead of a conventional double-action trigger the P7 K3 features a squeeze cocker. When the user grasps the pistol in the firing position, his fingers simultaneously depress the squeeze cocker, automatically cocking the firing pin for the first shot. This cocking grip remains engaged in the cocked position for subsequent shots, but as soon as the grip on the pistol is released it snaps forward and automatically uncocks the firing pin. If a round fails to fire, the user simply eases his grip, squeezes again to recock the firing pin and pulls the trigger. The pistol may be silently uncocked by pulling the slide approximately 10 mm to the rear, releasing the squeeze cocker, and manually letting the slide go forward.

This system gives light trigger pull from the first shot onward, contributing to accuracy. It eliminates the need for a conventional firing pin system incorporating a hammer, which is more expensive, complex and space-consuming. There is no need for a lateral slide catch and release lever; thus the P7 K3 is equally suited to right- and left-handed users. It is an important safety feature; should the pistol be dropped, it is uncocked before it strikes the ground.

Although the pistol grip is at the ergonomically favourable angle of 110° relative to the barrel, it has been possible to position the magazine almost vertically to the barrel. This provides optimum cartridge feed from magazine to chamber, even when special ammunition with unconventional bullet configurations is used. The P7 K3 is claimed to be much less susceptible to malfunctions than conventional pistols.

After the last round has been fired, the slide remains open. To continue firing, replace the empty magazine with a filled one and then either depress the squeeze cocker, thus snapping the slide forward and automatically cocking the firing pin, or pull the slide back and let it snap forward. The pistol is now ready to fire again. The P7 K3 may be unloaded by actuating the ambidextrous magazine release, removing the magazine and fully retracting the slide to eject the chambered cartridge. After ensuring that the chamber is clear, allow the slide to snap forward again. The catch lever at the rear of the trigger-guard may be used to keep open the action without a magazine being inserted.

DATA
Cartridge: 9 mm × 17 Short/.380 ACP (and see text)
Operation: blowback, self-loading, single-action
Feed: 8-round detachable box magazine
Weight, empty: 750 g
Length: 160 mm
Barrel: 96.5 mm
Sights: (fore) blade; (rear) notch
Sight radius: 139 mm

P7 K3 Heckler and Koch pistol

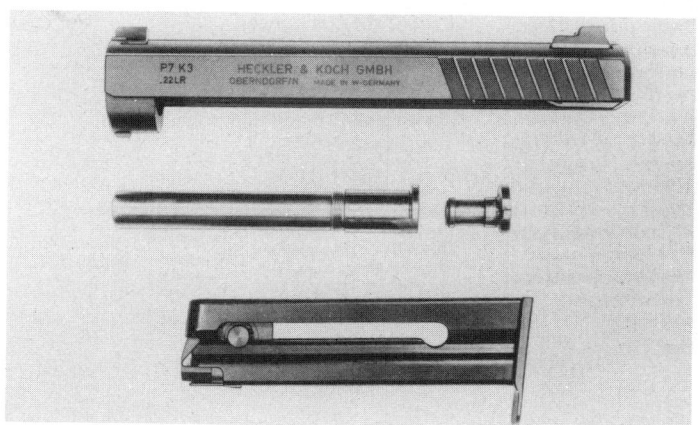
.22LR conversion kit for Heckler and Koch P7 K3 pistol. Similar conversion kit for .32/7.65 mm is also available

Manufacturer
Heckler and Koch GmbH, D-7238 Oberndorf-Neckar.
Status
Current. Production.
Service
Military and police forces in various countries and wide commercial sales.

GREECE

9 mm EP 7 pistol

The Hellenic Arms Industry (E.B.O.) SA have recently begun manufacture of several weapons, one of which is the Heckler and Koch P7 pistol.

Designated the EP 7 it is exactly the same as the West German pistol, to which reference should be made for specifications. It is marked Mod 'EP 7' with the Hellenic Arms Industry monogram (in the form of a diamond) and has been adopted by the Greek Armed Forces as well as being exported to various markets.

Manufacturer
Hellenic Arms Industry (E.B.O.) SA, 160 Kifissias Avenue, 11525 Athens, Greece.
Status
Current. Production.
Service
Greek military and security forces and for export.

9 mm Hellenic Arms Industry EP 7 pistol

HUNGARY

7.62 mm Model 48 pistol

The Soviet 7.62 mm Tokarev (TT-33) was manufactured in Hungary as the Model 48. The Hungarian pistol is identified by the crest on the grip (a star, wheatsheaf and a hammer surrounded by a wreath) and the uniform narrow vertical cuts on the slide for the firer to grip while cocking the pistol. All other details are the same as the Soviet Tokarev (TT-33). The pistol was produced by state arsenals and is still in service.

DATA
Cartridge: 7.62 mm × 25 Pistol Type P cartridge
Operation: short recoil, semi-automatic
Method of locking: projecting lugs
Feed: 8-round box magazine
Weight: 846 g
Length: 196 mm
Barrel: 116 mm
Rifling: 4 grooves rh
Sights: (foresight) blade; (rearsight) notch
Muzzle velocity: 420 m/s

Manufacturer
State arsenals.
Status
Current.

7.62 mm Model 48 pistol, Hungarian copy of Soviet Tokarev TT-33

Service
Hungarian armed forces.

9 mm (and 7.65 mm) PA-63 and AP pistols

Based on the West German Walther PP pistol, these are virtually the same weapon, the distinction being that the PA-63 is made only in 9 mm Makarov calibre and is the official sidearm of Hungarian military and police forces, while the Model AP is produced in both 7.65 ACP and 9 mm Short (.380 ACP) calibres for commercial sale. Both weapons are made with an aluminium frame and steel slide, the PA-63 having the frame left bright while the AP has the frame anodised black.

DATA
Calibre: 9 mm Makarov and 7.65 mm ACP
Operation: blowback, double-action, semi-automatic
Feed: (9 mm) 7- or (7.65 mm) 8-round detachable box magazine
Weight, empty: 595 g
Length: 175 mm
Barrel: 100 mm, 6 grooves rh
Sights: (fore) blade; (rear) notch
Muzzle velocity: 7.65 mm–ca 310 m/s

Manufacturer
Fegyver es Gazkeszuelekgyara NV, Soroksári út 158, H-1095 Budapest.
Status
Current. Production.
Service
Hungarian forces and export.

9 mm AP pistol

9 mm FEG Model FP9 pistol

This is an identical copy of the Browning High-power M1935GP, described under Belgium; the copy is so exact that many parts are completely interchangeable, though the Hungarian model adds a ventilated rib to the slide. It has not been adopted by Hungarian forces, because of the calibre, but is commercially available throughout Europe and is believed to have been adopted by some smaller police forces. Dimensions are almost identical with the GP35 (see Page 5) except that the FP9 weighs 950 g empty.

Manufacturer
Fegyver es Gazkeszuelekgyara NV, Soroksári út 158, H-1095 Budapest.
Status
Production.

7.65 mm FEG Model R pistol

This is a 7.65 mm pocket pistol which appears to be an original design, though the lockwork still carries traces of its Walther PP ancestry. It is a double-action blowback weapon, with a slide-mounted safety catch which drops the cocked hammer when applied. There is a stripping catch and lever on the left of the slide, ahead of the trigger-guard which allows the slide and barrel to be removed very easily. The slide and barrel are of steel, while the frame is of light alloy. The bore is chromium-plated.

DATA
Cartridge: 7.65 mm ACP
Operation: blowback
Weight: 450 g
Length overall: 140 mm
Barrel length: 72 mm
Magazine capacity: 6 rounds
Muzzle velocity: ca 300 m/s

Manufacturer
Fegyver es Gazkeszuelekgyara NV, Soroksári út 158, H-1095 Budapest.
Status
Production.

7.65 mm Model R pistol

9 mm FEG Model P9R and P9RA pistols

These pistols are derived from the FEG Model FP (above), but with locally-designed modifications to give them a more modern specification. The principal change is the adoption of double-action lockwork, together with a slide-mounted safety catch which lowers the hammer when applied. The operation of the safety catch locks the firing pin and interposes a positive stop between hammer and firing pin. In other respects the mechanism is identical to that of the Browning GP35 pistol.

The Model P9R has a steel frame; the Model P9RA has a light alloy frame. An interesting and unusual variation is the manufacture of a completely left-handed version of this pistol which has the safety catch, slide release and magazine catch on the right side of the frame.

DATA
Cartridge: 9 mm Parabellum
Operation: short recoil
Locking system: Browning cam
Weight: (empty) P9R–1.00 kg; P9RA–820 g; (loaded) P9R–1.17 kg; P9RA–980 g
Length overall: 203 mm
Barrel length: 118.5 mm
Magazine capacity: 14 rounds
Muzzle velocity: ca 380 m/s

Manufacturer
Fegyver es Gazkeszuelekgyara NV, Soroksári út 158, H-1095 Budapest.
Status
Production.

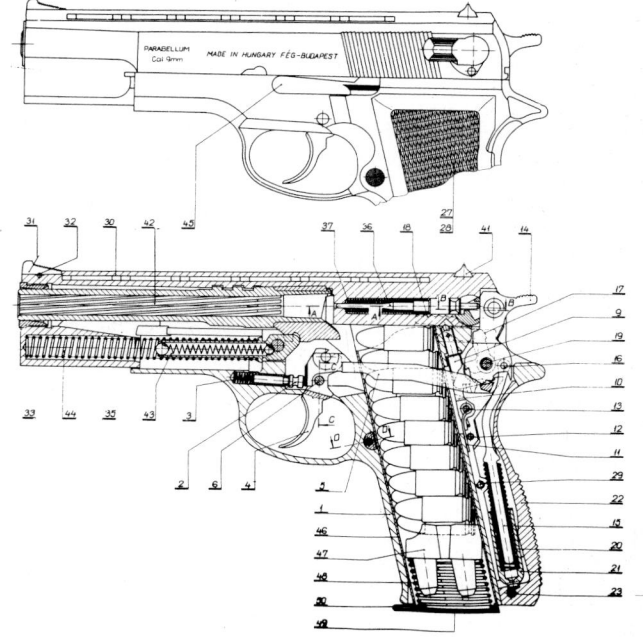

Mechanism of FEG Model P9R pistol

INDONESIA

9 mm Pindad pistol

The ordnance factory at Pindad manufactured the 9 mm High-power FN Browning pistol under the name 'Pindad'. It carries on the left side of the slide the marking 'Fabrik Sendjata Ringan' and below that 'Pindad' followed by 'P1A9mm'. The rearsight is a simple notch and there is no tangent ramp version in use.

DATA
Cartridge: 9 mm × 19 Parabellum
Operation: recoil, semi-automatic
Method of locking: projecting lug
Feed: 13-round detachable box magazine
Weight: 0.88kg
Length: 196 mm
Barrel: 112 mm
Rifling: 6 grooves rh
Sights: (fore) blade; (rear) notch
Muzzle velocity: 354 m/s

Manufacturer
Ordnance factory at Pindad.
Status
Obsolescent; being replaced by license-built Beretta 92S.
Service
Indonesian forces.

9 mm Pindad pistol

ISRAEL

Israeli Army pistols

The Israeli Army started with a wide variety of pistols. Some came from British sources and were mainly .455 Webley and .38 Enfield. There were also German P38 and Luger pistols and a few Pistols 9 mm High-power FN Browning. The Israeli 'Workers Industry for Arms' produced a six-chamber revolver of 9 mm calibre firing the Parabellum cartridge. This necessitates the use of three-round clips. The revolver is based on the Smith and Wesson .38 Military and Police revolver, weighs 0.9 kg, is 28 cm long, has a 155 mm barrel and is used by police units.

The military pistol which the Israeli Army adopted is the Italian 9 mm Model 1951 Beretta.

9 mm Model 1951 Beretta pistol currently used by Israeli armed forces

9 mm Uzi pistol

This is a shortened and lightened modification of the Uzi sub-machine gun, with a mechanism permitting semi-automatic fire only. Although appearing cumbersome by comparison with conventional pistols, it has the advantages of an exceptional magazine capacity for a pistol and a shape and size which allows a very firm two-handed grip. The bulk also helps to absorb recoil so that it is easy to control during the firing of a rapid succession of shots. It has been designed for civilian use, but there are obvious applications to military and security forces.

DATA
Cartridge: 9 mm Parabellum
Operation: blowback, closed breech
Weight: (empty) 1.65 kg; (loaded) 2.005 kg
Length overall: 240 mm
Barrel: 115 mm, 4 grooves, rh, one turn in 254 mm
Magazine capacity: 20 rounds
Muzzle velocity: ca 350 m/s
Effective range: 60 m

Manufacturer
Israel Military Industries, PO Box 1044, Ramat Hasharon, 47100.
Status
Production.

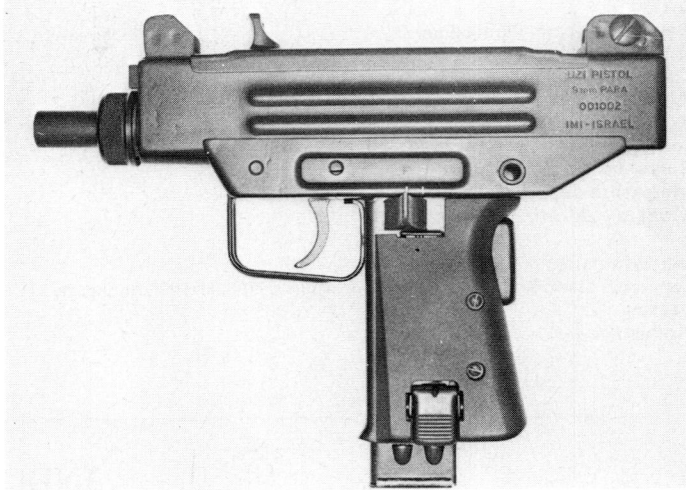

9 mm Uzi pistol

.357 Desert Eagle pistol

The Desert Eagle has been available for some time as a sporting pistol, but its undoubted utility has caused interest to be taken in other fields and it is now offered as a military weapon.

The design is unusual in that it is a gas-operated locked-breech pistol using a rotating bolt. The barrel is fixed to the frame and has a gas port just ahead of the chamber; this connects with a channel in the frame which runs forward to a point beneath the muzzle and there turns down to a gas cylinder inside which

is a short-stroke piston. The slide is formed into a receiver at the rear end, which contains the bolt, and into two side-arms which run forward below the barrel. On firing, a portion of the gas is diverted through the port and channel to drive the piston backwards. This drives the slide to the rear, against a spring. Movement of the slide first rotates the bolt, by a cam, to unlock it from the chamber, and then withdraws the bolt to extract and eject the fired case. The return spring, under the barrel, then drives the slide forward again; the face of the bolt chambers a fresh cartridge, and the final movement of the slide rotates the bolt and locks it into the chamber.

Another unusual feature of this weapon is that it fires the .357 Magnum rimmed cartridge, widely available as revolver ammunition. It will feed lead, semi-jacketed or full-jacketed cartridges with equal facility. A larger version in .44 Magnum calibre, also firing rimmed revolver cartridges, and the standard model chambered for the .41 Action Express cartridge are now available.

The pistol is equipped with an ambidextrous safety catch on the rear of the slide which locks the firing pin and also disconnects the trigger from the hammer mechanism. The standard sear assembly can be removed and replaced by a special assembly which allows adjustment of trigger pull for both length and weight. The pistol is normally fitted with combat sights, but an adjustable rearsight is available as an option, and the barrel is grooved to accept mounts for a sighting telescope.

.357 Desert Eagle pistol

DATA
Cartridge: .357 Magnum, .44 Magnum or .41 Action Express
Operation: Gas, semi-automatic, single action
Locking: rotating bolt
Feed: 9-round box magazine
Weight, empty: 1760 g (steel frame); 1466 g (aluminium frame)
Length overall: 260 mm
Barrel length: 152 mm (355 mm also available)
Rifling: 6 grooves rh, one turn in 355 mm
Sights: (fore) blade; (rear) U-notch
Sight radius: 225 mm

Manufacturer
Israel Military Industries, PO Box 1044, Ramat Hasharon 47100.

Status
Current. Production.

ITALY

7.65 mm Model 81 double-action Beretta pistol

Model 81 is one of three Beretta pistols which entered full-scale production in 1976. The other two are the 9 mm Short Model 84, which is identical to the Model 81 in most respects, and the larger Model 92 which fires the 9 mm Parabellum round and operates on the short recoil principle whereas the two smaller weapons are operated by blowback. All three pistols have a number of design features in common, however, and while the description that follows relates primarily to Model 81 it covers points that are relevant also to the other two weapons.

Operation
To load the weapon a loaded magazine is inserted in the butt and the slide operated manually to chamber a round. The firing pin is spring-loaded and shorter than the breech-block: it thus requires a sharp blow to cause it to overcome the spring resistance and fire the cartridge so that the hammer can be safely lowered under thumb restraint without firing a round. A manual safety, operable from either side of the weapon locks both the trigger mechanism and the slide in the closed position.

When the round is fired the pressure in the chamber drives the cartridge case, together with the slide assembly, back against the combined pressure of the recoil spring and hammer spring, the case being held on the face of the breech-block by the extractor until it strikes the ejector. If the magazine is empty the slide is held open. If the magazine is not empty the slide will move forward to chamber a round and when it has done so the extractor will protrude laterally, showing red, and can thus be seen and felt. The hammer will remain cocked and the next round can be fired by single action.

In the event of a misfire, there will be no blowback action and the hammer will be forward. A second attempt to fire the round can then be made by releasing the trigger and pulling it again.

Other features of the weapon include a 12-round magazine with a staggered loading arrangement, a reversible magazine release button to suit right- or left-handed firers, a stripping catch arrangement, which makes stripping easy

but guards against accidental disassembly, and an optional magazine safety which prevents the weapon from being fired when the magazine is removed and a cartridge remains in the chamber.

Stripping
To strip the pistol, first ensure that the chamber is empty and then remove the magazine. Hold the pistol in the left hand and with the left forefinger then

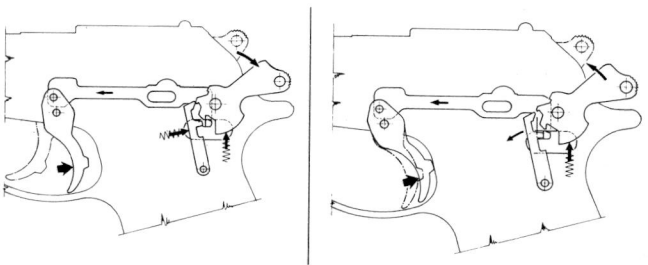

Trigger action

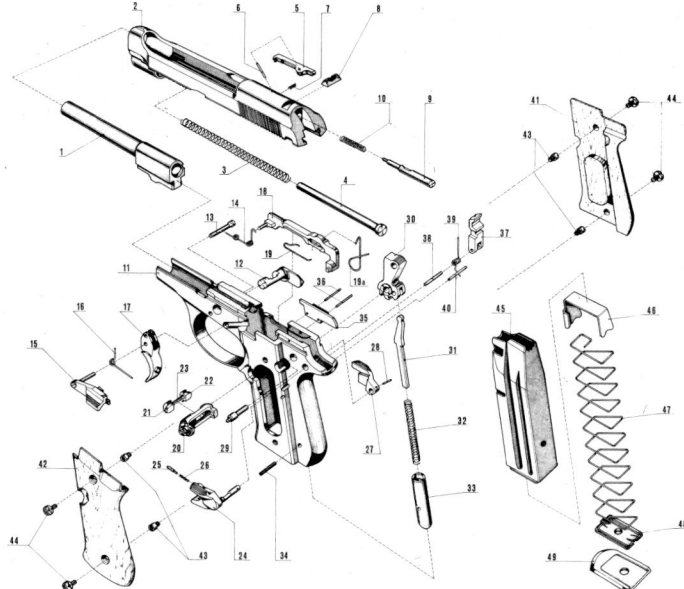

7.65 mm Models 81 and 84 double-action Beretta pistols
(1) *barrel* **(2)** *slide* **(3)** *recoil spring* **(4)** *recoil spring guide* **(5)** *extractor* **(6)** *extractor pin* **(7)** *extractor spring* **(8)** *rearsight* **(9)** *firing pin* **(10)** *firing pin spring* **(11)** *frame* **(12)** *disassembling latch* **(13)** *disassembling latch release button* **(14)** *trigger spring* **(15)** *slide catch* **(16)** *slide catch spring* **(17)** *trigger* **(18)** *trigger-bar* **(19)** *trigger-bar spring* **(19a)** *magazine safety spring* **(20)** *magazine release button* **(21)** *magazine release button spring bush* **(22)** *magazine release button spring bush* **(23)** *magazine release button spring* **(24)** *left safety* **(25)** *safety spring pin* **(26)** *safety spring* **(27)** *right safety* **(28)** *safety pin* **(29)** *hammer pin* **(30)** *hammer* **(31)** *hammer spring strut* **(32)** *hammer spring* **(33)** *hammer strut guide* **(34)** *hammer strut guide pin* **(35)** *ejector* **(36)** *ejector pin (2 pieces)* **(37)** *sear* **(38)** *sear pin* **(39)** *sear spring* **(40)** *sear spring pin* **(41)** *right grip* **(42)** *left grip* **(43)** *grip bush (4 pieces)* **(44)** *grip screw (4 pieces)* **(45)** *magazine box* **(46)** *magazine follower* **(47)** *magazine spring* **(48)** *magazine plate* **(49)** *magazine bottom*

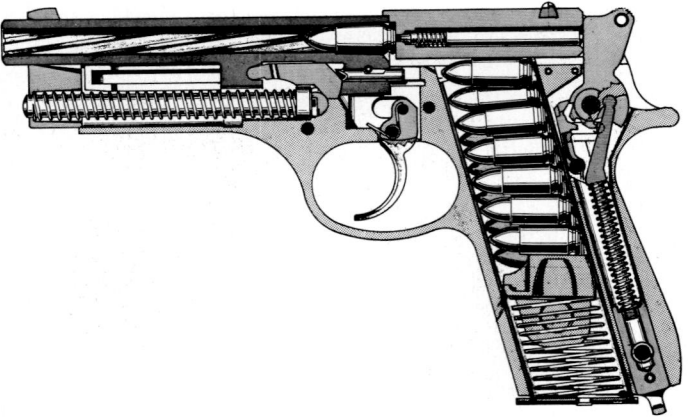

7.65 mm Model 81 double-action Beretta pistol

Sectional drawing of 7.65 mm Models 81 and 84 double-action Beretta pistols

press the disassembling latch release button, with the right thumb rotte the disassembling latch (on the right of the weapon) anti-clockwise as far as it will go. Still holding the pistol in the left hand, pull forward the slide barrel group with the recoil spring and guide. Remembering that the recoil spring and guide are under compression, slightly press the spring and guide to disengage them from the barrel then allow the spring to stretch slowly. Remove the spring and guide and take the barrel out of the slide. Further stripping is not normally required.

DATA
Calibre: 7.65 mm
Operation: blowback, single- or double-action
Feed: 12-round detachable box magazine
Weight: 670 g with empty magazine

Length: 172 mm
Barrel: 97 mm
Rifling: 6 grooves rh. 250 mm pitch
Sights: (foresight) blade, integral with slide; (rearsight) notched bar dovetailed to slide
Sight radius: 124 mm
Muzzle velocity: 300 m/s nominal
Safety devices: manual applied safety; inertia operated firing pin; half-cock position; optional magazine safety

Manufacturer
Pietro Beretta SpA, Gardone Valtrompia, Brescia.
Status
Current. No longer in production.
Service
Many law-enforcement agencies in and outside Europe. Also extensive commercial sales.

9 mm Model 1951 Beretta pistol

The Model 1951 is the standard pistol of the Italian armed forces and is is also used by both the Israeli and Egyptian Armies and in Nigeria.

The weapon has three main parts; the frame, barrel and slide. The frame holds the magazine, trigger and firing mechanism and has a forward extension to take the slide. The barrel carries a swinging locking piece pivoting from a lug on the underside, and the slide fits over the frame, sliding in grooves.

To load the pistol the magazine release in the lower left side of the grip is pressed and the magazine removed. When the loaded magazine is in place the pistol is cocked by pulling the slide fully back and then releasing it. The safety catch is a push-through type mounted at the top rear of the butt. 'Safe' comes from pushing from right to left.

When the pistol is fired, the breech-block, integral with the slide, goes back and the barrel which is locked to the slide goes back with it. After a short period of free travel of about 13 mm the unlocking plunger on the rear barrel lug strikes the receiver and stops. As the barrel and slide continue back the locking piece strikes the stationary plunger and is forced down into recesses in the slide in a manner similar to that of the Walther P38. The barrel comes to rest but the slide continues rearward for a further 5 cm. The hammer is

re-cocked and the empty cartridge case is extracted. The return spring then pushes the slide forward and the feed rib pushes the next cartridge into the chamber. The slide picks up the barrel and the locking piece on the barrel lug is lifted up by the receiver cam to lock the barrel to the slide. The forward motion of the slide and barrel stops when the barrel reaches the take-down lever spindle.

When the trigger is pulled to fire the first shot, the trigger bar moves back and rotates the sear on its pivot to release the hammer. The hammer spring rotates the hammer on to the firing pin and the cartridge is fired. If there is a misfire the hammer must be manually re-cocked for a second blow. There is no double-action.

Stripping
Remove the magazine, retract the slide and check that the chamber is empty by inspecting it.
Hold the slide to the rear until the take-down lever is aligned with the notch in the slide. Rotate the take-down lever forward and the slide will come forward off the receiver.
Disengage the return spring guide from the housing and then lift it up and back. Invert the slide, press on the unlocking plunger and then lift the rear of the barrel from the slide and pull it out backwards.

DATA
Cartridge: 9 mm × 19 Parabellum
Operation: short recoil, self-loading, single-action
Method of locking: swinging arm
Feed: 8-round box magazine
Weight, empty: 870 g with steel slide, 780 g with alloy slide
Length: 203.2 mm
Barrel: 114.2 mm
Rifling: 6 grooves, rh, one turn in 254 mm
Sights: (fore) blade; (rear) square notch adjustable for windage
Sight radius: 140 mm
Muzzle velocity: 350 m/s

Manufacturer
Pietro Beretta SpA, Gardone Valtrompia, Brescia.
Status
No longer in production.
Service
Italian armed forces. Also in use with Egyptian and Israeli forces, Nigerian police and in some other countries.

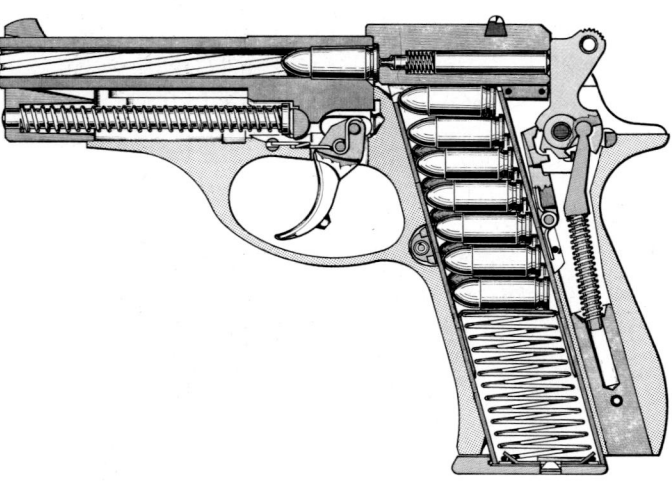

9 mm Model 1951 Beretta pistol

Italy/**PISTOLS** 35

9 mm Model 84 double-action Beretta pistol

This Beretta pistol resembles the 7.65 mm Model 81 described above in all respects save those that are relevant to the change of calibre to 9 mm Short. Components affected are the barrel, magazine box, magazine follower and magazine spring and it should be noted that the magazine capacity of the Model 84 is 13 rounds instead of 12. Details of the characteristics that distinguish the Model 84 from the Model 81 are given below: all other characteristics are the same and the drawings illustrating the entry for the Model 81 are relevant also to the Model 84. The accompanying photograph is of the right-hand side of the Model 84: apart from the markings this is indistinguishable from that of the Model 81 and the same is true of the left-hand illustration in the previous entry.

DATA
Calibre: 9 mm Short
Feed: 13-round detachable box magazine
Weight: 660 g with empty magazine
Muzzle velocity: 280 m/s nominal
Muzzle energy: 235 J

Manufacturer
Pietro Beretta SpA, Gardone Valtrompia, Brescia.
Status
Current. No longer in production.

9 mm Model 84 double-action Beretta pistol

Service
Law-enforcement agencies in many parts of the world and commercial sales.

Models 82BB, 85BB, 85F, 87BB and 87BB/LB Beretta pistols

These pistols are derivatives of the Models 81 and 84 and in most of the essential features they are the same. The Model 82BB is in 7.65 mm, the 85BB and 85F in 9 mm Short and the Models 87BB and 87BB/LB in .22 RF calibre. For the general method of operation and stripping reference should be made to the entry for the Model 81. The differences between these models and the 81 are as follows

BB Models:
Chamber loading indicator: a pin projects laterally from the slide to indicate when a round is in the chamber. This pin is coloured red and it gives both a visual and a tactile indication.
Magazine capacity: the magazine is smaller so allowing a lighter and thinner grip. This favours the smaller hand and also makes the pistols easier to conceal.

New safety system: this is really divided into four sections: the manual safety breaks the connection between the trigger and the sear; the firing pin is permanently locked until the last stage of the trigger pull when it is released; the firing pin is operated by inertia and so there is no direct contact with the primer; there is a half-cock position. It is worth noting that these pistols can be fired from the half-cock position by simply pulling the trigger to raise and drop the hammer.

F Models:
These have all the features of the BB models described above, with the addition of a hammer de-cocking facility built in to the applied safety system. To lower the hammer safely, the manual safety catch is applied; this drops the

7.65 mm Model 82BB Beretta pistol. Chamber loaded indicator is directly below and just forward of rearsight

.22RF Model 87BB Beretta pistol

.22LR Model 87BB/LB Beretta pistol

9 mm Short Model 85BB Beretta pistol, nickelled version

hammer against an interceptor, locks the slide in the closed position and interrupts the connection between trigger and sear. The half-cock hammer position is not available in these models. The F models also have the barrel and chamber chromium-plated.

Model 87BB/LB

This model differs from the remainder of the group in being single action only. The notation '/LB' indicates the use of a long (150 mm) barrel.

DATA
AS FOR MODELS 81 AND 84 EXCEPT:
DATA
Cartridge: Model 82BB: 7.65 mm ACP; Mods 85BB,85F: 9 mm Short;Models 87BB,87BB/LR: .22 Long Rifle RF
Feed: (82BB) 9-round magazine; (85BB,85F,87BB,87BB/LB) 8-roundmagazine
Weight, empty: 82BB: 630 g; 85BB,85F: 620 g; 87BB: 570 g;87BB/LB: 660 g

Manufacturer
Pietro Beretta SpA, Gardone Valtrompia, Brescia.
Status
Current. Production.
Service
Law enforcement agencies in many parts of the world, and commercialsales.

9 mm Short Model 85F Beretta pistol

Models 81BB, 84BB and 84F Beretta pistols

Although of greater magazine capacity than Models 82 and 85 described above, Models 81BB and 84BB are, like them, derivatives of Models 81 and 84 and resemble them in most essential features. Model 81BB is in 7.65 mm; Model 84BB in 9 mm Short. For the general method of operation and stripping, reference should be made to the entry for Model 81.

Models 81BB and 84BB are double-action pistols, incorporating staggered detachable box magazines (12-round capacity in the Model 81BB; 13-round capacity in the Model 84BB); chamber loading indicator; reversible magazine release button; stripping catch arrangement (as described for Model 81) and a manual safety operable from either side of the weapon. The safety system is as described for the Models 82BB and 85BB. A magazine safety, acting on the trigger mechanism when the magazine is removed, is optional.

The Model 84F is similar to the BB series but has the additional feature of a hammer de-cocking facility, operated by the safety catch. Pressing the safety catch will allow the hammer to fall safely on to an interceptor bar, lock the slide and disconnect trigger and sear.

The front and back straps of the grips are longitudinally grooved to ensure a firm hold in wet conditions or during rapid firing. Plastic grips are optional. It is also available in a limited de luxe series featuring gold-plated hammer and trigger, walnut grips and blued or gold-plated finish.

7.65 mm Model 81BB Beretta pistol

DATA
MODEL 81BB AS FOR MODEL 81, EXCEPT:
Operation: blowback, semi-automatic, double-action
Model 84BB as for Model 84

Manufacturer
Pietro Beretta SpA, Gardone Valtrompia, Brescia.
Status
Current; production.
Service
Law enforcement agencies in many parts of the world, and commercialsales.

9 mm Model 84F Beretta pistol

9 mm Model 86 Beretta pistol

Introduced in 1985, this is a pocket blowback pistol which would appear to have applications in police and security roles. It is unusual in employing a tip-up barrel, a design feature which was once common but which has not been seen for many years.

The Model 86 is of modern appearance and has double-action lockwork with an external hammer. There is a manual ambidextrous safety catch which acts to block the firing pin, and a hammer release lever which alows the hammer to be safely lowered, and a red chamber-loaded indicator. The barrel is pivoted beneath the muzzle, and by pressing a catch the breech end is released and allowed to rise, under spring pressure. This simplifies cleaning the barrel without having to dismantle the gun, and it also permits an additional

cartridge to be loaded into the chamber when a full magazine has been inserted into the butt. The pistol frame is of anodised light alloy, and walnut or plastic grips are available.

DATA
Cartridge: 9 × 17 mm Short/.380 ACP
Operation: blow-back, semi-automatic
Feed: 8-round box magazine
Weight, empty: 660 g
Length overall: 185 mm
Barrel length: 111 mm

9 mm Model 86 Beretta pistol

9 mm Model 86 Beretta pistol with barrel raised

Rifling: 6 grooves rh
Sights: (front) blade; (rear) U-notch
Sight radius: 127 mm
Muzzle velocity: ca. 310 m/s

Manufacturer
Pietro Beretta SpA, Gardone Valtrompia, Brescia.
Status
Current. Available.

.22LR Model 89 Beretta pistol

The Beretta Model 89 has been designed as a target pistol which, due to its general configuration, can also be used as a training pistol for heavier service weapons.

The Model 89 is a simple blowback semi-automatic with a heavy fixed barrel mounted on a light alloy frame. There is an external hammer, ambidextrous safety catch, a rearsight fully adjustable for both elevation and windage, and an adjustable trigger stop. The front sight is interchangeable, allowing infinite degrees of adjustment for zero. A magazine safety device is fitted.

DATA
Cartridge: .22 Long Rifle rimfire
Operation: blowback, semi-automatic
Feed: 8-round box magazine
Weight, empty: 1160 g
Length overall: 240 mm
Barrel length: 152 mm
Rifling: 6 grooves, rh, one turn in 350 mm
Sights: (front) interchangeable blade; (rear) fully adjustable U-notch
Sight radius: 185 mm
Muzzle velocity: 345 m/s nominal

Manufacturer
Pietro Beretta SpA, Gardone Valtrompia, Brescia.

.22LR Model 89 Beretta pistol

Status
Current. Production.

9 mm Model 92 double-action Beretta pistol

Firing the 9 mm Parabellum round, this third member of the family of Beretta pistols put into production in 1976 is both larger and more powerful than the Models 81 and 84: it also employs a short recoil operating system in place of the blowback system suited to the less powerful rounds of the smaller weapons.

In most general design respects, however, the Model 92 clearly resembles the other pistols. It has a double-action trigger system working on the same principles, a similar firing pin assembly and a similar arrangement for stripping. The short recoil system uses a falling locking block which is driven down to disengage the slide from the barrel and halt the rearward motion of the barrel but otherwise the extraction, cocking and loading operations are similar to those of the smaller weapons; the extractor provides the same loaded-chamber indication and the slide is held to the rear when the magazine is empty.

Stripping
Although the method of stripping generally resembles that of the smaller weapons there are some differences of detail. First remove the magazine by pressing the magazine release button and ensure that the chamber is empty. Hold the pistol in the right hand, press the disassembling latch release button with the left forefinger and rotate the latch, with the left thumb, anti-clockwise until it stops. Pull forward the slide barrel group with the locking block, recoil spring and guide. Remembering that they are under compression, press the recoil spring and guide sufficiently to disengage them from the barrel then lift them, allowing the spring to stretch slowly. Press the locking block plunger and remove the barrel and locking group from the slide. Further stripping requires the services of an expert. Assemble in reverse order,

and when rotating the disassembling latch ensure that the slide is aligned with the rear end of the frame.

DATA
Calibre: 9 mm Parabellum
Operation: short recoil, semi-automatic with single- or double-action

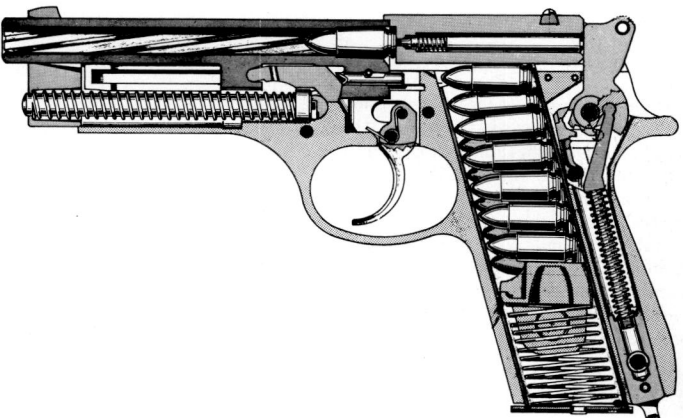

Sectional drawing of 9 mm Model 92 double-action Beretta pistol

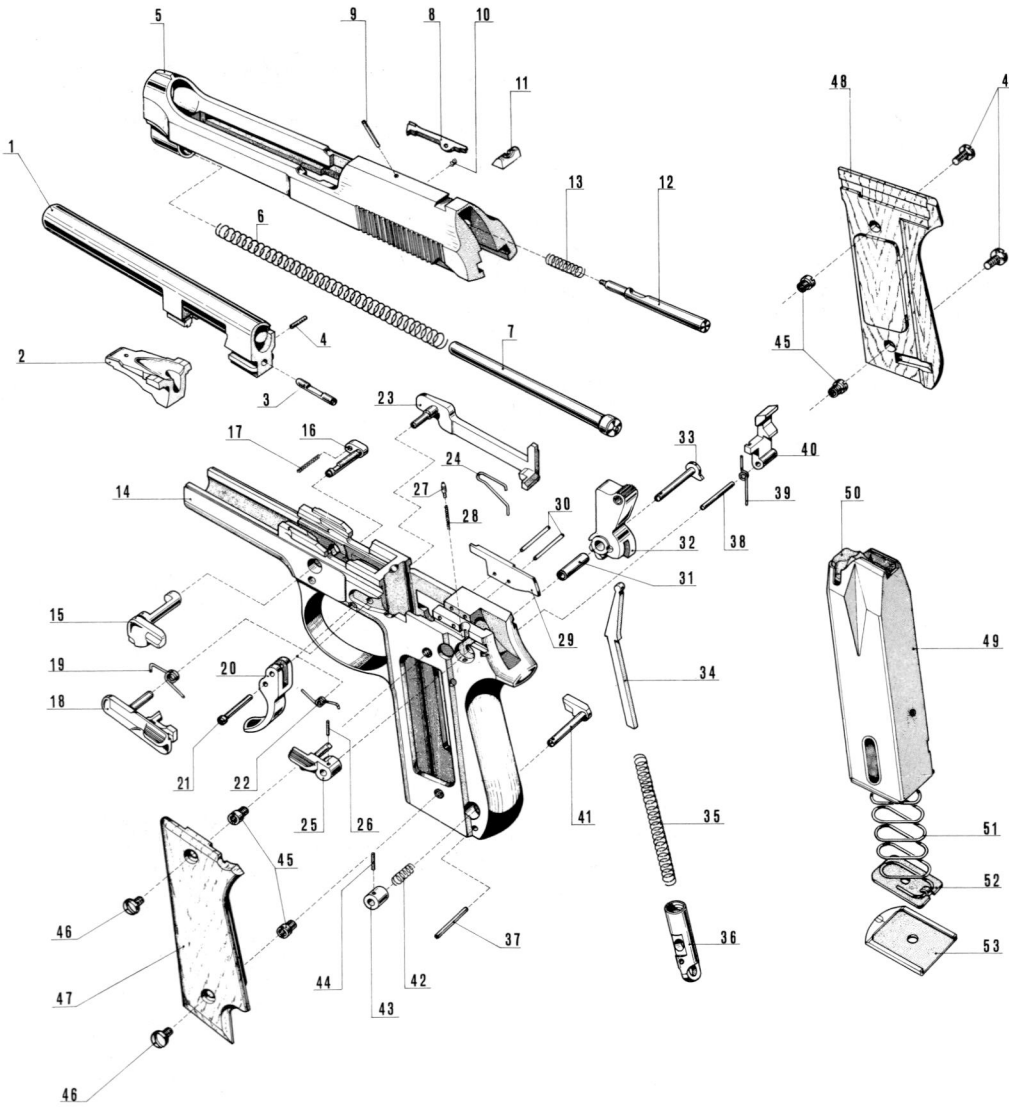

Model 92 double-action Beretta pistol
(1) barrel (2) locking block
(3) locking block plunger (4) locking block plunger pin (5) slide (6) recoil spring (7) recoil spring guide
(8) extractor (9) extractor pin
(10) extractor spring (11) rear sight
(12) firing pin (13) firing pin spring
(14) frame (15) disassembling latch
(16) disassembling latch release button (17) disassembling latch release spring (18) slide catch
(19) slide catch spring (20) trigger
(21) trigger pin (22) trigger spring
(23) trigger bar (24) trigger bar spring (25) safety (26) safety pin
(27) safety spring pin (28) safety spring (29) ejector (30) ejector pin
(2 pieces) (31) hammer bush
(32) hammer (33) safety pin
(34) hammer spring strut
(35) hammer spring (36) hammer strut guide (37) hammer strut guide pin (38) sear pin (39) sear spring
(40) sear (41) magazine release button (42) magazine release button spring (43) magazine release button bush (44) magazine release button pin (45) grip bush (4 pieces)
(46) grip screw (4 pieces) (47) left grip (48) right grip (49) magazine box (50) magazine follower
(51) magazine spring (52) magazine plate (53) magazine bottom

Method of locking: falling block
Feed: 15-round detachable magazine
Weight: 950 g with empty magazine
Length: 217 mm
Barrel: 125 mm
Rifling: 6 grooves rh. 250 mm pitch
Sights: (foresight) blade integral with slide; (rearsight) notched bar dovetailed to slide
Sight radius: 155 mm
Muzzle velocity: 390 m/s nominal
Safety devices: manual applied safety; intertia-operated firing pin; half-cock position

Manufacturer
Pietro Beretta SpA, Gardone Valtrompia, Brescia.
Status
Current. No longer in production.
Service
Italian forces and some foreign armies.

9 mm Model 92 double-action Beretta pistol

9 mm Model 92 S Beretta pistol

This closely resembles the Model 92 described above but has a modified safety mechanism. Whereas in the Model 92 the safety is mounted on the frame (as clearly shown in the illustrations in the preceding entry) the Model 92 S safety is mounted on the slide and provides a safe de-cocking facility. When applied it deflects the firing pin from the hammer head, releases the hammer and breaks the connection between the trigger bar and the sear.

If the weapon is cocked, therefore, application of the safety will allow the hammer to fall safely into the uncocked position; if the weapon is not cocked, operation of the trigger will not operate the hammer; and if the hammer should be inadvertently operated by some other means the weapon will not fire even though there may be a round in the chamber.

9 mm Model 92 S pistol showing manual safety mounted on slide

DATA
GENERALLY AS MODEL 92 (SEE ENTRY ABOVE) EXCEPT:
Weight: 980 g with empty magazine

Manufacturer
Pietro Beretta SpA, Gardone Valtrompia, Brescia.

Status
No longer in production.
Service
Italian armed forces and police and some foreign armies.

9 mm Model 92 SB Beretta pistol

The Model 92 SB is a direct development of the 92 S to which reference should be made for the main details. The SB differs in the following features: the safety lever is on both sides of the slide, allowing the pistol to be used by left-handers without alteration; the magazine release button has been moved to a position underneath the trigger-guard where it can be pressed without moving the hand from the grip. The button can be switched from left to right side to allow for left-handed firing.

A new series of safeties comprises the following items: the manual safety disengages the trigger from the sear; the firing pin is permanently locked until the last movement of the trigger on firing; the firing pin is inertia operated; there is a half-cock position.

The butt is grooved in the front and rear to improve the grip.
All data is identical with the Model 92 S pistol above.

Manufacturer
Pietro Beretta SpA, Gardone Valtrompia, Brescia.
Status
Current. Production.
Service
Italian Armed Forces (Special Units) and Police forces; some foreign armies.

9 mm Model 92 SB pistol

9 mm Model 92 SB pistol

9 mm Model 92 SB-C Beretta pistol

The SB-C is a smaller and handier version of the Model 92 SB. The main differences are in the size and the magazine capacity. All others features are the same as for the Model 92 SB above.

DATA
Magazine capacity: 13 rounds
Length overall: 197 mm
Height: 135 mm
Barrel length: 109 mm

Manufacturer
Pietro Beretta SpA, Gardone Valtrompia, Brescia.
Status
Current. Production.
Service
Police forces in Italy and abroad.

9 mm Model 92 SB-C pistol

9 mm Model 92 SB-C pistol

9 mm Model 92 SB-C Type M Beretta pistol

The SB-C Type M is almost the same as the SB-C but has an eight-round single-column magazine with a special base providing a small rest for the finger, and a slightly curved front face to the grip frame. Other features, and dimensions, are as for the SB-C.

Manufacturer
Pietro Beretta SpA, Gardone Valtrompia, Brescia.
Status
Current. Production.

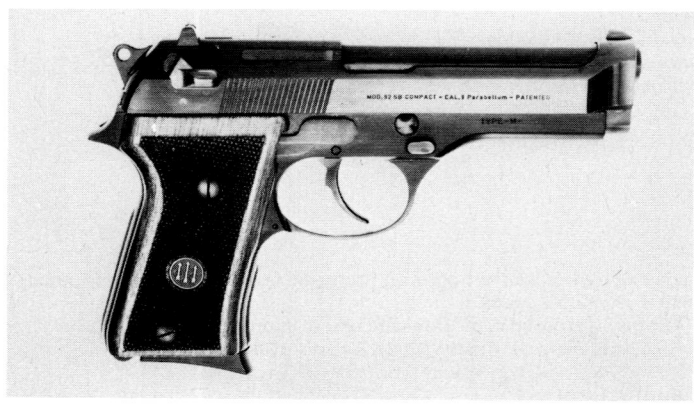

9 mm Model SB-C Type M pistol

9 mm Model SB-C Type M pistol

Service
Law enforcement agencies and commercial sales.

9 mm Model 92F Beretta pistol

The Model 92F is dimensionally and mechanically the same as the Model 92 SB. It differs in having the trigger-guard formed to suit a two-handed grip, an extended base to the magazine to improve the grip, a curved front edge to the grip frame, new grip plates and a new lanyard ring. The barrel is chromed internally, and the pistol is externally finished in 'Bruniton', a Teflon-type material.

Competition conversion kit
For competition purposes, the pistol can be converted by replacing the existing barrel with a 185 mm barrel, adding a counterweight, and fitting anatomical wooden grips and target sights. These items, together with a carrying case for the converted pistol, are available as a kit.

Manufacturer
Pietro Beretta SpA, Gardone Valtrompia, Brescia.
Status
Current. Production.
Service
US Army, Navy, Marine Corps, Air Force and Coastguard. French Gendarmerie Nationale. Law enforcement agencies worldwide and commercial sales.

9 mm Model 92F competition conversion

9 mm Model 92F pistol

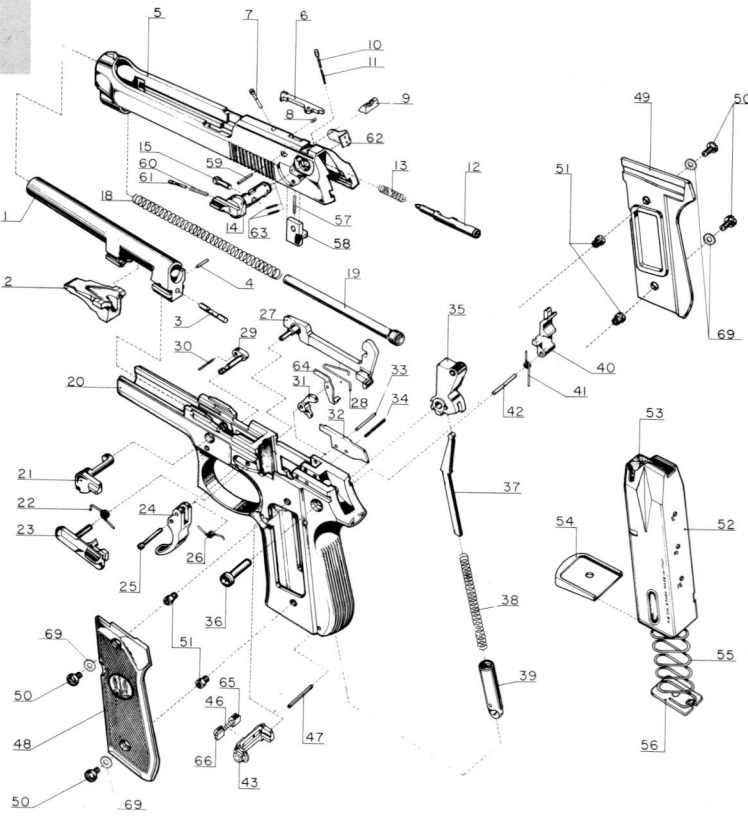

Exploded drawing of the components of Beretta 92F and 92F Compact pistols

9 mm Model 92F Compact Beretta pistol

The 92F Compact is the 92 SB-C modified in the same manner as the Model 92F above; that is, improvements designed to facilitate handling and shooting comfort. It is dimensionally the same as the Model 92 SB-C.

Manufacturer
Pietro Beretta SpA, Gardone Valtrompia, Brescia.

Status
Current. Production.
Service
Law enforcement agencies and commercial sales.

9 mm Model 92F Compact pistol

9 mm Model 92F Compact pistol

7.65 mm Model 98 Beretta pistol

This is the same as the Model SB-C but chambered for the 7.65 mm Parabellum cartridge. It is primarily intended for police use. Data differs from the SB-C only in the weight, which is approximately 900 g empty.

Manufacturer
Pietro Beretta SpA, Gardone Valtrompia, Brescia.
Status
No longer in production.
Service
Commercial sales.

7.65 mm Model 98 pistol

7.65mm Model 98F Beretta pistol

The Model 98F is identical to the Model 92F (above) but is chambered for either the 7.65 mm Parabellum or 9 × 21 mm cartridge. The only difference lies in the ballistic performance.

Manufacturer
Pietro Beretta SpA, Gardone Valtrompia, Brescia.
Status
Current. Production.

9 mm Model 98F Target Beretta pistol

Like the 98F, of which it is a variant model, the 98F Target is generally the same as the 92F but differs in being chambered for the 9 × 21 mm cartridge, thus permitting it to be sold commercially in several countries where 'military' calibres are prohibited. It is also fitted with a fully adjustable rear sight, wooden anatomical grips and a muzzle counterweight. Dimensions are as for the 92F except for overall length of 242 mm, barrel length of 150 mm, and an empty weight of 1100 g.

Manufacturer
Pietro Beretta SpA, Gardone Valtrompia, Brescia.
Status
Current. Production.

9 mm Model 98F Target Beretta pistol

7.65 mm Model 99 Beretta pistol

Also intended for police use, the Model 99 is the same as the SB-C Type M but chambered for the 7.65 mm Parabellum cartridge. Data as for the Type M except that the empty weight is approximately 900 g.

Manufacturer
Pietro Beretta SpA, Gardone Valtrompia, Brescia.

Status
Current. No longer in production.
Service
Commercial sales.

7.65 mm Model 99 pistol

---

9 mm Model 93R selective fire Beretta pistol

The Model 93R is an advanced self-loading pistol which can fire either single shots or three-round bursts and it thus falls more into the category of 'machine pistol' in the true sense of the word. However, it is meant to be carried and normally used in the same way as a single-handed pistol and as such it handles as a slightly large 9 mm self-loading pistol of conventional design. It can be used and fired just like any other pistol, but should the firer wish to engage a target beyond the normal pistol range, or even a difficult one close at hand, he can quickly fold down the front handgrip and hold the pistol with both hands. This hold is far steadier than the much-publicised modern method of clasping the butt with two hands for, with the forward handgrip, the two hands are a finite distance apart and able to exert a sensible control on the direction of the barrel.

For firing three-round bursts at any range it is essential that both hands are used, and if there is time, the firer is recommended to fit the folding carbine stock and take proper aim with the weapon in the shoulder. With both hands holding the pistol the right hand holds the butt in the normal way; the left hand grasps the forehand grip and the thumb is looped through the enlarged trigger-guard. To assist in holding the weapon on the target when firing bursts there is a small muzzle brake, which also acts as a flash-hider for night shooting.

The basic frame of the pistol is similar to that of the Model 92, but there is a burst-controlling mechanism in the right-hand butt grip. The fire selector lever is added above the left grip, and the lower frame forward of the trigger guard is deepened to carry the hinges of the forehand grip. The fire selector lever can be moved with the right thumb to select either single shots (one white dot) or three-round bursts (three white dots) without disturbing the aim, and the safety can be similarly applied or released.

The metal folding stock quickly clips on to the bottom of the butt without interfering with the magazine and provides a reasonably steady hold for burst fire.

Although normal cleaning can be carried out by the firer, it is recommended that stripping of the burst control mechanism be carried out by an armourer.

DATA
Calibre: 9 mm Parabellum
Operation: short recoil, single-shot or 3-round burst
Method of locking: hinged block
Feed: detachable box magazine, 15 or 20 rounds
Weight: (with 15-round magazine) 1.12 kg; (with 20-round magazine) 1.17 kg
Length: 240 mm
Barrel: 156 mm, including muzzle brake
Metal stock: (length folded) 195 mm; (extended) 368 mm
Weight: 270 g
Sights: as for Model 92, but sight radius 160 mm
Cyclic rate of fire: about 1100 rds/min
Selector: manual, single self-loading shots or 3-round bursts
Muzzle velocity: 375 m/s

Manufacturer
Pietro Beretta SpA, Gardone Valtrompia, Brescia.

9 mm Model 93R selective fire Beretta pistol with 20-round magazine, ready for single-handed firing. Stock is folded up for carriage

9 mm Model 93R selective fire Beretta pistol with extended stock fitted and forehand grip folded down

Status
Current. Production.
Service
Adopted by Italian and foreign Special Forces.

---

9 mm Model 951R semi- and full-automatic Beretta pistol

This weapon is directly derived from the standard pistol of the Italian armed forces, the 9 mm Model 1951 self-loading Beretta, and reference should be made to the entry for that weapon.

The major characteristics of the Model 951R are apparent from the accompanying photographs (note the forward handgrip and the fire selector switch) and data. The manufacturer states that data for rate of fire and muzzle velocity is dependent on ammunition quality and atmospheric conditions.

DATA
Calibre: 9 mm Parabellum
Operation: short recoil, semi- or full-automatic
Method of locking: hinged block
Feed: 10-round detachable box magazine
Weight: (with empty magazine) 1350 g
Length: 170 mm

9 mm Model 951R Beretta pistol

9 mm Model 951R Beretta pistol

Barrel: 125 mm
Sights: (foresight) blade; (rearsight) notch
Practical rate of fire: about 750 rds/min
Selector: automatic or semi-automatic
Muzzle velocity: 390 m/s

Manufacturer
Pietro Beretta SpA, Gardone Valtrompia, Brescia.
Status
Obsolete. No longer in production.
Service
Italian Special Forces.

P-018 Bernadelli pistol

The P-018 is a new military automatic pistol which joins functionality to strong design and convenient features.

The most important characteristics are an all-steel frame, double-action lockwork and a magazine capacity of 15 rounds. The P-018/9 in 9 mm Parabellum chambering has been specially built for use by military and police forces.

The pistol is a semi-automatic double-action design, with the breech locked on discharge. The outline is squared-off but compact, and the trigger-guard is shaped for a two-handled grip. The magazine holds 15 rounds in staggered column, plus one round in the chamber. The P-018 is said to handle well, being well-balanced and to have a relatively soft recoil due to the block locking system employed.

P-018 Compact
This model is generally the same as the P-018 but smaller in all dimensions. The magazine capacity is reduced by only one round.

P-018 Compact Bernadelli pistol

DATA*
Cartridge: 9 mm Parabellum
Operation: semi-automatic, double-action
Method of locking: block locking breech
Feed: 15-round box magazine (14)
Weight: (empty) 998 g (950 g)
Length: 213 mm (190 mm)
Barrel: 122 mm (102 mm)
Rifling: 6 grooves rh
Sights: (fore) blade; (rear) notch, adjustable for windage
Sight base: 160 mm (137 mm)

Manufacturer
Vincenzo Bernadelli SpA, I-25063, Gardone Valtrompia, Brescia.
Status
Current. Production.

(*)Variations of Model P-018 compact shown in brackets

P-018-9 Bernadelli pistol

P-060/A Bernadelli Pistol

This pistol is available in three calibres: 9 mm Short, 7.65 mm ACP and .22LR. It is a single-action semi-automatic employing a locked-breech, and is entirely made of steel. There is a loaded chamber indicator and provision for lowering the hammer on a loaded chamber. The rearsights are adjustable for windage and elevation.

DATA
Cartridge: 9 × 17 mm (9 mm Short/.380 Auto)
Operation: semi-automatic, single-action

Feed: 7-round magazine
Weight, empty: 690 g
Length: 164 mm
Barrel: 90 mm
Rifling: 6 grooves rh
Sights: rearsight adjustable
Sight base: 113 mm

Manufacturer
Bernadelli SpA, I-25063 Gardone Valtrompia, Brescia.
Status
Current. Production.

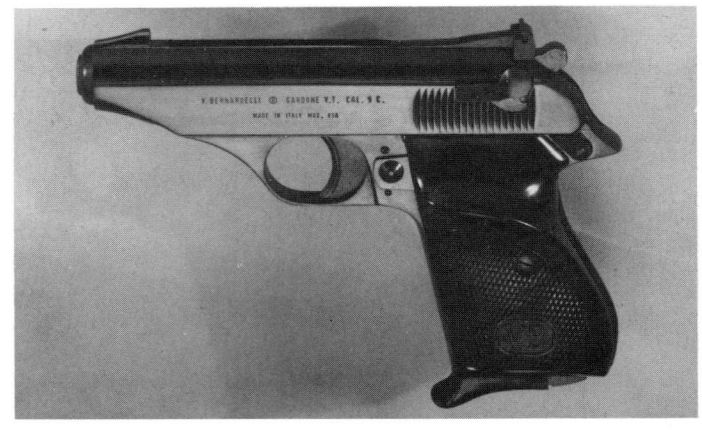

P-060/A Bernadelli pistol

9mm Model TA90 Tanfoglio pistol

The Model TA90 is a conventional double-action, locked-breech pistol firing the standard 9 mm Parabellum cartridge. Slide and frame are machined steel castings, and the barrel is of steel, rifled by cutting in the traditional manner. Breech locking is by a dropping barrel, on the well-known Colt/Browning system. The manual safety catch on the slide operates to lock the firing pin, hammer, sear and trigger, ensuring absolute safety at all times. The pistol is normally supplied in burnished black finish, but can be chromium plated or engraved if required. Rubber grips are available.

DATA
Cartridge: 9 × 19 mm Parabellum
Operation: Recoil, semi-automatic, double-action
Locking: Dropping barrel
Feed: 15-round box magazine
Weight, empty: 1015 g
Length overall: 202 mm
Barrel length: 120 mm
Rifling: 6 grooves, rh, one turn in 250 mm
Sights: (fore) blade; (rear) adjustable U-notch
Muzzle velocity: 350m/s nominal

9mm Tanfoglio Model TA90 pistol

Manufacturer
Fabbrica d'armi Fratelli Tanfoglio SpA, Via Valtrompia 39/41, Gardone Valtrompia, Brescia.

Status
Current. Production.

9 mm Models GT30, TA18 and GT21 Tanfoglio pistols

These three pistols are identical to the TA90 (above) but are chambered for different cartridges. The GT30 shoots the 7.65 mm Parabellum cartridge, the TA18 shoots the 9 x 18 mm Police cartridge, while the GT21 shoots the new 9 x 21 mm IMI cartridge. This cartridge has more power than the 9 mm Parabellum with a case which is 2 mm longer. Mechanically the pistols are the same as the TA90, being double-action locked-breech semi-automatics.
    Dimensions and weights are exactly the same as for the TA90.

Manufacturer
Fabbrica d'armi Fratelli Tanfoglio SpA, Via Valtrompia 39/41, Gardone Valtrompia, Brescia.
Status
Current. Production.

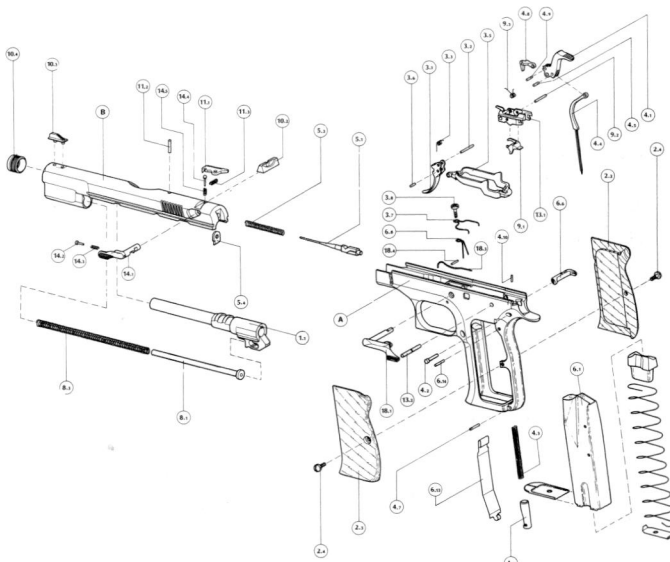

Exploded drawing of Tanfoglio TA90 pistol

9 mm 'Baby' Tanfoglio pistol

The Tanfoglio 'Baby' pistol is to the same general mechanical design as the TA90 but is of compact dimensions. It fits well in the hand, is easily concealable, delivers exceptional accuracy and is comfortable to shoot.

DATA
Cartridge: 9 × 19 mm Parabellum
Operation: recoil, semi-automatic, double-action

Locking: dropping barrel
Feed: 12-round box magazine
Weight, empty: 850 g
Length overall: 175 mm
Barrel length: 90 mm
Rifling: 6 grooves, rh, one turn in 250mm
Sights: (fore) blade; (rear) U-notch adjustable for windage

Manufacturer
Fabbrica d'armi Fratelli Tanfoglio SpA, Via Valtrompia 39/41, Gardone Valtrompia, Brescia.
Status
Current. Production.

9 mm Tanfoglio 'Baby' pistol

9 mm Tanfoglio GT21 series pistols

The GT21 is essentially the same as the TA90 (above) but it is chambered for the 9 × 21 mm IMI cartridge and there are some small changes in styling. These changes in outline (notably the trigger-guard and butt contour) are also applied to current production of the TA90 series. The GT21 has been produced principally for sale in those countries in which the normal 9 × 19 mm Parabellum cartridge is prohibited.

Variant Models
The **GT21 Combat** is of the same general appearance and dimensions as the standard GT21 but has the safety mounted on the frame, instead of on the slide, and has a less obtrusive slide stop lever.

The **GT21 Baby** has the same construction as the GT21 but is smaller in all dimensions and has a magazine of lesser capacity. The **GT21 Baby Combat** is of the same dimensions but has the safety on the frame and the smaller slide stop lever.

GT21 STANDARD and COMBAT:
DATA
Cartridge: 9 × 21 mm IMI
Operation: recoil, semi-automatic
Locking system: Browning cam
Feed: 15-round magazine
Weight, empty: 1015 g
Length: 202 mm
Barrel: 120 mm
Rifling: 6 grooves, rh, one turn in 27.7 calibres

GT21 BABY AND BABY COMBAT
As for GT21 except:-
Feed: 13-round magazine
Weight: 850 g
Length: 175 mm
Barrel: 90 mm

Manufacturer
Fabbrica d'armi Fratelli Tanfoglio SpA, Via Valtrompia 39/41, 25063 Gardone Val Trompia, Brescia.
Status
Current. Production.

9 mm Tanfoglio GT21 Baby pistol; Baby Combat is similar but with safety and slide stop as in the GT21 Combat

9 mm Tanfoglio GT21 Combat pistol

9 mm Tanfoglio GT21 pistol

.41 Tanfoglio GT41 pistol

The GT41 is to the same design as the GT21 and TA90 series but is chambered for the .41 Action Express cartridge. The safety is mounted on the frame, and the slide stop is smaller, as in the 'Combat' models described above. The dimensions, weights, etc, are exactly the same as for the GT21 except that the magazine holds only 11 rounds.

Manufacturer
Fabbrica d'armi Fratelli Tanfoglio SpA, Via Valtrompia 39/41, 25063 Gardone Val Trompia, Brescia.
Status
Current. Production.

.41 AE Tanfoglio GT41 pistol

.41 Tanfoglio 'Ultra' pistol

The 'Ultra' is a de luxe version of the GT41 with longer barrel and better finish; the frame is of polished steel, the slide is burnished black and the grips are of wood.

DATA
Cartridge: .41 Action Express
Operation: recoil, semi-automatic, double-action
Locking system: Browning cam
Feed: 11-round magazine
Weight, empty: 1040 g
Length: 214 mm
Barrel: 130 mm
Rifling: 6 grooves, rh, one turn in 38.5 calibres

Manufacturer
Fabbrica d'armi Fratelli Tanfoglio SpA, Via Valtrompia 39/41, 25063 Gardone Val Trompia, Brescia.
Status
Current. Production.

.41 AE Tanfoglio 'Ultra' pistol

9 mm Model B76 Benelli pistol

The Benelli B76 is an unusual delayed-blowback weapon firing the 9 mm Parabellum cartridge. It is conventional in outline but the slide conceals a separate breech-block in the rear end and the barrel is rigidly fixed to the frame. The breech-block is attached to the body of the slide by a toggle lever, and a recoil spring beneath the barrel completes the mechanical arrangements.

When the pistol is loaded and ready to fire the recoil spring holds the slide forward on the frame. In this position the rear of the slide, acting through the toggle lever, pushes on the rear of the breech-block; since this is tight against the base of the loaded cartridge, the effort is expended by forcing the rear of the bolt downwards and into a locking recess in the frame. Thus at the instant of firing the bolt and frame are locked together.

The pressure generated by firing attempts to drive the bolt back; this movement is resisted by the inertia of the slide, which also continues to exert a strong downward camming action on the block. Eventually the block overcomes the resistance, the toggle is straightened out, the rear of the block is lifted from its recess, and the slide and block are then free to move backwards to complete the reloading cycle.

The trigger unit is double-action, operating an external hammer. There is a manual safety catch on the frame which positively locks the hammer sear mechanism, when the hammer is cocked or uncocked. The firing pin is of the inertia type, and there is a loaded chamber indicator.

9 mm Model B76 Benelli pistol

DATA
Cartridge: 9 × 19 mm Parabellum
Operation: delayed blowback, semi-automatic, double action
Feed: 8-round box magazine
Weight, empty: 970 g
Length overall: 205 mm
Barrel length: 108 mm
Rifling: 6 grooves, rh, one turn in 250 mm
Sights: (fore) blade; (rear) square notch, white outlined

Manufacturer
Benelli Armi SpA, Via della Stazione 50, I-61029 Urbino.
Status
Current. Production.

9 mm Model MP3S Benelli pistol

The Benelli MP3S is a target version of the B76 which can also be used as a training weapon. It is available chambered either for the 9 mm Parabellum cartridge or for the .32 Smith and Wesson Wadcutter revolver cartridge, applicable to international shooting competitions.

The mechanism of the pistol is the same as that described for the B76 (above), a delayed blowback system using a toggle joint. The principal difference is that the MP3S uses a single-action trigger and is to an exceptionally high standard of finish both inside and out, is furnished with anatomical grips, and has the frame extended around the muzzle to form a counterweight.

DATA
Cartridge: 9 × 19 mm or .32 Smith and Wesson Wadcutter
Operation: delayed blowback, semi-automatic
Feed: 6-round box magazine
Weight, empty: 1175 g
Length overall: 237 mm
Barrel length: 140 mm
Sights: fully adjustable

Manufacturer
Benelli Armi SpA, Via della Stazione 50, I-61029 Urbino.
Status
Current. Production.

9 mm Benelli MP3S pistol

.38 RF83 Luigi Franchi revolver

The RF83 is a conventional solid-frame, double-action revolver using a swing-out cylinder. The entire weapon is made from nickel-chrome-molybdenum steel, and the firing pin is floating.
   The pistol is available in Compact, Standard, Service and Target versions, the various models differing only in dimensions and sights.

DATA (RF83 Service Model)
Cartridge: .38 Special
Operation: double-action
Feed: 6-chambered cylinder
Weight: 800 g
Barrel: 101 mm
Sights: fixed U-notch and blade, with optional adjustable rearsight available.
Manufacturer
Luigi Franchi SpA, Via del Serpente 12, 25020 Fornaci, Brescia.
Status
Current. Production.

.38 RF83 Luigi Franchi Service revolver with optional adjustable rearsight

JAPAN

.38 Model 60 New Nambu revolver

This Smith and Wesson type pistol in .38 Special has been the Japanese Police pistol since 1961 and 110 000 have been sold. The revolver is also issued to the Japanese Maritime Safety Guard.

DATA
Cartridge: .38 Special
Operation: manual – revolver, single- or double-action
Feed: 5-chamber cylinder
Weight: 680 g
Length: 197 mm
Barrel: 77 mm
Sights: (foresight) fixed half-tapered serrated ramp; (rearsight) square notch
Muzzle velocity: 220 m/s
Rate of fire: 15 rds/min
Effective range: 40 m

Manufacturer
Omori Factory, Division of Minebea Co Ltd (Tokyo), 18-18 Omori-nishi 4 Chome Ohta-ku, Tokyo 143.
Status
Current. Production.

.38 Model 60 New Nambu revolver

Service
Japanese police forces and Maritime Safety Guard.

KOREA, NORTH

7.62 mm Type 68 pistol

This is a much modified Tokarev TT-33. It is shorter and bulkier than either the Soviet TT-33 or the Chinese Type 51 or Type 54. It may further be distinguished from these pistols by the serrations on the rear of the slide intended to give a grip while the weapon is being cocked. The variations are: the Soviet TT-33: vertical, alternately wide and narrow; the Chinese Type 51 or 54: vertical, narrow; the North Korean Type 68: sloping forward, narrow.

Internally the Tokarev has been re-worked considerably. The link system, used to lift and lower the barrel ribs into and out of the grooves in the slide, has been replaced by a cam cut into a lug under the chamber in a manner similar to that used in the Browning 9 mm High-power pistol. The magazine catch has been relocated and is now at the heel of the magazine. The magazine from the Tokarev TT-33 will work in the Type 68 pistol but the converse is not

North Korean 7.62 mm Type 68 pistol

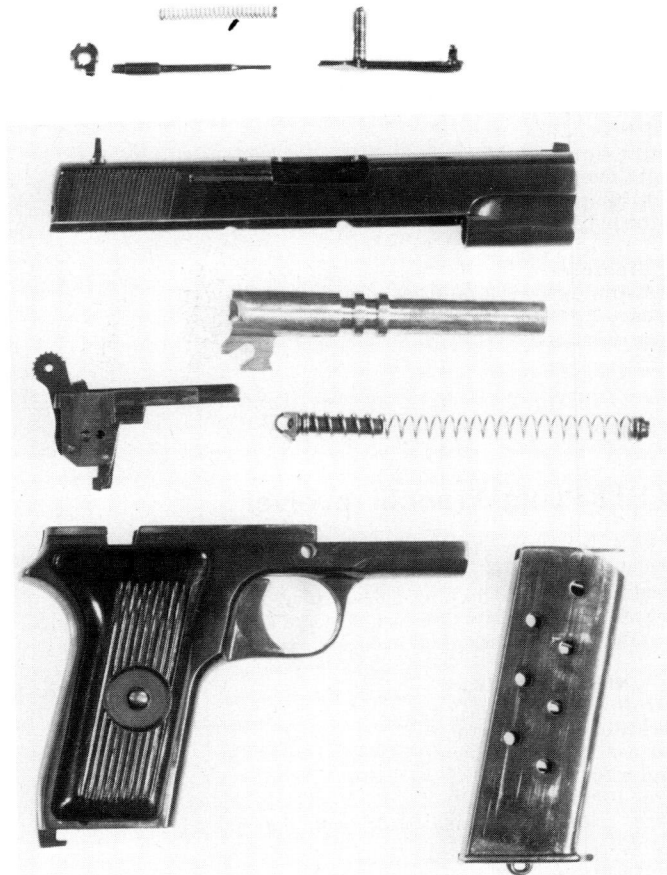

true as the Type 68 magazine lacks the necessary cut-out for the magazine catch. The firing pin is retained by a plate instead of a cross pin and the slide stop is a robust pin instead of the rather fragile clip in the TT-33.

A poor feature of the TT-33 which has had to be retained is the large-radius curve at the junction of butt and slide which presses into the web of the thumb when firing. It has not been possible to machine this into a smaller radius without a complete re-design of the hammer mechanism.

DATA
Cartridge: 7.62 mm × 25 Type; P 7.63 mm Mauser
Operation: short recoil, self-loading, single-action
Method of locking: projecting lug
Feed: 8-round box magazine
Weight, empty: 795 g
Length: 185 mm
Barrel: 108 mm
Rifling: 4 grooves, rh, one turn in 305 mm
Sights: (fore) blade; (rear) notch
Sight radius: 160 mm
Muzzle velocity: 395 m/s

North Korean 7.62 mm Type 68 pistol stripped showing FN Browning cam under barrel, note how it compares with Tokarev TT-33

Manufacturer
State factories.

Status
In service with North Korean forces.

7.65 mm Type 64 pistol

In introducing this pistol the North Korean design authority resurrected the Browning Model 1900. A photograph of Browning's pistol is shown.

The North Korean Type 64 has the stamping '1964 7.62' on the left side but in fact takes the 7.65 mm × 17SR cartridge which is the American Colt .32 Automatic Colt Pistol (ACP).

There is also a silenced version of the Type 64. It has a shortened slide to allow the muzzle to protrude and the end of the barrel is threaded to take the silencer attachment.

The basic parameters of the pistol are the same as the Browning Model 1900.

DATA
Cartridge: 7.65 mm (.32ACP)
Operation: blowback, semi-automatic

Method of locking: nil
Feed: 7-round detachable box magazine
Weight: 624 g
Length: 171 mm
Barrel: 102 mm
Rifling: 6 grooves rh
Sights: (fore) blade; (rear) notch
Muzzle velocity: approximately 290 m/s
Effective range: 30 m

Status
Current. Probably no longer manufactured.
Service
North Korean forces.

North Korean 7.65 mm Type 64 pistols

7.65 mm Model 1900 Browning pistol

NORWAY

Norwegian Army pistols

From 1883 to 1914 the Nagant revolver was used by the Norwegian Army. In 1914 the Colt .45 was adopted and manufactured at Kongsberg, and still remains in service in the Norwegian Army as does the 9 mm Walther P38 and other pistols. The Norwegian Colts .45 are direct copies of the American Model 1911 and Model 1911A1, the second having a lengthened finger piece on the slide stop.

In 1988 the Glock 9 mm pistol was adopted as the Norwegian army standard sidearm, but it will be several years before the older pistols are completely replaced.

POLAND

9 mm P-64 self-loading pistol

The Polish armed forces were at one time equipped with the 7.62 mm Pistolet TT which is identical to the Soviet Tokarev TT-33 pistol except for the handgrips. This pistol is now obsolete and has been replaced by the blowback-operated P-64 which, although an original Polish design, looks rather like the Soviet Makarov and has some design features which originated with the German Walther PP pistol. It has the inscription '9 mm P-64' on the left side of the slide.

The P-64 is a blowback-operated pistol of conventional form. It is prepared for firing by inserting a magazine, pulling back the slide and releasing it to chamber a round. The hammer is then cocked and the pistol may be fired. Alternatively, the safety catch may be pressed down; this rotates two lugs into place to protect the firing pin and allows the hammer to fall on to these lugs. The pistol can then be fired by pushing up the safety catch and pulling through on the trigger in double-action mode. When the magazine is emptied the slide is held by a slide stop.

Stripping
Remove the magazine and then check that the chamber and feedway are clear of live rounds. Pull down the front of the trigger-guard and push it to one side and rest it against the receiver. Pull the slide back, lift the rear end out of the receiver. Move the slide forward off the weapon. The return spring can then be moved from the barrel.

9 mm P-64 pistol

To re-assemble, the hammer is cocked and the safety lever is pushed up to the 'safe' position. The trigger-guard is placed in the position adopted for stripping. The return spring is placed over the barrel which is pushed through the hole in the front of the slide. The slide is pulled to the rear, dropped into the locating slots in the receiver, and forced forward by the return spring. The trigger-guard is re-located and the pistol is ready for use.

DATA
Cartridge: 9 mm × 18
Operation: blowback, self-loading, single- or double-action
Method of locking: nil
Feed: 6-round box magazine
Weight, empty: 0.636 kg
Length: 155 mm
Barrel: 84 mm
Sights: (fore) blade; (rear) notch
Sight radius: 114 mm
Muzzle velocity: 314 m/s
Effective range: 50 m

Manufacturer
State arsenals.
Status
Still produced.
Service
Polish forces.

9 mm P-64 pistol field-stripped

SOUTH AFRICA

9 mm Z-88 pistol

In 1985 the South African police expressed a requirement for a new pistol. Due to the arms embargo the acquisition of foreign pistols is very difficult in South Africa, and it was decided to investigate the possibility of local manufacture. Subsequent discussions between the police and Armscor led to the drawing up of a specification in April 1986, and the Lyttleton Engineering Works was instructed to proceed with the project, the aim being to start production within two years. The project plan proceeded according to schedule and by August 1988 200 pistols had been manufactured to confirm the production capability and to allow a quantity of pistols to be subjected to field testing. The testing involved the firing of some 146 000 rounds from 20 pistols, as well as the usual environmental tests.

The Z-88 pistol takes its name from the late Mr. T.D. Zeederberg, former general manager of LEW, who was instrumental in the successful handling of the project by the company, and from its year of introduction.

The pistol is a double-action locked-breech weapon in 9 mm Parabellum calibre. The barrel is locked by a floating wedge between barrel and frame;

indeed, as is obvious from the photograph, the pistol has been based on the Beretta 92 and it can be assumed that the dimensions and data are the same as for that pistol.

Manufacturer
Armscor, Private Bag X337, Pretoria 0001.
Status
Current. Production.
Service
South African police and security forces; approximately 500 pistols annually will be available to the commercial market.

9 mm Z-88 pistol

SPAIN

9 mm Llama M-82

This is a locked-breech 9 mm Parabellum pistol using a double-action lock. Breech locking is performed by a dropping block, similar to that used in the Walther P-38 pistol, a system which generally provides more reliable feed, firing and extraction since the barrel remains fixed in relation to the rest of the weapon. There is a slide-mounted safety catch which, when operated, conceals and locks the firing pin and disconnects the trigger bar.

To load, the safety is lowered to the 'safe' position, a magazine inserted, and the slide drawn back and released. This feeds a round into the chamber and as the slide closes so the hammer drops in a safe condition. To fire, all that is necessary is to move the safety up to the 'fire' position and pull the trigger; this will cock and release the hammer. Subsequent shots are fired in the single-action mode. After the last shot in the magazine has been fired the magazine follower rises and holds the slide in the rearward position.

Stripping
Remove the magazine, inspect the chamber and ensure the weapon is empty.
Draw the slide back until the inclined face of the slide, below the ejection port, aligns with a similar inclined face in front of the slide stop lever. Rotate the stripping catch forward and down.
Push the slide and barrel unit forwards off the frame. Remove the return spring and guide assembly.
Press in the locking block disconnector plunger, remove the locking block, and remove the barrel from the slide.
Re-assembly is simply a matter of proceeding in the reverse order.

9 mm Llama M-82 pistol

DATA
Cartridge: 9 × 19 mm Parabellum
Operation: recoil, semi-automatic
Locking system: dropping block
Feed: 15-round magazine
Weight, empty: 1110 g (steel frame); 875 g (alloy frame)
Length: 209 mm
Barrel: 114 mm
Rifling: 6 grooves rh
Muzzle velocity: 345 ± 3 m/s (V_{15})

Manufacturer
Llama Gabilondo y Cia SA, Portal de Gamarra 50, Apartado 290, 01080 Vitoria.
Status
Current. Production.
Service
Spanish Army.

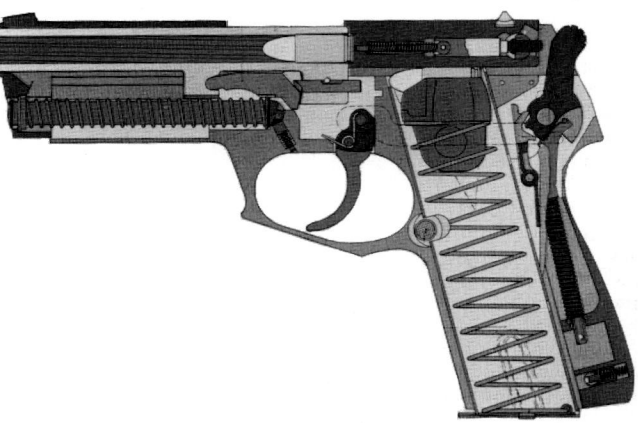

Llama M-82 pistol mechanism

9 mm Firestar pistol

The most recent design from Star Bonifacio Echeverria, the Firestar is among the smallest of 9 mm Parabellum pistols. It uses the well-tried Colt-Browning system of dropping barrel to lock the breech, with three locking lugs on the barrel and a shaped cam in the breech lump. The muzzle end of the barrel is conical and shaped so as to lock with the slide after each shot and render a barrel bushing redundant. The interaction of the shaped muzzle and the slide also improves the accuracy of the pistol. Operation is single-action only, there is an ambidextrous applied safety catch and an automatic firing pin safety.

DATA
Cartridge: 9 mm Parabellum
Operation: short recoil, semi-automatic
Method of locking: projecting lugs
Feed: 7-round detachable box magazine or 8 rounds with special extension magazine
Weight, empty: 798 g
Length: 163 mm

Barrel: 86 mm
Sights: (fore) blade; (rear) notch
Sight base: 115 mm
Muzzle velocity: 348 – 380 m/s

Manufacturer
Star Bonifacio Echeverria SA, PO Box 10, Eibar.
Status
Current. Production.

9 mm Firestar pistol

9 mm Model 30M and Model 30PK Star pistols

These two pistols are updated versions of the Models 28 and 28PK, descriptions of which will be found in the 1986/87 edition of *Jane's Infantry Weapons*; they are of the same construction and appearance, but incorporate an ambidextrous safety catch which locks the firing pin when applied. The Model 30M is made entirely of forged steel, while the Model 30PK has a light alloy frame. The pistol is somewhat unusual in having the slide running in internal frame rails; this gives excellent support throughout the slide movement, with a minimum bearing of over 110 mm. There is an ambidextrous safety catch on the slide which retracts the firing pin into its tunnel, out of reach of the hammer. The trigger and hammer action are quite unaffected by the safety catch action and it is possible to pull the trigger so as to drop the hammer after applying the safety, and also to pull the trigger to rise and drop the hammer for 'dry firing' practise, without needing to unload the weapon. Whether this is entirely a good thing is open to argument; it is quite possible to draw the pistol, set at safe, and pull the trigger through, expecting it to fire; the normal safety system at least prevents the trigger or hammer moving and thus gives a quick reminder that the safety is set. There is also a magazine safety which prevents firing if the magazine is removed, but in recognition of the fact that some users do not like these, the manufacturers have built it so that it is possible to remove this feature at will.

The trigger-guard fore-edge is shaped for the two-handed grip, and there is a loaded chamber indicator which stands proud of the slide when the weapon is loaded. The sights are clear, the rearsight being adjustable for windage, and the gun is very accurate.

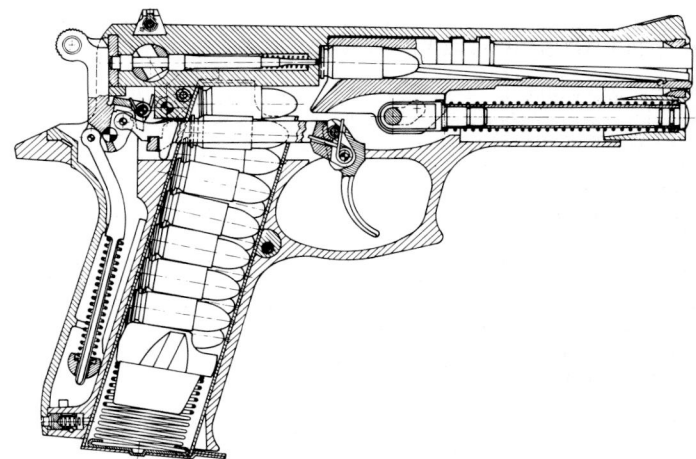

Sectioned view of Star 30M pistol

Stripping
Remove magazine, operate slide, check that chamber is empty.
Pull back the slide until white marks on slide and frame match, then push out the slide stop pin.
Push slide off front end of frame. Remove barrel, mainspring and guide from slide.
Depress plunger at bottom rear of butt and remove hammer unit through top of frame. This also releases the butt grips, which can be lifted from the butt frame.
The trigger and hammer units can be dismantled by driving out pins as necessary.

DATA
Cartridge: 9 mm Parabellum
Operation: short recoil, semi-automatic
Method of locking: Browning cam
Feed: 15-round box magazine
Weight: 1.14 kg (30M); 860 g (30PK)
Length: 205 mm (30M); 193 mm (30PK)
Barrel: 1190 mm (30M); 98 mm (30PK)
Sights: (fore) blade; (rear) notch, adjustable for windage
Muzzle velocity: ca 380 m/s

Manufacturer
Star Bonifacio Echeverria SA, PO Box 10, Eibar.

9 mm Star Model 30M double-action pistol

Status
Current. Production.
Service
Spanish armed and police forces; Peruvian police and security forces.

9 mm Model DKL Star pistol

The DKL is a small pistol which is rather unusual in employing a locked-breech in a calibre not usually thought worth such mechanical complication. This is though, less unusual among Spanish manufacturers than it is elsewhere in the world, and it certainly makes for a robust weapon which is normally easier to shoot than a comparative blowback design. The Star DKL has a light alloy frame, which is another good argument for using a locked-breech, and is a light and handy weapon for police and security forces. Breech locking is done by the usual Browning swinging link system, and, in general, the design is a scaled-down Colt M1911 in most respects.

DATA
Cartridge: 9 mm Short
Weight, empty: 420 g
Length overall: 145 mm
Barrel: 80.5 mm, 6 grooves, rh twist
Magazine capacity: 6 rounds
Muzzle velocity: ca 270 m/s

Manufacturer
Star Bonifacio Echeverria SA, PO Box 10, Eibar.
Status
Production.

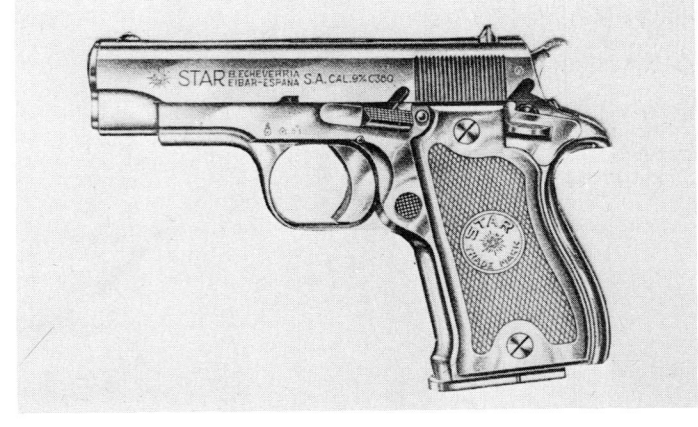

9 mm Star Model DKL

9 mm Models BM and BKM Star pistols

These are military-pattern pistols though of compact dimensions; the BK uses a steel frame, while the BKM has a light alloy frame. Both have locked breeches, using the Browning swinging link system and have a manual safety catch which locks both hammer and slide when the hammer is either cocked or uncocked; it will not lock if applied with the hammer at half-cock. There is also a magazine safety which prevents firing if the magazine is removed.

DATA
Cartridge: 9 mm Parabellum
Operation: short recoil, semi-automatic
Locking: Swinging link
Weight, empty: 965 g (BKM 725 g)
Length overall: 182 mm
Barrel: 99 mm, 6 grooves, rh one turn in 311 mm
Magazine capacity: 8 rounds
Muzzle velocity: ca 390 m/s

Manufacturer
Star Bonifacio Echeverria, SA, PO Box 10, Eibar.
Status
Production.

9 mm Star Model BM

.45in Model PD Star pistol

The Model PD was developed with the intention of providing a powerful pistol capable of being easily carried and concealed. It was among the first in this class and has remained one of the most successful. In order to bring the size down there have been some minor changes from the classic Browning system of locking; there is only one interlocking lug between barrel and slide, and the recoil spring and guide rod are an assembled sub-unit instead of being separate components. The frame is of light alloy and there is no grip safety. The foresight is a blade and the rearsight a fully adjustable notch.

The light weight makes the PD difficult to control, though it is not uncomfortable to shoot. Due to the short barrel the velocity and energy are somewhat less than normal with the .45 ACP cartridge.

DATA
Cartridge: .45 ACP
Operation: short recoil
Locking system: Browning swinging link
Weight, empty: 710 g
Length overall: 180 mm
Barrel: 100 mm, 6 grooves, rh 1 turn in 311 mm
Magazine capacity: 6 rounds
Muzzle velocity: ca 220 m/s

Manufacturer
Star Bonifacio Echeverria SA, PO Box 10, Eibar.

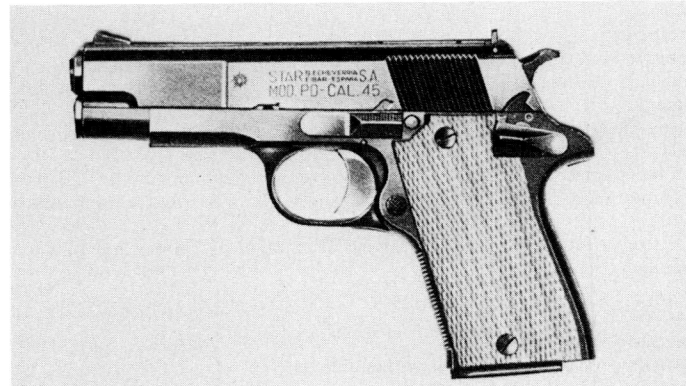

.45 Star Model PD

Status
Current. Production.

9 mm Model A-70 single-action Astra pistol

The Astra A-70 is a locked-breech single-action semi-automatic pistol. The powerful 9 mm Parabellum cartridge and small size make it ideally suitable for self-defence, police or military use.

The pistol employs three independent safety devices; there is an automatic firing pin safety which ensures that the firing pin cannot move unless the trigger is deliberately pressed, a manual safety applied with the thumb which, with the hammer cocked, blocks the trigger and slide, and a half-cock notch

on the hammer which prevents the hammer striking the firing pin should the hammer be allowed to slip during cocking or un-cocking the weapon.

DATA
Cartridge: 9 × 19 mm Parabellum
Operation: short recoil, semi-automatic
Locking system: Colt-Browning dropping barrel

Feed: 7-round box magazine
Weight: 840 g empty with empty magazine
Length: 166 mm
Barrel length: 89 mm

Manufacturer
Astra-Unceta y Ciá SA, Apartado 3, Guernica, Vizcaya.
Status
Current. Production.

9 mm Astra Model A-70 pistol

9 mm Model A-80 double-action Astra pistol

The A-80 is reminiscent of the SIG-Sauer P220 in its angular outlines and in the use of a de-cocking lever. When the gun has been loaded the de-cocking lever is pressed; this releases the hammer which falls into a first step on the de-cocking lever and goes down until arrested by a notch on the sear, preventing the hammer reaching the firing pin. The firing pin itself is restrained from movement by a spring-loaded plunger which is engaged at all times except when the trigger is pulled; when this is done the action of the sear in releasing the cocked hammer also forces the plunger out of engagement and permits the pin to move when struck by the hammer. Unless the trigger is consciously pulled, therefore, even an accidental fall of the hammer cannot drive the firing pin forward. In view of this, there is no manual safety catch on the A-80 pistol.

The de-cocking lever is on the left side of the frame where it can be conveniently operated by the (right-handed) firer's thumb; for left-handed firers it is possible to remove the lever and install a replacement lever on the right side of the frame.

The magazine holds 15 rounds and has holes which show when there are 5, 10 or 15 rounds in the magazine. Front and rearsights are of the Stavenhagen pattern with white inlays to provide an aiming mark in poor visibility. On test the pistol functioned well and displayed acceptable accuracy.

Stripping
Remove magazine, operate slide, ensure chamber is empty.
With slide held back, turn down the stripping catch on the right side of the frame above the trigger.
Depress the slide retainer latch (behind de-cocking lever on left side) and push the slide assembly off the frame.
Remove recoil spring assembly and barrel from slide.
No further dismantling is recommended. Assembly is undertaken in the reverse manner.

DATA
Cartridge: currently manufactured in 7.65 mm Parabellum, 9 mm Steyr, 9 mm Parabellum, .38 Super Auto and .45 ACP
Operation: short recoil, semi-automatic
Method of locking: Browning swinging link (cam)
Feed: 15-round box magazine
Weight, empty: 985 g
Length: 180 mm

9 mm Model A-80 double-action Astra pistol

Barrel: 96.5 mm
Sights: (fore) blade; (rear) notch, adjustable for windage
Muzzle velocity: 350 m/s

Manufacturer
Astra-Unceta y Ciá SA, Apartado 3, Guernica, Vizcaya.
Status
Current. Production.
Service
In service in several unspecified countries.

9 mm Model A-90 Astra pistol

The Astra A-90 was introduced in 1985 and is an updated version of the A-80 (above). It has an improved double-action mechanism, adjustable sights, large magazine capacity, and compact dimensions.

Safety in operation has been made a particular feature of the design, and the facilities on the A-90 permit the user to adopt a variety of safety procedures. There is firstly a hammer de-cocking lever; when pressed, this releases the sear to allow the hammer to drop, but the hammer stroke is arrested by the de-cocking lever and it does not contact the firing pin. There is also a manual safety catch on the slide which operates to rotate a portion of the two-piece firing pin out of the hammer path; this safety can be applied whatever the position of the hammer. Finally the major (front) portion of the firing pin is

locked by an automatic block which is only released when the trigger is pulled to its full extent, as when firing; at any other time the forward part of the firing pin cannot move, irrespective of what other safety is or is not in operation.

DATA
Cartridge: 9 × 19 mm Parabellum or .45 ACP
Operation: Recoil, semi-automatic, double-action
Locking: dropping barrel
Feed: 15-round (9 mm); 8-round (.45) box magazine
Weight, empty: 985 g (9 mm); 955 g (.45)
Length overall: 180 mm
Barrel length: 96.5 mm

Manufacturer
Astra-Unceta y Ciá SA, Apartado 3, Guernica, Vizcaya.
Status
Current. Production.

9mm Model A-90 Astra pistol

9 mm Astra A-50 pistol

The Astra A-50 is a compact pistol designed for personal defence and police use. It is of conventional blowback type, with the barrel fixed to the frame, and in spite of its appearance it is a single-action weapon. It is also available in 7.65 mm ACP chambering.

DATA
Cartridge: 9 mm Short (7.65 mm ACP)
Operation: blowback
Weight: 650 g (660 g)
Length overall: 168 mm
Barrel length: 89 mm
Magazine capacity: 7 rounds 9 mm (8 rounds 7.65 mm)

Manufacturer
Astra-Unceta y Ciá SA, Apartado 3, Guernica, Vizcaya.
Status
Production.

9 mm Astra A-50 pistol

9 mm Astra A-60 pistol

The A-60 pistol is a fixed-barrel blowback weapon of conventional pattern, using double-action lockwork. The innovative element of the design is that the safety catch is duplicated on both sides of the slide, so that it can be operated with equal facility by a right- or left-handed firer. In addition, the magazine catch, located in the forward edge of the butt, behind the trigger-guard on the left side, can be easily removed and refitted into the right side of the frame, so making it easier for left-handed firers to operate with their firing hand.

The Astra A-60 is available in either 9 mm Short/.380 Auto or 7.65 mm ACP chambering. The dimensions are the same in both calibres, the only change being in the magazine capacity.

DATA
Cartridge: 9 mm Short or 7.65 mm ACP
Operation: Blowback, semi-automatic
Feed: Box magazine; 12 shots (7.65 mm) or 13 shots (9 mm)
Weight, empty: 700 g
Length: 168 mm
Barrel: 89 mm
Sights: (fore) fixed blade; (rear) square notch, adjustable for windage

Manufacturer
Astra-Unceta y Ciá SA, Apartado 3, Guernica, Vizcaya.
Status
Current. Production.

9 mm Astra A-60 pistol

9 mm Astra Falcon pistol

This is the sole remaining example of the style which made Astra famous, a design which has its roots in the Campo-Giro pistol of 1913. Astra pistols in this distinctive 'round-barrel' pattern formed the Spanish service sidearm for many years and were widely distributed throughout Europe. The Falcon is available in 7.65 mm or 9 mm Short calibres for police or military use. The mechanism is simple and robust, with the return spring coiled around the removeable barrel and a visible hammer. The safety catch operates on the trigger, and there is a magazine safety incorporated.

DATA
Cartridge: 9 mm Short (7.65 mm ACP)
Operation: blowback
Weight: 646 g (668 g)
Length overall: 164 mm
Barrel length: 98.5 mm
Magazine capacity: 7 rounds 9 mm (8 rounds 7.65 mm)

Manufacturer
Astra-Unceta y Ciá SA, Apartado 3, Guernica, Vizcaya.
Status
Production.

9 mm Astra Falcon pistol

.38 Special Model 960 Astra revolver

The Astra company produce a number of service pattern revolvers, of which the Model 960 is a good example. It is a conventional solid-frame weapon with swing-out cylinder, simultaneous ejection and double-action lockwork. The rearsight is fully adjustable for elevation and windage and there is a regulator which permits adjustment of the pressure of the mainspring. Various barrel lengths are available, and while the standard finish is blue with checkered wooden grips, there are alternative finishes.

DATA
Cartridge: .38 Special
Operation: double-action revolver
Feed: 6-shot cylinder
Weight, empty: 1.15kg (with 102 mm barrel)
Length: 241 mm (with 102 mm barrel)
Barrel: 102 or 152 mm to choice
Sights: (fore) blade; (rear) notch, fully adjustable
Muzzle velocity: 265 m/s

Manufacturer
Astra-Unceta y Ciá SA, Apartado 3, Guernica, Vizcaya.
Status
Current. Production.

.38 Special Model 960 Astra revolver with 102 mm barrel

Service
In service in several unspecified countries.

Astra Model 44 and 45 revolvers

These are heavy military revolvers, the model number indicating the calibre of either .44 Magnum or .45 Colt. Both are solid-frame weapons with swing-out cylinders and simultaneous ejection. The rearsights are fully adjustable and there is an adjustment which allows the pressure of the mainspring to be varied. The cylinder is released by the usual thumb-catch on the left side of the frame, and the ejector rod is carried in a shrouded housing under the barrel.

DATA

	Model 44	Model 45
Cartridge	.44 Magnum	.45 Colt
Operation	double-action revolver	
Cylinder capacity	6 shots	
Length overall	293 mm or 356 mm	
Barrel length	152 mm or 216 mm	
Weight, empty:		
152 mm barrel	1.28 kg	1.24 kg
216 mm barrel	1.31 kg	1.27 kg
Muzzle velocity	ca 450 m/s	ca 270 m/s

Astra Model 44 revolver

Manufacturer
Astra-Unceta y Ciá SA, Apartado 3, Guernica, Vizcaya.

Status
Production.

Astra Police revolver

This is a conventional solid-frame, swing-out cylinder, double-action revolver chambered for the .357 Magnum cartridge and intended for use by police and security forces. The sights are fixed (the rearsight being merely a groove in the top strap) and the weapon has been designed with the rigours of service well in mind. The pistol is available in the three calibres enumerated below, and in addition it can be supplied with a spare cylinder in either 9 mm Parabellum or 9 mm Steyr calibre. Cylinders for rimless cartridges are provided with loading clips.

DATA
Cartridge: .357 Magnum, .38 Special, 9 mm Parabellum
Weight: 1040 g
Length overall: 212 mm
Barrel length: 77 mm
Cylinder capacity: 6 rounds

Manufacturer
Astra-Unceta y Ciá, Apartado 3, Guernica, Vizcaya.
Status
Production.

.357 Astra Police revolver

SWEDEN

9 mm pistol m/07

This is the Browning Model 1903 in 9 mm Browning Long chambering. First supplies were purchased from FN, after which it was manufactured by Husqvarna Vapenfabrik AB from 1908 to 1943, when it was partly replaced by the pistol m/40. In recent years it has been brought back into service and is currently the standard Swedish Army pistol.

9 mm revolver m/58

This is the commercial Smith and Wesson Model 12 military and police Airweight Special revolver with 2 in barrel and aluminium frame, chambered for the .38 Special cartridge.
It is used by flight crews in all Swedish services.

SWITZERLAND

P 210 and 9 mm Model 49 SIG pistols

Schweizerische Industrie-Gesellschaft (SIG), based at Neuhausen Rheinfalls, took up Charles Petter's patents from Société Alsacienne de Constructions Mécaniques (SACM). Development resulting in important technical improvements continued over the years 1938–46 and a series of weapons was produced. The 9 mm Model 44/16 held 16 rounds and the 44/8 had eight rounds but these were produced in small numbers only. The 9 mm Model 49 is the Swiss Army pistol (designated 9 mm Pistole 49) but is being replaced by the SIG-Sauer P 220 (see entry below). It is identical to the P 210-2. The SIG P 210 pistol has been produced in several versions:the P 210-1 (polished finish and wooden grip plates); P 210-2 sand-blast finish and plastic grip plates; P 210-4, a special production model for the West German Border Police; P 210-5, a target version with 150 mm barrel; and the P 210-6, also a target model but with a 120mm barrel.

The P 210-1, -2 and -6 are produced in either 9 mm or 7.65 mm Parabellum. The calibre can be changed by substituting the other barrel, of the alternative calibre, with its own return spring. The pistols can also be converted to .22 long rifle by changing the barrel, return spring, slide and magazine, and this of course reduces training costs.

Operation
The magazine is removed for loading by pressing rearward the magazine catch at the heel of the pistol grip. The magazine takes eight cartridges. With the magazine in place, pulling back the slide and then releasing it drives a cartridge into the chamber and pulling the trigger fires one round.

On firing the slide and barrel, locked together by lugs on the barrel mating with recesses in the slide, recoil together until the shaped cam beneath the chamber, acting against the slide stop pin, pulls the lugs out of engagement. The barrel then stops and the slide recoils, cocking the hammer at the end of its stroke. On returning, the slide collects a fresh round from the magazine and chambers it, then lifts the barrel back into engagement with the slide before coming to rest. When the ammunition is expended the slide is held to the rear by the slide stop which is forced into the notch on the slide by a lip on the magazine follower.

Stripping
Stripping the pistol is effected as follows: the magazine is removed and the slide pulled back about 6 mm and held back in this position. The slide stop pin

9 mm P 210-5 SIG pistol

9 mm P 210-6 SIG pistol

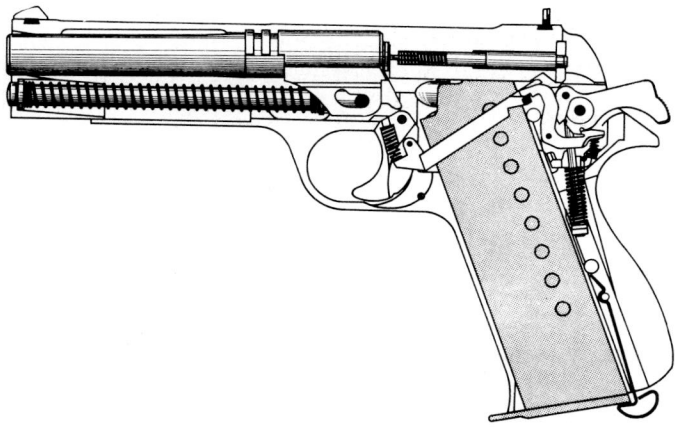

P 210-1 or P 210-2 SIG pistol

SP 44/8 SIG pistol

can then be pressed through from right to left until the stop is clear of the slide. The slide can then be released and the slide stop pulled out. Take the slide off the receiver. Invert it and lift the rear of the return spring guide out of engagement with the barrel. Hold the cam lug and pull the barrel up and back. The hammer mechanism will lift out as a single unit.

DATA
DATA BELOW IS COMMON TO 9 MM, 7.65 MM OR .22 VERSIONS OF P210-1 OR P210-2 UNLESS OTHERWISE SPECIFIED

Cartridge: 9 mm Parabellum (or 7.65 mm Parabellum or .22 longrifle)
Operation: short recoil, self-loading
Method of locking: projecting lug

Feed: 8-round box magazine
Weight, empty: 900 g (845 g for .22)
Length: 215 mm
Barrel length: 120 mm
Rifling: 6 grooves rh (4 for 7.65 mm); twist – 1 turn in 250 mm (450 mm for .22)
Sights: (fore) blade; (rear) notch
Sight radius: 165 mm

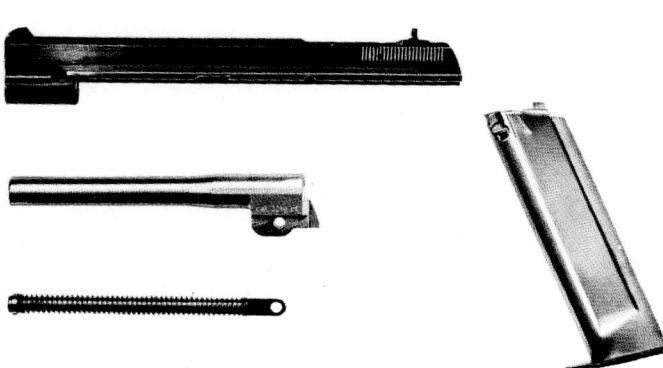

.22 long rifle conversion unit

Model 49 SIG pistol

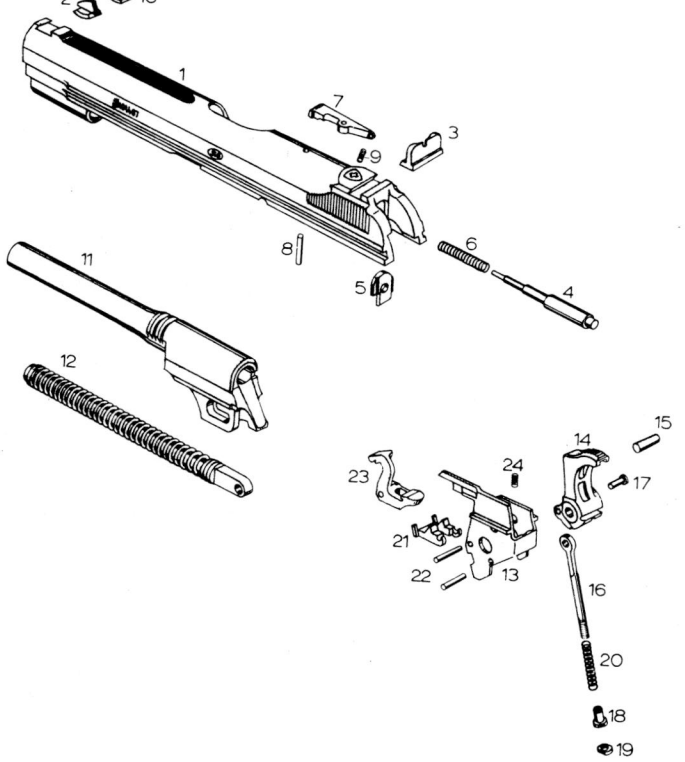

9 mm P 210 SIG pistol exploded view

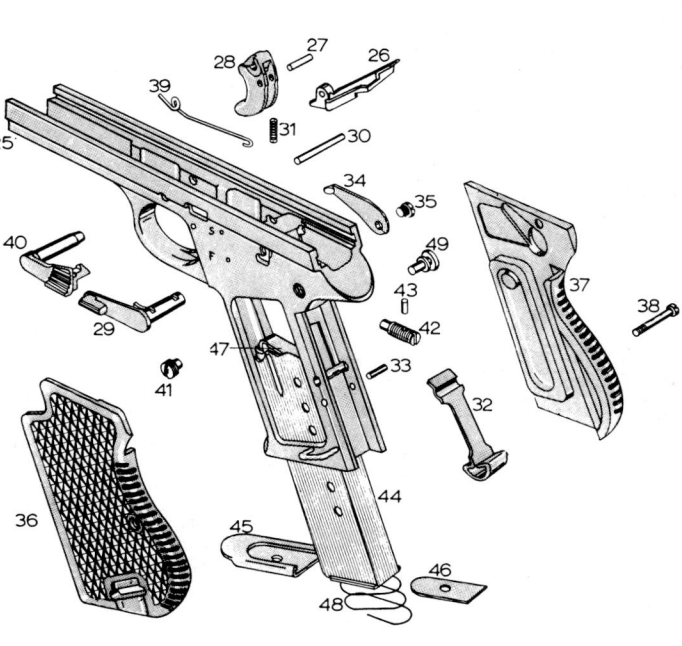

FIRING CHARACTERISTICS

	9 mm	**7.65 mm**	**.22**
Muzzle velocity:	335 m/s	385 m/s	330 m/s
Muzzle energy:	454 J	453 J	141 J

Status
Current. Production.
Service
Swiss Army has complete stocks and takes no more deliveries. Also in use by Danish Army. Commercial sales of target versions continue.

Manufacturer
SIG Swiss Industrial Company, CH-8212 Neuhausen-Rheinfalls Commercial target pistols are marketed through Hämmerli Ltd, Lenzburg.

P 220 (Model 75) SIG-Sauer pistol

This is a short-recoil operated, self-loading, single- or double-action pistol. The magazine catch is below the heel of the butt (except for the .45 ACP model, where it is a button in the front edge of the butt) and the magazine is loaded in the usual way by being pushed up into the butt until the catch clicks into place.

The slide is pulled to the rear and released to feed a round into the chamber. If it is not intended to fire the weapon immediately the cocked hammer should be lowered by pressing down on a hammer de-cocking lever above and slightly behind the trigger on the left side of the receiver. The effect of depressing this lever is to lift the sear out of engagement with the hammer which is rotated by its spring until the safety notch is caught by the sear and it comes to rest held clear of the firing pin. The firing pin itself is locked by a pin which is forced through it by a spring and cannot move even if the pistol is dropped.

The pistol can be fired double-action by a long pull on the trigger or it can be used as a single-action weapon by cocking the hammer by hand and then using a shorter, lighter, trigger pull.

When the round is fired the pressure in the chamber forces the cartridge case back against the breech face. The slide and barrel, locked together, recoil for about 3 mm, after which the barrel is unlocked. The locking system is a modification of the Browning dropping-barrel method; the barrel has a large shaped lug above the chamber which, in the locked position, engages in a recess formed around the ejection slot in the slide. A shaped lug beneath the chamber strikes a transom in the frame and this lowers the rear of the barrel, disengaging it from the slide.

The slide continues rearward and extracts and ejects the empty case. The return spring is compressed and the hammer is cocked. The slide comes to rest when it reaches a stop in the frame above the butt, and is then thrown forward. The cam formed beneath the chamber lifts the rear of the barrel into the recess in the slide. The cam slides forward another 3 mm along the flat top of a supporting ramp, and this keeps the rear end of the barrel firmly locked into the slide.

When the ammunition is expended the magazine follower rises and lifts the slide stop, located on the left of the receiver above the butt, into a recess in the slide, thus holding the slide in an open position.

When a loaded magazine is inserted the slide can be released either by pressing down on the slide catch with the thumb of the right hand or by gripping the slide, pulling back slightly and then releasing it. In both cases the slide goes forward and chambers a cartridge and the hammer remains cocked for immediate action.

Stripping
To disassemble the P 220 the magazine must be removed and the chamber and feedway checked. The slide is retracted until the cutaway lines up with the take-down pin which is on the receiver. This pin is rotated down and the slide can then move forward off the receiver. The barrel and return spring can be lifted out of the slide.

9 mm P 220 SIG-Sauer pistol

P-220-1 pistol in .45 ACP calibre, showing magazine release button

DATA

DATA IN 9 mm COLUMN IS COMMON TO ALL VERSIONS EXCEPT WHERE THE CONTRARY IS INDICATED

Cartridge	9 mm Parabellum	7.65 mm Parabellum	.45 ACP	.38 Super
Operation	short recoil self-loading, single- or double-action			
Method of locking	projecting lug			
Magazine capacity	9	9	7	9

WEIGHTS

Pistol without magazine	750 g	765 g	730 g	750 g
Empty magazine	80 g	80 g	70 g	70 g

LENGTHS

Pistol	198 mm
Barrel	112 mm

MECHANICAL FEATURES

Barrel: (rifling)	6 grooves rh	4 grooves	6 grooves	6 grooves
One turn in	250 mm	250 mm	400 mm	250 mm

Sights: (fore) blade 3 mm wide, white spot on surface; (rear) square notch 3 mm wide, white spot below notch; (graduation) nil; (zeroing) lateral, rearsight moves in dovetail. Elevation. Change rearsight. 8 sizes available in steps of 0.27 mm corresponding to 4.2 cm at 25 m or 8.4 cm at 50 m
Sight radius: 160 mm (Stavenhagen contrast sights)

FIRING CHARACTERISTICS

Muzzle velocity	345 m/s	365 m/s	245 m/s	355 m/s
Rate of fire:	single shots 40 rds/min			

Manufacturer
SIG-Sauer. Production at JP Sauer and Sohn GmbH, Eckernförde, Federal Republic of Germany.
Status
Current. Production.

Service
Orders for 35 000 of the 9 mm version have been placed with SIG by the Swiss Government and the weapon is now in service under the Swiss Army designation 9 mm Pistole 75. Also in service with some foreign police and special forces.

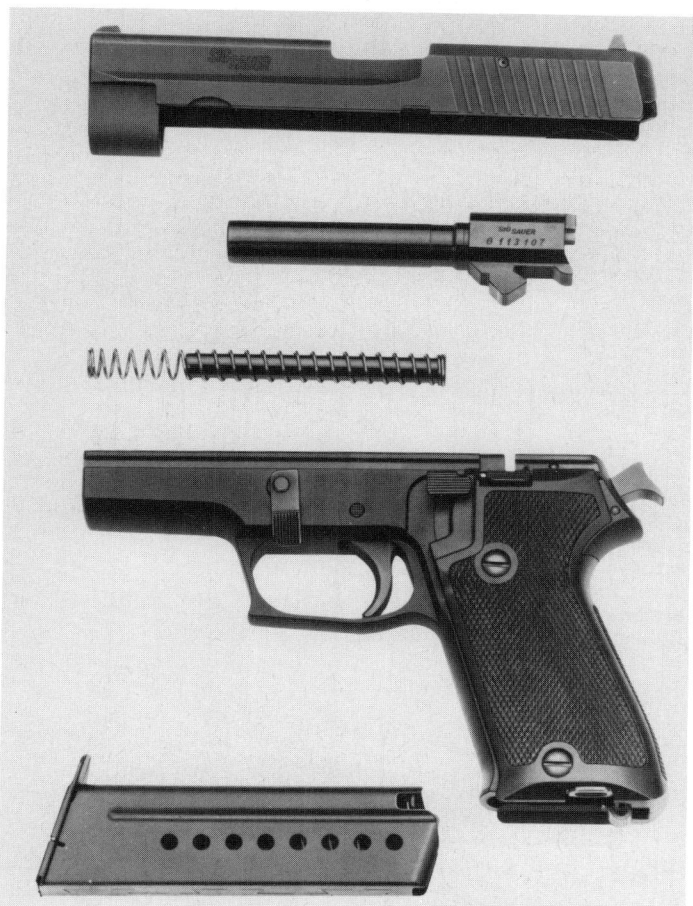

9 mm P 220 SIG-Sauer pistol stripped

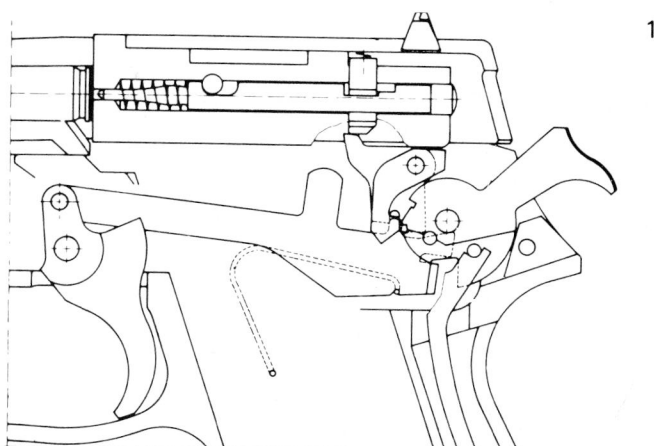

1

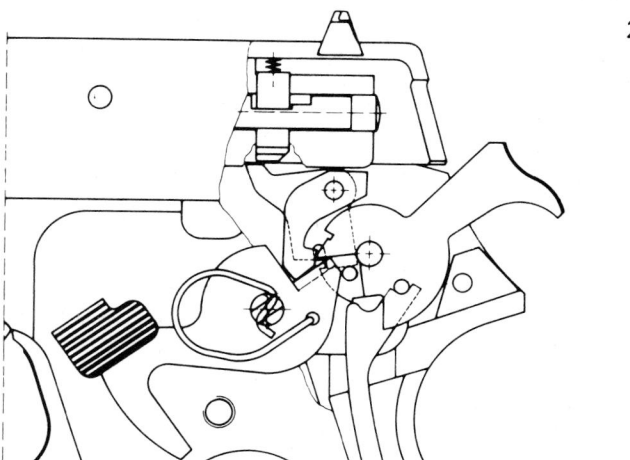

2

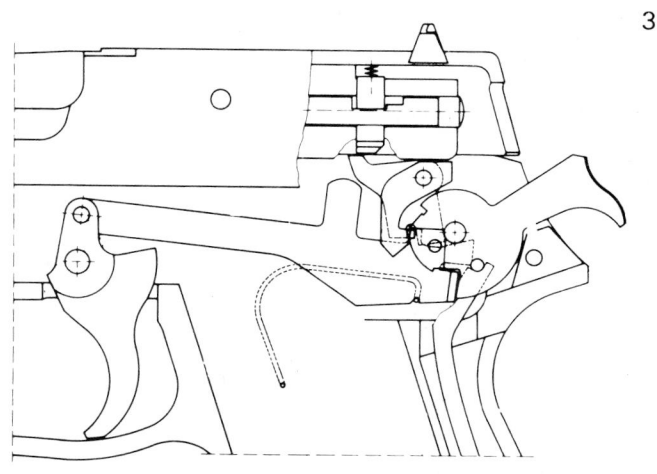

3

Action of 9 mm SIG-Sauer P 220 pistol
(1) double-action triggering. If the loaded weapon is not cocked, the shot can be fired directly via the trigger, by way of the double-action of the latter. The trigger is squeezed, whereby the hammer is cocked via the trigger rod. In the process the safety lever is pressed against the lock pin. The sear is moved away from the hammer, and the firing pin released by the lock pin. On squeezing further, the hammer lifts out of register and fires the shot (2) de-cocking lever releasing the hammer into the safety notch, so that the loaded weapon can be carried without danger. The safety notch is the position of rest for the hammer. The firing pin is always blocked during and after de-cocking. The weapon is therefore absolutely safe (3) firing pin safety. In order to achieve maximum safety, the firing pin is locked. Quick readiness for firing is always assured, as this safety pin is released automatically by the trigger action, without the manipulation of any lever. The pin is thus not released until the shot is about to be deliberately fired. Hence with this style of safety device, a loaded weapon is always safe, even if dropped with the hammer cocked

9 mm/.380 P230 SIG-Sauer pistol

The SIG P230 is a blowback pocket pistol which, although designed for police work, could well be carried by military personnel such as staff officers or second-line troops for personal defence.

It can be used as a single- or double-action pistol and has the same facility for lowering the hammer, by depressing the hammer de-cocking lever, as the P 220. Similarly the firing pin is permanently locked except when released by the trigger bar immediately before the hammer falls.

The pistol is available chambered for the 9 mm Short/.380ACP or the 7.65 mm ACP cartridges and both calibres are also available in stainless steel models.

The magazine platform forces the slide stop on the left of the receiver up when the ammunition is expended. When a loaded magazine has been inserted, the slide is pulled back to the stop and then allowed to snap forward. This cocks the hammer and feeds a cartridge into the chamber.

Stripping follows the same general pattern as the P 220. The magazine is removed, feedway checked, and the slide retracted. The take-down pin on the left of the receiver is rotated down and the slide pulled back to the stop and lifted at the back. It will then go forward off the receiver.

DATA
Cartridge: 9 mm Short, 7.65 mm ACP
Operation: blowback, semi-automatic, double-action
Feed: 7-shot box magazine (7.65 mm – 8-shot)
Weight, empty: 9 Short–460 g; 7.65–465 g; 9 Short stainless–590 g; 7.65 stainless–600 g
Trigger pull: single action 17N; double action 45N
Length: 168 mm
Barrel: 92 mm

Rifling: 6 grooves, rh, one turn in 250 mm
Sights: (fore) blade; (rear) notch; Stavenhagen pattern
Sight radius: 120 mm
Muzzle velocity: 9 Short–275 m/s; 7.65–290 m/s

Manufacturer
SIG-Sauer. Production at JP Sauer and Sohn GmbH, Eckernförde, Federal
Republic of Germany.
Status
Current. Production now limited to 9 mm Short in Aluminium and stainless
versions.
Service
Several Swiss State Police Corps and smaller Swiss Police units. Numerous
American police forces.

9 mm/.380 P230 SIG-Sauer pistol

9 mm P225 (P6) SIG-Sauer pistol

The P225, as may be seen from the data tables, is slightly smaller and lighter
than the P220, and it carries one round fewer in the magazine. It is similar in
its operation in that it is a mechanically-locked recoil-operated weapon with
an automatic firing-pin lock, double-action trigger, de-cocking lever and
external slide catch lever. An additional safety has been built in which
provides an absolute lock even if the pistol is accidentally dropped with the
hammer cocked, de-cocked, or half-way to being cocked.

Much thought has gone into ensuring that a shot can be fired only by
actually pulling the trigger and since there is no safety catch to be pushed off,
the weapon is remarkably quick to bring into action. In fact the firing of the
P225 is almost identical to that of a revolver, and it thus overcomes one of the
quoted advantages of the revolver against the automatic pistol, the fact that
the revolver does not need an applied safety and so can shoot the first round
more quickly.

A good deal of care has gone into the design of the grips and the positioning
of the centre of balance so as to ensure a good hold and positive control of the
weapon while firing. The finish is excellent, as are all SIG weapons, and all
parts are interchangeable between pistols of the same designation. For
training, the P 225 PT has been designed to fire the 9 x 19 mm PT (Plastic
Training) ammunition. In West Germany the P225 is known as the P6.

DATA
Cartridge: 9 × 19 mm (9 mm Parabellum)
Operation: recoil, single- or double-action trigger
Method of locking: projecting lug
Feed: 8-round magazine
Weight without magazine: 740 g
Length: 180 mm
Height: 131 mm
Barrel: 98 mm
Rifling: 6 grooves, rh, 1 turn in 250 mm
Sight base: 145 mm

Manufacturer
SIG-Sauer. Production at JP Sauer and Sohn GmbH, Eckernförde, Federal
Republic of Germany.
Status
Current. Production.
Service
Swiss and West German police forces.

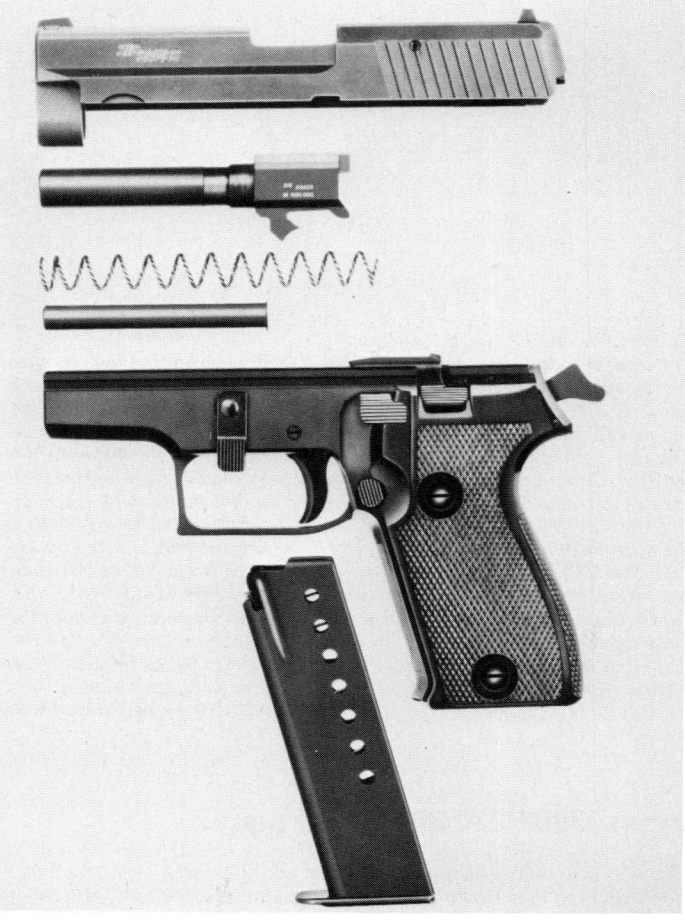

P225 SIG-Sauer pistol field-stripped. Further stripping requires an armourer

9 mm P225 SIG-Sauer pistol

9 mm P226 SIG-Sauer pistol

The P226 was conceived in late 1980 as SIG's candidate in the competition for
a new automatic pistol in 9 mm Parabellum for the US Armed Forces, in
which it ended up as a 'technically acceptable finalist'. About 80 per cent of its
parts come from current production P220 and P225 pistols and, like those

pistols, it is a positively-locked short-recoil weapon with automatic firing pin
lock, double-action trigger, de-cocking lever and external slide latch lever.

As will be seen from the data tables, the dimensions of the P226 are similar
to those of the P 220, from which it differs in magazine capacity and in the

provision of an ambidextrous magazine catch. The detailed technical description given above for the P 220 applies equally to the P226.

Constructed and finished to the standard of excellence that has come to be expected of SIG weapons, the P226 is stated by the company to be the best pistol it has made.

DATA
Cartridge: 9 × 19 mm (9 mm Parabellum)
Operation: short recoil, self-loading, single- or double-action
Method of locking: projecting lug
Magazine capacity: 15 or 20 rounds
Weight, empty: 750 g
Length: 196 mm
Barrel: 112 mm
Rifling: 6 grooves, rh, one turn in 250 mm
Sights: (fore) blade; (rear) square notch
Sight radius: 160 mm
Trigger pull: single-action 20N; double-action 55N
Muzzle velocity: 350 m/s

Manufacturer
SIG-Sauer. Production at JP Sauer and Sohn GmbH, Eckernförde, West Germany.
Status
Current. Production. Over 400 000 units of the P220 family have been produced to date.

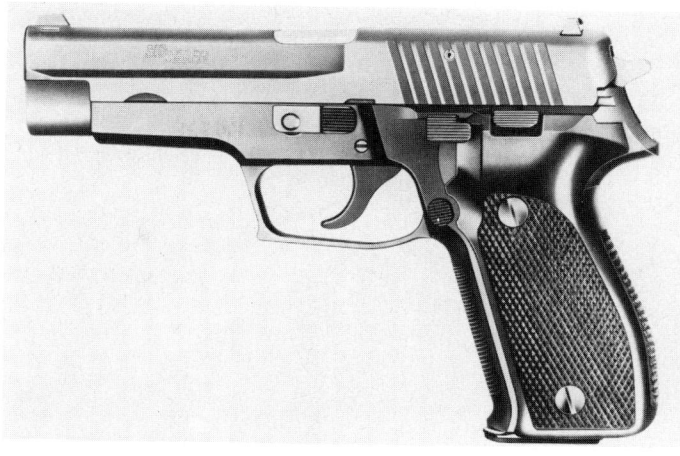

9 mm P226 SIG-Sauer pistol.

9 mm P228 SIG-Sauer pistol

The P228 has been designed to provide a compact pistol with large magazine capacity,. so as to round off and complement the range of SIG-Sauer pistols. It is a mechanically locked recoil-operated autoloader in 9 mm Parabellum calibre which is particularly suitable for concealed carrying and for individuals with smaller hands.

A double-action pistol, the P228 has a magazine capacity of 13 shots. It has the usual automatic firing pin safety system used on previous SIG-Sauer pistols, and the magazine catch can be mounted on the left or right side to suit the firer's preference. The magazine floorplate, made of high-impact synthetics, gives the magazine a considerable degree of protection against damage due to dropping.

An essentially closed design concept renders the P228 particularly resistant to dust and dirt. The majority of wearing parts are identical to those of the P225 and P226 pistols, and most of the accessories for these weapons are suitable for use with the P228.

DATA
Cartridge: 9 × 19 mm (9 mm Parabellum)
Operation: short recoil, self-loading, single- or double-action
Method of locking: projecting lug
Magazine capacity: 13 rounds
Weight, empty: 830 g
Length: 180 mm
Barrel: 98 mm
Rifling: 6 grooves, rh, one turn in 250 mm
Sights: (fore) blade; (rear) square notch
Sight radius: 145 mm
Trigger pull: single action 20N; double action 55N

Manufacturer
SIG-Sauer. Production at JP Sauer and Sohn GmbH, Eckernförde, West Germany.
Status
Current. Production.

SIG-Sauer 9 mm P228 pistol

Sig-Sauer 9 mm P228 pistol, right side

9 mm ITM AT-88S Pistol

The AT-88S pistol was originally the Czechoslovakian CZ75 manufactured under license but now, after many improvements developed by the Swiss manufacturers, it can be considered a new and independent design. The assembly tolerances and quality of finish are much improved and the barrel has been slightly changed in dimensions so that it is not interchangeable with that of the CZ75. The company have recently developed a Peters Stahl barrel of exceptional accuracy and long-wearing properties. The safety catch can now be applied whether the pistol is cocked or uncocked. In 1987 an automatic firing pin safety was introduced, which prevents movement of the firing pin except when the trigger is approaching the final movement of firing, and the most recent innovation is the fitting of an ambidextrous safety catch and, if desired, an ambidextrous slide stop pin.

The pistol is available chambered for either the 9 mm Parabellum cartridge or the .41 Action Express cartridge. A conversion kit allows changing a 9 mm pistol to .41 calibre by simply changing the barrel and magazine.

9 mm ITM AT-88S pistol

9 mm ITM AT-88S pistol in cutaway form

DATA
Cartridge: 9 mm Parabellum or .41 Action Express
Operation: recoil, semi-automatic, double-action
Method of locking: Browning cam
Feed: 15–round (9 mm) or 12-round (.41AE) box magazine
Weight, empty: 1.00 kg
Length: 206 mm
Barrel: 120 mm
Rifling: 6 grooves, rh, one turn in 250 mm
Sights: (fore) blade; (rear) square notch, adjustable
Sight radius: 161 mm
Muzzle velocity: 352 m/s

Manufacturer
ITM: Industrial Technology & Machines AG, Allmendstrasse 31B, Postfach 260, CH-4503 Solothurn.
Muller & Co. (England), High Street, Cleobury Mortimer, Worcs., England.
Status
Current. Production.
Service
Australian, Hong Kong, USA, Norway police forces. British forces in Hong Kong.

9 mm ITM AT-88P Pistol

Like the AT-88S (above) the AT-88P is based on the design of the Czechoslovakian CZ75, but it is a shortened and lightened version which has been designed by ITM. The company have also developed a Peters Stahl barrel for this model, and a version chambered for the .41 Action Express calibre is now in production. As with the AT-88S, an automatic safety firing pin, ambidextrous safety catch and ambidextrous slide stop pin have been added to this model, and there is also a similar conversion kit to allow changing a 9 mm pistol to .41AE calibre

DATA
Cartridge: 9 mm Parabellum or .41 Action Express
Operation: recoil, semi-automatic, double-action
Method of locking: Browning cam
Feed: 13-round magazine
Weight, empty: 910 g
Length: 184 mm
Barrel: 93 mm
Rifling: 6 grooves, rh, one turn in 250 mm
Sights: (fore) blade; (rear) square notch, adjustable
Sight radius: 150 mm

Manufacturer
ITM: Industrial Technology & Machines AG, Allmendstrasse 31B, Postfach 260, CH-4503 Solothurn.
Status
Current. Production.

ITM AT-88P pistol in .41 Action Express calibre

9 mm ITM AT-88H pistol

The AT-88H is the 'Hideaway' version of the AT-88 family. It is mechanically similar to the other models and exhibits the same safety features, but is smaller in all dimensions and has had the slide and barrel re-designed. It is normally supplied in 9 mm Parabellum chambering but can be converted to .41 Action Express. It can also be converted to the new 9 mm Action Express calibre, and a patented design of magazine has been developed which will accept any of these calibres without requiring modification.

DATA
Cartridge: 9 × 19 mm; .41 Action Express; 9 mm Action Express
Operation: recoil, semi-automatic, double-action

Method of locking: Browning cam
Feed: 10-round magazine
Weight, empty: 740 g
Length overall: 172 mm
Barrel length: 87 mm
Rifling: 6 grooves rh
Sight base: 140 mm

Manufacturer
ITM: Industrial Technology & Machines AG, Allmendstrasse 31B, Postfach 260, CH-4503 Solothurn.
Status
Current. Production.

9 mm ITM AT-88H pistol

9 mm ITM Sphinx .3AT pistol

This pistol has been developed primarily for police use, but it has undoubted applications for second-line military employment. It is chambered for the 9 mm Short (.380 Auto) cartridge or for the 9 × 18 mm Police cartridge. The action is self-cocking (or double-action only) with a light and very smooth trigger action. It is fitted with new and patented automatic decocking, automatic firing pin safety, ambidextrous magazine release and slide locking catch systems. The frame is of stainless steel and the remaining parts are of MnCrV or CrMo steels. The barrel is made by Lothar Walther of Germany.

DATA
Cartridge: 9 × 17 mm Short or 9 × 18 mm Police
Operation: blowback
Feed: 10-round box magazine
Length: 153 mm
Barrel: 82 mm
Muzzle velocity: 270 m/s

Manufacturer
ITM: Industrial Technology & Machines AG, Allmendstrasse 31B, Postfach 260, CH-4503 Solothurn.
Status
Current. Production.

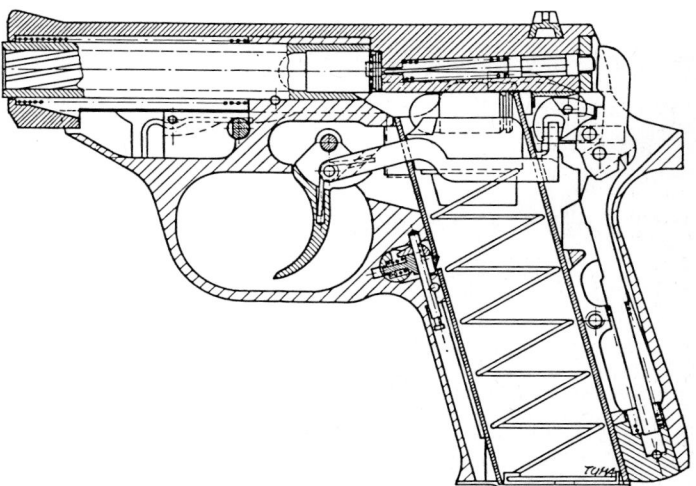

Sectioned drawing of the 9 mm ITM Sphinx pistol

TURKEY

9 mm MKE and 7.65 mm MKE pistols

These pistols are made by Makina ve Kimya Endüstrisi Kurumu at Kirikkale, Ankara. They are based on the Walther PP. The pistols are marked 'MKE' on the grip and the slide on the right side of the pistol and marked 'KÍRÍKKALE CAP 9 mm' (or '7.65 mm') on the left.
    The method of operation is exactly the same as the Walther PP and there are only minor external modifications such as the shape of the magazine finger rest.

DATA
Cartridge: 9 mm Short or 7.65 mm
Operation: blowback, double-action trigger
Method of locking: nil
Feed: 7-round detachable box magazine
Weight: 680 g empty
Length: 170 mm
Barrel: 98 mm
Rifling: 6 grooves rh
Sights: (foresight) blade; (rearsight) notch
Muzzle velocity: 260–280 m/s
Rate of fire: 35 rds/min
Effective range: 30 m

Manufacturer
Makina ve Kimya, Endüstrisi Kurumu, Ankara.
Status
Current. In production.

9 mm MKE pistol

Service
Turkish Army.

UNION OF SOVIET SOCIALIST REPUBLICS

5.45 mm PSM

This pistol is the standard sidearm of all Soviet police and internal security forces and also the security element of the armed forces. It resembles the Walther PP (see under West Germany) in size, appearance and general mechanical arrangements, being a fixed-barrel blowback weapon with double-action lock. There are some minor differences in the lock mechanism from that of the Walther, and the safety catch is fitted so as to protrude at the rear of the slide; this is in order to reduce the thickness of the pistol as much as possible, for concealed carrying.

The PSM, standing for 'Pistolet Samozaryadniy Malogabaritniy' ('pistol, self-loading, small'), fires the 5.45 mm × 18 bottle-necked cartridge described in the Ammunition section. The known features of the pistol and cartridge suggest low performance and an excess of complication for the chosen role.

5.45 mm PSM

DATA
Calibre: 5.45 mm
Operation: Blowback, double action
Feed: 8-shot box magazine
Weight: (empty) 460 g; (full magazine) 500 g
Length overall: 160 mm
Width: 17.5 mm
Barrel length: 85 mm
Rifling: 6 grooves rh
Muzzle velocity: 315 m/s

Manufacturer
State arsenals.

Status
Current. Production.
Service
Civil and military security forces.

7.62 mm TT-33 Tokarev pistol

The TT-33 Tokarev is now obsolete in the Warsaw Pact countries and production ceased in 1954 in its native country, though there is reason to believe that it continued for longer than that in Yugloslavia and China. There are therefore many of these pistols still to be met in the world, and it is no longer safe to say that their use is confined to Soviet and former Soviet states. It was derived from the Browning design in the 1920s at the Tula Arsenal by Feodor Tokarev and he simplified parts of the design and modified others, but the basis was the Model 1911 Colt. The main differences are the lock mechanism, the magazine, and the safety arrangements.

The Tokarev was fully described in *Jane's Infantry Weapons 1978*, page 53, but the stripping instructions are repeated here for the benefit of anyone faced with one of these pistols.

Stripping
The weapon is stripped as follows: the magazine is removed and the slide retracted to enable the chamber and feedway to be inspected. The recoil spring plunger, under the muzzle, is pressed in, using the nose of a cartridge, to unlock the barrel bushing. The barrel bushing is rotated and removed together with the return spring and plunger. The base of the magazine is used to move the clip retaining the slide stop, to the rear. This clip is on the right side of the receiver above the trigger. The slide stop can then be pulled to the left out of the receiver and when the pistol is inverted the slide can be moved forward off the receiver. When the slide is removed, the barrel can be freed by lifting the chamber end and sliding forward out of the slide. The hammer mechanism comes out of the receiver as one assembly.

DATA
Cartridge: 7.62 mm × 25 Type P
Operation: recoil, single-action
Feed: 8-round box magazine
Weight, empty: 0.85kg

TT-33 Tokarev pistol

Length: 196 mm
Barrel: 116 mm
Rifling: 4 grooves rh
Muzzle velocity: 420 m/s

Status
Obsolete. No longer in production.
Service
Anywhere in the Soviet Bloc and with many guerrilla forces.

9 mm Stechkin automatic pistol (APS)

The Stechkin pistol is now out of service with regular Soviet forces, and probably with all other forces, but it may well turn up in unlikely parts of the world where there is a supply of ammunition for it and where its peculiar mechanical attributes hold attraction for less astute users. As a pistol it is unusual in having the facility for full automatic fire. Naturally, such firing is an almost certain waste of ammunition since the weapon is too light to be controllable, even when fitted with its optional carbine-style wooden butt, and it would seem that the Soviets came to this conclusion and withdrew the weapon.

Mechanically the Stechkin is similar to the Makarov in that it is a blowback pistol firing the standard Soviet 9 mm × 18 cartridge. The main difference lies in the fact that it has a selector lever which permits single shots or automatic fire, and it is larger and heavier. In fact, like so many machine pistols, it is too large for a handy pistol, and too small for a useful sub-machine gun. The design is very complex and most expensive to make. The standards of finish

and fitting are of the highest and much above that usually found in Soviet weapons. It is fully described in *Jane's Infantry Weapons 1978*, page 54.

Stripping
Remove the magazine and inspect the chamber and feedway for live rounds. Pull the front of the trigger-guard down. Retract the slide, lift its rear end out of the receiver and ease it forward. Take off the return spring. To re-assemble: check that the hammer is cocked, the selector is not at safe, and the trigger-guard is pulled down.

Put the return spring over the barrel and insert into the slide. When the barrel goes through the hole in the front of the slide, pull the slide to the rear and press it down into the receiver and allow it to go forward. Press the trigger-guard up into position.

DATA
Cartridge: 9 mm × 18
Operation: blowback, single- or double-action, selective fire
Method of locking: nil
Feed: 20-round box magazine
Weight, empty: 1.03 kg; with holster stock 1.58 kg
Length: 225 mm; with stock attached 540 mm
Barrel: 127 mm
Rifling: 4 grooves rh
Sights: (fore) blade; (rear) notch, adjustable by rotating drum to 25, 50, 100 or 150 m
Muzzle velocity: 340 m/s
Rate of fire: 750 rds/min cyclic; 80 rds/min practical

Manufacturer
State arsenals.
Status
No longer produced.
Service
Out of service. May be occasionally encountered, but the pistol was not exported in quantity.

9 mm Stechkin automatic pistol

9 mm Makarov self-loading pistol (PM)

The Makarov dates from the early 1950s and is the standard pistol for the Soviet forces and for most of the countries in the Warsaw Pact as well. It has also appeared in some of the smaller countries which receive Soviet military aid. It is a copy of the Walther PP in its general size, shape and handling. It is well made and uses good quality steel, but the handling is a little awkward because of the very bulky grip. It is designed to give the best possible ballistics with an unlocked breech and the 9 mm × 18 pistol cartridge used by the Soviets. The different varieties in use at the moment are listed below;

USSR
Pistole Makarov (PM)
Five-pointed star on grips
Lanyard loop at heel of pistol grip

EAST GERMANY
Pistole M
No markings on grips
No lanyard loop

CHINA
Type 59 pistol
'59 SHI' on receiver

Operation

The magazine is inserted into the butt and the pistol is cocked by drawing the slide back and letting the return spring force it forward, feeding a round into the chamber. The safety catch is on the slide above the left-hand grip. When the selector is rotated upwards to the 'safe' position the red dot is covered. Setting the safety catch to 'safe' with the hammer cocked will drop the hammer, but the firing pin is blocked. When it is required to use the pistol the safety is moved to the 'fire' position, and the trigger can then be operated either in the single-action mode, by cocking the external hammer with the thumb, or in double-action mode by a long pull-through on the trigger which cocks and releases the hammer. When the magazine is empty the slide remains to the rear; it can be released by depressing the slide stop latch on the left side of the frame or by releasing the magazine, then pulling back and releasing the slide.

Trigger and firing mechanism

The trigger and firing mechanism is unusual. Assuming the pistol is fired at double-action, sustained pull on the trigger, pivoted about its mid-point, moves the trigger bar forward. The cocking lever is pivoted to enter a notch in the hammer and the further movement of the trigger bar forces the cocking lever round to rotate the hammer back. This movement continues until the cocking lever slips out of the notch in the hammer, taking the sear with it. The hammer is then free to go forward and drive the firing pin forward to fire the cap. This double-action pull is poor from the point of view of the shooter.

The recoiling slide forces the cocking lever sideways, clear of the sear. The sear is then impelled against the hammer by its spring; as the hammer is rocked back by the slide, the sear engages and holds the hammer notch. When the slide runs out the hammer remains held back. The slide completes its forward movement but the trigger is still pressed and must be released. When this occurs the cocking lever rotates forward and moves under the sear. When the trigger is pressed the cocking lever engages the sear and moves it out of engagement with the hammer which swings forward, under the influence of its own spring, to drive the firing pin into the cap.

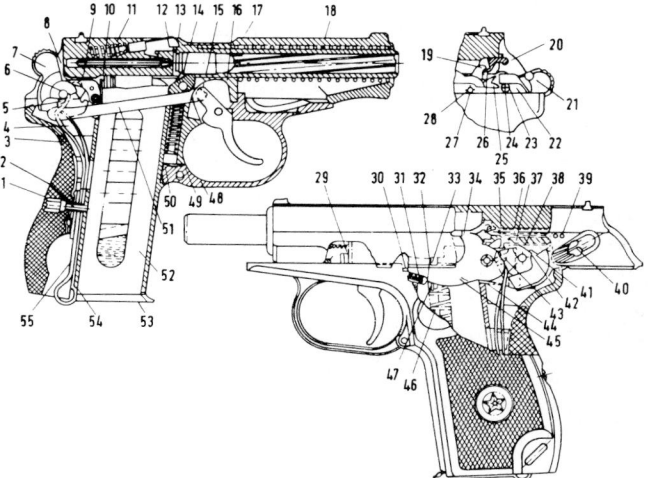

Sectioned view of 9 mm Makarov self-loading pistol
(1) *screw* **(2)** *hole for screw* **(3)** *mainspring wide leaf* **(4)** *mainspring narrow leaf* **(5)** *recess for sear lug* **(6)** *curved end of mainspring wide leaf* **(7)** *hammer* **(8)** *safety lug* **(9)** *sear lug* **(10)** *sear spring* **(11)** *extractor spring* **(12)** *extractor* **(13)** *extractor hook* **(14)** *bullet guide of barrel* **(15)** *trigger upper end* **(16)** *receiver curved slot* **(17)** *recoil spring turn (of less diameter)* **(18)** *recoil spring* **(19)** *sear tooth* **(20)** *hook for locking hammer* **(21)** *recess on hammer head* **(22)** *slide guiding slot* **(23)** *trunnion seat for hammer trunnion* **(24)** *hammer trunnion* **(25)** *hammer rebound tooth* **(26)** *shoulder of safety recess* **(27)** *sear trunnion* **(28)** *trunnion seat for sear trunnion* **(29)** *recess on trigger-guard lug* **(30)** *slide stop recess* **(31)** *slide stop catch knob* **(32)** *slide stop lug* **(33)** *slidetooth* **(34)** *slide stop ejector* **(35)** *disconnecting lug of cocking lever* **(36)** *sear* **(37)** *hammer cocking notch* **(38)** *slide rib* **(39)** *recess for safety thumb catch* **(40)** *safety thumb catch* **(41)** *lug for locking hammer* **(42)** *hammer safety notch* **(43)** *rebound lug of cocking lever* **(44)** *shoulder for trigger bar* **(45)** *slide stop* **(46)** *follower spring* **(47)** *follower claw* **(48)** *trigger-guard* **(49)** *trigger-guard spring* **(50)** *trigger-guard lug* **(51)** *trigger bar* **(52)** *magazine body* **(53)** *follower spring bent end* **(54)** *lug for magazine catch* **(55)** *mainspring lower end*

9 mm Makarov self-loading pistol

The safety lever is moved up to the 'safe' position and interposes a block between the hammer and pin; shortly afterwards a projection meets a tooth on the sear and lifts the sear from the hammer. The hammer falls and is locked in its forward position by the safety. When the safety is applied the slide is locked.

Stripping

To strip the pistol, the magazine is removed, the safety set to 'fire' to release the slide which is then retracted, and the chamber and feedway inspected.

If there is no cartridge in the pistol the slide is returned to the forward position. The front of the trigger-guard is pulled down out of the receiver, pressed to one side and rested against the receiver. The slide is pulled fully back and the rear end lifted out of the receiver. It is then eased forward over the barrel.

DATA

Cartridge: 9 mm × 18
Operation: blowback, self-loading, double-action
Method of locking: nil

Feed: 8-round box magazine
Weight, empty: 663 g
Length: 160 mm
Barrel: 91 mm
Rifling: 4 grooves rh
Sights: (fore) blade; (rear) fixed notch
Sight radius: 130 mm
Muzzle velocity: 315 m/s

Manufacturer
State factories in the USSR, East Germany and China.
Status
Current. Production in several Soviet Bloc countries.
Service
Most Warsaw Pact armies and several other Soviet forces.

UNITED STATES OF AMERICA

.45 Model 1911A1 automatic pistol

This weapon can trace its ancestry back to 1896 when Colt purchased four designs from Browning. In 1900 Colt produced the automatic Sporting Model using the then new .38 Colt Automatic Pistol Cartridge, thereafter known as the .38ACP, which had the unprecedentedly high velocity of 384 m/s. The pistol was recoil-operated and locked by grooves on the upper side of the barrel entering recesses in the ceiling of the slide. A link was provided at each end of the barrel to disconnect the barrel and slide after a short initial recoil. On this gun the slide was removed backwards and was held during gas pressure by a small metal slide stop.

In 1902 a modified model of the 1900 Sporting Model appeared and, more importantly, the .38ACP Military Model. The weight had gone up from 35 to 37 oz (992–1049 g), there were eight shots available instead of seven and there was a holding-open device when the ammunition was expended.

Following the criticisms of the .38 calibre noted above, Frankford Arsenal developed .45 and .41 cartridges; the former being used in the 1907 pistol trials. The Colt entry for these trials was a modified version of the 1902 ACP, retaining the double link arrangement and the slide stop as the only impediment preventing the slide from blowing back into the firer's face.

Between the 1907 trials, which narrowed the choice to the Colt and Savage pistols, and the troop trials in 1911, substantial modifications were made. The modified weapon had a grip safety and an applied safety, and the slide was mounted on the receiver from the front. The double link was replaced by a single link at the rear. In effect the 1911 pistol was a new pistol. It was adopted for the US Army and remained unaltered until the introduction of a new version in 1926. This resulted from a development begun in April 1923 by the Springfield Armoury and aimed at improving both sighting and weapon control in which the shape of the back of the handgrip was altered to a more arched form and checkered, a shorter, grooved, trigger was fitted, the receiver behind the trigger was chamfered to take the trigger finger, the grip safety was lengthened slightly and the hammer spur was shortened. The new version was designated the Model 1911A1.

Operation

The magazine is removed from the pistol by pressing the thumb release on the left side of the grip, behind the trigger. The loaded magazine is inserted into the handle and pushed in until the magazine locks. When the milled grips at the rear of the receiver are grasped in the left hand, the slide can be retracted. When the slide is released the compressed return spring, under the barrel, drives it forward picking up the top round from the magazine and loading it into the chamber. The extractor enters the extraction groove at the rear of the cartridge. The breech-block face strikes the barrel and pushes it forward. The barrel is connected to the receiver by a single link and as the barrel goes forward the link lifts the chamber, and two ribs on the top surface of the barrel enter recesses in the ceiling of the slide. The link rotates beyond the top dead centre position and the barrel is locked into the slide. When the slide is fully forward the disconnector rises into the recess and when the trigger is pulled the hammer flies forward and hits the firing pin. The firing pin is a notable feature of Browning's design. It is shorter than the length of its hole and is spring-retracted. The hammer strikes it forward, it strikes the cap and is immediately withdrawn into the breech-block.

The gas pressure drives the bullet up the bore. The breech face is forced back and the barrel is pulled back with it. The lower end of the link remains locked to the receiver and the top end rotates through the arc of a circle. The breech end of the barrel is rotated down and disconnected from the slide. The barrel is halted and the slide continues rearward. The empty case comes back on the breech face and is then struck by the ejector in the receiver and thrown out through the ejection port on the right of the slide. As the slide continues back, the return spring is compressed and the hammer is rotated back and held by the sear. The slide goes forward and if there is ammunition in the magazine, the cycle is repeated; if there is no ammunition, the stop on the left side of the magazine follower will lift the slide stop catch into a slot in the slide

.45 Model 1911A1 automatic pistol

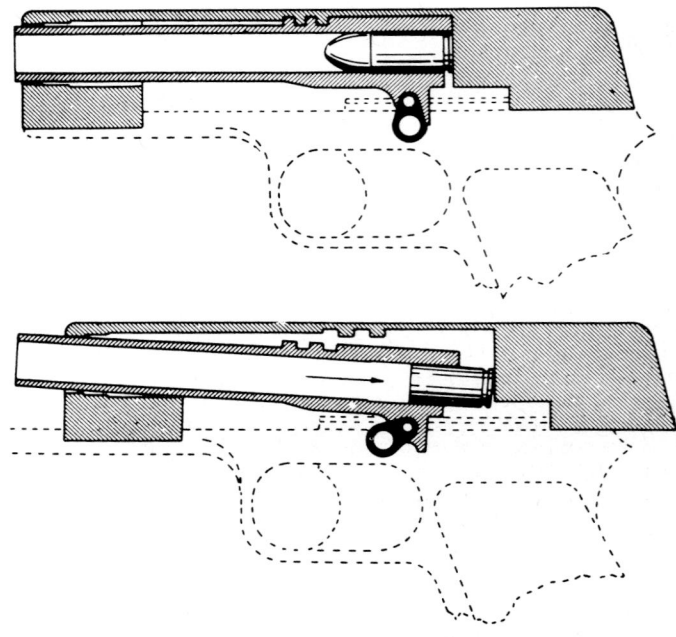

Locking system of .45 Model 1911A1 automatic pistol

and hold it to the rear. The barrel will go forward if the slide stop is manually depressed. If a loaded magazine is inserted, pressing down on the slide stop will enable the breech-block to feed a round out of the magazine lips and chamber it.

Stripping

Remove magazine, retract slide, check chamber and feedway are clear. Allow slide to go forward under control.
Press in return spring plug under the muzzle and rotate bushing clockwise. Allow plug to come out under control. Remove return spring. Remove barrel bushing.

Cock hammer. Pull slide back until rear of slide stop is aligned with rear recess in centre of slide.
Push slide stop out. Pull slide and barrel forward off receiver.
Remove return spring guide. Rotate barrel and remove from slide.
   Re-assembly is in natural reverse order but ensure barrel link is forward and link pin is not proud before sliding receiver forward into slide assembly.

DATA
Cartridge: .45 ball
Operation: short recoil, self-loading
Method of locking: projecting lug
Feed: 8-round box magazine
Weight, empty: 1.13 kg
Length: 219 mm
Barrel: 127 mm
Rifling: 6 grooves, lh, one turn in 406 mm
Sights: (fore) blade; (rear) U notch, adjustable for windage
Sight radius: 164.6 mm
Muzzle velocity: 253 m/s

.45 Model 1911 automatic pistol

Manufacturers
Colt's Patent Firearms Manufacturing Co, Hartford, Connecticut. Firearms Division, Colt Industries, Hartford, Connecticut. Ithaca Gun Co, Ithaca, New York. Remington Rand Inc, Syracuse, New York. Remington Arms-Union Metallic Cartridge Co, Bridgeport, Connecticut. Springfield Armoury, Springfield, Massachusetts. Union Switch & Signal Co, Swissvale, Pennsylvania. Also made under licence at various times in Argentina, Canada and Norway.
Status
Current for commercial production. Obsolescent for US military service.
Service
US Armed Forces and many others. One of the most widespread pistols in the world today.

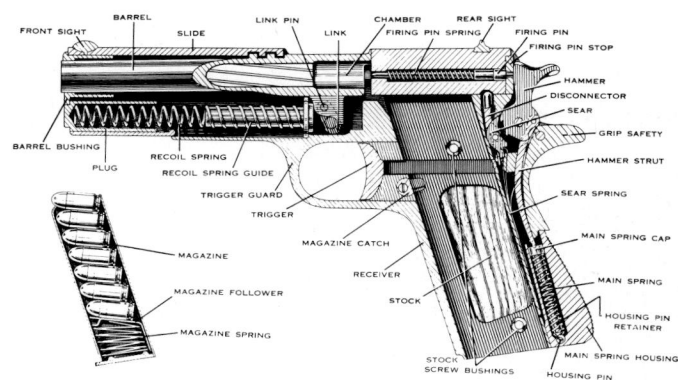

Section view of .45 Model 1911A1 automatic pistol

Pistol, 9 mm M9

This pistol, currently entering service, is the Beretta Model 92F, manufactured in Italy and by a Beretta subsidiary in the USA. Reference should be made to the entry under Italy for full details and dimensions.

.45 Model 15 General Officers' pistol

During the Second World War the US Army issued general officers with Colt Pocket Model automatic pistols in .380 Automatic Colt Pistol (ACP) calibre; when Colt took the pistol out of production in 1946 the Army still had sufficient stocks to last for several years, but in the early 1970s these stocks had dwindled and it became necessary to consider a new pistol for issue to senior ranks. The requirement was that it should be more compact than the issue Model 1911A1 but should have sufficient power to make it a practical combat weapon. A design developed by Rock Island Arsenal as their XM70 was selected and was standardised as the Model 15 in 1972.
   The Model 15 is, in essence, a .45 Model 1911A1 cut down in size and rebuilt. It operates in precisely the same way as the Model 1911A1 but is shorter in all dimensions. Due to the shorter barrel it tends to develop rather more flash and muzzle blast than the standard pistol, but this is felt to be acceptable since it is anticipated that the weapon will only be used in emergencies. It can be recognised by the excellent dark blue finish and the inscription 'General Officer Model RIA' on the slide. The left grip is inlet with a brass plate on which the individual officer's name is engraved, and the right grip has an inlaid Rock Island Arsenal medallion. The sights are raised rather more prominently from the slide than is the case with the Model 1911A1.

.45 Model 15 General Officers' pistol

DATA
Cartridge: .45 ACP
Operation: short recoil, semi-automatic
Method of locking: Browning swinging link
Feed: 7-round box magazine
Weight empty: 1.02kg
Length: 200 mm
Barrel: 106 mm
Rifling: 6 grooves lh
Sights: (foresight) blade on ramp; (rearsight) notch
Muzzle velocity: 245 m/s

Manufacturer
Rock Island Arsenal, Rock Island, Illinois.
Status
Current. Production capacity available.
Service
US Army general officers.

.45 Arminex Trifire

The Arminex Trifire is a modernised version of the US Government M1911 pistol, retaining all the proven features of that weapon but with additions which bring it to the optimum efficiency. The principal mechanical change is to place the safety catch on the slide where it actuates a positive firing pin lock; there is no hammer drop feature, and the grip safety has been eliminated, so that there is only one safety device to learn. The extractor and ejector are on pivoted mountings, giving better functioning and reliability, and the recoil spring guide system has been changed to eliminate the plug. The absence of the grip safety and re-shaping of the frame makes the weapon more comfortable to hold and easier to control.

The Trifire uses standard Government Model M1911 .45 calibre barrels, but conversion kits are available to change the calibre to either 9 mm Parabellum or .38 Super Auto. An ambidextrous safety catch is also available if required.

DATA
Calibre: .45 ACP, 9 × 19 mm or .38 Super Auto
Operation: short recoil, semi-automatic
Locking system: swinging link
Length overall: 219 mm
Barrel length: 127 mm
Weight: 1078 g
Sights: (fore) interchangeable blade; (rear) notch, adjustable for windage and elevation
Muzzle velocity: ca 255 m/s

Manufacturer
Arminex Ltd, 7882 East Gray Road, Scotsdale, AZ 85260.

Arminex Trifire .45 pistol

Status
Current. Production.

La France silenced Colt .45 pistol

This is a suppressor-equipped Colt .45 ACP Government Model pistol which functions reliably in the semi-automatic mode. The design uses the straight blowback system of operation by attaching the silencer directly to the slide, with a special fixed barrel. The mass of the suppressor/slide assembly has been engineered to produce sufficient dwell time so as to delay the opening of the breech, thereby gaining maximum bullet velocity while eliminating breech flash and inherent mechanical noise. Sights are incorporated directly on to the 800 g silencer, which is fabricated of aluminium and heat-resistant synthetic materials.

Since the design does not depend upon bullet wipes to achieve noise reduction, accuracy is excellent and noise level remains constant regardless of the number of rounds fired. The system has also been developed for the FN Browning GP35 Hi-Power and the Beretta M92 pistols.

DATA
Cartridge: .45 ACP
Operation: blowback
Feed: 7-round magazine
Weight, empty: 1.90 kg
Length overall: 485 mm
Barrel: 127 mm
Muzzle velocity: 259 m/sec
Effective range: 25 m

Manufacturer
La France Specialties, PO Box 178211, San Diego, CA 92117.
Status
Current. Production.

Firing the La France silenced .45 pistol

.45 Detonics Combat Master Mark VI pistol

The Detonics pistol has been designed with the object of providing the firepower of the .45 M1911 in dimensions more appropriate to a much smaller calibre. In essence it is a cut-down M1911, but in fact it is manufactured entirely from new material and is not a rebuild. The Mark VI is of polished stainless steel of high quality; the barrel is throated and has a polished feed ramp; the magazine aperture in the butt is polished and bevelled for rapid exchange of magazines; the trigger and sear have been carefully tuned to give the best pull-off; the rear of the slide is sloped to facilitate rapid thumb-cocking of the hammer; the magazine is of stainless steel; the ejection port has been relieved to prevent 'stove-piping' of empty cases; and the frame has been re-contoured so as to avoid the traditional 'bite' of the hammer tang into the web of the thumb on firing. Each barrel is carefully matched to its slide, and a patented V-block is used instead of a barrel bushing so as to retain concentricity of the barrel in spite of wear. The recoil spring assembly is built up from three separate and contra-rotated springs, which reduces recoil and allows the pistol to be brought back into the aim very rapidly.

Other models are similar except for finish; the Mark 1 is matt blue; Mark IV polished blue; Mark V matt stainless steel.

DATA
Cartridge: .45 ACP
Operation: short recoil
Locking system: Browning swinging link
Weight, empty: 820 g
Length overall: 171 mm
Barrel length: 89 mm
Magazine capacity: 6

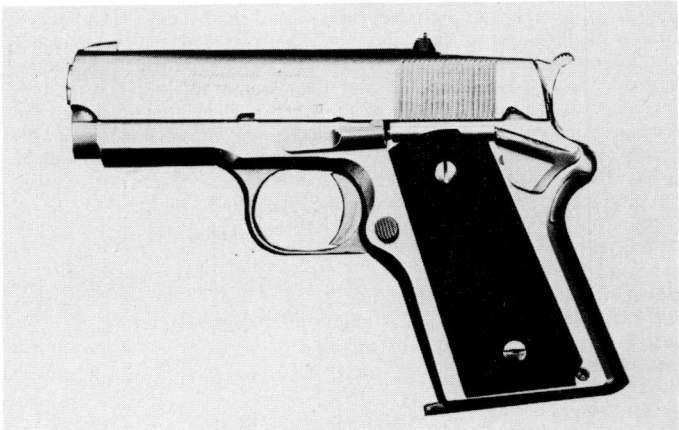
.45 Detonics Combat Master pistol

Manufacturer
Detonics Firearms Industries, 13456 SE 27th Place, Bellevue, WA 98005.
Status
Current. Production.

.45 Springfield Armory M1911A1 pistol

This new production by Springfield Armory is a standard M1911A1 pistol, the parts of which are completely interchangeable with any other model. In addition to being made in the usual .45 calibre, it is also available in 9 mm Parabellum chambering, and a conversion kit can be provided to convert existing M1911A1 pistols to 9 mm calibre.

Although of standard design, this version is entirely made of hardened and heat-treated steel machined forgings, and all parts have been surface treated for longer life and better wearing qualities.

DATA
Cartridge: .45 ACP or 9 mm Parabellum
Operation: recoil, semi-automatic
Locking: dropping barrel
Feed: 7-round box magazine (8 rounds in 9 mm)
Weight, empty: 1010 g
Length overall: 218 mm
Barrel length: 128 mm
Rifling: 6 grooves, rh, one turn in 406 mm
Sights: (fore) blade; (rear) square notch
Sight radius: 165 mm

Manufacturer
Springfield Armory, 420 West Main Street, Geneseo, IL 61254.

Status
Current. Production.

.45 Springfield Armory M1911A1 pistol

9 mm Springfield Armory P-9 pistol

The Springfield P-9 is based directly upon the CZ-75 design, though there appear to be some very minor differences, such as the sight contours and the adoption of a ring hammer rather than a spur pattern. The basic pistol is in 9 mm Parabellum chambering, but it is also available in .38 Super Auto and .45 ACP chamberings if desired. There is also a **P-9C Compact** model, with shorter slide and frame.

DATA
Cartridge: 9 × 19 mm; .38 Super Auto; .45ACP
Operation: recoil, semi-automatic
Method of locking: Browning lug
Feed: 16-round (P-9), 10-round (P-9C) magazine
Weight: 1.00 kg (P-9); 910 g (P-9C)
Length: 206 mm (P-9); 184 mm (P-9C)
Barrel: 120 mm (P-9); 93 mm (P-9C)
Rifling: 4 grooves, rh, one turn in 28 calibres

Manufacturer
Springfield Armory, 420 West Main Street, Geneseo, IL 61254.
Status
Current. Production.

Springfield 9 mm P-9 pistol

9 mm Ruger P-85 pistol

Announced early in 1987, this is the first military automatic pistol to be developed by Sturm, Ruger & Company. It is a conventional double-action pistol, using the familiar Browning type of swinging link to unlock the breech during recoil. Instead of using lugs on the top surface of the barrel, the chamber section is squared off and locks into the ejection opening in the slide. The frame is of lightweight aluminium alloy, is hardened to resist wear and is finished in matt black. The slide is of chrome-molybdenum steel and is similarly matt black. There is an external hammer and the safety catch is on

the rear of the slide. This catch is ambidextrous and locks the firing pin, blocks the hammer, and disconnects the trigger when applied.

The firing mechanism is double-action, the trigger-guard being proportioned so that the pistol can be fired by a gloved hand, and the forward edge of the trigger-guard is shaped to form a grip for the non-firing hand. The magazine release is in the forward edge of the butt and can be operated from either side of the pistol without adjustment. The sights are provided with white dot inserts to assist night firing.

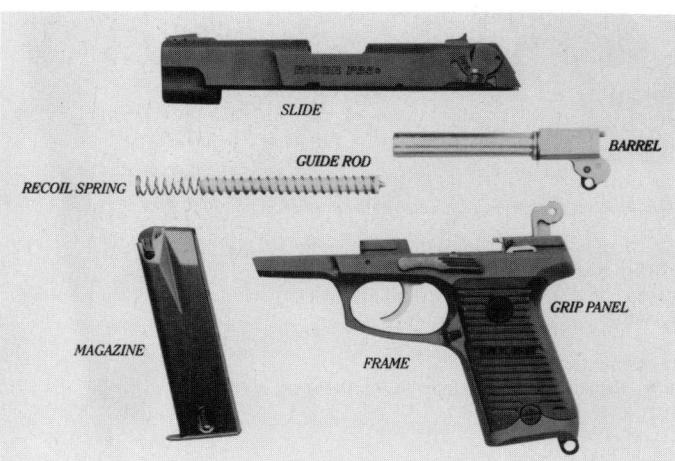

9 mm Ruger P-85 pistol dismantled

9 mm Ruger P-85 pistol

DATA
Cartridge: 9 mm Parabellum
Operation: Short recoil, semi-automatic, double action
Locking system: Browning link and tipping barrel
Feed: 15-shot box magazine
Weight, empty: 907 g
Length: 200 mm
Barrel: 114 mm

Sights: (fore) blade; (rear) square notch, adjustable for windage
Sight radius: 155 mm

Manufacturer
Sturm, Ruger & Co Inc., Ruger Road, Prescott, AZ 86301.
Status
Current. Production.

.38 Ruger SP101 small frame revolver

Following the introduction of the large frame .357 Magnum GP100 and the .44 Magnum Super Redhawk, Ruger completes the basic double-action family with this small frame, five-shot .38 calibre revolver. The all-stainless steel SP101 incorporates the engineering refinements of earlier Ruger models into a new compact revolver. This model completes the Ruger family of newly-designed revolvers for police, service and personal defence requirements.

Frame width has been increased in the critical areas which support the barrel, and both frame sidewalls are solid, to provide great strength and rigidity. The design of an offset ejector rod has allowed the building of a thicker and stronger frame in the forcing cone area, that which undergoes the most severe pressure.

The lock mechanism is contained within the trigger-guard, which is inserted into the frame as a single subassembly, without the need for frame-weakening sideplates.

The SP101 cylinder provides all the strength necessary to withstand the pressures of modern high-velocity .38 cartridges. The cylinder locking notches are offset, and are cut into the thick part of the cylinder walls, between the chamber centres. The crane and cylinder assembly swings out of the frame in the usual manner, but when the cylinder is in the firing position it is securely locked to the frame in two places: the traditional cylinder pin at the rear, and at the front of the crane by a large spring-loaded latch. Invented by Ruger, this forward lock ensures correct barrel and chamber alignment and also allows a larger thread diameter on the barrel and a thicker frame.

Barrels, cylinders and frames for the SP101 are made from ordnance quality 400-series stainless steel, as are the hammer, trigger and other internal parts.

The SP101 is available in two barrel lengths; 57 mm, the gun weighing 709 g; and 76 mm, the gun weighing 765 g. Although only offered in .38 calibre at present, it is anticipated that other chamberings will be made available in the near future.

Manufacturer
Sturm, Ruger & Co Inc., Ruger Road, Prescott, AZ 86301.
Status
Current. Production.

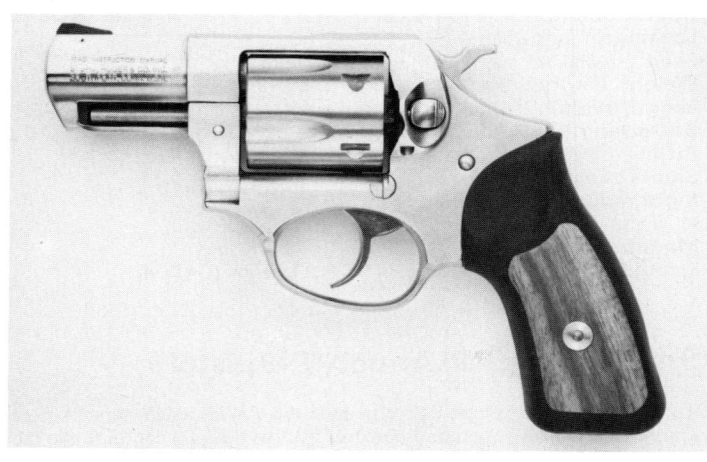

.38 Ruger SP101 double action revolver

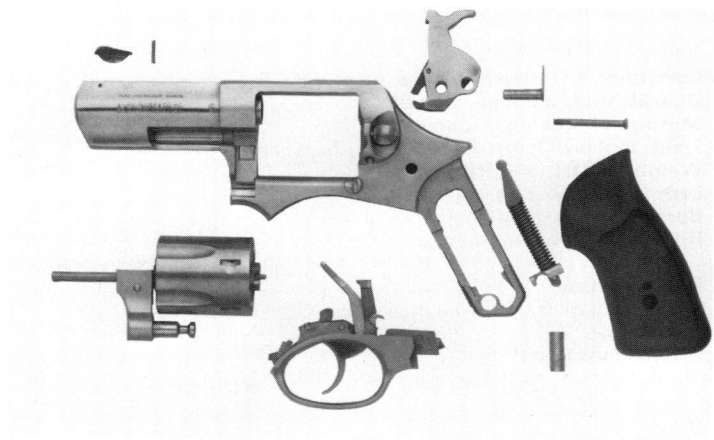

.38 Ruger SP101 revolver dismantled

.44 Magnum Ruger 'Super Redhawk' Revolver

The Ruger Super Redhawk revolver incorporates the mechanical design features and patented improvements of the GP-100 model, with a number of important additional features, the most significant of which is the massive extended frame and use of the exclusive Ruger Integral Scope Mounting System on the wide top strap, which provides a solid scope mounting surface. The extended frame also provides lengthened bearing surfaces and relocated barrel threads for greater strength and rigidity in barrel mounting.

The pistol is built of corrosion-resistant stainless steel in a brushed satin finish. It is available with 190 mm or 241 mm barrels. Both barrels are equipped with ramp front sight base with interchangeable insert blades. A steel adjustable rearsight with white outline square notch is standard.

Ruger Cushioned Grip panels are live rubber grip panels with Goncalo Alves wood inserts. The grip frame has been designed to allow installation of custom grips of a variety of shapes and sizes.

The pistol also incorporates a number of other Ruger features, including the floating firing pin mounted in the frame, transfer-bar safety, hammer and cylinder interlock, and the exclusive use of stainless steel springs throughout.

.44 Magnum Ruger Super Redhawk revolver

DATA
Cartridge: .44 Magnum
Operation: double-action revolver
Feed: 6-shot cylinder
Weight, empty: 1.502 kg
Length overall: 330 mm (with 190 mm barrel)
Barrel length: 190 mm or 241 mm

Rifling: 6 grooves, rh, one turn in 508 mm
Sights: (fore) ramp with insert blade; (rear) square notch with white outline, adjustable

Manufacturer
Sturm, Ruger & Co Inc, Ruger Road, Prescott, AZ 86301.
Status
Current. Production.

.41 and .44 Ruger 'Redhawk' Revolver

The Ruger Redhawk revolver is based on an entirely new mechanism and design philosophy. It encompasses a series of unique improvements and exclusive new features, making it oustanding in its field.

The frame has extra metal in the top strap and in critical areas below and surrounding the barrel threads. The frame has no side plate, so preserving both sidewalls intact as integral sections of the frame, resulting in increased strength and rigidity. The cylinder is locked in the firing position by a new, patented, Ruger cylinder locking system, which bolts the swinging crane directly into the frame.

With the new Ruger patented single-spring mechanism, the hammer and trigger are powered by opposite ends of the same coil spring, and the components which link the trigger and hammer to this spring function smoothly with minimum friction loss.

The Redhawk is available in blued or stainless steel finish. Two kits are available as accessories; the first includes four interchangeable front sights of glass-fibre reinforced nylon, coloured light blue, fluorescent orange, ivory and yellow. The second kit has a steel gold bead front sight with matching V-notch rearsight.

.44 Magnum Ruger Redhawk revolver with 140 mm barrel

DATA
Cartridge: .41 or .44 Magnum
Operation: double- or single-action
Feed: six-chambered cylinder, side-loading
Weight empty: 1474 g with 140 mm barrel
Length overall: 280 mm with 140 mm barrel
Barrel length: 140 mm or 190 mm
Rifling: 6 grooves, rh, one turn in 56 calibres
Sights: Interchangeable red insert front; interchangeable rear

Manufacturer
Sturm, Ruger and Co Inc, Ruger Road, Prescott, AZ 86301.
Status
Current. Production.

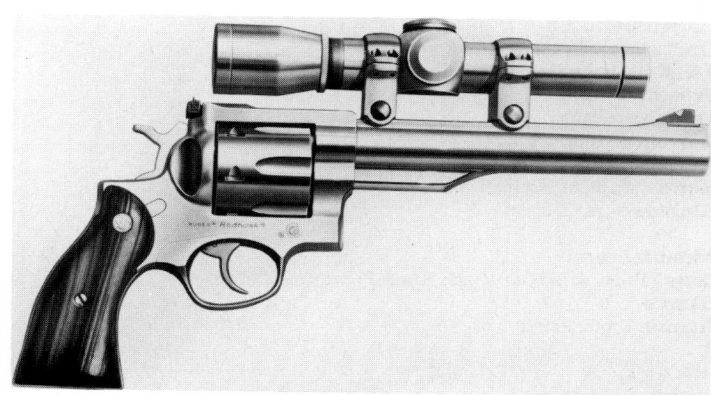

.44 Magnum Ruger Redhawk revolver with 190 mm barrel and Ruger Integral Scope Mounting System

.357 Ruger GP-100 Revolver

This is a new revolver in .357 Magnum calibre. The frame width has been increased in critical areas which support the barrel, and both frame sidewalls are solid and integral to provide strength and rigidity to the whole weapon.

The lock mechanism is contained within the trigger-guard which is inserted into the frame as a single subassembly. The cylinder locking notches are substantially offset and are located in the thickest part of the cylinder walls between the centres of the chambers. The crane and cylinder assembly swings out of the frame in the normal manner. When the cylinder is in the firing position it is securely locked into the frame by a unique new Ruger-invented mechanism.

The heavy 102 mm barrel with full-length ejector shroud is made from a hot-rolled section of ordnance-quality 4140 chrome-molybdenum alloy steel. The long shroud helps to achieve the slightly muzzle-heavy balance generally considered desirable by experienced users. The Ruger Cushioned Grips (Ruger patent) are of live rubber, with polished wood inserts.

The design incorporates a number of original Ruger innovations which have been in use for many years. These include the floating firing pin mounted in the frame, the transfer-bar safety system hammer and cylinder interlock, and the exclusive use of coil springs throughout the mechanism. The hammer, trigger, and most small internal parts are of durable, corrosion-proof stainless steel. The frame and cylinder are of ordnance-quality chrome-molybdenum steel alloys. The revolver is available in blued finish or stainless steel. The 102 mm barrel is available with a full shroud, the 152 mm barrel with full or short shroud and adjustable sights.

Ruger GP-100, Fixed Sight Model

GP-100 Fixed Sight Model
The 76 mm and 102 mm barrel models are available with full or short ejector shroud in .357 Magnum and .38 Special, with fixed sights, in blued or stainless steel finish. An optional red insert front sight is available from the factory. The Fixed Sight Model also uses a patented smaller round butt with one-piece cushioned grip.

.357 Ruger GP-100 revolver

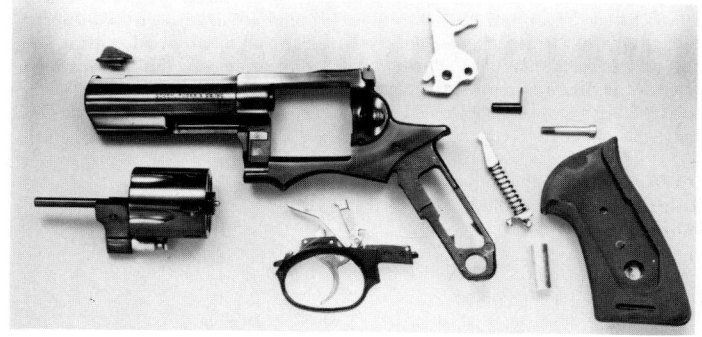

.357 Ruger GP-100 revolver dismantled

DATA
Cartridge: .357 Magnum or .38 Special
Operation: double-action revolver
Feed: 6-chambered cylinder
Weight, empty: 1247 g
Length overall: 238 mm
Barrel length: 102 mm
Rifling: 5 grooves, rh, one turn in 52 calibres

Sights: (fore) interchangeable blade; (rear) adjustable square notch with white outline

Manufacturer
Sturm, Ruger and Co Inc, Ruger Road, Prescott, AZ 86301.
Status
Current. Production.

.22LR Ruger Government Target Model pistol

This is a blowback automatic pistol firing the .22 rimfire cartridge and is well suited to training and target practice roles. This Mark II version of the well-known Ruger design retains the accuracy and handling characteristics of the earlier model.

DATA
Cartridge: .22LR
Operation: blowback, semi-automatic
Feed: 10-shot box magazine
Weight, empty: 1.247 kg
Length: 283 mm
Barrel: 175 mm
Rifling: 6 grooves, rh, one turn in 381 mm
Sights: (fore) blade; (rear) notch, adjustable for windage
Sight radius: 235 mm

Manufacturer
Sturm, Ruger & Co Inc, Ruger Road, Prescott, AZ 86301.
Status
Current. Production.

Ruger .22LR Government Target Model

.45 Model 645 Smith and Wesson pistol

Introduced in 1985 after several years of development and testing, the Smith & Wesson Model 645 is based on their successful Model 659 9 mm design but has a thicker and longer barrel. The mechanism uses the usual tipping barrel, controlled by a cam beneath the chamber, but the muzzle is located in a fixed barrel bushing which gives improved accuracy due to a tighter fit and simplfies field stripping. The magazine well is bevelled for easier and quicker reloading, and the magazine is numbered to show the remaining rounds. The pistol holds more rounds (eight in the magazine plus one chambered) than the M1911A1, yet weighs less. Trials have also shown that the gun will feed and fire all types of bullet without jamming.

There is a three-safety system with an ambidextrous manual safety catch on the slide which, when engaged, drops the hammer safely and interposes a steel bar between hammer and firing pin. There is also an automatic internal firing pin safety which prevents movement of the firing pin unless the trigger is fully pressed, and a magazine safety which prevents functioning of the trigger or hammer when there is no magazine in the weapon.

DATA
Cartridge: .45 ACP
Operation: recoil, semi-automatic, double-action
Locking: dropping barrel
Feed: 8-round box magazine
Weight, empty: 1063 g
Length overall: 219 mm
Barrel length: 127 mm
Rifling: 6 grooves, rh, one turn in 381 mm
Sights: (fore) ramp blade; (rear) square notch, adjustable for windage

.45 Model 645 Smith and Wesson pistol

Manufacturer
Smith and Wesson Inc, 2100 Roosevelt Avenue, Springfield, MA 01101.
Status
Current. Production ended.

Model 469 Smith and Wesson pistol

Announced in April 1983 the Smith and Wesson Model 469 is a shortened version of the existing Model 459, developed to meet a US Air Force specification. The pistol has a shortened Model 459 frame with contoured backstrap and serrated trigger-guard shaped to suit a two-handed grip. The butt carries roughened plastic grip plates and the magazine has a spur which extends the firer's grip area. The frame is of light alloy and the slide of steel, both having a sand-blasted blued finish. The double-action hammer has no spur.

DATA
Calibre: 9 mm Parabellum
Weight: 737.1 g

Length: 175 mm
Barrel length: 89 mm
Feed: 12-shot detachable box magazine; (Model 459 14-shot magazine may also be used)
Sights: (fore) ramped blade with yellow insert; (rear) U-notch, with white surround, adjustable for windage
Muzzle velocity: 360 m/s with NATO standard ammunition

Manufacturer
Smith and Wesson Inc, 2100 Roosevelt Avenue, Springfield, MA 01101.
Status
Current. Production ended.

9 mm Model 469 Smith and Wesson pistol

Smith & Wesson Third Generation semi-automatic pistols

This new series of semi-automatic pistols replaces earlier models in production and has been designed with the assistance of many US law enforcement agencies who were encouraged to make suggestions as to their requirements. Features incorporated in these new pistols include fixed barrel bushings for better accuracy and simpler dismantling; a greatly improved trigger pull; three-dot sights which allow a quicker sight picture in all conditions; improved wraparound grips; bevelled magazine aperture for quicker reloading; and a triple safety system comprising an ambidextrous manual safety catch, automatic firing pin safety system and magazine safety. Several other areas have been redesigned to improve service life.

9 mm 5900 series Smith & Wesson pistol

The 5900 series consists of two models, the 5904 and 5906. The 5904 has an aluminium alloy frame, carbon steel slide and stainless steel barrel and is finished in blue. The 5906 is entirely of stainless steel and is satin finished. Both are fitted with one-piece wraparound 'Delrin' grips with curved backstrap. The sights may be fixed or adjustable.

DATA
Cartridge: 9 × 19 mm Parabellum
Operation: Recoil, semi-automatic
Breech lock: Dropping barrel
Feed: 14-round magazine

Weight: 737 g (5904); 978 g (5906)
Length: 190.5 mm
Barrel: 101.6 mm
Sights: fore: post with white dot; rear U-notch fixed with 2 white dots, or adjustable for windage and elevation

Manufacturer
Smith and Wesson Inc, 2100 Roosevelt Avenue, Springfield, MA 01101.
Status
Current; production.

9 mm Smith & Wesson Model 5906, adjustable sight

9 mm Smith & Wesson Model 5904, fixed sight

9 mm 3900 series Smith & Wesson pistol

The 3900 series generally resembles the 5900 series but is slimmer, slightly lighter in weight, and has a lesser magazine capacity. Two models are manufactured, the 3904 with aluminium alloy slide, steel frame and stainless steel barrel, and the 3906 which is entirely of stainless steel.

DATA
Cartridge: 9 × 19 mm Parabellum
Operation: recoil, semi-automatic

Breech lock: dropping barrel
Feed: 8-round magazine
Weight: 723 g (3904); 964 g (3906)
Length: 190.5 mm
Barrel: 101.6 mm
Sights: As for 5900 series

Manufacturer
Smith and Wesson Inc, 2100 Roosevelt Avenue, Springfield, MA 01101.
Status
Current. Production.

9 mm Smith & Wesson Model 3904, fixed sight

9 mm Smith & Wesson 6900 series pistol

The 6900 series is a compact semi-automatic, using the same general design features as the rest of the Third Generation pistols but of smaller dimensions. Model 6904 has an aluminium alloy frame, carbon steel slide and stainless steel barrel, while Model 6906 is entirely of stainless steel. Both are available with fixed sights only.

DATA
Cartridge: 9 × 19 mm Parabellum
Operation: recoil, semi-automatic
Breech lock: dropping barrel
Feed: 12-round magazine
Weight: 666 g (6904)
Length: 174.6 mm
Barrel: 88.9 mm
Sights: Fore; laterally adjustable post with white dot: rear; fixed U-notch with two white dots

Manufacturer
Smith and Wesson Inc, 2100 Roosevelt Avenue, Springfield, MA 01101.
Status
Current. Production.

9 mm Smith & Wesson Model 6904 pistol

.45 Smith & Wesson Model 4506 pistol

The Model 4506 is chambered for the .45 ACP cartridge and provides the various modern features of the Third Generation with the power of the well known heavy calibre bullet. The 4506 is of stainless steel throughout, in satin stainless finish, and is available with fixed or adjustable sights.

DATA
Cartridge: .45 ACP
Operation: recoil, semi-automatic
Breech lock: dropping barrel
Feed: 8-round magazine
Weight: 1020 g
Length: 215.9 mm
Barrel: 127 mm
Sights: As for 5900 series

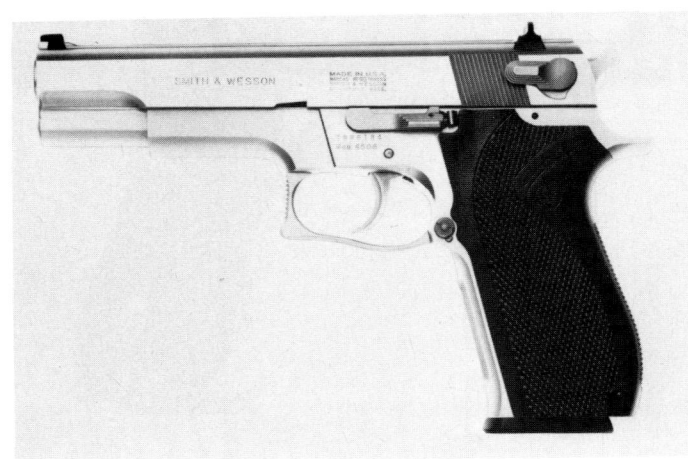

.45 Smith & Wesson Model 4506, adjustable sight

Manufacturer
Smith and Wesson Inc, 2100 Roosevelt Avenue, Springfield, MA 01101.
Status
Current. Production.

Rear Sight
Ejector Spring
Manual Safety (Fire Position)
Firing Pin Safety Lever
Hammer
Disconnector
Sear Release Lever
Drawbar
Sear
Sear Spring
Stirrup
Mainspring
Wraparound Grip
Mainspring Plunger
Magazine Butt Plate
Magazine Spring

Safety Lever Plunger Spring
Firing Pin Safety Plunger
Ambidextrous Manual Safety Lever
Extractor

Recoil Spring Guide Plunger
Recoil Spring Guide Plunger Spring

Front Sight
Barrel
Slide
Barrel Bushing
Recoil Spring Guide
Recoil Spring
Receiver

Drawbar Plunger Spring
Drawbar Plunger

Trigger
Extended Trigger Guard
Trigger Plunger Spring
Trigger Plunger
Magazine Catch
Magazine Follower

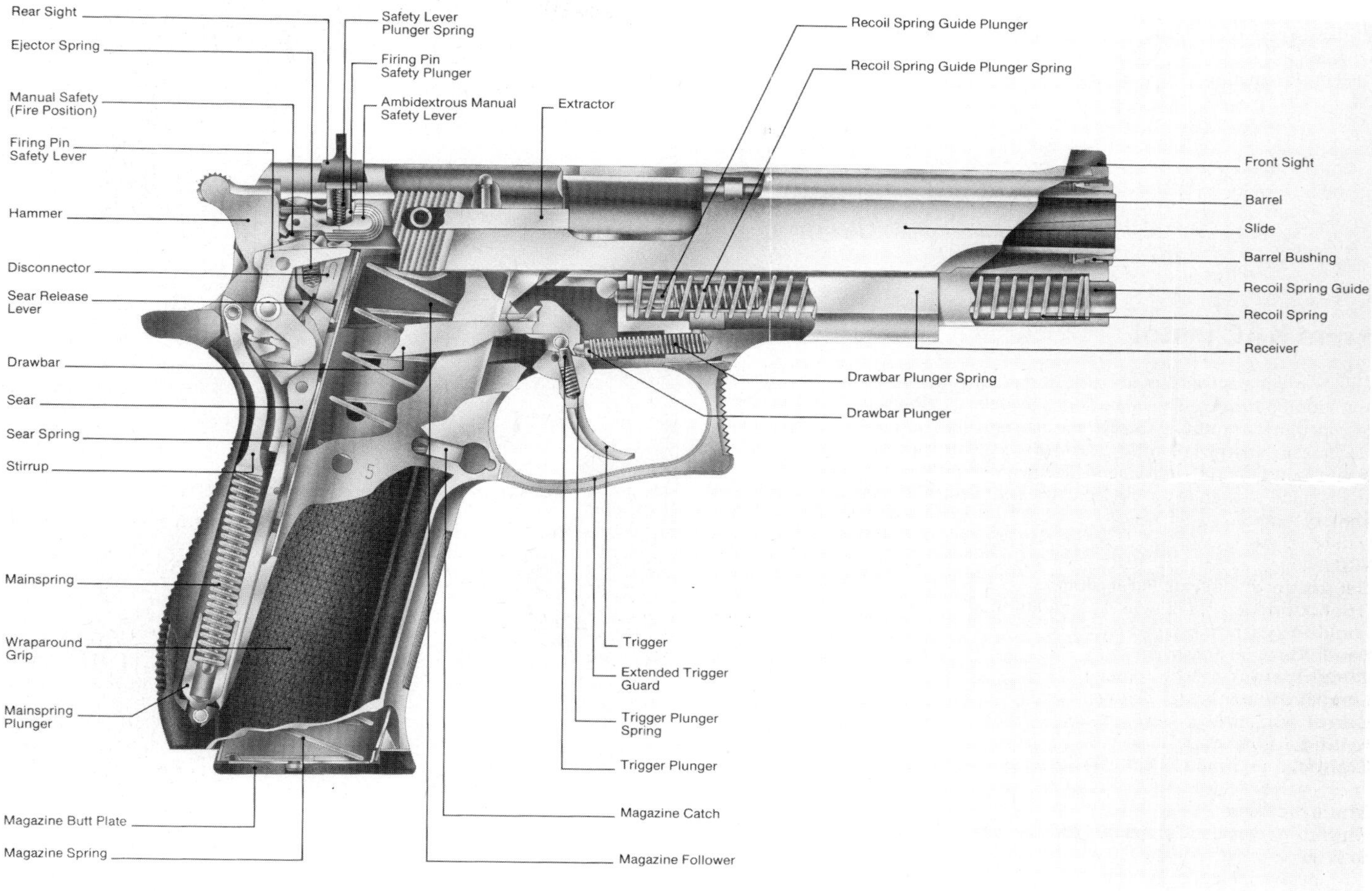

Sectioned drawing of the .45 Smith & Wesson Model 4506 pistol

.38 Model 64 Military & Police Stainless revolver

The Smith & Wesson Model 10 Military & Police revolver has been well-known and popular for many years; this is the stainless steel version which has obvious attractions for military service under arduous conditions. The metal is satin finished and the butt grips are of walnut. The standard 4 in (101 mm) barrel is designated as a 'heavy' barrel, being heavier in section than is usual in this calibre; a 2 in (50 mm) barrel of normal section is also available.

DATA
Cartridge: .38 Special
Operation: double-action revolver
Feed: 6-shot cylinder
Weight, empty: 865 g
Length: 235 mm
Barrel: 101.6 mm
Sights: (fore) serrated ramp; (rear) square notch

Manufacturer
Smith & Wesson, 2100 Roosevelt Avenue, Springfield, MA 01101.
Status
Current. Production.

.38 Smith & Wesson Model 64 Military & Police Stainless revolver

Service
Widely used by police and security forces.

Smith & Wesson .44 Magnum Model 29 revolver

Where the utmost power is required from a handgun, the Model 29 is the obvious answer, delivering upwards of 1600 J of muzzle energy with the longer barrels. The Model 29 is of the conventional Smith & Wesson design, a solid-frame revolver with swing-out cylinder and double-action lock, but the proportions are substantial in order to accommodate the powerful cartridge.

DATA
Cartridge: .44 Magnum
Operation: double-action revolver
Feed: 6-shot cylinder
Weight, empty: 1.332 kg with 165 mm barrel

Length: 302 mm with 165 mm barrel
Barrel: 101 mm, 165 mm or 212 mm
Sights: (fore) S&W Red Ramp; (rear) micrometer notch

Manufacturer
Smith & Wesson, 2100 Roosevelt Avenue, Springfield, MA 01101.
Status
Current. Production.

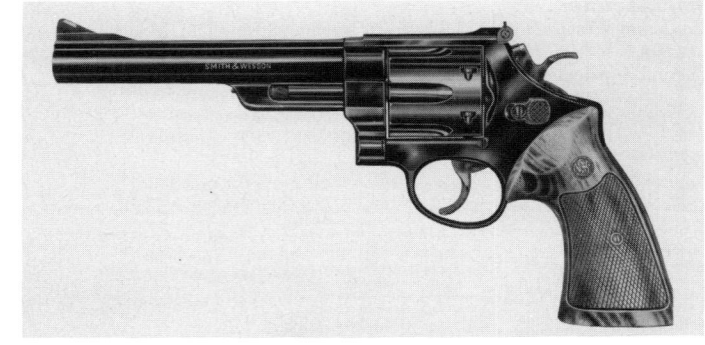

Smith & Wesson Model 29 .44 Magnum revolver

9 mm KAC pistol

This is a compact and innovative design which is currently available for evaluation purposes. The operation is by delayed blowback, the delay being provided by a rotating barrel mechanism, and the self-cocking trigger mechanism is carried on roller bearings so as to provide a constant 2.2 kg pull throughout the firing movement. The pistol is entirely of metal components and is small enough to be carried concealed, yet robust enough to serve as a primary weapon.

DATA
Cartridge: 9 × 19 mm Parabellum
Operation: delayed blowback, semi-automatic
Locking system: rotating barrel
Feed: 13-round magazine
Weight, empty: 822 g
Length: 171 mm
Barrel: 93.5 mm
Rifling: 6 grooves rh
Sights: (fore) fixed blade; (rear) notch, adjustable for windage

Manufacturer
Knight's Armament Company, 1306 29th Street, Vero Beach, FL 32960.
Status
Current; available.

9 mm KAC pistol

9 mm Göncz Assault Pistol

This weapon, though it resembles a sub-machine gun, is a semi-automatic pistol with a large magazine capacity and a long barrel giving excellent accuracy. The butt and frame are a one-piece casting of 4840 ordnance steel, while the receiver is also of high-tensile chrome-molybdenum steel drilled and ground to precise dimensions. The barrel is set well back in the receiver and the bolt surrounds the rear of the barrel in the firing position. There is a floating firing pin, and the weapon fires from a closed bolt by means of a patented sear mechanism. This mechanism has the unique advantage that if pressure has been taken up on the trigger in order to fire, and the firer changes his mind, releasing the trigger will return the firing mechanism to the rest position and does not leave the weapon in a 'hair-trigger' condition. The magazine fits into the butt, at the centre of balance, and the weapon balances and handles well in one hand, though the perforated barrel jacket can be used as a fore-grip should the firer prefer.

The standard model is the GA pistol, with 241 mm barrel; the Model GS Special pistol uses the same mechanism, frame and receiver, but dispenses with the barrel jacket and has a barrel only 127 mm long with the end screw-threaded to take a flash hider or sound suppressor. There is also the GAT Target pistol which is the same dimensions as the GA pistol but hand-finished and with a lapped barrel and adjustable trigger to give exceptional accuracy for this class of weapon.

These pistols are chambered for the 9 mm Parabellum cartridge as standard, but are also available in 7.63 mm Mauser, .38 Super and .45 ACP chamberings.

DATA
Cartridge: 9 mm Parabellum (and see text)
Operation: blowback, closed bolt
Feed: 18- or 36-round box magazines; (10 or 20 round in .45 ACP)
Weight, empty: 1.41 kg; (GS– 1.19 kg)
Length: 384 mm; (GS– 267 mm)
Barrel: 241 mm; (GS– 127 mm)
Sights: (fore) protected barleycorn; (rear) adjustable notch

Manufacturer
Göncz Company, 11526 Burbank Boulevard, North Hollywood, CA 91601.
Status
Current. Production.

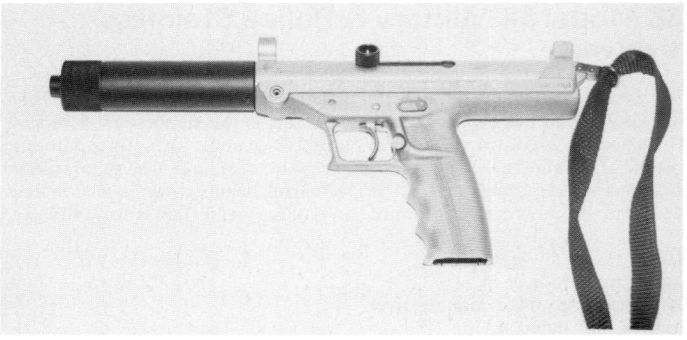

9 mm Göncz Model GA assault pistol

9 mm Göncz Model GS Special pistol with sound suppressor fitted

Service
Various police and special forces.

9 mm Calico M-950 pistol

This is a semi-automatic hand weapon with a unique Helical Feed system. The 50-round magazine can be loaded and stored without spring fatigue, and, using the Calico Speed Loader, it can be refilled from an open box in less than 15 seconds. There is very little muzzle climb, and in spite of its appearance the pistol sits well in the hand and is convenient to use.

The pistol operates by a roller-locking delayed blowback mechanism very similar to that used in the CETME and H&K rifles. The two-part bolt, return springs, guide rods, striker mechanism and buffer are in a self-contained unit which can be removed in one piece from the weapon for cleaning. The receiver is of cast aluminium, with glass-filled polymer furniture; the barrel is of CrMb steel and the breech mechanism is of stainless steel.

DATA
Cartridge: 9 × 19 mm Parabellum
Operation: semi-automatic
Locking: delayed blowback
Feed: 50- or 100-round helical magazine
Weight: 1.0 kg empty; 1.81 kg with 50-round loaded magazine
Length: 356 mm with 50-round magazine
Barrel: 152 mm
Rifling: 6 grooves, rh, one turn in 305 mm
Sights: (fore) post, adjustable for windage and elevation; (rear) fixed notch
Muzzle velocity: 393 m/sec

Manufacturer
Calico, 405 East 19th Street, Bakersfield, CA 93305.
Status
Current; production.

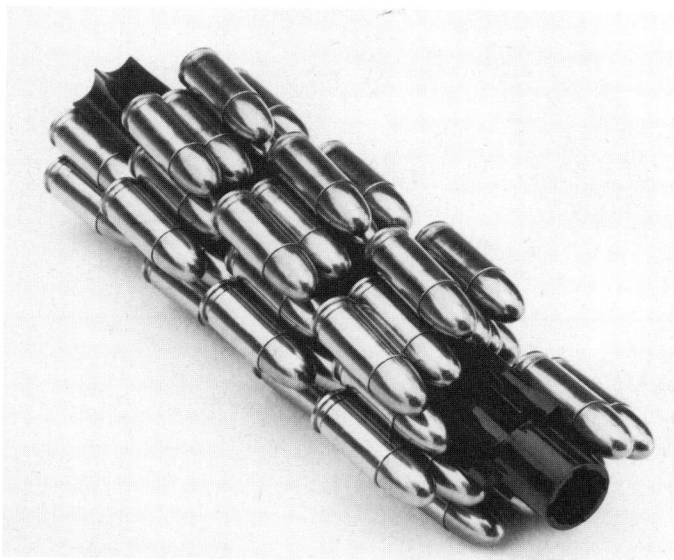

Model displaying the internal distribution of cartridges in Calico Helical Feed 50-round magazine

9 mm Calico M-950 pistol

9 mm Intratec TEC-9'A' pistol

Although resembling a sub-machine gun, this is a semi-automatic weapon classified by the makers as a pistol and generally in the same idiom as the Uzi and Göncz weapons. The TEC-9'A' family is based on a straightforward blowback design using a removable box magazine and with the magazine housing acting as a forward grip. The Basic model is the TEC-9, with matt blue-black finish, perforated barrel jacket, 36-round magazine and 127 mm barrel. The TEC-9S is to the same general specification but finished in matt stainless steel. The TEC-9M is a 'mini' version in blue-black, with a 76 mm barrel and no barrel jacket, and uses a 20-round magazine, and the TEC-9MS is a TEC-9M in matt stainless finish.

A variety of accessories, covering muzzle compensator, barrel extension and carrying cases are available.

DATA FOR STANDARD TEC-9 MODEL
DATA
Cartridge: 9 mm Parabellum
Operation: blowback, semi-automatic
Weight, empty: 1.417 kg
Length: 318 mm
Barrel: 127 mm

9 mm Intratec TEC-9 pistol

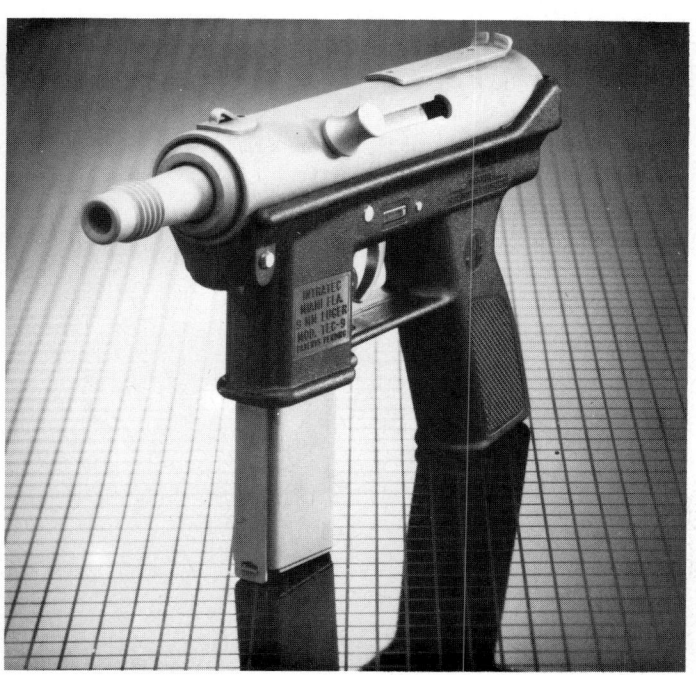

9 mm Intratec TEC-9M pistol

Sights: Fixed, fore blade, rear notch
Sight radius: 254 mm
Safety: firing pin block

Manufacturer
Intratec USA Inc., 12405 SW 130th Street, Miami, FL 33186.
Status
Current. Production.

.22 Intratec 'Scorpion' pistol

This weapon is suggested as a possible arm for covert forces where light weight and relative silence outweigh the limitations of the .22 bullet. It is on similar lines to the Intratec TEC-9 pistol (above), a blowback weapon using a removeable box magazine as its forward grip, though it can easily be used single-handed. It has ambidextrous safety and cocking and breaks down into only three component parts: the frame with cover, complete bolt assembly, and trigger housing assembly. Using a 30-round magazine, it will also accept any Ruger 10/22 type magazine without modification. At the time of writing, production has not begun and specifications are not yet available.

Manufacturer
Intratec USA Inc., 12405 SW 130th Street, Miami, FL 33186.
Status
Current. Production.

.22 Intratec 'Scorpion' pistol

YUGOSLAVIA

7.62 mm Model M57 and 9 mm Model M70 and M70A pistols

Pistol Model M57 is the Yugoslav model of the Soviet Tokarev TT-33 (which see) and can be distinguished from it by the maker's name on the slide and/or the emblem on the grip (see picture). The principle of operation and general mechanical details are similar to those of the TT-33 and reference may be made to the USSR entry for that weapon.

Model M70 is the M57 built to accept the 9 mm × 19 Parabellum cartridge. Apart from the change of calibre and the use of six-groove rifling instead of four grooves it is mechanically similar to the M57. Model M70A is an improved M70 which has had a slide-mounted safety catch added, which locks the firing pin when applied.

DATA

	M57	M70 and M70A
Cartridge	7.62 × 25mm	9 × 19mm Parabellum
Operation: short recoil		
Feed: 9-round box magazine		
Weight, empty: approx 900 g		
Length: 200 mm		
Barrel length: 116 mm		

	M57	M70 and M70A
Rifling	4 grooves	6 grooves
Muzzle velocity	450 m/s	330 m/s
Penetration at		
25 m: sand	35 cm	25 cm
Fir plank	6 cm	5 cm

Manufacturer
Zavodi Crvena Zastava, Beograd 29, Novembra 12.
Status
Current. Production.
Service
Yugoslav forces.

7.62 mm Model M57 pistol

9 mm Model 70A with added safety catch

9 mm Model M70(d) pistol

7.65 mm Model M70 and 9 mm Model M70(k) pistols

These two pistols, in 7.65 mm and 9 mm (Short) calibre are further developments of the Models M57 and M70(d) pistols (see entry above) with improvements in operation, handling and accuracy in addition to the changes in calibre. The distribution of masses have been improved to permit better control and consequent firing accuracy.

The side-mounted manual safety secures both the firing mechanism and the slide when applied and there is an automatic safety which blocks the sear when the magazine is removed.

The two weapons are chambered for 7.65 mm and 9 mm Browning cartridges and comparative data is given below.

9 mm Model M70(k) pistol

DATA

	M70	M70(k)
Cartridge	7.65 mm	9 mm (Short)
Operation: short recoil		
Feed: 8-round box magazine		
Weight (empty)	740 g	720 g
Length: 200 mm		
Barrel length: 94 mm		
Rifling: 6 grooves, 240 mm twist		
Muzzle velocity	300 m/s	260 m/s
Penetration at		
25 m: sand	25 cm	35 cm
Fir plank	10 cm	7 cm

Manufacturer
Zavodi Crvena Zastava, Beograd 29, Novembra 12.
Status
Current. Production.
Service
Yugoslav forces.

.357 Magnum Model 1983 revolver

In order to complete their production programme of weapons, Zastava has recently developed and manufactured the Model 1983 revolver, designed for the .357 Magnum cartridge. It can also chamber and fire the .38 Special cartridge, and with a change of cylinder and the use of a special adapter can fire rimless 9 mm Parabellum cartridges as well.

The revolver is a solid-frame, side-opening pattern with double-action lockwork. The barrel carries a ventilated rib and the extractor rod is fully shrouded. Cylinder release is by the usual thumb-operated latch on the left side of the frame.

In addition to the standard version, the revolver can be made with special handgrips or with adjustable sights, and sporting versions are available with 64 mm, 102 mm and 152 mm barrel lengths. On special request, the revolver can be made in a de luxe version, with engraving and chrome plating.

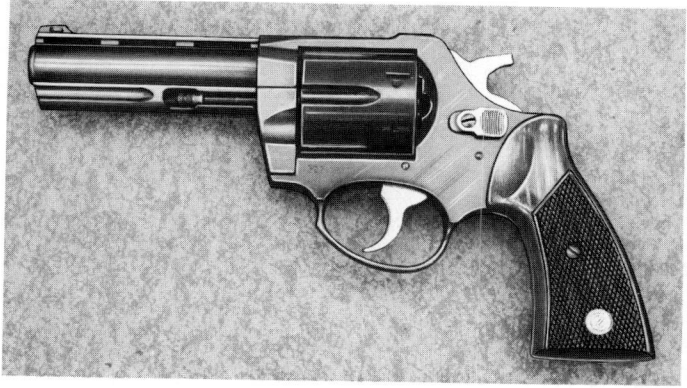

Zastava .357 Magnum Model 1983 revolver

DATA
Cartridge: .357 Magnum (and see text)
Operation: double-action revolver
Feed: 6 chambered cylinder
Weight, empty: 900 g
Length: 188 mm
Barrel: 64 mm (and see text)

Manufacturer
Zavodi Crvena Zastava, Beograd 29, Novembra 12.

Status
Current. Production.
Service
Yugoslav forces.

Sub-machine guns

ARGENTINA

9 mm FMK-3 Modif. 2 sub-machine gun

This weapon is the current sub-machine gun manufactured by Fabrica Militar de Armas Portatiles 'Domingo Matheu', Rosario. It is a blowback-operated weapon of modern design.

It was formerly produced in two models, one with a fixed plastic butt and the other with a sliding butt modelled on the US M3 (which see). The current 'Modification 2' model is produced only in sliding butt form. The body of the gun is a metal pressing and there is a screw-threaded cap at the front end to allow easy release of the barrel. A plastic fore-end grip is under the receiver. The 40-round magazine fits into the pistol grip which has a grip safety at the back. There is also a safety position on the selector.

The cocking handle is on the left side of the receiver, well forward, and there is a slide which covers the cocking slot to keep out dirt.

The FMK-3 is designed with a wrap-round bolt. The bolt encloses 180 mm of the barrel which itself is 290 mm long. This, the manufacturer claims, leads to good control and stability in firing, as well as reducing the length.

Stripping

The magazine is removed, the bolt retracted and the chamber checked.

The pin holding the butt in position is located at the rear of the receiver and can be pressed out with the nose of the round to allow the butt assembly to slide rearwards.

The bolt and return spring are withdrawn when the cocking handle is retracted.

The trigger group is released by pressing out the forward pin.

The barrel comes out of the receiver as soon as the front end cap is unscrewed.

DATA
Cartridge: 9 mm Parabellum
Operation: blowback, selective fire
Feed: 40-round box magazine
Weight, empty: 3.76 kg
Length: 690 mm butt extended; 520 mm butt retracted

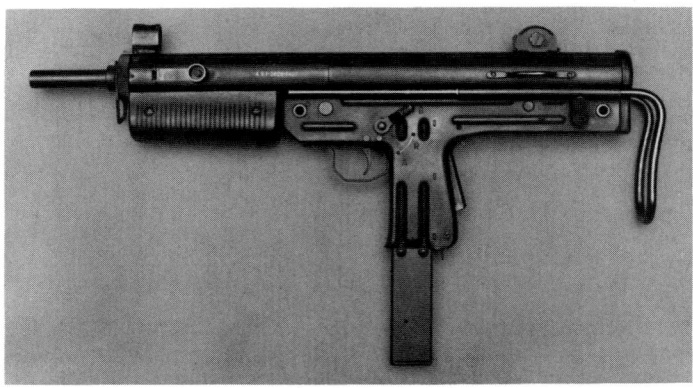

9 mm FMK-3 Modif. 2 sub-machine gun

Barrel: 290 mm
Rifling: 6 grooves, rh, one turn in 250 mm
Sights: (fore) pillar; (rear) flip aperture, 50 and 100 m
Sight radius: 320 mm
Muzzle velocity: 400 m/s
Rate of fire: (cyclic) 600 rds/min

Manufacturer
Fabrica Militar de Armas Portatiles 'Domingo Matheu', Avenida Ovidio Lagos 5250, 2000-Rosario.
Status
Current. Production.
Service
Argentinian forces.

AUSTRALIA

9 mm Owen sub-machine gun

The Owen sub-machine gun was designed by Pte. Evelyn Owen. It was adopted in November 1941 and manufacture was started shortly afterwards by Lysaghts in New South Wales. In all 45 000 were produced before production ended in September 1944.

The Owen is readily recognised by the forward sloping box magazine mounted on top of the receiver. It was extremely reliable and had a high reputation in the jungle of New Guinea. There were some unusual features such as a quick release finned barrel. The designer went to great pains to keep dirt out of the mechanism and the bolt carried a fibre washer to ensure that any dust and the like coming into the receiver through the cocking slot at the rear did not get forward to the bolt itself. The gun was usually painted in camouflage colouring.

The first model was called the Mark I/42. The Mark I/43 was lightened by having holes cut in the frame and the fins removed from the barrel. The Mark I/44 had a bayonet. The Mark II/43 was manufactured in very limited numbers for trials (about 200) but never went into large scale production.

The Owen was a good sturdy reliable weapon, but heavy and expensive to produce. The manufacturers, John Lysaght (Australia) Ltd, ceased production in September 1944 and passed the manufacturing machinery to the Small Arms Factory at Lithgow NSW.

DATA
MARK I/43
Cartridge: 9 mm Parabellum
Operation: blowback, selective fire
Feed: top mounted 33-round box magazine
Weight: (unloaded) 4.23 kg; (loaded) 4.86 kg
Length: 813 mm

9 mm Mark I/42 Owen sub-machine gun

Barrel: 250 mm
Rifling: 7 grooves rh
Muzzle velocity: 366 m/s
Rate of fire: (cyclic) 700 rds/min; (automatic) 120 rds/min; (single-shot) 40 rds/min

Status
Obsolete. Not in production.
Service
Australian reserve units. May be found in South-east Asia.

9 mm F1 sub-machine gun

The 9 mm F1 sub-machine gun has a cylindrical body extended forward over the barrel and perforated for cooling. The curved 34-round magazine is mounted over the receiver. The butt is a prolongation of the barrel and receiver and so there is no turning moment which will cause the muzzle to rise at full automatic. Since the magazine is mounted above the gun the sights must be offset. This feature, in common with the Owen (see previous entry), has the disadvantage that the magazine produces a blind area to the left front. The rearsight of the F1 is a plate which when lifted raises the aperture some 75 mm above the centre line of the barrel. This comes from the straight-through butt layout and the high sightline does tend to increase the exposure of the user when firing over cover.

The sling swivel at the nose of the weapon prevents the firer placing his hand or fingers over the muzzle and a handguard adjacent to the ejection port limits the rearward positioning of the fingers, thereby preventing personal injury by the breech bolt.

Magazine
The F1 has been designed to accept the UK and Canadian 9 mm 34-round magazines, thereby assuring interchangeability of magazines which are in current use and available from a number of sources.

Bayonet
The bayonet is the standard L1A2 as used on the 7.62 mm L1A1 rifle and is fitted to the right-hand side of the gun in a horizontal plane, cutting edge facing out.

Common components
A number of 7.62 mm L1A1 rifle components have been incorporated into the weapon. The pistol grip from the L1A1, complete with Arctic grip, has also been adapted to the F1.

DATA
Cartridge: 9 mm Parabellum
Operation: blowback, selective fire
Feed: top mounted 34-round box magazine
Weight: (unloaded) 3.27 kg; (loaded) 4.3 kg with bayonet
Length: 714 mm plus 203 mm with bayonet
Barrel: 213 mm
Sights: (fore) blade, offset to right; (rear) hinged plate, aperture. Offset
Muzzle velocity: 366 m/s
Rate of fire: (cyclic) 600-640 rds/min

Manufacturer
Australian Defence Industries, Lithgow Facility, New South Wales.
Status
Obsolete. No longer in production; limited spares available.
Service
Australian Defence Force.

9 mm F1 sub-machine gun with bayonet and sling

AUSTRIA

9 mm MPi 69 and MPi 81 Steyr sub-machine guns

The Steyr MPi 69 sub-machine gun is a simple and robust weapon. The receiver pressing is of light-gauge steel and is welded into a hollow box with two gaps on the right-hand side; one in the middle for ejection of the spent case and one approximately 76 mm from the front to accept an insert to take the barrel seating, barrel release catch and barrel securing nut. The cocking slide, a simple pressing, is on the front left-hand side of the receiver.

The ejector is a simple bent strip riveted in position in the middle of the base of the receiver to run in a groove in the bottom of the breech-block. Spot-welded under the receiver at the back is a small bracket which provides guides for the spring steel telescoping butt and spring-loaded release plungers. The strip-down catch is also at the rear. The moulded nylon receiver cover fits under the receiver and carries the trigger mechanism, the pistol grip and magazine housing.

The barrel is 250 mm long and is cold-forged on a rifling mandrel. This results in a cleaner groove than that obtained by the usual button rifling and both the inner and outer skins of the barrel are work-hardened. The breech-block has a fixed firing pin on the bolt face which is half-way along the bolt length. Thus the bolt wraps round the barrel and has a slot cut along the right-hand side for ejection. By this means the weight of the bolt which is required for blowback operation can be kept to the necessary level without being excessively long and at the same time the telescoping of the breech-block over the barrel enables the length of the latter to be increased with a resulting increase in muzzle velocity and accuracy. However, the long barrel maintains its pressure longer and so the bolt has to be somewhat heavier than it would be with a normal barrel length of about 20 cm. This ensures that the block does not blow back too rapidly and so allow the unsupported case to emerge from the chamber while the pressure is still high enough to destroy the relatively weak brass.

Trigger and firing mechanism
The weapon fires either single-shot or full automatic. The choice is controlled by finger pressure on the single trigger. A slight pull gives a single-shot, a long pull gives automatic fire. This is a similar arrangement to that used on the Steyr AUG rifle.

MPi 69 sub-machine gun

Cocking arrangements: **(top)** *MPi 69 is cocked by pulling on sling* **(bottom)** *MPi 81 uses conventional cocking handle*

The applied safety is a cross bolt which is pressed through the receiver. One side is marked 'S' in white and projects when the gun is safe. On the other end is marked 'F' in red and this projects from the receiver when the gun is set to fire. When the pin is in the middle position, the safe button is only half way through, the gun will fire single-shot only.

In order to prevent accidental discharge due to the bolt being jarred or mishandled during cocking, the breech-block is provided with three safety bents. The first bent is the front edge of the block and this engages the sear before the bolt has moved back enough to pass over the base of the round in the magazine. This is about 40 mm of backward travel. The second bent allows the bolt a further centimetre or so of travel and the same distance further back is the third bent. It is impossible to bounce the bent off the sear in any of the three positions.

The third bent is the normal working bent and the second bent is provided to prevent a runaway gun when using 9 mm ammunition of lower impulse than usual. The weapon cannot be accidentally discharged and is at least as safe as others with more complicated safety arrangements.

The cocking lever is a pressing running in a groove at the top of the left side of the receiver at the front. The method of cocking is unusual. The sling is attached to the cocking lever and the soldier cocks the gun by pulling back on the sling. To prevent this happening unintentionally a bracket is welded to the top of the receiver which prevents movement of the cocking lever backwards unless the sling is held out at right angles to the gun.

MPi 81
This is a variant model in which the sling is attached to a conventional swivel on the right side of the receiver and cocking is done by a normal cocking handle which protrudes at the top left side of the receiver. In addition the rate of fire has been increased to about 700 rounds per minute by slight internal improvements.

Silenced versions
Silenced versions of both the MPi 69 and MPi 81 are now available.

'Loop-hole' model
A special version of the MPi 81 has been developed for firing from APCs, IFVs and similar vehicles with loop-holes. The receiver is fitted with the optical sight of the AUG rifle carried in special brackets, and the barrel is extended and fitted with a locking collar. The rear positioning of the optical sight allows it to be used with a vision block, if the block position is suitable.

Stripping
To remove the bolt, press the button at the rear of the receiver and lift the catch. The bolt can be pulled straight out by the nylon plate attached to the return spring rod.

To remove the barrel, pull back the barrel securing catch, rotate and remove the barrel housing and the barrel is free.

To remove the nylon moulding holding the trigger mechanism, extend the shoulder stock, push the moulding forward, drop the rear and push straight off over the front. All this can be done in 15 seconds.

The weapon can be reassembled in reverse order in 15 seconds. There is only one point to watch. The barrel has a flange with a flat on each side. These flats are not parallel so the barrel will only go into the housing in one position.

DATA
Cartridge: 9 mm × 19 Parabellum
Operation: blowback, selective fire
Feed: 25- or 32-round box magazine
Weight, empty: 3.13 kg
Length: (butt extended) 670 mm; (butt retracted) 465 mm
Barrel: 260 mm
Rifling: 6 grooves, rh, one turn in 254 mm
Sights: (fore) blade; (rear) flip aperture, 100 and 200 m
Sight radius: 326 mm
Muzzle velocity: 381 m/s
Rate of fire: (cyclic) 550 rds/min; (automatic) 100 rds/min

Manufacturer
Steyr-Mannlicher GmbH, a member of the Steyr-Daimler-Puch AG group, Werke Steyr, Postfach 1000, A-4400 Steyr.
Status
Current. Available.
Service
Various military and police forces.

9 mm MPi 69 Steyr sub-machine gun

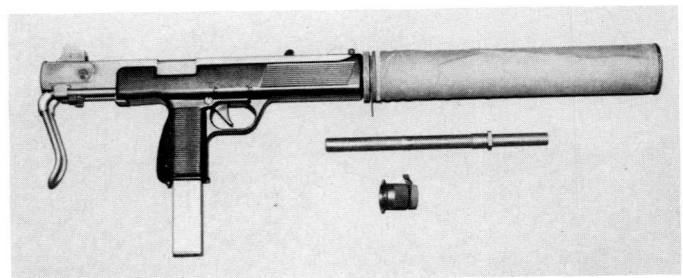
Mpi 69 with silencer. Standard barrel and retaining nut are below silencer. Changing from standard to silent version takes less than a minute

MPi 81 'Loop-hole' version with stock folded

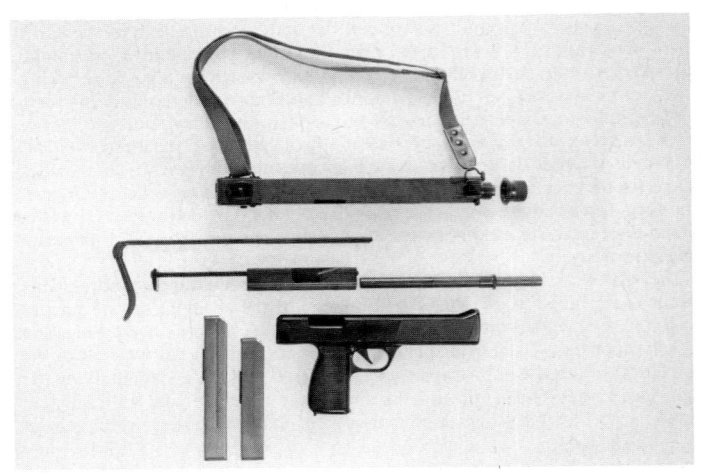
9 mm MPi 69 Steyr sub-machine gun field-stripped

Steyr AUG 9 mm Para

This is a sub-machine gun version of the standard AUG assault rifle. It uses the existing stock and receiver units but is fitted with a new barrel of 9 mm calibre, a special bolt group, a magazine adapter and a magazine. These components may be ordered in kit form in order to convert any existing AUG to the different calibre, or the weapon may be purchased in 9 mm form.

The change converts the weapon to blowback operation; the bolt is a one-piece unit rather than the carrier and bolt combination used with the rifle-calibre weapons, and firing takes place from a closed bolt.

Conversion of existing weapons is simply a matter of dismantling the rifle

and re-assembling it using the 9 mm component parts. It is, of course, necessary to re-zero the sight after the conversion has been made.

DATA
Cartridge: 9 mm Parabellum
Operation: blowback, selective fire
Feed: 25 or 32 round magazines
Weight, empty: 3.3 kg
Length: 665 mm

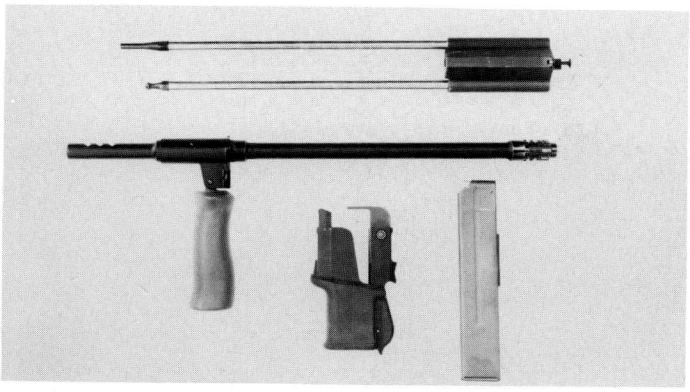

Conversion kit for Steyr 9 mm Para; bolt unit, barrel, magazine adapter and magazine

Steyr AUG 9 mm Para

Steyr AUG 9 mm sub-machine gun fitted with silencer

Barrel: 420 mm
Rifling: 6 grooves, rh, one turn in 250 mm
Sights: integral 1.5 × telescope
Rate of fire: 650–750 rds/min
Muzzle velocity: approx. 400 m/s

Manufacturer
Steyr-Mannlicher GmbH, a member of the Steyr-Daimler-Puch AG group,
Postfach 1000, A-4400 Steyr.
Status
Current. Production.

Steyr Tactical Machine Pistol

This weapon could equally be considered as a pistol, but its full-automatic
ability and general configuration places it into the new 'personal weapon'
category and gives it equal credence as a sub-machine gun.

The Tactical Machine Pistol (TMP) is a locked-breech weapon in 9 ×
19 mm Parabellum calibre. There are only 41 component parts, and the frame
and top cover are made from a plastic material which is sufficiently tough to
forego the use of inset steel guide rails for the bolt carrier. The locking system
is rotary, using a single lug. The cocking handle is at the rear of the weapon,
beneath the rearsight, and is pulled back to cock. Selection of single-shot or
semi-automatic fire is performed by a two-stage trigger, similar to the system
used in the AUG rifle; the first pressure on the trigger delivers single shots,
and further pressure brings automatic fire. A three-position cross-bolt safety
catch has a central semi-automatic-only setting, giving additional control.

The TMP is also to be produced in semi-automatic-only form, whereupon
it becomes the 'Special Purpose Pistol'. In TMP form there is a front hand
grip; in SPP form this is omitted. The TMP will also have facilities for
attaching a suppressor.

9 mm Steyr Special Purpose Pistol (SPP)

In addition to the standard 9 mm chambering, a modular system is planned
which will permit adaptation of the weapons to 9 mm Short, 7.65 mm
Parabellum, 9 mm Steyr, 10 mm Auto and .41 Action Express calibres.

DATA
Cartridge: 9 × 19 mm Parabellum (but see text)
Operation: recoil, semi-automatic (TMP selective fire)
Locking system: rotating bolt
Feed: 15-, 20- or 25-round magazines (9 mm Para.)
Weight, empty: 1.30 kg
Length: 270 mm
Barrel: 150 mm
Sights: (fore) blade; (rear) notch
Sight base: 190 mm

Manufacturer
Steyr-Mannlicher GmbH, Postfach 1000, A-4400 Steyr.
Status
Advanced development.

9 mm Steyr Tactical Machine Pistol (TMP)

BELGIUM

5.7 mm FN P 90 personal weapon

This unique weapon has been developed by FN Herstal to equip military personnel whose prime activity is not that of operating small arms — artillery, signals, transport and similar troops whose duty requires that they should be armed for self-protection but who do not wish to be burdened by a heavy weapon whilst performing their normal tasks. It will also be a suitable weapon for special forces who require very compact firepower. And finally, FN Herstal are of the opinion that the 9 mm Parabellum cartridge is reaching the end of its useful life and needs to be replaced with a round owing something to modern technology.

The P 90 is a blowback weapon firing from a closed bolt. The pistol grip, with a vestigial thumb-hole stock, is well forward on the receiver so that when gripped, the bulk of the receiver lies along the man's forearm. The controls are fully ambidextrous; a cocking handle is provided on each side, and the selector/safety switch is a rotary component located under the trigger. Even the forward sling swivel can be located on either side of the weapon as required.

The magazine lies along the top of the receiver, above the barrel, and the cartridges are aligned at 90° to the weapon axis. The 50 rounds lie in double-row configuration, and as they reach the mouth of the magazine they are ordered into a single-row by a fixed ramp. A spiral ramp then turns the round through 90° as it is being guided down into the feedway so that it arrives in front of the bolt correctly oriented for chambering. The magazine is of translucent plastic so that its contents can be checked at any time.

The sight unit is a reflex collimating sight, the graticule being a circle and dot; it can be used with both eyes open and allows very rapid acquisition of the target and accurate aiming. Should the sight be damaged it can be removed, whereupon two sets of iron sights are exposed, one on each side of the receiver; this duplication is another aspect of the ambidextrous design, allowing ease of use for both right- and left-handed firers.

The weapon strips easily into three basic groups for field maintenance. Much of the body and internal mechanism is of high-impact plastic material, only the bolt and barrel being of steel. The empty cartridge case is ejected downwards through the pistol grip, which is shaped so as to offer a grip for the disengaged hand when firing from the shoulder.

At the present time the P 90 is still undergoing development and refinement, and the data given below must be regarded as provisional.

DATA
Cartridge: 5.7 × 28 mm
Operation: blowback, closed bolt firing
Feed: 50-round magazine
Weight: 2.80 kg (empty); 3.70 kg (with full magazine)

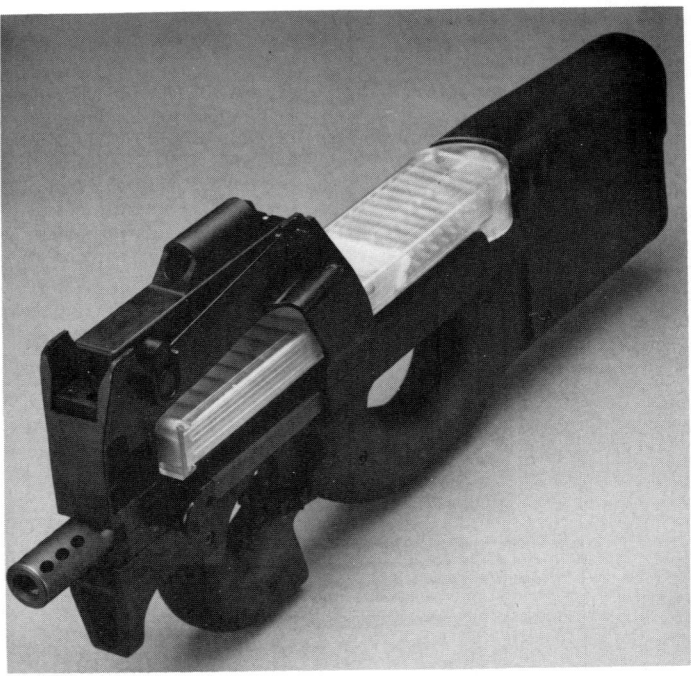

5.7 mm FN P 90 personal weapon

Length: 400 mm
Muzzle velocity: 850 m/s
Rate of fire: 800 – 1000 rds/min
Effective range: 150 – 200 m

Manufacturer
FN Herstal SA, B-4400 Herstal.
Status
Advanced development, which is expected to be complete by end 1989. Production possible by late 1990.

9 mm Mitraillette Vigneron M2 sub-machine gun

In 1953 the Belgian Army adopted the Mitraillette Vigneron M2. This was designed by Colonel Vigneron, a retired Belgian army officer, and manufactured by the Société Anonyme Précision Liègeoise at Herstal near Liège. This company no longer exists but spare parts and refurbishment services are available from FN Herstal SA.

The Vigneron M2 is made from sheet metal stampings. It is a blowback-operated design with a rather longer barrel than is customary in a weapon so operated. The cooling rings around the barrel and the compensator add to the cost of the barrel production without, apparently, significant improvement in performance. The stock is made of steel wire and telescopes along the receiver. There are three alternative positions. The pistol grip contains a grip safety which must be held in before the sear can release the bent of the bolt. The selector has three positions; safe, single and automatic.

When the lever is set to automatic it is still possible to obtain single shots by first pressure on the trigger and full automatic when the trigger is fully in.

In addition to the Belgian armed forces, the Vigneron was issued to the 'Force Publique' in the Belgian Congo (now Zaire) and when that country became independent some of the Vignerons remained with them. It was also used by the Portuguese Army as the 'Metralhadora M/961' for some years, with the result that numbers are still to be found in Angola and Mozambique.

DATA
Cartridge: 9 mm Parabellum
Operation: blowback, selective fire
Feed: 32-round box magazine
Weight: (unloaded) 3.29 kg; (loaded) 3.69 kg
Length: (butt extended) 886 mm; (butt retracted) 706 mm
Barrel: 305 mm with compensator
Rifling: 6 grooves rh
Sights: (foresight) blade; (rearsight) aperture fixed setting 50 m
Muzzle velocity: 381 m/s
Rate of fire: (cyclic) 620 rds/min; (automatic) 120 rds/min

Manufacturer
Société Anonyme Précision Liègeoise, Herstal.
Status
Obsolete. Not in production.
Service
Some Belgian units. May be found in Central Africa.

9 mm Mitraillette Vigneron M2 sub-machine gun

BRAZIL

Brazilian sub-machine guns

The first sub-machine gun to be adopted in Brazil was the 9 mm Bergmann MP.28, not by the armed forces, but by police forces. The first 'official' contact of the Brazilian Army with the sub-machine gun was in 1943-44, when US military advisers arrived in Brazil to familiarise Brazilian troops with the armament they would employ in combat operations in Europe. In the specific case of sub-machine guns, the .45 M3A1 was the standard Brazilian Expeditionary Force weapon, although the Thompson was also used. Both types equipped the Brazilian armed forces in the immediate post-war years, the US Reising Model 50 sub-machine gun having been briefly (and, for the record, unsuccessfully) evaluated for possible adoption.

In the early 1950s Indústria Nacional de Armas SA, of São Paulo, purchased all the manufacturing rights of the Danish Madsen M50 sub-machine gun, adapted to fire .45 calibre ammunition in place of the original 9 mm Parabellum round. For more than 10 years, the INA (as the Model 1950 was locally known) was the standard sub-machine gun of the Brazilian armed forces, police and para-military forces, though numbers of M3A1s and Thompsons remained in service.

In the early 1960s, the Brazilian Army decided to switch from .45 to 9 mm calibre for pistol and sub-machine gun use, and an international search for a new gun ended up with the choice of the Italian Beretta M12, which was eventually locally manufactured by Indústria e Comércio Beretta SA, in São Paulo. This weapon is being currently produced by Forjas Taurus SA. This is now in service with the Brazilian Army and Marine Corps.

In terms of local designs, a number of different prototypes have emerged in the last 10 years or so, but only one, the Uru sub-machine gun (which see), has reached production status, and is now in service with several police forces.

The old .45 INA has been relegated to use by paramilitary forces throughout Brazil, and is still seen (together with M3A1s) in Navy hands. Some police forces in Brazil also use the Walther MP-K 9 mm and Heckler and Koch MP5 sub-machine guns.

9 mm Mtr M 9 M1 – CEV

This sub-machine gun is, in fact, a modified Bergom BSM/9 M3 (see *Jane's Infantry Weapons, 1983-84*, page 77), which was never placed in production. The project was taken over in mid-1982 by Companhia de Explosivos Valparaiba, a well-established explosives manufacturer. The company says it is tooling up for series manufacture of the gun.

The Mtr M (Metralhadora de Mão, or 'Hand machinegun') 9 (millimetre) M (Model) 1 – CEV (the manufacturer's initials) is a pretty much conventional selective fire blowback-operated sub-machine gun firing from the open-bolt position. The receiver is a tubular structure with barrel-cooling perforations in the forward end, which also has a plastic handguard. The foresight is a protected post and the rearsight is a protected aperture, factory-set for 100 m. The cocking handle, which remains stationary when the gun is firing, is located at about 45° to the left at the forward end of the receiver, Beretta M12 style, though the handle proper is much smaller than the Italian sub-machine gun. The ejection port is located to the right, and is covered by a spring-loaded cover that snaps open when the weapon is cocked or fired.

The firing mechanism, together with the plastic pistol grip and trigger assembly, is an integral unit that fits under the receiver, just aft of the magazine housing. The magazine catch is a small blade at the rear of the housing, having to be pulled back (not easily in the prototype fired by *Jane's Infantry Weapons*) to release the empty magazine. The fire selector lever is on the left side, above the trigger, with the positions 'S' (Safety) forward, '30' (Automatic) middle, and '1' (Semi-automatic) rear. A grip safety is provided against accidental firing from drops or hits on the butt, blocking the bolt forward unless firmly squeezed by the firer.

The weapon has a wire-type telescoping stock, with the catch located under the receiver, making extension/retraction a two-hand operation. The stock will come out of the gun if pulled all the way.

To field strip the Mtr M 9 M1 – CEV, after removing the magazine and ensuring the gun is unloaded, the disassembly pin (above the pistol grip) is pulled out to either side, thus allowing the grip/firing mechanism assembly to be disengaged from under the receiver. This gives access to the working parts for normal cleaning. The barrel mounting nut, which in production weapons will also act as a flash hider, is then unscrewed, thus allowing the barrel to come out of the forward end of the tubular receiver, together with the cocking handle/rod/spring assembly, the bolt, and the two return springs with their respective guides. Re-assembly is, of course, a reversed procedure.

DATA
Cartridge: 9 mm × 19 Parabellum
Operation: blowback, selective fire
Feed: 30-round box magazine
Weight unloaded: 3 kg
Weight loaded: 3.7 kg
Barrel length: 228 mm
Sights: (fore) protected post; (rear) aperture, fixed for 100 m
Sight radius: 393 mm
Muzzle velocity: 400 m/s
Rate of fire: (cyclic) 600 rds/min

Manufacturer
Companhia de Explosivos Valparaiba, Praia do Flamengo, 200 – 20° andar 22210 Rio de Janeiro, RJ.
Status
Current. Final development.

Mtr M9 M1 – CEV

Mtr M9 M1 – CEV

Mtr M9 M1 – CEV

CEL 9 mm Madsen sub-machine gun

As noted in a later entry, the Madsen sub-machine gun was manufactured in Brazil, in .45 calibre, by the Indústria Nacional de Armas (INA).

Now IMBEL, at their Fábrica de Itajubá factory has developed and manufactured a conversion from .45 ACP calibre to 9 mm calibre to permit the use of the 9 × 19 mm Parabellum cartridge.

To improve the aiming and holding of the weapon during automatic fire, the IMBEL designers have added a highly efficient muzzle compensator. It has been found in practical testing that apart from the vibration of the weapon due to the reciprocating bolt there is no undue motion and the gun remains stable and under full control at all times.

The first lots of this 'new' weapon were delivered to Brazilian Army and police forces after approval of the prototypes at the Marambaia Military Proof Establishment; all proof and testing was in conformity with NATO standards.

DATA
Cartridge: 9 × 19 mm Parabellum
Operation: Blowback, selective fire
Feed: 30-round box magazine
Weight: 3.74 kg
Rifling: 6 groove rh
Rate of fire: 600 rds/min
Muzzle velocity: 392 m/s

Manufacturer
Indústria de Material Bélico do Brasil – IMBEL, Avenida dos Nacções Unidos 13.797, Bloco III, 1° Andar, CEP 04794 São Paulo SP.
Status
Current; production.
Service
Brazilian Armed Forces and police forces.

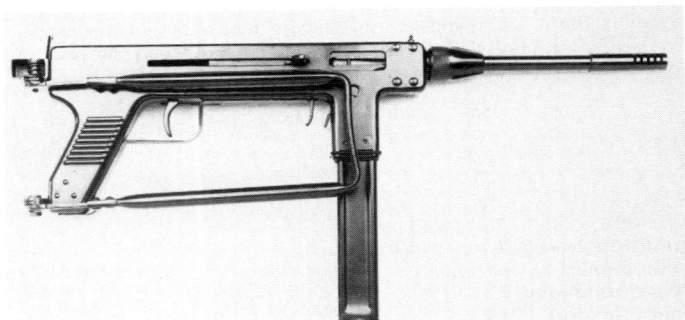

CEL 9 mm Madsen with butt folded

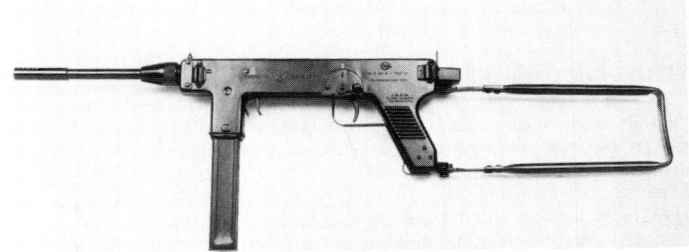

CEL 9 mm Madsen with butt extended

9 mm Uru Model II sub-machine gun

Named after a Brazilian jungle bird, the Uru sub-machine gun was designed by Olympio Vieira de Mello in mid-1974. Design work took him one month, and he was able to transform drawings into a working prototype in three months. When this was ready, in early 1975, the Mekanika company was formed to undertake manufacture and sales of the new gun.

The Uru was handed over to the Brazilian Army's Marambaia proving grounds, in Rio de Janeiro, for test and evaluation. Initial results were promising, though, as might be expected, some minor modifications were made: the shape of the removable stock was changed; the capacity of the box magazine was reduced from 32 to 30 rounds; the magazine guide was enlarged to act as a forward grip and better protect the magazine; the fire selector was enlarged and the wooden pistol grip was replaced by a plastic one.

Meanwhile, Mekanika had tooled up for series manufacture of the gun, and the official Brazilian Army Ministry production certification was granted in July 1977. Production of the Uru has since grown to meet local (Brazilian Army and State police forces) and foreign orders (Africa, Latin America and Middle Eastern countries). In 1988 Mekanika handed over all rights in the design to Bilbao SA Indústria e Comércio de Máquinas e Peças, Division FAU Guns, who developed the Mark II design and now produce the weapon.

The Uru sub-machine gun operates on the blowback principle, being largely built from tubular elements and stampings. The main parts are assembled by welding and spot welding. Heat-treated alloy steels are used in parts subject to wear or shocks. The body comprises a plastic rear pistol grip, trigger-guard (large enough to allow gloved use) and a metal forward grip, into which the 30-round box magazine is inserted. The magazine catch, at the rear of the forward grip, is pulled back to release the magazine.

The receiver is a tubular structure closed at the end, where there is a spring-loaded button on top, for the stock attachment. The forward end acts as a thermal jacket for barrel cooling and hand protection. The 175 mm barrel is fitted into place by its own mounting nut, which screws to the forward end of the receiver. A screw-on compensator/muzzle brake, adjustable by the firer for right- or left-hand use, is available.

The weapon is sighted for 50 m; the rearsight is of the aperture type, and the foresight is a flat blade, with a radius of 235 mm. The cocking handle is 45° to the right, and the rearward travel before firing is 60 mm. On firing, the bolt's rearward travel is 80 mm. The bolt remains closed after the last round is fired.

The fire selector is on the left side, above the plastic pistol grip, and has the three usual positions: 'S' (Safe) forward, 'SA' (Semi-Automatic) upward, and 'A' (Automatic) backward. In the 'S' position the trigger action is solidly blocked to prevent accidental firing when the gun is carried cocked. Accidental discharge caused by a fall or a hard blow, a common problem in free-bolt weapons, is guarded against by an inertial locking device which positively blocks the bolt at its forward position if the gun is dropped or struck on the butt. This device operates whatever the fire selector setting may be.

As the Uru has only 33 parts, including the magazine and the stock, it can be totally disassembled without tools in about 30 seconds. After the removal of the magazine and stock, the nose of a cartridge, or any other similar object, is used to push a button at the back of the body inwards, which allows the receiver to be lifted at the front and disengaged from the rear end. The barrel is then removed by unscrewing its integral mounting nut. After pulling out the cocking handle, the bolt slides out of the forward end of the receiver, and the return spring and its guide, as well as the extractor, can be removed from the bolt. Moving back to the body, the fire positioner is pressed down with one finger to allow the removal of the fire selector, which comes out sideways to the left. Removal is then made of the fire selector positioner, the trigger, the disassembling pin, and the fall lock and shock-absorber, together with its spring. By pressing the trigger spring down, the trigger shaft is removed. The last part to come out is the one-piece ejector/magazine catch/trigger spring. Assembling is a straightforward procedure, taking about the same time as stripping. Significantly, all parts of the Uru are interchangeable, and only the body of the weapon bears the manufacturer's number.

A silencer is offered to transform the Uru into a fully-silenced gun. The original barrel is retained and the silencer is merely screwed on to the muzzle. The gun can then be used with standard ammunition and at full automatic fire. This new silencer is smaller than the original combined barrel and silencer which had to be replaced as a complete unit.

Uru Model II sub-machine gun

9 mm Uru sub-machine gun completely stripped

DATA
Cartridge: 9 mm × 19 Parabellum
Operation: blowback, selective fire
Feed: 30-round box magazine
Weight: (without stock and magazine) 2.58 kg; (with stock, without magazine) 3.00 kg; (complete weapon, loaded) 3.69 kg
Magazine: (empty) 0.35 kg; (loaded) 0.69 kg
Length: (overall) 670 mm; (without stock) 432 mm
Barrel: 175 mm
Rifling: 6 grooves rh
Sights: (fore) blade; (rear) aperture, sighted for 50 m; (radius) 235 mm

Muzzle velocity: 389 m/s
Rate of fire: (cyclic) 750 rds/min

Manufacturer
Bilbao SA Indústria e Comércio de Máquinas e Peças, Division F.A.U.Guns, Rua Ribiero do Amaral 618, 04268 Sâo Paulo.
Status
Current. Production.
Service
Brazilian Army, Navy and some state police forces. Exported to unspecified countries.

.45 INA MB 50 and Model 953 sub-machine gun

The MB 50 is a copy of the Danish Madsen Model 1946 made by the Indústria Nacional de Armas SA, São Paulo, Brazil which acquired all the manufacturing rights. It has the Brazilian crest on the left side of the receiver and is marked 'Exercito Brasileiro'. The models used by the police are marked 'Policia Civil'.

The weapon remains a true copy of the Madsen.

The Model 953 has an enlarged magazine housing which has a wire clip around the mouth to keep the two halves together. The cocking handle has been removed from the top of the receiver to the right side.

The INA strips in the same way as the Danish Model 1946.

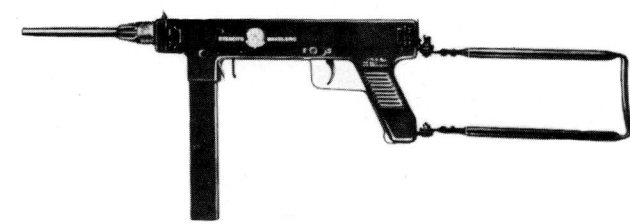

.45 INA MB 50 sub-machine gun

DATA
Cartridge: .45 Automatic Colt Pistol (ACP)
Operation: blowback, automatic
Feed: 30-round magazine
Weight: (unloaded) 3.4 kg; (loaded) 4.32 kg
Length: (stock extended) 794 mm; (stock retracted) 546 mm
Barrel: 213 mm
Rifling: 4 grooves rh
Sights: (fore) blade; (rear) aperture 100 m
Muzzel velocity: 280 m/s
Rate of fire: 650 rds/min

Manufacturer
Indústria Nacional de Armas SA, São Paulo.
Status
Obsolete. Not in production.
Service
Police and para-military forces. Also military reserve units.

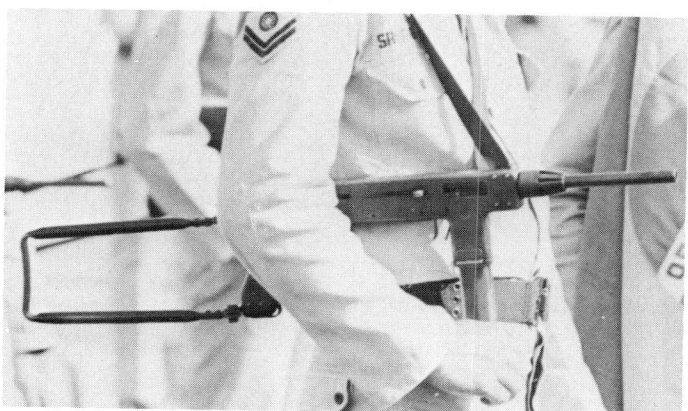

Model 50

9 mm LAPA SM Model 02 sub-machine gun

This sub-machine gun is the result of a project started in 1979 by Laboratório de Projetos de Armamento Automático (Automatic Armament Design Laboratory), a private research and development firm under the management of General (Retired) Fileto Pires Ferreira, Nalmo Suzano and Paulo Cochrane. A prototype, known as Model 01, was completed in about 90 days and, following initial testing conducted by the manufacturers, the feasibility of the design was proved, work then starting on Model 02, which took about a year to be completed. In September 1982, this was officially tested at the Brazilian Army's Marambaia proving grounds, successfully meeting all accuracy and working requirements under normal and adverse conditions, for example dust, sand, mud, rain, cold weather and sea water. After 6500 rounds were fired, only seven feed incidents (0.11 per cent) were recorded. The weapon has been cleared for production.

Operation
The LAPA SM (Sub-Metralhadora) Model 02 operates on the conventional blowback principle, firing from the open-bolt position. All its body is made of high-resistant, injected plastic material with smooth surfaces which makes the gun pleasant to handle and fire. The tubular receiver projects well into the stock hollow, thus giving the bolt a comparatively long (150 mm) rearward travel when firing and aiding stability. The prototype has a rate of fire of 485 rounds a minute, but this can be increased at the user's request.

The gun has an AR-10 type carrying handle, which also houses the two-position (50 and 100 m), flip-type rearsight, adjustable for windage. The frontsight, adjustable for elevation, is a protected post in an elevated position, this being dictated by the weapon's straight-line configuration. The cocking handle, which has a rearward travel of about 85 mm, is within the handle and remains stationary in the forward position when the gun is firing. The ejection port is on the right side, having a spring-loaded cover that snaps open as the weapon is cocked or when the bolt moves forwards. The magazine is inserted upwards into a housing foward of the trigger-guard, being a 30-round, staggered row, single-position feed, straight-line metal-box type in the prototype. Production models will use a 32-round, curved plastic magazine. The magazine catch is placed aft of the housing, being pressed forward to release the magazine from the gun.

The fire selector lever on the left side just above the pistol grip has three positions 'S' (safety), to the rear; '30' (automatic), upward; and '1' (semi-automatic), forward. The safety position blocks the trigger and firing mechanism; the gun possesses an additional safety in the form of an internal inertia-type locking device that automatically blocks the bolt at its forward position in case the weapon is accidentally dropped or hit on the butt, preventing unintentional and hazardous firing. The bolt remains closed after the last round is fired.

Stripping
Field-stripping the SM Model 02 LAPA is a straightforward procedure. The three metal pins which secure the body together (production guns will have only one) are first removed sideways to the right, with the aid of a round of ammunition or any other pointed object. This allows the buttplate to be pulled backwards and detached from the stock. The body can then be separated into two halves, the bottom one exposing all the firing mechanism for cleaning, making further disassembly unnecessary. The tubular metal receiver, with the integral sights, is pulled away from the top plastic cover/handle allowing, at the same time, the cocking handle and its rod to be detached from the bolt. The receiver may then be disassembled, which is done

9 mm LAPA SM Model 02 sub-machine gun

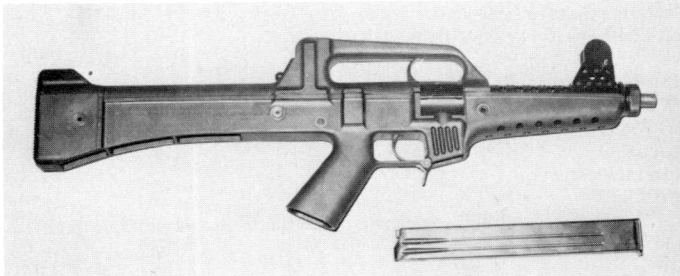

9 mm LAPA SM Model 02 sub-machine gun with magazine removed and ejection port closed

9 mm LAPA SM Model 02 sub-machine gun showing perforated upper part and frontsight assembly of forward end of tubular receiver

9 mm LAPA SM Model 02 sub-machine gun showing rearsight

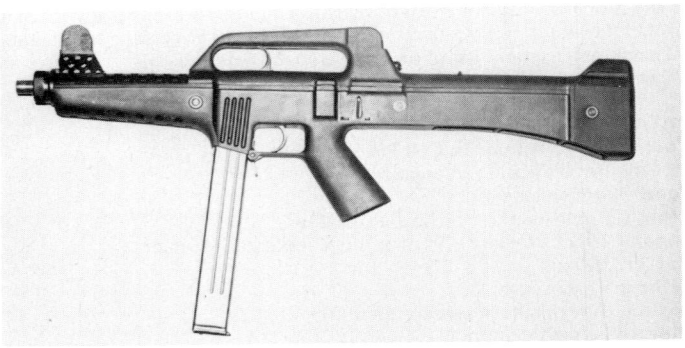

9 mm LAPA SM Model 02 sub-machine gun

by unscrewing the rear cap. This allows the return spring and the bolt to come out. The barrel is removed from the forward end by unscrewing its mounting nut.

DATA
Cartridge: 9 mm × 19 Parabellum
Operation: blowback, selective fire
Feed: 30- or 32-round straight or curved box magazine
Weight: (gun without magazine) 2.803 kg; (gun with 30-round magazine, loaded) 3.386 kg
Magazine: (empty) 0.21 kg; (30 round, loaded) 0.583 kg
Length: (overall) 623 mm
Barrel length: 202 mm

Rifling: 4 grooves rh
Sights: (fore) elevated protected post, adjustable for elevation; (rear) aperture, flip-type, 50 and 100 m, adjustable for windage
Sight radius: 290 mm
Rate of fire: 485 rds/min
Muzzle velocity: 405 m/s

Manufacturer
Laboratório de Projetos de Armamento Automático Ltda, Rua Homem de Melo 66, Grupo 201, Tijuca, 20510 Rio de Janeiro.
Status
Current. Development complete. Ready for production.

9 mm MSM sub-machine gun

The MSM (Mini Sub Metralhadora – mini sub-machine gun) is the result of a co-operative effort between LAPA, the research and development agency, and HAGA, a long-established metallurgical company. Design of the weapon began in February 1985 and a prototype was completed in December of the same year. Official testing was expected to take place early in 1986, with

full-scale production to follow about 10 months later under the responsibility of ENARM, a new research, development, production and marketing company.

The MSM is a compact weapon using conventional blowback operation and employing a telescoping bolt with a fixed firing pin. The receiver is a

9 mm ENARM sub-machine gun with single grip and folded stock

9 mm ENARM sub-machine gun with optional double grip and stock extended

box-type stamped structure to which may be attached one of two types of nylon pistol grip; one is a simple grip, the other has a forward hand-grip incorporated. The grip unit is retained by two pins and either type can be quickly removed and replaced by the other. The nylon magazine fits into the grip; it is a curved single-position-feed pattern holding 32 rounds. The fire selector lever is located on the left side of the receiver above the trigger, and has settings for automatic (marked '30') forward, semi-automatic ('1') centre, and safe ('S') rearward. The buttstock is retractable, with a locking catch on the lower rear end of the receiver. The cocking handle is on top of the receiver and as the first round is fired it runs forward and thereafter remains still while the weapon is being fired.

To strip the MSM the barrel retaining nut is unscrewed after pressing down on its locking latch. Pull the top receiver cover forward and up, removing it from the weapon. Press the barrel backwards, pull out the bolt together with the recoil spring and its guide rod. Press the ejector to the left and pull out the barrel. Re-assembly is the reverse procedure.

DATA
Cartridge: 9 mm × 19 Parabellum
Operation: Blowback, selective fire
Feed: 32-round curved box magazine
Weight: gun with empty magazine 2.70 kg
Length: stock extended 495 mm; stock retracted 290 mm
Barrel: 150 mm, 4 grooves, rh
Sights: (fore) fixed post; (rear) 100 m fixed aperture
Sight radius: 230 mm
Rate of fire: ca. 600 rds/min

Manufacturer
ENARM, Rua Homen de Mello 66, Apt. 201, Tijuca, Rio de Janiero.
Status
Final development.

CANADA

9 mm C1 sub-machine gun

When the British L2A1 sub-machine gun was developed the Canadian Army made some modifications and Canadian Arsenals Ltd has produced the C1 since 1958. The differences between the C1 and the British Sterling and L2 series are not great. The magazine has been changed both in capacity and design. It holds 30 rounds, having previously held 34. A 10-round magazine has also been produced. The design change involves the removal of the rollers and the substitution of an orthodox magazine follower. The bayonet for the C1 is that from the Fusil Automatique Légèr (FAL) rifle whereas the L2A3 takes the No 5 bayonet. The components are largely interchangeable between the two guns and the C1 is said to be cheaper to manufacture. The performance of the two weapons is much the same.

DATA
Cartridge: 9 mm Parabellum
Operation: blowback, selective fire
Feed: 30-round box magazine
Weight: (unloaded) 2.95 kg; (loaded) 3.46 kg
Length: (butt extended) 686 mm; (butt folded) 493 mm
Barrel: 198 mm
Rifling: 6 grooves rh

9 mm C1 sub-machine gun

Sights: (fore) blade; (rear) flip aperture 300 and 600 ft (91.4 and 182 m)
Muzzle velocity: 366 m/s
Rate of fire: (cyclic) 550 rds/min

Manufacturer
Canadian Arsenals Ltd, Long Branch, Ontario.
Status
Current. Not in production.
Service
Canadian forces.

CHILE

9 MM FAMAE sub-machine gun

This FAMAE design is a relatively simple blowback weapon which appears to have been influenced by the Sterling. The pistol grip has the same steep rake, the curved magazine fits into the left side of the cylindrical receiver, and the bolt has deep spiral dirt-clearing grooves. Beyond that though, there are features which have been adapted from other designs, such as the collapsing two-wire stock, similar to that of the M3 and a spoon-shaped muzzle intended to act as a compensator and counter the characteristic tendency of sub-machine guns to rise during automatic fire. The weapon appears to be a simple manufacturing and maintenance proposition and should be robust and reliable in service.

DATA
Cartridge: 9 × 19 mm Parabellum
Operation: blowback
Feed: 43-round box magazine
Weight: 2.44 kg
Length of barrel: 175 mm
Rifling: 4 grooves, rh, one turn in 27 calibres
Muzzle velocity: 400 m/s
Rate of fire: 800 – 875 rds/min
Dispersion at 100 m: 100 mm

Manufacturer
FAMAE Fabricaciones Militares, Av. Pedro Montt 1568/1606, Santiago.
Status
Current. Production.
Service
Chilean military & police forces.

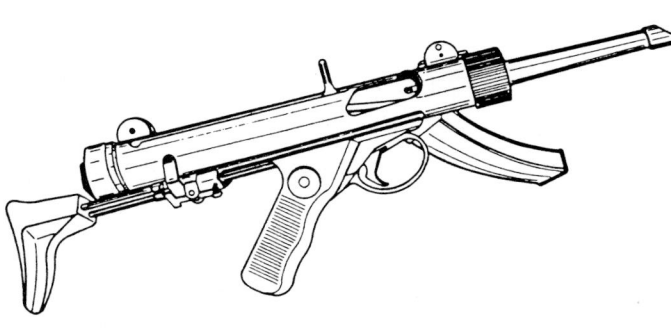

FAMAE 9 mm sub-machine gun

CHINA, PEOPLE'S REPUBLIC

7.62mm Type 79 light sub-machine gun

This is an extremely lightweight weapon, firing the standard 7.62 × 25mm pistol cartridge. The receiver is rectangular and made from steel stampings, and it has a safety lever and fire selector on the right side, above the pistol grip, which appears to have been modelled upon that of the AK series rifles. This is an interesting example of standardisation of operating characteristics between two different types of weapon; a man trained in using the AK system will find no difference if he picks one of these weapons up.

The operation is simple blowback, and the steel stock can be folded over the top of the receiver.

DATA
Cartridge: 7.62 × 25mm Pistol
Operation: blowback, selective fire
Feed: 20-round box magazine
Weight: gun with empty magazine, 1.90 kg
Length: stock extended 740 mm; stock folded 470 mm
Rate of fire: ca 650 rds/min
Muzzle velocity: 500 m/s

Manufacturer
China North Industries Corp, 7A Yue Tan Nan Jie, PO Box 2137, Beijing.
Status
Current. Production.
Service
Not known; available for export.

7.62 mm Chinese Type 79 light sub-machine gun

7.62 mm Light sub-machine gun Type 85

The Type 85 is a modified and simplified version of the Type 79. It is a plain blowback weapon, with a cylindrical receiver into which the barrel is fitted and which carries the bolt and return spring. There is a folding butt and the weapon uses the same magazine as the Type 79. It is notable that the manufacturers claim the ability to fire reduced-velocity Type 64 pistol ammunition as well as the standard Type 51 cartridge.

DATA
Cartridge: 7.62 × 25 mm pistol
Operation: blowback, selective fire
Feed: 30 round box magazine
Weight, empty: 1.90 kg
Length: (butt extended) 628 mm; (butt folded) 444 mm
Sights: (fore) blade; (rear) flip aperture
Muzzle velocity: 500 m/s
Rate of fire: 780 rds/min (cyclic)

Manufacturer
China North Industries Corp., 7A Yue Tan Nan Jie, PO Box 2137, Beijing.
Status
Current. Production.

7.62 mm Type 85 sub-machine gun

7.62mm Type 64 silenced sub-machine gun

This is a Chinese-designed and constructed sub-machine gun which combines a number of features taken from various European weapons. The bolt action is the same as that of the Type 43 copy sub-machine gun which was taken from the Soviet PPS-43. The trigger mechanism, giving selective fire, was taken from the Bren gun, numbers of which fell into Chinese hands during the Korean War. This mechanism was itself derived from that of the series of light machine guns purchased from Czechoslovakia by the pre-war mainland government. The weapon is blowback-actuated using the 7.62 mm × 25 pistol cartridge held in a curved magazine under the receiver. The chamber contains three flutes each 0.1 mm wide and 0.075 mm deep, extending from the commencement of the small cone to just beyond the mouth of the chamber: a total length of 10 mm. The suppressor is of the Maxim-type. The barrel is 200 mm long and is plain for the first 131 mm after which for a distance of 57 mm it is perforated by four rows of holes, each following the rifling groove, and each having nine holes of 3 mm diameter, making 36 holes in all. The tube surrounding the barrel continues forward for a further 165 mm and then there is a muzzle cap. Between the end of the barrel and the cap is a stack of baffles each of which is dished with a central hole of 9 mm diameter, and two rods pass down through the baffles keeping the stack together and properly lined up. The rods can be rotated and then are free to come out through the baffles, allowing ready disassembly.

7.62 mm Type 64 sub-machine gun with silencer removed

7.62 mm Type 64 silenced sub-machine gun

The suppressor is reasonably effective and also has the virtue of preventing any flash from either muzzle or breech.

This is a most unusual sub-machine gun, for it has been specifically designed and made as a silenced weapon; in all other cases the silencer is fitted to an existing gun, which at least reduces the cost and manufacturing effort.

DATA
Cartridge: 7.62 mm × 25 Type P Ball
Operation: blowback, selective fire
Feed: 30-round curved box magazine
Weight: (empty) 3.4 kg; (bolt) 0.39 kg
Length: (stock open) 843 mm; (stock closed) 635 mm

Barrel: 244 mm
Rifling: 4 grooves rh
Muzzle velocity: 513 m/s
Rate of fire: (cyclic) 1315 rds/min
Effective range: 135 m

Manufacturer
State factories.
Status
Current. Not in production.
Service
Chinese People's Army.

7.62 mm Type 85 silenced sub-machine gun

This is a simplified and lightened version of the silenced Type 64 sub machine gun, produced principally for export. It appears to be based on the simple mechanism of the Type 85 light sub-machine gun, but is of about the same size as the Type 64 and uses similar silencing arrangements. It is regulated for the Type 64 silenced cartridge but it is also possible to fire the standard Type 51 pistol cartridge, although the silencing effect will, of course, be much less and there will also be the problem of bullet noise. With the Type 64 cartridge, the sound of discharge is reduced to < 80 db.

DATA
Cartridge: 7.62 × 25 mm Type 64 (and see text)
Operation: blowback, selective fire
Feed: 30-round box magazine
Weight, empty: 2.50 kg
Length: (butt extended) 869 mm; (butt folded) 631 mm
Sights: (fore) blade; (rear) flip aperture
Muzzle velocity: 300 m/s
Rate of fire: 800 rds/min (cyclic)

7.62mm Type 85 silenced sub-machine gun

Effective range: 200 m

Manufacturer
China North Industries Corp., 7A Yue Tan Nan Jie, PO Box 2137, Beijing.
Status
Current. Production.

CZECHOSLOVAKIA

Models 23, 24, 25 and 26 sub-machine gun

These were all designed by Vaclav Holek and were produced from 1949 onwards. They differ only in cartridge and butt-stock.

The Model 23 fires 9 mm × 19 Parabellum cartridges and has a wooden stock.

The Model 25 fires 9 mm × 19 Parabellum cartridges and has a folding metal stock.

The Model 24 fires the 7.62 mm × 25 pistol cartridge and has a wooden stock.

The Model 26 fires the 7.62 mm × 25 pistol cartridge and has a folding metal stock.

All the guns are blowback-operated, selective fire weapons and use box magazines.

The 9 mm magazine capacity is 24 or 40 rounds and the 7.62 mm magazines hold 32 rounds.

The weapons offer selective fire with control exercised by trigger pull. A short pull produces single shots and a long pull produces automatic fire. The bolt is of the wrap-round variety with the firing pin and breech face well back from the front of the bolt which envelops the breech at the moment the firing pin reaches the chambered round.

All the weapons carry a magazine filler on the right of the foregrip. The Models 24 and 26 have magazines that slope forward whereas those of the Models 23 and 25 are vertical.

The Models 23 and 25, firing the 9 mm Parabellum cartridge, were in service with the Czechoslovak army in 1951 and 1952 and were then replaced by the 7.62 mm × 25 Models 24 and 26, which remained in service until 1962. This was but one example of many of the USSR forcing other Warsaw Pact countries to conform and use the standard Soviet cartridge.

DATA
MODELS 23 AND 25
Cartridge: 9 mm × Parabellum
Operation: blowback, selective fire
Feed: 24- or 40-round box magazine
Weight: (Model 23) 3.27 kg; (Model 25) 3.5 kg
Length: (stock extended) 686 mm; (stock folded) 445 mm
Barrel: 284 mm
Rifling: 6 grooves rh
Muzzle velocity: 381 m/s
Rate of fire: (cyclic) 650 rds/min

Manufacturer
Ceská Zbrojovka, Brno.
Status
No longer manufactured.

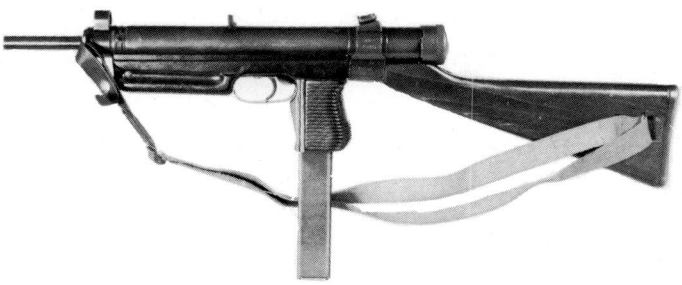

Czechoslovak 9 mm Model 25 sub-machine gun

Czechoslovak 9 mm Model 23 sub-machine gun

Service
No longer in Czechoslovak military service. Many used in Nigerian Civil War. Supplied to Cuba and Syria where some are probably still in use. Sold to South Africa.

MODELS 24 AND 26
As Models 23 and 25 except
Cartridge: 7.62 mm × 25 Pistol Type P
Magazine: 32 rounds
Weight: 3.41 kg
Rifling: 4 grooves rh
Muzzle velocity: 550 m/s

Status
Obsolescent. Not in production
Service
Cambodia, Cuba, Czechoslovakia, Guinea, Mozambique, Nicaragua, Nigeria, Romania, Somali, Syria, Tanzania

7.65 mm Model 61 Skorpion machine pistol

During the last few years of the 1950s Omnipol, which then controlled Czechoslovak arms production, hastened the pace of weapon development and the first post-war weapons were replaced. The Mark 52 pistol was superseded by a dual-purpose pistol/sub-machine gun called the Skorpion of unusual, but effective, design. Like the great majority of dual-purpose weapons it carries out neither role to perfection and is really neither a pistol nor a sub-machine gun but is best described as a 'machine pistol'. It fires the American .32 Automatic Colt Pistol (ACP) and this feature distinguishes it from other Soviet Bloc weapons. The .32 (7.65 mm) cartridge with its light bullet and low muzzle velocity produces a muzzle energy which is less than that of a high velocity .22 long-rifle cartridge; and it seems likely that commercial considerations dictated the choice of a cartridge which is readily available almost everywhere. There are now other versions firing alternative ammunition. The Model 63 is in .380ACP, Model 64 in 9 mm Makarov, and a Model 68 in 9 mm Parabellum is known to exist. A number of African states have purchased the Skorpion.

It is a weapon which is ideally suited to being carried by vehicle crews and others who have to work in a confined space, or who have to carry heavy and cumbersome loads, such as signallers. Despite the limitations of the ammunition the Skorpion enables its user to engage targets at greater ranges than a conventional pistol will allow, since it is easily held with two hands and, with shoulder stock extended, it shoots well.

Operation

The weapon can produce either single shots or full automatic fire. The change lever is marked '1' at the rear position for single-shot, '0' for safe and '20' for automatic, a system comprehensible to anyone familiar with Arabic numerals. The cartridge produces a very low recoil impulse and this enables simple blowback operation to be employed. Gas pressure drives the case back in the chamber against the resistance provided by the inertia of the bolt and its two driving springs. The block goes back, extracting the empty case which is ejected straight upwards and can hit the firer full in the face. Because the block is so light it is necessary to have a device to control the rate of fire which, otherwise, would be well over 1000 rds/min.

The rate reducer operates as follows: when the bolt reaches the end of its rearward stroke it strikes, and is caught by, a spring-loaded hook mounted on the back plate. At the same time it drives a light, spring-loaded plunger down into the pistol grip. The light plunger is easily accelerated and passes through a heavy weight which due to its inertia is left behind. The plunger, having compressed its spring, is driven up again and then meets the descending inertia pellet. This slows down the rising plunger which, when it reaches the top of its travel, rotates the hook, releasing the bolt which is driven forward by the compressed driving springs. This reduces the rate of fire to 840 rds/min. The bolt feeds a round from the 20 shot magazine, chambers it and supports the case in the chamber. Since it is a simple blowback-operated system there is no

7.65 mm Model 61 Skorpion machine pistol

bolting or delay device and the sole support to the cartridge is the inertia of the bolt and the strength of the driving springs. The magazine platform operates a plunger which retains the bolt in the rear position when the ammunition supply is exhausted.

DATA
Cartridge: .32ACP (7.65 mm), .380ACP, 9 mm × 18, 9 mm Parabellum
Operation: blowback with selective fire
Feed: 10- or 20-round box magazine
Weight, empty: 1.59 kg
Length: (butt extended) 513 mm; (butt retracted) 269 mm
Barrel: 112 mm
Rifling: 6 grooves, rh, one turn in 305 mm
Sights: (fore) post; (rear) flip aperture, 75 and 150 m
Sight radius: 171 mm
Muzzle velocity: 317 m/s; (with silencer) 274 m/s
Rate of fire: (normal) 840 rds/min; (silenced) 950 rds/min

Manufacturer
State ordnance factories. Also made in Yugoslavia where it is known as the Model 84.
Status
Current. No longer manufactured.
Service
Some Czechoslovak units. Angola, Libya, Uganda.

DENMARK

9 mm Model 49 Hovea sub-machine gun

This weapon was designed in Sweden at the Husqvarna arms factory and competed with the Carl Gustaf weapon in arms trials at which the latter weapon (the Model 45) was selected for Swedish use. The Hovea sub-machine gun was preferred by the Danish authorities, however, and they purchased a quantity together with a manufacturing licence; and Danish-built weapons are still in service with Danish forces.

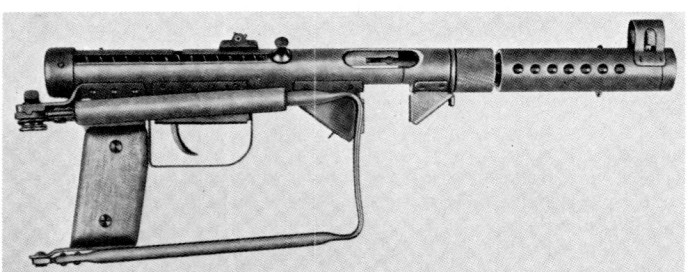

9 mm Model 49 Hovea sub-machine gun with butt folded and without magazine

The weapon was orginally designed to take the Finnish 50-round magazine, but this was subsequently replaced by the 36-round Carl Gustaf box. Fitted with a rectangular folding stock, the Model 49 Hovea bears a superficial resemblance to both the Carl Gustaf Model 45 and the Madsen weapons (see entries below).

DATA
Calibre: 9 mm Parabellum
Operation: blowback, automatic
Feed: 36-round detachable box (see text)
Weight: (empty) 3.4 kg; (loaded) 4 kg
Length: (stock extended) 810 mm; (stock retracted) 550 mm; (barrel) 215 mm
Sights: (foresight) blade; (rearsight) V-notch flips 100/200 m
Rate of fire: (cyclic) 600 rds/min

Manufacturer
Vabenarsenalet, Copenhagen. (Abolished 1970).
Status
Current. Not in production.
Service
Danish forces.

---

9 mm Madsen sub-machine guns

Madsen has produced a series of sub-machine guns of the same basic simple design, known as the Model 1946, Model 1950, Model 1953 and the Mark II. All have the same rectangular section folding stock. The Model 1946 had a shaped cocking handle which fitted over the receiver and down on each side; the Model 1950 has a small cylindrical knob on top of the receiver; the Model 1953 has a knurled cylindrical barrel nut that screws on to the barrel whereas

the 1946 and 1950 models have a grooved nut fitting on to the receiver. The magazines of the Model 1946 and Model 1950 are 32-round, single-position feed, straight flat-sided types. Those of the Model 1953 and Mark II are two-position feed types also holding 32 rounds. The Mark II is the only one giving selective fire: it may have a perforated barrel jacket. A feature of all these sub-machine guns is a grip safety behind the magazine wall. This has to be pressed forward before the weapon can be cocked or fired. It engages the bent in the bolt to hold it when the bolt is cocked and blocks the sear when the bolt is forward.

Operation
The 32-round magazine is most easily loaded by employing a filler which is carried inside the pistol grip. To get at it means opening up the two halves of the receiver shell and so putting the gun out of action for a short while. As a result the filler is not always used and hand filling is adopted which can be difficult as the magazine spring is progressively compressed. The filler fits over the top of the magazine and has a plunger to depress the magazine follower. The base of the cartridge is slid under the lips while the plunger takes the weight of the spring, and pushed fully home when the plunger is released.

When the grip safety is pushed forward, and the trigger operated, the sear will release the bolt which will go forward under the action of the compressed return spring. The feed rib on the bottom of the bolt picks up the top cartridge in the magazine and feeds it up over the bullet guide into the chamber. The cartridge is chambered, the extractor snaps into the groove and the fixed firing pin on the bolt-face fires the cap while the bolt is still moving forward. The gas pressure forces the cartridge case rearward and the bolt movement is halted and reversed. The extractor holds the case to the bolt-face until it hits the fixed ejector and is ejected to the right of the gun. The return spring is compressed. At the end of rearward travel the bolt is impelled forward again. If the trigger or grip safety has been released the bolt will be held back; otherwise the cycle is repeated until the magazine is empty.

Stripping
Basically the same procedure for stripping is used for all the Madsen sub-machine guns. There is a slight difference for the Model 1946 where the cocking handle must be pulled up and out of the bolt after releasing the spring-loaded detent.

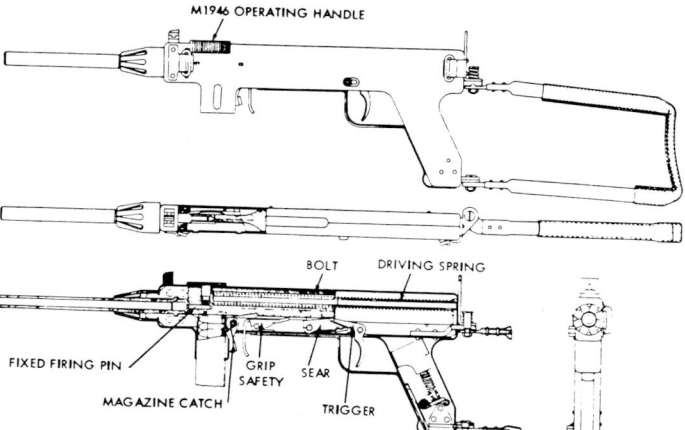

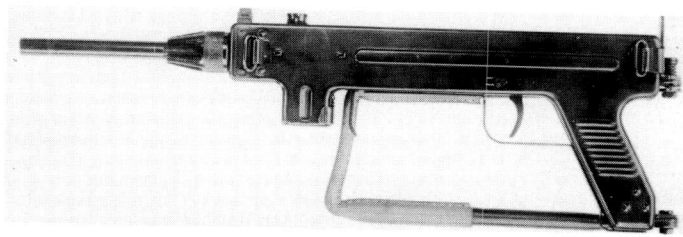

9 mm Model 1946 Madsen sub-machine gun

9 mm Model 1950 Madsen sub-machine gun

9 mm Model 1953 Mark II Madsen sub-machine gun with barrel jacket and bayonet

9 mm Model 1950 Madsen sub-machine gun

The magazine in each case is removed and the chamber and feedway inspected to see that they are empty. The barrel nut is unscrewed and removed. With the Mark II model fitted with a barrel jacket, unscrew the jacket which carries the barrel nut. The receiver can now be opened up like a suitcase, hinged at the rear. The barrel will lift out. The return spring and its guide rod come out after the rear end of the guide rod has been eased forward to unseat it. The bolt lifts out.

DATA
Cartridge: 9 mm Parabellum
Operation: blowback, automatic
Feed: 32-round detachable box magazine
Weight, empty: 3.2 kg
Length: (stock extended) 794 mm; (stock folded) 528 mm
Barrel length: 198 mm
Rifling: 4 grooves rh
Muzzle velocity: 390 m/s
Rate of fire: (cyclic) 550 rds/min

Manufacturer
Dansk Industri Syndikat, Madsen, Copenhagen.
Status
Current. Not in production.
Service
Danish police forces. Some South American and South-east Asian countries. Made under licence in Brazil.

EGYPT

Egyptian sub-machine guns

When Egypt obtained its independence a small arms industry was set up. The first weapon produced was a copy of the US Thompson. The barrels and the magazines were purchased and were originally produced by Auto Ordnance at Bridgeport. The weapon was altered in several respects. The cocking handle was removed from the top and, like that of the M1 and M1A1, placed on the side of the receiver. The rearsight was of local design and the cylindrical receiver had a screw-on end closure cap.

The weapon which was made in some quantity was the Port Said which was a very close copy of the Swedish Carl Gustaf Model 45 B. This weapon is still

in service. The photograph in the Egyptian Sub-machine Gun entry shows the Port Said carried by a soldier crossing the Suez Canal in the October 1973 war.

A simplified version of the Port Said was later manufactured as the Akaba. The barrel jacket was dispensed with and the barrel shortened to 150 mm; the foresight is a post on the receiver and the rearsight, moved back to the end of the receiver, is a fixed leaf with an aperture set to 100 m. The folding stock of the Port Said is replaced by a simple push-in wire stock based on that of the

Port Said sub-machine gun is a copy of Carl Gustaf

Akaba sub-machine gun, a simplified version of Port Said

American M3 weapon. The only identification is the marking 'AKABA Made in UAR' on the right side. Overall length with stock extended is 737 mm, with stock telescoped 482 mm. In other respects they conform to the basic Carl Gustav design.

Port Said sub-machine guns and Soviet Kalashnikov rifles used by Egyptian troops

Version of Thompson .45 sub-machine gun made in Egypt

FINLAND

9 mm Model 1931 sub-machine gun

Finland was a leader in Europe in the design of sub-machine guns after the First World War. The principal designer was Aimo Johannes Lahti who produced a number of Suomi (Finland) designs in the very early 1920s. The Model 1922, firing the 7.65 mm Parabellum was his first model, but was not produced and was followed by the Model 1926 which was adopted by the Finnish Army in the same calibre. Some guns were made in 9 mm Mauser and these had a large buffer housing behind the receiver. The Model 1926 had some unusual features; a compressed air buffer, a quick-change barrel and the first non-reciprocating cocking handle.

The Model 1931, which followed the Model 1926, was very successful. In Finland it was made by Oy Tikkakoski AB, in Sweden by Husqvarna and by

9 mm Model 1931 sub-machine gun

Hispano-Suiza in Switzerland. It was sold also to Norway. The Model 1931 was very heavy, extremely well made but as it was all machined from the solid it was very expensive. The magazine was original. It consisted of two compartments separated by a central partition extending to about 4 cm from the top, each with its own follower and spring. The rounds fed to a single position exit at the top.

There was also a 71-round drum magazine which was later copied by the Soviet PPD 34/38 sub-machine gun.

DATA
Cartridge: 9 mm Parabellum
Operation: blowback, selective fire
Feed: 50-round box magazine or 71-round drum magazine
Weight: (unloaded) 4.68 kg; (loaded with 71-round drum) 7.09 kg
Length: 870 mm
Barrel: 318 mm
Rifling: 4 grooves rh
Muzzle velocity: 399 m/s
Rate of fire: (cyclic) 900 rds/min; (automatic) 120 rds/min

Manufacturer
Oy Tikkakoski AB.
Status
Obsolescent. Not in production.
Service
Finnish forces.

9 mm Model 1944 sub-machine gun

This is a copy of the Soviet PPS-43 using the 9 mm Parabellum cartridge. A large number were made in Finland in 1944 before the Second World War ended and the weapon is still in service, having been modified to take the Carl Gustaf 36-round magazine as well as the 71-round drum magazine.

The Model 1944 was manufactured at Oy Tikkakoski AB and the manager of that factory, Willi Daugs, moved to Sweden at the end of the war, on to the Netherlands and then to Spain. He took the manufacturing drawings of the MP44 and subsequently the gun was produced at Oviedo as the Dux sub machine gun.

DATA
Cartridge: 9 mm × 19 Parabellum
Operation: blowback, automatic
Feed: 36-round box magazine or 71-round drum magazine
Weight: (unloaded) 2.9 kg; (loaded, 36-round box magazine) 3.6 kg; (71-round drum) 4.3 kg

Length: (stock extended) 831 mm; (stock folded) 622 mm
Barrel: 249 mm
Rifling: 4 grooves rh
Sights: (foresight) blade; (rearsight) notch
Muzzle velocity: 399 m/s
Rate of fire: (cyclic) 650 rds/min; (automatic) 120 rds/min
Effective range: 200 m

Manufacturer
Oy Tikkakoski AB.
Status
Obsolete. Not in production.
Service
Finnish forces.

9 mm Jati-Matic Sub-machine gun

The Jati-Matic is a design with some unusual features. The designers have aimed at producing a light weapon which is easy to use and control, and the weapon uses a patented system in which the bolt recoils up an inclined plane at an angle to the barrel. At the same time the bolt presses against the bottom of

the receiver. This upward movement of the bolt allows the pistol grip to be set higher than normal, so that the firer's hand lies on the axis of the bore. This, it is claimed, reduces the torque effects, easing the strain on the firer's wrist and also making the gun particularly steady when firing automatic.

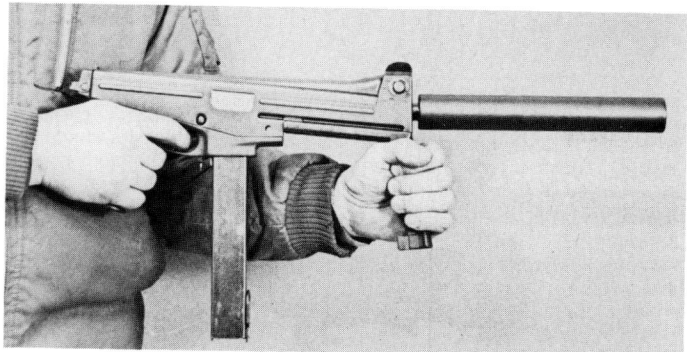

Jati-Matic sub-machine gun with sound suppressor

9 mm Jati-Matic with front grip folded down and 40-round magazine

9 mm Jati-Matic with 20-round magazine and front grip folded; note angle between axis of bore and axis of receiver

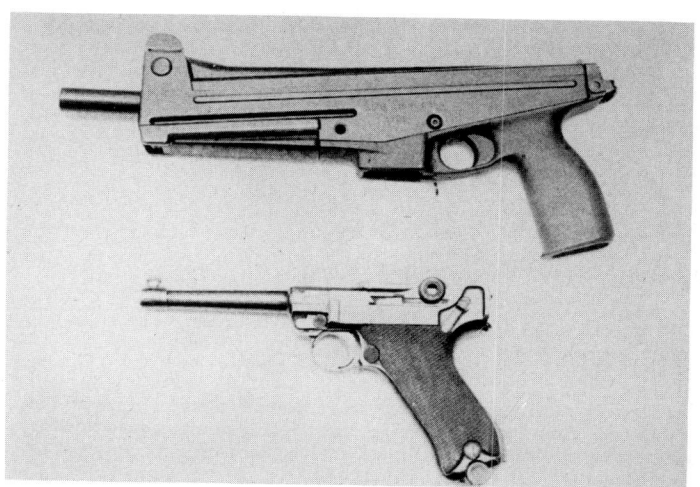

The Jati-Matic consists of a pressed steel receiver with a hinged top cover. Into this fits the bolt and barrel, retained by the top cover. A folding forward grip handle beneath the barrel acts as a steadying grip and is also the cocking handle. When closed, this handle acts as a positive bolt stop in either the cocked or uncocked positions and prevents inadvertent firing even if the trigger is pressed. The trigger is a two-stage type in which the first pressure fires single shots, and further pressure, against a spring stop, fires automatic.

The box magazine is inserted into the receiver ahead of the trigger; Smith and Wesson or Carl Gustaf magazines will also fit the Jati-Matic.

A range of accessories, including magazines, silencer, and a laser aiming spot are available. The weapon is optimised for use with 9 mm Parabellum ammunition having a bullet weight of 8 g and a muzzle velocity of 360-400 m/s. With other ammunition it may be necessary to modify the strength of the return spring.

Jati-Matic sub-machine gun with Parabellum pistol for size comparison

DATA
Calibre: 9 mm Parabellum
Operation: blowback, selective fire
Feed: detachable 20- or 40-round box
Weights: (gun empty) 1.65 kg; (20-round magazine) 300 g; (40-round magazine) 600 g; (gun and 20-round magazine) 1.95 kg

Length: 375 mm
Barrel length: 203 mm
Rifling: 6 grooves, rh. 1 turn in 250 mm
Sights: (fore) fixed blade; (rear) fixed U-notch; (zero) 100 m
Sight radius: 290 mm
Muzzle velocity: 360 – 400 m/s
Rate of fire: 600 – 650 rds/min
Effective range: 150 m

Manufacturer
Tampeeren Asepaja Oy, Hatanpäänvaltatie 32,33100 Tampere 10.
Status
Current. Negotiations in progress for licensed production outside Finland.

FRANCE

5.7 mm GIAT self-defence weapon

This design is a prototype weapon and is one of a potential family of small arms under the generic title of 'Armes de Défense Rapprochée' (ADR). The family is to consist of a semi-automatic pistol, a self-defence weapon (this weapon) and a small assault rifle. All will be chambered for a new 5.7 × 25 mm cartridge intended to replace the current 9 × 19 mm Parabellum round. GIAT claim that this round has an effective range four times that of the conventional 9 × 19 mm bullet, with a recoil impulse one-third less.

The prototype shown here is one of several ideas which have been developed; it is intended to be a compact and easily operated weapon which can be carried close to the body and which will be easy to manipulate in confined spaces. It is to weigh less than 2 kg and its envelope will be less than 300 × 120 × 30 mm. The magazine is inserted into the rear pistol grip. The reargrip and trigger are used when firing from the hip, the front grip then acting merely as a handgrip. For more deliberate fire, the reargrip can be placed to the shoulder like a conventional stock, and the fore grip and trigger are then the operative ones. No details of the mechanism have been divulged.

Manufacturer
GIAT: Groupement Industriel des Armements Terrestres, 10 place Georges Clemenceau, 92211 Saint-Cloud.

GIAT 5.7 mm self-defence weapon, showing 5.7 × 25 mm round compared to 9 × 19 mm

Status
Development.

7.65 mm Model 38 MAS sub-machine gun

This sub-machine gun followed an experimental model produced in 1935, the MAS 35, and used the 7.65 mm Long cartridge which was used only by France in the sub-machine gun role. The gun was manufactured by machining from the solid. It has a distinctive 'broken back' appearance because the barrel was angled to the receiver to avoid a high sight line. The breech-block travels at an angle to the line of the bore and its front face is angled back to fit flush as the round is chambered. The bolt travels well back beyond the sear and on its forward travel it has acquired a lot of kinetic energy when it strikes the sear. To prevent damage the sear is spring buffered – a feature found in no other sub-machine gun.

Efforts were made to keep out dirt by having a spring-loaded plate on the magazine well which closed it off when the magazine was withdrawn.

To lock the bolt in either the forward or cocked position, the trigger was pushed forward inside the trigger-guard, another unique feature.

When the gun was fully cocked the cocking handle opened the ejector port and was then disconnected from the bolt and retained to the rear, to prevent it reciprocating and at the same time to keep the ejection port open.

The MAS 38, although accurate, lacked penetration. It was produced in 1939 and production continued throughout the German occupation. It was used in the French Indo-China campaign and many fell into the hands of the Viet Cong and were used in that war.

A proportion of these captured weapons were re-barrelled to take the Soviet 7.62 mm pistol round and these may well still be in use in South-east Asia.

DATA
Cartridge: 7.65 mm Long
Operation: blowback, automatic
Feed: 32-round detachable box magazine
Weight: (unloaded) 2.87 kg; (loaded) 3.4 kg
Length: 734 mm
Barrel: 224 mm
Rifling: 4 grooves rh
Sights: (fore) blade; (rear) notch
Muzzle velocity: 351 m/s
Rate of fire: (cyclic) 600 rds/min

Manufacturer
Manufacture d'Armes de St Étienne.
Status
Obsolete. Not in production.
Service
May be found in South-east Asia.

7.65 mm Model 38 MAS sub-machine gun

9 mm MAT 49 sub-machine gun

The MAT 49 is a very solid, strongly-made weapon making considerable use of stampings from heavy gauge steel sheet. It is of conventional blowback design, firing the 9 mm × 19 cartridge, but it has a number of features which, although not original, are seldom met.

The magazine housing can be pivoted forward in front of the trigger to allow the magazine to lie along the underside of the barrel jacket. A housing release is forward of the trigger-guard; when this is operated the housing, holding the 32-round magazine, can be rotated forward. When under the barrel jacket it is retained by the housing catch. The magazine housing must be pulled manually into the firing position. It is shaped to afford a firing grip for the left hand. When the housing is under the barrel it seals off the feed opening into the receiver and this, together with the spring-loaded ejection port cover, makes the weapon proof against the ingress of sand, dirt and so on into the working parts. When the magazine is housed under the barrel the weapon is completely safe since forward movement of the bolt, whether accidental or otherwise, cannot feed, chamber or fire a round. When the magazine is swung into the firing position, safety is provided by the grip safety which has two functions. It locks the bolt in the forward position and locks the trigger when the bolt is cocked. This ensures that if the weapon is inadvertently dropped or jarred, the bolt cannot move back from its forward position, under its own inertia, to chamber and fire a round, nor can the cocked bolt go forward. When the grip safety is operated by the act of grasping the pistol grip, the trigger restraint is released and the lock preventing bolt movement is lowered. Since bolt movement is possible only when the grip safety is operated, no manual safety of the usual 'Fire' and 'Safe' type is required or provided.

shorten the overall length of the sub-machine gun by using a wrap-round bolt, enveloping the chamber, with a fixed firing pin placed towards the rear of the bolt. The head of the bolt of the MAT 49 enters an extension of the chamber thus producing a wrap-round barrel. The MAT 49 has a fixed firing pin and the weapon fires while the bolt is still moving forward; the design of the bolt-face and breech ensures that shielding is provided to minimise the effects of both early firing and a hang-fire.

The length of this French sub-machine gun can be adjusted by a sliding wire stock, similar to that of the US M3 sub-machine gun. There are two indentations in which the stock latch can engage: the stock may be locked in an intermediate position as well as in the fully extended position required for firing from the shoulder.

Stripping
Stripping is accomplished without the need of tools. The gun is made safe by removing the magazine and checking the chamber. The magazine housing is freed from the trigger frame but not latched to the barrel jacket. This permits pressing in the knurled take-down button under the rear end of the barrel

Bolt of MAT 49 sub-machine gun

9 mm MAT 49 sub-machine gun

9 mm MAT 49 sub-machine gun stripped

jacket and the barrel and receiver can then be swung up off the trigger frame. The bolt and its return spring can then be withdrawn to the rear to complete field stripping. Re-assembly is carried out logically by returning the bolt and spring into the receiver. The rear of the receiver goes into the cap at the rear of the trigger frame and the front is dropped down and pressed back until the receiver and trigger frame lock together.

DATA
Cartridge: 9 mm Parabellum
Operation: blowback, automatic
Feed: 32- or 20-round box magazine
Weight, empty: 3.5 kg
Length: (butt extended) 720 mm; (butt retracted) 460 mm

Barrel: 228 mm
Rifling: 4 grooves lh
Sights: (fore) hooded blade; (rear) flip aperture 100 and 200 m
Muzzle velocity: 390 m/s
Rate of fire: (cyclic) 600 rds/min

Manufacturer
Manufacture d'Armes de Tulle, Tulle.
Status
Current. Not in production.
Service
French forces and many former French colonies.

GERMANY, WEST

9 mm MP-K and MP-L Walther sub-machine guns

These two sub-machine guns differ only in the length of the barrel. The MP-L has a long and the MP-K a short barrel. The weapons are not used by the armies of any major power but were used by West German and some other police forces, by the armies of Brazil, Colombia and Venezuela, and by the Mexican Navy. Both have selective-fire systems and use 32-round box magazines.

The weapons are now out of production and are generally obsolescent; full details can be found in earlier editions of *Jane's Infantry Weapons*.

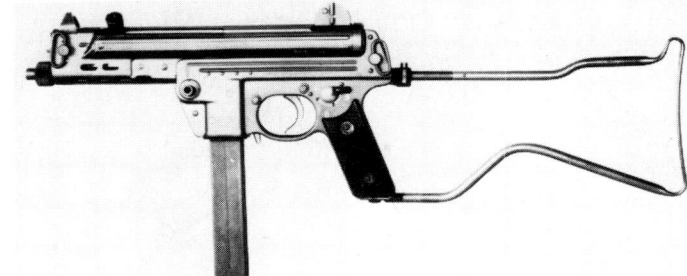

9 mm MP-K Walther sub-machine gun

9 mm MP5 Heckler and Koch sub-machine gun

The MP5 sub-machine gun was developed from the G3 rifle (which see) and has the same method of operation. Some parts are interchangeable with those used on the rifle. The sub-machine gun was adopted in late 1966 by the police forces of West Germany and by the border police. It has been purchased by military and police forces world-wide.

Operation
The weapon usually offers only a choice of single-shot or full automatic but a burst-fire device, allowing three-round bursts each time the trigger is operated, is offered by the manufacturers to those requiring this facility. In fact this trigger arrangement can be fitted to all automatic weapons in the Heckler and Koch series. The top position of the selector lever is 'safe' and the selector spindle lies over the trigger lug and prevents sufficient movement of the trigger to disengage the sear from the hammer notch.

The breech mechanism is of the same design as that used in the G3 rifle. It may be described, in brief, as a two-part bolt with rollers projecting from the bolt-head. The more massive bolt-body lies up against the bolt-head when the weapon is ready to fire and inclined planes on the front lie between the rollers and force them out into recesses in the barrel extension. The gas pressure forces back on the bolt-head which is unable to go back since the rollers are in the recesses in the barrel extension and must move in against the inclined planes of the heavy bolt-body. The selected angles of the recesses and the incline on the bolt-body produce a velocity ratio of about 4:1 between the bolt-body and the bolt-head. Thus the bolt-head moves back only about 1 mm while the bolt-body moves some 4 mm. As soon as the rollers are fully in, the

9 mm MP5A5E Heckler and Koch sub-machine gun with automatic and burst-fire facilities.

9 mm MP5 A3 Heckler and Koch sub-machine gun

9 mm MP5 A2 Heckler and Koch sub-machine gun with passive night sight

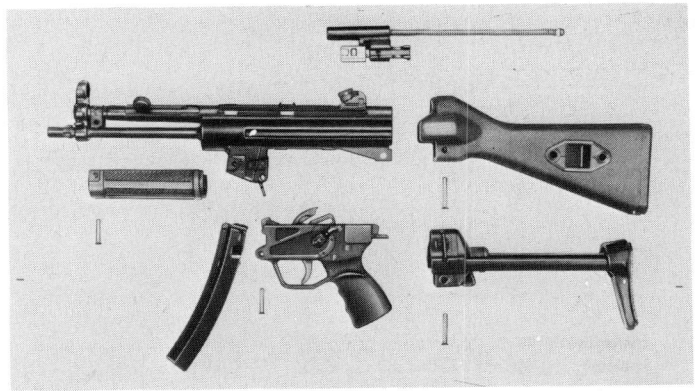

9 mm MP5 A2 Heckler and Koch sub-machine gun stripped

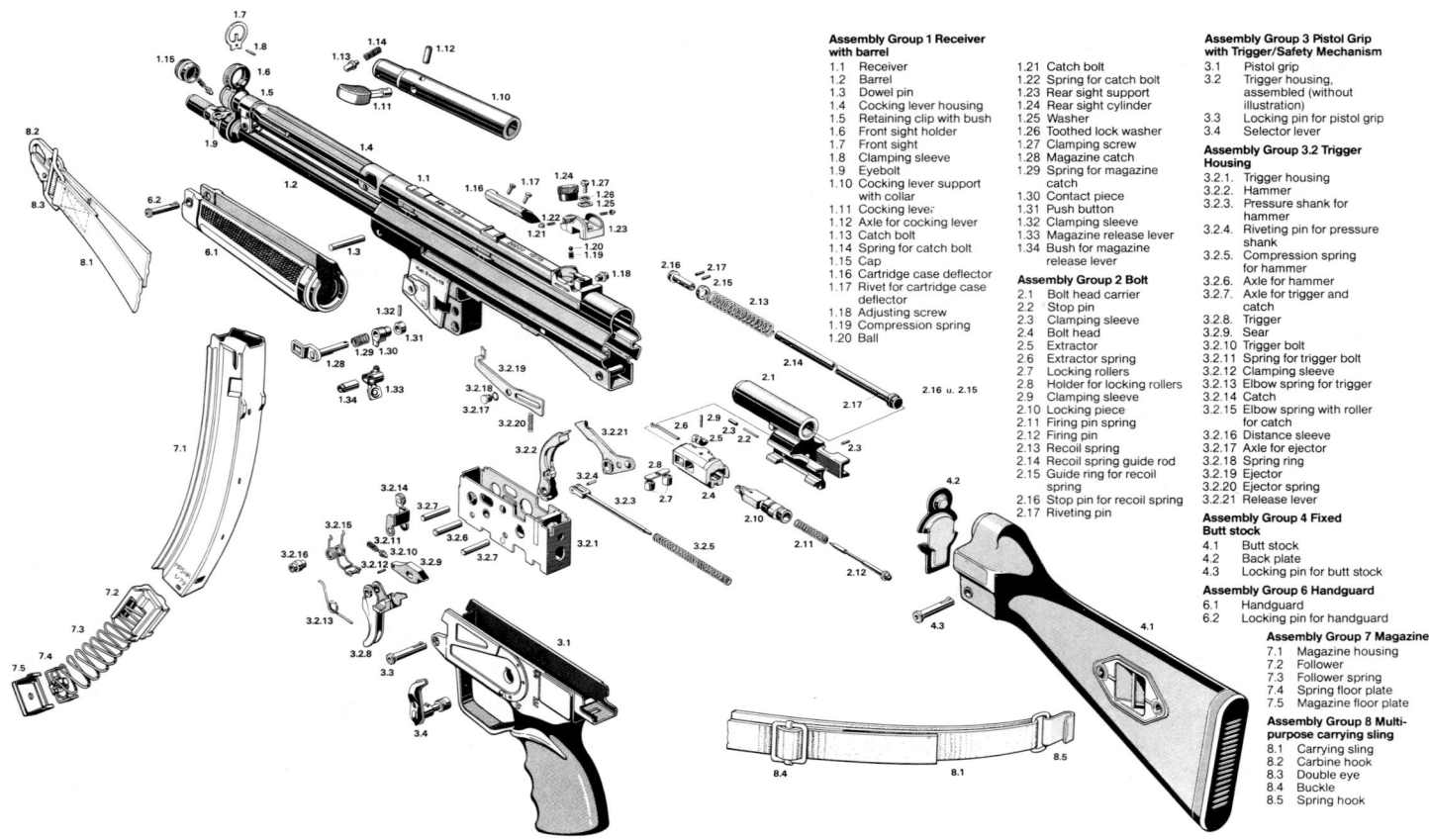

Assembly Group 1 Receiver with barrel				Assembly Group 3 Pistol Grip with Trigger/Safety Mechanism	
1.1	Receiver	1.21	Catch bolt	3.1	Pistol grip
1.2	Barrel	1.22	Spring for catch bolt	3.2	Trigger housing, assembled (without illustration)
1.3	Dowel pin	1.23	Rear sight support	3.3	Locking pin for pistol grip
1.4	Cocking lever housing	1.24	Rear sight cylinder	3.4	Selector lever
1.5	Retaining clip with bush	1.25	Washer		
1.6	Front sight holder	1.26	Toothed lock washer	**Assembly Group 3.2 Trigger Housing**	
1.7	Front sight	1.27	Clamping screw	3.2.1	Trigger housing
1.8	Clamping sleeve	1.28	Magazine catch	3.2.2	Hammer
1.9	Eyebolt	1.29	Spring for magazine catch	3.2.3	Pressure shank for hammer
1.10	Cocking lever support with collar	1.30	Contact piece	3.2.4	Riveting pin for pressure shank
1.11	Cocking lever	1.31	Push button	3.2.5	Compression spring for hammer
1.12	Axle for cocking lever	1.32	Clamping sleeve	3.2.6	Axle for hammer
1.13	Catch bolt	1.33	Magazine release lever	3.2.7	Axle for trigger and catch
1.14	Spring for catch bolt	1.34	Bush for magazine release lever	3.2.8	Trigger
1.15	Cap			3.2.9	Sear
1.16	Cartridge case deflector	**Assembly Group 2 Bolt**		3.2.10	Trigger bolt
1.17	Rivet for cartridge case deflector	2.1	Bolt head carrier	3.2.11	Spring for trigger bolt
1.18	Adjusting screw	2.2	Stop pin	3.2.12	Clamping sleeve
1.19	Compression spring	2.3	Clamping sleeve	3.2.13	Elbow spring for trigger
1.20	Ball	2.4	Bolt head	3.2.14	Catch
		2.5	Extractor	3.2.15	Elbow spring with roller for catch
		2.6	Extractor spring	3.2.16	Distance sleeve
		2.7	Locking rollers	3.2.17	Axle for ejector
		2.8	Holder for locking rollers	3.2.18	Spring ring
		2.9	Clamping sleeve	3.2.19	Ejector
		2.10	Locking piece	3.2.20	Ejector spring
		2.11	Firing pin spring	3.2.21	Release lever
		2.12	Firing pin		
		2.13	Recoil spring	**Assembly Group 4 Fixed Butt stock**	
		2.14	Recoil spring guide rod	4.1	Butt stock
		2.15	Guide ring for recoil spring	4.2	Back plate
		2.16	Stop pin for recoil spring	4.3	Locking pin for butt stock
		2.17	Riveting pin	**Assembly Group 6 Handguard**	
				6.1	Handguard
				6.2	Locking pin for handguard

Assembly Group 7 Magazine
7.1 Magazine housing
7.2 Follower
7.3 Follower spring
7.4 Spring floor plate
7.5 Magazine floor plate

Assembly Group 8 Multi-purpose carrying sling
8.1 Carrying sling
8.2 Carbine hook
8.3 Double eye
8.4 Buckle
8.5 Spring hook

9 mm MP5 A2 Heckler and Koch sub-machine gun

two parts of the breech are driven back together. The empty case is held to the breech face by the extractor until it strikes the ejector arm and is thrown out of the ejection port to the right of the gun. The return spring is compressed during the backward movement of the bolt and drives the bolt forward. A round is fed into the chamber and the bolt-face come to rest. The bolt-body continues to move forward and the inclined planes drive the rollers out again into the barrel extension recesses. The bolt-body closes up to the bolt-head and the weapon is ready to fire another round.

It can be seen from this explanation that the MP5 fires from a closed bolt and this makes it more accurate than the conventional blowback sub-machine gun.

Trigger and firing mechanism
If the selector is set to full automatic the trigger has moved up sufficiently for the nose of the sear to be depressed so far that it does not re-engage the hammer and the next round is fired as soon as the bolt is fully closed, and the safety sear is moved out of engagement with the hammer. If the selector is for a single-shot the trigger is unable to rise fully and the spring-loaded sear holds

the hammer until the trigger is released and re-pressed. Provided the bolt is fully closed the hammer will be released. The action is dealt with more fully in the description of the operation of the G3 rifle.

If a burst-fire control is fitted a ratchet counting device in the trigger mechanism holds the sear off the hammer until the allotted number of rounds have been fired. This device ensures the correct number of cartridges are discharged in a single burst and any interruption, for example the magazine is emptied, starts a fresh count. After the burst the trigger must be released to set the counter back to zero and another sustained pressure on the trigger fires another burst of three rounds.

The first models were fitted with a straight magazine, but in 1978 this was changed to one with a slight curve. It was found that the curved shape improved the feeding characteristics with the many different types of bullet and nose shapes in 9 mm Parabellum ammunition.

Stripping
Remove the magazine and then cock the weapon by pulling back on the cocking handle which is located well forward on the left hand side of the tube carrying the bolt extension and the return spring. After checking that the chamber and feedway are clear, the bolt may be allowed to travel forward. The butt-stock locking pin is behind the pistol grip and when it is pulled out the butt-stock can be withdrawn from the receiver and the trigger group allowed to hang down from its front locking pin. Retracting the cocking handle will pull the bolt assembly and return spring to the rear of the receiver where it can be gripped and extracted. The bolt head can be turned 90° clockwise and pulled out of the bolt-body. The firing pin and spring can then be taken out.

9 mm MP5 sub-machine gun
(1) receiver with cocking lever housing (2) grip assembly (3) safety catch (4) rotary rearsight (5) sight support (6) sight base (7) butt stock (8) sling holder (9) butt plate (10) eyebolt (11) locking pin for hand-guard (12) hand-guard (13) cylindrical pin (14) hook (15) locking pin for grip assembly (16) magazine catch (17) locking pin for butt stock (18) front sight holder (19) cap (20) clamping sleeve for front sight (21) front sight (22) catch bolt (23) stop ring (24) rivet for catch lever (25) catch lever (26) catch spring (27) axle for cocking lever (28) cocking lever support with collar (29) stop pin for recoil spring (30) guide ring (31) recoil spring guide rod (32) clamping sleeve (33) bolt head carrier (34) firing pin with firing pin spring (35) release lever (36) elbow spring for trigger (37) hammer (38) ejector spindle (39) ejector (40) ejector pressure spring (41) screw socket with rivet (42) binding screw with locking washer (43) recoil spring (44) rivet (45) back plate (46) barrel (47) barrel extension (48) follower (49) follower spring (50) magazine tube (51) magazine floor plate (52) safety cup with guide bush (53) magazine catch lever with roller (54) transmitting piece for magazine catch (55) magazine release lever (56) Elbow spring with roller (57) sear (58) trigger (59) safety pin (60) pressure shank and pressure spring (61) trigger housing (62) extractor with extractor spring (63) cocking lever (64) bolt head (65) locking roller (66) locking piece

9 mm MP5 A2 Heckler and Koch sub-machine gun

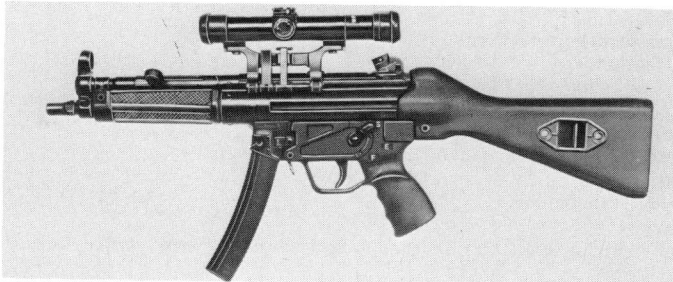

9 mm MP5 A2 Heckler and Koch sub-machine gun with telescopic sight

Variants

The basic MP5 mechanism is employed in six different sub-machine guns. Set out below is data relating to the MP5 A2 and MP5 A3 versions which differ only in that the former has a fixed butt stock whereas the MP5 A3 has a single metal strut stock which may be slid forward to give a considerable reduction to the overall length of the sub-machine gun. Three silenced variants with the general designation MP5 SD are described in the next entry; and the sixth weapon, the short-barrelled MP5 K is described in the following entry.

All versions are available with three-round burst control device.

DATA
MP5 A2, A3, A4 and A5
Cartridge: 9 mm × 19 Parabellum
Operation: delayed blowback, selective fire
Method of delay: rollers
Feed: 15- or 30-round box magazine
Weight, empty: 2.55 kg
Length: (fixed butt) 680 mm; (butt extended) 660 mm; (butt retracted) 490 mm
Barrel: 225 mm
Rifling: 6 grooves rh
Sights: (fore) fixed post; (rear) apertures for different eye relief adjustable for windage and elevation. A telescopic or night sight or an aiming projector may be fitted.
Sight radius: 340 mm
Muzzle velocity: 400 m/s
Rate of fire: (cyclic) 800 rds/min

Telescopic sight
The standard telescopic sight has × 4 magnification, has range settings for 15, 25, 50, 75 and 100 m and is adjustable for windage and elevation. A passive night sight may also be fitted.

Blank firing attachments
A firing attachment is available. It is a conventional device with a restricted gas flow. It is placed on the muzzle using the locking lugs and is retained by a catch. It is marked with a red ring to enable the firer to identify it as a training aid.

Manufacturer
Heckler and Koch GmbH, D-7238 Oberndorf-Neckar.
Status
Current. Production.
Service
West German police and border police. In use in USA, Switzerland, the Netherlands and many other police and military forces.

9 mm MP5 SD Heckler and Koch sub-machine gun

The MP5 SD is a silenced version of the MP5. Its mechanism is the same as that of the MP5 but the weapon differs in having a barrel in which 30 holes are drilled. The silencer on the barrel features two separate chambers, one of which is connected to the holes in the barrel and serves as an expansion chamber for the propulsive gases, thus reducing the gas pressure to slow down the acceleration of the projectile. The second chamber diverts the gases as they exit the muzzle, so muffling the exit report. The bullet leaves the muzzle at subsonic velocity, so that it does not generate a sonic shock wave in flight. The silencer requires no maintenance; only rinsing in an oil-free cleaning agent is prescribed.

There are six versions of the weapon. The MP5 SD1 has a receiver end cap and no butt stock; the SD2 has a fixed butt stock and the SD3 a retractable butt stock, the components being identical to those of the MP5 A2 and MP5 A3 respectively. The MP5SD4 resembles the SD1 but has a three-round burst facility in addition to single and automatic fire: the SD5 is the SD2 with three-round burst facility: and the SD6 is the SD3 with the three-round burst facility. It will be noted that the three latest models have a slightly changed contour of the pistol grip. Each may be used with iron sights, a telescopic sight, the Hensoldt Aiming-Point-Projector, or an image intensifier sight.

DATA
Cartridge: 9 mm × 19 Parabellum
Operation: delayed blowback, selective fire
Delay: rollers
Feed: 15- or 30-round curved box magazine
Weight, empty: (SD1) 2.9 kg; (SD2) 3.2 kg; (SD3) 3.5 kg
Length: (SD1) 550 mm; (SD2) 780 mm; (SD3) 610 or 780 mm
Barrel: 146 mm
Rifling: 6 grooves rh
Sights: as MP5
Sight radius: 340 mm
Muzzle velocity: 285 m/s
Rate of fire: (cyclic) 800 rds/min

Manufacturer
Heckler and Koch GmbH, D-7238 Oberndorf-Neckar.
Status
Current. Production.
Service
West German Special Forces and numerous military and police forces world-wide.

9 mm MP5 SD4 Heckler and Koch sub-machine gun with 15-round magazine and three-round burst facility

9 mm MP5 SD3 Heckler and Koch sub-machine gun

9 mm MP5 SD1 Heckler and Koch sub-machine gun

9 mm MP5 SD2 Heckler and Koch sub-machine gun

9 mm MP5K series Heckler and Koch sub-machine guns

Heckler and Koch introduced these weapons for use by special police and anti-terrorist squads. They are extra-short versions of the standard MP5 and are meant for carriage inside clothing, in the glove-pocket of a car, or any other limited space. They offer all the fire options of the MP5: the MP5K is fitted with adjustable iron sights, or a telescope if need be; the MP5K A1 has a smooth upper surface with a very small fore and rearsight so that there is little to catch in clothing or a holster as the gun is withdrawn. The MP5KA4 is similar to the MP5K but has a three-round burst facility as well as single and automatic fire. The MP5KA5 is similar to the A1, with the addition of the three-round burst facility. The prominent fore-hand grip gives the firer the best possible control for all types of fire. The butt stock is not fitted and both versions are meant to be fired from a hand-hold at all times, without any other support.

DATA
Calibre: 9 mm × 19 Parabellum
Operation: delayed blowback, selective fire
Delay: rollers
Feed: 15- or 30-round detachable box magazine
Weight: gun without magazine: 2 kg
15-round magazine, empty: 0.1 kg

30-round magazine, empty: 0.16 kg
30-round magazine, full: 0.52 kg
Length: 325 mm
Barrel: 115 mm
Rifling: 6 grooves rh
Sights: MP5K, MP5KA4: (fore) post; (rear) open rotary adjustable for windage and elevation, or × 4 telescopic
MP5K A1, MP5KA5: (foresight) fixed blade; (rearsight) fixed notch
Sight radius: MP5K, MP5KA4: 260 mm
MP5K A1, MP5KA5: 190 mm
Muzzle velocity: 375 m/s
Muzzle energy: 570 J
Rate of fire: (cyclic) 900 rds/min

Manufacturer
Heckler and Koch GmbH, D-7238 Oberndorf-Neckar.
Status
Current. Production.
Service
West German Special Forces and numerous military and police forces world-wide.

9 mm MP5K A5 Heckler and Koch sub-machine gun

9 mm MP5K Heckler and Koch sub-machine gun

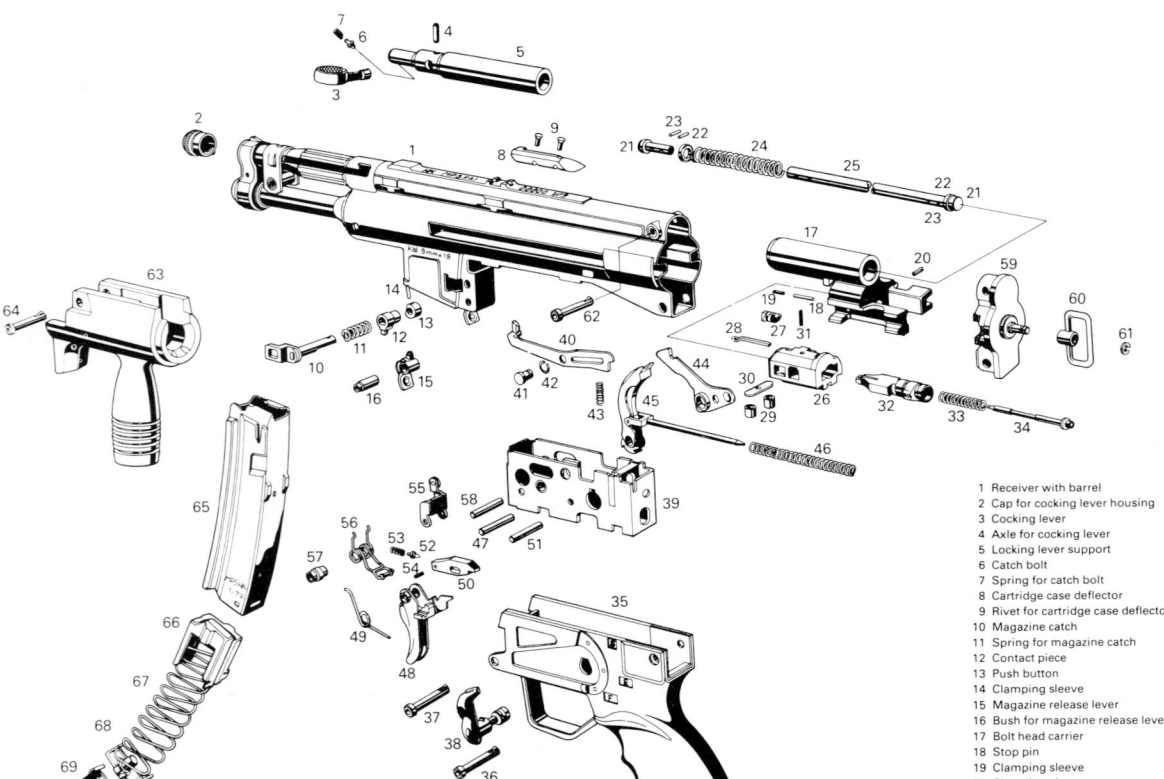

9 mm MP5K A1 Heckler and Koch sub-machine gun

1 Receiver with barrel
2 Cap for cocking lever housing
3 Cocking lever
4 Axle for cocking lever
5 Locking lever support
6 Catch bolt
7 Spring for catch bolt
8 Cartridge case deflector
9 Rivet for cartridge case deflector
10 Magazine catch
11 Spring for magazine catch
12 Contact piece
13 Push button
14 Clamping sleeve
15 Magazine release lever
16 Bush for magazine release lever
17 Bolt head carrier
18 Stop pin
19 Clamping sleeve
20 Clamping sleeve

21 Stop pin for recoil spring
22 Guide ring for recoil spring
23 Riveting pin
24 Recoil spring
25 Recoil spring guide rod
26 Bolt head
27 Extractor
28 Extractor spring
29 Locking rollers
30 Holder for locking rollers
31 Clamping sleeve
32 Locking piece
33 Firing pin spring
34 Firing pin
35 Pistol grip
36 Locking pin for receiver cap
37 Locking pin for pistol grip
38 Selector lever
39 Trigger housing
40 Ejector
41 Axle for ejector
42 Spring ring
43 Ejector spring
44 Release lever
45 Hammer with pressure shank
46 Compression spring for hammer
47 Axle for hammer
48 Trigger
49 Elbow spring for trigger
50 Sear
51 Axle for trigger
52 Trigger bolt
53 Spring for trigger bolt
54 Clamping sleeve
55 Catch
56 Spring for catch with roller
57 Distance sleeve
58 Axle for catch
59 Receiver cap
60 Clip with bush
61 Circlip
62 Locking pin for receiver
63 Handguard
64 Locking pin for handguard
65 Magazine
66 Follower
67 Follower spring
68 Spring floor plate
69 Magazine floor plate

9 mm MP2000 sub-machine gun

The MP2000 is something of a departure for Heckler and Koch since it is the first time that they have abandoned their well-tried roller-locked delayed blowback system and adopted a pure blowback design. The design of the MP 2000 is intended to reduce manufacturing and therefore sale costs, whilst providing a highly reliable and extremely versatile modular weapon system which can virtually be tailored to whatever role the user needs.

The gun is a fixed-barrel blowback, firing from a closed bolt. There is complete ambidextrous operation of stock, sling, selector/safety lever, bolt catch and magazine release. There is an automatic firing pin lock which prevents firing unless the trigger is deliberately pulled. The most unusual innovation is a chamber gas vent which is normally closed but which can be

opened when the gun is fitted with a sound suppressor and, in the absence of subsonic ammunition, has to fire standard 9 mm Parabellum cartridges. The open vent then bleeds a measured proportion of gas from the chamber, so reducing the muzzle velocity to a subsonic level. The sights are fully adjustable for subsonic or supersonic ammunition and are available with front and rear luminous inserts. The suppressor/barrel combination is simply interchanged for the standard barrel when required. Different magazines, a fore grip and a folding stock may be fitted or removed as desired, and as options it is possible to have the weapon fitted with a forward bolt lock for absolutely silent single shots, or with a three-round burst mechanism.

DATA
Cartridge: 9 × 19 mm Parabellum
Operation: blowback, selective fire
Feed: 30-round magazine
Length, standard: 565 mm (stock extended); 387 mm (stock closed)
Length, suppressed: 835 mm (stock extended); 657 mm (stock closed)
Weight: 2.78 kg (3.57 kg with suppressor)
Suppression level: 30 dB
Rate of fire: 880 rds/min

Manufacturer
Heckler and Koch GmbH, D-7238 Oberndorf-Neckar.
Status
Pre-production.
Service
Tests and evaluation in various countries.

Heckler & Koch MP 2000 sub-machine gun with suppressor

5.56 mm HK 53 Heckler and Koch sub-machine gun

The HK 53 sub-machine gun is a weapon chambered for the 5.56 mm × 45 cartridge which can be used either as a sub machine gun or as an assault rifle. The method of operation is similar to the other rifles and sub-machine guns made by Heckler and Koch and uses the system of rollers which delays the rearward movement of the bolt-head until the pressure has dropped sufficiently to allow the complete bolt unit to be blown back in safety.

In appearance it is very similar to the HK 33K rifle and it can be fitted with either a conventional plastic butt or a double strut telescoping butt stock. The HK 53 is the shortest of the 5.56 mm weapons. The length, with butt retracted, is only 56 cm whereas that of the HK 33K is 67 cm and the HK 33 A3 is 73 cm.

The HK 53 is capable of either single shots or full automatic, with a selector lever on the left of the receiver above the pistol grip. Optionally, it may be provided with a different grip with burst-fire control and an ambidextrous selector/safety lever. It is now available with a new type of flash suppressor which totally eliminates the muzzle flash which has previously been

experienced with this weapon. The new flash suppressor is seen in the accompanying illustration.

The HK 53 may be regarded as a scaled-down version of the G3 rifle, and many of the parts of the two weapons are interchangeable with each other and the HK 33. The 9 mm MP 5 sub-machine gun also has parts interchangeable with the G3 and HK 33 and it is extremely important that the bolts, (particularly the locking pieces), are not interchanged between weapons of different calibre and different recoil impulse.

DATA
Cartridge: 5.56 mm × 45
Operation: delayed blowback, selective fire
Method of delay: rollers
Feed: 25-round box magazine
Weight, empty: 3.05 kg
Length: (butt retracted) 563 mm; (butt extended) 755 mm
Barrel: 211 mm
Rifling: 6 grooves, rh, one turn in 178 mm or 305 mm as required
Sights: (fore) post; (rear) apertures for 200, 300, 400 m. V 100 m battle sight;
Sight radius: 390 mm
Muzzle velocity: 750 m/s
Rate of fire: (cyclic) 700 rds/min

Manufacturer
Heckler and Koch GmbH, D-7238 Oberndorf-Neckar.
Status
Production.
Service
Special military and police units in several countries.

5.56 mm HK 53 Heckler and Koch sub-machine gun with new flash suppressor

GREECE

9 mm EMP5 sub-machine gun

The Hellenic Arms Industry (E.B.O.) SA is currently manufacturing the Heckler and Koch MP5 as the EMP5, which has been issued to the Greek security forces and police. The weapon is also being offered for export. Details will be found in the West German section.

Manufacturer
Hellenic Arms Industry (E.B.O.) SA, 160 Kifissias Avenue, 11525 Athens.
Status
Current. Production.
Service
Greek security forces and police; exports.

9 mm Greek EMP5 sub-machine gun

HUNGARY

7.62mm 48M sub-machine gun

This is a Hungarian copy of the Soviet PPSh-41. It is well made with a good finish. Its physical characteristics and performance are identical to those of the Soviet weapon. It is believed to be now used only by police and border guards and may well be relegated to reserve forces.

The Hungarians also have a weapon, commonly classified as a sub-machine gun, which is based on the Soviet AKM assault rifle. Known as the AMD it uses rifle calibre ammunition (7.62 mm × 39) and is for that reason described in the rifle section.

7.62 mm 48M sub-machine gun

INDONESIA

9 mm P.M. Model VII sub-machine gun

An original sub-machine gun design by the Bandung Arsenal, Pabrik Sendjata Mesin, was produced in 1957. This was an orthodox blowback-operated sub-machine gun. The circular receiver was made from seamless steel tubing and the bolt was machined. The slotted-barrel jacket reached almost to the Beretta type compensator. The design owed some inspiration to the US M3 also the sliding wire butt stock. The magazine housing which is

9 mm Model VII sub-machine gun

also the forward grip was longer than is usual and had a distinctive sloping entry.

DATA
Cartridge: 9 mm × 19 Parabellum
Operation: blowback, automatic
Feed: 33-round box magazine
Weight: (unloaded) 3.29 kg; (loaded) 3.92 kg
Length: (stock extended) 840 mm; (stock retracted) 540 mm
Barrel: 274 mm (including compensator)
Muzzle velocity: 381 m/s
Rate of fire: (cyclic) 600 rds/min

Manufacturer
Pabrik Sendjata Dan Mesiu, Bandung.
Status
No longer manufactured. In limited use.

ISRAEL

9 mm Uzi sub-machine gun

During the Arab-Israeli War of 1948, immediately after the British Mandate in Palestine ended, the Israelis found themselves without a reliable sub-machine gun.

In 1949 Lieutenant (now Lieutenant-Colonel, Retired) Uziel Gal of the Israeli Army started work on a weapon of this type intended for use by all branches of the armed services of his country. He had previously studied the Czechoslovak pre-war sub-machine guns, particularly the 9 mm Models 23 and 25 and the designs produced for the Soviet 7.62 mm Type P cartridge which were the Models 24 and 26. These weapons and the post-war ZK 476 9 mm sub machine gun provided the foundation on which the Uzi was built and this weapon has retained many of the characteristics of its Czechoslovak parents. The bolt design shows this origin clearly.

Operation
The Uzi is a blowback-operated gun using the system of advanced primer ignition in which the round is fired while the bolt is still travelling forward. This produces a reduced impulse to the bolt and as a result this component can be designed to weigh less than half the amount that would be required for a static firing breech-block.

For any given weapon the minimum overall length is the sum of the barrel length plus the length of the breech-block behind the chamber plus the compressed return spring length. The Uzi is 445 mm long from the muzzle to the rear of the breech casing and has a very pronounced advantage in that in its short length it achieves a 260 mm barrel. This is managed by wrapping the bolt around the chamber and putting the breech face not on the front face of the bolt but 95 mm further back. Thus at the moment of firing the bolt completely surrounds the rear end of the barrel except for a cut-out section on the right-hand side which allows ejection of the fired case. A further advantage of this design is that if a round is fired early or a hang-fire occurs the soldier is protected from the effects of the bursting of the unsupported case by the wrap-round bolt.

The magazine is inserted into the pistol grip. This has the advantage of making magazine changing very easy in the dark and also giving positive support to the magazine over a greater length than is usual. The gun stops firing with the bolt to the rear and when the trigger is operated the bolt goes forward, collects a round from the top of the 25-round magazine and feeds it over the bullet guide into the chamber. The cartridge is held in the magazine at an angle with the nose slightly elevated so that it does not line up with the fixed firing pin on the breech face until the cartridge case enters the chamber. At any time thereafter it can be fired when the friction force between the case and the chamber wall produces the required resistance.

The change lever provides the three positions of Automatic (A), Single Shot (R) and Safe (S). Also there is a grip safety which must be fully depressed before the gun can be either cocked or fired. In addition extra safety is

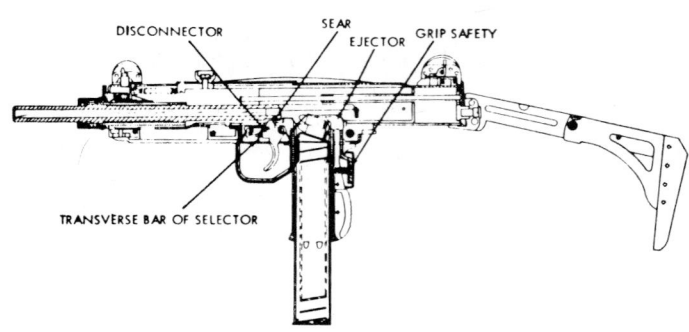

9 mm Uzi sub-machine gun

9 mm Uzi sub-machine gun stripped

Two versions of 9 mm Uzi sub-machine gun: with wooden stock and folding metal stock, both manufactured by IMI

Folding stock version of Uzi in use by West German soldier of ACE Mobile Force

provided by fitting a ratchet in the cocking handle slide to prevent accidental discharge if the hand cocking the gun should slip off the handle, after the breech-block has come back far enough to pass behind a round in the magazine. After the cocking handle has come back 475 mm, it cannot go forward again until it has been withdrawn fully to the rear which is a distance of 80 mm. Until the wire clip is pushed fully forward, that is the cocking handle is fully to the rear, the catch cannot go forward and the bolt is held up.

The method of operation of the grip safety is as follows. The projecting lug on the right-hand side of the trigger mechanism, alongside the disconnector, fits under the sear and prevents it from moving down. When the grip safety is operated the bar goes forward and the lug moving with it clears the way for the sear to drop.

Stripping
Remove the magazine, cock the action and check that the chamber is empty, return the bolt, under control, to the forward position.
Depress the catch in front of the rearsight and lift the top cover plate right off. Slide the bolt back 65 mm and lift out together with the return spring, rod and end plate.

Rotate the barrel nut off and withdraw the barrel forward.
Push out the transverse pin through the trigger group and drop out the trigger assembly.

DATA
Cartridge: 9 mm × 19
Operation: blowback, selective fire
Feed: 25- and 32-round box magazines
Weight, empty: (metal stock) 3.7 kg; (wood stock) 3.8 kg
Length: (wood butt, or metal extended) 650 mm; (retracted) 470 mm
Barrel: 260 mm
Rifling: 4 grooves, rh, one turn in 254 mm
Sights: (fore) post; (rear) flip aperture 100 and 200 m
Sight radius: 309 mm
Muzzle velocity: 400 m/s
Rate of fire: (cyclic) 600 rds/min

Manufacturer
Israel Military Industries, Ramat Ha Sharon 47100.
Status
Current. Production.
Service
Israeli forces, also Belgium, West Germany, Iran, Ireland, the Netherlands, Thailand, Venezuela and others.

9 mm Mini-Uzi sub-machine gun

Israel Military Industries has produced a new smaller version, designated the Mini-Uzi, of the 9 mm Uzi sub-machine gun described above. In operation it exactly resembles its larger 'parent', differing only in size, weight and firing characteristics as set out below. It will accept a 20-round magazine for its 9 mm Parabellum pistol ammunition, as well as 25- and 32-round magazines.

Since it can be easily concealed under ordinary clothing, and carried in the minimum of space in vehicles, the Mini-Uzi is particularly intended for security and law enforcement personnel and for use in commando operations. It can be fired full- or semi-automatic from the hip or, with stock extended, from the shoulder, and is said to maintain the high standards of reliability and accuracy set by the Uzi.

DATA
Cartridge: 9 mm Parabellum
Operation: blowback, selective fire

Feed: 20-, 25- and 32-round box magazines
Weight, empty: 2.70 kg
Length: (butt extended) 600 mm; (butt retracted) 360 mm
Barrel: 197 mm
Rifling: 4 grooves, rh, one turn in 254 mm
Sights: (fore) post; (rear) flip aperture 50 and 150 m;
Sight radius: 233 mm
Muzzle velocity: 352 m/s
Rate of fire: 950 rds/min

Manufacturer
Israel Military Industries, Ramat Ha Sharon 47100.
Status
Current. Production.

9 mm Mini-Uzi sub-machine gun field-stripped and with stock extended

9 mm Mini-Uzi sub-machine gun with stock folded

9 mm Micro-Uzi sub-machine gun

The Micro-Uzi sub-machine gun is the third and latest member of the Uzi family. In effect, it is the Uzi sub-machine gun reduced to its smallest practical size, and is marginally larger than the Uzi pistol. The operation is the same as

9 mm Micro-Uzi sub-machine gun with stock extended

that of the larger weapons, but the bolt has a tungsten insert which increases the mass and thus keeps the rate of fire down to a reasonably practical figure. The stock lies alongside the left of the receiver when folded. The Micro-Uzi is also available in .45ACP calibre, using a 16-shot magazine.

DATA
Cartridge: 9 mm × 19; also .45ACP
Operation: blowback, selective fire
Feed: 20-round box magazine
Weight, empty: 1.95 kg
Length: (butt extended) 460 mm; (butt folded) 250 mm
Barrel length: 117 mm
Rifling: 4 grooves rh
Rate of fire (cyclic): 1250 rds/min
Muzzle velocity: 350 m/s (9 mm): 240 m/s (.45ACP)

Manufacturer
Israel Military Industries, Ramat Ha Sharon 47100.
Status
Current. Production.

ITALY

9 mm Model 38/42 Beretta sub-machine gun

The Model 38/42 was issued to the Italian Army. It used stampings for the receiver and magazine housing. The bolt had a fixed firing pin and the return spring of the Model 1938A. The guide rod of the return spring projected to the rear of the end closure cap and provided a ready identification between this model and the 1938/44 which followed it.

The barrel jacket was removed and the barrel shortened. The prototype had a thick barrel, heavily fluted. The first production version had a fluted barrel but this was later supplanted by a barrel with a smooth exterior contour. This was often called 38/43. The compensator had two cuts above the muzzle. The cocking handle had an attached dust cover. After cocking was completed the

Production model of 9 mm Model 38/42 Beretta sub-machine gun with fluted barrel

handle was pushed forward and the cocking way sealed. The handle did not reciprocate with the bolt.

DATA
Cartridge: 9 mm Parabellum
Operation: blowback, selective fire
Feed: 20- and 40-round box magazine
Weight, empty: 3.27 kg
Length: 800 mm
Barrel: 213 mm
Rifling: 6 grooves rh
Sights: (fore) blade; (rear) flip, notch 100 and 200 m
Muzzle velocity: 381 m/s
Rate of fire: (cyclic) 550 rds/min

Manufacturer
Pietro Beretta SpA, Gardone Valtrompia, Brescia.
Status
Obsolete. Not in production.
Service
Not in military service but may be found anywhere. Many thousands were produced.

9 mm Model 38/44 Beretta sub-machine gun

This was a simplified version of the 38/42. The bolt length was reduced from 180 to 150 mm and the narrow diameter return spring used on all of Marengoni's designs thus far was replaced by a large spring and the end closure cap of the sub-machine gun was plain. The 38/44 went into production in February 1944 and production reached 3000 a month in December. These were sold to Syria, Iraq, Pakistan and Costa Rica.

Manufacturer
Pietro Beretta SpA, Gardone Valtrompia, Brescia.
Status
No longer in production.
Service
Italian and Nigerian police forces.

Smooth-barrelled version known as 9 mm Model 38/43 Beretta sub-machine gun distinguished from Model 38/44 which followed by guide rod recess in end cap

9 mm Model 38/49 (Model 4) Beretta sub-machine gun

The 38/49 was a modified version of the Model 38/44 with a cross-bolt safety above the forward of the double trigger. It locked the bolt in either the forward or cocked position. If the cross-bolt safety was in the midway position the sear was locked. In 1956 it was renamed the Model 4 and is still so designated as a standard sub-machine gun in the Italian Army. It was ordered by the West German Border Police in 1951 and designated MP1. With a folding bayonet attached it was sold to Egypt.

The Model 38/49 has been a very successful sub-machine gun both mechanically and financially. It showed that Marengoni's original concept of the Model 1938A was sound and could be modernised as users increased their requirements. In addition to the sales mentioned above it was sold to Costa Rica, Dominican Republic, Indonesia, Morocco, Thailand, Tunisia and North Yemen.

9 mm Model 38/49 (Model 4) Beretta sub-machine gun

Manufacturer
Pietro Beretta SpA, Gardone Valtrompia, Brescia.
Status
Obsolescent. Not in production.
Service
Italian Army and Navy. For others see text.

9 mm Model 12 Beretta sub-machine gun

A series of Beretta sub-machine gun developments in the mid-1950s culminated in the Model 12 which was first produced in 1958 and went into series production in 1959. It was adopted by the Italian Army as the Model 12 sub-machine gun. The Model 12 has been sold to Brazil, Gabon, Libya, Nigeria, Saudi Arabia and Venezuela. Heavy sheet metal stampings spot-welded together form the receiver and magazine housing. There are longitudinal grooves extending the full length of the receiver to ensure efficient functioning in conditions of dust, sand and snow. The receiver, forward pistol grip, magazine housing, trigger housing and pistol grip are all one unit. The breech-block is of the wrap-round type and envelops the barrel before firing. The fixed firing pin is well back and, of the total barrel length of 200 mm, 150 mm lies inside the breech-block at the moment of firing.

The principle of the wrap-round bolt does help in keeping the vibrations to a minimum. The gun is very steady at automatic fire and there seems little tendency for the muzzle to rise at full automatic.

The weapon has two safety systems. There is a grip safety in the front of the pistol grip, below the trigger, which locks the bolt, when released, in either the forward or cocked position. It must be held in before the action can be cocked. There is also a push-button safety just above the pistol grip and this locks the grip safety until pushed to the right. The selector lever is a push-through type. The weapon normally has a metal stock which folds laterally to the right; it can be fitted with a quickly detachable wooden butt.

DATA
Cartridge: 9 mm Parabellum
Operation: blowback, selective fire
Feed: detachable box magazine holding 20, 32 or 40 rounds
Weight unloaded: (folding metal stock) 3 kg; (wooden stock) 3.4 kg
Length: (butt extended) 645 mm; (butt folded) 418 mm; (wooden butt) 660 mm
Barrel: 200 mm
Rifling: 6 grooves rh
Sights: (fore) blade; (rear) flip aperture 150 and 250 m
Muzzle velocity: 381 m/s
Rate of fire: (cyclic) 550 rds/min

Manufacturer
Pietro Beretta SpA, Gardone Valtrompia, Brescia.
Status
Current. No longer manufactured in Italy.
Service
Italian Army, also Brazil, Gabon, Libya, Nigeria, Saudi Arabia and Venezuela. Made under licence in Indonesia and Brazil.

9 mm Model 12 Beretta sub-machine gun with stock folded

9 mm Model 12 Beretta sub-machine gun with stock extended

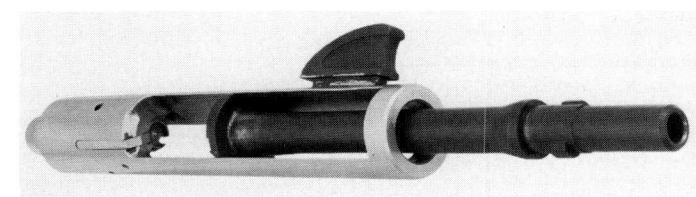

9 mm Model 12 Beretta sub-machine gun bolt wraps round barrel

9 mm Model 12S Beretta sub-machine gun

Generally similar to the Model 12, this is the latest in the sequence of Beretta sub-machine gun. The principal differences from the earlier model are a new design of the manual safety and fire selector and modifications to the sights: in addition the rear-cap catch has been modified, there is a new butt plate and an epoxy resin finish, resistant to corrosion and wear, has been introduced.

The manual safety and fire selector have been incorporated in a single-lever mechanism which can be operated by the right thumb without removing the hand from the grip. The three positions are marked (clockwise) 'S' (safe), 'I' (semi-automatic) and 'R' (automatic fire). When the manual safety is applied,

with the lever at 'S', both the trigger mechanism and the grip safety are blocked.

The front sight of the Model 12S is adjustable for elevation and windage and both front and rearsight supports have been strengthened.

The rear-cap catch has been moved to the top of the receiver, both to facilitate fastening and unfastening and to enable the user to see at a glance whether or not the catch is correctly fastened.

The butt plate of the metal stock has been modified by the addition of a catch which gives positive locking in the open and closed position. As with the Model 12 a detachable wooden stock is available.

9 mm Model 12S Beretta sub-machine gun with metal stock folded showing safety selector lever and modified rear cap catch

9 mm Model 12S Beretta sub-machine gun with metal stock extended

9 mm Model 12S Beretta sub-machine gun with detachable wooden stock

9 mm Model 12S Beretta sub-machine gun stripped

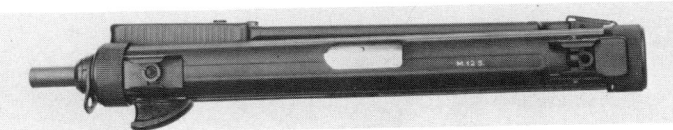

9 mm Model 12S Beretta sub-machine gun top view showing modified catch and foresight arrangement

DATA
Cartridge: 9 mm × 19 Parabellum
Operation: blowback, selective fire
Feed: 32-round detachable box magazine with options of 20 or 40
Weight, without magazine: (folding metal stock) 3.2 kg; (wooden stock) 3.6 kg
Length: (butt folded) 418 mm; (butt extended) 660 mm; (wooden butt) 660 mm
Barrel: 200 mm
Rifling: 6 grooves, rh, one turn in 250 mm
Sights: (fore) post, adjustable; (rear) flip aperture
Muzzle velocity: 380/430 m/s
Rate of fire: (cyclic) 500/550 rds/min

Manufacturer
Pietro Beretta SpA, Gardone Valtrompia, Brescia; also under licence by FN Herstal, Belgium.
Status
Current. Production.
Service
Italian and Tunisian armed forces and some other forces.

---

9 mm Model LF57 Franchi sub-machine gun

This gun was produced by the firm of Luigi Franchi of Brescia. The prototype, known as the Model LF56 appeared in 1956 and the production version, the Model LF57, was produced in the next year. The bolt is shaped so as to have a long arm overhanging the barrel and the firing pin and extractor, together with the feed horn, on the lower section, thus raising the centre of gravity of the gun up to the thrust line of the barrel and preventing muzzle rise at full automatic. The grip safety is at the front of the pistol grip. No other applied safety is fitted. The entire gun, except barrel and bolt, is produced from stampings of heavy sheet steel.

The Model LF57 was the first sub-machine gun manufactured in every respect by the Luigi Franchi Company and several thousands were taken by the Italian Army in 1962 and by some other countries.

The standard barrel is 205 mm long but in 1962 a semi-automatic version with a 406 mm barrel was made and exported to the USA as the Police Model 1962.

DATA
Cartridges: 9 mm × 19 Parabellum
Operation: blowback, selective fire
Feed: 10-, 20-, or 30-round detachable magazines
Weight: (overall) 3.3 kg; (loaded 40-round magazine) 0.73 kg

Length: (butt extended) 680 mm; (butt folded) 420 mm
Barrel: 205 mm
Rifling: 6 grooves rh. 1 turn in 250 mm
Sights: (fore) blade; (rear) notch, adjustable for windage
Sight radius: 292 mm

9 mm Model LF57 Franchi sub-machine gun

9 mm Model LF57 Franchi sub-machine gun field-stripped
(1) body **(2)** folding butt **(3)** bolt **(4)** barrel **(5)** return spring **(6)** return-spring guide **(7)** magazine **(8)** cocking handle spring **(9)** cocking handle

Muzzle velocity: 400 m/s
Rate of fire: 450-470 rds/min

Manufacturer
Luigi Franchi SpA, Via del Serpente 12, I-25020 Fornaci (Brescia).

Status
Current. Production.
Service
Italian Military Police and some African countries.

9 mm Socimi Type 821 sub-machine gun

The Socimi Type 821 is an automatic weapon, firing either single shots or full automatic, operating on advanced primer ignition by means of the blowback principle.

Because of its lightness, compactness, balance, simplicity and reliability, the Type 821 is suitable for combat in confined spaces and urban areas, use by mechanized and armoured troops, use by special forces and use by paramilitary security forces.

Optimum results are obtained within a range of 150 – 200 m, and the balance of the weapon is such that it can be fired one-handed with complete control.

DATA
Cartridge: 9 mm Parabellum
Operation: blowback, selective fire
Weight, empty: 2.45 kg

Weight, with full magazine: 2.65 kg
Feed: 32-round box magazine
Length: (butt folded) 400 mm; (butt extended) 600 mm
Barrel: 200 mm
Rifling: 6 grooves rh. 1 turn in 250 mm
Sights: (fore) adjustable post; (rear) flip aperture, 100 and 200 m
Sight radius: 288 mm
Rate of fire: (cyclic) 550/600 rds/min
Muzzle velocity: ca 380 m/s

Manufacturer
Socimi (Societa Constuzioni Industriali Milano SpA), Via Varesina 115, I-20156 Milano.
Status
Production.

9 mm Socimi Type 821

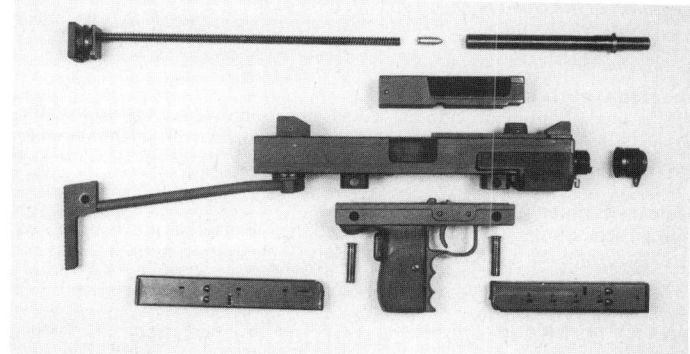
9 mm Type 821 dismantled

9 mm Spectre M-4 sub-machine gun

This weapon was announced at the 1983 AUSA Convention in Washington and has several features which support its claim to being a revolutionary design. In appearance it is a conventional enough weapon of its type, with the receiver and barrel shroud of pressed steel and with a folding steel butt stock which folds up and forward to lie along the top of the receiver when not in use. The first unusual feature is the patented four-column magazine which allows a capacity of 50 rounds in the length usually associated with 30 rounds. There is also a four-column 30-shot magazine which is only 160 mm long. The mechanism is blowback, but fires from a closed bolt and uses an independent hammer unit combined with a unique double-action trigger system. To fire, the magazine is inserted in the usual way and the cocking handle pulled back and released; this allows the bolt to close, chambering a cartridge. The hammer unit remains at the rear of the receiver; but depressing a de-cocking lever allows the hammer to move forward under control to stop a short distance behind the bolt. The weapon is now safe to carry, but when necessary to fire, a pressure on the trigger will retract the hammer from its rest position and allow it to go forward, impelled by its own spring, to strike the firing pin in the bolt and thus fire the chambered round. Thereafter the action is normal, the hammer being automatically released to go forward and fire each round after the bolt has closed.

A closed bolt system predisposes the weapon to heating of the barrel, and to counter this a forced draught system is employed; this is controlled by the movement of the bolt and ensures that air passes through and around the barrel during firing, thus keeping the barrel cool. It is claimed that the 'sinusoidal' rifling employed, which reduces bullet friction, also assists in keeping the barrel temperature down.

There is no applied safety: accidental firing is impossible due to a safety lock preventing the hammer striking the firing pin unless the trigger is pressed. The double-action trigger mechanism allows the user to open fire instantly without having to perform any operation on the weapon.

It can be appreciated from this description that the makers' claim, that the Spectre initiates a totally new concept in sub machine guns, has considerable validity. The weapon is based on an absolutely instinctive and automatic usage; the user can open fire instantly, without having to consider the safety condition or perform any manual action other than pulling the trigger. This is the most important feature of the weapon and makes it ideal for unconventional warfare, counter-terrorism and similar operations.

Variant models
The SITES company is currently manufacturing two variant models for the civil market, both derived from the Spectre sub-machine gun.

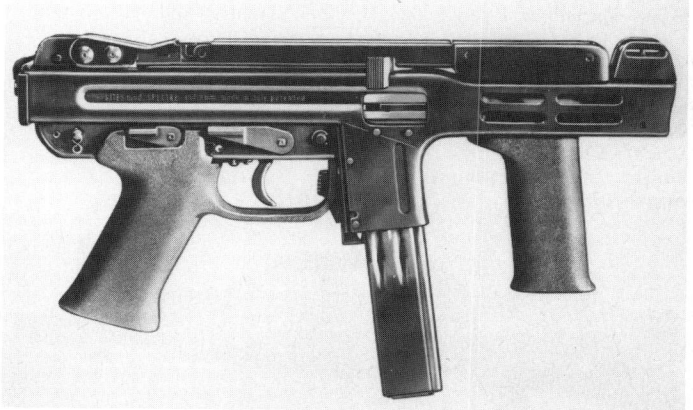
9 mm Spectre M-4 sub-machine gun

9 mm Spectre M-4 sub-machine gun with butt folded

9 mm Spectre M-4 sub-machine gun with stock extended

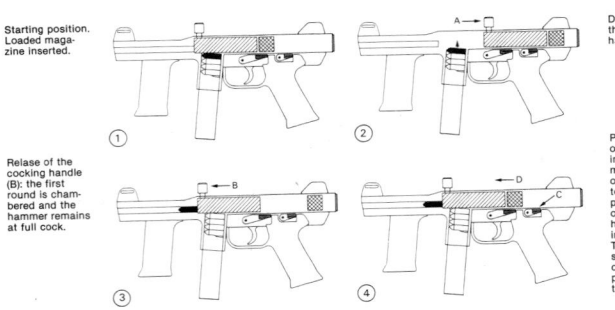

Starting position. Loaded magazine inserted.

Drawing back of the cocking handle (A).

Relase of the cocking handle (B): the first round is chambered and the hammer remains at full cock.

Pressing down on the de-cocking lever (C) permits the return of the hammer to its resting place (D), without having a round in the chamber. The weapon is set on «safe», but can fire by simply pressing the trigger.

Operation of the Spectre M-4 sub-machine gun

Spectre C carbine

This is a semi-automatic carbine; it has all the innovative features of the sub-machine gun but fires only single shots and has a barrel 420 mm long, to meet legal requirements.

Spectre P pistol

This is a semi-automatic pistol, also with all the sub machine gun features. As with the carbine the mechanism permits only single shots to be fired, and there

9 mm Spectre P pistol

is no stock and no fore-grip. It accepts the standard sub-machine gun magazines.

DATA
Cartridge: 9 mm Parabellum
Operation: blowback, closed breech, selective fire, double-action trigger mechanism, de-cocking lever
Weight, empty: 2.90 kg
Length: (butt folded) 350 mm; (butt extended) 580 mm
Barrel length: 130 mm
Feed: 30- or 50-round four-row box magazines
Rate of fire: (cyclic) 850 rds/min
Sights: (fore) post adjustable; (rear) U-notch
Muzzle velocity: ca 400 m/s

Manufacturer
SITES SpA, Via Magenta 36, 10128 Torino.
Status
Current. Production.

9 MM AGM-1 semi-automatic carbine

The AGM-1 is a semi-automatic weapon of bullpup design. It is configured for the 9 mm Parabellum cartridge as standard, but there are conversion kits which, by changing barrel, bolt, recoil springs and magazine, allow it to be changed to .45ACP or .22 LR chambering. The changeover takes about 10 minutes and can be done without the need for special tools or training.

The weapon fires from a closed bolt, and has a set trigger and a heavy target-grade barrel, giving the weapon a degree of accuracy above average for its class. The bullpup layout makes it extremely compact and well balanced.

DATA
Cartridge: 9 × 19mm Parabellum, or .45ACP or .22LR
Operation: blowback, semi-automatic
Feed: box magazines; 9 mm 13 or 20 rounds; .45ACP 6 or 10 rounds; .22LR 15 or 20 rounds
Weight: 9 mm version empty; 3.0 kg
Length: 670 mm

Barrel: 410 mm
Sights: (rear) two position flip aperture, adjustable for windage and elevation; (fore) post, adjustable for elevation

Manufacturer
ALGIMEC Srl, Via Melzi d'Eril 21, I-20154 Milan.
Status
Current. Production.

9 mm AGM-1 semi-automatic carbine

9 mm AGM-1 carbine field-stripped

MEXICO

9 mm Model HM-3 sub-machine gun

This sub-machine gun is manufactured by Productos Mendoza SA in Mexico. It is a lightweight weapon, of reduced overall length achieved by largely extending the wrap-round bolt forward around the barrel. A grip safety is provided to prevent accidental discharge. The stock is designed in a manner that makes folding and unfolding easy and provides a handy foregrip when the stock is folded. The selector lever is on the right-hand side of the weapon so that it can be operated by the right hand without releasing the weapon with either hand; similarly the stock can be folded or unfolded while gripping the weapon with both hands.

DATA
Cartridge: 9 × 19 mm Parabellum
Operation: blowback, selective fire
Feed: 32-round box magazine
Weight of weapon without magazine: 2.98 kg
Magazine: (empty) 0.275 kg; (full) 0.655 kg
Length: 635 mm stock open; 395 mm stock folded
Barrel: 255 mm
Rate of fire: approx 600 rds/min

Manufacturer
Productos Mendoza SA, Mexico DF.
Status
Current. Available.
Service
Mexican Forces.

9 mm Model HM-3 sub-machine gun with stock folded

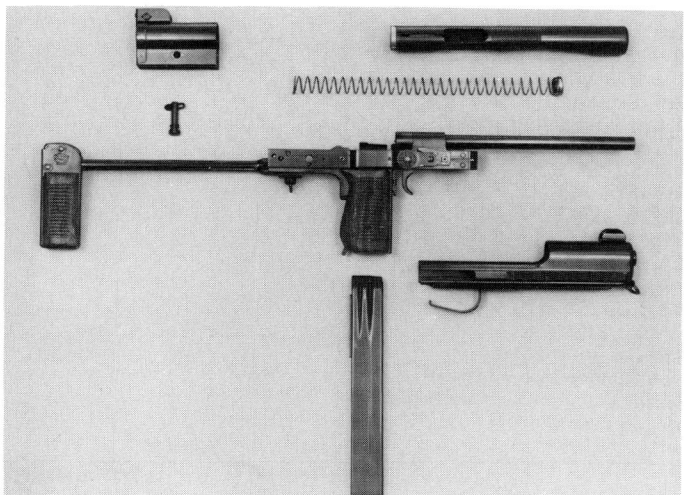

9 mm Model HM-3 sub-machine gun (stock extended) with magazines, accessories and magazine/accessory pack

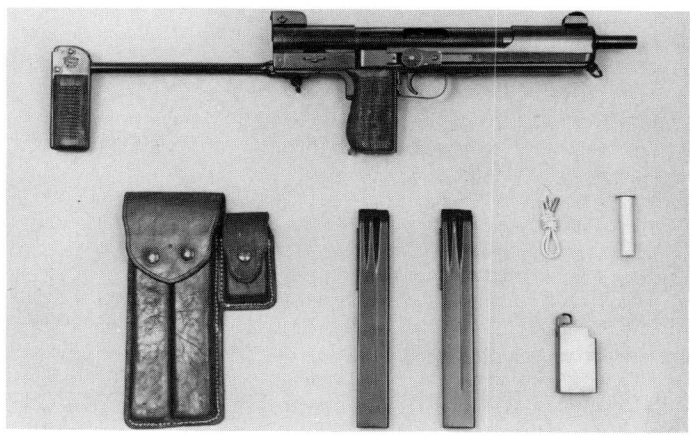

9 mm Model HM-3 sub-machine gun stripped

PERU

9 mm MGP-79A sub-machine gun

This weapon has been designed and manufactured in Peru and is a simple and robust blowback design. The butt is hinged at the rear of the receiver and folds sideways to lie along the right side of the weapon, allowing the shoulder pad to be gripped, together with the magazine, by the non-firing hand. The applied safety and fire selector switches are separated so that they can be operated by both hands. There is a perforated barrel jacket, allowing a different grip to be taken when deliberate aim is required, and the barrel and jacket can be replaced with a silencer assembly for special operations.

DATA
Cartridge: 9 × 19 mm Parabellum
Operation: blowback, selective fire
Feed: 20- or 32-round magazines
Weight, empty: 3.085 kg
Length overall: (butt extended) 809 mm; (butt folded) 544 mm
Barrel length: 237 mm
Rifling: microgroove, 12 grooves, rh
Sight radius: 258.5 mm
Sights: fore: adjustable blade; rear: two-position notch, 100 m and 200 m
Rate of fire: 600 – 850 rds/min
Muzzle velocity: 410 m/s

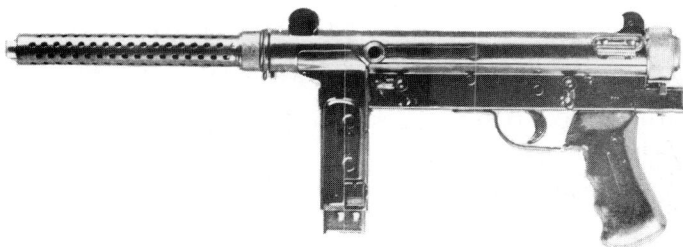

9 mm MGP-79A sub-machine gun

Manufacturer
SIMA-CEFAR, Av. Contralmirante Mora 1102, Base Naval, Callao.
Status
Current. Production.
Service
Peruvian armed forces.

9 mm MGP-87 sub-machine gun

This is built to the same basic design as the MGP-79A (above) but is shorter by virtue of a shorter barrel and shorter folding butt. There is no barrel jacket, and the cocking handle is turned up into the vertical position and made more prominent so that it can be more rapidly grasped and operated. The design is particularly intended for use in special operations and for counter-insurgency troops where rapid action is essential.

DATA
Cartridge: 9 × 19 mm Parabellum
Operation: blowback, selective fire
Feed: 20- or 32-round magazines
Weight, empty: 2.895 kg
Length overall: (butt extended) 702 mm; (butt folded) 500 mm

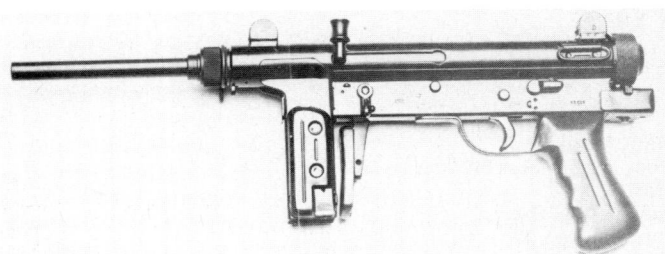

9 mm MGP-87 sub-machine gun

Barrel length: 194 mm
Rifling: microgroove, 12 grooves, rh
Sight radius: 258.5 mm
Sights: fore: adjustable blade; rear: fixed aperture 100 m
Rate of fire: 600 – 850 rds/min
Muzzle velocity: 410 m/s

Manufacturer
SIMA-CEFAR, Av. Contralmirante Mora 1102, Base Naval, Callao.
Status
Current. Production.
Service
Peruvian armed forces.

9 mm MGP-84 Mini sub-machine gun

This is a very small sub-machine gun designed for use by special forces and security guards who require compact firepower. It is a blowback weapon, the mechanism being the same as the larger weapons described above, but there is no exposed barrel and the magazine housing is incorporated into the pistol grip. There is a folding butt, hinged at the rear of the receiver and folding to the right side so that when folded the shoulder-piece can be utilised as a front grip. The safety and fire selector are incorporated into a single switch, located at the front end of the trigger unit in a position to be operated by the non-firing hand. The weapon is well-balanced and can be fired single-handed if necessary. The bolt is of the wrap-around type, allowing the maximum length of barrel, and, like the other SIMA-CEFAR designs, the magazine is compatible with the Uzi design so that Uzi magazines can be used interchangeably.

DATA
Cartridge: 9 × 19 mm Parabellum
Operation: blowback, selective fire
Feed: 20- or 32-round magazines
Weight, empty: 2.310 kg
Length overall: (butt extended) 503 mm; (butt folded) 284 mm
Barrel length: 166 mm
Rifling: microgroove, 12 grooves, rh
Sight radius: 232.5 mm
Sights: fore: adjustable blade; rear: two-position notch, 100 m and 200 m
Rate of fire: 650 – 700 rds/min
Muzzle velocity: 410 m/s

Manufacturer
SIMA-CEFAR, Av. Contralmirante Mora 1102, Base Naval, Callao.
Status
Current. Production.
Service
Peruvian armed forces.

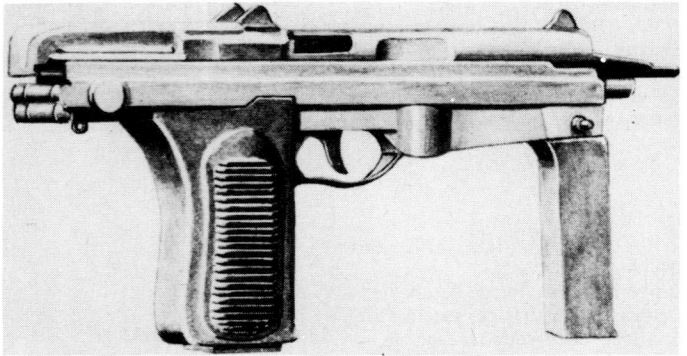

9 mm MGP-84 Mini sub-machine gun

POLAND

9 mm Wz63 (PM-63) machine pistol

This weapon has a combination of the characteristics of the self-loading pistol and the fully automatic sub-machine gun. It has the advantage that single shots can be fired using only one hand. If fully automatic fire is needed the shoulder stock can be pulled out and the fore-end dropped to provide a steady hold, which is necessary with the 9 mm × 18 cartridge.

In some ways the pistol resembles the Czechoslovak Skorpion (which see) but it has a few drawbacks which make it rather less attractive. It is difficult and expensive to manufacture the complex one-piece slide and the sights move during firing, so making it almost impossible to correct the fall of shot during a burst.

For a weapon which is intended to be fired one-handed, the use of an open-bolt system is questionable. Coupled to that is the balance and length of the whole weapon and it is interesting to find that the handbook only illustrates the two-handed hold.

Operation
The magazine catch is at the heel of the pistol grip and the magazine will hold 25 or 40 cartridges. Since the gun fires from the open-breech position, the slide must be pulled back to cock the action. Alternatively, the compensator can be placed against a vertical surface and the weapon pushed forward.

When the trigger is pulled the slide is released and driven forward, feeding a round from the magazine into the chamber. As soon as the cartridge is lined up with the chamber the extractor grips the cannelure and the gun fires while the slide is still going forward. The firing impulse halts the slide and drives it back against the return spring. The extractor grips the empty case until the ejector pushes it through the ejection port in the right of the slide. The slide continues to the rear and the return spring, under the barrel, is fully compressed. The slide rides over a retarder lever which snaps up and holds the slide to the rear. The retarder, an inertia pellet in the rear of the slide, continues rearward under its own momentum and compresses its own spring. When the spring is fully compressed it throws the retarder forward and this pushes the retarder lever down out of engagement with the slide and, provided the trigger is still depressed and ammunition remains in the magazine, the slide goes forward to repeat the cycle. The retarder keeps the cyclic rate of fire down to 600 rds/min from a natural frequency of about 840 rds/min.

Trigger and firing mechanism
The sear is forward of the trigger and is a long centrally-pivoted lever. The front end rises to engage a notch in the underside of the slide to hold it to the rear. The rear end is in contact with the disconnector mounted on the trigger. When a light pressure is exerted on the trigger the disconnector lifts the rear of the sear and the front end drops out of the slide notch. When the slide returns after the round is fired it impinges on the disconnector which frees the sear and the front end of the sear rises to hold up the slide. To fire another round the trigger must be released to allow the disconnector to move back under the tail of the sear.

When the trigger is pulled right back automatic fire results. The trigger revolves sufficiently to carry the disconnector clear of the recoiling slide and the gun continues firing at full automatic until the trigger is released.

After the ammunition is expended the back of the magazine follower lifts the slide stop up to intercept the slide and hold it to the rear.

The safety catch is on the left side of the receiver above the pistol grip. When it is rotated to the horizontal position the shaft rotates and a cam forces the slide stop up to lock the slide. The safety is only necessary when the slide is to the rear and cannot be applied at any other time.

9 mm Wz63 machine pistol

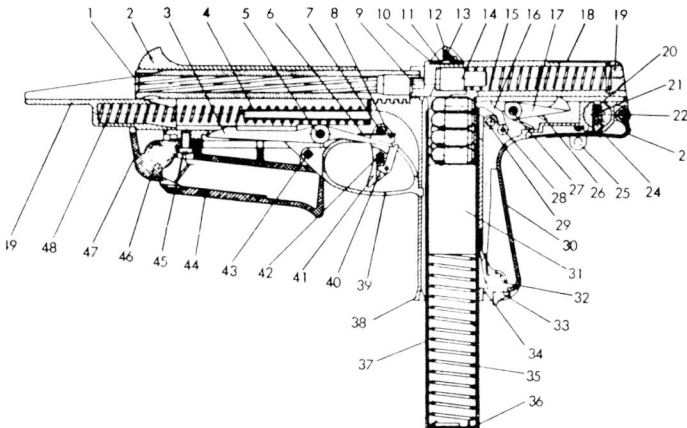

Sectioned view of the 9 mm Wz63 pistol
(1) *barrel* **(2)** *front sight* **(3)** *trigger lever* **(4)** *recoil spring guide* **(5)** *trigger lever (axis) pin* **(6)** *trigger spring* **(7)** *trigger pin* **(8)** *trigger (axis) pin* **(9)** *firing pin* **(10)** *sight-leaf* **(11)** *sight spring* **(12)** *rear sight* **(13)** *rear sight pin* **(14)** *retarder* **(15)** *retarder spring* **(16)** *slide stop* **(17)** *retarder lever* **(18)** *bearing latch* **(19)** *bearing* **(20)** *grip mount* **(21)** *butt-latch spring* **(22)** *butt-plate axis spring* **(23)** *butt plate* **(24)** *butt latch* **(25)** *lanyard loop* **(26)** *retarder lever spring* **(27)** *retarder lever (axis) pin* **(28)** *slide-stop (axis) pin* **(29)** *safety lock* **(30)** *stock* **(31)** *magazine follower* **(32)** *magazine-catch (axis) pin* **(33)** *magazine catch* **(34)** *magazine catch spring* **(35)** *magazine spring* **(36)** *magazine cover* **(37)** *magazine body* **(38)** *pistol grip (handle)* **(39)** *trigger-guard* **(40)** *trigger catch lever* **(41)** *trigger catch* **(42)** *trigger catch lever spring* **(43)** *back screw of the frame* **(44)** *grip (handle)* **(45)** *front screw of the frame* **(46)** *grip (handle) catch* **(47)** *frame* **(48)** *recoil spring* **(49)** *slide (compensator)*

Stripping

Remove the magazine, pull the slide back and check that the chamber and feedway are free. Leave the slide in the cocked position.

Grip the muzzle around the serrations and rotate the barrel anti-clockwise out of the receiver. Grip the slide, press the trigger and let the slide go forward under control, off the receiver. Take the barrel out of the slide, remove the return spring and its guide from the receiver.

To re-assemble the pistol place the return spring guide and then its spring into the receiver. Put the barrel in the slide and rotate so that the five lugs at the rear are in the recess in the slide. Fit the slide into the guide ribs in the receiver. Retract the slide and rotate the barrel to its locked position. The slide can then be moved forward by pulling the trigger.

DATA
Cartridge: 9 mm × 18
Operation: blowback, with selective fire
Feed: 25- or 40-round box magazine
Weight, empty: 1.8 kg
Length: (butt extended) 583 mm; (butt retracted) 333 mm
Barrel: 152 mm
Rifling: 6 grooves, rh, one turn in 203 mm
Sights: (fore) blade; (rear) flip aperture 100 and 200 m
Sight radius: 112 mm
Muzzle velocity: 323 m/s
Rate of fire: (cyclic) 600 rds/min

Manufacturer
State factories.
Status
No longer in production.
Service
Polish forces.

PORTUGAL

9 mm Indep 'Lusa' sub-machine gun

This is the most recent design to appear from INDEP and is a compact and robust weapon. It uses an overhung bolt, moving in a double-cylinder receiver. The magazine housing doubles as a front grip, and there is an extending stock of steel rod which slides into the recess between the two curves of the receiver. Two versions are manufactured; the standard model has a detachable barrel secured by a nut. The alternative model has a fixed-barrel surrounded by a perforated cooling jacket.

DATA
Cartridge: 9 mm Parabellum
Operation: Blowback, selective fire
Feed: 30-round box magazine
Weight, empty: 2.5 kg
Length: (stock extended) 600 mm; (stock folded) 445 mm
Barrel: 160 mm
Rifling: 6 grooves, rh, one turn in 250 mm
Sights: (fore) protected post; (rear) flipe aperture
Rate of fire: 900 rds/min
Muzzle velocity: 390 m/s

Manufacturer
INDEP; Fabrica de Braco de Prata, Rua Fernando Palha,1899 Lisboa-Codex.
Status
Current. Production.
Service
Portuguese Army and available for export.

9 mm INDEP 'Lusa' sub-machine gun; (a) with detachable barrel; (b) with fixed barrel and jacket

ROMANIA

9 mm Orita Model 1941 sub-machine gun

This sub-machine gun was designed by Leopold Jaska and manufactured at the Cugir Arsenal from 1941-44. It was issued to Romanian troops and used in the invasion of Soviet Russia in 1941. It remained the Romanian standard sub-machine gun. When Romania became part of the Soviet bloc it re-equipped with Soviet weapons and the Orita was relegated to the People's Militia and Police units and is still in service with these forces. It is much like the German MP 41 in appearance but major components were made by machining from the solid and consequently it is a very robust weapon. The butt is usually of wood but a folding type, again rather like that of the MP 40, is sometimes encountered. The magazine is straight, holding 25 rounds but it seems that a curved type holding 32 cartridges existed. The cocking handle is on the left of the receiver.

The weapon is a simple blowback-operated type but it has a bolt rather like the early Thompson, in which the hammer is carried in the bolt-head and

rotated to fire the cap by a projection standing proud of the floor of the bolt way. There is a choice of single shots or automatic fire. The safety is a push-through button in the front part of the trigger-guard with the button projecting on the right for 'safe'. The change lever is on the right of the receiver, forward of the trigger-guard. When it is rotated down, the gun produces automatic fire and single shots are selected when the lever is pushed up. The rearsight is large and sufficiently noticeable to provide a recognition feature.

DATA
Cartridge: 9 mm Parabellum
Operation: blowback, selective fire
Feed: 25-round box (a 32-round curved box was also made)
Weight: (empty) 3.46 kg; (loaded) 4 kg
Length: 894 mm
Barrel: 287 mm
Rifling: 6 grooves rh
Sights: (fore) blade; (rear) tangent notch 100 – 500 m
Muzzle velocity: 381 m/s
Rate of fire: (cyclic) 600 rds/min

Manufacturer
Cugir Arsenal, Cugir.
Status
Obsolescent. Not in production.
Service
Romanian police and militia.

9 mm Orita Model 1941 sub-machine gun

SOUTH AFRICA

9 mm BXP sub-machine gun

This is a conventional blowback weapon, constructed from stainless steel pressings and precision castings. It is simple to operate and fires from an open bolt. It is extremely compact and can be fired single-handed with the butt folded, like a pistol.

The bolt is of the 'telescoping' type which envelops the rear end of the barrel, thus allowing the overall length to be short whilst accommodating the maximum length of barrel. It also helps to keep the centre of gravity over the pistol grip and thus keep down the oscillations due to the movement of the bolt during automatic fire.

With the bolt in the forward position all apertures are closed and excessive dirt and dust cannot enter the receiver. The cocking handle is on top of the receiver but allows an unimpeded line of sight when taking deliberate aim. The barrel nut is externally threaded to accept a protector/compensator or a sound suppressor which can be used with standard or subsonic ammunition.

An ambidextrous change lever/safety catch is located on both sides of the receiver behind the trigger guard. In the horizontal position, marked with a green dot, the catch is on 'Safe', locking the trigger, sear and bolt, the latter being secured whether it is in the forward or rear (cocked) position. When the safety is moved to the red dot position the weapon is free to fire at automatic or single-shot; the first pull on the trigger gives single shots, a further pull against a felt resistance, gives automatic fire. There is an extra sear recess on the bolt to prevent accidental firing should the weapon be dropped on its butt, should the hand slip during cocking, or should the weapon be fired with a weak cartridge.

The exterior surfaces are coated with a rust-resisting finish which also acts as a life-long dry-film lubricant. The shoulder stock folds under the receiver when not required for aimed fire, and in this position the butt plate acts as a front grip and to deflect heat from the barrel away from the hand.

A 22- or 32-round magazine fits into the pistol grip. The rearsight is a combination of aperture and notch; the aperture is used for 100 m range, while the notch above it is used when firing at 200 m range. An occluded eye (collimating) sight may be fitted to improve shooting in poor light conditions.

Other optional accessories include a grenade launching tube, by means of which various grenades such as CS anti-riot grenades, may be fired; a muzzle compensator to reduce climb during automatic fired; a silencer; and a holster, carrying bag, carrying sling and magazine pouch.

Stripping and assembly of the BXP is simple and takes only a few seconds.

DATA
Cartridge: 9 × 19 mm Parabellum
Operation: blowback, selective fire
Weight, empty: 2.5kg
Length: (stock folded) 387 mm; (stock extended) 607 mm
Barrel length: 208 mm
Rifling: 6 grooves, rh, 1 turn in 254 mm
Feed: 22- or 32-round detachable magazine
Sights: (fore) adjustable cone; (rear) aperture and U-notch or OEG (collimating) sight
Rate of fire: 1000 – 1200 rds/min

9 mm BXP sub-machine gun with Occluded Eye Gunsight

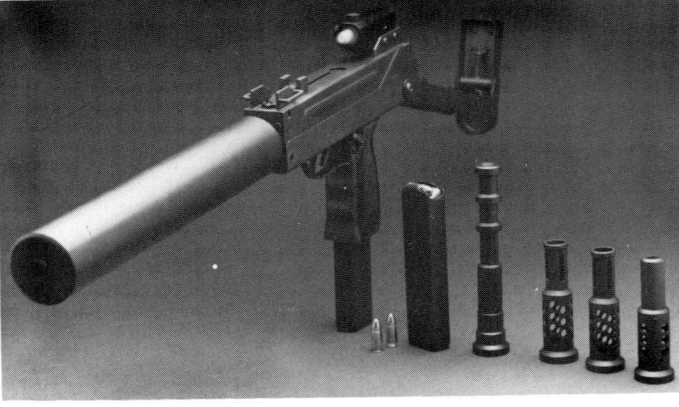

9 mm BXP sub-machine gun with silencer fitted, showing grenade launching barrel and alternative jacket/compensator units

Muzzle velocity: ca 370 m/s

Manufacturer
Armscor, Private Bag X337, Pretoria, 0001.
Status
Current. Production.
Service
South African Defence Force. South African Police.

SPAIN

9 mm Model Z45 Star sub-machine gun

The Model Z45 is based on the German MP 40 but differs in having the cocking handle on the right. It also has a protective perforated barrel jacket and an easily removable barrel which screws out.

There is a version of the Z45 with a wooden stock and both models have selective fire.

Like the MP 40 it is blowback operated and of simple design and construction. It has a compensator to keep the muzzle down at full automatic and it is by rotating this that the barrel can be removed.

The safety arrangements have been improved by fitting an extra catch to the cocking handle which secures the bolt in the forward position. The cocking handle and the catch must be pulled together to free the bolt. The trigger provides single shots when pulled halfway back and full automatic when fully back.

The Z45 was manufactured initially for sale to Germany but production did not get under way until July 1944. The gun was taken by the Spanish Civil Guard and then the Air Force. In June 1948 it was adopted by the Spanish Army. It has been sold to Chile, Cuba, Portugal and Saudi Arabia. The quantities involved have not been large and in no case has it become the standard sub-machine gun. The weapon was also one of those copied in Indonesia.

DATA
Cartridge: 9 mm Bergmann Bayard (9 mm Largo)
Operation: blowback, selective fire
Feed: 30-round box magazine
Weight: (unloaded) 3.86 kg; (loaded) 4.54 kg
Length: (stock extended) 838 mm; (stock folded) 579 mm
Barrel: 198 mm
Muzzle velocity: 381 m/s
Rate of fire: (cyclic) 450 rds/min

Manufacturer
Star Bonifacio Echeverria SA, Eibar. Also made in Indonesia.
Status
Obsolete. Not in production.
Service
Spanish forces. Sold to Chile, Cuba, Portugal and Saudi Arabia.

9 mm Model Z62 Star sub-machine gun

The Z62 is a blowback-operated weapon which replaced the Z45 as the standard sub-machine gun in the Spanish armed forces. It is of conventional design but has some features of interest. The problem of bolt setback when the weapon is dropped has always been difficult to solve and the Z62 incorporates a holding device which is independent of the firer.

The firing pin is operated by the hammer in the bolt. The hammer is rotated by a rod projecting from the bolt-face and driven back when the bolt closes. At the same time as the rod rotates the hammer it disengages a bolt lock which, at all other times, prevents hammer movement.

The trigger mechanism of the Z62 produces single shots from the lower part of the trigger and full automatic from the top half. The safety catch is a push-through type in the upper part of the pistol grip. A D-section normally allows the sear to move in the cut-out part of the safety but, when pushed across, the full diameter of the safety locks the sear and prevents bolt movement.

The weapon uses pressings, plastics and has a folding stock based on the Czechoslovak Model 25.

DATA
Cartridge: 9 mm Bergmann Bayard and 9 mm Parabellum
Operation: blowback, selective fire
Feed: 20-, 30-, or 40-round box magazine
Weight: (empty) 2.87 kg; (loaded, 30 rounds) 3.55 kg
Length: (stock extended) 701 mm; (stock folded) 480 mm
Barrel: 201 mm
Muzzle velocity: 380 m/s (9 mm Parabellum)
Rate of fire: (cyclic) 550 rds/min

Manufacturer
Star Bonifacio Echeverria SA, Eibar.
Status
Obsolete. No longer in production.
Service
Spanish forces.

9 mm Model Z62 Star sub-machine gun

9 mm Model Z-70/B Star sub-machine gun

The Z-70/B was introduced into the Spanish Army in 1971. It was produced because of difficulties with the trigger mechanism of the Z62. The basic mechanism of the Z62 remains unchanged but a new trigger mechanism has been designed. Instead of firing single-shot and automatic from the two halves of the trigger, a conventional system with a selector on the left of the receiver above the pistol grip has been installed. The safety catch lies under the trigger guard and is pulled back to the safe position and pushed forward to fire. This can be seen in the photograph.

The dimensions and weights remain unchanged from those of the Z62 above.

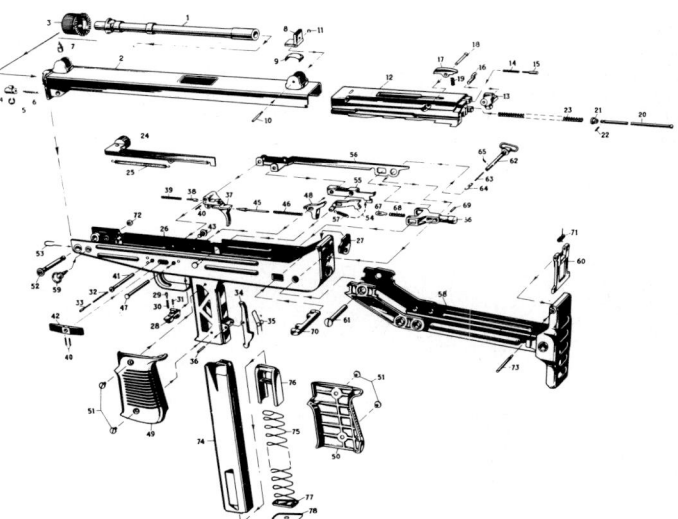

9 mm Model Z-70/B Star sub-machine gun

9 mm Model Z-70/B components

9 mm Star Model Z-84 sub-machine gun

The Star Model Z-84 sub-machine gun has been developed to meet the tactical requirements for a light, simple, reliable and effective personal weapon. Special attention has been paid during the design stage to permit manufacture with an extensive use of steel stampings and investment castings. Because of its specially-designed feed system it will fire soft-point and semi-jacketed bullets equally as well as full-jacketed military ammunition. Its underwater resistance and performance capabilities make it an adequate assault weapon for marine and commando units as well as for special services.

The Z-84 fires from an open bolt, using the well-known advanced primer ignition principle. The breech-block is recessed so that it wraps around the barrel when closed, allowing the maximum barrel length within a compact overall length. The centre of gravity is above the pistol grip, allowing the weapon to be fired single-handed with remarkable stability. The magazine housing is located in the pistol grip, allowing magazines to be replaced, even in darkness, with the utmost speed.

There are no external moving parts; after the breech-block has been cocked, the cocking handle returns to its forward location by spring power and remains still during firing. There is a dust cover in the cocking handle slot, and the interior of the weapon is only open when an empty case is being ejected. The breech-block is suspended on two rails, on four small contact points, and there is a minimum of 1 mm clearance on the other surfaces, so that any dust which finds its way into the weapon is unlikely to derange it; moreover there is ample space in the receiver for dirt to settle out of the way of any moving parts.

There is a sliding fire selector on the left side of the receiver; pushed forward, exposing one white dot, it gives single shots; pushed rearwards, exposing two white dots, gives automatic fire. The safety catch is a lateral sliding button inside the trigger-guard. In addition to the manual safety, the bolt is provided with three bents which allow it to be securely held in any position; should the hand slip while cocking, for example, the central safety bent will arrest the breech-block before it can fire a round. There is also an automatic inertia safety unit which locks the breech-block when it is in the closed position, so that a sudden shock will not cause the block to move. This mechanism is over-ridden by the cocking handle and is automatically put out of action when the weapon is being fired.

DATA
Cartridge: 9 × 19mm Parabellum
Operation: blowback, selective fire
Feed: 25 or 30-round box magazine
Weight: gun without magazine 3.00 kg
Length: Stock extended 615 mm; stock folded 410 mm (with short barrel)
Barrel: normal 215 mm; optional 270 mm
Sights: (rear) flip aperture for 100 m or 200 m; (front) post, adjustable for elevation
Sight radius: 325 mm
Rate of fire: 600 rds/min
Muzzle velocity: 400 m/s

Manufacturer
Star Bonifacio Echeverria SA, Eibar.

9 mm Star Z-84 sub-machine gun

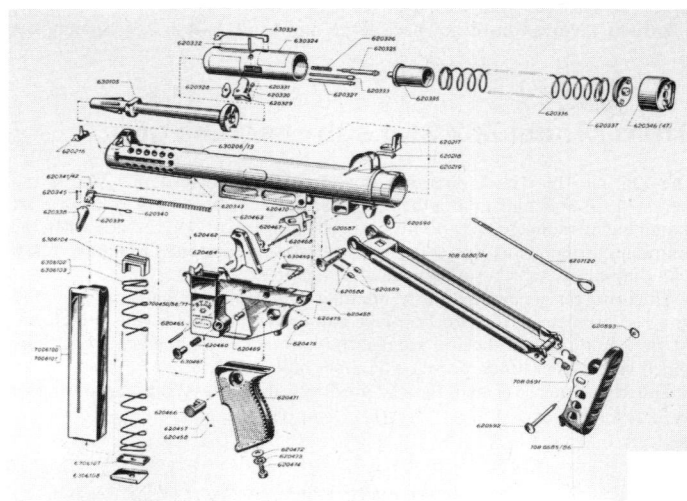

Component parts of Star Z-84 sub-machine gun

Status
Current. Production.
Service
Spanish Armed Forces (Special Units) and Security Forces. Military and Security services in various countries.

9 mm C2 sub-machine gun

This is a small and handy weapon which can be used with either the 9 mm Largo or 9 mm Parabellum round. Particular attention has been paid to the balance of the weapon which can be controlled with one hand.

Blowback-operated, the C2 differs from many weapons of similar size in not having a fixed firing pin, the system used being similar to that of the Z62, and incorporates a bolt-holding device as one of its three safeties. The bolt can also be locked forward for transport as in the Z45; and the third safety is a position on the selector lever.

Other characteristics of the weapon include a horizontal magazine holding 32 rounds in a double stack; a grooved bolt surface to reduce susceptibility to jamming by dust; a folding metal stock and a generally robust but economical construction.

DATA
Cartridge: 9 mm Parabellum or 9 mm Largo
Operation: blowback, selective fire
Feed: 32-round detachable magazine
Weight: (without magazine) 2.65 kg
Weight: (of empty magazine) 0.22 kg
Length: (with stock extended) 720 or 500 mm
Barrel: 212 mm
Rifling: 4 grooves rh
Sight radius: 400 mm
Muzzle velocity: 325 m/s (Parabellum) or 340 m/s (Largo)
Rate of fire: (cyclic) 600 rds/min
Designer: CETME

Manufacturer
Santa Barbara, Julian Camarillo 32, 28037 Madrid.
Status
Current. Available.

9 mm C2 sub-machine gun

SWEDEN

9 mm Model 45 sub-machine gun

Generally known as the Carl Gustaf, this is one of the standard weapons of the Swedish armed forces. Originally the gun used the Suomi 50-round box magazine from the Model 37-39 but in' 1948 an excellent two-column magazine holding 36 rounds was introduced. This was a wedge-shaped magazine which has been distinguished by its reliability and has been copied widely and been used, or a similiar type developed, in Czechoslovakia, other Scandinavian countries and in West Germany. This change of magazine led to the introduction of a detachable housing to allow both types of magazine to be used, the gun incorporating this feature being known as the Model 45. The M45B has an improved backplate lock, smaller holes in the barrel jacket, uses

only the 36-round magazine, and has a baked green enamel finish. The M45C is as the B model but with a bayonet lug, while the M45E has the option of selective fire.

Derivatives of the Model 45 include a silenced version which was used by US Special Forces in South-east Asia and the Model 49 Hovea weapon which is used by the Danish Army and resulted from further development of the Model 45 by Husqvarna. The weapon was also copied in Indonesia and was produced for a time in Egypt as the Port Said.

DATA
Cartridge: 9 mm M39B
Operation: blowback, automatic
Feed: 36-round box magazine
Weight: (loaded) 4.2kg; (empty) 3.9kg
Length: (stock extended) 808 mm; (stock folded) 552 mm
Barrel: 213 mm
Rifling: 6 grooves rh
Sights: (fore) post; (rear) flip
Muzzle velocity: 410 m/s
Rate of fire: 600 rds/min

Manufacturer
FFV Ordnance, Eskilstuna.
Status
Current. Not in production.
Service
Swedish forces. Also sold to Egypt, Indonesia and Ireland. Has also been made under licence in Egypt.

9 mm Model 45 sub-machine gun

SWITZERLAND

MP 310 SIG sub-machine gun

The MP 310 was developed in the mid-1950s from a previous design, the MP 48. The MP 48 had proved too expensive and the idea of the MP 310 was to bring down the cost and at the same time improve the design. The result was a very well-made sub-machine gun with some unusual features. The magazine folded forward under the barrel, although this was not an entire novelty at that time and the trigger had two pressures, the first for single-shot and the second for automatic. Expensive machined parts were discarded and

precision castings used instead, but the price was still high, and sales were not encouraging. Production started in 1958 but by 1972 only just over 1000 had been made and sold and the line closed. Some were exported and may be found in different parts of the world since this was a well-made weapon and with care it should last for many years.

The Swiss police took part of the output and are still using the gun.

DATA
Cartridge: 9 mm × 19 Parabellum
Feed: 40-round folding box magazine
Weight: (unloaded) 3.15 kg
Length: (stock extended) 735 mm; (stock folded) 610 mm
Barrel: 200 mm
Rate of fire: 900 rds/min
Muzzle velocity: 365 m/s

Manufacturer
Schweizerische Industrie Gesellschaft (SIG), CH-8212 Neuhausen Rheinfalls.
Service.
Swiss police. May be found in very small numbers anywhere in the world.

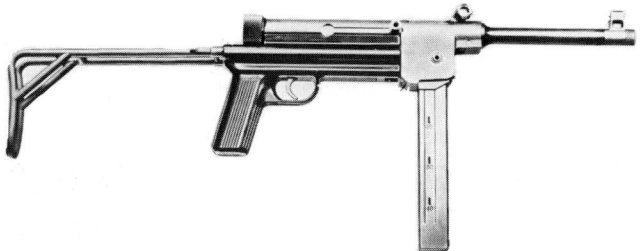

MP 310 SIG sub-machine gun with magazine in firing position

TURKEY

9 mm Model 1968 sub-machine gun

This weapon appears to have been derived from the Swiss Rexim-Favor; this was developed in the early 1950s and, according to some reports, was based on a design stolen from France. It was actually manufactured in Spain and marketed with little success until the Rexim company folded in 1957. Thereafter it was sold by the Spanish makers for some years as the La Coruna. The Turkish model is believed to have been manufactured in Turkey in relatively small numbers. It differs from the Swiss/Spanish design in having a longer receiver and barrel, a wooden butt and a short bayonet hinged below the muzzle.

The mechanism is unusual in that it fires from a closed breech. Releasing the cocking lever allows the bolt to go forward to chamber a round, after which pressing the trigger releases a hammer unit which is driven down the receiver by its own spring to strike the firing pin. Bolt and hammer are then blown back by recoil; if the weapon is set for single shots, the hammer is retained by the sear until the trigger is pressed; otherwise it follows the bolt and automatically fires after each round is chambered.

The mechanism is unnecessarily complicated and demands precision machining, though it appears to bestow better-than-average accuracy on the weapon.

DATA
Cartridge: 9 mm Parabellum
Operation: blowback, selective fire
Feed: 32-round box magazine
Weight: 4.69kg
Length: 880 mm
Barrel: 350 mm
Rifling: not known
Sights: (fore) blade; (rear) flip aperture

9 mm Model 1968 sub-machine gun

Muzzle velocity: 400 m/s
Rate of fire: 600 rds/min

Manufacturer
Not known.

Status
Service. No longer in production.
Service
Turkish Army.

9 mm MKE MP5 A3 sub-machine gun

The MKE MP5 A3 sub-machine gun is, as the name implies, the Heckler and Koch MP5 A3 manufactured under license by Makina ve Kimya Endustrisi. It is identical to the German-made model, to which reference should be made for operating information and data.

Manufacturer
Makina ve Kimya Endustrisi Kurumu, Tandogan Medyani, Ankara.
Status
Current. Production.
Service
Turkish forces, and for export.

MKE MP5A3 sub-machine gun

UNION OF SOVIET SOCIALIST REPUBLICS

5.45 mm AKSU-74 sub-machine gun

This weapon is a variant model of the AKS-74 rifle, but exhibits a number of interesting new features. Its construction varies from that of previous AK models by having the receiver top hinged to the gas tube retainer block so that it hinges forward on opening rather than lifting off. The short barrel is fitted with a cylindrical muzzle attachment which incorporates a bell-mouthed flash hider. It is believed that the body of the attachment forms an expansion chamber which will perform two functions; firstly it will give a sudden pressure drop before the bullet exits the muzzle, so reducing the otherwise high pressure in the gas tube and on the gas piston, caused by tapping off propellant gas very close to the chamber. Secondly it will also permit a degree of flame damping which will reduce flash and blast which would otherwise be inseparable from firing a full-charge rifle cartridge in such a short barrel.

The plastic magazine has been strengthened by moulded-in ribs on the front rebate. The rearsight is a simple flipover pattern with two U-notches, marked for 200 and 400 m. The skeleton stock folds to the left and locks into a spring-loaded lug on the side of the receiver. Apart from the shortness of the gas piston, return spring and spring guide rod, the internal mechanism is identical to that of the AK-74 rifle.

Full details are not yet available; the following has been derived from photographic analysis:

DATA
Cartridge: 5.45 × 39.5 mm Soviet
Length: (butt folded) 420 mm; (butt extended) 675 mm

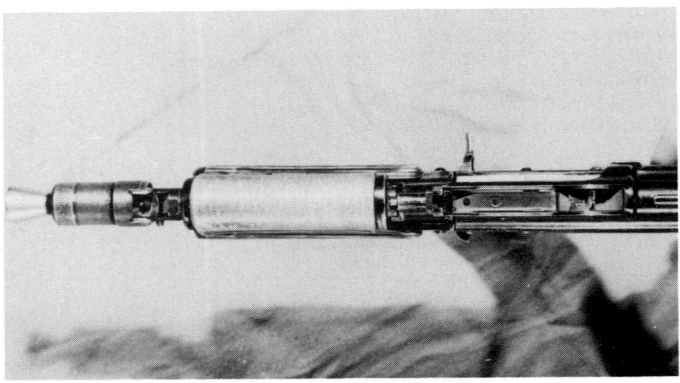

AKSU-74 sub-machine gun, top view; note flip-over sight

5.45 mm AKSU-74 sub-machine gun, left side

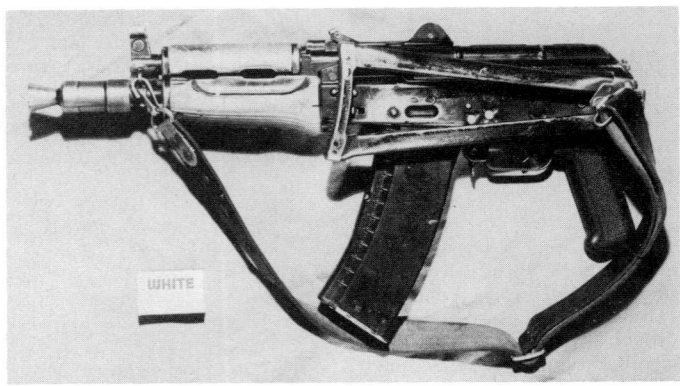

AKSU-74 sub-machine gun, left side, stock folded

5.45 mm AKSU-74 sub-machine gun, right side

Barrel length: 200 mm
Magazine capacity: 30 rounds
Muzzle velocity: ca 800 m/s
Rate of fire: (cyclic) ca 800 rds/min

Manufacturer
State Arsenals.
Status
Current. Production.
Service
Soviet forces.

AKSU-74 sub-machine gun field-stripped; note new method of attachment for receiver top

7.62 mm PPSh-41 sub-machine gun

The PPSh-41 was one of the main weapons of the Soviet infantry during the Second World War. Since then it has been widely spread throughout the world and for some years it was almost as potent a symbol of Soviet presence as the AK-47 is now. Although now largely supplanted in all Soviet-influenced states, so many were made up to 1945 that the gun is likely to be found in odd corners of the world for some years yet. Since late 1977 it has positively been identified in use by the two factions fighting in Beirut, and it is reported as being seen in Somalia and Ethiopia.

All details are given in *Jane's Infantry Weapons 1978*, page 89.

7.62 mm PPSh-41 sub-machine gun with 35-round box magazine. Change lever visible in front of trigger

7.62 mm PPSh-41 sub-machine gun with 71-round drum magazine, an early version with tangent rearsight

7.62 mm PPS-43 sub-machine gun

This gun, like the PPSh-41, is a relic of the Second World War which is still used in small numbers in various Soviet countries. It is no longer a first-line issue to any army, but it is carried by such men as border guards, factory militia and police. It was positively identified in Poland in 1978 in the hands of factory guards and it is still used by the Chinese Militia. Chinese versions supplied to the Viet Cong will probably still be found in Vietnam and neighbouring countries.

Full details of the PPS-43 were given in *Jane's Infantry Weapons 1978*, page 90.

DATA
Cartridge: 7.62 mm × 25 Type P pistol
Weight: (unloaded) 3.36 kg; (loaded) 3.93 kg
Length: (stock extended) 820 mm; (stock retracted) 615 mm
Rate of fire: (cyclic) 650 rds/min

7.62 mm PPS-43 sub-machine gun

UNITED KINGDOM

9 mm Sten sub-machine gun

The Sten is one of the weapons which has become so well-known that its name has become synonymous with sub-machine guns all over the English-speaking world and the title Sten is applied indiscriminately to any vaguely similar gun. The first models appeared in 1941 and were soon replaced with improved versions. By 1945 more than four million various marks had been made after which production virtually stopped apart from maintenance spares.

The Sten was supplied to all Commonwealth countries, some of which still have it in stock and some of which have sold it off to other nations. Probably half a million Mark II Stens were sent to the underground movements in Occupied Europe during the Second World War, and little or no attempt was made to recover them after 1945. There has been a steady trade in Sten guns of all marks and conditions through the reputable and also the less-reputable arms dealers since the late 1940s so that it is now quite impossible to say with accuracy where the gun may or may not be found. It was never made with any great precision, nor with good materials, so remaining examples are likely to be well-worn and should be treated with care until the effectiveness of the safeties are assured.

There is a full description of the Sten family in *Jane's Infantry Weapons*

1978, page 91 and the following list is intended only to identify any remaining specimens that may be found.

Mark I. So few of these were made that they may be disregarded for all practical purposes.

Mark II. This was the most prolific version. Over two million were made by 1945 and this was the model that was supplied to the underground movements. The short barrel jacket unscrewed and removed the barrel and with the stock also removed the gun can be packed into a small space such as a shopping bag. A small number of Mark IIs were fitted with a silencer (Mark IIS), but this could be used only with single-shot firing.

Mark III. A simplified version of the Mark II, with a fixed-barrel and the body extended over the whole length of the barrel. It usually had a single tube stock, but all the other versions could be fitted if desired.

Mark V. The standard British sub-machine gun until replaced by the Sterling in the 1960s. This mark is really a cosmetic attempt to make the Mark II look more like a modern sub-machine gun and it was little improvement over it. The wooden butt was heavy and the elaborate rifle foresight unnecessary. It is the one model most likely to have survived. A few were fitted with silencers and known as the Mark VI.

All Stens have certain features in common. The magazine fed from the left side and contained 32 rounds at full charge, but 30 rounds were better and led to fewer jams. All had fixed sights set for 100 m. All had a selector button which had to be pushed from one side to the other for automatic or single shots. Parts are interchangeable between different marks, so it is no positive identification that one is fitted with a certain type of stock or bolt. Few internal items changed during the production run and the quality standards were sufficient to ensure a life of one or two years in wartime use.

Wartime manufacture was restricted to factories in the UK and the Long Branch factory in Canada.

9 mm Sten Mark VI sub-machine gun

9 mm Sten Mark II submachine gun

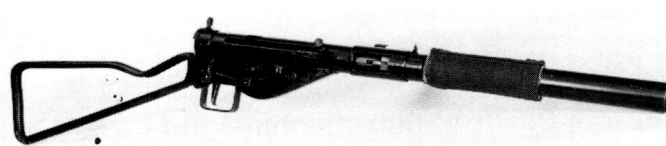

9 mm Sten Mark IIS submachine gun

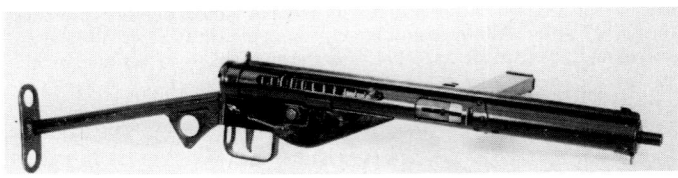

9 mm Sten Mark III submachine gun

DATA
Cartridge: 9 mm × 19 Parabellum
Operation: blowback, selective fire
Feed: 32-round box magazine

WEIGHTS	Mk II	Mk IIS	Mk III	Mk V	Mk VI
Gun with magazine					
unloaded	2.8kg	3.5kg	3.18kg	3.9kg	4.32kg
loaded	3.44kg	4.14kg	3.82kg	4.54kg	4.96kg
LENGTHS					
Overall	762 mm	857 mm	762 mm	762 mm	857 mm
Barrel	197 mm	91.4 mm	197 mm	198 mm	95 mm

MECHANICAL FEATURES
Barrel: (rifling) 6 grooves rh. 1 turn in 254 mm (Mk II has 2 grooves)
Sights: (foresight) blade; (rearsight) aperture

FIRING CHARACTERISTICS
Muzzle velocity: all 366 m/s except Mk IIS and VI 305 m/s
Rate of fire: (cyclic) 550 rds/min

Manufacturers
Birmingham Small Arms Co Ltd at Tyseley. Royal Ordnance Factory, Fazackerley. Long Branch Arsenal, Canada.
Status
Obsolete.

9 mm Sten Mark V submachine gun standard model

Service
See text ,

---

9 mm L2A3 Sterling sub-machine gun

Note on Sterling products: The Sterling Armament Company was purchased by Royal Ordnance plc in 1988 and the Dagenham factory closed down. Some of the tooling has been relocated at the Nottingham Small Arms Factory, but at the time of writing decisions have yet to be made on the future of the various items in the Sterling range. We have been informed that spares and repair facilities for all Sterling products exist at Nottingham, and that manufacture may take place in due course. Enquiries should be directed to Royal Ordnance Guns & Vehicles Division, King's Meadow Road, Nottingham NG2 1EQ, England.

The current sub-machine gun in the British Army is the L2A3. This gun is the military version of the Mark 4 Sterling sub-machine gun manufactured by the Sterling Armament Co at Dagenham. The gun was called the Sterling Mark 2 in 1953 and was adopted by the British Army as the L2A1. The L2A2 Mark 3 Sterling, which featured some small modifications, came into service in 1955 and the L2A3 in 1956.

The magazine fits horizontally into the left-hand side of the receiver. It holds 34 9 mm Parabellum cartridges. The magazines of the Australian F1 and Canadian C1 will fit the L2A3. The butt stock may be extended and the

Mark 4 Sterling sub-machine gun with fixed butt

weapon used from the shoulder. Alternatively it may be folded under the barrel and the gun fired from the hip. Fixed butts for both the Mark 4 and Mark 5 (see entry for 9 mm L34A1 sub-machine gun below) were introduced in 1981. The combination of a variable-length butt and a high cheek comb increases the efficiency of these weapons. The selector lever is directly above the trigger and the three positions are '34', '1' and 'SAFE'. The weapon is also available with single-shot (open bolt) capability only and in this version it is known as the Sterling Police Carbine Mark 4. Also available is a single-shot, closed bolt, floating firing pin version known as the Sterling CBS Mark 8.

The weapon consists of a tubular, perforated receiver, a cylindrical bolt and the stock.

9 mm L2A3 Sterling sub-machine gun stripped

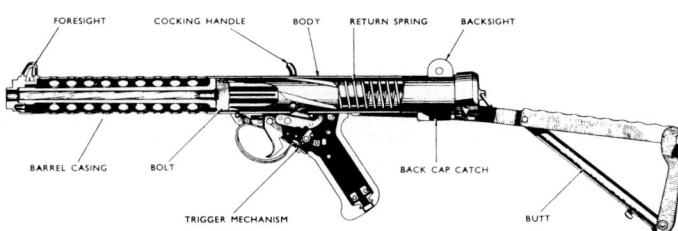

9 mm L2A3 Sterling sub-machine gun

9 mm L2A3 Sterling sub-machine gun

Mark 4 Sterling sub-machine gun with stock folded

9 mm Sterling sub-machine gun with normal trigger-guard removed and Arctic trigger installed

Operation

When the weapon is in the 'ready to fire' position, the bolt is held to the rear by the sear and the ammunition is in the magazine. When the trigger is pulled, the bolt goes forward under the influence of the return spring and the first round in the magazine is fed forward into the chamber. Since the magazine is on the side of the gun the round is angled inwards towards the chamber and the cap is not aligned with the fixed firing pin on the breech face. As the round enters the chamber it becomes lined up with the axis of the bore and as the extractor grips into the groove of the cartridge, the firing pin touches the cap of the cartridge case. The friction force between the cartridge and the chamber wall slows the cartridge and the fixed firing pin fires the cartridge while the bolt is still moving forward. The rapid development of the pressure quickly brings the bolt to a standstill and then reverses its direction and blows it back against the return spring. The weight and forward velocity of the bolt at the moment of cap initiation ensure that the rate of return of the bolt is not enough to allow the residual pressure to blow out the case. The cartridge case is forced back by the gas pressure and drives the bolt back. The extractor holds the case to the bolt-face until the cartridge hits the fixed ejector which throws it out of the ejection port to the right. The compressed return spring drives the bolt forward and if the selector is at '34' the automatic cycle continues; if the selector is at 'single shot', the bolt is held to the rear.

Trigger and firing mechanism

The trigger mechanism works as follows. The sear is fitted into a spring loaded sear carrier. The disconnector is ⌐ shaped and is pivoted at the junction of the two arms to the sear carrier. The nose of the disconnector is spring-loaded to bear against the step on the rear of the sear and the lower arm reaches down towards the spindle of the selector lever. The selector lever spindle has an inner arm which limits the movement of the lower end of the disconnector. The trigger is pivoted at the top and has a shoulder fitting under the sear carrier, which lifts the front of the carrier when the trigger is pulled.

When the selector is set for automatic the selector spindle is rotated forward so that the inner arm is not in contact with the lower end of the disconnector. When the trigger is pulled the sear carrier is rotated up at the front and the nose of the sear is lowered off the bent of the bolt which goes forward. The bolt reciprocation continues as long as the trigger is pulled and ammunition remains in the magazine.

When the selector is set to '1' the inner arm is rotated upwards toward the lower end of the disconnector. When the trigger is pulled the sear comes down as before and the bolt is released. The lower arm of the ⌐ shaped disconnector hits the inner arm of the selector lever and rotates clockwise. This causes the nose to slip off the spring-loaded sear which promptly rises to hold up the bolt. When the trigger is released the sear carrier rises and the top arm of the disconnector springs back over the shoulder of the sear.

When the selector is set to 'SAFE' the inner arm moves rearward under the lower arm of the disconnector. The disconnector is firmly held because it is pivoted on the sear carrier that also is held. Since the sear carrier does not descend the sear holds the sear notch of the bolt or, if the bolt is forward, the safety notch at the rear of the bolt.

Stripping

To strip the L2A3, press the magazine release catch above the magazine housing, and remove the magazine. Retract the bolt and check the feedway is clear. Allow the bolt forward under control. Press the release catch under the rear of the receiver and push the receiver cap forward and rotate anti-clockwise. The cap can then be drawn off. Pull the cocking handle back along its slot to the enlarged section where the cocking handle can be withdrawn. The return spring can then be withdrawn. If the gun is then elevated and the trigger pulled, the bolt will come out at the rear.

DATA
Cartridge: 9 mm × 19 Parabellum
Operation: blowback, selective fire
Feed: 34-round box magazine, 10- and 15-round magazines and twin stacked 10 × 2, 15 × 2 and 34 × 2 magazines are also available
Weight, empty: 2.72 kg
Length: (Butt extended) 690 mm; (butt retracted) 483 mm
Barrel: 198 mm
Rifling: 6 grooves, rh, one turn in 250 mm
Sights: (fore) blade; (rear) flip aperture
Sight radius: 410 mm
Muzzle velocity: 390 m/s
Rate of fire: (cyclic) 550 rds/min

Manufacturer
See entry for the L34A1 below.
Status
See L34A1 entry.

9 mm L34A1 Sterling sub-machine gun

This is the silenced version of the L2A3 sub-machine gun used by Infantry Battalions in the British Army. The commercial version is the Patchett/Sterling Mark 5 sub-machine gun. It is also available with single-shot (repetition) capability only. This weapon is known as the Sterling Police Carbine Mark 5.

The bore of the barrel has 72 radial holes drilled through it, allowing some of the propellant gas to escape and thus reduce the muzzle velocity of the bullet.

The gases which pass through the radial holes in the barrel enter a diffuser tube. This has a series of holes bored through it, and the gases pass into an expanded metal wrap. They emerge from this and are contained in the barrel casing. Eventually they seep forward through the barrel supporting plate or

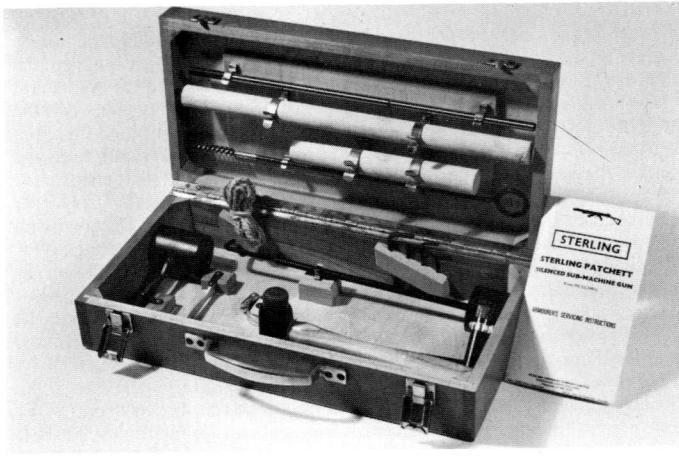

Armourer's kit for 9 mm Mark 5 silenced Sterling sub-machine gun

9 mm Patchett/Sterling Mark 5 silenced sub-machine gun, commercial version manufactured at Dagenham

return to the barrel. The silencer casing is extended forward of the barrel and contains a spiral diffuser. The bullet passes down the centre of this but the gases following behind are given a swirling action by the diffuser. The gases which follow closely behind the bullet are deflected back by the end cap and mingle with the gases coming forward through the diffuser. The result of this action is to ensure that the gas velocity on leaving the weapon is low.

At one regulated test, 60 000 rounds were fired in one weapon, after which the barrel was inspected and found to be within gauge tolerances and fit for further service.

As the effective pressure is reduced it is necessary to have a lightened bolt and only a single return spring to enable the blowback action to function efficiently.

Sterling makes available a wide variety of armourer's tooling and workshop equipment and, as shown in the accompanying photograph, a special armourer's kit is available for the L34A1 sub-machine gun.

The Sterling uses standard 9 × 19 mm Parabellum ammunition, unlike other silent weapons which require special subsonic ammunition.

DATA
Cartridge: 9 mm × 19 Parabellum
Operation: blowback, selective fire
Feed: 34-round box magazine
Weight, empty: 3.60 kg
Length: (butt extended) 864 mm; (butt retracted) 660 mm
Barrel: 198 mm
Rifling: 6 grooves, rh, one turn in 250 mm
Sights: (fore) blade; (rear) flip aperture
Sight radius: 502 mm
Muzzle velocity: 293-310 m/s
Rate of fire: (cyclic) 515-565 rds/min

Manufacturer
Both the L2A3 and the L34A1 were made by Sterling Armament Co at Dagenham. In addition the L2A3 was manufactured at the Royal Ordnance Factory at Fazackerley. The Mark 4 Sterling and the Patchett/Sterling Mark 5 were manufactured at Dagenham. The Canadian sub-machine gun, the C1, is basically the Sterling. The gun is also manufactured in India.
Status
Current. Production ended.
Service
The L2A3 and the L34A1 are in current service with the British Army. The Mark 4 Sterling has been sold abroad on a considerable scale, some 90 countries having purchased the gun in varying quantities. The principal purchasers have been Ghana, India (which now manufactures under licence), Libya, Malaysia, Nigeria and Tunisia. Many are also in service in the Arabian Gulf States.

Mark 6 Sterling carbine

This was produced for the American civil market and is a semi-automatic version of the standard Sterling, but with the barrel extended to conform with US regulations. The mechanism is so designed that it cannot be re-converted for automatic fire.

DATA
Cartridge: 9 mm Parabellum
Operation: blowback, semi-automatic

Weight, empty: 3.4kg
Length: (stock extended) 889 mm; (stock folded) 685 mm
Barrel length: 410 mm

Manufacturer
Sterling Armament Co Ltd, Dagenham, Essex.
Status
Production ended.

9 mm Mark 7 Sterling paratroopers pistol

Sterling designed this special-purpose weapon for use in confined spaces. It was available in four versions: Mark 7 A4 and Mark 7 A8, both open bolt, selective fire; Mark 7 C4 and Mark 7 C8, both closed bolt, semi-automatic fire. It should be noted that the closed bolt firing mode offers particularly high accuracy. Accessories include a light removeable plastic stock with three

add-on sections to adjust to the user's arm length; 10-, 15- and 34-round standard magazines and twin stacked 10 × 2, 15 × 2 and 34 × 2 magazines; a forward holding strap, long sling, and a suppressor (for the 4.25 in barrel only).

DATA
Cartridge: 9 mm Parabellum
Operation: (A4 and A8) open bolt, selective fire; (C4 and C8) closed bolt, semi-automatic fire

WEIGHTS
Gun: (empty) 2.2 kg (A4 and C4); 2.3 kg (A8 and C8)

LENGTHS
Gun overall: 355 mm (A4 and C4); 470 mm (A8 and C8)
Barrel: 89 mm (A4 and C4); 198 mm (A8 and C8)

Manufacturer
Sterling Armament Co, Dagenham, Essex.
Status
Current. Production ended.
Service
Three overseas navies and several US police forces.

9 mm Mark 7 Sterling paratroopers pistol

9 mm Sterling single shot Mark 8

Strictly speaking this is not a sub-machine gun, since it only fires single shots in the semi-automatic (self-loading) mode. It was produced for use by police and security forces requiring an accurate 9 mm weapon capable of being used out to ranges of 100-200 m, since the normal police pistol is inaccurate at such distances. The Mark 8 gun is externally the same as the normal L2A3 Sterling but the internal mechanism operates from a closed bolt. The usual 10- or 34-round magazines are used, and normal iron sights are fitted as standard. Other types of sight can be fitted to suit the purchaser and a rubber-armoured 3X or 9X zoom telescope sight is offered.

DATA
As for the L2A3 (above) except that only single shots can be fired

Manufacturer
Sterling Armament Co Ltd, Dagenham, Essex.
Status
Production ended.

UNITED STATES OF AMERICA

.45 Thompson sub-machine gun

The Thompson sub-machine gun is generally looked on as being the original sub-machine gun, which it is not, since it was not in existence in a practical form until after the First World War, but it gained much notoriety as a gangster gun in the 1920s and 30s. Manufacture ceased during the Second World War by which time about 1 400 000 had been made. Many survived the war to continue in scattered use throughout the world, but their numbers are

.45 Model 1928A1 Thompson sub-machine gun

Model M1 Thompson sub-machine gun

steadily declining since there are no longer any spares. Most that are found today are well worn and may not be safe to fire.

The original civilian Thompson is known for its unusual bolt delay by the use of an H-piece riding up and down inclined planes so that it slowed the opening of the bolt. Later wartime models dropped this component when redesigning for mass-production and went to a simple blowback system of operation. Most remaining Thompsons are the US Army Model M1 which used a vertical box magazine rather than the heavy and clumsy 50-round drum, and have a straight wooden fore-end instead of the elaborate carved handgrip of the civilian model.

The Thompson was heavy but reliable and well-liked by those who used it. It has not been used for some years now, though it was at one time popular with the IRA, the Chinese and the Egyptians. Full details are given in *Jane's Infantry Weapons 1978*, page 95.

DATA
Cartridge: .45 Automatic Colt Pistol (ACP)
Operation: blowback
Feed: 20- or 30-round vertical box magazine
Weight, empty: 4.8 kg
Length: 810 mm
Barrel: 267 mm
Muzzle velocity: 282 m/s
Rate of fire: (cyclic) 700 rds/min

Status
Obsolete. Likely to be found only with militia or irregular forces

.45 M3 sub-machine gun

When the USA entered the Second World War it had only one sub-machine gun in production. This was the Thompson which was heavy, expensive to produce and very demanding in labour and machine tools.

In October 1942 work was authorised on a new sub-machine gun. The requirement called for an all-metal weapon, easily disassembled and easily converted to accept the 9 mm Parabellum as well as the calibre .45 Automatic Colt Pistol (ACP) cartridge. The cost and performance were to be at least comparable to the British Sten gun. In December 1942 it was adopted as the sub-machine gun .45 calibre M3.

The M3 was manufactured by the Guide Lamp Division of General Motors Corporation at Anderson, Indiana. The bolts were made by the Buffalo Arms Company. Many difficulties were met in construction and the schedule fell back but in 1944 production was running at about 8000 a week.

A silencer was made for the M3 but only about 1000 silenced guns were made – all for the OSS. A flash hider was also developed and could be added to the weapon.

Over 600 000 were made, but it was reported in early 1944 that the bolt-retracting handle gave trouble and the opportunity was taken to improve the design still further to reduce manufacturing effort. The modified gun became the M3A1 and the chief difference lay in the fact that the bolt was pulled back by inserting the right forefinger into a hole in the bolt, and to allow for this the ejection port and its spring-loaded cover were both made

larger. A flash suppressor was fitted to some of the production run. There was a further order for M3A1s during the Korean War, but it was never completed and the total number made never exceeded 50 000.

Thus there was a grand total of roughly 650 000 of both types actually made in the USA, though from time to time copies have appeared in the Far East but there cannot have been many of these. Of the 650 000, 25 000 were specially produced in 9 mm with a Sten magazine and a replacement barrel. These were all supplied to various underground movements and it is possible that there are still some in use somewhere.

The M3 was a pleasant gun to shoot and handled as well as any other wartime gun of the same type. It was unusual in having a very slow rate of fire which allowed the firer to control the movement of the gun while firing bursts and it was also sufficiently slow to enable a single shot to be snatched off without too much disturbance of the aim. Despite its advantages the M3 was never popular in the US Army and in 1957 it was declared Substitute Standard and relegated to the reserve. It was little used after 1960, but may have been

.45 M3 sub-machine gun showing cocking handle

.45 M3A1 sub-machine gun. Cocking handle was removed and hole in bolt took firer's finger. Flash hider was optional screw-on feature

sold to some friendly countries under offshore aid programmes. Portugal bought components and perhaps complete guns to make the M948 FMBP.

A full description and history appears in *Jane's Infantry Weapons 1978*, page 97.

DATA

COMMON TO M3 AND M3A1 EXCEPT WHERE OTHERWISE INDICATED
Cartridge: .45 Automatic Colt Pistol (ACP)
Operation: blowback, automatic
Feed: 30-round box magazine
Weight, empty: (M3) 3.63 kg; (M3A1) 3.47 kg
Length: (stock extended) 757 mm; (stock retracted) 579 mm
Barrel: 203 mm

Rifling: 4 grooves rh
Muzzle velocity: 280 m/s
Rate of fire: (cyclic) 450 rds/min

Manufacturer
Guide Lamp Division, General Motors Corp, Anderson, Indiana. Ithaca Gun Co, Ithaca, NY. Buffalo Arms Co, Buffalo (bolts only).
Status
Obsolete.
Service
Present use problematical but may be found in South America and South-east Asia.

Ares FMG folding sub-machine gun

The Ares FMG has been designed to provide a compact weapon with high firepower which can be folded and carried in a volume no greater than that of a cigarette carton. It can also be carried concealed under clothing and unfolded and brought into action very rapidly. With practice it can be opened in two or three seconds and is immediately ready to fire.

The FMG consists of two metal shells, one containing the barrel and bolt assembly and the other, dimensioned so as to fold over the first, acting as the butt. A magazine housing is attached to the barrel/bolt casing in such a manner that it folds forward into the space beneath the barrel. There is a simple cocking handle below the muzzle, and the trigger and guard are folded as the magazine is moved. The weapon can be carried and folded with a loaded 20-round or 32-round magazine in place. At present it is made of metal, but it is anticipated that composite materials will be adopted in future development, resulting in a reduction in weight.

DATA

Calibre: 9 × 19 mm Parabellum
Operation: blowback, selective fire
Feed: 20- or 32-round box magazine
Length, folded: 262 mm
Length, opened: 503 mm
Depth, folded: 84 mm
Depth, opened, with 20-round magazine: 222 mm
Width, folded: 36 mm

Weight, with loaded 20-round magazine: 2.38 kg
Rate of fire: 650 rds/min

Manufacturer
Ares Inc, Building 818, Front Street, Erie Industrial Park, Port Clinton, OH 43452.
Status
Advanced development.

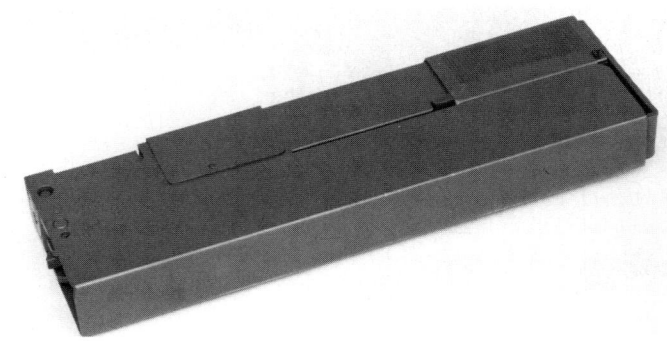

Ares FMG sub-machine gun folded

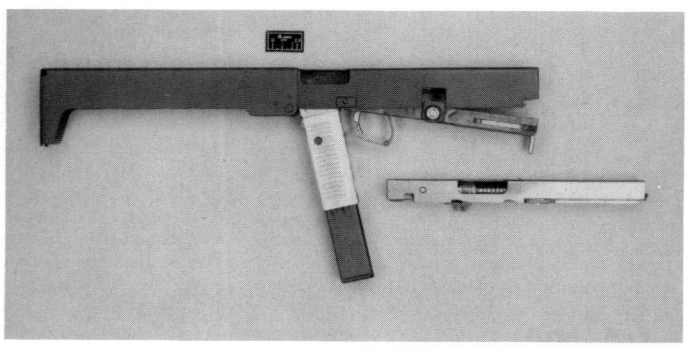

Ares FMG sub-machine gun about to fold, with barrel and bolt unit removed

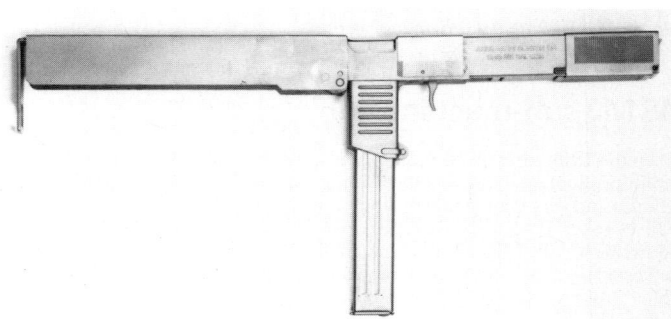

Ares FMG folding sub-machine gun ready to fire

9 mm Saco Model 683 sub-machine gun

This is a conventional blowback weapon designed with a firm eye on prime cost, ease of manufacture, and simplicity of operation and maintenance. The majority of parts are stamped steel, the butt is an unusual tubular telescoping design which slides over the receiver, and there is an unusual combined carrying handle and tubular sight. Castings, plastic mouldings and screw-machine parts are also used in the construction, and the whole design has been aimed at a weapon which can be easily manufactured in countries without high-grade weapon fabricating facilities.

DATA

Cartridge: 9 mm Parabellum
Operation: blowback, selective fire
Weight: (empty) 3.31 kg
Weight: (25-round magazine) 3.81 kg
Length: (butt folded) 520 mm; (butt extended) 699 mm
Barrel length: 203 mm
Feed: 25- and 32-round box magazines
Rate of fire: (cyclic) 650 rds/min
Muzzle velocity: ca 396 m/s

Manufacturer
Saco Defense Inc, a subsidiary of the Chamberlain Manufacturing Corporation, 291 North Street, Saco ME 04072.
Status
Development completed.

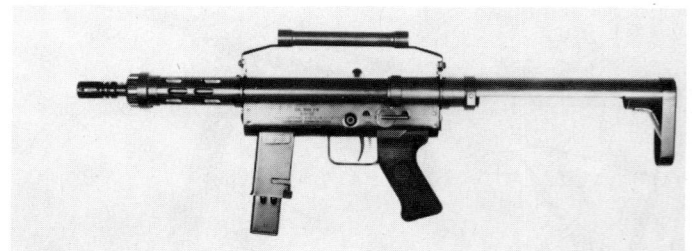

9 mm Saco Model 683

9 mm Colt sub-machine gun

The Colt 9 mm sub-machine gun is a light and compact delayed blowback weapon which embodies the same straight line construction and design as the well-known Colt M16A2 rifle. This straight line construction, coupled with the lower recoil impulse of the 9 mm Parabellum cartridge, provides highly accurate fire with less muzzle climb.

As with the M16 rifle, the sub-machine gun fires from a closed bolt and the chamber remains open after the last shot has been fired. This feature allows the user to replace magazines and re-open fire more rapidly. It is equipped with a rigid sliding buttstock and is readily field-stripped without the need for special tools. Operation and training is similar to the M16A2 rifle, Carbine or Commando, thus eliminating the need for extensive cross-training.

The Colt 9 mm SMG fires all standard 9 mm Parabellum ammunition, fed from 20- or 32-round magazines. It is extremely well suited for military and para-military organisations with a defined need for a lightweight compact weapon with sustained and controlled fire in close confrontations.

The US Drug Enforcement Agency (DEA) has adopted the weapon, after conducting extensive tests which included other 9 mm sub-machine guns. Other US armed forces and law enforcement agencies also use the weapon and it is under evaluation by various governmental agencies throughout the world.

DATA
Cartridge: 9 × 19 mm Parabellum
Operation: blowback, selective fire
Feed: 20- or 32-round box magazine
Weight: (gun without magazine) 2.59 kg
Length: (stock extended) 730 mm; (stock retracted) 650 mm
Barrel length: 260 mm
Sights: (Fore) adjustable post; (rear) flip aperture, adjustable for windage & elevation
Rate of fire: 800-1000 rds/min
Muzzle velocity: 397 m/s
Muzzle energy: 60:J
Max effective range: 150 m

Manufacturer
Colt Firearms, Hartford, CT 06101.
Status
Current. Production.
Service
US DEA, US Marine Corps and other countries.

9 mm Colt sub-machine gun, right side; note spent case deflector behind ejection port

9 mm Colt sub-machine gun, left side

Ingram sub-machine gun

Gordon B Ingram designed a series of sub-machine guns after he returned to the USA following the Second World War. All his weapons have been simple designs firing at automatic.

Description
The Models 10 and 11 are of the same basic design and differ in the weight and length dictated by the cartridge used. The Model 10 is chambered for the .45 Automatic Colt Pistol (ACP) or the 9 mm Parabellum. The Model 11 takes the 9 mm Short cartridge (.380 ACP). The weapons are very short, very compact, solidly built and made from steel pressings. The bolts are of the wrap-round type with the bolt-face and fixed firing pin placed well back to allow the greater part of the bolt to envelop the breech. With the bolt forward all openings are closed and no dirt can get into the action.

Both Models 10 and 11 are externally threaded at the muzzle to take the MAC suppressor. The suppressor differs from the conventional silencer in that the bullet is allowed to reach its full velocity and therefore becomes supersonic. The suppressor is added to the muzzle and is intended to reduce the emergent gas velocity to a subsonic level. The suppressor consists of helical channels going forward from the muzzle of the gun which meet similar channels coming back from the front of the screwed-on tube. The meeting of the two gas streams results in a dissipation of their energy inside the suppressor. The suppressor tube is covered with Nomex-A heat-resistant material.

The cocking handle is on top of the receiver and, like the early Thompson sub-machine guns, has a U-notch to allow an unimpeded line of sight. When the bolt is in the closed position, rotating the cocking handle through 90° locks it. The firer is, of course, warned that his bolt is locked because he can no longer use his sights. There is a second safety catch on the right of the trigger-guard, forward of the trigger. When it is pulled to the rear the bolt is locked in either the forward or cocked positions.

The shoulder stock pulls out for firing from the shoulder and pushes in when the gun is to be fired from the hip.

DATA	Model 10	Model 10	Model 11
Cartridge	.45 ACP	9 mm Parabellum	9 mm Short (.380 ACP) or 9 mm Parabellum
Operation	blowback	blowback	blowback
Feed, box magazine	30 rounds	32 rounds	16 or 32 rounds
Firing mode	selective	selective	selective
WEIGHTS			
Gun			
empty	2.84 kg	2.84 kg	1.59 kg
loaded, 16-round magazine	—	—	1.872 kg
30-round magazine	3.818 kg	—	—
32-round magazine	—	3.46 kg	2.10 kg
Suppressor	0.545 kg	0.545 kg	0.455 kg
LENGTHS			
Gun			
no stock	267 mm	267 mm	222 mm
stock telescoped	269 mm	269 mm	248 mm
stock extended	548 mm	548 mm	460 mm
Barrel	146 mm	146 mm	129 mm
Suppressor	291 mm	291 mm	224 mm
MECHANICAL FEATURES			
Barrel	6 grooves rh	6 grooves rh	6 grooves rh
	1 turn in 508 mm	1 turn in 305 mm	1 turn in 305 mm
FIRING CHARACTERISTICS			
Muzzle velocity	280 m/s	366 m/s	293 m/s
Rate of fire: (cyclic)	1145 rds/min	1090 rds/min	1200 rds/min

Accessories: long barrel with special receiver to increase the effective range and improve the power of the bullet. Full carrying equipment in leather.

Manufacturers
Various
Status
No longer in production as the Ingram. See below.
Service
Bolivia, Colombia, Guatemala, Honduras, Israel, Portugal, the UK, the USA and Venezuela

Ingram Models 10 and 11 sub-machine guns
(top) *Model 10 .45ACP* **(centre)** *9mm Parabellum* **(bottom)** *9mm Short*

Cobray M11 sub-machine guns

The Ingram M11 design, described above, is currently being manufactured as the 'Cobray M11'. Two types are produced, the M11 for the 9 mm Short (.380 Auto) cartridge and the M11/9 for the 9 mm Parabellum cartridge.

DATA
As for Ingram M11 above

Manufacturer
SWD Inc, 1872 Marietta Boulevard NW, Atlanta, GA 30318.
Status
Production.

5.56 mm Ruger AC-556F sub-machine gun

This is the mechanism of the Ruger AC-556 infantry rifle (for details of which see the Rifles section) but with a very short barrel and folding stock, reducing the dimensions so that it is compact enough to be used as a sub-machine gun. As with the rifle, the weapon has a selector which permits firing single shots, automatic fire, or three-round bursts.

DATA
Cartridge: 5.56 × 45 mm (.223)
Operation: short-stroke gas piston
Locking system: rotating bolt
Weight: (empty) 3.27kg; (with 20-round magazine) 3.50 kg
Length: (butt extended) 825.5 mm; (butt folded) 584 mm
Barrel length: 292 mm (330 mm with flash suppressor)
Feed: 20- or 30-round magazine

Manufacturer
Sturm, Ruger & Co Inc, Ruger Road, Prescott, AZ 86301.
Status
Current. Production.

5.56 mm Ruger AC-556F

KF-AMP assault machine pistols

The KF-AMP series of weapons have been designed as light, compact and accurate 'third generation' sub-machine guns having particular application to the needs of special, anti-terrorist, security and other forces engaged in urban, jungle and other close-quarter environments. The weapons are assembled from steel pressings and machined components and there is an extensive array of accessories which can be added to the basic weapon in order to fit it for specific tasks. These include sound suppressors, magazines of varying capacity, night, electronic, optical or iron sights, muzzle brakes, barrel extensions, and various holsters slings and straps for carriage. The muzzle is always threaded so as to accept a suppressor or muzzle brake if required. The cocking handle can be provided on the right or left side of the receiver, as required, and the weapon is made safe by engaging the cocking handle into a safety notch at the rear of the cocking handle slot. Operation is blowback, from an open bolt, and the rate of fire can be varied to order by substitution of the bolt return spring.

There are, currently, five models available:
KF-9-AMP   Standard 9 mm Parabellum, no forward grip
KF-59-AMP Similar to KF-9-AMP but with forward grip
KF-3-AMP   As KF-9-AMP but in 9 mm Short/.380ACP calibre
KF-11-AMP As KF-9-AMP but in .45ACP calibre
KF-54-AMP As KF-59-AMP but in .45ACP calibre

DATA
(KF-9-AMP)
Cartridge: 9 × 19 mm Parabellum
Operation: blowback, selective fire
Feed: 20-, 36-, 60- or 108-round box magazines
Weight, empty: 1.13 kg
Length: (stock extended) 603 mm; (stock folded) 273 mm
Barrel length: 76.2 mm, stainless or chrome-molybdenum steel
Rifling: polygonal, 6 grooves, one turn in 400 mm
Rate of fire: 800 rds/min

DATA
(KF-59-AMP) As above, except:
Weight: 1.27 kg

9mm KF-9-AMP assault machine pistol with stock folded

Length: (stock extended) 648 mm; (stock folded) 394 mm
Barrel length: 127 mm

DATA
(KF-3-AMP) As for KF-9-AMP except that calibre is 9 mm Short

DATA
(KF-11-AMP)
Cartridge: .45 ACP
Weight: 1.36 kg
Length: (stock extended) 626 mm; (stock folded) 356 mm
Barrel length: 89 mm

DATA
(KF-54-AMP) As above except:
Weight: 1.60 kg
Length: (stock extended) 668 mm; (stock folded) 406 mm
Barrel length: 127 mm

Manufacturer
Arms Research Associates, 1800 North Mannheim Road, Stone Park, IL 60165.
Status
Current. Available.

VIETNAM

7.62 mm K-50 Modified sub-machine gun

The Chinese copy of the Soviet PPSh-41 sub-machine gun is called the Type 50 sub-machine gun. The Viet Cong in Vietnam modified the Chinese weapon and named it the K-50 Modified (K-50 M) sub-machine gun. The major modifications are the removal of the upward folding butt and the substitution of a sliding metal butt stock of the type used in the French MAT 49 sub-machine gun (which see). The barrel jacket has been shortened and squeezed in at the second slot. The muzzle brake-cum-compensator has been dispensed with. The lower receiver has been reshaped and a pistol grip added. The foresight has been placed on the barrel.

The weapon is loaded and operates in exactly the same way as the Soviet PPSh-41 (which see).

DATA
Cartridge
7.62 mm × 25 Type P pistol
7.62 mm × 25 PRC Type pistol cartridge
7.63 mm Mauser
Operation: blowback
Feed: 35-round box magazine
Weight, empty: 3.4 kg
Length: (stock extended) 756 mm; (stock retracted) 571 mm
Barrel: 269 mm
Muzzle velocity: 488 m/s
Rate of fire: (cyclic) 700 rds/min

Manufacturer
Local workshops.
Status
Obsolescent.
Service
Not known, but likely to be still used.

7.62 mm K-50 Modified sub-machine gun

7.62 mm MAT 49 Modified sub-machine gun

The North Vietnamese had a number of French 9 mm MAT 49 sub-machine guns which they captured during the French Indo-China campaign. They decided to convert these to take the Soviet 7.62 mm × 25 pistol cartridge and did so by replacing the French 9 mm barrel with a much longer barrel. All the typical features of the MAT 49, the grip safety, the dust cover and the sliding rod stock, have been retained.

7.62 mm MAT 49 Modified sub-machine gun

YUGOSLAVIA

7.62 mm M49 and M49/57 sub-machine guns

The M49 and M49/57 sub-machine guns are based on the Soviet PPSh-41 sub machine gun. Although the general method of operation is the same there are several detail differences. The methods of construction are not the same.

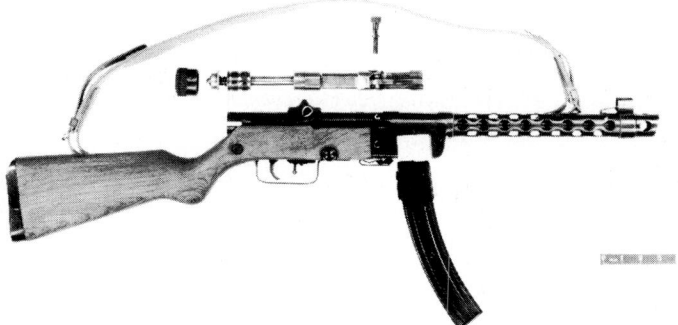

7.62 mm × 25 M49/57 sub-machine gun, direct copy of Soviet PPSh-41

Whereas the Soviet weapon used stamped parts, the Yugoslav gun is constructed from machined parts or drawn tube. The M49 and M49/57 have a large push through safety in the forestock in front of the trigger-guard; the barrel jacket has small circular holes rather than long slots. The buffer assembly, consisting of a separate buffer spring and split rings, is positioned at the end of the return spring guide. It resembles that used in the Beretta M38A.

There is no fitting for the Soviet 71-round drum magazine but the Yugoslav guns will not only take their own 35-round curved box magazine but will also take the Soviet magazine from the PPSh-41 and that from the Chinese Type 50 sub-machine gun (which see).

The M49 and M49/57 sub-machine guns differ only in unimportant details.

DATA
Cartridge: 7.62 mm × 25 Type P pistol or 7.63 mm Mauser
Operation: blowback, selective fire
Feed: 35-round curved box magazine

Weight: (unloaded) 3.95 kg; (loaded) 4.54 kg
Length: 870 mm
Barrel: 273 mm
Rifling: 4 grooves rh
Sights: (fore) blade; (rear) flip, notch 100 and 200 m
Muzzle velocity: 500 m/s
Rate of fire: (cyclic) 700 rds/min
Effective range: 200 m

Manufacturer
Zavodi Crvena Zastava, Kragujevac.
Status
Obsolete. Not in production.
Service
Yugoslav reserve and militia units.

7.62 mm M56 sub-machine gun

This sub-machine gun is similar in appearance to the German MP 40. The barrel looks very long and fragile and it takes a knife pattern bayonet. The cocking handle is on the right of the receiver and there are locking slots at each end of the cocking way allowing the bolt to be secured in either the forward or cocked positions. The folding stock is of the same pattern as the MP 40.

The internal mechanism is much simpler than was the MP 40 and there appears to be less insistence on keeping dirt out of the bolt way. A feature of the body is that there are several large cut-outs in the lower half of it, and the return spring is actually exposed for a part of its length. The bolt slot has no cover and it could be assumed that the entire gun is highly vulnerable to the effects of dirt and snow clogging the mechanism. The body is longer than with most contemporary sub-machine guns, with the magazine at the forward end of it. This makes it difficult to find a proper hold with the left hand, and the designer has incorporated the forward hand grip in the lower part of the body, just behind the magazine. The firing position is thereby distinctive and unusual.

DATA
Cartridge: 7.62 mm × 25 Type P pistol or 7.63 mm Mauser
Operation: blowback, selective fire
Feed: 35-round box magazine
Weight: 3 kg
Length: (stock extended) 870 mm; (stock folded) 591 mm
Barrel: 250 mm
Muzzle velocity: 500 m/s
Rate of fire: (cyclic) 600 rds/min

Manufacturer
Zavodi Crvena Zastava, Kragujevac.
Status
Current. Not in production.
Service
Yugoslav forces.

Yugoslav troops with 7.62 mm M56 sub-machine gun

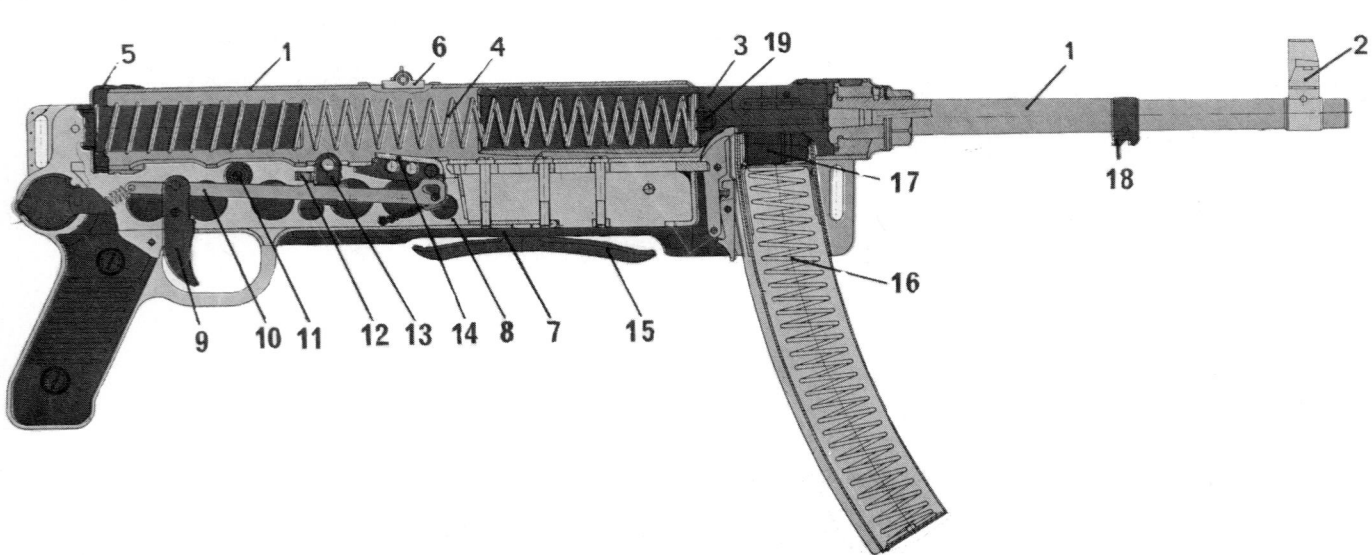

7.62 mm M56 sub-machine gun
(1) *receiver and barrel* **(2)** *foresight* **(3)** *bolt* **(4)** *return spring* **(5)** *rear cap* **(6)** *rearsight* **(7)** *left-hand grip* **(8)** *lower frame* **(9)** *trigger* **(10)** *trigger bar* **(11)** *safety button* **(12)** *selector catch* **(13)** *selector lever* **(14)** *sear* **(15)** *folded metal stock* **(16)** *30-round magazine* **(17)** *7.62 mm cartridges on the magazine platform* **(18)** *bayonet lug* **(19)** *cocking handle*

7.65 mm Model 84 machine pistol

Yugoslavia manufactures the Czechoslovak 7.65 mm Model 61 pistol (Skorpion) under licence. The pistol is identical with the Czechoslovak model (which see).

Manufacturer
Zavodi Crvena Zastava, Beograd 29, Novembra 12.
Status
Current. Production.
Service
Yugoslav forces.

7.65 mm Model 84 machine pistol

5.56 mm Zastava M85 sub-machine gun

This appears to be a variant of the M80 assault rifle which resembles the Soviet AKSU-74 in general appearance and intention. It uses the standard Kalashnikov-pattern rotating bolt gas action with a short barrel which is fitted with a tubular muzzle attachment which serves as a flash hider and also as a ballistic regulator to compensate for firing the 5.56 × 45 mm cartridge is a short barrel. The folding stock is a two-strut type which folds beneath the receiver to rest with its butt under the fore-end.

5.56 mm Zastava M85 sub-machine gun

According to the manufacturers, the M85 is intended for use by personnel in combat vehicles, helicopters and aircraft, and by special forces and internal security forces.
DATA
Cartridge: 5.56 × 45 mm
Operation: gas, selective fire
Locking system: rotating bolt
Feed: 20- or 30-round box magazine
Weight: 3.20 kg empty without magazine
Length: butt extended 790 mm; folded 570 mm
Barrel length: 315 mm with flash hider (estimated)
Sights: fore: hooded post; rear: flip U-notch 200 and 400 m. An optical sight may be fitted
Rate of fire: 650 – 750 rds/min
Muzzle velocity: ca 790 m/s
Recoil energy: 4 J

Manufacturer
Zavodi Crvena Zastava, Beograd 29, Novembra 12.
Status
Current. Production.

.22 MGV 176 sub-machine gun

This appears to be the latest incarnation of a design which first appeared in the USA in the 1960s as the Casull Carbine, then became the American 180, was briefly manufactured in the USA as a semi-automatic weapon, and in 1972 was licensed to Voere in Austria who changed it to a selective-fire weapon and produced about 5000 before ceasing manufacture.

The MGV 176 shows some minor changes, perhaps to facilitate manufacture and improve reliability, but it is basically the same blowback actuated weapon, chambered for a .22 rimfire high velocity cartridge with full-jacketed bullet or the standard .22 Long Rifle cartridge. There is a change lever on the left side and a grip safety. A simple wire stock folds beneath the receiver, where the butt section can act as a forward grip. The top-mounted drum magazine is of translucent plastic and contains 176 rounds; it takes about four minutes to load. A simple fixed aperture sight is used, with provision for mounting a laser spot marker.

The low recoil force of this weapon, even at full automatic fire, allows it to be held on target quite easily, and in spite of its small calibre it has an undoubted place in the armoury of special forces, internal security forces, and

as a personal defence weapon for mobile troops and aircrews. A sound suppressor may be fitted to the muzzle.

DATA
Cartridge: .22 rimfire, special or standard
Operation: blowback, selective fire
Feed: 176-round drum magazine
Weight: 2.78 kg with empty magazine; 3.45 kg with full magazine
Rate of fire: 1600 rds/min
Practical range: to 150 m

Manufacturer
Federal Directorate of Supply & Procurement, 9 Nemanjina Street, 11001 Beograd 9.
Status
Current. Production.

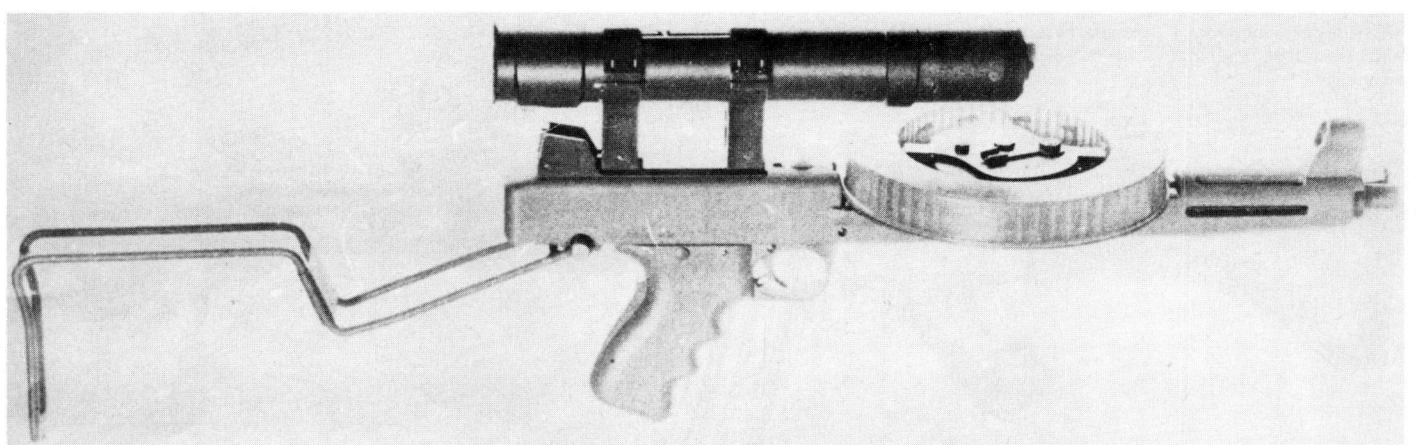

.22 MGV 176 sub-machine gun with stock folded and laser spot marker

Rifles

ARGENTINA

7.62 mm FN pattern rifles

Formerly equipped with Mauser-type 7.65 mm rifles, the Argentinian Army has for some time used the Fabrique Nationale Fusil Automatique Légér (FN-FAL) NATO-pattern rifle and the FN heavy-barrelled rifle which are made under FN licence at the ordnance factory at Rosario.

Three models are currently manufactured. The Fusil Automatico Liviano Modelo IV is the standard FN-FAL Model 50-00 and is identical in all respects other than being a few grammes heavier than the Belgian-manufactured weapon. The Fusil Automatico Liviano Modelo Para III is the FN-FAL Model 50-64 with side-folding butt and standard barrel. This differs slightly from the Belgian model, being 1092 mm long with extended butt, 897 mm long with butt folded, and 4.25 kg in weight when empty.

The third model is the Fusil Automatico Pesado Modelo II, which is the heavy-barrelled FN-FAL Model 50-41, with bipod. This also differs slightly in dimensions from the Belgian version, weighing 6.45 kg and being 1125 mm long.

The Fusil Automatiico Liviano Cal .22LR is a locally developed version of

the FAL which is chambered for the commercial .22 Long Rifle rimfire cartridge and is used for training. Details are given below.

DATA
Cartridge: .22 Long Rifle rimfire
Operation: blowback, selective fire
Feed: 20-round box magazine
Weight, empty: 4.60 kg
Length: 1090 mm
Barrel: 533 mm
Rifling: 6 grooves, rh, one turn in 400 mm
Sights: (fore) protected post; (rear) two-range flip, 50 and 100 m
Rate of fire: cyclic 800 rds/min
Muzzle velocity: 330 m/s

The Fabrica Militar also manufactures another .22 rifle, known as the Carabina Calibre .22 Modelo Militar. This generally resembles the well-known US M1 carbine and is a semi-automatic operating on blowback principles. It is apparently designed for training purposes but will also have application in the security and police fields. Details are given below.

DATA
Cartridge: .22 Long Rifle rimfire
Operation: blowback, semi-automatic
Feed: 10-round box magazine
Weight, empty: 2.5 kg
Length: 1020 mm
Barrel: 500 mm
Rifling: 6 grooves, rh, one turn in 400 mm
Sights: (fore) post; (rear) two-range flip, 50 and 100 m
Sight radius: 650 mm

.22 LR calibre Fusil Automatico Liviano

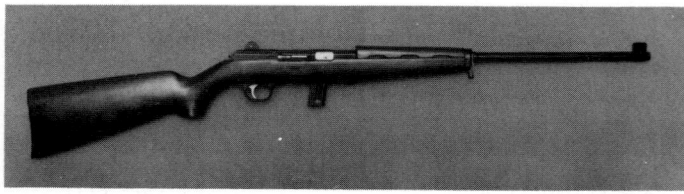
.22 LR Carabina Modelo Militar

Manufacturer
Fabrica Militar de Armas Portatiles 'Domingo Matheu', Avenida Ovidio Lagos 5250, Rosario 2000.
Status
Current. Production
Service
Argentinian forces and for export.

5.56 mm FARA 83 assault rifle

This is a gas-operated selective-fire rifle using a conventional gas piston arrangement to actuate a bolt carrier and two-lug rotating bolt. Manufacture is from steel forgings and metal pressings, the receiver being a welded sheet metal construction to which sub-assemblies are attached by rivets. The butt is of glass-reinforced plastic and is hinged so as to fold alongside the receiver. The fore-end is of plastic material.

The rear sight is a three-position drum, adjustable for windage by means of two screws. Two of the positions are for daylight firing, giving ranges of 200 m and 400 m, while the third position is a notch with Tritium inserts for firing at 100 m in poor light. There is also an auxiliary foresight fitted with a Tritium source for use in poor light, which can be folded down when not in use.

A bipod is available as an optional fitment; this can be attached below the gas block, and a special fore-end is fitted which has a recess to take the bipod legs when folded.

Issue of the FARA 83 rifle began in early 1984 but was halted after about 1000 rifles were made due to budgetary restrictions. Small numbers have been made since that time, and it will be generally issued as and when finances permit.

DATA
Cartridge: 5.56 × 45 mm M193 or SS109
Operation: gas, selective fire
Locking: rotating bolt
Feed: 30-round box magazine
Weight, empty: 3.95 kg
Length: 1000:mm butt extended; 745 mm butt folded
Barrel: 452 mm
Rifling: 6 grooves, rh, one turn in 229 mm
Sights: see text
Muzzle velocity: 965-1005 m/s depending upon ammunition
Cyclic rate: 750-800 rds/min

Manufacturer
Fabrica Militar de Armas Portatiles 'Domingo Matheu', Avenida Ovidio Lagos 5250, Rosario 2000.
Status
Current. Production.
Service
Argentinian forces.

5.56 mm FARA 83 assault rifle

AUSTRALIA

7.62 mm L1A1 and L1A1-F1 rifles

The standard Australian rifle is the L1A1 manufactured at Lithgow and conforming in most respects to the British 7.62 mm rifle. This is still current in spite of the use of the US 5.56 mm M16A1 which was employed by Australian troops in Vietnam only as a replacement for the F1 sub-machine gun. The L1A1 will be in service until 1993 and will be progressively phased out as the Australian Army introduces the F88 Australian version of the Steyr 5.56 mm AUG weapon. During this interim period limited spares will be available.

The L1A1-F1 is a modified version of the standard 7.62 mm self-loading rifle developed to meet a special requirement for a shorter personal weapon. It is a comfortable weapon particularly suited to personnel of smaller stature and for operating in close country.

The flash eliminator has been redesigned and the short butt is used effecting an overall reduction in weapon length of 70 mm (2¾ inches). The external profile of the flash eliminator has been retained and it is fitted over the muzzle as a sleeve and employed as a flash eliminator/muzzle brake. As a result there is a 20 per cent reduction in recoil energy and the modified eliminator is effective in reducing flash. The modification to the weapon does not affect its ability to accept the standard style of attachments, its reliability, its function or its accuracy.

DATA
L1A1
Cartridge: 7.62 mm × 51
Operation: gas, semi-automatic
Method of locking: tilting block
Feed: 20-round box magazine
Weight: (complete weapon) 5.443 kg
Length: 1136 mm
Barrel: 533 mm
Sights: (foresight) blade; (rearsight) aperture 200-600 m
Muzzle velocity: 823 m/s
Effective range: 600 m

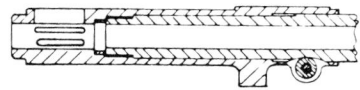
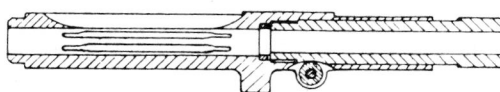

Standard and modified flash eliminator

L1A1-F1
As L1A1 except:
Weight: (complete weapon) 4.91 kg
Length: 1066 mm

Manufacturer
Australian Defence Industries, Lithgow Facility, New South Wales.
Status
Current. No longer in production; limited spares available.
Service
Australian Defence Force.

Australian soldier checking barrel of 7.62 mm L1A1 rifle

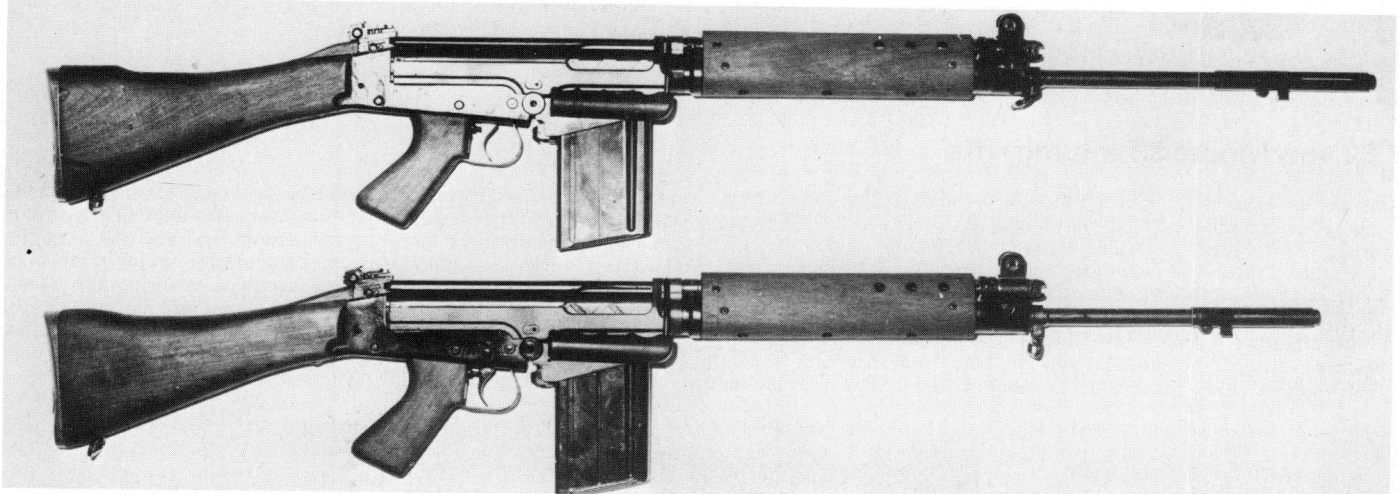

7.62 mm L1A1 and L1A1-F1 shortened version

7.62 mm L2A1 rifle

This is a heavy-barrelled version of the L1A1 rifle used to give some measure of supporting fire to units other than infantry.
Infantry are equipped with the 7.62 mm L4A4 Bren light machine gun.
This rifle is very similar to the standard Fabrique Nationale and Canadian C1 heavy-barrelled rifles.

DATA
Cartridge: 7.62 mm × 51
Operation: gas, selective fire
Feed: 30-round detachable box magazine
Weight: 6.9 kg

Length: 1137 mm
Barrel: 533 mm
Rifling: 6 grooves rh
Sights: (foresight) blade; (rearsight) aperture
Muzzle velocity: 838 m/s
Rate of fire: (cyclic) 675-750 rds/min
Effective range: 800 m max

Manufacturer
Australian Defence Industries, Lithgow Facility, New South Wales.
Status
Current. No longer in production. Spares available.
Service
Australian Defence Force.

7.62 mm L2A1 heavy-barrelled rifle

F1A1 magazine filler

To enable soldiers to load the magazines used with the L1A1 and L2A1 rifles more quickly and with less effort, the Australian Defence Engineering Development Establishment has developed a magazine filler which will accept either loose rounds or five-round chargers. It is said by using this device, the time to load a 20-round magazine is reduced from a minimum of 30 seconds to 10 seconds and the soldier has full control of the rounds during the loading process.

As is clearly indicated by the accompanying illustration the filler is fitted over the magazine and charged either with loose rounds, up to four at a time, or a five-round clip; the rounds are then pressed out of the filler and into the magazine. Smooth operation is obtained by coating the filler with Nylon II.

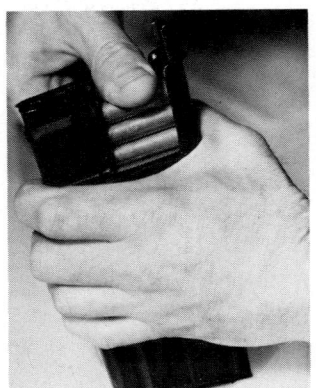

Magazine filler

Using magazine filler

DATA
Weight: 57 g
Dimensions: 79 × 69 × 25 mm

Manufacturer
Australian Defence Engineering Development Establishment, Maribyrnong, Victoria.
Status
Current. Production.
Service
Australian Army.

7.62 mm Model 82 sniping rifle

Australia has adopted the Parker Hale rifle for sniping. The rifle 7.62 mm sniper system comprises a Parker Hale Model 82 rifle with a 1200 TX barrel which has been matched to a Kahles Helia ZF 69 telescopic sight.

The scope is fitted with two bridge clamps dovetailed in the base to slide into the rear and front blocks of the receiver. It features a reticule pattern No 7, fixed magnification of 6 × 42 and visual and click adjustments for a range of 100 to 800 m in 50 m increments. Further click adjustments up to a range of 900 m are available although visual markings extend only to 800 m.

The Model 82 is also fitted with a Parker Hale 5E aperture rearsight and tunnel foresight. The 5E iron sights enable the weapon to be used as a sniping rifle in the event of the telescopic sight being damaged. The rear iron sight is mounted to the rearsight block on the receiver and contains a vernier adjustment for elevation and windage, which allows for zeroing at ranges of 100 to 1000 m. A carrying container is provided for the rearsight which cannot be attached to the rifle until the telescopic sight is removed.

The remaining details of this system are given in the section on the United Kingdom.

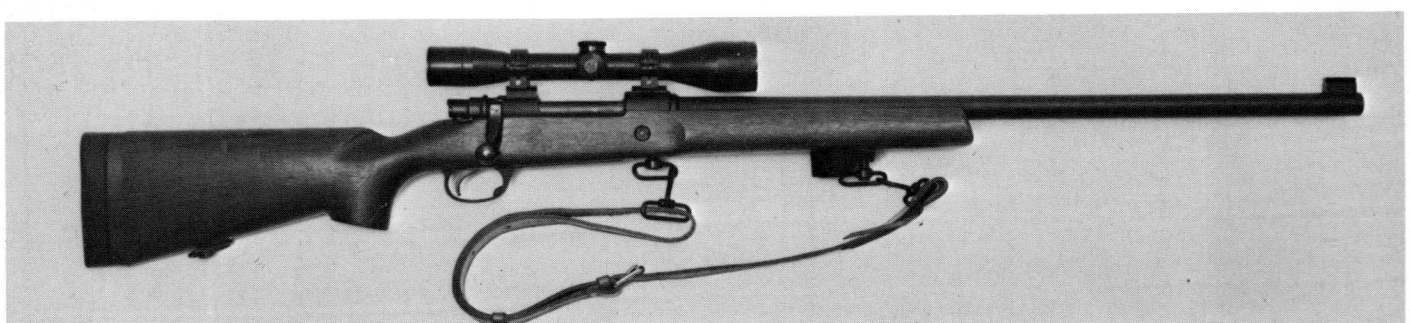

7.62 mm Model 82 sniping rifle

F88 Weapon System

The F88 System is the Australian-manufactured version of the Steyr AUG.

The F88 is an extremely simple weapon to operate. The large safety catch requires a simple but definite lateral movement to lock or unlock the trigger. The gun fires in semi-automatic mode when the trigger is pressed to a clearly felt point, then fully automatic when it is fully depressed. This gives the soldier total control over weapon function without the need to operate a change lever. Stripping for field cleaning and maintenance is performed without the need for any tools.

In all F88 models the barrel is connected to the receiver by eight locking lugs, an arrangement which gives maximum security of barrel location whilst retaining ease of barrel removal. Removal requires only a 22.5° rotation of the barrel in the housing assembly. This allows the weapon to be easily stripped and stowed in a knapsack. Its longest component is no more than 550 mm in length. All barrel styles are fully interchangeable with either receiver type. This allows a weapon to be instantly converted between the standard assault rifle, the machine carbine, the commando weapon or the heavy-barrelled light support weapon.

The F88 features an optical sight integrated into the weapon's carrying handle. This sight offers a magnification of 1.5 times, which allows a wide field-of-view (45 m at a range of 300 m) and allows the soldier to fire with both eyes open.

In adverse light conditions the optical sight will perform effectively long after open sights have ceased to be useful. The standard circular graticule covers a height of 1.8 m at a range of 300 m. This system allows for target acquisition in typically 1.5 s compared with 3 s for open sights. The circular graticule also provides an effective means for the soldier to estimate and compensate for target range without having to adjust the sight.

The 'bullpup' design makes the F88 an extremely short weapon. The overall length with the standard 508 mm barrel is only 709 mm. This is of tremendous advantage not only for mechanised troops but for any soldier since it makes him faster and therefore superior. With a 407 mm barrel fitted the overall length is reduced to 690 mm, and this eliminates the need for any consideration of a folding stock version of the weapon.

Full technical data on the F88 can be found in the Steyr AUG entry in the Austrian section below.

Manufacturer
Australian Defence Industries, Lithgow Facility, New South Wales (under licence from Steyr-Mannlicher GmbH, Austria).
Status
Current. Production.
Service
Australian and New Zealand Defence Forces.

5.56 mm F88 assault rifle, Australian manufacture

AUSTRIA

15mm AMR 5075 Anti-Material Rifle

This is a heavyweight precision rifle designed as a relatively inexpensive system for the long-range attack of vulnerable equipment such as light armoured vehicles, aircraft on the ground, fuel and supply dumps, radar installations and similar targets. It can be dismantled into two units for man carriage.

The rifle is a semi-automatic bullpup, using plastics and light metal to reduce the weight as far as is consistent with the strength demanded by its role. The mechanism employs the long recoil principle of operation, the barrel and bolt recoiling for about 20 cm, after which the bolt is unlocked and held while the barrel is returned to battery. The bolt is then released, collects a cartridge from the magazine, and chambers it, locking into the barrel by a rotary motion.

Recoil of the barrel is reduced by a multi-port muzzle brake of considerable efficiency, and is controlled by a hydro-pneumatic annular system carried in a ring cradle forming the front portion of the tubular receiver. The weapon is supported by a bipod, attached above the recoil cradle, and by an adjustable firing pedestal beneath the butt. A 10 × optical sight is fitted as standard.

The box magazine is inserted from the right side, at an angle of about 45° below the horizontal. Currently the magazine holds five rounds, but eight and ten round versions are being evaluated. Other possible future options include a low-rate automatic fire version, and the use of a rifled barrel so as to explore other ammunition design possibilities.

The complete round weighs 150 g. The cartridge case is of part-synthetic construction, of conventional bottle-necked form, and carries a 36 g 5.5 mm fin-stabilised tungsten dart projectile which has a muzzle velocity of 1500 m/s and a striking energy of 1200 N/mm² at 800 m range. It has a practical range of 1000 m and a probable range of 1500/2000 m depending upon the type of target. At 800 m range the current projectile has been demonstrated to pierce 40 mm of RHA plate, and the secondary fragmentation behind the plate is considerable. The high velocity bestows an exceptionally flat trajectory; the vertex at 1000 m range is no more than 80 cm above the line of sight.

DATA
Cartridge: 15mm Special
Operation: long recoil, semi-automatic

15 mm Steyr AMR 5075 anti-material rifle

Method of locking: rotating bolt
Feed: 5-round box magazine
Weight: approx. 20 kg
Length: n/a
Barrel: 1200 mm, smoothbore
Sight: 10 × telescope
Muzzle velocity: 1500 m/s

Manufacturer
Steyr-Mannlicher GmbH, Postfach 1000, A-4400 Steyr.
Status
Development.

7.62 mm SSG 69 rifle

This is the standard sniping rifle for the Austrian Army. It is called the Steyr-Mannlicher-Scharfschützen-Gewehr and is manufactured by Steyr-Mannlicher GmbH at its factory at Steyr. The Austrian Army calls it the Scharfschützengewehr 69 (SSG 69).

It has some features of interest. The barrel is made by the cold forging process, or hammering, in which the billet is placed on a mandrel with the rifling raised in relief. A series of hammers force the rifling on to the tube and form the external contour simultaneously. This process, originally developed by Steyr and now used by many barrel makers, results in work hardening both the bore and the outside surface of the barrel. It also allows, as in the case of the SSG 69, a slightly tapered bore.

The bolt is manually operated and moves through 60°. It has six rotating locking lugs, symmetrically arranged in pairs, set at the rear. Rear locking, although it allows a shorter bolt movement than the Mauser front locking system, has always been considered to be somewhat less desirable since the whole body of the bolt rather than the bolt-head is placed in compression. Also the body of the weapon is weakened by the cut-out on the right-hand side of the body in front of the locking shoulders. In the SSG 69 this has been offset to some extent by strengthening and lengthening the receiver to give the barrel seating a length of 5 cm which places the chamber within the receiver and so makes the barrel-receiver group very rigid.

The trigger mechanism gives a double pull. The length and weight of the pull are both adjustable externally. The safety catch on the top right of the rear of the receiver is of the sliding type. It locks the bolt and the firing pin.

The standard magazine, holding five rounds, is the rotating spool type which has been used on Mannlicher sporting and military rifles for many years. The rifle can, however, be used with a 10-round box magazine.

The stock on the sniping rifle is made of a synthetic material. It can be adjusted for length by the addition or removal of the butt pad giving a long, medium or short butt.

The receiver has a longitudinal rib machined on the top. Mounts, which are available for all types of sight, can be fixed to this rib. The standard sight is the Kahles ZF69. It is graduated up to 800 m and internally adjusted. The sight is attached by lever clamped rings on to the rib. There are also iron sights brazed on to the barrel for emergency use only. An infra-red night sighting system may also be fitted, or an image intensifier sight.

The same rifle with a heavier barrel and a differently shaped butt stock is available for competition shooting. The stock may be obtained either in the same material as the SSG 69 or in walnut. This rifle has match sights: Walther tunnel front and rear Walther micrometer.

Using RWS match rounds the manufacturer claims the following grouping figures:

   5 shots at 100 m 15 mm
   10 shots at 300 m 90 mm
   10 shots at 400 m 130 mm
   10 shots at 600 m 200 mm
   10 shots at 800 m 400 mm.

DATA
Cartridge: 7.62 mm × 51 or .243 Winchester
Operation: manual, single shots
Method of locking: rotating bolt
Feed: 5-round rotary magazine (or 10-round box)
Weight, empty; 3.9 kg; with telescope 4.6 kg
Length: 1140 mm

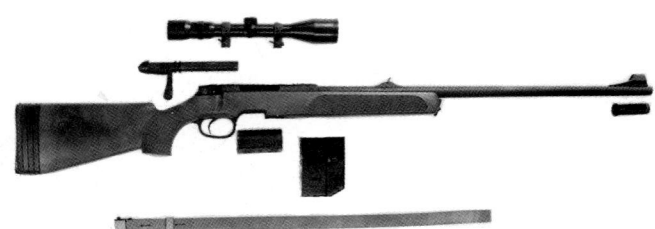

7.62 mm SSG 69 rifle showing alternate magazines and Kahles ZF69 sight

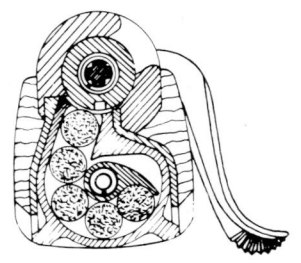

SSG 69 Mannlicher type five-round magazine

Steyr sniping rifle (Model SSG) calibre 7.62 × 51 (308) fitted with Star-Tron image intensifier

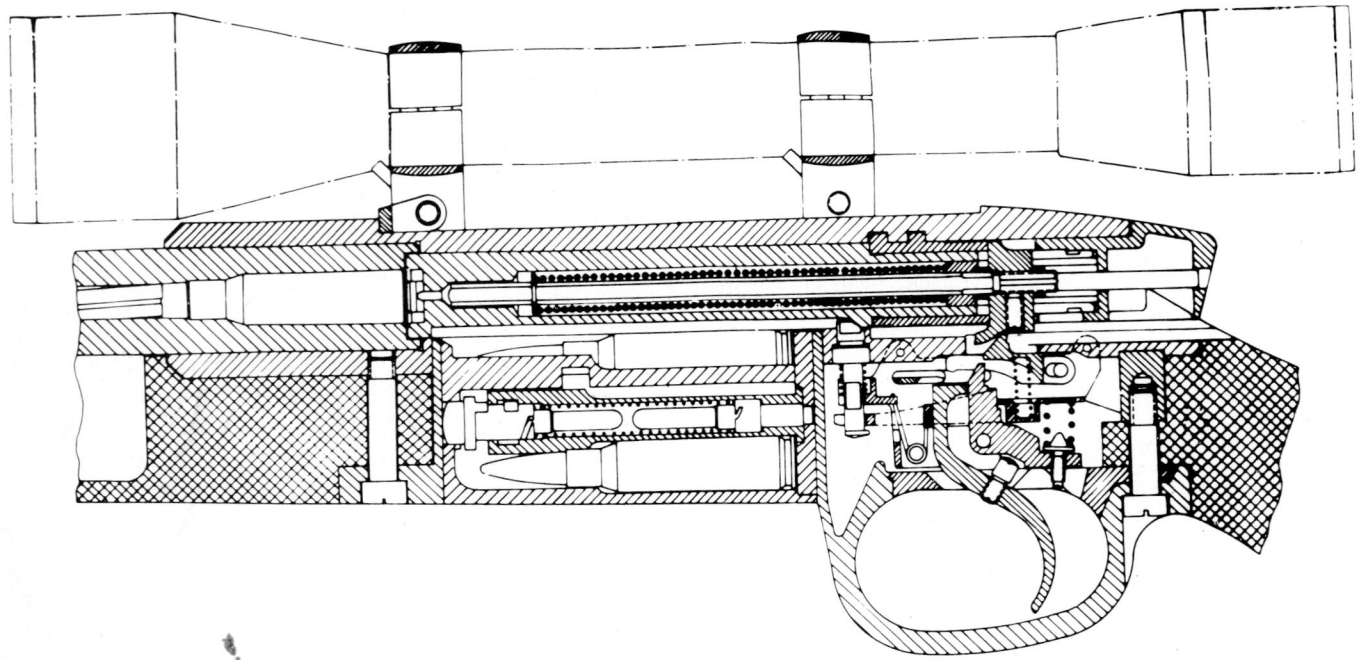

SSG 69 trigger mechanism

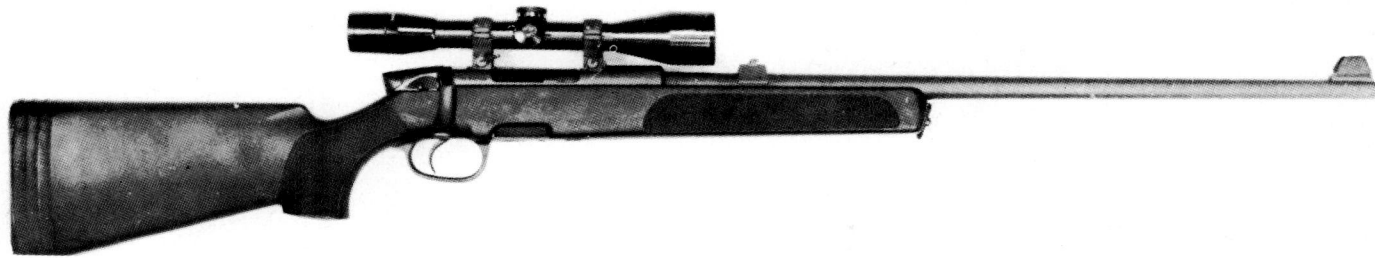

7.62 mm SSG 69 sniping rifle

Barrel: 650 mm
Rifling: 4 grooves rh
Sights: 6× telescope, internally adjustable to 800 m; graticule inverted pointer and broken cross–wires. Emergency iron sights (fore) blade, (rear) V–notch
Muzzle velocity: 860 m/s
Effective range: 800 m

Manufacturer
Steyr-Mannlicher GmbH, a member of the Steyr-Daimler-Puch AG group, Postfach 1000, A-4400 Steyr.
Status
Current. Production.
Service
Austrian Army and several foreign police and military forces.

7.62 mm Steyr Police rifle

This is a special version of the standard SSG rifle, designed for use by police and security forces. It is basically the same as the SSG but the stock is in matt black finish, the heavy barrel of the Steyr Match-UIT rifle is used, the match rifle bolt handle, with extra-large grip is fitted, a detachable shooting sling is mounted on a fore-end rail, and there is a sensitive adjustable set trigger.

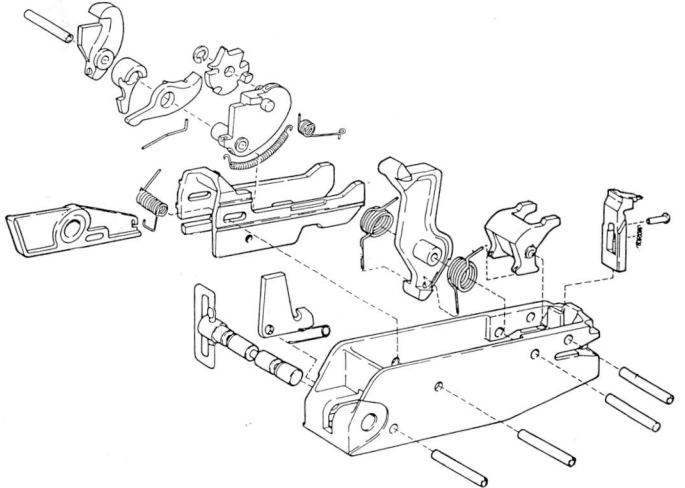

7.62 mm Steyr Police rifle

There are no iron sights on the barrel, a 6 × 42 telescope being provided as standard, although other telescopes can be fitted if preferred.

DATA
Cartridge: 7.62 × 51 mm NATO
Operation: turnbolt repeater
Feed: 5-round rotary or 10-round box magazines
Weight: 4.20 kg
Length: 1135 mm
Barrel: 650 mm
Sights: 6 × 42 telescope standard

Manufacturer
Steyr-Mannlicher GmbH, a member of the Steyr-Daimler-Puch AG group, Postfach 1000, A-4400 Steyr.
Status
Current. Production.
Service
Extensive overseas sales to security and police forces.

5.56 mm Steyr AUG rifle

This modern assault rifle was developed by Steyr in conjunction with the Austrian Army and the first production models came off the line early in 1978. Since then the rifle has been extensively adopted by armies in various parts of the world. The Armee Universal Gewehr (AUG) is designed to be convertible to different modes; it can be used as a parachutist's rifle or sub-machine gun, with a 350 mm barrel; as a carbine, with a 407 mm barrel; as a standard assault rifle, with a 508 mm barrel; and as a heavy-barrel automatic rifle in the light machine gun role, using a 621 mm barrel. The rifle versions all use a 30-round magazine, while the heavy-barrelled version uses a new 42-round magazine. It also has a light bipod. All types have an optical sight integral with the carrying handle, but customers may specify a sight-mounting bracket instead; this can be used to mount a variety of telescope or night-vision sights.

The AUG is a bullpup design of novel appearance. The long slender barrel strikes the eye at first glance, as does the raking optical sight in the integral carrying handle, and the prominent stubby butt. However the intention

behind the weapon is to have a light, handy gun with particular emphasis on use in and from vehicles and commonality of parts for the different modes of use. Every effort has been made to reduce weight and to keep all parts simple and as a result this is a most interesting and attractive weapon.

The AUG has six main groups of components: barrel; receiver; hammer; bolt; magazine; and stock group

Barrel group
The barrel is made from high quality steel and the internal profile is obtained by the cold-hammering process. Both chamber and barrel are chromed to reduce wear and fouling. An external sleeve is shrunk on to the barrel and carries the gas port and cylinder, gas regulator and forward handgrip hinge jaw. There is a short cylinder which contains a piston and its associated return spring. The gas regulator has two positions, normal and large, the latter to be used, when and if, the gun is slowed by fouling or dirt. A flash suppressor is

Exploded drawing of AUG selective hammer mechanism

AUG field-stripped, showing six basic groups

AUG Steyr rifle
(Top) *assault rifle version with standard receiver group* **(centre)** *carbine version, 407 mm barrel* **(bottom)** *commando carbine version (350 mm barrel) and fixed (non-folding) barrel grip*

AUG rifle with image-intensifying sight

5.56 mm AUG rifle with accessories

HBAR-T (621 mm barrel) and special receiver group with telescopic sight

AUG assault rifle with LS-Scope

screwed to the muzzle and it is internally threaded for taking a blank firing attachment; externally it is to the 22 mm standard to permit rifle grenades to be launched without the need for any accesories . This description applies to all three rifle barrels, which differ only in their dimensions. All barrels lock into the body by a system of multi-lugs; each barrel has eight lugs arranged around its chamber end, and these engage with the locking sleeve in the body. The bolt locks into the rear half of the same sleeve.

A forward handgrip is pivoted to the jaws in the gas regulator. This grip is normally used in the vertical position, but it can be folded forward to lie under the barrel for easy carriage or stowing. The grip also rotates the barrel to unlock it and lifts it clear, thus when changing barrels the hand does not touch any hot metal and there is sufficient leverage to release a barrel which has become fouled.

Receiver group
The receiver is made from an aluminium pressure die casting which holds the bearings for the barrel lugs and the guide rods. The carrying handle and sight are integral with the receiver casting. The cocking handle works in a slot on the left side of the receiver, and is operated by the firer's left hand. The forward sling swivel is anchored to the base of the sight casting at its front end. The rear end of the receiver carries a steel sleeve which has three rows of interrupted lugs; the rearmost row serves to lock the bolt, while the front two rows lock the barrel in place. When both are in the sleeve they turn and lock with their faces completely mating, making a gas-tight seal.

Hammer group
The hammer mechanism is contained in one group and is inserted into the weapon through an opening in the butt, covered by the synthetic rubber butt plate. An unusual feature of this hammer mechanism is that except for the springs and pins it is entirely made of plastic and it is contained in an open-topped plastic box. This box fits between the magazine and the butt-plate and the bolt group recoils over the top of it. Since the trigger is some distance away it transmits its pressure through a transfer rod which passes by the side of the magazine. The actual operation of the firing pin is by a plastic hammer driven by a coil spring. There is no change-lever for different modes of fire. Single shots are fired by pulling the trigger a short distance to the first sear action. Further movement operates the sear in the automatic mode. This trigger action requires a little practice to master, but it can be used with ease once it has been tried a few times. The automatic firing lever prevents firing until the bolt is locked.

The mode of operation of the weapon can be changed by changing the hammer mechanism unit; thus for police use a mechanism which permits only semi-automatic fire is available. For military use the standard mechanism offers the usual options of single-shot or full automatic fire, but the company has recently announced that a new mechanism has been developed. This has a selector switch which permits the user to withdraw the mechanism and select either full automatic fire or three-round burst fire as the alternative to single shots. The option can only be changed by withdrawing the mechanism to make the choice.

Bolt group
The bolt is a light multi-lug rotating component contained in a carrier and operated by a camway in that carrier. The carrier has two guide rods brazed to it and these rods run in steel bearings in the receiver. The guide rods are hollow and contain the return springs. The left-hand rod takes the pressure from the cocking handle and it can be locked to that handle by a catch so that the firer can push the bolt home, though in normal firing the cocking handle is free of the rod and does not reciprocate. The right-hand rod is the gas piston which operates the bolt. The left-hand rod can also be used as a clearing rod if the gas cylinder should become fouled. The bolt locks by seven lugs into the locking sleeve in the body and carries an extractor in place of an eighth lug.

Magazine group
The magazine is entirely plastic apart from the spring and platform. The body is translucent so that the ammunition stock can be seen, but the firm also supply darkened versions for better camouflage. The magazine is easily stripped for cleaning and it is not intended to be disposable. Two models are available, for 30 or 42 rounds.

Stock group
The stock group forms a substantial part of the actual weapon and is almost entirely plastic. At the forward end is the pistol grip with its large forward guard enclosing all the firing hand. The trigger is hung permanently on this pistol grip, together with its two operating rods which run in guides past the magazine housing. Immediately above the grip is a cross-bolt safety catch

which acts on the trigger rods. Behind that again, and just forward of the magazine housing, is the locking catch for the stock group. Pressing this to the right allows the receiver and stock to be separated. The magazine catch is behind the housing, on the underside of the stock. Above the housing are the two ejector openings, one of which is always covered by a movable piece of plastic. The rear of the stock is the actual shoulder butt and it contains the hammer group and the rear end of the bolt-way. The butt is closed by the butt plate which is held in place by the rear sling swivel. This swivel is in the form of a pin which pushes in across the butt and locks the plate.

Operation

The AUG works as a completely conventional gas-operated weapon and is unusual only in the gas cylinder's being off-set to the right and working on one of the two guide rods. For single-shot operation the gas piston drives the bolt unit back, cocking the hammer which is held by the disconnector lever. The return springs pull the bolt forward, loading a fresh cartridge, and the bolt is rotated to lock. For semi-automatic fire the trigger must be released between shots. For automatic fire the operation is the same, but the hammer is held by the automatic-fire lever which is released after the bolt has locked.

Sights

The optical sight has been optimised for battle ranges and the graticule is a black ring in the field-of-view. This can be quickly and easily placed around a man-sized target at ranges up to 300 m, though after that it may take some care to get the target in the centre. But for all normal infantry engagements it offers an easily-taught and rapidly-used sight with very good accuracy and performance in poor light. Although it appears that the sight is vulnerable to injury from rough handling, this is not so and it will withstand a good deal of abuse.

An alternative graticule with a fine dot in the centre of the aiming circle is also available, allowing a more precise aim. This graticule is supplied as standard on police models.

A special receiver group, with a flat mounting platform in place of the optical sight, permits the use of all types of optical, electro-optical or collimating sights.

Stripping

Before stripping check that the safety catch is on safe, the chamber clear and the magazine off. Cock the action and lock the bolt to the rear by turning the cocking handle up into a notch. To remove the barrel, press the barrel-locking stud down and right with the thumb, grasp the handle, rotate the barrel to the right until clear of the lugs and pull forward. The bolt is then moved to the forward position by turning the cocking handle down and out of the notch.

The housing lock is pushed to the right until it disengages and the housing group pulled out of the butt, with the bolt assembly. The bolt assembly is removed from the housing group. The butt cap is pressed with the thumb and the sling swivel pulled out to the left. The butt cap is removed. The trigger mechanism is removed by pulling it out backwards from the butt. No further stripping is necessary.

DATA
Cartridge: 5.56 mm × 45
Operation: gas, selective fire
Locking: rotating bolt
Magazine: 30- and 42-round detachable translucent plastic box
Rate of fire: 650 rds/min

DIMENSIONS	Sub-machine gun	Carbine	Rifle	Heavy-barrelled
Length				
overall	626 mm	690 mm	790 mm	900 mm
barrel	350 mm	407 mm	508 mm	621 mm
Weights				
unloaded	3.05 kg	3.3 kg	3.6 kg	4.9 kg
loaded magazine	0.49 kg	0.49 kg	0.49 kg	

SIGHTS
Optic sight: set in carrying handle. × 1.5 power

ACCESSORIES
Bayonet; blank-firing attachment; special receiver for sights; grenade launcher M203 (AUG-8); muzzle cap; carrying sling; cleaning kit; bipod with wire-cutter; fixed and adjustable bipod (HBAR only); rifle grenades

Manufacturer
Steyr-Manlicher GmbH, a member of the Steyr-Daimler-Puch AG group, Werke Steyr, Postfach 1000, A-4400 Steyr. Manufacture has also been licensed to the Australian government.
Status
Production.
Service
Austrian Army as the Stg 77. Adopted by the Australian, Moroccan, New Zealand, Omani, Saudi Arabian, Irish and other armed forces.

Steyr Advanced Combat Rifle

Details of the Steyr design submitted for the US Army Advanced Combat Rifle programme will be found, together with the other entrants, in the USA section on page 224. This course has been adopted so as to permit ready comparison of the various designs.

BELGIUM

7.92 mm M49 FN semi-automatic rifle

Before the Second World War, Fabrique Nationale (FN) was working on a self-loading rifle. The designer was DJ Saive, and when the Germans invaded Belgium in 1940 he and several of his colleagues went to England to work in the Design Department of the Royal Small Arms Factory at Enfield. After the war the rifle was produced as the SAFN 7.5 mm (Semi-Automatique FN) in 7.92 mm, 7.65 mm, .30-06 in and 7 mm calibres and was sold to several countries in some of which it is still in service.

A gas-operated weapon with a semi-automatic action, the M49 uses a tilting-block locking system of the Tokarev type. The mechanism is piston-operated and there is a gas regulator which is adjusted by a sleeve to which access is obtained by removing the front hand-guard. The rifle is fired from the closed breech position and the hammer spring guide, projecting from the trigger-guard, acts as a cocking indicator. A 10-round magazine is used and this is loaded with either five-round charges or loose rounds. The action is held to the rear when the magazine is empty; there is also a manual hold-open device to enable the rifle to be loaded with single rounds.

The standard design had a tangent aperture rearsight adjustable for elevation and windage but there was also a sniper version with a telescopic sight. A bayonet lug was a standard fitting and some models had a muzzle brake.

DATA
Calibre: 7.92 mm, 7.65 mm, .30-06, 7 mm or 7.5 mm
Operation: gas
Feed: 10-round charger-loaded magazine
Fire: semi-automatic
Weight: 4.3 kg
Length: 1110 mm
Barrel: 589 mm
Sight: (foresight) shielded post; (rearsight) tangent aperture
Effective range: 700 m

Manufacturer
FN Herstal SA, Branche Défense et Sécurité, B-4400 Herstal.
Status
No longer in production.
Service
Supplied to Argentina, Belgium, Brazil, Colombia, Egypt, Indonesia, Luxembourg, Venezuela and Zaïre and still in service in some of these countries.

7.92 mm M49 FN semi-automatic rifle

7.62 mm FN FAL rifle

This rifle has been adopted by more than 90 countries all over the world. It is, or has been, manufactured under licence in several countries and some of these have incorporated their own minor modifications to suit their own particular needs. In some instances these modifications have resulted in deviations from the original specification to the extent that the licensed weapons are not interchangeable with others. There are several versions of the basic FAL and they are listed at the end of this entry. One interesting version is the one with the heavy barrel, which is in fact a machine rifle. Several users equip their infantry with this for use as a light machine gun and it has the advantage that it has many components which are completely interchangeable with the standard FAL.

Operation

To load the FAL, the front end of a loaded magazine is placed in the housing and the rear end rotated back and up until the magazine catch clicks in to lock it in position.

The selector lever is above the pistol grip on the left of the receiver and is operated by the thumb of the right hand. The 'safe' position is at the top, 'repetition' lies below it and 'automatic' is further forward. In the British, Canadian, Indian and Netherlands versions, there is no automatic fire. The cocking handle is on the left side of the receiver. In some models it folds flat and must be pulled back to extend it at a right angle before it can be used. When the cocking handle is retracted fully and then released, the bolt picks up the top round from the magazine and chambers it. The weapon always fires from the closed breech position.

When the trigger is pulled the hammer strikes the firing pin and this, in turn, hits the cap to ignite the propellant charge. When the bullet passes the gas port in the top of the barrel, some of the propellant gas is diverted into the gas cylinder where it expands and drives back a light, spring-loaded piston. This piston strikes the top of the front face of the bolt carrier, transfers its energy and, after a short travel, is returned by its spring to its forward position. The carrier is driven rearwards and has about 6 mm of free travel during which the chamber pressure is dropping to an acceptably low level. After this idle motion, known as 'mechanical safety after firing', the unlocking cam of the carrier moves under the bolt lug and lifts the rear end of the bolt out of the locking recess in the floor of the receiver. The bolt and carrier travel back together, compressing the return spring. The extractor withdraws the empty case from the chamber and holds it on the face of the bolt until it strikes the fixed ejector and is thrown out of the rifle to the right, through the ejection port.

The return spring drives the carrier and bolt forward; the feed rib on the bolt pushes the top cartridge out of the magazine and over the bullet guide into the chamber. The extractor snaps into the extraction groove of the cartridge case and the bolt can travel forward no further. The carrier continues on and the locking cam rides over the bolt, forcing the rear of the bolt down into a recess in the receiver floor and holding it here. The final forward movement of the carrier trips the automatic sear. If the trigger is pressed, the hammer will drive the firing pin into the cartridge cap.

Trigger and firing mechanism

The trigger and firing mechanism of the FN FAL is ingenious, fool-proof and copied, in principle, by many other manufacturers. It consists of a sear mounted on a fixed pin which passes through an elongated slot. The sear is spring-loaded and moves forward, when it can, along its elongated slot. The trigger is pivoted at the top of the same fixed pin. When the trigger is pulled a shoulder rotates up and forces the rear end, the tail, of the sear up. The front end is depressed and moves out of the hammer notch. The hammer rotates slightly but is held by the automatic or 'safety' sear until the bolt carrier is fully home and bolt locking is completed. When the automatic sear is depressed by the carrier the hammer is free to rotate and drive the firing pin into the cartridge cap. If the selector is set to 'automatic' the trigger shoulder can rise sufficiently to hold the nose of the sear down clear of the hammer so that the hammer is controlled only by the automatic sear. If the selector is set to 'repetition' the trigger shoulder can rise only a short distance. The sear nose is depressed out of the notch and the hammer swings forward as before. The sear spring forces the sear forward along its elongated slot, the nose presses on the underside of the hammer, and the tail slips forward off the shoulder of the trigger. When the hammer is rotated back by the recoiling bolt carrier the nose of the sear is pressed into the hammer notch. The full force of the hammer spring overcomes the weaker sear spring and forces the tail of the sear back against the side of the shoulder and so another round cannot be fired. When the trigger is released the shoulder moves down and the hammer spring forces the sear further across and the tail moves on to the top of the shoulder of the trigger. Pulling the trigger lifts the trigger shoulder which in turn lifts the tail of the sear and the nose moves down out of engagement with the hammer notch. The hammer will be held up by the automatic sear if the bolt carrier is not fully forward.

When the selector is placed at 'safe' the trigger shoulder is prevented from rising sufficiently to lift the tail of the sear.

When the last round is fired a rib on the magazine platform lifts the bolt stop into the path of the bolt. On the UK version the last 6 mm of the bolt stop is ground off and thus the bolt is in the forward position when the ammunition is expended. By pushing up the bolt stop by hand when the bolt is to the rear, the bolt can be held open whenever required.

Gas regulation

The gas regulator works on the 'exhaust to atmosphere' principle. When the gun is clean and firing under ideal conditions, a large proportion of the gas is passed through the regulator and out into the atmosphere. When the need comes to increase the gas pressure on the piston-head, the regulator is screwed up and more gas is diverted into the cylinder.

The gas plug must be rotated when firing grenades to ensure all the gas pressure goes to drive the grenade out and none goes to the piston. To do this the plunger is depressed and the plug rotated until the marking is on the bottom. After each grenade the bolt must be retracted by hand and a fresh cartridge loaded into the propelling chamber. Some FN FAL rifles have a combined flash eliminator/grenade launcher which is a cylindrical fitting on

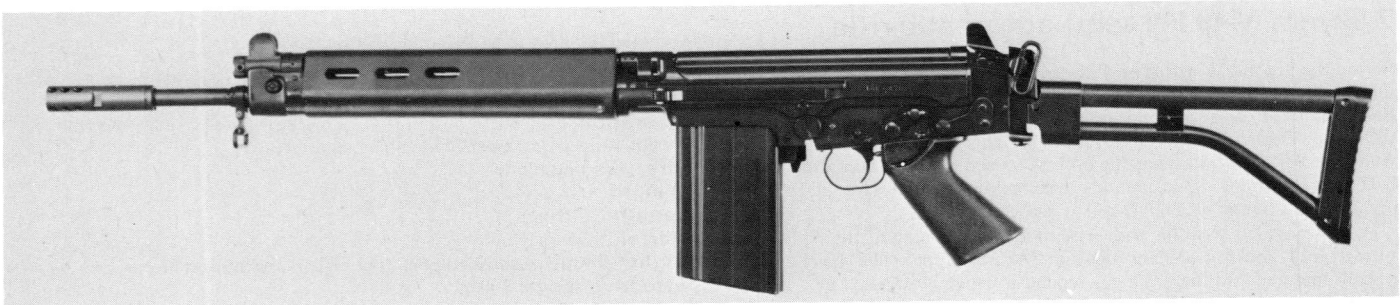

7.62 mm FN FAL Para rifle (50-63)

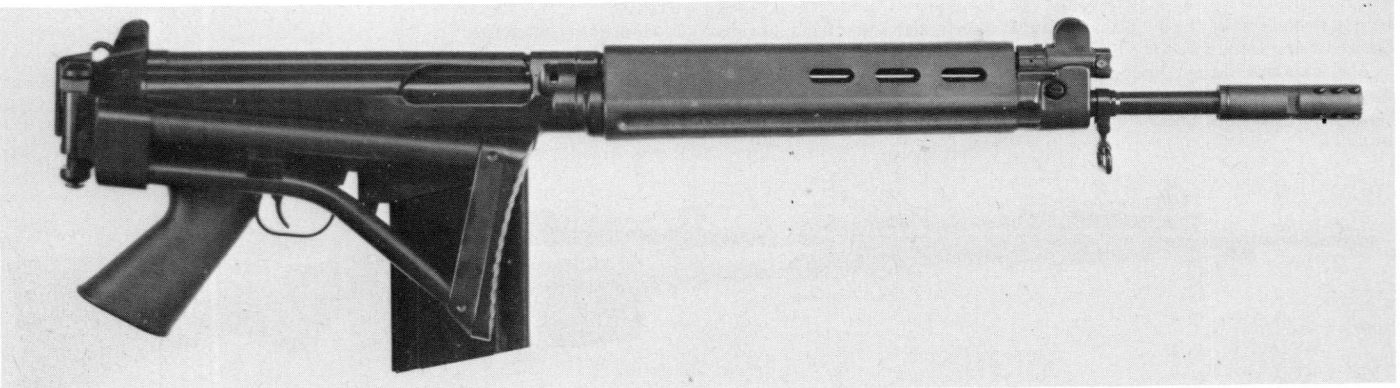

7.62 mm FN FAL Para rifle, butt folded (50-63)

7.62 mm FN FAL rifle heavy-barrelled version, with bipod folded for carrying (FAL HB)

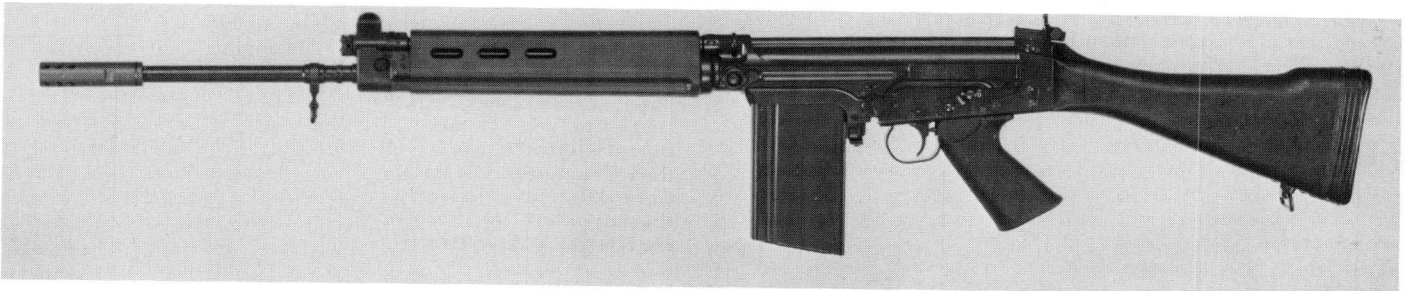

7.62 mm FN FAL rifle standard version (50-00)

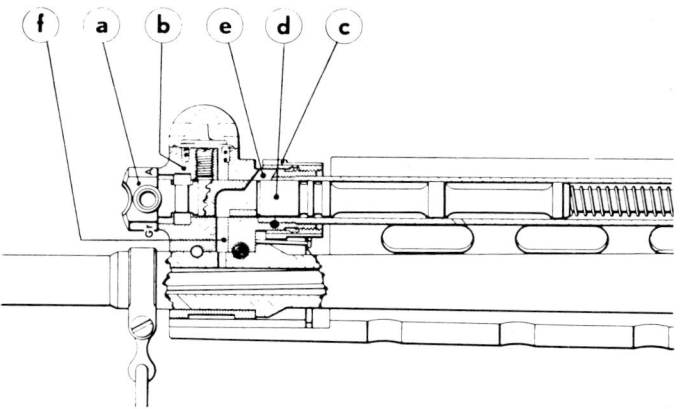

Gas regulation
(a) *plug* **(b)** *gas block* **(c)** *gas regulator* **(d)** *piston-head* **(e)** *gas outlet* **(f)** *gas port*

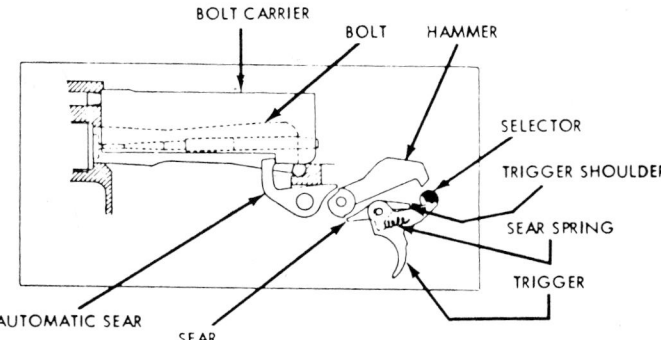

FAL firing mechanism

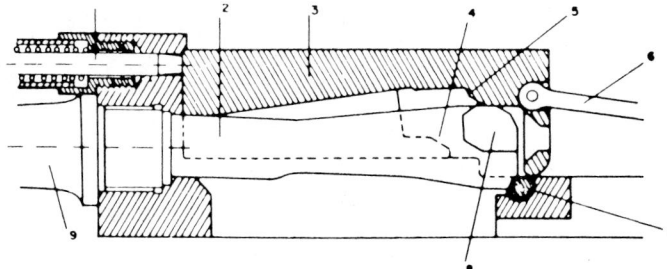

FAL bolt functioning
(1) *piston* **(2)** *bolt* **(3)** *bolt carrier* **(4)** *unlocking cam* **(5)** *locking cam* **(6)** *driving spring rod* **(7)** *locking seat* **(8)** *lugs* **(9)** *barrel*

the muzzle. Others have a prong-type flash eliminator and with these a spigot launcher must be placed over the muzzle. Some of these have a sight but most modern grenades have the sight mounted integrally.

Zeroing the sights is achieved by screwing the foresight up or down to adjust for elevation and the rearsight aperture can be moved across by loosening the screw underneath on the side to which movement is required and tightening the screw on the other side.

Stripping
Remove the magazine and check the chamber and feedway to ensure no cartridge is in the weapon. The bolt is then allowed forward under control, the hammer remaining cocked. The take-down lever is located on the left side of the receiver at the rear and should be rotated anti-clockwise and the butt stock swung down to open the rifle. The bolt cover is then pulled to the rear off the receiver. The bolt assembly can then be withdrawn by pulling on the return spring rod, and the bolt removed from the carrier.

DATA
General characteristics and special characteristics of four versions of the weapon are given below. The four versions, with their FN reference numbers are:
FN 50-00: Standard model with fixed butt stock and standard barrel
FN 50-63: Model with side-folding butt stock and short barrel
FN 50-64: Model with side-folding butt stock and standard barrel
FN 50-41: Model with fixed butt stock, bipod and heavy barrel

GENERAL CHARACTERISTICS
Cartridge: 7.62 mm × 51 NATO
Operation: gas, selective or semi-automatic fire
Method of locking: dropping bolt
Feed: 20-round steel or light box magazine
Weight: (empty, without magazine) 4.325 kg
Barrel: (regulator) exhaust to atmosphere system
Rifling: 4 grooves rh. 1 turn in 305 mm
Sights: (foresight) cylindrical post; (rearsight) aperture, sliding on bed, adjustable 200-600 m × 100 m; or flip aperture 150/250 m; or fixed battlesight 300 m (see table)
Rate of fire: (cyclic) 650-700 rds/min
Effective range: 650 m
Bayonet: (length) 290 mm; (weight) 221 g

SPECIAL CHARACTERISTICS

Type	50-00	50-64	50-63	50-41
Length				
without bayonet	1090 mm	—	—	1150 mm
stock extended	—	1095 mm	1020 mm	—
stock folded	—	845 mm	770 mm	—
Weight: (without bayonet or magazine)	4.25 kg	3.9 kg	3.75 kg	6 kg
Barrel length	533 mm	533 mm	436 mm	533 m
Rearsight type	Sliding	Flip	Fixed	Sliding
Sight radius	533 mm	549 mm	549 mm	553 mm
Muzzle velocity	840 m/s	840 m/s	810 m/s	840 m/s
Muzzle energy	34.17 J	34.17 J	31.62 J	34.17 J

Manufacturer
FN Herstal SA, B-4400 Herstal and licensees.
Status
Current. Production ceased in 1987.
Service
The FAL has been or is still in production not only in Belgium but also in a number of other countries such as Argentina, Australia, Austria, Canada, India, Israel, Mexico, Nigeria, South Africa, the UK and Venezuela. Included in the countries using the FN FAL are: Argentina, Australia, Austria, Barbados, Belgium, Brazil, Burundi, Cambodia, Chile, Cuba, Dominican Republic, Ecuador, Gambia, West Germany, Ghana, Guyana, India, Indonesia, Ireland, Israel, Kuwait, Liberia, Libya, Luxembourg, Malawi, Malaysia, Morocco, Mozambique, New Zealand, Norway, Oman, Paraguay, Peru, Portugal, Singapore, South Africa, United Arab Emirates and the UK.

7.62 mm Model 30-11 FN sniping rifle

The FN sniper rifle was introduced into the range of FN weapons to provide a precision instrument for either police or military use. It is foreseen that the main customers for this rifle will be police and gendarmerie forces, but it is equally applicable for an army and there is no intention that it shall be restricted to civilian use only.

The rifle is a high-precision bolt-action weapon, using a Mauser action and a carefully mounted heavy barrel. There is an integral magazine holding five rounds and an adjustable butt which can be altered in two planes by means of inserts so that it can accommodate any size of firer.

The accessories which are designed for this rifle make it one of the most comprehensively equipped rifles in the world. There is a wide variety of sighting systems, since it is the sighting which is so important in sniping, and the weapon can be supported with a shooting sling or the bipod from the MAG machine gun. The latter is recommended when using an electronic night sighting system. Finally there is a carrying case in which the rifle and all its accessories can be kept in safety.

DATA
Cartridge: 7.62 mm × 51 NATO
Operation: manual bolt
Feed: 10-round removable box
Overall length: 1117 mm
Barrel length: 502 mm
Barrel diameter: 24.4 mm
Bare weight of rifle: 4.85 kg
Trigger pull: 1.3 kg
Muzzle velocity: 850 m/s

ACCESSORIES
Sights: Anschutz aperture with adjustable dioptre; FN telescope 4 × 28 magnification; II night sight
bipod; stock extensions; stock inserts; sling

Manufacturer
FN Herstal SA, B-4400 Herstal.
Status
Current. Ceased production in 1986.
Service
Belgium and other armies.

7.62 mm sniping FN rifle

5.56 mm FN FNC rifle

The FNC is a light assault rifle intended for use by infantry who are operating without continuous logistical support, or who are in jungle, mountain or other difficult country. The construction is from steel, aluminium alloy and, for non-working parts, plastic, great use being made of stampings and pressings. There are two versions of the rifle; the first has a standard length barrel together with a folding tubular light alloy butt stock encased in a plastic coating and braced by a plastic strut; the second is similar but has a shorter barrel. Optional on both versions is a fixed plastic stock, and a special bracket to accept the US M7 bayonet is now standard. The layout of both models follows the general pattern of FN rifles and the body opens on a front pivot pin to allow the working parts to be taken out in much the same way as the Fusil Automatique Légèr (FAL).

Operation
The system of operation is by gas, using a conventional piston and cylinder mounted above the barrel. The breech is locked by a rotating bolt, with a two-lug head locking into the barrel extension. In the FNC the bolt and carrier are among the very few items which need precision machining in manufacture. The gas regulator acts directly on the gas passage at the port opening and the escape hole is open or shut by the gas cylinder. The open hole is the standard setting and is used for all normal firing. The closed hole allows full gas to flow and is meant for use only in adverse conditions.

Construction
The body is made from pressed steel, with the bearing surfaces for the bolt and carrier inserted. The trigger frame is of light alloy. The top and bottom of the body are held together at the front by a pin on the under side and they are locked by a pin which pushes in from the right side just above the pistol grip. When this pin is pushed out the lower body can be dropped away and the working parts are exposed and can then be lifted out. No tools are needed and in the field no further stripping is necessary.

The cocking handle is on the right side, working in a slot which is normally protected by a cover held shut by a spring. This cover remains closed at all times and ensures that the working parts are protected from mud, dirt and snow. The movement of the cocking handle opens the cover by a simple cam action, but it closes as soon as the handle reaches the forward position.

Sights
The sights are conventional in that they are an adjustable post for the foresight and an aperture for the rearsight. There is lateral adjustment for the rearsight and flip settings for range. For grenade firing a gas-tap is folded up over the foresight and this cuts off the flow of gas to the cylinder, thereby allowing the maximum pressure behind the grenade. This tap then acts as a sight and it is aligned along the top of the grenade to give the correct elevation for launching.

The skeleton tubular butt folds to the right, fitting below the cocking handle. This reduces the length of the standard version to 766 mm and the short version to 680 mm so allowing it to be used as a sub-machine gun. It is also particularly valuable in confined spaces such as vehicles or aircraft.

The receiver is provided with a telescope support capable of accommodating all sighting devices conforming to the NATO standard mounting plan. When firing with these sights it is recommended that a light bipod be clipped on to steady the weapon.

Accessories
The steel magazine is interchangeable with the US M16A1 rifle and both types of magazines can be used for the FNC, the Minimi light machine gun and the M16.

The manufacturer has anticipated that possible users may have different ammunition requirements and the rifle is offered with two possible barrel riflings. The first is optimised for the FN SS 109 cartridge which is now the recommended NATO standard. It has a twist of 1 in 178 mm (1:7 inches). The

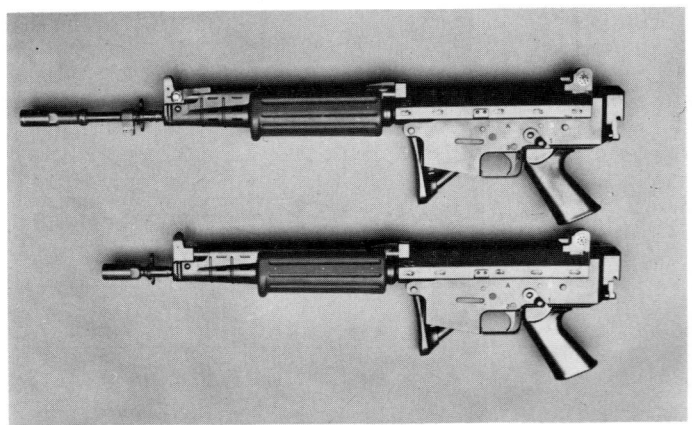

Standard and short-barrelled versions of 5.56 mm FNC rifle

second rifling is available on special order and is the same as the M16A1 with a twist of 1 in 305 mm (1:12 inches) and will fire the M193 or FN SS92 cartridge.

DATA
Cartridge: 5.56 mm × 45 (M193 or SS 109)

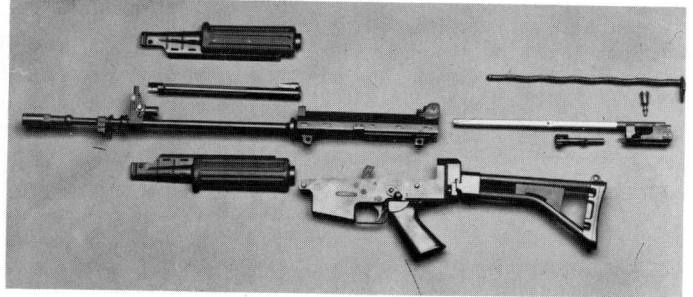

5.56 mm FNC rifle stripped

Operation: gas, selective fire with 3-round burst controller
Locking: rotating bolt
Feed: 30-round box magazine
Weight, empty: 3.8 kg
Length: (butt extended) 997 mm; (butt retracted) 766 mm. Short model: 911 mm and 680 mm
Barrel: (standard) 449 mm; (short) 363 mm
Rifling: 6 grooves, rh, one turn in 305 mm or 178 mm
Sights: (fore) cylindrical post; (rear) flip aperture 250 and 400 m
Sight radius: 513 mm
Rate of fire: (cyclic) 625-750 rds/min
Muzzle velocity: (M193) 965 m/s; (SS 109) 915 m/s
Accessories: bayonet, blank firing attachment, NATO bracket for night sight, optical sight, sling

Manufacturer
FN Herstal SA, B-4400 Herstal.
Status
Current. Production. Manufacture licensed to Indonesia.
Service
Military trials in many countries; adopted by Belgium, Indonesia, Nigeria, Sweden, Tonga, Zaïre and others.

BRAZIL

Brazilian military rifles

In the mid-1950s the Brazilian armed forces adopted the NATO 7.62 mm × 51 cartridge. Older rifles were modified to fire the new round, pending the service introduction of a new weapon. This was chosen to be the Belgian Fabrique Nationale's Fusil Automatique Légèr (FN FAL), which together with the heavy-barrelled variant (known in Brazil as the FAP (Fuzil Automático Pesado), or heavy automatic rifle) began to enter service in the early 1960s. An initial batch of FALs, FAPs and Para-FALs (foldable-butt) was purchased directly from FN, but the standard model was subsequently manufactured by the Army's Fábrica de Itajubá, Minas Gerais State. The Brazilian Marine Corps, having previously used the semi-automatic FN (SAFN) Model 49 rifle in 7.62 mm calibre, followed and is now using FALs, FAPs and Para-FALs. The Brazilian Air Force, after using the US M1 carbine with foldable butt for a long time, decided in the early 1970s to adopt the 5.56 mm calibre for its ground units, and selected the West German Heckler and Koch HK 33 rifle, which was purchased in substantial numbers in both the HK 33A2 (fixed-butt) and HK 33KA1 (shortened version with retracting butt) variants.

Brazilian-made FAL rifles in service

7.62 mm IMBEL semi-automatic rifle

IMBEL produces the standard FN-FAL rifle in semi-automatic standard and paratroop form, solely for supply to the armed forces of Brazil and other countries. The dimensions and data are exactly as for the Belgian original.

Manufacturer
Indústria de Material Bélico do Brasil – IMBEL, Avenida dos Nacções Unidos 13.797, Bloco III, 1° Andar, CEP 04794 São Paulo SP.
Status
Current. Production.
Service
Brazilian and other military forces.

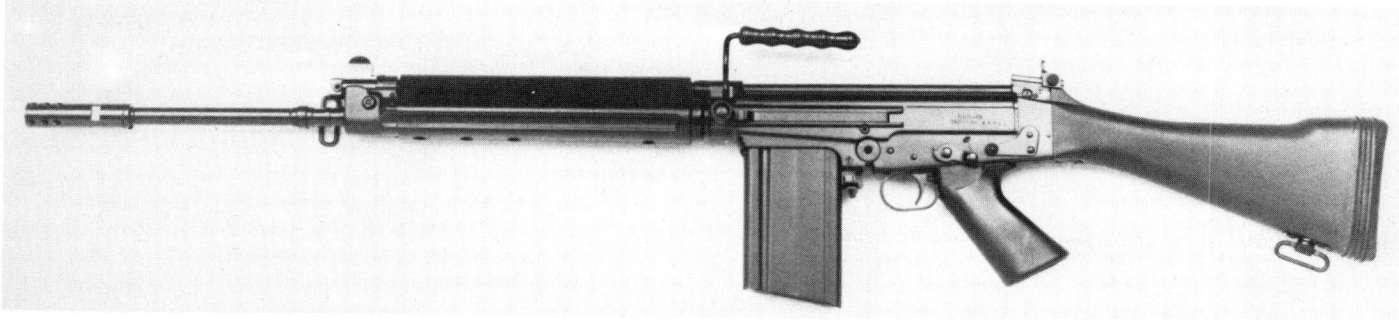

IMBEL 7.62 mm semi-automatic rifle

IMBEL 7.62 mm Light Automatic Rifle

The IMBEL Light Automatic Rifle in 7.62 × 51 mm calibre is an internationally accepted weapon, having been proved in use under the most severe conditions. It is the FN-FAL manufactured under licence and over 150 000 have been produced to date, 40 000 of which have been exported.

The rifle is gas-operated, with an adjustable regulator to ensure smooth and certain action without excessive recoil. The bolt is mechanically locked before firing and is not unlocked until the bullet has left the barrel.

Since the bolt is in the closed position when the trigger is pressed, there is no disturbance of aim due to movement of a heavy mass, a drawback to many automatic weapons.

DATA
Cartridge: 7.62 × 51 mm NATO
Operation: gas, selective fire
Method of locking: tilting bolt

Feed: 20-round box magazine
Weight, empty: 4.50 kg
Length: 1100 mm
Barrel: 533 mm
Rifling: 4 grooves rh
Rate of fire: 650-750 rds/min

Manufacturer
Indústria de Material Bélico do Brasil – IMBEL, Avenida dos Nacções Unidos 13.797, Bloco III, 1° Andar, CEP 04794 São Paulo SP.
Status
Current. Production.
Service
Brazilian and other military forces.

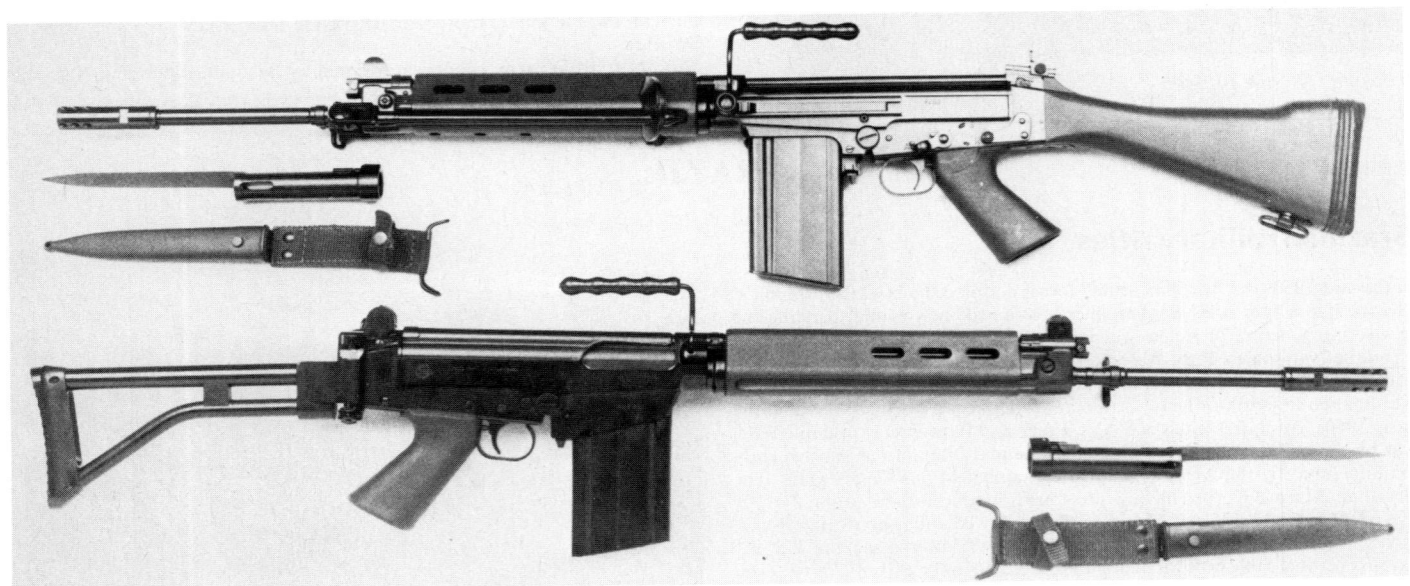

IMBEL Light Automatic Rifles with bayonets

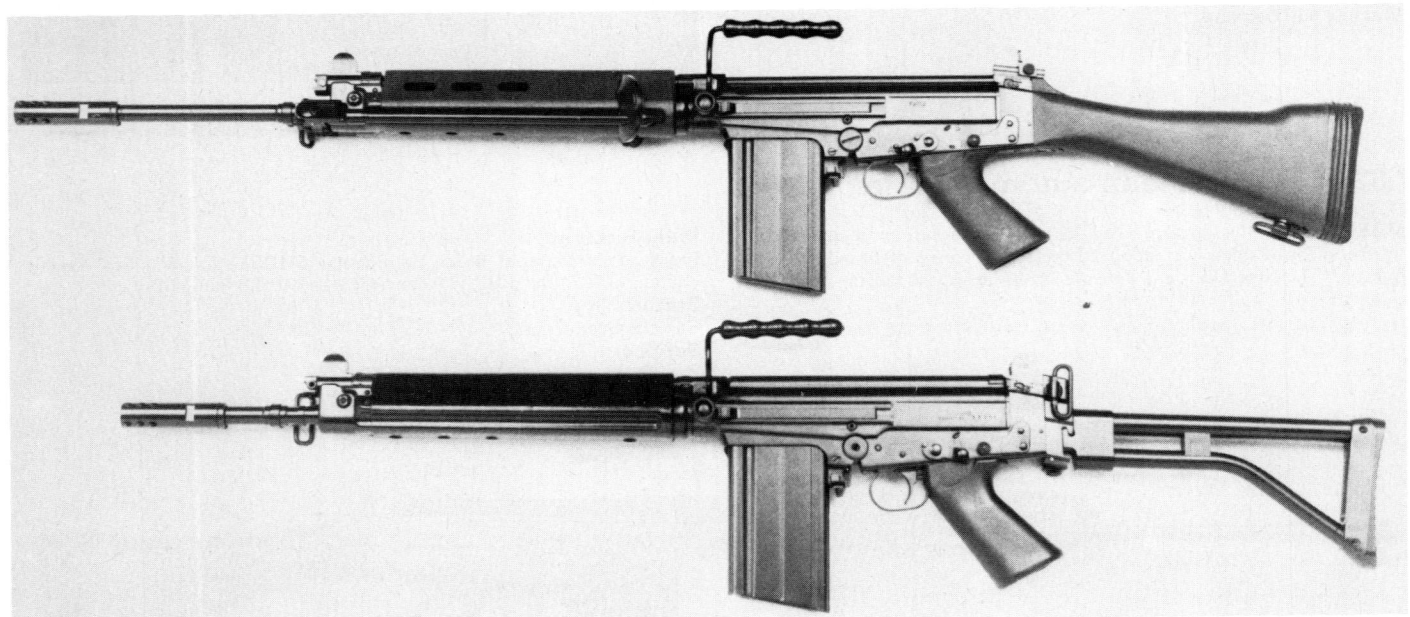

IMBEL standard and paratroop 7.62 mm Light Automatic Rifles

5.56 mm Imbel MD1 rifle

Using the 7.62 mm Light Automatic Rifle (above) as a starting point, the Fábrica de Itajubá has developed a 5.56 mm version, for use by the Brazilian Army and for export. The MD1 has a folding butt, is gas-operated, offers selective single-shot or automatic fire, and will fire either M193 or SS109 ammunition with equal facility.

The MD1 has been designed and developed in conformity with various international and NATO standards and has the advantage of a high proportion of parts interchangeable with the 7.62 mm L.A.R.

DATA
Cartridge: 5.56 × 45 mm M193 or SS109
Operation: gas, selective fire
Feed: 20-round box magazine or 30-round US M16 magazines
Sights:
foresight: post
rearsight: flip aperture 150 m and 250 m (paratroop versions) or 200 m and 600 m (standard version); adjustable for windage

5.56 mm Imbel MD1 rifle

Barrel length: 453 mm without flash hider
Rifling: 6 grooves rh
Rate of fire: 700-750 rds/min

Manufacturer
Indústria de Material Bélico do Brasil – IMBEL, Avenida dos Nacções Unidos 13.797, Bloco III, 1° Andar, CEP 04794 São Paulo SP.
Status
Current. Production.
Service
Brazilian and other armies.

5.56 mm OVM assault carbine

This weapon has been developed by Olympio Vieira de Mello, who designed the successful 9 mm Uru sub-machine gun and the 7.62 mm Uirapuru general-purpose machine gun. The OVM carbine is a gas-operated weapon using a rotating bolt locking system, no gas regulator being provided. The body is made of light alloys, and the butt, pistol grip and hand-guard are made of high-impact, heat-resistant plastic material. Fixed and foldable-butt versions are anticipated. The gun fires automatic and semi-automatic, being fed by 20- or 30-round box magazines. The muzzle is fitted with a novel flash-eliminator/compensator, which can easily be set by the firer for right- or left-handed use.

An as yet unnamed industrial concern has reportedly reached an agreement with Vieira de Mello to undertake the development and production programme of the OVM assault carbine.

DATA
Cartridge: 5.56 mm × 45
Operation: gas, selective fire
Method of locking: rotating bolt
Feed: 20- and 30-round box magazines
Length: (with fixed butt) 930 mm
Barrel: 464 mm
Barrel: (regulator) nil
Rifling: 6 grooves rh. 1 turn in 305 mm
Sights: adjustable to 600 m

Manufacturer
Olympio Vieira de Mello, Rio de Janeiro.
Status
Prototype construction, development.

5.56 mm LAPA FA Modelo 03 assault rifle

Work on this 5.56 mm assault rifle was started in late-1978 by the design team General Fileto Pires Ferreira, Nelmo Suzano and Paulo Cochrane of Laboratório de Pesquisa de Armamento Automático Ltda (Automatic Armament Research Laboratory), a Rio de Janeiro-based research and development concern. The technical feasibility prototype was completed in about a year, the promising characteristics of which led to the construction of a second prototype. This has undergone official testing at the Brazilian Army's Marambaia proving grounds, and LAPA is considering a number of offers from abroad for licenced manufacture of the weapon.

Operation
The Modelo 03 FA (Fuzil Automático) uses a bullpup configuration for compactness and is largely built of high-impact injected plastic material. The prototype uses M16-type metal box magazines for 20 and 30 rounds, but production examples will employ curved 20-, 30- and 40-round plastic magazines. The gun fires 5.56 mm × 45 M193 cartridges, but production examples will be rifled either for this cartridge or for the SS 109 NATO round. The magazine is inserted upwards into its guide inside the stock, the catch being at the rear and being pressed forward to release the magazine. The ejection port, fitted with a spring-loaded cover that snaps open when the weapon is cocked or fired, can be set for right- or left-handed firers.

The weapon has a carrying handle which also houses the flip-type 200 and 400 m rearsight, which is adjustable for windage. The protected post foresight is on top of a raised frame, a configuration resulting from the rifle's straightline configuration. The cocking handle is within the carrying handle, and the top end runs along a slot on the underside of the handle, giving it more rigidity.

The firing mechanism incorporates some novel features for a shoulder arm weapon, such as double-action for the hammer in place of the more traditional safety arrangements. This keeps the weapon ready for immediate use, and at the same time it is protected from accidental firing. The fire-selector lever on the left side, between the magazine housing and the butt plate, has three fire positions, ('1', '3' and '30') and two action settings ('DA' and 'SA'). Position '1' indicates semi-automatic fire, '3' is for controlled three-round bursts and '30' is full automatic.

Before or after cocking and loading the weapons, which is done by pulling the cocking handle backwards and releasing it, the firer sets the fire-selector lever at any of the three fire positions. If fired immediately, the rifle will be naturally set for single action operation, featuring a lighter and shorter trigger pull. After firing some rounds, or if no immediate use is anticipated, the firer may turn the lever to the 'DA' (double-action) mark

5.56 mm LAPA FA Modelo 03 assault rifle

5.56 mm LAPA FA Modelo 03 assault rifle with bayonet and magazine removed

and return it to any desired fire position. This will drop the hammer safely and allow the gun to be carried safely but ready for action: all that is needed is to pull the trigger which will have a heavier and longer pull. If more precision in single or first shots is required, the firer may move the lever to the 'SA' (single action) marking and return it to the chosen fire position which will make the weapon operate at single action. The Modelo 03 LAPA FA will be available with any required combination of action/fire selection.

Stripping
Field stripping is simple and requires no tools. After removing the magazine and checking that the rifle is unloaded the butt-plate pin is removed sideways to the right, which will allow the butt-plate together with the return spring and its guide to be pulled away from the stock. The take-down pin, behind the pistol grip, is now removed from the weapon sideways to the right, separating the body into two halves. The lower half, which includes the magazine guard, pistol grip and forward hand-guard, also houses all the firing mechanism, which becomes exposed for cleaning, no additional disassembly of this group being necessary. With the cocking handle pulled to the rear position, the top plastic cover may be removed, exposing the barrel/bolt assembly. The cap of the tubular metal receiver is unscrewed and this allows the bolt and bolt carrier group to slide out. The gas rod and piston may then be removed. To re-assemble the weapon reverse the procedure.

DATA
Cartridge: 5.56 mm × 45
Operation: gas, selective fire

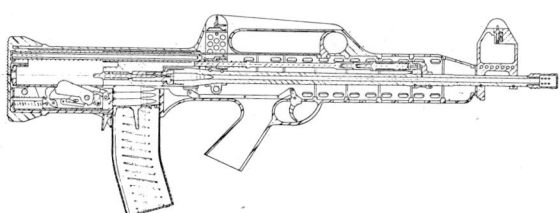

5.56 mm LAPA FA Modelo 03 assault rifle

Type of locking: rotating bolt
Feed: 20-, 30- or 40-round plastic box magazine (production)
Weight: (empty) 3.16 kg; (with loaded 20-round metal box magazine) 3.48 kg; (with loaded 30-round metal box magazine) 3.62 kg
Length overall: 738 mm
Barrel: (length) 489 mm; (rifling) 6 grooves rh. 1 turn in 305 mm
Sights: (fore) raised protected post, adjustable in elevation; (rear) flip aperture 200 and 400 m, adjustable for windage
Sight radius: 374 mm

Muzzle velocity: approx 1000 m/s
Rate of fire: (cyclic) 650-700 rds/min

Manufacturer
Laboratório de Pesquisa de Armamento Automático Ltda, Rua Homem de Melo 66, Grupo 201, Tijuca, 20510 Rio de Janeiro.
Status
Current. Final development.

.22 Imbel MD2 training rifle

This is virtually a replica of the Light 7.62 mm calibre automatic rifle (Fabrique Nationale's FN-FAL) in terms of size and weight but it fires the cheaper .22 rimfire cartridge. Rather than offering a conversion kit for the service rifle, the MD2 is factory-made in the smaller calibre. The 12-round .22 calibre non-removable magazine fits inside a false FAL magazine; the cocking handle is on the left of the receiver, and the rifle is used in exactly the same manner as the 7.62 mm version.

Prototypes have been evaluated by the Brazilian Army and the rifle is available for export.

DATA
Cartridge: .22LR rimfire
Operation: blowback, selective fire
Feed: 10-round box magazine
Weight: 4.14 kg

Manufacturer
Indústria de Material Bélico do Brasil – IMBEL, Avenida dos Nacções Unidos 13.797, Bloco III, 1° Andar, CEP 04794 São Paulo SP.
Status
Current. Production.

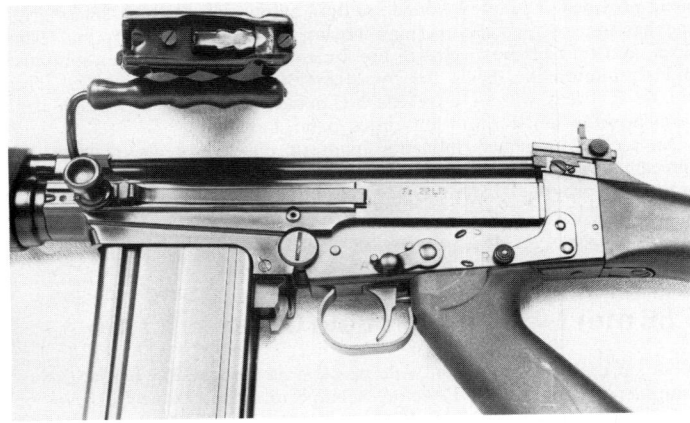

IMBEL .22 MD2 training rifle, showing cocking handle and a top view of the special magazine

.22 in Imbel MD2 training rifle

CANADA

7.62 mm C1 rifle

From the middle of the First World War, after the withdrawal of the Ross rifle, to the end of the Second World War the Canadian forces were equipped with Lee Enfield rifles, initially from UK sources but latterly from Long Branch Arsenal where some 962 000 rifles, mainly the Lee Enfield No 4, were manufactured during the Second World War.

After the war the Canadian authorities tried a number of variants on the Fabrique Nationale's Fusil Automatique Légèr (FN FAL) before settling on the C1 in June 1955. The FN rifles known as the X8E1 and X8E2 were tested in Canada as well as in UK.

The Canadian C1 differs from the L1A1 in having an opening in the feed cover through which the magazine can be re-charged using chargers and guides. It also has a rearsight which is a disc offering five different ranges, from 200 to 600 m, by 100 m intervals, by rotating the outside of the disc. The range appears in an aperture at the bottom of the front face of the disc.

Canadian 7.62 mm C1 rifle showing charger guides and disc rearsight

7.62 mm C1A1 modified rifle

In 1960 the British L1A1 was found to be firing rounds before the breech was closed. This was due to the long firing pin bending on firing and failing to withdraw into the breech-block. When the next round was chambered the projecting firing pin could fire the cap. The long firing pin was replaced by a two-part firing pin. In Canada this modification produced the C1A1 rifle which also has a plastic carrying handle.

DATA
Cartridge: 7.62 mm × 51
Operation: gas, semi-automatic
Feed: 20-round box magazine
Weight: 4.25 kg
Length: 1136 mm

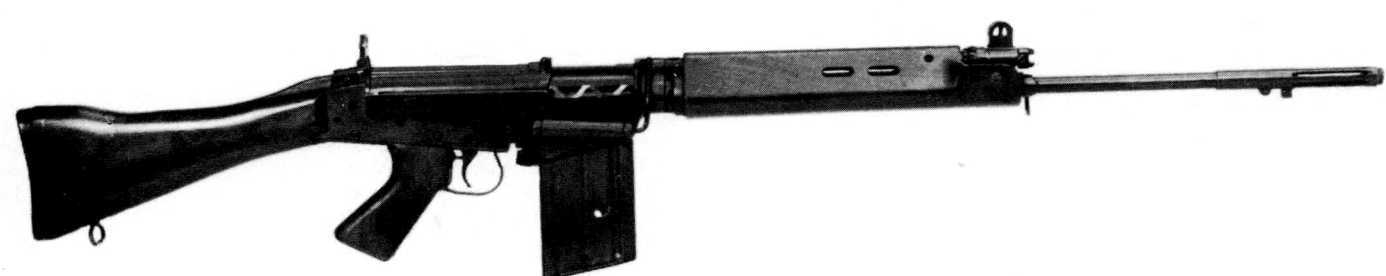

Canadian 7.62 mm C1 rifle

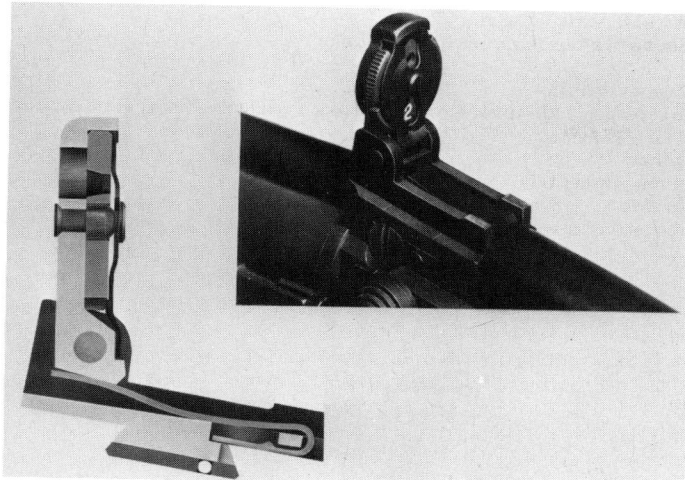

Canadian disc rearsight

Barrel: 533 mm
Sights: (foresight) post; (rearsight) aperture, disc
Muzzle velocity: 840 m/s

Manufacturer
Long Branch Arsenal, Ontario.
Status
No longer in production.
Service
Canadian forces.

7.62 mm C2 and C2A1 automatic rifles

The heavy-barrelled Fabrique Nationale (FN) rifle was taken by the Canadian Army as its Squad Light Automatic. It has selective fire allowing either single shots or full automatic. There are box magazines holding 20 and 30 rounds. The rearsight is the same type of rotating disk as used on the C1 rifle, which can be adjusted from 200 to 900 m.

The C2 rifle may readily be recognised by the uncovered gas cylinder over the forward grip. The gas regulator takes a light bipod.

The C2 rifle was modified by the substitution of a two-part firing pin and a plastic carrying handle, and known as the C2A1.

DATA
Cartridge: 7.62 mm × 51
Operation: gas, selective fire
Method of locking: tilting block
Feed: 20- or 30-round box magazine

Weight: 6.93 kg
Length: 1136 mm
Barrel: 533 mm
Sights: (foresight) post; (rearsight) tangent, aperture 200-1000 yds (182.8–914.4 m)
Muzzle velocity: 854 m/s
Rate of fire: (cyclic) 710 rds/min

Manufacturer
Long Branch Arsenal, Ontario.
Status
No longer in production.
Service
Canadian forces.

7.62 mm C3 sniping rifle

The standard Canadian sniper rifle is the British Parker Hale Model 82 modified slightly to meet Canadian requirements. The stock is fitted with four half-inch (12.7 mm) spacers to permit length adjustment. The rifle has two male dovetail blocks on the receiver, to accept either the Parker Hale 5E vernier rearsight, or the Austrian Kahles six-power telescope. All exposed metal parts of the rifle are non-reflective.

The Canadian designation for this rifle is Rifle 7.62 mm C3.

All data appears in the relevant entry in the section on the United Kingdom.

7.62 mm C3A1 sniping rifle

The Canadian Department of Defence has adopted a revised version of the C3 (M82) rifle which has been redesignated C3A1. The rifle incorporates a number of important changes from the earlier C3 model which entered Canadian service in the mid-1970s. The stock configuration remains unchanged but major alterations have been made to the action body which is greatly strengthened and equipped with a six-round capacity steel box magazine with release catch located in the leading edge of the trigger guard. The trigger mechanism has a redesigned safety catch and is of all-steel construction. Rifling specifications have been changed to provide optimum performance from both standard NATO 7.62 × 51 mm ball and heavier

bullets which have been developed for sniping and competition use. A Parker-Hale bipod is provided which attaches to a handstop assembly. The bipod has extendable legs, allows approximately 14° of swivel and cant, and can be folded either forward or back in addition to being hand detachable.

The telescope sight which has been specifically designed for use in conjunction with the C3A1 rifle is a 10 × magnification Unertl model, similar to that adopted by the US Marine Corps. The body and controls are of all-steel construction and the consequent mass requires specially reinforced mounts which have been designed to withstand a very high degree of stress.

The bolt handle has been modified by addition of an aluminium extension

7.62 mm C3A1 sniper rifle

knob which provides easier purchase particularly with a gloved hand, and greater clearance from the telescope sight than was the case with the C3 model.

DATA
Cartridge: 7.62 × 51 mm NATO
Operation: Bolt action repeater
Feed: 6-shot box magazine
Weight, empty: 6.30 kg with Unertl sight
Length: 1140 to 1210 mm

Barrel length: 660 mm
Rifling: 4 grooves, rh, one turn in 305 mm

Manufacturer
Parker-Hale Ltd., Bisley Works, Golden Hillock Road, Birmingham B11 2PZ, England.
Status
Current; production.
Service
Entered service with Canadian Army 1989.

5.56 mm C7 assault rifle

In 1984 the Canadian Forces adopted a new generation of 5.56 mm NATO calibre weapons for all services. The chosen weapons are based upon the M16 design and incorporate some of the features of the A1 and A2 versions of that family. The C7 retains the fully automatic firing mode rather than burst firing, as well as a single shot mode. It uses a simple two-position aperture rear sight with one large aperture for short range firing and a second smaller aperture for longer ranges. The rifle has a cartridge case deflector present to permit left-handed firing and the barrel is a cold-forged chrome lined design rifled to accept both SS109 and M193 type ammunition. A full range of accessories for the C7 have been developed, including bayonet, adjustable length stock, cleaning kits, BFA and sling.

DATA
Cartridge: 5.56 × 45 mm
Operation: direct gas, selective fire
Method of locking: rotating bolt
Feed: 30-round box magazine
Weight: (empty) 3.3 kg; (loaded) 3.9 kg
Length: 1020 mm (normal butt)
Barrel length: 510 mm
Rifling: 6 grooves, rh, one turn in 178 mm
Sights: (fore) post; (rear) 2-position aperture
Muzzle velocity: 920 m/s (SS109 at 24 m)
Rate of fire: 800 rds/min

5.56 mm C7 assault rifle and accessories

Manufacturer
Diemaco Inc., 1036 Wilson Avenue, Kitchener, Ontario N2C 1J3.
Status
Current. Production.
Service
Canadian forces.

5.56 mm C7A1 assault rifle

The C7A1 rifle is an improved version of the basic C7 combat rifle, incorporating a low mounted optical sight. The C7A1 eliminates the carrying handle of the C7 and substitutes an optical sight mounting rail which positions the selected optic in a natural position for rapid target acquisition. A fully capable backup iron sight fits into the butt trap of the C7A1, for use should the optic not be fitted. The C7A1 requires only five new components, plus the optic module, and will be introduced into service alongside the existing C7.

The C7A1 is available with either a Weaver type rail configuration or a RARDE type rail.

5.56 mm C7A1 rifle with backup iron sight fitted

DATA
AS FOR C7 RIFLE EXCEPT:
Weight: with Canadian sight: (empty) 3.9 kg; (loaded) 4.5 kg
Sights: optical or (fore) post; (rear) 2-position aperture

Manufacturer
Diemaco Inc., 1036 Wilson Avenue, Kitchener, Ontario N2C 1J3.
Status
Current. Production.
Service
Canadian forces.

5.56 mm C8 assault carbine

The C8 carbine is a compact version of the Canadian Forces standard C7 rifle. The carbine features a telescoping buttstock and a shortened barrel, while retaining all normal field replaced parts common with the C7 rifle. The carbine will be issued to armoured vehicle crews and may subsequently be issued to Special Forces and other users requiring a more compact personal weapon than the C7.

In spite of the shortened barrel the C8 has demonstrated accuracy nearly equivalent to the C7 in service, and it shares most of the firing characteristics of the rifle. The carbine uses the same hammer-forged barrel as the rifle, for superior life and accuracy, and it also accepts all C7 accessories.

DATA
AS FOR C7 RIFLE EXCEPT:-
Weight: (empty) 2.7 kg; (loaded) 3.2 kg
Length: (butt closed) 760 mm; (butt extended) 840 mm
Barrel length: 370 mm
Rate of fire: 900 rds/min

Manufacturer
Diemaco Inc., 1036 Wilson Avenue, Kitchener, Ontario N2C 1J3.
Status
Current. Production.

5.56 mm C8 assault carbine and accessories

Service
Canadian forces.

.22LR Small Bore Cadet rifle

The Small Bore Cadet rifle is an exact copy of the C7 assault rifle, manufactured to fire the .22 LR rimfire cartridge in the semi-automatic mode only. The Small Bore rifle is a primary marksmanship trainer for use by Cadets, Militia units and Regular Force units on indoor range facilities. The rifle features a barrel designed for the .22LR projectile and a 10 or 15-round capacity magazine which fits within a standard 5.56 mm magazine body for total weapon handling practice. Many components of the C7 have been redesigned in plastic, including the complete lower receiver and buttstock assembly, to make the Small Bore rifle less expensive to acquire.

The Small Bore rifle is also available as a kit to convert existing C7/M16 type rifles.

DATA
AS C7 RIFLE, EXCEPT:
Feed: 10- or 15-round box magazine
Rifling: rh twist, one turn in 406 mm

Manufacturer
Diemaco Inc., 1036 Wilson Avenue, Kitchener, Ontario N2C 1J3.
Status
Current. Production.

.22LR Small Bore Cadet Rifle and conversion kit

Service
Canadian forces.

CHINA, PEOPLE'S REPUBLIC

7.62 mm Type 56 carbine

This is a Chinese copy of the Soviet Simonov self-loading carbine (SKS). The rifle may be identified by the Chinese symbols on the left front of the receiver. The weapon is fashioned and functions in exactly the same way as the SKS

Chinese militia receiving instruction on 7.62 mm Type 56 carbine

(which see). Later versions of the Type 56 have a spike bayonet replacing the folding blade of conventional shape used on all other SKS variants.

The Type 56 is now manufactured for sale as a sporting rifle.

DATA
Cartridge: 7.62 mm × 39
Operation: gas, semi-automatic
Method of locking: tilting block
Feed: 10-round internal magazine
Weight: (unloaded) 3.86 kg
Length: (bayonet folded) 1021 mm
Barrel: 521 mm
Sights: (foresight) cylindrical post; (rearsight) tangent, notch
Muzzle velocity: 735 m/s
Effective range: 400 m

Manufacturer
State factories.
Status
Obsolescent. Manufactured for commercial sale.
Service
Reserve units only.

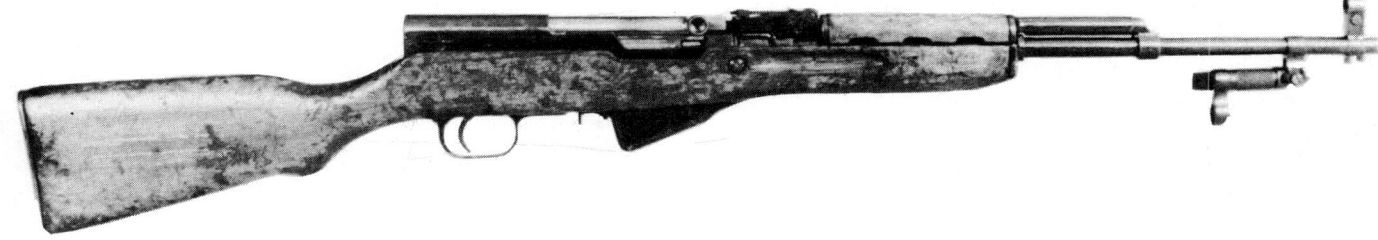

Chinese SKS 7.62 mm Type 56 carbine with knife bayonet

7.62 mm Types 56, 56-1 and 56-2 assault rifles

The Type 56 assault rifle is a copy of the later model of the Soviet AK-47 in which the rear end of the top surface of the receiver is straight. Earlier models had a down slope on the top rear surface of the receiver. This model is identified by the Chinese characters on the right-hand side of the receiver showing the positions of the selector lever, the upper and lower positions being for full automatic and single shots respectively. Later production models show 'L' for the full automatic marking and 'D' for single-shot. It is fitted with a folding bayonet which hinges down and back to lie beneath the fore-end.

There is also a Type 56-1 assault rifle which has a folding metal stock. This may be distinguished from the Soviet AK-47 not only by the Chinese markings or 'L' and 'D' for automatic and single-shot respectively, but also by the prominent rivets in the arms of the butt stock, which are not present in the Soviet version.

The Type 56-2 generally resembles the 56-1 but has a butt stock which folds sideways to lie along the right side of the receiver.

All three models are commercially available in semi-automatic form as Type 5.65 sporting rifles.

DATA
Cartridge: 7.62 mm × 39
Operation: gas, selective fire
Method of locking: rotating bolt
Feed: 30-round detachable box magazine
Weight: 3.8 kg

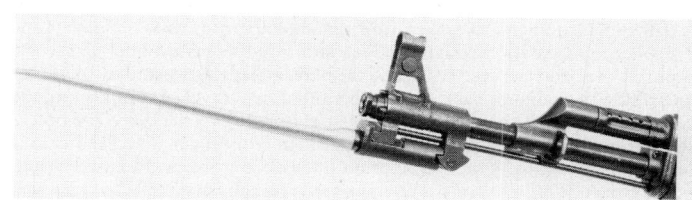

Chinese 7.62 mm Type 56 assault rifle with bayonet in place

Chinese 7.62 mm Type 56 rifle

Length: 874 mm
stock folded (Type 56-1) 645 mm; (Type 56-2) 654 mm
Barrel: 414 mm
Sights: (foresight) cylindrical post; (rearsight) tangent, notch
Muzzle velocity: 710 m/s
Rate of fire: (cyclic) 600 rds/min
Effective range: 300 m

Manufacturer
State factories.
Status
Current. In production.
Service
Chinese forces. Supplied to some South-east Asian countries.

Chinese 7.62 mm Type 56-1 assault rifle showing rivets in arms of butt stock

7.62 mm Type 68 rifle

This weapon is of Chinese design and manufacture. In general appearance it resembles the Type 56 (SKS) carbine but the barrel is longer, the bolt-action is based on that of the AK-47 and the rifle provides selective fire. It has a two-position gas regulator. It normally uses a 15-round box magazine but if the bolt stop is removed, or ground down, the 30-round magazines of the AK-47 and AKM can be used.

There are two versions of the Type 68. The earlier version has a receiver machined out of the solid whereas the later version has a pressed steel receiver. It can be recognised by the large rivets in the side of the receiver.

Operation

The Type 68 can be loaded in one of four ways. First, using an empty 15-round magazine, cocking the action holds the bolt to the rear; a 10-round SKS-type charger can then be put into the feed guides and the rounds forced down. Five further rounds can then be pressed in from the next charger. Secondly, if chargers are not available, 15 rounds can be forced down, one after the other, into the magazine. Thirdly, the 15-round magazine can be pre-loaded off the gun and then placed in position. Lastly if the bolt stop has been removed or ground down the 30-round magazine can be used but it must be pre-filled off the gun as the bolt will automatically close on the empty chamber and close off the top feed opening.

When the bolt is retracted over the loaded magazine and released, the top round is fed into the chamber as the bolt flies forward.

The selector lever is positioned directly in front of the trigger on the right. When the selector is pulled to the rear, reading 'O', the trigger is locked. The bolt however can be pulled back if required. The middle position, marked '1', with the lever vertical, provides semi-automatic fire and when the lever is fully forward, the gun will fire at full automatic. This position is marked '2'.

The sights are adjusted by pressing in on the side locks and moving the sight bar along the leaf to the required range. The position marked III is a battle sight position for all ranges up to 300 m.

Normally the gas regulator will be set with the smaller of the two holes nearer the barrel. If it is necessary to use the large hole to increase the gas flow the gas regulator retainer is pressed in towards the cylinder. When it disengages from the hand-guard, the retainer is pulled out of the gas cylinder and then the regulator is rotated to select the other hole. The retainer is replaced and rotated and the gun should then be ready to fire.

The bayonet is like that of the SKS. The handle is forced back against the spring and the bayonet then rotated forward and locked at the muzzle.

To fire the weapon, the selector lever is placed on '1' or '2'. When the trigger is pulled the hammer strikes the firing pin and the cap is crushed. Some of the propellant gas passes through the vent and into the gas cylinder above the barrel. The piston is forced back, and the bolt carrier starts to move to the rear and rotate the hammer back. After a period of free travel of about 6 mm, the cam cut in the bolt carrier reaches the operating lug of the bolt. The bolt is rotated, providing initial extraction, and then withdrawn. The fired case is held to the bolt-face by the extractor until the ejector pivots it out to the right of the gun. The return spring is compressed. The bolt carrier hits the rear wall of the receiver and rebounds. As it goes forward the next round is fed and chambered. The extractor clips over the cartridge rim and the bolt comes to rest. The carrier continues on and the cam forces the operating lug across and rotates the bolt into the locked position. The carrier, in its final foward motion, after the bolt is fully locked, releases the safety, or automatic, sear and then stops as it reaches the front of the receiver.

Trigger and firing mechanism

The automatic (or safety) sear is placed forward of the hammer and the latter is held back until the bolt is fully home and locked.

The trigger sear is to the right behind the hammer and moves with the trigger. The semi-automatic sear is on the left and functions only when the selector is set to '1'. The gun can never fire, regardless of the selected mode, until locking is completed. The safety (automatic) sear is then struck by the carrier and driven off the hammer. When the trigger is pulled the trigger sear releases the hammer. The round is fired and the recoiling carrier forces the hammer back. As soon as the carrier starts to move back the spring-loaded automatic sear is engaged and holds the hammer up. At single-shot the semi-automatic sear is engaged and holds the hammer back. Since the trigger is held back after the round has been fired it must be released before another shot. When this happens the semi-automatic sear moves off the hammer. The hammer starts to move forward but is caught by the trigger sear. When the trigger is pulled the hammer is released and if the bolt is locked the hammer will hit the firing pin. If the bolt is not locked the hammer is held back by the automatic sear.

When the selector lever is set for automatic fire, the semi-automatic sear is held clear of the hammer and plays no part. The trigger sear moves off the hammer and the round is fired. When the hammer comes back it is held by the automatic sear and only when the bolt is fully closed does the gun fire. When the trigger is released the trigger sear catches the hammer and holds it back.

When the selector is in the safe position 'O' the selector shaft blocks movement of the trigger sear and so the hammer is held.

Stripping

Remove the magazine, retract the bolt, inspect the chamber. Press in the catch on the left rear of the receiver cover and pull the cover back off the receiver.

If the gun is an early model, the return spring guide rod can be lifted out of the machined receiver. If the gun has the stamped receiver of the later model, the driving spring rod must first be eased forward before lifting out. Pull back the cocking handle. As the carrier reaches the rear end of the receiver, press down on the cocking handle and it will disengage from the receiver. The carrier can then be lifted out. The bolt can be rotated and removed from the carrier.

The piston comes out of the gas cylinder when the gas regulator retainer is removed. The hand-guard and heat shield are free to come off.

Rotate the catch at the rear of the trigger-guard and lift the trigger group out.

DATA
Cartridge: 7.62 mm × 39
Operation: gas, selective fire
Method of locking: rotating bolt

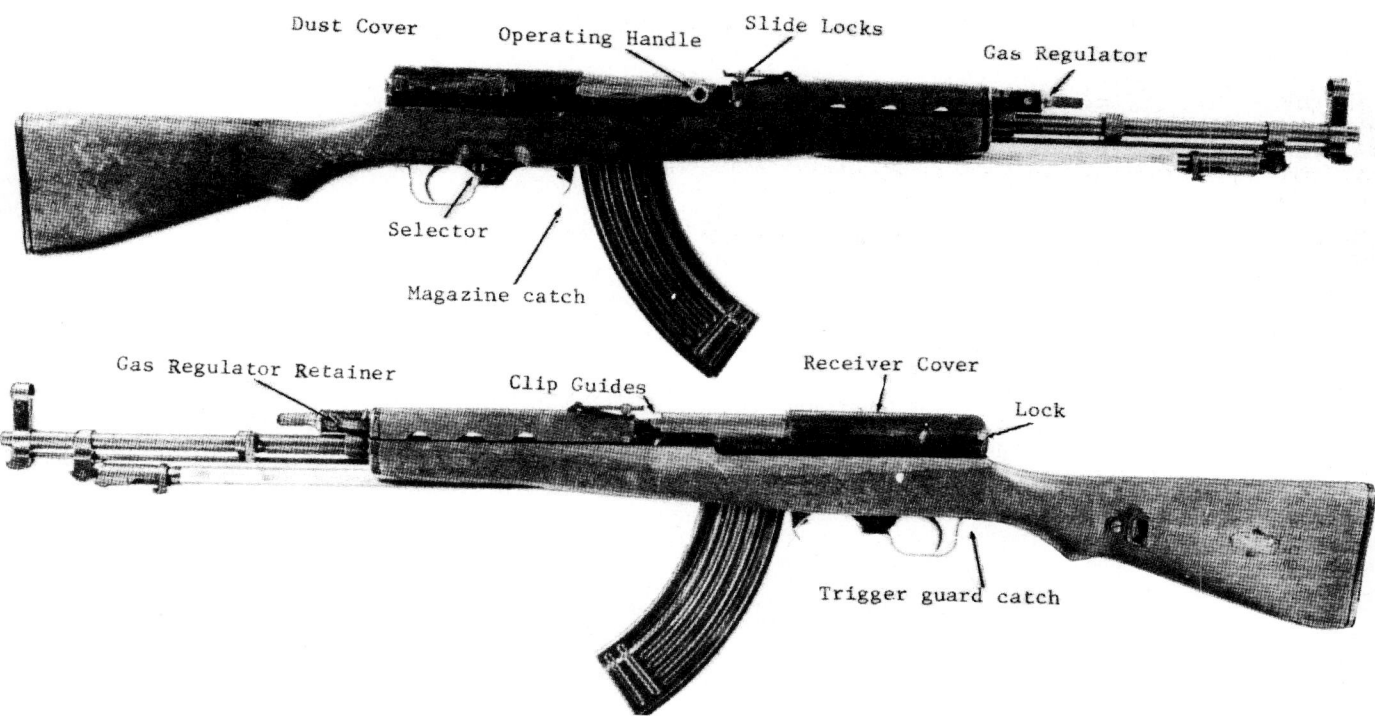

Chinese 7.62 mm Type 68 rifle (first version)

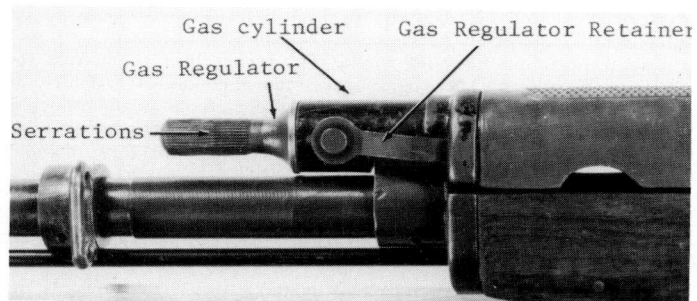

Gas regulator of Chinese 7.62 mm Type 68 rifle

Feed: 15-round detachable box magazine (30-round box magazine from the AK-47, AKM can be used only if the holding open stop is removed)
Weight: 3.49 kg
Length: 1029 mm
Barrel: 521 mm
Rifling: 4 grooves rh
Sights: (foresight) cylindrical pillar; (rearsight) tangent, notch
Muzzle velocity: 730 m/s
Rate of fire: (cyclic) 750 rds/min
Effective range: (single-shot) 400 m; (automatic) 200 m

Manufacturer
State factories.
Status
Current.
Service
Chinese forces.

7.62 mm Type 79 Sniping Rifle

This rifle is a precise copy of the Soviet SVD Dragunov sniping rifle, except that the butt is slightly shorter. It is equipped with an optical sight which is a copy of the Soviet PSO-1 and has the same ability to detect infra-red emissions. For details and data, refer to the SVD entry under USSR.

5.56 mm Type CQ automatic rifle

This rifle appears to have been based upon the American M16A1, from which it differs only in minor details. It is rifled to suit the 5.56 mm cartridge Type CJ, the Chinese equivalent of the M193. The construction of the rifle differs very slightly, in that in dismantling the butt and pistol grip group is removed entirely, rather than hinged to the receiver, but the rotating bolt and trigger mechanism appear to be the same.

DATA
Cartridge; 5.56 mm Type CJ (M193)
Operation: gas, selective fire
Locking: rotating bolt

Feed: 20-round box magazine
Weight, empty: 3.20 kg
Length: 987 mm
Barrel: 505 mm
Rifling: one turn in 305 mm
Muzzle velocity: 990 m/s

Manufacturer
China North Industries Corp., PO Box 2137, Beijing.
Status
Current. Production for export.

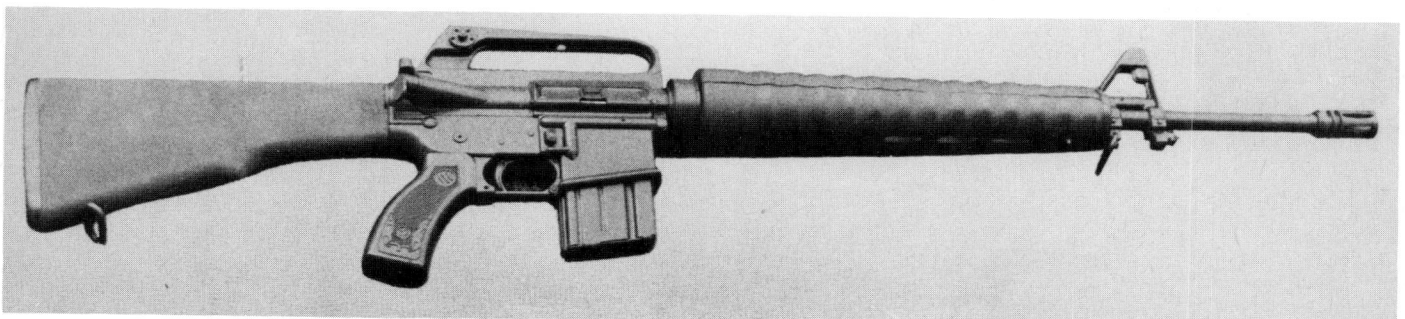

5.56mm Type CQ automatic rifle

CZECHOSLOVAKIA

7.62 mm Model 58 assault rifle

Most countries of the Warsaw Pact make weapons based on Soviet designs and the armies of most of them are equipped with locally made versions of the AK-47 or AKM rifles. Not so the Czechoslovak Army, the soldiers of which are equipped with the Model 58 which is an indigenous product of original Czechoslovak design. The earliest known versions had wooden butts, pistol grips and fore-ends, but most production weapons use a wood-fibre filled plastic for those parts. There are three standard versions:
Model 58 P which has a solid butt and is the normal infantry rifle;
Model 58 V in which the butt folds; and
Model 58 Pi which is a P version with a long dovetail bracket on the left side to accept a night-sight. This version is usually seen with a light bipod and an enlarged conical flash hider.

The Model 58 bears a superficial resemblance to the Soviet AK-47 but a closer look immediately reveals considerable differences.

Operation

The arm is gas-operated with a vent, 215 mm from the breech face, opening into a cylinder placed above the barrel. There is no gas regulator and the full gas force is exerted on the piston-head. The entire piston is chromium-plated so as to remain free of fouling. The gas pressure can drive the piston back only 19 mm and the shoulder on the shank then butts up against the seating and no further movement is possible. There is a light return spring held between the piston shoulder and the seating which returns the piston to its forward position. The cylinder is vented on the underside and the gas pressure gives the piston an impulsive blow before exhausting to atmosphere after the piston has gone back 16 mm. There is a protective metal heat shield over the piston which has a covering of the wood-filled plastic used throughout for the furniture. It is exactly the same shape as that in the AK-47 and it is removed by the withdrawal of a pin passing through the rearsight block in which the piston sits.

The short tappet-like stroke of the piston strikes the breech-block carrier and drives it rearwards. After 22 mm of free travel during which the chamber pressure falls to a very low level, an inclined plane on the carrier moves under the locking piece and lifts it out of engagement with the locking shoulders in the steel body. The locking piece swings and this movement provides the slow powerful leverage required for primary extraction. The breech-block is then carried rearwards extracting the empty case from the chamber. A fixed ejector in the receiver passes through a groove cut in the underside of the bolt and the case is rotated around the extractor and flung upwards clear of the gun. The continued rearward movement of the carrier and bolt compresses two double-coiled helical springs. The larger of these fits into the top hole of three drilled in the carrier and the smaller rests in the hollow steel tube which acts as a hammer. The carrier is driven forward and the feed horns on the underside of the bolt-face force a round out of the magazine and into the chamber. When the round is fully chambered the carrier still has 16 mm of travel and, as

it advances, a transverse cam face forces the locking piece down and the two lugs enter the locking shoulders in the body. There is a strong similarity in this locking system to that used in the Walther P38 self-loading pistol. It should be noted that the arm can be assembled and fired without the locking piece and this could lead to a serious accident.

Trigger and firing mechanism

Unlike the majority of self-loading and automatic rifles the Model 58 does not have a rotating hammer which strikes a firing pin; the hammer is a steel bar hollowed from one end almost throughout its full length to take the hammer spring. At the open end is welded a bent and there is a groove cut in each side of this to slide on the receiver guideways. This hammer enters the hollow bolt and drives forward a fully floating firing pin.

The trigger mechanism consists of two sears side-by-side with the left sear slightly in front. The right-hand sear is connected to a forward sear trip operated by the carrier. Unless the bolt carrier is fully forward (when locking is completed) the trip holds up the right-hand sear and the hammer is held back. The left-hand sear works with a disconnector to provide single-shot fire.

The selector is on the right-hand side of the receiver; single-shot is indicated by '1' and full automatic by '30'. When the selector is rotated forward to full automatic the disconnector is lowered and is disengaged from the semi-automatic sear. When the carrier is fully forward the right-hand sear is depressed and the hammer is held on the semi-automatic sear only. As soon as the trigger is operated the trigger bar is moved forward and the left-hand sear is depressed to release the hammer. So long as the trigger is pressed and there is ammunition in the supply, the firing of the gun is controlled by the bolt closure. Each time the bolt recoils the automatic sear rises and holds up the hammer and as soon as the locking is completed the right-hand sear drops and the gun fires.

When the selector is rotated to the rear to the semi-automatic position, the trigger bar is lowered and forced clear of the semi-automatic sear but the disconnector is allowed to rise to engage the semi-automatic sear.

Trigger operation fires one round. The carrier then recoils and the automatic sear rises. The carrier strikes the disconnector and lifts the semi-automatic sear. The hammer is held by the automatic sear which is released when the bolt carrier goes fully forward, and since the automatic sear is 1.5 mm behind the semi-automatic sear, the bolt then goes forward and is held up by the semi-automatic sear.

To fire another round it is necessary to release the trigger to allow the disconnector to move back and up to engage under the semi-automatic sear. The trigger can now be pressed and the hammer will go forward off the semi-automatic sear.

At the 'safe' position with the selector pointing vertically downwards, the trigger bar and the disconnector are lowered and so there is no connection between the trigger and the semi-automatic sear which holds the hammer.

7.62 mm Model 58 V rifle with butt stock folded and field-stripped

7.62 mm Model 58 V rifle

7.62 mm Model 58 Pi rifle with NSP-2 IR sight, bipod and enlarged flash hider

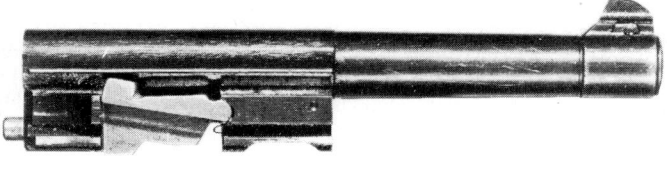

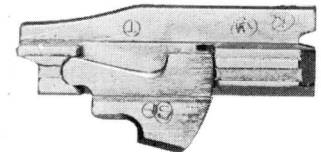

Comparison of locking systems of Model 58 and West German P38 pistol

Stripping

Remove the magazine, by rotating the release catch forward,operate the bolt and check that the chamber is empty. Ensure that the folding butt is extended.

The receiver cover is pinned to the receiver and is released by moving the locking pin, at the top rear, fully to the right. The cover is pressed forward and when the rear end is lifted it can be slid back and removed with the hammer spring and return spring.

Withdraw the cocking handle and at the end of the carrier travel rotate the cocking handle upwards. As the front of the carrier lifts, place the middle finger under the bolt and lock to prevent them falling out, and remove the assembly. The hammer can be taken out by pulling it fully to the rear and then moving it forward about 6 mm until it will rotate anti-clockwise. It can then be withdrawn.

The hand-guard over the gas piston can be released as soon as the pin under the rearsight is forced across to the right. The piston-head can be lifted when the piston is fully retracted and when tilted the piston and spring can be taken out.

Re-assembly is based on reverse order to stripping and presents no problems. Note that the hammer spring attached to the receiver cover goes into the hollow hammer and the return, or driving, spring goes into the top hole of the three drilled in the carrier.

Accessories

The usual accessories are available consisting of a sling, bayonet, cleaning rod, oil bottle and zeroing tool.

In addition to these however are the unusual adjuncts of a flash hider and a bipod. Either or both can be fitted. A blank firing attachment can be placed over the muzzle. This replaces the muzzle nut and has a small orifice which passes a limited amount of gas and allows the build-up of enough pressure to operate the gas system.

DATA
Cartridge: 7.62 mm × 39
Operation: gas, selective fire
Method of locking: pivoting locking piece
Feed: 30-round box magazine
Weight, empty: 3.14 kg
Length: (butt extended) 820 mm; (butt retracted) 635 mm
Barrel: 401 mm
Rifling: 4 grooves, rh, one turn in 240 mm
Sights: (fore) post; (rear) tangent leaf V notch
Sight radius: 356 mm
Muzzle velocity: 710 m/s
Rate of fire: (cyclic) 800 rds/min
Effective range: 400 m

Manufacturer
Uherský Brod ordnance factory.
Status
Current. Available.
Service
Czechoslovak Army. Sold on the commercial market.

DOMINICAN REPUBLIC

.30 Models 2 and 62 Cristobal light rifles

The Hungarian designer Pal Kiraly, who was responsible for the 39M and 43M Hungarian sub-machine guns, started up an arms factory at San Cristobal in the Dominican Republic in 1948 with extensive help from the Italian firm of Pietro Beretta and at first the Beretta sub-machine gun was produced. Kiraly then designed the Cristobal Model 2 carbine, using the .30 cartridge of the US M1 carbine and this was produced.

The Model 2 is a delayed blowback rifle using the two-part bolt delayed blowback system which Kiraly patented in 1912 and used in the Hungarian 39M and SIG MKMO sub-machine guns. It fires from an open breech. A lightweight bolt-head carries a lever, one end of which is engaged in the floor of the receiver and the other end bears against the heavy bolt body. The bolt-head is restrained by the lever when the pressure is high and the heavy bolt body is accelerated backwards. When the lever is rotated out of the receiver the entire bolt is blown back. On the feed stroke the two parts of the body must come together and rotate the lever into the receiver before the bolt body can drive the firing pin, in the bolt-head, into the cap of the cartridge.

Subsequently a Model 62 was made which was a gas-operated assault rifle firing the 7.62 mm × 51 cartridge. It was not manufactured in quantity.

DATA
MODEL 2
Cartridge: .30 M1 carbine
Operation: delayed blowback, selective fire
Method of delay: lever and 2-part bolt
Feed: 25- or 30-round detachable box magazine
Weight: 3.52 kg

.30 Model 2 Cristobal selective fire carbine

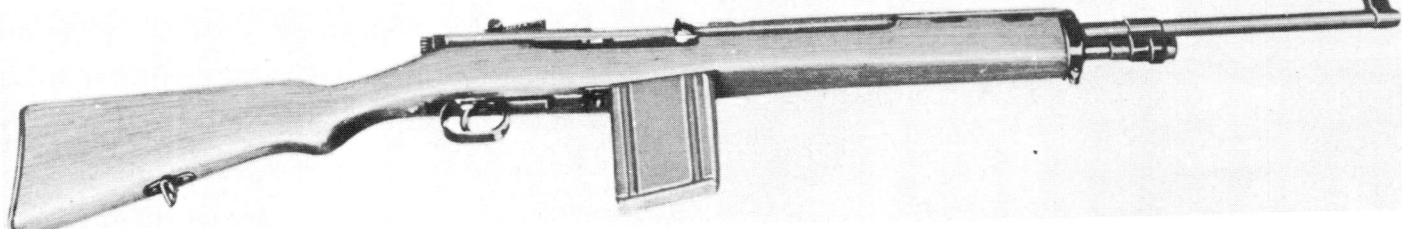

Dominican Model 62 rifle

Length: 945 mm
Barrel: 409 mm
Sights: (foresight) blade; (rearsight) notch
Muzzle velocity: 572 m/s
Rate of fire: (cyclic) 580 rds/min
Effective range: 300 m

Manufacturer
San Cristobal Arsenal.
Status
Obsolescent. Not in production.
Service
Dominican Republic. Some were supplied to Cuba.

EGYPT

7.92 mm Hakim rifle

The Egyptian Army historically used the same rifles as the British. In 1949, when the Egyptians became independent, they purchased a number of Saive-designed Fabrique Nationale (FN) self-loading rifles.

The 7.92 mm Hakim rifle was manufactured in Egypt. It was essentially a copy of the 6.5 mm AG42 Ljungman rifle, made by Carl Gustaf in Sweden, but adapted to take a 7.92 mm round. In 1954 Egypt started to obtain substantial assistance from the USSR and among the weapons so obtained were the Czechoslovak Model 52 self-loading rifle and the Soviet SKS self-loading carbine. These apparently appealed to the Egyptians who

manufactured a self-loading carbine with an attached folding bayonet, using the Soviet 7.62 mm × 39 cartridge, which they called the Raschid rifle. This was based on the Hakim rifle using the modified Ljungman system of direct gas action. Few were made and the weapon, if still in service, is not used by first-line troops. Some of the 7.92 mm Hakim rifles were also modified to take the 7.62 mm round, but these too are at best obsolete weapons.

The weapons most widely used by the Egyptian Army today are the Soviet AK-47 and AKM rifles.

7.92 mm Hakim rifle

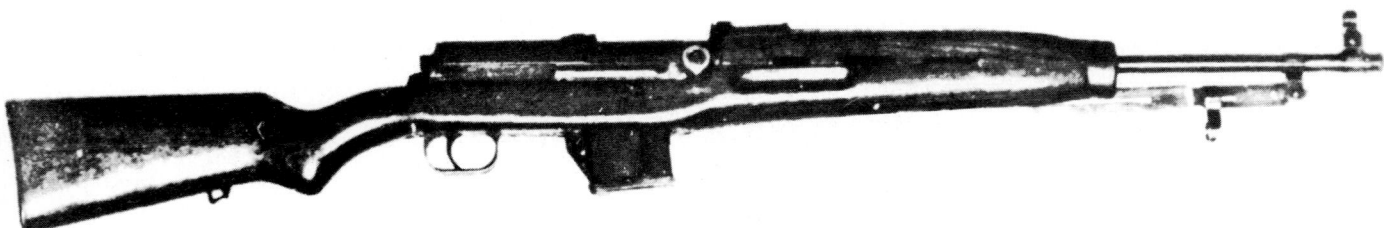

7.62 mm Raschid rifle

FINLAND

M60, M62 and M76 automatic Valmet rifles

The Finnish assault rifle is essentially a Soviet AK-47 manufactured at Sako's Tourula Works. The Finns adapted the design in the late 1950s and the first small pilot batch required for troop trials bore the designation M60. After some minor changes the Finnish Defence Forces adopted the weapon under the designation M62. The latest version is called the M76.

The M60 differed from the AK-47 in several respects. There was no wood on the M60. The metal fore-end and the pistol grip were both plastic covered. The butt stock was a cylindrical tube with a cross bar riveted on for the shoulder piece. The cleaning rod and brush were carried in the tube of the stock. The forearm of the Model 60 had 10 large and 10 small holes to allow dissipation of heat. The receiver was made from milled steel bar. The barrel had a three-prong flash eliminator. The foresight was a hooded post over the front end of the gas cylinder. The rearsight was a tangent aperture protected by two wings. Both foresight and rearsight had night sight option. Since, unlike the USSR practice, an aperture rearsight was used, it was placed at the back of the receiver cover. There was no trigger-guard, but a protective bar dropped down in front of the trigger to prevent accidental release of the magazine. The safety catch and selector lever on the right of the receiver was exactly the same as that of the AK-47. The semi-automatic marking was a single dot, and next came full automatic with three dots. The M60 did not go into production and after the trials the remaining examples were recalled.

The 1962 modifications were not of great significance: the fore-end was

made of plastic, ribbed, with 24 large cooling holes, a trigger-guard was fitted, the pistol grip was ribbed plastic and swept back further at the rear upper end and the fitting of the shoulder piece to the tubular stock was amended somewhat. A recent innovation is the addition of tritium light sources to the sights to facilitate aiming in poor light.

The M62, selective fire weapon, is strictly a military rifle. For commercial sale, however, the manufacturer makes a semi-automatic version, the M62/S, which is available with either a metal butt stock, like the M62, or a wooden stock. A version of the M62 with a folding butt is also available.

The following versions of the M76 have been made both in 7.62 mm and in 5.56 mm:
    M76T tubular butt stock;
    M76F folding butt stock;
    M76P plastic butt stock;
    M76W wooden butt stock.

DATA
M76W
Cartridge: 7.62 mm × 39 or 5.56 mm × 45
Operation: gas, selective fire
Method of locking: rotating bolt
Feed: 15-, 20-, or 30-round detachable box magazines
Weight: (empty) 3.6 kg; (loaded with 30 rounds 7.62 mm) 4.51 kg
Length: 914 mm
Barrel: 420 mm
Rifling: 4 grooves rh
Sights: (foresight) hooded blade; (rearsight) tangent aperture
Muzzle velocity: 719/960 m/s
Rate of fire: (cyclic) 650 rds/min
Effective range: 350-400 m

Manufacturer
Sako Ltd, Tourula Works, PO Box 60, SF-40101, Jyväskylä.
Status
Current. Production.
Service
Finnish and Qatar forces, and Indonesian security forces.

M76F automatic rifle

Finnish 7.*62 mm M76T automatic rifle*

M76W automatic rifle

Valmet 7.62 mm M78 long barrel automatic rifle

The M78 Valmet is a version of the M76 assault rifle. The two main differences are a longer and heavier barrel with a carrying handle, and a bipod fitted to the front of the barrel. Normal rifle magazines are used. The weapon is light and easily portable and good reliability is claimed, stemming from the considerable experience gained with the M76. The low weight makes it particularly suitable for areas in which infantry have to operate on their feet rather than from vehicles. It must therefore be attractive to those forces who expect to fight in jungle or mountain conditions.

DATA
Cartridge: 7.62 × 39 mm; 5.56 × 45 mm; 7.62 × 51 mm in semi-automatic form only
Operation: gas, selective fire

Feed: 15- and 30-round box magazines
Weight: (empty) 4.7 kg; (loaded with 30 rounds) 5.9 kg
Length: 1060 mm
Muzzle velocity: 719 m/s (7.62 × 39 mm)
Rate of fire: (cyclic) 650 rds/min

Manufacturer
Sako Ltd, Tourula Works, PO Box 60, SF-40101 Jyväskylä.
Status
Current. No longer in production.
Service
Military trials.

7.62 mm Valmet M78 long barrel rifle

7.62 mm Sako TRG-21 sniping rifle

This is a conventional bolt-action repeating rifle, Sako being of the opinion that for first-round effectiveness and accuracy the bolt action cannot be bettered.

The barrel is manufactured from tempered steel bar by the cold-hammering method. It is fitted with a detachable muzzle brake which also acts as an efficient flash hider. Alternatively, a silencer can be fitted.

The bolt and receiver are made from tempered steel, also by cold hammering, and are over-dimensioned so as to permit use of cartridges more powerful than the 7.62 × 51 mm in due course. The bolt has three forward locking lugs and is opened by a 60° movement. The safety catch, which is silent in operation, is on the right side of the receiver, and at the rear of the bolt is a cocked indicator which can be checked visually or by touch.

The two-stage trigger mechanism can be adjusted for trigger position and pull-off pressure between 1.0 and 4.0 kg. These adjustments can be carried out without dismantling the weapon. The trigger mechanism and trigger guard can be detached without dismantling the rifle.

The buttstock can be made of wood or glass fibre, and the action is bedded in a form of epoxy resin which has been selected to last for the operating life of the weapon. The stock may be adjusted for height and length, and there are

attachments for fitting a sling. The bipod is of steel and is articulated so as to allow movement of the weapon in aim without requiring the bipod position to be continually changed. It can be folded beneath the stock for carrying.

There is a 17 mm integral dovetail for attachment of optical day or night sights, and folding iron sights are provided for emergency use.

DATA
Cartridge: 7.62 × 51 mm
Operation: bolt-action, magazine repeater
Feed : 10-shot double-row magazine
Weight, empty: 5.3 kg without sights
Length: 1200 mm
Barrel: 660 mm
Rifling: 4 grooves, rh, one turn in 270 mm
Sights: telescope mount and iron emergency sights

Status
Current. Production

7.62 mm Sako TRG-21 sniping rifle

Vaime silencers

These attachment silencers are to a patented design and can be attached to almost any firearm. Three models are available; the 'Sniper' model is for application to military and sporting rifles and pistols from 4 mm to 11.6 mm calibre and is claimed to reduce the noise of discharge to the level of a .22 rimfire cartridge. It also eliminates muzzle flash and damps recoil by a considerable amount. The 'Marksman' model is smaller and intended for use on .22 rimfire weapons. A third model, the 'Varminter' is a fixed silencer/barrel assembly which replaces the standard barrel on .22 rimfire weapons; it is extremely efficient, eliminating the sound of discharge completely.

DATA

	Sniper	Marksman	Varminter
Diameter:	35 mm	25 mm	25 mm
Length:	350 mm	240 mm	500 mm
Weight:	500 g	150 g	

Manufacturer
Oy Vaimenninmetalli Ab, Sulantie 3, SF-04300 Hyrylä.
Status
Current. Production.

Vaime silenced sniping rifles

These rifles are built around standard bolt-action mechanisms but with integral silencer/barrel assemblies. The stock is of non-reflecting plastic material, and an articulated bipod with adjustable legs is standard. The rifles are not fitted with iron sights, but have a telescope mount which, by means of adapters, will accept any telescope sight or electro-optical sight required.

Two models are currently available; the SSR Mark 1 is chambered for the 7.62 × 51 mm NATO cartridge, while the SSR Mark 3 is chambered for the .22 Long Rifle rimfire cartridge.

DATA
SSR MARK 1
Length: 1180 mm
Weight: 4.1 kg

Magazine: 5-round box
Silencer: fixed, self-cleaning

SSR MARK 3
Length: 1010 mm
Weight: 3 kg
Magazine: 5-round box
Silencer: fixed, self-cleaning

Manufacturer
Oy Vaimenninmetalli Ab, Sulantie 3, SF-04300 Hyrylä.
Status
Current. Available.

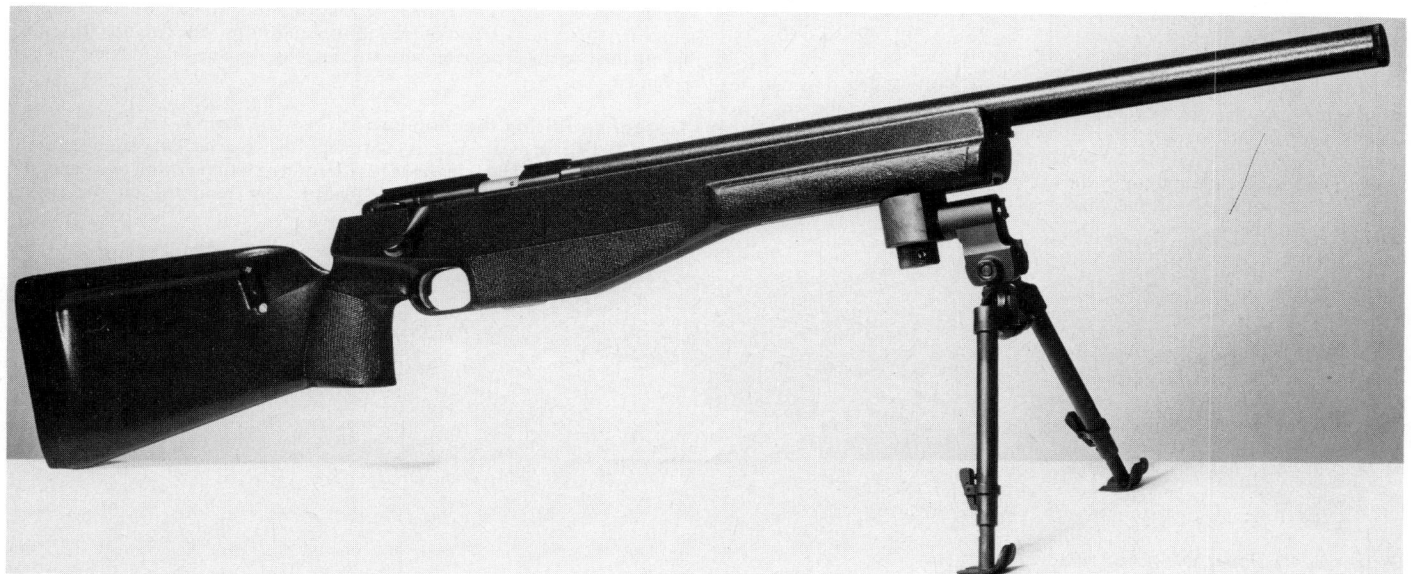

.22LR Vaime SSR Mark 3 silenced sniping rifle

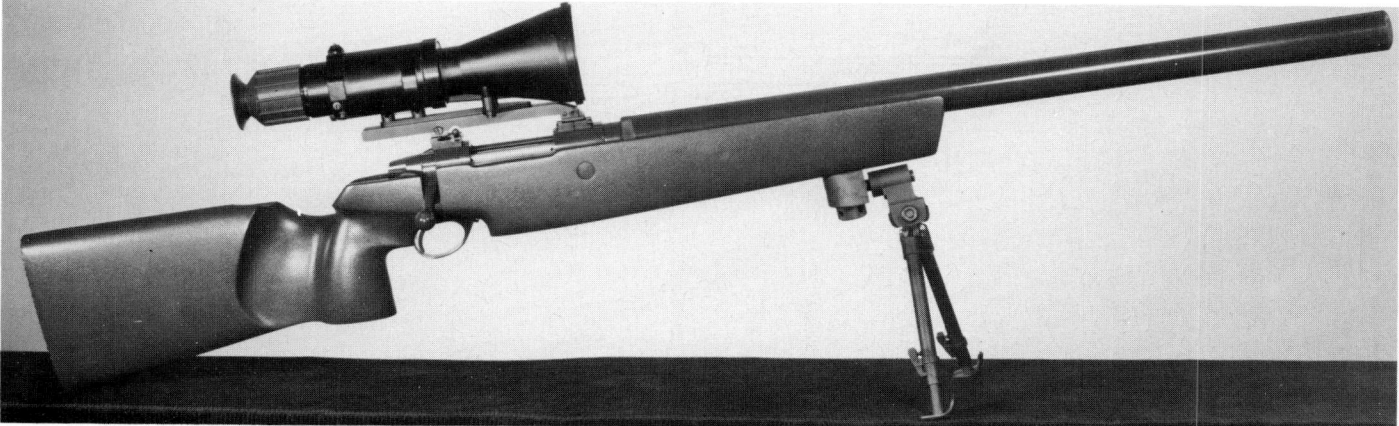

7.62 mm Vaime SSR Mark 1 silenced sniping rifle

FRANCE

7.5 mm MAS 49 rifle

The MAS 49 is a self-loading rifle using gas operation but dispensing with the gas piston and cylinder which are normally employed. The gas is diverted back, over the barrel, and impinges on the bolt carrier to drive it to the rear. The carrier moves back freely for about 6 mm and then lifts the back of the breech-block out of its locking recess in the receiver and carries it to the rear. This system reduces the overall weight of the weapon but is generally regarded as producing a lot of fouling. However it seems to be effective with the MAS 49.

The weapon is typical of its period using wooden furniture, a limited capacity magazine and a ball round ranging out well beyond the soldier's capacity to make use of it. There is a grenade sight mounted on the left side just behind the muzzle. A graduated spigot can be moved in and out and thus, by variation of the length of shot travel, the grenade range can be varied.

DATA
Cartridge: 7.5 mm Model 1929
Operation: gas, direct action; semi-automatic

Method of locking: tilting block
Feed: 10-round detachable box magazine
Weight: 4.7 kg
Length: 1100 mm
Barrel: 580 mm
Rifling: 4 grooves lh
Muzzle velocity: 823 m/s
Rate of fire: SA 30 rds/min
Effective range: 600 m

Status
Obsolete.
Service
Some former French colonies.

7.5 mm Model M1949/56 rifle

This rifle is a modified version of the MAS 49 which was adopted by the French Army after the Second World War to re-equip their infantry units. The Model M1949/56 differs from the 49 in that the wooden fore-end is shortened and the integral grenade launcher has been replaced by a combined muzzle brake/grenade launcher.

Operation
The MAS M1949/56 is a self-loading gas-operated rifle with a 10-round box magazine. Part of the propellant gas is deflected into a gas tube which lies along the top of the barrel, below the wooden furniture and emerges over the breech. The gas passes along the tube and is delivered into the bolt carrier inside which it expands and forces the carrier back some 9 mm. Cam grooves at the back of the carrier then contact lugs on the side of the bolt, the tilting

block is lifted out of engagement with the locking recess in the receiver and this rocking motion provides primary extraction of the case. The carrier collects the bolt and the assembly moves rearwards as an entity. The return spring is compressed, the empty case is withdrawn from the chamber and ejected, and the hammer is rotated back against its spring. The back end of the bolt carrier reaches the end of the receiver and stops. The compressed return spring then drives the carrier forward, loading a round and locking the bolt behind it.

Trigger and firing mechanism
The trigger/firing mechanism is in principle the same as that employed in the US Garand M1 and later in the M14 rifle, using a hammer with two bents and a primary and secondary, spring-loaded, sear mounted on the trigger

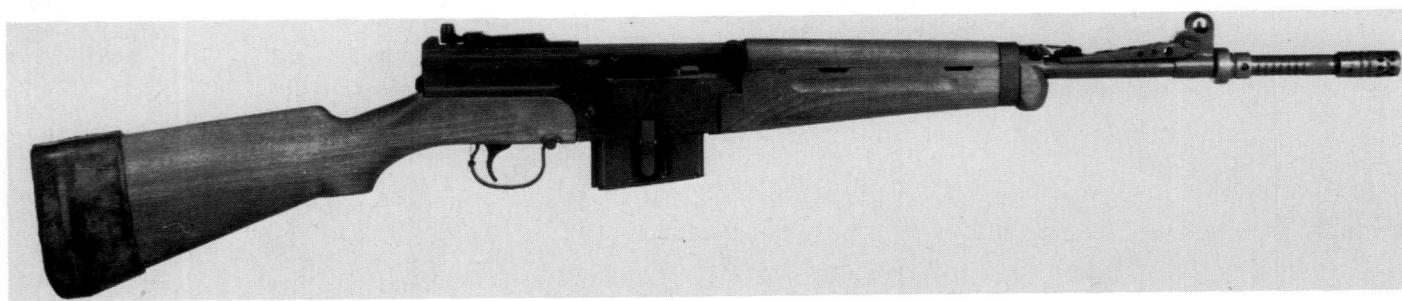

7.5 mm MAS Model M1949/56 rifle

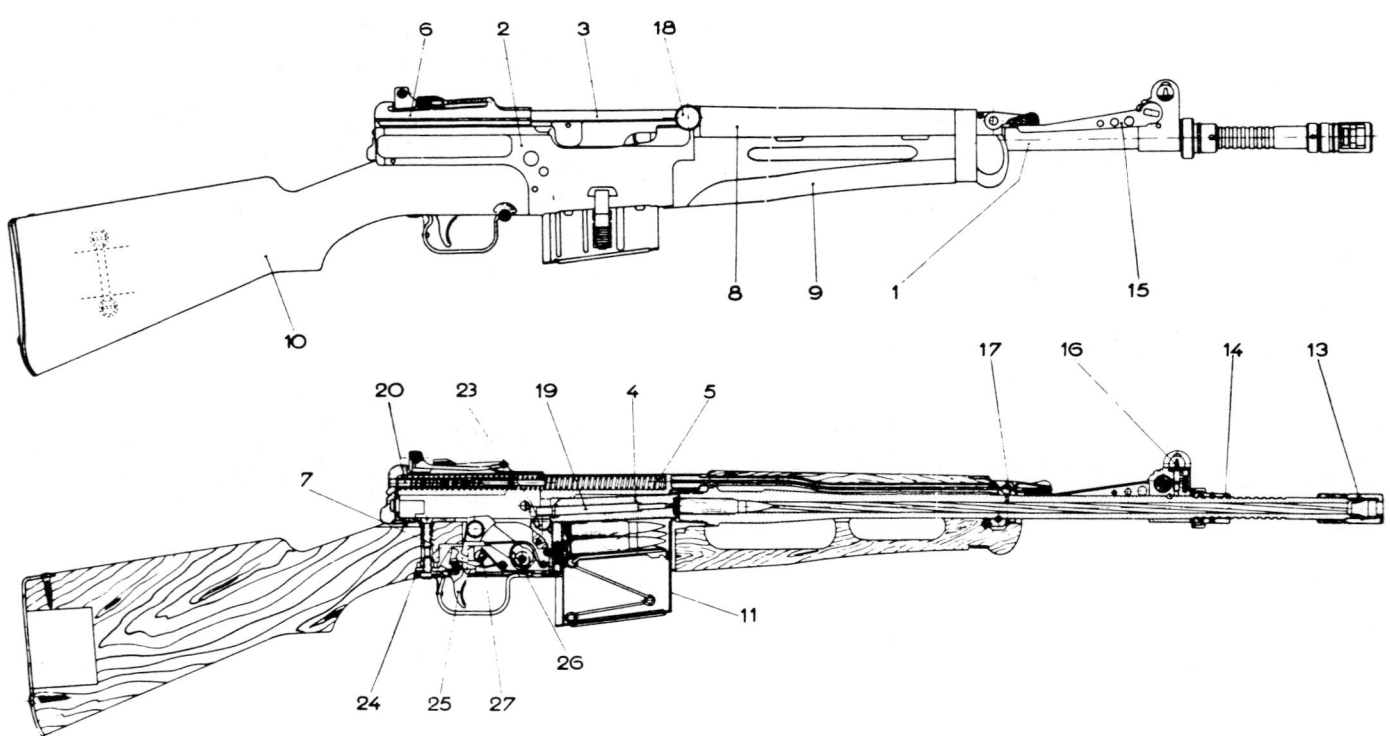

7.5 MAS Model M1949/56 rifle

extension. When the weapon is ready for firing the hammer bent is held by the primary sear. Pulling the trigger moves the primary sear out of the main bent of the hammer and the hammer spring rotates the hammer forward. When the hammer is brought back by the recoiling bolt carrier its rear hook is caught by the spring-loaded secondary sear and held. It is not possible to fire another round as the trigger is still held back. When the firer releases the trigger the upper end of the trigger revolves back and the spring-loaded secondary sear releases the hammer but before it can clear the trigger mechanism the primary bent is caught by the primary sear and the hammer is held. Pulling the trigger releases the hammer and the cycle starts again.

After the ammunition in the magazine is expended, the carrier will be held to the rear by the holding device operated by the magazine platform. The magazine catch on the right side of the magazine can be depressed, the magazine removed and replaced with a full one or alternatively two five-round chargers can be loaded via guides at the front of the bolt cover. The bolt can then be pulled back and when released will fly forward to chamber the top round.

Stripping
Stripping the M1949/56 rifle is straightforward. After the magazine has been removed and the chamber checked, the cover catch at the rear of the receiver is pressed and the receiver cover pushed forward until the rear end can be lifted out of the receiver. If it is then withdrawn the cover can be removed. The return spring is then exposed and can be lifted out. If the cocking handle is pulled back the bolt carrier assembly can be lifted out. The bolt can then be removed from its carrier.

Accessories include a telescopic sight and a night sight. The telescopic sight is the M53 bis. It fits on to the sight base on the left of the receiver, sliding on from rear to front, and is locked by swinging the adaptor lever forward. Further details of this sight are given in the FR-F1 sniping rifle entry below. The night sight is put on over the flash suppressor and foresight and the clamping wing nut is tightened.

DATA
Cartridge: 7.5 mm × 54 (some 7.62 mm × 51)
Operation: gas, direct action; self-loading
Method of locking: tilting block
Feed: 10-round box magazine
Weight, empty: 3.9 kg
Length: 1010 mm
Barrel: 521 mm
Rifling: 4 grooves, rh, one turn in 305 mm
Sights: (fore) blade; (rear) leaf tangent, aperture
Sight radius: 569 mm
Muzzle velocity: 817 m/s
Effective range: 600 m

Status
Obsolescent
Service
French forces; almost entirely replaced by FA MAS. Widely used in former French territories in Africa.

British soldier at French Commando School being instructed in grenade firing with 7.5 mm MAS Model M1949/56 rifle

FR-F1 sniping rifle

The French sniping rifle, the Fusil à Répétition Modèle F1, is a 10-shot, manually operated bolt-action rifle based on the action of the now obsolete 7.5 mm Model 1936. It has a detachable 10-round box magazine. The rifle was specifically designed for sniping but two other versions are in existence. The three models are known as: the Tireur d'Elite (sniping rifle); Tir Sportif (target rifle with target sights and a 1.5 to 1.9 kg trigger pull); Grande Chasse (which is described as being designed for big game hunting: it has an APX model 804 telescopic sight and 2 to 2.5 kg trigger pull).

Operation
The bolt is manually operated with a turn-down knob on the right-hand side. The locking lugs are rear mounted. As the bolt is forced forward it feeds the top round into the chamber and as it is closed the extractor grips the rim of the cartridge. At the same time the sear holds up the firing pin lug to retain the firing pin in the cocked position.

When the trigger is operated it rotates back around the trigger pin joining the trigger and sear until the trigger stud reaches the bottom of the receiver. Continued pressure pivots the sear and compresses its spring. When the top of the trigger adjusting screw reaches the receiver the first pressure is felt. When

Bolt and Modèle 53 bis telescopic sight

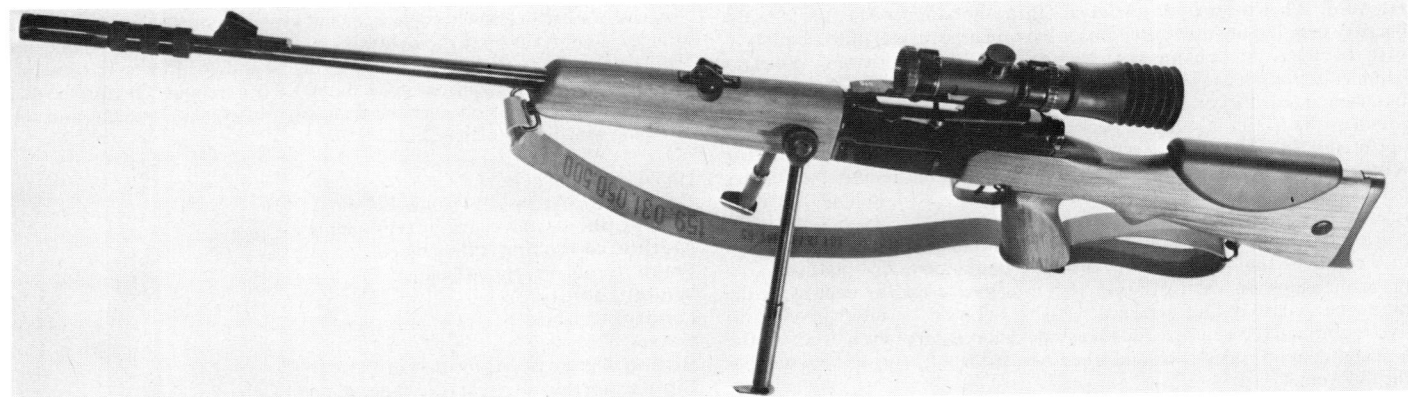

FR-F1 sniping rifle on bipod with telescopic sight fitted and iron sights folded down

the second pressure is taken the sear releases the lug of the firing pin which goes forward under its own spring to hit the cap. The lock time is very short. It should be noted that if the trigger is held back and the bolt operated, the bolt stop which pivots on the trigger pin is depressed against its spring and the bolt comes freely out of the receiver.

Telescopic sight

The FR-F1 sniping rifle uses the Modèle 53 bis telescopic sight which is carried, with its adjusting tool, in a transit case. The mount is placed with the locking lever to the rear and when in place it is secured by rotating the lever forward. To zero the telescope the screws at the bottom of the mount ring are loosened, the elevation knob set to '2' and the rifle is bore-sighted on a distant object. The plastic rings between the telescope and the mount rings are rotated until the telescope reticle is on the aiming mark with zero windage set. When the bore and graticule are aligned on the same aiming mark the screws on the mount rings are tightened. A three-round check firing at 200 m should then be carried out to ensure that the graticule is on the mean point of impact. Once the sight is correctly zeroed it may be dismounted, cased, and re-mounted without change of zero.

Firing preparation

The rifle is issued to one man in the sniping section and he fits the weapon to suit his own physical characteristics. The length of the butt stock can be increased by fitting one or both of the wooden spacers provided. The smaller is 20 mm and the other 40 mm thick. The butt plate screw is removed and the spacer fitted to match the contour of the butt stock. The butt plate is then screwed on again. Next the cheek pad is fitted. There are two heights, 8 or 17 mm. The cheek pad has pins on the underside which mate into holes on the comb of the stock. It should be noted that it is essential not to damage the pins on the cheek pad because this fitting has to be removed from the rifle before the bolt can be withdrawn.

The magazine catch on the right of the receiver is pressed to release the magazine. The magazine holds 10 rounds. If it is to be replaced in the rifle the rubber cap is left on the bottom but if the magazine is being carried as a loaded replacement the rubber cap is placed over the top to keep out dust and put back on the bottom, if time permits, when the magazine is placed in the rifle. It is possible to load single rounds through the bolt way to replenish a partially expended magazine but this procedure is not recommended in action.

The bipod is permanently fixed at the rear of the fore-end and the rifle is carried with the bipod legs swung forward and fitting into recesses on each side of the forearm. The bipod is always used unless firing over cover such as a low wall. The bipod legs can be lengthened by twisting the knurled collar above the spade and rotating it back when the spring-loaded lower leg has extended sufficiently. The legs should be fully retracted for stowing the bipod along the forearm.

There may be occasions when the telescope cannot be used. The fore- and rearsights are normally laid flat and must be raised. The rearsight is a square shouldered notch. The foresight is a flat topped pyramid. The shoulders of the rearsight and the centre of the foresight have three luminous green dots which at night are evenly spaced in a horizontal line and placed on the target. To use the iron sights the telescope must first be dismounted.

The FR-F1 uses match quality 7.5 mm × 54 ball ammunition and the use of tracer or armour piercing ammunition, which will damage the bore, is discouraged. The FR-F1 is also manufactured in 7.62 mm × 51 NATO and the marking 7.5 or 7.62 mm is inscribed on the left-hand side of the receiver.

DATA
Cartridge: 7.5 mm × 54 or 7.62 mm × 51
Operation: manual, single shots
Method of locking: rotating bolt
Feed: 10-round box magazine
Weight, empty: 5.2 kg
Length: 1138 mm without stock spacers
Barrel: 552 mm
Rifling: 4 grooves rh. 1 turn in 305 mm
Sights: (optical) × 4; (iron foresight) flat topped pyramid with luminous spot; (rear) square notch with luminous spots
Muzzle velocity: 852 m/s
Effective range: 800 m

Manufacturer
Groupement Industriel des Armements Terrestres, 10 place Georges Clémenceau, 92211 Saint-Cloud.
Status
Current. No longer being manufactured.
Service
French Army.

FR-F2 sniping rifle

Introduced in late 1984, the FR-F2 is an improved model of the FR-F1. The basic characteristics, action and dimensions are the same as the F1 model, the changes being in the nature of functional improvements. The fore-end is now of metal, covered in matt black plastic material; the bipod is more robust and has been moved from its location at the front end of the fore-end to a position just ahead of the receiver. This allows it to be adjusted more easily by the firer, and it is now suspended from a yoke around the rear of the barrel where it is less likely to affect the stability of the rifle when firing. The most innovative

7.62 mm FR-F2 sniping rifle

change is the enclosure of the barrel in a thick plastic thermal sleeve; this, it is claimed, reduces the possibility of heat haze interfering with the line of sight and it is also likely that it will reduce the infra-red signature of the weapon.

The FR-F2 is chambered only for the 7.62 × 51 mm NATO cartridge; otherwise there is no change in data from the F1, to which reference should be made.

5.56 mm FA MAS rifle

A bullpup design, the FA MAS can be fired from either shoulder without the difficulty inherent in most bullpup rifles of ejection of the spent case. This has been achieved by providing two extractor positions on the bolt-face. The rifle is issued with the extractor on the right and with this arrangement the spent cases are ejected to the right. By moving the extractor to the other position the direction of ejection is reversed, the cheek rest being removed and positioned on the other side of the butt stock where it closes off the ejection slot provided for operation on the other side. The ability to fire the weapon off either shoulder is an advantage that removes one of the major objections to the bullpup design, but in arranging for this change of shoulder other disadvantages have arisen. It is not a simple matter to change the ejector around, and it is not recommended to do it in the field as the small parts are easily dropped and lost.

The cocking handle is placed centrally above the receiver to permit operation by either hand and the centrally mounted sights can be used regardless of the shoulder the soldier uses.

The barrel is of plain steel, and has a fluted chamber. The forward end is formed as a grenade launcher and an adjustable collar controls the position of the grenade and so varies the velocity and hence the range. A flash hider is fitted to the muzzle.

The receiver is of light alloy and the other assemblies are pinned to it.

Working parts

The breech-block fits into the carrier. It is drilled to take the firing pin and the ejector rod and spring, and has a detachable front section carrying the extractor and a dummy extractor plug. This front section is easily removed by levering out a securing pin on top of the breech-block with the aid of a bullet.

The carrier has grooves at the bottom of the rear end which enable it to slide in the receiver. It is drilled at the top of the front face to take the cylinder containing the return spring. This is secured by a transverse pin which can be pushed out using the nose of a bullet.

The delay lever has two parallel, angled, arms joined by a cross-piece. The arms connect the breech-block to the carrier and the cross-piece controls the position of the firing pin. The lower ends of the arms bear against a hardened steel pin across the receiver and the upper ends rest against the back face of the breech-block carrier. It is the means of holding up the breech-face, while the chamber pressure is high, and transferring energy to the carrier. It also controls the trigger mechanism by operating the safety sear.

Trigger assembly

The mechanism is self-contained in a plastic box which is pinned to the receiver. It provides single shots, automatic fire and three-round bursts by using two selector controls; the usual 'safe', 'semi-automatic' and 'automatic'

5.56 mm FA MAS rifle fired from shoulder showing small size, short sight radius and general compactness

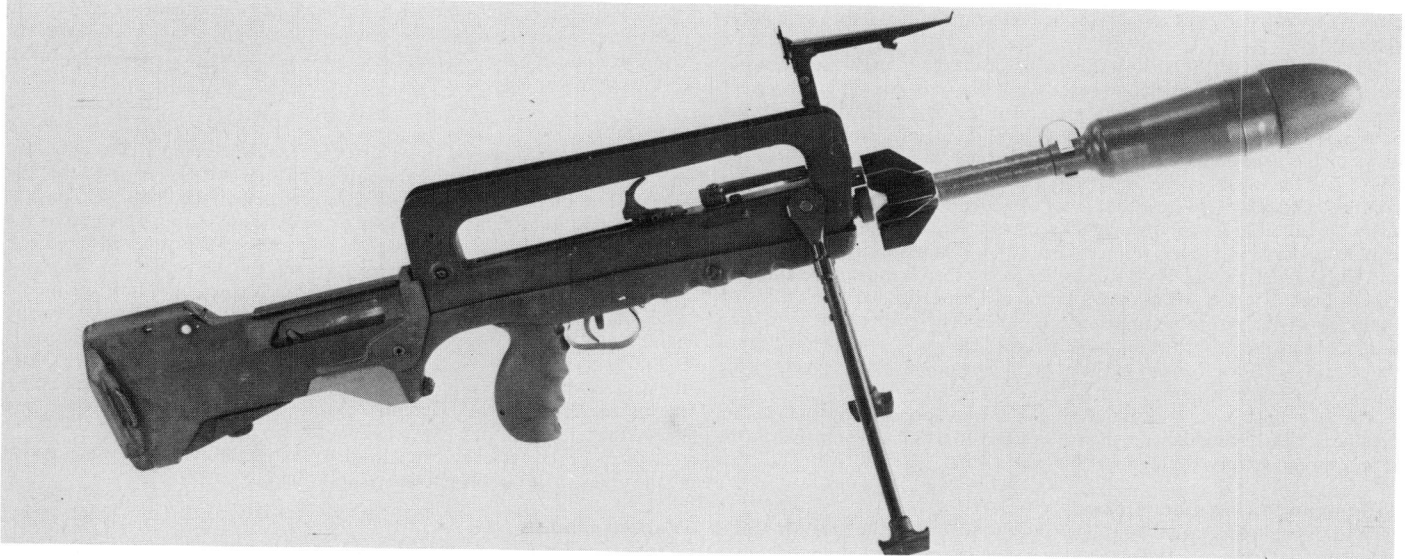

5.56 mm FA MAS rifle in grenade-launching configuration

control which is near the trigger, and the burst-fire controller under the trigger mechanism. This arrangement has been adopted because the burst-fire escapement is kept completely separate from the basic mechanism and should the burst-fire device be inoperable it will not affect the primary means of fire.

There are three types of fire available:
Semi-automatic: fire selector on '1'; burst selector on '0'
Automatic fire: fire selector on 'R'; burst selector on '0'
Three-round bursts: fire selector on 'R'; burst selector on '3'.

The weapon is 'safe' when the selector lever is on 'S', and parallel to the barrel axis.

The trigger-guard can be pulled down from the pistol grip and rotated through 180° to allow the use of Arctic mittens.

Sights
The raised foresight is mounted on a column pinned to the barrel. The foresight blade is mounted on a leaf spring and can be moved across the pillar, by using a notched screw. Each notch on the screw head moves the mpi 4 cm at 200 m (0.2 mils). There is a foresight cover with a luminous bead for night firing and a detachable open sight for direct laying of anti-tank and anti-personnel grenades.

The rearsight is also on a column, above the return spring cylinder. It is an aperture with the choice of two diameters selected by the use of two hinged plates, one on the front of the sight column with a small aperture for good light, and one on the rear with a large aperture for poor light. At night both shutters are lowered and the top of the pillar makes a large aperture which is used in conjunction with the luminous bead on the foresight support.

The rearsight is adjustable for elevation. There is a milled screw which compresses a spring to raise and lower the sight cradle. One turn of the screw moves the mpi 7 cm at 200 m (3.35 mils).

Furniture
The plastic butt stock carries a spring buffer in the top half. This cushions the blow of the working parts as they recoil and is compressed by about 25 mm. There is a rubber shoulder pad to reduce the impact on the firer's shoulder. The carrying handle is of plastic. It provides protection for the sights against accidental damage, for the return spring cylinder and against barrel heat. It carries the sight for the indirect fire of grenades at 45 or 75°, the detachable sight for direct grenade fire and the bipod legs.

Operation
The magazine holds 25 rounds, and is placed in the weapon by locating the front end in the well and rotating it rearwards to engage the magazine catch. The magazine has holes drilled in the sides to indicate that it contains (from top to bottom) 5, 10, 15, 20 or 25 rounds.

If the weapon is not to be fired immediately, the safety catch should be set to 'S'. The cocking handle is pulled to the rear as far as possible and then released, whereupon it flies forward under the force of the compressed return spring and the feed lug under the front face of the bolt pushes the top round into the chamber. As the bolt comes to rest the bolt carrier bears against the top of the delay lever, causing it to rotate forward, with the pin in the body acting as a fulcrum, until it assumes a vertical position. When it does this the bottom end of the left arm of the lever presses down on a spring-loaded rod which releases the safety sear and at the same time the centre cross bar, connecting the two arms, rotates out of a notch in the firing pin, which is then free and can be driven forward when struck by the hammer. When the trigger is operated (with the selector set to a fire position) the hammer flies forward and hits the firing pin which goes forward between the top arms of the delay lever and crushes the cap of the cartridge.

The pressure developed by the expanding gases forces the cartridge case back against the breech face and the bolt starts to move back. The lower arms of the delay lever lie below the bearing pin in the receiver and the force exerted by the breech-block causes the delay lever to rotate backwards. The top ends of the arms bear against the inside of the bolt carrier and since the top lever arms are longer than the bottom arms, a leverage differential exists and the carrier is accelerated back relative to the bolt which moves only a very short distance.

During the first 45° of backward rotation of the delay lever, the following occur: the bolt carrier is accelerated backwards; the cross bar of the delay lever engages in the notch in the firing pin and withdraws it into the bolt; the bottom of the left arm of the delay lever moves off the rod leading to the safety sear which is then released so that the hammer is held back regardless of the position of the trigger or the firing sear; the movement of the breech-block is retarded and this ensures that the cartridge case does not emerge, unsupported, from the chamber whilst the gas pressure is high.

When the delay lever has rotated some 45°, it clears the pin in the receiver and the residual pressure in the chamber forces the cartridge case and bolt rapidly back. The bolt and its carrier then travel back together with the piston compressing the return spring in the cylinder. The empty case is rotated around the extractor, by the centrally mounted ejector as soon as it is clear of the chamber. It then flies out of the ejection port on the side opposite to the cheek rest, clear of the gun.

The back of the bolt carrier hits the buffer and compresses it by some 25 mm and the bolt carrier and bolt are then reversed in direction and, impelled by the return spring, go forward to repeat the cycle.

Trigger and firing mechanism
Due to the bullpup design the trigger is well forward of the hammer and is connected by a long rod. When the trigger is pulled the connecting rod moves forward. When the selector is set to single shot '1' the forward movement of

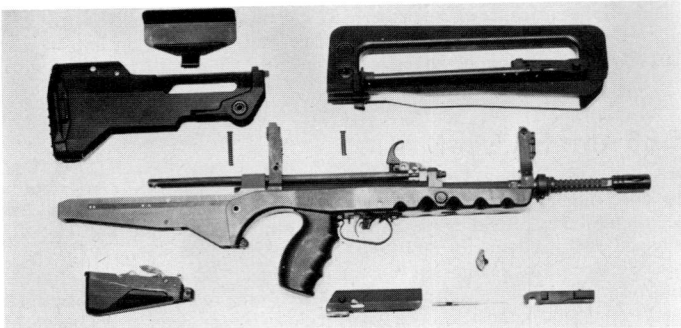

5.56 mm FA MAS rifle stripped

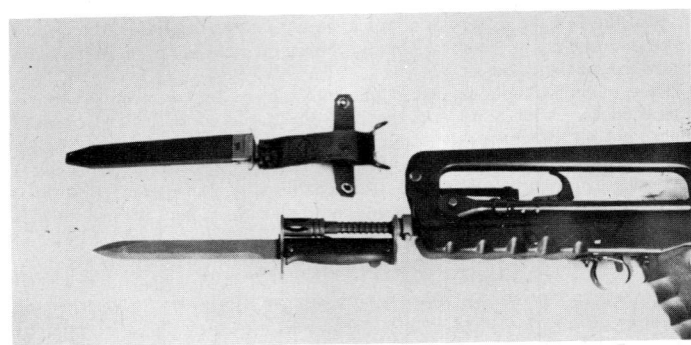

5.56 mm FA MAS rifle bayonet

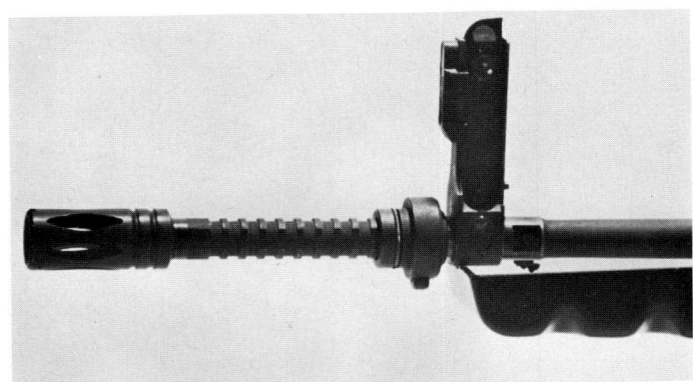

5.56 mm FA MAS rifle foresight

the connecting rod is limited. The rod carries a sleeve which rotates the manual sear off the hammer. If the delay lever is vertical it presses down the spring-loaded rod bearing on the safety sear and the safety sear is lifted off the hammer. The hammer flies forward and the round is fired. The recoiling bolt carrier goes back and rotates the hammer over. It is held by a third sear which is rotated into engagement by the spring. When the trigger is released the sleeve on the connecting rod moves back. Since the trigger spring is powerful the spring of the third sear is overcome and the hammer is then held by the manual sear. When the trigger is pulled another round is fired.

When the selector is set to automatic, the connector rod is able to move further forward. The manual sear is moved off the hammer and the increased travel of the sleeve holds the third sear back and the hammer is held by the safety sear. When the delay lever reaches the upright position it releases the safety sear and another round is fired. Thus at automatic the fire is controlled by the safety sear only.

Stripping
Remove the magazine and check that the chamber is empty. The only tool required is a cartridge; remove the butt by pushing out the pin in front of the magazine well and pulling the butt back; to remove the cheek rest, push it straight back; remove the hand-guard by pushing out the pin which enters the back sight column; the hand-guard is then rotated forward and lifted off; remove the trigger mechanism by pushing out the pin behind the magazine well; remove the working parts by pulling back the cocking handle and hold it while the pin securing the piston to the bolt carrier is pushed out; the cocking handle goes forward under control and the bolt carrier assembly is drawn off to the rear; separate the working parts by moving the block in the carrier until the delay lever is vertical and the bolt will separate from the carrier; remove the locking lever by rotating it until the upper arms are lined up with the centre line of the bolt. The firing pin comes out to the rear.

To remove the detachable bolt head, to change the extractor from one side to the other, prise out the pin on top of the breech-block.

Variant models
Variant models of the standard FAMAS assault rifle were announced in 1984. The **FAMAS Export** is the standard 5.56 mm rifle but is modified so as to

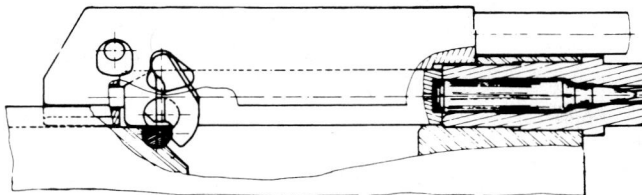

The bolt has just completed its forward movement and the cartridge is fully in the chamber. The bolt is up against the face of the breech and the bolt carrier has come forward to rotate the delay lever to the upright position. The lever can be seen to be locked over the pin, which is picked out in dark shading. As the lever comes to the fully locked position it releases the firing pin and also the sear

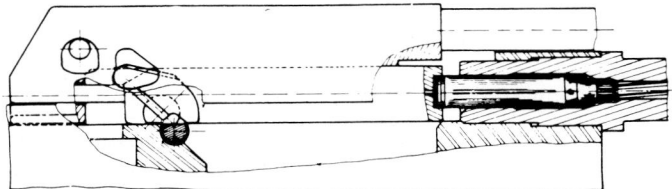

The cartridge has fired and the bullet has left the muzzle. The bolt has moved a short distance back and the cartridge is beginning to emerge from the chamber. The delay lever has acted on the carrier and given it a considerable push, it can be seen how much further it has moved in comparison with the bolt, and the carrier is now moving rapidly to the rear, compressing the spring on the way. The way in which the delay lever rotates to clear from the pin is apparent

permit only single-shot firing. There is no grenade launching capability, and the rifle is intended for the overseas commercial market. The **FAMAS Civil** is intended for the French commercial market, and in order to comply with French firearms laws has had its barrel lengthened to 570 mm and the calibre changed to .222 Remington chambering. It, too, can only fire single shots and has the grenade launching rings removed. The **FAMAS Commando** is a short version intended, as the name suggests, for use by Commando and similar special forces. The barrel has been shortened to 405 mm and there is no grenade launching facility, but in other respects it is the same as the service rifle and offers the full range of selective fire options.

DATA
Cartridge: 5.56 mm × 45
Operation: delayed blowback, selective fire and three-round burst facility
Method of delay: vertical delay lever

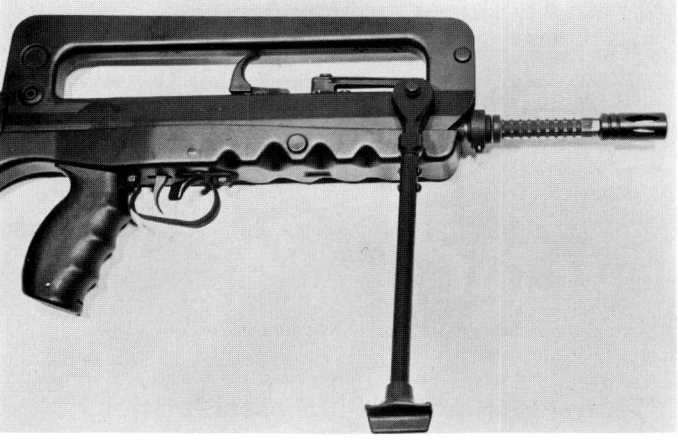

5.56 mm FA MAS rifle bipod

Feed: 25-round box magazine
Weight, empty: 3.61 kg
Length: 757 mm
Barrel: 488 mm
Rifling: 3 grooves rh. 1 turn in 305 mm
Sights: (fore) blade; (rear) aperture 0-300 m
Sight radius: 330 mm
Muzzle velocity: 960 m/s
Rate of fire: (cyclic) 900-1000 rds/min
Effective range: 300 m

GRENADE FIRING
Anti-tank: (normal operating range) 80 m
Anti-personnel
Direct fire: (normal operating range) 100 m
Indirect fire 45°: minimum 140 m; max 360 m
Indirect fire 75°: minimum 70 m; max 180 m
Muzzle velocity: 65 m/s
Grenade weight: 500 g

Manufacturer
Manufacture Nationale d'Armes de St Étienne, St Étienne.
Status
Current. Production.
Service
In service with the French Army. Also exported to Djibouti, Gabon, Senegal and the United Arab Emirates.

GERMANY, EAST

7.62 mm MPiKM assault rifle

This is the East German copy of the Soviet AKM assault rifle. It exists in two forms. The earlier has a wooden, fixed stock and a solid wooden non-laminated lower handgrip to the forearm. The rear pistol grip is plastic and the top half of the forearm is made of the same material.

The later version has a plastic stock, pistol grip and hand-guard, with the grip and the stock having prominent studs. It does not have the Soviet compensator. The other details will be found in the section on Soviet rifles.

7.62 mm MPiKM assault rifle with plastic furniture

7.62 mm Karabiner S carbine

This is a copy of the Soviet Simonov semi-automatic carbine (SKS). It differs from the Soviet original only in having a hole through the stock to attach the lower end of the sling. In all other respects it is the same and the details will be found in the entry covering Soviet equipment below.

7.62 mm Karabiner S carbine showing slot in butt for sling

7.62 mm MPiK and MPiKS assault rifles

The MPiK is the East German version of the Soviet AK-47 assault rifle; it has a fixed wooden stock and may be distinguished from the Soviet or any of the satellite weapons by distinctive fire-selector markings. The upper (full automatic) position is marked 'D' and the lower position which indicates single-shot is marked 'E'. The East German MPiK is the only AK-47 copy which does not have a built-in recess in the butt for cleaning kit, and neither does it have a cleaning rod under the barrel. The piston and bolt are finished to a very high standard and are chromium plated. Details of weight and performance are the same as the Soviet AK-47 and will be found in the Soviet section below.

The MPiKS is the copy of the AK-47 with the folding stock. Again it has the individualistic markings and the chromium-plated piston and breech-block. Like the MPiK it has neither cleaning rod nor provision for cleaning kit.

7.62 mm MPiKS assault rifle

7.62 mm MPiK assault rifle

7.62 mm MPiKMS-72 rifle

This rifle is in use in the East German Forces though it was originally intended for use by armoured troops and parachutists only. As with all the other folding versions of the AK series this rifle is identical to the MPiKM except for the butt. This is an unusual single-strut variety hinged at the rear of the receiver so that it swings alongside the right of the receiver. Locking is achieved using a press-button catch.

Dimensional differences between this and the MPiKM are very small.

7.62 mm MPiKMS-72 rifle

5.6 mm KKMPi69 rifle

This is an East German training rifle. It is chambered for the .22RF cartridge (5.6 mm × 16). It would appear to be used for all forms of field exercise and is carried by troops on assault courses, route marches, and all preliminary training on weapon handling and familiarisation is done on this rifle.

Note that the selector lever, sights, magazine and pistol grip are made as far as possible to resemble those of the full-bore rifle.

5.6 mm KKMPi69 rifle

GERMANY, WEST

Model 98 Mauser rifle

The Model 98 Mauser is probably the best-known bolt-action rifle in the world. It was in production for at least the first 50 years of this century, and is still in use with many military, paramilitary, police and private owners. It has been made in a wide variety of calibres and many have been changed from one calibre to another during their lifetime.

Perhaps the largest number of these rifles which are remaining are the ex-German Army stocks, but there have been so many different models that it is impossible to be sure. It is sufficient to say that Mauser rifles will continue to turn up in all parts of the world until the end of the century.

The rifle itself is generally well-known, and is not particularly distinguished in appearance. It gained its popularity from the robust and excellent locking arrangement which uses forward lugs engaging in recesses in the front end of the body, right against the barrel. The stress paths of the firing shock are thus kept short, and the bolt movement is also short. Mauser rifles were all orginally chambered for a rimless round, generally the 7.92 mm Mauser. Loading and feeding this ammunition from the five-round integral magazine was easy, though the bolt was never smooth to operate. Lifting the handle to open the breech cocked the action as well as causing the primary extraction, and this called for a hard upward push, causing the rifle to waver in the hand of the inexperienced firer. To assist this movement most models of the rifle have a straight bolt handle sticking out at right angles to the body, though some models, particularly those used for sniping, had a handle that was bent down.

The original rifle was long and fully stocked, but in 1935 the German Army adopted a short version, known as the Model 98K, which was made in very large numbers until 1945 and is the one most commonly found today.

DATA
Cartridge: 7.92 mm × 57
Operation: manual, bolt
Method of locking: rotating bolt
Feed: 5-round internal box
Weight: 3.89 kg
Length: 1103 mm
Barrel: 597 mm; (rifling) 4 grooves rh
Sights: (foresight) blade; (rearsight) tangent, notch
Muzzle velocity: 754 m/s
Effective range: 600 m

Manufacturer
Mauser-Werke AG and many others.
Status
Obsolete; no longer in production.
Service
To be found world-wide in many different calibres.

Model 98K Mauser rifle

7.62 mm Model SP 66 Mauser sniping rifle

The Model SP 66 is a heavy-barrelled bolt-action rifle of considerable accuracy which has been specifically designed and manufactured for use by law enforcement agencies and military snipers.

The Model SP 66 uses the well-known Mauser short-action bolt system in which the bolt handle is placed at the forward end of the bolt, just behind the locking lugs. By so doing the bolt movement is reduced by 90 mm and this length can then be added to the barrel. Since the body is correspondingly shorter, it is also lighter. Another advantage is that on opening the bolt there is less overhang behind the body and the firer does not have to move his head away from the sight.

Special emphasis has been placed on lock time and the operation of the sear/trigger combination. By using a very high rate on the firing pin spring and reducing the travel of that pin it has been possible to cut lock time to 50 per cent of that with the conventional bolt on the Model 98 (see entry above).

The trigger has a 'shoe' 10 mm wide so that the firer can feel it even when wearing gloves.

The stock is fully adjustable to suit all builds of firer and features the thumb hole which is becoming common on many good quality target rifles. All surfaces which are touched with the hands are roughened to offer the best grip and the broad fore-end is designed to allow the left hand to have a comfortable and steady grasp whatever the length of the firer's arm.

The muzzle is fitted with a combined flash hider and muzzle brake. The requirement to reduce flash is critical with this and similar rifles since they are intended to be fired with high-magnification optical sights or image-intensifiers and the presence of muzzle flash inhibits the firer with such magnified devices. Mauser claims that the chance of successive hits is high due to the lack of blinding from the muzzle.

Mauser offers the Model SP 66 rifle in standard form complete with

7.62 mm Model SP 66 Mauser sniping rifle with Oldelft night sight

7.62 mm Model SP 66 Mauser sniping rifle with Zeiss telescope

telescopic mount, Zeiss Diavari ZA 1.5-6 × 42 mm zoom telescopic sight, a second mount for night-vision devices and a case. It can be supplied with a night-vision sight on request.

DATA
Cartridge: 7.62 mm × 51 (selected sniper batches)
Operation: manual, Mauser short-action bolt
Method of locking: forward lugs
Feed: 3-round integral magazine
Barrel length: 650 mm without muzzle brake; 730 mm with muzzle brake

Manufacturer
Mauser-Werke Oberndorf GmbH, Postfach 1349, 7238 Oberndorf.
Status
Current. Production.
Service
West German forces and the forces of at least 12 other countries.

7.62 mm Mauser Model 86SR Sniping rifle

The Model 86SR is offered by Mauser as an alternative to the SP66 (above). It uses a newly-developed bolt-action, locking by forward lugs, has a ventilated stock to provide additional heat-dissipating area, and is fitted with an adjustable match trigger. The barrel is fitted with a muzzle brake. The laminated wood match stock has an adjustable recoil pad and a rail in the forearm allows use of a firing sling or bipod. The rifle is not fitted with iron sights but a special telescope mount allows the use of optical telescope sights or night sights and also the use of a combined scope with laser rangefinder.

DATA
Cartridge: 7.62 × 51 mm NATO (.308 Winchester)
Operation: bolt action, magazine repeater
Feed: 9-shot double-row box magazine
Weight: 4.90 kg without sight
Length: 1210 mm
Barrel: 730 mm with muzzle brake
Rifling: 4 grooves, rh, one turn in 305 mm
Trigger pull: adjustable between 0.8 and 1.4 kg
Stock: Match stock with thumb-hole optional

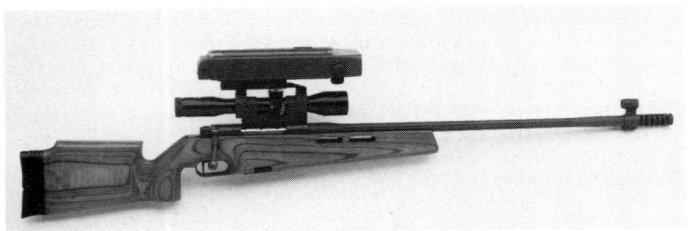

7.62 mm Mauser Model 86SR sniping rifle with 'Teleranger' combined laser rangefinder and telescope sight

Manufacturer
Mauser-Werke Oberndorf GmbH, Postfach 1349, 7238 Oberndorf-a-N.
Status
Current. Production.

7.62 mm G3 Heckler and Koch rifle

After the Second World War certain German designers went to the Centro de Estudios Tecnicos de Materiales Especiales (CETME), Madrid, and participated in the design of a delayed blowback-operated rifle known as the CETME. Subsequently, development of this rifle continued in Spain using an unusually long bullet in 7.92 mm calibre. The firm of NWM at s'Hertogenbosch was interested in developing the rifle and obtained rights to do so but the Bundeswehr considered the design had potential and the development of the weapon was transferred to Heckler and Koch at Oberndorf-Neckar. Work was continued there and in 1959 the Heckler and Koch G3 rifle became standard Bundeswehr equipment.

Description
The receiver is a steel pressing, grooved on each side to guide the bolt and seat the back plate, and carrying the barrel. Above the barrel a tubular extension, welded on to the receiver, houses the cocking lever and the forward extension of the bolt. The cocking lever runs in a slot cut in the left side of this tubular housing and can be held in the open position by a transverse recess. The barrel is screw-threaded at the muzzle with a serrated collar to engage the retaining spring of the flash eliminator or blank firing attachment. The rifling is orthodox and the chamber carries 12 longitudinal flutes extending back from the lead to 6 mm from the chamber face.

The bolt is ⌐ shaped. The long forward extension is hollow and takes the return spring. It extends into the tube above the barrel. The bolt-head carrier and bolt-head are placed on the barrel axis. The bolt-head carrier has long bearing surfaces on each side which slide in the grooves in the sides of the receiver. The bolt-head carries two rollers which project on each side and are forced out by the inclined front faces of what the manufacturer calls the 'locking piece'; this term will be used but, as will be seen, it is really a misnomer since the breech is never truly locked. When out, the rollers engage in recesses in the barrel extension. The bolt-head and locking piece seat in the bolt-head carrier and are held by the bolt-head locking lever to prevent bounce on chambering the cartridge. The trigger mechanism fits into the grip assembly, secured to the receiver by locking pins.

Operation
The weapon is carried ready for action with the cartridge in the chamber, the

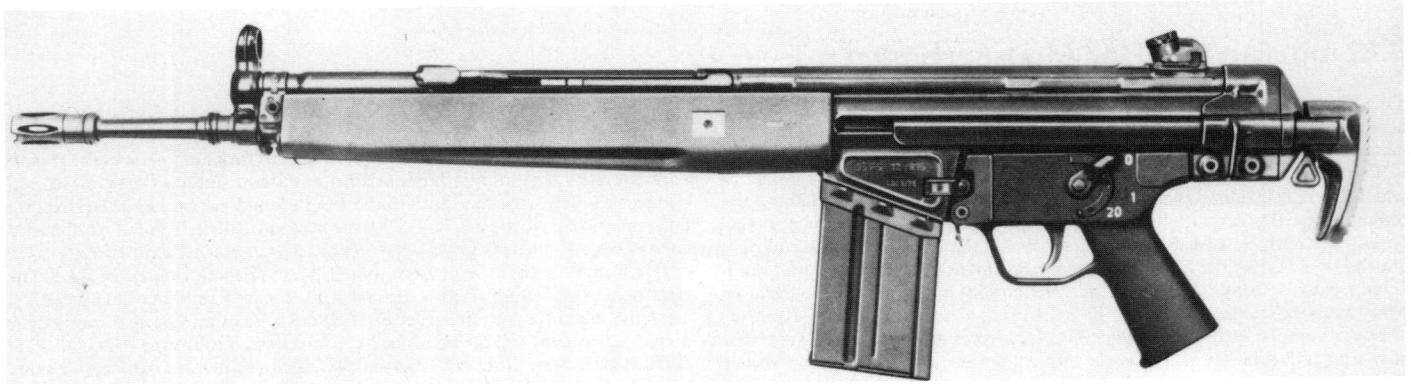

7.62 mm G3 Heckler and Koch rifle with butt retracted and plastic pistol grip

7.62 mm G3 A3ZF Heckler and Koch rifle (West German Army sniping version)

bolt-head supporting the base of the round, and the rollers held out in the recesses in the barrel extension by the front face of the locking piece. The hammer is cocked and held by the trigger sear.

When the trigger is squeezed the sear is lowered out of the hammer notch and the hammer flies forward and drives the firing pin through the hollow locking piece to fire the cap. The pressure generated in the chamber forces the cartridge case rearwards and exerts a force on the breech face which tends to drive the bolt-head back. Before the bolt-head can go back, the rollers must be driven inwards. The rollers, which are carried in the bolt-head, are pushed back and the angle of the recesses in the barrel extension is such that the rollers are forced inwards against the inclined planes on the front of the locking piece. In turn this inward force drives the locking piece back and with it the bolt-head carrier. The angle of the locking piece face is such that the velocity ratio between bolt-head carrier and bolt-head is 4:1. Thus while the rollers are moving backwards and inwards the bolt-head carrier is travelling back four

times as far as the breech face. As the carrier moves back the bolt-head locking lever is disengaged. After the bolt-face has moved back a little over a millimetre the rollers are clear of the recesses in the barrel extension, the entire bolt is blown back by the residual pressure with the bolt-head and bolt carrier maintaining their relative displacement of about 5 mm. The bolt-head carrier cocks the hammer and compresses the return spring. The cartridge case, held by the extractor, hits the ejector arm and is thrown out to the right. The rear end of the carrier hits the buffer. The buffer action, assisted by the return spring, drives the bolt carrier forward. The front face of the bolt-head strips a cartridge from the magazine and chambers it. The extractor springs over the extraction groove of the cartridge and the bolt-head comes to rest. The locking piece and bolt carrier then close up the gap of 5 mm and the rollers are pushed out into the recesses of the barrel extension. The bolt-head locking lever engages the bolt-head shoulder thus preventing bounce. The weapon is then ready to be fired again.

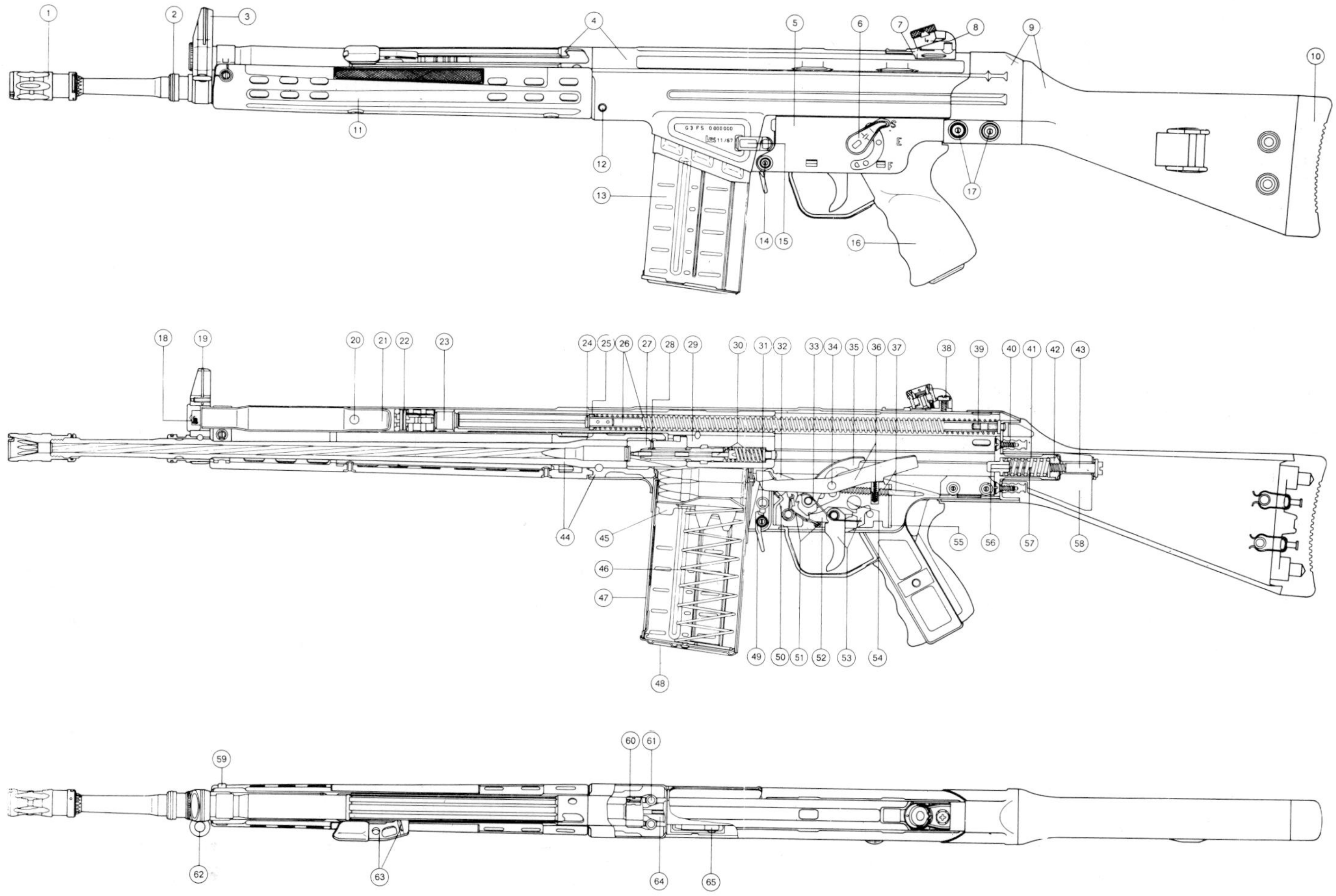

7.62 mm (NATO) G3 automatic Heckler and Koch rifle
(1) *flash suppressor (grenade launcher) with retaining spring for flash suppressor* **(2)** *snap ring to mount grenade* **(3)** *foresight holder* **(4)** *receiver and operating housing* **(5)** *grip assembly* **(6)** *safety* **(7)** *sight base* **(8)** *rotary rearsight* **(9)** *back plate with butt stock* **(10)** *butt plate* **(11)** *hand-guard* **(12)** *cylindrical pin* **(13)** *magazine* **(14)** *grip assembly locking pin* **(15)** *magazine catch* **(16)** *grip* **(17)** *butt stock locking pins* **(18)** *cap* **(19)** *foresight* **(20)** *stop pin* **(21)** *stop abutment* **(22)** *operating handle spindle* **(23)** *operating handle support* **(24)** *stop pin for recoil spring* **(25)** *recoil spring guide ring* **(26)** *recoil spring tube with recoil spring* **(27)** *bolt-head* **(28)** *clamping sleeve and holder for locking roller* **(29)** *bolt-body with recoil spring tube* **(30)** *firing pin with firing pin spring* **(31)** *contact piece* **(32)** *release lever* **(33)** *elbow spring for trigger* **(34)** *ejector spindle* **(35)** *hammer* **(36)** *ejector with ejector pressure spring* **(37)** *pressure shank and pressure spring* **(38)** *fixing screw* **(39)** *stop pin for spring guide* **(40)** *countersunk screw* **(41)** *buffer housing* **(42)** *buffer closure* **(43)** *screw for buffer* **(44)** *barrel with barrel extension* **(45)** *follower* **(46)** *follower spring* **(47)** *magazine tube* **(48)** *magazine floor plate* **(49)** *magazine release lever* **(50)** *catch* **(51)** *elbow spring with roller* **(52)** *sear* **(53)** *trigger* **(54)** *safety pin* **(55)** *trigger assembly* **(56)** *buffer pin* **(57)** *buffer spring* **(58)** *support for buffer housing* **(59)** *hand-guard locking pin* **(60)** *extractor with extractor spring* **(61)** *locking roller* **(62)** *eyebolt* **(63)** *operating handle with elbow spring* **(64)** *locking piece* **(65)** *bolt-head locking lever*

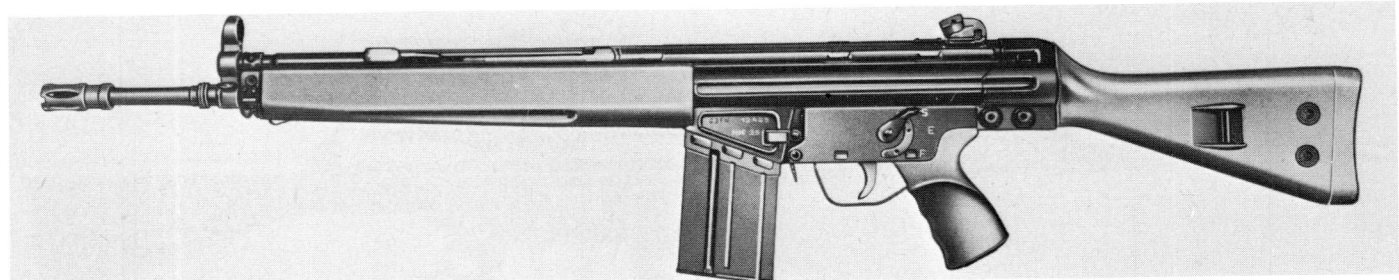

7.62 mm G3 A3 Heckler and Koch rifle

G3K carbine

7.62 mm G3 rifle with G41 hand-guard, bipod, and plastic pistol grip

Trigger and firing mechanism

The trigger mechanism is the same in principle as that employed in the Belgian 7.62 mm Fusil Automatique Légèr (FAL) rifle.

A spring-loaded sear has an elongated slot in which the trigger pin fits. The spring tries to drive the sear forward over the trigger. At the same time another spring holds up the nose of the sear. The rifle can never be fired unless

G3 with fitted silencer

7.62 mm G3 Heckler and Koch rifle with plastic pistol grip

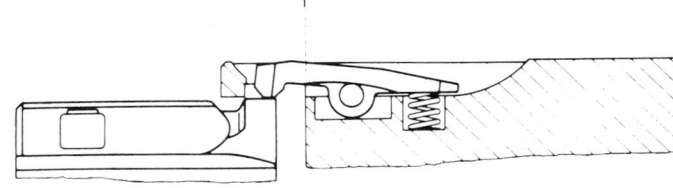

Disengagement of the bolt-head locking lever: as the bolt-head carrier travels backward, the bolt-head locking lever (1) is simultaneously pressed over the bolt-head shoulder against the pressure of its spring. After the round has been chambered, the bolt-head stops at the mouth of the barrel. The bolt-head carrier continues forward until the locking piece presses the locking rollers into engagement in the barrel extension. The bolt-head locking lever now engages the bolt-head shoulder thus preventing it from rebounding

the bolt carrier is fully forward. Completion of the forward movement moves a safety sear off the hammer bent.

In single-shot firing the bolt is closed and the trigger sear holds up the hammer. When the trigger is operated the sear is rotated down at the front and out of the hammer notch. When the hammer rotates to fire the round, the spring-loaded sear slips forward and its tail drops down off a fixed step called the pull-off surface; but the firer is still holding the trigger back. The hammer is rotated back by the bolt and catches the nose of the sear. The hammer spring overcomes the sear spring and pushes the sear back against the side of the pull-off surface. The hammer is held by the automatic sear until the bolt closes and releases it. The hammer is held by the sear but the trigger is already pulled and another shot cannot be fired. When the trigger is released the rear end, or tail, of the sear is allowed to rise and the spring forces it back and the elongated hole allows it to move on to the pull-off surface. When the trigger is pulled another round is fired.

When the change lever is put to automatic, the rotation of the change lever spindle allows the sear tail to rise so high that the nose of the sear does not engage the notch on the hammer at all and the hammer is held by the automatic sear only. As soon as the bolt carrier travels fully forward the automatic sear is released and the hammer is freed. When the change lever is set to 'safe' the spindle prevents any upward movement of the sear and so the nose cannot drop out of engagement with the hammer notch.

Accessories

The G3 rifle has a number of accessories and training aids. First there is a telescopic sight with a mount which clips over the receiver on prepared surfaces. The sight is basically for sniping out to 600 m, surveillance or observation of fire at longer ranges.

A blank-firing attachment may be screwed on the muzzle instead of the flash eliminator. It has a retaining circlip which grips the serrations on the muzzle and prevents the attachment working loose. The attachment consists of an open-ended cylinder with a cross-bolt which closes the opening entirely. A groove is cut across the bolt so that by rotating the bolt the amount of gas escaping can be regulated to produce correct functioning of the rifle. This device has a dull chromium-plated finish to prevent confusion with the flash hider.

A special bolt marked 'PT' replaces the normal bolt and is used for firing plastic training ammunition. This works on pure blowback and has no delay rollers. It is not suitable for use with service ammunition. The normal magazine is used for plastic ammunition.

To conserve ammunition and allow troops to train in confined and built-up areas, a sub-calibre training device is available. It consists of a sub-calibre tube with a special bolt and magazine. The ammunition used is 5.6 mm x 16 (.22 long rifle). The sub-calibre tube is inserted from the breech and locked with a spring ring. In addition the tube is secured by a bolt on the magazine engaging in the end of the chamber face of the tube. The extractor enters a groove in the chamber face. The bolt assembly carries out the normal

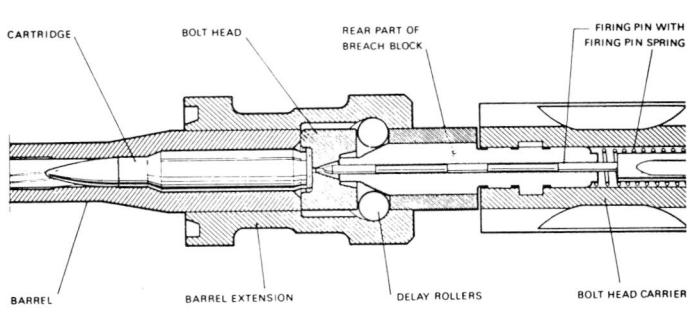

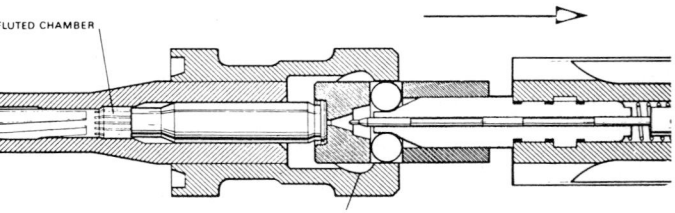

7.62 mm G3 Heckler and Koch rifle, method of operation

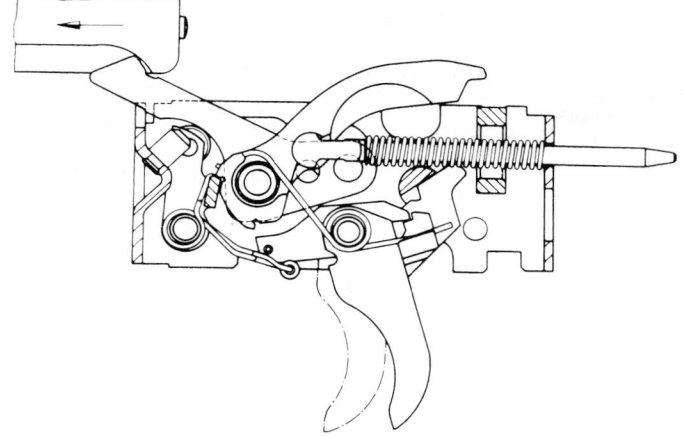

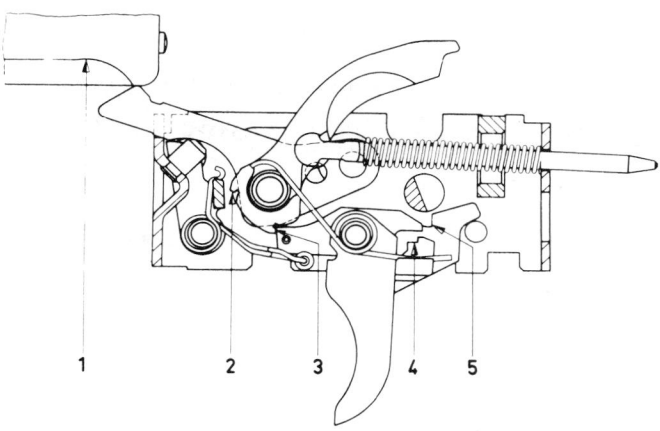

Burst function: setting the selective fire lever at 'F' automatically results in a longer trigger travel, which is required only for automatic fire. When the trigger is pulled, the first shot is fired in the same manner as if the selective fire lever were at 'E' single fire. However, the longer trigger travel swivels the sear so far downward that it can no longer catch the hammer. The hammer is now held only by the catch. As the bolt snaps forward, it pushes the safety sear downward, disengages the catch and releases the hammer. As this function repeats, it results in automatic fire. When the trigger is released, the front of the sear swivels upward again and engages the 'single fire' notch in the hammer

(1) cam surface (2) notch for burst (3) notch for single fire (4) notch for the sear (5) recess for single fire
The selective fire lever set at 'E' single fire. The safety pin permits a limited trigger pull

functions of chambering, supporting, firing, extracting and cocking the hammer. The magazine for 20 5.6 mm × 16 cartridges fits into the normal G3 magazine. The G3 rifle can be used to project rifle grenades.

Stripping
Stripping the G3 rifle is very simple. After removing the magazine and checking that the chamber is clear, press out the two locking pins holding on the back plate and remove the butt stock and its back plate. The trigger grip assembly can be dropped down and removed by pushing out the single holding pin. Retracting the cocking handle allows withdrawal of the bolt assembly. The flash hider and hand-guard can be removed at will. The bolt-head can be rotated from the carrier by holding the carrier and rotating the head. When the locking piece is removed from the carrier, the firing pin and spring can be taken out of the locking piece.

Silencer
A silencer may be screwed to the barrel in place of the flash hider and serves to reduce the muzzle report. It will fit all Heckler and Koch rifles with rifle grenade guides.

The use of subsonic ammunition is not recommended, in view of its reduced performance and the need for a special sight. The principal purpose of this silencer is for it to be used on firing ranges in order to reduce noise disturbance both for firers and local residents.

Variants
The standard rifle is known as the G3 A3 and has a plastic butt stock and plastic hand-guard.

When a telescopic sight is fitted to this rifle it is called the G3 A3ZF.

When the plastic butt stock is replaced by a retractable butt it is called the G3 A4.

DATA
Cartridge: 7.62 mm × 51
Operation: delayed blowback, selective fire
Method of delay: rollers

Feed: 20-round box magazine
Weight, empty: (fixed butt) 4.4 kg; (retractable butt) 4.7 kg
Length: (fixed butt) 1025 mm; (retracted butt) 840 mm
Barrel: 450 mm
Rifling: 4 grooves rh. 1 turn in 305 mm;
Sights: (fore) post; (rear) V battle sight at 100 m. Apertures for 200, 300 and 400 m
Sight radius: 572 mm
Muzzle velocity: 780-800 m/s
Rate of fire: (cyclic) 500-600 rds/min
Effective range: 400 m

Manufacturer
Heckler and Koch GmbH, D-7238 Oberndorf-Neckar.
Status
Current. Production.
Service
West German Army. The armed forces and/or police forces of the following countries have adopted HK rifles and are completely or partially equipped with them:

EUROPE	AFRICA	CENTRAL AND SOUTH AMERICA	MIDDLE EAST
Denmark	Burkina Faso	Bolivia	Iran*
France*	Chad	Brazil	Jordan
West Germany	Ghana	Chile	Qatar
Greece*	Ivory Coast	Columbia	Saudi Arabia*
Italy	Kenya	Dominican Republic	United Arab
Netherlands	Malawi	El Salvador	Emirates
Norway*	Morocco	Guyana	FAR EAST
Portugal*	Niger	Haiti	Bangladesh
Sweden*	Nigeria	Mexico*	Brunei
Switzerland	Senegal	Peru	Burma*
Turkey*	Sudan		Indonesia
	Tanzania		Malaysia
	Togo		Pakistan*
	Uganda		Philippines*
	Zambia		Thailand*

*Countries marked with an asterisk are producing Heckler and Koch rifles under licence. The rifles are also produced by the Royal Small Arms Factory, Enfield, England although they are not in UK service.

G3 SG/1 Heckler and Koch sniping rifle

The G3 SG/1 sniping rifle is used by the West German police. It is basically the G3 A3 but has some differences. During proof and testing of the standard rifles, those that demonstrate their ability to produce minimum groups are set aside for modification to sniper rifles. The sniper rifle is fitted with a special trigger unit. This incorporates a set trigger which has a pull-off which can be varied. Setting the trigger can be done only when the selector lever is set to 'E'. After firing one round the trigger can be re-set or the rifle can be fired again using the ordinary trigger setting which has been reduced to 2.6 kg. If the trigger has been set, it is automatically released as soon as the selector lever is changed from 'single-shot' to 'safe' or 'automatic'.

The sight is not that normally used by the G3 but is a special Zeiss or Schmidt & Bender telescope with variable power from 1.5× to 6× and with windage and range adjustment from 100 to 600 m. The mount employed fits over the receiver and allows use of the iron sights without dismounting the telescope. The graticule pattern is based on the use of the mil, for instance the angle subtended by one unit of length at a distance of 1000 units, for example 1 m at 1000 m subtends one mil. This system enables the firer to set the correct

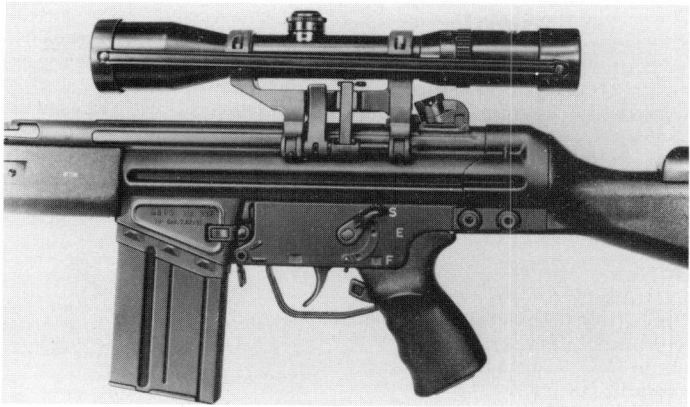

Sights and trigger of G3 SG/1 police sniping rifle

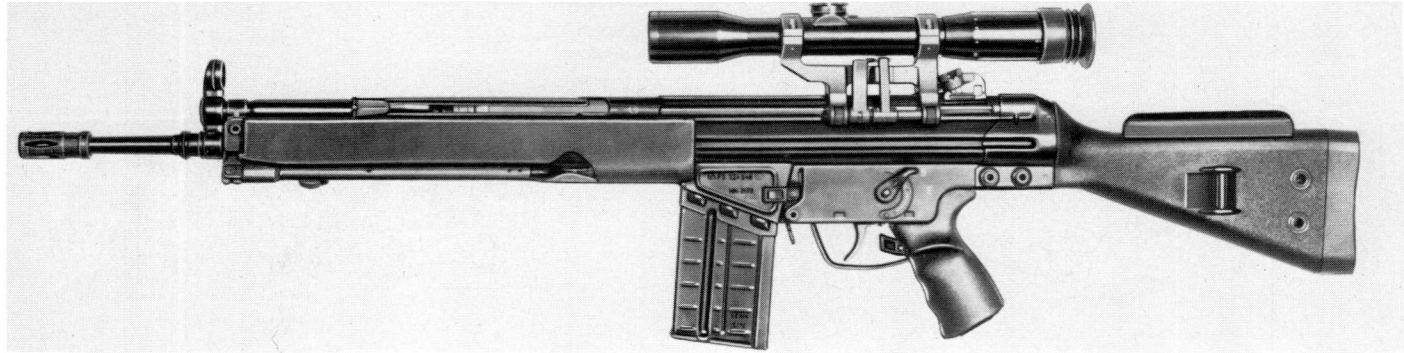

G3 SG/1 police sniping rifle

range provided he can align the telescope on an object of known size. For example a car of approximately 5 m length subtends 25 mils on the telescope. The range therefore is $^5/_{25} \times 1000 - 200$ m.

DATA
Weight: (unloaded but with sight) 5.54 kg
Trigger pull: (normal) 2.6 kg; (with set trigger) 0.9-1.5 kg
Sight: (optical) 1.5 × 6 telescope; (settings) 6 positions, 100-600 m

Manufacturer
Heckler and Koch GmbH, D-7238 Oberndorf-Neckar.
Status
Current. Production.
Service
West German police and Special Forces, Italian carabinieri and other police forces.

7.62 mm × 51 Heckler and Koch PSG 1 high-precision marksman's rifle

The Präzisionsschützengewehr (High-precision Marksman's Rifle) PSG 1 is manufactured for police and service use. It is a semi-automatic, single-shot weapon (with 5- or 20-round magazine option) incorporating the company's roller-locked bolt system. Superb accuracy is claimed: an average dispersion diameter of ⩽ 80 mm is quoted at a range of 300 m for test grouping (5 × 10 shots with Lapua .308 Winchester Match ammunition). A 6 × 42 telescopic sight with illuminated reticule is an integral part of the weapon. Windage and elevation adjustment is by moving lens, six settings from 100 to 600 m.

The manufacturer states that a special system provides for silent and positive bolt closing. A vertically adjustable trigger shoe enlarges the trigger width. The trigger pull is approximately 1.5 kg. Length of shoulder stock, vertical adjustment of cheek-piece and angular adjustment of butt to shoulder are all variable according to the firer's requirements. A precision tripod is available.

DATA
Cartridge: 7.62 mm × 51
Operation: delayed blowback, single-fire
Method of delay: rollers
Feed: single shot; option for 5- or 20-round magazine
Weight, empty: 8.1 kg
Length: 1208 mm
Barrel: 650 mm
Rifling: polygonal, 4 grooves rh
Sight: (optical) 6 × 42 telescope; (settings) 6 positions, 100-600 m

Manufacturer
Heckler and Koch GmbH, D-7238 Oberndorf-Neckar.
Status
Current. Production.
Service
West German and numerous other Police and Special Forces worldwide.

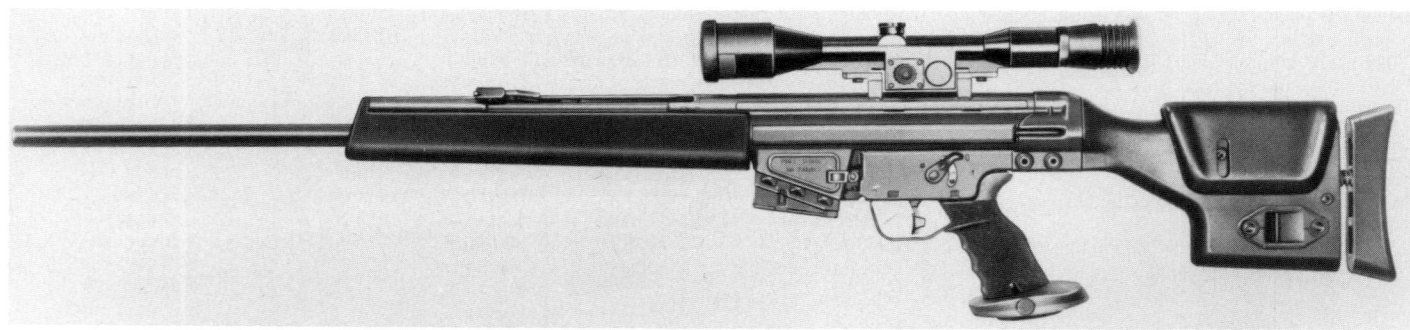

7.62 mm Heckler and Koch PSG 1 high-precision marksman's rifle

7.62 mm Heckler & Koch MSG90 sniping rifle

This sniping rifle, introduced in 1987, uses the same roller-locked delayed blowback system which is common to all Heckler & Koch rifles. Designed to military specifications, the MSG90 uses a special cold-forged and tempered barrel and has a pre-adjusted trigger with a shoe which enlarges the trigger width for better control. The trigger pull is approximately 1.5 kg. The stock is adjustable for length and has a vertically adjustable cheek-rest. There are no iron sights, the standard sight being a 12-power telescope with range settings from 100 to 800 m; the receiver is fitted with a STANAG mount which will accept any NATO-standard night-vision or other optical sight.

The fore-end is fitted with an internal T-rail which allows the attachment of a shooting sling or a bipod.

DATA
Cartridge: 7.62 × 51 mm
Operation: delayed blowback, semi-automatic only
Method of delay: rollers
Feed: 5- or 20-round box magazine
Weight, empty: 6.40 kg
Length: 1165 mm
Barrel: 600 mm

Manufacturer
Heckler and Koch GmbH, D-7238 Oberndorf-Neckar.
Status
Current. Production.

7.62 mm Heckler & Koch MSG90 sniping rifle

5.56 mm Heckler and Koch HK 33E rifle

The basic operating principle of the HK 33E rifle is that of delayed blowback, firing from a closed breech, using the Heckler and Koch system which employs a two-part bolt, the action of which has been described in the entry on the G3 rifle (above). The HK 33E is a scaled-down version of the 7.62 mm G3 rifle and uses exactly the same trigger and firing mechanism as the larger rifle. The mechanism provides automatic fire, single-shot fire and a 'safe' position.

The sights fitted to the .223 (5.56 mm) rifle have a V rearsight for battle setting and apertures for 200, 300 and 400 m. Zeroing for elevation and line is carried out on the rearsight. Optical sights can be fitted in a standard mount producing a four-power magnification with adjustment by hundreds of metres from 100 to 600. Infra-red and other special sights can also be accepted by the same mount.

There are five HK 33E variants: the standard rifle with fixed butt: rifle with retractable butt: rifle with bipod: sniper rifle with telescopic sight: and HK 33KE carbine version. All have the same operating system.

DATA
Cartridge: 5.56 mm × 45
Operation: delayed blowback, selective fire
Method of delay: rollers
Feed: 25-round box magazine
Weight, empty: (fixed butt) 3.65 kg; (retracting butt) 3.98 kg; (carbine) 3.89 kg
Length: (fixed butt) 920 mm; (extending butt model) 940 mm and 735 mm; (carbine model) 865 mm and 675 mm
Barrel: (rifle) 390 mm; (carbine) 322 mm
Rifling: 6 grooves rh. 1 turn in 178 mm or 305 mm as required
Sights: (fore) post; (rear) V battlesight 100 m. Apertures 200, 300 and 400 m; (optical) 4× telescope. 6 range increments from 100-600 m. Adjustable for windage and elevation
Sight radius: 480 mm

5.56 mm HK 33E with bipod

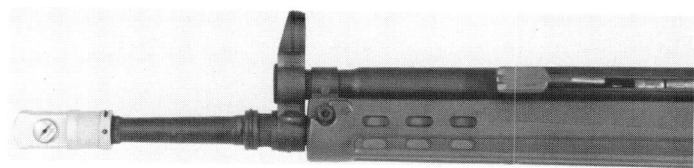

Standard blank attachment for G3, HK 33E and other rifles, here shown fitted to a G3 rifle

5.56 mm HK33SG1 sniper rifle

5.56 mm HK 33E rifle with retracting butt

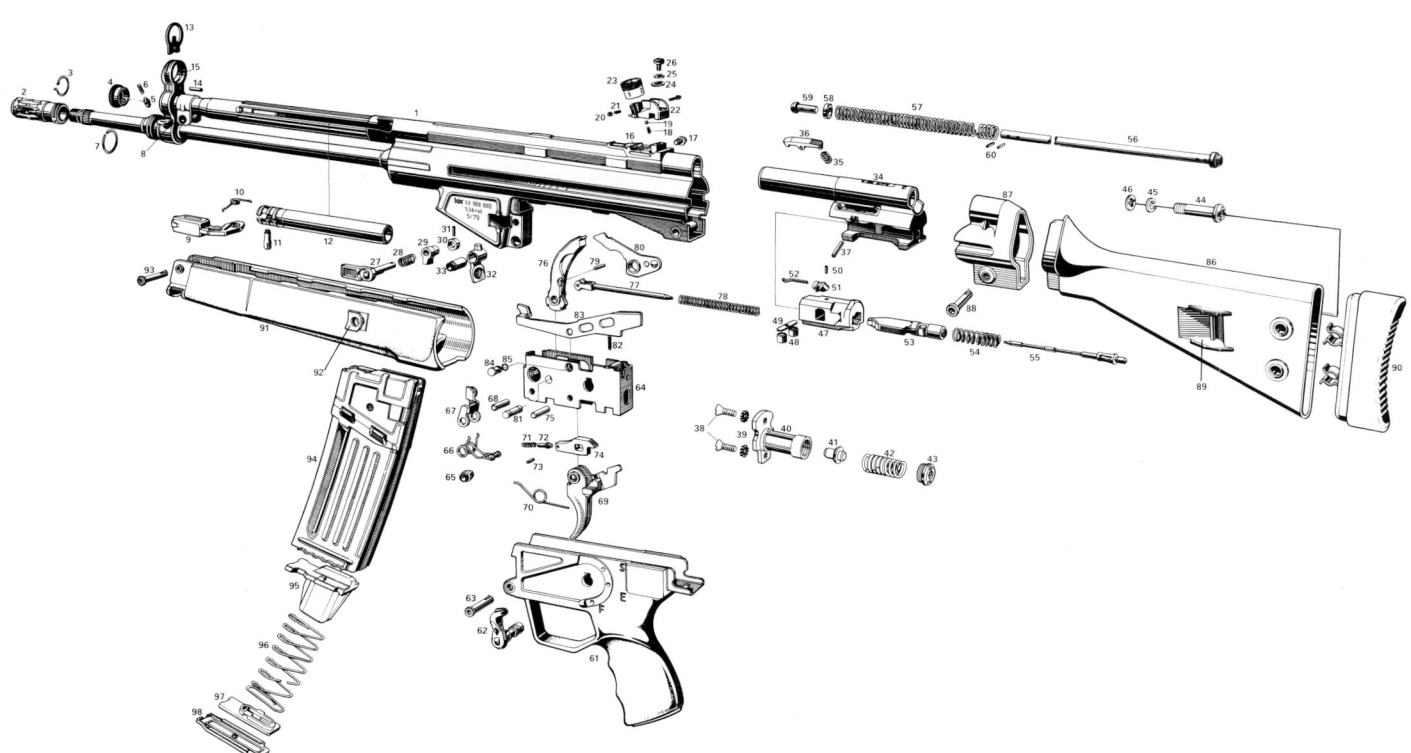

5.56 mm Heckler and Koch HK 33E rifle; exploded diagram

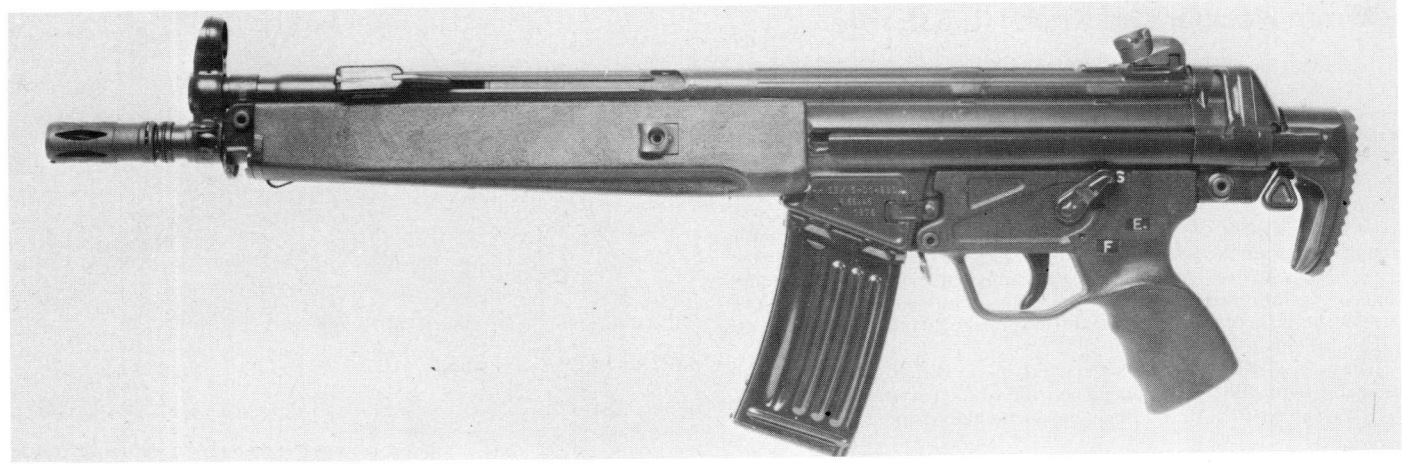

5.56 mm HK 33E Carbine

5.56 mm HK 33E Heckler and Koch rifle with fixed butt

Muzzle velocity: (rifles) ca 920 m/s; (carbine) ca 880 m/s
Rate of fire: (rifles) 750 rds/min; (carbine) 700 rds/min
Effective range: 400 m

Manufacturer
Heckler and Koch GmbH, D-7238 Oberndorf-Neckar.

Status
Production in West Germany and Thailand.
Service
Malaysia, Chile, Thailand, and Brazilian Air Force. Also sold in varying quantities in South-east Asia, Africa and South America.

5.56 mm G41 Heckler and Koch rifle

This design has been specifically developed to fire the new standard NATO cartridge 5.56 mm × 45 (standardisation agreement STANAG 4172) and is based on the 7.62 mm G3. As well as characteristics shared with that arm, including the 'roller-locked' bolt system (see entry for G3), the G41 is stated to incorporate several notable features: low-noise, positive-action bolt closing device; bolt catch, to keep open the bolt when the magazine has been emptied; dust-proof cover for cartridge case ejection port; facilities for employment of 30-round magazines standardised according to STANAG 4179 (M16); burst control trigger for firing three-round bursts; fitting for bipod; new telescopic sight mount according to STANAG 2324 for all daylight and night sighting devices; twilight sighting device; acceptance of G3 bayonets; robust design for a service life of more than 20 000 rounds.

5.56 mm G41A2 Heckler and Koch rifle

DATA
Cartridge: 5.56 mm × 45
Operation: delayed blowback, selective fire with three-round burst facility
Method of delay: rollers
Feed: 30-round box magazine
Weight, empty: (fixed butt) 4.1 kg; (retracting butt) 4.35 kg; (G41K) 4.25 kg
Length: (fixed butt) 997 mm; (retracting butt, long barrel) 996 and 806 mm; (retracting butt, short barrel) 930 and 740 mm
Barrel: 380 (G41K) or 450 mm
Rifling: 6 grooves rh. 1 turn in 178 mm
Sights: (fore) post; (rear) rotary. 4 settings: 200, 300, 400 m aperture V-notch for 100 m. Adjustable for windage and elevation
Rate of fire: approx 850 rds/min

G41K short rifle with butt retracted

Manufacturer
Heckler and Koch GmbH, D-7238 Oberndorf-Neckar.
Status
Production. In service and undergoing military evaluation trials.

5.56 mm G41 Heckler and Koch rifle

7.62 mm Heckler and Koch G8 rifle

This is a revised model of the pattern previously known as the HK11E; it has been specially designed for use by anti-terrorist police and security forces.

The mechanism is the familiar Heckler and Koch roller-locked retarded blowback system and the rifle permits semi- or full automatic fire and also has a three-round burst facility. Feed is by means of a standard box magazine, but a special 50-round drum magazine is available, and by means of an accessory feed mechanism outfit the weapon can be rapidly adapted to belt feed. There is a G8A1 variant which will not accept the belt feed mechanism and is restricted to use with magazines. The barrel is heavier than usual, precisely rifled, and capable of being changed rapidly when the weapon is being used in the automatic fire mode.

Iron sights are fitted, but the weapon is normally supplied with a four-power sighting telescope. The telescope mount is to STANAG specifications and will therefore accept all standard night-vision devices.

DATA
Cartridge: 7.62 × 51 mm NATO
Operation: delayed blowback, selective fire and 3-round burst facility
Method of delay: rollers
Weights
  Rifle empty, with bipod: 8.15 kg
  20-round magazine, empty: 280 g
  50-round magazine, empty: 690 g
  Barrel, with muzzle brake: 1.70 kg
Lengths
  Rifle (overall): 1030 mm
  Barrel, without muzzle brake: 450 mm
  Sight radius: 685 mm
Sights
  Mechanical: adjustable 100-1100 m in 100 m steps.
  Optical: 4 × telescope, graticules graduated 100-600 m, fully adjustable for elevation and windage
Performance
  Muzzle velocity: 800 m/s
  Muzzle energy: 3000J
  Rate of fire: ca 800 rds/min

Manufacturer
Hecker & Koch GmbH, D-7238 Oberndorf-Neckar.
Status
Current. Production.
Service
West German Border Police (Bundesgrenzschutz) and several other police forces.

7.62 mm G8 rifle

7.62 mm G8 rifle with accessories

Heckler & Koch Advanced Combat Rifle

Details of the Heckler & Koch design submitted for the US Army Advanced Combat Rifle programme will be found, together with the other entrants, in the USA section on page 225. This course has been adopted so as to permit ready comparison of the various designs.

4.7 mm G11 caseless Heckler and Koch rifle

In 1967 NATO initiated a general feasibility study into caseless small arms ammunition. At that time there were three main interested parties, a United States design, a Belgian semi-caseless design, and a West German one by Heckler and Koch. Within a few years only West Germany remained involved as the other designs failed and NATO interest faded. Shortly after that NATO began the long build-up to the 1979-80 small arms trials and caseless ammunition ceased to interest anyone. However, in West Germany the Government started its own feasibility study and let three contracts in 1969 to three different firms, Diehl, IWKA and Heckler and Koch. The terms of reference were very general, calling for an improved infantry weapon with a better chance of a hit (Ph) than any then in existence, yet fulfilling the FINABEL range and rate of fire characteristics. Designers were given a free hand as to the methods used, but it quickly became apparent that for any significant improvement a radical approach was necessary. Heckler and Koch continued with its caseless studies since this seemed to be a possible way of fulfilling the requirement. By mid-1974 the Bundeswehr selected Heckler and Koch to continue and the others stopped work. The NATO small arms trials interrupted this development since Heckler and Koch was persuaded to enter its incomplete and untried prototype and not surprisingly it failed, to be withdrawn amid much discouraging publicity.

Statistics showed that for the FINABEL battle range of 300 m a circular dispersion of three shots would hit the target allowing for normal aiming errors and in France this was shown to be true by building a special rig of three parallel barrels and firing them either simultaneously or in very rapid succession. However, one thing was plain, despite the low recoil of 5.56 mm rounds they were too large and heavy for any practical three-round firings. A necessary feature of the firing was that the firer must not affect subsequent shots once he had fired the first one, as happens with all normal automatic fire where the first shot goes to the target and the remainder stray off high and right. To overcome this all three must either be fired at the same time, an obvious impossibility, or fired in such rapid succession that the last one has left the muzzle before the firer can react. Experiments showed that to beat the firer's reactions required a rate of fire of at least 2000 rds/min, a virtual impossibility with conventional case ammunition but not totally impossible if

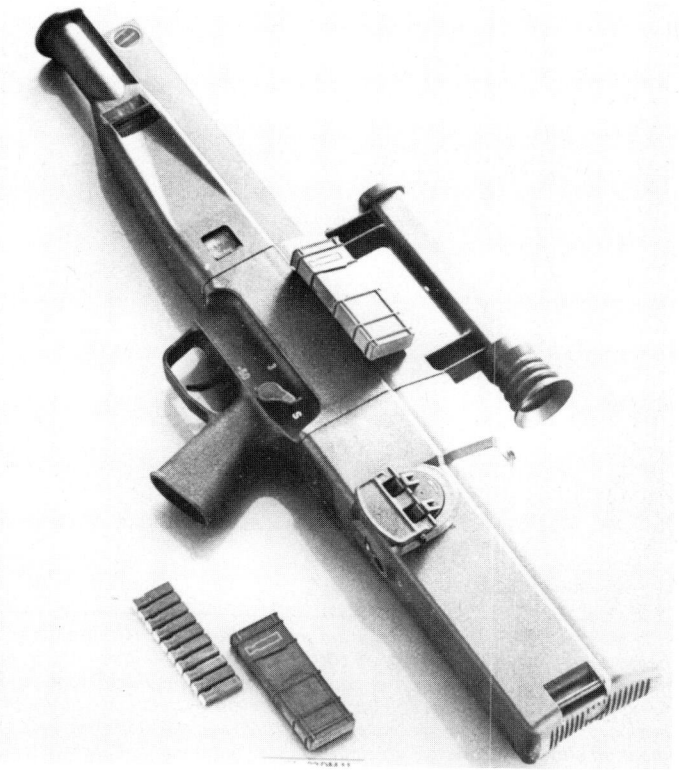

4.73 mm G11 caseless Heckler and Koch rifle

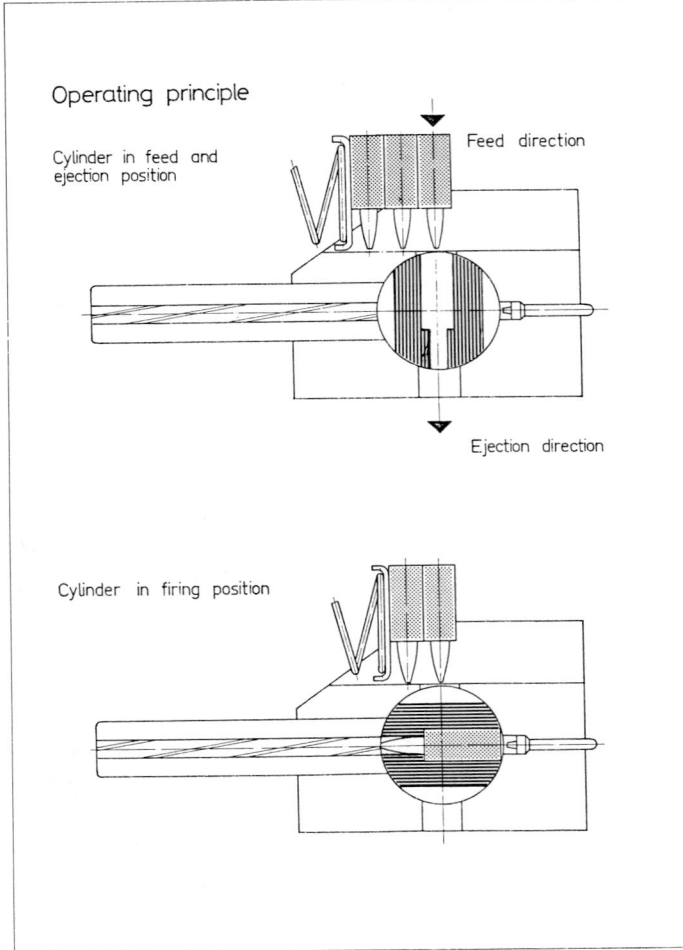

Operating principle

Cylinder in feed and ejection position

Feed direction

Ejection direction

Cylinder in firing position

Operating principle of the cylinder breech. In the upper drawing the cylinder is aligned vertically, ready to receive a fresh round which will be pushed down into it. In the second drawing the cylinder is aligned with the barrel and is ready to fire. Everything in the drawings moves together so that loading and firing can take place while the mechanism is recoiling

First shot fired

Second shot fired (3 round burst)

90ms   100ms

Third shot fired (3 round burst)

Buffer spring (3 round burst)

Reloading ended (3 round burst)

Second shot fired (normal automatic)

Shaded area is the graph of normal automatic fire at 600 rds/min each shot being separated by an interval of 100 milliseconds. The thick line graph above it shows the build-up of movement of the recoiling mass when firing a three-round burst. The first shot accelerates the mass at the same rate as for automatic, the second shot then fires and the motion speeds up, the third shot increases this still more, and when the bullet is clear of the barrel the mass reaches the buffer and is returned to battery, reloading on the way forward

caseless were used. It was this overwhelming need for a better Ph which drove the research into caseless ammunition and a suitable weapon to fire it.

Description

It can be seen from the illustration that the G11 is rather different from most existing designs. There is a single smooth outer casing, with virtually no protruberances or holes. The pistol grip is at the point of balance and the trigger pivots inside a rubber bellows which seals the slot. The selector lever is above the trigger where it can be reached by the thumb of the firing hand. Just behind the trigger is a large circular knob which is the nearest equivalent to a cocking handle. The optical sight is in the carrying handle above the receiver.

These are the only protruberances and there are only two holes, one being the muzzle, the other the ejection opening for clearing misfired rounds. While the muzzle is open at all times the ejection opening is sealed until it is necessary to clear a round. The casing provides complete protection for the mechanism against shocks, rough handling, immersion, frost, dust and dirt.

Ammunition

The crux of the weapon is the ammunition, since all else depends on this. In fact the development of it has been in the hands of Dynamit Nobel. The reasoning behind the choice of calibre is explained later and obviously much care has been taken with the bullet design to ensure a good ballistic shape and adequate stability.

The bullet is set into the block of propellant and is absolutely rigid in it. The propellant block is square in section and longer than the bullet itself so that the overall length of the complete round is just greater than the bullet length; certainly it is less than half the length of an equivalent round with a conventional case. The criteria for such caseless propellants are demanding and Dynamit Nobel has found that the normal types of nitro-cellulose are not adequate. It has developed a special propellant by taking Octagon, which is a high explosive, and treating it to burn as a propellant. This has apparently solved the problem of cook-off which has defeated every other caseless experiment. The primer is another area which has always been difficult and at least one previous version has a primer cap contained in a thin metal foil. Now the primer is a small pellet of some other explosive with a different figure of sensitivity, pushed into a cavity in the base. The complete round therefore is made up from only three components; the bullet, the propellant and the primer and for all practical purposes the propellant and primer can be considered as one.

Magazine

Caseless ammunition does not feed like conventional rounds and the demands for a very high rate of fire made it necessary to look for other ways of carrying the ammunition and presenting it to the breech.

The magazine positioned parallel to the bore axis, above the hand-guard, is removed from the front for reloading. It is a spring-powered, single row magazine which holds 50 cartridges. The approximate number of cartridges in the gun is shown on the outside by an indicator, even when the weapon is ready to fire.

The manufacturers claim that, because of its high capacity, the G11 rifle's weapon-integrated magazine ensures extremely quick, simple and safe reloading from a high-capacity reloading unit. This unit permits loading the magazine with 50 cartridges within a few seconds. These reloading units, designed as sealed air-tight packs, are only opened when required. The reloading units can be filled mechanically and can be used again, if required. The magazine can also be filled manually.

Mechanism

The mechanism is the most interesting of all its many novel features. The first step was to decide what was needed to deal with caseless rounds, then make a mechanism which would do it and finally make the mechanism work reliably. The resulting design is entirely different from anything that has been seen before in small arms, or in any other weapons for that matter.

The essential part of the breech is a cylinder rotating about an axis at right angles to the line of the barrel. The chamber is a hole bored in this cylinder, across the line of the axis. By rotating the cylinder the chamber comes into line with the barrel and, by a turn of 90°, the chamber is turned upside down and is ready to receive a fresh round of ammunition. Thus the first difference between this weapon and any previous one is that the breech motion has been reduced to partial rotation of a fixed cylinder. There is no axial movement of a bolt or a breech-block nor any complications of locking: the turning of the cylinder does it all. The breech cylinder is entirely contained within the barrel extension, and behind the cylinder is a further short extension containing the firing pin and presumably the hammer. When a round is fired the entire mechanism recoils with the barrel and does not separate from it at any part of the cycle. In the drawing this extension of the barrel is shown diagrammatically as a square block.

The cocking knob engages with the cylinder when that is in the forward position, but it disengages when the cylinder recoils and so there is no open slot as with most conventional systems. The trigger also stays still and the breech mechanism moves to and fro above it. When the mechanism comes to battery the trigger sear can then engage with whatever bent has been selected for the next shot. There is one exception to this rule, when the three-round burst is selected the trigger only fires the first round and the other two are controlled by the counter which is in the recoiling mass.

The next novelty in this design is that the magazine also moves with the recoiling mass. Thus when a shot is fired the barrel recoils, taking with it the breech mechanism, the firing pin mechanism, the three-round counter and also the magazine. There is no record of this having happened before in any small arm. One immediate effect of this is that the recoiling mass is maximised and so the impulse is considerably reduced. The return spring lies alongside the barrel so as to reduce overall length.

From this description it can be seen that the Heckler and Koch G11 is formed of two separate parts. There is the outside casing which is completely smooth and seals from all dust and dirt. The second part is the internal mechanism which slides back and forth in this casing and which comprises most of the components that make up a complete conventional gun. It might almost be described as being one gun inside another.

Firing sequence
Because the firing sequence is so different from any other in use today it is necessary to explain it in some detail, and this is best done by taking it step by step.

The first and most important aspect is that the mechanism only has to load and fire the round. There is no empty case to be pulled out of the breech and thrown clear of the gun and this fact significantly reduces the time and movement required for the firing cycle. There is no reason why a caseless design should not use the conventional bolt method, but to do so negates most of the advantages of the caseless round and the loading and ramming cycle imposes heavy strains on the block of propellant, as well as wasting time and space. With the G11 the first round is loaded by turning the cocking knob by hand. This cocks the firing pin and presents the chamber to the first waiting round at the end of the magazine. The round is fed downwards into the chamber by a feeding element under positive control and the cylinder rotates back into line with the barrel. On the outside, the firer rotates the cocking knob through 360°. The weapon is now loaded and cocked and if the selector lever is set at single-shot the trigger will fire this round. On firing the recoiling mass goes back and returns to battery. During this journey the cylinder rotates, a fresh round is loaded and the firing pin is cocked. A second shot can be taken at the instant that the recoiling mass is stationary again. The impulse to the firer is both low and long in duration, so that kick is very slight.

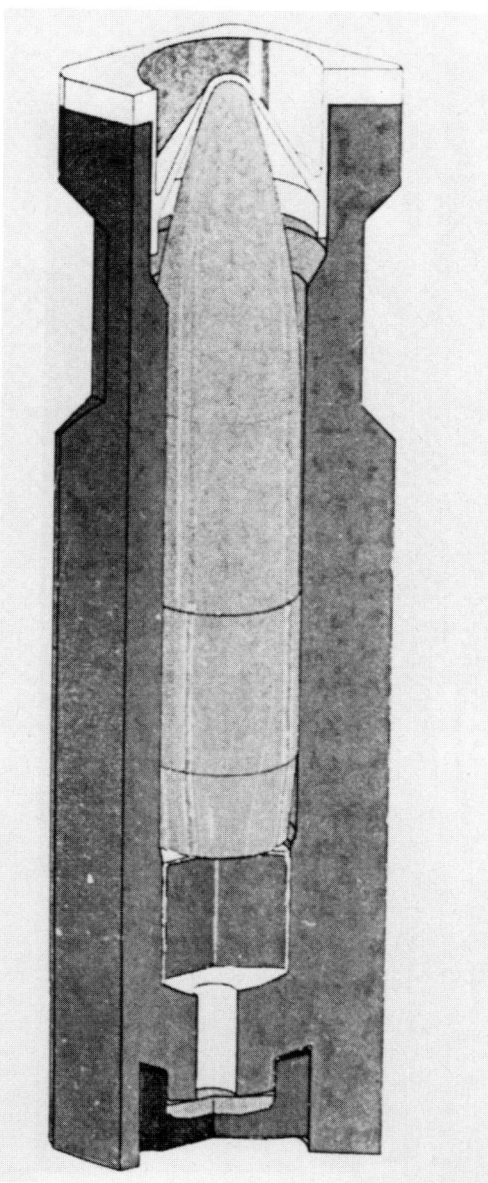

4.73 × 33 mm DM 11 caseless cartridge

When the selector is at automatic the procedure varies only slightly. The first round is obviously fed and fired in the same way. The hammer remains cocked after the first rotation of the cylinder until the rearward motion of the recoiling mass against the pressure of the return spring is completed and this spring has driven the recoiling mass back into its initial forward position (battery), where the second round is automatically ignited. There is no noticeable recoil at the firer's shoulder. The rate of fire is set at approximately 450 rds/min on the present model which is considered to be optimum for control of burst length and correction of aim.

With the selector lever set for three-round bursts a very different sequence is employed. The first shot is the same as for the others, but now the timing is not controlled by the trigger sear and the second and third shots are fired as soon as they are loaded. It is this action which demonstrates the true value of caseless ammunition for, with the short mechanical distance and low inertia loadings involved, the rate of fire is at least 2000 rds/min and all three shots are clear of the muzzle before the recoiling mass reaches the buffer. The mass then returns to battery. The diagram shows what happens but it may take some study to exactly distinguish each item. However, the main lesson which it does show is that all three shots of a three-round burst are fired within 60 milliseconds whereas the time between individual shots at the automatic rate of 450 rds/min is longer at 140 milliseconds. As the recoiling mass moves back, firing its three shots, it builds up a more or less steady load on the firer's shoulder, the peak coming after the last shot has left the muzzle, this being the impulse which the firer feels; but as the bullet has already gone it is too late for his reactions to affect it, and thus all three rounds are fired with a controlled dispersion set by the barrel/ammunition combination.

The final action in the sequence is the ejection of a round which has failed to fire. For this operation the cylinder tilts into the vertical position and the round is positively driven and ejected.

Calibre
Heckler and Koch made many experiments to find an ideal calibre and ranged from 4.3 to 4.9 mm, but finally decided on 4.7 mm as providing the best compromise. Other designers have found that these small bore sizes are prone to produce disastrous barrel wear but the G11 seems to have avoided this. There is no clear indication of the methods used to reduce barrel wear, but it might be that the use of progressive twist rifling contributes to it. Again, it could be that the bore is internally coated. Apparently the rifling is polygonal, which is known to reduce mechanical wear, and while no figures are yet quoted, Heckler and Koch predicts a barrel life equivalent to the 5.56 mm M16A1.

Sight
The optical sight is simple and designed with a view to mass production at the lowest cost. The lenses are toughened glass, protected by rubber at each end. Graticule adjustment is internal. It is intended that the firer can keep both eyes open if he wishes to.

Summary
The G11 can be said to have completed its development and in March 1988 was issued for technical trials. Troop trials with the weapon started in June 1988. Both trials are currently in progress and final results are expected in 1989. The rifle passed its 'cook-off test' and received Safety Certification in May 1988, and type classification is anticipated towards the end of 1989. The first production weapons are expected to be issued to West German Special Forces in 1990. Already sufficient data has been received from troop tests to show that the G11 gives average soldiers a considerable increase in their ability to hit targets when under simulated combat stress. A slightly modified version of the G11 is undergoing evaluation in the USA as part of the Advanced Combat Rifle Programme.

DATA
Cartridge: 4.73 × 33 mm DM11 caseless
Operating principle: gas, closed breech
Breech system: cylindircal drum
Magazine capacity: 50 rounds
Type of fire: selective + 3-round burst
Weight overall: (unloaded) 3.80 kg
Dimensions overall: (length) 752.5 mm; (width) 71.2 mm; (height) 317.5 mm
Length of barrel: 537.5 mm excluding chamber
Rifling: increasing twist rh, 1 turn in 29 calibres at the muzzle
Rifling profile: polygonal
Cyclic rate of fire: (automatic) 600 rds/min; (3-round burst) 2000 rds/min
Practical battle range: 300 m +
Penetration of steel helmet: 600 m +
Sights: optical 1:1 with automatic change to 1:3.5 starting at 300 m
Muzzle velocity: 930 m/s

Manufacturer
Heckler and Koch GmbH, D-7238 Oberndorf-Neckar.
Status
Final development. Technical and troop trials with West German Bundeswehr.

GREECE

Greek 7.62 mm G3A3 and G3A4 rifles

The Hellenic Arms Industry (E.B.O.) SA has been manufacturing the well-known G3A3 and G3A4 rifles for several years. Production is mainly for supply to the Greek Armed Forces for their operational needs and for export. In addition several accessories for the G3 rifle, such as bayonets and bipods, are also produced. The technical data for these weapons is exactly the same as that for the German-manufactured weapons, to which reference should be made.

Manufacturer
Hellenic Arms Industry (E.B.O.) SA, 160 Kifissias Avenue, 11525 Athens.
Status
Current. Production.
Service
Greek armed forces and exports.

Greek-manufactured 7.62 mm G3A3 and G3A4 rifles

HUNGARY

7.62 mm AKM-63 assault rifle

In accordance with standard Warsaw Pact policy the Hungarian Army is equipped with Soviet designs of weapons and for some years the AK-47 was built under licence, though with minor local modifications. In the 1960s the idea was considered of replacing the wooden furniture with plastic and this was incorporated in the production of the AKM. The resulting rifle was the AKM-63 in which the stock and fore-hand grip are made from polypropylene. The fore-hand grip is the distinctive feature of this weapon as it is a second pistol grip. It is attached to a perforated sheet steel fore-arm under the barrel. The resulting rifle is said to be 0.25 kg lighter than the Soviet standard AKM and cheaper to manufacture. All data, except for the reduced weight, is identical with the AKM (which see).

7.62 mm AKM-63 assault rifle

Hungarian 7.62 mm AK-47 assault rifle. The selector markings are '1' and '∞'

7.62 mm AMD-65 assault rifle

The Hungarians have taken the unusual step of modifying their AKM-63 to permit it to be used more easily inside vehicles and similar confined spaces. The modifications are a shorter barrel and a folding stock. The shorter barrel produces more muzzle flash and this is counteracted by a prominent flash hider with two large holes in each side. A magazine article has referred to this flash hider as a muzzle brake and claimed that it permits the rifle to be fired without using the stock at all.

There is also a version with a grenade launcher on the muzzle, and the length of this launcher is such that it almost takes the barrel back to the standard length. There is little detail available about the grenades, but it seems definite that a light anti-armour type is issued.

The folding stock of the AMD is supported with a single tubular strut and the strain on the small hinge must be considerable, however there are no signs of a support to overcome this. The grenade launching rifles have an extra shock-absorbing device on the stock and it may be that this is a spring-loaded plunger which allows some recoil movement. This same rifle has an optical sight, said to be specifically for firing grenades, and a folding fore-pistol grip. The idea of an optical sight for grenades is unusual, though it has to be admitted that most grenade sights are remarkably crude and there is plenty of room for improvement. The fore-pistol grip is actually described as being 'sliding' but this may be a mis-translation for 'folding', and the only photograph is not good enough to clarify this point. Whatever may be the precise facts, the AMD-65 is clearly an interesting variant of the AKM.

DATA
Cartridge: 7.62 mm × 39
Operation: gas
Method of locking: rotating bolt
Feed: 30-round detachable box magazine
Weight: 3.27 kg
Length: (butt extended) 851 mm; (butt folded) 648 mm
Barrel: (with muzzle brake) 378 mm; (no muzzle brake) 318 mm; (rifling) 4 grooves rh
Muzzle velocity: 700 m/s
Rate of fire: (cyclic) 600 rds/min

Manufacturer
State arsenals.
Status
Current. Production.
Service
Hungarian forces.

Hungarian troops using 7.62 mm AMD-65 assault rifle

Magazine photograph of 7.62 mm AMD-65 grenade launching version. The long grenade muzzle attachment can be seen, together with the optical sight on its large mounting plate. The shock absorber on the stock is contained within the thicker section by the shoulder rest. The folding fore-pistol grip is not easily discerned in this photo. There is an unexplained small projection on the gas regulator: this could be a cut-off for grenade firing

7.62 mm AMD-65 assault rifle with stock folded

5.56 mm NGM assault rifle

This weapon is officially known by the Hungarians as a sub-machine gun, but its calibre and proportions belie this description. It is more or less the Hungarian version of the Soviet AK-74, chambered for the 5.56 × 45 mm cartridge and manufactured for export.

The quality of manufacture is high. The barrel is cold forged from alloy steel and has 40 micron thick chromium plating in the bore. All other mechanical components are of high strength alloy steel. The barrel is rifled one turn in 200 mm (one turn in 36 calibres), a compromise value which produces good results with both M193 and SS109 types of bullet, and the makers claim that after 10 000 rounds the dispersion is no more than twice the proof figure.

The NGM is manufactured only in standard form, with wooden furniture and selective-fire capability.

DATA
Cartridge: 5.56 × 45 mm
Operation: gas, selective fire

Locking system: rotating bolt
Feed: 30-round box magazine
Length: 935 mm
Weight: 3.18 kg empty, without magazine
Weight of magazine: empty 330 g; loaded 660 g
Barrel length: 412 mm
Rifling: 4 grooves, rh, one turn in 200 mm
Sights: fire: hooded post; rear tangent U-notch
Muzzle velocity: 900 m/s
Rate of fire: 600 rds/min

Manufacturer
Technika, PO Box 125, Salgótarjáni u. 20, 1475 Budapest.
Status
Current. Production.
Service
Export.

5.56mm NGM assault rifle

INDIA

5.56 mm assault rifle

This is a gas-operated selective-fire weapon which shows an interesting blend of features culled from a variety of sources. The receiver and pistol grip show Kalashnikov influence, the butt, gas regulator and flash hider show FN-FAL influence, the fore-end appears to rely upon the AR-15, and the cocking handle is based on Heckler & Koch practice. The result is a well-balanced and attractive weapon which should prove reliable in service.

The gas system operates the usual rotating bolt, and the magazine housing has been made to accommodate standard M-16 magazines. The selector mechanism allows single shots and three-round bursts. A heavy-barrelled version has also been developed for use as a squad light machine gun.

DATA
Cartridge: 5.56 × 45 mm SS109
Operation: gas, selective fire (three-round bursts)

Locking: rotating bolt
Feed: 20- or 30-round box magazines
Weight, empty: 3.20 kg without magazine
Length: 990 mm (fixed butt)
Barrel: 464 mm
Sights: (fore) blade; (rear) flip aperture, 200 and 400 m
Muzzle velocity: 885 m/s

Manufacturer
Small Arms Factory, Kalpi Road, Kanpur 208009.
Status
Current; production.
Service
Indian armed forces.

5.56mm Indian assault rifle

INDONESIA

5.56 mm SS1-V1 and SS1-V2 assault rifles

The SS1 series assault rifles are licenced copies of the Belgian FN FNC rifle (see page 138) which are manufactured in Indonesia. The SS1-V1 is the standard assault rifle with 449 mm barrel, while the SS1-V2 is the carbine model with 363 mm barrel; both weapons have folding tubular metal butt stocks. The dimensions and data are identical with the information given in the FNC section except that the SS1-V2 is 910 mm long with butt extended and 666 mm long with butt folded. The muzzle velocities differ from the Belgian values; the SS1-V1 delivers 939 m/s and the SS1-V2 901 m/s, which probably reflects differences in the Indonesian propellant and environmental

standards. The Indonesian ammunition is based upon the SS109 bullet and both weapons are rifled 6 grooves, right-handed, one turn in 32 calibres.

Manufacturer
PT Pindad, BPPT Building, 8 Jl.M.H.Thamrin, Jakarta.
Status
Current. Production.
Service
Indonesian armed forces.

IRAQ

5.56 mm Tabuk assault rifle

The Tabuk rifle is simply a copy of the standard Kalashnikov AKM or AKMS chambered to fire the 5.56 × 45 mm cartridge. Details are scant, and we have no information as to the rifling, but presume that it will be rifled to suit the widely-distributed US M193 bullet. Versions are available with fixed wooden butt or with a light metal folding butt. The appearance is identical to the standard AKM rifle.

The data given below is that provided by the manufacturers; we are at a loss to account for the unusually low muzzle velocity quoted.

DATA
Cartridge: 5.56 × 45 mm

Operation: gas, selective fire
Method of locking: rotating bolt
Feed: 30-round box magazine
Weight: (wooden butt) 3.2 kg; (folding butt) 3.28 kg
Rate of fire: 600 ± 60 rds/min
Muzzle velocity: 700-710 m/s

Manufacturer
State arsenal.
Status
Current. Production.

ISRAEL

5.56 mm and 7.62 mm Galil assault rifle

The Galil ARM was designed to fill the roles of the sub-machine gun, assault rifle and light machine gun. It can also be used to project anti-tank and anti-personnel grenades from the shoulder and illuminating and signal grenades, with the butt on the ground. The folding metal stock is of rugged and lightweight construction.

In addition to the ARM there is an assault rifle equipped with a folding metal stock, without bipod and carrying handle, known as the AR (assault rifle), and a rifle with a shortened barrel and folding stock, without bipod and carrying handle, known as the SAR (short assault rifle).

There are three magazine capacities: 12, 35 and 50 rounds. The 12-round magazine is used only with the ballistite cartridges required for grenade launching, and the 35-round magazine is generally employed in the rifle role. When the weapon is employed as a light machine gun the 50-round magazine is used. The bipod is carried lying under the barrel and fitting into a groove under the foregrip. The powerful leverage that can be obtained when the bipod legs are swung down to the vertical position is used to provide a wire cutter. Any barbed wire placed under the hook at the bipod fulcrum can be sheared, usually in one movement of the bipod legs.

The sights enable the weapon to produce aimed fire out to 600 m. The foresight is a cylindrical post which can be screwed up or down for zeroing and the rearsight is a flip aperture with the choice of 0-300 m and 300-500 m. There are also night sights set for use at 100 m and these are normally folded, but when lifted produce a pattern of three luminous spots. The centre spot is that of the foresight, and three luminous aiming marks are lined up horizontally and placed on the target.

The Galil in many respects is modelled on the AK-47 series and most nearly approximates to the Finnish M62 assault rifle. It is a gas-operated rifle with no regulator. The gas block is pinned to the barrel and the gas track is drilled back at 30° into the gas cylinder. The piston-head and shank are chrome-plated. The bolt carrier is an extension of the piston end and is hollowed out over the bolt to take the return spring. The bolt has a cam pin operating in a slot in the carrier and has two locking lugs which are rotated in locking recesses to support the cartridge before firing. The cocking handle is attached to the bolt carrier to give positive bolt closure. It emerges from the right side of the receiver and is bent upwards to allow cocking with either hand. The change lever on the right of the receiver, like that of the AK-47, is a long pressing which when placed on 'safe' both closes up the cocking way against the entry of dirt and restricts the carrier movement. It is possible to retract the bolt sufficiently to inspect the chamber but not enough to feed a round.

Operation
The Galil fires from a closed bolt position. The magazine, fitting under the gun, is placed in position and held by the magazine catch in front of the trigger guard. The change lever is taken off 'safe' and the cocking handle pulled to the rear. When it is released the carrier is driven forward and the top round pushed out of the magazine into the chamber. The bolt comes to a halt and the

further forward travel of the carrier causes the cam pin, engaged in a cam slot in the carrier, to rotate the bolt. The bolt forces the cartridge forward, the extractor slips over the rim and the gun is ready for firing.

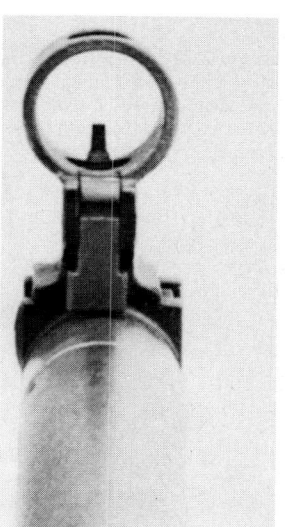

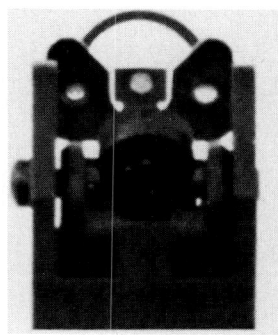

Sights of the Galil rifle consist of a post-type foresight and aperture 'L' flip-type rearsight, set for 300 and 500 m. Night sights are standard on all Galil models. They are normally folded, but when lifted they produce a pattern of three beta-light spots, the centre spot being that of the foresight. The three beta-lights are lined up horizontally and placed on the target

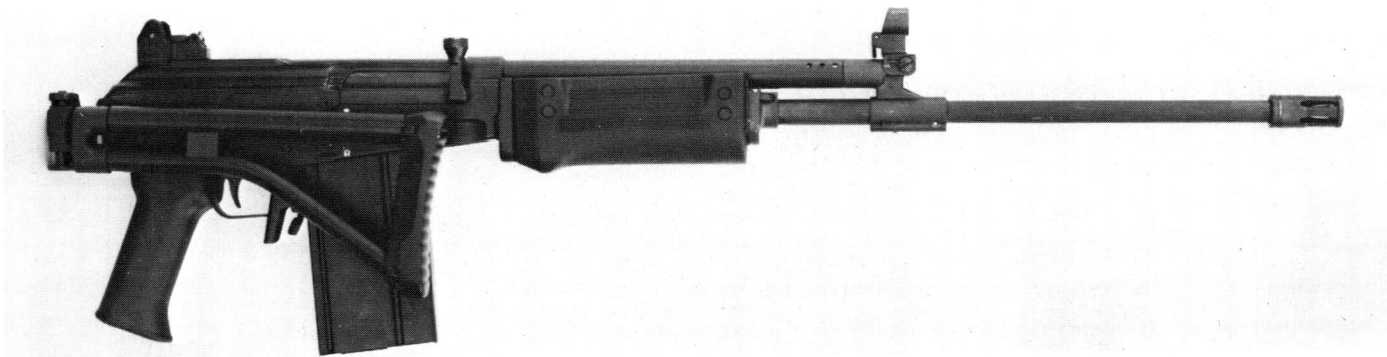

7.62 mm AR Galil assault rifle

5.56 mm AR Galil assault rifle with folding metal stock

When the trigger is pulled, the hammer drives the firing pin forward and the cap is crushed. Some of the propellant gas escapes through the gas vent into the cylinder and drives the piston rearwards. There is a brief period of free travel, while the gas pressure drops, as the width of the cam slot passes the cam pin. The slot then engages the pin which is rotated as the carrier proceeds back. The rotation of the bolt provides primary extraction and then the case is withdrawn as the bolt retracts. The return spring is compressed, the empty case is ejected and then the return-spring energy drives the carrier forward and the cycle starts again.

Trigger and firing mechanism
The system used is that employed in the M1 Garand rifle, the AK series and many others, in which the hammer has two bents and the trigger extension carries two sears. The hammer is initially held by the main sear and when the

trigger is pulled the trigger extension rotates forward and down and the hammer is freed. Provided that the carrier is fully forward the hammer can continue on to strike the firing pin. At semi-automatic the hammer is re-cocked by the recoiling carrier and is caught by the spring-loaded auxiliary sear. The bolt goes forward but since the trigger is already pulled nothing further happens. When the firer releases the trigger, the trigger extension moves back and up as it rotates and the hammer slips out of the auxiliary sear and is caught by the main sear. When the trigger is pulled another round is fired.

At full automatic the rear, spring-loaded, auxiliary sear is held back. The first round of the burst is fired off the front sear and the carrier comes back and rocks the hammer over. Until the carrier is fully forward the safety sear holds the hammer back. As soon as the carrier travels fully forward the safety sear is released and the hammer rotates to fire a further round. So after the

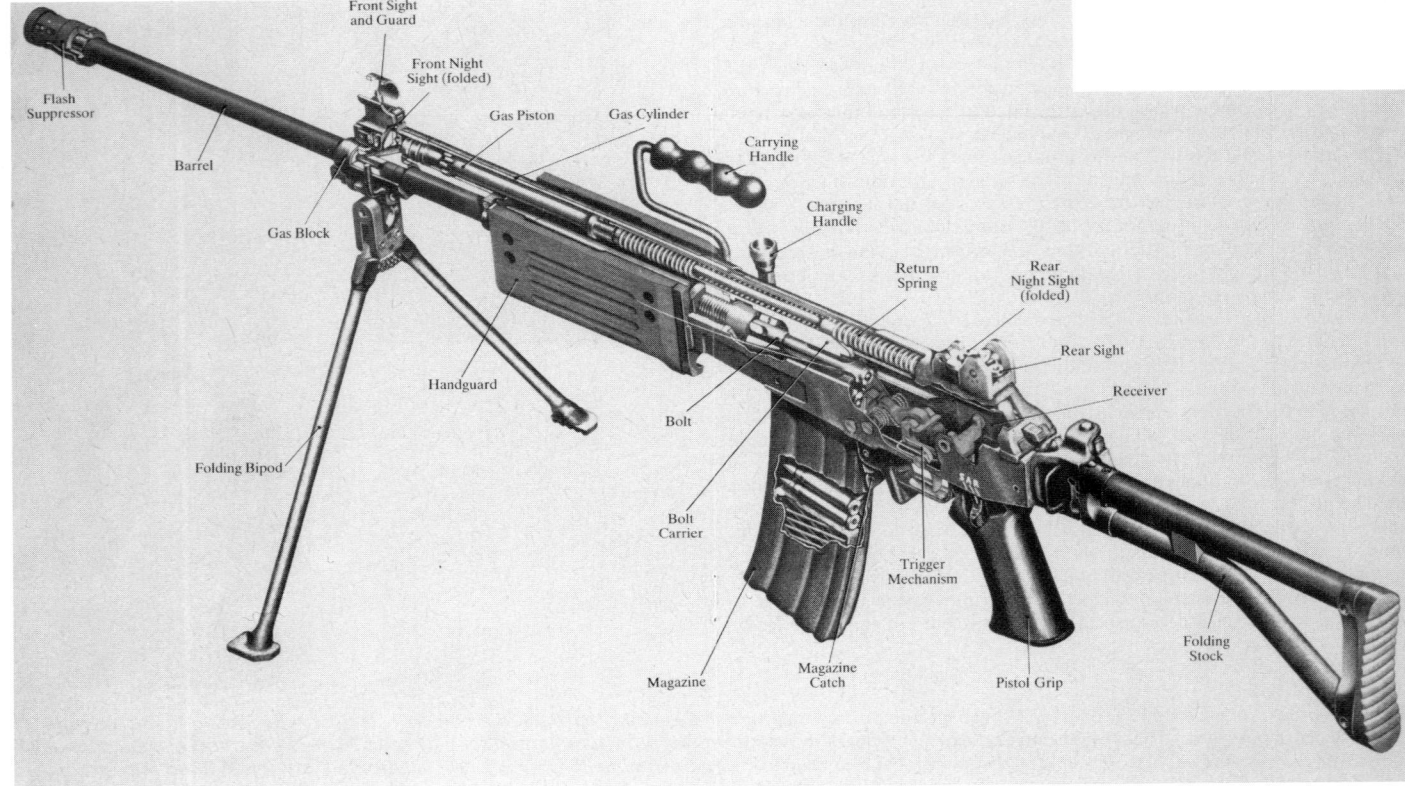

Phantom view of 5.56 mm Galil rifle

5.56 mm ARM Galil weapon system

7.62 mm SAR Galil short-barrelled rifle

first round the trigger sear plays no part. When the trigger is released the hammer is caught by the main sear and held back.

If the safety catch is applied the trigger extension is locked and neither sear can be depressed.

The fire-selector lever may be set from either the left or the right side of the rifle.

Stripping

The Galil breaks down into six parts for field stripping. The procedure follows that of the AK series. First the magazine is removed and the chamber checked; then the guide rod of the return spring is pushed forward and the receiver cover lifted off.

ARM Galil bipod

The return spring and rod are removed. The cocking handle is withdrawn and the carrier and bolt lifted up out of the receiver. The bolt is then rotated out of the carrier.

DATA
DATA REFERS TO 5.56mm ARM/AR EXCEPT AS NOTED
Cartridge: 5.56 × 45 mm
Operation: gas, selective fire
Locking system: rotating bolt
Feed: 35- or 50-round box magazine
Weight: (ARM)4.35 kg with bipod and carrying handle. (AR) 3.95 kg without bipod or handle. (SAR) 3.75 kg without bipod or handle
Length: (ARM/AR) 979 mm overall; 742 mm with stock folded. (SAR) 840 mm overall, 614 mm with stock folded
Barrel: (ARM/AR) 460 mm. (SAR) 332 mm
Rifling: 6 grooves, rh, one turn in 305 mm
Sights: (fore) post, with protector; (rear) flip aperture, 300 and 500 m. Night tritium sights
Sight radius: (ARM/AR) 475 mm; (SAR) 445 mm
Muzzle velocity: (ARM/AR) 950 m/s; (SAR) 900 m/s
Effective range: (ARM/AR) 500 m; (SAR) 400 m

DATA
DATA FOR 7.62mm RIFLES
Cartridge: 7.62 × 51 mm
Operation: gas, selective fire
Locking system: rotating bolt
Feed: 25-round box magazine
Weight: (ARM) 4.0 kg; (AR) 3.95 kg; (SAR) 3.75 kg; all without bipod or carrying handle.
Length: (ARM/AR) overall 1050 mm; stock folded 810 mm. (SAR) overall 915 mm; stock folded 675 mm
Barrel: (ARM/AR) 535 mm: (SAR) 400 mm
Rifling: 4 grooves, rh, one turn in 305 mm

Sights: (fore) post with protector; (rear) flip aperture, 300 and 500 m. Folding tritium night sights
Sight radius: (ARM/AR) 475 mm; (SAR) 445 mm
Muzzle velocity: (ARM/AR) 850 m/s; (SAR) 800 m/s
Rate of fire: (ARM/AR) 650 rds/min; (SAR) 750 rds/min
Effective range: (ARM/AR) 600 m; (SAR) 550 m

Manufacturer
Israel Military Industries, Export Division, PO Box 1044, Ramat Ha Sharon.
Status
Current. Production.
Service
Israeli forces and some other armies.

7.62 mm Galil sniping rifle

Developed in close co-operation with the Israeli Defence Forces, the Galil sniping rifle is a semi-automatic gas-operated rifle specially designed to meet the particular demands of sniping. It is chambered for the 7.62 × 51 mm NATO cartridge but will perform best with selected lots or with Match-quality ammunition. The accuracy requirement is considered to be a 12 to 15 cm circle at 300 m range and a 30 cm circle at 600 m range and the Galil sniper consistently exceeds these requirements.

The mechanism and general configuration is that of the standard Galil rifle, but there are a number of features peculiar to this model. A bipod is mounted behind the fore-end and attached to the receiver, so it can be adjusted by the firer and also relieves the barrel of supportive stress. The barrel is heavier than standard, which contributes to the accuracy. The telescope-sight mount is on the side of the receiver and is a precision cast, long-base unit giving particularly good support to the Nimrod 6 × 40 sight supplied as standard. The sight mount and telescope can be mounted and dismounted quickly and easily without disturbing the rifle's zero and any type of night sight can be fitted.

The barrel is fitted with a muzzle brake and compensator which reduces

jump and permits rapid re-alignment of the rifle after firing. A silencer can be substituted for the muzzle brake if desired; when in use, it is recommended that subsonic ammunition be used.

There is a two-stage trigger. The safety catch is above the pistol grip and there is no provision for automatic fire in this weapon. The wooden stock can be folded, to reduce the overall length of the rifle for carriage and when unfolded and locked is perfectly rigid. The butt stock is fitted with a cheek-piece and a rubber recoil pad, both of which are fully adjustable.

Each rifle is supplied in a specially designed case, which also contains the sight, two optical filters for use with the sight, a carrying and firing sling, two magazines and the cleaning kit.

DATA
Cartridge: 7.62 × 51 mm NATO
Operation: gas-operated, semi-automatic only
Method of locking: rotating bolt
Feed: 20-round magazine
Weight, empty: 6.4 kg including bipod and sling
Length: (butt extended) 1115 mm; (butt folded) 840 mm
Barrel: 508 mm without muzzle brake
Rifling: 4 grooves, rh, 1 turn in 305 mm
Muzzle velocity: (FN match ammunition) 10.9 g bullet, 815 m/s; (M118 match ammunition) 11.2 g bullet, 780 m/s
Sights: iron sights (fore) blade; (rear) aperture; (optical) Nimrod 6 × 40 telescope
Sight radius: 475 mm

Manufacturer
Israeli Military Industries, PO Box 1044, Ramat Ha Sharon.
Status
Current. Production.
Service
Israeli Defence Forces.

7.62 mm Galil sniping rifle

7.62 mm Sirkis M36 sniping rifle

The M36 is basically the action of the US M14 rifle but converted to bullpup form. This reduces the overall length of the weapon whilst retaining the full length of barrel. The mechanism and much of the weapon is encased in an in-line stock. The muzzle is fitted with an combined flash hider and muzzle brake, and there is provision for fitting a sound suppressor. The barrel is

free-floating and is bedded into a laminated or carbon-fibre stock. A lightweight bipod is provided, and there is a rubber butt-plate.

The M36 retains the rotary bolt of the M14 but some modifications to the mechanism have been made, including a two-stage trigger. Iron sights to 400 m are provided, but the primary sighting system will be a telescope.

DATA
Cartridge: 7.62 × 51 mm
Operation: gas, semi-automatic
Locking system: rotating bolt
Feed: 20-round magazine
Weight: 4.5 kg
Length: 850 mm; with suppressor 1030 mm
Barrel: 560 mm
Sight radius: 500 mm

Manufacturer
Sirkis Industries Ltd., Ramat Gan.
Status
Current. Production.

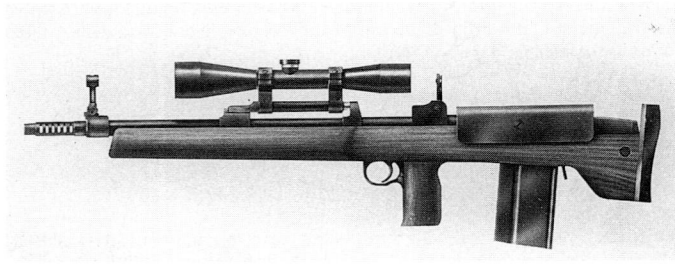

7.62 mm Sirkis M36 sniping rifle

Orlite plastic magazines for M16 and Galil rifles

The 30-round plastic magazine developed and manufactured by Orlite Engineering Company has been accepted for service on the M16 rifle by the Israeli Defence Forces. The 35-round plastic magazine for the Galil Assault Rifle has also been accepted for service by the IDF and is now in production. First deliveries to the IDF took place in February 1988.
Orlite magazines also received a favourable report from the US Army Research and Development Command (ARADCOM) after testing on the M16A1 rifle and M231 firing port weapon. ARADCOM stated that results of environmental extremes, solvent and functional tests were generally more favourable than those of drop tests; although the failure rate of plastic magazines was no worse than that of metal magazines.

The magazine is made from glass-reinforced nylon which offers not only inherent self-lubricating qualities but is also claimed to have greater strength than metal, high resistance to chemical attack and high and low temperature

stability. The injection moulding technique used for the magazines assures a smooth surface and accuracy in dimensions. The construction is in one piece and eliminates joints.

Maintenance is reduced with plastic magazines and they are cheaper to produce by about 40 per cent. In general, plastic is more resistant to damage as it does not distort under impact, but tends to deflect and re-assert itself.

Manufacturer
Orlite Engineering Co Ltd, Industrial Zone A, Ness-Ziona 70400.
Status
Current. Production.
Service
Israeli military forces; favourably evaluated by US Army.

ITALY

7.62 mm BM 59 Beretta rifle

At the end of the Second World War the firm of Pietro Beretta at Gardone Valtrompia began to manufacture the US M1 .30 rifle, the Garand, for the Italian Army. These rifles were later supplied to Indonesia and Denmark. Some 100 000 rifles were produced by 1961.

The need to compete with weapons of more modern design (the M1 rifle came into US service in 1936) led to the production of a modified M1 rifle firing the NATO 7.62 mm × 51 cartridge. Studies to convert the Garand into a modern selective fire weapon were commenced in 1959 under the direction of Domenico Salza and completed by Vittorio Valle.

The earliest versions were given letter suffixes. The first BM 59 was a Garand, lightened and shortened, modified to take a 20-round detachable box magazine.

Next came the BM 59 R which had a device to slow down the rate of fire. This was followed by the BM 59 D with a pistol grip and bipod and the BM 59 GL with a grenade launcher. The BM 60 CB had a three-round burst fire controller.

Of the BM 59 series the Mark E was the cheapest since it used the Garand barrel and gas cylinder. In all the others the barrel and gas cylinder was a Beretta design. The current series is as follows:

The BM 59 Mark I is a fixed wooden-stocked rifle with the gas cylinder under the barrel. There is a 20-round detachable box magazine which can be loaded and replenished using chargers and guides built into the receiver. The corresponding Italian Army version is known as the BM 59 Mark Ital and has a light bipod fitted to the gas cylinder. The BM 59 Mark II is like the Mark I but has a pistol grip, a winter trigger and a bipod.

The BM 59 Mark III has a folding-metal stock and two pistol grips.

The Italian Army BM 59 Mark Ital TA and the BM 59 Mark Ital Para were derived from the Mark III. The Mark Ital TA is for Alpine troops and the Mark Ital Para is for airborne use.

The BM 59 Mark IV is a heavy-barrelled version intended for the squad light automatic role. It has a plastic butt stock and a hinged butt-plate and a pistol grip. There is no forward pistol grip.

The BM 59 will fire any anti-personnel or anti-tank grenade with a tail tube having an inside diameter of 21 mm. Where the grenade launcher is fitted it is not intended that it should be removed, except for the Mark Ital TP, the parachutists' rifle. The erection of the grenade sight automatically cuts off the gas supply to the gas piston.

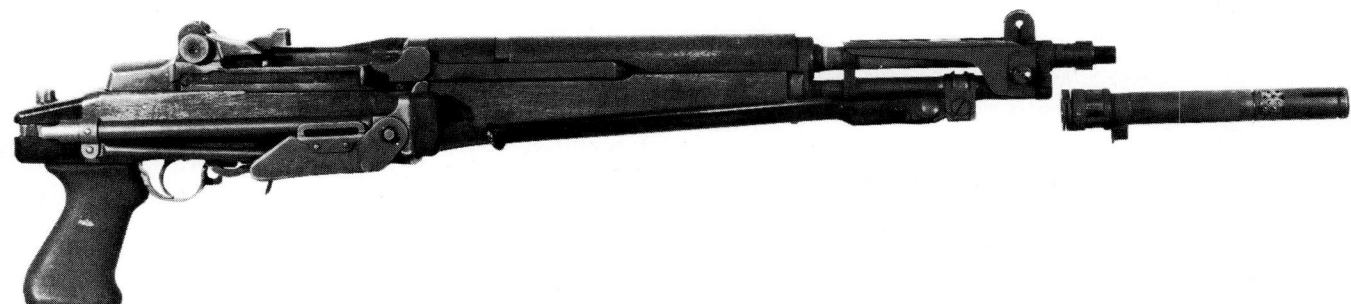

7.62 mm BM 59 Mark Ital TA Beretta rifle

7.62 mm BM 59 Mark Ital Beretta rifle

7.62 mm BM 59 Mark Ital Para Beretta rifle

7.62 mm BM 59 Mark I Beretta rifle

7.62 mm BM 59 GL Beretta rifle with permanently attached grenade launcher and sights showing winter trigger

DATA
BM 59 MARK ITAL
Cartridge: 7.62 mm × 51
Operation: gas
Method of locking: rotating bolt
Feed: 20-round detachable box magazine
Weight: 4.6 kg
Length: 1095 mm
Barrel: 490 mm
Rifling: 4 grooves rh
Sights: (fore) blade; (rear) aperture, adjustable for range and windage

Muzzle velocity: 823 m/s
Rate of fire: (cyclic) 750 rds/min
Effective range: 600 m

Manufacturer
Pietro Beretta SpA, Gardone Valtrompia, Brescia.
Status
Current. Not in production but spares available.
Service
Italian Army. Manufactured under licence in Indonesia and Morocco.

7.62 mm Socimi AR 832/FS assault rifle

The AR 832/FS is a gas-operated weapon using the conventional type of bolt carrier and rotating bolt which locks into the barrel extension. The receiver, grip frame and butt stock are of aluminium alloy, resulting in a very light weapon; due to careful matching of the return spring, the recoil, however, is not excessive. The butt stock folds to lie along the right side of the receiver. The gas regulator is adjustable and can be completely closed to permit firing grenades. Stripping is extremely simple; by removing only two pins the entire receiver and barrel group is separated from the grip and stock assembly, and the bolt, carrier and piston can be lifted out.

DATA
Cartridge: 7.62 × 51 mm NATO
Operation: gas, selective fire
Method of locking: rotating bolt
Feed: 30-round box magazine
Weight: 4.30 kg
Length: (stock extended) 1030 mm; (stock folded) 810 mm
Barrel: 450 mm
Rifling: 4 grooves, rh, 1 turn in 305 mm
Sights: (front) post; (rear) aperture, adjustable for elevation and windage
Rate of fire: 650-900 rds/min
Muzzle velocity: 820 m/s

Socimi AR 832/FS 7.62 mm rifle with butt extended

Manufacturer
Socimi SpA, Via Varesina 115, I-20156 Milano.
Status
Development completed.

5.56 mm AR70/223 Beretta rifle

In early 1968 Pietro Beretta at Brescia started an initial survey for a new assault rifle. The design team was headed by Vittorio Valle, head of the Beretta Research and Development Department, under the direct supervision of PC Beretta, the General Manager. Work included a detailed evaluation of existing 5.56 mm systems, including the AR-15 and the Stoner.

Eventually a gas-operated system was adopted. It was decided to have a conventional piston and cylinder system with the gas port as far forward as possible. The designers next looked at the type of bolt they were going to use and the forward locking, two-lug system of the Garand, .30 calibre M1 carbine and the Soviet Kalashnikov AK-47 was adopted. One of the advantages of forward locking, in modern design, is that the bolt can lock directly into the barrel extension. Thus only the barrel and bolt-head are stressed and must be of high tensile material but the rest of the bolt and the receiver need not be produced from such expensive steel. However Beretta has chosen to weld a sleeve into the receiver and the bolt lugs close into recesses cut in this.

Operation
The operating system of the AR70 is quite simple. On firing, the gas pressure drives the bullet forward up the bore. After a bullet travel of approximately 33 cm the gas passes through the vent into the cylinder and drives the piston rearward. There is a free travel of the piston for about 9 mm before the carrier starts to rotate the bolt and produce unlocking. This delay, mechanical safety, allows the pressure to drop. The bolt is cammed through 30° and is fully unlocked; it travels back with the carrier, extracts the spent case, compresses the operating spring around the piston and as the latter re-asserts itself the working parts go forward, feeding a new round into the chamber. When chambering is complete the bolt comes to a stop but the carrier continues forward and the cam rotates the bolt into the locked position. The trigger and firing mechanism are orthodox in their action. Single-shot is obtained using a disconnector between the trigger and the sear; the latter operates in a bent near the hammer axis. When firing at full automatic the hammer is controlled by a safety sear which is operated by the bolt carrier in its last forward motion.

5.56 mm SC70 Beretta short carbine with folding butt. Bayonet can be fitted to muzzle but it is not possible to launch grenades

5.56 mm AR70/223 Beretta rifle with grenade-firing gas lever folded forwards

5.56 mm SC70 Beretta folding-butt version of AR70

This ensures locking is completed before the hammer is released and it operates also in single-shot; if the bolt is not locked the hammer cannot move.

In the design of the weapon as an assault rifle it was considered, wisely, that all openings should be sealed to prevent the ingress of dirt when the rifle was ready to fire. A cover was provided over the slot in the receiver in which the cocking handle comes back (not the ejection opening) and this flies up as the firer initially cocks the weapon. This cocking handle is the only connection between the bolt carrier and the piston; if it is lost during stripping, the rifle becomes useless.

Stripping

The rifle strips easily. A take-down pin is placed on the left of the receiver above the rear surface of the pistol grip and is pushed through with the nose of a round. The butt group drops down and the cocking-handle catch is released and the cocking handle pulled out. The bolt and carrier come out to the rear. The fore-end is spring-retained. The flash eliminator can be removed and then the gas cylinder, piston and spring. The only snag in re-assembling the gun is ensuring that the slot in the piston, to connect it with the bolt carrier, is properly lined up.

It should be noted that the rear aperture sight and the foresight are both on the same side of the separating surfaces as the weapon breaks. Thus wear in the take-down pin or in the mating surfaces does not, unlike some current rifles, produce sloppiness in the sight. A telescopic sight can be fitted if required.

The rigid stock of the AR70 can easily be detached and replaced with the folding butt of the SC model.

The rifle fires any grenade of standard tail dimensions. To assemble the grenade over the flash eliminator a lever lying horizontally over the muzzle must be raised. This closes a valve, ensuring all the gas from the ballistite cartridge is used to drive the grenade, and the normal gas system is non-operative.

DATA
Cartridge: 5.56 mm × 45 (M193 and SS109)
Operation: gas, selective fire
Method of locking: rotating bolt
Feed: 30-round box magazine

	AR70	SC70	SC70 Short
Length			
overall	955 mm	960 mm	820 mm
butt folded	—	736 mm	596 mm
barrel	450 mm	450 mm	320 mm
Sight radius	507 mm	507 mm	455 mm
Rifling	4 grooves rh. 1 turn in 304 mm		
Weight			
empty magazine	3.8 kg	3.85 kg	3.7 kg
full magazine	4.15 kg	4.2 kg	4.05 kg
Rate of fire (cyclic)	650 rds/min	650 rds/min	600 rds/min
Muzzle velocity	950 m/s	950 m/s	885 m/s
Muzzle energy	167 kgm	167 kgm	142 kgm
Grenade	40 mm	40 mm	—

Manufacturer
Pietro Beretta SpA, Gardone Valtrompia, Brescia.
Status
Current. Production.
Service
Italian Special Forces, Jordan, Malaysia and some other countries.

5.56 mm Beretta 70/90 assault rifle

The Beretta 70/90 system has been developed primarily as a potential system for the Italian Army. It consists of four weapons: the assault rifle AR70/90 for infantry; the carbine SC70/90 for special forces; the special carbine (short) SCS70/90 for armoured troops; and the light machine gun AS70/90, for use as the squad automatic weapon and which is described in the appropriate section.

The AR70/90 is, broadly, an improved version of the 70/223 described in the previous entry, and its design has been influenced by service experience gained with the earlier weapon. In the 70/223, for example, the receiver was a pressed-steel rectangle in which the bolt moved on pressed-in rails; experience showed that this could distort under severe conditions, and thus the AR70/90 receiver is of trapezoidal section and has steel bolt guide rails welded in place. The method of operation is the same, using a gas piston mounted over the barrel which actuates a bolt carrier and a two-lug rotating bolt. A noticeable new feature is the carrying handle above the receiver. This clips into place by means of a spring-loaded catch and carries a luminescent source for use as an aiming aid in poor light conditions. Removing the handle reveals a dovetailed receiver cover which meets STANAG 2324 requirements for a sight mounting which will take any optical or electro-optical sight.

The trigger mechanism in the standard version permits selection of single shots, three-round bursts or full automatic fire. An optional mechanism is available which restricts the available modes to single shots and three-round bursts. However, the three-round burst mechanism can be removed and replaced by components which will permit automatic fire or semi-automatic action only.

The rifle feeds from a 30-round magazine, and the magazine housing interface is to STANAG 4179, allowing the use of any M16-type magazine.

The magazine is released by an ambidextrous button which is operated by the hand holding the weapon.

The gas cylinder is fitted with a regulator having three positions: open for normal use, further open for use in adverse conditions, or closed for grenade firing. The regulator is fitted with a lever which is raised for grenade firing; in this position it also impinges on the sight line, so reminding the firer that it is set. When lowered for normal firing, the lever will obstruct any attempt to load a grenade.

The normal iron sights consist of a blade foresight and a two-position aperture backsight. The supports of the carrying handle are open, so that the sight line passes through without obstruction, and the foresight can be adjusted for zeroing.

Accessories provided for the rifle include bayonet, bipod, slings, blank firing attachment and cleaning kit. The blank firing attachment is of thin metal and will rupture if a ball round is inadvertently fired. The trigger-guard can be hinged down to give adequate clearance when wearing arctic gloves.

The carbine SC70/90 differs from the rifle only in that it has a folding tubular metal butt. It is worth noting that when the butt is folded, the patented ambidextrous magazine release can still be conveniently operated.

The special carbine SCS70/90 has a shorter barrel; has no facility for launching grenades and hence no gas regulator lever; has no bayonet attachment; and has a folding tubular metal butt.

Particular features of the 70/90 system which are stressed by the manufacturers include a minimum number (105) of components; exceptionally easy field stripping, essentially the same as that of the 70/223; remarkable accessibility of the gas cylinder for maintenance – it can be field-stripped without tools, and is believed to be the only 5.56 mm rifle for which this claim

5.56 mm Beretta SC70/90 carbine with butt folded

5.56 mm Beretta SC70/90 carbine

5.56 mm Beretta SCS70/90 short carbine

5.56 mm Beretta AR70/90 infantry rifle

can be made; has no screws in its assembly; and has a unique method of attaching the barrel to the receiver. Normal practice is simply to screw the barrel into the receiver, but this means a certain amount of hand-fitting to obtain the correct cartridge head clearance, and also introduced problems when it is necessary to fit a new barrel. The system used with the 70/90 is to make a plain hole in the receiver, and a plain rear end to the barrel. There is a collar on the barrel, and a screw cap which fits over the barrel and, when screwed on to the front end of the receiver, clamps the barrel into place. During manufacture the headspace gauging datum is carefully located and the locating collar and chamber face are carefully ground, by computer-controlled machines, to give the correct headspace when the barrel retaining nut is tightened to a specified torque. Subsequent changes of barrel will thus always have the correct headspace.

DATA
Cartridge: 5.56 × 45 mm
Operation: gas, selective fire
Locking: rotating bolt
Feed: 30-round magazine

WEIGHTS
AR70/90: 3.99 kg
SC70/90: 3.99 kg
SCS70/90; 3.79 kg

LENGTHS
AR70/90: 998 mm
SC70/90: 986 mm
SCS70/90: (butt extended) 876 mm; (butt folded) 647 mm

BARREL
Lengths :450 mm or 352 mm
Rifling: 6 grooves, rh, one turn in 178 mm. Chrome plated

PERFORMANCE
No performance figures have been released by the manufacturers; it is anticipated that the AR70/90 will meet the NATO STANAG 4172 requirements with no difficulty.

Manufacturer
Pietro Beretta SpA, I-25063 Gardone Valtrompia, Brescia.
Status
Under trial by Italian Army.

7.62 mm Beretta sniper

The Beretta sniper is a conventional bolt-action repeating rifle chambered for the 7.62 × 51 mm NATO cartridge. It has a heavy free-floating barrel, with a flash eliminator on the muzzle and a tube in the fore-end acts as a locating point for a bipod; it is possible that it also contains a harmonic balancer to reduce barrel vibration.

The rifle is fitted with target-quality iron sights, the foresight being hooded to obviate reflection and glare and the rearsight being fully adjustable for elevation and windage. However, as a sniping rifle, it will normally be fitted with an optical sight; the sight recommended by the makers is the Zeiss Diavari-Z, a 1.5 to 6× zoom telescope. The sight mount is to NATO

7.62 mm Beretta sniper with bipod

7.62 mm Beretta sniper

STANAG 2324, and thus virtually any optical or electro-optical sight can be accommodated.

The thumb-hole pattern stock is of wood, and the recoil pad and cheek-piece can be adjusted to suit the firer; similarly, there is a hand stop under the fore-end which is adjustable and which also serves to locate the firing sling, should one be desired.

DATA
Cartridge: 7.62 × 51 mm NATO
Action: turnbolt
Weight
  empty: 5.55 kg
  bipod: 950 g
  telescope: 700 g

Length: 1165 mm
  barrel: 586 mm
Rifling: 4 grooves, rh, 1 turn in 305 mm
Feed: 5-round box magazine
Sights
  front: Hooded blade
  rear: Fully adjustable V-notch
  optical: Zeiss Diavari-Z 1.5 – 6 × 42T

Manufacturer
Pietro Beretta SpA, I-25063 Gardone Valtrompia, Brescia.
Status
Current. Production.

5.56 mm Franchi G41 assault rifle

This is the Heckler and Koch G41 rifle, manufactured under licence by Luigi Franchi SpA and submitted by it for possible adoption as the next Italian service rifle. The models available, G41, G41A2 and G41K are operationally identical to the Heckler and Koch versions, to which reference should be made for dimensions. There are, however, one or two minor differences, doubtless due to different manufacturing techniques. The barrel, which is chrome-plated throughout, is polygonally rifled with four grooves; the empty weight of the G41 version is 3.60 kg; the barrel is 480 mm long with its flash hider; and the cyclic rate of fire is 900 rds/min.

Manufacturer
Luigi Franchi SpA, Via del Serpente 12, I-25020 Fornaci, Brescia.
Status
Under trial by Italian Army.

5.56 mm Franchi G41A2 rifle

Bernadelli B2 and B2/S assault rifles

These rifles have been developed by Bernadelli for consideration by the Italian Army in its current rifle trial. Little information has been released, since the company informs us that the design is constantly under revision, and, since the final design is not settled, no pictures can be shown. However it is known that the design is based upon the well-known Israeli Galil. The B2 is the standard infantry rifle, while the B2/S is a short model for special troops. Lightweight and gas-operated, the rifle is capable of single shots or automatic fire and the muzzle is shaped into a NATO-standard grenade launcher which also acts as a flash suppressor. The weapon is easily operated by left- or right-handed firers, with both the fire-selector and magazine catch capable of operation from either side. A carrying handle is fitted and the rifle strips into six major components without the use of tools.

DATA

	B2	B2/S
Cartridge	5.56 × 45mm NATO	
Operation	gas, selective fire	
Locking system	rotating bolt	
Feed	30- or 50-round magazine	
Weight	3.9 kg	3.65 kg
Length, stock extended	979 mm	851 mm
Length, stock folded	742 mm	614 mm
Barrel length	460 mm	332 mm
Rifling	6 grooves, rh, one turn in 178 mm	
Muzzle velocity	915 m/s	840 m/s
Cyclic rate of fire	600/720 rds/min	

Manufacturer
Vincenzo Bernadelli SpA, I-25063, Gardone Val Trompia, Brescia.
Status
Advanced development.

.22 Bernadelli B/TR training rifle

The B/TR is a light semi-automatic rifle which has been specially designed to meet military training requirements. It is simple to use but has sufficient affinity with military weapons to allow it to be used for the recruit's first introduction to the rifle and thereafter to be used for marksmanship training. The mechanism is blowback, feeding from a box magazine, and the manufacture and finish appear to be to a high standard.

DATA
Cartridge: .22 LR rimfire
Action: semi-automatic, blowback
Feed: 5- or 10-round box magazines
Weight, empty: 2.45 kg

Length: 1020 mm
  barrel: 530 mm
Rifling: 6 grooves rh
Sights
  front: hooded blade
  rear: fully adjustable U-notch
Sight base: 440 mm

Manufacturer
Vincenzo Bernadelli SpA, I-25063 Gardone Valtrompia, Brescia.
Status
Current. Production.

.22 Bernadelli B/TR training rifle

5.56 mm Socimi AR 871 weapon system

The AR871 assault rifle is a selective fire weapon using a gas-operated rotating bolt which locks directly into the barrel. The rifle may be obtained with a selection system allowing single shots or automatic fire, single shots or three-round bursts, or single shots only. The weapon is laid out in a straight line configuration and this, together with the long recoil bolt stroke (110 mm), allows automatic fire with good control. The barrel is 450 mm long, allowing the bullet to develop its maximum velocity and optimum ballistic efficiency.

The rifle is light and strong, many of the components being made of

Socimi AR 871 5.56 mm rifle with fixed butt

Socimi AR 871 5.56 mm rifle dismantled

aluminium alloy or of plastic impact-resistant material. Other components are made of high tensile steel. The cocking handle is on the left side, allowing the rifle to be cocked without moving the grasp on the pistol grip. The magazine release is duplicated on both sides of the weapon for ambidextrous use. The barrel is chrome-lined and the muzzle is prepared for grenade launching.

Field stripping is quick and easy; after removing one of the two pins connecting the receiver to the butt stock, the receiver can be hinged open and the bolt carrier and bolt assembly can be easily removed.

DATA
Cartridge: 5.56 × 45 mm NATO
Operation: gas, selective fire
Locking system: rotating bolt
Feed system: 20- or 30- round box magazine, M16 interface to STANAG 4179
Weight, empty: 3.6 kg
Length: 970 mm
Barrel: 450 mm
Rifling: 6 grooves, rh, one turn in 178 mm (32 calibres)
Sights: (fore) blade; (rear) aperture, adjustable for elevation and windage. Mount for night sights to STANAG 2324
Rate of Fire: 700 – 750 rds/min
Muzzle velocity: 970 m/s

Manufacturer
Socimi SpA, Via Varesina 115, I-20156 Milano.
Status
Development completed.

Socimi AR 871 5.56 mm rifle with folding butt

JAPAN

7.62 mm Type 64 rifle

The formation of the Japanese Self-Defence Force led to the requirement for a new series of weapons and the Howa Machinery Company of Nagoya received a contract to develop a rifle. The project leader and designer was General K Iwashita. Several designs and modifications to those designs were produced until eventually the Type 64 emerged. This was the weapon accepted for the Japanese Self-Defence Force.

The cartridge used is a reduced load 7.62 mm × 51, the propellant charge having been reduced by 10 per cent below NATO normal. This leads to a reduced muzzle impulse and as no effort has been made to reduce the weight significantly, the recoil energy that the soldier has to absorb is less than that generated by the conventional 7.62 mm cartridge. Should it ever be necessary to employ the full charge NATO round then the gas regulator must be set to reduce the amount and pressure of the gases reaching the piston head. The rifle has been designed to launch rifle grenades and the gas regulator therefore includes a facility to cut off the gas supply to the piston entirely. The rifle is

orthodox in appearance and action. The gas cylinder and piston are positioned above the barrel. The locking system incorporates a tilting block, lifted into engagement and subsequently lowered and carried back by the bolt carrier.

The butt is a straight-through design and this, with the bipod under the front of the fore-end, and the muzzle brake, produces extremely accurate single-shot fire. The gun fires at 500 rds/min at full automatic and this relatively slow rate assists in maintaining the burst on the target. Three gas ports adjustable for load act as a regulator.

The very short length of the rifle of only 99 cm makes it distinctive and easily recognisable. This small overall dimension (13 cm less than the US M14) and the reduced velocity and recoil force make the Type 64 a very good weapon for the initiation of recruits and their subsequent advanced musketry training.

7.62 mm Type 64 rifle

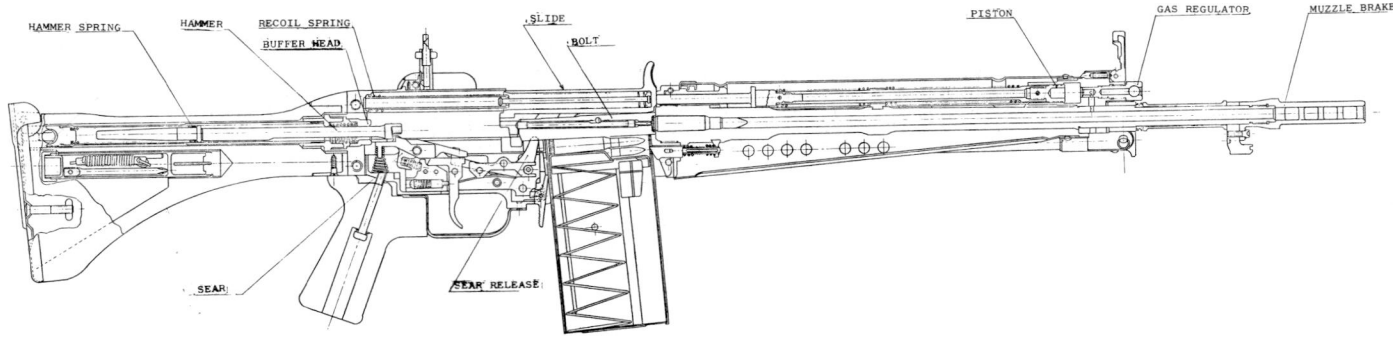

7.62 mm Type 64 rifle

DATA
Cartridge: 7.62 mm × 51 (reduced load)
Operation: gas, selective fire
Method of locking: tilting block
Feed: 20-round detachable box magazine
Weight, empty: 4.4 kg
Length: 990 mm
Barrel: 450 mm
Rifling: 4 grooves, rh
Sights: (fore) blade; (rear) aperture

Muzzle velocity: reduced charge 700 m/s; full charge 800 m/s
Rate of fire: (cyclic) 500 rds/min
Effective range: 400 m

Manufacturer
Howa Machinery Ltd, Shinkawacho, near Nagoya.
Status
Current.
Service
Japanese forces.

KOREA, NORTH

7.62 mm Type 68 assault rifle

The regular army of North Korea is largely equipped with the Type 68 assault rifle. This is based on the Soviet AKM but has distinct differences. Internally it does not have a rate of fire reducer but, like the AK-47, has no restraint on the hammer. Externally the forestock is not shaped with the two recesses, one on either side, for the fingers, that are a feature of all other AKM derivatives.

However the greatest source of difference is the folding stock which consists of two perforated rails joined at the shoulder piece. This makes the North Korean Type 68 the lightest of the AKM rifles.

The North Korean copy of the Soviet AK-47 is still retained for use with militia units and support troops.

North Korean AK-47 assault rifle

7.62 mm Type 68 assault rifle

7.62 mm Type 68 assault rifle showing perforated butt stock

KOREA, SOUTH

5.56 mm K2 assault rifle

The K2 is a gas-operated selective fire rifle with folding plastic stock. The operating system uses a long-stroke piston, similar to that of the AK series Kalashnikov rifles, operating a rotating bolt. The upper and lower receivers are machined from aluminium alloy forgings, and although the lower receiver may appear similar to that of the M16 there are considerable detail differences and the two are not interchangeable. The selector lever is rotatable in either direction and has four positions for safe, automatic fire, three-round burst fire and single shots. The burst control mechanism does not reset when the trigger is released; on pulling the trigger for a second time the burst picks up from where it stopped in the previous cycle.

5.56 mm Daewoo K2 assault rifle

The barrel is fitted with a muzzle brake/compensator which has no slot underneath so as to prevent dust being blown up. The folding stock is hinged to fold to the right of the receiver and uses a locking system similar to that of the FN FNC. The rearsight has a two-position flip aperture with additional cam-actuated elevation control. The normal position of the flip gives a small aperture for precise shooting. The other position of the flip provides two small spots for night shooting. Fore and rear sights are provided with luminous spots for firing in poor light. In either position, rotation of the elevation control raises the selected aperture, the maximum marked range being 600 m.

DATA
Cartridge: 5.56 × 45 mm M193 or SS109
Operation: gas, selective fire with 3-round burst
Locking system: rotating bolt
Feed: 30-round box magazine (M16 pattern)
Weight, empty: 3.26 kg
Length: 980 mm (butt extended): 730 mm (butt folded)
Barrel length: 465 mm without compensator
Rifling: 6 grooves rh
Muzzle velocity: (M193) 960 m/s: (SS109) 920 m/s

Manufacturer
Daewoo Precision Industries, PO Box 37, Dongrae, Pusan.
Status
Current. Production.

5.56mm K1A1 carbine

This weapon is variously referred to as a carbine or as a sub-machine gun, the manufacturers generally preferring the latter term. Although similar to the K2 assault rifle in general appearance, it differs internally, since it uses the gas tube and direct gas impingement system of operation used by the M16 rifle. The gas blast drives a bolt carrier with an eight-lug rotating bolt. The safety/selector switch is similar to that described under the K2, having four positions. The barrel has a much larger muzzle brake/compensator, largely in

5.56 mm Daewoo K1A1 carbine with stock extended

order to reduce muzzle flash and blast due to firing a rifle cartridge in a short barrel. The foresight is provided with a luminous spot for improving accuracy at night.

DATA
Cartridge: 5.56 × 45 mm M193
Operation: Gas, selective fire
Feed: 20- or 30-round box magazines
Weight, empty: 2.88 kg
Length: (butt folded) 645 mm; (butt extended) 830 mm
Locking system: rotating bolt
Feed: 30-round box magazine (M16 pattern)
Barrel length: 263 mm without compensator
Rate of fire (cyclic): 700 — 900 rds/min
Muzzle velocity: 820 m/s

Manufacturer
Daewoo Precision Industries, PO Box 37, Dong-Rae, Pusan.
Status
Current. Production.

NORWAY

Norwegian rifles

The Norwegian Army was equipped with the original 6.5 mm Krag-Jørgensen rifle with a rimless cartridge. These rifles were developed and made at Kongsberg, Norway. The rifles were largely lost during the German Occupation in the Second World War.

After the war the Norwegian Army decided to replace American, British and German rifles by the G3 A3. The G3 was also produced at Kongsberg. It differs from the standard Heckler and Koch by allowing silent cocking. When required, the bolt is allowed forward under control and then finally closed by finger pressure on the grooves cut in the side of the bolt.

Manufacturer
Norsk Forsvarsteknologi A/S, PO Box 1003, N-3601 Kongsberg.
Status
Current. No longer in production.

Norwegian G3 rifle bolt closure

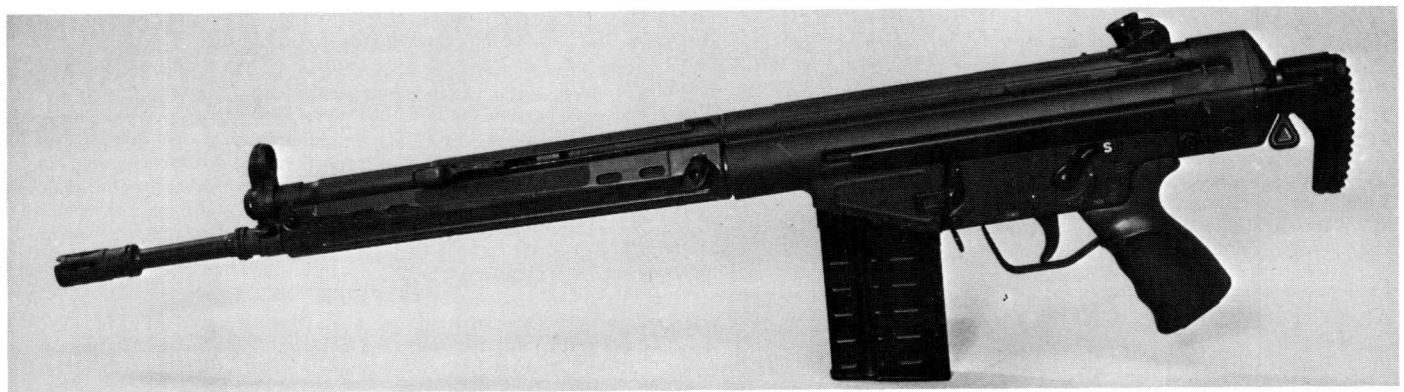

G3 A4 rifle manufactured by N.F.T.

G3 sniping rifle

7.62 mm NM149S sniping rifle

This rifle uses a Mauser M98 bolt action and has been developed in co-operation with the Norwegian Army and police forces. Intended for use out to a range of 800 m, it is fitted with a Schmidt & Bender 6 × 42 telescope sight which may be mounted and removed without altering the weapon's zero. Iron sights are also fitted, and the telescope sight may be replaced by a Simrad KN250 image-intensifying sight.

The stock is of impregnated and laminated beech veneer and may be adjusted for length by the use of butt-plate spacers; the Police model also has an adjustable cheek-piece. The rifle has a match trigger, adjusted to a 1.5 kg pull, and the weapon can be equipped with bipod and sound suppressor if required.

DATA
Cartridge: 7.62 × 51 mm NATO

Operation: bolt-action repeater
Feed: 5-round box magazine
Weight: 5.6 kg with telescope
Length overall: 1120 mm
Barrel length: 600 mm
Rifling: 4 groves, rh, one turn in 305 mm

Manufacturer
Våpensmia A/S, Box 86, 2871 Dokka.
Status
Current. Production.
Service
Norwegian Army and police forces.

7.62 mm NM149S sniping rifle

.22 VSM 004 Training rifle

The VSM 004 is a training and target practice rifle intended for use by the Youth Division of the Norwegian Home Guard.

The rifle is a bolt-action repeater, the receiver and action being the CBC 422 made by Companhia Brasileira de Cartuchoes of Brazil. This is fitted with a barrel and stocked-up by Våpensmia A/S. The stock is made from impregnated and laminated beech veneer and has been designed so that the appearance and shape approximate to the NM 149 sniping rifle (above). The butt-plate and spacer are equal to those used on the NM 149, and the sling is of 25 mm webbing and conforms to the requirements of the DFS (the Norwegian shooting organisation). The trigger is of match pattern adjusted to 1.4 – 1.7 Kp pull, and the safety catch is on the right side of the receiver.

DATA
Cartridge: .22 Long Rifle RF
Operation: bolt-action repeater
Feed: 5- or 10-round box magazine
Weight: 3.5 kg
Length: 1080 mm
Barrel length: 600 mm

Manufacturer
Våpensmia A/S, Box 86, 2871 Dokka.
Status
Current. Production.
Service
Norwegian Home Guard.

.22LR VSM 004 Våpensmia training rifle

POLAND

7.62 mm PMK assault rifle

The bolt-operated rifles used by Poland between the two World Wars were basically Mauser, and similar to the German Model 98 pattern weapons.

After the Second World War the Polish forces were equipped with the Mosin-Nagant Rifles but these were displaced by the Kalashnikov assault rifles as soon as these appeared.

The Polish-built copy of the Soviet AK-47, known as the PMK, does not differ significantly from the Soviet original. The variant described as the DGN-60 is described below.

Solid butt version of 7.62 mm PMK in training exercise. Man in background has PMKM

Polish PMK AK-47 folding butt version

7.62 mm PMK-DGN-60 assault rifle

This is the PMK modified for grenade launching. The muzzle has been made cone-shaped and threaded to take the LON-1 grenade launcher which is externally 20 mm in diameter.

The gas cylinder has been modified to take a gas cut-off valve which prevents gas from the grenade cartridge reaching the operating piston. A special grenade-launching sight is attached to the standard rearsight. A 10-round magazine, which will hold only grenade cartridges, is used for grenade launching. To absorb the considerable recoil the butt has been modified in shape and a boot fitted which goes into the firer's shoulder. The only mechanical change is the fitting to the rear of the breech cover of a securing catch to the return spring rod. This must be released before the rod can be pushed in to allow removal of the breech cover.

The Polish-made F1/N60 anti-personnel and the PGN-60 anti-tank grenades are used with the Polish PMK-DGN-60 rifle. It should be noted that these grenades have 20 mm internal diameters and will fit no other launcher. Similarly the conventional grenade with an internal diameter of 22 mm should never be used from the Polish rifle.

The head of the grenade screws on to the tail boom. The F1/N60 anti-personnel grenades have no fins on this boom but the PGN-60 does.

The grenade cartridge is white tipped and crimped. Under no circumstances should any other cartridge be used.

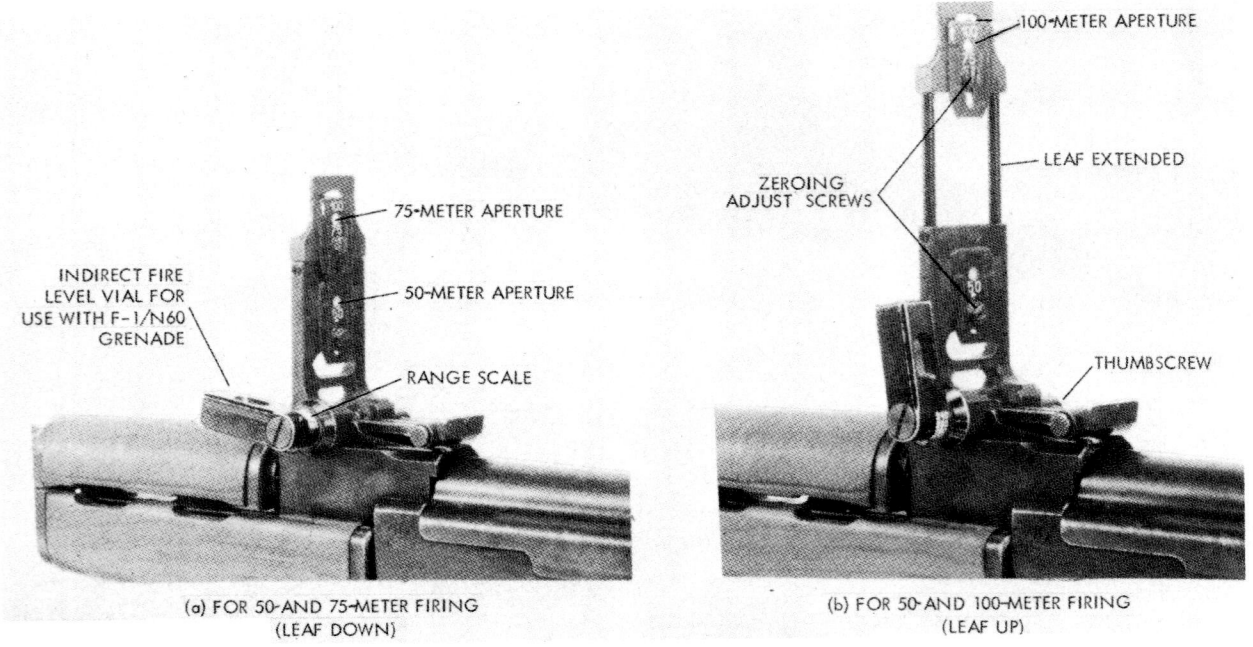

(a) FOR 50- AND 75-METER FIRING
(LEAF DOWN)

(b) FOR 50- AND 100-METER FIRING
(LEAF UP)

Grenade launching sights

Polish PMK-DGN-60 with LON-1 grenade launcher

Polish AK-47 Bulgarian rifle

The Polish arsenals produced a number of copies of the Soviet AK-47 which were manufactured for the Bulgarian Army. It is not known why this was done. It may have been purely commercial transaction or there may have been some political significance.

Bulgarian AK-47 rifle produced in Poland

7.62 mm PMKM assault rifle

This is the Polish copy of the Soviet AKM. Like the latest Soviet models it is fitted with a compensator at the muzzle. It cannot launch grenades and the muzzle is not coned.

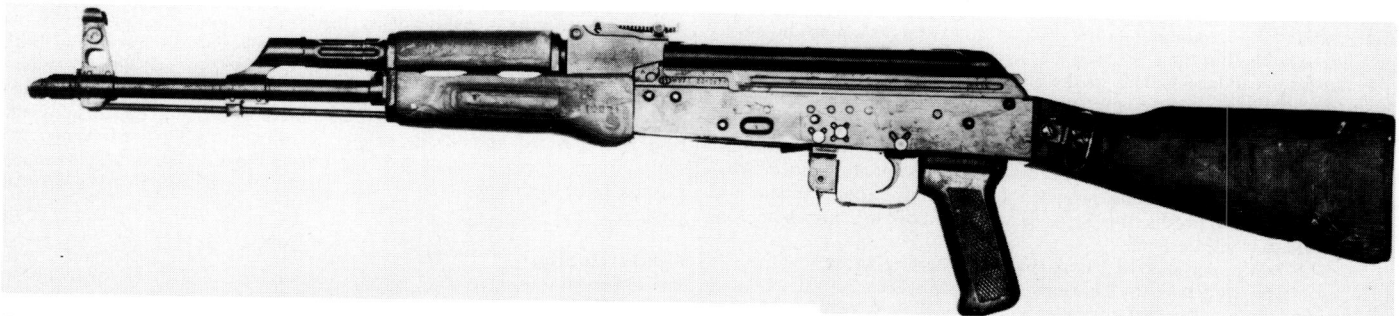

7.62 × 39 mm PMKM assault rifle copy of Soviet AKM showing compensator

PORTUGAL

Portuguese 7.62 mm G3A2, A3 and A4 Heckler and Koch rifles

Portugal manufactures the Heckler and Koch G3 rifles under licence. The rifles are identical with the West German models.

DATA
Cartridge: 7.62 × 51 mm NATO
Operation: delayed blowback, selective fire

Portuguese G3 A2 Heckler and Koch rifle

Feed: 20-round box magazine
Weight: (fixed butt) 4.30 kg; (retractable butt) 4.52 kg
Length: (fixed butt) 1020 mm: (retractable butt) 800 mm
Barrel: 450 mm, 4 grooves rh, one turn in 305 mm
Sights: (fore) post; (rear) V-notch 100 m, apertures for 200, 300 and 400m; adjustable for elevation and windage.
Sight radius: 572 mm
Rate of fire: 480-620 rds/min
Muzzle velocity: 780-800 m/s

Manufacturer
INDEP, Fabrica de Braco de Prata, Rua Fernando Palha, 1899 Lisboa-Codex.
Status
Current. Production.
Service
Portuguese and other armed forces.

ROMANIA

Romanian rifles

The Romanian Army has been equipped with the locally produced AK-47 copy, but this has now been relegated to militia and reserve use.

The current assault rifle is the copy of the Soviet AKM. This may be readily recognised because it has a wooden forward pistol grip made from laminated material. In addition the usual Romanian markings appear on the selector – 'FA' for full automatic and 'FF' on the lower position for single-shot.

Romanian AKM rifle

SINGAPORE

5.56 mm SAR 80 assault rifle

Chartered Industries of Singapore was founded in 1967 by the Ministry of Defence, Singapore to manufacture 5.56 mm small arms ammunition and to produce under licence the US 5.56 mm M16 assault rifle.

From its experience gained in the production of the M16, Chartered Industries of Singapore was appointed by the Singapore armed forces to develop a more cost-effective assault rifle. This rifle has to be simple, reliable and of rugged construction. It had to have performance characteristics at least equal to those of the M16 which had been used by the Singapore armed forces for over a decade. An additional proviso was that it must adopt modern and economical means of production. Sterling Armament of the United Kingdom was selected to be the designer.

The SAR 80 was designed by Frank Waters of Sterling Armament in early 1976. Prototypes were made by Chartered Industries of Singapore in 1978 and evaluated by the School of Infantry Weapons of the Singapore Armed Forces Training Institute. Modified models were then subjected to extensive troop trials and endurance evaluations to identify areas for further improvement. Design improvements were incorporated before commencement of full line production in 1980. To minimise cost of production, manufacturing concepts used in the production of the SAR 80 are based on the more economical production technologies known today. Approximately 45 per cent of components are made by sheet metal pressings. Another 40 per cent are standard parts readily available on the commercial market.

General design characteristics
The SAR 80 is a 5.56 mm magazine-fed, gas piston operated, air-cooled assault rifle. It is capable of both semi-automatic and automatic modes of fire from the closed bolt position. Both the US M16-type 20- and 30-round magazines can be used. The barrel is provided with a flash suppressor which serves as a grenade launcher and a front support for the bayonet.

The folding butt stock variant of the SAR 80 is easy to handle, especially for combat at close quarters and in confined areas. Combined with its low recoil and capability of automatic fire, the folded weapon is ideal for paratroopers and troops of armoured units.

The barrel, breech-block and butt stock are assembled in a straight line, and this construction improves controllability and reduces vertical dispersion in full automatic fire by minimising barrel climb. The SAR 80 is fitted with a spring-loaded firing pin to prevent firing if the rifle is accidentally jolted or dropped. Gas piston operation reduces fouling and this simplifies field maintenance. The gas regulator can be adjusted manually. This enables the SAR 80 to be operated under adverse conditions by releasing more return gas. Firing rifle grenades is also possible by sealing off the return gas completely.

Operation
The gas system uses a piston operating on to a breech-block carrier which carries the rotating breech-block. Rotation is achieved by means of a cam pin projecting out of the cam path in the breech-block carrier. The breech-block is prevented from rotating as the carrier is pushed forward to feed a round into the barrel chamber. However, when the feed is complete, the continued movement of the carrier rotates the breech-block and thus locks the breech-block to the barrel.

With the rifle armed, the hammer is controlled by a fire-control selector and an additional safety lever, the automatic sear. The automatic sear is so timed as to ensure that the breech-block is fully closed before the hammer can be released. In the unlikely event of a mechanical failure, and supposing the breech-block is not fully closed, the hammer will strike the rear of the carrier thereby dissipating its energy in pushing the carrier forward. No firing can therefore occur.

When firing a round, the carrier is forced backwards by the gas piston/push-rod. After a short delay, the cam slot on the carrier engages the cam pin in the breech-block. The breech-block unlocks, providing both primary extraction and loosening of the empty cartridge case in the chamber.

Once the carrier is forced back fully, the empty case is extracted and ejected and the return spring is compressed. The breech-block carrier runs out, the next round is fed and, if the trigger is still depressed, fired. When the ammunition supply is exhausted, the breech-block catch is automatically pushed up by the magazine follower to hold the breech-block carrier open at the rear.

Field stripping
The upper- and lower-receiver groups can be separated by disengaging the front and rear take-down pins. The soldier is not expected to further disassemble the lower receiver group. The breech-block carrier is removed from the upper receiver group by simply depressing the spring-loaded detent guide rod block with the thumb and first finger. The guide rod assembly is removed. The breech-block carrier group is then pushed towards the rear and the cocking handle removed. The top hand-guard can now be removed. The gas regulator and the gas piston, if desired, can be disassembled for cleaning. To remove the piston extension, the piston extension is depressed rearwards, tilted slightly to any direction away from the centre line and disengaged. This is the extent of field disassembly expected from a soldier.

Disassembly of the firing pin and the breech-block is only necessary on the SAR 80 when replacing a damaged component or when carrying out maintenance at the base. The firing pin is held in position by a locking pin. To disengage the firing pin and the breech-block, the locking pin is pushed across to free the firing pin freeing the breech-block at the same time.

Re-assembly of the SAR 80 is in the logical reverse sequence. For ease of re-assembly, the SAR 80 is so designed that incorrect re-assembly is impeded.

Accessories
A full range of accessories and options is available with the SAR 80: US M16-type ammunition magazines, 20 and 30 rounds; folding butt stock; telescopic sight; blank firing attachment; bayonet and scabbard; sling; and a cleaning kit.

DATA
Calibre: 5.56 mm
Ammunition: US M193 and M196
Operation: gas (piston, short stroke)
Method of locking: rotating bolt, multi-lug head
Feed: 20- or 30-round magazine
Weight: (unloaded) 3.7 kg
Length: (fixed butt) 970 mm; (butt folded) 738 mm
Barrel: 459 mm

5.56 mm SAR 80 assault rifle with 30-round box magazine and bayonet

5.56 mm SAR 80 assault rifle stripped

Sight radius: 541 mm
Rate of fire: (cyclic) 600-800 rds/min
Effective range: 400 m
Muzzle velocity: 970 m/s

Manufacturer
Chartered Firearms Industries Pte Ltd, 249 Jalan Boon Lay, Singapore 2261.
Status
Current. Production.
Service
Singapore armed forces.

5.56 mm SAR 80 assault rifle on bipod with 30-round box magazine

5.56 mm SR 88 assault rifle

The SR 88 is a lightweight gas-operated assault rifle which has been developed to meet all current military requirements. It has a four-position adjustable gas regulator which also serves as an anti-fouling device and which can be set to shut off gas action for firing grenades. Field maintenance is greatly simplified by the chrome-plated gas piston system which virtually eliminates carbon fouling of the bolt carrier. The straight-line layout allows excellent control and accuracy.

The receiver, which is an aluminium forging, houses the operating mechanism. The stock is of glass-reinforced nylon, and its length is adjustable by adding or removing the butt-pads. The rifle will function with the stock removed.

The bolt group assembly is in modular form and consists of the bolt, bolt carrier, buffer springs and rods. The bolt carrier and bolt are made from special high grade steel. The seven-lug bolt locks into the barrel, following a rotation of 22.5°, the rotation being controlled by a cam in the carrier. The bolt is always kept forward by the firing pin spring, a concept which permits rapid assembly after cleaning.

The upper receiver assembly comprises the upper receiver, barrel and carrying handle. The receiver is a high grade steel stamping and houses the bolt group assembly. The barrel is forged of high grade steel and features a chrome-plated chamber. The barrel is fitted with a flash suppressor that vents sideways and upwards and is internally threaded to accept a blank-firing attachment. The barrel is connected to the receiver by a locknut and detent, a system which simplifies barrel replacement.

The gas piston assembly consists of a chrome-plated steel regulator and piston-cum-pushrod. The pushrod and spring are shrouded by the hand-guards, leaving the venting of gas to the atmosphere. This keeps the hand-guard relatively cool and comfortable under sustained fire. The hand-guards are interchangeable — left and right are identical.

An effective safety feature is a spring-loaded firing pin which prevents inadvertent firing should the rifle be dropped or jarred. Other features include the provision for night aiming with a luminous front sight, the option of a folding butt-stock where compactness is desired, a carrying handle, an arctic trigger, and provision for mounting a sighting telescope. The cocking handle

Folding butt versions of the SR 88; (top) standard assault rifle, (lower) carbine

has an automatic lock to provide forward assist and silent bolt closure. A selector for full automatic or three-round burst fire can be fitted according to requirements. There is provision to mount a grenade launcher without the need to remove hand-guards.

The SR 88 is produced in fixed butt standard, folding butt standard and fixed and folding butt carbine versions.

DATA
Cartridge: 5.56 × 45 mm M193 or SS109
Operation: gas, selective fire and three-round burst
Method of locking: rotating bolt
Feed: 20- or 30-round magazine
Weight: (unloaded) 3.66 kg

Length: (fixed butt) 970 mm; (butt folded) 746 mm
Barrel: 459 mm
Rifling: 6-groove, rh, one turn in 32 or 55 calibres
Sight radius: 526 mm
Rate of fire: (cyclic) 650-850 rds/min
Effective range: 400 m
Muzzle velocity: 970 m/s

Manufacturer
Chartered Firearms Industries Pte Ltd, 249 Jalan Boon Lay, Singapore 2261.
Status
Current. Production.

SOUTH AFRICA

5.56 mm R-4 and R-5 assault rifles

The R-4 rifle, which is now replacing Fabrique Nationale's Fusil Automatique Léger (FN FAL) and G3 rifles as the South African Defence Force's standard infantry weapon, is a modified version of the Israeli Galil (which see). The modifications consist, in the main, of improvements in the strength of material and construction to better withstand the severe conditions of bush warfare. The butt has been lengthened, since South African soldiers are generally of larger stature than Israelis, and made from a reinforced synthetic material, and the hand-guard is similarly strengthened. A bipod is fitted as standard and there are some small changes in the receiver, gas piston and bolt assemblies in order to facilitate manufacture. The rifle is supplied with a 35-round magazine manufactured either from steel or from reinforced plastic material (nylon). The 50-round magazine is made from steel.

The sights incorporate tritium spots for night firing, set for 200 m engagement range.

Since its introduction the R-4 has been improved in various details, these improvements being due to experience gained in operations. The original butt was of aluminium covered with nylon; this has been replaced with a glass-filled nylon butt with a strengthening web halfway along its length, the result being stronger and more robust than the original. The metal magazine has been replaced by a nylon/glass-fibre design; this has little mechanical advantage but has reduced costs due to the abandonment of empty magazines in the heat of combat, and it is also much lighter than the metal version.

Other small changes include a redesigned gas tube lock, to prevent the tube shaking loose under the firing shock; the removal of the click adjustment from the foresight, thus allowing the mounting dovetail to be made stronger; and the introduction of a wider and stronger sear.

The most important improvement has been a redesign of the bolt. The old pattern was capable of firing a cartridge with a sensitive primer accidentally, by inertia of the firing pin as the bolt closed. A resilient retracting spring, of oil-proof polyurethene, now keeps the firing pin inside the bolt except when positively driven forward. It has the additional advantage of spreading the rotational torque on bolt closure, thus relieving the firing pin locating pin of excessive stress.

The R-4 assault rifle is also available in a short version, known as the R-5 assault rifle. It is identical to the R-4 except that it has a shorter barrel, no bipod and a shorter hand-guard.

The R-5 has been adopted by the South African Air Force and Marines. Semi-automatic versions of both the R-4 and R-5 are in production for use by police and paramilitary forces.

DATA
(DATA FOR R-5 IN PARENTHESES)
Cartridge: 5.56 × 45 mm
Operation: gas, selective fire
Locking system: rotating bolt
Feed: 35- or 50-round box magazines
Weight: (empty) 4.3 kg (3.7 kg); (with 35-round magazine) steel 4.6 kg (4.0 kg); nylon 4.44 kg (3.84 kg)
Length: (butt extended) 1005 mm (877 mm); (butt folded) 740 mm (615 mm)
Barrel: 460 mm (332 mm)
Rifling: 6 grooves. 1 turn in 305 mm
Rate of fire: (cyclic) 650 rds/min
Muzzle velocity: 980 m/s (920 m/s)

Manufacturer
Lyttleton Engineering Works (Pty) Ltd (Armscor subsidiary), Pretoria.
Status
Current. Production.
Service
South African Defence Force.

5.56 mm R4 assault rifle

SPAIN

7.62 mm Models A, B (58), C and E CETME assault rifles

When the NATO countries decided to standardise the 7.62 mm × 51 cartridge for small arms, the Spanish authorities decided that it would be advisable to use the same calibre for Spanish military weapons. It was considered, however, that the standard NATO round was unnecessarily powerful. Accordingly, a new cartridge was developed having the same external dimensions as the NATO round but having a lighter bullet, a smaller charge and thus a lower muzzle velocity and energy than the NATO round. The characteristics of this round are compared in the table below with those of the standard NATO round, the short Mauser round and the Compañia de Estudios Técnicos de Materiales Especiales (CETME) round developed for its 7.92 mm rifle. It should be noted, however, that the external dimensions of the two 7.62 mm cartridges were made the same despite the other differences.

DATA

	Light CETME	Short Mauser	CETME/ NATO	NATO
Calibre	7.92 mm	7.92 mm	7.62 mm	7.62 mm
Bullet				
weight	6.8 g	12.8 g	7.3 g	9.4 g
length	46 mm	35 mm	31 mm	32.5 mm
Weight				
charge	1.8 g	2.9 g	1.8 g	2.8 g
cartridge	18 g	27.4 g	19 g	24.5 g
Muzzle velocity	800 m/s	750 m/s	760 m/s	780 m/s
Recoil impulse	7.26 Ns	12.36 Ns	7.26 Ns	10.2 Ns

Models A and B
CETME developed a series of rifles in 7.62 mm calibre. The first of these, the Model A, entered production in 1956 and was identical in most respects with the 7.92 mm Model 2 described in *Jane's Infantry Weapons 1978*, page 164. It was followed, in 1958, by the Model B, also known as the Model 58, which incorporated several improvements including a grenade-launching fitment; the elimination of the hold-open feature whereby the gun was fired from the open-breech position (the objection to which was its mechanical complexity and the disturbance to the firer's aim as the action went forward); and the compensating additions of a metal jacket to assist cooling and serve as a foregrip in place of the folded bipod. Model B was taken into service with all Spanish armed forces in 1958, and in the same year the licence previously held by NWM in the Netherlands was transferred to West Germany and the development of what became the G3 rifle was put in hand there.

7.62 mm Model E CETME assault rifle

In 1964 it was decided that the Spanish forces should adopt the standard NATO 7.62 mm × 51 cartridge in place of the CETME/NATO round. Before this more powerful round could be used, however, certain modifications to the Model B/Model 58 rifle were necessary, resulting in the Model C.

Model C
This is a selective fire weapon using a 20-round detachable box magazine (five-round boxes are available for training) and using the roller-locked delayed blowback system developed for the earlier weapons. The rifle is fitted with a combined flash eliminator/grenade launcher attachment, may be fitted with a bayonet, a telescopic or night sight and has a case for cleaning tools in the cocking handle guide tube.

Accessories include a bipod, magazine filler, and a blank firing attachment which replaces the flash hider.

Operation
The weapon is loaded by inserting the leading edge of a loaded magazine into the front of the magazine-well and rotating it rearwards until the magazine catch engages (to remove the magazine the catch must be released either by pushing the magazine catch lever forward or by pressing the push-button behind the magazine-well). With the magazine in position set the selector lever to either 'T' (semi-automatic) or 'R' (automatic) and pull the cocking handle (on the left of the rifle) fully to the rear and release it. When it is released, the bolt moves forward, strips a round from the magazine, chambers it and cocks the action, and the weapon is ready to fire. To render the weapon safe move the selector lever to 'S'. When the weapon is thus cocked the cocking lever may be folded down.

If required the weapon may also be set in a safe condition with the bolt to the rear. To achieve this the first part of the cocking process is performed as before but then, instead of the cocking lever being released, it may be turned upwards into a recess at the rear of the guide tube, whereafter the action may be locked as before by setting the selector lever to 'S'.

The arrangement of the roller locking device is shown (in the locked position) in the accompanying drawing. When the round is fired, the pressure on the bolt-head is transmitted to the bolt carrier through the rollers. Until these rollers are clear of the recesses in the barrel extension the locking piece and bolt carrier are caused to move backwards faster than the bolt-head, thus giving a substantial mechanical advantage to the return spring (not shown) and introducing the required initial delay. When the rollers are clear of the recesses the bolt-head, locking piece and carrier move rearwards together, extracting the spent case, ejecting it and then returning to chamber a fresh round.

Stripping
Remove the magazine.

Draw the cocking handle to the rear, ensure that no round remains in the chamber and release the cocking handle.

Remove the two attaching pins from the butt stock and store them in the hole drilled in the butt stock for the purpose.

Withdraw the butt stock and return spring.

Tilt the handgrip forwards, draw the bolt to the rear by means of the cocking handle and withdraw the bolt through the rear of the receiver.

To strip the bolt, press the bolt-head back until it is stopped by the carrier, then rotate it 180° anti-clockwise and remove it.

Rotate and remove the locking piece.

Extract the firing pin and spring.

No further field stripping should be attempted. Assembly is in the reverse order. After re-assembly the action should be checked by cocking and firing on an empty chamber before inserting a magazine.

7.62 mm Model A CETME assault rifle

7.62 mm Model C CETME assault rifle

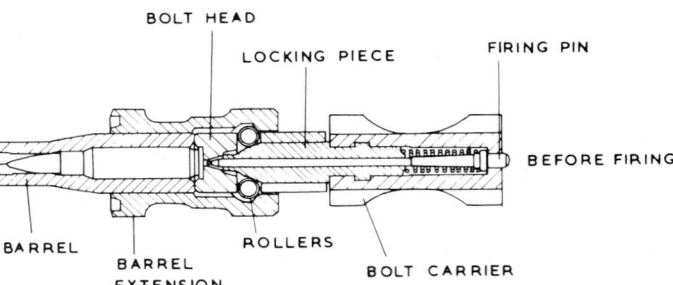

CETME delay action and locking piece

7.62 mm Model B (Model 58) CETME assault rifle

Model C
Cartridge: 7.62 mm × 51 NATO
Operation: delayed blowback using rollers
Feed: 20-round box magazine
Fire: selective
Weight: (with wooden hand-guard) 4.2 kg; (with bipod and metal hand-guard) 4.5 kg

Weight of empty magazine: 0.275 kg
Length: 1015 mm
Barrel: 450 mm
Rifling: 4 grooves rh.Twist 305 mm
Sights: (foresight) protected conical post; (rearsight) turning leaf with V-notch for 100 m and aperture leaves for 200, 300, and 400 m
Sight radius: 580 mm
Muzzle velocity: 780 m/s
Rate of fire: (cyclic) 550-650 rds/min
Effective range: 600 m

Model D
The designation Model D was applied to an intermediate development stage between Models C and E and was used only within the CETME organisation.

Model E
This was the last in the 7.62 mm CETME series and embodied several improvements over the Model C. These included improved ejection, improved sights and the introduction of a plastic handgrip. An experimental quantity was made but the weapon was not placed in production.

Development agency
Compañia de Estudios Técnicos de Materiales Especiales, (CETME), Julian Camarillo 32, 28037 Madrid
Manufacturer
Empresa Nacional 'Santa Barbara' de Industrias Militares SA, Julian Camarillo 32, 28037 Madrid
Status
Models A and B: no longer made and obsolete
Model C: production and service in Spain
Model D: never in production
Model E: experimental only
CETME rifles, among others, have been supplied at various times to Denmark, Norway, Pakistan, Portugal and Sweden for trials

7.62 mm Model R firing port weapon

This is a variant of the Model E (above) using the same basic mechanism but without a butt and with a shorter barrel. A flash hider is fitted to the muzzle, and there is a substantial collar forward of the receiver which locks into a firing port mounting (Rotula CETME RH 762/BMR) installed in the vehicle sidewall. The mounting allows a traverse of 50° and elevation from -20° to +30°. The position of the locking collar and mount means that the usual forward cocking handle of the Model E cannot be used, and a modified cocking device, using the MG3 machine gun cocking handle, is fitted. Fitting the weapon to the mount takes only four or five seconds.

The Model R fires automatic only, from a 20-round magazine.

DATA
Cartridge: 7.62 × 51 mm NATO
Operation: roller-delayed blowback

Feed: 20-round magazine
Weight: 6.40 kg
Length: 665 mm
Barrel length: 305 mm
Muzzle velocity: 690 m/s
Rate of fire: 500 – 600 rds/min

Manufacturer
Compañia de Estudios Técnicos de Materiales Especiales, (CETME), Julian Camarillo 32, 28037 Madrid.
Status
Development.

5.56 mm Models L and LC CETME assault rifle

This selective fire weapon in 5.56 mm calibre embodies Compañia de Estudios Técnicos de Materiales Especiales' (CETME's) extensive experience in developing its range of similar weapons in 7.92 mm and 7.62 mm calibres.

Two versions of the weapon have been designed: a standard model with a fixed butt stock (Model L) and a short-barrelled version with a telescopic

5.56 mm CETME Model L assault rifle

stock (Model LC). A 20-round magazine was first used, but the design was then modified to accept the standard 30-round M16 pattern magazine. Early rifles had a four-position sight graduated for 100, 200, 300 and 400 m ranges. Current models have a simple two-position flip-over aperture sight graduated for 200 and 400 m. Fore and rearsights are provided with luminous markers for firing in poor light.

Operation of the automatic mechanism is by a delayed blowback system substantially the same as that used on earlier CETME weapons, delay being achieved by the use of rollers as before.

Trigger mechanism
In general principle, the operation of the trigger mechanism for single shots (T) or normal automatic fire (R) is similar to that of the West German G3 rifle: similarly the safety setting (S) of the selector lever provides a conventional impediment to the operation of the trigger. Early models had a fourth (r) setting, giving a regulated three-round burst. This is no longer standard, since it has been found that the firer can easily regulate the length of the burst by using the trigger. However, if the three-round burst facility is desired, it can be fitted at the factory as an optional extra.

Stripping
The stripping procedure is similar to that employed with previous CETME designs, and applies to both L and LC models. Begin by removing the magazine and ensuring the weapon is unloaded.
Remove the two attaching pins in the butt stock, remove the butt stock and release the return spring and its guide.
Pull back the cocking handle and remove the bolt assembly through the rear of the receiver.
Withdraw the handgrip pin to its full extent (about 1.5 mm). Remove the safety/selector switch. Slide the handgrip down and to the rear and remove it.
Remove the hand-guard pin; slide the hand-guard forward and down to remove it.

5.56 mm CETME Model LC short assault rifle

5.56 mm CETME Model LC short rifle with stock telescoped

5.56 mm CETME Model L assault rifle

DATA
(Characteristics peculiar to the Model LC shown in parenthesis)
Cartridge: 5.56 × 45 mm NATO
Operation: delayed blowback, selective fire
Delay: rollers and locking lever
Feed: 12- or 30-round detachable box magazine
Weight: (unloaded) 3.40 kg
30-round magazine: (empty) 0.21 kg
Length: 925 mm (665 or 860 mm)
Barrel: 400 mm (320 mm)
Rifling: 6 grooves, rh, one turn in 178 mm
Sights: (foresight) protected conical post; (rearsight) flip-over aperture for 200 and 400 m; (sight radius) 440 mm
Muzzle velocity: 875 m/s (832 m/s)
Rate of fire: (cyclic) 600-750 (650-800) rds/min

Manufacturer
Santa Barbara, Julian Camarillo 32, 28037 Madrid.
Status
Current. Production.

To dismantle the bolt assembly, force the bolt-head towards the support until they touch, then give the bolt-head a half turn anti-clockwise and pull it away from the support. Turn the firing pin carrier through 90° and remove it, together with the firing pin and spring.
Re-assembly is performed in the reverse manner

7.62 mm C-75 special forces rifle

This is a conventional bolt-action rifle, using a Mauser action, intended for sniping, police and special forces tasks. Fitted with iron sights as standard, it also has telescope mounts machined into the receiver and can thus be equipped with virtually any optical or electro-optical sight. With a telescope, match ammunition and a skilled firer, it is capable of effective fire out to 1500 m range. It is also provided with a muzzle cup launcher for discharging various riot control devices, such as rubber balls, smoke canisters or CS gas grenades, by use of a standard grenade-discharging cartridge. It is not suited to firing combat types of grenade.

No data other than the weight, 3.7 kg, was available at the time of going to press.

Manufacturer
Santa Barbara, Julian Camarillo 32, 28037 Madrid.
Status
Production; available.

7.62 mm C-75 special forces rifle

SWEDEN

7.62 mm AK4 FFV rifle

The Swedish AK4 rifle, which is essentially the same as the standard West German G3, is manufactured by the FFV Ordnance Division at Eskilstuna.

Manufacturer
FFV Ordnance, Eskilstuna.
Status
Current. No longer in production.
Service
Swedish Army.

7.62 mm AK4 FFV rifle

5.56 mm AK5 assault rifle

The Swedish Army began looking for a new light, intermediate calibre, rifle in the mid-1970s, with a view to the eventual replacement of the AK4. Most existing 5.56 mm weapons were tested for reliability, endurance, accuracy, handling and other capabilities. The results of these studies led to the elimination of most designs until only two remained, the licence-built

FFV-890C (a slightly modified Israeli Galil – it will be found described in *Jane's Infantry Weapons 1984/85*, page 199 and the Belgian FN-FNC.

After comprehensive troop and technical trials in 1979/80 the FFV-890C was rejected in favour of the FN-FNC, which, it was considered, had 'developable characteristics'. Further tests were carried out in the ensuing

5.56 mm AK5 assault rifle, showing new hand-guard, gas block, arctic trigger-guard and cocking handle

5.56 mm AK5, showing modified butt, sights

years and the prototype FNC rifles were continuously modified to fill the particular Swedish requirements. In the course of this development period changes were made to the butt stock and butt stock lock, bolt, extractor, hand-guard, gas block, sights, cocking handle, magazine, selector switch, trigger-guard and sling swivels; the three-round burst facility was removed; and a surface finish of sand-blasting, Parkerising and finally baking on a dark green enamel was devised. However in all essential operating features (except the three-round burst) the AK5 remains similar to the FNC, and reference should be made to the description of that weapon on page 138.

DATA
Cartridge: 5.56 × 45 mm NATO
Operation: gas, selective fire
Method of locking: rotating bolt
Feed: 30-round box magazine

Weight, empty: 3.90 kg
Length: (stock extended) 1008 mm; (stock folded) 753 mm
Barrel: 450 mm
Rifling: 6 grooves rh, 1 turn in 178 mm
Sights
foresight: protected post
rear sight: flip aperture, 250 m and 400 m
Muzzle velocity: 930 m/s
Rate of fire: 650 rds/min
Accessories: bayonet, cleaning set, blank-firing adapter, optical sight

Manufacturer
FFV Ordnance, Eskilstuna.
Status
Current. Production.
Service
Swedish Army.

SWITZERLAND

7.62 mm SG 510-4 SIG rifle

In the early 1950s SIG's director Rudolf Amsler produced the AM55 which was a blowback-operated rifle using a roller-action delay and was based to some extent on the German Sturmgewehr 45 assault rifle. This rifle was adopted by the Swiss Army as the Stgw 57 and replaced the Schmidt Rubin bolt-operated rifle.

The SG 510 series was developed from the AM55 and consisted of:
the SG 510-1 firing the 7.62 mm × 51 NATO cartridge;
the SG 510-2 firing the same cartridge but lighter in construction with a correspondingly higher recoil;
the SG 510-3 firing the Soviet 7.62 mm × 39 cartridge and
the SG 510-4 using the 7.62 mm × 51 NATO round.
Only small quantities of the SG 510-3 were produced and this version is not discussed further here. A brief note on the differences between the SG 510-4 and the Stgw service rifle follows this entry.

Description
The SG 510-4 is a delayed blowback-operated rifle utilising a two-part block and delay rollers to hold up the breech face.

The body of the gun is made from pressings wrapped round and hard-soldered. The rear end of the receiver has been strengthened to take the butt, with the return spring, and the front end has a strengthened section to house the breech and the roller seatings. The roller seatings may be taken out and replaced when worn.

The barrel is made by the cold swaging or 'hammering' process. It has integral gas rings for grenade launching off the muzzle, and a muzzle brake, which also acts as a flash suppressor, is incorporated. The rear part of the barrel is surrounded by a perforated casing. A bipod is suspended behind the foresight and can be stowed by folding the legs and rotating the bipod until the legs lie along the top of the barrel.

The foresight is a pillar and the rearsight an aperture. The rearsight has press locks which allow the frame to be moved up or down the sight bed to the selected range. The front sight mount can be moved across in a dovetail for lateral zeroing. For elevation zeroing an Allen key is used to screw the pillar in or out.

A long arctic trigger is permanently fitted. It is also used for grenade firing. When not in use it hinges up flat under the receiver. When pushed down it has a short projecting arm which operates the normal trigger.

A blank firing device can be fitted over the muzzle for training. A bayonet can be fitted to the lug under the collar below the foresight. A ring fits over the muzzle brake.

The rifle will fire grenades using a blank cartridge. These are usually contained in a special magazine which takes the place of the normal one. This helps to ensure a ball round is not fired into the grenade.

Operation
A loaded magazine is placed in the magazine-well and pushed home until the catch clicks. The gun is loaded by pulling back and then releasing the T-shaped cocking handle; the bolt comes back, the hammer is cocked and held by the sear. On its forward travel the bolt chambers the cartridge and the gun is ready for action. The bolt is a two-part type with a light bolt-head and a heavier body. The bolt-head carries two light rollers. These are of unusual shape since they are not simply cylindrical in section but have small pivoting pieces. These are engaged in the bolt-head. When the round is chambered the bolt-head movement ceases and the bolt body, moving forward, is able to pass between the rollers. The front face of the bolt body is wedge-shaped and the rollers are forced out into recesses in the receiver.

When the cartridge is fired the bullet is forced up the bore and the cartridge case is driven back. The chamber is fluted and the case floats on a film of gas. A proportion of these flutes extend the full length of the chamber and appear at the chamber face, so that upon firing a small amount of high pressure gas is allowed to leak and strike the bolt-face. There are two small holes in the bolt-face through which this gas passes, through the bolt-head, to impinge on the bolt body. The pressure of the cartridge case and the gas on the breech face forces the rollers to the rear face of their recesses and the angle of the recesses forces the rollers inwards against the inclined faces of the bolt body. Due to the angle of the wedge and the pressure of the leaked gas, the bolt body is accelerated back and the two parts of the bolt are separated. When the rollers are clear of the recesses the entire bolt is drawn back with a displacement

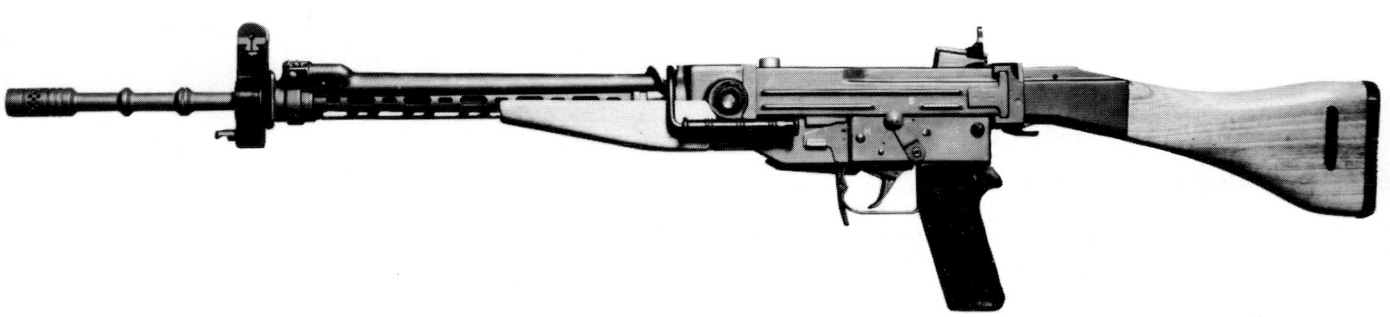

7.62 mm SG 510-4 SIG rifle

Sniping version

between the two parts. The empty case is held to the breech face by the extractor. The ejector is of an unusual type. It is a rocking system attached to the top of the bolt-head and as the bolt recoils the ejector contacts a ramp on the left wall of the receiver and the case is pushed through the ejection slot on the right of the receiver. This method is less violent than the usual fixed ejector system.

The return spring is compressed. The bolt comes up against the rear plate and the compressed return spring forces it forward. The bolt picks up the next round and chambers it.

Trigger and firing mechanism
The trigger mechanism is of simple design and unlike most modern rifles does not owe its origin to Garand's design for the M1 .30-06 rifle. It is based on the provision of a sear with an elongated slot through which passes an axis pin. A spring tends at all times to force the sear forward. The trigger is a pressing with a step at the rear. When the trigger is rotated the step lifts the rear end of the sear and the nose moves down out of the notch in the hammer and the hammer rotates forward to fire a round.

Semi-automatic fire
As soon as the hammer has gone forward the sear is pushed forward by its spring and it moves off the step of the trigger and lies on the flat portion in front of it. When the hammer is rotated it is caught by the sear engaging in the notch. The powerful hammer spring pushes the sear back against the front face of the step. Another round cannot be fired until the trigger is released. As soon as the trigger step drops, the sear is able to move back over the top of it. Pulling the trigger lifts the tail of the sear, depresses the nose, and another round is fired, provided the bolt is fully home and the rear part has closed fully up to the front part.

Automatic fire
When the change lever is rotated to 'automatic', the step at the rear of the trigger is able to rise considerably higher and this depresses the nose of the sear much further when it moves out of the hammer notch. As a result, when the hammer comes back it has no contact at all with the sear. It is caught by a quite separate safety sear and held back until the rear part of the two-part block comes forward to push the rollers out into their recesses. This final forward movement depresses the safety sear and releases the hammer. Thus the firing is controlled by the bolt closing position which satisfies the requirement for mechanical safety before firing. As soon as the trigger is released the sear moves up and catches the hammer on its next rearward rotation, and holds it back.

When the selector lever is rotated to 'safe', the shaft rotates over the rear of the trigger and prevents it from rising.

The hammer is mounted on the left of the receiver very close to the wall. It is therefore not operating down the centre line of the bore and it cannot reach the firing pin. Instead it hits the short end of an L-shaped bar which reaches right across the breech-block and is pivoted on the far side from the hammer. The centre of the transverse bar hits the firing pin and this in turn hits the cap of the case.

Stripping
Remove the magazine and check that the weapon is unloaded.

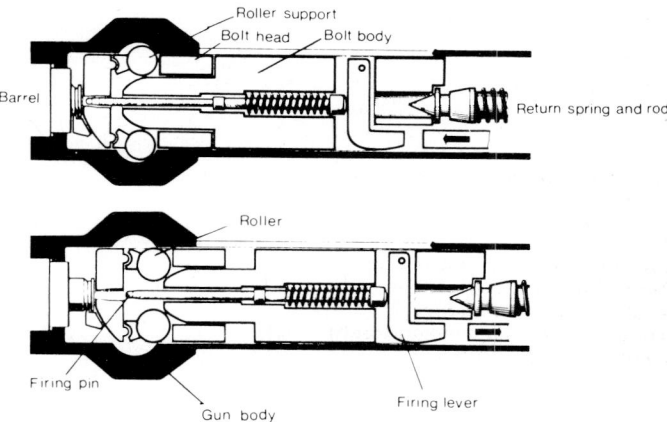

Before and after firing

Lift the catch under the receiver, behind the pistol grip, then rotate the butt anti-clockwise and pull it off to the rear.
Pull back the cocking handle and withdraw the bolt from the receiver.
Press the release pin to the right and remove the pistol grip and trigger assembly.
Separate the two parts of the bolt.

DATA
SG 510-4
Cartridge: 7.62 mm × 51 NATO
Operation: delayed blowback, selective fire
Feed: 20-round magazine
Weight, empty: 4.25 kg
Length: 1016 mm
Barrel: 505 mm
Rifling: 4 grooves rh. 1 turn in 305 mm
Sights: (fore) post; (rear) aperture
Sight radius: 540 mm
Muzzle velocity: 790 m/s
Rate of fire: (cyclic) 600 rds/min

ACCESSORIES
Telescopic sight, blank firing attachment, cleaning kit, night sights, ATk grenade sight

Manufacturer
SIG Swiss Industrial Company, CH-8212 Neuhausen Rheinfalls.
Status
No longer in production.
Service
In service as Stgw 57 with Swiss Army (see following entry), Chile and Bolivia.

SIG-Sauer SSG 2000 sniping rifle

The SIG-Sauer SSG 2000 has been purpose-built for military, law-enforcement and target marksmanship. The heart of this weapon is based on the design concept of the internationally proven bolt action of the Sauer 80/90 repeating rifle. This bolt-action uses hinged lugs at the rear end of the bolt which are driven outwards, to lock into the receiver, by the action of cams driven by the rotation of the bolt handle. The bolt body is non-rotating, giving very positive case extraction. This design results in a reduction in the angular travel of the bolt to only 65° and gives a fast and smooth loading action.

The heavy hammer-forged barrel is equipped with a combination flash hider and muzzle brake, providing excellent weapon recovery allowing a fast, controlled, follow-up shot.

The ergonomic thumb-hole stock can be optimally adjusted to suit the firer; right- and left-hand stocks are available.

The trigger is of the double set pattern. The sliding safety catch has a triple function: blocking of the sear, the sear pivot, and the bolt itself. The bolt can be opened when the safety catch is applied; the set trigger can be de-cocked by pulling the trigger when the weapon is in the safe condition, and it is automatically de-cocked when the bolt is opened. There is a signal pin to indicate when a round is in the chamber.

No iron sights are fitted; standard equipment includes a Schmidt & Bender 1½—6 × 42 or Zeiss Diatal ZA 8 × 56T telescope sight.

The SIG-Sauer SSG 2000 is available in 7.62 × 51 mm, .300 Weatherby Magnum, 5.56 × 45 mm and 7.5 × 55 Swiss calibres; the data which follows is for the 7.62 × 51 mm version, though the other calibres differ very little.

SIG-Sauer SSG 2000 sniping rifle

Detail of accessory bipod and adjustable sling swivel support, SSG 2000 rifle

DATA
Cartridge: 7.62 × 51 mm (see text)
Operation: Bolt action repeater
Feed: 4-round box magazine
Weight, with magazine and sight: 6.60 kg
Length overall: 1210 mm
Barrel: 610 mm excluding flash hider

Rifling: 4 grooves, rh, one turn in 40 calibres
Trigger pull: 18 N (standard); 3 N (set)
Muzzle velocity: 750 m/s depending upon ammunition

Manufacturer
SIG Swiss Industrial Company, CH-8212 Neuhausen Rheinfalls.
Service
Swiss police forces, Argentine Gendarmerie National, Police force of Kingdom of Jordan, Royal Hong Kong Police, Malaysian Police Kuala Lumpur, Combined Service Forces of Taiwan, some British police forces etc.

7.5 mm Stgw 57 service rifle

This is the Swiss Army version of the SG 510-4 SIG from which it differs in calibre and in several minor respects. The rifle is used with the Swiss 7.5 cartridge (Swiss Army designation GP 11) and it has a rubber butt stock, folding sights and a special bipod which can be used to support the weapon at either the rear or the front of the jacket.

For a general description of the weapon mechanism reference should be made to the preceding entry for the SG 510-4. The Stgw 57 has proved to be a

very reliable weapon even in the extreme conditions that can be encountered in Switzerland.

Manufacturer
SIG Swiss Industrial Company, CH-8212 Neuhausen Rheinfalls.
Service
Swiss forces.

7.5 mm Stgw 57 Swiss service rifle

7.62 mm Gruenel sniping rifle

This is a bolt-action repeating rifle specifically designed and built for long-range precision. It is the first rifle in the world to employ ball-bearing supports to aid bolt opening and closure, it uses a new and patented action bedding system, and is fitted with a fully adjustable electronic trigger system giving consistent pressures and crisp action. The butt-plate and cheek-piece are fully adjustable and the walnut stock is slotted for cooling. The accuracy life is approximately 10 000 rounds.

The barrel is drawn from special steel and precision lapped for accuracy. It is fitted with a muzzle brake to reduce the felt recoil. No iron sights are fitted, but the receiver is prepared for a telescope mount.

An adapter is supplied to convert the trigger to mechanical action if desired.

DATA
Cartridge: 7.62 × 51 mm NATO
Operation: bolt-action repeater
Feed: 10-round magazine
Weight: 5.30 kg without sight
Barrel length: 650 mm
Rifling: 4 grooves, rh, one turn in 46.5 calibres

Manufacturer
ITM; Industrial Technology & Machines AG, Allmendstrasse 31B, CH-4502 Solothurn.
Status
Current. Production.

7.62 mm Gruenel sniping rifle

540 Series SIG assault rifles

The French firm Manurhin was licensed to manufacture three types of SIG rifles, the SG540, 542 and 543. The SG540 and 543 are chambered for the 5.56 mm × 45 cartridge and the SG542 for the 7.62 mm × 51 NATO round; but all three are made to the same basic design thus giving them a high parts commonality. In 1988 the license operated by Manurhin was transferred to INDEP of Portugal, but in 1989 the only licensed production is in Chile, where FAMAE manufacture the SG542.

The rifles are gas-operated with a rotating bolt. The cocking handle reciprocates with the bolt and provides a means of closing the bolt if for some reason the return spring is unable to force it fully home. The gas regulator is mounted at the front of the cylinder and has a milled cap which can be rotated to one of the positions marked respectively 0, 1 and 2. At 'O' the valve is fully closed, no gas passes to the cylinder, and the entire gas force is used to project a grenade from the muzzle. The normal firing position is with the regulator set to '1', and '2' is an oversized port reserved for those occasions when the weapon is lacking in energy due to fouling or the entry of sand or snow, for example.

The weapon can be fired at single-shot, full automatic or with a three-round burst controller which provides for the release of three rounds at a cyclic rate of 725 rds/min from a single trigger operation. When firing at single-shot, accuracy is improved by the provision of a double pull pressure point trigger.

When the weapon is used in extremely cold conditions demanding the use of Arctic mittens, the trigger-guard can be rotated so that the hand has direct access to the trigger. The trigger-guard is similarly rotated when grenades are fired.

When the ammunition is expended the magazine platform operates a holding-open device. When a new, loaded, magazine is inserted the action can be released by a slight backward movement of the cocking handle and when this is released the bolt will chamber the next round. Alternatively the device may be released by pressing up a catch on the left of the receiver above and behind the magazine.

The foresight is a pillar which may be rotated up or down to allow for elevation zeroing. The rearsight is an aperture in a tilted drum which may be rotated to give ranges of 100 to 500 m for the 5.56 mm rifle and 100 to 600 m on the SG542. Each rifle has range alteration in increments of 100 m.

A bipod is mounted under the front of the barrel casing. This is normally stowed under the barrel and is pulled down when needed.

The flash suppressor is the closed prong-type and when grenades are used these are slid straight down on to the barrel. The spike bayonet has a tubular handle which fits over the flash suppressor.

A telescopic sight, an infra-red sight or an image intensifier can be fitted.

The use of plastics for the butt, trigger-guard, pistol grip and fore-end has reduced the weight and cost of the weapon without detracting from its efficiency in any way.

A conventional butt or a tubular one folding flat alongside the right side of the receiver can be fitted.

Stripping

The magazine is removed by pressing on the catch in front of the trigger-guard and the bolt retracted by pulling on the fixed cocking handle on the right. The chamber and feedway can then be inspected.

There is a take-down pin above the pistol grip just behind the change lever and when this is pressed through the butt can be forced down with the lower receiver and the trigger group. The cocking handle is pulled back to bring the bolt to the rear. The handle is then withdrawn sideways from its slot and the bolt withdrawn.

7.62 mm SG 542 assault rifle with 20-round magazine

From top: 5.56 mm SG 540 with 30-round magazine; SG 540 with folding stock and 20-round magazine; SG543 with folding stock and 20-round magazine

DATA
Cartridge: SG540 and 543, 5.56 mm × 45; SG542, 7.62 mm × 51
Operation: gas, selective fire
Method of locking: rotating bolt
Feed: magazine

WEIGHTS
Rifle: (with fixed butt) SG540, 3.26 kg; SG542, 3.55 kg; SG543, 2.95 kg; (with folding butt) SG540, 3.31 kg; SG542, 3.55 kg; SG543, 3 kg
Additional for bipod: SG540/542, 0.28 kg
Empty magazine: (20 rounds) SG540/543, 0.2 kg; SG542, 0.24 kg; (30 rounds) SG540/543, 0.24 kg; SG542, 0.35 kg
Full magazine: (20 rounds) SG540/543, 0.43 kg; SG542, 0.73 kg; (30 rounds) SG540/543, 0.585 kg; SG542, 1.085 kg

LENGTHS
Overall: (with fixed butt) SG540, 950 mm; SG542, 1000 mm; SG543, 805 mm; (with folding butt) SG540, 720 mm; SG542, 754 mm; SG543, 569 mm
Barrel: (without suppressor) SG540, 460 mm; SG542, 465 mm; SG543, 300 mm

MECHANICAL FEATURES
Barrel: (rifling) SG540/543, 6 grooves; SG542, 4 grooves rh. 1 twist in 305 mm. 2-position regulator
Sights: (foresight) cylindrical post; (rearsight) aperture; (sight radius) SG540, 495 mm; SG542, 528 mm; SG543, 425 mm

FIRING CHARACTERISTICS
Muzzle velocity: SG540, 980 m/s; SG542, 820 m/s; SG543, 875 m/s
Rate of fire: (cyclic) 650/800 rds/min
Effective range: 600 m

Manufacturer
SIG Swiss Industrial Company, CH-8212 Neuhausen Rheinfalls and licensed to Manurhin SA, France.
Status
Current. Production in Chile only (see text).
Service
Bolivia, Burkina Faso, Chad, Chile, Djibouti, Ecuador, France, Gabon, Indonesia, Ivory Coast, Lebanon, Mauritius, Nicaragua, Nigeria, Oman, Paraguay, Senegal, Seychelles, Swaziland.

5.56 mm SG550/551 SIG assault rifle

This rifle was developed to meet a Swiss Federal Army specification and was accepted as the Swiss Army assault rifle in early 1984 under the type designation Stgw 90.

In designing the SG550 SIG paid particular attention to weight saving, making extensive use of plastics for the butt, hand-guard and magazine. The magazine is of transparent plastic, allowing a check to be kept on the ammunition supply even when firing and it is provided with studs and lugs on the side which allow a number of magazines to be clipped together so that changing from an empty to a full magazine is greatly facilitated. The butt folds to one side and it is claimed that, even in this configuration, the weapon's optimum weight distribution and ergonomically convenient shape allow accurate firing in all positions. SIG states that accuracy at 300 m is equal to that of the Stgw 57.

Like the 540 series SIG-Manurhin the SG550 is provided with a three-round burst controller and a trigger-guard that can be rotated when rifle grenades are fired or so that a user wearing Arctic mittens can have direct access to the trigger. When the last round in the magazine is fired a hold-open device operates; a new magazine is inserted, the lock catch is actuated and the weapon is ready to fire once more.

The combined dioptre and alignment sight mounted on the breech housing is adjustable for traverse and elevation. The alignment sight has luminous spots for aiming during night firing, and when the daylight sight is adjusted the night firing sight is adjusted simultaneously. The foresight, with its tunnel and swing-up nightfiring sight with luminous spot, is permanently mounted. Telescopic and infra-red sights can be fitted to an integral telescope mount; this is dimensioned to Swiss Army standards, though a mount to NATO STANAG 2324 can be fitted if required.

SG 550/551 SP

This is a commercially available variation of the SG 550 or 551 which is also known in Switzerland as the Stgw 90 PE. It is intended as a sporting weapon, though it could have applications for security forces who do not require the full-automatic fire facility. It is dimensionally and ballistically the same as the SG 550 service weapon except that it is restricted to semi-automatic fire. Some 20 000 of these weapons have been sold to the civilian market.

DATA	SG550 Standard version	SG551 Short version
Cartridge	5.56 mm	5.56 mm
Operation	gas, selective fire	gas, selective fire

5.56 mm SIG SG550 field-stripped, with accessories

5.56 mm SIG SG550 assault rifle showing folded bipod and triple magazine assembly

5.56 mm SIG SG 550 right side showing folded bipod

5.56mm SIG SG 551 SP, short version. Note studs on magazine which allow two or more to be clipped together

5.56mm SIG SG 550 SP commercial version with telescope sight

Method of locking	rotating bolt	rotating bolt	Barrel	528 mm	372 mm
Feed	magazine	magazine	Rifling	6 grooves	6 grooves
			Sight radius	540 mm	466 mm
WEIGHTS			Rate of fire (cyclic)	approx 700	approx 700
Rifle				rds/min	rds/min

Manufacturer
SIG Swiss Industrial Company, CH-8212 Neuhausen Rheinfalls.
Status
Current. Production.
Service
Swiss Army (SP version, for target shooting).

WEIGHTS		
Rifle		
with empty magazine and bipod	4.1 kg	—
with empty magazine	—	3.5 kg
Empty magazine		
20 rounds	95 g	95 g
30 rounds	110 g	110 g
Full magazine		
20 rounds	340 g	340 g
30 rounds	475 g	475 g
LENGTHS		
Overall		
with butt extended	998 mm	827 mm
butt folded	772 mm	601 mm

5.56 mm SIG SSG 550 Sniping rifle

This has been developed from the standard SG550 assault rifle and is a semi-automatic version developed in close co-operation with special police units. The accuracy is due to the heavy hammer-forged barrel and low recoil, and there is a sensitive double-pull trigger. The semi-automatic action gives the sniper the ability to fire a very fast second shot should the occasion require it.

The rifle carries a fully adjustable bipod, a fully adjustable folding butt, and the pistol grip is adjustable for rake and has an adjustable hand-rest. The sighting telescope can be adjusted alongside and at right-angles to the sight base in such a way that the head leans naturally against the cheek rest and takes up the correct position relative to the line of sight. There are no iron sights fitted.

DATA
Cartridge: 5.56 × 45 mm
Operation: Gas, semi-automatic
Locking: rotating bolt
Feed: 20- or 30-round box magazine

SIG SG550 Sniper field-stripped, with accessories

5.56 mm SG550 Sniper, with telescope sight

SIG SG550 Sniper with night vision sight fitted

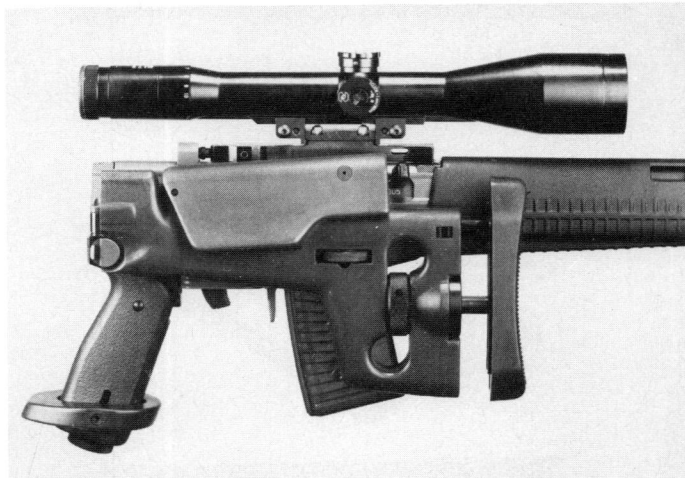

SIG SG550 Sniper, detail of folding adjustable butt

Weight: 7.02 kg empty, without magazine
Weight of magazine: empty 95 g; with 20 rounds 340 g
Length: 1130 mm butt extended; 905 mm butt folded
Barrel length: 650 mm
Rifling: 6 grooves, rh, one turn in 254 mm
Trigger pull: first pressure 800 g; second pressure 1.5 kg
Accessories: Spare magazine; sling; mirage band; bipod; carrying case; cleaning kit; sight cleaning equipment

SIG SG550 Sniper, with accessories, in carrying case

Manufacturer
SIG Swiss Industrial Company, Small Arms Division, CH-8212 Neuhausen Rheinfalls.
Status
Current. Production.
Service
Swiss police; Kingdom of Jordan police.

TAIWAN

5.56 mm Type 65 assault rifle

This generally resembles the M16 and the firing characteristics are markedly similar, as are many of the dimensions. The rifle is made entirely in Taiwan, probably on tools and machinery intended for the M16.

The lower receiver was taken directly from the M16A1, while the bolt, bolt carrier and piston-type gas system come from the AR-18. Initial Type 65 prototypes were made with stamped steel receivers, but later weapons were machined from blocks of aluminium.

The barrel is the same length as the M16 and gives a similar muzzle velocity with M193 ammunition. The receiver has the cocking handle of the original M16 without the bolt-closure device that was incorporated in the M16A1. The general shape of the receiver pressings are all but identical to the M16, even to the arctic trigger-guard. The missing feature is the carrying handle, and the rearsight is mounted on a substantial bracket in place of the rear part of the carrying handle. The foresight appears to be the same as the M16, as does the plastic butt. Twenty- or thirty-round M16 magazines are used.

A major recognition feature is the long plastic hand-guard which is somewhat larger than that on the M16. A light bipod can be fitted to the barrel under the foresight as required; this is not a normal fixture.

A modified version of the original Type 65 design incorporates the M16A1-type tube gas system and a more reliable, single-piece bolt assembly. It also has a higher rate of fire and can use transparent plastic magazines. This later model was previously (and incorrectly) referred to as the Type 68.

Although the Type 65 initially proved to be neither reliable nor easy to manufacture much effort was put into eradicating the various problems. The Type 65K1 incorporated modifications, notably improving the hardness of the aluminium buffer and the inadequacy of the heat insulation beneath the hand-guard. This was followed by the Type 65K2, which is rifled one turn in 178 mm to accept the SS109 bullet and also has a three-round burst facility in addition to full automatic fire. The gas system has also been redesigned in order to slow up the primary ejection of the spent case and thus overcome problems of case ejection which had occurred with the earlier models. The Type 65K2 has now undergone extensive testing and is in production. The bayonet used on the Type 65 rifle is different from the standard US M7. It is 8¼ in (209 mm) in overall length, 1½ in (38 mm) longer than the M7. Fully-bladed on the lower edge and 4¾ in (120-mm) bladed on the upper edge, the Taiwanese bayonet has a lined handle for a secure grip. It is designated the Type 65 bayonet.

DATA
Calibre: 5.56 mm × 45 M193 or SS109 (see text)
Feed: M16 20/30 round detachable box magazine
Weight: 3.17kg
Length: (overall) 990 mm
Barrel: 508 mm
Rifling: 4 grooves, rh. one turn in 178 mm or 304 mm (see text)
Rate of fire: (cyclic) 700-800 rds/min
Effective battle range: 400 m
Muzzle velocity: 990 m/s
Muzzle energy: 1745 J

Taiwan 5.56 mm Type 65 assault rifle demonstrating three-round burst firing

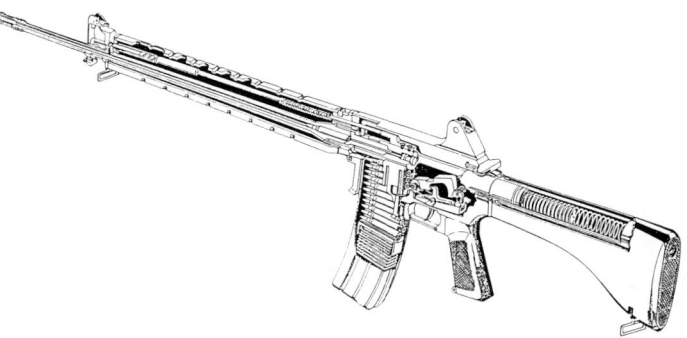

Sectioned view of Type 65 assault rifle

Manufacturer
Combined Service Forces, Hsing-Ho Arsenal, Kaohsuing.
Status
Current. Production.
Service
In service with Taiwan Army Airborne/Special Forces, military police, and Taiwan Marine Corps.

TURKEY

7.62 mm G3 rifle

This is the Heckler & Koch G3 rifle (described under West Germany) manufactured under licence in Turkey. From information supplied by makers it appears that different manufacturing methods have resulted in slight changes in dimensions. The Turkish G3 is issued with a locally designed bayonet which fits above the muzzle.

Licence-built G3 rifle by MKE

DATA
Cartridge: 7.62 mm × 51 NATO
Operation: delayed blowback, selective fire
Length overall: 1020 mm
Length of barrel: 450 mm
Weight: (empty) 4.25 kg
Weight: (with full magazine) 5.00 kg
Feed: 20-round box magazine
Muzzle velocity: ca 780-800 m/s
Rate of fire: 480-620 rds/min
Effective range: 400 m

Manufacturer
Makina ve Kimya Endustrisi, Kurumu, Ankara.
Status
Current. Production.

UNION OF SOVIET SOCIALIST REPUBLICS

7.62 mm Mosin-Nagant rifle

This rifle was designed by the Belgian Nagant brothers, Emile and Leon, and Colonel Sergei Mosin of the Russian artillery. The two Nagants and Colonel Mosin had both designed rifles of their own and the final Russian rifle of 1891 appears to be the action of Mosin and the magazine system of the Nagant brothers. The weapon was made to fire the Soviet three-line (7.62 mm) cartridge, the 'line' being a Soviet measurement of length equal to $1/10$ in or 2.54 mm.

The rifle was stocked almost to the muzzle and took a socket bayonet; it was originally sighted for the round-nosed 7.62 mm bullet and when the spitzer-type light bullet (L) was introduced in 1908, the sight was changed.

The bolt of the Mosin is of original and unique design. The removable bolt-head rotates with the bolt body and locking is achieved by two lugs on the bolt-head moving into recesses in the receiver. When the bolt is locked the two locking lugs are horizontal.

The cocking occurs as the cocking piece is held back by the sear as the bolt goes forward. Safety is achieved by pulling back on the cocking piece and rotating it anti-clockwise. This moves a lug into a recess in the bolt and also withdraws the striker point through a cylinder in the bolt-head; the striker point has two flats and so when turned it cannot pass between the flats in the inside of the cylinder to move towards the cartridge cap.

The magazine is very unusual in that when the top round has been chambered, the second round is held down and prevented from rising under the bolt until the final rotation of locking removes the feed interruptor and the cartridge is pressed up under the bolt.

There were a number of models following the original model of 1891, at least some of which may still be encountered, particularly in South-east Asia.

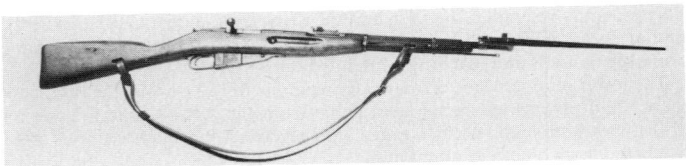

7.62 mm Mosin-Nagant M1944 carbine

DATA
Cartridge: 7.62 mm × 54R
Operation: manual, bolt
Method of locking: rotating bolt
Feed: 5-round internal box
Method of fire: single shots

Manufacturer
Various state arsenals, including Sestorets and Tula.
Status
Obsolete.
Service
Likely to be found anywhere, but not in first-line service with any regular army.

	Rifle M1891/30	Sniper rifle M1891/30	Carbine M1938	Carbine M1944
Weight	3.95 kg	5.05 kg	3.47 kg	4.03 kg
Length	1232 mm	1232 mm	1016 mm	1016 mm
Barrel	729 mm	729 mm	508 mm	518 mm
Sights				
foresight	post, tangent	post, tangent	post, tangent	post, tangent
rearsight	notch	notch	notch	notch
Muzzle velocity	811 m/s	811 m/s	766 m/s	766 m/s
Effective range	800 m	800 m	800 m	800 m

7.62 mm Simonov self-loading rifle (SKS)

Sergei Gavrilovich Simonov designed an anti-tank rifle, the PTRS, of 14.5 mm calibre in the Second World War and after the war produced a scaled-down version of it which was a light self-loading short rifle, or carbine, and was the first weapon to fire the new 7.62 mm × 39 M43 cartridge.

It is a gas-operated rifle of conventional design with a charger-loaded 10-round box magazine enclosed inside the receiver. There is a catch below the receiver, behind the magazine, which when pressed releases the bottom plate of the magazine to allow quick emptying. It has a permanently attached, folding blade bayonet.

It is now obsolescent in the Soviet Army, but has been manufactured by several communist countries and is still in use with several of these. In East Germany the rifle is known as the Karabiner-S; it can be identified by the year of manufacture and the serial number on the left front of the receiver and the fact that there is a hole through the butt for the sling. It does not carry the cleaning rod. In Yugoslavia it is called the M59/66 and is recognised by the permanently attached grenade launcher at the muzzle and the folding grenade sight behind the launcher. The Chinese PLA call it the Type 56 semi-automatic carbine and the later version has a triangular section bayonet in place of the flat blade. The symbol is carried on the left front of the receiver. The North Korean forces use the name 'Type 63' carbine and the receiver cover has '63' stamped in it.

Operation

To load the rifle the cocking handle on the right of the bolt is retracted. This is permanently attached to the bolt. If the magazine is empty the bolt will be held to the rear. The ammunition comes in 10-round chargers and is placed in the charger guides in the front of the bolt carrier and pressed fully down. When the 10 rounds are in the magazine the charger is removed; the bolt is pulled

slightly back and released and goes forward to chamber the top round. The magazine can, if necessary, be loaded or topped up as required, by using individual rounds.

The safety catch is along the rear of the trigger-guard and is pushed forward and up for 'safe' where it indicates its presence by obstructing the trigger finger as well as blocking the trigger.

Pulling the trigger releases the hammer which drives the firing pin into the cartridge cap. Some of the gases following the bullet up the bore are diverted through the gas port and impinge on the head of the piston. The piston is forced rearwards and the tappet strikes the bolt carrier. It is a short stroke action and the spring returns the tappet and piston to their forward position. The carrier is driven back and after about 8 mm of free travel, during which the gas pressure drops, it lifts up the rear end of the bolt out of engagement with the floor of the receiver. The bolt assembly then moves rearward as an entity. The hammer is rocked back to its cocked position and the return spring progressively compressed. The extractor holds the empty case to the bolt-face until it contacts the ejector and is expelled through the port on the right of the receiver.

The return spring drives the bolt assembly forward; the bolt picks up the top round in the magazine and chambers it. The extractor enters the cannelure of the cartridge case and bolt motion ceases. The carrier continues forward for about 8 mm and forces the rear of the bolt down into its recess in the receiver.

Trigger and firing mechanism

When the trigger is squeezed the trigger bar pushes the spring-loaded sear block forward and clear of the hammer. Provided the bolt carrier is fully forward, and the bolt is locked, the hammer rotates and hits the firing pin. The round is fired. The bolt moves rearward and rotates the hammer back. The

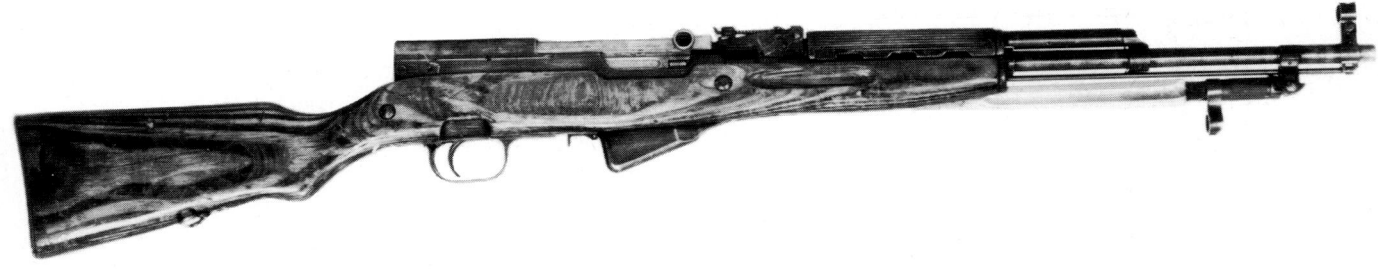

7.62 mm Simonov self-loading rifle

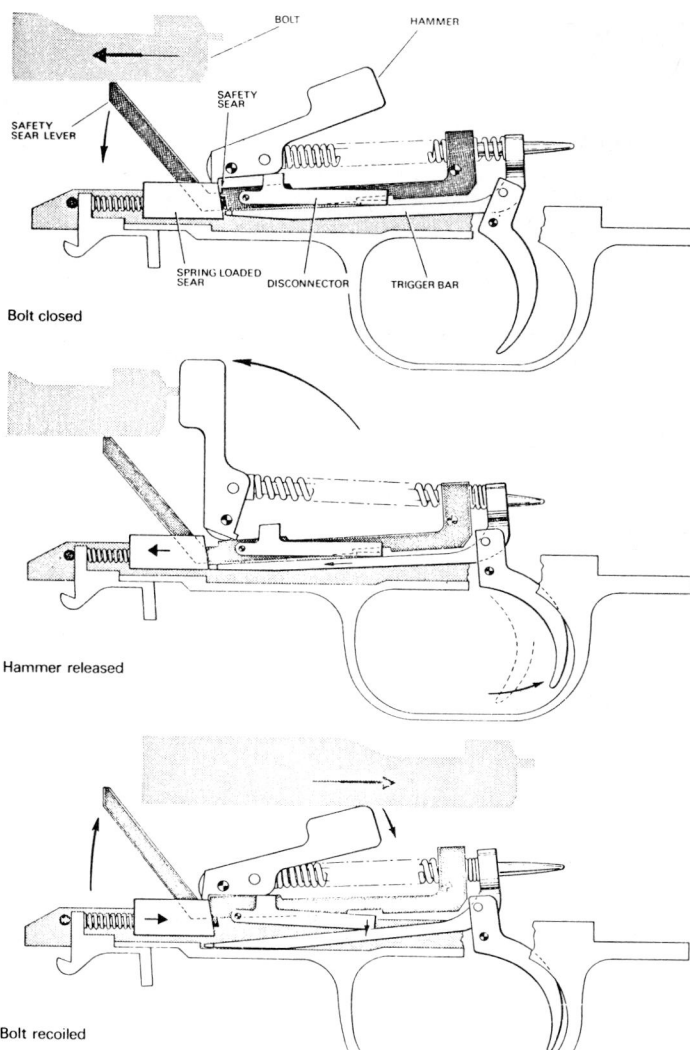

Bolt closed

Hammer released

Bolt recoiled

Trigger and firing mechanism of SKS

underside of the hammer forces down the disconnector which in turn pushes down the trigger bar to a level below the sear. The sear spring forces it back under the hammer and over the trigger bar. The hammer is held and cannot move. To fire another round the trigger must be released. The trigger bar is forced to the rear and the hammer/trigger spring forces it up so that it lies against the rear surface of the sear. When the trigger is pulled the trigger moves forward and pushes the sear forward against its spring, the hammer rotates and the cycle is repeated. Unless the bolt carrier is fully forward the safety sear holds up the hammer but the final closure of the carrier depresses the safety sear which is then moved clear of the hammer.

The rifle continues to fire single shots until the ammunition is expended. The magazine platform then rises and a small stud pushes up a bolt-retaining catch which holds the bolt to the rear. When the magazine platform is depressed by the insertion of ammunition, the catch continues to hold the bolt to the rear until the bolt is pulled slightly back, when the catch drops, and is released to feed the top round.

Stripping
The magazine catch is pulled back to swing the magazine open and so remove the cartridges. The bolt is retracted and the chamber checked.

The receiver cover pin on the right rear of the receiver is rotated to the vertical position and pulled as far as possible out of the rifle. The receiver cover can now be brought back and the return spring removed. The bolt carrier and bolt come back when the cocking handle is pulled rearwards and can be lifted out of the receiver and separated.

The gas cylinder tube retaining pin can be rotated upwards and the rear of the upper hand-guard lifted. The gas cylinder tube and piston can then be removed. The piston will drop out of the tube.

DATA
Cartridge: 7.62 mm × 39
Operation: gas, self-loading
Method of locking: tilting block
Feed: 10-round internal box magazine
Weight, empty: 3.85kg
Length: 1021 mm
Barrel: (rifling) 4 grooves rh
Sights: (foresight) post; (rearsight) tangent, notch
Muzzle velocity: 735 m/s
Effective range: 400 m

Manufacturer
State factories.
Status
Obsolete.
Service
No longer used in the USSR, except for ceremonial purposes, having been replaced in first-line service by the AK-47 (see entry below). Still used in some Asian countries.

7.62 mm AK-47 assault rifle

The Kalashnikov design has become one of the most important and widespread weapons in the world. It has been the standard rifle of the Soviet Army since the early 1950s and rapidly replaced all other rifles and sub-machine guns that were then in service. It was standard with all first-line units and was also adopted by the Soviet Navy and Air Force ground troops.

More significantly it has become the weapon of all the Eastern Bloc countries and most communist countries elsewhere. Practically all the communist-inspired guerrilla and nationalist movements seem to have been given stocks of it. Manufacture has taken place several of the Warsaw Pact arsenals in Eastern Europe and in China and it is impossible to estimate how many of these rifles have been made, nor how many are in use, but at a very rough guess it must be in excess of twenty million. Rifles manufactured outside Soviet Russia frequently exhibit slight detail changes, and reference should be made to the Warsaw Pact country entries for details of these.

General
There are some features of the AK series which are a departure from the conventional but in general it is a simple robust and handy weapon of sound design and generally excellent workmanship.

AK-47 night sight beneath rearsight notch

The AK-47 assault rifle is a compact weapon, capable of selective fire, robust and reliable. The Warsaw Pact countries have now largely replaced it with the AKM, a modernised and improved version, but it is still held for mobilisation in European Communist countries.

It is supplied in two configurations, one with a rigid butt and one with a double-strut folding metal butt stock controlled by a simple press button release above the pistol grip.

There are also early and late production models. The early model has a built-up receiver which has an angular shape to the rear end, sloping noticeably down to the butt. The later type has a straight receiver.

The various countries in the Warsaw Pact have produced a variety of materials for butt stocks and forearms ranging from laminated sheets of plywood to various types of plastic and all-metal construction. On the whole the standard of material employed for furniture is not up to that of American or West European weapons.

Operation
The magazine is loaded and positioned in the rifle by placing the front end into the magazine-well until the lug engages and then rotating the rear end back and up until the magazine catch engages. The cocking handle is permanently fixed to the right of the bolt carrier and reciprocates with it. This allows manual bolt closure if for some reason the bolt carrier does not go fully forward. If the cocking handle is retracted and released, the top round is fed into the chamber. When the trigger is operated the hammer hits the floating firing pin and drives it into the cap of the cartridge. Some of the propellant gases are diverted into the gas cylinder on top of the barrel. There is no gas regulator. The piston is driven back and the bolt carrier, built into the piston extension, has about 8.5 mm of free play while the gas pressure drops to a safe level. A cam slot in the bolt carrier engages the cam stud on the bolt and the bolt is rotated through 35° to unlock it from the receiver. There is no primary extraction during bolt rotation to unseat the case and so a large extractor claw is fitted which grips the empty case and holds it to the bolt-face until it contacts the fixed ejector formed in the guide rail and is thrown out of the right-hand side of the gun. As the bolt travels back it rocks the hammer over and also compresses the return spring. The bolt is brought to a halt by hitting

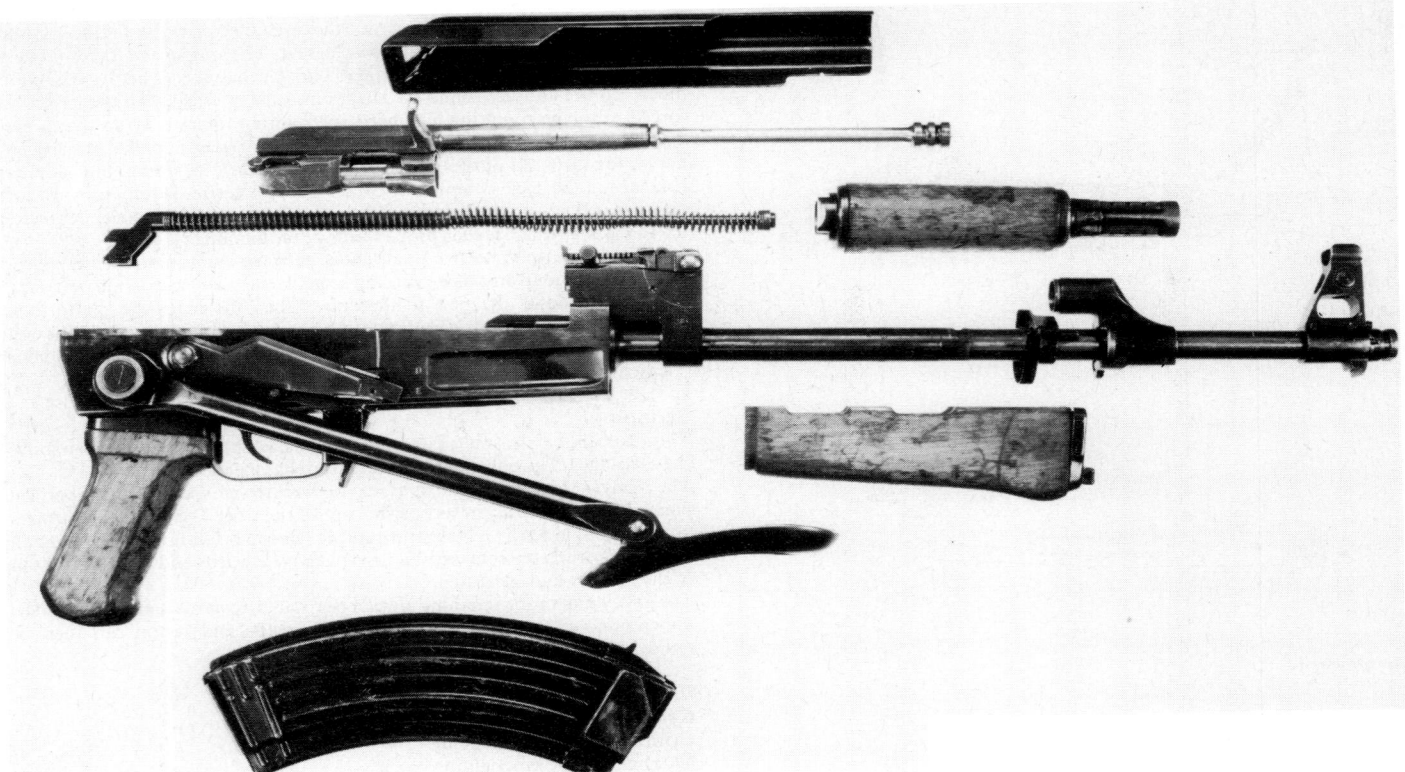

7.62 mm AK-47 assault rifle field-stripped

the solid rear end of the receiver. The return spring drives the bolt forward, another round is chambered, and the bolt comes to rest. The carrier continues on for about 5.5 mm after locking is completed. During this last forward movement of the carrier the safety sear is released and control of the hammer is returned to the trigger sear.

AK-47 night sight folded into place behind foresight blade

When the trigger is pressed, the hammer is released from the trigger sear and goes forward to fire the round. The recoiling carrier rotates the hammer back and it is held by the safety sear. As soon as the carrier is fully forward the safety sear is disengaged, the hammer is freed, and another round is fired. So long as the trigger is pressed, ammunition is available in the magazine, and the bolt carrier goes fully forward, the gun will continue to fire at its cyclic rate.

The change lever is mounted on the right-hand side of the receiver and is unique in its design. It is a long pressed-out bar pivoted at the rear and applied by the firer's thumb. The top position is 'safe'. Here it locks the trigger and physically prevents the bolt from coming back sufficiently to pass beyond the rear of a cartridge in the magazine, but does allow sufficient movement for checking that the chamber is clear.

The lever produces automatic fire at the centre position and single shots when fully depressed. It is invariably stiff to operate, noisy in functioning and extremely difficult to manipulate when wearing arctic mittens.

Trigger and firing mechanism

The trigger and firing mechanism is based on that of the US M1 Garand rifle. The hammer has two working surfaces on the bent and there are two sears which are hook shaped. When the hammer is cocked and the weapon is ready to fire, the main bent, which is the forward one, is held by the trigger sear which is part of the trigger lever. When the trigger is pulled the trigger lever rotates forward and the sear is disengaged from the hammer which is free to rotate, under the influence of its spring, into the firing pin.

When a single shot is set on the change lever, the hammer, when it is rocked back by the recoiling bolt carrier, is caught by the spring-loaded auxiliary sear. Since the trigger is already back another round cannot be fired. When the trigger is released the trigger lever carrying the main sear moves back and catches the hammer as it is released when the auxiliary is rotated clear of the secondary bent. Thus control of the hammer is restored to the main sear and provided the bolt carrier is fully forward, pulling the trigger will release the hammer to fire another round.

To provide mechanical safety, which implies that the bolt must be fully locked before the cartridge can be fired, a safety sear is fitted. Until this is disconnected by the carrier, in its final forward movement after locking is completed, it engages the hammer and holds it up.

When the change lever is set to automatic fire, a boss on the shaft presses against the spring-loaded auxiliary sear and it is forced back so far that it plays no part in controlling the hammer.

Soviet 7.62 mm AK-47 rifle, dated 1951

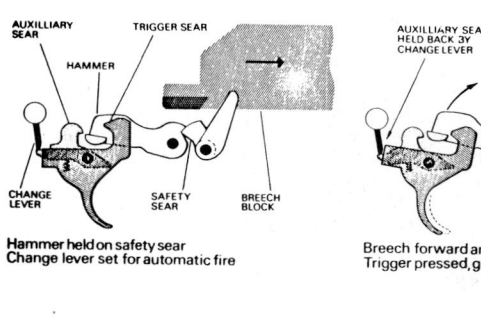

Hammer held on safety sear
Change lever set for automatic fire

Breech forward and locked
Trigger pressed, gun fires

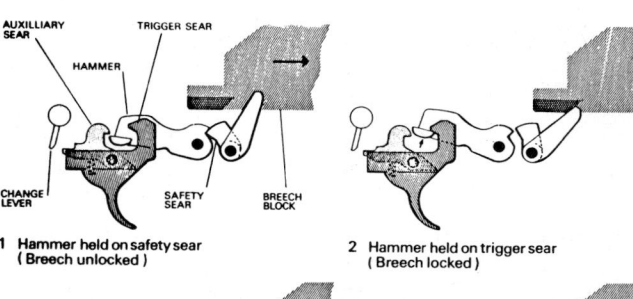

1 Hammer held on safety sear
(Breech unlocked)

2 Hammer held on trigger sear
(Breech locked)

3 Trigger pressed; hammer released,

4 Hammer held on auxilliary sear

Trigger and firing mechanism of 7.62 mm AK-47 assault rifle

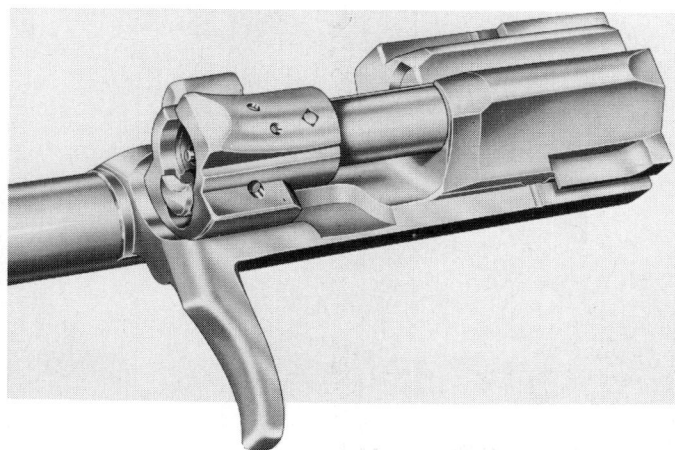

7.62 mm AK-47 assault rifle bolt

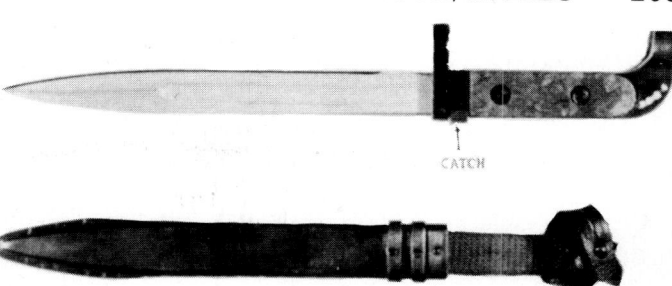

AK-47 bayonet

Sights

The foresight is a post, screw-threaded for adjustment when zeroing. The spanner in the combination tool kit is used for this. Lateral adjustment for zeroing is achieved by moving the foresight block in a dovetail.

The rearsight is an open U-shaped notch which will allow ranges up to 800 m by means of a slide and ramp. There is a battle sight setting for all ranges up to 200 m.

Stripping

Remove the magazine, check that the chamber and feedway are clear.
Press the end of the return spring guide into the rear end of the receiver cover and lift the receiver cover off the receiver.
Push the return spring guide forward to clear its rear housing and remove from the gun.
Pull the cocking handle to the rear and remove the bolt carrier and bolt.
Remove the bolt from the carrier.
Rotate the gas cylinder lock, mounted on the right of the rearsight block, and free the gas cylinder and upper hand-guard.

DATA
Cartridge: 7.62 mm × 39
Operation: gas, selective fire
Method of locking: rotating bolt
Feed: 30-round detachable box magazine
Weight: 4.30 kg
Length: (butt folded) 699 mm; (butt extended) 869 mm
Barrel: 414 mm
Rifling: 4 grooves, rh, one turn in 235 mm
Sights: (fore) post, adjustable; (rear) U-notch, tangent, adjustable to 800 m with battle sight for 200 mm
Sight radius: 376 mm
Muzzle velocity: 710 m/s
Rate of fire: 600 rds/min (cyclic)
Effective range: 300 m

Manufacturer
State factories.
Status
Current. Not in production.
Service
Has been replaced in first-line Warsaw Pact service by the AKM (see entry below). Widely used in Asian countries and in Egypt and Syria. Still used in reserve units in European Communist countries. The weapon and minor variants have been made in China (Type 56), Finland (M60 and M62), East Germany (MPiK, MPiKS), Hungary, North Korea (Type 58), Poland (PMK), Romania and Yugoslavia (M70 and M70A). See separate entries under country headings for variants.

7.62 mm AKM assault rifle

The AKM is a modernised version of the AK-47, produced in 1959. The forged and machined receiver of the AK-47 has been replaced with a body of pressed steel construction with riveting employed extensively to join the 1 mm thick U section to the inserts which house the locking recess, the barrel bearing, and the rearsight block at the front and the butt at the back. It should be noted that the bolt now locks into a sleeve and not directly into the barrel as in the AK-47. The slides on which the breech-block reciprocates are pressed out and spot welded inside the receiver walls. The results of these changes are reduced manufacturing costs and reduction in weight from 4.3 to 3.13 kg.

The AKM is produced with a wooden stock. It is also made with a folding butt stock and is then known as the AKMS.

There are several features which distinguish the AKM from the AK-47 and allow visual recognition (see illustration above). These are:
(a) There is a small recess in each side of the receiver centrally over the magazine. This is a magazine guide.
(b) The lower hand-guard has a groove for the firer's fingers.
(c) The receiver cover has transverse ribs.
(d) The bayonet lug under the gas tap-off point.
(e) The four gas escape holes on each side of the gas cylinder have been omitted.
(f) The sight is graduated to 1000 instead of 800 m, except in the case of the Hungarian short assault rifle.

(g) A small compensator is fitted to the muzzle in all but the earliest production models.

Versions of the weapon are made in several Warsaw Pact countries. They differ in furniture but in all essentials they are the same.

Operation
There is no essential difference between the functioning of the AKM and that of the AK-47. There is an additional assembly in the trigger mechanism which has, in the past, been referred to as a 'rate reducer' but which is simply a device to delay the fall of the hammer during automatic fire until the breech is closed and locked.

Accessories
The AKM and AK-47 carry the same accessories which are: bayonet, blank firing device, combination tool kit, magazine carrier, night firing sight, oil bottle and cleaning fluid and sling.

The bayonet of the AKM is quite different in shape from that of the AK-47. Whereas the AK-47 bayonet has an even taper to the point on each side, that of the AKM has an undercut reverse edge. The AKM bayonet has a slot in the blade into which a lug on the scabbard fits. The scabbard is electrically insulated and has a shearing edge so that, with the bayonet, an effective wire cutter is made. The blank firing device goes on to the muzzle thread which lies

Soviet 7.62 mm × 39 assault rifles AKM and AKMS

Soviet paratrooper with early model AKMS assault rifle. Note that there is no compensator on muzzle

7.62 mm AKM assault rifle stripped

Identification of 7.62 mm AKM assault rifle

under the muzzle nut. It produces sufficient pressure in the bore to operate the piston.

The combination tool kit fits under the butt-plate. It is contained in a case which also acts as a handle for the cleaning rod. The combined drift, screwdriver and spanner is used, among other things, to adjust the foresight.

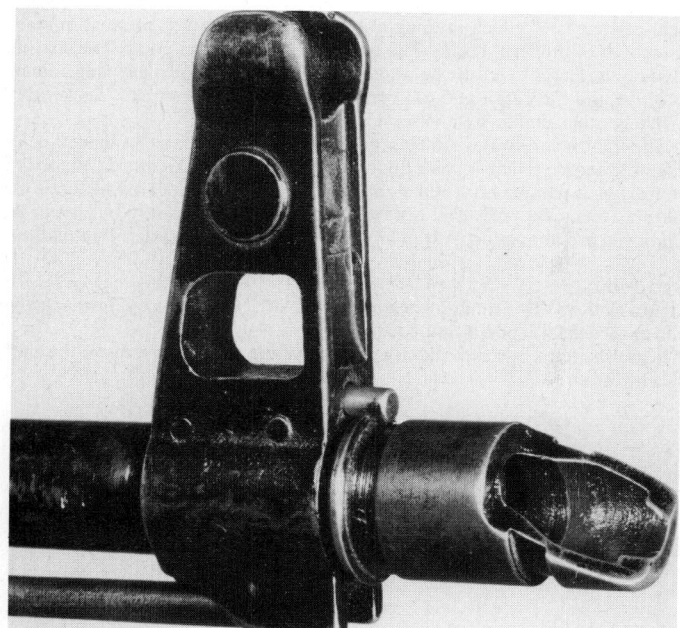

Compensator. It reduces muzzle climb and tendency to swing to right when using automatic fire

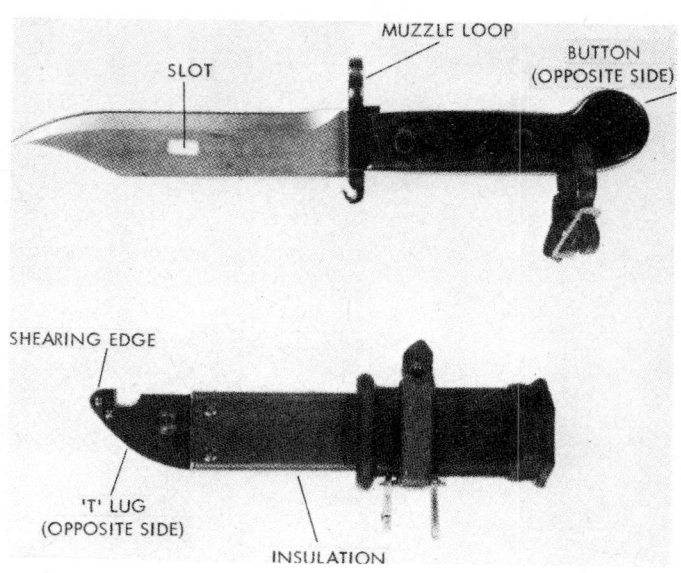

AKM bayonet

Stripping
The AKM strips in exactly the same way as the AK-47.

DATA
Cartridge: 7.62 mm × 39
Operation: gas, selective fire
Method of locking: rotating bolt
Feed: 30-round detachable box magazine
Weight: 3.15 kg
Length: 876 mm
Barrel: 414 mm; (rifling) 4 grooves rh
Sights: (foresight) pillar; (rearsight) U-notch
Muzzle velocity: 715 m/s

Rate of fire: (cyclic) 600 rds/min
Effective range: 300 m

Manufacturer
State factories.
Status
Current. Production.
Service
All Warsaw Pact countries since 1959. Supplied to Egypt and Syria and may now be found in many other countries. The weapon and variants (which are separately described under appropriate country headings) have been made in East Germany, Hungary, Poland and Romania as well as in the USSR.

7.62 mm Dragunov sniper rifle (SVD)

The Dragunov is a well-designed and well-built precision rifle with several interesting features. It is long, and not very easy to handle in the field, but snipers can be expected to be working on their own and thus able to take their time about movement. It is, of course, intended as a shooting instrument which will produce a first-round hit at ranges of 800 m or so. It is well suited for this task. It is a pleasant rifle to shoot and one which gives the firer immediate confidence in his equipment. It balances well and by using the sling there is little difficulty in holding it steady and taking a careful shot. The results are invariably encouraging, though, like all high-performance tools, it is intolerant of carelessness. The design has been copied in East Germany and China.

The rifle is a semi-automatic arm with a 10-round magazine, and is chambered for the rimmed 7.62 × 54R cartridge. It is gas-operated with a cylinder above the barrel. There is a two position gas regulator which may be adjusted using the rim of a cartridge case as a tool. The first position is employed in the usual operation of the rifle and the second is for extended use at a rapid rate or when conditions are adverse.

The bolt system is, in principle, exactly the same as that used in the AK-47, AKM and RPK, but the Dragunov bolt cannot be interchanged with that of the other weapons which fire the M43 7.62 × 39mm intermediate round. However the assault rifle and light machine gun are operated on a long-stroke piston principle which is somewhat heavy for a rifle since the movement of the fairly heavy mass with the attendant change in the centre of gravity militates against extreme accuracy. Therefore, in the Dragunov the designer has gone to a short-stroke piston system. The piston, of light weight, is driven back by the impulsive blow delivered by the gas force and transfers energy to the bolt carrier which moves back and a lug on the bolt, running in a cam path on the carrier, rotates the bolt to unlock it. The carrier and the bolt go back together; the return spring is compressed and the carrier comes forward and locks the bolt before firing can take place. Mechanical safety is produced by the continued movement of the carrier after bolting is completed. When the carrier is fully home a safety sear is released and this frees the hammer which, when the trigger is operated, can come forward to drive the firing pin into the cap.

Since the trigger mechanism has to provide only for single-shot fire, it is a simple design using the hammer, the safety sear controlled by the carrier and a disconnector. The disconnector ensures that the trigger must be released after each shot to re-connect the trigger bar with the sear.

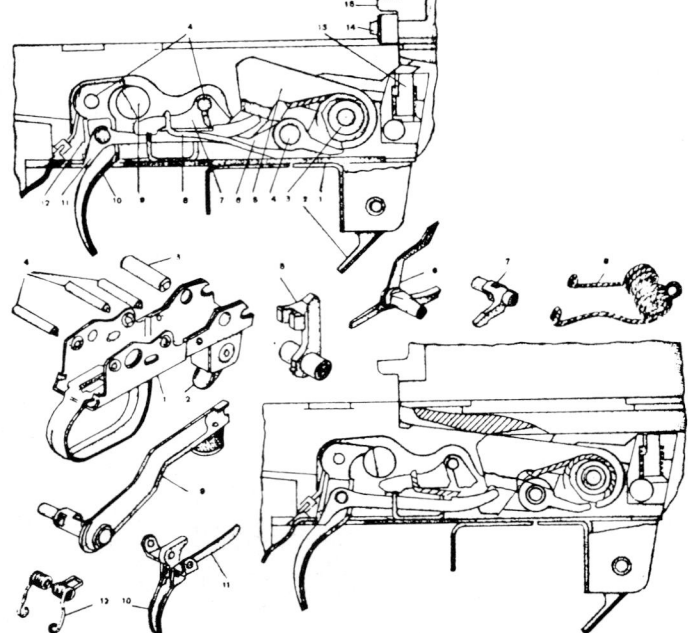

7.62 mm Dragunov sniper rifle (SVD). Parts of trigger mechanism: **(1)** *trigger housing* **(2)** *magazine catch* **(3)** *hammer pin* **(4)** *mechanism pins* **(5)** *hammer* **(6)** *safety sear* **(7)** *sear* **(8)** *hammer spring* **(9)** *safety* **(10)** *trigger* **(11)** *trigger bar* **(12)** *trigger spring* **(13)** *bolt stop* **(14)** *firing pin* **(15)** *bolt safety lug*

Sights
The Soviet sniper uses the PSO-1 sight. This is a telescopic sight of four magnification with power for graticule illumination supplied by a small battery. It is rather longer than most modern telescopic sights at 375 mm but a rubber eyepiece is included in this length. The firer's eye is in contact with this rubber which automatically gives the correct eye relief of 68 mm. The true field-of-view is 6° which is comparable with that obtained in most military telescopes of modern design. The optics of this sight are good. The coating used on the lenses to reduce light loss on the interchange surfaces is extremely effective and the depth and uniformity of the deposit compares favourably with any other similar sight in service.

The sight incorporates a metascope, meaning it is capable of detecting an infra-red source. There is some doubt whether this is sufficiently developed to be used as a passive infra-red sight but reports have appeared in the West German technical press that not only can it be used passively but it can be employed in conjunction with an infra-red light source for target illumination.

Stripping
Stripping the SVD is similar to stripping the AK series. Remove the magazine and ensure the weapon is unloaded.
Release the telescope sight latch and remove the telescope; remove the butt cheek-rest.

East German version of PSO-1 telescope. Note enlarged battery case (lower left) for graticule illumination and omission of long rubber eyepiece

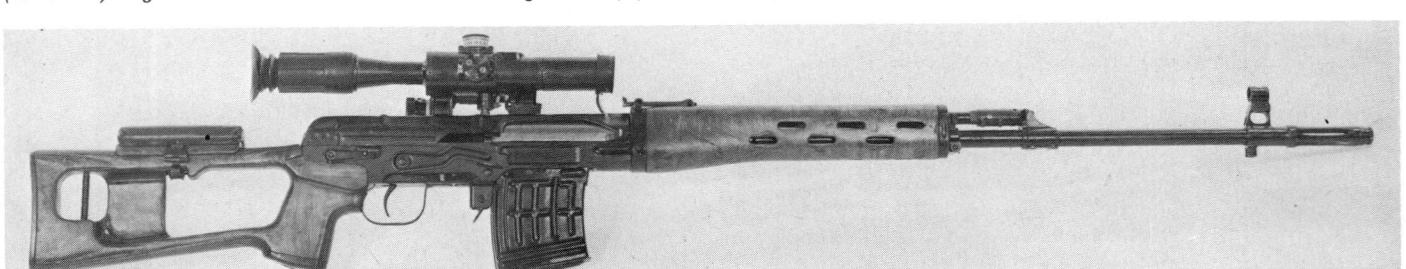

7.62 mm Dragunov sniper rifle (SVD)

Press up the receiver cover catch, pull the receiver cover and driving spring up and off the receiver. Pull the cocking handle to the rear, lift the bolt carrier and bolt from the rifle, and separate them.

Rotate the safety lever to the vertical and pull it to the right, out of the receiver. Remove the trigger group.

Press the hand-guard catch until it is free, then rotate it to the right. Push the hand-guard ferrule forward, then pull the hand-guard down and off.

Pull the operating rod to the rear, then move the front of the rod sideways. Pull the piston from the gas block, then ease the operating rod forward and remove it, together with its spring.

Re-assembly is the reverse procedure.

Left side of Dragunov rifle, showing PSO-1 telescope sight and mount

DATA
Cartridge: 7.62 mm × 54R
Operation: gas, short-stroke piston, self-loading
Method of locking: rotating bolt
Feed: magazine
Weight: 4.30 kg with telescope
Length: 1225 mm
Barrel: 622 mm
Rifling: 4 grooves, rh, one turn in 254 mm
Sights: (fore) adjustable post; (rear) U-notch, tangent.
PSO-1 telescope: 4 × 24, 68 mm eye relief, 6° field-of-view

Muzzle velocity: 830 m/s
Effective range: 800 m

Status
Current. Production
Service
Warsaw Pact armies

5.45 mm AK-74 and AKS-74 assault rifles

In the broadest terms the AK-74 is a small calibre version of the AKM. It is similar in size though slightly heavier. The effective range has decreased, but not below what is considered to be normal infantry fighting ranges. The magazine has the same capacity (30 rounds) as that of the AKM; the 40-round plastic magazine of the RPK-74 LMG is however interchangeable with that of the AK-74.

The AK-74 is one of the very few rifles to have been fitted with a successful muzzle brake. The reason seems to be to allow the firer to fire bursts without the muzzle moving away from the line of sight. The brake works by allowing

the emerging gases to strike a flat plate at the front of the assembly. This deflects the gas and produces a forward thrust. To counter the upward movement during automatic fire, gases escaping through three small ports on the upper part of the muzzle brake force the muzzle down, compensating for muzzle jump. The difficulty with all muzzle brakes is that the deflected gas comes back towards the firer; the AK-74 overcomes this, to a great extent, by means of narrow slits on the forward end of the muzzle attachment. These deflect gases forward from the muzzle brake instead of allowing it to flow back to the firer. A plan view of the muzzle would show a pattern of gas flowing out from the muzzle in a fan on either side, rather like a pair of butterfly wings. While this protects the firer from the effect of his own shots, it does nothing for anyone on either side of him, and there has already been an article in the Soviet military medical press expressing concern at the probable aural damage likely to occur on training ranges where firers are lined up on conventional firing points within a metre or two of each other.

However, despite the chance of deafness, the muzzle brake is highly effective and it allows the rifle to have no more recoil than is usually the case with an ordinary .22 long-rifle cartridge. This is not only an advantage in combat shooting, it is also a substantial step forward in training recruits.

Front view of muzzle brake from AK 74 rifle. Note upward vents which are intended to counter muzzle climb and vent slots pointing forward so as to deflect side blast from muzzle brake ports

Sectioned 5.45 mm bullets. Bright spots are test marks. Bullet on right is unfired and three sections of filling show well. Left bullet was fired into a soft catcher, but the nose bent over and the bullet tumbled

5.54 mm AK-74 assault rifle

There are defects however, the first being that a muzzle brake does not reduce flash and the AK-74 has a substantial muzzle flash of the order of three times the normal.

Construction

The AK-74 is built around the receiver of the AKM and there appears to be no difference between the receivers, leading to the belief that it is the same one from the same dies. The bolt carrier remains virtually the same, so that it runs in the same guide-ways, and it carries a smaller bolt. This bolt is lighter than the AKM version and so gives a better ratio of bolt to carrier mass and this leads to more efficient working. There is a slight modification to the AKM bolt and carrier design: the AK-74 bolt has a small flat, and the carrier a nib, which interact to prevent the bolt falling free of the carrier during disassembly of the weapon. The extractor has also been changed. The AKM extractor was always liable to breakage and for the AK-74 it has been substantially enlarged and strengthened. The magazine is made of a brownish plastic, which has been seen on some AKMs in recent years. It has thick walls and is extremely

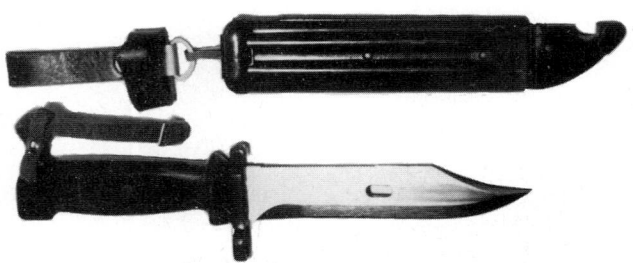

30-round plastic magazine of AK-74 (lower) compared with AKM 30-round magazine

AK-74 bayonet and scabbard

hard-wearing and strong and it fits into the existing housing, so saving a modification. This magazine holds 30 rounds.

The furniture closely resembles that of the AKM, the only obvious difference being that the AK-74 has a horizontal finger groove along each side of the butt. This is probably a recognition feature rather than an aid to holding and together with the muzzle brake it forms the quickest way of picking out a AK-74 from other AK rifles.

Variants

The most common variant is the folding butt version, the AKS-74. The folding butt is a tubular skeleton of a pattern not previously seen on Soviet rifles and it folds by swinging to the left and lying alongside the receiver.

A second variant is the light machine gun version, the RPK-74, which is described in the machine gun section. A third is the AKR sub-machine gun, described in the appropriate section. There are also reports of a version with integral silencer, but no details have yet been received on this.

Ammunition

The 5.45 mm cartridge has a number of novel features. The case is relatively conventional, using a double-based flake propellant. The case material is lacquered steel and the bullet has a high ratio of length to diameter and it requires a high rate of spin. It was at first thought that this spin rate would give good stability, but it is now known that the bullet is no more stable than that of an M16 and it will tumble on striking comparatively soft targets. The jacket is normal steel with a gliding metal cladding. The core is, however, mild steel and not lead. This steel core must be quite difficult to make since it tapers at the tail to make the boat-tail shape and it does not extend right to the nose. In fact it only goes two-thirds of the way up the jacket. In front of it is a small lead plug 3 mm long and in front of that is an air gap, a further 3 mm deep. The result of this variety is to bring the centre of gravity back and at the same time to ensure that the nose will deform easily. It is undoubtedly a clever way of extracting the maximum target effects from a small calibre bullet, but it must involve great difficulty in making and fitting the two metallic cores with the precision required. Tests have shown that this bullet, on striking, has a tendency to bend the nose (see the illustration appended) and this causes it to develop a curved track in soft targets, leading to severe wounding.

The rifling twist is steep (1:196 mm) and to avoid damaging the steel jacket the lands are bevelled on the leading edge, so allowing a gentler deformation of the metal. Without this bevelling it is likely that the jacket would rip.

DATA
Cartridge: 5.45 mm × 39
Feed: 30-round plastic box magazine
Weight: (unloaded AK-74 and AKS-74) 3.6 kg
Length: (AK-74 overall) 930 mm; (AKS-74 butt folded) 690 mm
Barrel: 400 mm; (rifling) 4 grooves rh. 1 turn in 196 mm
Sights: (foresight) post; (rearsight) U-notch
Muzzle velocity: 900 m/s
Rate of fire: (cyclic) 650 rds/min

Manufacturer
State arsenals.
Status
Current. Production.
Service
Soviet and some Warsaw Pact forces.

UNITED KINGDOM

.303 Lee-Enfield rifles

The first Lee-Enfield rifle came into British service in 1896. The well-known Short Magazine Lee-Enfield (SMLE) was adopted in 1902. It was used in the First World War.

After the First World War the British Army continued with the SMLE. Old models were converted to produce the Mark IV, similar to the Mark III.

The No 4 rifle was derived from a series of experiments in simplifying the manufacture of the Mark III SMLE during the 1920s. Volume production of the No 4 did not start until the late 1930s.

The No 4 rifle was used throughout the Second World War and saw service in every theatre in which British or Colonial troops were engaged. It differed from the No 1 rifle in having an exposed length of barrel at the muzzle. It was an accurate weapon and consistently placed a five-round diagram in a circle of 4 in (102 mm) diameter at 200 yds (183 m).

The No 5 Jungle Carbine was developed for use in the Far East.

DATA
NO 4 RIFLE
Cartridge: .303 Mk VII ball (11.3 g) or equivalent

Operation: manual, bolt-action
Feed: 10-round detachable box magazine
Weight: 4.1 kg loaded
Length: 1130 mm
Sights: (foresight) protected blade; (rearsight) adjustable aperture
Muzzle velocity: 751 m/s
Rate of fire: up to 20 rds/min
Effective range: 500 m

Manufacturer
Royal Small Arms Factory, Enfield Lock, Middlesex.
Status
Obsolete. Not in production.

Service
In the UK, cadet forces only. Still to be found in some former Commonwealth territories. Carried by many Afghan guerrillas.

.303 No 1 Mk III 1907 rifle

.22 No 8 rifle

.303 No 5 Mk 1 rifle

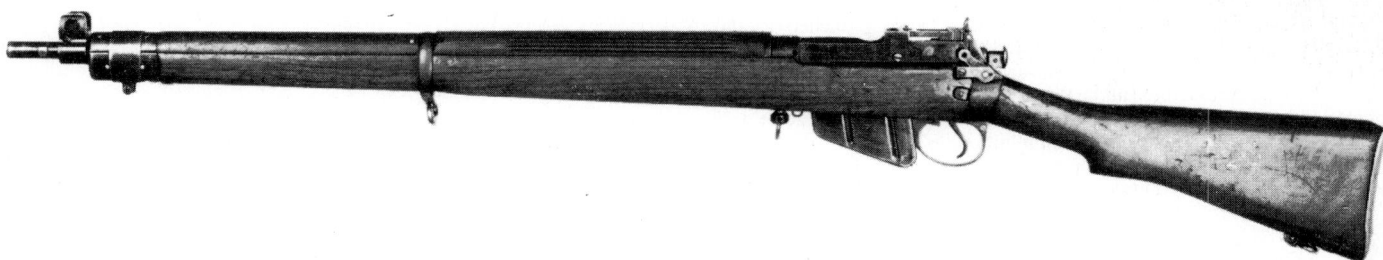

.303 No 4 rifle

7.62 mm Enfield L1A1 rifle

The 7.62 mm L1A1 is an easily handled, gas-operated self-loading rifle adapted from the very successful Belgium Fusil Automatique Légèr (FAL) rifle, with modifications to suit the special requirements of the British Forces. While the rifle is normally fitted for the self-loading single-shot mode, minor component changes in the trigger mechanism enable the rifle to be fully automatic. The bolt carrier is provided with oblique cuts on its outer surface which are designed to scour loose dirt from the inside of the receiver and eject it through the ejection port during firing. The cocking handle is fitted on the left-hand side of the weapon, allowing the right hand to remain on the trigger when cocking for firing. Both the cocking handle and the carrying handle fold down when not in use.

Robust, reliable and simple to maintain and operate, the L1A1 has continued as a leading design for over 20 years and currently remains the standard British infantry rifle. A trained man can reach a standard of rapid fire at a rate in excess of 20 accurate shots per minute firing at separate targets. All rifles are now fitted with plastic furniture: the butt may be varied in length by using one of four different butt-plates, thus enabling the weapon to be adjusted to suit the stature of the individual firer.

DATA
Cartridge: 7.62 mm × 51
Operation: gas, single shots

Method of locking: tilting block
Feed: 20-round box magazine
Weight: 4.30 kg empty; 5.00 kg with full magazine
Length: 1143 mm
Barrel: 554 mm
Rifling: 6 grooves, rh, one turn in 305 mm
Sights: (fore) blade; (rear) apertures; sight unit infantry trilux (SUIT) may be fitted
Sight radius: 554 mm
Muzzle velocity: 838 m/s
Recoil energy: 14.27 J
Effective range: 600 m with SUIT

Manufacturer
Royal Ordnance, Small Arms Division, Enfield Lock, Middlesex.
Status
Current. No longer manufactured, in process of replacement by L85A1.
Service
UK forces, also Australia, Barbados, Canada, Gambia, Guyana, Malaysia, New Zealand, Oman and Singapore. Made under licence in Australia and Canada.

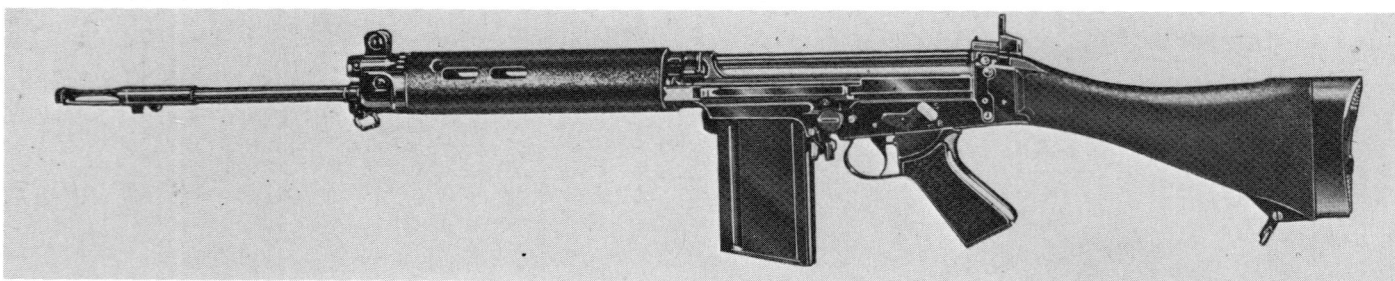

7.62 mm L1A1 rifle

7.62 mm Enfield L39A1 rifle

The L39A1 rifle was introduced to provide a satisfactory target rifle for competitive shooting for units of the British forces which normally would be equipped with the 7.62 mm L1A1 rifle. The L1A1 rifle is unsuitable for serious target shooting and the decision was made to produce target rifles from the .303 No 4 rifles held in ordnance depots. The No 4 rifles chosen for the conversion were the Marks 1/2 and 2. The choice of these marks of the No 4 rifles was made because the trigger is mounted on the receiver and not on the trigger-guard as on some earlier marks. With changes in the woodwork due to temperature variation and moisture content, a trigger attached to the trigger-guard does not give a consistent pull off.

The L39A1 is a manually operated bolt-action single-shot rifle with a heavy 7.62 mm barrel, manufactured at the Royal Ordnance Small Arms Division at Enfield, and the necessary modifications carried out on the extractor and receiver to permit the use of 7.62 mm ammunition.

The barrel is produced using the cold-forging process with a mandrel, carrying the rifling in relief, inserted in the drilled forging. The outside of the

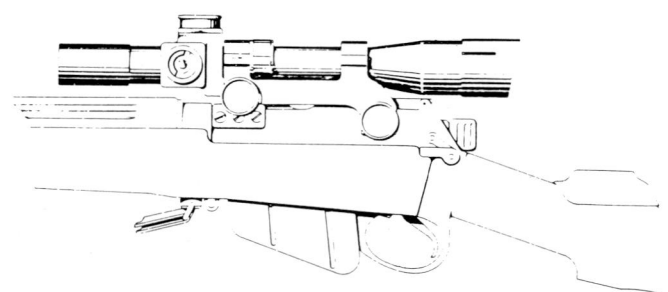

7.62 mm L39A1 rifle

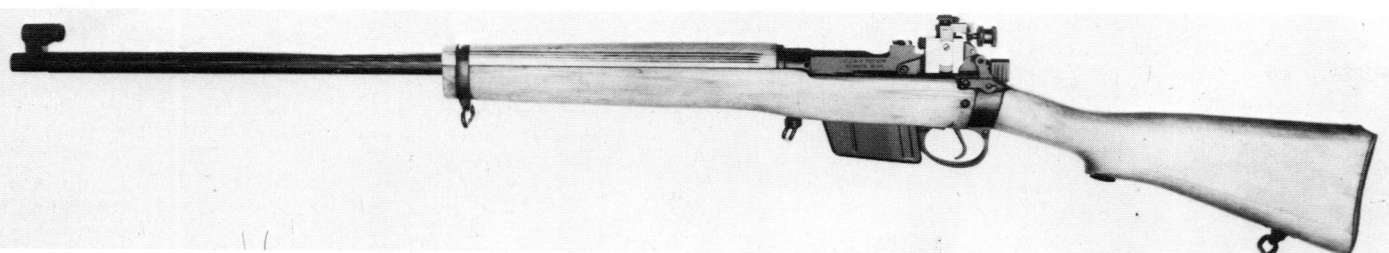

Telescope No 32 Mk 3 on No 4 rifle

barrel is hammered and as a result is elongated and the grooves impressed on the inner surface. Both inner and outer skins are work-hardened and the pattern of the hammering is clearly visible on the external surface of the barrel. A foresight block is soldered to the barrel but no foresight is fitted. Commercial target sights are fitted by the using unit.

The barrel projects from the wooden fore-end for 15 in (381 mm). The fore-end comes from the No 4 .303 rifle, cut down to half an inch of the lower band. The butt is unchanged from the No 4 rifle except that a recess has been machined under the knuckle to take a container holding spare foresight blades. The hand-guard is the same pattern as that on the No 8 .22 rifle.

The rifle is proved for 7.62 mm ammunition which produces a higher chamber pressure than the .303 Mark VII and, in addition to the usual military proof marks of crown, ER and crossed flags with the letter P, the receiver, bolt and bolt-head are stamped 19T. There are four sizes of bolt-head available, marked 0, 1, 2 and 3 in ascending size, for ensuring correct cartridge head space.

The prime purpose of the weapon is competition shooting, and since rounds are usually hand-fed into the chamber the only service required from the magazine is the provision of a platform to support the round while this is done. For this purpose the original .303 magazines are retained. There is no positive ejector when this magazine is used. A groove in the left of the receiver shallows towards the rear and the friction between the side of the case and the groove tips the mouth of the case to the right and the extractor loses its grip. As an alternative a 7.62 mm magazine holding 10 rounds can be provided. It has an ejector plate spot-welded to the lip at the left rear.

DATA
Cartridge: 7.62 mm × 51
Operation: manual, single-shot
Method of locking: rotating bolt
Feed: 10-round box magazine
Weight: 4.42 kg
Trigger pull: (first pull) 1.13-1.59 kg; (second pull) 1.81-2.04 kg
Length: 1180 mm
Barrel: 700 mm
Rifling: 4 grooves, rh, one turn in 305 mm
Sights: competition sights attached as required by user
Muzzle velocity: 841 m/s

Manufacturer
Royal Ordnance, Small Arms Division, Enfield Lock, Middlesex.
Status
Current. No longer manufactured.
Service
UK armed forces.

7.62 mm Enfield L42A1 rifle

The L42A1 rifle came into service to meet the need for a sniper's rifle.

It is a conversion from the .303 No 4 rifle using rifle No 4 Mark I(T) or Mark I*(T). These No 4 rifles were equipped originally with the telescope sighting No 32 Mark 3 and were used for sniping.

The conversion in all major aspects is similar to that used for the L39A1 rifle. However the trigger is pinned to the trigger-guard, not mounted on the receiver. There is an additional swivel, secured by the front trigger-guard screw.

The magazine on the L42A1 takes 10 7.62 × 51 mm cartridges and has an ejector plate spot welded to the left rear lip.

There are differences between weapons converted from the No 4 Mark I and the No 4 Mark I*. The bolt-head has a catch on the Mark I which must be depressed to permit the bolt-head to engage or disengage with the guide rib. The Mark I* has a break in the guide rib but no catch.

Telescope brackets are fitted to the left side of the body to take the Telescope, Straight, Sighting, L1A1 which is modified from the Telescope Sighting No 32 Mark III of the No 4(T) rifle. Those brackets also allow use of an image intensifier sight.

The open sights of the No 4 rifle have been retained and the Mark I rearsight is used. To allow for the different ammunition the datum line has been lowered by 0.07 in (1.78 mm) and the modified slide marked 'm' on the right side.

The foresight is a split block sweated to the barrel and it has an adjusting screw allowing the foresight to be clamped to the block. There are eight sizes of foresight from 0.03 to 0.075 in by increments of 0.015 in (0.762-1.905 mm × 0.381 mm) for zeroing.

Operation
The operation of the L42A1 is exactly the same as that of the L39A1 except for the ejector function.

DATA
Cartridge: 7.62 mm × 51
Operation: manual, single shots
Method of locking: rotating bolt
Feed: 10-round box magazine
Weight: 4.43 kg
Trigger pull: (first pull) 1.36-1.81 kg; (second pull) 2.27-2.95 kg
Length: 1181 mm
Barrel: 699 mm
Rifling: 4 grooves, rh, one turn in 305 mm
Sights: iron sights as described in text or telescope L1A1
Muzzle velocity: 838 m/s

Manufacturer
Royal Ordnance, Small Arms Division, Enfield Lock, Middlesex.
Status
Current. No longer in production.
Service
UK forces.

7.62 mm L42A1 rifle

5.56 mm Enfield L85A1 (Individual Weapon)

The Individual Weapon (IW) was designed from the start as a shooting weapon for soldiers in battle. For that reason no attempt has been made to conform to any conventional standards, apart from quality, and the result is the bullpup layout best suited to the sort of shooting that soldiers now have to undertake. It also allows the weapon to be easily handled and stowed inside vehicles and helicopters, a fact that is more important today and will become even greater in the next few years.

The resulting IW is a neat, handy, well-balanced weapon which is easy to fire, shoots well and gives little if any trouble in use. It is made from steel, using the modern processes of pressing and welding and the only machining is in the bolt, the carrier and the barrel. Almost all other parts are either stamped or pressed and pinned or welded together. The furniture is in plastic using high-impact nylon. Stripping and assembly is straightforward and can be done without using tools.

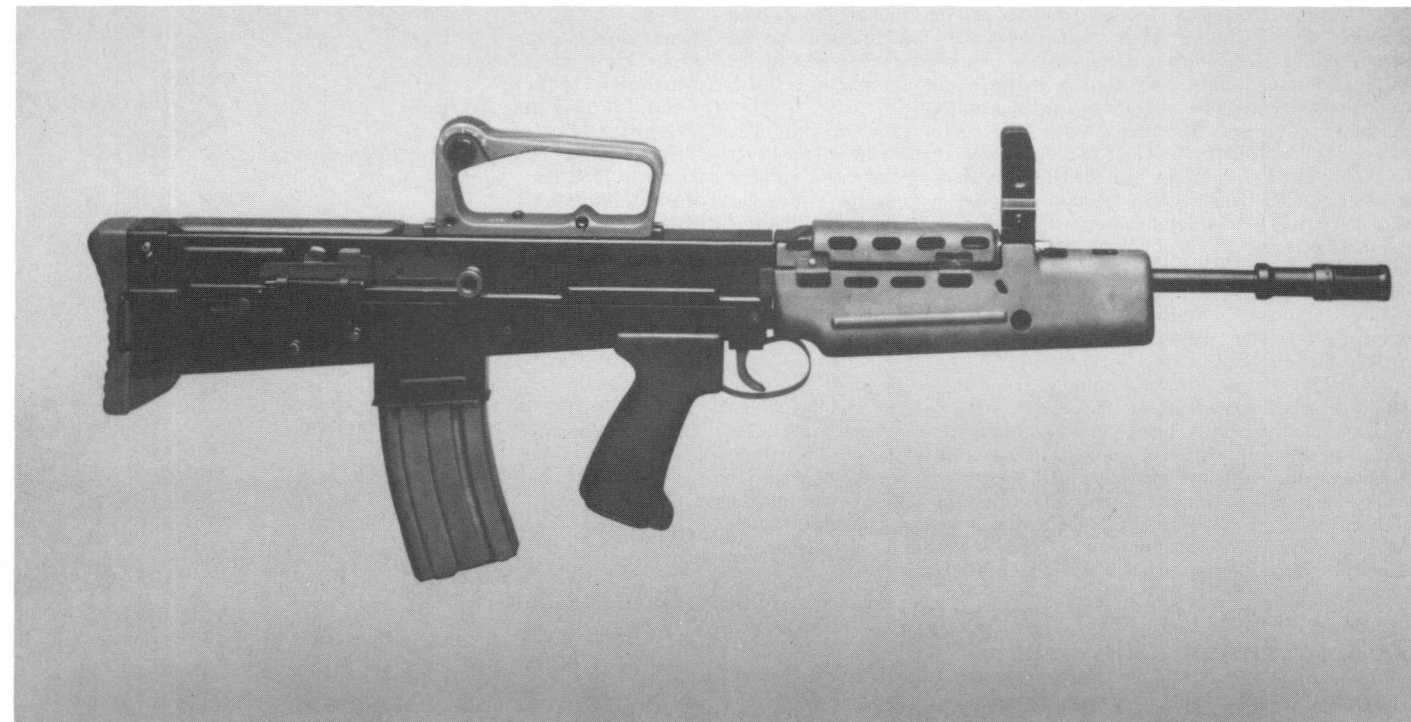

5.56 mm L85A1 individual weapon with secondary iron sight system, comprising rearsight/carrying handle (adjustable in azimuth) and front sight (adjustable for elevation)

Operation
The L85A1 is a conventional gas-operated rifle locked by a rotating bolt engaging in lugs behind the breech and carried in a machined carrier running on two guide rods; a third rod controls the return spring. The cocking handle is on the right side and has a cover which is spring-loaded to open. Moving the handle releases the catch and the cover flies open. The gas regulator has three positions, a normal opening for most firing, a large opening for use in adverse conditions and a closed position for grenade firing. Because the trigger has to be in front of the magazine there is a long connecting rod running to the mechanism, but this does not seem to have impaired the trigger pull-off. Finally there is a selector button which allows automatic fire, the first British rifle to incorporate this option.

The weapons
The two weapons in the Enfield Weapon System are the L85A1 IW (rifle/sub-machine gun) and the L86A1 light support weapon (LSW) which is a light machine gun. Both weapons fire common ammunition and also use some common components which gives increased flexibility, reduces spares requirements and simplifies maintenance. The recoil is very much less than with heavier calibre weapons which helps in recruit training and minimises weapon movement when fired. Both weapons are gas-operated, self-loading and magazine-fed with the facility for single-shot or automatic fire. The strength of construction enables muzzle launching of standard 21 mm grenades.

Sights
Both weapons are intended to be fitted with a robust, high-performance optical sight (SUSAT) of × 4 magnification which enables the weapons to be used operationally under poor light conditions and is also useful for surveillance. The sight is mounted on a bracket which incorporates range adjustment and zeroing. This bracket slides on to a dovetail base permanently fitted to the weapon body and which allows sight position adjustment.

An emergency open sighting system is also included, being permanently fitted to the body of the primary optical sight.

It is intended that the SUSAT sight will be provided with rifles used by infantry; other users will have a secondary sight, consisting of a double aperture rearsight housed in a carrying handle mounted above the receiver, with a foresight blade on the gas block.

The weapons can also be fitted with image-intensifying sights by dismounting the SUSAT from its dovetail base.

Construction and maintenance
The weapons are designed to be simple to dismantle without special tools into the main subassemblies for cleaning and maintenance. The trigger mechanism is a self-contained assembly in a pressed steel housing which is located to the main weapon body by two pins and a small butt-plate. The main body is a steel pressing which houses the bolt and carrier assembly and guide rods which locate in the barrel extension welded into the body and into which the barrel is screwed. Mass production techniques and ease of maintenance have been prime considerations throughout.

Accessories
A number of accessories are available including sling, bayonet, blank firing attachment, cleaning kit and multi-purpose tool.

Grenade launcher
A grenade launcher is in the final stages of developmnent. It fits on to the rifle without modification. It can fire standard 40 mm ammunition out to 350 m. The system fits within the length of the rifle and has automatic opening and ejection.

Variant models
The *Enfield Ensign Rifle* uses major components of the L85A1 but without the gas actuating system, thus converting the weapon into a manually operated

5.56 mm L85A1 individual weapon, infantry version with SUSAT sight

5.56 mm L85A1 individual weapon, showing major subassemblies and component parts, bayonet and scabbard, 30-round magazine and 10-round charger

single-shot rifle. It uses the secondary iron sight system as standard, has no provision for launching grenades, and can be fitted with an adapter to allow firing of .22 rimfire ammunition.

Enfield Carbine

The carbine is a more compact design suitable for light or specialist forces and for vehicle and helicopter crews. It has 80 per cent of parts interchangeable with the standard rifle, fires 5.56 × 45 mm NATO standard ammunition, and is accurate out to 300 m. Its length is 556 mm and it weighs (complete with sight and loaded magazine) 4.7 kg.

DATA
Cartridge: 5.56 mm NATO standard
Types: ball, tracer, blank, low power training
Operation: gas, selective fire

Type of locking: rotary bolt, forward locking
Feed: 30-round box magazine
Weight: (weapon without magazine and optical sight) 3.8 kg; (with loaded magazine and optical sight) 4.98 kg
Length: 785 mm
Barrel: 518 mm
Rifling: 6 grooves, rh, one turn in 180 mm (32 calibres)
Trigger pull: 3.12-4.5 kg
Muzzle velocity: 940 m/s
Rate of fire: (cyclic) 650-800 rds/min

Manufacturer
Royal Ordnance plc, Guns and Vehicles Divison, Nottingham.
Status
In service with British Army.

7.62 mm Model 82 Parker-Hale sniping rifle

The Model 82 is a bolt-action rifle with a four-round magazine and the option of a variety of sighting systems. It is intended for use by military and security forces as a precision weapon capable of engaging a point target with a 99 per cent chance of a first-round hit at all ranges out to 400 m in good light or to the range limits of the sights employed when fitted with a passive night vision infantry weapon sight. It is also designed for use as a marksman training and competition rifle when fitted with aperture sights, as illustrated.

The action is a Mauser 98-type with an internal magazine and a positive non-rotating extractor: twin locking lugs engage the front receiver ring and a third safety lug engages the rear receiver ring. The heavy, free-floating barrel, weighing almost 2 kg, is of chrome molybdenum steel, with rifling cold-forged to increase tensile strength by 5 to 10 per cent and improve wear characteristics.

The trigger mechanism is a separate self-contained assembly with full

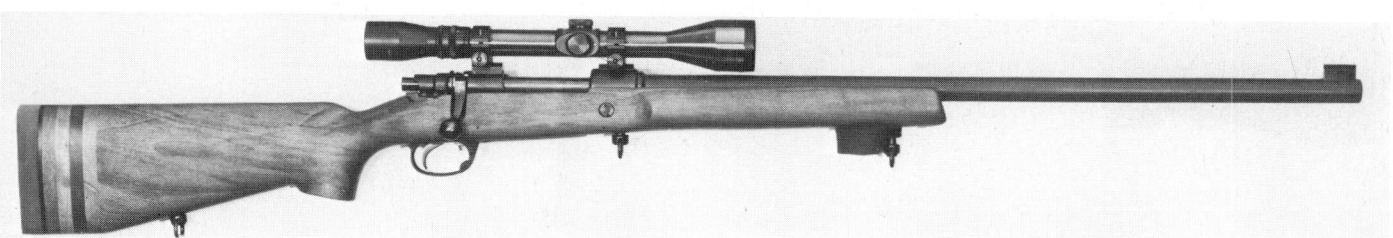

Parker-Hale 7.62 mm Model 82 sniper rifle

adjustment for alteration of trigger pull and wear. The safety locks the trigger, bolt and sear in a unique triple-action. The butt length is adjustable by detachable spacers (apparent in the illustration of the Model 82 in daytime sniper mode). A detachable, folding bipod with spring-loaded extending legs is available.

As with all sniper rifles the ultimate performance from the weapon depends to a great extent on using the correct ammunition.

DATA
Cartridge: 7.62 mm × 51, selected stock
Weight overall: (unloaded) 4.8 kg
Barrel weight: 1.93 kg

Length: 1162 mm
Barrel: 660 mm; (rifling) 4 grooves rh
Sights: fitted with iron target sights at the factory, optical may be fitted as desired to the machined dovetail on the receiver

Manufacturer
Parker-Hale Ltd, Bisley Works, Golden Hillock Road, Birmingham.
Status
No longer in production.
Service
In service with Australia, Canada and New Zealand.

7.62 mm Model 83 Parker-Hale target rifle

This rifle has been developed by Parker-Hale from their PH1200TX target rifle. It is a single-shot rifle, the absence of a magazine and feedway allowing the action to be stiffened and thus improving accuracy.

The rifling has been specially matched to the standard 144-grain 7.62 mm NATO bullet and is hand-bedded with Devcon F metal compound for the entire length of the action body. Accuracy is within half-minute of arc for 10 rounds, depending upon the quality of the ammunition.

The target trigger is single-stage and fully adjustable for weight, creep and back-lash. Lock time is relatively fast due to a striker travel of only 7 mm. The length of stock is adjustable to 317.5, 330 or 343 mm. The sights are fully adjustable and a variety of foresight elements are provided to suit individual users.

The Model 83 rifle has been adopted by the British Ministry of Defence as the Cadet Training Rifle L81A1 and entered service early in 1983.

DATA
Cartridge: 7.62 mm × 51 mm NATO
Operation: bolt, single-shot
Weight: 4.98 kg
Length: 1.187 m
Barrel: (weight) 2.04 kg; (rifling) 1 turn in 14 calibres
Sights: (foresight) tunnel with removable ring or blade; (rearsight) fully adjustable, aperture, 6 holes

Manufacturer
Parker-Hale Ltd, Bisley Works, Golden Hillock Road, Birmingham.
Status
Production.
Service
British cadet forces.

7.62 mm Model 83 Parker-Hale target rifle

7.62 mm M.85 Parker-Hale sniping rifle

The M.85 Sniper is a uniquely adaptable and robust high precision rifle designed to give 100 per cent first round hit capability at all ranges up to 600 m, and is sighted up to 900 m.

The rifle has been approved for service with the British Army, having met all the requirements of the NAGSR.

The specially designed action has a built-in aperture rearsight adjustable up to 900 m for emergency use, or when the use of optical sights would be impractical. The integral dovetail mounting is designed for rapid attachment and removal of telescope or passive night sights and features positive return-to-zero and recoil stop.

A suppressor can be fitted for use with supersonic or reduced velocity ammunition. The suppressor eliminates all muzzle flash and firing signature and significantly reduces the recoil energy. The muzzle is threaded to accept the suppressor and the front sight assembly, which is clamped to the barrel, may be removed and replaced with the aid of an Allen key.

Bracketing is available for fitting the Simrad KN250 night-vision sight to most daytime telescope sights, thus eliminating the need for an adjustable height butt-stock.

The M85 has an adjustable length butt-stock to suit all sizes of user, and is equipped with a quickly detachable militarised bipod with provision for both swivel and cant adjustment by the sniper when in the firing position.

A complete kit is offered, including high duty welded aluminium foam-lined transit and storage case containing all necessary accessories for

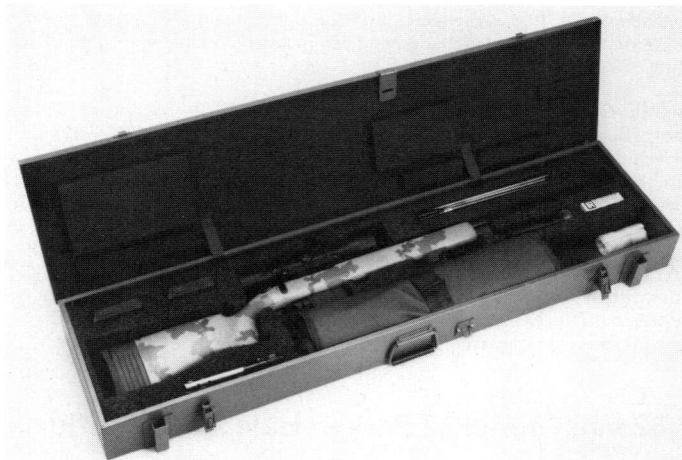

The M85, with desert camouflage stock, in aluminium transit and storage case

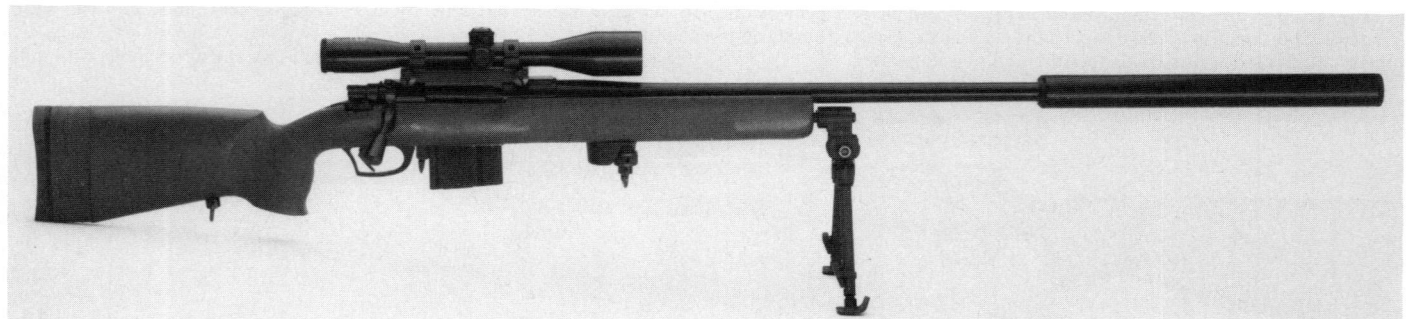

7.62 mm Model 85 Parker-Hale sniping rifle with current production fibreglass composite stock and optional suppressor

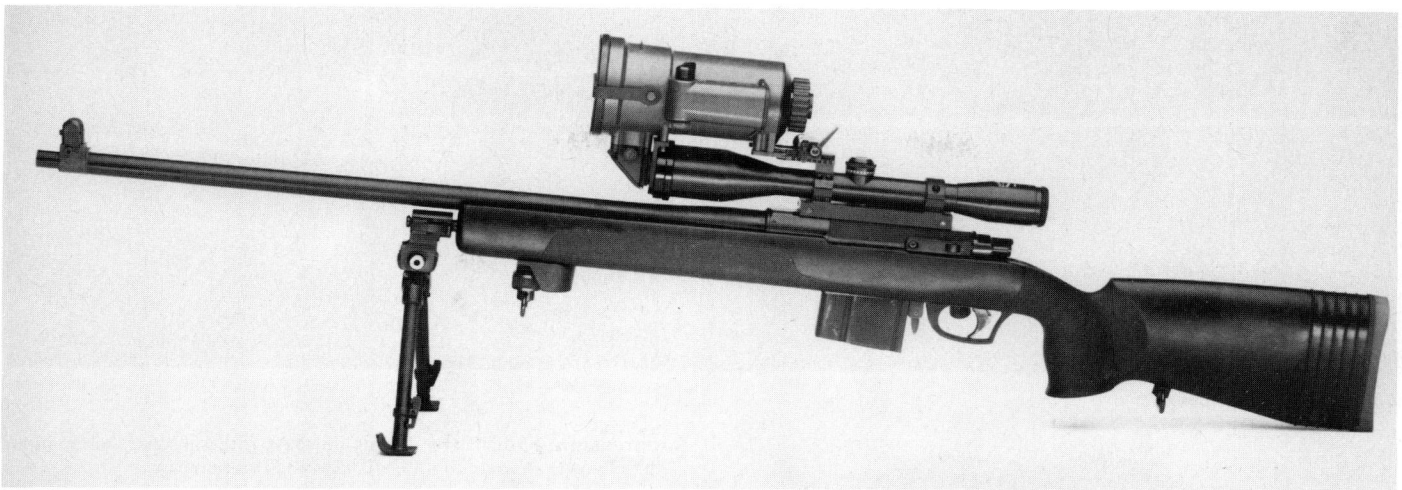

7.62 mm Parker-Hale M.85 Sniper Rifle fitted with Schmidt & Bender 6×42 telescope sight and Simrad KN250 night vision device

the sniper. These optional ancillary items may include passive night-vision individual weapon sight, soft padded cover, spotting telescope and bipod, cleaning and maintenance tools, spare magazine and sling. Final specifications are matched to the user's requirements.

DATA
Cartridge: 7.62 × 51 mm NATO
Operation: bolt-action, magazine repeater
Weight, empty: 5.70 kg with telescope sight

Length:(min) 1150 mm; (max) 1210 mm
Barrel: 700 mm
Rifling: 4 grooves, rh, one turn in 305 mm
Feed: 10-round box magazine

Manufacturer
Parker-Hale Ltd, Bisley Works, Golden Hillock Road, Birmingham B11 2PZ.
Status
Current. Production.

7.62 mm PM sniper rifle system

This weapon was designed from the start as a sniping rifle, intended to put its first shot on target, in any conditions, from a clean or fouled barrel. Developed by Accuracy International, it has been adopted by the British Army and designated L96A1.

The PM uses an aluminium frame to which the components are firmly attached. This is then clad in a high-impact plastic stock in which the stainless steel barrel floats freely. The bolt action is of conventional form, the bolt having three forward locking lugs and a safety lug at the handle. Bolt lift on opening is 60° and bolt throw is 107 mm allowing the firer to keep his head on

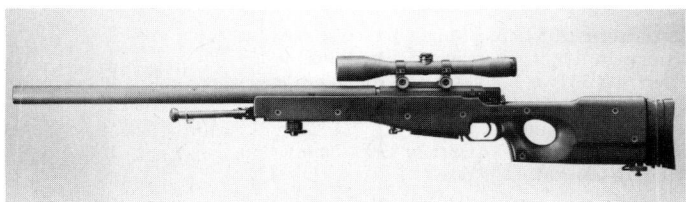

Accuracy International 7.62 mm suppressed sniper rifle with PM 6 × 42 sight

Accuracy International 7.62 mm 'PM Infantry' sniping rifle

the cheek-rest while operating the bolt and thus keep observation on his target while reloading. The rifle is equipped with a light alloy bipod which is fully adjustable.

The infantry version has fully adjustable iron sights for use out to 700 m, but is routinely fitted with a special design Schmidt und Bender 6 × 42 telescopic sight designated the L1A1. The accuracy requirement stated by the British Army was for a first round hit at 600 m range and accurate harassing fire out to 1000 m range. This has been achieved by the PM, which has an accuracy figure better than 0.75 minutes of arc.

The counter terrorist version is fitted with 12 ×, 2.5–10 × as well as the infantry 6 × 42 sight. A spring loaded monopod is usually fitted, which is concealed in the butt. This can be lowered and adjusted so that the rifle can be laid on the target and supported while the firer observes, without having to support the weight of the rifle for long periods. A spiral flash hider is fitted to the muzzle. Iron sights are not normally fitted to this version.

A suppressed rifle is also produced. Using special subsonic ammunition, this rifle is accurate out to 300 m without unreasonable trajectory or wind deflection. This increases the accepted 'state of the art' by at least 100 m.

The PM is now produced in single-shot Magnum calibres for long-range ..nti-terrorist work; calibres available now are .300 Winchester Magnum and 7 mm Remington Magnum for use up to 1000 m range, and an 8.6 mm cartridge is under development for use at ranges up to and exceeding 1000 m.

DATA
INFANTRY VERSION
Cartridge: 7.62 × 51 mm NATO
Operation: bolt-action
Feed: 10-round box magazine
Weight: 6.50 kg
Length: 1124-1194 mm
Barrel: 655 mm
Trigger: two-stage adjustable, 1-2 kg
Sights: Schmidt & Bender PM6X42

Manufacturer
Accuracy International, PO Box 81, Portsmouth PO3 5SJ.
Status
Current. Available to order.
Service
British Army, and several African, Middle East and Far East armies.

'Covert' Sniper rifle system

The latest variant of the PM System is the 'Covert PM'. The system consists of the suppressed PM rifle in a take-down version which, together with all its ancillaries, packs into an airline suitcase with fitted wheels and retractable handle.

The rifle is the 'Covert' folding PM bolt actioon repeating rifle fitted with either the PM 6 × 42, 10 × 42 or 12 × 42 Schmidt & Bender military sights, two 10-shot magazines, bipod, and one box of 20 rounds of subsonic ammunition.

DATA
Cartridge: 7.62 × 51 mm subsonic
Rifle weight: 6.5 kg
Length: 1250 mm
Trigger: 1-2 kg pull-off, two stage detachable
Muzzle velocity: 314-330 m/s
Range: 0-300 m

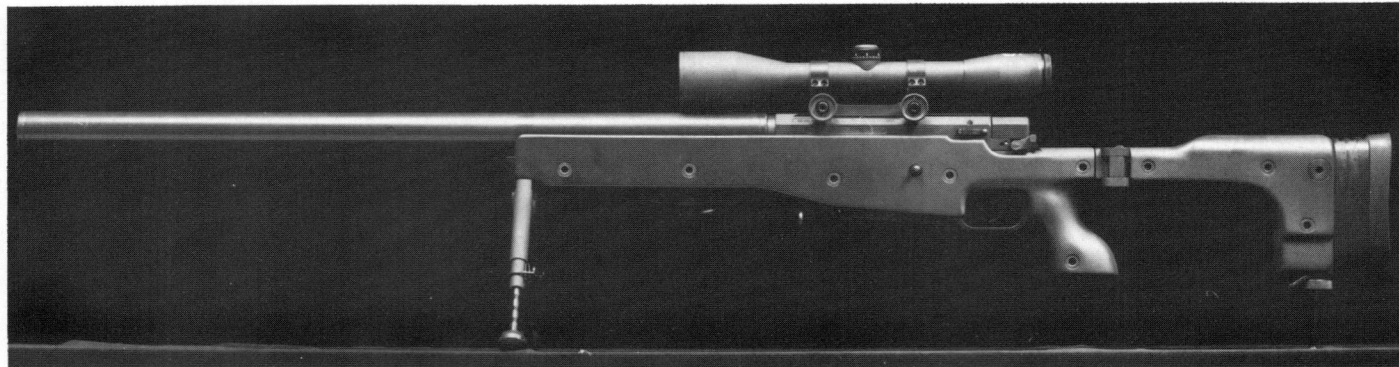

7.62 mm 'Covert' sniper rifle

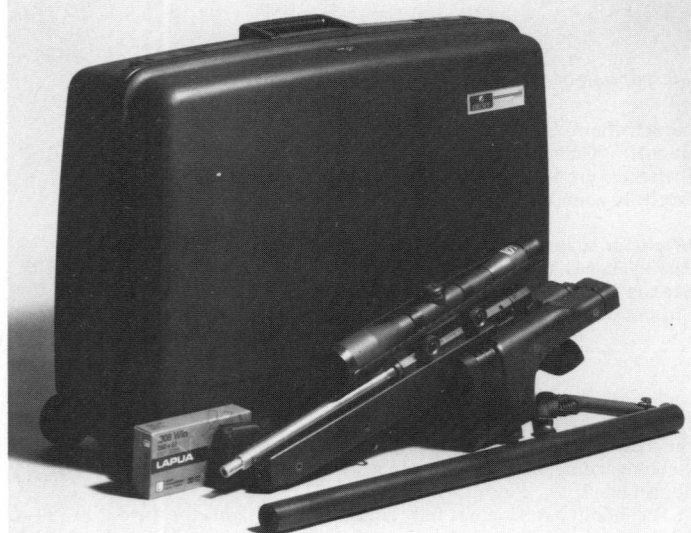

'Covert' sniper rifle dismantled, with carrying case

Suppression: 85 dBA over 125 ms timebase with subsonic ammunition. With full power ammunition, 109 dBA over 125 ms timebase.

Manufacturer
Accuracy International, PO Box 81, Portsmouth PO3 5SJ.
Status
Current. Available to order.

.338 Super Magnum sniper rifle

This has been designed as a dedicated sniping rifle giving guaranteed accuracy, ease of maintenance, reliability and military robustness. All the lessons learned during the development of the L96 sniping rifle, together with many new and innovative ideas, have been combined in this weapon.

Using the new 10 × sight and the new Lapua Magnum cartridge family, the weapon is able to meet requirements for equipment destruction and light armour penetration as well as the normal anti-personnel capability, to ranges well beyond 1400 m. The 16.2 g bullet is still supersonic at 1400 m, at which range there is still over 1000 J of energy remaining.

The rifle is to be supplied initially in three calibres: .338 Lapua Magnum, .300 Winchester Magnum and 7 mm Remington Magnum. Special accuracy ammunition is under development so as to provide a multi-projectile capability.

DATA
Cartridge: .338 Lapua Magnum (and see text)
Operation: bolt action magazine repeater
Feed: 4- (.338) or 5- (.300, 7 mm) round box magazine
Weight: 6.8 kg
Length: 2268 mm
Barrel: 686 mm (.338); 660 mm (.300, 7 mm)
Muzzle velocity: 914 m/s

Manufacturer
Accuracy International, PO Box 81, Portsmouth PO3 5SJ.
Status
Current. Available to order.

.338 Super Magnum sniping rifle

7.62 mm BGR sniping rifle

This is a carefully designed bolt-action rifle manufactured from components from a number of specialist makers. The bolt-action is a strengthened bench-rest action using three lugs and with a long and smooth bolt travel. The striker mechanism is carefully designed to have the shortest possible lock time and to ensure regular ignition even with the hardest primer caps. The match grade hammer forged barrel is fluted to achieve stiffness without excess weight and also to afford the maximum heat dissipation. It is fitted with a combination muzzle brake and flash suppressor, and a sound moderator may also be fitted. The action and barrel are bedded into the stock by means of a specially developed compound, and stocks are of wood, carbon fibre reinforced co-polymer material or, for the lightest possible weight, a special composite carbon fibre, Kevlar and GRP material.

The rifle is fitted with an Anschutz pattern accessory rail which will accept a bipod, hand rest or other accessories, and is equipped with a suitable mount for a variety of telescope sights. Iron sights may be fitted to special order.

The BGR rifle is normally supplied in 7.62 × 51 mm NATO chambering, but other chamberings from .243 Winchester to .300 Winchester Magnum can be provided.

7.62 mm BGR sniping rifle

DATA
Cartridge: 7.62 × 51 mm NATO and others
Operation: bolt-action repeater
Feed: detachable box magazines 5 to 20 rounds
Weight: 6.6 kg with bipod and telescope sight
Length overall: 1200 mm
Barrel length: 700 mm
Rifling: 4 grooves, rh, one turn in 280 mm

Manufacturer
Armalon Ltd, 44 Harrowby Street, London W1H 5HX.
Status
Current. Production.

UNITED STATES OF AMERICA

The U.S. Army Advanced Combat Rifle Programme

On 13 April 1989 the U.S.Army announced details of the various prototype rifles which have been put forward for consideration in the Advanced Combat Rifle (ACR) programme. Engineering and safety tests of the prototypes began in early April and were expected to last through May; a follow-on field experiment was then to be conducted at Fort Benning, commencing in August and extending to April 1990. During this field trial, the US Army and US Air Force are to test the weapons under simulated combat conditions on a specially designed range; the existing M16A2 rifle will be used as the control weapon during these tests. Descriptions of the four prototypes follow.

Colt Industries ACR Prototype

The Colt entrant is a gas-operated magazine fed rifle derived from the existing M16A2. The variations arise from ergonomic considerations and also from the use of an interchangeable duplex round of ammunition.

Human engineering considerations have led Colt to redesign the handguard, pistol grip and butt stock in order to improve the handling characteristics. The new handguard includes a heat resistant inner shield, air cooling vent holes, a front-end hand restricting ring, and an aiming/pointing rib on the upper surface for rapid target alignment. The pistol grip is similar to that of the M16A2 but with a different profile. The telescoping butt stock incorporates a cheek piece on both sides and allows for adjustment in length to suit the individual soldier.

The Colt ACR uses the basic M16A2 gas actuation system and rotating bolt, which engages the bolt to the barrel extension by means of locking lugs. The rifle has an advanced muzzle brake/compensator (MBC) and flash hider to improve controllability and reduce recoil. This MBC consists of concentric tubes positioned around the barrel so as to redirect gases which would otherwise adversely affect the accuracy of the weapon. Tests have shown that the MBC effectively reduces recoil and compensates for muzzle climb during firing. An oil-spring buffer assembly has been designed and installed in the lower receiver extension tubes to control the cyclic rate during automatic firing and help reduce felt recoil.

These changes in the system are said to reduce the total recoil effect by 40 per cent compared with the current M16A2.

The ACR uses the basic M16A2 trigger mechanism and has an ambidextrous fire control selector located on both sides of the weapon for safe, semi-automatic, and full-automatic firing modes. In the full-automatic mode, burst length is controlled by manual operation of the trigger.

The weapon has both iron and optical sights; the iron sight includes the carrying handle and a flip-type aperture sight for both short- and long-range targets. The recommended optical sight is of X3.5 power and is equipped with a self-powered illuminating device for periods of reduced light. Both sights have windage and elevation adjustments. The weapon is designed so that other standard optical devices may be exchanged or substituted for the recommended sight. For instinctive shooting in rapid target engagements there is a long, shotgun-type sighting rib along the top of the weapon to act as an aid in pointing.

Colt ACR, right side, iron sights

Colt ACR, left side, optical sight

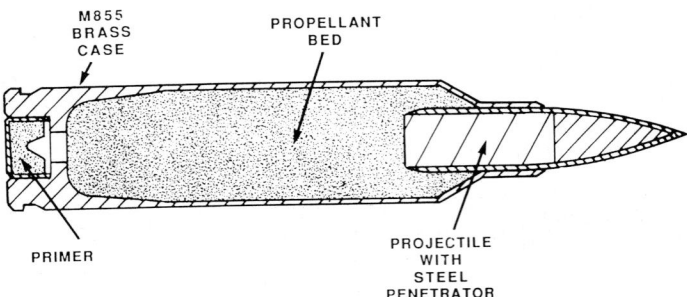

Standard 5.56 mm M855 cartridge

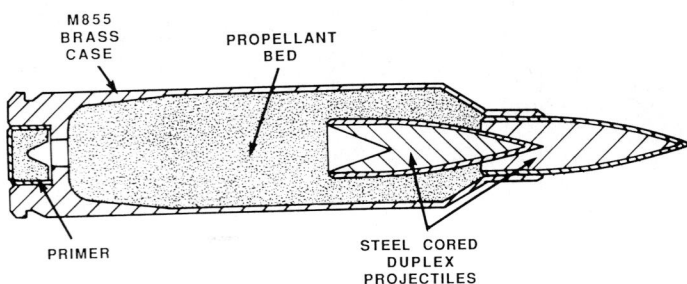

Colt ACR 5.56 mm Duplex cartridge

Ammunition

The Colt ACR fires NATO standard 5.56 × 45 mm cartridges interchangeably with new Colt/Olin developed duplex ammunition. Both rounds use the same 5.56 × 45 mm brass case.

The duplex round consists of two similar projectiles, one behind the other, each consisting of a hard steel core in a gilding metal jacket. The two projectiles weigh 2.26 g (front) and 2.14 g (rear), compared with the standard M855 ball bullet which weighs 4.02 g.

The duplex round has been developed in order to provide improved hit probability against shorter range targets, with a maximum effective range of 325 m. The standard M855 ball is used for longer range engagements. The theory behind the duplex round is that the leading bullet will go to the aimed point, while the trailing bullet will have a slight random dispersion around the point of aim in an attempt to compensate for human error.

Comment

This can be said to be evolutionary design at its best; Colt have taken a proven sound weapon and made improvements in the light of ergonomics and improved technology, and in a similar manner they have revived a cartridge concept which has been tried several times in the past but to which they have doubtless applied the results of many years of experiments and research. Thus the two concepts have been brought up to date and carefully mated together into a package which should meet the military requirements and which, because of its sound design background, is unlikely to provide any maintenance or reliability problems.

DATA

Cartridge: 5.56 × 45 mm M855 or Duplex
Operation: Gas, selective fire
Locking system: rotating bolt
Feed: 30-round magazine
Weight: 3.306 kg less magazine and sight
Length: 1031 mm butt extended; 933 mm butt retracted
Barrel: length not known
Rifling: one turn in 32 calibres
Sights: 3.5 × telescope or interchangeable iron sight
Muzzle velocity: 948 m/s (M855); 884 m/s (Duplex)
Chamber pressure: 50 750 psi (M855); 50 000 psi (Duplex)

Steyr-Mannlicher ACR Prototype

The Steyr-Mannlicher ACR is a bullpup weapon, its appearance having obvious affinities with the AUG rifle, using a rising chamber breech mechanism and a side-initiating flechette cartridge. There is a push-through firing mode selector giving safe, semi-automatic and three-round-burst modes of firing. The rifle has a long, shotgun-style rib/carrying handle along the top surface to aid as a pointing device for rapid target engagement. The peep sight is removeable and interchangeable with the preferred optical sight. An unusual feature is the provision of a recess in the stock around the muzzle, allowing muzzle-launched grenades to be fired in spite of the short exposed length of barrel.

The optical sight has no range adjustment; such adjustment is considered to be unnecessary because the the the flat trajectory of the flechette projectile, which has a muzzle velocity of 1480 m/s. The sight has two magnification settings: × 1.5 combat setting, and × 3.5 for ranges where the target is difficult to distinguish without magnification.

The rifle is gas actuated, using the barrel as a stationary piston and a sleeve around the barrel as a moving cylinder; an operating 'slide piece' is attached to the barrel sleeve. At the commencement of the operating cycle the slide piece is driven forward by a spring, stripping a round from the magazine and feeding it into the chamber, which is directly in front of the magazine. As the round enters the chamber, so it forces out the case of the previous round. A chamber guide pin is pushed out of its lower detent position and, by engagement of the pin in a cam surface on the slide piece, the chamber is raised by means of the block spring so that the chamber, containing the live round, is aligned with the barrel. The cartridge has an annular primer which is located beneath a hole in the chamber; as the chamber aligns with the barrel, so a fixed firing pin protrudes through this hole to strike the annular primer and fire the round. At the same instant the chamber guide pin snaps into its upper detent position and locks the chamber in alignment with the barrel axis. Gas pressure is tapped from the barrel into the annular space between barrel and sleeve, and the sleeve is thus driven backwards. The slide piece first disengages the chamber guide pin from the upper detent, then, by means of the cam surface, lowers the chamber to the loading position, and the cycle is ready to recommence.

The rifle can be disassembled to armourer maintenance level by the user without the use of tools.

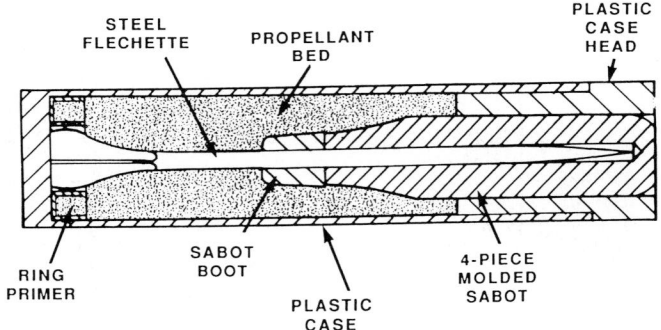

Steyr-Mannlicher ACR flechette cartridge

Steyr-Mannlicher ACR, right side, iron sights

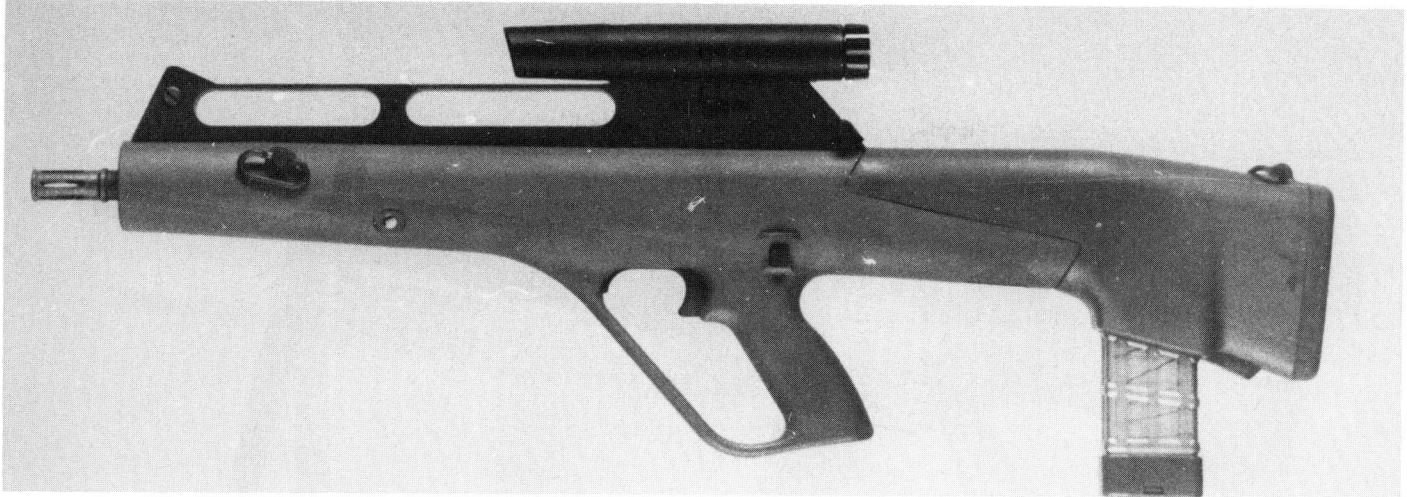

Steyr-Mannlicher ACR, left side, optical sight

Ammunition

The synthetic-cased flechette round is a simple cylinder 45 mm long and 10.4 mm diameter. At the rear end is an aluminium ring with a groove into which priming mixture is pressed; this is inserted into the plastic tube and fits into a prepared seat. The propellant, which is a relatively common but not military standard type, is then loaded. The flechette, pre-assembled with its four plastic sabot segments and held together with a plastic boot, is inserted into the head of the case so that its fins rest inside the annular primer ring. A plastic case head, which retains the forward end of the sabot, completes the assembly of the round.

The 0.66 g flechette is approximately 1.6 mm diameter and 41.25 mm long and has a slightly roughened surface to assist friction between the flechette and the sabot segments. These segments are made of liquid crystal polymer and are retained in place during shot travel; after muzzle exit they separate and fall clear, allowing the flechette to fly downrange in its most aerodynamic configuration.

Comment

This is an innovative mechanism, but one which appears to be simple enough to be reliable in use. There is obviously a question about how the moving chamber is sealed to the rear of the barrel, though it is possible that the plastic cartridge case head may have some function in this. The official Piccatinny Arsenal Fact Sheet observes that the primary drawback to this system is 'the energetic discard of sabot segments and their possible hazard to friendly

troops'; in peacetime the normal range safety precautions should obviate this problem, and in wartime one supposes that the friendly troops will have worse hazards to face. The flechette design is straightforward enough, and one can only assume that Steyr-Mannlicher have overcome the accuracy problems which plagued the single-flechette cartridges of the 1960s.

DATA
Cartridge: 5.56 mm SCF (synthetic cased flechette)
Operation: gas, semi-automatic and three-round bursts
Locking system: rising chamber block, detent locked
Feed: 24-round box magazine
Weight: 3.23 kg less magazine and sight
Length: 765 mm
Barrel: 540 mm
Rifling: one turn in 388 calibres (216 cm)
Sights: multi-power telescope, interchangeable iron sights
Muzzle velocity: 1480 m/s
Chamber pressure: approx. 63 000 psi
Weight of complete round: 5.1 g

Manufacturer
Steyr-Mannlicher GmbH, Postfach 1000, A-4400 Steyr.
Status
Advanced development.

Heckler & Koch ACR Prototype

As can be seen from the photographs, the Heckler & Koch ACR is obviously derived from the G11 rifle, although there are doubtless several internal differences. It is a gas-operated bullpup weapon firing caseless ammunition from a 50-round magazine which slides in above the barrel and feeds the round nose-downwards to the breech mechanism. There are four options on the safety/selector switch: safe, semi-automatic, full-automatic and three-round-burst.

The sight is a non-removable optical sight built into the carrying handle. It has two magnification settings, unit power for short ranges and 3.5 × for ranges beyond 300 m.

As with the G11, the ACR design uses a mechanism which is free to move inside an outer plastic cover which is fully sealed to prevent the ingress of water or dirt; this means that there is a certain amount of gas escape into the interior of the casing and there is a gas relief valve which prevents any dangerous build-up of pressure inside the casing.

The mechanism is best described as a radially reciprocating chamber. At the commencement of the operating cycle the chamber is vertically aligned, ready to be loaded. A feed arm pulls a round downward, bullet end first, into the chamber. The chamber then rotates through 90° into the horizontal position, where it is aligned behind the barrel. The firing pin strikes the cap and fires the

Heckler & Koch ACR, left side

Heckler & Koch ACR, right side

round, after which the chamber is rotated back through 90° to the original feed position and the cycle is complete. In the full-automatic mode this cycle repeats at about 450 times per minute; in the three-round burst mode the rate is stepped up to about 2000 times per minute. Initial cocking and charging is performed by a rotary cocking handle on the side of the weapon.

By withdrawing the magazine and rotating the cocking handle, the live round in the chamber is cleared, falling through a port in the bottom of the weapon casing which is opened at the appropriate moment.

There is no muzzle brake or compensator; absorption of recoil is performed by permitting the entire weapon mechanism to float, on a guide rail, inside the casing. The length of this floating stroke is dependent upon the firing mode selected. In the single-shot and automatic modes, the breech is actuated at the end of the recoil stroke, whereas in the three-round burst mode the operation is more complex, the mechanism recoiling after the opening shot, reloading and firing the second shot and continuing to recoil, reloading and firing the third shot and then completing the recoil movement. As a result, the firer does not feel the recoil blow until after the last shot has left the muzzle, there is no upward movement of the weapon between the three shots, and controlled projectile dispersion of a predetermined size is achieved.

The ACR has an unusual rotary firing pin, turning about its own axis to expose a spur which digs into the primer to initiate the propellant. The operation of this pin is dependent upon the special type of primer used with the ACR caseless cartridge, which is described below.

Ammunition

The 3.19 g bullet is of 4.92 mm calibre and is totally enclosed within the propellant body, secured in a machined cavity by a plastic end cap. The propellant casing has a contoured seating for the base of the bullet and the end cap an ogival seating for the bullet nose; this arrangement holds the bullet securely in place and also retains it in axial alignment with both the propellant body and the barrel axis.

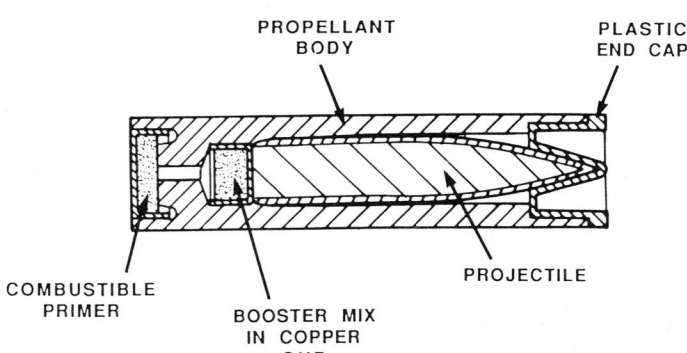

Heckler & Koch ACR caseless cartridge

Behind the bullet is a copper cup containing a booster compound based on lead styphnate. The open end of this cup faces rearward, to the primer, and is ignited by the primer. The burning of the booster composition ensures efficient ignition of the propellant and also launches the bullet into the commencement of rifling so as to engrave and form an efficient gas seal before propellant gas generation has reached a significant level. It is this initial sealing which is the key to ballistic consistency.

The primer cup is made of a combustible material based on nitramine, as is the propellant itself. The priming mixture incorporates ground glass to act as a friction device when struck by the rotating firing pin, in place of the conventional anvil. The cartridge is of square cross-section, approximately 7.9 mm side by 31.75 mm long, and the propellant body is extruded from solvent-wetted material, pressed into rough shape, dried for solvent removal and hardening of the body, and then finish machined. After assembly there are a variety of coating operations to provide waterproofing, low friction surfaces, heat resistance, and so on. In operation the remains of the copper booster cup and the plastic end cap are ejected from the muzzle in the wake of the bullet.

One of the constant objections raised to caseless ammunition is the 'exposed explosive' hazard. Tests have shown that there may be an advantage over conventional ammunition in that no metallic fragments are produced when caseless ammunition is hit by a bullet, the resultant fire is slow to generate at atmospheric pressures, and ignition will only occur if the primer is directly struck. The cook-off problem appears to have been solved by adoption of High Temperature Propellant which is far more resistant to induced heat than conventional nitro-cellulose propellants.

Comment

Whether or not the actual mechanics are the same or slightly different, what we are seeing here is essentially the G11 rifle and its ammunition. This has been exhaustively developed and tested over the past decade (for the latest information see the 'Summary' on page 171) and is within sight of adoption by the West German Army, so that there can be few problems left to be uncovered at Fort Benning. Unlike the other contestants though, approval of the H&K ACR involves acceptance of the caseless cartridge and in an army the size of the American, this would involve an enormous logistic upheaval which would need very careful preparation over a long period.

DATA

Cartridge: 4.92 × 34 mm caseless
Operation: gas, selective fire and three-round burst
Locking system: rotary breech
Feed: 50-round box magazine
Weight, empty: 3.90 kg
Length: 750 mm
Barrel length: not known
Rifling: one turn in 30.8 calibres
Sights: multi-powered fixed telescope, 1 × and 3.5 ×
Muzzle velocity: 914 m/s
Chamber pressure: 56 000 psi

AAI Corporation ACR Prototype

The AAI Corporation ACR is a gas-operated magazine-fed rifle firing a flechette cartridge. It is of the more conventional fully-stocked pattern. A long and unobstructed upper surface assists in rapid alignment of the rifle against fleeting targets. (Note that it is not the same weapon as that described on page 000.)

The ACR is a modified version of the 'serial bullet rifle' developed by AAI ca 1970 as part of the SPIW programme, so that there was an existing technology base upon which to build and the basic weapon mechanism was already proven. The major internal innovation is an entrapped gas operating system in which the propellant gas enters a cylinder and drives a piston to provide power for the operation. Benefits expected from this are a cleaner mechanism and the lower maintenance burden consequent upon cleanliness. Another modification worth noting is the addition of a muzzle compensator especially designed for long saboted projectiles.

The closed bolt mechanism has a two-position selector for semi-automatic and three-round burst modes. The safety catch is located forward of the trigger housing, and when 'safe' is selected a physical barrier prevents inserting the finger into the trigger-guard.

AAI ACR, right side, iron sights

AAI ACR, left side, optical sight

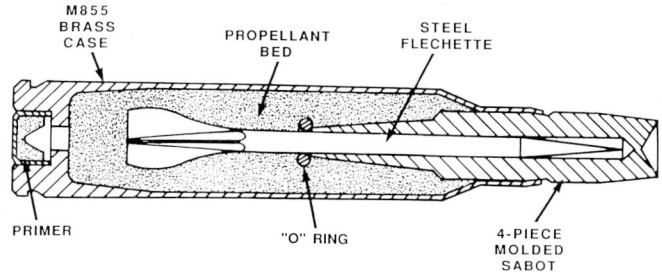

AAI ACR flechette cartridge

The flechette cartridge utilises the existing 5.56 × 45mm cartridge case, and the ballistic characteristics of this lightweight projectile demand that the gas port be located closer to the breech than is customary in conventional weapons. If a conventional bulletted 5.56mm cartridge was to be fired in the weapon the very different gas generation characteristics could cause a serious and dangerous malfunction. For safety purposes, therefore, the magazine interface has been designed so that a standard M16 magazine cannot be inserted, and the ACR magazine is designed so that a bulleted cartridge will not fit.

The rifle is provided with optical and iron sights which are interchangeable, with a quick-release lever to facilitate the change. The recommended optical sight has 4 × magnification and a tritium-powered graticule, but the mounting will accept any other standard optical device.

Ammunition
The AAI projectile is a 0.66g fin-stabilised flechette, approximately 1.6mm diameter and 41.27mm long, with a sharp point. The shaft is roughened to form a friction surface for the four plastic sabot segments, which are held in place around the flechette by a rubber 'O' ring at the rear of the sabot. This complete unit is inserted into a standard 5.56mm M855 cartridge case

When the complete assembly of flechette, sabot and ring leave the muzzle, air drag causes the segments to separate from the flechette and fall away, leaving the flechette to proceed to the target.

The weight of the complete assembly is about 1.36g, so that a low recoil impulse is delivered, allowing for controlled dispersion of the projectiles in a circular pattern around the point of aim.

The propellant is non-standard, having a different formulation than existing military propellants, but is well within the family of normally-produced propellant materials. The sabot segments are made from liquid crystal polymer. The development of this new plastic material has been a significant factor in the successful design of flechette rounds; earlier designs frequently failed due to unsuitable sabot materials.

As with the Steyr-Mannlicher design, the Fact Sheet on this weapon draws attention to the potential hazard to friendly troops from the discarded sabot segments.

Comment
It is noteworthy that the published information contains no description of this weapon's mechanism, other than the reference to it being based on the Serial Bullet Rifle. There were several weapons which went under that name in the late 1960s/early 1970s, as development succeeded development, and it is probable that the system referred to was the one in which three chambers were used and cycled rapidly in and out of alignment with the barrel to give the desired high rate of fire in the three-round burst mode. The AAI design of that period was well regarded, and its rejection was not upon mechanical grounds but rather upon the problems peculiar to the contemporary flechette ammunition. Among other things, it demanded a piston-primed cartridge and the sabot material generated dust which was considered a health hazard. The current round uses conventional priming and the new sabot material appears to do away with the dust problem, so we can see no real obstacles in front of the AAI design. The danger of hand-loading a conventional 5.56mm cartridge into the chamber is acknowledged in the Fact Sheet, but considered only attributable to deliberate sabotage. To avoid the incorrect ammunition being inserted into the chamber, more preventive steps should be considered.

DATA
Cartridge: 5.56mm sub-calibre flechette
Operation: gas, semi-automatic and three-round burst
Locking system: not known
Feed: 30-round magazine
Weight, empty: 3.53kg
Length: 1016mm
Barrel length: not known
Rifling: one turn in 388 calibres (85 in)
Sights: single power telescope, interchangeable iron sight
Muzzle velocity: 1402m/s
Chamber pressure: 55000psi

.30 M1 rifle

The M1 Garand rifle is one of the important milestones in the development of military small arms in this century, being the first military self-loading rifle to be adopted by and issued to an army in more than token quantities. Provision of the Garand was slow, but it began to be issued in small numbers in 1939. When the USA entered the Second World War in 1941 there were still not enough Garands, but production expanded enormously and by 1945 over four million had been turned out from the US arsenals and a further 500000

were made during the Korean war. The M1 remained in service with the US Army until replaced by the M14 in 1957. It was supplied to many countries in NATO after 1945 and a manufacturing base was set up in Italy in the factory of Pietro Beretta where more than 100000 M1s were made and more were converted into various modified versions of which the BM series are the best known. All told probably over 5½ million M1 rifles were made between 1939 and 1959 and they have spread throughout the world, either as a result of aid

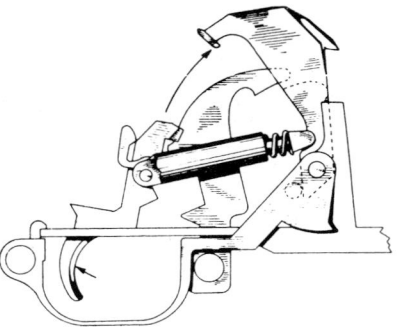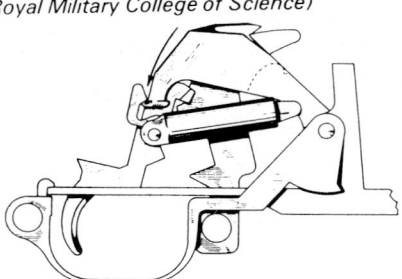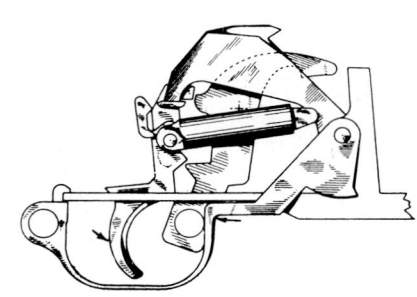

The trigger and firing mechanism *(Courtesy Royal Military College of Science)*

Trigger and firing mechanism
(1) *trigger is pulled, trigger extension goes forward and hammer is released* **(2)** *hammer is rotated by recoiling bolt. Rear bar catches in spring-loaded sear* **(3)** *when trigger is released hammer is transferred to main sear*

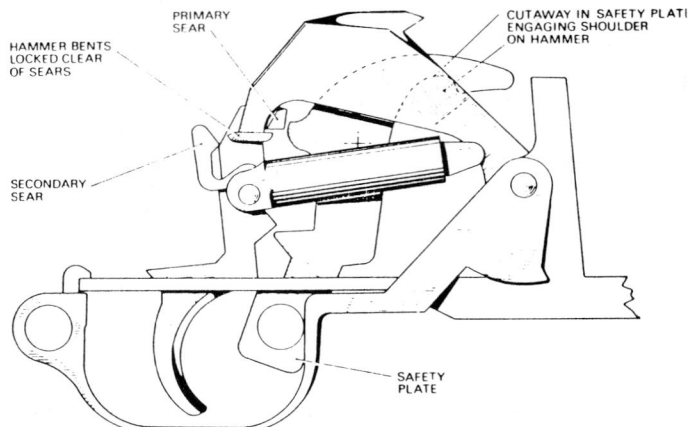

Applied safety mechanism

to friendly countries by the USA, or as a result of sales by both governments and private arms traders.

Operation

The cartridge clip takes eight .30 calibre cartridges. These are loaded into the clip so that the base of each round rests on the back wall of the clip and the extractor groove of each cartridge engages the inner rib of the clip. If the nose of any round is allowed to protrude, loading becomes difficult.

The operating rod handle is retracted and the bolt is held to the rear by a catch. The clip is placed on the magazine follower and pushed in with the thumb until it engages the clip latch. When this occurs the bolt is released and driven forward to engage the top cartridge and pushes it forward into the chamber. The bolt is rotated clockwise by the camming recess in the operating rod and the two lugs on the bolt engage in locking recesses in the receiver.

When the trigger is operated the hammer strikes the firing pin and this hits the cap of the cartridge. Some of the propellant gases are diverted through the gas port into the cylinder. The piston and operating rod are driven to the rear, compressing the return spring under the barrel. There is an initial free movement of 5/16 in (8 mm), allowing the gas pressure in the chamber to drop, until the cam surface of the recess in the operating rod contacts the bolt lug and lifts it up, rotating the bolt to the left and unlocking the lugs from the

receiver. The rotation and slight withdrawal of the bolt-face loosens the case in the chamber, forces the hammer away from the firing pin and withdraws the firing pin into the bolt.

The bolt is carried back by the operating rod, the empty case is pulled out of the chamber by the extractor and when the mouth of the case emerges from the chamber, the spring-loaded ejector emerges from the bolt-face to throw the case out of the rifle to the right front.

The hammer is cocked and the return spring fully compressed. The return spring also exerts pressure on the magazine follower arm which lifts the follower and forces the cartridges upwards in the clip. The operating rod is then forced forward by the pressure of the return spring, taking the bolt with it; the bolt collects the next round from the clip and loads it into the chamber. As the round is chambered the bolt stops, but the operating rod continues moving and the cam action rotates the bolt to lock it.

When the last round in the clip is fired the operating rod comes back and is caught by the operating rod catch and held to the rear. The clip is ejected by the clip ejector.

Trigger and firing mechanism

The Garand trigger and firing mechanism has been very widely copied and is now found in various forms in many other weapons.

The trigger has an extension, the primary sear, which is shaped like a hook and moves backward and forward as the trigger is operated. Mounted on this extension and behind it is a spring-loaded secondary sear.

The hammer carries two bars or bents and when it is rotated fully back these lie horizontally, one reaching forward the other back. The forward one is the primary bent, the rearward one the secondary bent.

When the hammer is cocked the primary sear engages the primary bent. When the trigger is rotated the primary sear goes forward and releases the hammer and the round is fired. The hammer is rocked back by the bolt and since the trigger is still pulled, the primary bent passes behind the primary sear but the secondary bent is caught by the spring-loaded secondary sear and is held to the rear. To fire another round the trigger must be released. The backward movement frees the secondary sear and the hammer starts to rotate but is immediately caught by the primary sear as it moves back.

When the trigger is operated again the primary sear moves forward, the hammer is free and rotates on to the firing pin.

However if the bolt is not fully rotated and locked, the tang of the firing pin is blocked by the receiver bridge. When locking is completed the tang is aligned with a slot in the bridge and can move forward. Also unless the bolt is fully locked the hammer cannot reach the firing pin but will drive the bolt forward and close it, and when it does reach the pin it will have insufficient energy to fire the cap and a misfire will result.

Garand .30 M1 sniper's version with M82 scope and flash hider

Standard M1 .30 rifle

The safety catch is in front of the trigger-guard. When it is pulled back to the 'safe' position a cutaway engages a shoulder in the hammer to lock it. The hammer is also forced back and disconnected from the sear. The safety plate, when it is fully back, also prevents any movements of the trigger.

Stripping

Remove the loaded clip from the receiver by pulling back the operating rod handle. Place the fingers of the left hand over the receiver, press in the clip latch with the left thumb and the clip will be ejected into the left hand.

Check no round remains in the chamber and then allow the operating rod to go forward under control.

Turn the rifle over, grasp the rear end of the trigger-guard and pull backwards and upwards on the guard until it comes free; then continue the pressure so as to withdraw the entire trigger group.

Separate the stock group from the receiver by pulling upwards on the butt end of the stock.

The magazine follower rod is pressed towards the muzzle and then lifted out to the side, together with the spring. Push the follower arm pin from its seat, remove the arm, bullet guide and operating rod catch assembly.

Pull the cocking handle slowly to the rear and upwards and outwards from the receiver to disengage it from the bolt. Once disengaged, withdraw the operating rod. Slide the bolt forward and remove it.

Further stripping is not usually necessary; re-assembly is the reverse procedure.

DATA

Cartridge: .30 M2
Operation: gas, semi-automatic

Method of locking: rotating bolt
Feed: 8-round clip
Weight: 4.30 kg
Length: 1106 mm
Barrel: 610 mm
Rifling: 4 grooves, rh, one turn in 254 mm
Sights: (fore) blade; (rear) aperture, adjustable for elevation and windage
Sight radius: 710 mm
Muzzle velocity: 865 m/s
Effective range: 600 m
Accessories: bayonet M1905; bayonet scabbard M3; bayonet scabbard M1910; sling

Manufacturer

Military production up to 1957 was by Springfield Armory (Government), Springfield, MA: Harrington and Richardson Arms Co, Worcester, MA: International Harvester. In 1974 Springfield Armory Inc, Geneseo, IL, acquired the rights to the Springfield Armory name and extensive parts, stocks and tooling from the US Government. Since 1979 the firm has manufactured the M1 Garand rifle for commercial sales world-wide, and currently offers factory service, replacement parts and accessories for all Government models of the rifle.

Status

Obsolescent. No longer made.

Service

US National Guard, Chile, Costa Rica, Denmark (m/50), Greece, Guatemala, Haiti, Honduras, Italy, Philippines, Taiwan, Tunisia and Turkey.

.30 M1 carbine

In 1940 the US Ordnance Department issued a specification for a light rifle not to exceed 5½ lb (2.5 kg) and capable of either self-loading or automatic action. This weapon was required to replace the pistol and sub machine gun in arms other than infantry.

The cartridge for this weapon was developed by Winchester Repeating Arms Co from its .32 self-loading rifle cartridge of 1905 from which it deviated only very slightly in external measurements. It was known as Calibre .30 SR M1 where SR stood for 'Short Rifle'.

The Winchester design was based on its experimental model rifle produced for test in 1940. This weighed 4.2 kg and incorporated a new form of gas operation, developed by David M Williams, and now universally referred to as 'short-stroke piston operation'.

It was accepted on 30 September 1941 as the US carbine calibre .30 M1 and eventually more of these carbines were produced than any other single US model. According to the best available figures, a total of 6 232 100 M1, M1A1, M2 and M3 carbines were made, of which 6 117 827 were accepted by the US government.

The M1 carbine was a self-loading model, of which 5 510 000 were made.

The M1A1 carbine, standardised in May 1942, had a side-folding stock. It was intended primarily for use by airborne troops and 150 000 were made.

The M2 carbine was standardised in November 1944 as a selective fire carbine and 570 000 were made.

The M3 was a M2 with no sights and a flash hider. It was intended specifically to carry the infra-red Sniperscope and only 2100 were made.

The carbine has been adopted, officially and unofficially, by many armies and it is still in service use although it was relegated to the reserve in the USA in 1957. There are still small quantities being produced by private manufacturers in the USA for sporting use. The carbine is popular with police forces because of its small size and relatively low power, making it a safer weapon to use in crowded urban areas.

Operation

The magazine is loaded and pushed up into the magazine-well. The cocking handle is on the right and is pulled back and released to chamber the top round in the magazine.

When the trigger is squeezed the hammer is released and the firing pin is driven into the cap of the cartridge. The gas pressure forces the bullet up the bore and some of the gas following it is diverted through a gas vent about 114 mm from the chamber. The pressure is high and it impinges on a very small piston which is driven back only 3.6 mm before its movement is stopped. It strikes the operating slide which acquires the momentum of the piston and moves back. There is a delay to allow the chamber pressure to fall while the slide travels back a little over 7.6 mm and then a cam recess engages the operating lug on the bolt. The bolt is rotated, providing primary extraction of the near parallel-sided case, and then unlocked. This rotation starts the cocking action on the hammer and also retracts the firing pin. The empty case is pulled out by the extractor and then the spring-loaded ejector in the bolt throws it forward to the right of the carbine. The hammer is now cocked and the return spring is compressed to drive the bolt forward.

The bolt chambers the round and rotates to the locked position, the operating slide pushes the piston forward inside its cylinder and the weapon is ready to fire again. Should the self-loading action fail, the cocking handle can be used to operate the cycle by hand.

The safety is a push-through type just forward of the trigger-guard. When it is pushed to the right the solid diameter of the plunger moves under the forward end of the trigger and prevents it from moving down.

Stripping

Remove the magazine, cock the action and inspect the chamber.

Push back the sling swivel against the fore-end. Loosen the front band screw

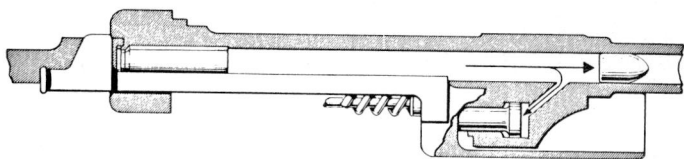

Short-stroke piston action

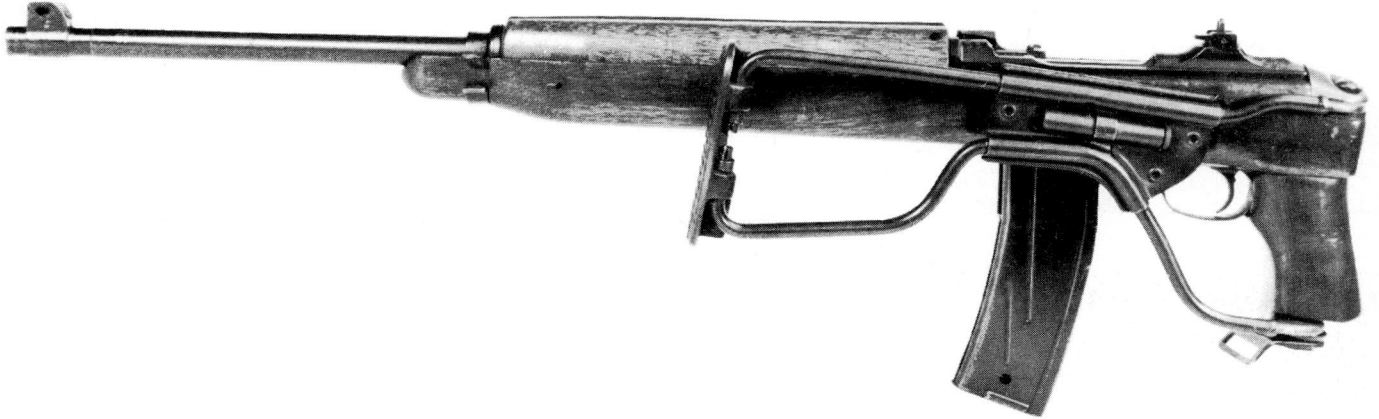

M1A1 folding stock carbine

.30 M2 US Army carbine with 15-round magazine

.30 M3 carbine with Sniperscope and flash hider

using a screwdriver or the rim of a cartridge case. Press the front end of the locking spring rearwards and pull the front band forward over the spring. Slide the upper fore-end forward and lift it clear of the barrel.

Lift the muzzle and pull the barrel receiver and trigger group forward and out of the stock.

Push out the pin at the front of the trigger-guard. Pull back the operating slide spring guide free of the operating slide. Pull it forward and to the right and the spring and guide are free.

Remove bolt by rotating it anti-clockwise, lift it at the front, rotate and remove.

Pull the trigger housing forward to clear the grooves in the receiver and lift it out.

Remove slide from barrel.

Re-assemble in reverse order.

DATA
Cartridge: .30 M1 Carbine
Operation: (M1) gas, self-loading; (M2 and M3) selective fire
Method of locking: rotating bolt
Feed: magazine with 15 or 30 rounds

WEIGHTS
M1 and M2 carbines: (with unloaded magazine) 2.36 kg; (with loaded magazine and sling) 2.63 kg
M1A1 carbine: (with unloaded magazine) 2.53 kg; (with loaded magazine and sling) 2.77 kg

LENGTHS
M1 carbine: 904 mm
M1A1 carbine: (stock extended) 905 mm; (stock folded) 648 mm
Barrel: 458 mm

MECHANICAL FEATURES
Barrel: no regulator; (rifling) 4 grooves rh. 1 turn in 508 mm
Sights: (foresight) blade; (rearsight) flip aperture 0-150, 150-300; leaf slide on M2 models; (sight radius) 546 mm

FIRING CHARACTERISTICS
Muzzle velocity: 607 m/s
Rate of fire: (cyclic) M2 and M3 only, 750 rds/min
Effective range: 300 m

Manufacturer
No longer manufactured but parts and repair service available from Springfield Armory Inc, Geneseo, IL.
Status
Obsolescent.
Service
US National Guard, Chile, Ethiopia, Honduras, Japan, South Korea, Mexico, Norway (in reserve), the Philippines, Taiwan and Tunisia. Made (M1 and M2) under licence by Beretta for the Italian Army and sold to Morocco and the Dominican Republic. Also adopted by many police forces throughout the world.

.30 Browning automatic rifle

The Browning Automatic Rifle (BAR) was produced by John M Browning in 1917 for giving covering fire to troops assaulting across no man's land against fixed machine gun fire. The idea was that the BAR could be carried and fired by one man as he walked forward, thus forcing the enemy to remain below cover until the assault was on top of them. The idea did not succeed, but the BAR was adopted by several countries during the 1920s and 1930s as an attractive and available light machine gun. It was manufactured in Europe as well as the USA and sold as a light machine gun with some minor modifications.

The BAR was a gas-operated weapon, firing from an open bolt. Bolt locking was performed by a link system which lifted the rear end of the bolt

into a recess in the roof of the receiver. It used a bottom-mounted 20-round box magazine and had a cyclic rate of fire of about 550 rds/min. The standard US Army model, used also in Britain during the 1939-45 war, was chambered for the .30-06 cartridge; weapons made in other countries can be found chambered for various other cartridges.

It is possible that small numbers of BAR may still be used in Central America and Africa, and may be held in reserve by some armies, but the weapon can be considered obsolete for all practical purposes. Full details of its operation, together with data, can be found in earlier editions of *Jane's Infantry Weapons*.

7.62 mm NATO M14 rifle

The M14 was adopted in 1957 as the successor to the M1 Garand and was the first weapon in the US Army to use the standardised NATO round. Basically the M14 is an evolution of the M1 and it has, for instance, the trigger mechanism of that rifle with virtually no changes. However it has incorporated some improvements and it could be fairly said that the M14

represents the ultimate development of the old M1. One major change is in the magazine: in the M14 the detachable box allows full magazines to be loaded on to the rifle without needing to use the unsatisfactory clip of the M1. Another feature is the gas cylinder which in the M1 was run right up to the muzzle. This interfered with the jump of the barrel and so affected accuracy

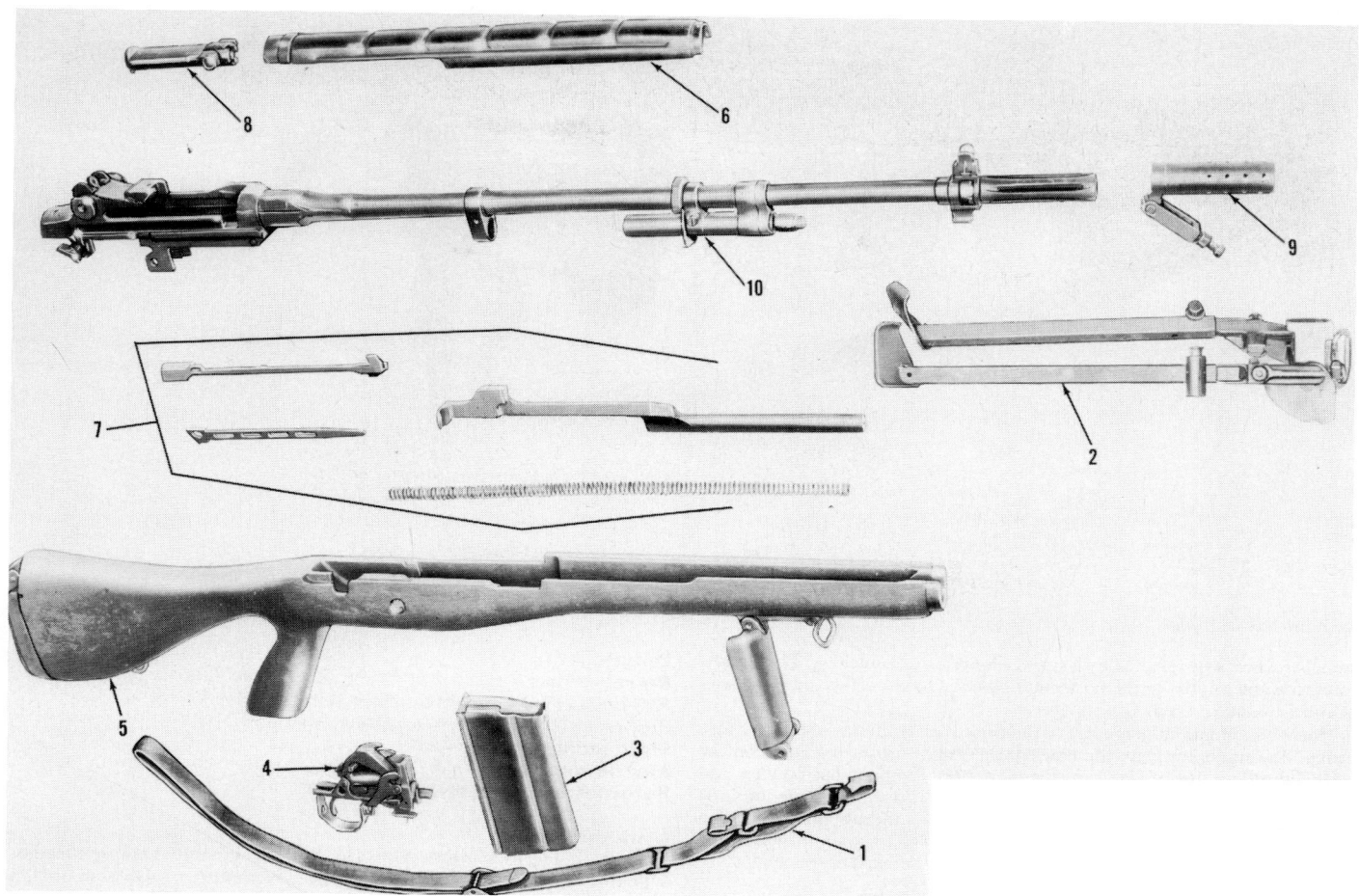

7.62 mm M14A1 rifle field-stripped
(1) *Sling;* **(2)** *M2 Bipod;* **(3)** *Magazine;* **(4)** *Trigger mechanism;* **(5)** *Stock;* **(6)** *Hand-guard;* **(7)** *Operating rod group;* **(8)** *Bolt;* **(9)** *M76 Grenade Launcher;* **(10)** *Gas cylinder*

and consistency and in the M14 the gas port has been moved back to a more normal place about two thirds of the way up the barrel.

Another change is in the gas system. The M1 had a direct-action piston which gave the operating rod a sharp and heavy blow. This was necessary to start the mass of the rod and its cam moving backwards. Although the M14 uses the same sort of rod and cam the gas port allows a more gentle and progressive push on the piston-head by using a different gas cut-off and expansion system. The result is that it is a more pleasant and steady rifle to fire and successive shots are made more quickly since the sights can be brought back on to the target a little faster.

There have been a number of variants on the basic M14, but only one was adopted in any numbers: the M14A1 which comes close to being a light machine gun. The basic rifle is fitted with a straight line stock, a shoulder strap, a rear pistol grip and a folding fore handgrip. There is a light folding bipod and a long sleeve over the muzzle which acts as a compensator to keep the barrel down when firing automatic.

US Government production of the M14 rifle ceased in 1964. Production was resumed on a commercial basis in 1974 by the Springfield Armory Inc.

Operation
The magazine is placed in the rifle by inserting the front end first and then rotating the rear until the magazine catch snaps into engagement.

When the cocking handle is pulled back and released, the bolt picks up the top round and chambers it. The safety catch is in the front of the trigger-guard and is pulled back for 'safe'.

The sights consist of a blade foresight and an aperture rearsight graduated from 200 to 1000 m. The elevation control is on the left of the backsight and is rotated clockwise to elevate. The windage control is on the right of the rearsight. The foresight can be moved across on its block in a dovetail for lateral zeroing. The range scale can be adjusted for elevation zero.

The normal M14 will fire at semi-automatic only. If a selector is fitted it must be rotated according to the type of fire required. When it is positioned with the face marked 'A' to the rear the rifle is set for automatic fire.

When the safety is pushed forward and the trigger squeezed, the hammer rotates forward and strikes the firing pin. The bullet passes up the bore and some of the gas passes through a port and into the hollow interior of the piston-head. This is filled with gas and the piston is forced rearwards, driving the operating rod and bolt with it. After the piston has travelled back slightly less than 4 mm, the gas ports are no longer aligned and no further gas can enter. This system requires no gas regulator as, by design, the pressure available to move the actuating rod back will automatically increase until it is sufficient to overcome the resistance to motion. The piston moves back 38 mm and the exhaust port in the bottom of the gas cylinder is then exposed to allow the expanded gases to escape into the atmosphere.

The operating rod has a free travel of about 9.5 mm to allow the chamber pressure to fall and then the camming surface inside the hump forces the bolt roller upward to rotate and unlock the bolt. The bolt is pushed back and the empty case is extracted and ejected. The compressed return spring forces the operating rod forward and the bolt chambers another round and is rotated into the locked position.

When the ammunition is expended the magazine follower operates the bolt lock and the bolt is held to the rear ready for a fresh magazine to be inserted.

The trigger and firing mechanism is essentially the same as that of the M1 Garand rifle, to which reference should be made.

Stripping
Stripping into basic units is done in the same manner as for the M1 Garand. With the barrel and receiver upside down, press the operating rod spring forward, disconnect the spring guide, and remove both units.
Retract the operating rod until the key on its lower surface coincides with the disassembly notch in the receiver. Lift the rod free and pull to the rear.
Grasp the bolt by its roller, lift up and forward to remove.
Re-assembly is in the reverse order.

Accessories
The blank firing attachment consists of the M12 muzzle attachment and the M3 breech shield. The tubular portion of the attachment is inserted into the muzzle opening of the flash suppressor and is secured by the bayonet lug and a spring clip. The shield is secured to the cartridge guide by a lug with a spring plunger.

The winter trigger kit M5 consists of an arctic trigger and an arctic safety installed to the pistol grip with screws. The winter trigger is a flexible grip lying under the small of the butt and attached to the trigger. Grasping the small of the butt with the right hand enables the firer to reach the flexible link and squeezing it up to the butt stock fires the rifle.

The winter safety is an extension piece reaching forward 38 mm outside the trigger-guard and long enough to be used with arctic mittens.

The M2 bipod is a light folding mount which slips on to the gas cylinder and gas cylinder lock and is secured using the combination tool to tighten the self-locking bolt.

The M6 bayonet knife and M8A1 bayonet knife scabbard. The groove of the bayonet handle slips over the bayonet lug on the flash suppressor and the ring goes over the flash suppressor. The bayonet is pushed back until the lugs of the latching lever snap over the bayonet lug.

The M76 grenade launcher is secured by sliding the launcher over the flash suppressor and pushing the clip latch rearward to secure it to the bayonet lug. The launcher has nine annular grooves numbered 6-1, 2A, 3A and 4A. These allow different ranges by placing the grenade at different positions on the

7.62 mm M14A1 rifle

launcher where it is retained by a spring clip prior to launching. The spindle valve must be rotated to the horizontal position to cut off gas to the piston before a grenade is fired.

The M15 grenade launcher sight can be used for both low- and high-angle firing. The mounting plate for this sight is installed by workshops on the left-hand side of the receiver over the magazine. The plate has notches into which the spring tips of the sight fit. The sight is turned clockwise until the index line of the sight is aligned with the 0° index on the mounting plate and the levelling bubble should then be central. The sight is then turned to the required elevation, and to level the bubble the muzzle must be elevated.

DATA
Cartridge: 7.62 mm × 51
Operation: gas
Method of locking: rotating bolt
Feed: magazine
Method of fire: (M14) self-loading, some selective; (M14A1) selective
Weight, loaded: (M14) 5.1 kg; (M14A1) 6.6 kg

Length: 1120 mm
Barrel: 559 mm
Rifling: 4 grooves, rh, one turn in 305 mm
Sights: (fore) fixed post; (rear) tangent aperture
Sight radius: 678 mm
Muzzle velocity: 853 m/s
Rate of fire: (cyclic) 700-750 rds/min

Manufacturer
Harrington and Richardson Arms Co, Worcester, MA: Thompson-Ramo-Wooldridge, Port Clinton, OH: Winchester-Western Arms Division of Olin Mathieson Corp, New Haven, CT: Springfield Armory (Government) Inc, Springfield, MA: Taiwan.
Status
Current. Manufactured commercially by Springfield Armory Inc.
Service
US Army, Taiwan, South Korea.

M14 (M1A, M1A-A1 and M1A-E2)
Springfield Armory rifles

Springfield Armory Inc is the only manufacturer of the M14 and its variants, designated the M1A, M1A-A1 and M1A-E2. When the US Government Springfield Armory in Massachusetts was closed in the late 1960s the tooling was sold. It has now been set up again in Illinois where the rifles are in production. Sales are largely to commercial users and the majority of these rifles are semi-automatic. However there are significant sales of full-automatic selective fire and semi-automatic models to military buyers.

The standard M1A model is a duplicate of the original general issue M14 rifle. It is also available with National Match and the even heavier Super

Match barrels. All three of these rifles, especially the two heavier barrel models, are consistent winners in civilian shooting competitions.

Springfield Armory also offers the M1A-A1 assault-paratrooper rifle.

The commercial version, designated M1A-E2, duplicates the M14-E2 military weapon previously manufactured, for those who prefer a pistol grip stock.

Springfield Armory makes and supplies a comprehensive range of parts and accessories for all models, military or commercial, and additionally it offers factory service for all Government models of the M14.

7.62 mm Springfield Armory M1A Super Match rifle

7.62 mm SAR-48 rifle

The SAR-48 rifle manufactured by Springfield Armory is an American-made copy of the Fabrique Nationale FAL rifle as originally made in Belgium. It has primarily been introduced, in semi-automatic form, to cater for the commercial market in the USA, where the military FAL was difficult to obtain and the commercial version was scarce and expensive. The rifle is manufactured to a very high standard, with a machined receiver and chrome-lined barrel.

Dimensions, data and ballistic performance of the Standard Model are all exactly the same as the standard Belgian weapon, to which reference should be made, and all parts are fully interchangeable with the original metric-version rifles.

In addition to the Standard Model, Springfield Armory has now developed variant models. These are:

The SAR-48 Bush Rifle. This is a compact version, with a 457 mm barrel but with the standard stock configuration. Apart from the different barrel, all other parts are the same as the Standard Model.

The SAR-48 Para Model. This is a compact, folding stock model. It is available with barrel lengths of 559 mm or 457 mm and may be had as a complete rifle or as a conversion kit which will allow conversion of almost any FAL or SAR-48 to Para specification. A 5.56 mm calibre version of the SAR-48 Para is under development and was expected to be in production by the end of 1987.

Manufacturer
Springfield Armory Inc, 420 West Main Street, Geneseo, IL 61254.
Status
Current. Production.

7.62 mm Springfield Armory SAR-48 rifle family

7.62 mm SAR-3 rifle

This, like the SAR-48 above, is an American-manufactured version of a standard service rifle, in this case the Heckler & Koch G3. The rifle is produced in semi-automatic form only, primarily for the commercial market. Dimensions and performance ar as for the standard German model, and Springfield manufacture both fixed and telescoping butt versions.

Manufacturer
Springfield Armory Inc, 420 West Main Street, Geneseo, IL 61254.
Status
Current. Production.

Standard Model

Retractable Stock Model

7.62 mm Springfield Armory SAR-3 rifles

7.62 mm M21 US rifle

This was originally called the US Rifle 7.62 mm M14 National Match (Accurised). It has been the standard US Army sniping rifle for several years but is now being phased out and replaced by the M24 Sniper Weapon System on page 235.

The differences between this rifle and the M14 are that the barrels are gauged and selected to ensure correct specification tolerances. Barrels are not chromium plated; the stock is of walnut and is impregnated with an epoxy resin; the receiver is individually fitted to the stock using a glass fibre compound; the firing mechanism is hand fitted and polished to provide a crisp hammer release. Trigger pull is between 2 and 2.15 kg; the gas cylinder and piston are hand-fitted and polished to improve operation and reduce carbon build up; and the gas cylinder and lower band are permanently attached to each other. The rifle must group consistently with an average extreme spread for a 10-round group not exceeding 15 cm at 300 m. A sound suppressor may be fitted to the muzzle. This does not affect the bullet velocity but reduces the velocity of the emerging gases to below that of sound. The suppressor is hand-fitted, and reamed to improve accuracy and eliminate misalignment.

Sniper's telescope
The telescope uses two stadia on the horizontal graticule which subtend

Suppressor on 7.62 mm M21 rifle

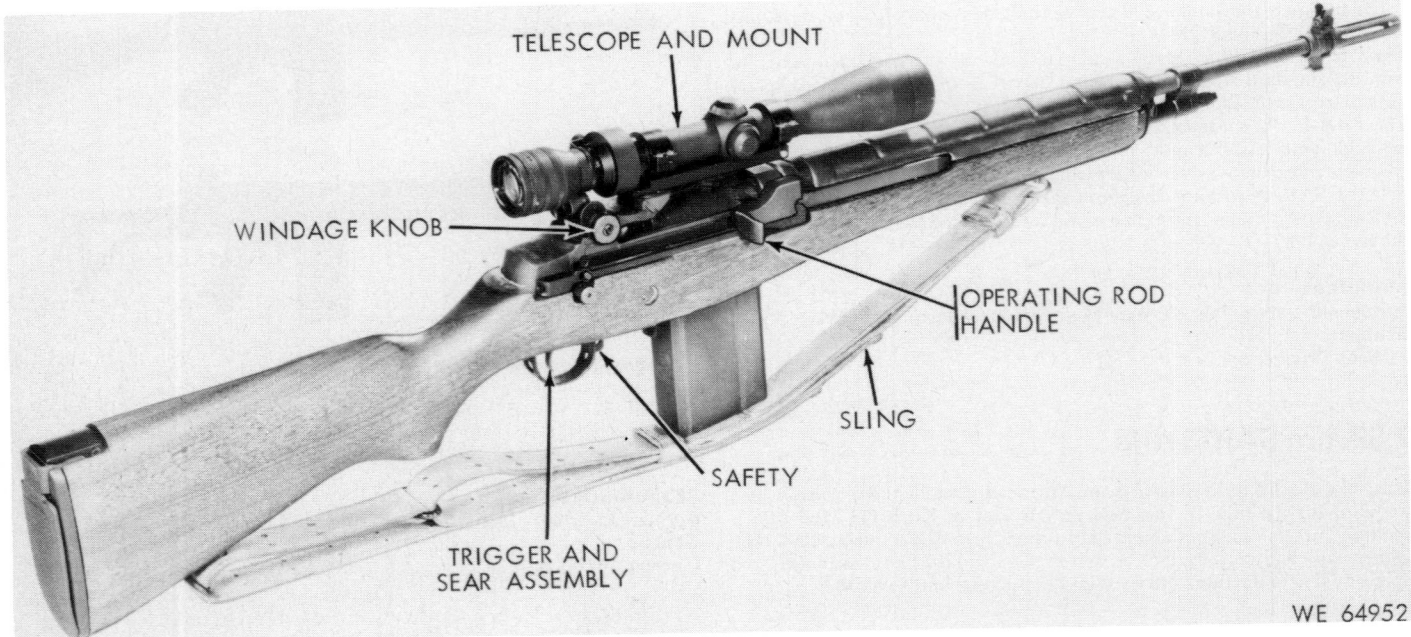

7.62 mm M21 rifle

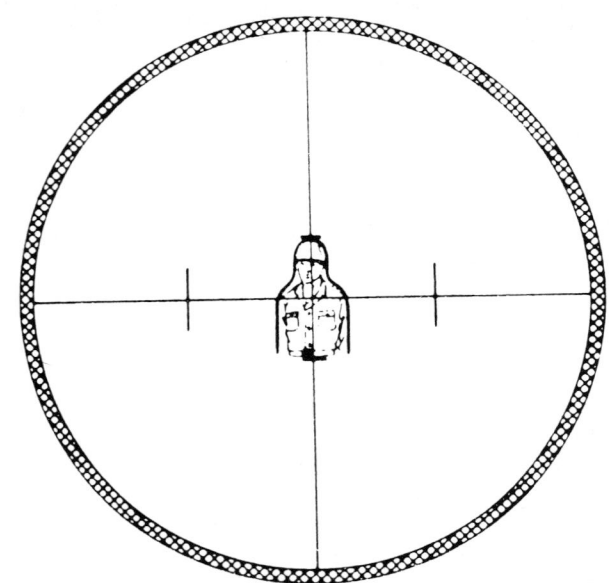

Sniper's sight picture-stadia correctly adjusted

152 cm at 300 m when viewed with the telescope variable magnification ring set to × 3 power. On the vertical graticule are two stadia which subtend 76 cm at 300 m at the same setting. As an illustration of this, the distance from the soldier's waist belt to the top of his helmet can be taken to be 76 cm. At 300 m

range the two stadia on the vertical graticule will, with a × 3 setting, rest on the waist belt and top of the steel helmet. If the range is greater than 300 m the power ring is used to increase the size of the picture until once again the two stadia lines rest on the waist belt and steel helmet. Clearly the range is proportional to the power used, for example since the × 3 magnification just places the stadia on belt and helmet at 300 m then the × 9 magnification will put the stadia on belt and helmet at 900 m.

This can be used to give the sniper a range read-out which he can use in reporting enemy positions, for example. In addition, however, a ballistic cam is attached to the telescope power ring. This cam is cut for the cartridge in use and so the act of placing the stadia in the correct position not only records the range but displaces the telescope axis and so automatically applies the correct tangent elevation to the rifle.

DATA
TELESCOPE
Weight: (with cam) 455 g
Length: 324 mm
Magnification: variable × 3 to × 9
Eye reliefs: 76.2-95.3 mm
Adjustments: internal; $\frac{1}{2}$ minute graduations for elevation and windage
Graticule: cross-hairs with stadia marks
Ballistic cam: for M118 match ammunition
Objective lens diameter: 46 mm
Eyepiece diameter: 34 mm
Finish: black matt anodised
Mount: (weight) 170 g; (material) aluminium; (operation) hand fixed, spring-loaded base; (finish) black matt anodised

Manufacturer
Redfield Inc, 5800 East Jewell Avenue, Denver, CO80224 (basic telescope); Leatherwood Industries (cam and mounting system).

Springfield Armory M21 Sniping Rifle

This is based on the M21 rifle (above) but with some improvements. The stock is of a new pattern, with adjustable cheek-piece and rubber recoil pad, and has the rifle action and barrel bedded in glass fibre resin. The bipod has been redesigned for ease of use and better stability. The heavy barrel is specially made by Douglas, air-gauged and has a twist of one turn in 254 mm; and the operating rod guide has been redesigned. The telescope mount is of Springfield's own design, and whilst the choice of sight is left to the purchaser, Springfield recommends the Leupold Stevens 3.5 × 10 variable-power sight.

Manufacturer
Springfield Armory Inc, 420 West Main Street, Geneseo, IL 61254.
Status
Available to special order only.

7.62 mm Springfield Armory M21 sniping rifle

7.62 mm M40A1 sniping rifle

The M40A1 is a conventional bolt-action magazine rifle with heavy barrel and wooden stock. There are no iron sights, a telescope sight being kept in place on the rifle at all times. A catch inset in the forward edge of the trigger guard allows the bolt to be removed, and a catch in front of the magazine floor plate enables the magazine spring and follower to be removed for cleaning.

DATA
Cartridge: 7.62 × 51 mm NATO
Operation: turnbolt, single-shot
Feed: 5-round integral magazine
Weight: 6.57 kg
Length: 1117 mm
Barrel: 610 mm
Rifling: right-hand, 1 turn in 305 mm
Muzzle velocity: 777 m/s
Sight: telescope, USMC sniper, 10 ×

Manufacturer
Remington Arms Co Inc., Bridgeport, CT 06601.
Status
Current. Production.
Service
US Marine Corps.

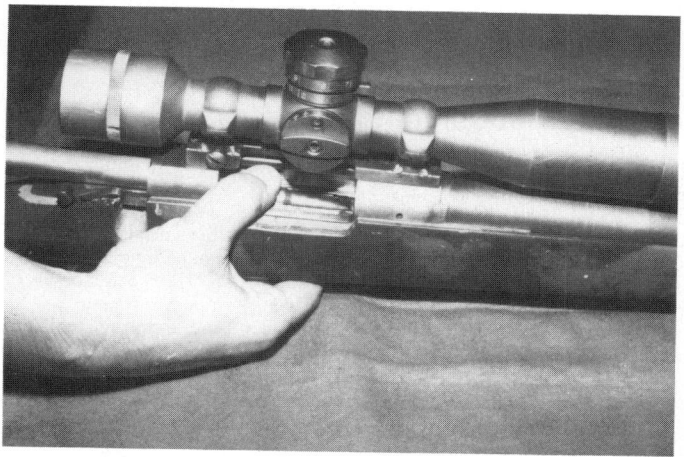

Loading M40A1 sniping rifle

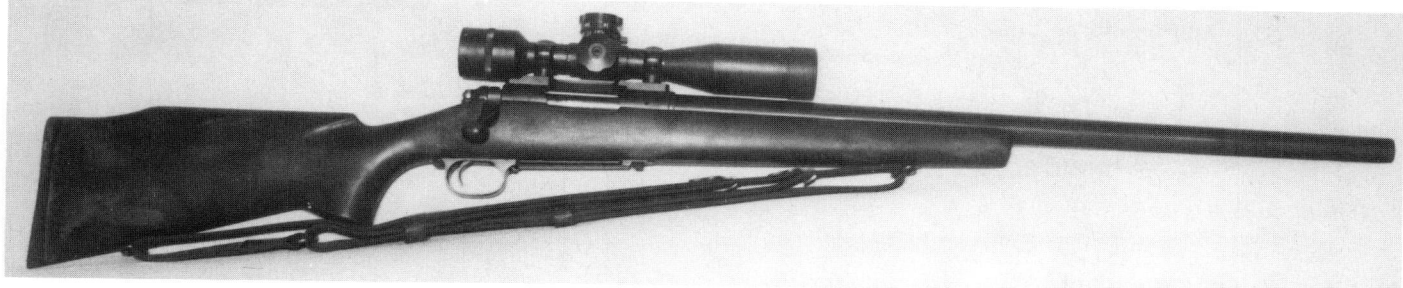

7.62 mm M40A1 sniping rifle

7.62 mm M24 Sniper Weapon System

The M24 is the US Army's first complete sniping system and will eventually replace all other service sniping rifles. First issues took place in November 1988 and the system will be issued to all infantry battalions, Special Forces and Ranger units. It has been stated that the procurement objective is 2510 systems.

The rifle M24 was developed by Remington, based upon their commercial M700 long bolt action and an M/40X custom trigger mechanism. The design was developed around the M118 special sniper ball cartridge, with the possibility of adapting to the .300 Winchester Magnum cartridge as a retrospective modification if required. The stock is synthetic, made of Kevlar-graphite, with an aluminium bedding block and adjustable butt plate.

The complete system consists of the six-shot bolt-action rifle, bipod, day optical sight, iron sights, deployment kit, cleaning kit, soft rifle carrying case, telescope carrying case and total system carrying case. The rifle, with its standard Leupold Ultra M3 10× telescope, weighs 6.35:kg; the complete system in its carrying case, weighs 25.4 kg.

Manufacturer
Remington Arms Co., 939 Barnum Avenue, Bridgeport, CT 06602.
Status
Current; production.
Service
US Army.

7.62 mm M24 Sniper rifle

7.62 mm Grendel SRT sniper rifle

The Grendel SRT is a modern sniping rifle, light and compact enough to complement 5.56 mm rifles under any conditions. It is a conventional bolt-action magazine rifle, with folding butt, integrated bipod mount and muzzle brake. There are no iron sights, but a tapered dovetail on the receiver allows the rigid mounting of any type of telescope sight or of emergency clip-on iron sights.

7.62 mm Grendel SRT sniper rifle

The action is the well-known Mauser bolt with two forward locking lugs and right side safety catch. The large diameter barrel is screwed into the action and then fluted to reduce weight. The muzzle is formed with a long taper and is normally fitted with a single-baffle muzzle brake which reduces the recoil by about 50 per cent. The muzzle taper will also accept sound suppressors in place of the brake.

The stock is injection moulded in DuPont 'Zytel' reinforced with 43 per cent glass fibre, giving outstanding impact strength and great structural rigidity. The butt stock folds down and forward; it is of hollow section, open underneath, so that when folded it encloses the magazine and trigger-guard. There is a rubber recoil pad fitted to the butt.

DATA
Cartridge: 7.62 × 51 mm NATO
Operation: bolt-action
Feed: 9-round box magazine
Weight: 3.00 kg
Length: (stock folded) 760 mm; (stock unfolded) 1035 mm
Barrel: 508 mm
Rifling: one turn in 254 mm

Manufacturer
Grendel Inc, PO Box 560908, Rockledge, FL 32956-0908.
Status
Current. Production.

7.62 mm M600 sniping rifle

This is a bolt-action rifle of conventional pattern, carefully manufactured so as to provide the utmost accuracy. There are three stock options; wood, glass fibre or metal. The wooden stock is of pistol grip type, with a high cheek-piece, and the action is bedded in by epoxy resin. The glass fibre stock is much lighter and thinner; if required, a special lightweight action can be supplied with this stock, reducing the weight of the rifle by about 1 kg. The metal stock is of aluminium alloy and is available in either fixed or folding versions.

The combination muzzle brake and flash hider serve to reduce recoil and muzzle flash. Universal sight mounting bases permit the use of virtually any type of telescope or image-intensifying sight. A laser spot device is also available.

Various types of ammunition are available in this calibre; the maker recommends a 168 grain (10.89 g) 'Medium Range' bullet for use up to 500 m, or the 190 grain (12.32 g) 'Long Range' bullet for use between 500 and 1000 m. To meet low-noise requirements a 'Subsonic' cartridge is available.

The M600 rifle is supplied with a pre-zeroed rangefinding optical telescope and cleaning equipment, in a foam-lined carrying case.

Manufacturer
Accuracy Systems Inc, 15203 North Cave Creek Road, Phoenix AZ 85032.
Status
Production.

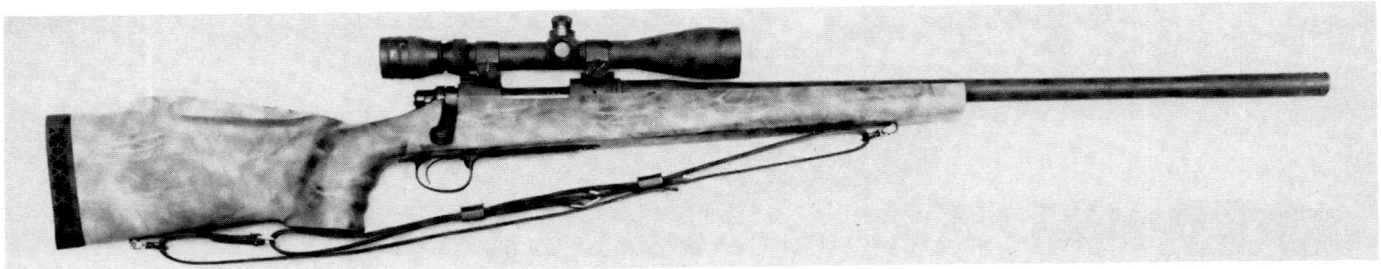

7.62 mm M600 sniping rifle

7.62 mm RAI convertible long-range rifle Model 300

Although of somewhat unconventional appearance, the Model 300 is a normal magazine-fed bolt-action rifle of considerable accuracy. It is normally chambered for the 7.62 × 51 mm NATO cartridge, but by changing the barrel and bolt-head it can fire 8.58 × 71 mm cartridges, a new round developed by Research Armament Industries for this weapon.

The heavy fluted barrel is free-floating, and vibration is damped out by an harmonic balancing assembly concealed within the fore-end. There are no iron sights, a telescope mount being fitted as standard. An adjustable bipod is attached to the fore-end, and the stock is fully adjustable for length, rake and height of cheek-piece.

7.62 mm RAI Model 300 convertible long-range rifle

DATA
Cartridge: 7.62 × 51 mm NATO, or 8.58 × 71 mm RAI
Operation: bolt-action repeater
Feed: 5-round box magazine (7.62 mm); 4-round box magazine (8.58 mm)
Weight: 5.67 kg
Length: (barrel) 610 mm
Rifling: 7.62 mm – 1 turn in 305 mm; 8.58 mm – 1 turn in 254 mm
Muzzle velocity: 7.62 mm – 800 m/s; 8.58 mm – 915 m/s

Manufacturer
Research Armament Industries, 1700 South First Street, Rogers, AR 71756.
Status
Current. Production.
Service
Not known.

7.62 mm Ruger M77V rifle

This is a bolt-action repeating rifle of commercial pattern which has been produced by Ruger for some years but which is now announced in a somewhat different form which has obvious attractions as a useful and economical military or paramilitary sniping rifle. The 'V' in the nomenclature stands for 'Varminter', meaning that the rifle is intended for use against small

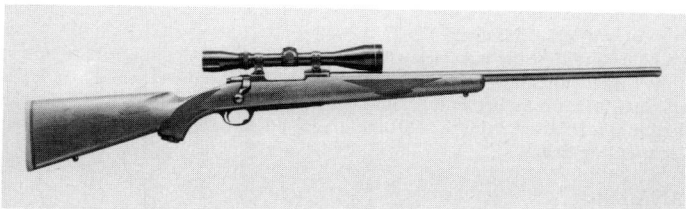

Ruger M77V bolt-action rifle

game, but it will be noted that it has a heavy barrel, no iron sights, and integral telescope mount. In addition to several American commercial calibres, it is available in 7.62 × 51 mm NATO chambering.

DATA
Cartridge: 7.62 × 51 mm NATO; also .22-250, 6 mm, .243, .25-06, .220 Swift
Operation: bolt-action repeater
Feed: 5-round integral magazine
Weight: approx 4.08 kg depending upon calibre
Length overall: 1118 mm
Barrel length: 610 mm

Manufacturer
Sturm, Ruger & Co Inc, Ruger Road, Prescott, AZ 86301.
Status
Current. Production.

7.62 mm Ruger Mini Thirty rifle

This is a modified version of the Ruger Mini-14 Ranch Rifle, chambered for the 7.62 × 39 mm Soviet M43 cartridge. The barrel, receiver and bolt have been engineered to accommodate the larger cartridge. The Mini Thirty is designed for use with telescope sights and features a low, compact telescope mounting which provides greater potential accuracy and carrying facility than found in other rifles of this calibre.

DATA
Cartridge: 7.62 × 39 mm M43
Operation: gas, semi-automatic
Locking: rotating bolt
Feed: 5-round box magazine

Weight: 3.26 kg
Length: 948 mm
Barrel: 470 mm
Rifling: 6 grooves, rh, one turn in 254 mm
Sights: (fore) bead on post; (rear) flip aperture; integral telescope mounts
Muzzle velocity: 713 m/s

Manufacturer
Sturm, Ruger & Co Inc, Ruger Road, Prescott, AZ 86301.
Status
Current. Production.

7.62 × 39 mm Ruger Mini Thirty rifle

5.56 mm AR-15 (M16) rifle

The .223 (5.56 mm) AR-15 was designed by Eugene Stoner, then an employee of Armalite Inc. Upon being standardised as a military weapon the AR-15 was designated the M16. The Air Force version still has this nomenclature but the Army rifle was modified and in 1967 became the M16A1. The differences are chiefly that the M16A1 has a bolt with serrations on the right-hand side and a plunger which protrudes from the body and can be used to force the bolt home if the return spring for some reason is unable to do so. This device, the forward assist assembly, allows the firer to close his bolt when a dirty cartridge produces a high friction force.

Operation
The rifle operates as follows: the cocking handle behind the carrying handle is withdrawn to cock the weapon. If the magazine is empty the magazine follower will rise under the force of the magazine spring and hold the bolt carrier to the rear. When a loaded magazine is in place the carrier will be driven forward by the return spring and the bolt will pick up a round from the magazine and feed it into the chamber.
The bolt motion stops when the cartridge is fully chambered but the carrier continues forward and the cam slot cut in the carrier rotates the bolt

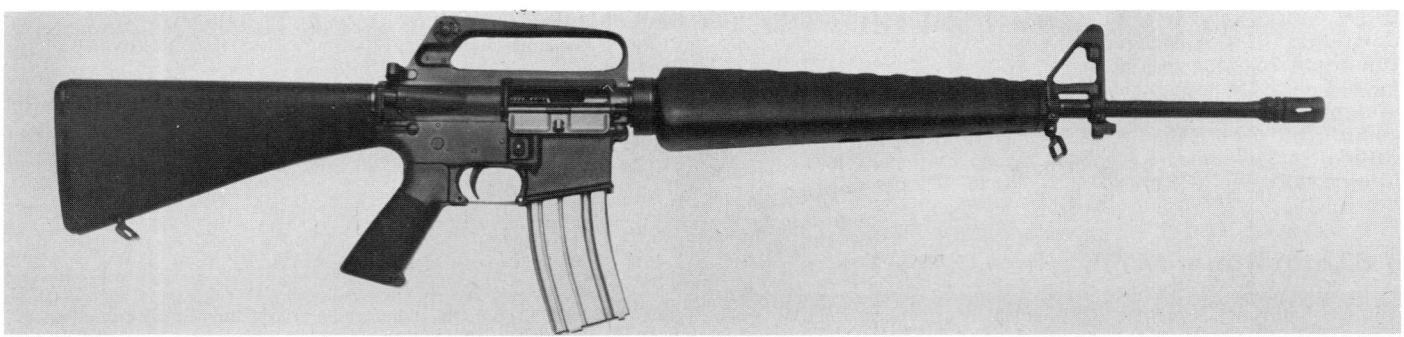

5.56 mm M16A1 rifle

anti-clockwise (viewed from the rear) and the eight locking lugs move behind abutments in the barrel extension. The rifle is now ready and if the fire control selection on the left side of the receiver is set either to automatic or semi-automatic, operation of the trigger will fire the round. When the trigger is pulled, the sear, extending forward of the trigger, is rotated down and moves out of the hammer notch; the hammer is then rotated forward by its spring and hits the firing pin which in turn strikes the cartridge cap and so fires the round. As the bullet passes the gas port some of the gas passes back along a stainless steel tube and through the bolt carrier key into the hollow interior of the carrier. The expanding gas forces the carrier back, the movement of the cam slot moves the cam pin, and the bolt rotates and unlocks. The momentum acquired by the carrier enables it to carry the bolt to the rear at a slightly reduced velocity. The extractor withdraws the cartridge from the chamber; the spring-loaded ejector rod emerges from the left of the bolt-face and rotates the case around the extractor as soon as the case clears the chamber, the case passing through the ejection port on the right side of the receiver. The carrier continues rearward compressing the return spring and cocking the hammer. The action of the buffer and the return spring force the carrier forward and the cycle starts again.

Trigger and firing mechanism

The trigger mechanism of the M16 is based on that of the M1 Garand. The hammer has a bent near its axis of rotation in which the trigger sear engages and a further bent on the underside which enables the hammer to be held by the spring-loaded disconnector sear, at semi-automatic fire. With the selector set to semi-automatic the carrier rotates the hammer back. The trigger is still held to the rear and the disconnector is therefore rotated forward by its spring, and its hook-shaped sear holds the upper inside bent of the hammer. To fire another round the firer must first release the trigger and the trigger sear moves back into its notch on the hammer. The release of the trigger moves the disconnector sear clear of the hammer which then is held only by the trigger sear. This process is repeated for each shot fired at semi-automatic.

Locking lugs of bolt-head, bolt-carrier and bolt cam pin

Bolt-closure device

When automatic is engaged and the trigger is operated, the hammer is released as before but the disconnector is prevented from moving forward to catch the hammer by a cam on the fire-selector lever and so plays no part in the control of the hammer.

To ensure that the hammer will not contact the firing pin until the bolt is fully locked, an automatic sear holds it up until released by the bolt carrier in its final forward movement. Should the trigger be released during firing, the hammer is released from the automatic sear but caught by the trigger sear so terminating firing.

Stripping

The M16 is field-stripped as follows: remove the magazine, cock the action and check that the chamber is empty. The take-down pin on the left rear of the receiver is pushed through and the butt and lower receiver dropped down. The bolt carrier assembly and cocking handle can then be slid out to the rear and the bolt separated from the carrier. The hand-guard is removed next, followed by the buffer and return spring.
Re-assembly is in reverse order.

Training aids

A blank-firing attachment, which restricts the gas flow from the muzzle and thus develops sufficient pressure inside the barrel to operate the bolt mechanism, is available.

Also available for training purposes is a calibre .22 long-rifle conversion unit consisting of a calibre .22 long-rifle bolt assembly and magazine insert.

DATA
Cartridge: 5.56 mm × 45
Operation: gas, direct action, selective fire; (M16A2) semi-automatic with 3-shot bursts
Method of locking: rotating bolt
Feed: 20- and 30-round box magazine

WEIGHTS
M16: (rifle without sling or cleaning equipment) 3.1 kg
M16A1: 3.18 kg
Magazine, empty: (20-round) 91 g; (30-round standard) 117 g; (30-round nylon) 113 g
Magazine, loaded: (20-round) 318 g; (30-round standard) 455 g
Sling: 182 g
M16: (rifle with sling and loaded 20-round magazine) 3.6 kg; (rifle with sling and loaded 30-round magazine) 3.73 kg
M16A1: (with sling and loaded 20-round magazine) 3.68 kg; (with sling and loaded 30-round magazine) 3.82 kg
Trigger pull: 2.3-3.8 kg

LENGTHS
Rifle: (with flash suppressor) 990 mm; (with bayonet knife M7) 1120 mm
Barrel: 508 mm; (with flash suppressor) 533 mm

MECHANICAL FEATURES
Rifling: (M16A1) 6 grooves rh. 1 turn in 305 mm
Sights: (fore) cylinder on threaded base; (rear) (M16A1) flip aperture

FIRING CHARACTERISTICS
Muzzle velocity: 1000 m/s
Rate of fire: (cyclic) 700-950 rds/min
Effective range: 400 m

Manufacturer
Colt Firearms, Hartford, CT 06101. Under licence in South Korea, the Philippines and Singapore.
Status
Current. Production.
Service
In service with US Forces. Also Chile, Dominican Republic, Haiti, Italy, Jordan, South Korea, Mexico, Nicaragua, Panama, the Philippines, the United Kingdom and Vietnam.

5.56 mm Colt M16A2 assault rifle

The Colt M16A2 is a lightweight, gas operated shoulder arm capable of semi-automatic, and either fully automatic or three shot burst control fire. Like all weapons in the M16 system, the M16A2 offers combat-proven performance, over eight million M16 rifles now having been produced and placed in military service throughout the world. When NATO adopted the 5.56 mm cartridge as standard, the Colt M16 rifle was the control weapon against which other entrants were judged.

The M16A2 system's straight line construction, with the barrel, bolt, recoil buffer unit and stock assembled in line, disperses recoil straight back to the shoulder, while keeping barrel climb to a minimum. This is critical to the need for accurate high-volume fire required by today's infantry. The bolt locks directly into the barrel, eliminating the need for a heavy steel receiver, while the direct gas operating system, eliminates the need for the conventional operating rod normally associated with gas operated weapons.

The M16A2 has a rifling twist of one turn in 178 mm, which allows it to accommodate the entire range of NATO 5.56 mm ammunition. The handguard, buttstock and pistol grip are made from new, significantly more impact-resistant, materials. The handguard is round, has an improved heat deflector, and is ribbed for greater grip control. The barrel is made heavier for added stiffness and better accuracy; it incorporates a new flash suppressor which further reduces muzzle climb and helps to eliminate dust dispersion when firing in the prone position. A cartridge case deflector has been added to the lower receiver, behind the ejection port, to guide the ejected case away from the face of left-handed shooters.

The rifle is equipped with a new target-style rear sight system, adjustable for both windage and elevation to 800 m. The M203 grenade launcher can be easily fitted to the M16A2, and the muzzle will accept any US or NATO-standard rifle grenade.

DATA
Cartridge: 5.56 × 45 mm NATO
Operation: gas, selective fire
Locking: rotating bolt
Feed: 20- or 30-round box magazine
Weight: 3.40 kg without magazine
Length: 1000 mm
Barrel: 510 mm
Rifling: 6 grooves, rh, one turn in 178 mm
Sights: (fore) adjustable post; (rear) aperture, adjustable for windage and elevation
Muzzle velocity: (M193) 991 m/s; (SS109) 948 m/s
Muzzle energy: (M193) 1722 J; (SS109) 1765 J
Rate of fire: 700-900 rds/min cyclic

Manufacturer
Colt Firearms, Hartford, CT 06101, and licensees.
Status
Current. Production.
Service
US Armed forces and 55 countries worldwide.

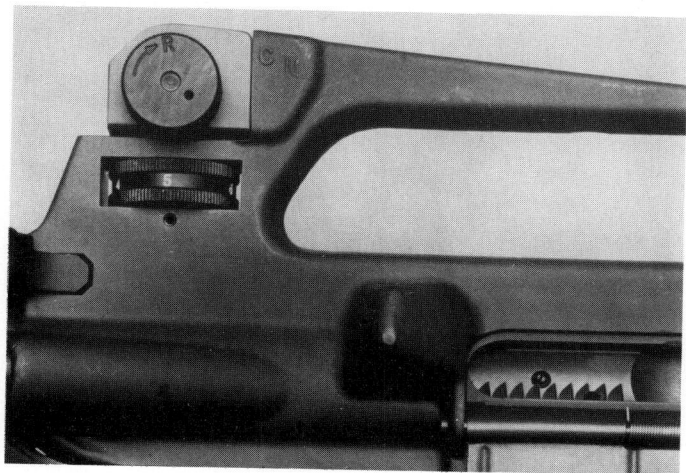

Adjustable rearsight of M16A2 rifle

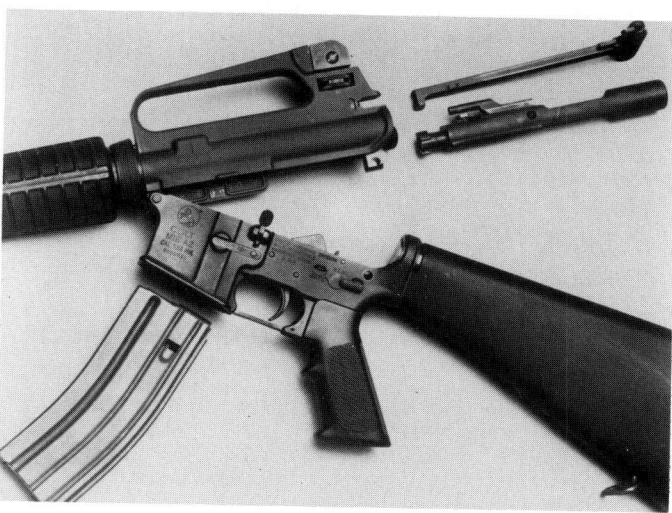

M16A2 rifle opened, bolt assembly and cocking handle removed

5.56 mm M16A2 rifle

5.56 mm Colt M4 (Model 727) Carbine

The Model 723 Carbine is a lighter and shorter version of the M16A2 rifle, intended for use where lightness and speed of action are paramount. It is an accurate and effective weapon under all practical field applications. The carbine is fitted with a new target-style rear sight, adjustable for both windage and elevation to a range of 800 m, which makes it possible to take advantage of the greater effective range and penetration available with the NATO standard rounds. The barrel is rifled one turn in 178 mm, the same as the M16A2 rifle.

All mechanical components are interchangeable with those of the M16A2 rifle, ensuring commonality of parts and reduced maintenance load. The carbine accepts all Colt M16 and NATO STANAG 4179 magazines. The sliding buttstock, when extended, allows the user to fire from the shoulder position, and when retracted to fire from the hip.

The M203 grenade launcher can be easily assembled to the M4 carbine, giving the user both point and area firing capabilities. All US and NATO standard rifle grenades can be fired from the muzzle without the need for supplementary equipment.

DATA
Cartridge: 5.56 × 45 mm NATO
Operation: gas, selective fire
Locking: rotating bolt
Feed: 20- or 30-round box magazine
Weight: 2.54 kg without magazine
Length: (butt extended) 840 mm; (butt retracted) 760 mm
Barrel: 368 mm
Rifling: 6 grooves, rh, one turn in 178 mm
Sights: (fore) adjustable post; (rear) aperture, adjustable for windage and elevation
Muzzle velocity: (M193) 921 m/s; (SS109) 906 m/s
Muzzle energy: (M193) 1509 J; (SS109) 1569 J
Rate of fire: 700-1000 rds/min cyclic
Max effective range: 600 m

5.56mm Colt M4 carbine, right side

5.56 mm Colt M4 carbine, with M203 grenade launcher attached

Manufacturer
Colt Firearms, Hartford, CT 06101.
Status
Current. Production.

Service
US Army; US Marine Corps; adopted in Canadian Army as C8 rifle; Honduras, UAE, Guatemala, El Salvador and various other countries.

5.56 mm Colt Model 733 Commando assault rifle

The Model 733 Commando is the smallest and most compact version of the M16A2 rifle, developed to provide a robust weapon with compact dimensions allowing speed of action. The Commando can be carried in the slung position with a minimum of inconvenience and bulk, much like a sub-machine gun. It can provide instant, accurate and effective fire under all combat conditions. It is ideal for a variety of tactical roles, including crews of cavalry fighting vehicles and other groups whose primary function is other than infantry operations.

The Commando's mechanical components are interchangeable with those of the M16A2 rifle, ensuring commonality of parts and reduced maintenance load. It accepts all Colt M16 and NATO STANAG 4179 magazines. Like the M4 carbine it has a sliding buttstock and the barrel is rifled one turn in 178 mm.

The Commando is used by US armed forces and Federal agencies, and several countries have adopted it as a 5.56 mm sub-machine gun.

DATA
Cartridge: 5.56 × 45 mm NATO
Operation: gas, selective fire
Method of locking: rotating bolt

5.56 mm Colt Model 733 Commando, right side

5.56 mm Colt Model 733 Commando, left side

Feed: 20- or 30-round box magazine
Weight: 2.44 kg without magazine
Length: (butt extended) 760 mm; (butt telescoped) 680 mm
Barrel: 290 mm
Rifling: 6 grooves, rh, one turn in 178 mm
Sights: (fore) cylindrical post; (rear) aperture, adjustable for windage and elevation
Muzzle velocity: (M193) 829 m/s; (SS109) 796 m/s
Muzzle energy: (M193) 1224 J; (SS109) 1270 J

Rate of fire: (cyclic) 700-1000 rds/min
Max effective range: 400 m

Manufacturer
Colt Firearms, Hartford, CT 06101.
Status
Current. Production.
Service
US Army, US Marine Corps, UAE, Guatemala and other countries.

5.56 mm Colt M231 firing port weapon

This is a gas-operated, air-cooled, magazine-fed weapon designed primarily to be mounted in armoured vehicles for offensive or defensive fire. It is, in most respects, an adaptation of the M16 rifle design to a specialised role for use in armoured personnel carriers and mechanised infantry combat vehicles (MICVs).

The M231 is used in the US M2 'Bradley' IFV, but is adaptable to a wide variety of armoured vehicles. Whenever required, it can be rapidly removed from the vehicle mount and used as a hand-held individual weapon.

DATA
Calibre: 5.56 mm × 45
Overall length: (stock retracted) 710 mm; (stock extended) 820 mm
Barrel length: 368 mm
Weight: (empty) 3.9 kg; (with full magazine) 4.32 kg
Feed: 30-round box magazine
Rate of fire: (cyclic) 1100/1300 rds/min; (practical) 50/60 rds/min
Muzzle velocity: 914 m/s
Effective range: 300 m

5.56 mm Colt M231 firing port weapon

Manufacturer
Colt's Industries, Firearms Division, Hartford, CT 06102.
Status
Current. Production.
Service
US Army.

La France M16K assault carbine

The smallest and lightest of the M16 variants, the La France M16K is ideal for carrying in the confined spaces of aircraft and motor vehicles. It is designed to give special operations units controllable firepower and high lethality in a weapon of minimum size and weight.

The specially-engineered gas system results in a cyclic rate of fire under 600 rds/min, unusually low for weapons of this calibre. The vortex flash suppressor is extremely effective, completely eliminating any trace of muzzle flash. The M16K utilises an aperture front sight in combination with the standard rear sight in an attempt to duplicate the effectiveness of the ring-type graticule in the optical sight of the Steyr AUG.

A molybdenum-disulphide dry-film lubricant is applied to all metal components. This maintenance-free finish prevents rust and corrosion and provides lubrication to moving parts in all extremes of climate and temperature.

DATA
Cartridge: 7.62 × 45 mm
Operation: gas, selective fire
Method of locking: rotating bolt
Feed: 20- or 30-round magazine
Weight, empty: 2.50 kg
Length: butt extended 686 mm; butt retracted 610 mm
Barrel: 213 mm; with flash suppressor 254 mm
Rifling: 6 grooves, rh, one turn in 32 calibres
Sights: (fore) aperture; (rear) flip aperture, adjustable for windage
Sight radius: 127 mm
Muzzle velocity: 762 m/s (M193)
Rate of fire: 550-600 rds/min

Manufacturer
La France Specialties, PO Box 178211, San Diego, CA 92117, USA.
Status
Current. Production.

La France M16K assault carbine

La France M14K assault rifle

The M14K is a successful attempt to improve the handling qualities and full-automatic controllability of the US 7.62 mm service rifle. The reduced overall length considerably improves the handling, while the shortened barrel causes a decrease in muzzle velocity and, in consequence, muzzle energy, thereby reducing muzzle climb. The high-efficiency muzzle brake aids in keeping the sights on target, as does the lower cyclic rate produced by the large-volume modified M60 gas system. Perceived recoil is little more than that of most 5.56 mm rifles.

The M14K is currently under evaluation by a US Marine Special Warfare unit.

DATA
Cartridge: 5.56 × 45 mm
Operation: gas, selective fire

Method of locking: rotating bolt
Feed: 20-round magazine
Weight, empty: 3.75 kg
Length overall: 902 mm
Barrel: 338 mm; with muzzle brake 411 mm
Rifling: 4 grooves, rh, one turn in 40 calibres
Sights: (fore) fixed post; (rear) tangent aperture
Sight radius: 445 mm
Muzzle velocity: 765 m/s
Rate of fire: 600-650 rds/min

Manufacturer
La France Specialties, PO Box 178211, San Diego, CA 92117, USA.
Status
Current. Production.

La France M14K field-stripped to three major groups

La France M14K compared with standard M14 (top)

Ciener AR-15/M16 belt-feed mechanism

The Ciener belt-feed mechanism is a factory-fitted conversion which allows the AR-15 or M16 rifles to be belt-fed in either semi-automatic or automatic fire. The feed mechanism is driven by the reciprocating bolt carrier, and it can be readily removed from the rifle by simply opening the action and lifting it out. With the unit removed, the rifle can be used with the normal magazine with no adjustment; when the magazine is in place, the holes in the receiver necessary for installation of the belt feed are closed off, and dirt cannot enter the weapon.

The belt is a disintegrating-link pattern. There is an accessory belt carrier which clips on below the rifle and carries a 100-round belt. In addition, it has a chute which deflects the empty links back into a partitioned section of the carrier, saving them for future use.

The manufacturer supplies the unit ready-assembled to new AR-15/M16 rifles or carbines, or will convert an existing weapon at its factory.

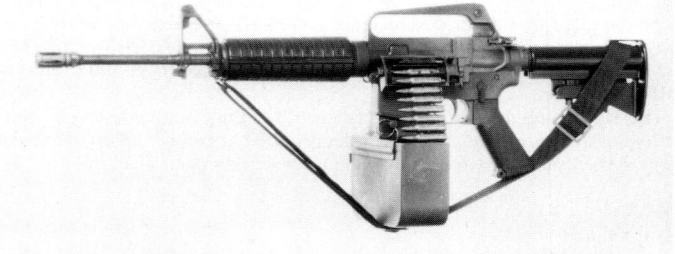

Ciener belt-feed mechanism, with belt box, on AR-15 carbine

Manufacturer
Jonathan Arthur Ciener Inc, 6850 Riveredge Drive, Titusville, FL 32780.

Status
Current. Production.

Atchisson .22 Conversion kit

The Atchisson .22 conversion kit permits conversion of Colt AR15, A1 and A2 firearms to fire .22 LR rimfire cartridges. The unit replaces the standard bolt in the Colt AR15/M16 family of firearms and allows the user top employ inexpensive .22LR ammunition instead of 5.56 mm. The unit is blowback operated.

The unit is made from chrome molybdenum steel, heat treated, and will fit all Colt AR15 A1 and A2 rifles. With the addition of the automatic trip and anti-bounce weight it will permit semi- and automatic fire from weapons with selective fire capability.

Manufacturer
Jonathan Arthur Ciener Inc., 6850 Riveredge Drive, Titusville, FL 32780.
Status
Current. Production.

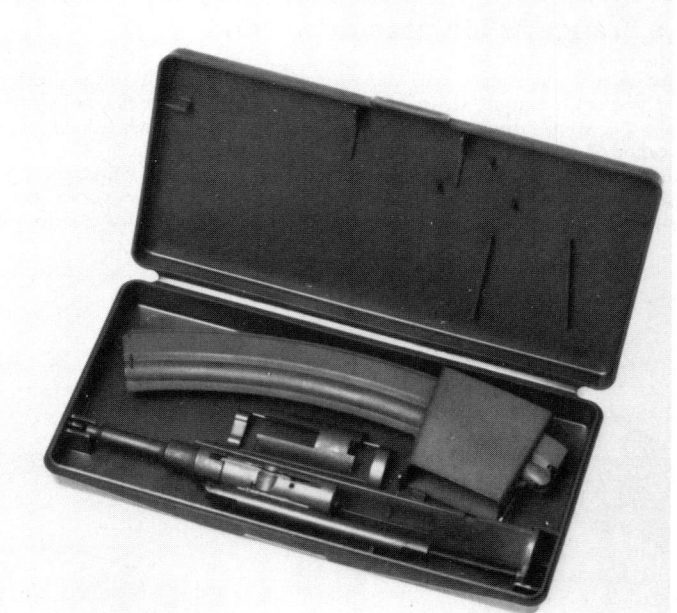

Atchisson .22 conversion kit

Colt AR15 rifle fitted with Atchisson .22 conversion kit

Ciener .22 Conversion kit

The Ciener .22 conversion kit permits conversion of AK series Kalashnikov rifles to fire .22LR rimfire cartridges. The conversion kit is available for both the 7.62 × 39 mm and 5.56 mm Kalashnikovs and copies. The unit consists of a sub-calibre chamber adapter with 228 mm barrel liner (without liner for 5.56 mm weapons), sub-calibre bolt, return spring and return spring guide rod locating block assembly, and 30-round .22LR magazine. These items are fitted into the dismantled weapon, replacing the standard parts, and the weapon re-assembled.

The unit is made from chrome molybdenum steel, heat treated, and will fit and operate in all AK-type rifles. With substitution of a special bolt and anti-bounce weight it will permit semi- and automatic fire from rifles with selective fire capability.

Manufacturer
Jonathan Arthur Ciener Inc., 6850 Riveredge Drive, Titusville, FL 32780.
Status
Current. Production.

Ciener .22 conversion kit

AK-type rifle with Ciener .22 conversion kit fitted

C-MAG 5.56 mm 100-round magazine

The C-MAG is a twin-drum high capacity ammunition magazine for rifles, light support weapons, firing port and other specialised weapons in 5.56 mm calibre. Covered by US and other patents, its main components are constructed of high-impact plastic. Avoiding trouble-prone compression springs, the C-MAG utilises a unique rotation device and low torsion stress springs which ensure positive control of each round and proper feeding. This advanced design makes the C-MAG the world's first magazine capable of providing a full 100-round capacity and yet which can be pre-loaded and stored indefinitely. The design provides for equal distribution of the ammunition in each of the twin drums located on either side of the weapon. This results in a fixed centre of gravity during firing, an important attribute during full automatic performance. The compact size allows for better ground clearance and light weight. The C-MAG loads and functions just like standard magazines. It may be re-loaded and re-used repeatedly.

Operation
Loading does not demand disassembly or special tools. The cartridges are loaded by pressing them into the feed aperture in the same way as a normal straight magazine. The cartridges will automatically feed alternately into each drum. Loading can be done by hand or by an appropriate mechanical loader. The springs are tensioned during the loading operation, without the need for any supplementary operations. The magazine does not need to be empty to be loaded and remains completely operational at all times; loading may be stopped at any point, the magazine inserted into the weapon and the C-MAG is ready for immediate operation.

The key feature of the C-MAG is the formation of the double rows of cartridges in each of the drums. The diagram illustrates how these double rows are formed in the interior of the magazine. Also shown are 15 solid spacer rounds, two pusher arm assemblies and the linked spacer round assembly. The solid spacer rounds occupy the feed clip when the magazine is

empty. The number required depends on the length and style of the feed clip, which is specific to an individual weapon system. The pusher arm assemblies transfer the spring energy to the cartridges while being loaded and unloaded. In addition, they serve to maintain the proper orientation of the leading solid spacer rounds. The linked spacer round assembly is very important; this assembly is positioned at the open end of the feed clip when the magazine is empty and is linked so that it cannot fall out of the feed clip. The upper half of the assembly is tapered so that the weapon bolt will pass over it (when the last round has been fired) and stop in the closed position.

During loading the live cartridges are pressed into the feed clip. The column of spacer rounds and live cartridges are, as loading progresses, pushed into the drums. The cartridge column splits at the point where the feed clip joins the

C-MAG 100-round 5.56 mm magazine

C-MAG 100-round magazine in place on M16 rifle

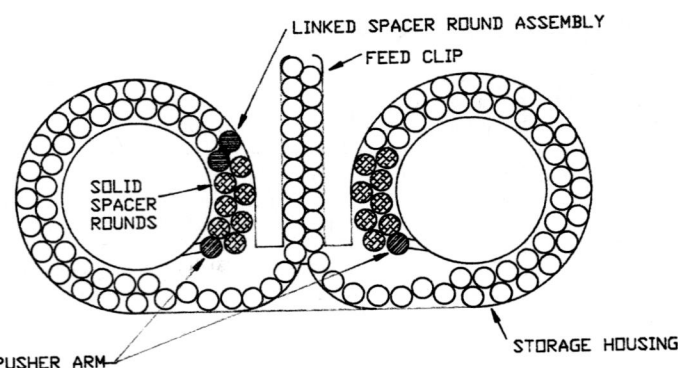

Diagram showing the feed system of the C-MAG magazine

storage housing; this allows for the concurrent division of the stored ammunition on each side of the magazine. Further into each drum the cartridges are allowed to divide into two concentric rows for storage. Loading stops when the maximum number of cartridges has been inserted.

Unloading during firing is the reverse of the above process. As cartridges are stripped from the feed clip, so the column of cartridges advances to fill the space. The springs in the C-MAG are designed to provide a virtually constant amount of tension, regardless of the number of cartridges remaining in the magazine.

Weapon compatibility
Due to the geometry of its construction, the C-MAG can be used on almost any modern 5.56 mm combat rifle or other weapon without modification to

the weapon being required. While the current production model employs the NATO STANAG 4179 (M16) magazine interface, it is easily adaptable by the manufacturer to any other 5.56 mm weapon by substituting a feed clip designed for the particular magazine housing. The feed clip is the centre piece of the C-MAG, detachable from the storage housing for maintenance or weapon compatibility purposes. The main storage housing is standard for all types of weapon; magazines can be ordered with compatible feed clips in place, or the changing can be done by the user. With the correct feed clip installed the C-MAG snaps into the standard magazine housing in the normal manner and is released by the weapon's magazine catch in the usual way.

Accessories
Mechanical and automatic loading devices holding 5, 20 and 100 cartridges for speed loading are available, as well as carrying pouches.

DATA
Capacity: 100 5.56 mm cartridges
Ammunition: ball, tracer, blank rounds
Weight, empty: 1.0 kg
Dimensions: 251 × 81 × 17 mm
Rate of fire: > 1300 rds/min
Materials: Main components: filled thermoplastic polyester. Minor components: non-corrosive steel, alloy materials
Shelf life, loaded: indefinite

Manufacturer
The Beta Company, 4200 Northside Parkway, Atlanta GA 30327.
Status
Current. Production.
Service
Under evaluation by various military and law enforcement agencies world wide, including US Army, US Navy, UK MoD.

MWG 5.56 mm 90-round magazine

The MWG Company offers a 90-round drum-type magazine for the M16/AR15 family of assault rifles including other assault rifles which can accept the same style magazine. The '90-Rounder' is a simple, reliable and low-cost magazine that has been selected as an optional accessory for the Colt M16 light machine gun.

The unit is constructed of high impact thermoplastics for durability and

The MWG 90-round magazine in place on M16 LMG

light weight. The rear panel of the drum is clear so that remaining ammunition is always in view. The unit interfaces with the weapon in the same manner as the standard 20-round box magazine. A single feed path with only one moving part ensures reliability. While the unit is designed for long-term repeated use it is inexpensive enough to be considered expendable in combat.

Loading is accomplished rapidly with a detachable plastic loading guide and a metal insertion tool included with each unit. An optional carrying case of heavy duty nylon is available; this can be worn over the shoulder or clipped to a belt or harness.

The '90-Rounder' does not compromise the feel or balance of the weapon in any firing position; the drum rests comfortably and solidly against the inside of the left arm. The unit has been shown to actually enhance weapon stability in both semi- and full-automatic firing modes.

Manufacturer
MWG Co Inc, 18689 SW, 103 Court, Miami, FL 33157.
Status
Current. Production.
Service
In service with or under evaluation by various military, government and police authorities worldwide.

Distributor
Calibre International Inc., 16608 Saddle Club Road, Fort Lauderdale, FL 33326

Redi-Mag speed loading system

This system allows two magazines to be carried on an M16 or AR-15 rifle, one in the feedway and the other alongside it, both being retained in place by a single magazine catch. In operation, the rifle is fired until the first magazine is empty; the firer then grasps the spare magazine with one hand and depresses the magazine catch with the other. The empty magazine drops out of the feedway and the second magazine is released, so that it can be withdrawn from its housing, moved sideways, and re-inserted into the feedway of the rifle. In this manner a fresh magazine can be inserted within two seconds of the last shot being fired. The housing clamps on to the existing magazine housing, does not interfere with the balance or operation of the rifle, and is of welded steel construction.

Manufacturer
Johnson Firearms Specialities, PO Box 12204, Salem, OR 97309.
Status
Current. Production.
Service
Under evaluation by US Marine Corps.

Redi-Mag speed loading system

JFS Safety Block

This device has been produced in response to a requirement for some method of protecting the interior of the M16 rifle when safety demands that it be carried with the bolt, ejection cover and magazine housing open during exercises. In such situations the rifle rapidly fills with dust, and subsequent live-firing exercises either fail or are delayed while weapons are cleaned.

The Safety Block is of hard plastic and fits into the weapon's magazine housing so as to block movement of the bolt and close off the apertures. It is currently under evaluation by the US Marines and the Ohio National Guard, each having 500 units for trial.

Manufacturer
Johnson Firearms Specialities Inc, PO Box 12204, Salem, OR 97309.
Status
Current. Available.
Service
Under evaluation by US forces.

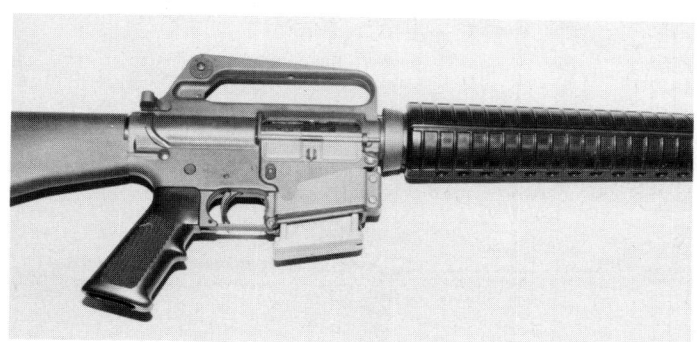
JFS Safety Block partly inserted into M16 magazine housing

5.56 mm Mini-14/20GB Ruger infantry rifle

The Mini-14/20GB infantry rifle is a special adaptation of the commercial model Mini-14 rifle intended for military applications. Changes from the basic Mini-14 configuration include the addition of a protected front sight which incorporates a bayonet lug, a heat-resistant glass fibre hand-guard, and a flash suppressor. Technical data and specifications for the Mini-14/20GB rifle are substantially the same as for those of the standard Mini-14 rifle. The Mini-14/20GB rifle is available in both blued and stainless steel versions.

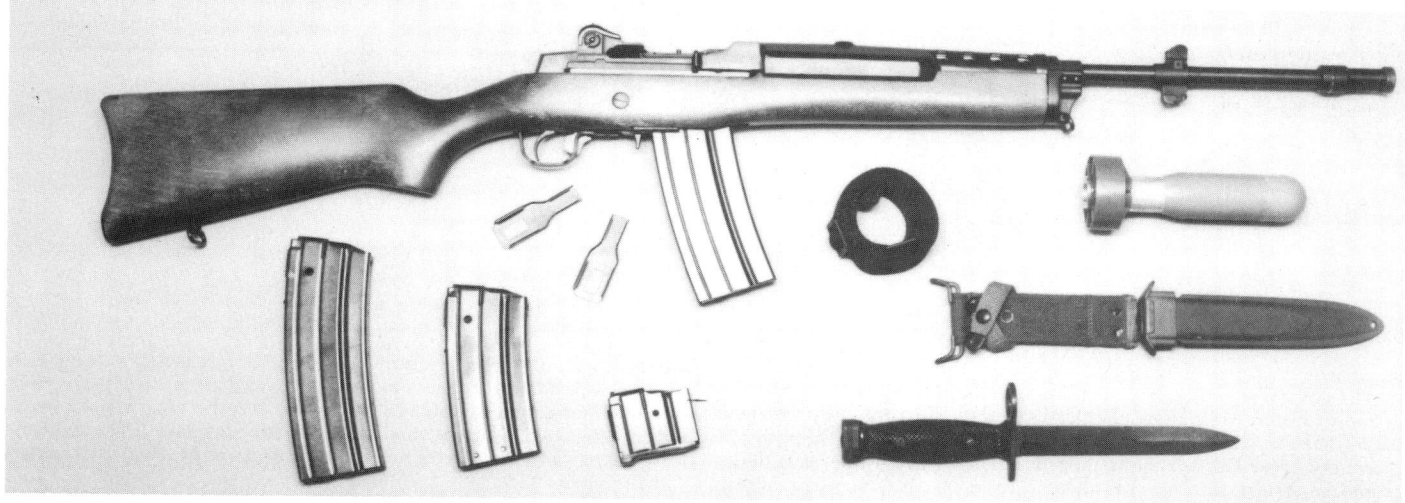
5.56 mm Mini-14/20GB Ruger rifle with full range of accessories and specimen grenade

5.56 mm Mini-14 Ruger rifle

This rifle was introduced commercially in 1973, since which time it has been adopted by various military and security forces. The mechanism is similar to that used on the M1 and M14 rifles in many respects, but the reductions made possible by the use of high tensile alloy steels have resulted in a light handy rifle with all the advantages conferred by the flat trajectory of the 5.56 mm cartridge.

Operation
To load the first round the cocking handle is pulled back and released. The compressed return spring drives it forward and the bottom face of the bolt picks up the top cartridge in the magazine and pushes it into the chamber. The operating slide carries a cam path in which the bolt roller engages. When the round is chambered the bolt comes to rest but the slide continues on and the bolt is rotated clockwise and locked. The slide continues on for approximately 6 mm over the bolt roller and when the end of the cam path contacts the roller all forward movement ceases and the gun is ready to fire.

When the trigger is pulled the hammer drives the firing pin against the cap and the round is fired. The gas pressure forces the bullet up the bore and the gas follows it up towards the muzzle. Some of the gas passes down through a vent drilled radially through the barrel wall and through a stationary piston to

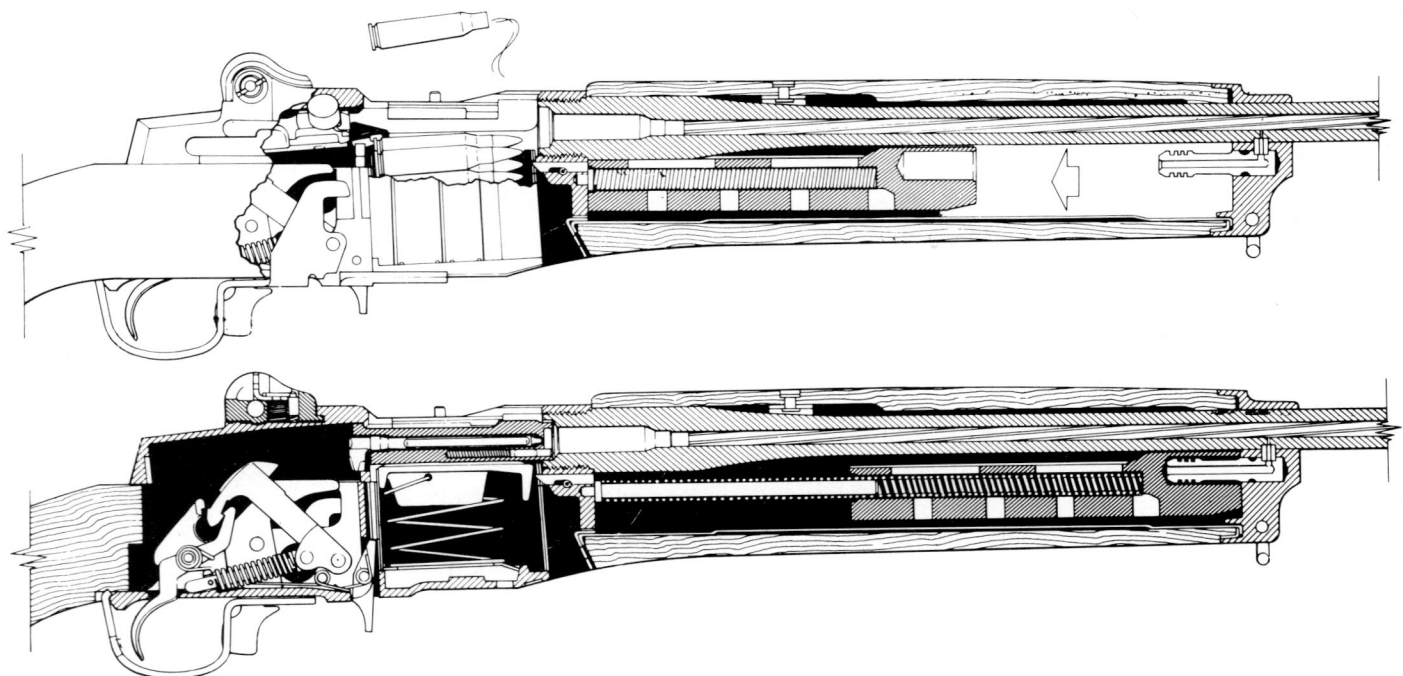

5.56 mm Mini-14 Ruger rifle
(above) *mechanism is full recoil; bolt will be driven forward under pressure of main spring to chamber bolt cartridge;* **(below)** *hammer cocked; bolt locked, chamber and magazine empty*

5.56 mm Mini-14 Ruger rifle with 30-round magazine

5.56 mm Mini-14 Ranch Rifle, folding stock model in stainless steel

impinge on the hollow interior face of the slide. The slide is driven back and the straight action of the cam path moves over the bolt roller. This free travel of approximately 6 mm allows the chamber pressure to drop to atmospheric before the roller is lifted by the cam path and the bolt is rotated. The firing pin is cammed back. The rotation of the bolt dislodges the cartridge case in the chamber and provides primary extraction. The bolt is then carried to the rear and the empty case is extracted and ejected to the right and forward. The bolt continues back over the rounds in the magazine and the return spring is fully compressed. The spring re-asserts itself and drives the slide forward to repeat the cycle.

If the ammunition is expended the magazine follower operates the bolt-holding open device. This can be released either by operating the bolt release on the left of the receiver, or by withdrawing the magazine, replacing with a full magazine and pulling back slightly on the cocking handle. When it is released the bolt will go forward and chamber the top cartridge.

The applied safety lies in the front of the trigger-guard and can be operated by the movement of the forefinger of the trigger hand. It is pushed forward to 'fire' and pulled back to 'safe'. When at 'safe' it blocks the hammer and pushes it down off the sear which is thus disconnected. When the weapon is set to 'safe' the slide can be retracted and the weapon loaded.

In 1977 changes were made to the Mini-14 rifle to simplify the mechanism and provide better protection of the action against dust and dirt. The later Mini-14 rifles can be distinguished from earlier models by their '181' serial

number prefix. In the later models, the bolt lock mechanism is completely enclosed and a manual bolt lock plunger is provided in the top of the receiver. In 1981 a flash suppressor of the old M14 type, incorporating radioluminous sights, was fitted. Current models also feature a glass fibre ventilated hand-guard which protects the firer's hand from barrel heat and from contact with the moving slide.

A stainless steel version of the Mini-14 rifle was introduced in 1978 and is in production. Data and specifications are substantially the same as for the standard model.

The **Mini-14 Ranch Rifle** is an additional model in the Mini-14 range. This incorporates an ideal telescope mounting system; integral bases in the receiver accept the Ruger 1 in Steel Scope rings. The Ranch Rifle is available in blued or stainless steel finish, with fixed or folding stock.

DATA
Cartridge: 5.56 × 45mm M193, SS109 or any commercial counterpart
Operation: gas, self-loading
Method of locking: rotating bolt
Feed: 5-, 20- and 30-round box magazine. 30-round magazine available only to government and law enforcement agencies
Weight: 2.9 kg
Length: 946 mm
Barrel: 470 mm

Rifling: 6 grooves, rh, one turn in 178 mm
Sights: (fore) bead on post; (rear) aperture. Adjustable for elevation and windage. Clicks of 1 minute of angle (1 minute is approx 3 cm at 100 m); (zeroing) elevation, rearsight. Line, rearsight
Sight radius: 561 mm
Muzzle velocity: 1005 m/s
Effective range: 300 m

Manufacturer
Sturm, Ruger and Co Inc, Ruger Road, Prescott, AZ 86301.
Status
Current. Production.

5.56 mm AC-556 Ruger selective fire weapon

Although similar in appearance to the standard Mini-14/20GB infantry rifle, the AC-556 Ruger is a selective fire weapon specifically designed for military and law-enforcement applications using standard 5.56 mm (.223 calibre) US military or commercial ammunition. The AC-556 is equipped with a heat-resistant, ventilated glass fibre hand-guard, protected front sight which incorporates a bayonet lug, and a flash suppressor. The fire control mechanism consists of a positive three-position selector lever which provides semi-automatic, three-shot burst or fully automatic fire. The selector lever is at the right rear of the receiver and is readily accessible to right- or left-handed shooters.

DATA
Cartridge: 5.56 × 45mm M193, SS109 or any commercial counterpart
Operation: gas, self-loading
Locking: rotating bolt

Feed: 5-, 20- or 30-round box magazines
Weight: 2.89 kg
Length: 984 mm
Barrel: 470 mm
Rifling: 6 grooves, rh, one turn in 305 mm
Sights: (fore) protected military type, with bayonet lug; (rear) aperture adjustable for windage and elevation; clicks of 1 minute of angle (1 minute is approx 3 cm at 100 m)
Muzzle velocity: 1058 m/s
Rate of fire: (cyclic) approx 750 rds/min

Manufacturer
Sturm, Ruger and Co Inc, 49 Lacey Place, Southport, CT 06490.
Status
Current. Production.

5.56 mm AC-556 Ruger selective fire weapon

5.56 mm AC-556F Ruger selective fire weapon

The Ruger AC-556F is a short barrel, selective fire weapon equipped with a steel folding stock. Employing the same basic mechanism as the Ruger AC-556 model, the AC-556F is adapted for aircraft and armour crews, paratroops, patrol vehicles, dignitary protection, law-enforcement and other

5.56 mm AC-556F Ruger selective fire weapon with stock folded

applications where a high rate of fire combined with compact configuration and short overall length are required. Except as listed below, technical data and specifications are the same as those for the AC-556 Ruger model. Both the Ruger AC-556 and AC-556F selective fire weapons are available in blued or stainless steel versions. In stainless steel they are designated the KAC-556 and KAC-556F respectively.

DATA
Weight: (basic weapon) 3.15 kg; (loaded, 20-rounds) 3.6 kg
Overall length: (stock folded) 603 mm with flash suppressor; (stock opened) 851 mm with flash suppressor
Barrel: 330 mm; (rifling) 6 grooves rh. 1 turn in 254 mm

Manufacturer
Sturm, Ruger and Co Inc, 49 Lacey Place, Southport, CT 06490.
Status
Current. Production.

5.56 mm Ruger M77 Mark II rifle

This design has evolved from the 7.62 mm M77V (see above) and is intended as a slender and compact hunting rifle in 5.56 mm calibre, though it has obvious applications for sniping and counter-insurgency roles. The rifle is a conventional bolt-action design, but has an entirely new trigger safety system using a three-position safety catch. In the rear position the bolt is locked and the gun will not fire; in the centre position the gun cannot fire but the bolt can be operated; in the forward position both bolt and trigger are free. The new trigger-guard incorporates a redesigned floor plate latch which prevents inadvertently losing the contents of the magazine. The magazine capacity has been reduced to four rounds, allowing the rifle to be more slender in contour.

The weight of this new rifle has been kept to a minimum of approximately 2.95 kg with the 508 mm ultra-light barrel. It will also be available with a 559 mm standard barrel. Additional calibres — .243 Winchester, 7.62 × 51 mm and .22-250 — and barrel configurations will also be available in the future.

Manufacturer
Sturm, Ruger and Co Inc, Ruger Road, Prescott, AZ 86301.
Status
Current. Production.

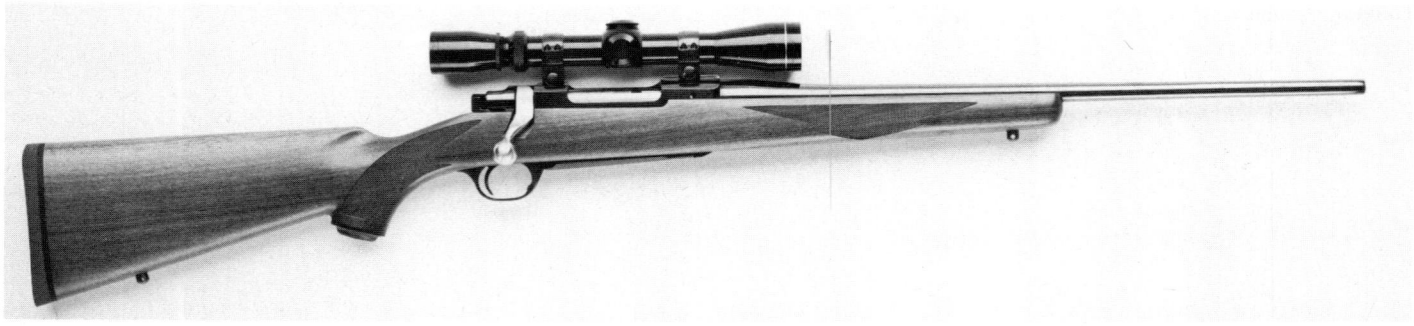

5.56 mm Ruger M77 Mark II rifle

Hohrein .22 conversion kit

The Hohrein .22 conversion kit permits conversion of the Ruger Mini-14 rifle to fire .22LR rimfire cartridges. The conversion kit is available for both the 5.56 mm (.223) and .222 calibre rifles. The unit consists of a sub-calibre chamber adapter which is placed in the chamber and secured by a set-screw; a sub-calibre bolt which replaces the normal bolt; a sub-calibre operating rod and a gas piston adapter. These items are assembled into the dismantled firearm, replacing the standard parts, and the firearm re-assembled. A 10- or 30-round .22 magazine, with an adapter, is fitted into the normal magazine housing and the firearm is then fully operative.

The unit is made from chrome-molybdenum steel, heat-treated, and will fit and operate in all Ruger Mini-14 rifles with serial numbers prefixed by '181' to present-day production.

Manufacturer
Jonathan Arthur Ciener Inc, 6850 Riveredge Drive, Titusville, FL 32780.

Status
Current. Production.

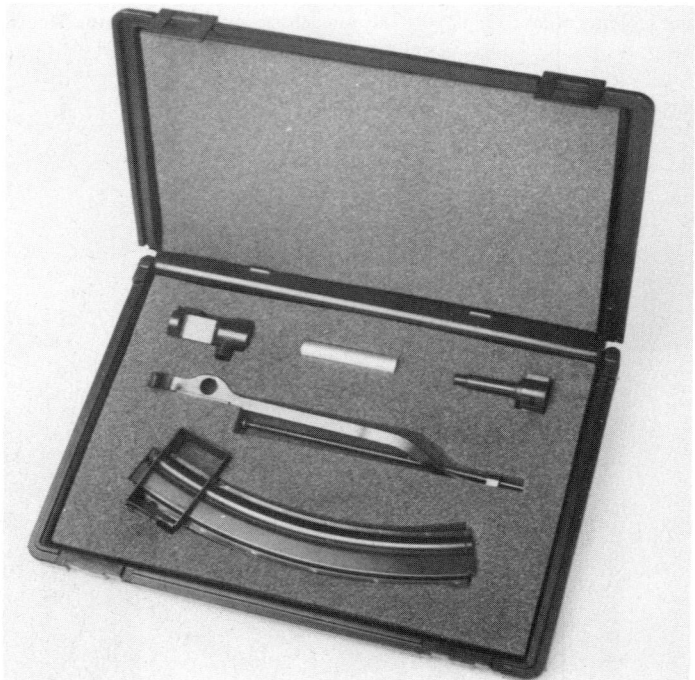

Hohrein conversion unit fitted to Mini-14 rifle

Hohrein .22 conversion kit

5.56 mm AAI Advanced Combat rifle

The AAI Advanced Combat rifle (ACR) has been developed in response to JSSAP's request for a 'technology base' which will lead to the design of the next generation of US military rifle. (See *Jane's Infantry Weapons 1984/85* page 129.) The AAI Corporation has considerable past experience in such projects as the XM19 SPIW and the 5.56 mm Caseless Rifle Programme, and

much of this experience has been utilised in the ACR project. Since much of the technology is state-of-the-art, the credibility of the design is maximised and difficulties in placing the weapon in production, should this be required, are few.

The principal feature of the ACR is that it is designed around a caseless

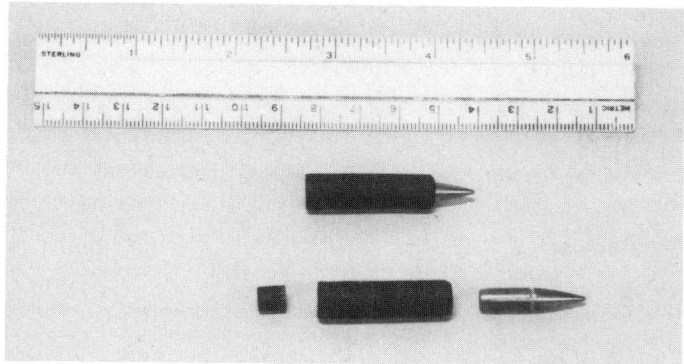

5.56 mm heavy bullet caseless cartridge

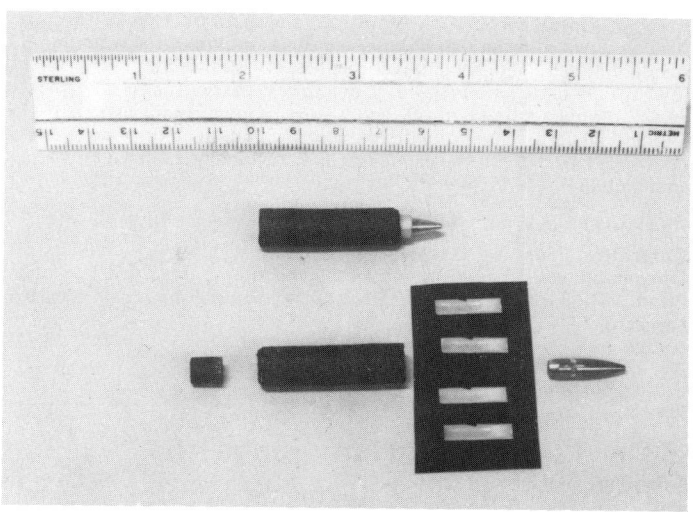

4.32 mm Saboted light bullet caseless cartridge

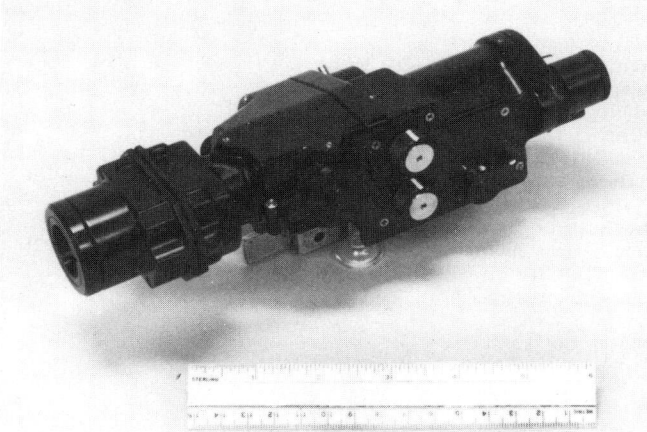

AAI optical sight for Advanced Combat rifle

cartridge; this, by removing the extraction-ejection phase from the operating cycle, allows the gun to be made with fewer apertures, removes the liability of the firer being struck by a hot case, and simplifies dust-proofing. The rifle is gas-actuated, though the precise method of breech closure is not disclosed; the rifle is capable of either open or closed bolt operation, depending upon the mode of fire selected. For single shots it fires from a closed bolt; at full-automatic fire, or when firing three-round bursts, it fires from an open bolt. In the three-round burst mode the cyclic rate of fire is about 2000 rds/min. Due to this controlled burst, the Hit Probability is expected to be in the order of 100 per cent improvement over current cartridge rifles at ranges out to 600 m.

The rifle is fitted with an advanced optical sight with a three-power optic, projected graticule, and immediate selection of the aiming point to suit the two different types of ammunition provided.

The caseless ammunition is of two types; a 5.56 mm conventional bullet and a 4.32 mm discarding sabot bullet, both of which are carried in cylindrical Compressed Propellant Charges. The rounds are fed to the weapon from

5.56 mm AAI Advanced Combat rifle

either a sealed 36-round two-row box magazine or from a 50-round weapon-powered drum magazine. There is also provision for mounting the standard M203 40 mm grenade launcher beneath the barrel.

The ACR has been proposed in two forms; the conventionally stocked model resembles a standard rifle and is perhaps more readily assimilable by soldiers, while a bullpup model reduces the overall dimensions without sacrificing barrel length and capitalises on the advantages offered by caseless ammunition.

DATA
Cartridge: 5.56 mm AAI caseless
Operation: gas; selective fire and 3-round burst
Feed: 36-round box magazine or 50-round weapon-powered drum magazine
Length
  Bullpup: 805 mm
  Standard: 1066 mm

Barrel: 457 mm
Weight (with sight and 36-round magazine)
  Bullpup: 3.35 kg
  Standard: 3.38 kg
Rifling: 1 turn in 178 mm
Muzzle velocity: 5.56 mm heavy bullet: 885 m/s; 4.32 mm light sabot bullet: 1158 m/s
Rates of fire (cyclic): 3-round burst: ca 2000 rds/min; automatic fire: 600 rds/min

Manufacturer
AAI Corporation, PO Box 6767, Baltimore, MD 21204.
Status
Development of ammunition has ceased. Rifle parts and assemblies are being supplied to the U.S.Army. Further development is in doubt.

.50 in (12.7 mm) RAI long-range rifle Model 500

The Model 500 is a long-range sniping rifle chambered for the .50 Browning (12.7 × 99 mm) cartridge. It is a bolt-action single-shot weapon; the bolt-action is unconventional in that the bolt is extremely short, consisting of little more than a breech plug with a handle, and it is entirely removed from the weapon for loading and unloading. To load, the bolt handle is turned up until the interrupted lugs disengage and the bolt is then withdrawn; a cartridge is then clipped into the bolt-face and held there by the extractor. The cartridge is then entered into the chamber, followed by the bolt, and the handle turned down to lock; inserting the bolt causes the striker to be cocked, and pressure on the trigger then releases it to fire the round. The bolt is then removed, bringing the spent case with it, and reloading takes place.

The heavy, fluted barrel is free-floating, and there is an harmonic balancer concealed in a tube within the fore-end which damps out harmonic vibrations of the barrel. There is a muzzle brake and a flash hider, but no iron sights are provided, the normal sighting system being a telescope, for which a Ranging Scope Base is provided.

The stock is adjustable for length and for height of cheek-piece.

DATA
Cartridge: .50 Browning (12.7 × 99 mm)
Operation: bolt-action, single-shot
Feed: single-shot, no magazine
Weight: 13.60 kg
Barrel: 840 mm
Muzzle velocity: 888 m/s

Manufacturer
Research Armament Industries, 1700 South First Street, Rogers, AR 72756.
Status
Current. Production.
Service
US Marine Corps and US Navy.

.50 in RAI Model 500 long-range rifle

.50 in Barrett 'Light Fifty' Model 82A1 sniping rifle

The Barrett 'Light Fifty' Model 82A1 is a semi-automatic rifle firing the .50 in (12.7 mm) Browning heavy machine gun cartridge. It is intended as a sniping and long-range interdiction weapon and is suggested as having uses in military and police employment and also as a suitable defensive weapon for ocean-going light craft.

The rifle operates on short recoil principles. Upon firing, the bullet is forced through the barrel, and considerable case-head thrust is delivered to the bolt face. This thrust transfers through the bolt lugs to the barrel extension and through the bolt body to the rear of the bolt carrier. The transfer to the carrier is unique to this design and dissipates much of the firing shock which might otherwise damage the locking assembly.

The barrel has recoiled approximately 13 mm when the bullet leaves the barrel. The barrel, bolt and bolt carrier continue to the rear under recoil, during which movement the cocking lever withdraws the firing pin and recocks it. At 25 mm of recoil travel an accelerator arm strikes an abutment in the receiver and swings forward, forcing the accelerator rod to separate the bolt carrier and the barrel extension. This action serves to rotate the bolt out of engagement with the barrel extension and to transfer energy from the barrel to the bolt carrier. The barrel continues to the rear for a total of 53 mm, stopping when a key, locked into a recess in the barrel, strikes the barrel stop. The bolt carrier continues to move under its own momentum and begins to separate from the barrel. A cam pin in the carrier, engaged in a helical cut in the bolt, causes the bolt to rotate 30° during the separation of carrier and barrel, thus moving the bolt locking lugs out of engagement with the barrel extension. At this point the bolt is fully extended from the carrier and the

locking recess, on the bolt body, has been rotated so that a bolt latch, mounted in the carrier, engages so as to lock the bolt in the forward position.

As the carrier pulls the bolt to the rear, the fired cartridge case is withdrawn from the chamber, and as it clears the barrel extension the case is expelled from the weapon by the ejector. As soon as the bolt is free of the barrel extension the barrel return springs restore the barrel to the forward battery position. The bolt carrier, having compressed the main return spring housed in the butt-stock, bottoms the buffer against the rear of the housing. The return spring then forces the bolt carrier forward, stripping a new cartridge from the magazine and chambering it. As the carrier moves forward the bolt latch is pivoted out of engagement with the bolt body by the pressure of the

.50 in Barrett Model 82A1 sniping rifle

bolt latch trip, allowing the bolt to retract and rotate when it meets the breech face, locking the lugs into engagement with the barrel extension.

The barrel is fitted with an efficient double-baffle muzzle brake, which reduces the recoil by some 36 per cent. There is an adjustable bipod, and the rifle can also be mounted on the standard M82 tripod or on any mounting which is compatible with the M60 machine gun.

Sighting
The Model 82A1 is fitted as standard with a Leupold & Stevens M3a Ultra 10X telescope, employing a ballistic graticule calibrated to the recommended ammunition types. The graticule allows adjustment of the point of impact from 500 to 1800 m, considered to be the maximum effective range when using standard Browning ammunition. In addition the rifle is fitted with flip-up iron sights for emergency use.

Ammunition
The manufacturers recommend using the APEI multi-purpose cartridge, but the rifle is capable of firing any standard 12.7 × 99 mm Browning machine gun ammunition. The preferred standard types are APHCI and AP M8.

DATA
Cartridge: .50 Browning (12.7 × 99 mm)
Operation: short recoil, semi-automatic
Method of locking: rotating bolt
Weight: 13.4 kg
Length: 1549 mm
Barrel: 737 mm
Feed: 11-round detachable box magazine
Sights: 10 × telescope; emergency iron sights
Muzzle velocity: 853 m/s (M33 ball)

Manufacturer
Barrett Firearms Manufacturing, Inc., PO Box 1077, Murfreesboro, TN 37133-1077.
Status
Current. Production.
Service
US Government and civil agencies and export.

.50 in Peregrine 50/12 Tactical Support Weapon

The Peregrine Industries 50/12 Tactical Support Weapon is a gas-operated semi-automatic rifle with interchangeable barrels which permit the alternative use of either 12.7 × 99 mm (.50 Browning) or 12.7 × 107 mm (12.7 mm Soviet) ammunition. The ability to quickly remove the barrel also makes the dismantled weapon relatively compact for transportation. The weapon is supported on a front bipod and rear monopod, is fitted with a Solothurn muzzle brake, and can accept a variety of sighting telescopes.

The data given below relates to the first prototype; development is continuing and the final dimensions may differ slightly.

DATA
Calibre: 12.7 mm

Operation: gas, semi-automatic
Locking: rotating bolt
Feed: 10-round side-feeding box magazine
Weight, empty: 15.2 kg with empty magazine
Length: 1410 mm
Barrel: 737 mm
Sights: iron sights; 10 × telescope

Manufacturer
Peregrine Industries Inc., PO Box 4199, Balboa, CA 92661.
Status
Development.

Springfield Armory M6 Survival Gun

During the Second World War the US Air Force developed a 'survival gun' for use by aircrews who made forced landings in deserted areas. The weapon was a combination of small-bore rifle and shotgun and was intended for killing small game for food in emergencies.

The M6 had now been put back into production by Springfield Armory Inc and consists of a .22 rifle barrel mounted above a .410 shotgun barrel. Both

barrels are single-shot, hammer-fired units and the gun breaks open for re-loading. It also folds around the break-open hinge to permit stowage in a small space or carriage in a pack. The butt stock incorporates a magazine which can hold 15 .22 cartridges and four shotgun rounds. Various types of signal flare are also available in the shotgun calibre. The M6 Survival Gun is of all-metal construction and is designed for quick disassembly and simple maintenance.

DATA
Cartridges: .22 long-rifle rim fire; .22 WMRF; .22 Hornet: .410 shot or flare
Operation: manual, single-shot
Weight: 1.81 kg
Length: 80 cm
Barrel length: 457 mm
Sights: (rifle) rear aperture, front blade; (shotgun) rear notch, front blade
Effective range: (rifle) 400 yards (365.7 m); (shotgun) 45 yards (41.1 m)
Variant models: may be obtained with barrels rifled to suit the various cartridges enumerated above

Manufacturer
Springfield Armory Inc, 420 West Main Street, Geneseo, IL 61254.
Status
Current. Production.

M6 Springfield Armory M6 Survival Gun folded for stowage

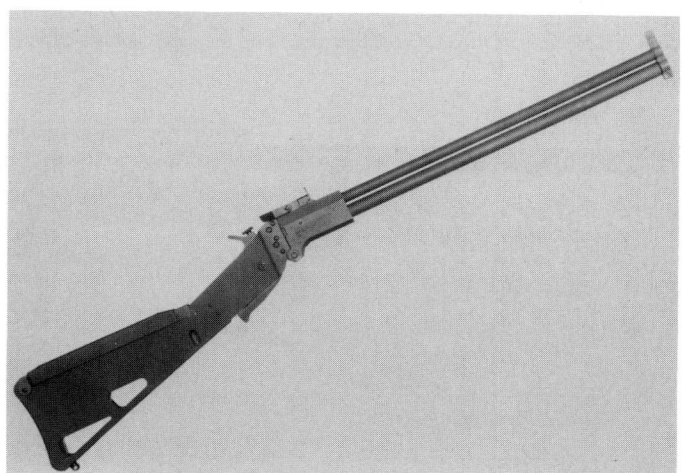

M6 Springfield Armory M6 Survival Gun

Ciener sound suppressors

This company manufactures sound suppressors for virtually any commercial or military weapon. Each is specifically designed to suit the particular weapon, and may be a suppressor complete with replacement barrel, or a device to fit an existing barrel, or a combination of suppressor and modified standard barrel as the particular design warrants. Suppressors to screw on to military weapons such as the AR-15/M16, M14, Heckler and Koch, Mini-14 and FN-FAL rifles and Uzi, M3, Sten and Carl Gustav sub-machine guns are readily available. All require minimal maintenance, as their designs do not demand periodical cleaning or rebuilding. Specifications cannot be given, since they differ according to the parent weapon, but suppressors fitted to semi-automatic weapons invariably reduce the noise of discharge to less than the mechanical noise of the weapon.

Ciener sound suppressor fitted to HK91 rifle

Manufacturer
Jonathan Arthur Ciener Inc, 6850 Riveredge Drive, Titusville, FL 32780.

Status
Current. Available.

9 mm Calico M-900 carbine

This is a semi-automatic shoulder weapon with a unique Helical Feed system. The 100-round magazine can be loaded and stored without spring fatigue, and, using the Calico Speed Loader, it can be refilled from an open box in less than 30 seconds. There is very little muzzle climb, and the carbine points easily and is well-balanced.

9 mm Calico M-900 Carbine with 100-round magazine, telescope sight and laser projector

The carbine uses a roller-locking delayed blowback mechanism very similar to that used in the CETME and H&K rifles. The two-part bolt, return springs, guide rods, striker mechanism and buffer are in a self-contained unit which can be removed in one piece from the weapon for cleaning. The receiver is of cast aluminium, with glass-filled polymer furniture; the barrel is of CrMb steel, with a muzzle compensator, and the breech mechanism is of stainless steel. There is a sliding adjustable shoulder-stock. The magazine slides on from the rear, above the receiver, and a special sight mount is designed to hold a telescope or other optical sight above the magazine. A laser projector can also be fitted, in front of the magazine.

DATA
Cartridge: 9 × 19 mm Parabellum
Operation: semi-automatic
Locking: delayed blowback
Feed: 50- or 100-round helical magazine
Weight: 1.68 kg empty; 2.49 kg with 50-round loaded magazine; 3.26 kg with 100-round loaded magazine
Length: 724 mm, stock folded; 884 mm stock extended
Barrel: 406 mm
Rifling: 6 grooves, rh, one turn in 305 mm
Sights: (fore) post, adjustable for windage and elevation; (rear) fixed notch
Muzzle velocity: 426 m/s

Manufacturer
Calico, 405 East 19th Street, Bakersfield, CA 93305.
Status
Current. Production.

9 mm Göncz carbines

These weapons are extended-barrel versions of the Göncz Assault Pistol, details of which will be found on page 70. They are blowback weapons, firing from a closed bolt, and are chambered as standard for the 9 mm Parabellum cartridge, though they are also manufactured in .38 Super, 7.63 mm Mauser and .45 ACP calibres.

The Model GC carbine uses the same mechanism as the pistol but has a 409 mm barrel and the receiver is extended to accommodate a walnut shoulder stock. It accepts the same magazines as the pistol, and has a sight radius of 510 mm. It is capable of highly accurate fire to ranges in excess of 200 m.

The Model GCH is the same weapon as the Model GC but has a housing cast into the frame, into which fits a halogen-iodide light source. The power pack for the lamp is fitted into the butt stock and the wiring and switching is built into the weapon. A three-position switch on the lamp housing gives 'on', 'auto' and 'off' positions; In the 'on' position the lamp gives a steady light of 25 000 candelas; when switched to 'auto' the light is actuated as the first pressure is taken on the trigger. The lamp beam is bore-sighted to the axis of the weapon and is focusable for varying ranges.

The Model GCL is similar to the Model GCH except that the housing contains a laser spot projector, similarly bore-sighted to the axis of the weapon. It has an effective range of 400 m.

Manufacturer
Göncz Co, 11526 Burbank Boulevard, North Hollywood, CA 91601.
Status
Current. Production.

9 mm Göncz carbine Model GCH

YUGOSLAVIA

7.62 mm M59/66A1 rifle

This gas-operated self-loading rifle closely resembles the Simonov SKS rifle (which see) of the USSR, from which it may be said to have been derived by way of the earlier M59 weapon which it has replaced.

Most noticeable of the differences between the M59/66A1 and the SKS is a spigot-type grenade launcher attached permanently to the muzzle. Associated with this is a folding grenade launching sight which is normally folded flat behind the front sight. To launch grenades it is necessary first to be sure that they are of the correct type with a boom of 22 mm internal diameter. The gas supply must be cut off from the gas piston by pressing in the gas cut-off valve catch and rotating it to the top of the gas cylinder. The grenade sight is then erected. The grenade cartridge is lodged in the tail boom of the Yugoslav grenade and is loaded individually into the chamber.

The sighting process is standard for this type of grenade, using the appropriate range arc aligned with the largest diameter of the grenade and placed on the target.

It is imperative that only a blank cartridge is used for grenade launching. If a ball round is used there is a very considerable chance that it will cause the grenade to explode on the launcher.

As with the SKS a bayonet is permanently attached to a pivot positioned below the foresight; and when not in use it is folded back under the barrel, the point being shielded by a recess in the stock. The presence of the grenade launcher slightly reduces the length of free blade when the bayonet is in the forward position.

DATA
Cartridge: 7.62 mm × 39
Operation: gas, self-loading
Method of locking: tilting block
Feed: 10-round internal box magazine
Weight: 4.1 kg approx
Length: (bayonet folded) 1120 mm; (bayonet forward) 1320 mm
Sights: (foresight) post; (rearsight) tangent notch, 100-1000 m; (sight radius) 444 mm
Muzzle velocity: 735 m/s
Rate of fire: 30-40 rds/min
Effective range: 500 m
Grenade launcher diameter: 22 mm

Manufacturer
Zavodi Crvena Zastava, Kragujevac.
Status
Current. Production.
Service
Yugoslav forces.

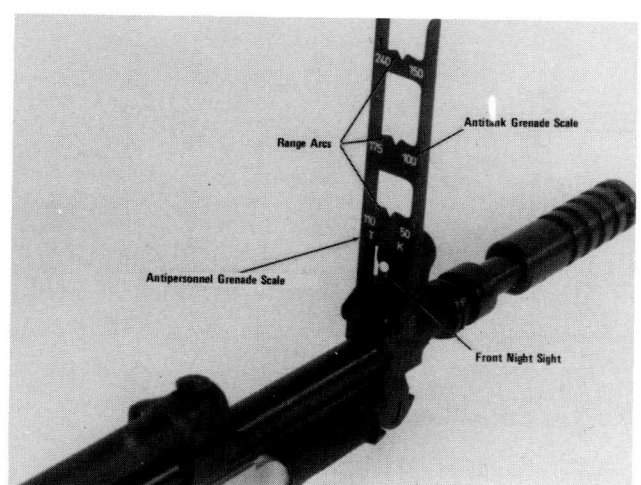

7.62 mm M59/66A1 rifle with grenade sight erected

7.62 mm M59/66A1 self-loading rifle

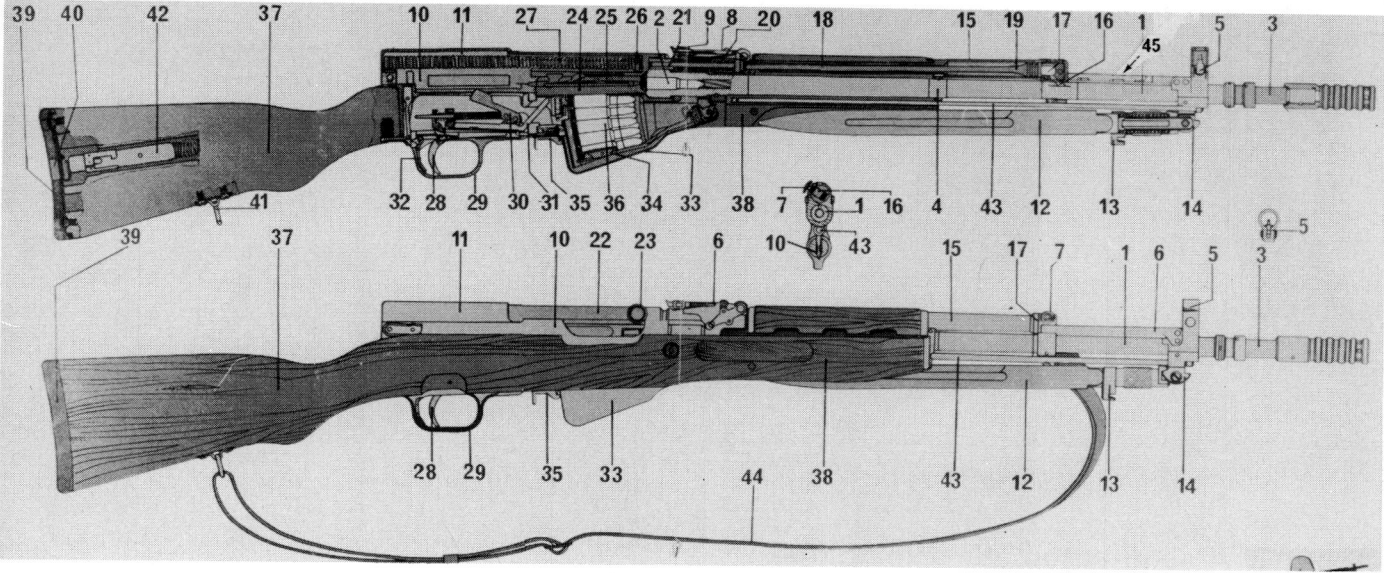

7.62 mm M59/66A1 self-loading rifle.
(3) *grenade launcher* **(45)** *grenade sight* **(16)** *gas cut-off valve* **(12)** *bayonet* **(14)** *bayonet pivot*

7.62 mm M70B1 and M70AB2 assault rifles

These two weapons are the rifle elements of the Yugoslav FAZ family of automatic arms based on the 7.62 mm × 39 cartridge, the other members of the family being the M72B1 and M72AB1 light machine guns (which see). They are the most recent developments of the M70 and M70A assault rifles (see *Jane's Infantry Weapons 1981/1982,* page 215). The M70B1 and M70AB2 differ only in that the former has a fixed wooden stock whereas the M70AB2 has a folding metal stock. Both are generally similar in design to the Soviet AK-47 and AKM (which see).

Like the AK-47 and AKM, the weapons are gas-operated and capable of selective fire. An important difference however is the provision of a grenade-launching sight which is permanently attached to the weapon at the gas port and which, when raised for use, cuts off the gas supply to the piston. At other times it lies flat on top of the gas cylinder, as shown in the accompanying overall views. The launcher itself is not permanently attached but is fitted to the muzzle when required.

Equipment provided includes a bayonet like that of the AKM, with a slot on the blade into which a lug on the scabbard fits to form an insulated wire cutter.

DATA
Cartridge: 7.62 mm × 39
Operation: gas, selective fire
Method of locking: rotating bolt
Feed: 30-round detachable curved box magazine
Weights: (M70B1) 3.7 kg; (empty magazine) 0.36 kg; (loaded magazine) 0.87 kg; (grenade launcher) 0.21 kg

Length: (M70B1) 900 mm; (M70AB2, stock retracted) 640 mm; (barrel) 415 mm; (rifling) 4 grooves rh
Sights: (foresight) pillar; (rearsight) tangent notch 100-1000 m; (sight radius) 395 mm; (night sights) tritium spots on iron sights
Muzzle velocity: 720 m/s
Rate of fire: (cyclic) 620-660 rds/min
Effective range: 400 m

Manufacturer
Zavodi Crvena Zastava, Kragujevac.
Status
Current. Production.
Service
Yugoslav forces.

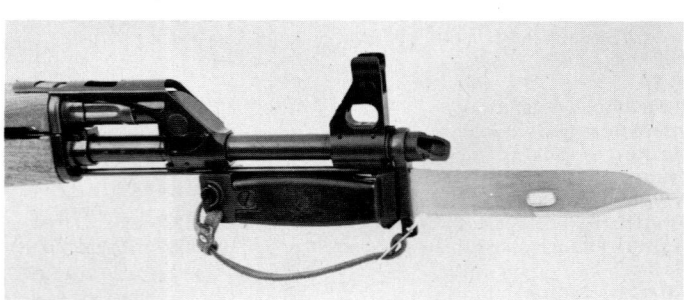

AKM-type bayonet on M70B1 rifle

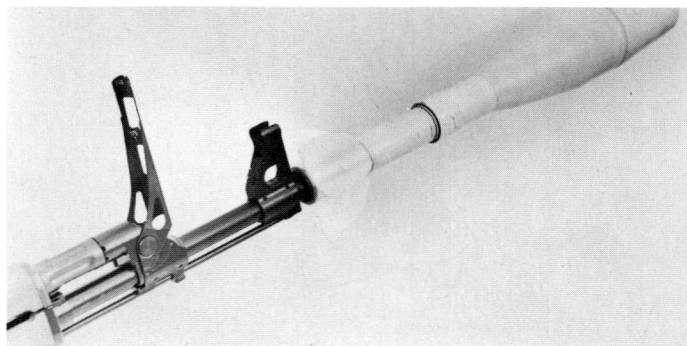

Grenade launcher with AT grenade on M70B1, with grenade sight raised

7.62 mm M70B1 assault rifle showing grenade sight pivoted at gas port

7.62 mm M70AB2 assault rifle with stock folded

7.92 mm M76 semi-automatic sniping rifle

Information received on the M76 semi-automatic sniping rifle is scanty, but it is said to be in production for the Yugoslav armed forces and to be based on the FAZ family of automatic arms. In the standard version, illustrated, the receiver is adapted to handle a 7.92 mm cartridge; variants have also been manufactured to accept the Soviet 7.62 mm × 54 and NATO 7.62 mm × 51 cartridges. A 10-round detachable box magazine is fitted.

The M76 is fitted with a telescopic sight of x4 magnification, apparently having much in common with the PSO-1 sight of the Soviet 7.62 mm Dragunov (SVD) (which see). It is stated to have a field-of-view (true) of 5° 10 mins and a sweeping range of 320 m for head silhouette (target height, 30 cm); 400 m for chest silhouette (50 cm); and 620 m for a running target

(150 cm). The most effective range is given as 800 m. As illustrated, the optical sight bracket is also designed to accept a passive optical night sight.

DATA
Cartridge: 7.92 mm
Operation: gas, self-loading
Method of locking: rotating bolt
Feed: 10-round detachable box magazine
Weight: 4.20 kg
Length: 1135 mm
Barrel: 550 mm

Rifling: 4 grooves, rh, one turn in 240 mm
Sights: (fore) blade; (rear) U-notch tangent. 4 × Telescope.
Muzzle velocity: 720 m/s
Effective range: 1000 m
Manufacturer
Zavodi Crvena Zastava, Kragujevac.
Status
Current. Production.
Service
Yugoslav forces.

7.92 mm M76 sniping rifle with passive optical night sight

7.92 mm M76 sniping rifle with telescopic sight

7.62 mm M77B1 assault rifle

This rifle, stated to be in production for export along with the M77B1 light machine gun (which see), closely resembles the M70B1 assault rifle. However it is based on the 7.62 mm × 51 NATO cartridge and, as illustrated, feeds from a 20-round detachable straight box magazine. It is provided with a grenade-launching sight which is not permanently attached (as is that of the M70B1) but is fitted to the muzzle, along with the launcher, when required. The AKM-type bayonet of the M70B1 is also used.

DATA
Cartridge: 7.62 mm × 51 NATO
Operation: gas, selective fire
Method of locking: rotating bolt
Feed: 20-round detachable straight box magazine
Weight: (rifle, with empty magazine) 4.8 kg

Length: 990 mm
Barrel: 500 mm
Rifling: 6 grooves, rh, one turn in 240 mm
Sights: (fore) pillar; (rear) tangent notch
Sight radius: 485 mm
Muzzle velocity: 840 m/s
Rate of fire: (cyclic) 600 rds/min
Effective range: 600 m

Manufacturer
Zavodi Crvena Zastava, Kragujevac.
Status
Current. Production for export.

7.62 mm M77B1 NATO assault rifle

5.56 mm M80 assault rifle

Development of a family of weapons for the 5.56 mm M193 and SS109 cartridges has now been completed. There are two rifles, the M80 with wooden stock and the M80A with folding metal stock, and two machine guns. All weapons are gas-operated and are capable of selective semi- or full-automatic fire. A special design of gas regulator ensures reliability of operation even with ammunition of varying energy levels. Both rifles are capable of firing rifle grenades; a launcher attachment and accessory grenade-launching sight is included in the equipment of each rifle.

5.56 mm M80 rifle with wooden stock and M80A with folding metal stock

DATA
Cartridge: 5.56 mm × 45 (M193)
Operation: gas, selective fire
Method of locking: rotating bolt
Feed: 30-round detachable curved box magazine
Weight: (rifle, without magazine) 3.5 kg; (magazine empty) 0.1 kg
Length: 990 mm
Barrel: 460 mm
Rifling: 6 grooves, rh, one turn in 178 mm

Sights: (fore) pillar; (rear) tangent notch
Sight radius: 439 mm
Muzzle velocity: 970 m/s
Effective range: 300 m

Manufacturer
Zavodi Crvena Zastava, Kragujevac.
Status
Available for export.

Light Support Weapons

AUSTRIA

M203 grenade launcher

The US M203 grenade launcher can be fitted to the barrel unit of the Steyr Army Universal Gun (AUG) rifle and, due to the interchangeable barrel system, can then be mounted to any variant of the AUG rifle family. In this way any AUG can be instantly converted into a grenade launcher without impairing its ability as a rifle.

DATA
Overall length of complete unit: 725 mm
Weight of complete unit: 3 kg
Rifle barrel length: 610 mm
Launcher barrel length: 305 mm
Launcher calibre: 40 mm
Max grenade range: 400 m

Manufacturer
Steyr-Mannlicher GmbH, a member of the Steyr-Daimler-Puch AG group, Postfach 1000, A-4400 Steyr.
Status
Current. Production.
Service
Not known.

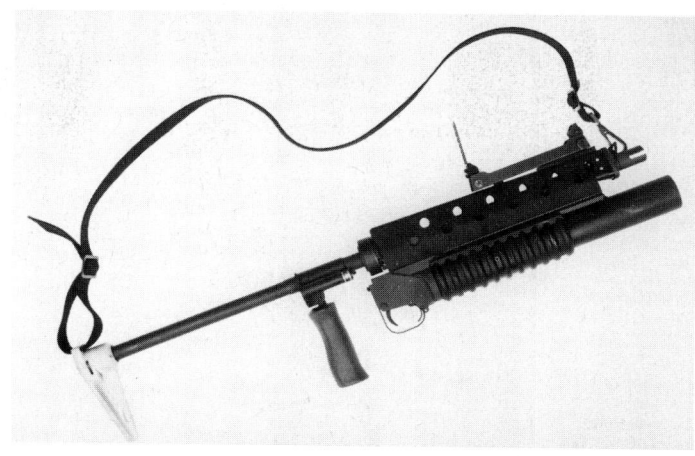

M203 grenade launcher fitted to AUG rifle-barrel assembly

BELGIUM

FLY-K PRB grenade launcher

There is a close resemblance between this launcher and the 52 mm PRB light mortar which is described in the mortar section (which see). In fact both use the same principle of propulsion and the same projectiles. The difference lies in the size and weight of launcher and the range and accuracy that it can achieve. The propulsion charges differ for each launcher, the grenade launcher giving a shorter maximum range.

The principle of operation is that of the spigot mortar. The tail boom of the projectile is hollow and it slides over a tubular spigot which has the firing pin in its top. The propellant charge is in the top of the tail boom, and when fired the projectile is forced clear of the spigot. This gives the projectile its acceleration as well as direction.

The feature of the FLY-K system is that it is flashless, smokeless and noiseless. The propellant gases are contained within the tail boom by a sliding piston which travels with the projectile, thus no propelling gases are released at the launcher. The sealed tail boom carries all the high pressure gas with it to the target and the firing of the projectile is without any signature.

NR 8111 launchers
The launcher is a simple high angle weapon, comprising a launch tube which merely has to locate the projectile correctly and give the firer something to hold; an internal spigot containing the firing pin and spring; a baseplate and a sling. There are two types of launcher, the NR8111A1 and the NR 8111A2. The A1 model has a spigot with three raised ridges on its surface; the A2 model has a smooth spigot and is designed to fire the projectiles used with the FLY-K 52 mm Mortar NR8113.

The launch tube, or guard tube, is a protective metal cylinder which encloses the spigot and is attached, at its lower end, to the baseplate by means of a ball and socket joint. The diameter of the NR8111A2 tube is slightly smaller than that of the NR8111A1, in order to fire mortar projectiles as previously noted. There is a simple bubble-level sight attached to the upper end of the tube.

The spigot is a cylindrical steel part with three external ridges set 120° apart. The upper end is coned, and it is bored internally to take the firing pin and spring. The spigot is firmly attached at its lower end to the guard tube. When set at 'Safe' the firing pin is retracted inside the spigot; at other times it floats freely, but so long as the weapon is held at angles of elevation greater than 40° it will not protrude from the spigot. An external sleeve actuates the firing pin and also acts as a safety device.

To fire, the grenade is introduced into the tube tail-first, and twisted to align three notches in the lower ring of the tail unit with the three ribs on the spigot. The grenade is then slid down until it stops. The launcher is aligned with the target by the firer lining up the sight unit with the distant object; he then elevates or depresses the launcher until the coloured liquid in the bubble sight is lined up against the estimated range. Holding the launcher with one hand the firer then pulls the sleeve back with the other, thus cocking and releasing the firing pin. This fires the propulsive cartridge inside the grenade; the explosion drives the internal piston back in the grenade tail unit and so launches the grenade. The piston is trapped at the end of the tail unit, so preventing the escape of gas, flame or noise.

Grenades
Three types of grenade are provided; NR 208 Fragmentation, NR 209 Illuminating and NR 210 Smoke. All consist of warhead, fuze, stabiliser and propulsion unit. In the case of the NR 208 the warhead is a thermo-plastic

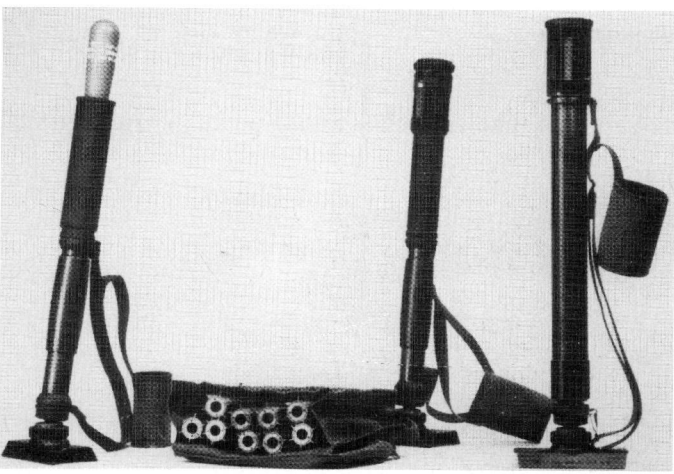

Hand launchers NR8111A1 and A2, with FLY-K 52 mm mortar and ammunition carrier

Sight unit, used with both hand launchers and 52 mm mortar

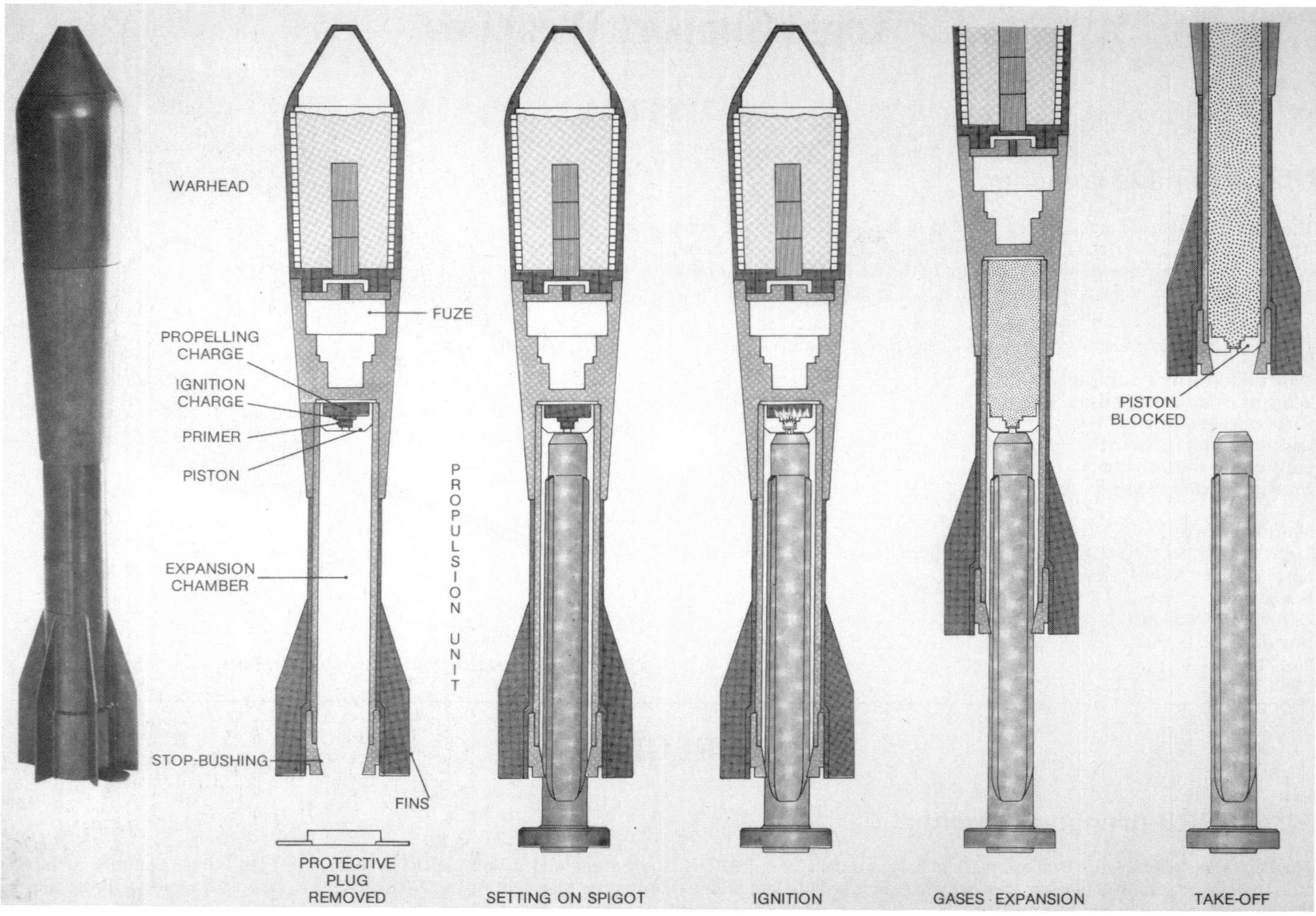

WARHEAD

FUZE

PROPELLING
CHARGE

IGNITION
CHARGE

PRIMER

PISTON

EXPANSION
CHAMBER

PROPULSION UNIT

STOP-BUSHING

FINS

PROTECTIVE
PLUG
REMOVED

SETTING ON SPIGOT

IGNITION

GASES EXPANSION

PISTON
BLOCKED

TAKE-OFF

FLY-K principle of operation

body moulded over a notched steel wire fragmenting sleeve, containing a bursting charge of Composition B (RDX/TNT) and closed by a plastic cap. The fuze is built in to the body of the grenade just behind the warhead; it is a superquick impact type with delayed arming which ensures safety in storage and handling and during the first 45 m of flight.

The stabiliser is a combination of light alloy tube and thermo-plastic fins, and the propulsion unit is built into the forward end of the stabiliser.

The illuminating grenade has a warhead containing a 200 000 candela star attached to a parachute. The head is closed by a nose cone of white plastic. The fuze is a gasless delay unit with a fixed burning time optimised for the best performance of the star unit. It is ignited on firing from the launcher and ignites both the star and an expelling charge at the vertex. The stabiliser and propulsion unit are as for the previous grenade.

The smoke grenade NR 210 has its warhead filled with white phosphorus and has the same delayed-arming impact fuze as the HE grenade. On impact the fuze causes a small explosion sufficient to break open the warhead and disperse the phosphorus, which then self-ignites to generate smoke.

DATA
LAUNCHER
Length overall: 605 mm
Weight: (unloaded) 4.5 kg

PROJECTILES

	NR 208 HE	NR 209 illuminating	NR 210 smoke
Calibre	52 mm	48 mm	52 mm
Overall length	330 mm	410 mm	330 mm
Weight	750 g	605 g	600 g
Muzzle velocity	90 m/s	100 m/s	85 m/s
Range	200–600 m	450 m range 220 m height	200–600 m
Fuze	Superquick	—	Superquick
Arming safety	45 m	—	45 m
Contents	135 g Comp B 580 fragments	Star 200 000 cd burning 25 s	60 g WP
Illumination	—	2 lux 225 m radius 4 lux 40 m radius	—

Manufacturer
Société Anonyme PRB, Département Défense, avenue de Tervueren 168, Bte 7, B-1150 Brussels.
Status
Current. Available.
Service
Foreign sales.

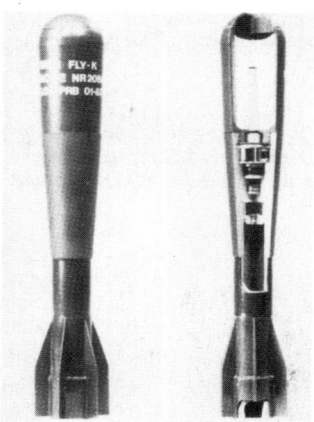

HE grenade NR 208A2

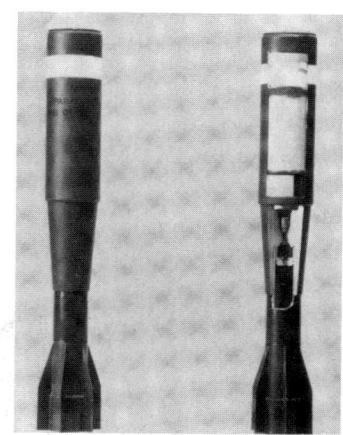

Illuminating grenade NR 209

NR 8464 PRB multiple grenade launcher

The NR 8464 multiple launcher is a modification of the FLY-K system which dispenses with the outer tube and merely carries 12 spigots on a rotating baseplate which can be attached to vehicles, light craft or used on a firm base as a positional defensive weapon. A level in the baseplate allows the launcher to be levelled independently of its vehicle, while the rotating plate permits firing through an arc of 359°. A lever-actuated retaining pawl allows adjustment of the elevation of the spigots to any one of five angles marked on a toothed sector – 75, 69, 62, 55 and 47.5° . These settings correspond, in the case of the standard fragmentation grenade NR 438, to ranges of 220, 360, 500, 640 and 780 m. The 12 spigots thus cover the entire range from 200 to 800 m; the adjusted range is based on the centre of the pattern, and the spigots are preset to divergent angles in both azimuth and elevation so as to give optimum dispersion of the 12 shots over an area of half a hectare with a front of about 100 m at maximum range. The divergence in elevation is gradually decreased as the elevation is raised so as to maintain approximately the same amount of dispersion at all ranges. The five zones, covered by the five fixed elevations, will overlap their splinter patterns.

The grenades are loaded by placing them over the spigot, and a spring hook at the base of each spigot will retain the grenade securely so that the vehicle can travel with the launcher fully loaded and ready to fire. The unit is fired electrically, via a multi-position switch; single shots can be fired, or any selected number up to the maximum load of 12, in rapid succession.

The grenades used with the multi-launcher are similar to those used with the hand launcher (above) but of different dimensions.

	NR 438 fragmentation	NR 440 illuminating
Overall diameter	58 mm	58 mm
Length	365 mm	410 mm
Weight	775 g	605 g
Muzzle velocity	100 m/s	100 m/s
Max range	800 m	450 m
Loaded	135 g Comp B	200 000cd star
Fragments	580	
Lethal radius	16 m	

The performance of the illuminating grenade is the same as that of the NR 209 (above). Both the NR 438 and NR 440 are fired electrically instead of by percussion.

DATA
LAUNCHER
Weight: (total) 95 kg
Height: 310–315 mm according to elevation
Base plate: 476 × 476 mm

Side view of multi-launcher loaded with NR 440 illuminating grenades: divergence in elevation between rows of spigots can be distinguished. On right is firing switch box

Multi-launcher unloaded: lever handle (upper left) adjusts elevation, set here for minimum range

Manufacturer
Société Anonyme PRB, Département Défense, avenue de Tervueren 168, Bte 7, B-1150 Brussels.

BRAZIL

LC T1 M1 flamethrower

The LC T1 M1 has been produced as a replacement for US models in use by the Brazilian Army and Marine Corps. It uses the principle of a single pressure tank containing air or nitrogen and two fuel tanks. The normal fuel is a mixture of gasoline and diesel oil, but other mixtures such as fish or vegetable oil with gasoline or thickened mixtures may also be used. The pressure in the air tank is sufficient for the entire fuel load without loss of range and power at the end.

The specially designed ignition system is energised by an electronic unit which is supplied by eight 1.5 V alkaline batteries. The electronic circuit converts this into about 20 000 V which provides an ignition arc across two electrodes mounted at the muzzle of the projector.

The high pressure cylinder of the portable flamethrower requires a charging unit that must be available during every operation, otherwise an excessive number of charged cylinders would be required. There is a mobile charging unit which brings logistic support to the front and permits weapon recharging

in a very short time. The unit is designed to fit any standard military trailer and consists of a diesel- or gasoline-driven compressor with two 8 m³ air storage cylinders, a flamethrower fuel tank of 600 litres capacity, fuel and oil tanks for the motor, and a kit of tools and spares.

The UR1 is a portable recharging unit consisting of one plastic container with 18 litres fuel capacity (the lid is used as a filling funnel) and one complete high pressure cylinder, attached to the carrier, a frame of highly resistant lightweight tubular construction.

Where a UM T1M1 mobile charging unit is not available to charge the high pressure flamethrower cylinder, commercial compressed air or nitrogen cylinders can be used. Hydroar can furnish a separate DC1 charging device to be connected to the commercial cylinders and to the high pressure flamethrowing cylinders.

LC T1 M1 flamethrowers in operation

LC T1 M1 flamethrower

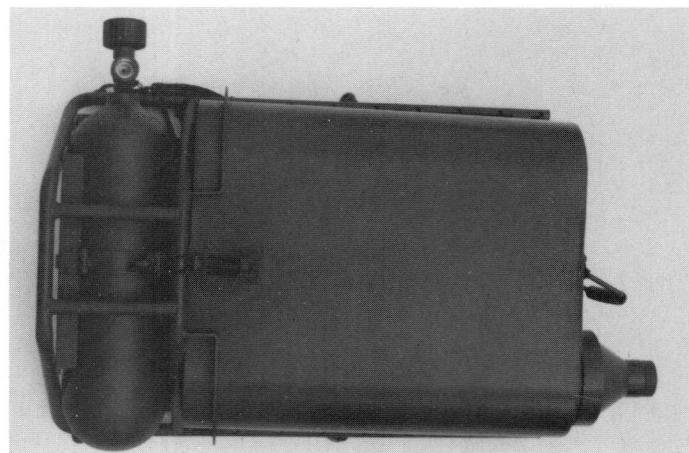

UR1 recharging unit

UM T1M1 Mobile charging unit

Flame gun of LC T1 M1 flamethrower, equipped with electronic ignition

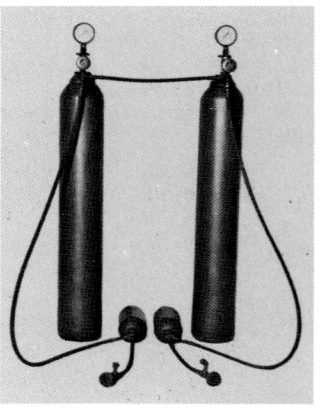

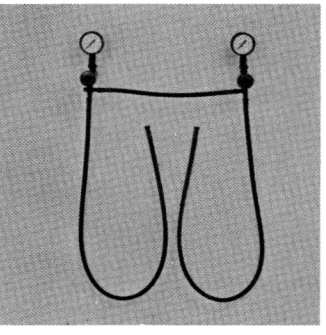

DC1 charging device

DATA
Weight: (unloaded) 21 kg; (with 15 l fuel) 32 kg; (with 18 l fuel) 34 kg
Max range: 70 m
Working pressure, fuel tank: 25 kg/cm²
Working pressure, air tank: 200 kg/cm²
Fuel hose length: 1.115 m
Weapon tube length: 635 mm

Manufacturer
Hydroar S/A, Rua do Rócio 196, CEP 04552 São Paulo, SP.
Status
Current. Production.
Service
Brazilian Army and Marines.

CHINA, PEOPLE'S REPUBLIC

30 mm grenade launcher

So far as can be determined, this is the Soviet AGS-17 grenade launcher manufactured in China. There appear to be some very minor dimensional differences, but in all major respects the weapon appears to be identical. The ammunition used is the standard Soviet 30 mm type, and HE and HEAT rounds are available.

DATA
Calibre: 30 mm
Operation: blowback, selective fire
Feed: belt
Length overall: 1280 mm
Height, zero elevation: 423 mm
Length of barrel: 300 mm
Muzzle velocity: 183 m/s
Maximum range: 1700 m
Rate of fire: 40 – 65 rds/min

Manufacturer
China North Industries Corporation, 7A Yue Tan Nan Jie, Beijing.
Status
Current. Available.
Chinese army.

Chinese troops with locally made 30 mm AGS-17 grenade launcher

35 mm grenade launcher Type W87

This is a new weapon of Chinese design and manufacture, firing a unique 35 mm round. Few details have been released, but the weapon is certainly smaller and lighter than the 30 mm AGS-17, even though it fires a larger projectile. The mechanism would appear to be a great deal simpler, and is probably a plain blowback; it is noticeable that the feed system is either a box magazine or a helical drum, so that no feed pawl complications are required and the operating system can be kept correspondingly simple. Also noteworthy is the obvious muzzle compensator, intended to keep the barrel climb under control during automatic fire, and the bipod folded beneath the barrel, suggesting a one-man light support role as well as the more common tripod-mounted role.

The ammunition available for this weapon has not been seen, but information from the manufacturers is that two rounds are provided, a high explosive/fragmentation projectile containing some 400 3 mm steel balls and with a lethal radius of 10 m; and a HEAT projectile capable of penetrating 80 mm of steel armour at 0° incidence. Ammunition is fed to the weapon from either six- or nine-round vertical box magazines or from a 12-round helical drum magazine.

DATA
Calibre: 35 mm
Operation: blowback, Selective fire
Feed: 6- or 9-round box, 12-round helical drum
Weight of launcher: 12 kg
Weight of tripod: 8 kg
Weight of projectile: 270 g
Muzzle velocity: 170 m/s
Effective range: 600 m
Maximum range: 1500 m
Rate of fire: 400 rds/min

Manufacturer
China North Industries Corporation, 7A Yue Tan Nan Jie, Beijing.
Status
Current. Available.
Chinese army.

Chinese 35mm Type W87 grenade launcher with helical drum magazine

EGYPT

Home Guard anti-personnel weapon

This is actually the Egyptian PG-7 anti-tank rocket launcher but employing a specially-developed anti-personnel warhead. It has been designed to fill various operational requirements for those ranges which fall between the hand grenade and larger weapons; it can be used to replace an infantry section mortar, and the rocket and warhead can be fired from the standard PG-7 launcher, thereby easing the logistic problem.

The warhead is cylindrical and consists of an explosive charge surrounded by controlled fragmentation steel rings. An all-ways fuze is fitted, highly sensitive for low angles of impact, allowing direct and indirect fire to ranges in excess of 1200 m.

DATA
Warhead calibre: 80 mm
Warhead length: 190 mm
Warhead weight: 2.5 kg
Projectile length: 520 mm
Projectile weight: 3.75 kg
Explosive content: 740 g
Fragmentation content: 800 g
Number of fragments: 980
Lethal radius: 30 m
Muzzle velocity: 90 m/s
Operating range: 100m flat trajectory fire; 1000 m at 10° elevation; 1200 m at 25° elevation
Time to self-destruction: 13 s

Manufacturer
Sakr Factory for Developed Industries, PO Box 33, Heliopolis, Cairo.
Status
Current. Production.
Service
Egyptian army.

Sakr Industries Home Guard light support weapon

FRANCE

Ruggieri D.I.P.S. area defence system

The D.I.P.S. (Défense Instantanée Position Stratégique) is, in effect, a ground application of the type of grenade discharger commonly found mounted upon armoured vehicles. The device consists of a control box with five double-barrelled launcher units fitted in a fan shape so as to disperse their projectiles over a wide area. The projectiles have a delay fuze which detonates them at pre-determined distances from the launcher, so that the entire area is covered. Each detonation releases 900 steel balls at a velocity of 1100 m/s, and is accompanied by a loud report, greater than 170 dB, which will add to the effect. The pattern of the detonating munitions is such that five launchers will completely protect an area of 2600 m² in a 120° arc; by adding more launch units it is possible to cover 360° around a point, an area of 7700 m². The danger area extends to 50 m from the launcher unit.

The D.I.P.S. system is suggested for the protection of vulnerable points and also as a portable method of protection for mobile units when deployed or when halted for the night. Other types of munition are available, including wounding and detonating projectiles with less range, neutralising projectiles with varying ranges, and tear gas and similar law enforcement projectiles.

Manufacturer
Ruggieri, Département Armement, 86 av. de Saint-Ouen, 75018 Paris.
Status
Current. Production.

Ruggieri D.I.P.S. area defence system

GERMANY, EAST

Model 41 flamethrower

The Model 41 flamethrower is carried on the infantryman's back and consists of two cylindrical tanks, one containing the fuel and the other being the pressure tank. The flamegun is connected to the tank by a hose and consists of a long cylinder with a smaller cylinder on the side. No details of the performance of the Model 41 have become available but its maximum range is probably about 20 m. It is unusual in that its tanks are carried horizontally.

Status
Current.
Service
East German Army.

GERMANY, WEST

HAFLA DM 34 hand flame cartridge launcher

Nomenclature: Flammpatrone, Hand, DM 34

This is in service with the West German Army. It is a single-shot, one-man operated, throw-away weapon designed to impel an incendiary smoke charge to a range of 70 to 80 m.

The weapon comprises an aluminium launch tube and three compressed sections of incendiary smoke composition contained in a projectile. A pivotal handgrip fuze and firing assembly is fitted at one end of the tube and a plastic cover at the other end. The launcher is watertight. Three units are individually packed and moisture sealed in a plastic carrying case.

Operation
The HAFLA DM 34 is delivered with the handgrip in the safe position. By depressing the safety button the grip can be moved outwards exposing the trigger. The trigger cannot be activated until the grip is moved completely to the rear and is locked in position. The weapon is then ready to fire.

The trigger is depressed and the firing pin strikes the primer which ignites the first propellant. Simultaneously the projectile is set in motion and the second propellant is ignited and the aluminium capsule is blown out of the tube.

As the second propellant burns, the delay fuze begins to smoulder. This action concurrently activates the incendiary smoke payload. After some two seconds, the burning mixture in the delay fuze reaches the dispersion charge. An immediate burst then demolishes the projectile spreading burning fragments of incendiary smoke composition.

The projectile bursts after travelling 70 to 80 m and burning particles will be spread over an area 10 m wide and 15 m long. If however the projectile strikes a hard surface within a range of 8 to 70 m, the incendiary smoke charge immediately scatters on impact in a brilliant flash of blinding smoke and fire covering an area of 5 to 8 m. In either case the combustion of hot (1300°C) fragments continues for at least two minutes.

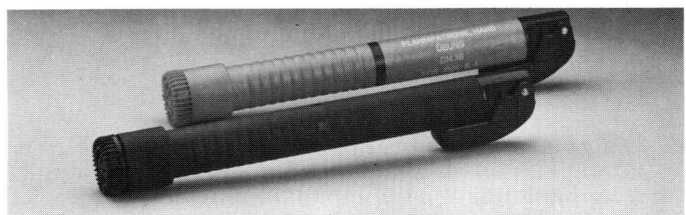
HAFLA DM 34 and DM 38 cartridge launchers

DATA
Weight: 625 G
Weight of composition: 240 g
Overall length: 445 mm
Calibre: 35 mm
Incendiary composition: red phosphorus
Packing: 3 HAFLAs per pouch; 51 HAFLAs per crate

Practice grenade DM 38
This has the same dimensions and weight as the HAFLA DM 34 but has an inert filling of lime and a smoke marker to indicate the point of impact.

Manufacturer
Buck Werke GmbH and Co, Postfach 2405, D-8230 Bad Reichenhall.
Status
Current. Production.
Service
West German armed forces.

---

40 mm Granatpistole grenade launcher

This is a light and handy weapon designed to fire grenades at either low or high angle and can project a 40 mm grenade up to a range of 350 m. It thus fills the gap between maximum hand throwing and minimum mortar range.

The grenade launcher is a single-shot, break-action weapon with a retractable butt stock, a fixed foresight and a folding ladder-pattern rear-sight. The barrel is rifled.

To load the weapon, pull the locking lever to the rear until the barrel pivots to its loading position; the striker does not cock during this movement. If there is a spent case in the breech it may be extracted manually, a recess cut in each side of the breech-end enabling the firer to grip and withdraw the case. When the next round has been inserted all the way into the chamber, the barrel is pivoted back to the receiver, where it is secured by the locking lever which engages in the locking slots on the barrel. The striker must now be cocked by hand, and may be secured by applying the safety lever positioned on either side of the pistol grip.

Grenades may be launched with the butt stock either extended or retracted. To extend it, rotate it 90° to the left or right, extract it all the way, and rotate again 90°. Retraction of the butt stock is carried out in the reverse manner. The ladder rearsight is raised for aiming if the range exceeds 100 m.

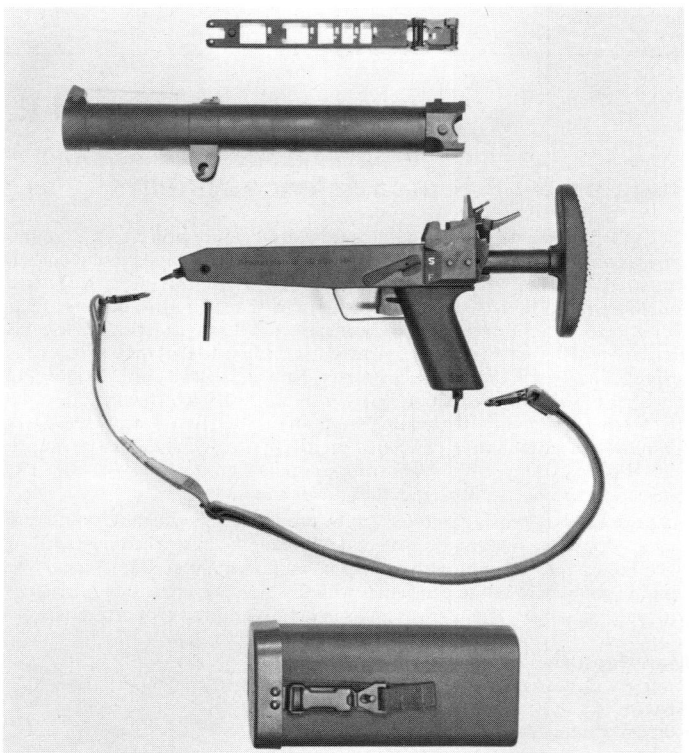
40 mm Granatpistole grenade launcher disassembled

40 mm Granatpistole grenade launcher

DATA
Calibre: 40 mm
Mode of fire: single-shot manually loaded
Length: 463 or 683 mm with stock extended
Height: 205 mm
Sights: (foresight) fixed barleycorn; (rearsight) folding ladder
Sight radius: 342 mm
Muzzle velocity: approx 75 m/s
Barrel length: 356 mm

Manufacturer
Heckler and Koch GmbH, 7238 Oberndorf-Neckar.
Status
Current. Production.
Service
West German Army. Several military and police forces worldwide.

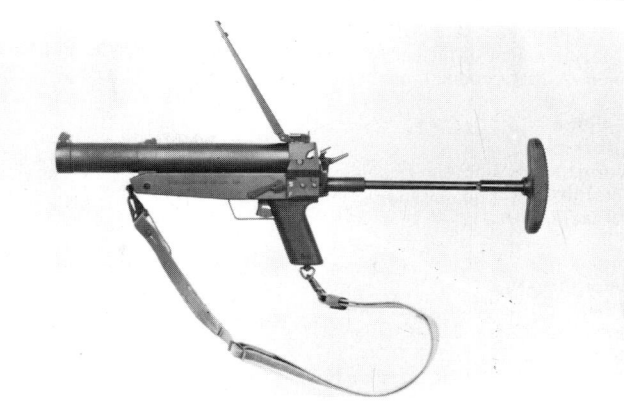

40 mm Granatpistole grenade launcher with butt stock extended and rear ladder sight raised

40 mm Heckler and Koch add-on grenade launcher HK79

The HK79 add-on launcher fits on all G3 and G41 rifles except the short K models. It is assembled to the weapon in place of the hand-guard. This does not affect the accuracy of the rifle, since the barrel is still free to oscillate. Handling and operating functions of the rifle are not affected by the presence of the launcher.

A versatile range of ammunition, the universally-employed 40 mm grenades based on the American M79 design, is available, enabling targets from 50 to 350 m to be engaged with various effects.

The HK79 launcher is a single-shot weapon with a dropping steel barrel. To load, the barrel is unlatched and drops down under its own weight; the round is inserted into the chamber and the barrel closed. The tilting barrel will accept rounds of any length. The firing pin is cocked by hand and the launcher is fired by pressing the trigger on the left-hand side of the hand-guard. At the rear of the launcher is a sliding safety catch, the status of which is shown by coloured rings – white for 'Safe' and red for 'Fire'; the weapon can be loaded and cocked with the safety set at either position. A mechanical tilting sight is fitted to the right side of the grenade launcher and permits range setting from 50 to 350 m.

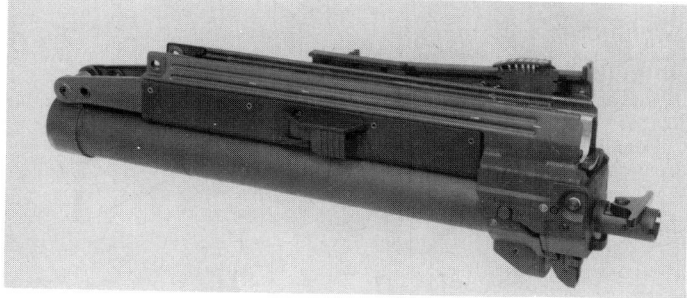

HK79 40 mm add-on launcher

DATA
Calibre: 40 mm
Weight, rifle and launcher: G3 – 5.6 kg; G-41 – 5.4 kg
Weight of launcher: 1.5 kg
Sight steps: 50 to 350m in 50m steps
Muzzle velocity: 76 m/s

Manufacturer
Heckler and Koch GmbH, D-7238 Oberndorf-Neckar.
Status
Current. Production.

HK79 40 mm add-on launcher; mechanical tilting sight

G3 rifle with Heckler & Koch HK79 40 mm add-on launcher

G41 rifle with HK79 add-on 40 mm launcher opened for loading

40 mm HE Heckler and Koch cartridge

When Heckler and Koch developed special anti-riot cartridges, they discovered that the existing 40 mm cartridges were somewhat lacking in performance. In particular the fuzes did not always come up to expectations and the decision was taken to develop their own rounds in this calibre.

The first improvement has been to the fragmentation effect. To achieve this a new production technique has been developed which allows the use of pre-formed ball fragments impregnated into the plastic body. This arrangement uses all the explosive power of the charge instead of dissipating some of it in bursting the steel body used in other designs. The pre-formed fragments can also be relied on to break clear and fly in predictable sizes, so precisely defining the velocity and pattern of the fragment spread. Thus, the fragments are lethal at 5 m but not at 12 m and this is reliable.

The greatest difficulty with the fuzes of these small grenades is to produce reliable detonation. They need to be of such small size, yet have to react to a slow speed of arrival coupled with a low angle of graze in many cases. Heckler and Koch overcame this difficulty by adding a self-destruction device to the fuze. The fuze still meets all the safety requirements and passes the severest environmental tests, yet it ensures that the grenade explodes reliably.

The propellant cartridge is unchanged.

Development of the cartridge was carried out by Heckler and Koch and Diehl GmbH; production is by Diehl.

DATA
CARTRIDGE
Length: (total) 99 mm
Weight: (overall) 230 g; (projectile) 180 g
Muzzle velocity: 76 m/s (at 21°C)

PROJECTILE
Weight: 180 g
Fragments: 700 steel pellets
Fragment size: 2.25 mm diameter
Explosive: 31 g Hexal 70/30
Performance: (5 m effective radius) penetration of 2 mm dural, about 2 lethal fragments/m²; (12 m effective radius) penetration of 0.5 mm dural, about 0.5 lethal fragments/m²; (15 m effective radius) penetration of 0.5 mm dural, about 0.2 wounding fragments/m²

FUZE
Muzzle safety: 9 m minimum
Arming distance: (max) 15 m
Sensitivity: 1 mm aluminium F40 within 0–60° (NATO standard)
Self-destruction time: 8 s (equal to 350 m range)

Manufacturer
Production by Diehl GmbH, Ammunition Division, Fischbachstrasse 14, D-8505 Rothenbach.
Export marketing by Heckler and Koch GmbH, 7238 Oberndorf-Neckar.
Status
Current. Production.
Service
West German armed forces and numerous other countries world-wide.

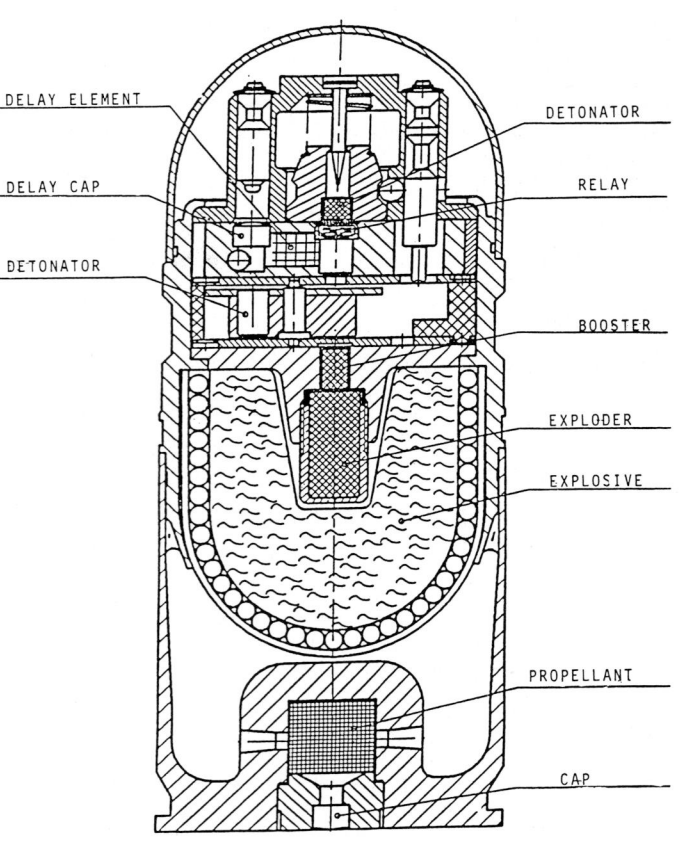

General arrangement drawing of 40 mm Heckler and Koch cartridge. Relative proportions of fuze to main charge are particularly well shown emphasising amount of space needed to fit in safe and reliable fuze

HUNGARY

Model 51 flamethrower

This consists of two vertical tanks carried on the man's back and connected to the flamegun by a hose. It is believed that the weapon is operated by first pressing the fuel lever towards the rear of the flamegun and then pressing a second lever to ignite the fuel. No details of the range of the Model 51 have become available but its maximum range is probably about 20 m.

Status
Current.
Service
Hungarian Army.

ITALY

12-gauge Franchi SPAS 12 shotgun

The firm of Luigi Franchi is well known in the field of shotguns where it has for many years produced excellent guns which are used all over the world by sportsmen. Franchi decided that it was time that a specific riot shotgun was produced since practically all the law enforcement shotguns in use are sporting guns with or without some modification. Since the requirements for a riot gun are rather different from those needed by a sports shooter it follows that there are some deficiencies in the present police and military weapons. Franchi set out to correct this by designing a gun from the requirements. The result is its Special Purpose Automatic Shotgun (SPAS) which was first produced in October 1979.

The gun is a short-barrelled semi-automatic shotgun with a skeleton butt and a special device enabling the firer to carry and fire the gun with one hand if need be. The receiver is made from light alloy and the barrel and gas cylinder are chromed to resist corrosion. All external metal parts are sand blasted and phosphated black.

The characteristic scattering effect of a smoothbored barrel spreads the pellets to about 900 mm diameter at 40 m and progressively greater at longer ranges so that all that is needed to hit the target is a quick, rough aim.

The automatic action will fire at about four shots a second. Using standard buckshot rounds it is possible to put 48 pellets a second on to a target 1 m² at

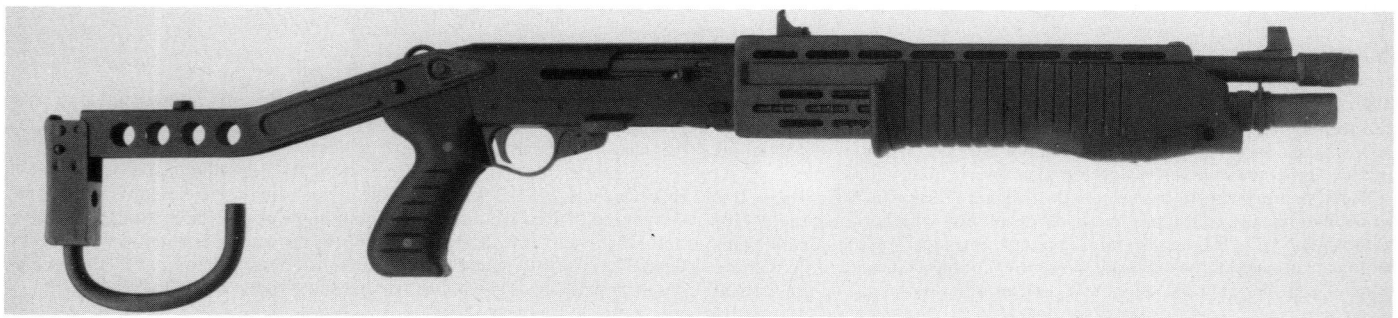

Franchi SPAS 12 shotgun

Entry hole of Franchi MPS shot in 8 mm steel plate

Four layers of bullet proof glass pierced by Franchi MPS metal-piercing shot from SPAS 12 shotgun

Exit hole of Franchi MPS shot in 8 mm steel plate

40 m. At that range the pellets each have a residual energy approximately 50 per cent greater than that of a 7.65 mm pistol round at the same range.

The wide range of ammunition offers a selection to the user which he can adapt to each particular situation, from buckshot and solid slug (with the same energy as a 7.62 mm NATO round) to small pellets and tear gas rounds which shoot a plastic container filled with CS gas up to a maximum range of 150 m.

When using the lighter ammunition the firer selects manual reloading, using the pump action, but for all other types of round the gun will fire automatically.

The weapon is stripped by unscrewing the tubular magazine and then sliding the entire barrel group away from the receiver and clear of the gun.

Two retaining pins hold the trigger mechanism in place. The breech-block is slid forward out of the body, having first removed the cocking handle. Further stripping is not necessary by the firer.

With the grenade launcher fitted to the muzzle, a grenade can be fired out to a maximum range of 150 m. There is also a special scattering device which fits to the muzzle and produces an instantaneous spread of pellets. This is very useful for indoor firing.

DATA

Model	11	12
Calibre	12 bore	12 bore
Length	900 mm	930 mm
(with folded stock)	—	710 mm
Barrel	500 mm	460 mm
Weight	3.2 kg	4.20 kg

Operation: gas, semi-automatic or hand pump
Rate of fire: (theoretical) 250 rds/min; (practical) 24-30 rds/min
Reloading time: 15s for 7 rounds into magazine
Sights: notch rearsight and ramp foresight

Manufacturer
Luigi Franchi SpA, Via del Serpente 12, 25020 Brescia (Fraz. Fornaci).
Status
Current. Available.

12-gauge Franchi SPAS 15 shotgun

Following the success of its SPAS 12 shotgun, the Franchi company, after discussions with American and Italian authorities, decided to transform the SPAS 12 by replacing the tubular magazine with a box magazine. The resulting weapon was the SPAS 14, which, though technically effective, had some ergonomic shortcomings and went no further than the prototype stage. In 1983 Franchi decided to approach the problem from a completely different angle and develop a new system. Prototypes of the SPAS 15 were built early in 1984 and these matched both the American JSSAP and Italian Ministry of the Interior requirements. The design has now been brought to completion and production has begun.

The SPAS 15 uses both semi-automatic and manual pump action. Pump action, while not required by military authorities, is nevertheless considered necessary for police employment where the ability to generate a large volume of fire is not so important. The semi-automatic functioning is performed by a gas piston which drives a bolt carrier and rotating bolt; the same components are manually operated by pulling back the sliding fore-end in the conventional pump action manner. The bolt remains open after the last shot has been fired, and replacing the empty magazine by a loaded one will automatically release the bolt, which then loads a fresh round into the chamber. Pump or semi-automatic action is selected by a button in the fore-end.

12-gauge Franchi SPAS 15 shotgun

The receiver is of die-formed and drawn nickel chromium steel; the external surface is sand-blasted and phosphated. The barrel is of tempered alloy steel, is chromed internally and phosphated externally, and has a screwed section at the muzzle for the attachment of either a shot concentrator or one of three available grenade launching attachments.

The stock is of tubular steel and can be folded. There is a carrying handle above the receiver, which protects the cocking handle (used when operating in the semi-automatic mode) and can also be used to mount a variety of sights.

DATA
Cartridge: 12-gauge 70 mm
Operation: Gas, semi-automatic, or manual pump action
Locking system: rotating bolt
Feed: 6-round box magazine

Weight: 3.80 kg without magazine
Length: 920 mm
Barrel: 400 mm

Manufacturer
Luigi Franchi SpA, Via del Serpente 12, 25020 Brescia (Fraz. Fornaci).
Status
Current. Production.

12-gauge Bernadelli B4 shotguns

The B4 shotgun family has been designed specifically for use by military and security forces. The frame is a single piece of aluminium alloy, giving high strength, the barrel is a high grade steel, and all external surfaces have an anti-corrosion and anti-glare finish. The breech mechanism employs a rotating bolt for additional security, and it is possible to launch grenades from the muzzle. The weapon feeds from a removable box magazine. There is a grip safety device which prevents the weapon being fired unless properly handled.

The B4 shotgun can be operated as a semi-automatic weapon or manually, as a pump action gun, the choice being selected by a simple lever. It was developed to meet the American JSSAP specification for a combat shotgun. The action is gas-operated in the semi-automatic mode. It can fire all normal plastic and brass-cased shotgun ammunition.

The B4/B model is derived from the B4 but dispenses with the semi-automatic action and is pump action only. Its dimensions are the same as those of the B4 except that it weighs 450 g less.

DATA
Cartridge: 12-gauge, 70 mm
Operation: gas, semi-automatic or manual pump action
Magazine capacity: 3, 5, or 8 cartridges
Length: stock extended 950 mm; folded 730 mm
Barrel: 460 mm, cylinder bored
Weight: 3.45 kg

Manufacturer
Bernadelli SpA, I-25063 Gardone Val Trompia, Brescia.

Bernadelli B4 (lower) and B4/B (upper) shotguns

Status
Current. Production.

Mod T-148 portable flamethrower

The flamethrower Mod T-148 is a light, efficient and easy to operate weapon. It consists of a tank assembly, a flexible hose and a flame gun. Operation is based on two fundamental concepts: an innovative electronic ignition system and the elimination of the high pressure tank from the tank assembly.

The electronic ignition system is far superior to the traditional ignition cartridges. The electronic ignition is silent and almost invisible, thus allowing surprise actions, whereas ignition cartridges may reveal to the enemy the presence of the operator, exposing him to danger.

The concept behind the operation of the tank assembly consists in pressurising the tank full of thickened fuel to the design value, and having the pressure decrease during projection. This way the traditional high pressure bottle, and the associated pressure reducer, have been eliminated from the tank assembly, making the design simpler and more rugged.

The main components
The flame gun: light and robust, this comprises the ejection nozzle and valves and the electronic ignition. The two handgrips have been designed to achieve optimum ease of operation.
The electronic ignition system is housed in a shock-resistant casing attached to the front handgrip. Its components are completely encapsulated in the casing, so that the ignition system will continue to function even if the flamethrower has been completely immersed in water. The system is powered either by commercial alkaline batteries or by Ni-Cd rechargeable cells.
The tank assembly consists of two cylindrical containers, made of weldless aluminium alloy, surface treated with thick anodic coating. It is equipped with a safety valve with a built-in pressure indicator that signals the pressurised condition.

Before operation the tank assembly is filled with thickened fuel for two-thirds of its capacity and, after closing the plugs, the tank is pressurised with air or nitrogen through the filling valve by simply connecting it to a suitable compressed gas supply.
The flexible hose connecting the tank assembly to the flamegun is provided, at both ends, with ball-bearing couplings. This allows easy handling of the weapon, even when it is pressurised, by allowing free relative rotation between the flamegun and the tank.

DATA
Weight of empty tanks: 8.5 kg
Weight of flame gun: 4.3 kg
Weight of hose: 1.0 kg
Weight of 15 l of fuel: 11.2 kg
Weight of batteries: 0.5 kg
Total weight: 25.5 kg
Operating range: in excess of 60 m
Working pressure: 28 kg/cm^2

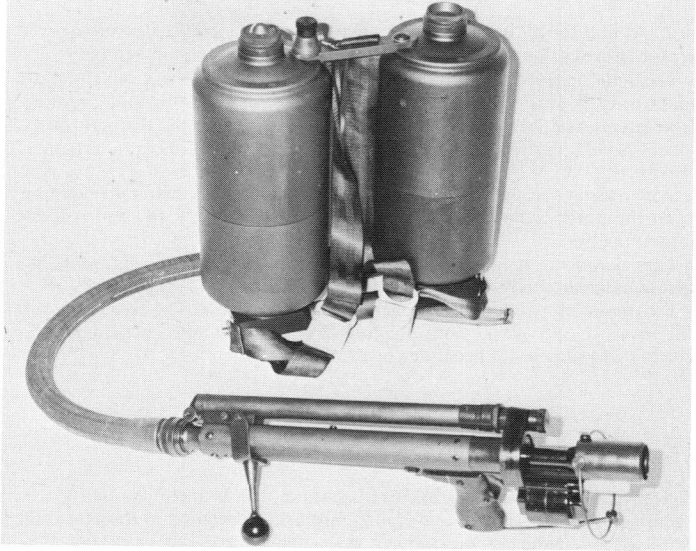

Mod T-148 portable flamethrower

Test pressure: 60 kg/cm^2
Operational temperature limits: -20 to +55°C

ACCESSORIES
Field Unit: steel container with accessories to prepare and transfer the thickened fuel. Capacity 45 litres
Pressure Reducer: pressurises the flamethrower directly from commercial cylinders up to 28 bar.
Double Scale Tester: to check battery and circuit efficiency
Haversack: contains one complete refill of fuel and nitrogen

Manufacturer
Tirrena SIPA, via Monte d'Oro, 00040 Pomezia (Roma).
Status
Current. Production.
Service
In service with unspecified armies.

ROMANIA

40 mm Automatic Grenade Launcher

This weapon is broadly based upon the Soviet AGL-17 but is of 40 mm calibre and is chambered for a special cartridge which is longer than the universal US-type 40mm cartridge and appears to be slightly bottle-necked. The launcher is normally mounted on a tripod, with elevation and traverse locks. Feed is from a drum containing 10 rounds, mounted high on the left side of the receiver. The launcher can be fired in the single-shot or full automatic modes, though the literature suggests restricting burst fire to no more than five rounds.

DATA
Cartridge: 40 mm special to weapon
Operation: blowback
Feed: 10-round drum magazine
Weight: gun 23:kg; with tripod, without magazine 35.5 kg
Length: 875 mm
Barrel: estimated 300 mm
Height of bore, on tripod: 600 mm
Muzzle velocity: 220 m/s
Maximum range: 1300 m

Manufacturer
State arsenals.

Romanian 40 mm automatic grenade launcher

Status
Current; production.
Service
Romanian armed forces.

122 mm support rocket launcher

This is a simple single-tube launcher normally mounted on a tripod, though it can also be vehicle-mounted. It can be man-carried for short distances fully assembled; for longer movement the launch tube is removed from the tripod. Firing is done by means of a magneto which can be located up to 35 m from the launcher and connected by cable. An interesting and unusual feature is the presence of what appears to be a muzzle compensator at the front of the launch tube; it is believed that the electrical firing linkage partly closes the rear of the tube and the muzzle compensator balances the small recoil of the tube due to this obstruction.

DATA
Calibre: 122 mm
Launcher length: 2.50 m
Launcher weight: 30 kg
Tripod weight: 25:kg
Rocket weight: 46.3 kg
Rocket length: 1927 mm
Maximum range: 13 400 m

Manufacturer
State arsenals.
Status
Current. Production.
Service
Romanian armed forces.

Romanian 122 mm rocket launcher

SINGAPORE

40 mm CIS-40GL grenade launcher

The CIS-40GL is a modular-construction, single-shot grenade launcher capable of firing all types of standard 40 mm grenades. The weapon is built up from four basic units: receiver, barrel, leaf sight and butt stock or rifle adapter. In normal form it is fitted with a butt stock as an individual weapon, but by removing the stock and replacing it with the rifle adapter it can be fitted beneath the barrel of a wide range of assault rifles, in the manner of the US M203 system.

To load, the lever at the left side of the receiver is pressed down; this simultaneously cocks the weapon, applies the safety catch and unlocks the barrel. The barrel can then be swung to the side and the grenade loaded, after which the barrel is swung back, when it automatically locks to the receiver. The safety catch must be manually released before the weapon can be fired.

The sight is a folding, graduated, leaf sight which can be adjusted for both elevation and azimuth. It is graduated to 350 m in steps of 50 m

DATA
Calibre: 40 mm
Length overall: 655 mm
Barrel: 305 mm
Weight, empty: 2.05 kg with stock; 1.70 kg without stock
Muzzle velocity: 76 m/s
Maximum range: 400 m

40 mm CIS-40GL grenade launcher

Manufacturer
Chartered Firearms Industries (Pte) Ltd, 249 Jalan Boon Lay, Singapore 2261.
Status
Current. Production.

40mm CIS 40-AGL Automatic grenade launcher

Development of this weapon began in 1986 and is still in progress. It is anticipated that initial production will commence in mid-1991.

The weapon is a blowback, firing from an open bolt. The bolt assembly is carried on a heavy central guide rod in the receiver, with two secondary guide rods, one on each side, which carry the return springs. The central rod carries a set of Belleville springs at the rear end which act as a bolt buffer. The weapon fires from an open bolt. The trigger is at the rear of the receiver and operates through a linkage carried above the receiver, to a sear which engages with a notch in the upper surface of the bolt. Ammunition is belt-fed, and a simple two-stage pawl unit inside the top belt cover indexes the belt forward. The feed is driven by the movement of the bolt unit; as the bolt moves rearward, the belt is moved one stage, and as the bolt moves forward the belt moves a further step, the cartridge is removed from its link and chambered.

The 40-AGL is currently chambered for the standard US-type 40mm grenade, but it is possible that CIS may develop a design of their own in order to provide better ballistic performance.

Manufacturer
Chartered Firearms Industries (Pte) Ltd, 249 Jalan Boon Lay, Singapore 2261.
Status
Development.

40 mm CIS 40-AGL automatic grenade launcher

SOUTH AFRICA

40mm Armscor grenade launcher

The 40mm low velocity grenade launcher is a single-shot, break-open, breech-loading, shoulder-fired weapon, fitted with a ranging occluded eye gunsight (OEG) system. It is an infantry small arm for firing low velocity grenades, whose tactical range lies between the maximum hand grenade throwing range and the minimum range of the mortar.

The high explosive grenades are effective against light-skinned vehicles and personnel up to 375 m. The grenade has an effective casualty radius of 5 m and it may normally be expected that 50 per cent of the exposed personnel will become casualties.

The launcher is capable of firing smoke, illuminating, flash, inert and high explosive fragmentation rounds. It is safe, reliable, accurate and economical to produce due to the utilisation of sheet-metal stampings, pressings and precision castings. The range quadrant is graduated in 25 m increments from 25 to 375 m and adaptable in poor light conditions.

Training, handling and maintenance under adverse conditions are made easier due to its simplicity.

DATA
Calibre: 40 mm
Weight, empty: 3.7 kg
Max length with butt extended: 665 mm
Max length with butt retracted: 475 mm
Muzzle velocity: 76 m/s
Barrel length: 340 mm
Minimum range
  Training: 100 m
  Combat: 30 m
Max effective range: (point targets) 150 m; (area targets) 375 m

40 mm Armscor grenade launcher

Manufacturer
Armscor, Private Bag X337, Pretoria 0001.
Status
Current. Production.
Service
South African Defence Force.

40mm Armscor MGL 6-shot grenade launcher

This weapon is a light, semi-automatic, low velocity, shoulder-fired 40 mm grenade launcher with a six-round capacity and a progressively-rifled steel barrel. It uses the well-proven revolver principle to achieve a high rate of accurate fire which can be rapidly brought to bear on a target. A variety of rounds such as HE, HEAT, anti-riot baton, irritant or pyrotechnic can be loaded and fired at a rate of one per second; the cylinder can be loaded or unloaded rapidly to maintain a high rate of fire. Although intended primarily for offensive/defensive use with high explosive rounds, with suitable ammunition the launcher is extremely useful in anti-riot and other security operations.

An adjustable occluded eye gunsight (single point) has been designed so that it can be used to determine the range to the target and be instantly adjusted accordingly. It enables the user to increase the hit probability at ranges up to 375 m. The range quadrant is graduated in 25 m increments and the aim is automatically compensated for drift.

The launcher consists of a light rifled barrel, sight assembly, frame with firing mechanism, spring-actuated revolving cylinder, and folding butt. The butt is adjustable to suit the eye relief and stance of the user, and the safety catch can be operated from either side. The launcher cannot be accidentally discharged if dropped.

The launcher is loaded by releasing the cylinder axis pin and swinging the frame away from the cylinder. By inserting the fingers into the empty chambers and rotating the cylinder it is then wound against its driving spring. The grenades are then inserted into the chambers, the frame closed, and the axis pin re-engaged to lock. When the trigger is pressed, a double-action takes place and the firing pin is cocked and released to fire the grenade. Gas pressure on a piston unlocks the cylinder and allows the spring to rotate it until the next chamber is aligned with the firing pin, whereupon the next round can be fired.

The occluded eye gunsight is a collimating sight which provides a single aiming spot; the firer sights with both eyes open and the effect is to see the

40 mm Armscor MGL 6-shot grenade launcher

aiming spot superimposed on the target, both target and spot being in sharp focus. The launcher is also fitted with an artificial bore-sight which can be used to zero the collimating sight. The collimating sight includes a radio-luminous lamp which provides the spot contrast and which has a life of approximately 10 years.

DATA
Calibre: 40 mm
Weight, empty: 5.3 kg
Weight, loaded: 6.50 kg
Length, butt extended: 788 mm
Length, butt folded: 566 mm
Barrel length: 300 mm
Rifling: 6 grooves, rh, increasing twist to one turn in 1200 mm at muzzle

Sustained rate of fire: as fast as the trigger can be pulled: approximately 20 rds/min
Muzzle velocity: 76 m/s
Max range: 400 m
Minimum range: 100 m (training); 30 m (combat)
Max effective range: (area targets) 375 m; (point targets) 150 m

Manufacturer
Armscor, Private Bag X337, Pretoria 0001.
Status
Current. Production.
Service
South African Defence Force.

SPAIN

40 mm AGL-40 automatic grenade launcher

The AGL-40 is a belt-fed automatic grenade launcher with an effective range of about 1500 m. The design is based on the long recoil principle, this being considered most appropriate in order to achieve the lightest possible weight for the weapon by reducing the recoil loadings. The light weight thus gives the launcher considerable versatility as a general purpose tactical weapon which can be readily carried in the field.

The long recoil system lends itself, in this calibre, to delivering the relatively low rate of fire of 200 rds/min which is tactically desirable and which allows the weapon to be pointed and controlled easily. This rate of fire also eases the logistic problem of ammunition resupply.

The AGL has been designed to be turret, tripod or pedestal mounted on wheeled or tracked vehicles, helicopters or boats and as a ground infantry weapon.

An important feature is the ability to shift the feed to the right or left side as desired, a change which can be effected very rapidly without the need for special tools. It is a simple weapon to operate and maintain. Field stripping can be done without tools and takes no more than 30 seconds.

The AGL-40 is designed to fire the 40 mm High Velocity grenades M383, M384, M385 and M430 and similar types.

The information given below is provisional, as small changes may be made in the course of preparing for series production.

DATA
Calibre: 40 × 53SR
Operation: long recoil, automatic
Feed system: 25- or 50-round linked belt
Weight: gun 30 kg; tripod 22.7 kg; cradle mount 10.5 kg; feed box with 25 grenades 10.4 kg
Length: 980 mm
Barrel: 400 mm
Rifling: 18 grooves, rh, one turn in 1219 mm
Sight: mechanical and optical panoramic
Muzzle velocity: 241 m/s
Maximum range: 2200 m

Manufacturer
Santa Barbara, Julian Camarillo 32, 28037 Madrid.
Status
Current. Pre-production.

40 mm AGL-40 showing left side ammunition feed and panoramic sight

40 mm AGL-40 automatic grenade launcher on tripod

SWITZERLAND

35 mm Arpad 600 close support weapon

Arpad 600 is a shoulder-fired multi-purpose close support weapon for infantry use. Lightweight and compact, it can be directed on targets very rapidly, can deal with stationary or moving targets, and can be fired from cover. Individual, flexible, and with multiple applications, the Arpad is meant to destroy point objectives such as machine gun posts, light vehicles or assault helicopters.

The rifled barrel is supported on a shoulder frame and is free to recoil, the movement being controlled by a hydro-pneumatic buffer mounted above the tube. High explosive and illuminating ammunition has been developed, and the rounds are light so that 20 to 30 can be carried by the firer. It is claimed that Arpad 600 is superior to conventional rifle grenades in both accuracy and range.

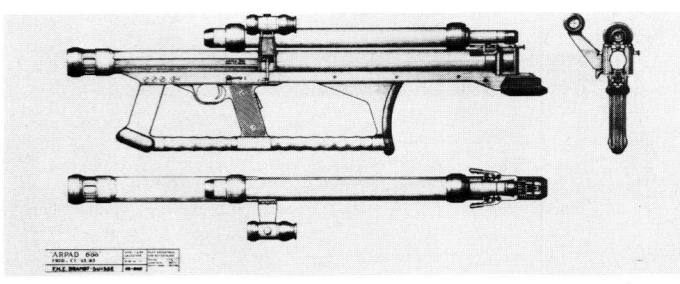

35 mm Arpad 600 close support weapon

DATA
WEAPON
Calibre: 35 mm
Length overall: 960 mm
Barrel length: 805 mm
Weight: 6.8 kg
Rifling: 7° constant twist rh
Chamber pressure: 500 bar max
Recoil stroke: 350 mm
Sights: optical, unit-power
Practical range: (stationary targets) 600 m; (moving targets) 400 m

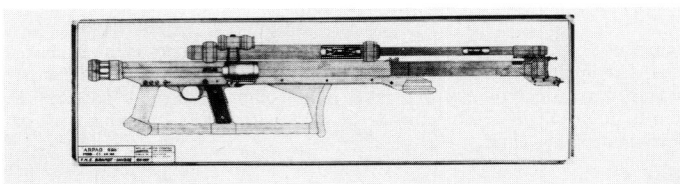

Arpad 600 close support weapon in fully-recoiled position

AMMUNITION
Calibre: 35 mm
Weight: 167 g
Length: 176 mm
Weight of bursting charge: 45 g
Weight of propellant: 17 g
Muzzle velocity: 600 m/s
Time to self-destruction: 2 s
Time of flight to 500 m: 1 s
Penetration: 50 mm steel plate

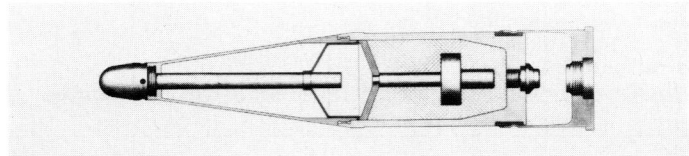

Sectioned view of ammunition for Arpad 600

Developer
FHE Brandt-Switzerland, PO Box 23, CH-1000 Lausanne 6.

Status
Design complete; available for license.

TAIWAN

Type 67 flamethrower

The Type 67 portable flamethrower consists of a fuel/pressure unit and the hose/gun unit. The fuel and pressure group comprises a tubular frame with two interconnected fuel tanks, a spherical pressure unit mounted beneath the smaller fuel tank and other accessories. The hose supplies fuel from the tanks to the gun group. Then the fuel is ignited by a replacable ignition cylinder in the nozzle end of the gun.

The principle of operation is similar to the US M9E1-7 flamethrower.

DATA
Weight: (net) 12.58 kg; (total with fuel) 23.51 kg
Height: 700 mm
Width: 530 mm
Length: 300 mm
Fuel: gasoline and napalm
Fuel capacity: 4.5 gallons (17 l)
Range: 40–55 m

Manufacturer
CSF; Hsing Hua Co Ltd, PO Box 8746, Taipei.
Status
Current. Production and available.
Service
Taiwan Army and Marine Corps.

Type 67 flamethrower

UNION OF SOVIET SOCIALIST REPUBLICS

30 mm AGS-17 grenade launcher

The AGS-17 'Plamya' ('Flame') was introduced into Soviet service in 1975 but did not achieve prominence until its employment in the Afghanistan campaign. It is issued down to infantry company level, a section of two AGS-17 being added to each infantry company. They appear to be employed as the company's base of fire during the advance. The equipment is also reputed to be installed in the Mi-8 'Hip-E' attack helicopter.

The weapon operates on the blowback system. The ammunition is belt-fed, into the right side of the receiver, and the bolt is cocked by pulling back on an operating handle at the rear of the weapon, which is attached by a steel cable to the bolt. On releasing this handle the bolt runs forward, propelled by two springs, and collects a round from the belt, loading it into the chamber. The firing pin is then released by means of a trigger mounted between the spade grips at the rear of the receiver. On firing, the breech pressure blows the bolt to the rear; its travel is damped by an hydraulic recoil buffer mounted on the left side of the receiver. The bolt then returns and chambers the next round. The empty case is extracted from the chamber by the rearward movement of the bolt and ejected through the bottom of the receiver. Belt feed is done by pawls actuated by a feed arm driven by the movement of the bolt.

A dial sight is normally fitted to the left rear of the receiver, and a plate on top of the receiver gives ranges and time of flight for low and high angle firing.

The standard infantry mounting is a tripod with elevation and traverse facilities. It is reported that it can also be mounted in most types of combat vehicle and photographs from Afghanistan have shown BMD reconnaissance vehicles which appear to have an AGS-17 mounted in place of the standard 73 mm gun. A helicopter-mounted version has recently been seen in an African country. This version appears to use the same receiver and mechanism but has a longer barrel and has the magazine inverted and held in a special carrier. The mounting is quite substantial and appears to have four recoil buffers. Elevation and traverse are free, controlled by a pair of widely-set grips at the rear of the weapon. The ammunition for this version is said to differ from that used with the ground model in having a longer cartridge case, though with the same overall length of round; this may indicate a strengthened design or it may indicate a more powerful propelling charge.

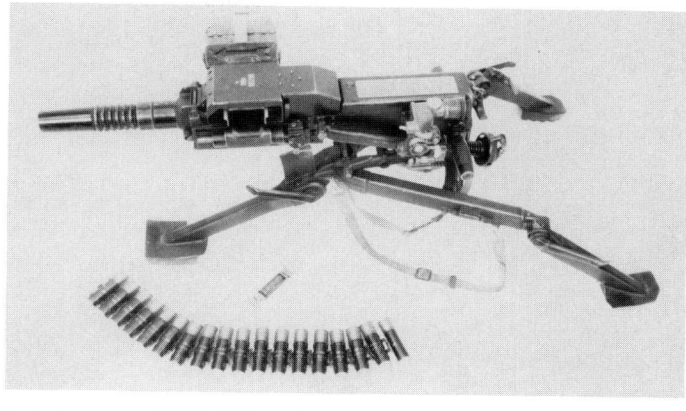

Left side view of AGS-17 launcher, with ammunition belt

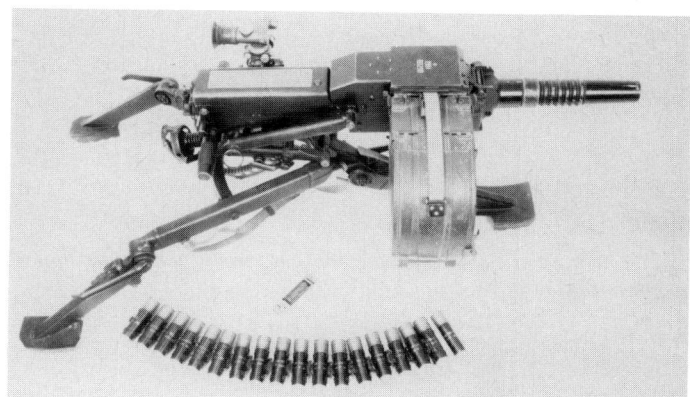

Right side view of AGS-17 launcher, showing belt magazine in position

AGS-17 launcher; belt feed and details of feed mechanism

Left side of AGS-17 launcher, showing details of optical sight

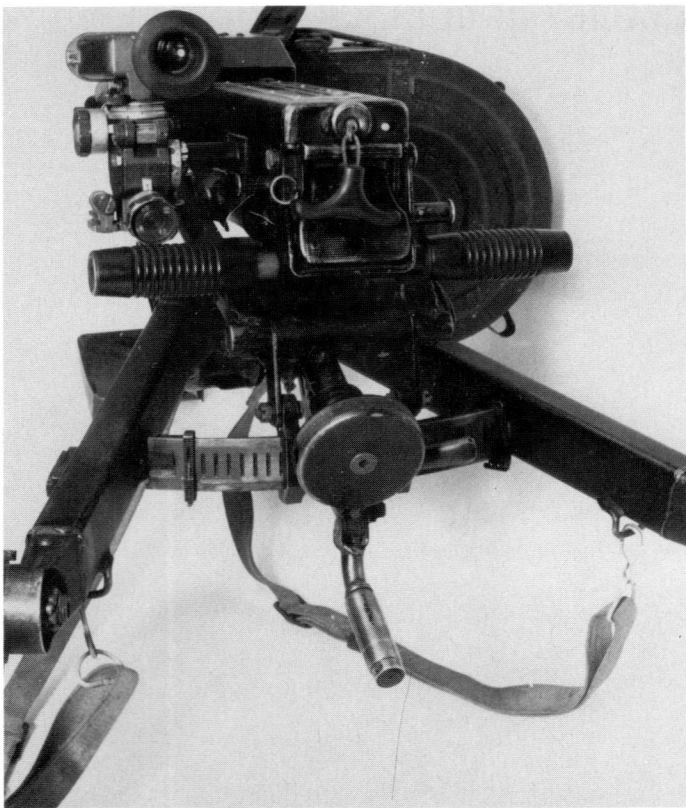

Rear of AGS-17 launcher, showing cocking handle, firing grips, trigger, and traverse control.

Helicopter-mounted version of AGS-17

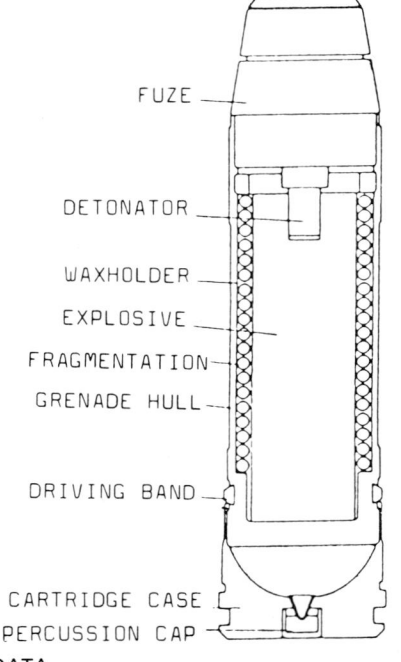

30 mm VOG-17M high explosive fragmentation grenade

DATA
Calibre: 30 mm
Operation: blowback
Feed: 29-round belt
Weights: (weapon) approx 18 kg; (tripod) approx 35 kg
Length overall: 840 mm
Barrel: 290 mm, 16 grooves rh
Range: (max) 1750 m; (effective) 8–1200 m
Rate of fire: approx 65 rds/min

PROJECTILE
Weight: 275 g
Filling: RDX/WX 94/6
Fuze: impact and self-destroying after 25 s
Cartridge case: belted, rimless, steel, 28.2 mm long

Status
Current. Production.
Service
Soviet Army only.

30 mm training dummy, with fired cartridge case

40 mm BG-15 rifle-mounted grenade launcher

This launcher was first seen in Afghanistan in 1984, mounted beneath an AK-74 rifle in the manner of the US M203 launcher. As can be seen, there is a trigger unit with a thumb-grip, and a grenade launching sight folded down on the left side of the barrel. The sight has three settings, marked '2', '3' and '42', which are understood to indicate fighting ranges of 200 and 300 m and the maximum range of 420 m. The launcher is short, no more than 350 mm overall, and is muzzle-loaded with a special grenade. These resemble the projectile used with the World War Two Japanese Type 89 50 mm mortar in that they have an integral propellant chamber at the rear of the projectile body, with a central percussion cap surrounded by gas vents. There is a gilding metal driving band surrounding the body of the grenade, and beneath this are more propellant gas ports. The grenade is thrust into the launcher base first and the self-cocking trigger pulled; this fires the percussion cap, the propellant explodes, and the flow of gas from the rear vents ejects the grenade. At the same time gas pressure forces the driving band outwards to engage in the 12-groove rifling of the barrel, so as to provide spin stabilisation.

Left side of launcher, showing folded sight

Two types of grenade have been reported, one of 118 mm length weighing 266 g and the other of 101 mm length weighing 250 g. Both are percussion nose fuzed and both appear to be simple HE/fragmentation types. Markings on grenades suggest that this weapon has been in production since at least 1978.

40 mm BG-15 grenades; one is inverted to show the cap and gas ports

Rifle-mounted grenade launcher showing method of firing

ROKS-2 flamethrower

Very few of these remain in service as they have been replaced by the ROKS-3 or the more recent LPO-50 flamethrowers. The ROKS-2 consists of a rectangular fuel container, with a smaller compressed air bottle attached to it, a hose and a flamegun which has the appearance of a rifle. Like the ROKS-3, the ignition cartridges are carried in a cylinder mounted on the muzzle of the flamegun and are activated by pressing the trigger. The fuel tank holds 10 litres of fuel, which is sufficient to give a maximum of seven to eight one-second bursts and a maximum range of some 35 m can be achieved in favourable conditions.

Status
Obsolescent but still in service in some Warsaw Pact armies.

ROKS-3 flamethrower

The ROKS-3 flamethrower has been replaced in almost all of the Warsaw Pact countries by the more recent LPO-50 flamethrower. It consists of a cylindrical fuel tank which holds 8 litres and has a filling aperture on top with the hose outlet at the base. The cylindrical compressed air tank is attached to the side of the fuel tank. The hose is connected to the gun which resembles a rifle in appearance. Mounted at the muzzle of the flamegun is the ignition cylinder containing 10 7.62 mm ignition cartridges: as the weapon is fired the cylinder advances automatically and brings a new cartridge into position. When the operator pulls the trigger a spring-loaded shut-off valve opens and allows the fuel to be ejected, a further pull releases the firing pin which ignites the ignition cartridge. When the trigger pressure is released the flamegun stops. There is sufficient fuel for 10 five-second bursts.

DATA
Weight filled: 23.5 kg
Max range: (with thickened fuel) 35 m; (with unthickened fuel) 15 m

Status
Obsolescent.
Service
Apparently still held by some armies of the Warsaw Pact.

LPO-50 flamethrower

The LPO-50 is currently the standard flamethrower in Warsaw Pact armies having largely replaced the earlier ROKS-2 and ROKS-3 flamethrowers. It consists of a tank group, hose and gun group. There are three tanks, on the top of each of which are a pressure relief valve and a cap for the filling aperture which also contains the chamber for the pressurising cartridge.

Wires from the three containers are combined in a harness which is fastened to the hose and attached to the gun group. Outputs from the three tanks are connected to a manifold, through one-way valves which prevent fuel flowing from one tank into another and this manifold is connected to the hose.

Ignition is by means of a slow burning pyrotechnic cartridge, three of which are grouped below the muzzle of the flamegun. A selector lever is mounted forward of the trigger-guard on the gun and, when the trigger is pressed, energy is supplied from a power pack of four 1.5 V cells to one of the ignition cartridges and simultaneously to one of the tank pressurising cartridges. Pressure from the latter drives fuel from the tank through the appropriate non-return valve into the manifold and then by way of the hose to the flamegun where it is ignited by the pyrotechnic cartridge. The firer can thus fire three shots, changing the selector lever position between shots. The tank capacity is 3.3 litres which is sufficient for a flame burst of two to three seconds. A trigger safety is fitted.

Firing LPO-50 flamethrower

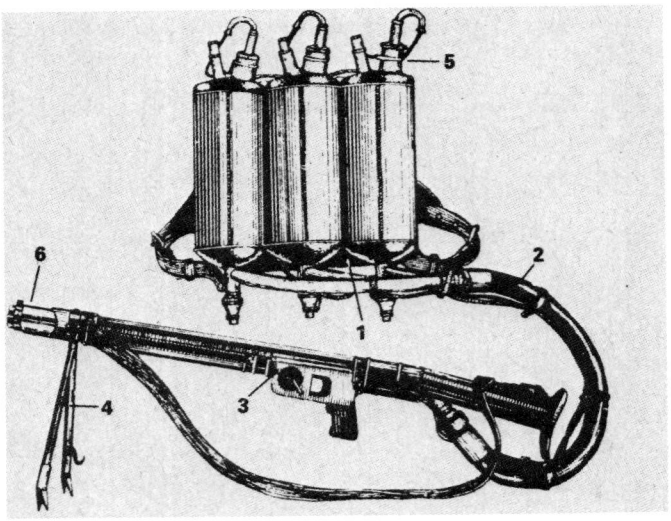

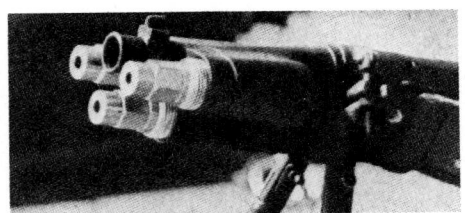

Close-up of muzzle showing 3 ignition cartridges

DATA
Fuel capacity: 3 × 3.3 l
Weight: (empty) 15 kg; (loaded) 23 kg
Max range: (with unthickened fuel) 20 m; (with thickened fuel) 70 m

Status
Current.
Service
Warsaw Pact armies.

Components of LPO-50 flamethrower
(1) *fuel manifold* **(2)** *hose* **(3)** *selector lever* **(4)** *bipod* **(5)** *container for pressure cartridge* **(6)** *ignition cartridges*

TPO-50 cart-mounted flamethrower

The TPO-50 consists of a two-wheeled handcart with three identical flamethrowers mounted horizontally on top and with a recoil spade at the rear. Each flamethrower holds 21 litres of fuel and they can be fired together or individually. If required the cylinders can be dismounted for ground use. Folding front and rearsights are provided on each cylinder and the cylinders can be elevated from +2 to +50°, traverse being carried out by moving the cart.

Each of the cylinders has a screw-on head and mounted on this is the discharge nozzle, cartridge holder and a relief valve. The fuel is forced out of the cylinder by gas pressure which is produced by an electrically ignited cartridge.

The ignition cartridge is the same as that used with the LPO-50 infantry flamethrower and ignites the fuel as it leaves the nozzle. Both the igniter and

pressure cartridges are simultaneously ignited electrically from a 6 V (4 × 1.5 V) power pack which can be located remotely to enable the TPO-50 to be fired from behind cover if required.

DATA
Crew: 2
Weight: (loaded) 170 kg; (empty) 130 kg
Max range: (with thickened fuel) 180 m; (with unthickened fuel) 65 m

Status
Obsolete
Service
Not known

UNITED KINGDOM

Series 50 shoulder-fired air launcher

This is a fully self-contained and versatile shoulder-fired weapon which can be used to fire a considerable variety of projectiles for a wide range of applications, including grapnel and line throwing, smoke and flare launching, HE projectiles and darts. It can also be used for riot control purposes.

The air supply is from a rechargeable cylinder which is easily fitted to and removed from the launcher. Depending upon the model, between 2 and 10 full pressure firings can be obtained from a small cylinder. Alternatively, should a scuba cylinder back-pack be used as the air supply, several hundred firings can be achieved.

The launcher is fitted with its own integrated control block, including charge valve, pressure limiter and dump valve to enable the launcher to be discharged at any time. The firing valve is operated by a conventional trigger mechanism which includes cocking lever and safety catch. In addition to the safety catch on the trigger mechanism, interlocks prevent the breech from being opened when the trigger is cocked. Additionally, if the breech is not closed and locked the trigger cannot be cocked.

DATA

Model	50A	50B	50C
Calibre	50 mm	50 mm	50 mm
Muzzle energy	2 kJ	2 kJ	1 kJ
Max Vo m/s	400	400	400
Max projectile mass, shoulder-fired	500 g	500 g	500 g
Max launcher pressure	21 MN/m²	21 MN/m²	21 MN/m²
Weight, depending upon equipment fitted	6.5 – 8.5 kg	6.5 – 8.5 kg	6.5 – 8.5 kg

Typical performance — muzzle velocities m/s:

Projectile mass	50A	50B	50C
50 g	280	280	200
100 g	200	200	140
250 g	125	125	89
500 g	89	89	63

Manufacturer
Ferranti International, Weapons Equipment Div, 10 Spring Lakes, Deadbrook Lane, Aldershot, Hants GU12 4HA.
Status
Current. Production.

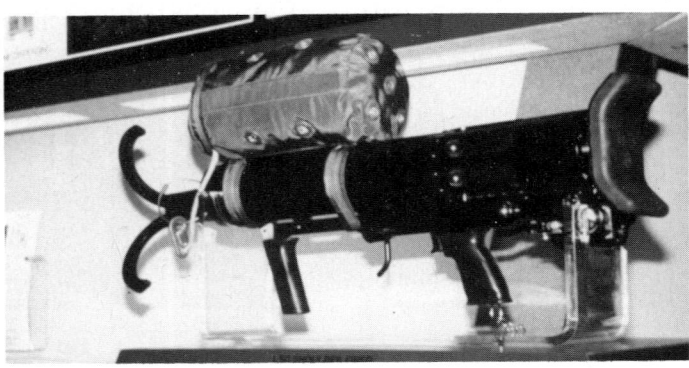

Ferranti air launcher

40 mm Hilton Multi-Purpose Gun

The Hilton Multi-Purpose Gun can be fitted with a rifled barrel for discharging the standard US series of 40 mm grenades. In this configuration the Gun is fitted with an adjustable Ring Sight collimating sight which can be set for elevation. Trials have shown that with an elevation of 10° the grenade reaches a range of 200 m and with 45° elevation 375-400 m can be reached, depending upon the particular grenade round in use. The 40 mm barrel is completely interchangeable with the other shot, pyrotechnic and rifle barrels provided for the Multi-Purpose Gun.

Manufacturer
Hilton Gun Co., 60-62 Station Road, Hatton, Derby DE6 5EL.
Status
Current. Production.

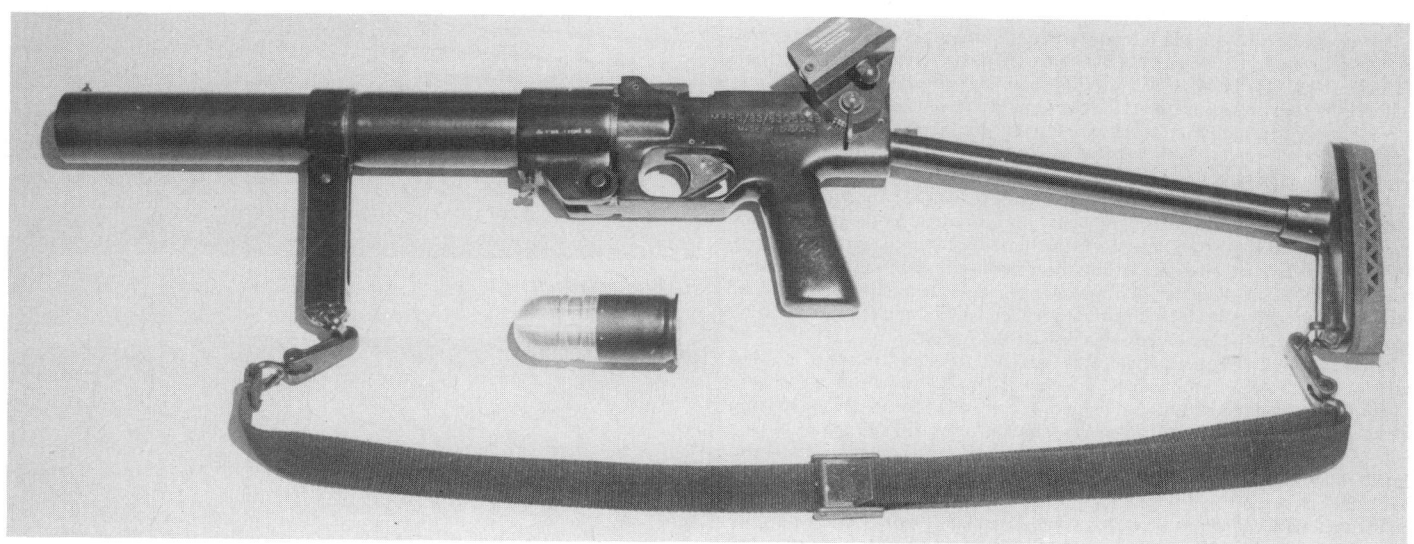

Hilton Multi-Purpose Gun fitted with 40 mm barrel

Hilton HG 40 grenade launcher

This is an add-on launcher for standard 40 mm grenades; originally designed for the UK L85 rifle, it has since been modified so that it can be fitted to the Steyr AUG, FN-FAL, AK47 and other rifles. The basic components are the fixing clamps; bracket; action housing; carrier and barrel; and sight unit. In the case of the L85 rifle the front clamp is in halves to allow easy fitting to the rifle barrel, and the rear clamp is designed to slide on to the barrel. The clamps are modified as necessary to suit fitting on other types of rifle.

The bracket is held to the front clamp by a hinge pin and to the rear clamp by a T-head supporting bolt. The action housing is attached to the bracket by screws, and the carrier and barrel, which are locked together during manufacture, are attached to the rifle at the muzzle end by means of the hinge pin. The breech-end is locked into position with a spring-loaded catch.

The barrel is rifled to provide the required rotational speed of the projectile, for accuracy and stability and for arming the grenade fuze. Sights of various types can be provided.

The launcher is loaded by releasing the barrel catch and allowing the breech end to drop, exposing the breech for loading. After loading the grenade, the barrel is swung up until the catch locks it in place in front of the action body. The firing pin is cocked, the weapon aimed and the launcher trigger pressed to fire.

DATA
Calibre: 40 mm
Weight: 1.50 kg
Length: 388 mm
Barrel length: 310 mm
Width: 70 mm
Height: 142 mm
Muzzle velocity: approx. 75 m/s
Maximum range: 375 m

Manufacturer
Hilton Gun Company, 60-62 Station Road, Hatton, Derby DE6 5EL.

Hilton HG40 40 mm grenade launcher on L85 rifle

Hilton HG40 grenade launcher open for loading

Status
Current; available.
Service
Under trial by British and other forces.

UNITED STATES OF AMERICA

140 mm Brunswick RAW HESH rocket

RAW is the acronym for Rifleman's Assault Weapon, developed by the Brunswick Corporation in conjunction with the US Army Missile Command. RAW, although rifle launched, is comparable with firepower normally associated with light artillery and heavier tube-launched assault weapons. It should be considered as a companion to smaller muzzle-launched grenades and used in a high/low mix of infantry rifle munitions.

The origin of RAW lies in the resurgence of interest in urban warfare and Fighting In Built Up Areas (FIBUA) (or Military Operations in Urban Terrain - MOUT). In street warfare the enemy is protected by strong walls and sandbagged firing positions, and the same is true of those occasions when it is necessary for infantry to engage bunkers in close country. In these conditions it is rarely possible to bring tanks or guns to where they can bring fire to bear on the target and the infantry have to take unnecessary casualties in the subsequent attack. In these conditions there is a need for some weapon which will blow sizeable holes in masonry and concrete. In the past this has been tried in two ways, by firing anti-tank rounds or by emplacing a charge by hand. The latter is usually suicidal and the former only results in a small hole a few centimetres in diameter. To make a big hole demands a payload of explosive beyond that which a normal grenade carries and it requires to be accurately placed and exploded. This is what RAW does.

RAW is a rocket-propelled projectile launched from the muzzle of an M16 rifle, or can be easily modified to fit other rifles. The warhead is a sphere 140 mm in diameter containing 1.27 kg of HE which is detonated on striking the target giving a 'squash-head' effect which is powerful enough to blow a hole 360 mm in diameter in 200 mm of double reinforced concrete. It will also defeat light armour, soft-skinned vehicles and a wide range of similar battlefield targets.

Another warhead option has been demonstrated; this has multi-purpose potential against both medium armour and structures. An Aerojet Ordnance shaped charge warhead and Motorola active optical fuze provide significantly improved armour penetration due to the standoff sensing capability of the fuze. By manually switching the optical fuze off, the warhead will then detonate on impact, giving the same effect as the high explosive warhead.

RAW after launch

RAW in flight showing flat trajectory

RAW mounted on M16A1 rifle, both weapons ready to fire without interfering with each other. Curved spin-up tubes are just behind projectile body

Projectile

The projectile is a light metal sphere to which is attached a small rocket nozzle. The sphere contains three elements, the warhead, its associated fuze and the rocket motor. It is spun prior to launch and flies at a constant angle to the line of sight, using the thrust from its motor both for propulsion and overcoming gravity. The balance of thrust ensures that it flies on a flat and undeviating path for at least 200 m, after which it begins to assume a ballistic trajectory of as much as 2000 m for indirect or interdiction fire. Within 200 m the firer can confidently aim directly at his target without having to make any allowances for range, and 200 m will encompass the overwhelming majority of urban targets. The average range in street fighting is about 45 m.

Launcher

The present launcher is designed for the M16 rifle, but there is no reason why launchers could not be made to fit any type of rifle, regardless of calibre. It is a metal bracket which slips over the muzzle and engages the bayonet lug. When fitted it does not affect the use of the rifle in any way, and the rifleman can continue to fire normal ball ammunition, the only restriction being the weight of the RAW on the end of his rifle. The bracket of the launcher holds beneath it a launch tube which is free to rotate on two simple bearings in the bracket. The tube has a turbine activated by escaping rocket gases emerging from it at right angles to the line of launch. Muzzle gases from the rifle flash suppressor will strike a firing pin to initiate the launch sequence. Although the rifle can continue to be fired after RAW is mounted on it, the launch sequence can begin only after the rifleman has set a valve allowing gases to reach the firing pin. Within the projectile is a matching ignition cap, aligned with the firing pin. The rotating tube is closed at its free end and there are gas escape slots cut into the walls. A safety pin passes through the firing pin and positively locks it in position.

Functioning

The sequence of events on firing the projectile is as follows. The firer's first action is to turn the arming valve. Having taken aim the firer then fires a normal ball round. The emergent gases at the muzzle are funnelled through the collar around the flash hider and some gas runs down a tube and drives the firing pin forward to strike the ignition cap in the wall of the projectile. This cap fires the rocket motor and the efflux blows into the launch tube, emerging through the slots in the side walls of the tube but also emerging through the two curved tubes. These tubes impart a spin to the tube and so to the projectile and before it has left the grip of the launch tube the spherical projectile is spin stablised for its flight. The projectile flies with a constant acceleration, reaching a velocity at 200 m of 173 m/s with a flight time of 1.9 seconds. A considerable advantage with the spun projectile is that the absence of aerodynamic fins eliminates the tendency to 'weathercock' into a cross wind. There is very little back-blast on launch and the only effect on the gunner is a slight recoil and a sudden change of centre of gravity in his rifle.

Tactical use

The tactical advantages of RAW are considerable. The launcher is an unmodified rifle and the ignition is provided by an unmodified ball round. Any man armed with a rifle can fire the projectile and the amount of training required is minimal. The 'soft' launch means that firing can be carried out in enclosed spaces without danger to the firer and the straight line of flight does away with the need to estimate range at all expected targets in urban warfare. The rocket efflux is hidden from the target during the flight and this, combined with the low launch signature, makes the whole weapon extremely difficult to spot. The manufacturers suggest that there could be alternative

RAW HESH installed on M16 rifle

Result of firing RAW HESH against a reinforced concrete wall

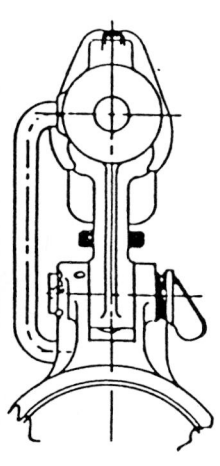
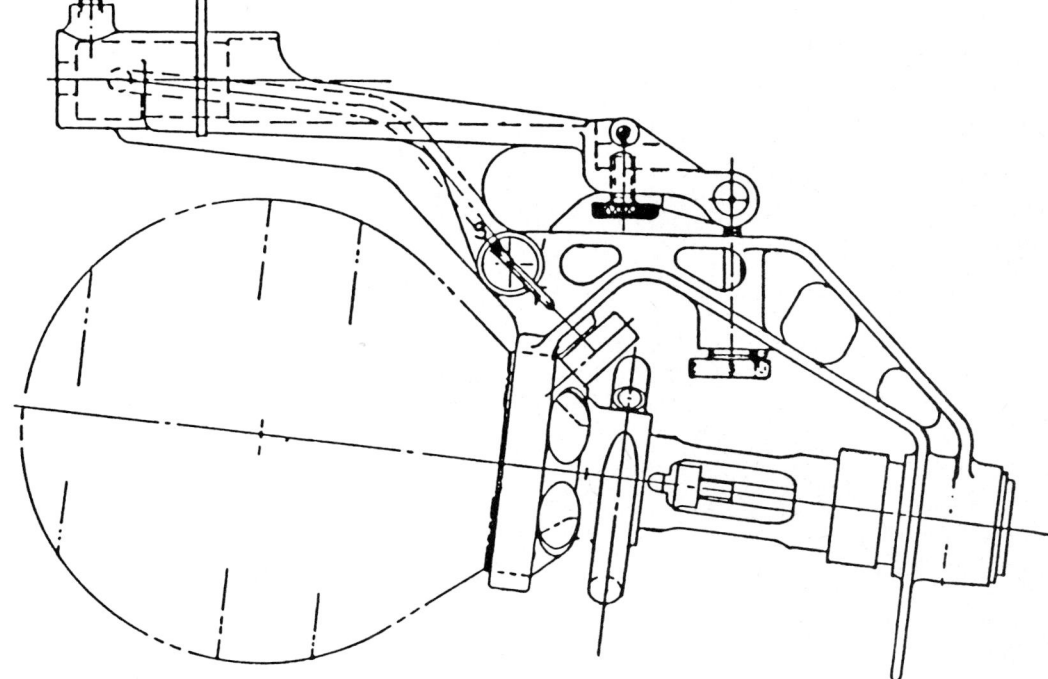

Diagram of RAW attachment and projectile. Note gas tube leading down from rifle muzzle to RAW firing pin

warhead loads such as anti-armour, fragmentation, smoke, flame or a chemical filling but it would seem that the simple HESH would be the most useful.

Status, 1990
Brunswick completed a company-funded NDI (Non-Developmental Item) qualification of the HESH RAW in late 1988, and the weapon is now being offered for procurement. The HESH RAW has been demonstrated as being capable of defeating most infantry targets with the exception of the most modern tanks and armoured vehicles, though it is capable of damaging the running gear and immobilising even these. RAW is also a candidate for the US Army Multi-Purpose Individual Munition (MPIM) competition, details of which were released in FY87 by the US Army Missile Command, Huntsville. The purpose of MPIM is to develop one weapon which can defeat existing and postulated light armour, field fortifications and sand-reinforced targets. RAW has demonstrated its capability to meet these multi-target requirements. MPIM competitors delivered Proof-of-Principle test units in September 1989; selection for full scale development is scheduled for the spring of 1990, with an in service date (IOC) of 1993.

DATA
Projectile and launcher: (overall weight)3.86kg
Payload: 1.27 kg
Length: (overall) 305 mm
Diameter: 140 mm
Propulsion: rocket motor
Range: (effective) 200 m; (max) approx 2000 m
Velocity at 200 m: 173 m/s
Time to 200 m: 1.9 s
Accuracy: 3 m
Possible payloads: HESH, smoke, flame, chemical, fragmentation, and anti-armour charges.

Result of firing HESH-RAW against an APC

Manufacturer
Brunswick Corporation, Defense Division, 3333 Harbor Boulevard, Costa Mesa, California 92628-2009.
Status
Completed company-funded qualification of HESH-RAW in February 1989. Delivered proof-of-principle rounds for MPIM tests in September 1989.
Service
Trials only.

---

83 mm McDonnell Douglas SMAW

The Shoulder-launched Multi-purpose Assault Weapon (SMAW) is a man portable shoulder-launched 83 mm system consisting of a re-usable launcher and individually encased High Explosive Dual Purpose (HEDP) and High Explosive Anti-Armour (HEAA) rockets. The re-usable launcher features a 3.8 power optical sight and a 9 mm spotting rifle that uses tracer cartridges ballistically matched to the tactical rounds. The cartridges are packaged in the end cap of each HEDP and HEAA rocket. The spotting cartridges enable the gunner to achieve a high degree of hit probability when firing the main rocket. The re-usable launcher is used for both tactical and training purposes, allowing for more realistic weapon system training.

The HEDP rocket was designed to defeat earth and timber bunkers, concrete and brick walls, and light armoured vehicles. Its unique warhead and fuzing allows point detonation for hard targets and delayed detonation for soft targets. With the addition of the HEAA rocket, which completed development testing and certification in 1988, SMAW is truly a multi-purpose weapon that will meet customer needs for a close-in anti-armour defeat capability and for urban fighting. The HEAA warhead was designed to defeat 600 mm of rolled homogenous armour (RHA) at angles of obliquity up to 70°. The SMAW gunner can now engage tank sized targets out to 500 m range using the new HEAA rocket.

In four years of service, SMAW has a demonstrated reliability of over 96 per cent.

83 mm SMAW in firing position

SMAW ready to fire

SMAW High Explosive Dual Purpose (HEDP) rocket

DATA
LAUNCHER
Weight: 7.5 kg
Length: 825 mm
Calibre: 83 mm
Optical sight: x3.8, 6° field-of-view

ENCASED ROCKETS
HEDP:
Length: 746 mm
Weight: 5.9 kg
Velocity: 220 m/s

HEAA:
Length: 842 mm
Weight: 6.3 kg
Velocity: 217 m/s

SYSTEM READY TO FIRE
Launcher with HEAA:
Weight: 13.9 kg
Length: 1372 mm

Launcher with HEDP:
Weight: 13.1 kg
Length: 1372 mm

EFFECTIVE RANGE
HEDP: against 1 m × 2 m target — 250m; against tank size or larger targets —500 m
HEAA: against 2.3 m × 2.3 m target — 500 m

Manufacturer
McDonnell Douglas Missile Systems Co., Combat Systems Division, PO Box 516, St Louis, MO 63166.
Status
Qualified for US military; more than 180 000 rockets produced to date.

SMAW High Explosive Anti-Armour (HEAA) rocket

SMAW HEDP rocket in storage package

SMAW HEAA rocket in storage package

Scorpion urban fighting weapon

Scorpion is the McDonnell Douglas candidate for the US Army's Multi-Purpose Individual Munition (MPIM) programme. McDonnell Douglas is one of three companies who are competing for a full-scale development contract for the MPIM.

Scorpion is an add-on rocket launcher which attaches below an assault rifle. It is zeroed to the rifle's 300 m battle sight and has its own trigger and firing mechanism. The rocket has folding fins, and the warhead, believed to be a tandem shaped charge, is capable of defeating 300 mm of brick or 200 mm of reinforced concrete. It will also defeat earth-and-rubble bunkers and light armoured vehicles.

Used from the rifle there is no recoil, and the attachment of Scorpion does not interfere with the use of the rifle. Although the rifle-mounted version is being developed first, development of a shoulder-fired disposable launcher is expected in the near future.

DATA
Calibre: 66 mm
Weight: 4 kg
Length: 775 mm
Maximum velocity: 400 m/s
Effective range: 300 m

Manufacturer
McDonnell Douglas Missile Systems Co., Combat Systems Division, PO Box 516, St Louis, MO 63166.
Status
Development.

Scorpion Multi-purpose Individual Munition

Marquardt Multi-Purpose Individual Munition (MPIM)

This has been developed under a joint Army/industry programme with the intention of providing the infantry soldier with a lightweight multi-purpose weapon capable of defeating the broad spectrum of targets likely to be encountered during military operations in urban environments. The MPIM uses a shaped charge follow-through (FT) warhead in which the shaped charge penetrates the target and the FT projectile is then launched through the hole and into the interior of the target so as to attack the materiel behind the protection rather than merely attacking the protection itself. The warhead is capable of defeating field fortifications, concrete and brick structures and light armoured vehicles.

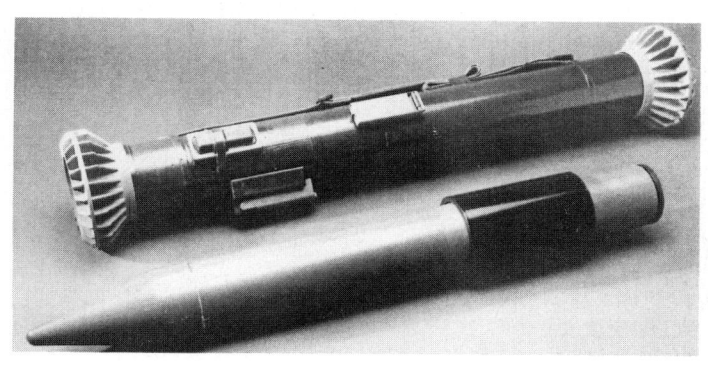

Marquardt MPIM projectile and launcher

The size weight and bulk of the MPIM have been minimised to permit employment by airborne, air assault, special operations and light infantry forces without adversely affecting their ability to manoeuvre, engage and defeat threat forces.

DATA
Calibre: 84 mm
Carry weight: 4.50 kg
Carry length: 813 mm

Effective range: 300 m
Practical range: > 500 m
Accuracy: CEP 1.3 mils

Manufacturer
The Marquardt Company, 16555 Saticoy Street, Van Nuys, CA 91409-9104.
Status
Advanced development.

AT8 multi-purpose weapon

The AT8 is a further development from the FFV AT4 anti-tank launcher. Development has been carried out by the Honeywell company in conjunction with the US Marine Corps, the object being to provide a light, disposable, multi-purpose direct fire weapon for urban warfare, capable of breaching walls, destroying fortified targets and defending against attack by light armour.

The weapon is the existing AT4 shoulder-fired launcher, loaded with a new projectile developed by the US Marine Corps. This is an explosive warhead device with a dual mode fuze which distinguishes between hard and soft targets. In use against hard targets the fuze detonates the warhead on impact, allowing the explosive to develop its full effect on the target surface. In use against a soft target the fuze will delay the detonation so as to allow the warhead to penetrate some distance into the target before it takes effect. This will have a considerable cratering effect should the warhead still be within the target structure, or, if it has passed through the protection, it will have an anti-materiel and anti-personnel effect within the protected structure.

Since the AT8 uses precisely the same external launch tube and controls as the existing AT4, the soldier familiar with the AT4 can operate the AT8 without requiring further training.

Honeywell AT8 multi-purpose weapon in firing position

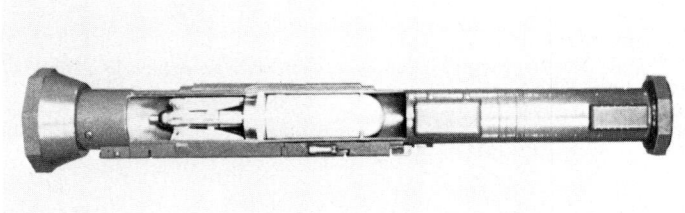

DATA
Calibre: 84 mm
Weight: 7.17 kg
Length: 1.00 m
Muzzle velocity: 219 m/s
Time of flight to 200 m: 0.99 s
Accuracy at 250 m: < 2 mils dispersion

Manufacturer
Honeywell Corp., Armament Systems Division, 7225 Northland Drive, Brooklyn Park, MN 55428.
Status
Development completed.

Honeywell AT8 multi-purpose weapon, sectioned

40 mm Colt M203 grenade launcher

The M203 grenade launcher was developed by the AAI Corporation of Cockeysville, Maryland, under the direction of the United States Army Weapons Command. Development started in May 1967 and it was type classified Standard 'A' in August 1969. The M203 replaced the M79 grenade launcher since it fulfilled the US Army requirement for a rifle/grenade launcher package, whereas the earlier M79 was only a grenade launcher.

The 40 mm Colt M203 grenade launcher is a lightweight, single-shot, breech-loaded, sliding barrel, shoulder-fired weapon designed especially for attachment to the M16A1 and M16A2 rifles. It allows the grenade launcher to fire a wide range of 40mm high explosive and special purpose ammunition, in addition to permitting 5.56 mm rifle power.

The launcher consists of a receiver assembly, barrel assembly, quadrant sight assembly, and handguard and sight assembly group (for M16A1 and M16A2). The quadrant sight assembly mounts on the left side of the carrying handle of the M16 rifle or M4 carbine without the need of tools. This permits the sight to pivot on the range quadrant to the desired range setting (to 400 m) in 25 m increments. A leaf sight on the receiver is adjustable in 50 m increments to 250 m.

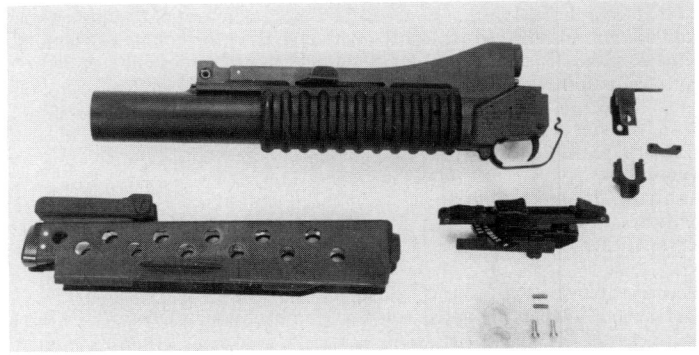
Colt 40 mm M203 grenade launcher component parts

Colt 40 mm M203 grenade launcher attached to M16A2 rifle

Colt M203 launcher attached to M4 carbine, showing quadrant sight

The M203 attaches directly to any standarde Colt M16 rifle or M4 carbine, utilising only two screws and without modification to the parent weapon. The rifle or carbine handguard is removed, and the M203 can then be easily installed in five minutes with the aid of a standard screwdriver.

DATA
Calibre: 40 mm
Operation: single shot
Feed: breech-loading, sliding barrel
Length of launcher: 380 mm
Length of barrel: 305 mm
Weight: (unloaded) approx 1.36 kg; (loaded) approx 1.63 kg
Weight: (loaded M16A1 and M203) 5.484 kg; (loaded M4 and M203) 4.624 kg
Muzzle velocity: 75 m/s with M406 grenade
Max range: 400 m
Types of ammunition: M406 HE; M433 HEAP; M576 buckshot; M407 and M781 practice; most other specialised 40 mm ammunition usable in the M79 launcher

Manufacturer
Colt Firearms, Hartford, CT 06101.
Status
Current. Production.
Service
US armed forces and several other countries.

Typical 40 mm grenades

PI-M203 40 mm grenade launcher

This is a product-improved version of the M203 grenade launcher which can be rapidly mounted or removed from the rifle. This is achieved by a unique locking mechanism on the launcher receiver which mates with an interbar which is secured to the barrel of the rifle. The interbar is attached to the rifle within the handguard and does not affect the operation of the weapon in any way.

Formerly the M203 has been available only on a limited range of rifles. Now with the production of interbars for any type of rifle, the M203PI launcher can be mounted on every assault rifle in use today,

For situations where only the launcher is required, the M203PI can be 'snapped on' to a pistol grip assembly with a folding stock. This results in a highly portable weapon weighing less than 2 kg and capable of delivering accurate high explosive firepower.

M203PI attached to Steyr AUG rifle

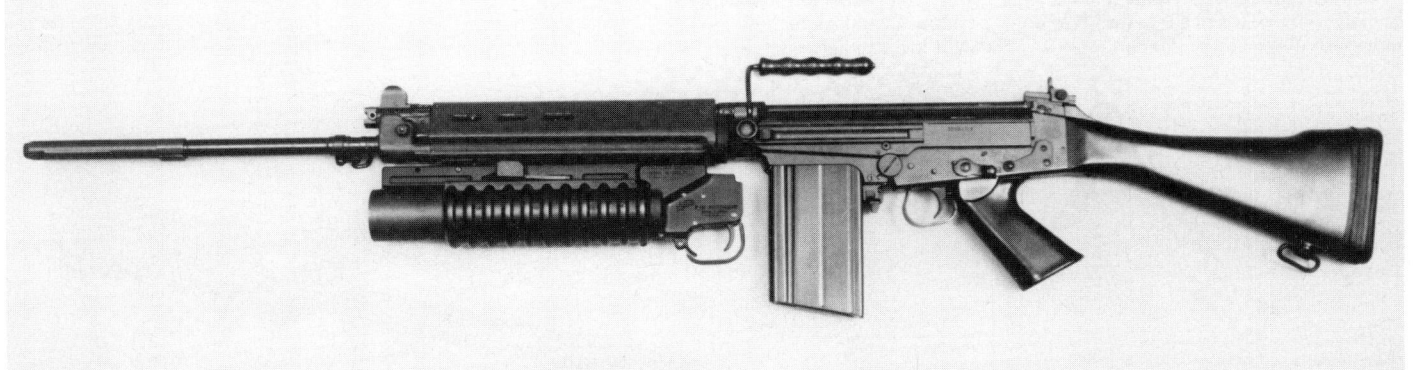

M203PI attached to FN-FAL rifle

Use of this system means that tactical requirements, rather than the availability of hardware, will govern the choice which commanders have in the distribution of rifles and grenade launchers

DATA

Cartridge: 40 mm all types
Weight, empty: 1.36 kg
Weight loaded: 1.63 kg
Length: 381 mm
Barrel length: 305 mm
Depth: 84 mm below rifle bore axis
Width: 84 mm max
Muzzle velocity: 74.7 m/s
Max range: 400 m

Manufacturer
RM Equipment Inc, 6975 NW 43rd Street, Miami, FL 33166.
Status
Current. Production.
Service
Under test for certification and acceptance by US Marine Corps, US Navy and US Army, and British, French, Pakistan, Swedish and Venezuelan armies. In service with various US law enforcement agencies.

M203PI attached to AK rifle

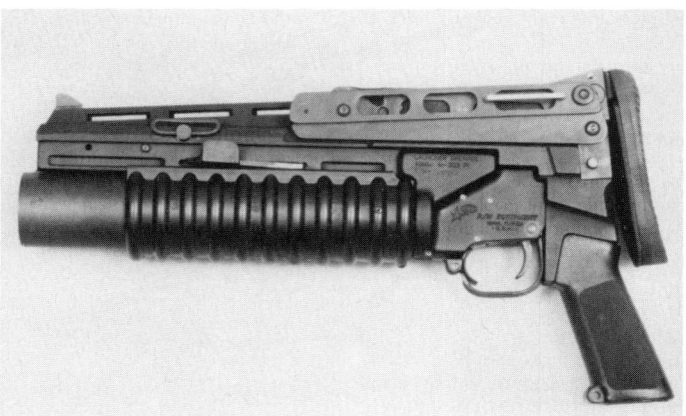

M203PI fitted to pistol grip unit, butt extended

M203PI fitted to pistol grip unit, butt folded

40 mm M79 grenade launcher

The 40 mm grenade launcher, M79, is a single-shot, break-open, breech-loading, shoulder-fired weapon. It consists of a receiver group, fore-end assembly, barrel group, sight assembly and stock assembly. A rubber recoil pad is attached to the butt of the stock to absorb part of the recoil. A sling is provided to carry the weapon.

The M79 was the first weapon to come into service which was specifically designed to fire spin-stabilised grenades. It was first issued to the US Army in 1961 and some 350 000 were made in the next 10 years before production ceased in 1971. It was replaced by the M203 (above) in the early 1970s and is now obsolete in US service though still widely used elsewhere. It is a light, handy weapon with an acceptable recoil and an adequate range. A trained man can put grenades through nominated windows in a house at 150 m range, although at greater ranges it is necessary to know the exact distance to the target since the round has a very high trajectory. This ability to place rounds on the target compensates to a great extent for the limited effective lethal radius of the grenade.

Safety
To fire the launcher the safety must be in the forward position. In this position the letter F is visible near the rear end of the safety. When the letter S is visible just forward of the safety, the launcher will not fire. The safety is automatically engaged when the barrel locking latch is operated to open the breech.

Rearsight assembly
The adjustable rearsight assembly consists of a rearsight lock, a windage screw and windage scale, an elevation scale and lock screw, a sight carrier and retainer lock-nut, an elevating screw wheel and elevating screw, and a rearsight frame and fixed sight.

The rearsight lock is spring loaded and permits the rearsight frame assembly to be locked in either the up or down position. To unlock the sight frame push down on the flat surface of the rearsight lock. By releasing the

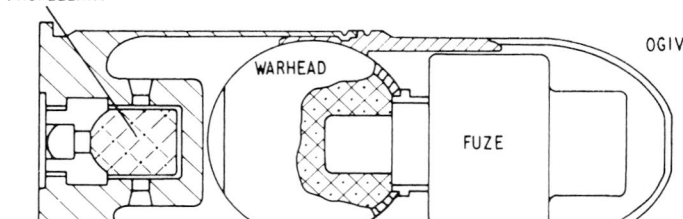

40 mm M79 HE grenade

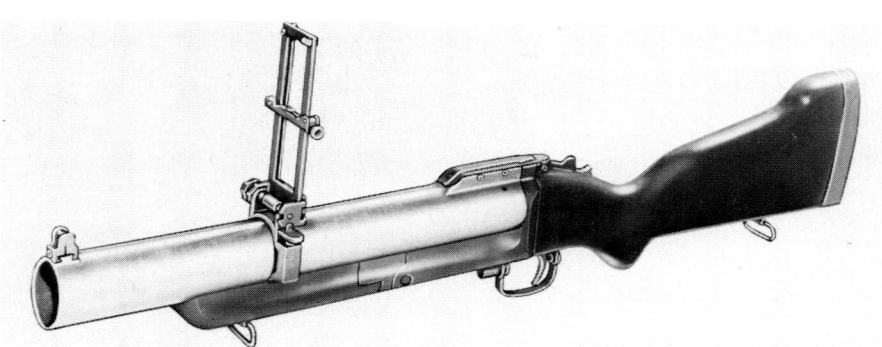

40 mm M79 grenade launcher

pressure the frame is locked in the desired position. A knob at its right end turns the windage screw to adjust the rearsight for deflection.

The elevation scale is graduated from 75 to 375 m in 25 m increments and numbered at 100, 200, 300 and 375 m. As the rearsight carrier is moved up the scale, the rearsight is cammed to the left compensating for the normal right-hand drift of the projectile. The lock screw holds the elevation scale in position. When the rearsight frame is in the down position, the fixed sight may be used to engage targets at ranges up to 100 m.

Front sight
The front sight consists of a tapered front sight blade and two front sight blade guards.

DATA
WEAPON
Length of launcher: (overall) 737 mm
Length of barrel: 356 mm
Weight of launcher: (loaded) 2.95 kg; (unloaded) 2.72 kg

OPERATIONAL CHARACTERISTICS
Operation: break-open, single-shot
Sights: (front) blade; (rear) folding leaf, adjustable
Chamber pressure: 210 kg/cm²
Muzzle velocity: 76 m/s
Max range: approx 400 m
Max effective range: (area targets) 350 m; (point targets) 150 m
Minimum safe firing range: (training) 80 m; (combat) 31 m

Manufacturer
Originally manufactured by Colt; now made under license by Daewoo Industries, South Korea.
Status
Obsolete and superseded in US service by M203.
Service
Several South American and Far Eastern countries.

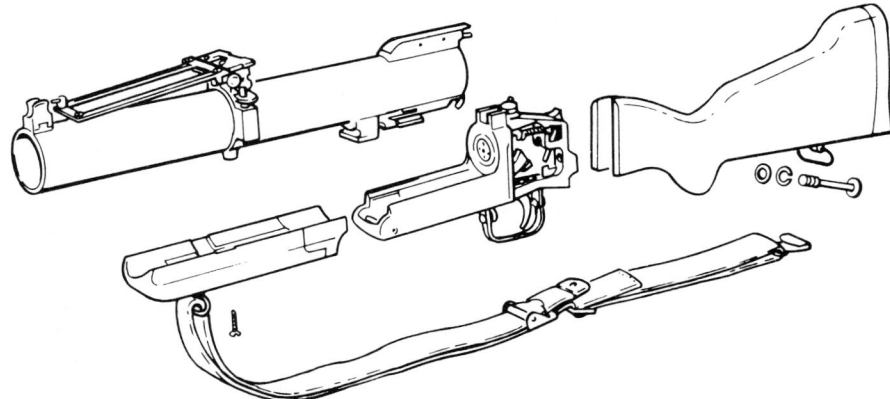

Major components of 40 mm M79 grenade launcher

40 mm Mark 19 Mod 3 automatic grenade launcher

The 40 mm Mark 19 machine gun was developed by the US Naval Ordnance Station, Louisville, Kentucky, in order to provide the US Navy with a suitable weapon for riverine patrol work in Vietnam. It was designed around the high velocity 40 mm M384 grenade round. Work began in July 1966 and the first functioning models were ready early in 1967. The Mark 19 was effectively used in combat in Vietnam and subsequently by Israel. A Product Improvement program initiated by the Navy in the late 1970's resulted in the Mark 19 Mod 3 machine gun which has significantly improved reliability over previous models. A contract for production of the Mark 19 Mod 3 weapon and the Mk 64 Mod 4 Mount was awarded by the US Government to Saco Defense Inc in October 1983. This and a subsequent contract awarded in December 1988 provides for deliveries of the weapon and mounts to the US Army, Air Force, Marine Corps and Navy through the early 1990s.

The Mark 19 is an air-cooled, blow back type automatic machine gun which fires a variety of 40 mm grenades at muzzle velocities of 790 ft/s. These include the M383 and M384 high explosive anti-personnel; M430 high explosive, dual purpose anti-personnel and armour piercing; the M918 Flash-Bang practice round, and the M385 practice inert round. With an effective range exceeding 1600 m, it is an excellent weapon against both personnel and light armoured vehicles. The system can be ground or vehicle mounted, and turret mounted applications are currently being explored.

Ammunition is belted by a unique link which stays with the cartridge case and is ejected with the case after firing. The gun is fired from the open bolt position with the ammunition feed occurring during recoil similar to the M2 machine gun. It can be fired manually or remotely by use of an electrical solenoid, a single round at a time, or fully automatic at 325-375 rds/min.

The round, which is belt-fed from ammunition containers holding either 20 or 50 rounds, is fed into the gun by pawls actuated by the movement of the bolt. The bolt recoils, withdrawing a round from the belt and, as the two travel rearward, a curved vertical cam forces the round down into a T-slot on the face of the bolt until it is aligned with the barrel chamber. When the bolt is released by the trigger mechanism, it is propelled forward by drive springs, chambering the round. Shortly before the cartridge case is fully closed in the chamber, the firing pin is released to impact the primer and ignite the round. The recoil force thus has to arrest the forward movement of the bolt and reverse it resulting in lower reaction forces and lower overall weight. The bolt is driven back by normal blowback action, extracting the empty case from the chamber. As it moves back and withdraws the next round from the belt, the vertical cam forces this round down into the T-slot and displaces the fired case, ejecting it through the bottom of the receiver.

40 mm Mark 19 Mod 3 automatic grenade launcher

The Mark 19 is comprised of five major subassemblies: the bolt and backplate, sear, top cover, feed slide and tray, and receiver. It is fitted with a sight, spade grips, and charger assembly similar to those on the M2 calibre .50 machine gun.

DATA
Calibre: 40 mm
Gun: (weight) 34 kg; (length) 1028 mm; (height) 206 mm
Muzzle velocity: 240 m/s
Effective range: 1600 m
Rate of fire: 325-375 rds/min
Mountings: turret, pedestal or tripod
Ammunition: 40 mm grenades M383, M384, M385, M385E4, M430 and M918

Manufacturers
Saco Defense Inc, a subsidiary of the Chamberlain Manufacturing Corporation, 291 North St, Saco, ME 04072.
Status
Adopted by all US Military forces.

40 mm MM-1 multiple grenade launcher

This is a revolver-type launcher which is capable of being operated by one man at a high rate of fire. It can be usefully deployed in a variety of tactical situations and forms a convenient and highly effective source of emergency firepower for boat, helicopter and tank crews in emergency.

The MM-1 is a 12-shot revolver-type weapon which can be easily and quickly loaded, using any US or foreign 40 mm grenades up to 101 mm in length. The weapon is manufactured from aluminium, steel and high strength plastic materials and, being simple, is extremely resistant to field conditions. It is easily maintained, requiring no more maintenance than a service revolver.

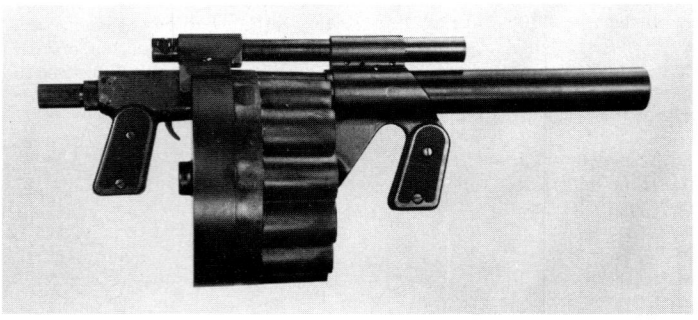

40 mm MM-1 grenade launcher

DATA
Cartridge: 40 mm grenades
Operation: self-cocking revolver
Feed: 12-shot cylinder, spring-assisted
Length: 635 mm
Weight, empty: 5.7 kg
Rate of fire: (cyclic) 144 rds/min; (practical) 30 rds/min
Range: approximately 350 m

Manufacturer
Hawk Engineering Inc, 42 Sherwood Terrace, Suite 101, Lake Bluff, IL 60044.

Status
Current. Production.
Service
With special warfare units in USA, Africa and Central America.

Olin/Heckler and Koch Close Assault Weapon System (CAWS)

This weapon has been developed by Olin Industries and Heckler and Koch in response to a JSSAP requirement for a Repeating Hand-held Improved Non-rifled Ordnance (RHINO) put forward in 1979. Other than overall cartridge length and the degree of recoil felt by the firer, there were virtually no restraints placed on design.

The Winchester ammunition group of Olin Industries began by developing the ammunition. The principal requirements were that it should have no greater recoil than a Remington 870P shotgun firing M162 and/or M257 buckshot; have a magazine capacity of at least 10 rounds; prevent the use of this military ammunition type in commercial 12-gauge shotguns; and have a hit incapacitation probability and penetration significantly better than the existing M162 or M257 12-gauge shotgun cartridge. These requirements were met by a 3 in (76 mm) belted brass cartridge case with a tungsten alloy load consisting of eight pellets. This cartridge carries eight tungsten alloy pellets, each weighing approximately 3.1 g, which are ejected at 538 m/s velocity at 1 m, penetrating a 20 mm pine or 1.5 mm mild steel plate barrier at 150 m. The cartridge case is also capable of being loaded with '000 Buck' pellets or with a flechette package.

The gun, designed around this cartridge by Heckler and Koch, has a distinct family resemblance to the G11 rifle and uses a similar outer casing which in this case is of high impact plastic. It is a semi-automatic weapon operated by recoil and due to the interior 'floating' system the received felt recoil has been significantly reduced. The gun is fed from a box magazine inserted behind the pistol grip. It fires from a closed bolt, has ambidextrous selection and safety levers, and can be fitted with choke or with a choked barrel extension if required.

The chamber is configured so that it will accept commercial-type 12-bore cartridges for training or service purposes as well as the special belted

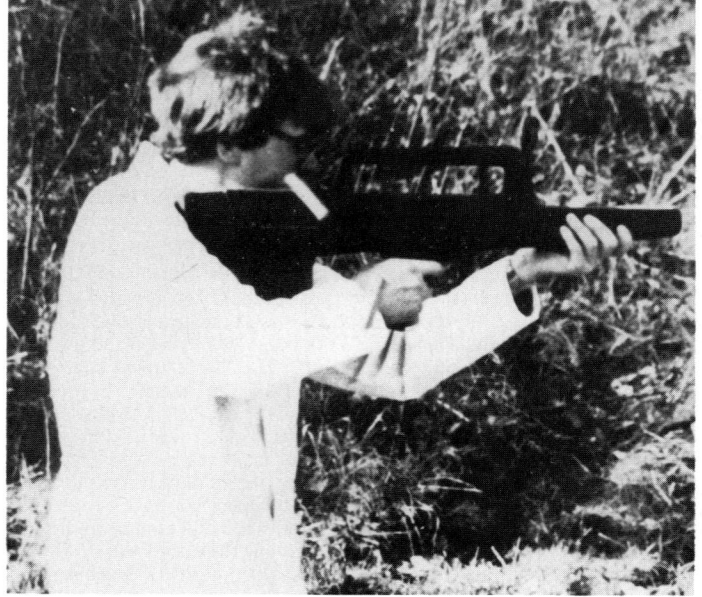

Firing Olin/HK CAWS

Olin/HK CAWS. Note new optical sight and attachment of choke to muzzle

cartridge developed for the gun. Both left- and right-handed firers can use the weapon, since it can be readily converted from left to right hand ejection by merely changing the bolt-head position by 180°. The carrying handle houses an optical sight or can be modified to a rail type sighting system, consisting of a front and rearsight similar to a match pistol sight, and beneath this handle is the cocking handle.

DATA
Cartridge: 12-bore 3 in belted (19.5 × 76 mm)
Operation: recoil, selective fire or semi-automatic
Weight, empty: 3.7 kg with magazine
Length overall: 764 mm. (With barrel extension 989 mm)
Barrel length: 460 mm. (With barrel extension 685 mm)
Magazine capacity: 10 rounds
Muzzle velocity: tungsten alloy ca. 538 m/s; '000 Buck' load ca. 488 m/s; flechette load ca. 900 m/s

Manufacturers
Olin Industries (Winchester Group), 707 Berkshire, East Alton IL 62024
Heckler and Koch Inc, 21480 Pacific Boulevard, Sterling, VA 22170-89093

Olin flechette cartridge

Status
Development suspended

AAI Close Assault Weapon System (CAWS)

The AAI Corporation's CAWS, developed for the US Naval Weapons Support Centre, is a shotgun-type weapon for use in quick fire, close combat situations. It has the ability to fire automatic at 450 rds/min, and a very low 'felt recoil' (slightly less than that of an M16 rifle) makes it easy to control in burst fire.

The weapon is recoil-operated and is laid out in straight-line fashion, which reduces muzzle climb. Butt and pistol grip are standard 16 components, and a short version, without butt, has been developed for airborne troops.

A special cartridge has been developed for use with the AAI CAWS. It is larger than the commercial 12-gauge cartridge, ensuring that it will not chamber in a sporting-type weapon, but CAWS will fire commercial 12-gauge ammunition by means of an adapter. The anti-personnel cartridge is loaded with eight drag-stabilised flechettes each weighing about 1 g. The accuracy of this round is such that all eight sub-projectiles will strike within a 4 m circle at 150 m range. At this range the flechettes have a remaining velocity of about 365 m/s and will penetrate 76 mm of pine or 3 mm of mild steel.

A complete family of ammunition, to include HE, AP, tear gas and other types is in course of development.

DATA
Cartridge: 12-gauge special
Operation: recoil, selective fire
Feed: 12-round box magazine
Weight: 4.08 kg
Length: 984 mm
Sights: day/night optical reflex with iron sight backup

Manufacturer
AAI Corporation, PO Box 6767, Baltimore, MD 21204.
Status
Development.

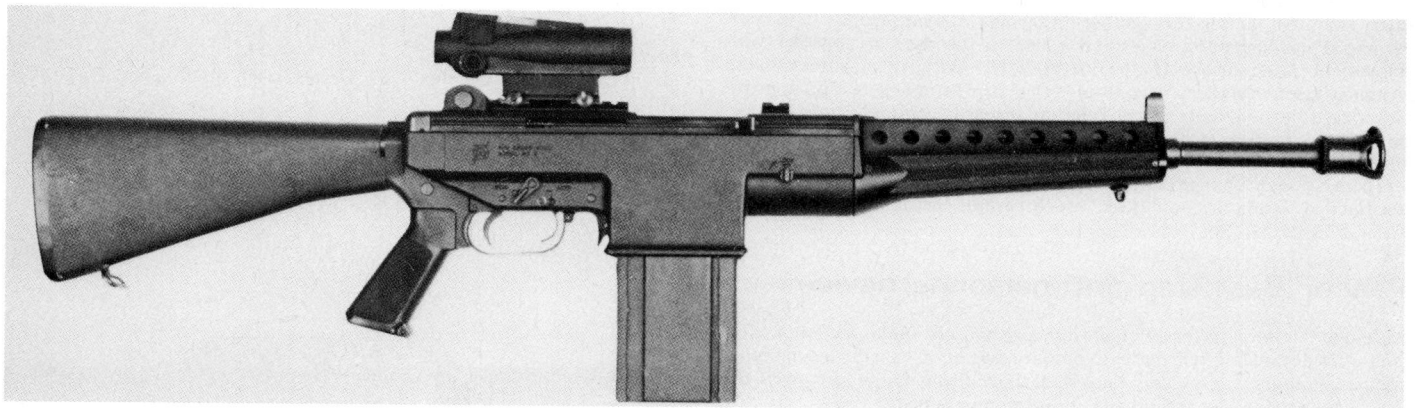

AAI Close Assault Weapon System (CAWS)

Pancor Jackhammer Mark 3-A2

Jackhammer is an automatic, gas-operated, 12-gauge shotgun which uses a pre-loaded rotating cylinder as its magazine. The cylinder has grooves incised on its outer surface which are engaged by a stud on an operating rod, so that as the rod oscillates it rotates the cylinder; it is very similar to the system of cylinder rotation used by the Webley-Fosbery automatic revolver. The Jackhammer barrel floats, and is driven forward by gas pressure after the shot is fired, then returned by a spring, this giving movement to the cylinder

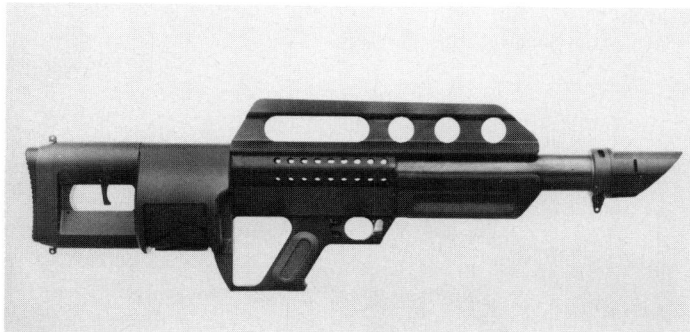

Pancor Jackhammer Mark 3-A2 combat shotgun

operating rod. The significant point about this movement of the barrel is that it disconnects the barrel from a gas-tight connection with the cylinder, so allowing the cylinder to be revolved to index the next round, and then, on the return stroke, re-seals barrel and cylinder together.

The barrel, flash eliminator, return spring, and 'Autobolt' (the Pancor patented name for the actuating rod which revolves the cylinder) are all of high quality steel; the remainder of the weapon is almost entirely plastic, a new product by Du Pont called Rynite SST which is an extremely strong and durable polyethylene terephthalate (PET) plastic toughened by the admixture of glass fibre.

The cylinder, called by Pancor the 'Ammo Cassette', is also of Rynite SST plastic, contains 10 shots, and is pre-loaded and sealed with a shrink-film plastic, colour-coded to indicate the type of ammunition loaded. The sealing is removed by a pull-strip and the cassette simply clips into the weapon and immediately engages with its operating system. The weapon is then cocked by sliding the fore-end, and it is ready to fire. It can fire at a rate of four shots per second, and once the cassette is emptied a simple movement of the fore-end and a hold-open latch allows it to be dropped out and a fresh cassette loaded. It is not possible to load single rounds into the gun and empty cases are not ejected whilst firing.

The Jackhammer Mark 3-A2 is the most recent step in the continuing development of this combat shotgun and is referred to by the company as its pre-production model. Whilst the Mark 3-A2 operates on the same principles as earlier models, described above, design modifications have been made

Rear view of Jackhammer Mark 3-A2, showing new butt stock and improved form of 'Ammo Cassette'

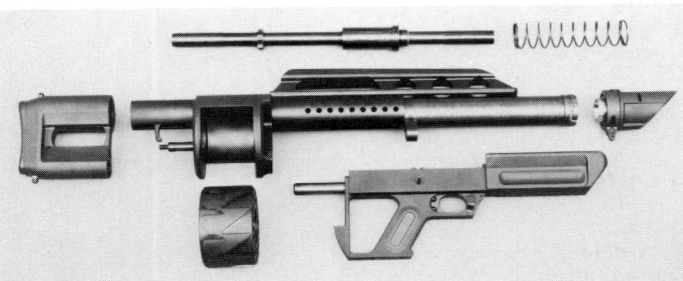

Jackhammer Mark 3-A2 dismantled

which facilitate production and which have improved the various elements and subassemblies. One of the principal concerns has been to keep the gun weight at an acceptable figure, and this has meant careful engineering in order to make the design compatible with the DuPont plastic materials which are being employed for many components.

The Mark 3-A2 differs in outward appearance from previous models of the Jackhammer. The butt stock assembly has been re-contoured to improve control and absorption of recoil; the receiver has been re-dimensioned to allow smoother action; the sighting bridge and carrying handle have been re-shaped; cooling ports have been added to the receiver, together with improved bearing surfaces for the floating barrel; the advancing track on the 'Ammo Cassette' has been re-contoured to reduce firing stresses and improve cassette ejection and replacement time; and the pistol grip and front grip have been re-designed, together with the general exterior contour, to give a smoother and more compact outline.

There is a slight change in the method of operation. A de-cocking lever has been added, inside the butt stock, which permits the hammer to be safely

de-cocked when the weapon has been loaded. It can then be transported safely in the loaded condition. When required for action, the lever is again operated, this time to cock the hammer, which is performed silently.

A particular feature is the rotating front sling swivel, mounted behind the flash guard. This allows the sling to take up the most convenient position according to the needs of the man carrying the weapon; it can be suspended in a firing position or slung across the back, in which position it automatically conforms to the body in the most comfortable manner.

In December 1987 Pancor Inc announced that the Jackhammer's receiver has been modified to accept and fire a special round that allows the payload or shot capacity of the standard shotgun cartridge (2.75 in or 70 mm) to be increased by 100 per cent and perform with elevated pressures in the 20 000 psi (1406 kg/cm²) region.

This cartridge, known as the 'Jack Shot', is a moulded cylindrical container of DuPont Rynite SST™ which holds a sized — cut back to 50 mm length — standard 12-gauge shotgun cartridge pressed in from the rear end, with the powder, carrier and payload inserted from the front end. This allows unusual payloads to be loaded, such as armour penetrators, flechettes, fragmentation/shrapnel loadings, canister loadings, liquid or solid chemicals, rocket-assisted projectiles or simply larger loads of conventional lead shot. Moreover, the construction of the 'Jack Shot' permits ready reloading in the field.

The operation of the Jackhammer weapon using the 'Jack Shot' cartridge is similar to the normal use of the Ammo Cassette, except that the floating barrel is locked in its firing position (to prevent damage to the normal cyclic gas system from the extra pressure) and the auxiliary silent cocking lever is used to cock the hammer.

The company is studying a number of other ammunition concepts, with a view to developing a range of ammunition specifically suited to the Jackhammer shotgun. It is also involved in the development of a sound suppressor which will be relatively inexpensive and which can be discarded after a limited life.

DATA
Cartridge: 12 gauge, 70 mm
Feed: 10-shot rotating cylinder
Overall length: 787 mm
Barrel length: 525 mm
Height: 230 mm
Width: 125 mm
Weight: 4.57 kg
Cyclic rate of fire: 240 rds/min

Manufacturer
Pancor Corporation, PO Box 7557, Ocean Beach Station, San Diego, CA 92107.
Status
Advanced development.

Pancor 'Bear Trap' anti-personnel device

Although this device is far different from anything else in this section, we feel it has to be considered here due to its unique association with the Jackhammer combat shotgun; it is the only known instance of the ability to convert the magazine of a shoulder arm into an anti-personnel mine.

Because of the functional shape of the Jackhammer shotgun Ammo Cassette, with its radial arrangement of 10 shotgun cartridges, it has been found feasible to develop the cassette into an anti-personnel mine or trap device. So configured it becomes the 'Bear Trap'™ and consists of three units: (A) the Ammo Cassette, loaded with from 1 to 10 rounds of 12-gauge shot cartridge; (B) the detonator base, containing the firing mechanism, which surrounds the lower portion of the Cassette; and (C) the detonator plunger or pressure plate which can be set to fire the contents of the cassette by direct mechanical action when pressed or by delayed reaction to a spring-loaded 12-hour timer which can be set at one-hour release intervals.

Once assembled, the 'Bear Trap' is placed in the ground or other position, with the open end of the cassette pointing in the desired direction. Upon pressure being applied to the detonator plunger, whether by pushing or pulling, the contents of the cassette will be discharged, according to the method of firing selected. The cassette is designed to withstand the blast effect of the discharge of its contents and be re-usable.

Pancor Jackhammer shotgun with the Ammo Cassette removed and converted into a 'Bear Trap' anti-personnel mine

Manufacturer
Pancor Corporation, PO Box 7557, Ocean Beach Station, San Diego, CA 92107.
Status
Advanced development.

Ciener Ultimate over-under combination

Rifles are particularly suited to long range use; shotguns provide close-in defence when reaction time is at a minimum. The Ceiner combination offers the opportunity to have both available as and when necessary. The Ultimate combination consists of a Remington 870 pump-action shotgun which attaches beneath the Colt AR15/M16 rifle, utilising the bayonet lug as the basic fixing point.

Ciener Ultimate over-under combination

A bayonet lug adapter is attached to the shotgun barrel and a yoke adapter to the receiver. This adapter engages the ends of a special upper receiver to lower receiver hinge pin. The bayonet lug adapter has two spring pressure clips which fix the shotgun in place, allowing removal of the shotgun by simply squeezing the clips.

Manufacturer
Jonathan Arthur Ciener Inc, 6850 Riveredge Drive, Titusville, FL 32780.
Status
Current. Production.

SAM Model 88 'Crossfire'

This is a slide-armed semi-automatic rifle/shotgun combination which is custom-designed rather than being merely a shotgun unit added to an existing weapon. The two barrels, an upper 12-ga shotgun barrel and a lower 7.62 × 51 mm rifle barrel, are mounted in a common receiver and are fed by two independent magazines; the rifle magazine feeds directly into the bottom of the rifle action, the shotgun magazine feeding into the side of the shotgun action. The weapon is loaded and cocked manually for the first shot by operating a fore-end slide; thereafter reloading is performed automatically by a dual gas piston system, leaving the weapon cocked ready to fire. Each unit can be independently reloaded while the other is in operation, thus permitting continuous fire to be sustained. A simple selector switch allows the user to select either the rifle or the shotgun or set the complete weapon to safe.

DATA
Cartridge: 12-ga 2¾in aand 7.62 × 51 mm NATO
Operation: semi-automatic, gas operation, manual initial cocking
Locking system: dual rotating bolts
Feed: 20-round box (rifle); 7-round box (shotgun)
Weight empty: 4.31 kg
Length: 1010 mm
Barrels: 508 mm (both)
Sights: Optional adjustable iron sights or optical

Manufacturer
SAM Inc, 405 Rabbit Trail, Edgefield, SC 29824-1547.
Status
Production to commence early 1990.

Remington M870 Mark 1 US Marine Corps shotgun

In 1966 the US Marine Corps conducted comparative trials of a number of shotguns and selected the Remington Model 870, slightly modified to suit the preferences of the Corps. The M870 Mark 1 is a rifle-sighted slide-action gun based on the commercial Remington 870 action, with a modified choke barrel. All metal parts are either Parkerised or with a black oxide finish, and there is a plain plastic butt-plate. The smooth extended fore-arm has prominent finger grooves. The magazine is extended to hold seven cartridges and the front end of the magazine tube is stepped to form a seating for the ring of a standard M7 bayonet; there is a bayonet locking lug forming part of the front sling swivel.

DATA
Cartridge: 12-gauge × 2¾ in
Operation: slide-action

Feed: 7-shot tubular magazine
Weight: 3.60 kg
Length: 1060 mm
Barrel: 533 mm modified choke
Sights: (front) blade; (rear) adjustable notch

Manufacturer
Remington Arms Inc, Bridgeport, CT 06601.
Status
Current.
Service
US Marine Corps.

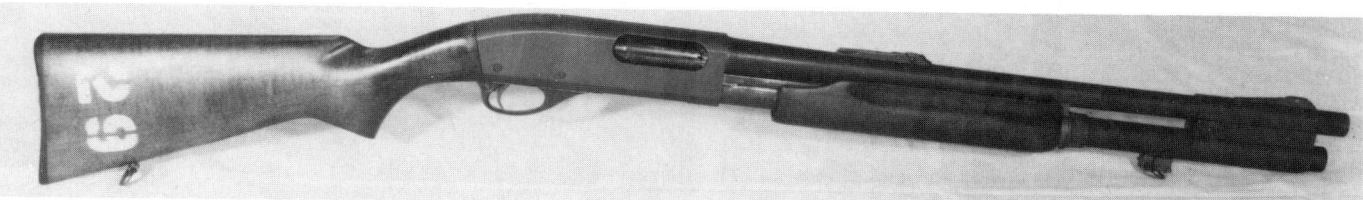

Remington M870 Mark 1 US Marine Corps shotgun

ABC-M9-7 portable flamethrower

The ABC-M9-7 flamethrower consists of the M9 fuel and pressure unit, M7 flamethrower gun and the M8 hose group which connects the other two units.

The fuel and pressure unit consists of two connected aluminium fuel tanks, a spherical steel pressure tank and valves, and a pressure regulator. Compressed air enters the tops of the fuel tanks and forces thickened fuel through the hose to the gun under 25 kg/cm² pressure. Four US gallons (15.14 litres) of fuel is ejected 40 to 55 m in 8 seconds in a continuous stream or in a maximum of five bursts. The fuel is ignited by an incendiary charge from an ignition cartridge as it leaves the nozzle of the gun barrel, five incendiary charges being contained in an ignition cylinder which fits over the forepart of the barrel.

DATA
Weight: (empty) 11.8 kg; (loaded) 22.7 kg
Fuel capacity: 15.14 l
Range: 40–55 m

Status
Obsolete.
Service
Probably only held in reserve stocks.

M2A1-7 portable flamethrower

The M2A1-7 portable flamethrower is carried on the soldier's back and consists of the M2A1 tank group, M7 flamethrower gun and M8 hose group. The latter connects the tank group to the gun.

The tank group consists of a pressure tank and two fuel tanks. Compressed air enters the tops of the fuel tanks and forces thickened fuel through the hose to the gun. Unthickened fuel can also be used but with considerable loss of range. Under pressure approximately 15 litres of fuel is ejected to a distance of between 40 and 55 m. The flame lasts for 10 seconds if fired in a continuous stream or can be delivered in five two second bursts. The fuel is ignited by an ignition cylinder as it passes from the nozzle of the gun barrel, there being five igniter cartridges in the ignition cylinder which fit over the forepart of the barrel. To fire the gun a lever on the reargrip must be pressed as well as the forward trigger.

M2A1-7 portable flamethrower in firing position

DATA
Weight: (unloaded) 18.7 kg; (loaded) 29.4–32.2 kg according to fuel
Fuel capacity: 15 l
Fire duration: 6–9 s
Range: (thickened fuel) 40–50 m; (unthickened) 20–25 m

Status
Obsolescent.
Service
US armed forces.

M2A1-7 portable flamethrower

M9E1-7 portable flamethrower

This is the replacement for both the M2A1-7 and the M9-7 flamethrowers. It consists of a tank group, a hose group, a gun group and a separate carrier group. The tank group comprises a tubular frame with two interconnected fuel tanks, a protective cover, a flame gun holster and carrying straps. A high pressure sphere is mounted beneath the smaller of the two fuel tanks. Air pressure from this sphere is distributed to the fuel tanks through a pressure regulator mounted between the tanks. The hose group supplies fuel from the tanks to the gun group. The fuel is then ignited by a replaceable ignition cylinder in the nozzle end of the gun group. Each cylinder contains five igniter cartridges.

The carrier group consists of a frame, protective cover and carrying straps. Four spare high pressure spheres and four ignition cylinders are stowed in the carrier group.

The principle of operation is the same as for the M2A1-7 and M9-7 flamethrowers.

DATA
Weight of complete flamethrower: (empty) 11.3 kg; (loaded) 21.7–23.6 kg according to fuel
Fuel capacity: approx 15 l
Range: (thickened fuel) 40–50 m; (unthickened) 20–25 m

Status
Current.
Service
US Army and Marine Corps.

US Marine wearing protective clothing and carrying M9E1-7 portable flamethrower

66 mm M202A2 multi-shot portable flame weapon

This consists of a light, shoulder-fired, four-tube rocket launcher complete with a sling and a pre-loaded four-round incendiary rocket clip which slides into the launcher tubes. Hinged protective covers are fitted at each end of the assembly, the forward one of which acts as a safety by inhibiting the firing system.

The weapon can be fired from the standing, sitting or kneeling positions at least one round in five seconds. The launcher is re-usable.

DATA
Weight, complete: 12 kg
Length, extended for firing: 82.26 cm
Range: (max effective) 750 m; (minimum safe) 20 m
Back blast area: 40 m

Status
Current.
Service
US Army and Marine Corps.

M202A2 multi-shot flame weapon in firing position

YUGOSLAVIA

128 mm light rocket launcher M71

The M71 is a lightweight single-shot launcher which can be man-carried. It is particularly recommended for use by commando and similar raiding forces.

The launcher consists of a lightweight tube hinged to a baseplate, and two telescopic legs which support the launch tube at any required angle. There is a sight bracket on top of the tube to take a panoramic collimating sight.

The M63 rocket is a spin-stabilised rocket, using tubular propellant in a rear compartment and with a high explosive warhead fitted with a UTI M63 point detonating fuze. Firing is carried out by an electric magneto firing unit which can be positioned up to 25 m from the launcher.

DATA
LAUNCHER
Calibre: 128 mm
Tube length: 1000 mm
Weight in firing position: 18.5 kg unloaded; 45 kg loaded
Elevation: 2° – 45°
Levelling limits: ± 10°
Detachment: 3 men

ROCKET
Weight: 23 kg
Weight of warhead: 7.55 kg
Bursting charge: 2.30 kg RDX/TNT
Motor weight: 4.78 kg
Length, with fuze: 814 mm
Max velocity: 444 m/s
Minimum range: 800 m
Max range: 8564 m
Dispersion at max range: 0.56% of range

Manufacturer
Federal Directorate of Supply and Procurement, 9 Nemanjina Street, 11001 Beograd.
Status
Current. Production.

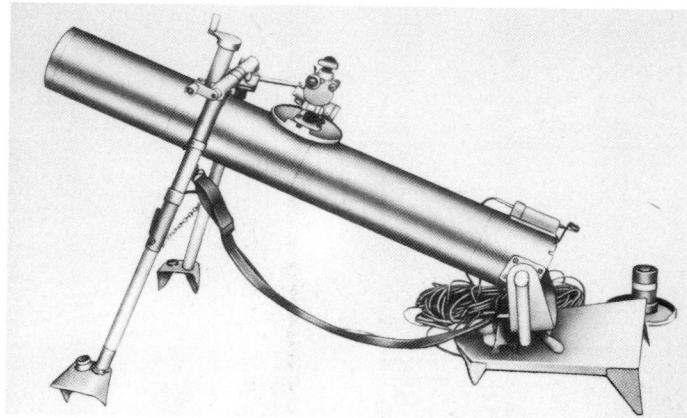

128 mm rocket launcher M71

Loading the M71 rocket launcher

Crew-served Weapons

Machine Guns
Cannon
Anti-Tank Weapons
Mortars and Mortar Fire Control

Manroy Engineering

MANROY .50" M2HB-QCB (QUICK CHANGE BARREL) MACHINE GUN

Manroy Engineering is the sole UK manufacturer of the .50″ M2HB QCB Machine Gun. The QCB gun combined with Softmount is the most comprehensive .50″ weapon system on the market, offering substantial cost savings over other .50″ and 20mm guns. By the use of a QCB Conversion Kit existing M2HB guns may easily be converted to the Manroy QCB system.

Manroy manufacture a full range of mounting systems for .50″ and 7.62mm machine guns including M3, M63, M31, Softmount and specialist mounts to customer requirements.

MANROY DEFENCE PRODUCTS
Machine Guns, Rifles, Pistols, Gauges, Tools, Accessories, Automatic Stud Gun, Signal Pistols and Gun Cleaning Equipment. The Company offers weapon repair and rebuild facilities to full Base Workshop standards.

Ammunition
Manufactured to US MIL Specifications. Available in calibres .50″ (12.7×99), 7.62 NATO, 5.56×45 and 9×19 Parabellum. Other natures are available on request.

Manroy Engineering Ltd, Beckley, East Sussex TN31 6TS
Tel: Beckley (079 726) 553   Telex: 95249   Fax: (079 726) 374

Machine Guns

ARGENTINA

Argentine machine guns

The Argentine armed forces have used a variety of machine guns in the past, including French AA-52, Browning .30 M3 converted to 7.62 × 51 mm chambering, and the .50 Browning M2HB. Numbers of these are still in use, particularly the .50 model, but the standard machine gun is now the locally manufactured version of the FN-MAG. This is almost identical with the Belgian-made version, to which reference should be made for details of operation, differing only in the length (1255 mm) and empty weight (10.80 kg). The locally manufactured tripod appears to be based on the FN Model 04-41.

Manufacturer
Fabrica Militar de Armas Portatiles 'Domingo Matheu', Avenida Ovidio Lagos 5250, 2000-Rosario.
Status
Current. Production.

Argentine-manufactured 7.62 mm FN-MAG machine gun

AUSTRALIA

7.62 mm M60 general-purpose machine gun

Currently in service in Australian infantry formations the 7.62 mm M60 general-purpose machine gun is purchased from the USA. The weapon itself is fully described in the appropriate US entry later in this section but two Australian developments, designed to provide more satisfactory methods of handling ammunition in what are considered to be the most likely combat conditions, are briefly described here.

Magazine belt box
This is a metal box which holds 40 rounds of 7.62 mm ammunition in standard belt form. It is so designed that it can be permanently attached to the gun without impeding the use of continuous belt ammunition, and so that the feed can be changed from external belt feed to boxed belt feed without opening the gun feed top cover.

The 40 rounds held by the magazine include those on the clipping platform of the magazine lid (see photographs); and the lid performs the additional function of a belt stop pawl, so that the belt can be loaded into the magazine

and retain these exposed rounds in position in one operation without having to thread the belt through a gate.

To load the magazine the box portion is simply pressed downwards to open the spring-loaded lid and a suitable length of belt is lowered into it. Holes in the rear side of the box enable the gunner to see roughly how many rounds there are in the box. To bring the magazine ammunition into operation, if the gun is empty, the magazine belt is simply drawn out of the box by the top clipping round and loaded in the normal way. If the gun has been firing rounds from an external belt, however, the top round on the clipping platform is clipped to the end of the belt protruding from the gun, any surplus length of the other belt having first been unclipped.

Operational use
Two important uses are envisaged. First, the firer can keep the belt box loaded with ready ammunition. If he then needs to move without his number two, either of them can break the external belt at a suitable length and clip it to the magazine belt. Alternatively, if the belt box is empty, the external belt can be broken at a suitable length and the loose end dropped into the box without either removing the belt from the gun or having to thread it through a gate.

The box is made of aluminium and is nylon coated to reduce noise, avoid corrosion and provide a dry lubricant. It weighs 343 g.

Bandolier
To provide a convenient method of carrying general-purpose machine gun belt ammunition, with adequate camouflaged protection for the belt, a special 50-round bandolier has been developed. Made of PVC-coated nylon it provides waterproof protection for the ammunition, is sufficiently durable to be re-used and cheap enough to be treated as expendable in appropriate circumstances. Three bandoliers will fit into an M19A1 ammunition box.

40-round box loaded and ready to attach to general-purpose machine gun showing three rounds on clipping platform and internal rounds visible through holes in side of box

40-round box mounted on general-purpose machine gun with its belt coupled to belt in gun

Alternative fastenings are provided which enable the effective length of the bandolier to be varied or two or more bandoliers to be joined together so that it or they can be worn in any convenient way.

To load the gun from the bandolier it is necessary to withdraw only a few rounds from its mouth (which can be identified by touch at night) thus having the bulk of the rounds in protective covering. There are then two ways of removing the remainder of the rounds for firing: either the bandolier cover can be pulled off like a sock, in which case it remains in a re-usable condition but the rounds are unprotected, or a gutting slip can be used to lay the whole

belt open quickly, the opened cover remaining under the rounds to protect them from picking dirt up from the ground.

Manufacturer
Ordnance Factory, Maribyrnong, PO Box 1, Ascot Vale, Victoria 3032.
Status
Current. Available.
Service
Australian forces.

7.62 mm converted Browning

The Engineering Development Department of the Australian Department of Defence at Maribyrnong, Victoria has produced a 7.62 mm version of the well-known and popular Browning .30 calibre Model 1919A1 machine gun. The gun is still used in quite large numbers in Australia and in other countries

Two feed systems compared. In both cases dummy rounds are in belts

but in recent years a criticism has been that the .30 ammunition is becoming more difficult to obtain, furthermore it is no longer standard with any first line army which complicates the supply organisation. Maribyrnong has succeeded in converting the gun to accept the NATO 7.62 mm round in the M13 linked belt.

Most of the conversion has been in the feed mechanism where it has been necessary to alter the arrangements for extracting the rounds from the belt. The Browning belt is a pocketed cloth one, and the .30 round is longer than the 7.62 mm. The working of the gun is not changed and the NATO round is withdrawn backwards from the linked belt to be fed in the normal Browning manner. Fortunately the muzzle energy of the two rounds is almost identical and so the functioning of the gun is not changed. Some minor changes have been made to the mechanism to increase the overall life and the rate of fire has been increased. Naturally, a new barrel has been fitted.

These modifications have resulted in a reliable and robust gun with all the Browning characteristics unchanged. Barrel life is now over 10 000 rounds and it looks likely that the old Browning could carry on giving good service for many years yet.

DATA
Weight: (overall) 13.6 kg
Barrel: 3.2 kg, chrome-plated; (rifling) 6 grooves, rh, 1 turn in 305 mm
Rate of fire: (cyclic) 650–750 rds/min
Muzzle velocity: 870 m/s

Status
Prototypes only.
Service
No firm orders yet.

5.56 mm F88 heavy barrel light support weapon

This is a 621 mm barrel length version of the F88 individual weapon, designed to be used as a light support weapon. The individual modules of the weapon are all fully interchangeable with all other F88 models. It differs from the rifle model in having a slightly heavier barrel to cope with the increased thermal stresses of a higher rate of fire, an adjustable fold-up bipod attached at the muzzle end of the barrel, and a 42-round magazine in place of the usual 30-round magazine.

Data and specifications for the F88 light support weapon are exactly as

given for the Steyr AUG light support weapon in the following section, to which reference should be made.

Manufacturer
Australian Defence Industries, Lithgow Facility, New South Wales (under licence from Steyr-Mannlicher GmbH).
Status
Current. Production.

5.56 mm F89 Minimi light support weapon

The F89 Minimi light machine gun has been adopted by the Australian Army as its future light support weapon, and is to be phased into service

commencing in 1990. Production capacity for the weapon is currently being established at the Australian Defence Industries Lithgow Facility.

Australian manufactured F89 Minimi machine gun, left side

Right side of Australian F89 Minimi machine gun

The F89 is a multi-purpose machine gun capable of sustaining high rates of fire. It can be tripod-mounted and has the same capabilities as heavier calibre weapons. It can also be fired from the bipod or from the hip.

A full technical description of the F89 Minimi, together with all data, can be found in the entry in the Belgian section.

Manufacturer
Australian Defence Industries, Lithgow Facility, New South Wales (under licence from Fabrique National Herstal).
Status
Current. Production.
Service
Australian Defence Force.

AUSTRIA

5.56 mm Steyr Army Universal Gun light support weapon (LSW)

One of the three versions of the new Army Universal Gun (AUG) is a light machine gun. All the details relevant to its action and operation are given in the section dealing with Austrian rifles and the data here refers to the light machine gun version only.

There are variant models of the LSW: the type firing only from a closed bolt (see below) is known as the Heavy Barrelled Automatic Rifle (HBAR). The model firing from an open bolt is known as the 'Light Machine Gun' (LMG). Both open and closed bolt types are generically known as the 'Light Support Weapon' (LSW). All types use a newly designed barrel group which is available with either 178, 228 or 305 mm pitch of rifling. A new type of muzzle attachment acts as a flash hider and reduces recoil and muzzle climb during automatic fire.

There are two basic versions of this weapon; the LMG has the usual carrying handle and built-in optical sight of the AUG family. The LMG-T has a different receiver assembly which has, instead of the carrying handle and sight, a flat bar upon which any telescope sight or night-vision device can be mounted. In addition both types can be furnished with modifications which change the operation of the weapon so that it fires from an open bolt. This modification involves substitution of a new hammer mechanism block for the standard type and changing the cocking piece in the bolt assembly; these components can also be supplied by the company to permit modification to be carried out on existing weapons, and the conversion can be done in less than one minute. In the open bolt mode there is no change in the firing characteristics of the weapon.

DATA
Cartridge: 5.56 mm × 45
Feed: 30- or 42-round box magazine
Weight: (unloaded) 4.9 kg
Length: 900 mm
Barrel: 621 mm

5.56 mm Steyr LMG-T, showing bipod and optional night sight

Sights: optical, integral with carrying handle
Muzzle velocity: 1000 m/s
Rate of fire: (cyclic) 680 rds/min

Manufacturer
Steyr-Mannlicher GmbH, a member of the Steyr-Daimler-Puch AG group, A-4400 Steyr.
Status
Current. Production.
Service
Commercial sales.

BELGIUM

15 mm FN BRG-15 heavy machine gun

Development of this weapon began in 1980 in order to fill the gap between the existing .50 Browning gun and the 20 mm and up group of cannon. A completely new cartridge, the 15 × 115 mm, was developed to meet a specified level of muzzle energy. A gun was then designed which would employ the latest proven techniques of operation, incorporate a dual-feed system, but still use, as far as possible, existing manufacturing technology, materials and equipment so as to keep the cost within reasonable limits. The resulting gun is a gas-operated weapon with simple mechanism, capable of being easily and rapidly dismantled for carriage or storage, and adaptable to any role including ground, vehicular or aerial employment. The dimensions were selected so that the gun could be adapted to existing mounts by means of accessory equipment, and could also be retrofitted so as to replace existing

weapons in various applications. The gun is belt-fed, so as to develop a high rate of fire suited to anti-aircraft applications, and the dual feed capability gives the firer the option of two types of ammunition at the flick of a lever. It fires from the open bolt mode, the bolt locking into the barrel extension, and has a selective fire capability.

In the course of final development, it was discovered that barrel wear in burst fire was excessive, and after investigation it was discovered that jacketed bullets were unsuitable at the desired velocity. New projectiles, based more upon artillery practice than upon small arms practice, and using a plastic driving band to provide spin and sealing, were developed, and it was discovered that the optimum calibre for this new projectile was 15.5 mm. This lead to development of a completely new cartridge (described in more detail in

15.5 mm FN BRG-15 on pintle mount

15.5 mm FN BRG-15 machine gun

15.5 mm BRG-15, showing dual feed, left belt engaged

FN BRG-15, showing gas regulator and piston, and recoil absorbers

the ammunition section) which led, in turn, to modification of various parts of the machine gun. As a result, a major re-design took place.

In its new form the weapon is slightly longer, due to development of a new quick-change barrel and modification of the feed mechanism. The barrel incorporates a gas port with short-stroke piston which, when the barrel is installed, interfaces with the mechanism operating rod in the receiver. The barrel is now grooved to dissipate heat more efficiently and is fitted with a flash eliminator. Dual recoil-absorbing buffers are fitted, which means that the recoil blow transmitted to any mounting will be approximately the same as that of the .50 Browning M2 machine gun.

The BRG-15 retains the dual belt feed, but the feed changing mechanism has been simplified and made more compact. Feed from right or left is available instantly, and a central 'neutral' position of the feed selector renders the weapon safe.

The gun can be rapidly dismantled into its principal component groups. A pintle mount, suitable for vehicle or boat mounting, is in an advanced stage of development and provides a compact and very stable mount.

DATA
Cartridge: 15.5 × 106 mm
Operation: gas, selective fire
Locking system: rotating bolt
Feed: dual disintegrating link belt
Ejection: cases and links downward
Weight, empty: 60 kg
Length: 2.150 m
Height: 245 mm
Width: 220 mm
Rate of fire: 600 rds/min
Muzzle velocity: 1055 m/s
Effective range: 2000 m

Manufacturer
FN Herstal SA, B-4400 Herstal.
Status
Advanced development.

.50 M2 heavy-barrel FN Browning machine gun

The M2HB (heavy barrel) Browning machine gun is similar in design and operation to other types of Browning machine gun, a recoil-operated, air-cooled, belt-fed weapon. The barrel is easily and quickly removed from the receiver in order to form loads of convenient size for packing and carrying.

Operation
The operating cycle begins when the firing pin strikes the cap and fires the cartridge. At this point the bolt is locked to the barrel/barrel extension assembly by the breech lock, a lug which is held up into a slot in the bolt by the locking cam fitted at the bottom of the receiver.

After the cartridge fires, the reaction to the explosion drives the moving parts to the rear. After about 20 mm of recoil the breech lock pin strikes the breech lock depressors, which push the pin and breech lock lug downward, out of engagement with the bolt. The bolt is thus unlocked, separates from the barrel/barrel extension, and travels to the rear. The barrel and barrel extension are completely halted by the barrel buffer body.

During rearward travel the barrel and extension assembly comes into contact with the accelerator, which pivots on the accelerator pin. The contact point between the barrel extension and the accelerator moves downwards progressively while the barrel extension slides backwards. This downward movement changes the ratio of leverage exerted by the tip of the accelerator against the bolt, so that the bolt is thrown backwards at an increased speed relative to the other recoiling parts. In addition, the gradual transfer of energy from the barrel extension to the bolt, via the accelerator, also damps the shock given by the barrel extension against the barrel buffer. A further advantage

accruing from the accelerator is that the increase in bolt speed assists in extracting the spent case from the chamber in a progressive manner.

The rearward travel of the bolt compresses the return spring and is finally halted when the bolt strikes the buffer plate. The buffer plate disks absorb the shock and, rebounding, start the bolt forward once more, this movement being continued by the energy stored in the return spring. The lower edge of the bolt strikes the tip of the accelerator, turns it forward, and thus unlocks the barrel and barrel extension assembly from the barrel buffer, whereupon the barrel buffer spring drives the barrel and barrel extension assembly forward. This forward motion is augmented by that of the bolt and its driving spring as the bolt reaches its foremost position in the barrel extension.

At the end of the forward movement the breech lock lug, driven by the barrel extension, rides up the inclined plane of the breech lock cam and enters the recess in the bottom of the bolt, thereby locking the bolt to the barrel/barrel extension assembly.

During the recoil of the bolt the spent case in the breech is withdrawn by the T-slot in the face of the bolt. At the same time, a fresh cartridge is pulled from the belt by the extractor-ejector. The extractor is gradually forced down by the cover extractor cam until the fresh cartridge is located on the barrel axis. This downward movement also forces out the spent case which is then ejected downwards. As the bolt returns, so the fresh cartridge is inserted into the chamber.

Also during the recoil movement of the bolt, the belt feed lever engages in a diagonal groove in the top of the bolt and is pivoted due to the bolt's movement. This forces the belt feed slide to move to the left in the slot in the feed cover. The belt feed pawl engages behind the next cartridge. As the bolt moves forward the belt feed lever is pivoted in the reverse direction and the belt feed slide moves to the right, ready to begin the next feed.

Headspace adjustment
The cartridge headspace is critical in this design; ideally, it should be adjusted by means of 'go-no go' gauges as described in the various field manuals on the machine gun, but in default of the correct gauges an approximation may be achieved by screwing the barrel in fully, then unscrewing it two clicks. Should the gun fire sluggishly, the barrel should be unscrewed by one further click.

FN .50 Browning M2HB machine gun

DATA
Cartridge: .50 Browning (12.7 × 99 mm)
Operation: short recoil, selective fire
Feed: disintegrating link belt, M2 or M9 links
Weight, empty: 38.15 kg
Length, overall: 1.656 m
Length of barrel: 1143 mm
Rifling: 8 grooves, rh

Rate of fire: 450-550 rds/min
Muzzle velocity: 916 m/s with M33 ball
Max range: 6800 m

Manufacturer
FN Herstal SA, B-4400 Herstal.
Status
Current. Production.

.50 M2HB/QCB FN machine gun

Since the late 1970s the demand for the .50 M2HB machine gun has been steadily growing, and this led FN to develop a quick-change barrel version. This new model eliminates tedious headspace adjustment, ending the chance of dangerous errors, saves time in training and in operations, and reduces the risk taken in action whilst adjusting headspace. To accompany this machine gun FN has also introduced new APEI and HEPI rounds.

All the M2HB and QCB machine guns made by FN are provided, as standard series production items, with a trigger safety, a height-adjustable front sight, and a timing requiring no adjustment on the user's part.

Conversion kit
Fabrique Nationale Herstal, the only company to manufacture the QCB system, also supplies a simple and inexpensive kit allowing any M2HB machine gun to be converted into the QCB version. This kit comprises:
a barrel*
a barrel extension
a set of breech locks
a set of shims for the barrel support sleeve
and an accelerator.

*The kit will not necessarily include the supply of a new QCB barrel, because it is possible to factory-modify the M2HB barrel to make it compatible with the QCB system.

A new accelerator has been developed to improve the performance of this machine gun.

With the exception of the parts specific to the QCB kit, all other parts are interchangeable with the M2HB model.

Combined recoil booster
This accessory enables the .50 machine gun to be fired in the automatic mode with blank cartridges having brass or plastic cases (star or American crimping) by effectively reproducing live round operating conditions.

It also allows the use of plastic bullet (PT) training ammunition by removal of the barrel muzzle plug.

Manufacturer
FN Herstal SA, B-4400, Herstal.
Status
Current. Production.

FN .50 M2HB/QCB machine gun

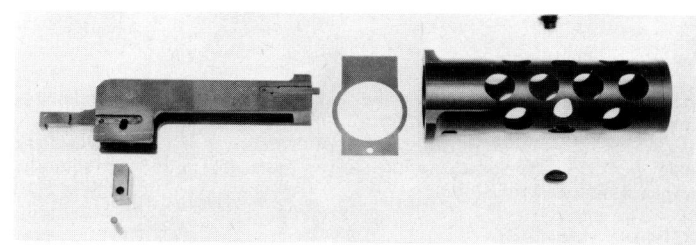

Conversion kit for M2HB to QCB machine gun

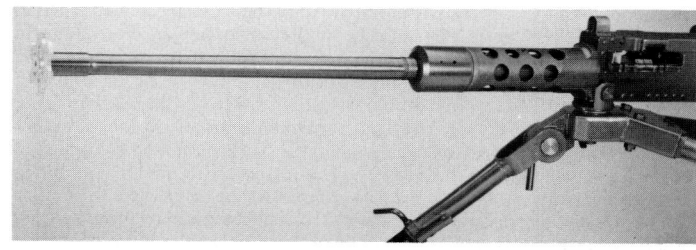

Combined recoil booster in place; with muzzle plug removed, plastic ammunition can be fired through barrel

7.62 mm FN MAG general-purpose machine gun

The FN MAG (Mitrailleuse à Gaz) is gas-operated, belt-fed and has a quick-change barrel. It is light enough to be carried by the infantrymen in the section and is capable of producing sustained fire over considerable periods when mounted on a tripod. It fires the standard NATO 7.62 mm cartridge from a disintegrating link belt of the US M13 type; alternatively the 50-round continuous articulated belt can be used but the two types of belt are not interchangeable. It has been made in other calibres, notably for Sweden in a calibre of 6.5 mm as the M58 general-purpose machine gun.

Receiver
The receiver is made of pressings riveted together to make a rectangular section of considerable strength. It is reinforced at the front to take the barrel nut and gas cylinder and at the rear end for the butt and buffer.

Along the inside of the receiver are ribs which support and guide the breech-block and piston extension in their reciprocating movement. The breech-guides are shaped to force down the locking lever when the breech-block is fully forward, and in the floor of the receiver is a substantial

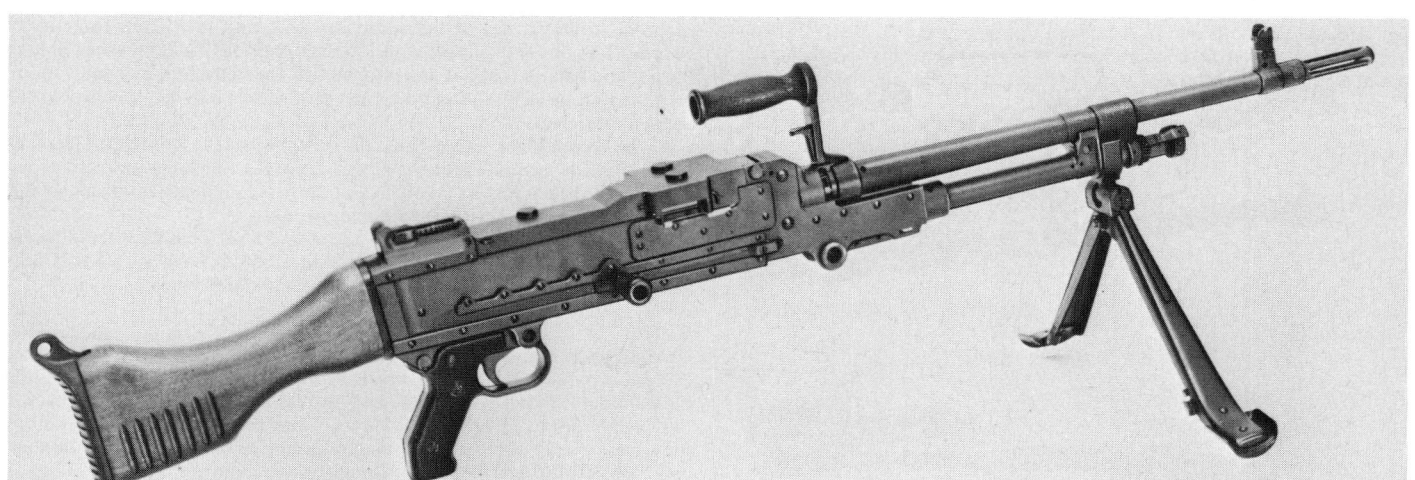

7.62 mm FN MAG general-purpose machine gun

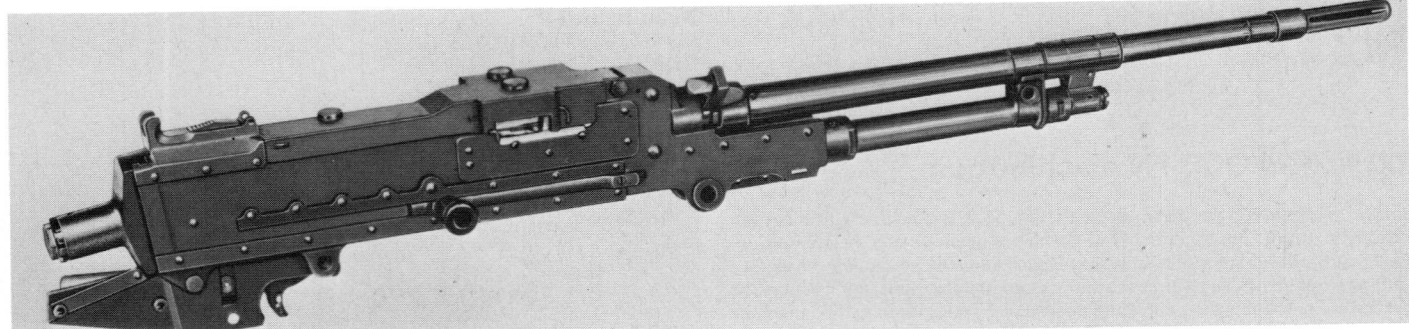

Coaxial version of MAG for mounting in armoured vehicles

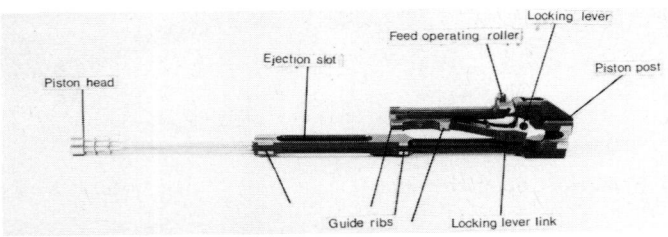

Piston and bolt assembly

locking shoulder against which the locking lever rests when locking is completed.

There is a cut-out section on the right of the receiver in which the cocking slide operates and the bottom of the receiver has a slot for the ejection of the empty case.

Barrel

The barrel is threaded externally at the rear end to fit into the barrel locking nut. This has an external interrupted buttress thread to engage into the receiver and is prevented from rotating by the barrel locking catch attached to the left side of the receiver. The carrying handle sleeve is engaged on flanges in the barrel locking nut and rotating the carrying handle locks the nut into the receiver. The handle is then clear of the line of sight.

A gas vent is drilled through the barrel leading down into the regulator. The gas comes into the regulator, which has a surrounding sleeve inside, which is a gas plug with three gas escape holes. When the gun is clean and cold most of the gas passes out through these three holes and only the minimum required to operate the gun passes back to the piston-head. As the need arises to increase the gas pressure to overcome the frictional resistance caused by the expansion of heated components, gas fouling or the ingress of sand and so on, the gas regulator knob is rotated, the gas regulator sleeve slides along the gas block and the three holes are progressively closed until eventually all the gas is diverted to the piston-head. This same arrangement can be used to vary the rate of fire within the limits of 600 to 1000 rds/min.

At the muzzle is the foresight block and the screw-threaded flash eliminator of the closed-prong type. The bore is chromium plated.

Barrel change

The gun stops firing in the open breech position, when the trigger is released, while ammunition remains in the belt. The breech is closed on an empty chamber when the last round is fired.

The safety catch is pressed through to the right to the 'safe' position and to do this the breech-block must be to the rear; so if the gun is empty the action must be cocked.

The barrel can be changed without unloading. The actuating catch of the handle is engaged in the recess of the barrel locking nut, the head of the barrel locking catch, on the left of the receiver, is pressed in, the carrying handle is rotated to the vertical position and pushed forward. The barrel is then moved forward and lifted off.

To replace the barrel the carrying handle is held vertically up, the barrel is rested on the V formed by the forepart of the gas cylinder and the barrel is slid

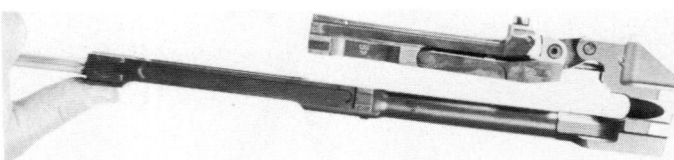

Bolt unlocked

Bolt locked. Locking lever is down and supported by locking face

back. The gas regulator goes into the gas cylinder and the interrupted thread of the barrel enters the barrel nut.

The carrying handle is rotated to the right to engage the threads of barrel and nut. The barrel is then secure.

Piston and breech-block

The piston is attached to the piston extension which has a cut-out section at the front to allow passage of the ejecting case from above. At the rear is a massive piston post, attached to which is the locking lever link, which in turn is connected to the locking lever. Finally the locking lever is pinned half way along the breech-block. When the block is travelling forward the locking lever lies in recesses in both sides of the block. The block is hollowed to take the firing pin which is mounted in the piston post. On the front face is the extractor. The return spring fits into the hollow interior of the piston extension.

Trigger mechanism

The trigger is centrally pivoted at its top edge so when it is pulled the rear end rises. Resting on the rear of the trigger is a long, centrally pivoted sear. When the trigger is pulled the front end of the sear, the tail, is forced up and the rear end, the nose, drops down out of engagement with a bent on the bottom of the piston extension. Also on the front top edge of the trigger is a pivoting tripping lever which projects up into the piston way. This tripping lever has a spring which pivots it forward. On its front face is a step.

When the trigger is pulled the central pivot causes the front, carrying the tripper, to go down and the back to rise, lifting the tail of the sear. As the tail of the sear rises, the spring-loaded tripping lever moves forward underneath it and holds it up.

When the trigger is released the tripping lever rises with it and holds up the tail of the sear so the nose cannot rise to engage the piston and hold it back. The tripping lever, as it rises, re-enters the bolt way. As the piston comes back it hits the top of the tripping lever and rotates it back against the spring. The tail of the sear flies down and its nose rises. The piston going back passes over the sear but when it comes forward again it is caught by the full width of the sear and the area of contact is large enough to prevent any chipping or bending.

The safety catch is a push-through plunger. When at 'safe' it rests under the sear nose and prevents it falling. At 'fire' a cut-out section allows full sear movement.

Feed mechanism

This is a two-stage system with the belt moving half way across on the forward motion of the bolt and moving the other half pitch during bolt recoil. The top of the breech-block carries a spring-loaded roller which engages in a curved feed channel in the feed cover over the receiver. This channel is pivoted near its rear end and at its front end is attached to the end of the short feed link. This link swings about its centre so that as one end goes in towards the centre line of the gun the other end moves out. It carries an inner feed pawl at one end and two outer feed pawls at the other. Thus as the breech-block travels forward the roller will first travel down a straight section of the feed channel, while the cartridge is forced out of the stationary belt, and then enter the curved portion. This forces the feed channel to the right. This swings the feed link to the left and rotates it about its centre so that the inner feed pawl moves out over the waiting cartridge and the outer pawls force it in half the distance to the gun centre line.

As the breech-block moves back, the roller swings the front of the channel to the left. The feed link swings to the right so that the inner pawl comes in, bringing the belt across half a pitch, and the first round comes up against the cartridge stop and is positioned for chambering.

The outer pawls move out over the next cartridge. Thus each set of pawls acts alternately as feed and stop pawls and the cartridge moves half way across for each forward and backward movement of the breech-block.

Buffer

The buffer assembly consists of a bush which receives the impact of the piston extension and moves back into a cone which it expands outwards. This grips the walls of the buffer cylinder and also moves back slightly. In moving back it flattens a series of 11 Belleville washers. These saucer-shaped washers store the kinetic energy of the piston and in returning to shape they drive the cone and bush forward and force the piston forward again.

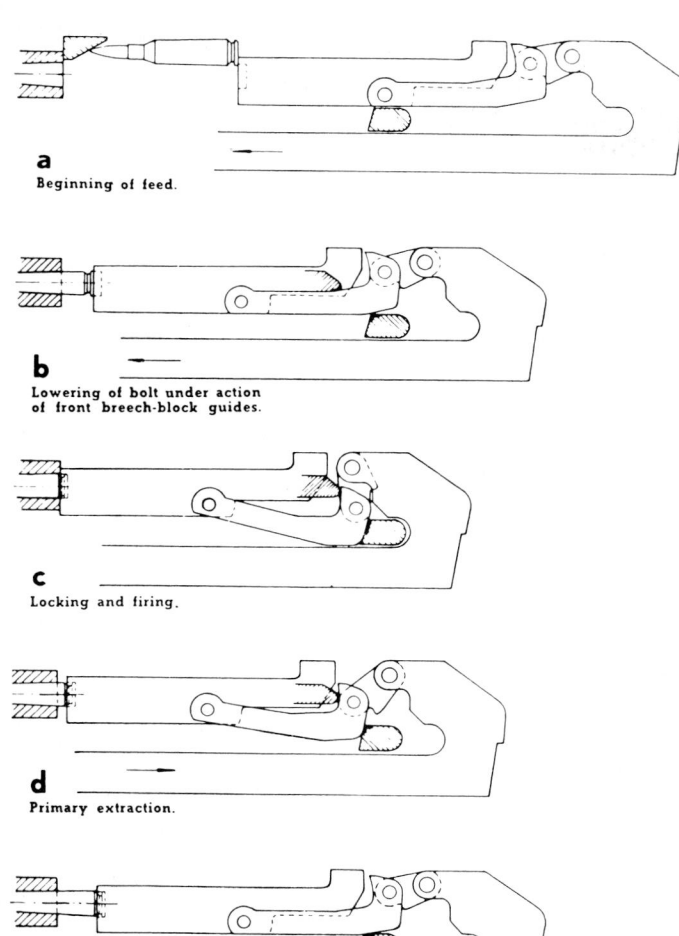

a
Beginning of feed.

b
Lowering of bolt under action
of front breech-block guides.

c
Locking and firing.

d
Primary extraction.

e
Unlocking.

Operating cycle

Sights
The foresight is a blade mounted on a screw-threaded base which fits into a block mounted on a transverse dovetail.

The rearsight is a leaf which can be used folded down for ranges marked from 200 to 800 m by 100 m increments. A slide with two spring catches allows the aperture rearsight to be set to the correct range.

When the leaf is raised ranges are marked from 800 to 1800 m at intervals of 100 m. There is a V rearsight on the slider.

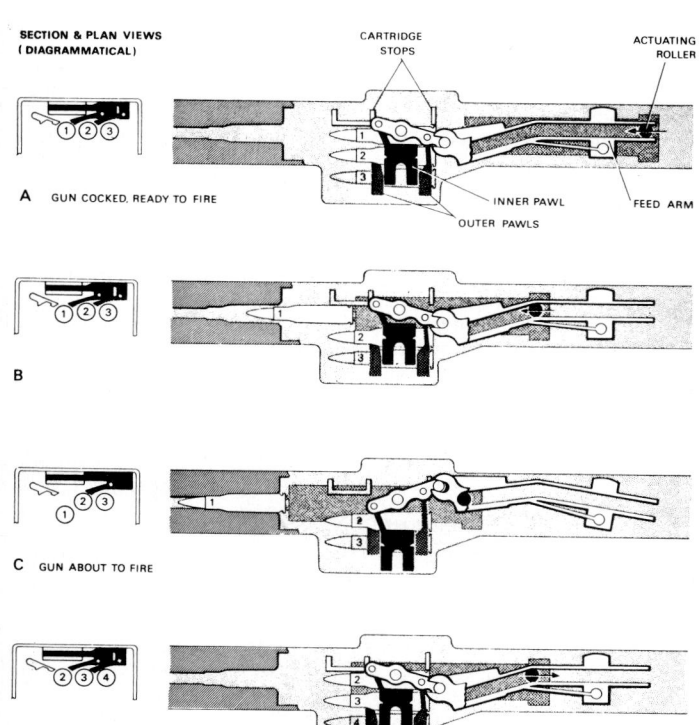

Feed mechanism of FN MAG general-purpose machine gun viewed from above

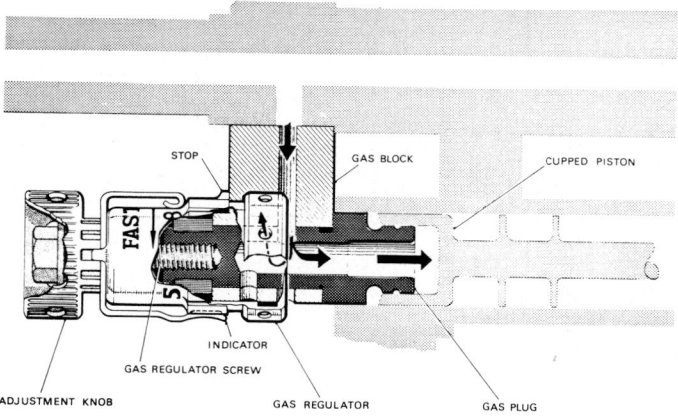

Gas regulator exhaust-to-atmosphere system of FN MAG (UK Model)

Elevation zeroing is carried out on the foresight by lifting the securing stirrup, rotating through multiples of 180° movement and then replacing the stirrup. A tool is needed.

Lateral zeroing is achieved by moving the foresight bodily in the dovetail. The securing screw is loosened on one side and tightened on the other.

Operation
Disintegrating belts come factory-packed and are not intended to be re-filled in the field. The continuous 50-round link belt can be re-filled by placing cartridges in the open pocket of the belt with the nib pressing into the extraction groove. Lengths of belt, of either type, can be joined to increase the ammunition available without stopping firing to change belts. The belts may be held in a 50-round belt box attached to the left of the gun or kept in a 250-round box placed beside the gun on the left.

The gun is cocked by pulling the cocking handle fully to the rear and then returning it to the forward position. When the gun is cocked, and only then, the safety catch can be set to 'safe' by pushing it from left to right through the gun body. The letter 'S' can then be read on the right side of the plunger facing the firer.

The top cover is opened by squeezing in the two catches at the back and lifting the cover to the vertical position. The loaded belt is then inserted, open side down, across the feed tray so that the leading cartridge rests against the cartridge stop on the right. The top cover is lowered. The safety is pushed through from right to left. The letter 'F' then shows, facing the firer, on the part of the catch projecting from the left side of the gun. When the trigger is pressed the nose of the sear drops and the piston extension is forced forwards by the compressed return spring, carrying the bolt with it. The feed horn on the top edge of the breech-block pushes the first round in the belt straight through the belt link and the bullet is directed down into the chamber by the bullet guide. The top surfaces of the locking lever are forced down when they hit the underside of the top breech-block guide. The cartridge is fully chambered, the extractor slips into the groove and the base of the cartridge seats into the recessed face of the breech-block. The ejector is forced back and the ejection spring is compressed. The breech-block stops.

The piston extension continues forward. The locking lever continues to move down and drops in front of the locking shoulder in the bottom of the receiver. Further forward movement of the piston extension causes the locking lever link to rotate forward past the vertical position. The final forward movement of the piston post pushes the long firing pin through the front face of the breech-block and the cartridge is fired. The forward motion of the piston extension ends when its shoulder hits the stop face of the gas cylinder. The gas pressure forces the bullet up the bore and some of the gas flows through the vent into the gas regulator and then to the piston-head which is driven back.

The piston post withdraws the firing pin into the breech-block. The locking lever link is rotated back to the vertical position. The continued movement of the piston post rotates the link further and this lifts the locking lever out of contact with the locking shoulder. The locking lever then pulls the breech-block rearward, the empty case is withdrawn and as soon as it is clear of the chamber the compressed ejection spring pushes forward the ejector which rotates the case down, around the extractor, through the slot in the piston extension and out through the bottom of the gun. The rearward motion of the piston extension compresses the return spring. The piston extension hits the buffer and is thrown forward. If the trigger is still pressed the cycle starts again.

Mounts
The bipod is mounted on the forward end of the gas cylinder and can rotate from side-to-side to allow firing across a slope, one leg higher than the other with the sights vertical. The legs are not adjustable for height.

The bipod can be folded back for carrying. A hook on each leg engages in a slot in the side of the receiver and is held securely in place by a sliding retainer catch.

The tripod mount is a spring-buffered assembly to which the gun is attached by two pins fitting into the rings on the underside of the receiver, and a push-through pin which enters a hole above the trigger-guard.

Stripping

Lift the top cover plate and remove the belt.

Cock the gun and check the chamber is clear. Return the working parts under control.

Remove the butt by depressing the butt catch and lifting the butt upwards.

Remove the return spring by pushing the rear end of the return spring rod forward and then lifting it. It is then pulled out of the rear end of the receiver.

Remove the piston, piston extension and breech-block by retracting the cocking handle and withdrawing these parts.

Remove the barrel.

Re-assemble in reverse order.

DATA

Cartridge: 7.62 mm × 51 NATO
Operation: gas, automatic
Method of locking: dropping locking lever
Feed: belt
Weight: 10.85 kg with butt & bipod
Length: 1255 mm
Barrel: 545 mm
Rifling: 4 grooves, rh, 1 turn in 305 mm

Sights: (fore) blade; (rear) aperture when leaf is lowered, U-notch when leaf is raised
Sight radius: sight folded down 848 mm, sight raised 785 mm
Muzzle velocity: 840 m/s
Rate of fire: (cyclic) 650–1000 rds/min
Effective range: 1200 m

ACCESSORIES
Combination tool; extractor remover; sight key; bore cleaning brush; chamber cleaning brush; gas regulator tool; set of drifts.

Manufacturer
FN Herstal SA, B-4400 Herstal.
Status
Current. Production in Belgium, India, Israel, Sweden, UK and USA.
Service
More than 150 000 guns have been bought by more than 75 countries. Known to be in service with the armed forces of Argentina, Belgium, Canada, Cuba, Ecuador, India, Indonesia, Israel, Kuwait, Libya, Malaysia, Netherlands, New Zealand, Peru, Qatar, Sierra Leone, Singapore, South Africa, Sweden, Tanzania, Uganda, UK, USA, Venezuela, Zimbabwe and others.

FN LGM 04-41 light ground mount

This is a light, fully adjustable tripod for use with the FN MAG general-purpose machine gun and FN Minimi machine gun in both ground and air defence roles. It consists of three basic units: a tripod head provided

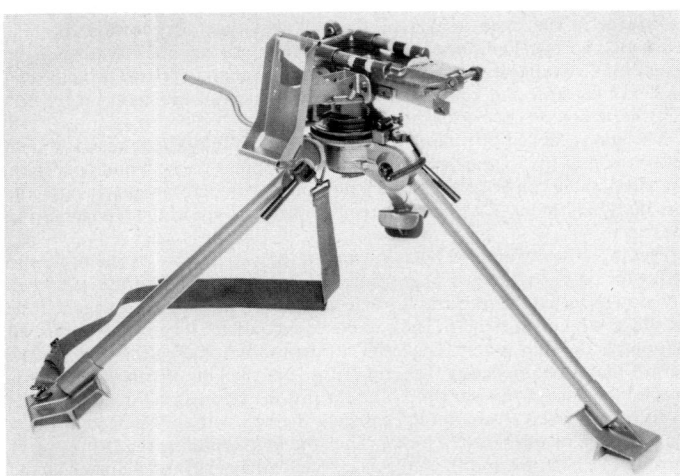

FN LGM light tripod

with three adjustable legs disposed at 120° intervals; an elastic cradle assembly, on to which fits the weapon and the ammunition box tray; and the ammunition box tray. There is also an accessory telescopic column for anti-aircraft firing which can be fitted to the tripod head in place of the elastic cradle. The top of this tube is adapted to receive the elastic cradle and the tube is telescopic to give a wide range of heights. There is also a sling for carrying the folded tripod. The tripod legs have a rack-gearing system of adjustment, and the most usual positions of the legs are indicated as '0-1-2-3' for quick positioning. Each leg is fitted with a ground anchoring shoe which is efficient at any angle of elevation of the leg. The azimuth angle can be restricted, for safety or other requirements, by adjustable clamps.

DATA

Weight: 11 kg
Dimensions, folded: 600 × 250 × 250 mm
Elevation (ground firing): 30°
Elevation (air firing): 80°
Depression (ground firing): 30°
Depression (air firing): 20°
Weight of AA column: 3.35 kg
Length of AA column: (max)1.315 m; (min) 795 mm

Manufacturer
FN Herstal SA, B-4400 Herstal.
Status
Current. Production.

5.56 mm NATO Minimi light machine gun

The Minimi is gas-operated, using gas tapped from the forward part of the barrel in conventional fashion. The rotary gas regulator is of a simple design, based on the earlier MAG general-purpose machine gun type, and having two basic settings (normal and adverse conditions). Adjustment is by hand, even with a hot barrel.

The breech-locking mechanism is an FN design, where the bolt is locked into the barrel extension by a rotational action. This action is initiated by a cam in the bolt carrier.

Normal gas operation, in which the gas piston is forced to the rear, moves the bolt carrier back, leaving the bolt still locked to the barrel extension. The residual chamber pressure has become virtually zero by the time the cam action referred to earlier unlocks the bolt. Primary extraction of the spent case begins with the rotation of the bolt before it unlocks. The extra care taken in this process has eliminated one of the common troubles with many other 5.56 mm weapons, the jamming of swelled or ruptured cases in the body.

Another feature of the Minimi is the fact that the bolt and carrier are controlled throughout their travel by two rails welded inside the body and

these rails make a significant contribution to the smoothness of the operation of the gun and its remarkable freedom from stoppages and breakage of the working parts.

In the ammunition feed system on the Minimi the disintegrating link belt is held in a box magazine of plastic material, which, apart from acting as an ammunition carrier when not on the gun, locks firmly to the gun and becomes virtually integral with it when in action. There is one type of magazine, of 200-round capacity.

The gun is most unusual in that it can accept either a magazine or a belt feed without any modification. The accompanying photograph shows the magazine projecting down from the left side of the body.

The gun is normally bipod-mounted but can, if required, be mounted on a tripod. It can also be used with either a fixed or a sliding stock. In addition, for use as a port-fire weapon in a mechanised infantry combat vehicle, the Minimi can be used with a backplate only.

Belgian 5.56 mm Minimi light machine gun, M249 model

Belgian 5.56 mm Minimi light machine gun showing belt feed

5.56 mm Minimi standard on FN 360° tripod and M249 model on US M122 tripod

For those users who need a gun which is lighter and shorter than the standard version there is a Para model with a sliding butt and a shorter barrel. The chief advantage of this version is that it is much easier to take in and out of vehicles, helicopters, and similar confined spaces.

DATA
Cartridge: 5.56 mm × 45 (FN SS 109 NATO or M193)
Operation: gas, firing fully automatic

5.56 mm Minimi, vehicle model

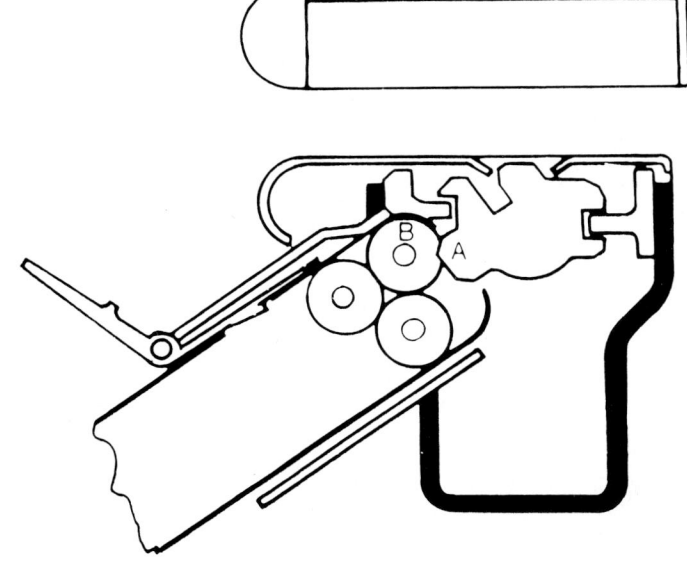

5.56 mm Minimi Para model

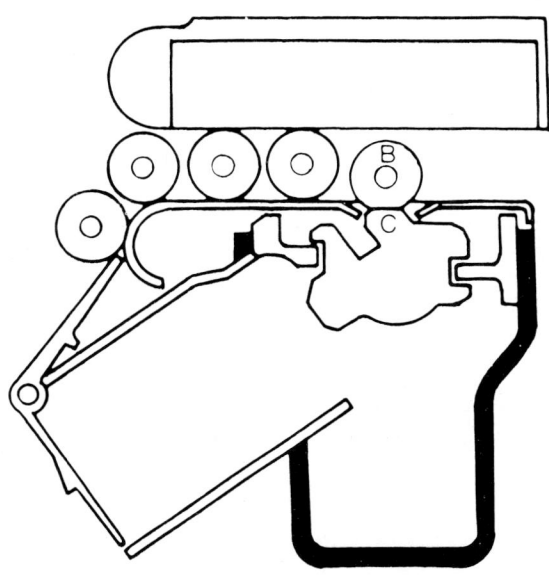

Alternative feed mechanism
(left) *magazine feed* **(right)** *belt feed showing bolt has two sets of feed horns for stripping rounds from two feed presentations*

5.56 mm Minimi, showing magazine feed

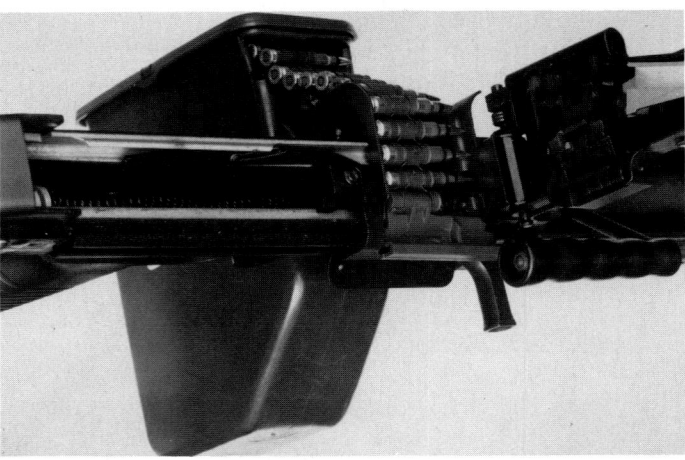

Plastic box magazine for disintegrating link belt, locked to Minimi

Method of locking: rotating bolt-head
Feed: 200-round belts or 30-round magazine (M16A1)
Weight: 6.850 kg (standard); 7.00 kg (Para)
Length: 1040 mm (standard); 910 mm (Para, stock extended); 755 mm (Para, stock folded)
Barrel: 465 mm (standard); 347 mm (Para)
Rifling: 6 grooves rh. 1 turn in 304 mm (M193) or 1 turn in 178 mm (SS 109)
Sights: (fore) semi-fixed hooded post, adjustable for windage and elevation; (rear) aperture, adjustable for elevation and windage

Muzzle velocity: M193 965 m/s, SS 109 915 m/s
Rate of fire: (cyclic) 700-1000 rds/min

Manufacturer
FN Herstal SA, B-4400 Herstal.
Status
Current. Production.
Service
Adopted by Australia, Belgium, Canada, Indonesia, Italy and USA. Military trials in New Zealand, Sweden, Tunisia and other countries.

BRAZIL

7.62 mm Uirapuru Mekanika general-purpose machine gun

This weapon began as a research project at the Brazilian Army Instituto Militar de Engenharia (IME) in Rio de Janeiro in November 1969. A design group consisting of Colonel Alcides Nasario Geurreiro Brito, Olympio Vieira de Mello and Nelmo Suzano built three prototypes of a gas-operated machine gun firing 7.62 mm × 51 ammunition, which received the name Maria Bonita. It worked satisfactorily but the lack of any previous design experience in this field by the Brazilian technicians led to a number of inherent malfunctions. It became clear that only a careful research programme with no strict deadlines would produce a reliable gun of 100 per cent indigenous design and, as other more urgent programmes were under way at the IME, the general-purpose machine gun project was handed over to a private concern in 1972 but no additional progress was made.

As a final resort, the Army placed the gun in the hands of Olympio Vieira de

Mello, one of the three members of the original group, who set out to solve the problems on his own. In 1976 he produced a substantially modified weapon which worked very well indeed. Demonstrated to the Army, it showed good potential for large-scale production and adoption, and the interest in the gun shifted from the study-orientated IME to the Instituto de Pesquisa e Desenvolvimento (IPD) (Research and Development Institute).

In December 1977 the IPD granted Mekanika Indústria e Comércio Ltda a contract to build two testing and evaluation prototypes of what had been called the Uirapuru general-purpose machine gun, named after a Brazilian jungle bird. The gun was fully tested and approved by the Brazilian Army Marambaia Proving Grounds in Rio de Janeiro and manufacture is in progress for the Brazilian armed forces and for export.

Firing standard 7.62 × 51 mm ammunition, the Uirapuru is a true

7.62 mm Uirapuru GPMG on bipod

Prototype stripped. Coil return spring lies above barrel, on which muzzle brake/flash hider is well shown. Barrel is unusual for gas-operated gun in that it has no gas block at point of gas take-off

Uirapuru GPMG on tripod mount

general-purpose machine gun, fulfilling a variety of roles, for example light machine gun (with butt and bipod), medium machine gun (without butt, with tripod), tank coaxial machine gun and aircraft machine gun (fitted with a solenoid), armoured vehicle machine gun (firing from pedestal), and so on. It is a belt-fed, gas-operated automatic weapon which fires from an open bolt. The carrying handle also acts as a barrel-changing handle, being integral with the barrel. Changing the barrel takes a matter of seconds, being recommended after the continuous firing of about 400 rounds. The barrel has a flash-eliminator/muzzle-brake at the forward end. The two-position fire-selector (safe/automatic) is above the plastic pistol grip.

DATA
Cartridge: 7.62 mm × 51
Operation: gas, automatic
Method of locking: dropping locking lever
Feed: belt (NATO standard)
Weight: 13.0 kg
Length: 1300 mm
Barrel: 600 mm
Rifling: 6 grooves, rh, 1 turn in 305 mm
Sights: (fore) adjustable for windage/elevation; (rear) adjustable from 200 to 600 m, and from 800 to 1400 m
Sight radius: 320 mm
Muzzle velocity: 850 m/s
Rate of fire: (cyclic) 650-700 rds/min

7.62 mm Uirapuru GPMG with feed cover open

Manufacturer
Mekanika Indústria e Comeŕcio Ltda, Rue Belisário Pena, 200-Penha-21020 Rio de Janeiro, RJ.
Status
Current. Production.

CANADA

5.56 mm C7 Light Support Weapon

The C7 light support weapon is a co-development between Diemaco and Colt Firearms (where the weapons is known as the M16 light machine gun). The C7 LSW features an open bolt firing mechanism, an hydraulic recoil buffer and an adjustable bipod. The weapon also features a heavy contour Diemaco hammer-forged barrel. The superior life and durability of the barrel in independent testing has eliminated the need for a quick-change barrel system, greatly simplifying the design and making the LSW accurate to the maximum effective range of the ammunition.

The C7 LSW is fed from any M16 compatible box magazine, including high capacity types, and it uses the same family of accessories as the C7 rifle.

The C7 LSW uses the M16A2 type rear sight as standard but is also available mounting an optical sight as the C7A1 LSW.

DATA
AS FOR C7 RIFLE EXCEPT:-
Weight: (empty) 5.8 kg; (loaded) 6.2 kg
Sights: (fore) post; (rear) adjustable aperture, 300 to 800 m
Rate of fire: 625 rds/min

Manufacturer
Diemaco Inc., 1036 Wilson Avenue, Kitchener, Ontario N2C 1J3.
Status
Current. Production.

5.56 mm C7 Light Support Weapon and accessories

Service
Under evaluation in several countries.

CHINA, PEOPLE'S REPUBLIC

Chinese machine guns

The armed forces of China have very largely been equipped with copies of weapons produced originally elsewhere. The weapons they hold are described individually in the sections devoted to the country of origin. They are:

Chinese Name	*Origin*
Type 53	Soviet 7.62 mm DPM light machine gun
Type 54	Soviet 12.7 mm Model 38/46 heavy machine gun
Type 56	Soviet 7.62 mm RPD light machine gun
Type 57	Soviet 7.62 mm SG43 medium machine gun
Type 58	Soviet 7.62 mm RP-46 Coy machine gun
Type 63	Soviet 7.62 mm SGM medium machine gun

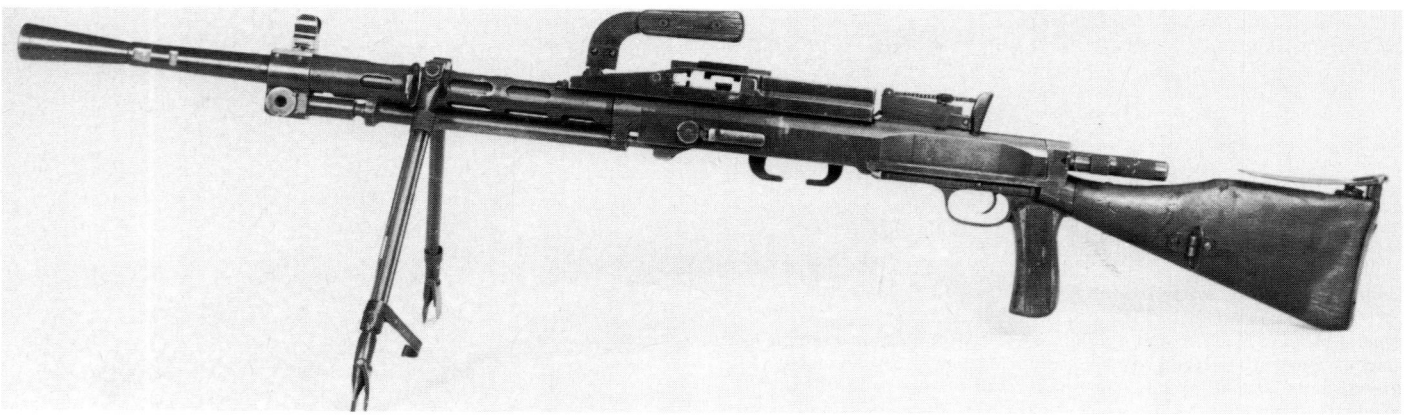

Chinese Type 58 machine gun

Chinese Type 53 light machine gun corresponding to Soviet DPM

Status

Types 53, 54, 56 and 58, at least, are believed to be manufactured in China. All appear to remain in service although some are probably used only by reserve units or the People's Militia.

Training with Chinese Type 56 light machine gun (Soviet 7.62 mm RPD)

7.62 mm Type 67 light machine gun

The type 67 is an indigenous Chinese design and has replaced the Types 53 and 58 in front-line units. It has been in production since the early 1970s and some of the early issues were given to North Vietnam. The gun is a sound design and bears all the hallmarks of being strong and reliable though perhaps a little heavy. It can be expected to remain in service for a long time.

The gun is gas-operated and belt-fed and may be used with either a bipod or a tripod. Its design is a mixture of features from the other guns in the previous entry and the part each has played in its design is as follows:

Type 24 (Maxim)	Feed mechanism
Type 26 (Zb26)	Bolt and piston
Type 53 (DPM)	Trigger mechanism
Type 56 (RPD)	Gas regulator
Type 57 (SG43)	Barrel-change system

The Type 67 uses the 7.62 mm × 54R cartridge, and has an open pocket metal belt with the pockets joined by spring metal coils. Each pocket carries a nib, extending to the rear, with a bent-over tab. The round is pressed into the link with the tab against the rear face of the cartridge. This type of belt allows the bolt to push the round out of the link and straight into the chamber.

Operation

To load the belt the feed cover catch at the rear of the receiver is pressed and the feed cover lifted about its front hinge. The cocking handle on the right of the receiver is normally folded vertically downwards and must be lifted through 45° before it can be pulled back to cock the action. Before cocking it is essential to check that the safety catch is back to 'fire'. If forward to 'safe' the sear will jam on the underside of the piston. The belt is placed in the feedway with the open side of the link facing down and the first round in the feedway slot. The cover is then closed. (Note that if the belt is inverted an immediate stoppage results.)

The rearsight, of similar shape to that of the Czechoslovak Model 59, has an elevation knob on the left and a windage knob on the right. There is a clicker system with each click on the elevation knob corresponding to 25 m up to 1000 m range and 20 m thereafter. The windage knob produces one click for each minute of angle, for instance 1 in (2.54 cm) at 100 yds (91.4 m).

The gun will fire always from the open bolt position and when the ammunition is expended the bolt closes on an empty chamber.

When the bolt goes forward the cartridge is pushed into the chamber. Bolt movement ceases but the piston continues forward and the rear of the bolt is lifted up to lock into the ceiling of the receiver. The piston continues forward after locking is completed and the flat face of the piston post drives the firing pin into the cartridge cap.

After firing, the bullet travels up the bore and some of the gas following it is diverted into the gas cylinder, below the barrel, where it drives the piston rearward. There is a period of free travel allowing the gas pressure to drop, before the ramp on the rear face of the piston post pulls the rear end of the bolt down out of its recess in the top of the receiver. The piston then carries the bolt to the rear; the empty case is extracted and then ejected downwards out of the gun. The return spring is compressed along its guide rod. When the piston comes to rest the return spring drives it forward and the cycle is repeated.

The belt is fed into the gun from the right and the feed mechanism is operated by having a cam track on the top side of the piston extension at the front, in a manner reminiscent of the Czechoslovak Model 52. This forces a roller on the bottom of the lower feed arm to move out and the rotation of the lower feed arm is transmitted by a vertical shaft to the upper feed arm. As the upper feed arm moves, a slot in it engages a roller on the feed slide and the feed slide moves outward to engage a round in the belt.

As the piston extension moves forward the bolt forces a cartridge through the belt and into the chamber. As the piston extension continues forward the lower feed arm is moved back to its original position and this in turn pulls in the upper feed arm and the next cartridge is pulled across from right to left by the feed slide and moved into the slot in the feed tray where it is pressed down by a pair of cartridge guides in the cover. A spring-loaded stop pawl on the feed tray prevents the belt from slipping out as the feed slide oscillates.

The barrel will normally be changed after two minutes of firing at the rapid rate. The procedure is to lift the top cover plate, remove the belt if any cartridges remain in it, and then press the barrel retaining catch to the left, as in the Soviet SGM. Grasp the carrying handle and push the barrel forward off the gun. Place the new barrel in position, with the gas cylinder in the gas

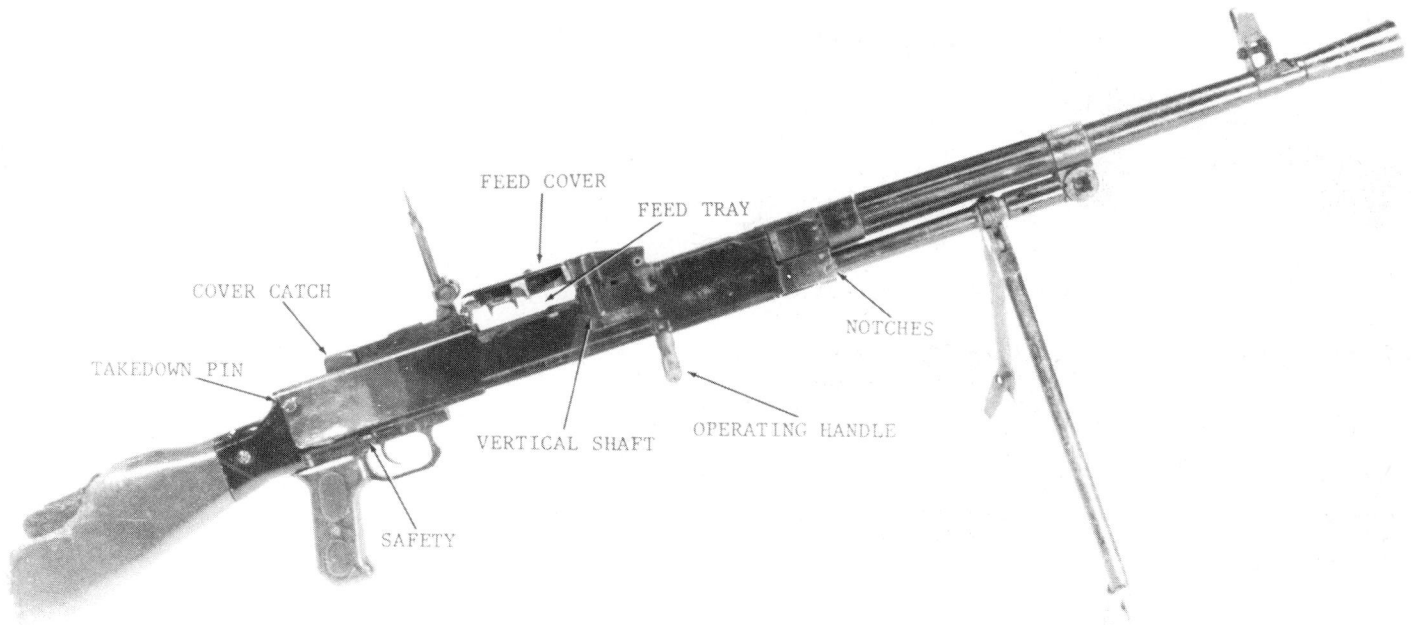

Chinese 7.62 mm Type 67 light machine gun

cylinder tube, and pull the barrel back. Push the retaining catch across to the right. Put in a new belt. Close the top cover plate and resume firing.

The same micrometer adjustment for cartridge head space is used in this gun as in the SGM.

The gas regulator works in the same way as that in the RPD. The nut on the left side of the regulator is loosened and then the regulator is pushed through to the right. This disengages it from the index pin and the regulator can then be rotated to the selected position. There are three settings, 1, 2 and 3, of which the first is usually used. The regulator is then pushed back and the nut tightened.

Stripping
Lift the top cover plate, remove the belt, check that the chamber and feedway are clear. Remove the barrel.
Press the take-down pin, at the rear of the receiver, to the right. Pull the butt to the rear. Remove the return spring and guide.
Pull the cocking handle back and remove the piston and bolt.
Re-assembly is in reverse order.

Anti-aircraft mounting
The anti-aircraft sights consist of a pillar permanently attached to the top of the stirrup-like rearsight and a speed ring foresight, which fits into dovetails in the top of the receiver and is held by a spring catch. The tripod is up-ended and

the gun attached by fitting the notches on the underside of the front of the receiver on to the pins in the mount and rotating the gun down until the catch locks on the front of the trigger-guard.

DATA
Cartridge: 7.62 mm × 54R
Operation: gas, automatic
Method of locking: tilting block
Feed: 100-round continuous open pocket metal belt
Weight: 9.9 kg
Length: 1143 mm
Barrel: 597 mm
Sights: (fore) pillar; (rear) leaf notch, adjustable for windage
Muzzle velocity: 835 m/s
Rate of fire: (cyclic) 650 rds/min
Effective range: 800 m

Manufacturer
State factories.
Status
Current. Production.
Service
China and Vietnam.

7.62 mm Type 74 light machine gun

This is a squad light machine gun, gas-operated and fed from a drum magazine; it is also possible to use the Type 56 rifle magazine in place of the drum. Details of the mechanism are not known, but there is a gas cylinder above the barrel and a cocking handle on the right side. It appears to be an entirely new design and not a made-over Kalashnikov.

DATA
Cartridge: 7.62 × 39 mm
Operation: gas
Feed: 101-round drum magazine
Weight: 6.20 kg
Length: 1070 mm
Muzzle velocity: 735 m/s
Rate of fire: 150 rds/min, effective
Effective range: 600 m

Manufacturer
China North Industries Corp, PO Box 2137, Beijing.
Status
Current. Production.

7.62 mm Type 74 light machine gun

7.62 mm Type 81 light machine gun

This, like the Type 74 above, is intended as a squad automatic, though it seems likely that it has been produced principally for export. It is gas-operated and appears to use a mechanism similar to that of the Type 68 rifle, a rotating bolt moving in a bolt carrier. Feed is normally from a drum magazine, though not

the same pattern as that used with the Type 74 machine gun, and the magazine of the Type 81 automatic rifle can be used interchangeably. In addition, most of the working parts of the Type 81 machine gun and rifle are interchangeable.

DATA
Cartridge: 7.62 × 39 mm
Operation: gas, automatic
Locking: rotating bolt
Feed: 75-round drum magazine
Weight: 5.3 kg
Length: 1024 mm
Rifling: 4 grooves, rh, one turn in 240 mm
Muzzle velocity: 735 m/s
Effective range: 600 m

Manufacturer
China North Industries Corp, PO Box 2137, Beijing.
Status
Current. Production.

7.62 mm Type 81 light machine gun

12.7mm anti-aircraft machine gun Type 77

The Type 77 is an automatic weapon of new design which is primarily intended for use for air defence, though it can also be used against ground targets. Full details have not been disclosed, but it appears to be recoil operated and is of light construction for this calibre. Feed is from a belt, carried in a belt box on the left-hand side of the gun, and a special optical sight is provided for air defence firing. The tripod can be set at various heights and is provided with geared traverse and elevation controls which can presumably be unlocked to pmerit free movement against aerial targets.

DATA
Cartridge: 12.7 × 107mm Type 54
Feed: 60-round belt
Weight, with tripod: 56.1 kg
Rate of fire: 650 – 750 rds/min cyclic
Muzzle velocity: 800 m/s
Effective range: (air) 1600 m; (ground) 1500 m

Manufacturer
China North Industries Corp., PO Box 2137, Beijing.
Status
Current. Production.

12.7mm Type 77 machine gun on low tripod, with air defence sight

12.7mm anti-aircraft machine gun Type W-85

This is a new design of battalion automatic gun principally intended for use against aerial targets but which can also be used as a heavy support machine gun against ground targets. The Type W-85 gun is gas-operated and belt-fed, and is exceptionally light, some 58 per cent lighter than the obsolescent Type 54, which is the Chinese version of the Soviet DShK. It is provided with an

adjustable tripod, for ground or aerial firing, and with a telescope sight. Ammunition provided for this weapon covers the usual range of AP, AP-T and AP-I with the addition of a bullet described as 'tungsten alloy cored' which may possibly be an APDS pattern.

DATA
Cartridge: 12.7 × 107 mm Type 54
Feed: 60-round link belt
Length: 1.995 m
Weight: 18.5 kg (gun); 15.5 kg (tripod)
Rate of fire: 80 – 100 rds/min practical
Muzzle velocity: 800 m/s (AP,AP-T,AP-I); 1150 m/s (tungsten alloy cored)
Effective range: (air) 1600 m; (ground) 1500 m

Manufacturer
China North Industries Corp, PO Box 2137, Beijing.
Status
Current. Production.

12.7 mm machine gun Type W-85 on low tripod

14.5 mm anti-aircraft machine gun Type 75-1

This is a Chinese version of the Soviet KPV machine gun, and differs from the original in certain details such as the arrangement of the belt feed and the provision of cooling fins on the barrel. It is mounted on a tripod with two small wheels, the tripod folding to become a lightweight trailer. Mechanical elevation and traversing gears are fitted, and there is a layer's seat and an optical course and speed sight mounted on a parallelogram arm so as to place the sight line in a convenient position. It will be noted that the picture shows an image-intensifying sight fitted; it is thought that this would have negligible value as an air defence sight and may possibly be employed in the ground role against light armoured vehicles.

DATA
Cartridge: 14.5 × 114 mm Type 56
Feed: 80-round link belt
Length: 2.39 m
Weight: 165 kg with mount
Elevation: -10 to +85°
Traverse: 360°
Rate of fire: 550 rds/min cyclic
Muzzle velocity: 995 m/s
Effective range: (air) 2000 m; (ground) 1000 m

Manufacturer
China North Industries Corp, PO Box 2137, Beijing.
Status
Current. Production.
Service
Chinese Army.

14.5 mm machine gun Type 75-1

CZECHOSLOVAKIA

7.92 mm Models 26, 27 and 30 light machine gun

The Czechoslovak armament firm of Ceskoslovenska Zbrojovka, Brno, produced a series of excellent light machine guns which were manufactured in large numbers and very widely exported. They were also produced under licence by a number of different countries.

The Model 26 is a gas-operated light machine gun with a 20-round box magazine mounted over the receiver. The gas cylinder carrying the bipod reaches right out to the muzzle of the finned barrel. There is no regulator. The barrel may be changed rapidly by rotating the barrel locking lever and using the carrying handle to push the barrel forward.

The gun has the bolt mounted on the piston extension. The piston extension carries a piston post which is seated in the hollow interior of the bolt and controls its position as the piston reciprocates. The bolt is raised at its rear end at the conclusion of the feed stroke, to lock into a recess in the ceiling of the receiver. When the chamber pressure is reduced sufficiently, the ramp on the rear of the piston post lowers the bolt and carries it to the rear. The gun fires from the open breech position.

The gun will fire at single-shot as well as automatic. The trigger mechanism is illustrated and explained in some detail in the description of the Bren gun (which see).

The Models 27 and 30 machine guns were virtually the same, but there were detail changes which prevented the interchange of parts. All made extensive use of machined parts and were expensive to produce.

The British Bren was a derivative of the Czechoslovak series: the relationship is described in the relevant UK entry.

DATA
Cartridge: 7.92 mm × 57
Operation: gas, selective fire
Method of locking: tilting block
Feed: 20-round detachable box magazine
Weight: (unloaded) 9.69 kg
Length: 1163 mm
Barrel: 602 mm; (rifling) 4 grooves rh; (cooling) air. Quick-change barrel
Sights: (foresight) blade; (rearsight) tangent aperture
Muzzle velocity: 762 m/s
Rate of fire: (cyclic) 500 rds/min
Effective range: 700 m

Manufacturer
Ceskoslovenska Zbrojovka, Brno.
Status
Obsolete. No longer manufactured.
Service
Not known to be in first-line service but held as reserve stock in many countries.

7.92 mm Model 30 light machine gun

7.92 mm Model 27 light machine gun

7.92 mm Model 37 medium machine gun

This gun was manufactured in 1937 and its nomenclature in the Czechoslovak Army was the Model 37. It was sold commercially as the Model 53.

It was an air-cooled, belt-fed, gas-operated gun. The action was much the same as that of the Model 26 but the bolt locked to the barrel extension and not into the receiver. The barrel and bolt could recoil together within the receiver before unlocking occurred. This in no way affected the method of operation of the gun nor the cycle of operations. It was intended to reduce the trunnion pull by dissipating energy before the whole gun recoiled and so reduce the load on the tripod or vehicle mount.

The Model 37 fed from the right and could use either a metal or a fabric-cum-metal continuous open pocket belt. It introduced the cocking system which was later used on Models 52 and 59, in which the trigger mechanism is pushed forward and the sear used to pull the piston and breech-block to the rear.

The Model 37 could be employed on its tripod as a medium machine gun or as an anti-aircraft gun. The British version, manufactured by BSA, was used as a tank weapon. The German Army used it both as a tripod-mounted medium machine gun and in their tanks as the MG 37 (t). It has two rates of fire, a buffer device, sometimes misleadingly called the 'accelerator' being interposed to shorten the recoil distance and to return the bolt forward at high speed. Without the device the rate of fire was about 450 rds/min and with it the firing rate went up to about 700 rds/min. In British manufacture the idea was dropped at an early stage to simplify production.

DATA
Cartridge: 7.92 mm × 57
Operation: gas, selective fire
Method of locking: tilting block
Feed: continuous open pocket belt
Weight: 18.82 kg
Length: 1105 mm
Barrel: 678 mm; (rifling) 4 grooves rh; (cooling) air. Quick-change barrel
Sights: (foresight) blade; (rearsight) leaf graduated from 300 to 2000 m, battlesight at 200 m
Muzzle velocity: 793 m/s
Rate of fire (cyclic) 450 or 700 rds/min
Effective range: 1000 m

Manufacturer
Ceskoslovenska Zbrojovka, Brno.
Status
Obsolete. No longer manufactured.
Service
Not in first-line service. May be held in reserve stocks. Probably in use in West Africa.

Czechoslovak 7.92 mm Model 37 medium machine gun

7.62 mm Model 59 general-purpose machine gun

This machine gun fires the Soviet 7.62 mm × 54R cartridge. As a squad automatic weapon with a light barrel and bipod or as a light machine gun with heavy barrel and bipod, it is known as the Model 59L.

As a medium machine gun with a heavy barrel and a light tripod it is called the Model 59. This tripod also enables the gun to fire in the air defence role. As a tank coaxial machine gun, fitted with a solenoid, it is referred to as the Model 59T.

The gun is also manufactured (for export) to use the NATO 7.62 mm × 51 cartridge and is then designated the Model 59N. The different cartridge contour leads to a different chamber and bolt face.

The gun fires only from the open pocket metal non-disintegrating Czechoslovak belt. The open pocket allows the Soviet rimmed cartridge to be pushed straight through the belt and the advantages obtained by having a push-through feed are clearly demonstrated when comparing this system with the complex arrangements for the Soviet PK general-purpose machine gun.

The continuous metal belt can be re-loaded as required, and a belt filler which can be clamped on an ammunition box is available.

In the light machine gun role a 50-round metal box can be hung from the right-hand side of the gun. This is used for firing on the move during the assault. In the medium machine gun role a box holding 250 rounds in five belts is available.

The rearsight for the early model light machine gun is a horse-shoe shaped bar with odd ranges on the left hand arm and even ranges on the right. The V notch is adjusted up and down by means of knobs on each side. The sight reads from 1 to 20 in hundreds of metres. Later models, and particularly the Model 59N, have a light folding frame rearsight of conventional design with spring thumb catches for setting the range.

The foresight is a cylinder mounted eccentrically on a screw threaded base and has a hooded protector. All zeroing is carried out on the foresight.

The adjustable bipod clamps into its seating below the foresight. It will pivot back along the barrel by closing the legs when they are hanging vertically below the gun and laying them under the barrel.

Operation
The belt is loaded by pressing the cartridges into the open links. Each link has a tab at its rear which fits up against the base of the cartridge – not into the cannelure if rimless NATO ammunition is being used.

The 50-round box has two grooves into which fit two lugs at the end of the

7.62 mm Model 59N general-purpose machine gun

Heavy-barrelled version on tripod with 250-round container

feed tray on the right of the gun. When the box is fitted into place the cover will spring open. Pull out the loop fitted at the leading end of the belt. Press the cover catch, on top of the cover, forward and lift the cover. Press the belt exit cover back and it will fly open. Place the belt in position with the open side of the link downwards. It will be retained in the gun by the stop pawl when the cover is closed.

To cock the gun pull down the trigger group latch which is the cone-shaped projection mounted on the left side of the receiver above the pistol grip. This releases the trigger grip from the receiver. When the trigger is pressed the sear goes down and the pistol grip can be pushed fully forward. The trigger is then released and the sear rises. It engages in the bent under the piston and when the pistol grip is pulled back the sear acts as a catch and pulls the piston and bolt rearwards compressing the return spring. The trigger group is reconnected to the receiver and the weapon is cocked.

The safety catch also lies on the left of the receiver behind the trigger group latch. When pressed up it locks the sear. When the safety catch is moved down and the trigger is pressed the sear is depressed and the piston goes forward carrying the breech-block. The feed ribs on top of the block force a cartridge through its link and into the chamber. The extractor snaps over the rim of the cartridge and the rim of the round hits the breech face and the breech-block can advance no further. The piston continues on and the two ramps force the locking piece to rotate upwards and into two recesses in the receiver side walls. They ride under the locking piece and hold the lock in position. When locking is complete there is a short delay providing mechanical safety before firing and then the piston post hits the rear of the breech-block, driving the firing pin into the cap.

The gun fires and some of the propellant gases are diverted into the gas port. The piston is driven to the rear. The piston post moves away from the breech-block and the spring-loaded firing pin is retracted. The central unlocking cam has an inclined plane which pulls the lock down out of its recesses in the receiver and the piston carries the bolt back with it. The empty case is held to the bolt-face and ejected downwards through the cut-away section in the piston. The return spring is compressed by the piston and rearward motion ceases when the back of the bolt hits the rather soft buffer.

If the trigger remains pressed the cycle will be repeated.

Feed system
The feed system owes its design to the Model 52 and is copied in the Soviet PK general-purpose machine gun. In this system the flat sides of the piston are raised to form cam paths on which runs a roller attached to a feed arm. The cam forces the roller outwards. The roller in turn pushes out the bottom end of a feed arm, mounted vertically on the receiver, which is pivoted at its centre point. As the bottom goes out, the top of the lever must go in and the feed pawl attached to it forces the belt in towards the feeding position on the centre line of the gun, until the leading cartridge comes up against the cartridge stop of the feed tray.

When the piston goes forward another cam path controls the feed lever and it is positively actuated and the top is pivoted out. The feed pawl, which is spring-loaded, slips over the next round in the belt. The belt cannot slip back because it is held up to the gun by a stop pawl on each side of the feed pawl. There is also another pair of stop pawls on the feed tray.

Trigger mechanism
The trigger mechanism, like that of most guns firing only at automatic, is simple. The trigger is ⌐ shaped and when it is pressed the top arm rises and lifts the front of the sear. The sear is centrally pivoted and the rear end falls, allowing the piston to go forward.

Incorporated in the trigger mechanism is a controlled sear device to ensure that the sear rises to present a full mating surface with the bent of the piston. It consists of a sear catch powered by the trigger spring. The catch lies beneath the top arm of the trigger and passes through it. It presses up on the front end of the sear and so keeps the nose down. When the trigger is pressed the spring pressure on the catch is increased. When the trigger is released the trigger arm rises but the sear catch holds the sear tail up (and the nose down) until the spring pressure is sufficiently reduced and the trigger levers the sear catch clear of the sear. The sear rises rapidly and there can be no question of a slow return even if the firer releases the trigger extremely slowly. In this way the chipping of the sear face resulting from partial engagement is reduced.

The safety catch, on the left of the receiver, is pushed up to engage and in

Model 59 tripod in anti-aircraft role

pivoting it blocks the sear which cannot fall; at the same time a flange at its upper edge slides in a slot in the receiver and locks the trigger group. This prevents the gun from being cocked.

Gas regulation
The gas regulator in all the Model 59 guns except that chambered for NATO ammunition has two positions. The normal position is marked '1' and this is changed to '2' using the combination tool.

The NATO guns have a four-position regulator. This is of unusual design in that movement is controlled by the carrying handle. To unlock the regulator press the carrying handle lock, which lies parallel to the barrel at the foot of the handle, and twist the carrying handle and then rotate it until the pointed lug at the front of the handle can be lined up with the notch at the top rear of the gas cylinder. Position '1' is used when the gun is functioning normally. The pointer on the right of the regulator is rotated to '1' and the regulator pressed fully in to the left. To get to position '2' the regulator is pressed to the right and rotated until the pointer is aligned with '2'. To get to '3' the regulator is rotated backwards still being held to the right and the pointer is at '3'. The position '4' is reached by pressing the regulator to the left and rotating it forward. To lock the regulator in the selected position, press the carrying handle lock, then twist and rotate the carrying handle. It should be noted that position '4' produces a very high rate of fire and was intended for anti-aircraft use.

Barrel changing
Press the cover catch forward, pivot the cover up about its front hinge and then push the cover sideways so that it rotates to the right. The cover is attached to the barrel nut and this is rotated out of engagement with the interrupted threads on the barrel.

Grasp the carrying handle and push the barrel forward out of the receiver. Remove the bipod by squeezing the legs together and swinging them forward until the legs are at right angles to the barrel. Put the bipod on a cool barrel and insert it into the gun.

Stripping
Set the safety catch up to 'safe', press the cover catch forward, lift the cover and remove the belt. Check no cartridge remains in the chamber or feedway. Set the safety to 'fire'. Depress the trigger group latch and control the forward movement of piston, breech-block and trigger group.

Push the take-down pin at the rear of the receiver fully to the left and remove the butt stock to the rear.

Pull the trigger group back and the group will come out of the gun together with the piston and breech-block.

Telescope
A telescopic sight is available and can be used with the gun mounted either on its bipod or a tripod. The telescope is attached to the receiver by means of a clamp. The sight can be adjusted for windage and elevation. The graticule can be illuminated at night and the telescope has a dovetail to accept the lamp housing. An active infra-red source and receiver can also be used with the gun.

Mounts
The bipod is standard with the light guns. To mount the gun on the tripod, the rearsight is lifted and the gun passed under the traversing arc and the mounting lugs on top of the rear of the receiver are then pinned to corresponding lugs under the main frame of the mounting. There is provision for free traverse and traversing stops are also provided. The tripod height can be adjusted.

If required the tripod can be used for anti-aircraft firing by using extension pieces.

DATA
Cartridge: 7.62 mm × 54R or 7.62 mm × 51
Operation: gas, automatic
Method of locking: swinging lock
Weight: (with bipod and light barrel, empty) 8.67 kg; (with tripod and heavy barrel) 19.24 kg
Length: (with heavy barrel) 1215 mm; (with light barrel) 1116 mm
Barrel: (heavy) 693 mm; (light) 593 mm
Rifling: 4 grooves, rh, 1 turn in 240 mm
Sights: (fore) pillar. Adjustable for zeroing both line and elevation; (rear) V notch. Adjustable from 100 to 2000 m by 100 m increments
Sight base: (heavy barrel) 744 mm
Muzzle velocity (light bullet): (heavy barrel) 830 m/s; (light barrel) 810 m/s
Muzzle velocity (heavy bullet): (heavy barrel) 790 m/s; (light barrel) 760 m/s
Rate of fire: (cyclic) 700-800 rds/min
Rounds before barrel change: (heavy barrel) 500 rounds; (light barrel) 350 rounds
Effective range: (tripod) 1500 m; (bipod) 1000 m
Max range: 4800 m

TRIPOD
Weight: 9.3 kg
Barrel height: (minimum) 300 m; (max) 500 m
Height of gun: (in anti-aircraft role) 1440 mm
Max elevation: 21° 36′
Max traverse between stops: 43° 12′

Manufacturer
Zbrojovka Brno, Narodni Podnik.
Status
Current. Production.
Service
Czechoslovak forces.

DENMARK

Madsen light machine guns

The Madsen light machine gun was manufactured in various calibres from 1903 until the middle 1950s. It was used by over 30 countries at various times and although no longer in service anywhere, specimens could still be encountered with irregular forces.

The Madsen used a unique mechanism based on the Martini-Peabody dropping breech-block; since there was no bolt as such, it required an independent rammer to load the cartridge into the breech and an extractor to remove the case. The intricate design meant expensive manufacture, one of the reasons the Madsen never appeared in numbers in any major army.

A full description of the Madsen machine gun will be found in earlier editions of *Jane's Infantry Weapons*.

Field and Vehicle Mountings for light automatic weapons

Dansk Industri Syndikat A/S (DISA) makes a range of special mountings to enable light or medium machine guns to be used effectively against ground or aircraft targets. Both field and vehicle mountings are included in the range, the design objective being the creation of a mount which is robust and extremely stable yet which permits the gunner, working single-handed, full freedom of weapon aiming. This requirement is met in the field mountings by a pivot mount on a tripod which, although light in weight, can be adjusted to suit any terrain and gunner's position whilst remaining highly stable.

The vehicle mountings employ a patented lever arm system comprising two arms. One of these is a rigid horizontal bar, vertically pivoted at one end on the mounting baseplate and joined by a vertical pivot at the other end to the second arm. The second (upper) arm is a spring-loaded vertical parallelogram which carries the gun on a combined horizontal and vertical pivot at the end remote from the junction with the lower lever arm.

The mounting can be locked in both elevation and azimuth. The whole arrangement provides a stable gun platform and makes it possible for the gunner to track an aircraft through 360°.

The vehicle mounting is fitted to the vehicle by a baseplate with four bolts. Currently the mounting is configured for the FN-MAG 58, the West German MG-3 and the US M60, but any light machine gun can be fitted, the only change required being a modified gun seat.

Mk 400 Anti-aircraft Mountings for light automatic weapons

DATA
Weight: 25.5 kg
Muzzle height: 520 mm (min); 920 mm (max)

Elevation: –10 to +85°
Dimensions in transport position: 900 mm long, 230 mm wide, 480 mm high

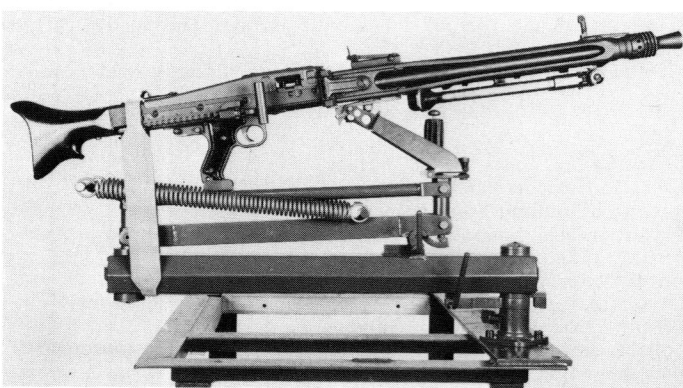

DISA vehicle mounting for West German MG3

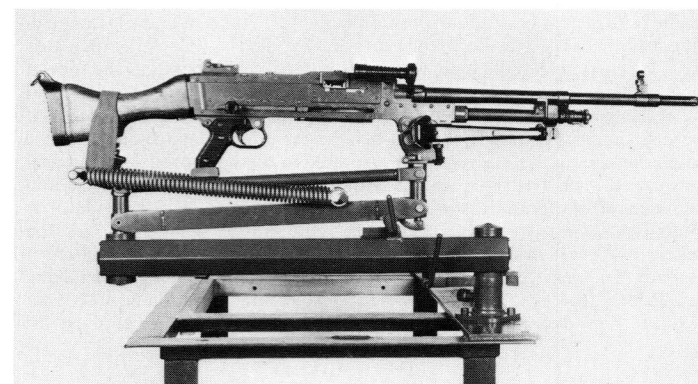

DISA vehicle mounting for FN-MAG

Madsen/DISA Field Tripod, Mk 200 series

This is a folding portable mount for the light machine gun in the sustained fire role. It can be used for aimed fire, barrage and searching fire, or for traversing fire, the mounting being vertically and laterally adjustable. There are graduated scales for both elevation and azimuth, and stops can be individually set so as to limit movement in both arcs. The gun recoil is fully absorbed by spring buffers, leading to low recoil loadings on the mounting, and the trigger can be operated by the gunner from a low or prone position. The gun can be fired without unfolding the rear tripod legs, giving a short tripod effect, and the mounting is provided with an adjustable sight bracket for optical sights.

DATA
Weight: 16 kg
Muzzle height: 700 mm (max); 345 mm (min)(short tripod)
Vertical adjustment: 200 mils
Lateral adjustment: (coarse) 400 mils; (fine) 20 mils
Dimensions in trasnport position (folded): 820 mm long, 660 mm wide, 225 mm high

Madsen/DISA Mk 200 field tripod

Madsen/DISA AA Field Tripod Mk 350 series

This is a folding and portable mount for light machine guns of all types in the air defence role which can rapidly be adjusted for firing in the ground role. Operable by one man, it is easy to erect and extremely stable in either position. The telescopic legs are individually adjustable, allowing the tripod to conform to any kind of terrain; non-retractable legs can be furnished if required. Movement in azimuth is unlimited; movement in elevation in the air defence role extends from +10 to +85°. Recoil is partially absorbed by spring buffers.

DATA
Weight: 9.5 kg
Transport dimensions (folded): 860 mm long, 290 mm wide, 260 mm high
Muzzle height: (air targets) 1400 mm; (ground targets) 390 mm

Manufacturer
Dansk Industri Syndikat A/S (DISA), Stationsvej 9-11, DK-3550 Slangerup.
Status
Current. No longer in production.
Service
In service with several NATO armies. Under trial in others including USA.

Madsen/DISA Mark 350 tripod in ground role

FINLAND

7.62 mm M62 Valmet machine gun

This is a gas-operated light machine gun which, although it bears little superficial resemblance to it, is based on the Czechoslovak Zb26 series and has a tilting block which locks into the roof of the receiver of the gun. It was started as a project in 1957 and the prototype was made in 1960. Trials were completed in 1962 and the M62 came into service with the Finnish Forces in 1966. It is belt-fed and uses the Soviet 7.62 mm M43 cartridge. It fires only at full automatic at a high rate of fire, and is unusual in modern machine guns in feeding from the right. Its empty weight of 8.3 kg suggests that, with the cartridge it uses and a quick-change barrel, it should be capable of a large volume of fire. It has been supplied to the armed forces of Qatar.

DATA
Cartridge: 7.62 mm × 39
Operation: gas, automatic
Method of locking: tilting block
Feed: 100-round continuous link belt

Weight: (empty) 8.3 kg; (with full belt) 10.6 kg
Length: 1085 mm
Barrel: 470 mm; (rifling) 4 grooves rh
Sights: (foresight) pillar; (rearsight) aperture
Range setting: 100–600 m by 100 m steps
Muzzle velocity: 730 m/s
Rate of fire: (cyclic) 1000-1100 rds/min; (practical) 300 rds/min (maker's figures)
Effective range: 350-450 m

Manufacturer
Sako Ltd, Tourula Works, PO Box 60, SF-40101 Jyväskylä 10.
Status
Obsolescent. No longer in production.
Service
Finnish and Qatar armed forces.

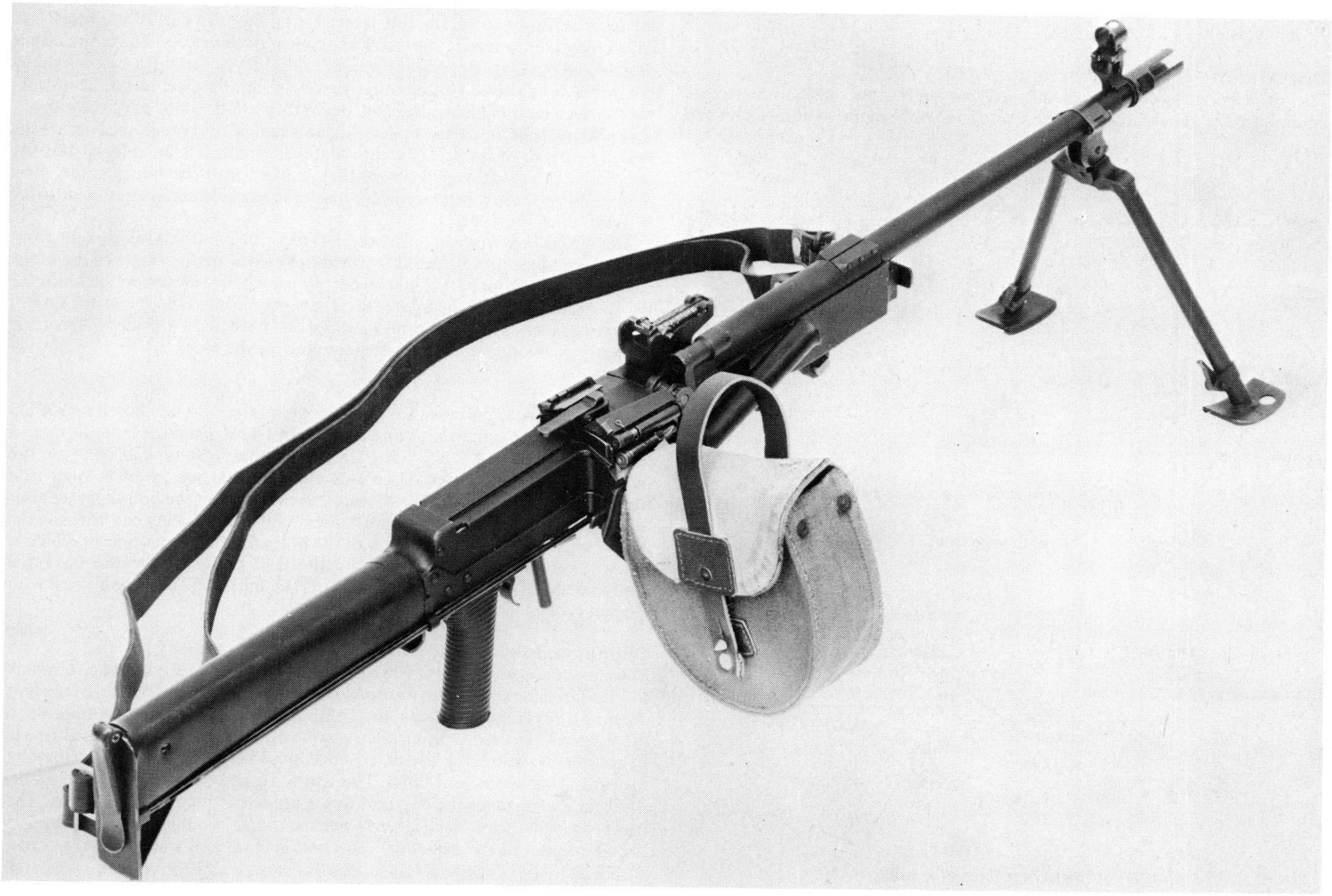

7.62 mm M62 Valmet machine gun

FRANCE

7.5 mm Model 1924/29 light machine gun

The Chatellerault was the standard French machine gun at the beginning of the Second World War and after that war it was put back into production to supply the French Army in Indo-China. It is no longer used by the French but was employed extensively by the Viet Cong in the Vietnamese war.

The gun is gas-operated with a top-mounted magazine holding 26 7.5 mm × 54 cartridges. The action is sturdy and reliable. The barrel cannot be removed without the use of a spanner to rotate it out of the receiver. Locking is performed by a tilting bolt. Two triggers are employed; the front trigger delivers single shots and the rear trigger full automatic fire.

A monopod can be fitted into the underside of the butt to make an improvised tripod.

A development of the Model 1924/29, the Model 1931A, may also be found, usually on a French Model 1945 or US M2 tripod: it has a heavier barrel, a side-mounted drum or box magazine and a prominent ejector housing on top of the receiver and is capable of automatic fire only using a single trigger.

French 7.5 mm Model 1924/29 light machine gun

DATA
Cartridge: 7.5 mm × 54
Operation: gas, selective fire
Method of locking: tilting block
Feed: 26-round top-mounted detachable box magazine
Weight: 9.2 kg
Length: 1082 mm
Barrel: 500 mm; (rifling) 4 grooves lh
Muzzle velocity: 850 m/s

Rate of fire: (cyclic) 500 rds/min
Effective range: 800 m

Status
Obsolete. No longer made.
Service
Not in first-line service, but likely to be in use by reserve and militia troops in South-east Asia and Africa.

7.5 mm AA 52 and AA 7.62 NF-1 general-purpose machine gun

The Arme Automatique Transformable Model 52, general-purpose machine gun Model 1952, was designed for ease of production, using stampings wherever possible. The receiver of the gun is made of fabricated semi-cylindrical shells welded together. As originally designed it fired the French 7.5 mm × 54 cartridge of 1929, and in this form it was known as the AA 52. The majority of present guns have been converted to fire the NATO 7.62 mm round and the 7.5 mm has been phased out. For both versions there is a choice of light and heavy barrels. The following description refers to the 7.5 mm version.

Operation
The principle of operation is delayed blowback, but since rifle calibre ammunition is used special arrangements have to be made to ensure safe and

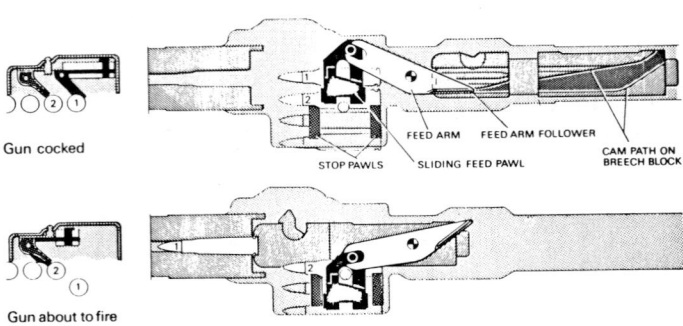

Gun cocked

Gun about to fire

7.5 mm AA 52 general-purpose machine gun feed system

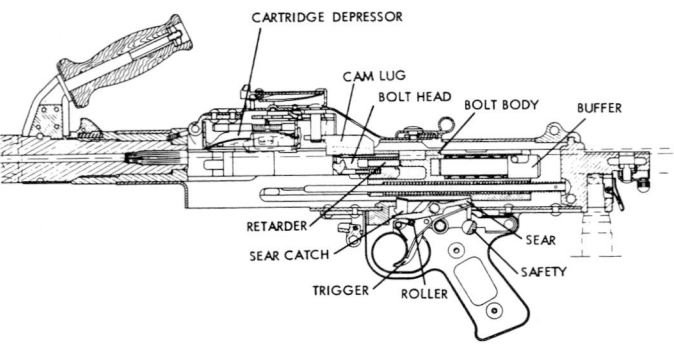

7.5 mm AA 52 general-purpose machine gun section

smooth extraction of the spent case. There are a number of longitudinal grooves in the neck of the chamber running out halfway towards the mouth. Gas enters these grooves and the case 'floats' with equal pressure on each side of the brass wall and can move comparatively freely.

To prevent premature breech opening while pressure is still high a two-part block is employed. The front part carries a lever. The short end of the lever rests in a recess in the side of the receiver and the long arm bears against the massive rear part of the bolt. Gas pressure exerts a force on the bolt-face and the lever is forced to rotate, thus accelerating the rear part of the block but restraining the front part until the lever is clear of the recess. The entire bolt is then blown back by the residual pressure. The empty cartridge case is held by the extractor to the bolt-face until it strikes the double ejectors on the bottom rear of the feed tray and is deflected down through the bottom of the gun.

It is found that the cartridge head space is critical and when wear occurs the bearing surface in the receiver can be quickly replaced. In spite of this the ejected cases are deformed where they expand into the bullet guide. Any relaxation of the tolerances in case manufacture and increase in headspace will result in a blowout.

The AA 52 fires from an open breech. The firing pin floats freely in the front part of the block and is forced forward only when the rear part of the block closes up to the front part. This can occur only when the lever, which holds the two parts separated, can rotate into the recess in the receiver at the end of forward travel. Thus the round cannot be fired until the delay device is properly positioned to hold up the front face of the block.

Feed
The 7.5 mm AA 52 gun uses a French disintegrating link belt based on the US M13 pattern but somewhat more flexible and with a smaller extension for a given load. The AA 7.62 NF-1 can use either this type of belt or the M13-NATO belt. The AA 52 cannot use the usual system of a roller in a feedway to operate the feed pawl since the two parts of the bolt separate after firing and remain with a gap between them until immediately before firing the next round. A cam groove on top of the bolt accommodates a lever which is swung from side to side to operate the feed pawl. Unlike the Fabrique Nationale MAG or the British L7A2 GPMG, the AA 52 must be carried with the gun cocked when a loaded belt is in place.

Trigger and firing mechanism
The gun is designed to fire at 900 rds/min at full automatic only, when clean. A soft buffer in the rear end of the bolt keeps the forward acceleration of the bolt down. To ensure that the sear and bent mate on full engaging surfaces when the trigger is released, a 'controlled sear' is used which incorporates a tripper to release the sear at a pre-determined position while the bolt is moving forward. This works as follows. The sear is an arm pivoted at its mid-point. The rear end engages the bolt and this is known as the nose of the sear. The front end (tail) is over the trigger. When the trigger is pulled the tail of the sear is pushed up and the nose falls. The bolt is freed and starts to reciprocate. Attached to the trigger is a sear catch, or tripper, which moves down when the

7.5 mm AA 52 general-purpose machine gun on US M2 tripod for sustained fire

7.5 mm AA 52 general-purpose machine gun

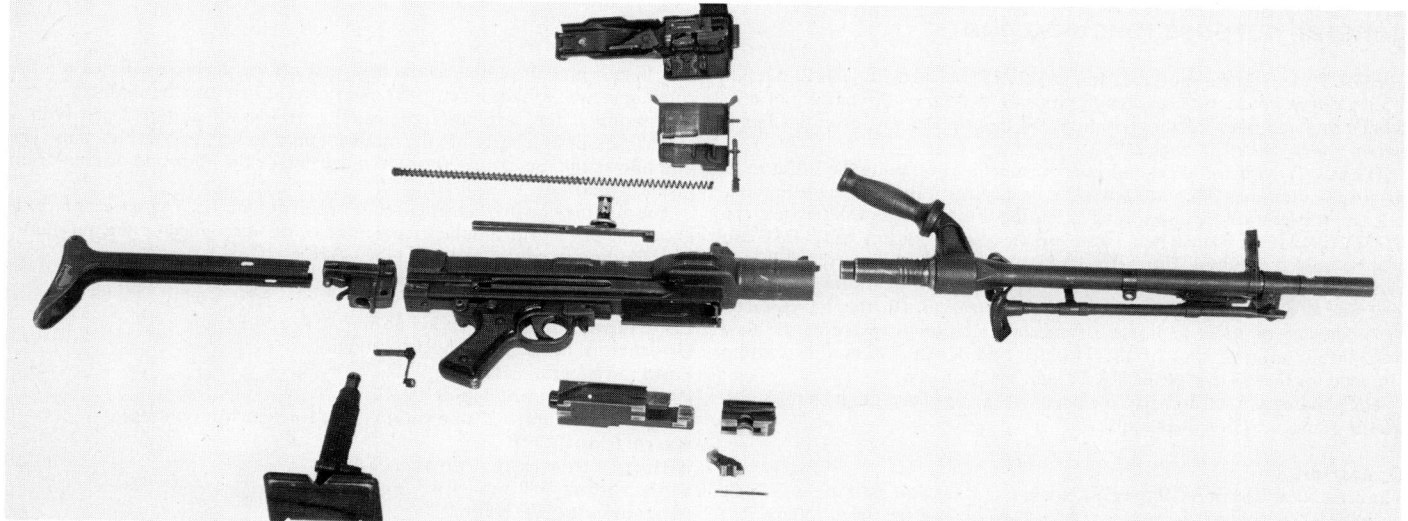

7.5 mm AA 52 general-purpose machine gun field-stripped

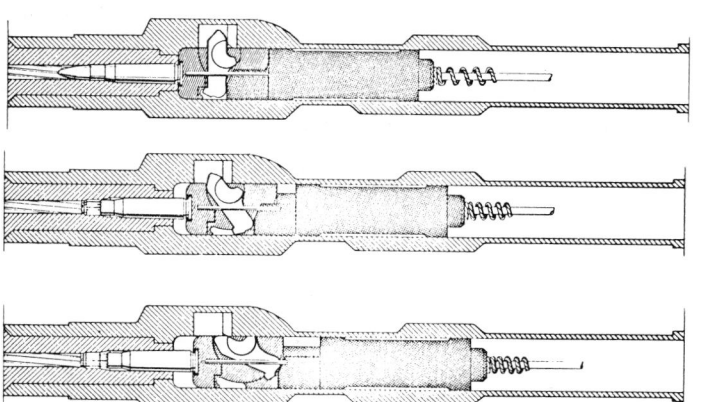

7.5 mm AA 52 general-purpose machine gun two-part bolt-action **(top)** *gun fires, lever engaged in receiver* **(centre)** *bolt body moving back* **(bottom)** *lever rotated clear of receiver*

trigger is pulled and has its own spring which rotates it under the tail of the sear. When the trigger is released the tripper goes up and prevents the tail of the sear from falling. When the bolt next comes back it hits the tripper, revolves it back against its spring, and frees the tail of the sear to fall. The nose of the sear rises, the bolt going back rides over it, and the full face of the sear is presented to the bent of the bolt as it comes forward. This prevents chipping of the sear due to contact of reduced mating areas.

Barrel change
The barrel change arrangements of the AA 52 are awkward if the gun is being fired off its bipod. A barrel-release catch must be pressed in (in earlier guns, pulled back) and then the wooden barrel-carrying handle rotated clockwise and pushed forward. Since the bipod is permanently attached to the barrel, once the barrel is drawn off there is no forward support for the gun and the gunner is left holding a very hot gun.

Sights
The foresight is a hinged block with a slot in the top. For firing by day the slot is used and at night the entire foresight block is aligned with the notch of the rearsight, the sights being luminescent.

Stripping
To strip, the gun is made safe and then the butt is removed by rotating a catch underneath. The monopod has a catch at the back. The rear plate of the receiver, and the spring rod, spring and bolt can be withdrawn by six revolutions of the actuating handle.

Variants
In addition to the dismounted infantry versions described above the two guns can also be supplied in versions without butt stock for vehicle mounting and arrangements can be made for them to be fired electrically.

DATA
COMMON TO BOTH VERSIONS UNLESS OTHERWISE STATED
Cartridge: 7.5 mm M/29 or 7.62 mm NATO
Operation: delayed blowback, automatic
Feed: disintegrating link belt (see text)

	Light Barrel	Heavy Barrel
Weight of gun		
(without bipod, empty)	9.15 kg	10.55 kg
Weight of barrel	2.85 kg	4.25 kg
Weight of bipod	0.82 kg	0.82 kg
Monopod	0.685 kg	0.685 kg
Weight of gun		
(with bipod, empty)	9.97 kg	11.37 kg
Weight of modified		
US M2 tripod	—	10.6 kg
Length of gun		
(butt extended)	1145 mm	1245 mm
(butt retracted)	980 mm	1080 mm
Length of barrel		
(without flash hider)	500 mm	600 mm

	AA 52	AA 7.62 NF-1
Number of grooves	4	4
Sights: (foresight) slit blade; (rearsight) leaf graduated 200-2000 m		
Muzzle velocity	840 m/s	830 m/s
Rate of fire (cyclic)	700 rds/min	900 rds/min
Practical range		
(light barrel)	800 m	800 m
(heavy barrel)	1200 m	1200 m

Manufacturer
Originally developed and produced by Manufacture Nationale d'Armes de Chatellerault. Production subsequently transferred to Manufacture Nationale d'Armes de Tulle. Enquiries to GIAT, 10 place Georges Clémenceau, 92211 Saint-Cloud.

Status
Current. No longer manufactured.
Service
French armed forces and some former French colonies, particularly in Africa.

GERMANY, EAST

7.62 mm LMGK

This is the East German version of the Soviet RPK light machine gun. It is identical to the Soviet gun and all details will be found in the USSR entry for that weapon.

Similarly, the East Germans use the PK as their general-purpose machine gun, and details of this will also be found under the USSR entry.

GERMANY, WEST

7.62 mm HK 21 Heckler and Koch general-purpose machine gun

The HK 21 is no longer produced and has been replaced by the HK 21A1, but it has been in production for some years and there are still numbers of it in service in some parts of the world. A full description was given in *Jane's Infantry Weapons 1978*, page 267.

The HK 21 is a belt-fed general-purpose machine gun using the 7.62 mm × 51 NATO cartridge. This will normally be carried in the disintegrating link belt and the gun will function using either the West German DM 60 belt, the US M13 belt or the French belt. However, the continuous link belt, DM1, can also be used if required. Furthermore by changing the barrel, the belt feed plate and the bolt, the gun can be converted to firing the 5.56 mm × 45 or the 7.62 mm × 39 cartridge. The usefulness of the weapon is further increased by the ability of the user to insert a magazine adaptor in place of the feed mechanism, to take any of the Heckler and Koch 7.62 mm magazines intended for the G3 rifle or the HK 11 light machine gun.

The gun has a practical and effective quick change barrel that enables it to produce sustained fire when required.

Operation
The method of operation of the HK 21 is the same as that used in the G3 rifle (which see) employing the two-part breech-block and the delay rollers. The delayed blowback system, with a fluted chamber for easy cartridge movement, operates from the closed breech position with a round in the chamber when the gun is ready to fire.

The feed system of the belt-fed gun functions as follows:

If the disintegrating belt has a feed tag this is pushed through the feed tray from left to right and pulled until the first round reaches the cartridge stop. Since the bolt will pass over the belt the open side of the belt links must be placed uppermost before the tag is inserted.

Where the belt has no tag the gun is first cocked and the cocking handle held to the rear in the recess of the cocking slideway. The feed mechanism catch is depressed and the mechanism moved over to the left. The first round is placed in the feed sprocket which is rotated to the right until it locks. The first

cartridge is now in place and the feed unit can be pushed in. Releasing the cocking handle will feed the first round from the belt into the chamber.

If required the belt feed unit can be withdrawn and replaced with a magazine unit which fits into the receiver and is held by two locking pins. The unit allows the use of the 20-round magazine or the 80-round double-drum plastic magazine.

The gun may be fired at single-shot or full automatic. The selector lever is above the pistol grip and the trigger mechanism, disconnector and automatic sear are the same as those used in the G3 rifle.

DATA
Cartridge: 7.62 mm × 51 or 5.56 mm × 45 or 7.62 mm × 39
Operation: delayed blowback, automatic
Feed: disintegrating link belt
Weight: 7.92 kg with bipod
Length: 1021 mm; (820 mm without butt for vehicle mounting)
Barrel length: 450 mm
Rifling: 4 grooves, rh, 1 turn in 305 mm
Sight radius: 589 mm
Muzzle velocity: 800 m/s
Rate of fire: (cyclic) 900 rds/min
Effective range: 1200 m

Manufacturer
Heckler and Koch GmbH, D-7238 Oberndorf-Neckar.
Status
Current. No longer produced by Heckler and Koch, but manufactured in 7.62 mm × 51 calibre under licence in Portugal by Industrias Nacionais de Defesa EP.
Service
Portugal and some African and South-east Asian countries.

7.62 mm HK 21 Heckler and Koch general-purpose machine gun with belt-feed system

7.62 mm HK 21A1 Heckler and Koch general-purpose machine gun

The HK 21A1 is a development of the HK 21, which has been in service for a number of years. One major change from the HK 21 is the abandoning of the option of using a magazine feed: the gun is now belt-fed only. It uses the

under-fed mechanism of its predecessor, but the mechanism can now be hinged down to allow the belt to be inserted which makes the operation of loading much quicker and easier.

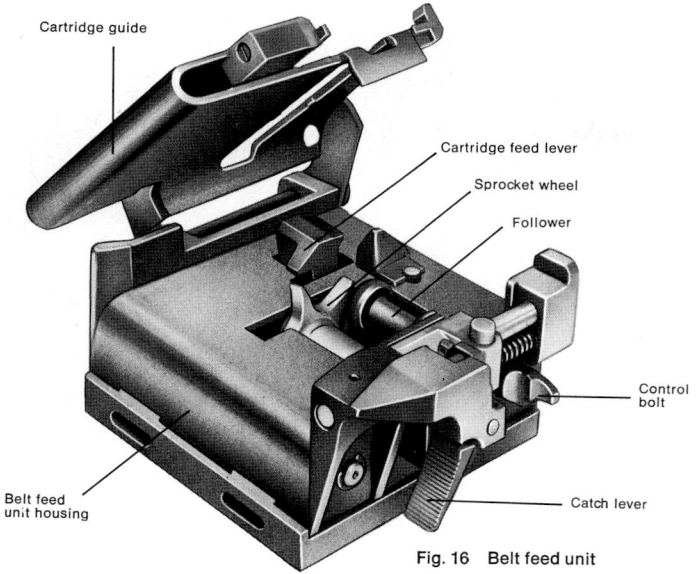

Cartridge guide

Cartridge feed lever

Sprocket wheel

Follower

Control bolt

Belt feed unit housing

Catch lever

Fig. 16   Belt feed unit

7.62 mm HK 21A1 Heckler and Koch general-purpose machine gun belt feed unit

Tactically the gun is intended as a squad light gun, but it is also adaptable as a general-purpose machine gun and there is the usual range of Heckler and Koch accessories which convert it to this role.

Operation

The HK 21A1 is a closed bolt automatic weapon operated by delayed blowback utilising the system of roller-locking of the bolt in the same way as the G3 rifle. The feed system is a left-hand belt feed mechanism which will accommodate a jointed or disintegrating link belt, and the gun will accept either 7.62 mm × 51 NATO ammunition, or 5.56 mm × 45. The latter requires that the barrel, bolt and feed mechanism be changed, but that is all. As with the other Heckler and Koch guns, when the weapon is cocked, the bolt is closed and there is a round in the chamber.

The belt feed unit is detachable from the gun if required for cleaning. It is in two parts, the cartridge guide and the belt feed housing. The belt feed housing is hinged to the receiver by a socket pin and is held in place by a spring catch. Actuating the catch lever releases the feed unit and allows it to swing down. At the same time two cams cut in the control bolt move the follower and cartridge feed lever downwards below the level of the feed plate, so making it easier to insert a new belt, or to remove the existing one if necessary. When the feed unit is swung back into place the follower and feed lever are forced back into position.

Cartridges are fed into position for the bolt to engage them by the combined actions of the feed sprocket, feed lever and follower. These are all driven by a camway cut in the bolt carrier which engages with a slide-bolt in the feed

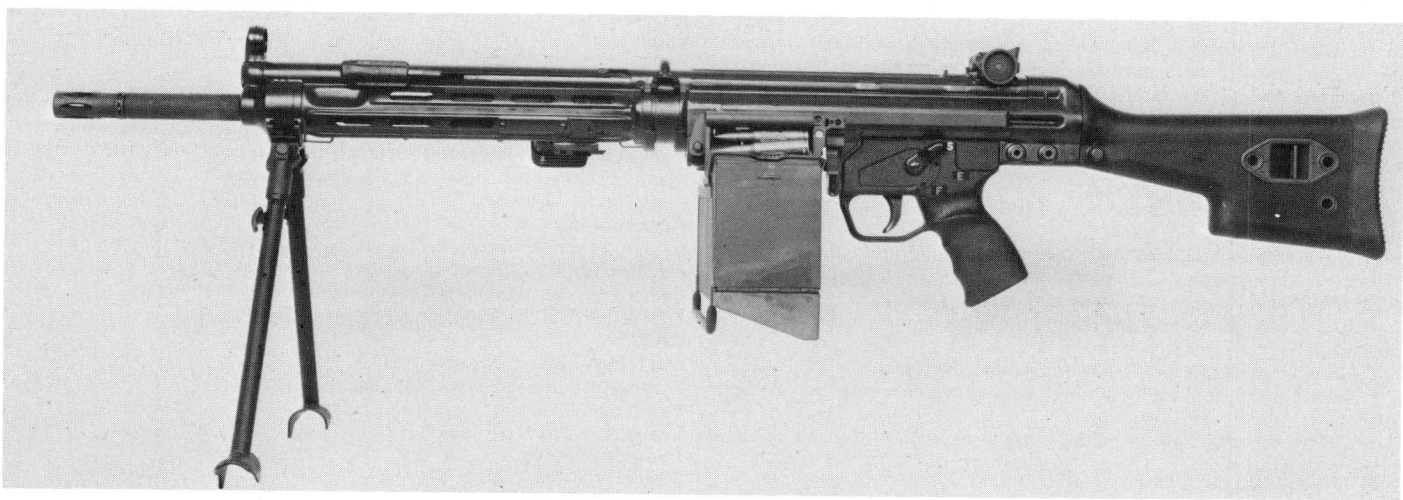

7.62 mm HK 21A1 Heckler and Koch general-purpose machine gun with loaded belt in carrying box

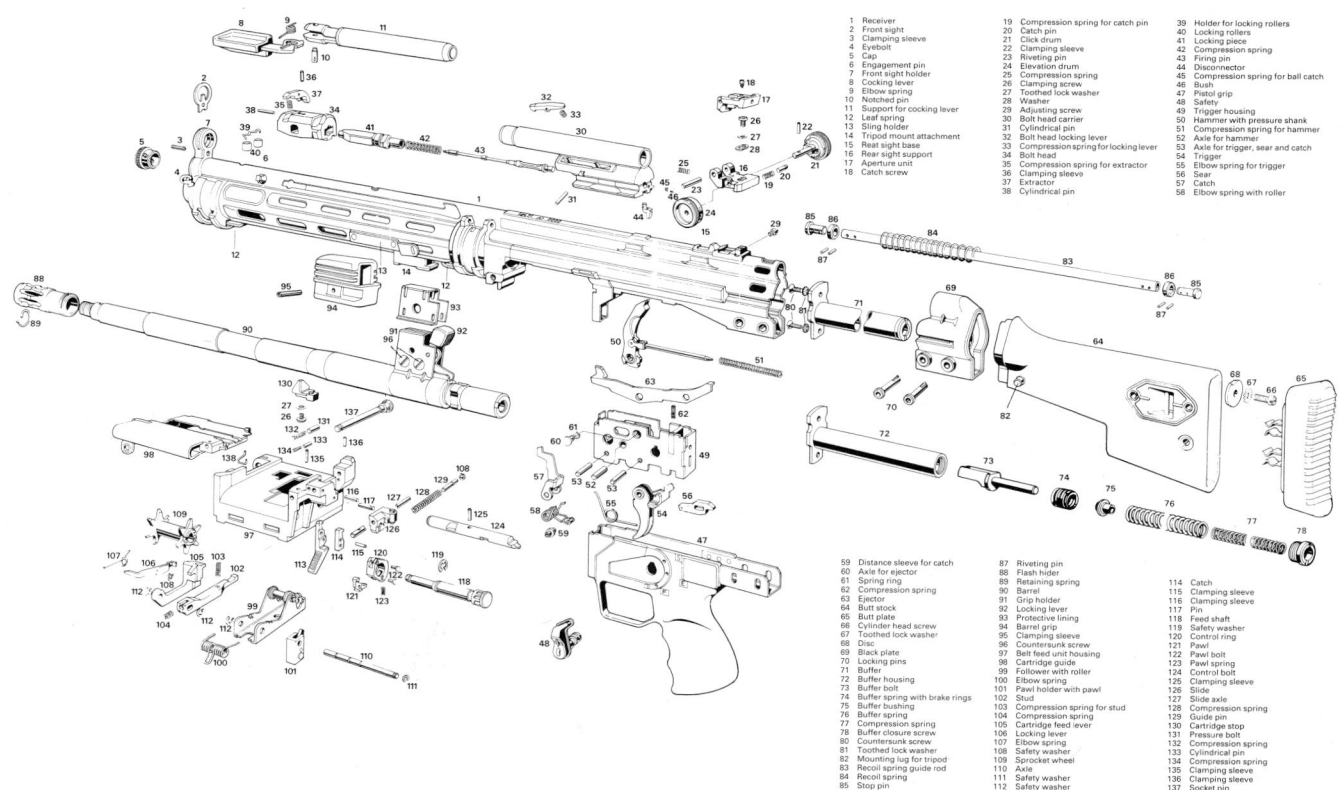

1	Receiver	19	Compression spring for catch pin	39	Holder for locking rollers
2	Front sight	20	Catch pin	40	Locking rollers
3	Clamping sleeve	21	Click drum	41	Locking piece
4	Eyebolt	22	Clamping sleeve	42	Compression spring
5	Cap	23	Riveting pin	43	Firing pin
6	Engagement pin	24	Elevation drum	44	Disconnector
7	Front sight holder	25	Compression spring	45	Compression spring for ball catch
8	Cocking lever	26	Clamping screw	46	Bush
9	Elbow spring	27	Toothed lock washer	47	Pistol grip
10	Notched pin	28	Washer	48	Safety
11	Support for cocking lever	29	Adjusting screw	49	Trigger housing
12	Leaf spring	30	Bolt head carrier	50	Hammer with pressure shank
13	Sling holder	31	Cylindrical pin	51	Compression spring for hammer
14	Tripod mount attachment	32	Bolt head locking lever	52	Axle for hammer
15	Rear sight base	33	Compression spring for locking lever	53	Axle for trigger, sear and catch
16	Rear sight support	34	Bolt head	54	Trigger
17	Aperture unit	35	Compression spring for extractor	55	Elbow spring for trigger
18	Catch screw	36	Clamping sleeve	56	Sear
		37	Extractor	57	Catch
		38	Cylindrical pin	58	Elbow spring with roller

59	Distance sleeve for catch	87	Riveting pin	114	Catch
60	Axle for ejector	88	Flash hider	115	Clamping sleeve
61	Spring ring	89	Retaining spring	116	Clamping sleeve
62	Compression spring	90	Barrel	117	Pin
63	Ejector	91	Grip holder	118	Feed shaft
64	Butt stock	92	Locking lever	119	Safety washer
65	Butt plate	93	Protective lining	120	Control ring
66	Cylinder head screw	94	Barrel grip	121	Pawl
67	Toothed lock washer	95	Clamping sleeve	122	Pawl bolt
68	Disc	96	Countersunk screw	123	Pawl spring
69	Black plate	97	Belt feed unit housing	124	Control bolt
70	Locking guns	98	Cartridge guide	125	Clamping sleeve
71	Buffer	99	Follower with roller	126	Slide
72	Buffer housing	100	Elbow spring	127	Slide axle
73	Buffer bolt	101	Pawl holder with pawl	128	Compression spring
74	Buffer spring with brake rings	102	Stud	129	Guide pin
75	Buffer bushing	103	Compression spring for stud	130	Cartridge stop
76	Buffer spring	104	Compression spring	131	Pressure bolt
77	Compression spring	105	Cartridge feed lever	132	Compression spring
78	Buffer closure screw	106	Locking lever	133	Cylindrical pin
80	Countersunk screw	107	Elbow spring	134	Compression spring
81	Toothed lock washer	108	Safety washer	135	Clamping sleeve
82	Mounting lug for tripod	109	Sprocket wheel	136	Clamping sleeve
83	Recoil spring guide rod	110	Axle	137	Socket pin
84	Recoil spring	111	Safety washer	138	Spring
85	Stop pin	112	Safety washer		
86	Guide ring	113	Catch lever		

7.62 mm HK 21A1 Heckler and Koch general-purpose machine gun

7.62 mm HK 21A1 Heckler and Koch general-purpose machine gun with belt feed system

housing and as the bolt travels backward and forward it moves the slide-bolt from side to side and this motion rotates the sprocket. The belt is housed in a box attached to the gun by a catch, and it hangs under the centre line of the gun.

Barrel changing

The gun is first cocked by pulling the cocking lever to the rear and engaging it in the recess. The barrel-locking lever is then depressed and the barrel grip turned upwards until the locking lugs disengage. The barrel is pushed forward to clear the locking recess and then pulled to the right and withdrawn backwards. Another barrel is inserted by reversing the procedure.

Stripping

MAJOR ASSEMBLIES

Unload and make sure that the weapon is clear and safe. Close the bolt onto the breech.

Remove both locking pins from the backplate and insert them into the tubular rivets in the stock.

Slide off the butt and pistol grip by pulling to the rear.

Pull back the cocking lever and remove the bolt assembly, recoil spring and its guide tube.

Push cocking lever forward again and remove the barrel.

Unlatch the belt feed unit and hinge it down together with the cartridge guide.

Withdraw the socket pin by pulling it out to the right of the feed unit and remove the entire unit.

Detach the bipod.

BOLT

Grasp the bolt-head with one hand and rotate it anti-clockwise to remove it from the locking piece. Remove the locking piece, firing pin and firing spring from the bolt-head carrier.

PISTOL GRIP

Let the hammer come forward.

Rotate the fire selection lever upwards and first lift out the fire selector lever and then the trigger housing from the pistol grip.

Any further stripping must be performed by an armourer.

Re-assembly

Attach the bipod and insert both barrel and bolt assembly and move bolt forward to meet breech.

Insert the feed guide and unit.

Attach pistol grip, making sure that the hammer is cocked before starting.

Attach butt and press in the locking pins from the left side.

Check for proper action by cocking and releasing the cocking lever several times.

Mountings

The usual mounting is the bipod, which has two positions on the gun, forward at the end of the receiver where it gives the maximum stability; or directly in front of the feed unit where it is near to the point of balance. The alternative mounting is the 1102 tripod, which is described separately.

Accessories

A comprehensive range of accessories is offered for the gun, enabling it to be used for several different purposes: belt box, carrying sling, anti-aircraft sights, telescopic sight, passive night sight and tripod mount.

DATA
Cartridge: 7.62 × 51 mm NATO
Operation: delayed blowback, selective fire
Feed: metal link belt
Weight, empty: (with bipod) 8.3 kg
Length: 1030 mm
Barrel: 450 mm
Sight radius: 590 mm
Rate of fire: (cyclic) 900 rds/min
Muzzle velocity: 800 m/s

Manufacturer

Heckler and Koch GmbH, D-7238 Oberndorf-Neckar.
Status
Production.

7.62 mm HK 21E and 5.56 mm HK 23E Heckler and Koch general-purpose machine gun

The 7.62 mm HK 21E and 5.56 mm HK 23E machine guns are the most recent developments of the HK 21A1 design, based on the results of stringent testing of that weapon. A number of technical modifications have been made

resulting, it is claimed, in higher efficiency and greater robustness and durability.

The HK 21E and HK 23E have the following significant modifications in

7.62 mm HK 21E Heckler and Koch general-purpose machine gun with belt-feed system

common: a 94mm extension of the receiver, giving a longer sight radius; reduced recoil, contributing to greater accuracy in all modes of fire; standard fitting of a burst-control trigger for three-round bursts, and provision of an attachable winter trigger; improved quick-change barrel grip; assault grip as standard fitting; rear drum sight – 100 to 1200 m (HK 21E); 100 to 1000 m (HK 23E) – with windage and elevation adjustment and side-wind correction. Other features include a special device for quiet closing of the bolt; carrying handle; cleaning kit housed in the grip; and replacement barrel for automatic firing of blank cartridges.

On the HK 21E the barrel has been lengthened to 560 mm. The belt transport system in the feed has been modified so that the belt is now transported in two operations, ensuring that the feed unit and the belt itself are subjected to lower stress. As the bolt travels forward, and following the ejection of a cartridge from the belt, one operation is effected for the following cartridge. When the bolt has opened and has moved to the rear, a second operation brings the cartridge fully into the feed position.

Both weapons are provided with a bipod with three elevation settings and the ability to traverse to 30° to either side and can also be mounted on the 1102 tripod and other Heckler and Koch mounts, described separately. The comprehensive range of accessories is the same as for the HK 21A1.

5.56 mm HK 23E Heckler and Koch general-purpose machine gun with belt-feed system

DATA

	HK21E	HK23E
Cartridge	7.62 mm × 51	5.56 mm × 45 (M193 or SS 109)
Operation	delayed blowback, automatic	delayed blowback, automatic
Feed	metal link belt	metal link belt
Modes of fire	automatic, 3-round burst, single-shot	automatic, 3-round burst, single-shot
WEIGHTS		
Gun (unloaded with bipod)	9.30 kg	8.75 kg
Barrel	2.2 kg	1.6 kg
Bipod	0.55 kg	0.55 kg
LENGTHS		
Length overall	1140 mm	1030 mm
Barrel	560 mm	450 mm
Rifling	1 turn in 305 mm	1 turn in 178 mm
Sight radius	685 mm	685 mm
FIRING CHARACTERISTICS		
Rate of fire (cyclic)	approx 800 rds/min	approx 750 rds/min
Muzzle velocity	approx 840 m/s	approx 950 m/s

Manufacturer
Heckler and Koch GmbH, D-7238 Oberndorf-Neckar.
Status
Production. Also manufactured under license in Mexico (HK21).

Service
Mexican Army and trials elsewhere.

7.62 mm HK 11A1 Heckler and Koch light machine gun

The HK 11A1 is a magazine-fed version of the HK 21A1 and employs all the same internal mechanisms. For the general system of operation reference should be made to the HK 21A1.

The major difference lies in the feed: instead of the belt feed unit a magazine unit is pinned to the receiver, using the same method of attachment, and 20-round box magazines from the G3 rifle are clipped into it using a conventional small catch.

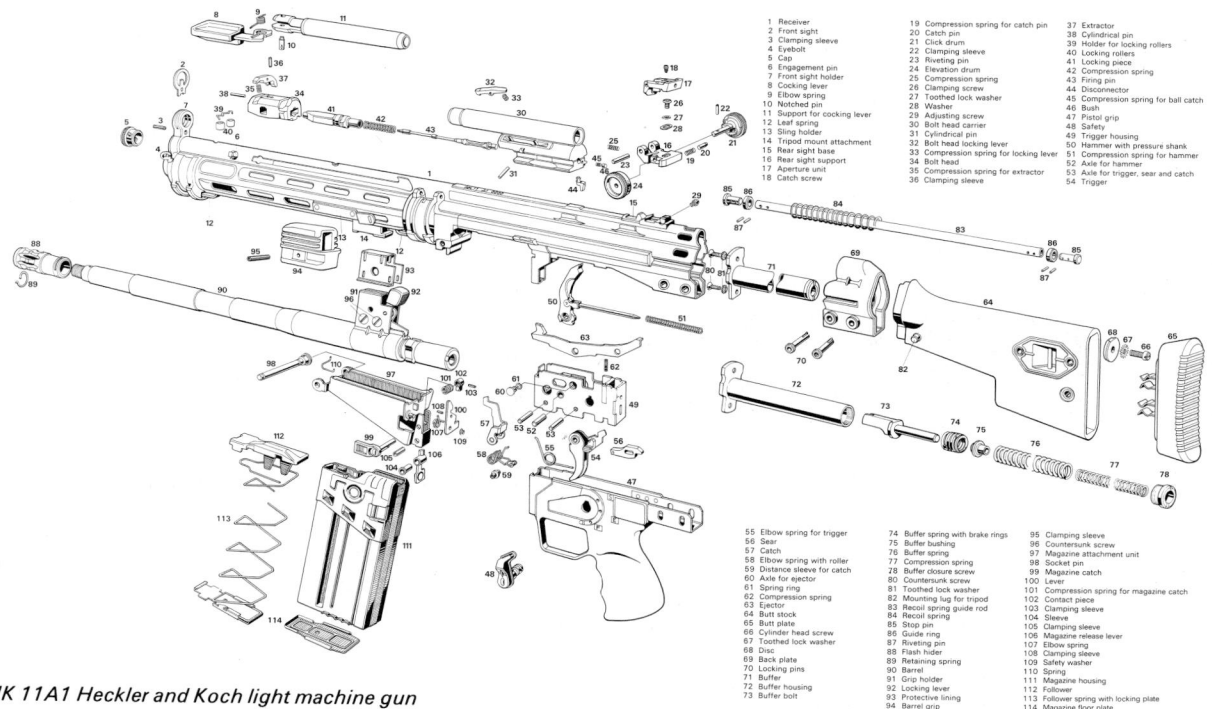

1 Receiver
2 Front sight
3 Clamping sleeve
4 Eyebolt
5 Cap
6 Engagement pin
7 Front sight holder
8 Cocking lever
9 Elbow spring
10 Notched pin
11 Support for cocking lever
12 Leaf spring
13 Sling holder
14 Tripod mount attachment
15 Rear sight base
16 Rear sight support
17 Aperture unit
18 Catch screw
19 Compression spring for catch pin
20 Catch pin
21 Click drum
22 Clamping sleeve
23 Riveting pin
24 Elevation drum
25 Compression spring
26 Clamping screw
27 Toothed lock washer
28 Washer
29 Adjusting screw
30 Bolt head carrier
31 Cylindrical pin
32 Bolt head locking lever
33 Compression spring for locking lever
34 Bolt head
35 Compression spring for extractor
36 Clamping sleeve
37 Extractor
38 Cylindrical pin
39 Holder for locking rollers
40 Locking rollers
41 Locking piece
42 Compression spring
43 Firing pin
44 Disconnector
45 Compression spring for ball catch
46 Bush
47 Pistol grip
48 Safety
49 Trigger housing
50 Hammer with pressure shank
51 Compression spring for hammer
52 Axle for hammer
53 Axle for trigger, sear and catch
54 Trigger

55 Elbow spring for trigger
56 Sear
57 Catch
58 Elbow spring with roller
59 Distance sleeve for catch
60 Axle for ejector
61 Spring ring
62 Compression spring
63 Ejector
64 Butt stock
65 Butt plate
66 Cylinder head screw
67 Toothed lock washer
68 Disc
69 Back plate
70 Locking pins
71 Buffer
72 Buffer housing
73 Buffer bolt
74 Buffer spring with brake rings
75 Buffer bushing
76 Buffer spring
77 Compression spring
78 Buffer closure screw
79 Countersunk screw
80 Toothed lock washer
81 Toothed lock washer
82 Mounting lug for tripod
83 Recoil spring guide rod
84 Recoil spring
85 Stop pin
86 Guide ring
87 Riveting pin
88 Flash hider
89 Retaining spring
90 Barrel
91 Grip holder
92 Locking lever
93 Protective lining
94 Barrel grip
95 Clamping sleeve
96 Countersunk screw
97 Magazine attachment unit
98 Socket pin
99 Magazine catch
100 Lever
101 Compression spring for magazine catch
102 Contact piece
103 Clamping sleeve
104 Sleeve
105 Clamping sleeve
106 Magazine release lever
107 Elbow spring
108 Clamping sleeve
109 Safety sleeve
110 Spring
111 Magazine housing
112 Follower
113 Follower spring with locking plate
114 Magazine floor plate

7.62 mm HK 11A1 Heckler and Koch light machine gun

7.62 mm HK 11A1 Heckler and Koch light machine gun

DATA
Cartridge: 7.62 mm × 51
Operation: delayed blowback, selective fire
Feed: 20-round box magazine
Weight: 7.7 kg
Barrel: 1.7 kg
Bipod weight: 600 g
Length: 1030 mm
Barrel length: 450 mm
Sights: (foresight) blade; (rearsight) aperture; (setting) 200-1200 m, click adjustment per 100 m
Muzzle velocity: 800 m/s
Rate of fire: (cyclic) 800 rds/min

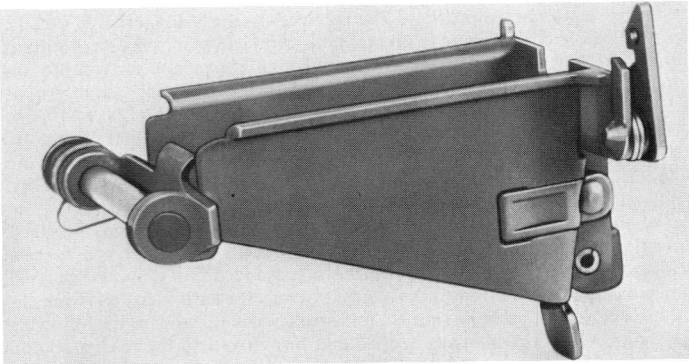

Magazine feed unit of 7.62 mm HK 11A1 Heckler and Koch light machine gun

Manufacturer
Heckler and Koch GmbH, D-7238, Oberndorf-Neckar.
Status
Production.
Service
Africa, Greece, South-east Asia.

Heckler & Koch Field Mount

The HK Field Mount is used for the deployment of H&K machine guns in the heavy support role. Designed as a tripod, the three legs can be individually adjusted in height. The mount is provided with elevation and azimuth adjusting devices which are set by means of clamping levers. A sprung recoil absorber enhances the rigidity of the mount during firing.

The HK Field Mount offers simple handling and training, excellent precision, low weight, portable design incorporating the means of carrying, and the quick insertion and removal of the weapon. It allows all H&K machine guns to be used with their attached belt case, and it permits fire on pin-point targets, long-range and short-range defensive fire, suppressive fire and area engagements.

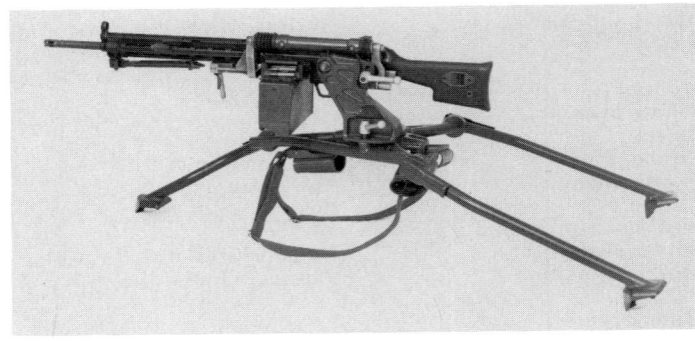

HK Field Mount with HK21E machine gun

DATA
Length, folded: 640 mm
Width, folded: 500 mm
Height, folded: approx 450 mm
Weight: 11.6 kg
Weight of carrying attachments: 1.3 kg
Muzzle height: from 350 to 550 mm
Azimuth angle: 40°side-to-side
Elevation angle: 20°

Manufacturer
Heckler and Koch GmbH, D-7238, Oberndorf-Neckar.
Status
Current. Production.

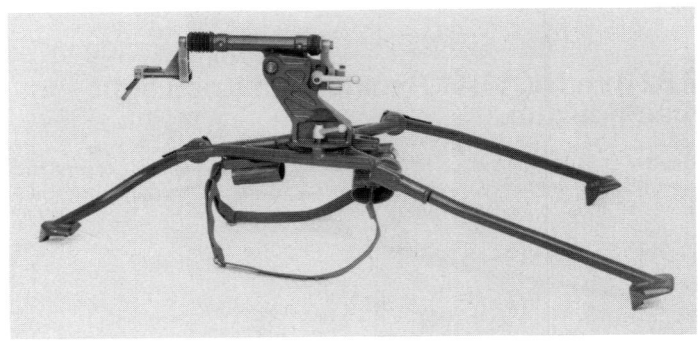

HK Field Mount without weapon

1102 tripod mount

The 1102 is a later development of the 1100 and it is specifically designed for the HK 21A1. It is a spring-buffered tripod which can be folded and carried into action on one man's shoulders. When set up its height (measured as bore centre line above ground) can be adjusted between 750 and 350 mm. This allows fire over low cover and where no cover is available the gun is sufficiently low to present a small target and also prevent any overturning moment. The rear legs of the tripod can be folded to allow use in a small weapon pit.

There is a traversing arc between the two rear legs which allows an arc of 43°. The rear attachment of the gun to the arc is by the elevating column which permits an elevation of 14°. If required the elevating column can rapidly be disconnected to allow a 360° traverse of the gun.

Attached to the elevating column housing is a dial sight with a periscope optical viewing eye piece allowing the gunner to operate the weapon without unduly revealing his position. The sight can be levelled by spirit level and, by reference to an aiming mark, it can be used to engage targets, the bearing and elevation of which have been previously recorded after ranging.

The sight also allows the gunner to produce traversing fire between set stops on the traversing arc and searching fire.

DATA
FOLDED DIMENSIONS
Length: 770 mm
Width: 570 mm
Height: 230 mm
Weight: 10.2 kg
Max traverse: 43°
Max elevation: 14°

1102 tripod mount for HK 21A1 general-purpose machine gun

2400 column mount

This mount is intended for use on a light vehicle. It is a spring-balanced system and with the butt of the weapon replaced by a receiver end cap the weapon can be traversed rapidly and elevated up to 75° or depressed to minus 15°. The traverse varies with the type of vehicle and the restrictions imposed by vehicle design, but with an open vehicle such as the Jeep or Land Rover, the gun can fire 90° on either side while the vehicle is moving. If the vehicle is stationary and the firer can leave his seat, the mount allows 360° traverse.

The same mounting can be used for a static anti-aircraft defence system.

2400 column mount ground target fire

2700 anti-aircraft/ground target mount

This system is designed for use on armoured vehicles with a top hatch. The weapon can be rotated rapidly through 360° and from minus 10° up to 75°. At any point it can be locked in position for the engagement of ground targets.

The gun is used without the butt stock and is so mounted that no turning moments are produced when firing.

2700 anti-aircraft/ground target mount

Circular track mount

This mount is designed to go on to unarmoured load-carrying vehicles. It can be fitted above the passenger seat in the cab of a lorry and allows the gun to deal with either ground or aerial targets.

The circular track rotates through 360° and allows rapid traverse. When it is locked in position the gun has a traverse of some 180° and can be elevated from minus 15° to 75°. The butt stock is removed and the rear of the gun is supported in prolongation of the barrel axis. Thus there is no turning moment when the gun is firing. The 100-round belt is in a box below the gun. To use the HK 21 in a dismounted role, the gun can be disengaged from the mounting, the butt stock fitted and the bipod positioned in a very short while.

Circular track mount. Anti-aircraft fire

Circular track mount. Ground target fire

DATA
Weight of the circular track mount: 33.5 kg
Diameter: (outside) 762 mm; (inside) 647 mm; (hole) 725 mm
Height of mount above roof: 310 mm; (with gun fitted) 325 mm

These mounts are also available with cradles for other machine guns.

5.56 mm HK 13 Heckler and Koch light machine gun

The HK 13 is a light machine gun using the 5.56 mm × 45 cartridge, is largely the same as the HK 33 rifle (which see) and operates in exactly the same way. Its physical dimensions are similar but it has the heavier quick-change barrel necessary for its light machine gun role.

The HK 13 will take the 25-round magazines used in the HK 33 rifle.

A bipod may be fitted either at the front of the barrel casing or at a centre point just in front of the magazine.

The light machine gun will produce selective fire. The change lever is above the pistol grip on the left of the receiver. The trigger and firing mechanism functions in the same way as that of the HK 33 rifle.

The HK 13 strips in the same manner as the HK 33.

DATA
Cartridge: 5.56 mm × 45
Operation: delayed blowback, selective fire
Method of delay: rollers
Feed: 25-round box magazine
Weight: 6.0 kg with bipod
Length: 980 mm
Barrel: 450 mm
Rifling: 4 grooves, rh, 1 turn in 305 mm
Sights: (fore) blade; (rear) 100 m V rearsight. Apertures for 200, 300 or 400 m. Adjustable for windage and elevation. Provision for fitting telescope sight
Sight radius: 541 mm
Muzzle velocity: 950 m/s
Rate of fire: (cyclic) 750 rds/min
Effective range: 400 m

5.56 mm HK 13 Heckler and Koch light machine gun field-stripped (25-round magazine)

Manufacturer
Heckler and Koch GmbH, D-7238 Oberndorf-Neckar.
Status
Current. No longer in production.
Service
Some South-east Asian forces.

5.56 mm HK 13 Heckler and Koch light machine gun with telescopic sight

7.62 mm HK 11E and 5.56 mm HK 13E Heckler and Koch light machine guns

The HK 11E and HK 13E light machine guns are developments of the HK 11A1 and HK 13 light machine guns described above and incorporate the technical modifications described in the entry for the Heckler and Koch HK 21E and HK 23E general-purpose machine guns including burst-control trigger, rear drum sight and assault grip. Both are primarily magazine-fed weapons, but they can be converted to belt feed by exchanging the bolt assembly with the magazine adaptor for bolt assembly with belt feed unit.

Like the HK 11A1, the HK 11E will accept the 20-round magazine of the G3 rifle and also the Heckler and Koch 50-round drum magazine. The HK 13E will accept the 20-round and the 30-round magazine of the G41 rifle, standardised according to STANAG 4179.

5.56mm Heckler & Koch HK11E light machine gun with 20-round box magazine

7.62 mm Heckler & Koch HK13E light machine gun with 20-round box magazine

DATA

	HK 11E	HK 13E
Cartridge	7.62 mm × 51	5.56 mm × 45
Operation	delayed blowback, automatic	delayed blowback, automatic
Feed	20-round box magazine and	30-round box magazine
	50-round drum magazine	20-round box magazine (M16 type)
Modes of fire	automatic, 3-round burst,	automatic, 3-round burst,
	single-shot	single-shot
WEIGHTS		
Gun (unloaded with bipod)	8.15 kg	8 kg
Barrel	1.7 kg	1.6 kg
Bipod	0.55 kg	0.55 kg
LENGTHS		
Length (overall)	1030 mm	1030 mm
Barrel	450 mm	450 mm
Rifling	1 turn in 305 mm	1 turn in 178 mm (SS 109)
Sight radius	685 mm	685 mm
FIRING CHARACTERISTICS		
Rate of fire (cyclic)	approx 800 rds/min	approx 750 rds/min
Muzzle velocity	approx 800 m/s	approx 950 m/s

Manufacturer
Heckler and Koch GmbH, D-7238 Oberndorf-Neckar.
Status
Production.

Service
Military trials.

7.92 mm MG42 and 7.62 mm MG1, MG2 and MG3 machine guns

The German Army entered the Second World War with the MG34 as its principal ground machine gun both in the infantry squad and in armoured vehicles. It was a slow and expensive gun to produce with forgings machined to close tolerances, and it had long been evident that a successor would be needed. Prototypes of a new gun, which subsequently became the highly successful MG42, were made in 1938 with the maximum use of stampings and the very minimum of forgings.

When the West German forces came into NATO they decided to modify

the MG42 design from 7.92 mm calibre to 7.62 mm × 51 and adopt it as their standard general-purpose machine gun. It was manufactured by Rheinmetall in 1959 which called it the MG42/59. The Bundeswehr called it the MG1. All MG1 versions (MG1, MG1A1, MG1A2 and MG1A3) were chambered only for the 7.62 × 51 mm cartridge in a West German 50-round continuous belt known as the DM1. The MG1A3 had some small changes intended to speed production including the rounded-muzzle booster.

In parallel with this development process, some of the original MG42

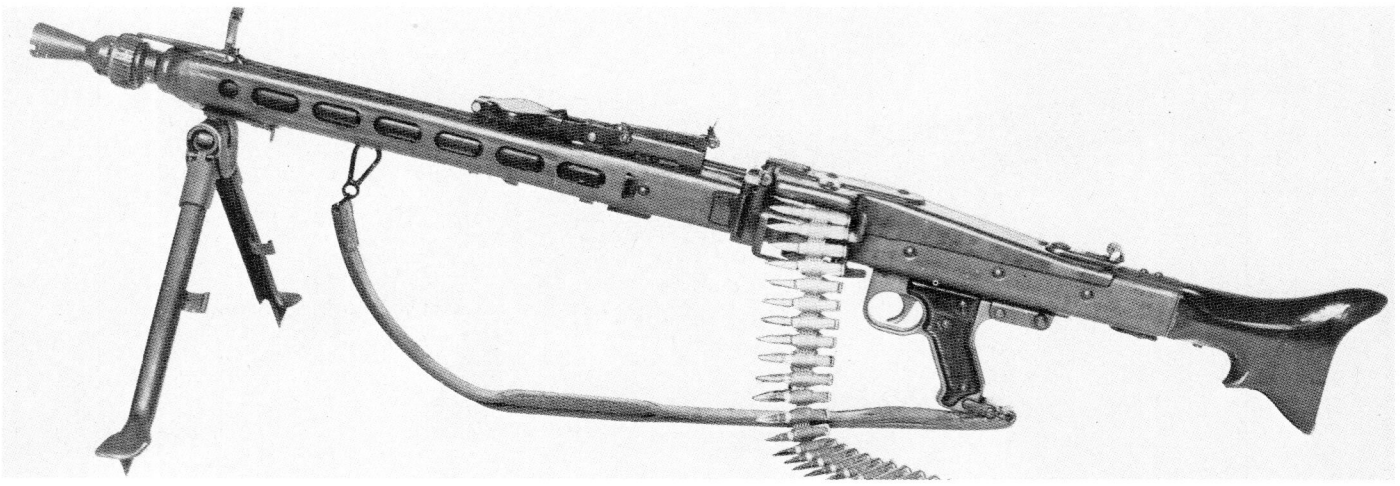

7.62 mm MG3 machine gun

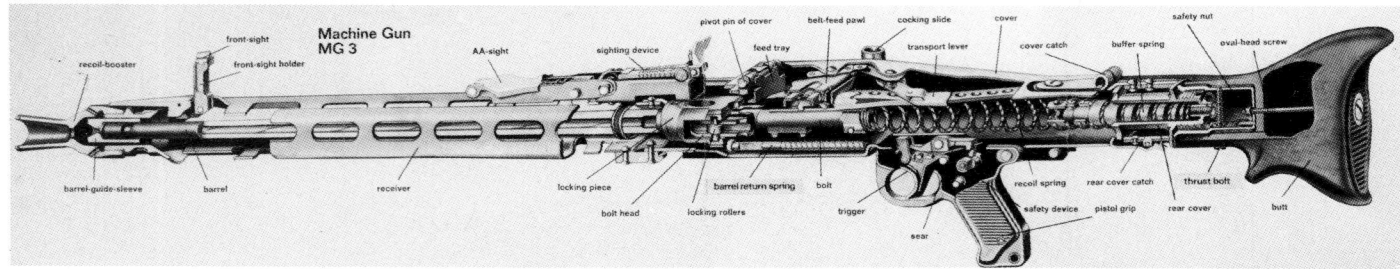

7.62 mm MG3 machine gun

weapons were converted from 7.92 to 7.62 mm calibre and were re-designated MG2. The current weapon, however, is a further development known as the MG3 which came into service in 1968; it has the external shape of the MG1A3 and can be fed from the West German DM1 continuous belt or either the West German DM6 or the US M13 disintegrating link belts. It has an AA sight and a belt retaining pawl to hold the belt up to the gun when the top cover plate is lifted.

Operation

The disintegrating link belt of the US M13 or the West German DM6-type can be used in the MG3. It is factory filled and is not intended to be refilled after use. The continuous belts of the West German DM1-type which are the only type used in the MG42, MG1 and MG1A3, and can also be used in the MG1A2 and MG3, can be refilled after use. To load these belts, cartridges are placed in each semi-circular link with the nib on the belt fitting into the extraction groove of the cartridge.

The T-shaped cocking handle is grasped and pulled straight back and the bolt cocked. When the bolt is to the rear, and only then, the safety catch above the pistol grip can be pushed through to the left to lock the sear.

The ammunition must be placed in the gun with the open side of the links downward so that the belt is above the cartridges. If it is put in with the link down, an immediate stoppage will result. If a feed tab is fitted this can be pushed straight through the gun. If not, the top cover plate must be lifted and the belt laid across the feed tray, cartridges downwards. When the rear of the cover plate is pressed down, the gun is ready for action. Pushing the safety catch from left to right with the side of the right thumb puts the gun in the

7.62 mm MG3 machine gun on anti-aircraft twin pedestal mount

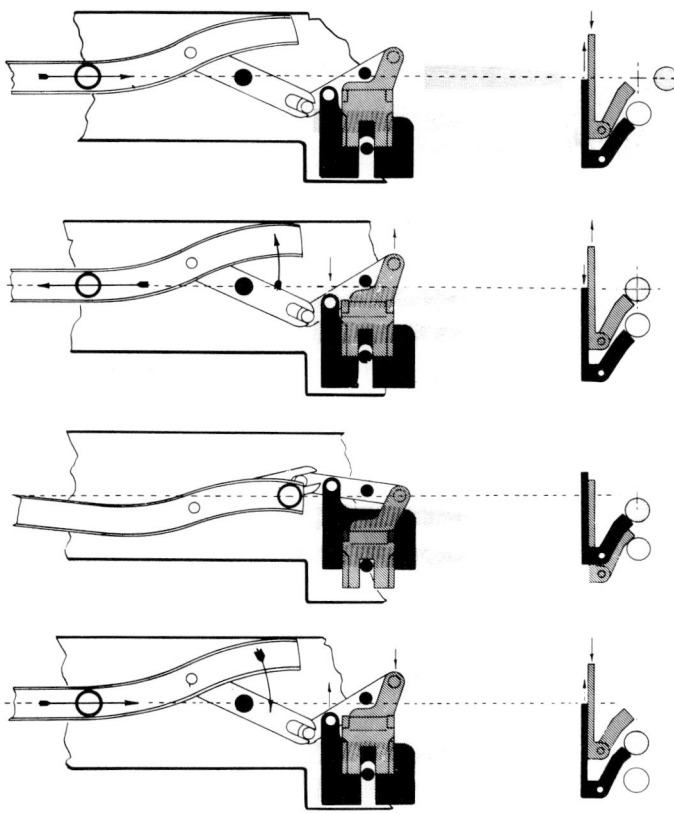

Bolt reciprocating motion pulls belt across gun (all versions)

'ready to fire' position. When the trigger is pressed, the sear, which engages the bent of the bolt, is lowered and the bolt goes forward, forcing a round out of the belt into the chamber.

The locking system of the gun is its most interesting and original feature. The bolt-head has an undercut slot on each side extending forward and

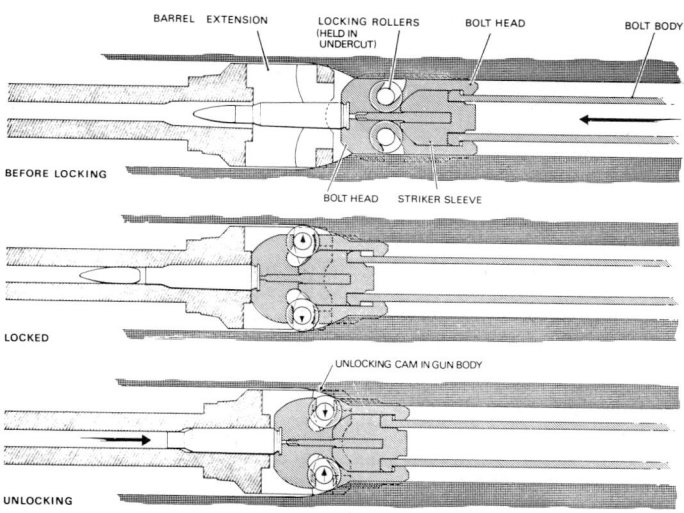

Gun locking and unlocking process (all versions)

inwards. In each of the undercuts rests a locking roller. The rollers are shaped like the wheel on a wheel barrow, ie a shaft and roller in one piece. As the bolt goes forward, the striker sleeve, impelled by the return spring, forces the rollers forward into the slots and completely clear of the receiver. There is therefore no contact between the rollers and the receiver, and no friction. The bolt-head enters the barrel extension and the rollers enter cammed slots which drive them back out of the undercut and then outwards. The outward motion is accelerated by the angled face of the striker sleeve which is driven forward by the full force of the return spring. The rollers move out, not only along the cam grooves in the barrel extension, but also along their grooves in the bolt-head and so the barrel extension and bolt-head are firmly locked together. As soon as the rollers are fully out the striker sleeve, carrying the striker, drives forward between them and the cap is struck.

The gas pressure developed in the chamber forces the breech face back but as the bolt-head is locked to the barrel, which is free to move, the barrel is pulled back with it. Gas pressure produced in a muzzle booster, or recoil intensifier, gives further energy to the barrel.

The shafts of the rollers hold the bolt-head to the barrel extension during firing and during the subsequent 8 mm of free recoil. The rollers then contact cam paths in the non-recoiling receiver and the rollers are forced inwards along the cam paths in the barrel extension and the bolt-head; the striker sleeve is forced back and the bolt is accelerated away from the barrel.

This acceleration of the bolt is an essential feature of all short-recoil operated machine guns of rifle calibre to impart sufficient energy to the bolt to carry out the cycle of operations. In the MG3 it is accomplished during the process of unlocking the bolt from the barrel. The roller shafts travel in both the barrel extension and the bolt-head and complete their travel in both components in the same time. However the length of their two paths is not the same and since the distance along the path in the barrel extension is greater than that in the bolt-head, the bolt-head must travel faster than the barrel and so is accelerated relative to it.

As the bolt is freed from the barrel extension, after a total travel of 21 mm, the barrel return spring drives the barrel forward to its run out position and the empty case in the chamber is held by the extractor on the bolt-face as the bolt continues rearwards. The ejector is a two-piece rod. The rear portion strikes the buffer spring and the front portion is forced out of the top of the breech face to pivot the case about the extractor and out of the bottom of the gun. The buffer spring is very powerful and returns the bolt to the forward position at a high velocity.

The belt movement is produced by a stud on top of the bolt riding in a curved feed channel in a feed arm. This feed arm rests under the top cover plate and is pivoted at its rear end. As the bolt reciprocates, the front of the feed arm moves across the receiver and operates a lever attached to the belt feed slide. This slide has two sets of spring-loaded pawls mounted one on each side of the centrally positioned slide pivot. Thus when one set of pawls is moving out and springing over the rounds in the belt, the other set is pulling the belt in. Each set of pawls in turn moves the belt across one half pitch. This sharing of the load reduces the forces on the belt and the feed mechanism. It produces a smooth belt flow rather than a series of jerky belt movements.

Trigger and firing mechanism

Due to the high rate of fire the bolt velocity is very considerable and the connection of the trigger sear and bent of the bolt must take place with a full face engagement. If the sear is only partly up when the bent hits it, there will be chipping of the mating surfaces. To prevent this the mechanism incorporates a controlled sear which rises at a predetermined point relative to the position of the bolt to ensure full area contact of the two parts.

The arrangement of the sear is shown in the diagram. The trigger carries a light, pressed-steel tripper which has its own spring forcing it forward. The tail of the sear passes through the tripper and a T-bar on its end limits the forward rotation of the tripper. The front face of the tripper has a step projecting forward. When the firer pulls the trigger, the trigger rotates and the tail of the sear is pushed up and the other end of the sear, the nose, pivots down and disengages from the bent of the bolt. The bolt is driven forward by the double-wound return spring. The spring of the tripper rotates it forward and the step rides under the T-bar of the raised tail of the sear. When the firer releases the trigger the tail of the sear can only fall as far as the step on the

7.92 mm MG42 machine gun showing shape of muzzle booster, compare with MG3

7.92 mm MG42 machine gun on its tripod mounting

7.92 mm MG42 and MG1 machine gun bolt

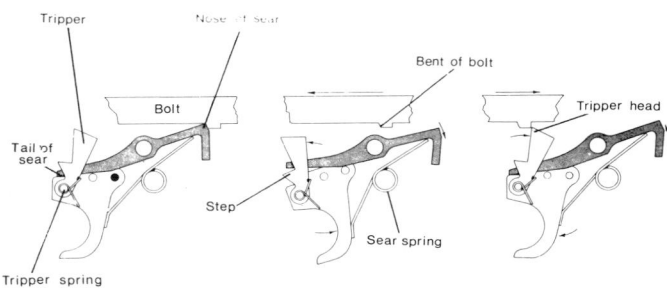

Trigger mechanism, all versions

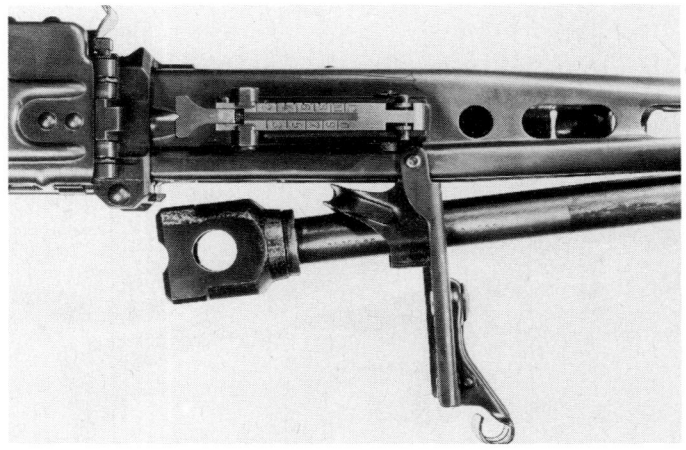

Barrel change, all versions

tripper and it comes to rest on the step. The nose of the sear therefore cannot rise to intercept the bolt. The rotation of the trigger pushes the tripper, which is attached to it, up into the path of the bolt which rides over it on the forward stroke and fires another round. On the recoil stroke the bolt hits the tripper head and rotates it back. This moves the step from under the tail of the sear which then falls and the nose of the sear rides up into the path of the bolt. The bolt rides over it as it moves back and continues back on its over-run. By the time the bolt comes forward again the nose of the sear has risen fully and presents its full frontal area to the bent. By this means the damage caused by high velocity contact and inadequate surface areas is avoided.

Rate of fire
The gun in its original form as the MG42 fired at about 1200 rds/min.
The standard bolt used in the MG1A1, MG1A3 and the MG3 weighs 550 g and produces a rate of fire of 1100 ± 150 rds/min. The MG1A2 uses a heavy bolt weighing 950 g and this produces rates of fire of about 900 rds/min. The West German Army uses the lighter bolt but the Italian MG42/59 uses the heavier bolt.

Sights
The foresight is mounted on the front end of the barrel casing and hinges flat. The rearsight is a U notch that is mounted on a slide moving on a ramp. Graduation is from 200 to 1200 m on the MG3. The MG42 was from 200 to 2000 m. Since the foresight is not on the barrel zeroing must be carried out on the rearsight.

Barrel change
The barrel must be changed at frequent intervals. Firing in short bursts at a rate of about 200-250 rds/min, the barrel should be changed after 150 rounds, that is three 50-round lengths. The barrel change is very quic k and simple. The gun is cocked, and the barrel catch on the right of the barrel casing is swung forward. The breech end of the hot barrel swings out and can be removed by elevating the gun. A cool barrel is pushed through the barrel catch and the muzzle bearing. When the catch is rotated back the barrel is locked.

Stripping
The process of stripping is uncomplicated. The bolt is retracted, the chamber checked and the bolt returned to the forward position. A catch below the receiver allows a 90° rotation of the butt and buffer spring out of the receiver. The large T-shaped cocking handle on the right of the receiver will withdraw the bolt and the return spring. The barrel catch moves outwards on the right of the breech to permit withdrawal of the barrel. The bolt can be further broken down by holding the bolt-head stationary with the rollers fully out and then rotating the bolt body through 90°. The bolt body and head can be separated and the ejector and spring-loaded insert removed.

Accessories
A buffered tripod to allow sustained fire is available. A dial sight allowing

engagement of unseen targets and recording of previously registered targets can be fitted to the tripod. A blank-firing attachment can be fitted in lieu of the normal-recoil booster at the muzzle.

DATA
Cartridge: 7.62 mm × 51
Operation: short recoil, automatic
Method of locking: roller locking
Feed: belt
Weight: 11.05 kg with bipod
Length: 1225 mm; (1097 mm without butt)
Barrel: (with extension) 565 mm; (without extension) 531 mm
Sights: (fore) barleycorn; (rear) notch. Also an AA sight
Sight radius: 430 mm

Muzzle velocity: 820 m/s
Rate of fire: (cyclic) 700-1300 rds/min
Effective range: bipod 800 m, tripod 2200 m

Manufacturer
Rheinmetall GmbH, Ulmenstrasse 125, D-4000 Dusseldorf. Licensed production in Italy by Beretta, Luigi Franchi and Whitehead Moto-Fides. Also production in Greece, Pakistan, Spain and Turkey.
Status
Current. Production.
Service
West German armed forces. Also with the forces of Austria, Chile, Denmark, Greece, Iran, Italy, Norway, Pakistan, Portugal, Spain, Sudan and Turkey.

MG3 belt drum

This belt drum has been developed by Heckler & Koch for the West German Army. Its purpose is to hold the normally loose belt of a length of approximately 80 cm. The belt, when loose, is exposed to fouling and is prone to become damaged, especially during a change of firing position, and these may cause malfunctions of the weapon due to belt twist or displaced cartridges.

The belt drum holds 100 belted rounds and is latched on the left of the weapon feed unit. The rear side of the drum is provided with a transparent cover serving as a visual indicator for the amount of ammunition available.

For transport a number of drums may be easily linked together, forming convenient stacks. The drum is made of highly durable synthetic material and consists of only two parts. It may be safely employed in temperature ranges between 63°C and -35°C. When empty it weighs only 200 g.

Manufacturer
Heckler & Koch GmbH, Postfach 1329, D-7238 Oberndorf/Neckar.
Status
Current. Production.
Service
West German Army.

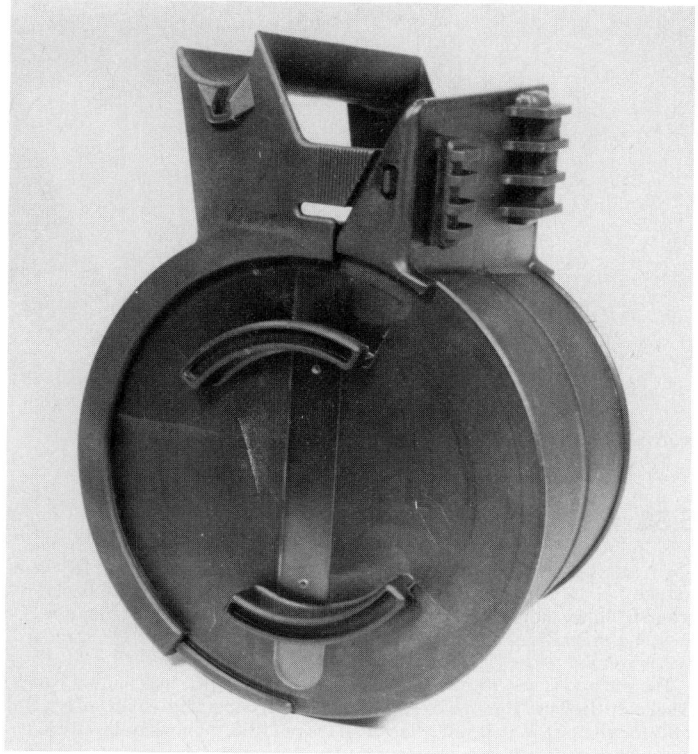

MG3 machine gun with belt drum in position

MG3 belt drum

GREECE

7.62 mm MG3 and HK11A1 machine guns

Hellenic Arms Industry (E.B.O.) SA is currently manufacturing the well-known 7.62 mm MG3 general-purpose machine gun as the principal machine gun for the Greek Army. In addition the company is also manufacturing the Heckler & Koch HK11A1 light machine gun, which has been adopted by the Greek Army as the light support weapon.

The relevant technical specifications can be found by referring to the West German section.

Manufacturer
Hellenic Arms Industry (E.B.O.) SA, 160 Kifissias Avenue, 11525 Athens.
Status
Current. Production.
Service
Greek Army.

7.62 mm HK11A1 machine gun manufactured by Hellenic Arms Industry

7.62 mm MG3 machine gun manufactured by Hellenic Arms Industry

ISRAEL

5.56 mm and 7.62 mm Model ARM Galil light machine gun

When the Israelis adopted the 5.56 mm Galil assault rifle family as standard weapons they did so with the intention that the new weapons should be used in the sub-machine gun, rifle and light machine gun roles. The ARM version of the weapon (described in the Rifles section) is fitted with a bipod and offers a choice of magazine sizes up to 50 rounds and is thus eminently suitable for use in the light machine gun role.

Details of the weapon will be found in the corresponding rifle entry.

Manufacturer
Israel Military Industries, PO Box 1044, Ramat Ha Sharon 47100.
Status
Current. Production.
Service
Israeli forces and other armies.

5.56 mm Model ARM Galil light machine gun

5.56 mm Negev light machine gun

The Negev light machine gun is a multi-purpose weapon that can be fed from standard belts, drums or magazines and fired from the hip, bipod, tripod or ground/vehicle mounts. It is designed to fire standard 5.56 × 45 mm NATO SS-109 ammunition, and with a replacement barrel will fire US standard M193 rounds.

The weapon is gas operated with a rotating bolt which locks into a barrel extension. It fires from an open bolt. It can be operated either as a light machine gun or as an assault rifle. As an assault rifle, manoeuvrability can be increased and weight reduced by removal of the bipod and installation of a short barrel and magazine. It can be fired in semi-automatic and automatic modes from standard magazines fitting beneath the weapon; Galil magazines or M16 magazines with adapters can be used.

The gas regulator has three positions; in normal conditions, position 1 gives a rate of fire of 650-800 rds/min; position 2 gives 800-950 rds/min; and position 3 shuts off gas supply to the piston so that the weapon can be used for launching grenades.

The Negev can be easily and quickly stripped into six sub-assemblies (including the bipod). All parts, including the quick-change barrels, are fully interchangeable. Various types of telescope and sight mounts can be installed on the receiver.

5.56 mm Negev light machine gun with belt feed unit

DATA
Cartridge: 5.56 × 45 mm SS109 or M193
Operation: gas, open bolt, selective fire
Method of locking: rotating bolt
Feed: 30-round box magazine, or belts, or drums
Weight: 7.2 kg empty, without magazine or belt
Length: (long barrel) (stock extended) 1020 mm; (stock folded) 780 mm: (short barrel) (stock extended) 890 mm; (stock folded) 650 mm
Barrel: (long) 460 mm; (short) 330 mm. Standard rifling 6 grooves, rh, one turn in 275 mm; optional one turn in 472 mm
Sights: (fore) post, adjustable for elevation and windage for zeroing; (rear) aperture, 300-1000 m; folding night sight with tritium illumination
Rate of fire: 650-950 rds/min (see text)

5.56 mm Negev machine gun stripped to its major units

Manufacturer
Israel Military Industries, PO Box 11044, Ramat Hasharon 47100.
Status
Current. Production.

Service
Israeli defence force.

ITALY

7.62 mm MG42/59 machine gun

This is the standard Italian general-purpose machine gun and is manufactured under licence from Rheinmetall.

Information on the method of operation is given under MG3 in the German section. However, there are a number of differences in dimensions and data as noted below.

DATA
Cartridge: 7.62 × 51 mm NATO
Operation: short recoil, automatic
Method of locking: rollers
Feed: belt
Weight, with bipod: 12.0 kg
Length overall: 1220 mm
Barrel: 567 mm with extension
Rifling: 4 grooves, right-hand twist
Sights: front post; rear notch, graduated 200-2200 m. Sight base 432 mm
Rate of fire: ca 800 rds/min
Muzzle velocity: 820 m/s
Effective range: bipod 400-500 m; tripod 800-1000 m
Maximum range: 3650 m at 33° elevation

Manufacturers
Pietro Beretta SpA, Brescia; Luigi Franchi SpA, Brescia; Whitehead Moto-Fides SpA, Livorno.
Status
Current. Production.
Service
Italian Army. Exported to Chile, Denmark, Mozambique, Nigeria, and Portugal.

7.62 mm 42/59 machine gun

5.56 mm Beretta AR70-84 light machine gun

The AR70-84 is a light machine gun firing 5.56 mm × 45 cartridges and is based on the AR-70/.223 Beretta rifle, with which it has a number of parts in common. It has superseded the earlier Model 70-78, which also formed part of this rifle family. It operates on the same gas system and the main alterations are a heavier barrel with a perforated steel hand-guard, fixing points for vehicle or other types of mounting, and a new butt with shoulder rest and facilities for gripping it more firmly.

The AR70-84 fires from an open bolt; otherwise the gas system and rotating bolt is the same as that employed with the rifles. The quick-change barrel of the earlier 70-78 model has been abandoned. Iron sights are fitted as standard,

and optical or electro-optical sights can be accommodated, provided that the receiver has been suitably modified during manufacture.

The bipod legs have a hinge in the middle of each leg to allow for coarse adjustment for height when using different cover.

The muzzle is machined to act as a grenade launcher, and the grenade sight folds down in front of the foresight when not in use.

DATA
Cartridge: 5.56 mm × 45
Operation: gas, selective fire
Locking: rotating bolt
Feed: 30-round box magazine
Weight: 5.3 kg
Length: 955 mm
Barrel: 450 mm
Rifling: 4 grooves, rh, one turn in 304 mm
Sight radius: 507 mm
Muzzle velocity: 970 m/s
Rate of fire: (cyclic) 670 rds/min

Manufacturer
Pietro Beretta SpA, 25063 Gardone Valtrompia, Brescia.
Status
Current. Available.

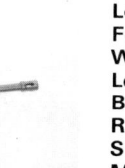

5.56 mm Beretta AR70/84 light machine gun

5.56 mm Beretta AS70/90 light machine gun

This is the squad automatic weapon designed to accompany the AR70/90 rifle, with which it forms the 70/90 system.

The AS70/90 uses the same gas-operated, rotating bolt system as the rifle, but fires from an open bolt and has a heavier barrel which canot be quick-changed. There is a prominent metal hand-guard, and an articulated bipod which can be adjusted to several positions. Grenades may be launched from the muzzle, though the launcher is of a different pattern to that used with the rifles. The same carrying handle is fitted, and like those on the rifles, can be removed to expose a mount for optical or electro-optical sights. The stock is of rather unusual design, but it gives good support at the shoulder and a good grip for the firer's disengaged hand.

DATA
Cartridge: 5.56 × 45 mm NATO
Operation: gas, selective fire
Locking system: rotating bolt
Feed: 30-round box to STANAG 4179 (M16 standard)
Weight: 5.34 kg without magazine and bipod
Length: 1000 mm
Barrel: 465 mm
Rifling: 6 grooves, rh, one turn in 178 mm
Sights: (front) post; (rear) two-position aperture, micrometer adjustment, for 300 and 800 m
Sight base: 555 mm
Rate of fire: ca 800 rds/min

Manufacturer
Pietro Beretta SpA, Via Pietro Beretta 18, I-25063 Gardone Val Trompia, Brescia.
Status
Current. Available.
Service
Undergoing trials by the Italian Army.

5.56 mm Beretta AS70/90 light machine gun

5.56 mm Franchi LF23/E light machine gun

This has been developed by Franchi in conjunction with Heckler & Koch and is essentially the HK 23E machine gun as described in the German section. There is a minor difference in dimension, the length being 1030 mm, and the chromed barrel is polygonally rifled one turn in 178 mm to suit the NATO standard 5.56 mm cartridge.

Manufacturer
Luigi Franchi SpA, Via del Serpente 12, I-25020 Fornaci, Brescia.
Status
Current. Production.
Service
Trials by Italian Army.

5.56 mm Franchi LF23/E light machine gun

JAPAN

7.62 mm Model 62 machine gun

This gun started as the Model 9M and was accepted into service with the Japanese Self-Defence Forces in 1962.

The gas cylinder and piston are below the barrel. The locking system is very unusual in that it is a tilting block with the front of the bolt forced up by cams on the piston extension (slide) to lock. Two wings level with the centre line move into recesses in the receiver and the bolt is held in position by the piston extension under it. The final movement of the piston, after locking is completed, carries the firing pin, fixed to the piston post, through the block, into the cartridge cap. Until the front of the block has risen there is no hole for the firing pin.

After the round has fired, some of the propellant gas passes through the gas port and drives the piston to the rear. The firing pin is withdrawn and the front of the bolt is first cammed down and then pulled back.

The extraction is unusual. There is no spring-loaded extractor hook but while the round is in the chamber, before firing, there is a spring-loaded plunger forced up into the cannelure from below. When the front end of the bolt is carried down, a fixed hook on the bolt-face above the firing pin hole, drops down and grips the cannelure of the cartridge head. When the bolt starts back the case is withdrawn. There appears to be no primary extraction.

The feed system uses a bolt-operated feed arm which oscillates as the bolt reciprocates and the belt is drawn across in two stages corresponding to the extraction and feed strokes.

The barrel-retaining catch is depressed by lifting the top cover plate. The carrying handle is rotated and then the barrel can be pushed forward. While the top cover plate is lifted it prevents the front of the bolt from rising into the locked position and so the firing-pin hole is blocked. This prevents the feed and firing of a round when this is unlocked or there is no barrel in position.

The Model 62 is a somewhat complex gun but has a reputation for reliability and accuracy.

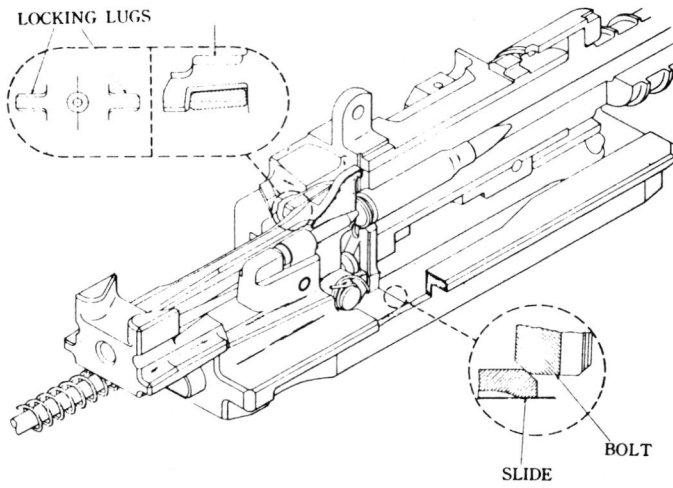

7.62 mm Model 62 machine gun locking mechanism

7.62 mm Model 62 general-purpose machine gun

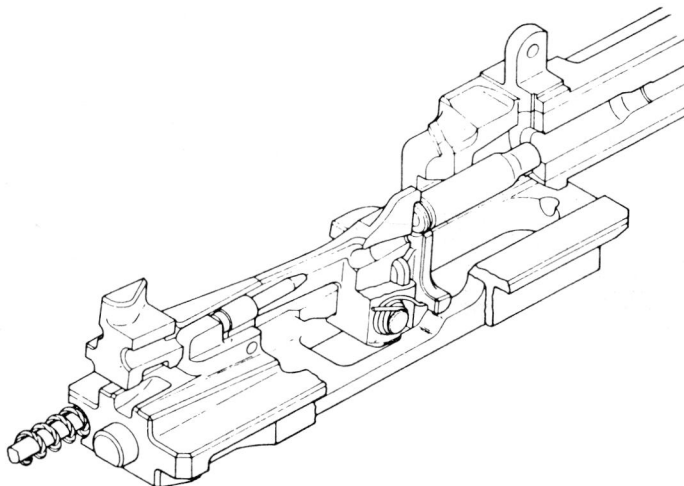

DATA
Cartridge: 7.62 mm × 51
Operation: gas, automatic
Method of locking: front tilting block
Feed: disintegrating link belt
Weight: 10.7 kg
Barrel weight: 2 kg
Length: 1200 mm
Barrel: 524 mm
Sight radius: 590 mm
Muzzle velocity: 855 m/s
Rate of fire: (cyclic) 600 rds/min

Manufacturer
Sumitomo Heavy Industries Ltd, 2-1-1 Yato-Cho, Tanashi-Shi, Tokyo.
Status
Current.
Service
Japanese Self-Defence Forces.

7.62 mm Model 62 machine gun positive extraction

7.62 mm Model 74 machine gun

The Model 74 is a variant of the standard Model 62 and was adopted by the Japanese Defence Agency in 1974 as a coaxial gun for armoured vehicles. It is mounted in the tanks and armoured personnel carriers of the Self-Defence Forces, the guns being in a special mount in the front glacis plate.

The gun is very similar to the Model 62, but uses more robust components to maintain reliability. At an average rate of fire of 300 rds/min the gun can fire without difficulty for three minutes or any need to change the barrel. The trigger is operated either by a solenoid or by the manual triggers at the rear of the body and the firing rate is adjustable by means of a selector lever on the right lower side of the body.

All guns can be dismounted from their vehicle if need be and mounted on the infantry tripod and used as support machine guns.

DATA
Cartridge: 7.62 mm × 51
Operation: gas

Locking: front tilting block
Feed: M14 link belt, left-hand side
Weight: 20.4 kg
Length: 1085 mm
Barrel: 625 mm
Sight radius: 306 mm
Rate of fire: (cyclic) 700-1000 rds/min
Muzzle velocity: 855 m/s

Manufacturer
Sumitomo Heavy Industries Ltd, 2-1-1 Yato-Cho, Tanashi-Shi, Tokyo.
Status
Current. Production.
Service
Japanese Self-Defence Forces.

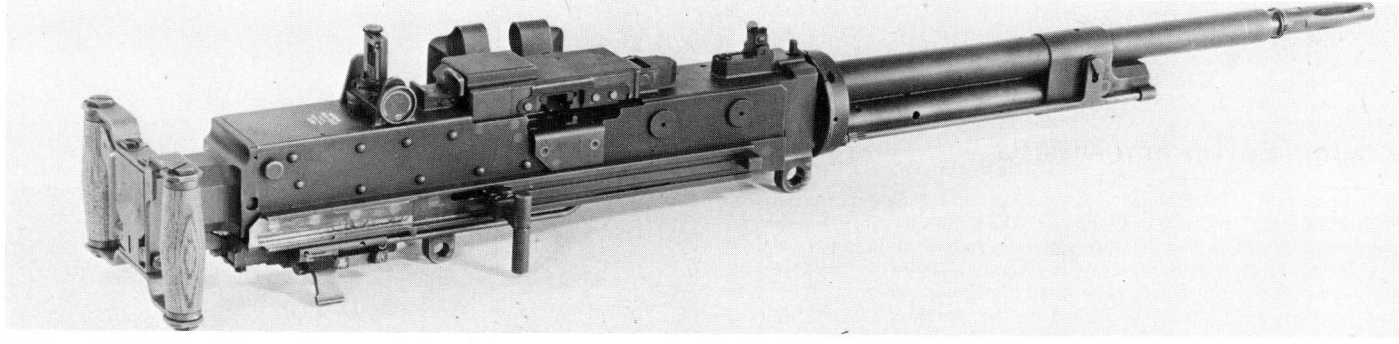

7.62 mm Model 74 vehicle machine gun. Trigger handles are unusual in modern gun, as are iron sights. This gun is not fitted with solenoid trigger

KOREA, SOUTH

5.56 mm K3 light machine gun

This is a gas-operated, air-cooled, belt-fed light machine gun with quick-change barrel. The gas system uses a piston, bolt carrier and rotating bolt and there is an adjustable gas regulator.

DATA
Calibre: 5.56 × 45 mm M198 or SS109
Operation: gas, selective fire
Locking: rotating bolt
Feed: 100-round link belt
Weight: 6.1 kg with bipod
Length: (butt folded) 830 mm; (butt extended) 1000 mm
Rifling: 6 grooves, rh
Muzzle velocity: 1000 m/s
Rate of fire: 700-900 rds/min

Manufacturer
Daewoo Precision Industries Ltd., PO Box 37, Dongrae, Pusan.
Status
Current. Production.

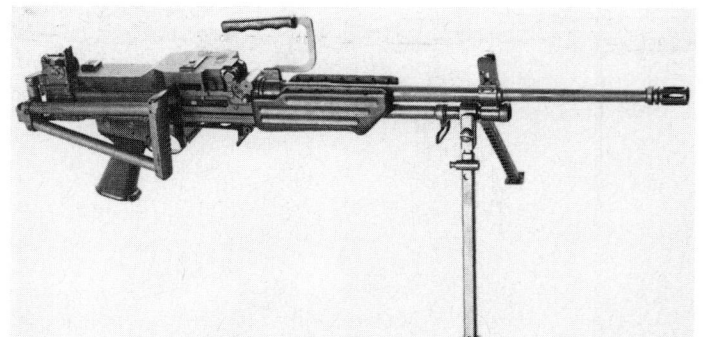

5.56 mm K3 light machine gun

Service
South Korean Army.

MEXICO

.30 RM2 Mendoza light machine gun

This is a very lightweight light machine gun, designed and built in Mexico by Productos Mendoza, and is an improved version of an earlier weapon known as the Model 45. The RM2 is currently in service with Mexican forces.

A simple gas-operated weapon, the RM2 has a top-mounted detachable box magazine, a permanently attached bipod and a muzzle brake. The action provides a selective fire option: the barrel can be changed quickly, using a cartridge as a releasing tool. The firing pin can be reversed in case of breakage; and access to the working parts, for this or any other reason, is obtained by hinging the butt and trigger mechanism down: all the working parts may then be drawn to the rear in much the same way as they can on the Type D BAR. The action is locked by a turning bolt.

DATA
Cartridge: 30-06
Operation: gas, selective fire
Method of locking: rotating bolt
Feed: 20- or 32-round detachable box magazine
Weight: 6.4 kg
Length: 1100 mm
Barrel: 610 mm. No barrel change
Rate of fire: (cyclic) can be regulated between 450 and 650 rds/min

Manufacturer
Productos Mendoza SA, Bartolache 1910, Mexico City 12.
Status
Obsolescent. No longer in production.
Service
Reserve stocks of Mexican armed forces.

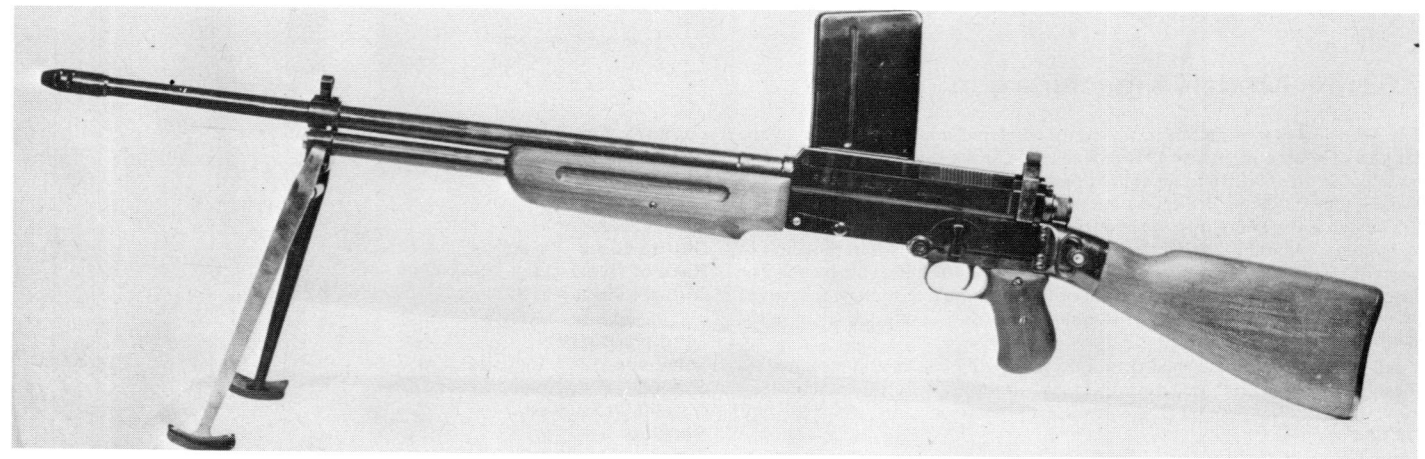

.30 RM2 Mendoza light machine gun

NORWAY

Vinghgs Softmount NM152

The Vinghgs NM152 Softmount is a recoil-absorbing cradle developed for the .50 Browning M2HB heavy machine gun. The Softmount supports the gun receiver and interfaces with the normal ground, vehicle and naval mounting. It can be fitted with shoulder supports and can be adapted to carry accessory sights. It has been designed with operation in extreme climatic conditions in mind, and has been successfuly tested by the Norwegian Navy Materiel Command.

DATA
Length: 850 mm
Length with shoulder rest: 1.20 m
Width, shoulder rest: 350 – 470 mm
Width hand grips: 500 mm
Weight: 17.5 kg

Softmount in use, with shoulder support

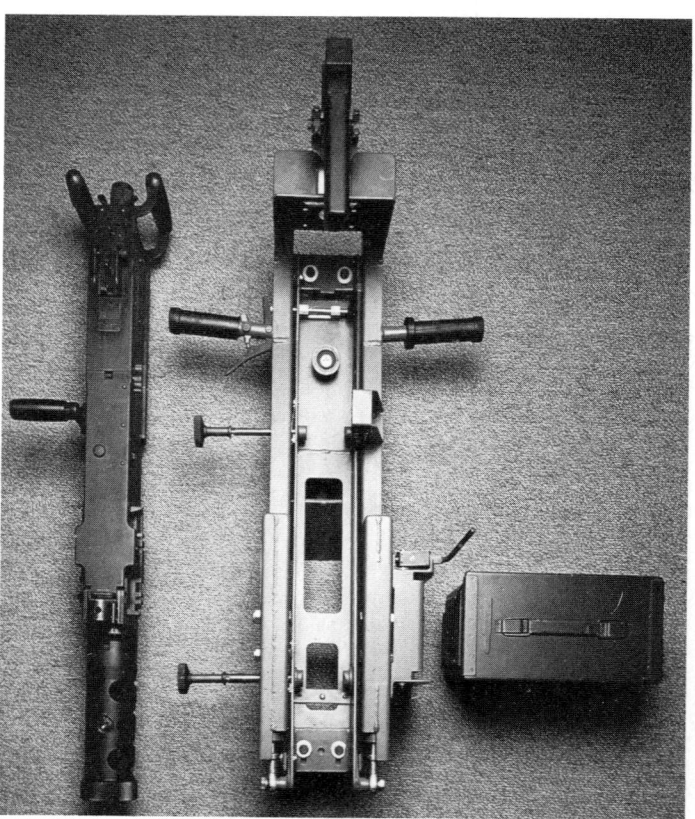

Top view of Softmount, with gun and ammunition box

Manufacturer
Vinghgs Mek. Verksted A/S, Lindholmvn. 14, N-3133 Duken.
Status
Current. Production.

Service
Norwegian armed forces.

PORTUGAL

The Portuguese company INDEP manufactures the Heckler and Koch 7.62 × 51 mm HK21 machine gun under licence. Full details of the weapon can be found in the German section.

Manufacturer
INDEP, Frabrica de Braco de Prata, Rua Fernando Palha, 1899 Lisboa-Codex.

Status
Current. Production.
Service
Portuguese Army.

SINGAPORE

5.56 mm Ultimax 100 light machine gun

Chartered Industries of Singapore (CIS) began designing the Ultimax 100 light machine gun in 1978 to meet a need for an automatic weapon for an infantry squad which would be light enough to be operated by one man moving with the rest of the squad in assault. Mobility, firepower and ease of handling were primary design considerations. Three years development resulted in a robust light machine gun which, it is claimed, handles like an assault rifle and weighs, when fully loaded with a 100-round magazine, less than any other 5.56 mm machine gun when empty.

Since accuracy in automatic fire depends on a weapon's controllability, the Ultimax 100 incorporates 'constant recoil': its dynamics ensure a smooth recoil action which enables the Ultimax to be fired under full control from the hip, even using only one arm. CIS states that no other machine gun or assault rifle can claim this capability.

Sustained firepower is provided by 100-round drum magazines that can be changed by a soldier on the move within five seconds. The use of standard 20- or 30-round box magazines is also possible. Rounds can be directly loaded into the drum magazines under finger pressure, with no need for accessories. The full-length 508 mm barrel, pre-zeroed, can be quickly changed, allowing fire to be sustained. The robustness and simplicity of the trigger and other operating mechanisms contribute to reliability. The loose fit of the bolt group within the receiver during its long recoil travel enables the gun to function even if dirt intrudes. A gas regulator is fitted. Safety against cook-off is provided by the open bolt operating feature.

The Ultimax 100, after extensive technical and troop trials, is in production to meet the total requirements of the Singapore armed forces. CIS has entered into agreements with two NATO manufacturers for the promotion and possible manufacture of the Ultimax in their respective countries.

Description
The Ultimax 100 is a magazine-fed, gas piston operated, rotating bolt, air-cooled light machine gun. It is a one-man weapon and can be fired from the shoulder, hip or with a bipod. It fires only in full automatic mode, from the open bolt position, at 520 rds/min. It feeds from 100-round drum magazines or standard 20- and 30-round flat magazines.

The heavy, air-cooled, quick-change (Mark III only) barrel can sustain 500 rounds of full automatic fire without heat damage. The handle on the barrel is carefully situated for balance and is also used in the quick-twist-action barrel change.

The rearsight is marked in 100 m increments to 1000 m and is fully adjustable by moving the sight slide. The foresight is also fully adjustable for elevation and windage; thus each quick-change barrel can be pre-zeroed with respect to the receiver.

The trigger system's lock-out mechanism prevents accidental firing if the weapon is improperly cocked or is dropped. The fire-selector has two positions: 'fire' and 'safe'. On 'safe' the trigger is blocked and the connection between trigger and sear moved out of position so that, if any part of the trigger mechanism fails, the gun will not fire.

Low recoil enables the gun to be fired without the easily detachable butt stock. This, in conjunction with the short para-barrel, makes it a handy weapon in confined spaces or for paratroopers.

The bipod legs are positively locked, either in the down position or up flush with the sides of the receiver and are quickly adjustable for length. The bipod allows a 15° sweep and roll to either side and the entire assembly is easily detachable without tools.

The cocking handle does not cycle with the bolt and is locked in the forward position during firing. It is on the left side of the receiver to enable the gun to be cocked quickly while leaving the right hand ready for firing.

The gas regulator has three apertures. Use of a high pressure gas system prevents fouling and reduces the need to clean the gas piston and regulator. Both piston and regulator can be removed for cleaning if desired.

Operation
The Ultimax 100 fires from the open bolt position: the bolt group is held back behind the feed area and the bolt carrier is engaged by a sear. The front end of the sear makes contact with the sear actuator in the fire-selector. Pulling the trigger releases the bolt carrier by means of the sear actuator and strips the top cartridge from the magazine. When chambering is complete, the continued forward movement of the bolt carrier rotates the bolt via a cam and locks it to the barrel. The forward momentum of the bolt carrier assembly drives the

5.56 mm Ultimax 100 Mark III light machine gun field-stripped

5.56 mm Ultimax 100 Mark III light machine gun with quick-change barrel and 100-round drum magazine; bipod legs extended

firing pin to ignite the cartridge. On firing, return gas is tapped off relatively close to the chamber, when barrel pressure is still high and the bolt carrier assembly is forced backwards by the gas piston, rotating the bolt via cam action. The bolt unlocks providing primary extraction and loosening the empty case in the chamber, which is extracted, ejected, and the return spring compressed. The bolt group has a long recoil travel and is designed to avoid impact with the receiver back plate. The bolt group then moves forward and if pressure on the trigger is maintained, chambers the next round and fires it. If the trigger is released, the bolt carrier assembly will be engaged by the sear. When the ammunition supply is exhausted, the bolt carrier assembly will remain in the forward closed position.

Stripping

The Ultimax 100 can be rapidly field-stripped without tools, since no small parts or pins are disassembled in normal field stripping. The weapon must be cleared and is then field-stripped into five main groups: receiver; bolt carrier assembly; butt stock; bipod assembly and quick-change barrel. To remove the butt stock, which must be detached from the receiver before the bolt carrier assembly can be removed, pull out the top and bottom take-down pins and pull the butt stock away from the receiver. To remove the bolt carrier assembly, push the take-down button forward with the thumb of the right hand: the backplate will slide down automatically. Eject the bolt carrier assembly through the rear opening by pulling back the cocking handle. Complete the action by pulling on the main spring guide rod. To disassemble the bolt carrier assembly further compress the main spring to release the guide rod. Press the firing pin forward to release the cross pin: the firing pin will now slide out freely. Push out the cam pin and remove the bolt.

To remove the bipod, lower one or both legs from the sides of the receiver, assuming the bipod is folded, and release the plunger in the bipod assembly.

This is the extent of field stripping expected of a soldier. Re-assembly follows the logical reverse sequence; the gun is designed to make incorrect re-assembly unfeasible.

Accessories

Accessories and options available with the Ultimax 100 are: 100-round magazines, either in single pouches or in packs of four; standard 20- and 30-round box magazines; bipod; blank-fire attachment; sling; cleaning kit; quick-change barrels; twin-gun mount for vehicles, helicopters or cupolas.

DATA

Cartridge: 5.56 mm × 45 (US M193 or SS 109)
Operation: gas, automatic
Method of locking: rotating bolt, multi-lug head
Feed: 100-round drum magazine; 20- or 30-round box magazine
Weight: 4.9 kg with bipod
Length: 1024 mm; (810 mm without butt)
Barrel: 508 mm
Rifling: 6 grooves, rh, 1 turn in 305 mm or 178 mm
Sights: (fore) adjustable for elevation and windage; (rear) adjustable from 100 to 1000 m by 100 m increments; rapid shift; fine click adjustment for range; windage screw
Muzzle velocity: 970 m/s
Rate of fire: 400-600 rds/min

Manufacturer

Chartered Industries of Singapore Pte Ltd, 249 Jalan Boon Lay, Singapore 2261.
Status
Current. Production.
Service
Singapore armed forces.

7.62 mm general-purpose machine gun

The 7.62 mm general-purpose machine gun is manufactured in Singapore in both infantry and coaxial versions. A tripod mount is also made, together with an anti-aircraft adaptor. The guns are identical in most respects with the Belgian-made product, to which entry reference should be made for dimensions etc.

Manufacturer

Ordnance Development & Engineering Co of Singapore (Pte) Ltd, 15 Chin Bee Drive, Postal District 2261.
Status
Production.

.50 CIS machine gun

The M2 Browning is a popular machine gun but the drawbacks of this weapon are well-known, namely adjustment of headspace, closed bolt firing system, complex design and high manufacturing cost. Knowing that there is still a need for such a weapon, one which can fill the gap between the 7.62 mm machine gun and the 20 mm cannon, Chartered Industries of Singapore (CIS) began designing a new .50 machine gun in late 1983. The objective for such a development was a simpler and lighter weapon than the M2 Browning in order to improve the system portability and ease the problems of field maintenance. The ability to fire the developmental Saboted Light Armor Penetrator (SLAP) ammunition was another objective.

The design of this new machine gun is simple and modular in construction, consisting of five basic assemblies. The modular construction allows ease of assembly and maintenance. Simplicity of design means reliability and inherently allows for relative ease of manufacture and thus lower cost.

Operation

The CIS .50 machine gun is gas-operated and fires from the open bolt position to prevent 'cook-off'. The bolt carrier group is held back behind the feed area and the bolt carrier is engaged by a sear. Activating the trigger releases the bolt carrier, and as the bolt moves forward it strips the round centred on the feed tray from the ammunition belt feeding in from either the left or the right hand side of the gun.

When chambering is completed, the combined forward movement of the bolt carrier rotates the bolt by means of a cam and locks it to the barrel. The forward momentum of the bolt carrier assembly drives the firing pin forward to fire the cartridge.

On firing, propellant gas is tapped off into a gas cylinder. The bolt carrier is driven back by the gas pistons, causing the bolt to rotate, unlock and extract the empty case from the chamber.

As the bolt carrier continues to move rearward, the empty case is ejected via an ejection port at the bottom of the receiver. When the return springs are almost fully compressed, the bolt carrier begins to move forward and, if the pressure on the trigger is maintained, the bolt will chamber the next round and fire it. If the trigger is released the bolt carrier assembly will be engaged by the sear. When the ammunition belt is exhausted the bolt carrier group will remain in the forward closed position.

Construction

The design of this new machine gun is simple and modular in construction, consisting of five basic groups:-
Receiver body: this is made of pressed steel, including two tubes at the front end to house the pistons and recoil rods.

Feed mechanism: this is located on top of the receiver body, using a single sprocket. It is designed to facilitate either left or right hand feed of ammunition belts.
Trigger module: the trigger module houses the trigger and sear mechanisms, with provision for safety lock. The selector has two modes, Automatic and Safe.
Barrel group: a quick barrel change can be accomplished within seconds, without any headspace problem. The CIS gun has a fixed headspace, unlike

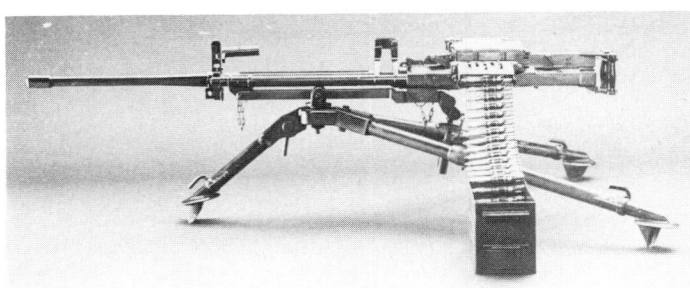

.50 CIS machine gun

.50 CIS machine gun mounted on armoured vehicle

the M2 Browning which requires adjustment whenever the barrel is replaced. The gas regulator has two positions to allow setting for normal and adverse environmental conditions.

Bolt carrier group: this group consists of a pair of pistons and recoil rods attached to the bolt carrier body by a pair of quick-release catches. The bolt is prevented from accidental firing out of battery by a 'sleeve lock' device between the bolt and bolt carrier body.

Stripping

The weapon can be rapidly field-stripped without the use of any tools, into the five main groups: barrel, bolt carrier group, receiver and trigger mechanism group. The feed mechanism, under normal circumstances, need not be field stripped.

Prior to stripping, ensure the bolt carrier group is in the forward position. To remove the barrel assembly, depress the latch and turn the barrel extension sleeve so that the latch engages the alternate detent hole in the sleeve: the lock lugs in the barrel extension are now completely disengaged from the barrel locking lugs. Slide the barrel forward to remove it.

Lift up the feed mechanism together with the feed tray, and remove the two catches on top of the bolt carrier body. The pistons, guide rods and return springs can now be removed from the front end of the receiver tubes. Remove the bolt carrier assembly through the top opening of the receiver. The trigger housing, together with the sear mechanism, can now be detached by removing the take-down pin attached.

Mounts
Two types of mount are available; a pintle mount for the M113 APC, and an adapter for the M3 tripod.

DATA
Cartridge: 12.7 × 99 mm (.50 Browning)
Operation: gas, automatic
Locking: rotating bolt
Feed: dual disintegrating link belt
Weights
    Gun: 30 kg
    Barrel: 9 kg
Length: 1670 mm
Barrel: 1143 mm
Rifling: 8 grooves, rh, one turn in 30 cal
Rate of fire: 400 – 600 rds/min
Muzzle velocity: M8: 890 m/s. APDS: 1200 m/s
Maximum effective range: 1830 m
Maximum range: 6800 m

Manufacturer
Chartered Industries of Singapore, 249 Jalan Boon Lay, Singapore 2261.
Status
Current. Production.

SOUTH AFRICA

7.62 mm SS-77 light machine gun

Development of this weapon began in 1977 and a limited quantity entered service in 1986. It is of conventional type, gas-operated, with quick-change barrel, bipod and folding butt.

Bolt locking is performed by swinging the rear end of the bolt sideways into a recess in the receiver wall, similar to the system used in the Soviet Goryunov machine gun. A gas piston beneath the barrel drives the bolt carrier back after firing; a post on the carrier, engaging in a cam path on the rectangular bolt, swings the bolt out of engagement with its locking recess and then withdraws it, extracting and ejecting the spent case. A post on the top rear of the carrier engages in a belt feed lever in the top cover of the receiver. Due to the shape of this lever it is turned around a pivot so that the forward end, carrying the feed pawls, moves the incoming round half a step towards the feed position. On the return stroke the feed lever moves the round the remaining distance, the bolt loads it into the chamber, and as the bolt comes to rest the carrier continues and the post, engaged in the bolt cam path, swings the rear of the bolt into the locking recess. The post then strikes the firing pin and the round is fired.

The barrel can be removed by depressing a locking lever and rotating the barrel to disengage its interrupted lugs from the receiver. The barrel is externally fluted to save weight, and this also increases the surface area and aids cooling.

A three-position gas regulator is fitted; this can be used to increase the supply of gas to overcome friction, or it can also be used to adjust the rate of fire. Position 3 on the regulator also closes down the exhaust and gives minimal emission of gas, allowing the gun to be safely fired in enclosed spaces or from inside vehicles.

DATA
Cartridge: 7.62 × 51 mm
Operation: gas, automatic
Locking system: transverse tilting block
Feed: disintegrating link belt

7.62 mm SS-77 light machine gun

Weight: 9.60 kg
Length: (butt folded) 940 mm; (butt extended) 1155 mm
Barrel: 550 mm without flash hider
Rifling: 4 grooves, rh, one turn in 305 mm
Sights: (front) post, with Tritium spot; (rear) tangent leaf with Tritium spot; 200-800 m aperture, 800-1800 m U-notch
Sight base: 747 mm (short range); 816 mm (long range)
Rate of fire: 600-900 rds/min, adjustable
Muzzle velocity: ca 840 m/s

Manufacturer
Lyttleton Engineering Works (Pty) Ltd, (Armscor subsidiary), Pretoria.
Status
Current. Production.
Service
South African Defence Force.

SPAIN

7.62 mm MG3 machine gun

This gun is described by its Spanish manufacturer as being the MG42 chambered for the 7.62 mm × 51 NATO cartridge. This makes it indistinguishable from the West German MG1 and it seems highly likely that it is in fact an MG1 made under licence in Spain. All data and details are apparently identical with the MG1 (which see).

Manufacturer
Empresa Nacional de 'Santa Bárbara'.

5.56 mm CETME Ameli assault machine gun

The 5.56 mm Ameli operates on the delayed blowback system with the use of rollers; the system used in all CETME's light weapons since 1956. Similarities with the CETME 7.62 mm Model C and 5.56 mm Model L rifles (which see) extend to the interchangeability of certain parts.

The Ameli is fed by a disintegrating-link belt, hanging free or from a 100- or 200-round box. The manufacturer states that the air-cooled barrel can be changed within five seconds and that the rate of fire, from 800 to 1200 rds/min, is dependent on 'the changing of one small part'.

Accessories include: an integral ammunition state indicator; an illuminated

sight and night-vision equipment; and a tripod as well as the bipod shown here.

Operation
The roller-locking device operates in the same way and gives the same advantages as described in the entry for the CETME 7.62 mm assault rifle: reference should be made to that entry and to the accompanying diagram illustrating delay action and locking piece.

The gun is loaded and cocked with the safety catch off, the sear holding the

5.56 mm Ameli light machine gun on bipod with 100-round belt box

5.56mm Ameli with 200-round belt box

5.56 mm Ameli with night-vision sight, showing method of loading belt

bolt immobile. When the trigger is pressed, the sear disengages and the bolt begins its forward travel, forcing the first round from the belt into the breech. Under the pressure of the bolt-head, the rollers move outward to the sides of the breech. The extractor (which acts as a guide throughout, rather than as a part of the ejector system) engages the grooves of the cartridge case. When the bolt completes its forward travel, the rollers completely lock the assembly: the bolt-head forces forward the firing pin and ignites the charge. As gas pressure falls, the recycled gases loosen the case and residual pressure in the barrel forces it back, to be ejected downward and forward.

DATA
Cartridge: 5.56 mm × 45 (M193 or SS 109)
Operation: delayed blowback using rollers
Feed: disintegrating-link belt or 100- or 200-round box
Mode of fire: automatic

WEIGHTS
Gun: 5.20 kg
Bipod: 0.20 kg
Barrel: 0.82 kg
200-round box magazine: 3 kg
100-round box magazine: 1.50 kg

LENGTHS
Gun: (overall) 970 mm
Barrel: 400 mm; (rifling) 6 grooves rh. 1 turn in 178 mm

FIRING CHARACTERISTICS
Muzzle velocity: 875 m/s
Muzzle energy: 1570 J
Rate of fire: 900 rds/min
Range: 1650 m max

Sights: 4 position rear aperture graduated 300, 600, 800 and 1000 m.
Sight radius: 340 mm

Manufacturer
Santa Barbara, Calle Julian Camarillo 32, 28037 Madrid.
Status
Current. Production.
Service
Adopted by Spanish Army.

SWITZERLAND

7.5 mm M25 light machine gun

This is an elderly recoil-operated weapon which is still in limited service in the Swiss Army. Compared with the post-war MG42-derived M51 weapon described below it has a very slow rate of fire, but it has the merit of being somewhat lighter and thus better-suited to the needs of the foot-soldier.

7.5 mm M25 light machine gun

DATA
Cartridge: 7.5 mm M11
Operation: recoil
Feed: 30-round detachable box magazine
Weight: 10.8 kg with bipod
Length: 1160 mm
Sights: (foresight) blade; (rearsight) tangent 100-2000 m
Rate of fire: (cyclic) 450 rds/min
Effective range: 800 m from bipod. Tripod also available

Status
Obsolete. No longer in production.
Service
Reserve for Swiss armed forces.

7.5 mm M51 light machine gun

Currently a standard light machine gun of the Swiss Army, this weapon was designed by the government establishment at Bern. The design is based on the German MG42 of Second World War fame but, largely because the design calls for machined parts in many places where the MG42 uses stampings, it is better made but it is also more expensive and more than 4 kg heavier.

Features of the design include a quick-change barrel arrangement and a belt feed system, both of which are similar to those of the MG42. The locking system is also based on that of the West German weapon but the Swiss design uses flaps instead of the rollers used on the MG42. Accessories include a tripod mount which can be fitted with an optical sight and a drum attachment which enables a 50-round belt to be carried conveniently.

DATA
Cartridge: 7.5 mm M11
Operation: gas-assisted recoil, automatic
Method of locking: flaps

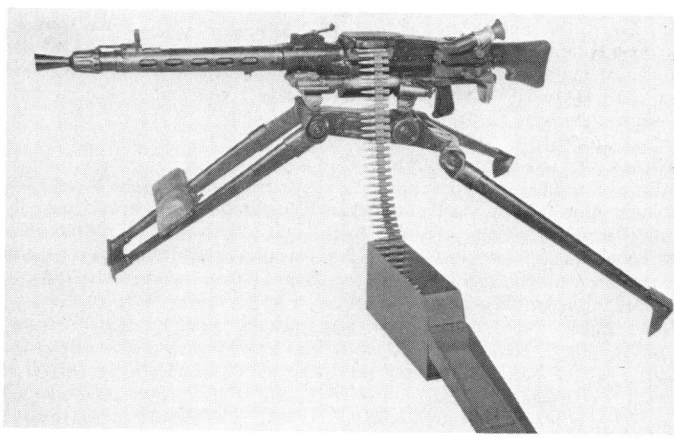

Feed: 50- or 250-round link belt
Weight: 16 kg with bipod, 27 kg with tripod
Length: 1270 mm
Barrel length: 564 mm
Sights: (foresight) folding blade; (rearsight) tangent
Rate of fire: (cyclic) 1000 rds/min
Effective range: (from bipod) 800 m; (from tripod) 1000 m

Manufacturer
Swiss Federal Arms Factory, Bern.
Status
Current.
Service
Swiss Army.

7.5 mm M51 light machine gun

7.62 mm SIG 710-3 general-purpose machine gun

The SIG 710-3 is a general-purpose machine gun designed to fire the NATO 7.62 mm round. It makes use of modern manufacturing techniques, using pressings wherever possible and it is probably one of the most advanced general-purpose machine guns today.

Operation
When the gun is fired the high pressure developed in the cartridge drives the case back hard against the bolt-face. Gas escapes from the mouth of the case when the bullet begins to move and this gas runs into the grooves of the fluted chamber, thus partially neutralising the internal pressure and floating the case in a film of gas.

The net effect is to allow the case to move back and maintain a force against the breech-block. This is the source of energy to carry out the cycle of operation.

The breech-block has three major parts: the forward part which contains

the rollers; the rear part which supports the anvil on which the rollers bear; and a sleeve which surrounds the rear part.

At the moment of firing the rollers in the forward part of the bolt are forced outwards into the recesses in the body of the gun by the anvil which itself is held forward by the rear part of the bolt.

The backward force exerted by the gas drives the cartridge rearwards and the front part of the bolt starts to move. This movement is transmitted to the body of the gun through the rollers. The reaction of the body to this applied force has a component which drives the rollers inwards. When they start to move in, they force the inclined plane of the anvil shoulders back and the rear part of the block is driven back. This movement is resisted by the inertia of the rear part of the block, including its sleeve, and the force exerted by the return spring. The rear part of the block moves back a greater distance than the front due to the angles of the anvil shoulder and is therefore accelerated with respect to the front part. When the rollers have moved in and are clear of the body the

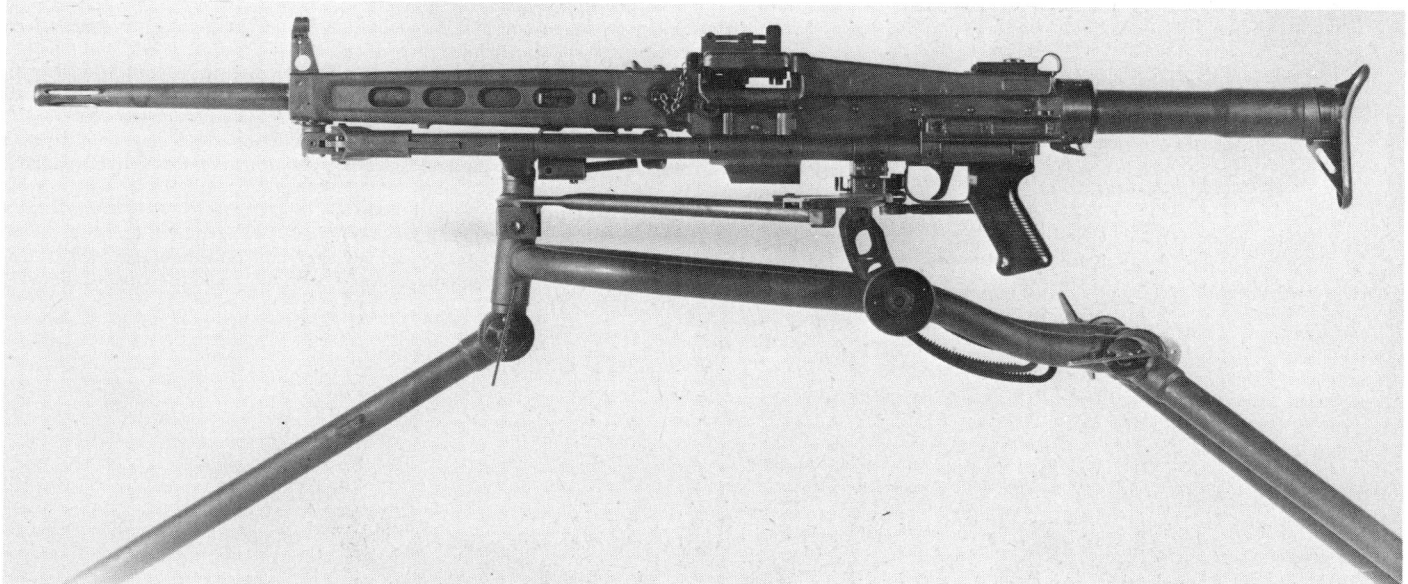
7.62 mm SIG 710-3 general-purpose machine gun on tripod SIG L810

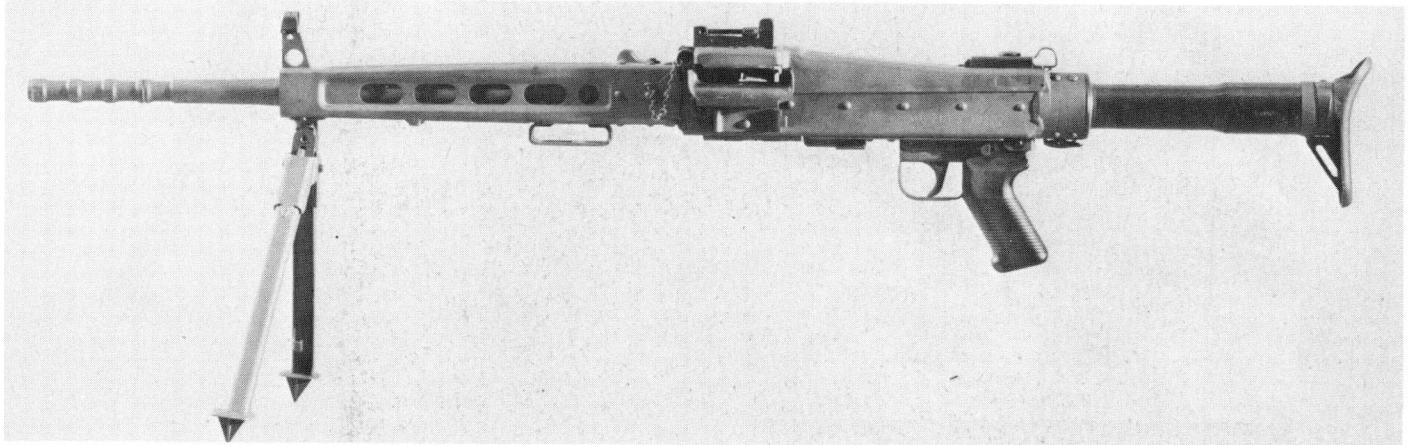
7.62 mm SIG 710-3 general-purpose machine gun

whole block moves back under the force exerted on the cartridge case by the residual pressure of the gases.

When the return spring is fully compressed the bolt hits the buffer. The bolt then travels forward with the two parts still separated because the rollers are held in by the body guide ribs. When the round is fully chambered the rollers come opposite the recesses cut into the body. The rear part of the bolt has its own momentum and is also being impelled by the return spring. The anvil is driven forward between the rollers which are forced out into the body. When they are fully out (and not until then) the anvil can go far enough forward to strike the free floating firing pin which then strikes the cap.

The part played by the sleeve around the rear part of the bolt is unique in this type of mechanism. When the cap is struck the rear part of the block has reached the limit of its forward travel but the sleeve has a further 5 mm of forward movement. The kinetic energy it contains is used for two purposes. First it prevents any bounce of the bolt which may occur if there is a slight

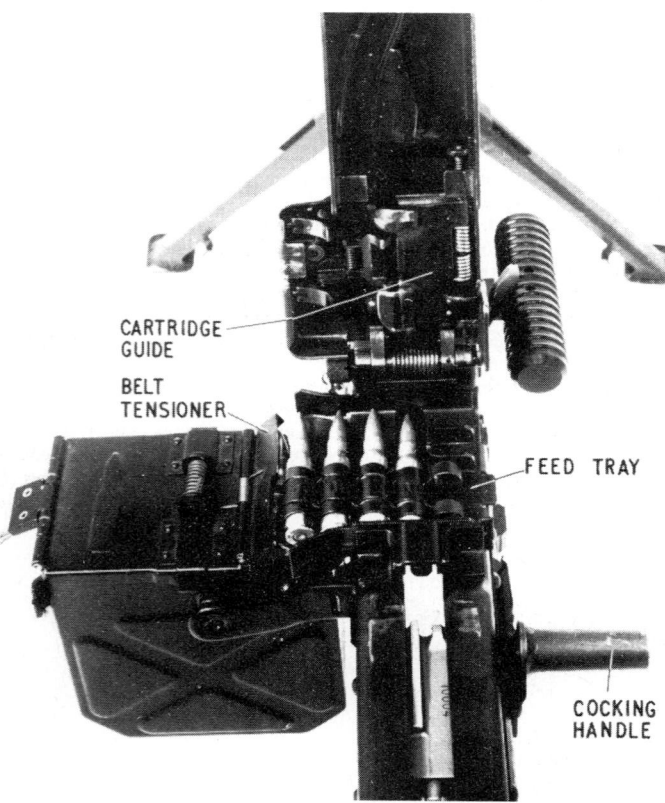

CARTRIDGE GUIDE
BELT TENSIONER
FEED TRAY
COCKING HANDLE

Loading 7.62 mm SIG 710-3 general-purpose machine gun

hang-fire in the ignition of the propellant and secondly it is still moving forwards when the backward thrust is developing in the chamber so that its forward momentum must be destroyed before there can be movement of the front part of the bolt. This provides another, although small, check to bolt movement when the pressure is high.

Trigger and firing mechanism

The trigger mechanism is developed directly from that used in the MG42. The moment at which the sear rises when firing is discontinued is carefully controlled to ensure that the bent of the bolt, which is moving quite fast, encounters the full face area of the sear. This prevents chipping and general wear of the bent and sear which is quite common on guns where the sear can rise into the bolt path at any moment.

The trigger carries a tripper which is forced forward by its own spring. When the trigger is rotated the tail of the sear goes up and the nose of the sear disengages from the bent of the bolt. The spring of the tripper drives it forward and a step rides under the tail of the sear. When the trigger is released the tail can fall only a little way because it is held up by the step on the tripper, and the nose of the sear cannot rise. The rotation of the trigger pushes the tripper up into the path of the bolt which rides over it on the forward stroke but on the rearward stroke (having fired another round) drives the tripper backwards. This moves the step from under the tail of the sear which then is free to fall and the nose of the sear rises into the bolt path. The bolt rides over the sear on its way back but the sear rises to present its full face to the bent on the next forward stroke. This mechanism is very successful in saving the bent and sear from damage but if the tripper is released it can produce a hazard to safety when the safety catch is set to 'safe' with bolt forward.

The applied safety consists of a bolt pushed across the body of the gun to prevent the nose of the sear from being depressed. Since the nose of the sear is held down until the tripper is pushed back, it follows that the gun must be cocked before the safety can be applied unless the trigger mechanism has been interrupted by a short jerk on the cocking handle before setting the safety catch. Pull the cocking handle back a short way and let it snap forward again before setting the safety catch.

Feed system

The feed system is the same in all essential details as that developed for the MG42. A roller is mounted on top of the sleeve over the rear of the bolt to engage in a rear pivoted track, or feedway, carried under the top of the feed cover. As the bolt reciprocates the feedway swings about its pivot and moves from side-to-side across the gun.

The gun has a considerable reserve of energy for feeding and under good conditions will lift a belt length of 1.2 m.

One very useful feature is the gun's ability to cater for rounds held either in NATO disintegrating links of the US M13 pattern or in the West German type continuous metal belt.

To change from one type of belt to the other requires a change of the feed tray and the cartridge guide. The whole operation takes less than 60 seconds.

Barrel and sights

The barrel-change arrangement is very good. The barrel-release catch on the right of the barrel casing is pushed and the barrel pulled out and back. It can be done with one hand and the change is definitely faster than that of the West

7.62 mm SIG 710-3 general-purpose machine gun showing method of barrel changing

Switzerland/**MACHINE GUNS** 339

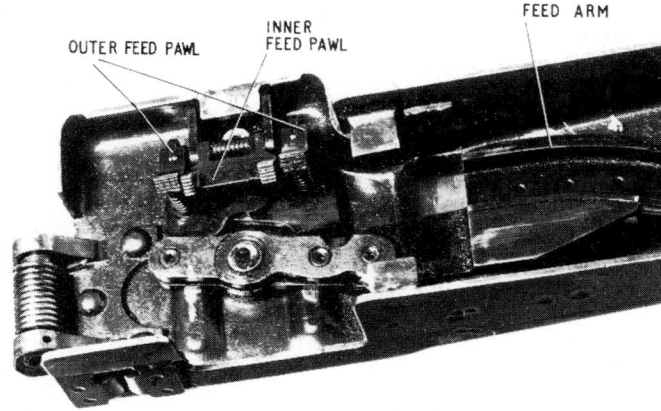

Feed system

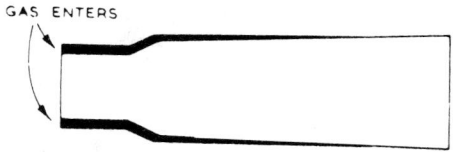

GAS ENTERS

SIDE ELEVATION OF THE CHAMBER

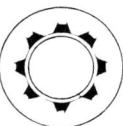

Fluted chamber

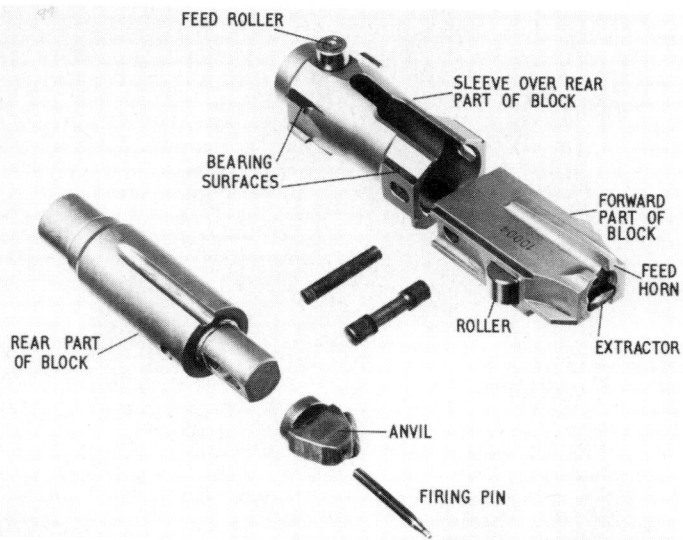

Breech-block

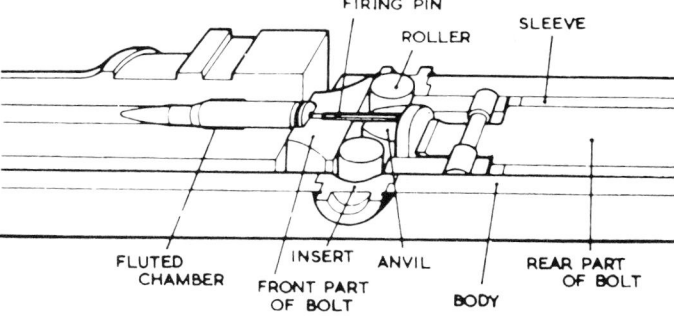

Breech in firing position

German MG3. This is probably the fastest change yet produced. The barrel handle is a good heat insulator and even when the barrel is cherry red the change is neat and simple.

The foresight is a blade fixed between protectors and the whole slides in a dovetail on the barrel casing. The rearsight is a leaf working on a ramp. It is graduated from 100 to 1200 m in hundreds of metres and has clear bold figures.

Stripping

Stripping is a simple process. With the bolt forward and the top cover plate lifted, the forward of the two catches below the neck of the butt is pressed and the butt rotated. The butt, buffer, and return spring, rod and casing are withdrawn. A sharp jerk on the cocking handle throws out the bolt. The trigger group requires the extraction of one pin. The barrel catch releases the barrel and the weapon is then field-stripped.

When the two pins in the bolt are pressed out the bolt is fully stripped. The rollers are held in by a low strength wire spring and can be released with the fingers. The top cover plate is held to the body by a pin incorporating a flat. When the punch marks on pin and top cover plate are lined up, the pin can be withdrawn.

The feed mechanism can be stripped by:
Removing cartridge guide. Push forward, rotate and remove hinge pin.
Remove feedway by compressing spring plate at the pivot end.
Remove feed pawls.

The trigger mechanism is easily stripped and assembled and the procedure is straightforward when the gun is seen.

Tripod

The tripod carries a dial sight weighing 2.5 kg which can incorporate a telescope of $\times 2\frac{1}{2}$ magnification or a night sight device. The tripod is the sprung SIG L810. The gun recoils when the first round is fired and the springs are compressed. Before the gun is fully forward again the next round is fired and thereafter the gun quivers in a state of near equilibrium.

There is an automatic device for dispersing fire in either elevation, line or both. This in principle shows a distinct similarity to that used on the MG42 during the Second World War.

The tripod can be mounted at any height between 305 and 700 mm and it weighs 10.3 kg. It folds up and can be carried by one man on his back.

The weapon has the usual cleaning kit but there are also one or two very useful ancillaries. There is a blank-firing barrel which operates the gun very well using standard 7.62 mm NATO blank of the British pattern. This is carried in a metal case. There is also an extremely good belt filler for the West German-type belt.

DATA
Cartridge: 7.62 mm × 51
Operation: delayed blowback, automatic
Method of delay: rollers
Feed: belt; disintegrating link type or continuous link belt

WEIGHTS
Gun: 9.25 kg
Heavy barrel: 2.5 kg
Standard barrel: 2.04 kg
Blank-firing barrel: 1.89 kg
Belt carrier: 0.8 kg

LENGTHS
Gun overall: 1143 mm
Barrel: 559 mm

MECHANICAL FEATURES
Barrel: (regulator) nil; (cooling) air; quick-change barrel
Sights: (foresight) blade; (rearsight) leaf notch; (sightbase) 412 mm; (telescopic sight) x2.5

FIRING CHARACTERISTICS
Muzzle velocity: 790 m/s
Rate of fire: (cyclic) 800-950 rds/min
Effective range: (bipod) 800 m; (tripod) 2200 m

TRIPOD CHARACTERISTICS
Designation: tripod mount L810
Minimum firing height: 305 mm
Max firing height: 700 mm
Weight: 10.3 kg
Max elevation: 500 mils
Max traverse: 800 mils
Dimensions: (folded) 82 × 46 × 27 cm
Weight of telescope: 1.45 kg
Weight of telescope mount: 400 g
Accessories: blank-firing barrel, belt carrier, ammunition box, sling, cleaning kit, belt filler

Manufacturer
SIG Swiss Industrial Company, CH-8212, Neuhausen Rheinfalls.
Status
Not in production.
Service
Supplied to the Bolivian Army, Brunei, and Chilean police.

7.5 mm Model 87 tank machine gun

The 7.5 mm tank machine gun has been designed specifically for installation into vehicle fighting compartments (for example battle tanks, armoured personnel carriers and so on) or as a casemate weapon for fortifications. It was developed from the 7.5 mm machine gun Model 51, described above.

The design of this gun decisively reduces the emission level in the fighting compartment. Special ventilation to remove propellant gases is not necessary. Measurement inside the Swiss tank 87 (Leopard 2), at a continuous fire rate of 900 rds/min, showed a COHb increase of one per cent at the loader's position.

By replacing only four component parts it is possible to adapt the gun to fire 7.62 × 51 mm NATO ammunition.

DATA
Cartridge: 7.5 × 55 mm Swiss GP11 or 7.62 mm NATO
Operation: recoil, automatic
Locking system: flaps
Feed: belt
Weight: 30.0 kg including mount
Length: 1175 mm
Barrel: 475 mm
Rate of fire: (full) 1000 rds/min: (reduced) 700 rds/min
Effective range: 1200 m

Manufacturer
Swiss Federal Arms Factory, Stauffacherstrasse 65, CH-3000 Berne 22.
Status
Current. Production.
Service
Swiss Army.

7.5 mm Model 87 tank machine gun

7.5 mm Model 87 tank machine gun installed

TAIWAN

Machine guns in Taiwan

The Chinese Nationalists carried to Taiwan a collection of machine guns, all of foreign origin. They included:

Type 24	—German Maxim MG'08 made in China
Type 26	—Czechoslovak Zb26 made in China
Type 30	—Czechoslovak Zb30 made in China
Madsen 7.92 mm	—Danish Madsen
Bren guns 7.92 mm	—Canadian manufacture
Browning .30 Model 1919A4	—US manufacture.

5.56 mm Type 75 squad automatic weapon

Subsequently the Bren was re-chambered for .30 calibre and manufactured in some quantity in Taiwan. Further supplies of Browning .30 Model 1919A4 and .50 M2HB machine guns were provided.

In 1968, Taiwan was provided with the necessary machinery to manufacture the M60 general-purpose machine gun and since then tens of thousands of these have been produced, known as the Type 57 machine gun. Around 1978, the Combined Service Forces (CSF) developed manufacturing capability for the 7.62 mm M134 Minigun. Whether this weapon was produced in any quantity is unknown.

More recently, an official statement suggested that the .50 M2HB machine gun may be in production in Taiwan for use as anti-aircraft armament on Taiwan-built M41 tanks. A 7.62 mm coaxial machine gun (of unspecified type) is also made.

Type 74 light machine gun
This has been developed from the FN-MAG general-purpose machine gun and will replace the Type 57 machine gun as the platoon support weapon. The Type 74 differs from the FN-MAG only in minor details, for example a slightly different shape of butt, different sights and a barrel with cooling ribs.

Type 75 Squad automatic weapon
This closely resembles the FN Minimi (M249) and will replace the M14A1 as the squad automatic weapon so that both rifles and support weapon will now use 5.56 mm ammunition. The Type 75 is rifled one turn in 178 mm, suggesting that the ROC army has standardised on the SS109 bullet.

Manufacturer
Combined Service Forces arsenals.
Status
The M60 is in production and service. Other older weapons are probably restricted to reserve units.

TURKEY

7.62 mm MG3

The West German MG3 is manufactured in Turkey for supply to the Turkish Army. It is identical in all respects with the original German weapon, to which reference should be made for dimensions etc.

Manufacturer
Makina ve Kimya Endustrisi, Kurumu, Ankara.
Status
Production.
Service
Turkish forces, and for export.

Turkish 7.62 mm MG3 machine gun on tripod mount

UNION OF SOVIET SOCIALIST REPUBLICS

7.62 mm Degtyarev DP and DPM light machine gun

The Degtyarev Pekhotnyy was designed by Vasily Alexeyevitch Degtyarev in the early 1920s and was completed in 1926.

In 1944 the DP was modified by W Shilin and in this form it became known as the DPM – DP modernised. The return spring was taken from under the barrel and placed in a cylindrical housing behind the receiver, projecting back over the small of the butt. A pistol grip was added. The grip safety was replaced by a safety lever mounted on the right of the receiver just above the trigger.

The two weapons differ only in these details and will be treated as one. The DPM has been manufactured in China as the Type 53 light machine gun.

Description

The DP has 65 parts, is of very simple construction and was intended for manufacture by unskilled labour. The design has many large flat bearing surfaces which mate during operation and the friction caused by the introduction of dirt causes stoppages in action. The barrel in the earliest versions had annular cooling fins but these were omitted during the war to speed production. The barrel of the earliest DP was fixed and so over-heated readily. To get the barrel out required the unscrewing of the flash hider and gas-cylinder nut. The cylinder was then slipped back and two pins holding the barrel in place knocked out with a drift. The barrel could then be rotated, using a tool on the flats at the muzzle, and pulled out.

In 1940 a replaceable barrel of more modern design was fitted. When the gun is cocked the barrel lock, on the left side of the receiver, is pressed in and an open-jawed spanner placed over the flats near the muzzle. The barrel is rotated through 60° and then pulled forward out of the barrel jacket.

The barrel casing is made of two tubes welded together. The top tube takes the barrel and the bottom one contains the gas piston and the return spring in the DP. The gas regulator is of the variable gas track type and has three passages of 2.7, 3.3 and 4 mm. To change the setting remove the cotter pin, unscrew the nut, move the regulator out and reset. Replace the regulator, tighten the nut and replace the cotter pin. The piston is a rod attached to the slide which carries the breech-block. The slide has a slot to allow passage of the extracted cartridge and is machined at the rear end to provide cams to force the locking lugs in to unlock the bolt. The cocking handle projects from the right side of the slide and reciprocates with it. The breech-block is rectangular in shape. Along each side is a locking lug. In the centre is the firing pin which is attached to the slide. Camming shoulders on the firing pin force the two locking lugs outward into recesses in the sides of the receiver as the slide continues forward after the breech-block has chambered the round and come to rest.

The foresight is a post with two foresight protectors. The rearsight is a leaf tangent V. It is adjusted by pressing in the side catches and moved along the bar to the appropriate range. The magazine catch is at the back of the rearsight block.

Operation

The drum magazine is loaded by inserting cartridges, nose towards the centre, into the feed mouthpiece, on top of the empty case which is used as a follower. The magazine is rotated by pulling round the leather tag. This will often have been torn off and rotation must be accomplished by pressure on the small stud that holds the leather tag. Separate positions are provided for each cartridge. Slide forward the dust cover and place the fork, at the front of the magazine, into the T slot on top of the barrel jacket. The rear of the magazine is then dropped down in front of the rearsight block. Check that the catch is holding the magazine. Pull the cocking handle to the rear. The magazine spring will place a round in the magazine lips.

The DP has a grip safety. The DPM has a rotating lever above the trigger on the right side of the gun and this is rotated to the rear for 'fire'.

When the trigger is pulled the return spring forces the slide forward carrying the breech-block. The feed piece of the breech-block drives a cartridge from the drum into the chamber. The cartridge is rimmed and comes to rest against the barrel face and the extractor on the bolt-face snaps over the rim as the bolt motion ceases. The slide continues on and the firing pin moves on with it and cams out the locking lugs on the sides of the bolt into recesses in the receiver sidewalls. The firing pin continues on and strikes the cap of the cartridge. It is prevented from emerging too far and piercing the cap when its shoulders strike the interior of the bolt.

The bullet passes up the bore and some of the gases following it pass through the gas port and impinge on the front face of the piston. The piston moves back and the slide, which is attached to it, carries the firing pin back. The slide has cams machined in its top surface and after a period of free travel while the gas pressure drops, these force the locking lugs out of the recesses in the receiver and into the sides of the bolt. The slide then carries the whole bolt rearwards compressing the return spring. The extractor holds the empty case to the breech face until the ejector revolves it down through the aperture in the slide and out of the bottom of the gun. The rear of the slide hits the rear wall of the trigger housing, stops and is driven forward again by the compressed return spring.

The bolt picks up another round and the cycle is repeated as long as the trigger is pressed and the drum contains ammunition.

Trigger and firing mechanism

Since the gun fires only at full automatic the trigger system is rudimentary. The trigger has a hook at its upper end, extending forwards over the sear.

Grip safety and trigger of DP

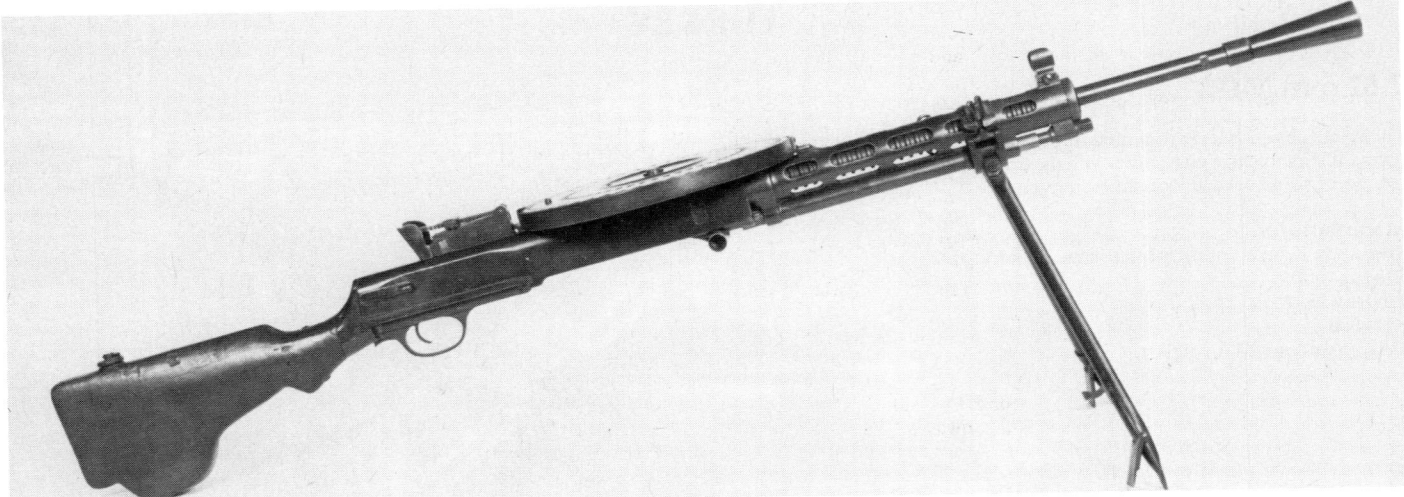

7.62 mm Degtyarev DP light machine gun

When the trigger is pulled the hook depresses the sear which drops out of engagement with a bent in the underside of the slide.

The grip safety of the DP is immediately behind the trigger and blocks its rotation until it is lifted by the firer's grip and frees the trigger.

Stripping
Remove the magazine by depressing the catch behind the rearsight block and lifting the rear end.

Check the chamber is clear.

Next the return spring must be disconnected or removed.

The DP has the return spring under the barrel and this must be unlocked by pressing the handle on the left front of the receiver forward and then down.

On the DPM the spring is housed in a tube behind the receiver. Pull back the detent on the tube, rotate the tube, remove it and withdraw the spring. Unscrew and remove the body locking pin at the right rear side of the receiver. Force down the butt.

Pull the cocking handle to the rear and withdraw the slide, bolt and firing pin. Lift the bolt up off the slide.

DATA
Cartridge: 7.62 mm × 54R
Operation: gas, automatic
Method of locking: projecting lug
Feed: 47-round pan magazine

WEIGHTS
Gun empty: 9.1 kg
Bipod: 0.7 kg
Flash hider: 0.2 kg
Barrel: 2 kg
Magazine: (empty) 1.6 kg; (loaded) 2.8 kg

LENGTHS
Gun overall: 1270 mm
Flash hider: 127 mm
Barrel: 605 mm

MECHANICAL FEATURES
Barrel: 3 position regulator
Sights: (foresight) cylindrical post; (rearsight) leaf tangent with a V backsight; (sight radius) 616 mm

FIRING CHARACTERISTICS
Muzzle velocity: 840 m/s
Rate of fire: (cyclic) 600 rds/min
Effective range: 800 m

Manufacturer
State arsenals.
Status
Obsolete.
Service
No longer in first-line units in Warsaw Pact countries. Likely to be found in any Soviet-influenced Third World country or with guerrilla movements.

7.62 mm RP-46 company machine gun

During the Second World War the USSR felt the need for a machine gun held at company level to produce a greater volume of fire than that supplied by the DPM, and yet light enough to be carried forward in the assault. During 1944 the Soviet designers W Shilin, P Poliakov and A Dibinin worked on the DPM and eventually produced a version which could be belt-fed for sustained fire and by changing the top cover could make use of the flat drum of the DPM during the assault. Although the design was finalised during the war it did not come into service until 1946 and the gun was known as the RP-46.

The RP-46 was officially adopted by the Soviet Army but it has been very little in evidence. The Chinese manufactured the gun and called it the Type 58.

The North Korean version is known as the Type 64. Apart from the markings the guns are the same.

In order to convert the DPM for belt feed the magazine has been removed and a feeder placed over the receiver by using the magazine lug at the front and the magazine catch at the rear. The belt feeder is operated by a somewhat crude use of the cocking handle which, as in the DP, is fixed to the right hand side of the slide and reciprocates with it. The handle fits into the claw-like feed operating slide and causes it to move backwards and forwards. The internal diameter of the claw is three times that of the cocking handle so that the bolt is able to accelerate to its full velocity in either direction before the feed slide is

RP-46 company machine gun

operated. This enables the full power of the gun to be used for unlocking the breech and accelerating the bolt back in one direction and similarly the full spring power is available to accelerate the bolt and push a round into the chamber.

The feed slide moves back, a diagonal feed groove cut in the top surface moves the belt feed pawl inward, and the first round in the belt is moved up to the cartridge stop. The round is held by a retaining pawl in the top feed cover when the feed slide moving forward forces the feed pawl out to the right to slip over the next round to be fed.

The feed slide carries a pair of jaws which grip the rim of the case and extract it from the belt as the feed slide travels back. The cartridge is pulled back under a spring-loaded depressor arm mounted under the top feed cover plate and this arm pushes the cartridge down into the feedway where the bolt feed piece can drive it forward into the chamber.

The RP-46 takes a 250-round continuous closed pocket metallic belt. The cartridge is simply placed in the conical pocket as far as it will go. The belt is usually carried in a metal box.

The dust cover over the feedway on the gun is lowered. The belt is loaded into the gun by inserting the feed tab into the feedway on the right of the gun and pulling the tab through until the first round reaches the stops. To load the gun the cocking handle must be pulled twice. The first time the bolt is retracted the jaws cannot grip the cartridge rim. The second time the leading round is pulled back and forced down into the feedway. The gun is then cocked and ready to fire.

If for any reason the belt feed device is not required it can readily be removed from on top of the receiver by pressing the magazine catch behind the rearsight block. The cocking handle must be pulled back until it lies in the centre of the claw. Tilting the feeder forward frees it from the cocking handle. The front end is held by the T lug and can be moved back to free the feeder completely from the gun. If required a DP or DPM 47-round drum magazine can then be used on the gun.

The RP-46 differs from the DPM in some other minor details. There is a barrel-release lever on the left of the receiver to depress the barrel lock. The gas regulator has a catch that engages one of the three grooves in the bottom of the gas block and can be moved from position '1' to '2' and then '3' as required.

The Chinese Type 58 originally had this type of regulator but it was later changed to a rotary type similar to that used in the RPD light machine gun (see entry below).

DATA
Cartridge: 7.62 mm × 54R
Operation: gas, automatic
Method of locking: projecting lugs
Feed: 250-round belt

WEIGHTS
Gun: (empty) 13 kg
Barrel: 3.2 kg

LENGTHS
Gun: 1283 mm
Barrel: 607 mm

MECHANICAL FEATURES
Barrel: 3 position regulator; (cooling) air. Quick-barrel change
Sights: (foresight) cylindrical post; (rearsight) leaf tangent with a V backsight; (sight radius) 635 mm

FIRING CHARACTERISTICS
Muzzle velocity: 840 m/s
Rate of fire: (cyclic) 600 rds/min
Effective range: 800 m

Manufacturer
State arsenals.
Status
Obsolete in Warsaw Pact countries.
Service
Likely to be found in parts of the world where there is a Soviet influence.

7.62 mm RPD light machine gun

Design work was started on the RPD by Degtyarev in 1943 to make use of the newly developed 7.62 mm × 39 cartridge. The urgent needs of the armed services, all of whom used Degtyarev machine guns firing the 7.62 mm × 54R cartridge, caused slow progress to be made but, shortly after the war ended, production of the new short cartridge was accelerated and the RPD was the first machine gun to use this cartridge.

The RPD was manufactured in large numbers and it formed the standard squad automatic for the Soviet Army and also its satellites. It was manufactured in China as the Type 56 and Type 56-1 and in North Korea as the Type 62 light machine gun.

It is now obsolescent in the Warsaw Pact countries but is still used in South-east Asia and by irregular forces in Africa.

Description
The RPD is a belt-fed automatic weapon. It is gas-operated and has a locking system which, in principle, is like that of the DP; it uses locking lugs of the same sort but they are pushed out to lock, not by the firing pin, but by a wedge on top of the slide. The belt is a continuous, metal, open-pocket design. The links are connected by a short metal spiral spring link. The cartridge goes into the pocket and the nib at the rear of the belt fits into the extraction groove at

Soviet Marines using 7.62 mm RPD light machine gun showing grip used by soldier

Soviet 7.62 mm RPD Model 2 light machine gun

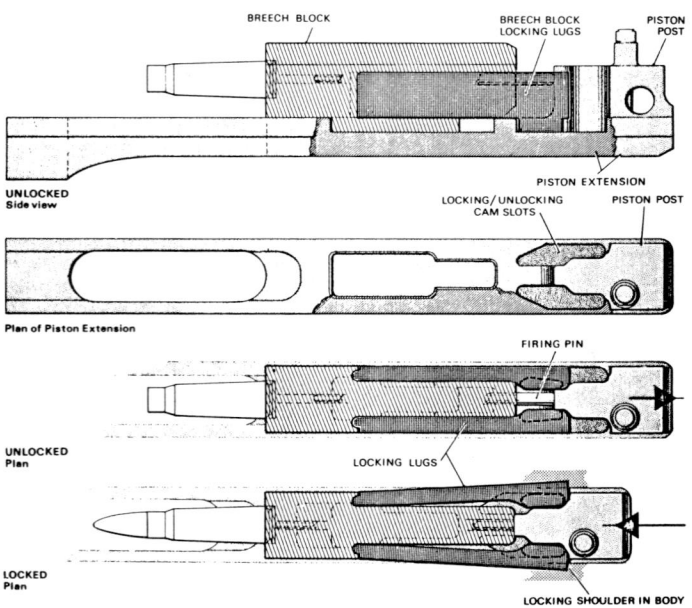

RPD locking system

the rear of the case. Some belts have no nib but a right-angled projection fits against the base of the cartridge. The 50-round belt can be joined to another by putting the end link of one belt against the joining link of the other. A cartridge is used to keep the two belts together. The cover on the belt carrier is opened and the belt placed in, from the left, with the loading tab outside. The belt carrier is slid onto the mounting bracket under the gun from the rear and the mounting bracket lock is swung down to secure the belt carrier in place. On the Chinese Type 56-1 the dust cover over the feedway must be swung down and locked into place to act as a belt carrier bracket.

The gun is fed by a feed arm oscillated across the receiver by a roller carried on the slide.

There is a regulator mounted in a block under the gas vent. It is adjusted by loosening the nut on the left side of the regulator and pressing it to the right. The regulator has three settings marked '1', '2' and '3' and is rotated to align the required number with the index pin. The regulator is then pushed back to the left and the nut tightened.

The barrel is fixed and cannot be changed.

The foresight is a post. The rearsight is a tangent U. The rearsight leaf registers from 100 to 900 m in 100 m steps.

For zeroing, the foresight is screwed in or out for elevation adjustment. For line, the lock nut must be loosened and the slide moved across its dovetail. The rearsight has a separate windage knob.

Variants

During its service life several small variations were made to the RPD and as a result there are five recognised versions of the gun. These are:

First version: Female gas piston fitting over male gas spigot. No dust cover. Rigid, reciprocating cocking handle. Windage knob on right of rearsight.
Second version: Male gas piston fitting into female cylinder. No dust cover. Rigid, reciprocating cocking handle. Windage knob on left of rearsight.
Third version: Male gas piston, female cylinder. Dust cover on feed mechanism. Folding non-reciprocating cocking handle. This is the version copied by the Chinese as their Type 56.
Fourth version: Male gas piston, female cylinder. Longer gas cylinder. Extra

friction roller on piston slide. Dust cover on feed mechanism. Buffer in butt. (Sometimes called RPDM)
Fifth version: As fourth version but with folding feed cover belt carrier. Multi-piece cleaning rod carried in butt trap. (Also Chinese Type 56-1.)

Operation

After loading the belts and placing one, or two joined together, in the belt carrier place the belt carrier under the gun. Push the loading tab through the feedway from left to right and pull through to seat the first cartridge under the retaining pawl. If no loading tab is fitted the feed cover must be lifted and the belt put, open side down, on the feedway. The belt has a nib, already mentioned, fitting into the extraction cannelure, and there is also another nib at the front end level with the shoulder of the case. This nib must fit over a guide which pushes the nose of the bullet down for locating, ready for feeding. If the belt is not so positioned a stoppage is inevitable.

When the belt is in place and the top cover replaced, the cocking handle can be retracted; the cocking handle stays to the rear on the first two versions of the gun but is pushed forward and folded upward on the later models. The gun is then ready to fire. When the trigger is pulled the slide is driven forward carrying the bolt which pushes the first round out of the belt and into the chamber. The extractor slips into the extraction groove and the bolt comes to rest. The slide goes on and the solid wedge at the rear drives between the locking lugs and forces them out into the locking recesses in the receiver walls. The bolt is now firmly secured to the receiver. The slide goes on and the wedge hits the rear of the light firing pin in the bolt. The cap is struck, the propellant ignited and the bullet driven up the bore. A proportion of the gas passes through the gas port, and after giving the piston-head a sharp blow, disperses to atmosphere from the open cylinder. The slide is attached to the piston and when it goes back there is a period of idle movement in which the breech-block remains firmly locked while the chamber pressure drops to a safe level. Then the locking lugs, which rest in cam slots cut in the top face of the slide, are forced inwards against the sides of the bolt and unlocking is completed. The slide carries the breech-block back. The empty case is held to the breech face and ejected downwards out of the gun when the ejector strikes the top of the base of the case and revolves it round the extractor. The return spring is fully compressed and the slide is driven forward again.

Feed mechanism

The slide carries a feed roller which, mounted on the left hand side, fits into the feed arm under the top cover plate.

Pivoted at the centre of the feed arm is a feed slide which carries a spring-loaded feed pawl. When the feed arm is moved to the right, the feed pawl is moved to the left.

Starting with the gun cocked and ready for action, the cartridge is positioned by the feed pawl against the cartridge guide. It is lined up with the bullet guide and is ready for chambering.

When the trigger is pulled the bolt goes forward and pushes the round out of the belt into the chamber. During this action the cartridge belt must be stationary so the roller runs down a straight portion of the feed arm. As soon as the cartridge is clear of the belt, the roller pushes the feed arm to the right and the feed slide moves out of the gun to the left. The stop pawl on the feed tray prevents the belt being pushed to the left out of the gun as the spring-loaded feed pawl rides over the round in the belt. At the end of the forward stroke of the bolt the round is fully chambered and the feed pawl is gripping the next round in the belt.

The round is fired. The piston is driven to the rear. The roller moves back and there is an initial free travel before it bears on the feed arm. The feed arm is then pulled to the left, the feed slide is pivoted to the right. The feed pawl pulls the round it is gripping to the centre line of the gun where it is held by the cartridge guide and the feed pawl.

In brief the gun brings the belt across during the recoil stroke and positions the feed pawl over the next round, ready for feeding, during the firing stroke.

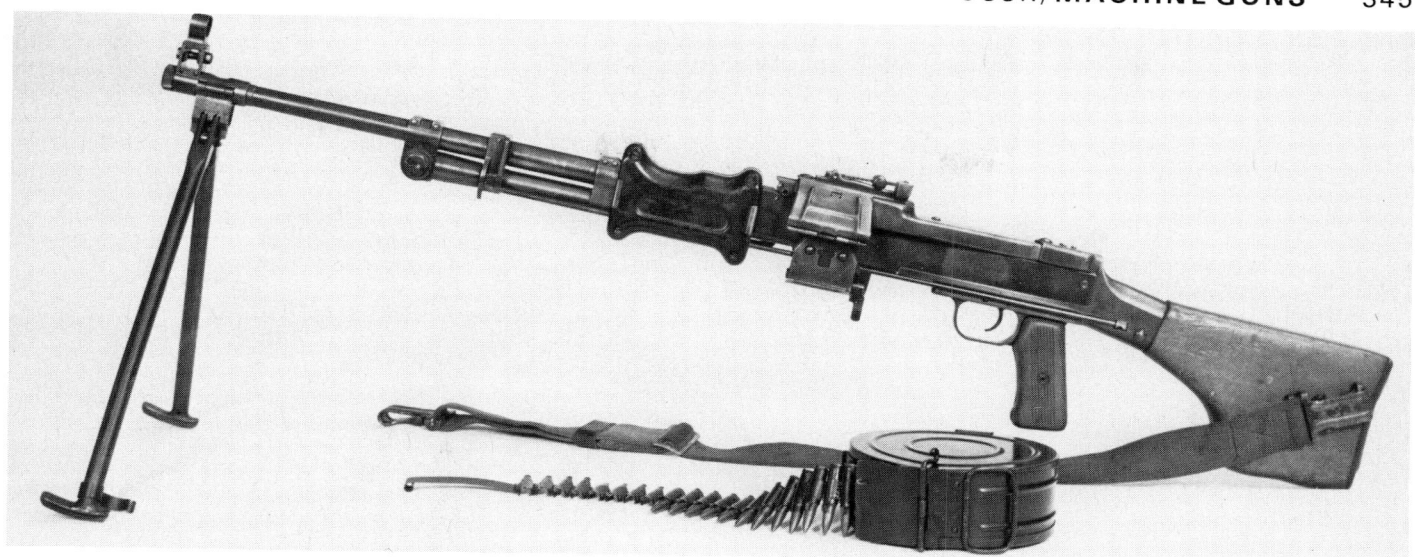

7.62 mm RPD Model 4 (Hungarian) light machine gun with anti-friction roller on back of bolt and enclosed gas cylinder

Trigger and firing mechanism

Since the gun fires only at automatic the trigger mechanism is not complex. The trigger has a hook at the front end which enters a window in the sear. When the trigger is pulled the hook pulls the sear down. The safety catch is on the right of the receiver, directly above the trigger. When it is applied (put forward), it locks the sear in the up position. If an attempt is made to cock the gun with the safety forward (at safe) the underside of the slide will jam on the immovable sear and remain locked. The gun will then be useless until the slide can be driven forward again. This presents no problem with the early version guns with fixed cocking handles but is not easy with non-reciprocating cocking handles.

Stripping

Lift the top cover plate by pushing forward on the catch at the rear end and then rotating the cover forward and up. It will only elevate about 45°.
Remove belt. Check chamber is clear. Return bolt to forward position.
Press out the locking pin at the rear of the receiver. Slide butt and trigger group rearwards.
Pull cocking handle rearwards and take slide and bolt out. Lift bolt off slide.

Re-assembly

Put locking lugs in bolt; place bolt on slide with feet of locking lugs in cam slots.
Push forward on slide; insert rails in grooves in receiver walls and push forward.
Slide butt and trigger group into the receiver. Replace locking pin.

DATA
Cartridge: 7.62 mm × 39

Operation: gas, automatic
Method of locking: projecting lugs
Feed: 100-round disintegrating link belt, carried in drum

WEIGHTS
Gun only: 7.1 kg

LENGTHS
Gun: 1036 mm
Barrel: 521 mm

MECHANICAL FEATURES
Cooling: air. Fixed barrel
Sights: (foresight) cylindrical post; (rearsight) leaf tangent with V backsight; (sight radius) 600 mm

FIRING CHARACTERISTICS
Muzzle velocity: 700 m/s
Rate of fire: (cyclic) 700 rds/min
Effective range: 800 m

Manufacturer

State factories of the Soviet Union and all European satellites. Made in China and North Korea.

Status

Obsolescent in Warsaw Pact countries. Current in many others.

Service

Known to be in service with China, Egypt, North Korea, Pakistan and Vietnam. Likely to be found in other parts of the world also.

7.62 mm SG43 and SGM (Goryunov) medium machine gun

The Soviet designer Peter Maximovitch Goryunov produced his SG43 to supplement the fire of the DP and DPM. The gun was modified and improved until there were eventually six versions of it with comparatively minor differences. These guns are:
SG43: This has a smooth barrel with no fins at all. It has the cocking handle lying horizontally between the two vertical spade grips of the firing gear. The sear is attached to the return spring guide. The barrel lock is a simple wedge which comes out to the side. There are no dust covers over the feed and ejection openings.
SG43B: This has a micrometer barrel lock and dust covers over feed and ejection openings.

SGM: This has longitudinal barrel fins and a separate sear housing. Dust covers are on late production models. The cocking handle is on the right hand side of the receiver.
SGMT: This is the tank version of SGM with a solenoid on the backplate of the receiver.
SGMB: Similar to SGM but has dust covers over the feed and ejection ports.
Hungarian SG general-purpose machine gun: This gun has an RPD butt stock with a pistol grip. It has an external resemblance to the PK general-purpose machine gun but there is no hole in the butt stock and the ejection slot is lozenge-shaped rather than rectangular.

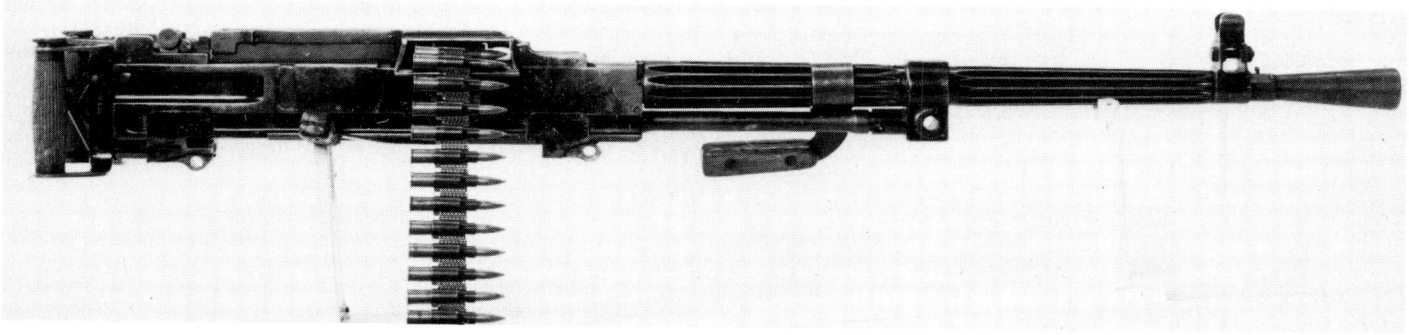

7.62 mm SGM medium machine gun showing cocking handle on right-hand side and fluted barrel

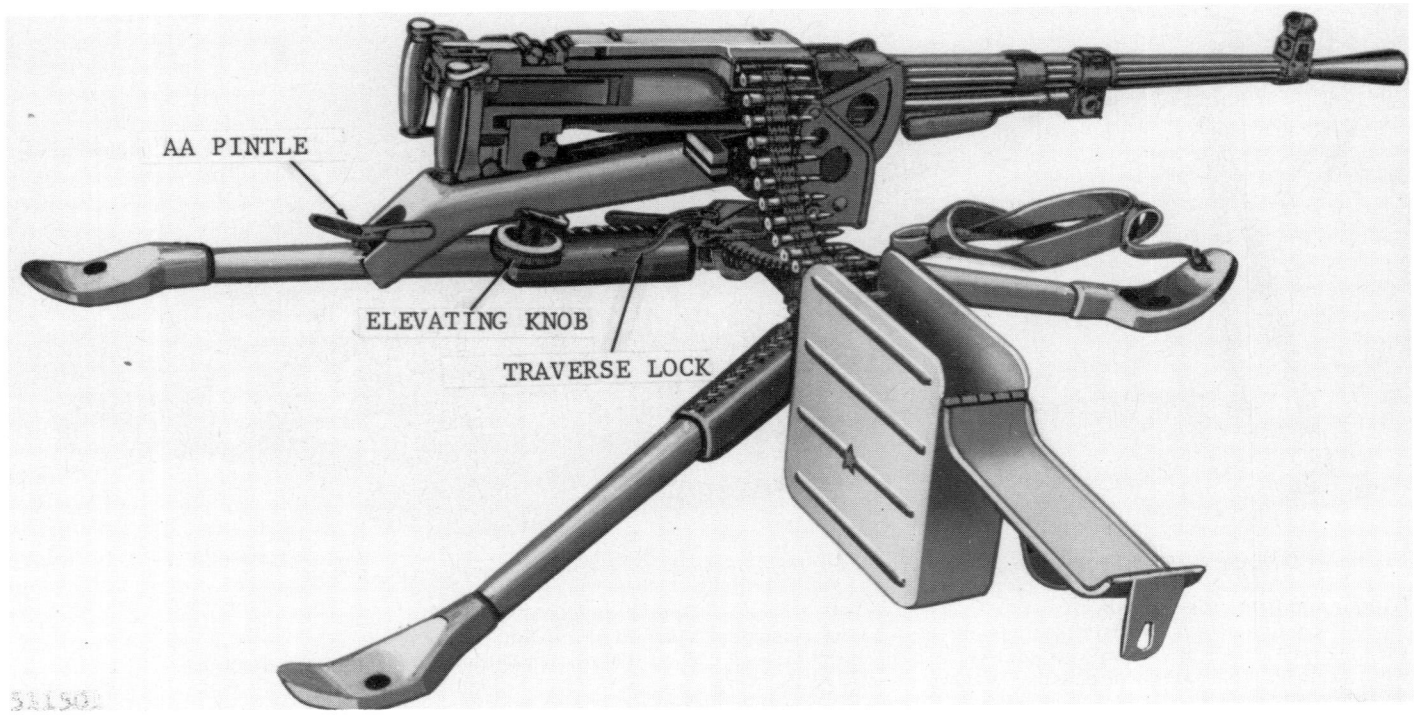

AA PINTLE

ELEVATING KNOB

TRAVERSE LOCK

7.62 mm SGM medium machine gun on Sidorenko-Malinovski tripod

The Goryunov guns have been used by client countries of the USSR in Europe, Asia and the Middle East.

The Chinese version of the SG43 is known as the Type 53 and the SGMB is called the Type 57 heavy machine gun. The Czechoslovak version has the marking 'vz.43' and the Polish gun is stamped 'Wz43'.

Description

The Goryunov is a sturdy, simple and reliable gun. It is gas-operated, belt-fed and fires from a wheeled tripod which can be manhandled or pulled by a draught animal or vehicle. It has a variable track gas regulator.

The barrel is changed readily. The SG43 has a simple wedge which drives into a dovetail slot on top of the barrel. All other Goryunov machine guns have a milled catch on the barrel lock which must be depressed before the lock can be withdrawn. The headspace of all these weapons, except the SG43, is adjusted by the micrometer barrel lock. The tool provided in the tool kit for this purpose is used to rotate the square socket screw. The gun is adjusted only at base workshops. The barrel is pulled back until the breech-block will just lock into its locking recess on the right side of the receiver wall.

The belt is the standard Soviet closed pocket type used in the RP-46, holding 250 7.62 mm × 54R cartridges. It is loaded by forcing the rounds into the metal pockets as far as they will go.

The gun fires from the open breech position and since it has the closed pocket belt, the cartridge is withdrawn from the belt, lowered to the level of the barrel and then rammed by the bolt.

It fires at automatic only and the massive barrel and the easy change make it possible to sustain a good rate of fire. The wheeled mounts incorporate coarse traverse and elevating gears and a fine elevating knob. The mounts have

separate lock levers for elevation and traverse which are normally secured before firing but the traverse lock lever can be released so that free traverse through 360° is possible.

The sights are simple and easily adjusted. The foresight is a cylinder mounted between two protectors. It is zeroed for elevation by screwing in and out, and for line by undoing the nut at the base and pushing the foresight on its dovetail into the required position. The rearsight is a U-notch mounted on a tangent leaf.

Operation

The feed cover is lifted to the vertical position and the belt is placed in the feedway from right to left with the first round positioned so that its rim lies inside the claws of the cartridge gripper. The cover is lowered and the gun cocked. The cocking handle on the SG43 and SG43M lies horizontally between the grips and on the others projects from the right of the receiver.

To fire the gun the trigger-locking lever must first be raised with the left thumb and then the trigger can be operated with the right thumb.

The return spring drives the slide forward carrying the bolt. The bolt picks up the cartridge which is lying above the feedway, with the nose slightly down, and pushes it into the chamber. As it enters the chamber the extractor on the bolt-face grips the rim. The bolt comes to rest but the slide continues on. The raised piston post moves up the hollow interior of the bolt and forces the rear end out into a locking recess in the right-hand wall of the receiver. The piston post then continues on and strikes the firing pin which is driven into the cap of the cartridge.

As the slide travels forward the diagonal grooves cut in its top surface engage ribs in the feed slide which is pushed out to the right and the feed pawl slips over the next cartridge in the belt.

Lying under the feed cover is the lower feed cover. Resting on top of this, with a lug which reaches down to the bolt, is the cartridge gripper. The lug engages in the bolt so the cartridge gripper reciprocates above it. The cartridge gripper has a pair of spring-loaded claws which grip over the rim of the cartridge in the belt when the bolt goes fully forward.

The gases push the bullet up the bore and some are diverted into the gas cylinder and force the piston and slide rearwards. The piston post on the slide moves back along the hollow interior of the bolt. There is a period of free travel while the chamber pressure drops to a safe level and then the piston post forces the rear end of the bolt out of the locking recess and carries it back. The empty case is held to the breech face and is ejected through the lozenge-shaped port on the left of the gun.

As the bolt goes back the cartridge grippers pull the round out of the belt and withdraw it. The spring-loaded depressor arm projecting down from the underside of the feed cover forces the round down into the lower feed cover where it rests, nose slightly down, in the feedway where it can be fed by the next forward movement of the bolt.

As the slide goes back the grooves on the top surface pull the feed pawl inwards and the next round is moved over into position for the cartridge grippers to seize it at the end of the next feed stroke.

Stripping

Lift the feed cover. Remove the belt. Ensure the chamber is clear.

With the SG43 or SG43M pull the take-down pin on the lower right rear of the receiver to the right and remove the backplate.

With the SGM, SGMB or SGMT depress the detent in the latch at the top of the backplate. Slide the latch back. Rotate the backplate clockwise and remove it.

Withdraw the return spring.

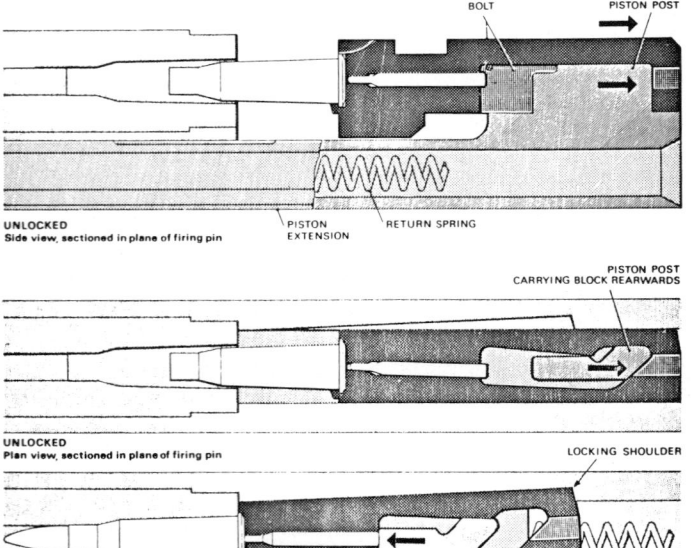

BOLT   PISTON POST

UNLOCKED
Side view, sectioned in plane of firing pin

PISTON
EXTENSION   RETURN SPRING

PISTON POST
CARRYING BLOCK REARWARDS

UNLOCKED
Plan view, sectioned in plane of firing pin

LOCKING SHOULDER

LOCKED

Locking action

7.62 mm SG43 medium machine gun showing smooth barrel and cocking handle between spade grips

With the SGM, SGMB and SGMT remove the sear housing out of the back of the receiver.

Grip the cocking handle and pull it rearwards. Remove the bolt and slide. The bolt can be lifted off the slide. The belt feed slide comes out of the receiver to the right. The cartridge gripper can be slid along the lower feed cover until it reaches the cutaway section and can then be removed.

Re-assembly

Put the cartridge gripper back in the lower feed cover with the lug at the back projecting down.

Put the belt feed slide in the receiver. Push the ejector forward in the bolt.

Put the bolt as far forward as possible over the piston post on the slide.

Put the cocking handle forward and place piston and bolt in the receiver.

Slip the sear housing of the SGM, SGMB and SGMT, sear forward, into the sear.

Put the return spring into the slide. Put the guide in the spring. Close up the backplate.

Close the lower feed cover and adjust the lug of the cartridge gripper in the bolt. Close the feed cover.

Pull the cocking handle and ensure everything works.

DATA
Cartridge: 7.62 mm × 54R
Operation: gas, automatic
Method of locking: tilting block
Feed: 250-round pocketed belt

WEIGHTS
Gun: (empty) 13.6 kg
Barrel: 4.8 kg

LENGTHS
Gun: 1120 mm
Barrel: 719 mm

MECHANICAL FEATURES
Barrel: (cooling) air. Quick change
Sights: (foresight) cylinder on dovetail; (rearsight) tangent leaf with U backsight; (sight radius) 850 mm

FIRING CHARACTERISTICS
Muzzle velocity: 800 m/s
Rate of fire: (cyclic) 650 rds/min
Range: (max effective) 1000 m

Manufacturer
State factories in the USSR, Czechoslovakia and Poland.
Status
Obsolete. No longer made.
Service
Not known to be in first-line service with any army, but likely to be found in the Middle East, Africa, China and South-east Asia in general.

7.62 mm RPK light machine gun

Since the AK-47 assault rifle was introduced, Soviet policy has been to replace the older weapons with modern counterparts designed by Kalashnikov. The light machine gun in the Soviet infantry section has been, for several years, the RPD, a Degtyarev design. This has now been replaced by the RPK, the Ruchnoi Pulemet Kalashnikov. It was first seen in large numbers carried in the May Day Parade in Red Square in Moscow in 1966 although its existence was known before this. The picture here shows a model stamped '1964' and there are no indications that it is other than a standard gun. The RPKS differs from the RPK in having a folding butt.

The RPK is basically an AKM with a longer and sturdier barrel. It will take the magazines from the AK-47 or the AKM. It also has a 40-round box magazine of its own and a 75-round drum magazine. The 75-round drum magazine takes some time to fill and it appears that it would be used initially in action but thereafter replaced, during the assault or the early stages of defence, by the 40-round magazine. The box magazines are loaded by placing a cartridge between the feed lips and pressing down. The drum magazine has a loading lever on the front which is used to compress the magazine spring before each round is inserted into the feed lips. It is released after each round is loaded. The drum has an extension piece which is inserted in the magazine-well under the gun, in the same way as a box magazine. The magazine rests across the gun, sloping forward at about 45°.

Operation

The RPK is gas-operated with a vent 21.5 cm from the commencement of rifling and 28 cm from the muzzle, opening into a cylinder placed above the barrel. There is no gas regulator, which results in a slightly harsh action and a vigorous ejection to the right rear when the gun is clean, and a gradual slowing down of the rate of fire and force of ejection as firing progresses and fouling accumulates. The piston is forced back by gas pressure in the cylinder and there is a short period of free travel while the width of the cam slot cut in the piston extension travels across the unlocking lug of the bolt-head and then the two locking lugs are caused to rotate anti-clockwise out of engagement with the locking shoulders. After unlocking is completed the piston carries the bolt to the rear.

The loading stroke of the bolt chambers the round; the bolt movement then ceases and further forward movement of the piston rotates the bolt to enable the locking lugs to engage in front of the locking shoulders.

The firing pin floats freely in the bolt. It cannot be reached by the hammer which comes up against a stop until such time as locking is completed. If the hammer is released early the bolt is forcibly closed and the reduced blow on the firing pin produces a misfire.

The cocking handle is integral with the bolt carrier and reciprocates as the

75-round drum magazine of 7.62 mm RPK light machine gun

7.62 mm RPK light machine gun with 40-round box magazine

gun fires. This allows positive closure if adverse conditions of sand or mud prevent completion of locking.

There is a single extractor which is located on the right-hand side of the bolt face which itself is recessed so that the base of the round is always fully supported in the chamber. A fixed ejector in the body runs along a groove in the bolt producing a strong ejection to the right rear of the gun. The barrel and chamber are both chromium plated.

There is no holding-open device so the mechanism always stops in the forward position.

Fire selection

The selector is on the right side of the machined receiver and is a long lever of similar shape to that in the rifle. There are three positions: fully automatic at the middle position of travel, single-shot at the bottom and safe at the top.

The selector arm in the 'safe' position carries out two functions. Its spindle carries an arm which, when rotated to 'safe', covers the trigger extension and prevents trigger movement. The selector itself comes up to block the cocking lever and thus prevents the weapon being cocked while set at 'safe'. The limited bolt movement does not pass over the base of the round in the magazine but is sufficient for the firer to check that the chamber is empty either by visual inspection by day or by touch at night.

Trigger mechanism

The weapon employs a system embodying a hammer with a cross bar at the top which forms two bents, one on the left arm and one in the centre. The function of the right extension on the cross bar will be dealt with later.

The nose of the trigger carries a single-hook sear which engages in front of the left arm of the hammer bar, and a secondary spring-loaded sear is rotated with the trigger.

The mechanism, in principle, is the same as that employed in the US M1 Garand and the M14. It is derived from that used in the Soviet AK-47 but is considerably modified. There is a safety sear which always holds back the

hammer until the bolt carrier is fully forward when this rotates the safety sear clear of the hammer notch near the hammer axis.

When single-shot is selected, ie the bottom position of the selector lever, and the weapon is loaded and locking is completed, the hammer is held back only by the nose of the trigger. Rotation of the trigger moves the nose forward, the hammer bent is disengaged and one round is fired. The hammer is rotated back by the rearward movement of the bolt and the T bar at the top of the hammer is held back by the spring-loaded secondary sear engaging the centre of the cross bar of the T. Since the trigger is still held back by the firer another shot cannot be fired until the trigger is released. The nose of the trigger moves up and holds the left-hand arm of the hammer head and at the same time releasing the trigger moves the secondary sear clear, leaving the firer free to rotate the trigger which will, if locking is complete, allow the hammer to go forward under the influence of the two-strand multiple wire spring.

When automatic is selected the spindle on the selector lever is rotated. This carries an attachment which forces the secondary sear back and it is no longer contacted by the hammer in its rearward rotation. The fire is controlled entirely by the safety sear. When the bolt carrier is fully forward it rotates the safety sear extension which releases the hammer to fire another round. The process continues until the ammunition is expended (in which case the hammer remains forward) or the trigger is released and the hammer is held back by the nose of the trigger and the safety sear.

The trigger-firing mechanism is obviously based on the Kalashnikov AK-47 but the following changes taken from the AKM have been incorporated. Instead of a double-hook, as on the AK-47, the nose of the trigger has only the one on the left-hand side. Mounted on the trigger axis and lying on the right side of the receiver is a spring-loaded pawl which is struck by the right arm of the T of the hammer cross bar every time the hammer rises. The pawl is linked to another hook which moves over the hammer when the pawl is struck and holds the hammer down. The long curved face of the pawl results in the hook holding the hammer for an appreciable time. When the

7.62 mm RPK light machine gun showing firing position

spring snaps the pawl back and releases the hook, the hammer rises but the delay reduces the cyclic rate.

Sights

The rearsight is a leaf sliding on a ramp. It has, like all Soviet hand-held weapons, a U rearsight. There is also a night sight with an enlarged U with a fluorescent white spot. The rearsight is graduated in hundreds of metres marked 1 to 10. There is a windage scale.

The foresight is a cylindrical post which can be screwed up or down for elevation adjustment, when zeroing.

There is a battle sight which is a coarse U covering 100 to 300 m.

Stripping

The RPK strips in exactly the same way as the AK-47 and AKM.

Accessories

The same accessories are carried as the AK-47 and AKM but there is no bayonet. A bipod is mounted well forward which can be folded under the barrel and clipped in place.

Some RPKs mount an infra-red night sight.

DATA

Cartridge: 7.62 mm × 39
Operation: gas, selective fire
Feed: 40-round box or 75-round drum magazine. Can also use 30-round rifle magazine

WEIGHTS
Weapon: 5 kg
30-round magazine: (empty) 340 g; (loaded) 850 g
40-round magazine: (empty) 368 g; (loaded) 1.13 kg
75-round drum magazine: (empty) 0.9 kg; (loaded) 2.1 kg

LENGTHS
Weapon: 1035 mm
Barrel: 591 mm

MECHANICAL FEATURES
Barrel: (regulator) nil; (cooling) air. No barrel change
Sights: (foresight) cylindrical post; (rearsight) leaf notch; (sight radius) 570 mm

FIRING CHARACTERISTICS
Muzzle velocity: 732 m/s
Rate of fire: (cyclic) 660 rds/min
Effective range: 800 m

Manufacturer
State factories.
Status
Current. Production.
Service
USSR, East Germany and other Warsaw Pact countries.

5.45 mm RPK-74 light machine gun

The RPK-74 is the light machine gun version of the AK-74 rifle and it follows the same logical design trend as the RPK. It is a machine rifle with a longer and heavier barrel and a larger magazine. A light bipod is clipped to the muzzle. The butt and fore-end appear to be identical with those of the RPK and, as would follow from experience with the rifle which uses the maximum possible components of the AKM, the RPK-74 seems to do the same with the RPK.

There are no firm details of the dimensions, but it can be calculated from photographs that the barrel of the RPK-74 is roughly 1.6 times longer than that on the rifle.

5.45 mm RPK-74 light machine gun with bipod folded

7.62 mm PK machine gun family

As the family of Kalashnikov weapons replaced those of Degtyarev the time came for the replacement of the RP-46 and the SGM series by a Kalashnikov design. The weapon selected for this was the PK which is a mixture of components and ideas from other weapons. The rotating bolt comes from the other Kalashnikov weapons, the AK-47 and RPK, the cartridge gripper comes from the Goryunov SGM and so does the barrel change. The system of feed operation using the piston to drive the feed pawls comes from the Czechoslovak model 52 light machine gun. The trigger comes from the Degtyarev RPD.

Description

The Soviet 7.62 mm PK machine gun was first seen in 1964. Since then it has been modified and improved so that it has become a true general-purpose machine gun and has replaced the RP-46 and SGM. The following versions exist:

PK: the basic gun with a heavy fluted barrel, feed cover constructed from both machined and stamped components and a plain butt plate. The PK weighs about 9 kg.
PKS: the basic gun mounted on a tripod. The lightweight tripod (4.75 kg) not only provides a stable mount for long-range ground fire, it also can be quickly opened up to elevate the gun for anti-aircraft fire.
PKT: the PK as altered for coaxial installation in an armoured vehicle. The sights, stock, tripod, and trigger mechanism have been removed, a longer heavy barrel is installed, and a solenoid is fitted to the receiver back plate for remote triggering. An emergency manual trigger and safety is fitted.
PKM: a product improved PK, with a lighter, unfluted barrel, a feed cover constructed wholly from stampings, and a hinged butt rest fitted to the butt plate. Excess metal has been machined away wherever possible to reduce the weight to about 8.4 kg.
PKMS: PKM mounted on a tripod (similar to the PKS).

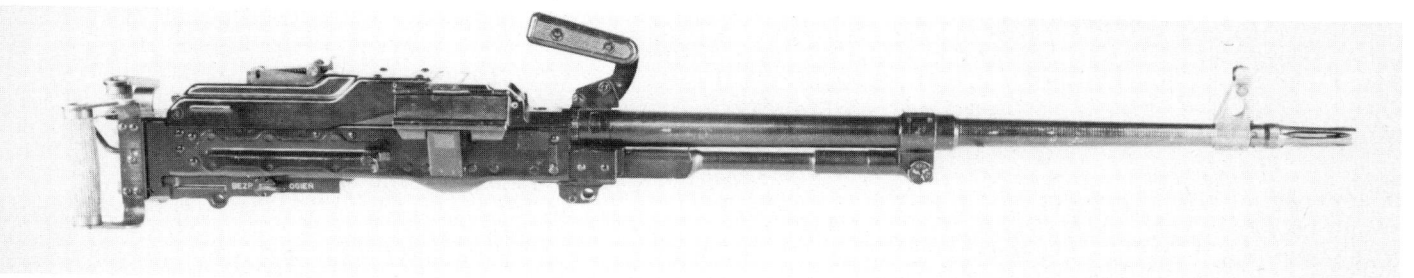

7.62 mm PKB light machine gun

Interesting line-up of small arms on Polish exercise. Left to right: Dragunov, PK with belt box and AKM

PKB: The PKM with the tripod, butt stock, and trigger mechanism removed and replaced by twin spade grips and a butterfly trigger similar to those on the SGMB. This gun may possibly be known as the PKMB.

The PK and PKM machine guns are infantry support guns normally fired from their bipod mounts; however both versions can be installed in the front firing ports of the Soviet BMP infantry combat vehicle. The PKS also is an infantry gun and is used in the classic role of a heavy machine gun to provide long-range area fire, cover final protective lines, and, in addition, provide anti-aircraft fire. The PKT is used in coaxial installations on most modern Soviet tanks and armoured personnel carriers. The PKM, although about 450 g lighter than the PK, is used in the same role as the PK. The PKB (PKMB) is probably used as a pintle-mounted gun on older APCs such as the BRDM, BTR-50, and BTR-60 similar to the manner in which the old SGMB was used.

The PK family of machine guns are gas-operated, rotary bolt locked (Kalashnikov system), open bolt fired, fully automatic, belt-fed machine guns firing the 7.62 mm × 54 rimmed cartridge. The ammunition is fed by non-disintegrating metallic belts; current belts are composed of joined 25-round sections but earlier feed belts were made of one 250-round length. The belts are held either in 250-round ammunition boxes, in special large capacity boxes on tanks (for the PKT) or in a 100-round assault magazine attached to the bottom of the gun's receiver. The PK, PKM, and PKB guns are effective against area targets at ranges to 1000 m. They have a cyclic rate of 650 shots per minute. Limited US firing tests indicate that the PK has a very mild recoil effect and practically no muzzle climb which makes the weapons easily controllable by the gunner. The component parts of the PK are individually simple.

The receiver is constructed of stampings riveted up. It carries the very simple, automatic only, trigger and the belt feed mechanism.

The barrel is chrome-plated and has a change system which is not as quick as that employed in most modern guns. To operate it, the gun must be

unloaded; the feed cover comes up when the return spring guide is pressed and the feedway pivots up on the same pin. The barrel lock comes out to the side in the same way as the SGM and the barrel can be pulled forward out of the receiver, presumably with an asbestos glove. The new barrel goes in, the lock is replaced, the feedway is lowered and feed cover lowered and the gun is ready to fire. A gas regulator is fitted. It is a variable track-type with three numbered positions, similar to that in the Goryunov, and the head of the regulator is disengaged from its detent and moved to the selected number.

The gas piston is attached to the slide. Running along the length of the slide is a cam path on which bears the roller of the belt feed lever. The roller follows the cam as the slide goes back and the feed lever pivots outward causing the feed pawl attached to it to move in with the next cartridge in the belt. This is placed in position ready for the cartridge gripper. As the bolt goes forward a second cam path forces the feed lever in at the bottom, out at the top and the spring-loaded feed pawl moves out over the next cartridge.

On the top of the slide is the bolt carrier. The relationship between this carrier and the bolt is exactly the same as that in the AK-47. As soon as the bolt stops moving forward, having chambered the round, the cam path on the carrier engages the cam pin on the bolt and rotates it to lock into recesses.

At the rear of the slide, projecting forward over the bolt, is the cartridge gripper. When the bolt is fully forward the two spring-loaded claws of the gripper slip over the rim of the cartridge in the belt. When the bolt recoils the cartridge is pulled out of the belt and back. The spring-loaded cartridge depressor arm pushes the cartridge down out of the claws and it is held in the feed lips waiting for the bolt to drive it forward into the chamber.

Operation

The 250-round belt is loaded in the same way as the RP-46 or the Goryunov. The feed cover of the gun can be lifted when the return spring guide is pressed. The belt is placed in the feedway, coming in from the right, with the first round gripped in the claws of the cartridge gripper. The feed cover is closed, the

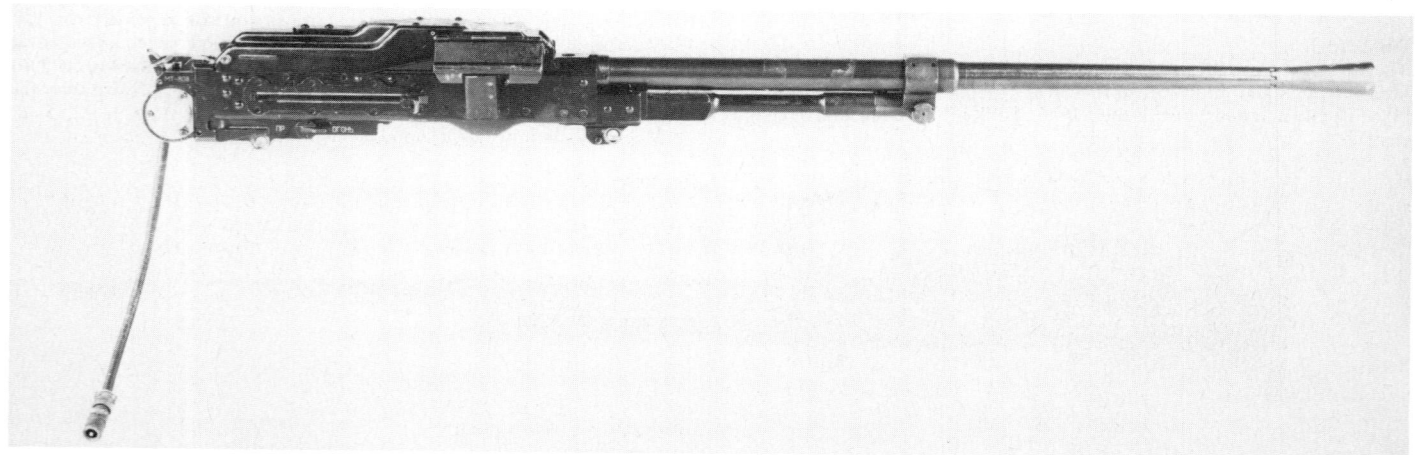

7.62 mm PKT light machine gun

7.62 mm PKM light machine gun

NO BUTT REST

FLUTED BARREL

7.62 mm PK general-purpose machine gun

safety catch set to 'fire', and the cocking handle on the right of the receiver is retracted and returned to the forward position. If the safety catch is left in the rear position at 'safe' the gun cannot be cocked.

With the safety at 'fire', pulling the trigger releases the sear and the compressed return spring drives the slide forward. The bolt picks up the round in the feed lips and chambers it. The extractor grips the rim and bolt movement ceases. As the slide goes forward the cam path along the side forces the feed pawl outwards over the next cartridge in the belt. The bolt carrier moves forward with the slide after the round is chambered and the bolt is rotated 30° to the locked position. The firing pin is carried on the bolt carrier and it comes through the bolt to fire the cap.

The gases force the bullet up the bore. Some of them pass through the gas vent and drive the piston and slide back. The bolt carrier moves back and the firing pin is retracted. There is a period of free travel while the pressure drops and then the cam slot on the bolt carrier engages the cam pin on the bolt and the bolt rotates to become unlocked. The bolt recoils. The empty case is held on the bolt-face and the slide compresses the return spring. The cartridge gripper pulls the cartridge out of the belt and carries it back. The cam path on the side of the bolt carrier forces the feed roller outwards and the upper end of the pivoted feed arm moves inwards so the feed pawl pushes the belt into the gun and the next round reaches the cartridge stop where it remains for the gripper to come forward again.

The empty case is ejected by the fixed ejector. The depressor arm forces the live round down into the feed lips, waiting for the bolt to go forward again. The cycle is repeated as long as the trigger is pulled and ammunition remains in the gun.

Stripping
Press forward the return spring guide and lift the feed cover. Remove the belt

and check the chamber. Remove the barrel by slipping the lock out and pulling the barrel forward.

Press in the return spring guide and ease the return spring with its guide upwards, out of the gun. Hold the cartridge grippers and pull the bolt carrier, bolt slide and piston out of the gun.

Pull the bolt forward and rotate it free of the carrier. Lift the firing pin out of its recess in the carrier.

Re-assembly
Put the firing pin in its recess in the bolt carrier. Put the bolt back over the firing pin onto the bolt carrier.

Pull the bolt forward on the slide, place the piston in the gas cylinder. Pull the trigger and push the bolt carrier home. Put the return spring into its seating in the slide and press the guide forward until it seats against the rear wall of the receiver. Replace the barrel and lock it. Close the feed cover.

Accessories
The largest accessory is the tripod which although primarily designed for general use, can be quickly adapted for anti-aircraft fire.

Belt boxes for 100-, 200- and 250-round belts are issued. The 100-round box can be attached to the underside of the PK for the assault role.

Spare parts and the usual combination tool and the oil-solvent container are carried.

DATA
PKS
Cartridge: 7.62 mm × 54R
Operation: gas, automatic
Method of locking: rotating bolt
Feed: belt; 100, 200 and 250 rounds

WEIGHTS
Gun: (empty) 9kg
Tripod: 7.5kg
100-round belt: 2.44kg

LENGTHS
Gun: 1160 mm; (on tripod) 1267 mm
Barrel: 658 mm

MECHANICAL FEATURES
Barrel: (bore) chromium plated; (regulator) 3 position; (cooling) air. Quick-change barrel
Sights: (foresight) cylindrical post; (rearsight) vertical leaf. Windage scale

FIRING CHARACTERISTICS
Muzzle velocity: 825 m/s
Rate of fire: (cyclic) 690-720 rds/min
Effective range: 1000 m

Manufacturer
State factories.
Status
Current. Production.
Service
Soviet and allied armies.

12.7 mm Degtyarev (DShK-38 and Model 38/46) heavy machine gun

The DShK-38 was designed by Degtyarev and was based on an earlier model of his, the DK, which was produced in very limited numbers in 1934. The feed was designed by G S Shpagin and was a rotary type in which the rounds in the belt were successively removed from the links and then fed through a feed plate and collected by the bolt in its forward travel. This was a complicated process and required skill and training if stoppages were to be avoided.

In 1946 the rotary feed was replaced by the conventional shuttle-type feed used on the Degtyarev Model RP-46. This difference produces a ready recognition feature as the DShK-38 has the large circular drum-like feed mechanism and the Degtyarev Model 38/46 has a flat rectangular feed cover. There is no interchangeability between the parts of the two guns.

The 12.7 mm guns have been used extensively by Soviet-influenced European and Asian countries. The European countries who originally used the gun on a tripod now fit it into a number of vehicles for anti-aircraft and anti-vehicle duties. It is still used on the ground mount by Asian countries. The Chinese version is a copy of the Model 38/46 known as the Type 54 heavy machine gun and can be identified by the Chinese characters on the receiver behind the feeder. The Czechoslovaks make a towed four-gun anti-aircraft assembly.

The DShK-38 feeds from the left and has a fixed barrel. The Model 38/46 can readily be adapted for feed from either side, by changing some parts in the feed mechanism, and it has a quick-change barrel.

Operation
The belt is a continuous metallic-link type holding 50 rounds. Each link takes a cartridge which is located by the nib on the belt entering the groove of the cartridge.

The cover latch of the DShK-38 is at the rear of the cover and it permits the cover to be rotated about the hinge at the front.

The belt is placed in the gun from the left with the first link under the link stripper and the first cartridge in the top compartment of the rotating feed drum. The weight of the ammunition is taken off the feed drum and the ammunition and drum pushed round through some 120°. When the drum will rotate no further the feed cover is closed. The cocking handle between the spade grips is retracted and the slide held to the rear on the sear. The compartments of the rotating feed drum are now full and the first round is pressed through the feed plate into the path of the bolt.

The Model 38/46 should be regarded as a large version of the RP-46. The belt tab is inserted into the feed guide and pulled through until the first round passes over the stop pawl which will prevent it from falling out. When the cocking handle is fully retracted the gun is ready to fire.

When the trigger is pressed it raises the rear end of the sear release lever which pivots and forces down the sear out of engagement with the slide. The

compressed return spring drives the slide and bolt forward. The feed rib on the bolt passes under the feed drum of the DShK-38 and picks up the round which is projecting through the feed plate. In the Model 38/46 the round is held up to the catch stop. The round is pushed forward and into the chamber. The extractor grips into the cannelure at the base of the cartridge and the bolt comes to rest. On each side of the bolt is a long locking lug which is pivoted at the front end into a recess in the side of the bolt and opens out from the bolt at the rear end, like a flap, as the firing pin, which is attached to the slide, drives forward. The projecting lugs are cammed out by shoulders on the firing pin and engage in recesses in the side walls of the receiver. The firing pin goes in a further 16 mm after locking is completed and the cap is fired.

Some of the propellant gases are diverted into the cylinder and drive the piston back. The slide carrying the firing pin is retracted and there is about 16 mm of free play while the chamber pressure drops. The recesses cut into the top surface of the slide, bolt and firing pin go back as one unit. The empty case comes out on the bolt-face and is ejected through the bottom of the receiver. The return spring is compressed to provide energy for the next forward stroke.

As the operating stud on the slide comes back it enters the open stirrup of the feed lever and rotates it back. On the Model 38/46 this operates the feed slide as in the RP-46. A cam path moves the belt feed slide inward and the feed pawl brings the next round up to the cartridge guide and over the stop pawl. As the feed slide goes forward with the operating stud, the feed pawl is moved out to engage the next round in the belt.

On the DShK-38 the backward movement of the operating stud rotates the feed lever and a pawl bears on a ratchet in the feed drum causing it to rotate (in a manner not unlike the revolving chamber of a revolver pistol). At the same time another pawl prevents counter rotation. The rotation of the drum draws the belt into the gun. The first cartridge is pressed along the lips in the feed plate and is subsequently forced forward on the next feed stroke of the bolt.

Stripping
Lift the feed cover and remove the belt.
Check the feed drum of the DShK-38 is empty.
Check the chamber is clear.
Push out the locking pin at the rear of the receiver and detach the back plate from the receiver.
Remove the sear through the back of the receiver.
Force the gas cylinder forward and rotate it clockwise to free it from the barrel.
Retract the cocking handle. Remove gas piston and slide with bolt and firing pin. Remove gas cylinder.
On the Model 38/46 unscrew the barrel lock securing nut. Remove barrel lock to the side. Pull barrel forward out of the receiver.

Soviet 12.7 mm DShK-38 heavy machine gun on Sokolov mount

Re-assembly

Assemble the firing pin in the bolt and place locking lugs on the sides. Place bolt on slide and push into receiver.

Pull gas cylinder forward and rotate anti-clockwise to re-connect it with the barrel.

Slide sear mechanisms into receiver. Drop backplate. Replace locking pin. On Model 38/46 replace barrel, barrel lock and securing nut.

Accessories

The following are issued with the 12.7 mm heavy machine guns; cleaning rod; chamber cleaning brush; oil and solvent container; combination tool; punch and separated case extractor.

Sights

The ground sights are of orthodox configuration. The foresight is a pillar which screws up and down for zeroing. The bolt through the base of the foresight can be loosened and the sight moved totally for zeroing.

The rearsight has vertical twin pillars with a U backsight between them. When the sight is upright the elevating screw is at the top. Rough setting of the sight is achieved by pressing the slide catch and moving the slide connecting the pillars, up or down until the upper edge is aligned with the required range reading which is shown in hundreds of metres. The elevating screw gives fine adjustment. There is a windage knob at the base of the sight.

The anti-aircraft sight Model 1943 requires the concerted action of two men. The number two rotates his sight and lines it up along the fuselage of the target and so indicates the approach angle. He does this by rotating the target course hand crank. The drive shaft also rotates the number one's sight. The number one lines up the rear, peep sight with an aim off lead determined by his estimate of the target speed.

Mountings

The basic mount for both the DShK-38 and the Model 38/46 is the Model 1938. This is a two-wheeled mount that can be moved by man, mule or motor. A shield is sometimes provided for ground use. The traverse and elevation are coarse with a fine elevation adjustment. The gun is locked in position before firing. The mount can be converted for anti-aircraft use by removing the gun and shield and then taking off the wheels and axle. The three legs are spread, the gun mounted above them on the saddle and the sight inserted in the dovetail on the left side of the receiver.

DATA

Cartridge: 12.7 mm × 108
Operation: gas, automatic
Method of locking: projecting lugs
Feed: belt

WEIGHTS
Gun: 35.7 kg
Barrel: 12.7 kg

LENGTHS
Gun: 1588 mm
Barrel: 1070 mm

MECHANICAL FEATURES
Barrel: (regulator) 3 position; (cooling) air. Model 38/46 has changeable barrel
Sights: (foresight) cylindrical post; (rearsight) vertical leaf with U backsight
Sight radius: 1111 mm

FIRING CHARACTERISTICS
Muzzle velocity: 860 m/s
Rate of fire: (cyclic) 575 rds/min
Effective range: 2000 m

Manufacturer
State factories.
Status
Current. No longer in production.
Service
Soviet and allied armies as well as several guerrilla forces.

12.7 mm NSV heavy machine gun

Development of this weapon began in 1969 and was carried out by a team of three designers, GI Nikitin, JM Sokolov and VI Volkov. It is employed variously in the ground role as a heavy support weapon or as an air defence weapon, and as a tank air defence machine gun. In the ground role it is provided with a shoulder stock and pistol grip, tripod and optical sight; in the air defence role it is provided with the collimating optical sight K10-T, and it is also fitted with normal iron sights.

The NSV is a gas-operated weapon using a rotating bolt and not, as previously thought, a sliding block breech. Feed is from a 50-round disintegrating metal link belt. The barrel is interchangeable and it is recommended that it be changed for a cool one after firing 1000 rounds.

DATA
Cartridge: 12.7 × 107 mm
Operation: gas, automatic only
Method of locking: rotating bolt
Weight: 25.0 kg
Weight of barrel: 9.0 kg
Weight of 50-round belt: 7.70 kg
Length overall: 1.560 m
Rate of fire cyclic: 700 – 800 rds/min
Rate of fire practical: 80 – 100 rds/min
Sights: see text; sighted to 1500 m for aerial targets, to 2000 m for ground targets
Effective range: 1500 – 2000 m
Max range: 6000 m
Muzzle velocity: 845 m/s

Manufacturer
State arsenals.

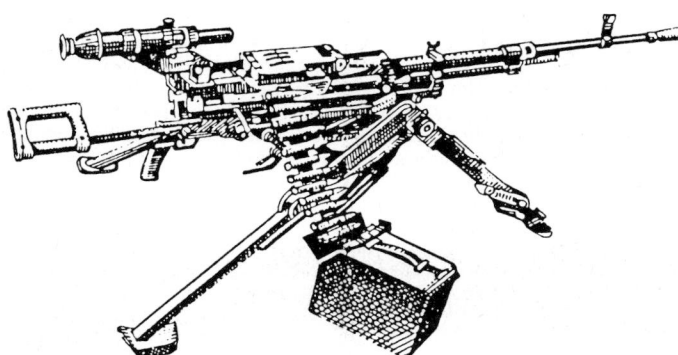

12.7 mm NSV machine gun in ground role

12.7 mm NSVT tank machine gun showing feed cover, barrel and conical flash hider

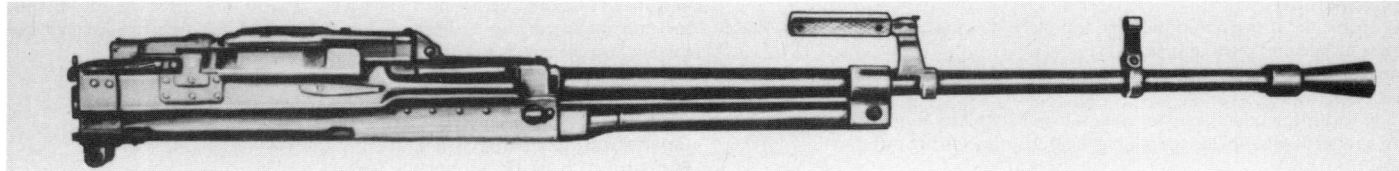

12.7 mm NSV machine gun

Status
Current. Production.

Service
Warsaw Pact armies; also Yugoslavia (qv).

14.5 mm KPV heavy machine gun

This gun was designed immediately after the Second World War by Vladimirov expressly to fire the high velocity round produced for the PTRD-41, Degtyarev, anti-tank rifle. It is a very well made, solid weapon with some characteristics and ideas new to Soviet weaponry.

The gun was designed for simplicity of manufacture. The body is a simple metal cylinder to which the various attachments are riveted. There is some welding. There is considerable use of stampings, for example, the feed tray.

There is no lack of strength and all the components are both robust and well finished. Apart from the ejection opening it is well sealed against ingress of dirt and dust.

While the gun was designed initially as an anti-aircraft gun it is extremely suitable as an armoured fighting vehicle gun. It has a very short inboard length (60 cm) which is less than the .50 Browning, the belt feed system, which allows breakdown in 10-round lengths, is very suitable for vehicle use and so is the alternative left- or right-hand feed but the forward barrel change is not an advantage in this role although it is difficult with a 135 cm barrel to devise any suitable alternative.

Operation
The gun is short-recoil operated with gas assistance supplied by a muzzle booster. Before firing the bolt assembly is securely locked to the barrel. The locking lugs consist of eight rows of castellated form each 1.65 mm deep which are rotated to lock into corresponding recesses on the barrel extension. The pressure built up forces the breech-block rearwards and this in turn draws the barrel back with it against the resistance offered by the inertia of the assembly and the force of the return spring. Since the barrel has its own run out spring this also opposes the rearward movement. When the projectile reaches the hole in the muzzle cap the gases are briefly sealed and the resulting pressure build-up inside the booster drives the barrel backwards to give increased momentum to the recoiling parts. The breech-block and barrel have a free recoil of slightly less than 5 mm which provides the mechanical safety after firing and allows the bore pressure to fall to a safe level. The bolt-head carries a roller on each side running in a curved cam path in the bolt body and these rollers come up against curved cam paths in the body of the gun. The rotation imposed on the bolt-head unlocks it from the barrel. There is a 6 mm pitch on the locking lugs and while the bolt-head is being rotated it is drawn back this distance which provides the slow powerful leverage required for the initial extraction of the cartridge case. At the same time the firing pin is withdrawn. The unlocking is completed when the barrel and bolt have recoiled together for 18 mm. As the rollers are rotated by the cams in the body they are running in the slots in the bolt body and thus accelerate the bolt body to the rear. The relative displacement between the bolt-head and bolt body is slightly less than 18 mm.

KPV bolt. Rotation of roller in cam slot rotates bolt-head which locks into interrupted threads in barrel

The fired case is held in a T-slot in the bolt-head which grips the rim and extractor groove, and the rearward movement of the unlocked bolt withdraws the expended case from the chamber.

The next round to be fed is held by two claws which move back with the bolt and extract this round from the closed-pocket metal belt. When the bolt-head is rotated to unlock, the T-slot is then lined up to receive the next round which is forced down by a cam operated arm and displaces the case of the previous fired round which falls out of the bottom of the gun. However this system cannot operate when the gun fires single-shot or the last round of the belt is fired. To cater for these circumstances a spring-loaded arm is fitted behind the cam which depresses the cartridge feed arm down far enough to eject the empty case.

During bolt recoil the belt is fed across and a live round positioned ready to be gripped by the claws above the bolt, as the latter comes forward.

When the bolt has reached its rearmost position energy is stored in the return spring and the buffer. This carries the bolt forward, driving a round into the chamber and slipping the feed claws over the waiting round in the belt. The bolt-head, as with all rotating bolt recoil operated machine guns, must be correctly positioned to allow the locking lugs to enter the recesses in the barrel extension. This is accomplished in the KPV by locking the bolt-head and releasing it to be free for rotation when the locking latch hits a cam in the body. The two bolt-head rollers run on to locking cams in the side walls of the body and are rotated to lock the bolt-head to the barrel. The bolt body then closes up to the bolt-head and the firing pin is driven through into the cap and the round is fired. As the bolt goes forward the feed pawl is moved outwards to slip over the next round in the belt ready to draw it towards the barrel on recoil.

The system is not as complex as it may sound and the weapon is obviously designed for ease of manufacture.

Loading and unloading
To load the gun it is necessary to allow the claws that withdraw the round from the belt to grip the rim. Since the claws are mounted 51 mm behind the bolt-face and the round is held in the belt at the shoulder it is necessary to withdraw the bolt about 63 mm before inserting the belt and pulling it fully through. The bolt is then released and the return spring will force it forward and allow the extractor claws to ride over the rim of the positioned round.

The bolt is then withdrawn fully to the rear until held by the sear. This cocking movement must be continuous. If the bolt is checked half way back the live round will fall off the T-slot and out of the gun.

To unload, cock the gun and lift the feed tray.

Pull back the round which is in line with the bolt, release the belt stop pawls and withdraw the belt.

Allow the bolt forward under control until the round on the bolt-face reaches the extraction position where it will fall out of the T-slot. In a hot dirty gun it will probably need to be pushed down and out of the gun. Continue to control the bolt until it is fully forward.

Removing barrel
Unload.
Cock the gun.
Rotate the triangular barrel catch 45° clockwise. This rotates the locking sleeve and the barrel is then taken out forward using the handle provided.

Stripping
The gun is designed to be field-stripped without the use of any special tools. The procedure is simple: raise the top cover; cock the gun; remove the barrel; rotate the cover through 60° and lift the front end up; lift the feed tray out of the body; control the bolt until it is fully forward to release the compressed return spring; release the catch and rotate the body end plate controlling it against the return spring which will then come out; depress the sear and withdraw the bolt. The bolt-head can then be removed from the body.

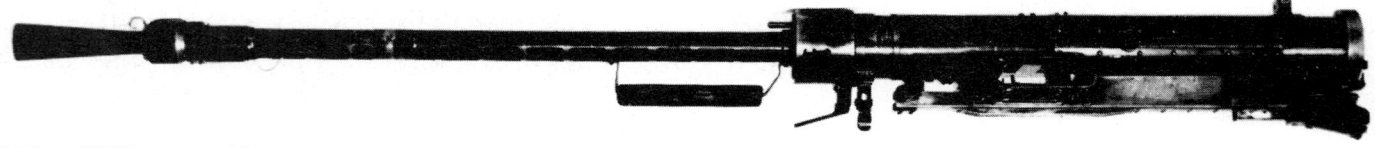

14.5 mm KPV heavy machine gun

Mountings

Towed-carriage mountings exist for one, two or four guns: they are known as ZPU-1, ZPU-2 and ZPU-4 respectively. These have all been very successful and the twin-towed and four-gun assemblies were used extensively in North Vietnam. The mountings are also found in China and possibly manufactured there.

DATA
Cartridge: 14.5 mm
Operation: gas-assisted short recoil
Feed: belt
Gun: (length) 2006 mm; (weight) 49.1 kg
Barrel: (length) 1346 mm; (weight of barrel with jacket) 19.5 kg
Inboard length: 597 mm

SIGHTS
Front: cylindrical post
Rear: tangent leaf U. 200-2000 m by 100 m intervals
Radius: 736 mm

BARREL
Chromium plated. Fitted muzzle booster and flash hider. Air-cooled, quick-change

FEED
Left or right hand
Ammunition: AP-I or HEIT
Belt: continuous metallic closed pocket; breaks down into groups of 10 rounds; can be fed either way up
Bullet weight: 923 grains (60 g)
Muzzle velocity: 1000 m/s
Rate of fire: (cyclic) 600 rds/min

Manufacturer
State factories.
Status
Current. No longer in production.
Service
USSR, other Soviet influenced countries and China.

UNITED KINGDOM

.303 Vickers machine gun

The Vickers medium machine gun was declared obsolete in British service in 1968. After that time a small number were retained by some former Commonwealth countries, but these have now gone out of service and so far as can be discovered there are no Vickers guns in use anywhere in the world by any regular forces. A full description of the gun can be found in earlier editions of *Jane's Infantry Weapons*.

Bren guns

At the end of the First World War the British Army possessed large quantities of the Vickers Mark 1 medium machine gun and the Lewis gun. In 1930 exhaustive trials were conducted to find a replacement for the Lewis. The choice finally fell on the Czechoslovak Zb26, modified to fire .303 in ammunition and with a slightly improved gas system. This became the ZGB30. The trials and modifications were completed in 1934 and the gun put forward to the War Office Acceptance Committee.

.303 in Bren series

When the acceptance committee were satisfied, arrangements were made for the production of the gun at the Royal Small Arms Factory at Enfield. The gun was named the Bren from BRno and ENfield. The gun was made by conventional machining from the solid and the first gun was finished in September 1937. By July 1938 production was 300 a week, rising to 400 a week in September 1939. By June 1940 more than 30 000 guns had been produced and issued. The gun was manufactured only at Enfield.

The original gun was designated the Mark I. The Mark II had the same length barrel, a simplified rearsight, and the flash hider/gas regulator/front sight, which was originally a single stainless-steel fabrication, was produced

as three separate units with only the regulator in stainless steel. The bipod was made with non-telescoping legs and the handle below the butt was omitted. These changes to assist production increased the weight from 10.04 to 10.52 kg. The Mark III simplified production, reduced barrel length, and also the gun weight was reduced to 8.76 kg. The Mark IV had the shorter barrel and weight was reduced to the minimum compatible with the stresses imposed by the .303 Mark VII cartridge.

Remarkably, .303 in Brens still appear and the Royal Small Arms Factory convert them to L4A4 or L4A3 according to the needs of the customer. The .303 in guns are stripped, cleaned and gauged and any defective components are replaced. The gun is rebarrelled, tested and proof-fired. While spares are made for the 7.62 mm guns, none are made for the .303 in version.

7.62 mm L4 Bren series

When the decision was made to adopt the 7.62 mm NATO round various conversions of the .303 in Bren gun were made to adapt it for 7.62 mm. These generally employed the breech-block made for the Canadian 7.92 mm guns, with new barrels.

Bren team in Hong Kong. Compare gunner's method of holding with illustration of Soviet soldier using RPK

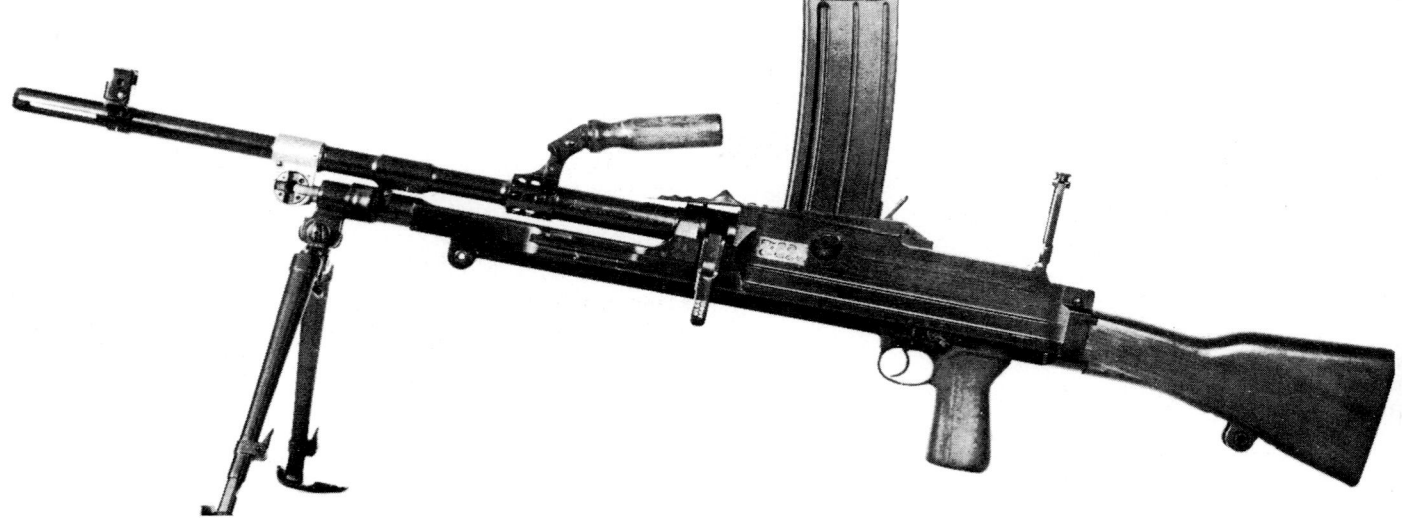

7.62 mm L4A3 Bren converted from .303 in Mark II Bren light machine gun

Camouflaged L4A3 Bren light machine gun on exercise in Norway

A brief summary of the L4 series follows:

L4A1: Converted Mark III .303 in Bren. First known as the X 10 E1. Two steel barrels. Bipod Mark 1. Obsolescent.

L4A2: Converted Mark III .303 in Bren. First known as the X 10 E2. Two steel barrels. Light bipod. Land and naval use. Obsolescent.

L4A3: Converted Mark II .303 in Bren. One chromium plated barrel. Obsolescent for land service.

L4A4: Converted Mark III .303 in Bren. One chromium plated barrel. Current weapon all services.

L4A5: Converted Mark II .303 in Bren. Two steel barrels. Obsolescent for land and air service. Still in naval service.

L4A6: Converted L4A1. One chromium plated barrel. Introduced only for land service.

L4A7: Conversion of Mark I .303 in Bren. None made but drawings prepared for the Indian Army. One chromium plated barrel.

Operation

The Bren light machine gun is a magazine-fed, gas-operated gun using a tilting-block locking system lifting the rear end of the breech-block into a locking recess in the top of the body.

During the period of initial pressure build up, the body, barrel, breech-block gas cylinder and bipod recoil on the butt slide approximately 6 mm. The movement is buffered by the piston buffer and spring. When the energy has been absorbed the piston buffer spring reasserts itself and returns the body, barrel, cylinder and bipod to their normal positions on the butt slide. This recoil and run out of these assemblies reduces the shock experienced by the firer and makes for fewer breakages in the affected components.

When the gun is fired the gas forces the bullet up the bore and a small proportion of it is diverted through a tapping in the barrel, passes through the regulator and impinges on the piston-head. The piston is driven back.

Attached to the piston by a flexible joint is the piston extension on which is supported the breech-block.

A piston post on the extension fits into the hollow interior of the breech-block and two ramps hold the rear of the block up into the locked position engaged in the locking recess at the top of the body.

When the piston extension moves back there is a movement of about 32 mm during which the bolt remains fully locked. Further movement removes the ramp support under the block and then an inclined surface on the rear of the piston post forces the back end of the bolt down and unlocking is completed.

The tilting motion of the breech-block provides primary extraction and the cartridge case is first unseated in the chamber and then withdrawn by the extractor claw as the breech-block moves back. A fixed ejector rides in a groove on top of the block and it is chisel-shaped so that as it strikes the brass of the cartridge case above the primer cap, brass is burred over the cap to prevent the latter falling out and causing a stoppage. The empty case is pushed through a cut-away section in the piston extension and thrown downwards out of the gun. As the piston goes back the return spring is compressed, storing energy and this, plus the action of the soft buffer, throws the piston forward again. The soft buffer has a low co-efficient of restitution and so the piston speed forward is not excessive and this keeps the cyclic rate to about 500 rds/min. The feed horns on top of the front of the block push a round out of the 30-round box magazine mounted vertically above the gun and the bullet is guided downwards into the chamber. As the cartridge goes forward the extractor claw clips over the rim of the round. When the round is fully chambered bolt movement ceases. The piston continues forward under its own momentum, and the remaining force in the return spring, and the two ramps at the rear end lift the rear of the breech-block so that the locking surface on top of the rear of the block rises into the locking recess in the body. The ramps remain under the block and hold it locked. The forward movement of the piston continues for another 32 mm and the front face of the piston post

7.62 mm L4A1 Bren light machine gun

.303 in Mark III Bren light machine gun

acts as a hammer to drive the spring-retracted firing pin into the cap at the base of the cartridge. The system is extremely simple. The applied safety disconnects the trigger from the sear by holding the trigger lever in the middle of the sear window. In theory this is not a satisfactory arrangement as a heavy jar caused by dropping the gun could dislodge the sear from the piston bent.

The gas regulator was installed in the Bren gun to give greater flexibility when the gun is firing under adverse conditions. The regulator has four tracks and a larger diameter gas track can be rotated into position as required. It should be noted that the gas impulse is applied only for a very short distance and then the gas escapes to atmosphere through vents bored in the cylinder walls. If excessive fouling occurs the bipod can be twisted and this cuts away any build up of carbon which is then dispersed by the next blast of gas. This feature produces an extremely reliable gun even after prolonged firing.

The barrel can be changed in a matter of seconds by raising the barrel latch and pulling the barrel forward using the carrying handle. With the gun fired at 120 rds/min, four magazines, the barrel requires changing every $2\frac{1}{2}$ minutes. The hot barrel can be cooled by air after removal from the gun or, as often happened in action, by laying it in wet grass or even in a stream.

Trigger mechanism
The weapon can be fired either at full automatic or at single-shot. The sear has a window through which projects the tripping lever. When the change lever is rotated to 'single-shot' the tripping lever bears against the upper surface of the window in the sear and its tripping head is raised into the path of the gas piston which depresses the tripping head as it comes forward. This forces the tripping lever down away from the sear window and the sear is released to rise and hold the piston to the rear. Releasing the trigger re-positions the tripping lever against the top surface of the sear window and operating the trigger fires one more shot. When the change lever is set to 'automatic' the tripping lever is forced down to bear on the bottom side of the sear window and the tripping lever head is pulled down clear of the piston. The gun continues firing as long as ammunition remains in the magazine and the trigger is depressed.

Stripping
Remove the magazine. Cock the action. Check chamber and feedway clear.
Push body locking pin, at rear of body from left to right.
Pull back the butt and trigger group.
Pull back the cocking handle and remove the piston and breech-block.
Re-assemble in reverse order.

DATA
MARK I
Cartridge: .303 in Mark 7
Operation: gas, selective fire
Method of locking: tilting block
Feed: 30-round box magazine

WEIGHTS
Gun: 10.04 kg
Magazine: (empty) 0.48 kg; (full (30 rounds) 1.25 kg; (drum (100 rounds) empty) 2.92 kg; (drum (100 rounds) full) 5.41 kg
Barrel: (assembled) 2.85 kg

LENGTHS
Gun: 1156 mm
Barrel: 635 mm

MECHANICAL FEATURES
Barrel: (regulator) 4 position; (cooling) air. Quick-change
Sights: (foresight) blade; (rearsight) aperture, drum-operated arm
Sight radius: 788 mm

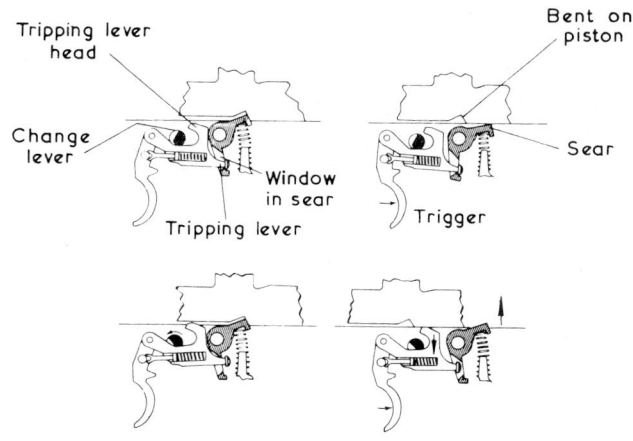

Fire selector of Bren gun

FIRING CHARACTERISTICS
Muzzle velocity: 744 m/s
Rate of fire: (cyclic) 500 rds/min
Effective range: 600 m

MARK II
As Mark I except:

WEIGHTS
Gun: 10.52 kg
Barrel: 2.93 kg

MECHANICAL FEATURES
Rearsight: tangent leaf, 200–1800 yds (1646 m) × 50 yds
Sight radius: 782 mm
Rate of fire: (cyclic) 540 rds/min

MARK III
As Mark I except:

WEIGHTS
Gun: 8.76 kg
Barrel: 2.31 kg

LENGTHS
Gun: 1090 mm
Barrel: (Mk IV) 565 mm

MECHANICAL FEATURES
Rearsight: as Mark II
Sight radius: 694 mm
Rate of fire: (cyclic) 480 rds/min

MARK IV
As Mark III except:

WEIGHTS
Gun: 8.68 kg
Barrel: 2.27 kg

MECHANICAL FEATURES
Rate of fire: (cyclic) 520 rds/min

L4A4
As Mark I except:
Cartridge: 7.62 mm × 51

WEIGHTS
Gun: 9.53 kg
Magazine: (empty) 0.45 kg; (with 30 rounds) 1.18 kg; no drum magazine
Barrel: 2.72 kg

LENGTHS
Gun: 1133 mm
Barrel: 536 mm

MECHANICAL FEATURES
Rearsight: as Mark III
Sight radius: 743 mm
Muzzle velocity: 823 m/s
Rate of fire: (cyclic) 500 rds/min

Manufacturer
Royal Small Arms Factory, Enfield; John Inglis Ltd, Toronto, Canada; Small Arms Factory, Lithgow, New South Wales, Australia.
Status
Current. No longer manufactured.
Service
British forces and many Commonwealth and former Commonwealth countries.

7.62 mm L7A1 and L7A2 general-purpose machine gun

The British Army ended the Second World War with the Vickers Mark 1 machine gun as its sustained fire weapon and the Bren gun as the section light machine gun. Both guns used the .303 in cartridge, the Vickers employing the Mark 8Z and the Bren the Mark 7.

When the NATO 7.62 mm round was introduced the opportunity was taken to select a general-purpose machine gun. After long trials the Belgian Mitrailleuse d'Appui Général (MAG) was chosen and built under licence at the Royal Ordnance, Small Arms Division, Enfield. Several minor changes were made to the design to accept British manufacturing methods and material specifications, but the essential dimensions, features and functioning of the gun are identical to those of the Belgian MAG. However, parts are not interchangeable between Belgian and British guns.

A drill purpose (inoperative) version of the weapon has been produced and is designated L46A1.

DATA
L7A2
Cartridge: 7.62 mm × 51
Operation: gas, automatic
Method of locking: dropping locking lever
Feed: belt

WEIGHTS
Gun: (in light machine gun role) 10.9 kg
Barrel: 2.73 kg
Ammunition: (100 rounds linked) 2.95 kg
Wallet: (spare parts and tools) 1.02 kg

LENGTHS
Light machine gun: 1232 mm
Sustained-fire machine gun: 1048 mm
Barrel: including 50 mm overhang of carrying handle (with flash hider) 679 mm; (without flash hider) 597 mm

MECHANICAL FEATURES
Barrel: (regulator) exhaust-to-atmosphere type. 10 positions; (cooling) air. Quick-change
Sights: (foresight) blade; (rearsight) aperture
Sight radius: light machine gun role (sight down) 851 mm; sustained-fire machine gun role (sight up) 787 mm

FIRING CHARACTERISTICS
Muzzle velocity: 838 m/s
Rate of fire: (cyclic) 750–1000 rds/min

ACCESSORIES
50-round, belt feed box. Blank-firing attachment L3A1 with blank-ammunition guide plate. Cover muzzle 7.62 mm machine gun L1A3. Wallet, spare parts, filled. Chest, carrying, 7.62 mm Mark 1.

7.62 mm L7A2 general-purpose machine gun in light role

General-purpose machine gun in sustained-fire role, mounted on L4A1 tripod. C2 dial sight is on gun, and gunner does not need to look along iron sights while firing. The detachment commander is using binoculars to observe fall of shot. This is a posed photograph. The gun crew would normally be well under cover

Manufacturer
Royal Ordnance PLC, Guns & Vehicles Division, Nottingham.
Status
Current. Production complete.
Service
British and some Commonwealth forces.

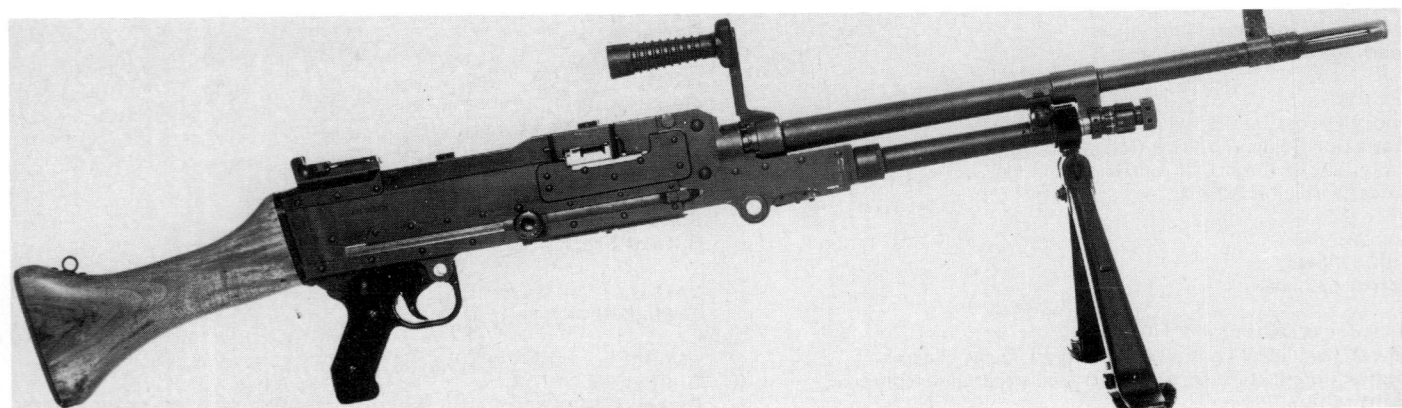

7.62 mm L7A2 general-purpose machine gun

General-purpose machine gun buffered tripod mounting and sustained fire kit

To provide a strong and stable yet small and light mounting for the L7 general-purpose machine gun the tripod mounting, 7.62 mm L4A1 machine gun was adopted at the Royal Ordnance, Small Arms Division at Enfield. Details of this mounting are given below; but it should be noted that the manufacturers have now produced basically similar mountings adapted for use with a wide variety of other machine guns.

The mounting incorporates its own recoil buffer unit, permitting all round traverse and has a quick-release mechanism allowing free traverse, elevation and depression.

The tripod mounting consists of: three legs – two short, one long; tripod head; cradle and recoil unit; elevating and traversing mechanisms.

The three tubular legs support the tripod head. At the foot each leg has a shoe to give an improved ground grip. Each leg is locked by a cam lever which when pushed inwards secures the leg in any desired position. When the locking levers are pulled outwards the clutch faces are forced out by coil springs and the leg is free. The tripod can be set to 'high' or 'low' and these settings are notched on the legs and on the tripod head at the junction of the clutches. There are three lugs on the tripod head to take legs. The dial ring is held in position by friction and can be rotated by hand. It has a traverse scale of 0–3200 mils in each direction with graduations every 250 mils, numbered every thousand mils. The indicator moves with the head and a line on it indicates the traverse on the scale. The cradle and recoil unit is mounted on the tripod head. The cradle is attached to the tripod head by a ball joint which can be locked; when free it allows 600 mils elevation and all round traverse. The gun is secured to the recoil unit by front and rear mounting pins. The cradle has an enclosed buffer system of tubes, buffers and springs. When the first round is fired, the whole gun recoils and then the buffer returns it to battery. Before run out is completed the next round is fired and thereafter the gun movement is extremely small as it rests in equilibrium imposed between the recoil force and the buffer unit.

The elevation mechanism provides a fine adjustment using the handwheel on the left. It is controlled and locked by an eccentric cam tightened by a thumb lever. The fine traversing mechanism is controlled by the handwheel on the right. The range of controlled traverse is 200 mils and one turn of the handwheel traverses the gun through 6.6 mils. A clicker control operated by a sliding sleeve gives 2.2 mils to each click; three clicks per turn. If desired the sleeve may be pushed in to give a smooth, silent control.

The dial sight is the Sight unit C2, Trilux which is also used with the 81 mm mortar. It has a right-angled telescope with a magnification of x1.7 and a field-of-view of 180 mils. The eyepiece may be rotated 1600 mils (90°) right or left from the vertical. The bearing assembly comprises the coarse and fine bearing scale rings and a worm gear. The coarse bearing ring has a scale reading from 0 to 6400 mils (0 to 360°) and is numbered every 200 mils clockwise. The fine bearing scale ring is assembled on the worm spindle and has a slipping scale graduated in single mils from 0–100 and numbered every 10 mils. The two-bearing scales allow the gun to be laid on to a selected zero line and then bearings read off to the nearest mil from that zero line. The sight can then be locked if so required.

The elevation assembly has a coarse scale plate and a fine scale ring. A worm gear allows the completed sight unit to be rotated up or down and the angle of elevation can be read to the nearest mil. The worm gear can be locked in a similar way to that for bearing.

The elevation and cross levelling bubbles are mounted below their corresponding coarse scales.

The sight unit is fitted with perspex bearing and elevation scales which, with the telescope graticule and the levelling bubbles, are Trilux illuminated by sealed in beta sources, using tritium gas, which lose half their brightness in 10 years.

When the L7 machine gun is used on the tripod the butt is removed and the recoil buffer is used in its place. This carries out the same function as the buffer assembly and carries the same braking cone and ring together with the Belleville compression washers as those on the buffer assembly of the gun in the light machine gun role.

DATA
Mounting: tripod L4A1

WEIGHTS
Tripod: 13.61kg
Sight unit: (cased) 2.58kg
Conversion kit: (complete) 32.66kg

LENGTHS
Legs spread
Distance: (across short legs) 1118 mm; (front to rear) 1118 mm
Height of sight line above ground: 330–635 mm
Folded dimensions: 813 × 191 × 191 mm

ADAPTOR ANTI-PARALLAX

PERSPEX BEARING AND ELEVATION SCALES

Sight unit C2, Trilux and adaptor, anti-parallax

7.62 mm L7A2 machine gun mounted on mounting tripod 7.62 mm L4A1 machine gun. Note butt has been removed and buffer, recoil, inserted

FRONT MOUNTING PIN

SIGHT BRACKET

WING NUT

REAR MOUNTING BRACKET

CRADLE

LEVER

DIAL RING

TRIPOD HEAD

TRAVERSING MECHANISM

ELEVATING MECHANISM

MECHANICAL FEATURES
Traverse: (free) 360° (6400 mils); (mechanical traverse (max)) 11° (200 mils)
Clicker control: (angle/click) 7' (2.2 mils)
Elevation: (max) 22° (400 mils); (mechanical elevation (max)) 2° 48' (50 mils)
Depression: (max) 11° (200 mils); (mechanical depression (max)) 2° 48' (50 mils)

Manufacturer
Royal Ordnance plc, Guns & Vehicles Division, Nottingham.
Status
Current.
Service
British, some Commonwealth and other forces, principally those which have adopted the British-built general-purpose machine gun. The L5A1 tripod (Australian and New Zealand service) is essentially the L4A1 adapted for the US M60 machine gun.

Cradle and recoil unit

7.62 mm L8A2 tank machine gun

During the Second World War the standard British tank gun was the 7.92 mm BESA which was developed from the Czechoslovak Model 37. After the war it was decided to abandon the 7.92 mm cartridge and British armoured fighting vehicles were mainly equipped with the .30 calibre Browning Model 1919A4 tank machine gun. Before the introduction of the Chieftain tank it was decided to adopt a machine gun using the 7.62 mm × 51 NATO cartridge and the decision was made to adapt the L7 machine gun for tank use. The resultant weapon was designated L8A1: there is also a drill purpose (inoperative) version designated L41A1.

The L7 general-purpose machine gun can be fed only from the left; its gas system is of the exhaust-to-atmosphere type, and to remove the belt from the gun it is necessary to lift the top cover plate, housing the feed mechanism.

All these factors make it unsuitable for tank use. The necessity to feed from the left dictates the position of a coaxial machine gun relative to the main armament and this is not always the position the designer or the user would prefer.

The L8 machine gun embodies the necessary changes to improve the L7 for tank use but embodies the same basic design features. In 1978 the L8A1 was updated to the A2 version in order to introduce changes which would improve performance in the vehicle.

Gas system
The gas regulator now has a single gas port fitting into a conical seating to eliminate any gas leakage. The diameter of the gas port was chosen to limit the rate of fire so as to prevent a build-up of links and spent cases in the vehicle collector system, which would cause a gun stoppage.

Since a feature of this weapon is the requirement for an interchangeable barrel, the inevitable small escape of gas from the close clearance fit of the gas plug in the front end of the gas cylinder is dealt with by a fume extractor which has been improved on the A2 version by fitting a tubular venturi flash suppressor and fume extractor to the muzzle. Also, to exclude gas from the interior of the vehicle, the usual gas escape holes are omitted from the gas cylinder.

Feed system
An investment-cast feed pawl operating arm, actuated by a breech-block roller, with improved retention of the roller and stronger spring, reduces wear and improves feed operation. A one-piece feed tray is lighter than the previous design and incorporates a positive cartridge stop to allow the use of both disintegrating and non-disintegrating ammunition belts.

Feed pawl depressor
This is a sheet-metal plate which is fitted over the feed pawls. When the depressor is moved to the left, the spring-loaded feed pawls are free to work in the normal manner and the belt cannot move to the left. When the depressor is moved to the right the feed pawls are pushed up against their springs and the belt can be pulled to the left out of the gun.

Ground role
Dismounting of the L8A2 for use in the ground role is not contemplated except in extreme emergency.

In all other respects the L8A2 is similar to the L7A2 and operates in the same way as the Fabrique Nationale's Mitrailleuse d'Appui Général (which see).

Changing barrel of L8A2 tank machine gun in Chieftain tank. The whole gun must be withdrawn into tank to allow barrel to go forward

DATA
Cartridge: 7.62 mm × 51
Operation: gas, automatic
Method of locking: dropping locking lever
Feed: belt

WEIGHTS
Gun: (complete) 10.43kg
Barrel: 3.06kg

LENGTHS
Gun: 1099 mm
Barrel: (including flash suppressor and 50 mm overhang of carrying handle) 737 mm

MECHANICAL FEATURES
Barrel: (regulator) 3 position; (cooling) air; quick-change

FIRING CHARACTERISTICS
Muzzle velocity: 838 m/s
Rate of fire: (cyclic) 625–750 rds/min
Accessories: wallet S.A. spare parts and tools No 2 Mark 1 filled; spare parts box containing: spare barrel, bipod, butt, breech-block and piston assembly, return spring assembly

Manufacturer
Royal Ordnance plc, Guns & Vehicles Division, Nottingham.
Status
Current. Production completed.
Service
British and some Commonwealth armies. Iran.

L8A2 tank machine gun showing short barrel locking handle to facilitate mounting in restricted space

7.62 mm L37A2 tank machine gun

This is a machine gun based on the L8 and designed to be used in armoured fighting vehicles, scout cars and armoured personnel carriers and also have a capability for use in a dismounted, ground role.

The gun is basically the L8 machine gun and to enable it to carry out its ground role it carries L7 machine gun subassemblies. The gun is made up of the following items: L8 machine gun body assembly. This normally has a trigger mechanism operated by a solenoid which is installed adjacent to the weapon in the tank turret; L8 or L37 machine gun barrel assembly. This barrel, the L6A1, does not have a foresight, or an insulated barrel carrying handle; L7 machine gun trigger mechanism. This replaces the normal L8 trigger; L7/L8 machine gun buffer, recoil assembly Mark I.

To convert the gun to its ground role, parts of the gun equipment needed are: L7 machine gun barrel assembly. This barrel, the L1A2, has a foresight and a carrying handle.

L7 bipod assembly.

L7 butt assembly.

DATA
WHERE DIFFERING FROM THE L8 MACHINE GUN
Gun
Weight: 9.53kg
Length: 1075 mm

BARREL L6A1 ARMOURED FIGHTING VEHICLE ROLE
Weight: 2.7kg
Length: 663 mm

BARREL L1A2 GROUND ROLE
Weight: 2.7kg
Length: 711 mm

SIGHTS GROUND ROLE
Foresight: adjustable blade
Rearsight: aperture

Status
Approved for service. In production.
Service
UK and some foreign armies. Commander's gun of Chieftain and Challenger tanks. Coaxial in Fox, Scimitar and Scorpion CVRs.

L37 machine gun with attachments for converting it to L7 in ground role, including L1A2 barrel

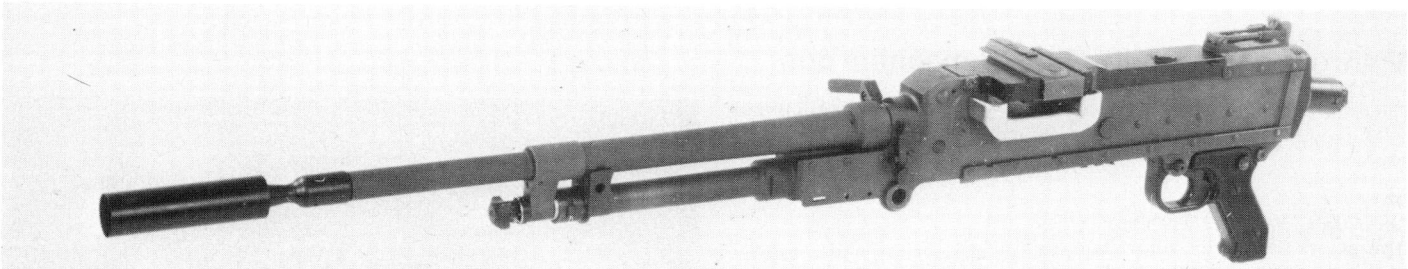

7.62 mm L37A2 tank machine gun in tank configuration with short barrel locking handle

7.62 mm L20A1 machine gun

This gun is a modification of the L8 tank machine gun to enable it to be used in a helicopter. The trigger is electrically controlled, there is no sighting system and the barrel does not have the fume extractor of the L6A1 barrel. It has the gas system of that barrel but uses a prong-type flash eliminator. The gun can be fed from either side by changing the feed cover with the feed pawls and feed arm and the feed plate.

A duct is installed to convey the expended links into a bag.

7.62 mm L43A1 ranging machine gun

Until recently it was not possible to combine the role of the ranging tank machine gun with that of the coaxial machine gun. To some extent barrel wear suffered by the coaxial machine gun resulted in loss of consistency, essential to a ranging gun, but in addition a problem existed with the movement of the Mean Point of Impact (MPI) of a group of shots from a cold to a hot gun. This was critical since effective ranging is essential with the gun either cold or hot. The L43A1, with its barrel bearing, was specially developed to reduce this MPI shift to an acceptable level. This bearing is between the gas block and the muzzle, and supports the barrel at the forward end. The L43A1 does not have the flash hider and gas tube of the L3A2 barrel. A barrel clamp which effectively 'breeches up' the barrel, to improve further the consistency of the grouping, is also fitted.

The L43A1 has been specifically developed as a ranging gun for the Scorpion tracked reconnaissance vehicle.

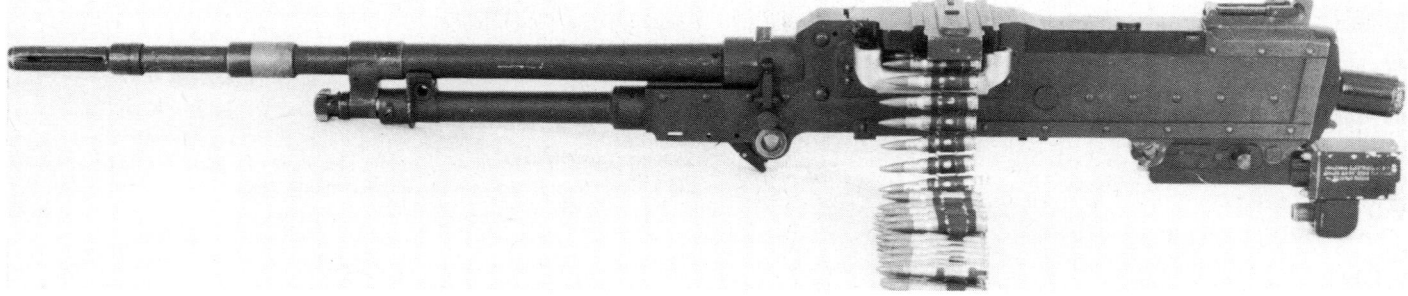

7.62 mm L43A1 ranging machine gun

5.56 mm L86A1 (light support weapon) Enfield Weapon System

Most of the information on this weapon is already contained in the entry dealing with the Individual Weapon (IW) in the Rifles section. The light support weapon is essentially a heavy-barrelled version of the IW and should more properly be called a machine rifle; it shares the same design concept. Although a change-barrel version was investigated it has not been followed up.

The gain from the light support weapon is in the longer and heavier barrel and the greater muzzle velocity and accuracy that this gives. When fired from the bipod it becomes an impressively accurate and consistent weapon which is able to take full advantage of the SUSAT sight which is a standard fitting.

DATA
Cartridge: 5.56 mm × 45

5.56 mm L86A1 light support weapon in latest form with forward mounted bipod, rear grip and shoulder strap

Operation: gas
Type of locking: rotary bolt, forward locking
Feed: 30-round box magazine

WEIGHTS
Weapon: (less magazine and optical sight) 5.4 kg; (complete with loaded magazine) 6.58 kg
Optical sight: 0.7 kg
Magazine: (empty) 0.12 kg; (loaded) 0.48 kg (30 rounds)

LENGTHS
Barrel: (including flash hider) 646 mm
Weapon: 900 mm

MECHANICAL FEATURES
Rifling: 6 grooves, rh, 1 turn in 180 mm
Trigger pull: 3.12–5kg

FIRING CHARACTERISTICS
Muzzle velocity: 970 m/s
Rate of fire: (cyclic) 700–850 rds/min

Manufacturer
Royal Ordnance plc, Guns and Vehicles Division, Nottingham.
Status
Current. Production.
Service
In service with the British Army.

Manroy .50 Browning M2HB machine gun

This is the standard .50 M2HB heavy machine gun, manufactured in Britain by Manroy Engineering. It matches the standard US specification in every respect.

DATA
Cartridge: .50in (12.7 × 99mm)
Operation: short recoil, automatic
Locking system: projecting lug
Feed: disintegrating link belt
Weight: 38.5 kg
Length overall: 1651 mm
Barrel length: 1143 mm
Rifling: 8 grooves, rh, one turn in 381 mm
Muzzle velocity: 893 m/s
Rate of fire: 450 – 500 rds/min
Maximum range (M2 ball): 6766 m
Effective range: 1850 m

Manufacturer
Manroy Engineering Ltd, Hobbs Lane, Beckley, E. Sussex TN31 6TS.

Manroy .50 M2HB machine gun

Status
Current. Production.
Service
NATO and other forces worldwide.

Manroy .50 M2HB-QCB machine gun

In addition to manufacturing the standard .50 Browning M2HB heavy machine gun, Manroy Engineering also produce a quick-change barrel version. As with other QCB systems the barrel can be changed by one man in under 10 seconds and there is no need to adjust headspace at any time. The

Manroy .50M2HB-QCB machine gun

DATA
Cartridge: .50in (12.7 × 99mm)
Operation: short recoil, automatic
Locking system: projecting lug
Feed: disintegrating link belt
Weight: 38.0 kg
Length overall: 1650 mm
Barrel length: 1140 mm
Rifling: 8 grooves, rh, one turn in 381 mm
Muzzle velocity: 893 m/s
Rate of fire: 450 – 500 rds/min
Maximum range (M2 ball): 6766 m
Effective range: 1850 m

Manufacturer
Manroy Engineering Ltd, Hobbs Lane, Beckley, E. Sussex TN31 6TS.
Status
Current. Production.
Service
NATO and other forces worldwide.

change is made by rotating the cocking handle rearward and holding it, rotating the barrel, withdrawing it, inserting the new barrel, rotating it to stop and releasing the cocking handle.

Should it be necessary, the standard barrel can still be used with the gun, though in this case the normal headspace adjustment will have to be made.

Manroy M3 Mount for .50 M2HB machine guns

This is the standard US M3 tripod, manufactured by Manroy in UK. The design is robust and well-proven and can be improved by the addition of the Vinghøgs Softmount (see p.332) which is manufactured under licence by Manroy.

DATA
Weight: 20 kg
Length, extended: 1890 mm
Length, folded: 1155 mm
Height: 355 mm
Width, fully open: 1562 mm
Traversing range: 800 mils, bar fixed; 6400 mils free
Traverse graduations: 800 mils

Manufacturer
Manroy Engineering Ltd, Hobbs Lane, Beckley, E. Sussex TN31 6TS.
Status
Current. Production.
Service
NATO and other forces worldwide.

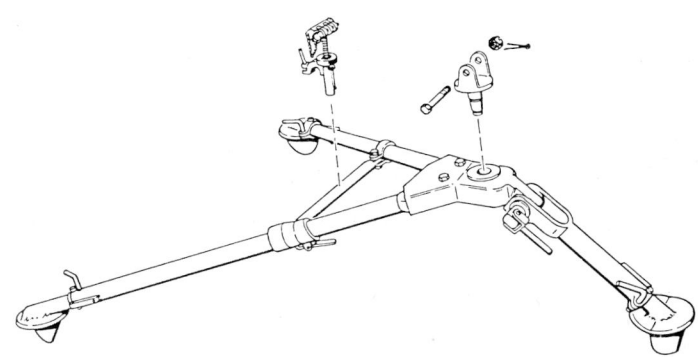

Manroy M3 Mount for M2HB machine gun

Manroy M63 anti-aircraft mount

This is a pedestal mount designed to accommodate the .50 Browning M2HB machine gun for air defence use. It consists of a four-legged base, central pedestal, and a traversing and elevating cradle controlled by handgrips with firing lever. A support tray for the standard ammunition belt box fits on the side. The mount can be rapidly and easily dismantled into its four basic units for transportation and storage.

DATA
Weights:-
Complete mounting: 65 kg
Base and legs: 35.3 kg
Pedestal: 5.5 kg
Cradle assembly: 20 kg
Ammunition tray: 4.5 kg

Height overall: 1065 mm
Base diameter: 1320 mm
Max elevation: 85°
Max depression: 29°
Traverse: 360°

Manufacturer
Manroy Engineering Ltd, Hobbs Lane, Beckley, E. Sussex TN31 6TS.
Status
Current. Production.
Service
NATO forces.

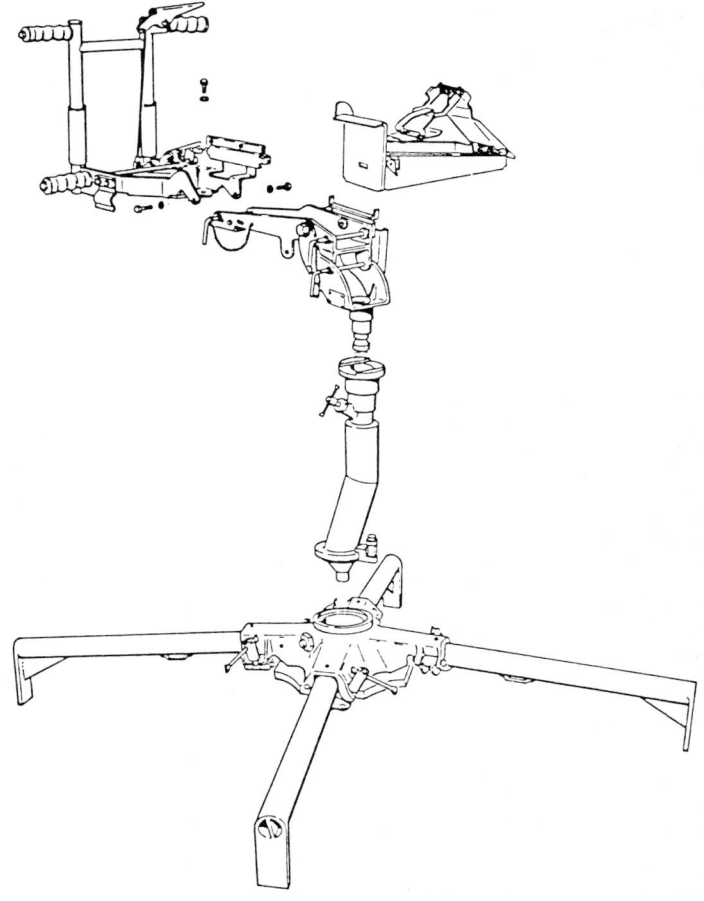

Manroy M63 anti-aircraft mount

Manroy lightweight AA mount adapter

The Manroy lightweight anti-aircraft adapter has been specially designed to convert the M3 ground mount into a dual role ground/AA mount which can be quickly and easily switched between configurations. The light weight makes it suitable for use in temporary defence tasks where there is a demand for both air defence and ground cover using standard .50 M2HB machine guns. It provides a cost-effective and viable alternative to the much heavier and dedicated combination of M3 ground and M63 anti-aircraft mounts.

To convert the standard M3 tripod to the air defence role the rear mounting pin is removed and the pintle catch released to allow the gun and pintle to be lifted out of the mount. The three support arms are then positioned on the legs of the M3 mount and locked in place. The gun is then lifted back into the cradle, the pintle locating in a bush identical to that of the M3 mount, and locked in position by the catch. The rear mounting holes are aligned with the corresponding holes in the cradle and the rear mounting pin pushed through and locked.

To ensure complete stability throughout 360°, particularly on soft ground, two stainless steel straining wires may be pegged into the ground on either side of the mount.

Manufacturer
Manroy Engineering Ltd, Hobbs Lane, Beckley, E. Sussex TN31 6TS.
Status
Current. Production.
Service
NATO forces.

Manroy lightweight AA adapter attached to M3 tripod

UNITED STATES OF AMERICA

.30 calibre M1919A4 Browning machine gun

The M1919A4 was used as a fixed gun in tanks in the Second World War and on many post-war armoured fighting vehicles. It was also used by infantry as a company level weapon, mounted on the M2 tripod, with the flash hider M6 and a detachable carrying handle. Its evolution can be traced back to the .30 calibre Browning tank machine gun which itself was derived from Browning's air-cooled aircraft machine gun designated the M1918.

The mechanism of the M1919A4 is that of the original M1917 Browning; it operates by short recoil of the barrel, barrel extension and bolt locked together. After 8 mm of movement the bolt is unlocked by cams in the receiver and is then accelerated to the rear while the barrel and extension move back into battery. During the rearward movement of the bolt a fresh round is extracted from the belt and fed into the T-slot in the bolt-face, displacing the empty case just extracted and ejecting it downwards. On the return movement of the bolt the new cartridge is aligned and fed into the breech and the bolt is once more locked to the barrel extension.

For a more complete description of the basic Browning machine gun mechanism, readers are referred to earlier editions of *Jane's Infantry Weapons*.

DATA
Cartridge: .30 M1 or M2 (also used in 7.62 mm in Canada)
Operation: short recoil, automatic
Method of locking: projecting lug
Feed: 250-round belt

WEIGHTS
Gun: 14.06kg
Barrel: 3.33kg
M2 tripod: 6.35kg

LENGTHS
Gun: 1044 mm
Barrel: 610 mm

MECHANICAL FEATURES
Barrel: (cooling) air
Sights: (foresight) blade; (rearsight) leaf aperture; windage scale; (sight radius) 354 mm

.30 calibre M1919A4 Browning machine gun on tripod M2

FIRING CHARACTERISTICS
Muzzle velocity: 860 m/s
Rate of fire: (cyclic) 400–500 rds/min
Effective range: 1000 m

Status
Obsolescent. No longer in production.

Service
Used by many countries as a vehicle gun. Some have rechambered it to accept 7.62 mm × 51 NATO. In service with Canada, Denmark, Dominican Republic, Greece, Guatemala, Haiti, Iran, Israel, Italy, South Korea, Liberia, Mexico, South Africa, Spain, Taiwan, United Kingdom, Vietnam. In reserve in the USA.

.30 calibre M1919A6 Browning machine gun

The M1919A6 is a belt-fed, air-cooled, recoil-operated machine gun. It can have a muzzle-mounted bipod or be mounted on the M2 tripod. It is similar to the M1919A4 except for the following modifications.
  A removable metal shoulder stock has been added.
  The cover latch has been modified to provide for easier opening.
  The return spring has been lightened to make bolt retraction easier.
  A removable handle has been added to enable easier carrying of a hot gun.
  The barrel jacket has been modified to mount a front barrel bearing, a removable bipod assembly and a lock ring.
  The bolt latch has been removed.
  The M7 flash-hider has been incorporated.
  The M1919A6 has a barrel which can be changed when hot by grasping the muzzle and (preferably with asbestos mittens) unscrewing the barrel and sliding it forward out of the jacket. The new barrel is slid into the jacket and screwed into the barrel extension. The gun is a poor attempt at an infantry general-purpose machine gun and has never been popular.

DATA
WHERE DIFFERENT FROM M1919A4
Weight: (with metal stock and bipod) 14.74kg; (of metal stock) 0.79 kg; (of barrel) 2.11 kg
Trigger pull: 3.86kg
Length overall: 826 mm

Status
Obsolete. No longer in production.
Service
Likely to be found only in reserve stocks of countries that had US armaments after 1945.

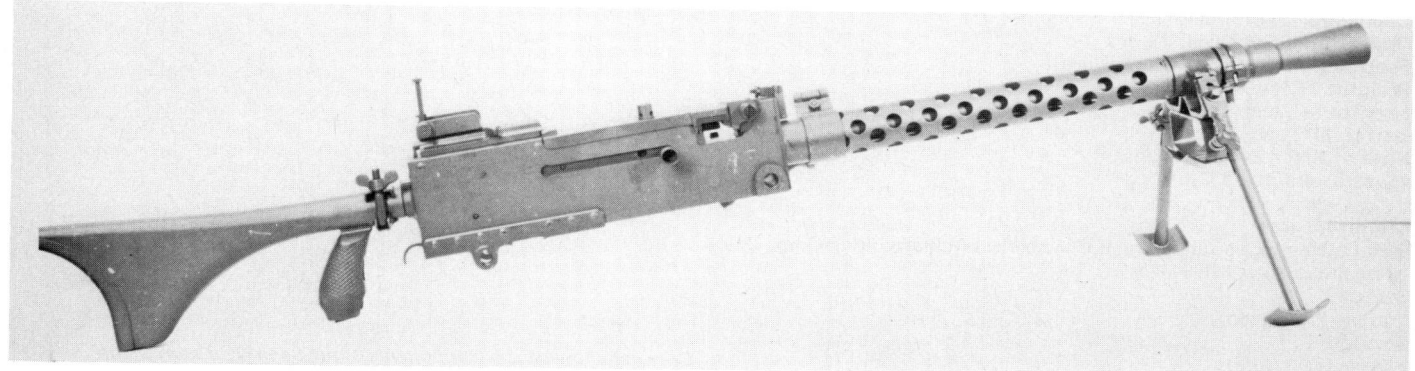

.30 calibre Browning M1919A6

.50 calibre Browning machine guns

The air-cooled .50 Browning was developed for aircraft use and was adopted in 1923 as the Model 1921. In 1933 it was renamed the M2 and subsequently it was found necessary to increase the mass of the barrel and replace the oil buffer with a simpler design that eliminated the oil and improved service life and maintenance. It was then renamed the M2HB (heavy barrel) and in this form has been used and is still used extensively on ground movements and on vehicles. It is one of the most widespread and successful heavy machine guns in service and looks like continuing as such for many years.
  Production has been resumed in Belgium and the USA and in both countries the output is primarily for export. The demand does not appear to be slackening.
  The US Marine Corps has recently expressed a requirement for a light

weight .50 machine gun, and three solutions have been offered. One is the GPHMG (see below) while the others are redesigns of the M2.

DATA
(M2HB)
Cartridge: .50 M2 Ball, (12.7 × 99 mm)
Operation: short recoil
Method of locking: projecting lug
Feed: disintegrating link belt
Weight: 39.1 kg
Length of gun: 1653 mm
Length of barrel: 1143 mm

.50 calibre M2 HB machine gun on M3 tripod

Rifling: 8 grooves, rh, 1 turn in 381 mm
Sights: (foresight) blade; (rearsight) leaf aperture
Rate of fire: 450–600 rds/min
Muzzle velocity: 810 m/s

Manufacturers
Saco Defense Inc., a subsidiary of Maremont Corporation, 291 North Street, Saco, Maine 04072.

Status
Current. Production.
Service
US forces and at least 20 other countries.

Saco Defense .50 M2HB QCB machine gun (or kit)

Saco's most recent improvement to the .50 M2HB machine gun provides fixed headspace and a quick-change barrel. All the other features of the weapon have been retained, and ammunition may be fed to the gun from either side with a simple adjustment. The barrel is Stellite lined and chromium plated for long service life.

The new design is an improvement on the latest US military specifications developed in 1978 by Saco Defense under contract with the US Government. Almost all the weapon components are unchanged; the only new parts are the barrel, barrel extension and barrel support. It is possible to convert a standard M2HB gun to the fixed headspace configuration by substitution of new for old parts. Fitting by a technician, followed by infrequent inspection, is sufficient to ensure continuous safe and reliable operation.

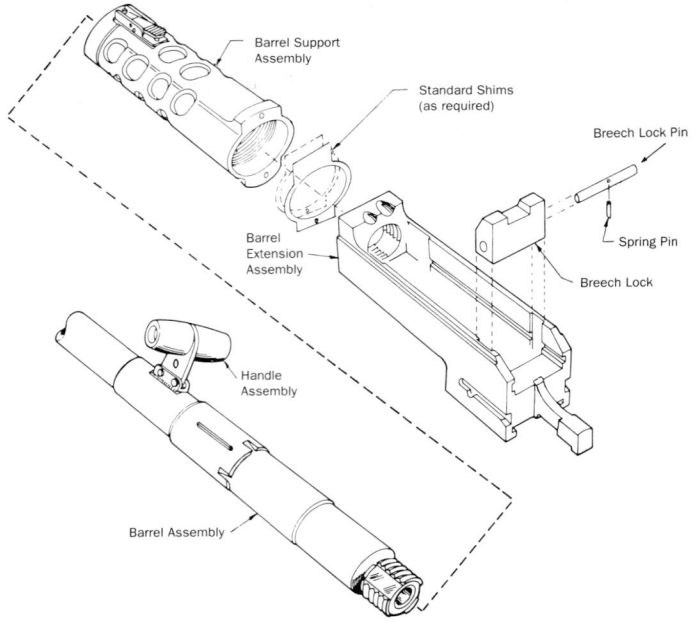

DATA
Cartridge: .50 Browning (12.7 × 99mm)
Operation: Short recoil, selective fire
Feed system: Disintegrating link belt
Weight: 38.2 kg
Length: 1651 mm
Barrel: 1143 mm
Rate of fire: 450-600 rds/min
Muzzle velocity: 853 m/s

Manufacturer
Saco Defense Inc., a subsidiary of Chamberlain Manufacturing Corp., 291 North Street, Saco, ME 04072.
Status
Current. Production.
Service
Foreign armed forces.

Drawing showing the components of the Saco QCB conversion

Saco Defense 'Fifty/.50' .50 in heavy machine gun

The 'Fifty/.50' is Saco Defense's 25 kg .50 in general-purpose machine gun. This model is the latest product-improved version of the Browning .50 M2HB and has been refined over the earlier Saco model described in previous editions of *Jane's Infantry Weapons*. This development was carried out by Saco Defense Inc as a private venture, taking over three years, and reflects numerous user-desired features.

The new weapon offers fixed headspace with quick interchangeable barrels, an adjustable rate of fire capability from 500 to 750 rds/min, welded receiver construction, a new charging system, and a new ergonomic backplate assembly. It retains the Browning operating mechanism, whilst weighing some 35 per cent less than the M2HB. Interest in new .50 calibre applications

have emerged from users of light airframes and small boats, where more firepower is desired but is restricted by system weight.

The gun is capable of firing all existing M2 machine gun ammunition, including higher and lower impulse rounds which are under development. Although many components have been changed, in most cases standard M2HB spare parts can be used, though they will carry an added weight penalty.

DATA
Cartridge: 12.7 × 99 mm (.50 Browning)
Operation: recoil
Method of locking: projecting lug
Weight: 25 kg
Barrel weight: 7 kg
Length overall: 1560 mm
Rate of fire cyclic: 500 – 725 rds/min (adjustable)
Muzzle velocity: 850 m/s
Range: max 6800 m; effective 1800 m

Manufacturer
Saco Defense Inc., a subsidiary of Chamberlain Manufacturing Corporation, 291 North Street, Saco, Maine 04072.
Status
Current. Production.

Saco Defense 'Fifty/.50' .50 in machine gun

.50 RAMO M2 lightweight machine gun

The RAMO .50 calibre M2 Lightweight is a new development weapon designed as a response to both ground and air requirements for a reduced weight .50 calibre weapon. Utilising the M2 Browning principles, the weapon has been upgraded to include an adjustable buffer to vary the rate of fire, a patented quick-change Stellite-lined chrome bore barrel with flash suppressor, 'Max Safe' charging system, and a trigger safety switch. The weapon is constructed of the same basic material as the standard M2 but incorporates a protective finish to further enhance the maintainability of the weapon, especially in marine environments.

The gun will fire all existing types of .50 (12.7 × 99 mm) ammunition,

including multi-purpose explosive projectiles and the SLAP saboted light armour penetrator. The rate of fire is adjustable from 550 to 750 rds/min so as to adapt the weapon to a ground support or an air defence role. The lined, chrome bore barrel is manufactured using the latest technology so as to maximize wear life and barrel accuracy.

The Lightweight M2 retains 75 per cent commonality of parts with the standard .50 M2 weapon. Consequently all these parts are logistically supportable through routine channels. The remaining items peculiar to the Lightweight are manufactured by RAMO and are available for world-wide distribution.

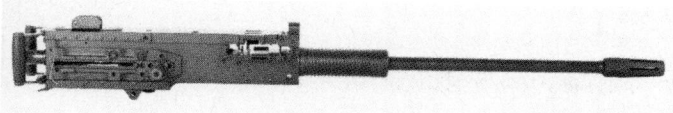

RAMO .50 Lightweight M2 machine gun

DATA
Cartridge: .50 Browning (12.7 × 99 mm)
Operation: Recoil, automatic
Method of locking: projecting lug

Feed: belt
Weight, complete: 26.72 kg
Weight of barrel: 7.48 kg
Length of gun: 1.524 m
Length of barrel: 914.4 mm
Muzzle velocity: 866 m/s

Manufacturer
RAMO Manufacturing Inc., Building D, 412 Space Park South, Antioch
Pike, Nashville, TN 37211.
Status
Current. Production.

.50 RAMO M2HB Quick-Change barrel kit

The RAMO Quick-Change barrel kit allows the advantage of a quick-change barrel to be retrofitted to any Browning .50 M2HB machine gun, regardless of

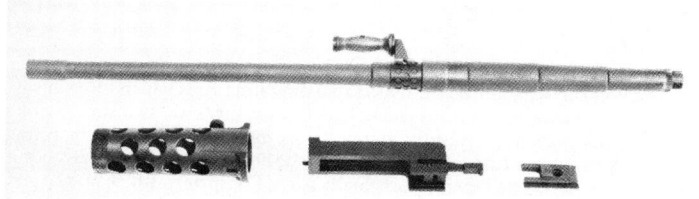

RAMO .50 M2HB Quick-Change barrel kit

manufacturer. The kit may be installed without major tools and increases the effectiveness of the M2HB machine gun by eliminating the time-consuming complications of headspace adjustment.

The system allows for the modification of new conventional M2HB barrels to the quick change design, with consequent savings to the customer. The barrel supplied in the kit can be either standard (unlined, without chrome bore), unlined with chrome bore, or Stellite lined with chrome bore which complies with US Government specifications.

Manufacturer
RAMO Manufacturing Inc., Building D, 412 Space Park South, Antioch
Pike, Nashville, TN 37211.
Status
Current. Production.

.50 RAMO M2HB-QCB machine gun

This is an otherwise standard M2HB .50 Browning machine gun but with a quick-change barrel, allowing the gunner to change the barrel in a fraction of the time required by the conventional design and without the need to adjust for cartridge headspace clearance. A unique feature of the RAMO system is that the gun can still be fitted with the standard M2 heavy barrel should the need arise, though the quick-change ability is no longer available. Conversely, the QCB barrel can be used in a standard M2HB weapon.

The QCB can be supplied in a new gun or the feature can be retrofitted to an existing weapon. Only the barrel, barrel extension and barrel support assembly differ from the standard design while still maintaining 100 per cent interchangeability.

RAMO .50 M2HB-QCB machine gun

DATA
Cartridge: .50 Browning (12.7 × 99 mm)
Operation: Recoil, selective fire
Method of locking: projecting lug
Weight complete: 38.20 kg
Weight, barrel: 10.91 kg
Length overall: 1.651 m
Length of barrel: 1.143 m
Muzzle velocity: 929 m/s
Maximum range: 6800 m
Effective range: 1800 m
Rate of fire: 450 – 600 rds/min

Manufacturer
RAMO Manufacturing Inc., Building D, 412 Space Park South, Antioch
Pike, Nashville, TN 37211.
Status
Current. Production.

Mount, Machine Gun, RM2 (Flex)

The RM2 (Flex) is a mount designed for adapting the M2HB, M2 aircraft, or the RAMO M2 Lightweight .50 machine guns to either vehicle, marine, ground or aircraft installations. It employs a damping system to provide low

RAMO RM2 (Flex) machine gun mount on M3 tripod

recoil forces on the basic mounting and increase the accuracy of the weapon. The mount is constructed of stainless steel and other corrosion-resistant materials, making it particularly suitable for marine use. The M2 machine gun can be installed without modification to the weapon.

DATA
Weight: 14.52 kg
Length: 869 mm
Height: 216 mm
Width (hand grips): 489 mm
Ammunition capacity: 100 rounds on mount
Radius of rear turning circle: 551 mm
Radius of front rurning circle: 267 mm

Manufacturer
RAMO Manufacturing Inc., Building D, 412 Space Park South, Antioch
Pike, Nashville, TN 37211.
Status
Current. Production.

.50 General-purpose heavy machine gun (GPHMG) programme

The US forces have indicated a requirement for a .50 machine gun, lighter than the existing Browning M2HB (above) which could replace the existing M2HB and M85 guns in all roles. Two designs were developed in the early 1980s, but most recent information indicates that the project is in abeyance.

ARRADCOM design
This was developed by the US Army Armament Research and Development Command (ARRADCOM) at Dover, New Jersey, and was the final stage in the 'Dover Devil' design programme which originated as a cannon project. (See *Jane's Infantry Weapons 1983/84*, page 352.)

.50 GPHMG, AAI design

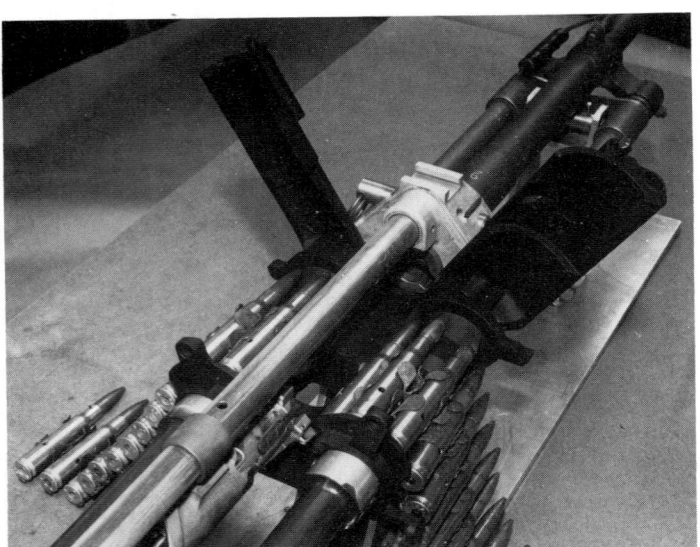

.50 GPHMG, AAI design, showing dual feed

The GPHMG developed at Dover is a gas-operated, fully automatic, belt-fed machine gun which fires from the open bolt position. The weapon is modular in construction, consisting of six basic assemblies. This modular construction and simple design allows ease in stripping, assembly and maintenance; it can be field-stripped in seconds and re-assembled in under one minute without the use of tools, and all component parts requiring operator maintenance, cleaning or inspection are readily accessible. This simplicity of design also means ease of manufacture.

The gun is compatible with all existing .50 (12.7 × 99 mm) ammunition as well as with the Saboted Light Armor Penetrator (SLAP) .50 ammunition currently under development. It may be fired from the M122 (7.62 mm) tripod mount, the M3 heavy machine gun tripod, or in a variety of vehicular applications using an adaptor to the standard Browning Pintle mounts. The ability to use the M122 mount, normally employed with a 7.62 mm weapon, is due to the reduced weight and gas operation of the GPHMG.

The receiver consists of two aircraft-quality steel tubes mounted in a vertical plane and with an aluminium end cap. Included in this assembly is the barrel release mechanism, the two gas pistons which drive the operating group, the cartridge feed ramp, the feed door and the fixed foresight. The barrel release mechanism doubles as the forward mounting point of the weapon.

The barrel assembly is an M85 machine gun barrel modified to permit fixed headspace and the employment of a rotary locking bolt. There is a handle to permit changing a hot barrel and a flash hider. When installed in the receiver the barrel assembly is located at three points, these being the receiver end cap and the two receiver tubes. A 180° rotation of the release mechanism securely locks the barrel to the receiver.

The operating group consists of the bolt carrier, charging handle, bolt assembly with firing pin, drive pin and feed cam roller, two operating rods and two drive springs. Machined in the bolt carrier is a linear cam path which is linked to the bolt by the bolt cam pin. Movement of this pin in the cam path, due to the motion of the bolt carrier, rotates and locks the bolt at the end of the return stroke and rotates and unlocks the bolt during the recoil cycle. The bolt has three locking lugs which engage in the barrel extension. A rammer strips the cartridge from its link during counter-recoil, and the extractor removes the fired case during the recoil movement; two spring-loaded ejectors then eject the case to the right of the gun. The firing pin is fixed to the bolt carrier and is taken forward through the bolt-head to fire the cartridge; a positive stop ensures that the firing pin cannot strike the cartridge cap unless the bolt has been rotated and locked.

The bolt carrier is propelled by the two operating rods which move inside the receiver tubes; at the front end of the receiver tubes are two short-stroke gas pistons which, impelled by gas tapped from the barrel, strike the operating rods and drive them rearwards. This carries the bolt carrier back, rotating and opening the bolt. The moving mass loads two drive springs which, at the end of the recoil stroke, drive the bolt carrier and operating rods back into battery. The use of two gas pistons and two operating rods evenly distributes the loads to both the upper and lower rods, resulting in a balanced loading of the moving components.

The GPHMG is fed from the left side, employing a sprocket-driven M15A1 disintegrating link belt. There is a neutral selector lever at the rear of the feed box assembly which, when operated, disengages the feed mechanism, allowing the weapon to be fired in the single shot mode; it also assists in unloading and provides a means of stopping a runaway gun.

During the recoil cycle the feed cam roller, on the operating group drive pin, rides in a linear cam path machined in the feed cam, causing the cam to rotate in a clockwise direction, driving the feed sprocket; this feeds the next round into position and ejects the previous link. During counter-recoil the feed cam rotates in the opposite direction but is disengaged from the sprocket by a ratchet. The round in the feed position is removed from the belt by the rammer, travels up the feed ramp and into the chamber. Loading is done by first cocking the weapon, leading the belt through the feed aperture and rotating the sprocket by hand until the first round is held by the cartridge stop.

The dust cover over the receiver is not essential to the operation of the weapon; it merely provides protection from the elements.

DATA
Calibre: .50 Browning (12.7 × 99 mm)
Operation: gas
Locking: rotating bolt
Weights
   Weapon: 21.32 kg
   Spare barrel: 6.35 kg
   M122 tripod: 8.6 kg
   Gun and tripod: 29.94 kg
Length, overall: 1829 mm
Barrel, length: 914 mm
Number of parts: 188
Feed: disintegrating link belt
Rate of fire: 400 rds/min
Muzzle velocity: 865 m/s
Max range: 6650 m

AAI Corporation design
This has also been influenced by the 'Dover Devil' modular design theories

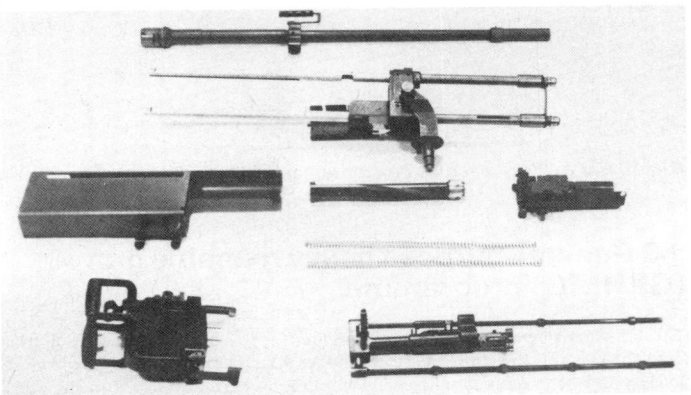

.50 GPHMG, ARRADCOM design, dismantled into major component groups

.50 GPHMG, ARRADCOM design

and in some respects resembles the ARRADCOM design described above. The AAI GPHMG, however, uses a three-tube configuration for the receiver which increases inherent strength and also makes it easier to incorporate a dual feed mechanism. There are two operating rods, driven by gas pistons and controlling the movement of a bolt carrier with a rotating three-lug bolt. Feed, by disintegrating link belts, is performed from both sides of the gun simultaneously, and a selector allows the firer to select which belt to use; this selector clutches in the appropriate feed sprocket. It can also be set in a neutral position, preventing either belt from feeding, and this position is used during loading for positive safety. There is a quick-change barrel, requiring no headspace adjustment. The manufacturers claim that this gun can be manufactured for 30 per cent less cost than the current M2HB machine gun.

DATA
Calibre: .50 Browning (12.7 × 99 mm)
Operation: gas
Locking: rotating bolt
Weight: 24.95 kg
Length, overall: 1549 mm
Number of parts: 190
Feed: disintegrating link belt
Rate of fire: 400 rds/min

Manufacturer
AAI Corporation, PO Box 6767, Baltimore, Maryland 21204.
Status
Development has ceased and further work is not anticipated.

.50 Ares Telescoped Ammunition Revolver Gun (TARG)

The TARG is a gas-operated self-powered gun incorporating a revolving four-chambered cylinder. The ammunition for this gun is a special telescoped round which is cylindrical and which allows the adoption of push-through feeding and ejection, removing the need for a separate extraction phase in the operating cycle. The feed system is linkless, the rounds being supplied to the gun and then transferred to the cylinder by a feed rotor. The round is chambered, presented to the barrel and fired, and the empty case is then driven forward to be ejected from the gun. The four-chambered cylinder allows various portions of the feed-fire-eject cycle to be performed simultaneously rather than serially, and hence the gun has an extremely high rate of fire.

The Plastic Cased Telescoped Ammunition (PCTA) developed for this gun has the bullet in the centre of the cartridge, surrounded by solid propellant. A starter charge ejects the bullet from its casing and enters it into the barrel, after which the propellant charge explodes and drives the bullet from the gun. This sequence, by ejecting the bullet from the cartridge, leaves air space within the propellant which gives more efficient combustion and allows a high velocity to be developed. Moreover, by ensuring that the bullet is entering the barrel before the main charge ignites, various gas erosion problems are avoided.

The TARG is still in the initial development phase, but a working weapon has been built and fired, and development is continuing. The gun is considerably lighter, and has fewer parts, that the .50 M2HB Browning, and it is estimated that the plastic ammunition will cost about 20 percent less than conventional brass-cased ammunition.

DATA
Calibre: .50in

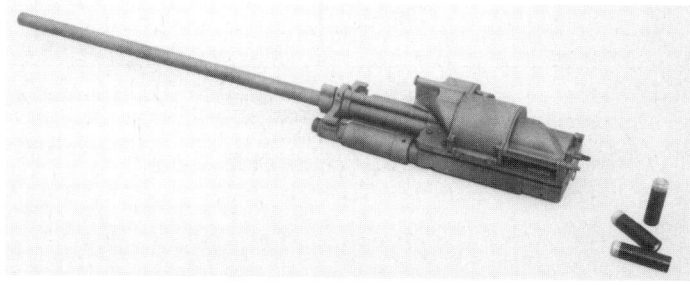

Ares .50 TARG machine gun

Operation: gas, automatic, revolver
Feed: linkless
Weight: 20.4 kg
Length: 1128 mm
Barrel: 927 mm
Muzzle velocity: 900 m/s
Rate of fire: 1400 rds/min proven; 2000 rds/min projected

Manufacturer
Ares Inc, 818 Front Street, Erie Industrial Park, Port Clinton, Ohio 43452.
Status
Development.

7.62 mm M60 general-purpose machine gun

The M60 is the US Army general-purpose machine gun and it came into service in the late 1950s. The prime producer has been Saco Defense Inc, a subsidiary of the Chamberlain Manufacturing Corporation.

The original design had some interesting features. The straight line layout allows the operating rod and buffer to run right back into the butt and reduce some of the overall length. The large fore-hand grip is most convenient for carrying at the hip, and the folded bipod legs continue the hand protection almost up to the muzzle. The gun can be stripped using a live round as a tool.

An undesirable feature of the first production guns was the barrel having the bipod and gas cylinder permanently attached to it. Removing the barrel meant that these extra components were taken off too, and a barrel change was a cylinder and bipod change also. This was quite unnecessary and led to additional weight being carried with the spare barrels which is objectionable in the field. Removing the barrel also takes the supporting bipod away and means that the gun has to be laid on the ground while reaching for the new barrel, or it has to be held by the gunner while his number two tries to fit a new barrel. Neither operation is at all easy in the dark. To overcome these failings the M60E1 was developed, and with it the opportunity was taken to

incorporate other changes which service in Vietnam had established as necessary. The 12 alterations to the basic design are: the barrel does not have the cylinder and bipod attached to it; the bipod is on the fixed gas cylinder, and so supports the gun when the barrel is off; simplified gas cylinder; simplified attachment for gas cylinder; modified gas piston; modified rearsight; pressure die-cast feed cover with fewer parts; modified feed tray; modified belt box attachment; pressure die-cast fore-hand grip allowing easier changing of the barrel; relocated swing swivels for easier carrying and larger carrying handle.

The M60E1 was not introduced into service, but the improvements were incorporated in the lightweight M60E3 design described in the next entry.

Operation
The M60 is a gas-operated weapon. The barrel is drilled radially downwards at some 20 cm from the muzzle and after the bullet has passed this point a small proportion of the propellant gas passes through the vent.

The gas enters the gas cylinder, passes through a drilling in the side wall of the piston and expands to fill the interior of the piston and the forward section

7.62 mm M60 general-purpose machine gun fitted with gas evacuator tube for use in armoured fighting vehicles. M60E2 armoured fighting vehicle version of weapon has similar tube

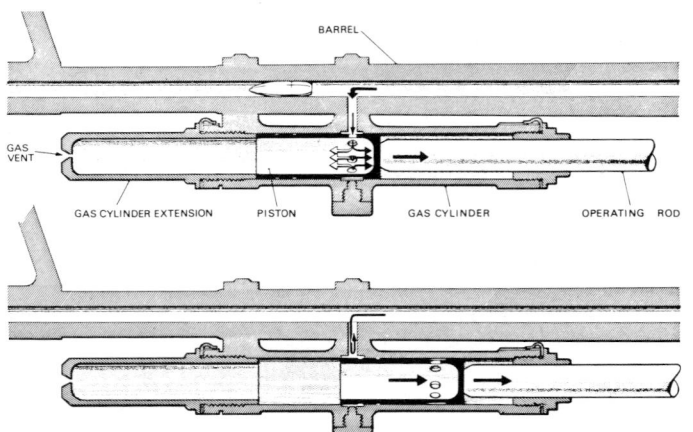

Constant energy gas regulator

of the gas cylinder. When sufficient pressure has built up the piston is forced rearwards. As soon as it moves the radial drillings through the piston are moved out of alignment with the gas vent and no further gas enters the cylinder. This is shown diagramatically below.

The piston moves back and drives the operating rod, which carries the bolt rearwards. The action is that of a short-stroke piston since the piston travel is limited to 60 mm, and a sharp impulsive blow is given to the operating rod which imparts enough energy to carry out the complete cycle of operations.

The M60 has no method of gas adjustment. There is no regulator of any sort adjustable by the firer. The general theory of operation is that the piston will move back when enough energy has been supplied to overcome fouling and external friction. When the piston moves, it automatically cuts off its own gas supply and can be said to be self-regulating. It is often referred to as a constant volume system.

The operating rod assembly has a post which rides in the hollow interior of the bolt and it is noteworthy that this post carries an anti-friction roller which

is the bearing surface against the camway cut in the bolt. The post rides initially in the curved portion of the cam slot and tries to rotate the bolt which is restrained by the forward locking lugs riding in longitudinal grooves in the gun body. When the bolt is fully forward it can rotate clockwise and lock in the curved cam path cut in the barrel extension. When the rotation is completed the post is able to enter the last inch of the cam slot, which is parallel to the longitudinal axis, and travel straight forward. The firing pin, which is carried on the post, then strikes the cap.

After the bullet has passed the gas port the pressure starts to build up in constant volume cylinder. Before there is sufficient pressure to move the piston the bullet has left the muzzle. The piston impinges on the operating rod and this moves back 21 mm before unlocking commences.

The M60 is designed to fire full automatic only, at a cyclic rate of 550 rds/min. This is a slow enough rate for an accomplished firer to get off a single round. The tactical rate 'rapid' is 200 rds/min.

Feed system
The M60 started out with the German MG42 feed system. The MG42 was the first gun to come into large scale use employing a system of inner and outer feed pawls driven by a bolt-operated feed lever, which were arranged to move in opposite directions and move the belt in two stages as the bolt moved back and forward again.

As the M60 was progressively developed the feed system was changed. The friction roller at the back of the bolt was retained as was the feed arm on the top of the feed cover in which it operated. The best feature, the inner and outer pawl system, was abandoned and a single pawl system substituted. In this the roller carried by the bolt moving forward swings the feed arm to the right. The feed arm is pivoted so the front end carrying the feed pawl slips to the left over the next round to be fed. As the bolt goes back the pawl moves to the right and lifts the entire weight of the belt through one complete pitch.

Barrel assembly
On the M60 the gas cylinder and the bipod are both permanently attached to the barrel. Thus the member of the gun crew who carries the spare barrel has to cope with a kilogram or so of weight which could be on the body of the gun. The M60E3 has a barrel with gas cylinder only, as the bipod is attached to the front of the receiver assembly. The barrel is chromium plated and also has a stellite liner in the 6 in (152 mm) of barrel forward of the chamber.

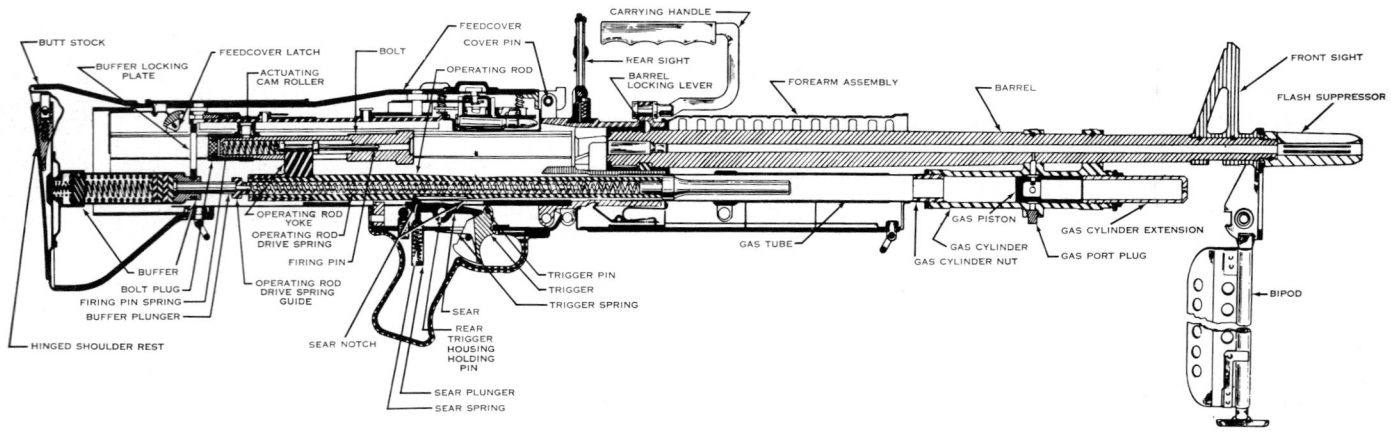

7.62 mm M60 general-purpose machine gun

British soldier on exchange visit to USA demonstrating prone firing position with 7.62 mm M60 general-purpose machine gun showing barrel assembly

Sights

The M60, having a fixed foresight, makes adjustment for zeroing on the rearsight. This means that the number one on the gun must first recognise which barrel is in the gun and secondly know the correct zero setting for both elevation and line for that barrel. In practice this inevitably means all barrels are fired from a common zero and the consequent loss of accuracy is accepted. An adjustable front sight has been provided with the M60E3 model.

Stripping

The technical manual lists 20 stages in stripping the weapon and re-assembling it. In practice it is a fairly straightforward procedure although care must be taken in removing the bolt operating rod.

The general procedure for stripping is as follows: check the gun is safe and the bolt forward; lift the shoulder strap, depress the butt stock latch with the point of a cartridge and remove the butt stock; rotate the latch lever assembly to lift the top cover plate; remove the buffer yoke; slide the bolt and operating rod out sufficiently far to insert the yoke into the bolt; this holds the bolt spring in place; remove the bolt and operating rod; rotate the barrel catch and remove the barrel; the trigger group can be removed by displacing the leaf spring and pushing the retaining pin through the body.

The above order of events is not that taught officially but it does have the advantage that taking the barrel off at the end allows the bipod to support the front end of the weapon while the bolt, and so on, is being removed.

Variants

Five versions of the weapon are distinguished by separate designations. In addition to the basic M60 and M60E3 infantry weapons there are;
M60C: remotely fired, for external mounting on helicopter. No longer in production.
M60D: helicopter and vehicle weapon, pintle-mounted in helicopter doorways, vehicle platforms etc. Current, and in production.
M60E2: armour weapon for internal mounting in armoured fighting vehicles. Current and in production.

7.62 mm M60 general-purpose machine gun fired from hip

7.62 mm M60D general-purpose machine gun used as door gun in helicopter

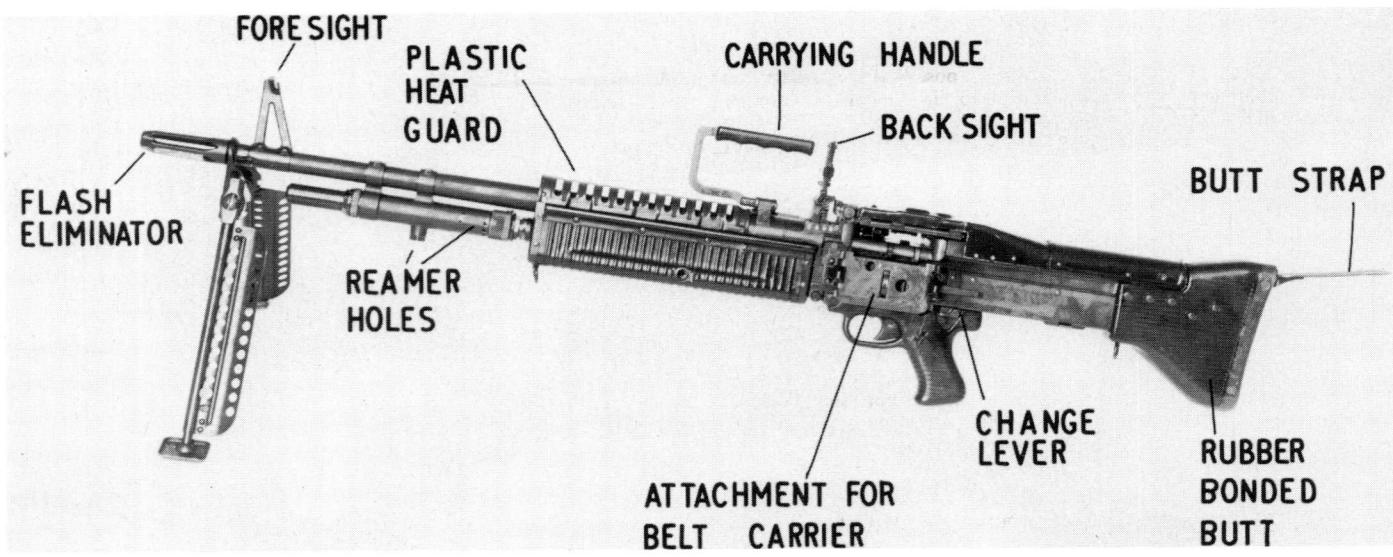

FORE SIGHT

FLASH ELIMINATOR

PLASTIC HEAT GUARD

REAMER HOLES

CARRYING HANDLE

BACK SIGHT

BUTT STRAP

ATTACHMENT FOR BELT CARRIER

CHANGE LEVER

RUBBER BONDED BUTT

7.62 mm M60 general-purpose machine gun

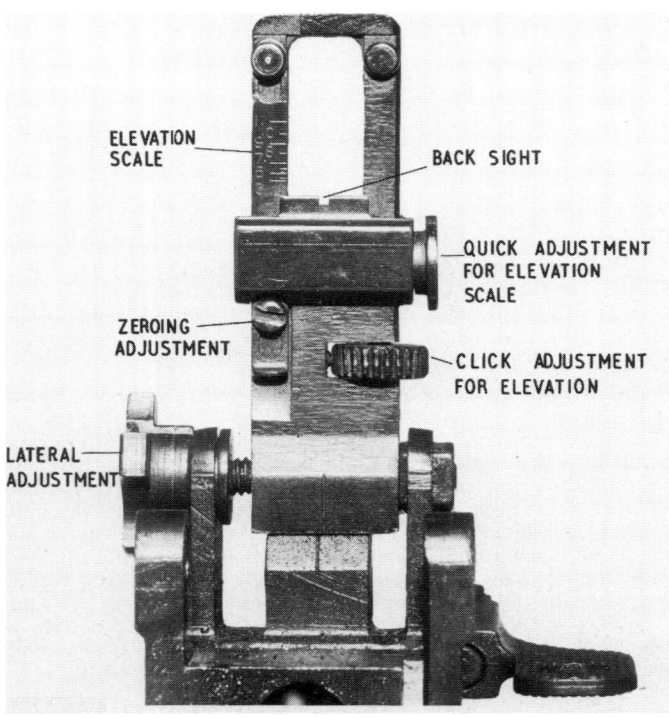

ELEVATION SCALE

BACK SIGHT

QUICK ADJUSTMENT FOR ELEVATION SCALE

ZEROING ADJUSTMENT

CLICK ADJUSTMENT FOR ELEVATION

LATERAL ADJUSTMENT

M60 rearsight. Figures on scale are very small

DATA
Cartridge: 7.62 mm × 51
Operation: gas, automatic

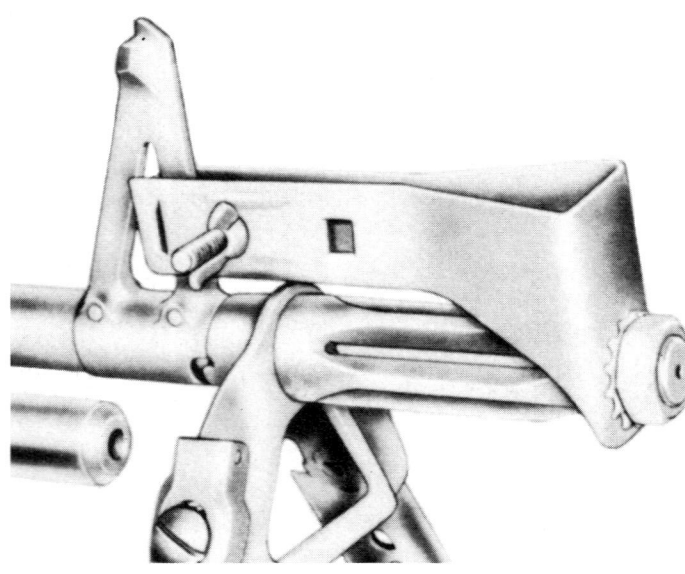

M13BF attachment

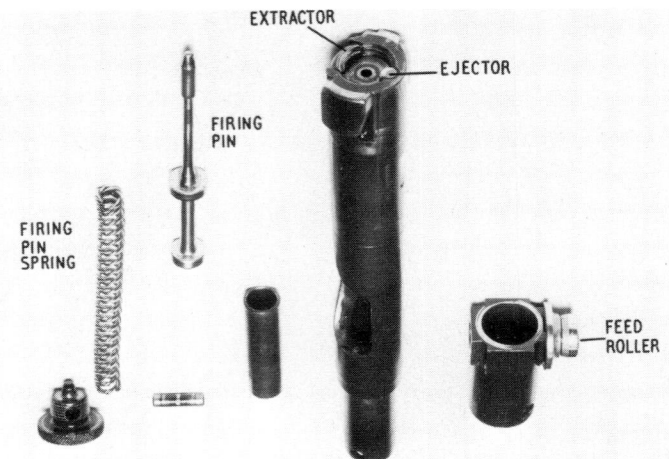

EXTRACTOR

EJECTOR

FIRING PIN

FIRING PIN SPRING

FEED ROLLER

Bolt assembly

Method of locking: rotating bolt
Feed: disintegrating link belt

WEIGHTS
Gun: 10.51 kg
Barrel: (including bipod and gas cylinder) 3.74 kg

LENGTHS
Gun: (overall) 1105 mm
Barrel: (excluding flash hider) 560 mm

MECHANICAL FEATURES
Barrel: (regulator) self-regulating; (cooling) air. Quick-change barrel
Sights: (foresight) fixed blade; (rearsight) U notch; (sight radius) 540 mm

FIRING CHARACTERISTICS
Muzzle velocity: 855 m/s
Remaining velocity: (at 500 m) 492 m/s; (at 3700 m) 108 m/s
Rate of fire: (cyclic) 550 rds/min
Effective range: (bipod) 1000 m; (tripod) 1800 m

MOUNT
Weight: 6.8 kg
Length: (extended) 825 mm; (folded for travel) 685 mm
Spread of rear legs: 762 mm
Height: 358 mm
Traversing range: (using traversing bar) 50°
Free: 360° (6400 mils)
Traversing bar: (graduated) 875 mils
Elevating range free: +28°30′ +28°45′
Locked: +14° -12°35′
Least increment: 1 mil
Elevating handwheel: (graduated) 1 mil

Manufacturer
Saco Defense Inc, a subsidiary of Chamberlain Manufacturing Corporation, 291 North Street, Saco, Maine 04072.
Status
Current. Production.
Service
US forces. Australia, South Korea, Taiwan and many other countries.

7.62 mm M60E2 fixed machine gun

This gun is a modification of the M60 general-purpose machine gun for use as a coaxial machine gun in armour. This weapon has undergone extensive evaluation testing by the US Army Armor and Engineer Board. The US Marine Corps has adopted the M60E2 as standard coaxial machine gun in their M60A1 tanks.

A barrel extension tube with integral flash suppressor and an evacuator tube on the gas system vent the combustion gases outboard through the firing port. A collar on the operating rod and reduced clearance at the rear of the gas cylinder virtually eliminate the discharge of combustion gases into the fighting compartment through the receiver. Length of the barrel extension tube and evacuator tube depend on the installation. Arrangements are provided to support the barrel extension at the gun port.

A short backcap is fitted in place of the butt stock. A chain-type charger with return spring replaces the standard charger. The forestock is removed. Because of the confined clearance above the gun at full depression, feedcover opening is normally limited by a stop to 35° with override provisions to allow full opening for rapid clearance of malfunctions. The trigger assembly is

similar to that used in the M60C helicopter version, but has a rotating safety and push type manual trigger.

Saco has developed mounting hardware for the M60A1 tank installation. The mount is rigid and contains traverse and elevation adjustments for zeroing the coaxial machine gun to the main gun. Attachment points existing in the standard receiver are used. A quick disconnect feature allows rapid dismounting of the gun for barrel changes. Mounting hardware for other tank installations is under development.

Manufacturer
Saco Defense Inc, a subsidiary of Chamberlain Manufacturing Corporation, 291 North Street, Saco, Maine 04072.
Status
Current.
Service
US, South Korean and other armies.

7.62 mm M60E2 machine gun fitted with evacuator tube

7.62 mm M60E3 lightweight machine gun

The M60E3 lightweight machine gun was developed by Saco Defense Inc, a subsidiary of Chamberlain Manufacturing Corporation. It was designed to provide a lighter, more versatile calibre 7.62 mm machine gun maintaining all the capabilities of its parent weapon, with several additional features of its own. A lightweight bipod is mounted on the receiver eliminating the bipod on the spare barrel. The carrying handle is attached to the barrel for ease of changing and handling when hot without the need for asbestos mittens. The forearm has been eliminated and a second pistol grip with heat shield mounted forward under the bipod for better control of the weapon when firing. The feed system has been modified to permit charging of the weapon after the feed cover has been closed. A double sear-notch prevents uncontrolled fire. A winter trigger-guard allows the user to fire while wearing heavy gloves or mittens. The foresight is adjustable for both windage and elevation allowing any barrel assembled to the weapon to be zeroed without adjustment of the rearsight. The gas system has been simplified and provided with interlocking cylinder nuts. Except for the gas system, field stripping of the M60E3 lightweight machine gun is the same as it is for the standard M60. All the operating group is common and major assemblies are interchangeable.

DATA
Calibre: 7.62 mm × 51
Feed: link belt
Weight: 8.61 kg
Length: 1067 mm
Length of barrel: 560 mm
Muzzle velocity: 860 m/s
Rate of fire: (cyclic) 550 rds/min
Max effective range: 1000 m

Manufacturer
Saco Defense Inc, a subsidiary of Chamberlain Manufacturing Corporation, 291 North Street, Saco, Maine 04072.
Service
Adopted by US Marine Corps: in service with US Navy and Air Force, and numerous foreign armed forces.

7.62 mm M60E3 Saco Defense lightweight machine gun

Saco M60 to M60E3 conversion kit

The M60E3 Conversion Kit can be used to convert any serviceable, unwelded, M60 receiver to the lightweight M60E3 version. The Conversion Kit has been procured by the US Marine Corps, prior to the acquisition of complete M60E3 machine guns. The kit conversion can be completed at the Saco Defense factory in the USA or technical personnel can be trained to perform the conversion as a qualified facility.

Manufacturer
Saco Defense Inc, a subsidiary of Chamberlain Manufacturing Corporation, 291 North Street, Saco, Maine 04072.
Status
Current. Production.
Service
US and foreign armed forces.

5.56 mm M249 Squad Automatic Weapon (SAW)

The M249 Squad Automatic Weapon, employed as the section light machine gun in US service, is the FN 'Minimi' light machine gun with small modifications to suit American military specifications and US manufacturing techniques. It does not differ in significant detail from the Belgian Weapon which is fully described and illustrated on p300.

7.62 mm Ares externally powered machine gun (EPG)

This has been developed by Ares Inc, for use as a secondary weapon on armoured vehicles. It is adaptable to existing coaxial machine gun mounts, and its small size and weight make it an excellent choice for use in small turrets. Electrically driven by a cam system which actuates the bolt, the EPG has a variable rate of fire which allows it be be adapted to a variety of tactical demands. The mechanism is simple, reliable, and easily maintained. The barrel can be rapidly changed, and feed can be easily changed from left to right side as required. It is fed by the normal type of disintegrating link belt; the empty cases are ejected downward from the receiver, and the spent links are ejected below the feed chute. In the event of power failure the weapon can be operated by a hand crank.

DATA
Cartridge: 7.62 × 51 mm NATO
Operation: powered cam
Weight: 12.52 kg
Barrel weight: 2.36 kg
Length overall: 808 mm
Barrel: 597 mm
Rate of fire: variable up to 650 rds/min
Power requirements: 18 – 32 V; 9 A at 24 V

Manufacturer
Ares Inc, 818 Front Street, Erie Industrial Park, Port Clinton, Ohio 43452.
Status
Development completed.

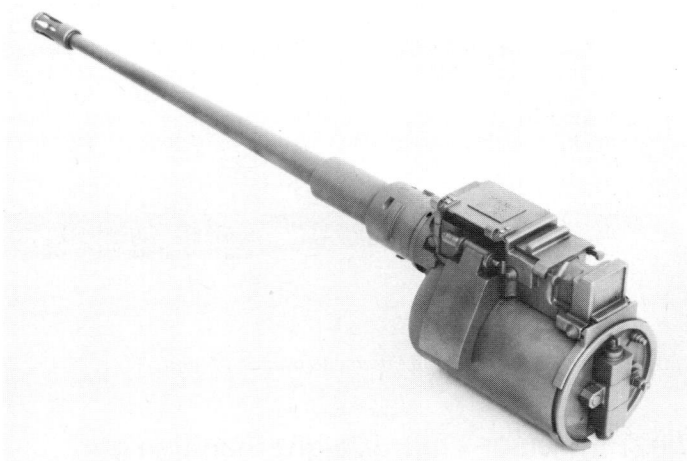

7.62 mm Ares EPG

7.62 mm M134 Minigun General Electric machine gun

General Electric (USA) of Burlington, Vermont, came into the machine gun field with the production of the Minigun which was specifically planned for use in aircraft and helicopters in the Vietnam conflict. It was based on the 20 mm Vulcan gun which armed many types of American aircraft.

This weapon is based on the Gatling gun principle in which a high rate of fire was achieved by having a number of rotating barrels which fired in turn when the 12 o'clock position was reached.

The first Gatling was made in 1862 so possibly the progress made in the last 100 plus years is not as great as some would believe. The original Gatling gun was hand-cranked and could produce a cyclic rate of 1000 rds/min for a very brief period. The delays in loading etc normally meant about 250 rds/min was the optimum.

The Minigun is driven by a 28 V DC or 115 V AC electric motor and produces a steady rate of fire which varies according to the type from between 2000 and 6000 rds/min as a top rate, down to 300 rds/min as the slowest rate of fire. At a steady 6000 rds/min the drive motor draws 130 A. The gun itself consists of four groups: the barrel group; gun housing; rotor assembly and bolt assembly.

Barrel group
The six 7.62 mm barrels fit into the rotor assembly and each is locked by a 180° turn. They pass through the barrel clamp which grips the barrel using a special tool to secure the bolt. The barrels are normally parallel but there are other clamps available which are designed to produce varying dispersion patterns by ensuring that the barrels converge at some selected range.

Gun housing
This is a one piece casting which holds the rotor assembly and provides a mounting for the 'safety sector' and the guide bar. The function of these two components will be mentioned later. There is an elliptical cam path on the inner surface in which the bolt bearing roller runs.

Rotor assembly
This is the main structural component of the gun and is supported in the gun housing by ball bearings. The front part of the rotor assembly holds the six barrels. Six bolt tracks are cut into the rotor and six removable tracks are bolted on to the ribs on the rotor. Each bolt track carries an S-shaped triggering cam which cocks and fires the firing pin of one bolt.

Bolt assembly
There are six bolts each of which is mated to a barrel and lock into the barrel by means of a rotating head: when each is unlocked and withdrawn from the barrel the empty case is pulled out by a fixed single extractor. The firing pin is spring-operated and carries a projection or 'tang' at the rear end. This tang engages the S-shaped triggering cam on the rotor. The side slots on the bolt engage in the bolt tracks of the rotor and this causes the bolt to move round with the rotor.

Operation
The red master switch cover must be elevated to expose the switch which controls all the electrics. The firing button on the left-hand trigger grip starts the rotor assembly and barrel assembly rotating in an anti-clockwise direction viewed from the rear. As the rotor assembly turns, the roller on each bolt follows the elliptical cam path on the inner surface of the gun housing. Each bolt in turn picks up a live round from the guide fingers of the guide bar.

As the bolt is carried round, the roller, working in the groove, moves the bolt forward to chamber the round. The bolt-head is now rotated by the cam path in the bolt body and locks into the barrel. The firing pin has been cocked by the tang action in the triggering cam in the rotor and at the top position of the bolt the S-shaped groove releases the firing pin and the round is fired.

The elliptical cam path in the gun housing has a flat profile, or dwell profile, which holds the bolt locked until the bullet has gone and the pressure has dropped to a safe level. The bolt assembly roller enters the reverse segment of the path and the bolt-head is unlocked. The cam path carries the bolt to the rear, extracting the empty case which is pushed away by the guide bar and

7.62 mm Minigun General Electric machine gun

ejected. The bolt has now completed a 360° cycle and is ready to pick up another live round. All six bolts repeat the same procedure in sequence.

Ammunition boxes carry a normal load of 4000 linked rounds. This is pulled up a plastic chute to the gun. If the length of this chuting exceeds 1.5 m, or the radius of the curve is small, it is advisable to have another motor on the top of the ammunition boxes driving a sprocket which helps by pushing the belted ammunition up to the gun.

It will be noted that since the entire gun is driven by an external source, a misfired round is simply ejected and thrown down the large section hose with the empties either into a box or more usually out over the side.

It is equally possible to use unlinked ammunition and feed it up from the boxes.

When the gun ceases firing the bolts cease to pick up ammunition from the guide bar and the bolts close on an empty chamber. Thus there is no danger of a cook-off during the second it takes the barrels to come to rest.

DATA
Cartridge: 7.62 mm × 51
Operation: external power, Gatling-type action, automatic
Method of locking: rotating bolt
Feed: linked belt or linkless

WEIGHTS
Basic gun: 16.3 kg
Drive motor: 3.4 kg
Recoil adaptors: (2) 1.36 kg
Barrels: (6) 1.09 kg each
Feeder: 4.8 kg

LENGTHS
Gun: (overall) 801.6 mm
Barrel: 559 mm

MECHANICAL FEATURES
Barrel: (regulator) nil; (cooling) air
Sights: vary with employment
Power required: at 6000 rds/min (steady rate) 130 A
Starting load: 260 A for 100 m/s
Reliability: (gun only) not less than 200 000 MRBF
Gun life: 1 500 000 rounds minimum

FIRING CHARACTERISTICS
Muzzle velocity: 869 m/s
Rate of fire: up to 6000 rds/min
Dispersion: 6 mRad maximum
Average recoil force: 0.5 kN

Manufacturer
General Electric Co., Armament Systems Department, Lakeside Avenue, Burlington, VT 05401.
Status
Current. Production.
Service
US Army, Air Force, Marine Corps and Navy, and other countries.

5.56 mm Minigun

This is very similar to the 7.62 mm weapon described above but has several design innovations. These include an access cover lever/safety lever which allows the gun to be made incapable of firing; bolts that are removeable from the gun without tools; a clutch mechanism that stops the feeder drive gear upon trigger release; and a case ejection sprocket.

The Minigun has been produced as a complete lightweight system which can be aircraft or vehicle mounted or fired from a ground tripod. With 1000 rounds of ammunition the system weighs 38.6 kg and can be quickly broken down into two equal loads for carrying.

Particular attention has been paid to ease of maintenance. The access cover, gun bolts and feeder can be removed without tools, and the feeder drive gear is a self-timing design which provides correct feeder-gun timing.

DATA
Cartridge: 5.56 × 45 mm
Operation: Gatling action, 6 revolving barrels
Feed: linked belt
Weight: 10.2 kg
Length: 731.5 mm
Rate of fire: 400 to 6000 rds/min, selectable
Reliability: 240 000 MRBF
Dispersion: 8 mRad
Gun life: 500 000 rounds
Average recoil: 0.5 kN

Manufacturer
General Electric Co., Armament Systems Dept., Lakeside Avenue, Burlington, VT 05401.

GE 5.56 mm Minigun

Status
Not currently in production.
Service
US Air Force, not currently employed.

5.56 mm Ares light machine gun

The Ares 5.56 mm LMG has been designed using a minimum number of parts so as to provide a reliable and lightweight weapon. The design uses components from the Stoner 63 system, as well as accessories from weapons such as the M60, M249 and M16. The modular components are made of corrosion-resistant material, stainless steel, and aluminium with protective finishes. The weapon's modular design enables quick field stripping without the use of tools.

The 5.56 mm LMG is adaptable to both belt and magazine feeds. No variation in firing rates occurs due to environmental conditions, feed type or belt length. The use of a long bolt stroke provides low recoil forces, eliminating the need for a buffer and gas adjustments. The interchangeability of components provides versatility and enables rapid conversion to lightweight and assault gun variants.

An assault gun variant (weighing approximately 4.2 kg) of the Ares 5.56 mm light machine gun is quickly created by removing the bipod and butt stock and replacing the standard barrel with a lighter and shorter barrel.

Ares 5.56 mm light machine gun

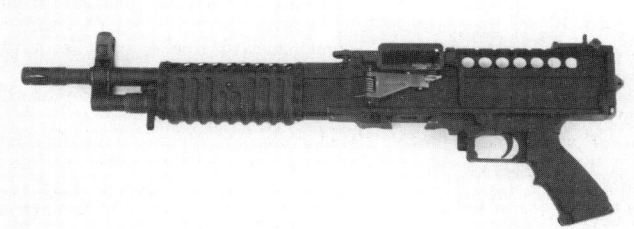

Ares 5.56 mm light machine gun, assault gun variant

DATA
Cartridge: 5.56 × 45 mm
Operation: gas, automatic
Method of Locking: rotating bolt
Weights:
  Gun, empty, without bipod: 4.91 kg
  Gun, without bipod, with 100 rds: 6.32 kg
  Gun, empty, with bipod: 5.47 kg
  Barrel assembly: 1.73 kg
Length, butt extended: 1073 mm
  butt retracted: 971 mm
  butt removed: 813 mm

Barrel length: 550 mm
Rate of fire: 600 rds/min
Muzzle velocity: 945 m/s with SS109 ammunition

Manufacturer
Ares Inc., Bldg 818, Front Street, Erie Industrial Park, Port Clinton, OH 43452-9399.
Status
Current. Low rate initial production.

5.56 mm Colt M16A2 light machine gun

The Colt M16A2 LMG is a gas operated, air cooled, magazine fed, fully automatic weapon which fires from the open bolt position. The heavier barrel increases stability, accuracy and heat dissipation. The M16A2 LMG features a rugged bipod design which deploys and retracts rapidly and easily with the use of a single hand. The various features of the weapon combine to yield an unprecedented degree of control and accuracy in a light machine gun at extended ranges.

The M16A2 LMG shares common features with the M16A2 rifle. The LMG operating controls and handling characteristics are immediately familar to soldiers trained in operating the M16 family of weapons. The LMG uses the same wide variety of ammunition as the rifle.

The Colt M16A2 LMG has wider and more square handguards, and a forward grip on the bottom part of the handguard. It utilises the M16's standard magazine and other large-capacity magazines available today with 90 to 100 round capacities. The LMG's magazine-fed design reduces the number of parts and the ammunition weight carried by the soldier when the weapon is operated, while it provides the ability to provide ammunition for sustained fire faster than most belt fed weapons.

DATA
Cartridge: 5.56 × 45 mm NATO
Operation: gas, fully automatic
Method of locking: rotating bolt
Feed: 30-round magazine standard; will accept any M16 interfacing magazine
Weight: empty, without magazine, 5.78 kg. With loaded magazine, 6.234 kg
Length: 1000 mm
Barrel length: 510 mm
Rifling: 6 grooves, rh, one turn in 178 mm
Sights: (fore) adjustable post; (rear) aperture, adjustable for windage and elevation
Muzzle velocity: (M193) 991 m/s. (SS109) 948 m/s
Muzzle energy: (M193) 1722 J. (SS109) 1765 J
Rate of fire: 600-750 rds/min
Max effective range: 800 m

Manufacturer
Colt Firearms, Hartford, CT 06101.
Status
Current. Production.
Service
US Marine Corps, El Salvador, Brazil and various other countries.

5.56 mm Colt M16A2 light machine gun, left side

5.56 mm Colt M16A2 light machine gun, right side

GECAL 50 machine gun

GECAL 50 is a six-barrel lightweight .50 calibre Gatling gun capable of firing standard 12.7 × 99 mm ammunition at variable firing rates up to a maximum of 8000 rds/min. A three-barrel variant model can fire at rates up to 4000 rds/min. Drive power may be derived from electric, pneumatic or hydraulic sources, and both models have the add-on capability of a self-starting gas drive for applications where external power is either unavailable or undesirable. Barrels may be had in lengths from 914 mm to 1.295 m in order to permit the ballistic performance to be matched to the role. A controlled burst firing rate control gives the gunner the capability of firing a volley of approximately 10 rounds at the high rate of fire, resulting in nearly simultaneous impact of the shots.

The gun, when provided with a de-linking feeder, will handle the standard M9 link belt. It can also be equipped with a linkless feed system for high reliability at the high firing rates. Gun clearing is accomplished by a declutching feeder in linked systems and through gun reversal in linkless systems.

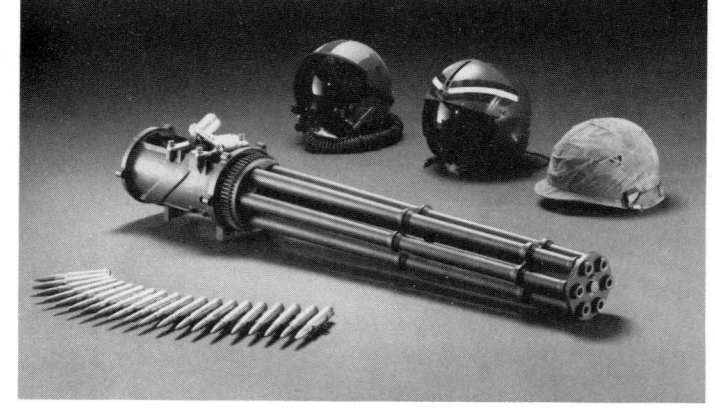

.50 GECAL 50 machine gun, six-barrel version

DATA
Cartridge: 12.7 × 99 mm (.50 Browning)
Operation: rotary, automatic
Locking: rotating bolt
Feed: linked or linkless belts
Weight: 43.6 kg (6-barrel); 30 kg (3-barrel)
Length: 1181 mm
Barrel: 914 mm to 1295 mm to choice
Rate of fire: 8000 rds/min max (6-barrel); 4000 rds/min (3-barrel) (linkless feed)

Time to rate: 0.3 seconds
Muzzle velocity: 884 m/s

Manufacturer
General Electric (USA), Lakeside Avenue, Burlington, VT 05402 and Astra Defence Systems Ltd, Grantham, Lincs., England.
Status
Undergoing qualification in 1990, production scheduled for 1991.

7.62 mm EX-34 Chain Gun automatic weapon

This 7.62 mm automatic weapon, specifically developed for coaxial use and the smallest member of the Chain Gun family, is readily adaptable to a variety of armoured vehicle and aircraft gun roles. For coaxial installation, the weapon is so configured as to permit barrel change in less than 10 seconds without removing the belt feed. It incorporates the basic features of the proven 30 and 25 mm weapons and addresses the critical problems of gas, brass and feed normally associated with all enclosed gun installations. Like the M242, this weapon also ejects fired cases forward (overboard) and capitalises on long dwell time to eliminate gas build-up in cupola or turreted installations.

Compact and lightweight, the 7.62 mm machine gun is physically interchangeable with the coaxial weapon in M60 tanks and no mounting modification is required to replace the presently installed machine guns.

The coaxial version for armoured fighting vehicles has a long barrel jacket running right to the muzzle. A venturi system draws air forward through the jacket to cool the barrel and also acts as a fume extractor. Empty cases are thrown forward clear of the vehicle.

The weapon has been tested by both the US Army and the US Naval Surface Weapons Center, Dahlgren, Virginia. The Navy tests were to determine reliability, durability, performance capabilities, safety and ease of handling. The Navy report stated that the performance of the EX-34 gun during all phases of testing was outstanding.

As a result of extensive, highly successful tests conducted by the Royal Small Arms Factory, Enfield in 1978 and 1979, the British Army purchased a small number of EX-34 Chain Guns. Maintenance and operational tests were scheduled to provide additional data on the suitability of the weapon for use in vehicle installations. Following the evaluation of these tests, the UK Ministry of Defence adopted the EX-34 for coaxial installation in the Warrior MICV and for use on all future British Army combat vehicles including Challenger II. The gun is now in serial production in the UK under licence to McDonnell Douglas Helicopter Company.

Canadian forces are also evaluating the EX-34 for use in both their Leopard tank and their armoured vehicle, general-purpose, (AVGP) Cougar and Grizzly fleets. Toxicity tests conducted by the UK and Canada have shown that the EX-34 is essentially a gasless weapon.

Although initially developed as an armoured vehicle weapon, the EX34 has also been configured to a lightweight helicopter armament subsystem, designated HGS-55, for installation on the McDonnell Douglas 500MD 'Defender' light attack helicopter. The HGS-55 system provides accurate suppressive fire at ranges out to 900 m and is extremely effective against non-armoured targets. The HGS-55 system is now in limited production for sale throughout the free world and interest has been generated in the US Army Aviation community for specialised military applications.

Kenya and Colombia purchased a quantity of 7.62 mm Chain Gun weapons for installation in armament kits on their 500MD helicopters.

DATA
Ammunition: 7.62 mm NATO (MB linked)
Muzzle velocity: 856 m/s

WEIGHTS
Unloaded: 13.70 kg

LENGTHS
Overall: 940 mm
Barrel: 559 mm
Behind rear feed: 114 mm

MECHANICAL FEATURES
Rate of fire: 570 rds/min
Time to full rate: 0.15 s
Power required: 0.3 hp (0.22 kW)

Manufacturer
McDonnell Douglas Helicopter Co, Mesa, AZ 85205-9797.
Status
In production in UK.
Service
UK and trials elsewhere.

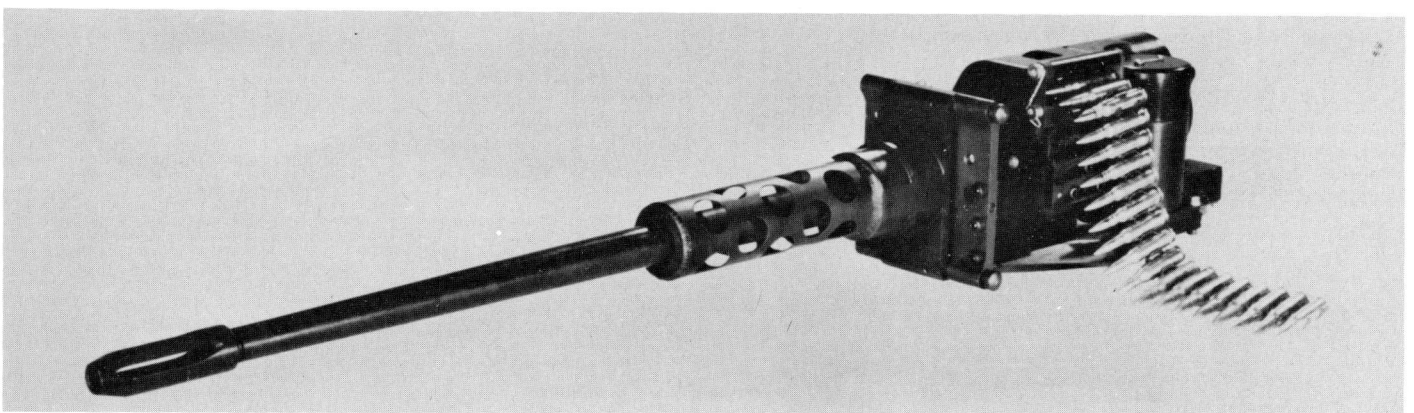

7.62 mm EX-34 Chain Gun automatic weapon

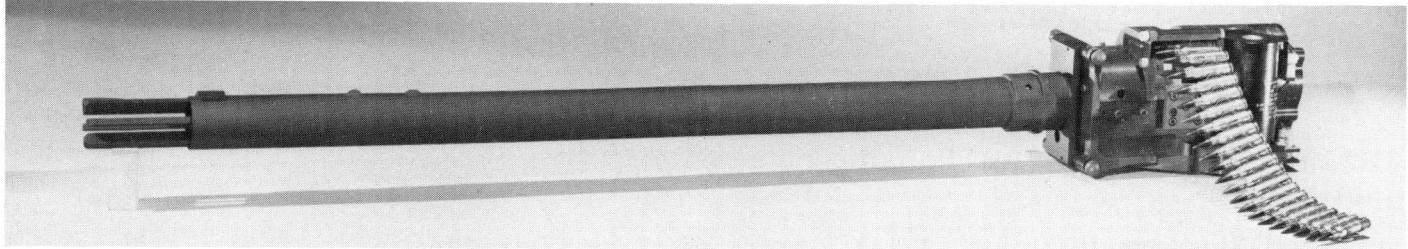

Coaxial version of 7.62 mm EX-34 Chain Gun automatic weapon

VIETNAM

7.62 mm TUL-1 light machine gun

This light machine gun looks very much like the Soviet RPK and carries a 75-round drum magazine below the receiver in the same way as the RPK. It is in fact a weapon produced within Vietnam and is based on the Chinese Type 56 assault rifle which itself is built with the characteristics of the Soviet AK-47.

The TUL-1 therefore has the receiver of the AK-47 and the same trigger mechanism though without the integral firing safety which is a feature of the AKM and RPK designs.

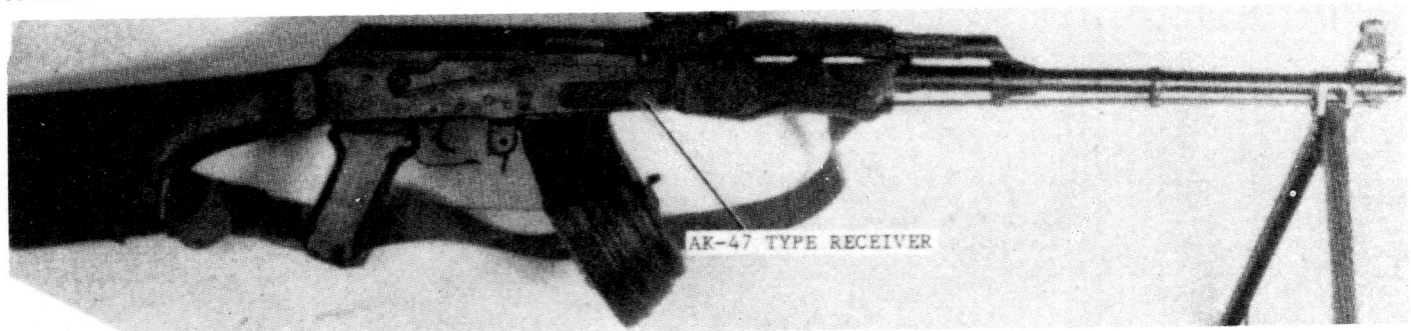

Vietnamese 7.62 mm TUL-1 light machine gun

YUGOSLAVIA

12.7 mm machine gun NSV-12,7

This is the Soviet NSV heavy machine gun (see p353) manufactured under licence for service with the Yugoslavian armed forces. It is precisely the same as the Soviet weapon, to which reference should be made for dimensions and photograph.

Manufacturer
Zavodi Crvena Zastava, Kragujevac.
Status
Current. Production.
Service
Yugoslavian armed forces.

7.62 mm PKT tank machine gun

This is the Soviet-designed PKT (see p328) manufactured under licence in Yugoslavia for supply to the Yugoslav armed forces. It is the only model of the PK family to be made in Yugoslavia and is entirely for use in tanks and light armoured vehicles; it is not fitted with sights, being directed on to the target by the tank fire control system. The PKT uses the rimmed 7.62 × 54R Soviet cartridge and fires in the automatic mode only. Feed is by a 250-round belt, and firing is performed by an electric solenoid. The barrel is interchangeable, and is recommended to be changed after firing 500 rounds.

DATA
Cartridge: 7.62 × 54R
Operation: gas, automatic
Method of locking: rotating bolt
Weight: 10.5kg
Weight of barrel: 3.23 kg

Length overall: 1098 mm
Barrel: (with flash hider): 793 mm
Rifling: 4 grooves, rh, 1 turn in 240 mm
Rate of fire cyclic: 700-800 rds/min
Rate of fire practical: 250 rds/min
Effective range: 1000m
Muzzle velocity: 855 m/s

Manufacturer
Zavodi Crvena Zastava, Kragujevac.
Status
Current. Production.
Service
Yugoslavian armed forces.

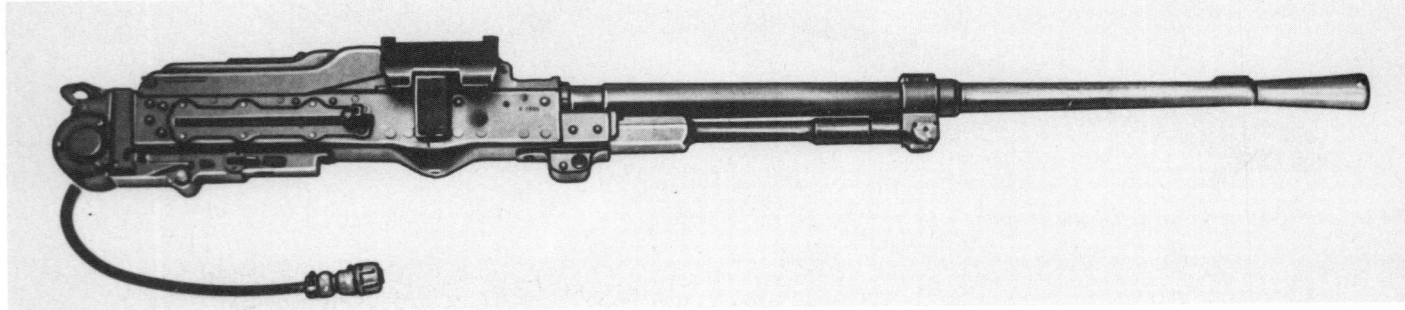

7.62 mm PKT machine gun

7.62 mm M72B1 and M72AB1 light machine guns

These two light machine guns, together with the M70B1 and M70AB2 assault rifles, belong to the Yugoslav FAZ family of automatic weapons. The M72B1 and M72AB1 are the most recent developments of the M72 light machine gun (see *Jane's Infantry Weapons 1981 – 82*, pages 296–297). They differ only in that the M72B1 has a fixed wooden stock whereas the M72AB1 has an easily-detachable folding metal stock and is intended to be used by armoured or parachute units. All the weapons of this family have interchangeable magazines, are all gas-operated and have many parts in common. The main

points of difference lie in the fins along the short stretch of the barrel under the gas cylinder and in the shape of the butt. The fins probably help slightly in cooling the hottest part of the bore and the shape of the butt is more a matter of individual preference which is not likely to have a great effect on the performance of the weapon. In fact it is possible to deduce the national attitudes to this light machine gun through the butt shape. The Soviet version has a layout which allows the gunner to hold the underside of the stock with his left hand, the traditional grip of a machine-gunner in the Soviet Army.

7.62 mm M72B1 light machine gun with bipod open showing finned barrel. Compare butt with Soviet RPK

7.62 mm M72AB1 light machine gun with metal stock folded and bipod open

One could say from this that the RPK is looked on as a definite machine gun. The Yugoslavs use a rifle butt, which cannot possibly be held in the same way, so the left hand has to go forward to the fore-hand guard, just like a rifle. It could be taken from this that the Yugoslavs treat the RPK as a machine rifle and if this is true then they are correct to do so, for the light barrel and awkward magazine do not lend the gun to being used as a proper light machine gun. It is interesting that the Yugoslavs do not have a larger magazine than the 30-round box, which is something that tends to support this view.

In the Yugoslav Army the light machine gun is carried in the infantry squad and provides extra fire-power. Support machine gun fire comes from the M53. The light machine gun can fire either single shots or automatically, the change lever being on the right of the weapon above the trigger-guard and incorporating a safety position. Structurally, although there are many points of difference, the M72B1 and M72AB1 resemble the Kalashnikov RPK (which see) and have a similar performance.

Tactically the weapon is designed for use against ground targets up to 800 m and air targets up to 500 m. The rearsight is graduated from 100 to 1000 m in 100 m steps and incorporates an engraved windage scale with two mil divisions. Both sights are marked with tritium spots for firing in conditions of poor visibility.

The M72B1 may now be equipped with a drum magazine. This fits into the normal magazine housing and holds 75 rounds. It weighs 2.175 kg when loaded.

When fitted with a bracket to carry the PN5 × 80 night-vision sight, the M72B1 machine gun takes the designation M72B1 N-PN.

DATA
Cartridge: 7.62 mm × 39
Operation: gas, selective fire
Method of locking: rotating bolt
Feed: 30-round curved box magazine or 75-round drum

WEIGHTS
M72B1: 5 kg
Empty magazine: 0.36 kg
Full magazine: 0.87 kg

LENGTHS
M72B1: 1025 mm
Barrel: 540 mm

MECHANICAL FEATURES
Sights: (foresight) cylindrical post; (rearsight) leaf, notch, 100–1000 × 100 m; (night sight) tritium marks; (sight radius) 525 mm

FIRING CHARACTERISTICS
Muzzle velocity: 745 m/s
Rate of fire: (cyclic) 600 rds/min
Effective range: 800 m; (anti-aircraft) 500 m

Manufacturer
Zavodi Crvena Zastava, Kragujevac.
Status
Current. Production.
Service
Yugoslav forces.

7.62 mm M77B1 light machine gun

In production for export, this light machine gun, like the M77B1 assault rifle (which see), is based on the 7.62 mm × 51 NATO cartridge. Otherwise it closely resembles the M72B1 light machine gun, although it appears to be supplied only with a 20-round detachable straight box magazine.

7.62 mm M77B1 NATO light machine gun

DATA
Cartridge: 7.62 mm × 51 NATO
Operation: gas, selective fire
Method of locking: rotating bolt
Feed: 20-round detachable straight box magazine
Weight: (gun with empty magazine) 5.1 kg; (magazine, empty) 0.25 kg
Length: (gun) 1025 mm; (barrel) 535 mm
Rifling: 6 grooves, rh, 1 turn in 240 mm
Sights: (foresight) pillar; (rearsight) tangent notch; (sight radius) 525 mm

Muzzle velocity: 840 m/s
Rate of fire: (cyclic) 600 rds/min
Effective range: 800 m

Manufacturer
Zavodi Crvena Zastava, Kragujevac.
Status
Current. In production for export.

5.56 mm M82 and M82A light machine guns

These machine guns have been developed as part of a 5.56 mm 'family', the other members of which are the M80 and M80A rifles. The design is obviously derived from the Kalashnikov-inspired M72 gun and is gas-operated. The manufacturers claim that a specially designed gas regulator ensures faultless operation even though ammunition of varying energy levels may be loaded into the magazine.

DATA
Cartridge: 5.56 × 45 mm (M193 or SS109)
Operation: gas, selective fire
Method of locking: rotating bolt
Feed: 30-round detachable box magazine

Weight, empty: 4.0 kg
Length: 1020 mm
Barrel length: 542 mm
Rifling: 6 grooves, rh, 1 turn in 178 mm
Sight radius: 498 mm
Muzzle velocity: 1000 m/s
Effective range: 400 m

Manufacturer
Zavodi Crvena Zastava, Kragujevac.
Status
Available for export.

5.56 mm M82 machine gun

5.56 mm M82A machine gun

7.62 mm M84 general-purpose machine gun

In production for the Yugoslav armed forces, the M84 general-purpose machine gun is designed to use the Soviet 7.62 mm × 54 cartridge. It is a rotary bolt-locked, belt-fed gun and all available data suggests that it closely resembles, both in appearance and mechanically, the PKM version of the Soviet 7.62 mm PK machine gun family (which see).

The M84 fires either from a bipod, as shown, or from a tripod and is claimed to have an effective range of up to 1000 m against both ground and aerial targets. The air-cooled barrel is capable of firing 500 rounds continuously before changing. It is fed by 100- or 250-round belts for which belt boxes are issued.

DATA
Cartridge: 7.62 mm × 54
Operation: gas, automatic
Method of locking: rotating bolt
Feed: 100- or 250-round belt

WEIGHTS
Gun: 10 kg
Tripod: 5 kg
Belt: (100 rounds) 3.9 kg; (250 rounds) 9.4 kg

7.62 mm M84 machine gun on tripod, with optical sight ON-M80

7.62 mm M84 general-purpose machine gun with bipod open and belt box fitted. Compare with Soviet PKM

LENGTHS
Gun: 1175 mm
Barrel: (with flash hider) 658 mm

MECHANICAL FEATURES
Rifling: 4 grooves, rh, 1 turn in 240 mm
Sights: (foresight) cylindrical post; (rearsight) vertical leaf; (sight radius) 663 mm

FIRING CHARACTERISTICS
Muzzle velocity: 825 m/s
Rate of fire: cyclic: approx 700 rds/min; (practical) up to 250 rds/min
Effective range: 1000 m

Manufacturer
Zavodi Crvena Zastava, Kragujevac.
Status
Current. Production.

7.92 mm M53 general-purpose machine gun

Designed for use either as a light machine gun, fired from its bipod, or as a sustained fire machine gun using a tripod, the M53 general-purpose machine gun is generally similar to the West German MG42 machine gun (which see).

Operation
Like the MG42, the M53 operates on the short-recoil principle with bolt acceleration and recoil intensification by a muzzle booster. A full description will be found in the West German entry in this section.

Application
As a light machine gun the weapon is intended for use against ground targets up to 800 m with maximum effect up to 500 m. Corresponding ranges for the weapon on its tripod are 1000 and 600 m, but two or more such weapons can produce effective fire against multiple targets at ranges up to 1500 m.

The cyclic rate of fire is 800 to 1050 rds/min with a practical combat rate of 300 to 400 rds/min. Unless special cooling arrangements are made the barrel must be changed after 150 rounds of sustained fire. The gun may be used either with a 50-round drum magazine or with 50-round belts.

Tripod
The buffered tripod (illustrated here) is fitted with an automatic recoil-operated search attachment. Another attachment enables it to be used for anti-aircraft fire.

DATA
Cartridge: 7.92 mm × 57
Operation: short recoil, gas-assisted, automatic, air-cooled, quick-change barrel
Method of locking: projecting lugs
Feed: 50-round drum magazine or 50-round continuous metal-link belt

7.92 mm M53 general-purpose machine gun in firing position

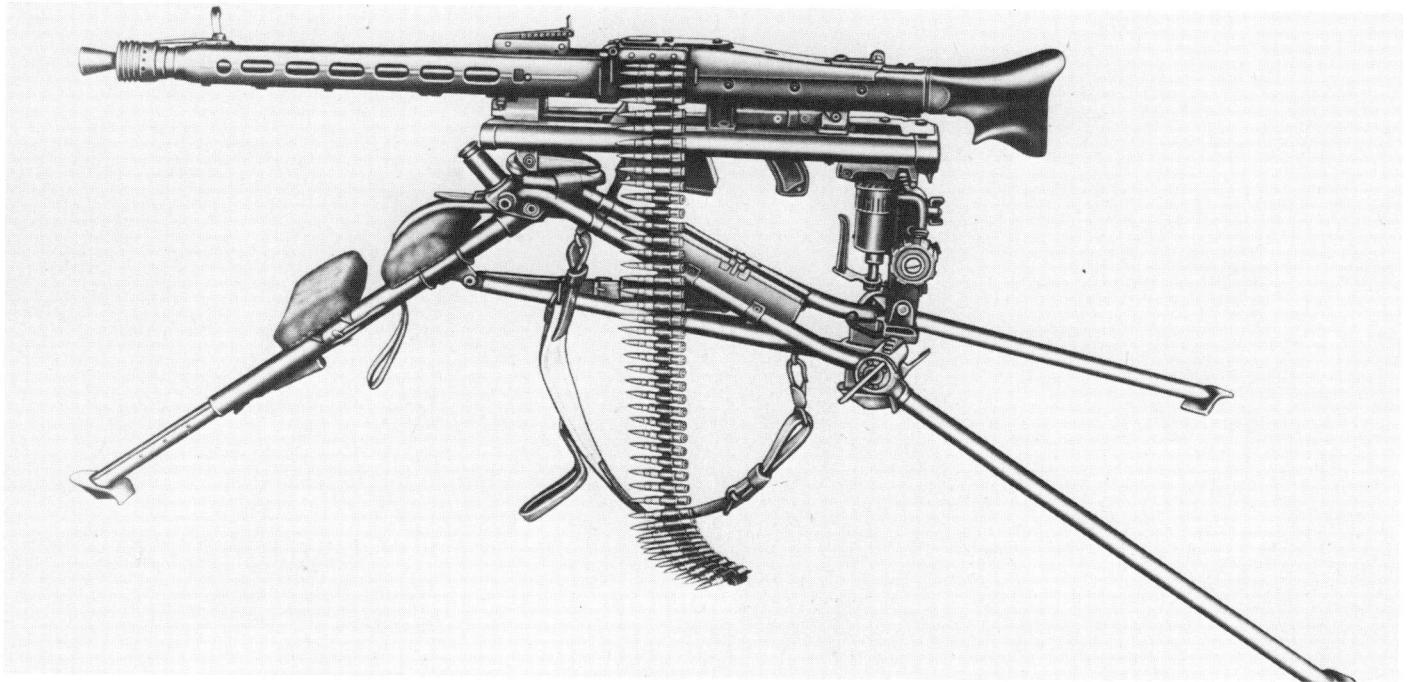

7.92 mm M53 general-purpose machine gun on buffered tripod

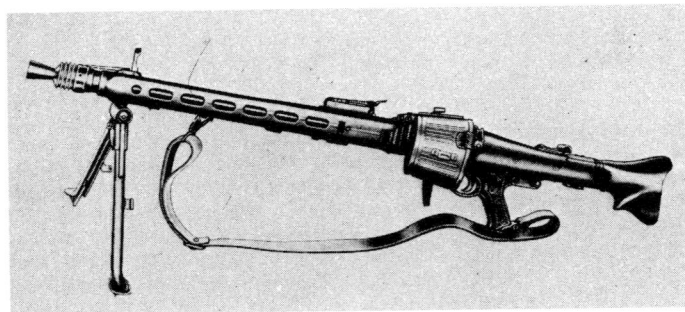

M53 in light machine gun role on bipod

WEIGHTS
Gun only: 11.5 kg
Bipod: 1 kg
Tripod: approx 22 kg
Anti-aircraft attachment: approx 1.7 kg
Band: approx 1.75 kg
Empty ammunition box: approx 2.5 kg

Full ammunition box: (300 rounds) approx 10 kg
Drum magazine: (50 rounds) approx 2.3 kg
Belt loading device with box: approx 7.5 kg

LENGTHS
Gun: approx 1210 mm
Barrel: 560 mm

MECHANICAL FEATURES
Sights: (foresight) barleycorn; (rearsight) tangent, notch, 200–2000 m; (sight radius) 430 mm

FIRING CHARACTERISTICS
Muzzle velocity: 715 m/s
Rate of fire: (cyclic) 800–1050 rds/min
Effective range: (bipod) 500–800 m (see text); (tripod) 600–1000 m (see text)

Manufacturer
Zavodi Crvena Zastava, Kragujevac.
Status
Current.
Service
Yugoslav forces.

Cannon

CZECHOSLOVAKIA

30 mm anti-aircraft cannon

This weapon is known to exist in two twin mountings, M53 and M59, which are briefly described in the Anti-aircraft Weapon section. It is a conventional recoil-operated weapon fed horizontally by 10-round clips of 30 mm ammunition and there is provision for a quick-barrel change.

Not normally an infantry weapon, the gun is found in divisional anti-aircraft regiments and anti-aircraft companies of armoured regiments of the Czechoslovak Army. Although primarily intended as an anti-aircraft weapon it can also be used, firing API ammunition, in an anti-armour role.

DATA
Cartridge: 30 × 220 mm
Operation: recoil, automatic
Feed: 10-round clips

Barrel length: (including muzzle brake) 2430 mm
Muzzle velocity: 1000 m/s
Rate of fire: (cyclic) 420-450 rds/min; (practical) 100 rds/min
Projectile weight: 0.435 kg
Round weight: approx 0.9 kg
Effective range: 2000 m
Max range: (horizontal) 9700 m; (vertical) 6300 m
Penetration: 55 mm armour at 500 m with API

Status
Current.
Service
Czechoslovak and Yugoslav forces.

FRANCE

20 mm M621 cannon

This is an electrically controlled cannon designed for use on small ground vehicles but which also has applications in any field where its two major characteristics of light weight and low recoil are at a premium. It is therefore found in helicopters, light aircraft and rivercraft.

The locking system of the gun makes use of two swinging locking pieces, one in each side of the bolt, projecting outwards into the bolt way and engaging in recesses in the body. After the round is fired these locking pieces are retracted by gas action. There is a gas vent on each side of the barrel some 650 mm from the breech face: gas from these drives back two pistons which force back the supports that secure the locking pieces: these are then driven inwards and the residual gas pressure enables the bolt to move back. When the locks are fully in, the block is blown to the rear.

The gun can fire at single-shot from a closed breech or at 300 rounds or 740 rds/min. These rates of fire are controlled electrically. In the event of a misfire there is an automatic re-cocking system which operates after a delay of 0.3 seconds to allow for a hang-fire. It is also possible to re-cock the gun by hand using a wire cable provided with a T-handle.

The gun uses mainly the US 20 mm M56 cartridge (or the M55 or M53), all of which are electrically fired and allow ready variation of the rate of fire and also permit both open- and closed breech firing. These rounds are used in the US 20 mm M39 gun and in the 20 mm M61 Vulcan gun. The US M12 link is used.

The feed systems of the gun provide its chief technical interest. The first of the different modes is the usual flat feed with the belt coming in from one side and the empty links ejected on the opposite side. The direction of feed can be reversed by the reversal of the sprocket feed and by changing a few parts. The second method is referred to as the 'enveloping feed mechanism'. It uses the same arrangements except that the axis of the sprocket drive has been raised so that the rounds enter the gun from a slightly elevated position and the links are ejected on the same side as the feed and are flung out under the incoming belt. With both methods of feed, the electrical control box enables the gunner to select single-shot fire with a closed breech operation, or automatic fire from an open breech. At automatic the length of the burst can be fixed and a record maintained of the rounds fired.

By selection of the various available mounts, the gun can be fitted to helicopters (Pod Mount 20-621; Puma Mount 19A; Gazelle Mount 22A or Ecuriel Mount 23A); trucks (Mount 15A); boats (Mount 15A) or in a number of turrets.

DATA
Cartridge: 20 × 102 mm
Operation: delayed blowback with locked breech
Weight of weapon: 45.5 kg
Weight of cradle: 12.5 kg
Length of gun: 2207 mm
Width of gun cradle: 202 mm
Height of gun in cradle: 245 mm
Muzzle velocity: 990-1026 m/s according to ammunition
Rate of fire: 740 or 300 rds/min
Voltages: (firing) 250 V; (re-cocking) 24 V; (feed) 24 V

Manufacturer
GIAT, 10 place Georges-Clémenceau, 92211 Saint-Cloud.
Status
Current. Production.
Service
French forces.

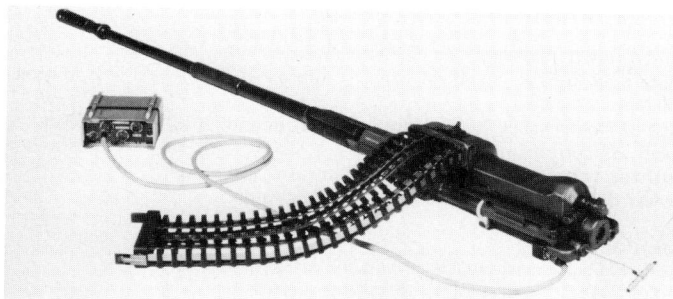

20 mm M621 gun showing feed, which can be from left or right side

20 mm 20F2 (M693) cannon

The M693 cannon embodies many features of the M621 weapon from which it is derived but differs principally in that it fires cartridges of the more powerful HS820 family instead of the electrically primed US M56 and similar cartridges fired by the M621. Because the HS820 cartridges are percussion primed an electrical power supply is not required for the firing operation; and although provision has been made in the M693 design for remote electrical operation, a feature which is obviously useful in many vehicle-borne applications of the weapon, it can be fired under local control without electricity.

M693 is the commercial designation of the weapon which was first adopted by the French Army as the Canon Mitrailleur 20 F1 and which, following changes of detail, is now known as the F2 (CN-MIT-20 F2). In French Army use it forms the main armament of some of the Toucan vehicle air defence mountings.

Operation
The principle of operation is delayed blowback with positive breech locking. Immediately before firing, with the breech-block fully forward and a round in

the chamber, the breech-block is locked in position by two swinging locking pieces (shown in the accompanying illustration) which act as struts between the block and the main body of the gun. When the round is fired, some propellant gas is diverted through two vents a short distance from the breech

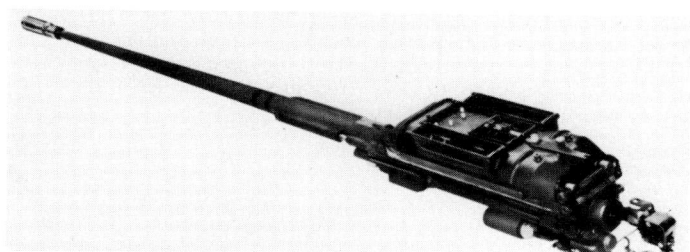

20 mm 20F2 cannon

and impinges on the faces of two pistons disposed symmetrically on either side of the chamber causing them to move rearwards (middle illustration). As they do so they rotate the breech-block locking pieces about their pivots; and when the locking pieces have been disengaged from the breech-block the residual gas pressure drives it rearwards to carry out the mechanical cycle of operations.

While the breech remains locked, the gas pressure inside the bore causes the whole barrel and body assembly to recoil, moving rearwards in its mounting cradle. When the breech opens, the gas pressure ceases to accelerate the gun body directly and accelerates the breech-block instead; and this produces a rearward force on the gun body which rises to a peak when the breech-block is at its rearmost point. Countering these rearward pressures, gases expanding through the muzzle brake after the bullet has left the barrel exert a forward pull on the barrel. The net recoil energy to which these factors contribute is absorbed by the recuperator assembly in the gun cradle: moreover the actual period of recuperation is arranged to be somewhat longer than the cycle time of the firing mechanism, so that the barrel and body are moving forwards in the counter-recoil motion when the next round is fired. With this floating firing arrangement much of the rearwards force produced by the propellant gas while the breech is locked is dissipated in extinguishing the counter-recoil momentum of the gun assembly, and the resultant pull of the gun on the mounting trunnions is very small. Furthermore the symmetrical arrangement of the moving parts about the centre-line of the barrel ensures that lateral stresses which might deflect the gunner's aim are kept to a minimum.

Feed mechanism

Two ammunition belts, one on each side, can be coupled to the feed mechanism through flexible feed chutes, between which there is sufficient room for the ejection of empty links. Two side sprocket shafts are driven independently through ratchet and pawl assemblies, and these are driven by a rack and pinion arrangement operated by the breech-block as it reciprocates. A simple change-over lever, remotely actuated through a flexible cable, engages the rack with the left- or right-hand pinion according to the gunner's ammunition requirements. The feed mechanism is well to the rear of the gun, thus reducing vehicle installation problems; the short inboard length requirement is apparent from the accompanying photographs.

In addition to the choice of ammunition type from the two belts, the gunner can choose single-shot or burst fire, by using a selector switch which also incorporates a safety position. The gun will normally be fired remotely, the trigger being solenoid-operated, and in this case the firing mode is selected by a control on a remote-control box. With the gun selector switch set for automatic fire, the control provides for single-shot limited burst or full-automatic fire and incorporates a safety position.

Sub-calibre ammunition

In addition to the standard HS820 range of ammunition there is a special sub-calibre armour-piercing (APDS) round which can be used with the 20F2 weapon. This will penetrate the French double plate target (a 1.5 mm mild steel plate 500 mm behind a 20 mm armour plate) at 1000 m and 60° incidence and comprises a 12.5 mm sub-projectile with a tungsten carbide core in a 20 mm sabot.

DATA
Cartridge: 20 × 139 mm
Operation: delayed blowback with locked breech, selective fire (see text)
Feed: dual-selectable belt
Belt traction: 2 m vertical

WEIGHTS
Gun only: 70.5 kg
Cradle: 10.5 kg
Control box: 1.25 kg
Barrel: (with muzzle brake) 25 kg

DIMENSIONS
Length: (gun with cradle) 2695 mm; (barrel with muzzle brake) 2065 mm
Width: (gun with cradle) 205 mm
Height: (gun only) 236 mm; (gun with cradle) 266 mm

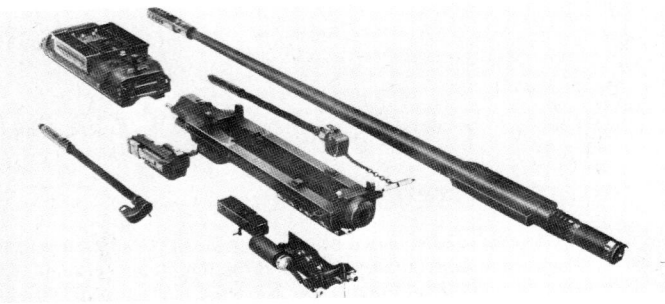

20 mm 20F2 cannon basic subassemblies

20 mm 20F2 cannon with flexible ammunition feed chutes

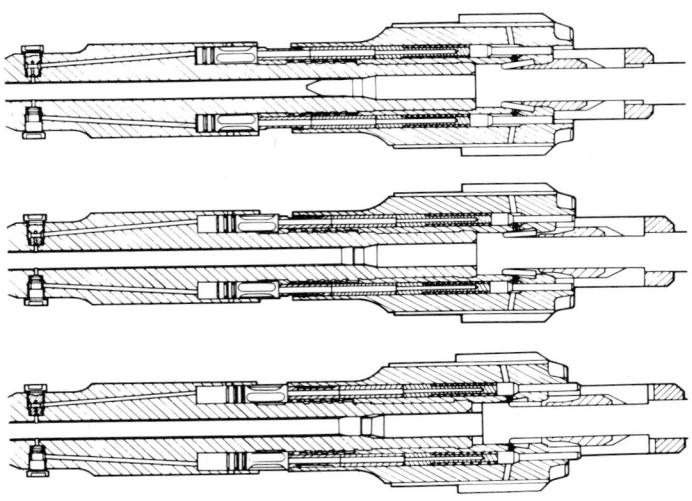

Locking arrangements of 20 mm Model F2 gun
(top) *gun ready to fire, breech locked* **(centre)** *gun fired, gas moving pistons back* **(bottom)** *breech unlocked*

FIRING CHARACTERISTICS
Muzzle velocity: (standard rounds) approx 1050 m/s; (APDS) approx 1300 m/s
Rate of fire: (cyclic) 740 rds/min, or 900 rds/min for 20F2 ACA cannon
Recoil distance: approx 40 mm
Counter-recoil run-out: approx 20 mm after last round
Power drain: 24V DC, 6-7 A for remote operation

Manufacturer
GIAT, 10 place Georges-Clémenceau, 92211 Saint-Cloud.
Status
Current. Production.
Service
French forces.

25 mm 25M811 cannon

This is a mechanical cannon with single or dual feed, intended for mounting in armoured or soft vehicles and on air defence mountings. It was first shown at the Satory Exhibition in 1983.

Operation is by an electric motor which drives a camshaft lying at the side of the receiver. This shaft has a spiral cam groove which engages with a lug on the bolt, so that as the shaft revolves, so the bolt is moved back and forth. The shaft is also geared to the feed mechanism, so that feed is in strict synchrony with the bolt's movements; this is of importance since the design allows for various rates of fire as well as for single shots.

25M811 cannon

DATA
Cartridge: 25 × 137 mm
Operation: Mechanical, selective fire
Length: (with cradle) 2680 mm
Weight: (with cradle) ca 120 kg including 8 kg of electronic controls
Feed: dual belt
Rates of fire: 150, 400 or 650 rds/min

Manufacturer
GIAT, 10 place Georges-Clémenceau, 92211 Saint-Cloud.
Status
Current. Production.

30 mm Model 781 cannon

This 30 mm cannon uses a similar system of operation to the 25 mm 25M811 (above), an electrically-driven camshaft which drives the bolt back and forth. It has been designed principally with helicopter applications in mind, but it could well find employment in a variety of ground and vehicle roles.

The Model 781 uses single feed and has a rate of fire which is variable up to a maximum of 750 rds/min. Single shots can also be fired, and the length of burst can be precisely limited.

The mechanical design makes it possible to actuate the gun for training or maintenance without the need to fire any ammunition.

DATA
Cartridge: 30 × 150B
Operation: mechanical, selective fire
Length: 1870 mm
Weight: 65 kg including 10 kg of fire control electronics
Recoil thrust: 600 daN
Rate of fire: variable to 750 rds/min

Manufacturer
GIAT, 10 place Georges-Clémenceau, 92211 Saint-Cloud.
Status
Prototype.

30 mm Model 781 cannon

GERMANY, WEST

MK 30 mm × 173 Mauser Model F cannon

The Mauser MK 30 mm × 173 Model F cannon was developed to meet the emerging threats for light air defence and combat vehicle armament systems, and is suitable for all applications.

The MK 30 is a fully gas-operated gun in which not only the movement of the bolt but also the ammunition feed, which is absolutely independent of the bolt, is operated by means of a gas piston. In the loaded condition, or during interruption of fire, there is no cartridge in the chamber, therefore the danger of cook-off in a hot barrel has been eliminated.

All movable parts, such as return springs, gas piston and buffer, are installed along the gun axis to eliminate transverse loads that adversely affect gun motion. This guarantees excellent dispersion. The breech system consists of dual locking flaps with a long rigid lock.

The receiver assembly is forged and its special configuration ensures reliability in most adverse environmental conditions. The gun itself is installed in a very simple lightweight cradle in which it can recoil.

Mechanical simplicity was the basic design goal for the MK 30 and this ensures easy maintainability and the possibility of dismantling and re-assembling the gun without special tools.

DATA
Cartridge: 30 × 173 mm
Operation: gas, automatic
Breech locking: twin locking flaps; modified Friberg-Kjellman system
Feed: single belt, dual belt, dual linkless

WEIGHTS
Gun: (complete) 146.5 kg (single belt): 155 kg (dual belt)
Barrel: 61 kg
Feeder: 18.5 kg (single belt): 29.5 kg (dual belt)

LENGTHS
Gun: 3.35 m with muzzle brake

FIRING CHARACTERISTICS
Muzzle velocity: HEI/SD-T 1025 m/s: APDS 1195 m/s
Rate of fire: 800 rds/min
Recoil force: 18 kN (peak): 16 kN (during bursts)

30 mm Mauser Model F cannon with single feeder

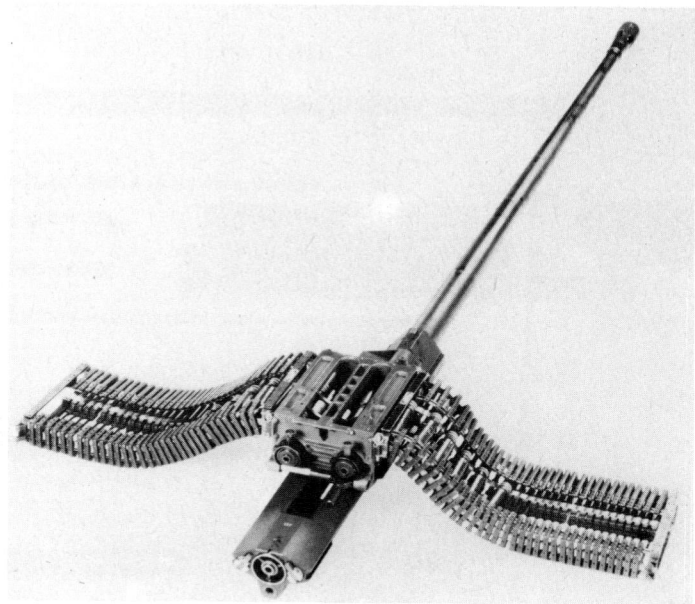

30mm Mauser Model F cannon with dual feeder

Manufacturer
Mauser-Werke Oberndorf GmbH, Geschäftsbereich Waffensysteme, 7238 Oberndorf-Neckar.
Status
Current. Production.

Service
Wildcat twin SPAA gun prototype; Artemis twin field mount (Greece); Breda twin field and naval mounts (Italy); Defence Equipment & Systems Naval Mount DS30F; Arrow ADS field mount; and others.

MK 25 mm × 137 Mauser Model E cannon

The Mauser MK 25 mm × 137 Model E has the same design and operating principles as the Mauser Model F (above).

The Model E was developed to replace the 20 mm cannon in the West German Marder Infantry Fighting Vehicle, and it is suitable to all existing mounts of 20 × 139 mm guns after small modifications have been made.

The Model E fires the standard 25 × 137 mm ammunition and uses a dual belt feeder. The gun can be stripped down into its five main components (barrel, receiver, feeder, recoil system and bolt) without requiring any special tools.

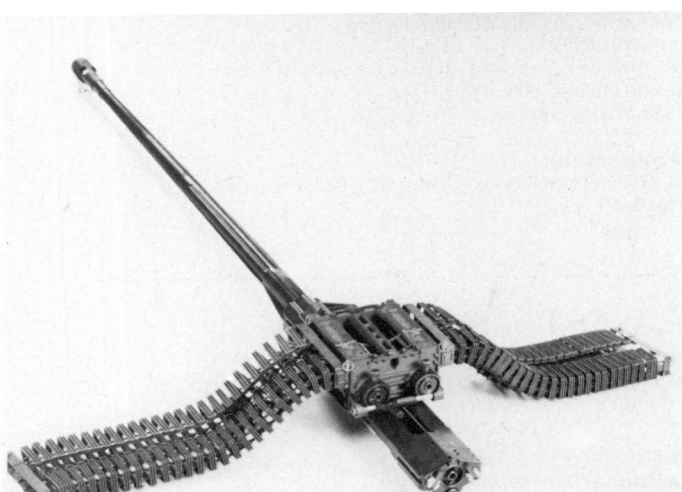

DATA
Cartridge: 25 × 137 mm
Operation: gas, automatic
Breech locking: twin locking flaps; modified Friberg-Kjellman system
Feed: dual-selectable disintegrating link belts

WEIGHTS
Gun, complete: 109 kg
Barrel: 38 kg
Dual feed mechanism: 25 kg

LENGTHS
Gun: 2850 mm
Barrel: 2100 m (without muzzle brake)

FIRING CHARACTERISTICS
Muzzle velocity: 1100 m/s
Rate of fire: 900-1000 rds/min
Recoil force: 9 kN (peak)

Manufacturer
Mauser-Werke Oberndorf GmbH, Geschäftsbereich Waffensysteme, 7238 Oberndorf-Neckar.
Status
Ready for series production.
Service
For installation in Marder IFV Prototype, Krauss-Maffei PUMA ACV, MaK Wiesel (West Germany); Hagglund PBV302 (Sweden); Kongsberg towed field AA gun systems (Norway); and others.

25 mm Mauser Model E cannon

20 mm MK20 Rh 202 automatic cannon

The MK20 Rh 202 has been designed and developed to a broad spectrum of applications. It has a high rate of fire yet low recoil forces so that it can be adapted to numerous mounts which up to now have not been suitable for a 20 mm high performance weapon.

The gun is gas-operated and has a rigid bolt: a noteworthy feature is that the bolt is locked symmetrically by two locking pieces so that the forces are absorbed centrally. Recoil forces are reduced by the use of a muzzle brake and by firing each round before the recoil travel from the previous round is complete. The ammunition feed is gas-operated and does not depend on the movement of the belt and weapon.

The Rheinmetall designed MK20 Rh 202 uses the NATO standard 20 x 139 mm cartridge with disintegrating belt. Two different belt feed mechanisms are available. In one (Type 2) two standard cartridge belts can be introduced in parallel simultaneously from above, and the operational belt can be selected by a simple lever control. In the other (Type 3) a single standard cartridge belt is used but can be introduced from the left, from the

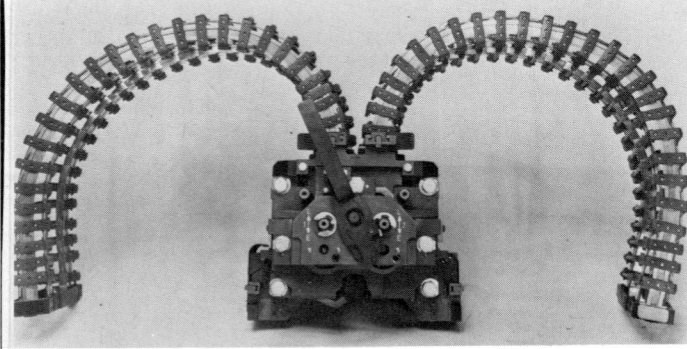

Type 2 belt feeder

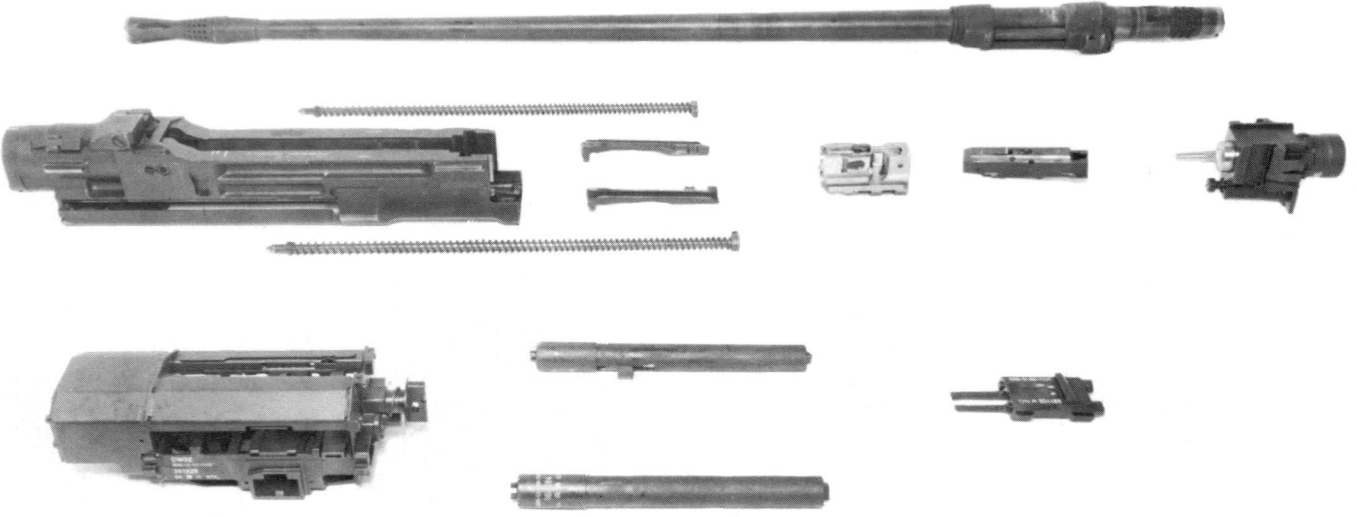

20 mm MK20 Rh 202 automatic cannon stripped to its component parts

Type 3 belt feeder

right or from above without the necessity of making any mechanical changes.

The gun has been designed to operate satisfactorily under the most arduous physical conditions including temperatures below -54°C and exposure to water and heavy contamination, and can be dismantled and re-assembled without the use of any tools.

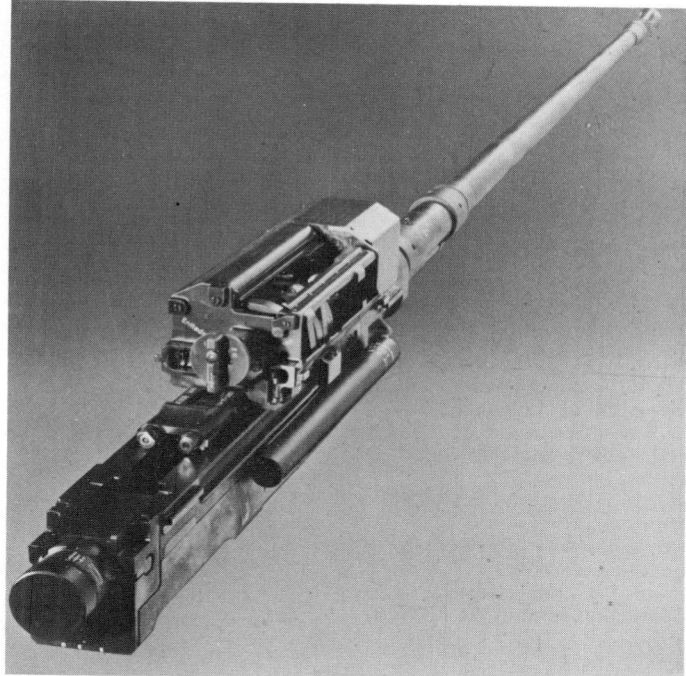

20 mm MK20 Rh 202 automatic cannon with Type 3 belt feeder

DATA
Calibre: 20 × 139 mm NATO
Weight: approx 75 kg
Length: 2612 mm

FEEDER
Type 2: 2 belts attached parallel, selector, gas-operated
Type 3: feeding of one belt only from left, right or top (depending on mounting characteristics), gas-operated

FIRING CHARACTERISTICS
Muzzle velocities: (HEI-T, TP-T and Disintegrating Projectile) 1050 m/s; (API-T) 1100 m/s; (APDS-T) 1250 m/s

Rate of fire: 1000 rds/min
Recoil force: 550-750 kgf
Tactical range: 2000 m

Manufacturer
Rheinmetall GmbH, Ulmenstrasse 125, 4000 Düsseldorf 30.
Status
Current. Production.
Service
Armed forces of Argentina, West Germany, Greece, Indonesia, Italy, Nigeria, Norway, Pakistan, Portugal, Saudi Arabia, Spain, Thailand and others.

GREECE

Component manufacture

The Hellenic Arms Industry (E.B.O.) SA is currently manufacturing cannon barrels, using an advanced cold-forging technique which can be applied to any calibre between 20 and 40 mm.

The company is currently manufacturing barrels and bolts for the 20 mm Rheinmetall cannon and the 30 mm Mauser Model F cannon as well as spare parts for its range of small arms. For information on these weapons reference should be made to the relevant entries under West Germany.

E.B.O. has recently established a separate and independent tool-making workshop within the Aeghion Plant. This unit is equipped with special

machining equipment of high producttion fidelity for precision tool manufacturing. The plant has the capacity to undertake specialist tooling and component manufacture for the Greek and foreign markets.

Manufacturer
Hellenic Arms Industry (E.B.O.) SA, 160 Kifissias Avenue, 11525 Athens.
Status
Current. Production.

SOUTH AFRICA

20 mm GA1 cannon

The GA1 is an improved and updated version of the GermaMG151/20 cannon, which orignated as a 15 mm weapon in 1936 and was later reworked into 20 mm calibre during World War Two.

The GA1 operates on short recoil, the barrel and bolt moving back a few millimetres in the receiver upon firing. The bolt is a two-piece assembly, the front section being free to revolve and being furnished with lugs to lock into the chamber. On the front section of the bolt are two rollers which run on cam tracks in the receiver; as the bolt closes, so the rollers meet the cam tracks and are constrained to turn, so rotating the bolt-head into engagement with the chamber lugs and locking the bolt. After firing, as the barrel and bolt move back, so the rollers follow the cam tracks in the opposite direction and so unlock the breech. As the bolt reaches the unlocked position, an accelerator in the receiver is struck by the recoiling barrel and transfers energy to the bolt, giving it an additional impulse to the rear.

Feed is from a belt, using a reciprocating feed pawl unit driven by the fore-and-aft movement of the bolt. The gun can be quickly adapted to accept the belt from either side.

The cartridge (20 × 83.5mm) is relatively low powered by today's standards but this gives the weapon a low recoil impulse which permits it to be

GA1 cannon and aircraft sighting system on proof mount

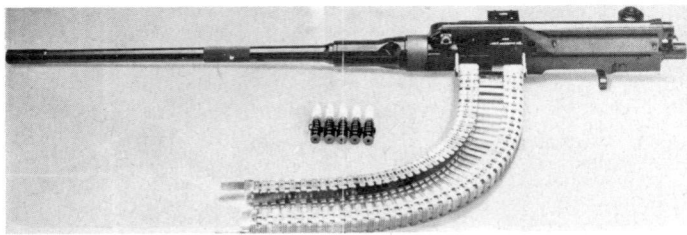

20 mm GA1 cannon arranged for left-hand feed

placed on a simple pintle mount in a light vehicle or AFV, or fired from light helicopters.

DATA
Cartridge: 20 × 83.5 mm
Operation: short recoil, automatic
Method of locking: roller-operated rotating bolt-head
Feed: disintegrating link belt
Weight, empty: 39 kg
Length: 1.760 m
Rifling: 8 grooves rh
Muzzle velocity: 720 m/s
Recoil force: 4000 N
Rate of fire: 700 rds/min
Power requirement: 22 – 29 V DC

Manufacturer
Armscor, Private Bag X337, Pretoria 0001.

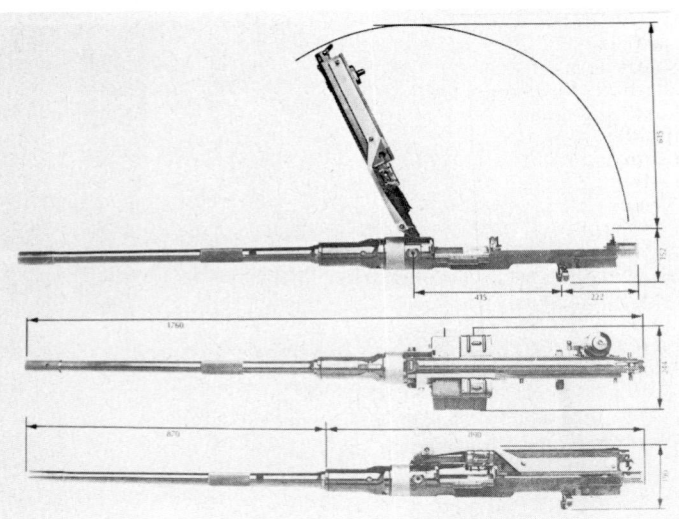

Dimensions of the 20 mm GA1 cannon

Status
Current. Production.
Service
South African armed forces.

20 mm G12 cannon

The G12 cannon is a development from the French M693, used in South Africa for some years. In comparison with other cannon it is of reduced overall dimensions and weight and is therefore well suited for infantry applications, as well as for vehicle and helicopter armament.

The G12 is a gas-assisted blowback weapon. The bolt is locked in place by two side flaps; after firing, gas pressure drives two pistons back to rotate these flaps and unlock the bolt, after which residual chamber pressure forces the bolt backwards to complete the functioning cycle. There is a dual feed mechanism, with two ammunition belts permanently in place; firing from one or other belt can be selected by a locally or remotely controlled selector.

The complete gun slides in a cradle, along the receiver guideways. The receiver is machined from a solid forging for maximum strength and stability, and the bore is nitrided to resist wear and prolong accuracy life.

An electrical fire control device is connected to the weapon by means of a cable; this allows the selection of single shots, limited bursts or sustained automatic fire.

The G12 has a symmetrical arrangement of moving parts around its axis, which is instrumental in giving the weapon a very low dispersion figure. It can be installed upright, on its side or upside down without affecting its performance.

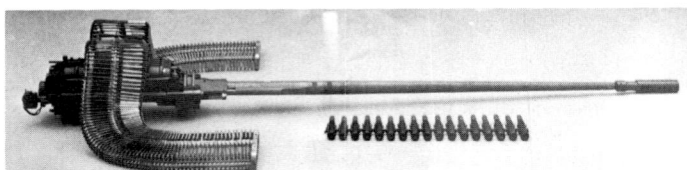

G12 cannon showing dual feed

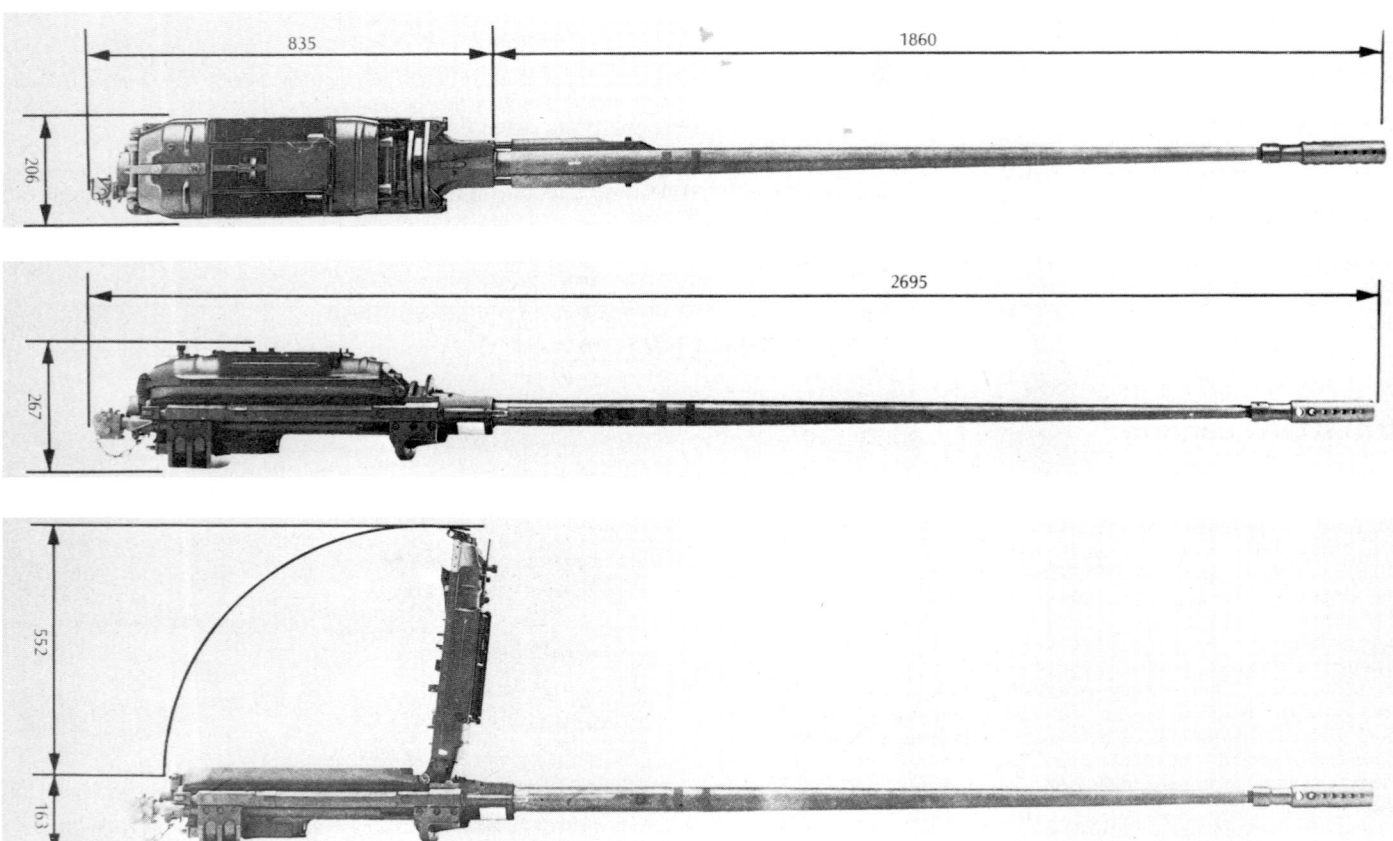

Dimensions of the 20 mm G12 cannon

DATA
Cartridge: 20 × 139 mm
Operation: gas assisted blowback
Method of locking: flaps
Feed: dual disintegrating link belts
Weight: 73.5 kg
Length: 2.695 m
Rifling: 15 grooves rh
Muzzle velocity: 1050 m/s
Recoil force: 4500 N
Rate of fire: 740 rds/min
Power requirement: 24 VDC
Belt traction: 2 m vertical

Manufacturer
Armscor, Private Bag X337, Pretoria 0001.
Status
Current. Production.
Service
South African armed forces.

20 mm G12 cannon

SWEDEN

20 mm M40 anti-aircraft cannon

Used by infantry brigade anti-aircraft companies in the Swedish Army, this light anti-aircraft weapon is the latest in a development series extending back to before the Second World War. It is mounted on a two-wheel trailer for towing but is supported on a three-part trail with its wheels raised when firing.

DATA
Calibre: 20 × 145 R
Operation: recoil, automatic
Feed: drum magazine
Weight in combat order: 500 kg
Barrel length: 70 cal

Traverse: 360°
Elevation: –5° to +85°
Muzzle velocity: 815 m/s
Rate of fire: (cyclic) 360 rds/min
Max range: 7500 m
Practical anti-aircraft range: 1600 m

Status
Current. No longer in production.
Service
Swedish forces.

SWITZERLAND

20 mm KAA (Type 204 GK) belt-fed cannon

This is a gas-operated gun firing from a locked breech.

Operation
The gun fires from the open breech position. When the trigger is pressed the sear holding the breech-block assembly is disengaged and the return spring drives the breech-block forward. The head of the breech-block forces a round through the link of the belt and into the chamber, stopping its movement when the front face of breech-block head strikes the stop insert. The breech-block tail continues forward under force from the return spring and pushes out two pivoting locking pieces which engage with locking inserts in the receiver body. The breech-block tail is prevented from rebounding by a spring-loaded detent holding it to the breech-block head. The firing pin, attached to the tail, strikes the cap and the round is fired. In automatic firing, after the first round is fired, this occurs when the receiver and barrel are still moving forward in the cradle in counter-recoil from firing the previous round. The gas pressure forces the projectile up the bore and as it passes the gas port some of the gas forces back a piston, which in turn drives the return spring housing back. A lug on the return spring housing is engaged with the breech-block tail, which is pulled to the rear withdrawing the firing pin and releasing the pivoting locks. The breech-block assembly is then blown back by the residual pressure in the bore. The chamber is fluted to reduce extraction friction. Each empty case is held by the extractor on the breech-block face and ejected down out of the gun as the breech-block passes under the fixed ejector.

The breech-block assembly moves back and is cushioned by a pneumatic/oil buffer. If the change lever is at 'single-shot' the sear retains the breech-block assembly to the rear, the sear being spring-loaded to cushion the shock of arresting the breech-block. If the change lever is at 'automatic' the breech-block, under the influence of the return spring and breech-block buffer, goes forward to fire again.

The feed mechanism is operated by the return spring housing. A lug on the housing engages in a spiral slot in the feed cylinder, rotating the cylinder in one direction during recoil and the other direction during counter-recoil. At the front end of the cylinder teeth are cut which engage in the feed slide and move the slide backwards and forwards across the gun. As the slide moves in, the spring-loaded feed pawls carried on it bring the next round in front of the breech-block. When the feed slide moves out to collect the next round the ammunition belt is prevented from dropping back by spring-loaded retaining pawls.

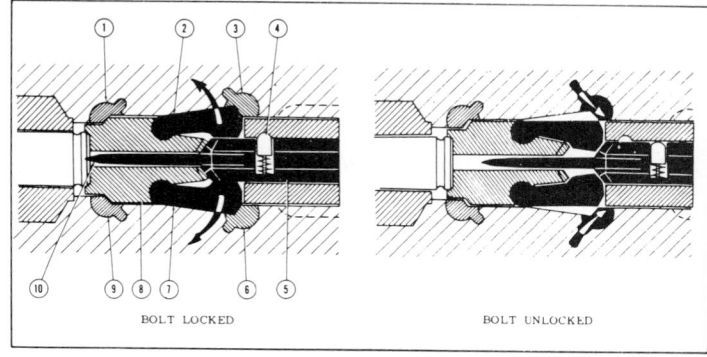

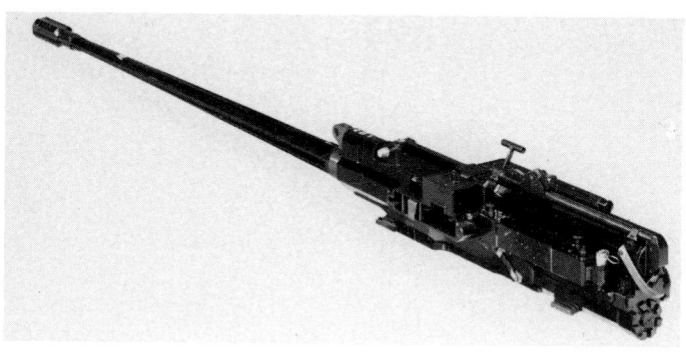

Oerlikon 20 mm KAA belt-fed cannon

Bolt action of 20 mm KAA belt-fed cannon
(1) stop insert (2) pivoting lock (3) lock insert (4) detent (5) bolt tail (6) lock insert (7) pivoting lock (8) bolt-head (9) stop insert (10) firing pin

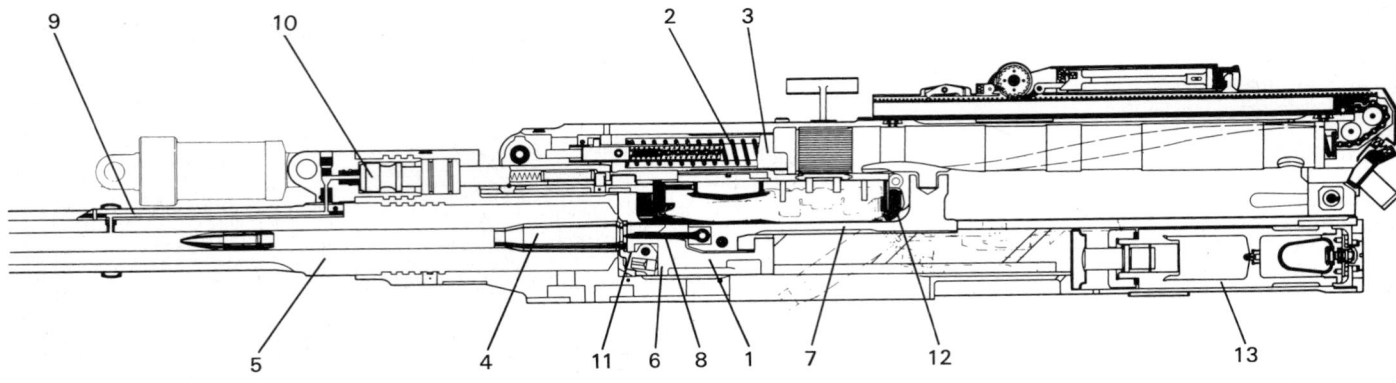

20 mm KAA belt-fed cannon
(1) *breech-block* **(2)** *charging spring* **(3)** *spring housing* **(4)** *cartridge* **(5)** *barrel* **(6)** *breech-block head* **(7)** *breech-block tail* **(8)** *firing pin* **(9)** *gas port* **(10)** *gas piston* **(11)** *extractor* **(12)** *ejector* **(13)** *breech-block buffer*

DATA
Cartridge: 20 × 128 mm
Operation: gas; positively locked
Method of locking: pivoting locks
Feed: disintegrating link belt
Fire: selective
Barrel length: 85 cal (1700 mm)

WEIGHTS
Barrel with muzzle brake: 26.6 kg
Receiver: 26.4 kg
Bolt: 4.1 kg
Belt feed plate: 4.5 kg
Cover group: 20 kg
Trigger mechanism: 5.4 kg
Complete gun: 87 kg

MECHANICAL FEATURES
Rifling: 12 grooves. Increasing twist 0°–6° 30′
Force required to cock gun: 7–10 kg
Trunnion pull: 2400 kg

FIRING CHARACTERISTICS
Muzzle velocity: 1050 m/s
Rate of fire: (cyclic) 1000 rds/min

Manufacturer
Machine Tool Works Oerlikon-Bührle Ltd, CH-8050 Zurich and Astra Defence, Grantham, Lincolnshire, England.

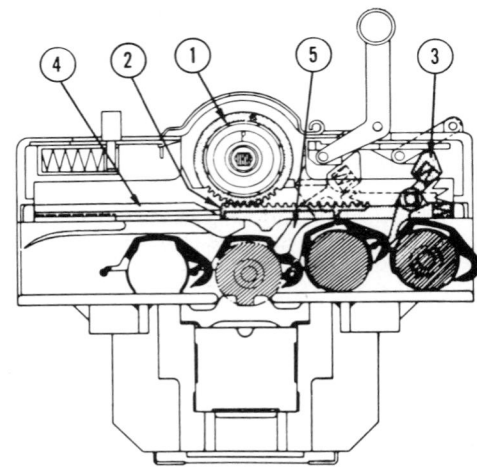

Ammunition feed of 20 mm KAA belt-fed cannon
(1) *feed roller* **(2)** *belt feed slide* **(3)** *feed pawls* **(4)** *stop* **(5)** *holding pawls*

Status
Current. Production.
Service
Various countries.

20 mm KAB-001 (5TG) automatic cannon

This is a positively locked gas-operated gun suitable for anti-aircraft or ground target applications. It is mechanically fired and uses a mechanical trigger, for either single-shot or automatic operation, with a trigger locking device which prevents the bolt from moving forward without feeding a round into the chamber. Ammunition feed is by either drum or box magazine. The barrel has Oerlikon progressive rifling and is fitted with a four-stage muzzle brake. The gun may be field-stripped without tools.

DATA
Cartridge: 20 × 128 mm
Barrel length: 2400 mm (120 cal)

WEIGHTS
Cannon: (including barrel) 109 kg
Barrel: 51.6 kg
Magazine: (20-round drum, empty) 17 kg; (50-round drum, empty) 24.5 kg; (8-round box, empty) 4.5 kg; (single round) 320-340 g according to type

MECHANICAL FEATURES
Rifling: 12 grooves rh, progressive 0–6°
Mean recoil force: 800 kg
Recoil barrel: 10 mm

FIRING CHARACTERISTICS
Muzzle velocity: 1100–1200 m/s
Rate of fire: approx 1000 rds/min

Manufacturer
Machine Tool Works Oerlikon-Bührle Ltd, CH-8050 Zurich and Astra Defence, Grantham, Lincolnshire, England.
Status
Current. Production.
Service
Various countries.

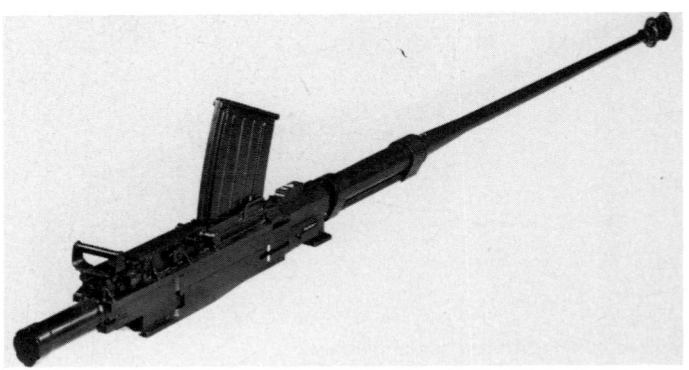

20 mm KAB-001 automatic cannon with box magazine

20 mm KAB-001 automatic cannon with drum magazine

25 mm KBA automatic cannon

Application

The NATO nominated Oerlikon 25 mm automatic cannon type KBA was developed as the close range multi-purpose weapon for the modern battle field.

Due to its firepower, various available types of ammunition and its unique instant ammunition selection device, the KBA enables a very effective engagement of all lightly armoured vehicles, infantry and anti-tank gunner positions, helicopters and combat aircraft.

Description

The Oerlikon KBA 25 mm cannon is a fully automatic, positively locked, gas-operated weapon with a rotating bolt-head and double belt feed. This feature guarantees a high reliability and safety, even under the most extreme environmental conditions.

The KBA offers a wide range of firing modes as single-shot, programmable rapid single-shot with a rate of fire of up to 200 rds/min and full automatic fire with 600 rds/min. The cannon functions such as cocking and firing are electrically actuated via remote-control from the gunner's control box and in auxiliary mode mechanically by hand crank and a trigger pedal.

Small size and low weight offer various integration possibilities such as one and two man gun turrets on APC's and AA-tanks, field, naval and helicopter mounts.

For maintenance purposes the cannon can be stripped to its main assemblies without tools.

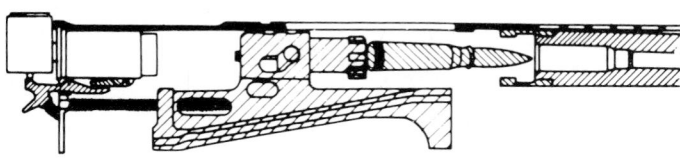

25 mm KBA automatic cannon showing sloping ramp which controls belt movement

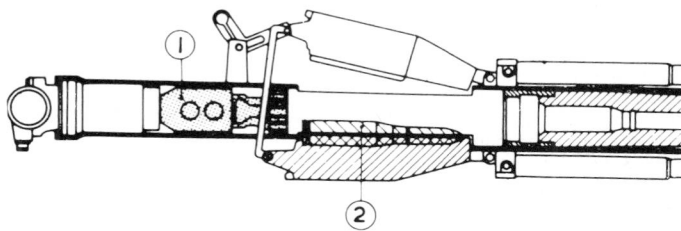

25 mm KBA automatic cannon showing how feed is switched from one belt to that on other side (**1**) bolt (**2**) round in belt

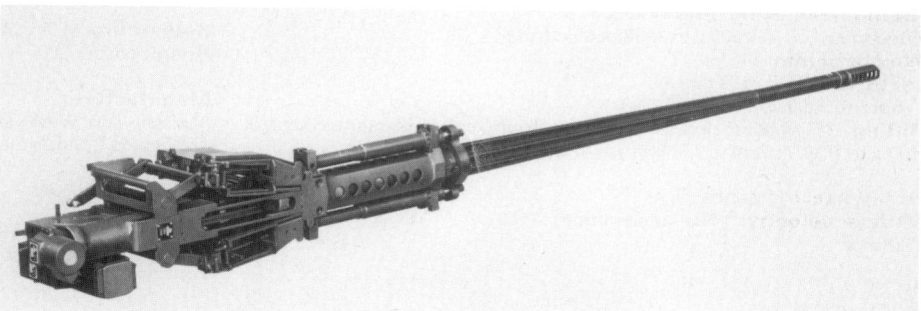

25 mm KBA automatic cannon

Oerlikon 25 mm KBA cannon installed in Oerlikon one-man GBD-COA turret on Mowag Piranha 6 × 6 APC

DATA
Cartridge: 25 × 137 mm NATO standardised
Operation: gas, selective fire
Method of locking: Rotating bolt
Feed: dual, selective belt, instantaneous change
Weight of gun: 112 kg
  barrel: 37 kg
Length overall: 2888 mm
  barrel: 2173 mm
Rifling: 18 grooves, increasing twist from zero at commencement of rifling to 7° (1 turn in 25.5 cal) at muzzle

Rate of fire: 600 rds/min
Muzzle velocity: HEI/T, SAP/HEI/T, TP/T: 1100 m/s; APDS/T: 1335 m/s

Manufacturer
Machine Tool Works Oerlikon-Bührle Ltd, CH-8050 Zurich.
Status
Current. In production for NATO and other armies. Installed in Oerlikon GBD-AOA turret on M113 C + R; in Oerlikon GBD-COA on Mowag Piranha 6 × 6, 8 × 8 and M113; in FMC EWS on AIFV; four-barrel OTO-Melara AA gun turret; OTO-Melara T25; Creusot-Loire 25 mm turret; Hagglunds 25 mm turret; Helio Mirror 25 mm turret; Oerlikon twin field mounting GBF-AOA and others.

25 mm KBB cannon

The KBB cannon is a development by Oerlikon intended for use as an anti-aircraft weapon or as vehicle armament. It is a gas-operated locked-breech weapon using two positively actuated locking flaps to secure the breech-block on firing. It has two parallel direct-feed belt mechanisms which permit two types of ammunition to be held ready and selection made instantly, without needing to fire off rounds in the feedway before the new ammunition is loaded. Changeover is made simply by swinging the appropriate feed unit into place.

The KBB is currently used in a four-barrel Oerlikon naval mounting and in the twin-gun 'Diana' air defence equipment. It has obvious applications as armament for light armoured vehicles and armoured personnel carriers.

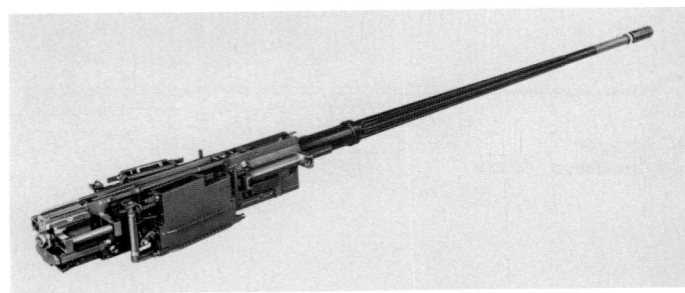

25 mm KBB cannon

DATA
Cartridge: 25 × 184 mm
Operation: gas, automatic, selective fire
Method of locking: twin locking bars
Feed: dual belt, selective, instantaneous change
Weight of gun: 146 kg
Length overall: 3190 mm
  barrel: 2300 mm (92 cal)
Rifling: 18 grooves, rh twist increasing from zero at commencement of rifling to 7°30⁸ (1 turn in 23.86 cal) at muzzle

FIRING CHARACTERISTICS
Muzzle velocity: (HE ammunition) 1160 m/s; (APDS ammunition) 1355 m/s

Rate of fire: 800 rds/min
Recoil force: 8000 N

Manufacturer
Machine Tool Works Oerlikon-Bührle Ltd, CH-8050 Zurich; Astra Defence, Springfield Road, Grantham, Lincs NG31 7JB, England.
Status
Current. Production.
Service
Not known.

30 mm KCB cannon

The 30 mm KCB cannon (formerly known as the HS831) feeds from a left- or right-hand belt feed mechanism, and fires at 600-650 rds/min. Early versions fed by means of five-round clips, but this system is no longer manufactured.

Operation
The gun fires from the open breech position. When the weapon is cocked and ready to fire, the breech-block is held to the rear by the sear engaging in the locking piece. When the trigger is operated the bolt is driven onward by the compressed return spring and the feed horn on top of the front face forces the first round out of the feedway and it is guided into the chamber. As the block reaches the end of its travel, the locking piece is forced down by guides on the inside of the receiver and the locking piece drops in front of a locking shoulder in the bottom of the gun. As soon as the locking piece moves down it frees two locking plates which are driven forward by their springs to ride over the locking piece and ensure it cannot rise. The bolt is halted. The locking piece acts as a strut and holds the breech-block firmly in place. The firing pin goes forward and the round is fired.

The whole gun recoils as the projectile goes up the bore. After a short travel the shell passes under a gas vent and the gas pressure is used to force back a tappet which pushes the locking plates back off the top of the locking piece. The angles between the locking piece and the locking shoulder are such that once this has happened the locking piece is lifted and the bolt is freed from the receiver. The residual pressure drives back the bolt and the empty case is held to the breech face until it hits the spring-loaded ejector and is thrown down out of the gun. The rear end of the breech-block strikes the buffer and is thrown forward. If the trigger is still depressed, the cycle is repeated. Release of the trigger allows the buffered sear to rise and hold the bolt to the rear.

DATA
Calibre: 30 × 170 mm

WEIGHTS
Weapon: 135.5 kg
Gun barrel: 61.5 kg
Feed mechanism: 21 kg

LENGTHS
Bore: 75 cal
Overall: 3524 mm
Gun barrel: 2555 mm
Rifling: 18 grooves, rh constant, 1 turn in 29.9 cal

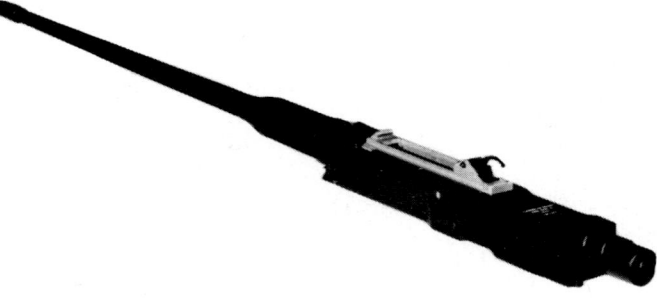

30 mm KCB cannon formerly designated HS831

30 mm KCB cannon with feeder

FIRING CHARACTERISTICS
Muzzle velocity of projectile: 1080 m/s
Rate of fire: 650 rds/min
Recoil: 50 mm

Manufacturer
Astra Defence, Grantham, Lincs NG31 7JB, England.
Status
Current. Production.

UNION OF SOVIET SOCIALIST REPUBLICS

23 mm ZU-23 cannon

The Zu-23 is one of the most frequently used cannon in the armoury of the Soviet ground forces and it has gained an enviable reputation as an anti-aircraft weapon. In general concept it dates from a gun that appeared on Soviet aircraft just after the Second World War and the first specimens to be examined in the West came from fighters shot down in the Korean War. Since then it has remained in service with what appear to be very few alterations to either gun or ammunition, but some advance in the mounting and tactical use. In Vietnam it showed itself to be a useful anti-aircraft gun when mounted in pairs and fours and its most formidable features were its rate of fire and range. It is a gas-operated gun with a vertically moving breech-block locking system which drops to unlock. The breech-block is raised and lowered in guideways in the body of the gun and its position is controlled by cams on the piston extension.

Operation

The gun stops firing with the working parts held to the rear on the sear. When the trigger is operated and the sear is lowered, the return spring drives the working parts forward. A round is rammed by the separate rammer and extractor through the link and into the chamber. As the piston extension moves forward three diagonal grooves cut across the top surface move a cross-feed slide outwards and the feed pawls are retracted. The extractor on the rammer engages the rim of the round in the chamber and the breech-block is raised and pushed slightly forward by the cams on the piston extension raising it in the inclined guideways in the gun body. The firing pin is released during the last 20 mm of breech-block movement. The piston extension holds the block in position and is prevented from bouncing. The gun fires and some of the propellant gases are diverted through the gas regulator at a point 368 mm from the commencement of rifling. The regulator is chromium plated and has two ports of 3.2 and 3.4 mm diameter. The gases enter the gas cylinder which is also chromium plated. The gas piston (chromium plated) is driven back, and after 10 mm of free piston movement the breech-block is lowered by cams on the piston extension. The block moves back as it goes down and moves the extractor back to provide primary extraction. As the piston continues back, the extractor withdraws the empty case which is forced down through guides and out of the bottom of the gun. The further movement of the piston draws the feed slide across and the one-piece spring-loaded feed pawl

Receiver and feed of 23 mm ZU-23 cannon

pulls the belt across. The piston extension is checked by the buffer and thrown forward. If the ammunition is depleted a last round lever, operated by the linked round and connected to the firing mechanism, ensures that the last round is left on the feed tray ready for ramming; so that the mechanism need not be re-cocked before a new belt is placed in the feed tray.

The gun can be fed from either side and, to effect the change, only the feed pawl, the three-tracked cam plate on the piston extension and the retaining pawls have to be changed.

DATA
Cartridge: 23 × 152B
Operation: gas, automatic
Method of locking: vertically rising block
Feed: disintegrating link belt

WEIGHTS
Gun: 75 kg
Barrel with muzzle attachment: 27.2 kg
Recoiling parts: 6.14 kg

LENGTHS
Gun overall: 2555 mm
Barrel and muzzle attachment: 2010 mm
Barrel only: 1880 mm
Receiver: 953 mm
Inboard length: (front face of feed opening to rear of weapon) 565 mm

23 mm ZU-23 cannon stripped

23 mm ZU-23 cannon

MECHANICAL FEATURES
Regulator: 2 position
Rifling: rh, 10 grooves increasing over 920 mm from 1 turn in 1150 mm to 1 turn in 575 mm. Then constant to muzzle
Chamber: chromium plated. 12 flutes
Cooling: air. Quick-change

FIRING CHARACTERISTICS
Muzzle velocity: 970 m/s
Chamber pressure: 3100 kg/cm²

Rate of fire: 800-1000 rds/min
Max range: 2500 m (horizontal); 1500 m (vertical)

Manufacturer
State factories.
Status
Current.
Service
USSR and almost all other Soviet influenced countries.

UNITED KINGDOM

30 mm (L21A1) RARDEN cannon

The concept of the RARDEN cannon dates from the early 1960s. The initial study concentrated on the defeat of armoured personnel carriers at a range of 1000 m and a deterrent effect against low and slow flying aircraft. The then current Soviet armoured personnel carrier, the BTR-50P, had 14 mm thick armour-plated sloping back at 45° from the vertical. It was considered that by the time the project came to fruition, the armour plate of comparable vehicles could have been increased to 40 mm and it was on this basis that the RARDEN calibre was decided. Examination of current weapons and their penetration led to the view that 30 mm was the minimum acceptable calibre. The principle requirements influencing the design of the weapon were accuracy, low trunnion pull, short inboard length, low weight and low toxicity. The decision was made to design a completely new 30 mm gun to meet these requirements and the design work was done at the Royal Small Arms Factory at Enfield. RARDEN was designed by Norman Brint; the name RARDEN derives from the initials of the Royal Armament Research and Development Establishment and Enfield.

Since the primary requirement was accuracy, the weapon was designed as an automatic, aimed, single-shot weapon which had a resulting automaticity of some 90 rds/min and a correspondingly low trunnion pull. No attempt was made to degrade the accuracy to achieve a higher rate of fire. To cut down the inboard length it was decided to feed from the side with clips. The requirement for low toxicity was met by arranging for forward external ejection of the empty case and by enclosing the entire gun mechanism.

The final design was a gun operated on a long-recoil system with a sliding block to reduce the length behind the breech and clip fed, rear-loaded ammunition.

Mechanism
The gun operates on the long-recoil system. When the gun fires, the breech-block, breech ring, rammer, barrel and recuperator cylinder all move back some 330 mm inside the non-recoiling casing, against the buffer, compressing the air in the recuperator.

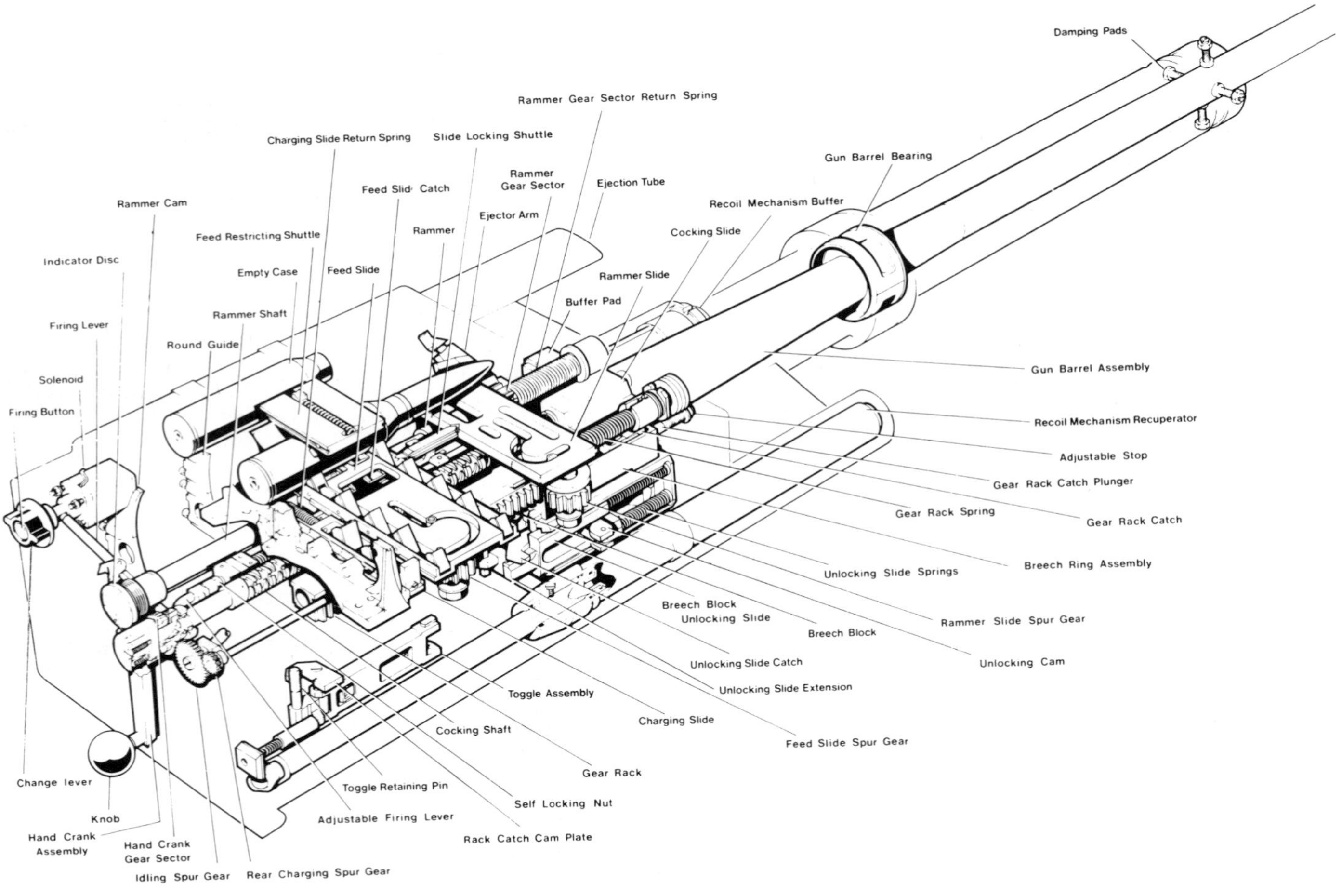

30 mm L21A1 cannon

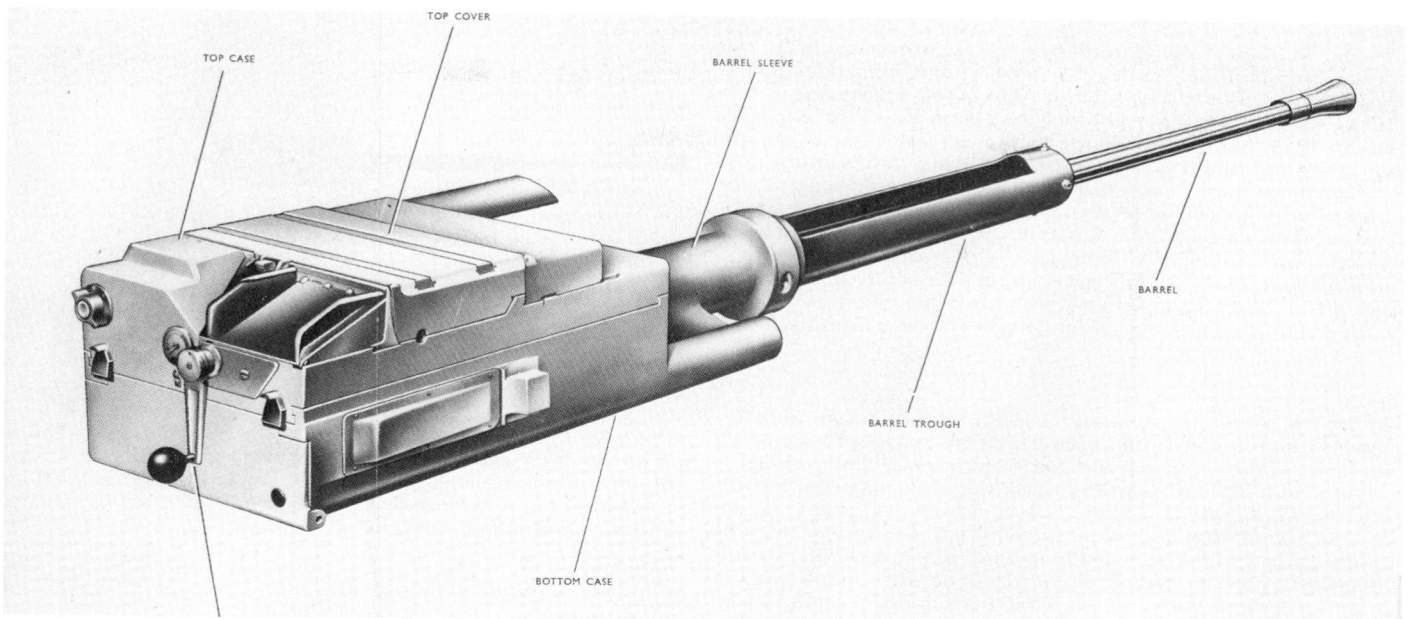

TOP CASE · TOP COVER · BARREL SLEEVE · BARREL · BARREL TROUGH · BOTTOM CASE

30 mm RARDEN cannon: handle at rear cocks action; clips are inserted from back and empty cases are ejected from chute on left. Gun casing is sealed and no fumes come into crew compartment

The breech-block remains locked for the first 229 mm of recoil and is then opened by a cam attached to the block rolling round a radial wall of twice its own diameter, in the breech ring. This produces a movement of the breech-block at right angles to the barrel axis, and at the same time the rammer is drawn down to grip the empty case.

When the recuperator drives the barrel and ring forward, the empty case is left on the rammer at the rear. The breech-block is held open. After 203 mm of run out, the empty case is clear of the breech and the rammer arm is rotated so that it carries the case upwards until it comes to rest in line with the feed slide, above the mechanism. The next round is now fed across and it pushes the empty case out of the rammer jaws and into the ejection chute, itself being gripped by the rammer.

The final forward movement of the breech ring ejects the empty case from the previous round forward out of the chute, out of the gun. The rammer is rotated to carry the live round down into line with the bore and then forward to feed the round into the chamber. The rammer releases the sliding block which pushes it into a recess in the breech ring. The gun is now ready to fire and the firing hammer is released from the sear by the gunner operating either the firing button or a solenoid.

Mechanical safety before firing is controlled by the breech-block which pushes out a safety plunger when it is fully home. This releases the safety sear and leaves the hammer held by the sear. After firing, the free recoil of 230 mm ensures that the pressure is down at atmospheric before the breech opens.

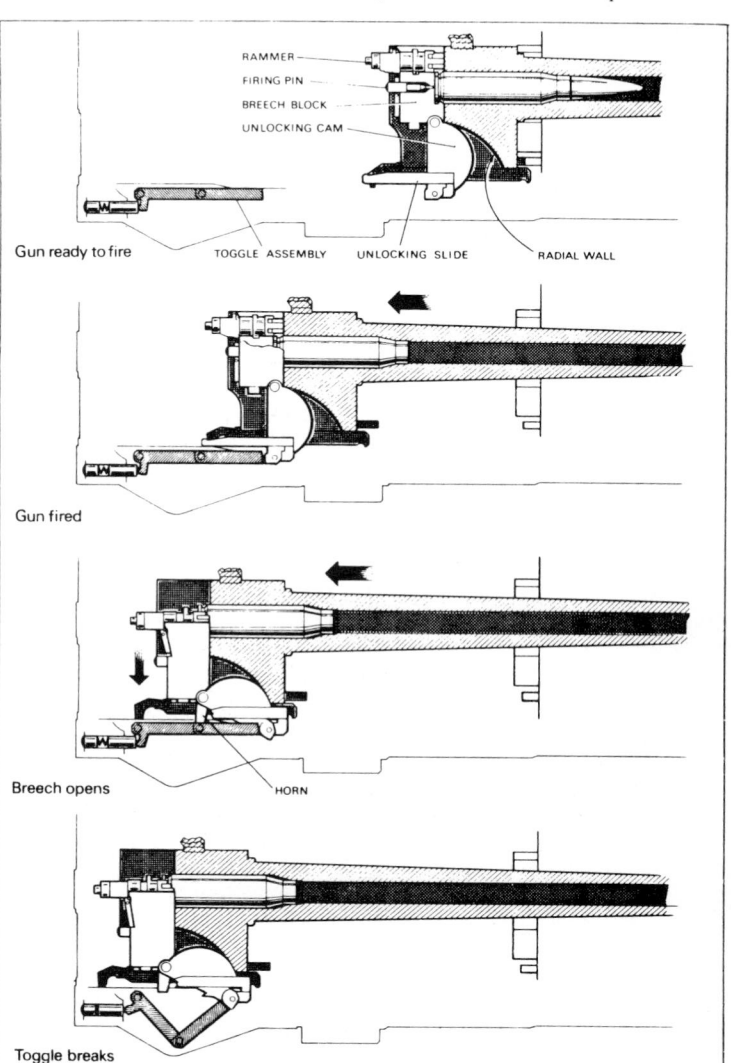

RAMMER · FIRING PIN · BREECH BLOCK · UNLOCKING CAM

Gun ready to fire — TOGGLE ASSEMBLY · UNLOCKING SLIDE · RADIAL WALL

Gun fired

Breech opens — HORN

Toggle breaks

Sequence of events to completion of recoil is shown. Note that when toggle assembly makes contact with unlocking slide, unlocking cam is rotated around radial wall which moves block across at right angle to barrel axis

EMPTY OASE IN CHAMBER · EJECTOR · CASE BEING EJECTED · EJECTION CHUTE

RAMMER · BREECH BLOCK CATCH

Commencement of run-out

RAMMER CLAW AT FEED SLIDE LEVEL

FEED SLIDE

Rammer rotated in line with feed slide

Rack operates feed slide · RACK

Run-out sequence and positioning of next round for firing

Feed

The feed mechanism is designed to take two three-round clips. The clip releases the rounds as soon as they are pushed in from the rear. For lightness the gun body is made of cast aluminium alloy in two sections. The bottom casing contains the recoiling masses and the top case houses the feed, firing and cocking mechanisms. The loading action is unusual in that the clips are pushed into the feed-way from the rear, but the rounds travel sideways to be fed. There is insufficient room for ammunition to come into the body from the side, so there is an opening in the top of the body which just accepts a three-round clip. Once the clip is in, it is taken by feed pawls across the top of the mechanism towards the left. Six rounds can be accommodated in this way. The rammer arm rotates up at the end of each cycle and collects a fresh round from this row, and takes it down into the body interior to present it to the breech. The length of time required for this activity is the main reason for the low rate of automatic fire.

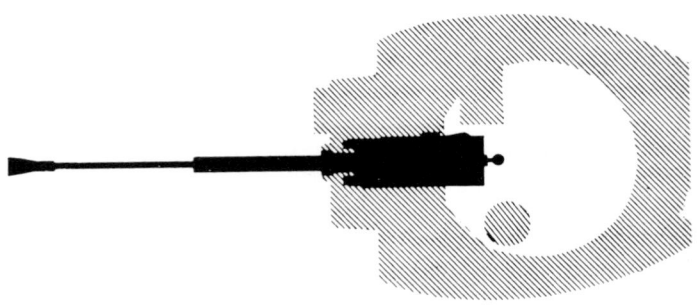

Short inboard length of gun inside Fox turret

Barrel

The barrel is of monobloc construction and it is made of high-yield steel, threaded at the breech-end for attachment to the breech ring and at the muzzle for a flash suppressor. A barrel catch engages serrations at the breech-end to prevent barrel rotation. The use of high quality steel has produced a very light barrel of only 24.5 kg with a length of 2.44 m. As with all light, long barrels, the paramount problem in obtaining accuracy and consistency is to control the vibration pattern on loading and firing. This is done in the RARDEN with a front bearing positioned about one quarter of the barrel length from the breech. This slides in the barrel sleeve. The vibrations are damped by four damping pads at the end of the barrel sleeve. Clearance between pads and the barrel is 0.076 mm.

Changing the barrel is a simple operation. The flash suppressor is screwed off, the barrel catch is lifted and the barrel screwed out. The breech-end is then pushed forward, lifted and then pulled back. The barrel will normally be removed during long, non-operational moves and stowed on the hull of the vehicle.

Service use

Applications of the gun in British Army service include the Fox armoured car, the Warrior Infantry Fighting Vehicle, the Scimitar combat reconnaissance vehicle and some of the AFV 432 armoured personnel carriers. The gun can also be mounted in a two man turret in the US M113A1. In all these vehicles the gun is compatible with day- and night-sighting systems including image intensifiers and infra-red sights.

Ammunition

For RARDEN the basic design was the HS831L series, and the HE, and Practice are little changed from the Swiss design. There is a special armour-piercing round for the specific attack of armoured personnel carriers, designed to operate against light armour only. For the attack of heavy armour an APDS round has been developed by the Royal Ordnance Factories in conjunction with the firm of PATEC in California.

DATA
Calibre: 30 × 170 mm
Operation: long recoil
Feed: 3-round clips
Trunnion pull: 1360 kg

WEIGHTS
Overall: 113 kg
Barrel: 24.5 kg; (with flash hider) approx 26.8 kg

LENGTHS
Overall: 2959 mm
Barrel: 2440 mm

Manufacturer
Royal Ordnance plc, Guns & Vehicles Division, Nottingham; Astra Defence, Grantham, Lincs NG31 7JB.
Status
Current. Production.
Service
UK forces, also Belgium, Honduras, Malawi and Nigeria.

UNITED STATES OF AMERICA

30 mm M230 Chain Gun® weapon

Developed by McDonnell Douglas Helicopter Co. (MDHC), the 30 mm M230 Chain Gun cannon is a single-barrel, externally powered weapon which incorporates a rotating bolt mechanism driven by a simple and reliable chain drive. The entire gun weighs 57.6 kg and has a firing rate of 625 ± 25 rds/min. The chain-drive principle permits a simplified gun cycle which operates safely from an open bolt without requirement for chargers, declutching feeders, or other special devices. Since the M230 is motor-driven, it provides inherent high belt pull, eliminating the need for a powered booster. Long bolt-lock time ensures an essentially gasless action and the resultant long dwell following firing provides for hangfire-safe weapons.

As all the moving parts are keyed together, every motion is precisely timed and fully controlled. Further, complete round control within the M230's ballistically independent system eliminates malfunctions due to conventional ammunition problems. These factors coupled with the smooth power flow from an electric-drive motor ensure the highest reliability at all rates of fire.

MDHC initiated development of a 30 mm Chain Gun cannon in December 1972 and after an accelerated engineering and fabricating effort, the weapon was first fired in April 1973 and burst-fired in May of that year. A 2500-round feasibility trial sponsored by the US Army was successfully completed in September 1973 and by December 1973 4000 rounds had been fired.

Subsequent models of M230s were built and tested and a quarter of a million rounds of M552/M639 ammunition were fired through these several Chain Gun weapons.

Early in 1976 the Department of Defense directed the M230 to be rechambered to fire ADEN/DEFA type ammunition for NATO interoperability. To ensure interoperability MDHC was given the responsibility for development of a family of US 30 mm rounds: M788 (TP), M789 (HEDP) and M799 (HE). The modified M230 was first fired in March 1978. While the basic M230 gun can fire either linked or linkless ammunition through the use of a lightweight transfer unit, the AH-64 attack helicopter uses a linear linkless conveyor system.

The M230 is in production for installation in the AH-64 Apache attack helicopter.

DATA
Calibre: 30 × 113 B

WEIGHTS
Receiver: 38 kg including motor
Barrel: 14.5 kg

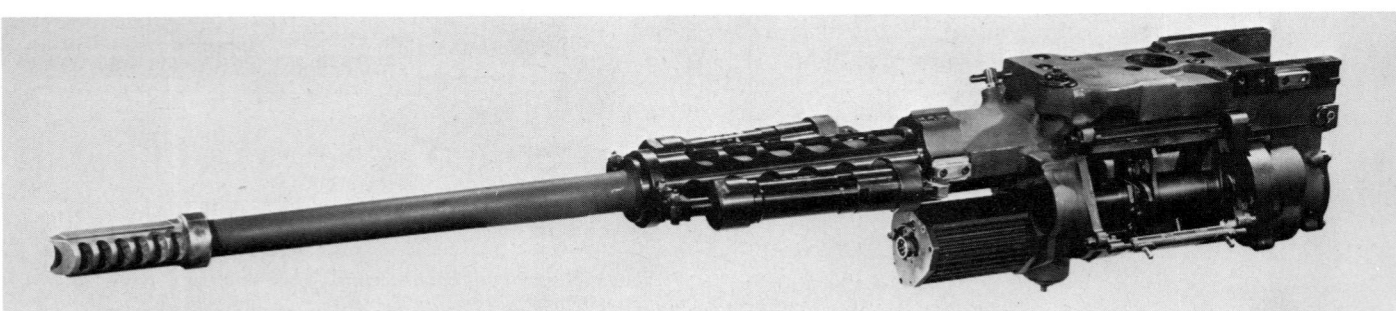

30 mm M230E1 Chain Gun® cannon

Recoil adaptors: 5.0 kg
Linkless transfer unit: 3.6 kg
Total gun weight: 57.5 kg

DIMENSIONS
Length: 1676 mm
Width: 267 mm
Height: 290 mm
Barrel life: 10 000 rounds
Rate of fire: 625 ± 25 rds/min
Time to rate: 0.2 s
Time to stop: 0.1 s
Clearing method: open bolt
Effective impulse: (with muzzle brake) 20.4 kg/s

Dispersion: < 2 mil
Power required: 3.0 hp
Reliability predicted: 15 000 minimum rounds between failures

AMMUNITION
Combat: ADEN/DEFA/M789/M799
Practice: M788

Manufacturer
McDonnell Douglas Helicopter Co., Mesa, AZ 85205-9797.
Status
Current. In production.
Service
US Army.

30 mm ASP-30® Infantry Support Weapon

The McDonnell Douglas Helicopter Co. ASP-30 (Automatic, Self-Powered) cannon has been developed as a combat support weapon which is readily interchangeable with .50 in heavy machine guns. The gun has built-in dual recoil adapters and can be used in virtually any existing vehicle mount or dismounted and placed on a standard M3 heavy machine gun tripod for use in the direct infantry support role.

The gun is gas-operated, using a rotating bolt to lock the breech securely during the firing phase. The feed is single, from the left side, and the short length of the receiver behind the feed allows simple and easily controlled spade grip operation from any cupola mounting structures

Ammunition is the standard ADEN/DEFA M789 pattern, widely used throughout the world, carried in a disintegrating link belt.

DATA
Cartridge: 30 × 113B
Operation: gas, selective fire
Method of locking: rotating bolt
Length overall: 2027 mm
Length behind feed: 292 mm
Width: 203 mm
Height: 241 mm
Barrel: 1321 mm (1473 mm with blast suppressor)
Weight: 47.62 kg
Mounting provision: identical to .50 Browning M2HB
Rate of fire: 400 – 450 rds/min, and single shots
Muzzle velocity: 838 m/s

30 mm ASP-30 cannon on M3 tripod

30 mm ASP-30 cannon mounted on M113 APC

Manufacturer
McDonnell Douglas Helicopter Co., Mesa, AZ 85205-9797. Royal Ordnance plc, Guns & Vehicles Division, Nottingham, UK.
Status
Current. Production.

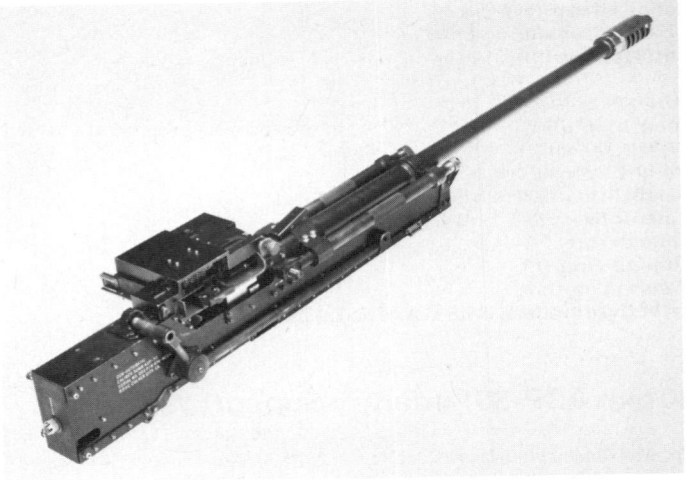

30 mm ASP-30® cannon

25 mm M242® Chain Gun weapon

The McDonnell Douglas Helicopter Company's 25 mm M242 dual feed Chain Gun weapon is in production for the US Army as the primary weapon for installation in the M2 and M3 Bradley Fighting Vehicles, on the US Marine Corps Light Armoured Vehicle (LAV-25), and in deck mountings on US Navy patrol boats. It has also been selected by the Spanish government for installation in armoured personnel carriers. McDonnell Douglas has delivered 208 guns to Spain. To date, over 6000 M242 guns have been delivered to the US military since production began in 1981 and a multi-year production contract from the US Army calls for a further 2000 guns over the next two years.

The entire gun, with integral dual feed, remote feed select and internal recoil mechanism, weighs 110 kg and provides for semi-automatic and automatic fires of 100, 200 and 500 rds/min.

The M242 effectively fires both European and US 25 mm ammunition, ensuring interchangeability and logistic commonality on the modern battlefield. Fired cases are ejected forward (overboard) and long dwell after firing eliminates gas build-up in turreted installations.

DATA
Calibre: 25 × 137 mm

WEIGHTS
Receiver assembly: 40.8 kg
Barrel assembly: 43.0 kg
Feeder assembly: 26.7 kg
Total gun system: 110.5 kg

DIMENSIONS
Length overall: 2760 mm
Width: 330 mm
Height: 380 mm
Barrel length: 2032 mm

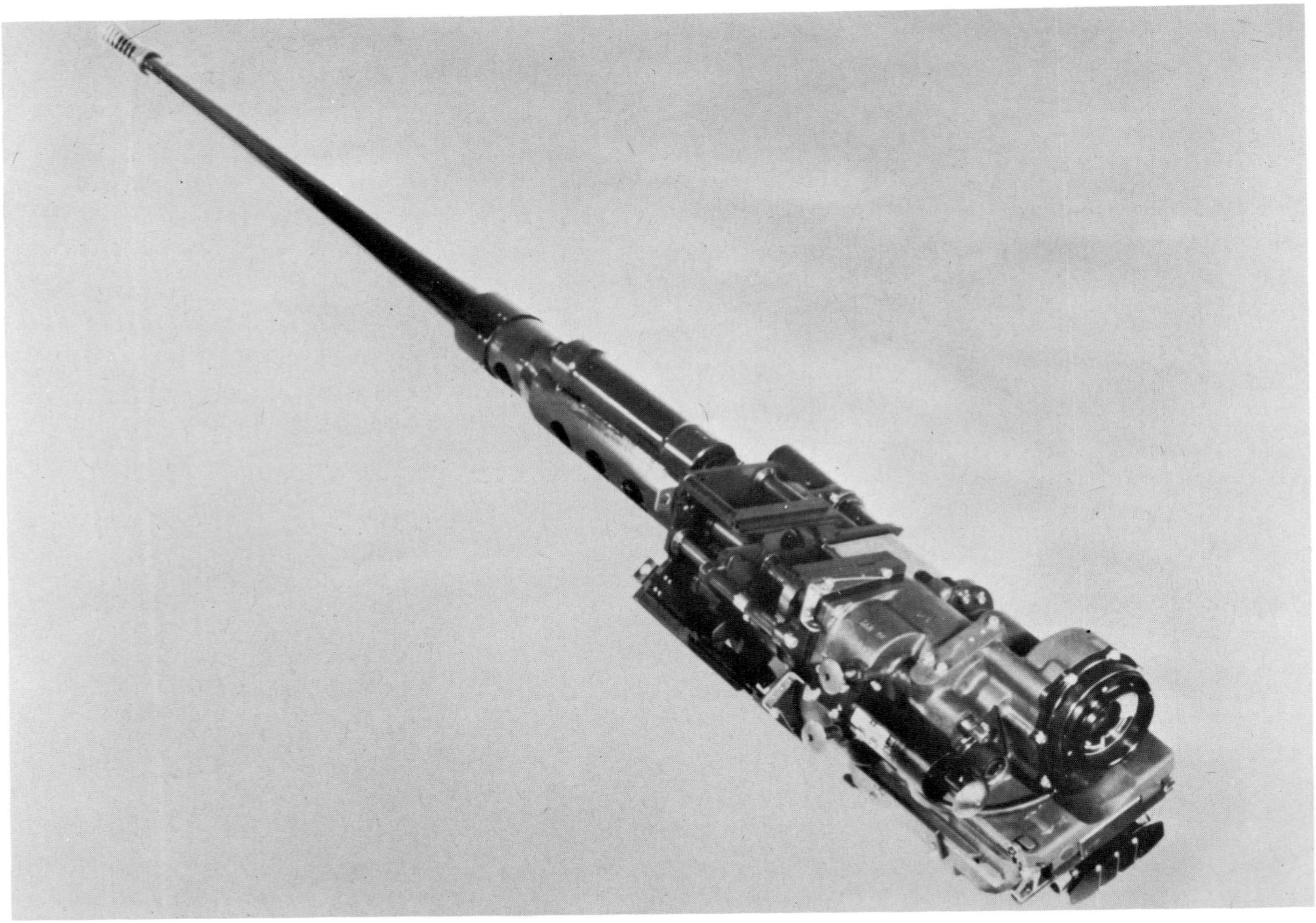

25 mm M242 Bushmaster cannon

FIRING CHARACTERISTICS
Barrel life: surpasses requirements
Rate of fire: (single-shot) 100, 200 or 500 rds/min
Time to rate: 0.15 s
Time to stop: 0.12 s
Power required: 1.5 hp for 200 rds/min; 8 hp for 500 rds/min
Clearing method: (cook off safe) open bolt
Safety: absolute hang-fire protection
Case ejection: forward
Peak recoil force: 3175 kg

Dispersion: 0.5 mil
Demonstrated reliability: 18 000 minimum rounds between stoppages

Manufacturer
McDonnell Douglas Helicopter Co.,5000 East McDowell Road, Mesa, AZ 85205-9797.
Status
Production. Fitted in US Army M2 and M3 Bradley Fighting Vehicles, in US Marine Corps Light Armoured Vehicle (LAV-25) and deck mountings on US Navy patrol boats. Spanish Army.

30 mm Bushmaster II cannon

This weapon has been developed from the M242, to the extent that some 70 per cent of the components are common to both guns. The Bushmaster II fires standard 30 × 173 mm GAU-8 ammunition, using a side-stripping link developed by McDonnell Douglas. It can also fire Rarden and Oerlikon KCB (30 × 170 mm) ammunition by changing the barrel, bolt and aft feed plate. Like the M242, the Bushmaster II has a dual feed capability.

DATA
Calibre: 30 mm (see text)
Operation: mechanical (Chain Gun)
Feed: dual, disintegrating link belt
Weight: 147.5 kg
Length: 3.499 m

Width: 335 mm
Height: 396 mm
Peak recoil: 5443 kg
Power required: to 200 rds/min 1.5 hp; at 400 rds/min 8 hp
Rate of fire: single shot, 200 or 400 rds/min
Dispersion: <0.5 mil
Muzzle velocity: HEI, API, TP 1036 m/s; APDS-T 1220 m/s

Manufacturer
McDonnell Douglas Helicopter Co.,5000 East McDowell Road, Mesa, AZ 85205-9797.
Status
Development completed.

25 mm General Electric GE225 lightweight gun

This interesting weapon has revived the operating principle of the German Gast machine gun of 1918, two side-by-side mechanisms arranged so that the recoil impulse of one side provides the motive power for loading and firing the other. The two barrels are firmly secured in a common receiver, while the bolts are inter-connected by a reciprocating lever. Thus while one bolt is extracting a fired case, the lever transfers the motion to the other bolt so that it is chambering the next round to be fired. The two barrels thus fire alternately. Power is derived from short-stroke gas pistons, though it is also possible to power the mechanism externally. Feed is simple, requiring only one shaft and sprocket to serve both gun bolts from a single belt; the belt feed can be installed to feed from either side of the weapon, and the empty cases are ejected downwards from both sides.

The gun has been designed principally with helicopter applications in mind, but it could also be applied to various vehicle and ground mounts and would appear to have advantages in the anti-aircraft role.

DATA
Cartridge: 25 × 137 mm
Operation: gas or external power
Length overall: 2209 mm
Width: 216 mm
Height: 203 mm
Weight: 81.6 kg with feeder and external drive
Rate of fire: variable to 2000 rds/min
Power requirement: 25 hp at 2000 rds/min; 3 hp at 750 rds/min (externally powered gun only)

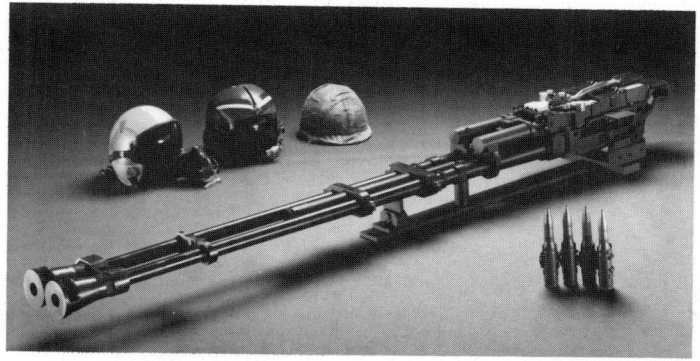
25 mm GE225 lightweight gun

Feed: disintegrating link belt

Manufacturer
General Electric Co, Armament Systems Dept, Lakeside Avenue, Burlington, VT 05401.
Status
Under development.

20 mm M168 six-barrelled gun

This is an M61A1 gun modified by the omission of the automatic clearing function and the addition of a fixed casing.Like the M61A1 it is externally powered with a cluster of six barrels but the rate of fire has been reduced: whereas the airborne weapon fires 6000 rds/min the M168 is capable of firing at 1000 or 3000 rds/min. The gun is basically a Gatling-type mechanism in which each of the six barrels fires only once during each revolution of the barrel cluster. Barrels are attached to the gun rotor by interrupted threads and no headspace adjustment is required. The muzzles are held in a muzzle clamp which allows the dispersion of shot to be spread into a flattened ellipse.

The mechanical arrangement and method of firing are similar to those of the 7.62 mm M134 Minigun described and illustrated in the section on machine guns. The gun rotor rests on bearings inside the stationary outer housing and contains the six-gun bolts. As the bolts rotate around the rotor, a cam follower on each bolt follows a stationary cam path fixed to the housing and this causes the bolt to reciprocate and carry out the functions of feeding, chambering, locking, firing, unlocking, and extraction. Since each barrel fires only once for each revolution, a maximum of 500 rds/min/barrel, there is no chance of cook-off. Any misfires are thrown out of the gun and so are hang-fires. A declutching feeder is used to permit firing of all ammunition actually in the gun at the end of a burst.

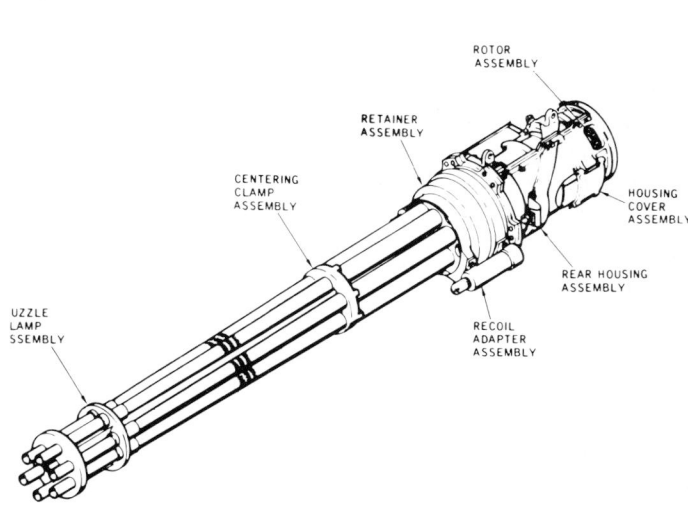
20 mm M168 cannon

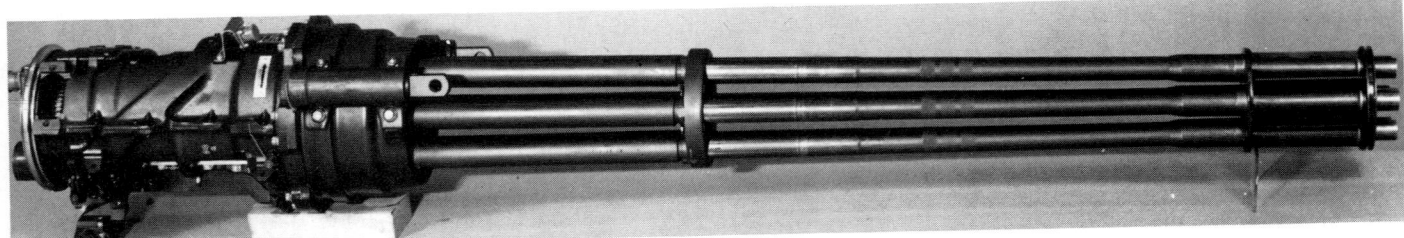

20 mm M168 six-barrelled cannon, modified M61A1 Vulcan gun

DATA
Calibre: 20 × 102 mm
Operation: externally powered. Gatling type automatic
Feed: boxed linked rounds
Weight: (gun) 136 kg
Barrel length: 1524 mm
Rate of fire: 1000 or 3000 rds/min selectable

Manufacturer
General Electric Co, Armament Systems Dept, Lakeside Avenue, Burlington, VT 05401.
Status
Current. Production.
Service
US forces, Belgium, Japan, Saudi Arabia, South Korea and others.

20 mm M197 three-barrelled gun

The M197, 20 mm gun is a three-barrel, externally powered, lightweight version of the M61A1 Vulcan gun. It is currently in the US military inventory for use in applications requiring a lightweight, highly reliable gun capable of firing at rates of up to 1500 shots per minute.

The gun operation is based on an externally powered, rotating cluster of barrels. Each barrel has its own bolt which sequentially rams, locks, fires, unlocks and extracts rounds during one revolution of the barrel cluster. The barrels are installed in the gun rotor with a 'twist'; no headspace adjustment is necessary. The gun rotor rotates within a stationary gun housing which contains the main cam. The bolts, which slide within tracks on the gun rotor, are driven fore and aft by the controlling action of the main cam. Each barrel fires only once during each revolution of the gun rotor. This reduced rate for each barrel contributes to a long weapon life and high reliability.

A total peak recoil load of less than 1181 kg allows the weapon to be mounted in many configurations including turret, pod and pintle.

The gun fires standard M50 series ammunition. It is at present used only in the air application but can form the basis of a lightweight infantry anti-aircraft system.

DATA
Cartridge: 20 × 102 mm
Operation: externally powered, Gatling type
Feed system: linked or linkless
Gun: including delinking feeder: (length) 1829 mm; (width) 405 mm; (height) 342 mm; (weight) 66.5 kg
Number of barrels: 3
Gun life: 75 000 rounds
Barrel: (length) 1524 mm; (weight) 8.18 kg
Barrel life: (per set of 3) 15 000 rounds
Rate of fire: 400-1500 rds/min
Muzzle velocity: 1030 m/s
Power required: 3 hp at 1500 rds/min
Reliability: 30 000 minimum rounds before failure

Manufacturer
General Electric Co, Armament Systems Dept, Lakeside Avenue, Burlington, VT 05401.
Status
Current. Production.
Service
US Army, US Marine Corps, and forces of other countries.

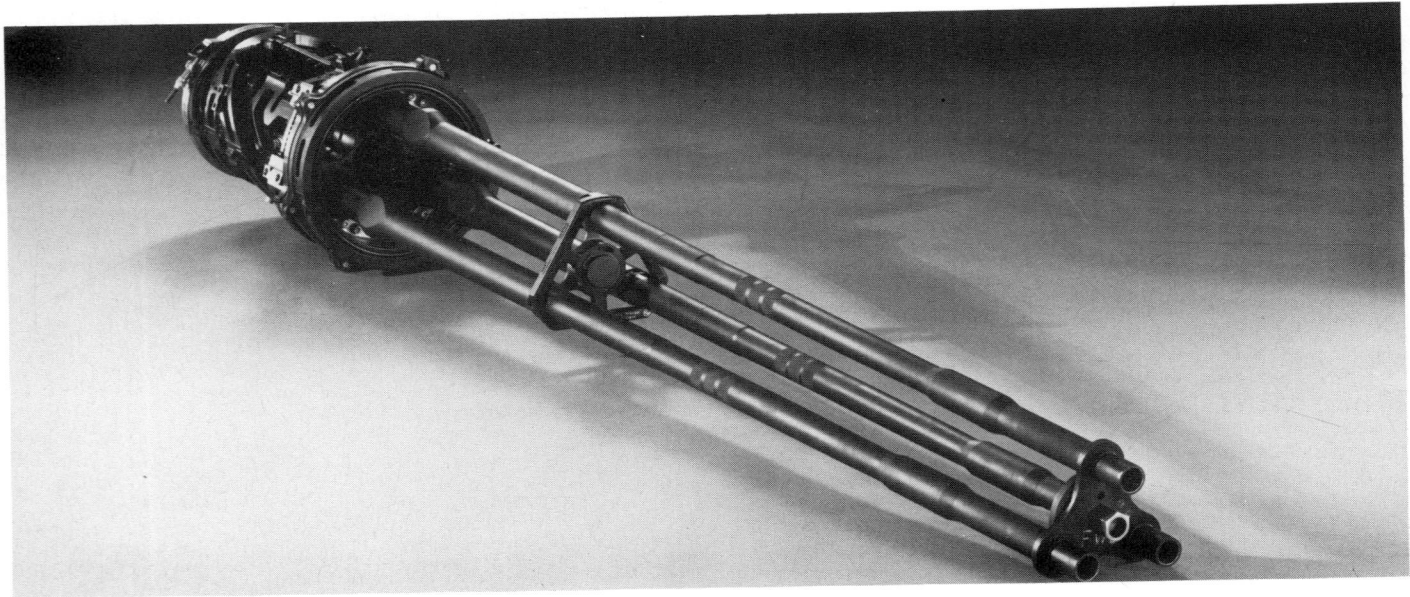

20 mm M197 three-barrelled gun

YUGOSLAVIA

20 mm M1955 cannon

Designed and manufactured in Yugoslavia this weapon closely resembles the Hispano-Suiza (later Oerlikon) HS804 cannon and is believed to be substantially similar in design.

Long familiar in a triple air defence mounting used by both land and naval forces the gun has recently been seen also on a single, two-wheel, towed mounting.

DATA
Cartridge: 20 × 110 mm
Operation: blowback with positive breech locking, automatic
Feed: 60-round drum magazine
Barrel: 70 cal

Rate of fire: (cyclic) 800 rds/min
Muzzle velocity: 835 m/s
Practical range: (ground) 2000 m; (anti-aircraft) 1200 m
Armour penetration: 25 mm at 1000 m

Manufacturer
Zavodi Crvena Zastava, Kragujevac.
Status
Current.
Service
Yugoslav forces.

30 mm CZ automatic cannon

This weapon is of Yugoslavian design and is intended for mounting on armoured vehicles and for use in anti-aircraft systems. It is a gas-operated, belt-fed gun which can feed from either side and ejects the spent case forwards. There is an integral gas buffer and spring recuperator which reduces the recoil load on the mounting. The barrel is easily removed and replaced, and the feed system is powered by recoil energy and is capable of lifting a 40-round length of belt. Bolt cocking and sear release are both performed by hydraulic mechanisms, and firing is controlled by a 24 V electric circuit.

DATA
Cartridge: 30 × 210B
Operation: Gas, automatic
Feed: belt
Weight: 150 kg
Length: 3065 mm
Height: 264 mm
Width: 242 mm
Barrel: 2100 mm; 60 kg
Rifling: 12 grooves, rh, increasing twist to 6° 24'
Gas pressure: 3000 bar at 300 MPa
Recoil force: 2500 daN
Recoil distance: 45 mm maximum
Muzzle velocity: 1050-1100 m/s
Rate of fire: 650-750 rds/min
Ammunition: Mine HEI, HEI/T; APHEI/T; Prac, Prac/T

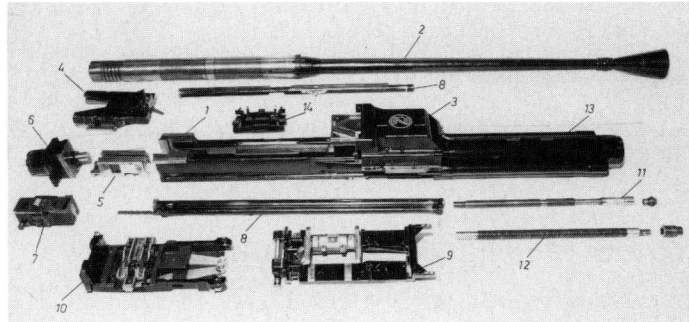

Component parts of 30 mm CZ automatic gun
(1) *breech casing;* **(2)** *barrel assembly;* **(3)** *front seat;* **(4)** *rear seat;* **(5)** *breech-block;* **(6)** *buffer;* **(7)** *trigger mechanism;* **(8)** *Recoil brake and recuperator mechanisms;* **(9)** *Lower feeder;* **(10)** *upper feeder;* **(11)** *gas piston;* **(12)** *shock absorber;* **(13)** *barrel support;* **(14)** *link chute*

Manufacturer
Zavodi Crvena Zastava, Kragujevac.
Status
Current. Production.

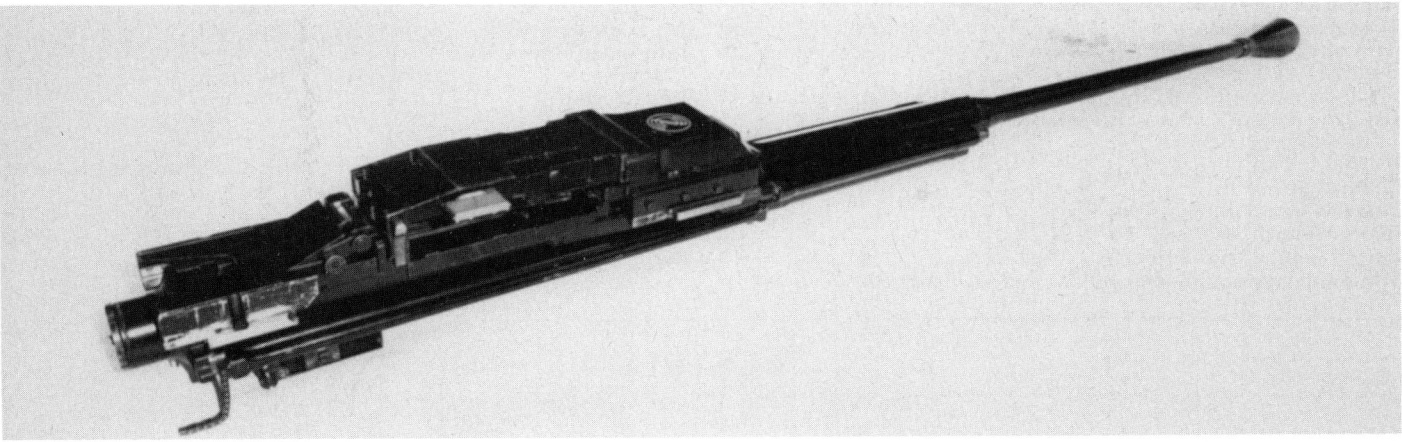

30 mm CZ automatic gun

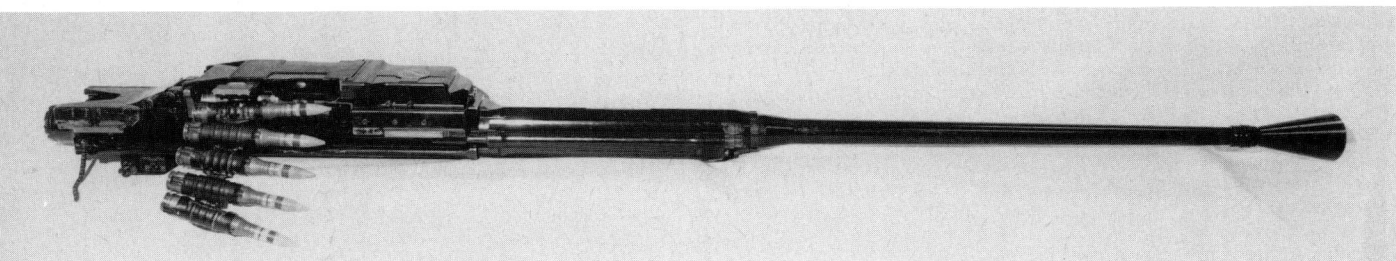

30 mm CZ automatic gun showing right-hand belt feed

Anti-Tank Weapons

ARGENTINA

Mathogo anti-tank missile

Mathogo was developed in the late 1970s by CITEFA to meet a specification from the Argentine Army. It is a first-generation wire-guided missile, reminiscent of the British Vigilant in general appearance. As well as being manportable, it can also be vehicle- or helicopter-mounted, though its accuracy in the latter field must be considered doubtful due to the guidance system used.

The missile has a shaped charge warhead and is carried in and launched from a simple water-tight container. A control unit with telescope sight can be set up and connected with up to four launchers up to 50 m away. After firing, the operator gathers it into his field-of-view and then commands it in flight to impact with the target. Commands are transmitted down the wire link and are effected by means of spoilers in the wings. A boost rocket launches the missile, after which a sustainer motor takes over.

Two versions of this missile are known to exist; one has a range of 2000 m, the other a range of 3000 m. Both use the same control unit.

DATA
Length: 998 mm
Diameter: 102 mm
Wingspan: 470 mm
Weight: 11.3:kg
Warhead: 2.80 kg
Launcher: 8.2 kg
Cruise velocity: 90 m/s
Minimum range: 400:m
Maximum range: 2000 or 3000 m (see text)
Penetration: 400 mm

Manufacturer
CITEFA: Instituto de Investigaciones Cientificas y Tecnicas de las Fuerzas Armadas), Zufriategui y Varela, 1603 Villa Martelli, Buenos Aires.
Status
Current; production.
Service
Argentine armed forces.

105 mm Model 1968 recoilless gun

The gun has a cylindrical perforated combustion chamber and a centrally hinged breech-block. The backplate has a venturi mounted on it in such a way as to be easily replaced, and there are two spiral vanes in the body of the diffuser to compensate for the torque force resulting from the interaction of the barrel rifling and the shell. An additional peripheral chamber permits the use of a lower-density charge and gives optimum combustion of the powder and efficient use of the resultant gases. Either HE or hollow-charge (ECH) ammunition can be fired.

Several different firing positions are available to meet different operational requirements. When mounted on its towing carriage the gun can be fired with the axle high, half-lowered or low; while still on its wheels it can also be fired (with its muzzle forward) from a standard Unimog 421 ½-ton carrying vehicle: it can also be fired from a tripod. Minimum height to the centre of the barrel in the low-axle position is 62 cm.

Aiming is by means of a telescope which is calibrated up to 1800 m and fitted with a stadiametric ladder anti-tank sight. An auxiliary spotting rifle (7.62 mm FAP – heavy automatic rifle) is fitted.

The two-wheeled gun-carriage can be fitted with conventional or puncture-proof tyres. It has a torsion bar suspension.

DATA
Calibre: 105 mm
Barrel length: 3000 mm
Overall length: 4020 mm
Height in road trim: 1070 mm
Minimum height of barrel: 620 mm (low angle position)

Weapon with carriage and normal tyres: 397 kg
Spotting rifle weight: 6.4 kg
Additional for puncture-proof tyres: 56 kg
Vertical field of fire: -7 to +40°30′
Max range: 9200 m
Range of spotting rifle: 1200 m effective
Sight field-of-view: 12°
Magnification: 4 ×
Rear danger zone: 40 m (90°)
Crew: 4
Rate of fire: 3–5 rds/min

Ammunition	HE shell	Hollow-charge (ECH) shell
Weight	15.6 kg	11.1 kg
Muzzle velocity	400 m/s	400 m/s
Internal pressure at breech	930 kg/cm²	830 kg/cm²
Penetration	—	200 mm

Manufacturer
Rio Tercero Military Factory, Direccion General de Fabricaciones Militares, Cabildo 65, Buenos Aires.
Status
Current. Available.
Service
Argentinian forces.

AUSTRIA

106 mm M40A1 (towed) recoilless rifle

The US M40A1 recoilless rifle is in service with the Austrian Army in two forms. One of these is as a vehicle-mounted weapon – as it is in its US Army applications; the other is as a towed weapon on an Austrian-designed two-wheeled carriage. The carriage is notable for having two stable firing positions, the barrel being 26 cm higher in one position than it is in the other.

DATA
Calibre: 106 mm
Barrel length: 3400 mm
Elevation: –17 to +65°
Traverse: 360°
Weight of rifle in combat order: 113.9 kg
Weight of shell: 7.71 kg
Weight of carriage: 170 kg
Muzzle velocity: 503 m/s
Max range: 6900 m
Rate of fire: 5 rds/min

Gun fully raised

Status
Current.
Service
Austrian Army.

Loading gun in lowest position

LAT 500 anti-tank weapon

The LAT (Light Anti-Tank) 500 is a disposable shoulder-fired weapon for infantry use. It is completely self-contained and does not require any prior assembly or loading routines on the part of the user. It is provided with a shoulder rest and forward handgrip, and there are simple pop-up sights. The weapon is recoilless and may be fired from confined spaces such as rooms or pillboxes. The firing signature has been kept to a minimum.

DATA
Calibre: 82 mm
Overall length: 900 mm
Total weight: 7 kg
Weight of projectile: 1.6 kg
Weight of explosive: 0.5 kg
Velocity: 255 m/s
Penetration, armour: > 500 mm
Penetration, concrete: > 1000 mm

Manufacturer
Intertechnik GmbH, Industriezelle 56, Postfach 100, A-4040 Linz.
Status
Undergoing trials by Austrian Army. Ready for mass production.

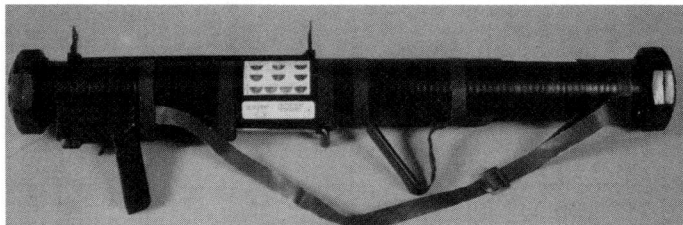
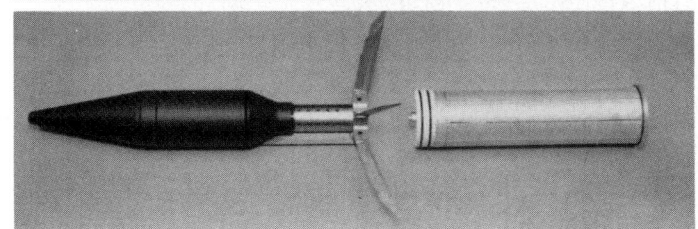

LAT 500 launcher (above) and complete round (below)

RAT 700 106 mm projectile

This is a shaped-charge projectile for use with the 106 mm recoilless rifle M40A1. By careful design the internal ballistics of the weapon have been improved, increasing the muzzle velocity and thus the hit probability. The warhead is claimed to perform well against all types of armour, including laminated and reactive.

DATA
Calibre: 106 mm
Length, with fins: 575 mm
Weight: 5 kg
Explosive weight: 1.1 kg
Penetration: 700 mm homogeneous rolled armour

Manufacturer
Intertechnik GmbH, Industriezelle 56, Postfach 100, A-4040 Linz.
Status
Advanced development.

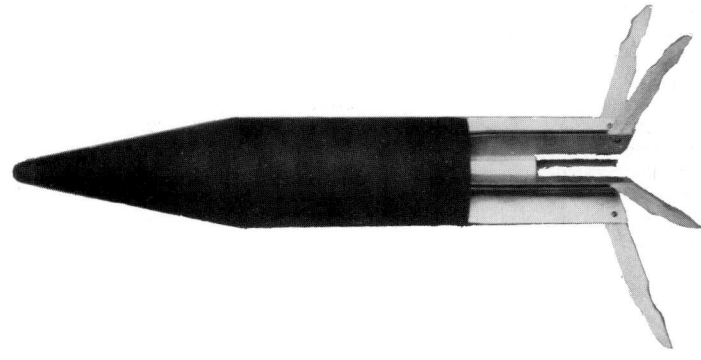

RAT 700 106 mm anti-armour projectile

BELGIUM

3.5 in PRB NR 415 HE anti-tank and anti-personnel rocket

Among a wide range of military products (mostly rockets, ammunition and explosive devices) made by the PRB Group is a series of ground-launched rockets. In 2.36 in (60 mm) and 3.5 in (89 mm) calibres it makes HEAT, smoke and incendiary rounds. It also makes dummy, blank and sub-calibre practice rounds.

An interesting development in the 3.5 in range is the dual purpose PRB 415 replacement for the standard American M28A2 super-bazooka rocket. The novel feature of the Belgian round is that its warhead is designed to be effective in both anti-personnel and anti-tank roles. This has been achieved by slightly reducing the size of the normal hollow-charge arrangement and incorporating an anti-personnel fragmentation sleeve made of spirally-wound steel wire which has been pre-notched by a patented process. At the same time

the rocket propellant and igniter have been improved. The same modification has been developed for the 2.36 in (60 mm) rocket and could be easily adapted to other HEAT heads.

Sub-calibre training device
PRB also manufactures a training round which is externally identical in appearance to the 3.5 in rocket but fires a 20 mm projectile. It can be used with M20B1 or M20A1B1 launchers. The sub-calibre launch tube is rifled and the projectile is a tracer round having ballistics similar to those of the real rocket. Handling and firing procedures are the same as for the real rocket and the dummy rocket is ejected by recoil from the rear of the launcher.

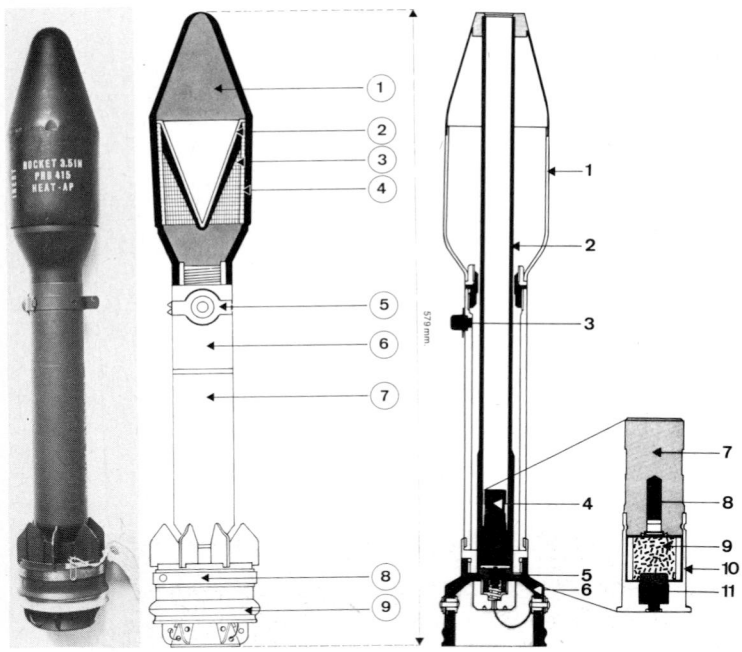

DATA
OPERATIONAL WEAPON
Calibre: 89 mm
Weight of rocket: 4 kg
Weight of explosive charge: 0.6 kg
Type of explosive charge: Composition B
Lethal radius (50%): 20 m
Penetration: 200 mm of armour
Range at 45°: 735 m
Muzzle velocity: 95 m/s
Fragmentation sleeve: 1250 fragments

Manufacturer
PRB SA, Département Défense, avenue de Tervueren 168, B-1150 Brussels.
Status
Obsolescent, but still produced.
Service
Belgian forces and many others.

Left to right 3.5 in PRB NR 415 dual purpose rocket; drawing showing construction of PRB 415; drawings showing construction and arrangement of sub-calibre training rocket

(1) head (2) copper cone (3) explosive charge (4) fragmentation sleeve (5) safety band (6) base fuze M404A1 (7) motor (8) contact ring (9) groove

Training rocket drawing
(1) body (2) 20 mm barrel (3) dummy safety (4) cartridge (5) electrode (6) breech (7) bullet (8) tracer (9) powder (10) cartridge case (11) electric primer

90 mm Mecar light gun

This weapon is designed primarily for anti-tank defence but fires a variety of other ammunition types in addition to the HEAT-T round. It is supplied with or without muzzle brake to meet customer requirements. Due to its low recoil force and light construction, the gun is particularly suited to mounting on light armoured vehicles and on small gun carriages for employment in mountainous and or difficult terrain. Two versions of the gun exist, the heavier CAN-90H having a lower recoil velocity than the lighter CAN-90L.

DATA
GUN WITHOUT CARRIAGE
Weight: (90L) 285 kg (90H) 416 kg
Calibre: 90 mm
Height overall: (90L) 350 mm (90H) 400 mm
Width overall: (90L) 290 mm (90H) 450 mm
Barrel length: 2.9 m
Length of chamber: 368 mm
Chamber volume: 2560 ml
Max pressure: 1200 kg/cm^2
Max recoil distance: 420 mm
Max recoil velocity: (90L) 10.2 m/s (90H) 7.8 m/s
Barrel life: up to 10 000 rounds

AMMUNITION
HEAT-T round
Weight: (projectile) 2.28 kg; (round) 3.54 kg
Length: 635 mm
Muzzle velocity: 633 m/s
Range: (max) 3500 m; (effective) 1000 m
Arming distance: 16 m
Penetration: (armour) 350 mm; (concrete) 1200 mm

HE-T
Weight: (projectile) 4.1 kg; (round) 5.21 kg
Length: 520 mm
Muzzle velocity: 338 m/s
Range: 4200 m max
Arming distance: 16 m
Lethal radius: 115 m
Effective radius: 25 m

Smoke WP or FM (TTC) Weight: (projectile) 4.1 kg; (round) 5.21 kg
Length: 520 mm
Muzzle velocity: 338 m/s
Maximum range: 4200 m
Arming distance: 16 m

Canister round
Weight: (projectile) 4.1 kg; (round) 5.95 kg
Length: 373 mm
Contents: 1120 balls
Muzzle velocity: 338 m/s
Effective range: 200 m
Cone angle forms at: approx 7 m

Manufacturer
Mecar SA, 6522 Petit-Roeulx-lez-Nivelles.
Status
Current. Available.
Service
Sold to 15 countries.

BRAZIL

57 mm M18A1 recoilless rifle

This is a copy of the American M18A1, manufactured in Brazil by Hydroar. Due to its portability and ease of operation, the weapon is still a viable defence against light armour. HE, HE/AP and HE/AT ammunition is readily available. The full data is given in the entry in the United States section, but Hydroar claims a barrel life of 2500 rounds and a life for the breech-block and throat of 500 rounds.

A 7.62 mm sub-calibre training adapter is available.

Ammunition is made by IMBEL.

Manufacturer
Hydroar SA, Rua do Rócio 196, CEP 04552, São Paulo.
Status
Current. Available.
Service
Brazilian Army.

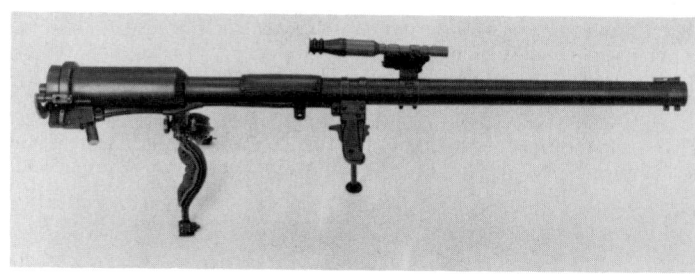
57 mm Hydroar RCL rifle

Hydroar 3.5in M20A1B1 rocket launcher with PE-1 electro-electronic grip

Hydroar manufacture a copy of the American 3.5in rocket launcher which is fitted with an electro-electronic grip of their own design. This replaces the original battery-powered M9 grip and the magneto-type M20 grip with a more modern circuit using two AA batteries and micro-electronics to produce the required power impulse. In other respects the launcher remains the same as the original US model, details of which will be found in the USA section.

In addition, the same grip can be supplied as a replacement kit for fitting to the 2.36in M9A1 and 3.5in M20A1B1 launchers. It will fit either launcher, and replacement is very simply performed.

Manufacturer
Hydroar SA, Rua do Rócio 196, CEP 04552, São Paulo.
Status
Current. Production.

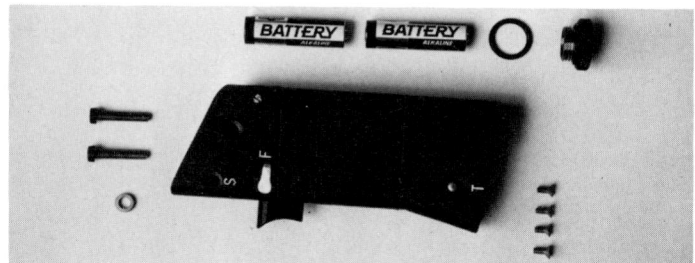
Grip kit PE-1 for modernisation of rocket launchers

Hydroar 3.5in M20A1B1 launcher with PE-1 electro-electronic grip

3.5 in Rocket (Rj 3,5 AE AP AC M1-CEV)

This rocket is for use with M20A1, M20A1B1 or similar launchers and combines high explosive, anti-personnel and anti-tank performance in one round. Muzzle safety extending to 15 m from the launcher is standard, and a training rocket with inert head is also available.

DATA
Calibre: 88.9 mm (3.5 in)
Weight: 4 kg
Weight of explosive: 860 g Composition B
Length: 600 mm
Muzzle velocity: 98 m/s
Max range (45°): 800 m
Number of fragments: 1300 minimum
Lethal radius: 20 m
Penetration: 20 mm homogeneous armour minimum

Manufacturer
Companhia de Explosivos Valparaiba, Praia do Flamengo 200, 20° Andar, 20000 Rio de Janiero, RJ.
Status
Current. Production.
Service
Brazilian and other armies.

Brazilian Marines firing 3.5 in rocket launcher

3.5 in M1-CEV rocket

CHINA, PEOPLE'S REPUBLIC

Red Arrow 8 guided weapon system

Red Arrow 8 is a second-generation guided missile intended for use by infantry against tanks and other armoured targets with a range of 100 to 3000 m. It is a crewportable weapon, fired from a ground tripod mount; it can also be configured for mounting in a variety of wheeled and tracked vehicles.

The system uses a tube-launched, optically tracked, wire guided missile which is controlled by a SACLOS system based on an infra-red flare in the tail of the missile. The position of this flare is detected by the sight unit and

corrections are automatically generated and signalled down the wire so as to fly the missile into the axis of the line of sight.

In general appearance the system is similar to Milan, having a sight unit on to which the missile transport and launch tube is attached before firing. The tripod resembles that used with the TOW system. A full range of first and second echelon maintenance equipment and a training simulator are also manufactured to accompany the system in service.

Red Arrow 8 missile installed on light wheeled vehicle

DATA
Operation: SACLOS, wire guided
Warhead diameter: 120 mm
Missile length: 875 mm
Missile weight: 11.2 kg
Wingspan: 320 mm
Launch tube diameter: 255 mm
Launch tube length: 1566 mm
Launch tube weight, with missile: 24.5 kg (carry); 22.5 kg (firing)
Sight unit weight: 24 kg
Tripod weight: 23 kg
Effective range: 100 to 3000 m
Penetration: > 800 mm
Flight velocity: 200 – 240 m/s
Hit probability: ⩾90%

Manufacturer
China North Industries Corp, PO Box 2137, Beijing.
Status
Current. Production.
Service
Chinese army.

40 mm Type 56 anti-tank grenade launcher

This is a copy of the Soviet RPG-2 launcher and has the same weight and dimensions. It fires the Chinese Type 50 grenade which was designed and

produced in that country. This is a better round as far as penetration is concerned than the Soviet PG-2 HEAT but its superiority is apparent only at normal impact; at 45° it may be less efficient.

DATA
Calibre of launcher: 40 mm
Calibre of warhead: 80 mm
Overall length of launcher: 1194 mm
Weight, empty: 2.83 kg
Max effective range: 150 m
Type of ammunition: Type 50 HEAT
Projectile weight: 1.84 kg
Rate of fire: 4-6 rds/min
Armour penetration: Type 50 HEAT 265 mm at normal (0° obliquity) (USSR PG-2 HEAT 150-175 mm at normal)

Manufacturer
State factories.
Status
Obsolete.
Service
Known to be used by militia and presumably a reserve item.

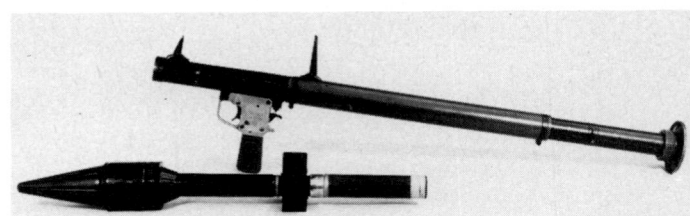

RPG-2 launcher. This has been manufactured in China and is copy of Soviet weapon. It is shown with practice grenade of Chinese origin

Sectioned drawing showing dimensions of Type 56 HEAT round

40 mm Type 69 anti-tank grenade launcher

This is a copy of the Soviet RPG-7 launcher and was not seen until 1972. It has the same performance as the Soviet version.

DATA
Calibre of launcher: 40 mm
Calibre of projectile: 85 mm
Length of launcher: 990 mm
Weight of launcher: 7 kg
Weight of grenade: 2.25 kg
Muzzle velocity: 300 m/s
Range, static target: 500 m
Range, moving target: 300 m
Range, to self-destruction: 920 m
Penetration of armour: 120 mm/65°

Manufacturer
State factories.
Status
Current. Production.
Service
Chinese forces.

Chinese Type 69 launcher

57 mm Type 36 recoilless rifle

This is a copy of the US M18A1. The United States sent the drawings to Nationalist China and gave technical assistance. When the Nationalists were forced off the mainland the Communists took over the factory and produced the 57 mm Type 36 RCL rifle. This is a breech loading single-shot weapon using a HEAT round against armour and an HE or canister round against personnel. The Chinese weapon will fire the American 57 mm round but the converse is not true and the Chinese ammunition can be used by no one else. The weapon can be fired from the shoulder but is usually fired from the rear bipod – front monopod system. It can also be mounted on a tripod, usually a Czechoslovak-pattern Zb26 or Zb30.

DATA
Calibre: 57 mm
Method of operation: recoilless
Overall length in firing position: 1549 mm
Weight: 35.4 kg
Max HE range: 450 m
Ammunition: HE and HEAT
Armour penetration: 63.5 mm

Manufacturer
State factories.
Status
Obsolete.

Service
Probably no longer in service in China, but supplied to many other countries. Recently identified in use by Tanzania.

57 mm Type 36 tripod-mounted recoilless rifle

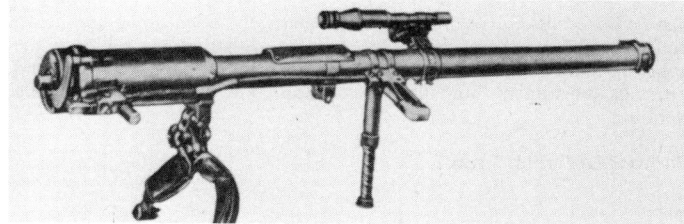

57 mm Type 36 recoilless rifle

57 mm Type 36 recoilless rifle in action

75 mm Type 52 recoilless rifle

This is a copy of the obsolete US M20 recoilless rifle and no improvement over the American weapon has been obtained. The gun can fire either Chinese or US ammunition.

The weapon is towed on a pair of solid wheels and a tripod with extensible legs is used for firing.

There is also a Type 56, which is a slightly improved version of the Type 52 and will fire only the Chinese ammunition.

DATA
Calibre: 75 mm
Method of operation: recoilless
Overall length in the firing position: 2132 mm
Weight with US type mount: 85.1 kg
Type of ammunition: HE or HEAT

Chinese 75 mm Type 56 recoilless rifle

American troops using M20

Max range of HE shell: 6675 m
Max effective anti-tank range: 800 m
Armour penetration: 228 mm

Status
Obsolete.
Service
Not known.

82 mm Type 65 recoilless gun

This is a copy of the Soviet B-10 recoilless gun. The performance of the Chinese version is the same as that of the Soviet weapon.

DATA
Cartridge: 82 mm
Method of operation: recoilless, with multi-vented breech-block with enlarged chamber section
Barrel length: 1659 mm
Overall length: 1677 mm
Weight in travelling position: 87.6 kg
Weight of HE round: 4.5 kg
Weight of HEAT round: 3.6 kg
Muzzle velocity: 320 m/s
Max range HE round: 4470 m
Max effective anti-tank range: 390 m
Armour penetration: 240 mm
Rate of fire: 6-7 rds/min

Status
Obsolete.
Service
Not known, but likely to be found in South-east Asia and Africa.

82 mm Type 65 gun. This is copy of Soviet B-10

62 mm portable rocket launcher

This is a manportable launcher similar in principle to the US M72 light anti-armour weapon. It consists of a telescoping tubular launcher into which a HEAT rocket is pre-assembled. The unit is carried in the telescoped condition and extended prior to firing; extending the launch tube automatically cocks the firing mechanism. After firing the launcher is discarded.

DATA
Length: (in firing position) approx 1.7 m
Calibre: 62 mm
Projectile length: 543 mm

Projectile weight: 1.18 kg
Effective range: 150 m
Penetration: 100 mm at 65° angle of incidence

Manufacturer
China North Industries Corp, PO Box 2137, Beijing.
Status
Current. Production.
Service
Chinese forces.

Ammunition for recoilless guns

Chinese projectiles for recoilless guns are generally based upon, and hence similar to, the projectiles originally acquired with the weapon from the supplying country. Various manufacturing changes may have been made and materials substituted, but the construction and operation show little alteration.

57 mm HEAT shell Type 7

DATA
Calibre: 57 mm
Type: HEAT
Weight, fuzed: 1.23 kg
Bursting charge: 159 g TNT
Fuze: Unknown, but appears to be a spitback PIBD
Using weapon: Type 36 RCL
Remarks: The steel pre-rifled driving band is forged as part of the body. General design based on US M307 shell

75 mm HEAT projectile

DATA
Calibre: 75 mm
Type: HEAT
Weight, fuzed: 2.81 kg
Bursting charge: 341 g RDX/TNT
Fuze: Types 1, TS-1 and TS-2 PIBD
Using weapon: Types 52 and 56 RCL

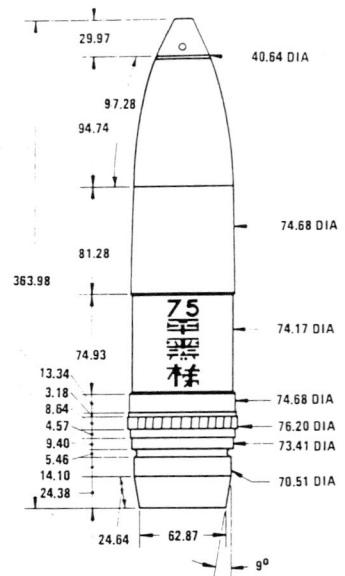
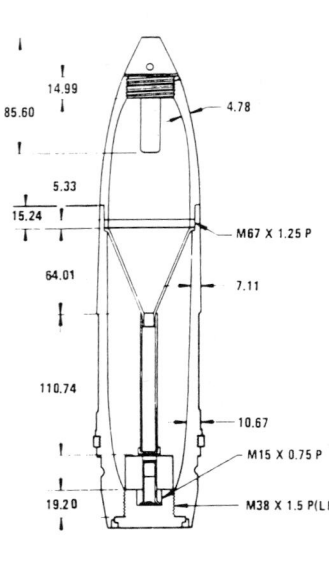

75 mm HEAT shell

75 mm HEAT shell

DATA
Calibre: 75 mm
Type: HEAT
Weight, fuzed: 5.39 kg
Bursting charge: 622 g RDX/TNT
Fuze: Unknown, but appears to be a spitback PIBD
Using weapon: Types 52 and 56 RCL
Remarks: General design based on US M310A1 shell

75 mm HE Shell

DATA
Calibre: 75 mm
Type: HE fragmentation
Weight, fuzed: 5.94 kg
Bursting charge: 712 g TNT
Fuze: Types 1, 3, 53 or M6 point detonating
Using weapon: Types 52 and 56 RCL
Remarks: General design based on US M309 shell. Frequently found using Soviet M6 mortar fuze

82 mm HEAT projectile Type 65

DATA
Calibre: 82 mm
Type: HEAT
Weight, fuzed: 2.95 kg
Bursting charge: 422 g RDX/Wax
Fuze: Base detonating Type 4
Using weapon: Type 65 RCL

82 mm HEAT projectile, Variant 1

DATA
Calibre: 82 mm
Type: HEAT
Weight, fuzed: 3.54 kg
Bursting charge: 454 g RDX/TNT
Fuze: Type 3 spitback PIBD
Using weapon: Type 65 RCL

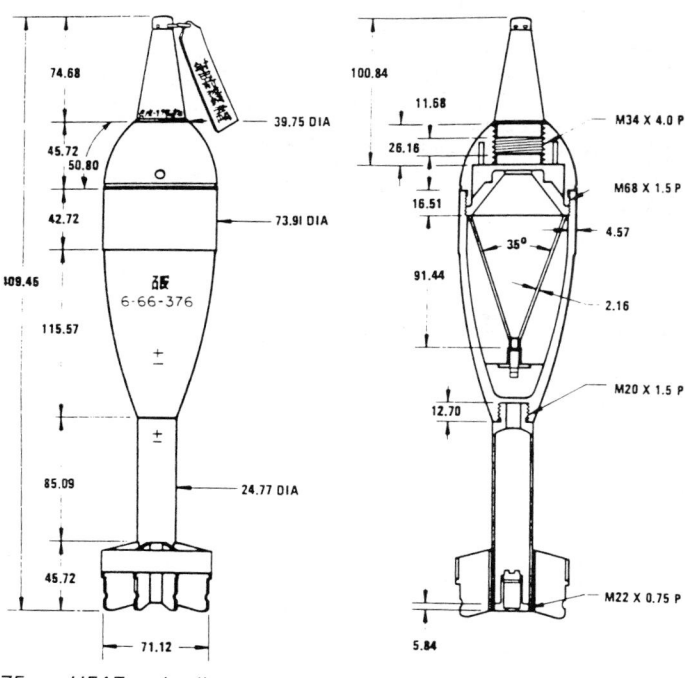

75 mm HEAT projectile

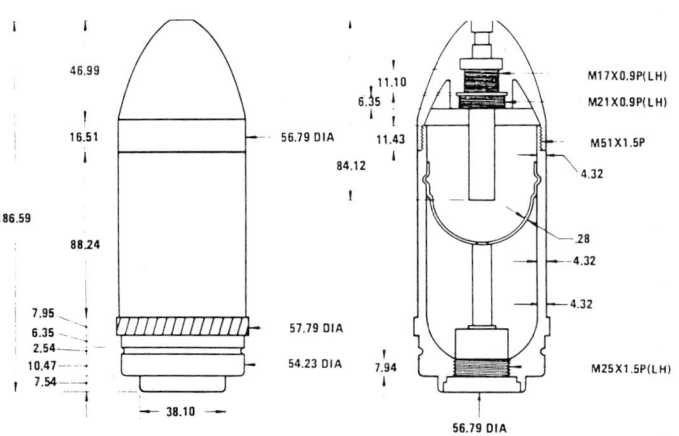

57 mm HEAT shell

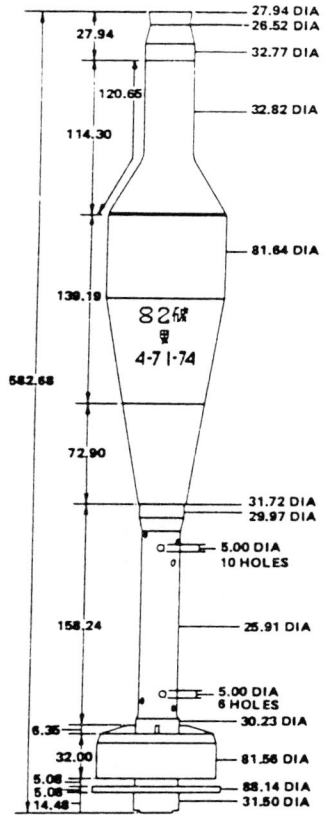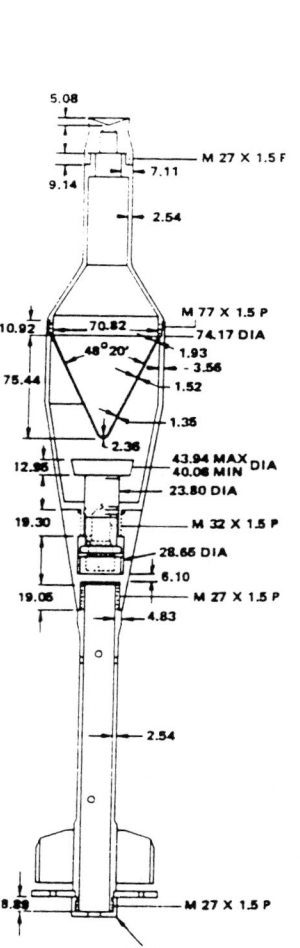

82 mm HEAT projectile Type 65

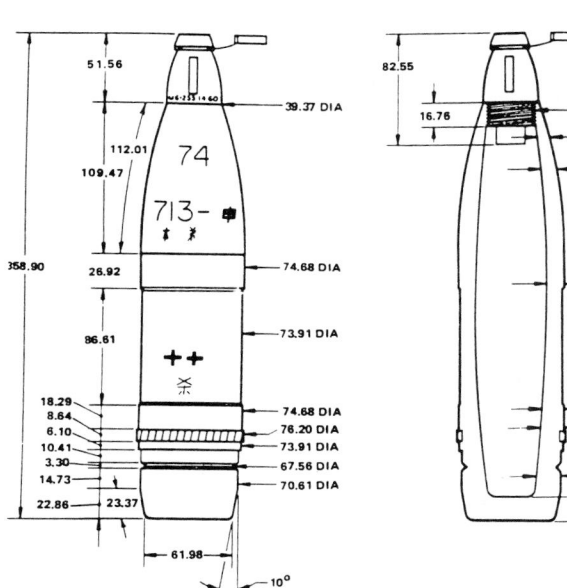

75 mm HE fragmentation shell

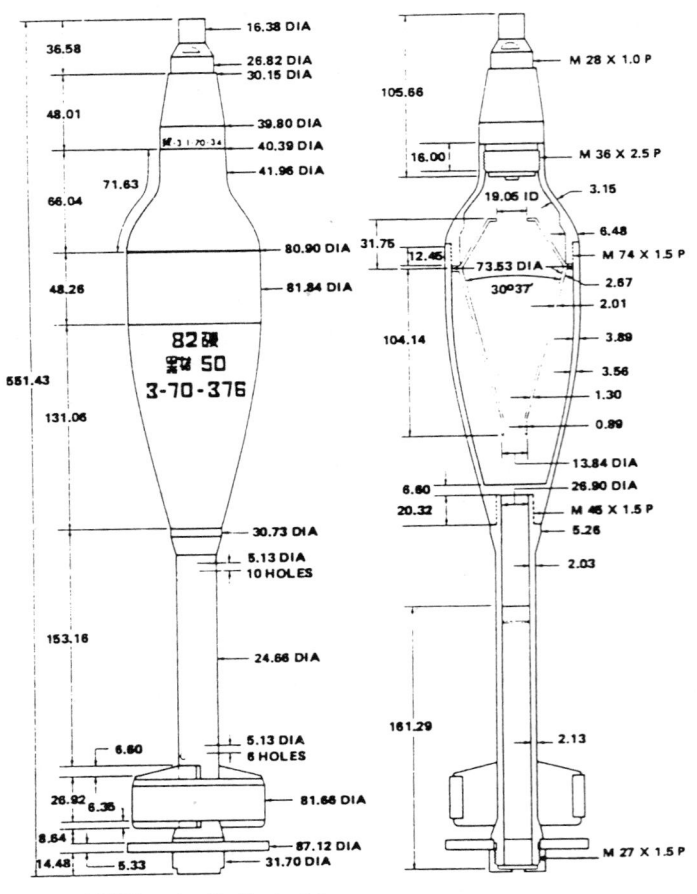

82 mm HEAT projectile, Variant 1

82 mm HEAT projectile, Variant 2

DATA
Calbre: 82 mm
Type: HEAT
Weight, fuzed: 2.94 kg

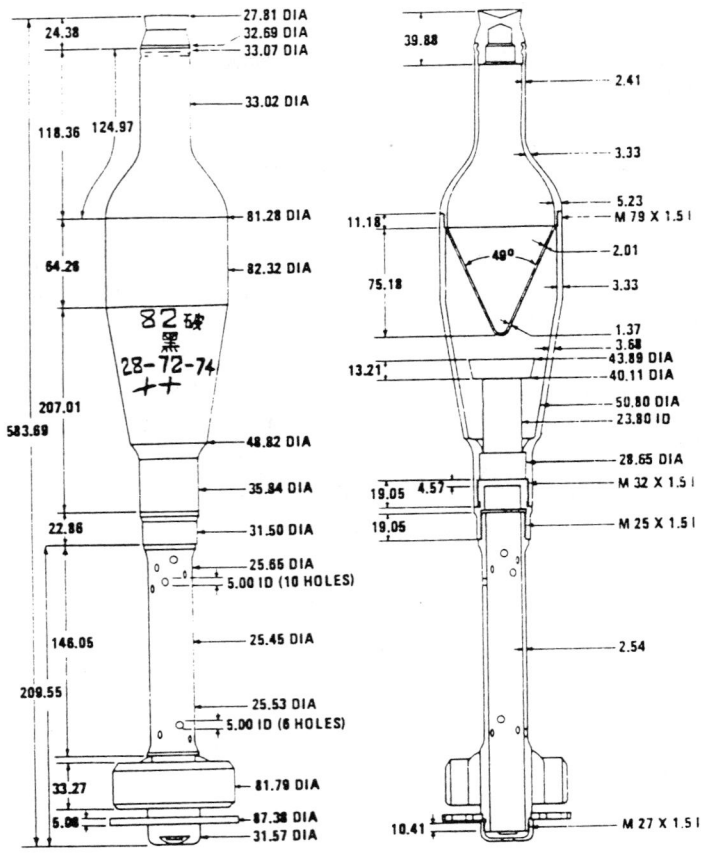

82 mm HEAT projectile, Variant 2

Bursting charge: 454 g RDX/Wax/PETN
Fuze: Base detonating Type 4
Using weapon: Type 65 RCL
Remarks: The charge is in two layers; the forward layer, around the cone, is RDX/Wax; the rear layer is PETN.

CZECHOSLOVAKIA

Type P-27 ('Pancerovka') anti-tank grenade launcher

This is a Czechoslovak version of the Soviet RPG-2 grenade launcher. The ammunition is not interchangeable with that of the Soviet weapon and the launcher has a folding bipod under the muzzle. It is more substantial and heavier than the Soviet weapon. The propellant contains iron filings which increase the momentum of the gases leaving the rear end.

DATA
Calibres: (launcher) 45 mm; (projectile) 120 mm
Method of operation: rocket
Overall length of projector: 1092 mm
Weight of launcher: 6.39 kg
Type of ammunition: HEAT
Projectile weight: 3.3 kg
Max effective range: 100 m
Armour penetration: 250 mm
Rate of fire: 3-4 rds/min

Manufacturer
Skoda.

45 mm Type P-27 ('Pancerovka') anti-tank grenade launcher

Status
Obsolete.
Service
Some service in Asia and Africa.

RPG-75 light anti-armour weapon

This is a shoulder-fired rocket launcher which appears to be a local variant of the Soviet RPG-18, and is one of several existing designs which have been developed from the American M72 66 mm launcher.

The RPG-75 is made of an extruded light alloy tube. It telescopes and is 633 mm long when in the carrying position. The two halves are held together with a bayonet catch which has to be rotated to release the halves. The rear portion has a bell-shaped venturi not unlike the RPG-7. There are rubber end-caps to protect the tube and the projectile but these caps are not attached to the sling, as they are for the M72. The firing mechanism is mechanical and

not unlike that on the M72. The sights are simple pop-up frames made of plastic; the foresight is marked for ranges of 200, 250 and 300 m and there is a simple rangefinder in the form of a stepped slot, indicating the apparent length of an average tank at the same ranges. The rearsight has three apertures on a revolving disc, marked 'N', '+' and '-' which adjust the sight line to compensate for temperature effects on the rocket motor.

Each launcher has a series of simple drawings showing the firer how it is used. In this it follows the style of M72. Each drawing has a few instructions. The series translates as follows:

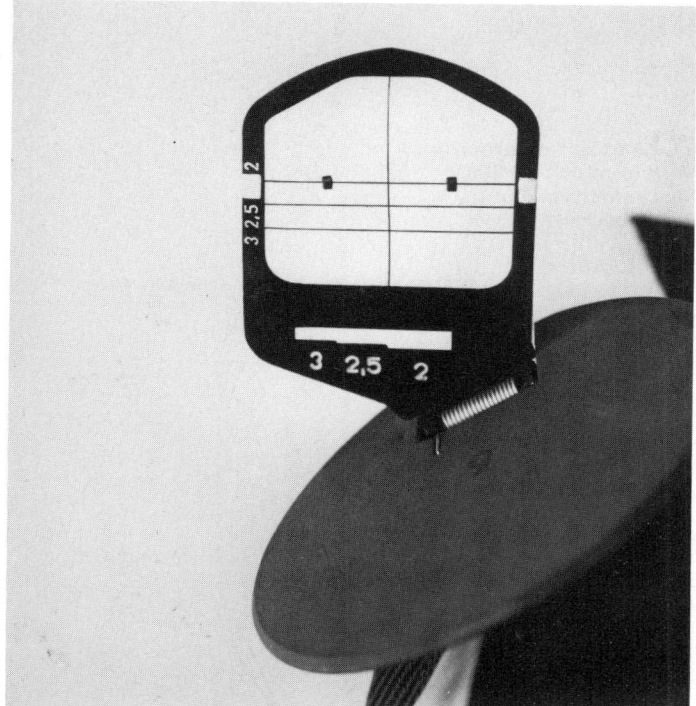

Front sight of RPG-75

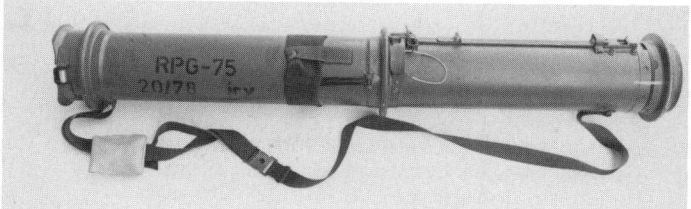

RPG-75 in carrying mode

RPG-75 in firing position

1 Remove rubber cap from muzzle
2 Tear off seal
3 Turn chamber as far as possible
4 Slide chamber as far as possible
5 Turn chamber as far as possible
6 Release catch of rearsight
7 Erect front sight by pressure of button
8 Erect rearsight and adjust correction of temperature
9 Pull out carrying safety
10 Squeeze the firing safety with your forefinger as far as possible and hold it
11 Fire with pressure of your thumb.

   Actions 3, 4 and 5 relate to the extending of the launch tube. Action 8 is again a similar one to the preparations involved in M72 and shows that the motor gives a different performance in cold weather. There is no shoulder stop and the firer has to judge where to put his head and eye when firing. One training instruction warns that the venturi should not be within 2 m of a wall or similar solid obstruction.

DATA
Calibre: 68 mm
Length: (folded) 633 mm; (extended) 890 mm
Weight, firing: 3.1 kg
Launch velocity: 189 m/s
Effective range
Moving targets: 200 m
Stationary targets: 300 m
Time to self-destruction: 5 s
Penetration: 300 mm
Warhead: shaped charge 0.32 kg

Status
Current. Apparently in production.
Service
Warsaw Pact armies.

82 mm M59 and M59A recoilless guns

These are two versions of a recoilless gun, smooth bored, breech-loaded, firing fin-stabilised ammunition. The original M59 had a smooth exterior and the M59A differs only in having a radially finned section over the chamber to dissipate the heat. Each is capable of being towed behind an armoured personnel carrier (usually OT-810) or any other suitable vehicle. It can be mounted on top of an OT-62 armoured personnel carrier and has been seen inside. It can be manhandled over quite long distances with two men, with a towing harness, pulling on the bar across the muzzle. In addition to the HEAT

anti-tank round it fires an HE shell in either direct or indirect roles. It is the only Warsaw Pact gun to use the ranging rifle technique.

DATA
Calibre: 82 mm
Method of operation: recoilless
Overall length: (firing position) 4597 mm
Weight: 385 kg
Type of ammunition: HE and HEAT
Projectile weight: 6 kg
Max effective anti-tank range: 1200 m
Max HE range: 6657 m
Armour penetration: 250 mm

Manufacturer
State factories.
Status
Obsolescent. Not in production.
Service
Probably with reserve units in Czechoslovakia.

82 mm M59 recoilless gun

82 mm M59A recoilless gun

EGYPT

PG-7 weapon system

This is essentially the Soviet RPG-7 copied and manufactured in Egypt, wih some modifications to suit Egyptian manufacturing methods.

DATA
Calibre: (launcher) 40 mm; (warhead) 85 mm
Launcher length: 950 mm
Rocket length: 930 mm
Launcher weight: 6.8 kg
Rocket weight: 2.25 kg
Rocket length: 230 mm
Warhead weight: 900 g
Effective range: (AFVs) 300 m; (other) 500 m

Range to self-destruction: 900 m
Muzzle velocity: 120 m/s
Max velocity: 300 m/s
Penetration: > 260 mm

Manufacturer
Sakr Factory for Developed Industries, PO Box 33, Heliopolis, Cairo.
Status
Current. Production.
Service
Egyptian Army.

Cobra anti-tank weapon system

This is an improved warhead for the standard PG-77 weapon system described above. The shaped charge is of Octol, and has a new pattern liner manufactured by a specially developed process, giving an approximately 20 percent improvement in performance compared to standard copper cones. The fuzing sytem is electro-magnetic and is initiated by contact between the double ogives of the warhead. It is capable of functioning at angles of 80° from normal and has a response time less that 10 µs.

The Cobra warhead has been demonstrated to penetrate over 500 mm of armour steel at 3.5 calibre stand-off distance in static tests.

Manufacturer
Sakr Factory for Developed Industries, PO Box 33, Heliopolis, Cairo.
Status
Current. Production.
Service
Egyptian Army.

FINLAND

55 mm M-55 recoilless anti-tank grenade launcher

This weapon, which is the current light anti-tank weapon of the Finnish Army, is of rather more elaborate construction than many such manportable recoilless launchers. The weapons are issued on a scale of six per company in the motorised battalions. The weapon was developed in Finland.

Finnish 55 mm M-55 recoilless anti-tank grenade launcher

DATA
Calibre: 55 mm
Launcher length: (unloaded) 940 mm; (with grenade in place) 1240 mm
Weapon weight: 8.5 kg
Grenade weight: 2.5 kg
Effective anti-tank range: 200 m
Rate of fire: 3-5 rds/min
Penetration: 200 mm

Status
Current. Production complete.
Service
Finnish forces.

95 mm 95 SM58-61 recoilless anti-tank gun

We are informed that this weapon is no longer in service with the Finnish Army, and since it was never employed elsewhere it is unlikely to be encountered. Details can be found in earlier editions of *Jane's Infantry Weapons*.

Raikka recoilless guns

The firm of Raikka developed an interesting series of anti-tank guns using the original recoilless principle of the counter-mass. In this arrangement the barrel is a plain tube and the cartridge is inserted into the centre of it. On firing, the shell goes forward and an equivalent weight of another substance is blown backwards, thus balancing the recoil.

The principle has not been used in this manner for many years, largely because of the extra weight of the counter-mass and the danger to troops behind. However, Raikka has found that guns of this type are substantially cheaper to produce than normal recoilless models and by using water, sand or iron filings as the counter-mass the back-blast danger is not excessive. Raikka offers a range of such guns in the following calibres: 41, 55, 81, 120 and 150 mm.

Barrels are either smoothbored or rifled and a most unusual feature of the ammunition is that the 120 mm gun has a fin-stabilised APDS round

120 mm recoilless gun with 3.5 m barrel and round

120 mm recoilless gun with 6 m barrel

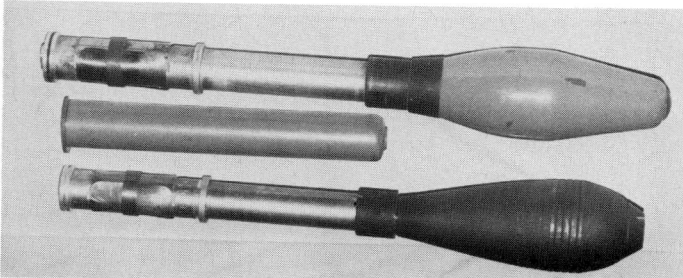

Two types of projectile with propellant cartridge between them. Note that with this size of launcher warhead is outside barrel and stabilising fins are wrapped around propellant tube. Upper projectile appears to have warhead made from 81 mm mortar bomb

Raikka 41 mm counter-mass light anti-armour weapon. Note no shoulder stop to hold launcher in correct position for aiming

41 mm launcher with sights extended. Tube beneath barrel may have some connection with trigger mechanism

(HVAPDS(FS)). It is claimed that this projectile can achieve 1500 m/s and so is effective beyond 1000 m against main battle tanks.

The entire system is novel and interesting. It might appear that the firm is a little late in producing these guns now that the recoilless principle is being phased out in most armies as a main launcher of anti-tank projectiles, but it may be that there is more to these guns than at first appears.

The accompanying illustrations show what are obviously prototype weapons, but they do show that this interesting principle has been explored in some detail.

Manufacturer
Raikka Oy, PO Box 30, SF-00251 Helsinki.
Status
Development completed.

DATA

Weapon VEHICLE MOUNTED	Calibre (mm)	Length (mm)	Weight (kg)	Ammunition weight (kg)	Muzzle velocity (m/s)	Effective range (km)
Raikka 150	150	4500	1200	42	400	over 10
Raikka 120	120	6000	1500	6	1500	1.5
	120	6000	1500	15	900	over 15
	120	3500	500	5.3	1000	over 10
Raikka 81	81	3000	250	3.5	1000	over 10
MANPORTABLE						
Raikka 81	81	1150	15	2	350	0.4
	81	1150	15	3.3	250	3
Raikka 55	55	900	4.5	2	170	0.2
Raikka 41	41	760	3	1	170	0.2

81mm double-barrelled recoilless gun. Probably prototype as no sights fitted

FRANCE

89 mm LRAC 89 anti-tank rocket launcher

The Lance-Roquette Anti-Char de 89 mm (LRAC 89) is a lightweight waterproof sub-sonic rocket system capable of very accurate anti-tank fire out to 400 m. It can also be used, with a different round, against personnel and light vehicles out to 1000 m.

The launching tube is made of glass fibre and plastic. It carries an adjustable shoulder piece which can be moved fore and aft to fit the firer and from the bottom of this comes a small bipod which may be pulled out and secured by a spring-loaded catch. The fore-hand grip may also be moved forward or back at will. The pistol grip contains the firing mechanism. This has a mechanical safety catch and when this is off, the action of pulling the trigger generates the current to launch the rocket. The telescopic sight is carried on the left side of the launcher.

The APX M 309 sight is graduated in ranges of 100 to 1000 m. The NATO requirement is for a maximum ordinate of 1.8 m and a corresponding time of flight for this is 1.25 seconds and a range of 315 m. The sight is arranged with the horizontal cross wire corresponding to this range. The horizontal cross wire is marked 10, 20 and 30 on each side of the vertical centre line, giving the correct lead for a tank travelling at 10, 20 or 30 km/h. Two stadia lines are provided to allow range assessment, based on the subtension angle produced by a 6 m tank.

An image-intensification night sight can be fitted in place of the M 309.

Anti-tank rocket
The rocket is contained within the cylindrical carrying tube. The front cap is removed and the container is then attached to the rear end of the launcher to complete the firing assembly. The rear plug is left in position to waterproof the rocket until ready for use. Until it is removed the rocket cannot be fired but when it is taken off the firing circuit is completed.

The propulsion system uses a number of long sticks of tubular propellant and they give a constant pressure burning and therefore a constant acceleration of the missile. Burning is complete before the tail of the rocket leaves the tube. There are nine tail vanes folded forward along the body and

these emerge into the airstream as soon as the missile leaves the launcher.

The warhead contains a shaped charge which is initiated by a piezo-electric sensor fitted in the front of the ballistic cap and connected to the base fuze. The fuze cannot function until the rocket container is screwed into the launcher, and it is then armed by the propellant gas when the motor is fired. The system provides a bore safety distance of 10 m.

The penetrative effect of the warhead enables it to meet the NATO heavy tank requirement of penetrating a single plate of 120 mm set at 65° and a double plate of 40 mm followed by a 110 mm plate, both set at 60°, and

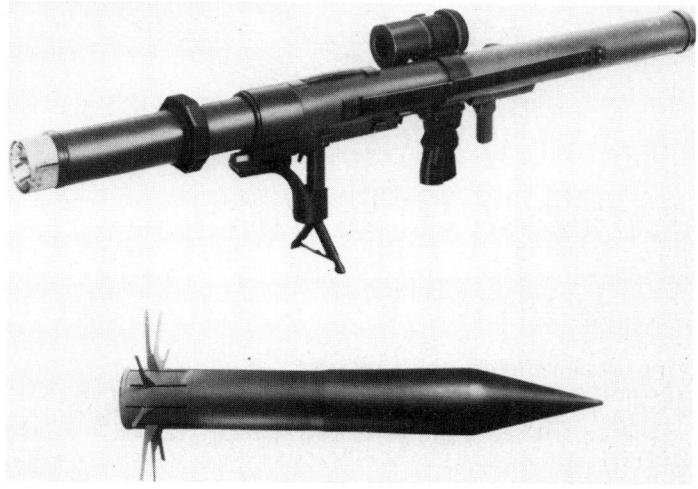

89 mm anti-tank rocket launcher and round

French Army anti-tank team loading LRAC 89

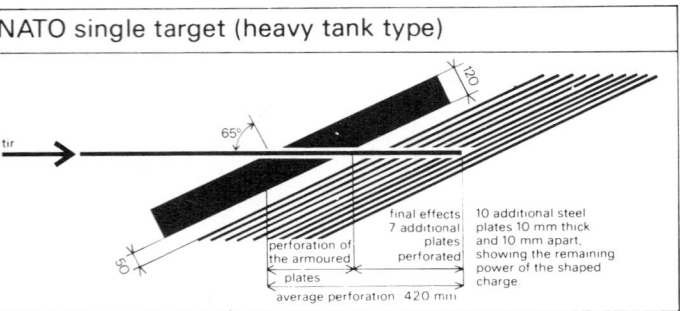

NATO single target (heavy tank type)

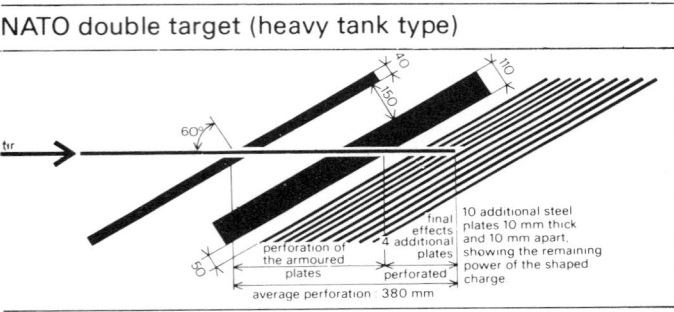

NATO double target (heavy tank type)

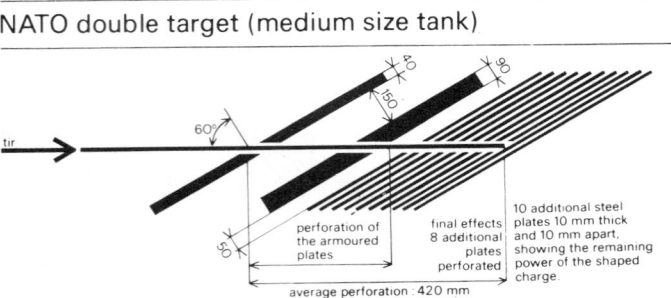

NATO double target (medium size tank)

Effects of 89 mm missile against standard NATO tank targets

150 mm apart. The head has been demonstrated penetrating 400 mm of plate at normal (0° obliquity).

Anti-personnel/anti-vehicle rocket
Although originally designed as an anti-tank weapon, the LRAC 89 can be used effectively against personnel and light vehicles using a special round. This is generally similar to the anti-tank round and has a dual purpose warhead comprising moulded steel balls which become about 1600 projectiles and are lethal up to at least 20 m from point of burst, and a hollow-charge which, while less effective than that of the anti-tank rocket, gives the round

considerable anti-armour capability. Handling and firing are the same as for the anti-tank round.

Smoke and illuminating rockets
Two smoke rockets are available for use with the LRAC 89. One, with a TT4 smoke head, contains a liquid smoke generating composition which produces dense smoke immediately on bursting on any surface: the other, with a P (phosphorus) head, produces smoke more slowly but also has an incendiary effect.

There is also an illuminating rocket containing a parachute flare which produces a light intensity of 300 000 candela for 30 seconds. It has present ranges of 400, 800, 1200 and 1800 m and is aimed using an accessory sight integral with the main (M 309) sight support.

Training devices
The training devices consist of: dummy rocket; practice rocket with an inert head and impact marker; blank shot trainer; and sub-calibre barrel.

The dummy rocket is intended to instruct soldiers in loading and unloading.

The practice rocket is similar in all respects to the live round except that it is inert and marks the point of strike by the emission of talc. It can be fired against a tank without danger to the crew or damage to the vehicle.

The blank shot trainer consists of a dummy rocket fitted in a container identical to that of the anti-tank rocket. A cartridge is fitted into the electrical firing circuit and this produces the noise and smoke of discharge.

The sub-calibre barrel is inside a dummy container which is handled and loaded in exactly the same way as with a live round. It is loaded with a 7.5 mm small arms cartridge and it offers realistic training at minimum cost.

DATA
WEAPON
Calibre: 88.9 mm
Length: (during transport) 1.168 m; (in firing position) 1.6 m
Weight: (in transport condition) 5.5 kg including sight; (loaded) 8.2 kg

ANTI-TANK ROCKET
Calibre: 88.9 mm
Length: 600 mm
Length of container: 626 mm
Weight: (without container) 2.2 kg; (with container) 3.2 kg
Weight of propellant: 300 g
Diameter of shaped charge: 80 mm
Muzzle velocity: 300 m/s
Combat range for time of flight: 1.25 s (max ordinate 1.9 m), 330 m; 1.36 s (max ordinate 2.2 m), 360 m; 1.55 s, 400 m
Max effective anti-tank range: 600 m
Max range at 45°: 2300 m
Penetration: steel plate at normal (0° obliquity) more than 400 mm; concrete (0° obliquity) 1300 mm

OTHER ROCKETS
Dimensions generally as for the anti-tank rocket except:
Weight (in container): (AP/AV) 3.8 kg; (smoke) 3.95 kg; (illuminating) 3.3 kg
AP/AV round: (penetration) 100 mm steel plate; (lethal range of fragments) 20 m plus

Smoke round: (duration) 35 s approx
Illuminating round: (intensity) 300 000 candela; (duration) 30 s

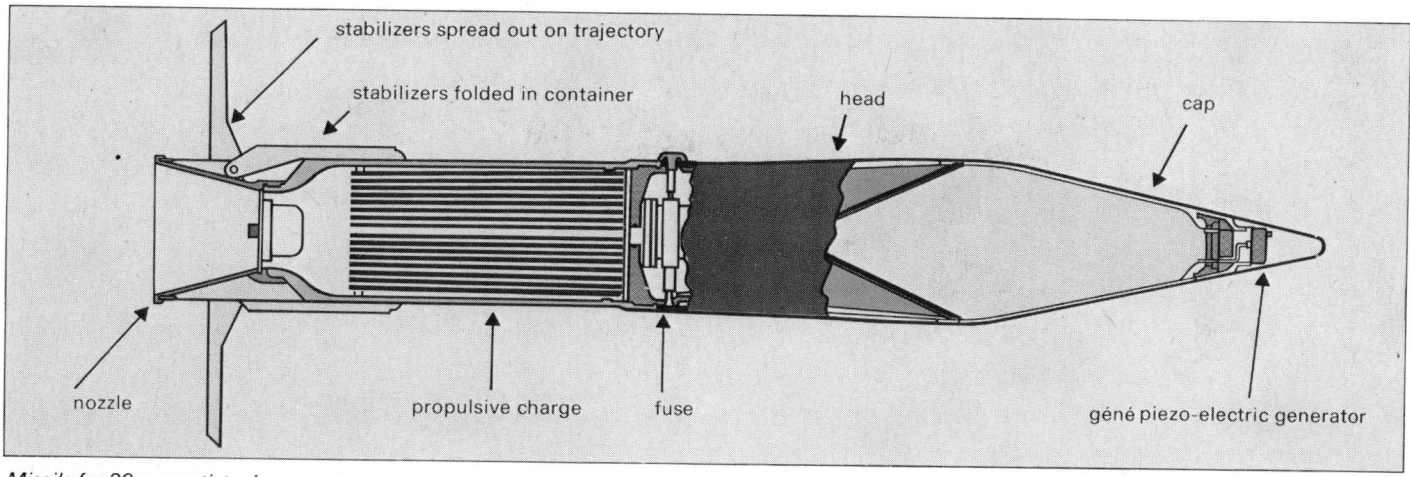

Missile for 89 mm anti-tank weapon

Manufacturer
Luchaire Defense, 180 Boulevard Haussmann, 75382 Paris Cedex 08.
Status
Current. Production.

Service
French Army, which calls it LRAC de 89 mm Mle F1. Commercial sales to foreign armies.

Luchaire Defense Wasp 58 assault rocket launcher

Specially designed for the direct attack of light armoured vehicles at short range (250 m), the Wasp 58 is a light and disposable weapon system that combines the principles of the recoilless gun and of the counter-mass rocket launcher. Efficient, compact and easily handled, it may be fired with minimal signature, including firing from a confined space.

The Wasp 58 is composed of a fully-equipped launching tube and a 58 mm

Luchaire Wasp 58 assault rocket launcher

hollow-charge warhead with significant piercing capabilities. Care has been taken in the design to facilitate bringing the weapon into use instantly, without lengthy preparation.

Due to its versatility the Wasp 58 is an ideal weapon system for small infantry units. It gives any soldier substantial firepower capability, enabling him to defeat a wide spectrum of targets. In addition to infantry use, it is also perfectly suited to commando, paratroop or heliborne operations in the enemy's rear areas.

DATA
Length: 800 mm
Weight, complete: < 3.5 kg
Projectile weight: 620 g
Projectile calibre: 58 mm
Countermass weight: 650 g
Muzzle safety: 10 m
Minimum operating range: 20 m
Maximum ordinate (at practical range): 1.90 m
Muzzle velocity: 250 m/s
Combat range: 250 m
Penetration: 300 mm steel armour at 0° incidence
Hit probability Ph 20-250 m: Stationary target 0.95; moving target 0.88
Accuracy Standard Deviation: 0.30 m

Manufacturer
Luchaire Défense, 180 Boulevard Haussmann, 75382 Paris Cedex 08.
Status
Advanced development; production at end of 1989.

SEP DARD 120 close anti-armour weapon

DARD 120 was designed to defeat main battle tanks head-on at ranges up to 600 m, with a combat range of 300 m when using the standard simple optical sight.

The configuration of the weapon is similar to the B-300 or the LRAC 89 (STRIM) and the projectile is in a 1.2 m disposable kevlar launch tube which is then attached to a re-usable firing and sighting unit which could have a life of up to 100 firings.

The projectile is launched from the tube by a charge of double-based propellent, giving it an initial velocity of 280 m/s. At the same time, a countershot of compressed plastic flakes is ejected to the rear to balance the recoil forces. The manufacturer claims that recoil is at a minimum with noise levels less than 180 dB. It gives little signature on launch and it can possibly be fired from within some buildings.

After launch, the projectile's six spring-loaded penknife fins are erected and a small sustainer motor ignites to maintain a velocity of 280 m/s. The fins give sufficient spin to maintain stability but not enough to lower the performance of the warhead which has been developed by GIAT.

The manufacturer claims that the operation of the weapon is quick and simple in the following sequence: lock the container/launch tube to the firing unit; arm the weapon; select the point of aim having assessed range and any necessary lead angle; fire; unlock the empty container/launch tube and continue the cycle.

The forward pistol grip can be pivoted through 180° allowing DARD 120 to be fired by either left- or right-handed soldiers.

The large calibre, 120 mm, should give the weapon a good performance against current main battle tanks and possibly against future main battle tanks. Details of the warhead's functioning mode have not been given other than that it contains 1.95 kg of explosive and that it can penetrate the NATO Heavy Triple Target. The inference is that it is a HEAT type warhead with a standoff of about two and a half times the weapon's calibre.

DARD 120 assembled with container/launch tube

DARD 120 firing unit

DARD 120 projectile with sustainer motor in tail

DARD 120 firing unit and launch tube

DARD 120 ready to be fired by right-handed firer

With an overall length when ready to fire of 1.6 m, a weight of 14 kg with a simple sight, a crew of two is required to operate DARD 120.

DATA
Calibre: 120 mm

LENGTHS
Container/launch tube: 1.2 m
Firing unit: 0.76 m
Ready to fire: 1.6 m

WEIGHTS
Projectile in tube: 8.9 kg
Firing unit (simple sight): 4.5 kg
Countermass: 1.7 kg
Weapon ready to fire: 14 kg
Warhead: 3.3 kg

Muzzle velocity: 280 m/s
Max range: 600 m
Combat range: 300 m
Sights: simple sight (rangefinder or FCU would add up to 6 kg to the weight but extend the combat range to 600 m)
Penetration: 820 mm
Fuze graze angle: 75°
Warhead filling: 1.95 kg
Propellant: DB cordite
Noise level: 180 dB

Manufacturer
Société Européenne de Propulsion (SEP), Tour Roussel-Nobel, Cedex 3, F92080 Paris La Défense.
Status
Development.
Service
Nil.

Eryx short-range anti-tank missile

Eryx was designed for use by forward infantry, providing them with a weapon effective against new types of armour to ranges up to 600 m. The objects in view during development were to use the most simple and compact solutions to obtain high penetration power on all armour likely to be in service during the remainder of this century, to have extreme precision, to be able to fire from enclosed spaces, and at the same time to keep the weight down to 11 kg and the unit cost compatible with broad distribution at section level.

The equipment consists of a missile pre-packed in a container tube in which it is transported and stored and from which it is launched. The compact firing unit contains the ignition, detection and timing systems, together with a remote-control. The missile can be emplaced and made ready to fire in less than five seconds. During flight (which lasts less than four seconds) the firer is required only to keep his sight on the target. The missile carries an infra-red beacon which is detected by the sight unit, corrections derived, and steering commands sent to the missile by means of a wire link which is unspooled as it flies. The application of a new concept, the direct thrust flight control, efficient even at low speed, allows the launching to be achieved using a small propulsion unit. The missile can thus be used in confined spaces. After launch the rocket motor accelerates to the flight speed of 300 m/s.

In 1989 it was announced that a Memorandum of Understanding has been signed by France and Canada with the intention of producing Eryx as a co-operative venture; industrial agreements are in course of being finalised.

DATA
Length: 925 mm
Diameter: 160 mm
Weight: 11 kg
Weight of firing post: 3.4 kg
Weight of explosive charge: 3.6 kg
Range: 40 – 600 m
Time of flight: 3.6 s to 600 m
Max velocity: 300 m/s
Penetration: 900 mm homogeneous armour

Manufacturer
Aérospatiale, Division Engins Tactiques, 2 rue Béranger, 92322 Chatillon-Cedex.

Eryx anti-tank missile in firing position

Status
Pre-series production.
Service
Adopted for French Army; initial deliveries in 1991.

Eryx anti-tank missile being fired inside building

MILAN portable anti-tank weapon

The Missile d'Infanterie Léger Anti-char (MILAN) is a weapon which a was introduced by Euromissile, an international consortium of the French Aérospatiale Group and the West German MBB Group. The design originated in France, which explains its presence in the French section, but it is now a truly international weapon and is built simultaneously by both concerns and under licence in the United Kingdom.

MILAN is designed to be used by infantry from a defensive position and the emphasis is placed on this use rather than on vehicle mounting. MILAN is a SACLOS wire controlled missile and the task of the operator is to keep his cross-hairs on the target. The flare on the missile emits an infra-red signature which enables the computer to measure the error between its position and the line of sight. The velocity is twice that of most early portable missiles and allows the MILAN missile to reach 1500 m in 10 seconds and 2 km in 12.5 seconds.

The complete weapon system is made up of two units as follows: a round of ammunition, consisting of a missile, factory loaded into a sealed launcher/container tube; a combined launching and guidance unit, consisting of a launcher combined with a periscopic optical sight and an infra-red tracking and guidance system, the whole being mounted on a tripod.

The round of ammunition comprises an assembled missile factory loaded with wings folded into a sealed tube which serves the dual purpose of storage/transport container and launching tube.

The container/launcher tube is fitted with mechanical and electrical quick-connection fittings and a self-activating battery is mounted on the outside to provide electrical power for the firing installation.

The MILAN now has a night firing capability through the addition of the MIRA-thermal imaging device adopted by the French, West German and British armies. This consists of a case weighing 9 kg which can be mounted on the standard firing post. Target detection is possible at a range of over 3 km, and firing at 2 km.

Missile

The missile is an assembly of the following main components: an ogival head containing a shaped charge and fuze; a two-stage solid propellant motor discharging through an exhaust tube to a central nozzle located at the rear of the missile; a rear part containing the jet spoiler control system and guidance components. The guidance components include: a gas driven, turbine operated gyro; an infra-red flare; a spool carrying the two guidance wires in one cable; a decoder unit; a self-activating battery for internal power supply.

The missile is launched from its tube by a booster charge gas generator which is contained in the tube and burns for 45 milliseconds. Initial velocity is 75 m/s.

The recoil effect is compensated but a part of it is used to eject the tube to the rear of the gunner to a distance of 2-3 m.

The two-stage propulsion motor burns for 12.5 seconds and increases the velocity of the missile, at first rapidly, then more slowly to 200 m/ s. The

operator must keep his sight cross-hairs on the target throughout the engagement.

Guidance is achieved by means of a single jet spoiler operating in the sustainer motor exhaust jet. The jet spoiler operates on guidance command signals generated automatically by the launcher/sight unit (by measurement of the angular departure of the missile from the reference directions of the infra-red tracker in the sight unit) and transmitted to the missile via the guidance wires which unwind from the missile.

MILAN being fired from light vehicle

MILAN tactically sited

MILAN fitted with MIRA night firing device

MILAN being unloaded from helicopter

The guidance commands are decoded by a transistorised decoder unit within the missile. The self-activating battery which provides internal power is designed for long-term storage and use in world-wide temperature conditions.

MILAN 2, introduced in 1984, is a warhead assembly intended to improve performance against the front face of main battle tanks. The warhead has been increased in diameter from 103 mm to 115 mm, the explosive weight has

been increased from 1.2 kg to 1.8 kg and the fuze initiating switch is located at the tip of a probe which adds about 145 mm to the length of the warhead. These modifications have considerably improved the piercing performance, largely by improving the standoff distance and the explosive effectiveness of the shaped charge. The improvement is calculated to be some 65 per cent, and penetrations into steel armour of 1060 mm have been recorded. The MILAN 2 missile does not affect the interface with the firing post, the overall missile weight remains the same and the tubular container is unchanged. Existing firing posts can therefore fire both MILAN 1 and MILAN 2 missiles.

MILAN 2T is an improved weapon using a tandem warhead and extended nose probe. The probe contains the primary shaped charge which is designed to detonate reactive armour protection and thus clear the line of attack for the main shaped charge. No modification will be required to the existing launchers, and it is anticipated that MILAN 2T will be fully operational in 1992.

Safety arrangements

The missile is locked inside its tube and the solid propellant gas generator cannot be ignited until the missile is unlocked by the gunner.

The sustainer motor ignites when the missile is released from the tube and the wings have unfolded.

The fuze cannot arm until the sustainer motor is ignited and an electrical safety device functions when the missile has flown approximately 20 m.

DATA
MISSILE
Type: wire guided SACLOS system
Weight: (ready to fire) 6.65 kg
Length: 769 mm
Mid-body diameter: 90 mm
Warhead diameter: 115 mm
Wingspan: (wings unfolded) 265 mm
Warhead weight: (with fuze) 2.98 kg
Warhead filling: 1.80 kg
Cone diameter: 115 mm
Fuze: 0.3 kg

AMMUNITION (TACTICAL PACKAGE)
Weight in carrying condition: 11.8 kg
Weight, ready to fire: 11.3 kg
Length: 1260 mm
Body diameter: 133 mm

LAUNCHING/GUIDANCE UNIT (FOLDED)
Weight: 16.4 kg
Length: 900 mm
Height: 540 mm
Width: 420 mm
Traverse: 360°
Elevation: up to 20°

PERFORMANCE
Velocity: (at launch) 75 m/s; (at 2 km) 200 m/s
Time of flight to max range: 12.5 s
Effective range: 25-2000 m
Chance of a hit: (0-250 m) average 75%; (250-2000 m) greater than 98% (maker's figures)
Warhead: shaped charge. Detonated by electrical connection produced by crush-up of ogive when missile hits. Minimum performance is the triple penetration of NATO heavy tank plate

Manufacturer
Aérospatiale, Division Engins Tactiques, 2 rue Béranger, 92322 Chatillon Cedex, France.
Messerschmitt-Bölkow-Blohm, 8012 Ottobrunn bei München, West Germany
Licensed manufacture in the UK by British Aerospace, Dynamics Division, Six Hills Way, Stevenage, Hertfordshire SG1 2DA, England
Marketed by Euromissile, 12 rue de la Redoute, 92260 Fontenay-aux-Roses, France.
Status
Current. Production.
Service
In service in France, West Germany and the UK. Also being supplied to other NATO and non-NATO countries.

APILAS light anti-armour weapon

APILAS (Armour Piercing Infantry Light Arm System) is a product of the Matra Manurhin Défense Company. It is a disposable, manportable launcher containing a projectile carrying a HEAT warhead propelled by a fast burning rocket motor. APILAS has several interesting characteristics.

The rocket is supplied pre-packed in the aramid-fibre launch tube which has a retractable sight; this is bore-sighted in the factory during assembly and needs no field adjustment. A separate box containing two lithium batteries is clipped to the launcher, and in order to fire three successive actions are required: rotation of a security switch for mechanical arming, pressure on an arming switch, and finally pressure on the trigger.

The rocket motor cone is of wound aramid-fibre construction, with an aluminium venturi. The shaped charge warhead is initiated by an electrical fuze system powered and armed by gas pressure released when the rocket is launched. There is a mechanical delayed-arming system which guarantees the warhead to be unarmed at 10 m from the launcher and fully armed from 25 m onward. The warhead has a long tapered nose cone which permits normal functioning up to incidence angles of 80°, and the fuzing system will ensure detonation even on graze contact.

The training system comprises a sub-calibre firing weapon using special 7.5 mm tracer bullet cartridges, and a drill weapon.

APILAS in launching position

APILAS light anti-armour weapon launcher

WEIGHTS
Launcher: 4.7 kg
Projectile: 4.3 kg
Weight of explosive: 1.5 kg
Complete weapon: 9 kg

Effective range: 330 m
Time of flight: 1.2 s to 330 m; 1.9 s to 500 m
Muzzle velocity: 293 m/s
Penetration: (into RHA) more than 700 mm; (into reinforced concrete) more than 2000 mm

Manufacturer
Matra Defense, 21 Avenue Louis Breguet, BP 60, 78145 Vélizy-Villacoublay Cedex.
Status
Current. Production.
Service
France, Finland, Italy, Jordan.

DATA
Calibre: 112 mm

LENGTHS
Launcher: 1.29 m
Projectile: 925 mm
Warhead stand-off distance: 3 cal

Apajax autonomous anti-tank system

This is a combination of the Matra Apilas anti-tank launcher and the British Aerospace Ajax sensor and control system, which together form an autonomous and controlled anti-armour system. The infra-red detection system is alerted by acoustic and seismic sensors by the approach of an armoured vehicle. The system is programmed to engage armoured vehicles at ranges up to 150m. As soon as a target is detected, the computer in the system determines the optimum firing point, and automatically fires the weapon at the appropriate time. The system is capable of destroying any existing tank whilst moving at up to 80 km/h in lateral attack, from -45° to +45° to the line of fire. The system can engage targets within the range bracket 3 m to 150 m from the launcher.

Further refinements of the control system permit the system to be active or dormant at will; it can be programmed to be active in four-hour steps up to 96 hours from installation and thereafter up to 40 days. The system discriminates between types of target and will not respond to marching troops, light vehicles, animals or other false alarms. It is also possible to programme the system to engage a specified vehicle among the first six in a column. The direction of the target is programmable, so that, for example, the system can be instructed to ignore targets proceeding from right to left, but engage those moving left to right.

The system can be emplaced in any terrain by one man in less than five minutes by day and 10 by night, and is self-neutralising by the loss of power in the system battery at the end of the programmed active period. Optionally, it can be arranged to self-sterilise the electronic circuitry at the end of the programmed period, and can be fitted with anti-handling devices which result in the device self-sterilising itself in the event of any attempt to move it.

A training version firing a 7.5 mm sub-calibre bullet is available.

DATA
Length of folded system: 1.00 m
Height of deployed system: 0.65 m
Total weight: 12 kg

Manufacturer
Matra Defense, 21 av Louis Bréguet, BP 60, 78145 Vélizy-Villacoublay Cedex.
British Aerospace Dynamics Division, Downshire Way, Bracknell, Berks RG12 1QL, England
Status
Current; available.

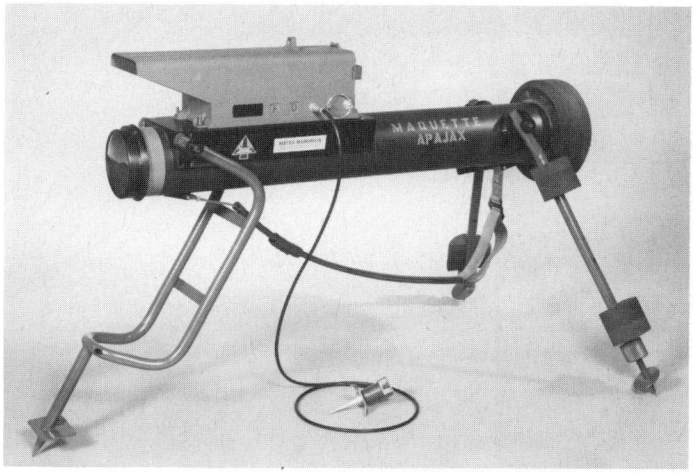

Matra/BAe Apajax system

Matra/BAe Apajax system in operation

GERMANY, WEST

Panzerfaust 3

The Panzerfaust 3 (Pzf 3) is a compact, lightweight, manportable, shoulder-fired, unguided weapon system providing effective engagement capabilities against current and future Main Battle Tanks (MBTs). The Pzf 3 system, which was developed under contractual agreement with the West German Army, consists of a disposable cartridge with 110 mm warhead, and a re-usable firing and sighting device.

The Pzf 3 concept was selected by the West German Army because it fulfils the following basic tactical requirements: high kill probability against current and future MBTs, even when frontally engaged; capability of being fired from enclosed spaces; low cost, so that it can be widely distributed to all types of military formation.

The weapon system is based on the Davis countermass principle; adoption of this principle permits firing the weapon in enclosed spaces, vitally necessary for effective anti-tank defence in built-up areas. Moreover the system admits potential growth, since it is possible to modify the warhead shape and calibre to meet different tactical requirements without affecting the launch tube. To emphasise the versatility of the system, development of HEAT, multi-purpose fragmentation, smoke and illuminating warheads has formed part of the project.

Night combat capability is achieved by use of an infra-red target marker fitted to the telescope sight mount and used in conjunction with infra-red goggles. Details will be found in the Sighting Equipment section.

An autonomous firing mount known as 'Fire Salamander' has been developed. This carries four Pzf 3 equipments around a central command module which can contain a variety of sensors for automatic firing. It can use a remote-controlled TV camera as an aiming device.

DATA
Calibres: (launcher) 60 mm; (projectile) 110 mm
Weight: 12 kg
Length: 1.2 m
Projectile weight: 3.8 kg
Weight of firing device: 2.4 kg
Range: (moving targets) 300 m; (stationary targets) 500 m
Velocity: (initial) 170 m/s; (max) 250 m/s
Vertex: (300 m range) 2 m
Time of flight: (to 300 m) 1.3 s
Safety distance: (rear) 10 m
Penetration: in excess of 700 mm

Manufacturer
Dynamit Nobel AG, Postfach 1261, D-5210 Troisdorf.
Status
Adopted by West German Army June 1987. Volume production 1990.
Service
West German Army.

'Fire Salamander' autonomous firing mount with TV sensor

Warheads for Panzerfaust 3 (left to right): HEAT 125 mm; HEAT 110 mm; HEAT 90 mm; practice; HESH; IR screening smoke; screening smoke; multi-purpose projectile; illuminating

Panzerfaust 3 in firing position

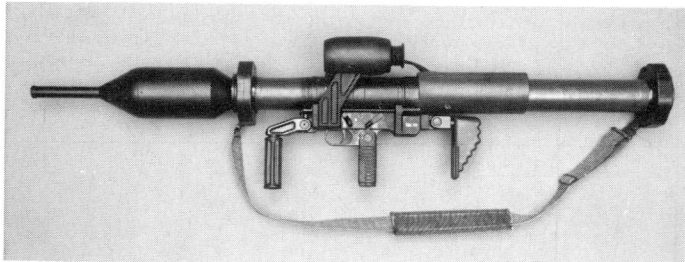

Panzerfaust 3

Panzerfaust Off-route Mine System

The Panzerfaust 3 can be adapted for use as an unattended sensor-operated off-route mine by the addition of the SIRA acoustic-infrared sensor manufactured by Honeywell. The SIRA unit is fitted to the weapon by means of the dovetail on the Panzerfaust firing device.

The sensor will detect targets approaching the position by means of its acoustic function and activates the infra-red sensor. Once activated, the infra-red unit will precisely determine the target position by evaluation of the speed, direction of movement, temperature and distance. This evaluation is performed by a microprocessor which then fires the weapon at the optimum moment. The system has an all-weather, day and night capability, and the multi-sensor principle provides for a high degree of functional reliability. Background noise and countermeasures will be rejected by a sophisticated algorithm especially developed for this application.

DATA
Target speed range: 30-60 km/h
Effective range: up to 150 m
Dimensions of sensor: 100 × 150 mm
Operational time: up to 40 days
Power supply: 2 × 3.4 V lithium cells
IR sensor: passive, two-colour sensitive
IR optics: double parabolic, off-axis
Acoustic sensor: capacitative microphone
Processor: 8-bit CMOS
Special functions: programmable target counting; direction of movement discrimination

Manufacturer
Dynamit-Nobel AG, Defence Division, D-5210 Troisdorf.
Status
Advanced development.

Panzerfaust 3 off-route mine with SIRA sensor

PZF 44 2A1 (Lanze) portable anti-tank weapon

The PZF 44 2A1, also known as Lanze and often simply referred to as the Panzerfaust, is a recoilless, one-man weapon designed for close-range anti-tank engagement.

The idea of such a weapon dates back to the closing stages of the Second World War and the PZF 44 has been in service since the early 1960s. Dynamit Nobel has recently produced a greatly improved ammunition which has extended the useful life of the Panzerfaust.

The weapon consists of a tube, open at both ends, with a pistol grip and firing mechanism underneath at the point of balance. At the front end is a forward pistol grip. The telescopic sight is mounted on the left side just forward of the rear pistol grip.

The ammunition is in two parts. It consists of the warhead and rocket motor, and the propelling charge and counterweight system contained in a tubular cartridge.

The propellant cartridge is inserted into the muzzle of the launcher and pushed down until only the top is projecting. The rear end of the rocket motor housing is then pushed down into a detent on the top of the propellant cylinder and the two are locked together. The rocket body is then pushed down until only the warhead is protruding. The precise position is determined by a stop under the warhead.

When the weapon is fired the propellant charge is ignited and a genuine recoilless effect is obtained by the missile being discharged in one direction and a nearly equal mass of powdered iron being driven to the rear. This iron consists of a very large number of small particles with little mass and soon drops to the ground. As soon as the missile is free of the tube the six fins are extended, driven by a spring-loaded sleeve. The propellant ignites the rocket motor which burns when it is clear of the firer and the rocket is accelerated as it moves towards the target. The acceleration forces arm the detonator which is fired by a piezo-electric voltage induced when the nose strikes the target. It can be fired from closed rooms. HEAT, multi-purpose and practice warheads are available.

DATA
WEAPON
Calibre: 44 mm
Length: (in travelling position) 880 mm; (in firing position) 1162 mm
Weight: (of telescope) 0.75 kg; (of weapon with telescope and carrying case) 7.8 kg; (in firing position) 10.3 kg
Life: in excess of 2000 firings
GRENADE
Diameter of warhead: 67 mm
Length: 550 mm
Weight: 1.5 kg
PROPELLANT CHARGE
Diameter: 44 mm
Length: 538 mm
Weight: 1 kg
FIRING CHARACTERISTICS
Velocity: (initial) 168 m/s; (max) 210 m/s
Penetration at normal: 0° obliquity 370 mm
Combat range: (moving targets) 300 m; (stationary targets) 400 m

PZF 44 2A1 fired from shoulder

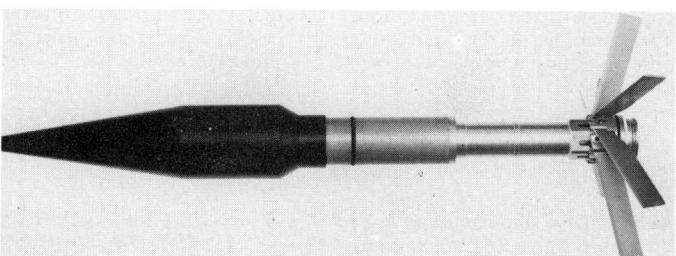

Projectile with fins unfolded

Manufacturers
Dynamit Nobel AG, D-5210 Troisdorf; Heckler and Koch GmbH, 7238 Oberndorf/Neckar
Status
Obsolescent. No longer in production.
Service
West German Army.

Warhead and propelling charge of PZF 44 2A1

Armbrust short-range anti-armour and self-defence weapon

Armbrust has unique features which set it apart from most other short-range anti-armour weapons. Armbrust has no firing signature, emits neither smoke nor blast from the muzzle nor flash from the rear, is quieter than a pistol shot, can be fired from small enclosures or roofed foxholes without danger or discomfort to the firer, has no recoil, requires no maintenance and weighs only 6.3 kg.

Primarily, Armbrust is a support weapon for all levels of command, and is particularly suited for combat under special conditions. Its characteristics provide Armbrust with a wide scope of applicability not attainable with any other short-range armour weapon.

The anti-armour warhead penetrates 300 mm of rolled homogeneous armour at 0°. In addition it also pierces such materials as stone, reinforced concrete and so on. Armbrust is ideally suited to targeting a vehicle at its weaker points; thus a gunner can engage a vehicle from an upstairs window without endangering himself. Furthermore it can generate a higher rate of fire since there is no disclosure of the firing location to the target.

Armbrust is a manportable, shoulder-fired, expendable weapon with a maximum range of about 1500 m and an operational range against tanks and armoured vehicles of up to 300 m. Its main characteristics are: (1) low

signature; no flash, no smoke, no blast and low noise, (2) low IR detectability and (3) a small rear danger area; it can be safely fired from small enclosed rooms, requiring a minimum distance between weapon and wall of 0.8 m only. For firing it requires no additional appliances, no testing or maintenance. The launcher is thrown away once the weapon has been fired. Armbrust offers the soldier safe concealment; he has a very low silhouette due to the side-mounted reflex sight, needs no ear protection, and does not flinch when pulling the trigger, so that he can concentrate fully on aiming.

DATA
Length of weapon: 850 mm
Weight: approx 6.3 kg
Time of flight to 300 m: approx 1.5 s
Normal penetration: 300 mm homogenous armour at 0°
Fuze ignition at impact angles: up to 78°

Manufacturer
Messerschmitt-Bölkow-Blohm GmbH, Defence Systems Group, PO Box 80 11 49, D-8000 Munich 80.
Status
Current. Production.

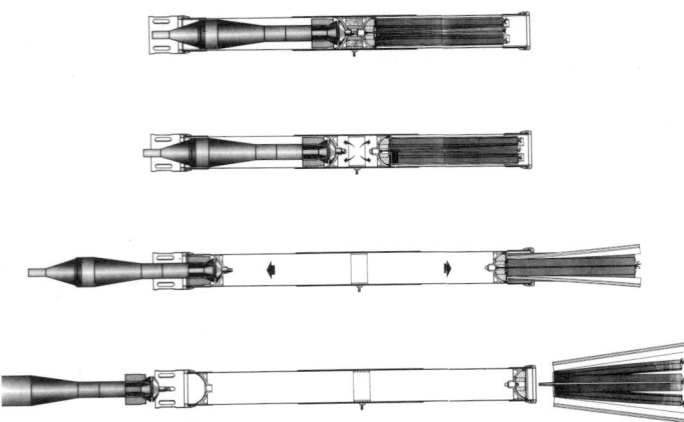

Firing sequence for Armbrust from pre-ignition (top) to expulsion of missile and counter-mass with pistons locked (bottom)

Firing Armbrust from enclosed position

Diehl 106 mm fragmentation shell

Although this is not an anti-tank projectile, it is included in this section since it affords a useful extension to the utility of the 106 mm RCL gun, giving it the ability to engage soft targets.

The Diehl projectile consists of a flat-based shell with pre-engraved driving band, fitted with a nose fuze. Inside the shell is a central tube containing high explosive, and the annular space between this tube and the shell wall is filled with some 10 000 steel balls held in a polyester resin matrix. The explosive charge is RDX/WX five per cent, with a tetryl exploder beneath the fuze.

The fuze is a highly sensitive impact type which can be set for instantaneous action or delay. A graze mechanism makes it sensitive to shoulder impacts, and this permits the fuze, when set to delay, to be used for ricochet firing. In this method the shell is aimed to strike the ground a short distance in front of the target; the impact initiates the fuze delay action, which takes place as the shell ricochets from the ground, so that detonation takes place when the shell is several feet in the air over the target. This ensures the optimum distribution of fragments across the target area and is a highly effective anti-personnel tactic. A skilful gunner can also 'bounce' the projectile into doors, caves, and beneath cover.

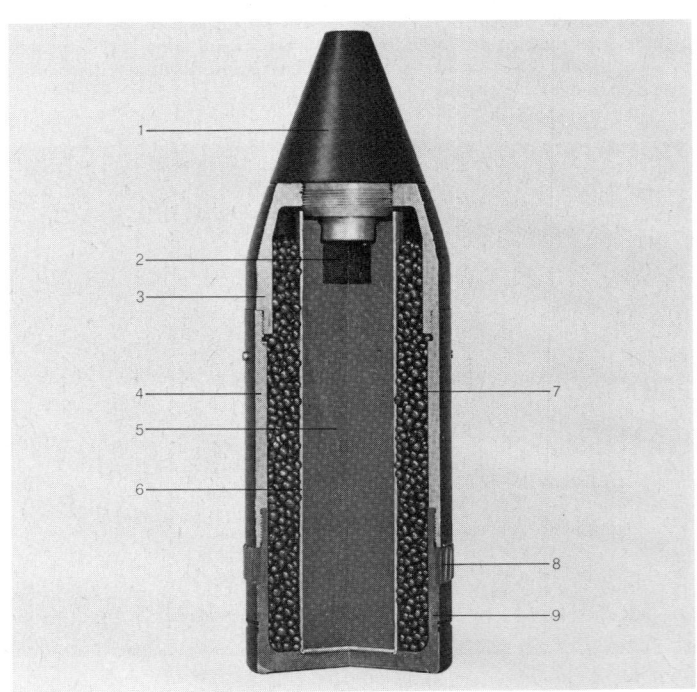

Diehl 106 mm fragmentation projectile. (1) fuze (2) tetryl exploder (3) aluminium ogive (4) aluminium shell wall (5) RDX/WX burster (6) aluminium burster tube (7) steel balls in resin (8) copper driving band (9) steel base

DATA
Calibre: 106 mm
Weight of shell: 8 kg
Weight of explosive: 770 g
Weight of fragments: 3.5 kg
Fuze arming bracket: 30 – 70 m
Fuze delay setting: 0.06 s

Manufacturer
Diehl GmbH & Co, Wehrtechnik, Fischbachstrasse 16, D-8505 Röthenbach.
Status
Current. Production on request.
Service
Austrian army.

Fuze MTSQ 504

This fuze is a modification of the DM42MT nose fuze and is intended for use with the 84 mm Carl Gustaf HE/Fragmentation projectile, whereas the DM42MT is for use with the illuminating shell. The fuze is set according to the range in metres, from a minimum of 60 m to a maximum of 1250 m.

Manufacturer
Junghans Feinwerktechnik, PO Box 110, D-7230 Schramberg.
Status
Current. Production.

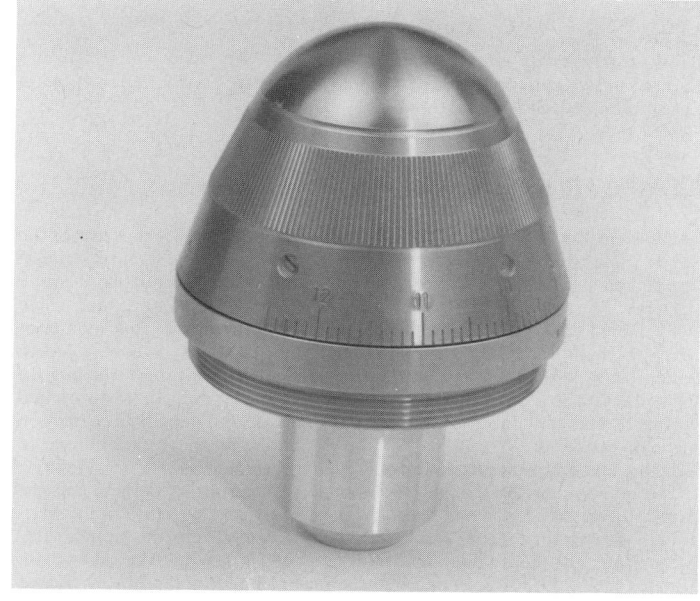

Junghans MTSQ 504 fuze

ISRAEL

B-300 light anti-armour weapon

The B-300 is a manportable, shoulder-fired, semi-disposable anti-armour system with an effective range of 400 m. It is carried, loaded and fired by one man. The launcher and three rounds weigh 19 kg.

The system has a calibre of 82 mm and is very similar to the French LRAC 89. It breaks down into two major sections: the launcher, with a variety of sight options, which is re-usable; and a projectile in a sealed container that is disposed of after firing. The grp container acts as an extension to the launcher during firing and is coupled quickly to the launcher making all the necessary mechanical and electrical connections. The life of the launcher is not known.

In addition to the basic Mk 1 HEAT round there are a Mark 2 HEAT round, slightly heavier than the Mark 1 but with significantly better penetration (ca 550 mm); a High Explosive Follow-through (HEFT) round which blows a hole in the target by means of a shaped charge and then launches a secondary charge through the hole and into the interior of the target; an illuminating round of 600 000 candle-power and a range of 1700 m; practice rounds with an impact marker for day or night use; and dummy

Four illustrations showing B-300 light anti-tank system in field use

rounds with a noise cartridge for loading and firing practice. A further training aid is a 9 mm sub-calibre device. The projectile has eight penknife fins folded forward around the motor nozzle. No details have been given of the motor which accelerates the projectile after an initial launch velocity of about 270 m/s.

The launcher is provided with an in-built battle sight although a stadia telescopic sight is also available. The stadia sight is equipped with a betalight for use in poor visibility. This can be switched in as required. For night operation, where white light is not available, a starlight telescopic sight can be mounted using an adaptor provided with each launcher.

The launcher is equipped with a folding bipod and shoulder rest and a broad sling. The B-300 is claimed to have a good hit probability.

DATA
Calibre: 82 mm
Launcher: weight: (unloaded) 3.65 kg; (loaded) 8 kg; (in wooden box) 14 kg. Length: (unloaded) 775 mm; (loaded) 1400 mm

Round: weight: 4.5 kg; (projected mass) 3.1 kg. Length: 725 mm
Weapon weight plus 3 rounds: 19 kg
Initial muzzle velocity: approx 270 m/s
Range: 400 m
Temperature range: –10 to +60°C
Penetration: in excess of 400 mm at a graze angle of 65°
Into action time: 20 s
Packaging, launcher: wooden box 400 × 860 × 190 mm weighing 14 kg
Ammunition: 6 rounds in a wooden box 450 × 840 × 300 mm weighing 30 kg
Ammunition backpack: 3 rounds weighing 15.5 kg

Manufacturer
Israel Military Industries (IMI), PO Box 1044, Ramat Ha Sharon 47100.
Status
Development.

MAPATS anti-armour missile system

This system, developed by Israel Military Industries, exhibits elements of the Soviet Sagger and US TOW systems, so much so that during its early stages it was codenamed 'Toger'. The system consists of a beam-riding missile and a launcher. The missile is packed in a glass fibre launch and storage tube which is assembled to the launching unit. After launch, the missile follows a laser beam which is directed at the target and maintained there by the operator. An optical sight, with the laser slaved to its axis, is used for target tracking. In flight the missile detects the presence of the beam by using a rear-mounted sensor, detects any deviation from the beam axis, and generates correction manoeuvres so as to stay on the line of sight. The system is immune to jamming and entirely autonomous.

The components of the MAPATS system are launch tube and missile, tripod, traverse and guidance unit, and the night-vision system. The missile is housed in a factory-sealed container which is loaded into the launch tube. After firing, the empty container is discarded. Setting up takes about two minutes, including an automatic self-test routine. Loading subsequent rounds takes only a few seconds. On firing, the missile leaves the launch tube, the ejector motor falls away, and the laser beam modulator starts to transmit signals to the sensor in the missile. The laser beam cross-section is kept constant by a zoom system, creating a constant corridor which coincides with the line of sight. Beam intensity on the target is initially very low, increasing as the missile approaches.

DATA
Calibre: 148 mm
Missile length: 1.45 m
Missile weight: 18.5 kg
Launch velocity: 70 m/s
Flight velocity: 315 m/s
Time of flight to 4 km: 19.5 s
Max range: 5000 m
Firing angles: Traverse 360°; elevation -20—+30°
Warhead: 3.6 kg shaped charge
Penetration: in excess of 800 mm
Weight of tripod and battery: 14.5 kg
Weight of traversing unit: 25 kg
Weight of guidance unit: 21 kg
Weight of launch tube: 5.5 kg
Weight of night-vision system: 6 kg
Weight of missile in container: 29.5 kg

Manufacturer
Israel Military Industries, PO Box 1044, Ramat Ha Sharon 47100.
Status
Current. Production.
Service
Israeli army.

Electro-optical stand-off sensor

This device, which has been developed by Reshef Technologies in conjunction with International Technologies (Lasers) Ltd, is an electro-optical sensor which detects a target and triggers an output at a predetermined distance (presently 1 m) before reaching the target. An accuracy of better than 10 per cent has been achieved in laboratory tests, with approach velocities between 100 and 1000 m/s.

The sensor uses a low-cost infra-red LED. Distance sensing is obtained by the triangulation of narrow transmitting and receiving beams. Additional resolution within the common overlap of the beams is obtained by signal processing. This and signal modulation provide high immunity to noise, background illumination and interference, and provide the sensor with a very wide dynamic range. Tests have demonstrated performances within

specification for target inclination between 0° and 80° for all practical target surfaces.

The sensor detection algorithm also provides immunity to volume scattering, such as that due to smoke and dust. Such attenuation, typical of a 2 m visibility condition, does not degrade the sensor's performance.

A sensor for a 1 m stand-off distance requires an overall diameter of about 70 mm.

Manufacturer
Reshef Technologies Ltd, 16 Hacharoshet Street, 60375 Or-Yehuda.
Status
Advanced development.

IMI 127 mm Tandem warhead

This is a warhead design which is claimed to be highly effective against reactive armour and composite armour. It uses a front probe which carries a small precursor shaped charge of explosive which activates the explosive charge of the reactive armour. This also initiates a time delay which then detonates the main charge. This delay ensures that the explosion of the reactive armour is completed and that there is no debris to deflect or hinder the main shaped charge jet. The probe is normally recessed into the warhead and locked in place by a bore-riding pin; on leaving the launch tube, this pin ejects and allows a spring to drive the probe out to its functioning position.

The principal charge is initiated at the correct stand-off distance by an electro-optical proximity fuze.

Although the first design is for a 127 mm warhead, IMI say that the technology can be equally well applied to other calibres. In 127 mm calibre the warhead is capable of defeating 1020 mm of conventional armour protected by reactive armour.

Manufacturer
Israel Military Industries, PO Box 1044, Ramat Ha Sharon 47100.

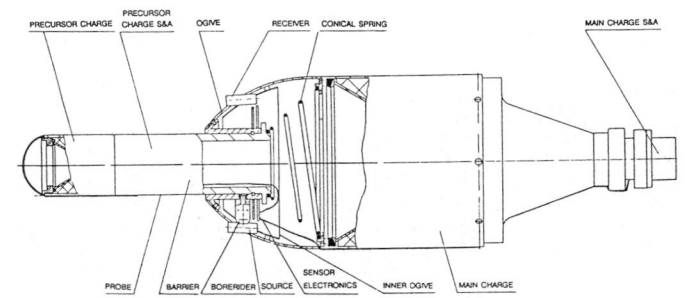

IMI 127 mm tandem warhead

Status
Development.

ITALY

Mosquito

Mosquito was produced by Contraves Italiana SpA as a one man infantry anti-tank weapon. A relatively early manual command to line of sight system, it is optically-aimed, wire guided and roll-stabilised.

It entered production in 1961 and has been in service in the Italian Army for many years. Production ceased some years ago, however, and the missile must now be regarded as obsolescent. A description appeared in *Jane's Infantry Weapons 1976*.

Mosquito manportable anti-tank missile

Folgore anti-tank weapon

Folgore is the name given to a light, recoilless, anti-tank rocket projector by Breda Meccanica Bresciana. It has a maximum range of 4500 m, and a maximum effective anti-tank range of 1000 m. Folgore can be fired from the shoulder or from a tripod.

In the tripod version, two men perform the function of aimer and loader, and the aimer is aided in his task by an optical-electronic system fixed to the launcher, which enables him to estimate target range, target speed and elevation angle in a few seconds. In the shoulder version, the weapon is provided with a lighter optical device and with a bipod and it can be carried, loaded and fired by one man. The ammunition, the same for both versions, is composed of a rocket and a launching charge; the rocket is fin-stabilised and has a hollow-charge warhead.

DATA
Calibre: 80 mm
Muzzle velocity: 385 m/s
Max velocity: 500 m/s
Max trajectory height: (at 500m) 2.2 m; (at 700 m) less than 3 m
Range: 50 m (minimum); 4500 m (max); 1000 m (effective anti-tank)

TRIPOD VERSION
Launcher (recoilless gun, tripod and optical-electronic system)
Length: 1.80 m
Weight: (including optronic sight) 27 kg
Range: (minimum) 50 m; (max) 1000 m

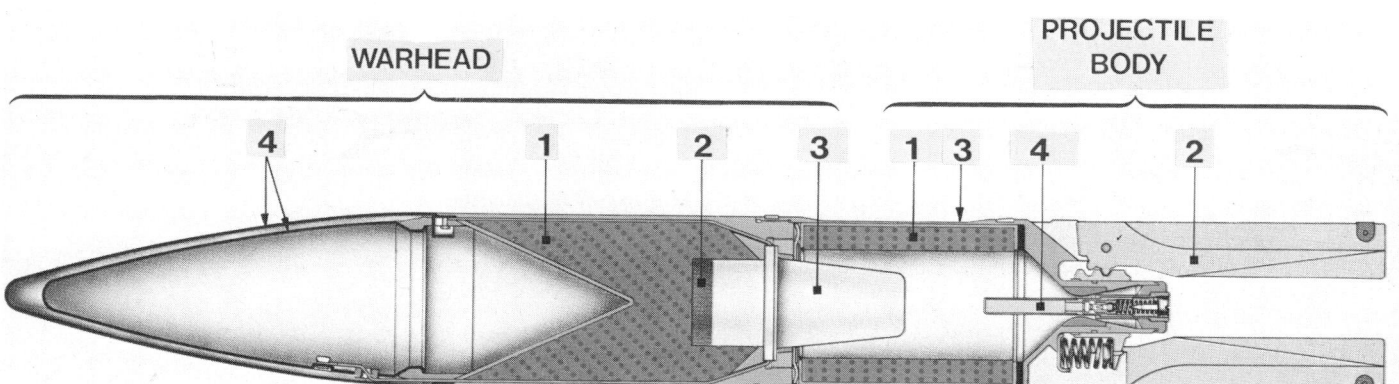

Folgore projectile
Warhead: **(1)** *Hollow-charge explosive, with Teflon wave-shaping lens* **(2)** *Cone is made of pure copper, flow-turned.* **(3)** *Fuze. Electric system using coil induced current with double safety. First safety is released only when projectile reaches fixed level of acceleration. Second will only fire on impact. Nose cone* **(4)** *is double skin which crushes on impact and closes firing circuit.*
Projectile body: **(1)** *is double-base extruded cylindrical motor, firmly held in steel motor casing* **(3)** *Pyrotechnic igniter* **(4)** *ensures even ignition of motor so that burning is regular and build up of pressure rapid. Motor is not lit until projectile is safe distance from launcher. Aerodynamic fins* **(2)** *are spring-loaded and open when projectile leaves muzzle. They are canted to give slow roll during flight and reduce errors in trajectory.*

Folgore on tripod mount. Gunner is using optronic sight

Folgore can be carried by one man. In this version, without tripod and optronic sight, Folgore is effective out to 700 m

BIPOD VERSION (also used for shoulder firing)
Launcher (recoilless gun and simplified optical system)
Length: 1.80 m
Weight: 18.0 kg

AMMUNITION
Round: (length) 740 mm; (weight) 5.2 kg
Rocket weight: 3 kg

Manufacturers
Launcher: Breda Meccanica Bresciana, Via Lunga 2, 25126 Brescia
Ammunition: Breda Meccanica Bresciana in co-operation with BPD Difesa e Spazio
Status
Current. Production.
Service
Adopted by Italian Army.

MAF manportable missile system

MAF is the name given to a long-range anti-tank weapon system of advanced SACLOS type now under final development by OTO-Melara, in co-operation with ORBITA of Brazil, in order to meet a requirement stated by the Brazilian Army.

The system consists of a missile, in its container launcher, and a firing post. Guidance is by laser beam riding, the missile being subsonic. The firing post consists of a tripod, projector unit and sights.

The missile is provided with a two-stage solid fuel rocket motor. The first stage is all-burnt inside the launch tube; the second stage ignites at a safe distance from the launcher and boosts the missile to maximum speed.

The guidance beam is obtained from a laser modulating emitter operating in the near-IR band. The beam is generated in the launcher unit and the receiver is in the rear of the missile body, giving a very high degree of protection against jamming. Missile guidance is ensured by a servo loop which corrects the missile trajectory as a function of the deviations from the line of sight. The sighting system in fair weather is by means of an optical sight, and by night through a thermal imaging camera.

The missile configuration is of the canard type (roll free) with conical nose and tail. Folding stabilisers are mounted at the rear to provide high manoeuvring capabilities and aerodynamic damping. The warhead is a shaped charge with very high penetrative ability.

The maximum range is in excess of 3 km. The guidance system ensures a processable signal at more than 2000 m in the worst conditions of optical visibility, humidity, temperature and turbulence.

The present system is manportable; vehicle and helicopter-borne applications are being studied.

DATA
Launcher post weight: 23 kg
Missile weight: 14.5 kg
Length of the launcher/container: 1380 mm
Weight of missile and container: 20 kg
Missile length: 1380 mm
Body diameter: 130 mm
Warhead weight: 4.1 kg
Range: (minimum) 70 m; (max) 3000 m
Max speed: 290 m/s
Time of flight to 3 km: 16 s

Manufacturer
OTO-Melara SpA, Via Valdilocchi 15, Casella Postale 337, 19100 La Spezia.
Status
Advanced development.

MAF missile and firing post

MAF Missile with operating crew

JAPAN

KAM-3D anti-tank system

Work started on this manual command to line of sight anti-tank guided missile in 1957. The prime contractor was Kawasaki Heavy Industries, which was responsible to the Technical Research and Development Institute of the Japanese Defence Agency. The missile was finally called the Type 64 ATM and after prolonged trials it was adopted as standard equipment in the Japanese Ground Self-Defence Forces. It now equips all 13 of the Divisions.

The missile has a cylindrical body with cruciform wings incorporating full width spoilers for control. It carries a flare which enables the operator to track it by day using an optical system with a thumb-button control box. By night the sustainer rocket exhaust provides the required illumination. The missile is placed on a metal frame launcher, which gives an angle of 15°, by the two man

crew. When launched, a booster motor accelerates the missile to cruising speed (85 m/s) in 0.8 second, after which a sustainer motor maintains this speed.

Missiles can be fired singly or in multiple units by infantry or from Jeeps or helicopters.

During trials it was found that three out of four unskilled operators were able to hit a target with their first round after completion of a two-week training course. Skilled operators could score 19 hits in 20 firings.

A field test set (KAM-3TE) and a simulator (KAM-3TP) are made by the main contractor.

KAM-3D missile on its ground launcher

Firing KAM-3D from Jeep mounting

DATA
Length: 1015 mm
Body diameter: 120 mm
Wing span: 600 mm
Launch weight: 15.7 kg
Cruising speed: 85 m/s
Range: (minimum) 350 m; (max) 1800 m
Crew: 3

Manufacturer
Prime contractor, Kawasaki Heavy Industries Ltd, World Trade Center Building, 4-1 Hamamatsu-cho, 2-chome, Minato-ku, Tokyo.
Status
Current. Production.
Service
Japanese forces.

KAM-9 (TAN-SSM) anti-tank missile system

KAM-9 is an extended range SACLOS anti-tank weapon with a better performance than the KAM-3D and suitable for use against armoured targets on both land and water.

The missile is launched from a tubular container which is also used for transport and storage. A solid-propellant launch motor ejects the missile from the container to a safe distance from the operator; the flight motor then ignites and accelerates the missile to its cruising speed in a few seconds.

Prior to firing the container is placed on the launcher which comprises firing control device, missile check-out device, tracking mechanism and installs sight unit. The optical sighting device is designed to be operated by one man who, during the missile's flight, keeps his sight trained on the target. Sensors in the sight unit translate any course deviation by the missile into electrical signals which are fed into a computer; and this calculates and generates course correction signals which are communicated to the missile by way of the guidance cable.

The launching system has two modes: direct and separate fire.

Full operational details have not yet been released. It is believed, however, that the following data are substantially correct.

DATA
Missile: (length) 1500 mm; (diameter) 150 mm; (span) 330 mm
Container: (length) 1700 mm; (diameter) 200 mm
Propulsion: solid propellant booster and sustainer
Guidance: SACLOS. Optical sighting; probably infra-red tracking; wire command
Warhead: hollow charge with piezo-electric fuze or semi-armour piercing HE with contact/electro-magnetic fuze
Range: not released. Presumably in the 3-4000 m region

Manufacturer
Kawasaki Heavy Industries Ltd, World Trade Center Building, 4-1 Hamamatsu-cho, 2-chome, Minato-ku, Tokyo.
Status
Current. Production.
Service
Japanese Ground Self Defence Force.

KAM-9 launcher

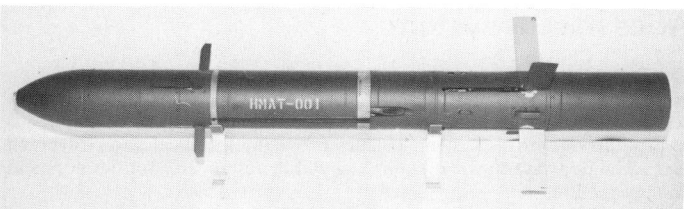

KAM-9 missile

PAKISTAN

RPG-7 Weapon System

The Soviet RPG-7 anti-tank rocket launcher is now manufactured in Pakistan. The design, operation and basic dimensions are precisely the same as the Soviet model, to which further reference should be made. The performance figures given below are those quoted by the makers and differ somewhat from those associated with the Soviet weapon.

DATA
Calibre: 40 mm
Length of launcher: 950 mm
Weight, with sight: 6.3 kg
Maximum aimed range: 500 m
Rate of fire: 4 – 6 rds/min
Armour penetration: 260 mm

Manufacturer
Pakistan Machine Tool Factory Ltd., Landhi, Karachi 34.
Status
Current. Production.
Service
Pakistan armed forces.

RPG-7 rocket launcher in use by Pakistan forces

106 mm recoilless gun M40A1

This is a locally manufactured version of the standard US design, and the entire system consists of the Gun M40A1, Mount M79, Spotting Rifle M8C, Elbow Telescope M92F, Telescope Mount M90 and Instrument Light M42. The Spotting Rifle is bought-in from Empresa Nacional Santa Barbara of Spain and the sight assembly is manufactued in Pakistan by Mogul Industries.
    The gun may be mounted on and fired from a light vehicle or it can be easily dismounted and brought into action on its own carriage.

DATA
Calibre: 105 mm
Weight of gun: 115 kg
Weight of Mount M79: 82 kg
Maximum range: 2012 m direct fire, 7680 m indirect fire
Muzzle velocity: 507 m/s

Manufacturer
Pakistan Machine Tool Factory Ltd., Landhi, Karachi 34.
Status
Current. Production.
Service
Pakistan armed forces.

106 mm recoilless gun in use by Pakistan forces

SPAIN

Aries missile system

This is a development by a consortium of three Spanish companies, Union Explosivos Espanola, Instalaza and Ceselsa, under the name Esprodesa. The consortium has formed a joint company with Hughes Aircraft Co of America to produce the missile system, Hughes providing technical knowledge and management expertise while Esprodesa will be responsible for development and production.
    The system is to be manportable, fired from a small tripod mount, and operates in the SACLOS mode, the operator keeping his sight on the target and the missile steering itself into the sightline to impact. It is promised that the missile will have a high kill probability, a low launch signature and will be highly resistant to ECM.

DATA
Calibre: 148.6 mm
Length: 1016 mm
Weight: 15.9 kg with disposable launch tube
Launch unit weight: 11.4 kg
Min range: 65 m
Max range: 2000 m
Time of flight: (1000 m) 4.5 s; (2000 m) 8.5 s

Status
Development.

Full-scale mock-up of the Aries missile with fins extended

88.9 mm M-65 anti-tank rocket launcher

This is a Spanish-developed weapon, with an electro-magnetic firing mechanism and a sight unit fitted with an adjustable light source which lights up the graticule for night sighting. The electrical connection between the round and launcher is established automatically when loading.

There are three types of ammunition available, CHM81 L anti-tank, MB66 anti-personnel/anti-armour and FIM66 smoke, plus their practice rounds; the launcher and its ammunition are made by the same manufacturer.

DATA

LAUNCHER
Calibre: 88.9 mm
Length: (folded) 850 mm; (in firing order) 1640 mm
Weight: 6 kg

AMMUNITION

	CHM81L	MB66	FIM66
Rocket weight	2.3 kg	2.9 kg	2.7 kg
Initial velocity	215 m/s	145 m/s	155 m/s
Effective anti-tank range	450 m	300 m	—
Max range	—	1000 m	1300 m
Armour penetration	400 mm	250 mm	—

Manufacturer
Instalaza SA, Monreal 27, Zaragoza.

88.9 mm rocket launcher M-65

Status
Current. Production.
Service
Spanish and other armed forces.

C-90 C light anti-tank weapon

This light, disposable, anti-tank weapon system, developed and manufactured by Instalaza SA, has been designed to provide the infantryman with a very simple and effective anti-tank weapon at ranges of up to 250 m for moving targets or 400 m for stationary targets.

A glass reinforced resin container-launcher holds the warhead and rocket-motor and also supports the firing mechanism, optical sight and carrying strap, and, optionally, a pistol-grip and a shoulder-rest.

The round consists of a shaped-charge warhead, an instantaneous fuze placed behind the warhead which makes it effective over its full calibre, a rocket motor and a fin stabilising unit.

The firing mechanism is pyrotechnic, thus dispensing with the need for batteries or any other electric device. The whole system is completely maintenance-free.

The optical sight has a two times magnification and a reticle with the appropriate markings for distance and lateral prediction. A permanent light source fitted inside the sight permits the gunner to see the markings even in the night.

There are a number of variant models of the C-90 system which differ in length and weight or in the nature of the pre-packed rocket. Details are given in the data tables below.

A specific training system, called TR-90, has also been developed. Based on the principle of the recoilless gun, it uses a single cartridge to launch an aluminium alloy arrow and to compensate for the recoil by the backwards ejection of a jet of gas. The aluminium arrow can be re-used many times provided that it is fired against a soft target.

Using this training system, a thorough and extremely cheap training programme can be carried out, leaving the firing of inert rockets for the final stage.

C-90 C light anti-armour weapon in firing position

C-90 C light anti-armour weapon in carrying position

DATA

Version	C90-C	C90-C-AM	C90-CR	C90-CR-RB	C90-CR-AM	C-90-CR-FIM
Function	A/Tank	A/Tank-A/Pers	A/Tank	A/Tank	A/Tank-A/Pers	Smoke/Incendiary
Calibre	90 mm	90 mm	90 mm	90 mm	90 mm	90 mm
Length	840 mm	840 mm	940 mm	940 mm	940 mm	940 mm
Weight	4.2 kg	4.2 kg	4.8 kg	4.8 kg	4.8 kg	5.35 kg
Penetration, steel	400 mm	220 mm	400 mm	500 mm	220 mm	—
Penetration, concrete	1 m	0.65 m	1 m	1.2 m	0.65 m	—
Anti-tank range	200 m	200 m	300 m	300 m	300 m	—
A/Personnel range	—	600 m	—	—	800 m	750 m
Sight graduations	0-350 m	150-600 m	0-450 m	0-450 m	200-800 m	200-750 m
A/Pers fragments	—	>1000	—	—	>1000	—
Lethal radius	—	21 m	—	—	21 m	—

Manufacturer
Instalaza S/A, Monreal 27, 50002 Zaragoza.
Status
Current. Production.

Service
Spanish and other armed forces.

106 mm recoilless rifle

This is a Spanish-made version of the American M40A1 RCL rifle and is identical in all respects. It is fitted with an aiming rifle and optical sight and is provided with the usual range of HEAT, TP and anti-personnel ammunition. The standard mounting is the tripod ground mount, but it can also be installed into any appropriate light vehicle.

DATA
Length, overall: 3.40 m
Height on mount: 1.13 m
Rifling: 36 grooves rh
Weight: 219 kg
Max elevation: 27° over tripod, 65° between legs
Max depression: −17°
Traverse on mount: 360°
Muzzle velocity: 503 m/s
Max range: 7640 m

Manufacturer
Santa Barbara, Calle Julian Camarillo 32, 28037 Madrid.
Status
Current. Production.
Service
Spanish Army.

106 mm recoilless rifle on ground mount

SWEDEN

Miniman light anti-armour weapon

The Miniman is a one shot, throw-away, recoilless gun intended to be issued to infantry to enable them to have an effective defence against close-in tanks. It arrives in the forward area with the projectile already in place and the soldier has simply to estimate range and speed, cock his firing mechanism, lay on the target and fire.

The barrel is made of filament-wound GRP. Attached to the outside is a descriptive label giving simple illustrated instructions which make easy the process of applying the correct lead to a moving target and demand from the firer only that he judges the range, out to 150 m, to the nearest 25 m. The speed of the target is classified as very slow (6 km/h), slow (20 km/h) or fast which is 30 km/h, and the firer is called on to place his target in one of these categories, and that at a range not exceeding 150 m is not excessively difficult.

The HEAT shell in the barrel is linked to the igniter by a breakable joint and the igniter is at the front of the combustion chamber which is made of high strength aluminium alloy and contains the igniting and propellant charges. The combustion chamber is cylindrical and perforated. Behind the combustion chamber the barrel is opened out to form a venturi. The firer cocks the weapon by moving the cocking lever over to the right. When he pushes the firing button forward the firing rod goes rearward and the pin ignites the primer. The flame travels down the ignition transmission line and reaches the igniting and propelling charges. The system works on the high-low pressure system. The pressure in the combustion chamber is at a high level and the gases escape through the holes into the barrel where the pressure is reduced. The pressure in the barrel breaks the joint between the combustion chamber and the shell and the shell moves forward while the gases emerge from the rear and the combustion chamber remains in the barrel. The

momentum of the gases escaping through the venturi is equal to the momentum of the shell and gases going forward and so the gun remains stationary.

The HEAT shell consists of:
the distance tube at the front, made of alloy and responsible for establishing the standoff distance
the shell body with copper liner and shaped charge of octol and
the stabilising tube of light alloy. In the rear of this tube are four H-shaped slots and these form eight flaps which are forced out by gas pressure to form four fins to stabilise the shell.

When the shell strikes the target the body is compressed and the piezo crystal produces a voltage which initiates the electric detonator. The shaped charge is initiated from the rear and the resulting jet penetrates the armour plate.

There is also a sub-calibre device using a 9 mm tracer round which is used for training or, with standard 9 mm Parabellum, it can be used against miniature targets and for indoor firing against projected film targets.

DATA
Weight of weapon loaded: 2.9 kg

HEAT SHELL
Calibre: 74 mm
Weight: 880 g
Length: 325 mm
Weight of explosive: 300 g

Firing position with Miniman. Firer's right thumb is on trigger-catch. Immediately to left of his thumb is cocking handle

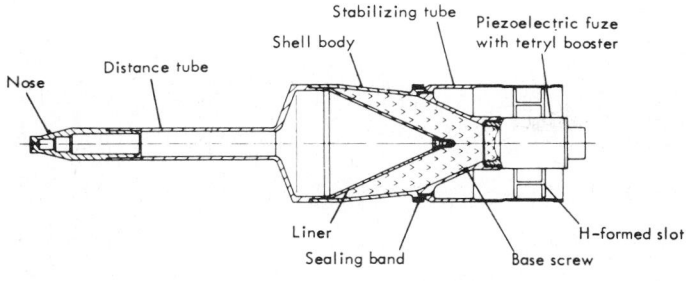

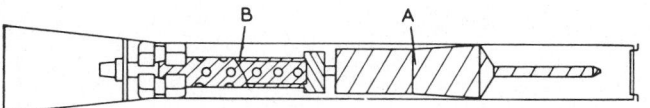

Projectile and propellant in tube
(a) projectile (b) propellant, contained in metal case. Note breakable link
between projectile and propellant case

Miniman launching tube is of filament-wound fibre and projectile and charge are inside

Length of weapon: 900 mm
Weight of package containing two weapons: 7.1 kg

PERFORMANCE
Muzzle velocity: 160 m/s
Armour penetration: 340 mm
Range: (moving target) 150 m; (stationary target) 250 m
Time of flight for 150 m: 1.2s

Manufacturer
FFV Ordnance, S-631 87 Eskilstuna.
Status
Current. Production complete.
Service
Swedish, Finnish and Austrian forces.

84 mm light anti-armour weapon AT4

The AT4 is a preloaded, disposable and recoilless anti-armour weapon, intended for firing one round, after which the barrel is discarded. It is light in weight and very easy to handle.

The main parts of the weapon are:
barrel, made of glass-fibre reinforced plastic and fitted with an aluminium venturi to the rear
shock absorbers and muzzle cover
firing mechanism, placed on top of the barrel
sights and
HEAT round.

On the barrel are labels with instructions for handling, safety etc.

The HEAT round consists of a cartridge case assembly, similar to that used with Carl Gustaf ammunition, and a HEAT shell. The main parts of the HEAT shell are the fin assembly, base fuze, standoff cap and hollow-charge. The fins pop out after the shell has left the muzzle and stabilize the shell in flight. The fuze has an out-of-line detonator safety device to prevent accidental initiation. On impact, the fuze detonates the hollow-charge, even at angles of impact as shallow as 80° to the normal. The special design of the hollow-charge causes behind-armour damage by over-pressure, extensive spalling and intense heat.

The sights are protected by sliding covers during transport. The rear sight is adjustable and factory-set to 200 m. The front sight has a centre post and two lead posts to be used depending upon target speeds.

The standard package is a handy and rugged plywood box with carrying handles, containing four weapons sealed in moisture-proof plastic bags.

AT4 Training Devices
There is a complete range of training devices for the AT4: for marksmanship training there is the 9 mm sub-calibre AT4, a weapon with an integral sub-calibre barrel firing 9 mm tracer ammunition matched to the ballistics of the HEAT round. For realistic training there is the back-blast weapon, simulating the bang and flash of the live HEAT weapon. For field exercises there is the AT4 TPT weapon, firing a full-calibre inert filled projectile.

There is also a completely inert handling weapon for drill purposes.

84 mm LMAW AT4
FFV Ordnance is currently developing the Light Multi-purpose Assault Weapon (LMAW) AT4. This is a recoilless one man portable weapon intended for fighting in built-up areas. The LMAW uses a slightly modified 84 mm HEDP round from the Carl Gustav system.

The launcher is similar to that of the normal AT4 but is fitted with a fuze mode selector switch with two positions, 'D' for Delay burst and 'I' for Instantaneous burst on impact. The 'I' mode is used when engaging armoured vehicles, the 'D' mode being used when engaging troops protected by field fortifications or light cover. In the anti-armour 'I' mode the warhead causes special behind-armour damage.

Development is currently proceeding on a new version, the AT4CS (Confined Space) which will permit the weapon to be fired in enclosed spaces.

AT4 light anti-armour weapon in firing position

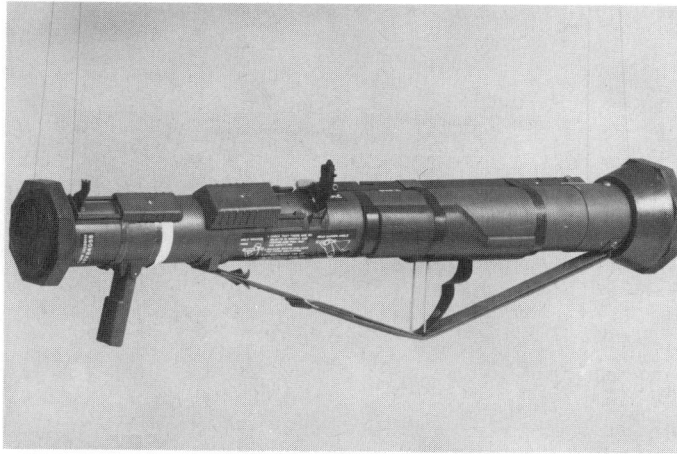

AT4 light anti-armour weapon

DATA
Calibre: 84 mm
Weight: 6.7kg
Weight of HEAT round: 3 kg
Length: 1000 mm
Effective range: 300 m
Muzzle velocity: 290 m/s
Armour penetration: greater than 400 mm

Manufacturer
FFV Ordnance, S-631 87 Eskilstuna.
Status
Current. Production.
Service
Swedish Army; US Army and Navy. Venezuelan Army.

120 mm AT 12-T anti-tank weapon

This is a shoulder-fired rocket anti-tank weapon under development for the Swedish Army. Feasibility studies aree expected to be completed in 1991 and series production to commence in 1995. It is similar in general form to the AT-4 and a folding mount will allow it to be fired from the prone, kneeling or standing positions.

The weapon uses a tandem warhead designed to defeat reactive armour and penetrate over 950 mm of conventional armour. The company claim that it will defeat all tanks likely to be fielded in the coming decade, throughout their frontal arc.

DATA
Calibre: 120 mm
Length: 1.2 m
Weight: approx 14 kg
Effective range: > 300 m
Penetration: 950 mm

Manufacturer
FFV Ordnance, S-631 87 Eskilstuna.
Status
Development.

FFV 120 mm AT 12-T anti-tank weapon

The Carl-Gustav System
84 mm RCL Carl-Gustav M2

The Carl-Gustav is a one man portable, recoilless gun, originally conceived for the anti-tank role but in later years upgraded to a true multi-purpose capability. It is a reliable and rugged design, tailored for long operational life under adverse conditions. Recoilless functioning is obtained by allowing a proportion of the propellant gases to escape to the rear of the weapon under control.The barrel is rifled and is fitted with a cone-shaped venturi funnel at the rear.

The weapon is best served by a two man detachment. One fires the gun and the other carries ammunition and loads. The gun is breech loaded and is opened by releasing the venturi fastening strap and rotating the venturi sideways. The empty case is removed and the next round can be loaded straight in. Until the breech is rotated to the fully closed position the firing mechanism is inoperative. Since the round is percussion fired by a mechanism on the side of the chamber, it is necessary to index it into the chamber by a tongue and notch.

The firing mechanism is contained in a tube on the right side of the barrel and is cocked when the cocking lever, behind the pistol grip, is pushed forward to compress the main spring. The safety catch is on the right side of the pistol grip. The gun can be fired from the shoulder or from the prone position or it can be rested on the edge of a trench or fired from a mount on an armoured personnel carrier. When fired off a flat surface the weapon is supported on a flexible bipod immediately in front of the shoulder piece.

There is an open sight but the usual sighting system is a × 2 telescope with a 17° field-of-view. This is fitted with a temperature correction device and luminous front and rear adaptors are available for night work.

DATA
WEAPON
Calibre: 84 mm
Length: 1130 mm
Weight: 14.2 kg

TELESCOPE
Weight: 1 kg
Magnification: × 2
Field-of-view: 17°
HEAT AMMUNITION
Weight of complete round: 2.6 kg
Weight of shell: 1.7 kg
Muzzle velocity: 310 m/s
Armour penetration: 400 mm
RANGES
HEAT: 450 m
HE: 1000 m
Smoke: 1300 m
Illuminating: 2300 m

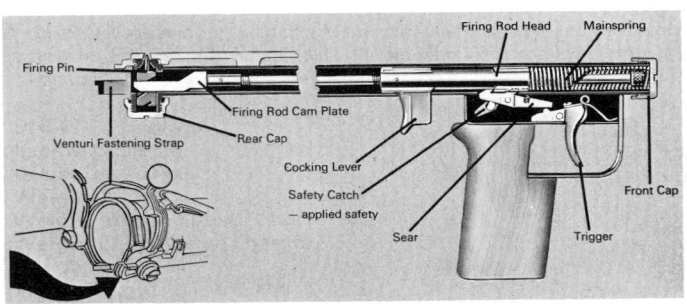

Components of the gun. Breech open, ready for loading

84 mm Carl-Gustav M2 gun

Firing mechanism

84 mm RCL Carl-Gustav M3

The Mark 3 gun is essentially a lightweight version of the Carl-Gustav M2 described above. The barrel consists of a rifled steel liner, around which is wound a laminate of carbon fibre and epoxy. The venturi is made of steel, all other external parts being of aluminium or plastic.

The gun is fitted with a carrying handle to facilitate rapid moving of the firing position.

The M3 will accept and fire all existing types of 84 mm Carl-Gustav ammunition.

The M3 is in use by the Danish and Swedish Armies and the Venezuelan Army and Navy.

DATA
Length: 1070 mm
Weight: 8.5 kg
Mount weight: 500 g
Packed gun, with accessories: 21.5 kg

Carl-Gustav M3 gun

84 mm Carl-Gustav ammunition

The multi-purpose capability of the Carl-Gustav system is provided by light weight, compact dimensions, easy handing of the gun, and a complete range of ammunition.

84 mm HEAT round FFV 751
This new round is currently under development and is scheduled for production in 1993. It is a rocket-assisted tandem warhead projectile which is intended to defeat reactive armour and penetrate over 500 mm of conventional armour.

84 mm HEAT round FFV 551
The High Explosive Anti-Tank round FFV 551 is effective against all types of armoured vehicles, even those fitted with standoff or 'bazooka' plates. The shell is fin-stabilised and has a rocket motor sustainer to flatten the trajectory and extend the range. It is fitted with a piezo-electric fuze system which will reliably detonate the hollow-charge even at oblique impact.

The HEAT round consists of a HEAT shell and a cartridge assembly. The latter is a common feature of the entire ammunition range and is of a reliable and well-proven design. It has a case of light alloy and a propelling charge of double-base strip propellant ignited by a lateral percussion primer. The rear end of the case is closed by a plastic blowout disc, one of the vital precision components instrumental in making the Carl-Gustav system recoilless.

The HEAT shell consists of a nose, with ogive and standoff tube; the shell body with Octol bursting charge, copper liner, CE booster and electric fuze system; the rocket motor of double-base smokeless propellant; a Teflon slipping driving band to reduce shell spin; and a stabilising unit with pop-out fins.

The rocket motor has a delay unit which is ignited by the propellant gases on firing. When the delay composition has burned for 45 ms it ignites the rocket propellant, which delivers thrust of 325 N for 1.5 seconds.

The FFV 551 comes in handy, rugged and durable twin containers of high-density polyethylene. Three containers — six rounds — are packed together in a plywood transport box which is pressure impregnated to resist rot and termite attack.

DATA
Weight, complete round: 3.2 kg
Weight of shell: 2.4 kg
Muzzle velocity: 255 m/s
Maximum velocity: approx 330 m/s

84 mm Carl-Gustav amunition: from left, HE FFV 441B; HEDP FFV 502; Smoke FFV 469B; Illuminating FFV 545; and HEAT FFV 551.

Armour penetration: > 400 mm
Arming range: 5 – 15 m
Effective range: 700 m

84 mm HE round FFV 441B
The high explosive shell FFV 441B is intended for use against unprotected troops, entrenched troops, troops in machine gun posts, soft-skinned vehicles and similar targets.

The main parts of the shell are the mechanical time and impact fuze FFV 447 and the shell body with ball inserts and an RDX/TNT bursting charge.

Fuze FFV 447 has the safety features demanded in modern fuzes and complies with MIL-STD 331. It is set by hand, without tools; setting is made between 40 and 1250 m and is stepless. The impact mechanism functions at angles of impact as shallow as 85° to normal.

DATA
Weight of complete round: 3.1 kg
Weight of shell: 2.3 kg
Muzzle velocity: 240 m/s
Arming range: 20 – 70 m
Practical range: to 1100 m

84 mm smoke round FFV 469B
Smoke shell FFV 469B is intended for rapid laying of blinding and screening smoke. It is also suitable for marking targets for artillery and close support aircraft.

The shell consists of the direct action and graze fuze, and the shell body with smoke composition and a central burster tube. The fuze impact mechanism functions at angles of impact as shallow as 85° to normal. The smoke composition consists of titanium tetrachloride (FM) adsorbed by powdered calcium silicate. It is non-toxic and leaves no harmful residues.

DATA
Weight of complete round: 3.1 kg
Weight of shell: 2.2 kg
Weight of smoke composition: 800 g
Muzzle velocity: 240 m/s
Practical range: up to 1300 m

84 mm illuminating round FFV545
The FFV 545 is intended for quick battlefield illumination in support of direct fire anti-tank weapons and missiles. It allows sub-units to supply their own illumination of targets as required.

The shell consists of a pyrotechnic time fuze, a shell body of light alloy, a canister with illuminating composition, and a nylon parachute.

The fuze is fitted with a setting ring, the graduations of which range from 200 to 2300 m, with subdivisions of 50 m. The canister is ejected in mid-air and falls slowly, braked by the parachute. It produces a sodium light of some 650 000 candela and has a burning time of about 30 seconds.

DATA
Weight of complete round: 3.1 kg
Weight of shell: 2.2 kg
Weight of illuminating composition: 500 g
Muzzle velocity: 260 m/s
Practical range: from 300 – 2100 m
Illuminated area, diameter: 400 – 500 m
Burning time: 30 s

84 MM HEDP round FFV 502
The High Explosive Dual Purpose (HEDP) shell FFV 502 is tailored for fighting in built-up areas (FIBUA). It makes it possible to shift quickly

between HE and HEAT fire. The 502 is suitable for deployment with Rapid Deployment Forces, airborne trops and commando units.

The HEDP has two modes, selectable when loading. The impact mode is used when engaging armoured vehicles — the HEAT role — and the delay mode is for use when engaging troops protected by light cover and field fortifications. In the HEAT role the shell causes special behind-armour damage.

The shell consists of a nose-cap, providing standoff; a hollow charge specially designed to produce behind-armour damage in the HEAT role and optimum fragmentation in the FIBUA role; a base fuze; and a fin assembly with pop-out fins.

The cartridge case contains esentially the same components as the other Carl-Gustav rounds. It is, however, fitted with two lateral percussion primers; the two primers are required since the fuze mode is set by selective orientation of the round on loading.

When using the FFV 502 HEDP against earth or timber bunkers or similar targets, an effect is obtained from both the hollow charge and the blast which destroys the target.

DATA
Weight of complete round: 3.3 kg
Weight of shell: 2.3 kg
Muzzle velocity: 230 m/s
Arming range: 15 – 40 m
Armour penetration: > 150 mm
Effective range: moving targets 300 m; bunkers 500 m
Practical range, troops in open: up to 1000 m

FFV 553B Training System

FFV Ordnance has developed a complete training system for the Carl-Gustav. The 553B Training System makes it possible to conduct marksmanship training and drill under realistic, near-live conditions, including battle noise and stress.

The core of the 553B system is the 7.62 mm FFV 553B sub-calibre adapter, externally similar in shape to the HEAT round FFV 551. It is fitted with a sub-calibre barrel firing 7.62 mm tracer ammunition matching the ballistics of the 551 round.

The sub-calibre adapter is loaded into the gun and the gun is fired in the normal manner, the firing pin striking a percussion cap which in turn operates a hammer mechanism to fire the 7.62 mm cartridge.

The sub-calibre adapter weighs 3.7 kg and is 60 cm in length.

There is also an optional back-blast charge to simulate the bang and flash of full-calibre Carl-Gustav ammunition. There are three ways of combining the ammunition components:

tracer round plus back-blast charge and percussion cap for live-fire exercises;
tracer round plus percussion cap for marksmanship training;
back-blast charge plus percussion cap for combat exercises.

84 mm TP round FFV 552

Target Practice projectile FFV 552 is intended to be used in training. The shell contains no fuze, booster or main bursting charge; it is fitted with a rocket motor to provide the same ballistic performance as the HEAT round FFV 551. The cartridge case contains a propelling charge.

Manufacturer
FFV Ordnance, S-631 87 Eskilstuna
Status
Current. Production.
Service
In service with Swedish forces and Australia, Austria, Burma, Canada, Denmark, West Germany, Ghana, India, Ireland, Japan, Malaysia, The Netherlands, New Zealand, Nigeria, Norway, Singapore, United Arab Emirates and the UK.

106 mm 3A-HEAT-T anti-armour round

This round has been developed by Bofors in order to defeat reactive armour and ensure penetration of main armour protection. It is a fin-stabilised shaped-charge projectile with an electric contact fuze and has a considerable behind-armour effect. It has been demonstrated to defeat a reactive armour protected sloping MBT glacis at 60° angle of impact, and will defeat more than 700 mm of plate after detonating the reactive protection. In general, the performance of the 3A-HEAT-T projectile is double that of a conventional 106 mm HEAT round under any conditions.

DATA
Projectile weight: 5.5 kg
Complete round weight: 13.0 kg
Explosive filling: 1.0 kg Octol
Muzzle velocity: 570 m/s
Tracer duration: 55 s
Operating temperature: -40 to +55°C

Manufacturer
AB Bofors, S-69180 Bofors.
Status
Current. Production.
Service
Swedish Army and under evaluation elsewhere.

Bofors 106 mm 3A-HEAT-T projectile (front) with practice round

90 mm PV-1110 recoilless rifle

This weapon is employed either mounted on a light wheeled vehicle, on the tracked vehicle Bv2062, or on a two-wheel trailer. In the latter version the rifle is mounted on a turntable which forms part of the carriage and serves as an arm-rest for the gunner who fires from a kneeling position. There is a 7.62 mm spotting rifle mounted above the main barrel.

DATA
Calibre: 90 mm
Barrel length: 3.7 m
Elevation: –10 to +15°

Traverse: 75 – 115° according to elevation
Weight in combat order: 260 kg as a trailer
Weight of shell: 3.1 kg
Muzzle velocity: 650 m/s
Effective range: 800 m
Penetration: 550 mm armour at 0.5 m standoff
Rate of fire: 6 rds/min
Crew: 3

PV-1110 recoilless gun on trailer mount

PV-1110 recoilless gun mounted on Bv 2062 tracked carrier

Manufacturer
AB Bofors, S-69180 Bofors.
Status
Obsolete. Not in production.
Service
Swedish forces and Ireland.

Bantam portable anti-tank missile

This is a small, one man, wire guided, manual command to line of sight (MCLOS) anti-tank missile. The missile container holds the missile and a 20 m control cable. The missile is launched directly from the container and the cable connects the missile to the control unit. The control unit will take up to three missiles but it is also possible to connect a distribution box with outlets for six missiles to each of the control cables from the control unit which can thus be used to fire up to 18 missiles. The control unit consists of test facilities, a missile selector, safety and firing buttons, a monocular optical sight and a joystick to control the missile in flight.

The missile container has a carrying harness and is carried on the soldier's back. To bring it into action the container is placed on the ground and the control unit connected. If necessary extra cable can be used to displace the control unit by 120 m from the missile. The operator carries out his tests of the battery and transmitter and then selects the missile. Each missile carries four guidance flares and according to the degree of visibility, the firer selects the

required number of flares, one or all four. He then turns the safety to the off position and the transmitter starts operating. The firing button is pressed. The missile battery is made operative by the bursting of the electrolyte container. The gyro starts and the booster is ignited. After the gyro has run up to speed, it is uncaged after the missile has travelled about 30 m. The spools on the missile each contain 2000 m of wire which is reeled out as the missile moves forward. A microswitch is mounted in each spool. One of the microswitches ignites the sustainer motor and the tracers after 40 m of wire has reeled out and the other arms the warhead fuze after 230 m.

When the missile leaves the container, the wings, which are folded in the launcher, open and the angled corners cause the missile to roll. The signals transmitted by the operator along the wires are communicated to the spoilers on each pair of wings. Since the missile is rolling, the spoilers on each pair of wings switch from transverse guidance to elevation guidance and vice versa at every quarter revolution.

DATA
Type: infantry, one man anti-tank missile
Type of guidance: wire guidance (MCLOS)
Method of guidance: manual commands transmitted to spoilers on wings
Propulsion: booster and sustainer solid motors
Warhead: hollow-charge
Penetration: 500 mm
Weights: (container with missile) 11.5 kg; (container with missile, 20 m of cable and harness) 14 kg; (control unit with monocular sight) 5 kg; (reel with missile selector and 100 m cable) 10 kg
Range: less than 300–2000 m
Cruising speed: 85 m/s

Manufacturer
AB Bofors, S-69180 Bofors.
Status
Obsolescent. No longer manufactured.
Service
Swedish and Swiss forces.

Bantam emerging from launching/carrying box

RBS 56 BILL light anti-armour missile

This has been developed by Bofors under a contract from FMV (the Swedish Defence Materiel Administration) in July 1979.

BILL (the acronym stands for Bofors, Infantry, Light and Lethal) fills the need for a portable, easily operated, long-range anti-tank weapon capable of combating any known or projected armoured vehicle. BILL is basically a SACLOS wire guided missile, using coded signals between missile and sight unit for guidance. The missile body is made from wound Kevlar/epoxy filament, is about 1 m long, and contains, from front to rear; a dual impact/proximity fuze, a four-nozzle IMI rocket motor, warhead, four folding wings, electronics package, wire bobbin and four folding guidance fins.

The launcher tube is also of Kevlar/epoxy filament and is mounted on a stand which permits a 90° traverse and 10° elevation. The operator controls the traverse by the left hand grip, which also carries the thumb-actuated trigger. Elevation is controlled by twisting the right hand grip. A mobile mounting has been designed for fitting to vehicles; two or three missiles can be carried in a vehicle with the launcher and three man crew and another 10 missiles can be carried in a trailer.

The missile warhead is designed for top attack by means of a hollow-charge

Vehicle-mounted RBS 56

RBS 56 in firing position

RBS 56 squad on move

RBS 56 attack profile

unit which is canted 30° downward and triggered by the proximity fuze; the missile is kept in the appropriate attitude by means of gyro-roll stabilisation. The fuzing incorporates a delay time so that the missile will pass over the more heavily armoured areas of the target and strike at the less well protected roof areas. In order to ensure correct action and simple operation, the missile is programmed to fly 1 m above the sight line. If, therefore, the operator aims for the hull/turret joint, he will automatically place the missile over the top of the tank and in position for the most effective attack. This elevated flight path adds an important element of ground clearance to that of the line of sight, thus avoiding many terrain obstacles which might otherwise be limiting factors. The proximity fuze has the effect of extending the target upwards, to the limit of the fuze sensitivity, so that it now becomes possible to inflict damage on targets which display little or no hit area; for example, a tank in a hull-down position will give the operator a sufficient aiming mark, and the proximity fuze will ensure an effective hit as the missile passes over.

DATA
Length: 900 mm
Diameter: 150 mm
Weight: (firing unit) 27 kg; (missile) 16 kg; (sight and stand) 11 kg
Range: (stationary target) 150-2000 m; (moving target) 300-2000 m
Time of flight: (to 1200 m) 6 s; (to 2000 m) 10 s
Setting-up time: < 20 s
Reloading time: < 7 s

Manufacturer
AB Bofors, S-69180 Bofors.
Status
Swedish and Austrian armies and under trial in several countries.

SWITZERLAND

BB-65 anti-tank missile

This is the Swiss designation of the Swedish Bantam anti-tank guided weapon (ATGW) which is the only weapon of its kind currently in service with the Swiss Army. No indigenously-designed ATGW has been deployed in Switzerland but some of the early work on the Italian Mosquito missile was done by Contraves AG in Switzerland before the project was transferred to Contraves Italiana SpA.

25mm Oerlikon Iltis infantry gun

The Iltis infantry gun is a 25mm Oerlikon KBB automatic cannon on a special ground mount which permits it to be used against ground targets or helicopters. Development of this weapon was initiated in order to fill the need for a weapon not easily saturated by multiple targets and effective against helicopters, neither of which demands are satisfied by the present generation of anti-tank missiles. Iltis can be deployed in a ground mount, can be fitted to a two-wheeled mount, or can be carried on a wide variety of all-terrain vehicles. The equipment is light enough to be man-carried when dismantled.

The mounting is made from composite materials to combine strength with lightness; the dismantled mounting can be carried by one man. The gun is the standard KBB cannon which has had the feed system modified so that instead of dual belt feed it now uses two 15-round box magazines, one inserted into each side of the receiver.

The unique feature of the Iltis gun is the method of operation. Instead of the gunner sitting upright behind it, affording a good target to the enemy, he lies prone behind the weapon, well below the axis of the bore. Sighting is performed by having an electro-optical sensor – a TV camera, thermal imager or other type – mounted on the gun and bore-sighted to it, and connected to a

Rear view of Iltis showing gunner's position

Front view of 25 mm Iltis infantry gun

visual display unit by means of a fibre optic cable. The VDU is at the foot of the mounting, in front of the prone gunner's eyes, and the handle for controlling the movement of the gun, and the trigger, are extended down and to the rear so that the gunner can grasp them whilst observing the VDU. He therefore sights the gun via the VDU, controlling it in accordance with the sight picture he sees.

There is a basic iron sight, for emergency use and for collimation purposes. A parallax-free reflex sight is available as an intermediate step between the iron sight and the principal electro-optical sight. A laser rangefinder can also be incorporated into the sighting system if required.

DATA
Calibre: 25 mm
Weight of gun: 117 kg
Weight of complete equipment: 240 kg
Rate of fire: 800 rds/min
Elevation: -10° to +45°
Traverse: 360°
Muzzle velocity: 1160 – 1355 m/s depending upon ammunition

Manufacturer
Machine Tool Works Oerlikon-Bührle Ltd, CH-8050 Zurich.
Status
Advanced development.

90 mm PAK 50 and 57 anti-tank guns

These two light towed anti-tank guns are no longer in production but both are still in service with infantry formations of the Swiss Army. The more recently developed of the two (Model 57) is heavier and has a somewhat higher performance; but the two are functionally similar and both fire a HEAT round. The PAK 57 carries a 12.7mm spotting rifle. Comparative data is given below.

DATA

	PAK 50	PAK 57
Calibre	90 mm	90 mm
Length of barrel	2.9 m	2.92 m
Length of rifling	2530 mm	2550 mm
Weight, emplaced	550 kg	530 kg
Weight, travelling	600 kg	580 kg
Elevation	-10 – +32°	-15 – +23°
Traverse	34 – 66°*	70°
Weight of HEAT shell	1.95 kg	1.95 kg
Weight of propellant	250 g	400 g
Muzzle velocity	600 m/s	650 m/s
Maximum range	4000 m	4000 m
Effective range	700 m	900 m
Max rate of fire	8-10 rds/min	8-10 rds/min
Crew	6	6

Sights (both models): sighting telescope for daylight. Infra-red night sight
*depending upon elevation

Manufacturer
Eidgenössische Konstruktionswerkstätte, CH-3602 Thun.
Status
Obsolescent. Not in production but spares are still manufactured.

PAK 50 anti-tank gun in firing position

PAK 57 gun in firing position

Service
Swiss forces.

83 mm Rocket Launcher 58/80

The Rocket Launcher 58/80 portable anti-tank weapon system was developed by the Swiss Federal Arms Factory, Berne, as a successor to the 83 mm Rocket Launcher 58.

It can fire every type of 83 mm rocket ammunition in the Swiss Army.

DATA
Calibre of launcher: 83 mm
Length: 1300 mm
Weight: 8.5 kg
Effective range: (moving targets) 200 m; (stationary targets) 300 m

Manufacturer
Swiss Federal Arms Factory, Berne.
Status
Current.
Service
Swiss Army.

83 mm Rocket Launcher 58/80

TAIWAN

The Taiwan Army uses the M20 3.5 in rocket launcher and the M72 66 mm light anti-armour weapon (LAW) in the infantry anti-tank role. The TOW anti-tank guided weapon (ATGW) and 106 mm recoilless rifle are used on vehicular installations. The M20, M72 and their projectiles are all manufactured in Taiwan. The 106 mm recoilless is locally produced as the Type 51.

Kuen Wu 1 anti-tank guided missile

Kuen Wu 1 is a manual command to line of sight (MCLOS) missile developed in 1974-78 by the Sun Yat-sen Scientific Research Institute of Taiwan. It closely resembles the Soviet AT-3 Sagger in general appearance and in many aspects of dimensions, weight and performance. It differs from Sagger in having a somewhat larger and differently shaped warhead.

Kuen Wu is powered by two solid rocket motors and control is by thrust vector control (TVC) jetavator nozzles. Guidance is by a joystick, correcting to the line of sight, via twin wires. The warhead is of the shaped-charge type.

The missile is stored and carried in the same glass fibre case as Sagger and launch preparations are similar. Normally the weapon system comprises a quadruple launcher on a M-151 Jeep mount, and a maximum of eight missiles can be carried by the launch vehicle. The missile could also be fired by infantry and, presumably, from helicopters. The present system is designated Kuen Wu 1, which suggests that newer versions may be in the process of development.

A computer-based missile simulator and field test set are made by the Sun Yat-sen Scientific Research Institute.

DATA
Type: surface-to-surface, anti-tank guided missile
Guidance: optical tracking, wire guidance, MCLOS
Control: TVC; roll rate stabilised
Propulsion: 2 solid rocket motors
Warhead: (HEAT) 2.72 kg
Length: approx 880 mm
Diameter: approx 119 mm
Weight: (missile) approx 11.3 kg; (missile and container) 15 kg
Effective range: 500–3000 m
Velocity: average 120 m/s
Penetration: 400–500 mm

Manufacturer
Sun Yat-sen Scientific Research Institute, PO Box 2, Lung Tan, Taiwan 325.
Status
Current. Production.
Service
Taiwan Army.

Kuen Wu 1 ATGW on M-151 Jeep mount

UNION OF SOVIET SOCIALIST REPUBLICS

RPG-2 portable rocket launcher

The RPG-2 was developed from the German Panzerfaust of the Second World War. It fires a fin-stabilised HEAT projectile which is loaded into the muzzle of the launcher. The fins are made of a stiff flexible sheet and they can be rolled around the cylindrical container of the motor and retained in place with a ring. As the fins are pushed down into the barrel the ring is slid off. The fins spring out and are held against the inside of the launcher. As soon as the missile leaves the muzzle the fins extend fully. The metal tube of the launcher has a wood-encased section in the middle to protect the firer against heat and also to make the weapon easier to hold when it is used in cold climates. The firing mechanism is a spring-loaded hammer and this means that the round has to be indexed into position to ensure that the hammer and the cap are in line. The diameter of the head of the rocket is twice that of the launcher tube.

The rear end of the tube is usually a plain cylinder but some have been seen with a flange-like blast deflector at the rear of the launcher.

The Chinese have manufactured the RPG-2 as the anti-tank grenade launcher (ATGW) Type 56 and they have also a round of their own known as the Type 50. This round has an excellent performance when it hits at normal but at 45° it is said to be less efficient than the Soviet round.

DATA
Calibres: (launcher) 40 mm; (warhead) 82 mm
Overall length of launcher: 1494 mm
Weight, empty: 2.83 kg
Max effective range: 150 m
Type of ammunition: (Soviet) PG-2 HEAT; (Chinese) Type 50 HEAT
Projectile weight: 1.84 kg
Rate of fire: 4–6 rds/min
Armour penetration: (Soviet round) 150–180 mm; (Chinese round) 265 mm

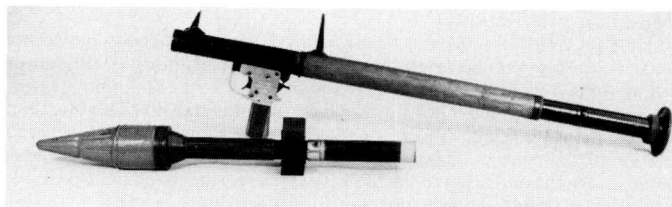

Manufacturer
Soviet state arsenals. Chinese state factories.
Status
Obsolete. Not in production.
Service
Still to be found with some militia and irregular forces in South-east Asia and
the Middle East.

RPG-2 rocket launcher with PG-2 grenade

RPG-7 portable rocket launchers

The RPG-7 is the standard manportable short-range anti-tank weapon of the
Warsaw Pact countries and their allies. It has had wide operational use in
several wars since it was introduced in 1962 and despite its unconventional
appearance it is an effective and efficient weapon.

It is similar to the RPG-2 in having a tube of 40 mm but the maximum
diameter of the PG-7 grenade is 85 mm. The RPG-7 has a stadia line,
sub-tension type, rangefinding optical sight for day use and an image
intensifier sight for night firing. It is percussion fired with a positively indexed
round. The gases emerge from the convergent-divergent nozzle at high
velocity and the weapon rests on the firer's shoulder. The original version of
the weapon, referred to simply as RPG-7 was rather larger than the current
model and fired a smaller projectile. The current model is known as the
RPG-7V and has a variant (see below) called the RPG-7D.

The grenade has large knife-like fins which spring out when the projectile
emerges from the tube. At the rear end of the missile are some small offset fins
which give a slow rate of roll to improve stability. After a travel of 10 m the
rocket motor fires. This gives a small increase to the velocity then sustains it
out to 500 m. The point at which the rocket assistance comes in is very
consistent and this regularity is a major factor in obtaining round-to-round
matching of trajectory.

The penetration of the PG-7 grenade is good. However the fuze is a
piezo-electric type which produces a voltage when the nose of the round is
crushed against an inner skin. This provides a means of rendering the missile
inoperative by shorting out the fuze – for example, by placing wire in front of
the target so that the nose cone, in touching two strands of the mesh, shorts
out the fuze. The HEAT warhead of the missile will then not function. If it is
fired into sandbags the self destruction element of the fuze will function but
the main HEAT charge will by then have broken up and the target effect is
virtually nil.

The procedure for firing is as follows. The user screws the cardboard
cylinder containing the propellant to the missile. He then inserts the grenade
into the muzzle of the launcher with the small projection mating with the
notch in the muzzle to line up the cap with the percussion hammer. The
nosecap is then removed and the safety pin extracted. The cocked hammer is
released when the trigger is fired and the missile is launched.

The missile has reasonable accuracy if there is no cross wind but with a
cross wind it is erratic. While the rocket motor is burning and producing an up
wind thrust greater than the down wind drag, the missile heads into wind and
moves up wind. When the motor produces less up wind thrust than the down
wind drag, or has burnt out, the rocket heads into wind, as would be expected

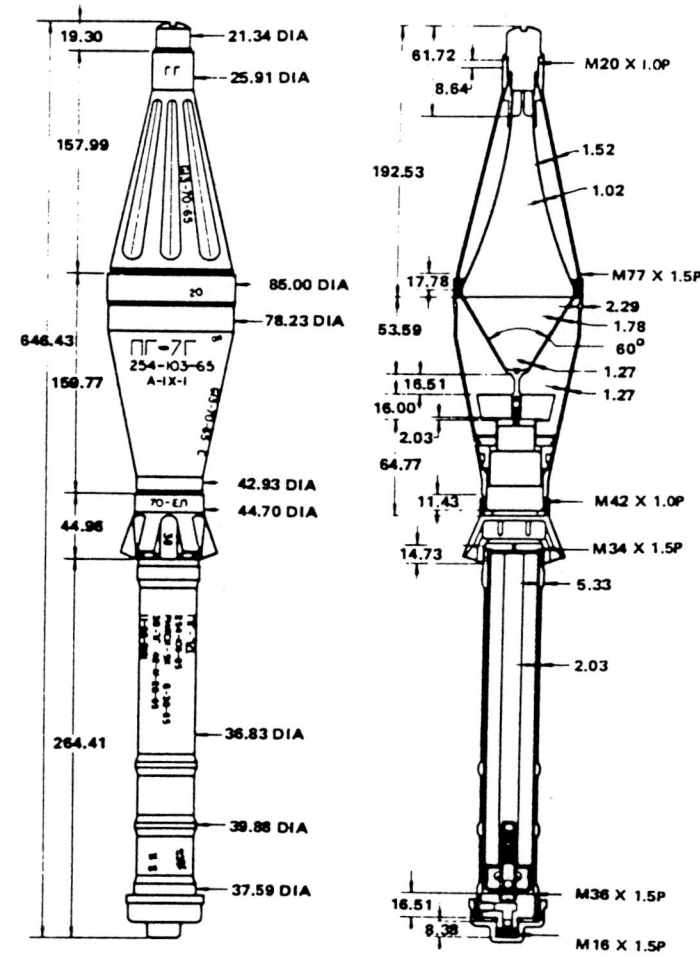

Dimensioned drawing of PG-7 projectile

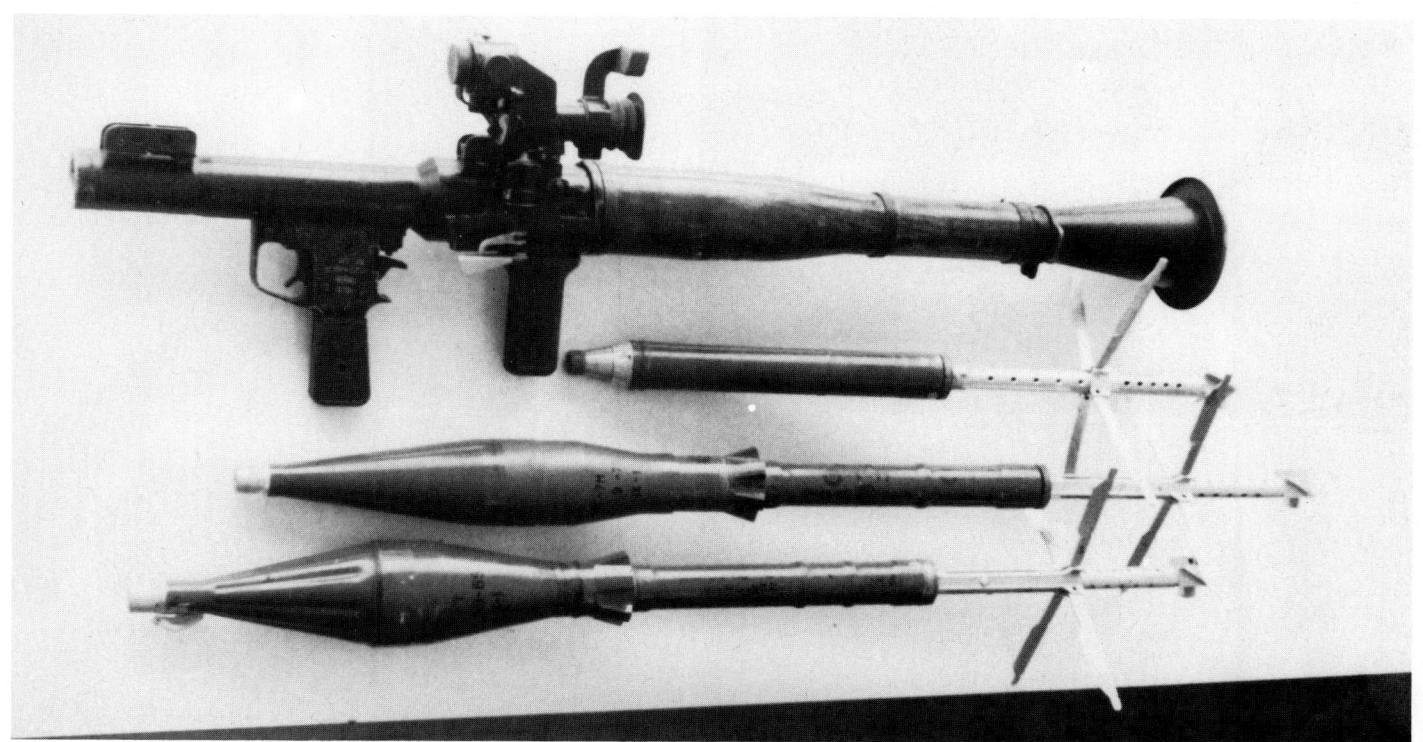

RPG-7 launcher with PGO-7 optical sight, OG-7 anti-personnel rocket, PG-7M HEAT rocket and original PG-7 HEAT rocket

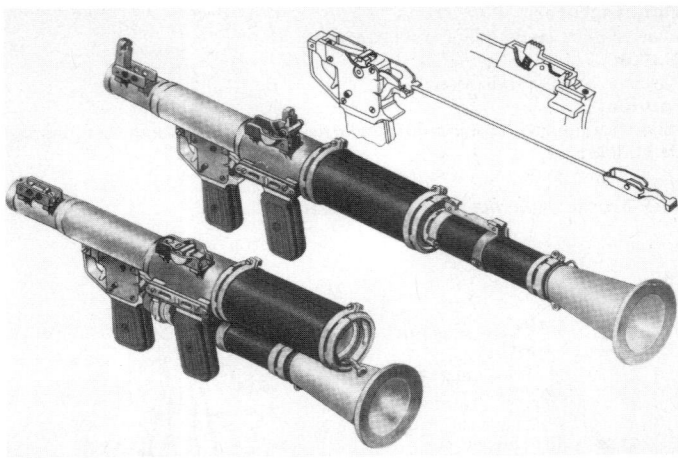

RPG-7D in folded and firing modes, and showing firing mechanism

RPG-7V rocket launcher. This has rangefinding optical sight

with a fin-stabilised projectile, but is blown bodily down wind. Another operational problem is that the weapon is very noisy.

Folding version

In 1968 a folding version of the RPG-7V was seen. This was initially taken to be a new launcher and was tentatively named the RPG-8 but then it was realised it was a variant of RPG-7V and it was named the RPG-7D. Originally used by airborne troops, it is now on general distribution throughout Warsaw Pact forces.

Product improvements

We have previously reported a long-range version with a two-stage rocket motor. It seems probable that this was a normal product improved (PI) motor with marginal increases in motor performance and it has undoubtedly now been incorporated in the RPG-16 (which see). A more positive identification of a PI is in the warhead. The drawing on this page shows the older warhead with a cone diameter of 73 mm and a fluted nose cone. A newer warhead is identical with the one fitted to the SPG-9 (which see) and the gun fitted in the turret of BMP. It has a cone diameter of 70 mm, but due to better design of the fuzing and firing arrangements it has a better performance. This new warhead has a slightly longer nose cone so that it appears to be slimmer.

A high explosive projectile has been introduced, designated the OG-7. This is ballistically identical to the standard PG-7 HEAT rocket but uses the 0-4M impact fuze, that used with the 82 mm mortar HE bomb. This new round is for anti-personnel use and has been so used in Afghanistan; reports from there speak of rounds date-marked from the late 1970s, pointing to that period as being the time of first introduction into service.

Sights

The PGO-7 and PGO-7V optical sights are marked with ranges from 200 to 500 m at intervals of 100 m. They have a 13° field-of-view with a × 2.5 magnification. They have, as already mentioned, a rangefinding stadia type sight. The NSP-2 infra-red night sight can also be used.

DATA
(RPG-7V)
Calibres: (launcher) 40 mm; (projectile) 85 mm
Length of launcher: 950 mm
Weights: (launcher) 7.9 kg; (grenade) 2.25 kg
Muzzle velocity: 120 m/s
Max flight velocity: 300 m/s
Range: (moving target) 300 m; (stationary target) 500 m
Range to self-destruction: 920 m
Penetration of armour at normal: 330 mm

Manufacturer
State factories.
Status
Current. Production.
Service
All Warsaw Pact countries and fairly general use in Africa, Asia and Middle East.

RPG-16 portable rocket launcher

The RPG-16 appears to be a product improved RPG-7 and it appears to retain most of the characteristics of that weapon.

The launcher appears to be the same in most details as for RPG-7, but the warhead is of unknown calibre, though still of the over-calibre type. Some sources have suggested that it may be a tandem warhead, utilising two shaped charges, one to make the initial penetration and the second to enlarge the hole and do more damage behind the armour. A bipod appears to be fitted to the front of the tube, probably to aid steadiness of aim when there is scope for its use.

RPG-16 rocket launcher being fired. Picture suggests that general principles of launcher are unchanged

DATA
Calibre: 58.3 mm
Launch tube length: 1.10 m
Launcher weight, empty: 10.3 kg
Grenade weight: 3 kg
Muzzle velocity: 130 m/s
Max flight velocity: 350 m/s
Practical range: 500–800 m
Armour penetration: 375 mm
Sights; optical, with IR night sight available

RPG-16 rocket launcher being carried on exercise. Bipod on muzzle can be identified

RPG-18 light anti-armour weapon

The RPG-18 has been confirmed as being a small rocket with a motor which is all-burnt on launch. The warhead is a shaped charge with a calibre of 64 mm. The weapon is disposable.

The launcher is made of an extruded light alloy tube. It telescopes and the two halves are held together with a bayonet catch which has to be rotated to release the halves. The launcher is extended before firing. The firing mechanism is mechanical and not unlike that on the M72. The sights are simple pop-up frames, made of plastic and are graduated for ranges of 200, 250 and 300 m. There is also a stepped cut-out slot which forms a rudimentary rangefinder; an average length tank will fit inside one step of the slot at the three ranges noted above.

Each launcher has a series of simple drawings showing the firer how to use it. In this it follows the style of M72. Each drawing has a few instructions with it. There is no shoulder stop and the firer has to judge where to put his head when firing. One of the training instructions warns that the venturi should not be within 2 m of a wall or similar solid obstruction. The simple diagrams indicate that the weapon can be used by anyone in a defensive position.

DATA
Calibre: 64 mm
Length: (folded) 705 mm
Weight: 2.7 kg
Projectile weight: 1.4 kg
Penetration: 375 mm armour
Combat range: 200 m
Muzzle velocity: 115 m/s
Sight markings: at 50, 100, 150 and 200 m

Status
Current. Apparently in production.
Service
Soviet and some Warsaw Pact armies.

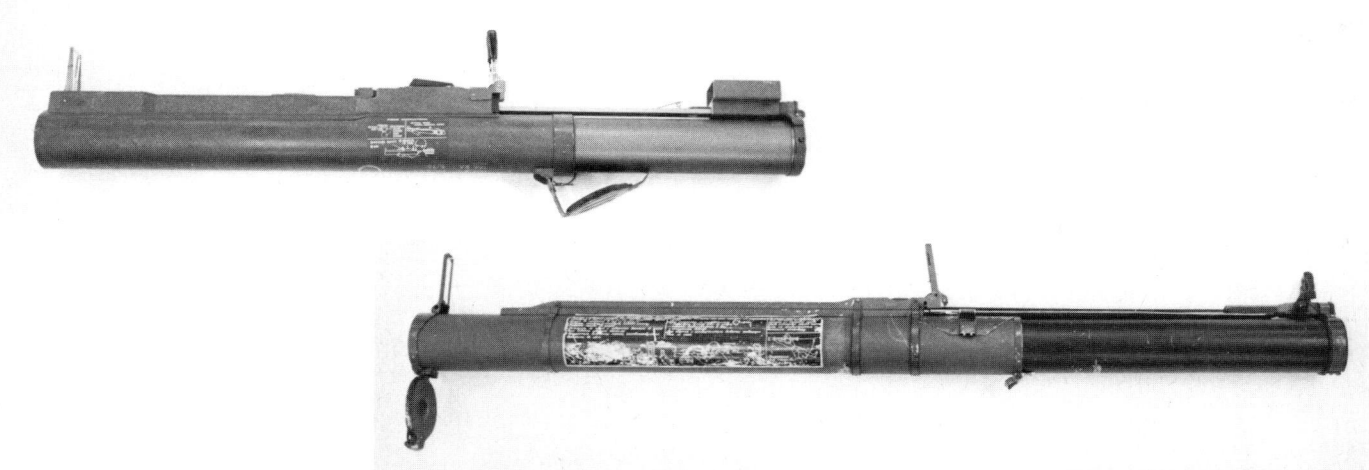

RPG-18 and M72 LAW extended to firing condition

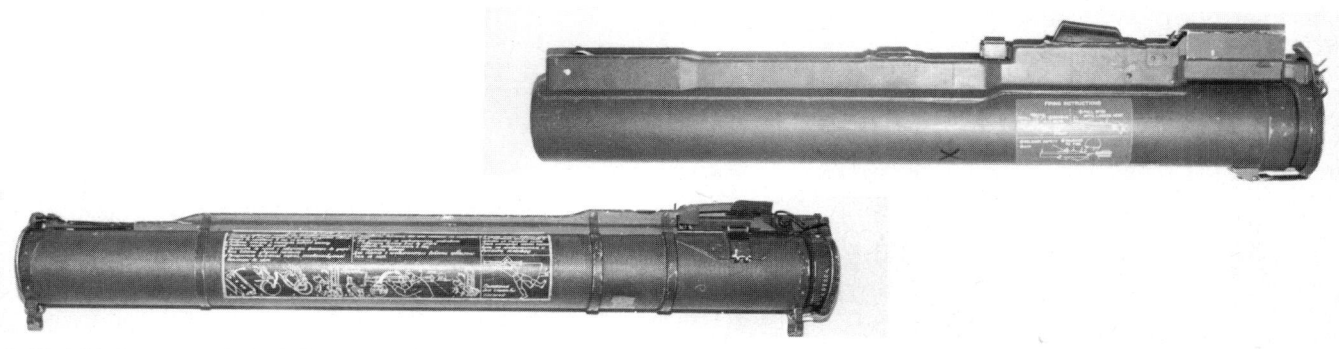

RPG-18, folded, compared with M72 LAW

RPG-22 light anti-tank weapon

Reports from Afghanistan in early 1987 indicate the issue of an improved LAW known as the RPG-22. This is of the same general construction and appearance as the RPG-18 but fires a HEAT rocket of 80 mm calibre. The weapon has simple pop-up sights graduated for ranges of 50, 150 and 250 m, together with a temperature-compensating rearsight. The launcher tube is about 850 mm long when extended. Reports also speak of a similar launcher firing a 76 mm rocket, and it is thought that this may be a variant manufactured elsewhere within the Warsaw Pact.

57 mm M-43 (ZIS-2) anti-tank gun

This is an elderly gun which is probably no longer used by first-line troops in the USSR but continues in service in some other Warsaw Pact countries and has been copied in China as the Type 55. Its penetration capability is no longer sufficient for use against modern battle tanks, but light tanks and other armoured vehicles would be vulnerable to it. A version with an auxiliary power unit and known as the M-55 probably still exists.

DATA
Calibre: 57 mm
Weight: (M-43) 1150 kg; (M-55) 1250 kg with auxiliary motor
Elevation: 5–25°

Traverse: 54°
Ammunition: HE, APHE, AP, HVAP with shell weights from 1.75–3.74 kg
Muzzle velocity: (HE) 700 m/s; (HVAP) 1270 m/s
Penetration: (HVAP) 100 mm armour at 500 m at 30°

Status
Obsolete.
Service
Thought to be in reserve units in some countries. Has been identified in Yugoslavia.

85 mm M-1945 (D-44) anti-tank gun

This gun was developed towards the end of the Second World War. Two models are in service, one being a towed weapon with split trails and the other a self-propelled version with a trail-mounted motor. Firing characteristics are the same for both weapons but the SD-44 self-propelled version weighs 525 kg more than the D-44. The weapon is used in many countries and is standard equipment of Soviet parachute regiments.

Manhandling 85 mm D-44 anti-tank gun

DATA
Calibre: 85 mm
Weight: 1725 kg
Length: 1.78 m
Elevation: -7 to +35°
Traverse: 54°
Projectile weight: (HE) 9.5 kg; (APHE) 9.3 kg; (HVAP) 5 kg
Muzzle velocity: (HE, APHE) 792 m/s; (HVAP) 1030 m/s
Range: 1600 m
Penetration: (APHE) 91 mm at 500 m at 30°; (HVAP) 113 mm at 500 m at 30°

Status
Obsolescent.
Service
USSR and at least 20 other countries.

73 mm SPG-9 recoilless gun

The SPG-9 is a lightweight anti-tank gun, normally carried by two men, crewed by four men and mounted on a tripod for firing. It can be towed using a small two-wheel carriage. It is now used not only by the Motor Rifle Battalions of the Soviet Army but is also in service in Bulgaria, East Germany, Hungary and Poland.

It fires a fin-stabilised round with a HEAT warhead. The propellant charge is carried in a case attached behind the fins and so it makes a very long round. However not only does this system produce a high muzzle velocity but the projectile is also rocket assisted so that the velocity is further increased. The projectile is given a slow rate of spin inside the barrel by means of offset holes in the launching charge. Once clear of the muzzle, the main rocket motor lights up after 20 m.

The device sometimes seen mounted over the barrel, and resembling a spotting rifle, is a sub-calibre training device.

DATA
Calibre: 73 mm
Weight of launcher: 47.5 kg
Weight of bipod: 12 kg
Length of launcher: 2110 mm
Height of launcher: 800 mm
Ammunition: HEAT, rocket assisted
Muzzle velocity: 435 m/s

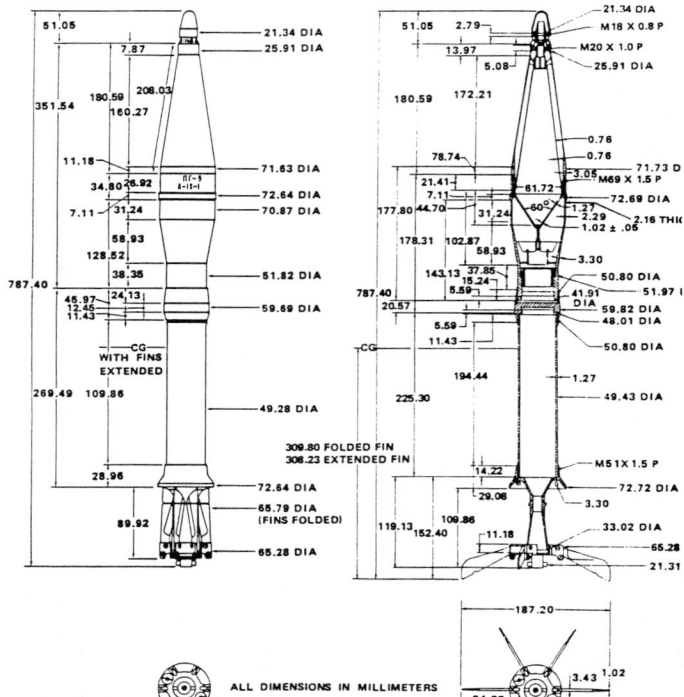

73 mm HEAT/FS projectile PG-9

Velocity with rocket assistance: 700 m/s
Max range: 1300 m
Penetration: more than 390 mm

73 mm HEAT/FS projectile Model PG-9

DATA
Calibre: 73 mm
Type: HEAT/FS
Weight, fuzed: 2.53 kg

73 mm SPG-9 recoilless gun

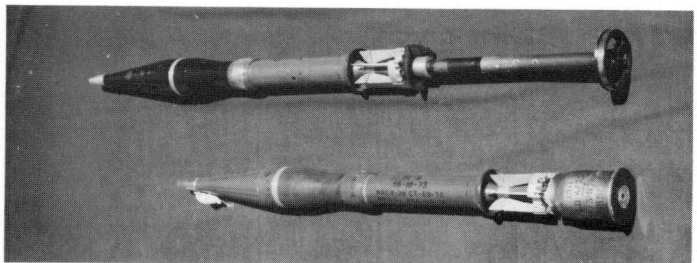

Bursting charge: 320 g RDX/Wax
Fuze: VP-9 piezo-electric PIBD

Manufacturer
State factories.
Status
Current.
Service
Warsaw Pact armies.

Ammunition for SPG-9 gun; top round is for towed gun, lower round for version mounted in BMP and BMD vehicles

82 mm B-10 recoilless gun

This is a smoothbore recoilless gun. The projectile is fin-stabilised and at first sight can easily be taken for a mortar bomb. It has a HEAT warhead. The weapon is towed on two wheels which are removed for firing, and the gun is fired off a tripod which is folded under the barrel in the travelling position and lowered when the gun is emplaced. If necessary the weapon can be fired from the wheels. A bar attached to the muzzle is used to enable the weapon to be dragged into position. The gun has an optical sight without rangefinding capability.

It was standard equipment in the Soviet Army until the early 1960s and then it was gradually phased out until now it is used only in the parachute battalions of the airborne divisions. It has been used for some years by the Egyptian and Syrian armies, was copied by the Chinese as the Type 65 recoilless gun and is to be found in the armies of several Asian countries.

DATA
Cartridge: 82 mm
Method of operation: recoilless, with multi-vented breech-block with enlarged breech section
Barrel length: 1659 mm
Overall length: 1677 mm
Weight: (in travelling position) 87.6 kg
Weight of round: (HE) 4.5 kg; (HEAT) 3.6 kg
Muzzle velocity: (HE round) 320 m/s
Armour penetration: 240 mm
Max range: (HE round) 4500 m
Max effective anti-tank range: 400 m
Rate of fire: 6–7 rds/min

Status
Obsolescent.
Service
In limited European service but widely used in the Middle and Far East.

82 mm HEAT projectile BK-881

DATA
Calibre: 82 mm
Type: HEAT/FS

O-881A fin-stabilised fragmentation bomb for 82 mm B-10 gun

82 mm B-10 recoilless gun

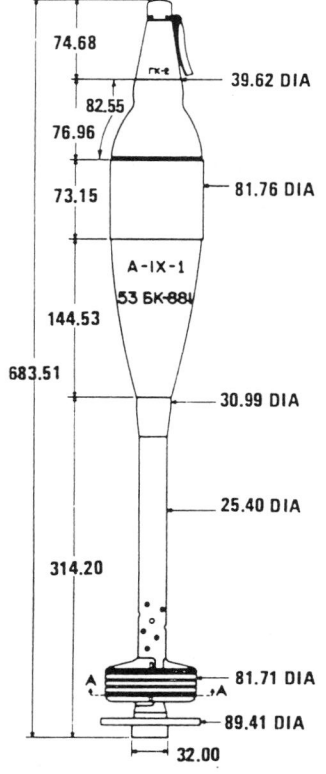

82 mm HEAT projectile BK-881

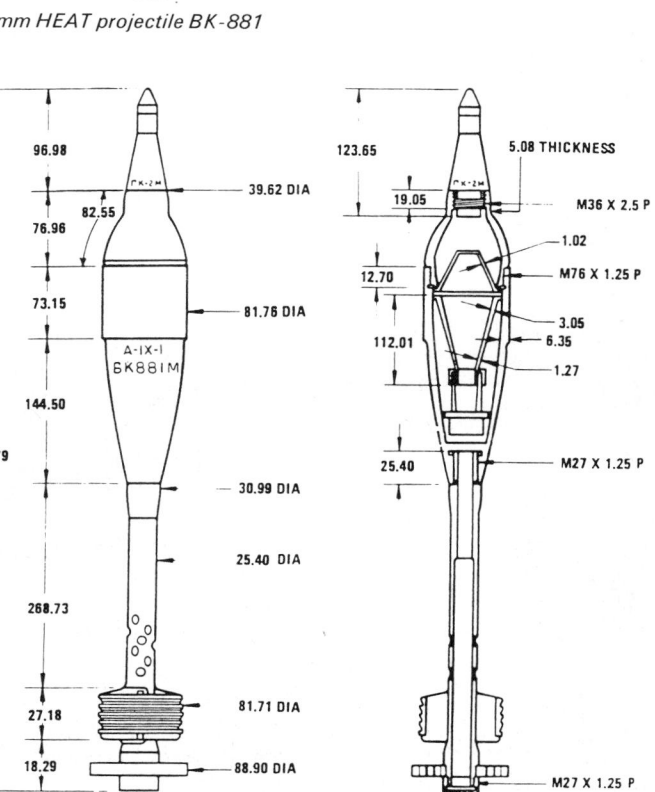

82 mm HEAT projectile BK-881M

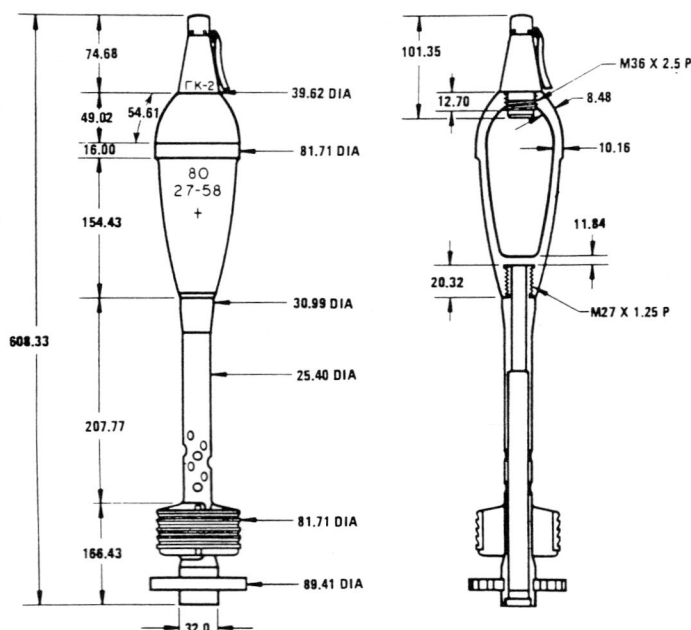

82 mm HE fragmentation projectile O-881A

Weight, fuzed: 3.87 kg
Bursting charge: 463 g RDX
Fuze: GK-2 PIBD

82 mm B-10 recoilless gun on wheels. It is easily pulled for short distances by two men

82 mm HEAT projectile BK-881M

DATA
Calibre: 82 mm
Type: HEAT/FS
Weight, fuzed: 4.11 kg
Bursting charge: 545 g RDX
Fuze: GK-2M PIBD

82 mm HE fragmentation projectile O-881A

DATA
Calibre: 82 mm
Type: HE fragmentation
Weight, fuzed: 3.90 kg
Bursting charge: 465 g TNT/dinitronaphthalene
Fuze: GK-2 point detonating

107 mm B-11 recoilless gun

Like the B-10 this is another smoothbore recoilless gun firing a fin-stabilised mortar-like round. It is towed by the muzzle from a vehicle. The tripod legs are folded under the barrel in the travelling position and it is brought into action and fired normally off the tripod but, in an emergency, it can be fired off the wheels with reduced accuracy. It can be employed as an anti-tank weapon but the sights also allow it to be used in the indirect fire role with an HE round.

DATA
Cartridge: 107 mm
Method of operation: recoilless, with multi-vented breech-block with enlarged chamber section
Barrel length: 2718 mm
Overall length: 3314 mm
Weight: 305 kg
Weight of round: (HE) 13.6 kg; (HEAT) 9 kg
Max range HE round: 6650 m
Effective anti-tank range: 450 m
Armour penetration: 380 mm
Rate of fire: 6 rds/min

Status
Obsolescent.

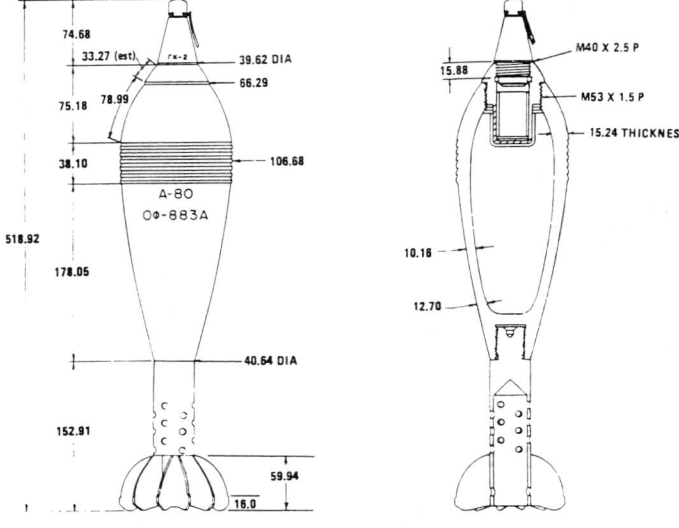

107 mm HE fragmentation projectile O-883A

107 mm B-11 recoilless gun. This is a bigger equipment than the B-10 and is normally vehicle-towed from the muzzle. Larger wheels are a readily distinguishable feature

Service

No longer in use in the Warsaw Pact armies. May be found in some Middle East and South-east Asian forces.

107 mm HEAT projectile BK-883

DATA
Calibre: 107 mm
Type: HEAT/FS
Weight, fuzed: 7.51 kg
Bursting charge: 1058 g RDX/Aluminium
Fuze: GK-2 PIBD

107 mm HE fragmentation projectile O-883A

DATA
Calibre: 107 mm
Type: HE fragmentation
Weight, fuzed: 8.5 kg
Bursting charge: 2.088 kg Amatol 80/20
Fuze: GK-2 point detonating

107 mm HEAT projectile BK-883

3M6 Snapper anti-tank guided missile

Snapper is the NATO codename given to the earliest of the Soviet anti-tank guided missiles. It is also known by the US alphanumeric designation AT-1 and is believed to be designated 3M6 by the Soviet authorities.

Generally similar in function and layout to the French SS10 missile it has, like its counterpart, been almost completely displaced by later weapons and is now unlikely to be encountered in Europe at least. One of the weapons supplied to the Arab armies in the early days of Soviet involvement in the Arab-Israeli contest, it was employed during the Six Day War.

Snapper is a manual command to line of sight (MCLOS) wire guided missile and has always been seen mounted on a vehicle but can be controlled by an operator located in a remote position. Early mountings were on the GAZ-69 vehicle, but later the BRDM-2 was used.

DATA
Type: infantry anti-tank system
Guidance principle: MCLOS. Optical tracking of rear-mounted flares. Joystick control
Method of guidance: wire control using spoilers in the trailing edge of the wings. Roll stabilised
Length: 1130 mm
Body diameter: 140 mm
Wing span: 775 mm
Launch weight: 22.25 kg
Cruising speed: 53.6 m/s
Range: (minimum) 370 m; (max) 2700 m
Warhead: hollow-charge weighing 5.25 kg. Direct action fuze
Penetration: 380 mm

Manufacturer
State factories.
Status
Obsolescent. No longer manufactured.
Service
Probably phasing out of service in all countries. Has been supplied to all Warsaw Pact armies and Egypt, Syria and Yugoslavia.

Yugoslav troops loading Snapper missiles

Snapper fired from GAZ-69

Snapper's pointed nose and wide wing span make it easily recognised

Swatter anti-tank guided missile

Swatter is the NATO codename for the Soviet anti-tank missile which is also known by the US alphanumeric designation AT-2. It is a manual command to line of sight (MCLOS) vehicle-mounted system and is unusual in that radio command guidance is used, there being two radio frequencies on the command link as an ECCM measure.

The missile has two pairs of fins at the rear and two canard fins at the front. The canard fins are lined up with one pair of rear fins and these are parallel to

the ground when the missile is in flight. The fins have controllable surfaces. Two flares are attached to one pair of fins. Propulsion is from a solid rocket motor which both boosts and sustains.

There have been two versions of the missile. The earlier version known as Swatter A had two sustainer nozzles, one on each side. The Swatter B missile does not have the sustainer nozzles and has larger flares – presumably indicating a longer time of flight.

BRDM installation
The BRDM reconnaissance vehicle has been modified to take Swatter. The crew compartment has been widened and an optical sight has been installed. The missiles have been placed in the crew compartment which has three light

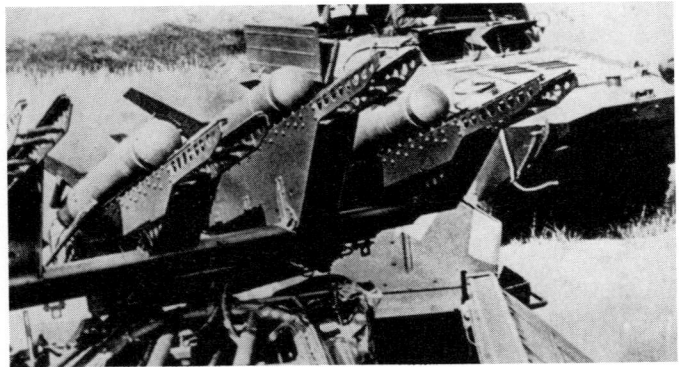

Swatter on BRDM in foreground

covers over the top. There are four missiles with launchers in the missile compartment together with the erector gear. The missiles are on rails which can be elevated and depressed.

DATA
Type: radio command guided MCLOS missile system
Weight at launch: 26.5 kg
Length: 1130 mm
Diameter: 130 mm
Rocket motor: solid
Guidance: manual command. Radio link
Control: control surfaces on fins
Warhead: HEAT. Direct action fuze
Range: 600–2500 m
Penetration: 400 mm +

Manufacturer
State factories.
Status
Current.
Service
USSR and Warsaw Pact armies. Supplied to Egypt and Yugoslavia.

Sagger anti-tank guided missile

Sagger is the NATO codename for the Soviet manual command to line of sight (MCLOS) anti-tank missile, also known by the US alphanumeric designation AT-3, which is now extensively deployed in Europe and elsewhere. Currently it is known to be used in at least three ways: as a manportable missile; mounted on BRDM; mounted on BMP.

Manportable system
This is used in the motor rifle and airborne units. The detachment consists of an NCO and two men. The missile is carried into action in a glass fibre case in which the warhead is separated from the motor. The lid of the case forms a base for launching the missile which has a launch rail on the underside of the motor section. The two parts are taken out of the case and the front legs of the rail on the motor section are slotted into the lid so that there is a small angle of elevation. This is done to ensure that the missile enters the operator's field-of-view when launched. The warhead is then clipped to the body. The missile is aligned on the target or the centre of the primary arc, and the launcher is strapped to stakes driven into the ground. The leads from four missiles are taken to the control box and connected up. At the back of the control box are the batteries. The periscopic sight emerges from the centre and the control stick is in front of it.

The NCO is the operator. If the target is under 1000 m distant he does not use the sight but guides the missile to the target by eye. If the target is further away he gathers the missile to a line above the target and then controls it through the × 10 magnification periscopic sight.

The method of guidance is line of sight command and the corrections are transmitted through a multi-core cable fed out from the missile. There is a flare between the two upper fins. The missile has two solid propellant motors. The front annular motor is a booster. The rear motor controls the missile through two jetavator nozzles at the rear which swivel according to the commands transmitted through the wire. The four plastic fins are folded when the missile is towed in the case. The warhead is a HEAT design.

BRDM mounted system
The BRDM reconnaissance vehicle is modified to take a rectangular hatch lifted by a centrally mounted hydraulically operated column. Six launch rails are attached to the underside of the hatch and missiles are clipped to these rails. Elevation, depression and swivelling are possible. The sight and control unit is between the two forward hatches. Control of the missiles is carried out in the same way as with the manportable system. Eight missiles are carried as a reserve in the vehicle.

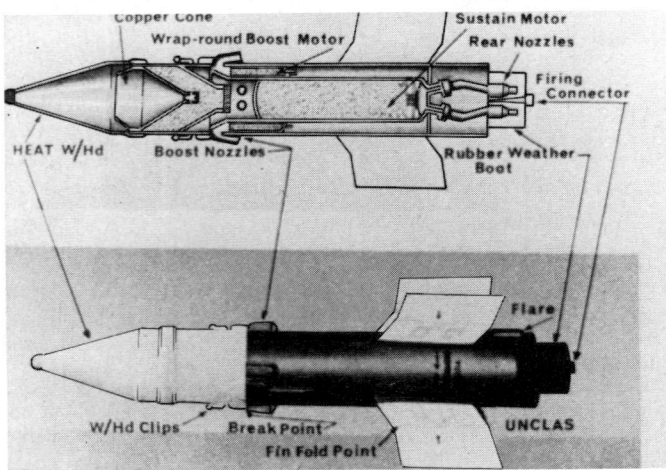

Sectioned view of Sagger missile. Note wrap-round booster motor

BMP mounted system
In this single mounting the guide rail of the missile is fitted into a bracket above the 73 mm gun. The guide rail can be elevated and traversed at will. The missile is fired from inside the vehicle and controlled in the same way as the manportable version.

Putting warhead on body

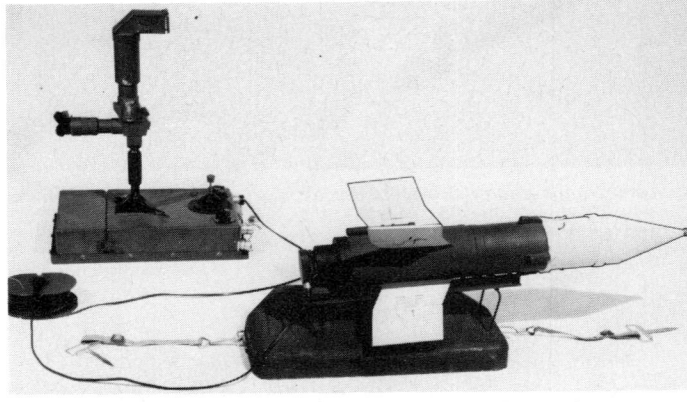

Sagger set up on its carrying case — launching ramp, connected to control box. Control stick and periscopic sight are clearly seen

Sagger on BRDM

DATA
Type: wireguided MCLOS system
Weight at launch: 11.3 kg
Length: 880 mm
Diameter: 120 mm
Propulsion: 2 solid rocket motors
Guidance system: MCLOS wire link
Control: jetavator nozzles. Roll rate stabilised
Warhead: HEAT. Piezo-electric fuze

Range: 500–3000 m
Penetration: 400 mm +

Manufacturer
State factories.
Status
Current. Production.
Service
USSR and Warsaw Pact armies, Egypt, Syria and Yugoslavia.

AT-4 Spigot anti-tank guided missile

There had been reports for some years of a Soviet semi-automatic command to line of sight (SACLOS) system but it was not until 1980 that photographs and a little information appeared in some Warsaw Pact military journals. The name is not certain, but in the West it is known as AT-4 and it has been given the NATO reporting name of Spigot. It may be known to the Soviets as Fagot. The missile is uncannily similar to MILAN and almost certainly uses the same principles. The chief outward difference lies in the fact that the sight is smaller than for MILAN and the computing mechanism and goniometer are contained in a box below the launch rail, whereas in MILAN the sight container is larger and has the goniometer inside it. The tactical advantages of such detailed placings are very small, though it may be that AT-4 has a slightly smaller silhouette from the front when in the firing position.

The introduction of AT-4 to the Warsaw Pact greatly improves the capability of the Pact infantry to defend itself against armoured attacks, and it can be assumed that the older manual command to line of sight systems will now be phased out.

DATA
ESTIMATED
Operation: SACLOS, wire guided
Overall weight: 40 kg
Warhead diameter: 120 mm
Range: 2000–2500 m
Flight time: 11 s
Speed: 180–200 m/s
Penetration: 500–600 mm armour at 90°

Status
Current. Presumed in production.
Service
Soviet, East German and perhaps other Warsaw Pact armies.

AT-4 Spigot SACLOS system

AT-4 anti-tank guided missile in East Germany. Compare with MILAN

AT-7 Saxhorn guided missile

This weapon, generally considered to be in the same category as the US Dragon missile, is believed to have been placed on general issue in the early 1980s, but so far little hard information has appeared in the West. The photograph reproduced here shows AT-7 in use by Polish troops, and it is believed to be deployed at company level in the Soviet and some Warsaw Pact armies.

The weapon appears to be the usual type of pre-packed launch tube with missile, mounted upon a firing post with optical and electro-optical sights. The method of guidance is not known, though it is possibly a laser beam rider. The missile is believed to weigh about 6 kg and has a shaped charge warhead capable of defeating better than 500 mm of armour. The maximum range is probably 1500 m.

AT-7 Saxhorn anti-tank missile

UNITED KINGDOM

120 mm Wombat battalion anti-tank gun

This equipment has now been declared obsolete in British service, though some may remain in use in other countries. It is known to have been supplied to Jordan, Kenya, Malaysia and New Zealand.

Full details of the gun and its performance can be found in previous editions of *Jane's Infantry Weapons*.

Swingfire anti-tank guided missile

Swingfire is a long-range command-controlled (manual command to line of sight (MCLOS)), wire commanded, anti-tank weapon system. It is designed for use by armoured formations and is intended to be mounted on a vehicle. It is capable of destroying the armour on any known tank. It is, however, also suitable for infantry use.

The controller can place his vehicle in a totally concealed position and site himself in a vantage position, up to 100 m away. Alternatively the controller can be 23 m above the launch vehicle. The target can be up to 45° on either side of the launch bearing and up to 20° above or below the horizontal axis of the vehicle. For use with armoured forces, the system can be mounted in a variety of armoured vehicles. In the British Army the Striker vehicle is used. The Striker is from the Combat Vehicle Reconnaissance (Tracked) (CVR(T)) family produced by Alvis Ltd. Striker is also in service in the Belgian Army. Both vehicles permit Swingfire engagements in the direct and separated fire modes. The Swingfire system is adaptable to almost any vehicle and can readily be installed as a removable crew-portable launcher for infantry use.

Swingfire is fired from its launching box which forms a primary package for the missile and seals it up to the moment of launch. In the armoured fighting vehicles variants, the launcher boxes are housed in armoured bins for protection against shell splinters and small arms fire. The launch box weight is 10 kg. No launcher aiming is required. The launcher bins are elevated to the constant launching angle of 35°. If the vehicle is tilted, this is compensated for by a special vehicle tilt compensating unit.

The Swingfire missile is gathered automatically, after launch, into the controller's field-of-view, close to the line of sight to the target, by a programme generator in the ground equipment. The data which is fed into the programme generator depends on the relative positions of the launch vehicle, controller and target bearing. When the system is operated in the separated fire mode, the controller pays out the separation cable and sets up the sight with its reference mark parallel to the line of launch. At the vehicle additional data is passed to the programme generator: vehicle tilt angle (automatic compensation); the angle of elevation to the point of the crest over which the missile will pass to the target; azimuth angle between the line of the launcher box and the separation sight; separation distance from launcher to sight.

The angles of elevation and azimuth to the target from the separation sight are read off and passed automatically to the programmer when the firing button is pressed.

Whether in the direct or separated fire mode, the automatic programmer brings the missile on to the line of sight to the target and into the controller's field-of-view; and thereafter he controls the missile manually.

The Swingfire missile command system is one of velocity control. In the missile autopilot there are two gyros, one for pitch and one for yaw. In the absence of commands from the controller the autopilot maintains the missile on a constant course. When the joystick is moved from the central position, the missile alters course and continues to do so until the stick is centralised. It then maintains its current heading. So the controller moves the joystick until he sees the missile framed in the target and moving towards it. He then centres the stick and the missile continues on its course to the target.

The controller's demands, communicated by the trailing wire link, from the missile, are interpreted in the missile by altering the direction of thrust of the main motor.

Warhead lethality

At a demonstration carried out in 1979 to show the power of the warhead, a live missile was fired at a Conqueror tank at a range of almost 4000 m. Conqueror was the heaviest and best-protected tank ever built in the UK and it still serves as an adequate yardstick of the best armour that can be expected to appear on the battlefield in the 1980s. The missile struck the commander's cupola on the top of the turret and the jet went right through the cupola, leaving two holes in the sides. The explosion blew off all the hatch covers and caused extensive damage to the interior. There is no doubt that the Conqueror would have been completely knocked out, even by a hit on the cupola.

The latest Swingfire improvements include: the introduction of a clip-on thermal imaging sight for use with the current separated sight, enabling targets to be identified and successfully engaged at night and in poor visibility. This is a lighter and cheaper alternative to the combined sight used by the British Army. There is also a new compact solid-state micro-miniaturised missile control equipment leading to improved reliability and simplified maintenance in the field and a new lightweight palletised launcher.

Thermal imaging sight

Using the existing separated sight the missile operator can control an engagement at a distance from the launcher, which itself can be concealed. The separation sight optics comprise an open ring sight for initially gathering the missile into the line of sight to the target, and a × 10 telescope used when controlling the missile during the final flight phase. The thermal imaging sight, made by Thorn-EMI Electronics Ltd and based on its Multi-Role Thermal Imager (MRTI), carries a visual gather-sight and can be clipped on to the separation sight in place of the existing optical telescope. The MRTI

Entry hole made by Swingfire missile in Conqueror tank cupola. At top of picture is hatch, which was blown open

Simplified launcher and micro-miniaturised missile control equipment. Also shown, in foreground, is Thorn-EMI MRTI

Thorn-EMI MRTI with visual gather-sight mounted on separation sight in place of optical telescope

Firing Swingfire from Egyptian Army CJ6 Jeep during desert exercises

Swingfire launched from CVR(T) vehicle 'Striker'. Combat reconnaissance vehicle and missile are also in service with Belgian Army

operates in the 8-14 micron wavelength far infra-red band. It can be used, if desired, with all Swingfire systems currently in service, as no modification is required to the separation sight itself or to the launcher system.

Besides weapon aiming, the MRTI sight can also be used for surveillance.

Micro-miniaturised missile control electronics

There have been considerable advances in electronic technology since Swingfire entered service with the British Army in 1969. Advantage has been taken of new techniques to reduce both the size and weight of the Swingfire ground control equipment which, in turn, has led to significant improvements in overall system reliability and maintainability. Two compact units, together weighing 20 kg, replace the present five units, giving a volume reduction to about one-fifth. The new units contain hybrid solid-state integrated-circuit packages mounted on single-sided Eurocard (160 × 100 mm) printed circuit boards. Because all the electronics are now housed on six boards, fault diagnosis and repair-by-replacement in the field is a much easier and quicker process. Spares holdings can also be reduced and logistic support greatly simplified. The new micro-miniaturised units are suitable for installation in armoured vehicles such as Striker, as well as for use with palletised Swingfire systems.

Lightweight launcher

A new fixed-elevation launcher of simplified construction has been designed. When equipped with the micro-miniaturised ground control electronics, the overall weight of the system has been reduced by 50 per cent compared with the original palletised version.

DATA
Length overall: 1060 mm
Max body diameter: 170 mm
Span with wings open: 373 mm
Range minimum: less than 150 m at direct fire, up to 300 m with 100 m separation; (dependent on separation and launch azimuth angle) (max) 4000 m
Guidance principle: optical sighting. Programmed gathering to line of sight. Wire guidance (MCLOS)
Guidance method: thrust deflection of main motor-jetavator
Propulsion: integral boost and sustainer motor
Warhead: shaped charge

Manufacturer
British Aerospace plc, Dynamics Division, Six Hills Way, Stevenage, Hertfordshire SG1 2DA.
Status
Current. Available.
Service
British Army. Supplied to Belgium and Egypt.

LAW 80 light anti-tank weapon system

LAW 80 is a one-shot low cost disposable short-range anti-tank weapon developed to replace existing weapons which are either light and ineffective or of a size which demands crew operation to achieve lethality. It provides a capability for the individual soldier to engage current and future Main Battle Tanks at ranges out to 500 m and achieve a high probability of hitting the target. The unique performance of the HEAT warhead (armour penetration in excess of 650 mm) ensures that Main Battle Tanks both current and future, can be defeated from any aspect, including frontal attack. LAW 80 is stored and transported in Unit Load Containers (ULC) holding 24 launchers. LAW 80 is issued from the ULC direct to the infantryman and is fully manportable together with his own personal weapons and pack, each launcher being provided with a carrying handle and shoulder sling.

The tactical use of LAW 80 in the British forces is to provide defence against an armoured threat at ranges out to 500 m. In this, it complements the cover provided by medium- and long-range anti-tank guided weapons which suffer a minimum range problem at shorter distances. The weapon will be operated not only by infantry but also by Royal Marines and the RAF Regiment. The system is simple to operate and does not require a dedicated gunner or specialist training.

Much attention has been placed on the need for the firer to obtain a good hit on the target, since the firing of almost all types of LAW projectile is likely to reveal the firing position. It is dangerous and time consuming to engage with a second weapon. To achieve this accuracy with operators drawn from combat and support arms, some of whom may not be expert in the estimation of range and lead for moving targets, a spotting rifle is used. This built in spotting rifle contains five pre-loaded rounds, any number of which may be fired, without revealing the position of the firer. The 9 mm spotting rifle ammunition is ballistically matched to the main projectile marked by a tracer and by a flash head to record a hit on a hard target. At any time the operator can select and fire the main projectile. Trials have proved the large improvement in hit probability obtained using the spotting rifle principle.

The weapon is issued as a round of ammunition, complete with its projectile and the integrated spotting rifle pre-loaded and pre-cocked. Safety is assured by the provision of detents and links which require two failures before any hazard would be caused. In deploying and operating the weapon the safety links are progressively removed automatically up to the point of firing the rocket projectile. The safety features are listed under data.

Shock absorbent end caps protect the weapon for carriage and storage. These also provide some side protection, in conjunction with the resilient carrying handle. The end caps provide sealing for the tubes to protect the projectile against the effects of water immersion, even though the projectile itself is sealed. The end caps also support the projectile against the effects of mishandling such as end drops.

After removal of the end caps, the tube containing the HEAT projectile is extended rearwards from the outer tube. The launch tube is automatically locked into position. This moves the centre of gravity of the weapon from the carrying handle to the shoulder rest to provide an ideal balance suitable for any firing position.

The spotting rifle is an integral part of the outer tube. Tests have shown that the five rounds of spotting ammunition sealed in the rifle are sufficient for two engagements. Also included in the external furniture is the firing handgrip/carrying handle and a folding shoulder rest. The sight is separately bonded to the front tube and a forward sliding protective cover allows the sighting prism to erect. The sight is of unit magnification with the graticule projected through the sighting prism into the firer's line of sight, like a

LAW 80 with Adder remote firing apparatus

LAW 80 in firing position

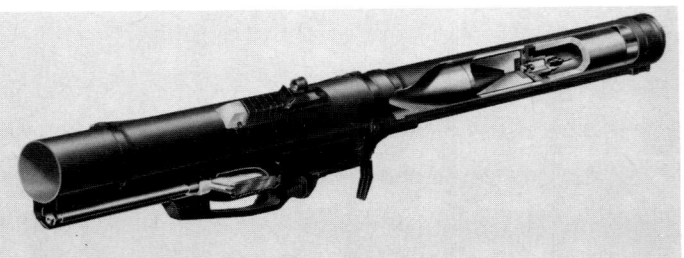

Sectioned view of LAW 80

head-up display. The sight can be used with one or both eyes open and for lowlight use the graticule can be illuminated with a tritium light source selected by the firer, making the weapon effective down to starlight levels of illumination.

With the weapon extended and cocked and with the sight erected, the gunner only has to select 'Arm' on the Safe/Arm lever and use the trigger to fire either the spotting rifle or the rocket projectile. When the rocket projectile is to be fired the change lever is moved forward by the thumb of the firing hand. The system is completely reversible if no target is engaged with the rocket projectile. As part of the safety design, the main round cocking lever covers the detent to unlock and close the weapon so that the weapon must be uncocked and in its spotting rifle mode before it can be closed down.

The projectile is initiated with a totally non-electric (and therefore totally EMC proof) system comprising a percussion cap in the launcher connected by a flash tube to the rocket igniter. The rapid burn of the rocket motor, which is completed before the rocket leaves the launcher, ensures that the efflux is contained and directed rearwards by the launch and blast tubes, leaving the projectile to coast from the muzzle to the target. No debris of afterburn is projected at the firer, the recoil is not discernible, and the noise level at the firer's ear is significantly less than 180 dB, which meets the US Surgeon General's requirements, unlike many other weapons of this class.

The forward part of the projectile consists of the HEAT warhead and its fuzing unit, and a double ogive nose switch which also provides the optimum stand-off distance. The fuzing unit generates electrical energy to fire the warhead by means of peizo crystals and contains various safety devices to ensure that the warhead does not arm until safe separation from the firer is achieved.

At the rear of the projectile the composite aluminium and filament wound motor case has an extruded vane propellant. Four wrap-around fins are mounted on the rear of the motor. These are spring-loaded to erect at muzzle exit to provide stability and to spin the projectile as it coasts to the target.

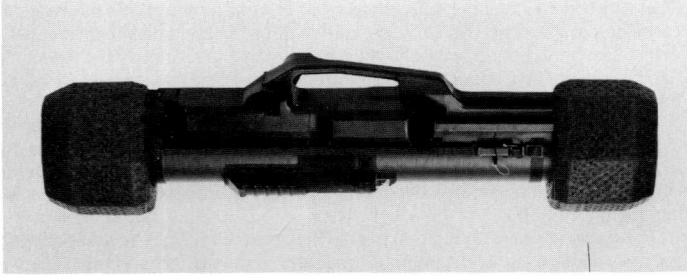

LAW 80 ready for carriage

A complete and simple training package is available. This includes:
drill weapon
indoor trainer
outdoor trainer.
Training courses are available to suit customers' requirements, both in the UK and overseas.

The LAW 80 weapon system is designed to kill modern armour at all aspects, including head-on attacks, out to 500 m. It is simple and cheap, and is designed for wide scale issue to all arms. It has been developed under the direction of the prime contractor, Hunting Engineering Ltd. The system has been ordered by the UK Ministry of Defence to replace the currently used 84 mm Carl Gustaf and 66 mm M72 weapons. Deliveries commenced in 1987. Export enquiries should be directed to Hunting Engineering Ltd.

Associated equipment
The British Army night sight is fully compatible with LAW 80. After the projectile has been fired the sight is removed and the weapon discarded in the normal way.

A number of systems have been developed to fire LAW 80 remotely. A standard weapon and identical tripod and firing unit have been incorporated into the following four systems:
    Adder: the weapon is command fired from a distance of up to 200 m.
    Addermine: suitable for use as an off-route mine. The weapon is fired when its trip or breakwire is disturbed.
    Addermine Ajax: this is a fully autonomous off-route mine system. Its programmable sensor is able to select a particular target before firing.
    Adderlaze: a coded laser pulse fires the weapon.

DATA
Calibre: 94 mm
Penetration: in excess of 650 mm
Length: 1 m (folded) 1.5 m (extended)
Weight:
Carry: 10 kg
Shoulder: 9 kg
Effective range: 20 m – 500 m
Safety and performance: double failure logic throughout; motor burn completed well before exit from launcher; imperceptible shoulder recoil; spotting rifle allows aim correction before disclosing fire position with three fold increase in hit probability at some ranges; ergonomic design allows operating sequence to be done only in correct way; noise levels at firer's ear significantly less than 180 dB and meet UK MIL STD 1474A
Construction: filament wound tube, high proportion of plastic mouldings, ensuring low UPC for very high performance.
Shelf life: 10 years

Prime contractor
Hunting Engineering Ltd, Reddings Wood, Ampthill, Bedfordshire MK45 2HD
Status
Current. Production.
Service
British Army, Royal Marines, RAF Regiment.

UNITED STATES OF AMERICA

57 mm M18A1 recoilless rifle

No longer in service with US forces but still to be found in several other armies, the M18A1 is by today's standards a relatively cumbersome and ineffective arm. Although American ammunition has not been obtainable for many years, suitable rounds have been manufactured in China and South America and may well still be in circulation.

DATA
Calibre: 57 mm
Total weight: (with cover) 25 kg
Weight: (with bipod and sight) 22.04 kg; (with bipod only) 20.15 kg
Barrel weight: 18.2 kg
Overall length: 1564 mm
Barrel length: 1219 mm

Length of rifling: 1181 mm (20.7 cal)
Rifling: rh 1 turn in 30 cal
Breech-block: interrupted lug
Firing mechanism: percussion
Muzzle velocity: 365 m/s
Range: (effective) 450 m; (max) 3976 m
Ammunition: HEAT, HE, WP or TP
Crew: 2

Status
Obsolete.
Service
Occasionally seen in the hands of guerrilla forces.

Olin RAAM Rifle-launched Anti-Armour Munition

This is a rocket-boosted bullet-trap rifle grenade which can be launched from any unmodified M16A2 rifle by means of an expendable plastic launch adapter. It has a direct-fire engagement range of 250m, can defeat 400 mm of armour, and is safe during transport and handling, arming only after travelling 10 m from the rifle. The combination of rifle launch and rocket boost means that the firer is in no danger from back-blast and can fire the weapon from inside a confined space in perfect safety.

DATA
Weight: 1.65 kg
Length: 564 mm
Range: 250 m
Accuracy: 4 mrad
Penetration: 400 mm RHA

Manufacturer
Olin Ordnance, 10101 Ninth Street North, St Petersburg, FL 33716.
Status
Advanced development.

Olin RAAM launched from M16A2 rifle

66 mm M72A2 and M72A3 HEAT rocket launcher

The M72 series are one man, throw-away type rocket launchers. Each consists of two concentric tubes. The outer tube carries the trigger housing assembly on the top, the trigger assembly, trigger safety handle, rear and foresight assemblies and the rear cover. The inner tube of aluminium will extend telescopically along a channel assembly which rides in an alignment slot in the trigger housing assembly. The channel assembly houses the firing pin rod assembly and locks the launcher in the extended position. The firing pin rod assembly locks under the trigger assembly and cocks the weapon when the inner tube is extended.

Firing mechanism
The trigger, on the top rear of the outer tube, is a bar which must be depressed to release the tension on the firing pin rod assembly which then strikes the centre of the primer.

The firing pin housing is fixed to the top rearmost portion of the inner tube. Closely associated with the housing is the firing pin rod assembly, firing pin rod spring, primer block and primer. The primer is in line with the firing pin. The trigger safety handle must be pushed forward to the release or 'arm' position before the trigger bar can be depressed. When the safety is at 'safe' the firing pin rod assembly cannot move to the rear to strike the primer. Cocking is accomplished only in the last inch of travel of the inner tube as it is extended and so the inner tube must be pushed in at least 1 in and pulled out again if re-cocking is necessary – for example after a misfire.

Sights
The front sight of the M72A2 and M72A3 consists of a central vertical range line with ranges from 50 to 350 m by 25 m increments. There is a rangefinding device consisting of two stadia lines which subtend the length of a 20 ft (6 m) tank at that range. The tank is assumed to be twice as long as it is wide.

On either side of the central range line is a series of crosses which give the correct aim-off, or lead, for a 15 mph (24 km/h) directly crossing target. The rearsight of the M72A2 launcher consists of the bracket, a plastic rearsight and a rubber rearsight cover. Inside the plastic rearsight is an aperture plate which is attached to a spring that automatically compensates for a temperature change.

Rocket
The rocket is made up of the 66 mm HEAT warhead M18, the point-initiating base-detonating fuze M412 and the rocket motor M54. Attached to the nozzle of the rocket motor are six spring-loaded fins which are folded forward along the motor when the rocket is in the launcher. When the rocket emerges these fins spring out to stabilise the rocket in flight. The rocket motor is designed to ensure that all the propellant is fully burnt before the rocket leaves the launcher, even under Arctic conditions where the propellant burns more slowly.

The HEAT rocket warhead M18 consists of a tapered lightweight steel body, cylindrically shaped, containing ³/₄ lb (340 grams) of octol. The cone is of copper. The nose cap contains a piezo-electric crystal which is crushed on impact and generates a small electric charge which is led to the detonator of the base fuze.

The rocket and launcher are identified by colour markings. The HEAT rocket is black with yellow stencilling.

Five rocket launchers, complete, are packed in a fibre board inner package and three such packs are contained in a wire-bound wooden box. The inner pack weighs 12.5 kg and the outer pack containing 15 rocket systems weighs 54.4 kg.

Training device
This device was introduced for instructional purposes. It consists of the M190 sub-calibre launcher and the M73 sub-calibre rocket. It is a lightweight shoulder-fired rocket launcher used to simulate the M73A1, M73A2 series weapons. Although much smaller in calibre, length, and weight than the M73A2 and the M73A3 it simulates the noise, smoke, blast and flight trajectory of the tactical rocket.

The M190 sub-calibre launcher is a tubular, telescoping, smoothbore, open

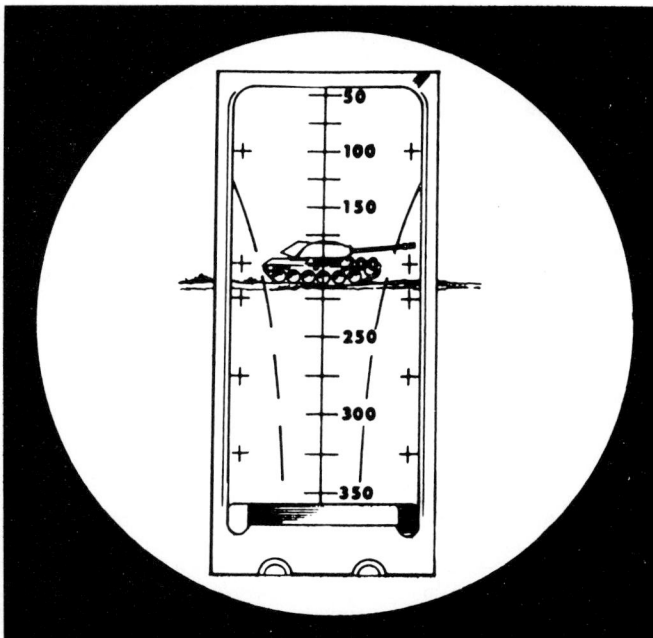
Firing position with fingers on trigger

Front sight using stadia lines for range estimation. When tank is straddled by stadia lines, range is marked. Here tank is at 200 m

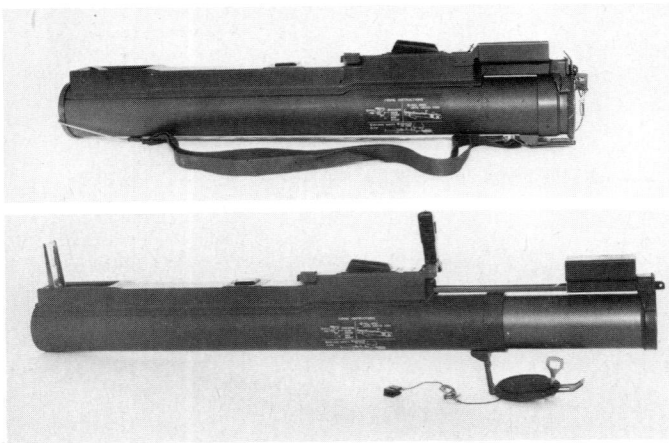

Launcher in its extended and retracted positions

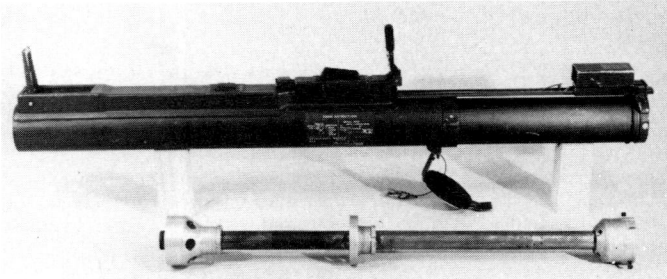

Sub-calibre training device

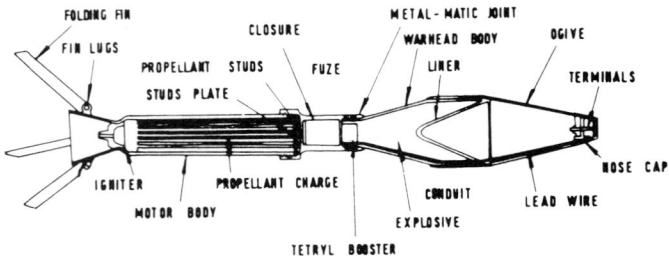

66 mm HEAT rocket

breech weapon. Its external appearance resembles the M72A2 launcher. It has a re-usable sub-calibre inner tube for launching the M73 sub-calibre rocket. The sub-calibre inner tube is re-usable.

The M73 sub-calibre rocket is 35 mm calibre and consists of a detonating head, a motor closure, a rocket motor and an igniter assembly. The detonating head, made of rigid plastic, contains 1.5 g of M80 composition mix. The forward end of the motor closure contains a base detonating fuze and an M26 stab primer. The steel motor casing contains three tubular grains of M7 propellant. The rocket is stabilised by six moulded plastic fins.

The M73 sub-calibre rocket is launched in exactly the same way as the 66 mm rocket. When the head strikes a target an inertia driven firing pin sets off the M26 stab primer which sets off the spotting head, producing noise, flash and white smoke.

The launcher controls, sights, and operation are the same as for the M73A2 and M73A3 system.

The training device can be used against stationary or moving targets. It penetrates 3 mm of steel plate or 20 mm of soft wood.

DATA
M72A2 AND M72A3
Rocket: M72A2 with 66 mm HEAT warhead M18 PIBD fuze M412; motor M54
Launcher: M72A2
Method of operation: rocket
Method of feed: single-shot and discard
Method of fire: percussion

WEIGHTS
Launcher only: 1.36 kg
Rocket: 1 kg
Complete assembly: 2.36 kg

LENGTHS
Launcher: (closed) 655 mm; (extended) 893 mm
Rocket: 508 mm

MECHANICAL FEATURES
Launcher: smooth bore
Sights
Front: reticle, graduated
Rear: peepsight. Adjusts automatically to temperature change
Graduations of front sight: 50–350 m by 25 m increments
Zeroing: nil; lead; 15 mph (24 km/h) marking
Sight radius: 490 mm

FIRING CHARACTERISTICS
Muzzle velocity: 145 m/s at 21°C
Recoil energy: nil
Range: max effective: (moving targets) 150 m; (stationary targets) 300 m. Max practical: 1000 m
Penetration: approx 305 mm of armour (RHA)

Manufacturer
Developed by the Hesse Eastern Company, Brockton, Massachusetts. Prime contractor, US Army Munitions Command. Made under licence by A/S Raufoss Ammunisjonsfabrikker, Norway.
Status
M72, M72A1 obsolescent: M72A2, M72A3 in production.
Service
United States forces and most NATO armies. It is also widely used in many countries all over the world.

M72 ILAW

The well-known 66 mm M72A3 anti-tank weapon has now passed through considerable improvement and modification. The external appearance and function is the same, the improvements being inside the launcher – to the rocket and warhead – and in the training system. Longer effective range, greater penetration, higher velocity and improved aerodynamics have given the weapon a significant increase in hit probability.

DATA
Calibre: 66 mm
Length: (carry) 777 mm; (firing) 983 mm
Weight: 3.44 kg
Muzzle velocity: 230 m/s
Effective range: 250 m

Operational range: 350 m
Penetration: 355 mm at 0°
Dispersion: 1.78 mils

Manufacturers
NI Industries Inc, Los Angeles, California (launcher); Talley Defense Systems, 4551 E McKellips, Mesa, Arizona,(System integrator and production source); Tracor Aerospace, San Ramon, California, (warheads); A/S Raufoss Ammunisjonsfabrikker, 2831 Raufoss, Norway (NATO systems integrator and production source).
Status
Under development.

3.5 in M20 rocket launcher

The M20 rocket launcher (sometimes called the Super Bazooka) followed the 2.36 in (60 mm) M9A1 rocket launcher. The 3.5 in (89 mm) calibre launcher is a two-piece tube, open at both ends and with a smoothbore. Its function is to ignite the rocket which it launches and to give initial direction to its flight. Ignition is electrical.

To reduce the weight, the tube is made of aluminium and breaks into two sections for ease of carrying. The barrels of the M20 and M20A1 models are made from aluminium tube and the component parts are fastened by means of screws. The barrels of the M20A1B1 and M20B1 are aluminium castings and

many component parts are cast integrally with the barrel. The sight is attached to the left side of the rear tube. It consists of a single lens and a graticule fitted into a housing. The graticule consists of a single broken vertical line up the centre of the lens and five broken horizontal lines. Each section of the vertical line and each space between sections represents 50 yards (45.7 m). The horizontal lines are marked in hundreds of yards and the appropriate range is laid on the target. Each section of the horizontal line and each space between the sections, represents 5 mph (2.2 m/s). The appropriate aim-off is applied when firing at a moving target.

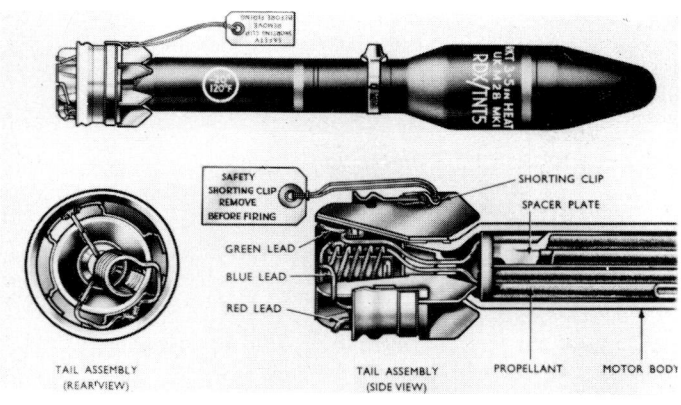

Rocket and ignition system

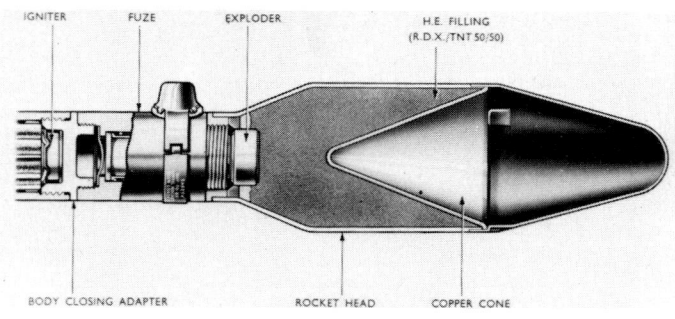

Warhead, filling and igniter

The current to fire the rocket is provided by a magneto, the movement of the armature being produced when the trigger is pulled and also when it is released. Thus a firing impulse is produced for both movements of the trigger and a rocket that fails to fire when the trigger is pulled, may fire when the trigger is released.

The rocket HEAT 3.5 in is issued assembled and is loaded into the launcher as a unit. It consists of a rocket head, a fuze and a motor to which is attached a tail assembly. The head is cylindrical with a ballistic head made of steel. The fuze is a base percussion type. The motor is a steel tube with propellant held in spacer tubes. The tail assembly is attached to the rear.

The igniter is at the front of the propellant with two leads passing out through the nozzle at the rear. These convey the firing current from the magneto-trigger when connected up. When the rocket is fired, there is a flow of high velocity high temperature gas from the venturi at the rear and there is a triangular danger area with a base and height of 25 yards (23 m). The firer should wear a face mask to protect his eyes from particles of unburnt propellant.

Although a very elderly design, the 3.5 in rocket launcher is still listed by several countries as being in service and the firm of Instalaza SA in Spain

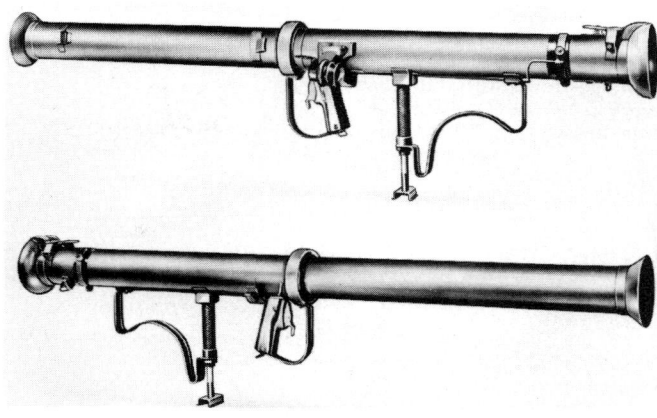

3.5 in M20 rocket launcher used during Korean War

manufactures a close copy of the launcher and also makes ammunition. It is possible that it is still made in China, though now it would only be for use by the militia. In Italy it is shown as being in service and the firm of Tirrena SpA makes and sells a practice head for training. How much longer this old equipment can survive is not clear. Although the launcher is unlikely to wear out, the remaining ones must be suffering from damage by handling, and it would not seem worthwhile making more when there are so many better weapons available. The ammunition is not as effective as modern designs, and its battle range is short, despite the claims of some manufacturers.

DATA

3.5 IN M20 ROCKET LAUNCHER
Calibre: 89 mm
Length of launcher assembled: 1549 mm; (front tube) 768 mm; (rear tube) 803 mm
Weight of launcher: 5.5 kg; (front tube) 1.6 kg; (rear tube) 3.9 kg
Range: (max against armour) 110 m; (max practical) 1200 m

ROCKET HEAT 3.5IN M28, 28A1, 28A2
Length: 598 mm
Weight: 4.04 kg

ROCKET WARHEAD ·
Filling: comp B
Weight of bursting charge: 0.87 kg
Rocket motor propellant: M7
Weight of propellant: 163 g
Fuze: fuze rocket BD M404, percussion, non-delay

Status
Obsolete, but still made outside USA.
Service
Still in service in many countries.

90 mm M67 recoilless rifle

The 90 mm M67 recoilless rifle is a lightweight, portable, crew-served weapon intended primarily as an anti-tank weapon although it can also be employed against pill boxes or sandbag emplacements. It is designed to be fired from the ground off its bipod and monopod but may be fired from the shoulder. It is an air-cooled, breech loaded single-shot weapon firing the high explosive anti-tank round M371E1. The rifle is equipped with a manually operated breech mechanism and a centrally located percussion type firing mechanism. It is designed for direct fire only. The M103 sight has a fixed focus, × 3 telescope with a field-of-view of 10°. Its lead lines and stadia lines are provided for ranging on targets having a 10 ft (3 m) width and a 20 ft (6 m) length.

Some rifles used in the Persian Gulf have the M103A1 sight. This is a sight modified for the experimental XM591 round which was only issued on a limited scale for a short while. The ranges on the left-hand side of the sight remain unchanged but those on the right have been modified.

Operation
The breech-block is hinged open to load the round but before this can be done the lock ring must be rotated clockwise. This withdraws the sear and forces the firing pin rearwards. The breech-block can then be unlocked and swung clear of the breech opening. The firing pin and spring are housed in a bar crossing the hollow body of the breech-block. The round is placed in the breech and the block closed behind it. When the firing mechanism is operated the sear is withdrawn by a cable and the firing pin goes forward under the influence of the spring, and this strikes the cap and fires the round. The rear end of the cartridge case is made of frangible material and is blown out when the round fires. The gases pass out to the rear and flow over the firing pin housing, through the breech-block.

The shell is effective out to 400 m. There is a 7.62 mm sub-calibre training device which has a case blowout arrangement with six equally spaced holes in

the chamber shoulder to keep the pressure down and allow the trajectory to approximate to that of the 90 mm round.

Ammunition
The ammunition of the M67 is a fixed round. The HEAT round M371E1 is a 3.06 kg fin-stabilised projectile with a shaped charge head. It has fuze PIBD M530. This is a point-initiating, base-detonating fuze with an inertially operated graze system. There is also a practice round with a small charge called TP M371. The complete HEAT round weighs 4.2 kg and the shell has a muzzle velocity of 213 m/s.

90 mm recoilless rifle. This has now been replaced in US service by Dragon

DATA
Weight of complete system: (unloaded) 16 kg
Overall length: 1346 mm
Height, ground mounted: 432 mm
Range: (max effective) 400 m; (max sighted) 800 m; (max practical) 2100 m
Rates of fire: (rapid) 10 rds/min (max of 5 rounds); (sustained) 1 rds/min. When firing at the rapid rate, a 15 mins cooling period must be observed after every 5 rounds

Manufacturer
Now produced by Kohema Industries Ltd, 446 Shindorim-Yong, Kuro-Ku, Seoul, South Korea.
Status
Obsolescent.
Service
US Army in Berlin. Otherwise in reserve.

106 mm M40, M40A2 and M40A4 recoilless rifle

The 106 mm M40 recoilless rifle is a lightweight weapon designed for both anti-tank and anti-personnel roles. It is air cooled, breech loaded and fires fixed ammunition. The marks of the gun are made up as follows and it will be seen that there is a substantial commonality between them:

the piece (cannon) M206
mount M79 (M40A2)
mount M92 with tripod M27 (M40A4)
spotting rifle M8C
elbow telescope M92F
telescope mount M90
cover, telescope and mount and
instrument light M42.

The 106 mm recoilless rifle is one of the most successful examples of its type. It came into service in the mid-1950s and is still used in substantial numbers in many parts of the world. It is lighter than its contemporaries, yet it is little reduced in performance, and it was the first recoilless gun to have a spotting rifle for the gunner. It is the spotting rifle which has made it so popular, since the chance of a hit with the first main armament is much higher than without it.

One of the lesser known peculiarities of the 106 mm is that it is actually 105 mm and is an improved version of the M27 105 mm. To avoid confusion at a time when both types were in service it was found convenient to measure the bore of the newer weapon from the bottom of the grooves and so give it a new calibre.

The barrel assembly consists of the tube, an enlarged reaction chamber and a mounting bracket. The tube is made of alloy steel, screw threaded at the back to take the reaction chamber. The breech-block is a short cylinder with an interrupted screw thread to mate with the segmental thread in the interior of the recoil chamber. The breech-block is hinged on the left side of the breech and is drilled to take the percussion firing mechanism. The mount M79 is basically a tripod. The two rear legs have carrying handles and clamps and the front wheel has a hard rubber tyre. The mount has the traversing and elevating gear and part of the firing mechanism. The two rear legs of the mount are adjustable for lateral expansion. The mount can be lifted at the rear by two men and pushed forward on the single wheel, like a wheelbarrow. The mount can also be clamped in position on a vehicle. The traversing mechanism allows 360° of controlled or free traverse using the traverse knob. The elevating mechanism on the left of the mount provides slow or rapid rates of elevation. The handwheel gives rapid elevation (16 mils a turn) while the firing knob gives slow elevation (3.8 mils a turn). With the breech between the legs the gun can be elevated to +65° and depressed to −17° but over the legs the maximum elevation is 27°. The .50 spotting rifle (Cal. .50 Spotting Gun M8C) is a gas-operated, magazine-fed, self-loading rifle, firing a bullet whose trajectory matches that of the recoilless rifle. The sight allows engagement up to 2200 m and has stadia lines based on a target 20 ft (6 m) long and 10 ft (3 m) wide.

Operation
The breech-block is opened by depressing and rotating the breech operating lever and the round can then be loaded. When the breech operating lever is rotated back the breech-block is locked. This connects the firing mechanism

to the trigger. The M8C spotting rifle is used to determine range. The firer estimates range using, normally, the stadia lines and then when the target is thus considered to be in range the spotting rifle is fired. When a hit is obtained, the main armament is fired. Provided the same point of aim is used, the trajectory of the HEAT round will be the same as that of the spotting rifle and a hit will ensue. Note that the correct lead to allow for the target speed must be applied to the spotting rifle before a hit can be obtained.

The sub-calibre device for training personnel on the 106 mm recoilless rifle consists of a case fitting into the chamber, carrying a .30 calibre barrel. This

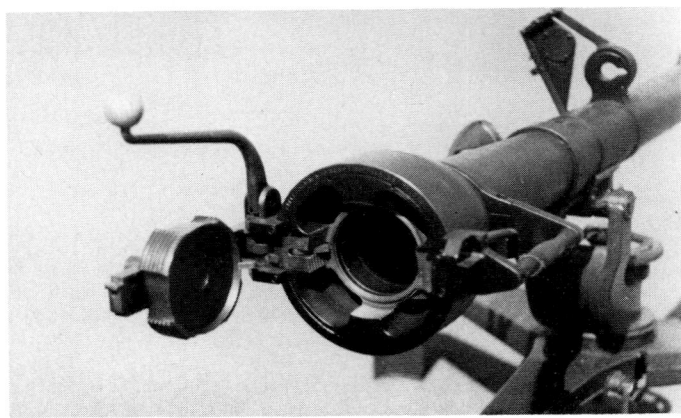

Breech arrangement of M40

M40A4 rifle, complete with spotting rifle, mounted on M274 Mechanical Mule

106 mm M40 recoilless rifle

barrel has 24 holes drilled in it to reduce the velocity of the sub-calibre round and, at the same time, this creates a back-blast which simulates that of normal discharge. The sub-calibre device is held in place by the breech-block and fired by the gun firing pin. The cartridge case is extracted manually.

Ammunition

Ammunition for the 106 mm rifle is issued in complete, fixed cartridges. Propellant is contained in a plastic bag within the cartridge case. The cartridge case is perforated to permit expanding propellant gases to escape into the enlarged reaction chamber. The cartridge case has a well around the primer to permit the breech to be closed. Proper headspace is automatically provided for by the seating of the flange of the cartridge case against the vent bushing.

Depending on the type of projectile, ammunition for the 106 mm rifle is classified as high explosive anti-tank (HEAT), or high explosive plastic with tracer (HEP-T). Both projectiles are painted olive drab with yellow markings. These identifying colours are repeated on the sealing tape around the fibreboard container. Both rounds have a maximum range of 7700 m.

There is also an anti-personnel round with tracer (APERS-T), which has a maximum range of 3300 m.

The spotting rifle employs a spotter-tracer cartridge. The projectile contains a tracer element and an incendiary filler. On impact, the incendiary filler produces a puff of white smoke which aids in adjusting fire. The tracer burnout point is between 1500 and 1600 m. The cartridge can be identified by the red and yellow marking on the nose of the projectile.

DATA
Calibre: 105 mm
Weight, empty: 209.5 kg
Overall length: 3404 mm
Height on M79 mount: 1118 mm
Width of M79 mount: (legs spread) 1524 mm; (legs closed) 800 mm
Length of bore: 2692 mm
Type of breech-block: interrupted thread
Max range: 7700 m
Max effective range: 1100 m
Rate of fire: (sustained) 1 rds/min
Ammunition: HEAT, HEP-T or APERS-T
Weight of HEAT: (round) 16.9 kg; (projectile) 7.96 kg
Weight of HEP-T: (round) 17.25 kg; (projectile) 7.96 kg
Weight of APERS-T: (round) 18.6 kg; (projectile) 9.89 kg

Manufacturer
Watervliet Arsenal, Watervliet, New York.
Status
Obsolescent. Not in production.
Service
In reserve in USA. Has been supplied to 32 countries and is in service with some. Has been built under licence in Brazil, Israel, Japan and Spain and may now be met almost anywhere in the world.

M136 (AT4) light anti-armour weapon

M136 is the US designation for the FFV AT4 weapon, which is more fully described in the Swedish section (page 431). The M136 is manufactured by Honeywell in the USA under licence from FFV, and the current contract is for over 200 000 units.

In order to remain ahead of armour improvements, Honeywell is developing new warheads for the M136. Their AT4E1 programme is focussed upon defeating the large number of older tanks which remain in service throughout the world and modern tanks not protected by reactive or composite armour. The present warhead can defeat over 600 mm of RHA and should therefore be capable of defeating this class of tank.

To provide a multi-mission capability and to defeat IFVs with advanced armour protection, buildings or field fortifications, Honeywell is developing AT4E2. This has an improved shaped charge warhead which is optimised to defeat reactive or ceramic armour, with enough residual penetration to ensure target destruction. Used against soft targets such as bunkers or trench lines, the AT4E2 will have a delay fuze that allows the projectile to penetrate into the target structure before detonating, ensuring that the maximum energy is released at the protected enclosure.

All improved systems, AT8 (see page 280), AT4E1 and AT4E2, use the same launcher as the basic AT4 and will maintain AT4's reputation for reliability and accuracy. Other developments currently in progress include weight reduction by using composite materials, and providing a 'Fire from Enclosure' capability to AT4 and its family of associated weapons.

Manufacturer
Honeywell Inc., Defense Systems Group, Armament Systems Division, 7225 Northland Drive, Brooklyn Park, MN 55428.
Status
Current; production.
Service
US Army, Navy, Air Force and Marine Corps.

M47 Dragon II Medium Anti-tank Assault Weapon

The Dragon II weapon system is a product-improved version of the M47 Dragon fielded by the US Army in 1970. The Product Improvement Programme (PIP) was initiated by the US Marine Corps in 1985 and managed by the Naval Surface Warfare Center (NSWC), Dahlgren, Virginia and contracted to McDonnell Douglas.

History

The first generation Dragon is a 1000 m system with an 11.2 seconds time of flight. This system can be fired at stationary or moving targets within this range. Operation is simple and automatic; all the firer is required to do is keep the sight cross-wires on the target until the missile hits. The warhead power of

Dragon II in action

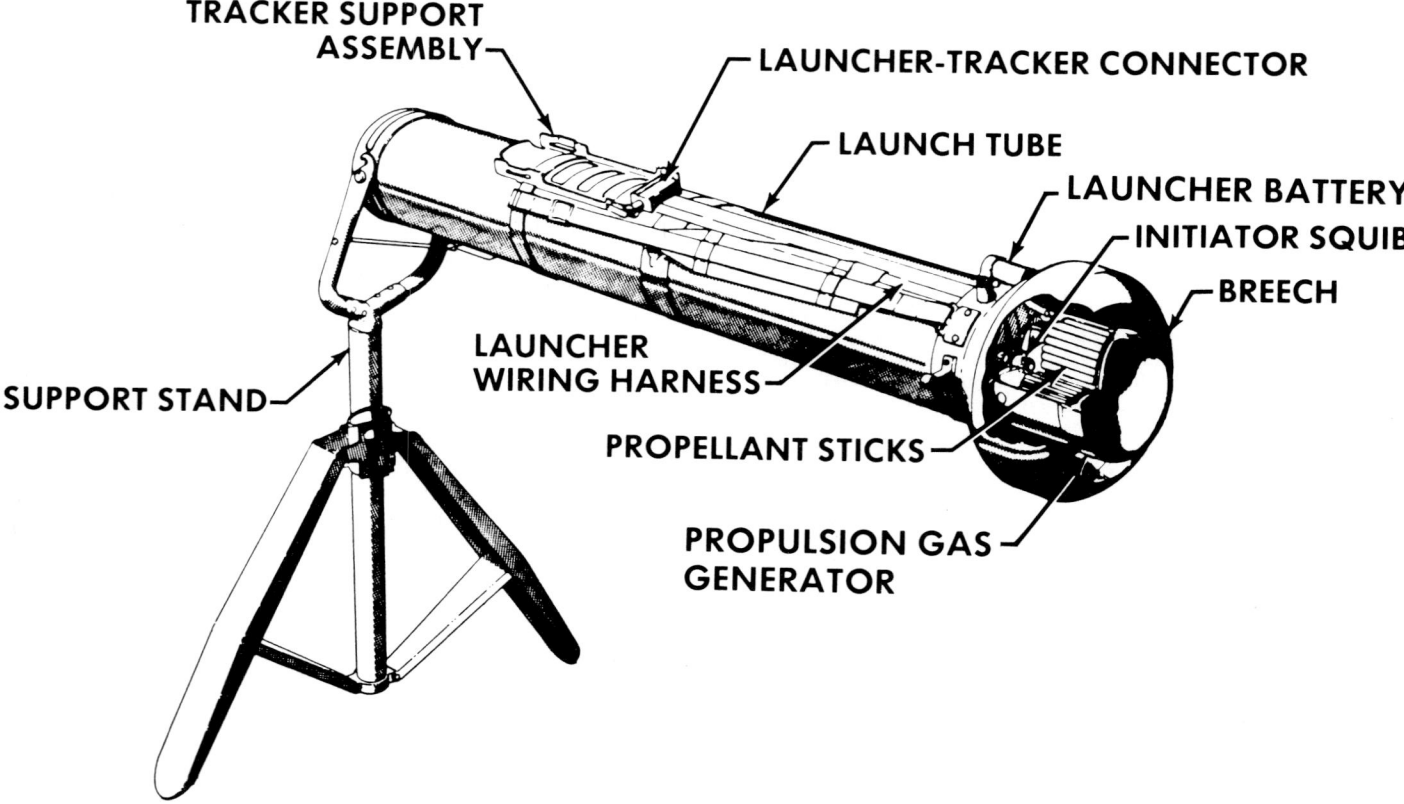

TRACKER SUPPORT ASSEMBLY

LAUNCHER-TRACKER CONNECTOR

LAUNCH TUBE

LAUNCHER BATTERY

INITIATOR SQUIB

BREECH

SUPPORT STAND

LAUNCHER WIRING HARNESS

PROPELLANT STICKS

PROPULSION GAS GENERATOR

Dragon launcher

Dragon makes it possible for a single infantry soldier to challenge armoured vehicles, fortified bunkers, concrete gun emplacements or other hard targets. Dragon I was designed to be light in weight and the ready to fire system weighs approximately 14 kg including the round and the day tracker.

The round consists of the launcher and the missile. The missile is installed in the launcher during final assembly and is shipped to the customer in a ready to fire condition. The launcher consists of a smooth bore glass fibre tube, breech/gas generator, tracker support, bipod, tracker battery, sling, and forward and aft shock absorbers. The launcher serves as the storage and carrying case for the missile, from the time of delivery until launch. For a missile firing, the tracker is mounted on the launcher. Upon completion of missile flight the tracker is removed and the launcher is discarded. Dragon I utilises either a day tracker ot a separate night tracker to acquire its target.

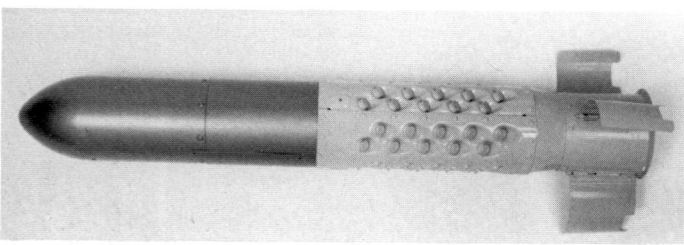

Dragon II missile

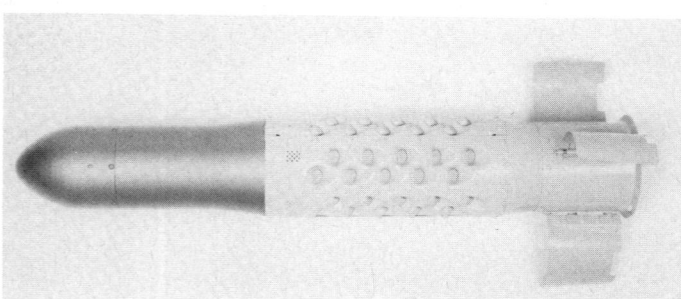

Dragon I missile

Dragon II
Dragon II is a warhead improvement programme that provides an 85 per cent increase in penetration performance when compared to the currently fielded Dragon I. The warhead is now in production and plans for retrofitting US Army and USMC rounds are currently underway at the Florida Missile Production facility in Titusville. The current day and night trackers are used without modification. No changes were made to the outside dimensions of the launchers.

Dragon II+
An additional improvement to the Dragon II programme is being considered for the international market. Upgrades would include a sustainer motor to increase range and shorten time of flight, and an extendable probe to increase warhead effectiveness.

DATA

	Dragon I	Dragon II
WEAPON		
Guidance	SACLOS	SACLOS
Length	1154 mm	1154 mm
Carry weight	14.6 kg	16.2 kg
Firing weight	13.8 kg	14.8 kg
MISSILE		
Length	774 mm	846 mm
Weight	6.2 kg	6.97 kg
Propulsion	Gas launched, then by 30 pairs of side thruster solid fuel motors	
Electronics	Solid-state	Solid-state
Battery	Thermal	Thermal
Flare	Tungsten lamps	Tungsten lamps lamps
Gyroscope	Roll reference	Roll reference
Range	1000 m	1000 m

Manufacturer
McDonnell Douglas Systems Co., Combat Systems Division, PO Box 516, St Louis, MO 63166.
Status
Dragon I no longer in production; Dragon II in production.
Service
Dragon I with US forces, Israel, Jordan, Morocco, Netherlands, Saudi Arabia, Spain, Switzerland, Thailand and Yugoslavia.

TOW heavy anti-tank weapon system

The TOW weapon system is a crew-portable or vehicle-mounted, heavy anti-tank weapon (HAW). The name is derived from its description as a Tube-launched, Optically-tracked, Wire command-link guided missile. Design work started in 1962, the first firings were carried out in 1968 and the weapon entered service with the US Army in 1970.

Since then it has been proved in action in Vietnam and the Middle East and has been sold to more than 37 countries. It is now well on its way to being the most successful and widely used anti-tank missile system in the world. According to US Army statistics, the missile has a cumulative reliability record of over 93 per cent in nearly 12 000 test and training firings conducted

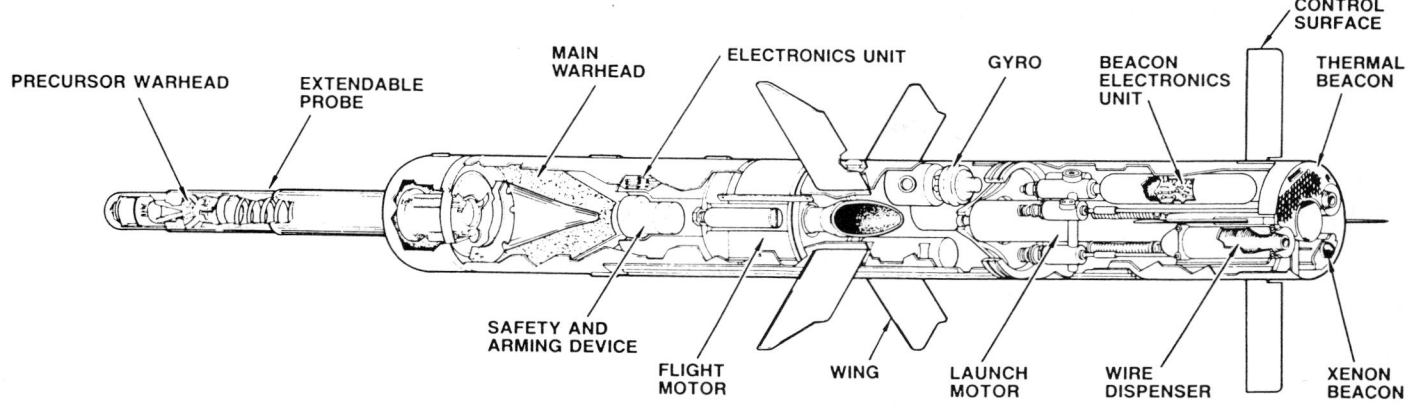

TOW 2A missile

TOW 2 missile launcher

since 1970. One contributory factor to its success is the fact that the training of the operator is far less complicated than with so many previous types, and the excellent simulator enables training shots to be carried out under realistic conditions. Another useful feature of TOW is that the same missile is fired from the ground and helicopter mountings which simplifies logistics.

The TOW weapon system, M220A1, consists of six major units. These are: tripod, traversing unit, launch tube, optical sight, missile guidance set and battery assembly, housed in the missile guidance set.

The tripod provides a stable mounting base for the traversing unit and allows levelling on ground sloping up to 30°. The traversing unit is an electro-mechanical assembly attached to the tripod and is the mounting base for the optical sight and the launch tube. The launch tube is constructed of a lightweight honeycomb material, covered with laminated glass fibre. It provides initial guidance and stability to the missile and protects the gun crew from the missile launch blast on firing. The optical sight is used to track the target, and to detect the infra-red signal from the missile in flight. The sight contains a 13 power telescope, bore-sighted to an infra-red tracker, with a field-of-view of 5.5°. The cross-hairs in the eyepiece may be illuminated and the task of the gunner is to keep these cross-hairs on the target. The missile guidance set consists of two rechargeable 50 V batteries and one rechargeable 24 V battery. These are nickel cadmium using potassium hydroxide electrolyte.

The basic TOW missile is the BGM 71A. It consists of the launch motor, the flight motor, four control surfaces, the infra-red (IR) source, two wire dispensers, three batteries, a gyro, safety and arming devices and the shaped charge warhead. The launch motor is at the rear of the missile and consists of 1.2 pounds of M7 double-based propellant which is completely expended before the missile leaves the tube. The flight motor uses a solid propellant and is near the centre of gravity of the missile. The burning gases emerge from a pair of nozzles at 30° to the horizontal axis of the missile. This configuration eliminates interference with the wire link and minimises changes in the centre of gravity of the missile as the motor burns. The flight motor ignites about 12 m in front of the launch tube to protect the gunner. The wire command link is a two-wire system dispensed from two spools at the back of the missile. These wires carry steering commands from the launcher to the missile, and they are applied, together with signals from the gyro, to the four control surfaces. The IR source provides a beacon which is detected by the IR tracker in the optical sight to determine the missile's position. The IR lamp current is modulated to allow discrimination against a background of other strong IR emitters such as the sun.

The TOW 2 launcher is a modified version of the original launcher, with improvements aimed at permitting guidance through obscurants such as smoke or dust and at night. Changes include converting the launcher's

AN/TAS-4 sight (the upper of three mounted sights) used to track targets at night, to function as a totally independent fire control sensor. Other new or modified components include the post amplifier electronics in the small rectangular box above the AN/TAS-4 sight, and the digital missile guidance set (shown beneath the tripod in the accompanying picture). All types of TOW missile – basic TOW, ITOW, TOW 2, TOW 2A or any future TOW missile contemplated by the US Army can be fired from this modified launcher.

Operation
When the missile is installed in the launcher, the gunner uses the self-test facility. This is a switch with seven positions and is used to check the circuitry associated with weapon functioning. When this is completed the operator connects the weapon system to the encased missile by raising the arming lever. When a target appears the gunner moves the sight to place the cross-hairs on the target. When the target is in range the trigger is pressed. This activates the missile batteries and gas is released to spin the gyro up to speed. About 1.5 seconds later the launch motor is fired. This burns out completely within the tube but the missile acquires sufficient momentum to coast until the flight motor ignites. The missile wings and control surfaces are extended from the missile body. The IR sources on the missile start to operate and the two command-link wires are dispensed from the internal spools. The first stage of arming the warhead occurs. The flight motor is activated at the end of the 12 m coasting period and the warhead is fully armed. The missile has then travelled about 60 m.

Guidance
The gunner operates the traversing unit and keeps the cross-hairs of the optical sight on the target. The IR sensor tracks the signal from the modulated lamp in the missile and detects any deviations from the line of sight path to the target. It provides continuous information over the wire link and to the missile guidance set which produces signals which are delivered, with those from the gyro, to the control surface to correct the flight path and bring the missile back to the line of sight.

Warhead improvements
The US Army has conducted a number of programmes to improve the performance of the TOW warhead against advanced enemy armour. The first was an improved 5 in (127 mm) warhead (ITOW), which is roughly the same size and weight as the basic TOW warhead but has better performance against armour. A feature of this warhead is a telescopic nose probe which is retracted into the missile while in the launch tube but which extends out when the missile has cleared the launcher and is in flight. The probe gives a longer standoff distance and enhances the jet so as to be better formed before striking the target plate. The second upgrade includes a heavier, 6 in (152 mm) warhead which occupies the entire diameter of the missile body. With this warhead, the guidance system has been improved by introducing a digital guidance set based on a microprocessor which provides greater flexibility in programming and higher precision. The warhead increases the weight and, to compensate for this, the flight motor is loaded with an improved propellant.

This second step, referred to as TOW 2, can be retrofitted into the earlier versions, but requires some fairly sophisticated procedures, including reloading the motors. However, the earlier missiles may be fired from the TOW 2 launcher.

TOW 2A, a third warhead improvement, adds a small warhead to the missile probe. This detonates reactive armour and so clears a path for the primary warhead.

TOW 2B, the most recent improvement, is designed for top attack. It features a dual mode sensor and a new armament section with two warheads substantially different from those used in other TOW versions. The first of these tandem warheads will detonate the explosives in reactive armour, and the second will achieve penetration of the exposed target. The missile is programmed to fly over the target and the sensor will trigger the two warheads to shoot downwards at the proper time.

Part of the development programme will include launcher software changes to enable TOW 2B to be compatible with launchers and equipment already in use. The missile itself will look similar to TOW 2A, with a 152 mm diameter warhead. First production deliveries are expected in the early 1990s.

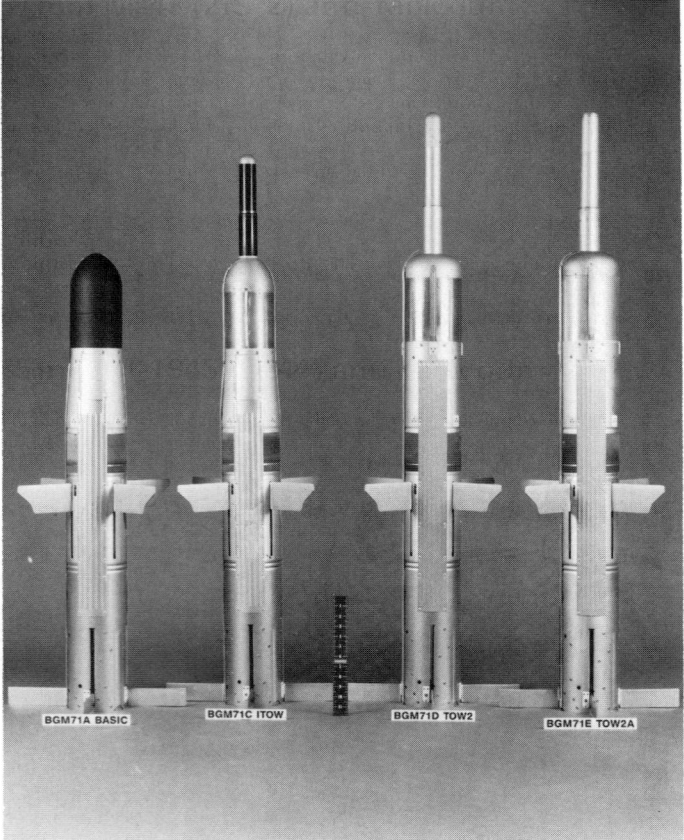

TOW family
(Left to right) *basic TOW: ITOW: TOW 2; TOW 2A*

TOW 2 in operation
(From right) *ignition of launch motor; probe, wings and tail partly deployed; probe, wings and tail fully deployed; ignition of flight motor*

FITOW

In late 1987 Thorn-EMI were awarded a major contract by the UK MoD for Further Improvements to the TOW system (FITOW). As prime contractor the company will lead a team comprising Royal Ordnance, Hughes Aircraft and Westland Helicopters. The heart of the FITOW project is a new proximity fuze which, when combined with a new type of warhead, will provide TOW with a top-attack capability. The work is based on development work carried out over several years by Thorn-EMI Electronics and Royal Ordnance in conjunction with the Royal Armament Research and Development Establishment (RARDE), and it is anticipated that the enhanced performance of FITOW will make it capable of defeating the next generation of armour and maintaining TOW as a viable defence system well into the next century.

Swiss co-production

The Swiss Parliament, after studying the results of extensive evaluation and

A US Marine fires a TOW 2 missile from a ground launcher.

testing, has approved procurement of the TOW 2 system. Swiss industry will co-produce the missile under license from Hughes Aircraft. The missile will be carried on the Piranha 6 × 6 armoured vehicle and will be fired from a turret mount designed by Thune-Eureka of Norway, also to be built under license in Switzerland. The turret will provide under-armour protection for the gunner and fire-unit equipment.

Helicopter launchings

The airborne TOW system used by the US Army is designated the M65 and it is the anti-tank system for the AH-1 Cobra series attack helicopters of the US Army and Marine Corps, as well as several other nations. The M65 system is also installed on the Hughes Helicopters 500MD, Agusta A-109, the Westland Lynx, the MBB B0105C, and the Bell 206-L-1. It can also be installed on the Agusta A129, the Sikorsky S-76, and the Aérospatiale Gazelle, Dauphin, and Ecureuil-2.

The system is available with chin sight, roof-mounted sight, and mast-mounted sight configurations.

An airborne TOW system including roof-mounted sight is licensed for production in the UK by British Aerospace Group. All TOW missiles are produced by Hughes in the United States.

DATA

Designation: BGM 71A (basic TOW); BGM 71C (improved TOW/ITOW); BGM 71D (TOW 2); BGM 71E (TOW 2A)
Type: HAW
Guidance principle: automatic missile tracking and command to line of sight guidance from optical target tracker (SACLOS)
Guidance method: wire command link controlling aerodynamic surfaces
Propulsion: 2 stage, solid propellant motor
Warhead: HEAT
Missile velocity: 200 m/s
Range: (minimum) 65 m; (max) 3750 m
Crew: 4

	Length	Width	Height	Weight
Launcher, tubular, guided missile, M151E2	2210 mm	1143 mm	1118 mm	78.5 kg
Launcher, with AN/TAS-4				87.5 kg
Launcher, with TOW 2 Mods				93 kg
Tripod (retracted)	1064 mm	645 mm	569 mm	9.5 kg
Traversing unit	297 mm	518 mm	511 mm	24.5 kg
Optical sight	544 mm	295 mm	315 mm	14.5 kg
Launch tube	1675 mm	191 mm	191 mm	5.9 kg
Missile guidance set (with battery)	406 mm	406 mm	254 mm	24 kg
Battery assembly	394 mm	117 mm	178 mm	10.9 kg
Guided missile, basic TOW, BGM 71A	1174 mm	221 mm	221 mm	22.5 kg
Guided missile ITOW, BGM 71C	1174 mm	221 mm	221 mm	25.7 kg
Guided missile TOW 2, BGM 71D	1174 mm	221 mm	221 mm	28.1 kg
Guided missile TOW 2A, BGM 71E	1174 mm	221 mm	221 mm	28.1 kg
Guided missile practice (inert warhead, live motor)				
BTM 71A	1285 mm	221 mm	221 mm	24.5 kg

Manufacturer
Hughes Aircraft Co, Canoga Park, CA91304.
Status
Current. Production. Over 460 000 have been made.

Service
US Army, Marine Corps. Purchasers include Bahrain, Botswana, Cameroon, Canada, Chad, China, Colombia, Denmark, Egypt, Finland, West Germany, Greece, Israel, Italy, Japan, Jordan, Kenya, Kuwait, Lebanon, Luxembourg, Morocco, Netherlands, Norway, Oman, Pakistan, Portugal, South Korea, Saudi Arabia, Somalia, Spain, Sweden, Thailand, Tunisia, Turkey, United Kingdom and South Yemen.

Advanced Anti-tank Weapon System – Medium (AAWS-M)

This is a development programme aimed at the acquisition of an advanced, light, multi-purpose, medium anti-tank weapon to replace the M72A3 LAW and Dragon with a more cost-effective system having better Ph and Pk values.

The US Army originated a programme in 1980 known as IMAAWS (Infantry Manportable Anti-armor/Assault Weapon System) but then cancelled it. In 1981 they again requested industry proposals, but lack of funding aborted this programme. A third programme, Rattler, began in 1982 and was cancelled in 1983. In the wake of growing complaints about the relative ineffectiveness of current weapons against the projected next generation of tanks, in 1986 the AAWS-M programme was launched, with contracts to the value of $30 million, calling for the three contractors to produce viable weapons for evaluation and competitive testing. Ford

Aerospace, Hughes and Texas Instruments were given contracts for the Technology Demonstration/Proof-of-Principle phases of the AAWS-M programme. In early 1989 it was announced that the Texas Instruments/ Martin Marietta submission had been selected for futher development.

Texas Instruments AAWS-M
This company began working on a fire-and-forget anti-armour weapon in the early 1980s, using as a basis a design first devised in DARPA's 'Tank Breaker' programme. Information so far released shows that the TI design uses infra-red homing, an advanced, top-attack, warhead and a high performance airframe. In February 1988 it was announced that TI were linking with Martin Marietta to continue the development of this missile system, and since the 1989 award both companies have continued working on the project.

Manufacturers:
Texas Instruments Inc., PO Box 660246, Dallas TX 75266
Martin Marietta Inc, Electronics & Missiles Group, PO Box 5837, Orlando FL 32855

Texas Instruments AAWS-M deployed

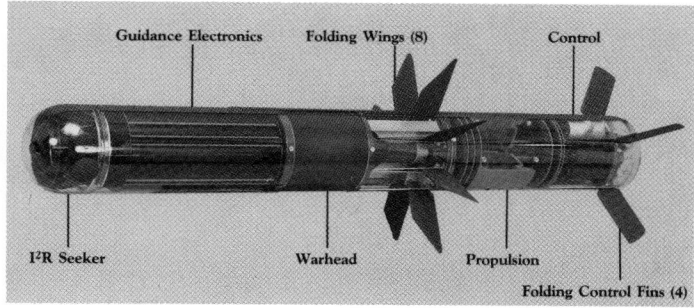

Principal components of the Texas Instruments AAWS-M

Saber dual-purpose missile system

This is a shoulder-fired, Laser Beam Rider (LBR) guided, dual purpose (anti-aircraft and anti-tank) missile system currently under development by Ford Aerospace. It consists of three elements: missile, launcher assembly and a re-usable guidance unit which can be quickly attached to the launcher assembly. The missile is a simple body/tail configuration which is attitude

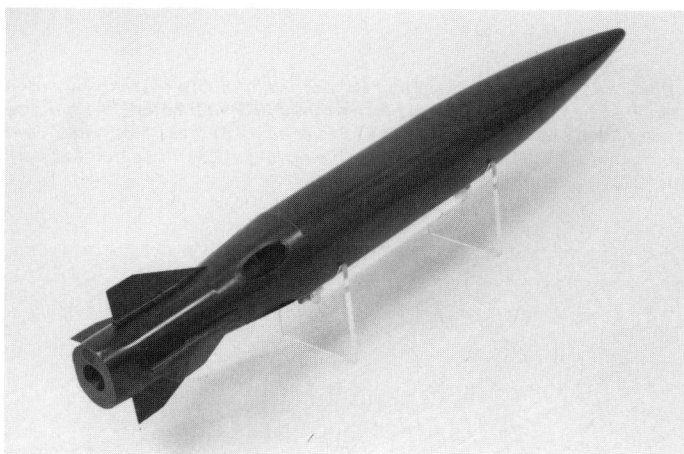

The Ford Aerospace manportable Saber dual-purpose missile

stabilised about all three axes. It is ejected from the launcher at low speed to minimise flashback and noise, and the flight motor ignites after the missile has travelled a short distance from the launcher. The launch and in-flight signature is very low, making it extremely difficult to detect from the air or by adjacent ground troops. Guidance is performed by the re-usable Command Launch Unit which incorporates a stabilised optical sight for target tracking and a low intensity laser beam projector which provides guidance data to the missile's laser receiver. The misile receiver looks backwards and flies the laser beam to the target.

The system has inherently low susceptibility to countermeasures and can be used with a minimum of operator training. The missile can be produced at low cost since it does not require a seeker. System reaction time is short. In addition to its shoulder-fired configuration, Saber can be adapted to vehicle-mounted multiple launcher systems.

Saber is designed for the defeat of all types of subsonic air threats; the quick-reaction point-and-shoot characteristic is particularly effective against high speed incoming and pop-up targets, since there is no delay while a seeker locks on. The missile can be configured with a fuze/warhead combination that is optimised for either the air defence or anti-armour role. The Saber missile is 120 mm in diameter and 1092 mm in length. The total system weight, including missile, launcher and guidance unit, is approximately 16 kg.

Manufacturer
Ford Aerospace, Aeronutronic Division, Ford Road, Newport Beach, California 92658-8900.
Status
Development.

Long Range Saber missile system

Long Range Saber is designed for the defeat of all types of air threats such as fighter-bombers, helicopters and gunships. The quick reaction point-and-shoot characteristic is particularly effective against high speed incoming and pop-up targets since no seeker lock-on is required. The missile can also be adapted to carry a dual purpose shaped charge warhead of a size adequate to defeat armoured and hard point ground targets. With this dual capability, a

vehicle equipped with Long Range Saber becomes a formidable air or ground weapon system.

The system is composed of three elements: Missile, Launcher Pack Assembly and a ground- or vehicle-mounted Guidance Unit. The missile is a simple body/tail configuration which is attitude-stabilised about all three axes. It is ejected from the launcher at low speed to minimise flashback and

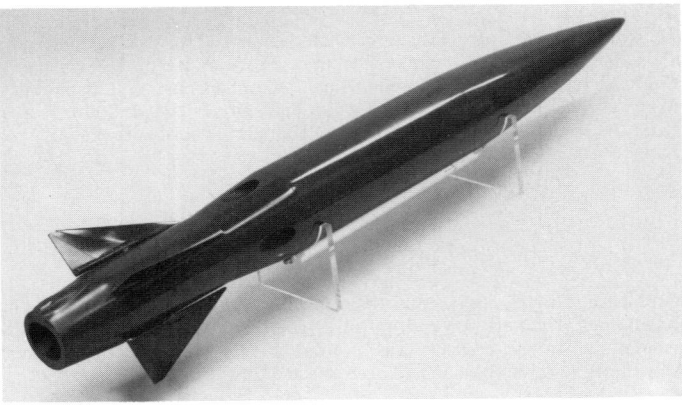

noise, with the flight motor being ignited after the missile has travelled a safe distance. The missile launch and in-flight signatures are very low, making it extremely difficult to detect from the air or by adjacent ground forces. The Guidance Unit incorporates a stabilised sight line for target tracking and a guidance beam projector which delivers guidance data to the missile receiver.

A number of significant operational and cost advantages are inherent in this system. The missile system has an extremely low susceptibility to countermeasures and can be used with a minimum of operator training. The missile can be produced at very low cost because it does not require a seeker. System reaction time is short, and outstanding self-defence capability is provided. Long Range Saber can be adapted to vehicle carry in multiple-launcher configurations, utilising existing TOW launchers or new launch boxes or packs. The Long Range Saber missile is 152 mm in diameter, 1.783 m long, and weighs approximately 32 kg.

Manufacturer
Ford Aerospace, Aeronutronic Division, Ford Road, Newport Beach, CA 92658-8900.
Status
Development.

Ford Aerospace Long Range Saber missile

YUGOSLAVIA

44 mm M57 (RB 57) anti-tank launcher

This is basically derived from the German Panzerfaust and is generally similar to the Czechoslovak P-27. The M57 also has a permanently attached bipod and is served by a two man crew. The firer lies behind the weapon and the loader lies to his right.

DATA
Calibres: (launcher) 44 mm; (projectile) 90 mm
Length of launcher: 960 mm
Weights: (launcher) 8.2 kg; (projectile) 2.4 kg
Type of projectile: HEAT
Max effective range: 200 m
Penetration: 300 mm

Status
Current.
Service
Yugoslav Army.

44 mm M57 anti-tank launcher in firing position

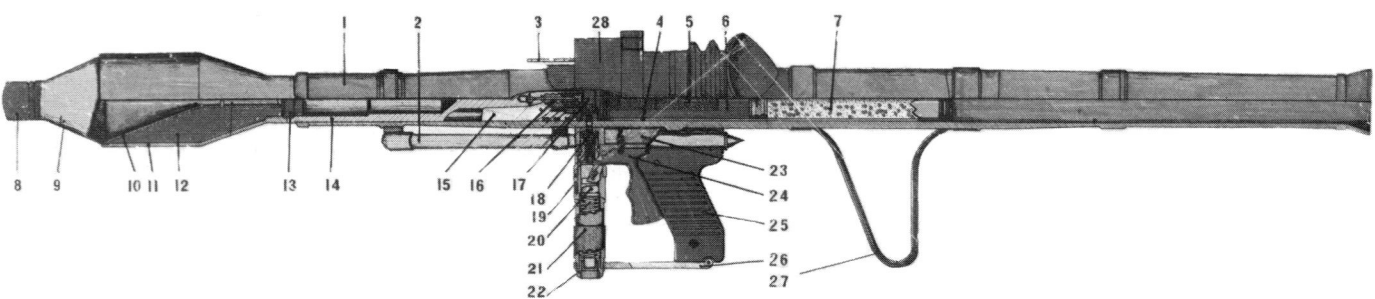

M57 rocket launcher
(1) *launcher tube (barrel)* **(2)** *support legs* **(3)** *mechanical sight* **(4)** *propellant capsule (cartridge)* **(5)** *igniter* **(6)** *propellant filling* **(7)** *recoil absorber* **(8)** *UTI M.61 fuze* **(9)** *ballistic nose cap* **(10)** *funnel-shaped hollow-charge former* **(11)** *mine shell* **(12)** *explosive filling* **(13)** *detonator* **(14)** *stabiliser* **(15)** *stabiliser wings* **(16)** *stamping with shaft and spring* **(17)** *small cap with propellant charge and striker* **(18)** *striker pin* **(19)** *brake (restrainer) with stopper/cap and spring* **(20)** *striker* **(21)** *fairing (cover) around the firing mechanism* **(22)** *turnable brace/clamp/clip* **(23)** *frame for handgrip* **(24)** *trigger with spring* **(25)** *handgrip* **(26)** *brace/clamp* **(27)** *shoulder support* **(28)** *optical sight*

82 mm M60 recoilless gun

This gun was designed and developed in Yugoslavia. It is a lightweight anti-tank gun, firing a fin-stabilised round of native design. It can be towed behind a Jeep-like vehicle or dragged into position by hand.

DATA
Calibre: 82 mm
Length of weapon: 2200 mm
Overall weight: 122 kg
Traverse: 360°
Elevation: −20 to +35°
Weight of projectile: 4.3 kg
Muzzle velocity: 380 m/s
Max effective range (HEAT): (against moving targets) 1000 m; (against stationary targets) 1500 m
Armour penetration: 220 mm
Crew: 5

Training with 82 mm M60 recoilless gun

Status
Current.
Service
Yugoslav forces.

82 mm M60 recoilless gun

82 mm M79 recoilless gun

This is a lightweight weapon of high efficiency; it may have had its initial inspiration from the Soviet B-10, but the weapon eventually developed shows considerable differences. The gun is made from high-tensile steel and special alloys and uses the characteristic perforated case and annular chamber system. There is a single venturi at the rear of the breech-block. A telescopic sight is mounted on the left side of the gun, and there are also open sights for emergency use.

The mounting is a tripod with a flat top to which the gun is coupled; there is a rear pivot and a front support which screws up and down to alter elevation and can move laterally in a slot to give limited traverse. The gunlayer can thus control both movements of the gun with one hand while keeping the other on the firing lever. The firing mechanism is electro-mechanical and is interlocked so the firing is impossible unless the breech is closed and locked.

The round of ammunition is fixed, with a perforated cartridge case. The projectile is fin-stabilised by six jack-knifed fins which flip out after shot ejection. There are two projectiles; the anti-tank HEAT shell is rocket-boosted by a sustainer which fires after the projectile has left the muzzle. The HE anti-personnel projectile has no assistance.

82 mm M79 recoilless gun with HEAT ammunition

DATA
Calibre: 82 mm
Weight in action: 27.5 kg with tripod
Length overall: 1.785 m
Muzzle velocity: (HEAT) 340 m/s; (HE) 320 m/s
Max flight velocity (HEAT): 460 m/s
Max range (HE): 2000 m
Operational range (HEAT): 670 m
Penetration (HEAT): 350 mm

Manufacturer
Federal Directorate of Supply and Procurement, 9 Nemanjina Street, 11001 Beograd.
Status
Current. Production.

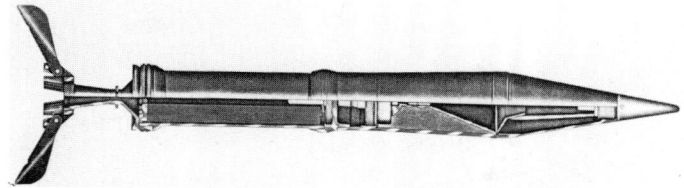

Sectioned M72 HEAT projectile for M79 RCL gun

Service
Yugoslav Army.

90 mm M79 rocket launcher

This is a rocket launcher which, like the French LRAC-89 and others, is loaded by coupling a pre-packed rocket and combustion chamber to the rear end of the launch tube. After firing, the empty combustion tube is removed and discarded and another pre-loaded unit is attached. Connection of the combustion tube automatically connects the electrical firing circuits.

The projectile is a rocket carrying a shaped charge warhead, initiated by a piezo-electric impact fuze capable of acting at angles up to 70° from normal to the target.

DATA
Calibre: 90 mm
Weight of launcher: 6.2 kg
Weight of launcher, loaded: 10.7 kg
Length of launcher: 1.432 m
Length of launcher, loaded; 1.91 m
Weight of rocket: 3.5 kg
Length of rocket: 672 mm

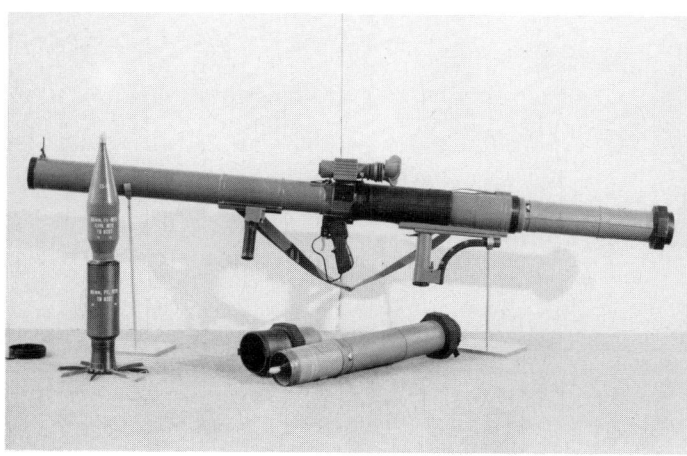

90 mm M79 rocket launcher showing rocket in combustion tube

90 mm M79 rocket launcher

Launch velocity: 250 m/s
Max range: 1960 m
Operational range: (AFVs) 350 m; (other) 600 m
Armour penetration: 400 mm

Manufacturer
Federal Directorate of Supply and Procurement, 9 Nemanjina Street, 11001 Beograd.
Status
Current. Production.
Service
Yugoslav Army.

64 mm RBR-M80 LAW

This is another introduction into Yugoslav service and appears to be a near-copy of the well-known M72 light anti-armour weapon (LAW). The launch tube is of telescopic construction, with the HEAT rocket packed inside and sealed. The rocket consists of a motor using tubular propellant, a percussion igniter, and a shaped charge warhead initiated by a piezo-electric fuze. The shaped charge uses phlegmatised octol (HMX), and the fuze is bore-safe, arming between 6 and 20 m from launch, with self-destruction after six seconds of flight should it miss the target.

DATA
Calibre: 64 mm
Weight complete: 3 kg
Weight of rocket: 1.58 kg
Length, travelling: 860 mm
Length, firing: 1.2 m
Length of rocket: 664 mm
Max range: 1280 m
Operational range: 250 m
Muzzle velocity: 190 m/s
Penetration: 300 mm

Manufacturer
Federal Directorate of Supply and Procurement, 9 Nemanjina Street, 11001 Beograd.
Status
Current. Production.
Service
Yugoslav Army.

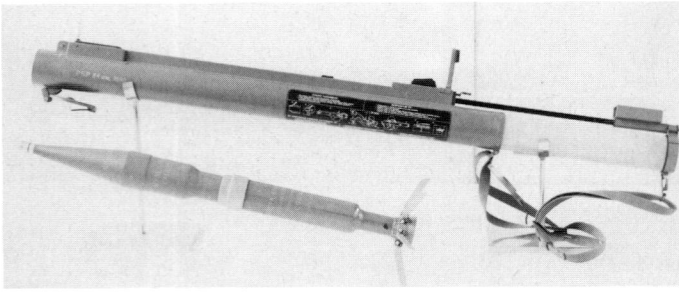

64 mm RBR-M80 LAW, with rocket

64 mm RBR-M80 LAW in firing position

Mortars

ARGENTINA

Argentinian mortars

Mortars in service in Argentina include 81 and 120 mm weapons of Hotchkiss-Brandt design. These were originally purchased from the French manufacturer but recent supplies have been locally manufactured under the auspices of Fabricaciones Militares.

Manufacturer
Controlled by Fabricaciones Militares, Cabildo 65, Buenos Aires.

AUSTRIA

81 mm SMI mortar

Computer-aided design and low weight alloys used in the aircraft industry have enabled Südsteirische Metallindustrie (SMI) to keep the weight of this 81 mm mortar down to 40 kg, making it among the lightest currently available.

The barrel is made of forged steel, reinforced at the rear end, and the lower part of the barrel is finned to assist in cooling. The breech, which holds the fixed firing pin, is screwed to the barrel and contains the ball by which the barrel is positioned on the baseplate. Two flats on the ball allow the barrel to be inserted and removed from the socket on the base plate.

The mounting is a new design which enables the SMI mortar to be emplaced on very uneven ground and yet be folded to a compact 700 mm for transport. It comprises bipod, elevating gear, traversing gear, cross-levelling gear and the barrel clamp. The elevating gear is completely enclosed in a tube which forms the centre shaft of the bipod.

Three radial ribs on the underside of the baseplate provide the strength to prevent the baseplate from buckling under firing stresses and also anchor the mortar to the ground.

The mortar can be broken down into three manpack loads and can be carried in any suitable vehicle.

DATA
BARREL
Calibre: 81 mm
Weight: 12.25 kg
Length with breech plug: 1330 mm

MOUNTING
Weight: 13.25 kg
Length, folded: 1000 mm
Elevation range: 45–85° (800–1510 mils)
Traverse range: 100 mils left and right

BASEPLATE
Weight: 13.75 kg
Diameter: 570 mm

OPTICAL SIGHT
Weight: 1.3 kg
Field-of-view: 160 mils
Magnification: × 2.5
Elevation range: 900 mils
Traverse: 6400 mils

PERFORMANCE
Weight of HE long-range bomb: 4.15 kg
Max velocity: 300 m/s
Max range: 5800 m
Max pressure: 750 bar
Accuracy: at max range, 50% of bombs fall within a zone 40 m long and 20 m wide based on the target
Ammunition: The SMI 81 mm mortar is designed for the HE long-range bomb described below. However, it can fire all 81 mm mortar bombs of current manufacture

Manufacturer
Südsteirische Metallindustrie GmbH, A-8430 Leibnitz.
Status
Current. Available.
Service
Various countries have placed orders or have evaluated the mortar.

81 mm SMI mortar

Design of 81 mm SMI mortar mounting enables mortar to be emplaced in difficult terrain

60 mm Böhler C6 Commando mortar

The C6 mortar is a light and reliable mortar designed for easy carriage, especially in difficult terrain such as mountains or jungle. There are two models: the C6-110 has a fixed firing pin, while the C6-210 has a self-cocking firing pin controlled by a trigger built into a special handgrip.

The C6 mortar consists of the following parts: barrel with breech piece and firing pin assembly; baseplate; handgrip and trigger mechanism; and a set of auxiliaries and spare parts.

The barrel is of high strength forged special steel, is provided with a collar at the muzzle, and is externally threaded at the rear end to receive the breech piece. The muzzle is bell-mouthed to facilitate loading. There is a protective sleeve of heat-resistant plastic, about 15 cm in length, shrunk on to the barrel behind the muzzle.

The breech piece is of steel, is internally threaded to mate with the barrel, and has a tapered sealing face in order to ensure a gastight seal. There is a central hole into which the firing pin assembly is mounted, and four threaded holes for attachment of the trigger mechanism. During factory assembly of the mortar the breech piece is screwed tightly to the barrel, the threads being treated with a special adhesive, and the two cannot be separated in the field.

The C6-110 firing pin is a solid assembly fitted into the hole in the breech piece and retained by a cover. The C6-210 firing pin is a spring assembly which is retained in the breech piece by the trigger mechanism housing. The trigger assembly consists of an aluminium housing into which the stainless steel (and hence maintenance-free) mechanism is fitted. Operation of the firing pin is controlled by a thumb-actuated trigger set into the carrying handle part of the mechanism housing. There is a cross-bolt safety catch which, when applied, locks the mechanism. An advantage of this trigger firing system is that the mortar can be fired at flat trajectories.

The baseplate is circular and made of high tensile aluminium alloy. It carries a central stainless steel spike which serves to retain the baseplate to the breech piece as well as anchoring the mortar securely on hard ground.

The carrying strap is attached to the barrel and breech piece. It serves as a carrying sling and is marked with ranges; it is used by allowing the sling to drop on the ground when the mortar is supported in the firing position. The firer then kneels with his boot on the desired range mark and elevates the mortar until the vertical portion of the strap is taut. There is also an optional elevation indicator which can be clamped to the barrel; this can be seen in the picture of the mortar in its carrying frame.

The auxiliaries and spares consist of cleaning rod and brush, screwdrivers, spanners, oil can and a spare firing pin assembly of the appropriate type.

DATA
Calibre: 60.70 mm
Overall length: 815 mm
Length of bore: 640 mm
Outer diameter: 72 mm
Weight of mortar: 5.10 kg (C6-210): 4.30 kg (C6-110)
Weight of auxiliaries: 1.20 kg
Maximum range: 1600 m
Maximum rate of fire: 30 rds/min

Manufacturer
Böhler Special Products, PO Box 188, A-1232 Vienna.
Status
Current. Production.
Service
Overseas armies; to enter Austrian Army service in 1990.

Böhler 60 mm C6-210 mortar in firing position

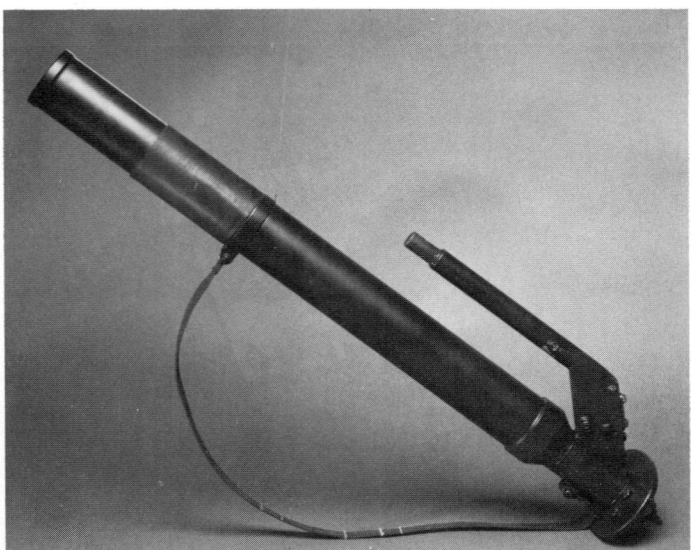

Böhler 60 mm C6-210 mortar

Böhler 60mm C6-210 mortar in carrying frame

81 mm Böhler M8-111 medium mortar

This mortar was developed for the Austrian Army, with whom it is in service as the mGRW (mitteler Granatwerfer) 82. It is also in service with other armies. It is available in a number of variant models: M8-111 81 mm standard version; M8-211 81 mm long-range; M8-313 82 mm standard; and M8-413 82 mm long-range. These differ only in length of barrel and calibre.

The barrel is forged from ESR steel, is smoothbored with a bottleneck at the breech end, and has the muzzle bell-mouthed to facilitate loading. There are radial fins on the rear section which add strength and aid cooling. The bipod barrel clamp is mounted on the cylindrical part of the barrel between the cooling ribs and an annular collar. The stainless steel breech piece seals the rear of the barrel and supports the barrel on the baseplate. A copper washer between barrel and breech piece provides a gastight seal. To facilitate assembly and disassembly for cleaning, by means of the appropriate tools, the barrel has two flats formed on it and the breech piece has a transverse hole. The breech piece is bored with an oblique threaded hole with tapered sealing face, to hold the firing pin. Due to this inclined position the firing pin can be removed and replaced when the mortar is in the firing position or can be completely retracted in the event of a misfire.

The bipod assembly consists of a barrel clamp, twin shock absorbers, traversing mechanism, sight bracket, cross-levelling mechanism, elevating mechanism and bipod legs. One leg of the asymetrical bipod can be adjusted and locked in any desired position. The bipod is surmounted by the elevation mechanism, which is a screw device operated by a handwheel. By adjustment of the position of the barrel clamp on the barrel, in conjunction with the elevating mechanism, any desired elevation can be easily achieved.

The baseplate is a die-forged circular aluminium alloy unit, with ribs and spades on its under-surface. A stepped recess in the upper side holds the pivoted ball socket, which is secured by a retaining ring. Beneath the ball are damping rings and discs which absorb some of the recoil blow. A lateral opening in the ball socket allows insertion of the barrel and also permits lowering the barrel to an elevation of about 700 mils (39°). When the barrel is moved in azimuth the ball socket moves with it, and thus all-round firing is possible without repositioning the baseplate.

The auxiliaries and spare parts kit contains cleaning and maintenance equipment and tools, spare firing pin, breech piece washer and baseplate components, two aiming posts and two aiming post lamps.

All standard types of 81 mm and 82 mm bombs can be fired from these mortars, though optimum results will be obtained by using the HE 70 range of bombs designed for it by Hirtenberger.

81 mm Böhler M8-111 mortar

DATA
BARREL
Calibre: 81.0 mm (M8-111/211); 82.0 mm (M8-313/413)
Length: 1280 mm (M8-111/313); 1480 mm (M8-211/413)
Weight: 12 kg (M8-111); 14 kg (M8-211); 12.6 kg (M8-313); 14.5 kg (M8-413)

BASEPLATE
Diameter: 546 mm
Depth: 130 mm
Weight: 12 kg

BIPOD (folded)
Length: 1050 mm
Width: 630 mm
Depth: 220 mm
Weight: 11.6 kg

PERFORMANCE
Elevation: 700 – 1500 mils (39 – 84°)
Minimum range: 100 m all models
Maximum range: 6100 m standard, 6500 m long-range, both calibres
Maximum rate of fire: 30 rds/min

Manufacturer
Böhler Special Products, PO Box 188, A-1232 Vienna.
Status
Current. Production.
Service
Austrian and other armies.

Laying the Böhler 81 mm M8-111 mortar

81 mm Böhler M8-522 medium mortar

This mortar offers the same performance as the 81 mm M8-111 described above but is of more conventional and somewhat heavier construction. As with the M8-111 series there are variants: M8-522 81 mm standard; M8-622 81 mm long-range; M8-722 82 mm standard; and M8-822 82 mm long-range.

The barrel is externally smooth and without cooling fins and is made of special ESR steel. The breech piece is screwed and glued to the barrel but can

be removed in the field if necessary. The bipod is of symmetrical pattern with the elevating gear mounted in a central tube and the traversing gear attached to the top, working on the barrel clamp. Two shock absorbers prevent the transmission of recoil force to the bipod during firing. The baseplate is similar to that of the M8-111 series. Auxiliaries and spares to the same specification as the M8-111 are provided.

DATA
Barrel: As for M8-111 above except weights: M8-522 13.5 kg; M8-622 14.8 kg; M8-722 12.5 kg; M8-822 13.8 kg
Baseplate: as for M8-111 series except weight 13 kg
Bipod: length 1050 mm; width 440 mm; weight 15.6 kg
Performance: as for M8-111 series above.

Böhler 81 mm M8-522 medium mortar

120 mm Böhler M12-1111 heavy mortar

This mortar was jointly developed by Böhler and the Austrian Army, with whom it entered service in February 1985 as the sGRW (schwere Granatwerfer) 86. It is normally mounted on a towed carriage but can also be fitted into armoured vehicles of the M113/M125 type, when it is known as the M12-2330. Almost all types of 120 mm bomb can be fired from this mortar.

The barrel is of ESR steel. The muzzle is bell-mouthed to facilitate loading and the barrel clamp mounts between two collars. There is a carrying handle mounted towards the rear end. The breech piece is screwed into the barrel and prevented from rotating by a safety lock; a patented sealing washer between barrel and breech piece ensures a gastight seal. The firing mechanism, which is contained in the breech piece, can be adjusted to fire in fixed or trigger-actuated modes. For safety reasons the change to the fixed firing pin mode requires the use of two hands. There is an external trigger lever to which

a firing lanyard can be attached, and there is also an external safety lever which withdraws and locks the firing pin when applied.

The bipod is of symmetrical type with a central elevating screw and a screw traversing mechanism attached to the barrel clamp. There are two shock absorbers, using hollow rubber springs, which attenuate the recoil blow to both bipod and baseplate.

The baseplate is of sectional welded pattern. Handles on the upper side serve also as mounting devices for the carriage lifting gear. The bottom side has six radial spade ribs which provide stability on any type of ground. A rapid lock permits quick removal of the barrel from the baseplate.

The carriage is a suspended single-axle carriage of solid-frame construction. By use of a cable winch and tilting frame, the complete mortar can be either lowered into the firing position or pulled on to the carriage for transport. There is a locking mechanism which fastens the barrel and bipod in the travelling position. The drawbar is adjustable for height and there is a front castor wheel which aids manoeuvring and can be retracted when towing. The carriage frame is provided with hooks to facilitate helicopter lifting and handgrips for manhandling.

The auxiliaries and spare parts kit contains all the necessary cleaning and maintenance equipment and tools, spare firing pin and sealing washer, and two aiming posts and lamps.

M12-2330 APC Version
This version uses a shorter barrel and instead of a bipod has a special 'support' which consists of two tubular legs which carry the normal elevating gear in the centre but which end in rollers travelling on an arcuate racer set into the floor of the vehicle. Clamps are provided for locking the support at any desired point in the traverse. The rear end of the barrel fits into a ball socket of the usual type which is assembled into a heavy tubular member bolted to the vehicle sidewalls. There is sufficient freedom of movement in the ball and socket to accommodate the full range of elevation and traverse. Sighting, elevation and traversing mechanisms are the same as those for the towed mortar, though traverse is restricted by the length of the racer to about 45°.

The mortar is supplied with a bipod and standard baseplate, and can be removed from the vehicle and brought into action on the ground if desired. The bipod differs from standard in having no elevating gear; that from the APC mounting is removed with the barrel and fitted into a prepared fork in the bipod.

DATA
M12-1111 towed version:
WEIGHTS
Mortar, complete with carriage: 670 kg
Mortar in firing position: 277 kg
Barrel with breech piece: 100 kg
Bipod: 60 kg
Baseplate: 117 kg

120 mm Böhler M 12-2330 mortar on APC mounting

120 mm Böhler M12-1111 mortar on travelling carriage

120 mm Böhler M12-1111 mortar in firing position

PERFORMANCE
Elevation: 700 – 1500 mils (39 – 83°)
Traverse on bipod: 300 mils (16.8°)
Minimum range: 400 m
Maximum range: 9000 m
Maximum rate of fire: 15 rds/min
Sustained rate of fire: 10 rds/min

DATA
M12-2330 APC Version
WEIGHTS
Barrel complete: 92 kg
Support: 50 kg
Baseplate: 85 kg

DIMENSIONS
Barrel overall length: 1750 mm

PERFORMANCE
Maximum range: 8600 m (from baseplate)

Manufacturer
Böhler Special Products, PO Box 188, A-1232 Vienna.
Status
Current. Production.
Service
Austrian Army.

DIMENSIONS
Calibre: 120.2 mm
Barrel length overall: 1900 mm
Bipod spread: 1340 mm
Baseplate diameter: 1100 mm
Baseplate depth: 350 mm
Carriage length: 3700 mm
Carriage height: 1400 mm
Carriage width: 1710 mm
Carriage track: 1450 mm
Ground clearance: 350 mm

BELGIUM

FLY-K PRB mortar NR8113A1

The FLY-K mortar uses a form of spigot, combined with a propulsion unit sealed within the bomb, to produce effective mortar fire with very little noise, no emission of explosive gases, no flash, and therefore no firing signature capable of detection. In addition the design ensures that the mortar barrel does not get hot, therefore there is no detectable infra-red emission.

The FLY-K mortar consists of a guide tube and a truncated A-shaped baseplate fitted with fixed firing pin automatic percussion. The projectiles are equipped with a propulsion unit of new design which is integrated into the tail fin and stabiliser tube of the bomb. On loading, the bomb drops down the barrel in the usual way, but the fixed spigot inside the guide tube passes into

the bomb tail unit to ignite the propelling charge by percussion. The charge explodes and drives aa captive piston down the tail tube, so thrusting the bomb off the spigot and into the air. As the piston reaches the end of the tail tube it is locked, so retaining the explosive gases inside the tube; since these gases do not reach the atmosphere they do not generate the usual noise therefore FLY-K mortar is virtually silent.

A completely new and patented system of sighting has been adapted to the FLY-K mortar. A column of coloured liquid inside a transparent tube indicates te firing angle or the equivalent range against graduations marked on the tube. By simply tilting the mortar tube until the surface of the liquid aligns with the required graduation, the mortar is accurately laid for range. The sight is all plastic and is fixed on the tube near the muzzle, allowing the firer to take directional aim and check the elevation at the same time. A Trilux illuminator makes it possible to read the sight in any light condition. The system is particularly stable since there is no temperature rise in the tube due to firing.

DATA
Calibre: 52 mm
Overall length: 710 mm
Weights
barrel: 2.15 kg
baseplate: 2.07 kg
sight: 120 g
total: 4.50 kg
Elevation: 40°–82°
Range: 200–700 m
Ammunition
fragmentation: NR235
practice: NR234
illuminating: NR236
smoke: WP NR233
smoke: RP NR237 (under development)

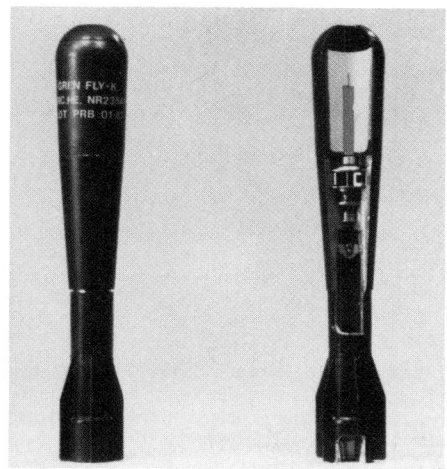

52 mm PRB FLY-K fragmentation bomb NR235A1

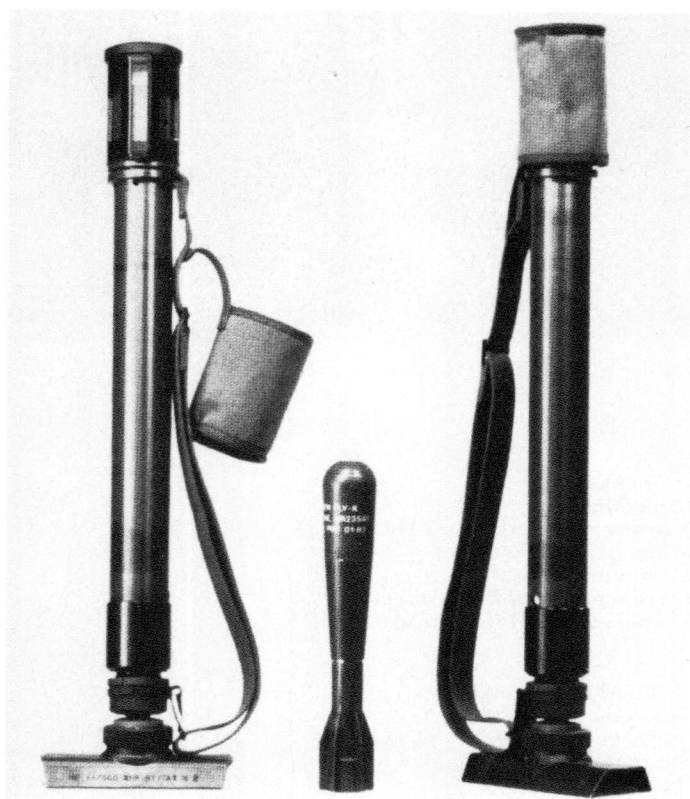

Manufacturer
Société Anonyme PRB, Département Défense, Avenue de Tervueren 168, Bte
7, B-1150 Brussels.
Status
Current. Production.

52 mm PRB FLY-K mortar NR8113A1

NR 8120 PRB grapnel launcher

This is a short mortar using a variation of the FLY-K system in which the
propelling charge is fired in the mortar. This drives a piston up the barrel to
eject the grapnel. The piston is trapped at the muzzle, sealing the propulsive
gases inside the weapon. The grapnel is a strong four-armed device with arms
that fold back for firing, but which open out when the rope is pulled. The rope
is kept in a sealed bag until the grapnel is fired.

The launcher is a small firing support with adjustable elevation. It can be
fired by a trigger or a remote cord and it can be mounted on a wide variety of
surfaces, including boats and vehicles. The normal crew for firing is two men.

DATA
GRAPNEL
Length: (closed) 520 mm; (open) 550 mm
Span: (closed) 250 mm; (open) 430 mm
Weight: 2.4 kg

PROPULSION CARTRIDGE
Diameter: 60 mm
Weight: 2.7 kg

MOUNTING
Length: 360 mm
Weight: 9.3 kg
Muzzle velocity: 150 m/s

Manufacturer
Société Anonyme PRB, Département Défense, Avenue de Tervueren 168, Bte
7, B-1150 Brussels.
Service
Not known.

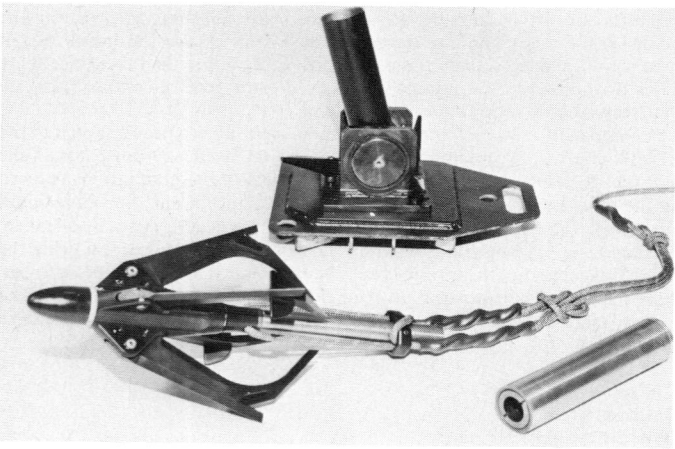

Grapnel and launcher with propulsion cartridge in foreground

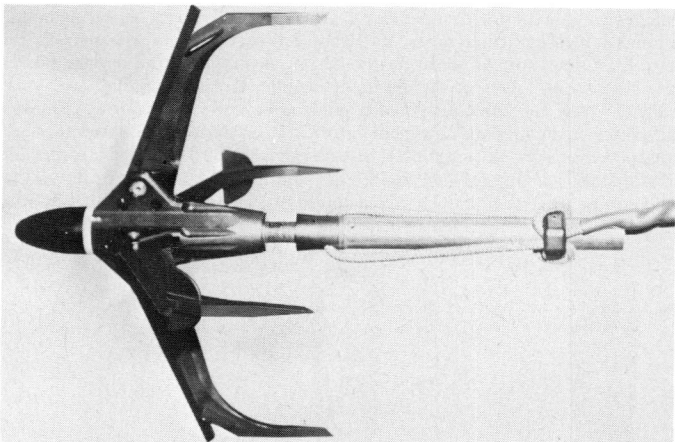

Grapnel open

60 mm NR 493 PRB mortar

This is a muzzle-loaded smoothbore mortar of fairly conventional design with
a fixed firing pin. It can fire all standard 60 mm ammunition but is particularly
effective when used with the special NR 431 round (see entry below).

Barrel
Made of special steel this is a smoothbore tube of 60.7 mm diameter closed at
one end by the breech with its fixed firing pin. The breech has a spherical

projection which mates with the baseplate socket and there is a carrying belt
attached at one end to this projection and at the other to the barrel near the
muzzle. The belt carries a protective cap which is placed on the muzzle when
the mortar is not in use.

Bipod
This carries the elevation and traverse mechanisms and the sight mounting

NR 493 PRB mortar

and is fitted with two mechanical shock absorbers and a strap fastener. It is robustly made and the sliding parts ar enclosed in sealed protective sheaths to prevent corrosion and provide permanent lubrication.

Baseplate
This is circular and dish-shaped. A socket on the upper face accepts the breech projection and is locked by a lever. On the underside are three ribs, one of which serves as an anchoring spade. The rim of the plate is perforated to ease extraction from the ground after firing.

Sight unit
This comprises a collimator sight, and an open sight for use in emergency, coss-levelling gauges and sighting controls. The deflection scale has 64 intervals of 100 mils each and the deflection knob is graduated in 50 intervals of 2 mils with the 10 mil intervals numbered. The elevation scale has nine 100 mil graduations (numbered every 200 mils) and the elevation knob is graduated in 50 intervals of 2 mils with the 10 mil intervals marked.

Crew
In normal service the mortar will be carried and operated by three men. In emergency, however, it is possible for one man to carry and fire the weapon.

DATA
Calibre: 60 mm
Weight: (total) 22.1 kg; (barrel) 6 kg; (bipod) 10.4 kg; (baseplate) 5.1 kg
Length of barrel: 780 mm
Diameter of baseplate: 320 mm
Traverse on bipod: 135 mils
Range: (with NR 431 bomb) 1800 m
Crew: 1–3

Manufacturer
Société Anonyme PRB, Département Défense, Avenue de Tervueren 168, Bte 7, B-1150 Brussels.
Service
Belgian Army.

81 mm NR 475 A1 PRB medium mortar

This mortar has been designed for use by infantry with either US or PRB bombs and especially with the highly lethal NR 436 round. It is suitable for transport by land vehicle or helicopter or by pack animal. It may also be carried by three soldiers and is robust enough to be dropped by parachute.

Barrel
The barrel is a smoothbore tube with radial fins on the lower half and is terminated by a breech piece in which is inserted a fixed firing pin. The

81 mm NR 475 A1 PRB medium mortar

exterior of the breech piece is terminated by a ball which engages in a socket in the baseplate.

Baseplate
This is circular and has four welded ribs and five spikes on the underside. Holes are provided to ease extraction from the ground. A socket in the centre of the plate accepts the ball projection from the breech piece and has a clamping device with a securing screw. Carrying handles are attached to the sides.

Bipod
The bipod legs have spiked feet and are distanced by an adjustable spring-loaded chain. They are connected to the elevating mechanism which consists of an elevating screw rotating in a nut and driven through a gear mechanism. The elevating rod is terminated by the traversing mechanism which comprises the yoke, the traversing drive and two shock absorbers. Cross-levelling is effected by roughly adjusting the cross-levelling clamp on the left bipod leg and then completing the adjustment using the screw mechanism between the clamp and the elevating mechanism housing.

BR 1 sight unit
This is mounted at the left end of the yoke by a dovetail bracket and comprises a collimator, elevating and deflecting mechanisms and longitudinal and cross-levels. The elevation scale is graduated in seven intervals of 100 mils, numbered at the even numbers from 8(00) to 16(00), and the elevation knob is graduated in 2 mil intervals and numbered every 10 mils from 0 to 90. The deflection scale is graduated in 100 mil intervals, and numbered from 0 to 64(00), and the deflection knob is graduated in the same way as the elevation knob.
Standard accessories for the mortar include two aiming posts and aiming-post lamps.

DATA
Calibre: 81 mm

WEIGHT
Total: 43 kg
Barrel: 15.3 kg
Bipod: 12.5 kg
Baseplate: 14.6 kg
Sight unit: 0.6 kg
Length of barrel: 1350 mm
Diameter of baseplate: 500 mm
Elevation ange: 700–1515 mils
Traverse on bipod: 140 mils at 60° elevation
Operating range: (minimum) 300 m; (max) 5500 m using NR 436 bomb and max charge
Rate of fire: 15–20 rds/min

Manufacturer
Société Anonyme PRB, Département Défense, Avenue de Tervueren 168, Bte 7, B-1150 Brussels.

Service
Belgian Army.

CANADA

Mortar sight unit C2

The mortar sight unit C2 is designed for use with a variety of different types of medium mortar. It has also proved to be effective with heavy mortars and can be used to lay a variety of medium and heavy machine guns in the indirect and sustained fire roles.

The sight unit is a fixed focus type optical instrument using external light sources to illuminate the elevation and azimuth scales, graticule and level bubbles. It is manufactured in accordance with all the relevant military specifications. The elbow telescope incorporates a dovetail for accepting a periscope extension. This raises the line of sight so that the mortar can be fired from a pit or a modified weapons carrier.

DATA
Magnification: × 1.8
Field-of-view: 10°
Weight, with case: 4.9 kg
Illumination: Instrument light with batteries
Scales:
Azimuth: 0 – 6400 mils
Elevation: mortar: 600 – 1600 mils
Elevation: machine gun: –200— + 600 mils
Accuracy: 1 mil
NATO Stock Number: 1240-21-111-2454 (with case and instrument light)

Manufacturer
Ernst Leitz Canada Ltd, 328 Ellen Street, Midland, Ontario L4R 2H2.
Status
Current. Production.
Service
Worldwide.

Mortar sight unit C2

Mortar sight unit C2A1

The sight unit C2A1 was designed and developed by Ernst Leitz Canada Ltd. It was designed for use with the 81 mm mortar and it can also be used to lay a variety of machine guns when employed in the indirect fire role.

The sight unit is a fixed focus type optical instrument which uses tritium light sources to illuminate the elevation and azimuth scales, graticule and level vials.

The sight unit elbow telescope incorporates a dovetail to permit a periscope extension (below) to be fitted so as to raise the line of sight so that the mortar can be operated from a pit or from a modified weapons carrier.

DATA
Magnification: × 1.8
Field-of-view: 10°
Weight, with case: 1.4 kg
Illumination: Tritium (hydrogen 3)
Scales:
Azimuth: 0 to 6400 mils
Elevation: (mortar) 600-1600 mils; (machine gun) -200 to 600 mils
Accuracy: 1 mil
NATO stock number: 1240-21-876-5058 (with case)

Manufacturer
Ernst Leitz Canada Ltd, 328 Ellen Street, Midland, Ontario L4R 2H2.
Status
Current. Production.
Service
Military forces worldwide.

Leitz sight unit C2A1

Periscope extension C2/C2A1

The periscope extension is used with the C2/C2A1 family of 81 mm mortar sight units in order to raise the line of sight by 355 mm so that the mortar can be employed from weapon pits or in modified weapons carrying vehicles. The periscope extension consists of a periscope optical system in a tubular aluminium body. The extension is designed to be quickly and easily installed and removed from the lamp housing assembly, which is fitted to a dovetailed slot on the elbow telescope of both the C2 and C2A1 sight units.

DATA
Weight, including case: 270 g
Length: 379.5 mm
Diameter: 19 mm
Magnification: × 1
Field-of-view: 8°
Exit pupil: 9 mm
NATO Stock Number: 1240-21-851-6968 (with case)

Manufacturer
Ernst Leitz Canada Ltd, 328 Ellen Street, Midland, Ontario L4R 2H2.
Status
Current. Production.
Service
Military forces worldwide.

Leitz periscope extension mounted on C2A1 sight unit

Mortar Sight unit M53A1

The Sight unit M53A1 is the primary medium mortar sight of the US Army and is used to lay the 81 mm mortar. It can be used with other weapons, including rocket launchers.

The M53A1 consists of the Mount, Telescope M128A1 and the Telescope, Elbow, M109. The M109 Elbow Telescope is a fixed focus optical instrument. Illumination of the M128A1 scales and the M109 graticule is provided by the Light Instrument M53E1.

DATA
Length: 152 mm
Width: 127 mm
Height: 254 mm
Elevation: 1600 mils
Azimuth: 0 to 3200/6400 mils
Accuracy: 1 mil
Magnification: 4 ×
Field-of-view: 10°
NATO Stock Number: 1240-00-856-9452

Manufacturer
Ernst Leitz Canada Ltd, 328 Ellen Street, Midland, Ontario L4R 2H2.
Status
Current. Production.

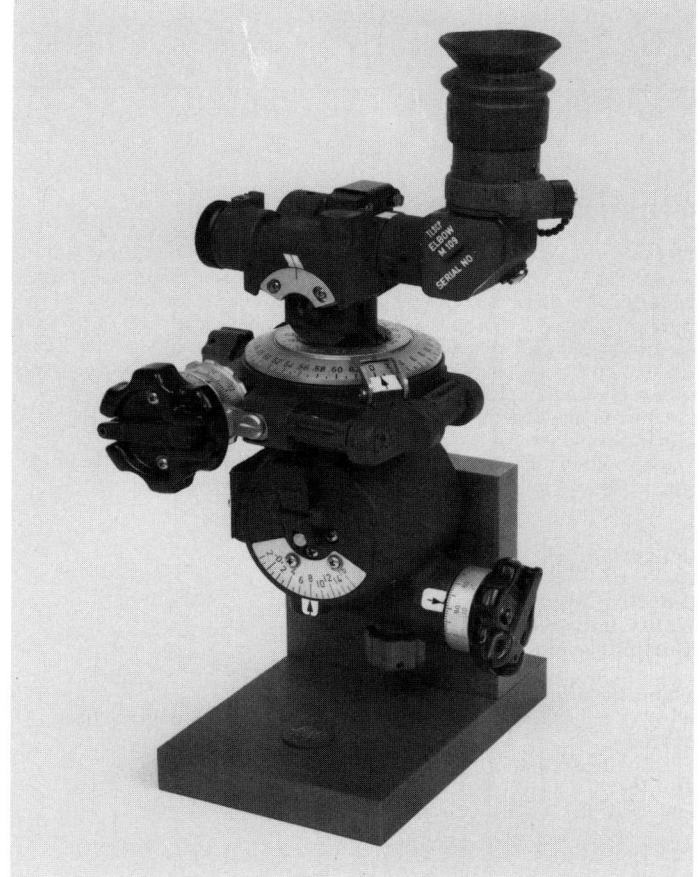

Leitz M53A1 mortar sight unit

Mortar Sight Unit M64A1

The Sight Unit M64A1 was designed to lay the 60 mm M224 lightweight company mortar of the US Army. It can also be used with the 81 mm M252 mortar and with a variety of other lightweight and medium mortars.

The M64A1 Sight Unit consists of the Mount, Telescope, M64A1; the Telescope, Elbow, M64A1; and the Ancillary Sighting Equipment Set. The Sight Unit is a fixed focus opto-mechanical instrument which uses tritium light sources to illuminate all elevation, azimuth and graticule scales and level vials. Ancillary equipment is also available as part of a complete logistic support package. The M64A1 is built to all applicable military specifications.

DATA
Length: 111 mm
Width: 111 mm
Height: 187 mm
Weight: 1.1 kg
Elevation: 700 to 1600 mils
Azimuth: 0 to 6400 mils
Accuracy: 1 mil
Magnification: Unity

Field-of-view: 8°
Illumination: Tritium
NATO Stock Number: (Mount, Telescope) 1240-01-201-8299; (Elbow Telescope) 6650-01-211-3608

Manufacturer
Ernst Leitz Canada Ltd, 328 Ellen Street, Midland, Ontario L4R 2H2.
Status
Current. Production.

Leitz M64A1 mortar sight unit

CHILE

60 mm Commando mortar

This is a conventional smoothbore drop-fired 60 mm mortar which appears to be of the Brandt Commando Type V pattern. The firing pin is fixed. Laying for direction is done visually, by means of an axis line painted on the mortar barrel, and laying for range is done either by a simple level sight or by using the carrying sling. By laying the sling on the ground and trapping it there with one foot on a range marker attached to the sling, the mortar can be given a reasonably accurate elevation sufficient for an opening round.

A variant model of this mortar has a trip-operated firing pin which can be withdrawn and set to safe if required.

Ammunition for this mortar is a licence-manufactured copy of the Brandt 60 mm Mark 61 series, to which reference should be made for details.

DATA
Calibre: 60 mm
Weight: 7.7 kg
Length of barrel: 650 mm
Total length: 680 mm
Maximum range: 1050 m

Manufacturer
FAMAE Fabricaciones Militares, Av Pedro Montt 1568/1606, Santiago.
Status
Current. Production.
Service
Chilean Army.

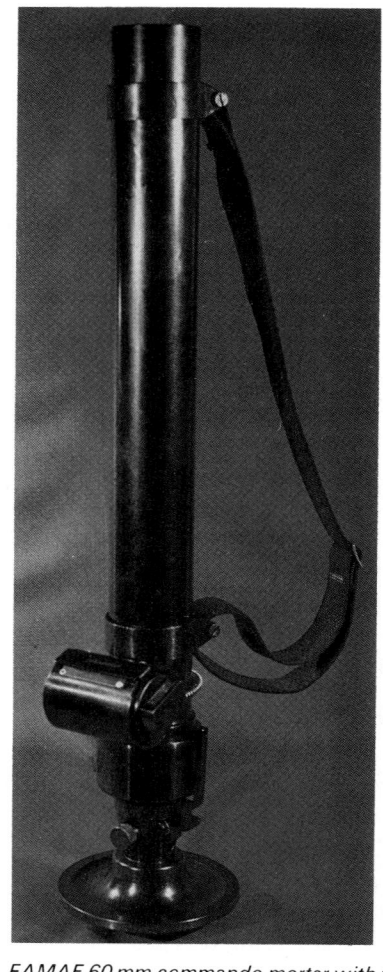

FAMAE 60 mm commando mortar with trip firing pin

81 mm FAMAE mortar

Although referred to by the manufacturer as 'Brandt type' this has some features which appear to be of Chilean design, though there are distinct leanings towards the Esperanza form of tripod. This is a conventional smoothbore mortar with fixed firing pin. The baseplate is a distinctive hexagonal shape with prominent spades and the whole equipment is rather heavier than is usual for this class of weapon.

The ammunition fired with his mortar is a licence-manufactured copy of the Brandt 81 mm M57 series, to which reference should be made for details.

DATA
Barrel weight: 20.6 kg
Baseplate weight: 28 kg
Tripod weight: 17.2 kg
Maximum range: 4200 m
Minimum range: 100 m

Manufacturer
FAMAE Fabricaciones Militares, Av Pedro Montt 1568/1606, Santiago.
Status
Current. Production.
Service
Chilean Army.

FAMAE 81 mm mortar

FAMAE 81 mm mortar dismantled and manpacked

FAMAE 120 mm mortar

This is a FAMAE-designed light 120 mm mortar which has some features reminiscent of the Esperanza Model L, though it is considerably lighter. The mortar is a conventional drop-fired smoothbore supported by a circular baseplate and a tripod. It is moved by dismantling and fitting the three basic components on to a light two-wheeled trailer which can be vehicle-drawn or pulled by the mortar crew.

The ammunition used with this mortar is a licence-manufactured copy of the Brandt 120 mm M44 series of bombs, to which reference should be made for details.

DATA
Barrel weight: 40 kg
Tripod weight: 25 kg
Baseplate weight: 44 kg
Trailer weight: 80 kg
Total weight, march order: 189 kg
Maximum range: 6650 m
Minimum range: 500 m

Manufacturer
FAMAE Fabricaciones Militares, Av Pedro Montt 1568/1606, Santiago.
Status
Current. Production.
Service
Chilean Army.

FAMAE 120 mm mortar

FAMAE 120 mm mortar on transport trailer

CHINA, PEOPLE'S REPUBLIC

60 mm Type 31 and variants

This was originally a Chinese Nationalist weapon and is a copy of the US M2 mortar. Both the M2 and the Type 31 weapons resemble the pre-war French Stokes-Brandt M1935 mortar. All three have a square baseplate, a handcrank on the end of the elevating screw housing and a cross-levelling mechanism of two-piece construction. Recognition features are brass feet on the Chinese Nationalist weapon and a folding handcrank on the traversing knob of the French and US versions which can be distinguished from each other by the markings.

DATA	Type 31	M2	M1935
Calibre	60 mm	60 mm	60 mm
Barrel length	675 mm	726 mm	700 mm
Firing weight	20.2 kg	19 kg	17.6 kg
Max range	1530 m	1820 m	1760 m
Rate of fire	20–30 rds/min		

Status
Obsolete. No longer made.
Service
May still be found in Asia.

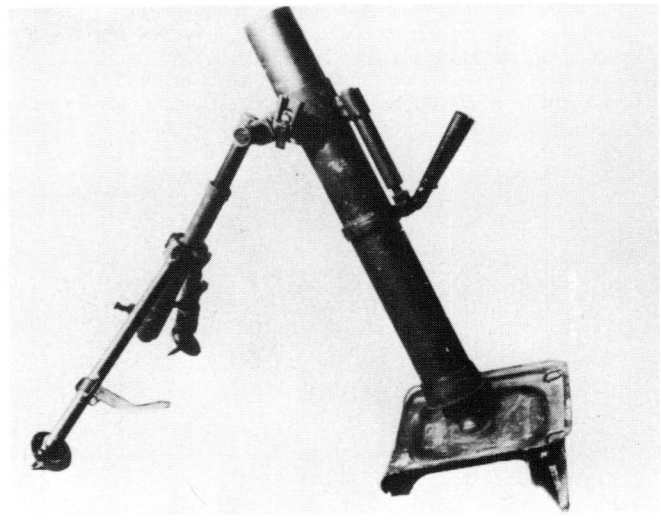

Chinese 60 mm Type 31 mortar

60 mm Type 63 mortar

The Type 63 light mortar is essentially an updated model of the earlier Type 31 mortar. It is lighter and has only one recoil cylinder, its base plate is smaller and rectangular in shape (on the Type 31 the baseplate is square), the bipod and mortar barrel are shorter and a carrying handle is provided on the barrel, just below the recoil system. It has the same operational range as the Type 31.

DATA
Calibre: 60 mm
Weight in firing position: 12.39 kg

Length of barrel: 610 mm
Rate of fire: 15–20 rds/min
Range: 1530 m
Ammunition: HE
Muzzle velocity: 158 m/s

Status
Current.
Service
China. Also Albania, Vietnam and some parts of Africa.

82 mm Type 53 mortar

This is a copy of the Soviet M1937 which is still the current mortar of this calibre in the USSR.

Status
Current.
Service
China. Also Pakistan, Tanzania, Uganda, Vietnam.

120 mm Type 53 mortar

This is a copy of the Soviet M1943.

Status
Obsolescent.
Service
China. Also in Pakistan, Tanzania and Vietnam.

160 mm mortar

This is a copy of the Soviet M1943.

Status
Current.
Service
China.

CZECHOSLOVAKIA

Czechoslovakian mortars

During and shortly after the Second World War the Skoda Works produced a series of interesting and unusual mortars including some of very considerable size. A 305 mm mortar with breech loading obturation was tried but work ceased at the end of the war. The B24 was a 120 mm mortar which, until recently at any rate, was still in use with some reserve formations but not with first-line troops. It appears to be similar to the Brandt mortar and was towed into action on a very light carriage. It fired a 13 kg bomb out to some 7 km, and in the firing position weighed approximately 250 kg.

In 1948 a short 120 mm mortar was produced. This was intended for airborne use and to equip troops engaged in fighting in the mountains or over other terrain where weight assumed major importance. The bipod is of the same shape as the B24 with the curved members joining at the back of the elevation gear case and pinched in together at their junction.

The 120 mm M1948 mortar is no longer in front-line service, although quantities are probably held in reserve. It fires a HE bomb weighing 16.3 kg to a maximum range of 5900 m. Ammunition is the same as that used for the 120 mm B24 mortar. Total weight of the M1948 is 317.5 kg and maximum rate of fire is stated as 6 rds/min.

Only two Czechoslovakian mortars are in current service with first-line troops and both of these are 81 mm calibre. Both are said to have performances much the same as the Soviet M37 and they use the same ammunition. These two are the M1948 and the M1952.

Short 120 mm M1948 mortar

Current 81 mm M1948 mortar

The mortar breaks down into three manpack loads for easy handling; barrel, baseplate and bipod. The M1948 fires a 3.63 kg HE bomb to a maximum range of 3700 m. Total weight of the mortar is 63.5 kg and maximum rate of fire is 10 to 12 rds/min.

B24 mortar on travelling carriage

DENMARK

Danish mortars

Three mortars are in service with Danish forces. One of them, the 60 mm M/51, is the US M2 mortar and has been supplied by the USA. The other two are made by the Vabenarsenalet in Denmark. The 81 mm M/57 is of indigenous design and the 120 mm M/50 is a licence-built copy of a Hotchkiss-Brandt design.

Manufacturer
Vabenarsenalet, Copenhagen (Abolished in 1970).

EGYPT

Egyptian mortars

Three mortars are known to be in use by the Egyptian Army. The 60 mm is a copy of the Chinese Type 63, weighing 12.3 kg and can reach a maximum range of 1530 m.

The 82 mm appears to be the Soviet M1937 barrel and bipod with a new baseplate based on a Thomson-Brandt design. The only other difference between this and the M1937 is a slight change in the positioning of the elevating handle. The mortar fires a 3.05 kg bomb to a range of 3045 m, weighs 45 kg in the firing position, and has a muzzle velocity of 205 m/s.

The 120 mm mortar appears to be a copy of the Soviet M1943. It weighs 282 kg in the firing position, 500 kg in the travelling mode on the usual two-wheeled trailer, and fires a 15 kg bomb at 272 m/s to a maximum range of 5520 m.

Manufacturer
Helwan Machine Tools Co, 23 Talaat Harb Street, PO Box 1582 Cairo.
Status
Current. Production.
Service
Egyptian Army.

FINLAND

60 mm TAM 18 Tampella mortar

This lightest Tampella mortar is intended to serve as a support weapon for troops, companies and guerrillas when fighting in wide forest areas. The equipment is so light that a patrol of three men can carry it along for several kilometres even in roadless conditions, both summer and winter. The effect of fire as well as the wide range and accuracy offer various different tactical applications.

DATA
Calibre: 60.75 mm
Barrel length: 800 mm
Weight in firing position: 18 kg
Weight of bomb: 1.8 kg
Max range: 4000 m
Rate of fire: 25 rds/min

Manufacturer
Oy Tampella Ab, PO Box 267, SF-33100 Tampere 10.
Status
Current. Available.
Service
Finnish Army.

60 mm TAM 18 Tampella mortar, showing firing table marked around barrel

60 mm TAM 18 Tampella mortar bomb

81 mm M71 Tampella mortar

This weapon has been developed by Tampella to meet the requirements of the Finnish Army. As can be seen from the accompanying picture it has a smooth, uniform barrel, a circular dished baseplate, a bipod mounting with a distancing chain, a screw-jack elevation mechanism, a cross-levelling device with coarse and fine adjustment, a sight unit mounted on the yoke and shock absorbers. The weapon is clearly of robust construction and is said by the manufacturer to be considerably stronger than other weapons of the same calibre.

DATA
Calibre: 81.4 mm
Weight in firing position: 55 kg
Weight of bomb: 4.2 kg
Max range: 5000 m
Rate of fire: 20 rds/min

Manufacturer
Oy Tampella Ab, PO Box 267, SF-33100 Tampere 10.

81 mm M71 Tampella mortar

81 mm TAM 4.2 Tampella mortar bomb

120 mm M73 Tampella mortar

Like the 81 mm M71 weapon (which see) this mortar has been developed by Tampella to meet Finnish Army requirements. Inspection of the accompanying picture shows that its design is generally similar to that of the 81 mm mortar. It is of robust construction and special design attention has been paid to accuracy.

DATA
Calibre: 120.25 mm
Weight in firing position: 236 kg
Weight of bomb: 12.8 kg
Max range: 8000 m
Rate of fire: 15 rds/min

Manufacturer
Oy Tampella Ab, PO Box 267, SF-33100 Tampere 10.

120 mm TAM 12.8 Tampella mortar bomb

120 mm M73 Tampella mortar

160 mm M58 Tampella mortar

160 mm mortars have been developed only in the USSR and Finland. The Soviet mortars are both large and heavy and have to be loaded from the breech. This is both expensive and complicated since it requires balancing springs to allow the barrel to lower. In the Tampella mortar the traditional muzzle-loading has been retained. This does not affect the rate of fire since the time required to lift a bomb to the muzzle is much the same as that needed to open a breech and close it again.

The Tampella mortar makes use of a new series of HE bombs, which considerably extend the range and the terminal effectiveness. The use of a proximity fuze improves this effectiveness by up to three times, and is therefore much to be recommended.

The Finnish Army has not yet accepted the 160 mm mortar, but it is expected to do so in the near future.

DATA
Calibre: 160.4 mm
Weight in firing position: 1450 kg
Weight of HE bomb: 40 kg
Max range: 10 000 m
Rate of fire: 4–8 rds/min

Manufacturer
Oy Tampella Ab, PO Box 267, SF-33100 Tampere 10.
Status
Current. Available.
Service
Under consideration by the Finnish Army.

160 mm M58 Tampella mortar in firing position. Two-wheeled carriage assists in traversing weapon

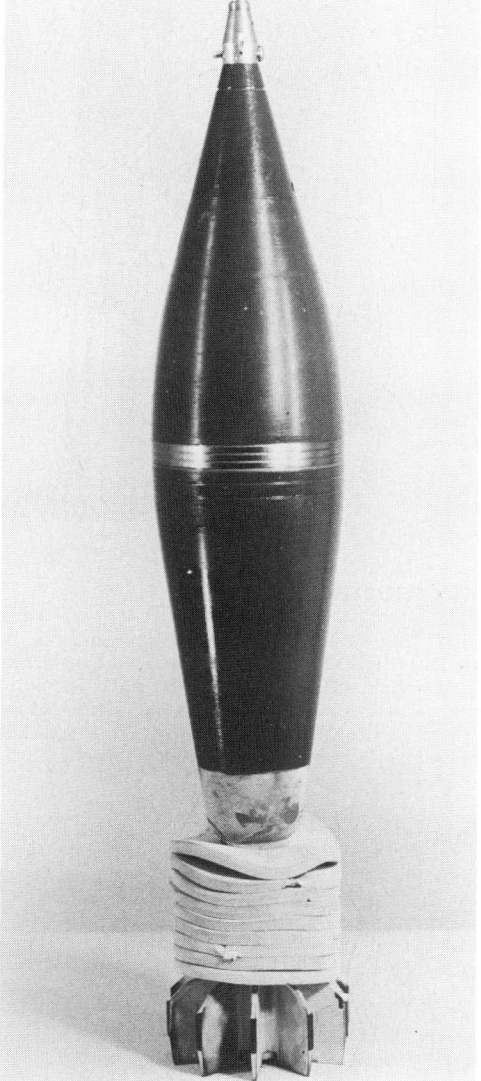

160 mm M58 HE Tampella mortar bomb

FRANCE

60 mm Commando mortar

This is the lightest and most portable of the Brandt mortars. It is produced in two variants, types V and A. Both have small spades and do not use a mounting. They are controlled for both line and elevation directly by the firer.

The type V has a short barrel and a fixed firing pin. As soon as the muzzle-loaded bomb reaches the firing pin it is automatically fired. The sight is a simple clamp-on type with a bubble fitted on the right-hand side. The firer sets the range on a drum and when the bubble is levelled the sight is elevated for the correct tangent elevation for the selected range. A white line is scribed along the front of the barrel and the azimuth is applied by lining up this mark with the target. A broad webbing strap is around the barrel to provide a grip for the firer's left hand which would otherwise find it difficult to control the hot barrel. A webbing sling is provided for carriage of the mortar and has a number of brass bars rivetted to the fabric. These are graduated in ranges according to the firing charge involved – from left to right, Charge 0, Charge 1, Charge 2. If the gunner places his foot on the sling on the selected range bar and then pulls the sling tight by pulling on the barrel, the mortar will be at the correct elevation, provided the ground is horizontal. A muzzle cover is supplied to keep out the dirt, sand and so on which greatly increase barrel wear. The spade is small and button shaped.

The mortar may be fired at a rate of 12 to 20 rds/min. There is a two charge

system providing a maximum range of 1050 m and a minimum range of 100 m. The overall length of the mortar is 680 mm and the barrel itself is 650 mm. The total weight is 7.7 kg.

The type A is longer and a breech piece below the barrel holds the firing mechanism and is connected to a small trough-shaped spade. The mortar is controlled by a firing lever which has a trip-over action. The bomb is dropped in from the muzzle and rests on the firing pin plug. The user pulls back on the lanyard attached to the firing lever and the firing pin is retracted, the spring is compressed and then the pin trips over and is driven forward to fire the round. There is no sight fitted and the soldier lying behind the mortar lines up the barrel to control his fire for azimuth and estimates the correct quadrant elevation for his first round.

The type A mortar has an overall length of 861 mm but the barrel length is the same as that of the type V mortar. The weight is 10 kg.

Ammunition

The Commando mortars have a range of 60 mm ammunition. Primary cartridges for all bombs are 24 mm in diameter and 65 mm long, containing 4.2 g of ballistite. Secondary charges are generally of the horseshoe-type, although earlier bombs used secondary charges fitted between the fins. The supercharge is horseshoe-shaped and contains 5 g of ballistite.

Manufacturer

Thomson-Brandt Armements, Tour Chenonceaux, 204 Rond-Point du Pont de Sevres, F-92516 Boulogne-Billancourt.

Status

Current. Production.

Service

The armed forces of 20 countries. Most use the type A.

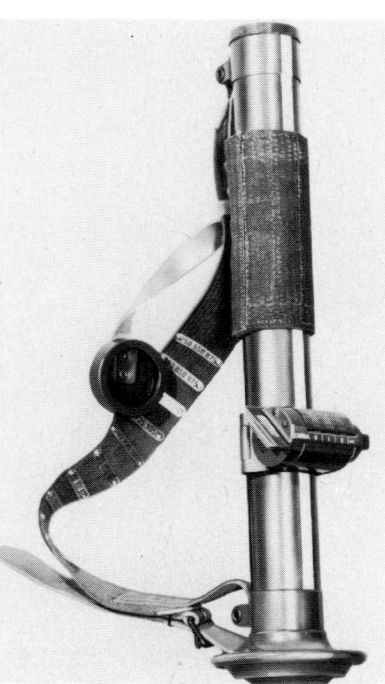

60 mm type V Commando mortar. This has fixed firing pin and fires bomb as soon as it drops. Note button-shaped spade

Firing 60 mm type A Commando mortar. Note boot-and-sling method of adjusting elevation

60 mm proximity mortar

This weapon derives its name from its intended use – in close proximity to the enemy by forward troops. Due to its lightness and handiness it can be carried easily and thus gives the advanced infantry squad accurate and efficient fire support. It can be carried in a shoulder bag by one man, and is adjusted for elevation by sliding the bipod sleeve up and down the barrel; the curved ends of the bipod permit this movement to be made easily on any nature of ground. The baseplate permits all-round firing.

DATA
Total weight: 6.0 kg
Overall length: 860 mm
Elevation: 45° – 82°
Range: (HE, smoke, practice) 110–650 m; (illuminating) 350–900 m

All standard 60 mm bombs can be fired from this mortar.

Manufacturer

Thomson-Brandt Armements, Tour Chenonceaux, 204 Rond-Point du Pont de Sevres, F-92516 Boulogne-Billancourt.

Status

Current. Production.

Service

Swiss Army.

Thomson-Brandt 60 mm proximity mortar

60 mm light mortar

This mortar was produced as a private venture by Hotchkiss-Brandt which started work on it in 1963. It has proved to be a very successful light mortar and is still in production.

The weapon is of orthodox configuration with a baseplate, barrel and bipod mounting. The barrel is smoothbored and made of nickel chrome steel. The base of the barrel is threaded to take the breech piece. The breech piece is made of steel and is screwed into the barrel. It is pinned in place and can be removed only by an armourer. The striker is dome headed and screws into the breech piece from the rear. It protrudes into the base of the chamber and projects about 1.5mm. The bombs are dropped on the striker.

The baseplate is an equilateral triangle reinforced by three webs forming a star round the central cone. On the top surface there is a socket in which the breech piece fits.

The barrel is locked into position by rotating it in the socket. The baseplate allows all-round traverse and the webs give good stability in very wet soil. A lifting handle is fitted. The bipod supports the barrel. It has two legs made of aluminium alloy hinged to a cross piece at the top. The legs can be moved out or in and the width between the feet is controlled by an adjustable chain. Passing upwards from the centre of the cross piece, in a tube connecting the two bipod legs, is the elevating screw. This is attached to the yoke. The tube containing the elevating screw is connected to the left bipod leg by a rod. This rod can be slid up and down the tube and in so doing the tube is tilted. This provides the cross-levelling.

The yoke carries the traversing gear and at the end of it is the traversing handle. On the top of the yoke is the cradle into which the barrel fits, and at the other end to the traversing handle is the sight bracket. There is a shock-absorber which connects the yoke to the cradle and provides some isolation from the vibrations of firing.

The sight in use on the 60 mm light mortar is the F9. This has an elevating scale graduated from 40 to 60° at 1° intervals. The deflection micrometer is graduated from 0 to 8° 30 on the right and from 360 to 351° 30 on the left. Other graduation systems are available.

The mortar can be brought into action very quickly even on an unprepared platform. The baseplate should be level. If possible the first round should be fired at an angle greater than 60° and serves to bed the plate in.

If the ground is soft the baseplate will sink in but using the barrel as a lever it can be withdrawn speedily. The detachment can move on foot carrying the mortar over considerable distances.

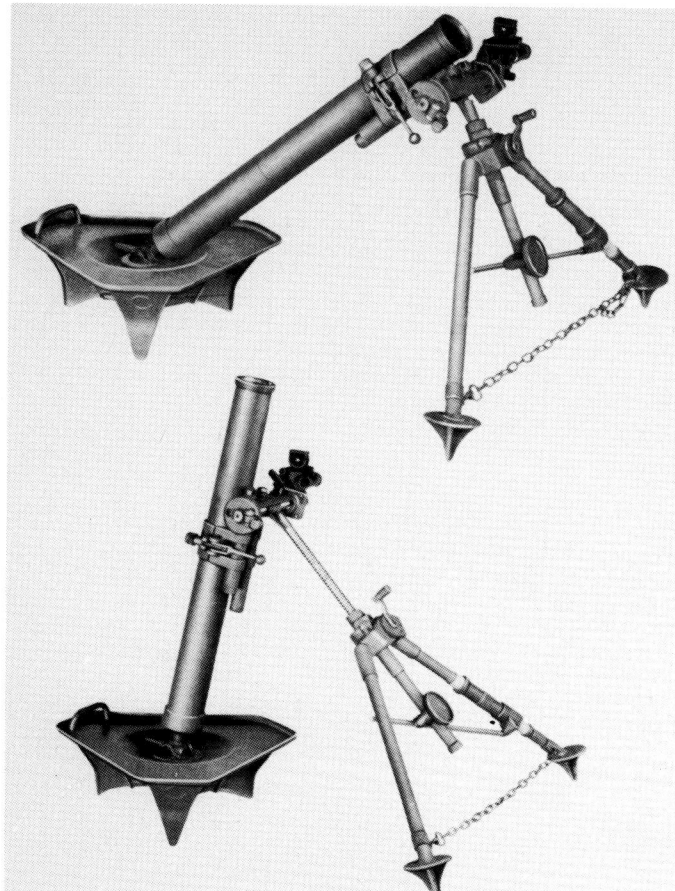

60 mm light mortar showing the maximum and minimum angle of depression

Sight of the 60 mm mortar
(1) *bearing scale* **(2)** *sighting tube* **(3)** *cross-level bubble* **(4)** *sight catch* **(5)** *elevation scale* **(6)** *elevation fine scale* **(7)** *elevation bubble*

DATA
Calibre: 60 mm
Weight: 14.8 kg
Length of barrel: (with breech piece) 724 mm
Weight of barrel: 3.8 kg
Weight of baseplate: 6 kg
Weight of bipod: 5 kg
Top traverse: 300 mils
Elevation: 40–85°
Range: (minimum) 100 m; (max) 2050 m

Mark 61 ammunition range showing shape and position of secondary charges

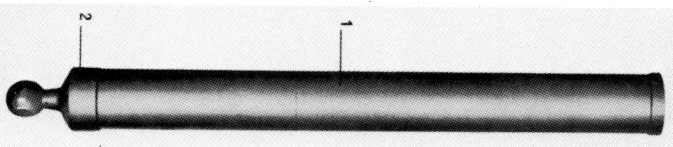

Barrel of 60 mm light mortar. Breech piece (2) is pinned in place and carries striker

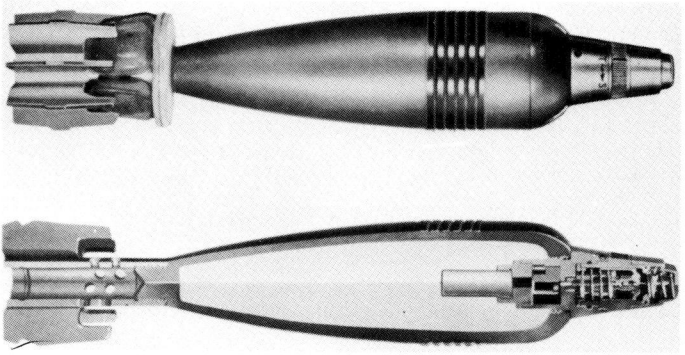

Mark 60/61 bomb with V9 fuze

60 mm light mortar in action

Ammunition

The 60 mm light mortar was designed for the Brandt 60 mm Mark 61 bomb (which see).

The mortar will also fire the US 60 mm bomb (old type), the old Brandt Mark 35/47 (no longer manufactured by Thomson-Brandt), the M61 coloured marker, smoke and practice bombs and the M63 illuminating bomb described in the entry for the 60 mm Commando mortar (which see).

Manufacturer

Thomson-Brandt Armements, Tour Chenonceaux, 204 Rond-Point du Pont de Sevres, F-92516 Boulogne-Billancourt.

Status

Current. Production.

Service

French and several other armies.

60 mm long-range mortar

The Thomson-Brandt long-range mortar has been developed to combine the flexibility of the 60 mm light mortar and the firepower of the 81 mm mortar. For ease of transport it can be broken down into three components: barrel, baseplate and bipod.

It can fire the standard Thomson-Brandt M61 and M72 ranges of HE, smoke, practice and illuminating bombs. In addition a new long-range projectile called the LR has been developed for this mortar. The LR is of malleable pearlitic cast iron and has a fuze which detonates under any angle of impact, including grazing fire. The LR bomb weighs 2.2 kg and has a total length of 381 mm. It has a TNT charge which gives the bomb comparable efficiency to the 81 mm mortar bomb.

DATA
Calibre: 60 mm

WEIGHTS
Total: 23 kg
Barrel: 8.4 kg
Baseplate: 8.8 kg
Bipod: 5.8 kg

Length of barrel: (including breech) 1350 mm
Traverse: 200 mils
Elevation: 40–85°
Range: (max) 5000 m

Manufacturer

Thomson-Brandt Armements, Tour Chenonceaux, 204 Rond-Point du Pont de Sevres, F-92516 Boulogne-Billancourt.

Status

Pre-production.

60 mm Thomson-Brandt long-range light mortar

60 mm Type LR long-range bomb

60 mm MCB gun-mortar

Development of the 60 mm MCB began in 1970 with the aim of producing a mortar for use in an armoured personnel carrier. It has been fitted to several vehicles, for example the Panhard M3 series of wheeled armoured vehicles.

The weapon can be muzzle-loaded from outside the vehicle. The firer drops the bombs down the smoothbore tube. When this procedure is followed the mortar can be regarded as conventional and the bomb drops on to a fixed firing pin. It then can be fired in the upper register (45 to 90°) at a rate of 12 to 20 rds/min.

However, the mortar has an opening-breech which is locked by a falling block much like an artillery piece. It can therefore be loaded and fired from under cover and used in the same way as any other conventional vehicle gun. There are different ways of opening the breech to fit in with the available space in different vehicles. One is by using a lever-breech mechanism, or alternatively a simple hand bar may be positioned across the rear of the block and then rotated back and down to operate the breech.

As soon as the breech-block is unlocked the firing pin is withdrawn and will not emerge again until the block is locked. This ensures that the mortar cannot be loaded from the muzzle and fired while the breech-block is unlocked.

When the firing is controlled from within the vehicle, either electrical or mechanical firing can be provided through a trigger lever.

Since the recoil force must be taken on the trunnions, a hydraulic buffer is provided. This allows a recoil of 135 mm. The maximum recoil force is kept

Thomson-Brandt gun-mortar being used as light support weapon

60 mm MCB gun-mortar

Installation of 60 mm gun-mortar in Thomson-Brandt Armements turret. This turret can also accommodate 60 mm long-range gun-mortar described below

down to 1700 kg. The recoiling mass is 42 kg and this of course makes a major contribution to keeping down the velocity of recoil.

Ammunition

The gun-mortar will fire the old Mark 35/47, the standard M61 and the M72 bomb.

The M63 illuminating bomb is used for target illumination, and the M61 smoke bomb for the provision of immediate cover.

The bombs used in this mortar have the V9 fuze (except the illuminating bomb which has a clockwork time fuze) which has a safe and fire setting. In addition a modified V9 fuze can be provided to give arming closer to the muzzle. This provides functioning at 20 m and will also give graze functioning for horizontal firing.

DATA

Length of weapon: 1210 mm
Bomb travel: 905 mm
Elevation: –11 to +75°
Max range: (Mark 72 bomb) 2600 m; (Mark 61 bomb) 2050 m
Flat trajectory: 500 m

Manufacturer

Thomson-Brandt Armements, Tour Chenonceaux, 204 Rond-Point du Pont de Sevres, F-92516 Boulogne-Billancourt.
Status
Current. Production.
Service
Export sales, probably to African states.

60 mm Model LR gun-mortar

This mortar can be turret mounted in a variety of vehicles such as the Panhard M3 armoured personnel carrier, in addition it can also be mounted in riverine or coastal patrol craft. It can be breech- or muzzle-loaded. It has a lever-breech mechanism with an automatic withdrawal of the firing pin during breech opening. It incorporates a safety mechanism which ensures that the firing pin is clear from the front face as long as the breech is not locked.

It can be fired either electrically or mechanically through a trigger mechanism or by gravity when muzzle loaded.

It can fire the LR (long-range) bomb, standard M61, M72 and Mark 35/47 bombs and the M63 illuminating bomb.

DATA

Calibre: 60 mm
Total length of barrel: 1800 mm
Total length of bomb travel: 1500 mm
Length of recoil: 170 mm
Max recoil thrust: 2800 kg
Weight of recoiling mass: 75 kg
Elevation: –11 to +75°
Max ranges as a mortar: (LR bomb) 5000 m; (Mark 72 bomb) 3000 m; (Mark 61 bomb) 2300 m
Max ranges with flat trajectory: (LR bomb) 500 m; (Mark 72 bomb) 400 m

Manufacturer

Thomson-Brandt Armements, Tour Chenonceaux, 204 Rond-Point du Pont de Sevres, F-92516 Boulogne-Billancourt.
Status
Production.

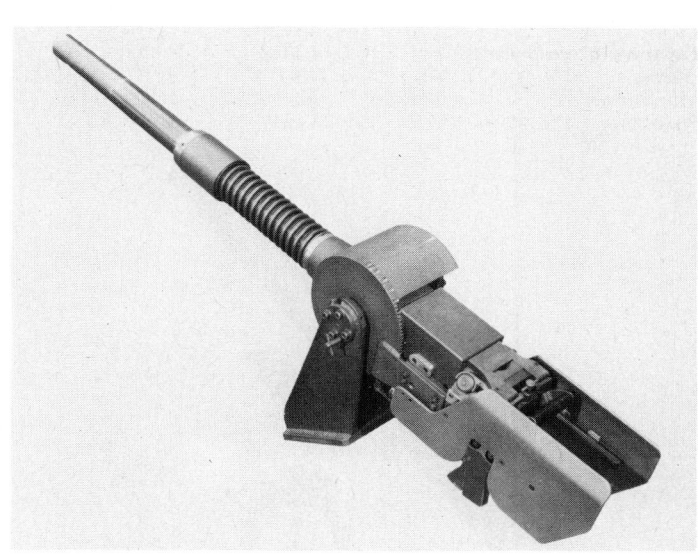

60 mm Model LR gun-mortar

81 mm light mortar

This mortar was designed in 1961 and has been in service for several years. It exists in two versions. These are known as MO 81-61 C and MO 81-61 L. The two versions have barrels of 1150 and 1450 mm respectively. The mortar is typical of its type and consists of barrel, mounting and base plate. The barrel is made of steel, reinforced at the rear end. At the rear end is screwed on the breech piece which holds the firing lock. This allows the user to retract the firing pin when required in the safe position or allow it to protrude in the firing position. The end of the breech piece contains the ball which fits into the socket on the baseplate. It has two flats which allow for the insertion of the knob and subsequently its rotation.

The mounting consists of the bipod, elevating gear, traversing gear, cross-levelling gear and the barrel clamp. The bipod legs can be splayed at will and the distance between them is controlled by the chain connecting the feet. The elevating screw thread is enclosed in a tube. The tube is connected by a rod to the left bipod leg. By sliding the tube along the rod the elevating screw column is tilted and so the mortar is cross-levelled. The traversing gear is enclosed in a tube. The traversing handle is at one end of the tube and the sight mounting is at the other.

The baseplate is made of chrome molybdenum steel. It has, on the top surface, a socket in which the knob of the breech piece fits. This socket is movable to allow the equipment to rotate around the baseplate. The underside of the baseplate has three ribs which radiate outwards from under the socket and provide the strength to prevent the baseplate from buckling as the impulsive loading is applied.

The mortar may be carried in any suitable vehicle. It can also be broken down into mule loads or even carried by the detachment in framed rucksacks.

The sustained rate of fire for the mortar is 12 to 15 rds/min.

In addition to the standard HE bombs, M57D and M61, the following variants are available in both series: coloured, HE, target marker bomb: (green, yellow or red) filled with TNT and colouring material. Smoke bomb: filled with liquid titanium tetrachloride, also white phosphorus. Practice bomb: filled with sulphur/naphthalene mixture and with dummy or live fuze.

V19 P fuze

The V19 fuze may be set for instantaneous action or to function with a delay. The mechanism is arranged so that there is complete safety in the bore. The fuze cannot detonate the main filling until 0.8 seconds after firing. This means that on charge 0 the bomb will have travelled 55 m and at charge 6 it will have travelled 200 m before the main filling can be detonated. In the event of double loading the bombs have been proved safe.

DATA

	Short barrel	Long barrel
Total weight	39.4 kg	41.5 kg
Length of barrel	1.15 m	1.45 m
Weight of barrel	12.4 kg	14.5 kg
Weight of baseplate	14.8 kg	14.8 kg
Weight of mount	12.2 kg	12.2 kg

Ammunition

HE BOMB

Body	M57D	M61	M82
	steel	malleable pearlitic cast iron	malleable pearlitic cast iron
Total weight of round	3.3 kg	4.20 kg	4.45 kg
Filling	TNT	TNT	TNT
Max range	4550 m	5000 m	5600 m
Fuze	V19 Pa	V19 P	V19 P

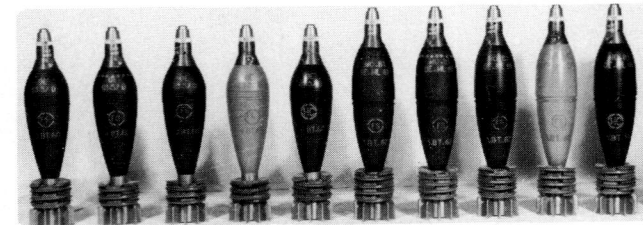

HE bombs, smoke and illuminating. Larger bombs on right are M61

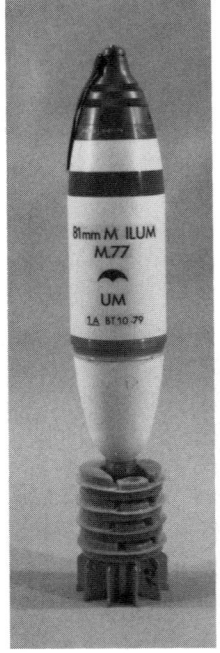

81 mm M77 illuminating bomb

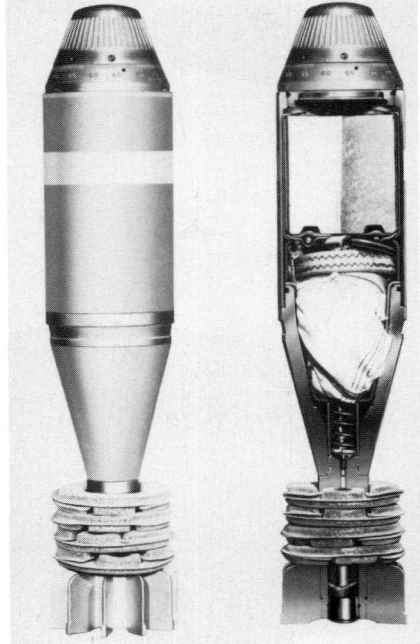

81 mm M68 illuminating bomb

81 mm Thomson-Brandt light mortar

81 mm Thomson-Brandt light mortar with short barrel

COLOURED HE BOMB
Filling   TNT and colouring (green, yellow or red), same general characteristics as for HE bomb

SMOKE BOMB
Filling   Liquid titanium tetrachloride or white phosphorus, same general characteristics as for HE bomb

PRACTICE BOMB
Filling   Mixture of sulphur and naphthalene provided with dummy or live fuze

ILLUMINATING BOMB

	M68	M77
Overall length	417 mm	478 mm
Total weight of round	3.5 kg	4 kg
Filling	magnesium-base compound	
Light intensity	600 000 candela	800 000 candela
Minimum burning time	35 s	40 s
Illuminating radius	250 m	300 m
Fuze	clockwork time	

Manufacturer
Thomson-Brandt Armements, Tour Chenonceaux, 204 Rond-Point du Pont de Sevres, F-92516 Boulogne-Billancourt.
Status
Current. Production.
Service
French Army (short barrel) also several other armies.

81 mm MCB 81 mortar-cannon

Similar in principle to the two 60 mm gun-mortars (which see) made by the same company, the Thomson-Brandt 81 mm weapon is designed for installation in armoured vehicles, riverine craft or small patrol boats for close support missions. It can be mounted in a turret or on a ring mounting. It was originally designated 'Canon Lisse de 81 mm' by Brandt, indicating that it was considered more as a smoothbore gun than a mortar. If mounted in a turret an automatic cross-levelling device can be incorporated in the installation to ensure that the trunnion mounting is horizontal.

Like the 60 mm weapons, the 81 mm mortar-cannon can be loaded from either the breech or the muzzle and will fire all 81 mm Thomson-Brandt projectiles including a long-range projectile with a maximum range of 8000 m, and an armour-piercing shot with a range of 1000 m, the SP 81 APFSDS cartridge.

DATA
Calibre: 81 mm
Weight: 500 kg
Recoiling mass: 400 kg
Length: 2300 mm
Elevation limits: −10° to +70°
Breech: sliding

Initiation: electrical or manual
Recoil system: mechanical
Max range: (mortar with LR bomb) 8000 m; (gun with AP shell) 1000 m

SP 81 (APFSDS)
Muzzle velocity: 1000 m/s
Armour piercing capacity at 1000 m: 90 mm with a nil angle of incidence; 30 mm with a 60° angle of incidence

Manufacturer
Thomson-Brandt Armements, Tour Chenonceaux, 204 Rond-Point du Pont de Sevres, F-92516 Boulogne-Billancourt.
Status
Production.
Service
Export sales.

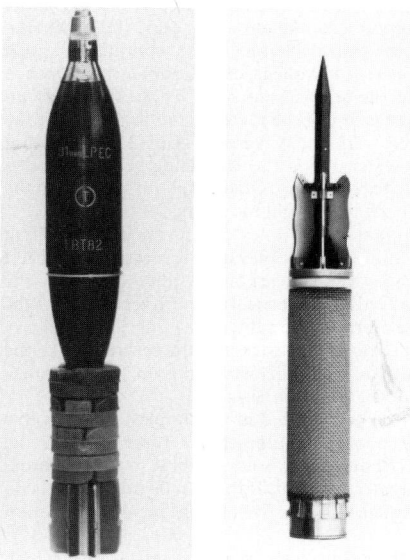

81 mm mortar-cannon on AMX10-PAC

Long-range HE bomb and SP 81 APFSDS cartridge

81 mm mortar-cannon with its circular mounting plate in position

81 mm long-range mortar

This has been designed in order to combine the flexibility of employment of the light 81 mm mortar with the range and efficiency of the 81 mm mortar-cannon CL 81 and to approach the capabilities of the 120 mm mortar.

It is of conventional design and can be broken down into three loads, barrel, baseplate and bipod – for man-carriage; it can also be carried by any vehicle, dropped by parachute, or transported by helicopter. It fires the 81 mm long-range bomb to a range of 7600 m or the 81 mm M82 bomb to a range of 5800 m.

DATA
Barrel length: 1700 mm
Weights
barrel: 35 kg
baseplate: 36 kg
bipod: 14.3 kg
sight: 700 g
Total: 86 kg in action
Top traverse: 520 mils
Elevation: 533–1511 mils (30–85°)

Manufacturer
Thomson-Brandt Armements, Tour Chenonceaux, 204 Rond-Point du Pont de Sevres, F-92516 Boulogne-Billancourt.
Status
Current. Production.

81 mm long-range mortar

120 mm MO-120-60 light mortar

The 120 mm light mortar is a simple mortar in which mobility and firepower have been allied. The mortar is made up of the barrel, mounting and base-plate and it fires a variety of ammunition providing all-round fire from a minimum range of 600 m out to 6610 m.

The barrel is smoothbore and has a breech piece screwed over the end. The firing mechanism is enclosed in the breech piece and the firing pin can be set to protrude, thus firing a bomb dropped on it under gravity, or alternatively the pin can be retracted into a safe position. The bottom of the breech piece has a spherical ball, with two flats, which enables it to be secured in the baseplate. The mounting consists of the bipod legs, the elevating gear, the traversing gear, cross-levelling gear and the barrel clamp, which contains a shock absorbing buffer. The bipod consists of two legs, each terminating in a shoe with a spike, and the spread of the legs is controlled by the length of an adjustable chain. The elevating gear in its tube is connected to the top of the bipod and reaches up to connect with the traversing gear. The traversing mechanism consists of a yoke which holds the traversing screw and also connects the elevating gear to the barrel clamp. On the right of the yoke is the traversing handwheel and on the left is the sight mounting.

To enable the sight to be upright at all times, there is a cross-levelling mechanism made up of the rod connecting the elevating screw tube to the left bipod leg. This can be moved across by a control on the bipod leg and so the elevating screw column can be moved to the vertical position regardless of the slope of the ground on which the mortar is standing.

The baseplate is triangular in shape with a socket in the centre to take the breech piece ball and underneath are three ribs which provide the complete rigidity required and also prevent the baseplate slipping.

The mortar has been designed to be operated in the simplest way possible and can be brought into action and maintained by three men in the detachment. The baseplate must be dug in if it is not possible to fire the initial round at an angle of elevation greater than 60°. If this can be done there is no need for a bedding in round but there will be a five to eight per cent reduction in range in this first round.

In the event of a misfire, the firing mechanism can be set to 'safe' which withdraws the firing pin and then the bomb extractor can be used to remove the bomb from the tube.

The 120 mm light mortar can be moved by vehicle, mule, manpack or by air.

DATA
WEIGHT
Barrel: 34 kg
Bipod: 24 kg
Baseplate: 36 kg
Total weight of equipment: 94 kg
Length of barrel with breech piece: 1632 mm
Elevation: 40–85° (711–1511 mils)
Traverse: 360° (6400 mils)
Rate of fire: normal 8 rds/min for 3 mins; 15 rds/min for 1 min

Manufacturer
Thomson-Brandt Armements, Tour Chenonceaux, 204 Rond-Point du Pont de Sevres, F-92516 Boulogne-Billancourt.
Status
Current. Production.

120 mm Thomson-Brandt light mortar

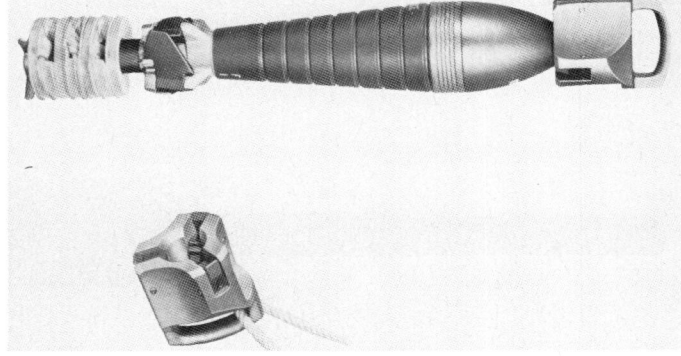

Bomb extractor used to remove bomb from barrel in event of misfire. It is lowered from muzzle and grips fuze

Service
French Army and several others.

120 mm MO-120-M65 strengthened light mortar

This 120 mm mortar uses the baseplate and mount of the 120 mm light mortar but the barrel is reinforced and weighs some 10 kg more than that of the light mortar. There is also a difference in the method of moving this mortar. A trolley with pneumatic tyres and elastic suspension is used on which the entire mortar rests. A muzzle cap with a towing eye is used to enable the mortar to be pulled by a light vehicle of the jeep type. Alternatively if the necessity arises, the mortar can be dragged along by the detachment, using the towing handle which is also part of the muzzle cap. When the mortar is brought into action, the baseplate is lowered on to the required position and the bipod is used to connect the barrel to the baseplate. The mortar cannot be fired from the wheels.

The mechanism of the mortar allows conventional gravity firing but there is a spring-controlled firing pin which allows the bomb to be loaded into position and fired by means of a lanyard attached to the firing lever. The firing mechanism also allows the firing pin to be withdrawn as a safety measure. The mortar has an all round field of fire without need to move the baseplate and with the barrel clamp giving an angle of elevation of 60°, there is a top traverse of 300 mils.

120 mm MO-120-M65 strengthened light mortar

Ammunition
The mortar can use the following bombs: HE bomb 120 M44/66 AT; illuminating bomb Mark 62 ED; smoke, practice, and marker bombs of the M44 series; PEPA bomb; PEPA/LP bomb.

DATA
MORTAR
Total weight: (equipment mounted on carriage) 144 kg; (in firing position) 104 kg
Barrel and breech piece: 44 kg
Mount: 24 kg
Baseplate: 36 kg
Trolley and muzzle cap: 40 kg
Length of barrel and breech piece: 1640 mm
Rate of fire: 8 rds/min for 3 mins; 12 rds/min for 1 min

AMMUNITION
Total length with tail assembly: 758 mm
Total weight of complete round: 13.42 kg
Filling: RDX/TNT
Minimum range: 500 m
Max range using rocket assistance: 9000 m

Manufacturer
Thomson-Brandt Armements, Tour Chenonceaux, 204 Rond-Point du Pont de Sevres, F-92516 Boulogne-Billancourt.
Status
No longer manufactured. Replaced in production by MO-120-LT (see entry below).
Service
Unspecified armies.

Mortar in motion

PEPA/LP sectioned

PEPA/LP bomb

120 mm MO-120-LT Brandt mortar

This mortar was introduced to replace the MO-120-AM 50. It has a massive baseplate and is transported on a wheeled carriage. The carriage is substantial, with pneumatic tyres and torsion bar suspension, but the mortar is fired off the bipod mount. It is claimed that the mortar on its carriage can be pulled by two men but this can only be over short distances on favourable terrain.

Ammunition
This mortar fires all the ammunition used by the AM-50: PEPA/LP long-range rocket-assisted bomb; HE bomb M44/85 (range 7000 m); illuminating bomb M62-ED; smoke, practice and marker bombs corresponding to the above.

DATA
WEIGHTS
Total weight: (equipment with the wheeled carriage) 247 kg; (in firing position) 167 kg
Barrel: 62 kg

120 mm MO-120-LT Thomson-Brandt mortar in action

120 mm MO-120-LT Thomson-Brandt mortar in travelling mode

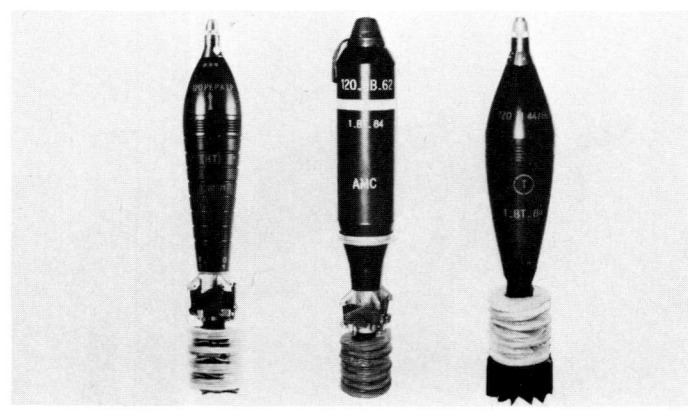

Rocket-assisted, illuminating and HE bombs for Thomson-Brandt MO-120-LT mortar

Mount: 25 kg
Baseplate: 80 kg
Travelling carriage: 80 kg including towing attachment
Length of barrel and breech piece: 1.7 m
Elevation: 45–85° (711–1511 mils)
Top traverse at 60° elevation: 17° (300 mils)
Firing: gravity or controlled firing; pin can be withdrawn for safety
Rate of fire: (normal) 8 rds/min; (max) 20 rds/min

Manufacturer
Thomson-Brandt Armements, Tour Chenonceaux, 204 Rond-Point du Pont de Sevres, F-92516 Boulogne-Billancourt.
Status
Current. Production.
Service
Uncertain.

120 mm MO-120-RT-61 rifled mortar

The 120 mm rifled mortar is probably the most complex of modern mortars and in some aspects approaches very closely to the gun. It can be fired only off its wheels and can be deployed only in areas to which the towing vehicle has access. It is a massive piece of equipment which fires a heavy bomb out to 13 km. It has a rifled barrel and is muzzle loading. To cope with the windage problem it has a pre-engraved driving band.

Main components are the barrel, cradle and undercarriage and the base-plate. The barrel is a substantial forging equipped at the muzzle to take the towing eye by which it is attached to the vehicle. The outside is radially finned to increase the surface area for heat dissipation. The interior is rifled and with the pre-engraved driving band imparts rotation to the shell to produce stability throughout the trajectory.

The cradle consists of a steel tube connecting the two wheels and carrying the torsion-bar suspension. The traversing gears are totally enclosed to exclude foreign matter. The elevating handwheel rotates a worm and gear assembly which transmits motion, through a multiple disc clutch, to a pinion meshing with the rack formed by the barrel finning. The collar sliding along the barrel produces the necessary change of elevation. The cross-levelling shaft, actuated by the upper left handwheel, tilts the traversing assembly, together with the collar to which the sight is attached.

The baseplate is very heavy with massive webs on the underside. After a prolonged period of firing the baseplate can be extricated by using the barrel as a lever and employing the towing vehicle to pull the baseplate up.

DATA
Total weight of the equipment: 582 kg

BARREL
Length of barrel: 2080 mm
Weight of barrel: 114 kg
Weight of towing eye: 17 kg

MOUNT
Weight of mount: 257 kg
Elevation: 30°–85°
Top traverse at 60° elevation: 250 mils

BASEPLATE
Weight of baseplate: 190 kg
Clamping collar: 4 kg

DIMENSIONS OF THE EQUIPMENT
Wheel base: 1.73 m
Overall width: 1.93 m
Overall length: 3.01 m
Height in running position: 1.33 m
Ground clearance: 0.32 m
Wheel diameter: 0.7 m
Time: (into action) 1.5 mins; (out of action) 2 mins
Rate of fire: (normal) 10–12 rds/min; (max) 15–20 rds/min – for a very limited period

Ammunition
Although this is a rifled mortar, it will fire smoothbore bombs except those types having spring-loaded tail fin assemblies with straight fins. Smoothbore bombs are frequently used for bedding in the baseplate (1 round charge 3, 1

120 mm MO-120-RT-61 rifled mortar

120 mm MO-120-RT-61 rifled mortar in action

charge 5, 1 charge 7) and also for cheaper training. Bombs for the MO-120-RT-61 are equipped with a tail tube carrying the primary and secondary cartridges. This tube is ejected just after the bomb has left the mortar and falls about 100 m from the muzzle. An anti-armour bomb has also been developed and is described in the Mortar Ammunition section.

Manufacturer
Thomson-Brandt Armements, Tour Chenonceaux, 204 Rond-Point du Pont de Sevres, F-92516 Boulogne-Billancourt.
Status
Current. Production.
Service
French Army. Under evaluation by Japanese Army. Also exported to several countries.

Lacroix grapnel launcher

This is a self-contained unit which can be rapidly emplaced and used to fire a grapnel either vertically (for scaling cliff faces) or horizontally (river crossings, removing wire, wreck rescue). The size of the grapnel and the weight and length of cable can be varied according to the task in hand. The capabilities of the system are inversely proportional to the length of cable deployed. Due to its low weight it can be easily carried by commandos or raiding parties.

The grapnel is prepared for launching by first driving in the base at a suitable angle. The rocket unit is then attached to the grapnel and the grapnel to the cable. The rocket is then ignited electrically from a distance and tows the cable to its destination.

The launcher unit, ready for operation, weighs 1.7 kg and the grapnel 1 kg. It has a horizontal range of 250 to 350 m depending on the diameter and weight of cable used.

The unit can also be used to carry a scrambling net weighing 50 kg; in this case the maximum range is no more than 50 m.

Manufacturer
Ste E Lacroix, BP 213, 31601 Muret.
Status
Current. Production.
Service
French Army.

Lacroix grapnel launcher deployed for firing

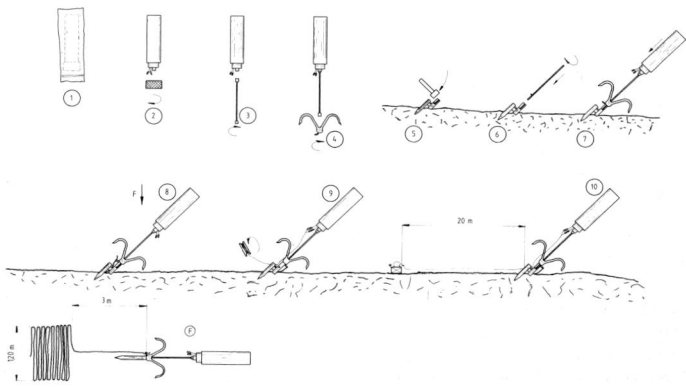

Sequence of operation of Lacroix grapnel launcher

GERMANY, WEST

Bundeswehr mortars

The German Army used the mortar with considerable success during both world wars; the British Army regarded the German mortars, and the way they were used, with great respect. Surprisingly, however, the Germans looked to the Soviets as the masters in this field and they copied the Soviet 120 mm mortar and also shared the same design of 82 mm mortar.

After West Germany came into NATO they again adopted the mortar. They did not accept any kind of light mortar (50 to 60 mm) or a medium mortar (81 mm) but purchased, from France and from Israel, 120 mm mortars made, respectively, by Hotchkiss-Brandt and Soltam. The Israeli mortar is made to a design produced by the Finnish firm of Tampella (which see). The current mortars of the Bundeswehr are illustrated below. Complete details are given in the entries on the originating countries.

Panzermörser M113: this US armoured personnel carrier has been adapted by the Bundeswehr to carry the 120 mm Tampella mortar. This mortar is intended to be used, as far as possible, directly from the vehicle. It can, if necessary, be deployed dismounted from the armoured carrier. It has a range when mounted in the vehicle of 450 to 6350 m. It is used by armoured, mountain and light infantry, armoured reconnaissance units and tank regiments.

120 mm Tampella mortar: can be broken down to mule loads for deployment in mountainous terrain. The barrel and bipod make up one mule load and the heavy baseplate together with the sight and two rounds make up another.

When the mortar is deployed in mountainous country a special baseplate is available for rocky surfaces. This makes, with the simple bipod, a very basic mortar.

120 mm Tampella mountain mortar

M113 carrier with 120 mm Tampella mortar

120 mm Brandt mortar 120-AM50 towed by 1.5 tonne truck

Panzermörser HS 30: has been modified from its primary role as an armoured personnel carrier and has been fitted with the 120 mm Brandt mortar. It has the ability to engage targets between 400 and 6150 m.

The 120 mm Tampella light mortar is used by mortar detachments of the airborne infantry regiment. Its range zone is 450 to 6350 m.

The 120 mm AM50 Brandt mortar is used by the light infantry units of the Territorial Army. It is carried on the mortar trailer which is towed by a 1.5 tonne truck. The mortar can engage targets between 425 and 6150 m.

GREECE

81 mm mortar Type E44

This light and versatile 81 mm mortar has been designed by Hellenic Arms Industry (E.B.O.) SA to provide quick, accurate and effective firepower at battalion or company level in every phase of battle and in every type of terrain. The features which it incorporates make it an important area weapon in the conduct of land operations.

The simplicity of its employment, versatility, speed and accuracy, coupled with the lightness of the equipment and the resulting ease of carriage, provides small units, operating on wide fronts, with effective fire support whilst remaining highly mobile. The E44 mortar can fire most types of 81 mm projectile.

DATA
Calibre: 81 mm
Range: 100–6000 m
Rate of fire
max: 30 rds/min
sustained: 16 rds/min
Elevation: 45°–85°
Arc of fire: 360°
Traverse at 45° elevation (without moving bipod): ± 5°
Max operating gas pressure: 850 bar
Design gas pressure: 1250 bar
Total weight of mortar: 39.5 kg
   Barrel weight: 13.8 kg
Barrel length: 1240 mm
Bipod weight: 12.5 kg
Baseplate weight: 11.2 kg
Sight: weight: 1.3 kg
   magnification: 2.5 ×
   vertical tilt: ± 25°
   eyepiece axis: 45°

Manufacturer
Hellenic Arms Industry (E.B.O.) SA, 160 Kifissias Avenue, 11525 Athens.
Status
Current. Production.

81 mm mortar Type E44

120 mm mortar Type E 56

The E 56 mortar incorporates features of modern requirements which make it an important weapon in the conduct of land operations. It is mounted on a two-wheel carriage which can be towed by any suitable vehicle. The mortar supported on its special mounting can fire from any type of ground or from an armoured personnel carrier. High accuracy, long-range, high rate of fire and robust construction are some of the main features of this mortar.

DATA
Calibre: 120 mm
Range: 300–9000 m
Rate of fire
max: 12 rds/min
sustained: 8 rds/min

Max operating gas pressure: 1800 bar
Design gas pressure: 2400 bar
Total weight (without carriage): 260 kg
Elevation: 40°–85°
Arc of fire: 360°
Traverse at 45° elevation: ± 8°

Manufacturer
Hellenic Arms Industry (E.B.O.) SA, 160 Kifissias Avenue, 11525 Athens.
Status
Under development.

INDIA

81 mm Mortar E1

The 81 mm mortar E1 is designed to provide quick, accurate and heavy sustained fire together with operational and logistic economy. It is the only weapon system offering range, accuracy, lethality and weight of fire with low weight for easy portability. Because of its exceptional range, this weapon can be used for tasks which would otherwise demand field artillery.

The mortar consists of the usual three basic components, barrel with breech-piece, baseplate and bipod. Elevation, traverse and cross-level controls are caried on the bipod, which assembles to the barrel by means of a split collar. The triangular baseplate is formed with spade surfaces on the underside.

DATA
Calibre: 81 mm
Barrel weight: 14.5 kg
Bipod weight: 11.7 kg
Baseplate weight: 14.4 kg
Transit weight: 135 kg including cases, sight and accessories
Barrel length: 1280 mm
Elevation range: 30° to 85°
Traverse range: 7°30 to 15°3
Maximum range: 5000 m
Accuracy: L 70 m × B 26 m
Rate of fire: 6-8 rds/min with relaying; 20 rds/min without

Manufacturer
Indian Ordnance Factory Board, 10A Auckland Road, Calcutta 700 001.
Status
Current. Production.

Service
Indian Army and export.

120 mm Mortar E1

The 120mm Mortar E1 has been designed to provide heavy and sustained firepower in all phases of battle in various types of terrain. It can provide support to amphibious forces immediately upon landing, and can be effectively deployed in mountainous country.

The mortar is a smoothbore, muzzle-loading weapon which can be easily transported by men, mule, boat, vehicle or aircraft. It can also be parachute-dropped. It consists of the usual three principal components, barrel, bipod and baseplate, and is also provided with a light two-wheeled carriage with pneumatic tyres. The bipod carries the elevating, traverse and cross-level controls and connects to a twin-cylinder recoil buffer mechanism which attaches to the barrel by means of two split collars.

DATA
Calibre: 120 mm
Total weight: 421 kg with sight and accessories
Barrel length: 1750 mm
Elevation range: 45° to 85°
Traverse range: 16° to 31°
Pack load: 5 mules; 3 mules without carriage
Maximum range: 6650 m

Manufacturer
Indian Ordnance Factory Board, 10A Auckland Road, Calcutta 700 001.
Status
Current. Production.
Service
Indian Army and export.

IRAQ

60 mm 'Al-Jaleel' light mortar

This is a conventional drop-fired mortar. The bipod carries the cross-level, elevation and traversing apparatus and is connected to a twin recoil buffer unit which holds the barrel in a quick-release clamp. The barrel has a removeable breech-piece which is fitted with a ball unit engaging in the rectangular baseplate.

The mortar is fitted with an NSB-3 sight unit which is of the usual collimating pattern with Betalight illumination.

DATA
Calibre: 60 mm
Weight, in firing order: 22 kg
Elevation: +45 to +85°
Max range: 2500 m
Rate of fire: 25-30 rds/min

Manufacturer
State arsenal.
Status
Current. Production.
Service
Iraq army and export.

60 mm 'Al-Jaleel' mortar

82 mm 'Al-Jaleel' mortar

This is a conventional drop-fired mortar and is generally similar in its arrangements to the 60 mm model described above. The NSB-4A sight unit is fitted with an optical elbow telescope.

DATA
Calibre: 82 mm
Weight in firing order: 63 kg
Elevation: +45 to +85°
Max range: 4900 m
Rate of fire: 20-25 rds/min

Manufacturer
State arsenal.
Status
Current. Production.
Service
Iraq army and export.

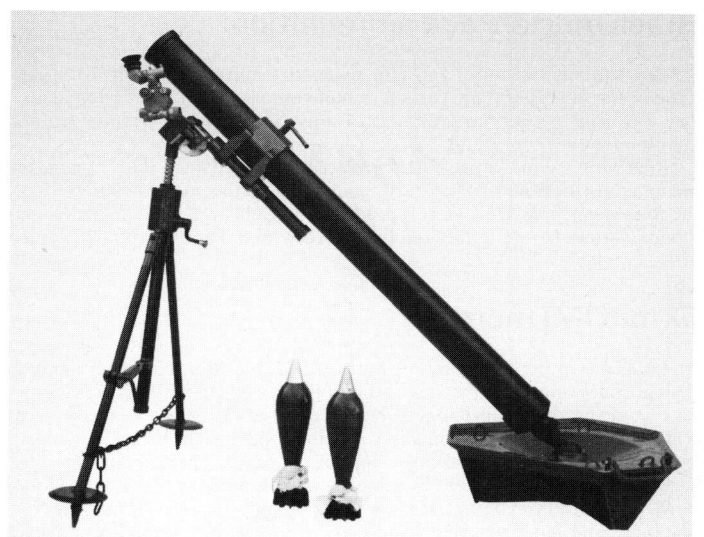

82 mm 'Al-Jaleel' mortar

120 mm 'Al-Jaleel' mortar

This is a conventional pattern of drop-fired heavy company mortar. The barrel has a removeable breech piece which fits into the rectangular baseplate. The bipod carries the cross-level, elevation and traverse mechanisms and attaches to a twin recoil buffer unit which holds the barrel in a quick-release clamp. The NSB-4A sight unit employs an optical elbow telescope.

DATA
Calibre: 120 mm
Weight in firing order: 148 kg
Elevation: +45 to +85°

Max range: 5400 m
Rate of fire: 5-8 rds/min

Manufacturer
State arsenal.
Status
Current. Production.
Service
Iraq army and export.

120 mm 'Al-Jaleel' mortar

ISRAEL

Israeli mortars and ammunition

Soltam Ltd manufactures mortars made basically to Finnish Tampella designs but incorporating various modifications and new developments. These are sold to a very large number of countries. Soltam also manufactures ammunition for its mortars.

Israel Military Industries (IMI) was set up by the Israeli Government to manufacture those items of military equipment that the resources and skiils of the country are able to provide. It produces the Galil rifle, hand grenades and a wide variety of munitions for the Artillery and Tank Corps. However, unnecessary duplication of efforts is avoided and agreements with Soltam and other manufacturing concerns have been made so that IMI produces stores which are not made elsewhere in the country. The entries that follow list the IMI products relevant to this section of the book.

Care should be taken, when using the abbreviation IMI, to avoid confusion with the former Imperial Metal Industries in the UK which is also known by that abbreviation.

52 mm IMI mortar

This is a conventional smoothbore muzzle-loaded mortar. It is of very simple design, consisting of a tube and a trough-like baseplate mounted across the base of the barrel. Connecting the barrel to the baseplate is the breech piece which contains the firing mechanism. The firer operates the firing mechanism by rotating the small wheel on the right of the barrel. There are no sights as such and aiming is carried out by using the white line on the barrel to align the bore. The mortar can be used for low- or high-angle fire, is easily managed by one man, and has a carrying handle and a muzzle cover. High explosive, illuminating and smoke bombs are all fired.

DATA
Type: smoothbore, muzzle-loaded
Range: 130–420 m
Rate of fire: 20–35 rds/min
Elevation: 20–85°
Traverse: 360°
Ammunition: HE, illuminating, smoke
Barrel length: 490 mm
Length, with baseplate: 673 mm

Baseplate dimensions: 150 × 85 × 35 mm
Baseplate weight: 1.3 kg
Overall weight with baseplate: 7.9 kg
Attachments: carrying handle, leather muzzle cover

Manufacturer
Israel Military Industries (IMI), PO Box 1044, Ramat Ha Sharon 47100.
Status
Current. Production.
Service
Israeli forces.

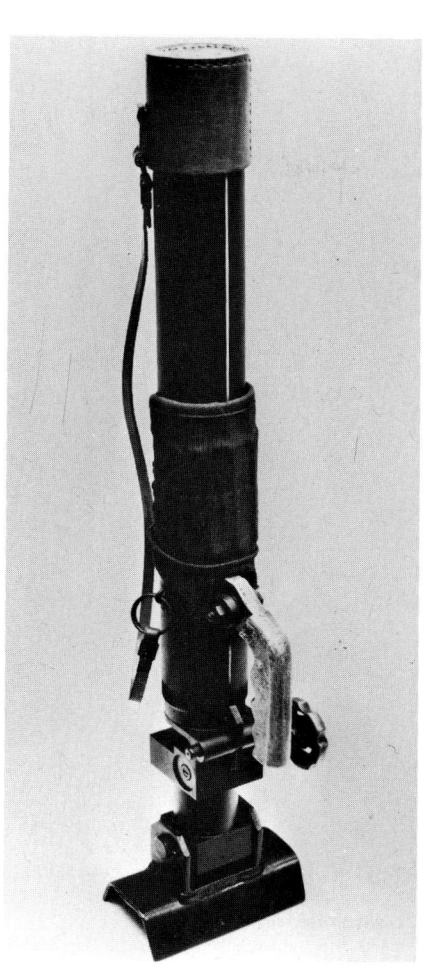

52 mm IMI mortar

60 mm Soltam mortars

Soltam makes four versions of a 60 mm mortar which it describes as 'Type Tampella' mortars because they are based on designs by the Finnish firm. The four weapons, which are designated Long-Range, Standard, Commando and Armoured Mounted have many features in common. The Standard mortar can be used with or without its bipod; the Commando is a short-range weapon, is always used without a bipod and has a firing mechanism instead of a fixed pin. The Armour Mounted mortar is a unique muzzle loaded mortar that is mounted inside a tank's turret ot an APC fighting compartment. This mortar is operated while the operator is completely protected. The main description which follows relates to the Standard and it is followed by notes on the other weapons and a table of comparative data.

Standard mortar
This mortar was designed to fulfil three roles. Equipped with a bipod it was intended to be used in the normal way as a company support weapon; alternatively, without the bipod, it could be used as a light mortar in the assault role, or externally mounted on a tank turret or APC roof.

In the standard role, the mortar consists of three elements, barrel, bipod and baseplate. The barrel is made of alloy steel and the bore is given a particularly smooth finish. Screwed to the bottom of the mortar tube is the breech piece. This is rounded off at its lower end into a ball which fits into the socket of the baseplate. The striker pin is contained in the breech piece. It is of the fixed, non-retractable type and has no control mechanism. The impact of the bomb on the fixed striker is enough to set off the cap of the primary cartridge. A canvas sleeve is placed around the barrel to act as a heat shield and to allow the attachment of a carrying handle. Around the muzzle is a simple drum sight which is used only in the attack or 'Commando' role.

The bipod is of conventional design with the spread of the legs limited by the construction of flanges in the plates where they are joined at the top of the legs. Passing up through the bipod is the elevating screw and the elevating column which is joined to the yoke. This carries the traversing screw. At the left end is the sight bracket, while the right-hand end has the traversing handwheel. The cross-levelling gear connects the left leg of the bipod to the bottom of the elevating gear. Thus when the mortar is brought into action on uneven ground, the elevating screw and column can be swung to allow the sight to be brought to the upright position. Two recoil cylinders are interposed between the barrel collar and the yoke. The sight has a clamp which will normally be placed around the tubular yoke but can also be placed around the barrel when so required. There is a bearing scale which runs from 0 to 6400 mils. The target or aiming post is viewed over a collimator. The elevation scale is set out in five columns corresponding to the ranges achieved by the primer and the four charges and the range reader is scribed on a perspex plate attached to the body of the sight. When the range is applied, the body of the sight is thrown out of level and to centre the elevation bubble the barrel must be elevated or depressed.

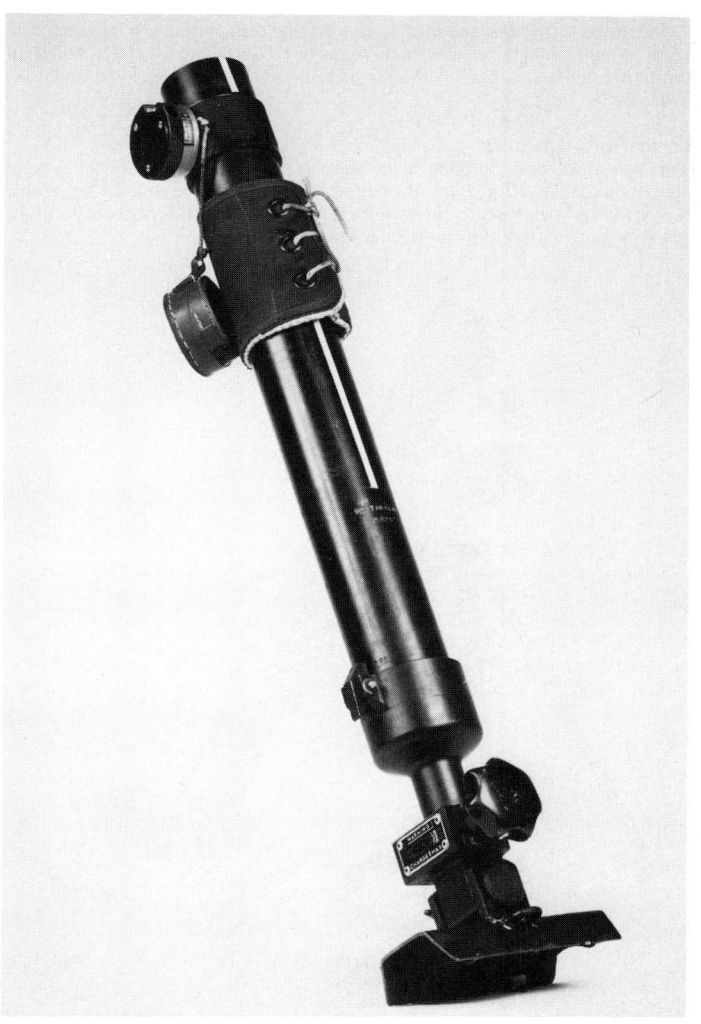

60 mm Commando Soltam mortar

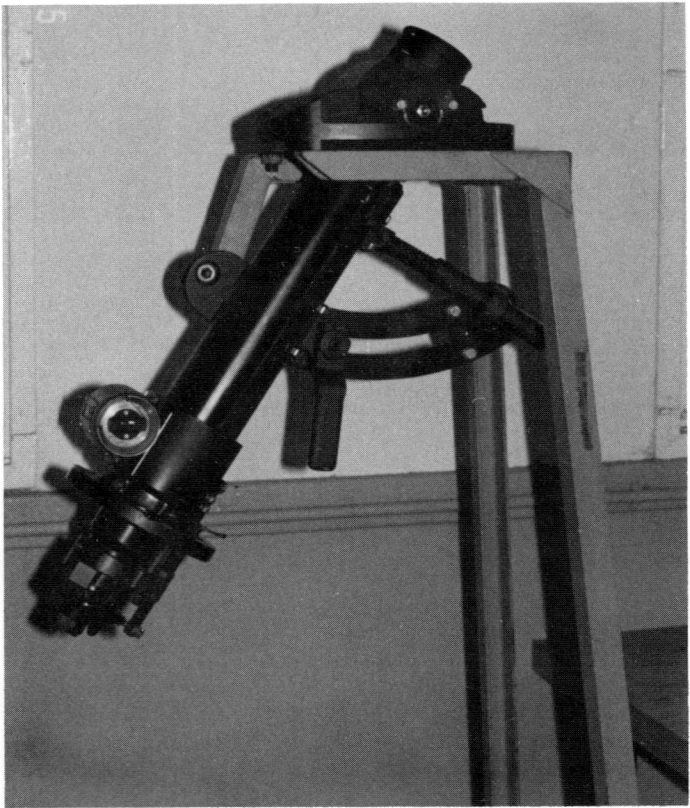

60 mm Under Armour mortar

The baseplate is of welded construction with a flat, dished top plate and three webbed ribs which enable the baseplate to take the recoil force without slipping. In the centre of the baseplate is a socket to take the ball at the bottom of the breech piece. The socket has a clamping ring with a securing screw that can be used to clamp the barrel rigidly to the base plate. The construction of the baseplate is such that a 360° traverse can be obtained. The second way of using the mortar is in the assault role. Here the bipod is not used. The normal sight can be secured to the muzzle or a simple drum sight can be used. The barrel is controlled by hand until it is required to go to fire for effect, and then the clamp on the baseplate is used to lock the barrel at the required elevation in the desired line.

Long-Range mortar
This mortar has previously been described as the Type A but is now known by its new name. It differs from the Standard version by having a longer and heavier barrel designed to throw a bomb to 4000 m with a rate of fire of 20 rds/min. For this it requires a stronger breech also.

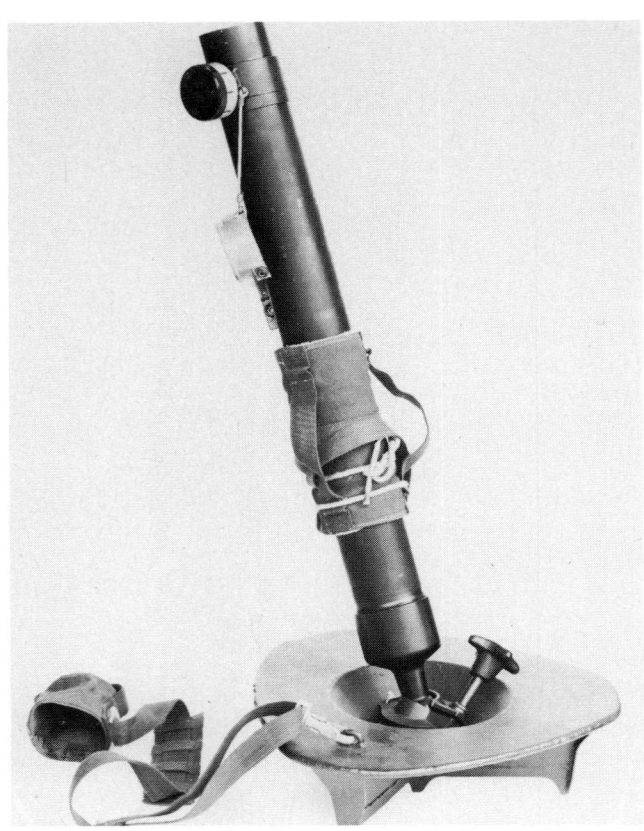

60 mm Standard Soltam mortar used without bipod in assault role

Commando mortar
This is the lightweight version of the Standard mortar and it is a much simplified design. The barrel is considerably shorter than the Standard and it has no bipod. There is a simple trigger mechanism incorporated in the breech, and this is rigidly attached to a small baseplate. The whole mortar can be conveniently carried and fired by one man. A sight is clipped to the barrel, and is capable of indicating elevation and azimuth, but it is probable that the majority of firing would be undertaken by rough alignment and judgement.

Under Armour mortar
The 60 mm under-armour mortar is a far reaching weapon system that enables the user to conduct a complete mortar firing mission while being completely protected inside a tank turret or other fighting compartment.

The system covers ranges up to 3500 m and provides screening, illumination and the advantage of hitting soft targets and threatened personnel with an effective, accurate, low cost ammunition. The smoothbore muzzle loading mortar fires all types of existing 60 mm mortar rounds, HE, smoke and illuminating.

The system, mounted on the turret or vehicle roof, pivots on trunnions to a bolted base which transmits the firing loads to the vehicle body.

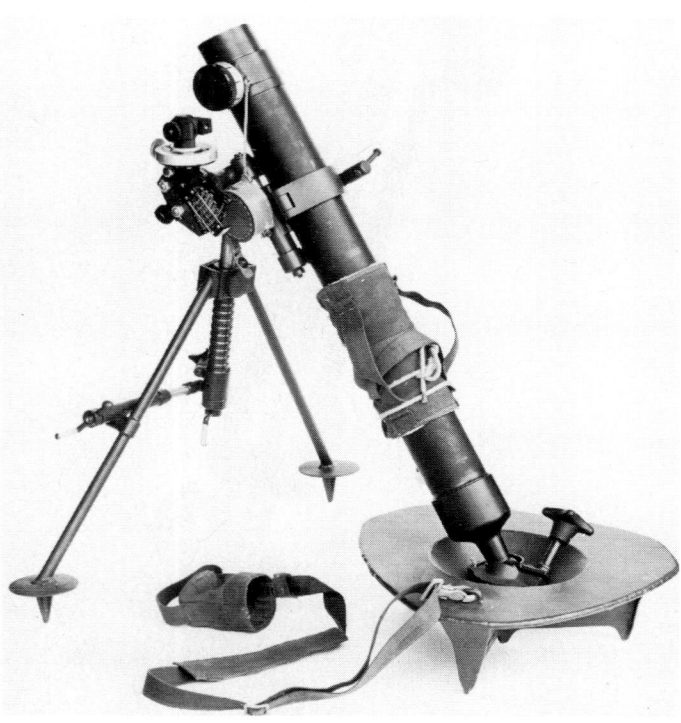

60 mm Standard Soltam mortar set up as conventional company support weapon showing sight on bipod and drum sight at muzzle

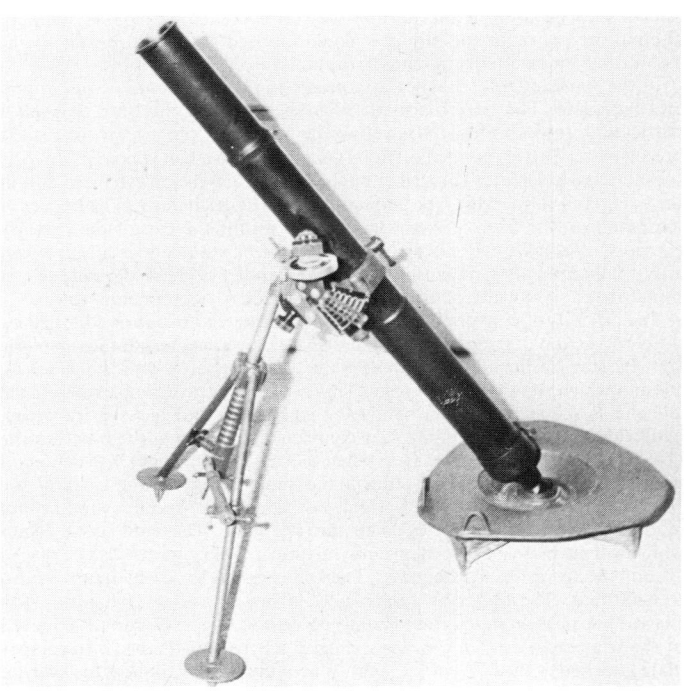

60 mm Long Range Soltam mortar

Laying: the mortar mount is parallel to the main armament. Deflection is controlled through the gunner's control handle, while the elevation is set by the mortar operator. The mortar is equipped with a quadrant elevation indicator which indicates the ranges according to the charge selected.

Loading: for loading, the barrel must be unlocked, and the round can then be inserted into the muzzle. The barrel is then pushed forward and locked.

Firing: the mortar is fitted with a trip-action firing mechanism, and is also equipped with safety devices to prevent accidental double loading or firing when the barrel is in the loading position.

DATA

Type	"Under-armour" mounted C-04	Long-range C-06	Standard C-08	Commando Regular C-03	Light C-576
Calibre	60 mm	60 mm	60 mm	60 mm	60 mm
Max range	3500 m	4000 m	2550 m	1000 m	1600 m
Min range	150 m	150 m	150 m	100 m	100 m

WEIGHT

Total in					
firing position	62 kg	18.3 kg	16.3 kg	6.8 kg	6.3 kg
Barrel	10 kg	7 kg	5.7 kg	5.7 kg	4.6 kg
Bipod	—	4.7 kg	4.5 kg	—	—
Baseplate	—	5.5 kg	5.1 kg	1.1 kg	1.7 kg
Sight	—	1.1 kg	1.0 kg	0.5 kg	0.5 kg

DIMENSIONS

Barrel length	800 mm	940 mm	740 mm	725 mm	675 mm
Bipod length, folded		640 mm	540 mm		
Baseplate diam.	—	350 mm	350 mm	—	200 mm

Manufacturer
Soltam Ltd, PO Box 371, Haifa.
Status
All are current and available.
Service
Standard mortar in service with Israeli forces. All types supplied to many foreign countries.

81 mm Soltam mortars

As with the 60 mm versions, the intention with the 81 mm mortars made by Soltam is to provide variations in the design to suit all requirements and to do this with the least possible alteration to the basic characteristics and manufacturing processes. The mortars are of a conventional pattern, based on the Tampella designs and using as many common items as possible. There are four versions of the 81 mm, namely Standard type with either a one-piece or split barrel, and Long Range type with a reinforced barrel similarly having a one-piece or a split barrel.

In each case the barrel is made from a high grade steel with a honed finish to the bore and a breech piece which fits into a socket in the base plate. This breech piece carries a fixed firing pin, though the firm can fit a controllable firing pin for those who need it.

The variety of barrels provides a flexibility of use which is unusual in modern mortars. The long-range version is obviously less mobile than the others due to its extra weight, but it can provide an impressive range of 6500 m and for many purposes this range will be worth the slight extra load of the barrel. The standard mortar could be looked on as the basic version, and its performance is similar to others of the same size and weight. The split barrelled mortar is intended for greater mobility when carrying on foot or in small vehicles, though the penalty for the smaller bulk is a small increase in weight due to the screw collar in the middle of the barrel.

The bipod is of orthodox pattern but has some interesting features. The two legs of the bipod have spiked feet and the distance between them is controlled by an adjustable chain. The elevating column is attached to the yoke and has a cross-levelling gear connecting it to the left bipod leg. The elevating screw thread is totally enclosed within the elevating tube and so is protected from the ingress of dirt, sand and snow. The yoke carries the traversing screw and protects it. The traversing handwheel is on the left and the sight mounting on the right. There is a shock-absorbing unit consisting of two cylinders joining the yoke to the barrel clamp.

The baseplate is of circular shape with welded ribs on the underside. The socket takes the ball of the breech piece and allows a full 360° traverse without need to move the baseplate. A carrying handle is attached and is needed to get the baseplate out when coming out of action.

The sight is graduated in mils and covers the full 6400 traverse. There is also a slipping scale. This enables the sight to be used on an aiming post and the target recorded in terms of a bearing rather than a switch from the point of aim. This permits registration of the target and subsequent engagement of the target without further ranging. If meteorological information is available, then the target data can be corrected and stored in a true form. The elevation scale and the cross-levelling are controlled by bubbles and a collimator is used for sighting. A quick-release is available for the traversing scale.

81 mm Soltam mortar with split barrel

81 mm Soltam mortar

Three manpack loads of split barrel 81 mm Soltam mortar

Sight

Weight:				
Total	42 kg	44 kg	49 kg	51 kg
Barrel	15.5 kg	17.5 kg	22.5 kg	24.5 kg
Bipod	14 kg	14 kg	14 kg	14 kg
Baseplate	12.5 kg	12.5 kg	12.5 kg	12.5 kg
Dimensions:				
Barrel length	1560 mm	1560 mm	1583 mm	1583 mm
Bipod length folded	960 mm	960 mm	960 mm	960 mm
Baseplate diameter	518 mm	518 mm	518 mm	518 mm
Sight				
Weight	1.55 kg	1.55 kg	1.55 kg	1.55 kg
Traversing scale	6400 mils	6400 mils	6400 mils	6400 mils
Elevating scale	700-1600 mils	700-1600 mils	700-1600 mils	700-1600 mils

Manufacturer
Soltam Ltd, PO Box 371, Haifa.
Status
Current. Available.
Service
Israeli forces. Exported to other countries and can be found in East Africa. Believed to have been evaluated by US Army.

DATA

Mortar	Standard Solid Barrel	Split Barrel	Long Range Solid Barrel	Split Barrel
Max Range	4900 m	4900 m	6500 m	6500 m
Min Range	150 m	150 m	150 m	150 m
Elevation	43-85°	42-85°	43-85	43-85°
Traverse	360°	360°	360°	360°

81 mm Soltam mortar on M113 APC

Soltam has developed armoured vehicle mountings for its 81 mm and 120 mm (light) mortars for use on US M113 pattern vehicles accompanying motorised infantry formations.

The mountings, which involve the use of a modified bipod assembly and a vehicle-mounted baseplate fitting, are similar for the two mortars. The accompanying illustration shows the 81 mm mounting: general data will be found in the entry for the 120 mm mounting below.

Manufacturer
Soltam Ltd, PO Box 371, Haifa.

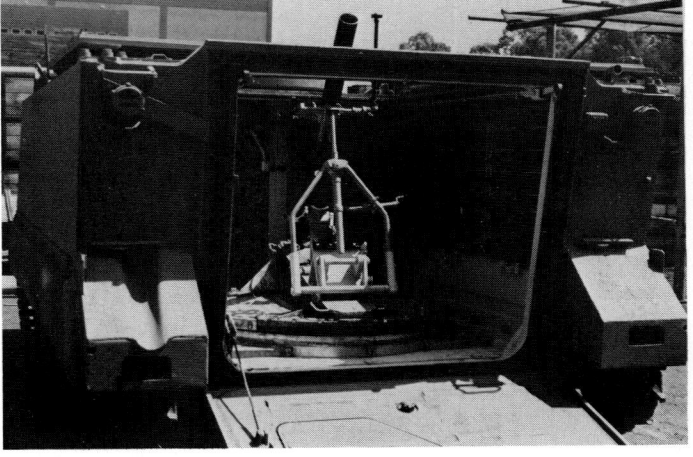

81 mm Soltam long barrel mortar on M113 APC mounting

120 mm Soltam light mortar and K6 mortar

This mortar was designed to provide infantry with a mortar capable of firing the very lethal 120 mm ammunition but sufficiently light to enable it to be moved by a vehicle of the 'Jeep' type, transported by a mule, carried by a detachment of three soldiers, moved in a helicopter or dropped by parachute. The light mortar can fire all the 120 mm bombs available including the rocket-assisted bomb. The K6 is an updated version of the older light mortar and has been adopted by the US Army.

Travelling carriage of 120 mm Soltam light mortar. Containers on axle tree carry six rounds ready for use

Barrel
The barrel is an alloy steel tube with a honed interior. The lower end is externally screw threaded to take the breech piece. The breech piece holds the striker. This is a fixed stud on which the bomb falls under gravity. However, to ensure safety while dealing with a misfire, there is a safety catch which, when rotated, draws the pin back into the interior of the breech piece. The lower end of the breech piece is shaped into a ball which enters a socket in the baseplate. The barrel can be carried by a single man.

Bipod
The two legs have spikes at the bottom and are joined at the elevation gear housing. The distance between the legs is controlled by an adjustable length chain. The lower end of the elevation column, which contains the elevation screw thread is attached to the left leg of the bipod by a cylinder containing a screw thread. The rotation of the handle at the end of this thread moves the elevation column out of the vertical and so allows the mortar to be cross levelled to allow for irregularities in the ground. At the top end of the elevating column is the yoke which holds the sight at the left and the traversing handle at the right. The thread of the traversing gear is, like all the threads in this mortar, completely enclosed. The yoke is connected to the barrel clamp by a pair of shock-absorbing units which ensure that the barrel-yoke position does not change as the barrel recoils and then returns to its original firing position.

Baseplate
The baseplate is a heavy, welded, steel dish which has a socket at the centre to take the breech piece. The mortar can traverse through 360° without need to

120 mm Soltam light mortar in firing position

Carriage

This is a lightweight two-wheeled vehicle with a torsion-bar suspension. It enables the mortar to be pulled by troops. It is parachutable and can be towed behind any vehicle with a towing hook of the right height and size. In some roles the carriage is not used and the mortar is brought into action without it. In addition to supporting the mortar, the carriage carries the spare parts and tools for the weapon and also has six metal containers set across the axle, allowing the carriage of that number of ready to use bombs.

Ammunition

The mortar takes the M48 and M57 series of rounds.

K6 mortar

The K6 is a new version of the light mortar and it incorporates a new and reinforced barrel. The firm now quotes this barrel as being the one for use with the long-range M57 ammunition and it is this barrel which gives a maximum range of 7200 m. All other dimensions and performance parameters are unaltered.

DATA

	Standard	Long Range
Type	K-5	K-6
Weight in march order	266 kg	322 kg
Max Range	6250 m	7200 m
	with M48 round	with M57 round
Min range	250 m	200 m
Elevation	43-85°	40-85°
Traverse	6400 mils	6400 mils
Weights:		
Total, firing position	140 kg	144 kg
Barrel	46.5 kg	50 kg
Bipod	31.5 kg	32 kg
Baseplate	62 kg	62 kg
Dimensions		
Barrel length	1758 mm	1790 mm
Bipod length, folded	1040 mm	1230 mm
Baseplate diameter	880 mm	880 mm

Manufacturer
Soltam Ltd, PO Box 371, Haifa.
Status
Current. Available.
Service
Israeli forces and some other armies.

Manufacturer
Soltam Ltd, PO Box 371, Haifa.
Status
Current. Available.
Service
Israeli forces and some other armies.

shift the baseplate. The ribs welded on to the underside of the plate prevent the baseplate from slipping sideways.

Sight

This incorporates a 360° (6400 mils) traverse scale and a slipping scale. Thus when the zero line has been recorded the slipping scale can be adjusted to zero. The mortar can therefore engage targets, register at the conclusion of the action and re-engage without recourse to further ranging. Large switches are made easier by the incorporation of a quick-release lever which enables the sight head to be rotated rapidly on to a new bearing without laborious rotation of low geared handwheels. Elevation and cross-level are controlled by bubbles and a collimator is used to lay on the aiming post.

Tools

The tools supplied include one which can be inserted into the muzzle to engage around the fuze of a bomb and is used to withdraw it in the event of a misfire.

120 mm Soltam light mortar and K6 mortar on APC mounting

Soltam has developed an armoured vehicle mounting for its 120 mm light mortars for use in US M113 APC or M106/125 mortar carriers. There are two construction variants: a rigid mounting, which uses a bolted traverse beam to anchor the baseplate, and a turntable mounting.

The latter is an adaptation of the towed configuration (K6) to the M106A2 vehicle. Consequently, its general performance duplicates that of the K6 while minimising additional logistic support requirements.

The adaption kit is similar to that of the 4.2 in mortar. The turntable is modified to include a bipod support and breech piece socket. The barrel and the bipod are those of the K6, the bipod having removable ground spikes.

In both cases the mortar is assembled in the vehicle with the muzzle facing backwards and is fired in this position. Alternatively, it can be rapidly dismantled from its mount for conventional ground deployment.

DATA
VEHICLE WITH MORTAR
Crew: 5

DIMENSIONS
Length: 4830 mm
Width: 2450 mm
Height: 2020 mm
Ground clearance: (front) 435 mm; (rear) 395 mm
Track width: 380 mm

WEIGHTS
Combat ready: 11.3 tons
For swimming: 10.68 tons

POWER PACK
Engine type: Chrysler 75 M
Power: 212 PS SAE
Engine revolutions: 4100 rpm
Gear: Allison TX200-2A

PERFORMANCE
Max speed: 64 km/h
Speed in water: 5.6 km/h
Max trench: 1680 mm
Fuel consumption: 94 l/100 km
Fuel capacity: 305 l
Cruising range: 320 km
Power/weight ratio: 18.4 hp/ton
Ground pressure: 0.51 kg/cm²
Electrical system: 24 V
Communication: wireless and intercom
Armament: (light mortar) 120 mm; (machine gun) 7.62 mm

120 mm Soltam light mortar in M113 APC

MORTAR
Traverse: (left) 18.5°; (right) 25° for M113 APC: 45° right and left for M106/M125 carrier
Elevation: (minimum) 45°; (max) 80° for M113 APC; (min) 45°, (max) 85° for M106/M125 carrier
120 mm ammunition: (swimming) 23 rounds; (on land) 54 rounds

Rate of fire: (max) 15 rds/min
Freeboard: (front left) 305 mm; (front right) 295 mm; (rear left) 370 mm; (rear right) 365 mm
Roof opening for mortar: (length) 1580 mm; (width) 1150 mm

Manufacturer
Soltam Ltd, PO Box 371, Haifa.

RMS-6 120 mm Recoiling Mortar System

This is a vehicular adaptation of the 120 mm K6 mortar, enabling it to be mounted in tracked or wheeled armoured vehicles. The new system is primarily intended to replace 81 mm and 4.2 in mortars mounted on M125 and M106 tracked armoured carriers as well as to be mounted on wheeled armoured vehicles. The system can be mounted in the M125 and M106 carriers without any structural changes to the carrier. Although it fires 120 mm bombs to a range of 7 200 m, its recoil forces do not exceed those of an 81 mm mortar and allow a 360° arc of fire.

The unit consists of a saddle mounted on a bearing which is bolted to the structure of the vehicle. A ring cradle is trunnioned to the saddle and carries the 120 mm K6 barrel running in a concentric recoil system inside the cradle. The barrel is inserted from the rear of the cradle and can be withdrawn for ground use within one minute.

The system is fitted with manual elevating and traversing controls and an extended sight bracket lifting the sight line above the hull of the vehicle if required. The barrel can be elevated between 43° and 85°.

'Breaking' the elevating screw allows the barrel to be lowered instantly below roof level, so that roof doors can be closed without delay.

Manufacturer
Soltam Ltd, PO Box 371, Haifa.
Status
Entering service after completion of testing.

RMS-6 recoiling mortar mounted in M125/M106 carrier

120 mm recoiling mortar mounted on LAV-25 MOWAG Piranha

120 mm A4 Soltam heavy mortar

The heavy 120 mm mortars follow the usual Soltam practice of being slightly larger and heavier versions of the standard mortar and differ in only minor aspects. The heavy mortar has been in service for some years now, and the A4 is the latest version of it. Both mortars use a longer and heavier barrel than the standard, giving an increased range using the Tampella M59 ammunition. Both have a heavier and larger baseplate also to absorb the increased recoil forces. Strangely, the carriage is a little lighter than for the other versions.

Manufacturer
Soltam Ltd, PO Box 371, Haifa.
Status
Current. Available.
Service
Not known.

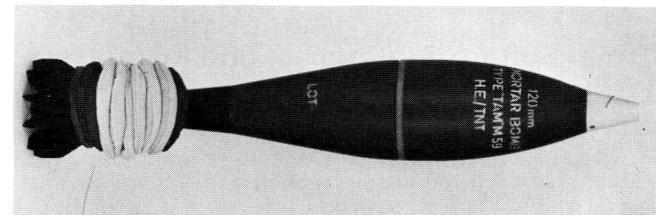

120 mm M59 Soltam HE mortar bomb

120 mm Soltam heavy mortar A7

This new 120 mm heavy mortar has been designed for rapid deployment and for operation by fewer men than the older models.

It uses the same ammunition and has the same ballistic performance as the A4 described above, but it can be handled by only two or three men. The

Soltam 120 mm heavy mortar in travelling mode

Soltam 120 mm heavy mortar in firing position

carriage has a lightweight torsion bar suspension from which the weapon is not dismounted for use although when being fired the mortar does not rest on its wheels.

If necessary, the carriage can be removed if the weapon is to be carried by sleigh or by pack animals.

Manufacturer
Soltam Ltd, PO Box 371, Haifa.
Status
Current. Production.

120 mm M-65 Soltam standard mortar

The standard 120 mm is a substantial mortar which can be used in the traditional way and can also be mounted in an armoured personnel carrier or similar vehicle. Such a vehicle should be able to carry sufficient ammunition to make full use of the very considerable firepower of the weapon.

Barrel and breech piece
The barrel is made of high tensile alloy steel and is given a very good internal finish. The breech piece screws into the bottom of the barrel and holds the firing mechanism. The spring-loaded striker is fired by pulling a firing lever which cocks the spring before allowing it to trip over to fire the bomb. The firing pin can be retracted to a safe position and this is of particular value when clearing a misfire. The ball-shaped end of the breech piece fits into the socket of the baseplate. At the front end of the barrel is a bayonet-type catch into which the towing eye is secured. This is sprung to reduce the shocks of travelling.

Bipod
The bipod is, in principle, similar to that of the light 120 mm mortar. All exposed and sliding components are either chromium plated or made of stainless steel to resist wear or corrosion and all the gears and screw threaded columns are totally enclosed.

Baseplate
This is a circular dish-shaped plate of steel with strengthening ribs welded on. There are also carrying handles and eyes for securing the baseplate to the carriage welded onto the upper side of the baseplate.

Sight
The sight employed on the standard mortar is the same as that used on the 120 mm light mortar.

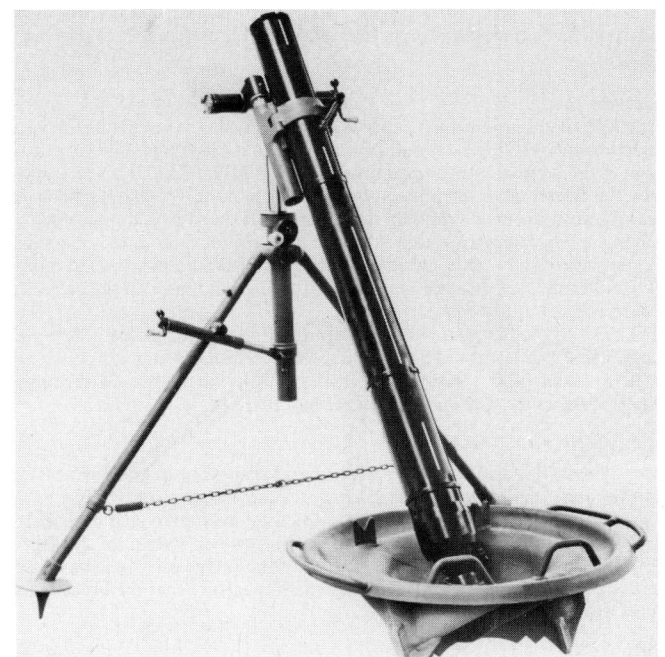

120 mm M-65 Soltam standard mortar in firing position

Carriage
This is a more substantial carriage than that used with the light mortar. It is of box construction made up of welded steel sheet. The top portion carries the barrel clamping collar and baseplate clamping hooks. The wheels are of standard jeep size and type. Drag rings are mounted on the wheel hubs for man-handling the mortar across country. The frame of the cross member is boxed in to take tools and has a lid. There are welded attachments to which can be secured the cleaning rods, carriage stay and baseplate levering rods during transit.

120 mm M-65 Soltam standard mortar in travelling position

DATA

Type	Standard M65	Heavy A-4	Heavy A-7
Weight, march order	351 kg	364 kg	431 kg
Max range	6500 m with M48 bomb	8500 m with M59 bomb	8500 m with M59 bomb
Min range	400 m	400 m	400 m
Elevation	43-85°	43-85°	43-85°
Traverse	360°	360°	360°
Crew	6	6	6
Weights			
Total, firing position	231 klg	244 kg	257 kg
Barrel	82 kg	95 kg	102 kg
Bipod	69 kg	69 kg	70 kg
Baseplate	80 kg	80 kg	85 kg
Dimensions			
Barrel length	1950 mm	2154 mm	2154 mm
Bipod length folded	1675 mm	1675 mm	1520 mm
Baseplate diameter	900 mm	900 mm	900 mm
Sight			
Weight	1.5 kg	1.5 kg	1.5 kg
Traversing scale	6400 mils	6400 mils	6400 mils
Elevating scale	700-1600 mils	700-1600 mils	700-1600 mils

Manufacturer
Soltam Ltd, PO Box 371, Haifa.
Status
Current. Available.
Service
Israeli forces and some other armies.

120 mm M-65 Soltam standard mortar mounted in Israeli half-track

160 mm Soltam M-66 mortar

The 160 mm mortar fires a 40 kg bomb out to 9600 m. To do this requires a heavy weapon (1700 kg in the firing position) and a detachment of six to eight men.

Barrel

This is a high tensile strength, alloy steel tube. At the bottom end is the breech piece which contains the firing mechanism. This is operated by a firing lever which allows the spring-loaded striker to trip off the sear and go forward to fire. The striker can be withdrawn to provide safety, particularly in the case of a misfire.

Carriage

There is no bipod mounting for this weapon and instead the barrel is elevated and depressed by a single column which is part of the carriage. The very heavy weight of the bomb makes conventional loading difficult and it is necessary to lower the barrel to a loading position. This is done by folding back the elevating strut from a hinge at its mid-point. There is a counterbalance mechanism which makes it easy to elevate the barrel each time it is loaded. This consists of steel cables attached to the barrel some 2 m from the muzzle and led over sheaves into the lower pair of tubes that make up the chassis of the carriage.

The axles allow the road wheels to be turned in and locked by a cam plate attached to the axle tube. The offside road wheel incorporates a clutch and handle to allow slow traverse. The carriage rolls on its wheels through a complete 360° circle without moving the baseplate.

Baseplate

This is a heavy flat disc welded up with a number of webs to give stiffness and to prevent the tendency to slide sideways. Four handles are welded to the top-plate to allow carrying the baseplate and they are used to get the baseplate out of the ground. There is a central socket into which the tail of the breech piece fits. This allows a full 360° traverse without movement of the baseplate. There is a spring-loaded locking arrangement to ensure that the breech piece cannot leave the baseplate.

DATA
Calibre: 160 mm
Weight of mortar in travelling position: (excluding baseplate) 1450 kg
Weight of baseplate: 250 kg
Total weight of mortar in firing position: 1700 kg
Length of barrel: 2850 mm
Range: 600–9600 m
Elevation: 43–70° (764–1244 mils)
Traverse: 360° (6400 mils)

160 mm Soltam M-66 mortar in firing position

SIGHT
Weight: 1.57 kg
Traversing scale: 360° (6400 mils)

Manufacturer
Soltam Ltd, PO Box 371, Haifa.
Status
Current. Available.
Service
Israeli Army.

160 mm Soltam M-66 mortar on modified Sherman chassis as used by Israeli Army

M-66 mortar in travelling position

ITALY

Italian mortars

Currently, the Italian Army is equipped with 81 mm and 120 mm mortars of French (Brandt) design. The 81 mm mortar is the MO-81-61-L, the long barrel version of the Brandt light mortar (which see) and supplies for the Italian Army were manufactured in Italy by OTO Melara some years ago.

Otherwise development and production of mortars has not been a significant feature of the Italian defence scene in the years since the Second World War.

There is, however, significant activity in the manufacture of mortar ammunition in Italy. Both SNIA-BPD and Simmel make complete rounds of mortar ammunition and Borletti makes fuzes.

Mortar ammunition in the SNIA-BPD and Simmel current range comprise the 81 mm and 120 mm families and the 107 mm (4.2 in). Further information on these items will be found in the Ammunition section.

Manufacturers
OTO Melara SpA, Via Valdilocchi 15, 19100 La Spezia. SNIA-BPD SpA, Via Sicilia 162, 00187 Rome. Simmel SpA, Borgo Padova 2, 31033 Castelfranco Veneto (TV). Borletti FB, s.r.l., Via Verdi 33, I-20010 S.Giorgio su Legnano (MI).

81 mm Breda light mortar

The well-known firm of Breda has produced a light mortar in 81 mm calibre, probably with the intention of gaining the contract to replace the fairly elderly types currently in service with the Italian army. The Breda mortar follows most other types in having a circular baseplate to permit all-round traverse without having to re-bed the plate and a sturdy bipod showing its ancestry to the Brandt. A special feature however is the long barrel, specifically intended to give long-range without having to resort to special bombs or unconventional means of propulsion.

Another feature, and one which is much needed on mortars is a means of withdrawing the striker when there is a misfire. The striker can be pulled out from the breech, leaving the barrel entirely safe, the bomb can then be loosened and withdrawn without the danger of it slipping back on to the striker and exploding.

Two small recoil cylinders have been interspersed between the bipod clamp and the legs to reduce the shock load on the legs and the sight. This arrangement improves the consistency and reduces the number of times that the mortar has to be relaid. The cradle is made of aluminium alloy and it is held to the barrel by a half-ring locked by an over-centre clamp. The sight fits on to a bayonet fitting.

Traversing is by a horizontal screw on the lower side of the cradle and a nut on the top of the bipod legs travels along this screw, turned by a small wheel on the right-hand side.

Cross-levelling is by means of a screw adjuster running from one leg to the central column.

The mortar breaks down for carriage into three basic loads, each about 15 kg, which allows the weapon to be carried by three men, though more would be needed for the ammunition.

DATA
Calibre: 81 mm

WEIGHTS
Complete mortar: 43 kg
Barrel: 14.5 kg
Baseplate: 13 kg
Mount: (without sight) 15.5 kg

Barrel length: 1455 mm
Range: (minimum) 75 m; (max) 5000 m
Elevation: 35-85°
Rate of fire: (max) 20 rds/min

Manufacturer
Breda Meccanica Bresciana SpA, Via Lunga 2, 25126 Brescia.
Status
Development completed.
Service
Trials with Italian Army.

81 mm Breda light mortar. Sight is not fitted, but box for it is standing behind baseplate

JAPAN

Japanese mortars

The Japanese Self-Defence Force uses both 81 and 107 mm mortars. The 81 mm mortar Type 64 is a modified version of the US M29A1 mortar, manufactured by the Howa Machinery company. The 107 mm is the MS M2 supplied from American military sources.

KOREA, SOUTH

South Korean Army mortars

The South Korean Army, while equipped mainly with weapons of US origin, is making considerable efforts towards self-sufficiency. Daewoo Heavy Industries Ltd produces versions of the 60 mm mortar M19 and the 81 mm mortar M29A1 designated KM19 and KM29A1 respectively. Details of the items produced differ slightly from those of the original US product and are therefore given below.

DATA
60 MM MORTAR KM19
Max range: (full charge at 45° elevation) 1975 m

81 MM MORTAR KM29A1
Max range: (full charge at 45° elevation) 4730 m

Manufacturers
Daewoo Heavy Industries Ltd (mortars); Yangji Metal Industrial Company Ltd, Dae Chang Building 605, 42 Jan Gyo-Dong, Chung-Gu, Seoul (81 mm base plate)
Status
In service with South Korean Army.

60 mm Mortar KM181

This is a light company mortar, similar in design to the US M224. It is capable of firing any current 60 mm ammunition, and has a non-selective fixed firing pin for drop-firing only. The CN81 barrel is finned for cooling, permitting a sustained rate of fire of 20 rds/min. The bipod is symmetrical, with a traversing screw at the top and a cross-levelling strut at the bottom of the central elevating tube. Cross-levelling is performed by screwing a sleeve on the left tripod leg up or down. The sight unit is provided with a self-powered radioactive luminous source to permit operation at night.

The mortar can be carried by the two man crew; one man carries the entire mortar, a load weighing 21.20 kg, while the second man carries eight bombs weighing 17.40 kg.

DATA
Calibre: 60 mm
Barrel length: 1.00 m
Weights: Barrel 5.50 kg
Bipod 6.00 kg
Baseplate: 6.50 kg
Sight: 1.50 kg
Weight in action: 19.5 kg
Elevation: 800 – 1511 mils
Traverse: 250 mils
Maximum range: 3590 m

Manufacturer
Kia Machine Tool, KIA Building, 15 Yoido-dong, Youngdeungpo-gu, Seoul.
Status
Current. Production.

Man-carrying the 60 mm KM181 light mortar

60 mm KM181 light company mortar

PAKISTAN

60 mm Light mortar

This light mortar is based on the Chinese Type 63 but has been improved in various details, notably by an increase in the weight and strength of the barrel, which permits a more powerful propelling charge to be used and thus provides greater range.

The light mortar is of conventional drop-fired pattern, using a rectangular baseplate and a symmetrical bipod. The bipod collar fits around the barrel and is connected to it by means of a recoil buffer, behind which is a carrying handle. The bipod has a simple screw traverse mechanism and cross-level is catered for by one telescopic leg on the bipod.

DATA
Calibre: 60 mm
Barrel weight: 4.5 kg
Baseplate weight: 5.7 kg
Bipod weight: 4.6 kg
Weight in action: 14.8 kg
Length of barrel: 623 mm
Elevation: +45— +80°
Rate of fire: 8 rds/min
Range: 50 – 2000 m

Manufacturer
Pakistan Machine Tool Factory Ltd., Landhi, Karachi 34.
Status
Current. Production.

Service
Pakistan armed forces.

81 mm Light mortar

The design of this mortar appears to be based upon that of the
Thomson-Brandt 81 mm MO-81-61L; it is a conventional drop-fired weapon
using a triangular baseplate and a symmetrical bipod. Elevation is provided
by a central screw mechanism in the bipod, traverse by a cross-screw at the top
of the bipod, and cross-level by a telescoping leg connected to the central,
elevating screw casing. An optical panoramic sight is provided on the left side.

The mortar can be rapidly broken down into its three basic components
and can be pack-loaded, man-carried or parachute dropped.

81 mm Pakistan Machine Tool light mortar

DATA
Calibre: 81 mm
Length of barrel: 1450 mm
Weight of barrel: 14.5 kg
Weight of baseplate: 14.8 kg
Weight of bipod: 12.2 kg
Weight in action: 41.5 kg
Shipping weight: 108 kg
Elevation: 45 – 85°
Rate of fire: 15 rds/min
Range: 75 – 5000 m

Manufacturer
Pakistan Machine Tool Factory Ltd., Landhi, Karachi 34.
Status
Current. Production.
Service
Pakistan armed forces.

120 mm Heavy mortar

Like the 81 mm weapon, the 120 mm mortar appears to be based upon a
Thomson-Brandt original, though with small changes in detail to suit local
military preferences. It is a conventional type of smoothbore mortar which
can be drop- or trigger-fired and is transported by means of a light
two-wheeled trailer. The mortar can be brought into action without
disconnecting it from the trailer, by simply lowering the baseplate to the
ground and then running the wheels forward until the bipod can be lowered
into contact with the ground; alternatively the entire mortar can be removed
from the trailer and placed in action on its bipod in the usual manner. The
former method is preferred when manpower is restricted, but the latter
method can be easily accomplished by only four men.

The bipod is connected to the barrel through a two-cylinder recoil buffer,
and the usual traverse, elevation and cross-level facilities are provided.
Towing is performed by attaching a towing eye to the muzzle by means of an
interrupted collar.

120 mm Pakistan Machine Tool heavy mortar

DATA
Calibre: 120 mm
Barrel length: 1746 mm with breech cap
Traverse: 17°
Elevation: 45 – 80°
Weight of barrel: 76 kg with muzzle towing attachment
Weight of baseplate: 80 kg
Weight of bipod: 86 kg
Weight of carriage: 137 kg
Weight in march order: 402 kg
Maximum range: 8950 m
Rate of fire: 12 rds/min

Manufacturer
Pakistan Machine Tool Factory Ltd., Landhi, Karachi 34.
Status
Current. Production.
Service
Pakistan armed forces.

PHILIPPINES

Philippine army mortars

The Philippine Army is equipped with two mortars of local design and
manufacture. The 60 mm M75 uses a tripod and a thick-gauge round,
stamped and welded baseplate. It is fitted with a locally manufactured M4
sight. The M53 60 mm bomb uses a locally designed aluminium alloy tail fin
unit, and the percussion fuze is also locally made. The 81 mm M2 mortar is
another Philippine design; the lower portion of the barrel is radially finned,
and the circular baseplate is stamped from heavy gauge steel.

Manufacturers
Mortars: Philippine United Machinery & Foundry Co
Bombs: Dayton Metals Corp, PO Box 435, Araneta Center Post Office,
Fiesta Carnibal Building, Cubao, Quezon City

PORTUGAL

Portuguese mortars

The former Fábrica Militar de Braço de Prata (FMBP) has been merged with another Portuguese military factory, FNMAL, specialising in small arms ammunition. The resulting company, Indústrias Nacionais de Defesa EP (INDEP), which is wholly owned by the Portuguese Government, has continued the production lines of the former factories and intends to start study and production of new items on a gradual basis, most of the production being intended for export.

INDEP makes most of the mortars for the Portuguese Army and the types

produced are: 60 mm light mortar, M/965; 60 mm commando mortar, M/968 and 81 mm medium mortar, M/937.

In addition to these, the company has designed and developed a new model in 81 mm calibre (model HP), currently in production, which is to replace the M/937 which will be dropped from production.

All models (including the HP) are offered and have been sold for export. The company also makes HE ammunition in both calibres, while production of smoke ammunition is under consideration.

60 mm M/965 light mortar

The M/965 mortar is a standard configuration weapon, designed for close support of infantry and airborne troops. It is composed of three main groups: barrel, bipod mounting and baseplate.

The barrel, of high-quality alloy steel, carries a breech piece with a fixed firing pin, easily replaceable through the back of the breech piece. The connection between the barrel and the baseplate is made by a ball and socket arrangement, which provides 360° traverse without moving the base plate. The barrel may, however, be locked on any desired azimuth to give the best accuracy. The baseplate, made of welded steel plate, is designed to provide good stability even in soft soil. The bipod mounting carries the elevation and traverse mechanism, the sight and the shock absorbers.

DATA
Calibre: 60 mm
Barrel length: 650 mm
Weight: 15.5 kg
Range: (with M49A2 bomb) 50-1820 m

Manufacturer
Indústrias Nacionais de Defesa EP (INDEP), Rua Fernando Palha, 1899 Lisbon-Codex.
Status
Production.
Service
Portuguese armed forces and other countries.

60 mm M/965 light mortar

60 mm M/968 Commando mortar

The M/968 is a light and simple model, with emphasis placed on extreme portability. It mainly consists of a tube and a button-shaped baseplate which carries a fixed firing pin. Sighting is effected by a white line painted on the barrel and a series of engraved plates fixed to a sling which doubles as a carrying sling. For firing, the gunner aligns the white line with the target, steps on the plate corresponding to the desired range and pulls the sling taut, which accounts for elevation.

An alternative model is available with a small rectangular baseplate, trip firing mechanism, and a simple clamp-on sight.

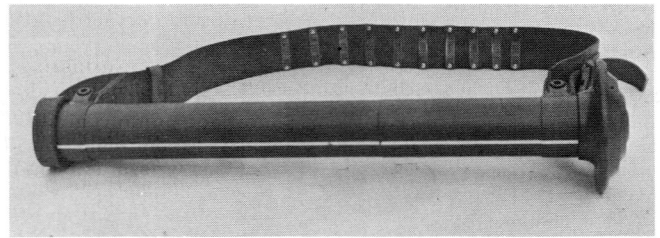

60 mm M/968 Commando mortar

Although the barrel can withstand higher pressures (being manufactured to the same specifications as the one used for the 60 mm light mortar) the manufacturer recommends that, for the tactical situations for which this mortar is designed, a maximum of only two increments (of a possible four currently supplied with the ammunition) should be used. This will provide range coverage between 50 and 1050 m.

The mortar is supplied with a muzzle cover and a protecting sleeve to provide a grip for the hand supporting the barrel.

DATA
Calibre: 60 mm
Barrel length: 650 mm
Weight: 6.6 kg
Range: 50–1050 m

Manufacturer
Indústrias Nacionais de Defesa EP (INDEP), Rua Fernando Palha, 1899 Lisbon-Codex.
Status
Production.
Service
Portuguese armed forces and other countries.

81 mm HP (FBP) mortar

The 81 mm HP (FBP) mortar is a medium mortar designed for use by infantry units. The letters HP stand for High Pressure meaning that this mortar is prepared to accept high performance ammunition of the latest types. The development of this weapon was completed in 1979. The main goals were firstly to obtain a better performance compared with that of the mortar then in production (M/937) and secondly to reduce weight as much as possible. These were successfully achieved and this model has replaced the M/937 in

production. The general configuration is quite orthodox. The barrel of high grade alloy steel, may be supplied in two different lengths, the longer one giving a slightly better performance at the cost of a little more bulk. There is a reinforcement in the lower third of the barrel, near the breech. Ignition is by a fixed firing pin, which was designed so as to be easily replaceable through the back of the breech piece.

As with the 60 mm mortar, the design allows 360° traverse and, although

the gun has been lightened as much as possible, all the usual controls are included. The elevating, traversing and levelling mechanisms, the double-shock absorbers and the sight are mounted on the bipod assembly. The weapon is supplied with two baseplates, a standard one of embossed, welded and reinforced steel plate and another, smaller one for use over hard ground.

An adaptation of this gun for firing from Jeeps has been studied and made in prototype form, but is not in production.

DATA
Calibre: 81 mm
Barrel length: 1155 mm (short): 1455 mm (long)
Weights
   Firing position: 42.5 kg (Short): 46 kg (Long)
   Barrel assembly: 15 kg (Short): 18.5 kg (Long)
   Bipod assembly: 13.5 kg
   Baseplate: 13 kg (Normal): 2.8 kg (Small)
   Sight unit: 0.85 kg
Elevation: +40 to +87°
Rate of fire: 25 rds/min
Min range: 87 m (Short); 100 m (Long barrel)
Max range: 3517 m (Short); 3837 m (Long barrel)

Manufacturer
Indústrias Nacionais de Defesa EP (INDEP), Rua Fernando Palha, 1899 Lisbon-Codex.
Status
Production. Export.

81 mm HP (FBP) mortar

SINGAPORE

60 mm ODE mortar

This is a conventional drop-loaded mortar which appears to be produced to meet high specifications. It is notable that there are spring shock buffers on the bipod collar, a collimating dial sight with elevation scales, and a wheel-protected elevating screw. The baseplate is well ribbed for soft ground and of a triangular shape.

DATA
Calibre: 60 mm
Weights
Barrel: 5.4 kg
Bipod: 4.8 kg
Baseplate: 5.4 kg
Sight: 1.2 kg
In action: 15.5 kg

Barrel length: 740 mm
Elevation: 800–1420 mils
Top traverse: 116 mils
Max range: 2555 m
Minimum range: 150 m
Ammunition: HE 1.68 kg; FM Smoke 1.68 kg; Illuminating 1.50 kg

Manufacturer
Ordnance Development and Engineering Co of Singapore (Pte) Ltd, 15 Chin Bee Drive, Postal District 2261.
Status
Current. Production.

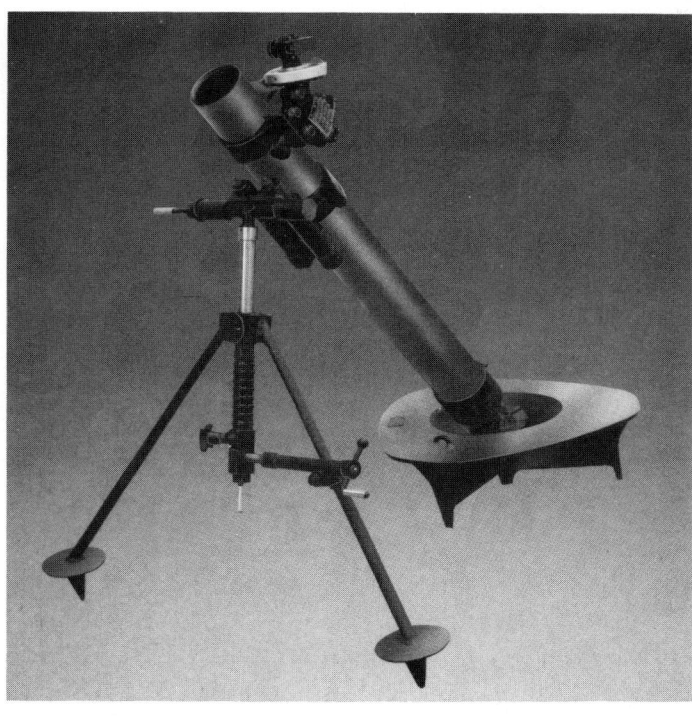

60 mm ODE mortar

81 mm ODE mortar

The 81 mm ODE mortar is of conventional drop-fired pattern. The barrel is of high-strength forged steel, the rear section being reinforced. The screwed-on breech piece contains the retractable firing pin and terminates in a ball unit which engages in the baseplate.

The baseplate is circular, of heavy duty steel alloy and is designed so as to facilitate extraction from the ground even in wet conditions. The bipod has been ergonometrically designed, with all the controls at the top for ease of operation.

The mortar can be carried in any convenient vehicle or can be broken down for man-carriage; manpack carriers are available.

DATA
Calibre: 81 mm
Weights
Barrel: 14.8 kg
Baseplate: 13.6 kg
Bipod: 13.5 kg
Sight: 1.7 kg
In action: 43.6 kg

Barrel length: 1319 mm
Elevation: 800–1550 mils
Top traverse: 200 mils
Max range: 6200 m

Manufacturer
Ordnance Development and Engineering Co of Singapore (Pte) Ltd, 15 Chin Bee Drive, Postal District 2261.

Status
Current. Production.

81 mm ODE mortar

120 mm ODE mortar

The 120 mm mortar is of conventional drop-fired design and is carried into action on a light two-wheeled trailer which also has provision for a small supply of ammunition. The mortar consists of the usual barrel, base plate and bipod and is clamped to the trailer without being dismantled.

DATA
Calibre: 120 mm
Weights
Barrel: 80 kg
Bipod: 70 kg
Baseplate: 85 kg
Sight: 1.7 kg
In action: 236.7 kg
Travelling: 512 kg (with trailer and accessories)

Barrel length: 1940 mm
Elevation: 800–1422 mils
Top traverse: 240 mils
Max range: 6500 m
Minimum range: 400 m
Ammunition: HE 13.2 kg; WP Smoke 12.8 kg; Illuminating 12.6 kg

Manufacturer
Ordnance Development and Engineering Co of Singapore (Pte) Ltd, 15 Chin Bee Drive, Postal District 2261.
Status
Current. Production.

120 mm ODE mortar

SOUTH AFRICA

Recent developments

In order to improve the performance of the existing 60 mm mortars which were built under license from Hotchkiss-Brandt some 30 years ago, a new long-range 60 mm mortar is under development. This will have a longer barrel and a reinforced breech piece but will be otherwise of standard pattern. The bomb will be more streamlined and will use an obturating ring. Using a 1.4 m barrel the new mortar will have a range of over 5000 m, and at 5 km will have a 50 per cent zone of 250 × 100 m. It is anticipated that the new mortar will replace current 81 mm designs rather than replace 60 mm weapons; the advantages of lightweight and improved lethality mean that the performance

of an 81 mm mortar can now be achieved with a more convenient weapon which saves weight and simplifies transport and logistic problems.

By present-day standards the current 120 mm mortar lacks both range and target effectiveness. Two approaches are being made to remedy this, firstly the design of a new bomb for the existing mortar and secondly an entirely new mortar to give a range of 10 km.

Information on ammunition developments will be found in the mortar ammunition section.

60 mm M4 series Commando mortars

The majority of portable and lightweight mortars made in South Africa are the 60 mm M4 series of Commando-type mortars of which the M4 and the M4 Mark 1 proved to be of the most popular designs. These mortars are of completely South African design and form an essential piece of weaponry used by infantry troops in bush war conditions.

The two models are identical except for the breech pieces in that the M4 model's breech piece is equipped with a trigger mechanism which is pulled by a lanyard, with a toggle for better grip. As soon as the lanyard is released the trigger mechanism automatically recocks, being ready for the next shot. With this configuration foot patrols may walk with a bomb in the barrel to enable quick firing in contact situations. The M4 model can also be equipped with a fixed firing pin which is issued with the mortar as part of the accessories.

The M4 Mark 1 model is only issued with a fixed firing pin. Both models are muzzle loaded.

Both models are issued with a dome-type baseplate with an elevating action limiting elevation to one plane only. Sighting is by a simple clamp-on type of sight unit. All particulars, for example distance in metres, elevation in degrees or mils as well as the charge to be used are engraved on the curved face plate. The face plate also houses a curved bubble for elevation and a small bubble for setting the weapon horizontal. The handgrip of the sight is used to elevate the barrel till the bubble corresponds with the required distance on the face plate; the charge to be used as well as the elevation can now be read off the face plate. This handle type grip is very convenient in handling or carrying the weapon.

Among a full range of accessories and equipment, some of which are optional, is a webbing sling for carriage of the mortar. This carriage sling is issued with a muzzle cover to prevent dirt and sand from entering the barrel when not in use.

DATA

	M4	M4 Mk 1
Calibre	60 mm	60 mm
Total mass of weapon as carried	7.6 kg	7.0 kg
Length of barrel	650 mm	650 mm
Mass of barrel	3.2 kg	3.2 kg
Mass of baseplate	1.8 kg	1.8 kg
Mass of breech-block assembly	1.5 kg	0.9 kg
Mass of sight unit	1.1 kg	1.1 kg
Sight illumination	3 Betalights for night use	

Manufacturer
Armscor, Private Bag X337, Pretoria, 0001.
Status
Current. Production.
Service
South African Defence Force and some other armies.

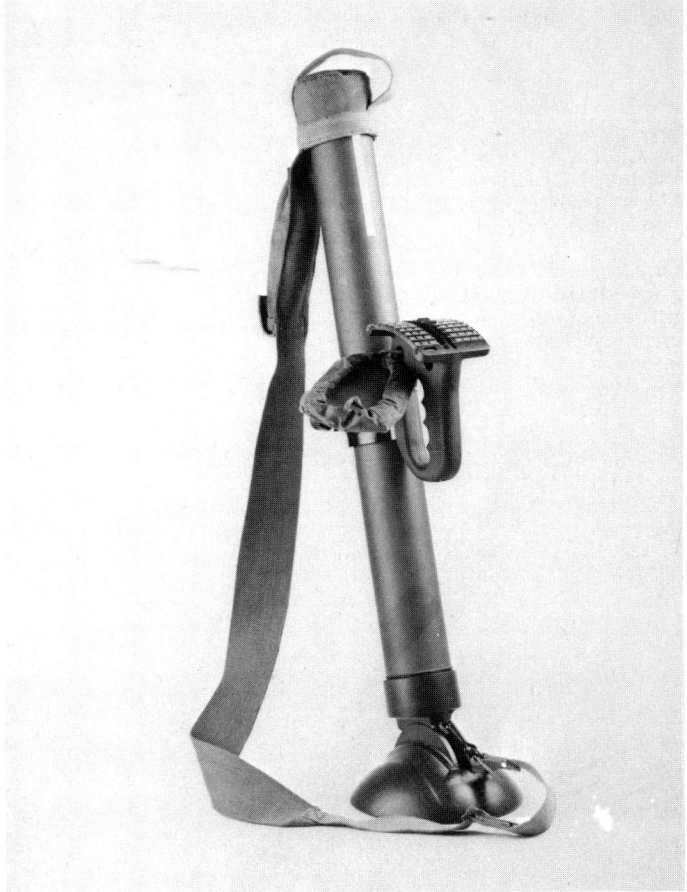

60 mm M4 Mark 1 mortar

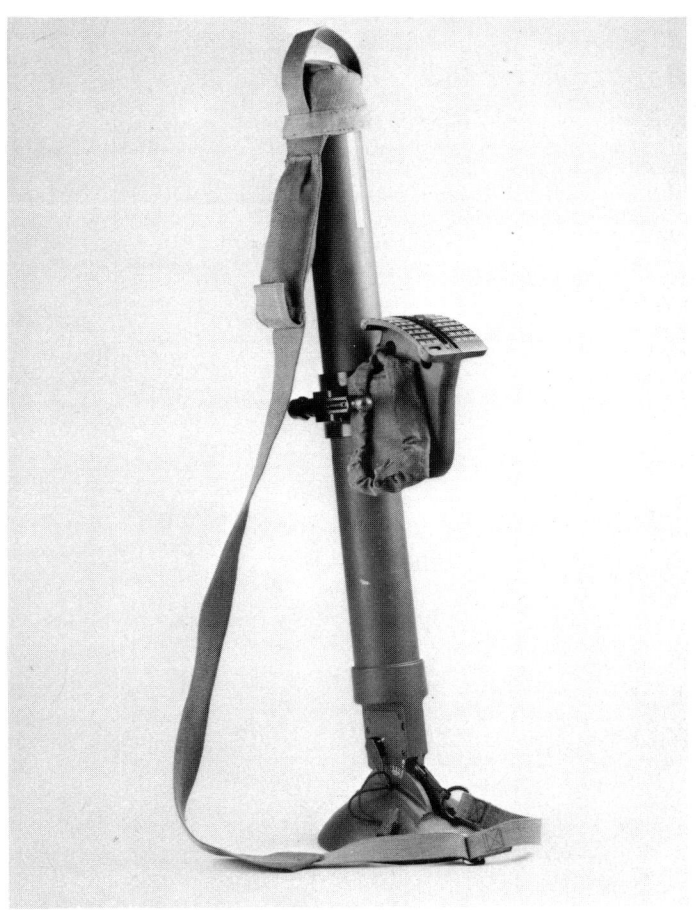

60 mm M4 mortar

60 mm mortar M1

This originated as a Hotchkiss-Brandt design built in South Africa under licence. It continued in production after the expiry of the licence agreement with some very small changes, largely intended to facilitate manufacture.

DATA
Calibre: 60 mm

WEIGHTS
Barrel: 3.2 kg
Baseplate: 6 kg
Bipod: 5 kg

Sight: 1.5 kg
Breech piece: 0.9 kg
In action: 16.6 kg

Length of barrel: 650 mm
Traverse: 300 mils
Elevation: 710–1510 mils
Minimum range: 100 m
Max range: 2100 m
Muzzle velocity: 171 m/s

Manufacturer
Armscor, Private Bag X337, Pretoria, 0001.
Status
Current. Production.
Service
South African Defence Force.

60 mm M1 mortar

81 mm mortar M3

As with the 60 mm this is based on a Hotchkiss-Brandt design, in this case the MO-81-61 light mortar and the current locally manufactured version is almost identical to the French original.

It is a conventional drop-fired smoothbore mortar, supported by bipod and baseplate and with a lensatic dial sight.

DATA
Calibre: 81 mm

WEIGHTS
Barrel: 14.5 kg
Baseplate: 14.8 kg
Sight: 1.5 kg
Bipod: 12 kg
In action: 42.8 kg

Length of barrel: 1450 mm
Traverse: 300 mils
Elevation: 800–1510 mils
Minimum range: 75 m
Max range: 4856 m
Muzzle velocity: 279 m/s

Manufacturer
Armscor, Private Bag X337, Pretoria, 0001.
Status
Current. Production.
Service
South African Defence Force.

81 mm mortar

81 mm mortar

H-019 mortar sight

The sight and sight mounting is designed for use on various weapons which includes the 120 mm M5 mortar, 81 mm M3 mortar, 60 mm M1 mortar and the 7.62 mm Vickers machine gun.

The sight is of modular South African design to ensure interchangeability, adaptability and ease of maintenance. Two variations in sighting methods are available: a lensatic type for use on the 81 mm M3 mortar, 60 mm M1 mortar

H-019 mortar sight

and the 7.62 mm Vickers machine gun, and an optical telescope for more accurate sighting for use on the 120 mm M5 artillery mortar.

Mechanically the sight comprises the following:
a dovetail connection with a quick decoupling mechanism
an elevation scale and azimuth scale graduated in 1 mil intervals. The elevation range is between + 800 and + 1600 mils for mortar applications and between - 400 and + 800 mils for machine gun applications
double levelling bubbles are provided for setting in each plane
a decoupling for fast adjustment is provided on the azimuth scales.

The azimuth scales comprise a main and a secondary scale for ease of correction adjustment, both with a 6400 mil range.

All scales and levelling bubbles are illuminated by Betalights for night use; a cover is also supplied to hide Betalights, when not in use at night.

The optical sighting mechanism is adjustable in the vertical plane and is provided with Betalights for night use.

Worm gears are used inside the sight with a built-in self-eliminating play system, to ensure minimum maintenance and accuracy at all times.

DATA
Mass of sight: 1.5 kg; (in transit case complete with accessories) 3.4 kg
Accuracy: within 3 mils
Ilumination: 11 Betalights

Manufacturer
Armscor, Private Bag X337, Pretoria, 0001.
Status
Current. Production.
Service
South African Defence Force and some other armies.

SPAIN

60 mm Models L and LL M-86 mortars

All the mortars used by the Spanish Army are manufactured by Esperanza y Cia SA, under the general trade designation ECIA. These mortars, and the ammunition manufactured by the same company, are in current use by the armies of many countries.

The Models L and M86 are easily carried by one soldier and capable of producing high explosive fire, smoke or illumination. The Model M86 is heavier and can fire to a greater range, but otherwise the two are similar and the description which follows is appropriate to both models.

The barrel of the mortar is of steel with a thread turned on the outside of the lower end, over which the breech piece fits. At the bottom of the breech piece is a ball shaped extension which fits into the socket of the baseplate. The baseplate is circular; webs are welded on to the bottom to stiffen the body and prevent any slipping when the mortar is fired at a low angle of elevation. In the centre is the socket which takes the ball of the breech piece. This socket can revolve and so the mortar has a complete 360° movement.

In the L Model the breech piece contains the trigger mechanism. This allows the firer the choice of gravity firing or trigger operation. There is a safety device which withdraws the pin when required.

In the M86 Model the firing pin is fixed for drop-firing and it is externally removable.

The sight has a telescope for viewing the aiming point and elevating and traverse scales.

The Model L mortar is provided with a tripod; the M86 uses an asymmetrical bipod.

DATA
MORTAR MODEL L
Tripod System
Calibre: 60 mm
Barrel length: 650 mm
Weight in firing position: 12 kg
Max rate of fire: 30 rds/min
Max range: Bomb Model N 1975 m; bomb Model AE 3800 m
Number of charges: 5

MORTAR LL M-86
Calibre: 60 mm
Barrel Length: 1000 mm
Weight in firing position: 17.75 kg
Max rate of fire: 30 rds/min
Max range: Bomb Model N 2350 m; bomb Model AE 4600 m
Number of charges: 5

60 mm LL M-86 mortar in firing position

Manufacturer
Esperanza y Cia SA, Marquina (Vizcaya).
Status
Current. Production.
Service
Spanish forces and several others.

60 mm Commando mortar

This is a light mortar operated 'and carried by one man. The barrel is a plain
tube with a small circular button type baseplate at the bottom. There is no
mount. The soldier supports the barrel with his left hand, grasping the canvas
sleeve placed around it to protect his hand. A very simple sight is held on the
right of the tube by a 'hose clip' fastener.

Of all the light mortars in use today, this is one of the simplest.

DATA
Calibre: 60 mm
Barrel length: 650 mm
Total weight: 6.5 kg
Firing system: drop firing
Max range: Bomb Model N 1060 m; bomb Model AE 1290 m
Number of charges: Bomb Model N 2' bomb Model AE 1

Manufacturer
Esperanza y Cia SA, Marquina (Vizcaya).
Status
Current. Production.
Service
Spanish forces and several others.

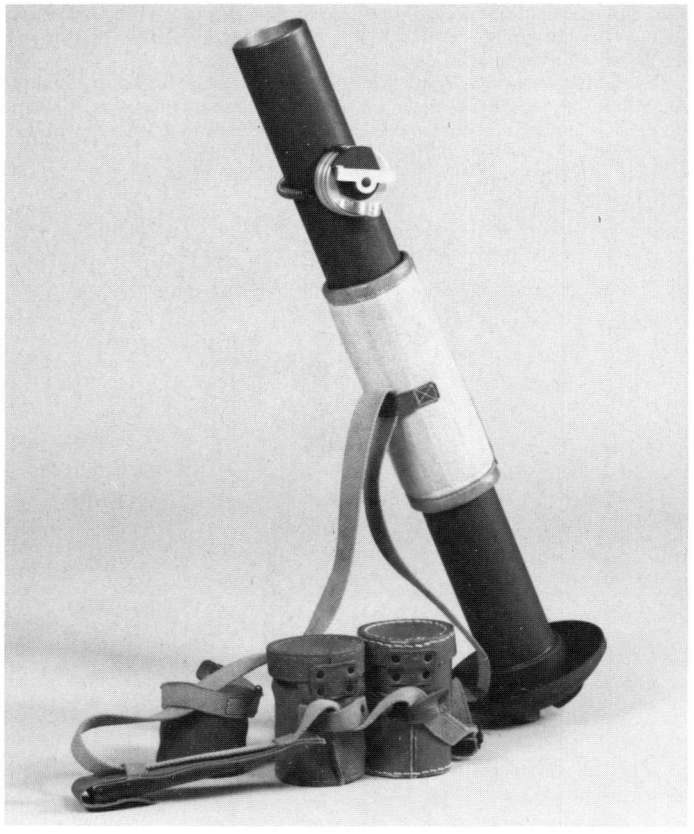

60 mm Commando mortar with accessories

60 mm Model MC-2 gun-mortar

Still only in prototype form, the ECIA 60 mm gun-mortar is similar to the
Brandt 60 mm gun-mortar in concept and execution but was developed
entirely by ECIA. It has a smoothbore and may be either muzzle- or
breech-loaded. The manufacturer anticipates that it will be used to arm light
armoured vehicles and patrol boats. The firing mechanism allows the firing
pin to be fixed, permitting 'automatic fire', or manually cocked and fired. The
breech is locked by a rotary latch mechanism and the firing pin is
automatically retracted while the breech is open. Provision is made for a
mechanical firing linkage from the vehicle installation. The mounting for the
mortar incorporates a hydraulic buffer with automatic run-out.

DATA
Overall length: 1396 mm
Shot travel: 905 mm
Weight, excluding mounting: 85 kg
Weight of recoiling mass: 17 kg
Recoil distance: 76 mm
Elevation: -10 to +80°

Manufacturer
Esperanza y Cia SA, Marquina (Vizcaya).
Status
Development.

60 mm Model MC-2 gun-mortar

81 mm Models L-N and L-L mortar

The differences between these models are very slight and amount to little more
than the length of the tube.

The barrel is a steel tube, strengthened at the rear end, and screw threaded
on the outside of the bottom to take the breech piece. This contains the firing
mechanism and allows either drop firing or firing by trigger operation. The
ball extension of the breech piece fits into the socket of the circular baseplate
and is held by a spring latch. Two different barrel lengths are available.

The tripod follows the same pattern as that of the 60 mm mortar, with the
three legs arranged one forward and two to the rear, fitting into the fabricated
tripod head. The mortar is designed to be manportable in three loads.

DATA
MORTAR MODELS L-N AND L-L
Tripod System
Calibre: 81 mm
Barrel length: L-N 1150 mm; L-L 1450 mm
Weight in firing position: L-N 43 kg; L-L 47 kg
Firing system: drop or trigger
Max rate of fire: 15 rds/min
Max range, Model L-N: Bomb Model NA 4125 m; bomb Model AE
6200 m
Max range, Model L-L: Bomb Model NA 4680 m; bomb Model AE
6900 m

Manufacturer
Esperanza y Cia SA, Marquina (Vizcaya).
Status
Current. Production.
Service
Spanish forces and several others.

81 mm Model L-L mortar in firing position

81 mm Model LN M86 and LL M86 mortars

This is the latest version and introduces an asymmetrical bipod. It is manufactured in two different barrel lengths, 1150 mm and 1450 mm. The barrel is made of special high alloy steel and manufactured from a solid bar, and it is strengthened at the breech end. The ball extension fits into the socket of the baseplate, allowing a 6400 mil movement. It has a drop-firing system, the firing pin being externally removable.

The mortar is equipped with a new sight with night vision and, like the old model, is manportable in three loads.

DATA
MORTAR MODELS LN M-86 AND LL M-86
Bipod System
Calibre: 81 mm
Barrel length: LN 1150 mm; LL 1450 mm
Weight in firing position: LN 41 kg; LL 43.5 kg
Firing system: drop firing
Max rate of fire: 15 rds/min
Max range, Model LN: Bomb Model NA 4125 m; bomb Model AE 6200 m
Max range, Model LL: Bomb Model NA 4680 m; bomb Model AE 6900 m

Manufacturer
Esperanza y Cia SA, Marquina (Vizcaya).
Status
Current. Production.
Service
Spanish forces and several others.

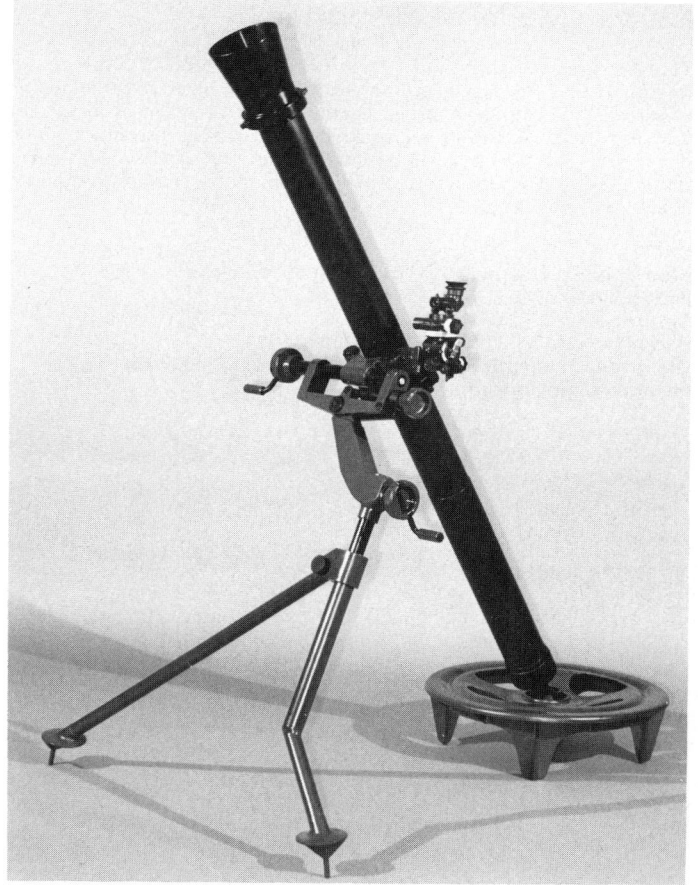

81 mm Model LL M-86 mortar in firing position

120 mm Model L mortar

The Model L has all the characteristics of the other ECIA mortars with the tripod mount, simple barrel and circular base-plate. It can be fired either by gravity drop or by trigger control. The mortar is transported on a carriage towed by a light vehicle and it is also capable of being mounted on armoured personnel carriers.

The baseplate lies over the axle and is held in position by a number of clips. The barrel and mount rest on a frame above the baseplate and are held in position by straps.

DATA
MORTAR
Barrel and breech piece: 61 kg
Baseplate: 100 kg
Mount: 40 kg
Sight: 2 kg
Carriage: 165 kg with tools and accessories
Weight in firing position: 203 kg

120 mm Model L mortar in travelling position

Length of barrel: 1600 mm
Max rate of fire: 20 rds/min
Max range: Bomb Model L 6725 m; Bomb Model AE 8000 m
Manufacturer
Esperanza y Cia SA, Marquina (Vizcaya).
Status
Current. Available.
Service
Spanish forces and several others.

120 mm Model L mortar in firing position showing large baseplate

120 mm Model M-86 mortar

The M-86 is a new design of mortar, lighter than the Model L and with two lengths of barrel. The design incorporates a light alloy bipod and an alloy steel barrel which is chromium plated internally. The firing pin is locked for drop-firing and is externally removable. The mortar, when dismantled, fits on to a very light but strong two-wheeled trailer, the frame of which is of tubular steel with brackets for the retention of the mortar unit. It can also be mounted in several types of APC.

DATA
MORTAR MODELS M-86 (M120-13 AND M120-15)
Bipod System
Calibre: 120 mm
Barrel length: M120-13 1600 mm; M120-15 1800 mm
Weight in firing position: M120-13 155 kg; M120-15 160 kg
Firing system: drop firing

Max rate of fire: 20 rds/min
Max range M120-13: Bomb Model L 6725 m; bomb Model AE 8000 m
Max range M120-15: Bomb Model L 7000 m; bomb Model AE 8250 m

Manufacturer
Esperanza y Cia SA, Marquina (Vizcaya).
Status
Current. Production.
Service
Spanish Army; South American, Middle Eastern and African armies.

120 mm Model M-86 (M120-13) in firing position

120 mm M-86 (M120-13) barrel adapted for use in APCs

SWEDEN

81 mm m/29 (1929) mortar

The 81 mm m/29(1929) mortar is the standard light mortar of the Swedish Army and has been in service for over 50 years. It is basically the French 81 mm Stokes/Brandt Model 1917 mortar produced under licence in Sweden. In 1934 the French sights were replaced by sights of Swedish design. In appearance the mortar is similar to most other mortars of that period: rectangular base with carrying handle, bipod with chains connecting the legs and an anti-cant device on the left leg.

DATA
Calibre: 81.4 mm
Weight in firing position: 60 kg
Barrel length: 1000 mm

Elevation: 45-80°
Traverse: 90°
HE bomb weight: 3.5 kg
Muzzle velocity: (minimum) 70 m/s; (max) 190 m/s
Max range: 2600 m
Rate of fire: 15-18 rds/min

Status
Current. Not in production.
Service
Swedish forces.

120 mm m/41 D (1941) mortar

The 120 mm m/41D (1941) mortar is the standard heavy mortar of the Swedish Army. It is basically the Finnish Tampella 120 mm Model 1940 mortar built in Sweden. In 1956 Sweden built a number of Hotchkiss-Brandt M-56 baseplates under licence to replace the earlier Tampella baseplates. More recently the sights have been replaced. The mortar is issued on the scale eight per mortar company of the infantry brigades and six per mortar company of the Norrland brigades.

DATA
Calibre: 120.25 mm
Weight travelling: 600 kg
Weight in firing position: 285 kg
Barrel length: 2000 mm
Elevation: 45-80°
Traverse: 360°
HE bomb weight: 13.3 kg
Muzzle velocity: (minimum) 125 m/s; (max) 317 m/s
Max range: 6400 m
Rate of fire: 12-15 rds/min

Status
Current. No longer in production.
Service
Swedish forces and Ireland.

120 mm m/41D mortar

SWITZERLAND

60 mm Model 87 mortar

This 60 mm mortar has been accepted for service with the Swiss Army and entered production in 1988. Some elements of the design and ammunition are licensed from Thomson-Brandt Armaments of France.

The mortar is a conventional drop-loaded smoothbore with fixed firing pin. A simple bipod is attached to the barrel by means of a collar, and the rear of

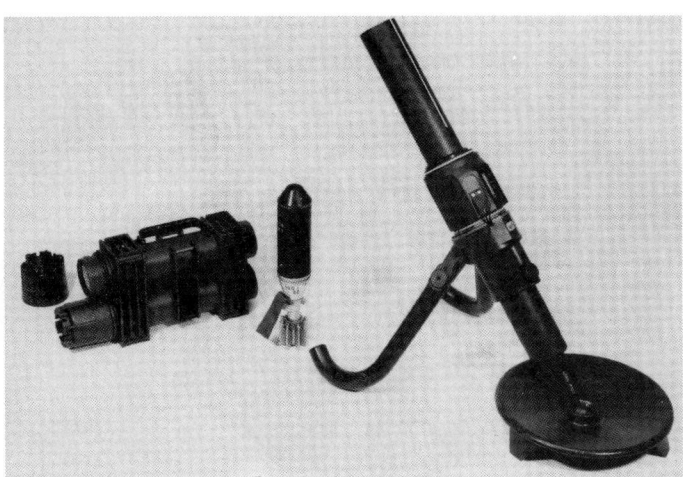

60 mm Model 87 mortar with bomb and 'Duopack'

Loading the 60 mm Model 87 mortar

the barrel rests on a small circular baseplate. Elevation of the barrel is done manually, by slipping the collar up or down the barrel, and the amount of

60 mm Model 87 mortar elevation scale, showing setting for illuminating bomb

elevation can be determined by a bubble-type unit mounted in the bipod collar and visible from behind the mortar. This unit can be adjusted to provide elevation data for different types of ammunition. Direction is assessed by the firer by using a white line marked on the axis of the barrel, aligning this with his target.

The ammunition is packed and carried in a special 'Duopack' of plastic, containing two bombs. The Duopack is designed so that any number can be clipped together for transport or stacking as a single unit.

DATA
Calibre: 60 mm
Barrel length: 845 mm
Weight in action: 7.6 kg
Traverse: 6400 mils
Elevation:
    Illuminating bombs: 800–1100 mils
    HE and smoke bombs: 800–1520 mils
Operational range: up to 1000 m
Rate of fire: 15 rds/min (intense); 3 rds/min (normal)

Manufacturer
Swiss Federal Arms Factory, Postfach 3000, Stauffacherstrasse 65, CH-3000 Berne 22.
Status
Current. Production.
Service
Swiss Army.

81 mm Model 1933 mortar

This mortar is very similar to the German, US and Soviet mortars produced in this calibre in the 1930s. In the shape of the baseplate, the rather complicated arrangements for the socket annd the layout of the elevating and traversing screws, it has much in common with the US M1 mortar. The bipod legs, with their spiked feet, are positioned by the chin connecting them. The cross-levelling gear is operated from the left leg of the bipod and the elevating gear tube is displaced as the connecting rod is moved across. The tube passes through the junction of the bipod legs and there is a gear case at the junction from which projects the elevating handle. The top of the elevating column is attached to the nut through which passes the traversing screw. The wheel on the right-hand end of the traversing screw rotates the thread which moves the nut along and with it goes the barrel. The sight is connected to the left end of the yoke. There are two shock absorbers interposed between the barrel latch and the yoke and they ensure that the yoke is correctly positioned before a round is fired.

In Swiss service the mortar is known as the 8.1 cm Mw 33.

DATA
MORTAR
Weights: (all up) 62 kg; (barrel) 21 kg; (mounting) 18 kg; (baseplate) 21 kg; (sight mount and sight) 2 kg
Lengths: (barrel) 1265 mm; (shot travel) 1155 mm; (mount) 940 mm; (base-plate) 680 × 410 mm
Top traverse: 8°
Traverse: (moving bipod) 45-56°
Elevation: 45-90°

Manufacturer
Swiss Federal Arms Factory, Postfach 3000, Stauffacherstrasse 65, CH-3000 Berne 22.
Status
Obsolescent.
Service
Swiss Army.

81 mm Model 1933 mortar

81 mm Model 1972 mortar

This is an improved Mw 33. The baseplate is a flat disc with webs welded on to the under side. The bipod has been cleaned up by having the screw threads fully enclosed and the cross-levelling gear, although working in exactly the same way, is much easier to operate. A very considerable weight saving has been made. Like its predecessor, the mortar breaks down into three manportable loads. Its military designation is 8.1 cm Mw 72.

DATA
Weights: (all up) 45.5 kg; (barrel) 12 kg; (mounting) 15 kg; (baseplate) 16.5 kg; (sight mounting and sight) 2 kg
Lengths: (barrel) 1280 mm; (shot travel) 1154 mm; (mount) 990 mm; (base-plate diameter) 550 mm
Top traverse: 10°
Traverse: (moving bipod) 360°
Elevation: 45-85°
Max range: 4100 m

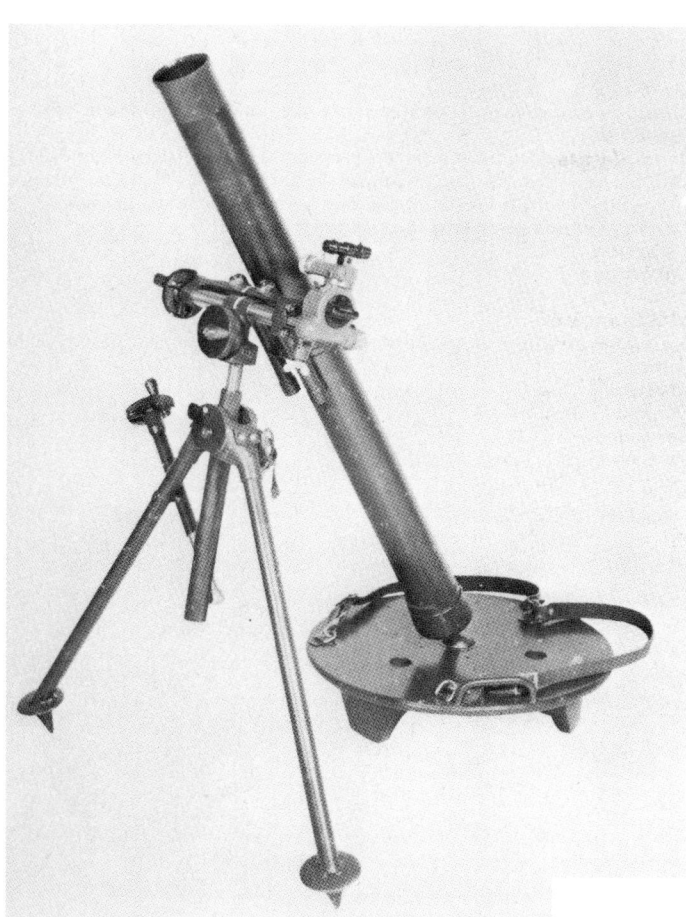

Manufacturer
Swiss Federal Arms Factory, Postfach 3000, Stauffacherstrasse 65, CH- 3000 Berne 22.
Status
Current.
Service
Swiss Army.

81 mm Model 1972 mortar

120 mm Model 64 mortar in M106 mortar carrier

This mortar is normally fired from inside the carrier using a special baseplate built into the structure of the vehicle. A conventional base plate is carried externally on the side of the vehicle and if necessary the mortar may be dismounted and brought into action as a conventional infantry mortar.

An 81 mm sub-calibre training device is used which consists of a tube inserted in the barrel and enables firings to be carried out on smaller ranges and at reduced cost.

The 120 mm mortar is muzzle loaded and the firing mechanism is manually controlled using an external firing lever.

The mortar is known by the Swiss Army as the 12 cm Mw Pz 64.

DATA
MORTAR
Weights: (equipment overall) 239 kg; (barrel) 87 kg; (mounting) 44 kg; (baseplate) 95 kg; (sight mounting and sight) 4 kg
Lengths: (barrel length) 1770 mm; (shot travel) 1524 mm; (mount) 1400 mm; (baseplate) 950 mm
Top traverse: 10°
Traverse: (moving bipod) 60°
Elevation: 45-85°
Max range: 7.5 km

Manufacturer
Swiss Federal Arms Factory, Postfach 3000, Stauffacherstrasse 65, CH-3000 Berne 22.
Status
Current.
Service
Swiss Army.

120 mm M106 mortar carrier in action

Interior of carrier showing 120 mm M64 mortar, and ammunition ready for use

120 mm Model 74 mortar on 2-wheel carriage

This mortar in this form was produced from 1974 – 78. The mortar itself is the same as that used in the carrier, that is the Model 64. The equipment is designed to be towed by a light vehicle or carried as an underslung load on a helicopter. The carriage is of a simple design with a cross axle on which the weight of the mortar is taken. Alongside the mortar is an ammunition box allowing six ready for use rounds to be carried on the mounting.

The mortar itself is of orthodox construction and is a scaled-up version of the 81 mm weapon, Model 1972.

DATA
Weights: (baseplate) 96 kg; (barrel) 87 kg; (mount, with sight) 56 kg; (limber) 250 kg
Lengths: (length with mortar on carriage) 2350 mm; (length unloaded) 1140 mm; (overall width) 1510 mm; (height loaded) 1130 mm; (height unloaded) 970 mm; (wheel base) 1285 mm; (ground clearance) 240 mm
Towing speeds on roads: (max) 80 km/h
Elevation: 45–85°
Max range: 7.5 km

Manufacturer
Swiss Federal Arms Factory, Postfach 3000, Stauffacherstrasse 65, CH-3000 Berne 22.
Status
Current. Production.
Service
Swiss Army.

120 mm Model 74 mortar in action position

120 mm Model 74 mortar on two-wheel carriage

TAIWAN

Mortars

Taiwan currently produces a full range of mortars and mortar bombs for both home and export markets. These include: 60 mm Type 44 mortar (M-19); 81 mm Type 44 mortar (M-29); 107 mm Type 62 mortar (M-30); 120 mm Type 63 mortar.

Manufacturer
Combined Service Forces; Hsing Hua Co Ltd, PO Box 8746, Taipei.
Status
Current. Production. Available.

120 mm Type 63 mortar made by Combined Service Forces

TURKEY

81 mm MKE UT1 mortar

DATA
Calibre: 81 mm
Weight: (barrel) 28.1 kg; (baseplate) 19.6 kg; (bipod) 23.2 kg
Length of barrel: 1453 mm
Rate of fire: (normal) 16 rds/min
Max number of charges: 6
Weight of MKE Mod 214 HE bomb: 4.68 kg

RANGE (WITH MKE MOD 214 HE BOMB)
Max range: 6000 m
Minimum range: 400 m
Max pressure: 625 kg/cm²

MOUNTING
Traverse: 90 mils left and right
Elevation: 40-85°

81 mm MKE UT1 mortar

Manufacturer
Makina ve Kimya Endustrisi Kurumu, Tandoğan Meydani, Ankara.

Status
Turkish Army.

81 mm MKE NTI mortar

DATA
Calibre: 81 mm

WEIGHTS
Barrel: 21.4 kg
Baseplate: 14.6 kg
Bipod: 22.6 kg
Length of barrel: 1150 mm
Rate of fire: (normal) 16 rds/min
Max number of charges: 6
Weight of M43A1B1 HE bomb: 3.2 kg
Range: (with US M43A1B1 HE bomb) (minimum) 200 m; (max) 3200 m
Max pressure: 425 kg/cm²

MOUNTING
Traverse: 90 mils left and right
Elevation: 40–85°

Manufacturer
Makina ve Kimya Endustrisi Kurumu, Tandoğan Meydani, Ankara.
Status
Turkish Army.

81 mm MKE NTI mortar

MKE 120 mm Tosam HY 12 DI mortar

DATA
Calibre: 120 mm

WEIGHT
Barrel: 143 kg
Baseplate: 152 kg
Bipod: 275 kg
In action: 570 kg
Length of barrel: 1900 mm
Rifling: 40 grooves. 1 turn in 16.95 calibres
Range: (with MKE Mod 209 HE) (max) 8000 m; (minimum) 1500 m
Rate of fire: (normal) 5 rds/min
Max number of charges: 13
Weight of MKE Mod 209 HE bomb: 16 kg
Traverse: 160 mils left and right
Elevation: 35–80°

Manufacturer
Makina ve Kimya Endustrisi Kurumu, Tandoğan Meydani, Ankara.
Status
Current. Production.
Service
Turkish Army.

MKE Tosam HY 12 DI mortar

UNION OF SOVIET SOCIALIST REPUBLICS

82 mm M-36 mortar

This was the first of the Soviet 82 mm mortars. It was very similar to the US 81 mm M1 mortar and the German 82 mm was based on this design. The ammunition was conventional with a primary cartridge and six increments. Three different types of sight were used, the MP-1 and MP-82 optical sights and a sheet metal aiming circle-clinometer.

DATA
Calibre: 82 mm
Total firing weight: 57.3 kg
Total length of barrel: 1288 mm
Max range: 3100 m
Ammunition: HE and smoke

Manufacturer
State factories.
Status
Obsolescent. Not in production.
Service
Not in Soviet or Warsaw Pact armies, but likely to be found in Asia and Africa.

Soviet 82 mm M-36 mortar

82 mm M-37 mortar

This was a modification of the M-36 having a circular baseplate. It may be distinguished by the cross-levelling and connecting rod on the right leg rather than the left. However when using this as an identifying feature the position of the elevating handle should be noted. This should emerge from the gear case at the rear. If not, the bipod gear will be on the left. The circular baseplate was a great success and although a rectangular plate has been used in the mountain divisions, it is rarely seen. The sight is the MPM-44.

DATA
Calibre: 82 mm
Length of barrel: 1220 mm
Weight: 56 kg
Elevation: (max) 85°; (minimum) 45°
Top traverse: 6°
Rate of fire: 15–25 rds/min
Bomb weight: 3.05 kg
Muzzle velocity: 211 m/s
Max range: 3000 m
Minimum range: 100 m
Detachment: 5

Manufacturer
State factories.
Status
Current.
Service
Most Warsaw Pact armies. Produced in China as the Type 53.

82 mm M-37 mortar was restored to favour after 1941–45 war

82 mm 'new' M-37 mortar

This is the current 82 mm mortar. It is the same as the original M-37 but has a lightened tripod and baseplate. The muzzle is fitted with a double-loading stop which prevents the possibility of dropping a second bomb down before the first has cleared the muzzle.

The ammunition for the 82 mm mortar has remained unchanged since its inception in 1936. However the Vietnamese used a bomb fitted with a chemical delay fuze. The firing pin was held back against the spring by a plastic arrestor. A corrosive solution attacked the plastic and when it was sufficiently weakened the firing pin drove forward into the cap and the filling was detonated. This was achieved by having the acid solution in a glass tube loaded into the detonator just before firing. When the detonator was tightened in its housing the glass tube was broken.

Status
Current.
Service
Soviet Army. Also, with various variations in: Albania, Bulgaria, China, Congo, Cuba, Czechoslovakia, Egypt, East Germany, Ghana, Indonesia, Iraq, North Korea, Syria, Vietnam, Yugoslavia.

82 mm 'new' M-37 mortar showing muzzle safety device fitted to prevent double loading

82 mm M-41 mortar

This was introduced to improve the performance and mobility of the M-37 mortar. Instead of the conventional bipod and yoke, two short legs supported the long elevation column and the traversing gear. Cross-levelling was accomplished in the same way as on the 50 mm M-40 mortar, with a linkage between the elevating shaft and the traversing screw. At the foot of each bipod leg was a stub axle to which a wheel could be fitted. When the mortar came out of action the bipod was folded back and clamped to the circular base-plate. The wheels were attached and the mortar was towed from the muzzle end by whatever type of transport was available – man, mule or motor. However, the equipment was found to be less successful and less stable than the M-37 when firing and the slight advantages of greater mobility were discounted against a much poorer ballistic performance. It is no longer in first line service, but has been supplied to many Soviet-influenced countries and may be met almost anywhere.

DATA
Calibre: 82 mm
Length of barrel: 1220 mm
Weight: (in firing position) 52 kg; (in travelling position) 58 kg
Elevation: (minimum) 45°; (max) 85°
Top traverse: 5°
Rate of fire: 15–25 rds/min
Bomb weight: 3.05kg
Muzzle velocity: 211 m/s
Max range: 2550 m
Minimum range: 100 m
Detachment: 5

Manufacturer
State factories.
Status
Obsolescent. No longer in production.
Service
Some Soviet-influenced and Third World countries.

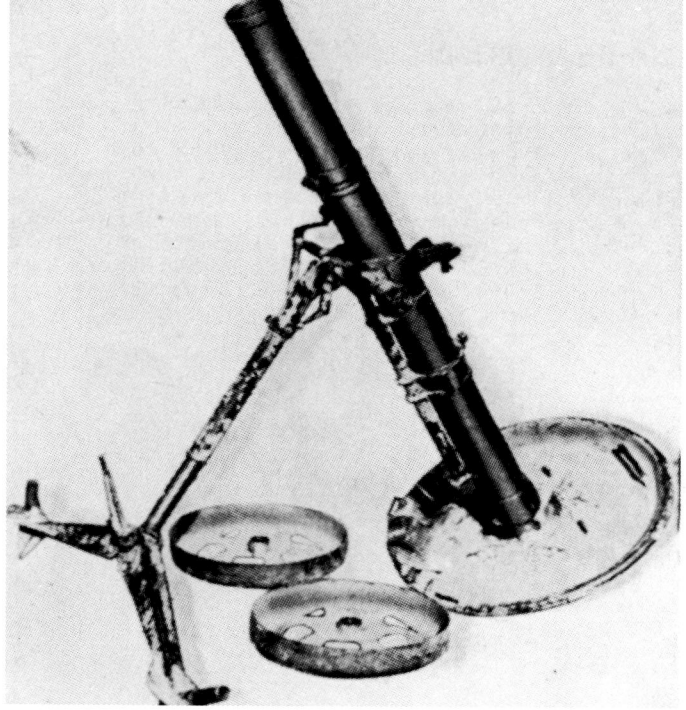

82 mm M-41 mortar has unconventional mount with short legs and long elevation column. Wheels are removed before firing

82 mm automatic mortar AM 'Vasilek'

This weapon was apparently introduced into Soviet service some time in 1971 but its existence was not revealed until late in 1983. It is officially designated AM for 'Avtomaticheskiy Minomet' or automatic mortar; its full nomenclature is not known, but it also bears the familiar name of 'Vasilek', in line with the current Soviet practice of giving nicknames to equipment.

The only available photograph of the weapon (not reproduced here due to its poor quality) shows it to be mounted on a light two-wheeled split-trail carriage resembling that of the 76 mm mountain gun M1969. It is fitted with some form of recoil system and a firing jack and can fire through the full range from point-blank to about 80° elevation. This allows the weapon to function either as a conventional mortar or in the direct-fire role as a support or anti-tank weapon.

No details are known of the ammunition apart from the fact that high explosive fragmentation and shaped charge anti-tank projectiles are available. The round is breech loaded, but details of the automatic mechanism and feed system are not known.

It is understood that at least four versions can be found mounted on different vehicles; reports from Afghanistan speak of a turreted mounting in either BTR-60 or BTR-70 APCs, and other reports refer to similar mounting in BMP-1 and BMD-1 IFVs.

The towed weapon is issued on a scale of one battery of six mortars per motor rifle battalion; airborne units and those equipped with IFVs use self-propelled versions.

107 mm M-38 mortar

This mortar was produced after the 120 mm mortar and is a reduced size copy of that weapon designed for mountain use and animal transport. The weapon is carried complete on a two-wheeled trolley. It has now been replaced with the M-107 which is an improved version.

DATA
Calibre: 107 mm
Length of barrel: 1670 mm
Weight: (in firing position) 170 kg; (in travelling position) 340 kg
Elevation: (minimum) 45°; (max) 80°
Top traverse: 3°
Rate of fire: 15 rds/min
Bomb weight: (HE heavy) 9 kg; (HE light) 7.9 kg
Muzzle velocity: (HE heavy) 302 m/s; (HE light) 263 m/s
Minimum range: (HE heavy) 800 m
Max range: (HE heavy) 5150 m; (HE light) 6300 m
Detachment: 5

Manufacturer
State factories.
Status
Obsolescent.
Service
Soviet reserves and in China, North Korea and possibly Vietnam.

107 mm M-38 mortar

120 mm M-38 mortar

The Soviet 120 mm mortar occupied the same place in the Infantry Division as the 107 mm did in the Mountain Division. During the Second World War and afterwards the 120 mm mortar was found in the divisional artillery as well as in the separate non-divisional artillery units. In recent years motor rifle battalions have each been issued with six of these mortars.

The M-38 was a great deal more mobile than most mortars of that calibre. It was lifted bodily on a two-wheeled transporter and towed behind any suitable vehicle. It can be broken down to three loads for animal pack transport. It can be fired directly by dropping the bomb down on a protruding firing pin, or alternatively a trigger device can be used.

120 mm M-38 in travelling order

120 mm M-38 mortar fitted with muzzle safety device

120 mm M-43 mortar

Although the M-38 was a very successful mortar and was copied by the Wehrmacht, it was modified into the M-43. This has longer shock absorber cylinders but retains the very efficient baseplate of stamped circular form and the two-wheeled carriage. The M-43 is produced in China as the Type 53.

DATA
Calibre: 120 mm
Length of barrel: (overall) 1854 mm
Width: (in travelling position) 1548 mm
Height: (in travelling position) 1206 mm
Weight: (in firing position) 274.8 kg; (in travelling position) 500 kg
Track: 1210 mm
Elevation: 45–80°
Top traverse: 8°
Rate of fire: 12–15 rds/min
Bomb weight: 15.4 kg

120 mm M-43 mortar in travelling order

120 mm M-43 mortar

Muzzle velocity: 272 m/s
Max range: 5700 m
Minimum range: 460 m
Detachment: 6

Manufacturer
State factories.

120 mm M-43 mortar

Status
Obsolescent. Not in production.
Service
Soviet Army, also: Albania, China, Czechoslovakia, Egypt, East Germany, Iraq, North Korea, Romania, Syria, Vietnam, South Yemen, Yugoslavia.

160 mm mortars

The Soviet Army introduced the 160 mm M1943 to provide the infantry divisions with a weapon producing a heavy weight of high explosive but not making undue demands on the hard pressed manufacturing resources. It was the heaviest mortar used by the Soviet Army during the Second World War. Because of the long barrel it had to be a breech loading weapon and the barrel pivoted for loading about trunnions placed not far from the centre point. It was towed by the muzzle, using either an armoured personnel carrier or a

heavy lorry. After the war it was used by Poland, Romania and Bulgaria and deployed in troops of four mortars. It has now been replaced throughout the Warsaw Pact countries by the 160 mm M-160 mortar. This is very similar in design and the method of breech loading has been preserved but it has a longer barrel and a greater range. It was originally used as a divisional mortar in all types of division but is now employed with the mountain divisions where its range, explosive shell capacity and high angle of fire are very useful.

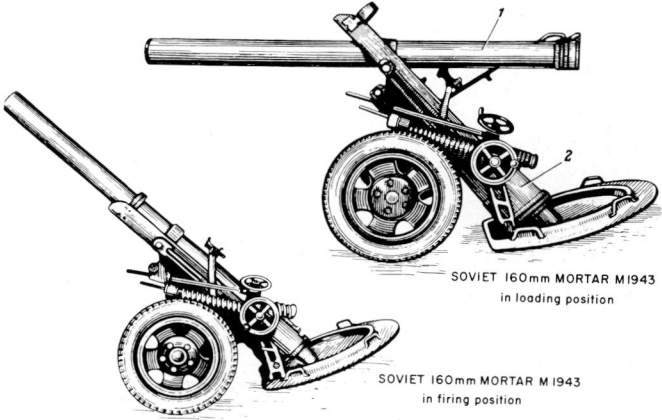

160 mm M-43 mortar in loading and firing positions

Detail of trunnions about which barrel is lowered for loading

160 mm M-43 mortar

160 mm M-160 mortar

DATA

	M-43	M-160
Calibre	160 mm	160 mm
Length of barrel	3030 mm	4550 mm
Length in travelling position	3985 mm	4860 mm
Width in travelling position	1770 mm	2030 mm
Height in travelling position	1414 mm	1690 mm
Weight		
in firing position	1170 kg	1300 kg
in travelling position	1270 kg	1470 kg
Track	1750 mm	1750 mm
Elevation		
minimum	45°	50°
max	80°	80°
Traverse	25°	24°
Rate of fire	3 rds/min	2–3 rds/min
Bomb weight	40.8 kg	41.5 kg
Muzzle velocity	245 m/s	343 m/s
Max range	5150 m	8040 m
Minimum range	630 m	750 m
Detachment	7	7

Manufacturer
Current, but no longer manufactured.
Status
Soviet Army M-160, also used by China and Egypt.
M-43, Albania, China, Czechoslovakia, Egypt, East Germany, North Korea, and Vietnam.

UNITED KINGDOM

51 mm mortar

The 51 mm mortar system is in service with the British Army.

The barrel is a steel tube, bell mouthed at the muzzle to assist in loading bombs and also to add some strength to the vulnerable end of the barrel. The barrel is held to the breech by four spring-loaded plungers which lock into four holes in the barrel. The breech assembly is a reversion to the well-tried and proven 2 in (5 cm) mechanism with several detailed improvements. It is trigger operated with a short firing lever tripping the firing pin when pulled over centre.

A maximum range of 800 m can be achieved for all ammunition types. This range is realistically the furthest that a simple mortar of this type can be expected to fire with acceptable accuracy and consistency. The mortar provides the infantry platoon with a means of immediate close-range illumination, screening smoke, and anti-personnel capability under the direct control of the platoon commander.

The most interesting part of the design is probably the means adopted to achieve the minimum range of 50 m. This is a short-range insert (SRI) which is dropped into the barrel from the muzzle. It is a rod with a long firing pin along its longitudinal axis. The lower end of the firing pin rests on the firing pin in the breech piece and is driven forward when the mortar is fired. Thus the bomb does not reach the bottom of the tube and the short-range is achieved by increasing the chamber volume and so reducing the working chamber pressure; at the same time there is a reduction in shot travel. When not in use the SRI is held in a clip attached to the muzzle cover.

DATA
51 MM MORTAR
Barrel
calibre: 51.25 mm
weight: 2.6 kg
length overall: 750 mm
outside diameter: 55 mm
Breech piece assembly: (weight including baseplate) 3.05 kg

51 mm mortar firing in lower register

51 mm mortar dismantled, with accessories, showing L3A1 Trilux sight and short-range insert

Infantryman carrying 51 mm mortar

Total weight: (including carrying harness) 6.275 kg
Accuracy: 2% probable error in range; 3 mils in line
Rate of fire: 3 rds/min for 5 mins; 8 rds/min for 2 mins

51 mm mortar firing in upper register

Manufacturer
Royal Ordanance plc, Guns & Vehicles Division, Nottingham.
Status
Current. Production.
Service
In service with the British Army.

3 in mortar

The 3 in (7.62 cm) mortar was introduced into British Service in 1936. It has now been replaced by the 81 mm mortar L16 in the British Army but remains in service in other countries. It was fully described in *Jane's Infantry Weapons 1978*, page 547, and the reader is referred to that for all background detail.

The reduced data table here is sufficient to identify the mortar should it be encountered.

DATA
MARK 5 EXCEPT WHERE SPECIFIED
MORTAR
Barrel
calibre: 81.48 mm
muzzle to front of striker stud: 1150 mm
total weight: 20 kg (Mark 4, 22.2 kg)
Mounting
weight including cradle: 20.5 kg
traverse, angle of left and right: 5.5°
means of firing: striker stud
Baseplate: No 5 (Mark 4 has No 6)
depth: 203 mm
length: 521 mm
width: 368 mm
weight: 19.5 kg (Mark 4, 15.54 kg)
Range: 450–2560 m
Rate of fire: up to 15 rds/min

AMMUNITION
Mark 4 high explosive: (max weight, complete) 4.42 kg

Manufacturer
Royal Ordnance plc.
Status
Obsolete. No longer made.
Service
Some former Commonwealth countries.

3 in mortar with Mark 5 barrel, No 6 baseplate and Mark 5 bipod

81 mm L16 ML mortar

The 81 mm mortar is normally deployed as part of a section used conventionally in a prepared position on the ground. It has also been developed to be fired from the FV 432 which greatly increases its immediate mobility and enables speedy movement to another site at the conclusion of an engagement. It can be carried in a Land Rover and when necessary it can be broken down into three loads of 11.3, 11.8 and 12.25 kg which can be carried by the mortar section.

Barrel
To enable the mortar to fire bombs with the hot propellants used by some NATO countries, a barrel slightly heavier than that originally specified was adopted. Firing the British bomb, the mortar can fire 15 rds/min indefinitely, the barrel temperature reaching an equilibrium value of 1000°F (540°C).

The barrel is made from a forged steel tube which has been reduced in diameter at the bottom and to save weight it has been screw threaded internally for the insertion of a breech plug. The lower half of the barrel has been finned to increase the surface area available for heat dissipation, and the top half of the barrel has been left plain. There is a collar at the mouth of the barrel to strengthen the section there and a small internal taper is provided to assist in loading the bomb.

Breech plug
This fits into the barrel at one end and has a ball shape to fit into the socket of the baseplate. It carries a longitudinal hole screw threaded to take the firing pin.

Mounting
This is of unusual shape and has been referred to as a 'K' mount. The shape was adopted because with the elevating screw incorporated in one of the legs there is a significant weight saving and no loss of function. All the screw threads associated with elevation and traverse of the mortar have been enclosed to reduce wear and increase life.

Baseplate
This is of Canadian design and is of forged aluminium. It produces an adequate flotation area and the design of the web prevents any tendency for the plate to slip sideways. The circular baseplate allows all round traverse without need to disturb the plate.

Sight
This is the Canadian C2 sight. It fits not only the mortar but also the general-purpose machine gun for use in the sustained fire role. It allows either direct laying or indirect laying using a 45° angled telescope of some × 1.7 magnification. The sight is illuminated for night use with a tritium source.

DATA
BARREL: ORDNANCE, MUZZLE-LOADING, L16
Weight: 12.7 kg
Overall length: 1280 mm
Outside diameter: (muzzle) 86 mm; (breech) 94 mm
Calibre: 81 mm
Construction: steel monobloc forging

MOUNTING, L5A5
Weight: 12.3 kg
Overall length: (folded) 1143 mm
Construction: steel and light alloy
Traverse: 100 mils left and right at 800 mils elevation
Elevation: over 125 mils left and right at elevations of greater than 950 mils; (minimum) 45° (800 mils); (max) 80° (1422 mils)

BASEPLATE, CANADIAN MK 1
Weight: 11.6 kg
Diameter: (overall) 546 mm
Socket size: 50.8 mm
Construction: forged aluminium alloy

SIGHT UNIT, C2
Weight: 1.25 kg

Manufacturer
Royal Ordnance plc, Guns & Vehicles Division, Nottingham.
Status
Current. Production.

81 mm L16 ML mortar

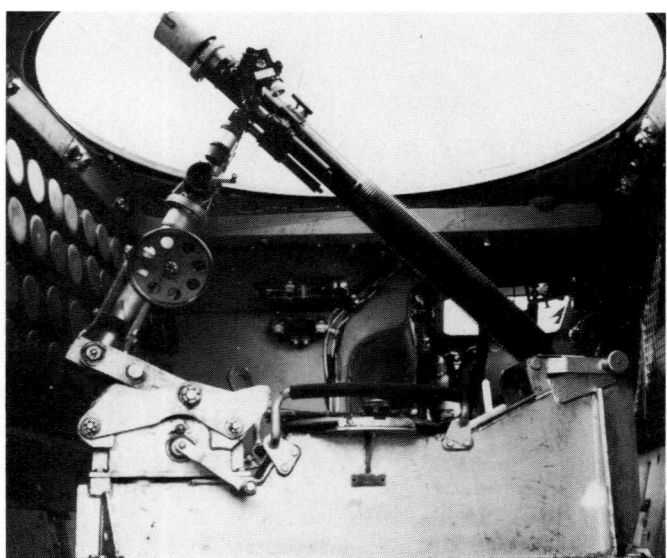

81 mm L16 ML mortar mounted in FV 432

81 mm L16 ML mortar in firing position showing HE bomb being loaded

Service
British Army, also: Austria (designation 8 cm Granatwerfer 70), Bahrain, Canada, Guyana, India, Kenya, Malawi, Malaysia, New Zealand, Nigeria, Norway, Oman, Qatar, United Arab Emirates and North Yemen. Under the designation M252, the L16 has been accepted by the US Army.

120 mm Breech Loading Mortar

This is a new mortar system developed by Royal Ordnance and intended to be fitted into armoured vehicle turrets as an advanced infantry support weapon. The system comprises a smoothbore breech loading mortar and a ring cradle carrying a hydro-pneumatic recoil system, which can be retrofitted into light wheeled or tracked armoured vehicles or incorporated in the design of new vehicles. The mortar is designed to fire current types of conventional ammunition and will be capable of firing future types of precision guided and terminally guided submunition projectiles.

DATA
Calibre: 120 mm
Weight, with cradle: 850 kg
Breech mechanism: Front conical, rear cylindrical screw with Crossley obturating pad
Ammunition: HE, Smoke, Illuminating, Training and Drill

Manufacturer
Royal Ordnance plc, Guns & Vehicles Division, Nottingham.

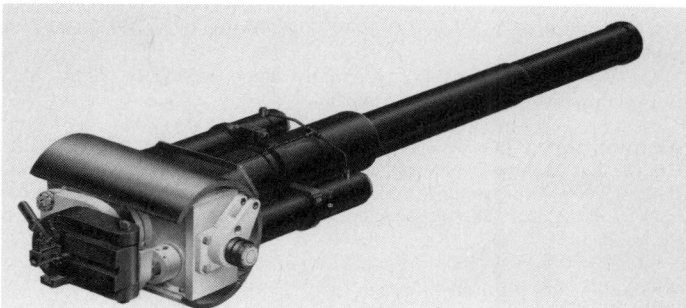
Royal Ordnance 120 mm breech loading mortar

Status
Advanced development.

UNITED STATES OF AMERICA

60 mm M2 mortar

Originally designed and made by Edgar Brandt in France, this mortar was licence-built in the USA from 1938 onwards. The designation M1 was applied to the samples purchased from France and US production mortars were type-designated M2 and were used as the standard light infantry mortar for US forces in the Second World War. It was used mainly with HE ammunition (particularly M49A2) but also fired the M83 illuminating round. It had a rectangular baseplate and a bipod of characteristic Brandt design.

Long withdrawn from US combat units, the M2 mortar can still be found in many US-supplied countries.

DATA
Calibre: 60 mm
Barrel length: 726 mm
Weight in firing position: 19.07 kg
Elevation: 40–85°

Top traverse: 14°
Bomb weight: (HE) 1.36 kg
Muzzle velocity: 158 m/s
Minimum range: 91 m
Max range: 1816 m

Manufacturer
US arsenals.
Status
Obsolete. Out of production.
Service
No longer in US service. May be found in many countries supplied by the USA and ammunition is still manufactured by FMBP in Portugal. Probably still in service in Denmark, Greece, Guatemala, Haiti, Indonesia, South Korea, Morocco, Taiwan and several South and Central American countries.

60 mm M19 mortar

The M19 mortar is a conventional smoothbore, muzzle-loading weapon designed to produce high angle fire. The equipment consists of the barrel M19, bipod M2, sight M4 and baseplate M1 or M5. The bipod and base plate together make up the M5 mount.

It is a development of the M2 mortar, and the main difference lies in the longer and heavier barrel. The weight is increased by roughly 2 kg, but the range and accuracy are improved.

The mortar is fired by dropping a bomb down from the muzzle. The setting of the selector lever determines the method by which the firing takes place. If the firing pin is locked out, the primer drops on it and the bomb is fired. If the firing pin is retracted, the firing lever must be pulled back to compress the firing spring so that when the firing pin trips off the lever, it is driven forward to fire the round. At high angles of elevation either method may be used. At low angles only lever fire will provide the necessary impulse to fire the cap.

M19 barrel
The barrel is a steel tube with an external thread at the lower end to take the base cap. The firing mechanism is contained in a housing which is attached to the base cap by a threaded adaptor. The barrel is fitted to the baseplate by a spherical projection which fits into a socket and is locked in place. The firing mechanism consists of the firing pin, a spring, a trigger and pawl and a firing lever. A firing selector acts as a cam on the rear end of the firing pin and allows the mortar to be fired with or without the firing lever.

M2 bipod
This consists of the two legs, the elevating mechanism assembly and the traversing gear. The bipod legs are connected by a clevis joint which limits the spread of the legs. The elevating screw tube passes through the junction of the legs and connects up with the yoke. The left leg of the bipod carries the sleeve of the cross-levelling gear. When the sleeve is moved up or down the leg the connecting link which is attached to the elevating tube forces the elevating tube out of the vertical. This in turn moves the yoke out of the horizontal. Thus, when the mortar comes into action on uneven ground and the two spiked feet on the bipod legs are on different levels, the cross-levelling gear enables the yoke to be made level and the sight which is fitted into a dovetail slot in the yoke to be upright.

The elevating gear consists of the screw thread in the tube with a handle at the bottom. This enables the yoke to be pushed up or down and the barrel clamp, mounted on the yoke, raises or lowers the muzzle of the barrel.

The traversing gear consists of a nut which is moved across inside a tube by the rotation of the traversing handwheel. This displaces the yoke to one side or the other. The lower half of the barrel clamp has two shock absorber cylinders which permit movement between the yoke and clamp assembly during firing.

Baseplate
The standard component is the M5 baseplate which is a rectangular plate with

Checking lay before HE bomb is dropped

ribs on the underside and a centrally mounted socket to take the spherical end piece of the barrel. The barrel is held in place by a locking lever.

There is alternatively a small M1 baseplate which may be attached to the barrel and the weapon can then be hand-held and fired without the bipod.

M4 sight
The M4 sight has an elevating scale covering the arc from 40 to 90°, by 10° divisions. The micrometer scale is divided up into 40 divisions each one representing $\frac{1}{4}$ degree. The traversing scale has 60 graduations each of 5 mils. The graduations are numbered every 10 mils from 0 to 150 on each side of the zero position.

The M4 sight weighs 0.57 kg. Alternative sights M34A2 (1.8 kg) or M53 (2.3 kg) may also be used with the mortar.

Ammunition
HE: the high explosive round is the M49A4. This has a steel body containing 154 g of TNT. A fin assembly is screwed on the rear of the body and a fuze is inserted in the nose. The fins envelop four secondary cartridges. The fuze is the M525 or M525A1.

Smoke (WP): the M302A2 is used as a screening, signalling or incendiary round. It is filled with white phosphorus (WP) and has a burster charge. The fuze is the PD M527B1.

Practice: the M50A2 is the practice round. It differs from the HE round in colour and filling. It has the same ballistic characteristics and has a small black powder charge to indicate the point of fall.

Illumination: bomb M83A3 is the standard round. The illuminant assembly is expelled after 14.5 seconds of flight and drifts down on its parachute to illuminate the ground below.

The illumination lasts for a minimum of 25 seconds with a minimum of 330 000 candela.

Training: the M69 training bomb has a solid cast iron body and the normal fin assembly. Only the primary cartridge is used and the maximum range is 225 m.

DATA
MORTAR
Weight: mortar (complete with M5 baseplate) 21.03 kg; (with M1 baseplate and no bipod) 9.3 kg
Weight of barrel: 7.25 kg
Weight of bipod: 7.42 kg
Baseplate: (M5) 5.79 kg; (M1) 2.03 kg
Overall length: 819 mm
Elevation: (with M5 mount) 40–85° (710–1510 mils); (with M1 baseplate) 0–85° (0–1510 mils)
Traverse right or left: 129 mils
Rate of fire: (max rds/min) 30; (sustained rds/min) 18 (for 4 mins) or 8 (indefinitely)

Manufacturer
Watervliet Arsenal, Watervliet, New York 12189.

60 mm M19 mortar

Status
Obsolete. Out of production.
Service
In USA, reserve stocks only. Supplied to many countries in South-east Asia and South and Central America which may well still have it in first-line service. It is also used in Belgium (licence-built by PRB), Canada (licence-built by Turnbull Elevator Co), Chile, Iran and Japan.

60 mm M224 lightweight company mortar

During the Vietnam campaign, the US Infantry found the 81 mm mortar to be excessively heavy and unwieldy for use outside the firebases. As a result patrols had to rely for their own support on the 60 mm M19 mortar which was then obsolescent. In December 1970 the XM224 mortar was approved for development to succeed the 81 mm mortar in the Infantry, Airmobile Infantry and Airborne Infantry. It was intended to be a light portable weapon capable of providing the volume of fire required by the infantry and of operation at either high or low angle fire.

The mortar was developed at Watervliet and Picatinny arsenals: the advanced development objective was approved in 1971, engineering tests were

completed at Aberdeen Proving Ground in 1972 and the engineering design was approved in 1973. The weapon was tested at Fort Benning and it was type classified standard in July 1977 and placed in production.

To achieve the required range a barrel 254 mm longer than the M19's and with double the working pressure was developed. The lower part of the tube has been radially finned to increase the surface for heat dissipation and the wall thickness has been kept to much the same value as that of its predecessor by the use of a higher tensile steel. The baseplate is circular, with a pronounced

60 mm lightweight company mortar. This is an early prototype

Comparison of M53 sight used on 81 mm mortar with M64 used on 60 mm M224 mortar

web on the underside, and is an aluminium forging. There is also a light rectangular auxiliary baseplate, the M8, which is used when the weapon is hand-held. In this method of employment the weight of the mortar is only 7.8 kg. The firer grips the barrel towards the muzzle and at the rear there is a handle which holds the trigger mechanism. The mortar can be fired by either gravity or a spring-loaded firing pin controlled by a firing lever which cocks and fires the pin in one movement. The bipod has alloy legs and the elevating screw thread is contained in an alloy elevating tube. Attached to this tube is the connecting rod of the cross-levelling gear which is moved across by a sleeve on the left leg of the bipod. The M64 sight developed for the M224 mortar weighs only 1.1 kg, and is also self-illuminated for night operations. For easy transport the mortar breaks down into two man loads, the heavier being 11.8 kg.

Ammunition for the M224 comprises the M720 HE round with M734 fuze and M702 ignition cartridge; the M721 illuminating; M722 smoke; M723 smoke (white phosphorus); M816 and M840 practice rounds.

To reduce the number of different sorts of fuze required, running at 15 different types with the 81 mm mortar, a multi-option fuze, designated M734, was developed by the Harry Diamond Laboratories. This allows the user to select one of four options: high airburst, low airburst, point detonating and delay. Should the fuze fail to function on the chosen option it will automatically select the next one and function on that: if low airburst is selected and the fuze does not fire, it will fire on point detonating, and should that not function then it will fire on delay. The fuze uses microcircuitry and generates the required electrical power from a tiny turbine in the nose. This

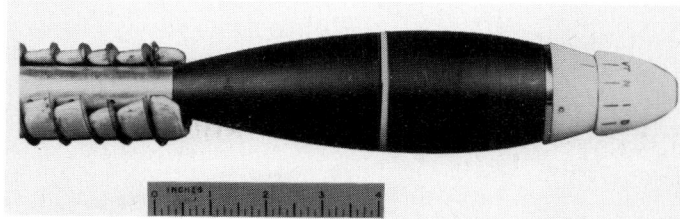

M720 high explosive bomb fitted with multi-option fuze M734

turbine unlocks with the setback of firing and air is fed through a hole in the extreme nose of the fuze. The spinning turbine operates a miniature generator and it takes a finite time before the current has built up to the required level. This time provides a safety, in that the bomb is well clear of the baseplate position before the fuze is active. A plastic cover over the fuze must be removed before the round is fired but if it is inadvertently left in place, air pressure tears it off. The nose ring of the fuze (see photograph) provides the method of selection.

Manufacturer
US Arsenals.
Status
Service.

81 mm M1 mortar

This mortar was declared obsolete in March 1970 and the supply of spares ceased soon after that time. Any of these mortars which are still being used will therefore have a limited life, unless the user is prepared to go to the trouble and expense of making his own replacement parts.

The 81 mm M1 mortar consists of the barrel M1 and the mount M1. The barrel consists of a tube, base cap and a firing pin. The base cap is hollowed and threaded to screw on the tube. It ends in a spherical projection flattened on two sides, which fits into and locks in the socket of the baseplate. The firing pin is screwed tightly into the base cap against a shoulder. The firing pin protrudes a little over a millimetre through the base cap into the barrel.

The mount consists of a bipod and a baseplate. The bipod consists of the leg assembly, the elevating mechanism assembly and the travelling mechanism assembly. The baseplate is a rectangular pressed steel body to which are welded a series of ribs and braces, a front flange, three loops, a link, two handles and the socket. The sight fitted is the M34A2. The mortar can also be mounted in a variety of vehicles including the M4 and M21 half-tracks.

DATA
Calibre: 81 mm
Weight of barrel: 20.18 kg
Weight of mount: 19.27 kg
Weight of baseplate: 20.41 kg
Total weight: 59.87 kg
Barrel length overall: 1266 mm
Length of barrel: 1155 mm
Elevation: 40–85°
Traverse: 5° left and right
Type of firing mechanism: fixed firing pin
Rate of fire: (normal) 18 rds/min; (max) 30 rds/min
Ammunition types: HE, illuminating, smoke – FS, WP and TP

RANGE AT 45° ELEVATION
HE (M43A1) and TP (M43A1): 3016 m
HE (M56 and M56A1): 2317 m
HE (M362 – 5 increment max): 2467 m
Illuminating (M301A1 and M301A2): 51¼°, 2102 mm
Smoke FS (M57 and M57A1): 2216 m
Smoke WP (M57 and M57A1): 2568 m
M68 training, ignition charged only: 283 m

MUZZLE VELOCITY AT 45° ELEVATION
HE (M43A1) and TP (M43A1): 211 m/s
HE (M56 and M56A): 174 m/s
HE (M362): 234 m/s
Illuminating (M301A1 and M301A2): 51¼°, 174 m/s
Smoke FS (M57 and M57A1): 174 m/s
Smoke WP (M57 and M57A1): 174 m/s
M68 training, ignition charge only: 53 m/s

Status
Obsolete. Out of production.
Service
No longer in use in the US Army. Has been supplied to many countries and built under licence by PRB in Belgium. It may be found in almost any country outside the Eastern Bloc and it is likely that stocks are being sold off as more modern mortars become available.

81 mm M29 and M29A1 mortar

The 81 mm mortar is a smoothbore, muzzle-loaded weapon. It consists of the barrel, the mount and the baseplate. The standard M29 mortar comprises the M29A1 barrel assembly, the M23A3 mount, the M53 sight and the M3 baseplate. It may be used in the conventional manner as a ground weapon or mounted into the Carrier Mortar M125A1.

The barrel is made up of the tube, externally threaded at the rear to take a base plug which has a ball-shaped projection on its lower end to fit into the socket of the baseplate. There are two white lines painted 432 and 533 mm from the muzzle for the location of the mount attachment ring. The exterior of the barrel is radially finned to increase the cooling area. The mount comprises the bipod legs, elevating mechanism and the traversing mechanism. The bipod has two tubular steel legs hinged at the sides of the elevating mechanism: they have spiked feet and their spread is limited by an adjustable chain with a spring to relieve the shock on the legs during firing. The left leg carries the cross-level mechanism on a sliding bracket mounted on the leg with a locking sleeve and an adjusting nut. The sliding bracket is connected to the elevating housing by a connecting rod. When the mount is on uneven or sloping ground, the sliding bracket is moved across by rotation of the sleeve on the bipod leg and this in turn moves the elevating mechanism assembly across taking the barrel with it. This enables the sight, at the left end of the yoke, to be moved into an upright position. The elevating mechanism assembly includes a vertical elevating screw moving inside the elevating housing assembly. There is a handle projecting back from the gear case at the junction of the bipod legs, and rotation of this handle elevates or depresses the mortar barrel.

The traversing assembly consists of the yoke assembly, traversing

mechanism and shock absorber. The yoke body supports the upper end of the barrel when the mortar is assembled. The older models of the mortar had a yoke with a levelling bubble. The sight unit is mounted in the dovetail sight slot on the left side of the yoke. The traversing mechanism is an internal screw shaft operating within a nut and tube. The handwheel turns a screw which forces the nut to traverse the yoke and take the barrel with it. The tube over the nut is connected to the elevating shaft which protrudes from the gear case of the bipod. The shock absorber is a compression spring, mounted in the yoke. When the barrel is located to the yoke by the barrel clamp the shock absorber connects the barrel to the bipod.

The M3 standard baseplate is a one-piece aluminium alloy forging. In the centre is the barrel socket which rotates through 360°.

Sight units M53 and M34A2 can be used with the mortar. The M53 is now the standard sight and has very largely replaced the M34A2. Each incorporates a telescope mount and an elbow telescope fastened together into one unit. Both have fixed and slipping scales.

The mortar was originally given the service designation M29, its development designation being T106. The baseplate M3 was developed by the Canadian Armaments Research and Development Establishment (now DREV). A product-improved version, of the mortar, with a barrel (designated M29E1 instead of M29) capable of sustaining a higher rate of fire, was standardised in 1970 with the designation M29A1. The M125A1 mortar carrier mounts the M29 mortar in a modified M113 carrier, originally designated T257E1.

81 mm mortar with sight M53 and boresight fitted

DATA
Calibre: 81 mm
Weight of barrel: 12.68 kg
Weight of mount: 18.12 kg
Weight of baseplate M3: 11.3 kg
Sight unit: (M53) 2.37 kg; (M43A2) 1.81 kg
Elevation: 800–1500 mils
Top traverse: (right or left from centre) 95 mils
Length: 1295 mm
Method of fire: muzzle-loaded, drop-fired
Max range: varies with ammunition. See notes on ammunition below

81 mm M29A1 mortar, ground emplaced with M53 sight and M3 aluminium baseplate

Ammunition: HE, smoke, illuminating and practice
Crew: 5

RATE OF FIRE

Bomb	Mortar	Max rds/min	Sustained rate of fire
M362	M29	15/2	4
		27/1	
M362	M29A1	20/2	5
		30/1	
M374 and M375	M29	18/2	5
M374 and M375	M29A1	25/2	8
		30/1	
M374A3	M29	22/2	7
	M29A1	25/3	13

Manufacturer
Watervliet Arsenal, Watervliet, New York 12189.
Status
Current. Production.
Service
US forces. Also many other countries including Australia, Austria and Italy.

81 mm M252 mortar

US Army interest in the British 81 mm mortar L16 began in 1975; in 1978 18 mortars were provided for test, together with a supply of ammunition. Subsequent tests showed various areas where improvement was desirable, among them problems with cracked baseplates and with wet efficiency of the ammunition. A new bomb, based on the British design, was developed as the M821, followed by the M889. A WP smoke bomb and an illuminating bomb have also been developed.

An initial order for the supply of mortars from Royal Ordnance plc was placed in December 1984, and a further order for 400 tubes and 350 000 bombs was placed in September 1985. Some sources place the potential US order at 2000 complete mortars, though a large proportion of these will be built in the USA.

For a general description of the 81 mm system, reference should be made to the UK section. The data given refers specifically to the US M252 version and shows some small differences in dimensions and performance.

DATA
Calibre: 81 mm
Length of barrel: 1277 mm
Weight of barrel: 12.25 kg
Weight of bipod: 11.79 kg
Weight of baseplate: 11.34 kg
Elevation: 45° to 85°
Traverse: 5.62° (100 mils)
Minimum range: 80 m
Max range: 5675 m

107 mm M30 rifled mortar

The unusual feature of the US 107 mm mortar is that it fires spin-stabilised projectiles which do not therefore have tail-fins, although the propellant charges can be adjusted as on any other mortar.

The 107 mm mortar, M30 (T104), is a rifled, muzzle-loaded, drop-fired weapon which can be hand-carried for short distances when disassembled into five loads. The complete weapon comprises a barrel, a standard, a bridge assembly, a rotator assembly, a baseplate and a sight unit. The design is such that the recoil forces are absorbed by the main mechanical assemblies as a whole.

The barrel is a rifled tube 152.4 cm long with an inside diameter of 107 mm. The rifling consists of 24 lands and grooves of which the first 22.86 cm as measured from the base inside the barrel, are straight. The twist increases to the right from zero at this point to one turn in 19.98 calibres at the muzzle.

The standard assembly, consisting of the elevating, traversing and recoil mechanism, connects the barrel and the bridge assembly.

The traversing mechanism consists of an enclosed screw and bearing located at the top of the standard assembly. It is operated by turning the traversing crank on the left side of the mechanism.

The recoil mechanism consists of a series of springs mounted in the lower section of the standard assembly. These are designed to ease the downward shock of firing and return the mechanism to its pre-firing position.

The bridge assembly consists of two pieces joined by a swivel point. One end receives the cap at the base of the barrel: the other is fitted with a spade which facilitates the digging action of the bridge assembly during firing.

The rotator assembly is approximately 50.8 cm in diameter. It carries the barrel end of the bridge piece on its upper side and pivots on the baseplate.

The baseplate assembly is approximately 96.5 cm in diameter. A recess in the centre receives the bottom insert of the rotator assembly. The lower surface carries six ribs to increase the area in contact with the ground. Each rib has a depth of 16.51 cm. Two carrying handles are provided.

M30 mortar in M106 carrier

Firing 107 mm mortar from M106A1 carrier

M30 mortar ground emplaced with M53 sight

Sight units, M53 and M34A2

The M53 sight unit is the standard sighting device. It consists of an M128 telescope mount and an M109 elbow telescope fastened together into one unit for operation.

The M109 telescope is a lightweight four-power, fixed focus instrument, with a 10° field-of-view that provides the optical line of sight for aiming the weapon for line and elevation. There is a coarse elevation scale on the left side of the sight body with 18 graduations, each of 100 mils, numbered every 200 mils. The elevating knob has an adjustable micrometer with 100 graduations, each of 1 mil and numbered every 10 mils from 0 to 90. The deflection micrometer scale has 100 graduations numbered from 0 to 90. It is fastened to the deflection knob. There is a coarse deflection slip scale and an adjustable micrometer deflection slipping scale. There are two levelling bubbles at 90° to each other on the main sight housing.

Ammunition

Ammunition for the 107 mm M30 mortar is issued in the form of complete rounds. The propelling charge consists of an ignition cartridge and 41 propellant increments assembled in a bag and sheets. To adjust the charge, individual increments are removed. Information on the standard rounds authorised for the 107 mm mortar M30 will be found in the Ammunition section.

Development

At the request of the Army Development and Employment Agency, which is testing various light division concepts, a programme has begun to improve the mobility, sights and fuzing of the M30 mortar. This will include development of a terminally-guided anti-armour bomb; a trailer or wheeled carriage; adaptation of the M734 multi-option fuze to 4.2 in ammunition; and fitting the M64 sight and fire control system. Information received in late 1986 suggests that this programme may be in abeyance and that a 120 mm mortar may be the preferred solution.

DATA
Calibre: 107 mm (4.2 in)

WEIGHTS
Barrel: (M30) 70.89 kg
Bridge assembly: 76.55 kg
Baseplate assembly: (M24A1) 87.42 kg; (M24, inner baseplate) 53.45 kg; (outer baseplate ring assembly) 46.66 kg
Standard assembly: 26.95 kg
Rotator assembly: (cast magnesium) 26.05 kg; (welded steel) 40.32 kg
Sight equipment: (M34A2) 1.81 kg; (M53) 2.48 kg
Mortar complete: 305 kg with welded steel (latest model) rotator, M24A1 baseplate and M53 sight

Length of barrel: 1524 mm
Rifling: 24 grooves. Straight to 228.6mm, thereafter increasing rh twist to 1 turn in 19.96 cal

Elevation	Low range	High range
Minimum	706 mils	919 mils
Max	933 mils	1156 mils
Firing elevations	800 or 900 mils	1065 mils

Per turn of elevating handle: approx 13 mils
Shifts: (per turn of traversing crank) approx 10 mils; (on traversing mechanism, right or left from centre) 125 mils
Rate of fire: the maximum rate of fire is 18 rds/min for the first minute and 9 rds/min for the next 5 mins. This can be followed by a sustained rate of fire of 3 rds/min for prolonged periods.

Status
Current. Out of production.
Service
US Forces. Also Austria, Belgium, Canada, Greece, Iran, South Korea, Liberia, Netherlands, Norway, Oman, Turkey, Zaïre.

YUGOSLAVIA

50 mm M-8 mortar

This is very similar to the British 2 in (50 mm) mortar but differs in having a supporting and carrying handle at the point of balance. The barrel is a plain steel tube internally screw threaded at the bottom to take the breech piece which holds the trigger and firing mechanism. The weapon is fired by a lever on the right of the breech piece. The sight is a simple pointer with a bubble to indicate when it is horizontal and a range scale. The baseplate is trough shaped and lies across the barrel axis. The mortar M-8 fires an HE bomb weighing about 1 kg, a smoke bomb and an illuminating bomb.

DATA
Calibre: 50 mm
Weight: 7.3 kg
Max range: 480 m
Minimum range: 135 m
Rate of fire: 25–30 rds/min
Muzzle velocity: 80 m/s

Manufacturer
State factories.
Status
Current.
Service
Yugoslav Army.

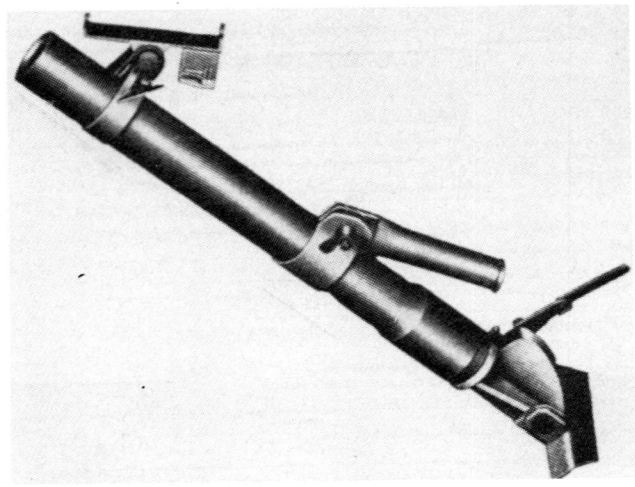

50 mm M-8 mortar

60 mm M-57 mortar

This was developed from the US M2 60 mm which is now obsolete. The breech piece is screw threaded to fit over the bottom of the barrel, and has a ball-shaped end to fit into the socket of the rectangular baseplate. The bipod has a cross-levelling gear operated by a sleeve on the left leg and connected to a rod attached to the elevating gear. This is attached to the yoke which carries the traversing screw and the shock absorbing spring cylinders. At the right-hand end of the yoke is the traversing handwheel and on the left is the sight. This has no slipping scale but allows switches from a chosen aiming point.
The mortar fires HE, smoke and illuminating bombs.

DATA
Calibre: 60.75 mm
Weight of barrel: 5.5 kg
Weight of bipod: 4.5 kg
Weight of baseplate: 8.85 kg
Weight of sight: 1 kg
Muzzle velocity: 159 m/s
Minimum range: 75 m
Max range: 1700 m
Rate of fire: 25–30 rds/min
Weight of bomb: 1.35 kg

Manufacturer
State factories.
Status
Current.
Service
Yugoslav Army.

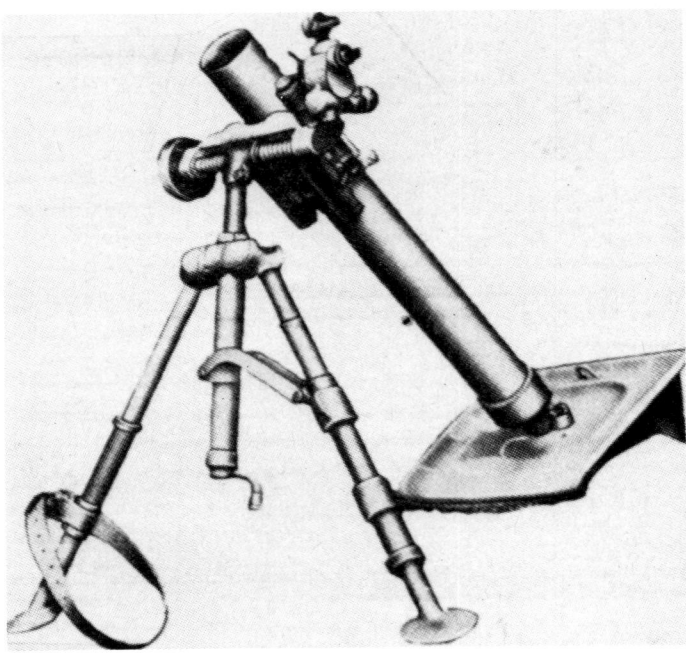

60 mm M-57 mortar

81 mm M-31 mortar

This mortar was the Brandt Model 30 Export, manufactured under licence by the Kragujevac State Arsenal after firing trials in 1930. It has a plain steel tube, a rectangular baseplate and a simple bipod with the spread of the legs controlled by a chain. The cross-levelling gear is controlled from the right leg of the bipod.

DATA
MORTAR
Calibre: 81 mm
Length of barrel: 1310 mm
Weight of barrel: 21 kg
Weight of bipod: 19 kg
Weight of baseplate: 20 kg
Weight of sight: 1.5 kg

Rate of fire: 20 rds/min
Max range: 4100 m
Minimum range: 85 m
Muzzle velocity: 300 m/s

AMMUNITION
Weight of light HE bomb: 3.2 kg
Weight of TNT filling: 650 g
Weight of heavy HE bomb: 4.2 kg
Weight of TNT filling: 750 g
Smoke bomb: as heavy HE bomb: produces smoke for 2–4 mins
Illuminating bomb: as heavy HE bomb: produces 300 000 candela for 30 s and illuminates a circle of diameter 250–300 m

Manufacturer
State factories.
Status
Current.
Service
Yugoslav Army.

81 mm M-31 mortar

81 mm M-68 mortar

Although similar in appearance to the Brandt MO-81-61L mortar, the Yugoslav M-68 is of Yugoslav design and construction.

DATA
MORTAR
Calibre: 81 mm
Length of barrel: 1640 mm
Weight of barrel: 16 kg
Weight of bipod: 13 kg
Weight of baseplate: 11 kg
Weight of sight: 1.5 kg
Rate of fire: 20 rds/min
Max range: 5000 m
Minimum range: 90 m
Muzzle velocity: 300 m/s

AMMUNITION
HE bomb
Weight: (fuzed) 3.3 kg
Bursting charge: 463 g TNT
Fuze: Model UT M45P1 point detonating

Manufacturer
State factories.
Status
Current.
Service
Yugoslav Army.

81 mm M-68 mortar

Firing 81 mm M-68 mortar

82 mm M-69 mortar

Although nominally of different calibre this is a shortened version of the M-38 and fires the same ammunition. It is of a more modern design and owes little or nothing to foreign models. The bipod uses telescoping legs and has the elevating handle on the rear side. There are two small recoil buffer cylinders forming part of the barrel clamp and a small circular baseplate. Maximum range is said to be 3150 m and the weight in action 62 kg, though we think this latter figure excessive.

120 mm UBM 52 mortar

The 120 mm UBM 52 is of Yugoslav design and construction. Some reports have suggested that the mortar is in fact the French 120 mm Brandt MO-120-AM50 built in Yugoslavia but both Brandt and the Yugoslav Ministry of Defence have informed *Jane's Infantry Weapons* that this is not the case. It is apparent from the photographs that there is little resemblance between the Yugoslav and French mortars.

The UBM 52 has been designed for both field and mountain deployment. For field use it is normally towed by its muzzle by a 4 × 4 truck, for example the Zastava AR-51. For transport in rough country it can be quickly broken down into five loads and carried by animals.

When in the firing position the wheels are retained and the hydro-elastic recoil system permits immediate commencement of fire once in position. The mortar, which is smoothbore, can also fire the Soviet 120 mm mortar round. It can be fired either by conventional drop-firing or by a trigger. The mortar has a crew of five.

DATA
MORTAR
Calibre: 120 mm
Weight of complete equipment: 400 kg
Range: heavy bomb: (minimum) 195 m; (max) 4760 m. Light bomb: (minimum) 225 m; (max) 6010 m

AMMUNITION
Light bomb, HE Model 62
Weight: (fuzed) 12.2 kg
Bursting charge: 2.5 kg
Fuze: Model 45UTU point detonating

Heavy bomb, HE Model 49P1
Weight: (fuzed) 15.1 kg

120 mm UBM 52 mortar

Bursting charge: 3.1 kg
Fuze: Model 45TU point detonating
Smoke bomb M64: produces smoke for 2–4 mins
Illuminating bomb: burns for 30 s and produces 1 100 000 candela

120 mm M74 light mortar

This is a light, portable 120 mm mortar which is particularly suited to operations in mountainous terrain or in other areas where vehicular access is restricted or difficult. The M74 is a conventional drop-fired weapon with bipod and triangular baseplate. The three basic components can be separated for carrying, or can remain attached and the mortar transported on a light two-wheeled trailer which can be vehicle-towed or pulled by hand.

The M74 fires the standard 120 mm family of ammunition, which includes light and heavy HE bombs, a rocket-assisted bomb which can be fired with or without rocket boost, smoke, illuminating and practice bombs. Details of the ammunition will be found in the relevant section.
DATA
Calibre: 120 mm
Barrel length: 1690 mm
Weight in firing order: 120 kg
Weight in travelling order: 208 kg with trailer
Weight of barrel: 45 kg
Weight of bipod: 25 kg
Weight of baseplate: 49 kg
Minimum range (light shell): 267 m

Max range (light shell): 6213 m
Muzzle velocity (light shell): 266 m/s
Minimum range (heavy shell): 300 m
Maximum range (heavy shell): 5374
Maximum range (rocket-assisted shell): 9056 m
Elevation: 45° to 85°
Traverse on bipod: 6°

Range coverage diagram for 120 mm M74 and M75b light mortars

120 mm M74 light mortar in travelling mode, with FUG-M77 bomb

Detachment: 4
Maximum rate of fire: 12 rds/min
Sustained rate of fire: 35 rds/10 mins

Manufacturer
Federal Directorate of Supply and Procurement, 9 Nemanjina Street, 11001 Beograd 9.
Status
Current. Production.
Service
Yugoslavian Army.

120 mm M74 light mortar in firing position; rocket-assisted bomb in foreground

120 mm M75 light mortar

Although still called 'light', the M75 is somewhat heavier than the M74 and uses a much more substantial circular baseplate in order to withstand the recoil stresses of the heavier bombs. All types of 120 mm ammunition may be fired, as can a rocket-assisted bomb, giving a useful increase in range.

The M75 is of conventional pattern and is transported on the same lightweight trailer as the M74; a harness rig for towing by four horses is available. It can also be mule-packed, helicopter-lifted or parachute dropped.

DATA
Calibre: 120 mm
Weight in firing order: 177 kg
Weight in travelling order: 263 kg with trailer
Weight of barrel: 65 kg
Weight of bipod: 25 kg
Weight of baseplate: 86 kg
Length of barrel: 1690 mm
Minimum range: (light shell) 275 m; (heavy shell) 300 m
Max range: (light shell) 6340 m; (heavy shell) 5551 m; (rocket-assisted shell) 9056 m
Elevation: 45° to 85°
Traverse on bipod: 6°
Max rate of fire: 15 rds/min
Detachment: 4

Manufacturer
Federal Directorate of Supply and Procurement, 9 Nemanjina Street, 110011 Beograd 9.
Status
Current. Production.
Service
Yugoslavian Army.

120 mm M75 light mortar in travelling position

120 mm M75 light mortar in firing position

Mortar Fire Control

CHINA, PEOPLE'S REPUBLIC

AMC-II mortar fire control computer

The AMC-II computer is a self-contained unit incorporating a computer, keyboard for data entry, display, batteries, illuminator and leather case. The computer is suitable for all types of mortar at all altitudes, can perform routine topographical surveying functions, has high computation speed and short response time, and requires minimum maintenance. It is powered by rechargeable Ni-Cd batteries which can be re-used not less than 800 times.

Operation of the computer is simple and similar to that of an ordinary calculator. Without previous knowledge, a soldier can be instructed to a competent level in no more than two days.

DATA
Dimensions: 112 × 243 × 40 mm
Weight, with batteries: 1.2 kg
Power requirement: 5 V DC
Power supply: 4 × 1.5V Ni-Cd batteries
Survey accuracy: ± 1 m
Range accuracy: less than 8 m
Azimuth accuracy: 1 mil
Firing data calculation time: 4 s

Manufacturer
China North Industries Corp, PO Box 2137, Beijing.
Status
Current. Production.
Service
Chinese Army.

AMC-II mortar fire control computer

FRANCE

Thomson-Brandt ATAC computer

This is capable of being used with virtually any artillery piece or mortar, since it uses a plug-in ballistic module which is available in eight weapon/ammunition combinations. There is an alphanumeric display panel and keyboard, and the display primes the operator to each step in a 'dialogue' form. A warning signal is emitted automatically should the operator attempt to calculate an impossible or dangerous mission.

The ATAC (Aide au Tir de l'Artillerie de Campagne) can perform the usual functions: survey computations, change of co-ordinates between rectangular and polar; intersections, resections, traverses and triangulation. Conduct of fire programs include adjustment and computation of initial data, corrections during fire, computation and application of meteor corrections, reduction to gun data at the end of mission and target recording. Fire plans and engagement of moving targets are also possible. It is particularly well adapted

to the trajectories of rocket-assisted projectiles (for which it was originally designed).

The computer can store 9 gun positions, 99 targets, 20 observer locations, 10 survey base points, 10 protected areas, and a STANAG 4082 meteor message up to Line 21. There is a full self-testing facility.

DATA
Dimensions: 240 × 160 × 80 mm
Weight: 4 kg with batteries
Power supply: 17 V DC–35 V DC (24 V DC nominal) (ensured by 12 commercial 1.5 V R14 cells)
Consumption: 1 W
Battery life: at ambient temperature of 20°C battery life depends on type of cell in use
Ni-Cd 30 h continuous
Alkaline 100 h continuous
Lithium 138 h continuous

Thomson-Brandt ATAC computer in use

Thomson-Brandt ATAC computer

Manufacturer
Thomson-Brandt Armements, Tour Chenonceaux, 204 Rond-Point du Pont de Sevres, F-92516 Boulogne-Billancourt.

Status
Current. No longer in production.
Service
French and other armies.

Thomson-Brandt FAC 101 computer

This was designed for use with 155 mm guns and howitzers and with heavy mortars such as the Brandt 120 mm models. It incorporates plug-in ballistic modules for each gun/ammunition combination. In general features it is similar to the ATAC model described above but is larger and has a smaller memory; it can store 6 gun positions, 10 targets, 10 observer locations, 10 protected areas and can also store the STANAG 4082 meteor message up to Line 21. All memories are protected against accidental erasure.

DATA
Dimensions: 370 × 355 × 213 mm
Weight: 18.5 kg
Power supply: 19 V – 32 V (24 V nominal)
Consumption: 50 W
Battery life: 2 h continuous

Manufacturer
Thomson-Brandt Armements, Tour Chenonceaux, 204 Rond-Point du Pont de Sevres, F-92516 Boulogne-Billancourt.
Status
No longer in production. Replaced by ATAC.
Service
French Army.

Thomson-Brandt FAC 101 computer

ESD fire control computer

This is a portable computer for fire control of mortars or guns and can be supplied with ballistic programming for any desired weapon/ammunition combination. It is provided with a shoulder harness and can be operated while on the move if necessary.

The usual topographical and ballistic calculations are provided, and the memory can contain 10 weapon locations, 50 targets, 10 observer locations, 10 protected areas and 2 meteor telegrams to STANAG 4082. All memory functions are protected against accidental erasure and are retained while the instrument is switched off. There are also programs covering the usual safety routines for range firing and built-in test facilities.

It is possible to link the computer with a suitable radio for data transmission and reception, and links are available for various peripheral equipment such as printers or VDUs.

DATA
Dimensions: 220 × 220 × 100 mm
Weight: approx 3 kg
Power supply: internal 12 V or 24 V NiCd battery or from vehicle battery
Battery life: 1 Ah battery–2 h; 4 Ah battery–8 h

Manufacturer
Electronique Serge Dassault, 55 quai Marcel Dassault, 92214 Saint-Cloud.

Status
Current. Production.

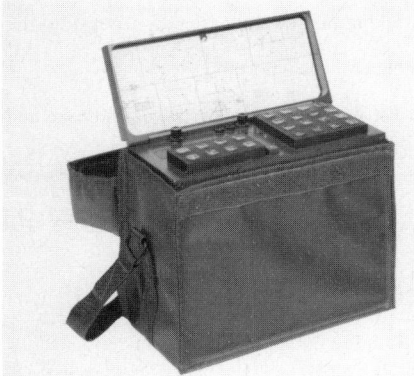

ESD fire control computer

TIRAC fire control computer

TIRAC has been developed by SECRE in co-operation with the French Direction Technique des Armements Terrestres. For single operator use, the computer can calculate all trajectory data required for gun batteries or mortar platoons.

Firing data is implemented from (1) ballistic data applicable to specific types of ammunition (by means of plug-in ballistic modules); (2) non-standard gun conditions including charge temperature, shell weight and muzzle velocity; (3) survey data covering locations of guns, targets and observers; (4) meteorological data to STANAG 4082, or from datum shooting, or an estimated met message; and (5) possible stored residual corrections following adjustment of fire. Firing data (azimuth, elevation and fuze setting) is determined for each gun of the battery, and the charge is selected by the operator on the basis of data displayed by the computer.

The computer can handle up to 6 guns, 10 targets and 10 registration points. Computation accuracy is to within 1 mil for azimuth and 0.5 PE for range.

In addition TIRAC can calculate observer's adjustments, reduce fired data to map data, determine target co-ordinates from firing data, and perform all routine survey computations.

The equipment is supplied from a DC power source, either a vehicle battery or an internal power pack. Selection of the appropriate ammunition is simply a matter of inserting the correct ballistic module. There are output facilities for data transmission to individual guns or a teleprinter link.

SECRE TIRAC gun and mortar computer

DATA
Dimensions: 400 × 330 × 180 mm
Weight: 12 kg
Power supply: 19 – 32 V DC external; 2 × BA58 batteries back-up
Consumption: 25 W
Battery life: 72 hours

Manufacturer
SECRE, 214-216 Faubourg Saint-Martin, 75483 Paris-Cedex 10.
Status
Current. Production.
Service
French Army.

ITALY

CMB-8 mortar fire control computer

The CMB-8 is a military, hand portable computer which automates the procedures needed to compute firing data for a battery deployed as three platoons, each of three or four mortars.

It makes use of all the information stored in its memory to compute firing data for the different mortars of the battery, and it can be connected to a Galileo military hand-portable printer for recording of the executed procedures.

An internal rechargeable battery and the possibility of connection to an external power supply give the CMB-8 a long combat autonomy.

Compared with conventional methods of fire control the CMB-8 offers many advantages:

a drastic reduction in time of calculation, together with increased precision

a reduction in the number of members in the fire direction centre and in working time, since a menu sequence directs the operator step by step to completion of the procedures

a possibility of automatic data transfer through a digital link with other units of the battery (forward observers) or with higher echelons of command.

The most important functions performed by CMB-8 are:
(1) fire preparation, including
  (a) survey preparation, with several programs such as:
    polar co-ordinates replot, closed traverse replot, change of grid, intersection and resection
  (b) the introduction of positional data for all the mortars in the battery
  (c) ballistic corrections from meteor messages, charge temperature and weight and muzzle velocities
(2) evaluation of engagements taking into account friendly troops, no-fire areas and crest clearance
(3) computation of initial firing data for each mortar
(4) fire adjustment
(5) fire for effect, including fire distribution for smoke screens, barrages and area targets.

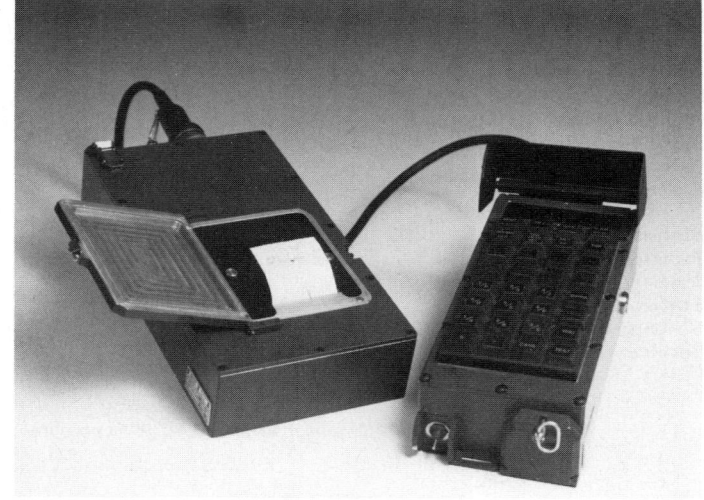

CMB-8 mortar computer with optional printer

Manufacturer
Officine Galileo SpA, Via Einstein 35, 50013 Campo Bisenzio, Firenza.
Status
Current. Production.
Service
Italian Army.

NEW ZEALAND

MERE mortar fire control computer

MERE (Mortar Elevation and Ranging Equipment) has been developed by the New Zealand Department of Scientific and Industrial Research in conjunction with the New Zealand Army and is intended for use with the 81 mm L16 mortar.

MERE consists of a keyboard and display panel, computing circuitry and a clip-on rechargeable battery pack. It will store up to 99 targets, 24 observers or friendly positions, and 9 mortar lines. Positions can be specified by grid reference or by azimuth and range from known points. Altitudes can be in feet or metres and NATO format meteor data can be entered and stored. Charge temperature, ammunition variations and similar corrections can also be entered. After storing the mortar location and entering target information, the computer will produce the complete fire mission data of range, charge, azimuth, elevation, fuze length and time of flight. Fall of shot corrections notified by observers can then be inserted to produce corrections to the fire, and at the end of the mission the target can be recorded. Initial firing data is produced in about 12 seconds and corrections in about six seconds. When used in training it is possible to enter range safety information, and the display will indicate any safety hazard; similarly, with crest and friendly force information inserted, the display will produce warnings covering safety of own troops.

DATA
Dimensions: 265 × 80 × 125 mm
Weight: under 4 kg
Power supply: 6 V battery, rechargeable in field
Battery life: 70 h continuous, 1000 h intermittent use

Manufacturer
Wormald Vigilant Ltd, 211 Maces Road, PO Box 19545, Christchurch.
Status
Current. Production.

MERE mortar computer

Service
New Zealand Army.

NORWAY

Hugin Mortar Fire Control System

This system is currently under development by NFT for the Norwegian Army. The programme involves the development of the KMT 400 Multi-purpose Terminal for military use (the hardware) and the Mortar Fire Control Application (the software). Twenty sets of the pre-production series are currently undergoing field trials with the Army.

The KMT 400 is a compact, ruggedised, modular computer/terminal for field use. The unit has a fully graphic, monochromatic LCD screen with touch data entry based on a menu technique. It is EMP-resistant and operates over the full -40—+55°C temperature range. It can be easily operated by personnel wearing NBC or winter equipment.

Standard interfaces are to handset, field line, radio, RS232C and digital communications networks. These interfaces are flexible and can be tailored to meet any specification. The applications software is designed to be readily adaptable to different mortar calibres and types of ammunition. Ballistic calculations are performed using the Mass Point Model, and necessary software extensions for field artillery applications are available.

DATA
Processors: 80C88
Memory: EPROM 512 kb; RAM 96 kb; EEPROM 8 kb
Modulation: FSK for built-in modem
Speed: 600 bps
Weight: 1.7 kg
Size: 70 × 120 × 200 mm
Power source: lithium BA-5052N; or 10 – 32 V DC external
Battery life: 50 h

Manufacturer
Norsk Forsvarsteknologi AS, PO Box 1003, N-3601 Kongsberg.

Hugin mortar fire control system in use

Status
Advanced development.
Service
Troop trials with Norwegian Army.

SPAIN

Seimor mortar data computer

This is a compact hand-held computer with high resistance to drop, damage and water. Operation is reliable and simple, the keyboard being colour-coded to facilitate operation.

The display prompts the operator during standard operations, and there is automatic switch-off to conserve battery power. When the computer is switched off, the data is stored and can be recalled. Batteries can be changed without memory loss.

The computer has capacity for 80 targets, 12 observers, 21 mortar locations, up to 21 prohibited areas, 7 survey programmes and 9 fire for effect procedures. Two simultaneous missions can be computed and there are five initial corrections and five modes to commence firing missions available. Charge selection can be done manually or can be automatic, and the display will produce projectile, range, charge, elevation, azimuth, individual corrections and time of flight. There is the facility to attach a printer for hard copy.

DATA
Dimensions: 90 × 60 × 180 mm
Weight: 500 g
Operating temperature range: -20° to +60°C
NATO Stock Number: 1220-33-0003093

Manufacturer
Seidef SA, Pso. de la Castellana 140, 28046 Madrid.
Status
Current. Production.
Service
Spanish Army.

Seimor mortar data computer

UNITED KINGDOM

MORCOS

MORCOS is a self-contained, hand-held unit which incorporates the entry keyboard, visual display, batteries and computing circuitry. The unit is in a polycarbonate case of convenient size and is fully militarised to withstand all battlefield environmental conditions. The keyboard has 24 keys, 10 for the numerical digits and the rest for various functions. After initiating a function, the display then leads the operator through the required sequences for the various operations, either by displaying the parameters for which the operator must insert values or by presenting, on the display, the names of the various routines which the operator can select. Every entry is displayed and, if incorrect, can be cancelled. The ballistic program can be changed in seconds by removing a plug-in module and substituting another; modules for most weapon/ammunition combinations are available or can be produced to order. The modules can be changed in the field without impairing the weather-sealing of the instrument. There is a built-in self-test facility.

The memory can accommodate 10 mortar positions, 50 targets, 10 observer locations and 10 locations of own troops. Calculations include normal survey routines, laser rangefinder calculations, firing data, corrections to fall of shot, meteor data, prediction for three parallel missions, and close target routines.

DATA
Dimensions: 230 × 110 × 54 mm
Weight: 1.35 kg
Power supply: 9 V DC integral batteries (6 AA cells) or 12 V DC external supply

Manufacturer
Marconi Command and Control Systems Ltd, Chobham Road, Frimley, Camberley, Surrey, GU16 5PE.
Status
Current. Production.

MORCOS fire control computer

Service
Many overseas countries.

MORZEN II

MORZEN II is a second-generation instrument which complements the increased sophistication of the latest weapon technology and provides fast, accurate firing data for any mortar, using any ammunition and any sighting system, in the indirect-fire role. It is in full service with the British Army which uses it with the L16 81 mm mortar, and with an increasing number of other countries.

Lightweight, hand-held, robust and simple to operate, MORZEN II uses standard prompts and procedures familar to each user. With only a few hours training it can be operated sucessfully in the field to produce faster and more accurate data than any other proven system.

The main features of MORZEN II can be summarized as all computations to 10-digit accuracy from 6-, 8- or 10-digit reference points; stores up to 314 target locations and 9 observation post locations, plus 14 mortar lines for normal or 'shoot and scoot' operations; polar co-ordinates for use with laser

MORZEN II computer in use

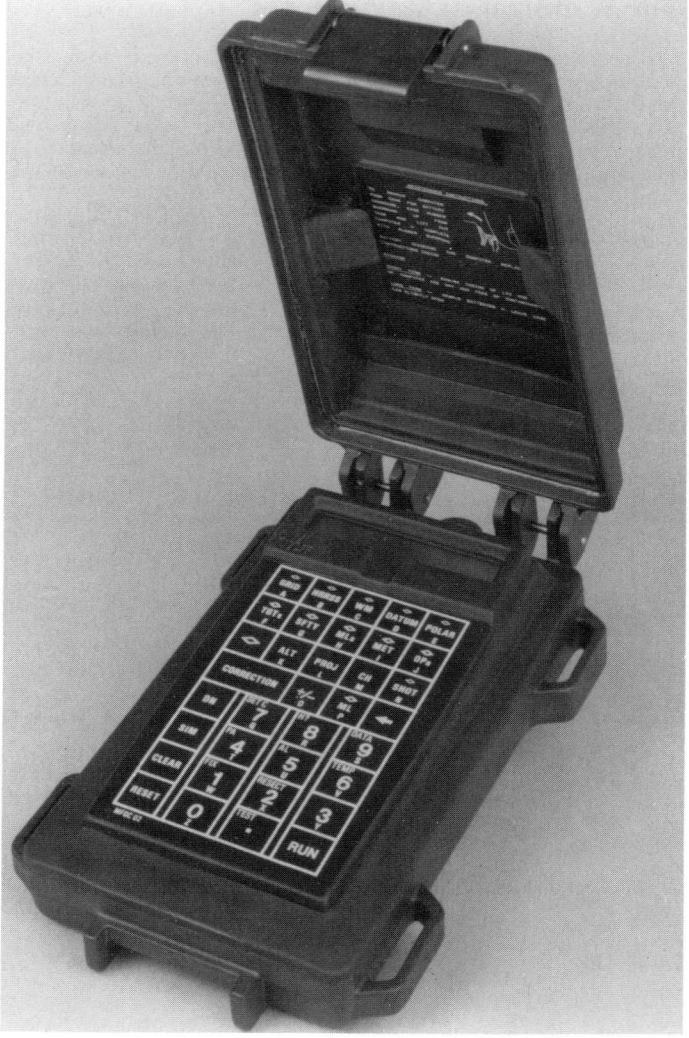

MORZEN II mortar fire control computer

rangefinders; easily used datum procedure for accurate location of new mortar lines; without-maps procedures; on-board clock for real-time, time relative to H-Hour, splash and co-ordinated illumination; and easily understood error messages generated by miskeying or invalid data input.

MORZEN II has a battery life in excess of 80 hours of continuous use, with advance warning of battery failure being given. It has been designed for use with NBC or arctic clothing and operates under a wide range of environmental conditions.

The computer has passed the British Army testing for high and low temperature (-30° to +55°C), dust implosion, drop testing, vibration, immersion, bounce, nuclear flash burn and electro-magnetic compatibility. It is qualified to DEFSTAN 00-1 and DEFSTAN 07-55. Zengrange is qualified to AQAP1 and AQAP13.

DATA
Dimensions: 200 × 118 × 58 mm
Weight: 625 g
Target store: 314 max
ML store: 14 max
OP store: 9 max
Power supply: 4 × 1.5 V dry cell
Battery life: > 80 h
Accuracy: Range ±1 m; bearing and elevation ±1 mil; fuze setting ±0.5 units; time of flight ±0.5 s; co-ordinates ±1 m
NSN: V5/1220-99-967-4824

Manufacturer
Zengrange Ltd, Greenfield Road, Leeds LS9 8DB.
Status
Current. Production.
Service
British and other armies.

UNITED STATES OF AMERICA

M23 Mortar Ballistic Computer

The M23 is a portable computer which can be adapted, by using appropriate software, to the US 60, 81 and 107 mm mortars in current use. It can deal with 3 concurrent fire missions, 3 firing locations, 18 individual weapon locations and 12 observer locations. All the usual facilities for ballistic, meteorological and survey computation are available. It is possible to interface with higher-level computer systems such as TACFIRE and ACS, and it can receive inputs from most in-service digital message devices.

DATA
Dimensions: 267 × 182 × 60 mm
Weight: 3.2 kg
Power supply: 20 – 32 V DC external; or battery pack

Manufacturer
Magnavox Electronic Systems Co, Fort Wayne, Indiana.
Status
Current. Production.
Service
US Army and US Marine Corps.

Ammunition

Small Arms and Cannon Ammunition
Combat Grenades
Mortar Ammunition
Pyrotechnics

Small Arms and Cannon Ammunition

STANDARD AMMUNITION

The list of cartridges adopted for military service since the coming of the central-fire cartridge is almost endless, but the last 30 years have seen a considerable amount of standardisation, either by agreement or by simple economic hard facts, and thus the multiplicity of cartridges in use has considerably diminished. Nevertheless, there are still a fair number about, as the following tabulation shows. In an endeavour to condense, only those rounds known to be in first-line service or in reserve have been listed. Doubtless there are soldiers walking around parts of the Balkans carrying 8 mm Roth automatic pistols and 10.15 mm Serbian Mauser rifles, but we regret that we cannot cater for them.

Pistol

5.45 × 18 mm Soviet

V_o 315 m/s
E_o 129J

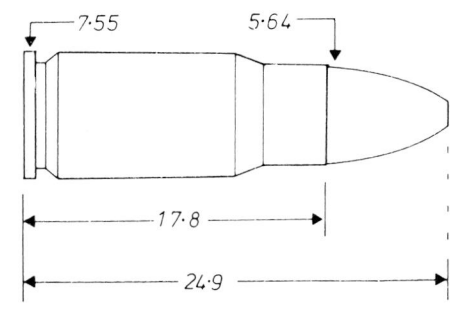

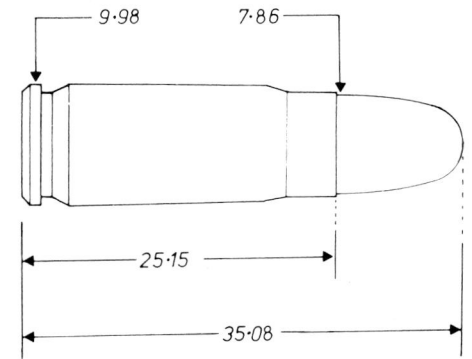

5.45 × 18 mm Soviet cartridge compared with 9 mm Makarov, 9 mm Parabellum and .45 ACP cartridges

This has existed, by evidence of headstamps, since the mid-1970s, but apart from the fact that it existed little else was known until early in 1983. It is solely for use with the PSM pistol (which see) and is a bottle-necked, rimless cartridge weighing 5 g and with a 2.6 g jacketed-compound bullet of non-streamlined form. The bullet is unusually long for a pistol, being approximately 14.5 mm or 2.58 calibres long, where the usual pistol bullet is approximately 1.5 calibres long. The shape and the performance figures suggest poor stopping power but probably good accuracy, and it is possible that target effect relies on bullet tumble due to deliberate instability. The bullet has a gilding metal jacket surrounding a core which has its front half of steel and its rear half of lead, like modern small-calibre rifle bullets.

7.63 mm Mauser

7.63 mm × 25
V_o 455 m/s
E_o 576J

This cartridge was originally introduced for use with the Mauser military pistol in 1896 and is still commercially manufactured for that purpose. The Mauser was widely used in Russia and, from this, the cartridge came to be adopted as the standard Soviet pistol and sub-machine gun round, under the name 7.62 mm Tokarev. Although there are minute differences, they are often swallowed up in the makers' tolerances and for all practical purposes the Mauser and Tokarev cartridges are interchangeable. Mauser cartridges can be found with a wide variety of bullet types, though the military standard is a full-jacketed bullet of 5.57 g.

The Soviet service cartridges are as follows:

Ball, Type P	5.57 g full-jacketed bullet in brass or brass-coated steel case
AP/Incendiary Type P-41	developed for sub-machine guns, uses steel-jacketed bullet with hard steel core and filling of incendiary composition in nose. Weighs 4.79 g
Tracer, Type P-T1	also for sub-machine guns, uses lead-cored compound bullet with recess at base filled with tracer composition

7.65 mm Parabellum

7.65 mm × 21
V_o 368 m/s
E_o 407J

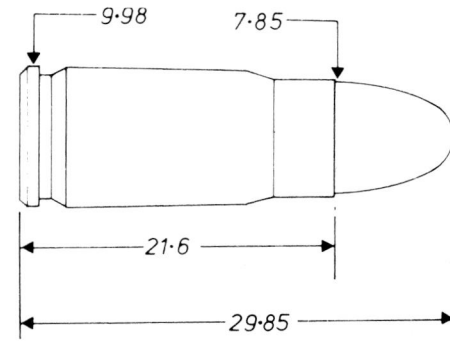

This cartridge was designed by George Luger for use with the original Parabellum pistol and it was, for several years, the service pistol cartridge of the Brazilian, Bulgarian, Portuguese, Swiss and other armies. A few sub-machine guns were also adapted to its use. It is not currently in first-line service with any army, but will be found in the hands of various police and security forces. SIG of Switzerland, Beretta of Italy and Walther of West Germany still produce variants of their pistols chambered for this cartridge.

An unusual and dangerous variant which may occasionally be encountered is a high powered round for use in the Parabellum pistol-carbine. These, if fired from a pistol, could produce dangerous pressures and cause mechanical damage, if nothing worse, but they can easily be recognised by their cases being chemically blacked.

As with most commercially popular cartridges, several weights and types of bullet exist. The military standard is a 6.02 g full-jacketed type with a round nose.

7.65 mm Browning

7.65 mm ACP
7.65 mm × 17
V₀ 300 m/s
E₀ 216J

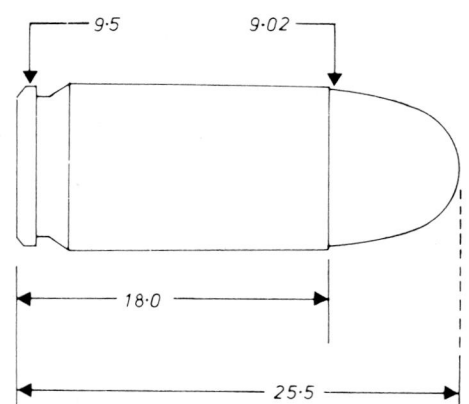

Developed by John Browning for his early blowback automatic pistol in 1897, this cartridge has since become the world standard for pocket pistols: one authority has estimated that some 75 per cent of the automatic pistols made since 1900 have been chambered for this cartridge. In military service it has usually been restricted to staff officers' pistols and to substitute standard pistols issued to second-line troops. It is, though, widely used as a police and security force sidearm calibre and the Czechoslovaks have even produced a sub-machine gun chambered for it.

The usual military loading is a 4.6- to 4.8 g full-jacketed bullet. Note that this case, like most of the cases designed by John Browning, is actually a semi-rimmed type and not a true rimless.

7.65 mm Longue

7.65 mm × 19.5
V₀ 366 m/s
E₀ 368J

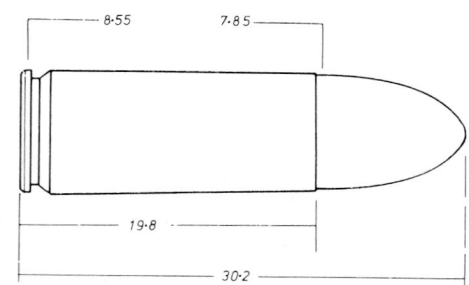

This cartridge is unique to French forces, having been designed for the French SACM Model 1935A automatic pistol and subsequently used in the Manufacture d'Armes de St Etienne (MAS) sub-machine gun. It was replaced in military use by the 9 mm × 19 cartridge in 1950, but it is still used by French police forces and may still be encountered in military reserve.

Some authorities argue that it was derived from the abortive US .30 Pedersen round, but the connection seems tenuous. It is more probable that it was a French designer's attempt to produce a 7.65 mm cartridge of comparable power to the contemporary 9 mm Browning Long cartridge.

The French standard round was:
Ball Type L s 5.5 g lead-cored compound bullet in brass or steel case.

9 mm Short

9 mm × 17
V₀ 270 m/s
E₀ 224J

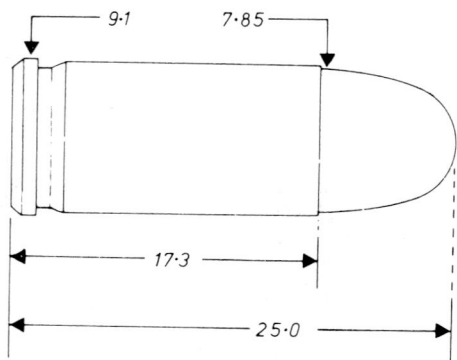

This cartridge was originated by Colt as the .380 automatic pistol in 1908, and in 1910 was adopted by Fabrique Nationale of Liège as the 9 mm Short Browning. It has since come to be known throughout Europe as the '9 mm Short' and in the USA as the '.380 Auto'.

It was widely adopted in Central Europe as a police and military cartridge in the 1920-40 period. At present it does not feature in any military inventory but will be found in police use and in staff officers' pistols.

The standard form of this cartridge is with a full-jacketed compound bullet weighing about 6.15 g. Although theoretically low powered, the 9 mm Short is one of the more effective pocket-pistol cartridges in that it combines reasonable stopping power with a low velocity which reduces the risk of ricochet.

9 mm Police

9 mm × 18
V₀ 345 m/s
E₀ 363J

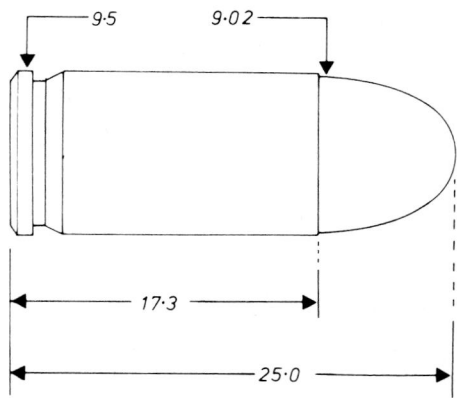

This cartridge is a recent introduction, developed in West Germany in the early 1970s. It appears to have been designed to replace the 9 mm Short as the most powerful cartridge capable of being fired safely from a weapon with an unlocked breech. In appearance and in philosophy it resembles the Soviet 9 mm × 18 Makarov, but there are small dimensional differences which ensure that the two are not completely interchangeable. It can be seen from the performance figures above that it falls a little short of the 9 mm × 19 Parabellum cartridge in energy and, as the title implies, it is being taken up by several European police authorities. Both Mauser and Heckler and Koch have produced suitable pistols.

The service loading is a full-jacketed ogival or truncated conical bullet weighing 6.1 g. The cartridge is also known in West Germany as the 9 × 18 Ultra and this marking will be found in the headstamp of cartridges made by Dynamit-Genschow; other makers, notably Hirtenberger, stamp '9 mm Police' on the base of the case.

9 mm Makarov

9 mm × 18 Soviet
V₀ 340 m/s
E₀ 348J

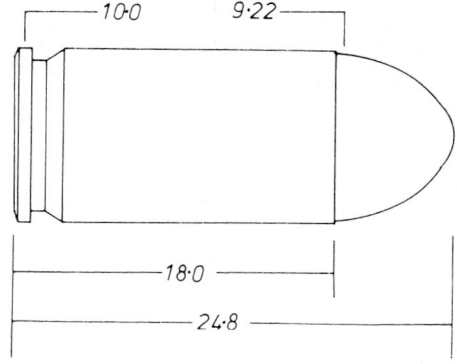

This cartridge appeared from the Soviet Union in the early 1960s, allied with the Makarov automatic pistol. It is understood to have been based on the pre-Second World War German experimental Ultra cartridge and it fills the same niche, namely to be more powerful than the 9 mm Short but not so powerful as to require a locked breech. The metric notation, 9 mm × 18, is the same as that of the recently introduced West German 9 mm Police cartridge (see above) but there are dimensional differences.

The 9 mm × 18 Makarov is now manufactured in other Warsaw Pact countries and also in China, where it is known as the Type 59. So far as is known there is only one loading, a ball round carrying a 6.02 g full-jacketed round-nosed bullet.

9 mm Parabellum

9 mm × 19
V₀ 396 m/s
E₀ 584J

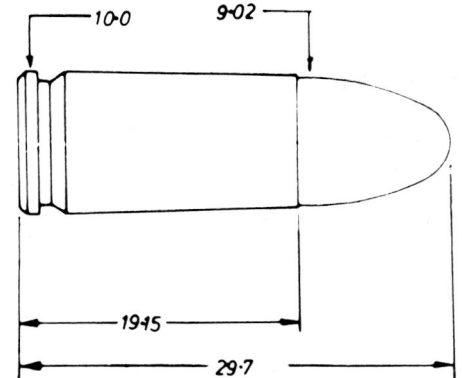

This cartridge was developed in 1902 by George Luger and Deutsche Waffen and Munitionsfabrik (DWM) in order to improve the stopping power of the Parabellum pistol: it is, in effect, the 7.65 mm Parabellum cartridge opened up and fitted with a 9 mm bullet. This bullet was originally cylindro-conoidal with a flat nose, a similar shape to its 7.65 mm ancestor, but in 1915 the German Army switched to an ogival bullet which has since become the standard type, though some commercial manufacturers continued to offer the original bullet until the 1930s, and indeed the bullet alone is still available, for re-loading purposes.

Since its introduction the 9 mm × 19 has become a universal military cartridge, and it has been manufactured by innumerable firms all over the world (with the possible exception of the Soviet Union). As a result it can be found with brass or steel cases and with every sort of bullet, including plastic. The NATO standard specification calls for a 7.45 g full-jacketed bullet loaded to deliver a velocity of 396 m/s; mean pressure to be 2050 kg/cm²; and the Accuracy Figure of Merit 76 mm at 50 m. Other existing variant designs used by military forces include:

BELGIUM (FN)
Ball   8 g bullet, V₀ 356 m/s
Tracer   7.65 g bullet, V₀ 350 m/s

FRANCE
'Balle D'   8 g bullet, V₀ 396 m/s
'Balle T' Tracer   8 g bullet, V₀ 396 m/s

WEST GERMANY (DYNAMIT-NOBEL)
Ball   8 g bullet, V₀ 360 m/s
Tracer   7.5 g bullet, V₀ 375 m/s
Action   5.6 g bullet, V₀ 430 m/s

ISRAEL
Ball   7.45 g bullet, V₀ 343 m/s
Carbine Ball +P   7.45 g bullet, V₀ 420 m/s
Carbine Ball +P   7.45 g hollow point bullet, V₀ 430 m/s
Subsonic Ball   8.03 g bullet, V₀ 314 m/s
Subsonic Ball   10.24 g bullet, V₀ 286 m/s

ITALY
M38   8 g bullet, V₀ 305 m/s

SOUTH AFRICA
Ball   7.6 g bullet, V₅ 401 m/s
Tracer   7.1 g bullet, V₅ 386 m/s, trace to 183 m

SWEDEN
M39B   6.8 g bullet, V₀ 420 m/s
M39 (Parabellum)   7.6 g bullet, V₀ 395 m/s
Tracer   no details
Gallery practice   plastic bullet with steel ball
Blank   plastic bullet; functions at full automatic

UNITED KINGDOM
Mk 2Z   7.45 g bullet, V₀ 414 m/s; E₀ 637J
ROTA (Training)   4.8 g bullet, V₀ 395 m/s; E₀ 375J, frangible bullet;
                    max range 850 m

UNITED STATES OF AMERICA
M882   8.03 g bullet, V₀ 375 m/s ± 15; E₀ 563J
Olin OSM   9.52 g bullet, V₀ 308 m/s; E₀ 450

9 mm Action Express

9 × 23RB

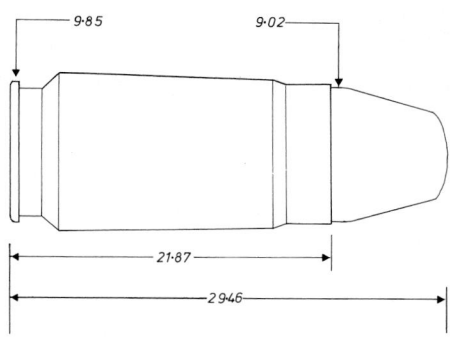

This cartridge is so new that no reliable performance figures can yet be given, those so far quoted by various sources being contradictory. Basically the 9 mm Action Express is the case of the .41 Action Express (below) necked down to take a standard 8 g 9 mm bullet. This should provide more chamber space and thus yield better performance within the 9 mm parameter, and it would appear to have military potential.

9 mm Bergmann-Bayard

9 mm Largo
9 mm × 23
V₀ 340 m/s
E₀ 505J

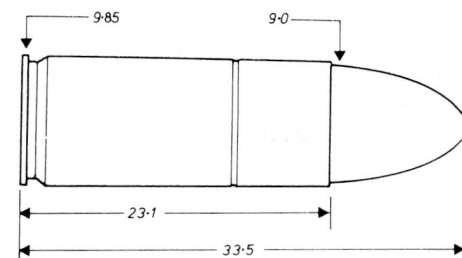

This was developed in 1903 for a Bergmann automatic pistol subsequently made under licence by Pieper of Liège (whose trademark was 'Bayard'). This pistol was adopted by the Danish and Spanish armies, and though the pistol has been long abandoned by both, the Spaniards retained the cartridge as their standard pistol round until comparatively recently. It has now been replaced in military service by the 9 mm Parabellum but various pistols and sub-machine guns are still in use by Spanish police and other civil authorities and the 9 mm Largo (as it is known in Spain) will be in use for a long time to come.

It is quite powerful yet for many years was used in the Spanish Astra pistol with an unlocked breech. The normal military loading is with a 8.75 g full-jacketed bullet.

9 mm Winchester Magnum

9 mm × 29
V₀ 450 m/s
E₀ 754J

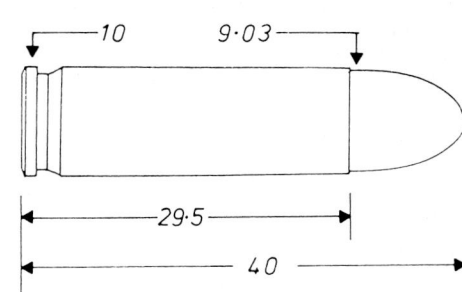

This cartridge was developed in the USA in 1979 for use with the Wildey gas-operated automatic pistol, a weapon intended for long-range target shooting. Though some 5 mm longer, its performance is very close to that of the 9 mm Mauser Export round; indeed, the similarities are such that one is inclined to wonder why anyone bothered. Although designed purely as a commercial venture, the 9 mm Winchester Magnum appears to have considerable military potential as a sub-machine gun or carbine cartridge, and for this reason it is included here.

The current commercial loading is a jacketed ogival 7.45 g bullet, which would be satisfactory for military purposes.

.357 Magnum

V₀ 436 m/s
E₀ 972J

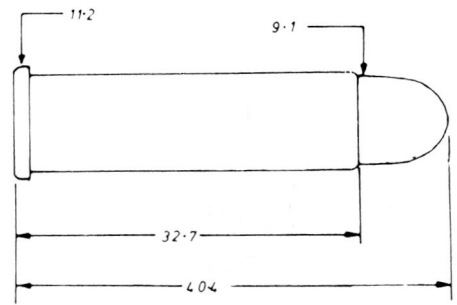

This cartridge was introduced in 1935 by Smith and Wesson and has since become accepted as a police cartridge in the USA. The calibre is the same as the common .38 revolver cartridge, but the notation is deliberately changed to distinguish this more powerful loading, and the case is longer so that it will not chamber in ordinary .38 revolvers. The standard factory loading is a 10.23 g bullet but numerous variations exist.

.38 British Service

.38/200
V₀ 180 m/s
E₀ 186 J

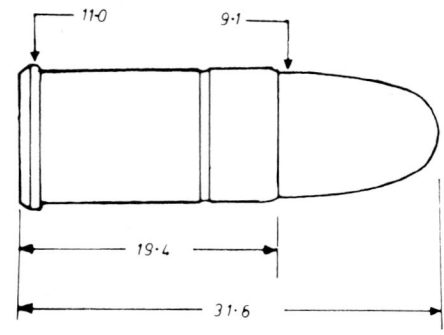

This was introduced in 1930 when the British Army adopted a .38 revolver in place of the .455 which had been standard for many years. The theory expounded at the time was that the smaller calibre would be easier to shoot and, by using a 12.96 g bullet, would still deliver a blow comparable to that of the .455 cartridge. It is difficult to understand that argument, since the .455 delivered 306 J while with the 12.96 g bullet the .38 gave only 210 J; but since 75 J was considered sufficient force to produce a casualty and, since the .38 was certainly easier to handle, the argument went through. The chosen bullet was a blunt, lubricated-lead article and, in view of its dubious legality in the eyes of the Hague and other agreements, a 11.53 g full-jacketed bullet was introduced in 1938 and has remained standard ever since. This substitution reduced the power to a point rather better than a 7.65 mm Browning round.

Although no longer in British service, numbers of ex-Imperial forces retained the revolver in reserve and it is widely used by police forces. The authorised service rounds were:

Ball Mk 1   12.96 g lead with cordite charge
Ball Mk 1Z   as for Mk 1 but with nitro-cellulose charge
Ball Mk2   11.53 g jacketed bullet, cordite charge
Ball Mk 2Z   as Mk 2 but with nitro-cellulose charge
SOUTH AFRICA (Armscor) manufacture:
Ball   9.40 g jacketed bullet, V₅ 240 m/s
               with double-base propellant

.38 Special

.38 Smith & Wesson Special
9 × 29.5R
V₀ 260 m/s
E₀ 346 J

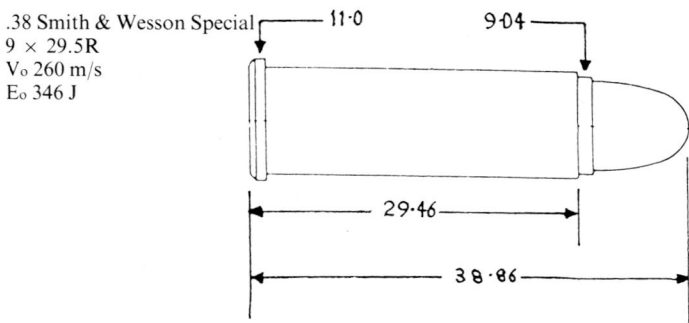

Developed by Smith & Wesson in about 1900 as a potential military cartridge, before the US Army decided that .45 was the minimum it would consider for pistols, this cartridge went on commercial sale in 1902 and has prospered ever since, becoming popular with police and security forces as well as being an accurate target and sporting cartridge. It combines good ballistic performance with high accuracy and controllable recoil.

Innumerable different bullets and loadings can be found; the commercial standard is a 10.23 g round-nosed lead or jacketed bullet, while a 12.96 g jacketed bullet is widely sold as a police loading. Tracer bullets have been developed, as well as specially hardened 'metal-piercing' bullets for police use against cars. The performance figures quoted above are based on the REM-UMC cartridge with 10.23 g bullet, as procured by the US Government.

The Société Française de Munitions (SFM) manufactures a special high-velocity round in this calibre, known as THV (Tres Haute Vitesse). This uses a lightweight bullet with an internal cavity and a concave ogive shape, achieving remarkable penetration performance. The bullet weighs 2.9 g and from a 102 mm barrel gives a muzzle velocity of 690 m/s, giving it a muzzle energy of 688 J, twice the energy of the conventional round. Similar rounds are also available in 9 × 19 mm, .357 Magnum and .45 ACP calibres.

.41 Action Express

10.4 × 22 mm
V₀ 352 m/s
E₀ 772 J

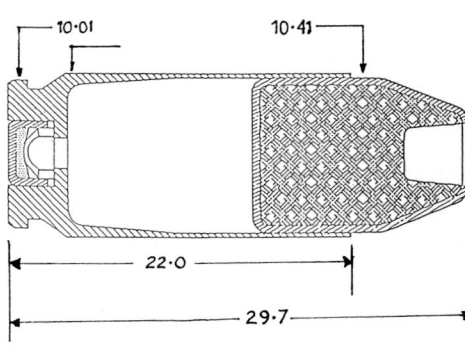

This cartridge has been developed as a means of upgrading 9 mm Parabellum weapons with a minimum amount of rebuilding. The significant feature is that the dimensions of the base and extraction rim are identical to those of the 9 × 19 mm Parabellum cartridge, so that the dimensions of the weapon's breech face remain unchanged. All that is required is to change the weapon's barrel and magazine, make adjustments to the return springs and recalibrate the sights. The case is somewhat longer than that of the 9 mm round but not so much as to cause major problems in redimensioning magazines or feedways. The larger bullet and charge produce a significant increase in energy. A number of loads have been developed, but the only one suitable for military purposes uses a 12.46 g full metal jacketed bullet, upon which the data given above has been derived. Loads with 11 g soft-point bullets, with various weights of propelling charge, develop muzzle velocities from 355 to 396 m/s.

.45 Automatic Colt Pistol (ACP)

.45 M1911
V₀ 250 m/s
E₀ 474 J

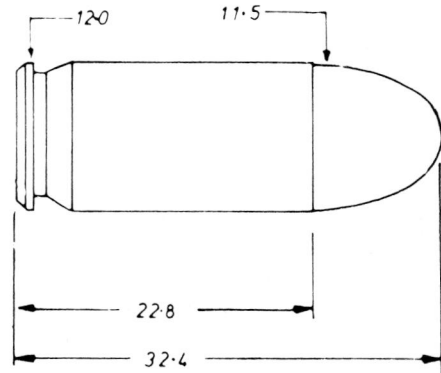

The .45 ACP has been the standard American service pistol cartridge since 1911, though its inception dates from 1907. It is one of the most accurate and certainly the most powerful of service-pistol rounds and it is widely used in other countries, particularly in Central and South America. In British service two types have been authorised, the Mark 1Z, made briefly in Australia in 1940/42 and the Mark 2Z which is simply the standard American-made product. Two types are currently in American service:

Ball   M1911   15.16 g jacketed bullet
Tracer   M26   13.48 g jacketed bullet with tracer in base
Steel-cased cartridges were extensively produced during the Second World War and may still be encountered.

Rifle and Machine Gun

4.73 × 33 mm Caseless

Patr. 4.73 mm DM11
V₀ 930 m/s
E₀ 1380 J

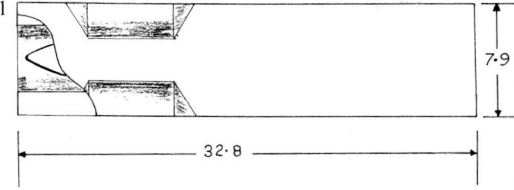

This cartridge was developed by Dynamit Nobel in collaboration with Heckler and Koch and is for use solely with the Heckler and Koch G11 rifle. The original version, used in early models of the G11 during the NATO small arms trials of 1977-79, gave rise to problems, notably with heat build-up and cook-offs. This version used a conventional nitro-cellulose propellant with a binder material to form it into a solid mass with adequate mechanical strength. This propellant mass was shaped into an 8 × 9 × 21 mm rectangle; the bullet was embedded in the front end, while at the rear was a recess for a tin-foil percussion cap and an 'intermediary' or 'booster' charge which transferred ignition from cap to propellant.

The cartridge was subsequently redesigned. In its present form the propellant is a moderated high explosive, probably based on a RDX derivative, mixed with a suitable binder and formed into an 8 × 8 × 32.8 mm block. The bullet is completely embedded in the block, its tip showing at the front end of a central hole, and the percussion cap is now made of thin leaves of propellant so that there is no danger of tin-foil residues in the chamber after firing. This new propellant, known as HITP (high ignition temperature propellant) can withstand chamber temperatures some 100°C above the cook-off point of the earlier nitro-cellulose material and has great mechanical strength.

The bullet is of conventional streamlined (boat-tailed) form and consists of a lead core, steel jacket and gilding metal envelope. It shows no fragmentation on impact and will pierce 6 mm of mild steel plate at 300 m range and a standard steel helmet at 600 m range. The bullet weighs 3.2 g and the complete cartridge 5.2 g.

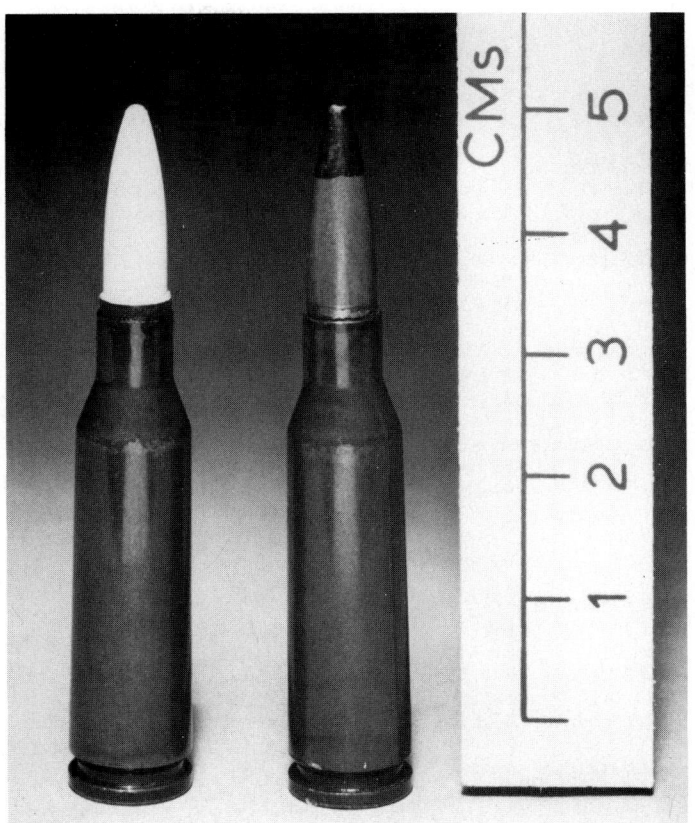

Soviet 5.45 × 39.5 mm Plastic Practice and Tracer cartridges

5.45 mm Soviet

5.45 mm × 39.5
V₀ 900 m/s
E₀ 1383 J

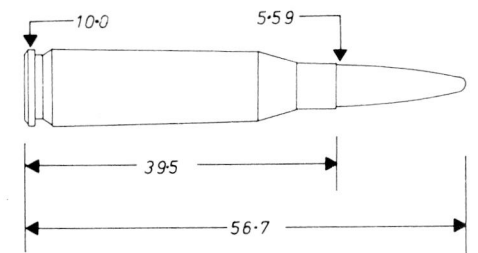

This cartridge was developed in the early 1970s to accompany the AK-74 rifle. The case is slightly fatter and shorter than the 5.56 mm × 45 and, as is usual with Soviet ammunition, the dimensions are such that no Western weapon will be able to chamber the 5.45, nor will the AK-74 chamber the 5.56 mm. The bullet weighs 3.415 g and is of the full-jacketed streamlined pattern, 4.47 calibres long. The cartridge case is of lacquered steel and a stripe of red lacquer seals the case-bullet joint. The Berdan pattern primer is of brass and there are two fireholes in the primer pocket. Matching AP, tracer and plastic-bulletted practice cartridges are known to exist.

It should be noted that this cartridge bears no relationship to the earlier 5.56 mm Soviet 'Biathlon' round which was a 5.56 mm × 39 developed in the Soviet Union in the 1960s based on the 7.62 mm × 39 case necked-down and was developed primarily as a sporting and target cartridge, though it doubtless played its part in the development of the 5.45 mm round.

5.56 × 45 mm

.223 Armalite
V₀ 1005 m/s
E₀ 1798 J

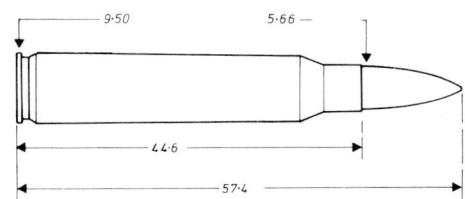

This was developed from a commercial cartridge, the .222 Remington, in the mid-1950s and was first used with the Armalite AR-15 assault rifle in 1957. In the 1960s this became the M16 and was extensively used in Vietnam, and in the late 1960s the US Army in Europe changed over completely to the 5.56 mm cartridge, abandoning the 7.62 mm × 51 except for machine gun use. This precipitated a rush to evaluate this calibre and several rifle makers produced suitable weapons. The AR-15 has also been adopted to a greater or lesser degree by several countries and thus the manufacture of 5.56 mm ammunition has now become widespread. The NATO trial of 1977-79 came to the conclusion (which many observers thought was fairly predictable) that the 5.56 mm cartridge was to be the next NATO standard small-arms round and that the Belgian SS 109 bullet, developed by Fabrique Nationale, was to supersede the US M193 ball which had hitherto been the standard projectile.

Several variant bullets are still on offer for military adoption. In addition, it is produced commercially, since a number of sporting rifles have been made in this calibre, and in sporting cartridges the bullet is almost always a non-streamlined soft-point type which tends to deform easily when fed through a military semi-automatic rifle and cause malfunctioning.

Royal Ordnance plc have produced two calibres of ROTA (Royal Ordnance Training Ammunition). It is available now in 5.56 × 45 mm and in 9 × 19 mm by June 1989. The ROTA ammunition uses a frangible bullet which is effective out to combat ranges, after which its performance deteriorates rapidly so that range safety templates can be drastically reduced. The bullet has no lead content and is non-toxic so it is ideal for use in indoor training facilities. It minimises range damage. The ammunition can be fired without requiring any weapon modification.

Standard types of 5.56 mm × 45 cartridge include:

AUSTRIA (SMI and HIRTENBERGER)
Ball   3.6 g bullet, V_o 980 m/s
Ball   4.0 g bullet, V_o 900 m/s (SS109 equivalent)

BELGIUM (FN)
Ball SS92   3.56 g streamlined bullet, V_o 965 m/s, E_o 1657 J
Tracer L95   3.3 g semi-streamlined bullet
AP P96   3.43 g steel-cored streamlined bullet
Ball SS109   3.95 g streamlined bullet, V_o 922 m/s, E_o 1680 J
Tracer L110   4.14 g semi-streamlined bullet, V_o 860 m/s
AP P112   3.95 g steel-cored streamlined bullet, V_o 922 m/s

FRANCE (GIAT)
Ball Type O   3.56 g streamlined bullet, V_o 950 m/s, E_o 1605 J
Tracer Type T   3.36 g semi-streamlined bullet, V_o 950 m/s

WEST GERMANY (DYNAMIT-NOBEL)
Ball   3.56 g bullet, V_o 1010 m/s, E_o 1815 J
Tracer   3.5 g bullet, V_o 1010 m/s, E_o 1785 J

PORTUGAL (FNM)
Ball E1   3.55 g streamlined bullet, V_o 990 m/s, E_o 1740 J
Tracer E2   3.35 g semi-streamlined bullet, V_o 975 m/s

SOUTH AFRICA (ARMSCOR)
Ball   3.52 g bullet, V_s 965 m/s
Tracer   3.44 g bullet, V_s 950 m/s, trace to 457 m

SWEDEN
Ball m/5   4.0 g, streamlined, V_o 930 m/s

UNITED KINGDOM
Ball L2A1   4.0 g, streamlined, V_o 944 m/s, E_o 1782J
Tracer L1A2   4.2 g, V_o 865 m/s; dark ignition, trace to 600 m
Blank L1A1   7.87 g, full brass casing — no muzzle debris
Drill L5A1   10.85 g, representing Ball L2A2 and Tracer L1A2
ROTA (Training)1.4 g, frangible bullet; V_o 920 m/s, E_o 1744J; max range 800 m

UNITED STATES OF AMERICA
Service
Ball M193   3.56 g streamlined bullet
Tracer M196   3.43 g semi-streamlined bullet with trace to 400 m
Developments
Ball M777   3.46 g streamlined bullet; velocity not disclosed
Tracer M778   3.46 g semi-streamlined bullet, trace to 725 m

Winchester 'Penetrator' M855
This 5.56 mm round uses a bullet with compound steel/lead core and is optimised for rifles with 173 mm twist rifling. The bullet is 23.2 mm long, weighs 4.02 g, and has a V_o of 945 m/s . It can penetrate 19 mm aluminium armour at 300 m or 3.5 mm NATO plate at 750 m.

5.7 mm FN P90

5.7 × 28 mm
V_o 850 m/s
E_o 540 J

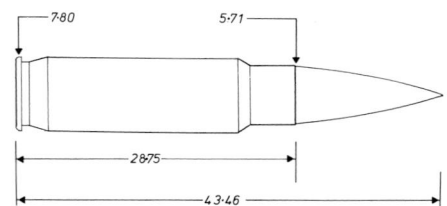

An entirely new cartridge, the 5.7 × 28 mm has been developed by Fabrique Nationale Herstal for their P90 'personal weapon'. The bullet is streamlined and jacketed and weighs 1.5 g and is claimed to be capable of penetrating a steel helmet at 150 m and 48 layers of Kevlar at over 50 m. In spite of this, the recoil impulse is one-third that of a 5.56 × 45mm round and less than two-thirds that of a 9 mm Parabellum round. The stopping power is claimed to be three times that of the 9 mm Parabellum, and this is largely due to the tumbling movement of the bullet after it strikes; it does not fragment, even after ricochet, and gives up its latent energy very rapidly. FN Herstal intend to develop other weapons to fire this round, and consider it to be a replacement for the 9 mm Parabellum, which they consider to be technically obsolescent.

6.5 mm Swedish Mauser

6.5 mm × 55
V_o 793 m/s
E_o 2852 J

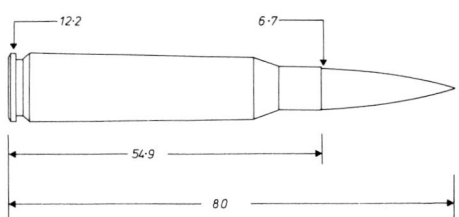

This was adopted by the Swedes in 1894 and remained in first-line service until replaced by the 7.62 mm × 51 in the 1950s, after which it was relegated to the reserve. It was also used by the Norwegians in suitably chambered Krag-Jorgensen rifles. The cartridge was well-designed, and it had a reputation for delivering superb accuracy and for this reason it is still widely used for target and sporting activities throughout Scandinavia.

It should be noted that although this, the 6.5 mm × 57, the 6.5 mm × 58 Mauser and the 6.5 mm × 54 Mannlicher all look very similar, none of them is interchangeable, even though some will load into the wrong rifle.

Standard Swedish loadings in this calibre were:

Ball PRJM41   9.07 g streamlined compound bullet
Tracer SLPRJ M41   7.78 g non-streamlined bullet
AP PPRJ M41   non-streamlined bullet steel-cored
The cartridge is currently manufactured by Hirtenberger of Austria, with a 9.25 g ball streamlined bullet giving V_{10} 800 m/s, and a 7.12 g tracer bullet giving V_{10} 800 m/s, and tracing to 800 m.

7 mm Spanish Mauser

7 mm × 57
V_o 700 m/s
E_o 2744 J
Original specification
with 11.2 g bullet

The 7 mm Spanish Mauser was introduced in 1892 for an improved Belgian rifle, but gained its name by being adopted in the following year by the Spanish Army. It was subsequently adopted by the Brazilian, Chinese, Colombian, Mexican, Transvaal and other forces and it attained fame in the Spanish-American War. Although no longer a first-line weapon, it still serves in several countries as a police and security force weapon and ammunition is currently manufactured in several countries. Typical military loadings, these by Fabrique Nationale, are:

Ball   9 g streamlined, full-jacketed bullet, V_o 840 m/s, E_o 3175 J
Tracer   8.95 g semi-streamlined bullet

7.5 mm French MAS

7.5 mm × 54
V_o 793 m/s
E_o 2852 J

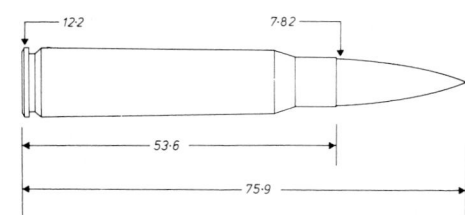

This cartridge was developed in the 1920s, primarily as a machine gun round, a suitable rifle not appearing until 1934. In broad terms it was based on studies of the 7.92 mm × 57 Mauser, with a change of calibre to produce the required ballistic performance. The 7.5 mm has never been produced outside France and weapons in this chambering have been used only by France and ex-French possessions. It is still a service calibre with the French Army and ammunition is currently produced by government arsenals and commercially by Gévelot and GIAT.

Current French service ammunition includes:

Ball	Mle1929 'O'	flat-based, pointed, lead core, 9.07 g, bullet V_o 793 m/s
Tracer	Ml1942 'TO'	steel core, lead sleeve red tip, 9.4 g bullet
Tracer	Mle1949A 'TO'	steel core, lead sleeve red tip, 9.4 g bullet
Tracer	Mle1950 'TO'	lead core, green tip, 9.07 g bullet
Tracer	Mle1958A 'TO'	lead core, plastic tracer cover, violet tip, 9.07 g bullet
Tracer	MleG.59 'TO'	lead core. Has longer trace than Mle 58, red tip, 9.07 g bullet
AP	Mle1949 'P'	hardened steel core, 9.4 g bullet
AP	Mle1949A 'P'	hardened steel core, 9.4 g bullet

APT   Mle1949 'TP'   originally white band, black tip bullet
APT   Mle1949A 'TP'   now red band, black tip bullet

Commercially available ammunition:

FABRIQUE NATIONALE
Ball   9 g streamlined bullet, V_o 845 m/s
Tracer   8.93 g semi-streamlined bullet, V_o 840 m/s

GÉVELOT SA
Ball   weight not stated, V_o 810 m/s
GIAT
Ball   LGF1

7.5 mm Swiss

7.5 mm × 55
V_o 780 m/s
E_o 3450 J

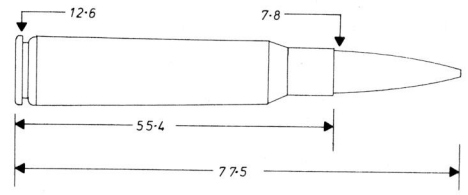

This was one of the first small-calibre rimless cartridges to be developed and was adopted by the Swiss in 1889. After various improvements it had settled down, by 1911, to being an 11.34 g streamlined bullet which was actually of 7.8 mm calibre, and remained the standard Swiss round thereafter. It is still a service calibre used in rifles and machine guns. Both brass- and steel-cased cartridges may be found and the Swiss also experimented with light alloy cases in the 1940s. Military ammunition in this calibre is made only in Switzerland, but there is a small output of sporting ammunition, mainly by Norma of Sweden, since large numbers of earlier pattern Swiss rifles were disposed of as surplus.

Current military loadings include:

Ball Type GP11   11.34 g streamlined bullet
Tracer L11   10.11 g non-streamlined bullet
AA tracer FLAB11   9.45 g streamlined bullet
AP StK11   11.34 g steel-cored streamlined bullet

7.62 mm Soviet Model 1943

7.62 mm × 39
V_o 710 m/s
E_o 1993 J

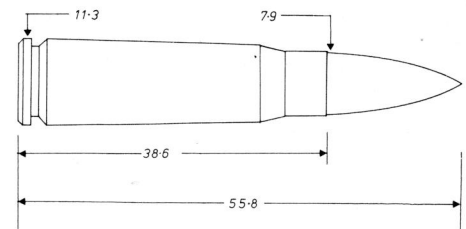

This cartridge is said to have been developed in 1943 but no weapon for it appeared until after the Second World War. It was doubtless influenced by the German 7.92 mm × 33 Kurz cartridge and rapidly became the standard infantry cartridge of the Soviet Army. It is one of the most successful military calibres, giving good performance from small bulk and has become a universal cartridge with the widespread adoption of Soviet weapons by Third World countries. Original manufacture was, of course, confined to the USSR, but this was soon followed by production in other eastern bloc countries and it has since been taken up by manufacturers in Finland, the Netherlands, Austria and West Germany. Brass, lacquered steel or brass-coated cases may be found.

Current Soviet service cartridges are as follows:

Ball Type PS   7.91 g streamlined steel-cored bullet
Tracer Type T45   7.45 g non-streamlined bullet
AP1 Type BZ   7.77 g streamlined bullet, steel-cored with incendiary pellet
Incendiary/ranging Type ZP   6.61 g non-streamlined bullet with incendiary and tracer elements

Service ammunition in this calibre includes:

AUSTRIA (HIRTENBERGER)
Ball   7.95 g non-streamlined bullet, V_o 710 m/s

FINLAND
Ball Type S309   8.04 g non-streamlined bullet, lead-cored
Tracer Type VJ313   8.04 g non-streamlined bullet
Both these bullets have a different contour to the Soviet bullet

Commercially available ammunition includes:

BELGIUM
FN ball   8 g streamlined bullet, V_o 695 m/s

7.62 mm NATO

7.62 mm × 51
V_o 838 m/s
E_o 3276 J

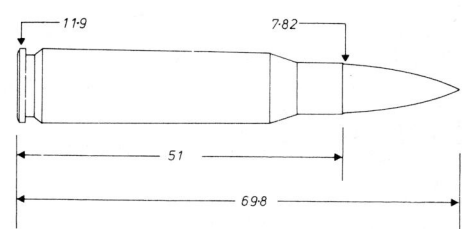

This cartridge, a triumph of political expediency over ballistics, became the NATO standard round in 1952. On the one hand the more percipient designers wanted a short-cased 'intermediate' cartridge similar to the Soviet 7.62 mm × 39; on the other, the Americans were adamant about the need to have full-sized performance comparable to their .30-06. The result was, in effect, the .30-06 in a shorter case, a compromise which satisfied nobody except the financial wizards. Such, however, was the mystique of NATO acceptance that the 7.62 mm × 51, and the various weapons made for it, have found wide acceptance outside NATO and are in use throughout the world.

The majority of 7.62 mm × 51 will be found with brass cases, though steel-cased rounds have been produced from time to time. As well as being manufactured by government arsenals, most commercial makers produce military cartridges in this calibre, generally based on the NATO specification, for open sale. Those cartridges produced by member countries of NATO to the agreed specification will be marked on the base with the NATO symbol, a cross within a circle, signifying interchangeability. The 7.62 mm × 51 is also sold as a sporting cartridge, known as the .308 Winchester.

Current standard cartridges used by various countries are as follows:

BELGIUM (FN)
Ball SS71/1   9.3 g streamlined bullet, V_o 837 m/s
Tracer L78   8.93 g semi-streamlined bullet, V_o 834 m/s
AP P80   9.75 g streamlined bullet, V_o 845 m/s

FRANCE (SFM)
The Société Française de Munitions (SFM) manufactures a unique piercing round known as PPI. This has a steel core fitted into a self-tightening non-discarding sabot of brass. The hardened steel makes no contact with the barrel, the brass sabot engaging with the rifling; it has a special profile which reduces friction and this, together with the lighter bullet, admits of a higher velocity than is attainable with conventional AP bullets. In 7.62mm calibre the bullet weighs 9.07 g and has a V25 of 884 m/s, with a penetration of 10 mm of armour steel at 300 m range. It will pierce 15 mm of aluminium alloy at 700 m.

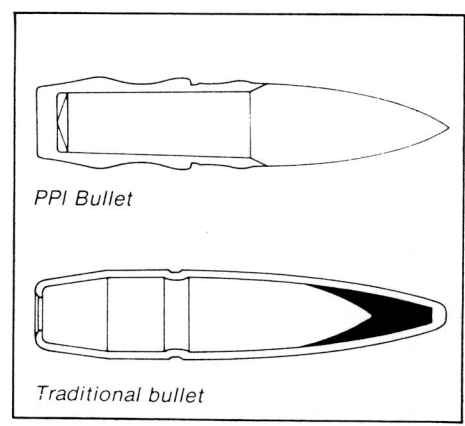

SFM 7.62 mm PPI bullet

WEST GERMANY (DYNAMIT-NOBEL)
Ball   9.45 g streamlined bullet, V_o 850 m/s
Tracer   8.95 g bullet, V_o 850 m/s
AP   9.55 g bullet, V_o 850 m/s

PORTUGAL (FNM)
Ball   9.3 g streamlined bullet, V_o 837 m/s
Tracer   8.8 g semi-streamlined bullet, V_o 834 m/s
AP   no details of bullet, V_o 845 m/s

SOUTH AFRICA (ARMSCOR)
Ball   9.27 g streamlined bullet, V_s 850 m/s
Tracer   9.1 g semi-streamlined bullet, V_s 827 m/s
Plastic   1.25 g non-streamlined plastic bullet, V_s 810 m/s

548 **AMMUNITION**

SWEDEN
Ball Type 10PRJ   9.4 g streamlined bullet, Vo 830 m/s
Tracer Type
10SLPRJ   no details of bullet
FFV AP   8.4 g streamlined bullet with heavy metal core, Vo 950 m/s; penetrates 15 mm of armour at 300 m
Blank   wooden bullet

UNITED KINGDOM
Ball L2A2   9.33 g streamlined bullet, lead core
Blank L14A1   brass case, crimped round
Tracer L5A3   8.75 g non-streamlined bullet, lead tip filler, tracer element in base
Practice L14A1   plastic bullet
Drill L1A2   inert round for training purposes
Green Spot   for target/sniping use; accuracy better than mean radius of 100 mm at 600 m range
Optimised machine gun ammunition: heavier bullet for improved range

UNITED STATES OF AMERICA
Ball M59   7 g non-streamlined bullet, lead core
Tracer M16   6.93 g semi-streamlined bullet, lead tip filler, tracer element in base
Tracer M27   6.54 g semi-streamlined bullet, dark ignition

7.62 mm Czechoslovak M1952

7.62 mm × 45
Vo 744 m/s
Eo 2325 J

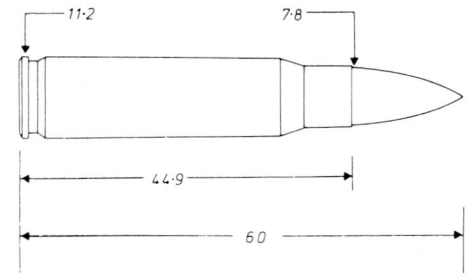

This was developed independently by the Czechoslovaks in the late 1940s and was adopted in 1952 for use in their assault rifle. It appears to have been an excellent cartridge, but was abandoned in the early 1960s in favour of the Soviet 7.62 mm × 39, in the interests of standardisation within the Warsaw Pact countries. It can be found with brass or with lacquered-steel cases, but it is doubtful whether it is even retained in reserve. So far as is known, the only service issues were:

Ball M52   8.4 g streamlined bullet, steel core
Tracer M52   8.4 g non-streamlined bullet

.30-06 Springfield

.30 US Service
7.62 mm × 63
Vo 837 m/s
Eo 3450 J

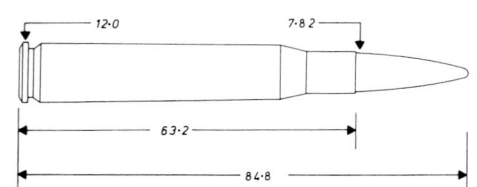

Another venerable cartridge, this first appeared in 1903 to accompany the Springfield Model 1903 rifle. It was improved in 1906 by substituting a pointed, non-streamlined, bullet for the original round-nosed type and it remained in this form until the adoption of a streamlined bullet, the M1, in 1926. This, though, led to problems in maintaining the desired ballistics, and in 1940 a new 9.7 g non-streamlined bullet was standardised as the M2. This has remained in service ever since and has been supplemented by other types of bullet from time to time.

The .30-06 was adopted by many other countries either for general military use or, as for example, in the case of the United Kingdom, in connection with the use of Browning machine guns in tanks or aircraft. It is, therefore, a widely-distributed cartridge and has been manufactured in many countries. In addition to its military variants, it will be found in considerable quantity as a sporting cartridge.

Current .30-06 loadings include:

UNITED KINGDOM
Ball   Mk 4Z   flat-based, pointed, lead core. 9.72 g bullet, Vo 855 m/s
Tracer   G Mk1Z   flat-based, pointed, lead tip filler. Unlike US production, trace composition is held in metal canister. 9.72 g bullet (dim ignition tracer)
Incendiary   B Mk2Z   flat-based, pointed, steel sleeve. 0.45 g composition. 9.98 g bullet, Vo 870 m/s

UNITED STATES OF AMERICA
Ball   M2   flat-based, pointed, lead core, 9.85 g bullet, Vo 837 m/s
AP   M2   flat-based, pointed, steel core, 10.69 g bullet, Vo 829 m/s
AP1   M14   flat-based, pointed, steel core, 9.72 g bullet, Vo 849 m/s
Incendiary   M1   flat-based, pointed, steel sleeve, 9.07 g bullet, Vo 901 m/s
Tracer   M25   flat-based, pointed, lead point filler, dim ignition, 9.46 g bullet, Vo 814 m/s
Frangible   M22   flat-based, solid construction of powdered lead and bakelite, 7 g bullet, Vo 396 m/s

Commercial loadings:
FABRIQUE NATIONALE
Ball   9.75 g bullet, Vo 845 m/s, Eo 3480 J
AP   10.51 g bullet, Vo 830 m/s, Eo 3620 J
Tracer   9.14 g bullet, Vo 830 m/s, Eo 3148 J

Other companies offering .30-06 include:

Fabrica Nacional de Municoes de Armas Ligeiras, Moscavide, Portugal
Gévelot SA, rue Ampere, Paris, France
Oy Sako AB, Riihimaki, Finland
Società Metallurgica Italiana SpA, Florence, Italy
Hirtenberger AG, Hirtenberg, Austria
All these firms produce ammunition based on the US standard specifications.

.30 Carbine

7.62 mm × 33
Vo 579 m/s
Eo 1173 J

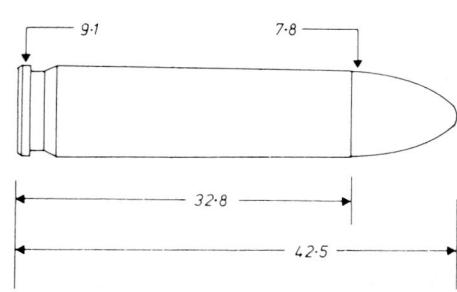

This cartridge was developed in 1940 from a commercial automatic-rifle cartridge and was tailored to the specification of the US carbine M1. It is a moderately effective short-range cartridge which has since been adopted in one or two South American semi-automatic carbine and sub-machine gun designs; it has also, at various times, been toyed with as a pistol cartridge, though rarely with any success since its ballistic characteristics are not well-suited to short barrels. It is interesting to see that the energy figure for this cartridge is close to that of the Soviet 5.45 mm round, but the poor ballistic shape of the .30 carbine bullet ensures that the muzzle energy falls off very rapidly; by 300 m the .30 has dropped below 300 m/s and the E_{300} 317 J, while the Soviet bullet is still in the sonic zone and delivering in excess of 700 J.

Military .30 carbine cartridges can be found in steel or brass-cased versions, and there are a number of commercial sporting rounds available.
Current military standard rounds are in service as follows:

UNITED KINGDOM
Ball Mark 1   7.19 g bullet, to US pattern, Vo 600 m/s, Eo 1294 J

UNITED STATES OF AMERICA
Ball M1   7 g round-nosed, non-streamlined bullet, with lead-antimony core exposed at base
Tracer M16   6.93 g flat-based bullet, with lead-antimony point filler and red tracer composition
Tracer M27   6.54 g bullet, similar to M16 but with dark ignition

Commercial manufacture:

FABRIQUE NATIONALE
Ball   7.1 g bullet, to US pattern, Vo 590 m/s, Eo 1235 J
Tracer   6.53 g bullet, bright ignition, red tracer, Vo 565 m/s, Eo 1042 J

HIRTENBERGER
Ball   7.15 g bullet to US pattern, Vo 580 m/s

7.62 mm Mosin-Nagant

7.62 mm × 54R
V₀ 818 m/s
E₀ 4008 J

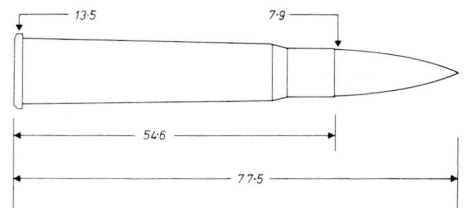

Another cartridge dating from the 1890s this was introduced into Russian service with the M1891 Mosin-Nagant rifle and, with minor modifications, has survived ever since. It is the last of the old-style, rimmed-case, large-capacity cartridges to survive in military service, but still performs more than adequately in the support machine gun role. It also survives in other countries which adopted the Mosin-Nagant rifle at various times, for example Finland and China. Although most of the ammunition has, of course, been made in Soviet state factories, other countries have produced it at various times and some quite surprising headstamps can be found. Cases are usually brass or copper-washed steel and are instantly recognisable by their unique part-convex base contour. Some commercial sporting ammunition is also made in this calibre.

An enormous number of variant cartridge designs have, of course, existed since its inception, but the current service rounds are as follows:

FINLAND
Ball Type D166	12.96 g streamlined bullet, V₀ 700 m/s, E₀ 3175 J
Tracer Type D278	
AP Type D277	no details of bullet available
Incendiary Type S276	

UNION OF SOVIET SOCIALIST REPUBLICS
Ball, heavy, Type D	11.98 g streamlined bullet, lead core
Ball, light, Type LPS	9.65 g streamlined bullet, steel core, V₀ 870 m/s
API Type B32	10.04 g streamlined bullet, steel core with incendiary composition, V₀ 870 m/s
API Type BS-40	12.11 g non-streamlined bullet, tungsten core, incendiary composition, V₀ 805 m/s
Tracer Type T46	9.65 g, non-streamlined bullet
API-T Type BZT	9.2 g, non-streamlined bullet, steel core with incendiary and tracer composition
Incendiary/Obs Type ZP	10.36 g, non-streamlined bullet, steel jacket with incendiary composition, explosive charge and striker in nose

7.65 mm Argentinian Mauser

7.65 mm × 54
V₀ 825 m/s
E₀ 3403 J

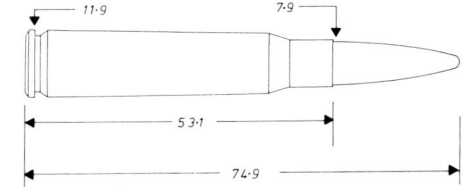

This cartridge, designed in 1889, was widely adopted in South America, notably by the Argentinian, Bolivian, Colombian, Ecuadorian and Peruvian armies, as well as being used elsewhere in the world. As a result, it can be met all over South America and may well still be in reserve stocks. Sporting rifles in this chambering were once popular and sporting ammunition is still made.

The standard military loadings currently manufactured by Fabrique Nationale of Liège are as follows:

Ball Type S	10 g streamlined bullet, lead core
Ball Type SS	11.25 g non-streamlined bullet, V₀ 750 m/s
AP	10.95 g bullet, V₀ 730 m/s
Tracer	10.45 g bullet, V₀ 725 m/s

.303 British

7.7 mm × 56R
V₀ 731 m/s
E₀ 3011 J

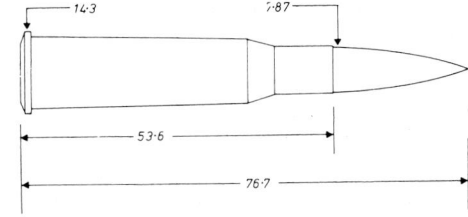

This was the standard British and Imperial cartridge from 1889 to the 1960s and accompanied the Lee-Enfield rifle and the Vickers, Vickers-Berthier and Bren machine guns. It is still in considerable use throughout the world and is

currently manufactured by a number of commercial firms in both military and commercial (sporting) loadings. To the best of our knowledge it has never been made in steel-cased form.

Principal standard loadings are in service as follows:

UNITED KINGDOM
Ball Mk7	11.27 g non-streamlined bullet
Ball Mk8Z	11.34 g streamlined bullet, for use in machine guns only
AP W Mk1	11.27 g non-streamlined bullet with hard steel core, V₀ 722 m/s. Penetration 10 mm at 100 m
Tracer G Mk3	9.98 g non-streamlined bullet, traces to 800 m, V₀ 701 m/s
Tracer G Mk8	10.95 g non-streamlined bullet, dark ignition, traces to 1000 m, V₀ 722 m/s
Incendiary B Mk7	10.75 g non-streamlined bullet, V₀ 722 m/s

Commercially available:

FABRIQUE NATIONALE
Ball Type S	10 g bullet, V₀ 760 m/s, E₀ 2888 J
Ball Type SS	11.25 g bullet, V₀ 745 m/s, E₀ 3122 J
AP	11.6 g bullet, V₀ 735 m/s
Tracer	10.45 g bullet, V₀ 755 m/s

7.92 mm Mauser

7.92 mm × 57
V₀ 750 m/s
E₀ 3600 J

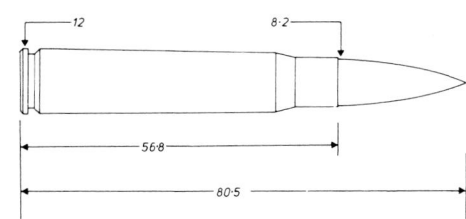

Probably the most widely-distributed military rifle cartridge in history, the 7.92 mm Mauser has appeared in innumerable variant forms and has been manufactured in every major country at some time or other. But, since it is a full-sized, old-style, powerful cartridge, its use in rifles is declining, though all sorts of border guards, police, customs guards and similar para military forces throughout the world are still furnished with Mauser rifles and carbines. Its major military application today is in belt-fed machine guns, eg the Yugoslav M53, but even this is being eroded in favour of the 7.62 mm × 51 cartridge. Nevertheless, millions of suitable weapons exist, and they appear with unfailing regularity in any coup d'état or civil commotion. Vast amounts of ammunition must exist and the cartridge is still in production by several companies. Brass or steel-cased military ammunition can be found, with a wide selection of bullet types. Much ex-Second World War German and British (for the Besa machine gun) ammunition is still to be found, though a lot of it is quite unsuited to use with infantry weapons. British 7.92 mm cartridges should never be used in automatic rifles, while German ammunition with a green ring on the bullet is a high-velocity type for use only in aircraft machine guns.

The more common service loadings are as follows:

GERMANY (EX-SECOND WORLD WAR)
Ball SmE lang	11.53 g streamlined bullet, mild steel core, V₀ 837 m/s
Heavy Ball, sS	12.83 g streamlined bullet, V₀ 765 m/s
Semi-AP SmE	11.53 g streamlined bullet, V₀ 770 m/s
AP SmK	11.53 g, non-streamlined bullet, steel core, V₀ 798 m/s
AP SmK(G)	12.57 g, non-streamlined bullet, tungsten carbide core, V₀ 911 m/s. Penetrates 19 mm at 100 m
AP-Incendiary PmK	10.11 g streamlined bullet, steel core, white phosphorus incendiary compound, V₀ 835 m/s
AP-Tracer SmK L	10.12 g streamlined bullet, trace to 1000 m, changing from green to red at 600 m, V₀ 832 m/s
Incendiary/Obs SmK B	10.82 g streamlined bullet, V₀ 814 m/s

UNITED KINGDOM
Ball	Mk 2Z	12.83 g streamlined bullet, V₀ 756 m/s, E₀ 3666 J
AP W	Mk 2Z	11.53 g streamlined bullet, V₀ 786 m/s. Penetrates 11 mm of plate at 100 m
Tracer G	Mk 3Z	10.49 g non-streamlined bullet, dark ignition, traces to 1000 m, V₀ 756 m/s
Incendiary B	Mk 2Z	11.66 g streamlined bullet, V₀ 771 m/s

Commercial (typical):

FABRIQUE NATIONALE
Ball	12.8 g streamlined bullet, V₀ 750 m/s, E₀ 3600 J
AP	12 g non-streamlined bullet, steel core, V₀ 735 m/s
Tracer	11.5 g semi-streamlined bullet, V₀ 735 m/s

12.7 mm Soviet

12.7 mm × 107
V_o 840 m/s
E_o 16 920 J

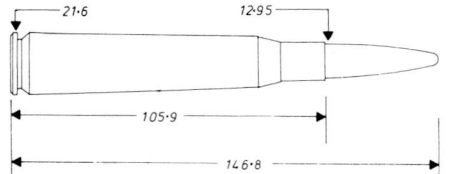

This cartridge was developed in the late 1920s and was influenced by the German First World War 13 mm TuF (Tank und Flieger) cartridge. It first saw service with the Degtyarev DK heavy machine gun in the mid-1930s and has remained in constant use ever since. It has been widely distributed throughout Soviet satellite and Third World countries, but manufacture has been confined to China, Czechoslovakia, Poland and the Soviet Union. Cases are usually of brass, but steel is known to have been used in small batches.

Current types on issue include:

AP-1 Type BZ	47.95 g non-streamlined bullet, steel core	
API-T Type BZT	44.06 g non-streamlined bullet, steel core	

.50 Spotting Rifle

12.7 mm × 77
V_o 532 m/s
E_o 7585 J

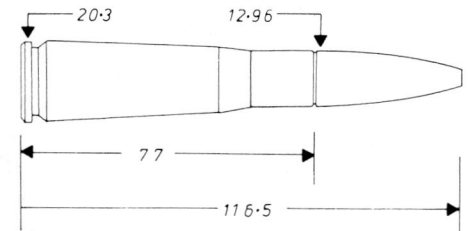

This cartridge, a shortened version of the .50 Browning, was developed solely for use with the spotting rifle M8 used with various recoilless guns and the ballistic performance is expressly matched to that of the parent guns so that there is a specified relationship between the trajectory of the .50 bullet and the parent-gun projectile. The bullet is a special design of observing-tracer which carries a spotting charge of incendiary material in the front section so that the strike of the bullet is signalled by a flash and a puff of smoke.

Tracer M48A1   53.6 g bullet

.50 Browning

12.7 mm × 99
V_o 888 m/s
E_o 16 916 J

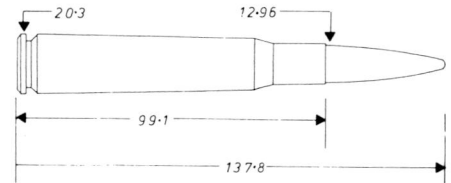

This is another cartridge which was influenced by the German 13 mm TuF round. It was developed in the 1920s and came into prominence as an aircraft machine gun cartridge during the Second World War. It was widely adopted as a ground machine gun during the war, and in post-war years became international standard, adopted by almost every army outside the Soviet bloc. Ammunition has been manufactured in several countries, but invariably to the US standards prevailing at the time, so that inter-changeability has always been assured. Steel-cased cartridges have been made, but brass-cased types are by far the most common.

Current standard rounds are as follows:

UNITED KINGDOM
Ball	Mk 3Z	boat-tailed, pointed, steel core, 46.01 g bullet V_o 823 m/s
Tracer	G Mk 6Z	flat-based, pointed, lead tip filler, dark ignition tracer, 44.39 g bullet V_o 823 m/s
Incendiary	B Mk 2Z	flat-based, pointed, steel sleeve, 46.01 g bullet V_o 793 m/s
API	Mk 1Z	boat-tailed, pointed, steel core, 42.77 g bullet V_o 870 m/s
MG observing	L11A1	boat-tailed, pointed, steel core, 40.56 g bullet V_o 914 m/s
MG observing	L11A2	boat-tailed, pointed, steel core, 40.56 g bullet V_o 903 m/s
MG observing	L13A1	boat-tailed, pointed, steel core, 62.53 g bullet

UNITED STATES OF AMERICA
Ball	M2	boat-tailed, pointed, 46.01/46.79 g bullet V_o 858 m/s
Ball	M33	boat-tailed, pointed, steel core, 42.9 g bullet V_o 888 m/s
AP	M2	boat-tailed, pointed, steel core, 45.88/46.53 g bullet V_o 885 m/s
Tracer	M10	flat-based, pointed, lead tip filler, 41.67 g bullet V_o 873 m/s
Tracer	M17	flat-based, pointed, lead tip filler, 41.67 g bullet V_o 873 m/s
Tracer	M21	flat-based, pointed, lead tip filler, 45.3 g bullet V_o 867 m/s
API	M8	boat-tailed, pointed, steel core, 42.06 g bullet V_o 888 m/s
APIT	M20	boat-tailed, pointed, steel core, 39.66 g bullet V_o 888 m/s
Incendiary	M1	boat-tailed, pointed, steel sleeve, 41.02 g bullet V_o 901 m/s
Incendiary	M23	flat-based, pointed, steel incendiary container, 33.18 g bullet V_o 1036 m/s

Other manufacturers are as follows:

BELGIUM (FABRIQUE NATIONALE)
Ball M33	43 g bullet, V_{25} 910 m/s
Tracer M17	41.7 g bullet, V_{25} 910 m/s
Tracer M33T	no details
AP M8	43 g bullet, V_{25} 910 m/s
API M8	43 g bullet, V_{25} 910 m/s
API/T M20	40.4 g bullet, V_{25} 910 m/s
APEI 169	43 g streamlined bullet with heavy alloy core and explosive/incendiary filler. V_{25} 890 m/s; E_{25} 17 030 J

CHINA (CHINA NORTH INDUSTRIAL CORP)
API	48.2 g bullet, steel core and incendiary agent in gilding metal jacket, V_{25} 810 m/s
APIT	44 g bullet as above, with tracer element V_{25} 810 m/s

FRANCE (MANURHIN)
Manurhin of France have developed a special 'anti-ricochet' bullet, filled with a loose inert compound, which disintegrates on impact. It is intended for training use, particularly in ranges with small safety areas. It can also be used as an impact marker.

FRANCE (SFM)
The Société Françcaise de Munitions (SFM) manufactures a unique piercing round known as PPI. This has a steel core fitted into a self-tightening non-discarding sabot of brass. The hardened steel makes no contact with the barrel, the brass sabot engaging with the rifling; it has a special profile which reduces friction and this, together with the lighter bullet, admits of a higher velocity than is attainable with conventional AP bullets. In 12.7 × 99mm calibre the bullet weighs 47.82 g and has a V25 of 890 m/s, with a penetration of over 25 mm of armour steel at 300 m range and better than 13 mm at 1200 m range. The .50 PPI round is in current supply to the French Army.

SOUTH AFRICA (ARMSCOR)
Ball	45 g bullet, V_{27} 823 m/s
API	41.18 g bullet, V_{27} 868 m/s
Tracer	43.7 g bullet V_{27} 823 m/s

UNITED KINGDOM (ELEY-KYNOCH)
Eley manufacture ball, tracer, observing, AP, API, API-T and blank cartridges which conform generally to the US standard patterns. The ball cartridge has a muzzle velocity of 914 m/s and at 550 m range the mean radius of a 20-shot group will not exceed 228 mm. Other types of bullet are matched to the ball at 600 m range.

14.5 mm Soviet

14.5 mm × 114
V_o 976 m/s
E_o 30 675 J

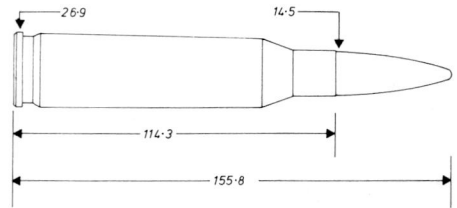

This originally appeared in the late 1930s as the cartridge for the PTRD and PTRS anti-tank rifles, and it proved itself so efficient that when the anti-tank rifles were scrapped, the cartridge was retained and a heavy machine gun, the KPV, was designed around it. Ammunition has been manufactured in the USSR and China and is usually brass-cased.

Current Soviet standard rounds are:

API Type BS41	64.4 g non-streamlined bullet with tungsten core
APIT Type BZT	59.6 g non-streamlined bullet with steel core and tracer element
Incendiary-T Type ZP	59.6 g non-streamlined bullet, steel core, incendiary and tracer elements

Chinese manufactured rounds by China North Industrial Corp:

API   steel core and incendiary agent in gilding metal jacket, V_{25} 988 m/s
APIT   as above but with tracer element, V_{25} 1005 m/s

14.5 mm Soviet cartridge (left) and .50 (12.7 mm × 99) Browning (right)

Comparison of (left to right) 15.5 × 106, 14.5 × 114, 12.7 × 107 and 12.7 × 99 (.50) cartridges

15.5 × 106 mm FN BRG-15 system

During the development of the 15mm BRG-15 machine gun, it was discovered that firing heavy jacketed bullets at very high velocities caused unacceptable barrel wear, and Fabrique National Herstal have therefore

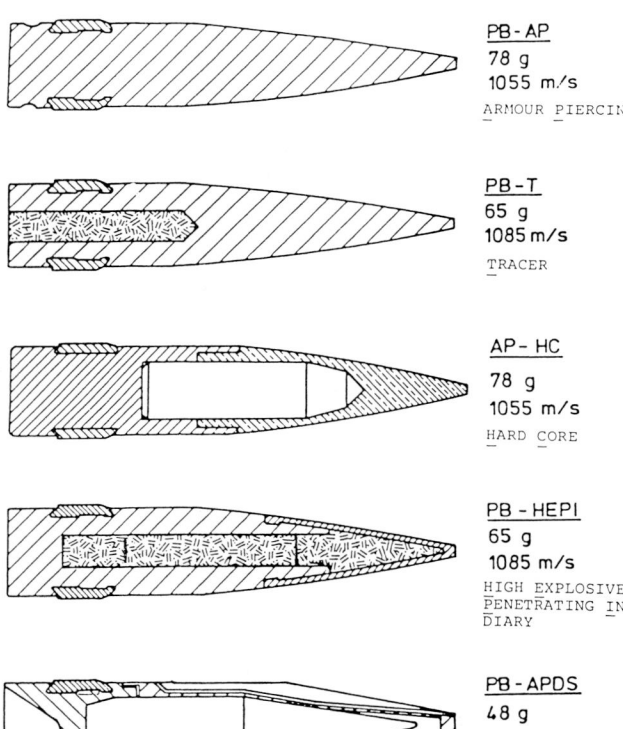

PB - AP
78 g
1055 m/s
ARMOUR PIERCING

PB - T
65 g
1085 m/s
TRACER

AP - HC
78 g
1055 m/s
HARD CORE

PB - HEPI
65 g
1085 m/s
HIGH EXPLOSIVE PENETRATING INCENDIARY

PB - APDS
48 g
1340 m/s
ARMOUR PIERCING DISCARDING SABOT

15.5 mm bullet types; (representative only and not purporting to be accurate depictions of construction)

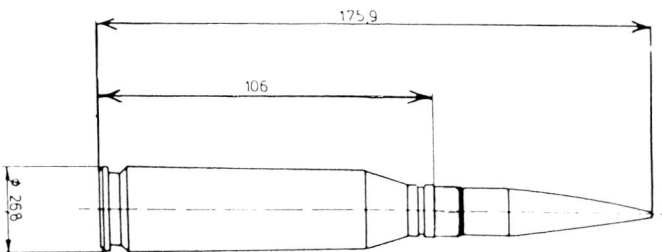

Dimensions of 15.5 × 106 mm round

re-designed their 15mm range of ammunition. In consequence, the calibre and cartridge case length have changed from 15 × 115mm to 15.5 x 106mm.

The new design abandons the use of jacketed bullets and, instead, bases the design on artillery practice rather than upon small arms conventions. The resulting standard projectile is a steel slug with a considerable parallel-walled body, a long tapering ogival nose, and a plastic driving band pressed into the usual type of undercut seat. The long-walled body gives the projectile excellent support and axial stability within the barrel.

The changed contour of the cartridge case has given a 10 per cent increase in volume, which will permit greater latitude in selecting propellants.

The standard projectile is known as 'PB-AP' for Plastic Banded, Armour Piercing. In addition, tracer, hardcore, HE Penetrating Incendiary and APDS designs are in advanced states of development, as well as blank and practice ammunition.

The PB-AP bullet has a muzzle velocity of 1055 m/s, developing 43300J muzzle energy. It is capable of defeating 19.1mm of Rolled Homogeneous Armour of 360 Brinell Hardness at 800 m range at an impact angle of 45°; by comparison the Soviet 14.5 × 114mm bullet cannot defeat this target beyond 250 m range.

CANNON AMMUNITION

Cannon Ammunition Related to Guns

The matching of cartridges to cannon is becoming daily more complicated; not so long ago the number of cannon rounds could be counted on the fingers, and it was comparatively easy to memorise the related weapons. Today the situation has changed and it looks as if it will continue to change for some time to come. In the hope that it might prove as useful to others as it has to us, we append here a list of cannon together with their cartridges. Several of these weapons and rounds are not detailed in the following section, since they are not ground weapons; moreover, some are obsolescent or even obsolete. But for the sake of completeness we have cast the net widely. Unfortunately, as will be seen, there are one or two weapons for which we have been unable to identify the correct cartridge; we hope that this will be remedied in due course.

Gun type	Ammunition type
20 mm 5TG (Switzerland)	20 × 128
20 mm Breda (Italy)	20 × 138B
20 mm EX-29 (USA)	20 × 102
20 mm Flak 30	20 × 138B
20 mm Flak 38	20 × 138B
20 mm GAU-4 (USA)	20 × 102
20 mm GAM-BO1 (Switzerland)	20 × 128
20 mm GE-120 Mk 29 (USA)	20 × 102
20 mm Hispano HS404	20 × 110
20 mm Hispano HS804	20 × 110
20 mm Hispano HS820	20 × 139
20 mm KAA (204GK) (Switzerland)	20 × 128
20 mm KAB (5TG) (Switzerland)	20 × 128
20 mm KAD (HS-820) (Switzerland)	20 × 139
20 mm KAD-B (HS-820) (Switzerland)	20 × 139
20 mm M3 (USA)	20 × 110
20 mm M24 (USA)	20 × 110
20 mm M39 (USA)	20 × 102
20 mm M40 (Sweden)	20 × 145R
20 mm M50 (USA)	20 × 102
20 mm M61A1 (USA)	20 × 102
20 mm M139 (USA)	20 × 139
20 mm M168 (USA)	20 × 102
20 mm M197 (USA)	20 × 102
20 mm M621 (F1) (France)	20 × 102
20 mm M693 (F2) (France)	20 × 139
20 mm M1955 (Yugoslavia)	20 × 110
20 mm Madsen (Denmark)	20 × 120
20 mm Mark 2* (HS-404) (UK)	20 × 110
20 mm Mark 4 (US Navy)	20 × 110RB
20 mm Mark 5 (HS-404) (UK)	20 × 110
20 mm Mark 11 Mod 5 (USA)	20 × 110USN
20 mm Mark 12 (USA)	20 × 110USN
20 mm Mark 22 Mod 2 (US Navy)	20 × 102
20 mm Mark 100 (US Navy)	20 × 110USN
20 mm Mauser Model B (Germany)	20 × 139
20 mm Meroka (Spain)	20 × 128
20 mm MG 151/20 (Germany)	20 × 82
20 mm MK20 (Rh202 or DM5) (Germany)	20 × 139
20 mm Oerlikon FF (Switzerland)	20 × 72RB
20 mm Oerlikon FFL (Switzerland)	20 × 101RB
20 mm Oerlikon FFM (Switzerland)	20 × 80RB
20 mm Oerlikon Marks 2,3 and 4 (UK)	20 × 110RB
20 mm Oerlikon 'S' (Switzerland)	20 × 110RB
20 mm Polsten (UK)	20 × 110RB
20 mm Rh 202 (Germany)	20 × 139
20 mm RK 206 (Switzerland)	20 × 128
20 mm RK 251 (Switzerland)	20 × 128
20 mm ShVAK (USSR)	20 × 99R
20 mm Solothurn (Switzerland)	20 × 138B
20 mm T20E2 (USA)	20 × 102
20 mm TCM-20 (HS-404) (Israel)	20 × 110
20 mm Type 99/1 (Japan)	20 × 72RB
20 mm Type 99/2 (Japan)	20 × 101RB
20 mm Vulcan Air Defence System (USA)	20 × 102
20 mm XM168 (USA)	20 × 102
20 mm XM195 (USA)	
23 mm AM-23 (USSR)	23 × 115
23 mm GSh-23 (USSR)	23 × 115
23 mm NS-23KM (USSR)	23 × 115
23 mm NR-23 (USSR)	23 × 115
23 mm VYa (USSR)	23 × 152B
23 mm ZU-23 (USSR)	23 × 152B
23 mm Twin (China)	23 × 152B
23 mm Type 2H (China)	23 × 115
25 mm 2-M3 and 2-M8 (USSR)	25 × 218
25 mm GAU-12/U (Bushmaster) (GE USA)	25 × 137
25 mm KBA (Swiss)	25 × 137
25 mm KBB (Swiss)	25 × 184
25 mm M242 Bushmaster (Hughes USA)	25 × 137
25 mm M1940 (USSR)	25 × 205SR
25 mm Mauser Model E (Germany)	25 × 137
25 mm MK25 Rh205 (Germany)	25 × 137

Gun type	Ammunition type
27 mm Mauser MRCA (Germany)	27 × 145B
30 mm Aden Type 3M (UK)	30 × 86B
30 mm Aden Mks 4 & 5 (UK)	30 × 113B
30 mm CZ (Yugoslavia)	30 × 210B
30 mm DEFA 541 (France)	30 × 97B
30 mm DEFA 551 (France)	30 × 113B
30 mm DEFA 552A (France)	30 × 113B
30 mm DEFA 553 (France)	30 × 113B
30 mm DEFA 554 (France)	30 × 113B
30 mm GAU-8/A (USA)	30 × 173
30 mm GAU-13/A (USA)	30 × 173
30 mm GCM-A (Switzerland)	30 × 170
30 mm GPU-5 (USA)	30 × 137
30 mm HSS-831A (Switzerland)	30 × 170
30 mm KCA (Switzerland)	30 × 173
30 mm KCB (HSS-831) (Switzerland)	30 × 170
30 mm M53 (Czechoslovakia)	30 × 220
30 mm M230E1 (USA)	30 × 113B
30 mm Mauser MK30F (Germany)	30 × 173
30 mm NR-30 (USSR)	30 × 155B
30 mm NN-30 (USSR)	30 × 210B
30 mm NR-20 (USSR)	
30 mm 2NR (USSR)	
30 mm 6-barrel Gatling (USSR)	
30 mm Rarden (UK)	30 × 170
30 mm T121 (USA)	30 × 86B
30 mm TCM-30G (Israel)	
30 mm XM188E1 (USA)	30 × 113B

20 mm MG 151

20 mm × 82

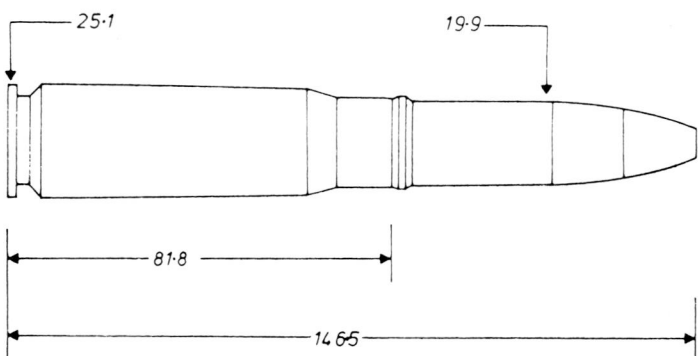

This was developed by the Mauser company in the 1930s as a possible anti-tank projectile but was eventually used in the aircraft cannon and light anti-aircraft guns used by both Germany and Japan. In post-war years the cannon was manufactured in France and was additionally used to arm light armoured vehicles. Manufacture ceased in about 1970 but numbers of guns still exist, notably in various African states and the ammunition is still manufactured by Manurhin.

All rounds are brass-cased. The current range of projectiles comprises:

HE-I	filled Hexal 70/30, percussion fuze Mle 61, weight 112 g, V₀ 720 m/s
HEI-T	filled Hexal 70/30, percussion fuze Mle 61, trace to 1200 m, weight 112 g, V₀ 720 m/s
SAP-I	filled incendiary composition, weight 115 g, V₀ 720 m/s
API-T	filled incendiary composition, weight 120 g, trace to 1500 m, penetration 20 mm, V₀ 700 m/s
I-T	filled incendiary composition, unfuzed, trace to 1500 m, V₀ 720 m/s
Prac-T	filled inert, weight 112 g, trace to 1500 m, V₀ 720 m/s

20 mm Type 151

20 × 83.5 mm

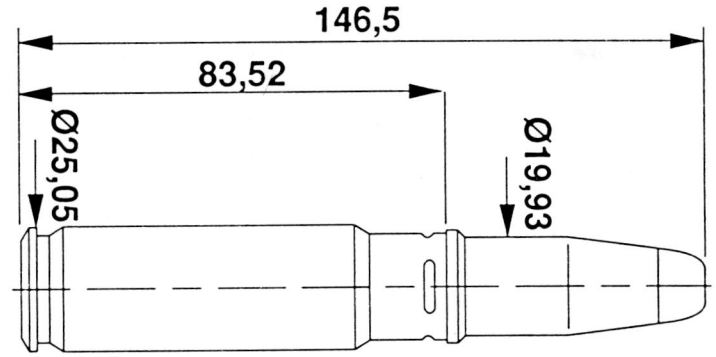

Although this round is nominally the same as that for the MG 151 discussed above, it does in fact differ in dimensions and contour as can be seen from the drawings. The 20 × 83.5 mm was developed in South Africa in the early 1980s to provide a cartridge for the 20 mm GA1 cannon, itself a development from the MG 151 design. While the overall length of the round remains 146.5 mm it can be seen that there is a slight lengthening of the case and also a much different shoulder contour, so that it would seem unlikely that standard 20 × 82 mm cartridges would fit the South African weapon.

The ammunition is manufactured only in South Africa; all rounds are steel-cased, and the standard HEI-T specification calls for a 110 g shell with a muzzle velocity of 720 m/s. It is fitted with an impact fuze and 2 s trace. Other types of ammunition produced are HEI, SAPHEI (with base fuze), TP and TP-T. The HEI and SAPHEI shells are filled Hexal 30 (HEI) and Hexal P18 (SAPHEI) ensuring a strong blast and powerful incendiary effect. The SAPHEI shell will penetrate 15 mm armour plate (110 kg/m²) at 100 m range with adequate incendiary effect behind the plate. All projectiles are fitted with soft iron driving bands designed to ensure long barrel life and minimum fouling.

20 mm US M39

20 × 102 mm

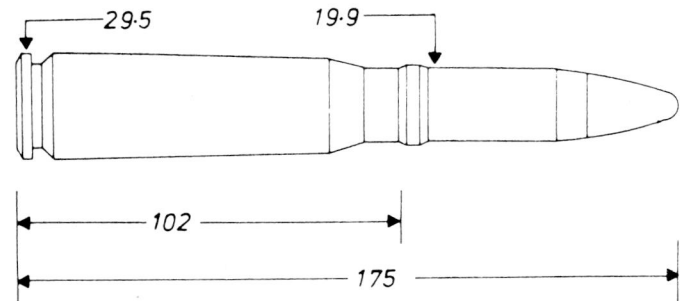

After experience with Oerlikon and Hispano guns during the Second World War the Americans set out in the post-war years to develop a 20 mm cannon of their own. The resulting weapon was based on the German Mauser 312 revolver gun and was taken into service as the M39. It was later supplemented by the M39A1 and then by the M61 and M168 Vulcan Gatling-pattern guns. A Naval version, the Mark 22, was also developed for use in fast patrol boats but was not put into service. The guns and ammunition were used in several NATO aircraft and the cartridge was adopted by the French for their M621 ground gun.

The cartridge cases may be of brass or steel and use electric primers, though cases with percussion primers were produced for various prototype weapons and may be encountered.

US SERVICE ROUNDS
Ball/TP M55A1   steel inert projectile with dummy fuze
HE/I M56A3   steel shell loaded with RDX and incendiary composition, and with Fuze, Point Detonating, M505. Weight 101 g
AP/I M53   steel pointed shell with alloy ballistic cap and filling of incendiary material

FRENCH SERVICE ROUNDS
AP/T   steel pointed projectile with ballistic cap and with base tracer functioning to 1200 m range
HE/T/SD   HE shell with self-destroying nose fuze and base tracer functioning to 1400 m range
HE/SD   as above but without tracer

BELGIUM (FABRIQUE NATIONALE)
API FN144   101 g projectile, V_{25} 1030 m/s
HEI FN143A1   101 g shell with impact fuze, V_{25} 1030 m/s
HEIT-SD FN142A1 101 g shell, impact fuze, tracer self-destruction. V_{25} 1030 m/s
TP FN140   99 g shell, inert, V_{25} 1030 m/s
TPT FN141(71)   99 g shell, inert, with tracer. V_{25} 1030 m/s

BRAZIL (COMPANHIA BRASILEIRA DE CARTOUCHOS)
HEI   No details
TP   No details
TP-T   No details

20 mm × 110 RB Oerlikon Type S

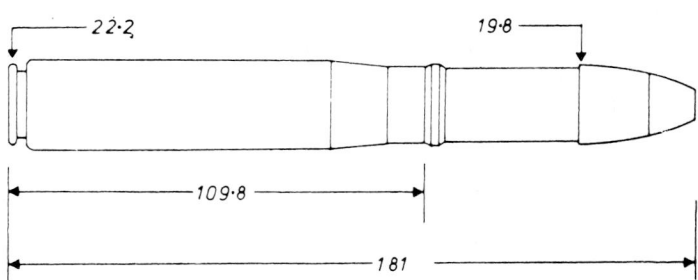

This was developed by Oerlikon in the 1930s and was subsequently adopted by Britain and the USA during the Second World War. Several other countries took Oerlikon guns of this chambering into service, and ammunition of this pattern is still manufactured in Belgium, Italy and Yugoslavia.

Cartridge cases of steel or brass may be found, though the former is not common. Ignition is by a conventional percussion cap pressed into the base of the case. Projectiles vary from solid shot to explosive and incendiary projectiles with impact fuzes. Service patterns which may be encountered include designs from:

UNITED KINGDOM
Ball Mk1   hollow steel projectile weighing 131.5 g and completely inert. Normally used only for training and practice
AP Mk2   solid steel pointed shot with moulded plastic ballistic cap
Tracer   hollow steel inert projectile with tracer in base
HEI-T   shell filled with Tetryl, fuzed No 253, 254 or 258 in nose; trace in base

UNITED STATES OF AMERICA
HE Mk3   shell filled Pentolite or Tetryl, with fuze Mk 26
HE-I   shell filled Pentolite and incendiary composition, nose fuze Mk 26
HEI-T Mk 7 as for HE-I but with tracer in shell base and reduced filling of high explosive
AP   solid steel shot

Current   ~cture (typical):

FABRIQ~
TP FN65A2   ⌐ inert shell, weight 122 g, V_o 830 m/s, for ₁ practice use
TP-T FN63A1   simi… .bove, but with addition of tracer unit in shell base
HEI FN60   shell filled high explosive, with compression-ignition impact nose fuze, no self-destruction and with rear half of cavity carrying incendiary charge. Weight 122 g, V_{25} 830 m/s
HEI-T FN62   similar to HE-I but with tracer. Weight 120.4 g
HEI-T/SD FN71 HEI, with tracer, with pyrotechnic self-destruction; as tracer burns out it ignites relay unit which, in turn, ignites explosive filling and causes self-destruction of shell. Weight 122 g
SAPI-T FN74   Perforating head, body filled with incendiary composition, red tracer. Weight 122 g, V_{25} 830 m/s

20 mm × 110 Hispano HS404

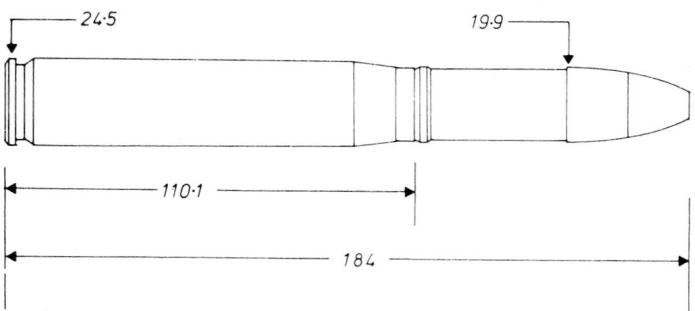

Types of projectile available include:

HE-I	Model SSB/K	filled 12 g Hexal (RDX/TNT/Al) with self-destroying nose fuze
HEI-T	Model SSBL/K	as for SSB/K but with 2.5 s tracer
Mine	Model MSB/K	filled 18 g Hexal, and self-destroying nose fuze
Mine/T	Model MSBL/K	as MSB/K but with only 12 g Hexal and 2.5 s tracer
APHE/I	Model BSBH/B	pointed shell with base fuze and filling of 5.4 g Hexal
APSV/T	Model PKLH	composite rigid shot; tungsten carbide core in light alloy body, with 1.7 s tracer in base
Practice	Model SU	malleable iron inert projectile
Prac-T	Model SUL	as Model SU but with tracer

This cartridge was extensively used by Britain and the USA during the Second World War and large numbers of these guns are still in use. The ammunition is almost identical with that of the 20 mm × 110RB Oerlikon, but the case is a rimless type rather than a rebated rimless. In post-war years the US Air Force developed electrically-primed ammunition in this chambering: it is prominently marked 'ELECTRIC' on the side of the cartridge case, and the primer cap exhibits a ring of black insulating material around it.

Both Britain and the USA produced the same range of cartridges in this chambering as for their Oerlikon guns; in fact, the projectiles were exactly the same, only the cartridge case and propelling charge differed. Details given under the Oerlikon cartridge can thus be applied to this cartridge. The standard muzzle velocity was 844 m/s.

Similarly, the ammunition currently offered by commercial concerns uses identical projectiles; the details given for the Fabrique Nationale (FN) cartridges under Oerlikon can be applied to this cartridge in the same way. The muzzle velocity of FN Hispano ammunition is 845 m/s.

20 mm × 99R ShVAK

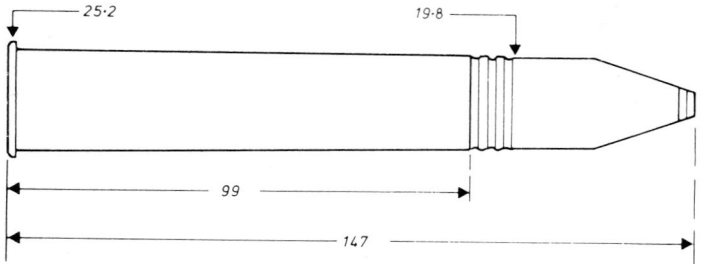

This round was developed in the late 1930s for the Shpitalny-Vladimirov ShVAK aircraft cannon used by the Soviet Air Force. It has long since ceased to be used as air armament, but the cartridge has survived to be used in a sub-calibre barrel insert for training with the various 122 mm howitzers used by the Soviet Army. The cartridge is a straight, rimmed, case carrying an explosive shell with percussion fuze. The shell weighs 97 g, and in its cannon role had a velocity of 860 m/s. Its performance in the sub-calibre device is not known.

20 mm × 128 Oerlikon KAA

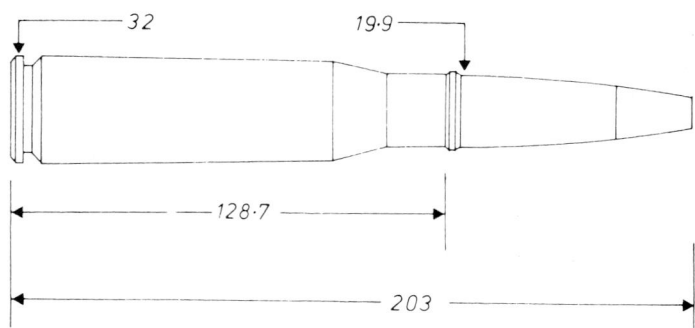

This round was adopted by Oerlikon in the 1950s for use in various designs of ground and aircraft guns; rounds for ground guns used percussion primers, while those for aircraft use were electrically primed. In 1972, when Oerlikon and Hispano-Suiza combined, the terminology of various guns was changed but production continued. Ammunition has been made in Switzerland, France and Spain.

The case is of lacquered steel, with a screwed-in percussion primer. Projectile weights vary between 120 and 130 g, but charges are regulated to give a normal muzzle velocity of 1050 m/s in an 85-calibre-length barrel.

20 mm Long Solothurn

20 × 138B

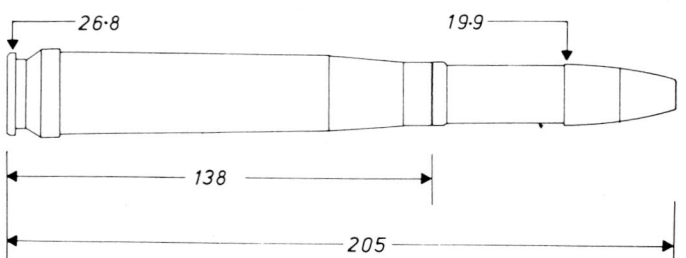

This was developed by Rheinmetall in the early 1930s and marketed through their subsidiary, the Waffenfabrik Solothurn AG of Switzerland. In addition to being used by Rheinmetall/Solothurn as an AA gun (the German Flak 30 and 38 series) it was also adopted by the Italians for use in a number of Breda and Isotta-Fraschini AA and air service guns, and by the Swiss in the Solothurn S18-1000 tank gun and S18-1100 anti-tank rifle. Ammunition with brass or steel belted cases and with percussion primers, has been made in Finland, Germany, Greece, Italy, Sweden and Switzerland at various times. It is currently made in Italy and Yugoslavia, since numbers of guns are still to be found in smaller nations.

GERMAN STANDARD ROUNDS

AP/T	pointed steel shell of 146.5 g with inert filling and 2.5 s green tracer
AP/I/T	as above but the body cavity holds 2.7 g white phosphorus in a light alloy capsule
HE/T/SD (1)	same projectile as used with 20 × 80RB Oerlikon, fuzed AZ5045
HE/T/SD (2)	123 g streamlined shell, Penthrite/Wax in front section, 6 s green tracer in rear section, with self-destroying relay. Fuzed AZ5045
APSV/T	Tungsten carbide core in light alloy body. 1.5 s tracer, white changing to red. Weight 101 g, V₀ 975 m/s

ITALIAN STANDARD ROUNDS

HE/T/SD	116 g two-section shell, filled 2.8 g TNT in front with impact fuze T74, and 5 s white tracer in rear, with a self-destruction relay
AP/HE/T	151 g pointed shell filled 1.2 g Penthrite/Wax, with base fuze and white tracer

Ammunition produced in other countries was usually based on the above designs.

20 mm × 139 Hispano HS820

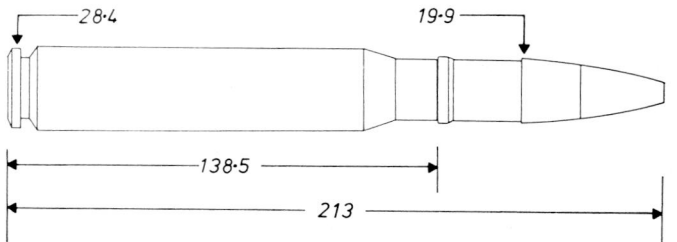

This round was developed in the 1940s to improve on the wartime 20 mm × 110 Hispano cartridge, and it was widely adopted. The West German Rheinmetall company developed a number of suitable guns, as did the French, and ammunition has been made in several countries. The case is usually of lacquered steel, though brass is occasionally seen, and it uses a screwed-in percussion primer. Conventional projectiles weigh about 120 g and charges are regulated to produce a muzzle velocity of 1050 m/s.

Manufacturers include:

FRANCE

HE-I	Model OEI	shell filled 16 g Hexal and fitted with self-destroying nose fuze
HEI-T	Model OEIT	similar to OE but with filling reduced to 6 g Hexal to provide space for tracer
APDS	Mod OPTSOC	discarding sabot, total weight 90 g. No details revealed
AP-T	Mod OPT	unfuzed piercing shot with tracer
TP	Mod OX	inert filled steel body
TP-T	Mod OXT	as OX but with tracer

WEST GERMANY

HEI-T	DM51A2	shell filled Hexal, and fitted self-destroying nose fuze DM281. V₀ 1045 m/s
HEI-T	DM81	as above but fuze DM301
HEI	M599	122 g shell with impact fuze and self-destruction. V₀ 1045 m/s
HEI	DM101	120 g shell with base fuze. V₀ 1055 m/s
HEI	M-DN71	Multi-purpose shell; HE in body, incendiary under ballistic cap, base fuze. Shell 98 g, V₀ 1030 m/s
Shrapnel	DM111	118 g shell loaded with 120 heavy metal balls which are ejected after a delay of about 40 ms. For airfield defence
HVAP/T	DM43	tungsten core in alloy body, with tracer at base. V₀ 1100 m/s
API	DM23A1	100 g shell filled incendiary mixture in body and under ballistic cap. V₀ 1033 m/s
TP	DM48A1	steel body with dummy fuze. Weight 122 g, V₀ 1045 m/s
TP-T	DM48A1	as DM48 but with tracer
APDS/T	DM63	tungsten core with discarding sabot. V₀ 1150 m/s

NETHERLANDS

APDS/T	Mod DM63	discarding sabot with tungsten core. No details revealed
Breakup	Mod DM78	plastic projectile casing with filling of powdered lead shot. On firing, projectile causes gun to recoil but centrifugal force causes lead shot to fracture plastic casing outside muzzle and 'projectile' disintegrates. Used for gun testing and practice in confined spaces. V₀ 1045m/s

SOUTH AFRICA

HE-I	filled 9 g of Hexal P30. Weight 120 g with KZA 348 impact fuze. V₀ 1050 m/s. Self-destruction time 4-9 s
HEI-T	as above, with 2.5 s trace
APC-T	pointed, unfuzed, filled 1.3 g HE, V₀ 1100 m/s. Tracer 2.5 s. Penetration 40 mm/30°/100 m
TP	inert. V₀ 1050 m/s
TP-T	as TP but with 2.5 s trace

SWITZERLAND

Mine HE-I	Type MSA	shell filled Hexal, with self-destroying nose fuze. Weight 125 g
Mine HEI-T	Type MLA	as MSA but with 3 s tracer
APHE	Type PSA	steel pointed shell, with aluminium ballistic cap; base fuze with self-destruction
APHE/T	Type PLA	as PSA but with 1.2 s tracer
APSV/T	Type HLA	tungsten core in alloy body, with 1.5 s tracer. V₀ 1100 m/s

20 mm Bofors

20 × 145R

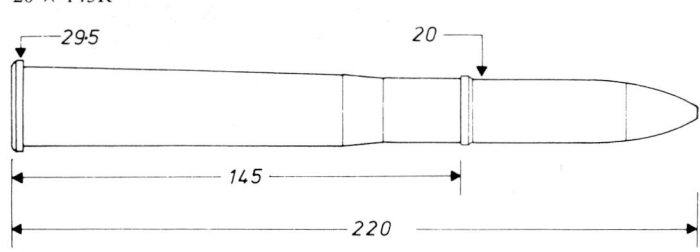

This cartridge was developed by AB Bofors of Sweden for their M/1940 anti-aircraft cannon, and its use was carried over to the later m/40/70 (70-calibre) improved version. Both these guns are obsolescent, if not obsolete, but the cartridge remains in use with various sub-calibre adapters for 57 mm and 75 mm anti-tank guns and 75 mm and 105 mm armoured vehicle guns.

Little is known of the range of projectiles fired by the anti-aircraft weapon; it is presumed that HE/T or HE/I/T would be the standard projectile, and a weight of 80 g has been quoted. No information is available about the projectiles used in the sub-calibre application. The data given above is based on the performance in the anti-aircraft role in the m/40/70 gun.

23 mm Soviet NS

23 × 115mm

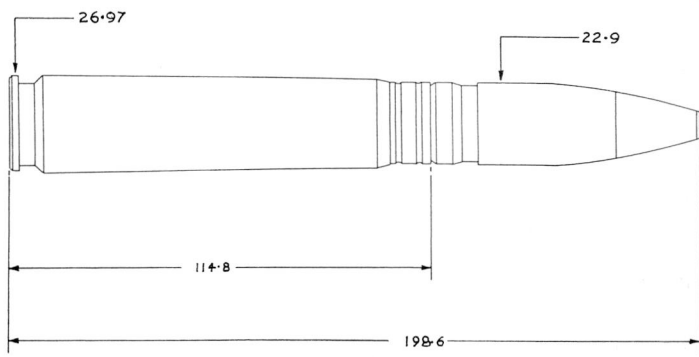

This round was developed during the Second World War for use in the Nudelman-Suranov aircraft cannon, which in use by Soviet ground attack aircraft. The gun was recoil-operated and was scaled-down from an earlier 37 mm design. After the war it was replaced by the gas-operated AM-23 and GSh-23 guns which have a higher rate of fire.

Although the ammunition is similar, cartridges for the two types of guns differed and were not functionally interchangeable. Those for the older NS gun were loaded to a lower velocity (690 m/s), while those for the AM and GSh guns used a more powerful primer and improved propellant to achieve 740 m/s. In order to differentiate between the two, cartridges for the AM/GSh series have their projectiles marked with a white band.

Cartridge cases were of brass, percussion primed, and can be found with Soviet or Czech headstamps.

Inevitably, the guns and their ammunition were exported as part of the equipment of aircraft, and the cartridge has now been adopted by the People's Republic of China as a land service cartridge for use in the Type 2 automatic gun, used in the light air defence role.

Known service rounds are:

HE/I/T Type OZT	HE filling, K-20M impact fuze. Weight about 175 g. Copper driving band. Soviet made.
HE/I/T Type OZT	As above but with N2231A fuze and sintered iron driving band. Czech manufacture.
AP/I Type BZ	Pointed steel shell with incendiary filling and light alloy ballistic cap. Weight about 200 g. Copper driving band.
China Type 2	174 g AP/I shell, base fuzed, filled 5 g explosive. V₀ 720 m/s. Steel cartridge case

23 mm Soviet VYa and ZU-23

23 mm × 152B

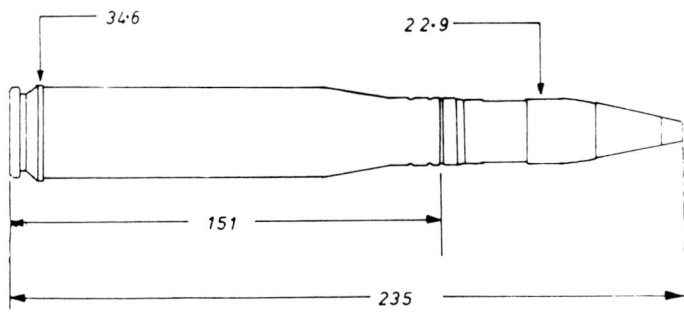

HEI-T Type OZI for ZU-23 gun

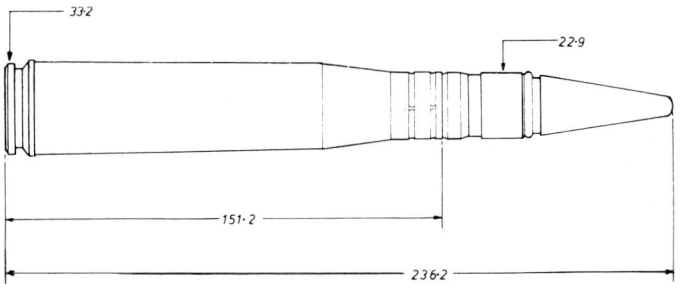

API Type BZ for VYa gun

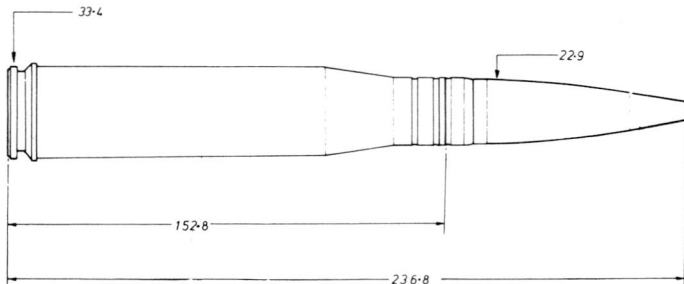

HEI-T Type OZI for VYa gun

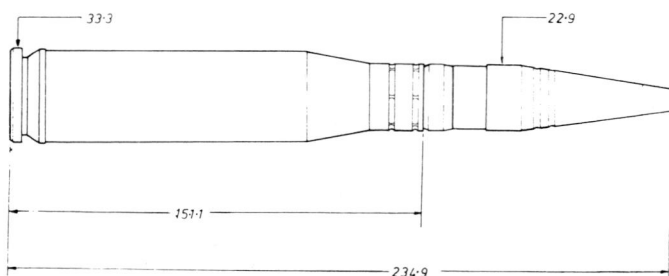

HE Type OZT for ZU-23 gun

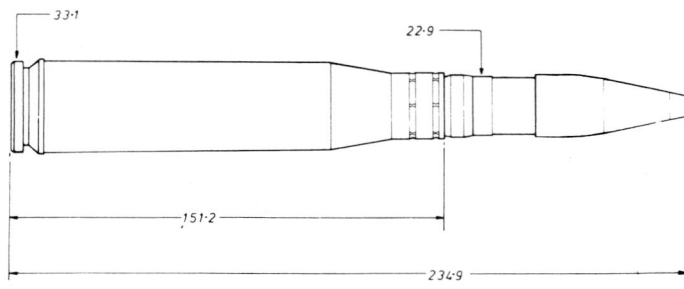

API-T Type BZT for ZU-23 gun

This cartridge was originally developed for the VYa aircraft cannon, a Soviet Air Force weapon produced by simply scaling up the existing 12.7 mm Beresin machine gun. Both gun and cartridge proved successful and were extensively used during the Second World War. When the VYa cannon was retired from air service, the cartridge was retained and used for the ZU-23 anti-aircraft gun. In this role it has been widely distributed among Soviet satellite countries. In adapting the cartridge to the ZU gun some very slight dimensional changes were made, and thus the steel-cased ZU cartridge is not interchangeable with the brass-cased VYa cartridge. Both may still be found, since the VYa cartridge is used on a tank gun sub-calibre training device. Both rounds are percussion fired, but the primer in the ZU cartridge is an improved design, capable of withstanding higher pressures than the original VYa type.

VYa AMMUNITION
API	Type BZ	shot with carbon steel core and incendiary pellet
HEI-T	Type OZI	HE shell with impact fuze
TP-T		as Type BZ but without incendiary pellet

ZU-23 AMMUNITION
HE	Type OZT	HE shell with MG-25 impact fuze
HEI-T	Type OZI	HE shell with MG-25 fuze and incendiary pellet
API-T	Type BZT	steel shot with incendiary filler in ballistic cap

25 mm Oerlikon KBA

25 mm × 137

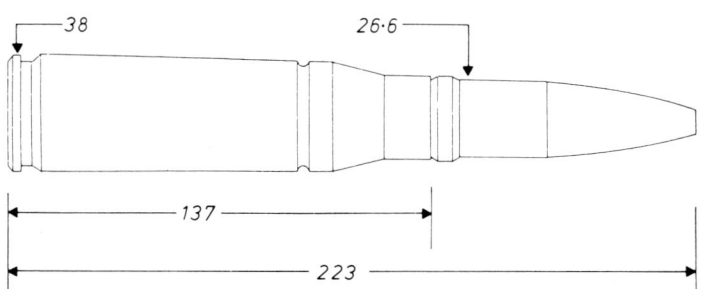

This was developed by Oerlikon in the late 1960s and has also been adopted for the American Hughes Chain Gun. The lacquered steel case is readily identified by the unusual belt-locating groove just below the shoulder; this is expanded on firing, due to internal pressure, but leaves a conspicuous mark on the fired case. Swiss-, American- and French-manufactured ammunition is available.

Current Swiss ammunition includes:

Frag HE/I Type SLB 050	180 g steel shell filled Hexal, with self-destroying nose fuze and tracer
Frag HE/I Type SSB 051	as SLB 050 but without tracer
AP/HE/I Type PSB	180 g pointed shell, filled Hexal, with ballistic cap and self-destroying base fuze
AP/HE/I/T Type PLB	as PSB with the addition of tracer
APDS Type PKHT	tungsten carbide core in plastic sabot, with light alloy ballistic cap. V_o 1463 m/s
APDS/T Type TLB 044-2	improved PKHT with tracer. Weight 128 g. V_o 1360 m/s

Current American ammunition includes:-
HE/I/T M792	185 g steel shell loaded 30.2 g HE and fitted with M758 self-destroying nose fuze and 4.7 s tracer
TP/T M793	185 g inert-filled shell with Plug Representing Fuze in nose and 4.7 s tracer
APDS M?	depleted uranium core. No details available

Current French (Manurhin) ammunition includes:-
HE/I/T/SD	183 g shell filled Hexal, with Manurhin MR25 impact and SD nose fuze or US M758 fuze, and tracer. V_o 1100 m/s
TP	simulates the HE/I shell
TP/T	as for TP but with tracer

25 mm Oerlikon KBB

25 mm × 184

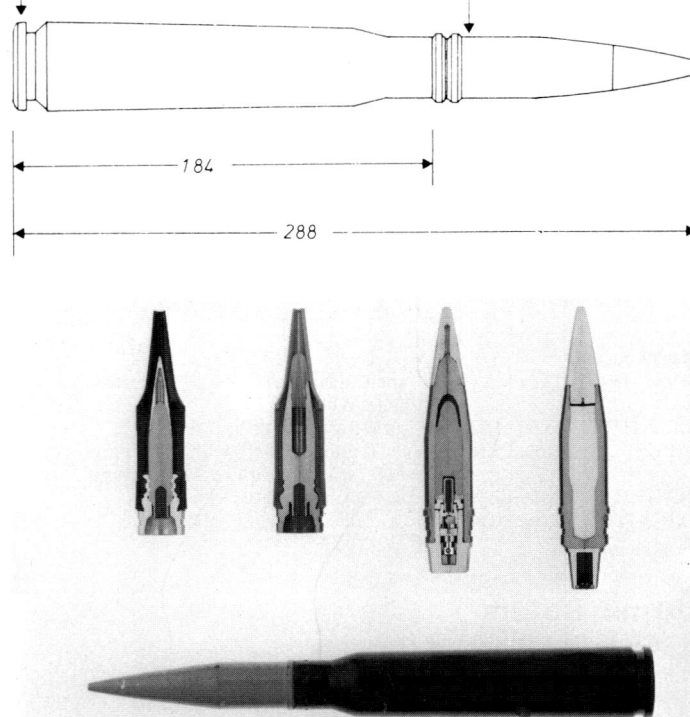

25 mm Oerlikon KBB

This round of ammunition has been developed by Oerlikon specifically for the 25 mm KBB cannon in order to develop a higher muzzle velocity than the 25 mm KBA cartridge.

The cartridge case is steel and has a screw-in percussion primer. The projectile is secured to the case by an eight-point crimp. The cartridge is usually supplied belted, using a disintegrating link belt of special form which longitudinally locates the cartridge by means of the extractor groove.

The HE-I shell is provided with a base percussion fuze which incorporates self-destruction by spin decay. There is also an arming delay giving about 10 m safety distance from the muzzle before the fuze can function. The fuze is a 'thinking' fuze insofar as it senses the resistance, and thus the thickness, of the structure it is penetrating and arranges for the delay between first impact and detonation to be proportional to the resistance. In this way detonation behind the structure is assured and maximum effect is achieved. Should the projectile miss, then the self-destruction element comes into play after about seven seconds time of flight.

The following ammunition types are currently available:

SSB 064	HE-I	this is a fragmentation high explosive/incendiary shell weighing 230 g and with an explosive filling of 22 g. V₀ 1160 m/s
TLB	APDS/T	discarding sabot shot with hard core and plastic sabot. Weight of shot 180 g; weight of core 150 g. V₀ 1400 m/s. Penetration 30 mm at 1000 m at 30° impact angle
ULB 066	APP/T	target practice projectile simulating the APDS/T shot at ranges up to 1000 m. Weight of shot 150 g. V₀ 1460 m/s. Tracer burns about 1.5 s, to a range of about 2000 m
ULB 057	TP-T	target practice shell simulating the HE-I shell. The shell weighs 230 g and is fitted with an inert material. V₀ 1160 m/s. Trace to 3.5 s or 2500 m

25 mm Soviet M1940

25 × 205 SR

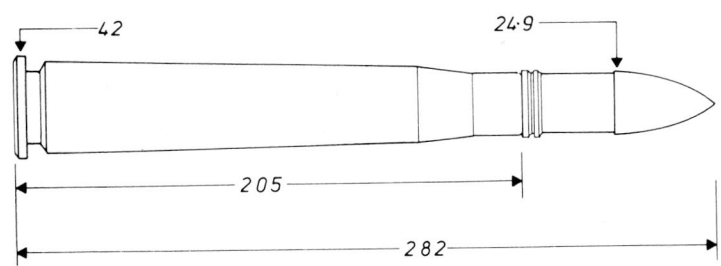

The cartridge appears to have been based on a 25 mm Bofors design, adopted by the Soviets in 1933, and it will be noted that the characteristic Bofors clip groove has been retained. It was developed for use with the 25 mm field anti-aircraft cannon M1940 and the projectiles are also used with the 25 × 218 naval AA gun.

Standard projectiles included the UOZR-132 HE/I/T shell weighing 288 g and the UZR-132 AP/T shot weighing 295 g. Both attained the same muzzle velocity of 900 m/s.

It is doubtful if any M1940 guns remain in Soviet service, but they are likely to have been distributed to smaller nations, particularly on the African continent and the rounds could still be met.

30 mm ADEN/DEFA

30 × 113B

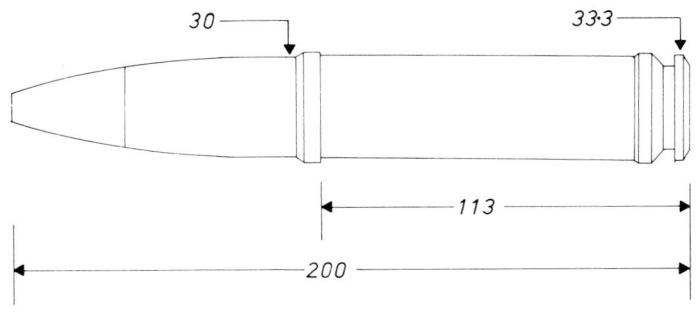

Britain and France worked on revolver cannon for some years after the war, each pursuing their own cartridge design. Subsequent shared development led to the adoption by both countries of a new 30 × 113B round for use in improved models of both the ADEN and DEFA guns.

Cases for these cartridges may be of brass or steel and are fitted with electric primers. French cases may be recognised by their steel cases, French headstamps and stencilled markings. British cases are usually of brass and have their identifying marks stamped into the bottom of the extraction groove. Ballistic performance is slightly different, so that the rounds are not entirely interchangeable. French ammunition is regulated to produce velocities of 760/820 m/s, while British ammunition produces velocities between 600 and 800 m/s. Projectile weights are similar. In addition, ammunition for both ADEN and DEFA guns has been made in Belgium, West Germany, Israel, the Netherlands, Sweden and Switzerland. Ammunition has also been developed in the USA for use in the M230 Chain Gun.

DEFA AMMUNITION (BELGIUM — FABRIQUE NATIONALE)

TP FN45	270 g hollow shell, inert, dummy fuze. V₀ 765 m/s
TP FN47	244 g hollow shell, inert, dummy fuze. V₀ 820 m/s
TPT FN46A1	244 g hollow shell with tracer. V₀ 820 m/s
HE/I FN24	244 g shell with aluminised explosive filling and impact nose fuze. V₀ 815 m/s
HE/I/SD FN23	243 g high-capacity shell filled 44 g aluminised explosive and with SD nose fuze. V₀ 820 m/s
AP/HE/I/SD FN21	270 g pointed shell with HE/Incendiary filling, SD base fuze and ballistic cap. V₀ 765 m/s

DEFA AMMUNITION (BRAZIL — COMPANHIA BRASILEIRA DE CARTOUCHOS)

| TP | No details |
| TP-T | No details |

DEFA AMMUNITION (FRANCE — GIAT)

HE/I/SD/T Type 6522	238 g shell with 50 g HE/Incendiary filling, SD nose fuze and tracer. For air-to-air combat
HE/I/T Type 5432	244 g shell with 22 g Hexal filling and percussion fuze. For air-to-ground use
AP/I/T Type 5970	275 g steel shot with ballistic cap and tracer. For air-to-ground use against hard targets
TP Type 2102	238 g inert projectile

DEFA AMMUNITION (FRANCE— MANURHIN)

Mine HE/I/SD F7570	thin-walled 245 g shell with 50 g of Hexal
TP F2570	simulates the HE Mine shell for practice
TP/T F3170	as F2570 but with tracer
AP/I/T F5970	275 g pointed shell with incendiary filling and dark ignition tracer. For air-to-ground attack of hard targets
HE/I F 5270	275 g fragmentation shell with nose fuze for air-to-ground anti-personnel attack

DEFA AMMUNITION (SOUTH AFRICA — ARMSCOR)

| HE/I Type 3060 | 247 g shell filled 50 g HE/Incendiary composition, with SD nose impact fuze. V₀ 810 m/s |
| APC/I | 238 g composite shot containing a 130 g tungsten carbide core with pyrophoric composition to give incendiary effect. V₀ 820 m/s |

ADEN AMMUNITION (BELGIUM — FABRIQUE NATIONALE)

TP FN2	220 g hemispherical-based mine shell with inert filling and dummy fuze. V₀ 780 m/s
FN-HE	shell as above but with aluminised explosive filling and impact nose fuze
HEI FN3	shell as above but with HE/Incendiary filling

ADEN AMMUNITION (FRANCE — MANURHIN)

HE/I Type 5469	246 g shell containing 22 g Hexal explosive and fitted with impact nose fuze with pyrotechnic arming delay. V₀ 765 m/s
AP/I Type 5969	246 g shell carrying a 127 g penetrating head and an incendiary filling in the body. V₀ 765 m/s
TP Type 2469	246 g inert projectile and dummy fuze

ADEN/DEFA AMMUNITION (SWEDEN —FFV)

Mine/HE M55	219.5 g shell with 52.5 g Torpex. V₀ 795 m/s
TP M55B	219.5 g inert projectile. V₀ 795 m/s
TP Type DEFA	inert projectile

ADEN/DEFA AMMUNITION (USA —HONEYWELL)

HEDP M789	Dual-purpose HEAT/fragmentation, nose fuzed
HEI M799	Hemispherical base, high-capacity, nose fuzed
TP M788	Inert, hollow projectile, with plug representing fuze

30 × 150B GIAT

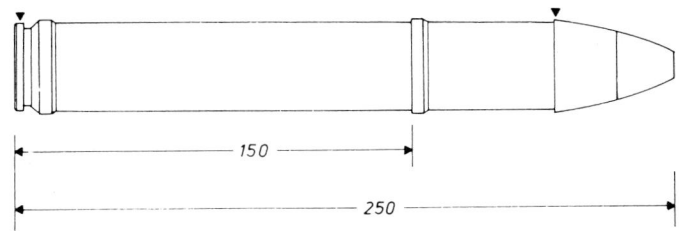

This cartridge was announced in October 1983 and is intended to go into French service in 1985 together with two 30 mm cannon, the Mle 30-791B and the Mle 30-792. The former is a high-rate revolver gun firing at 2500 rds/min and intended for the armament of aircraft and anti-missile systems. It will replace the present 30 mm ADEN and DEFA cannon in French service. The latter is a mechanically actuated, external power supply, gun firing at 650 rds/min and intended for use in vehicles and helicopters and as a ground anti-aircraft weapon, and will replace the various 20 mm cannon presently used by ground troops. The standard projectile (for which the data above is valid) is a Mine HE/I shell weighing 275 g. Other natures which have been designed include a 170 g APDS/T which has a muzzle velocity of 1350 m/s and a 275 g AP/T with a velocity of 1050 m/s.

The drawing above is provisional, being based on incomplete information.

30 mm RARDEN/HSS 831

30 mm × 170

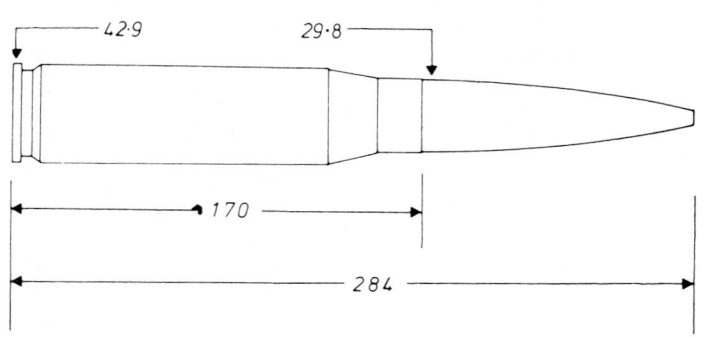

This was originally developed by Hispano-Suiza for their 831L cannon which, after their amalgamation with Oerlikon, became the Model KCB. The cartridge was also adopted by the British as the basic round for their RARDEN cannon, though the RARDEN ammunition was separately designed in Britain. This was not a very wise move, since it entailed redesigning the steel case into brass, developing new projectiles and fuzes, and a great deal of time was wasted in getting these new designs into production. The long-heralded APDS for the RARDEN has still not materialised: in-house development by RARDEN was abandoned and is now being undertaken by an American firm of consultant engineers. The Swiss ammunition, on the other hand, appears to function quite satisfactorily.

Current ammunition includes:

FRANCE
HE-I	Type OE	steel shell, filled Hexal, and fitted with self-destroying nose fuze
HEI-T	Type OET	as OE but with smaller explosive filling to allow space for tracer
TP	Type OX	hollow steel projectile, inert
TP-T	Type OXT	as OX but with tracer

SWITZERLAND
Mine HE-I	Type MSC	360 g shell, filled Torpex, with self-destroying nose fuze, V₀ 1080 m/s
Mine HEI-T	Type MLC	as MSC but with 4 s tracer
APHEI	Type PSC	360 g pointed steel shell with aluminium ballistic cap, Torpex filling, and self-destroying base fuze
TP	Type UGC	360 g hollow steel shell, inert
TP-T	Type ULC	as UGC but with 4 s tracer

UNITED KINGDOM (RARDEN)
HE-T	L8A2	357 g shell filled Torpex, fitted with impact nose fuze L86A2 and tracer. V₀ 1070 m/s. No self-destruction
APSE-T	L5A2	357 g pointed steel shell with aluminium ballistic cap. Filled with explosive and smoke mixture CS5390. Unfuzed. Tracer
Practice	L7A4	357 g hollow shell with dummy fuze and tracer

30 mm Oerlikon KCA

30 × 173 mm

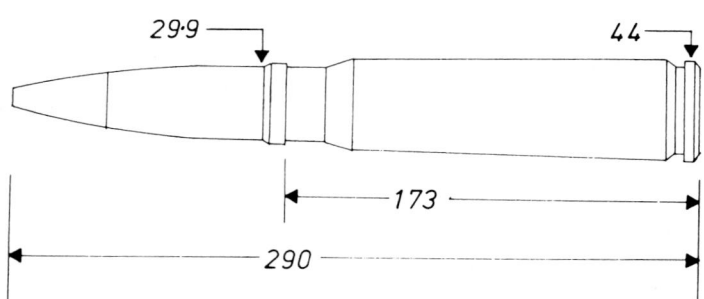

The revolver cannon KCA is the most recent Oerlikon aircraft cannon and is the most powerful of the class. It has a rate of fire of 1350 rds/min. Ammunition includes AP/I, HE/I and SAP/HE/I projectiles, and the steel cartridge cases are electrically primed. Projectile weights average 410 g.

The cartridge has also been adopted for the Mauser Model F cannon and for the American GAU-8A multiple-barrelled Gatling-pattern gun (with percussion priming) and the GAU-9A Gatling-pattern gun (with electric priming) used in the A10 close support aircraft. American ammunition types include HE/I, AP/I and TP.

30 × 210B

30 mm Soviet NN-30

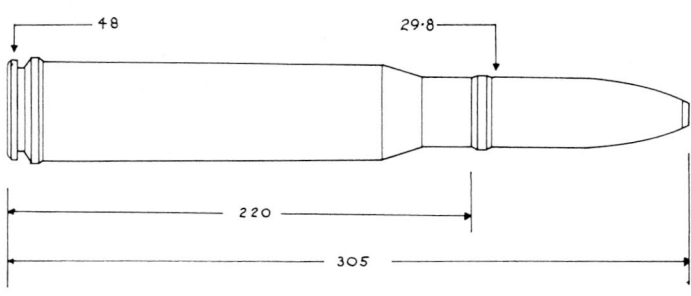

This was developed for the Soviet NN-30 naval anti-aircraft gun, which was introduced in 1960 and is widely employed in many classes of ship, including missile boats and other small craft. It is a two-barrelled, remote-controlled weapon with a rate of fire of about 500 rds/min. The gun was exported, as the equipment of ships, to a number of countries; among them is Yugoslavia, who have now adopted the cartridge for use with a land service automatic cannon. It is probable that other countries will follow suit, since the round has good performance.

The standard Soviet round is the UF-83, carrying a 435 g high explosive shell which is fired at 1000 m/s muzzle velocity. The Yugoslavian standard HE-Incendiary projectile weighs 356 g and has a muzzle velocity of 1100 m/s. The complete round weighs 1066 g and is 307 mm long. In addition, the Yugoslavian gun is provided with APHE-I-T and Practice projectiles with and without tracer.

30 mm Czech M53

30 × 220 mm

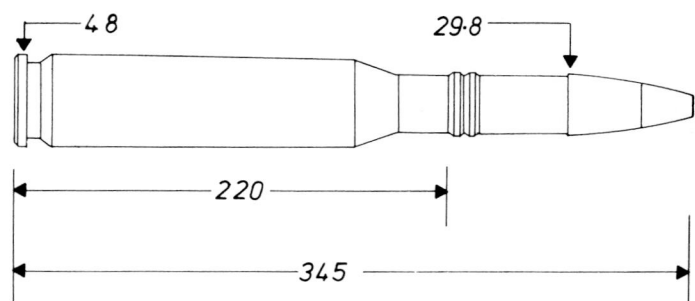

This cartridge is used with the Czechoslovak twin-barrelled M53 anti-aircraft cannon, which can be encountered either in towed or self-propelled form with the Czech, Cuban, Romanian, Vietnamese and Yugoslavian armies. The weapon has a cyclic rate of fire of about 500 rds/min per barrel.

Two types of projectile are known; an AP/I shell weighing 450 g and capable of penetrating 55 mm of plate at 500 m range, and an HE/I shell also of 450 g.

DM381 25 mm PDSD cannon fuze

The DM381 25 mm fuze is a mechanical, explosive nose fuze with spin-controlled self-destruction (SD) function. In the safe configuration the primer carrier (rotor) retains its detonator out-of-line; this configuration is guaranteed by two elements independent of each other.

Due to physical and environmental influences occuring during firing and on the trajectory, arming is performed. One of the safety elements is responsible for delayed arming, thus controlling muzzle safety.

The fuze is distinguished by two separate functions. One is the highly sensitive impact mechanism which also has high angular sensitivity. The other feature is the low-dispersion SD function which causes self-destruction after a certain period of time, should the projectile have missed the target.

Manufacturer
Junghans Feinwerktechnik, PO Box 110, D-7230 Schramberg, Federal Republic of Germany.
Status
Current. Production.

DM381 25 mm PDSD cannon fuze

No. 569 30 mm PDSD cannon fuze

The No. 569 fuze is identical in its operation to the DM381 described above, but is dimensioned to fit 30 mm cannon projectiles. It has the same safety elements, impact action and self-destruction function as described above.

Manufacturer
Junghans Feinwerktechnik, PO Box 110, D-7230 Schramberg, Federal Republic of Germany.
Status
Current. Production.

No 569 30 mm PDSD cannon fuze

AMMUNITION EMPLOYMENT, MANUFACTURE AND IDENTIFICATION

This section lists the types of ammunition in current military use, the national manufacturing establishments, with examples of their headstamps, and details the systems of colour coding in use on small arms and cannon ammunition, where these are known.

Argentina

Types used: 9 mm × 19; 7.62 mm × 51; 7.65 mm × 54; .45 ACP; .30-06

Principal manufacturers
Fabrica Militar de Munitiones de Armas Portatiles, Borghi **(a) (b)**.
Fabrica Militar de Munitiones de Armas Portatiles, San Francisco, Cordoba **(c) (d)**.
Fabrica Militar de Cartouchoes 'San Lorenzo', Borghi **(e)**.
Fabrica Militar 'Frey Luis Beltran', Borghi **(f)**.
Cartoucheria Orbea Argentina, Buenos Aires **(g)**.

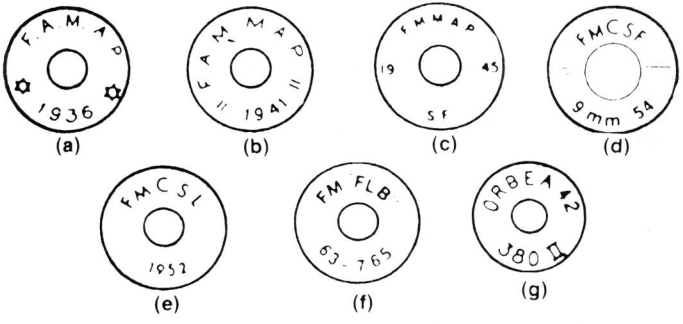

Colour code: small-arms ammunition: coding by coloured bullet tips, as follows:
Tracer (light)   blue
Tracer (smoke)   yellow

AP-tracer   green
AP   red
Observing   black
Incendiary   white
Cannon ammunition, above 20 mm: projectile bodies coloured as follows:
AP   white
Practice   red
HE   green
Incendiary   grey
AP-incendiary   white body, grey tip
Tips of projectiles are additionally coloured when fitted with tracers:
Flame tracer   blue
Smoke tracer   yellow

Australia

Types used: 9 mm × 19; 5.56 mm × 45; 7.62 mm × 51; 12.7 mm × 99

Principal manufacturers
Small Arms Ammunition Factory No 1, Footscray, Victoria **(a)**.

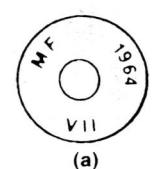

(a)

Colour code: originally used British; now uses NATO standard.

Austria

Types used: 9 mm × 18; 9 mm × 19; 5.56 mm × 45; 7.62 mm × 51

Principal manufacturers
Hirtenberger AG, Hirtenberg **(a)**.
Österreichische Jagdpatronfabrik, Kremsach, Tirol **(b)**.
Südsteirische Metallindustrie GesmbH, Leibnitz **(c)**.

(a)
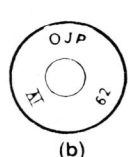
(b)
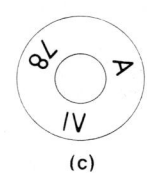
(c)

Colour code: NATO standard, plus one national code:
Sniper ball   purple bullet tip

Belgium

Types used: 9 mm × 19; 5.56 mm × 45; 7.62 mm × 51; 12.7 mm × 99

Principal manufacturers
Fabrique Nationale Herstal SA, Herstal.

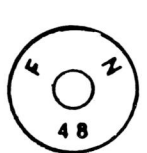

Colour code: small-arms ammunition: NATO standard 20 mm and upwards: body colours:
HE   yellow
HE1   yellow with red band superimposed
AP1   black with red band superimposed
Practice   light blue

Bolivia

Types used: 9 mm × 19; 7.62 mm × 51; 12.7 mm × 99

Manufacturers: none known.

Colour code: believed to use NATO standard.

Brazil

Types used: 9 mm × 19; 5.56 mm × 45; .45 ACP; 7 mm × 57; 7.62 mm × 51; 12.7 mm × 99

Principal manufacturers
Fabrica Nacional de Cartuchos e Municoes, São Paolo **(a)**.
Companhia Brasileira de Cartuchos, São Paolo **(b) (e)**.
Fabrica Realengo, Rio de Janeiro **(c) (d)**.

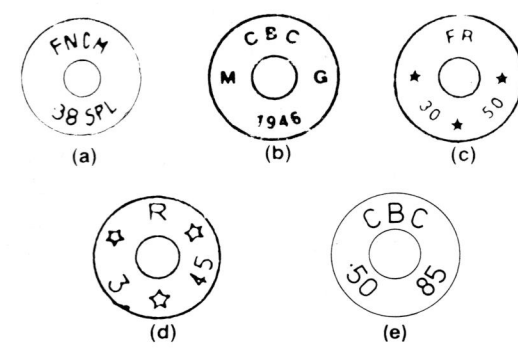

(a)   (b)   (c)

(d)   (e)

Colour code: believed to use US standard.

Bulgaria

Types used: all Soviet standards. 7.62 mm × 25; 7.62 mm × 39; 7.62 mm × 54R; 12.7 mm × 107; 14.5 mm × 114

Manufacturers
State factory, identified only by the number '10'.

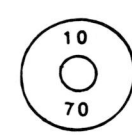

Colour code: Soviet standard.

Burkina Faso

Types used: 7.62 mm × 51; 7.5 mm × 54; 5.56 mm × 45; 9 mm × 19

Manufacturers
Cartoucherie Voltaique, Ouagadougou.

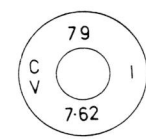

Colour code: not known.

Burma

Types used: 7.62 mm × 51; 9 mm × 19; .303

Manufacturers
A modern factory, erected by a West German contractor exists, location unknown.

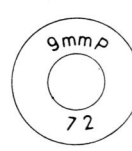

Colour code: not known.

Cambodia

Types used: 5.56 mm × 45; 9 mm × 19; 7.5 mm × 54; .30-06

Manufacturers
State-operated factory at Stung Chral operated 1969-70 under Czechoslovak supervision. Current manufacturing capacity unknown.

Colour code: not known.

Cameroon

Types used: 7.62 mm × 51; 7.5 mm × 54; 9 mm × 19

Manufacturers
Manufacture Camerounaise de Munitions (Manucam).

Colour code: not known.

Canada

Types used: 9 mm × 19; .45 ACP; 5.56 mm × 45; 7.62 mm × 51; 12.7 mm x 99; .303; .30-06

Manufacturers
Dominion Arsenal, Quebec **(a) (b)**.
Industries Valcartier Inc, Valcartier, Quebec **(c)**.

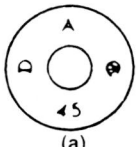 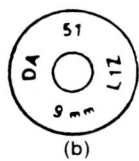 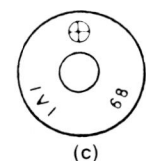

(a) (b) (c)

Colour code: British system on .303; US system on .30-06; NATO on other calibres.

Chile

Types used: 9 mm × 19; 5.56 mm × 45; 7 mm × 57; 7.62 mm × 51; 12.7 mm × 99

Manufacturers
Fabrica de Materiel del Ejercito, Santiago.

Colour code: not known.

China, People's Republic

Types used: 7.62 mm × 25; 7.62 mm × 39; 7.62 mm × 54R; 12.7 mm × 107; 14.5 mm × 114

Manufacturers
State factories, identified only by the numbers 11, 31, 41, 61, 71, 81, 321 and 661.

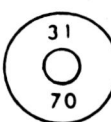

Colour code: China adopted the Soviet code in the 1950s but later modified it since they did not require the diversity of bullet types for which the Soviet code caters. In 1967 a new system was adopted. Both systems are explained below:

7.62 mm × 39
Tracer   green bullet tip
AP/UI   black bullet tip over red ring (pre-1967), black bullet tip, black primer annulus (post-1967)
Incendiary/T red bullet tip (post-1967)

7.62 mm × 54R
Ball   white bullet tip (pre-1967. No marking after 1967)
AP1   black bullet tip over red ring (pre-1967), black bullet tip (post-1967)

12.7 mm × 107
AP1   black bullet tip over red ring (pre-1967), black bullet tip (post-1967)
AP1-T   violet bullet tip over red ring (pre-1967), violet bullet tip (post-1967)

14.5 mm × 114
AP1   black bullet tip over red ring (pre-1967), black bullet tip (post-1967)
Incendiary/T red bullet tip (post 1967)

Colombia

Types used: 9 mm × 19; 7.62 mm × 51; 12.7 mm × 99

Manufacturers
Talleres Central de Ejercito, Bogota **(a)**.
Industrias Militar, Bogota **(b)**.

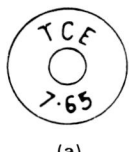 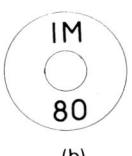

(a) (b)

Colour code: not known.

Cuba

Types used: 9 mm × 19; 7.62 mm × 39; 7.62 mm × 51; 12.7 mm × 99; 12.7 mm × 107; 14.5 mm × 114

Manufacturer: none known; believed all ammunition imported.

Colour code: not known; probably adheres to the Soviet system.

Czechoslovakia

Types used: 7.62 mm × 25; 7.62 mm × 39; 7.62 mm × 45; 7.62 mm × 54R; 7.92 mm × 57; 12.7 mm × 107; 14.5 mm × 114

Manufacturers
Sellier and Bellot, Vlasim **(b) (d)**.
Unknown factories using codes **(a) (c)**.

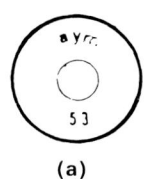 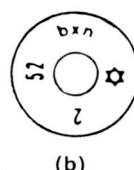 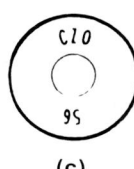 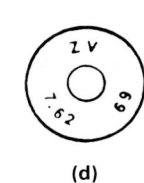

(a) (b) (c) (d)

Colour code: uses Soviet standard system, with two national additions:
Ranging   green bullet tip over white ring
Incendiary   yellow tip
Sniping

Denmark

Types used: 6.5 mm × 55; 9 mm × 19; 7.62 mm × 51; .30-06; 12.7 mm × 99

Manufacturer
(Haerens) Ammunitionsarsenalet, Copenhagen.

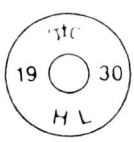 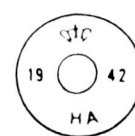  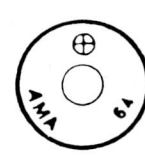

Colour code: NATO standard.

Dominican Republic

Types used: 9 mm × 19; .45 ACP; .30 carbine; .30-06; 5.56 mm × 45; 7.62 mm × 51; 12.7 mm × 99

Manufacturer
Armeria de Fuerzas Armatas, San Cristobal.

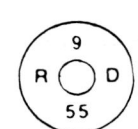

Colour code: not known; believed to be US standard.

Ecuador

Types used: 9 mm × 19; 7.62 mm × 51; 12.7 mm × 99

Manufacturers: none known.

Colour code: not known.

Egypt

Types used: 9 mm × 19; 7.62 mm × 39; 7.62 mm × 54R; 7.92 mm × 57; .303; 12.7 mm × 107; 14.5 mm × 114

Manufacturers
Various state factories exist but are not necessarily identified in cartridge markings; all cartridges carry Arabic symbols indicating 'Egypt' or 'UAR'.

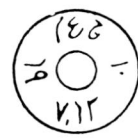

Colour code: Soviet standard.

Ethiopia

Types used: 7.62 mm × 39; .30-06; 9 mm × 19

Manufacturers
None known. Cases with Ethiopian headstamps exist but it is not known whether they are locally made or imported.

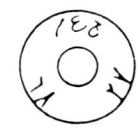

Colour code: not known.

Finland

Types used: 9 mm × 19; 7.65 mm × 21; 7.62 mm × 39; 7.62 mm × 54R

Manufacturers
Lapuuan Patruunatehdas, Lapua (a).
Sako AB, Riihimaki (b) (c).
(Sako = Suojeluskuntain Ase-Ja Konepaja Osakeyhtiö = Company for the manufacture of armaments for the Civil Guard).
Valtion Patruunatehdas, Lapua (d) (e).

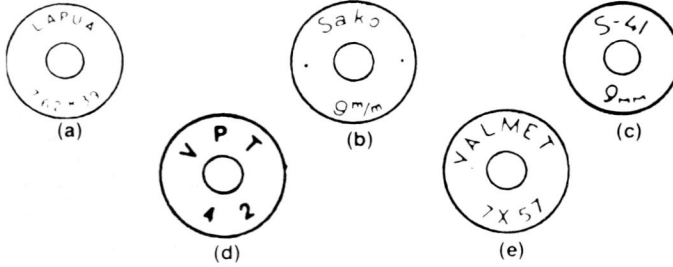

Colour code: bullet natures are identified by a coloured tip:

Tracer	white
AP	blue
AP1	black over red ring
Observing	red
Incendiary	yellow

France

Types used: 7.65 mm × 19; 9 mm × 19; 5.56 mm × 45; 7.5 mm × 54; 7.62 mm × 51; 8 mm × 51R; 12.7 mm × 99

Manufacturers
MILITARY ESTABLISHMENTS
Atelier de Construction de Tarbes (a).
Atelier de Construction du Mans (b).
Atelier de Construction de Paris.
Atelier de Construction de Puteaux (c).
Atelier de Construction de Rennes (d).
Atelier de Construction de Toulouse.
Atelier de Construction de Tarbes (e).
Cartoucherie de la Seine (f).

Atelier de Construction de Valence (g).
Atelier de Construction de Vincennes.
Atelier de Construction de Versaille (h).

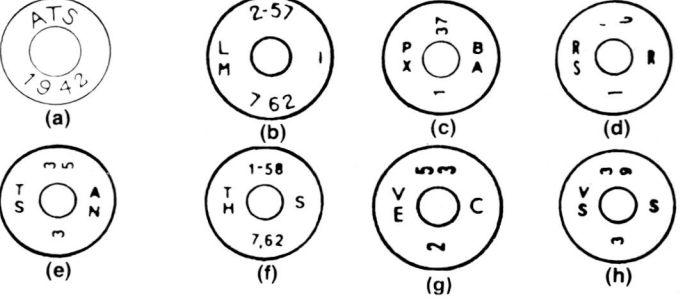

PRIVATE COMPANIES
Etablissements Luchaire SA, Evreaux (a).
Manufacture de Machines du Haut Rhin, Mulhouse (b) (c).
Société Française des Munitions, Paris (d) (e).
Société Méridionale d'Industrie R Paulet and Cie, Marseille.
Gévelot SA, Paris (f).

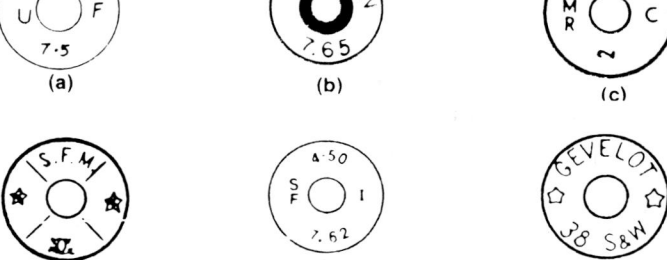

Colour codes: the French adopted a variety of colour identification systems after the introduction of the 7.5 mm × 54 cartridge in 1929, and the system underwent considerable change in 1945 and again in 1958. Since both the 7.5 mm × 54 and the 8 mm × 51R cartridges may still be met outside France, it is necessary to tabulate both pre- and post-Second World War systems.

PRE-1939
7.5 mm × 54

Light ball	no colour
Heavy ball	violet cap annulus and case-mouth seal
AP	brass-washed bullet
APT	green bullet tip, green cap annulus
Tracer	black bullet tip, black cap annulus
Incendiary	blue bullet tip, blue cap annulus

8 mm × 51R

Solid ball	bronze-coloured bullet
Lead-cored ball	nickel-plated bullet
Tracer	tinned bullet
AP	chemically blackened bullet

1945-58
A general code was instituted

Heavy ball	violet bullet tip
Ball (12.7 mm × 99)	grey bullet tip
Incendiary	red bullet tip
Tracer	white bullet tip
APT	black bullet tip over white ring

1958 ONWARDS
A new general code was instituted

Tracer	red bullet tip
AP	black bullet tip
Incendiary	blue bullet tip
AP1	silver bullet tip
APT	black bullet tip over red ring
APIT	silver bullet over red ring
Observing	yellow bullet tip
Incendiary-T	blue bullet tip over red ring
Observing-T	yellow bullet tip over red ring

Germany (pre-1945)

Types used: 9 mm × 19; 7.92 mm × 57; 7.92 mm × 33

Manufacturers
No practical purpose would be served by attempting to tabulate the scores of factories which manufactured military ammunition between 1934 and 1945.

Colour code: since German wartime ammunition continues to appear from time to time, a resumé of the standard pre-1945 colour coding system may be of value:

Light ball	blue primer cap or cap annulus
Heavy ball	green cap annulus
Long ball	zinc primer cap
APT	black bullet tip, red annulus
AP	red cap or cap annulus
Explosive/incendiary	black cap or annulus, black bullet with plain tip
Practice ball	green stripe across cartridge base
Practice tracer	green stripe across base, black bullet tip
High velocity	green ring on bullet

Germany, East

Types used: 9 mm × 18; 7.62 mm × 25; 7.62 mm × 39; 7.62 mm × 54R; 12.7 mm × 107; 14.5 mm × 114

Manufacturers
State factories identified only by the numeral codes '04' and '05'. It is highly likely that these represent pre-war factories at Schönebeck-a-d-Elbe and Magdeburg.

Colour code: Soviet standard.

Germany, West

Types used: 9 mm × 19; 7.62 mm × 51

Manufacturers
Dynamit AG, Nuremberg (a).
Dynamit-Nobel-Genschow GmbH, West Berlin (b) (c).
Industrie-Werke Karlsruhe Augsburg AG, Karlsruhe (d).
Metallwerk Elisenhütte GmbH, Nassau (e) (f).
Manusaar, Budingen (g).

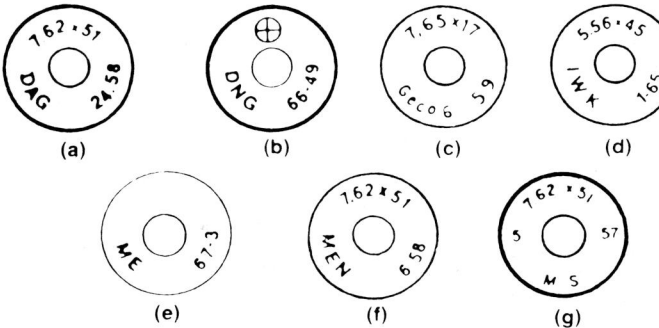

Colour code: small-arms ammunition: NATO standard system 20 mm and upward: body colours:

High explosive	yellow
AP	black, white lettering
API	black, red lettering
APHE	black, yellow lettering
Practice	blue

Greece

Types used: 9 mm × 19; .45 ACP; 7.62 mm × 51; .30-06; 12.7 mm × 99

Manufacturers
Greek Powder and Cartridge Co, Athens.

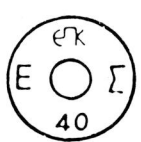 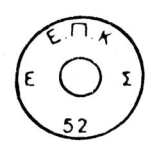 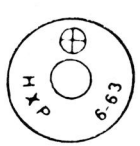

Colour codes: NATO standard.

Hungary

Types used: 9 mm × 18; 9 mm × 19; 7.62 mm × 25; 7.62 mm × 39; 7.62 mm × 54R; 12.7 mm × 107; 14.5 mm × 114

Manufacturers
State factories identified only by the code numbers '21' or '23' (a).

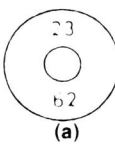  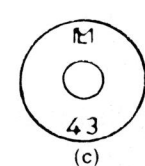

Earlier manufacturers
Czepel Arsenal, Budapest (b)
Magyar Loszervermuk, Veszprem (c)

Colour code: Soviet standard.

India

Types used: 9 mm × 19; 7.62 mm × 51; .303; 12.7 mm × 99

Manufacturers
Kirkee Factory, Kirkee (a).
Ordnance Factory, Khamaria Jubbulpore (b).
Ordnance Factory, Varangoan (c).

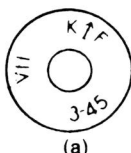 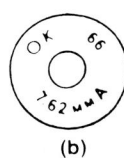 

Colour code: NATO standard.

Indonesia

Types used: 12.7 mm × 99; 7.62 mm × 51; 5.56 mm × 45; 9 mm × 19

Manufacturers
Pabrik Sendjasta Mesin, Bandung (state-owned); state factory at Turen.

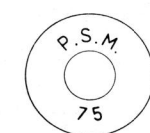

Colour code: not known.

Iran

Types used: 9 mm × 19; .45 ACP; .30-06; 7.62 mm × 51; 12.7 mm × 99

Manufacturers
National Ammunition Factory, Tehran.

 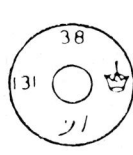

Colour code: not known.

Iraq

Types used: 7.62 mm × 25; 7.62 mm × 39; 7.62 mm × 54R; 12.7 mm × 107; 14.5 mm × 114

Manufacturers
Al Yarmouk Factory, Baghdad.

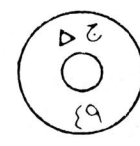

Colour code: not known, presumably Soviet standard system.

Israel

Types used: 9 mm × 19; 5.56 mm × 45; 7.62 mm × 51; 7.92 mm × 57; .30-06; 12.7 mm × 99

Manufacturers
Israel Military Industries, Ramat Ha Sharon.

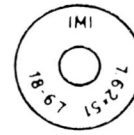 

Colour code: uses a mixture of cap annulus and bullet tip colours
7.92 mm × 57

Ball	purple cap annulus
Tracer	red bullet tip, green annulus
AP	black bullet tip, green annulus
API	black bullet tip, red annulus
Incendiary	blue bullet tip, green annulus

OTHER CALIBRES

Tracer	red bullet tip
AP	black bullet tip
API	blue bullet tip over black ring
APIT	blue bullet tip over black ring over red ring

Italy

Types used: 9 mm × 19; 5.56 mm × 45; 7.62 mm × 51; .30-06; 12.7 mm × 99

Manufacturers
Bombrini, Parodi e Delfino SpA, Rome **(a)**.
Pirotecnico de Esercito, Capua **(b) (c)**.
Guilio Fiocchi, Lecco **(d)**.
Leon Beaux and Co, Milan **(e) (f)**.
Società Metallurgica Italiana, Florence **(g)**.

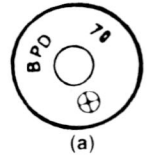 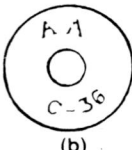 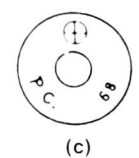 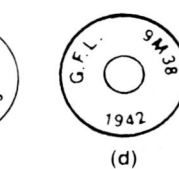

(a)　　　　(b)　　　　(c)　　　　(d)

  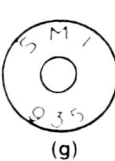

(e)　　　　(f)　　　　(g)

Colour code: US standard system until replaced by NATO system in the mid-1950s.

Japan

Types used: .38 Special; .45 ACP; .30 carbine; 5.56 mm × 45; 7.62 mm × 51; 12.7 mm × 99

Manufacturers
Asahi-Okuma Corp, Asahi **(a) (b) (c) (d)**.
Toyo Seiki Kogaku, Amagasaki **(e)**.
Chuo Kagaku Kako Keisha, Nagoya **(f)**.
Showa Kagaku, Totsuka **(g)**.

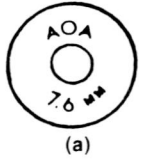 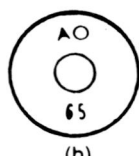 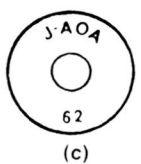 

(a)　　　　(b)　　　　(c)　　　　(d)

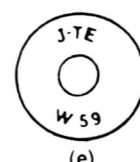 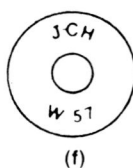 

(e)　　　　(f)　　　　(g)

Colour code: US standard system until 1960, thereafter NATO.

Jordan

Types used: 9 mm × 19; 7.62 mm × 51; 12.7 mm × 99

Manufacturers: not known.

Colour code: NATO standard.

Korea, North

Types used: 7.62 mm × 25; 7.62 mm × 39; 7.62 mm × 54R; 12.7 mm × 107; 14.5 mm × 114

Manufacturers
State factories, identified only by symbols or number codes.

Colour code: Soviet standard.

Korea, South

Types used: .45 ACP; .30 carbine; .30-06; 7.62 mm × 51; 5.56 mm × 45; 12.7 mm × 99

Manufacturers
Pusan Arsenal, Pusan **(a)**.
Poonsang Metal Mfg Co, Seoul **(b)**.
Hyosung Corp., Pusan **(c)**.

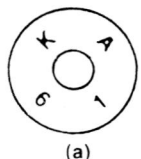 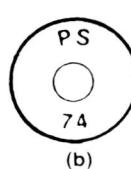 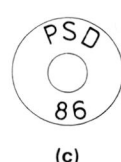

(a)　　　　(b)　　　　(c)

Colour code: US standard system.

Lebanon

Types used: 9 mm × 19; 5.56 mm × 45; 7.5 mm × 54; 7.62 mm × 45; 7.62 mm × 51; 7.92 mm × 57; 12.7 mm × 99

Manufacturers
None known to exist; ammunition is made elsewhere, and can be identified as Lebanese by the stylised 'Cedar of Lebanon' symbol, plus two Arabic script symbols indicating 'Republic of Lebanon'.

Colour code: not known.

Libya

Types used: 9 mm × 19; 7.62 mm × 51

Manufacturers: none known.

Colour code: not known.

Malaysia

Types used: 9 mm × 19; 5.56 mm × 45; 7.62 mm × 51

Manufacturers
Syarikat Malaysia Explosives Ltd, Batu Arang.

Colour code: not known.

Mexico

Types used: .45 ACP; .30 carbine; 5.56 mm × 45; .30-06; 12.7 mm × 99

Manufacturers
Fabrica National de Cartouchoes, Mexico City **(a)**.
Fabrica National de Munitiones, Mexico City **(b)**.

Colour code: not known.　　　(a)　　　　(b)

Morocco

Types used: 9 mm × 19; .30 carbine; 7.5 mm × 54; 7.62 mm × 39; 7.62 mm × 51; 12.7 mm × 99

Manufacturers
Manufacture Nationale d'Armes et Munitions.

Colour code: not known.

Netherlands

Types used: 9 mm × 19; 7.62 mm × 51; 12.7 mm × 99

Manufacturers
Eurometaal, Zaandam (a).
Nederland Wapen and Munitiefabrik, s'Hertogenbosch (b).
Artillerie Inrichtingen, Hembrug (now EMZ) (c).

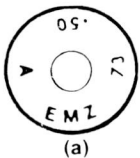 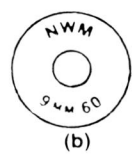 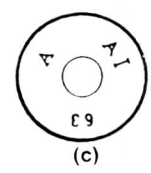

(a) (b) (c)

Colour code: NATO standard.

New Zealand

Types used: 9 mm × 19; 5.56 mm × 45; 7.62 mm × 51

Manufacturers
CAC Industries Ltd (late Colonial Ammunition Co), Auckland.

Colour code: NATO standard.

Nicaragua

Types used: 9 mm × 19; 5.56 mm × 45; 7.62 mm × 51

Manufacturers: none known.

Colour code: not known.

Nigeria

Types used: 7.62 mm × 51; 9 mm × 19

Manufacturers
Nigerian Ordnance Factory.

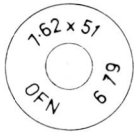

Colour code: NATO standard.

North Atlantic Treaty Organisation

The NATO colour code was based on a modified US system: it is used not only by the member nations of NATO but many other countries which buy NATO-types of ammunition in Europe and, for convenience, accept the standard markings. Ammunition standardised within NATO is head-stamped with a cross-in-a-circle symbol; unfortunately, this symbol has come to be applied somewhat indiscriminately and can be found on ammunition which is not to NATO specifications, which is not a NATO standard calibre and even ammunition which has never been seen in a NATO country.

Colour code:

Tracer	red bullet tip
AP	black bullet tip
AP1	silver bullet tip
Incendiary	blue bullet tip
Tracer (Dark Ignition)	orange bullet tip
Observation	yellow bullet tip
Observation/T	yellow tip over red band

Norway

Types used: 9 mm × 19; .45 ACP; 7.62 mm × 51; 12.7 mm × 99

Manufacturers
Raufoss Arsenal, Raufoss (a) (b).
Norma Projectilfabrik, Oslo (c).

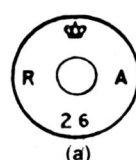 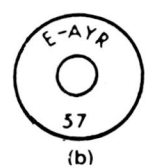 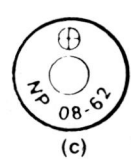

(a) (b) (c)

Colour code: NATO standard.

Pakistan

Types used: 9 mm × 19; .303; .30-06; 7.62 mm × 39; 7.62 mm × 51; 12.7 mm × 99

Manufacturers
Pakistan Ordnance Factory, Wah Cantonment, Rawalpindi.

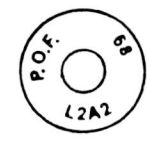

Colour code: not known.

Paraguay

Types used: 9 mm × 19; 7.62 mm × 51

Manufacturers: none known.

Colour code: not known.

Peru

Types used: 9 mm × 19; 7.62 mm × 39; 7.62 mm × 51

Manufacturers
Fabrica de Armas y Municiones de Guerra, Lima (a) (b).
Arsenal de Guerra, Lima (c).

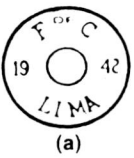 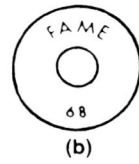 

(a) (b) (c)

Colour code: not known.

Philippines

Types used: .45 ACP; 7.62 mm × 51; 5.56 mm × 45; .30 carbine; .30-06; 12.7 mm × 99

Manufacturers
Republic of the Philippines State Arsenal, Manila.

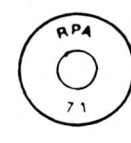

Colour code: not known.

Poland

Types used: 9 mm × 18; 7.62 mm × 25; 7.62 mm × 39; 7.62 mm × 54R; 12.7 mm × 107; 14.5 mm × 114

Manufacturers
Current manufacture is by state factories identified only by the code numbers 21 and 343.

Colour code: Soviet standard.

(a)

Portugal

Types used: 9 mm × 19; 7.62 mm × 51; 7.92 mm × 57; 12.7 mm × 99

Manufacturers
Fábrica Nacional de Municoes de Armas Ligeiras, Moscavide **(a)**.
Fábrica Cartouchoes e Pulveras Quimicas, Lisbon **(b)**.
Arsenal do Ejercito, Lisbon **(c)**.

  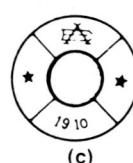

    (a)   **(b)**   **(c)**

Colour Code: NATO standard.

Romania

Types used: 7.62 mm × 25; 9 mm × 18; 9 mm × 19; 7.62 mm × 39; 7.62 mm × 54R; 12.7 mm × 107; 14.5 mm × 114

Manufacturers
State factories, identified only by the code numbers 21RPR and 22RPR (RPR = Romanian People's Republic).

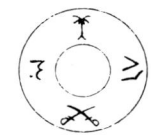

Colour code: Soviet standard.

Saudi Arabia

Types used: 12.7 mm × 99; 7.62 mm × 51; 5.56 mm × 45; 9 mm × 19

Manufacturers
State factory at Riyadh.

Colour code:
AP	purple tip
APT	green tip
Incendiary	orange tip

Singapore

Types used: 9 mm × 19; 5.56 mm × 45; 7.62 mm × 51; .303

Manufacturers
Chartered Industries of Singapore **(a) (b) (c)**.

    (a)   **(b)**   **(c)**

Colour code: NATO Standard.

South Africa

Types used: 9 mm × 19; 7.62 mm × 51; .303; 12.7 mm × 99

Manufacturers
Pretoria Metal Pressings (Pty) Ltd, Pretoria **(a) (b) (c)**.
South African Mint, Pretoria **(d) (e)**.
South African Mint, Kimberley **(f)**.
Musgrave and Sons, Bloemfontein (manufacture by PMP) **(g)**.

   (a)   **(b)**   **(c)**   **(d)**

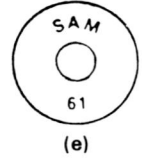 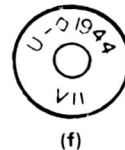 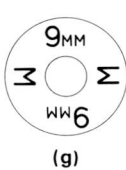

    (e)   **(f)**   **(g)**

Colour code: uses both UK and NATO codes, depending on the original source of the ammunition.

Spain

Types used: 9 mm × 19; 9 mm × 23; 7.62 mm × 51; .30-06; 12.7 mm × 99

Manufacturers
Consortio de Industrias Militares, Toledo **(a)**.
Consortio de Industrias Militares, Seville **(b)**.
Fábrica Nacional de Palencia **(d)**.
Fábrica Nacional de Toledo **(e)**.
Manufacturas Metálicas Madrileñas, Madrid **(f)**.
Empresa Nacional de Santa Bárbara, Palencia **(g) (i)**.
Pirotecnia Militar de Sevilla **(c) (h)**.

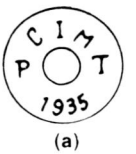  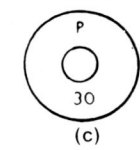

    (a)   **(b)**   **(c)**

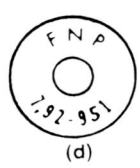 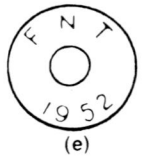 

    (d)   **(e)**   **(f)**

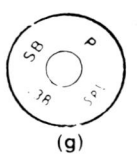 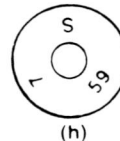 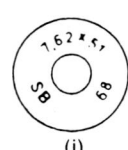

    (g)   **(h)**   **(i)**

Colour code: not known.

Sudan

Types used: 7.62 mm × 51; 7.62 mm × 39; 9 mm × 19

Manufacturers
State factory, Khartoum.

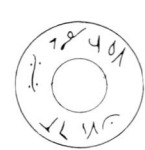

Colour code: not known.

Sweden

Types used: 9 mm × 19; 9 mm × 20SR; 9 mm × 29R; 5.56 mm × 45; 6.5 mm × 55; 7.62 mm × 39; 7.62 mm × 51; 8 mm × 63; 12.7 mm × 99

Manufacturers
Norma Projectilfabrik, Amotfors **(a) (b) (h)**.
Karlskrona Naval Arsenal, Karlskrona **(c)**.
Marieborg Arsenal, Stockholm **(d)**.
Svenska Metallverken AB, Västerås **(e) (f)**.
FFV Ordnance, Eskilstuna **(g)**.

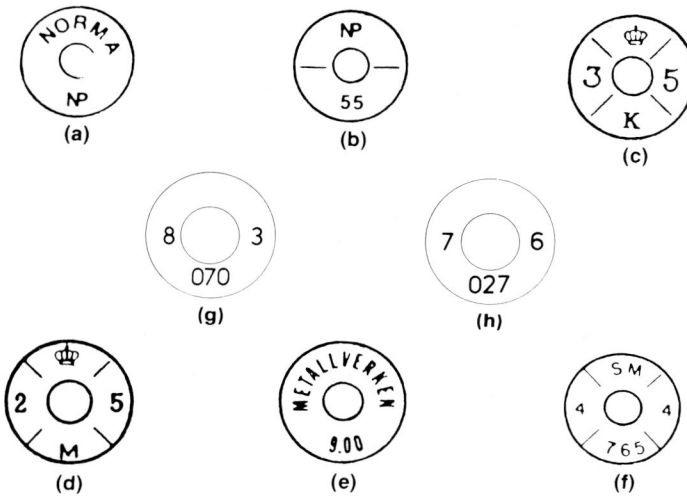

(a) (b) (c)

(g) (h)

(d) (e) (f)

Colour code: based on coloured bullet tips, but varies with the calibre:

Tracer	white
Incendiary	orange
API	yellow (on 12.7 mm × 99 only)
AP	black

Switzerland

Types used: 7.65 mm × 21; 9 mm × 19; 7.5 mm × 55; 12.7 mm × 77; 12.7 mm × 99

Manufacturers
Eidgenossische Munitionsfabrik, Altdorf **(a)**.
Eidgenossische Munitionsfabrik, Thun **(b)**.

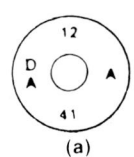 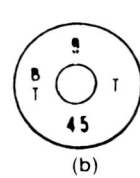

(a) (b)

Colour code: small-arms ammunition: based on colouring the base of the cartridge case:

AP	violet
Tracer	red

Cannon ammunition, 20 mm and over: the system consists of two parts, a basic code applied to both projectile bodies and to package labels, and a secondary code applied to the projectile bodies.

BASIC CODE: packages and projectiles

Training ammunition	black
Blank ammunition	green
Pyrotechnics	pale blue
Drill ammunition	brown
Service ammunition	grey

SECONDARY CODE: projectiles only

High explosive	yellow
Incendiary	pink
Tracer	red
Smoke	white

The conjunction of these two codes demands the application of some common sense. Thus a high explosive projectile will be yellow, but will be packed in a box identified by a grey label; an HE-incendiary shell will be coloured half yellow and half pink; an inert projectile (eg shot) but carrying a tracer would be grey with a red ring.

Syria

Types used: 9 mm × 19; 7.62 mm × 39; 12.7 mm × 107

Manufacturers
Unidentified state factories; note that ammunition is also bought abroad but marked with Syrian headstamps, so that it is not easy to identify ammunition of Syrian manufacture.
Defence Industries Syndicate, Damascus.

Colour code: Soviet standard.

Taiwan

Types used: .30 carbine; .30-06; .45 ACP; 9 mm × 19; 5.56 mm × 45; 7.62 mm × 51; 7.92 mm × 57; 12.7 mm × 99

Manufacturers
It is known that a state factory has been set up in Taiwan for the production of 7.62 mm × 51 and possibly 9 mm × 19 cartridges: it is identified only by the marking '60A'. It should be noted that dates are calculated from 1912, the year of the Republic, and that therefore the indicated date should have 12 added to it to produce the final digits of the Gregorian calendar. Ammunition of military calibres is also produced commercially by the Hsing Hua Company of Taipei.

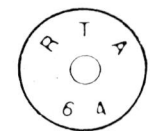

Colour code: US standard.

Thailand

Types used: 9 mm × 19; 5.56 mm × 45; 7.62 mm × 51; 12.7 mm × 99

Manufacturers
Royal Thai Arsenal, Bangkok.

Colour code: not known.

Turkey

Types used: 7.65 mm × 17; 9 mm × 17; .45 ACP; 7.62 mm × 51; .30-06; 12.7 mm × 99

Manufacturers
Makina ve Kimya Endustrisi, Kurumu **(a)**.
Turkish ammunition can be additionally identified by the initials 'TC' (for 'Turkiye Cumhuriyeti' – Turkish Republic) which usually appear in some form in the headstamp **(b)**.

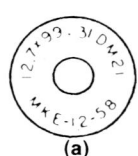 

(a) (b)

Colour code: NATO standard.

Union of Soviet Socialist Republics

Types used: 9 mm × 18; 7.62 mm × 25; 7.62 mm × 38R; 5.45 mm × 39.5; 7.62 mm × 39; 7.62 mm × 54R; 12.7 mm × 107; 14.5 mm × 114

Manufacturers
All Soviet (and Soviet bloc generally) ammunition is marked with a code number to indicate the factory of origin and the last two digits of the year of manufacture. Symbols, triangles and stars, also appear but their significance is not known.

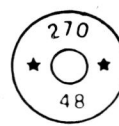  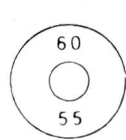

Colour code: small-arms ammunition: the Soviet colour code is used by all Warsaw Pact countries and also by several other countries which rely on the USSR for their ammunition. The system relies on coloured bullet tips:

Heavy ball	yellow
Light ball	silver or white
Tracer	green
AP	black
API	black tip, red ring or red bullet
API (tungsten core)	red over black ring (14.5 mm × 114 only)
API-T	purple tip, red ring or red bullet
Incendiary/T	red
Reduced velocity ball	black over green ring
HE/incendiary	red bullet, with fuze (14.5 mm × 114 only)

20 mm ammunition and above

API-T	black projectile with yellow tip
HEI-T	black projectile with magenta tip

United Kingdom

Types used: 9 mm × 19; .38/200; .303; 7.62 mm × 51; 5.56 mm × 45; .30-06; 12.7 mm × 77; 12.7 mm × 99

Manufacturer
Current: Royal Ordnance plc, Ammunition Division, Radway Green, Cheshire.

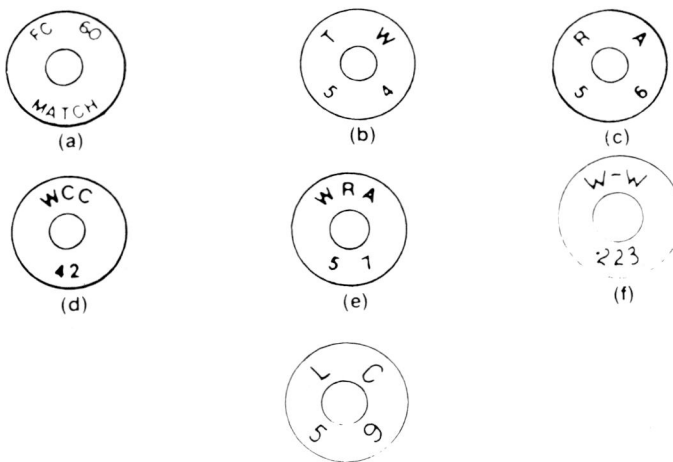

Colour and other codes
British ammunition types introduced before 1957 use a letter identifier on the cartridge base, preceding the mark number, to specify the bullet type, for example 'B II' or 'G8Z'. (Roman numerals were used until t he end of 1944). These identifying code letters are:

B incendiary
D drill
E smoke bomb projector
F semi-armour-piercing
G tracer
H grenade discharger
L blank
O observing
P practice
Q proof
R explosive
U dummy
W armour-piercing

The absence of any identifying letter indicated ball.

Where a bullet has two functions, two letters can be used, for example 'PG' for practice tracer.

With the introduction of the 7.62 mm × 51 cartridge, the system was changed, each model of cartridge being given a unique model number, for example L1 for 'Land Service, Model 1'. Subsequent modifications to this basic model attracted suffixes, to produce L1A1, L1A2 etc. No type prefix letter is used, since bullets of different design took totally different model numbers; thus the 7.62 mm × 51 ball is L2, the tracer L5. But these model numbers only apply in that specific calibre; eg the 7.62 mm × 51 L11 is a ball cartridge, whereas the 12.7 mm × 99 L11 is an observing bullet cartridge. Note also that designs pre-dating 1957 retain their old nomenclature, eg the 9 mm × 19 ball is still 'Mark 2Z'.

The major British colour coding system before the adoption of the NATO code in 1957 was based on the use of coloured varnish in the primer cap annulus, and this system is still in use on some calibres of ammunition today, in addition to the NATO code. At the same time a bullet tip colour code also existed, though its application tended to be haphazard and it is not as reliable a guide as the cap annulus code.

Colour code
CAP ANNULUS
Ball	purple
Tracer	red
AP	green
Incendiary	blue
Explosive	black
Proof	yellow

PRE-NATO BULLET TIP
Tracer (short-range)	white
Tracer (long-range)	red
Tracer (dark ignition)	grey
AP	green
Explosive	black
Incendiary	blue

CANNON AMMUNITION BODY
20 mm Oerlikon and Hispano:
HEI	blue
HEI-T	light green
APHE-I	red body with white tip
Tracer	dark green
Practice	olive drab

30 mm RARDEN:
AP	black
API	black with red ring
APSE	black with green ring
HE	yellow
Practice	light blue

United States of America

Types used: .45 ACP; 9 mm × 19; 5.56 mm × 45; .30-06; 7.62 mm × 51; 12.7 mm × 77; 12.7 mm × 99

Manufacturers
GOVERNMENT FACILITIES
Current
Lake City Ordnance Plant, Independence, Missouri **(g)**.
Twin Cities Ordnance Plant, Minneapolis, Minnesota **(b)**.

COMMERCIAL SUPPLIERS OF MILITARY AMMUNITION
Federal Cartridge Co, Anoka, Wisconsin **(a)**.
Remington Arms Co, Bridgeport, Connecticut **(c)**.
Western Cartridge Co, East Alton, Illinois **(d)**.
Winchester Repeating Arms Co, New Haven, Connecticut **(e)**.
Winchester-Western, New Haven, Connecticut **(f)**.

Colour code: small-arms ammunition: before the adoption of the NATO code in 1957 a bullet tip colour code was in use:
AP	black
API	silver
Incendiary	blue
Tracer	brown
Tracer	red
Tracer	maroon
Frangible ball	green tip over white ring
API-T	red tip over silver ring
Incendiary	dark blue tip over light blue ring (12.7 mm × 99 only)
Spotter/T	yellow tip over red ring (12.7 mm × 77 only)
Duplex ball	green

CANNON AMMUNITION BODY
20 mm Oerlikon and Hispano:
Practice ball	black
AP	black with white lettering
HE (filled Tetryl)	white
HE (Pentolite)	yellow
HEI (Tetryl)	red
HEI (Pentolite)	pink
HE-T (Tetryl)	grey
HE-T (Pentolite)	dark blue
Practice inert	greenish-grey
Practice/T	greenish-grey with yellow ring

CURRENT 20mm AND UPWARD
Practice ball	blue
AP	black
API	black with red ring
HEI	yellow with red ring

Venezuela

Types used: 9 mm × 19; 7.62 mm × 51; 12.7 mm × 99

Manufacturers: CAVIM (Companhia Anonima Venezolana de Industrias Militares), Caracas. **(a)**.

(a)

Colour code: not known.

Vietnam

Types used: 7.62 mm × 25; 7.62 mm × 39; 7.62 mm × 54R; 12.7 mm × 107; 14.5 mm × 114

Manufacturers: none known.

Colour code: Soviet standard.

Yugoslavia

Types used: 7.62 mm × 25; 9 mm × 19; 7.62 mm × 39; 7.62 mm × 51; 7.92 mm × 57; 12.7 mm × 99; 12.7 mm × 107

Manufacturers
Igman Zavod, Konjic **(a) (b) (c)**.
Voini Techniki Zavod, Kragujevac **(d) (e)**.
Prvi Partizanki Zavod, Titovo Uzice **(f) (g)**.

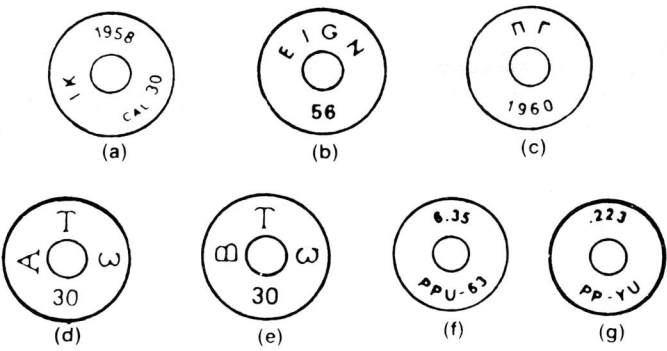

Colour code: a mixture of US and Soviet codes depending on the origin of the cartridge.

Combat Grenades

ARGENTINA

GME-FMK2-MO hand grenade

The GME-FMK2-MO hand grenade was designed by Fabricaciones Militares. The body design is based on a nodular-iron technique which is claimed by the company to be unique. Its final characteristics are the result of its metallic structure (Fe, C, Mg, Mn, etc) and the post-casting heat treatment used.

The fuze body consists of one piece in aluminium which can be produced by drop forging or injection moulding. It incorporates the detonator and firing mechanism and can be removed from the grenade body as a single unit.

On explosion the fuze body is reduced to fragments, in most cases of a weight of 3 to 5 g. These produce a lethal area of about 5m radius and will not endanger the thrower whose throw in the prone position is limited to about 10 or 30 m if kneeling.

Firing is initiated by percussion and the powder train produces a delay of between 3.6 and 4.5 seconds between the throwing of the grenade and the moment of explosion.

The bursting charge consists of a mixture of recrystallised hexogene and flaked TNT in proportions which can be varied to suit fragmentation requirements. The net weight of the charge varies between 75 and 79 g.

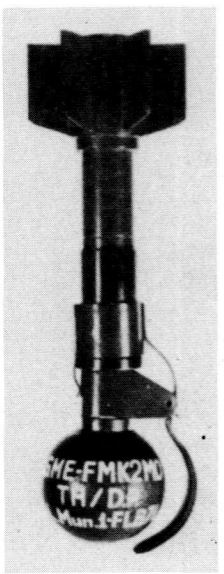

GME-FMK2-MO hand grenade

Grenade and tail for rifle launching

Safety devices
The hand grenade has two safety devices. One is a specially-shaped spring-steel wire securing the grenade in transit, and the other consists of two split pins on the safety pin ring. The first can be removed at any time and the grenade can continue to be handled with perfect safety. It is released by firm thumb pressure and must always be removed before issuing the grenades.

The second split pin is the last safety device on the grenade and once it is removed the grenade is ready for throwing. Should the safety lever come away on pulling out the pin, the grenade will explode.

Grenade used with rifle launcher
If a launching device is used with the grenade, it can be fired to a distance of between 350 and 400 m, to explode either at ground level or at varying heights, according to the degree of skill acquired by the soldier in practice.

The launcher is provided separately, complete with its propellant cartridge. To fire: make sure the grenade is firmly in its seating and that the safety lever is in place; place grenade and launcher in the muzzle of the automatic light rifle (FAL). There is no need to use a muzzle adaptor; rotate gas vent plug to close; load rifle with the propellant cartridge provided with launcher; remove safety split pin; aim and fire.

Practice and drill grenades
The grenade and fuze bodies FMK 2 are used, and have the following features: inert detonator and weighted body; supplied with five spare safety levers, creep springs and safety pins in case of loss or damage to those fitted; painted in white; markings as on original with addition of the letters EJ after letters GM; fitted with safety wire for transport.

DATA
Weight: (grenade body) 165 g; (fuze body) 40.8 g; (charge) 75/79 g
Length overall: approx 120 mm
Body diameter: approx 55 mm
Delay: 3.6–4.5 s

Manufacturer
Fabrica Militar de Armas Portatiles, Domingo Matheu, Cabildo 65, Buenos Aires.
Status
Current. Production.
Service
Argentinian forces.

GMINS-FMK4-MO instructional hand grenade

This is similar to the operational FMK2 grenade but has the following special characteristics: smoke emitting powder train which, on ignition (3.6 to 4.5 seconds) indicates the moment of explosion of the grenade; grenade body perforated to allow smoke emission; a set of interchangeable smoke emitting powder trains in quantities established by the instruction programme requirements; a spare fuze body.

The fundamental feature of this grenade is that, though the powder train must be changed each time, the grenade body can be used indefinitely while in good condition, giving it unlimited life.

Manufacturer
Fabrica Militar de Armas Portatiles, Domingo Matheu, Cabildo 65, Buenos Aires.
Status
Current. Production.
Service
Argentinian forces.

AUSTRIA

Type HG77 fragmentation hand grenade

The HG77 is a spherical grenade designed for defensive use in open spaces. The construction is to the usual Arges pattern, using pre-formed fragments, and the spherical shape gives a particularly even distribution of fragments around the point of burst. There is also the advantage that, from its shape, the grenade will roll when it hits the ground.

If required, the HG77 grenade can be supplied as an offensive grenade (OFF HG77).

DATA
Weight: 480 ± 20 g
Length: 96 ± 2 mm
Diameter: 65 ± 1 mm
Weight of charge: 70 ± 5 g
Number of fragments: approx 5500
Diameter of fragments: 2 – 2.3 mm
Delay time: 4 s + 1.5/-0.5 s

Manufacturer
Arges, Armaturen-GmbH, A-4690 Schwanenstadt-Rüstorf.
Status
Current. Production.

Arges HG77 fragmentation hand grenade

Type HG78 fragmentation hand grenade

The HG78 is an improved version of earlier Arges designs; the same system of pre-fragmentation is used, together with a wave shaper which controls the detonation of the main filling so as to achieve a more even distribution of the fragments. The safety lever must sweep through 50° before allowing the firing pin to rotate and strike the detonator, and a plastic moulding keeps the safety pin ring in place so that it cannot be accidentally pulled free. The standard HG78 is a defensive grenade, but it can also be supplied as an offensive grenade, the 'OFF HG78' which has no fragmenting capacity and is identified by a smooth body.

DATA
Length: 115 ± 2 mm
Diameter: 60 ± 1 mm
Weight: 520 ± 20 g
Charge weight: approximately 70 ± 5 g
Number of fragments: approx 5500
Diameter of fragments: 2 – 2.3 mm
Delay time: 4.0 s nominal
OFF HG78 is of the same dimensions but weighs 230 g

Manufacturer
Arges, Armaturen-GmbH, A-4690 Schwanenstadt-Rüstorf.
Status
Current. No longer in production.
Service
Supply to Middle and Far East, Africa, America and NATO countries.

HG78 defensive grenade

OFF HG78 offensive grenade

Type ÜbHG84 practice grenade

This resembles the HG84 service grenade but is solely for training troops in handling and throwing. The body is of cast iron, with a hole at the top into which the practice igniter fits, and a hole at the bottom to release the gas pressure and noise of the igniter when it fires. The grenade body is covered with black, high-impact, plastic material with a 12 mm blue strip moulded in, giving permanent identification.

The practice igniter unit corresponds to a service igniter, except that instead of a detonator it carries a pyrotechnic charge giving a loud report; the noise, escaping from the open end of the grenade, gives ample evidence of its functioning and is most realistic. Having once been thrown, the grenade can then be salvaged, cleaned, and a new practice igniter set fitted; due to the robust construction the life of the grenade body is virtually limitless.

DATA
Length: 115 ± 2 mm
Diameter: 60 ± 1 mm
Weight: 520 ± 20 g
Delay time: 4.0 s + 1.5/-0.5 s

Manufacturer
Arges, Armaturen-GmbH, A-4690 Schwanenstadt-Rüstorf.
Status
Current. Production.
Service
Supply to Middle and Far East countries and Africa.

ÜbHG84 practice grenade

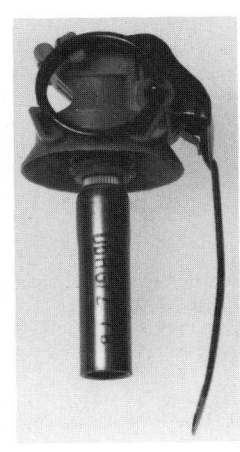

ÜbHG84 complete practice igniter set ready for insertion in grenade

ÜbHG84 practice igniter sectioned

Type HG84 fragmentation grenade

The HG84 is the largest type in the Arges HG range and is an improved version of the Arges HG78. The special design of the fragmentation body gives an optimum distribution of fragments and thus is effective for close combat operations in open terrain. In spite of being an egg-shaped grenade, development has resulted in the same symmetrical fragment distribution as is found in grenades of spherical shape. Detonated in a lying position, 1 m above the ground, at a distance of 5 m from the target, the HG84 gives an average fragment density of 14.3 per m^2 in 20 mm soft wood.

The Offensive Hand Grenade 84 (OFF HG 84) differs from the HG84 by having a plastic core without fragments, a smooth 'lemon skin' surface, and being somewhat lighter in weight. In all other respects it resembles the HG84.

DATA
Weight: 480 ± 20 g
Length: 115 ± 2 mm
Largest diameter: 60 ± 1 mm
Charge weight: 95 ± 5 g
Fragment diameter: 2.0 – 2.3 mm
Number of fragments: approx 5000
Delay time: 4.0 s + 1.5/-0.5 s at 21°C

Arges HG84 fragmentation grenade

Manufacturer
Arges, Armaturen-GmbH, A-4690 Schwanenstadt-Rüstorf.
Status
Current. Production.

Service
Austrian Army.

Type HG85 fragmentation hand grenade

The Arges HG85 is the medium size in the Arges HG range and is a further development of the HG79 design. As with the earlier design it consists of a plastic body, an inner liner carrying the pre-formed fragments, and a filling of Pentolit high explosive. Fragment distribution covers the full 360° area around the grenade, and an HG85, detonated 1 m above the ground and 5 m from a 20 mm soft wood target gives an average fragment density of 8.1 per m^2.

An offensive pattern, OFF HG85, is available and differs from the HG85 by having a plastic core without fragments, a smooth 'lemon skin' surface, and being slightly less in weight.

DATA
Weight: 340 ± 20 g
Length: 96 ± 2 mm
Diameter: 57 ± 1 mm
Weight of explosive charge: 50 ± 3 g
Diameter of fragments: 2.0 – 2.3 mm
Number of fragments: approximately 3500
Delay time: 4 s + 1.5/-0.5 s at 21°C

Manufacturer
Arges, Armaturen-GmbH, A-4690 Schwanenstadt/Rüstorf.
Status
Current. Production.
Service
Supply to Far East countries.

HG85 fragmentation hand grenade

Type HG86 Mini fragmentation hand grenade

The Arges HG86 (Mini 86) is the smallest type in the Arges range and is a further development of the Arges HG80 Mini. The principal improvement is in the control of fragment distribution. An HG86 detonated 1 m above the ground, 5 m from a 20 mm soft wood target, averages a fragment density of 3.05 per m^2.

The HG86 Mini can also be supplied as an offensive grenade (OFF HG86) in which case it has a plastic core without fragments and a smooth surface.

DATA
Weight: 180 ±20 g
Length: 76 ±1 mm
Diameter: 43 ±1 mm
Weight of explosive charge: approximately 17 g
Diameter of fragments: 2.0 – 2.3 mm
Number of fragments: approximately 1600
Delay time: 4 s + 1.5/-0.5 s at 21°C

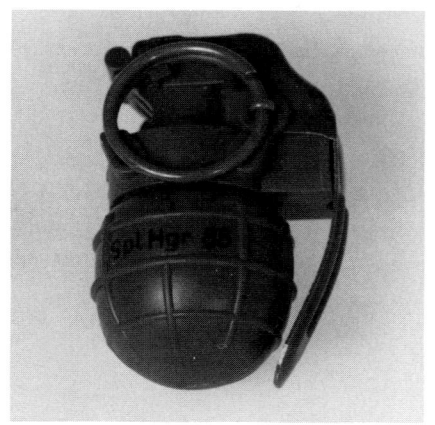

Manufacturer
Arges, Armaturen-GmbH, A-4690 Schwanenstadt-Rüstorf.
Status
Current. Production.

HG86 Mini fragmentation grenade

Arges 40 mm grenades for M79/M203 pattern launchers

These grenades are manufactured by Arges to their own designs but are dimensioned to suit the standard types of launcher such as the American M79 and the several others which have since been developed on the same pattern. However, they differ from compasable designs in that they contain more explosive, individually embedded fragments, and a newly developed propellant system for greater accuracy.

Types available are as follows:-
No 91: fragmentation airburst at user's choice from 1.5 to 5 s.
No 112: Fragmentation with impact and self-destruction after 8 s. (HE/PDSD = High Explosive Point Detonating Self-Destruction).
No 102: Dual purpose AP and fragmentation with impact and self-destruction after 8 s. (HE/DPSD = High Explosive, Dual Purpose, Self-Destruction).
No 92: Incendiary grenades, bursting on impact to give instantaneous incendiary and smoke effect.
No 93: Smoke, instantaneous on impact or self-destruction after 1.5 s.
Training ammunition is available on request.

 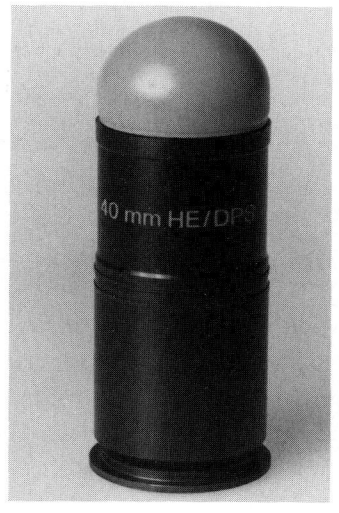

Arges 40 mm Grenade No 91   *Arges 40 mm Grenade No 102*

DATA

40 mm ammunition	No 91 (delay)	No 112	No 102	No 92	No 93
Total length	102 mm	102 mm	112 mm	122 mm	122 mm
Total weight	262 g	230 g	244 g	234 g	260 g
Weight of projectile	210 g	195 g	192 g	182 g	202 g
Weight of explosive charge	40 g RDX	26 g RDX	32 g RDX	85 g RP	—
Number of fragments (without projectile fragments)	approx 900	approx 700	approx 600	—	—
Diameter of fragments	2.0 – 2.3 mm	2.0 – 2.3 mm	2.0 – 2.3 mm	—	—
Perforation of steel plates	—	—	50 mm	—	—
Max range	400 m	400 m	400 m	400 m	100 m

Manufacturer
Arges, Armaturen-GmbH, A-4690 Schwanenstadt-Rüstorf.

Status
Current. Production.

Rifle-launched HG84, 85 and 86 Mini fragmentation grenades (SGG)

These rifle-launched grenades, consist of the standard grenades previously described, fitted with longer delay fuzes in the igniter sets, and attached to a launching unit which consists of a hollow tail with fins and a set of claws to hold the grenade as a 'warhead'.

When inserted into the claws the grenade's firing lever is secured by an inertia clip; on firing, the shock of launch causes this clip to set back and thus release the lever, allowing the normal ignition sequence to be followed. An ingenious, patent-protected design of the holder permits safe removal of the safety pin before firing. Since the time of flight is greater than with hand throwing, the fuze delay is set at a nominal 5.5 seconds.

The tail unit can be provided with a sight; it can also be supplied with a bullet trap and also with a booster cartridge inside the tail unit to give an increase in the maximum range.

The SGG launching assembly can be used with 5.56 mm or 7.62 mm rifles of any type, provided they have the requisite launching rings on the muzzle; the maximum range achieved depends upon the type of weapon and launching cartridge.

SGG 84 and 85 rifle-launched grenades

DATA

	without bullet trap and without booster fired with propellent cartridge			with bullet trap and with booster fired with cal 5.56 mm only	
	SGG 84	SGG 85	SGG 86	SGG 85	SGG 86
Total diameter	78 ± 1 mm	71 ± 1 mm	61 ± 1 mm	71 ± 1 mm	61 ± 1 mm
Total length	260 ± 2 mm	245 ± 2 mm	220 ± 2 mm	335 ± 2 mm	310 ± 2 mm
Weight of complete grenade	640 ± 20 g	495 ± 20 g	295 ± 10 g	695 ± 20 g	495 ± 10 g
Weight of explosive charge	90 ± 5 g	50 ± 3 g	17 ± 1 g	50 ± 3 g	17 ± 1 g
Delay-action fuze	5.5 + 1.5 s − 0.5 s	5.5 + 1.5 s − 0.5 s	5.5 + 1.5 s − 0.5 s	5.5 + 1.5 s − 0.5 s	5.5 + 1.5 s − 0.5 s
Range for calibre 5.56 × 45	135 m	200 m	280 m	220 m	300 m
Range for calibre 7.62 × 51	200 m	250 m	380 m	—	—

Manufacturer
Arges, Armaturen-GmbH, A-4690 Schwanenstadt-Rüstorf.
Status
Current. Production.
Service
Some armies in the Far East and Africa.

SHG 60 DNW hand grenade

DNWs SHG 60 grenade is manufactured from non-corroding components and is impervious to humidity and temperatures in the range from –40 to +70°C. Even contact with sand will not affect the operation. The body is plastic, and the high explosive filling is surrounded by some 4000 steel balls to provide the fragmentation effect. A special 'splinter disc' is interposed between the fuze detonator and the high explosive filling which, by carefully controlling the propagation of detonation, ensures an even distribution of fragments and, on test delivers a minimum of 20 hits per m² on a wall-circle of 5 m diameter.

DATA
Height: 115 mm
Diameter: 60 mm
Weight: 575 g
Fragments: 4000 steel balls 2.5 mm diameter
Fuze delay: 4.5 ± 0.5 s

Manufacturer
Dynamit Nobel Wien GmbH, Postfach 74, Opernring 3-5, A-1015 Wien.
Status
Current. Production.

DNW SHG 60 grenade

Sp1HGr 80/DNW hand grenade

This is a high explosive grenade consisting of a plastic body into which approximately 4000 pre-formed fragments are embedded, and inside this is the plastic explosive bursting charge. A percussion ignition set is installed in the centre of the grenade, controlled by the usual safety pin and fly-off lever. Upon detonation the pre-formed fragments are effective in all directions, with a danger area of 100 m radius. The Sp1HGr 80/DNW can be considered as both an offensive and a defensive grenade, since 90 per cent of the fragments are effective within a radius of 20 m.

DATA
Weight: 585g
Bursting charge: 75g
Number of fragments: 4000 approx
Diameter of fragments: 2.5 mm
Length: 115 mm
Diameter: 60 mm
Delay time: 4.5 s nominal
Temperature range: –40° to +70°C

Manufacturer
Dynamit Nobel Wien GmbH, Postfach 74, Opernring 3-5, A-1015 Wien.
Status
Production.

Sp1HGr 80/DNW hand grenade

Smoke grenade HC-75

The HC-75 consists of a green varnished tin plate can, fitted with a fly-off lever type of igniter and loaded with a metal oxide/hexachloroethane smoke compound. The grenade is thrown in the usual manner, and a cloud of whitish-grey smoke of high obscuring power is generated in about 10 seconds. It is not recommended that this grenade be used in confined spaces, nor where there is a risk of fire.

DATA
Weight: 570 g
Height: 108 mm
Diameter: 73 mm
Delay time: 4.5 s nominal
Duration of smoke: 2 mins.

Manufacturer
Dynamit Nobel Wien GmbH, Postfach 74, Opernring 3-5, A-1015 Wien.
Status
Production.
Service
Austrian Army.

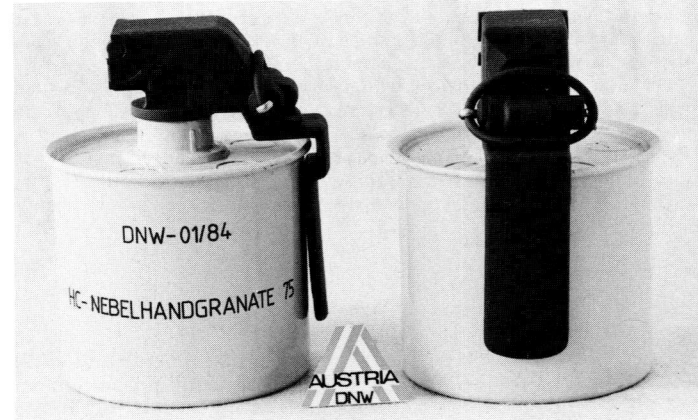

Smoke grenade HC-75

Illuminating hand grenade LHG-40

This can be used either as a close-area illuminating device or as an incendiary grenade in urban areas. The intensity of illumination is such that most optical and electro-optical devices are dazzled. In order to obtain maximum effectiveness in fog, rain or other poor weather conditions, the emitted light is yellow rather than pure white. The grenade consists of a metal canister containing the flare composition, fitted with the standard type of fly-off lever grenade igniter set.

DATA
Weight: 500 g
Height: 145 mm
Diameter: 55 mm
Ignition delay: 4.5 s nominal
Duration of flare: 35 s nominal
Temperature of flare: 1800°C
Luminous power: 280 000 candelas over radius of 150 m

Manufacturer
Dynamit Nobel Wien GmbH, Postfach 74, Opernring 3-5, A-1015 Wien.
Status
Production.
Service
Austrian Army.

Illuminating hand grenade LHG-40

SMI incendiary smoke hand grenade

This is an oval plastic casing containing a charge of stabilised red phosphorus composition which gives both incendiary and smoke-producing effects. The grenade is designed as a close-combat weapon for short-range applications against moving or stationary targets. It is particularly effective against fortified positions, houses, in street fighting, tanks and other armoured vehicles and against concentrations of soft vehicles. Instead of the customary spring-actuated percussion ignition system, this grenade uses a friction igniter. Removing the cap reveals a cord attached to the cap and to the central pin of the igniter; pulling the cord withdraws this pin, which has a roughened surface, through a friction-sensitive substance which is thus ignited. This, in turn, ignites a delay fuze which, after 1.9 seconds, ignites the combustible mass in the body of the grenade. On striking the ground or target, the body of the grenade is shattered, exposing the smouldering composition to the air; this sudden exposure causes the mass to explode with a bright flash, after which it will burn for up to five minutes, giving off dense smoke.

A practice and drill grenade, identical in form, shape and weight but without fuze and charged with an inert lime dust compound for visual target effect, is also available.

DATA
Length: 135 mm
Diameter: 67 mm
Weight: 320 g
Bursting charge: 260 g stabilised red phosphorus
Delay time: 1.9 s nominal
Temperature range: –40° to +50°C
Burning temperature: 1200°C

Manufacturer
Sudsteirische Metallindustrie GmbH, A-8430 Leibnitz.
Status
Production.

Glock SHG Fragmentation hand grenade

The body of this grenade contains approximately 6000 ball-shaped fragments of equal size (diameter approximately 2.3 mm) made of hardened steel and moulded into a polymer body and fragmentation top plate. The body's coating, a durable polymer material, guarantees full protection against corrosion, even if exposed to seawater, due to its firm adhesion even under the most severe conditions. The construction of the inner body results in an equal distribution of the fragments and a lethal area of approximately 10 m diameter from the point of burst.

The firing mechanism is also made of polymer synthetic material (and is therefore corrosion-free) with a safety pin and spring made of stainless steel and a hammer made of brass. A special safety feature is the employment of a safety pin with polymer ring which rests in the polymer structure of the safety lever.

The grenade is delivered without fuze or explosive; Composition B is recommended as a filler.

DATA
Length: 114 mm
Width: 62 mm
Weight, empty: 428 g

Manufacturer
Glock GmbH, Hausfeldstrasse 17, A-2232 Deutsch-Wagram.
Status
Current. Production.

Glock UHG Training hand grenade

This training grenade features a synthetic coated seamless steel tube body (wall thickness 5 mm) to provide complete safety in use. The body coating is proof against any form of corrosion and the rigid polymer material retains its firm adhesion under all climatic conditions. The firing mechanism is also made of polymer synthetic material, with the safety pin and firing spring of stainless steel and the hammer of brass. As with the fragmentation grenade, the safety pin, with polymer pull-ring, rests in a specially-formed portion of the safety lever. The training grenade is delivered fully inert, without detonator or delay fuze.

DATA
Length: 117 mm
Width: 66 mm
Weight, empty: 500 g

Manufacturer
Glock GmbH, Hausfeldstrasse 17, A-2232 Deutsch-Wagram.
Status
Curent, Production.

Glock UHG (left) and SHG hand grenades

BELGIUM

FN TELGREN Telescopic rifle grenade

The FN Telgren telescopic rifle grenade is a totally new concept for which the manufacturers claim four distinct advantages:

1. the grenade can be launched with any type of ammunition, including armour-piercing and tracer bullets. Instead of the conventional bullet trap, the Telgren is made with a clear passage through the axis of the grenade, through which the bullet passes without impact. A specially designed gas trap ensures that the propellant gas is retained and utilised in projecting the grenade

2. having a low mass, it can be comfortably fired from the shoulder in any combat position, thus increasing accuracy and reducing the vulnerable exposure time for the soldier

3. due to the compact method of construction, the rifleman can carry several grenades. Moreover, since the grenades are small, the soldier is more likely to carry them into combat and less likely to discard them as excess weight

4. again due to the compact construction, grenades can be stocked in large numbers in reduced areas and transportation is utilised to better effect.

The Telgren consists of two basic units, the head and the tail. During shipping and transportation the tail is telescoped into the head, giving an overall length of only 19 cm. In this configuration the detonator is out of line with the firing pin and a safety shutter isolates the detonator from the explosive train.

When required for use, the grenade is extended by simply pulling out the tail unit. This gives an overall length of 29 cm, and safety is maintained since the firing pin is in the extended tail while the detonator is in the forebody. The extended tail unit also carries the fragmenting sleeve, and this is now withdrawn from proximity with the central explosive charge.

The grenade is placed on the muzzle of the rifle and launched in the usual way, using whatever ammunition is in the rifle chamber. After leaving the muzzle a spring causes the tail unit to retract once again into the body; while doing so it rotates so that the firing pin and detonator are brought into

FN Telgren in extended and retracted positions

FN Telgren in shipping configuration, compared with conventional rifle grenade

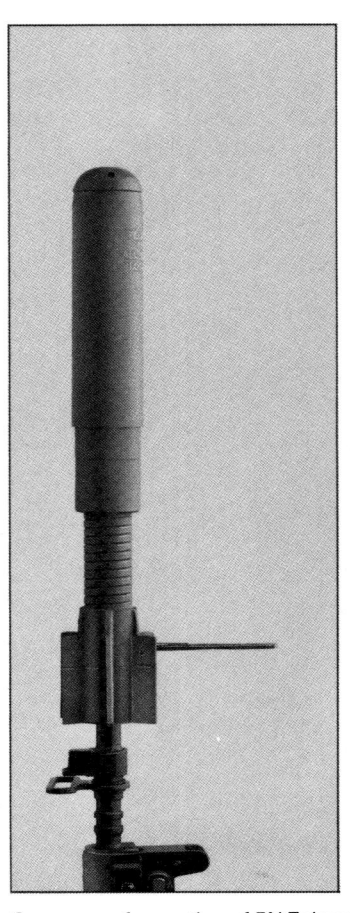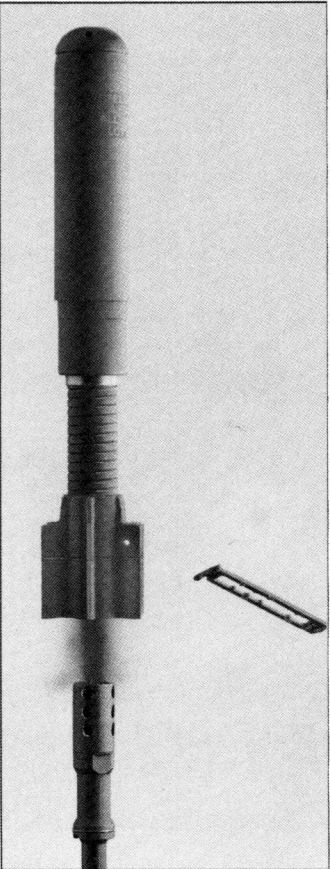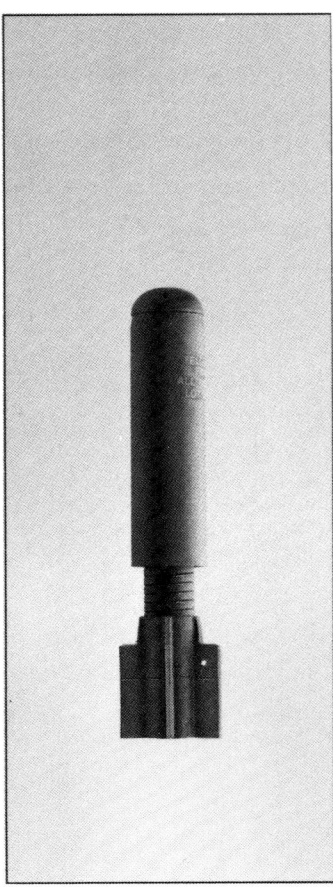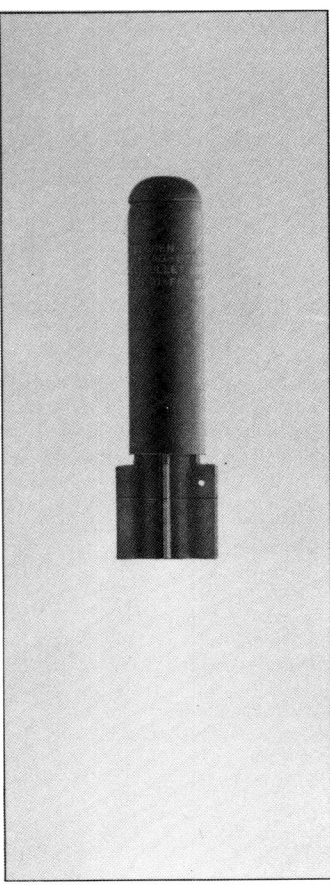

Sequence of operation of FN Telgren when fired

alignment. As the two parts come together, so the fragmentation sleeve takes up position around the central explosive charge. During flight the safety devices in the fuze function and the fuze is armed after about 10 m of flight so as to detonate on impact.

DATA
Calibre: 37 mm
Length retracted: 190 mm
Length extended: 290 mm
Weight: 295 g
Maximum range: 300 m (5.56 mm rifle); 400 m (7.62 mm rifle)
Recoil energy: 45 J (5.56 mm rifle); 60 J (7.62 mm rifle)
Lethal radius: 10 m

Manufacturer
FN Herstal SA, Branche Défense et Sécurité, B-4400 Herstal.
Status
Current. Production.

Firing FN Telgren; note the carrying case on the soldier's belt

MECAR rifle grenade designations

MECAR grenade designations indicate the designed function of the rifle grenades. Abbreviations are:
AR – anti-riot
ARP – anti-armour
BTU – bullet trap universal
CN – tear gas
CS – anti-riot (irritant gas)

DS – disorientation (stun)
LR – long-range
D – delay
PRAC – practice inert BTU grenade PFL – parachute flare
RFL – rifle-launched
SMK – smoke
35, 40 or 55 – warhead diameter in millimetres.

MECAR rifle grenades

The MECAR 35, 40 and 55 mm rifle grenades are the latest series from the factory. These rifle grenades have been designed to permit each soldier to act as a grenadier without special equipment other than his own rifle. They are light in weight and effective in almost every combat situation.

The MECAR rifle grenade system was designed to fill requirements for those ranges between hand grenades and mortars and rocket launchers. While there is some overlap with the latter, because of their versatility, range, lightness, accuracy and lower cost, the MECAR system solves the logistic problems created by heavier weapons without sacrificing efficiency.

The MECAR system includes blast and fragmentation, armour-defeating, smoke, illumination and riot control rifle grenades. The grenades fit over the 22 mm diameter muzzle of any modern infantry rifle and do not require an attachment. Each rifle grenade in the MECAR system is provided with the MECAR patented Bullet Trap Universal (BTU) to permit grenades to be launched using standard issue ball rounds.

Grenades fitted with the BTU in their tail can be fired from any standard military rifle of 5.56 or 7.62 mm calibre with all regular ball ammunition including M43 and SS 109 using the aiming grid provided. The BTU has been specifically designed to accept soft steel-cored as well as lead-cored bullets.

Adapters are available for rifles with a muzzle diameter other than 22 mm. Muzzle velocity and recoil energy vary between the rifles used to fire the grenades.

The HE blast and fragmentation rifle grenade has a triple-safety, super sensitive fuze and can be used in any combat situation where a quick response to eliminate opposition is required. The grenade is effective from 15 to 350 m in both direct and indirect fire against light structures, material targets and personnel. The armour-defeating type is effective against soft and armoured vehicles, small bunkers and structures. Safe arming distances can be varied according to the requirements of the buyer.

The illuminating, WP smoke/incendiary and non-WP smoke grenades enable the soldier to deal with a range of specific battlefield situations.

Practice and practice inert grenades, ballistically matched to the service grenades, are available for use in training. When used with MECAR-supplied ballistite cartridges, the practice inert grenades may be recovered and re-used a number of times.

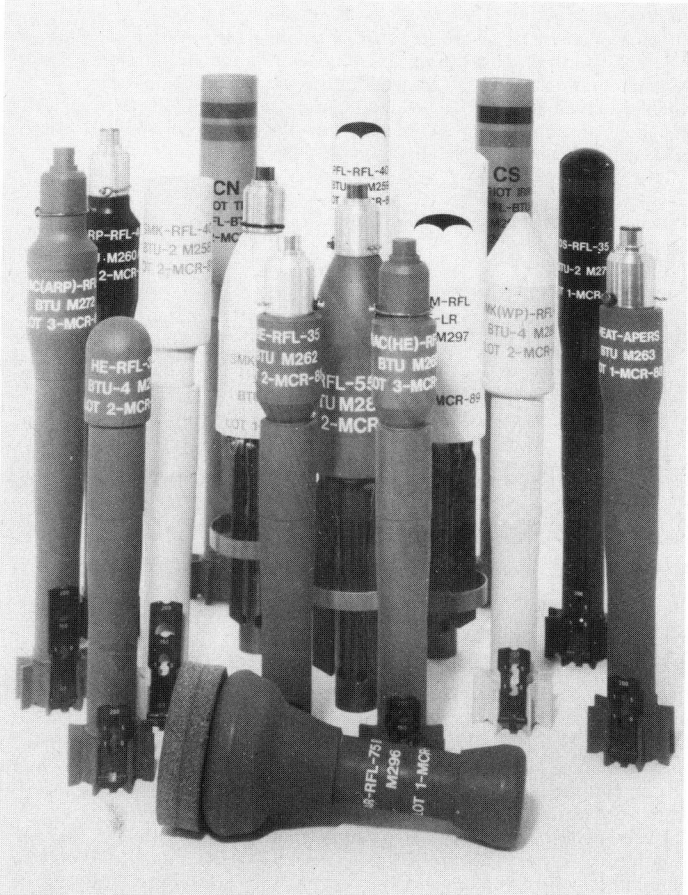

A selection of MECAR rifle grenades

MECAR Anti-Armour BTU rifle grenade: ARP-RFL-40 BTU M260 series

The ARP-RFL-40 BTU M260 series rifle grenade is used for the defeat of soft and armoured targets by means of a shaped charge HEAT warhead.

The grenade may be launched from any standard military 5.56 or 7.62 mm assault rifle having a 22 mm diameter muzzle, using standard issue ball ammunition, or may be adapted for use with other rifle models. The grenade is launched with the rifle at the shoulder or any other conventional position.

The patented MECAR Bullet Trap Universal (BTU) has been specifically designed to accept soft steel core bullets as well as lead core bullets. Muzzle velocities and recoil energy values vary between rifles.

DATA
Type: 40 mm HE shaped charge, anti-armour
Length: 330 mm
Weight: 390 g
Fuze: PIBD F55
Sight: weapon related aiming grid with each grenade

Arming distance: 15 m
Max range: 300–400 m
Operational range: 25–200 m
Probable error at 100 m: H + W 1.25 × 1.25 m
Impact function: 70° obliquity
Penetration: (RHA HB220): > 160 mm; concrete: > 300 mm
Recoil energy: eg FNC 6 kg/m; FAL 9 kg/m; M16 9 kg/m
Operating temperature range: –32 to +52°C

Manufacturer
Mecar SA, B-6522 Petit-Roeulx-lez-Nivelles.
Status
Current. Production.
Service
More than 35 countries including NATO forces.

MECAR anti-armour BTU rifle grenade, ARP-RFL-40 BTU M260 series

---

MECAR blast and fragmentation BTU rifle grenade, HE-RFL-35 BTU M262 Series

The shoulder-launched MECAR HE-RFL-35 BTU M262 series blast and fragmentation rifle grenades provide riflemen with the ability to engage and defeat, in both direct and indirect fire roles, light structures, material targets and personnel.

The grenade may be launched from any standard military 5.56 mm or 7.62 mm assault rifle having a 22 mm diameter muzzle, using standard issue ball ammunition, or may be adapted for use with other rifle models.

The patented MECAR Bullet Trap Universal (BTU) has been specifically designed to accept soft steel core bullets as well as lead core bullets. Muzzle velocities and recoil energy values vary between rifles.

DATA
Type: 35 mm, HE, blast and fragmentation
Length: 288 mm
Weight: 400 g
Fuze: PD F85

Sight: weapon related aiming grid with each grenade
Arming distance: 15 m
Max range: 300–400 m
Operational range: 25 – 200 m
Impact function: 70° of obliquity
Lethal radius 10 m
Effective radius: 18 m
Number of fragments: > 300
Operating temperature range: –32 to +52°C

Manufacturer
Mecar SA, B-6522 Petit-Roeulx-lez-Nivelles.
Status
Current. Production.
Service
More than 35 countries including NATO forces.

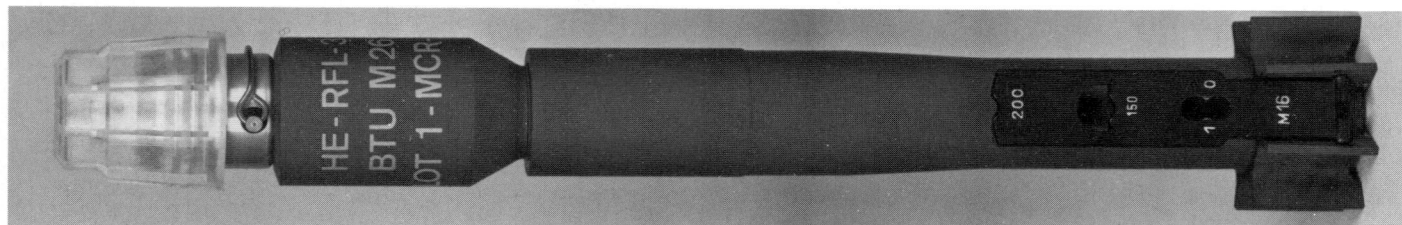

MECAR blast & fragmentation rifle grenade HE-RFL-35 BTU M262 series

---

MECAR dual purpose BTU rifle grenade HEAT/APERS-RFL-35 BTU M263 series

The HEAT/APERS-RFL-35 BTU M263 series dual purpose anti-armour/blast and fragmentation rifle grenade enables riflemen to simultaneously defeat armoured and material targets and personnel by means of an HE shaped charge and a pre-notched fragmenting body.

The grenade may be launched from any standard military 5.56 or 7.62 mm assault rifle having a 22 mm diameter muzzle, using standard issue ball ammunition, or may be adapted for use with other rifle models. The grenade is launched with the rifle at the shoulder or any other conventional position.

The patented MECAR Bullet Trap Universal (BTU) has been specifically designed to accept soft steel core bullets as well as lead core bullets. Muzzle velocities and recoil energy values vary between rifles.

DATA
Type: 35 mm HE shaped charge, anti-armour
Length: 287 mm
Weight: 438 g
Fuze: PIBD F100
Sight: weapon related aiming grid with each grenade
Arming distance: 15 m
Max range: 300–400 m
Operational range: 25–200 m
Impact function: 68° of obliquity
Probable error at 100 m: H + W 0.20 × 0.10 m

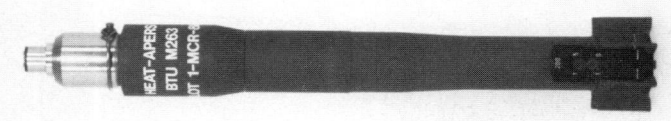

MECAR dual purpose rifle grenade HEAT/APERS-RFL-35 BTU M263

Penetration: (RHA HB220): >65 mm; concrete: >160 mm
Number of fragments: >300
Lethal radius: 8 m
Effective radius: 15 m
Operating temperature range: –32 to +52°C

Manufacturer
Mecar SA, B-6522 Petit-Roeulx-lez-Nivelles.
Status
Current. Production.
Service
More than 35 countries including NATO forces.

MECAR blast and fragmentation BTU rifle grenade with delay, HE-RFL-35 BTU M261 (Delay) series

The MECAR HE-RFL-35 BTU M261 (Delay) series grenades differ from other rifle grenades in that they have a pyrotechnic delay of four seconds, which allows them to be launched through light structures such as walls, doors, windows and sandbag emplacements, and to detonate within the enclosure. If fired into the air at an angle of 45° the grenades detonate at a height of 80 m and in areas of dense vegetation will penetrate brush and scrub before functioning.

The grenade may be launched from any standard military 5.56 or 7.62 mm assault rifle having a 22 mm diameter muzzle, using standard issue ball ammunition, or may be adapted for use with other rifle models.

The patented MECAR Bullet Trap Universal (BTU) has been specifically designed to accept soft steel core bullets as well as lead core bullets. Muzzle velocities and recoil energy values vary between rifles.

DATA
Type: 35 mm HE blast & fragmentation
Length: 243 mm
Weight: 320 g

Fuze: pyrotechnic with 4 second delay
Sight: weapon related aiming grid with each grenade
Max range: to 300 m
Operational range: 200 m
Lethal radius: 10 m
Effective radius: 18 m
Number of fragments: >300
Operating temperature range: –32 to +52°C

Manufacturer
Mecar SA, B-6522 Petit-Roeulx-lez-Nivelles.
Status
Current. Production.
Service
More than 35 countries including NATO forces.

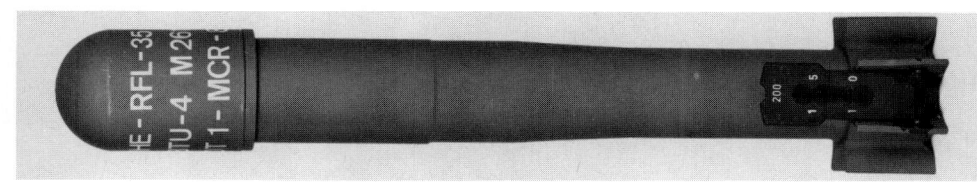

MECAR blast and fragmentation rifle grenade HE-RFL-35 BTU M261 (delay)

MECAR parachute flare rifle grenade PFL-RFL-40 BTU M259 series

The MECAR PFL-RFL-40 BTU M259 series rifle grenade is used when a specific area of operations is required to be illuminated. The flare is ejected from the grenade four seconds after launch and gives an intense yellow illumination for some 30 seconds during its parachute descent. When fired at an angle of 80° the grenade reaches a height of at least 100 m to light up an area of 200 m radius.

The grenade may be launched from any standard military 5.56 or 7.62 mm assault rifle having a 22 mm diameter muzzle, using standard issue ball ammunition, or may be adapted for use with other rifle models. The grenade is launched with the rifle at the shoulder or any other conventional position.

The patented MECAR Bullet Trap Universal (BTU) has been specifically designed to accept soft steel core bullets as well as lead core bullets. Muzzle velocities and recoil energy values vary between rifles.

DATA
Type: 40 mm parachute illuminating
Length: 359 mm
Weight: 420 g
Fuze: pyrotechnic with 4 s delay
Operational range: 175 m
Intensity of illumination: 60 000 candela
Duration of illumination: 30 s
Operating temperature range: –32 to +52°C

Manufacturer
Mecar SA, B-6522 Petit-Roeulx-lez-Nivelles.
Status
Current. Production.
Service
More than 35 countries including NATO forces.

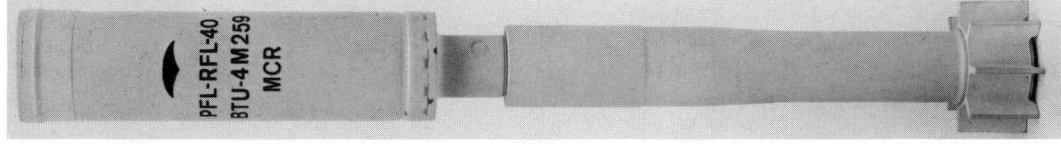

MECAR parachute flare rifle grenade PFL-RFL-40 BTU M259

MECAR smoke emission rifle grenade SMK-RFL-40 BTU M258 series

The MECAR SMK-RFL-40 BTU M259 series rifle grenade is used for screening and or signalling purposes. The grenade produces a dense opaque and persistent smoke commencing two seconds after the grenade leaves the rifle muzzle and continuing for a period of at least 80 seconds.

The grenade may be launched from any standard military 5.56 or 7.62 mm assault rifle having a 22 mm diameter muzzle, using standard issue ball ammunition, or may be adapted for use with other rifle models. The grenade is launched with the rifle at the shoulder or any other conventional position.

The patented MECAR Bullet Trap Universal (BTU) has been specifically designed to accept soft steel core bullets as well as lead core bullets. Muzzle velocities and recoil energy values vary between rifles.

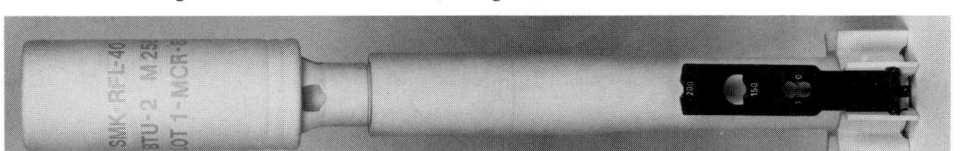

MECAR smoke emission rifle grenade SMK-RFL-40 BTU M258

DATA
Type: 40 mm smoke emission (white, red, yellow, blue, green etc.)
Length: 308 mm
Weight: 480 g
Fuze: pyrotechnic with 2 s delay
Sight: weapon related aiming grid with each grenade
Max range: 250 m
Operational range: 175 m
Operating temperature range: −32 to +52°C

Manufacturer
Mecar SA, B-6522 Petit-Roeulx-lez-Nivelles.
Status
Current. Production.
Service
More than 35 countries including NATO forces.

MECAR smoke and incendiary rifle grenade with delay, SMK(WP)-RFL-40 BTU M288 (Delay) series

The MECAR SMK(WP)-RFL-40 BTU M288 (Delay) rifle grenades are used when smoke and or incendiary effects are required. With their four second delay pyrotechnic fuzes they are particularly suited to use in urban situations where they may be used to penetrate light structures such as walls, doors, windows and emplacements to function inside the enclosure, producing a fireball of 6 m diameter with a combustion temperature of some 1000°C.

The grenade may be launched from any standard military 5.56 or 7.62 mm assault rifle having a 22 mm diameter muzzle, using standard issue ball ammunition, or may be adapted for use with other rifle models.

The patented MECAR Bullet Trap Universal (BTU) has been specifically designed to accept soft steel core bullets as well as lead core bullets. Muzzle velocities and recoil energy values vary between rifles.

DATA
Type: 40 mm smoke & incendiary
Length: 289 mm
Weight: 423 g
Loading: White phosphorus
Fuze: pyrotechnic with 4 s delay
Sight: weapon related aiming grid with each grenade
Max range: to 300 m
Operational range: 200 m
Effective radius: > 3 m
Operating temperature range: −32 to +52°C

Manufacturer
Mecar SA, B-6522 Petit-Roeulx-lez-Nivelles.
Status
Current. Production.
Service
More than 35 countries including NATO forces.

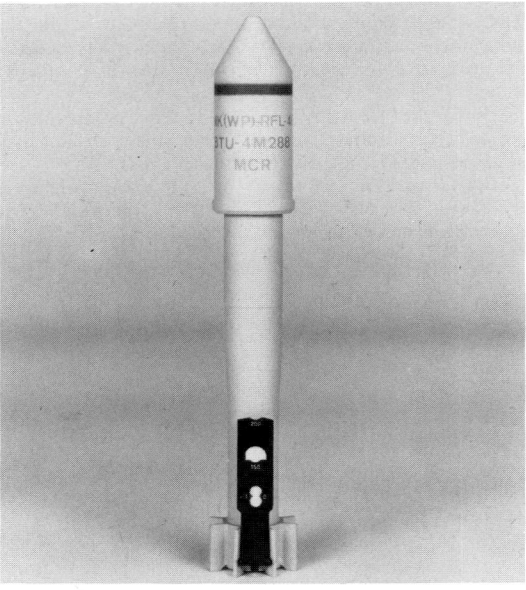

MECAR smoke and incendiary rifle grenade SMK(WP)-RFL-40 BTU M288 (Delay)

The MECAR 55 mm Long-Range BTU rifle grenade family

The MECAR family of long-range, rocket assisted rifle grenades are an effective replacement for the squad light mortar and are able to successfully engage targets out to 600 or 700 m range.

The grenade may be launched from any standard military 5.56 or 7.62 mm assault rifle having a 22 mm diameter muzzle, using standard issue ball ammunition, or may be adapted for use with other rifle models. The grenade is launched with the rifle in the under-arm firing position for direct or indirect fire, or with the heel of the butt placed firmly on the ground for indirect fire.

The patented MECAR Bullet Trap Universal (BTU) incorporated in the tail tubes of these grenades has been specifically designed to accept soft steel

core bullets as well as lead core bullets. Muzzle velocities, recoil energy values and maximum achievable ranges are obviously dependent upon the model of rifle used to launch the grenades.

Aiming, in both direct and indirect fire roles, is performed using a quadrant aiming device attached to the exposed rifle barrel.

A practice round, designated PRAC-RFL-55 LR BTU M292 series, is available and is ballistically matched to the service HE M287, SMK(WP) M282 and SMK(TTC) M283 rounds, differing from the service rounds in that the service fillings are replaced by a compound which provides a sight and sound impact signature.

MECAR blast and fragmentation long-range rifle grenade HE-RFL-55 LR BTU M287 series

The MECAR HE-RFL-55 LR BTU M287 rifle grenade is used effectively as a squad light mortar enabling riflemen to engage and defeat targets through the effects of blast and fragmentation at extended ranges out to 600 – 700 m. These ranges are attained by the use of a rocket motor which is ignited immediately after grenade launch.

Grenades may be launched from any standard 5.56 mm or 7.62 mm assault rifle having a 22 mm diameter muzzle, using standard issue ball ammunition. The patented MECAR Bullet Trap Universal (BTU) has been designed to accept soft steel core bullets as well as lead core bullets.

Aiming, in both direct and indirect fire modes, is achieved using a quadrant aiming device attached to the exposed rifle barrel.

DATA
Type: 55 mm, HE, blast & fragmentation
Weight: 790 g
Length: 304 mm
Max range: 600 – 700 m
Operational range: 25 – 700 m
Fuze: PD F80
Arming distance: 25 m
Impact dispersion: 30 × 30 m at 650 m
Impact function: 70° of obliquity
Lethal radius: 12 m
Effective radius: 20 m
Number of fragments: > 300
Operating temperature range: −32 to +52°C

Manufacturer
MECAR SA, B-6522 Petit-Roeulx-lez-Nivelles.
Status
Current. Production.
Service
More than 35 countries worldwide, including NATO forces.

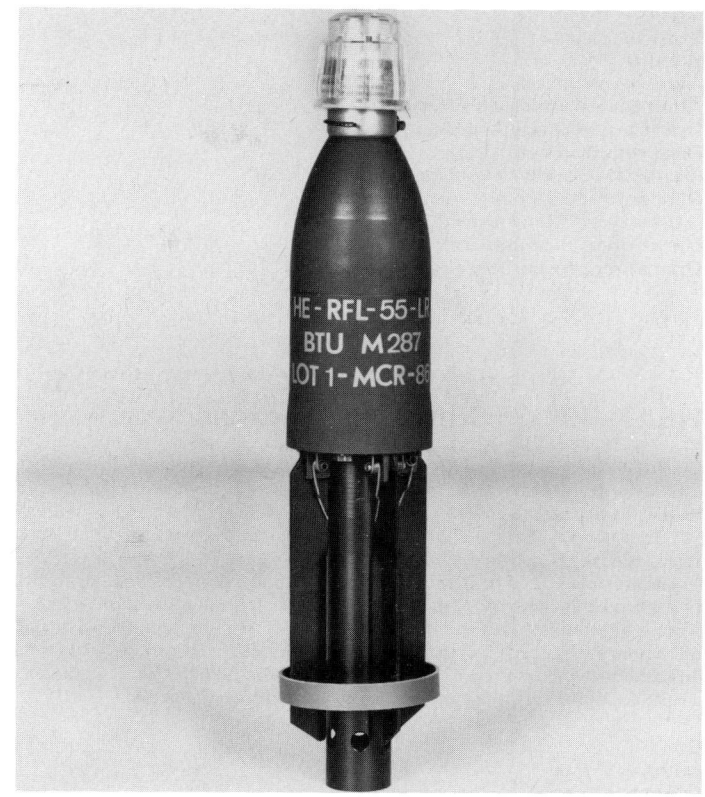

MECAR blast and fragmentation long-range rifle grenade HE-RFL-55 LR
BTU M287 series

MECAR smoke and incendiary long-range rifle grenade SMK(WP)-RFL-55 BTU M280 series
MECAR smoke bursting rifle grenade SMK(TTC)-RFL-55 BTU M281 series

The MECAR SMK(WP)-RFL-55 BTU M280 rifle grenades are used to provide signalling or screening smoke, or to use the incendiary effects of the white phosphorus filling aganist specified targets. Riflemen in the squad armed with this model of grenade are able to engage targets out to ranges of 600 – 700 m.

A second model of grenade, SMK(TTC)-RFL-55 BTU M281 series, having the same ballistics as the SMK(WP) M280, is also available. This has a non-phosphorus smoke composition Titanium Tetrachloride (FM) and is used to provide signalling or screening smoke.

Both models of grenade may be launched from any standard 5.56 mm or 7.62 mm assault rifle having a 22 mm diameter muzzle, using standard issue ball ammunition. The patented MECAR Bullet Trap Universal (BTU) has been designed to accept soft steel core bullets as well as lead core bullets.

Aiming, in both direct and indirect fire modes, is achieved using a quadrant aiming device attached to the exposed rifle barrel.

DATA
Type: 55 mm (WP), screening or signalling WP smoke and incendiary 55 mm (TTC), screening or signalling smoke
Weight: 790 g
Length: 304 mm
Fuze: PD F80
Aiming: weapon related quadrant aiming device
Max range: 600 – 700 m
Operational range: 25 – 700 m
Fuze: PD F80
Arming distance: 25 m
Dispersion at 650 m: 30 × 30 m
Impact function: 70° of obliquity
Effective radius: 20 m
Operating temperature range: –32 to +52°C

Manufacturer
MECAR SA, B-6522 Petit-Roeulx-lez-Nivelles.
Status
Current. Production.

MECAR smoke and incendiary long-range rifle grenade SMK(WP)-RFL-55 BTU M280

MECAR Parachute illuminating Long-Range rifle grenade ILLUM-RFL-55 BTU M297

The MECAR ILLUM-RFL-55 BTU series long-range rifle grenades enable riflemen to illuminate specific areas of operations, to ranges in excess of 500 m. The parachute flare delivers a yellow light with an intensity of illumination of 200 000 candelas for approximately 30 seconds. Grenades are launched from 5.56 mm or 7.62 mm assault rifles having a 22 mm diameter muzzle, using standard issue ball ammunition. The patented MECAR Bullet Trap Universal (BTU) accommodates both soft steel and lead core bullets.

Aiming is achieved by using a quadrant aiming device attached to the exposed rifle barrel.

DATA
Length: 321 mm
Weight: 790 g
Fuze: pyrotechnic delay
Aiming: weapon-related aiming quadrant
Operational range: to 550 m
Flare ejection height: 200 m
Illuminated area: 300 m diameter
Descent rate: 2.5 m/s
Intensity of illumination: 200 000 candela
Duration of illumination: 30 s
Operating temperature range: –32°C to +52°C

Manufacturer
MECAR SA, B-6522 Petit-Roeulx-lez-Nivelles.
Status
Current. Production.

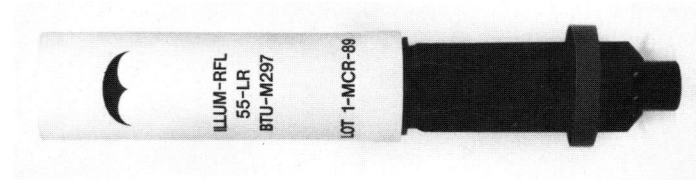

MECAR parachute illuminating rifle grenade ILLUM-RFL-55 BTU M297

HEAT-RFL-75 Super Energa grenade

This is a rocket-boosted counterpart of the familiar Energa rifle grenade. It can be fired with accuracy at targets up to 200 m away and the shaped charge warhead will penetrate over 275 mm of armour plate at impact angles of up to 70° from normal. Concrete, brick and sandbag emplacements up to 600 mm thick are easily penetrated.

The grenade is fired from the rifle held in the under-arm position, by means of a special launch cartridge. Inside the tail unit is a small rocket motor which is ignited at the instant of launch and which burns for approximately 40 ms, so increasing the grenade's velocity and maximum range. It can be fired from any rifle provided with a 22 mm diameter muzzle launcher; special launch cartridges of the requisite calibre are provided with each grenade.

DATA
Calibre: 75 mm
Length: 425 mm
Weight: 765 g
Fuze: impact, superquick
Range: (moving targets) 150 m; (stationary targets) 200 m
Max range: (45°) 550 m
Penetration: (steel) 275 mm (Brinell hardness 220); (concrete) 600 mm
Launch velocity: ca 75 m/s
Arming distance: 6 m
Recoil energy: ca 30 kg

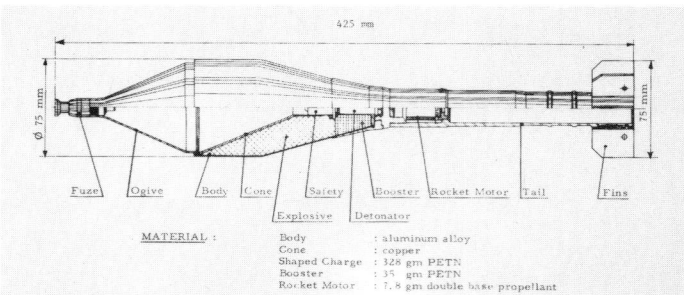

HEAT-RFL-75 Super Energa Mecar grenade

Manufacturer
MECAR SA, 6522 Petit-Roeulx-lez-Nivelles.
Status
Production.
Service
Several armies.

PRB NR 423 hand grenade

The hand grenade PRB NR 423 produces a controlled fragmentation due to the number and shape of the splinters and to the weight and type of the explosive used.

The very high efficiency at close-range and the very short safety radius are the two main characteristics of the PRB NR 423 grenade.

The dispersion of the effective splinters is obtained in a homogeneous manner and is uniform in the space around the point of impact whatever the position of the grenade may be at the moment of explosion.

The safety radius is such that a man at 20 m from the point of burst will be unharmed.

This results in an important logistic advantage as the same grenade can be used offensively and defensively.

Grenade body
The egg-shaped grenade is composed of five elements: an outer plastic case of olive green colour with transversal and longitudinal ribs ensuring a good grip; a fragmentation sleeve lining the inside of the case; this sleeve is made of a spirally wound steel wire of rectangular section. This wire is pre-notched (according to a PRB patent) at regular intervals, inside the coil formed, so as to fragment into about 900 splinters during the explosion; 22 steel balls placed in the bottom closing plug and 30 in the upper part of the case to reduce the dead angles when the grenade explodes in horizontal position; a very powerful explosive filler and a plastic closing plug of olive green colour.

Fuze
The PRB NR 423 time fuze has a four second gasless delay. This delay is sufficient to allow the soldier to throw the grenade without danger, but is too short for an enemy to throw it back.

DATA
Weight of grenade body: 180 g
Weight of time fuze: 50 g
Weight of complete grenade: 230 g
Explosive filler weight: Comp B, 60 g
Fragmentation sleeve weight: 65 g
Number of fragments: 900 (average weight of one splinter) 0.105 g
Number of steel balls: 52
Average weight of each steel ball: 0.1 g
Lethal radius: 9 m
Safety radius: 20 m

Manufacturer
Société Anonyme PRB, Defence Department, Avenue de Tervueren 168, Bte 7, B-1150 Brussels.
Status
Current. Available.

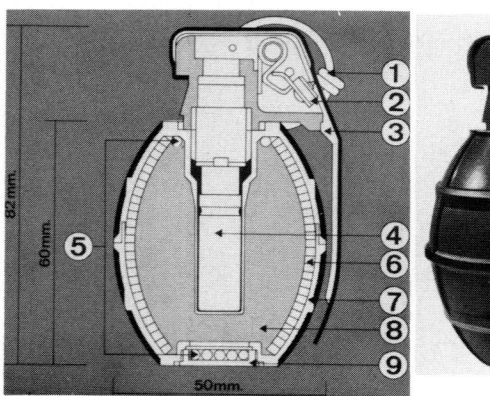

PRB NR 423 grenade
(1) safety clip (2) striker (3) lever (4) detonator (5) balls (6) fragmentation sleeve (7) plastic shell (8) explosive (9) plug

PRB NR 423 practice hand grenade

The practice hand grenade is intended for training soldiers in the use of the live hand grenade PRB NR 423.

It has the same weight (230 g) and shape and the same outside aspect. However it looks different from the live grenades due to its colour, marking, materials and elements used.

The effect produced is noise, simulating an explosion, and a small quantity of smoke.

The functioning does not produce any projection of metallic elements, consequently its use does not require any special safety provisions.

This grenade has two separate components, the grenade body and the time fuze.

Grenade body
The body is of aluminium alloy. This part has a through channel in which is screwed the fuze with its deflagrator and which permits the evacuation of the gases produced when functioning.

This body can be re-used (minimum 100 times) without other maintenance than simple cleaning with a rag.

Time fuze
The practice time fuze is identical in all respects to the PRB NR 432 (four seconds) gasless delay time fuze used with the PRB NR 423 live grenade; its use and functioning are the same. The only difference is that the detonator has been replaced by a deflagrator which produces noise and smoke. No dangerous projections are produced by its functioning. Like the PRB NR 432 time fuze, the practice time fuze cannot be re-used as it is destroyed on functioning.

Manufacturer
Sociéte Anonyme PRB, Defence Department, Avenue de Tervueren 168, Bte 7, B-1150 Brussels.

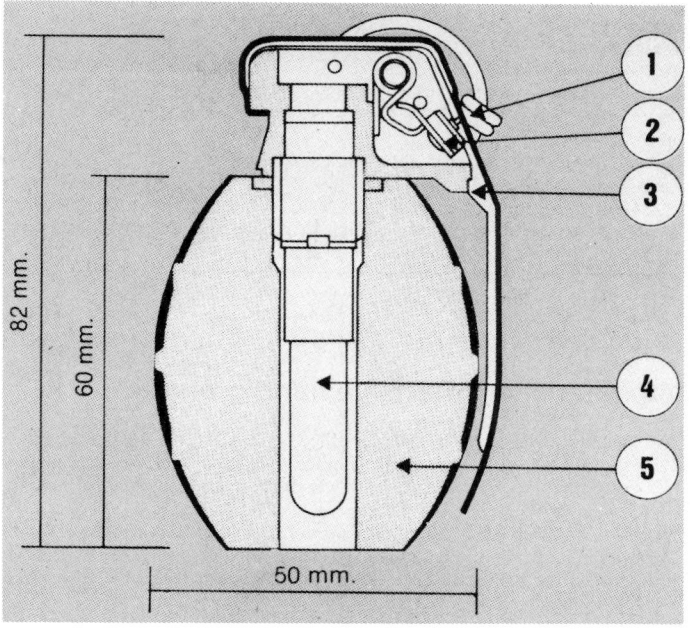

PRB NR 423 practice hand grenade
(1) *safety clip* **(2)** *striker* **(3)** *lever* **(4)** *deflagrator* **(5)** *grenade body*

PRB NR 446 offensive hand grenade

The purpose of the offensive hand grenade PRB NR 446 is to produce a powerful blast effect without fragments.

It has been developed with elements of the controlled fragmentation hand grenade PRB NR 423. Its outside shape is identical, the only difference being the colour and marking as well as a lighter weight; 165 g instead of 230 g.

All the components that can produce dangerous fragments, particularly the fragmentation sleeve and the balls, have been removed.

The same explosive is used, Composition B, but in larger quantity, 85 g instead of 60 g, in order to increase the power of the shock wave (same power as that of 110 g of cast TNT). This grenade uses the PRB NR 432 time fuze of four seconds gasless delay which is also used with the PRB NR 423 controlled fragmentation hand grenade.

DATA
Weight of grenade body: 115 g
Weight of time fuze: 50 g
Weight of explosive: Comp B, 85 g
Total weight of grenade: 250 g

Manufacturer
Société Anonyme PRB, Defence Department, Avenue de Tervueren 168, Bte 7, B-1150 Brussels.
Status
Current. Available.

PRB NR 446 offensive hand grenade
(1) *safety clip* **(2)** *striker* **(3)** *lever* **(4)** *detonator* **(5)** *plastic shell* **(6)** *explosive* **(7)** *plug*

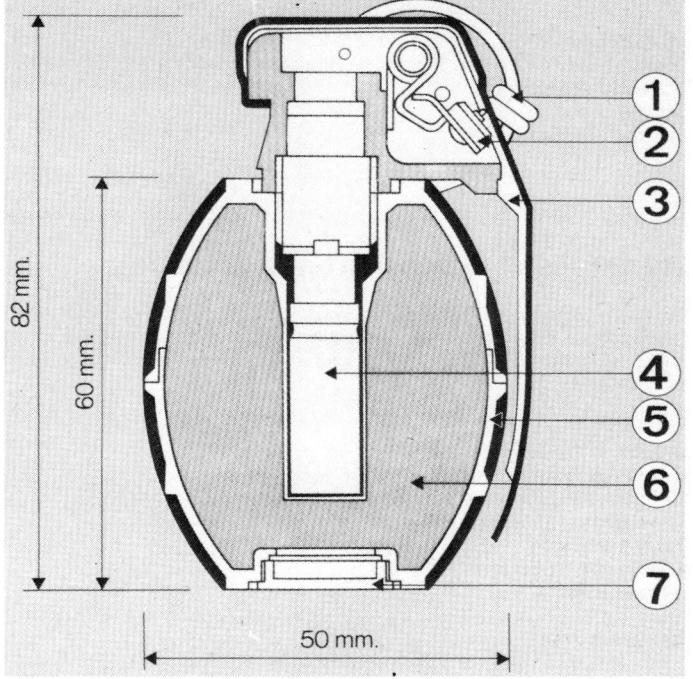

PRB smoke grenades

Two smoke grenades are currently made by PRB. One, type PRB 405, is a 'bursting' type of white phosphorus grenade creating a dense cloud of white smoke and having an associated incendiary effect and an anti-personnel effect within a 15 m radius of the explosives.

The charge is contained in a waterproof can and is detonated by a 4.5 second time fuze initiated by a conventional spring-loaded plunger secured by a lever and split pin with pull ring.

Type HC is a hexachlorethane grenade containing some 200 g of composition and emitting a cloud of dense, white, clinging smoke. Non-explosive, it can be fitted either with a ring and plunger mechanism with a 4.5 second delay to ignition or with a friction igniter with a four second delay.

DATA	PRB 405	HC
Diameter	50 mm	63 mm
Height	103 mm	125 mm
Weight		
Complete	330 g	320 g
Time fuze assembly	50 g	50 g
Smoke emission	c 45 s	c 90 s
Danger zone	15 m radius	—
Packing		
Grenades	54/case	50/case
Fuzes	108/case	—
Average packed weight	500 g	520 g

Manufacturer
Société Anonyme PRB, Defence Department, Avenue de Tervueren 168, Bte 7, B-1150 Brussels.
Status
Current. Available.

Smoke grenades PRB 405 (left) and HC

BRAZIL

M3 defensive/offensive hand grenade

This grenade is basically a cylinder filled with HE and detonated by a pyrotechnic delay fuze which is itself ignited by a percussion cap and striker lever. A fragmenting sleeve can be slid over the body of the grenade.

As a plain explosive charge the grenade can be used in the offensive role relying on its blast effect to demoralise the enemy. It can also be a small demolition charge for immediate action in the field. With the fragmentation sleeve in place the M3 becomes an effective defensive grenade which produces over 240 fragments.

DATA
Length: 96 mm
Diameter: 40 mm; (with sleeve) 47 mm
Weight: 415 g; (fragmentation sleeve) 200 g; (explosive charge) Comp B, 90 g
Delay: 4.5 s
Number of fragments: 240+
Fragment velocity: 2000 m/s
Lethal radius: 8 m

Manufacturer
Companhia de Explosivos Valparaiba, Praia do Flamengo, 200, 20° Andar, 22210 Rio de Janeiro, RJ.
Status
Current. Production.

M3 grenade with fragmenting sleeve in place

M4 defensive/offensive hand grenade

This is a conventional time-fuzed hand grenade using a plastic body shell lined with notched wire which provides controlled fragmentation. Being lighter in weight than older types of cast-iron grenade, it can be thrown further and with better accuracy.

DATA
Weight of fuzed grenade: 241 g
Weight of fragmentation wire: 75 g
Weight of explosive: Comp B, 75 g
Dimensions: 87 × 50.5 mm
Delay time: 4.5 s
Number of fragments: 800+
Lethal radius: 9.5 m

Manufacturer
Companhia de Explosivos Valparaiba, Praia do Flamengo, 200, 20° Andar, 22210 Rio de Janeiro, RJ.
Status
Current. Production.
Service
Brazilian army.

M4 hand grenade

M2-CEV rifle grenade

The M2 grenade is an anti-personnel HE type, fired from the muzzle of a 7.62 mm rifle, using a ballistite launching cartridge. It produces fragments on exploding, but also has a useful explosive effect which will penetrate concrete and light armour making it a general-purpose munition. Each grenade is packed in a waterproof container with a launching cartridge and an attachable sight for the rifle.

When fitted with Fuze SIG the nomenclature of this grenade changes to M4-CEV.

DATA
Length: 323 mm
Diameter, max: 40 mm
Weight: (total) 550 g; (explosive) 85 g Pentolite or Comp B
Max range: (42°) 380 m
Muzzle velocity: 70 m/s
Fragments: 450+
Penetration of armour: 50 mm

Manufacturer
Companhia de Explosivos Valparaiba, Praia do Flamengo, 200, 20° Andar, 22210 Rio de Janeiro, RJ.
Status
Current. Production.

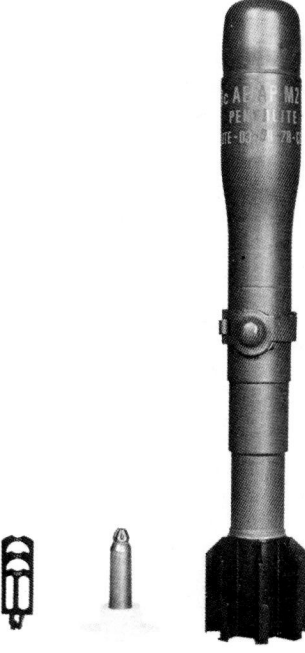

M2-CEV rifle grenade

M3-CEV anti-tank rifle grenade

The M3 is a larger and heavier version of the M2, containing a larger HE charge specifically for use against armoured vehicles. It is a muzzle-launched grenade in the same way as the M2, but due to its greater weight the velocity and range are less. This, however, is not important since it can be expected that engagement ranges will be short. As with the M2, the grenade is intended for use with the FAL rifle which is in use with the Brazilian forces.

When fitted with Fuze SIG the nomenclature of this grenade changes to M5-CEV.

DATA
Length: 410 mm
Diameter: 65 mm
Weight: (total) 770 g; (explosive) 265 g (Comp B or Pentolite)
Max range: (42°) 260 m
Muzzle velocity: 60 m/s
Penetration of armour: 76 mm

Manufacturer
Companhia de Explosivos Valparaiba, Praia do Flamengo, 200, 20° Andar, 22210 Rio de Janeiro, RJ.
Status
Current. Production.

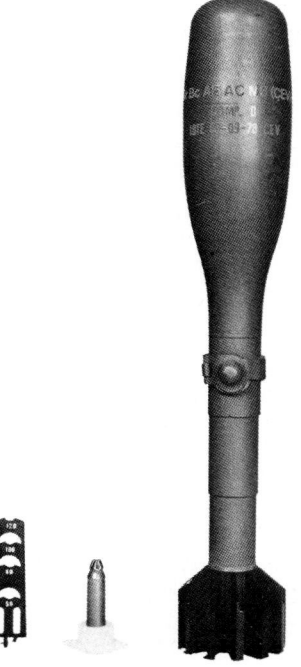

M3-CEV anti-armour grenade

MB-315 (GR M QT/1) defensive/offensive hand grenade

The MB-315 (GR M QT/1) is a typical dual-purpose hand grenade consisting of two major elements: HE hand grenade body filled with a main charge of Pentolite 50/50, and a fragmentation sleeve, the use of which depends on the operational requirements of the user. This grenade is supplied in 50-unit 29 × 29 × 26 cm wooden boxes weighing 27 kg. A separate wooden box, measuring 34 × 44 × 26 cm and weighing 16 kg, contains 100 delay fuzes.

DATA
Length: 97 mm
Diameter: 39 mm; (with sleeve) 47 mm
Weight: (total) 480 g; (fragmentation sleeve) 240 g; (explosive charge) 90 g
Delay: 4.5 s
Number of fragments: 300+
Lethal radius: 9 m

Manufacturer
Condor SA Industria Quimica, Av Rio Branco 26, 16° Andar, CEP 20090 Rio de Janeiro, RJ.
Status
Current. Production.

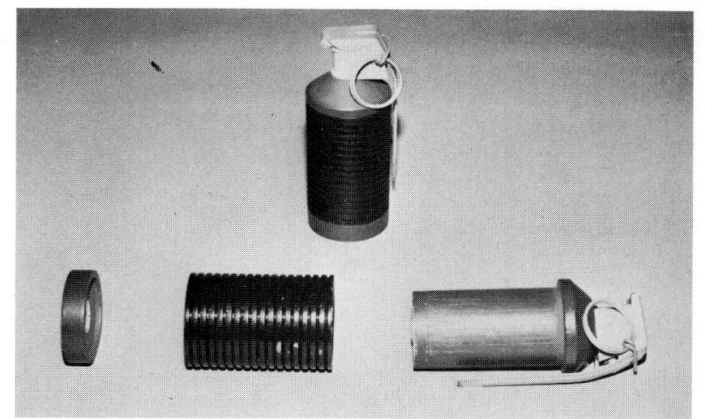

MB-315 defensive/offensive hand grenade

MB-309/T1 smoke grenade

The MB-309/T1 produces a dense cloud of grey smoke, which may be used for screening or signalling purposes. It is packed in 50-unit wooden boxes measuring 42.5 × 30 × 32.7 cm and weighing 4.8 kg.

DATA
Weight, total: 550 g
Weight of filling: 350 g
Diameter: 67 mm
Length: 175 mm
Delay: 4–5 s
Smoke emission: 120 s

Manufacturer
Condor SA Industria Quimica, Av Rio Branco 26, 16° Andar, CEP 20090 Rio de Janeiro, RJ.
Status
Current. Production.

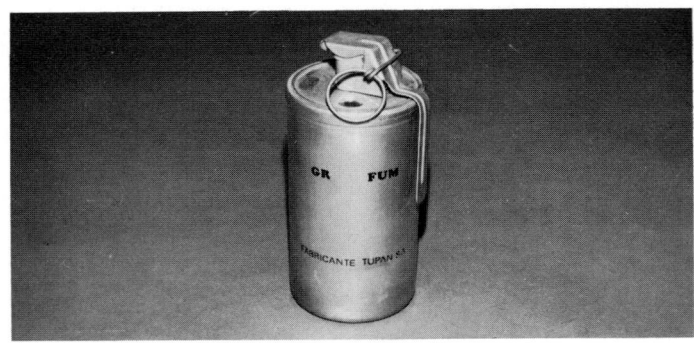

MB-309/T1 smoke grenade

MB-306/T1 vehicular smoke grenade

This is designed to be electrically fired from vehicle smoke dischargers. Loaded with a zinc/hexachloroethane mixture, it emits a dense smoke cloud in order to screen the vehicle. Models are available with ranges of 60 or 100 m.

DATA
Length: 182 mm
Diameter: 79.5 mm
Weight: 1.30 kg

Delay time: 1 s nominal
Emission time: 2.5 to 4.5 mins

Manufacturer
Condor SA – Industria Quimica, Av Rio Branco 26, 16° Andar, CEP 20090, Rio de Janiero.
Status
Current. Production.

MB-502 smoke hand grenade

This grenade has an aluminium cylindrical body with a cap perforated to allow the emission of smoke. It is fitted with a fly-off lever type of igniter and produces a dense grey smoke.

DATA
Length: 146 mm
Diameter: 67 mm
Weight: 510 g

Delay time: 2 – 4 s
Emission time: 100 to 135 s

Manufacturer
Condor SA – Industria Quimica, Av Rio Branco 26, 16° Andar, CEP 20090, Rio de Janiero.
Status
Current. Production.

CHILE

Offensive/defensive hand grenade

This grenade is a cylindrical sheet steel casing containing HE, detonated by a short pyrotechnic delay which is ignited by the usual striker and fly-off safety lever. A steel fragmenting case can be slid over the body for the defensive role, while for offensive use the grenade is thrown without it, relying on blast and shock for its effect.

A useful feature of this grenade is that the lower surface of the main cylinder is a hollow cone. If the grenade is stood upright on a flat surface it acts as a

Offensive/defensive grenade. Sectioned grenade on right shows the hollow cone in the lower surface

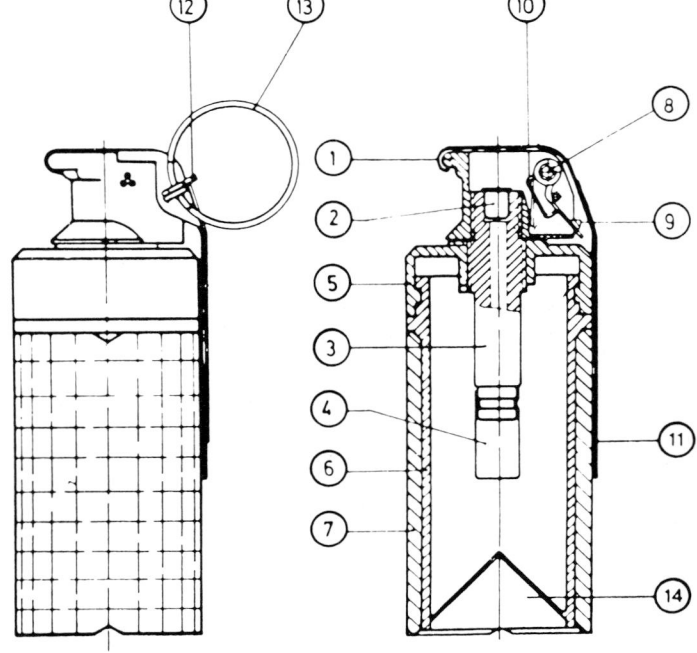

Cardoen offensive/defensive grenades
(1) *fuze head* **(2)** *cap* **(3)** *delay column* **(4)** *detonator* **(5)** *plastic head* **(6)** *interior body* **(7)** *removeable fragmentation case* **(8)** *spring axis* **(9)** *striker* **(10)** *spring* **(11)** *safety lever* **(12)** *safety pin* **(13)** *pull ring* **(14)** *shaped charge cone*

small hollow-charge demolition munition. The manufacturer supplies a special igniter assembly for use when the grenade is used as a booby trap or an anti-personnel mine. This igniter assembly operates on a push-pull principle and is intended to be connected to a trip wire.

DATA
Length: 121.5 mm
Diameter: 46.5 mm
Weight: 210 g (offensive); 500 g (defensive)
Weight of HE filling: 100 g TNT
Fuze delay: 5 s

Manufacturer
Industrias Cardoen SA, Division Defensa, Av Providencia 2237, 6° Piso, Santiago.
Status
Production.
Service
Chilean and other forces.

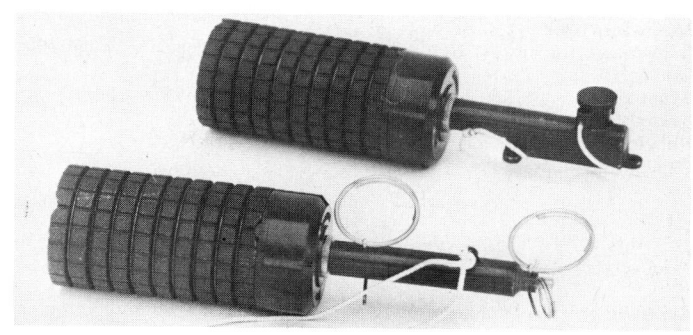

Two offensive/defensive grenades set up with pressure and pull igniters for use as booby traps

Type MK-2 hand grenade

This grenade bears a very close resemblance to the US M2, which is no longer used in the US forces and is described in *Jane's Infantry Weapons 1978*, page 444. It is a pear-shaped cast iron body with prominent serrations. The igniter mechanism is the same as that of the offensive/defensive grenade described in the previous entry. This grenade can also be fitted with the Cardoen special igniter for use as a mine or booby trap.

DATA
Weight: 662 g
Weight of HE filling: 115 g
Fuze delay: 5 s

Manufacturer
Industrias Cardoen SA, Division Defensa, Av Providencia 2237, 6° Piso, Santiago.
Status
Current. Production.
Service
Chilean and other forces.

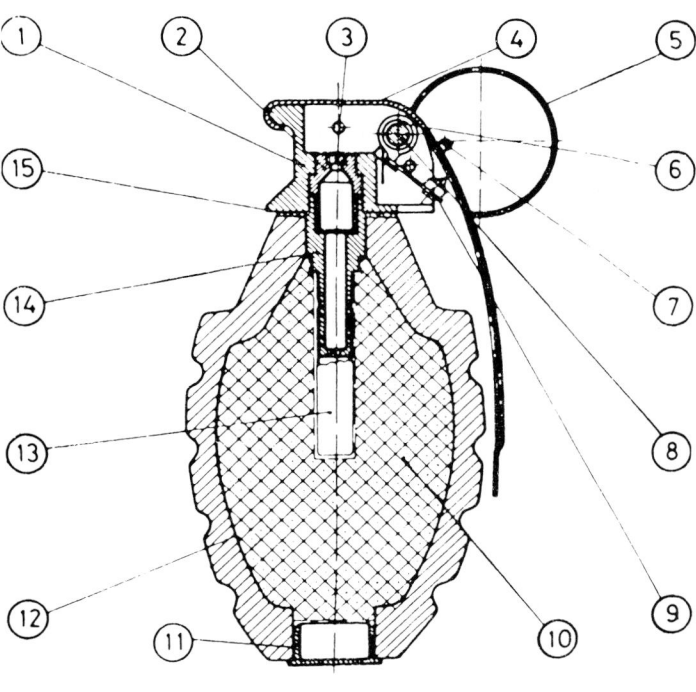

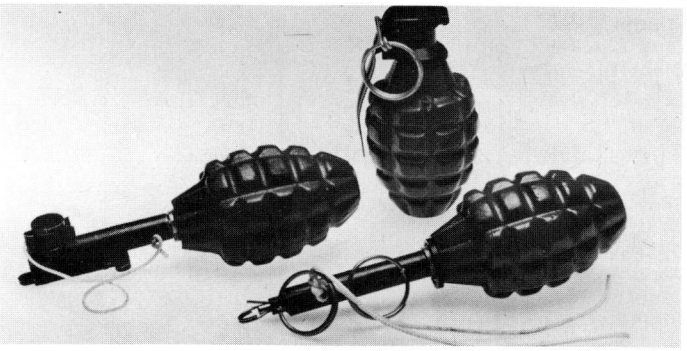

Type MK-2 grenades ready for hand throwing and with pressure and pull igniters for use as booby traps

Cardoen Mk-2 grenade
(1) *fuze head* **(2)** *fulcrum* **(3)** *cap* **(4)** *safety lever* **(5)** *pull ring* **(6)** *striker axis* **(7)** *safety pin* **(8)** *spring* **(9)** *striker* **(10)** *explosive filling* **(11)** *base plug* **(12)** *casing* **(13)** *detonator* **(14)** *delay column* **(15)** *sealing ring*

Mini hand grenade

This is a small grenade designed by Cardoen with a view to obtaining the maximum throw range consistent with good fragmentation effect. The cast iron body is loaded with TNT and provides all-round lethal coverage. The safety ring on the time fuze must be twisted and then pulled to remove; it cannot be withdrawn accidentally. The fuze unit is silent and smokeless in operation; it can be removed and pull or pressure switches inserted so as to convert the grenade into a booby trap or anti-personnel mine.

DATA
Dimensions: 52.8 × 79.3 mm
Weight: 332 g
Explosive filling: 77 g TNT
Delay: 4 s

Mini hand grenade
(1) *fuze head* **(2)** *interior fuze head* **(3)** *primer cap* **(4)** *lever* **(5)** *safety pin pull ring* **(6)** *striker spring* **(7)** *safety device* **(8)** *spring bolt* **(9)** *striker* **(10)** *high explosive* **(11)** *base plug* **(12)** *body* **(13)** *detonator* **(14)** *delay column* **(15)** *rubber*

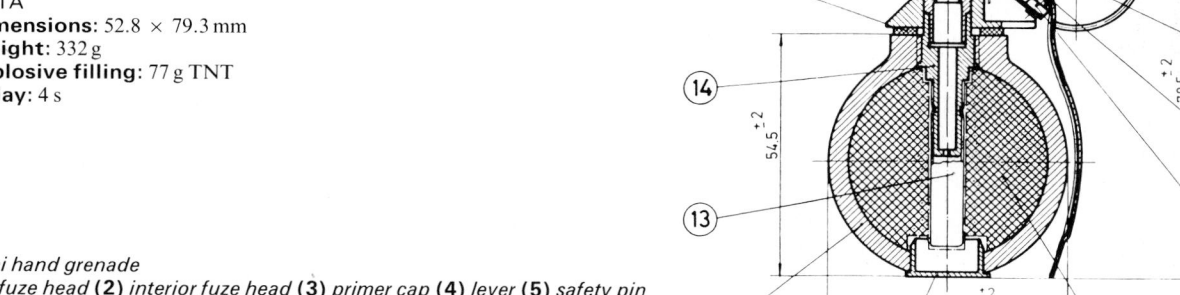

Manufacturer
Industrias Cardoen SA, Division Defensa, Av Providencia 2237, 6° Piso, Santiago.
Status
Current. Production.
Service
Chilean and other armies.

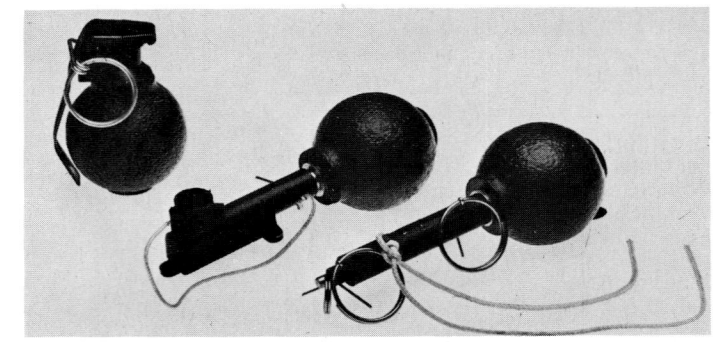

Mini hand grenades ready for hand throwing and with pressure and pull igniters for use as booby traps

CHINA, PEOPLE'S REPUBLIC

Chinese grenades

The armed forces of the People's Republic of China use a variety of grenades of both Chinese and Soviet manufacture. The Chinese grenades are similar in design to the Soviet models, except that in most cases they have different dimensions and, as the same design is often made by a number of factories, there are minor differences between copies of the same type of grenade. Listed below are some copies or near-copies of Chinese manufacture whose designations are known, together with a note on Chinese stick grenades. Another Soviet grenade known to have been copied in China is the RPG-6 (which see).

Type 42 (offensive/defensive) hand grenade

This is a direct copy of the Soviet RG-42 grenade and its method of operation is identical.

DATA
Weight: 385 g
Length: 127 mm
Diameter: 58 mm

Weight of filling: 110 g
Type of filling: pressed TNT
Fuze delay: 3–4 s
Effective fragmentation radius: 15 m
Unit pack: 20 grenades and 20 fuzes
Weight of pack: 14.97 kg

Type 1 (defensive) hand grenade

This is a copy of the Soviet F1 grenade.

DATA
Weight: 581 g
Length: 125 mm
Diameter: 57 mm

Weight of filling: 55 g
Type of filling: cast TNT
Fuze delay: 3–4 s
Effective fragmentation radius: 15 m
Unit pack: 20 grenades and 20 fuzes
Weight of pack: 19.5 kg

Type 59 (defensive) hand grenade

This is similar in design to the Soviet RGD-5 grenade and operates in the same manner.

DATA
Weight: 308 g
Length: 115 mm
Diameter: 54 mm

Weight of filling: 110 g
Type of filling: TNT
Fuze delay: 3–4 s
Effective fragmentation radius: 20 m
Unit pack: 20 grenades and 20 fuzes
Weight of pack: 19.5 kg

Type 73 pre-fragmented mini-grenade

This is a smaller version of the Type 59, using an internally pre-notched body to which is attached a prominent percussion ignition set similar in operation to that used with the Type 59 and Soviet RGD-5.

DATA
Diameter: 42 mm
Weight: 190 g
Lethal radius: 8.5 m

Manufacturer
China North Industries Corp, PO Box 2137, Beijing.
Status
Current. Production.

Stick grenades

The Chinese have manufactured a wide variety of stick grenades for defensive operations. Scored, serrated and plain types have been encountered. Their contents have included picric acid, mixtures of TNT or nitroglycerin with potassium nitrate or sawdust and schneiderite. Method of operation is that the cord of the pull-friction fuze, which is underneath the cap at the end of the throwing handle, is pulled. This ignites the delay element which lasts between 2.5 and 5 seconds, after which the detonator explodes the main charge. These grenades are generally packed in boxes of 20 already fuzed.

A typical example of a defensive stick grenade is illustrated here and is known to be still in use. It is a fragmenting type with a serrated head made of grey cast iron. This produces a small number of large fragments and a very large number of fragments so small that they could well be described as "dust". The filling is picric acid which was discarded as an explosive filling in the West many years ago, principally because it forms dangerous and unstable compounds. In the UK this was the HE Lyddite which was used in the First World War. It is essential that the inside of the container is varnished.

Chinese smooth fragmentation grenade with cap removed to show pull cord

DATA
Weight: 500 g
Length: 228 mm
Diameter: 50 mm
Weight of filling: 99 g
Type of filling: picric acid
Fuze delay: 2.5–5 s
Effective fragmentation radius: 10 m

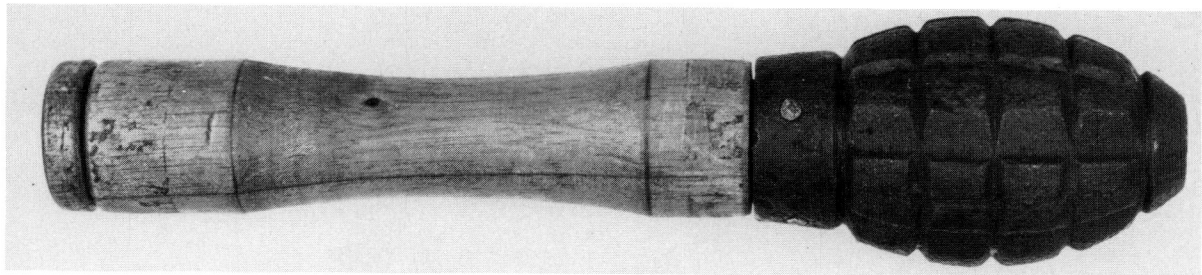

Chinese, serrated, fragmenting, defensive stick grenade

Type 77-1 stick grenade

This is commercially available through Norinco, as well as being on issue to the Chinese Army. Like the stick grenades described above, the 77-1 uses a friction pull igniter, revealed by removing a screwed cap from the handle. The head is of cast metal, smooth and ovoid.

DATA
Length: 170 – 173 mm
Weight: 360 – 380 g

Diameter: 48.5 mm
Delay time: 2.8 – 4 s
Lethal radius: 7 m

Manufacturer
China North Industries Corp, PO Box 2137, Beijing.
Status
Current. Production.

Type 82 hand grenades

These grenades are of recent introduction and local design. They are basically of the offensive pattern though the makers claim they can be used in either role. The body is internally serrated to give fragmentation control, and ignition is by the usual percussion 'mousetrap' igniter system. The **Type 82-1** has a smooth, slightly irregular, oval shape and is fitted with a friction pull igniter beneath a removable cap. The **Type 82-2** (illustrated) is built up from two halves and has a prominent joint around the centre. There is a close-fitting protective cap which covers the working parts and retains the safety pin and ring whilst being shipped or carried; after removing this cap the pin is pulled and the grenade thrown in the usual way. Light and compact, it can be thrown for a considerable distance, and several can be carried. The **Type 83-3** uses the same body as the **Type 83-1** but has a percussion igniter of somewhat different type to that used on the **82-2**.

DATA
Type 82-2:-
Length: 85 mm
Diameter: 48 mm
Weight: 260 g
Explosive charge: 62 g TNT
Effective fragments: in excess of 280
Lethal radius: 6 m
Safety radius: 30 m
Delay time: 2.8 – 3.8 s

Manufacturer
China North Industries Corp, PO Box 2137, Beijing.
Status
Current. Production for export.

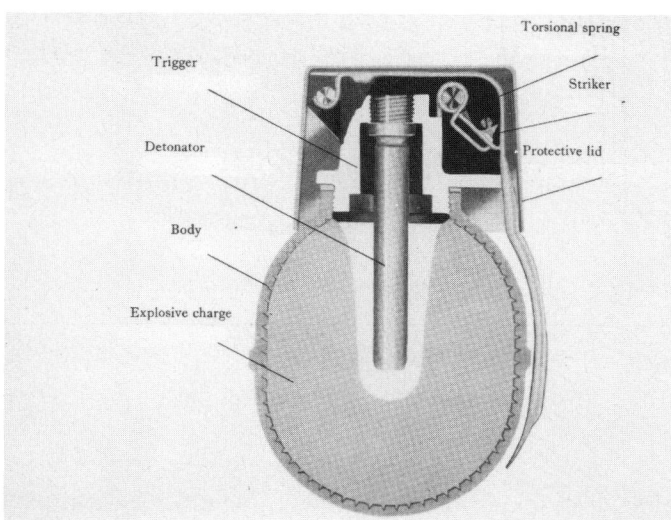

Chinese Type 82-2 hand grenade

CZECHOSLOVAKIA

RG34 and RG4 anti-personnel hand grenades

The Soviet hand grenades in current service are also used in Czechoslovakia. In addition two native grenades have been produced. These are both impact anti-personnel grenades. They are described as offensive hand grenades but the sheet metal bodies also produce some fragmentation. If required, each of these grenades can be fitted with an external fragmentation sleeve which improves the lethality.

RG34
The RG34 is a steel bodied cylindrical grenade, distinguished by the fluting around the mid-section.

DATA
Type: blast
Weight: (without fragmentation sleeve) 340 g
Length: 76 mm
Max diameter: 64 mm
Body material: steel
Filler weight: 100 g
Filler material: TNT
Fuze type: impact. All ways
Range thrown: 35 m
Effective fragment radius: 13 m; (with fragmentation sleeve) 25 m

Status
Obsolete, but still in stock.
Service
Czechoslovak forces.

RG4
The RG4 has replaced the RG34 in the Czechoslovak Army. It has a cylindrical steel body which is completely smooth. It can be converted to a defensive grenade by adding a fragmentation jacket. This grenade is unusual in containing an upper and lower bursting charge.

DATA
Type: blast
Weight: (without fragmentation sleeve) 320 g
Length: 84 mm
Max diameter: 53 mm
Body material: steel
Filler weight: 105 g
Filler material: TNT
Fuze type: impact
Fragmentation radius: 13 m; (with fragmentation sleeve) 25 m

Status
Current.
Service
Czechoslovak forces and some guerrilla forces in Africa.

Czechoslovak RG4 hand grenade

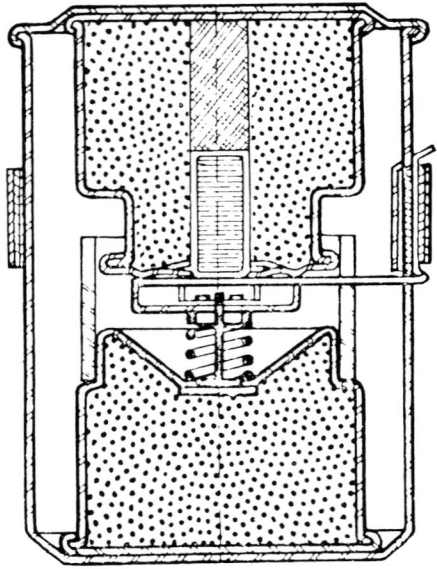

Czechoslovak Model RG34 hand grenade

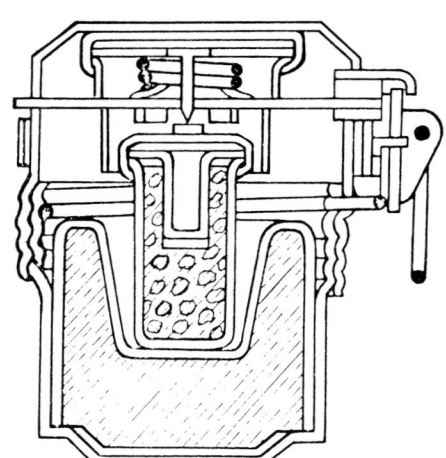

Czechoslovak RG4 hand grenade

Czechoslovak Model RG34 hand grenade

EGYPT

Hossam anti-tank grenade

This has been designed and manufactured in Egypt and uses a shaped charge to effect penetration. The grenade is stabilised in flight by a drag parachute which ensures that the impact fuze strikes the target and the shaped charge is correctly oriented.

DATA
Diameter: 63 mm
Length: 192 mm
Weight: 575 g
Filling: 149 g Hexogen 90%
Arming distance: 5 m
Penetration: 120 mm (armour); 400 mm (concrete)

Manufacturer
Sakr Factory for Developed Industries, PO Box 33, Heliopolis, Cairo.
Status
Current. Production.
Service
Egyptian army.

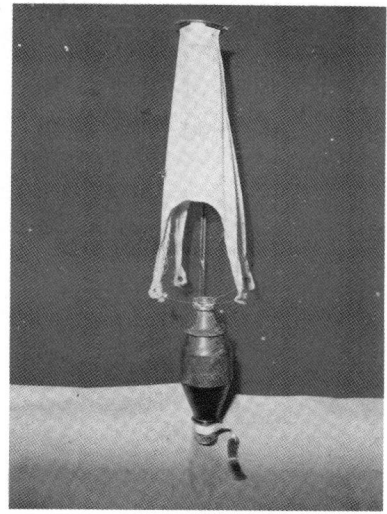

Hossam anti-tank grenade with parachute deployed

FINLAND

Wildcat plastic hand grenade

This grenade is of frangible plastic material lined with about 9600 2 mm diameter steel pellets. Inside this, and surrounding the central tube, is a bursting charge of 55 g of Hexotol high explosive. The igniter set is of the usual spring-powered type, released by a fly-off handle, which fires a detonator by releasing it to fly down and strike a fixed pin at the bottom of the central tube. This ignites the delay, which burns for about 3.5 seconds before detonating the filling. Whilst the standard delay is 3.5 seconds, other timings may be provided as required and are colour-coded for ready identification.

Wildcat has been tested and proofed against strikes by rifle bullets, dropping and is not sensitive to sympathetic detonation. The grenade is watertight and will function under water.

Manufacturer
Tampereen Asepaja Oy, Hatanpäänvaltatie 32, SF-33100 Tampere 10.
Status
Production.

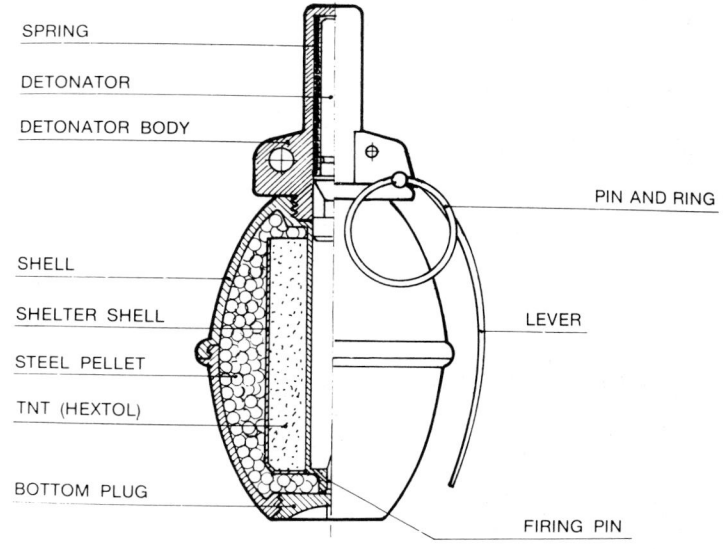

Wildcat hand grenade

FRANCE

58 mm Luchaire Defense anti-tank rifle grenade

The Luchaire 58 mm anti-tank rifle grenade is notable for a high level of performance and the ability to be shoulder-fired using any 5.56 mm rifle equipped with a 22 mm muzzle sleeve.

The grenade is fitted with a bullet trap which allows safe firing with any normal 5.56mm ball cartridge. Extremely effective, this grenade considerably increases the firepower of any soldier carrying a rifle.

A training version of the grenade is also available.

DATA
System: HEAT
Weight: 550 g
Length: 400 mm
Effective range: 75 m with SD of 0.12 m
Muzzle safety: > 5 m
Positively armed: 15 m
Recoil: < 35Ns
Penetration: Steel armour BH > 241, 320 mm; concrete 800 mm

Manufacturer
Luchaire Défense, 180 Boulevard Haussmann, 75382 Paris Cedex 08.
Status
Current. Production.
Service
French Army.

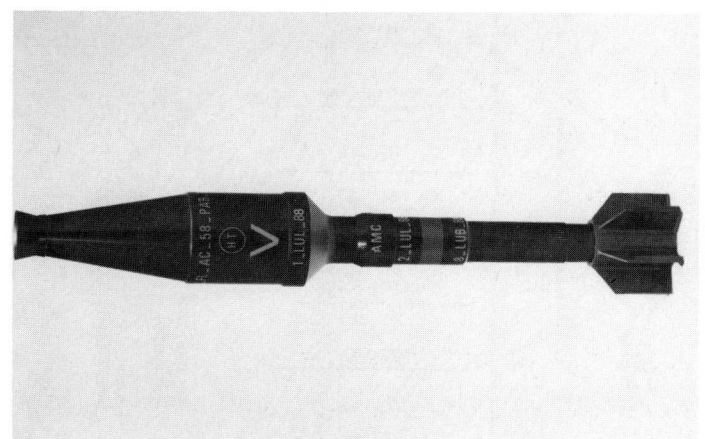

Luchaire Défense 58 mm anti-tank rifle grenade

40 mm Luchaire bullet trap rifle grenades

These rifle grenades combine the accuracy of very short-range flat firing and the efficiency of curved fire beyond 400 m range, thus providing an individual weapon system designed for anti-armoured vehicle and anti-personnel applications.

The 40 mm calibre is the result of a technical compromise between ergonomics, carrying power and optimal performance.

Owing to its bullet trap the Luchaire 40 mm rifle grenade may be safely and directly launched using any combat cartridge. The infantryman's rifle remains ready for instant use before, during and after launching a grenade.

Being without seal or safety pin the grenade is permanently ready for use. Versatility is assured, since the range comprises six combat warheads and two training rounds, all using the same fuze and tailtube assembly which is compatible with most modern 5.56 mm, 7.5 mm or 7.62 mm rifles.

Combat grenades
-anti-tank (HEAT)
-anti-personnel/anti-vehicle (HEAT-APERS-FRAG)
-anti-personnel (HE-APERS-FRAG)
-anti-personnel delay (HE-APERS-FRAG-Delay)
-smoke (SMOKE)
-illuminating (ILL)

Training/practice grenades
-training practice (TP)
-one-piece training practice with target marker (TPM)

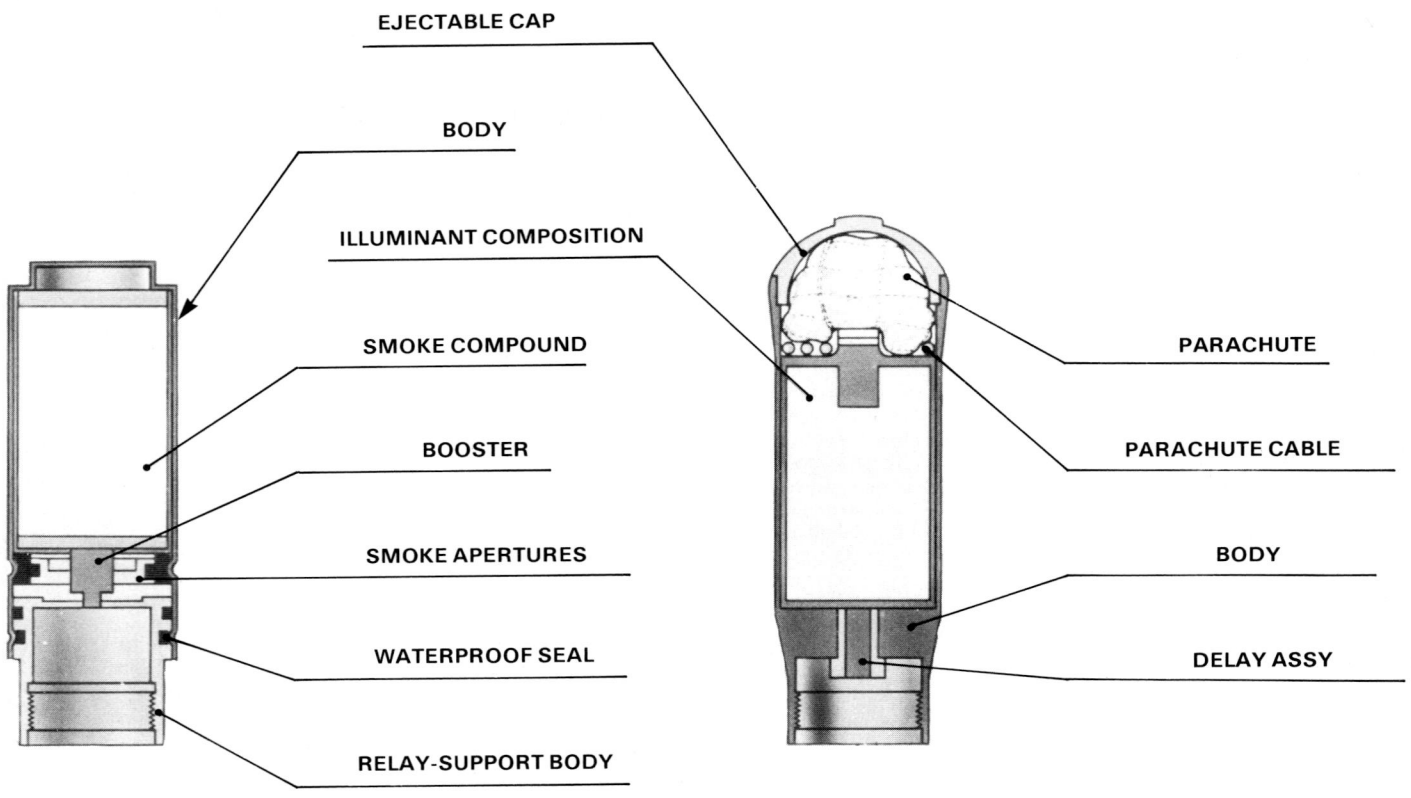

Warheads for the Luchaire AC, AP/AV and AP rifle grenades

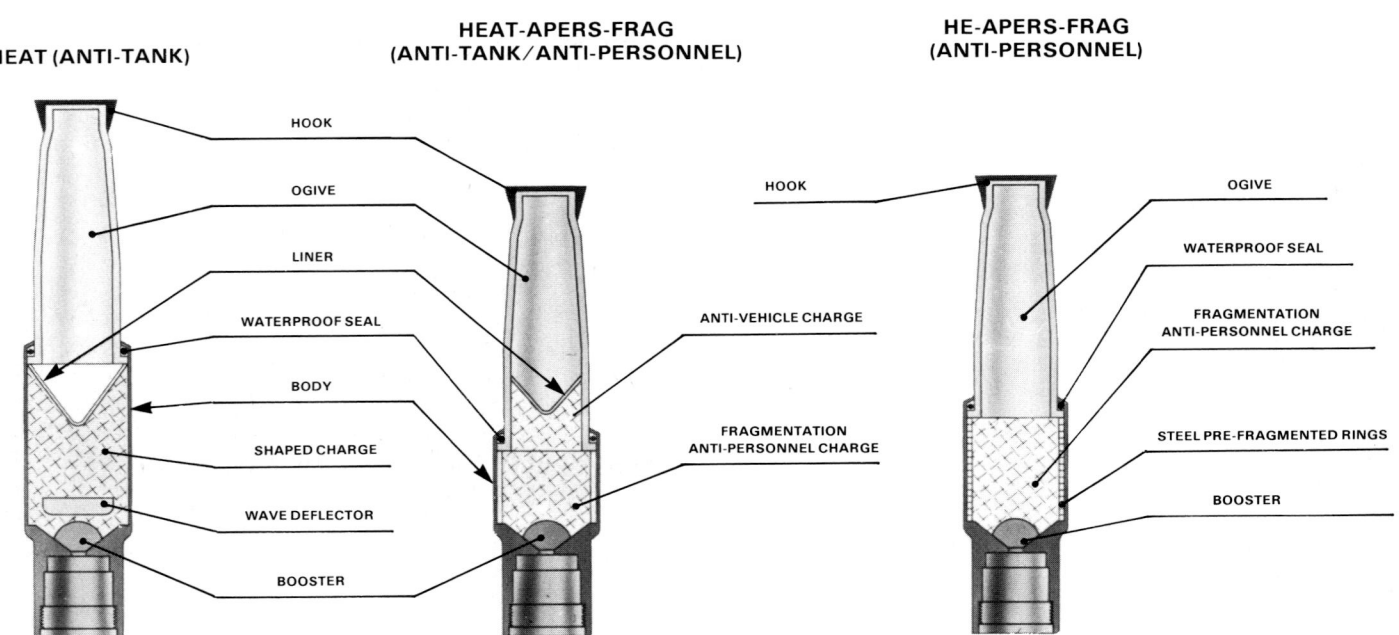

Warheads for the Luchaire smoke and illuminating rifle grenades

DATA	HEAT	HEAT-APERS FRAG	HE-APERS FRAG	HE-APERS FRAG-Delay	SMOKE	ILL	TP	TPM
External calibre, mm:	40	40	40	40	40	40	40	40
Total length, mm:	383	352	357	305	292	306	383	351
Weight, g:	412	405	412	410	410	390	415	405
Fuze type:	SQ	SQ	SQ	Delay	SQ	Delay		
Pre-fragmented splinters:		396	468	468				
Splinter weight, g:		0.17	0.17	0.17				
Muzzle velocity, m/s:	68	68	68	68	70	73	68	68
Muzzle safety:			5 or 12 m depending upon timer					
Safe operating range:			8 or 15 m depending upon timer					
Ranges:-								
effective combat range,								
flat trajectory:	100	100		100			100	100
max range, curved trajectory	360	360	360	360	(a)		360	360
Accuracy; H+at 100 m:	<2	<2	<2	<2			<2	<2
Lethal radius:		12	12	12				
Penetration, RHA, mm:	200	100						
Smoke emission time:					25 s			
Smoke screen duration:					60 s			
Illumination duration:						30-40 s		
Illumination intensity:						90 000 cd		

(Note (a) – smoke pot opening height and distance for a 45° firing angle is about 200 m

Manufacturer
Luchaire Défense, 180 Boulevard Haussmann, 75382 Paris Cedex 08.

Status
Current. Production.

Luchaire LU 213 HE-APERS-FRAG hand grenade

The LU 213 is a controlled fragmentation anti-personnel hand grenade in which particular attention has been paid to the fragmentation process. Perfect reliability and homegeneous fragment distribution is obtained by a coiled sleeve fragment generator and the use of small steel ball pads on the upper and lower parts of the grenade.

At the time of the grenade's detonation, whatever its position in flight, the steel balls and the fragments of the pre-notched sleeve are projected at very high speed, providing a dense and uniform pattern all around the point of detonation and beyond the lethal radius.

The safety radius has been studied so as to allow the thrower to remain unprotected within a radius of 20 m without being in danger.

DATA
Weight, with fuze: 280 g
Fuze weight: 55 g
Weight of pre-fragmented sleeve: 100 g
Explosive weight: 90 g (Comp. B or equivalent)
Length, with fuze: 94 mm
Diameter: 52 mm
Number of fragments: >1100 g
Average fragment weight: 0.085 g
Lethal radius: 9 m

Manufacturer
Luchaire Défense, 180 Boulevard Haussmann, 75382 Paris Cedex 08.
Status
Current. Production.

Luchaire LU 213 Inert ballasted hand grenade

This grenade is for demonstration and classroom instruction purposes. The exterior shape is identical with the LU 213 anti-personnel grenade, but the inert grenade is coloured orange and equipped with an inert fuze assembly. A safety device is placed at the base of the priming system so as prevent fitting a live fuze into the inert grenade body.

DATA
Weight, with fuze: 280 g
Weight of dummy fuze: 55 g
Length with fuze: 94 g
Diameter: 52 g

Manufacturer
Luchaire Défense, 180 Boulevard Haussmann, 75382 Paris Cedex 08.
Status
Current. Production.

Luchaire LU 213 (left) and LU 216 hand grenades

Luchaire LU 216 HE blast effect hand grenade

For this version of the Luchaire grenade, design was concentrated on producing maximum blast effect. Of ovoid form, the LU 216 HE differs from the LU 213 by the absence of the pre-notched fragmentation sleeve and steel ball pads. It is fitted with the same fuze as that of the LU 213 hand grenade.

DATA
Weight, with fuze: 190 g
Weight of fuze: 55 g
Weight of explosive: 110 g (Comp B or equivalent)
Diameter: 52 mm

Manufacturer
Luchaire Défense, 180 Boulevard Haussmann, 75382 Paris Cedex 08.
Status
Current. Production.

LU 219 PRAC practice hand grenade

Of blue colour and of identical form to the LU 213 hand grenade, the LU 219 PRAC is made of light aluminium alloy. Completely inert, it may be used up to 100 times. This version is equipped with an expendable practice fuze which simulates the functions of the real fuze (handling, firing, functioning and delay) and the terminal effects of the grenade (noise, flash and smoke).

The practice hand grenade fuze functioning system derives from that of the LU 213. Mounted on the grenade body it presents no danger to the user at any time.

DATA
Weight with fuze: 280 g
Length with fuze: 94 mm
Diameter: 52 mm

Manufacturer
Luchaire Défense, 180 Boulevard Haussmann, 75382 Paris Cedex 08.
Status
Current. Production.

63 mm type Grafac SERAT AC rifle grenade

This anti-tank hollow charge rifle grenade is designed for close-combat against armoured vehicles.

Very light, it has been specially designed to be launched from the 5.56 mm FA MAS rifles now in service with the French armed forces. However, it can also be launched from those 7.5 mm rifles still in service.

In both cases, the rifles will be provided with the standard 22 mm launching sleeve. Special launching cartridges, without bullet, are made for each type of rifle. The penetration is the same as that of the most recent grenades in service with the French armed forces and with the armed forces of many other countries.

Characteristics
The effective range for a 2.5 m vertex height (the tank height) is equal to approximately 100 m.

The fuze includes an electro-magnetic generator and the muzzle safety exceeds 6 m. It complies with the transport and handling safety requirements of the NATO specifications (safety device in case of accidental fall or cocking, igniter train cut-out device, etc). The grenade can fall from 5 m without any risk of burst. It is hermetically sealed. Properly packed, it can be parachuted.

The grenade is supplied with a ballistite cartridge without bullet, corresponding to the calibre of the rifle for which it is provided.

DATA
Max diameter of body: 63 mm
Total length, when stored: 365 mm
Weight of grenade: approx 515 g
Muzzle velocity: approx 65 m/s
Effective range with a flat-trajectory fire: approx 100 m
Max range with a 45° angle: approx 350 m
Penetrating power with a zero angle of incidence: 300 mm

Manufacturer
Société d'Etudes, de Réalisations et d'Applications Techniques, 180 boulevard Haussmann, 75382 Paris.
Status
Development.

63 mm Grafac SERAT AC rifle grenade

Losfeld fragmentation grenade

This hand grenade is transformable from defensive to offensive roles by simply removing the fragmentation sleeve which fits around the plastic body of the grenade. A conventional fly-off lever igniter fuze is screwed into the body, the lever being retained by a quick-release safety pin. The detonation train is positively interrupted before arming, and arming is carried out dynamically after throwing, the grenade being some 5 m from the thrower before arming takes place. Thus, should the grenade be dropped in the act of throwing it remains safe. The fragmentation sleeve breaks up into about 480 0.2 g splinters which are effective to a radius of 20 m around the point of burst.

DATA
Weight: 322 g with fragmentation sleeve; 174 g without
Length: 101 mm
Diameter: 47 mm
Explosive content: 50 g
Delay: 5 s nominal

Manufacturer
Losfeld-Industries, 15 rue Thiébault, 94220 Charenton.
Status
Current. Production.

Losfeld fragmentation grenade, with fragmenting sleeve   *Losfeld fragmentation grenade, offensive form*

Losfeld Multi-Purpose Anti-personnel grenade

This is a hand grenade similar to the Losfeld fragmentation grenade described above but with a selective fuze, and which can be attached to one of two bullet-trap tail units so as to be fired as a rifle grenade.

The basic grenade is of plastic, and has a fragmentation sleeve which can be slipped over the body. The fuze has a selector head which can be turned to any one of three positions: delay, giving a nominal 5 second delay after throwing, using the normal fly-off lever system; delay and impact, used for rifle firing; and impact only, which can be used with either hand throwing or rifle firing.

The tail units are finned and are dimensioned so as to fit over the NATO standard 22 mm rifle muzzle or launcher. One is for 5.56 mm rifles, the other

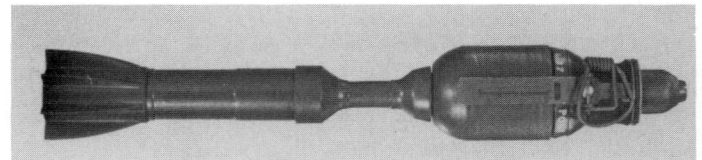

Losfeld multi-purpose grenade, hand offensive role

can be used with any calibre; the difference lies in the bullet trap, that of the 5.56 mm being only sufficient for that calibre while the 'universal' model will operate with any bullet. The design of the trap allows the bullet to impart some degree of spin to the grenade during its launch, and it is claimed that this roll stabilisation improves the accuracy. The rifle launched grenades have a maximum range of 300 to 400 m, depending upon the type of rifle and cartridge used to launch.

In either mode the grenade is safe before launch or throw due to physical interruption of the detonating train. Arming is completed after launch, at a distance of about 5 m from the thrower or 10 m from the rifle launcher.

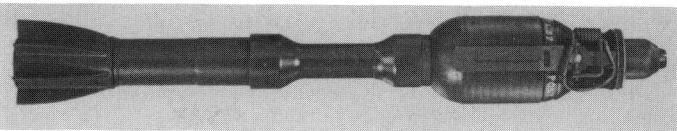

Losfeld multi-purpose grenade, 5.56mm rifle, fragmentation role

DATA
Weights: hand 212 g without sleeve, 360 g with sleeve. Rifle; with 5.56mm tail unit, without sleeve 329 g, with sleeve 477 g; with universal tail unit, without sleeve 379 g, with sleeve 527 g.
Lengths: Hand 124 mm; 5.56 mm rifle 315 mm; universal rifle 337 mm
Diameter: 47 mm
Explosive content: 50 g

Manufacturer
Losfeld-Industries, 15 rue Thiébault, 94220 Charenton.
Status
Current. Production.

ALSETEX defensive and offensive grenades

ALSETEX manufactures offensive and defensive combat grenades. The grenades are made with either a thin casing for offensive use, or a thicker fragmenting case for use as a defensive munition.

These grenades are provided with the Alsetex Mk F5A fuze. This safety fuze, with fly-off lever and compressed pyrotechnical delay, includes a pyrotechnic-chain interrupting device, allowing greater safety in use and storage.

Both models are filled with compressed Tolite (TNT) explosive. There are also practice models containing inert materials.

DATA

Grenade	No 1	No 2
Type	Offensive	Defensive
Case	Metal	Metal
Diameter	60 mm	55 mm
Height	95 mm	100 mm
Weight	140 g	540 g
Filling weight	90 g	56 g

Manufacturer
Société d'Armament et d'Etudes Alsetex, 4 rue du Castellane, 75008 Paris.
Status
Current. Production.
Service
In service with the French Army.

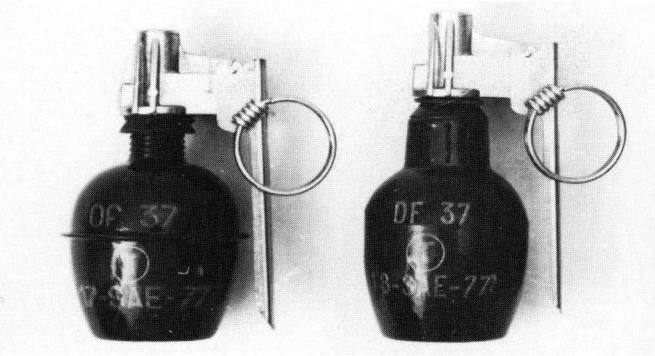

Alsetex defensive and offensive grenades

Alsetex AP/AV rifle grenade

This is a rifle grenade with bullet trap, intended for use against personnel or against armoured personnel carriers. It can be fired using bulleted cartridges, including those with armour-piercing bullets. It can be fired from any weapon with the standard 22 mm sleeves at the muzzle, though it is optimised for use with 5.56 mm rifles.

The warhead combines a shaped charge with a pre-fragmented surround to give the optimum anti-personnel effect around the point of detonation. A mechanical delayed-arming fuze, armed by the gas pressure of the propelling cartridge, provides safety until the grenade is en route to the target. The fuze incorporates a safety indicator, allowing a visual check of the arming state prior to firing.

DATA
Weight: 520 g
Length: 360 mm
Head diameter: 38 mm
Max range: 380 m
Max launch velocity: 67 m/s
Number of fragments: approx 900
Penetration: 100 mm homogeneous steel armour
Accuracy: ≤ 30 cm at 100 m
Arming delay: 5 – 14 m

Manufacturer
Société d'Armement et d'Etudes Alsetex, 4 rue de Castellane, 75008 Paris.
Status
Current. Production.
Service
In course of formal approval by French Army.

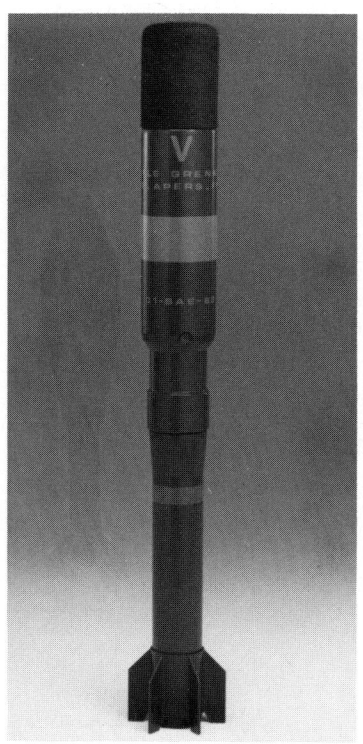

Alsetex AP/AV 5.56 mm rifle grenade

Offensive hand grenade Model 1937

The offensive hand grenade Model 1937 is made from a stamped aluminium body divided into two swaged parts. One of these parts bears a hole into which is fitted a light metal exploder tube, screwed and stuck to the body. The body is filled with nitrate explosive. An Igniter Plug Model F3 Type 1719 screws on to the top of the grenade, the delay and detonator unit being contained in the exploder tube. For transit and storage the Igniter Plug is not present, the tube being closed by a plastic plug.

DATA
Diameter: 60 mm
Height, fuzed: 100 mm
Weight of explosive: 78 g
Total weight, fuzed: 180 g
Delay time: 4 – 7 s

Manufacturer
Ruggieri, Département Armement, 86 av. de Saint-Ouen, 75018 Paris.
Status
Current. Production.
Service
French and other armies.

Ruggieri offensive hand grenade Model 1937

Defensive hand grenade Model 1937

The defensive hand grenade Model 1937 is made of cast iron. The body has a hole at the upper end into which a light metal exploder tube is screwed and stuck. The Igniter Plug Model F3 Type 1719 then screws into this hole, the detonator and delay assembly fitting into the exploder tube. Upon detonation, the cast iron body shatters to produce a cloud of dangerous fragments.

DATA
Diameter: 55 mm
Height, fuzed: 95 mm
Weight, fuzed: 540 g
Delay time: 4 – 7 s

Manufacturer
Ruggieri, Département Armement, 86 av. de Saint-Ouen, 75018 Paris.
Status
Current. Production.
Service
French and other armies.

Ruggieri defensive hand grenade Model 1937

GERMANY, EAST

Grenades in East German forces

East German troops are equipped almost entirely with grenades of Soviet design but two smoke grenades of East German design and manufacture, described below, are still in use.

S-53 smoke hand grenade

The S-53 smoke grenade is of similar construction to the S-32 but differs in having a handle like a conventional stick grenade. The grenade weighs about 300 g and burns for 30 seconds producing a white smoke cloud. It will continue to burn even if submerged in water.

S-53 smoke hand grenade

S-32 smoke hand grenade

This is a cardboard cylinder. It is ignited by the user rubbing the ignition pellet with a friction cap which activates the seven-second delay fuze. The grenade continues to emit black smoke for 1.5 to 2 minutes. The grenade weighs 250 g.

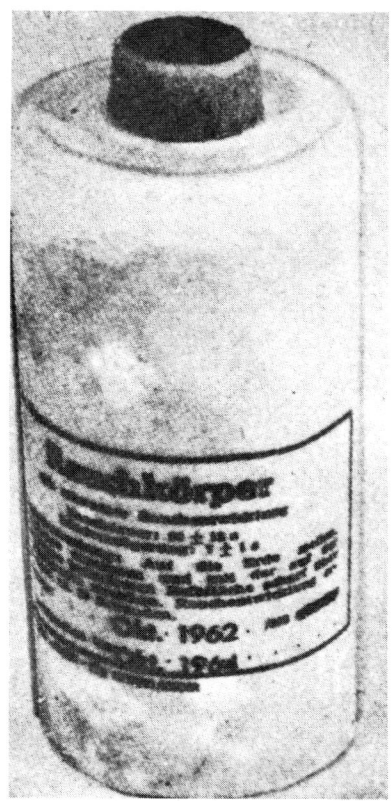

S-32 smoke hand grenade

GERMANY, WEST

M-DN 11 hand grenade

One of four defensive hand grenades made by Diehl (the others are the M-DN 21, 31 and 61) the M-DN 11 grenade is barrel shaped and made of plastic. The thick wall contains 3800 steel balls which are embedded in the plastic. The outside of the grenade body has longitudinal and transverse ribs which are raised from the surface to improve the user's grip and generally to improve handling under adverse conditions such as sub-zero temperatures or in mud. The interior is filled with plasticised nitropenta.

At the top a fuze thread is cut into the plastic material, and in this the grenade takes the DM 82A1B1 or M-DN 42 fuze. This has the same external appearance as previous fuzes made by Diehl and consists of a spring actuated hammer, held back by a long safety lever from hitting the primer cap, and a delay pellet above a detonator and a booster in the bursting charge.

However, the main feature of the fuze is that it incorporates a series of safeties. The first one is that the detonator is physically separated from the booster until the grenade is thrown. Additional safety is provided by a flap valve at the bottom of the delay tube. Should the cap be struck for whatever unlikely reason, or should the delay pellet be missing, this valve ensures that the flash cannot reach the detonator. After 2.5 seconds the soldered joint in the delay tube melts and allows the spring to force the tube into contact with the detonator. The flash material is now close enough to the detonator to ignite it and the flap valve is clear of the throat which held it shut, so that the flash can open it and pass through the fire holes. The entire arrangement is most ingenious and certainly safer than any fuzing that has been in service before. The sequence of events is explained in the following description.

When the safety pin **(1)** is withdrawn the torsion spring, **(2)** in the accompanying drawing, forces the hammer **(3)** round and it hits the DM 1024A1B1 primer cap **(4)**. The spring force throws off the protective cap **(5)** and the safety lever **(6)**. The flash of the percussion primer ignites the delay pellet **(7)** in its delay tube **(8)**. After about 2.5 seconds burning time the heat melts a solder ring **(9)** and so disconnects the delay tube *(8)* from the detonator holder **(10)**. The pressure spring **(11)** then forces the detonator holder and DM 1066B1 detonator **(12)** down on to the DM 1034 booster **(13)**. After four seconds burning time the flash from the delay pellet passes through two holes in the cup **(14)** over the detonator and strikes the flap valve **(15)** bending it up and passing it to reach the detonator *(12)*. The detonator sets off the booster which in turn ignites the main charge.

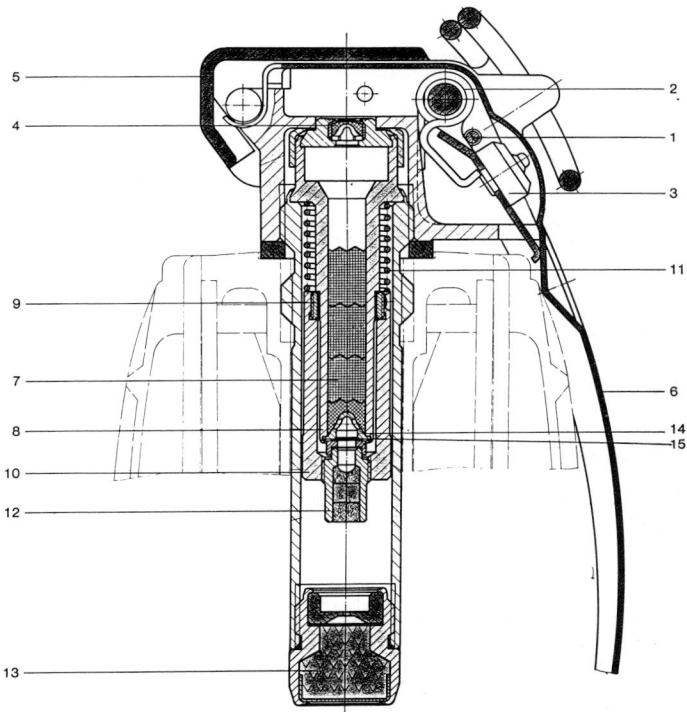

DM 82A1B1 fuze assembly (see text)

DATA
Weight, with fuze: 467 g
Weight of fuze: 61 g
Height, with fuze: 97 mm
Diameter: 60 mm
Grenade body: (empty) 364 g
HE filling: 42.5 g
Ball diameter: 2.5–3 mm
Number of balls: 3800
Danger area: ca. 100 m

Manufacturer
Diehl Ordnance Division, Fischbachstrasse 16, 8505 Röthenbach/Pegnitz.
Status
Current. Production.

M-DN 11 hand grenade

M-DN 21 hand grenade

This is a smaller edition of the M-DN 11. It functions in exactly the same way.

DATA
Weight of grenade with fuze: 224 g
Weight of fuze: 61 g
Height, fuzed: 85 mm
Diameter: 50 mm
Grenade body: (empty) 118 g
HE filling: 45 g
Ball diameter: 2–2.3 mm
Number of balls: 2200
Danger area: 45 m

Manufacturer
Diehl Ordnance Division, Fischbachstrasse 16, 8505 Röthenbach/Pegnitz.
Status
Current. Available.

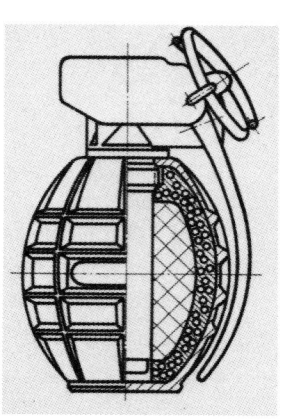
M-DN 21 hand grenade

M-DN 31 hand grenade

This grenade is similar in size and function to the M-DN 21 grenade but has more fragments and a greater danger area.

DATA
Weight of grenade with fuze: 247 g
Weight of fuze: 61 g
Grenade body: (empty) 151 g
Length, fuzed: 85 mm
Diameter: 50 mm
HE filling: 35 g
Ball diameter: 2–2.3 mm
Number of balls: 3000
Danger area: 60 m

Manufacturer
Diehl Ordnance Division, Fischbachstrasse 16, 8505 Röthenbach/Pegnitz.
Status
Current. Available.

M-DN 31 hand grenade

M-DN 61 hand grenade

The M-DN 61 is the last of the series of defensive grenades and closely resembles the others except for its size, which is intermediate between the M-DN 21 and the M-DN 11.

DATA
Weight of grenade with fuze: 317 g
Height, fuzed: 88 mm
Diameter: 57 mm
Grenade body: (empty) 206 g
HE filling: 50 g
Ball diameter: 2–2.3 mm
Number of balls: 4300
Danger area: 75 m

Manufacturer
Diehl Ordnance Division, Fischbachstrasse 16, 8505 Röthenbach/Pegnitz.
Status
Current. Available.

M-DN 61 hand grenade

DM 51 offensive-defensive hand grenade

This dual purpose grenade consists of two major elements: a high explosive hand grenade body and a fragmentation jacket. The body can be used separately as an offensive grenade, and when the jacket is placed around it the two parts together make up a defensive grenade. The parts are connected and held by a bayonet fixing in which the outer jacket is turned through 90° and the base portion of the jacket locks on to the central body. The HE grenade body has the shape of a hexagonal prism. It consists of a water-tight plastic container filled with compressed nitropenta. The top end of the container is screw threaded to take a fuze. The fragmentation surround is cylindrical and is made of plastic, with fragmentation inserts, in the same way as the defensive grenades described above. The two major parts, HE centre and jacket, can be joined and separated as often as required. Several grenade bodies can be joined to make up a cluster charge or connected end to end to make a Bangalore torpedo for assault engineer applications.

The offensive-defensive DM 51 hand grenade has been adopted by the West German Army under the nomenclature 'Handgranate Spreng/Splitter DM 51 mit Handgranaten-Zünder DM 82A1B1'. It is now the standard grenade for all units.

DM 51 offensive-defensive hand grenade

DATA
DM 51 hand grenade: 435 g
DM 82A1B1 fuze: 64 g
Height, fuzed: 107 mm
Diameter: 57 mm
Fragmenting jacket: 260 g
HE filling: 60 g
Ball diameter: 2–2.3 mm
Number of balls: 6500
Danger area: 35 m

Manufacturer
Diehl Ordnance Division, Fischbachstrasse 16, 8505 Röthenbach/Pegnitz.
Status
Current. Production.
Service
West German forces.

76 mm fragmentation grenade

This grenade is designed to be fired from the usual type of 76 mm smoke launcher fitted to most armoured vehicles, though a special launcher unit has been designed and produced by Wegmann and Company for fitting to vehicles requiring it. Since the standard type of launcher is designed to accept the low pressures associated with firing smoke grenades, this fragmentation grenade has a special propelling charge which, whilst maintaining low chamber pressure, ensures regularity of performance and bursts the grenade at the optimum point.

Prior to loading the safety pin (1) is removed. On firing, the ignition current flows via contact rings (2) and the cable (3) to the igniter cap (4) which fires the propellant (5). The gas pressure generated by the combustion of the propellant expands through the gas ports (6) into the remaining space in the launcher, thus accelerating the grenade to a velocity of about 24 m/s. The initial pressure generated by the propellant shears off a collar (9) on the striker (7) which then activates the fuze (8) and ignites the cap in the pyrotechnic time fuze; this initiates the explosive charge after a flight time of 3.8 seconds. The pyrotechnic fuze gives a delay in arming of 1.3 seconds.

The grenade is intended for use in the close-in defence of armoured vehicles, against infantry who are so close that the vehicle weapons cannot be depressed sufficiently to engage them at short-range. In addition, it can also be used against men in foxholes or behind cover protecting them against direct fire from the vehicle.

DATA
Length: 200 mm
Diameter: 76 mm
Weight: 1.7 kg
Warhead: RDX/TNT surrounded by steel fragments
Arming delay: 1.3 s
Firing delay: 3.8 s
Optimum range: 40 m at 63° elevation
Optimum burst height: 12 m at 63° elevation
Lethal area: approx 40 m² below optimum burst height

Manufacturer
Diehl Ordnance Division, Fischbachstrasse 16, D-8505 Röthenbach/Pegnitz.
Status
Current. Production.

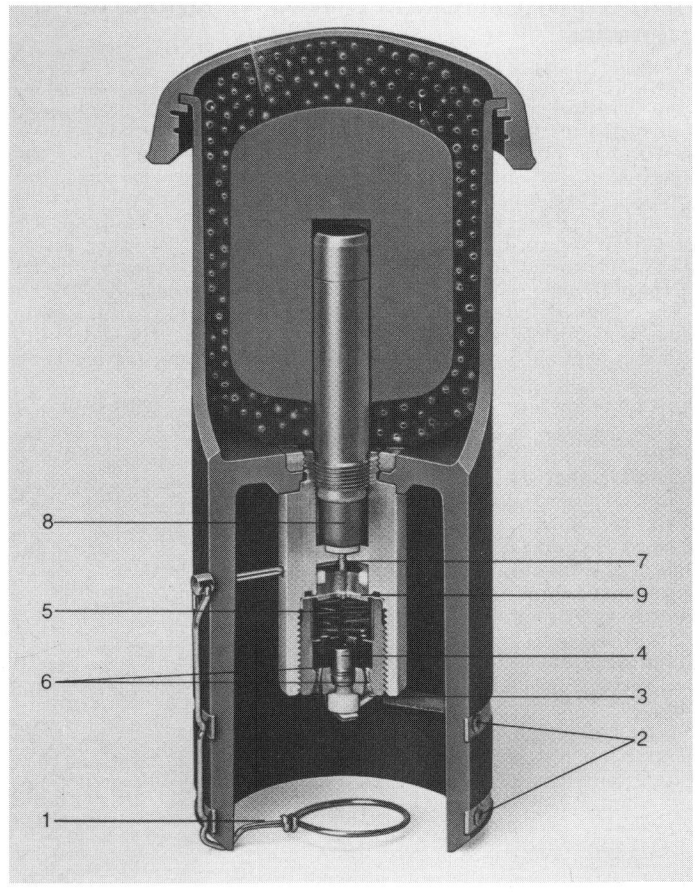

76 mm fragmentation grenade for vehicle launching

DM 24 incendiary smoke hand grenade

The DM 24 incendiary smoke hand grenade is the improved successor to the DM 19 and is designed to burst in a brilliant flash of fire on impact and simultaneously produce a cloud of dense smoke. The flash is blinding to the enemy and the smoke induces a severe irritating cough. The incendiary mass burns for about five minutes at a temperature of approximately 1200°C. This great heat ignites any combustible material the burning mass touches.

The grenade is absolutely safe to handle and to use because it does not contain an explosive charge. The incendiary mass is only activated by the ignition system. When striking a hard surface or object the plastic body breaks up so that the activated incendiary mass on being exposed to the oxygen of the air instantaneously bursts into a blaze of fire and smoke.

For transport the grenade is packed inside a cylindrical casing.

The training version of the DM 24 has the designation DM 68 and is handled in exactly the same way as the live grenade. The fuze does not have any live elements and, instead of the incendiary charge, it contains an inert lime dust compound for simulation on the target.

DATA
Weight: 340 g
Length: 133 mm
Diameter: 67 mm
Weight of incendiary charge: 255 g
Incendiary composition: red phosphorus

Manufacturer
Buck Werke GmbH & Co, Postfach 2405, 8230 Bad Reichenhall.
Status
Current. Production.
Service
Several armies.

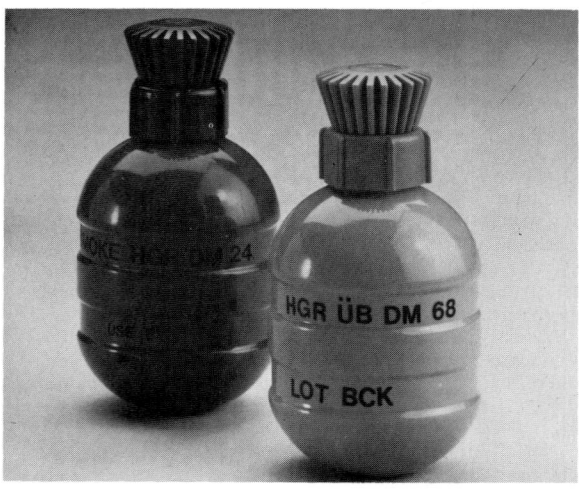

DM 24 and 68 incendiary smoke grenades

76 mm DM 15 HC and DM 35 RP smoke hand grenades

These smoke grenades are designed to be fired from launching tubes on an armoured vehicle as well as to be thrown by hand depending on the situation encountered. In either case some 2.5 seconds after ignition it starts to produce smoke which forms a dense screen preventing the opposing forces from observing any rearrangements of vehicles and personnel or any other operation. The emission of smoke continues for some 2.5 mins.

The smoke grenade has two separate ignition systems, the electrical and the mechanical system. Electrical ignition is used with the launching device allowing the grenade to be remotely fired from inside the armoured vehicle. Mechanical ignition is used when throwing the grenade by hand.

The cylindrical, thin-walled, metal body of the DM 15 HC grenade is filled with a smoke charge consisting mainly of hexachlorethane. The DM 35 RP grenade is filled with a red phosphorus composition and has the advantage of producing full cover more rapidly. Both grenades are moisture and shock-proof and will function within a temperature range of –40 to + 50°C.

DATA
Weight: 1.2 kg (DM 35 1.1 kg)
Length: 175 mm (DM 35 169.5 mm)
Diameter: 76 mm
Weight of smoke charge: 880 g
Range of projection under 45°: 40–70 m
Duration of smoke emission: 150 + 50 s; (DM 35 120 + 30 s)
Time to full cover: 6 – 8 s; (DM 35 0.5 s)

Manufacturer
Buck Werke GmbH & Co, Postfach 2405, 8230 Bad Reichenhall.
Status
Current. Production.
Service
West German Army, also other NATO armies and some other countries.

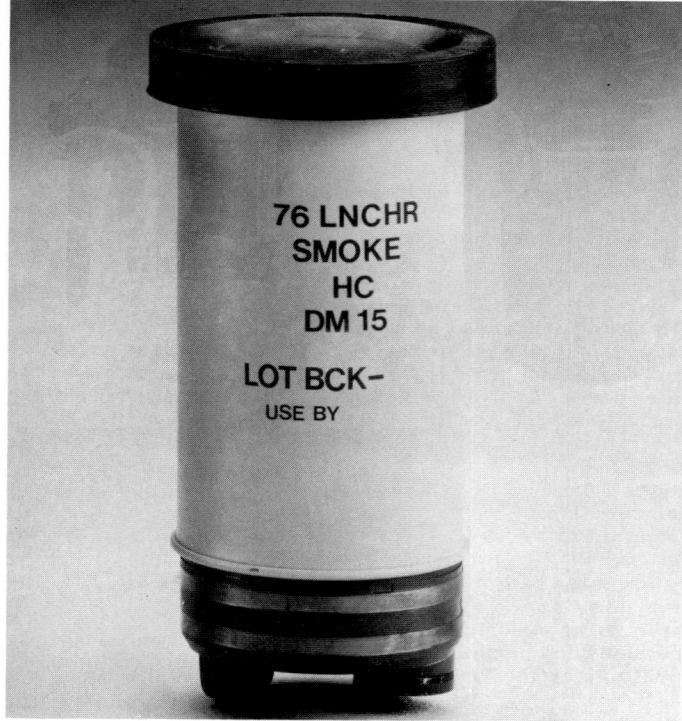

HC smoke grenade

Piepenbrock 76 mm Instant Tank Smoke grenade

The Instant Tank Smoke grenade is suitable for use with 76 mm Wegmann launchers as used on many armoured vehicles. It is designed for rapid production of defensive screening smoke, behind which the vehicle can manoeuvre.

The grenade is ignited from inside the vehicle by an electrical remote-control switchboard. After ignition a rocket motor propels the grenade from the launcher; after about a 5 m flight it begins to rotate, which effects a rapid and well-dispersed distribution of the smoke.

DATA
Length: 170 mm
Diameter: 76 mm
Weight: approx 1.95 kg
Emission time: approx 2.5 mins
Range: approx 35 m

Manufacturer
Piepenbrock Pyrotechnik GmbH, Ruhweg 21, D-6719 Göllheim/Pfalz.
Status
Development completed.

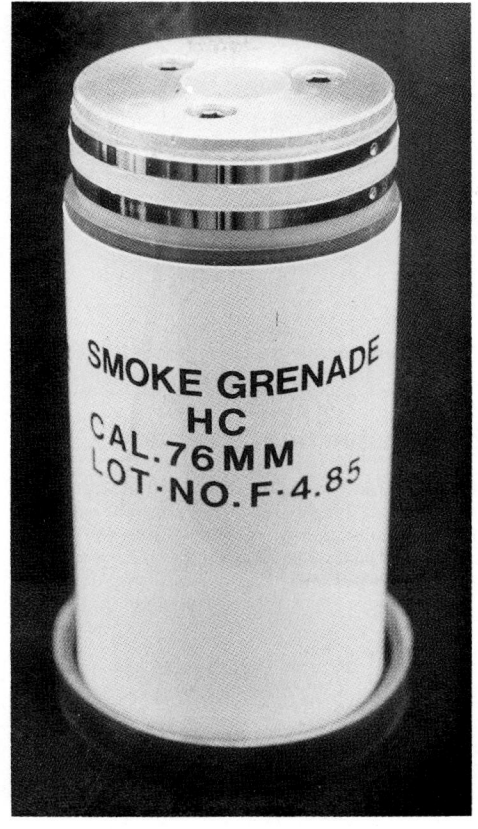

Piepenbrock 76 mm Instant Tank Smoke grenade

Piepenbrock 76 mm I R Screening tank smoke grenade

This resembles the Instant Smoke grenade described above and is used in the same manner, being fired from the standard 76 mm Wegmann launcher. On firing, however, it is ejected from the launcher by a propelling charge and at about 30 m from the vehicle, at about a 10 m height, bursts, releasing a quantity of pellets which fall to the ground and emit smoke. The smoke is effective in screening both optically and across the electro-optical spectrum, so preventing the concealed vehicle being seen by electro-optical surveillance or aiming devices or detected by target seeking sensors. The smoke itself emits radiation in the thermal band, so confusing thermal imaging sensors.

DATA
Length: 170 mm
Diameter: 76 mm
Weight: approx 1.20 kg
Emission time: 30 – 60 s
Time to full screening: 1 – 5 s

Manufacturer
Piepenbrock Pyrotechnik GmbH, Ruhweg 21, D-6719 Göllheim/Pfalz.
Status
Development completed.

GREECE

Elviemek EM01 defensive hand grenade

Elviemek SA was entrusted with the responsibility for manufacturing grenades for the Greek forces early in 1980. The defensive model EM01 has a plastic body containing pre-formed fragments in a matrix. The size, weight and configuration of the pellets, together with the amount of explosive, gives a lethal radius of 15 m. The igniter is largely made of plastic. It has a plastic head containing the metal striker, safety pin and lever and the percussion cap. The delay fuze inside the body is carried in a light metallic tube. The explosive filling is 'Pentaplastit' a PETN derivative exclusively produced by Elviemek.

DATA
Length: 91 mm
Diameter: 57 mm
Weight: 355 g
Weight of filling: 37 g
Number of pellets: 2600 ± 50
Diameter of pellets: 2.5 mm
Delay time: 4.0 s nominal

Elviemek EM01 defensive hand grenades

Manufacturer
Elviemek SA, Hellenic Explosives and Ammunition Industry, Atrina Centre,
32 Kifissias Avenue, Athens.
Status
Current. Production.
Service
Greek forces.

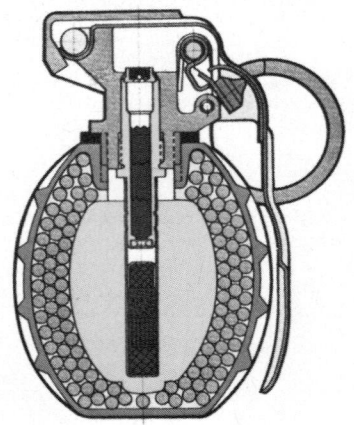

Sectioned drawing of Elviemek EM01 grenade

Elviemek EM02 offensive hand grenade

This is designed to produce blast with minimum fragment hazard and is
constructed entirely of plastic. It has a danger area of 2 to 3 m. The explosive
filling is 'Pentaplastit'. The fuze consists of a plastic housing with metal
spring, striker and delay tube, and is held safe by a metal fly-off lever retained
by a safety pin. The body of the grenade is smooth so as to differentiate it from
the defensive grenade.

DATA
Length: 92 mm
Diameter: 57 mm
Weight: 140 g
Weight of filling: 37 g
Delay time: 4.0 s nominal

Manufacturer
Elviemek SA, Hellenic Explosives and Ammunition Industry, Atrina Centre,
32 Kifissias Avenue, Athens.
Status
Current. Production.
Service
Greek forces.

Elviemek EM02 offensive hand grenades

Elviemek EM10 rifle grenade

The EM10 is an anti-personnel rifle grenade which consists of the standard
EM01 defensive grenade attached to a tail unit designed to be fired from a
standard rifle adapter or from any rifle with the muzzle conforming to NATO
standard dimensions. The grenade is equipped with the normal type of fly-off
lever igniter set but has a special delay element giving a longer delay time. A
simple inertia catch secures the fly-off lever until set-back, on firing, releases
it.

DATA
Length: 235 mm
Weight: 555 g
Diameter: 62 mm
Weight of filling: 37 g
Number of pellets: 2600
Delay time: 6.5 s nominal
Maximum range: 300 m

Manufacturer
Elviemek SA, Hellenic Explosives and Ammunition Industry, Atrina Centre,
32 Kifissias Avenue, Athens.
Status
Current. Production.
Service
Greek forces.

Elviemek EM10 rifle grenade

Elviemek EM14 rifle grenade

The EM14 rifle grenade uses the body of the EM01 defensive grenade
attached to a tail unit containing the fuze, a bullet deflector and tail fins. The
tail unit will fit any standard discharger attachment or the muzzle of any
NATO standard rifle. The grenade body contains an explosive charge and
pre-formed pellet fragments inside a plastic body. The arming mechanism and
fuze are located in the tail unit behind the grenade. A pyrotechnic train
initiated on firing ensures that arming does not take place until the grenade is
well clear of the rifle. Once armed, detonation takes place immediately on
impact and the grenade will function even on soft ground and at low angles of
incidence. A visual indicator in the arming mechanism shows whether the
grenade is safe or armed.

The bullet deflector is not intended to permit normal firing with bulleted
ammunition. Blank grenade discharger cartridges are used for launching, but
should a ball cartridge be used in error, the deflector will prevent detonation

of the grenade and will cause the grenade to fail safe and fall to the ground harmlessly a few metres in front of the rifle with the arming device locked in the safe condition.

DATA
Length: 275 mm
Weight: 510 g
Weight of explosive: 37 g
Number of pellets: 2600
Arming distance: 15 m
Maximum range: 360 m

Manufacturer
Elviemek SA, Hellenic Explosives and Ammunition Industry, Atrina Centre, 32 Kifissias Avenue, Athens.
Status
Current. Production.
Service
Greek forces.

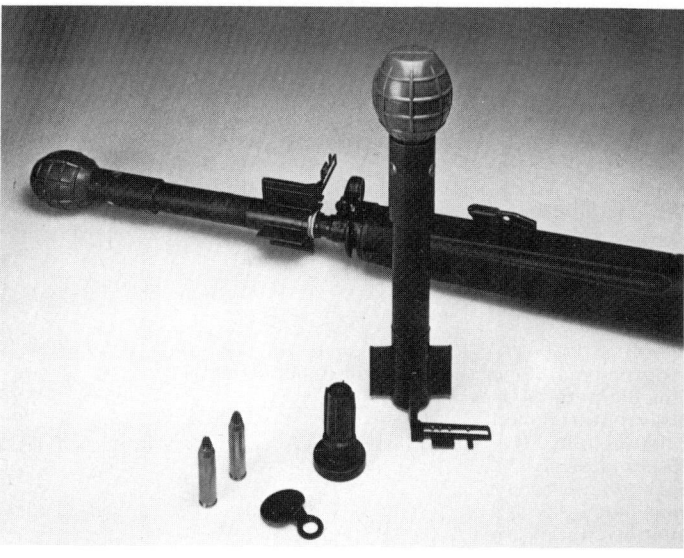

Elviemek EM14 rifle grenade

Elviemek EM04 smoke grenades

The smoke grenade is made of a metallic cylindrical pot containing the smoke mixture. The igniter system, the same as in the other hand grenade models, is screwed on top of the pot. Elviemek manufactures seven similar models in the following colours: white, red, blue, yellow, green, violet and fog.

DATA
Length: 145 mm
Weight: 650 g (fog); 450 g (colours)
Diameter: 66 mm (fog); 60 mm (colours)
Delay time: 2.5 s nominal
Emission time: 90 – 150 s (fog); 100 – 140 s (colour)

Manufacturer
Elviemek SA, Hellenic Explosives and Ammunition Industry, Atrina Centre, 32 Kifissias Avenue, Athens.
Status
Current. Production.
Service
Greek forces.

Elviemek EM04 smoke grenades

Elviemek EM12 screening smoke rifle grenade

The EM12 screening smoke grenade is the standard 'fog' EM04 grenade attached to a rifle launching adapter. This is a simple tail unit with an inertia device which holds the grenade safety lever securely until released by set-back when the grenade is fired.

DATA
Length: 290 mm
Weight: 560 g
Diameter: 66 mm
Delay time: 5 s nominal
Maximum range: 150 m
Emission time: 60 s minimum

Manufacturer
Elviemek SA, Hellenic Explosives and Ammunition Industry, Atrina Centre, 32 Kifissias Avenue, Athens.
Status
Current. Production.
Service
Greek forces.

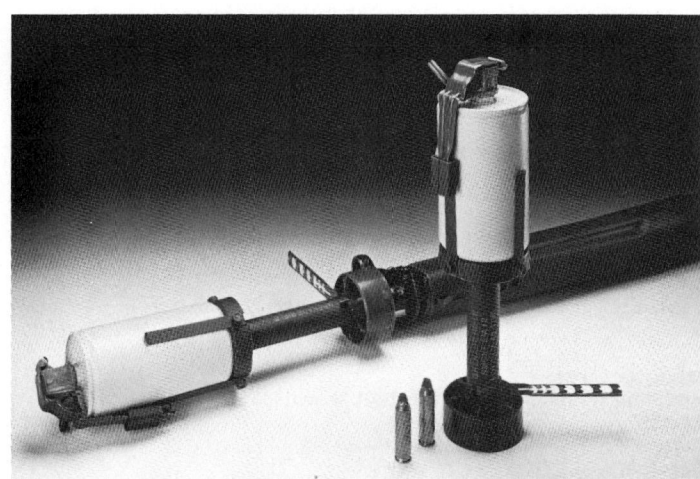

Elviemek EM12 screening smoke grenade

HUNGARY

M42 hand grenade

The only grenade of Hungarian design in current service is the M42. This is an offensive stick-type hand grenade. It employs a delay fuze. One unusual feature of M42 is the provision of a male thread at the top of the grenade and a female thread at the bottom which permits the junction of several grenades to provide a small demolition charge. There are three $\frac{1}{2}$ in (13 mm) red bands around the body.

DATA
Type: blast
Weight: 310 g
Length: (head) 76 mm; (stick) 118 mm; (total) 194 mm
Max diameter: 48 mm
Body material: steel
Filler weight: 134 g
Filler material: TNT
Fuze type: delay
Fuze delay: 3.5–4.5 s
Range: (thrown) 30 m

Manufacturer
State arsenals.
Status
Current.
Service
Believed still in service, but the design is old.

Hungarian M42 grenade with additional head

Hungarian M42 grenade with additional head screwed on

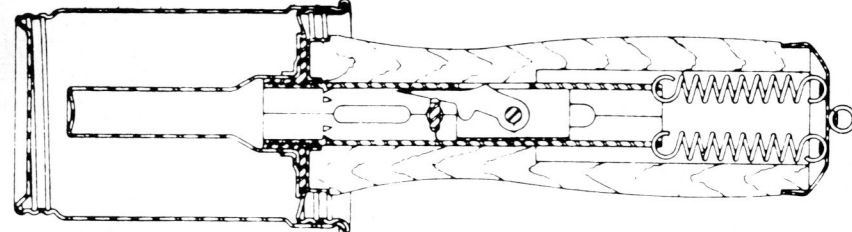

Sectioned view of Hungarian M42 offensive hand grenade, showing percussion fuze mechanism

ISRAEL

No 5 white smoke hand grenade

This grenade is used for signalling and also to produce a local smoke screen. The grey smoke is emitted for slightly less than two minutes and, being of hexachlorethane, has less tendency to pillar than some of the phosphorus smokes. The grenade consists of a tin-plated cylinder and is operated in the conventional fashion with a fly-off lever restrained by a safety pin.

DATA
Smoke colour: white
Smoke mixture: hexachlorethane
Mixture weight: 610 g
Body material: steel sheet
Functioning: delay fuze 2 s nominal
Height: 150 mm
Diameter: 63 mm
Weight: 800 g
Actuation: on removal of safety pin
Application: local smoke screen and signalling
Smoke emission: 90 – 130 s
Packing: 8 grenades in a sealed metal container, 2 containers in a wooden box

Manufacturer
Israel Military Industries, PO Box 1044, Ramat Ha Sharon 47100.
Status
Current. Production.
Service
Israeli forces.

No 5 white smoke hand grenade

No 14 offensive hand grenade

This grenade is used by assaulting infantry who need to close with the enemy immediately after the blast and so require the radius of effect of the grenade to be comparatively small without the production of splinters. The grenade operates in the usual way with a fly-off lever and safety pin.

DATA
High explosive filling: TNT flakes
Weight of HE filling: 200 g
Body material: laminated paper with sheet metal ends
Functioning: delay fuze 4.5 s ± 0.5 s
Height: (without fuze) 110 mm; (with fuze) 135 mm
Diameter: 64 mm
Weight: 325 g
Actuation: removal of safety pin
Application: offensive hand grenade
Packing: 24 grenades in metal container, 8 containers to a box

Manufacturer
Israel Military Industries, PO Box 1044, Ramat Ha Sharon 47100.
Status
Current. Production.
Service
Israeli forces.

No 14 offensive hand grenade

M26 A2 fragmentation hand grenade

This is a copy of the American grenade. It is a fragmentation grenade with a notched coil inside the thin-wall sheet steel body.

DATA
High explosive filling: Comp B, cast
Weight of HE filling: 155 g
Body material: sheet steel
Fragmenting material: spirally wound steel coil, pre-notched
Number of fragments: 1000
Functioning: delay fuze 4.5 s ± 0.5 s
Height: 107 mm
Diameter: 61 mm
Weight: 425 g
Actuation: removal of the safety pin
Application: defensive
Lethality: 50% chance of a hit – lying man – at 10 m

Manufacturer
Israel Military Industries, PO Box 1044, Ramat Ha Sharon 47100.
Status
Current. Production.
Service
Israeli forces.

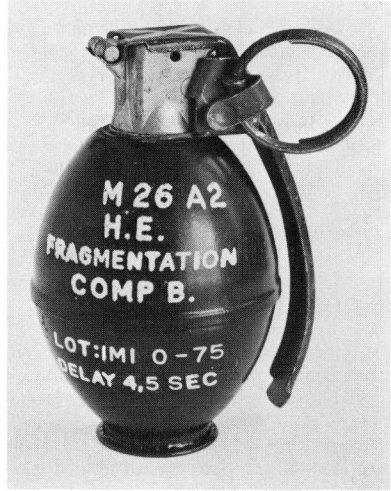

M26 A2 fragmentation hand grenade

No 5 coloured smoke hand grenade

This grenade is employed for ground position indication or ground to air indication. It is also used for the production of local smoke screens.

DATA
Smoke colours: red, yellow, green
Body material: sheet metal
Height: 150 mm

Diameter: 63 mm
Weight: 480 g
Smoke charge: 290 g
Type of fuze: delayed ignition
Delay time: 2 s nominal
Smoke emission: 45–85 s
Application: position marking and local smoke screen

Manufacturer
Israel Military Industries, PO Box 1044, Ramat Ha Sharon 47100.
Status
Current. Production.
Service
Israeli forces.

No 5 coloured smoke hand grenade

IMI Bullet-trap rifle grenades

These new grenades are propelled by a combination of the gases generated by the cartridge and the impact of the bullet on the bullet trap. In the remote event of failure of the bullet trap, the bullet is deflected and cannot, under any circumstance, cause the grenade to detonate or explode. The presence of a bullet trap or a bullet deflector is now commonplace in modern grenades, but only the IMI design offers both.

Further safety features of these grenades include the striker being out of alignment with the detonator until the fuze is completely armed; arming of the fuze by gas pressure from the propelling cartridge, with the option of adding a set-back safety element to work in conjunction with the gas pressure arming; and final arming by the movement of a spring-driven rotor governed by two independent set-back detents. Two indicator windows in the grenade show immediately the state of the grenade; if it is armed, these windows show red; if the grenade is safe, they show white. In addition a reflector disc inside the tail tube is present and visible if the grenade has not been fired. Arming takes

place during flight, and the grenade must travel a minimum distance of 7, 14 or 21 m, as specified by the customer. Should the grenade strike an obstacle or fail to fly before the minimum distance is reached, it remains perfectly safe, since the striker and detonator will not be aligned, so that disposal offers no danger.

Disposable plastic sights are provided with each grenade, the type of sight being appropriate to the rifle specified by the purchaser. Special night sights are also available as optional extras; these are marked with luminous lithium light sources and are designed for firing at ranges up to 100 m at night.

Inert grenades, for training purposes, are available; these match the shape, dimensions and ballistic performance of the service grenades. Two types are provided; one is completely inert and may be re-used; the other carries a flash and bang element in the head to allow identification of the point of impact by day or night.

BT/AP 30 APERS rifle grenade

This is for use against troops in the open and against armoured personnel carriers; it will penetrate 10 mm of steel armour plate. The grenade detonates on impact, projecting approximately 400 fragments at high velocity to a lethal radius of 9 m.

The grenade consists of three basic parts; the steel fragmentation body with high explosive charge, the tail tube and stabiliser assembly with bullet trap and bullet deflector, and a base detonating fuze with high graze sensitivity capable of detonating the grenade at angles of impact as low as 15°.

The grenade can be launched from 5.56 mm rifles, using ball or AP ammunition, or it can be launched with a grenade blank cartridge. The tail

unit fits all rifles with the standard 22 mm muzzle dimension, and can be launched from other rifles with appropriate adapters. A disposable plastic sight is supplied with each grenade.

DATA
Weight: 490 g
Length: 320 mm
Diameter: 30 mm
Range: 300 – 350 m depending upon rifle

IMI BT/AP 30 APERS rifle grenade

BT/AP 65 APERS rifle grenade

This grenade is for use against troops in the open or against armoured personnel carriers; it will penetrate 13 mm of steel armour plate.

It consists of a pre-scored steel fragmentation body with high explosive filling, a tail tube and stabiliser with bullet trap and deflector, and a base

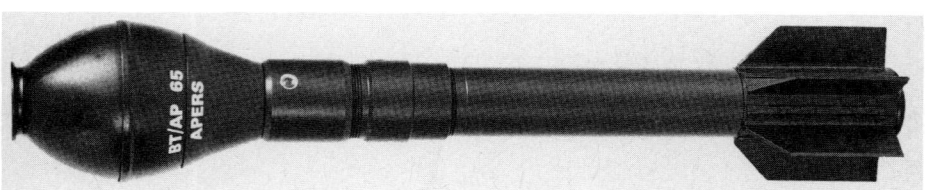

IMI BT/AP 65 APERS rifle grenade

detonating fuze capable of operating at angles of impact as low as 15°. It can be fired from 5.56 mm rifles using ball, AP or grenade blank ammunition, and the tail units fits the standard 22 mm muzzle launcher. Adapters will permit it to be fired from rifles of non-standard type. A disposable plastic sight is provided with each grenade.

DATA
Weight: 630 g
Length: 320 mm
Diameter: 60 mm
Range: 200 – 250 m depending upon rifle
Lethal radius: 18 m

BT/AP 67 APERS rifle grenade

This grenade is designed for use against troops in the open, under cover, or behind shelter, and in urban areas against troops in light structures, behind doors or light armoured cover. It will penetrate 15 mm of steel armour plate and thus is effective against troops in light armoured vehicles. The grenade detonates on impact, projecting over 1000 fragments over a lethal radius of 25 m.

The grenade can be launched from 22 mm muzzle launchers on 5.56 mm rifles, using ball, AP or grenade blank ammunition; it can be launched from non-standard rifles by means of an adapter. A disposable plastic sight is provided with each grenade.

DATA
Weight: 635 g
Length: 310 mm
Diameter: 64 mm
Range: 190 – 240 m depending upon rifle

IMI BT/AP 67 APERS rifle grenade

BT/AT 52 HEAT rifle grenade

This is for use against light armoured vehicles, bunkers, and light armour protection. It can penetrate 150 mm of steel armour plate.

The grenade consists of a body containing a shaped charge filled with Composition B, a tail tube and stabiliser assembly fitted with bullet trap and deflector, and a base detonating fuze capable of functioning at angles of impact as low as 15°. It can be fired from 5.56 mm assault rifles having the usual 22 mm muzzle dimension, and from non-standard rifles by means of an adapter. Ball, AP or grenade blank ammunition can be used. A disposable sight is provided with each grenade.

DATA
Weight: 510 g
Length: 400 mm
Diameter: 50 mm
Range: 270 – 320 m depending upon rifle

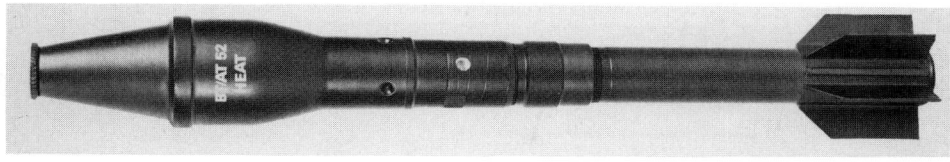

IMI BT/AT 52 HEAT rifle grenade

BT/AP-AT 39 APERS-HEAT rifle grenade

This model can be used against light armour, troops in the open or behind cover, in bunkers, or in urban areas. It will penetrate 140 mm of steel armour plate and the pre-fragmented head breaks into about 600 high velocity fragments giving a lethal radius of 12 m.

The grenade has a pre-scored fragmenting head surrounding a shaped charge element. The remaining features and details of employment are as for the other grenades in this series noted above.

DATA
Weight: 520 g
Length: 400 mm
Diameter: 39 mm
Range: 290 – 340 m depending upon rifle

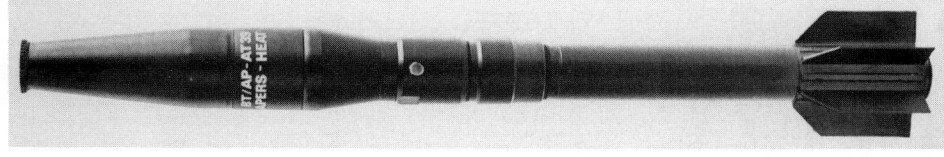

IMI BT/AP-AT 39 APERS-HEAT rifle grenade

BT/AP-AT 68 APERS-HEAT rifle grenade

This is similar to the previous grenade but is biased more to the anti-personnel performance; it delivers over 1000 fragments, but the armour penetration is reduced to 85 mm of steel armour plate. It has a lethal radius of 20 m.

DATA
Weight: 660 g
Length: 380 mm
Diameter: 64 mm
Range: 180 – 220 m depending upon rifle

Manufacturer
Israel Military Industries, PO Box 1044, Ramat Hasharon, 47100.
Status
Current. Production.
Service
Israel Defence Force.

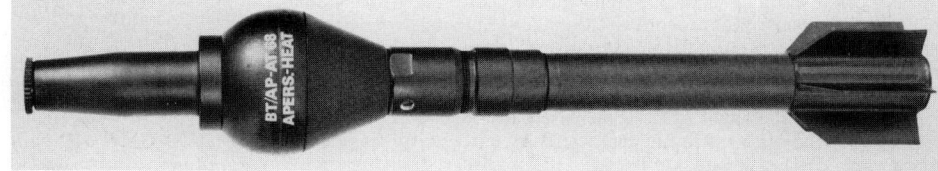

IMI BT/AP-AT 68 APERS-HEAT rifle grenade

MA/AP 30 anti-personnel rifle grenade

This is a general-purpose rifle grenade which also has the capability of penetrating up to 8 mm of armour and can thus be used against troops in armoured personnel carriers. It consists of three basic units: the steel body with Composition B filling; tail tube and fin unit; and a base percussion fuze which will function on impact or on graze down to a 15° impact angle.

The grenade can be launched from the Galil, FN-FAL or M16 rifles, and each grenade is packed with a ballistite cartridge and a disposable sight appropriate to the launching weapon. With an appropriate adaptor and cartridge the grenade can be used with other models of rifle.

DATA
Weight: 430 g
Length: 300 mm
Diameter: 30 mm
Max range: 300–400 m depending on rifle

Manufacturer
Israeli Military Industries, PO Box 1044, Ramat Ha Sharon 47100.
Status
Current. Production.
Service
Israeli and several other armies.

MA/AP 65 anti-personnel rifle grenade

This is also a general-purpose grenade, heavier than the MA/AP 30 and thus with greater effect: it can pierce up to 13 mm of armour plate. Its range is rather less, though. It can be used with the Galil or FN-FAL rifles or, with an appropriate adaptor, with other rifles. Each grenade is provided with a ballistite cartridge and disposable sight unit. A useful feature is the provision of two sight windows in the tail tube which give visual indication of the state of arming of the grenade.

DATA
Weight: 600 g
Length: 300 mm
Diameter: 60 mm
Max range: 250–300 m depending on rifle

Manufacturer
Israeli Military Industries, PO Box 1044, Ramat Ha Sharon 47100.
Status
Current. Production.
Service
Israeli and other armies.

MA/AT 52 HE/AT rifle grenade

This is for use against light armoured vehicles and against personnel behind light armour cover or in bunkers. The hollow-charge warhead will penetrate 150 mm of steel armour and is initiated by a base percussion fuze. In standard form the grenade can be fired from the Galil, FN-FAL and M16 rifles by means of a ballistite cartridge. With appropriate adaptors it can be fired by other types of rifle.

DATA
Weight: 450 g
Length: 380 mm
Diameter: 50 mm
Max range: 275–380 m depending on rifle

Manufacturer
Israeli Military Industries, PO Box 1044, Ramat Ha Sharon 47100.
Status
Current. Production.
Service
Israeli and other armies.

SGF 40 smoke rifle grenade

This grenade can be used for screening purposes or for target marking. It consists of two parts: the cylindrical body, containing a hexachlorothane/zinc mixture; and a tail tube and fin assembly. There is no fuze, a four-second pyrotechnic delay being ignited on firing. As issued it can be fired from the Galil, FN-FAL or M16 rifles, and with appropriate adaptors from other types.

Like all IMI rifle grenades the SGF 40 is provided with a bullet trap in the tail unit which ensures that even if a ball bullet is inadvertently fired instead of the proper ballistite cartridge, the grenade is undamaged and the firer is not at risk.

DATA
Weight: 440 g
Length: 250 mm
Diameter: 50 mm
Max range: 275–380 m depending on rifle

Manufacturer
Israeli Military Industries, PO Box 1044, Ramat Ha Sharon 47100.
Status
Current. Production.
Service
Israeli and other armies.

ITALY

MU-50/G MISAR hand grenade

The MISAR MU-50/G is described by its manufacturers as a controlled effects munition. It is a small hand-thrown grenade with a plastic shell containing a matrix of fragments which can be varied in size to obtain the desired effect. Thus, the lethal range can be altered by different sizes of fragment and different quantities and types of explosive.

The grenades are fitted with a silent, flashless and smokeless igniter and delay which leave no trace behind them when thrown. The igniter is activated by the usual spring-loaded striker held down by a safety lever and secured by a pin.

The grenade is quite small and light and this, together with its almost spherical shape make it easy to throw. The makers claim that it can be thrown with greater precision and to a greater range than any other grenade of the traditional size and shape.

MISAR also supplies other versions of this grenade with different sizes of fragment so that the lethal range is increased. The MU-50/E is a practice grenade with the same operating characteristics but with no fragments. It produces noise, flash and smoke. The MU-50/I is an inert drill grenade.

MU-50/G MISAR hand grenade on Franchi SPAS 12 special-purpose shotgun

DATA
Length: 70 mm
Diameter: 46 mm
Weight: (grenade) 140 g; (explosive) 46 g compressed TNT
Lethal range: 5 m
Safety distance: 20 m
Fuze delay: 4 s ± 0.5 s
Number of fragments: 1500 steel balls

Manufacturer
MISAR SPA, SS 236 Goitese, Loc. Fascia d'Oro, I-25018 Montichiari (Brescia).
Status
Production, in service.

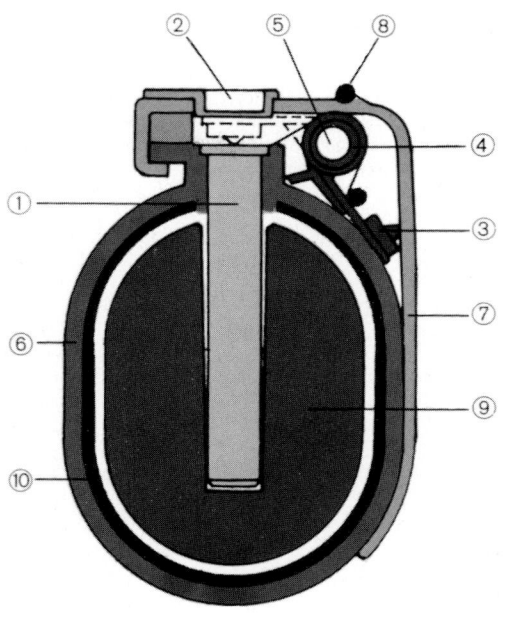

MU-50/G MISAR hand grenade
(1) detonator (2) plug (3) striker (4) spring (5) spring rod (6) grenade body (7) safety lever (8) safety (9) charge (10) fragmentation

OD/82 hand grenade

This grenade may be used as an offensive or defensive grenade without adding to or removing any component parts, since its fragmentation pattern and density is closely controlled by its design.

DATA
Weight: 286 ± 5 g
Length: 83 mm
Diameter: 59 mm
Bursting charge: 112 g Composition B
Weight of one fragment: 0.05 g
Efficacy at 5 m: 85%
Safety distance: 20 m

Manufacturer
La Precisa – Stabilimenti di Teano SpA, 81057 Teano (Caserta).
Status
Current. Production.

La Precisa OD/82 hand grenade

KOREA, SOUTH

K400 fragmentation hand grenade

This has been developed in South Korea and appears to be based on the American M67 design. The body is of steel, internally embossed to give break-up for optimum fragmentation, and with the usual type of igniter and fuze screwed into the top. Tests indicate that in excess of 1300 fragments may be expected from the detonation of one grenade.

DATA
Weight: 405 g
Bursting charge: 130 g Composition B
Height: 90 mm
Diameter: 60 mm
Fuze: K402
Delay: 4-5 s
Lethal radius: 15 m
Range: 40 m hand-thrown

Manufacturer
Korea Explosives Co Ltd, 34 Seosomoon-Dong, Chung-Ku, Seoul.
Status
Production.
Service
South Korean Army.

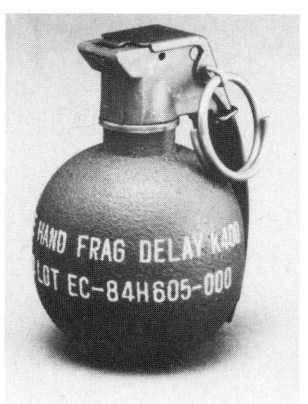

K400 fragmentation hand grenade

Smoke hand grenade K-M8

As might be gathered from the nomenclature this is based on old American design. The K-M8 is a screening smoke grenade, filled with HC mixture and fitted with the usual Bouchon igniter set. An aluminium cup under the igniter contains a starter composition, and four holes in the top of the grenade, normally tape-sealed, allow the smoke to escape.

DATA
Weight: 710 g
Height: 11 mm
Fuze delay: 1.5 s nominal
Emission time: 105–150 s

Manufacturer
Korea Explosives Co Ltd, 34 Seosomoon-Dong, Chung-Ku, Seoul.
Status
Production.
Service
South Korean Army.

K-M8 smoke grenade

40 mm grenades

These are based on the American M79/M203 launcher design of projected grenade. They consist of a cartridge case, detonating unit, fuze assembly and sheet-metal ogive. The detonating unit is a spherical grenade body carried in a parallel-walled skirt which, together with the ogive unit, form a ballistic projectile.

In addition to the high explosive grenade K200 there is also a practice grenade using the same impact fuze K502 and with a smoke pellet to indicate the strike instead of a high explosive bursting charge.

DATA
Weight: 226 g
Length: 98.8 mm
Filling: (HE) Composition B; (TP) plastic ball and smoke pellet
Max range: 400 m
Muzzle velocity: ca 75 m/s

Manufacturer
Korea Explosives Co Ltd, 34 Seosomoon-Dong, Chung-Ku, Seoul.
Status
Production.
Service
South Korean Army.

K200 40 mm fragmentation grenade

40 mm multiple projectile cartridge

The multiple projectile cartridge is fired from all standard 40 mm grenade launchers and is intended for close defence and rapid response to ambushes. The cartridge case carries a plastic sabot holding a pellet cup loaded with 20 metal pellets. On firing, the sabot and cup are driven up the bore of the launcher; setback causes the cup to move slightly back in the sabot and thus removes the covering cap. At the muzzle, air resistance causes the sabot and cup to discard, leaving the pellets to fly freely to the target.

DATA
Round weight: 115 g
Round length: 67.2 mm
Projectile weight: 24 g
Loading: 20 metal pellets
Effective range: 30 m
Muzzle velocity: 269 m/s

Manufacturer
Korea Explosives Co Ltd, 34 Seosomoon-Dong, Chung-ku, Seoul.
Status
Current. Production.

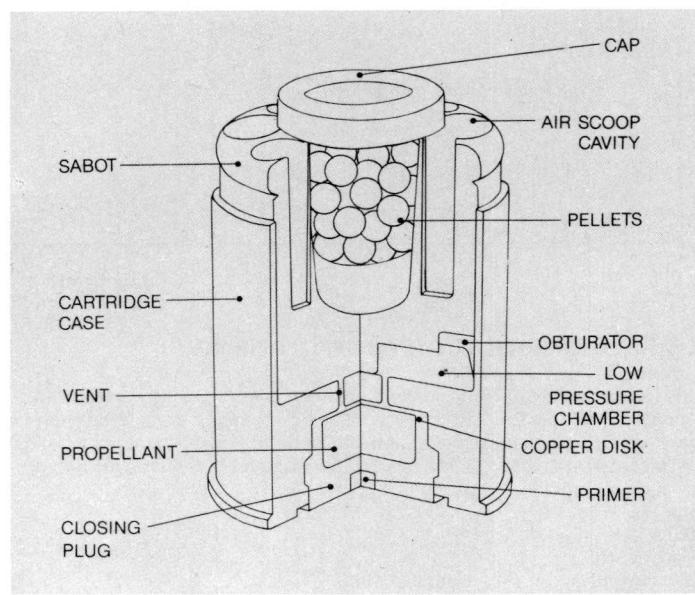

Interior arrangement of M576E1 multiple projectile cartridge

NETHERLANDS

NR 1C1 fragmentation hand grenade

This is similar to the earlier NR 1 grenade but differs in having the fuze assembly 19C1 or 19C2.

The NR 19 fuze mechanism is intended to be the standard Dutch fuze for all future grenades adopted by the Armed Forces. The original fuze is the NR 19. The first modification was numbered 19C1 and the second 19C2. A typical designation for a fuze assembly, that is the fuze plus detonator, is 19C1/20. This indicates that the Fuze NR 19 including the first modification is being used with detonator NR 20. The fuze/detonator combinations are used as follows: 19C1/20 for offensive, defensive and explosive smoke grenades; 19C2/17 for smoke, incendiary, and tear gas grenades; 19/9 for practice fragmentation grenades.

The NR 19 fuze is very orthodox in its functioning. When the safety pin is withdrawn the striker rotates and the lever is forced off the grenade body. The striker ignites the detonator cap which in turn ignites a pyrotechnic delay which takes about three seconds before the main filling is detonated.

There is no indication that the NR 19 fuze has been initiated and neither smoke, flame nor sound is apparent.

DATA
Body: cast iron
Filling: 55 g TNT powdered
Diameter: 56 mm
Height: 122 mm
Weight: 670 g

Status
Obsolescent.
Service
Dutch armed forces.

Fragmentation hand grenade NR 1C1 with fuze assembly NR 19C2/20

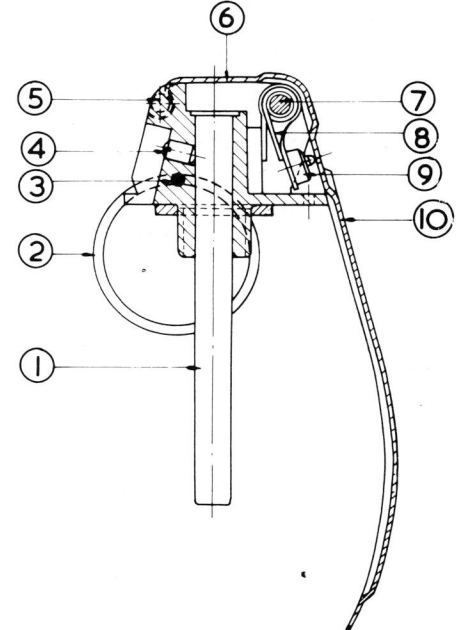

Fuze NR 19
(1) *delay train* (2) *safety pin with ring* (3) *safety pin* (4) *retaining screw* (5) *safety lever axis* (6) *fuze 19C1* (7) *firing pin axis* (8) *firing pin spring* (9) *firing pin* (10) *safety lever*

Rifle grenades

The Royal Netherlands Army uses the Rifle Grenade HEAT NR 4. This is based on the Belgian Mecar Energa Grenade.

Practice grenades
There are two practice rifle grenades. These are the practice rifle grenades NR 5 and NR 18. The latter has a white powder marker head.

Status
Current.
Service
Royal Netherlands Army.

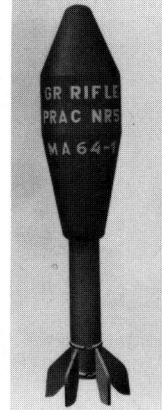

NR 5 practice rifle grenade

NR20 C1 Eurometaal hand grenade

The Eurometaal NR20 C1 hand grenade, developed for the Royal Netherlands Army, consists of a plastic body with an inner lining of steel balls and a HE charge. It is fitted with the standard 19C3 mechanical fuze with pyrotechnic delay, which is a member of what has been the standard range of fuzes of the Netherlands Army for some years. The safety pin blocks the lever. This lever in turn blocks the firing pin, which, when released, ignites the percussion primer. The primer then ignites the rest of the explosive train.

During trials it was found that the grenade exploded into about 2100 fragments at an initial velocity of 1600 m/s and that it is highly lethal up to a distance of 5 m (almost 100 per cent against personnel in the prone position) and not lethal above 20 m from the point of detonation.

Using targets of mild steel 1.5 mm thick and spruce 25 mm thick the following perforation data were obtained.

NR20 C1 Eurometaal grenade

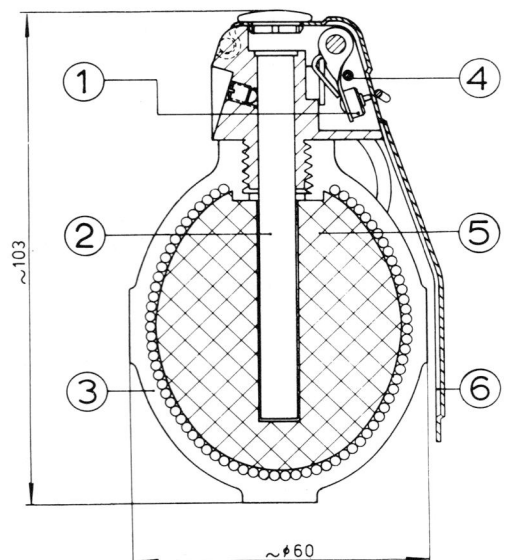

Components of the NR20 C1 Eurometaal hand grenade
(1) striker (2) detonator with pyrotechnic delay (3) fragmentation body
(4) safety pin (5) explosive filling (6) safety lever

	3 m	4 m	5 m	20 m
Mild steel 1.5 mm	18	10	6	0
Spruce 25 mm	18	10	6.5	0

DATA
Length: 104.5 mm
Diameter: 61 mm
Weight of hand grenade: (with filling) 390 g
Weight of fragmentation body: 160 g
Weight of filling: Comp B, 150 g
Fuze: 19C3

Delay: 3.5 s ± 0.5 s
Packing: 18 grenades per box
Weight of box: approx 11.7 kg
Volume of box: approx 0.018 m³

Manufacturer
Eurometaal NV, PO Box 419, 1500 EK Zaandam.
Status
Current. Production.
Service
Royal Netherlands Army.

POLAND

Polish hand grenades

The Polish Army still uses the Model 31 hand grenade. This is a fragmentation grenade of similar appearance and function to the Soviet F1 hand grenade. The latter is used by the Polish Army too, as is the Soviet RC-42 anti-personnel hand grenade.

Poland and Hungary are the only Warsaw Pact countries to equip their forces with rifle grenades. These are launched from the Polish model of the Soviet AK-47 assault rifle, the PMK-DGN-60. The muzzle of the rifle is coned to take the Polish LON-1 grenade launcher and the rifle has a gas cut-off to ensure that the gas energy is devoted to grenade launching and at the same time the piston is not subjected to excessive pressure from the grenade launching cartridge. The LON-1 has an external diameter of 20 mm and therefore should not be used with the more normal 22 mm grenades usually projected from rifles. The two grenades, the F1/N60 anti-personnel grenade and the PGN-60 anti-tank grenade, with their internal diameter tubes of 20 mm cannot be launched from any Western rifle, all of which have 22 mm launchers.

The F1/N60 rifle grenade is a Polish adaptation of the Soviet F1 fragmentation grenade (which see). The delay fuze has been replaced with an impact fuze and an unfinned stabilising boom has been added. The sights of the PMK-DGN-60 rifle show ranges of 100–240 m.

The PGN-60 HEAT anti-tank rifle grenade is of Polish design and the shaped charge warhead has a finned stabilising boom, and is employed out to ranges of 100 m.

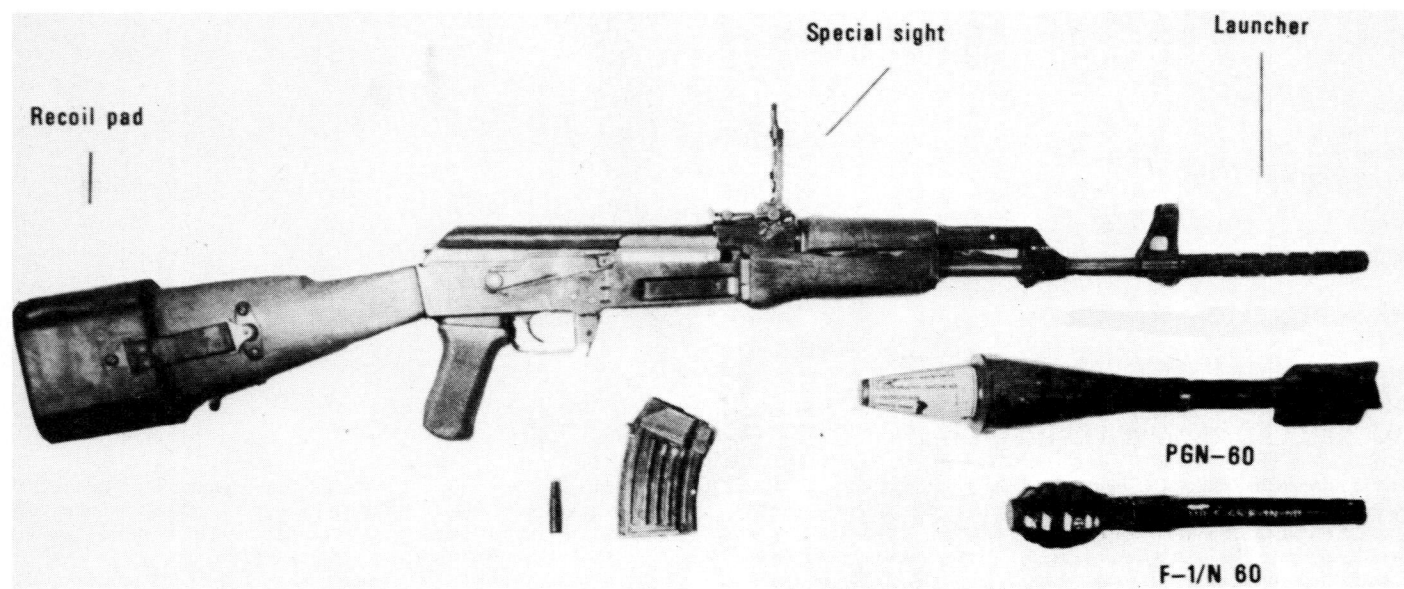

Polish PMK-DGN-60 rifle with associated grenades, PGN-60 and F1/N60

DATA

	F1/N60	PGN-60
Weight, fuzed	632 g	580 g
Length	270 mm	405 mm
Diameter	55 mm	67.5 mm
Weight of HE filler	45 g	218 g
Fuze type	impact	impact
Max range	40 m	100 m
Fragment radius	15–20 m	—
Penetration	—	100 mm

Status
Current.
Service
Polish armed forces.

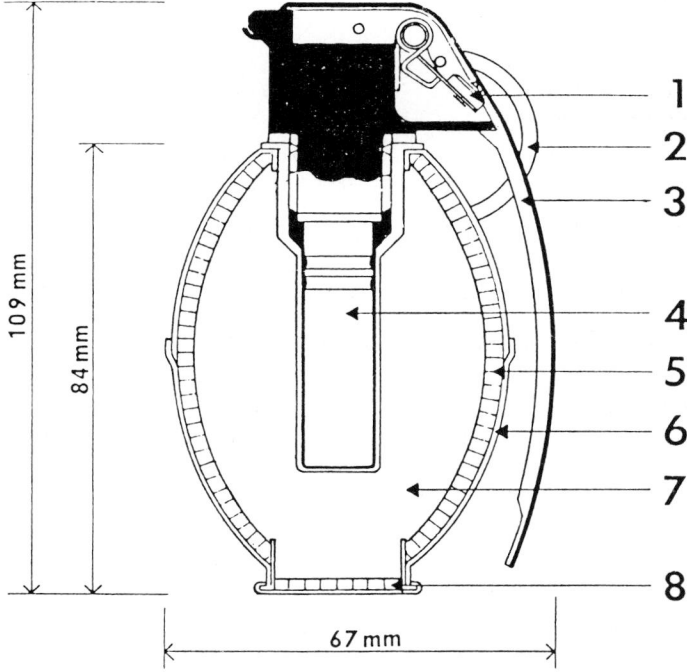

Polish Model 31, defensive, fragmenting, hand grenade

PORTUGAL

M312 fragmentation hand grenade

This is similar in design to the US M26 grenade, a thin sheet metal body lined with a notched wire fragmentation sleeve. There are certain detail differences in the shape of the detonator pocket and in the base plug being lined with a pre-fragmented plate. Ignition is by the usual type of fly-off lever percussion unit. Average fragmentation over a 20 m radius is 756 perforations of plywood screens set at varying distances.

DATA
Weight: 450 g
Weight of filling: 150 g Composition B
Weight of fragmentation sleeve: 150 g
Number of fragments: approx 900
Average weight of fragment: 0.13 g
Delay time: 4 – 5 s
Effective radius: 10.5 m
Safety radius: 30 m

Manufacturer
Sociedade Portugesa de Explosivos SA, Av. Infante Santo 76, 5°, 1300 Lisbon.
Status
Current. Production.

M312 fragmentation grenade. **(1)** *fuze;* **(2)** *safety pin;* **(3)** *safety lever;* **(4)** *detonator;* **(5)** *fragmentation sleeve;* **(6)** *metal body;* **(7)** *explosive filling;* **(8)** *plug*

M313 fragmentation hand grenade

The M313 grenade uses a plastic body inside which is a double layer of steel balls packed into a resin matrix surrounding the plastic explosive filling. There is the usual threaded detonator well, into which a conventional fly-off lever percusion igniter set is screwed. Average fragmentation of the grenade produces over 900 perforations in plywood screens arranged at distances from 2 to 15 m from the point of burst.

DATA
Weight: 370 g
Weight of filling: 51 g plastic HE
Weight of fragmentation liner: 230 g
Number of fragments: approx 1800
Average fragment weight: 0.11 g
Delay time: 4 – 5 s
Lethal radius: 9 m
Safety radius: 20 m

Manufacturer
Sociedade Portugesa de Explosivos SA, Av. Infante Santo 76, 5°, 1300 Lisbon.
Status
Current. Production.

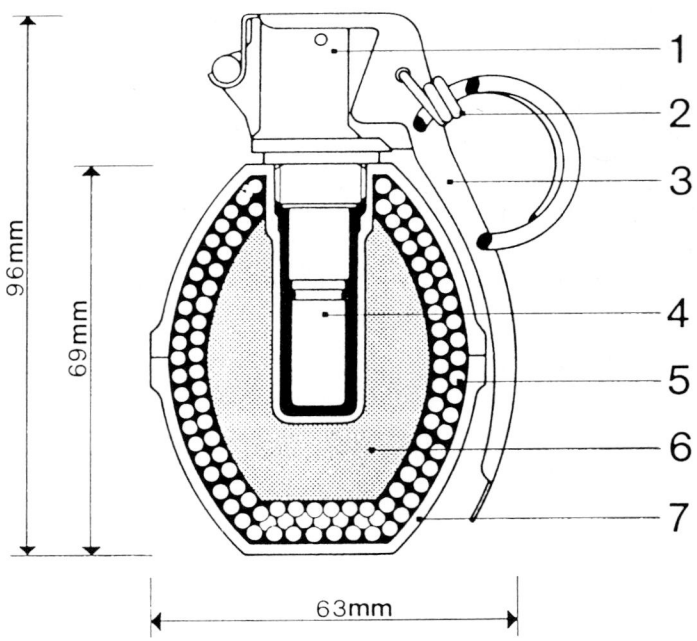

M313 fragmentation grenade. **(1)** *fuze;* **(2)** *safety pin;* **(3)** *safety lever;* **(4)** *detonator;* **(5)** *fragmentation liner* **(6)** *explosive filling;* **(7)** *plastic body*

M314 fragmentation hand grenade

The M314 is of similar design to the M313 (above) but larger and containing a heavier explosive charge and more fragments, a third layer of smaller steel balls being incorporated into the liner. As a ready means of identification this grenade has a smooth exterior finish, whereas the M313 has a roughened finish.

DATA
Height: 103 mm
Diameter: 74 mm
Weight: 575 g
Weight of filling: 58 g
Weight of fragmentation liner: 400 g

Number of fragments: approx 6500
Average fragment weight: 0.06 g
Delay time: 4 – 5 s
Lethal radius: 8 m
Safety radius: 20 m

Manufacturer
Sociedade Portugesa de Explosivos SA, Av. Infante Santo 76, 5°, 1300 Lisbon.
Status
Current. Production.

SINGAPORE

SFG 87 fragmentation hand grenade

This is a pre-fragmented hand grenade, the body of which is of high strength plastic material. The filling is Composition B (RDX/TNT) embedded with steel balls. The lethal radius is 10 m, the safety radius 25 m. Initiation is by the conventional fly-off lever percussion mechanism.

DATA
Weight: 300 g
Diameter: 54 mm
Height: 68 mm
Explosive content: 80 g
Fragments: 2800 steel balls of 2 mm diamater
Delay time: 4.5 s nominal

Manufacturer
Chartered Industries of Singapore, 249 Jalan Boon Lay, Singapore 2261.
Status
Current. Production.

Anti-Frogman grenade (AFG)

This is a powerful blast grenade made from high strength cross-linked plastic and filled with Composition B (RDX/TNT) so as to develop a powerful blast overpressure in water. The fuze and grenade are entirely watertight and the grenade will detonate at a depth of between 3 and 8 m below the surface. The safety radius is 30 m.

DATA
Weight: 320 g
Diameter: 54 mm
Height: 138 mm
Explosive content: 180 g
Delay time: 4.5 s nominal

Manufacturer
Chartered Industries of Singapore, 249 Jalan Boon Lay, Singapore 2261.
Status
Current. Production.

SOUTH AFRICA

All South African grenades are marketed by Armscor, Private Bag X337, Pretoria, 0001.

Fragmentation hand grenade

This is a copy of the American M26 grenade, using the same construction of sheet metal body and notched wire fragmentation unit, with a RDX/TNT filling. The fuze unit contains a striker mechanism and 4.5 second pyrotechnic delay, actuated by the usual type of pin and fly-off lever.

Inert drill grenades and re-usable practice grenades are also available.

DATA
Dimensions: 60 × 113 mm
Weight fuzed: 465 g
Explosive filling: 160 g RDX/TNT
Delay time: 4.5 s
Fragments: 1000 +
Lethal radius: 15 m with 50% lethality chance

Status
Current. Production.
Service
South African Defence Force.

Fragmentation hand grenade

M791 Anti-personnel rifle grenade

This grenade is constructed by using the body of the M26 hand grenade and attaching it to a special base fuze unit and tail unit. The fuze unit contains a pyrotechnic delay train which is initiated by setback on firing and ensures that

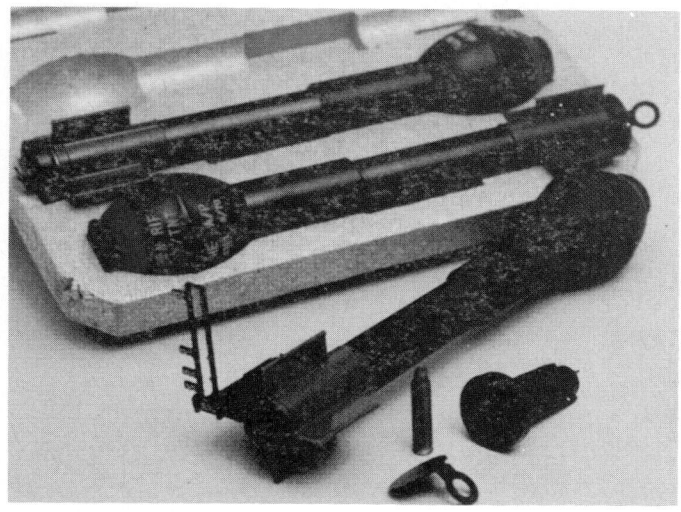

Anti-personnel rifle grenade

arming only takes place at a safe distance from the rifle (15 m). Until this point the firing pin is securely locked and the grenade cannot explode should it strike any object. Once armed, detonation takes place on impact. The fuze is exceptionally sensitive and detonation will take place even on soft ground or at slight glancing angles of impact. A visual indicator is provided to show whether the grenade is in the safe or armed condition.

The tail unit contains a bullet deflector so that the grenade will fail safe should it be fired with a ball round instead of the correct ballistite cartridge. In such a case the grenade will not explode, but will travel a few metres from the rifle before falling to the ground with the arming device locked in the safe position.

The grenade can be fired from any suitable 5.56 or 7.62 mm rifle, and is provided with a disposable plastic sight.

DATA
Length: 297 mm
Weight: 610 g
Explosive filling: 150 g RDX/TNT
Fragments: 1000 +
Average fragment weight: 0.2 g
Max range: (5.56 mm rifle) 210 m; (7.62 mm rifle) 300 m
Arming distance: 15 m
Lethal radius: 15 m with 50% lethality chance

Status
Current. In service.
Service
South African Defence Force.

75 mm HE/AT rifle grenade

This is a shaped-charge grenade, based on the Energa design and designed to be fired from most 7.62 mm rifles. The tail unit carries a 1 g incremental charge which augments the propulsive effort derived from the ballistite launching cartridge; the tail unit is pressure-tested to 750 kg/cm² to ensure safety during firing.

The grenade is fitted with a spit-back impact fuze which initiates a detonator at the rear of the shaped charge; this detonator is carried externally in a plastic case fitted to the tail tube and is installed in the grenade before firing by simply unscrewing the tail tube. The fuze is protected by a plastic nose cap which is resistant to dropping or other rough treatment and is removed immediately before firing. There is also an internal safety unit which seals the spit-back from the fuze away from the detonator unless unlocked by setback due to actual firing.

DATA
Weight fuzed: 720 g
Explosive filling: RDX/WX 93/7 main charge and CH6 booster
Fuze: DA and Graze Type BS02
Max velocity: 62 m/s
Accuracy: H + L 2.5 m at 100 m

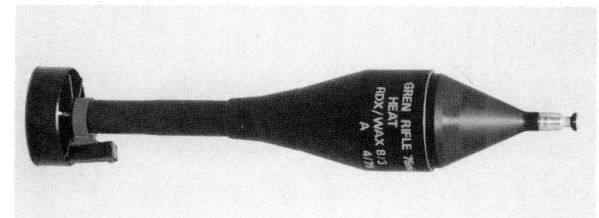

75 mm HE/AT rifle grenade

Range: (max) 375 m; (optimum) 75 m
Penetration: 275 mm

Status
Current. No longer in production.
Service
South African Defence Force.

Practice rifle grenade

This resembles the anti-personnel grenade (see entry above) but is completely inert. The arming mechanism of the live grenade is replaced by a metal spacer, but the bullet deflector is retained so the firer is protected if accidentally using a ball cartridge for launching. The grenade can be fired from most suitable 5.56 and 7.62 mm rifles and the head can be re-used up to 10 times, after recovery, by fitting a new tail unit.

DATA
Length: 297 mm
Weight: 610 g
Max range: (5.56 mm rifle) 210 m; (7.62 mm rifle) 300 m

Status
Current. Production.
Service
South African Defence Force.

Practice rifle grenade

SPAIN

M5 hand grenade

This cylindrical grenade has an ABS plastic body in its basic offensive pattern, and can be fitted with a fragmentation sleeve to convert it into a defensive grenade. The fly-off lever mechanism is somewhat different to the usual percussion delay type, since it incorporates a one-second arming delay during which time the detonating train becomes aligned, after which it will detonate on impact. If, for any reason, it should fail to detonate there is a self-destroying explosion after eight seconds.

DATA
Diameter: 62 mm
Height: 105 mm
Weight: 290 g
Weight of fragmentation sleeve: 190 g
Weight of TNT charge: 130 g

Manufacturer
Explosivos Alaveses, Apdo 198, 01080 Vitoria, Alava.
Status
Current. Production.

M5 cylindrical hand grenade

M6 hand grenade

This is a spherical grenade having an ABS plastic body lined with several hundred steel balls in a resin matrix around the hexolite filling. The igniter is of the conventional fly-off lever percussion type, and initiates a gasless delay composition. The construction of the grenade ensures very symmetrical fragment distribution around the point of detonation.

DATA
Diameter: 65 mm
Height: 98 mm
Weight: 325 g

Manufacturer
Explosivos Alaveses, Apdo 198, 01080 Vitoria, Alava.
Status
Current. Production.

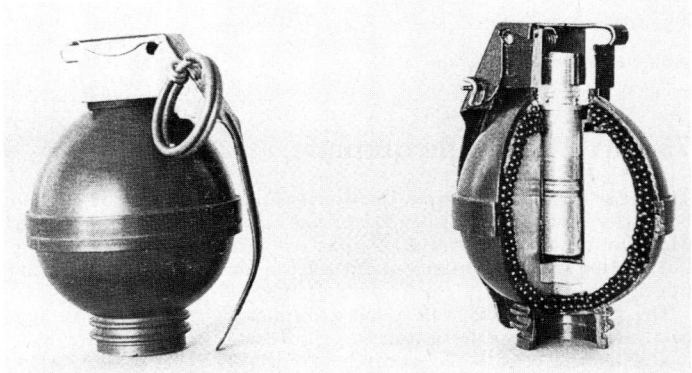

M6 spherical hand grenade

MB-8 hand grenade

This is a controlled fragmentation defensive grenade which can easily be converted into an offensive pattern by removing the metal sleeve. Safety, simplicity and effectiveness have been the fundamental objectives to be met in the design of this grenade.

Safety is provided by the delayed action fuze with detonation occurring at 4 ± 0.5 s after release of the striker. Simplicity is imparted by the ease of conversion from defensive to offensive and vice versa. Effectiveness is achieved by the type and uniform distribution of fragments as well as the ratio of high explosive to fragment material.

A practice version of this grenade is also produced.

MB-8 hand grenade dismantled to show ignition set, offensive unit and defensive fragmentation sleeve.

Hand Grenade MBS-9

This grenade is externally the same as the MB-8 but differs in that it has a trajectory safety device incorporated.

DATA
Length: (defensive) 103 mm; (offensive) 95 mm
Diameter: (defensive) 61 mm; (offensive) 50 mm
Weight: (defensive 440 g; (offensive) 171 g

Manufacturer
Explosivos Alaveses, Apdo 198, 01080 Vitoria, Alava.
Status
Current. Production.

MB-8 hand grenade in defensive form

Expal Incendiary hand grenades

Explosivos Alaveses manufacture a family of incendiary hand grenades which differ in their filling agent and thus are capable of being matched to different tactical requirements.

The basic 'GWP' grenade is filled with white phosphorus and therefore has applications as a smoke producer, an anti-personnel weapon or as an incendiary grenade. When detonated the WP particles are distributed over an area some 40 m in diameter. Ignition is by the usual type of fly-off lever igniter, which has a nominal five second delay. The grenade is 140 mm high, 55 mm in diameter and weighs 750 g.

The 'GRP' grenade is filled with red phosphorus; it thus has a primary role as a smoke producer but it will also act as an incendiary device with easily-ignited substances. When detonated, the red phosphorus particles are distributed over an area of 30 m diameter, and the phosphorus continues to burn for about 30 seconds. The same igniter set is used as in the 'GWP' grenade above. The grenade is 142 mm high, 55 mm in diameter and weighs 560 g.

The 'CTE' grenade is filled with thermite and is therefore purely an incendiary device which will ignite anything capable of being burned, since the thermite combustion reaches a temperature of 3000°C. Burning continues for 40 seconds. The igniter set is of similar pattern to the previous grenades, but the delay time is only two seconds. The grenade is 140 mm high, 55 mm in diameter and weighs 580 g.

Manufacturer
Explosivos Alaveses SA, Apartado 198, 01080 Vitoria, Alava.
Status
Current. Production.

Expal GWP Incendiary hand grenade

R-41 Oramil delay hand grenade

This model has replaced the earlier R-1 in production. It is an offensive pattern grenade, using a plastic casing, a charge of high explosive, and the normal Oramil pattern of fly-off lever igniter set. The igniter includes the Oramil delayed arming system in which a safety ring, holding the detonator away from the rest of the explosive train, requires to be melted by heat generated by the ignition system, so giving an arming delay of about 1.2 seconds.

When required in the defensive role, the R-41 is enclosed by a separate fragmentation container which carries 3500 steel balls of 2 mm diameter. It is possible to load the container with balls of different sizes, and accordingly the quantity of fragments will be varied. The container is locked to the body of the grenade by a simple bayonet catch.

In 1987 the R-41 grenade, in competition with several other designs of Spanish and foreign grenades, won a contest organised by the Spanish Ministry of Defence to select a hand grenade for provision to the Spanish Army.

DATA
Length: 114 mm(defensive): 110 mm (offensive)
Diameter: 53 mm (offensive); 61.5 mm (defensive)
Weight: 242 g (offensive); 400 g (defensive)
Charge weight: 125 g
Delay time: 4.0 s

Manufacturer
Plasticas Oramil SA, Division AM, Apdo Correos 192, San Sebastian.

R-41 Oramil hand grenade in offensive and defensive forms

Status
Current. Production.

POI and POMI Oramil hand grenades

The Oramil POI grenade was among the earliest plastic grenades and has been in production for over 20 years. It consists of an ABS plastic body containing a fuze, detonator and bursting charge of 110 g of TNT. The percussion fuze is in the upper part of the grenade and is of the 'all-ways' pattern. The fuze is held safe during transit by a safety pin attached to a length of tape and a plastic 'apron', all of which are secured in place by the safety cap. The cap is removed and the grenade thrown, whereupon the air drag on the 'apron' causes the safety tape to unwind and pull out the safety pin, leaving the grenade armed. The process of unwinding the tape ensures that the grenade is 10 to 15 m away from the thrower before it is armed. Should the grenade fail to detonate on impact, the firing pin is automatically returned to a locked position and the grenade is then safe to handle.

The POMI grenade is a development from the POI and incorporates both percussion and delayed-action fuzing. When the safety cap is removed a small lever is exposed; moving this will set the fuze for either impact or time functioning. At the latter setting, which is shown by a 'V' in a small window, the fuze detonates four seconds after impact. Should both fuzes fail, the firing pin is locked in a safe condition and the grenade may be handled.

A metal splinter ring, made of steel wire, is available and can be slid over the bodies of both the POI and POMI grenades to enhance their effect.

DATA
Weight: (POI) 285 g; (POMI) 335 g; (splinter ring) 190 g

Manufacturer
Plasticas Oramil SA, Division AM, Apdo Correos 192, San Sebastian.
Status
POI: Current. Production.
POMI: Current. Not in production but capacity available.
Service
POI: Spanish and other armies.

POI and POMI Oramil hand grenades

PO plastic grenade launching cartridges

PO plastic projection ammunition is similar in construction to the plastic blank cartridges manufactured by Plasticas Oramil but are loaded with a propelling charge suited to the launching of grenades of all types. Although made of plastic (except for the head and extraction rim) these are considered to be war-grade ammunition, since it fulfils the ballistic requirement and is capable of withstanding all the environmental and other service rigours.

The cartridge is currently manufactured in 5.56 × 45 mm and 7.62 × 51 mm NATO calibres and consists of a one-piece plastic casing, a 2.3 g powder charge, a metal base and a percussion cap. The plastic material is self-coloured green in order to distinguish this type of cartridge from practice or blank ammunition. On firing, the charge splits the plastic head of the cartridge, releasing the gas which then projects the grenade. There is sufficient power to project a 473 g grenade to a range of 207 m.

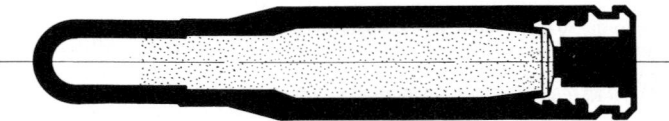

PO grenade launching cartridge

Manufacturer
Plasticas Oramil SA, Division AM, Apdo Correos 192, San Sebastian.
Status
Current. Production.

GL Type I, Model 61 rifle grenade

This grenade, manufactured by Instalaza, is an anti-tank grenade with a shaped charge warhead. It can penetrate up to 250 mm of armour plate and 650 mm of concrete. It can be employed with any rifle with a launcher or flash eliminator of 22 mm.

A practice grenade is produced for training purposes. The tail is made of high-strength alloy steel and allows the practice grenade to be used repeatedly.

DATA
COMBAT GRENADE
Diameter: 64 mm
Length: 395 mm
Internal diameter of fin assembly: 22 mm
Weight: (ready for firing) 700 g; (in carrying case) 850 g
Weight of projecting charge: 1.9 g
Max range: 300 m
Muzzle velocity: 54 m/s
Effective range: 100 m
Projecting cartridge: 7.62 mm contour with 5 pointed crimped ends
Minimum range: 5 m

Manufacturer
Instalaza SA, Monreal 27, Zaragoza, 50002.
Status
Current. Available.

GL Type 1, Model 61, rifle grenades with transit containers

Service
Spanish and other armed forces.

GL Type II, Model 63B rifle grenade

This grenade is manufactured by Instalaza. It is a shaped charge, fragmentation grenade for use against tanks and personnel which will produce penetration of armour plate of up to 130 mm and up to 350 mm in concrete. It may be used with any rifle with a 22 mm flash eliminator, or launcher of that dimension.

The sights are graduated from 75 to 200 m for direct fire and, with indirect fire, read 150, 200, 250, 300, 350, 400 and 425 m.

A re-usable practice grenade is also made.

DATA
Calibre: 40 mm
Overall length: 333 mm
Internal diameter of fin assembly: 22 mm
Weight: (grenade) 500 g; (grenade in case) 600 g
Weight of projection charge: 1.9 g
Weight of HE filling: 88 g
Max range: 425 m
Muzzle velocity: 75 m/s
Effective radius against personnel: 30 m
Lethal radius: 6 m
Number of fragments: approx 600
Fragment velocity: 1300 m/s

Manufacturer
Instalaza SA, Monreal 27, Zaragoza, 50002.

Firing the GL Type II, Model 63B rifle grenade

Status
Current. Production.
Service
Spanish and other armed forces.

TV dual-purpose rifle grenade

The TV rifle grenade can be launched from any 5.56 mm rifle with the standard 22 mm flash suppressor and using any kind of live ammunition, even the SS109 bullet.

This grenade features a shaped charge warhead inside a fragmentation casing, thus combining anti-armour capability with the anti-personnel effect. It can penetrate more than 110 mm of armour steel and breaks up into more than 550 lethal fragments. A density of at least one lethal fragment per square metre is obtained up to a distance of 8 m, and the fragments are still lethal at 20 m distance.

Two different practice grenades are also available, the TV-INS-1 and the TV-INS-20. The former is a one-shot grenade, and the latter is re-usable, being supplied with the appropriate number of spare launch cartridges and sights.

DATA
Calibre: 36 mm
Length: 345 mm
Weight: 440 g
Weight, cased: 520 g
Range, maximum: 300 m
Muzzle velocity: 65 m/s
Lethal radius: 8 m
Penetration: steel 110 mm; concrete 280 mm

Manufacturer
Instalaza SA, Monreal 27, Zaragoza, 50002.

TV dual purpose rifle grenade prepared for firing, with expendable sight

Status
Current. Production commenced 1989.
Service
Spanish armed forces.

SWEDEN

FFV 915 smoke rifle grenade

The FFV 915 smoke grenade has been developed to meet the tactical requirements of infantry where a quick and effective smoke screen is called for. The grenade is simple and safe to handle, operates in a wide temperature range, gives an effective screen within seconds of launch and is not smothered by snow or soft ground.

It is fired from the standard Swedish 7.62 mm rifle, or any other rifle with the standard 22 mm grenade launcher, and is projected by a special grenade cartridge. To ensure that ordinary ball rounds are not fired with the grenades, the launch cartridges are issued in a 20-round magazine which will not accept ball rounds.

The grenade is normally intended for low-angle fire and if fired to a range greater than 250 m it will begin to emit smoke while still in flight. The Swedish requirement is for two grenades to produce a screen 20 m wide and 4 m high at 250 m range in a wind speed of 5 m/s in open terrain at normal temperature within 10 seconds, which the 915 achieves easily.

The smoke composition consists of titanium dioxide and hexachlorethane ignited by black powder. Emission begins on the instant of ignition and dense and persistent smoke is produced which is emitted through the rear of the tail. Thus smoke can be released even if the grenade lands in soft ground.

Smoke screen after 10 s in a wind speed of 5 m/s

Smoke screen after 25 s

DATA
Weight: 570 g
Weight of smoke composition: 320 g
Length: 272 mm
Smoke emission: 59 s
Fuze: firing pin actuated by gas pressure
Delay: 3.8 s
Launch velocity: 65 m/s
Range: 300 m
Operating temperature range: –40 to +60°C

Manufacturer
FFV Ordnance, Eskilstuna.
Status
Current. No longer in production.
Service
Swedish Army.

Firing the FFV 915

FFV smoke hand grenade

FFV Ordnance manufactures a smoke grenade for exercise purposes. The smoke producing composition is of low toxicity. The grenade is of cylindrical metal canister type with a centrally located igniter. It can also be fitted with a plastic cone, making it possible to use the grenade in deep snow.

DATA
Diameter: 55 mm
Length: 106 mm
Weight: 560 g

Weight of smoke composition: 530 g
Delay: 6 s
Duration of smoke: 120 s

Manufacturer
FFV Ordnance, S-63187 Eskilstuna.
Status
Current. Production.

TAIWAN

The Nationalist Chinese forces use a variety of combat grenades, almost all of US origin. The Hsing Hua Company also manufactures a number of types in substantial quantities for use by the Taiwan army and for export. These include:

Type 67: A small grenade resembling the M67/68 types
Fragmentation grenade: based on the obsolete American Mk IIA1
Offensive grenade: Cylindrical body
Smoke grenade: Various colours, for screening and marking
Rifle grenade: Resembles the 40 mm Mecar type (see Belgium entry) and has a range of 300 m
Tear gas grenade: Filled with CS, CN or CN/DM mixture

Manufacturer
Hsing Hua Co Ltd, PO Box 8746, Taipei.
Status
Current. Production.
Service
Chinese Nationalist Army and others.

Fragmentation grenades manufactured by Hsing Hua Company

UNION OF SOVIET SOCIALIST REPUBLICS

Anti-personnel hand grenades

The Soviet Union has a number of anti-personnel hand grenades. Those produced before the Second World War are all officially obsolete but some of them have found their way to the Far Eastern area and have been used in Vietnam. Among these very early grenades are the RDG-33 and RPG-40.

RDG-33

This is a hand grenade with a delay detonator, functioning in both offensive and defensive roles. In the offensive mode it is a seamed tin plate cylindrical grenade with a screw thread to allow the attachment of a metal throwing handle. As an offensive grenade it has a lethal radius of 5 m. When converted to the defensive role a metal sleeve is placed over the cylindrical body holding the 85 g TNT filling and a stud on the metal cylinder engages in a recess in the sleeve to retain it in position. The sleeve is pre-notched in a diamond pattern. It then has a lethality radius of 25 m. Without the jacket it weighs 508 g and with the jacket 722 g. The delay time is 3.2 to 3.8 seconds. The grenade is obsolete in the USSR but was used in Vietnam and it still appears in odd corners of the world.

RPG-40

This is an anti-tank hand grenade, using an HE filling to produce a blast effect. It has a separate primer detonator fitted into a recess in the head of the grenade. It is stabilised in flight by a ribbon which streams behind it. It is effective against soft skinned and load carrying vehicles but has little chance of success with armoured vehicles of modern construction. It weighs 1105 g and has an instantaneous all-ways fuze. The 794 g of TNT give an effective fragmentation radius of 20 m. The Soviet wartime grenades are still encountered with militia units of the Warsaw Pact countries.

RPG-40 hand grenade

RDG-33 hand grenade

RG-42 anti-personnel hand grenade

This is a fragmentation concussion hand grenade which was used in the Second World War and was retained for some years in the Soviet Army. It was adopted by the Chinese Army as the Hand Grenade Type 42. Since then it has been taken by all the satellite countries and used by them for several years. It has now been relegated to the various militia bodies of these countries. It is obsolete in the USSR.

The grenade body is a plain steel, light gauge, cylinder with no serrations. It encloses a separate fragmentation sheet that is formed into a pre-grooved diamond shaped pattern. The grenade is employed in much the same way as any other delay fuzed type and should be thrown from behind cover. The UZRG fuze gives a delay of three to four seconds and the pressed TNT filling drives the fragments out to an effective radius of 25 m.

The grenades are packed, unfuzed, 20 to a wooden box, with the fuzes packed in a separate container in the same box. The packed box weighs 16 kg.

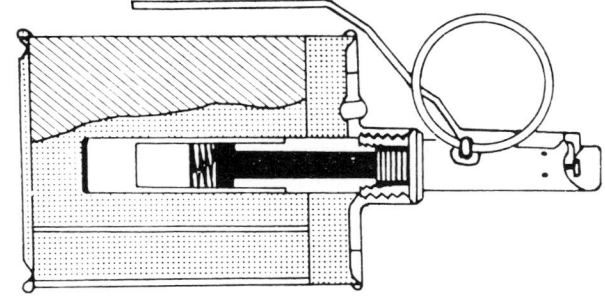

RG-42 anti-personnel hand grenade

DATA
Type: fragmentation
Weight: 436 g
Length: 121 mm
Max diameter: 54 mm
Body material: steel
Filler weight: 118 g
Filler material: pressed TNT
Fuze type: delay
Fuze delay: 3–4 s
Identifying markings: P-42
Range thrown: 35 m
Effective fragment radius: 25 m

Status
Obsolete, but still encountered in Africa and Far East.

RG-42 anti-personnel hand grenade

RGD-5 anti-personnel hand grenade

This is an egg-shaped anti-personnel fragmentation grenade with a smooth exterior surface to the two piece steel body and with a serrated fragmentation liner. It is a compact, easily handled grenade which can be thrown slightly further than the earlier Soviet defensive hand grenades. It uses the fuze UZRGM. The detonator assembly protrudes.

DATA
Type: fragmentation
Weight: 310 g
Length: 114 mm
Diameter: 57 mm
Weight of filling: 110 g
Type of filling: TNT
Fuze: 3.2–4.2 s delay UZRGM
Range thrown: 30 m
Effective fragment radius: 15–20 m

Status
Current.

RGD-5 anti-personnel hand grenade

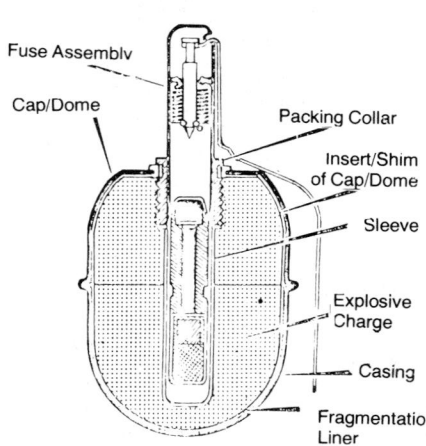

RGD-5 anti-personnel hand grenade

F1 anti-personnel hand grenade

This grenade was introduced during the Second World War. It is a fragmentation grenade with a cast iron body notched into cubes on the outside surface in a manner reminiscent of the US Mark II or the British Mills 36 grenade. It suffers from the same defects as the latter and produces a number of fragments from the base plug and filler which can be lethal out to 200 m and so the thrower would be well advised to throw the grenade from under cover.

The Polish Army has also introduced this grenade in a modified form with an impact fuze as the F-1/N 60 rifle grenade.

A later pattern of F1 has been seen; this has a fuze assembly of the 'mousetrap' type, similar to that on the US M26 or British L2A2 grenades. This new model is painted buff but otherwise resembles the old pattern F1.

Information previously given under this head, relating to varying delay times, has been found to be incorrect, and there is only one nominal delay time. Nonetheless, it is no more than prudence to examine the detonator assembly of any unfamiliar grenade before using it.

DATA
Type: fragmentation, defensive
Weight: 600 g
Length: 124 mm
Max diameter: 55 mm
Body material: cast iron
Filler weight: 60 g
Filler material: TNT
Fuze type: delay
Fuze delay: 3.2–4.2 s
Range thrown: 30 m
Effective fragment radius: 15–20 m

Status
Current.

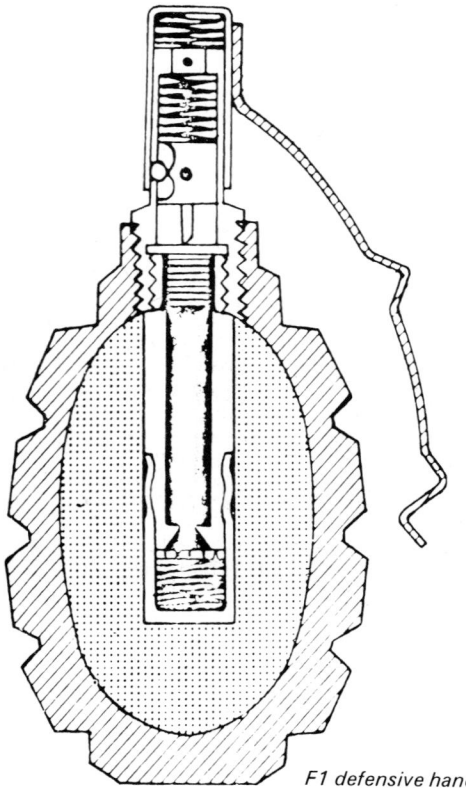

F1 defensive hand grenade

F1 fragmentation grenade. This grenade is usually painted olive green

M75 hand grenade

This is a grenade of modern pattern, using a plastic outer casing and pre-formed fragments, which appears to have been copied from the Arges HG79. Most specimens examined have been filled with Bulgarian explosive and fitted with Bulgarian fuzes.

DATA
Weight: 375 g
Length: 97 mm
Diameter: 57 mm
Delay time: 4 s nominal

Status
Current. Production.

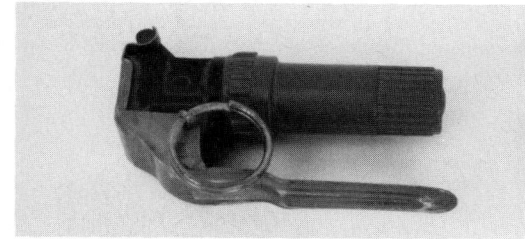

M75 fragmentation hand grenade

Igniter set removed from M75 grenade

RGO fragmentation hand grenade

The RGO (Ruchnaya Granata Oboronitel'naya) is a defensive hand grenade which has recently been introduced into Soviet service. It consists of a spherical pre-fragmented steel body formed of four hemispheres; the two inner hemispheres are internally segmented for fragment control. The outer hemispheres are also serrated to give regular fragments; the lower half has external serrations while the upper section is internally serrated and displays a smooth exterior to the top of the grenade body. There is a central detonator well which is threaded to accept the fuze assembly. This unit is of the usual 'mousetrap' type, with a fly-off lever and safety pin holding down a spring-actuated firing hammer. The complete grenade weighs 530 g, with an explosive filling of 92 g. The fuze has a nominal delay of 3.8 ± 0.5 seconds, and has a delay feature which prevents the grenade arming for between 1 and 1.8 seconds after release of the lever.

Manufacturer
State arsenals.
Status
Current. Production.

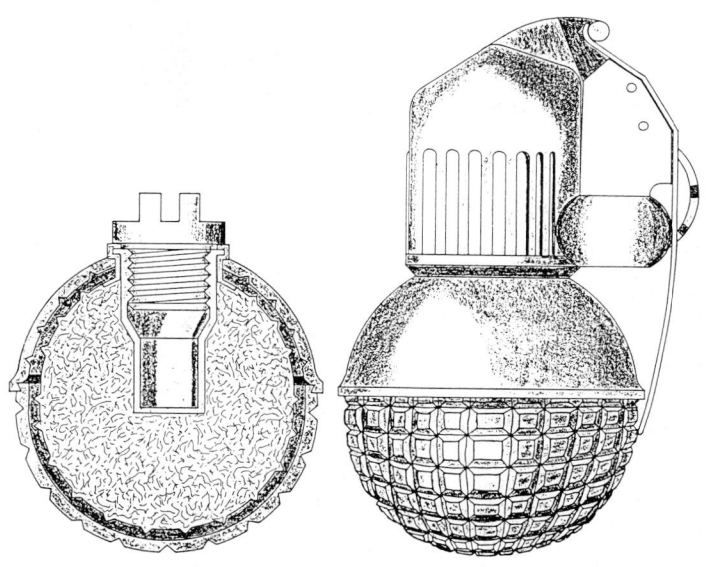

External and section drawings of RGO hand grenade

RGN fragmentation hand grenade

The RGN (Ruchnaya Granata Nastupatel'naya) is an offensive type hand grenade using a spherical pre-fragmented aluminium casing containing the high explosive bursting charge. As can be seen from the drawing, the grenade body consists of upper and lower hemispheres, both of which are internally serrated to give the desired fragmentation control. In similar fashion to the RGO there is the same central detonator well threaded to accept the same type of fly-off lever fuze mechanism. The two grenades can be told apart in darkness by the exposed serrated lower section of the RGO, which contrasts with the entirely smooth surface of the RGN. The RGN weighs 310 g, with an explosive filling of 114 g. The fuze timing and arming delay is the same as for the RGO grenade.

Manufacturer
State arsenals.
Status
Current. Production.

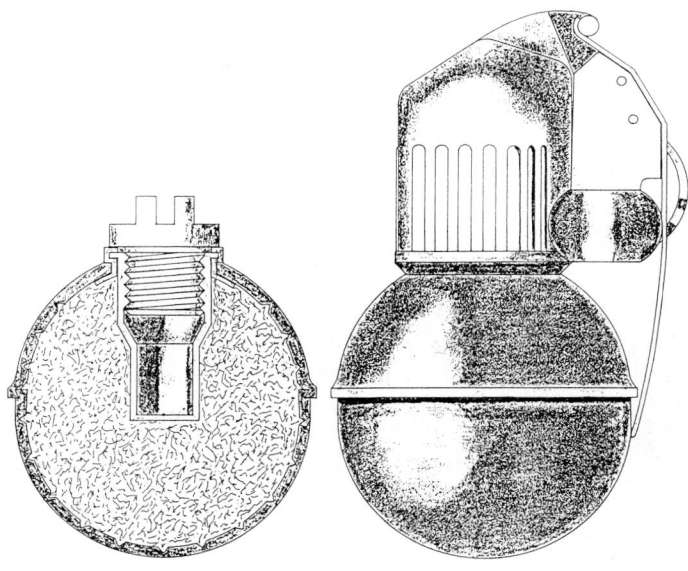

Soviet RGN hand grenade

RPG-6 anti-tank hand grenade

This anti-tank grenade was used in the Second World War by the Soviet Infantry and was subsequently adopted by most of the Eastern Bloc including China. Remarkably, it is still used for training and is identified from time to time in the hands of troops. It has absolutely no place on the modern battlefield, but it must be presumed that the Soviets and their allies look on it as a worthwhile weapon for guerillas and possibly for street-fighting by militia units.

It has several features in common with the RPG-43. It has four trailing cloth strips to stabilise it in flight. These are dragged out by the safety lever from the throwing handle, when the grenade is thrown. The RPG-6 can be distinguished by its hemispherical head on the cone shaped casing over the shaped charge. It is a more efficient anti-armour grenade than its predecessor, and because of its pronounced fragmentation effect it can be used as an anti-personnel grenade.

DATA
Type: anti-tank
Weight of grenade with fuze: 1100 g
Weight of HE filling: 562 g
Type of HE filling: TNT
Penetration: 100 mm
Type of fuze: instantaneous impact
Effective fragment radius: 20 m
Throwing distance: 15–20 m
Length: 343 mm
Diameter: 102 mm

Status
Long obsolete.
Service
Still found in training. See text.

RPG-6 HEAT hand grenade

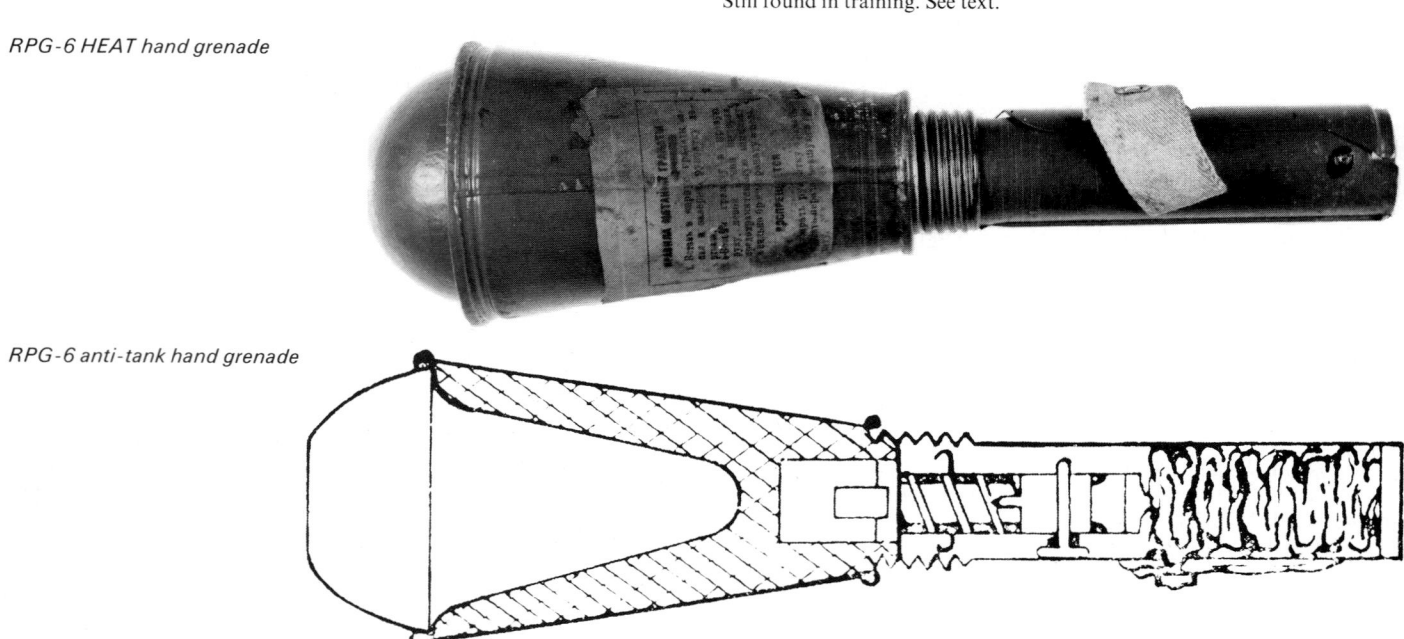

RPG-6 anti-tank hand grenade

RPG-43 anti-tank hand grenade

This is the earliest of the Soviet anti-tank hand grenades. It was intended for employment at very close ranges, against armoured vehicles and defensive, protected positions. To ensure that the head of the grenade strikes the target there is a stabiliser consisting of two fabric strips which pull a conical metal sleeve open at both ends, which trails behind the grenade during its flight through the air.

The fuze is an instantaneous impact type. The grenade was used by the Egyptian troops in the October War of 1973 with the Israelis. It is still seen occasionally in the hands of troops, particularly in training units. The remarks made of RPG-6 (see entry above) apply to this also.

DATA
Type: anti-tank
Weight of grenade with fuze: 1200 g
Weight of HE filling: 612 g
Type of HE filling: TNT
Penetration: 75 mm
Type of fuze: instantaneous impact
Effective fragment radius: 20 m
Length: 279 mm
Diameter: 102 mm

RPG-43 anti-tank hand grenade

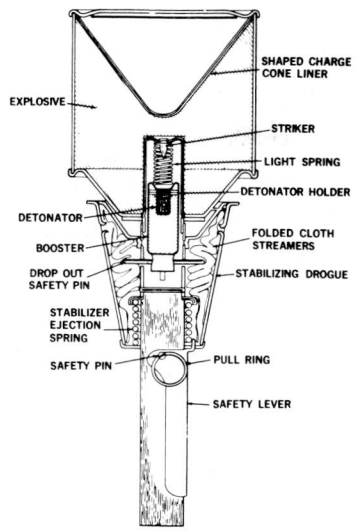

Soviet RPG-43 anti-tank hand grenade

RKG-3 anti-tank grenades

The family of RKG-3 grenades has now replaced the RPG-40, RPG-43 and RPG-6 anti-tank grenades in the Eastern Bloc and China.

The RKG-3 is a more modern type of grenade and is more efficient than those grenades it replaces. It is stabilised in flight by a four-panelled fabric drogue which is pulled out from the handle when the grenade is thrown. The drogue development completes the arming of the grenade. The drogue also ensures that it is possible to drop the grenade on the top of an armoured vehicle.

The earliest member of the family was the RKG-3 and this was capable of penetrating just under 5 in. Two further versions, the 3M and the 3T, have been produced. The RKG-3M has a slightly larger warhead with a copper cone instead of the steel liner of the first version and the 3T also has a different cone liner. The RKG-3M was used extensively in the 1973 October War and was shown to penetrate 165 mm of armour.

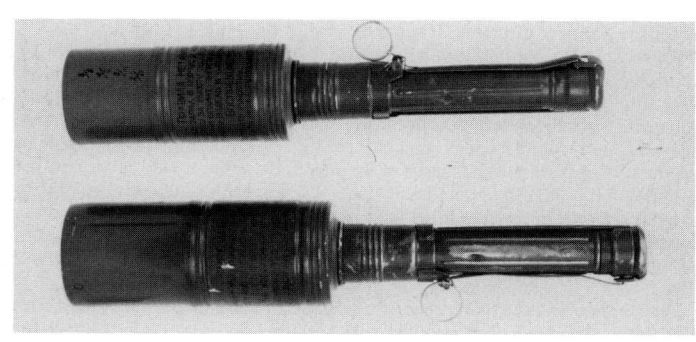

RKG-3 (top) and (lower) RKG-3M anti-tank grenades

DATA
Type: anti-tank
Weight of grenade with fuze: 1.07 kg
Weight of HE filling: 567 g
Type of HE filling: TNT/RDX
Penetration: (RKG-3) 125 mm; (RKG-3M) 165 mm
Type of fuze: instantaneous impact
Effective fragment radius: 20 m
Length: 362 mm
Max diameter: 55.6 mm

UPG-8 practice grenade
This is a practice anti-tank grenade used in training in place of the RKG-3 series of anti-tank grenades.

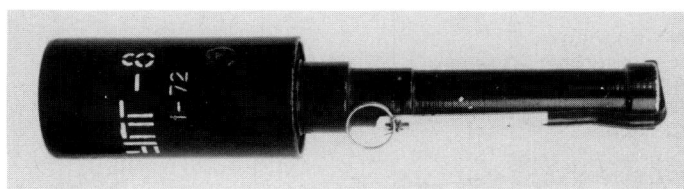

The UPG-8 practice grenade

RDG-1 smoke hand grenade

This is a stick type grenade. The earliest version, adopted about 1948, was made of grey moulded cardboard. Later versions are said to have a wooden handle. There is an inner cardboard tube with an igniter at each end and in the middle. The friction igniter is rubbed on the match head fuze at the wide end of the grenade to ignite the filling which can be either a white or a black smoke mixture. The grenade will float and can be used to produce screening smoke during the progress of a water crossing operation.

DATA
Weight: 500 g
Length: 222 mm
Throwing distance: 35 m
Average smoke area: 460 m²
Burning time: 1–1½ mins

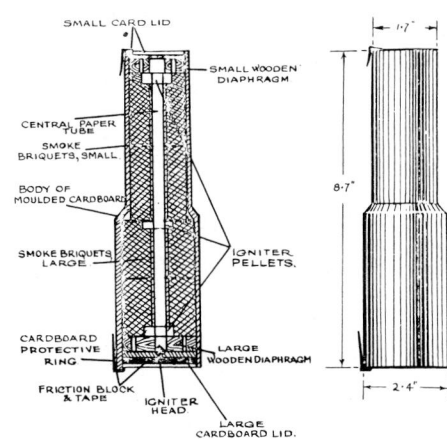

RDG-1 smoke hand grenade

RDG-2 smoke hand grenade

This grenade has been in service since the early 1950s and has been adopted by all the Soviet Bloc countries. It is a tactical grenade and is used to conceal the movements of small bodies of infantry or engineers. It is made of cardboard, with a cardboard tube down the centre. It is filled with a burning type filler and a friction fuze. It is waxed and is damp proof but it cannot be used to produce smoke over water. It produces a dense white screen of smoke approximately 20 m long and 8 m wide.

DATA
Length: 240 mm
Diameter: 46 mm
Weight: 500 g
Throwing range: 35 m
Smoke area: 160 m^2
Duration of smoke: 1$^1/_2$ mins

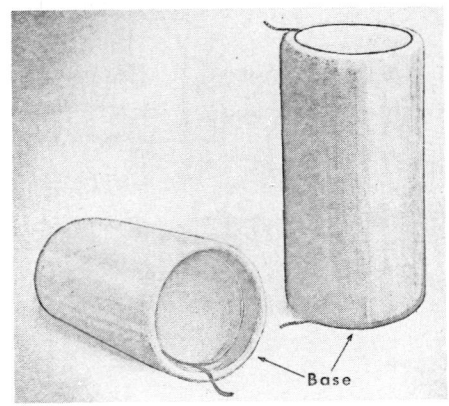

RDG-2 smoke hand grenade

UNITED KINGDOM

No 36M hand grenade

Nomenclature: Grenade, Hand, No 36M
The No 36M hand grenade is an improved form of the Grenade, Hand, No 36, modified for service in Mesopotamia in 1918 and especially sealed against moisture. It is an anti-personnel grenade which is projected by hand. It has a thick, cast iron body which is notched on the outside. The original idea was that the body would break up into pieces corresponding to the notched segments but it was found that the fragment size is in no way related to the notches and in fact the grenade produces a relatively limited number of very large fragments and also a lot of cast iron dust. The large fragments and the base plug give rise to a danger area which extends about 275 m from the point of burst and the thrower must be under cover. Weight of the filled grenade is 774 g.
    The grenade was also used as a rifle grenade and in this role was fitted with a gas check plate of steel. A cup launcher was used on the rifle. In this role the grenade had a range of about 180 m and was fitted with a seven second delay fuze. The components used in the rifle projected version have long been declared obsolete.
    The No 36M hand grenade has now been replaced by the Grenade, Hand/Rifle, Anti-Personnel L2A1.

Manufacturer
Royal Ordnance plc,Ammunition Division, Chorley, Lancs, PR7 6AD.
Status
Obsolete.
Service
No longer in the UK, but used by many Commonwealth and ex-Commonwealth countries.

No 36M hand grenade

No 80 WP Mark 1 and Mark 1/1 grenade

Nomenclature: Grenade Hand/Discharger No 80 WP
The No 80 WP is intended to produce screening smoke and may be either thrown or projected from a multi-barrel discharger on an armoured fighting vehicle, using a Fuze Electric No F 103. It has a tin plate body filled with white phosphorus, Detonator No 75 and Striker Mechanism Grenade No 2.

Striker mechanism grenade No 2
This consists of an adaptor with a screwed-in housing for a spring operated striker which is retained in the cocked position by a fly-off lever and safety pin.

Detonator No 75
The Detonator No 75 consists of a .22 in rimfire cap attached to a 1.5 in (38 mm) length of Fuze Grenade No 1 and a cap chamber. The fuze gives a delay of 2.5 to 4 seconds and then initiates a Detonator No 63 Mk 2 or No 78 Mk 1.

Fuze electric No F 103
This has a brass gunpowder magazine, containing G 20 gunpowder, closed by a brass cover into which is set a Fuze Electric No 53 Mark 2.

Drill grenade hand/discharger No 80 Mark 1
This is the body of a service grenade with an inert filling.

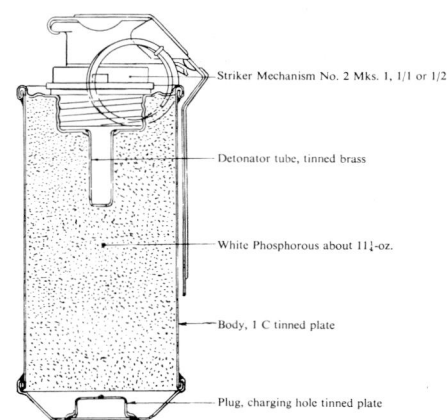

No 80 WP Mark 1 grenade

Drill detonator No 75 Mark 1
This is an empty .22 rimfire cap in a cap chamber and connected to a dummy fuze and an empty detonator tube. It is painted white with a hole through the tube of the detonator.

Manufacturer
Royal Ordnance plc, Bishopton, Strathclyde, Scotland.
Status
Obsolete. Stocks still exist.

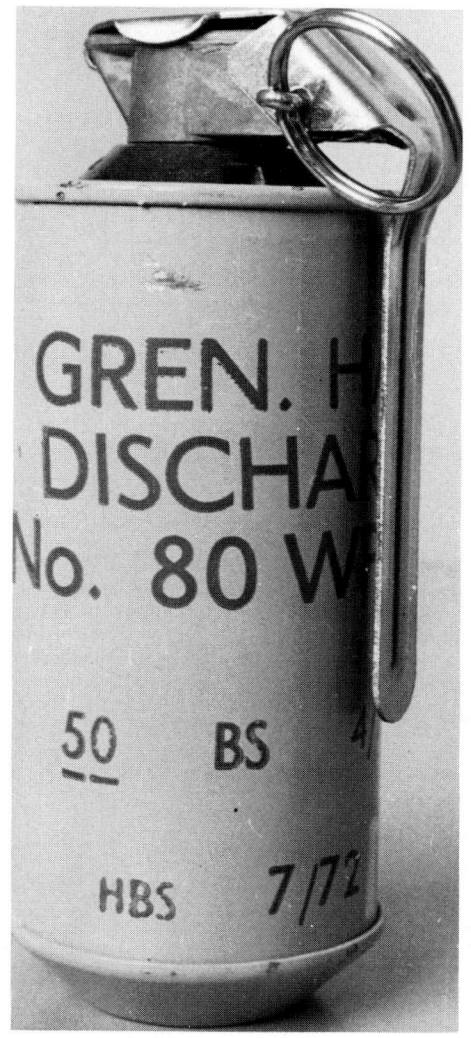

No 80 WP Mark 1 hand/discharger grenade

L2A2 anti-personnel hand grenade

The grenade L2 is a high explosive anti-personnel grenade based on the US M26 grenade and closely resembling it in appearance and performance. It differs in having a separate fuze assembly. It was designed to be thrown by hand or projected from a rifle but the British army abandoned the use of rifle-launched grenades.

The L2 grenade superseded the No 36M hand grenade. It exists in two variants, L2A1 and L2A2, but the only significant difference between the two is in the design of the fuze holder which was modified to assist production.

The grenade consists of a two part, tinned plate outer casing, a coil of notched wire, a fuze (L25), holder, cap and HE filling.

The body assembly (upper and lower) holds the fuze holder in the upper part and the two parts are 40.6 mm (upper) and 36.8 mm (lower) in length and circular in section. The coil is made from steel wire 2.4 mm in section and notched at intervals at 3.2 mm. The HE filling consists of 170 g of RDX/TNT.

There are a number of models of the L25 fuze, current issue being the L25A6. The fuze consists of a striker assembly and an adaptor assembly with detonator, delay pellet and cap housing.

Practice and drill grenades
Practice grenades L3A1, 2 and 3 and drill grenades L4A1 and 2 are similar in construction and appearance to the operational grenades. Practice grenades are inert and are painted light blue. Drill grenades are fitted with Drill Fuze, Grenade, Percussion L30 and are painted dark blue.

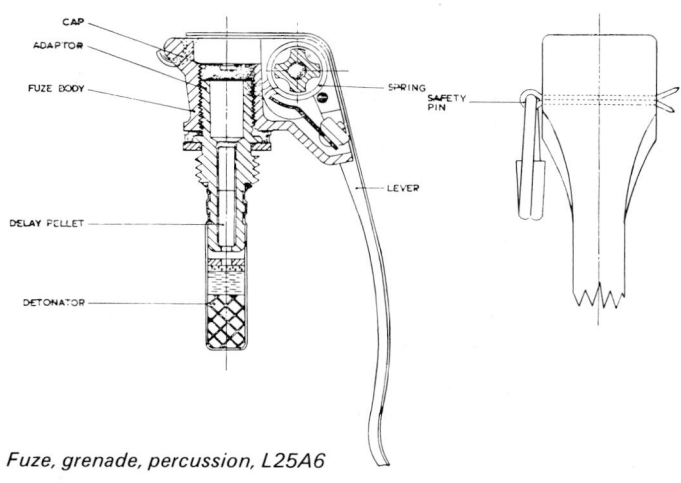

Fuze, grenade, percussion, L25A6

L2A2 anti-personnel hand grenade

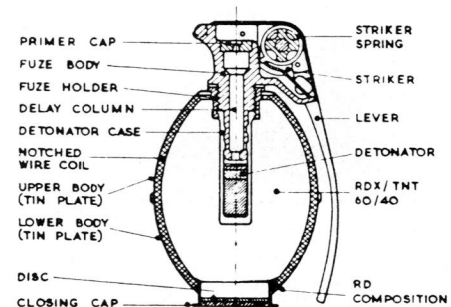

L2A2 anti-personnel
hand grenade

DATA
Length: 84 mm
Weight: 395 g
Max diameter: 60 mm
Explosive filling: RDX/TNT 55/45
Weight of explosive: 170 g
Fuze: L25A6
Delay: 4.4 s ± 0.5 s
Lethal radius: 10 m

Manufacturer
Royal Ordnance plc, Ammunition Division,Chorley, Lancs, PR7 6AD.
Status
Current. Production.
Service
British forces.

L8A4 smoke grenade

L8A4 grenades are electrically initiated smoke grenades for the screening and protection of armoured vehicles. They are fired from fixed launcher tubes attached to the hull or turret.

The grenade consists of a black rubber body, a closing plug and a metal housing containing the propelling charge and the fuze. The rubber body is filled with a mixture of 95/5 red phosphorus and butyl rubber. A plastic burster tube containing gunpowder passes through the body. At its top, in the metal housing, is the delay and this is ignited by the propellant. In the metal housing is a small propelling charge of gunpowder which throws the complete grenade clear of the vehicle.

When fired, the grenades burst clear of the vehicle and explosively disseminate the ignited RP pellets which then produce an almost instantaneous cloud of white smoke round a frontal arc, entirely concealing the vehicle.

DATA
Length: 185 mm
Diameter: 66.3 mm
Weight: 680 g
Range: 20–30 m
Height of burst: 6 m
Delay time: 0.75 s
Screen width: approx 35 m
Minimum duration of smoke cloud: (wind velocity at 24 km/h) 3 mins
Operating voltage: 3 V

Manufacturer
Royal Ordnance plc, Ammunition Division.
Status
Current. Production.
Service
UK and US armies.

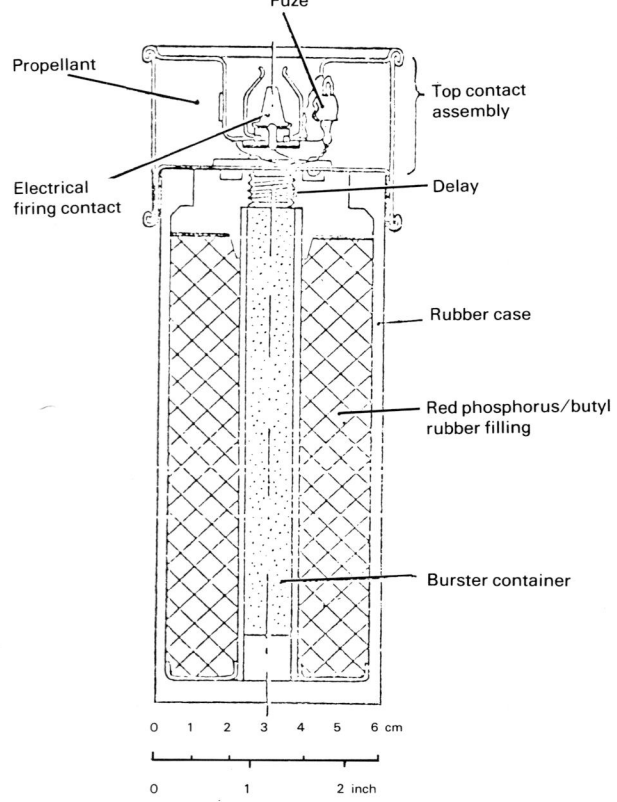

L8A4 grenade, sectioned to show interior arrangements

Haley and Weller V130 66 mm fragmentation grenade

This is an anti-personnel grenade designed to be launched from the standard 66 mm vehicle-mounted launchers, originally intended for discharging smoke grenades. The fragmentation grenade contains several thousand steel ball bearings embedded in a plastic matrix surrounding a central tube containing the bursting charge. The grenade is electrically fired and launched in the same manner as a smoke grenade, but bursts in the air, about 2 m above the ground, and distributes the fragments in an all-round pattern. It is an ideal anti-ambush device and is also valuable for defending vehicles against close-in infantry at ranges where the vehicle weapons cannot be brought to bear.

DATA
Length: 182 mm
Diameter: 66 mm
Weight: 1.48 kg
Explosive content: 14 g
Range: 100 m
Lethal area: 15 m
Danger area: 50 m

Manufacturer
Haley and Weller Ltd, Wilne, Draycott, Derbyshire, DE7 3QJ.
Status
Current. Production.

Haley and Weller 66 mm fragmentation grenade

Haley and Weller E105 fragmentation hand grenade

This is a high explosive fragmentation grenade for infantry use. It is supplied in two separate parts, one being the body with explosive charge and pre-formed fragments, the other being the electrical initiation set. The initiation set is of Haley and Weller's patented design which is completely silent in operation. The manipulation of the grenade is the same as for a conventional type, the thrower removing the safety pin and releasing the lever as he throws; but the lever does not fly off, therefore cannot make a noise when striking the ground, and there is no percussion element to be struck and cause noise. Initiation and timing is entirely electrical and silent, and the timing is therefore extremely accurate and consistent.

In addition to its normal use for throwing, the grenade can also be fitted with the tripwire mechanism E190 for use in perimeter defence, and it can be wired into a grenade necklace without requiring alternative detonators. The grenade is completely waterproof and has a shelf life of five years.

DATA
Length: 130 mm
Diameter: 50 mm
Weight: 580 g
Explosive content: 51 g
Number of fragments: approximately 2 000
Delay time: 4.0 – 4.5 s

Manufacturer
Haley and Weller Ltd, Wilne, Draycott, Derbyshire, DE7 3QJ.
Status
Current. Production.
Service
Middle and Far Eastern countries.

Haley and Weller E105 fragmentation hand grenade

Haley and Weller Incendiary grenade E108

This grenade employs the same silent ignition as the other grenades in this series. It has been developed for use as a sabotage and destruction weapon. The grenade is functioned in the standard manner, but in addition can be fired remotely by means of the external terminals. It burns in excess of 2700°C and will melt through 3 mm of steel.

DATA
Length: 134 mm
Diameter: 50 mm
Weight: 530 g
Explosive content: 400 g
Burn time: 60 s

Manufacturer
Haley and Weller Ltd, Wilne, Draycott, Derbyshire, DE7 3QJ.
Status
Current. Production.
Service
Various European countries.

Haley and Weller E108 incendiary grenade

Schermuly Mk 4 screening smoke grenade

This design is a hand-thrown grenade consisting of a metal body containing a solid block of smoke-producing composition (HCE) and a fly-off lever ignition system. Smoke emission is through apertures at the top of the body; the smoke is grey-white for maximum obscuration and is formulated to give a rapid build-up. The grenade has a strengthened base to allow it to be fired from a rifle using a discharger cup and launching cartridge.

Ignition of the grenade is instantaneous so as to contribute to the rapid screening effect.

The screening smoke grenade is used for the rapid concealment of troops, vehicles or buildings in any situation where it is necessary to degrade the line of sight.

The product is approved by the UK Ordnance Board and will shortly enter service with the UK armed forces.

A new low-toxicity screening smoke grenade is now available from Pains-Wessex. This product, while still maintaining the dense obscuration requirement, is safe for troops to use in training without the effects of conventional screening smokes which are considered to be potentially hazardous to health and the environment.

DATA
Length: 138 mm
Diameter: 62 mm
Weight: 650 g (HCE): 463 g (low toxicity)
Pyrotechnic weight: 450 g HCE; 284 g low toxicity
Burning times: 120 s or 60 s HCE; 60 s low toxicity
Product codes: (120 s HCE) 1719; (60 s HCE) 1720; (60 s low-toxicity) 1722

Manufacturer
Schermuly (Pains-Wessex Ltd), High Post, Salisbury, Wilts SP4 6AS (a member of the Chemring Group plc).
Status
Current. Available.
Service
Worldwide.

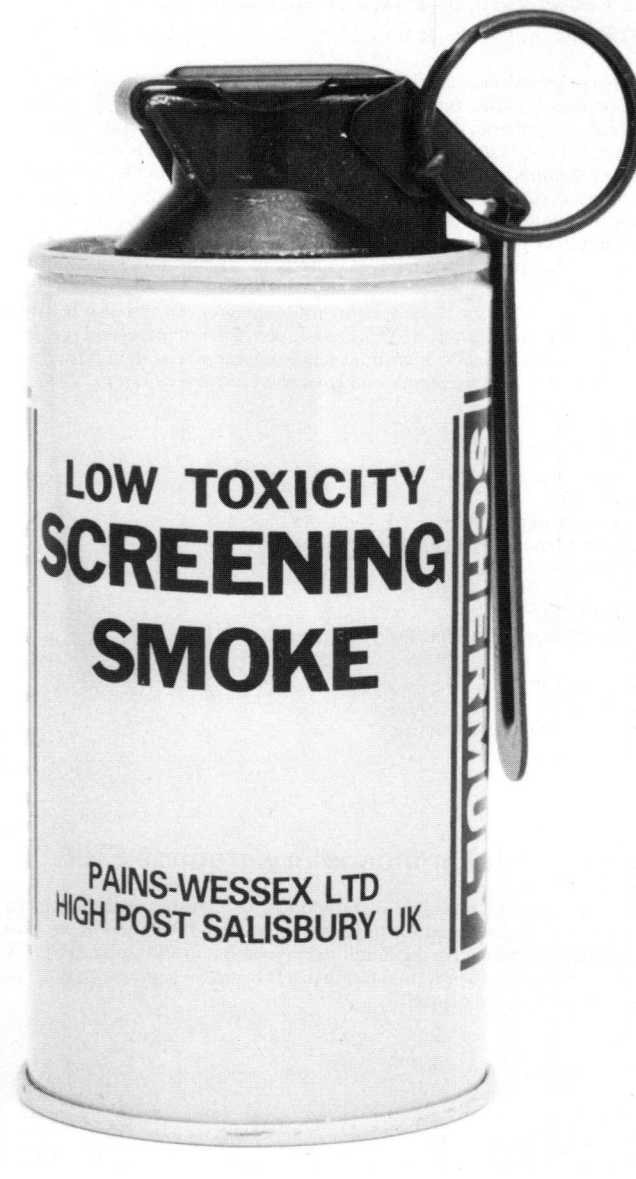

Schermuly screening smoke grenade Mk 4

66 mm Schermuly vehicle discharged grenade — screening

This product has been developed to allow a well-proven grenade design to be used from a vehicle. It is intended to be fired from any 66 mm discharger common to most armoured fighting vehicles.

The grenade consists of a discharger base fitted with the electrical jackplug contact and the grenade payload screwed to the base. Both parts of the grenade are ejected from the discharger. A pull ring is incorporated in the top of the grenade to facilitate withdrawal from the launcher in the event of non-use.

Smoke emission begins on ejection and full volume is achieved by the time the grenade has landed. The screen is produced by a dense greyish-white opaque smoke which develops rapidly. Unlike phosphorus screening smokes, this smoke will not burn exposed parts of the body or start fires and is therefore suitable in both operational and training roles.

The grenades are used in situations where an effective screen is required at a distance from the user to prevent accurate enemy fire and conceal one's own withdrawal.

DATA
Operating voltage: 1.5 V minimum
Range: 30 m nominal
Delay: instantaneous ejection on firing, 5 s from firing to full volume
Burning time: 30 s nominal
Composition: HCE type
All-up total weight: 745 g in primary pack
Pyrotechnic weight: 300 g
Length: 151 mm
Diameter: 66 mm

Manufacturer
Schermuly (Pains-Wessex) Ltd, High Post, Salisbury, Wiltshire SP4 6AS (a member of the Chemring Group plc).
Status
Current. Available. Product code 4028.

66 mm Schermuly vehicle discharged screening smoke grenade

Service
Middle East.

76 mm and 66 mm Vehicle Discharged screening smoke grenade, red phosphorus

The 76 mm screening smoke grenade is based on a red phosphorus composition which gives excellent visual screening with a very rapid build-up of smoke. It is fired from existing 76 mm launcher systems fitted with side electrical connectors and gives an almost instantaneous smoke screen.

In light wind conditions of up to 15 mph (24 km/h), a salvo of eight smoke grenades produces an effective visual screen within two seconds of being fired and over an arc of 80°.

In addition there is a low toxicity practice version for training.

Both these grenades are also available in 66 mm calibre

DATA
Operating voltage: 1.5 V minimum
Operating current: 0.76 A minimum
No-fire current: 0.3 A maximum
Screen duration: 90 s nominal
Screen height: 5 m minimum
Range: 20 – 30 m from vehicle
Length: 274 mm
Diameter: 76 mm
Schermuly Product Codes: 76 mm VDSS – RP: 1823; 76 mm VDSS – low toxicity: 1824

Manufacturer
Schermuly (Pains-Wessex) Ltd, High Post, Salisbury, Wilts SP4 6AS (a member of the Chemring Group plc).
Status
Current. Available.

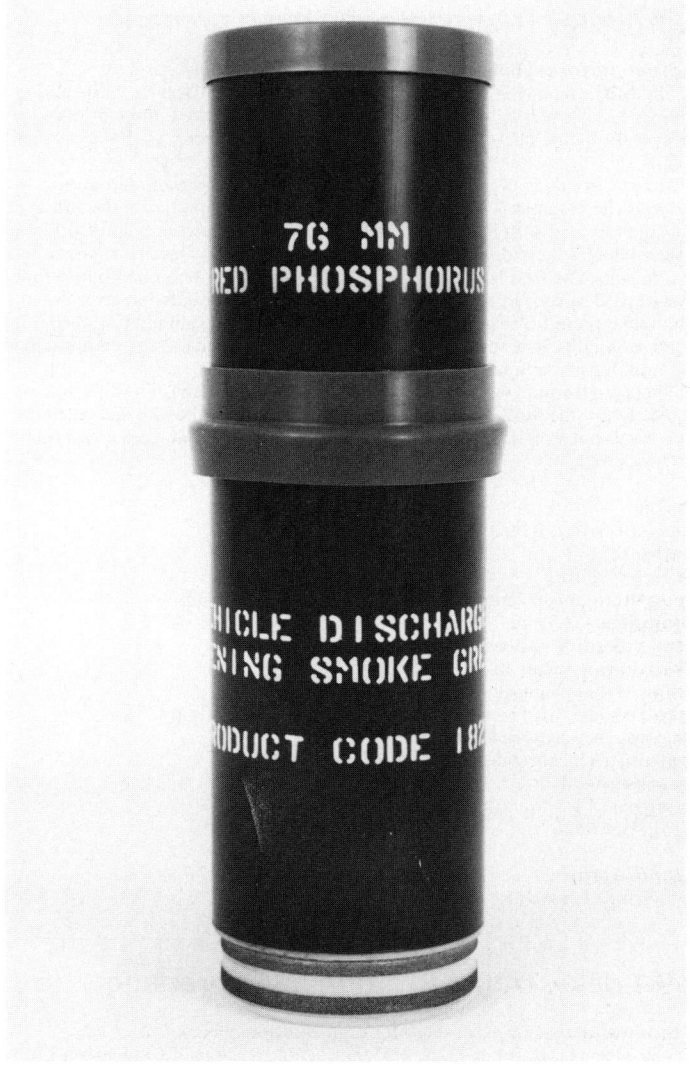

Schermuly 76 mm vehicle discharged screening smoke grenade

Brocks screening smoke grenade

This is a hand grenade, using the conventional type of pin-and-lever ignition set, intended to be used for screening purposes. It emits a dense cloud of grey smoke about one second after initiation, and is ideal for screening troop movements for a period of about one minute without the hazards often attendant upon the use of white phosphorus in this role. Robust and waterproof, it can be used in any climatic conditions.

DATA
Length: 106 mm
Diameter: 70 mm
Delay: 1 s
Duration: 60 s

Manufacturer
Brocks Pyrotechnics, Sanquhar, Dumfries and Galloway DG4 6JP, Scotland.

Ferranti EDNON electronic grenade fuze

EDNON is an electronic delay fuze for chemical hand grenades which provides a safer more reliable fuze than current pyrotechnic fuzes with more precise control of the time delay within the range 0.55 to 4.05 s. Operation is compatible with conventional grenades, the thrower removes the safety pin, retaining pressure on the safety lever. On release the safety lever operates the magnetic power source and initiates the time delay circuitry. The lever does not fly off on throwing, making the operation silent.

There are numerous safety features in the fuze including: ring pull and split pin lock, safety pin and safety lever, the power source is shorted out until the safety pin is released, electrical energy is generated only after release of the safety lever and safety pin, early impact sterilisation option available to ensure safety in the event of the grenade being accidentally dropped.

The fuze assembly is 55 mm in diameter and 33 mm high, it weighs 80 g.

Manufacturer
Ferranti International, Weapons Equipment Department, Moston, Manchester M10 0BE.
Status
Advanced development.

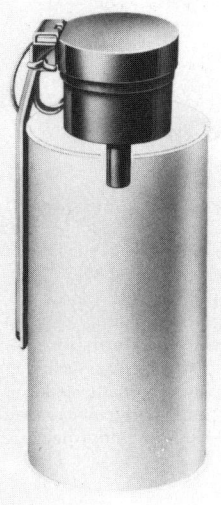

Ferranti EDNON electronic hand grenade fuze

UNITED STATES OF AMERICA

M67 delay fragmentation hand grenade

Nomenclature: Grenade, Hand: Fragmentation, Delay, M67

The M67 grenade is one of the standard grenades in service with the US Army. It is a small spherical grenade which is time-fuzed and can only be thrown by hand. There are no arrangements for projecting it by any other means.

The steel body is 63.5 mm in diameter and it breaks up on detonation to provide the fragments. The M213 fuze is integral with the body and the grenade is issued with its fuze *in situ*. The fuze is activated by a conventional striker which is held down by the safety lever. The safety lever is retained by two means. The first is the usual split pin which must be pulled out before throwing. The second is a small wire clip, which also holds the lever down. This clip is intended to act as a second safety should the split pin be pulled out unintentionally. In throwing, it is usual to remove the clip first, and leave the pin until the last moment.

There is a training version of the M67, known as the M69. The difference is in the colour, the M69 is blue with a brown band, and in the fact that when the fuze burns out it emits a small, sharp crack like a firework and releases a small puff of smoke.

DATA
GRENADE (WITH FUZE)
Body: steel
Weight: 0.39 kg
Length: (max) 89.7 mm
Diameter: 63.5 mm
Max casualty radius: 15 m
Throwing range: 40 m
Filler: (type) Comp B; (weight) 0.18 kg
Fuze: (model) M213; (type) pyrotechnic delay-detonating
Primer: (percussion) M42
Detonator: lead azide, lead styphnate, and RDX
Delay time: 4–5 s
Weight: 71 g
Length: 85 mm

Manufacturer
Government arsenals.

M67 fragmentation hand grenade

Status
Current. In production.
Service
US armed forces.

M61 delay fragmentation hand grenade

Nomenclature: Grenade, Hand: Fragmentation, Delay, M61

The M61 is a standard issue HE fragmentation grenade which is a little larger than the M67. It has been in service for some years and it formed the original pattern for the British L2 model. The major improvement of this grenade over the M67 is that it has a notched coil liner and so has regular and predictable fragments in the pattern. It is slightly heavier than the M67 and so cannot be thrown quite as far, but in overall effectiveness it ought to be a good deal better.

The fuzing and safety arrangements are the same as for the M67.

DATA
Body: thin-wall sheet steel with inner fragmentation coil
Weight: 0.45 kg
Length: (max) 99 mm
Diameter: 57 mm
Effective casualty radius: 15 m
Filler: 156 g Comp B and 8 g tetryl pellets
Fuze: (models) M204A1, M204A2; (type) pyrotechnic delay-detonating
Primer: (percussion) M42
Detonator: lead azide, lead styphnate, and RDX
Delay time: 4–5 s
Weight: 73 g
Length: 102 mm

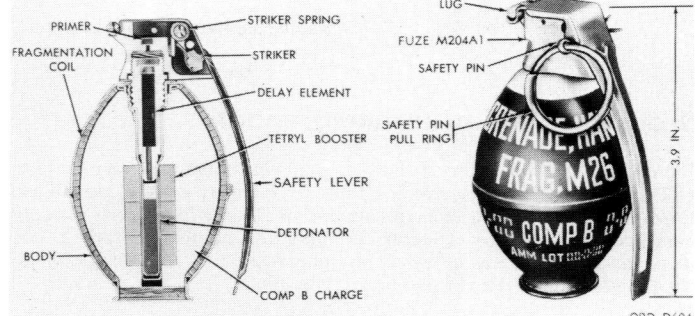

M61 fragmentation hand grenade

Manufacturer
US Army Material Readiness Command.
Status
Current. Production.
Service
US armed forces.

M68 impact fragmentation hand grenade

Nomenclature: Grenade, Hand: Fragmentation, Impact, M68

Essentially, the M68 is the M67 fitted with a different fuze. The bodies and fillings are the same, as are the lethal radii and throwing range.

The fuze is the M217 and is activated in exactly the same way as the fuze in the M67: by releasing a safety lever as the grenade is thrown. The M217 fuze, however, is a rarity in that it is an impact fuze with a time delay as a back-up. The impact part of the fuze acts by an electrical detonator and a tiny thermal power supply which is started by a powder train set off by the percussion cap. This thermal power supply requires a second or two to generate sufficient electricity, but after that the detonator will be fired if the grenade strikes a hard surface or is sharply jolted. Should the impact action fail for some reason the powder train continues burning and sets off the detonator by pyrotechnic action within seven seconds.

This complicated and clever fuze is contained within a very small space in the body and is apparently perfectly reliable. The only effect of the size of the fuze is that temperature alters the delay times which are quoted as lying between three seconds at +52°C and seven seconds at –40°C, with a rough mean of 4½ seconds at normal ambient of 20°C.

DATA
Body: steel
Weight: 0.39 kg
Length: (max) 126 mm
Diameter: 64 mm
Filler: (type) Comp B; (weight) 184 g

Fuze: (model) M217; (type) electrical impact with overriding delay function feature
Primer: M42
Detonator: lead azide, lead styphnate, PETN
Delay time: 3–7 s
Weight: 169 g
Length: 76.2 mm

Manufacturer
US Army Material Readiness Command.
Status
Current. Production.
Service
US armed forces.

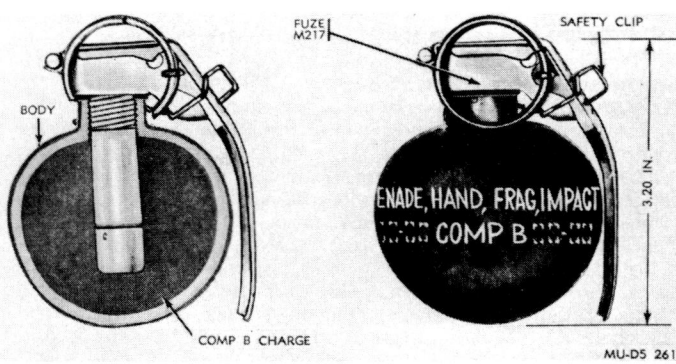

M68 fragmentation hand grenade

M57 and M26A2 impact fragmentation hand grenade

The M57 hand grenade is the M26A2 hand grenade with a safety clip. Each grenade is assembled with an electrical impact fuze.

Grenade body: bodies of the M61, M26A1 and M26 are identical to the M26A2 except the fuze threads are different. The body, constructed of two pieces of thin-wall sheet steel, has a notched fragmentation coil liner. Bodies contain a high explosive filler. Bodies of the M61, M26A1 and M26 contain booster pellets and are longer and narrower than those of the M26A2 and M57. Bodies of M26A2 and M57 do not contain booster pellets.

Fuze, hand grenade, M217: see description in the entry for the M68 grenade above.

Safety clip: the hand grenade safety clip is designed to keep the safety lever in place, should the safety pin be unintentionally removed from the grenade. It is an additional safety device used in conjunction with the safety pin.

Clips of spring steel wire consist of a loop, which fits around the neck of the grenade, and a clamp, which fits over the safety lever.

The safety clips on the hand grenades M68, M61, M67 and M57 are not interchangeable.

DATA
GRENADE (WITH FUZE)
Models: M57, M26A2
Body: thin-wall sheet steel with notched fragmentation coil
Weight: 454 g
Length: (max) 99 mm
Diameter: 57 mm
Filler type: Comp B with tetryl pellets
Comp B weight: 156 g
Tetryl pellets weight: 8 g
Fuze: (model) M217; (type) electrical impact with overriding delay function feature
Primer: M42

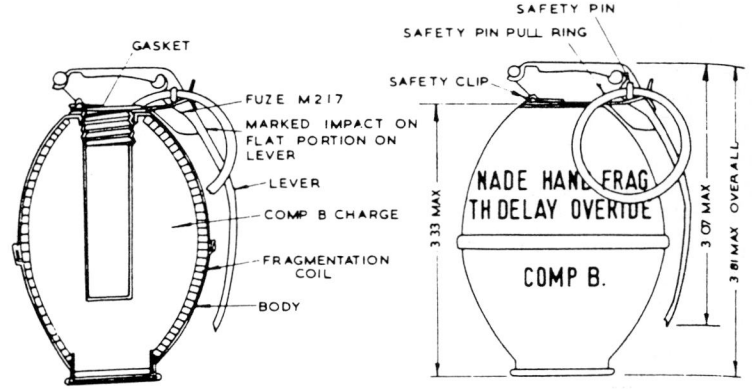

M26 hand grenade

Detonator: lead azide, lead styphnate, PETN
Delay time: 3–7 s
Weight: 76 g
Length: 76.2 mm

Manufacturer
US Army Material Readiness Command.
Status
Current. Production.
Service
US armed forces.

AN-M8 HC smoke hand grenade

The AN-M8 HC smoke hand grenade is a burning type grenade used to generate white smoke for screening activities of small units. It is also used for ground-to-air signalling. The duration of smoke screen or signal is 105 to 150 seconds. Throwing distance for an average soldier is said to be 30 m.

Grenade body: the grenade body is a cylinder of thin sheet metal. It is filled with HC smoke mixture topped with a starter mixture directly under the fuze opening.

Fuze, hand grenade, M201A1: fuze M201A1 is a pyrotechnic delay-igniting fuze. The body contains a primer, first-fire mixture, pyrotechnic delay column, and ignition mixture. Assembled to the body are a striker, striker spring, safety lever and safety pin with pull ring. The split end of the safety pin has an angular spread.

Safety clips: safety clips are not required with these grenades.

DATA
GRENADE (WITH FUZE)
Model: AN-M8
Body: sheet metal
Weight: 681 g
Length: 145 mm
Diameter: 63.5 mm
Burning time: 105–150 s
Filler: (type) HC (type C); (weight) 539 g
Fuze: (model) M201A1; (type) pyrotechnic delay-igniting
Primer: M39A1
Ignition mixture: iron oxide, titanium, zirconium
Delay time: 0.7–2 s
Weight: 43 g
Length: 99 mm

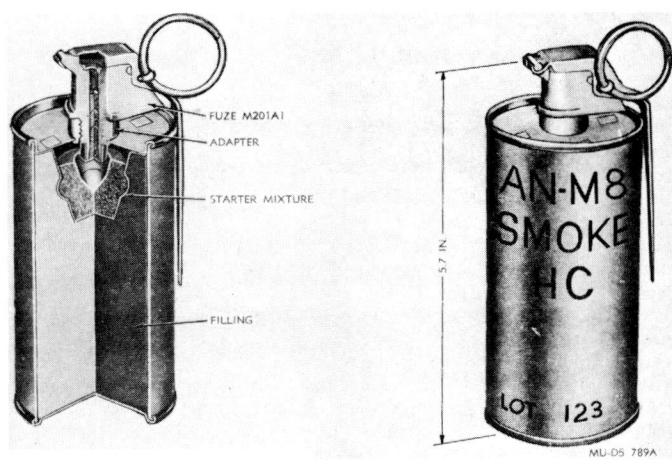

AN-M8 HC smoke hand grenade

Manufacturer
US Army Material Readiness Command.
Status
Current. Production.
Service
US armed forces.

AN-M14 TH3 incendiary hand grenade

Nomenclature: Grenade, Hand: Incendiary, TH3, AN-M14
The AN-M14 TH3 incendiary hand grenade is used primarily to provide a source for intense heat to destroy equipment. It generates heat to 2200°C. The grenade filler will burn from 30 to 45 seconds. This was developed by the Chemical Warfare Service in response to a military requirement for a hand held incendiary device that would allow the individual soldier to destroy equipment. Over 800 000 M14s were made, with final deliveries by Ordnance Products Incorporated in 1970. The grenade is normally hand thrown, although it may be rifle launched by using a special M2 series projection adaptor. Throwing distance for an average soldier is said to be 25 m.
Grenade body: the grenade body, of thin sheet metal, is cylindrical in shape. It is filled with an incendiary mixture, Thermite TH3 and First Fire Mixture VII.
Fuze, hand grenade, M201A1: see description in the entry for the AN-M8 grenade above.
Safety clips: safety clips are not required with these grenades.

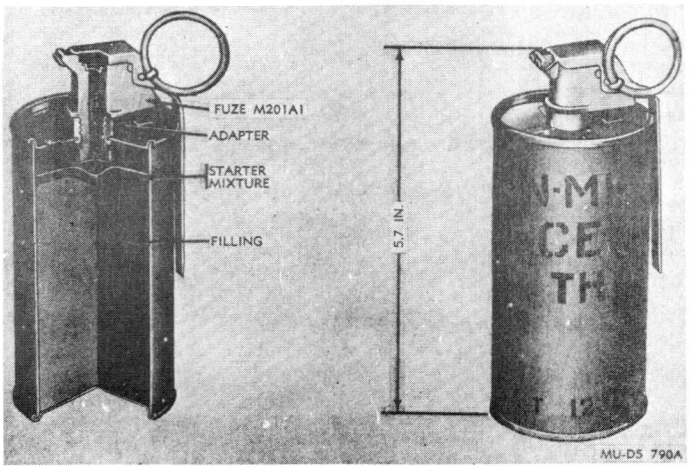

AN-M14 TH3 incendiary hand grenade

DATA
GRENADE (WITH FUZE)
Model: AN-M14
Body: sheet metal
Weight: 0.9 kg
Length: 145 mm
Diameter: 63.5 mm
Filler: (type) igniter mixture III, delay mixture V, FF mixture VII; incendiary mixture; Thermite TH3 and Thermite plain; (weight) 752 g
Fuze: (model) M201A1; (type) pyrotechnic delay-igniting
Primer: M39A1
Ignition mixture: iron oxide, titanium, zirconium
Delay time: 0.7–2 s
Weight: 42 g
Length: 99 mm

Manufacturer
US Army Material Readiness Command.
Status
Current. Probably no longer made.
Service
US Army.

M19A1 WP smoke rifle grenade

The M19A1 is filled with white phosphorus (WP). This chemical agent ignites spontaneously when exposed to air, producing a yellow-white flame and giving off a dense cloud of white smoke. When used as an anti-personnel weapon, the M19A1 grenade has an effective casualty radius of 10 m but will throw burning phosphorus up to 20 m. It has a maximum range of approximately 195 m.
The M19A1 consists of three basic parts: a steel stabiliser tube assembly, an integral fuze and a body. After the grenade is launched, the fuze functions on impact. It bursts the body and scatters particles of burning WP over a large area.
Grenade and fuze function as follows:
The grenade ogive strikes the ground or other resistant object.
Inertia of the firing pin overcomes spring tension and the firing pin strikes the primer.
The primer emits a small, intense spit of flame.
Flame from the primer explodes the detonator.
Explosion of the detonator ruptures the body. Fragments of the body and particles of WP scatter over an area with a radius of approximately 10 m.
Particles of WP ignite upon coming into contact with air and produce a dense cloud of white smoke.

DATA
Type: smoke (WP)
Weight: 681 g
Diameter: 51 mm
Height: 287 mm
Charge: (WP) 241 g
Fuze: integral
Type: mechanical impact detonating
Max range: 195 m
Effective casualty radius: 10 m

Manufacturer
US Army Material Readiness Command.
Status
Current, but Standard B. No longer made.
Service
US Army. Reserve stocks.

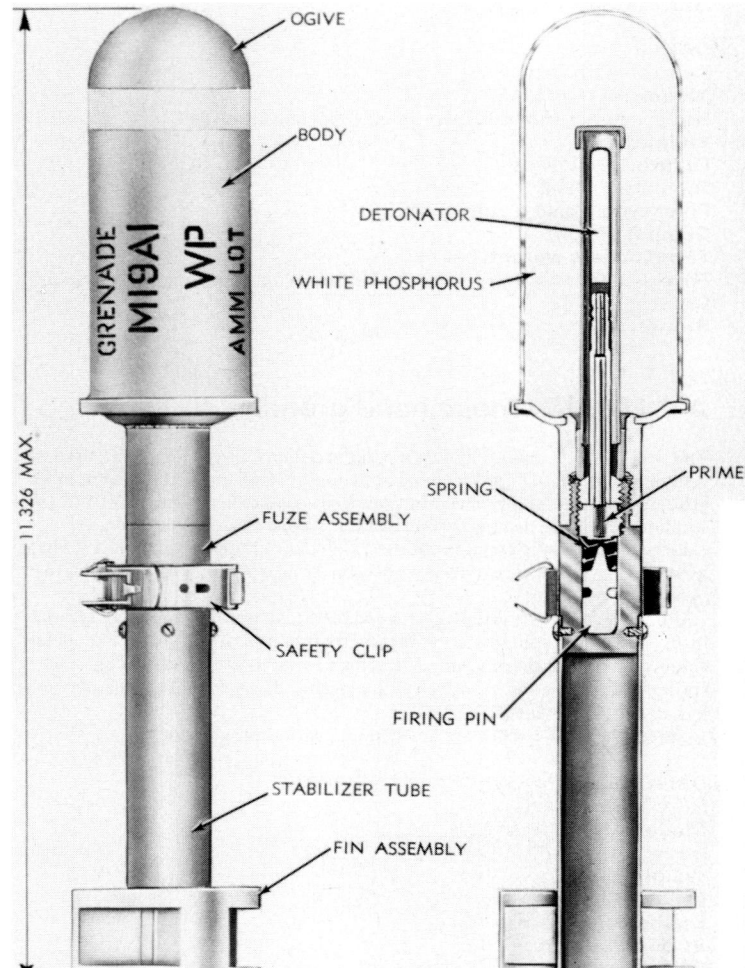

M19A1 WP smoke rifle grenade (dimensions in inches)

M76 vehicle smoke grenade

The M76 grenade has been developed by the AAI Corporation under contract to the US Army. It protects armoured vehicles and personnel with an instantaneous smoke screen which is effective against visual and infra-red observation. Though primarily developed for use with the M1 Abrams tank, the M76 can be fired from any system currently using the L8 grenade with either the six-tube M250 or four-tube M243 launcher.

Within two seconds of firing the M76 grenade forms a screen over a 100° arc, 7 m high and 30 m forward of the launcher. The screen remains effective for a minimum of 45 seconds with winds up to 20 km/h.

DATA
Length: 238 mm
Diameter: 66.5 mm
Weight: 1.81 kg

Manufacturer
AAI Corporation, PO Box 126, Hunt Valley, MD 21030-0126.
Status
Current. Production.
Service
US Army.

M545 Universal fragmentation grenade

The M545 is a thin 50 mm diameter spherical steel ball loaded to provide a number of different ballistic functions, including HE/fragmentation, HE concussion, HE/incendiary/concussion and fragmentation/incendiary. Three sizes of steel ball are available:
3.0 mm diameter, providing approximately 1300 0.12 g fragments per grenade
4.8 mm diameter, providing approximately 225 0.5 g fragments per grenade
6.5 mm diameter, providing approximately 150 1.0 g fragments per grenade.

The grenade is filled with the maker' 'Magdex Hi-Shock' explosive and is designed to provide the maximum amount of blast with very limited fragmentation spread. Alternatively, it can be filled with an explosive/incendiary composition (M545IC) or with an entirely incendiary composition (M545I).

Manufacturer
Accuracy Systems Inc, 15205 North Cave Creek Road, Phoenix, AZ 85032.
Status
Current. Production.

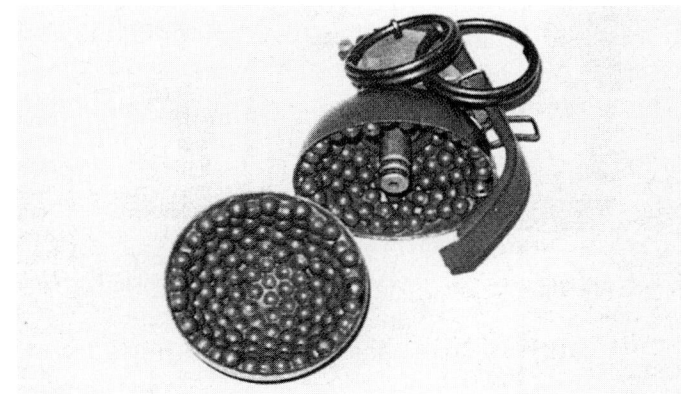

M545 Universal fragmentation grenade

M560 series anti-personnel fragmentation grenades

These grenades come in three sizes: 10 × 5 cm diameter; 11 × 6 cm diameter; and 7.5 × 4.3 cm diameter. All are manufactured with a special 'Hi-Frag' coined metal fragmentation shell. The smaller grenade is especially adaptable for room-to-room use by counter-terrorist forces in situations where the explosive blast and fragmentation of larger grenades would be unacceptable.

Manufacturer
Accuracy Systems Inc, 15205 North Cave Creek Road, Phoenix, AZ 85032.
Status
Current. Production.

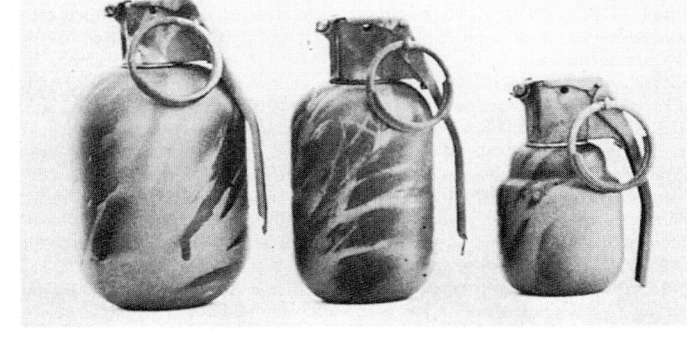

M560 series anti-personnel fragmentation grenades

40 mm grenade cartridges

There is a wide range of 40 mm grenade cartridges suitable for use with M79, M203, HK79 and other type of grenade launcher. The 40 mm cartridge is a fixed munition consisting of a cartridge case and a projectile.

The cartridge case is made of aluminium or plastic and has an integral propellant retainer into which is inserted a thin-walled brass cup, containing the propellant, followed by an aluminium base plug which seals the base of the cartridge case. This arrangement is the basis of the high-low propulsion system required to propel a 40 mm projectile from a shoulder-fired weapon. When the firing pin strikes the cartridge primer the propellant in the brass

powder cap is ignited and generates a pressure in the region of 2500 kg/cm². This causes the brass case to rupture at a ring of vent holes in the propellant retainer, allowing gas to flow into the remainder of the cartridge case which thus forms a low pressure chamber (around 200 kg/cm²). This pressure is adequate to propel the projectile and does not produce excessive recoil. The grenade leaves the launcher with a muzzle velocity of 76 m/s and is spin stabilised by the launcher rifling. The spin also provides the rotational force necessary to arm the fuze.

Grenades known to be in the series are listed below and it will be seen that they include high-explosive, riot control, practice and a variety of smoke, signalling and illuminating types.

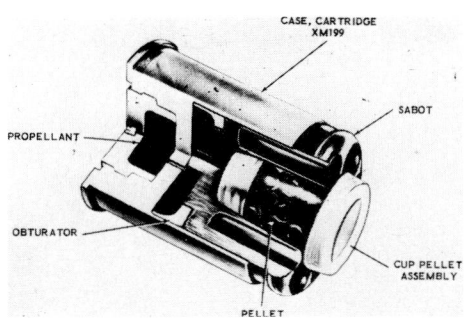

40 mm M576E1 multiple projectile cartridge

40 mm cartridge case and projectile

Fuzes

The M552 and M551 impact detonating fuzes are used with the HE and the TP rounds. The M552 fuze arms by a spin action and is armed about 3 m from the muzzle. The M551 fuze arms by a spin and setback action and must travel between 14 and 28 m before being armed.

Effective casualty radius

The high explosive grenade has an effective casualty radius of 5 m. The effective casualty radius is defined as the radius of a circle about the point of detonation in which it may be expected that 50 per cent of exposed troops will become casualties.

Combat load

The recommended minimum combat load is 36 rounds of HE 40 mm ammunition.

40 MM CARTRIDGES

Model	Type	Fuze	Remarks
M381	HE	M552	Obsolete
M386	HE	M551	Obsolete
M397A1	HE	M536E1	Airburst; obsolete
M406	HE	M551	Current production
M441	HE	M551	Obsolete
M433	HEDP	M550	Fragmentation/shaped charge
M576E1	Multi-projectile	None	Contains 20 balls
M583A1	White star	None	Parachute; obsolete
M585	White star	None	Cluster; obsolete
M661	Green star	None	Parachute
M662	Red star	None	Parachute
M675	Red smoke	None	Current
M676	Yellow smoke	None	Canopy smoke; obsolete
M680	White smoke	None	Canopy smoke: obsolete
M682	Red smoke	None	Canopy smoke: obsolete
M715	Green smoke	None	Current
M717	Yellow smoke	None	Current
M674	CS	None	Current
M382	Practice	M552	Filled yellow dye
M407A1	Practice	M551	Yellow smoke and dye
M781	Practice	None	Yellow or orange dye

Ring Airfoil Grenade (RAG)

The Ring Airfoil Grenade is, as the name implies, a grenade made in the form of an airfoil section wrapped round to make a ring-like projectile. The outer surface of the ring is a pre-fragmented metal layer, while the inner area contains a charge of high explosive. The RAG is intended as a subsonic launched projectile.

The RAG posesses flight performance characteristics which offer new potential for future weapon systems. Because of its specific shaping and physical properties, the RAG generates aerodynamic lift as it flies through the air. This lift retards the drop due to gravity and, together with the low-drag shaping, results in a considerable range increase coupled with a relatively flat trajectory. These performance characteristics are obtained for a reasonably low muzzle velocity and projectile weight, resulting in reduced recoil loads which make the principle especially applicable to a shoulder-fired grenade system.

Moreover the three-dimensional shaping of the RAG results in excellent terminal ballistic performance for its size and weight. The performance can be tailored, to some degree, to meet specific operational requirements. To this end two RAG shoulder-fired grenade systems were considered, RAG(A) and RAG(B). Sufficient exploratory development was undertaken to show that both systems were feasible for further development.

RAG(B) represents the current approach to a shoulder-fired system, and was a direct evolutionary result of RAG(A), and a description of RAG(A) is given since it provided the basis from which RAG(B) was evolved.

The RAG(A) projectile was of 63.5 mm diameter and had a length of 35.5 mm. The projectile weighed 142 g and the total weight of the round (cartridge, sabot and projectile) was 181 g. It had a fragmenting steel warhead. The lethal area of the RAG(A) is at least 1.4 times that of the 40 mm M406 grenade. The projectile could be fired from a shoulder launcher at a muzzle velocity of 76 m/s and a spin rate of 6000 rds/min. Targets could be reached out to ranges of 1000 m, and even this range was achieved at a relatively low launcher angle of elevation; thus sighting and firing are greatly simplified as compared to the high elevation angles and awkward sighting arrangements necessary with the 40 mm M79 launcher.

The RAG(B) projectile is of 54 mm maximum diameter with a length of 29.2 mm. Projectile weight is 90 g and the total weight of the round is between 113 and 160 g. The lethal area is equal to the current 40 mm M406 grenade.

A multi-shot launcher holding a clip of three rounds has been considered. The projectile would be launched at a muzzle velocity of 91 to 137 m/s, with a spin rate between 10 000 and 12 000 rpm. The flat trajectory allows RAG(B) to be aimed and fired like a rifle, and ranges in excess of 1400 m are possible.

The RAG(B) system was conceived and designed to provide an area-fire munition for small arms application where a short reaction time is the primary operational performance demand. Toward this end the RAG(B) represents the smallest possible warhead which will still retain a lethal area equal to that of current projected grenades whilst having less bulk and weight. This decrease in weight permits the muzzle velocity of the RAG(B) to be increased to a maximum of 137 m/s without exceeding the recoil limits for a shoulder-fired weapon.

Concepts have also been evaluated for launching RAG(B) at a higher subsonic velocity similar to that of the 40 mm grenade from the Mk 19 Grenade Machine Gun. The potential system would be significantly lighter than the Mk 19, which would allow the high velocity RAG(B) to be launched from the ground, from vehicles or from helicopters.

At the present time the RAG configuration has been adopted for a riot control system, details of which will be found in the relevant section; development of an explosive grenade for combat use is understood to be continuing. Recently there has been renewed interest by the U.S. Army in a shoulder-fired fragmenting projectile that can be useful at ranges exceeding that of the current 40 mm grenade. Hence additional attention is being given to the RAG performance potential.

YUGOSLAVIA

Yugoslav grenades

Yugoslavia uses both hand and rifle grenades. The hand grenades are egg shaped and they are basically of the same design. The earliest ones had an impact fuze but the later ones are all fitted with delay fuzes. Typical of the series is the M69 which weighs 600 g and has a 4.5 second delay fuze.

The older M52R grenade is also still in service. There is also a stick smoke grenade which weighs 600 g and emits smoke for up to two minutes.

The Mecar Energa rifle grenade was used in Yugoslavia for several years and was launched from a spigot attached to the muzzle of the M48 rifle, which was bolt operated.

The M59/66 semi-automatic rifle is now used to project the new grenades which were developed in the country. The Yugoslav M64 assault rifle, which is equipped with a grenade sight, takes a grenade launcher when the compensator is removed. Erecting the grenade sight automatically cuts off the gas supply to the piston. There are four Yugoslav rifle grenades, all of which are characterised by their rather small diameters. In addition to the anti-tank and anti-personnel grenades, there is an illuminating and a smoke grenade. The smoke grenade is 330.5 mm and 40 mm in diameter. It generates smoke for 90 seconds.

Yugoslav M69 anti-personnel hand grenade

DATA

	Anti-tank	Anti-personnel
Weight, fuzed	0.602 kg	0.52 kg
Length	390 mm	307 mm
Diameter	60 mm	30 mm
Weight of HE	235 g	67 g
Fuze type	Impact	Impact
Max range	150 m	400 m
Effect range	—	20–50 m
Penetration	—	—

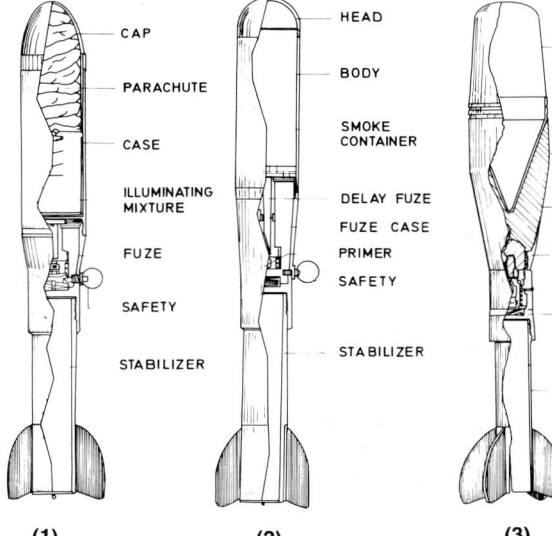

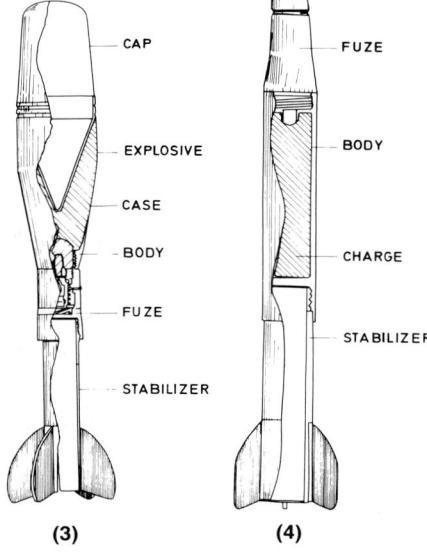

(1)   (2)   (3)   (4)

(1) *M62 illuminating rifle grenade* **(2)** *M62 smoke rifle grenade* **(3)** *M60 anti-tank rifle grenade* **(4)** *M60 anti-personnel rifle grenade*

M75 hand grenade

This is a defensive grenade using pre-fragmented steel balls packed in a plastic body around a high explosive bursting charge. The usual type of fly-off lever fuze is fitted centrally in the grenade. The grenade has a storage life of at least 10 years, will function in temperatures between –30° and +60°C, and is drop safe from a 10 m height.

DATA
Weight: 355 g
Height: 89 mm
Diameter: 57 mm
Bursting charge: 36 g plastic explosive
Steel balls: approx 3 000 of 2.5 – 2.9 mm diameter
Delay: 3 – 4 s

Manufacturer
Federal Directorate of Supply and Procurement, PO Box 308, 9 Nemanjina Street, 11105 Beograd.
Status
Current. Available.

M75 hand grenade

M79 anti-tank grenade

This is a hand-thrown, drag-stabilised shaped charge grenade for the attack of armour at close-range. The body contains the shaped charge unit and also incorporates the desired standoff distance. The handle carries the ignition system and also a drag parachute which keeps the grenade in a nose-forward attitude to ensure correct operation on impact. There is an impact fuze which is armed by a fly-off lever and by the action of the drag parachute after throwing.

M79 shaped charge anti-tank grenade

DATA
Weight: 1.10 kg
Length: 397 mm
Diameter: 76 mm
Charge: 400 g Hexolite
Throwing range: approx 25 m
Penetration: 220 mm homogeneous armour

Manufacturer
Federal Directorate of Supply and Procurement, PO Box 308, 9 Nemanjina Street, 11105 Beograd.
Status
Current. Available.

BRD M83 smoke grenades

The BRTD M83 series of smoke hand grenades are all of similar appearance and include white, red, yellow or green smokes, the contents being indicated by the colour of the base of the grenade body. The coloured smoke grenades are somewhat smaller and lighter than the white smoke model. There are two types of ignition system available, either the conventional fly-off lever percussion system or a pull-type friction igniter. In the latter model the cap is removed, revealing a ring and cord; the ring is pulled so as to drag the friction bar through the igniting composition, and the grenade is then thrown.

DATA
WHITE SMOKE M83:
Diameter: 80 mm
Length: 130 mm
Weight: 1000 g
Weight of HC smoke charge: 700 g
Smoke emission time: 3 mins
Shelf life: 5 years

COLOURED SMOKE M83:
Diameter: 60 mm
Length: 148 mm
Weight: 400 g
Weight of smoke composition: 240 g
Fuze delay: 2.5 – 3.5 s
Smoke emission time: 50 – 70 s
Shelf life: 3 years

Manufacturer
Federal Directorate of Supply and Procurement, PO Box 308, 9 Nemanjina Street, 11105 Beograd.
Status
Current. Available.

M83 smoke grenades; (left) pull-igniter type, (right) fly-off lever igniter type

Mortar Ammunition

AUSTRIA

81 mm SMI mortar ammunition

The firm of Südsteirische Metallindustrie (SMI) manufactures both the 81 mm L70 mortar and the 120 mm 60 and also the ammunition.

The 81 mm bombs are a licensed version of the British L15A1 and are made to close tolerances and high standards. As a result, their fragmentation pattern is very even and their flight regular and predictable. Distribution of the fragments is claimed to be precisely even, so producing the optimal effect from the weight of the bomb.

DATA
Overall weight: 4.15 kg
Weight of casing: 2.9 kg
Weight of explosive: 0.75 kg (TNT)
Number of fragments: approx 1600 of at least 0.5 g; approx 2000 of 0.3–0.5 g
Overall length of bomb: 487 mm
Muzzle velocity: 75–300 m/s
Max range: 5800 m
Max barrel pressure: 750 bars
Fuzes: percussion, M125A1, DM111A2 or equivalent

81 mm SMI HE bomb, without secondary charges

Manufacturer
Südsteirische Metallindustrie GmbH, A-8430 Leibnitz.
Status
Current. Production.
Service
Austrian Army.

120 mm SMI HE long-range mortar bomb

The 120 mm SMI bomb is a streamlined projectile much resembling the 81 mm in general shape and possessing a good aerodynamic efficiency which gives a long range with minimum dispersion. Unlike the 81 mm bomb the 120 mm does not rely on a plastic sealing ring for obturation, but uses a bourrelet section in the centre of the body with six machined grooves cut in it.

The fins are carried on a boom; the primary cartridge is carried inside this boom in the usual manner. Augmenting charges are flat plates covered with coloured silk, and these plates have a slot cut in them to allow them to be slipped over the tail boom as required. These charges run up to number 9, but the normal firing is with charge 6 since the higher charges can only be used with heavy mortars. The body is made from ferritic cast steel, machined all over and carefully controlled for weight and all dimensions. The normal fuze is the DM111 series but any one of several electronic proximity fuzes will also fit the nose adaptor.

DATA
Length overall: 747 mm
Weight: (overall) 14.5 kg; (explosive filling) 2.3 kg (TNT); (shell body) 10.5 kg (spheroidal graphite cast iron)

Charge	Muzzle velocity	Range	Pressure
1	135 m/s		
6	322 m/s	7040 m	800 bar
9	403 m/s	9010 m	1500 bar

Fragments: approx 6000 effective: 2800 between 0.3–0.5 g; 3200 between 0.5–30 g

Manufacturer
Südsteirische Metallindustrie GmbH, A-8430 Leibnitz.
Status
Current. Available.

Hirtenberger mortar ammunition

The Austrian company Hirtenberger manufactures ammunition of many small and large calibres, including four sizes of mortar bomb: 60, 81, 82 and 120 mm.

All bombs are of modern design; their outstanding maximum firing ranges and excellent accuracy are due to their aerodynamic shape and favourable ratio of weight to cross-section.

The stabiliser, made from extruded aluminium alloy, is screwed firmly into the body and contains the waterproof propelling cartridge. The horseshoe-shaped augmenting charges are clipped to the tail boom.

The bodies of the high explosive bombs are made of spheroidal graphite cast iron. All high explosive bombs are available also with inert filling and impact fuze or fuze plug as Practice versions for training purposes, with identical ballistic performance.

A new design is the Hirtenberger 60 mm LD bomb which, due to increased length and weight, offers a 25 per cent better fragmentation than does the standard version, though the maximum ranges of both types are the same.

For the smoke versions, Hirtenberger uses red phosphorus in the 60 mm calibre, both red and white phosphorus in the 81 mm and 82 mm calibres and white phosphorus or hexachloroethane (HC) in the 120 mm calibre.

The incendiary versions use red phosphorus, which achieves a much higher combustion temperature than conventional white phosphorus and offers an additional long-lasting smoke effect.

All Hirtenberger mortar bombs are suitable for all usual mortar types with corresponding calibres and gas pressure.

Manufacturer
Hirtenberger AG, A-2552 Hirtenberg.
Status
Current. Production.

60 mm mortar bombs

DATA

	60 mm HE 80	60 mm HE LD	60 mm ILL	60 mm RPS	60 mm RPI
Body	cast iron	cast iron	iron	iron	iron
Length	300 mm	360 mm	470 mm	470 mm	470 mm
Weight	1.6 kg	1.9 kg	2.3 kg	2.3 kg	2.3 kg
No of propelling charges	4 (5)*	4 (5)*	4	4	4
Max range	2900 (4400) m	3000 (4400) m	1900 m	2050 m	2050 m
Gas pressure	450 (550) bar	500 (550) bar	450 bar	450 bar	450 bar
Initial velocity	199 (276) m/s	210 (275) m/s	163 m/s	163 m/s	163 m/s
Illuminating power	—	—	400 000 cd	—	—
Burning time	—	—	35 s	90 s	90 s
Rate of descent	—	—	3–4 m/s	—	—
Fuze	impact fuze	impact fuze	time fuze	impact fuze	impact fuze

With one or two propelling charges only, the 60 mm HE 80 mortar bomb is suitable for commando mortars.
*Values in parentheses are for a barrel length of 1000 mm

Hirtenberger 60 mm mortar bombs

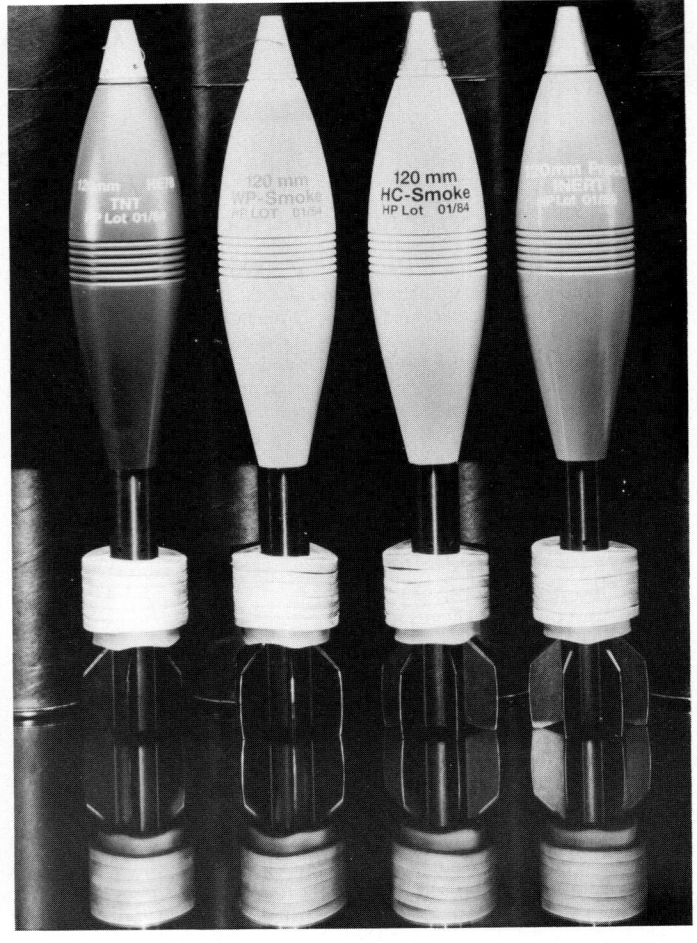

Hirtenberger 120 mm mortar bombs

81 mm mortar bombs

DATA

	81 mm HE 70	81 mm WP	81 mm ILL Mk 3	81 mm RPS Mk 3	81 mm RPI Mk 3
Body	cast iron	cast iron	aluminium	aluminium	aluminium
Length	487 mm	487 mm	635 mm	635 mm	635 mm
Weight	4.15 kg	4.17 kg	3.8 kg	3.8 kg	3.8 kg
No of propelling charges	6	6	6	6	6
Max range Standard mortar	5800 m	5800 m	5200 m	5500 m	5500 m
Max range Long-range mortar	6300 m	6300 m	5450 m	5700 m	5700 m
Gas pressure	850 bar	850 bar	850 bar	850 bar	850 bar
Initial velocity Standard mortar	295 m/s	295 m/s	311 m/s	311 m/s	311 m/s
Initial velocity Long-range mortar	303 m/s	303 m/s	320 m/s	320 m/s	320 m/s
Illuminating power	—	—	1 200 000 cd		
Burning time	—	—	35 s	90 s	90 s
Rate of descent	—	—	4–5 m	—	—
Fuze	impact or proximity fuze	impact or proximity fuze	time fuze	impact or proximity fuze	impact fuze

The above 81 mm mortar bombs are also available in 82 mm calibre with practically identical ballistic data.

The 81 mm HE 70 can also be equipped with a different set of propelling charges (three big ones, three small ones) and thus offers the same ballistic data as the L 15 HE Bomb.

120 mm mortar bombs

DATA

	120 mm HE 78	120 mm WP	120 mm HC
Body	cast iron	cast iron	cast iron
Length	747 mm	747 mm	747 mm
Maximum range			
Charge 6 (Standard mortar)	7100 m	7100 m	7100 m
Charge 7 (Standard mortar)	7750 m	7750 m	7750 m
Charge 7 (Long-range mortar)	8640 m	8640 m	8640 m
Charge 9 (Long-range mortar)	9670 m	—	—
Gas pressure			
Charge 6	800 bar	800 bar	800 bar
Charge 7	1000 bar	1000 bar	1000 bar
Charge 9	1500 bar	—	—
Initial velocity			
Charge 6 (Standard mortar)	320 m/s	320 m/s	320 m/s
Charge 7 (Standard mortar)	340 m/s	340 m/s	340 m/s
Charge 7 (Long-range mortar)	370 m/s	370 m/s	370 m/s
Charge 9 (Long-range mortar)	430 m/s	—	—
Duration of smoke effect	—	—	210 s
Height of burst			100–200 m
Fuze	impact or proximity fuze	impact or proximity fuze	time fuze

BELGIUM

60 mm NR 431 PRB HE mortar bomb

PRB manufactures 60 mm, 81 mm and 4.2 in (107 mm) mortar bombs. These are exported all over the world.

The NR 431 mortar bomb has a watertight thin metallic casing which contains a patented fragmentation sleeve and explosive charge. PRB claims that the NR 431 is twice as effective as the American M49A2 mortar bomb.

The fins, which are made of light alloy, hold a primary cartridge with an 'improved' ignition cap. The supplementary charges are held by spring blades fixed on the fin tube and are shorter than the fin blades.

The NR 431 can use all fuzes standardised on the American 1.5 in (38 mm) fuze well. It is usually fitted with the US M52A2 fuze but may be fitted optionally with the PRB NR 444 fuze which is exceptionally robust and has a muzzle safety. The bomb was originally designed for firing from the American M19 mortar, which was subsequently produced in a slightly modified form by PRB, and it is also suitable for use with the NR 493 mortar. According to PRB, it can also be fired from breech loaded mortars.

In addition, the loading of the mortar bomb can be modified so that the smoke produced by the explosion is red, yellow or green.

DATA
Total weight: 1.37 kg
Weight of TNT explosive: 150 g
Length of bomb: (with US M52A2 fuze) 255 mm; (without fuze) 195 mm
Lethal radius: 13.5 m
Propelling charge: 4 increments (powder M8) and 1 primary cartridge
Max range (M19 mortar): 1800 m
Packing: 1 shell per fibre container, 10 containers per wooden case

Manufacturer
Société Anonyme PRB, Département Défense, Avenue de Tervueren 168, Bte 7, B-1150 Brussels.
Service
Belgian Army.

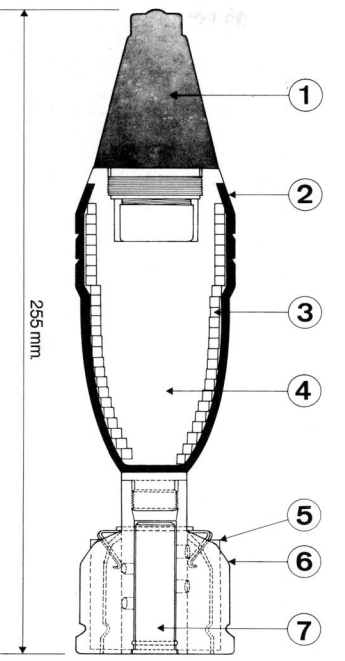
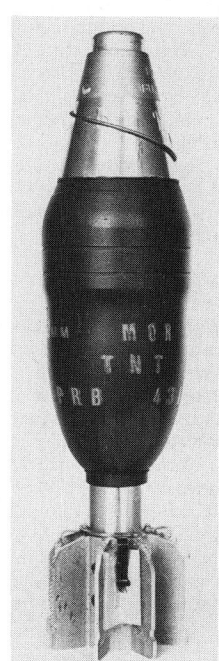

60 mm NR 431 PRB HE mortar bomb
(1) fuze (2) shell case (3) fragmentation body (4) explosive (5) increment charges (6) fin assembly (7) primary cartridge

60 mm NR 161 PRB WP smoke bomb

This round is similar to the US M302 bomb. It can be used for screening or spotting and, being filled with white phosphorus (WP), also has an incendiary effect.

DATA
Weight, complete: 1.83 kg
Weight of WP filling: 350 g
Max range: (charge 4) 1470 m

Manufacturer
Société Anonyme PRB, Département Défense, Avenue de Tervueren 168, Bte 7, B-1150 Brussels.
Service
Belgian Army.

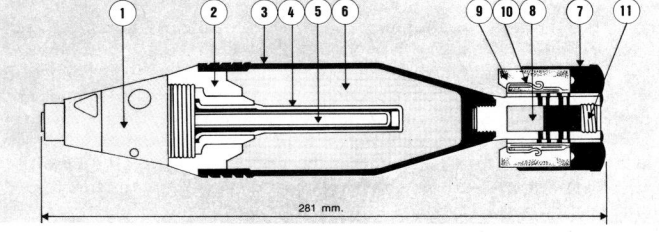

60 mm NR 161 PRB WP smoke bomb
(1) detonating fuze (2) adaptor (3) steel shell body (4) burster casing (5) burster charge (6) WP filling (7) fin assembly (8) ignition cartridge (9) propellant increments (10) increment holder (11) percussion primer

60 mm NR 162 PRB parachute illuminating bomb

This round is similar to the US M83A1 illuminating bomb. The fin assembly can be fitted with up to four incremental charges to give a maximum range of 1000 m at a parachute deployment height of 170 m. It is fitted with a M65A1 time fuze which explodes a charge of black powder which in turn expels the parachute and ignites the illuminating mixture.

DATA
Weight: 1.88 kg
Muzzle velocity: (charge 4) 132 m/s
Range: (charge 4) 1000 m
Deployment height: (charge 4) 170 m
Illuminating time: 25–30 s
Rate of descent: 3 m/s
Illuminating power: 300 000 cd
Bomb markings: grey with black markings

Manufacturer
Société Anonyme PRB, Département Défense, Avenue de Tervueren 168, Bte 7, B-110 Brussels.

Service
Belgian Army.

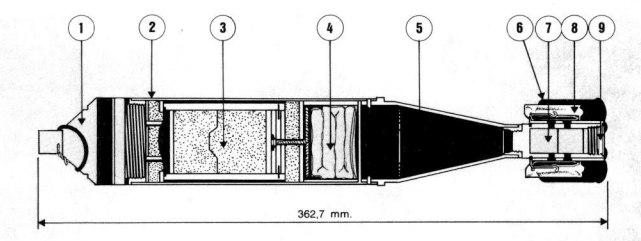

60 mm NR 162 PRB parachute illuminating bomb
(1) time fuze (2) steel shell body (3) illuminating charge (4) parachute (5) nose cone (6) fin assembly (7) ignition increments (8) propellant increments (9) percussion primer

81 mm NR 414 PRB HE bomb

The NR 414 PRB mortar bomb is made of a watertight, thin metallic sheath which contains a patented fragmentation sleeve and an explosive charge. The light alloy fins and the use of a hunting type primary cartridge have enabled the manufacturer to give the bomb maximum HE content.

The supplementary charges are held by spring blades staked on the fin tube and are shorter than the fin blades. The NR 414 PRB can use all standard American fuzes as it has a 1.5 in fuze-well. The loading of the mortar bomb can be modified so that the smoke produced by the explosion is red, green or yellow.

The mortar bomb has been designed to be fired from the American M1 or M29 mortars (or the PRB equivalent), as well as being a replacement for the American M43A1 bomb. PRB claims that its bomb is twice as effective as the American bomb.

DATA
Total weight: 3.25 kg
Weight of TNT explosive: 500 g
Length of bomb: (with US M54A2 fuze) 338 mm; (without fuze) 275 mm
Lethal radius: 21 m
Propelling charge: 6 increments (powder M8) and 1 primary cartridge
Muzzle velocity: (at max charge) 220 m/s
Max range: (at max charge in standard mortar) 3200 m

The ranges of the NR 414 PRB shell, when usedd with the American mortars M1 and M29 or any foreign mortars of the same geometrical characteristics (105 cm, low pressure 600 bars), are given by the American firing table for the M43A1 shell with the fuze M52A2. PR can also supply super charges for use in medium pressure (800 bars) and high pressure (1000 bars) mortars.
Packing: 1 shell per fibre container, 10 containers per wooden case

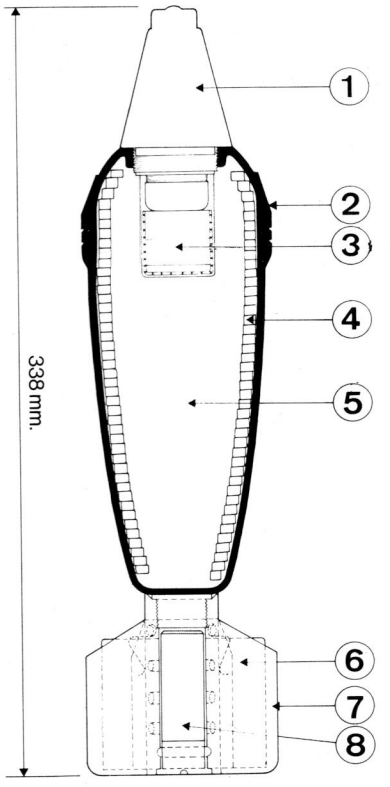
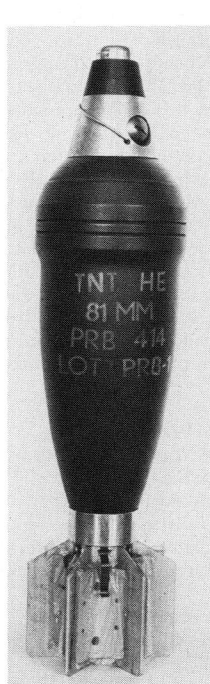

81 mm NR 414 PRB HE mortar bomb
(1) *fuze* **(2)** *body* **(3)** *booster* **(4)** *fragmentation body* **(5)** *explosive*
(6) *increment charges* **(7)** *fin assembly* **(8)** *primary cartridge*

81 mm PRB HE bombs

The PRB long-range mortar bomb has now been developed in three different types to suit mortars of differing characteristics. They may be fired in all types of mortars within the classes indicated by the following examples: the high pressure bomb NR 436 may be fired with seven charge increments in the PRB NR 8475A1 mortar; the medium pressure bomb NR 520 may be fired with five charge increments in Brandt-Armements 81 mm Types 61C and 61L mortars; and the low pressure bomb NR 254 may be fired with four increments in the US 81 mm M1 and M29 mortars. The manufacturer will calculate firing tables for the appropriate bomb for any other type of mortar.

The shape is slightly unusual, the ogive being long and conical to give good flight characteristics. There is a plastic obturating ring, and the interior of the bomb is lined with pre-notched wire to provide ample fragmentation. The bomb is fitted with a superquick impact fuze as standard.

DATA
Weight: 4.20 kg
Length: 543 mm
Maximum muzzle velocity: 321 m/s
Max range: 5820 m
Lethal radius: 25 m average

Manufacturer
Société Anonyme PRB, Département Défense, Avenue de Tervueren 168, Bte 7, B-1150 Brussels.
Status
Current. Production.

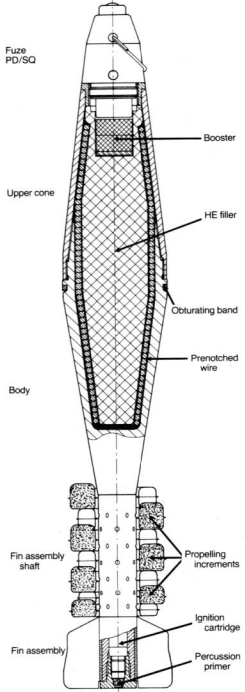

NR 436 PRB HE bomb

81 mm NR 163 PRB WP smoke bomb

Similar to the US M57A1 bomb this smoke bomb has a white phosphorus (WP) filling and thus has an incendiary effect in addition to its ability to produce a thick smoke screen. Construction details can be seen in the accompanying diagram.

DATA
Weight of bomb: 5.8 kg
Weight of WP filling: 1.86 kg
Length of bomb: 619.5 mm
Muzzle velocity: (with 4 incremental charges) 163 m/s
Max range: (4 increments) 2130 m
Packing: 1 bomb per fibre container, 2 containers in wooden box, 21.3 kg

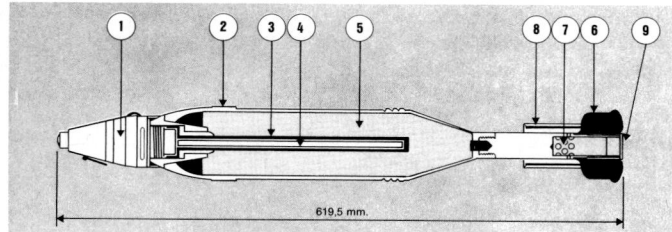

Manufacturer
Société Anonyme PRB, Département Défense, Avenue de Tervueren 168, Bte 7, B-1150 Brussels.
Service
Belgian Army.

81 mm NR 163 PRB WP smoke bomb
(1) *time and superquick fuze* **(2)** *steel body* **(3)** *burster casing* **(4)** *burster charge* **(5)** *WP filling* **(6)** *fin assembly* **(7)** *ignition cartridge* **(8)** *incremental charges* **(9)** *percwssion primer*

81 mm PRB illuminating bombs

The PRB illuminating bomb has been re-designed and is now available in three distinct types for use in different mortars. The high pressure bomb NR 544 may be used with seven incremental charges in the PRB Long-Range Mortar NR 8475A1; the medium pressure bomb NR 253 is to be used with five charge increments in the Brandt-Armement 81 mm mortars Models 61C and 61L; the low pressure bomb NR 240 is for use with four charge increments in the US 81 mm Mortars M1 and M29. The bombs may be used in other mortars having characteristics similar to those given as examples.

The functioning of the bomb is conventional; a mechanical time and percussion fuze ignites an expelling charge at the requisite point on the trajectory. This charge blows the nose of the bomb off, allowing the spring to thrust out the star unit and its parachute, the star having been ignited by the explosion of the expelling charge. Optimum expulsion height is 450 m, but different star units can be supplied so as to give different light and time characteristics; either 500 000 candela for 55 seconds or 900 000 candela for 30 seconds.

Fired at maximum charge, the NR 544 attains 321 m/s muzzle velocity and a range of 5600 m.

Manufacturer
Société Anonyme PRB, Département Défense, Avenue de Tervueren 168, Bte 7, B-1150 Brussels.
Status
Current. Production.

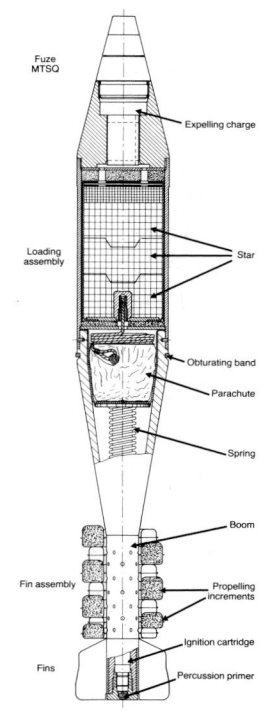

81 mm PRB illuminating bomb

MECAR 81 mm A1 Series mortar bomb family

The MECAR A1 series of 81 mm mortar bombs are used in M1 low pressure, M29/29A1 medium pressure and M252 and L16 series high pressure mortar systems and their equivalents. Five models of bomb are available and may be divided into two groups of ballistically matched bombs, ie M511A1 to M513A1 using a common Firing Table and M514A1 and M515A1 fired to a second Firing Table. The bombs available are:-
Smoke (FM) (TTC) M511A1
HE (TNT) M512A1
HE/ICM M514A1
Smoke (WP) M513A1 and
Illuminating M515A1.

Bombs are packed in three-round NATO approved waterproof polymer containers. HE and smoke bombs have a maximum range of 5500 m in high pressure mortars, 4500 m in medium pressure and > 2500 m in low pressure mortar systems.

The A1 propulsion system employs a primary cartridge comprising a screw-threaded primer and shotgun type ignition cartridge, and up to five 'horseshoe' type augmenting charges, the number of charges permitted depending upon the pressure capacity of the mortar. The base augmenting charge is coloured blue, and there are three translucent cased charges and a red coloured supercharge.
Charge 0 = Primary cartridge only
Charge 1 = Primary and base augmenting charge
Charge 2 = Primary, base augmenting charge and one translucent case charge
(Charge 2 is the MAXIMUM for low pressure mortars)
Charge 3 = Primary, base augmenting charge and 2 translucent case charges
Charge 4 = Primary, base augmenting charge and 3 translucent case charges
(Charge 4 is MAXIMUM for medium pressure mortars)
Charge 5 = Primary, base augmenting, 3 translucent cased charges and the red cased supercharge.
(Charge 5 is used ONLY in high pressure mortar systems.)

DATA
Primary and Augmenting Charges
Primary Cartridge: M519; Augmenting Charges (3) M521
Base Augmenting Charge M520; Supercharge Increment M522

81 mm High Explosive Bomb M512A1
Length (fuzed): 516 mm
Weight (fuzed): 4.10 kg
Filling 1.03 kg TNT
Body: nodular cast iron
Ranges: 100 to 4500 m (Ch 0 to 4); to 5500 m with Supercharge M522

81mm Smoke (TTC) (FM) bomb M511A1
Length (fuzed):516 mm
Weight (fuzed): 4.10 kg
Filling: 880 g Titanium Tetrachloride (FM)
Body: nodular cast iron
Ranges: 100 – 4500 m (Ch 0 to 4); to 5500 m with Supercharge M522

Smoke (WP) bomb M513A1
Length (fuzed): 516 mm
Weight (fuzed): 4.10 kg
Filling: 880 g white phosphorus (WP)
Body: nodular cast iron
Ranges: 100 – 4500 m (Ch 0 to 4); to 5500 m with Supercharge M522

81 mm HE/ICM bomb M514A1
Length (fuzed): 630 mm
Weight (fuzed): 4.135 kg
Filling: 12 HEAT/fragmentation sub-munitions
Body: steel
Ranges: 250 – 4100 m (Ch 0 to 4); to 4600 m with Supercharge M522

81 mm Illuminating bomb M515A1
Length (fuzed): 625 mm
Weight (fuzed): 4.20 kg Filling: parachute and flare canister
Body: steel
Ranges: 250 – 4100 m (ch 0 to 4); to 4600 m, with Supercharge M522

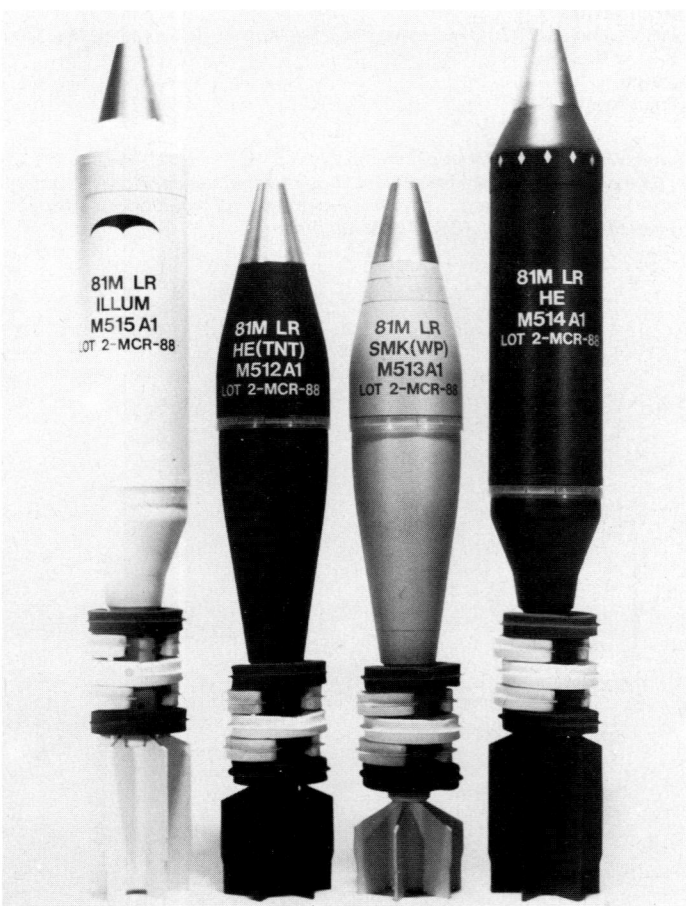

MECAR A1 family of 81 mm mortar bombs

Manufacturer
MECAR SA, B-6522 Petit-Roeulx-lez-Nivelles.
Status
Current. Production.

BRAZIL

60 mm HE bomb TIR 60 AE M3

This bomb can be used with all types of smoothbore mortars of corresponding calibre, being designed for exploding instantaneously on impact. Its fuze offers total safety during handling and, by means of a safety wire that does not allow the alignment of the explosive train when the round is in the bore, it also ensures complete safety in case of a double feed. The bombs are supplied in a reinforced wooden box with 20 rounds in individual waterproof paper containers (gross weight: 32 kg).

DATA
Total length: 245 mm
Total weight: 1.37 kg
Filling: 150 g of TNT or Comp B
Range: 1800 m

Propellant increments: 1–4
Number of fragments: 600+
Packing: (fuzes) 80 units each, separated boxes, gross weight 39 kg; (ignition cartridges, charge 0) 20 boxes with 25 cartridges each in reinforced wooden box, gross weight 53 kg; (propellant increments) 18 boxes with 240 charges each in reinforced wooden box, gross weight 43 kg

Manufacturer
Companhia de Explosivos Valparaiba, Praia do Flamengo, 200-20° Andar, 22210 Rio de Janeiro, RJ.
Status
Production. Service.

81 mm HE bomb TIR 81 AE M4

This round is for all types of smoothbore 81 mm mortars, incorporating the same safety items of the same manufacturer's 60 mm round. Grenades, fuzes, ignition cartridges, and propellant increments come in separate packages.

DATA
Total length: 335 mm
Total weight: 3.345 kg
Filling: 500 g of TNT or Comp B
Range: 4050 m

Propellant increments: 1–6
Number of fragments: 1000+

Manufacturer
Companhia de Explosivos Valparaiba, Praia do Flamengo, 200-20° Andar, 22210 Rio de Janeiro, RJ.
Status
Production. Service.

81 mm HE bomb TIR 81 AE M7

This is a more modern design than the AE M4, being longer and better streamlined. It uses the same EOP M4-CEV point detonating fuze as other 60 and 81 mm bombs, but the improved weight and shape give it a better ballistic performance and an improved lethal radius of burst.

DATA
Total length: 400 mm
Total weight: 3.875 kg
Filling: 600 g of TNT

Range: 5200 m
Propellant increments: 6

Manufacturer
Companhia de Explosivos Valparaiba, Praia do Flamengo 200-20° Andar, 22210 Rio de Janiero, RJ.
Status
Current. Production.

CHINA, PEOPLE'S REPUBLIC

81 mm fragmentation projectiles

There are two 81 mm bombs in service. Both are based on the US M43A1 projectile, one being a direct copy and the other a variant. Although marked 81 mm both bombs are of the same diameter as the 82 mm types and are fired from 82 mm mortars.

DATA
Calibre: 81 mm
Type: fragmentation
Weight: (fuzed) 3.22 kg (variant: 3.2 kg)
Bursting charge: 0.56 kg TNT (variant: 0.35 kg)
Fuze: point detonating
Known using weapons: mortars, Models 20 and 53

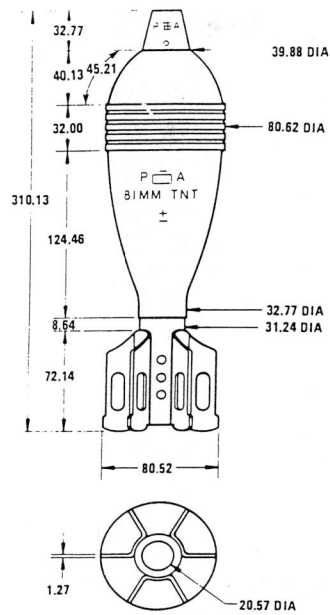
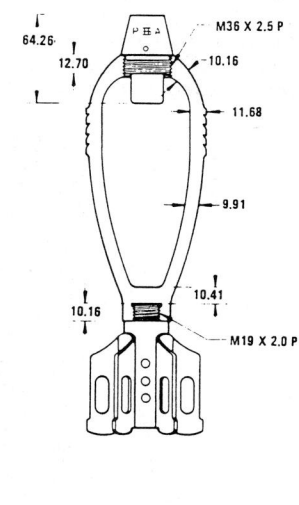

Chinese 81 mm HE bomb, M43A1 variant

82 mm HE bomb Type 53

DATA
Calibre: 82 mm
Type: HE
Weight: (fuzed) 3.87 kg
Bursting charge: 0.38 kg TNT
Fuze: point detonating
Known using weapons: mortars, Models 20 and 53

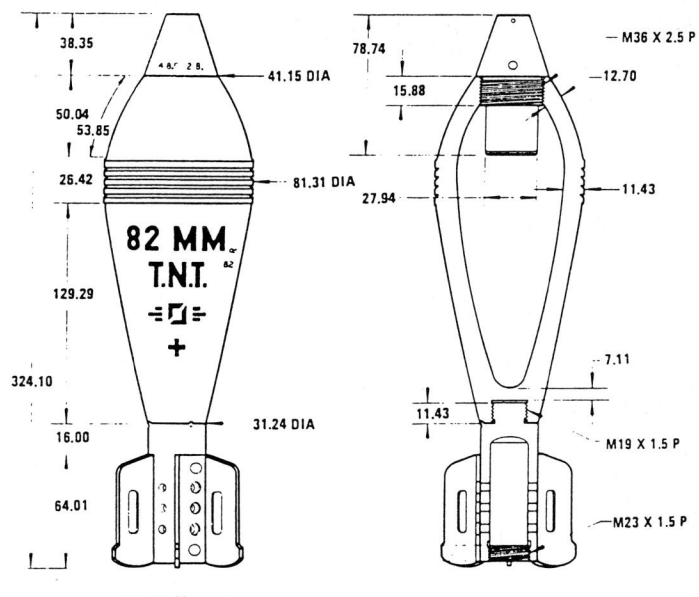

Chinese 82 mm HE Bomb Type 53

82 mm M30 HE fragmentation bomb

DATA
Calibre: 82 mm
Type: HE fragmentation
Weight: (fuzed) 3.15 kg
Bursting charge: 0.42 kg TNT/dinitronapthalene
Fuze: Type 6 point detonating
Known using weapons: mortars, Type 53
Remarks: the bomb is a copy of the Soviet O-832, the fuze a copy of the Soviet Model M-6

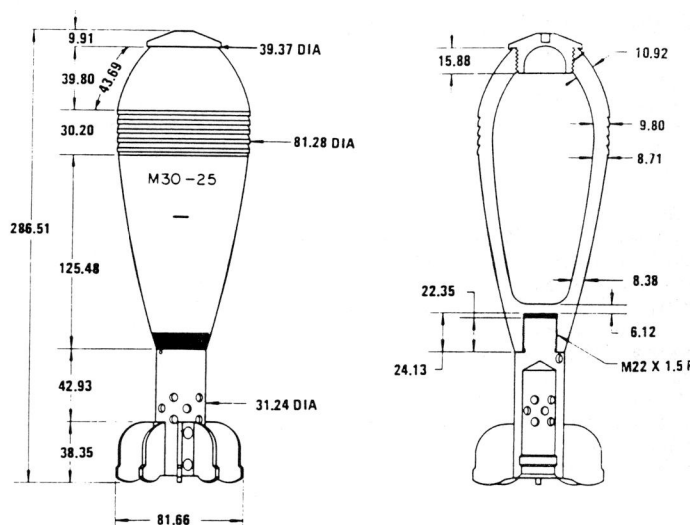

Chinese 82 mm M30 HE fragmentation bomb

82 mm HE fragmentation bomb Type 20

DATA
Calibre: 82 mm
Type: HE fragmentation
Weight: (fuzed) 3.82 kg
Bursting charge: 0.14 kg commercial dynamite
Fuze: Type 9 point detonating
Known using weapons: mortar Type 53

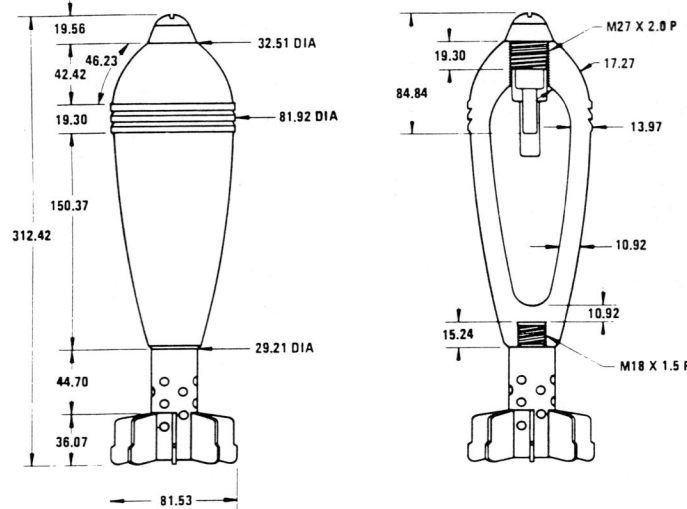

Chinese 82 mm HE fragmentation bomb Type 20

CZECHOSLOVAKIA

120 mm HE bomb Model OF-A

DATA
Calibre: 120 mm
Type: HE fragmentation
Weight: 15.33 kg (fuzed)
Bursting charge: 2.043 kg TNT
Fuze: MZ30AV point detonating
Using weapons: Soviet M1938 and M1943 mortars

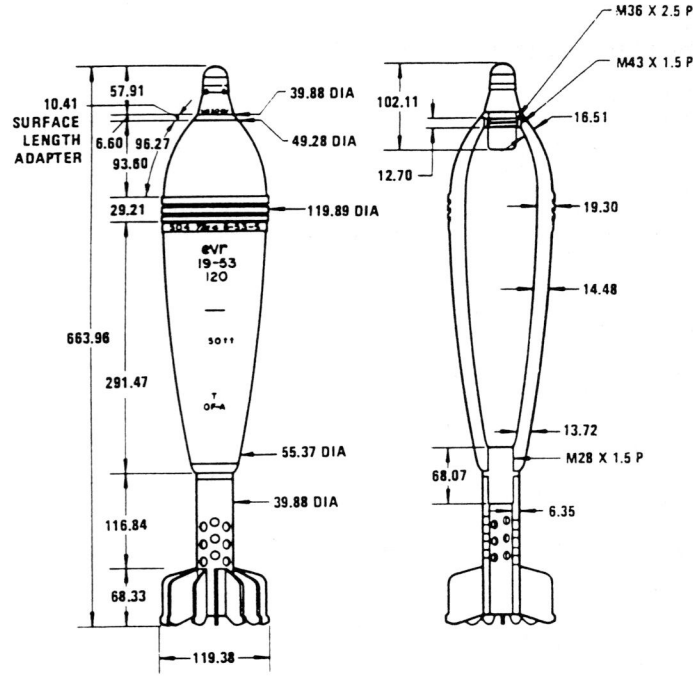

Czech 120 mm HE bomb Model OF-A

FRANCE

Ammunition for 60mm Mortars

Primary cartridges for all 60 mm bombs are 24 mm in diameter and 65 mm long, containing 4.2 g of ballistite. Secondary charges are generally of the horseshoe type, although earlier bombs used secondary charges fitted between the fins. The supercharge is horseshoe shaped and contains 5 g of ballistite.

Mark 61 HE bomb
The body of this mortar bomb is made of pearlitic, malleable cast-iron. It is filled with 260 g pressed TNT and is fitted with the V9 fuze. The total length of the bomb is 317 mm, and it weighs 1.72 kg. Using Charge 2 it is propelled to 1050 m.

Mark 61 HE colour marker bomb
This bomb is used for ranging and target indication and contains the normal HE filling but around it is colouring matter which may be green, yellow, red or black. The bombs have the same physical characteristics as the normal HE bombs.

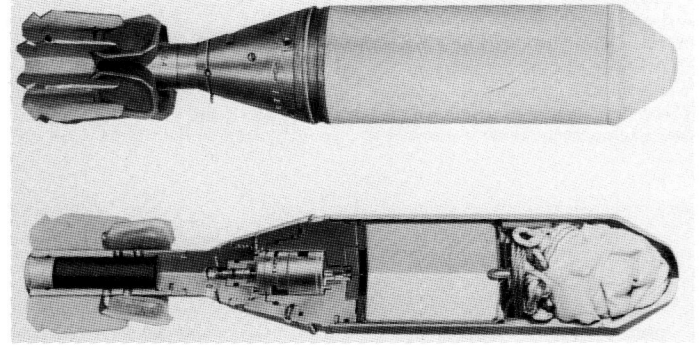

60 mm Mark 63 parachute illuminating bomb

Mark 61 ammunition, showing shape and position of secondary charges

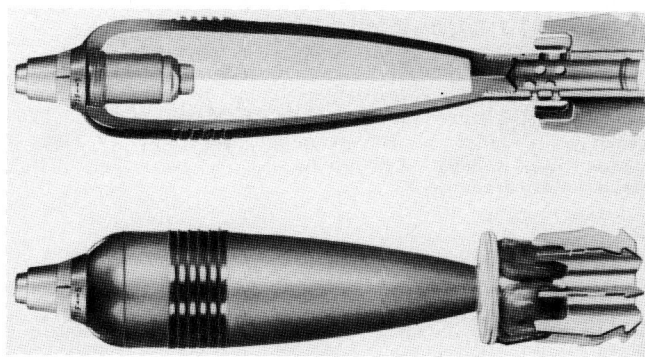

60 mm HE bomb Mark 61 with fuze V9

Mark 61 smoke bomb

This bomb, filled with liquid titanium chloride or white phosphorus, produces an effective smoke screen. The weight, length and so on are the same as those of the HE bombs. The same fuze is used.

Mark 61 practice bombs

This bomb is ballasted with a dummy head. Again it has the same ballistics as the HE bomb. Either live or inert V9 fuzes may be fitted. Some users fire the Mark 61 colour marker (black) as a practice bomb.

Mark 63 illuminating bomb

This bomb has a magnesium based filling and is hung on a parachute. The light persists for a minimum of 30 seconds and produces 180 000 candela. It provides illumination over an area with a radius of 150 m. The round is 333 mm long, weighs 1.55 kg and has a clockwork fuze, graduated from 7 to 35 seconds in half-seconds.

Manufacturer
Thomson-Brandt Armements, Tour Chenonceaux, 204 Rond-Point du Pont de Sevres, F-92516 Boulogne-Billancourt.
Status
Current. Production.
Service
The armed forces of 20 countries. Most use the type A.

60 mm CC anti-tank mortar bomb

This bomb has been developed by Brandt to be fired from the 60 mm gun-mortars, thus giving these weapons a full range of ammunition; ie HE, illuminating, smoke, practice and anti-tank. The mortar bomb has a piezo-electric fuze which gives it a 10 m muzzle safety; the bomb is not armed until it is 10 m away from the weapon. This fuze is the same type as is fitted to the 68 mm rocket, with the accelerometer modified for mortar loadings.

Although the hollow-charge will penetrate almost 200 mm of armour or concrete, it still has a good anti-personnel capability.

60 mm CC anti-tank mortar bomb

DATA
Weight of complete round: 1.54 kg
Total length: 343 mm
Charge: Hexolite (RDX/TNT)
Muzzle velocity: 200 m/s
Range: (stationary target) 500 m; (moving target) 300 m

Manufacturer
Thomson-Brandt Armements, Tour Chenonceaux, 204 Rond-Point du Pont de Sevres, F-92516 Boulogne-Billancourt.
Status
Production.

60 mm canister round

This has been designed for the close defence of armoured vehicles equipped with the 60 mm gun-mortar; it is also an expedient method of clearing attacking infantry from a nearby armoured vehicle. The round is a one-piece unit which is breech loaded into the gun-mortar and it contains a propelling charge and 132 hardened lead shot, each 8.7 mm in diameter. Upon firing, the crimped mouth of the case is blown open and the shot ejected. Once the shot has left the gun the empty casing is ejected through the muzzle. At 50 m range the shot charge covers an area of about 25 m², and most have sufficient remaining velocity to pierce 27 mm pine boards. The shot will not perforate 5 mm thickness of mild steel at 5 m range, so that it may be fired at a friendly vehicle without risk to the crew.

DATA
Length: 217 mm
Diameter: 60 mm
Weight: 1.125 kg
Filled: 132 hardened lead shot

Manufacturer
Thomson-Brandt Armements, Tour Chenonceaux, 204 Rond-Point du Pont de Sevres, F-92516 Boulogne-Billancourt.
Status
Current. Production.

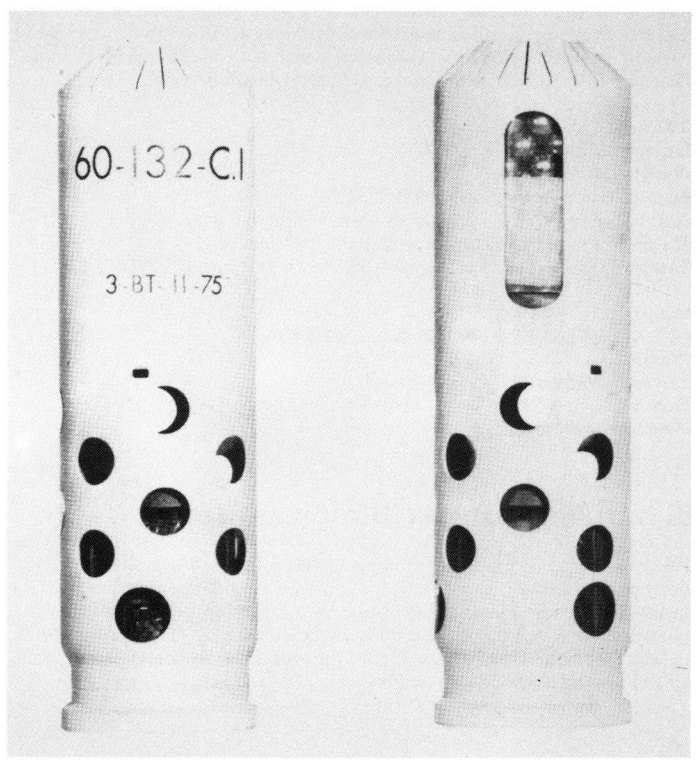

60 mm canister round

60 mm Lacroix canister shot

This has been developed for use by light armoured vehicles equipped with the 60CS gun-mortar. It consists of a casing containing a propelling charge and a loading of steel balls. The canister is loaded into the breech of the mortar and fired; the balls are ejected shotgun-fashion and form an effective anti-personnel screen out to a range of about 60 m. The empty canister is then expelled through the muzzle of the gun.

DATA
Diameter: 60 mm
Length: 235 mm
Weight: 1.455 kg
Weight of balls: 520 g
Diameter of balls: 8/9 mm
Number of balls: 135
Effective range: 10–60 m

Manufacturer
Ste E Lacroix, BP 213, 31601 Muret.
Status
Current. Production.
Service
French Army.

60 mm Lacroix canister shot

60 mm Alsetex mortar bombs

Alsetex makes a range of bombs suitable for use with 60 mm mortars of all types, including those that are breech loaded. The bombs are supplied as complete rounds with fuzes and cartridges in position and packed in individual containers which can be coupled for carriage. There are six models shown in the accompanying illustration.

DATA
Calibre: 60 mm
Weight: 1.35 kg
Body: French Army type FA Mark 47
Fuze: SNEM Mark F1 with double safety (except ref. **F** – see caption)
Propellant: central cartridge with 4 increments
Range: 100–1000 m

Manufacturer
SAE ALSETEX, 4 rue de Castellane, 75008 Paris.
Status
Current. Production.
Service
French armed forces.

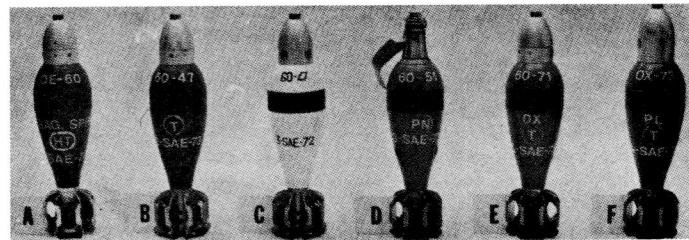

60 mm Alsetex mortar bombs
(A) *high-efficiency fragmentation, approximately 150 g of RDX producing about 350 fragments;* **(B)** *HE (Mark 47) 150 g of TNT;* **(C)** *inert (Mark 47) 150 g inert ballast;* **(D)** *smoke (Mark 51);* **(E)** *OX 60 PLT Mark F1 practice, inert marking ballast dispersed on impact by small TNT charge, SNEM Mark F1 fuze;* **(F)** *OX 60 PLT Mark F1 (as* **E** *but with 21 × 28 fuze)*

81 mm Alsetex mortar bombs

A range of bombs suitable for use with all types of 81 mm mortar is made by Alsetex. They are supplied as complete rounds, with fuze and cartridge in position, packed in individual containers which may be coupled for carriage. There are five models, shown in the accompanying illustration.

DATA
Calibre: 81 mm
Weight: approx 3.5 kg
Body: French Army type FA Mark 32
Fuze: SNEM Mark F1 with double safety (feed and trajectory)
Propellant: central cartridge with 3 or 6 increments
Range: (average practical) approx 2000 m; (max approx) 3000 m

Manufacturer
SAE ALSETEX, 4 rue de Castellane, 75008 Paris.
Status
Current. Production.
Service
French armed forces.

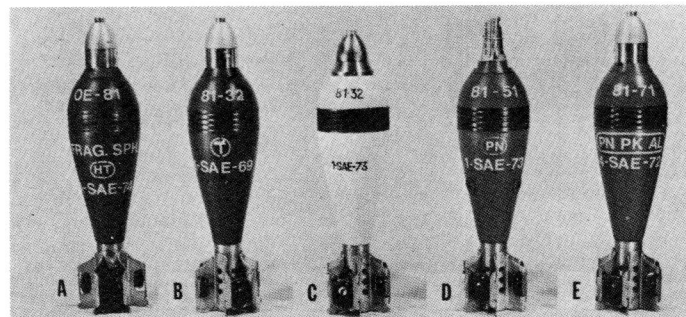

81 mm Alsetex mortar bombs
(A) *high-efficiency fragmentation, RDX filling producing about 300 splinters* **(B)** *HE, containing 550 g cast TNT* **(C)** *inert (Mark 32)* **(D)** *smoke practice (Mark 51)* **(E)** *practice PL PN Mark F1, inert marking, ballast dispersed by small charge*

81 mm Alsetex Alta 81 mortar fuze

The Alta 81 is a point detonating impact fuse for use with smoothbore 81 mm mortar bombs. Highly sensitive, it will function on any type of ground. Safety during storage and handling is ensured by an interruptor in the firing train and by positive locking of the arming elements. Arming is delayed on firing and the fuze cannot arm within 40 m of the muzzle. A visual indication of the state of arming is provided.

DATA
Length: 75 mm
Diameter: 48 mm
Weight: 150 g
Arming impulse: 800 to 8500 g acceleration
Operating temperature: −40° to +60°C

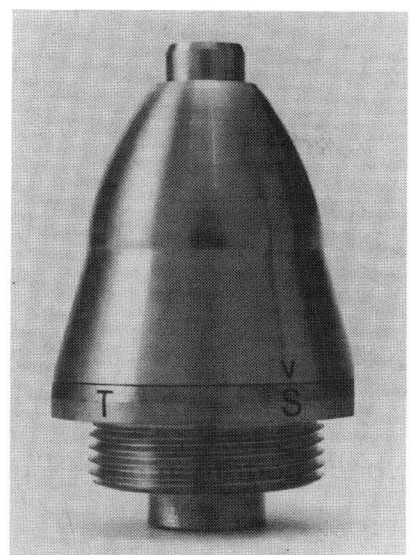

Manufacturer
SAE ALSETEX, 4 rue de Castellane, 75008 Paris.
Status
Prototype production.

Alsetex Alta 81 mortar fuze

Ammunition for 120mm Brandt mortars

M44 HE bomb range

This is a conventional high explosive bomb, fuzed V19, weighing 13 kg and filled TNT. It is 679 mm long, minimum range is 500 m and the maximum is 7000 m. There is a practice bomb, a smoke bomb filled WP or FM, and an HE marker bomb with the same weight and performance as the M44 bomb. The M44 bomb was produced up to 1966 by Brandt. Since that time it has been manufactured for the French Army by other contractors. The current Brandt HE bomb is designated M44/66 HE and fuzed V19P.

Smoke bomb
The standard smoke bomb is the Mark 44/67 which is filled with white phosphorus.

Mark 62-ED illuminating bomb
This has a mechanical time fuze, type FH81-B and spring-loaded tail fins.

DATA
Total length: 800 mm
Total weight of complete round: 13.65 kg
Filling: illuminating compound, magnesium based
Power: 700 000 cd
Burning time: 1 min
Radius of illumination: 300 m

Manufacturer
Thomson-Brandt Armements, Tour Chenonceaux, 204 Rond-Point du Pont de Sevres, F-92516 Boulogne-Billancourt.
Status
Current. Production.
Service
French Army and several others.

120 mm M44 series mortar bombs

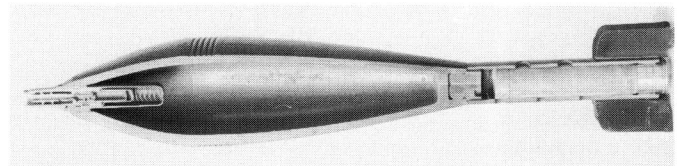

M44 bomb sectioned

120 mm PEPA rocket-assisted mortar bomb

This bomb consists of an steel body, a HE filling, an internal solid fuel rocket motor and the V19 P fuze. The propelling charge, made up of the primary and secondary charges, is at the rear. The body (3) is made in two parts, screwed together, with the tail assembly (10) screwed on to the rear part of the body.

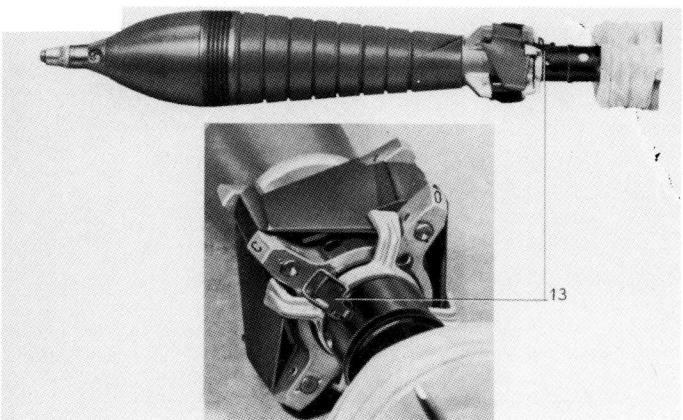

120 mm PEPA bomb
(1) *fuze V19 P* **(2)** *HE filling* **(3)** *steel body* **(4)** *tube containing rocket assistance* **(5)** *rocket motor* **(6)** *venturi* **(7)** *optional delay* **(8)** *selection lock*
(9) *spring-opened tail fin* **(10)** *tail assembly* **(11)** *primary cartridge* **(12)** *secondary charges* **(13)** *spring catch holding selection lock in place*

The HE content (2) is 2 kg of hexogen (RDX/TNT) cast inside the cavity and around the solid fuel motor. The solid fuel motor (5) of star-shaped section, is housed in a tube (4), externally inhibited, with a venturi at the end (6). The venturi is secured at the tail end by an obturator containing an optional delay (7) and a selection lock (8) which allows the use of the solid fuel propulsion unit if required. A spring catch (13) holds the selection lock in the required position.

The propulsion charge consists of the primary cartridge (11) and up to seven augmenting cartridges. The tube (10) holding the primary cartridge contains the flash holes communicating with the secondary charges. This tube is ejected after the bomb is fired.

In use the order of events is as follows. The bomb is removed from the container and the charge system adjusted by the removal of unrequired secondaries. If the rocket assistance is required the tube containing the primary cartridge is rotated clockwise. The fuze V19 P is then set for instantaneous action or delay.

The ranges obtainable with the PEPA are as follows. 'R' indicates rocket assistance.

Charge	Elevation (max)	Range (minimum)	Elevation (minimum)	Range (max)
1	74°	600 m	45°	1350 m
2	72°	1000 m	45°	1900 m
3	69°	1700 m	45°	2560 m
4	67°	2200 m	45°	3175 m
5	70°	2500 m	45°	3560 m
6	69°	2800 m	45°	4250 m
3R	66°	4000 m	55°	4800 m
4R	66°	4600 m	55°	5500 m
5R	63°	5300 m	50°	6100 m
6R	62°	5700 m	45°	6550 m

From this table it can be seen that the maximum range without rocket assistance is 4250 m. The muzzle velocity is 240 m/s. The maximum range with rocket assistance is 6550 m. The muzzle velocity is increased by 110 m/s. The PEPA/LP bomb was developed for use with the MO-120-M65 and 120 AM 50 mortars. It has a wider range bracket.

Manufacturer
Thomson-Brandt Armements, Tour Chenonceaux, 204 Rond-Point du Pont de Sevres, F-92516 Boulogne-Billancourt.
Status
Current. Production.
Service
French and other armies.

Ammunition for 120 mm rifled Brandt mortars

Although the MO-120-RT-61 is a rifled mortar, it will fire smoothbore bombs except those types having spring-loaded tail fin assemblies with straight fins. Smoothbore bombs are frequently used for bedding in the baseplate (1 round charge 3, 1 charge 5, 1 charge 7) and also for economy in training. Bombs designed for the MO-120-RT-61 are equipped with a tail tube carrying the primary and secondary cartridges. This tube is ejected just after the bomb has left the mortar and falls about 100 m from the muzzle.

DATA
PRPA BOMB
The projectile rayé à propulsion additionelle (PRPA) has rocket assistance which comes into action after a delay of 10 seconds and a pre-engraved driving band
Length with tail tube: 918 mm
Weight, complete round: 18.7 kg
Filling: RDX/TNT
Range: (max with rocket assistance) 13 000 m; (minimum without rocket assistance) 1100 m
Fuze: PD, M557 or Brandt VG-29. The latter has the normal Brandt pneumatic delayed arming features but makes use of setback, creep and centrifugal forces in the initiation of this delay

PR 14 BOMB
No rocket assistance. Pre-engraved driving band
Length including tail tube: 897 mm
Weight of complete round: 18.7 kg
Weight in flight: 15.7 kg

Filling: RDX/TNT
Max range: 8135 m
Minimum range: 1100 m
Fuze: PD, M557 or Brandt VG-29. The latter has the normal Brandt pneumatic delayed arming features but makes use of setback, creep and centrifugal forces in the initiation of this delay.

ILLUMINATING BOMB PRECLAIR
Overall length: 886.5 mm
Total weight of complete round: 18.3 kg
Weight in flight: 15.3 kg
Filling: magnesium-based compound
Fuze: FR 55 A clockwork
Fuze setting: 2–55 s
Optimum height of functioning: 300 m
Rate of fall of illuminant: 5 m/s
Radius of illumination: 500 m
Minimum burning time: 40 s producing 1 600 000 cd

Manufacturer
Thomson-Brandt Armements, Tour Chenonceaux, 204 Rond-Point du Pont de Sevres, F-92516 Boulogne-Billancourt.
Status
Current. Production.
Service
French Army. Also exported to several countries.

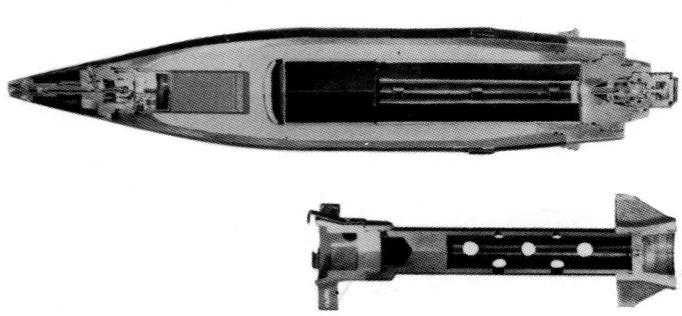

Sectioned PRPA bomb

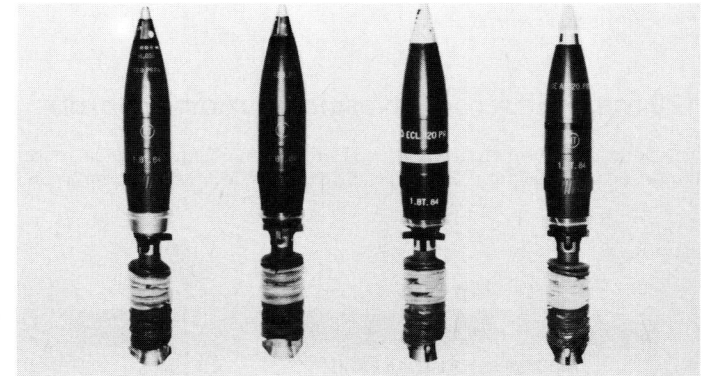

120 mm PRPA, PR 14, illuminating and anti-armour bombs

120 mm rifled anti-armour bomb

This has been developed by Thomson-Brandt to extend the range of 120 mm mortar ammunition. The basic idea is that the mortar bomb hits the ground almost vertically and sprays a large number of fragments out sideways. According to Brandt, if the new mortar bomb lands at an angle of 70° or more and 15 m or less away from the target, the fragments will penetrate 12 mm of armour to a height of approximately 1.5 m. Although these fragments might not affect a closed-down main battle tank, they can be extremely effective against armoured personnel carriers and similar lightly-armoured vehicles. The bomb can be fitted with either a graze fuze or a proximity fuze, the latter offering an enhanced performance.

DATA
Weight of complete round: 18.7 kg
Weight in flight: 15.7 kg
Total length including tail: 897 mm
Charge: Hexolite (RDX/TNT)
Range: (max) 8135 m; (minimum) 1100 m

Manufacturer
Thomson-Brandt Armements, Tour Chenonceaux, 204 Rond-Point du Pont de Sevres, F-92516 Boulogne-Billancourt.
Status
Current. Production.

120 mm Alsetex mortar bombs

The Alsetex company manufactures a range of 120 mm bombs suitable for use in any conventional mortar. They are of steel, loaded with cast TNT, and supplied complete with fuze, primary cartridge, and seven augmenting cartridges in an airtight container. The basic design is that of the standard French Army Mk 44 bomb, and the fuze is the SNEM Mark F1 with delayed arming and double-loading safety features.

Type A is a high capacity HE bomb with hexolite filling and a lining of spherical pre-formed fragments. Type B is a standard HE bomb with a 2.5 kg TNT filling. Type C is an inert practice bomb, and Type D is a practice bomb with a small pyrotechnic burster charge to indicate the point of impact. All types weigh approximately 13 kg and have a maximum range of about 7000 m.

Manufacturer
SAE ALSETEX, 4 rue de Castellane, 75008 Paris.
Status
Current. Production.

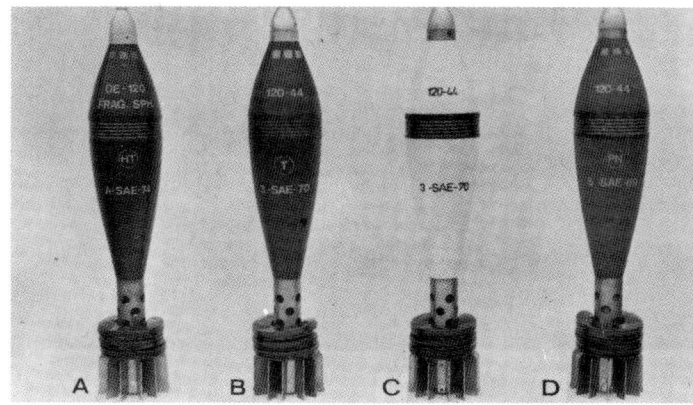

Alsetex 120 mm mortar ammunition
(A) *High explosive, high capacity* **(B)** *High explosive, standard* **(C)** *Inert practice* **(D)** *Practice with burster charge*

GERMANY, WEST

Buck 60 mm mortar ammunition

This is a new generation of 60 mm smoke, smoke/incendiary and illuminating mortar bombs. The design has produced a highly efficient bomb which will function with various types of 60 mm mortar including the US M224. Their enhanced payloads permit the achievement of the tactical function with less expenditure of ammunition.

Technical data	60 mm Smoke	60 mm Incdy/ Smoke	60 mm Illuminating
Weight	2.3 kg	2.3 kg	2.3 kg
Length	467 mm	467 mm	467 mm
Range	1900 m	1900 m	1600 m
Emission time	2 min	—	—
Luminance	—	—	450 000 cd for 35 s
Payload	400 g	Submunition	430 g

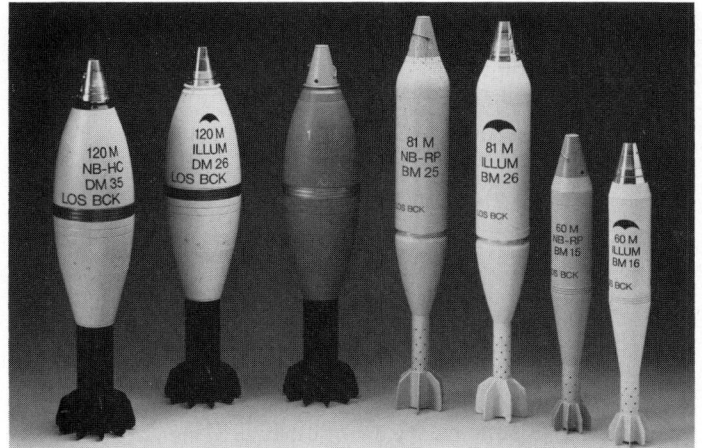

Buck family of mortar ammunition

Buck 81 mm mortar ammunition

As with the 60 mm rounds, these 81 mm smoke and smoke/incendiary bombs are of modern design and intended to produce the desired effects for less expenditure of ammunition. They will function in any modern 81 mm mortar including the US M29 and UK L16A1 series.

Technical data	81 mm Smoke	81 mm Incdy/Smoke	81 mm Illuminating
Weight	3.9 kg	3.9 kg	4.4 kg
Length	630 mm	630 mm	635 mm
Range	5600 m	5600 m	5400 m
Emission time	2.5 mins	—	—
Luminance	—	—	600 000 cd or 1 million cd for 35 s
Payload	1.4 kg	submunition	1.4 kg

Buck 120 mm mortar ammunition

These new-generation mortar bombs will function with all types of modern 120 mm mortar including Tampella, Thomson-Brandt and Noricum/VEW types. As with the smaller calibres, the object has been to produce bombs which are capable of delivering the requisite tactical effect with less expenditure of ammunition.

Technical data	120 mm Smoke	120 mm Incdy/Smoke	120 mm Illuminating	120 mm Practice
Weight	12.9 kg	12.9 kg	12.9 kg	12.9 kg
Length	590 mm	590 mm	583 mm	582 mm
Range	6000 m	—	—	—
Emission time	3 mins	—	—	—
Luminance	—	—	1.2 M cd/50 s	—
Payload	3.65 kg	submunition	1.71 kg	Spotting charge

Manufacturer
Buck Werke GmbH & Co., Postfach 2405, D-8230 Bad Reichenall.

Status
Current. Production.

120 mm Diehl 'Bussard' terminally guided mortar projectile

In 1975 Diehl was awarded a contract by the West German Ministry of Defence to perform a feasibility study into a terminally guided mortar projectile. Bodenseewerk Geratetechnik, AEG and Eltro were also involved in the project. The projectile which has resulted from this study is the Diehl 'Bussard', the first West German contribution to this group of munitions.

The objective of this study was to combine the advantages of traditional gun-launched projectiles with those of guided missiles so that the resulting munition would be used from existing ordnance, would have a high first-round hit probability, even against moving targets, and would permit engaging armoured targets in the 'top attack' mode. The Bussard projectile complements existing ammunition inventories and is compatible with the 120 mm mortar currently used by the Bundeswehr.

The projectile consists of a shaped charge warhead, a central structure with power supply and guidance controls, a seeker head, and a tail unit. It is loaded and fired in the same manner as any 120 mm ammunition. During the ascending portion of the trajectory the thermal battery is activated and the gyro is run up. Shortly after the vertex the wings are deployed, the gas generator begins to function so as to provide power for the controls, and the laser recognition system in the seeker head is activated. The selected target is illuminated by an observer using a Laser Target Designator. The Bussard seeker recognises the reflected laser signal and from this develops the commands necessary to actuate the control surfaces and steer the projectile to impact with the target.

The concept permits the employment of alternative types of seeker; thus the semi-active laser system can be replaced by an active millimetric-wave radar seeker or by a passive infra-red seeker. Both these alternatives would dispense with the need for laser illumination and make the projectile a fire and forget system.

Successful firings of the first development models took place in 1983.

DATA
Calibre: 120 mm
Length: approx 1 m
Weight: approx 17 kg
Range: 800–5000 m

Diehl Bussard terminally guided mortar projectile

Propulsion: standard mortar cartridges
Guidance: proportional navigation
Control: aerodynamic control fins driven by hot gas generator
Seeker: semi-active laser (Silicide photo-diode) or active millimetric-wave radar, or passive infra-red
Acquisition altitude: between 2000 m and 600 m
Accuracy: 0.5 m when laser designated
Warhead: shaped charge

Manufacturer
Diehl GmbH & Co, Fischbachstrasse 16, D8505 Röthenbach/Pegnitz.
Status
Advanced development.

Diehl optronic proximity mortar fuze

This fuze is suitable for use in 120 mm or larger non-spin-stabilised mortar bombs. The principle can also be adapted to use in cluster bombs or other similar types of ammunition. The optronic sensor relies upon a new principle in connection with a laser diode; when approaching a target, detonation occurs at a selected height with a narrow tolerance which is independent of the nature of the target. The height of burst, within the range 1 to 12 m, is set during manufacture.

The fuze has a small field-of-view and optic and electronic filtering, which results in a high resistance to electronic countermeasures. The fuze does not react to fog, smoke, rain, clouds or snow. The electronic components are few and are of monolithic and hybrid structure. A reliable lead/lead dioxide battery and rugged mechanical safety devices contribute to the high reliability of the fuze in service. Should the proximity function not be required, the protective cap is left on the nose and the fuze acts on impact; the impact feature will also function should the proximity circuits fail to work.

Development of the fuze has been carried out as a co-operative venture by:
Eltro GmbH, Heidelberg
Silberkraft GmbH, Duisburg
Junghans GmbH, Schramberg
Diehl GmbH, Röthenbach.

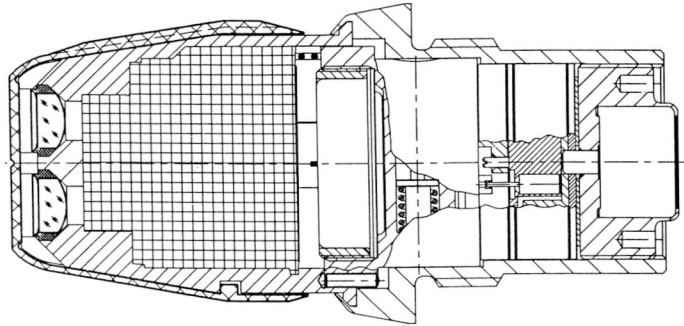

Diehl optronic proximity mortar fuze

Manufacturer
Junghans Feinwerktechnik GmbH, PO Box 110, D-7230 Schramberg.
Status
Advanced development; production in 1990.

DM111A4 point detonating mortar fuze

The DM111A4 is a mechanical nose fuze used for unspun HE, smoke and TP mortar ammunition of 51 to 160 mm calibre. The fuze is highly sensitive so that it will function even upon impact on fresh snow, marshy ground and water or against shrubs and bushes. Two modes can be set, Superquick (SQ) or Delay (D). In the safe position the rotor with two detonators is locked out of line by a safety pin and a pulse safety which is only released by firing. The muzzle safety distance is $\geqslant 40$ m.

Manufacturer
Junghans Feinwerktechnik, PO Box 110, D-7230 Schramberg.
Status
Current. Production.

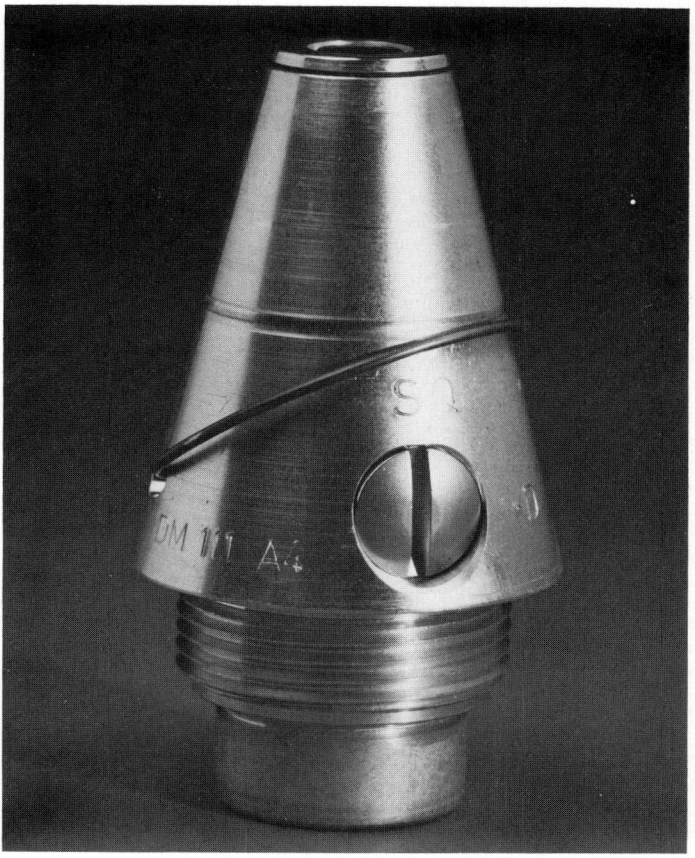

DM111A4 PD mortar fuze

DM111A5 point detonating mortar fuze

The DM111A5 is a mechanical nose fuze for unspun HE, smoke and TP mortar ammunition of calibres from 51 to 160 mm. This fuze is not as sensitive as the DM111A4, so that it can be used in wooded terrain (for example, in jungle areas).

Two modes can be set: superquick (SQ) or delay (D). In the safe position the rotor with two detonators is locked out of line by a safety pin and a pulse safety which is only released by firing. The muzzle safety distance is $\geqslant 40$ m

Manufacturer
Junghans Feinwerktechnik, PO Box 110, D-7230 Schramberg.
Status
Current. Production.

DM111A5 point detonating mortar fuze

DM111 AZ-W point detonating mortar fuze

This mechanical PD nose fuze provides a vane type arming mechanism in addition to the usual setback detent system. It thus has two entirely independent safe and arming systems activated by two different physical properties, acceleration and rotation. Due to the vane feature this fuze does not require a safety pin which must be removed before firing, as seen on many conventional PD fuzes. After the bomb has left the muzzle of the mortar, the wind streaming past the fuze rotates the vane and releases the vane safe and arming device.

The DM 111 AZ-W fuze can be used with smoothbore 51 mm to 120 mm mortar HE, smoke and practice bombs. Release criterion is a translational acceleration of 500 g or better. The arming distance is at least 40 m using the lowest charge.

All requirements concerning handling, transport, firing and muzzle safety, as well as the provisions of STANAG 3525 and MIL-STD-1316B are fulfilled.

Manufacturer
Junghans Feinwerktechnik, PO Box 110, D-7230 Schramberg.
Status
Current. Production.

DM 111 AZ-W wind vane arming impact fuze

DM 113A2 point detonating fuze

This is a mechanical nose fuze for use with mortars from 81 mm to 120 mm calibre. In the unarmed state the detonators are out of line due to the position of the safety rotor. The rotor is locked by two elements, a safety pin and a detent which is released upon firing. Movement of the rotor is then controlled by a delayed arming mechanism, upon firing. This ensures a safe distance of at least 40 m with the lowest charge. The rotor travels to the position set before firing, either delay or non-delay. The fuze has a sensitive impact mechanism which initiates the firing train upon striking the target.

Manufacturer
Junghans Feinwerktechnik, PO Box 110, D-7230 Schramberg.
Status
Current. Production.
Service
In British Army service as the Fuze, Nose, Percussion, DA, with optional delay, L127A3.

DM93/M776 MTSQ mortar fuze

The DM93 is a mechanical time fuze with an additional impact device, for use in unspun smoke and illuminating mortar projectiles of 51 to 160 mm calibre. It can be set from 6 to 54 seconds with a possible extension up to 67 seconds. At the zero setting the fuze functions upon impact. In the safe position the rotor is out of line and the clockwork is locked by a safety pin and a pulse safety which is released only by firing. The muzzle safety distance is ⩾40 m.

DM123 MTSQ mortar fuze
The DM123 is a modification of the DM93. By changing the detonation train this fuze can be used with HE ammunition.

Manufacturer
Junghans Feinwerktechnik, PO Box 110, D-7230 Schramberg.
Status
Current. Production.

DM93 time and percussion mortar fuze

M772 MTSQ mortar fuze

The M772 is a mechanical time fuze with an additional impact device, for unspun smoke and illuminating bombs in 81 mm calibre. It can be set from 4 to 55 seconds. The M772 is a derivation of the DM93 especially developed for the US 81 mm ammunition (M853 and M819) used by the US Army and US Marine Corps. At the zero setting the fuze functions upon impact. In the safe position, the rotor is out of line and the clockwork is locked by a safety pin and a pulse safety which is released only by firing. The muzzle safety distance is ≥40 m.

Manufacturer
Junghans Feinwerktechnik, PO Box 110, D-7230 Schramberg.
Status
Current. Production.

M772 MTSQ mortar fuze

GREECE

81 mm M374A2 HE bomb

This is the standard US pattern 81 mm HE bomb. It consists of a steel body with plastic obturating ring, filled with Composition B. The fuse is the PD M524A6 with selectable superquick or delay functioning; any other suitable fuze can be used. The fins are of aluminium alloy and a primary cartridge M285 is fitted into the forward section of the tail tube, with a percussion primer M71A2 screwed into the base of the tail unit. The flash from the primer passes up the centre of the tail unit to ignite the primary, which then vents through holes in the tail tube to ignite the secondary cartridges. There are nine secondaries in the form of long bags filled with flake M9 powder; these are retained by two increment holders. The secondaries are of two types: one Charge A unit of 12 g weight, and eight Charge B units of 10.875 g weight.

DATA
Calibre: 81 mm
Weight: 4.24 kg
Length: 529 mm
Explosive content: 945 g Comp B
Muzzle velocity: (Charge 0) 64 m/s; (Charge 9) 261 m/s
Max range: (Charge 0) 403 m; (Charge 9) 4500 m

Manufacturer
Greek Powder and Cartridge Co (PYRKAL), 1 Ilioupoleos Avenue, GR-17236 Hymettus, Athens.
Status
Current. Production.
Service
Greek Army.

81 mm HE bomb M374A2

107 mm HE GR M20 grenade carrier bomb

The Greek Powder & Cartridge Company has developed and is producing an improved mortar bomb for use with existing weapon systems, the 107 mm (4.2 in) HE GR M20 for M2 and M30 mortars. The bomb carries 20 grenades M20G which contain a 30 g shaped charge for penetration and fragmentation effects. Each grenade is equipped with a GR M3A2 fuze and a delay parachute and functions upon impact with the target.

Targets include personnel and light armoured vehicles, with a penetrating ability of 60 mm in steel plate, a dispersion radius of approximately 50 m, an effective radius of 15 m around the impact point of each grenade, and a total effective area of some 7000 m².

Functioning
The fuze of the projectile (MTSQ M577A1 or Electronic Time Fuze AXπ-1) is set to function at 300 m above the target area. At this predetermined point in flight the entire grenade load is ejected from the rear of the projectile and is radially dispersed over the target area.

The inherent superiority of the rifled 107 mm mortar over conventional smoothbore mortars accounts for the bomb's greater grenade dispersion capabilities.

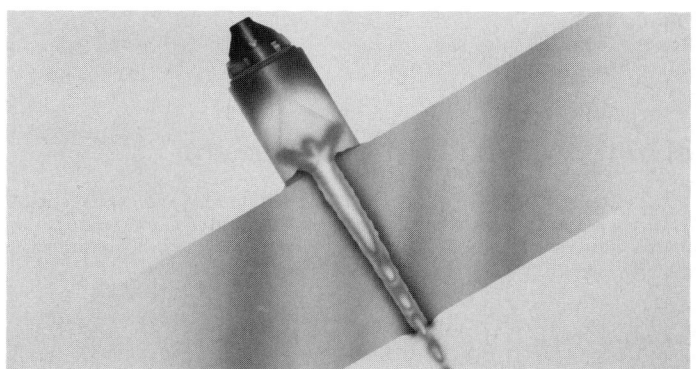

Action of M20G grenade sub-munition

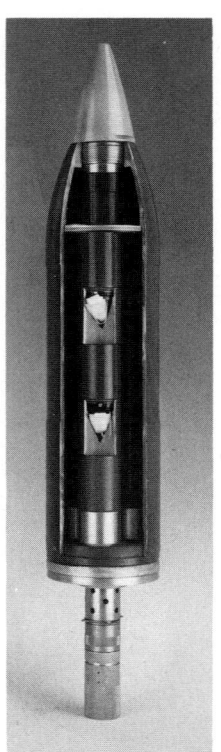

DATA
Calibre: 107 mm
Weight: 12.0 kg
Length, fuzed: 655 mm
Dispersion radius: 70 – 140 m around MPI
Max range: 5500 m

Manufacturer
Greek Powder and Cartridge Co (PYRKAL), 1 Ilioupoleos Avenue, GR-17236 Hymettus, Athens.
Status
Current. Production.
Service
Greek Army and Middle East countries.

107 mm HE GRM 20 grenade carrier bomb

INDONESIA

60 mm GMO-6 PE A1 mortar bomb

This bomb is intended for use in short 'Commando' type 60 mm mortars and is conventional pattern with gas-check rings around the waist. It has one propellant charge, composed of the usual ignition cartridge and a single incremental charge fitted around the tail boom. The tail fins are slightly canted to give some degree of roll stabilisation. The design is licensed from Tampella.

DATA
Calibre: 60 mm
Weight: 1.62 kg
Length: 300 mm
Body: forged steel
Filling: TNT
Range: 80 – 800 m

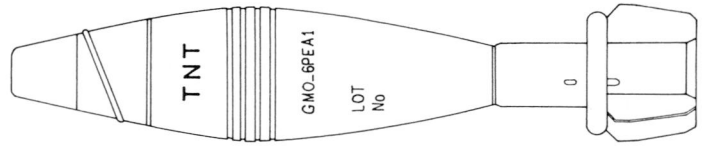

60:mm GMO-6 PE A1 mortar bomb

Muzzle velocities: 57 – 94.4 m/s
Fuze: M111B1 or DM111A2

Manufacturer
P T Pindad, Thamrin, Jakarta.

60 mm GMO-6 PE A2 mortar bomb

This bomb is for use in standard types of 60 mm mortars and is of modern pattern, with streamlined body, canted fins and an obturating ring around the waist. It is, in fact the Soltam M38 produced in Indonesia under license. The propelling charge consists of a primary cartridge and up to six secondary increments which fit around the tail boom.

DATA
Calibre: 60 mm
Weight: 1.84 kg
Length: 351 mm
Body: forged steel
Filling: TNT
Range: 100 – 4000 m
Muzzle velocities: 89 – 258 m/s

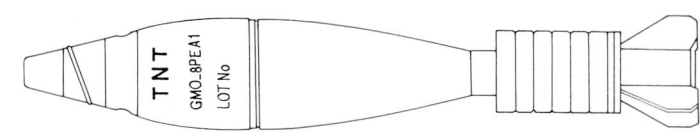

60:mm GMO-6 PE A2 mortar bomb

Fuze: M111B1 or DM111A2
PE at medium/maximum range: 0.6 × 0.3% of range

Manufacturer
P T Pindad, Thamrin, Jakarta.

81 mm GMO-8 PE A1 mortar bomb

This is another modern design, based on the Soltam M64 and manufactured under license in Indonesia. The streamlined bomb uses an obturating ring around the waist and has a propelling charge composed of a primary cartridge and up to eight secondary charges fitted around the tail boom.

DATA
Calibre: 81 mm
Weight: 4.6 kg
Length: 490 mm
Body: forged steel
Filling: TNT

81:mm GMO-8 PE A1 mortar bomb

Range: 150 – 6500 m
Muzzle velocities: 67 – 353 m/s

Fuze: M111B1 or DM111A2
PE at medium/maximum ranges: 0.5 × 0.3% of range

Manufacturer
P T Pindad, Thamrin, Jakarta.

INDIA

51 mm HE-1A HE bomb

This bomb is for the 51 mm mortar now being manufactured in India and is generally similar to the UK L127 bomb. The body is of light steel and is lined with a notched wire coil to provide controlled fragmentation. The filling is RDX/Wax. A percussion fuze, which appears similar to the UK Fuze Percussion 152 is fitted and there is a primary and one secondary cartridge.

DATA
Calibre: 51 mm
Weight: 950 g nominal
Length: 269 mm
Muzzle velocity: 107 m/s
Min range: 200 m
Max range: 850 m

Manufacturer
Indian Ordnance Factory Board, 10A Auckland Road, Calcutta 700 001.
Status
Current. Production.
Service
Indian Army and export.

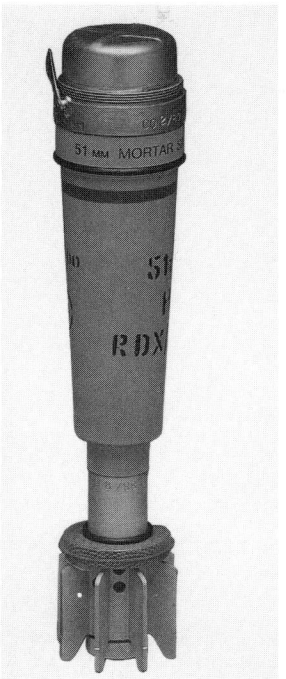

Indian 51 mm HE-1A bomb

81 mm HE Bomb

This is very similar to the old British 3 in HE bomb but has a raised bourrelet band at the waist and a plastic obturating ring. The tail unit is of machined light alloy and is similar in pattern to that used with bombs for the L16 mortar. The bomb is fitted with a locally manufactured version of the British Fuze Percussion No 162, an impact fuze which carries a safety pin beneath the cap and which is armed on firing. The propelling charge consists of a primary cartridge and eight 'horse-shoe' secondaries clipped around the tail unit.

DATA
Calibre: 81 mm
Weight: 4.2 kg
Weight of filling: 705 g TNT
Muzzle velocity: 295 m/s
Max range: 5000 m

Manufacturer
Indian Ordnance Factory Board, 10A Auckland Road, Calcutta 700 001.
Status
Current. Production.
Service
Indian Army and export.

Indian 81 mm HE bomb

81mm smoke bomb

This is essentially the same as the HE bomb described above, except that the filling is Plasticized White Phosphorus (PWP). The same impact fuze is fitted, and the eight-unit propelling charge is similar but adjusted to give the bomb the same ballistic performance as the HE bomb.

DATA
Calibre: 81 mm
Weight: 4.4 kg
Muzzle velocity: 295 m/s
Max range: 5000 m

Manufacturer
Indian Ordnance Factory Board, 10A Auckland Road, Calcutta 700 001.
Status
Current. Production.
Service
Indian Army and export.

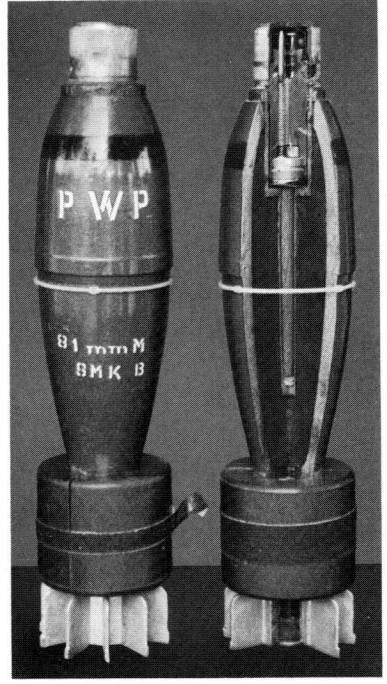

Indian 81 mm PWP Smoke bomb

120 mm HE bomb

This resembles the Brandt M44 HE bomb and is of forged steel with a series of obturation grooves around the waist. It is fitted with the Fuze DA4A, a locally manufactured version of the Brandt V19P impact fuze. The propelling charge consists of a primary cartridge in the tail unit, plus four secondaries in 'horse-shoe' containers clipped around the tail boom.

DATA
Calibre: 120 mm
Weight: 13.03 kg
Weight of filling: 448 kg TNT
Number of charges: 8
Muzzle velocity: Charge 0: 118 m/s. Charge 7: 331 m/s
Max range: 6650 m

Manufacturer
Indian Ordnance Factory Board, 10A Auckland Road, Calcutta 700 001.
Status
Current. Production.
Service
Indian Army and export.

Indian 120 mm HE bomb

ISRAEL

52 mm IMI smoke mortar bomb

DATA
Length: 240 mm
Weight: 940 g
Charge weight: 400 g
Filling material: Hexachlorethane, zinc oxide and aluminium powder
Propelling charge: primary cartridge with 4 g of ballistite
Muzzle velocity: 78 m/s
Max range: 450 m
Time delay: 7.5 s
Time of smoke generation: 100 s

Manufacturer
Israel Military Industries (IMI), PO Box 1044, Ramat Ha Sharon 47100.
Status
Current. Production.

Service
Israeli forces.

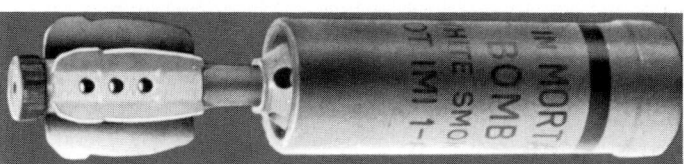

52 mm IMI smoke mortar bomb

52 mm IMI HE mortar fragmentation bomb

DATA
Length: 250 mm
Weight: 1.15 kg
Charge weight (TNT): 105 g
Propelling charge: base ignition cartridge containing 4 g of ballistite
Fuze: No 161
Muzzle velocity: 75 m/s
Max range at 45° elevation: 450 m

Manufacturer
Israel Military Industries (IMI), PO Box 1044, Ramat Ha Sharon 47100.
Status
Current. Production.
Service
Israeli forces.

52 mm IMI HE mortar fragmentation bomb

52 mm IMI illuminating mortar bomb

DATA
Overall length: 271 mm
Weight of complete round: 830 g
Illuminant charge: 150 g
Nature of illuminant charge: pyrotechnic composition
Propelling charge: primary cartridge with 5.5 g ballistite
Muzzle velocity: 95 m/s
Parachute opening: range 350 m; height 100 m
Delay time to burst: 6 s approx
Illuminating power: 100 000 cd
Minimum burning time: 30 s

Manufacturer
Israel Military Industries (IMI), PO Box 1044, Ramat Ha Sharon 47100.

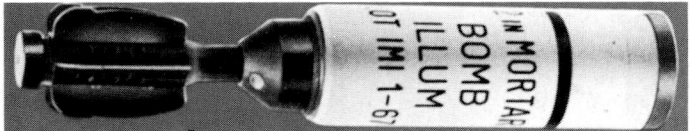

52 mm IMI illuminating mortar bomb

Status
Current. Production.
Service
Israeli forces.

Ammunition for 60 mm Soltam mortars

Bombs for 60 mm mortars are of the conventional pattern. The HE bomb is made from a steel forging and machined to shape. The tail unit is made from an extruded aluminium tube. The secondaries are arranged around the tail boom above the fins and when the bomb is dropped into a hot barrel the secondaries do not contact the walls of the tube, and there is no danger of a cook-off.

The ballistics of the smoke bomb are the same as the HE. There are two smoke fillings, white phosphorus and titanium tetrachloride. The white phosphorus is used also for ranging at night.

Weight, all bombs: 1.61 kg
Charges: primary charge; secondaries, 1, 2, 3, 4

HE BOMB, WITH M111 FUZE
Lethal area: 490 m²
Maximum range: 4000 m

ILLUMINATING BOMB
Illuminating power: 250 000 cd
Burning time: 40 s

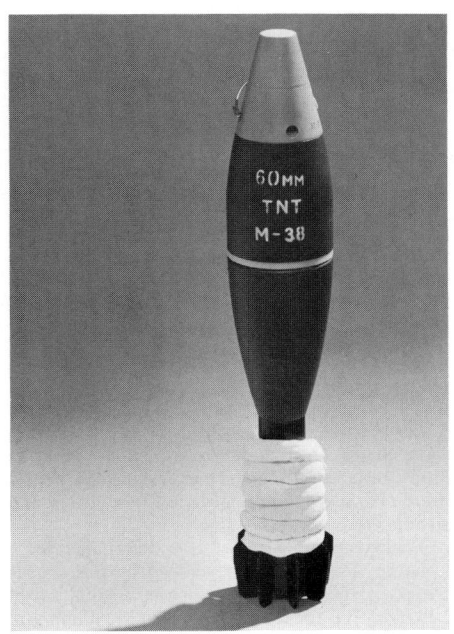

60 mm Long-Range Soltam HE bomb

60 mm long range illuminating bomb

This has been developed in order to provide effective illumination at double the range of previously available 60 mm bombs. The increased range has been achieved by reduction of weight and by increasing the propellant charge to take advantage of the maximum permissible barrel pressure in the mortar. Illumination power has been increased by 83 per cent.

The long range bomb incorporates two original design concepts: the parachute assembly is located in the rear section, permitting dedication of the entire interior to the illuminating assembly; and the acceleration/deceleration sequence is designed to permit opening of the main parachute only after the discarded tail section is well clear, making collision impossible.

The functioning of the bomb is initiated by a time fuze which ignites a small burster charge. This separates the two parts of the bomb by shearing the connecting pins, ignites the illuminant, and also ignites a small rocket motor. This motor gives a forward thrust to the illuminant assembly, ejecting it from the nose of the bomb and causing the small drag parachute to deploy. This drag parachute slows down the illuminant assembly, allowing the empty bomb body to pass it in flight. The drag parachute then extracts the main parachute from its container, which is still attached to the bomb body by a line. The main parachute then deploys, to lower the illuminant to the ground; the drag parachute has by now been burned by the illuminant, and the bomb body, trailing the empty main parachute bag, has continued on its own trajectory and cannot interfere with the descending illuminant unit.

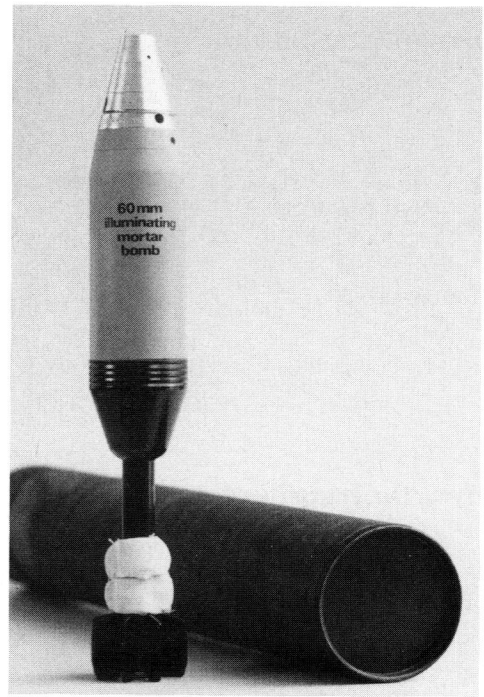

60 mm long range illuminating bomb

DATA
Calibre: 60 mm
Weight: 1.44 kg
Length: 432 mm
Propelling charge: primary + 2 secondaries
Max range: 2270 m
Illuminance: 500 000 cd
Duration: 38 s
Rate of descent: 4 m/s

Manufacturer
Israel Military Industries (IMI), PO Box 1044, Ramat Ha Sharon 47100.
Status
Current. Production.

Service
Israeli forces and overseas sales.

Ammunition for 81 mm mortars

HE BOMBS	Long-range	Standard
Type of mortar bombs	M64	M21
Max range	6500 m	4660 m
Minimum range	200 m	150 m
Max MV	350 m/s	285 m/s
Minimum MV	70 m/s	66 m/s
Bomb weight	4.6 kg	3.9 kg
Filling (TNT)	0.74 kg	0.54 kg
No of propelling charges	8	7

Fuze: DM111; point detonating superquick and delay
M25; proximity, with PD option
Body: forged steel

SMOKE BOMBS
Smoke bombs are filled with 0.54 kg of titanium tetrachloride, white phosphorus or plastic white phosphorus. All other data is as for the HE bomb for the long mortar.

PRACTICE BOMB
Weight of smoke charge: (to mark point of impact) 100 g
Weight of inert filling: 440 g
All other data as the HE round.

Manufacturer
Soltam Ltd, PO Box 371, Haifa.
Status
Current. Available.
Service
Israeli forces. Exported to other countries and can be found in East Africa. Believed to have been evaluated by US Army.

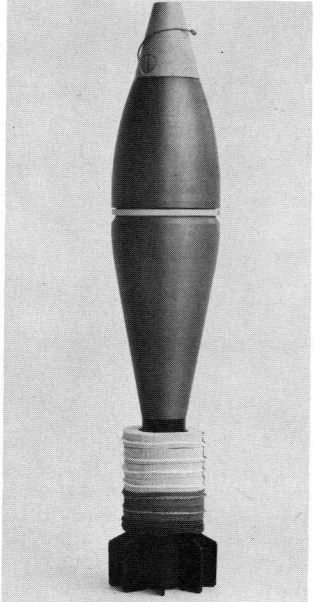

81 mm Soltam HE bomb

81 mm M-64 Soltam HE long-range bomb

81 mm M2A1 IMI illuminating mortar shell

The M2A1 illuminating mortar bomb embodies two original concepts in design and construction. The parachute assembly is located in the rear section, permitting dedication of the entire interior to the illuminating assembly, and the acceleration-deceleration sequence is designed to permit opening of the main parachute only after the discarded tail section is well clear, making collision impossible.

The bomb functions with a clockwork fuze which at a pre-determined point initiates a small propelling charge. The pressure developed by this charge shears the pins holding the two parts of the bomb body, ignites a small rocket motor and the pyrotechnic charge. The rocket motor drives the illuminant

forward and this pulls the auxiliary parachute out of the rear section of the bomb. The main parachute is also pulled out but remains attached to the rear section of the bomb. The auxiliary parachute slows the illuminant assembly and the rear part of the bomb goes past and pulls the main parachute from its bag. The main parachute opens and the illuminant descends gently.

Two versions of this bomb exist; one for use with the US pattern M29A1 mortar and the other for use with the British L16A3 (XM252) mortar. Whilst the bomb itself is the same, the propellant increments differ so as to take advantage of the maximum permissible barrel pressure of the two mortars, with increased maximum range as a result.

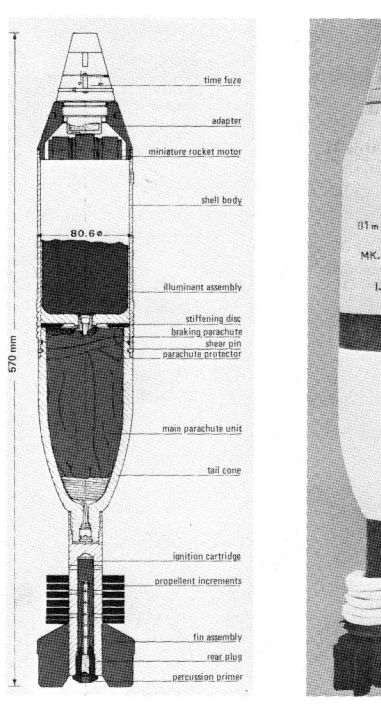

The illuminating power can be increased by adjusting the composition – with a corresponding decrease in burning time – in accordance with the operational requirements of the user. For example, it can be adapted to provide one million candela for a burning time of up to 40 seconds.

DATA
Overall length: 570 mm
Weight of complete round: 4.0 kg
Weight of illuminant: 800 g
Propellant weight: 132 g (M29A1 mortar); 147.5 g (L16A3 mortar)
Range (M29A1): from 400 m (Ch 1) to 4500 m (Ch 4)
Range (L16A3): from 400 m (Ch 1) to 5000 m (Ch 4)
Rate of descent: 5 m/s
Illuminating power: 700 000 cd for 55–60 s

Manufacturer
Israel Military Industries (IMI), PO Box 1044, Ramat Ha Sharon 47100.
Status
Current. Production.
Service
Israeli forces and overseas sales.

81 mm Mark 2 IMI illuminating mortar shell

Ammunition for 120 mm mortars

Ammunition for 120 mm Soltam mortars is of conventional type, using a streamlined bomb with alloy tail unit. Secondary charges are in the form of cloth-wrapped split rings which can be attached around the tail boom. The usual types of fuze are available.

Model	M48	M84	M57	M68	M59	M69
Type	HE	Smoke	HE	Smoke	HE	Smoke
Fuze	M111	M111	M111	M111	M111	M111
Weights (kg)						
Total	12.6	12.6	13.2	13.2	14.75	14.75
Bomb body	9.18	8.72	9.68	9.68	1.086	1.015
Filling	2.2	2.0(WP)	2.25	2.03(WP)	2.49	2.46(WP)
Fuze	0.205	0.205	0.205	0.205	0.205	0.205
Dimensions (mm)						
Length overall	581	581	664	664	703	703
Body length	333	321	350	350	377	377

Ballistic Data			
Mortar model	K5,K6	K6	A4,A7
Max prop charges	8	10	15
Range	250-6250	200-7200	200-8500
Muzzle velocities	113-310	102-320	95-308

For all types of ammunition, illuminating and practice rounds are available.

HIGH EXPLOSIVE ROUND M48
Weight of round complete: 12.6 kg
Body: forged steel 9.3 kg
Tail unit: aluminium alloy 700 g
Fuze: superquick/delay; arming safety 0.8 s; weight 0.2 kg. The performance of the PD fuze is improved upon by using the M25 proximity fuze.
Filling: TNT 2.3 kg
Charge system: primary cartridge + additional charge + 4 secondary charges
Overall length of round: 580 mm
Muzzle velocity: (minimum) 115 m/s; (max) 310 m/s
Max chamber pressure: 900 kg/cm²
Range: (minimum) 400 m; (max) 6500 m

SMOKE ROUND M48
FM round. Filling: titanium tetrachloride. Weight of filling 2.3 kg
WP round. Filling: white phosphorus. Weight of filling 2.3 kg
PWP round. Filling: plastic white phosphorus. Weight of filling 2.3 kg
All other data are the same as for the M48 HE round

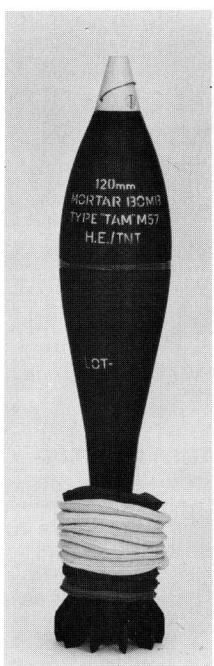
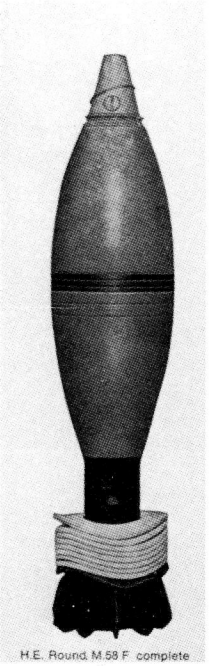

120 mm M57 HE and M58F HE rounds

PRACTICE ROUND M48
Weight of smoke indicator charge: 200 g
Weight of inert filling: 2.1 kg
All other data are the same as for the M48 HE round

Manufacturer
Soltam Ltd, PO Box 371, Haifa.
Status
Current. Available.
Service
Israeli forces and some other armies.

120 mm IMI HE rocket-assisted mortar shell

This is a rocket-assisted round which will reach 10 500 m. It can be fired from any current 120 mm mortar.

DATA
Overall length: 744 mm
Complete round weight: 16.7 kg

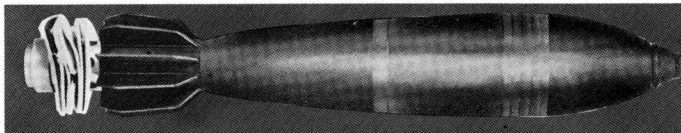

120 mm IMI HE rocket-assisted mortar shell

Propelling charge: primary and 9 increments
Rocket propellant: 1.15 kg
High explosive filling: 2.15 kg
Nature of filling: Comp B
Max range: 10 500 m
Muzzle velocity: (max) 280 m/s
Packing: 2 rounds to a wood box, each round in fibre container; separate fibre container holding 2 propellant charges

Manufacturer
Israel Military Industries (IMI), PO Box 1044, Ramat Ha Sharon 47100.
Status
Current. Production.
Service
Israeli forces and several other armies.

120 mm M3 IMI illuminating bomb

The M3 illuminating bomb is fired from a Tampella, Soltam or Brandt 120 mm mortar. Ballistics of the bomb are identical to those of the comparable HE bomb and the fuze is set in accordance with the standard firing tables. The M3 bomb works in the same manner as the 81 mm M2 bomb described above.

The M3 120 mm shell embodies two original concepts in design and construction. The parachute assembly is located in the rear section, permitting dedication of the entire interior to the illuminating assembly, and the acceleration-deceleration sequence is designed to permit opening of the main parachute only after the discarded tail section is well clear, making collision impossible.

DATA
Length: 580 mm including fuze
Weight: 12 kg including fuze
Weight of illuminant: 1.2 kg
Propelling charge: increments charges 0–9
Range: from 1100 m at charge 1–6100 m
Illuminating power: 1 250 000 cd minimum
Illuminating time: 45 s minimum
Rate of descent: 5–6 m/s

Manufacturer
Israel Military Industries (IMI), PO Box 1044, Ramat Ha Sharon 47100.
Status
Current. Production.

120 mm M3 IMI illuminating bomb

Ammunition for the 160 mm Soltam mortar

The HE bomb is made from forged steel. The tail unit is made from extruded aluminium. The secondaries, which are flat discs, do not protrude beyond the diameter of the fins and so do not come into contact with the barrel wall. This reduces the chance of a cook-off, even with a barrel temperature as high as 650°C. The high explosive filling is 5 kg of TNT. The propellant system is made up of the primary cartridge and nine secondaries. The Diehl fuze gives a direct action or a slight delay.

There are two kinds of smoke bomb. One has a titanium tetrachloride filling and the other plastic white phosphorus (PWP). The PWP bomb can be used also as a night-ranging bomb.

Weight of bomb complete: 40 kg
Weight of HE filling: 5 kg
Rate of fire: 5–8 rds/min

Manufacturer
Soltam Ltd, PO Box 371, Haifa.
Status
Current. Available.
Service
Israeli Army.

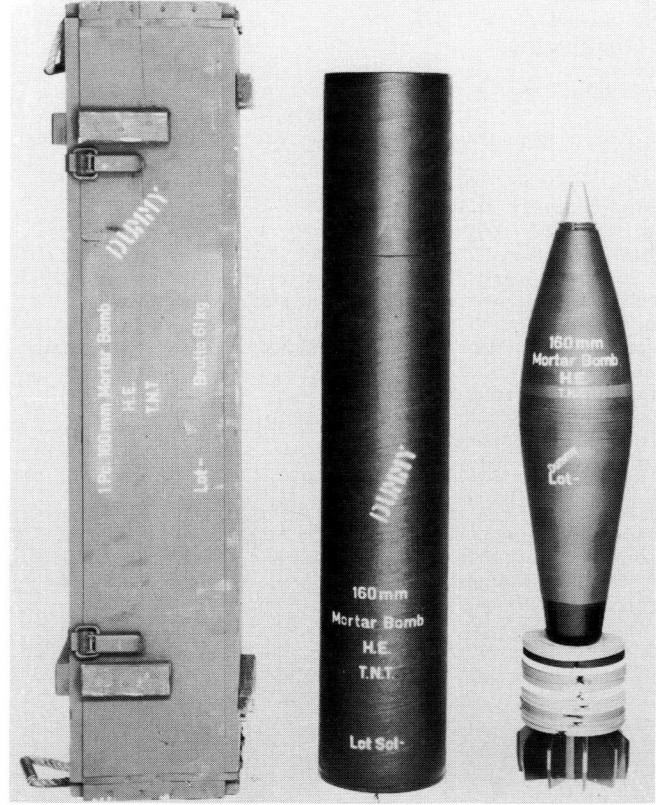

160 mm Soltam HE bomb

Alpha M787 dual-option PPD mortar fuze

This is a dual-option proximity/point detonating fuze compatible with all 60 mm, 81 mm, 82 mm, 120 mm and 160 mm mortar bombs, irrespective of their charges, muzzle or terminal velocities.

The M787 fuze is powered by a wind-driven alternator which serves as a second safety device in addition to the more usual setback-operated integrator. The safe arming distance is 200 m. The circuitry includes a peak trajectory sensor which activates the fuze only after the vertex has been passed and the bomb is on the downward leg of its trajectory. This feature gives additional own-troops safety and crest clearance safety. A fail-safe feature prevents malfunction; should air speed precede setback, the fuze will lock in the safe mode.

The fuze was designed and manufactured in accordance with Israel Defence Forces specifications, as well as all relevant Western safety and performance criteria.

Manufacturer
Reshef Technologies Ltd, 16 Hacharoshet Street, Or-Yehuda 60375.
Status
Current. Production.
Service
Israel Defence Force and other countries.

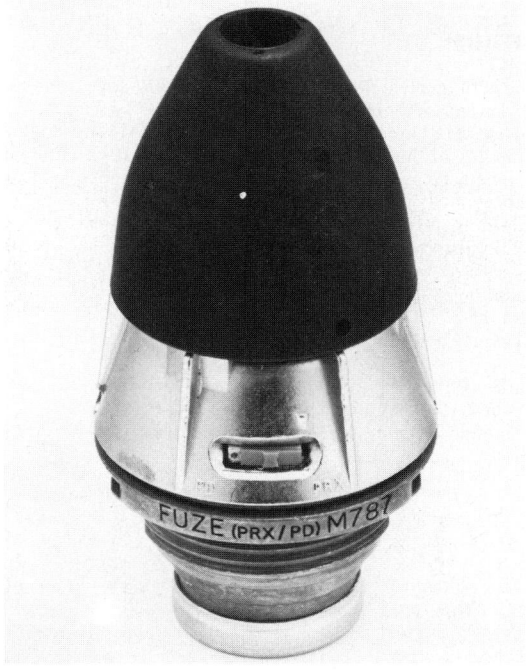

Reshef M787 Alpha dual-option mortar fuze

Lambda M760 electronic time mortar fuze (US-PTF XM 778)

This is an electronic time fuze compatible with all 60 mm, 81 mm and 120 mm mortar bombs, irrespective of their charges, muzzle or terminal velocities, for use with smoke, illuminating and carrier bombs (in expelling mode) or for HE and WP bombs (in detonating mode).

The fuze is powered by a wind-driven alternator which serves as a second safety device in addition to the more usual setback-operated integrator. The safe arming distance is 100 m. A fail-safe feature prevents malfunction; should air speed precede setback, the fuze will lock in the safe condition.

Lambda has an electronic time delay of 3 to 99.8 seconds in 0.1 second increments. The standard deviation of the delay error is less than 0.1 second. Point detonating is settable when 99.9 s is selected. The system setting is digital, manually operated (no tools are required) by three setting rings with positive locking.

The fuze was designed in accordance with the relevant Western safety and performance criteria.

Manufacturer
Reshef Technologies Ltd, 16 Hacharoshet Street, Or-Yehuda 60375.
Status
Current. Initial production.

Reshef Lambda M760 electronic time fuze

ITALY

Mortar ammunition

Mortar ammunition in the current SNIA-BPD and SIMMEL range comprise the 81 mm and 120 mm smoothbore and the 107 mm rifled (US 4.2 in).

Most interesting are the 81 mm and 120 mm families of SIMMEL design. These can be fired in any of the above calibres now existing. Each family is perfectly identical in ballistic properties and therefore employ a unique firing table for each calibre.

The 81 mm family consists of eight types of bomb as follows:-

S1A1   HE high capacity PFF (2600 splinters)
S2A1   WP smoke
S3A3   Illuminating (1.4 mega-cd over 35 s)
S5A1   HC smoke
S6A2   Cargo SA (9 bomblets)
S7A1   IC
S8A1   Cargo-SA-TP
S9A1   FSS (Flash-sound-smoke) TP.

The cargo bomb, containing nine bomblets with the ability to penetrate 60 mm of steel armour, has aroused considerable interest.

The 120 mm family consists of seven types of bomb as follows:-

S10B   HE
S11B   WP smoke
S23B   Illuminating (1.2 mega-cd over 45 s)
S23B   HC smoke
S12B   Cargo SA (12 bomblets)
S15B   Cargo SA TP
S14B   FSS-TP.

In addition, Simmel produces another 120 mm cargo bomb, having the same standard weight of 13 kg and containing 35 bomblets with high penetrating capability. It can be fired from all 120 mm mortars.

Manufacturers: SNIA-BPD, Corso Garibaldi 22, I-00034 Colleferro (Roma)
Simmel, Borgo Padova 2, I-31033 Castelfranco Veneto (Treviso)

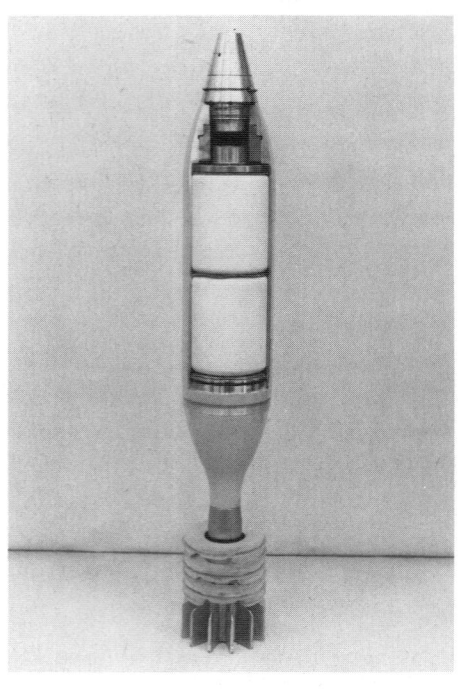
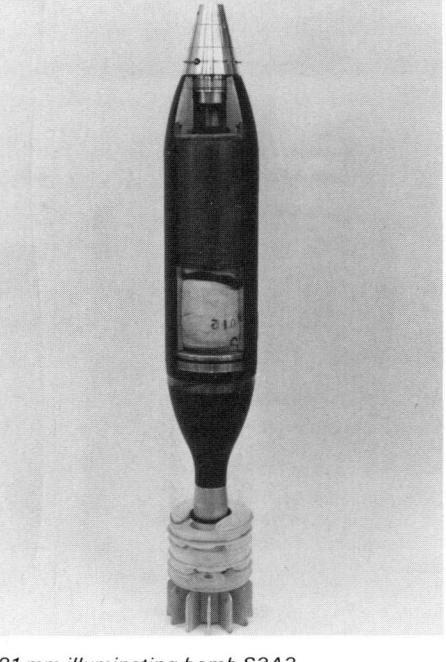
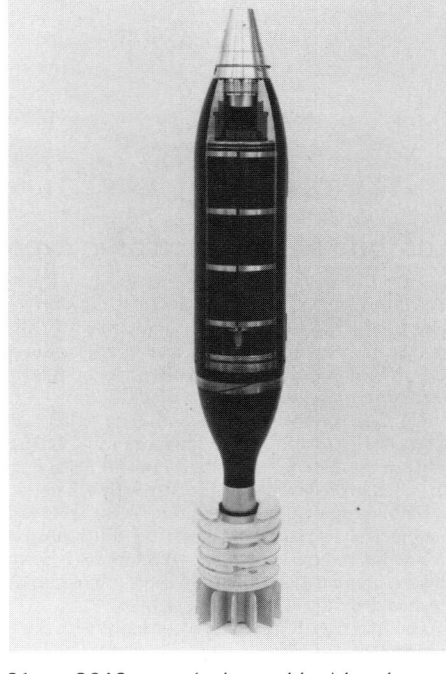

81 mm S5A1 HC smoke bomb   *81 mm illuminating bomb S3A3*   *81 mm S6A2 cargo (sub-munition) bomb*

Borletti FB 332 point detonating fuze

This is a conventional superquick impact fuze for use with all 60 mm, 81 mm and 120 mm mortars. The mechanism is bore-safe by means of an escapement-controlled delayed-arming shutter and safety conforms with MIL-STD-1316. There is a pull wire which locks the setback detent regulating shutter movement.

Manufacturer
Borletti FB SpA, Via Verdi 33/35, I-20010 S. Giorgio su Legnano (MI).
Status
Current. Production.

Borletti FB 332 point detonating fuze

Borletti FB-391 mortar proximity fuze

This is a radio-frequency proximity fuze for use with 60, 81 and 120 mm mortar bombs. Circuitry involving frequency agility and advanced electronic processing techniques ensures that electronic countermeasures are relatively ineffective against the fuze, whilst it is immune to mutual interference and battlefield disturbances such as radio emissions, dust and smoke. The signal processing techniques used ensure proximity function with sharper altitude discrimination than is obtainable with an unmodulated RF carrier. The Doppler processor triggers the fuze at a typical height of 4 m, and burst height is virtually independent of soil nature, approach angles and terminal velocities. Proximity action is not affected when the fuze is used with rocket-assisted bombs.

The fuze has a dual option selector giving the user the choice of either proximity with impact back-up or point-detonating effect. Electric power is provided by a wind-driven alternator. The safe and arming system complies with MIL-STD-1316C (the fuze is provided with an inspection window). Prior to firing, the electric detonator is out of line and short-circuited. A setback integration and several hundred revolutions of the turbine are necessary before arming can take place. An electronic timer, set during manufacture, inhibits fuze action during the initial part of the trajectory, ensuring delayed arming.

Two variant models are made: FB391A has intrusion and thread to comply with MIL-STD-331A; FB391B has intrusion and thread interchangeable with the V19P fuze. Other configurations of intrusion and thread are possible on request.

Manufacturer
Borletti FB SpA, Via Verdi 33/35, I-20010 S. Giorgio su Legnano (MI).
Status
Current. Production.
Service
Undergoing qualification by Italian Army.

Borletti FB-391 mortar proximity fuze with nose protection

Borletti FB282 Mortar point detonating fuze

The FB 282 fuze is a dual purpose superquick or 0.07 s delay type, designed for use with 81 mm or 120 mm mortars. The fuze uses a direct action striker for superquick functioning, and a graze pellet and pyrotechnic element for delay functioning which is designed to operate down to very low impact angles.

Safety is in accordance with MIL-STD-1316, and there is a pull-wire for safety in handling, shipment and para-dropping. The fuze has an external Safe and Armed indicator.

The fuze has a plastic nosecap, removeable to increase the PD sensitivity for operations on snow or soft surfaces.

Manufacturer
Borletti FB SpA, Via Verdi 33/35, I-20010 S. Giorgio su Legnano (MI).
Status
Current. Production.

Borletti FB 282 point detonating fuze

Borletti FB 392 Mortar electronic time fuze

The FB 392 is an electronic time (ET) fuze intended for use in smoke, illuminating and cargo rounds for mortar systems of 60 mm, 81 mm and 120 mm calibre.

Although it is understood that some types of rounds may require some adaptation of the basic configuration (especially at the mechanical interface between fuze and bomb — the fuze-hole bush) the basic configuration has been conceived in such a way as to permit the greatest possible compatibility.

The fuze has two operating modes: time, with impact back-up, or PD function only. The fuze is set by hand to whatever mode is desired.

Major subassemblies are the time setting unit; electronic module; turbine alternator power supply; Safe and Arming; detonating elements; and three plastic rings which allow selection of the required time setting from 3 to 99.9 seconds, in increments of 0.1 second. Setting is performed by aligning the selected number of each ring with a fixed index. When the PD-only mode is required, the fuze is set to zero.

Manufacturer
Borletti FB SpA, Via Verdi 33/35, I-20010 S. Giorgio su Legnano (MI).
Status
Advanced development.

Borletti FB 392 electronic time fuze with nose protection

KOREA, SOUTH

60 mm M83A3 illuminating mortar bomb

Designed to be fired from 60 mm M2 or M19 mortars, this is based on the American M83 design. The fixed-time fuze M65A1 operates about 15 seconds after firing, igniting the expelling charge which blows off the tail and ignites the illuminant charge. As the contents are ejected from the bomb body, so the parachute deploys and supports the illuminant during its fall.

DATA
Length, fuzed: 363 mm
Weight, fuzed: 1.88 kg
Burning time: 32 s
Intensity: 320 000 cd
Max range: 1006 m

Manufacturer
Korea Explosives Co Ltd, 34 Seosomoon-Dong, Chung-Ku, Seoul.
Status
Current. Production.
Service
South Korean Army.

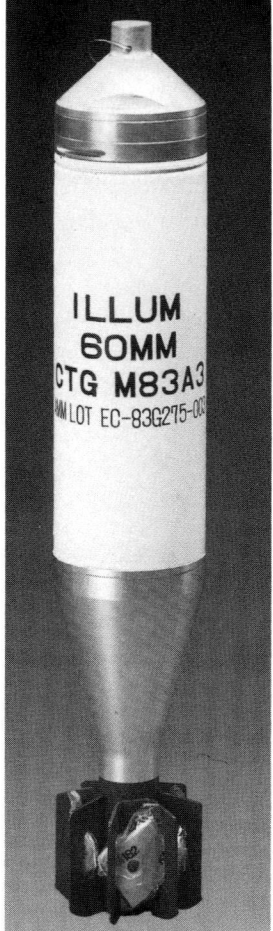

60 mm M83A3 illuminating bomb

60 mm M49A4 HE mortar bomb

This is a conventional type of mortar bomb, based generally on an American original but with an improved tail unit and extension. The primary cartridge is in the tail boom, while the four augmenting charges, sealed in plastic, are clipped between the fins. The bomb body is of pearlitic malleable cast iron, filled with Composition B, and the bomb is fitted with fuze PD M525

DATA
Length, fuzed: 295 mm
Weight: 1.47 kg
Range:
Charge 0   256 m
Charge 1   640 m
Charge 2 1063 m
Charge 3 1451 m
Charge 4 1795 m

Manufacturer
Korea Explosives Co Ltd, 34 Seosomoon-Dong, Chung-Ku, Seoul.
Status
Current. Production.
Service
South Korean Army.

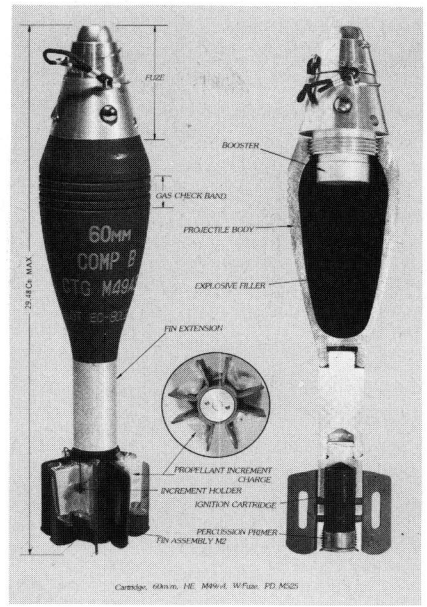

60 mm M49A4 HE bomb

81 mm M374 HE bomb

This is based on the US design and may be fired from 81 mm M1, M29 or M29A1 mortars. It consists of a bomb body with obturating ring, a point detonating or proximity fuze, a fin assembly which incorporates the primary cartridge housing, and two types of propellant charge enclosed in fabric bags. The body is filled with Composition B, and the fins are canted 5° to produce spin.

DATA
Weight, fuzed: 4.35 kg
Length: 529 mm
Maximum range, Charge 9: 4500 m

Manufacturer
Korea Explosives Co Ltd, 34 Seosomoon-Dong, Chung-Ku, Seoul.
Status
Current. Production.
Service
South Korean Army.

81mm M374 HE mortar bomb

81 mm M301A3 illuminating mortar bomb

This is based on the American M301 pattern and is for use in any 81 mm mortar. The bomb carries a time fuze which is set before loading in accordance with the firing tables. At the set time the tail is blown off and the bomb, the illuminant and parachute are expelled. This bomb is not fired at charges less than Charge 3.

DATA
Length: 628 mm
Weight: 4.58 kg
Burning time: 60 s minimum
Intensity: 500 000 cd
Range max: 3150 m

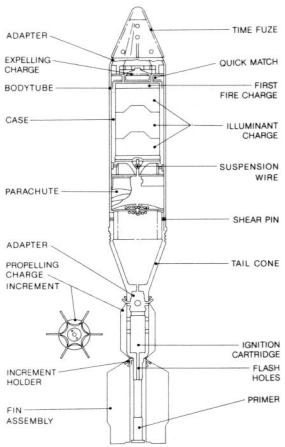

81 mm M301A3 illuminating bomb

Loading 81 mm M301A3 bomb

Manufacturer
Korea Explosives Co Ltd, 34 Seosomoon-Dong, Chung-Ku, Seoul.
Status
Current. Production.
Service
South Korean Army.

NORWAY

81 mm NM 123A1 HE mortar bomb

The 81 mm mortar bomb NM 123A1 HE has been developed by A/S Raufoss Ammunisjonsfabrikker in co-operation with Norwegian Army Material Command. The bomb is made from ductile cast iron with an extruded aluminium alloy tail unit. The bomb produces approximately 10 000 effective fragments. The NM 123A1 has a maximum range of more than 5800 m.

The bomb can be used with any type of mortar bomb fuze at the option of the user, for example the Norwegian Kongsberg fuze PPD 323. Normally the bomb is fitted with 1.5 in (3.8 cm) threads, but may, if desired, be produced with 2 in (5 cm) threads, or 2 in threads with an adaptor for 1.5 in threads. The NM 123A1 has six augmenting charges of equal weight. Identical charges allow simple and safer handling during night operations. The propellant charges are made from ballistite. The bomb has low loading densities and excellent ballistic performances. Probable error at maximum range is 4 m in azimuth and 22 m in range. The bomb is filled with 800 g of Composition B.

The NM 123A1 has a plastic obturating ring which reduces the spread in muzzle velocity. Significantly the ring has also been tested at excess pressure of 150 per cent at -40°C. Each container in triplex medium density polyethylene has three separate spaces, one bomb in each.

DATA
Calibre: 81 mm
Length overall: 522 mm (fuzed PPD 323)
Weight complete: 4.34 kg
Weight of body: 3.04 kg
Filling weight: 0.8 kg
Filling: 60/40 RDX/TNT

CONSTRUCTION
Body: ductile cast iron
Tail unit: extruded aluminium alloy
Propellant: ballistite in rayon cloth bag

Manufacturer
Raufoss A/S, N-2831 Raufoss.

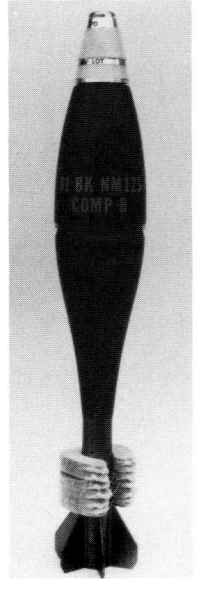

81 mm NM 123A1 HE mortar bomb

Status
Production.
Service
Norwegian, Swedish and Canadian Armies.

NVT-224 proximity fuze

Proximity fuzes for mortar bombs have been in production since 1965 in Norway. The fuzes, of the NVT family, are adaptable to all commonly-used mortar calibres, such as 81, 107, 120 mm etc. All the fuzes are of virtually the same design and the following description of the NVT-224 for 120 mm mortar bombs applies to all the other fuzes.

Description
The NVT-224 is a transistorised proximity fuze for 120 mm mortar bombs which causes the bomb to detonate at a predetermined height above the target area. By virtue of its external dimensions it is fully interchangeable with commonly-used point detonating fuzes.

A wind-driven turbine generator provides the required electrical power for the electronic circuits. By avoiding active chemical components in the electrical energy source, the best possible storage and reliability performance has been ensured. In its safe position the safety and arming mechanism keeps the detonator shorted and the explosive chain interrupted. The high-speed turbine also serves the function of providing mechanical energy to operate the safety and arming mechanism. Depending on the bomb velocity the fuze is

mechanically armed 0.6 to 1.5 seconds after firing which corresponds to a minimum safe distance of 150 m from the muzzle.

The NVT incorporates an electronic Apex Sensor which ensures that the fuze is kept inactive until the bomb has passed the apex of the trajectory, or until at least 60 per cent of the time of flight has elapsed.

A neoprene nosecap protects the fuze against moisture and dirt and also prevents the turbine from rotation. The magnetic coupling in the generator is, however, sufficient to keep the turbine locked during transport and handling.

An electro-mechanical impact switch connects the detonator to the firing capacitor if the proximity function of the fuze should fail.

Burst height
The burst height of the fuze is set during production for an optimum performance of the bomb, or according to the customer's requirement.

The fuze together with the shell forms a skew dipole antenna, which is fed by a Doppler sensitive oscillator. The radiation pattern or sensitivity of the fuze deviates only slightly from that of an ideal dipole antenna, depending on outer dimensions and form of the shell. This implies that the burst height

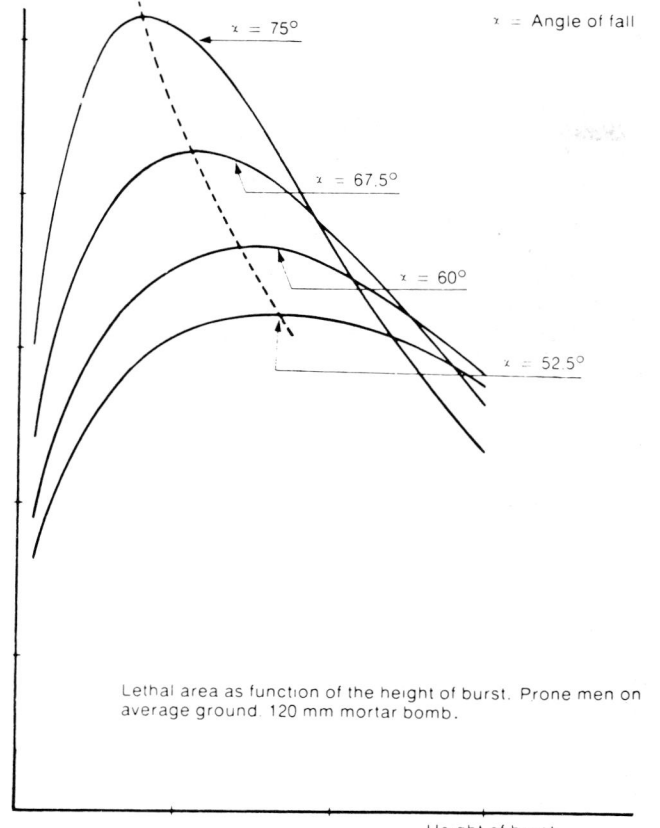

NVT-224 proximity fuze burst height variations

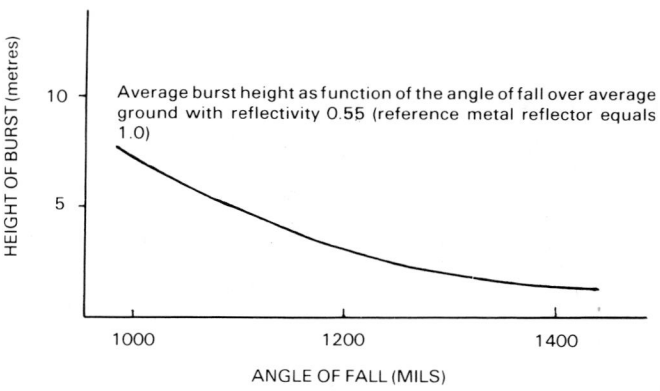

NVT-224 proximity fuze. Wind-driven turbine is clearly visible in extreme nose

DATA
Weight of fuze: 0.25 kg

Manufacturer
Norsk Forsvarsteknologi A/S, Postbox 1003, N-3601 Kongsberg.
Status
Current. No longer in production.
Service
Norwegian and other armies.

decreases with increasing angle of fall. In fact, this variation corresponds well with the burst height desirable for various types of shell to get a maximum effect in the target.

PPD-323 and PPD-324 proximity fuzes

This is a multi-role fuze for use in mortar bombs, PPD-323 for 81 mm bombs and PPD-324 for 120 mm bombs, providing selectable impact or proximity functions. Power for the proximity function is furnished by a turbo-generator, and the action of the turbine also provides mechanical arming which, in connection with an integrated setback detent, ensures full compliance with the safety requirements of STANAG 3525 (MIL-STD-1316B). Mechanical arming does not take place until after 250 m of flight, and electrical emission by the proximity circuits does not take place until the bomb has passed the top of the trajectory. The electronic circuits are fully protected against countermeasures. This is accomplished by high radiated power, radiation pattern, frequency modulation and protective features in the signal processor.

The burst height may be specified by the customer. The PPD-323 and 324 meet all environmental requirements of MIL-STD-331A.

Manufacturer
Norsk Forsvarsteknologi A/S, PO Box 1003, N-3601 Kongsberg.
Status
Current. Production.
Service
Norwegian, Swiss, West German and other armies

Operating characteristics of PPD-323 and PPD-324 fuzes

PPD-323/324 multi-role fuze

PAKISTAN

V19P impact and delay fuze

The V19P is a Thomson-Brandt design manufactured in Pakistan under license. The fuze is a point detonating impact fuze with selectable delay and has comprehensive safety arrangements which cover parachute-dropping, bore, muzzle and double-loading safeties.

The fuze has a firing pin set in its tip; below this is a pneumatically braked delayed arming system which, until the bomb has travelled some distance from the mortar, does not release the cylindrical bore-safety shutter. Once released, this shutter rotates through 60 or 120° depending upon the setting of the fuze for superquick action or delay action. Upon impact, the firing pin strikes a detonator and the subsequent flash initiates a further detonator, through a delay if so set, and this, in turn, initiates the RDX/WX filling of the fuze magazine.

The V19P fuze is for use with 120 mm mortar bombs; a similar fuze, known as V19PA, is for use with 81 mm bombs.

Manufacturer
Alsons Industries (Pvt) Ltd, PO Box 4853, Karachi 74000.
Status
Current. Production.
Service
Pakistan army.

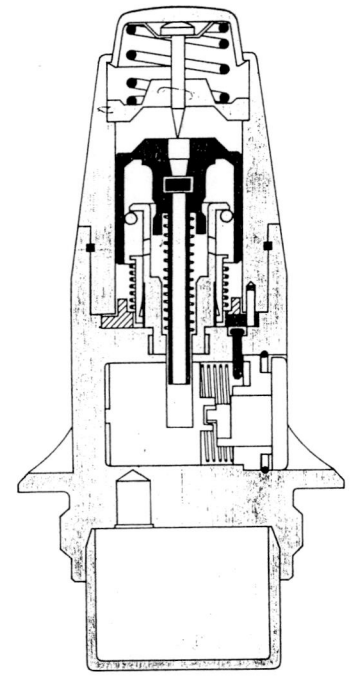

V19P mortar fuze

60mm HE bomb

This is a standard iron bomb, apparently based on a Brandt design, filled TNT and fitted with the impact fuze PAI-1A. Obturation is performed by four grooves around the waist of the bomb. The tail unit is of steel with welded fins; it carries the primary cartridge and three secondary cartridges in horse-shoe form, clipped round the tail boom.

DATA
Calibre: 60 mm
Weight: 1.36 kg
Muzzle velocity: 134 m/s
Maximum range: 1410 m

Manufacturer
Alsons Industries (Pvt) Ltd, PO Box 4853, Karachi 74000.
Status
Current. Production.
Service
Pakistan army.

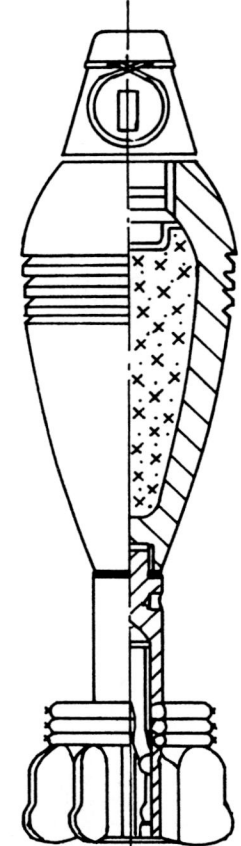

60 mm HE mortar bomb

60 mm smoke bomb

This matches the HE bomb described above and is essentially the same bomb but with a filling of white phosphorus. It is fuzed with the impact fuze PAI-1A.

DATA
Calibre: 60 mm
Weight: 1.50 kg
Muzzle velocity: 134 m/s
Maximum range: 1410 m

Manufacturer
Alsons Industries (Pvt) Ltd, PO Box 4853, Karachi 74000.
Status
Current. Production.
Service
Pakistan army.

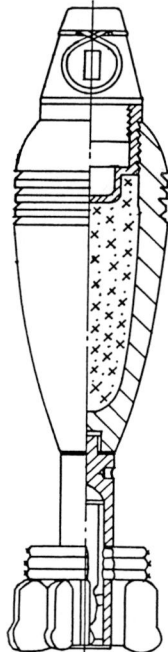

60 mm mortar smoke bomb

60 mm illuminating and signal bombs

This is a cylindrical fuzeless bomb which comes in three variant forms; a parachute illuminating bomb or a red or green signal bomb. The basic bomb is a canister attached to a tail unit which carries the primary cartridge. At the base of the bomb is a chamber containing a fixed time delay unit which is ignited by the propellant flash. After burning through, the delay ignites an expelling charge in the base of the body; this ignites the pyrotechnic payload and then ejects it, blowing off the front cap of the body to permit exit. In the case of the illuminating bomb, the parachute deploys and lowers the flare unit to the ground. In the case of the signal bombs, the flare unit is thrown clear and falls free, the duration of its fall being sufficient for signalling purposes.

DATA
Calibre: 60 mm
Length: 246 mm
Weight: 1.00 kg
Filling: Illuminating composition SR 562 and parachute; or green flare composition SR 429AM; or red flare composition SR 406
Delay time: 3.5 – 5.5 s
Duration: illumination 35 – 45 s; signal 8 – 12 s
Ejection height: approx 200 m

Manufacturer
Alsons Industries (Pvt) Ltd, PO Box 4853, Karachi 74000.
Status
Current. Production.
Service
Pakistan army.

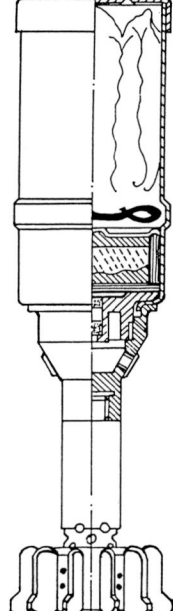

60 mm illuminating bomb

81 mm HE bomb M-57D Mk 1

This is a copy of the Brandt Mod 57D bomb, with steel body and a very short tail boom. The fuze is the V19 PA described above.

DATA
Calibre: 81 mm
Weight: 3.338 kg
Filling: TNT
Muzzle velocity: 291 m/s (Charge 7)
Maximum range: 4550 m (Charge 7)

Manufacturer
Alsons Industries (Pvt) Ltd, PO Box 4853, Karachi 74000.
Status
Current. Production.
Service
Pakistan army.

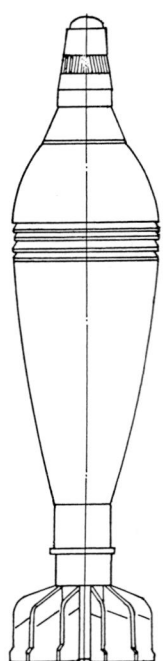

81 mm HE bomb M-57D Mk 1

81 mm smoke bomb

This is the partner to the HE bomb described above and is of the same general appearance, differing in being somewhat longer and more slender and in the use of white phosphorus as the filling. It uses the same V19 PA impact fuze and is ballistically matched to the HE bomb.

DATA
Calibre: 81 mm
Weight: 3.2 kg
Muzzle velocity: 291 m/s (Charge 7)
Maximum range: 4550 m (Charge 7)

Manufacturer
Alsons Industries (Pvt) Ltd, PO Box 4853, Karachi 74000.
Status
Current. Production.
Service
Pakistan army.

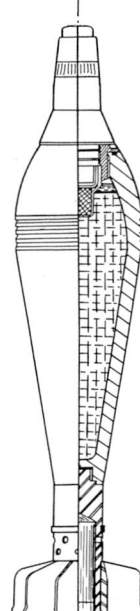

81 mm WP smoke bomb

120 mm HE bomb

This appears to be a locally designed bomb, of conventional construction though with an unusually long tail boom; this, though, probably contributes to the very good accuracy. The bomb is fuzed with the V19 P impact fuze and has a rather large exploder system which probably gives the bomb extremely efficient detonation and fragmentation characteristics. The primary cartridge is held inside the tail boom and the secondary charges are clipped around the boom in horse-shoe units.

DATA
Calibre: 120 mm
Weight: 13.00 kg
Weight of filling: 2.6 kg TNT
Muzzle velocity: 331 m/s (Charge 8)
Maximum range: 6745 m (Charge 8)

Manufacturer
Alsons Industries (Pvt) Ltd, PO Box 4853, Karachi 74000.
Status
Current. Production.
Service
Pakistan army.

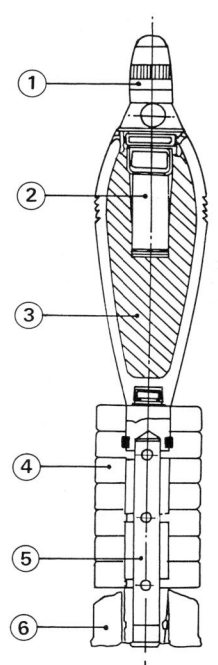

120 mm HE bomb; **(1)** *fuze V19 P;* **(2)** *exploder;* **(3)** *filling;* **(4)** *secondary charges;* **(5)** *primary cartridge;* **(6)** *tail fins*

PORTUGAL

Mortar ammunition

INDEP manufactures three types of mortar ammunition, two in 60 mm and one in 81 mm, all of them high explosive. These are usually supplied with the Fuze PD M525 or the Diehl AZ DM111A2 fuze manufactured in Portugal under licence. They may, however, be equipped with other fuzes. Both 60 mm rounds may be fired from Brandt breech loading mortars of the type used in armoured personnel carriers.

81 MM HE
This is similar to the US M43A1B1

DATA
Weight: 3250 g
Minimum range: 75 m
Max range: with 6 increments 3500 m; with 8 increments 4200 m
Filling: TNT or Comp B, 560 g

60 MM HE
This is similar to the US M49A2

DATA
Weight: 1340 g
Minimum range: 46 m
Max range: 1820 m
Filling: TNT, 155 g

60 MM HE, APERS, TYPE NR431A1 (UNDER PRB LICENCE)
DATA
Weight: 1360 g
Minimum range: 50 m
Max range: 2100 m
Muzzle velocity, Charge 4: 177 m/s
Filling: TNT, 156 g

Manufacturer
Indústrias Nacionais de Defesa EP (INDEP), Rua Fernando Palha, 1899 Lisbon-Codex.

SOUTH AFRICA

Recent developments

As noted in the entry on mortars, a programme is in progress to improve the current inventory by developing new mortars and by developing new types of ammunition.

The forthcoming 60 mm mortar bomb will be of streamlined form and fitted with an obturating ring. It will be TNT filled and fitted with the PDM 8807 direct action fuze and fragmentation trials have shown that it will deliver five lethal fragments per square metre over a 25 m radius around the point of burst. At 5 m range the bomb will have a 50 per cent zone of 250 × 100 m.

A new bomb is also under development for existing 120 mm mortars. This draws on the same sort of aerodynamic technology as most contemporary designs and is a well-streamlined bomb using a central obturating ring and with a long tail boom carrying machined fins. The result of improved shape, improved material and filling and modern fuzing, should produce a bomb which will offer a useful extension of range together with increased lethality.

A more advanced concept also under development is a 'smart' sub-munition bomb for the 120 mm mortar. This carries 21 shaped-charge/ fragmentation bomblets inside its cylindrical body and is fitted with a millimetric-wave seeker head. This has a 300 m radius of sensitivity in which it will detect targets and steer the bomb towards them. The bomb is fired in the normal way, and steerable fins are opened into the airstream after the bomb passes the vertex of the trajectory. The seeker switches on some 2500 m from

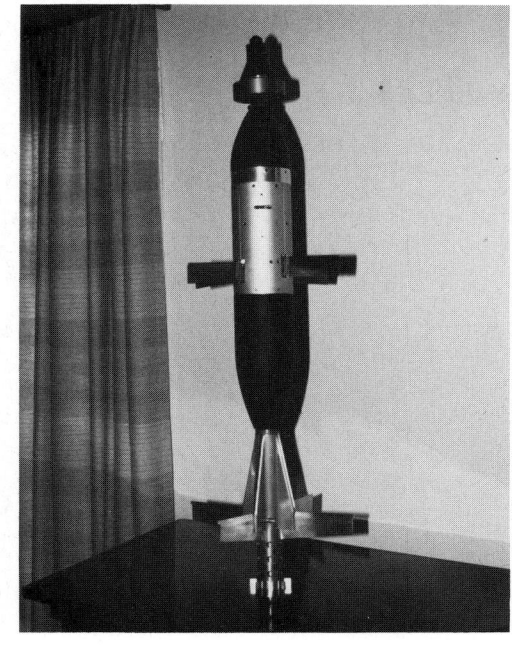

120 mm terminally guided sub-munition bomb under development by Armscor

the target and once it has identified its objective it steers toward it and ejects the bomblets at the optimum height.

Trial versions of the sub-munition bomb have been tested using laser guidance, and these have all homed to within 7 m of their targets. The technology for using either millimetric wave radar or laser guidance is equally suitable for the task, and the design could adopt either system, depending upon which offers the best cost-effectiveness.

60 mm mortar ammunition

As with the South African 60 mm mortars, this ammunition is based on Hotchkiss-Brandt designs but with some modifications and improvements in order to facilitate manufacture and to improve reliability.

60 mm M61 HE bomb

This can be used in either the standard infantry drop-fired mortars or in the 60 mm gun-mortar used with armoured vehicles. It uses a forged-steel body and is filled with RDX/TNT 40/60. A primary cartridge fits into the tail tube and four secondaries fit in between the fins while a fifth can be clipped around the tail tube ahead of the fins. It can be supplied fuzed with the V9 direct-action impact fuze or the SC12B direct action and graze impact fuze. The former is based on the Hotchkiss-Brandt V9 design but has been somewhat modified in the light of combat experience. The latter incorporates long- or short-arming times which can be selected by the user and is of French design.

DATA
Weight: 1.8 kg

VELOCITY
Charge 0: 62 m/s
Charge 1: 88 m/s
Charge 2: 111 m/s
Charge 3: 132 m/s
Charge 4: 149 m/s
Charge super: 171 m/s
Range: (charge super) 2100 m; (charge primary only) 358 m

Manufacturer
Armscor, Private Bag X337, Pretoria 0001.

60 mm M61 HE bomb

60 mm M61 bursting smoke bomb

This is ballistically matched to the HE bomb. It is charged with titanium tetrachloride (FM) which is preferred to white phosphorus (WP) due to its lower fire risk in bush country. The FM is distributed by a small explosive burster and generates dense white smoke by reaction with water vapour in the air. The direct action impact fuze V9 is fitted as standard.

DATA
As for M61 HE bomb (above)

Manufacturer
Armscor, Private Bag X337, Pretoria, 0001.

60 mm M61 bursting smoke bomb

60 mm M61 practice bomb

This is ballistically matched to the M61 HE bomb so that it may be used for practice when non-explosive bombs are necessary. The bomb is filled with an inert substance but carries a small marker charge of TNT/aluminium which detonates on impact to give an indication of the fall of shot. The direct action impact fuze V9 is fitted as standard.

DATA
As for M61 HE bomb (above)

Manufacturer
Armscor, Private Bag X337, Pretoria, 0001.

60 mm illuminating bomb

This is of South African design and is of the conventional pattern using a parachute-suspended star unit to provide illumination. A time fuze is fitted and set before firing; at the end of the set time the fuze ignites an expelling charge which blows off the tail unit and ejects the star and parachute.

DATA
Length: 377 mm
Weight: 1.6 kg
Illuminating power: 180 000 cd minimum
Burning time: 30 s minimum
Rate of fall: approx 4.5 m/s
Max effective range: 2000 m

Manufacturer
Armscor, Private Bag X337, Pretoria, 0001.

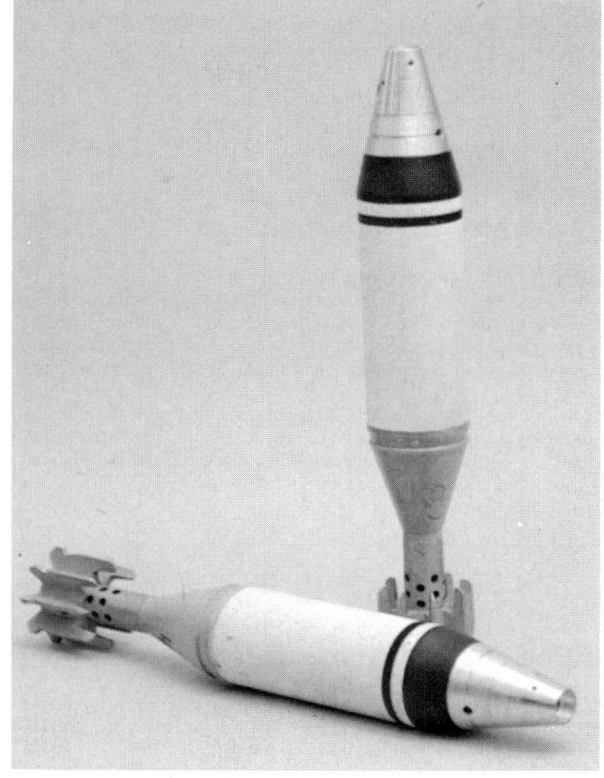

60 mm illuminating bomb

81 mm mortar ammunition

This range of ammunition is also based on an original Hotchkiss-Brandt design, though modified to suit local manufacturing methods and improved in various respects in the light of tactical experience.

81 mm M61 HE bomb

This uses a forged steel body with a Rilsan obturating ring and a light alloy tail tube and fin assembly. It is filled RDX/TNT 40/60 and fuzed with the direct action V19P fuze which has provision for instantaneous or delay action, selectable before firing. A primary cartridge fits into the tail tube, and secondaries can be clipped around the tube; there are eight charges.

DATA
Weight: 4.43 kg
Velocity: (charge primary) 75 m/s; (charge 8) 279 m/s
Max range: 4856 m
Minimum range: 75 m

Manufacturer
Armscor, Private Bag X337, Pretoria, 0001.

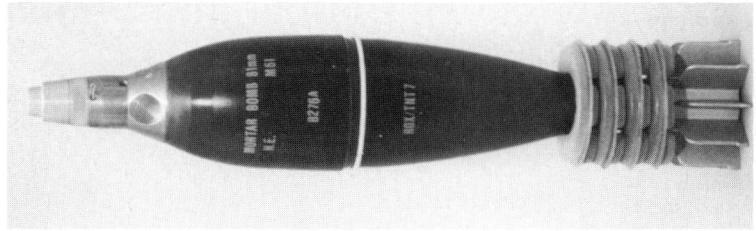

81 mm M61 HE bomb

81 mm M61 bursting smoke bomb

This is ballistically matched to the M61 HE bomb and is filled with titanium tetrachloride (FM). On impact a small burster charge of explosive ruptures the steel body and distributes the FM, which generates smoke by reaction with the water vapour in the air. It produces a dense smoke for screening or signalling purposes.

DATA
As for M61 HE bomb (above)

Manufacturer
Armscor, Private Bag X337, Pretoria, 0001.

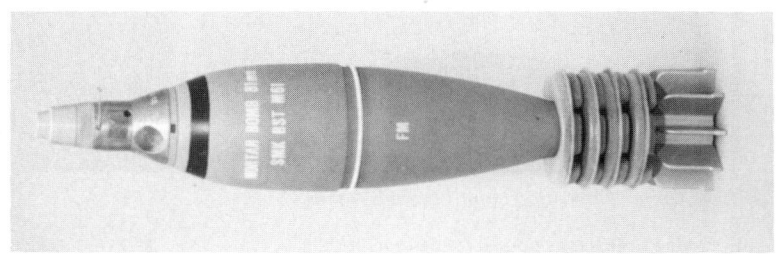

81 mm M61 bursting smoke bomb

81 mm M61 practice bomb

This is ballistically matched to the M61 HE bomb and is filled with an inert substance. There is a small indicating charge of TNT/aluminium which is detonated on impact by the V19 P fuze to give indication of fall of shot. It can be used at any charge up to Charge 8.

DATA
Weight: 4.43 kg
Velocity: (charge primary) 75 m/s; (charge 4) 196 m/s; (charge 8) 279 m/s
Max range: 4856 m (charge 8)

Manufacturer
Armscor, Private Bag X337, Pretoria, 0001.

Merlin proximity/impact fuze

Merlin is a selectable proximity or direct action impact fuze for mortars; it is suitable for 60, 81 or 120 mm mortars when fitted with appropriate adaptors to match the fuze-hole thread. It will work with all charges of all three systems and has been experimentally fired from other types of mortar with complete success. Using a loop antenna, it is 99 per cent independent of bomb size for its proximity functioning.

Power for the fuze is generated by a wind turbine inside the body, which is capable of producing adequate power even at the low velocities found in the 60 mm gun-mortar. The near-vertical slope detection circuitry reduces the possibility of jamming and also gives a good spread in burst height. There is a double-acting safe-and-arming device.

The fuze can be set to give proximity or direct impact action as desired, and can be set and re-set an infinite number of times without damage. In the proximity mode the DA action remains as insurance in case of electronic failure.

Manufacturer
Armscor, Private Bag X337, Pretoria 0001.
Status
Current. Production.

SPAIN

Ammunition for 60 mm mortars

Ecia manufactures two families of mortar ammunition, a conventional one and the aerodynamical (AE-84) type. Both types include the whole range (HE, Smoke HC or WP, Illuminating and practice rounds).

The ammunition manufactured by Ecia is obtained from forging, thus giving a knife-like fragment with higher velocity and therefore higher effectiveness. Upon request, Ecia can provide 60 mm bombs made from cast iron.

The Model AE-84 was introduced in 1984 and uses a highly streamlined body to obtain a good weight-to-calibre ratio and a plastic obturating ring to achieve good gas sealing and ballistic regularity. The tail unit, consisting of stabiliser tube and fins, is of forged aluminium and is attached to the body by a screwed section. The primary cartridge is inserted into the tail tube, and the incremental charges are in horse-shoe shaped containers which clip round the stabiliser tube.

HE bomb Model N
Length: 263 mm
Weight: 1.43 kg
Filling: TNT 232 g
Max range: 1975 m

HE Bomb Model AE
Length: 396 mm
Weight: 2.05 kg
Filling: 355 g TNT
Max range: 4600 m

Illuminating bomb Model N
Length with fuze: 368 mm
Weight: 1.97 kg
Max range: 1575 m
Intensity: 250 000 cd
Burning time: 23 s
Rate of descent: 4 m/s

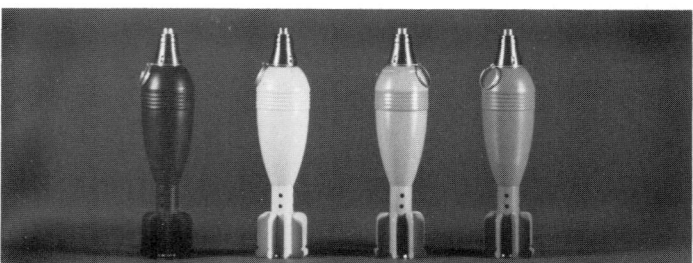
Model N series bombs for 60 mm mortars

Illuminating bomb Model AE
Weight: 1.55 kg (without fuze)
Max range: 4200 m
Intensity: 250 000 cd
Burning time: 25 s
Rate of descent: 4 m/s

Manufacturer
Esperanza y Cia SA, Marquina (Vizcaya).
Status
Current. Available.
Service
Spanish Army and several others.

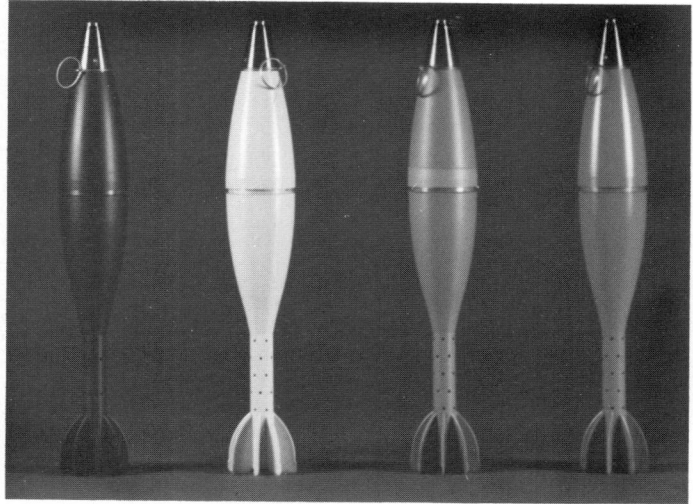

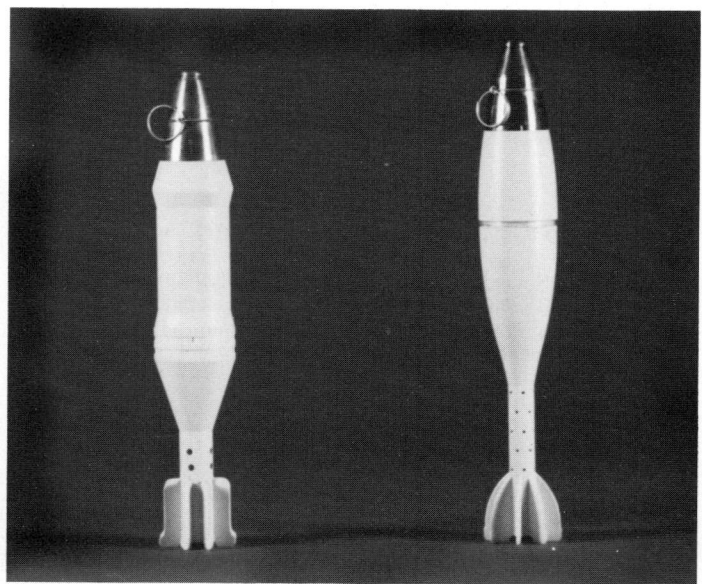
60 mm Illuminating bombs Models N and AE

Model AE series bombs for 60 mm mortars

Ammunition for 81 mm mortars

Ecia manufactures two families of mortar ammunition, a conventional one and the aerodynamical (AE-84) type. Both types include the whole range (HE, Smoke HC or WP, Illuminating and practice rounds).

The ammunition manufactured by Ecia is obtained from forging, thus giving a knife-like fragment with higher velocity and therefore higher effectiveness. Upon request, Ecia can provide 81 mm bombs made from cast iron.

The Model AE-84 was introduced in 1984 and uses a highly streamlined body to obtain a good weight-to-calibre ratio and a plastic obturating ring to achieve good gas sealing and ballistic regularity. The tail unit, consisting of stabiliser tube and fins, is of forged aluminium and is attached to the body by a screwed section. The primary cartridge is inserted into the tail tube, and the incremental charges are in horse-shoe shaped containers which clip round the stabiliser tube.

82 mm mortar rounds are also available.

81 mm HE bomb Model NA
Length overall: 343 mm
Weight: 3.2 kg
TNT filling: 496 g
Max range: mortar L-N 4125 m; mortar L-L 4680 m

81 mm HE bomb Model AE
Weight: 4.5 kg
TNT filling: 1009 g
Max range: mortar LN M-86 6200 m; mortar LL M-86 6900 m

81 mm Illuminating bomb Model N
Length overall: 474 mm
Weight: 3.93 kg
Intensity: 550 000 cd
Burning time: 30 s
Rate of descent: 4 m/s
Max range: mortar L-N 3350 m; mortar L-L 4000 m

81 mm Illuminating bomb Model AE
Weight: 4.25 kg without fuze
Intensity: 550 000 or 700 000 cd
Burning time: 35 s
Rate of descent: 4 m/s
Max range: Mortar LN M-86 5800 m; mortar LL M-86 6500 m

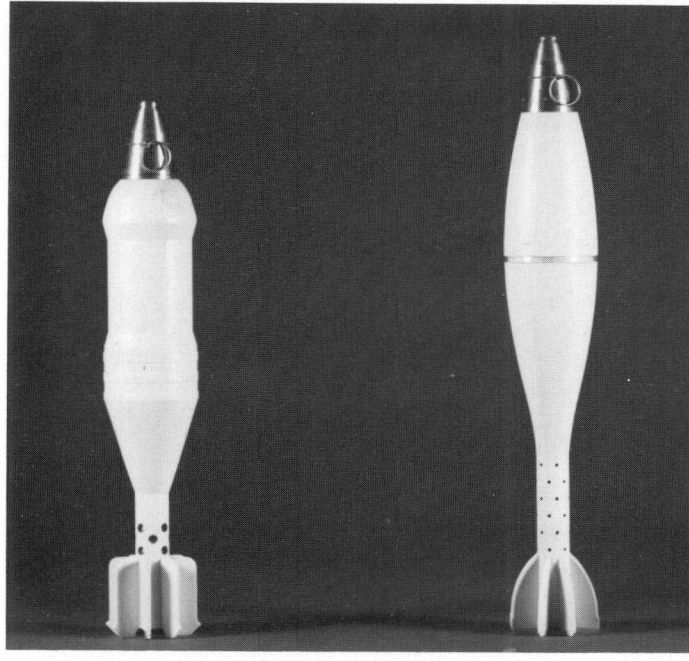
81 mm Illuminating bombs Models N and AE

Manufacturer
Esperanza y Cia SA, Marquina (Vizcaya).
Status
Current. Available.
Service
Spanish Army and several other armies.

Model NA series bombs for 81 mm mortars

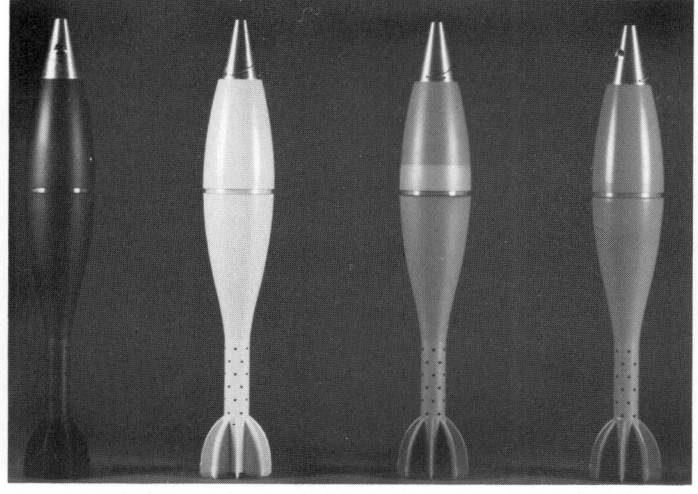
Model AE series bombs for 81 mm mortars

Ammunition for 120 mm mortars

Ecia manufactures two families of 120 mm mortar ammunition, a conventional one and the aerodynamical (AE-84) type. Both types include the whole range (HE, Smoke HC or WP, Illuminating and practice rounds).

The ammunition manufactured by Ecia is obtained from forging, thus giving a knife-like fragment with higher velocity and therefore higher effectiveness. Upon request, Ecia can provide 81 mm bombs made from cast iron.

The Model AE-84 was introduced in 1984 and uses a highly streamlined body to obtain a good weight-to-calibre ratio and a plastic obturating ring to achieve good gas sealing and ballistic regularity. The tail unit, consisting of stabiliser tube and fins, is of forged aluminium and is attached to the body by a screwed section. The primary cartridge is inserted into the tail tube, and the incremental charges are in horse-shoe shaped containers which clip round the stabiliser tube.

Ecia also manufactures 122 mm rounds.

HE bomb Model L
Weight: 13.0 kg
Filling: 2.391 kg TNT
Max range: with 1.6 m barrel 6725 m; with 1.8 m barrel 7000 m

HE bomb Model AE
Weight: 14.75 kg
Filling: 3.148 kg TNT
Max range: Mortar M-86 (M120-13) 8000 m; mortar M-86 (M120-15) 8250 m

Illuminating bomb Model N
Weight: 14.087 kg
Max range: 5450 m
Intensity: 1 000 000 cd
Burning time: 60 s
Rate of descent: 4 m/s

Illuminating bomb Model AE
Weight: 14.0 kg without fuze
Max range: Mortar M86 (M120-13) 7300 m; mortar M86 (M120-15) 7500 m
Intensity: 1 000 000 or 1 500 000 cd
Burning time: 60 s
Rate of descent: 4 m/s

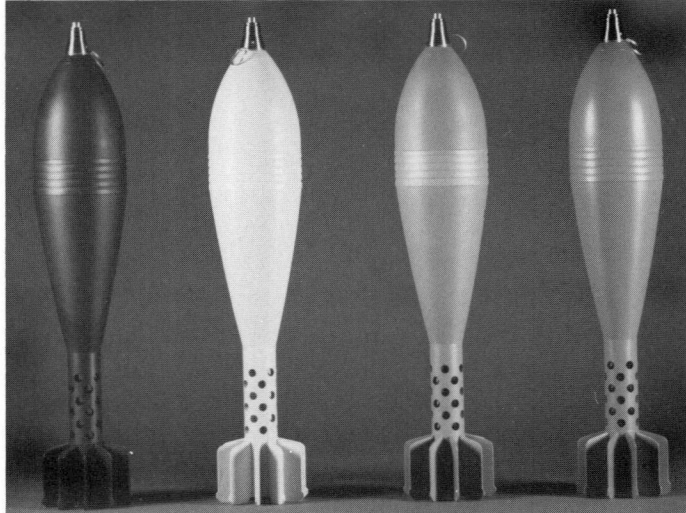

Model L series bombs for 120 mm mortars

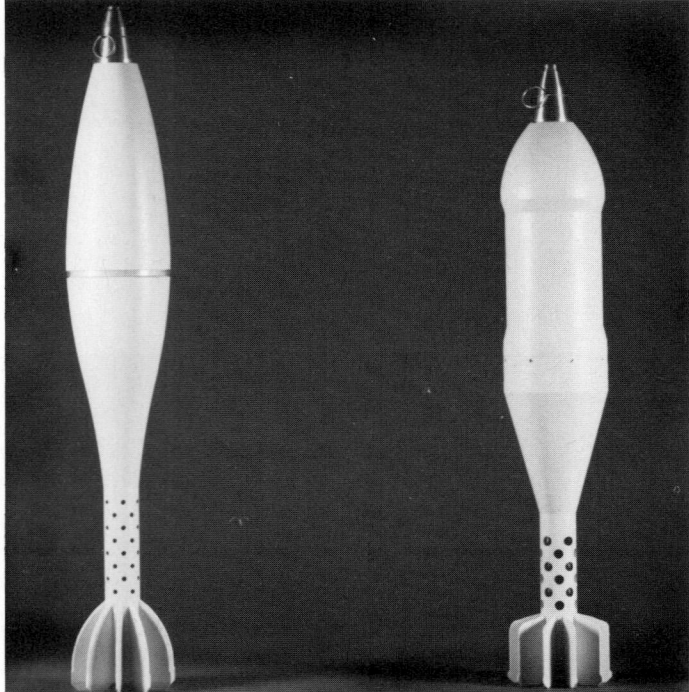

120 mm Illuminating bombs Models N and AE

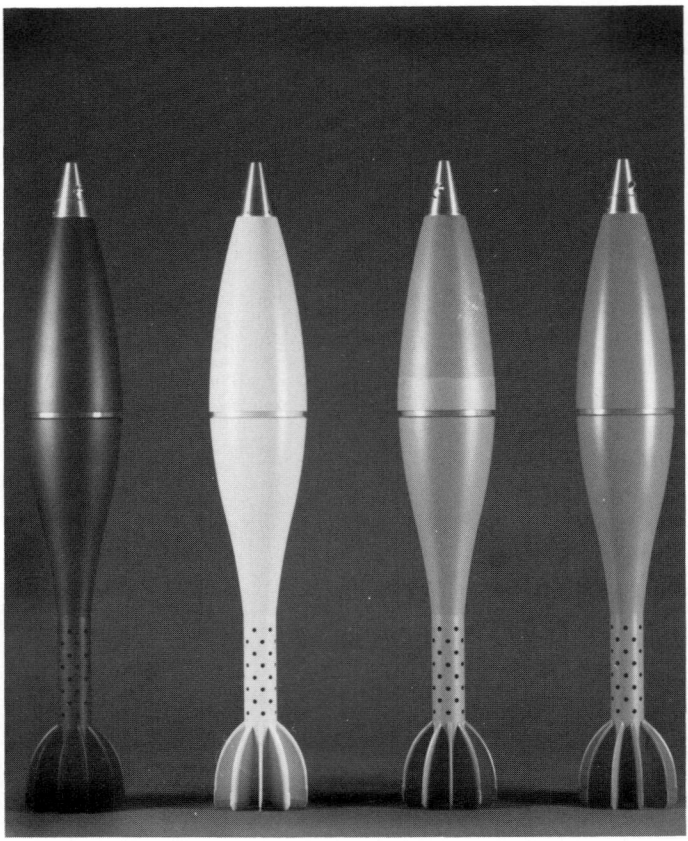

Model AE series bombs for 120 mm mortars

Manufacturer
Esperanza y Cia SA, Marquina (Vizcaya).
Status
Current. Production.
Service
Spanish forces and several others.

120 mm Espin multi-purpose ammunition

This is a new type of ammunition for the 120 mm mortar which provides it with the opportunity to widen its performance.

Each round of Espin ammunition contains 21 sub-munitions which are effective both against armour and personnel. Each sub-munition is provided with a shaped charge and also has an anti-personnel fragmenting capability. The covered area can extend from a few hundred to 4000 m² depending upon the height of burst selected by means of a time fuze, and can be effected at ranges of up to 6500 m from the mortar.

Espin ammunition is thus able to engage any type of troop concentration, tanks, armoured or soft vehicles, or any combination of them.

DATA
Number of sub-munitions: 21
Weight of bomb: 18.7 kg
Length of bomb, without fuze: 796 mm
Total number of fragments produced: approx. 13 600
Max range: 6500 m
Approximate area covered: up to 4000 m²

SUB-MUNITIONS
Diameter: 37 mm
Total weight: 285 g
Piercing capability in steel: 150 mm
Number of fragments: approx 650
Lethal area radius: 8 m
Action radius: 21 m

Espin 21 120 mm multi-purpose bomb

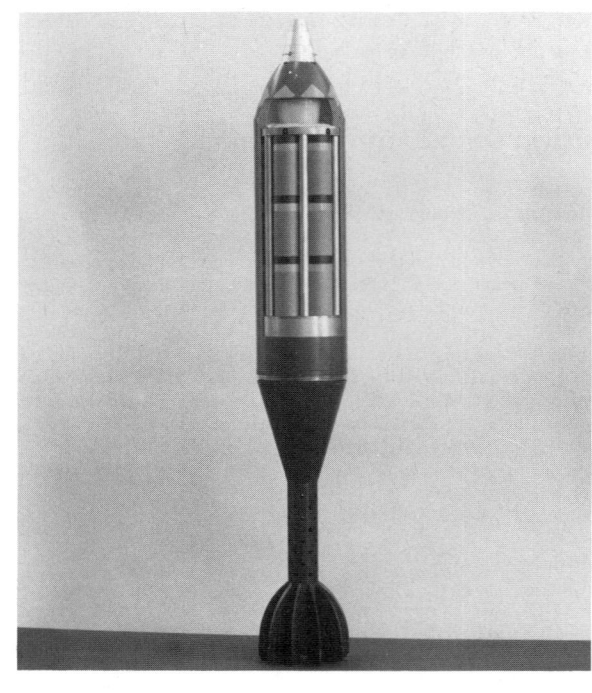

Manufacturer
Esperanza y Cia SA, Marquina (Vizcaya).
Status
Current. Production.

Service
Spanish forces and several others.

SWEDEN

120 mm FFV 029 illuminating bomb

The FFV 029 has been designed to meet the requirement for an easily handled and reliable illuminating bomb for heavy mortars such as the Hotchkiss-Brandt AM 50 and the Tampella types. It is fin-stabilised with a fin assembly similar to that used on Tampella ammunition. A mechanical time fuze is fitted. The bomb illuminates an area of 500 m diameter with not less than 5 lux.

DATA
Weight of bomb: 13.3 kg
Max muzzle velocity: 317 m/s
Max range: 6300 m
Height of burst: 350 m

Illumination: 1 000 000 cd
Burning time: 40 s
Speed of descent: 6 m/s

Manufacturer
FFV Ordnance, S-631 87 Eskilstuna.
Status
FFV 029 illuminating bomb was designed to match the ballistics of a HE bomb under development by FFV; at the time it was stated that the shape and weight of the FFV 029 could be changed to obtain a ballistic match with other HE bombs, and that the weight could be reduced to 12.8 kg if necessary. The present status of the FFV 029 bomb is uncertain.

120 mm FFV Strix guided mortar projectile

FFV, in collaboration with Saab, is developing a 120 mm mortar projectile terminally-guided by an infra-red target seeker. This projectile, designated Strix, is completely autonomous in operation, requiring no other information than that normally provided by a Forward Observation Officer. In this way it differs from the Diehl/Martin Marietta Bussard guided projectile, which depends on laser target designation.

The Strix projectile is handled exactly like any other mortar round except that it is fitted with a special tail unit which separates from the bomb after approximately 20 m of flight. The bomb follows a normal ballistic trajectory, sustained if required by a rocket motor which increases range and reduces the effect of wind. Only in the last stages of flight does the IR target seeker and guidance system come into operation to direct the bomb onto the upper surfaces of its target, and it is these surfaces which are generally the least protected.

The Strix projectile comprises a hollow-charge warhead, a guidance system, an electronic unit, the IR seeker, the afterbody with the fins and an optional rocket motor. Terminal guidance is performed by side-thruster units in the afterbody.

The high descent angle implicit in the use of a mortar projectile enables the Strix projectile to overcome several attack problems. For example, it is possible to attack the lighter top armour of armoured fighting vehicles rather than attempting to defeat the front and side armour. The near-vertical final trajectory also allows the IR seeker to overcome concealment methods such as camouflage and smoke screening. The application of terminal guidance to mortar bombs is also simplified by the relatively light setback forces on firing and the lack of spin in most 120 mm mortar systems.

DATA
Calibre: 120 mm
Length: 810 mm
Weight: 17.5 kg
Range: 1700-7500 m

Manufacturer
FFV Ordnance, S-631 87 Eskilstuna.
Status
Development.

Loading the 120 mm Strix projectile

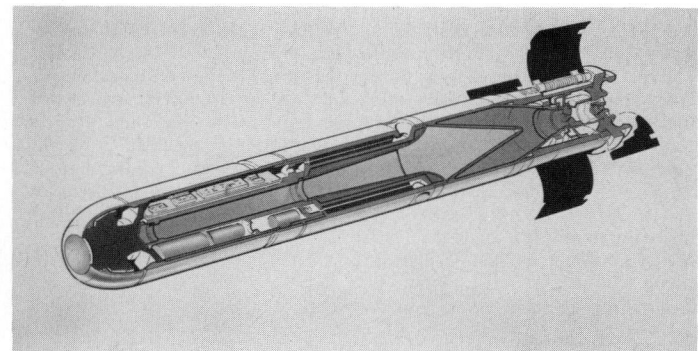

FFV Strix homing projectile

SWITZERLAND

Ammunition for 81 mm mortars

Ammunition for the Models 1933 and 1972 Swiss mortars is of the conventional streamlined pattern.

Manufacturer
Swiss Federal Arms Factory, Postfach 3000, Stauffacherstrasse 65, CH-3000 Berne 22.
Service
Swiss Army.

Types	HE	Smoke	High capacity fragmentation bomb
Fuzing	PD or DA	percussion	PD
Weight	3.17 kg	3.67 kg	6.89 kg
Filling	TNT	smoke composition	TNT
Charges	0-6	0-6	1-4
Muzzle velocity	70-260 m/s	70-210 m/s	64-110 m/s
Max range	4100 m	3000 m	1070 m

81 mm Model 73 illuminating bomb

The Model 73 illuminating bomb is of Swedish design and the first batch supplied to the Swiss Army was manufactured in Sweden; however the bomb is now in production at the Swiss Federal Arsenal, Thun. The mechanical time fuze and propelling cartridges are of Swiss design and manufacture.

The Model 73 is a conventional parachute illuminating bomb in which the flare is ejected at about 300 m above the target area and then illuminates an area of 650 m diameter for up to 30 seconds.

DATA
Weight: 3.5 kg
Range: 500-3250 m
Rate of descent of flare: 4 m/s

Flight duration at max range: 30 s
Time fuze setting: infinitely variable between 5 and 60 s
Duration of illumination: 30 s
Charges: 0-6

Manufacturer
Eidg Munitionsfabrik, Thun.
Status
Current. Production.
Service
Swiss Army.

Ammunition for 120 mm mortars

Ammunition for the Models 64 and 74 120 mm mortars is of conventional type.

Fuze: point detonation and slight delay (HE); point detonation (incendiary/smoke)
Weight: 14.33 kg
Filling: trotyl (HE); white phosphorus (incendiary/smoke)
Charges: 1-8
Muzzle velocity: 128-420 m/s
Max range: 7500 m

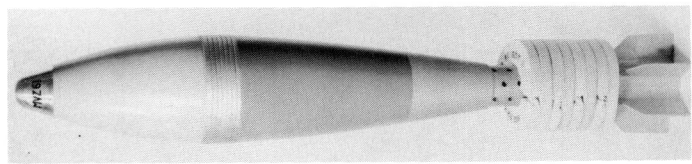

120 mm mortar ammunition (HE)

120 mm Model 74 illuminating bomb

The Model 74 is a conventional parachute illuminating bomb, the functioning of which is controlled by a mechanical time fuze of Swiss manufacture. The body of the bomb, including the flare, is manufactured in Sweden, but the tail assembly, propelling charges and fuze are produced in Switzerland. The Model 74 can be fired from the 120 mm mortars Mw 64 and Mw 74 as well as the statically emplaced Fest Mw 59.

The flare is ejected at about 500 m above the target and illuminates an area of approximately 1600 m in diameter.

DATA
Weight: 15 kg
Rate of descent of flare: 5 m/s
Max range: 7000 m
Flight duration at max range: 50 s
Fuze time setting: infinitely variable from 5-60 s

Manufacturer
Eidg Munitionsfabrik, Thun.
Status
Current.
Service
Swiss Army.

Degen Fuze K-85

The K-85 is a single-action point detonating fuze. It can be used with mortar bombs of 60, 81, 82 and 120 mm calibre. It is similar to fuzes M52, M525 and DM111. Safety is ensured by a shutter which holds the detonator out of alignment prior to firing, and by a safety pull wire which remains in place until removed immediately before loading the bomb. The minimum arming distance is 30 m from the muzzle. A fully supported and guided firing pin ensures detonation at low angles of impact. The fuze meets all requirements of MIL-STD 331.

Manufacturer
Degen & Co AG, CH-4435 Niederdorf.
Status
Current. Production.

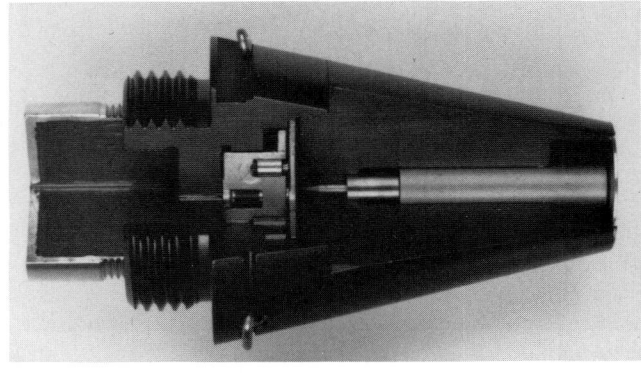

Degen K-85 mortar fuze

UNION OF SOVIET SOCIALIST REPUBLICS

82 mm Model 0-832 DU fragmentation bomb

DATA
Calibre: 82 mm
Type: HE fragmentation
Weight: (fuzed) 3.23 kg
Bursting charge: 0.436 kg TNT/dinitronapthalene
Fuze: M-6 point detonating
Known using weapon: mortar M-37 (1942–43 version)
Remarks: also found with M-1, M-2, M-3, M-4, M-5, and MP-82 fuzes

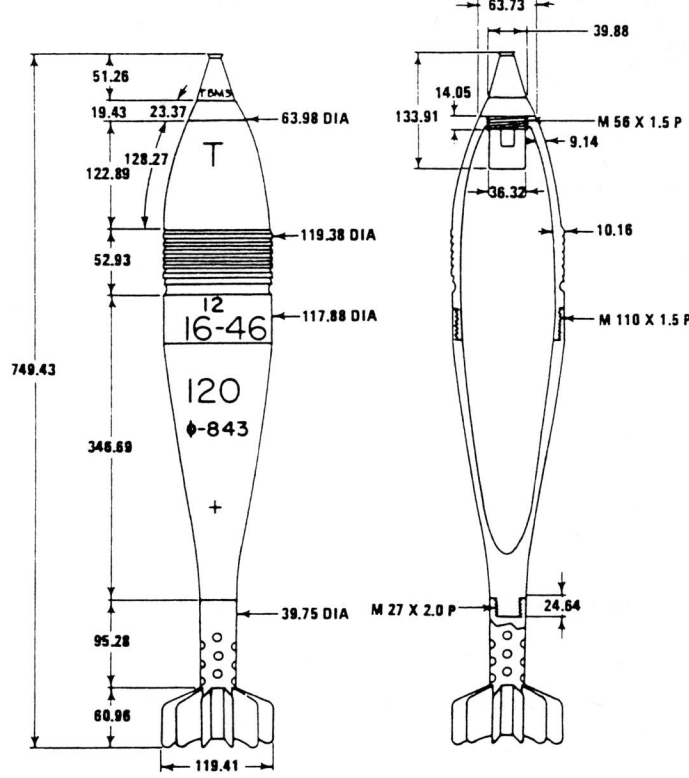

82 mm Model 0-832DU fragmentation bomb

107 mm Model OF-841A HE fragmentation bomb

DATA
Calibre: 107 mm
Type: HE fragmentation
Weight: (fuzed) 9.1 kg
Bursting charge: 1 kg TNT or Amatol
Fuze: GVMZ-7 point detonating
Known using weapon: mountain-pack regimental mortar M-38
Remarks: also uses Models GVMZ and GVMZ-1 point detonating fuzes
Fuze is shown with safety cap installed

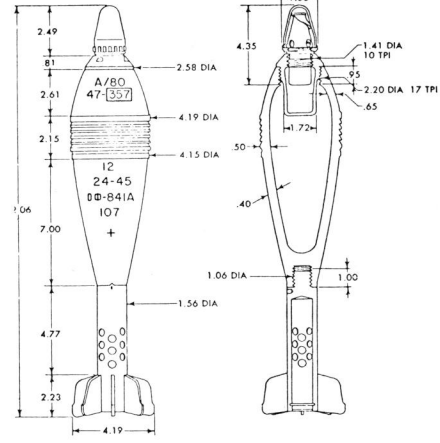

107 mm Model OF-841A HE fragmentation bomb; dimensions are in inches

120 mm Model OF-843 HE bomb

DATA
Calibre: 120 mm
Type: HE Fragmentation
Weight: (fuzed) 16.02 kg
Bursting charge: 2.68 kg TNT
Fuze: Model GVMZ point detonating
Known using weapons: regimental mortars M-38 and M-43
Remarks: also found with M-1 or M-4 PD fuzes

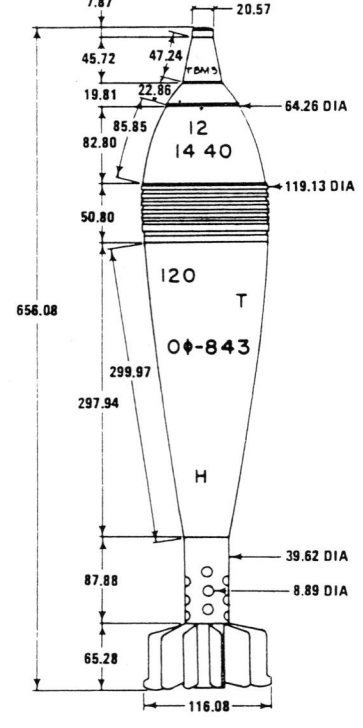
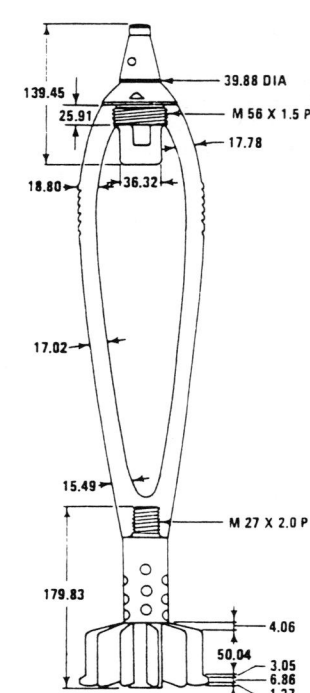

120 mm Model OF-843 HE bomb

120 mm Model OF-843A HE fragmentation bomb

DATA
Calibre: 120 mm
Type: HE fragmentation
Weight: (fuzed) 15.98 kg
Bursting charge: 1.58 kg Amatol 80/20
Fuze: Model GVMZ-1 point detonating
Known using weapons: regimental mortars M-38 and M-43
Remarks: also found with M-4 point detonating fuze

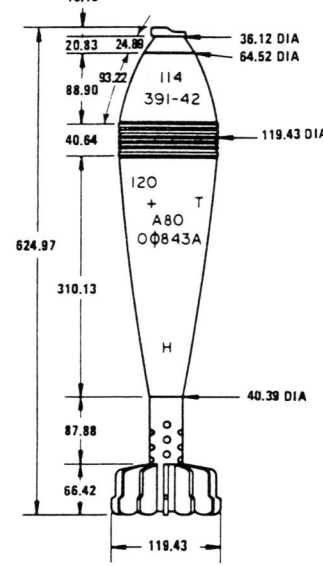
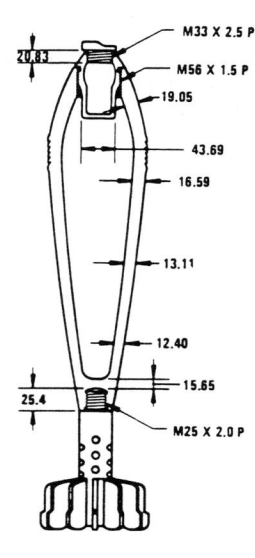

120 mm Model OF-843A HE fragmentation bomb

160 mm Model F-852 HE bomb

DATA
Calibre: 160 mm
Type: HE
Weight: 40 kg
Bursting charge: 7.39 kg TNT
Fuze: Model GVMZ-7 point detonating
Known using weapon: mortar M-43

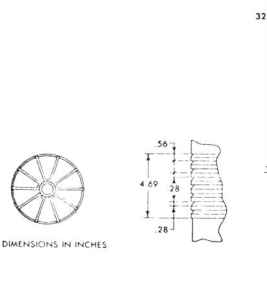
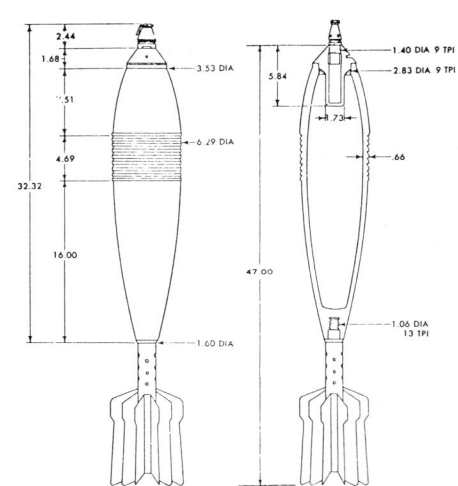

Soviet 160 mm Model F-852 HE bomb

160 mm Model F-853A HE bomb

DATA
Calibre: 160 mm
Type: HE
Weight: 41.18 kg
Bursting charge: 7.73 kg Amatol 80/20
Fuze: Model GVMZ-7 point detonating
Known using weapon: mortar M-160
Remarks: Cast iron bomb; fuze shown here with safety cap fitted.

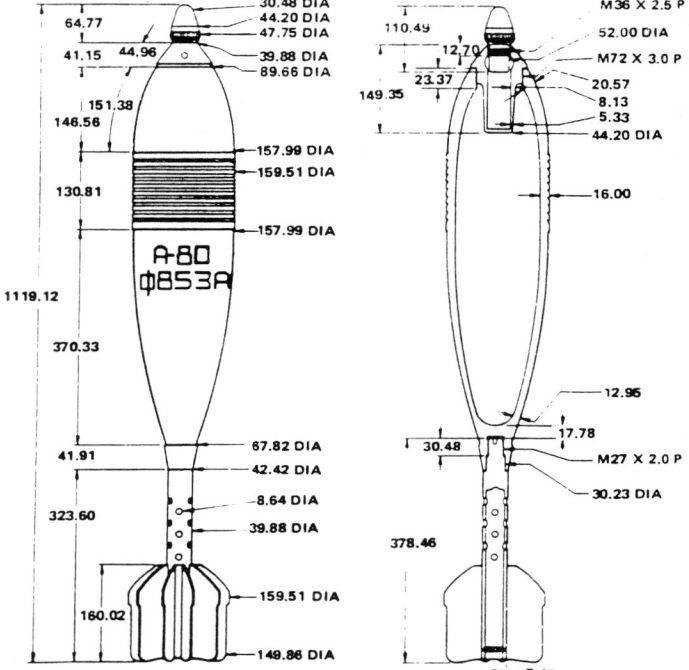

160 mm Model F-853A HE bomb

160 mm Model F-853U HE bomb

DATA
Calibre: 160 mm
Type: HE
Weight: 41.18 kg
Bursting charge: 8.989 kg TNT
Fuze: Model GVMZ-7 point detonating
Known using weapon: mortar M-160
Remarks: Steel bomb

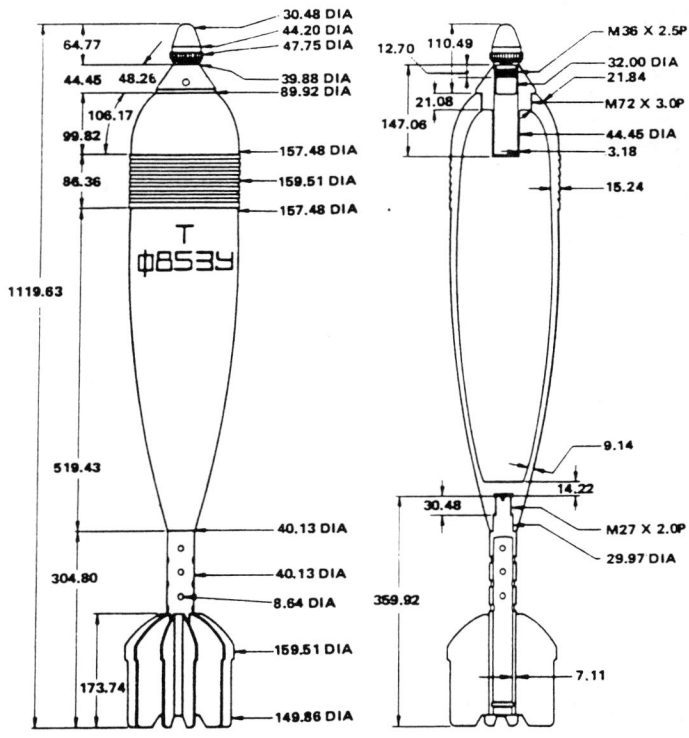

160 mm Model F-853U HE bomb

UNITED KINGDOM

Ammunition for the 51 mm Assault mortar

The ammunition for the 51 mm comprises high explosive, smoke and illuminating bombs; training and drill bombs are also available. All types have a common aluminium alloy casing of advanced design and the HE bomb body contains a mild steel fragmentation coil which is notched around its internal diameter to produce fragments of optimum size. It uses the L127A3 delayed arming fuze which has impact and delay operating modes. All types use a plastic obturating ring to give the most efficient ballistic performance. The cartridge has an all-metal cover and is completely waterproof. A number of packaging options are available, but for the British Army the HE bombs are supplied in the L17A1 steel box (10 bombs per box). Inside each box are two satchels, folded on top of the ammunition. In the field the ammunition is transferred from the box to the satchels, which are fitted with long carrying straps to permit shoulder slinging. The smoke and illuminating bombs are packed six to an H83 steel box. They are pre-packed in individual tubes, the whole being enclosed in a webbing satchel for ease of transportation in the field.

HE BOMB L1A1
Weight: 920 g
Filling: 60/40 RDX/TNT
Fuze: point detonating, L127A2
Overall length: (fuzed) 290 mm
Max range: 800 m
Minimum range: 50 m
Lethality: claimed to be 5 times greater than 2 in HE bomb
Muzzle velocity: 103 m/s

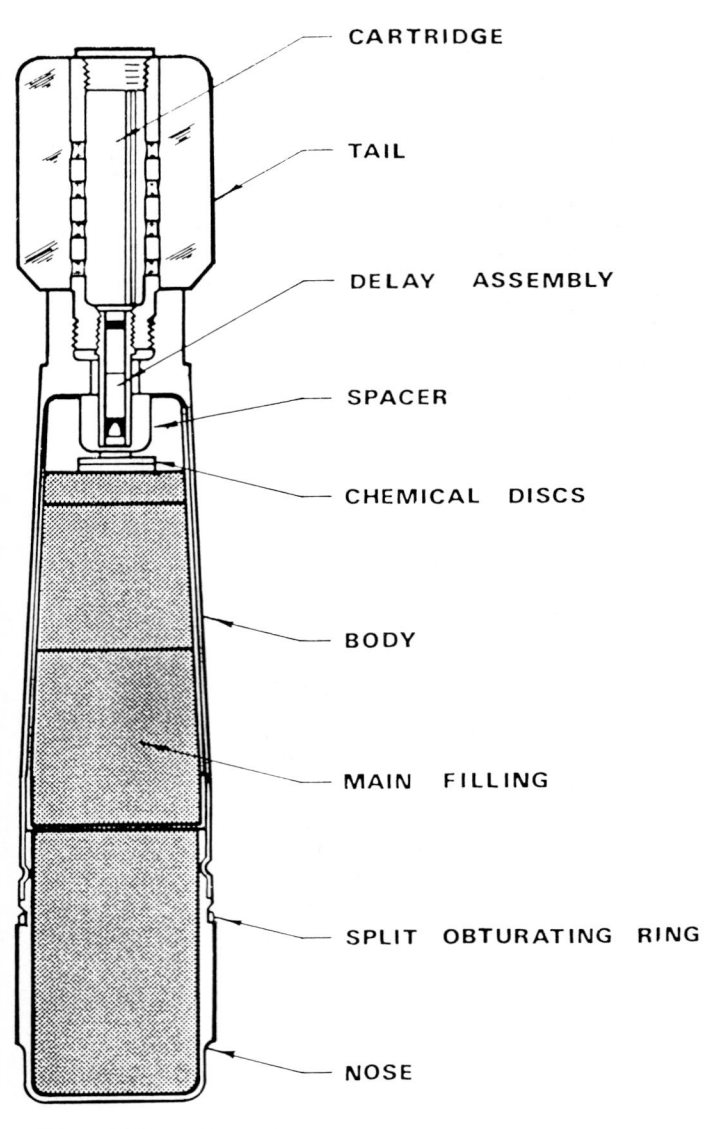

51 mm L2A1 smoke bomb

Labels for L2A1 smoke bomb:
- CARTRIDGE
- TAIL
- DELAY ASSEMBLY
- SPACER
- CHEMICAL DISCS
- BODY
- MAIN FILLING
- SPLIT OBTURATING RING
- NOSE

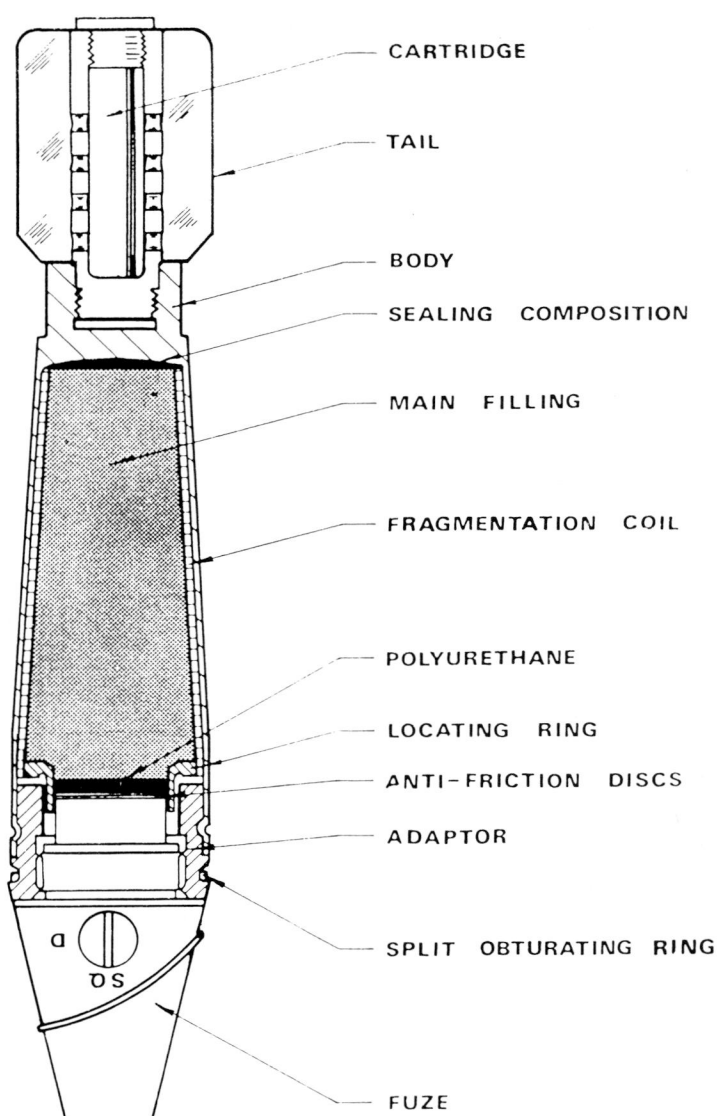

51 mm HE bomb fuzed L127A2

Labels for HE bomb L127A2:
- CARTRIDGE
- TAIL
- BODY
- SEALING COMPOSITION
- MAIN FILLING
- FRAGMENTATION COIL
- POLYURETHANE
- LOCATING RING
- ANTI-FRICTION DISCS
- ADAPTOR
- SPLIT OBTURATING RING
- FUZE

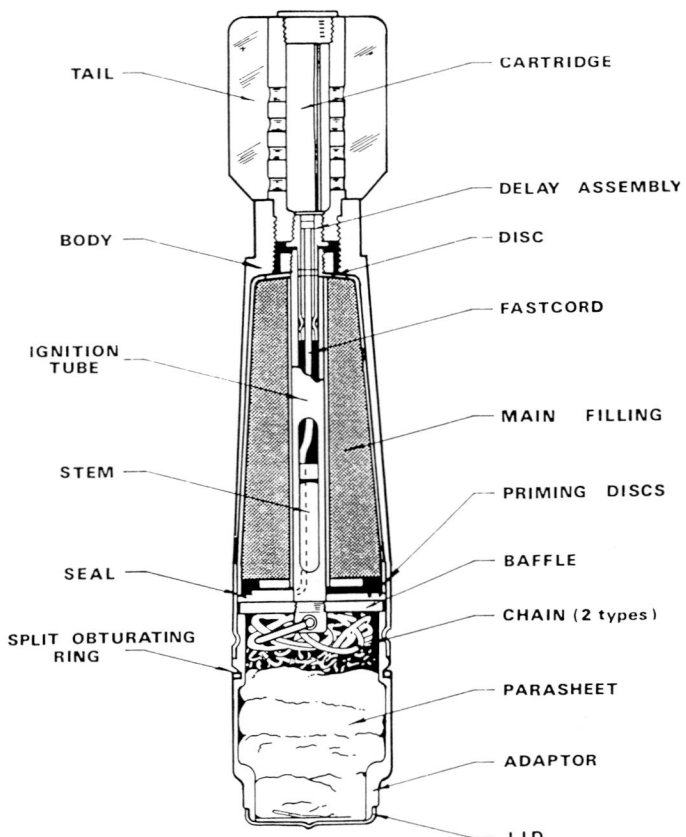

51 mm L3A2 illuminating bomb

Labels for L3A2 illuminating bomb:
- TAIL
- CARTRIDGE
- DELAY ASSEMBLY
- DISC
- BODY
- FASTCORD
- IGNITION TUBE
- MAIN FILLING
- STEM
- PRIMING DISCS
- SEAL
- BAFFLE
- CHAIN (2 types)
- SPLIT OBTURATING RING
- PARASHEET
- ADAPTOR
- LID

SMOKE BOMB L2A1
Weight: 0.90 kg
Filling: hexachlorethane (HCE)
Body: as for HE bomb, without fragmentation coil
Fuze: fixed time delay (4 s minimum)
Duration of emission: 100 s minimum
Muzzle velocity: 103 m/s

ILLUMINATING BOMB L3A2
Weight: 0.800 kg
Body: as for HE bomb without fragmentation coil
Fuze: 12–14 s pyrotechnic delay
Burst height: 325 m at 775 m
Rate of descent: 4.65 m/s
Illumination: average 135 000 cd
Duration of burning: 44 s approx
Muzzle velocity: 127 m/s

Manufacturer
Royal Ordnance plc.
Status
Current. Production.
Service
In service with the British Army.

Ammunition for the 81 mm Mortar

The high explosive round L36A2, fuzed L127A3, is streamlined in shape and has been designed to produce the maximum number of fragments of the optimum size. A study of cast iron, forged steel and spheroidal graphite cast iron, showed the third to be the best material. The 81 mm bomb has also gained in velocity and consistency from the incorporation of a sealing ring. This allows adequate windage as the bomb drops down the tube but the pressure produced by the burning propellant forces the polycarbon ring outwards against the interior wall of the mortar tube and so prevents gas leakage. The ring also centres the bomb to reduce the yaw at the muzzle associated with conventional mortars and which leads to both inaccuracy and inconsistency.

Royal Ordnance has developed two new charge systems, the Marks 4 and 5. The Mark 4 system has been chosen for introduction into service with the British Army, and the Mark 5 with the US Army.

Both systems offer the following improvements over the Mark 2 system:
Full wet efficiency
More robust construction
Reduced blast overpressure
No visible signature on firing
Equal size secondary charges for easier use.

The Mark 4 has six equal secondary charges, the Mark 5 has four equal secondary charges. The maximum charge gives the same range with either system.

The white phosphorus smoke bomb is ballistically matched to the HE bomb and can be fitted with any of the charge systems Marks 2, 4 or 5. Common fuzes can be fitted to both the HE and WP bombs.

Ammunition for the 81 mm mortar is supplied in a strong waterproof container holding two rounds, two containers fitting into a steel box. This provides a robust four-round package weighing 31 kg.

HE bomb L36A2
Fuze: L127A2, US M734, US M935 and other NATO fuzes
Length: (overall) 472.4 mm
Weight: (complete) 4.2 kg
Filling: 0.68 kg 60/40 RDX/TNT
Body: ductile cast iron
Tail: extruded aluminium alloy
Charge system: Marks 2, 4 or 5

Smoke bomb L40A1
The body of the smoke bomb is similar to that of the HE bomb and is ballistically matched to it. It is filled with white phosphorus.
Weight: 4.5 kg
Overall length: 472.4 mm

Practice bomb L27A1
The practice bomb is a re-usable item designed to fire to a maximum range of 80 m when an obturating ring is fitted.

Drill bomb
A dummy round, not intended for firing, but inert filled to the correct weight for training purposes.

Illuminating bomb
An illuminating bomb is available which delivers 1 million candela for 35 seconds. It is ballistically matched to the HE and smoke bombs.

Range
All charge systems permit the bombs to achieve consistent ranges of 100 m minimum and 5650 m maximum under IACO air conditions.

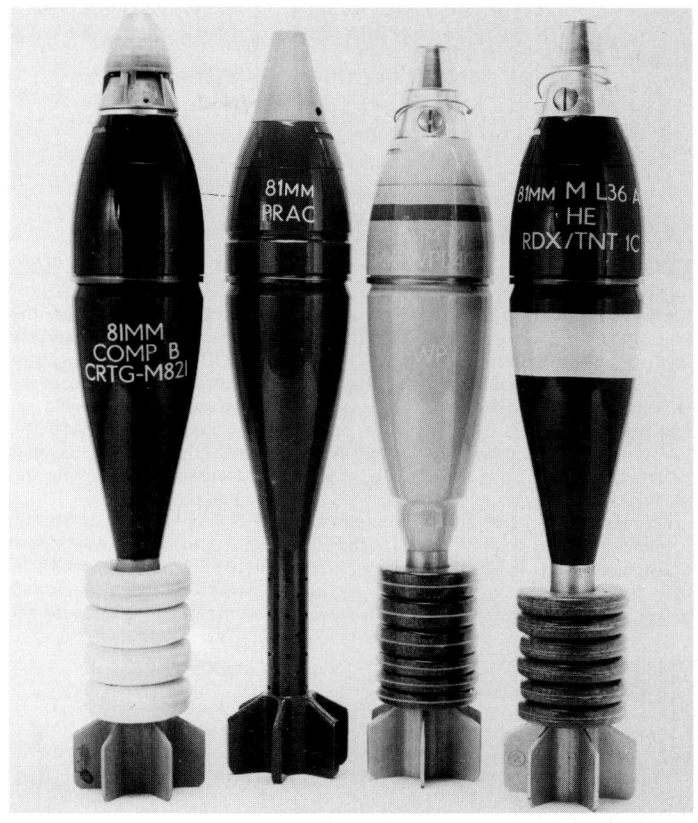

81 mm L16 mortar ammunition and various charge systems. From left to right: HE bomb (US variant) with Mark 5 charge system; recoverable practice bomb; smoke bomb with Mark 2 charge system; HE bomb (UK variant) with Mark 4 charge system. (Royal Ordnance plc, Glascoed)

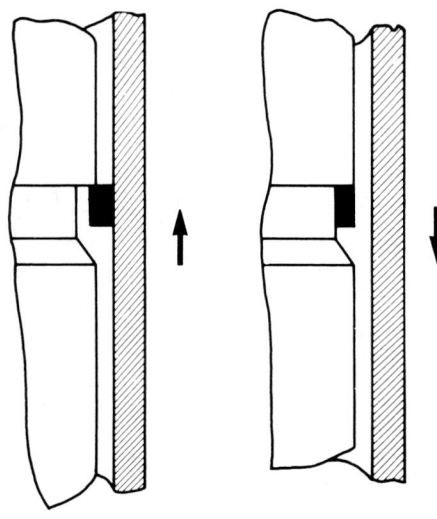

Sealing ring is driven out to close windage gap when pressure is developed

PERFORMANCE DATA, HE ROUND L36A2, MARK 2 CHARGE SYSTEM

	Augmenting cartridges			Muzzle velocity	Range		
Charge	Small	Intermediate	Large	L5 mounting & FV 432	minimum	max	50% zone
Primary	0	0	0	73 m/s	180 m	520 m	15 × 2 m
Charge 1	1	0	0	110 m/s	390 m	1120 m	30 × 5 m
Charge 2	2	0	0	137 m/s	580 m	1710 m	30 × 5 m
Charge 3	3	0	0	162 m/s	780 m	2265 m	30 × 5 m
Charge 4	3	1	0	195 m/s	1070 m	3080 m	40 × 10 m
Charge 5	3	2	0	224 m/s	1340 m	3850 m	50 × 15 m
Charge 6	3	3	0	250 m/s	1700 m	4680 m	60 × 20 m
Charge 7	0	0	4		1900 m	5255 m	
Charge 8	1	0	4		2100 m	5650 m	

PERFORMANCE DATA, HE ROUND L36A2, MARK 5 CHARGE SYSTEM

	Augmenting cartridges			Muzzle velocity	Range		
Charge	Small	Intermediate	Large	L5 mounting & FV 432	minimum	max	50% zone
Primary		0		70 m/s	166 m	497 m	15 × 2 m
Charge 1		1		133 m/s	525 m	1564 m	30 × 5 m
Charge 2		2		175 m/s	850 m	2500 m	30 × 5 m
Charge 3		3		212 m/s	1175 m	3425 m	40 × 5 m
Charge 4		4		243 m/s	1450 m	4200 m	50 × 10 m
Charge 5		5		272 m/s	1700 m	4975 m	60 × 20 m
Charge 6		6		297 m/s	1950 m	5650 m	60 × 30 m

PERFORMANCE DATA, HE ROUND L36A2, MARK 4 CHARGE SYSTEM

	Augmenting cartridges			Muzzle velocity	Range		
Charge	Small	Intermediate	Large	L5 mounting & FV 432	minimum	max	50% zone
Primary		0		70 m/s	166 m	497 m	15 × 2 m
Charge 1		1		154 m/s	675 m	2060 m	30 × 5 m
Charge 2		2		209 m/s	1125 m	3400 m	40 × 10 m
Charge 3		3		257 m/s	1525 m	4620 m	60 × 20 m
Charge 4		4		297 m/s	1837 m	5650 m	60 × 30 m

Merlin 81 mm terminally guided mortar projectile

Merlin is a terminally guided armour-piercing mortar bomb for use with any 81 mm mortar. It offers a highly effective fire and forget capability to infantry confronted with armoured vehicles.

Firing is carried out in the normal way, using target information from the Mortar Fire Controller and using standard procedures. After launch, six rear-mounted fins are deployed to provide basic aerodynamic stability, followed by four canard fins to provide directional control. The seeker is switched on as Merlin approaches the vertex and searches first for moving and then for stationary targets.

The miniature active millimetric seeker, which is a key feature of Merlin, provides information after launch to the control system to accomplish de-spin and attitude control. The seeker then carries out a search over a footprint area of 300 m by 300 m, giving priority to moving targets. Having acquired a target, the seeker provides the necessary angular error information to the

guidance system to ensure impact with the most vulnerable areas on the top of the armoured vehicle.

The effectiveness of Merlin has now been demonstrated in a series of successful firings against targets representing armoured vehicles. These have proved the ability of Merlin to carry out the full sequence of acquiring and locking on to moving and stationary targets and then to steer itself out of its ballistic trajectory to attack them.

Manufacturers
British Aerospace (Dynamics) Ltd, Six Hills Way, Stevenage, Hertfordshire SG1 2DA
Status
Advanced development.

Loading Merlin into 81 mm L16 mortar

AM52A3 point detonating mortar fuze

This fuze is designed for use on 60 mm, 81 mm and 120 mm mortar bombs. Functioning is in the superquick mode only, and safety is ensured by having the elements of the explosive train out of alignment until fired. A special feature is the 'Armed Fuze Indicator Pin'; should the fuze become armed prior to firing, the slider pin will protrude from the body of the fuze and indicate that it is unsafe to fire.

This fuze is interchangeable with DM111 and M525 type fuzes.

DATA
Length: 89 mm
Weight: 254 g
Arming impulse: 300 g
Intrusion: 28 mm
Thread: 1.15-12-NF1A

AM52A3 point detonating fuze

Manufacturer
Allivane International, 15 John Street, London WC1N 2EB.
Status
Current. Production.

UNITED STATES OF AMERICA

Ammunition for the 60 mm M19 mortar

The bombs for the 60 mm mortar are of conventional cast iron pattern, with obturating grooves around the body and steel tail units screwed in. The primary cartridge is of shotgun type, inserted into the centre of the tail unit; the secondary charges are in the form of leaves of smokeless powder stitched together and are sprung into place between the tail fins as required.

Maximum ranges given below are applicable only to the mortar used with the M5 baseplate. Using the M1 baseplate no more than one incremental charge should be added to the bomb.

M49A4 HE bomb
Weight: 1.46 kg
Filling: 154 g TNT
Propelling increments: 4
Minimum range: 45 m
Max range: 1814 m (charge 4)
Bursting area: 9 × 18 m

M302A2 (WP) smoke bomb
Weight: 2.26 kg
Filling: white phosphorus
Minimum range: 91 m
Max range: 1465 m (charge 4)
Bursting area: 10 m in diameter

M83A3 illuminating bomb
Weight: 2.27 kg
Minimum range: 375 m
Max range: 1000 m (charge 4)
Illuminated area: 600 m diameter
Illumination duration: 25 s

M50A2 practice bomb
Weight: 1.45 kg
Minimum range: 45 m
Max range: 1814 m (charge 4)

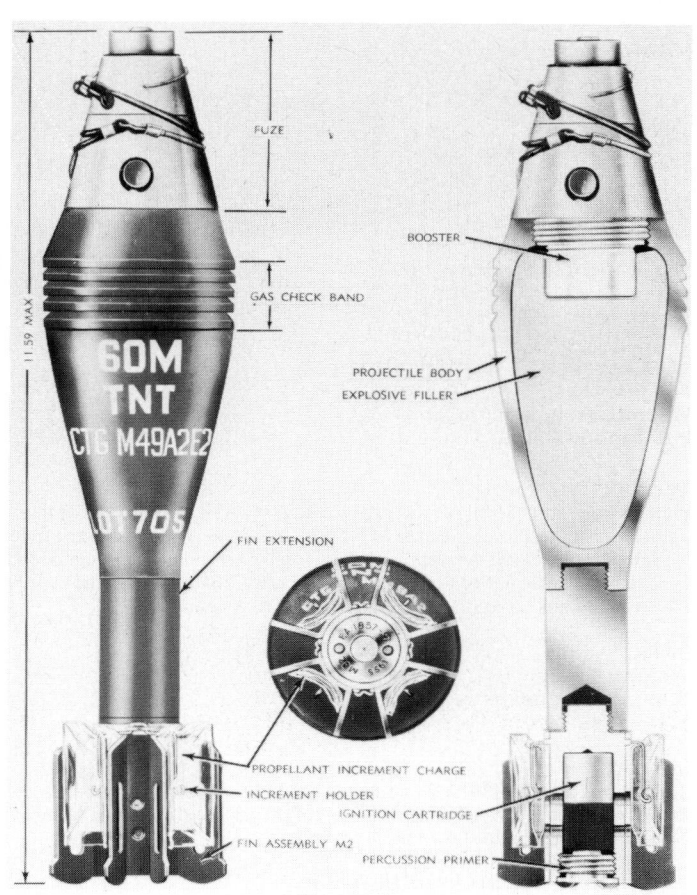

60 mm M49A2E2 bomb; this has since been improved into the M49A4 model, though the basic details remain the same

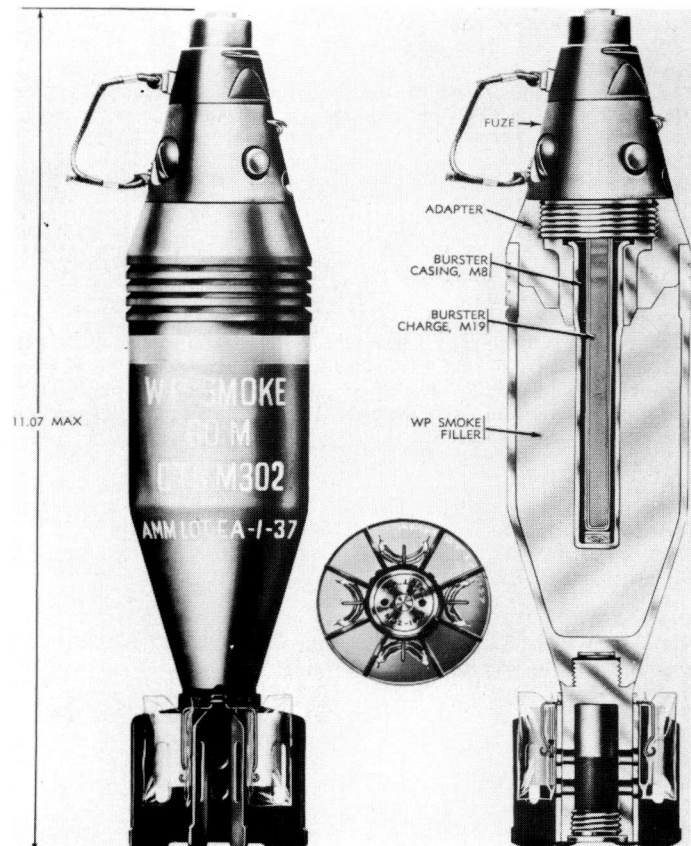

60 mm M302 white phosphorus smoke bomb

Ammunition for the 81 mm mortars M29 and M29A1

The latest ammunition for these mortars is based upon the original British designs with modifications to suit American systems of manufacture and operation. The principal obvious change is the use of cloth bags for the secondary charges, rather than the British plastic 'horseshoe' container, and the canting of the tail fins to produce a slight degree of roll stabilisation. Older designs of ammunition utilise simple forged steel bombs with steel tail units and the bombs have obturating grooves around the body rather than plastic obturating rings.

Authorised rounds:
High explosive, M43A1B1, M362A1, M374A2
White phosphorus, M370, M375A2
Illuminating, M301A1, M301A2, M301A3
Training practice, M43A1
Training (concrete), M798
White phosphorus, M784
Illuminating, M512E1, M652.

High explosive
HE M374A2
The pearlitic malleable iron projectile is loaded with approximately 0.95 kg of Composition B. The rear of the bourrelet section of the projectile is fitted with an obturator ring with a circumferential groove. The aluminium fin assembly, M170, consists of an ignition cartridge housing and six extruding fins canted counterclockwise 5° at the rear to stabilise the round in flight.

DATA
Weight: 4.23 kg
Minimum range: 72 m
Max range: 4595 m
Bursting area: 34 m diameter

HE M362A1
The steel forged projectile is loaded with approximately 0.95 kg of Composition B. The aluminium fin assembly is the M141. This bomb is not classified Standard A.

DATA
Weight: 4.25 kg
Minimum range: 46 m
Max range: 3987 m
Bursting area: 25 × 20 m

HE M43A1B1
The complete round consists of a relatively thin-walled shallow-cavity steel projectile containing a TNT bursting charge and PD fuze.

DATA
Weight: 3.22 kg
Minimum range: 69 m
Max range: 3890 m
Bursting area: 20 × 15 m

White phosphorus
WP M375A2
The M375A2 WP round is ballistically and otherwise similar to the HE M374A2 round except that it is loaded with approximately 725 g of white phosphorus, and contains a one-piece aluminium burster casing (M158) prefilled to the forward end of the body. The burster casing houses a central burster tube containing RDX.

DATA
Weight: 4.23 kg
Minimum range: 72 m
Max range: 4737 m
Bursting area: 20 m diameter

WP M370
The M370 WP round is ballistically and otherwise similar to the HE M362A1 round except for the white phosphorus filler. This round is not classified Standard A.

DATA
Weight: 4.23 kg
Minimum range: 52 m
Max range: 3987 m
Bursting area: 20 m diameter

Illuminating ammunition
M301A3
The complete round consists of a time fuze, a thin-walled steel body-tube containing the parachute and illuminant assembly, and a steel tail cone and fin assembly.

This round is designed to be fired with a minimum of two propelling charge increments and not more than eight. It has a burst height of 600 m and will illuminate a 1200 m area for a minimum of 60 seconds.

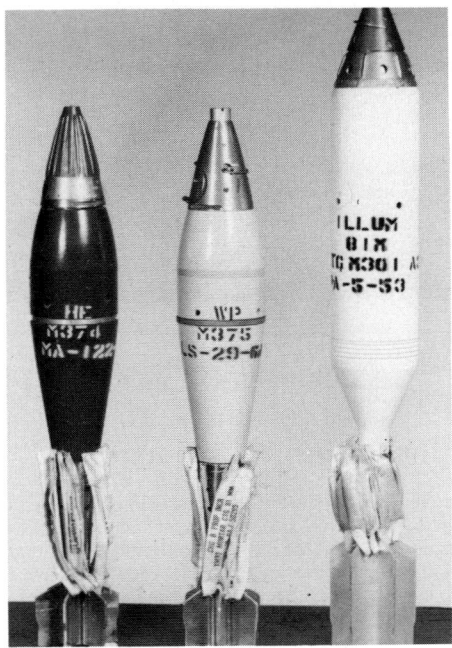

81 mm bombs
(left) *HE bomb M374* **(centre)** *white phosphorus smoke bomb M375* **(right)** *illuminating bomb M301A3*

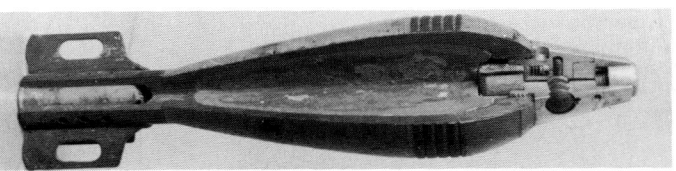

M43A1 bomb

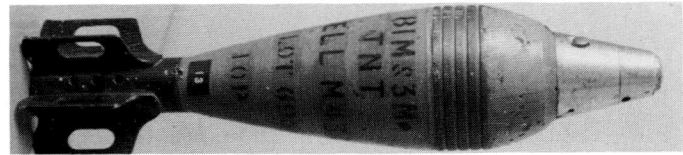

81 mm M43A1 HE bomb

DATA
Weight: 4.89 kg
Minimum range: 100 m
Max range: 3150 m

M301A2
Illuminating round M301A2 is similar to the M301A3 except that it has a tail fin that is 57 mm shorter than the M301A3. The round is designed to be fired with a minimum of two propelling charge increments and not more than four increments. The height of burst is 400 m and it will illuminate a 1100 m area for a minimum of 60 seconds. It is not a Standard A round.

DATA
Weight: 4.89 kg
Minimum range: 100 m
Max range: 2150 m

M301A1
Illuminating round M301A1 is similar to the M301A2 except that it has gas check bourrelet grooves and some minor dimensional differences in metal parts. It is not a Standard A round.

Target practice ammunition (TP)
M43A1
Target practice round M43A1 is intended for use in training only and is similar to the M43A1 HE except for the projectile filler and its colour. The target practice projectile is loaded with an inert material (plaster of Paris and stearic acid) and a 0.05 pound pack powder pellet. On impact, the black powder pellet and the fuze booster charge provide a spotting charge for observation purposes. The projectile is loaded to simulate the weight of the high explosive projectile and has the same ballistic characteristics.

Fuzes
Point detonating (PD) fuzes
M524 series. Dual purpose fuze M524A5, superquick or 0.05 second delay, is used with M362 series HE cartridges and M374 or with WP cartridges M370 and M375.

M525, M525A1. These fuzes are modifications of M52 series fuzes. The modification consists of the substitution of a head assembly containing a delayed arming device in addition to the firing pin mechanism. Those fuzes are used for M43 HE and TP series ammunition.

M526 series. These fuzes, which are replacing PD fuze M519, consist of the former M52 series fuzes modified, as in the M525 series, with an arming delay and, in addition, fitted with an adaptor containing booster pellets to adapt to a newer design ammunition. This fuze may be used instead of PD fuze M524A1 in cartridges M362 and M374 HE and M370 and M375 WP.

PD fuze M519, which is a combination of PD fuze M525A1 and a fuze adaptor, is a single-action type with a direct action firing device for use with cartridges M362 HE and M370 WP.

Proximity fuzes
M532
The M532 is a radio-Doppler fuze which is standard for the M374 HE round and may also be used on the M362 HE or the M375 WP round. It provides an airburst at or near a height for optimum effectiveness by employment of the radio-Doppler principle of target detection. A clock mechanism provides a nominal nine seconds of safe air travel (610 to 2340 m travel along trajectory

for charges 0 to 9, respectively). It can also be set to superquick (point detonating) to eliminate the proximity function.

The proximity fuze can be converted to a point detonating fuze action by rotating the top of the fuze more than 120° in either direction.

M517
This proximity fuze is provided for use with the M362 high-explosive round. It will not function in the M374 or M375 rounds because of the spin. The M517 fuze's operating principles are similar to the M532, differing primarily in the arming system. The minimum time (after firing) to arm for an impact function is in excess of 1.5 seconds. The minimum time (after firing) to arm for a proximity function is 3.2 seconds. This fuze does not provide a PDSQ option.

Mechanical time fuze
M84
This fuze is a single purpose, powder train, selective time type used with the 81 mm illuminating cartridges M301A1 and M301A2. It has a time setting of up to 25 seconds.

M84A1
This fuze is a single-purpose, tungsten ring, selective time type used with the 81 mm illumination cartridge M301A3. It has a time setting of up to 50 seconds.

Ammunition for the 107mm Mortar M30

The 107 mm mortar M30 is a rifled weapon and the bombs are fin-stabilised. The projectiles are fitted with a malleable copper plate, slightly dished, at the base, with a flat steel plate behind it. Both plates cover the base of the bomb and surround the cartridge container. The copper plate is of bomb diameter when loaded, and therefore the bomb slides down the barrel quite easily. On firing, the pressure of the propellant gases drives the steel plate forward, so flattening the copper plate and forcing its circumference to protrude beyond the body of the bomb so that it is forced to engrave in the rifling of the barrel. The spin so given is transferred to the bomb during its passage up the bore of the mortar.

Ammunition for the 107 mm M30 mortar is issued in the form of complete rounds. The propelling charge consists of an ignition cartridge carried inside the cartridge container and 41 propellant increments assembled in a bag and

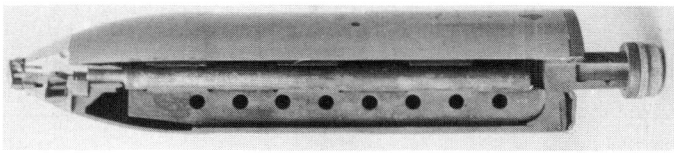

107 mm M328A1 white phosphorus smoke bomb; the construction, apart from the internal baffles, is standard for the 107 mm range of bombs.

sheets, which are fitted around the body of the cartridge container. To adjust the charge, individual increments are removed. The table below contains information on the standard rounds authorised for the 107 mm mortar M30.

Type	Range minimum	max	Weight	Effective area	Filler and designation	Notes
HE M329A1	920 m	5650 m	12.26 kg	40 × 20 m	TNT	—
HE M329A2	770 m	6800 m	10 kg	—	—	—
HE M34A1	870 m	4620 m	12.21 kg	40 × 20 m	TNT	—
Smoke M328A1	920 m	5650 m	12.98 kg	n/a	WP	—
Smoke M2	870 m	4620 m	11.32 kg	n/a	WP	burning time 90 s
Illumination M335A2	400 m	5490 m	12.09 kg	1500 m dia	—	rate of descent 5 m/s
Gas M2A1	870 m	4540 m	11.17 kg	n/a	HD	—
Tactical CS XM630	1540 m	5650 m	11.64 kg	n/a	CS	—

Range: (minimum) (M329A1) 920 m; (M329A2) 770 m; (max) (M329A1) 5650 m; (M329A2) 6800 m

Boeing Fiber Optic Mortar Projectile (FOMP)

This is a 120 mm mortar projectile which is fired from a normal mortar in the usual manner. Shortly after launch wings and fins are deployed, and as the bomb approaches the vertex of the trajectory, a rocket sustainer motor is ignited. During the flight, the bomb dispenses a fibre-optic link which is connected to a control station at the mortar position. The bomb is fitted with a television camera, the view from which is relayed down the fibre-optic link to be displayed at the control station. The operator, with the bomb's picture before him, can then select his target and steer the bomb to impact.

The project has reached the stage where feasibility of the various components has been demonstrated and actual fibre-optic links have survived live firing.

Manufacturer
Boeing Military Airplanes, PO Box 1470, MS JD-30, Huntsville, AL 35807.
Status
Development.

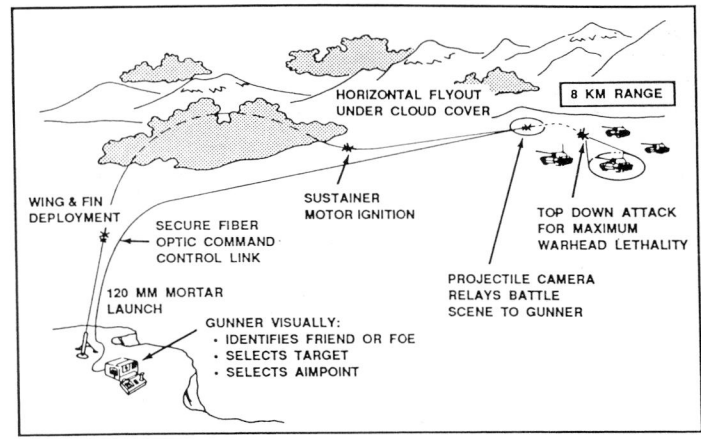

Operational principle of FOMP

VIETNAM

82 mm shrapnel bomb Model 81

This is an extremely unusual bomb of local design and manufacture and is a very potent weapon against unprotected personnel. The bomb contains a charge of high explosive covered by a dished plate, above which are 240 irregularly shaped steel fragments. The HE is initiated by a base time fuze, which appears to be armed by the mortar discharge and which then probably burns for a pre-set time, since there does not appear to be any means of setting the fuze. On detonation of the HE the steel fragments are blown out forward, giving an airburst effect.

DATA
Calibre: 82 mm
Type: combined HE and shrapnel
Weight: 3.23 kg
Bursting charge: 600 g TNT
Fuze: base, time, Type II
Using weapons: Chinese Types 20 and 53, Soviet M1937

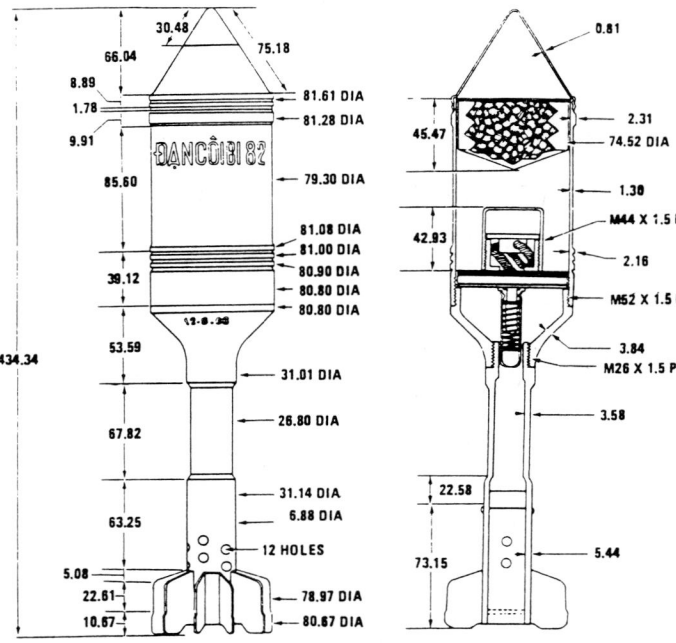

Vietnamese 82 mm shrapnel bomb Model 81

YUGOSLAVIA

Ammunition for 50 mm mortars

The round currently manufactured for 50 mm mortars is the **HE Bomb M82,** a modern design of streamlined shape. The bursting charge consists of TNT and the bomb is fitted with the direct action impact fuze TK-135. The propelling charge is a single 5 g primary cartridge, which is sufficient to send the 950 g bomb to a range of 1000 m.

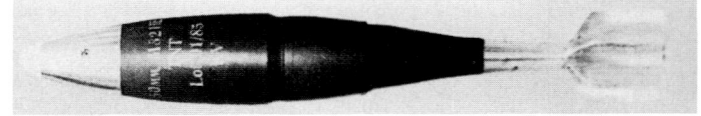

50 mm HE mortar bomb M82

Ammunition for 60 mm mortars

60 mm HE Bomb M73

This is a conventional 'tear-drop' bomb with a tail unit around which the supplementary charges are fitted in horseshoe form celluloid containers. The TNT-filled bomb is fitted with the direct action impact fuze UT M68P1. A primary cartridge fits into the tail tube to form Charge 0, and four supplementary charges can be fitted, except when using 'Commando' type light mortars, when only two supplementaries are permitted. Each supplementary charge contains about 4.2 g of nitro-cellulose powder.

DATA
Total weight: 1.35 kg
Weight of TNT filling: 220 g
Muzzle safety distance: 8 m
Performance:

Charge	Muzzle Velocity m/s	Minimum Range m	Maximum Range m
0	74	94	525
1	111	184	1072
2	143	283	1632
3	170	365	2136
4	193	440	2537

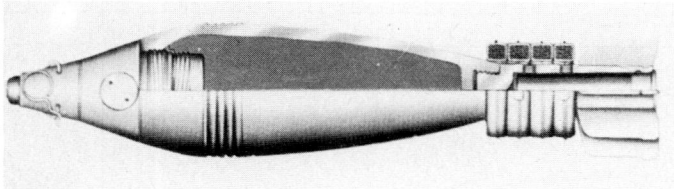

60 mm HE mortar bomb M73

60 mm WP smoke bomb M73

60 mm Smoke Bomb M73
The M73 smoke bomb is of the white phosphorus bursting type, using a central exploder container and fited with the direct action impact fuze UT M70P1. It is ballistically matched to the HE bomb M73 and uses the same propelling charges to obtain the same ranges.

DATA
Weight, total: 1.35 kg
Weight of WP: 190 g
Muzzle safety distance: 8 m

60 mm Illuminating Bomb M67
The M67 illuminating bomb contains the usual type of magnesium flare unit suspended by a parachute. The bomb is fuzed with the pyrotechnic time fuze M67 which, at the end of the set time, ignites an expelling charge and ejects the parachute and flare unit from the nose of the bomb. The M67 bomb cannot be fired with the Charge 0 (primary cartridge only) and is restricted to Charges 1 and 2 when fired from light 'Commando' type mortars.

DATA
Weight, total: 1.27 kg
Illumination intensity: 180 000 cd
Illuminated area: 500 m diameter

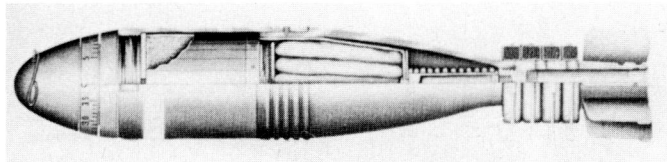

60 mm Illuminating bomb M67

Illumination duration: 35 s
Mean height of parachute opening: 180 m
Rate of descent: 2.5 m/s
Performance:

Charge	Muzzle Velocity m/s	Minimum Range m	Maximum Range m
1	117	200	950
2	154	300	1600
3	185		2100
4	210		2450

Ammunition for 81 mm mortars

81 mm HE bomb M72

This bomb might well be called a 'transitional' design, since it tends towards the older tear-drop shape but with a tapered ogive, giving it a more symmetrical look, and with a single gas-check groove around the waist. It is fitted with the direct action impact fuze UT M68P1, and the propelling charge consists of a primary cartridge and six combinations of 8 and 13 g secondary increments which clip around the tail boom.

DATA
Weight, total: 3.05 kg
Weight of TNT bursting charge: 680 g
Muzzle safety distance: 8 m
Performance:

Charge	Muzzle Velocity m/s	Minimum Range m	Maximum Range m
0	72	88	513
1	120	230	1302
2	167	380	2135
3	197	505	2932
4	229	650	3685
5	257	737	4307
6	283	858	4876

These values apply when fired from a barrel of 1150— 1200 mm length; when fired from a 1450 mm barrel the maximum range with Charge 6 is 5070 m.

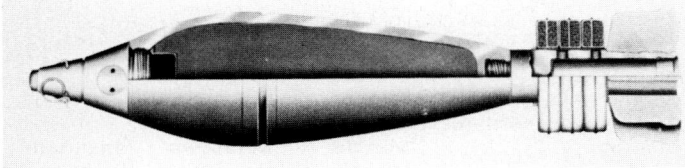

81 mm HE bomb M72

81 mm HE Bomb M72P1

This is similar to the M72 bomb described above but is fitted as standard with the direct action and delay fuze UTU M67. Data and performance are exactly the same as those for the M72.

81 mm HE bomb M71

The M71 bomb is of modern design, being symmetrically streamlined and with a plastic obturating ring fitted into a gas-check groove at the waist. The bomb is fitted with the direct action impact fuze UT M68P1, and the propelling charge consists of a primary cartridge (Charge 0) and eight combinations of 8 or 14 g supplementary increments which fit around the tail boom to give Charges One to Eight.

DATA
Weight, total: 4.10 kg
Weight of TNT bursting charge: 690 g
Muzzle safety distance: 8 m
Performance:

Charge	Muzzle Velocity m/s	Minimum Range m	Maximum Range m
0	73	90	520
1	109	190	1100
2	143	350	1790
3	174	700	2490
4	204	1200	3210
5	230	1500	3875
6	256	1900	4500
7	277	2300	5000
8	297	2600	5426

These values apply when fired from a barrel of 1150 – 1200 mm length.

81 mm HE bomb M71

81 mm HE bomb M84

This is to the same general design as the M71 bomb described above, but is lighter and therefore reaches to a greater range. It is fitted with the direct action impact fuze UT M84P1 and the propelling charge consists of a primary cartridge (Charge 0) and six combinations of 8 and 25 g secondary increments giving Charges One to Six. The increased bursting charge is of RDX/TNT (Composition B), giving the bomb a better lethality than previous designs.

DATA
Weight, total: 3.80 kg
Weight of RDX/TNT bursting charge: 890 g
Muzzle safety distance: 8 m
Performance:

Charge	Muzzle Velocity m/s	Maximum Range m
1	122	2320
2	169	3210
3	212	4030
4	255	4800
5	292	5500
6	318	6050

These figures apply when fired from a barrel of 1150 – 1200 mm length.

81 mm HE bomb M84

81 mm WP smoke bomb M72

The M72 smoke bomb is ballistically matched to the M72 HE bomb and delivers the same range performance with the same charges. The bomb is charged with white phosphorus and has a central HE burster, initiated by a direct action impact fuze UT M70P1. Except for the weight of the smoke charge (600 g), the data and performance figures are exactly as for the HE M72 bomb.

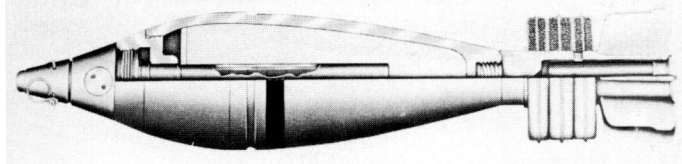

81 mm WP smoke bomb M72

81 mm WP smoke bomb M71

This is a ballistic match for the HE bomb M71 described above, data figures and performance being exactly the same except for the weight of white phosphorus of 630 g. The bomb is fuzed with the direct action impact fuze UT M70P1 which detonates a central high explosive burster to open the bomb and disperse the WP filling.

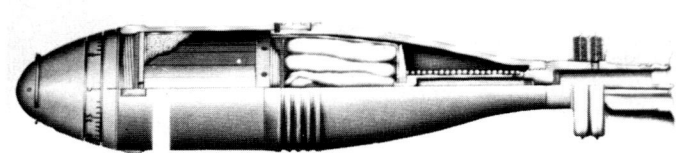

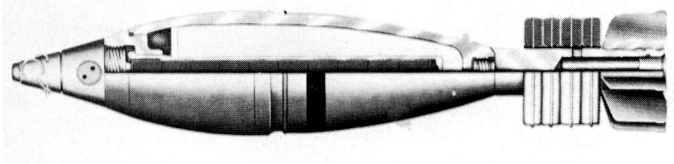

81 mm WP smoke bomb M71

81 mm illuminating bomb M67

The M67 illuminating bomb is of conventional nose ejection pattern, being fitted with the pyrotechnic time fuze M67. This ignites an expelling charge after the set time, removing the nose of the bomb and causing the parachute and flare unit to be ejected. The propelling charge consists of a primary cartridge and four combinations of 7.6 and 13.8 g secondary increments to give Charges One to Four. The bomb cannot be fired with the primary cartridge alone.

81 mm Illuminating bomb M67

DATA
Weight, total: 2.95 kg
Illumination intensity: 500 000 cd
Illuminated area: 800 m diameter
Illumination duration: 40 s
Mean height of parachute opening: 270 m
Rate of descent: 2.4 m/s
Performance:

Charge	Muzzle Velocity m/s	Minimum Range m	Maximum Range m
2	175	600	2050
3	212	1400	2770
4	244	210	3380

These values apply when fired from a 1150 – 1200 mm barrel; when fired from a 1450 mm barrel the maximum range with Charge Four is 3645 m

Ammunition for 82 mm mortars

The range of bombs for use in 82 mm mortars is much the same as those described above for the 81 mm mortar, but there are some small differences in nomenclature. Thus, the 82 mm HE bomb M74 is the same as the 81 mm HE M72, but the maximum range is 4943 m, there being small differences in the ranges achieved with each charge. The 82 mm HE bombs M72P1, M71 and

M84 are exactly the same of the similarly-numbered 81 mm bombs and have the same data and performance. The 82 mm smoke bomb M74 is similar to the 81 mm smoke bomb M72 but has a maximum range of 4943 m. The 82 mm smoke bomb M71 and illuminating bomb M67 are the same as the corresponding 81 mm bomb and have the same data and performance.

Ammunition for 120 mm mortars

120 mm HE bomb, light, M62P1

This is a conventional tear-drop bomb with multiple gas-check rings at the waist, and is filled with TNT. It is fitted as standard with a direct action and delay impact fuze, and the propelling charge consists of a primary cartridge (not fired alone) and six combinations of 35 and 75 g secondary increments providing Charges One to Six.

DATA
Weight, total: 12.60 kg
Weight of TNT bursting charge: 2.25 kg
Muzzle safety distance: 10 m
Performance:

Charge	Muzzle Velocity m/s	Minimum Range m	Maximum Range m
1	121	400	1400
2	162	800	2360
3	200	1500	3370
4	236	2000	4400
5	267	2500	5280
6	297	3000	6050

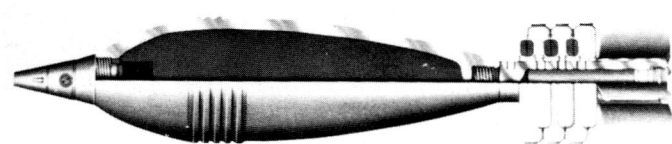

120 mm HE bomb M62P1

120 mm rocket-assisted HE bomb M77

This is a cylindrical bomb with an ogival head and a long tail boom which has a set of four folding fins ahead of the charge container. The bomb body is divided into two sections, the foward section containing the HE bursting charge and the rear section containing the rocket motor. A pre-firing adjustment permits the rocket motor to be switched in or out of action as required. The propelling charge consists of the usual primary cartridge (not fired alone) and six 71 g secondary increments giving Charges One to Six.

The rocket motor is ignited by means of a pyrotechnic delay train from the primary cartridge, and the initial ignition of the rocket motor blows away the cartridge container. This permits the rocket efflux to escape; the forward section of the tail boom remains in place to support the fins, which unfold into the slipstream as the bomb leaves the muzzle. These fins are skewed, so as to develop a degree of spin stabilisation during flight.

DATA
Weight, total: 13.65 kg
Weight of bursting charge: 2.91 kg
Performance, without rocket assistance:

Charge	Muzzle Velocity m/s	Minimum Range m	Maximum Range m
1	133	300	1500
2	180	800	2500
3	217	1600	3400
4	251	2600	4100
5	280	3500	4700
6	307	4200	5300
with rocket assistance:			
3	217	5100	7900
4	251	5100	8500
5	251	5100	8900
6	307	5100	9400

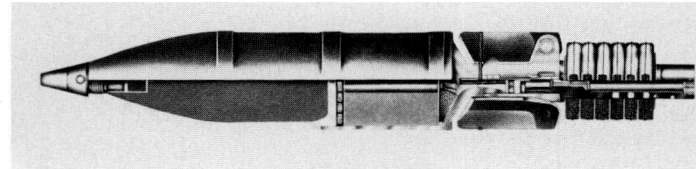

120 mm rocket-assisted HE bomb M77

120 mm smoke bomb M64P1

This is a conventional white phosphorus bursting smoke bomb, fitted with the direct action impact fuze UT M70P1 and a central HE burster. The propelling charge consists of a primary cartridge and up to six 76 g secondary incremental charges clipped around the tail boom, giving Charges One to Six.

120 mm WP smoke bomb M64P1

DATA
Weight, total: 12.40 kg
Weight of WP charge: 2.45 kg
Muzzle safety distance: 8 m
Performance:

Charge	Muzzle Velocity m/s	Minimum Range m	Maximum Range m
1	123	255	1410
2	165	435	2375
3	212	625	3400
4	240	810	4400
5	271	970	5250
6	302	1100	6010

These figures apply when fired from the 120 mm mortar M52; when fired from the 120 mm mortar M75 the maximum range is 6464 m.

120 mm smoke bomb M84

This is a nose ejection bomb using a single container of hexachloroethane-based persistent smoke mixture. The bomb is fitted with the pyrotechnic time fuze M84 which, at the set time, ignites an expelling charge which blows off the bomb nose and ejects the ignited smoke canister. The propelling charge consists of a primary cartridge and up to five 76 g secondary increments in silk bags which are tied around the tail boom, giving Charges One to Five.

DATA
Weight, total: 10.35 kg
Weight of smoke mixture: 1.20 kg
Duration of smoke emission: > 3 mins
Performance (in mortar with 1.5 m barrel):

Charge	Muzzle Velocity m/s	Minimum Range m	Maximum Range m
1	137	230	1260
2	185	900	2560
3	229	1300	3850
4	266	1700	4850
5	303	2000	5850

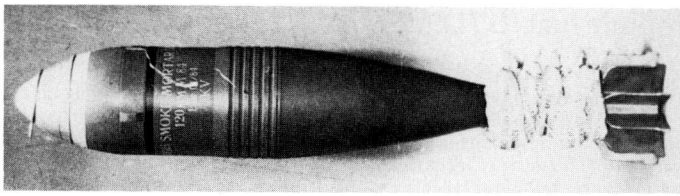

120 mm smoke bomb M84

120 mm illuminating bomb M84

As the type number suggests, this is almost identical to the screening smoke bomb, but contains the usual parachute and flare assembly. The bomb is fitted with the pyrotechnic time fuze M84 which, at the set time, blows off the nose and ejects the burning flare and parachute. The propelling charge consists of a primary and five secondary incremental charges in silk bags.

DATA
Weight, total: 10.35 kg
Illumination intensity: > 900 000 cd
Area of illumination: 1800 m diameter
Rate of descent: 3 m/s
Performance: as for smoke bomb M84 above

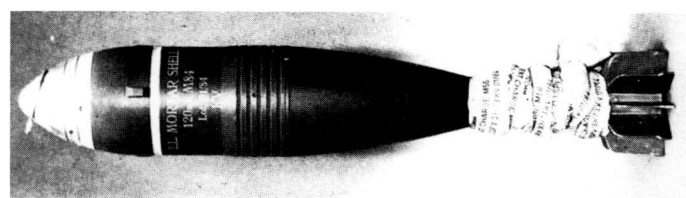

120 mm illuminating bomb M84

Impact Fuze UT M68P1

This is a bore-safe direct action fuze intended for use in unspun mortar bombs from 60 to 82 mm calibre. Muzzle safety is ensured by means of a spring-driven axial rotor which is locked by the firing pin in the safe condition. On firing, setback releases the firing pin, allowing the rotor to turn and slide until it aligns its detonator with the armed firing pin.

DATA
Weight: 175 g
Length: 85 mm
Drop safety height: 3 m

Impact fuze UT M68P1

Impact fuze UTU M67

The M67 fuze is an impact, superquick and delay pattern for use with 81 mm and 82 mm unspun mortar bombs. Bore safety is achieved by a rotor, locked in the safe condition by the firing pin. The fuze may be set before firing for delay or instant action; doing so sets or releases a block on the movement of the sliding rotor. In the instant setting, the firing pin is released by set back upon firing, after which the rotor moves across to align a detonator with the firing pin. On striking the target the pin is driven in, ignites the detonator, and this in turn fires the fuze magazine and the bursting charge of the bomb. At delay setting, the rotor moves so as to bring a second detonator, with pyrotechnic delay, under the firing pin. When the fuze strikes, the action is as before but the pyrotechnic delay has to burn through before the fuze magazine can be initiated.

DATA
Weight: 285 g
Length: 91 mm
Drop safety height: 3 m

Impact SQ and delay fuze UTU M67

Impact fuze UTU M78 (AU-29)

This fuze is for use in the 120 mm rocket-assisted bomb, where spin is available to assist in functioning. It uses a similar method of obtaining delay to the US M48 series, and has a Semple rotor locked by a centrifugal detent to provide bore-safety. The fuze can be set, before loading, for delay or non-delay; this closes or opens a central channel between an impact detonator and the rotor detonator. If this channel is open, the flash from the impact detonator, struck by the impact of the firing pin, passes through to fire the rotor detonator and the fuze magazine. If the channel is closed, then the flash from the impact detonator is channelled via a delay unit to a delay detonator which then fires the rotor detonator.

DATA
Weight: 433 g
Length: 105 mm

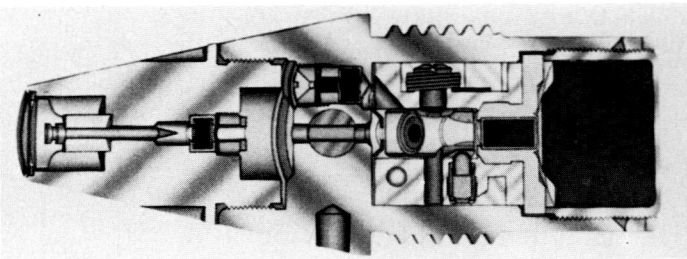

Impact, SQ and delay, fuze UTU M78

Impact fuze UT M70P1

This is precisely the same fuze as the M68P1 described above except that the magazine and lower portion of the fuze body are of differing dimensions so as to suit the internal arrangements of WP smoke bombs. It is for use with 60, 81 and 82 mm mortars. The fuze weighs 160 g and is 78 mm long.

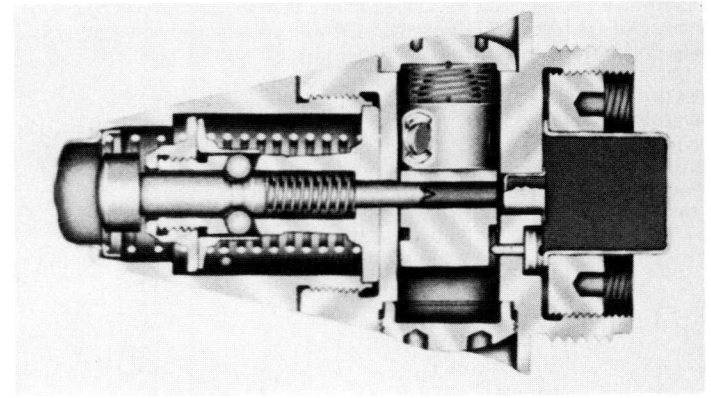

Impact fuze M70P1 for WP smoke bombs

Fuze, Time, TP M66

This is identical in form to the M67 fuze described below but is larger and offers timing from 5 to 50 seconds, being intended for use in 120 mm illuminating bombs. The fuze weighs 1.265 kg and is 108 mm long.

Fuze, time, TP M67

This is a single-banked, tensioned, igniferous, time fuze for use in nose ejection bombs in 81 mm and 82 mm mortars. The single time ring offers timing from 5 to 38 seconds and is filled with a zirconium-based powder. A central hammer, suspended on a shear-wire, drops on setback to strike a detonator which lights the time composition. This burns round until it fires a flash channel leading to the fuze magazine, which then explodes and initiates the bomb's functioning.

DATA
Weight: 580 g
Length: 85 mm

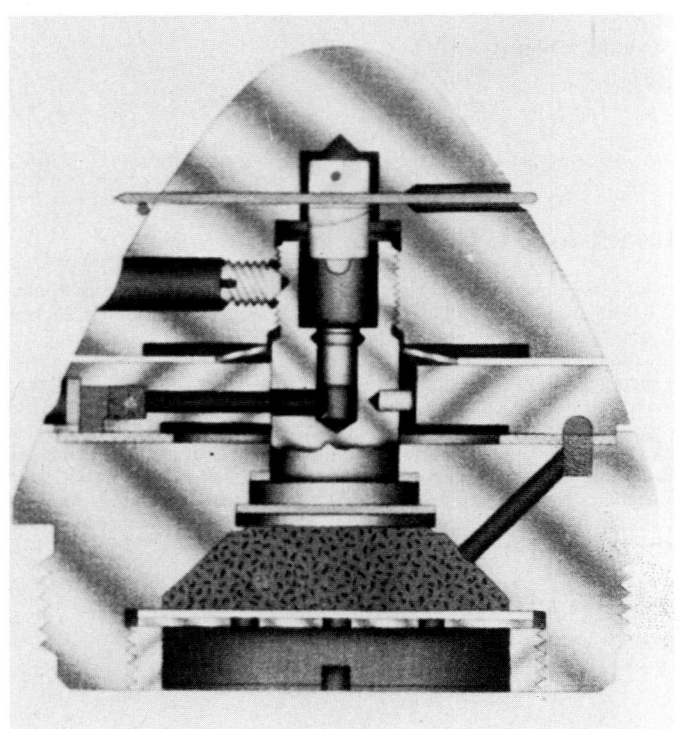

Time fuze TP M67

Pyrotechnics

AUSTRIA

Defence smoke cartridge

This is a screening smoke cartridge which can be fired from ground or vehicle-mounted electrically fired projectors. It is designed so that the smoke rises, concealing the site from aerial or superior observation while leaving the fields-of-view and fire clear close to the ground. The cartridge is a tinned-plate canister with propulsion charge and a smoke loading of about 1 kg of composition. Electrical ignition fires the propulsion charge to eject the canister to a range of about 25 m from the launcher and also ignites a pyrotechnic delay which in turn ignites the smoke composition. Emission of smoke begins during flight and continues for about 40 seconds.

DATA
Diameter: 80 mm
Length: 180 mm
Weight: 2.30 kg
Range: 25 ± 5 m
Duration: 40 ± 10 s

Manufacturer
Dynamit Nobel Wien GmbH, Postfach 74, Opernring 3-5, A-1015 Wien.
Status
Production.

Defence smoke cartridge

Smoke Cartridge HC-72

This is a vehicle smoke screening device intended to be electrically fired from the Discharger Equipment 69. It consists of a metal canister containing the hexachloroethane/zinc smoke composition in the front end and a propellant charge and electrical igniter in the rear. After loading, a protection cover closes the discharger so that the cartridge is protected against the influence of weather. To ensure safety there are two parallel firing circuits and both are safe against stray currents up to 360 mA. On firing, the grenade is thrown to a distance of about 50 m and the smoke composition is ignited. In about five to seven seconds after firing the smoke has reached full development.

DATA
Length: 322 mm
Diameter: 80 mm
Weight: 5.0 kg
Duration of smoke: approx 70 s
Colour of smoke: grey-white

Manufacturer
Dynamit-Nobel Wien GmbH, Opernring 3-5, A-1010 Wien.
Status
Current. Production.

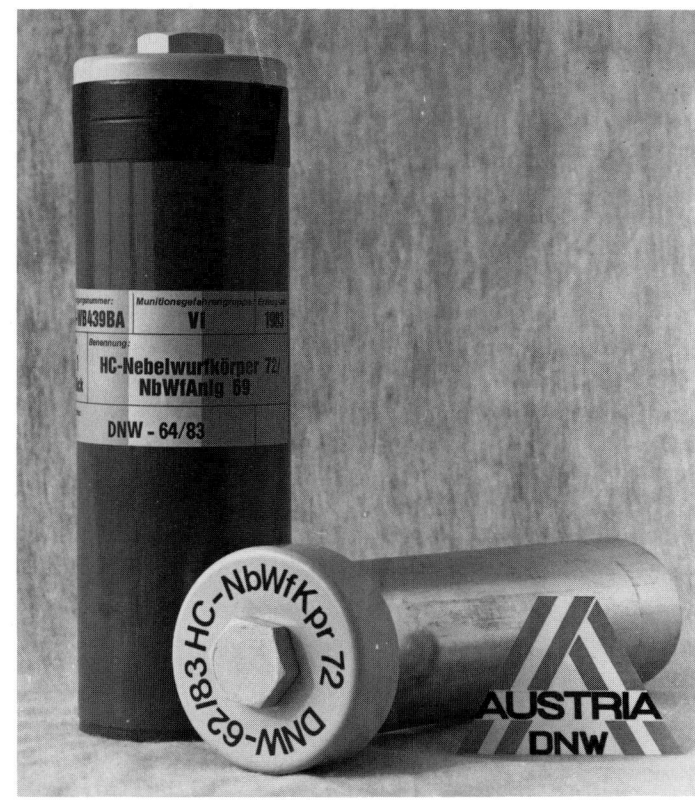

Smoke cartridge HC-72

Smoke generator HC-81

This generator is designed for the production of screening smoke for periods which can be as long as required. This variable time is achieved by making the generators with mating surfaces so that two or more may be screwed together to make up whatever size is desired. Each individual generator contains sufficient HC composition to burn for about 5 mins and where another generator has been connected, transfer of the burning is automatically carried out by a connector which is permanently installed in the base of the generator. It is even possible to add further generators to a group already ignited, provided the last generator of the group has not yet begun to burn.

Ignition is by friction or electric igniter, two of each being provided with each generator.

DATA
Weight: 4.5 kg
Diameter: 182 mm
Height: 140 mm
Igniter delay: 4.5 s nominal
Duration of smoke: 5 mins

Smoke generators HC-81

Manufacturer
Dynamit Nobel Wien GmbH, Postfach 74, Opernring 3-5, A-1015 Wien.
Status
Production.
Service
Austrian Army.

Smoke generator HC-75

This is a large generator developing screening smoke for periods of up to 22 mins. It is loaded with HC mixture and can be ignited by either a friction or an electric igniter, one of each being supplied with each generator. Emission of smoke is intense, and this equipment should not be used in confined spaces; there is also a certain amount of spark emission which should be taken into account when sitting in dry grass areas in training.

DATA
Weight: 10.5 kg
Duration: 18-22 mins
Ignition delay: (friction igniter) 9 s nominal; (electric igniter) 4.5 s nominal

Manufacturer
Dynamit Nobel Wien GmbH, Postfach 74, Opernring 3-5, A-1015 Wien.
Status
Production.
Service
Austrian Army.

Smoke generator HC-75

Illuminating mine LK-40

This is a trip-operated flare unit. The aluminium body is closed by a cap threaded to accept the Universal Fuze DDZ78 which functions in the usual way, by pull, pressure, or the cutting of the tripwires. Once the fuze fires, the ignition of the flare composition blows off the top cap so that the burning mass is exposed, illuminating the surrounding area.

The mine is packed with a mounting support, Fuze DDZ78, two rolls of tripwire, six pickets and two nails. A hammer is provided with every 10 mines.

DATA
Height: (without fuze) 120 mm
Diameter: 55 mm
Weight: 450 g
Weight of flare composition: 350 g
Duration of flare: 40 s nominal
Luminous power: 280 000 cd
Colour of flare: yellow or white

Manufacturer
Dynamit Nobel Wien GmbH, Postfach 74, Opernring 3-5, A-1015 Wien.
Status
Production.
Service
Austrian Army.

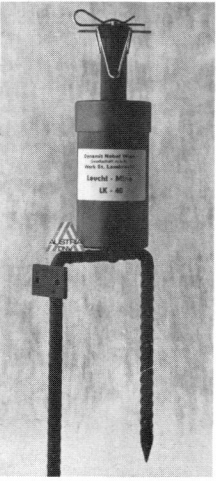

Illuminating mine LK-40

BELGIUM

PRB NR 403 trip flare

This device is intended to give visual warning of enemy approach in poor visibility conditions and provide illumination of sufficient intensity and duration to permit defensive action. It produces yellowish light with a minimum intensity of 60 000 candela, sufficient to illuminate an area with a radius of about 300 m, and burns for a minimum of 60 seconds. It responds to a pull on any of up to four tripwires which can be attached to a rifle flare.

Operationally, the flare is mounted vertically on a picket driven into firm ground. It consists of a cylindrical body, containing the burning compound, to the top of which is secured an igniter mechanism protected by a plastic cap, the latter also providing external storage for the tripwires. The operative part of the igniter mechanism is a striker, surrounded by a spring and coupled to a slide by a ball and groove arrangement. Four 'antennae', attached to the slide,

pass through a hole in the top of the igniter and are terminated by pull-rings. Traction applied to any one of these rings draws the slide and igniter upwards, compressing the spring, until the ball and groove coupling reaches a recess in the igniter head at which point the spring pressure forces the balls outwards into the recess, thus permitting the striker to move downwards and strike a percussion cap in the base of the igniter. This ignites the flare compound; and the gas pressure produced by the combustion blows the igniter head clear of the main body leaving the flare compound to burn in the open cylinder.

In the rest position the striker is held clear of the percussion cap by the slide and the slide is locked in place by a safety clip. The pull-rings are hooked round projections from the igniter head.

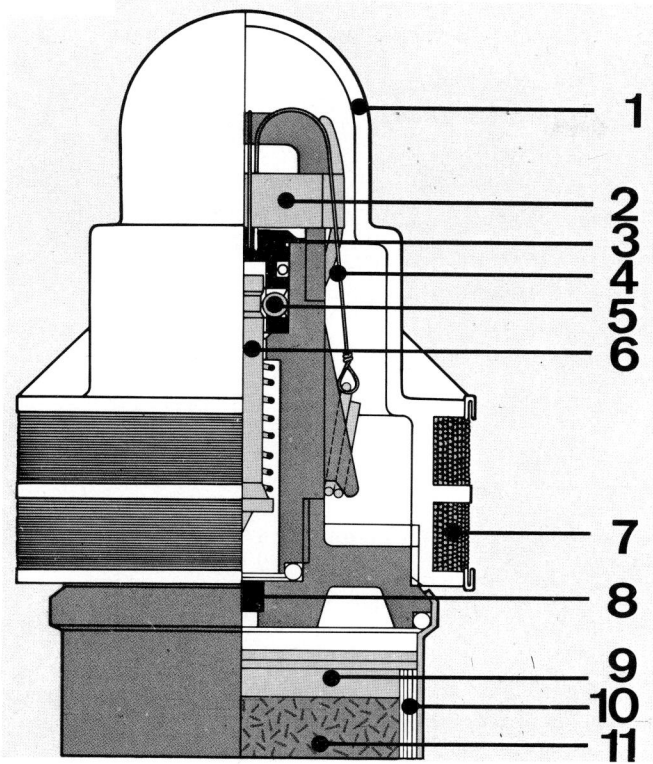

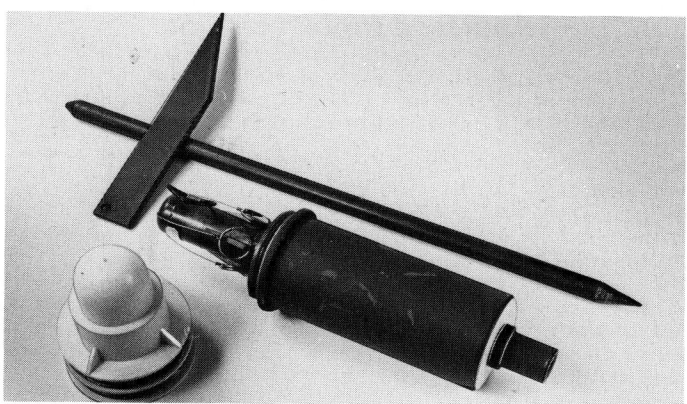

PRB NR 403 picket, flare assembly (with safety clips in place) and protective cap with tripwires wound in its grooves

Igniter mechanism of PRB NR 403
(1) *plastic cover* **(2)** *safety clip* **(3)** *slider* **(4)** *flexible antenna with pull-ring* **(5)** *steel ball* **(6)** *striker and spring* **(7)** *trip wire* **(8)** *cap* **(9)** *first-fire composition* **(10)** *aluminium case* **(11)** *illuminant charge*

Repeat this operation for other pull-rings as required. Finally remove (but retain) the safety clip. To dismantle, proceed in reverse order.

DATA
Weight of flare with cover: 39 kg
Weight of picket: 0.23 kg
Weight of complete assembly: 0.62 kg
Weight of 24-flare pack: 24.5 kg
Height of flare with cover: 225 mm
Diameter: 57 mm
Height of picket: 300 mm
Operating traction: 2–5 kg
Illuminating power: approx 60 000 cd
Burning time: 60 s minimum

Manufacturer
Société Anonyme (PRB), Département Défense, Avenue de Tervueren 168, Bte 7, 1150 Brussels.
Status
Current. Production.
Service
Belgian Army.

Installation sequence
First, drive the picket into the ground and mount the flare assembly on it. Remove the protective cap and attach a tripwire to one of the pull-rings, fastening the other end to some rigid object and drawing the wire tight.

PRB NR169 trip-flare

This trip-flare serves the same operational purpose as the PRB NR 403 (see entry above) but has the additional function of responding to an attempt to neutralise it by cutting the tripwire. This is achieved by mounting an extension arm on the picket carrying the flare and attaching one end of the tripwire to a spring-loaded slider on the arm, the open end of the wire being attached to a second picket. Thus tensioned the wire is attached to the flare igniter mechanism, so that traction on the wire between the pickets will operate the

igniter in the normal way; if the wire is cut however, the spring-loaded slider will apply traction in the opposite direction and operate the igniter. The igniter mechanism of this flare operates from the base of the flare assembly.

DATA
Flare assembly: (weight) approx 1.5 kg; (casing diameter) 75 mm; (height) 135 mm
Tripwire: 19 m
Illuminating power: approx 40 000 cd
Illumination uration: approx 60 s

Manufacturer
Société Anonyme (PRB), Département Défense, Avenue de Tervueren 168, Bte 7, 1150 Brussels.
Status
Current. Production.
Service
Belgian and other armies.

PRB NR169 trip-flare installed

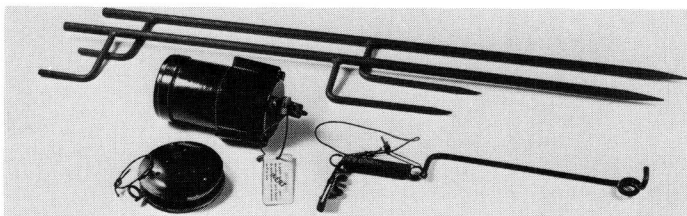

Components of PRB NR169 trip-flare installations

NR170 PRB surface aircraft beacon

This is a powerful long-burning flare intended for the illumination of aircraft landing areas and similar purposes. It has a spike at its base and can be ignited either directly using a special match or remotely by a traction fuze.

DATA
Weight: 0.5 kg
Length: approx 500 mm
Diameter: 42.5 mm
Illuminating power: 100 000 cd minimum
Burning time: 3.5–5 mins

Manufacturer
Société Anonyme (PRB), Département Défense, Avenue de Tervueren 168, Bte 7, 1150 Brussels.

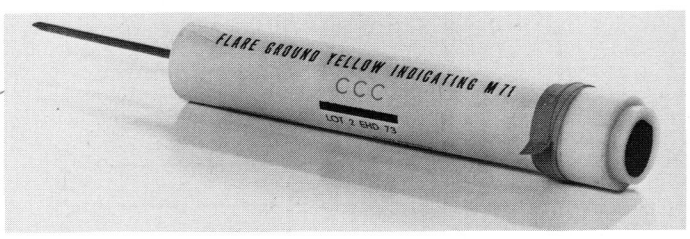

NR170 PRB surface aircraft beacon

PRB signal and illuminating cartridges

PRB make signal cartridges for use with the British 1 in signal (Very) pistol and the 1.5 in signal pistols, the smaller cartridges being suitable also for use with the West German calibre 4 (26.5 mm) pistol. They also make a special cartridge, for use with a service rifle, which can be launched using standard ball, blank or ballistite rounds.

1 in signal cartridge for British Very pistol

DATA

APPROX	1 in pistol	1.5 in pistol	1.5 in rifle
Weight	50 g	110/170 g	90 g
Diameter	26.5 mm	38 mm	38 mm
Length	69/80 mm	98 mm	60 mm
Range of star	100 m	80 m	80 m
Burning time	6–8 s	6–8 s	5.5–7.5 s

Colours available: white (illuminating), red, green, yellow or orange. Double stars on 1.5 in cartridges can be any combination of these colours. Night identification is by marks on the sealing cap.

Manufacturer
Société Anonyme (PRB), Département Défense, Avenue de Tervueren 168, Bte 7, 1150 Brussels.
Status
Current. Production.
Service
Belgian and other armies.

PRB coloured smoke grenades

PRB make a range of smoke signalling grenades which can be hand-thrown, launched from helicopters or ground vehicles or used by parachutists. Each type is available in versions producing any of six different smoke colours and the three types differ only in the method of ignition. The hand-thrown type is fitted with a lever-release mechanism which has an igniter instead of a detonator; the helicopter-launched type is fitted with an instantaneous electrical igniter and can be launched from any vehicle equipped with the appropriate type of discharger; the parachutist grenade has a delay fuze and igniter mechanism similar to that of the hand-thrown grenade but without the lever release. The grenade is attached to the parachutist's boot and the igniter mechanism functions as soon as the safety pin is withdrawn.

DATA
COMMON TO ALL
Weight: 0.3 kg
Diameter: 63 mm
Height: 125 mm
Emission time: 60 s
Colours: blue, green, orange, red, violet and yellow

Manufacturer
Société Anonyme (PRB), Département Défense, Avenue de Tervueren 168, Bte 7, 1150 Brussels.
Status
Current. Production.

PRB coloured smoke grenades
(left to right): *hand-thrown, helicopter-launched and parachutist smoke signalling grenades*

PRB coloured smoke generators

Available in the same range of smoke colours as the PRB coloured smoke grenades and in white, these generators are long-burning signalling devices producing dense coloured smoke for about 5 mins. The smoke generating composition is contained in a cylindrical can with vents in the top and is ignited by a friction igniter with a waterproof friction cap. There is a delay of 2.5 seconds between the operation of the igniter and the emission of smoke.

DATA
Weight: (complete) 445–595 g according to colour
Weight of composition: 290–410 g according to colour
Diameter of canister: 92 mm
Height: 125 mm
Emission duration: approx 5 mins
Ignition delay: 2.5 s
Colours: blue, green, orange, red, violet, white or yellow

Manufacturer
Société Anonyme (PRB), Département Défense, Avenue de Tervueren 168, Bte 7, 1150 Brussels.
Status
Current. Production.

PRB coloured smoke generator

PRB HC white smoke generator

This is a long-burning screening or signalling smoke generator loaded with a burning composition based on hexachlorethane which produces a dense and persistent cloud of white smoke. The composition is contained in a cylindrical canister, vented at the top, and is ignited by a built-in friction igniter through a 3 to 5 second pyrotechnic delay.

DATA
Weight: (complete) approx 1 kg
Weight of composition: approx 0.9 kg
Diameter of canister: 70 mm
Height: 170 mm
Emission duration: 4–6 mins
Ignition delay: 3–5 s

Manufacturer
Société Anonyme (PRB), Département Défense, Avenue de Tervueren 168, Bte 7, 1150 Brussels.
Status
Current. Production.

PRB HC white smoke generator

BRAZIL

SS-601 coloured smoke hand grenade

This grenade consists of an aluminium body with an opening in the cap through which the smoke can escape. A removeable plastic cap, which is of the same colour as the emitted smoke, conceals the friction igniter. To operate, the plastic cap is removed, exposing a loop of nylon cord; pulling this cord activates the friction igniter, after which the grenade is thrown.

DATA
Weight: 350 – 450 g depending upon colour
Length: 146 mm
Diameter: 67 mm
Delay time: 2 – 4 s
Emission time: 90–120 s
Colours available: red, orange, blue, yellow, green, white

Manufacturer
Condor SA – Industria Quimica, Rue México 148, Gr. 806 – 808, Rio de Janiero CEP 20031.
Status
Current. Production.

SS-601 coloured smoke grenade

Condor SS-602 floating smoke generator

This is a long-lasting generator for maritime use and releases a dense cloud of orange smoke for about 4 mins. A plastic lid covers the pull-ring and fuze assembly. Each smoke generator is protected by a plastic bag.

DATA
Length: 163 mm
Diameter: 76 mm
Weight of charge: 560 g
Delay: 6 s
Emission time: 3 mins

Manufacturer
Condor SA – Industria Quimica, Rue México 148, Gr. 806 – 808, Rio de Janiero CEP 20031.
Status
Current. Production.

SS-602 floating smoke generator

SS-603 Hand-held signal star

The SS-603 consists of a hand-held launching tube containing a propelling charge which fires into the air a yellow, green or red bright star; the colour of the star is indicated by the colour of the tube. Firing is simply performed by grasping the lower end, removing the red cap, and pulling the cord which operates a friction igniter. The signal device fires immediately.

DATA
Length: 218 mm
Diameter: 33 mm
Weight: 140 g
Burning time: 6 ± 1 s
Power: 20 000 cd
Altitude: 80 ± 10 m

Manufacturer
Condor SA – Industria Quimica, Rue México 148, Gr. 806 – 808, Rio de Janiero CEP 20031.

SS-603 hand-held signal star

Status
Current. Production.

SS-604 Hand-held signal flare

This is a simple non-ejecting flare for signal or distress use. It consists of a plastic body containing a pyrotechnic composition which is ignited by a friction igniter in the handle. Once ignited, the device produces a high-intensity red flare for about one minute.

DATA
Length: 242 mm
Diameter: 44 mm
Weight: 580 g
Power: 15 000 cd

Manufacturer
Condor SA – Industria Quimica, Rue México 148, Gr. 806 – 808, Rio de Janiero CEP 20031.
Status
Current. Producton.

SS-604 hand-held signal flare

SS-605 day and night signal

The SS-605 combines night and day signal functions in a single device. It consists of two plastic tubes connected together by a red tape, each having individual signal characteristics. Each unit has a friction igniter incorporated. Upon being separated and fired, the day signal emits a dense cloud of orange smoke, while the night signal produces a high-intensity red flare. After separation the unused flare can be retained for later use.

SS-605 day and night signal separated

DATA
Length: 251 mm
Diameter: 44 mm
Weight: 513 g
Burning time: 25 s either unit
Power of night flare: 15 000 cd

Manufacturer
Condor SA – Industria Quimica, Rue México 148, Gr. 806 – 808, Rio de Janiero CEP 20031.

Status
Current. Production.

SS-606 Hand-held parachute flare

The SS-606 is the usual type of hand-launched parachute flare, fired by a friction igniter concealed in the base of the plastic launch tube. The igniter has a delay of about two seconds, sufficient to allow the firer to pull the ignition cord and then grasp the launch tube with both hands before the rocket is fired.

DATA
Length: 324 mm
Diameter: 44 mm
Weight: 490 g
Altitude: 300 m minimum
Burning time: 40 s minimum
Power: 30 000 cd

Manufacturer
Condor SA – Industria Quimica, Rue México 148, Gr. 806 – 808, Rio de Janiero CEP 20031.
Status
Current. Production.

SS-606 hand-held parachute flare

SS-607 Hand-held five-star rocket

The SS-607 consists of an aluminium launching tube containing a rocket carrying five stars of a specified colour, which is indicated by the colour of the front closing cap. The rocket is fired by a friction igniter concealed under the rear cap. The rocket bursts at the vertex of the trajectory to display the five individual stars.

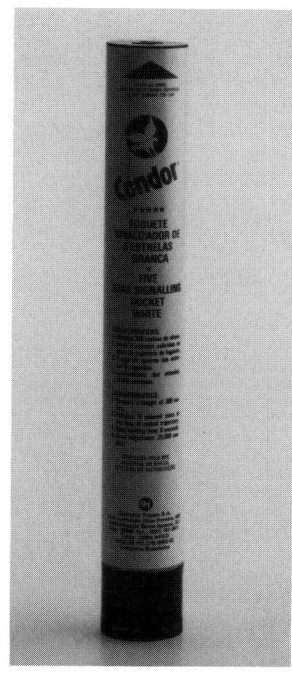

DATA
Length: 324 mm
Diameter: 44 mm
Weight: 490 g
Firing delay: 1.5 to 2 s
Star burning time: 6 s maximum
Power: 25 000 cd
Altitude: 300 m minimum
Star colours: red, green, white

Manufacturer
Condor SA – Industria Quimica, Rue México 148, Gr. 806 – 808, Rio de Janiero CEP 20031.
Status
Current. Production.

SS-607 hand-held five-star signal rocket

SS-608 40 mm signalling cartridges

This is a conventional signal cartridge of the 37/38/40mm class intended for rescue, survival or military signalling. Available in green, white or red colours.

DATA
Length: 96 mm
Diameter: 40 mm
Weight: 134 g
Burning time: 6 s
Power: 15 000 cd
Altitude: 80 m

Manufacturer
Condor SA – Industria Quimica, Rue México 148, Gr. 806 – 808, Rio de Janiero CEP 20031.
Status
Current. Production.

SS-608 40 mm signal cartridge

MB-306/T1 vehicular smoke grenade

This is designed to be electrically fired from vehicle smoke dischargers. Loaded with a zinc/hexachloroethane mixture, it emits a dense smoke cloud so as to screen the vehicle. Models are available with ranges of 10 or 60 m.

DATA
Length: 182 mm
Diameter: 80 mm
Weight: 1.30 kg
Delay time: 1.0 s nominal
Emission time: 2.5 to 4.5 minutes

Manufacturer
Condor SA – Industria Quimica, Rue México 148, Gr. 806 – 808, Rio de Janiero CEP 20031.
Status
Current. Production.

MB-306/T1 vehicular smoke grenade

M5-CEV coloured smoke grenade

The M5 is a general-purpose grenade intended mainly for daylight signalling. The coloured smoke is visible for more than 1 km in reasonable weather and conditions. Ignition is by a percussion cap and spring striker, held down by a safety lever and pin. It is available in white, red, brick, yellow, blue or green colours.

DATA
Length: 157 mm
Diameter: 63 mm
Weight: 600 g
Delay: 4 ± 1.5 s
Emission time: 60 s

Manufacturer
Companhia de Explosivos Valparaiba, Praia do Flamengo, 200-20° andar-22210, Rio de Janeiro, RJ.
Status
Current. Production.

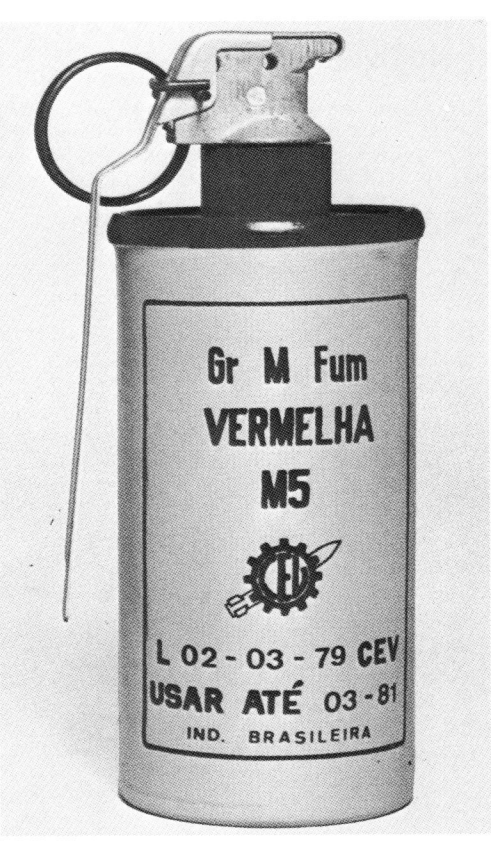

M5-CEV coloured smoke grenade

CHILE

FAMAE Infiltration alarm

This is a trip-flare which when activated delivers an intense white light for a minimum of 20 seconds, followed by a loud detonation. It consists of an illuminating grenade with its base formed into a socket so that it can be fitted to an upright stake. A tripwire is then attached to the pull-ring of the igniter safety pin. When the wire is tripped, the safety pin is pulled out and the fly-off lever ignition mechanism acts to fire the grenade.

Manufacturer
FAMAE Fabricaciones Militares, Av. Pedro Montt 1568/1606, Santiago.
Status
Current. Production.

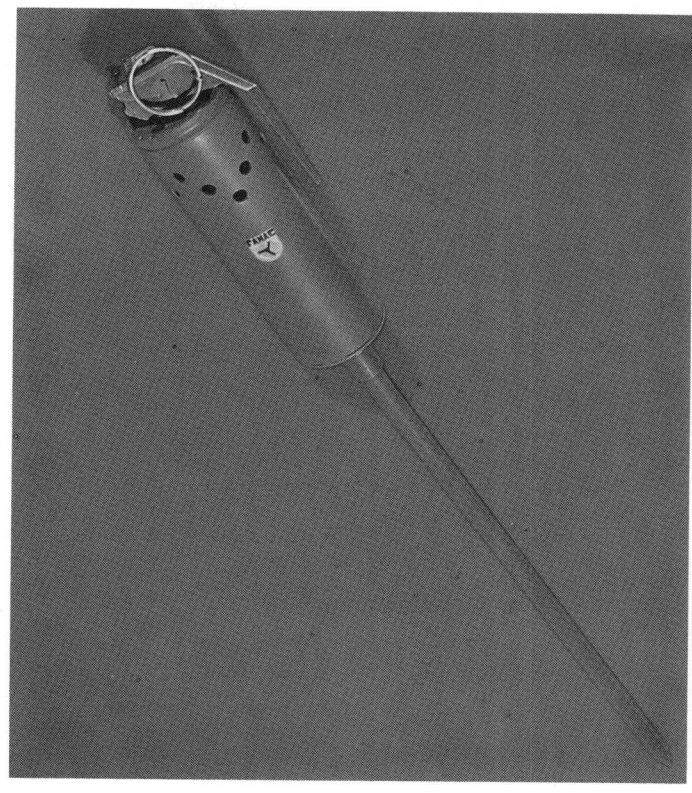

FAMAE infiltration alarm

CHINA, PEOPLE'S REPUBLIC

27 mm Type 57 signal pistol

This is a single-shot, manually operated pistol for firing the conventional 27 mm pyrotechnic cartridges. It is built of eight assemblies and contains only 30 component parts. The design is simple and robust, and the pistol is easy to operate and maintain.

DATA
Calibre: 26.65 mm
Length: 220 mm
Weight: 900 g
Rate of fire: (possible) 12 rds/min

Manufacturer
China North Industries Corp, PO Box 2137, Beijing.
Status
Current. Available.

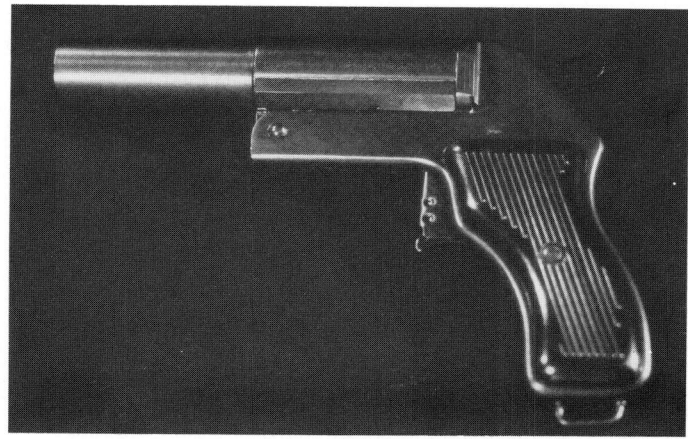

Type 57 signal pistol

27 mm signal cartridges

These signal cartridges are of the conventional 27 mm form, using a brass and plastic case to contain the pyrotechnic loading. They are optimised for use with the Type 57 pistol (above).

DATA
Calibre: 26.65 mm
Weight of complete round: 55.3 g
Length of cartridge: 79 mm
Colours: red, green, yellow, white

Burning time: 6.5 s
Candlepower: approx 50 000 cd
Vertex height: approx 90 m

Manufacturer
China North Industries Corp, PO Box 2137, Beijing.
Status
Current. Available.

CZECHOSLOVAKIA

DA25 coloured smoke generator

This is the smaller of the two similar devices. It consists of a metal canister, the top plate of which has a number of emission holes. This is normally covered by a white plastic cap. On removing the cap a central igniter socket is exposed, and any standard igniter can be inserted; the friction igniter 'ČAROZ' is the most common type, but electrical igniters can be used in cases where several generators are to be simultaneously fired. White, black, red, violet, blue, orange, green and yellow smokes are available, the colour of the smoke being indicated by a coloured stripe on the body of the generator. The DA25 emits smoke for about 25 seconds.

(1) *light alloy canister* **(2)** *white plastic cover* **(3)** *emission holes* **(4)** *sealing cover* **(5)** *smoke composition* **(6)** *priming charge*

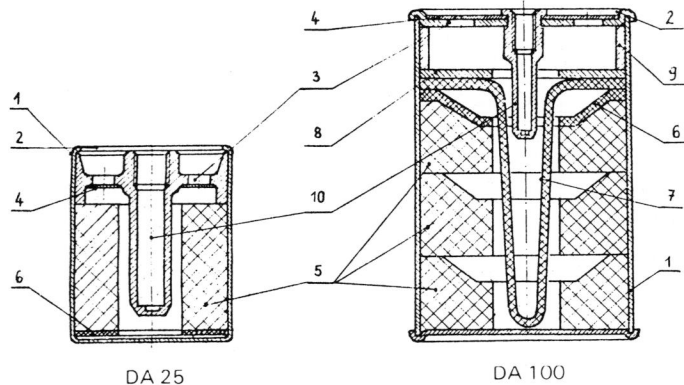

DA 25   DA 100

Czechoslovakian smoke generators DA25 and DA100

DA100 coloured smoke generator

This is similar to the DA25 but larger, emitting smoke for about one minute. The action is the same as the DA25, but the filling is in layered form to improve burning. White, black, red, blue and orange smokes are available.

Czechoslovakian ČAROZ general-purpose friction igniter

FRANCE

Coloured Smoke hand grenade Type 188

The Type 188 grenade consists of a light alloy cylindrical body filled with coloured smoke composition. The grenade is fuzed by a Model F8 or F12 igniter set, both being of the fly-off lever type. The grenade is used for daytime signalling, target indication, and for indicating wind direction to helicopters and light aircraft.

DATA
Diameter: 55 mm
Height, fuzed: 142 mm
Weight: 400 g
Delay time: 2.5 s nominal
Emission time: approx. 2 mins
Colours: red, green, yellow, blue, white, orange, violet

Manufacturer
Ruggieri, Département Armement, 86 ave de Saint-Ouen, 75018 Paris.
Status
Current. Production.
Service
French Army.

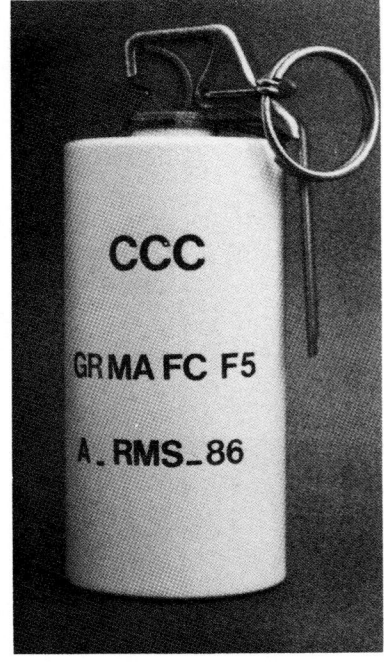

Ruggieri coloured smoke grenade Type 188

Flare Mine with sound effect Type 424

This is a form of trip-flare, actuated by a traction-type ignition device. The mine is buried in the ground up to its igniter and fitted with tripwires. When the wire is pulled, the igniter initiates the mine, causing a detonating charge to be ejected into the air to a height of about 10 m, where it detonates with a very loud report. At the same time the flare pot, in the mine body, is ignited and pushed upwards by a spring so that the burning flare comes partially out of the mine body and burns just above the ground surface.

DATA
Diameter: 60 mm
Height, without igniter: 373 mm
Height, with igniter: 420 mm
Weight, without igniter: 1.65 kg
Weight of igniter: 50 g
Luminosity: 40 000 cd

Manufacturer
Ruggieri, Département Armement, 86 ave de Saint-Ouen, 75018 Paris.
Status
Current. Production.
Service
French Army.

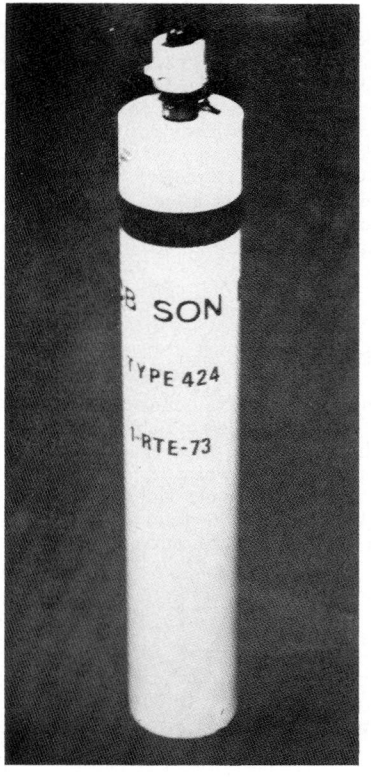

Ruggieri flare mine with sound effect Type 424

Ruggieri day and night distress signal Type 252

This is designed to provide a single system which is efficacious in providing an easily seen signal by day or night, on land or sea. It consists of a cylindrical body with a percussion igniter at each end. One end contains an orange smoke composition for use by day, the other a red Bengal Light for use by night. The daylight signal develops a cloud of orange smoke for 15 – 30 seconds, which can be seen up to 10 km away in clear weather. The night signal produces a red flame which can be seen from an aircraft 48 km away in clear weather. After using either end of the signal, the other end remains ready for subsequent use.

DATA
Diameter: 38 mm
Height: 126 mm

Manufacturer
Ruggieri, Département Armement, 86 ave de Saint-Ouen, 75018 Paris.
Status
Current. Production.
Service
French Army.

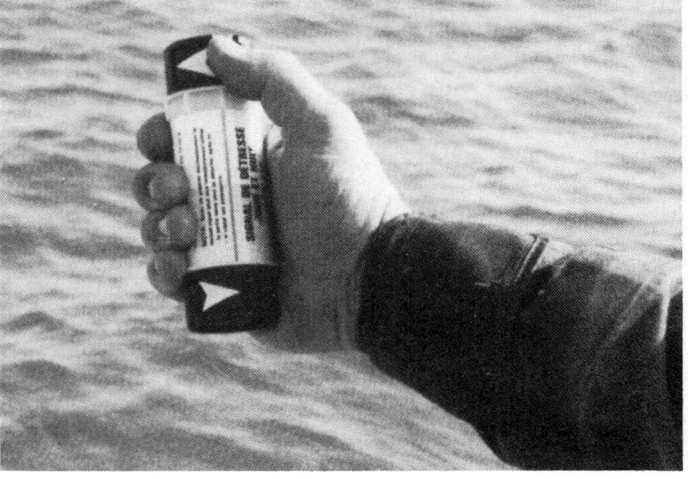

Ruggieri day and night distress signal Type 252

Model 50 trip-flare

The Model 50 trip-flare is designed to operate in the same way as most other similar flares. It is set up with one or more tripwires running from the igniter and when these are pulled or cut the igniter is activated and the flare burns brightly, illuminating the surrounding ground up to a radius of 100 m.

DATA
Weight: 465 g
Height: 110 mm (170 mm with igniter)
Diameter: 55 mm
30 m tripwire: 110 g
Illumination: 40 000 cd
Burning time: 45 s

Manufacturer
Ruggieri, Direction Commerciale Armement, 86 ave de Saint-Ouen, 75018 Paris.
Status
Current. Production.
Service
French Army.

Model 50 trip-flare

Alsetex audio-visual chemical alarm

This device is designed to transmit a chemical warfare alarm over an area of some 1 km radius by means of audible and luminous signals.

The device consists of a hand-held container-launcher with a percussion firing mechanism in its base. When the launcher is pointed upwards and the percussion unit fired, a rocket is discharged. This rocket carries a sound generator in its forward part, which emits a powerful modulated noise during the upward part of the rocket's trajectory. On reaching the vertex a small charge ejects three light signals (yellow – red – yellow) which ignite simultaneously and are braked in their descent by a parachute.

This combination of audible and visible effects attracts maximum attention and ensures perception over a wide area.

DATA
Length: 385 mm
Diameter: 58 mm
Weight: 1 kg
Duration of audible signal: 10 s
Duration of visible signal: 20 s
Vertex height: 600 m

Manufacturer
SAE Alsetex, 4 rue de Castellane, 75008 Paris.
Status
Advanced development.

Lacroix LXT signal pistol

This is a 40 mm smoothbore signal pistol with a self-cocking trigger mechanism and a barrel which swings sideways, to left or right, for loading. By means of an adapter it can also fire standard 26.5 mm signal cartridges. The breech face is mounted on a sprung recoil buffer which, on firing, absorbs much of the recoil, making the pistol very easy to handle. The exterior of the barrel is formed with interrupted lugs which allow the pistol to be clamped into a 'firing chimney' permanently fitted to vehicles, aircraft or control towers.

DATA
Calibre: 40 mm
Weight: 1.645 kg

Manufacturer
Ste E Lacroix, Route de Toulouse, 31600 Muret.
Status
Current. Production.

Lacroix LXT signal pistol

Lacroix signal cartridges

These cartridges can be fired from any 40 mm (1.5 in) signal pistol. The coloured star ignites during the upward part of its trajectory and extinguishes about 15 m above the ground.

DATA
Diameter: 40 mm
Length: 100 mm
Weight: 132 g
Illumination time: 6.5 s
Illuminating power: red 40 000 cd; green 25 000 cd; white 70 000 cd
Ignition height: 100 m +

Manufacturer
Ste E Lacroix, Route de Toulouse, 31600 Muret.
Status
Current. Production.

Lacroix signal cartridges

Lacroix audio-visual chemical alarm

This is a hand-held rocket which emits visual and audible warning of sufficient intensity to alert a platoon or company sized unit under combat conditions. The device is fired by holding it near vertical, turning the hand-grip to the left, and pulling it sharply. After a 1 to 1.5 s delay, so that the firer can ensure the device is pointed correctly, the rocket motor ignites and provides soft ejection of the payload. The signals are fully deployed in about five seconds.

DATA
Length: 347 mm
Diameter: 47.5 mm
Weight: 690 g
Altitude: 200 m
Radius of warning: 500 m
Audible signal intensity: 139 dB at 0.5 m from source
Duration of audible signal: 10 s
Duration of visual signal: 24 s

Manufacturer
Ste E Lacroix, Route de Toulouse, 31600 Muret.
Status
Current. Production.

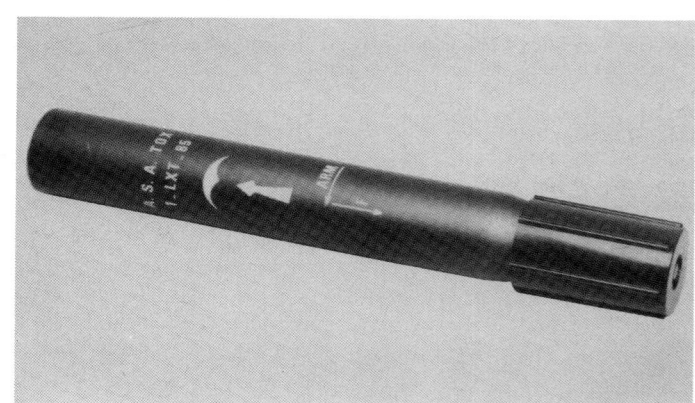

Lacroix audio-visual chemical alarm

Service
French Army.

Lacroix short-range illuminating rocket

This is a hand-held illuminating rocket designed to produce a minimum illumination on the ground of 5 lux within a 300 m diameter area; this is considered to be a sufficient level of illumination to permit identification of a target at night and engagement with any infantry weapon, more particularly anti-tank weapons, without the need for night-vision devices.

The device is fired by simply turning and pulling the handle-grip at the rear end, while holding the tubular body in the desired direction of launch. A rocket is discharged which, at its vertex, deploys a parachute-supported star unit. The recoil stress is short and weak, so that the rocket can be launched without danger and without the need for special training and there is no visible trail to disclose the firer's position.

DATA
Calibre: 40 mm
Length: 323 mm
Weight of rocket: 350 g
Weight: (complete) 580 g
Range: 300–350 m or 600 m
Height: (ejection) 140 m; (burnout) 40 m
Rate of descent: 3.5 s
Duration: 25 s
Power: 200 000 cd

Manufacturer
Ste E Lacroix, BP 213, 31601 Muret.
Status
Current. Production.
Service
French Army, in which it is known as 'Type F2'.

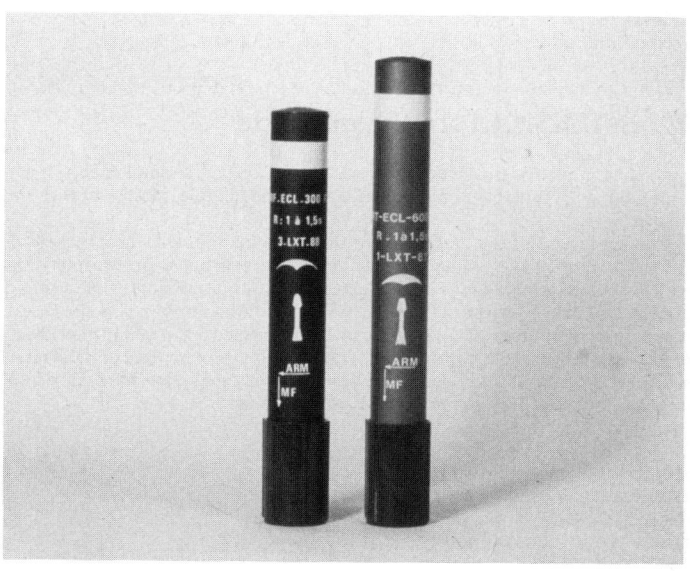
Lacroix short-range illuminating rocket

Lacroix stabilised illuminating device

This is a multiple star discharger which can be set on the ground or partly buried and can then be operated by conventional trip or pull switches as a trip-flare or can be operated by remote-control at distances up to 2 km from the operator. Once triggered, it fires a vertical star every second until the entire 16 stars have been used. As a result it delivers almost constant illumination over a period of about 23 seconds, sufficient to permit the aiming and firing of any infantry weapon. The device is called 'stabilised' because atmospheric conditions (particularly wind) have no effect, as they do in the case of the usual parachute-suspended star devices. The centre of the illuminated zone therefore remains constant during the period of operation. All the stars have identical trajectories and power; they burn out at approximately 40 m from the ground. It would be possible to slightly increase the operational time to suit specific requirements.

DATA
Weight of unit: 2.2 kg
Dimensions: 180 × 135 × 110 mm
Number of stars: 16
Illumination level: 5 lux
Radius: 200 m
Duration: 23 s

Manufacturer
Ste E Lacroix, BP 213, 31601 Muret.

Lacroix stabilised illuminating device

Status
Current. Production.
Service
French Army.

Long emission time signal mine Type F1B

This is a trip or command type ground flare specially designed to give almost constant light output for a period of 6 mins. It will illuminate an area of about 10 000 m². This operation time is achieved by a spring device which keeps moving the illuminant upwards in its container so as to keep the burning area exposed and thus emitting light in all directions. The mine is ignited by a pull-type switch 'Fusée F2'.

DATA
Length: 397 mm
Diameter: 65 mm
Weight: 2.36 kg
Duration: 6 mins
Light level: average 4500 cd; max 7500 cd

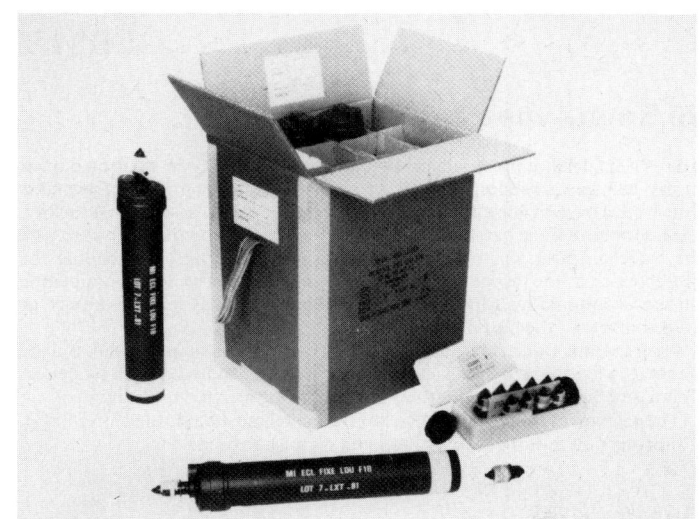
Lacroix F1B signal mine

Manufacturer
Ste E Lacroix, BP 213, 31601 Muret.
Status
Production.

Service
French Army.

Model 56 hand smoke grenade

This device is designed for the rapid emission of a smoke cloud for concealment. It is highly efficient, and operation is simple, rapid and without danger.

The upper end of the grenade is formed into an adapter which accepts a standard fly-off lever igniter set, and the lower end has a protruding tube which contains a smokeless powder relay unit. It is thus possible to use a single grenade by simply withdrawing the safety pin and throwing in the usual manner, or a number of grenades can be connected by pushing the lower protruding tube into the top adapter of another grenade, in place of the igniter set. Several grenades can be connected end-to-end, and as one burns down so it ignites the next, prolonging the emission of smoke.

DATA
Height: 140 mm without igniter set
Diameter: 60 mm
Weight: 400 g
Delay time: 2.5 s
Emission time: 2 mins

Manufacturer
SNPE, 12 Quai Henri IV, 75181 Paris Cedex 04.
Status
Current. Production.
Service
French Army.

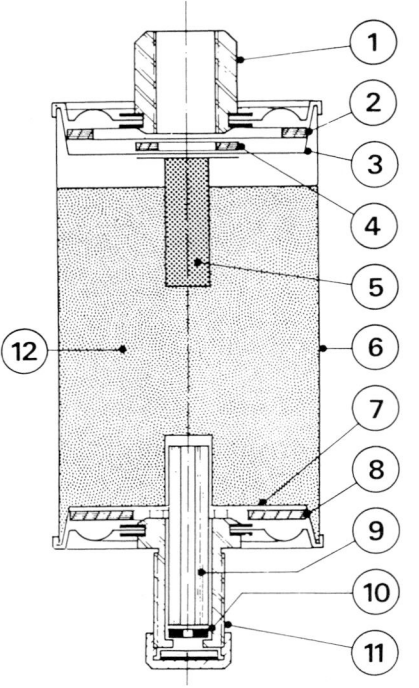

Model 56 smoke grenade; **(1)** *igniter adapter;* **(5)** *ignition relay* **(9)** *smokeless powder relay* **(10)** *black powder pellet* **(12)** *smoke composition.*

Type 428 distress signal kit

The kit consists of a launcher and six red flares which, when fired vertically, reach a height of 400 m and can be seen up to 60 km on a clear night. There is no recoil or spark from the launcher.

DATA
Weight of kit: 250 g
Dimensions of kit: 140 × 70 × 15 mm
Culminating height: 400 m
Duration of illumination: 11 s
Illumination: 5000 cd
Visibility: 60 km on a clear night
Packaging: 200 kits packed in a wooden crate

Manufacturer
Ruggieri, Direction Commerciale Armement, 86 ave de Saint-Ouen, 75018 Paris.
Status
Available.

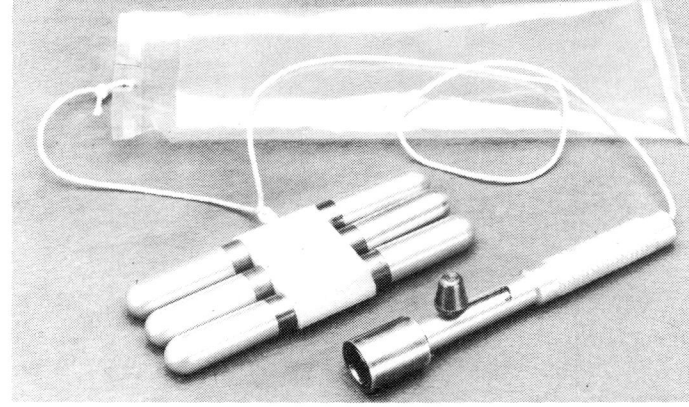

Type 428 distress signal kit

GERMANY, WEST

26.5 mm Type P2A1 signal pistol

This is a single-shot break-action signalling pistol which can be also adapted to fire tear gas grenades. In its basic form the pistol comprises a seamless drawn steel barrel which is pivoted on a receiver and grip assembly. A locking stud at the rear of the barrel engages with a breech catch on the receiver to lock the barrel in position; to break the weapon for loading or unloading the breech catch at the top of the grip must be depressed. The action of opening the breech automatically partially ejects an empty case by means of the ejector at the bottom of the barrel.

After loading a cartridge in the chamber the barrel is snapped shut and the weapon is loaded and safe. To fire it the hammer must be thumb-cocked. There is no applied safety but the hammer may be lowered under control.

The weapon is easily maintained but the barrel must be cleaned after firing. For protection in store any common gun oil may be used.

P2A1 signal pistol

DATA
Calibre: 26.5 mm
Weight: 0.52 kg
Length: 200 mm
Barrel: 155 mm

Manufacturer
Heckler and Koch GmbH, 7238 Oberndorf-Neckar.
Status
Current. Production.
Service
West German and Swiss armed forces, and numerous other countries world-wide.

Emergency flare kit

This flare launcher was specifically developed for use in emergency. The compact launcher is magazine loaded with up to five 19 mm DM13 signal cartridges, usually single star red although other colours are available. The flares reach a height of 65 m.

It was designed for fast, single-hand operation with the hammer moved to the rear by the thumb, the safety pushed up to fire, again by the thumb, and then the trigger squeezed with the index finger.

DATA
Calibre: 19 mm
Overall length: 80 mm
Height: 146 mm
Width: 37 mm

Loaded weight: 445 g
Launcher weight: 225 g
Filled magazine weight: 220 g

Manufacturer
Heckler and Koch GmbH, D-7238 Oberndorf-Neckar.
Status
Current. Production.
Service
West German armed forces; military and police forces in other countries and wide commercial sales.

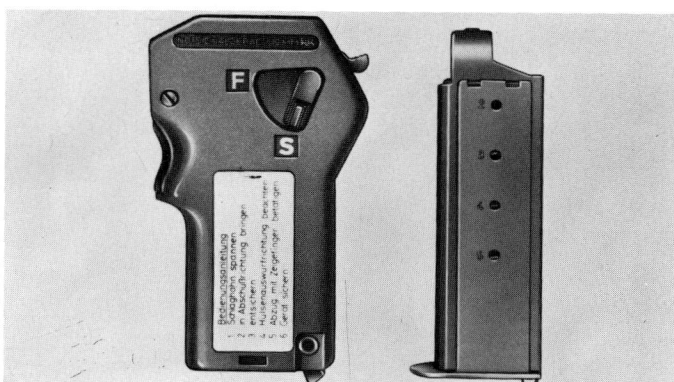

Emergency flare kit launcher and 5-round magazine

19 mm DM13 cartridge

NICO smoke grenades

NICO manufactures several different 44 mm and 60 mm smoke grenades. Two are activated by fly-off levers and a third version by a pull or friction device. All are available in red, orange, blue, yellow, violet, green, grey, white or black.

DATA

Igniter	Pull/friction	Lever	Lever
Diameter	44 mm	44 mm	44 mm
Length	137 mm	132 mm	132 mm
Emission time	120 s	120 s	60 s
Diameter	60 mm	60 mm	60 mm
Length	164 mm	132 mm	132 mm
Emission time	90 s	60 s	90 s

Manufacturer
NICO Pyrotechnik Hanns-Jürgen Diederichs GmbH and Co KG, Bei der Feuerwerkerei 4, PO Box 1227, D-2077 Trittau/Hamburg.

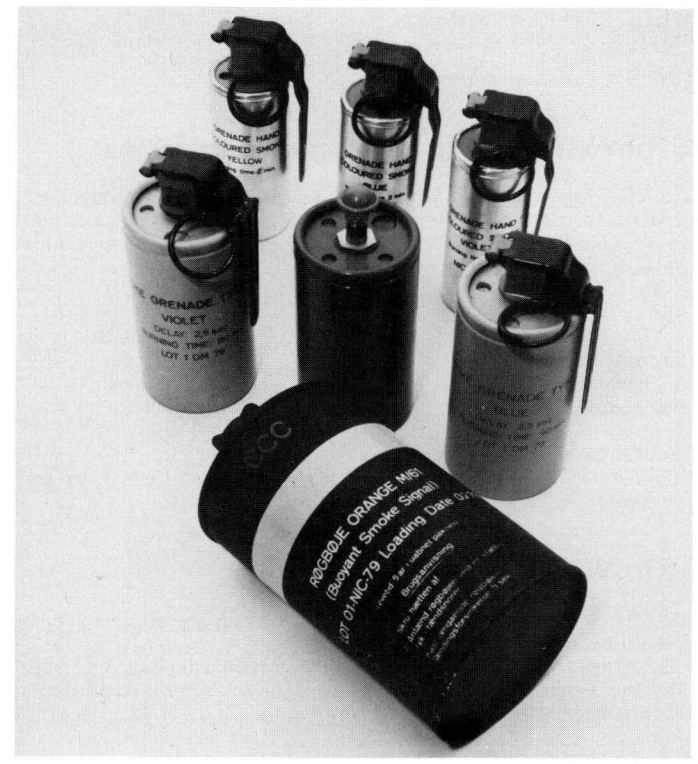

NICO smoke grenades

NICO flare cartridges

The parachute flare cartridge and the rocket-assisted parachute cartridges are conventional flare cartridges with all the characteristics to be expected of such a system.

However, the new Battle Line Illuminating Flare is different in that it is projected without visible trajectory and emits a high intensity light for seven to eight seconds. As it is not equipped with a conventional parachute, it does not drift off target in adverse wind conditions. There are two versions of this flare which achieve heights of 200 and 400 m respectively.

All these flare cartridges have a 25 mm calibre.

DATA

	Parachute flare	Rocket-assisted flare	Battle line illuminating (low level)	Battle line illuminating (high level)
Diameter	26.5 mm	26.5 mm	26.5 mm	26.5 mm
Length	Not known	Not known	110 mm	150 mm
Max height	70 m	300 m	200 m	500 m
Burning time	15 s	30 s	7–8 s	7–8 s
Colour and luminosity	Red 30 000 candela White 40 000 candela	Red 30 000 candela White 40 000 candela	150 000 candela leading to 10 lux 70–270 m forwards	150 000 candela leading to 10 lux 275–425 m forwards

Manufacturer
NICO Pyrotechnik Hanns-Jürgen Diederichs GmbH and Co KG, Bei der Feuerwerkerei 4, PO Box 1227, D-2077 Trittau/Hamburg.

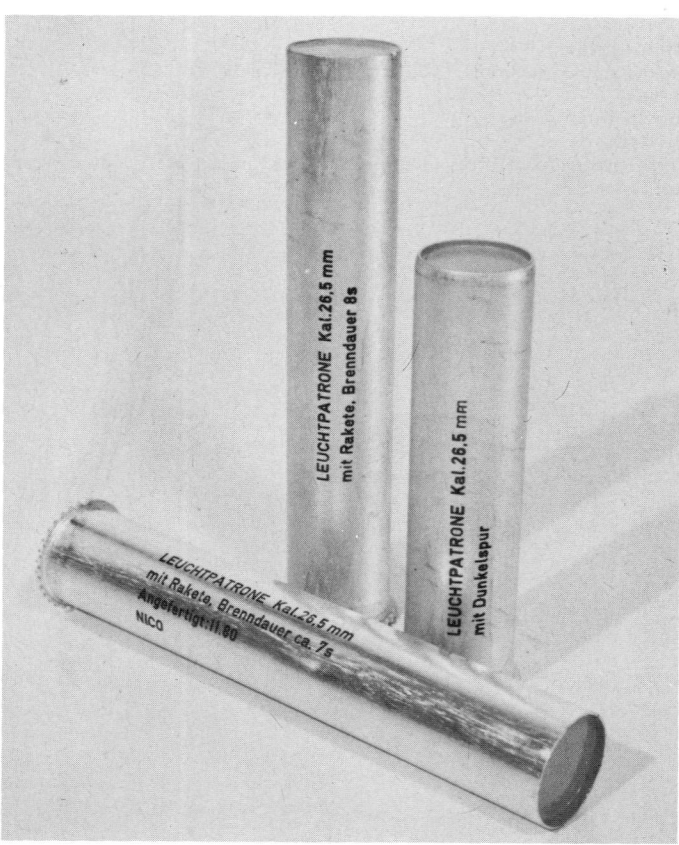

NICO rocket-assisted flare cartridges

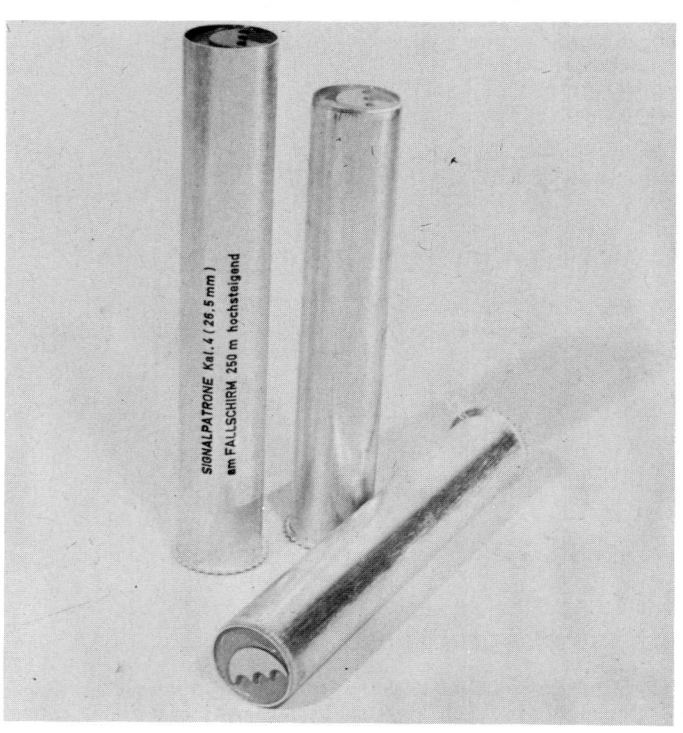

NICO parachute flare cartridges

28 mm NICO hand-held signal cartridges

Type A is designed for single-handed operation and Type B for two-handed operation. There are parachute and non-parachute versions of each cartridge operating at different heights. The standard cartridge is red although other colours are available.

DATA

	Type A	Type B
Diameter	28 mm	28 mm
Non-parachute height	70 m	70 m
Parachute height	300 m	300 m

Manufacturer
NICO Pyrotechnik Hanns-Jürgen Diederichs GmbH and Co KG, Bei der Feuerwerkerei 4, PO Box 1227, D-2077 Trittau/Hamburg.

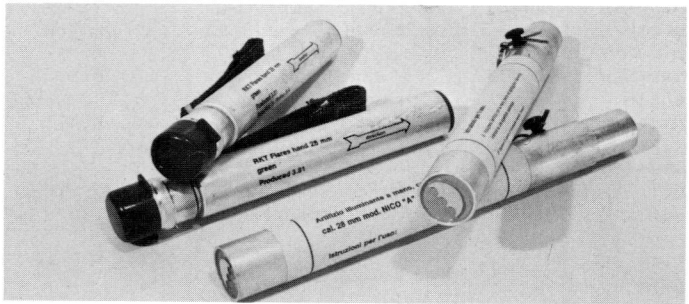

28 mm NICO hand-held signal cartridges

1.5 in NICO signal cartridges

NICO Pyrotechnik has developed a wide range of signal cartridges in both 1 in and 1.5 in calibres.

The cartridges produce a single star in the standard colours of red, green and white. Double, treble and quadruple stars can be produced in both calibres if required.

Manufacturer
NICO Pyrotechnik Hanns-Jürgen Diederichs GmbH and Co KG, Bei der Feuerwerkerei 4, PO Box 1227, D-2077 Trittau/Hamburg.

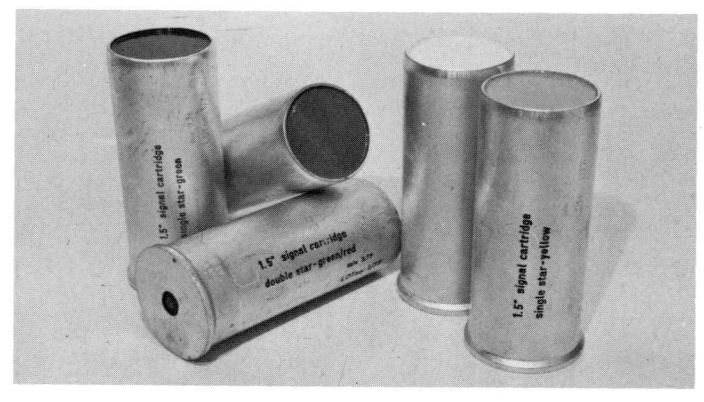

1.5 in NICO signal cartridges

Nicosignal flare kit

This is a hand-held repeating flare discharger manufactured from stainless steel and plastic materials to ensure resistance to sea water and corrosion. It consists of a handle with security loop, safety catch, trigger and firing mechanism, to which a six-shot revolving flare magazine is clipped. The magazine can be rapidly fired and a fresh magazine can be quickly installed. Cartridges are available in red, green and white, giving various signalling options; the intensity and duration of the ejected flare is in conformity with accepted military standards.

Nicosignal flare kit

DATA
DISCHARGER
Weight, with magazine: 280 g
Weight, replacement magazine: 90 g
Dimensions: 170 × 55 × 50 mm

AMMUNITION
Colours: red, green, white
Cartridge dimensions: 45 mm long × 16 mm diameter
Functioning height: ca 80 m
Burning time: ca 6 s
Intensity: red – 10 000 cd
Storage life: 3 years

Manufacturer
NICO Pyrotechnik, Hanns-Jürgen Diederichs GmbH & Co KG, Bei der Feuerwerkerei 4, PO Box 1227, D-2077 Trittau/Hamburg.
Status
Current. Production.

NICO NT screening smoke

This is a new and patented smoke composition developed by the NICO Pyrotechnik company which replaces the standard hexachloroethane/zinc composition. A particular advantage of this composition is that it has high mechanical strength and resists damage, so that it can be formed into any required shape without requiring any container or canister. The chemical characteristics of NT smoke are such that it has better storage life than HC smokes and the burning rate can be controlled by the method of use, so that it becomes possible to generate a large volume of smoke to start the cloud, then reduce the output to keep the cloud fed and maintain the cover. The absence of a metal container means that less weight needs to be transported to provide a given volume of smoke and there is no corrosion problem in store.

The smoke produced is very white, has a high TOP (total obscuring power) value, is not so weather-dependent as HC or FS smokes, and is chemically neutral (pH 5.6 to 6.2) so that it does not irritate soldiers operating in the screen.

The solid material is zinc oxide and ammonium salt based, is chemically neutral and has a virtually unlimited store life. Combustion is at ca 800/850°C and is flameless. Burning continues even under water spray conditions. The rate of combustion depends upon the free surface area, and the material can be ignited by pyrotechnic inserts, electrical squibs, storm matches, ordinary matches, percussion fuze or even a lighted cigarette.

Standard sizes are 0.5, 1 and 10 kg blocks, but any desired shape or size can be manufactured.

Manufacturer
NICO Pyrotechnik, Hanns-Jürgen Diederichs GmbH & Co KG, Bei der Feuerwerkerei 4, PO Box 1227, D-2077 Trittau/Hamburg.
Status
Current. Production.

NICO Super Mini Signal Cartridges (SMSC)

In order to meet the requirements of special forces and those who have a need for lightweight signal cartridges, NICO has developed SMSCs which are small and compact but have a good performance.

They can be fired single-handed from a lightweight pen-type launcher.

Two sizes are available.

NICO Super Mini Signal Cartridges

DATA

	Standard	Long
Length	32 mm	50 mm
Diameter	17 mm	17 mm
Height	60–90 m	60–110 m
Casing	Waterproofed aluminium assembly	Waterproofed aluminium assembly
Burning time	4–6 s	4–6 s
Colours	Red, green or white	Red, green or white

Manufacturer
NICO Pyrotechnik Hanns-Jürgen Diederichs GmbH and Co KG, Bei der Feuerwerkerei 4, PO Box 1227, D-2077 Trittau/Hamburg.

NICO area warning devices

Two specific warning devices have been developed by NICO for perimeter protection; the alarm flare and alarm mine. The alarm flare gives a high level of illumination to allow observation of an intruder together with an audible warning.

The alarm mine has been optimised to give a much louder report and a bright light when triggered by an intruder. It is small and inconspicuous.

Both alarms are activated by tripwires.

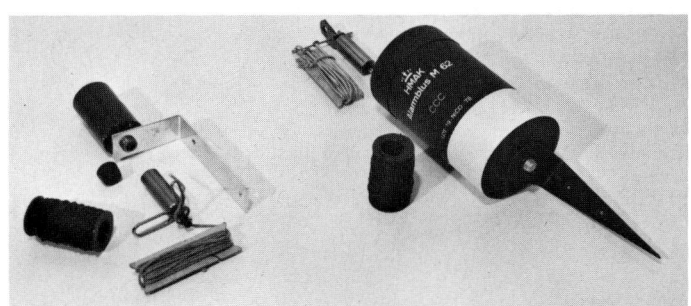

NICO alarm mines

DATA
ALARM FLARE
Length: 110 mm
Diameter: 60 mm
Composition weight: 600 g
Luminosity: 200 000 cd
Burning time: 35 s

ALARM MINE
Luminosity: 140 000 cd
Burning time: 10 s

Manufacturer
NICO Pyrotechnik Hanns-Jürgen Diederichs GmbH and Co KG, Bei der Feuerwerkerei 4, PO Box 1227, D-2077 Trittau/Hamburg.

Comet 26.5 mm signal cartridges

These are the usual type of cartridge for use in signal pistols. They are aluminium-cased and are provided in four colours (white, yellow, red and green) and also as a fulminating cartridge which delivers a flash and smoke signal accompanied by a loud report. Individual types are identified by notches in the cartridge rim, coloured end closures and by the model number on the outside of the case.

DATA
Length: 80 mm
Diameter: 26.5 mm
Weight: 55 g
Altitude: ca 120 m fired at 90°
Burning time: ca 8 s
Brilliance: ca 25 000 cd
Types and numbers: red No 1220; white No 1221; green No 1222; yellow No 1223; fulminating No 1225

Manufacturer
Comet GmbH Pyrotechnik Apparatebau, Postfach 100267, D-2850 Bremerhaven 1.
Status
Current. Production.

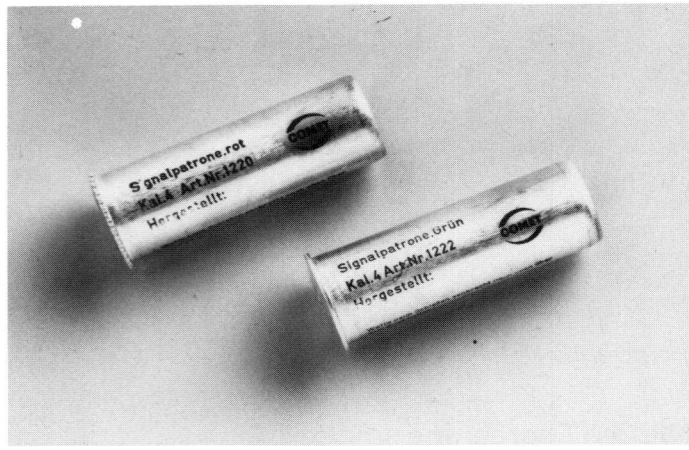

Comet 26.5 mm signal cartridges

Comet 26.5 mm parachute signal cartridges

These are elongated cartridges, for use in signal pistols, containing a parachute star unit. They are intended for close-in illumination of the battlefield or for use in search and rescue operations at sea. White and red flares are available.

This cartridge is also available with a rocket motor, using smokeless propellant, attached to the star unit. This gives a considerable increase in performance, approximating to that of the Type 1260 parachute signal rockets (below), and the smokeless propellant means that the trajectory and firing point cannot be detected.

DATA
Length: 170 mm
Diameter: 26.5 mm
Weight: 120 g
Altitude: 300 m fired at 90°
Burning time: 15 s
Brilliance: 120 000 cd
Illumination: 1 lux at 100 m altitude

Manufacturer
Comet GmbH Pyrotechnik Apparatebau, Postfach 100267, D-2850 Bremerhaven 1.

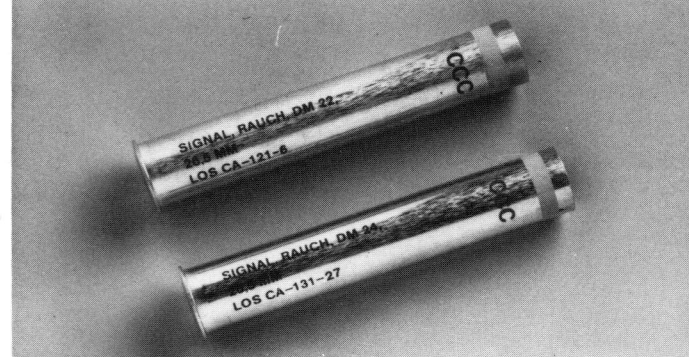

Comet 26.5 mm parachute signal cartridges

Status
Current. Production.

Comet 26.5 mm smoke signal cartridge

This is similar to the usual type of signal cartridge except that instead of firing a coloured star it fires a smoke-generating unit which takes effect at the vertex and then continues to burn while falling. It is thus useful for indicating wind direction and velocity to aircraft and helicopters, as well as for target indication and marking. Two colours, violet and orange, are available, while other colours may be ordered.

DATA
Length: 148 mm
Diameter: 26.5 mm
Weight: 85 g
Altitude: 120 m fired at 90°
Colours: violet No 1250; orange No 1251
Smoke duration: 7 s

Manufacturer
Comet GmbH Pyrotechnik Apparatebau, Postfach 100267, D-2850 Bremerhaven 1.
Status
Current. Production.

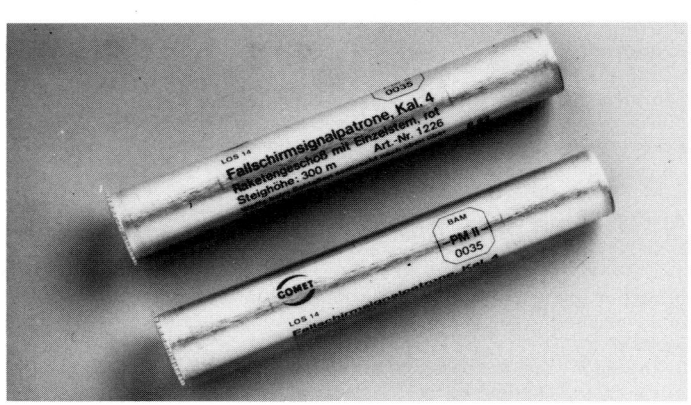

Comet 26.5 mm smoke signal cartridges

Comet coloured signal rockets

These are hand-held rockets which, at their vertex, eject a parachute and a single coloured flare unit or a flare cluster. They can be used for signalling, and the red version is approved for use as a marine distress signal.

DATA
Length: 285 mm
Diameter: 40 mm
Weight: 435 g
Altitude: 300 m fired at 90°
Types: red No 1232; green No 1233
Burning times: Type 1232–40 s; Type 1233–30 s
Brilliance: Type 1232–40 000 cd; Type 1233–80 000 cd

Manufacturer
Comet GmbH Pyrotechnik Apparatebau, Postfach 100267, D-2850 Bremerhaven 1.
Status
Current. Production.

Comet coloured signal rockets

Comet white parachute signal rockets

These are hand-held rockets, similar to the coloured star types described above, but intended primarily for illumination purposes. Two types exist; the Type 1234 is less powerful and is launched by a pull-wire igniter housed under

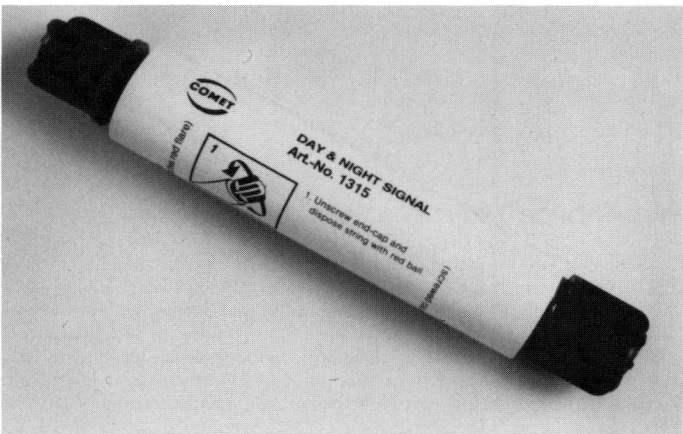

Comet parachute signal rocket Type 1260

a removable cap in the base; Type 1260 is more powerful and is launched by a new system operated by simply twisting the plastic end cap after removal of a safety pin. This model also incorporates smokeless propellant so that there is no firing signature and the launch point cannot be readily identified.

DATA

	Type 1234	Type 1260
Length	315 mm	312 mm
Diameter	40 mm	40 mm
Weight	470 g	660 g
Altitude fired at 90°	300 m	500 m
Burning time	30 s	30 s
Brilliance*	100 000 cd	160 000 cd
Illumination at 100 m	8 lux	16 lux

***Illumination up to 250 000 cd with 30 s burning time is possible**

Manufacturer
Comet GmbH Pyrotechnik Apparatebau, Postfach 100267, D-2850 Bremerhaven 1.
Status
Current. Production.

Comet Day and Night signal

This is a dual-purpose signal or distress device. By removing the brown end cap and pulling the igniter wire so exposed, a cloud of orange smoke is released for signalling by day. By removing the red cap, at the other end, and pulling the exposed wire, a red flare is initiated for use by night. This store is watertight to a depth of 30 m and is thus suitable for use by divers or other waterborne forces.

DATA
Length: 202 mm
Diameter: 34 mm
Weight: 270 g
Ignition: friction igniter
Burning time: 20 s
Light intensity: 20 000 cd
Smoke duration: 19 s

Manufacturer
Comet GmbH Pyrotechnik Apparatebau, Postfach 100267, D-2850 Bremerhaven 1.
Status
Current. Production.

Comet Day and Night signal

Comet ground smoke signal

This is a small smoke generator, ignited by a pull-wire friction igniter concealed under the removeable cap. It can be placed on the ground for use as a marker for aircraft or helicopters, as a wind indicator, or it can be thrown or dropped as a target marker. It is normally available in orange, though other colours can be provided on request.

DATA
Length: 95 mm
Diameter: 50 mm
Ignition delay: 5 s
Smoke duration: 60 s
Weight: 180 g

Manufacturer
Comet GmbH Pyrotechnik Apparatebau, Postfach 100267, D-2850 Bremer-haven 1.
Status
Current. Production.

Comet ground smoke signal

Comet signal pistols

Comet GmbH distribute three signal pistols. All are of 26.5 mm calibre and are single-shot, drop-barrel, hammer-fired types which differ principally in the method and materials of their construction.

DATA

	Type Comet		Type Diana	Type Mondial
Calibre	26.5 mm	26.5 mm	26.5 mm	26.5 mm
Length overall	235 mm	215 mm	235 mm	235 mm
Barrel length	150 mm	155 mm	150 mm	150 mm
Weight	1000 g	650 g	870 g	870 g

Manufacturer
Comet GmbH Pyrotechnik Apparatebau, Postfach 100267, D-2850 Bremerhaven 1.
Status
Current. Available.

Diana 26.5 mm signal pistol

Mondial 26.5 mm signal pistol

Comet 26.5 mm signal pistol

Piepenbrock 40 mm signal cartridges

These are designed to be fired from 40 mm grenade launchers such as the US M79 or West German HK69A1. The single star cartridge is available in red, white, green or yellow, operates at 100 m altitude and burns for eight seconds. The parachute flare cartridge can be used for signalling or illumination; it ejects a parachute and white flare at 350 m altitude, and burns for approximately 30 seconds, giving illumination of 100 000 candelas. The single star cartridges are 100 mm long, while the parachute flare cartridge is 250 mm long.

It should be noted that all Piepenbrock cartridge types produced in 1.5 in (37.5 mm) calibre are also available in 40 mm form.

Manufacturer
Piepenbrock Pyrotechnik GmbH, Ruhweg 21, D-6719 Göllheim/Pfalz.
Status
Current. Production.

Piepenbrock 37.5 mm signal cartridges

These are single star signals for use with standard 37.5 mm (1.5 in) signal pistols. All operate at an altitude of about 90 m and burn for seven seconds. The red and white stars are approximately 50 000 candela, the green star 30 000 candela and the yellow star 70 000 candela.

Manufacturer
Piepenbrock Pyrotechnik GmbH, Ruhweg 21, D-6719 Göllheim/Pfalz.
Status
Current. Production.

Piepenbrock 26.5 mm signal cartridges

This company manufactures a wide range of signal cartridges for the standard type of 26.5 mm signal pistol. The following are the most important types for infantry use:
Single star, red, white or green
Multi-star, giving four stars in red, white or green
Multi-star giving various combinations of stars; 3 red; 3 green; 2 red/1 green; 2 green/1 red; 2 white/1 green; 2 green/1 white; 2 red/1 white; 2 white/1 red.
Other combinations are possible and can be provided to order:
Yellow flare, with or without parachute
Smoke trail, orange or violet
Flash-report, giving a bright flash and loud explosion
Parachute signal, giving parachute-suspended flares in red, green or white.
   Non-parachute stars operate at 100-120 m altitude and burn for 6-8 s. Parachute types operate at about 75 m and burn for up to 12 s.

Manufacturer
Piepenbrock Pyrotechnik GmbH, Ruhweg 21, D-6719 Göllheim/Pfalz.
Status
Current. Production.

Piepenbrock 26.5 mm multi-star signal cartridges

Piepenbrock 16.5 mm signal projector

These are hand-actuated firing devices which are used with 16.5 mm screw-on cartridges. The Model 11010 is a simple spring device operated by pulling back the striker until it is retained by a trigger, after which the trigger is pressed to fire the cartridge. The Model 11009 is equally simple, being fired by pulling back on a cocking piece and releasing it. This model is furnished as part of a kit containing a selection of signal cartridges.
   The cartridges available include red, white, yellow and green single stars, and an 'audible effect' cartridge which emits a whistle during its upward flight and then bursts with a loud report.

Manufacturer
Piepenbrock Pyrotechnik GmbH, Ruhweg 21, D-6719 Göllheim/Pfalz.
Status
Current. Production.

Piepenbrock hand signal cartridges

These produce the same effect as a conventional signal pistol but are hand-held devices which are discarded after firing. They consist of a metal tube with wrist-strap and removable end cap. The forward end is closed with a coloured and notched plastic cap indicating the contents. To operate, the strap is looped around the wrist and the tube grasped firmly; the end cap is removed, whereupon a trigger ring is exposed; pulling this fires the signal cartridge from the tube.
   Types available are single star in red, green, white or yellow, which reach 100 m altitude and burn for about eight seconds: Multi-star, with four red, green or white stars, reaching 75 m and burning for six seconds: and parachute single star in red, green, white or yellow, reaching 70 m and burning for about 12 seconds. All cartridges are 28 mm diameter and about 200 mm long.

Manufacturer
Piepenbrock Pyrotechnik GmbH, Ruhweg 21, D-6719 Göllheim/Pfalz.
Status
Current. Production.

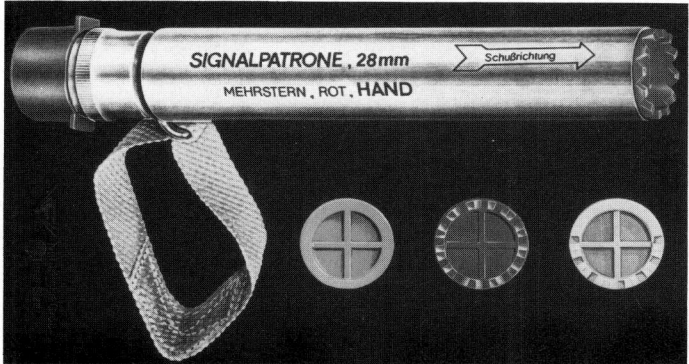

Piepenbrock hand signal cartridge

Piepenbrock 26.5 mm rocket-assisted illuminating cartridges

These are similar in effect to normal illuminating cartridges but are rocket-boosted during flight so as to give greater range and permit the deployment of the illuminating unit to points beyond the reach of normal signal pistol cartridges. Two types are available; one without parachute and one with parachute. The non-parachute model is 173 mm long, reaches a range of about 400 m when fired at 45° elevation, burns for eight seconds and gives about 130 000 candelas. The parachute model is 210 mm long, ranges to about 300 m when fired at 45° elevation, burns for about 15 seconds and gives 160 000 candelas.

Manufacturer
Piepenbrock Pyrotechnik GmbH, Ruhweg 21, D-6719 Göllheim/Pfalz.
Status
Current. Production.

Piepenbrock DM22 smoke generator

The DM22 generator is for marking landing areas, signalling, or indicating wind strength and direction. It consists of a light metal canister with friction igniter concealed under a plastic cap.

DATA
Length: 95 mm
Diameter: 50 mm
Weight: 185 g
Ignition delay: 4 s nominal
Duration of smoke: 60 s nominal
Colour of smoke: orange

Manufacturer
Piepenbrock Pyrotechnik GmbH, Ruhweg 21, D-6719 Göllheim/Pfalz.
Status
Current. Production.

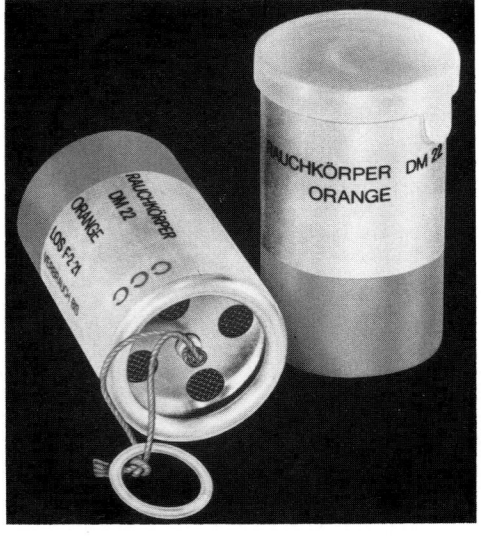

Piepenbrock DM22 smoke generator

Piepenbrock 40 mm screening smoke cartridge

This cartridge is designed to be fired from standard 40 mm grenade launchers such as the US M79/M293 or the HK69A1. On discharge, it launches a rocket-boosted projectile to a range of up to 200 m, which then generates light grey smoke for screening purposes.

DATA
Length: 250 mm
Calibre: 40 mm

Range: 200 mm
Emission time: 30 s

Manufacturer
Piepenbrock Pyrotechnik GmbH, Ruhweg 21, D-6719 Göllheim/Pfalz.
Status
Current. Production.

Piepenbrock alarm mine

This is a tripwire operated device which, upon being triggered, emits a loud report and a bright flash. It is intended for perimeter alarm, and the signal can be heard and seen for a considerable distance. The device is small and easily concealed.

DATA
Weight of mine: 50 g

Time of illumination: 10 s
Candlepower: 15 000 cd

Manufacturer
Piepenbrock Pyrotechnik GmbH, Ruhweg 21, D-6719 Göllheim/Pfalz.
Status
Current. Production.

Piepenbrock trip flare

The Piepenbrock trip-flare is a conventional alarm device which can be quickly emplaced and which functions when the tripwire is either pulled or cut. On firing it produces a vivid yellow flare; two types are available, of different light intensity and duration. In addition to its function as an alarm, it is possible to emplace this flare as an ambush illuminant and ignite it remotely by using an electric igniter.

DATA
Weight of flare: 600 g
Candlepower: type 1 250 000 cd; type 2 100 000 cd
Duration of flare: type 1, 35 s; type 2, 60 s

Manufacturer
Piepenbrock Pyrotechnik GmbH, Ruhweg 21, D-6719 Göllheim/Pfalz.
Status
Current. Production.

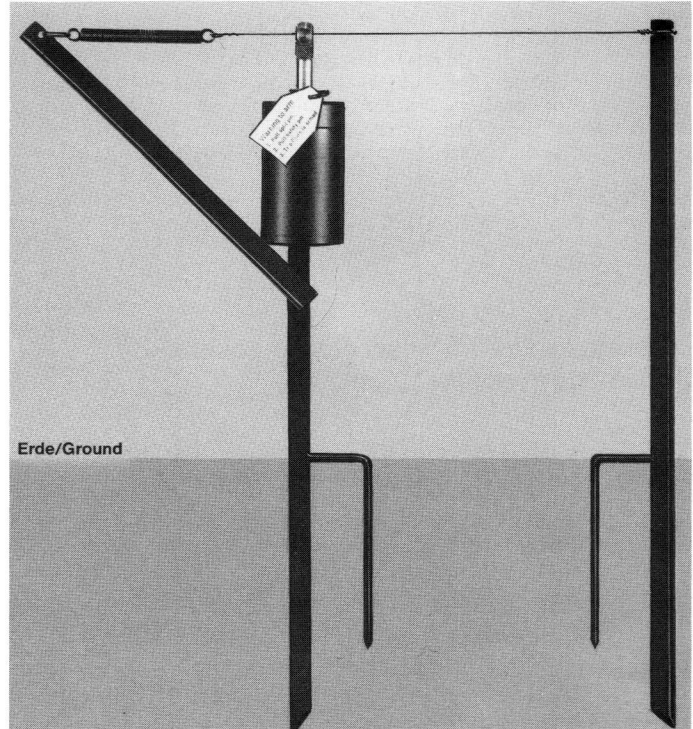

Piepenbrock trip-flare

GREECE

Hand-held distress flare EM53

For use as an emergency flare or for signalling and locating purposes, this consists of a steel tube containing the flare composition and a match-type ignition system under a waterproof cap. It conforms to standard maritime specifications, and has a minimum burning time of 55 seconds, emitting 15 000 candela.

Manufacturer
Elviemek SA, 32 Kifissias Avenue, Atrina Center, GR-151 25 Amaroussio, Athens.
Status
Current. Production.

Elviemek EM53 hand-held flare

Parachute signal or distress rocket EM51

This is the usual sort of hand-held discharger which fires a spin-stabilised rocket by means of a trigger unit at the bottom end. When fired vertically the rocket will reach an altitude of 300 m, releasing a bright red flare of 40 000 candela. The flare will burn for a minimum of 40 seconds.

Manufacturer
Elviemek SA, 32 Kifissias Avenue, Atrina Center, GR-151 25 Amaroussio, Athens.
Status
Current. Production.

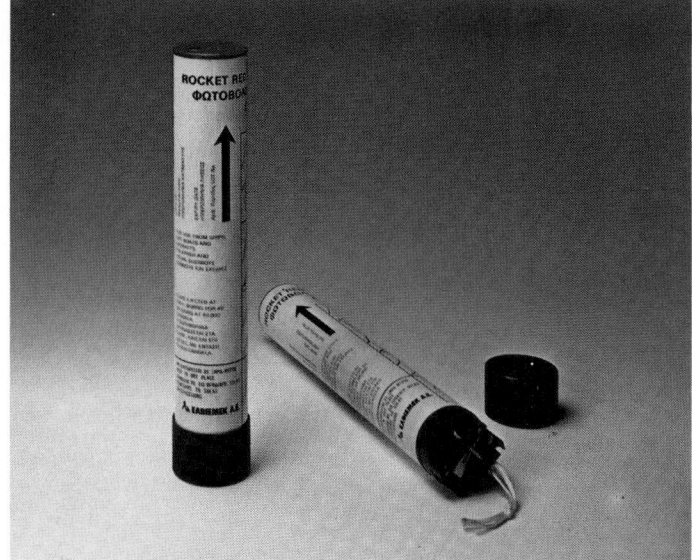

Elviemek EM51 parachute signal rocket

Trip flare EM55

This is a conventional trip-flare which is supplied with two pickets and a length of steel wire. One picket is driven into the ground and the flare grenade mounted on it; the fly-off lever of the grenade is held in a special clamp, to which the steel wire is attached and run out to the other picket some 19 m distant. Once adjusted, any pull or release of the tripwire will release the grenade igniter mechanism and illuminate the surrounding area. Full light output is almost instantaneous. In addition, the grenade unit can be thrown as a hand grenade if desired, and there is also provision for firing by means of an electric igniter.

DATA
Height of flare unit: 165 mm
Diameter of flare unit: 80 mm
Overall weight: 900 g
Pyrotechnic weight: 280 g
Burning time: 80 – 110 s
Luminance: 80 000 cd minimum

Manufacturer
Elviemek SA, 32 Kifissias Avenue, Atrina Center, GR-151 25 Amaroussio, Athens.
Status
Current. Production.

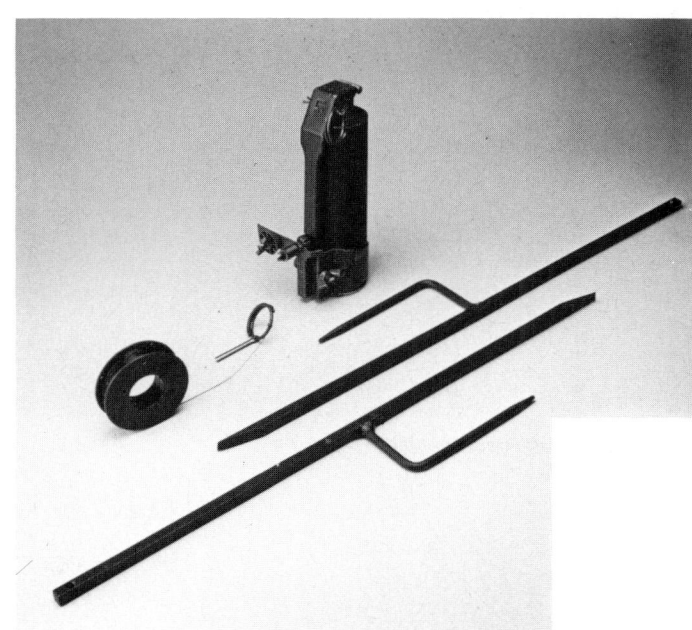

Elviemek EM55 trip-flare

ITALY

1 in Beretta signal pistol

The basic design of this pistol is a single-action pistol with a hammer cocked by the thumb and a trigger-guard which acts as a barrel-lock and safety. It can fire any 1 in (25 mm) signal ammunition.

The present version is an improvement on the one which has been in service for some years. The barrel is longer and it can be fitted with a sub-calibre device to enable the pistol to shoot the smaller 15 and 22 mm flares.

A novel feature is that by using the sub-calibre barrel the pistol can become a line-thrower, using a .32 blank cartridge. An ordinary fishing reel is attached to the barrel and a line can be thrown out to 50 m and then reeled in if necessary. The weight is a nylon muzzle cap and the line is normal 0.45 mm fishing line.

DATA
Calibre: 1 in (25 mm)
Weight: 590 g
Length: 213 mm
Height: 150 mm
Weight with line-thrower attached: 920 g
Barrel length: 145 mm
Range with line-thrower: 50 m

Manufacturer
Armi Beretta SpA, 25063 Gardone Valtrompia, Brescia.
Status
Current. Available.
Service
Not known.

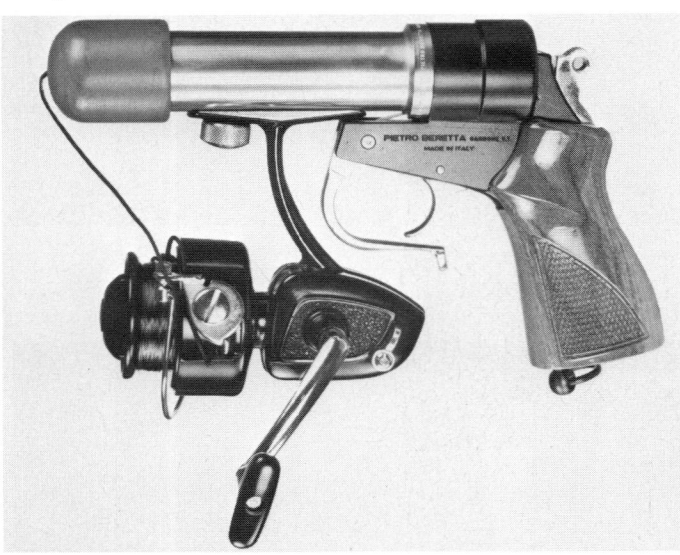

Beretta flare pistol with line-thrower attached

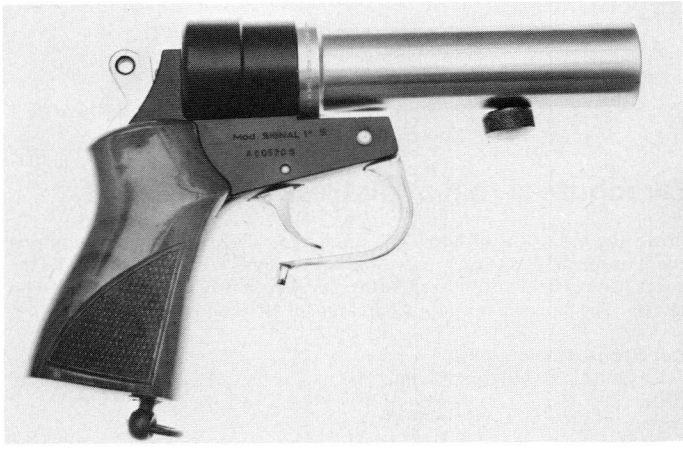

1 in Beretta flare pistol. Stud under barrel is attachment point for line-thrower

37 mm Bernadelli PS 023 signal pistol

The Bernadelli PS 023 is a pyrotechnic flare launcher manufactured in conformity to military standards. It can fire any standard 37 mm pyrotechnic cartridge. It can be used in ground operations or, with an adapter and Mount M1, in aircraft, for both signalling and illumination tasks. The PS 023 pistol is a single-shot, manually breech loaded weapon of conventional type.

DATA
Cartridge: 37 mm (1.5 in) pyrotechnic
Operation: manual, single-shot

Method of locking: manual top lever
Weight: 1 kg
Length: 209 mm
Height: 215 mm

Manufacturer
Bernadelli SpA, I-25063 Gardone Valtrompia, Brescia.
Status
Current. Production.

PS 023 signal pistol open for loading

Bernadelli PS 023 signal pistol

Valsella VS-T Pull-Release illuminating device

The Valsella VS-T illuminating device consists of a cylindrical plastic container, housing a flare, fitted with a tripwire fuze mounted on the top. The device can be planted in the ground or snapped on to a special plastic stake which can be driven into the ground. Actuation is by pressure on the fuze rods or by traction on tripwires, which may be up to 15 m long. Actuation by release can be obtained by the fitting of an optional spring.

DATA
Weight: 470 g
Diameter: 70 mm
Height: 210 mm
Weight of illuminant compound: 350 g

Flare intensity: at least 15 lux at 57 m distance
Illumination period: minimum 40 s
Operating force: (pressure) 4 – 10 kg; (pull) 3 – 7 kg
Operating temperature range: –32°C to +60°C

Manufacturer
Valsella Meccanotecnica SpA, 25014 Castenedolo, Brescia.
Status
Current. Production.
Service
Approved by Italian Army; in service with several countries.

 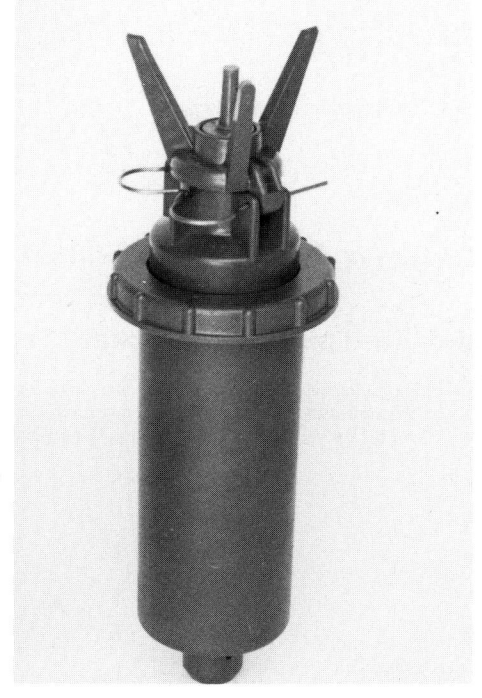

Valsella VS-T illuminating device and plastic mounting stake

Valsella VS-TA-90 Multi-function illuminating device

The Valsella VS-T-A is a combined ground-air, tripwire activated, warning illuminating device intended to give acoustic and visual warning of intrusion into a protected area. It is composed of a ground flare; an in-air flare; one plastic rest picket for each flare; two metal pickets for the tripwires of the ground flare; and two spools with appropriate tripwires. The fuze of the ground flare is triggered by pulling one of its tripwires or by a pressure and/or shock directly applied to the fuze prongs.

The in-air flare is automatically activated, with a proper delay, by the ground flare at the end of its illuminating period.

Several in-air flares could be connected in series so as to repeat the sequence, or in parallel, by means of a special connector, so as to increase the brilliance or the illuminated area.

Optional attachments are available to allow functioning of the device by release, functioning of the ground flare on command by remote-control, or conversion of the ground flare from illuminating to smoke discharging.

Valsella VS-TA-90 Multi-function illuminating device

DATA
Weights: ground flare 423 g; air flare 269 g
Diameter: ground flare 80 mm; air flare 60 mm
Height: ground flare 180 mm; air flare 140 mm
Illumination data: ground flare 15 lux for > 50 s at 57 m; in-air flare 8 lux for > 10 s at 33 m
Operating force: traction 5 kg: pressure 9 kg
Temperature limits: operating -32 to +44°C; storage -33 to +70°C

Manufacturer
Valsella Meccanotecnica SpA, 25014 Castenedolo, Brescia.
Status
Current. Production.

Valsella VS-MK-83 smoke hand grenade

The VS-MK-83 is a signalling device intended for manual activation and positioning. It is fitted with a friction igniter activated by pulling a cord protected by a removeable safety cap. Emission of smoke begins approximately three seconds after the igniter is pulled. The grenade can be placed on the ground and ignited, or the igniter can be fired and the grenade then thrown in the normal way.

The VS-MK-83 grenade has been developed in accordance with requirements resulting from the environmental and operational conditions foreseen by the Italian Army.

DATA
Height: 140 mm
Diameter: 60 mm
Weight: 250 – 350 g depending upon colour
Emission time: approx 70 s
Delay time: approx 3 s
Colours available: white, red, green, yellow, orange, violet, grey

Manufacturer
Valsella Meccanotecnica SpA, 25014 Castenedolo, Brescia.
Status
Current. Production.

Valsella VS-MK-83 signalling smoke grenade

STA-RIP/100 hand-fired parachute rocket

This is a conventional hand-held launcher which discharges a rocket carrying a parachute flare. It consists of an aluminium launch tube with a percussion firing device and a solid-propellant rocket carrying an illuminating flare with parachute assembly.

DATA
Diameter: 45 mm
Length: 240 mm
Weight: 360 g
Intensity: 100 000 cd
Duration of light: 30 s
Height: 400 m when fired at 30° elevation
Illuminated area: 5 lux over a circle of 50 m radius

Manufacturer
Stacchini Sud SpA, Via Sicilia 50, 00187 Roma.
Status
Current. Production.

STA-RIP/100 hand-fired parachute rocket

Stacchini M60F Smoke signal grenade

This smoke signal has been developed to meet Italian Army requirements for several applications including ground-to-ground signalling, ground-to-air signalling, ground recognition marking, wind indication, target identification and tactical screening.

The basic grenade is a hand-thrown canister; a plastic cap is removed before throwing which exposes the pull-type friction igniter cord and also permits the smoke to be emitted from the internal pyrotechnic unit. There is a delay of about 3.5 seconds before smoke is emitted.

If required to be rifle launched, the end cap is removed and a tail tube and vane unit is screwed on in its place. This unit has vents which permit the escape of the smoke. A suitable blank cartridge is packed with the tail unit; this is loaded into the rifle and the grenade is slipped over the muzzle. Any rifle with the NATO standard 22 mm muzzle configuration will accept the grenade.

DATA
Diameter: 64 mm
Length: (hand) 140 mm; (rifle) 340 mm
Weight: (hand) 300 g with coloured smoke, 420 g with white smoke; (rifle) 450 g with coloured smoke, 570 g with white smoke
Burning time: 90 s
Range, rifle launched: 180 m
Colours: white, grey, red, orange, green, blue, black

Manufacturer
Stacchini Sud SpA, Via Sicilia 50, 00187 Roma.
Status
Current. Production.

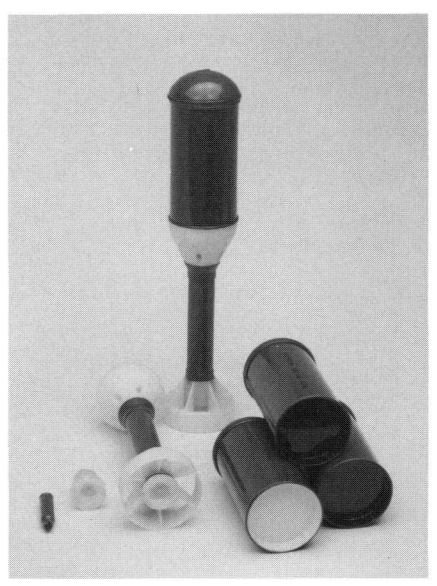

Stacchini M60F smoke grenade, rifle version

Stacchini 1.57 in signal cartridges

These cartridges are designed for the US Pistol, Signal, Mod 8, and are suitable for all military signalling purposes. The cartridge carries two coloured stars and a propellant charge, in an aluminium cartridge case sealed with a weatherproof disc. The colour of the signal is indicated by two coloured stripes on the case and by indentations on the base which can be distinguished in darkness.

DATA
Calibre: 1.57 in (39.5 mm)
Length: 97.5 mm
Altitude: 80 m
Intensity: ca 25 000 cd
Burning time: 5 to 7 s
Colours: white, red, green, yellow

Manufacturer
Stacchini Sud SpA, Via Sicilia 50, 00187 Roma.
Status
Current. Production.

Stacchini 1.57 in signal cartridges

Stacchini 1 in signal cartridges

These are designed for the common 1 in (nominal 27 mm) 'Very' signal pistol and are suitable for all military signal purposes. The round consists of an aluminium cartridge case which contains a single coloured star and a propellant charge. The case is sealed with a waterproof closing disc which is coloured to indicate the colour of the star; the base rim is also indented to identify the colour in darkness.

In addition to coloured stars, similar cartridges loaded with an illuminating flare, a single smoke pellet, two illuminating flares, or a combination of one flare and one smoke pellet can be provided.

DATA
Calibre: 1 in (27 mm nominal)
Altitude: 80 m
Duration: 5 s
Intensity: 10 000 to 15 000 cd according to colour
Colours: red, white, green, yellow

Manufacturer
Stacchini Sud SpA, Via Sicilia 50, 00187 Roma.
Status
Current. Production.

Stacchini 1 in signal cartridges

KOREA, SOUTH

Smoke hand grenades K-M8 and K-M18

As might be gathered from the nomenclature these are based on old American designs. The K-M8 is a screening smoke grenade, details of which will be found in the Grenades section, while the K-M18 is a coloured smoke grenade filled with the usual chlorate/dye composition and available in red, yellow, green and violet. The top plate of the grenade is coloured appropriately. The filling has a central hole which is lined with starter composition, and a layer of starter composition also covers the top of the smoke mixture. There is a hole in the bottom of the grenade body, normally tape-sealed, through which the smoke can escape.

DATA
Weight: 520 g
Height: 118 mm
Fuze delay: 1.5 s nominal
Emission time: 50 – 90 s

Manufacturer
Korea Explosives Co Ltd, 34 Seosomoon-Dong, Chung-Ku, Seoul.
Status
Production.
Service
South Korean Army.

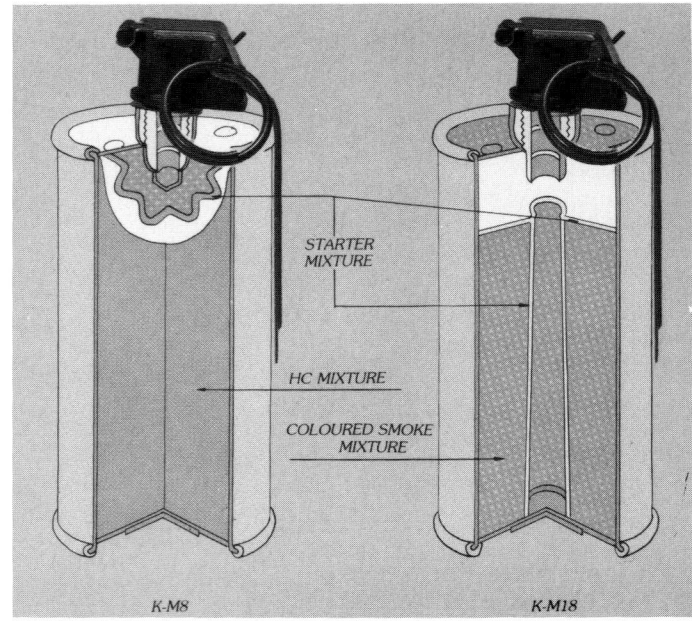

Internal arrangement of K-M8 and K-M18 grenades

40 mm illuminating parachute cartridge

These are pyrotechnic cartridges designed to be fired from standard 40 mm grenade launchers such as the US M79/M203 and similar types. They provide a convenient method of signalling and illuminating with less weight and bulk and greater accuracy than comparable hand-held signal devices.

There are three cartridge models: KM583A1 white star, KM661 green star and KM662 red star. Each comprises a flare candle attached to a parachute contained in a projectile unit which has a pyrotechnic delay lit by the propellant explosion. The delay burns through and fires an ejection charge which ignites the flare candle and blows both candle and parachute free from the projectile.

DATA
Cartridge weight: 220 g
Cartridge length: 134 mm
Candlepower: white 90 000 cd; red 20 000 cd; green 8000 cd
Burst height: 183 m at 85° elevation
Muzzle velocity: 76 m/s
Burning time: approx 40 s

Manufacturer
Korea Explosives Co Ltd, 34 Seosomoon-Dong, Chung-Ku, Seoul.
Status
Current. Production.

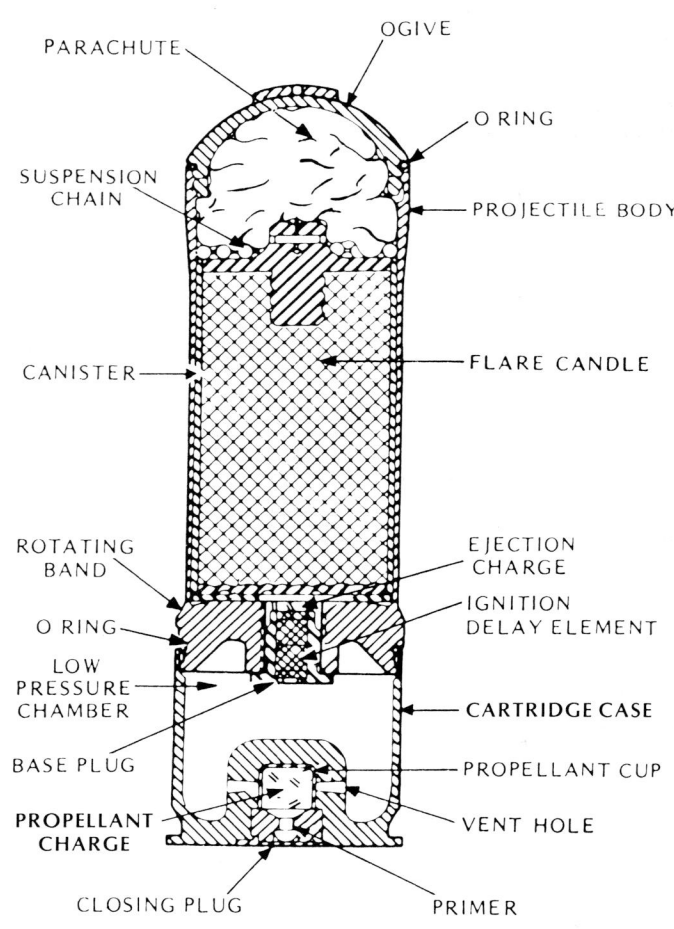

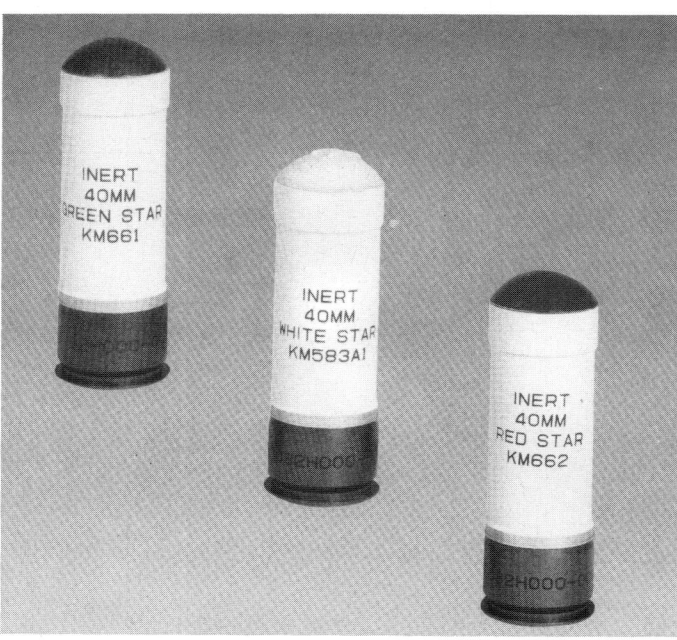

40 mm illuminating parachute cartridge

Internal arrangements of 40 mm parachute cartridge

PERU

Convertible signal pistol MGP-S2

This a rather unusual signal pistol on two counts; firstly because it is designed for 12-gauge signal cartridges rather than the more usual 26/27 mm type, and secondly because it is supplied with a drop-in auxiliary barrel which converts it into a .38 Special calibre single-shot survival pistol. The weapon uses a machined steel barrel, hinged to the frame so as to drop down for loading; there is an automatic ejector beneath the chamber. The barrel is locked in the firing position by an external catch which resembles the Webley stirrup lock, after which the hammer is manually cocked and the trigger pressed to fire. With the barrel open, the auxiliary .38 calibre barrel can be dropped into the chamber and secured at the muzzle by a screwed collar. The automatic extractor works with this barrel also.

DATA
Cartridge: 12-gauge pyrotechnic; .38 Special (see text)
Action: single action, single-shot
Weight, empty: 604 g (12-ga); 805 g (.38 Special)
Length: 206.5 mm
Barrel length: 150 mm (12-ga); 162.5 mm (.38 Special)

Manufacturer
SIMA-CEFAR, Av. Contralmirante Mora 1102, Base Naval, Callao.
Status
Current. Production.
Service
Peruvian armed forces.

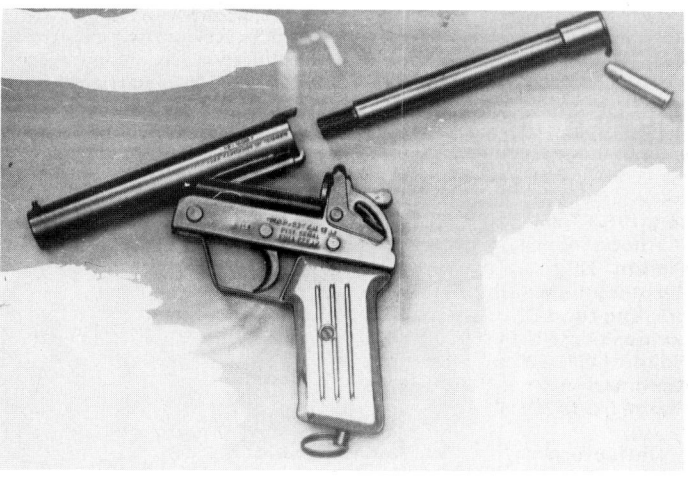

SIMA MGP-S2 convertible signal pistol

SOUTH AFRICA

38 mm signal and illuminating cartridges

These resemble the 25 mm but are of larger calibre. Red, green or yellow stars or an illuminating flare are available and are identified by a coloured identification band and, at night, by knurling on the case rim. The cartridges are fired from a standard 38 mm signal pistol.

DATA
Length: 100 mm
Diameter: 38 mm
Pyrotechnic weight: 70 g
Burning time: 6 s
Vertex (90°): approx. 100 m

Manufacturer
Armscor, Private Bag X337, Pretoria 0001.
Status
Current. Production.
Service
South African Defence Force.

38 mm signal and illuminating cartridges

15 mm signal cartridge

This consists of an anodised aluminium case fitted with a .22 rimfire cap. The case is threaded and will fit most standard launchers of the 'Pengun' or 'Miniflare' types. There are three colours available, identifiable by coloured closing cups and, at night, by raised ribs on the case: red star, two ribs; green star, one rib; white star, smooth case.

DATA
Length: 40 mm
Diameter: 20 mm
Burning time: 6 s
Max height: 75 m
Weight: 11.2 g
Pyrotechnic weight: 3.8 g

Manufacturer
Armscor, Private Bag X337, Pretoria 0001.
Status
Current. Production.

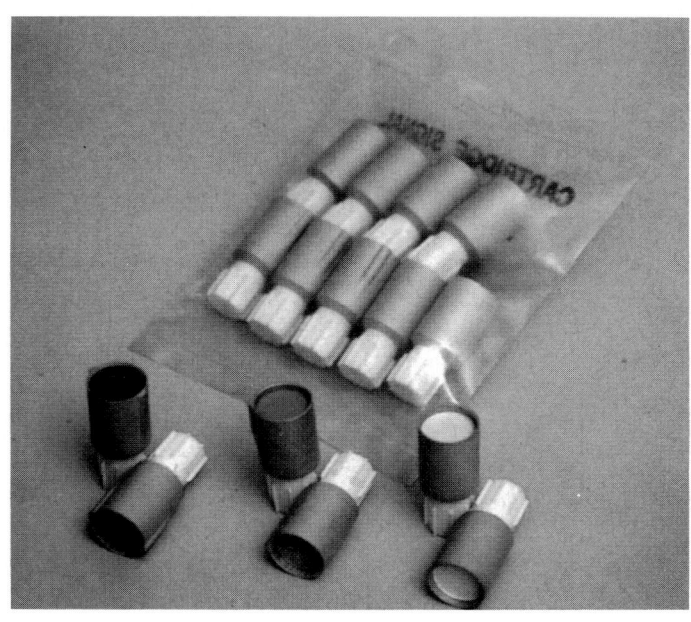

15 mm signal cartridge

Hand rocket flare

This is a hand-held rocket discharger. The rocket has a solid-fuel propellant motor and is spin-stabilised for greater accuracy. After firing it ejects a parachute-suspended illuminating flare at the vertex of its flight. When fired vertically it can be used to illuminate a large area or, alternatively, when fired at an angle of 45° a more concentrated light can be achieved over a restricted area. It is therefore suitable for target identification, counter-insurgency operations, search and rescue operations, perimeter defence and signalling.

The flare is designed for ease of operation. On removal of the bottom cap and safety pin, a trigger lever swings free. While holding the rocket firmly this trigger is pressed, resulting in instant ignition.

DATA
Length: 267 mm
Diameter: 48 mm
Weight: 350 g
Pyrotechnic weight: 83 g
Burning time: 25–35 s
Luminosity: 80–100 000 cd
Height (45°): 200 m
Illumination area: 200 m diameter
Range (45°): 300 m

Manufacturer
Armscor, Private Bag X337, Pretoria 0001.
Status
Current. Production.

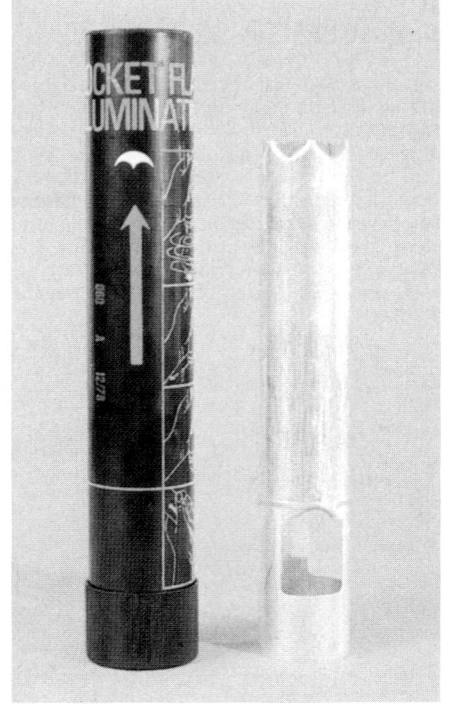

Hand rocket flare

Hand rocket signal

This is a tubular casing containing a solid-fuel rocket which is spin-stabilised for greater accuracy. After firing it ejects a free-falling coloured star at the vertex of its flight. It can be fired vertically or at an angle and is suitable for use as a position marker or for various signalling applications.

DATA
Length: 267 mm
Diameter: 48 mm
Weight: 300 g
Pyrotechnic weight: 83 g
Burning time: 6–10 s
Ejection height (45°): 250 m
Colours available: red, green, yellow

Manufacturer
Armscor, Private Bag X337, Pretoria 0001.
Status
Current. Production.

Hand rocket signal

Tripwire flare

The tripwire flare is used in temporary perimeter protection to give warning of intruders. The flare with its integral striker mechanism is mounted on a short steel picket. The tripwire is attached to the release mechanism and tensioning device, and stretched to another picket some 19 m distant. Either pulling or releasing the tripwire causes the flare to be ignited, thus illuminating the surrounding area. Full light output is almost instantaneous. An adaptor is available for electric ignition. The flare assembly can also be used as a hand-thrown illuminating grenade in an emergency.

DATA
Height of flare assembly: 173 mm
Diameter of flare assembly: 43 mm
Overall weight: 480 g
Weight of pyrotechnic: 250 g
Length of tripwire: 19 m
Length of electric wire: 50 m
Burning time: 80 – 110 s
Luminosity: 80 000 cd minimum

Manufacturer
Armscor, Private Bag X337, Pretoria 0001.
Status
Current. Production.
Service
South African forces.

Armscor tripwire flare

Illuminating hand grenade

The grenade consists of an aluminium case containing the illuminating composition, to which is fitted a conventional fly-off lever striker mechanism. The striker mechanism contains the integral delay element which provides the pre-functioning delay. Intended for use by ground forces, the illuminating grenade provides sufficient light for target identification and attack. The grenade can also be used as a light source for emergency conditions when other lights are not available.

DATA
Diameter: 43 mm
Length: 173 mm
Weight overall: 480 g
Weight of illuminating composition: 250 g
Delay time: 1.5 s
Burning time: 80-110 s minimum
Luminosity: 80 000 cd minimum

Manufacturer
Armscor, Private Bag X337, Pretoria 0001.
Status
Current. Production.
Service
South African forces.

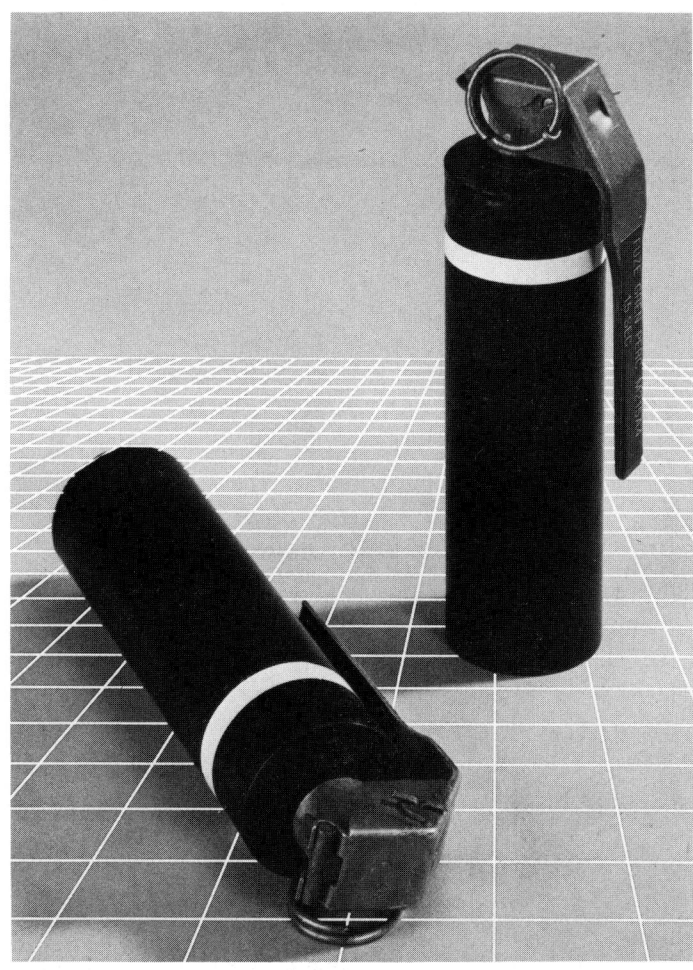

Armscor illuminating hand grenade

81 mm vehicle-launched smoke generator

This is an electrically initiated munition which is fired from the standard 81 mm launcher fitted to various types of armoured vehicle. It is waterproof and resistant to bumps and vibration, so that it can remain permanently in the launcher ready for instant use in an emergency.

The device consists of three major parts: the propulsion assembly and the primary and secondary smoke canisters. On firing, the launching charge propels the generator away from the vehicle and simultaneously lights a delay element which controls the time before smoke emission begins. A few metres above the ground a detonator bursts the primary canister, causing an instant cloud of smoke formed by a composition containing red phosphorus. Simultaneously, the secondary canister is ignited so that it is already emitting a dense cloud of HC smoke when it reaches the ground.

The generator is intended to provide an instant smoke screen in front of a vehicle subjected to attack. The RP cloud gives instant cover and the HC cloud provides sufficient time for either the vehicle to withdraw or at least for the crew to make their escape.

DATA
Length: 315 mm
Diameter: 80.5 mm
Weight: 2.50 kg
Weight of smoke compositions: 1.47 kg
Delay time: approx. 3 s
Range: 40 – 60 m
Time of smoke emission: 35 – 50 s

Manufacturer
Armscor, Private Bag X337, Pretoria 0001.
Status
Current. Production.
Service
South African forces.

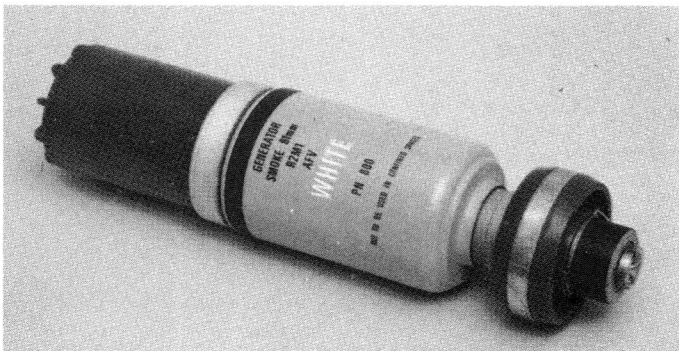

81 mm vehicle-launched smoke generator

SPAIN

Expal ME-1 penrocket flare

The ME-1 consists of a pen-shaped discharger fitted with a spring-driven firing pin, and a screwed socket. The flares are screwed into the socket and the firing pin released to fire the flare, which can reach an altitude of 70 m. The flare burns with a coloured light of 6 000 candela for eight seconds and can be used for identification or signalling purposes. The discharger and six flares, two each of green, white and red, are contained in a soft case.

Manufacturer
Explosivos Alaveses SA, Apartado 198, 01080 Vitoria (Alava).
Status
Current. Production.

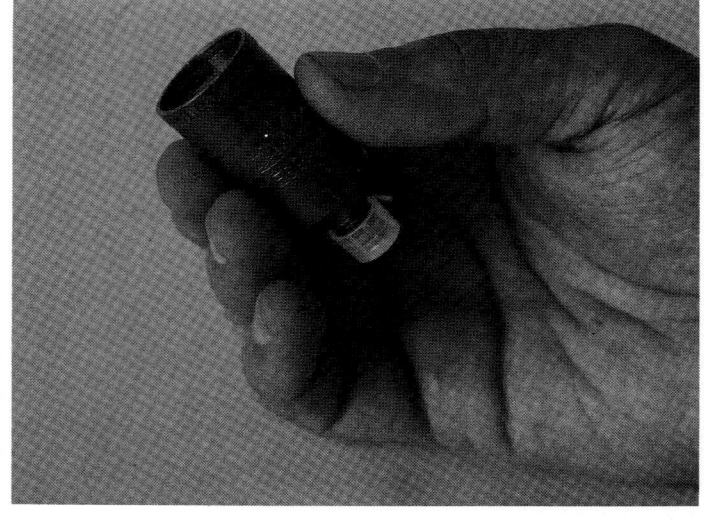

Expal penrocket flare

Expal MRH-1 signal smoke generator

This is a hand grenade type smoke generator which can be used for signalling purposes, for target indication or for indicating wind direction to helicopters. It is available in red, yellow, green or blue colours, burns for 1.4 mins, and has a delay of 12 seconds after releasing the fly-off lever igniter to allow the generator to be carefully set and for the operator to move clear.

Manufacturer
Explosivos Alaveses SA, Apartado 198, 01080 Vitoria (Alava).
Status
Current. Production.

Expal signal smoke generators

Expal day and night signal

This is a double hand-held signal consisting of an aluminium tube with a signal unit at each end with independent ignition. The day unit emits an intense orange smoke cloud for more than 30 seconds; the night unit emits a brilliant red light of over 10 000 candela for more than 30 seconds. Initiation is by percussion, having removed the screwed cap on the appropriate end.

Manufacturer
Explosivos Alaveses SA, Apartado 198, 01080 Vitoria (Alava).
Status
Current. Production.

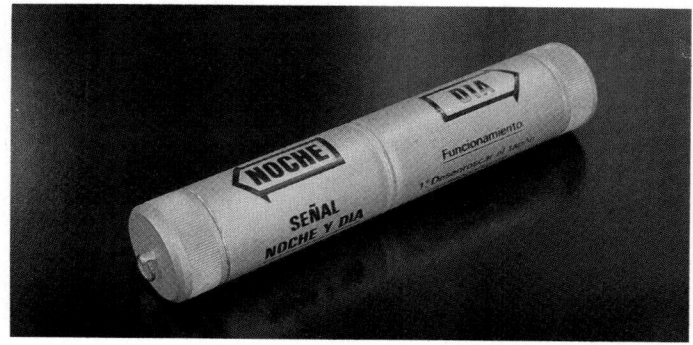

Expal day and night signal

Expal red parachute flare

This is a hand-launched rocket carrying a red parachute flare for use as a distress or other signal. The rocket is gyroscopically stabilised, and can be initiated and then thrown into water if necessary since it can float and will launch the rocket from the water. The canister is of plastic, and removal of the cap exposes the percussion firing device which has a two second delay.

DATA
Length: 245 mm
Base diameter: 65 mm
Body diameter: 45 mm
Weight: 420 g
Altitude reached: 200 m
Luminance: 25 000 cd
Burning time: 45 s

Manufacturer
Explosivos Alaveses SA, Apartado 198, 01080 Vitoria (Alava).
Status
Current. Production.

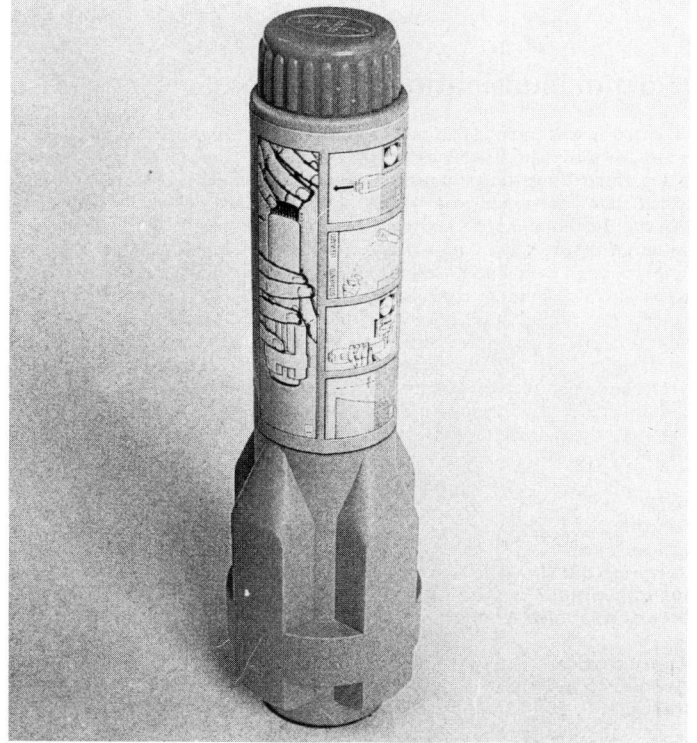

Expal red parachute flare

Expal red hand flare

This is a very compact red flare for emergency signalling. To use, the plastic tubular cover is unscrewed to expose the flare body, and is then reversed and screwed to the base of the body so that it forms a handle. The flare is then ignited by pulling the percussion igniter ring, exposed when the cover was removed.

DATA
Length: 122 mm
Diameter: 35 mm
Weight: 110 g
Luminance: 13 000 cd
Burning time: 45 s

Manufacturer
Explosivos Alaveses SA, Apartado 198, 01080 Vitoria (Alava).
Status
Current. Production.

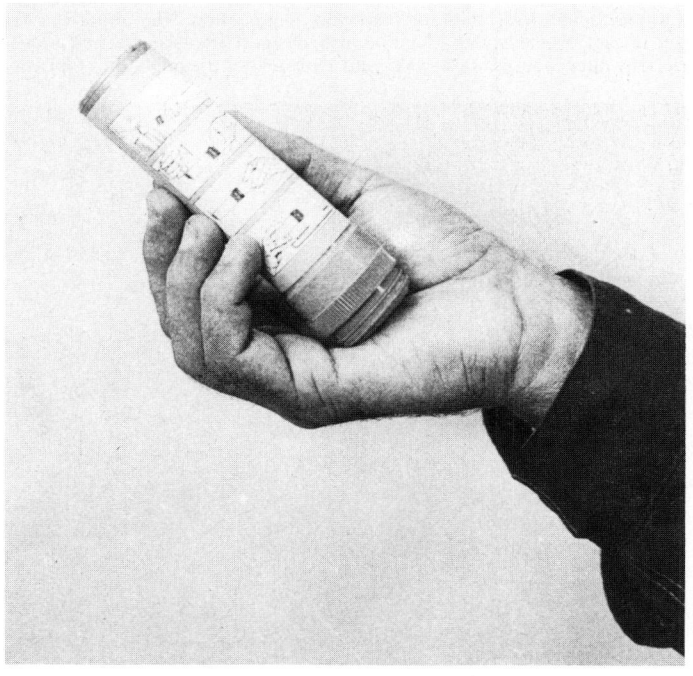

Expal red hand flare

Expal smoke generators

Expal manufacture three standard smoke generators. All are filled with hexachloroethane mixtures for screening or signalling use.
Model HC2 is a hand grenade pattern, weighing 580 g and giving off smoke for 2 mins.
Model HC4 is similar to the HC2 but weighs 1.70 kg and emits smoke for 4 mins.
Model HC9 is percussion fired, weighs 5.40 kg and emits smoke for 9 mins.

Manufacturer
Explosivos Alaveses SA, Apartado 198, 01080 Vitoria (Alava).
Status
Current. Production.

Expal smoke generators

SWEDEN

26.5 mm illuminating cartridge

This cartridge is intended to be fired from most standard 26.5 mm signal pistols and differs from other cartridges of the type by providing illumination lasting about three times longer due to modern parachute techniques. The ejected flare is attached to a parachute with a very slow rate of descent, allowing the illuminant to become totally extinguished before reaching the ground. Because of the long burning time it is possible for one man to fire rapidly enough to maintain continuous illumination of an area.

The cartridge contains a projectile with a metal front body containing the propellant, priming and illuminating charges. The plastic rear body with stabilising fins contains the parachute. When fired, the projectile leaves the pistol rear end first and then, during flight, turns 180° along the trajectory. A pyrotechnic delay charge, which burns for three to four seconds, then ignites a separation charge which splits the front and rear bodies. The illuminating unit is ignited, the parachute is deployed, and the unit then falls slowly to the ground.

DATA
Weight: 160 g
Range: 240 m
Illumination: 90 000 cd
Burning time: 15 s
Rate of descent: 2.5 m/s

Manufacturer
AB Bofors, S 69180 Bofors.
Status
Current. Production.

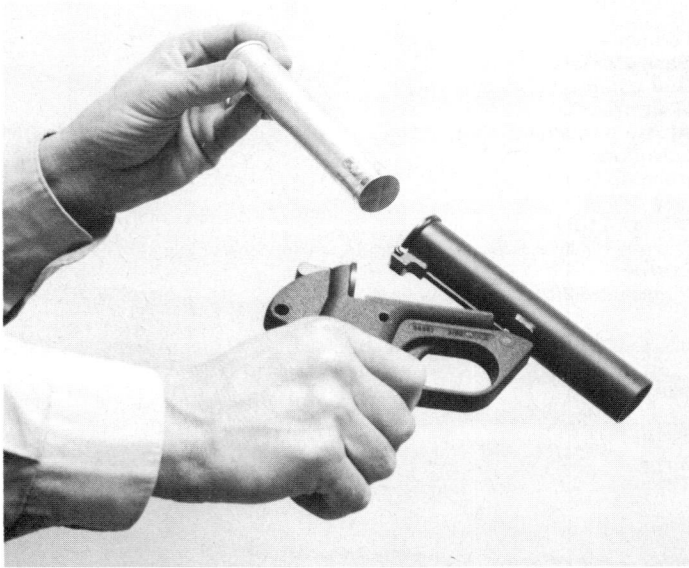

Loading Bofors 26.5 mm illuminating cartridge

71 mm Lyran illuminating system

This illuminating system has been developed in order to meet the primary requirements of small units for battlefield illumination. The objective has been, using only one type of ammunition, to design a simple, portable version, primarily intended for infantry use, and a version for use on various types of combat vehicles, an essential requirement then being that it should be possible to mount the system on existing types of combat vehicles, without any major vehicle modifications.

Infantry version
The infantry version consists of two packs, comprising a launcher and six mortar flare shells.

Pack 1 consists of a plastic mounting containing the barrel and two mortar flare shells. Before firing, the barrel is erected on the mounting. The elevation of the barrel is set with the aid of a spirit level. When firing, the operator is seated on the mounting.

Pack 2 consists of four mortar flare shells, individually packed in plastic tubes which can be connected together.

The mortar flare shell is provided with a pyrotechnical fuze which can be set for ranges of 400, 800 and 1300 m (the horizontal range to flare).

DATA
INFANTRY VERSION
Total weight: (pack 1) 9 kg; (pack 2) 8 kg
Ranges: (elevation 47°) 400 and 800 m

MORTAR FLARE SHELL
Weight: 1.17 kg
Length: 340 mm
Muzzle velocity: 115 m/s
Luminous intensity: mean 600 000 cd
Descending speed: mean 3 m/s
Burning time: 30 s
5-lux illumination: diameter 500 m

Operating Lyran launcher

Twin Lyran installation on Swedish Centurion main battle tank

Two Lyran packs can be carried by one man

Combat vehicle version

The combat vehicle version consists of a launcher intended for permanent mounting, firing device and ammunition packs.

The launcher, which is fired from the crew compartment of the vehicle, is provided with an electro-mechanical firing device. The launcher is connected via the firing device to the electric system of the vehicle. It is fastened with four screws, either directly to the top of the vehicle, or on a special fastening device and it is so positioned that, during combat, it can be served by the crew without anyone having to leave the crew compartment. Before combat

operations commence, the launcher is loaded, the elevation (45°) is set, and the safety device on the launcher is released.

In the transport position the barrel is swung down to protect it from being damaged.

A control unit is used for firing the launcher from inside the vehicle. Its design depends on the number of launchers to be mounted on the vehicle; there is a firing button for each launcher. When the mechanical safety of a launcher has been released, firing can take place immediately; the firing buttons are provided with covers to prevent accidental firing.

The vehicle ammunition pack is the same as pack 2 for the infantry version except that the mortar flare shells are provided with extra propellant charges. These charges are used to obtain a range of 1300 m. The connection devices on the individual packing tubes can be used for attaching extra ammunition on or in the combat vehicle.

DATA
COMBAT VEHICLE VERSION
Weight of launcher: 17 kg
Power requirements: 24V DC
Elevation settings: 5° steps
Ranges: (elevation 45°) 400, 800 and 1300 m
External dimensions of baseplate: 205 × 110 mm
Total length: 885 mm
Firing elevation: 45°
Mortar flare shell: as for infantry version
Muzzle velocity: (with extra propellant charge) 153 m/s

Manufacturer
AB Bofors, S-69180 Bofors.
Status
Current. Production.
Service
Swedish, Swiss, Belgian and Norwegian forces.

Horizon illuminating rocket

This is similar to the Helios rocket (see over) but is longer and heavier and differs in internal construction since it is intended for long-range firing.

DATA
Length: 400 mm
Diameter: 45 mm
Weight: 750 g
Operational range: 1000 m

Light output: 450 000 cd
Duration: 30 s

Manufacturer
Norabel AB, Box 1133, S-43600 Askim.
Status
Current. Production.

Helios illuminating rocket

This is a self-contained hand-fired rocket for the illumination of target areas. It will provide 5 lux illumination at ranges up to 400 m when burst at the optimum height of 170 m above the target area. This level of light is adequate for well-aimed effective fire on any type of target.

The Helios rocket is fired by removing the end cap, elevating the tube, and pressing the trigger lever against the outer case.

DATA
Length: 270 mm
Diameter: 45 mm
Weight: 550 g
Light output: 380 000 cd
Light duration: 30 s
Rate of descent: 5 m/s

Manufacturer
Norabel AB, PO Box 1133, S-43600 Askim.
Status
Current. Production.

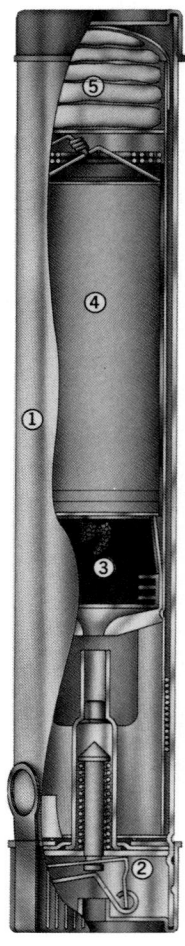

Helios illuminating rocket. (**1**) *launching tube;* (**2**) *firing mechanism;* (**3**) *rocket motor;* (**4**) *flare;* (**5**) *parachute*

UNION OF SOVIET SOCIALIST REPUBLICS

Soviet illuminating stores

26.5 mm signal pistol

This is the conventional type of single-shot signal pistol and is in wide use throughout Soviet forces. The illuminating cartridge has a range of about 120 m, burns for six seconds, and can illuminate an area of about 250 m². There is also the usual variety of coloured star signals available.

Hand-fired illuminating rocket PG 431

This is of 30 mm calibre and is comparable to similar devices used in the West. It has a range of about 450 m, emits a parachute star unit which burns for about nine seconds, and can illuminate an area of some 600 m². There are two known variant models; the illuminating pattern, and a special signal for warning of NBC attacks. This develops a piercing whistle, audible over long distances, during its upward flight, and at the top of its trajectory bursts to display a number of red stars.

To use either of these rockets, the rear cap is removed to expose a pull-ring and lanyard; the device is then grasped and pointed upwards at 45° and the lanyard pulled sharply to ignite the rocket. It can be fired free-hand, but the recommended method is to use the AK rifle as an improvised support; the rocket unit is held alongside the rifle barrel, with its rear end braced against the front of the handguard with the left hand, the butt of the rifle being rested on the ground. The lanyard is then pulled.

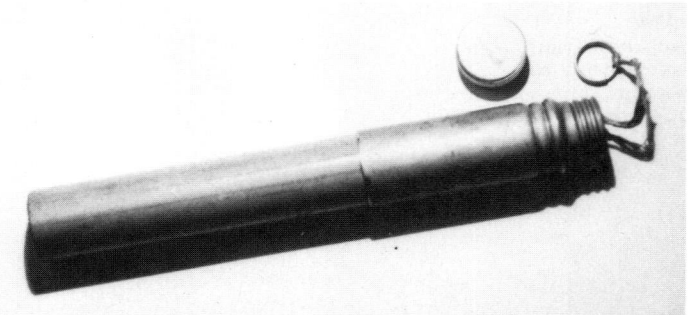

Soviet 30 mm hand-fired illuminating rocket PG 431

DATA
Calibre: 30 mm
Length: 225 mm
Weight: 190 g

Rocket illuminator

This is a larger version of the hand-fired device, of 40 mm calibre. It has a range of 300 m, burns for 20 seconds, and will illuminate an area of 400 m². It is presumably fired using the rifle as a support, as outlined above.

Two-stage rocket illuminator

This is of 50 mm calibre and is provided with a two-stage solid propellant rocket to give greater range. It will reach to 1200 m, burns for 25-30 seconds and can cover an area of 600 m². There is no information on how this device is supported or fired.

UNITED KINGDOM

Astra hand-held coloured flare

This is a high-intensity magnesium flare available in red, green, yellow, white and illuminating forms for use as hand-held signals, either tactical or distress.

The flare consists of a coated-steel tube filled with pyrotechnic composition and fitted with a pull ignition system protected by a removable plastic cap. The flare is fully waterproofed. The end cap is coded by colour and by raised symbols to indicate the contents by day or night, and removal of the cap releases the firing cord.

DATA
Length: 205 mm
Diameter: 32 mm
Weight: 300 g
Burning time: (signals) 60 s; (illuminating) 45 s
Intensity: (red) 20 000 cd; (green) 7000 cd; (yellow) 16 000 cd; (white) 5000 cd; (illuminating) 70 000 cd

Manufacturer
Astra Holdings plc, 43 Old Dover Road, Canterbury, Kent CT1 3DE.

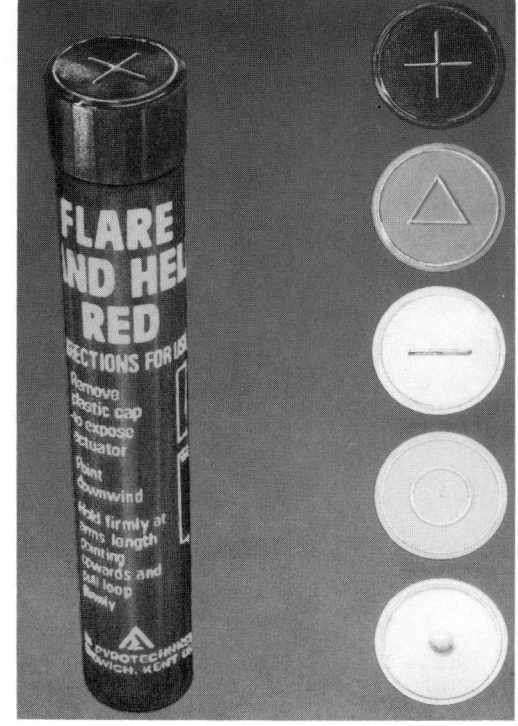

Astra hand-held coloured flare

Coloured smoke hand grenade Mark 2

This is a coloured smoke producing unit suitable for dropping from a light aircraft or helicopter as well as being thrown by hand. It operates in the conventional manner to produce a dense and highly coloured smoke in red, yellow, blue, green, orange or violet. The smoke composition does not contain phosphorus and is entirely non-toxic.

DATA
Height overall: 107 mm
Height of body: 85 mm
Diameter: 63.5 mm
Burning time: 30–60 s to customer's requirement

Manufacturer
Astra Holdings plc, 43 Old Dover Road, Canterbury, Kent CT1 3DE.
Status
Current. Production.

Astra coloured smoke hand grenade Mark 2

Astra signal flares

Astra produces a range of signal flares which are made to meet UK, US and Canadian defence specification requirements. These hand-held devices can be used for tactical or distress signalling purposes.

Coloured flares give up to 60 seconds of burn time and include red, green, yellow and white versions. The illuminating flare gives a much brighter light for approximately 45 seconds. There is also a hand-held rocket system which propels a parachute flare up to 350 m and illuminates a wide area for up to 40 seconds.

All flares have a sturdy waterproof casing; a removeable plastic cap protects the simple pull-ignition system. Ground marker variants are fitted with small spikes to fix them in place. These are used to mark landing areas and drop zones for aircraft and helicopters, and provide illumination for up to three minutes.

DATA
HAND FLARE
Length: 205 mm
Diameter: 32 mm
Weight: 300 g
Burn time: 60 s (coloured): 45 s (illuminating)

ROCKET FLARE
Length: 280 mm
Diameter: 47 mm
Burn time: 35 – 40 s
Altitude: 350 m

Manufacturer
Astra Holdings plc, 43 Old Dover Road, Canterbury, Kent CT1 3DE.
Status
Current. Production.

Astra rocket illuminating flare

Astra signal flare used as ground marker

Astra trip-flares

Trip-flares are used by ambush forces or as warning and protection devices in defensive operations. A trip-flare can additionally be used around the perimeter of a key point or static installation. Astra provides trip-flares to both UK MoD specifications and to the company's own design.

A trip-flare consists of two metal spikes or pickets, a reel of tripwire, a tensioning spring and a metal body containing the flare and the percussion cap igniter. The two pickets can be positioned up to 20 m apart, the tripwire fitted and the spring tensioned. The flare and striker mechanism is fitted to one picket, and the striker arm clamped to the wire. Once the safety pin has been removed the flare will ignite if the tripwire is pulled or cut.

An alternative means of firing is by use of an electric igniter, which allows remote-controlled operation to be performed.

DATA
Wire length: 20 m
Pickets: 500 mm
Burn time: 60 s
Light intensity: 150 000 cd
Total weight: 1.40 kg

Manufacturer
Astra Holdings plc, 43 Old Dover Road, Canterbury, Kent CT1 3DE.
Status
Current. Production.

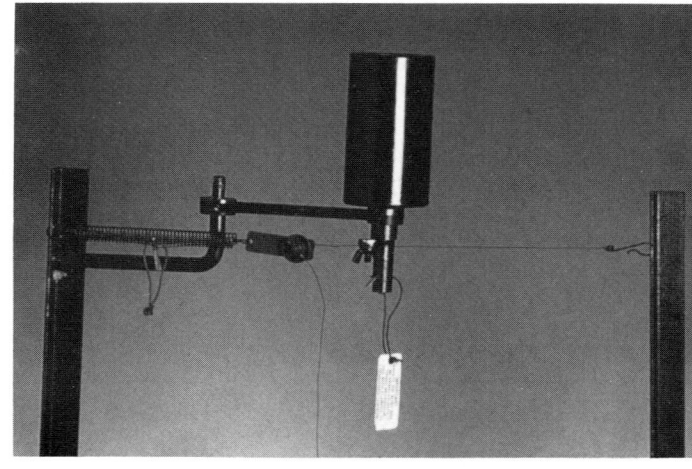

Astra trip-flare

Astra smoke generators

Smoke generators are used to screen static installations or troop locations during tactical operations, or to train personnel to operate in smoke-filled areas.

Astra produces a range of generators to meet various requirements, with differing smoke outputs and burn times. Each device consists of a metal body filled with smoke-producing composition. The ignition system is protected by a removeable plastic cap and can either be a match, friction or electrical device. Smoke generators are produced to UK, US and Canadian specifications.

The electrically ignited variants enable remote firing to be carried out. This allows smoke generators to be operated by devices such as the 'Battlemaster' control system, as part of a training exercise.

Data is given below for two examples of the range.

DATA
AP/3 SERIES
Length: 110 mm
Diameter: 87 mm
Weight: 300 g
Burn time: 3 mins

AP/6 SERIES
Length: 160 mm
Diameter: 87 mm
Weight: 780 g
Burn time: 6 mins

Manufacturer
Astra Holdings plc, 43 Old Dover Road, Canterbury, Kent CT1 3DE.
Status
Current. Production.

Astra smoke generators

36 mm ML hand-fired rocket

The 36 mm ML hand-fired rocket has, as its primary role, white light battlefield illumination, though it can also carry coloured signals smoke, chaff or IR payloads. Small and light, it can easily be carried by the soldier. The whole unit is sealed and protected against all weather conditions.

The rocket is fired simply by pulling a lanyard, possible even when wearing arctic gloves. It is spin stabilised for accuracy and, apart from a minimal flash at ignition, is virtually signatureless. The motor burns out in 0.2 seconds, so there is no tell tale smoke trail.

The illumination provided is a minimum 90 000 candela for 30 seconds nominal at ranges of 300 m or 600 m.

DATA
Unit weight: ca 360 g dependent on payload
Diameter: 37 mm o/d
Length: 215 mm
Range: 300 m or 600 m at 40° QE
Time of flight: 5.5–10 s nominal
Burning time: 30 s
Flare intensity: 90 000 cd minimum

Manufacturer
ML Lifeguard Ltd, Arkay House, Weyhill Road, Andover, Hampshire.
Status
Current. Production.
Service
NATO, Finland, Middle East and Africa.

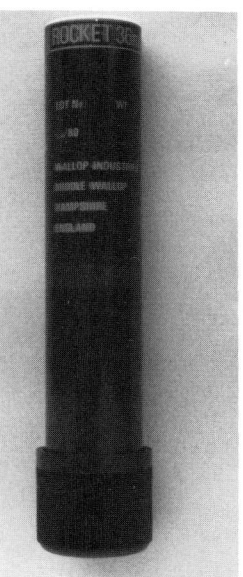

Firing ML 36 mm rocket

ML 36 mm hand-fired rocket

38 mm hand-held rockets

Schermuly offers a range of hand-held rockets, all of which are manufactured to the same basic design, have identical firing sequences and differ only in payload. These stores have been well proven in operational service with British and overseas armed forces. The parachute illuminating rocket described below is the one most commonly encountered for military applications. Other versions have payloads of parachute flares, free-falling stars, explosive maroons or radar-reflective dipoles.

The Satellite low-signature parachute illuminating rocket is a new product developed by Schermuly and is designed to be simpler to operate with a twist fire mechanism, a reduced signature and a higher output from the payload. The reduced signature will prevent a soldier being pinpointed as a target when firing a rocket and the increased light output will give better illumination of the enemy.

Each self-contained rocket consists of an environmentally protected launch tube containing the firing mechanism – uncocked for maximum safety, the rocket assembly and the payload. The rocket assembly itself has a solid fuel motor and is spun for improved accuracy.

Fired vertically the illuminating rocket can be used to illuminate a large area, or alternatively when fired at an angle of 45°, a more concentrated light can be achieved over a smaller area. It is therefore suitable for a wide range of illuminating applications from battle areas to specific targets, for night reconnaissance and for search and rescue.

38 mm hand-held parachute illuminating rocket

Operation
The unit has been designed for maximum ease of operation. Top and bottom transit caps and the safety pin are removed thus allowing the trigger lever at the base to swing free. While holding the rocket firmly, the trigger lever is pressed against the outer case. Ignition is instantaneous.

Other payloads
In addition to the parachute illuminating rocket the following payloads are available: multiple free-falling stars – red, green or white, for short duration general or distress signalling purposes; parachute suspended flares - red or green, for longer duration general or distress signalling purposes; maroon, flash and loud report; radar reflective dipoles which enable position location by radar under adverse weather conditions. The signal carries packs of dipoles to provide a radar echo and a free-falling star for visual sighting under more favourable conditions. Paratarget provides an airborne orange target for anti-aircraft practice with small arms.

DATA
PARACHUTE ILLUMINATING ROCKET L5A4
Complete rocket and launcher
Weight: 345 g
Length: 267 mm
Diameter: 48 mm
Flare composition: 120 g
Illuminating power: 80 000 cd minimum
Ejection altitude: (vertical firing) 300 m; (45° firing) 200 m
Burning time: 30 s nominal
NATO Stock Number: 1340-99-966-2984
Schermuly Product Code: 0803

SATELLITE ROCKET
Complete rocket and launcher
Weight: 0.35 kg
Length: 266 mm
Diameter: 46 mm
Flare composition: 80 g
Illuminating power: 135 000 cd average
Ejection altitude at 90° firing, 300 m: **at 45° firing**, 200 m
Burning time: 30 s minimum
Schermuly Product Code: 0903

Manufacturer
Schermuly (Pains-Wessex) Ltd, High Post, Salisbury, Wiltshire SP4 6AS (a member of the Chemring Group plc).
Status
Current; available.
Service
British and other armies in America, Africa, Europe, Middle East and Far East.

Effect of ground illumination by hand rocket

38 mm hand-held rocket Mk3 Radasound/Radaflare

Each self-contained rocket consists of an environmentally protective tube containing the uncocked firing mechanism, rocket assembly and radar reflective payload. The rocket assembly has a solid fuel motor, it is tail stabilised and rotates on its axis for greater accuracy.

For Radasound the typical payload is five packs of 'I' band chaff but this may be varied to meet customer requirements. For Radaflare, the payload comprises typically of four packs of 'I' band chaff and one free-falling red star. The red star gives a visual indication of the location of the chaff cloud. Both rockets are fired vertically and are operated in the same way as the 38 mm parachute illuminating Mk3 rocket.

Radaflare and Radasound can be used in a variety of situations including: distress signalling (position location in adverse weather conditions); decoy; radar training and testing; terrain mapping; radar range calibration; location aid to assist supply dropping.

The maximum theoretical radar cross-section return averages are 1100 m² for Radasound and 880 m² for Radaflare, and tracking is possible for 12-25 mins dependent on wind conditions. Both rockets exclusively use CHEM-CHAFF aluminised glass fibre.

DATA
RADASOUND
Length: 268 mm
Diameter: 48 mm
Overall weight: 345 g
Explosive content: 46 g
Deployment height: (vertical) 300 m typical; (45°) 200 m typical

Radar cross-section: 1100 m² at 9.3 GHz ('I' Band)
MoD No: L8A1
NATO Stock No: 1340-99-965-2469
Product Code: 0852

RADAFLARE
Length: 268 mm
Diameter: 48 mm
Overall weight: 390 g
Explosive content: 100 g
Deployment height: 300 m typical
Flare burn time: 11 s nominal
Light intensity: 30 000 cd minimum
Radar cross-section: 880 m² at 9.3 GHz ('I' Band)
Product code: 0851
NATO Stock number: 1370-99-966-0352

Manufacturer
Schermuly (Pains-Wessex Ltd), High Post, Salisbury, Wiltshire SP4 6AS and Chemring Ltd, Alchem Works, Fratton Trading Estate, Portsmouth PO4 8SX, members of Chemring Group plc.
Status
Current, available.
Service
Europe, Far East, UK.

Schermuly Day & Night Distress signal

The Day & Night signal was developed to military specifications for a compact personal survival and liferaft signal. It consists of a plastic outer case incorporating a red flare at one end for night signalling and an orange smoke at the other end for daylight use.

Both end caps are coded for easy night identification and sealed with 'O' rings to make the signal watertight to 30 m depth.

Either end can be fired independently by means of a pull-ring percussion striker mechanism, and the unused end stored until it is required.

DATA
Length: 139 mm
Width: 42.5 mm
Weight: 228 g
Flare: burn time 20 s nominal; > 10 000 cd average
Smoke: burn time 18 s nominal
NATO Stock Number: 1370-99-254-4254
Schermuly Product Code: 3031. CAA Approved

Manufacturer
Schermuly (Pains-Wessex) Ltd, High Post, Salisbury, Wilts SP4 6AS (a member of the Chemring Group plc).
Status
Current. Production.
Service
UK MoD and other countries worldwide.

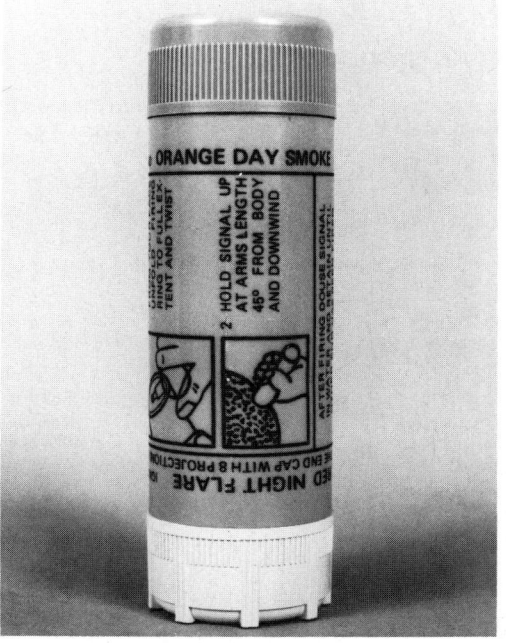

Schermuly Day & Night distress signal

Manroy 25 mm signal pistol

This is the standard Very pattern single shot, drop barrel signal pistol, using a thumb-cocked hammer single action mechanism. It is of all steel construction with bakelite grips and will accept and fire all types of standard 25mm (1 in) signal pistol ammunition.

The same pistol is also available in 26.5mm calibre to suit certain other patterns of signal cartridge.

DATA
Calibre: 25mm (1.0in)
Weight: 820g
Length: 250mm
Barrel: 145mm smoothbore

Manufacturer
Manroy Engineering, Hobbs Lane, Beckley, E Sussex TN31 6TS.
Status
Current. Production.
Service
British army.

Manroy 25 mm signal pistol

Haley and Weller screening smoke grenade E101

This grenade ejects six sub-munitions to give a very rapid build-up of dense smoke for screening purposes. It has been developed to supplant the white phosphorus grenade with a non-incendiary device which will provide smoke with equal speed. The ignition is performed by the unique Haley and Weller silent electrical system; this resembles the conventional fly-off lever pattern so far as the user is concerned but the lever does not leave the grenade and there is no noise of a cap being struck. The movement of the lever after throwing permits an electrical circuit to close and send current from a built-in battery to an electrical delay system which then ignites the smoke composition.

DATA
Length: 134 mm
Diameter: 51 mm
Weight: 460 g
Explosive content: 270 g
Fully effective screen: less than 3 seconds
Duration of smoke: 25 s

Manufacturer
Haley and Weller Ltd, Wilne, Draycott, Derbyshire DE7 3QJ.
Status
Current. Production.

Haley and Weller screening smoke grenade

Signalling smoke grenade E120

This resembles the screening smoke grenade described above but is filled with coloured smoke composition for signalling and indicating purposes. It uses the same electrical ignition system.

DATA
Length: 134 mm
Diameter: 50 mm
Weight: 300 g

Explosive content: 150 g
Colours: red, blue, yellow, green, orange, purple, white
Burning time: 45 to 55 s

Manufacturer
Haley and Weller Ltd, Wilne, Draycott, Derbyshire DE7 3QJ.
Status
Current. Production.

Illuminating hand grenade E140

This is designed to provide an instantaneous bright white light either for illuminating purposes or to blind electro-optical night sights by overloading their circuitry. It uses the same silent electrical ignition system as the smoke grenades previously described, making it a particularly effective countermeasure against snipers with night sights.

DATA
Length: 114 mm
Diameter: 50 mm
Weight: 292 g
Explosive content: 100 g
Intensity: 80 000 cd
Duration: 12 – 15 s
Delay time: 1 – 1.5 s

Manufacturer
Haley and Weller Ltd, Wilne, Draycott, Derbyshire DE7 3QJ.
Status
Current. Production.

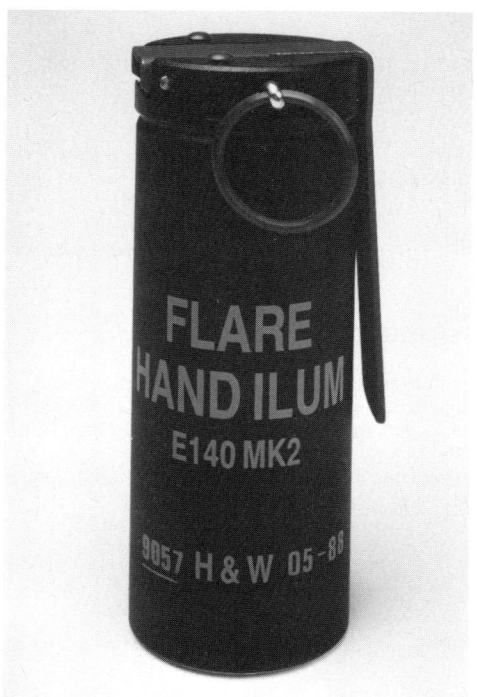

Haley and Weller illuminating hand grenade

Tripwire mechanism E190

This is an all-purpose trip mechanism into which various Haley and Weller grenades can be fitted; for example, the use of an illuminating grenade provides a trip-flare, while using a fragmentation grenade would provide an ambush device. The mechanism is reusable several times, except, of course, when a fragmentation grenade has been used. The trip mechanism will function if the wire is pulled or cut, and there is provision for remote electrical firing. The mechanism is of non-corroding metal, and the tripwire mechanism has a two-position spring which allows the tension to be accurately set whether the wire is long or short. There are also a number of unique safety features which make the device easier and safer to set.

The mechanism is so arranged that it can be emplaced in the ground in the usual way or can be adjusted so that it can be hammered into a tree. By turning the shaft through 90° a surface is exposed which can be hammered, thus avoiding damage to the working parts, and when folded the entire device fits into a convenient pouch.

DATA
Height of picket: 580 mm
Length of tripwire: 20 m
Total weight: 1.32 kg

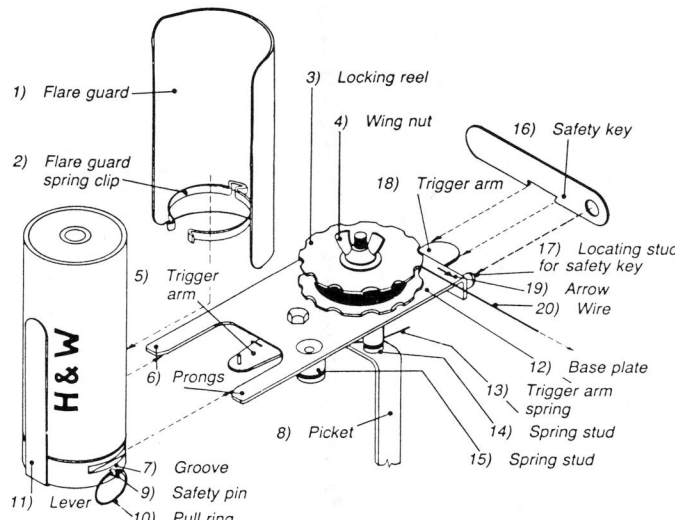

1) Flare guard
2) Flare guard spring clip
3) Locking reel
4) Wing nut
5) Trigger arm
6) Prongs
7) Groove
8) Picket
9) Safety pin
10) Pull ring
11) Lever
12) Base plate
13) Trigger arm spring
14) Spring stud
15) Spring stud
16) Safety key
17) Locating stud for safety key
18) Trigger arm
19) Arrow
20) Wire

Details of Haley and Weller tripwire mechanism

Manufacturer
Haley and Weller Ltd, Wilne, Draycott, Derbyshire DE7 3QJ.
Status
Current. Production.

Haley and Weller tripwire mechanism

66 mm screening smoke grenades

These grenades are designed to be fired from standard 66 mm vehicle mountings. There are two different types; V108 loaded with red phosphorus and V101 loaded with H & W composition CC144. Both grenades have a metal base and rubber body and are electrically fired. Their effect varies slightly; the CC144 mixture is to be preferred where the risk of fire is considered important, and the difference in functioning between the pellet and standard types shows in the nature of the screen.

DATA
Length: 185 mm
Diameter: 65.5 mm
Weight: 550 g
Explosive content: standard 445 g; pellet type 200 g
Range: 20 – 35 m
Duration of smoke: 30 – 40 s
Operating voltage: 1.5 V minimum

Manufacturer
Haley and Weller Ltd, Wilne, Draycott, Derbyshire DE7 3QJ.
Status
Current. Production.

Haley and Weller 66 mm screening smoke grenade

Ground indicating flare

A pyrotechnic flare for indicating landing or dropping zones for parachute troops, or for other indicating or illuminating functions. It is fitted to a spike so that it can be driven into the ground easily, and is then ignited by means of a portfire or other flame.

Two types are manufactured. The 'Flare, Ground Indicating, No 1' is to British Ministry of Defence pattern and uses a metal case; Product Code Y900. The commercial pattern 'Flare, Ground Indicating, Haley & Weller Y920, Yellow or White', uses a plastic casing. All characteristics are similar except that the commercial pattern is less expensive and, since the plastic is consumed during burning, leaves no litter on training grounds.

DATA
Length: 340 mm
Diameter: 35 mm
Weight: 340 g
Explosive content: 300 g

Haley and Weller Type Y920 Ground Indicating flare

Duration: 3 mins minimum
Intensity: 80 000 cd

Manufacturer
Haley and Weller Ltd, Wilne, Draycott, Derbyshire DE7 3QJ.
Status
Current. Production.

Schermuly 38 mm signalling system

The system consists of the 38 mm (1.5 in) signal pistol and 38 mm signal cartridges.

The pistol is constructed principally of high tensile aluminium alloy of rugged construction, giving light weight and durability. The mechanism is of the self-cocking type requiring a long firm trigger pull, the striker having automatic rebound to avoid accidental discharge when closing. The design also incorporates a safety interlock to ensure that the barrel catch is fully engaged before the weapon can be fired.

To load the weapon it is necessary only to press down the serrated part of the barrel catch and allow the barrel to swing open through 90°; enter the round as far as the extractor will permit, and close the pistol until the barrel catch fully engages.

Schermuly also supply the cartridges which are used for a wide range of signalling for all ground, sea and air applications and short duration illuminating purposes. Each consists of a rimmed aluminium case into which the propellant charge and signalling or illuminating star are inserted and sealed with a weatherproof closing disc. The colour of the star is indicated on the side of the case and by means of the NATO night identification markings embossed on the cap. Colours available are red, green, white and illuminating.

DATA
CARTRIDGES
Calibre: 38 mm
Length: 70 mm
Weight: 86 g
Trajectory height: 78 m nominal
Burn time: 6 s nominal (coloured stars)
Illuminating time: 5 s nominal
Candela: Red > 55 000: Green > 33 000: White > 55 000: Illg > 120 000; (average values)
Product codes: (red) 2511: (green) 2512: (white) 2513: (Illg) 2514
NATO Stock number: Illuminating: 1370-99-966-6664

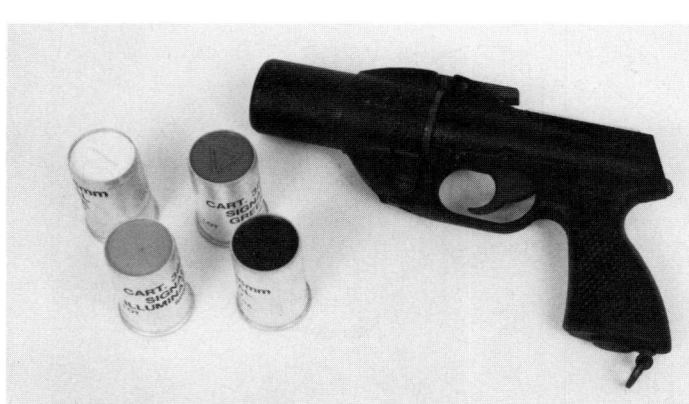

Schermuly 38 mm signal pistol and cartridges

SIGNAL PISTOL
Calibre: 38 mm
Length: 265 mm
Weight: 1.2 kg
Product code: 4092

Manufacturer
Schermuly (Pains-Wessex Ltd), High Post, Salisbury, Wilts. SP4 6AS (a member of the Chemring Group plc).
Status
Current. Available.
Service
World-wide.

Miniflare No 1 Personal Signalling Kit

This kit comprises a pen-sized projector and eight screw-on cartridges in a weatherproof plastic pack. Each cartridge, a self-contained waterproof unit, has a propellant and a star emitting an intense light. For signalling use, red, green and white colour-coded cartridges are available. An important feature of the Miniflare is that when fired vertically as a distress signal, the star burns out well above ground or sea level. With an effective range of 50 to 100 m this system is an ideal illuminator for small patrol and ambush operations.

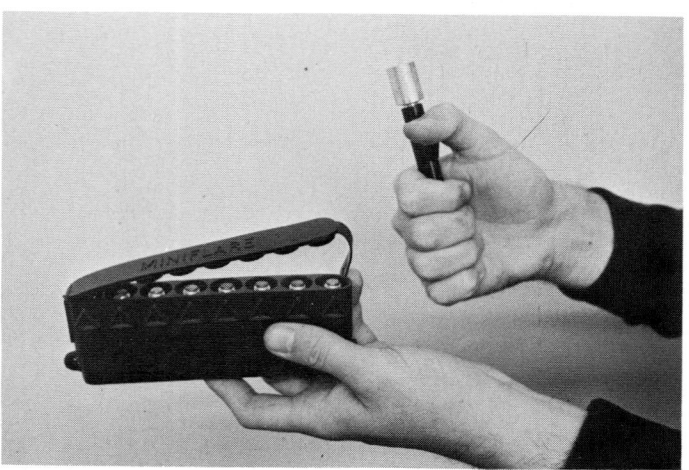

16 mm Miniflare signal kit

The cartridge consists of an aluminium case containing the propellant and star, the case also acting as a projection cup. A waterproofing disc seals the top of the cartridge and the threaded projection at the base contains the percussion cap.

The projector is anodised light alloy with a female thread at the top to take the male threaded cartridge. A trigger is attached directly to the spring-loaded striker.

The plastic pouch has NATO tactile markings for night identification.

DATA
Altitude: 85 m typical
Burning time: 5–7 s nominal

DIMENSIONS
Projector: 120 × 22 mm
Cartridge: 32.5 × 16.7 mm
Pack: 150 × 54 × 19 mm
Weight: (complete) 216 g
Colours and approx intensity: (red) 3000 cd; (green) 1500 cd; (white) 3000 cd
Signal Kit Product codes:
    Red: NSN 1370-99-965-5622
    Green: NSN 1370-99-965-5621
    White: NSN 1370-99-965-5620

Manufacturer
Schermuly (Pains-Wessex) Ltd, High Post, Salisbury, Wiltshire SP4 6AS (a member of the Chemring Group plc).

Status
Current. Production.
Service
UK MoD and world-wide.

Schermuly signal smoke grenade Mk 4

The latest version of the Schermuly signal smoke grenade incorporates substantial design improvements, including a fly-off lever ignition system which provides immediate generation of smoke though apertures in the top of the grenade body. The body is more compact and has been designed and tested for use under all climatic conditions.

Dense smoke is produced for 45 seconds, providing a visible signal for a minimum of 60 seconds. It is designed for use wherever colour definition and good smoke density are vital.

A FOD (Foreign Object Damage) free version is available for use from helicopters etc. A plastic loop is fitted to the grenade which attaches the fly-off lever and safety pin to the body of the grenade.

DATA
Weight: 365 g
Length: 115 mm
Diameter: 62 mm
Weight of smoke composition: 180 g
Burning time: 45 – 50 s
Colours: red 1781; orange 1782; yellow 1783; green 1784; blue/violet 1785; white 1786

Manufacturer
Schermuly (Pains-Wessex Ltd), High Post, Salisbury, Wilts SP4 6AS (a member of the Chemring Group plc).
Status
Current. Available.
Service
World-wide.

Schermuly Mk 4 signal smoke grenade.

Schermuly trip-flare kit

The trip-flare kit is an essential element of perimeter security at night. An immediate benefit of the device is the illumination of infiltrating enemy troops by the instantaneous ignition of the flare which produces a high candela for approximately one minute.

The kit basically consists of a flare with a fly-off lever (FOL) mechanism, two pickets, a spool of wire and a spring. Additional product benefits included in the kit are an anchorage spade to ensure picket stability in light soils and an anti-glare shield which, in its original state, doubles as a packing tin. The flare can also be fitted with an electrical connection for remote firing from a strike detonator. If required, the flare can be used as a hand-thrown incendiary device.

Operating instructions
Remove one kit consisting of two pickets, one spring and one box containing flare and spool.

Force one picket into the ground in the required position. Use the second picket as an axle to deploy wire from the spool. Fit anchorage spades as necessary.

Remove flare from packing. Hook one end of the spring assembly under the leg of the picket and slide the picket arms into the grooves on the side of the flare. Slacken the wing nut and bend wire under and around bolt. Tension wire until locking tongue is in the 12 o'clock position (securing the fly-off lever) when the wing nut is tightened. Re-adjust as necessary.

Remove safety pin ensuring trigger is still in the set position. Keep the safety pin for disarming. The flare is now armed. Fit the anti-glare shield as required.

DATA
Burn time: 45 s nominal
Candela: 60 000 average cd requirements
Total weight of flare body: 360 g
Pyrotechnic weight: 115 g

Manufacturer
Schermuly (Pains-Wessex Ltd), High Post, Salisbury, Wiltshire SP4 6AS (a member of the Chemring Group plc).
Status
Current. Available. Product code: standard kit 2341; with electric ignition 2342; anti-glare shield 2346; spade 2348.
Service
World-wide.

Schermuly trip-flare

51 mm illuminating rocket and launcher

This store has been designed to meet the requirement for an illuminating light source operating at a ground range of 900 m. The 51 mm rocket has a light intensity of 350 000 candela and burns for 30 seconds; deployed at a height of 250 m the parachute flare will adequately illuminate an area some 300 m in diameter.

The rocket, complete with payload, is supplied in an aluminium container/launch tube incorporating the ignition system and is completely weatherproof. It can be supplied with either a percussion or an electrical ignition device, and for both versions a tripod-mounted launcher weighing only 4 kg is available. The optimum launch angle is 30°.

Apart from the illuminating versions the 51 mm rocket can be produced with other payloads such as explosive maroons, radar-reflective chaff or para-target.

DATA
Weight of container/launch tube and rocket: 1.02 kg
Diameter: 55 mm
Length: 385 mm
Light intensity: 350 000 cd average
Burning time: 30 s
Deployment height: 250 m typical at 32°
Ground range for 32° launch angle: 900 m
Weight of infantry-type launcher: 3.9 kg

Manufacturer
Schermuly (Pains-Wessex) Ltd, High Post, Salisbury, Wiltshire SP4 6AS (a member of the Chemring Group plc).
Status
Current. Available. Product codes: Illuminating, electrical ignition 0561; percussion ignition 0562; maroon 0563; para-target 0564; radar chaff 0565.
Service
Armies in Africa, the Americas, Europe and the Middle East.

Loading infantry launcher with 51 mm Parachute Illuminating rocket

Portable Speedline 250 linethrowing unit

Although primarily intended for naval and coastguard use the Speedline 250 linethrower has applications for land forces where the ability to put a line across a difficult obstacle is valuable. Designed for satisfactory operation in extremely rough weather at sea, it is a robust and reliable equipment and is simple to use. The whole unit is completely waterproof and the rocket can be fitted with a rapid-fit buoyant head or a grapnel.

To operate the linethrower the protective front cap is removed and the free end of line attached to a strong point. Then, with the thrower held in one hand with the trigger-guard horizontal and the unit pointed in the right direction, the safety pin is removed and the trigger squeezed. The container is then held until the line is paid out.

To ensure reliable operation the rocket and striker must be replaced at stated intervals. The line can be re-used, if desired, and a special line-former is available to reload it into the container.

The Speedline 250 is now UK Ordnance Board approved and is carried on UK submarines.

DATA
Protechnic content: Rocket 170 g
Line length: 275 m
Line diameter: 4 mm
Breaking strain: 2000 newtons
Operational range: 250 m, nominal, in calm weather
Weight: complete unit 4.6 kg
Length: 334 mm
Container diameter: 190 mm
NATO stock numbers: Speedline 250 body & line 1370-99-743-4384: Rocket 1340-99-734-8853: Striker 1370-99-745-3920
Product code: 0013

Manufacturer
Schermuly (Pains-Wessex) Ltd, High Post, Salisbury, Wiltshire SP4 6AS (a member of the Chemring Group plc).

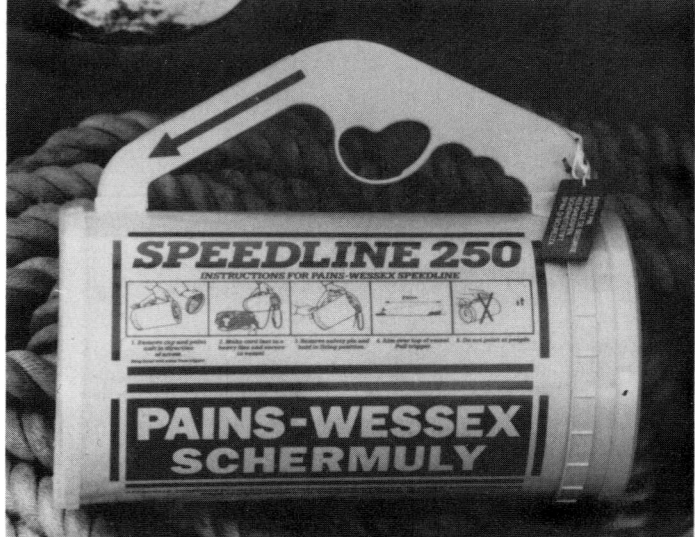

Speedline 250 Portable linethrowing unit

Status
Current. Available.
Service
Merchant Marine world-wide. UK Royal Navy. Armies in Africa, Far and Middle East.

57 mm ML rocket parachute illuminator

The 57 mm ML illuminating rocket system is used for target identification in combat areas, where a degree of high mobility is essential. With its lightweight single- or double-barrelled launcher carried in a rugged man-pack, it is immediately deployable in all roles where short term white light support is required, for example defence perimeters, infantry and armoured support.

It is spin-stabilised, highly accurate and, apart from a minimum flash at launch, has the advantage of being signatureless in flight. Rockets can be supplied with fixed ranges, eg 650, 1000, 1200 and 1800 m or with an infinitely variable electrical fuze to allow range selection within the 650 to 1800 m range band. The flare provides an illumination of 250 000 candela when deployed at a height of 250 m.

The rocket is fired from a recoilless, compact launcher weighing only 2 to 3 kg, ideally suited for infantry use, allowing simple one-man operation in both offensive and defensive situations. The launcher systems are mounted on a tripod which folds away for easy carriage. Luminous levelling and sighting devices are provided which enable the firer to set up the launcher very quickly.

DATA
Rocket weight: 1.8 kg
Diameter: 57 mm
Length: 300 mm
Range at 40° QE: 650–1800 m
Time of flight: approx 7–18 s

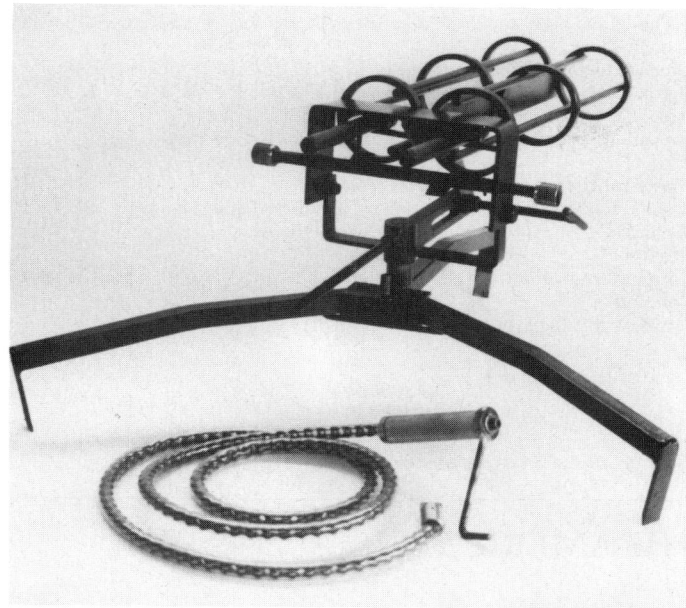

ML lightweight launcher, twin-barrelled version

Loading lightweight launcher

Nominal light intensity: 250 000 cd
Burning time: 30 s
Launch ignition: remote mechanical by Bowden cable, electric, electro-mechanical, or contactless induction.

Manufacturer
ML Lifeguard Ltd, Arkay House, Weyhill Road, Andover, Hampshire.
Status
Current. Production.
Service
NATO, Middle East, Africa.

Brocks Ground Indicating Markers

Brocks Explosives Ltd., a member of the Explosives Development Group, manufacture a range of signal devices designed to mark an area or light up a temporary landing site in situations where normal signal grenades are ineffective.

The **Arctic Grenade** is designed as a wind reference indicator and ground reference point by helicopter or light aircraft crews in situations where the terrain is covered in snow and in temperatures between –46°C and +20°C. By the use of expanding flaps, the grenade lies on the surface of soft snow emitting coloured smoke for a minimum of 25 seconds, whilst the ejected smoke pellets dye the surface of the snow to provide a ground reference mark in white-out conditions. The grenades can be manufactured in alternative colours.

The **Buoyant Grenade** is similar in purpose but is designed to float upon water. In addition to delivering a coloured smoke plume (various colours being available) the emitted dyestuff also stains the surface of the adjacent water making a prominent signal mark.

The **Ground Flare Indicating** is used for indicating landing or dropping areas. It consists of a tin plate body, holding a flare candle, which has a spike on one end to support the flare when in use. The flare is ignited by a portfire.

Manufacturer
Brocks Explosives Ltd, Sanquhar, Dumfries and Galloway DG4 6JP, Scotland.
Status
Current. Production.

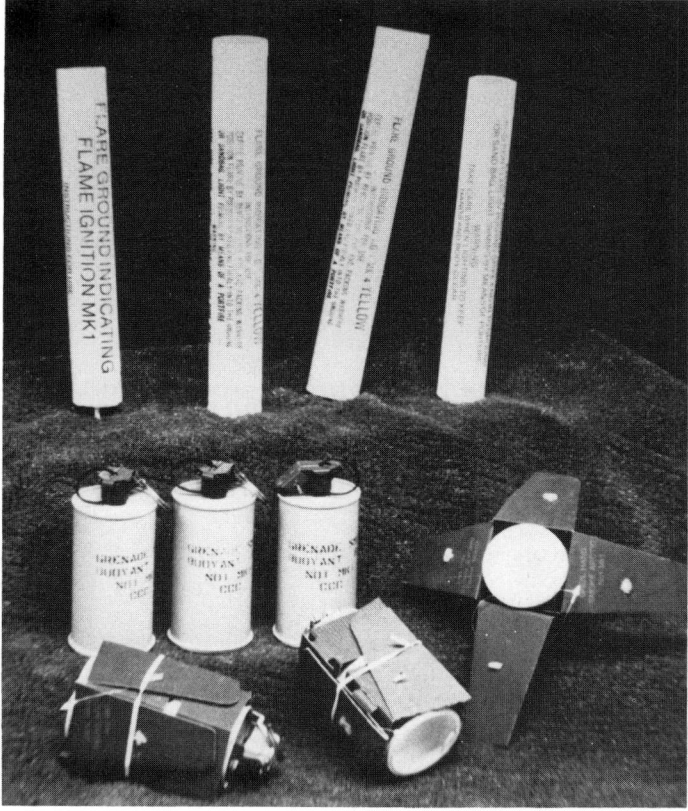

Brocks Ground Indication Markers

Brocks Smoke Grenades and Generators

Brocks manufacture smoke products for both training and operational purposes. The smoke generator and screening grenade produce a dense could of non-toxic grey smoke, screening the movements of infantry and vehicles without the risks associated with white phosphorus.

The signal grenades are manufactured in a range of colours to produce a distinctive plume of coloured smoke, allowing troops to carry oout manoeuvres without radio contact.

The **Smoke Generator** produces a dense cloud of grey/white smoke and burns for approximately 4 to 6 mins after ignition. The HCE smoke carries none of the risk sometimes associated with white phosphorus. The unit is electrically ignited and produces smoke almost immediately. It is robust and waterproof and is suitable for use in any climatic conditions.

The **Screening Smoke Grenade** is more fully described in the Combat Grenades section, page 570.

The **Signal Grenades** are small, lightweight, hand-thrown munitions which produce a dense cloud of coloured smoke shortly after release.

Brocks Smoke Grenades and Generators

They have been approved for British Army service under the following nomenclature:
Grenade, Hand, Signal, Smoke, Blue, L52A2
Grenade, Hand, Signal, Smoke, Green, L53A2
Grenade, Hand, Signal, Smoke, Red, L54A2
Grenade, Hand, Signal, Smoke, Orange, L55A2.

Manufacturer
Brocks Explosives Ltd., Sanquhar, Dumfries and Galloway DG4 6JP, Scotland.
Status
Current. Production.
Service
British Army and export sales.

UNITED STATES OF AMERICA

M18 coloured smoke hand grenade

The M18 coloured smoke hand grenade is used for ground-to-air or ground-to-ground signalling. Grenades may be filled with any one of four smoke colours: red, green, yellow or violet. Each grenade will emit smoke for 50 to 90 seconds.
  Grenade body: the body, of thin sheet metal, is filled with red, green, yellow or violet smoke composition. The filler is topped with a starter mixture.
  Fuze, hand grenade, M201A1: fuze M201A1 is a pyrotechnic delay-igniting fuze. The body contains a primer, first-fire mixture, pyrotechnic delay column and ignition mixture. Assembled to the body are a striker, striker spring, safety lever and safety pin with pull-ring. The split end of the safety pin has an angular spread.

DATA
Weight: 539 g
Length: 146 mm
Diameter: 64 mm
Colour: light green with black markings
Weight of filling: 326 g
Ignition mixture: iron oxide, titanium, zirconium
Delay time: 0.7–2 s

Status
Current.

M18 coloured smoke hand grenade

Service
US Army and others.

HC 110 Federal Laboratories smoke grenade

This small grenade is mainly intended for screening, and the manufacturer claims that it has sufficient capacity to ensure a satisfactory screen for troop use. It emits a dense white cloud for up to 150 seconds after ignition.

DATA
Dimensions: 60 mm diameter × 148 mm length
Weight: 480 g
Delay: 2 s
Discharge time: 90 – 150 s
Fuze type: U.S. MIL SPEC
Filling: HC smoke composition
Weight of composition: 290 g

Manufacturer
Federal Laboratories Inc, Saltsburg, PA 15681.
Status
Current. Available.
Service
US police and military.

HC 110 Federal Laboratories smoke grenade

F18-A Federal Laboratories hand smoke grenade

This coloured smoke hand grenade is used for ground-to-air or ground-to-ground signalling. Grenades are available in any of seven colours: red, green, yellow, orange, violet, blue and black.

DATA
Dimensions: 60 mm diameter × 148 mm length
Weight: 240 g
Delay: 2 s
Discharge time: 40 – 50 s
Fuze type: M201A1
Type of discharge: smoke, pyrotechnic burning

Manufacturer
Federal Laboratories Inc, Saltsburg, PA 15681.
Status
Current. Available.
Service
Foreign military.

Federal F18-A coloured smoke grenades

Federal 26.5 mm signal cartridges

These cartridges are used for both illuminating and signalling and can be fired from the 26.5 mm gas and flare pistol or any 1 in Very pistol, to an ejection altitude of 250 ft (77 m).

There are cartridges with parachute illuminating flare; parachute red distress flare; or meteor flare in red, green, white and yellow.

DATA
Dimensions: 26.5 × 79.6 mm or 26.5 × 140.8 mm
Weight: parachute flare 85 g; meteor flare 57 g
Height of ejection: 77 m
Burning time: 5 – 6 s
Intensity: varies from 35 000 cd (illuminating) to 18 000 cd (yellow meteor) depending upon colour

Manufacturer
Federal Laboratories Inc, Saltsburg, PA 15681.
Status
Current. Available.
Service
Foreign military.

Federal 26.5 mm signal cartridges

Cyalume Lightsticks

Though not strictly pyrotechnics, since they do not burn, these devices must be considered under this heading since they perform the same function as pyrotechnic flares. The Cyalume Lightstick is a sealed plastic tube containing a liquid chemical composition which is inert until the tube is bent between the fingers and shaken. It then emits light without heat, flames or sparks; depending upon the type of light in use, this emission will last from 30 mins to 12 h. The lights are provided in various colours so that they may be used for warning, signalling or illumination purposes. In military use they can be used as minefield, route or perimeter markers, drop-zone markers, river-crossing guides, or for map reading or illuminating the interior of command posts or communication centres. The lights work equally well under water and can be used by demolition frogmen. A rugged plastic carrying case (the Combat Light Device) has a clip by which it can be attached to clothes or equipment and is fitted with an adjustable shutter by means of which the light can be regulated or occluded as desired. There is also a Personnel Marker Light which can be attached to flotation devices; this clips to the clothing or life-jacket and carries a permanently-fitted Cyalume Lightstick which can be activated by squeezing it, after which it acts as a marker visible up to 1 mile and lasting for 8 h.

Types of Lightstick available include:
4 in General-Purpose (green)
6 in 30-mins High Intensity (yellow)
6 in 12-h General-Purpose (green)
6 in 12-h Red
6 in 8-h Blue
6 in 12-h Yellow
6 in 12-h Orange
6 in Non-visible Infra-Red
Personnel Marker Light (green-yellow, 8 h)
Combat Light Device (holder for 6 in lights).

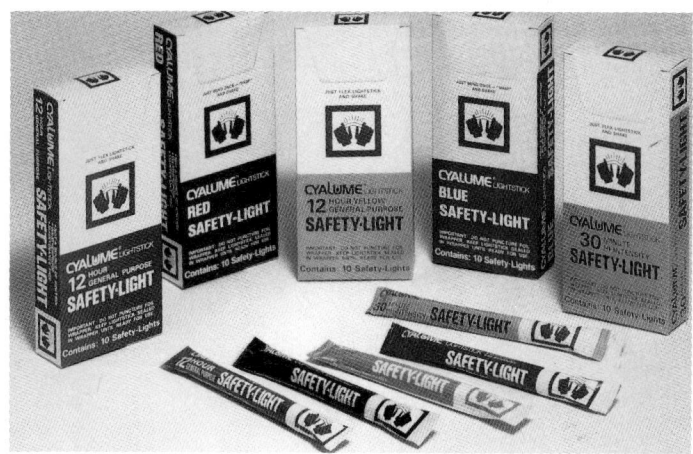

Cyalume Lightsticks

Manufacturer
American Cyanamid Company, Chemical Light Department, Wayne, New Jersey 07470.
Status
Current. Production.
Service
US Armed Forces and Coast Guard.

Ancillary Equipment

Viewing and surveillance devices
Sighting equipment
Personal Protection

Viewing and Surveillance Devices

FRANCE

OB-44 SOPELEM night observation binoculars

The OB-44 low light-level sight is designed for general night observation purposes and is capable of use from a moving vehicle. The latter feature is assisted by the eyepiece design which requires no interpupillary adjustment (between 58 and 72 mm), has no exit pupil and no eyepiece focussing, which enables an observer to retain spectacles. A special high performance refractive lens, designed for low light-level applications, is used and an inverter type micro-channel intensifier tube with manual and automatic gain control is employed.

DATA
Magnification: × 2.5
Field-of-view: 11°
Resolution: $\geqslant 0.7$ mrad at 10^{-1} lux
Focus range: 20 m to infinity
Lens: dioptre 120 mm focal length

Aperture: f1.6
Image tube: 25 mm type S-20ER cathode
Resolution: 25 lp/mm
Range performance: 600 m for human targets and 1200 m for tank targets at 10^{-1} lux; 450 m and 900 m at 10^{-3} lux
Dimensions: 90 × 120 × 330 mm
Weight: 2.25 kg
Power: 2 × 1.5 V commercial standard batteries

Manufacturer
SOPELEM, 19 Boulevard Ney, PO Box 264, 75866 Paris Cedex 18.
Status
Production.
Service
French Army and foreign forces.

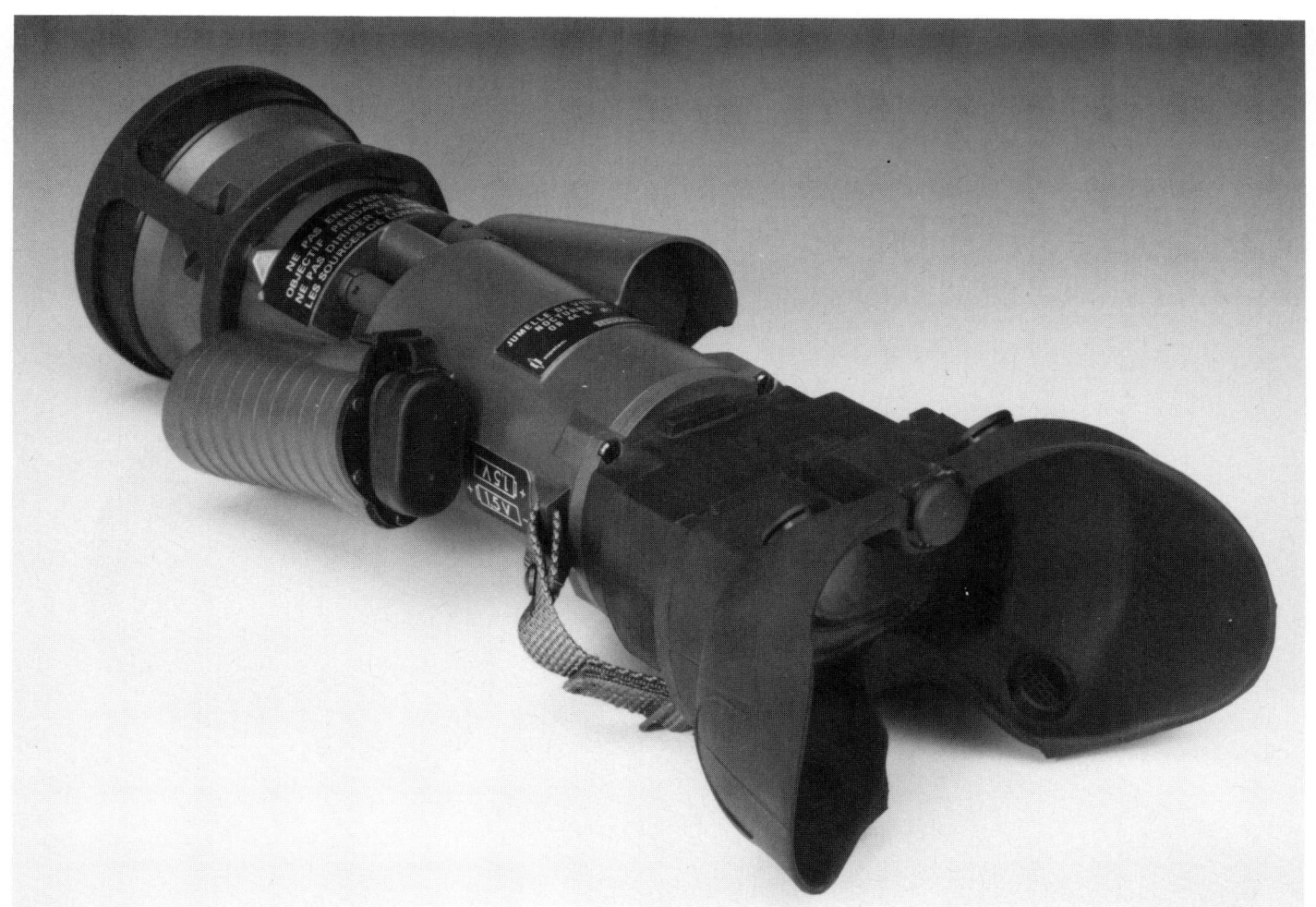

OB-44 night observation binocular viewer

JUNON SOPELEM night flying goggles

The JUNON goggles are designed for night observation, night flying, and instrument and map reading in helicopters and aircraft. They incorporate two second or third generation micro-channel amplifiers associated with the appropriate optics.

The goggles can be mounted in a fixed position on the pilot's helmet by using a special support. The power pack can be fitted at the back of the helmet. Image intensification is performed by two third generation micro-channel tubes with double proximity focus; they can be replaced by second-generation tubes if required.

DATA
Magnification: × 1
Field-of-view: 40°
Weight: 550 g (binoculars only); 910 g (complete equipment)
Power supply: on-board system (28 V) or 1 × PS31 3.5 V battery

Manufacturer
SOPELEM, 19 Boulevard Ney, PO Box 264, 75866 Paris Cedex 18.
Status
Current. Production.
Service
French Army and Air Force, and foreign forces.

JUNON night observation goggles

LISP day-night sight

The LISP sight is designed for use in both day and night observation and can be used for night driving or command and surveillance functions, operating on the light intensification principle. With the addition of an afocal module the sight can also be employed for night observation. The afocal module is designed to provide a monocular daylight observation function.

DATA
Day Path: monocular:
Magnification: 5.5 ×
Other features: graticule; anti-laser protection

Night path: binocular monotube:
Magnification: 1 × ; 3 × ; 5.5 ×
Field-of-view: 40° (1 ×); 12° (3 ×); 7° (5.5 ×)
II Tube: 18 mm 2nd or 3rd generation
Weight: 430 g (1 × power)
Power supply: 2 × AA batteries 1.5 v

Manufacturer
SOPELEM, 19 Boulevard Ney, PO Box 264, 75866 Paris Cedex 18.
Status
Prototype.

Sopelem LISP day-night sight with afocal module

OB-41 night driving binoculars

The OB-41 binoculars have two separate vision paths, each consisting of a micro-channel intensifying tube with double-proximity focalisation. In this way comfortable stereoscopic vision is achieved. The two body elements of the binoculars are identical and completely interchangeable. The unit is mounted on a close-fitting face mask and interpupillary adjustment is carried out by rotating the body elements in the interior of the mask. The battery container, switch and an additional light source (AsGa diode) are incorporated in the mask; the additional light source is faint but permits the OB-41 to be used in very bad light conditions (less than 0.1 millilux) up to a distance of about 10 m.

The OB-41 is compatible with the various types of helmet used by infantry, tank drivers, helicopter pilots and other servicemen; and it can be used for driving and other night activities such as minelaying, bridging and vehicle repair, even at very low light levels.

DATA
Magnification: × 1
Field-of-view: 33°
Focus range: 0.3 m to infinity
Weight: 900 g
Power supply: Mallory 2.7 V battery

Manufacturer
Télécommunications Radioélectriques et Téléphoniques (TRT), 88 rue Brillat-Savarin, 75640 Paris-Cedex 13.

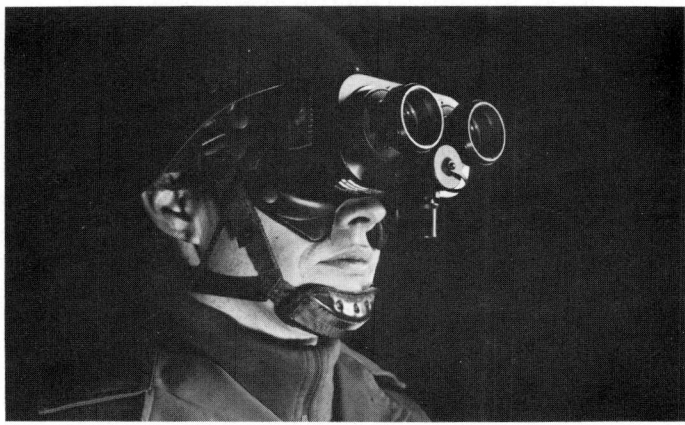

OB-41 night driving binoculars

Status
Current. Production.
Service
In several countries with military, police and customs forces.

OB-42 passive binoculars

The OB-42 binoculars comprise two independent light paths, each having an object lens, micro-channel intensifier tube and eyepiece. The right-hand unit is fitted with an illuminated reticle graduated in mils in elevation and azimuth; the degree of illumination is adjustable by the operator.

OB-42 passive binoculars

The binoculars are intended for surveillance tasks; they are small enough to be used conveniently from a tank turret or can be tripod-mounted. They are sufficiently sensitive to identify a tank at 600 m distance in a light level of one millilux, and their two separate light paths ensure comfortable stereoscopic vision. Automatic regulation of luminance gain, incorporated in the power supply, enables the binoculars to be used in light conditions from 0.1 to 100 000 millilux.

DATA
Magnification: × 4
Field-of-view: 8°
Focus: 20 m to infinity
Power supply: 2.7 V battery
Weight: 2.2 kg

Manufacturer
Télécommunications Radioélectriques et Téléphoniques (TRT), 88 rue Brillat-Savarin, 75640 Paris-Cedex 13.
Status
Current. Production.
Service
French Army and Navy and other countries.

Caliope thermal imaging device

Caliope is a passive thermal imager, designed for various needs: battlefield surveillance, night capability for weapon systems, protection for battle ships, surveillance for sensitive targets.

This imager operates in the 8 – 13 μm infra-red band, and the visible image is given either by a monocular or on TV display.

DATA
Weight: 8 kg
Overall dimensions: 520 × 220 × 190 mm
Field-of-view: 6° × 3°
Magnification: GX7
Scanning: serial-parallel
Detectors: detectors' array Cd Hg Te
Cooling process: Joule-Thompson relaxation or cooling machine
Display: direct view by monocular, or TV display
Power consumption: 7 W (in Joule-Thompson version)

Manufacturer
Télécommunications Radioélectriques et Téléphoniques (TRT), 88 rue Brillat-Savarin, 75640 Paris-Cedex 13.

Caliope thermal imaging device

CHINA, PEOPLE'S REPUBLIC

Type 1985 passive night vision goggles

These are small image-intensifying goggles of the usual type, based on a single stage intensifying tube. They are suitable for most night activities involving the use of the hands, such as driving, loading and unloading, operating weapons, and for map and document reading with a suitable IR source.

DATA
Magnification: × 1
Field-of-view: 44°
Resolution: 2.8 mRad at 1 × 10^{-2} lux; 4 mRad at 1 × 10^{-3} lux
Focus range: 0.3 m − ∞
Dioptre setting: ± 5
Weight: 960 g
Power supply: 0.5 Ah NiCd cell

Manufacturer
China North Industries Corp, PO Box 2137, Beijing.
Status
Current. Production.
Service
Chinese army.

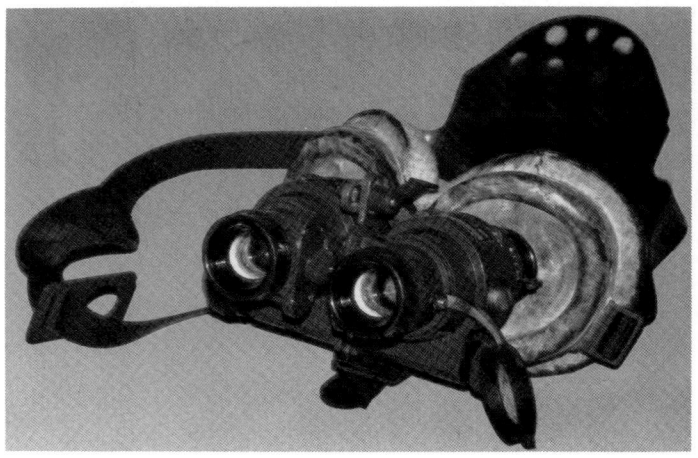

Type 1985 passive night vision goggles

GERMANY, WEST

Euroscope H II night viewing device

The Euroscope H II is a handy low-weight viewer for several applications. It uses a 25 mm second-generation micro-channel image intensifier tube. A wide range of accessories is available covering several different objective lenses, eyepieces, adapters for SLR cameras, motion picture or video cameras and laser illuminators.

DATA
Standard objective: 75 mm f/1.4
Magnification: 3 × (for 75 mm)
Field-of-view: 19°
Dimensions: 280 × 80 × 255 mm

Weight: 1.68 kg complete
Power supply: 2 × AA/LR6 batteries
Battery life: Approximately 50 h
Alternative lenses: 50 mm f/0.95; 135 mm f/1.8; 17.5 — 108 mm f/1.8 zoom

Manufacturer
Euroatlas GmbH, Zum Panrepel 2, D-2800 Bremen 45.
Status
Current. Available.

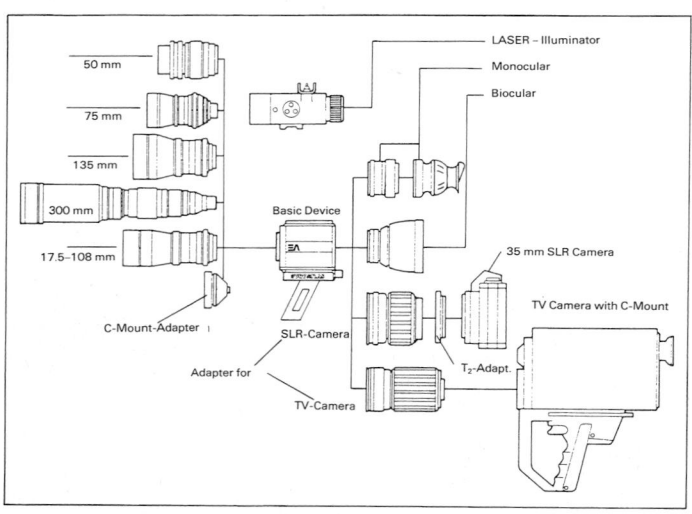

Euroscope H II accessories

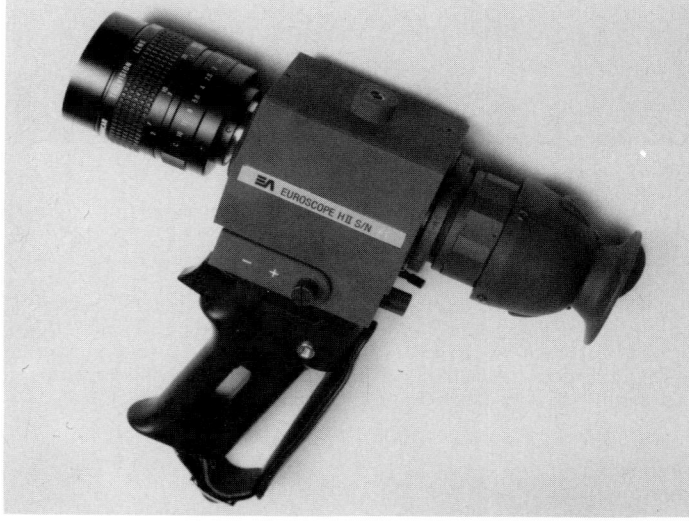

Euroscope H II night viewing device

Eurolook B II night vision binoculars

The Eurolook B II are lightweight general-purpose night vision binoculars for long-range observation and surveillance. They are designed for hand-held use by reconnaissance patrols, vehicle commanders, artillery observers and paratroops, or for law enforcement and police forces. The low weight allows permanent use over long periods without fatigue.

The device offers the proven performance of a second-generation 18 mm image intensifier MCP tube with a fast focussing optical system from 3.2 m to infinity. It is extremely easy to operate and fulfils military standards due to the stringent use of NATO-introduced assemblies.

DATA
Magnification: 3.8 ×
Field-of-view: 10.8°
Objective lens: 95 mm, f/1.3
Focus range: 3.2 m to ∞
Brightness gain: > 1250
Weight: 1.2 kg
Power supply: 2 × AA/LR6 batteries
Battery life: Approximately 65 h (at 20°C)

Manufacturer
Euroatlas GmbH, Zum Panrepel 2, D-2800 Bremen 45.
Status
Current. Available.

Eurolook B II night vision binoculars

Single-tube night vision goggles

These low-weight goggles use a single object lens, single amplifier and binocular eyepiece. They can be mounted on a face mask or helmet, or used as a hand-held viewer. The design of the mount permits the wearer to swing them out of his line of vision if necessary. The operating switch gives on/off positions and also switches on a laser diode-illuminator for map or message reading. The eyepiece has been designed to avoid fogging at low temperatures.

The image intensifier is an 18 mm micro-channel second-generation tube; it can be replaced by a third-generation amplifier, by modification of the object lens assembly, in the future.

DATA
Magnification: unity
Field-of-view: 40°
Object lens: 26.6 mm f/1.0
Focussing range: 25 cm to infinity
Weight: (goggles) 450 g; (face mask) 195 g
Power supply: 2 × AA/LR6 batteries
Battery life: approximately 65 h (at 20°C)

Manufacturer
Euroatlas GmbH, Zum Panrepel 2, D-2800 Bremen 45.
Status
Current. Available.

Single-tube night vision goggles

Eurocat day/night TV camera

The DNTV 2001 'Eurocat' is a newly-developed day and night CCD TV camera using approved assemblies combined in a unique modular system which is capable of operating 24 hours continuously, controlled by intelligent electronic circuits. The scenery illumination of 10^5 (brightest sunlight) to 5×10^{-5} lux (clouded night sky) is easily handled by the camera without any loss of image quality.

Hand-held, mobile or stationary versions offer a wide field-of-observation applications including the possibility of video tape documentation.

Various accessories such as different lenses (also motorised zoom lenses), weatherproof housing, remote-control, different power supply adaptors or sources etc, serve a modularity realising most of the tactical requirements.

DATA
Scenery illumination: 5×10^{-5} lux to 10^5 lux, automatically controlled
Spectral response: 400 nm — 920 nm
Resolution: 320 (H) × 380 (V) TV lines
Video output: 1 Vpp/75Ω, CCIR
Objective lens: 50 mm f/1.4 (standard), C-mount
Length: 380 mm without lens
Width: 160 mm
Height: 120 mm without handgrip
Weight: 3.9 kg without lens and grip
Power supply: 12 VDC, 6.5 W

Manufacturer
Euroatlas GmbH, Zum Panrepel 2, D-2800 Bremen 45.
Status
Current. Available.

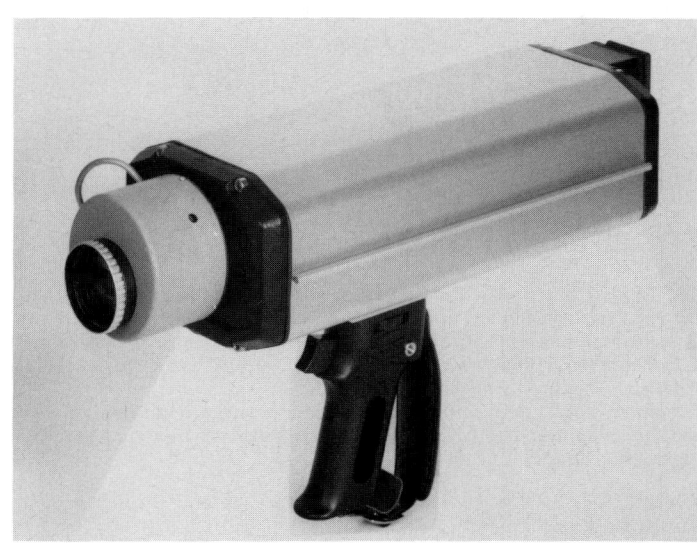

Eurocat day/night TV camera

ISRAEL

IL-7 Laser IR Illuminator

Image intensifying viewers and goggles rely entirely upon amplifying the contrast of extremely low light levels. In absolute darkness they cannot function, and there are occasions, when using night vision devices, when the area to be viewed falls into this category and vision fails. The IL-7 illuminator is a small lightweight clip-on unit which can be attached to any night vision device and which projects a beam of infra-red light into the field-of-view of the device. This beam can be varied from a point, for designating targets, to a 40° spread for overall illumination of a scene.

DATA
Weight: 116 g including battery
Dimensions: 60 × 45 × 20 mm
Wavelength: 850 nM
Output power: 2 mW average (standard version); 15 mW average (long range version)
Power supply: 3.5 V lithium battery, AA size
Battery life: 15 – 20 h continuous operation

Manufacturer
International Technologies (Lasers) Ltd, 12 Hachoma Street, PO Box 4099, 75140 Riion.
Status
Current; available.

IL-7 Mini-Laser IR Illuminator

NVG-7 single eye night vision goggle

The principal disadvantage of conventional night vision goggles is that there is a transitional period of over a minute while the pupil of the operator's eyes dilate after removing the goggles, a period in which he is effectively blind. This new and exclusive concept of single-eye night vision is based on independent use of each eye. One eye is provided with night vision capability, while the other is free and uncovered. The user can elect which eye to use in varying circumstances, and the design of the goggle allows it to be used with either eye, as the user prefers. The design also allows the optical unit to be flipped up and out of the field of vision, so that the user is immediately ready for any activity.

There are two versions; one (shown here) is for wear without helmets, while the other is designed to fit on a standard military helmet.

DATA
Magnification: 1 ×
Field-of-view: 40°
Dioptre adjustment: + 2 to -6
Power supply: 2.7 v mercury standard; others optional
Weight: 675 g (700 g helmet fitting type)

Manufacturer
International Technologies (Lasers) Ltd, 12 Hachoma Street, PO Box 4099, 75140 Rishon-Letzion.
Status
Current; available.

NVG-1 single-eye night vision goggle

LPL-30 long range laser pointer

The LPL-30 is a pocket-size, infra-red laser, long range pointer specially designed for commanders and soldiers on night missions. The pointer will operate in conjunction with any type of night vision system, allowing the user to point out targets for distances up to 4 km. It can also be used for signalling between two points up to 40 km apart.

For more local use, an optional beam expander can be fitted, which spreads the illumination so that the device can be used as a torch in conjunction with night vision goggles.

DATA
Output power: 15 mW
Beam divergence: 0.3 mRad
Wavelength: 830 nM
Power supply: 2 × 1.5 V AA alkaline or equivalent
Operating time: 5 to 10 hours
Dimensions: 124 × 40 × 20 mm
Weight: 195 g without batteries

Manufacturer
International Technologies (Lasers) Ltd, 12 Hachoma Street, PO Box 4099, 75140 Rishon-Letzion.
Status
Current; available.

LPL-30 long range laser pointer

ORTEK Type 5157 night vision goggles

The Type 5157 is a passive night viewing device which uses moon or starlight as its source of illumination. The optical system consists of two focusable object lenses, two image intensifier tubes and two adjustable eyepieces. Each tube contains a built-in high voltage power supply with automatic brightness contol and includes bright source protection. Two AA batteries are housed in the goggles body.

For near viewing tasks, such as control panel or map reading, a built-in auxiliary infra-red light source is provided. For maximum safety this source is only active when the appropriate switch is pressed.

The instrument is fixed to a face-mask secured to the wearer's head by means of self-tightening straps. After switching on and adjusting the instrument, the wearer has his hands free for other tasks.

DATA
Weight: 1.0 kg
Magnification: × 1
Field-of-view: 40°
Focus range: 25 cm to infinity
Resolution: 1.5 mRad
Eyepiece adjustment: +2 to -6 dioptres
Interpupillary distance: 60 to 72 mm
Object lens: 26.8 mm
Luminance gain: 7500 fl/fc
Power supply: 2 × AA batteries
Battery life: approx 50 h

Manufacturer
ORTEK Ltd, PO Box 388, Sderot 80100.

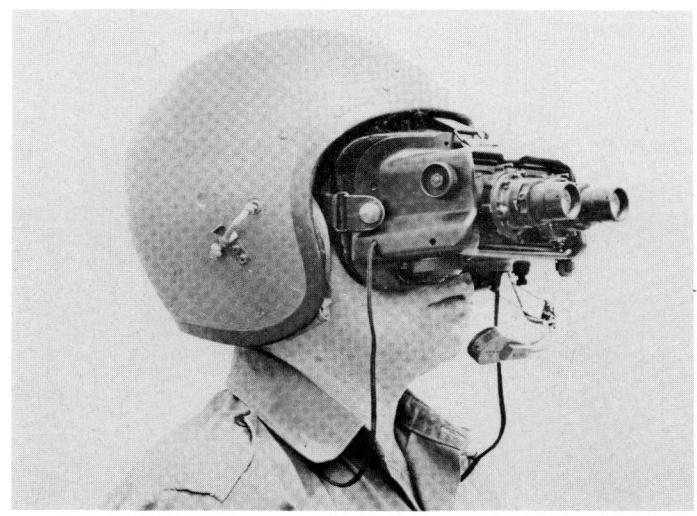
ORTEK Type 5157 night vision goggles

Status
Current. Production.

ORTEK ORT-PS1 pocket scope

The Pocket Scope is a small night viewing device suitable for a variety of night surveillance activities. It can be used as a self-contained passive night viewer (one-channel goggle) with the advantages of low weight and very small size. It can also be attached to day aiming devices (rifle telescopes, gun sights) and converts them within seconds into passive night aiming telescopes. It can be attached to ordinary day binoculars, turning them into a second-generation night viewer with × 7 magnification. It is supplied with adapters for standard 7 × 50 and 8 × 30 binoculars and for the ORTEK 'Hit Eye 1500' rifle telescope, according to requirements.

DATA
Magnification: x1
Field-of-view: 40°
Resolution: 1.5 mRad
Dynamic range: from 10^{-5} to 10^{-2} FC
Luminance gain: 7 500 fl/fc
Focus range: 25 cm to infinity
Objective lens: 26.8 mm
Dioptre adjustment: -2 to +6
Power supply: DL 123A 2.8 V lithium type battery
Battery life: 50 h minimum
Weight: 400 g

ORTEK ORT-PS1 pocket scope

Manufacturer
ORTEK Ltd, PO Box 388, Sderot 80100.
Status
Current. Production.

MILOS miniature long-range TV observation system

The ELOP MILOS is a lightweight, portable, long-range television observation device. The system has two TV fields-of-view with an integrated laser rangefinder and a very accurate rotary tilt pedestal. It is a stand-alone system with a night vision option. It can also be integrated with other systems.

MILOS can be operated on a tripod, from a vehicle, or from land or marine stations, for various applications such as forward observation, direction and co-ordination, or surveillance and border guarding.

DATA
Rangefinder coverage: at least 40 km to any visually recognisable target
Narrow field-of-view: f = 2000 mm (6.5 × 4.5 mRad)
Wide field-of-view: from 200 – 600 mm zoom (42 × 32 to 14 × 10 mRad)
Rangefinder resolution: 0.3 mRad
Sensitivity: 53 dB
Electrical pedestal resolution: 50 mrad
Power supply: 24 V vehicle battery, NiCd or lithium battery pack
Observation/rangefinder head weight: 13 kg
Dimensions: 230 × 250 × 530 mm

Manufacturer
ELOP – Electro-Optics Industries Ltd, Advanced Technology Park, Kiryat
Weizmann, PO Box 1165, Rehovot 76110.
Status
Current. Production.

MILOS miniature long-range TV observation system

ITALY

M167 weapon sight

The M167 is a longer-range instrument fitted with a second-generation tube.
It is intended for medium-range surveillance at the forward edge of the
battlefield and for mounting on crew-served support weapons. In the
surveillance role it is mounted on a damped tripod or on a vehicle mount,
however, it can also be used with a handgrip, though this requires to be rested
to maintain a clear picture free from shake.

The mounting bracket on the sight complies with STANAG 2324.

M167 weapon sight

DATA
Weight: 2.2 kg
Length: 371 mm
Diameter: 141 mm
Lens: 136 mm
Relative aperture: 1.2
Magnification: × 7.5
Focus range: 15 m to infinity
Tube: 2nd generation 20/30 mm tube
Battery: 3 V lithium cell
Battery life: 30–40 h continuous running
Operating temperature range: -45 to +60°C

Manufacturer
Aeritalia SpA, Avionic Systems and Equipment Group, Viale Europa, 20014
Nerviano (Milan).

Status
Current. Available.
Service
Not known.

OGVN6 night vision binoculars

This optical system, utilising mirrors with high characteristic, is equipped
with a second-generation micro-channel light-amplifier tube which has
Automatic Brightness Control (ABC) and a system for automatic switch-off
when the brightness of the viewing field rises above a certain level.

The instrument may be used for night surveillance, for naval manoeuvres,
for battlefield observation or for police surveillance and patrolling.

OGVN6 night vision binoculars

DATA
Dimensions: 360 × 130 × 110 mm
Weight: 1.95 kg
Magnification: × 4
Field-of-view: 11°
Dioptre adjustment: ± 5 dioptres
Interpupillary adjustment: 58 – 72 mm
Focus: Adjustable 12 m to infinity
Resolution: 0.8 mRad at 10^{-3} lux
Power supply: 2 manganese alkali batteries, 1.5 V LR14
Battery life: 100 h at 25°C

Manufacturer
Officine Galileo, Via Albert Einstein 35, 50013 Campi Bisenzio, Firenze.

Status
Current. Production.
Service
Italian Naval Special Forces.

TURBO thermal observation unit with rangefinder

TURBO has been specifically designed in order to allow round-the-clock capability for surveillance, observation, target spotting and target or impact point accurate location.

TURBO consists of a modified Galileo VTG 120 thermal sight integrated with a laser rangefinder on an accurate goniometric head. The unit is manportable and can be easily assembled and deployed by means of the built-in levelling and north-alignment devices.

The position data are given in real time by the laser rangefinder and by the goniometric head with great accuracy and reliability. The thermal sight gives significant improvement when observation with conventional optical equipment would be hampered by camouflage, smoke or night-time conditions.

DATA
Laser: NdYag
Range: 300 — 20 000 m
Range resolution: ±10 m
Pulse repetition rate: 2 sec
Display: 5 digits
Goniometric range: elevation ±400 mils; azimuth 6400 mils
Goniometric accuracy: ±1 mil
Thermal operating band: 8-12 μm
Field-of-view: wide 3.3 × 6.6°; narrow 1.1 × 2.2°
Frame frequency: 25 Hz
Typical detection range: tank >3500 m; 105 mm shell burst >6000 m

Manufacturer
Officine Galileo SpA, Via Albert Einstein 35, 50013 Campi Bisenzio, Firenze (FI).
Status
Pre-production.

TURBO thermal observation unit with laser rangefinder

JAPAN

Stabiscope Fujinon binoculars

The Stabiscope is a high power binocular with an integral gyro-stabilising system which allows it to be used when the viewing platform is moving in different planes. It provides ×10 magnification, at which level very small vibrations or movements by the operator will blur the image. The Stabiscope not only compensates for the normal body tremors which cause poor images, but it will also maintain a stable image in a vehicle, boat or aircraft.

When not used for stabilised viewing, the Stabiscope becomes a normal pair of optical binoculars. When a stabilised image is required the 'on-off' switch is put to 'on' and this starts a small internal gyroscope. Within one minute the gyro has reached its operating speed of 12 000 rpm. It is attached to a double gimbal assembly which carries the two prisms and at fully gyro speed this assembly 'floats' and stabilises the image. The binocular casing can shake and vibrate, but it does not affect the line of sight between the operator's eye and the object. The prisms control the line of sight and, if they remain steady, then so does the image presented to the eye. It does not matter that the lenses move for they have no effect on the image until they move so far as no longer to be pointing at the object at all, but that presupposes an abnormal move of the operator.

Power for the gyros comes from an internal battery, but for prolonged viewing the Stabiscope can be connected to any 12- or 28 V external supply.

The main effect of the stabilisation is in the vertical plane because this is the more common one for movement and it is the one that is more difficult for the operator to control. Horizontal movement is controlled by the gyro, but to a lesser extent; however this horizontal control limits rotational tracking to a rate of 5°/s. With full stabilisation in operation the Stabiscope cannot be used for high speed tracking in azimuth.

The outer casing is rubber-coated to reduce shock damage and resist weather while the optics are coated to permit maximum light transmission.

DATA
Dimensions: 200 × 180 × 96 mm
Weight: 2 kg
Magnification: × 10
Objective lens: 40 mm
Exit pupil: 4 mm
Field-of-view: 90 mil (5°)
Focus range: 30 m to infinity
Battery: 6 AA Type NiCd
External power: 12-28 V DC
Gyro motor speed: 12 000 rpm

Manufacturer
Fuji Optical Company Ltd, Omiya City.
Status
Current. Available.
Service
Not known.

Stabiscope Fujinon binoculars

NETHERLANDS

HNV-1 holographic night vision goggles

The HNV-1 lightweight single tube goggles give the user the unique opportunity of always having an image of the real world, irrespective of possible flares or flashes, as well as a large peripheral vision, during the night. Owing to the use of Holographic Optical Elements (HOEs) the HNV-1 goggles provide the unique feature of a see-through image and allow a compact optical layout. This use of holographic technology differentiates the HNV-1 goggles from all other night vision goggles currently obtainable. Wearing HNV-1 goggles allows the operator to perform night tasks under even better conditions than could ever be achieved with classical goggles: driving vehicles, flying helicopters and low-speed aircraft, map reading, maintenance, loading and unloading at night, mine laying, night patrols and surveillance. HNV-1 goggles are perfectly suited for Special Task Forces and also find many applications in non-military areas.

Oldelft have been granted substantial orders from the Belgian Army and from a Special Task Force of one NATO country for series production and supply of their holographic navigation equipment.

Oldelft HNV-1 Holographic goggles

The HNV-1 goggles see-through image give the user the opportunity of always having an image of the real world, without having to tilt or remove the goggles. Owing to this dual view the HNV-1 goggles can even cope with the fast fluctuations of light conditions such as flares or gun flashes. The transparent visor and the open frame construction give the user a large peripheral vision. Moreover the visor completely blocks the stray light of the intensified image and guarantees the undetectability of the user even at very low light levels.

The HNV-1 goggles fit directly on the head and are compatible with tank drivers' helmets. Adjustable fastening straps assure perfect adaptability to any head size or shape. Special care has been taken to locate the centre of gravity as close as possible to the user's face, ensuring maximum comfort. Moreover, a specially designed chin support provides maximum stability.

Because of the large exit pupil diameter no accurate positioning of the goggles on the head is necessary, and a large range of interpupillary distances is covered without any adjustment. Sharp focus is easily achieved by simple adjustment of one knob. For close work such as map and document reading, there is a built-in auxiliary light source emitting light in the IR spectrum so that it cannot be detected by the naked eye.

The goggles incorporate 14 modular parts; this modular construction provides significant logistic benefits and simple maintenance procedures.

The quality control system applied by Oldelft in the development and construction of their night vision equipment is in accordance with the NATO AQAP-1 directives.

DATA
Optical power: x1
Distortion: less than 7%
Field-of-view: night image: vertical 30°, horizontal 40°, overlapping L + R > 16°. See-through image; unobstructed peripheral vision.
Focus: continuously adjustable from 25 cm to infinity
Interpupillary distances: from 57 to 69 mm without adjustment
Resolution: full moon (10^{-1} lux): better than 3 mRad. Starlight (10^{-3} lux): better than 5 mRad
Tube: 18 mm 2nd generation wafer with narrow band phosphor
Batteries: 2×1.5 V standard AA size
Weight: approx. 1.0 kg
Centre of gravity: less than 50 mm from the forehead

Manufacturer
Oldelft (B.V. Optische Industrie 'De Oude Delft'), PO Box 72, 2600 MD Delft.
Status
Current. Production.
Service
Belgian and other armies.

GS6TV low light television system

This system consists of a high quality military image intensifying sight integrated with a solid state TV camera, a control box, a TV monitor (optional) and cabling. The GS6TV provides the operator with a monochrome monitor image including a ballistic graticule, giving complete control of graticule brightness and focussing from the control box.

The combined night sight and camera is intended for use in adverse environments, and in most applications the control box and monitor are placed in a protected area. The system can thus be used on armoured vehicles for aiming and observation, or in buildings and tents for observation and surveillance. Other applications can be considered, including those where a

GS6TV LLTV system

video distribution network is required. It can also be used as a stand-alone observation device with an optional tripod.

DATA
Tube: 2nd generation image intensifier
Field-of-view: 5.9° horizontal × 4.4° vertical; 3.7°H × 2.8°V optional, giving better resolution
Resolution: Better than 1 mrad at 1 mlux and 85 per cent contrast
Sensor: CCD, 576V × 604H pixels, integrated with tube
Video output: CCIR standard; RS170 optional
Remote-controls: on/off, graticule brightness and focus

MTBF: 1500 hours minimum
Dimensions, sight: 440 mm long, 180 mm diameter, nominal
Weight: Sight and camera approx 3.5 kg without mounting; control box approx 1 kg
Power supply: 24 vDC, 15 W

Manufacturer
Oldelft (B.V. Optische Industrie 'De Oude Delft'), PO Box 72, 2600 MD Delft.
Status
Current. Production.

Type Cyclop PC1MC night vision goggles

The Cyclop night vision goggles type PC1MC are a small lightweight night viewing device based on the use of a single second-generation image intensifier tube, lightweight housing materials and high speed optics.

These goggles make it possible to carry out night-time tasks without artificial illumination and with both hands, such as driving of vehicles,

patrolling, surveillance, first aid, map reading, repair duties, loading and unloading of vehicles, night flying and landing helicopters and low-speed aircraft, and working in bunkers and tunnels.

DATA
Length: 170 mm
Width: 131 mm
Height: 82 mm
Weight: approx 600 g without head support
Magnification: × 1
Field-of-view: 40°
Power supply: 2 × AA cells or 2 × NiCd AA cells or, with special adapter, one Lithium AA cell

Manufacturer
Oldelft (B.V. Optische Industrie 'De Oude Delft'), PO Box 72, 2600 MD Delft.
Status
Current. Production.
Service
Several armies.

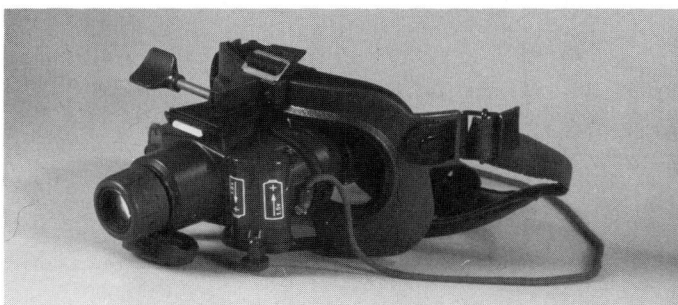

Type PC1MC Cyclop night vision goggles

Type PG1MS night vision goggles

The night vision goggles Type PG1MS are a small lightweight night viewing device based on the use of special image intensifier tubes, of lightweight housing materials, and of high-speed optics.

The goggles make it possible to undertake night-time tasks without illumination and with both hands free, such as driving, patrolling, surveillance, map reading, repair work, loading and unloading, and working in bunkers and tunnels.

DATA
Magnification: × 0.9
Field-of-view: minimum 40°
Power supply: one NiCd cell size AA or one normal AA dry cell
Weight: 1.0 kg
Dimensions: 120 × 138 × 90 mm

Manufacturer
Oldelft (B.V. Optische Industrie 'De Oude Delft'), PO Box 72, 2600 MD Delft.
Status
Current. Production.
Service
Several armies.

Type PG1MS night vision goggles

Type PB4SL passive lightweight binoculars

The lightweight passive binoculars Type PB4SL are designed for general-purpose night surveillance and reconnaissance, for example at airfields and frontiers. They are also in large-scale use by battery and troop commanders, for patrol operations, security and anti-terrorist operations, in boats, tanks and helicopters.

The binoculars are of rugged construction and stand up to hard use in the field. They incorporate extremely high-speed mirror entrance optics and a one-stage second-generation image intensifier assembly. The very minimum of light available at night at the scene to be observed is intensified thousands of times, so that a distinct image is obtained with maximum security from detection. The binoculars will also detect infra-red light sources.

DATA
Weight: 1.1 kg
Dimensions: 241 × 125 × 101 mm
Magnification: × 3.75
Field-of-view: 10.5°
Battery: 2 × normal AA cells; or 2 × NiCd AA cells; or, with special adapter, one lithium AA cell

Type PB4SL passive binoculars

Manufacturer
Oldelft (B.V. Optische Industrie 'De Oude Delft'), PO Box 72, 2600 MD Delft.
Status
Current. Production.

Service
Several armies.

Type PB4MP passive lightweight binoculars

The PB4MP binoculars are of a highly ergonomic design, and are suitable for general surveillance and reconnaissance activities at night. They meet full military specifications and have an overall magnification of × 4. The PB4MP can, without changes to the design, be fitted with a single stage second generation, a two stage second generation or a single stage third generation image intensifier tube.

The user-friendly design incorporates such features as an on-off switchable illuminated graticule for estimating distances (optional); an automatic shut-off feature to avoid unnecessary power consumption; low battery indication; tripod mounting; leak-proof battery compartment and simple battery replacement, with the possibility of using extra-long (2AA) lithium cells for better performance especially at low temperature; and protection against reversed polarity placement of the batteries.

Due to the design, maintenance procedures are very simple. The objective can be changed with only minor adjustment procedures and no special tools or measuring equipment, other than standard types, are needed to maintain the PB4MP at higher levels. Focussing is accomplished by simply turning the complete objective, and since the interpupillary adjustment is internal, the body has a fixed shape.

DATA
Dimensions: 265 × 170 × 112 mm
Weight: 1.50 kg with batteries
Magnification: × 4
Field-of-view: 9.5°
Diopter adjustment: -5/ + 5 diopters
Interpupillary adjustment: 58 – 72mm
Focussing range: 10 m to infinity
Identification range: 2nd gen: 1.0 mLux 375 m. 2nd gen, 2-stage:1.0 mLux 450 m. 3rd gen: 1.0 mLux 500 m
Power supply: 2 × AA NiCd or conventional cells; or one size 2AA lithium cell

Manufacturer
Oldelft (B.V. Optische Industrie 'De Oude Delft'), PO Box 72, 2600 MD Delft.
Status
Current. Available.

Type PB4MP passive binoculars

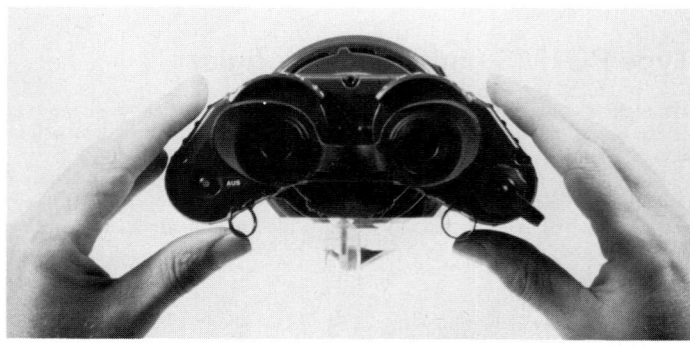

Type PB4MP passive binoculars.

Type HV7 × 200 AT night observation device

The night observation device Type HV7 × 200TG is an image intensifier system for field and battle surveillance at night. It may be additionally used as an aiming sight for crew-served weapons.

Its passive operation without artificial light gives maximum security from detection. The high-sensitivity second-generation tube used in this instrument enables detection of enemy infra-red sources.

This passive device is used for medium-range FEBA surveillance; it enables forward combat and reconnaissance units to detect and engage enemy force at night without the help of artificial light sources.

DATA
Magnification: × 7
Field-of-view: minimum 6.5°
Battery type: NiCd or, with adapter, four 1.5V AA cells
Weight: approx 12.8 kg
Dimensions: length 555 mm, width 270 mm

Manufacturer
Oldelft (B.V. Optische Industrie 'De Oude Delft'), PO Box 72, 2600 MD Delft.
Status
Current. Production.
Service
Various armies.

Type HV7 × 200AT second generation night observation device

LUNOS lightweight universal night observation system

LUNOS is a very versatile modular night vision system built to give optimal performance for a wide range of night-time tasks.

Unlike other systems, the LUNOS system allows exchange of objective lenses on the spot in any field situation. Switching from one objective lens to another is comparable with changing the lenses on a SLR camera. The driver of a reconnaissance vehicle, for example, can, after having used the LUNOS × 1 lens for night driving, easily switch to observation tasks as there is a × 4 lens for medium-range or a × 6 lens with handgrip for long-range observation.

DATA
Magnification: × 1, × 4 or × 6
Field-of-view: 40°, 10°, or 6.5°
Relative aperture: f/1.1; f/1.2; f/1.5
Weight, with batteries: 570 g; 1100 g; 2100 g
Dimensions:
    LUNOS 1: 160 × 72 × 131 mm
    LUNOS 2: 230 × 104 × 136 mm
    LUNOS 3: 300 × 150 × 136 mm

LUNOS Lightweight Universal Night Observation System

Manufacturer
Oldelft (B.V. Optische Industrie 'De Oude Delft'), PO Box 72, 2600 MD Delft.
Status
Current. Production.

Type UA 1242 night observation device

The UA 1242 is a biocular night observation device which has been designed to combine advanced visual sensor technology with ease of operation.

It has a × 5 magnifying second-generation micro-channel tube with an integrated power supply, automatic brightness control and point highlight suppression. Both objective and eyepiece lenses are adjustable, as is the inter-ocular distance. The general layout of the device allows it to be held without difficulty and to be used for long periods.

DATA
Weight: 2 kg
Magnification: × 5
Field-of-view: 7.5°
Resolution: 1 mil at 1 mlux (starlight) at a target contrast of 30%
Focus range: 25 m to infinity
Power supply: 2.2–3.4 V DC

Manufacturer
SIGNAAL-Usfa, P.O.Box 6034, 5600 HA, Eindhoven.
Status
Current. Production.
Service
NATO and Middle East armed forces.

Type UA 1242 night observation device

UA 1242/02 observation night sight with video output

The UA 1242/02 second-generation biocular night sight has a built-in CCD camera for use with standard CCIR video equipment. This enables the image viewed by the user tobe displayed on remote monitors and recorded for training or evaluation.

Incorporating automatic brightness control and bright source protection, the night sight weighs 3.5 kg and can be operated with gloved hands. The image intensifier is powered by two internal batteries, or by one lithium battery plus a dummy.

DATA
Image Intensifer Sight:
Magnification: x5.2
Field-of-view: 7.5°
Focus range: 25m to infinity, adjustable
Eyepiece adjustment: -5 to +3 dioptres
Operational temperature: -20 to +55°C

CCD Camera:
Field-of-view: 6.0° (hor.) × 4.5° (vert.)
Resolution: 450 TV lines horizontally
Pixels: 604 (hor.) × 588 (vert.)
Gain control: automatic
Operational temperature: -10 to +55°C
Power supply: 9 to 32 vDC external

Manufacturer
SIGNAAL-Usfa, P.O.Box 6034, 5600 HA, Eindhoven.
Status
Current. Production.

UA 1242/02 observation night sight with video output

NORWAY

Simrad KN150 second-generation night observation sight

The KN150 is a hand-held passive night observation sight which combines high performance with simple operation and low weight. Extensive tests have shown that tanks can be detected at ranges up to 1500 m, and other applications include general surveillance, battlefield night observation and target location.

The use of binocular eyepieces minimises operator fatigue compared with monocular viewing and make it possible to maintain continuous surveillance for long periods. To simplify handling, only two controls are fitted: an

KN 150 second-generation night observation device

on-switch and twist-grip focussing. An electronic off-switch turns the instrument off automatically after about one minute's operation.

An automatic brightness control reduces the amplification of sudden bright lights, thus preventing the operator from losing his night-adapted vision. Normal operation is restored less than one second after a sudden flash.

The KN150 incorporates a large aperture catadioptric object lens and a second generation, high resolution microchannel image intensifier tube. Two standard 1.5 V dry cells provide more than 40 hours of operation.

The KN150 can also be combined with a Simrad Optronics laser rangefinder for night observation and location of targets.

DATA
Magnification: × 4
Field-of-view: 8°
Max detection range: 500 m
Focussing range: 25 m to infinity
Resolution: 0.65 mrad
Object lens: catadioptric, 135 mm f/1.3 (T/1.9)
Eyepiece exit pupil: two, 9 mm
Dioptre setting: -1 D (fixed)
Operating voltage: 2.6 V DC nominal
Batteries: 2 standard C-cells
Battery life: approx 40 h at +25°C
Weight: 2.6 kg

Manufacturer
Simrad Optronics A/S, PO Box 6114, Etterstad, N-0602 Oslo 6.
Status
Current. Production.
Service
Norwegian Army.

Simrad GN1 night vision goggles

The Simrad GN1 night vision goggles are based on a new and unique patented design incorporating a folded optical path. This results in an extremely compact and light piece of equipment. The depth is half that of other existing goggles, which reduces strain on the neck muscles dramatically.

The GN1 is worn with a lightweight face mask. A dovetail enables the GN1 to be clipped on to the face mask or helmet. This mounting system also ensures that the distance between the face and the goggles can be adjusted for optimum viewing over the entire field-of-view.

The GN1 is simple to operate, having only an on/off/IR switch, focus control and dioptric setting. Because of the large exit pupils, adjustment of interocular distance is not necessary.

The goggles can focus down to 20 cm for tasks such as map reading. An integral infra-red diode provides extra illumination for close work.

The GN1 has been designed for both 2nd and 3rd generation image intensifier tubes; 3rd generation tubes can be introduced without any modification of the design. The tube used is the ANVIS type, available from a number of manufacturers.

DATA
GENERAL:
Width: 157.5 mm
Height: 72 mm
Depth: 58 mm
Weight: 390 g including batteries

Simrad GN1 night vision goggles

OPTICAL:
Objective lens: 26 mm f/1.0
Magnification: 1 ×
Field-of-view: 40°
Focussing range: 20 cm to infinity
Dioptre adjustment: + 2 to -4 dioptres
Eye relief: 20 mm
High contrast resolution: 1.5 mRad
Interocular distance: 66 mm
Exit pupil diameter: 9 mm

ELECTRICAL
Power source: 2 standard AA cells
IR source: LED
Battery life: 40 hours
Battery life with IR source on: 20 hours

Image intensifier tube: 2nd gen high performance inverting tube with glass input (ANVIS type) or 3rd gen ANVIS tube

ENVIRONMENTAL:
Operating temperature: -40 to + 52°C
Storage temperature: -40 to + 65°C
Immersion: 0.9 m for 2 hours

Manufacturer
Simrad Optronics A/S, PO Box 6114, Etterstad, N-0602 Oslo 6.
Status
Current. Series production from July 1990.

PAKISTAN

GP/NVB-4A, GP/NVB-5A night vision binocular

The GP/NVB-4A and 5A are general-purpose night vision binoculars for long-range observation and surveillance. The GP/NVB-4A has a 95 mm focal length objective and offers × 4 magnification, while the -5A has a 155 mm lens and provides × 6 magnification. Both systems offer the proven performance and reliability of a second generation image intensifier tube that has been designed especially to match the binocular optics. The eyepieces can be individually focussed to accommodate various eye prescriptions. A built-in objective focus mechanism, allows the operator to refocus on a new target quickly.

The units have been engineered to minimise weight and cost without sacrificing performance and quality.

GP/NVB-5A night vision binoculars

DATA

	GP/NVB-4A	GP/NVB-5A
Object lens	95 mm f/1.2	155 mm f/1.2
Magnification	× 4	× 6
Field-of-view	10.5°	6.5°
Resolution	0.4 mRad	0.25 mRad
Brightness gain	1000	1000
Focussing range	10 m—infinity	20 m—infinity
Interpupillary adjustment	55 – 72 mm	55 – 72 mm
Weight	1.13 kg	3.18 kg
Length	225 mm	305 mm
Width	145 mm	155 mm
Power source	2 × AA cells	2 × AA cells

Manufacturer
Institute of Optronics Pakistan, PO Box 1596, Rawalpindi.
Status
Current. Production.

AN/PVS-5A night vision goggles

These high performance goggles use two 18 mm micro-channel wafer type image intensifier tubes whose performance exceeds that of other tubes of a similar type. The matched precision optics are faster than those on any other night vision goggle. It is powered by two inexpensive AA type batteries which are available all over the world.

Specially designed for night driving, patrols, security duties, detection of IR sources and low level helicopter operations, it can also be used for maintenance of equipment and map reading.

DATA
Dimensions: 117 × 173 × 165 mm
Weight: 850 g
Magnification: unity
Field-of-view: 40°
Resolution: 27 lp/mm
Brightness gain: 2000 nominal
Object lens: EFL 27 mm f/1.0

Manufacturer
Institute of Optronics Pakistan, PO Box 1596, Rawalpindi.
Status
Current. Production.

AN/PVS-5A night vision goggles

SWEDEN

Bofors NK-25 night observation binoculars

The Bofors NK-25 is a passive, light weight observation device for a wide range of military applications. The binocular eyepiece permits long duration surveillance and reduces operator fatigue.

The NK-25 is equipped with a large aperture objective lens which, together with the second generation 18 mm micro-channel tube and the binocular eyepiece, enables detection of a 2.3 m × 2.3 m object at 500 m under starlight conditions (3×10^{-3} lux). The second generation tube may be exchanged for a third generation one.

A light weight but extremely rugged design with a minimum of movable parts and controls ensures simple and reliable operation under severe battlefield conditions.

DATA
Field-of-view: 8°
Overall magnification: 4×
Focussing range: 20 m to infinity
Resolution: 0.65 mrad
Objective: f/1.3 catadioptric
Front lens diameter: 110 mm
Eyepiece exit pupils: 9 mm
Dioptre setting: fixed
Operating voltage: 2.6 V DC nominal
Power source: 2 × C cells or rechargeable NiCd cells
Battery life: approx. 40 hours at +25°C
Operating temperature range: -40°C to +55°C
Weight: 1.9 kg

Manufacturer
Bofors Aerotronics AB, S-18184 Lidingö.
Status
Current. Production.

Bofors NK-25 night vision binoculars

NK-23/KN150 night vision instrument

The NK-23/KN150 is a night vision observation binocular with very short reaction time due to easy handling and swift focussing procedure. The binocular viewing permits long observation time with a minimum of operator fatigue.

The NK-23/KN150 includes a 20/30 mm second-generation image intensifier tube as standard, assuring very good light amplification and image resolution. It has a superior range capacity in all light conditions due to the light gathering ability of the 110 mm front lens. The weight is kept low because of an advanced optical construction. The NK-23/KN150 has been developed as a joint venture between Bofors Aerotronics AB, Sweden and Simrad Optronics A/S, Norway. It is approved by NATO.

DATA
Magnification: × 4
Field-of-view: 8°
Focus range: 25 m to infinity
Resolution: 0.65 mrad
Objective: 110 mm f1.3 catadioptric
Operating voltage: 2.6 V DC
Batteries: 2 × C-cells
Battery life: 40 h
Weight: 2.5 kg

Manufacturer
Bofors Aerotronics AB, S-181 84 Lidingö.

NK-23/KN150 night vision instrument

Status
Series production.

JAI Nighthawk low light level TV cameras

The JAI Nighthawk is a camera system which can be adapted for almost any surveillance task. The system is built around two cameras, the 733 SIT and the 743 ISIT, to which can be added various power supplies, fixed or zoom lenses, image intensifying lenses, control and transmission equipment and other options for special tasks.

All cameras are built over a closed form, as extremely robust units in dark colours for discreet employment. The cameras will operate under all light conditions; the 733 SIT from full sunlight to moonlight, and the 743 ISIT down to starlight conditions. The special design of the JAI EHT unit provides ultimate protection for the camera tube by means of automatic control of the camera's shutdown, enhanced by the unique JAI neutral density spot filter in

the lens. For increased picture stability the cameras incorporate an outstanding back-focus design which efficiently negates adjustments resulting from changes in the ambient temperature. By using the new JAI service unit, fault analysis and re-adjustment can be performed instantly, cutting down time to an absolute minimum.

Manufacturer
Jorgen Andersen Ingeniorfirma A/S, DK-2660 Glostrup, Copenhagen.
Status
Current. Production.

SWITZERLAND

Wild BIM25/35 night pocketscope

The Wild night pocketscope is a general purpose monocular night viewing device. It exists in two versions, differing only in the objective utilised: BIM25 has 1× magnification, whereas BIM35 has 3× magnification. Both are specifically designed for military applications and feature a watertight light metal housing purged with nitrogen to prevent fogging of the optics.

DATA

	BIM25	BIM35
Magnification	1 ×	3 ×
Field-of-view	41°	13°
Resolution	⩾0.6 lp/mrad	⩾2.3 lp/mrad
Dioptric setting	–6 to +2	–6 to +2
Objective	24.9 mm, f/1.12	75 mm, f/1.17
Focus range	25 cm to ∞	10 m to ∞
Image intensifier	18 mm MCP, 2nd or 3rd generation	
Power supply	1 × 2.7 V lithium or 2 × 1/5 V alkaline	
Dimensions	140 × 70 × 80 mm	200 × 100 × 100mm
Weight	510 g	1100 g

Manufacturer
Wild Leitz Ltd, CH-9435 Heerbrugg.
Status
Under development.

Wild BIM25 night pocketscope in use

Wild BIM235 night pocketscope (top) and BIM25 pocketscope below

Wild BIG2 night vision goggles

These goggles are intended for use by vehicle drivers, patrol leaders, medical corpsmen and others requiring unobstructed night vision. They use a single image intensifying tube presenting an image to binocular viewing lenses. The second-generation image intensifier incorporates automatic brightness control and bright-point protection and is designed so as to permit eventual upgrading to third-generation standard.

A pinhole in the lens cap allows the user to practice with the Wild BIG2 in daylight. There is an auxiliary led infra-red light source which can be switched on to provide close illumination for reading maps or messages or for operating instruments or equipment. When this source is in use there is an indication in the field-of-view to act as a reminder.

DATA
Magnification: × 0.98
Field-of-view: 41°
Resolution: ⩾0.6 lp/mrad
Objective: 24.9 mm f/1.3
Focus range: 25 cm to infinity
Weight: 580 g
Power supply: 1 × 2.7 V lithium cell or 2 × 1.5 V alkaline cells

Manufacturer
Wild Leitz Ltd, CH-9435 Heerbrugg.
Status
Current. Production.

Wild BIG2 night vision goggles

Wild BIG3 night binoculars

Wild BIG3 night binoculars are a hand-held passive instrument for night surveillance. They produce a high-resolution image at × 3 magnification; the recognition range depends upon a number of factors, but in average conditions a man-sized target can be recognised at a range of 500 to 700 m.

The binoculars are rubber-covered, ensuring a good grip and protection against shock. The light-alloy housing is water-tight and moisture-proof. A lens cap, neck strap, cleaning kit and pouch are standard accessories, and soft eyecups are provided in order to prevent light being reflected from the operator's face and also to ensure proper positioning of the instrument.

DATA
Dimensions: 260 × 140 × 99 mm
Weight: 1.5 kg
Magnification: × 3
Field-of-view: 12°
Limit of resolution: 2.3 lp/mrad
Objective lens: Catadioptric, 75 mm, f/1.17
Focussing range: 10 m to infinity
Power supply: 1 × 2.7 V Lithium AA cell; or 2 × 1.5 V alkaline
Battery life: ⩾40 h

Manufacturer
Wild Leitz Ltd, CH-9435 Heerbrugg.

Wild BIG3 night binoculars

Status
Current. Production.

Wild BIF image intensifying telescopes

These telescopes are image intensifying surveillance devices suited to all the usual night observations tasks in military, police or security roles. There are three patterns of telescope; BIF2-6M is a monocular instrument; BIF2-6B is a binocular instrument; and BIF1-6P is adapted for photography, allowing the attachment of various types of camera by means of adapters. They may be fitted to tripods, surveying or measuring instruments, or to weapons.

DATA

Image intensifier telescope	BIF2-6M	BIF2-6B	BIF1-6P
General			
Field-of-view	7.2°	7.2°	8.9°
Magnification	× 6.9	× 6.9	× 3.9
Focussing range	25 m to infinity	25 m to infinity	25 m to infinity
Reticle illumination	adjustable	adjustable	adjustable
Reticle adjustment range	−20 mil to +20 mil	−20 mil to +20 mil	−20 mil to +20 mil
Horizon mask	adjustable down to middle of field-of-view	adjustable down to middle of field-of-view	adjustable down to middle of field-of-view
Battery	1 × 2.7 V Lithium, size AA	1 × 2.7 V Lithium, size AA	1 × 2.7 V Lithium, size AA
Length/width/height (mm)	370/150/150	420/150/150	320/150/150
Weight	4.1 kg	4.9 kg	4.0 kg
Image intensifier tube, type	20/30	20/30	25/25
with automatic brightness control (ABC) and bright point protection (BPP)			
Objective, type	catadioptric	catadioptric	catadioptric
Focal length	160 mm	160 mm	160 mm
Clear objective aperture	150 mm	150 mm	150 mm
F number, effective	1.46	1.46	1.46
T number, effective	1.58	1.58	1.58
Eyepiece, type	monocular	binocular	monocular
Focal length	35 mm	35 mm	41.5 mm
Dioptric setting	−4 dptr to +4 dptr	−5 dptr to +5 dptr	−3 dptr to +2 dptr
Exit pupil diameter	7 mm	6 mm	28 mm
Eye relief	35 mm	11 mm	11 mm
Eye base adjustment	—	56 mm to 72 mm	—

Manufacturer
Wild Leitz Ltd, CH-8435 Heerbrugg.
Status
Current. Production.

Wild BIF2-6B image-intensifying telescope

Wild SPE2 trench periscope

The SPE2 is a lightweight optical periscope with a 400 mm offset from the line of sight, allowing it to be used for observation over or around obstacles and cover. It can be fixed in the ground by means of a peg, or mounted on a mini-tripod. With a covering in foam material and a matt black finish, it is easily hand-held, well protected against knocks and abrasion, and does not reflect light or attract the eye.

The instrument is of fixed focus but the eyepiece has dioptric adjustment so that observation can be maintained for long periods without eyestrain. A graticule divided into 5 mil intervals permits measurement of angles and estimation of range.

DATA
Length overall: 600 mm
Weight: 1.5 kg
Magnification: × 5
Field-of-view: 170 mils (9.5°)
Focus: 15 m to infinity

Manufacturer
Wild Leitz Ltd, CH-9435 Heerbrugg.
Status
Current. Production.

Wild SPE2 trench periscope

UNITED KINGDOM

IR26 Thermal Imaging Sensor Head (TISH)

The IR26 Thermal Imaging Sensor Head (TISH) is the latest member of the IR18 family, described in previous editions of *Jane's Infantry Weapons*. It has been designed and built round the reliable and proven IR18 scan head with its processing electronics on plug-in printed circuit boards and housed in the Thermal Imaging Processing Unit (TIPU). The sensor head incorporates a dual field-of-view telescope which is currently in service on the Chieftain/Challenger tank fire control system.

The TISH has been designed to fit externally any AFV, IFV, tracking platform or guided weapon system.

IR26 is a rugged TI sensor using cryogenically cooled SPRITE detector technology operating in the 8 – 13 micrometer IR waveband. It is fitted with a

IR26 Thermal Imaging Sensor Head (TISH)

wash nozzle and wiping facility to remove mud and/or water from the objective lens. The objective lens is coated with an externally durable anti-reflection coating which renders it immune to damage through continuous wiping in adverse environments.

IR26 utilises a dual field-of-view telescope giving a wide field-of-view suitable for surveillance and target detection, and a narrow field-of-view suitable for target recognition/identification and fire control or missile guidance.

The equipment is packaged in two units to ease installation and maintenance. The TISH contains the optics, detector and essential processing electronics while the TI Processing Unit (TIPU) contains the remainder of the processing electronics and power supplies. The TIPU can be mounted remotely with separation up to 10 m.

The video output of IR26 is fully compatible with normal European or US TV standards and can, therefore, be fed to any standard TV display or video recorder and can also be transmitted over a microwave link.

DATA
Weight: TISH 19.9 kg; TIPU 12.0 kg
Fields-of-view: Wide 13.6° horizontal by 9.1° vertical; narrow 4.75° horizontal by 3.18° vertical
Resolution: Wide field 0.51 mRad; narrow field 0.18 mRad
MRTD: Wide field: Spatial Frequency 0.4 cycles/mRad, MRTD 0.25 K. Narrow field: 1.15 cycles/mRad, MRTD 0.13K
Spectral bandwidth: 8 – 13 μm
Video output: CCIR system (625 lines, 50 Hz) or EIA (525 lines 60 Hz)
Power supply: (nominal) 55 W at 28 V DC

Manufacturer
Barr & Stroud Ltd, Caxton Street, Glasgow G13 1HZ, Scotland.
Status
Current. Available.

Steadyscope monocular viewer

The Steadyscope is a hand-held gyro-stabilised sight designed to be used in place of conventional binoculars and gives the user a steady picture, unaffected by hand tremor, vehicle vibration or other movement. It is a monocular instrument similar in shape and size to binoculars, but one eyepiece is blanked off. A flexible eyeshield is fitted to the instrument to enable the user to operate in harsh environments. The instrument can be reversed for use with either eye.

Stabilisation is accomplished by a gimbal-mounted mirror controlled by a gyroscope, powered by a battery-driven motor. An internal steering device provides effective tracking and the instrument may be held in any attitude, the user requires no steadying support.

Controls are available for power by a push button on/off switch, a push button for stabilisation and a focus ring with ± 5 dioptre adjustment.

The viewer is in service in fixed and rotary wing aircraft, fast patrol boats, hovercraft and some land vehicles.

The latest development is a night vision adapter incorporating a second generation image intensifier. The power supply for the image intensifier (II) system is a battery fitted in the inactive eyepiece. The entire instrument can be converted from day to night operation by the operator in approximately one minute. A conversion kit for existing users to adapt for day or night viewing is

in production. Customers can purchase the day version, complete with night viewer adaptor fitted, and buy the N/V module later if required, or in the complete dual-purpose kit.

DATA
Magnification: × 7
Field-of-view: 7.4°
Focus adjustment: ± 5 dioptres
Weight: (with battery, day) 2.4 kg; (night) 2.8 kg
Length: (day) 230 mm; (night) 270 mm
Width: 225 mm
Height: 96 mm
Power supply: 1.5 V (D Type) manganese alkaline cell for gyro, 1.5 V Mallory UK 402380 cell for II or equivalent; running time approx 10 h

Manufacturer
British Aerospace Systems & Equipment, Clittaford Road, Plymouth, Devon PL6 6DE.
Status
Current. Quantity production.

Service
In service with UK armed forces and 30 overseas armed forces and other civil organisations.

Steadyscope viewer in use

THORN EMI Multi-Role Thermal Imager

The Multi-Role Thermal Imager (MRTI) is one of a number of thermal imagers developed by THORN EMI under the UK MoD Thermal Imaging Common Module (TICM) programme. The MRTI uses Class 1 modules from this and has been developed as indirect and direct view systems. Both systems are now in use with the British Army, the Royal Navy, and other NATO forces.

Thermal radiation is detected by the cadmium mercury telluride detector array which is scanned across the scene in a series-parallel mode. The detector array operates at 80°K and can be cooled to its working temperature by either a Joule-Thompson cooler using compressed air or by a closed-cycle cooling engine. MRTI is configured to use a 0.6 litre air bottle which, when charged to a pressure of 300 Atm will cool the detector for up to 4.5 hours. The alternative closed-cycle cooler has a power comsumption of about 40 W. Power to the rest of the imager can be supplied at voltages between 10 and 30, with a consumption of less than 7 W.

The system operates in the 8 to 12 μm band and two telescope fields-of-view are provided. Alternative telescopes are available to give enhanced performance. The equipment maintains focus when switched from one field-of-view to another. Controls for temperature offset and temperature window are provided and enable the operator to emphasise different temperature ranges within the scene under view. Hot-white or cold-white is switch selectable.

When used in the direct role the thermal scene is reconstructed using a scanned LED array. By adding another module to the direct view system a TV Compatible (CCIR) output to a remote display can be obtained to form the indirect view system. The TV monitor can be either mounted on the imager or remoted.

The versatile Multi-Role Thermal Imager in field use

DATA
Dimensions: 500 × 250 × 200 mm
Weight: thermal imager 10.4 kg including bottle and battery; remote viewing unit 0.75 kg
Waveband: 8–12 µm
Temperature sensitivity: 0.1 K
Field-of-view: 4.9° × 3.2° narrow; 12.9° × 7.9° wide
Focussing range: 10 m – ∞ (wide): 30 m – ∞ (narrow)
Display: red LED raster with illuminated graticule and warning symbols for low battery and low air supply

Manufacturer
THORN EMI Electronics, Electro Optics Division, Forest Road, Feltham, Middlesex TW13 7HE.
Status
Current. Production.
Service
British Army, Royal Navy, Royal Thai Navy, US Air Force, Danish Army, Norwegian Navy, and other countries in the Middle and Far East.

Observer Thermal Imaging System (OTIS)

OTIS is an adaptation of the Multi-Role Thermal Imager described above, together with a laser rangefinder and angulation head. The MRTI and the laser rangefinder are mutually bore-sighted. The equipment is manportable and is intended for application in battlefield surveillance, observation and control of artillery and mortar fire, and target tracking for laser guided weapons.

   Azimuth and elevation readings are available as mechanical slipping scales or digital readouts. All other technical characteristics are as given for the MRTI.

Manufacturer
THORN EMI Electronics, Electro Optics Division, Forest Road, Feltham, Middlesex TW13 7HE.
Status
Current. Production.
Service
British Army; Danish and Jordanian Armies.

Observer Thermal Imaging System in use in field

THORN EMI hand-held thermal imager

The THORN EMI hand-held thermal imager is a compact lightweight unit intended for use by reconnaissance troops and forward observation officers, as well as a wide range of civil applications. The equipment uses modules developed from the UK Class 1 Thermal Imaging Common Module programme and employs a great deal of the technology which led to the MRTI and OTIS thermal imagers (above).

   It is, however, considerably lighter than either of these equipments, with the electronics having been hybridised to reduce the weight, but maintains considerable commonality to reduce spares holdings and ease maintenance.

DATA
Dimensions: 470 × 210 × 160 mm
Weight: 5 kg including cooling air bottle and battery
Waveband: 8 – 12 µm
Detector: series/parallel cadmium mercury telluride
Field-of-view: wide: 20° × 8.5° (× 2 magnification); narrow: 8° × 3.4° (× 5 magnification)
Focussing range: 5 m – ∞
Display: Red LED raster with illuminated graticule and warning symbols for low battery and low air bottle supply
Minimum resolvable temperature difference: 0.3 K

THORN EMI hand-held thermal imager

Manufacturer
THORN EMI Electronics, Electro Optics Division, Forest Road, Feltham, Middlesex TW13 7HE.
Status
Current. Production.

Service
British Army, Royal Navy, Royal Air Force, Royal Marines, Royal Danish Air Force, Royal Netherlands Marines, US Air Force, US Marine Corps, US Navy, and other countries in the Middle and Far East.

HHI-8 hand-held thermal imager

Development of this instrument began in 1980 against a requirement stated by the Royal Signals and Radar Establishment and was brought to production status as a private venture project. The HHI-8 has been designed to offer a self-contained, fully portable thermal imager that gives high quality pictures in total darkness, through battlefield smoke and most mist and fog.

Operation is simple and calls for the minimum of training; compressed air and electrical circuits are activated by a single control and the imager is ready for use in a few seconds. The image is displayed on a 1 in green CRT and can be selected to show a white-hot or black-hot image as desired. Focussing is by means of a front lens ring. Two different fields-of-view and degrees of magnification can be selected. Supplies are provided from a readily re-chargeable 0.6 litre compressed air bottle and a replacement 12 V lithium battery pack. These provide five hours of continuous running and replacement of either can be done in seconds.

In addition to being hand-held, the HHI-8 can be tripod-mounted together with a laser rangefinder to form a complete forward observer's station. Alternatively the image display unit can be detached from the body of the instrument for remote viewing, and it can be linked by a scan converter to a TV monitor.

Marconi HHI-8 thermal imager

DATA
Dimensions: 410 × 140 × 140 mm
Weight: 4.5 kg (without bottle and battery)
Spectral band: 8–12 m
Field-of-view
Narrow: 70 m × 40 m (4° × 2.3°) magnification × 4.9
Wide: 170 m × 100 m (9.6° × 5.6°) magnification × 2
Cooling system: Joule-Thompson mini-cooler

Manufacturer
Marconi Command and Control Systems Ltd, Chobham Road, Frimley, Camberley, Surrey, GU16 5PE.
Status
Current. Production.

Lowlight Miniscope viewer

The Miniscope is a second generation surveillance viewer designed to give distortion-free resolution in starlight conditions. The Miniscope uses a channel-plate image intensifier which results in a small enough bulk to be easily held in one hand and carried without difficulty. It incorporates a high sensitivity which extends into the red zone of the EM spectrum and it is relatively insensitive to highlights and flare.

It is therefore most suitable for a wide range of security applications and there is a wide range of lenses which extend the possible applications.

DATA
Dimensions: 140 × 63 mm; (with lens) 235 × 63 mm
Weight: 600 g; (with lens) 825 g
Magnification: × 9

Gain: (typical) variable 6000 to 60 000
Resolution (screen centre): 32 lp/mm
Lens: f2.8 135 mm standard
Batteries: 2 × RM401 mercury cells
Battery life: 50 h intermittent use

Manufacturer
Phychem Ltd, Turnpike Road Industrial Estate, Newbury, Berkshire RG13 2NS.
Status
Current. Available.
Service
Not known.

Night Hawk surveillance system

This is a versatile system built around a low-distortion, second generation image intensifier which, by means of a range of adapters, can be used for almost any surveillance task. The standard system comprises a main body, containing an 18 mm micro-channel amplifier and fitted with a 'C' mount lens thread; a CCTV lens adapter which couples the main unit to standard Newvicon or Saticon TV cameras; a battery pack; an eyepiece which fits the main unit and converts it into a hand-held viewer; a camera relay lens which adapts the main body to any single-lens reflex 35 mm camera; various cables, brackets, a carrying case and so on. Further lenses and special-purpose adapters are available as optional extras, which allow camera zoom lenses to be fitted as the object lens on the main unit. The carrying case measures 470 × 370 × 165 mm and weighs 7 kg when filled with the standard range of items.

Manufacturer
Phychem Ltd, Turnpike Road Industrial Estate, Newbury, Berkshire RG13 2NS.
Status
Current. Production.

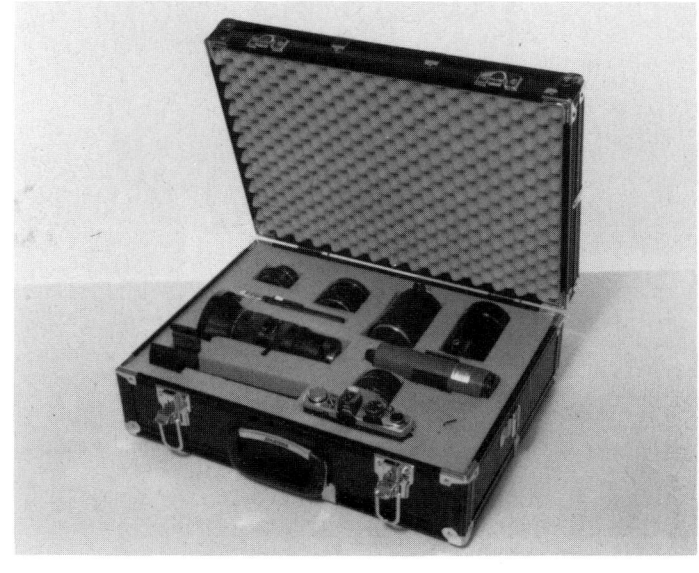

Night Hawk surveillance system (camera and zoom lens are optional extras)

Nova night vision goggles

The Nova general-purpose night vision goggles were designed and developed by Pilkington Optronics in response to a specific demand by the UK Ministry of Defence for a low cost compact head-mounted passive night surveillance device. Weighing only 600 g including the headmount, the goggles incorporate two eyepieces for user comfort and a single second generation image intensifier tube for cost effectiveness. Designed to be simple to operate, Nova has only three controls, namely an on/off switch, focus control and an adjustment for the position of the eyepieces. Power is provided by a single 2.7 V battery which provides over 60 hours continuous use.

Nova is ideal for a number of short-range night-time duties such as maintenance and map reading, and a tiny infra-red diode is incorporated in the system to illuminate objects at close quarters. The goggles are also designed for foot patrol commanders, drivers of soft skinned vehicles and a general passive (undetectable) look around capability. Affording the user a 40° field-of-view, Nova can recognise a standing man at 300 m in starlight (10⁻³ Lux) conditions. Good performance characteristics are supported by simple repair by virtue of modular subassembly design.

DATA
Field-of-view: 40° horizontal (instantaneous) 30° vertical
Overall magnification: × 1
Eye relief: 28 mm
Weight: 500 g

Manufacturer
Pilkington Optronics, Glascoed Road, St Asaph, Clwyd LL17 0LL.
Status
Current. Production.
Service
British Army; Middle Eastern and South American armed forces.

Nova night vision goggles

Eagle long-range night surveillance viewer

Eagle represents a new generation of night vision equipment developed to meet an ever-increasing demand for a long-range hand-held device which gives an increased performance over first generation equipment.

Eagle employs a binocular eyepiece and incorporates an illuminated graticule suitable for weapon aiming and adjustment of fire. In addition to acting as a long-range surveillance sight, Eagle can be used for a wide variety of tasks including its use as a sight in artillery and air defence fire control. It may also be mounted on a tripod, and is fully compatible with NBC defence equipment.

DATA
Magnification: × 8
Eye relief: 29 mm nominal
Power supply: 2 × 1.5 V batteries
Field-of-view: 5°
Weight: under 3.5 kg

Manufacturer
Pilkington Optronics, Glascoed Road, St Asaph, Clwyd LL17 0LL.
Status
Current. Production.

Eagle long-range night vision device in typical hand-held role

Modulux modular night vision equipment

Modulux was developed for the Police Scientific Development Branch of the British Home Office to satisfy requirements for night photography, closed-circuit television and direct surveillance. The Modulux system consists of a set of modules which allow the user to assemble quickly the most suitable system for any particular operation.

The basic component is the main body which houses the image intensifier tube and its battery power supply. Adapters permit the mounting of almost any standard 35 mm camera lens on to the receiving end of the main body; the options available for mounting on the output end depend upon the role in which the equipment is to be used. For photography and observation, a relay lens unit is mounted on the main body, together with an adapter to fit the lens throat of the camera to be used. This relay lens unit magnifies the image 1.5 times which effectively allows it to fill the standard 35 mm frame. The camera can then be used as an observation instrument, and the operator can take whatever photographs he requires for record. The low-distortion image intensifier ensures the highest possible quality of definition in the photographic image.

For CCTV purposes the relay lens unit is reversed and fitted with a standard 'C' mount, compatible with most TV cameras; the magnification is now 0.65, which reduces the image to TV frame size. The scene can then be viewed on a monitor (which can be remote from the surveillance equipment) and recorded on videotape, with date/time and time-lapse facilities if required.

For observation only, a monocular magnifier is provided, which gives the operator a magnified image of the scene.

DATA
Image intensifier: 2nd generation
Gain: 70 000 minimum; 100 000 typical
Distortion: 6% max
Lens adapters
Standard: Nikon, Pentax (M42) screw
Optional: Olympus, Canon, Pentax bayonet, Minolta, 'C' mount

Modulux night vision system in photographic role

Modulux night vision system in use for direct observation

Relay lens unit
Magnification: (35 mm) × 1.4; (CCTV) × 0.65
Aperture: f/1.1
Weights
Main body: 2.1 kg
Relay lens unit: 640 g
Pistol grip: 220 g

Manufacturer
Solartron Defence Systems, Optronics Division, 580 Great Cambridge Road, Enfield, Middlesex EN1 3RX.
Status
No longer in production.

OE8010 miniature surveillance device

This pocketscope is a compact, lightweight and versatile device ideally suited to a number of security-related applications. The standard unit is supplied with a 50 mm f/1.4 objective, second generation image intensifier, focussing eyepiece and pistol grip. A range of adapters is available to fit most SLR and C-mount lenses, including a high performance 6:1 zoom objective.

Manufacturer
Osprey Electronics Ltd, E27 Wellhead Industrial Centre, Dyce, Aberdeen AB2 0GD, Scotland.
Status
Current. Production.

Osprey OE8010 miniature surveillance device

OE8000 ruggedised zoom surveillance device

This system is completely sealed and ruggedised to withstand military environments. Target identification is achieved through a high performance 4:1 zoom objective that can accept second or third generation image intensifiers. An integral pistol grip and switching eyeguard both enhance the system's ergonomics and significantly increase the image tube and battery life.

Manufacturer
Osprey Electronics Ltd, E27 Wellhead Industrial Centre, Dyce, Aberdeen AB2 0GD, Scotland.
Status
Current. Production.

Osprey OE8000 surveillance device

OE1065 hand-held thermal imager

The Osprey OE1065 hand-held thermal imager is a compact and lightweight device designed to meet the diverse needs of the industrial, civil, security and military markets. Responding to heat radiation in the 3- to 5-micron waveband, the OE1065 imager is designed to indicate temperature differences of less than 0.5°C. A small portable battery pack provides up to eight hours of operation.

Manufacturer
Osprey Electronics Ltd, E27 Wellhead Industrial Centre, Dyce, Aberdeen AB2 0GD, Scotland.
Status
Current. Production.

Osprey OE1065 thermal imager

Davin Modulux night vision equipment

Extensively used by police forces and other security services, Davin Optical's Modulux night vision equipment is immensely versatile and can be used for night time photography, observation or for linking with closed circuit television (CCTV). The system operates under ambient light levels down to starlight illumination. Where the situation allows, additional image enhancement can be obtained using an infra-red flash gun or light source.

Offering unequalled performance based on a high resolution, low distortion image intensifier tube which amplifies light by a factor of at least × 250 000, Modulux is compatible with all popular commercially available cameras and lenses. As its name implies it is a flexible, modular system allowing the user to assemble the best package for any given set of circumstances.

Modulux can operate in any one of three modes:

Photography

For night time photography, the Modulux can be fitted to any 35 mm SLR camera and the basic kit includes a relay lens which magnifies the end-image to fill the 35 mm frame. The camera view finder can then be used for observation and the camera operated in the normal way using the built-in exposure metering. Identification from Modulux photographs is made easy by the low distortion image intensifier tube.

Observation

The Modulux system can be fitted with either a biocular or monocular lens to provide the night time observer with a sharp, magnified image of the scene.

Closed circuit TV

By using a 'C' mount adaptor, the Modulux can be fitted, via the relay lens, to any standard 2/3 in or 1 in CCTV camera. The scene may then be viewed on a remote monitor or recorded with a video cassette recorder.

Interchangeable bodies – first and second generation

Modulux is available with either a first or second generation image intensifier tube; the systems being designated Modulux 130 and Modulux 225 respectively. Modulux 130 produces the brighter end-image and is, therefore, recommended when there is little or no local lighting. It is also thought the better system for photography or filming.

Modulux 225, on the other hand is lighter and smaller and its second generation tube can ignore bright lights or sudden illumination in the field-of-view.

DATA

	Modulux 130	Modulux 225
Image intensifier tube	P8079HPDC	XX1470
Gain	250 000	variable up to × 70 000
Distortion	6% max	2% max
Battery	TR 132N/133N	Mallory TR 132N

Tripod fittings: ¼", ⅜" Whit. threaded holes
Weight: (incl image intensifier tube battery) 2.1 kg
Lens adaptor: Nikon, Pentax screw etc

BIOCULAR LENS
Magnification: × 3.5
Aperture: 72 mm
Weight: 900 g

RELAY LENS UNIT
Magnification: for 35 mm × 1.54: for CCTV × 0.65
Aperture: F/1.1 (infinity)
Weight: 640 g
Camera adaptors: Nikon, Pentax etc

Weight: (case with complete set of modules but without camera or lenses) 10 kg

Manufacturer

Davin Optical Ltd, 13 Alston Works, Alston Road, Barnet, Hertfordshire EN5 4EL.
Status
Current. Available.
Service
British Police Force, Home Office, Ministry of Defence and worldwide.

Davin Modulux 225

Davin Modulux 130

MH 218 dual-role lightweight military viewer

Davin designed the MH 218 to fulfil a dual military role: firstly as a fully sealed compact pocketscope for night-time observation and secondly as a robust lightweight rifle sight.

Equipped with a high performance F/1.4 objective lens, the MH 218 is a distortion-free viewer which can provide a variable light gain up to × 70 000. It features a fixed shadow graticule on the output screen and is supplied with a pistol grip and lanyard. It measures a compact 204 mm long and weighs only 980 g. A single battery gives up to 40 hours operation.

DATA
SIGHT CHARACTERISTICS
Overall length: 204 mm (excluding eyecup)
Max diameter: 62 mm
Overall weight: 980 g
Magnification: × 2.8
Field-of-view: 14.6°

INTENSIFIER TUBE
Type: Mullard XX1500 micro-channel plate
Gain: variable up to × 70 000
Power supply: Mallory TR 132N 2.7 V
Battery life: average 40 h operation

OBJECTIVE LENS
Focal length: 70 mm
Aperture: F/1.4
Type: 6 – element refractor
Focussing: 10 m – infinity

EYEPIECE
Focal length: 25 mm
Type: 5 – element distortion free
Focussing: fixed at – 1.75 dioptres

AIMING BRACKET
Weight: 300 g
Length: 87 mm
Width: 40 mm (excluding control knobs)
Height: 32 mm (excluding control knobs)
Azimuth and elevation adjustment in steps of 0.4 mrads; (equivalent to 40 mm at 100 m)

Manufacturer
Davin Optical Ltd, 13 Alston Works, Alston Road, Barnet, Hertfordshire EN5 4EL.
Status
Current. Available.
Service
British Police Force.

Davin MH 218 viewer

SH 218 hand-held night viewer

Davin Optical's SH 218 is a hand-held security night viewer which has been designed to accept any standard C-Mount lenses including wide angle and zoom lenses. This provides a small, versatile, lightweight hand-held viewer for general security use.

DATA
SIGHT CHARACTERISTICS
Overall length: 204 mm (excluding eyecup)
Max diameter: 62 mm
Overall weight: 980 g (as MH 218)
Magnification: × 2.8
Field-of-view: 14.6°

INTENSIFIER TUBE
Type: Mullard XX1500 micro-channel plate
Gain: variable up to × 70 000

Power supply: Mallory TR 132N 2.7 V
Battery life: average 40 h operation

EYEPIECE
Focal length: 25 mm
Type: 5 – element distortion free
Focussing: fixed at – 1.75 dioptres

Manufacturer
Davin Optical Ltd, 13 Alston Works, Alston Road, Barnet, Hertfordshire EN5 4EL.
Status
Current. Available.
Service
British Police Forces and Ministry of Defence.

Maxilux M and Maxilux P night vision system

Maxilux is an exceptionally high performance night vision system using first and second generation image intensifier tubes. There are two versions - Maxilux M for military use and Maxilux P for night photography – each using a 170 mm F/1.0 lens specially developed by Davin to provide extra-high magnification and resolution. Weight and size have been reduced to a minimum to aid portability and ease of use.

Maxilux M
The primary purpose of Maxilux M is long-range target acquisition and recognition in conditions down to starlight level. Easily portable by one man, the unit is fully sealed and ruggedised for field use. A single battery gives approximately 50 hours operation under average low light conditions. The high gain image intensifier tube incorporates an automatic brightness control and a shuttered eyecup is fitted to minimise the chance of user detection.

It is provided with a lens hood and an aperture stop as standard. The support bar has built-in tripod fittings.

Maxilux P
Using the same lens and main body unit as Maxilux M, this system is compatible with Modulux photographic accessories. This flexibility enables the Maxilux M to be used for long term surveillance with a biocular viewer lens; for night photography using 35 mm SLR cameras; for closed circuit television with miniature cameras and for 16 mm cine photography. For best photographic results, the Maxilux M uses a distortion controlled, high gain image intensifier.

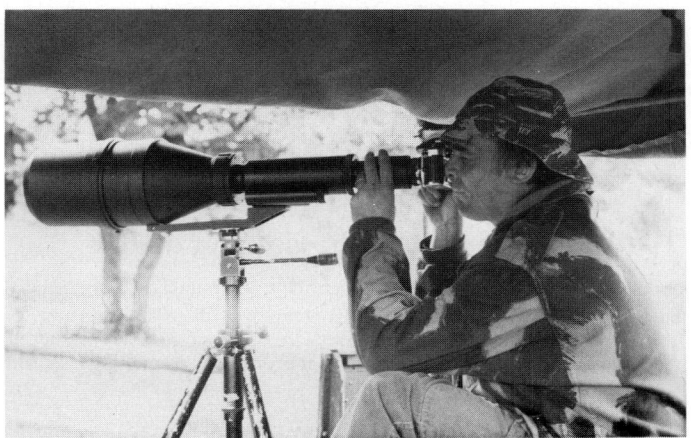

Davin Maxilux P with relay lens and 35 mm camera

DATA

	Maxilux P	Maxilux M
OBJECTIVE LENS		
Focal length	170 mm	170 mm
Focussing	20 m – infinity	20 m – infinity
Aperture	1.05	1.05
Field-of-view	8.4°	8.4°
Length	285 mm	285 mm
Diameter	172 mm	172 mm
Weight	5.5 kg	5.5 kg
MAIN BODY		
Intensifier tube	EEV 3-stage cascade P8079 HP DC with ABC	EEV 3-stage cascade P8079 HP DC with ABC
Length	217 mm	217 mm
Weight	2.0 kg	2.0 kg
Battery	Mallory TR132N/133N	Mallory TR132N

MONOCULAR EYEPIECE (Optional)		
Focal length	35 mm	35 mm
Magnification (eyepiece)	× 7	× 7
Magnification overall	× 4.5	× 4.5
Focussing	± 4 dioptres	± 4 dioptres
Sealing	yes	yes
Secure eyecup	yes	yes
Weight	0.3 kg	0.3 kg
BIOCULAR LENS (Optional)		
Magnification (biocular)	× 3.5	× 3.5
Magnification overall	× 2.2	× 2.2
Aperture	72 mm	72 mm
Weight	0.9 kg	0.9 kg
RELAY LENS (Optional)		
Magnification	for 35 mm photography × 1.54; for CCTV × 0.65	not used

Aperture	F/1.2 (at infinity)
Weight	0.48 kg

SYSTEM SPECIFICATION
For Maxilux-M the overall length is 540 mm (not including eyecup) with an overall weight of 7.8 kilos. A support bar is supplied for tripod mounting of the system.

ADAPTORS
Camera adapters can be supplied for fitting Maxilux P to Nikon, Pentax Canon and many other cameras.

Manufacturer
Davin Optical Ltd, 13 Alston Works, Alston Road, Barnet, Hertfordshire EN5 4EL.
Status
Current. Available.
Service
British Ministry of Defence.

Davin Spylux personal night viewer

Davin Optical has designed and manufactured a new personal night viewer which will sell for about one-half the cost of current image intensifier equipment; called the 'Spylux', this scope meets the needs of police and security forces for an inexpensive but effective night viewer.

Spylux is rugged and compact, giving high contrast, good resolution images at low light levels. It incorporates a focussing eyepiece, a finger-operated fully sealed push-button on/off switch and a C-mount lens adapter to allow the fitting of any suitable lens. An f/1.4 75 mm lens is supplied with the Spylux.

DATA
Weight: 0.6 kg
Resolution: 100 lp/mm
Luminance gain: 400 cd/m²/LX
Supply voltage: 2.0 to 5.0 V
Input current: 15 mA

Manufacturer
Davin Optical Ltd, 13 Alston Works, Alston Road, Barnet, Hertfordshire EN5 4EL.
Status
Current. Production.
Service
UK and overseas police forces and UK Ministry of Defence.

Davin Spylux in use

Davin Spylux, with optional 75 mm lens

Type SS32 Twiggy night viewing device

The Twiggy sight is a first generation device used extensively for general battlefield surveillance and fire control operations with artillery and mortars. Twiggy is also recommended for aerial reconnaissance using a suitably damped platform; naval gunfire control from turrets and night marine navigation and identification from ships' wheelhouses; internal security and police surveillance.

The large aperture lens produces a very high resolution image, giving exceptional performance down to the lowest ambient light levels.

The sight has been designed and manufactured to British Ministry of Defence standards and specifications and is now in service with the British Army where it is known as Telescope Straight Image Intensified L5A1.

Optional extras include an altazimuth mounting unit, laser rangefinder, high level tripod, and low level tripod with pan/tilt head.

DATA
Dimensions: 610 × 230 mm max diameter
Weight: 10 kg
Magnification: × 5
Field-of-view: (7.5°) 129 mils
Objective field: 20 m to infinity
Power supply: 6.75 V mercury cell (100 h) or rechargeable cell

Manufacturer
GEC Sensors, Pullin Controls Division, Langston Road, Debden, Loughton, Essex IG10 3TW.
Status
Current. Production.
Service
British and foreign armies.

SS32 Twiggy mounted on high level tripod

SS69 second generation Twiggy night viewing device

The SS69 night viewing device is a second generation version of the SS32 'Twiggy' and provides approximately the same performance. Mechanical construction is very similar but the SS69 offers a reduction in both size and weight over its first generation predecessor.

DATA
Dimensions: 510 × 238 mm max diameter
Weight: 9 kg
Magnification: × 5

Field-of-view: 129 mils (7.5°)
Objective field: 20 m to infinity
Power supply: 2.7 V mercury cell (40 h)

Manufacturer
GEC Sensors, Pullin Controls Division, Langston Road, Loughton, Essex IG10 3TW.
Status
Current. Production.

SS67 altazimuth head for Twiggy and laser rangefinder

The altazimuth mounting unit together with laser rangefinder makes Twiggy a versatile manportable instrument for a range of uses. The unit has its own adjustable feet, intended for use in low cover where tripod mounting is impracticable. It achieves fast and accurate target location due to its quick release clutch and five adjustment screws. A north-seeking compass is also built in for common reference for observer and artillery battery.

Manufacturer
GEC Sensors, Pullin Controls Division, Langston Road, Debden, Loughton, Essex IG10 3TW.
Status
Current. Production.
Service
British Army.

SS67 altazimuth head fitted with SS32 (Twiggy) and laser rangefinder

Type SS600 thermal surveillance device

The SS600 thermal surveillance device is a compact system for long-range thermal detection. It offers high thermal and spatial resolution and can be used for general battlefield surveillance and for fire control operations with artillery and mortars. The system is fully passive and exhibits superior performance at greater ranges against comparable image intensified systems, coupled with the ability to see through smoke, haze, mist and into shadow. In addition it is not blinded by intense light or heat sources such as flares, searchlights, impact flashes and fires.

The SS600 is based on the established Pullin Control's thermal imaging system, with the main camera unit comprising the following basic elements:
Scanner unit – consists of scan mechanism, optics, SPRITE detector, cryogenic cooling head, scanner drive and detector head amplifier
Telescope unit – single and dual field-of-view options available
Processing electronics unit (PEU) – converts infra-red signals to a video signal for TV type display(s).
Viewing is by either a 1 in display unit mounted on the camera body, or remote, and viewed directly through an eyepiece, or a remote 9 in diagonal screen display unit. Controls can be provided within the display unit or through a remote hand-held control box. In addition, an autocontrol option can be incorporated within the PEU. This module provides automatic computer optimisation of picture and scene.

An Image Processing Unit (IPU) is available which provides an output of 484 TV lines on standard 625-line format, for display on standard video monitors, together with the facilities of Electronic Magnification and freeze frame. An additional optional plug-in module provides the facility of noise reduction by frame integration to enhance the displayed picture during adverse thermal conditions.

Cooling for the system is provided by high pressure gas bottles, but compressor units and cooling engines are available as options.

Ancillary equipments available include a range of rugged tripod mounts, manual and electrically driven pan and tilt heads, laser rangefinders and telemetry interfaces.

DATA
Field-of-view: (basic Scanner) 40° × 30°
MRTD: typically better than 0.15°C
Detector: Mullard SPRITE
Power supply: 28 V DC nominal (voltage range 19–32 V DC)
Power consumption: less than 36 W at 28 V DC nominal

Type SS600 thermal surveillance device

Pupil diameter: 10.4 mm
Resolution: 2.1 mr
Bandwidth: 8 – 13 µm

Manufacturer
GEC Sensors, Pullin Controls Division, Langston Road, Debden, Loughton, Essex IG10 3TW.
Status
Current. Available.
Service
Commercial and military sales to UK and overseas.

W401 Hall and Watts night observation device

This instrument was announced early in 1983 by Hall and Watts, a company well known for optical and surveying instruments, but a newcomer to the electro-optical field.

W401 Hall and Watts night observation device

The W401 is a long-range surveillance device using a 25 mm second generation micro-channel amplifier tube. The objective lens is 200 mm in diameter and the optical design incorporates the latest lens coating techniques. The instrument has been designed and manufactured to British Ministry of Defence standards and specifications.

The W401 is supplied with shock-absorbent carrying case and carrying bag; optional accessories include various types of tripod, a bracket for mounting a laser rangefinder and an adaptor to fit a camera in place of the eyepiece.

DATA
Magnification: × 7
Field-of-view: (7° 45⁸) 138 mils
Focus range: 20 m to infinity
Graticule: available if required
Image intensifier: 25 mm second generation
Length: 420 mm
Diameter: 240 mm
Weight: 8.5 kg
Power supply: 3.5 V lithium dry cell
Range: (detection) 2000 m; (identification) 1500 m

Manufacturer
Hall and Watts Ltd, 266 Hatfield Road, St Albans, Hertfordshire AL1 4UN.
Status
Current. Production.

Betalight illuminated defile/route marker

The Betalight defile/route marker is a compact, robust, self-powered marker which permits rapid and reliable marking of routes, bridges and minefields. It will retain its luminous properties for up to 15 years. The plate carrying the luminous arrow is mounted on the body of the device so that it can be oriented to any of eight positions at 45° intervals. The arrow is clearly visible by day or night, and in starlight conditions can be seen at a distance of 100 m.

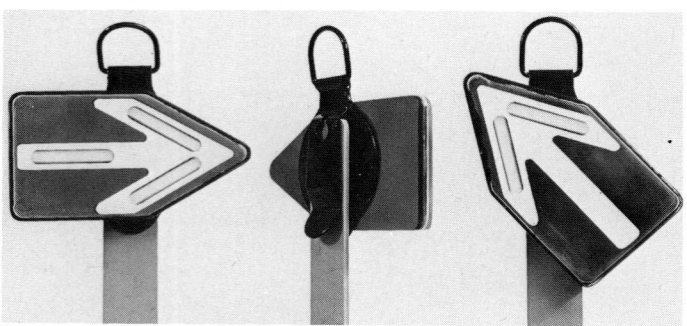

Betalight illuminated defile markers

The body is designed to be clipped to a standard British Military Police mounting pole or other narrow support, attached by screw to wood or threaded metal, or suspended by its loop from any convenient support such as wires, trees or bushes.

The Betalight illuminated route marker is similar in design to the defile marker but contains a third Betalight light source. This additional Betalight is coloured to correspond to the colour of the body, which is available in five colours. Twelve markers can be carried in a standard 7.62 mm ammunition box.

DATA
Length: 130 mm
Weight: 170 g
Width: 70 mm
Thickness: 27 mm
Betalight colours: green; blue; yellow; orange; white

Manufacturer
Saunders-Roe Developments Ltd, Millington Road, Hayes, Middlesex UB3 4NB.
Status
Current. Production.
Service
British forces and other armies worldwide.

Betalight illuminated torch

Another Betalight device is a small pocket torch which gives enough light to read a map or the graduations on fuzes and sights yet is small enough to be easily concealed without encumbering the user. A ring at the rear of the torch is attached to a neck-cord, and will fit over most knife handles for use as an aiming post or marker.

The torch is robustly made and housed in a black rubber moulding. The hinged front cover can be used to obscure the light in order to ensure maximum security in forward areas. With its Betalight self-powered light source this torch is completely independent, requiring no batteries, wires or maintenance.

DATA
Length: 75 mm
Diameter: 35 mm
Weight: 45 g
Betalight colour: green

Manufacturer
Saunders-Roe Developments Ltd, Millington Road, Hayes, Middlesex UB3 4NB.
Status
Current. Production.
Service
British and allied armies worldwide.

Betalight illuminated torch

Betalight illuminated aiming post lamp

Betalight (Tri-lux) aiming post lamps, which are in widespread service, are designed to be used to align mortars or artillery weapons during night firing. Their advantage over earlier types of aiming post lamp is that the Betalight source used is self-powered, reliable, weighs very little and lasts up to 15 years.

The lamps are supplied with either green or orange illumination and standard displays include a cross, a bar and an arrow. The lamps, which are small and light enough to be carried in the pocket, are designed to clamp on either square or round posts.

Using a combination of green and orange lamps it is possible to set up a series of mortars by planting a lamp to the front and rear of each. By using a C2 or similar sight each mortar can then be accurately aligned. By alternating orange and green lamps in both front and rear rows confusion between own and adjoining mortar lamps can be avoided.

Although designed primarily for weapon aiming, the lamp can be used for many other purposes either as a routine or in an emergency. Possible applications include route marking, map reading and minefield marking and the lamp can be used as a torch for close work or as a convoy tailboard lamp.

DATA
Length: 72 or 75 mm including mask
Width: 70 mm including clamp
Diameter: 45 mm
Clamping range: round or square section posts from 13–38 mm diameter
Weight: (complete) 230 g
Normal working range: up to 200 m
Angle-of-view: 7.5° standard, 45° version available
Operating temperature: -60 to +70°C

Tri-lux self-powered aiming post lamp

Manufacturer
Saunders-Roe Developments Ltd, Millington Road, Hayes, Middlesex UB3 4NB.
Status
Current. Production.
Service
British Army and many foreign armies. Standard equipment with the L16 81 mm mortar.

Betalight illuminated Peglight

Peglight is a development of the Betalights and is intended to combine several of the functions in one item. Specifically it is meant to be a route and minefield marker, gun position indicator or arc of fire indicator. The light is constructed by placing a long Betalight source inside a tube with three apertures. At night these three apertures produce a row of three strips of light which are easily and quickly picked out.

The advantage of these and similar light sources is that they require no maintenance and are only visible over a restricted arc so that they are quite secure. They emit no heat and give no infra-red emissions.

DATA
Length: 203 mm
Weight: 60 g
Diameter: 12 mm
Betalight colour: green
Viewing range: 25 m in clear starlight
Viewing angle: 90°

Manufacturer
Saunders-Roe Developments Ltd, Millington Road, Hayes, Middlesex UB3 4NB.

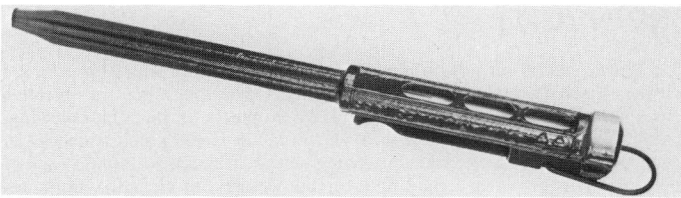

Betalight illuminated Peglight

Status
Current. Available.
Service
British and foreign armies world-wide.

Betalight illuminated map reader/magnifier

The Betalight illuminated map reader, requiring no battery, or wiring gives sufficient light output to illuminate a route map, without affecting night vision, while the recessed light source reduces the chance of enemy detection.

The compact, robust black rubber housing of the Betalight map reader/magnifier protects the sturdy plastic lens and side panel from scratching and breakage. This product is ideal for accurate map reading in all climatic conditions and where maximum security in forward areas is essential. It is perfectly suited to artillery and armoured vehicle manoeuvres.

DATA
Length: 85 mm
Weight: 200 g
Width: 65 mm
Thickness: 50 mm
Betalight colour: white

Manufacturer
Saunders-Roe Developments Limited, Millington Road, Hayes, Middlesex UB3 4NB.
Status
Current. Production.
Service
British and foreign armies world-wide.

Betalight illuminated map reader

Betalight SPI

The Betalight SPI has been designed in conjunction with the British MoD to provide the soldier with a multi-purpose self-powered light source. Supplied complete with neck cord and clip attachment, the robust and lightweight design makes SPI easy to use. By rotating the outer sleeve, five different configurations of light source can be obtained: torch, map reading light, personal marker, route marker or aiming post.

SPI has a low level light source ideal for reading maps and signs, or fuze and sight graduations, without losing the advantage of night vision. The light is derived from a controlled source ensuring maximum security against detection. Illumination is provided by Betalight (GTLS), a glass capsule internally coated with phosphor and filled with Tritium gas, activating the phosphor particles. The device can be used in all climatic conditions and has a life expectancy in excess of five years.

DATA
Diameter: 30 mm
Length: 67 mm
Weight: 40 g
Betalight: green
Body colour: black
NSN: 6260-99-781-4147

Manufacturer
Saunders-Roe Developments Ltd., Millington Road, Hayes, Middlesex UB3 4NB.

Betalight SPI self-powered light source

Status
Current. Production.
Service
British Army.

UNITED STATES OF AMERICA

Baird general-purpose night vision goggles (GP/NVG-1)

The Baird general-purpose night vision goggles are a product programme between Baird Corporation and Litton Systems. The goggle has one objective lens, one second generation II tube, and two eyepieces for binocular viewing. The unit is held to the eyes with a plastic facial mask which touches the forehead, temples, cheeks and chin, being held with straps round the back of the head. This provides a rigid base for the viewer and virtually eliminates shake in the field-of-view. The viewer is held to the mask by one quick-release clamp so that there is the minimum of fiddling required in donning or removing the viewer. The goggle can also be hand-held or mounted on a helmet.

DATA
Weight: 450 g
Magnification: × 1
Field-of-view: 40°
Gain: × 2500
Angular resolution: 1.4 mrad
Focus range: 250 mm to infinity
Batteries: 2 × AA alkaline or equivalent

Manufacturer
Baird Corporation, 125 Middlesex Turnpike, Bedford, Massachusetts 01730.
Status
Current. Available. In production.
Service
US and foreign military.

Baird GP/NVG-1A general-purpose night vision goggles and mask

Baird 'Maglens' night vision goggle magnifying lens

The Baird Maglens is an optical assembly designed as an attachment for use with night vision goggles. It will increase the scene magnification from × 1 (unity) to × 2.5, and it can also be used independently for daylight observation. The Maglens can be used with a wide variety of night vision goggle systems, including the GP/NVG-1, AN/PVS-5 (all versions), AN/PVS-6 (ANVIS) and AN/PVS-7B.

The Maglens provides a quick and easy way to bring a scene closer for detailed observation; it provides a compact and lightweight attachment to provide temporary magnification without altering the operational intent of a unity-powered goggle; and it is an economical approach to obtaining magnified scene viewing without the added size and weight of normal magnified night vision binoculars.

DATA
Magnification: × 2.5 ± 0.5
Aperture: entrance 50 mm; exit 20 mm
Effective aperture: f/1.3 based on a 26 mm FL objective
Distortion: not to exceed 4%
Weight: 165 g
Diameter: 57 mm
Length: 80 mm

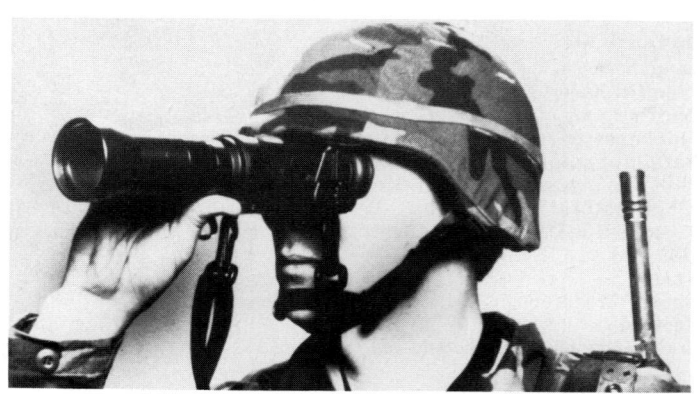
Baird Maglens night vision goggle magnifying attachment

Field-of-view: 20° on GP/NVG; 15 – 20° on others, depending on type
Focussing distance: 2 m to infinity

Manufacturer
Baird Corporation, 125 Middlesex Turnpike, Bedford, Massachusetts 01730.

Status
Current. Available. In production.
Service
US and foreign military.

General-purpose night vision binocular (GP/NVB-4, GP/NVB-6)

The Baird GP/NVB-4 and GP/NVB-6 are general-purpose night vision binoculars for long-range observation and surveillance. The GP/NVB-4 has a 95 mm focal length objective lens and offers four-power magnification. The GP/NVB-6 has a 155 mm focal length objective lens and provides six-power magnification. Both systems offer the proven performance and reliability of a second generation image intensifier tube that has been designed specially to match the Baird optics. The eyepieces can be individually focussed to accommodate various eye prescriptions. A built-in objective focus mechanism lets the operator refocus on a new target quickly.

The units have been engineered to minimise weight and cost without sacrificing performance and quality. They have a rugged construction and are easy to operate and maintain.

Specific applications include long-range night observation and surveillance for tank commanders, forward area observers, border patrols, and special forces.

DATA

	GP/NVB-4	GP/NVB-6
	95 mm focal length	155 mm focal length
	T 1.7, F 1.2	T 1.7, F 1.2
Magnification	× 4	× 6
Field-of-view	10.5°	6.5°
Resolution	0.4 mrad	0.25 mrad
Brightness gain	1000	1000
Focus adjust range	10 m to infinity	20 m to infinity
Image intensifier tube	18 mm, 2nd generation	18 mm, 2nd generation
Individual eyepiece focus	+2 to -6 dioptres	+2 to -6 dioptres
Interpupillary adjustment	55 to 72 mm	55 to 72 mm
Weight	1.13 kg (2.5 lbs)	3.18 kg (7.0 lbs)
Length	225 mm	305 mm
Width	145 mm	155 mm
Power source	Two AA alkaline batteries (or equivalent)	Two AA alkaline batteries (or equivalent)

Manufacturer
Baird Corporation, 125 Middlesex Turnpike, Bedford, Massachusetts 01730.
Status
Current. Production.

Baird GP/NVB-6 binocular

Baird GP/NVB-4 binocular

M-802 night vision goggles

The M-802 goggles have been developed under the auspices of the US Army to provide passive night viewing in minimal light conditions. They are used for such tasks as driving, helicopter flying, air, sea search and rescue, map reading and close handwork such as vehicle repairing and mechanical assistance.

The II tube is a second generation channel-plate and the whole unit is entirely self-contained and hands-free in operation.

The goggles are furnished with a universal battery cap, and use either the BA 1657/U or E-132 commercial batteries, or the lithium type 4405-BT for cold weather use.

The M-802 is functionally equivalent to the US Army standard AN/PVS-5A goggles.

DATA
Weight: 850 g (including batteries)
Dimensions: 119 × 172 × 165 mm
Magnification: × 1 (unity)
Gain: 500 × nominal
Resolution: 0.67 lp/mrad (nominal)
Field-of-view: 40°
Battery: 2.7 V, 40 mA mercury battery
Focus range: 250 mm to infinity

M-908 goggles, modified version for use with standard AA-type batteries

M-802 night vision goggles

Manufacturer
Litton Systems Inc., Electron Devices Division, 1215 South 52nd Street, Tempe, AZ 85281-6987.
Status
Current. Available.

Service
Military and paramilitary service.

Litton M975/976 night vision binoculars

The M975/976 night vision binoculars are based on the AN/PVS-7 night vision goggle. The binocular utilises the PVS-7 body and substitutes an x4 objective lens for the x1 lens normally used with the PVS-7. Field conversion between x4 and x1 power requires no special tools and can be accomplished in minutes.

The M975 and 976 differ only in the image intensifier; the M975 is standard with the GEN II Plus second-generation intensifier, while the 976 uses the GEN III third-generation for higher gain and increased resolution under conditions of extremely low light levels.

DATA
Field-of-view: 8.5°
Magnification: × 4
Brightness gain: × 1000 (M975); x1200 (M976)
Resolution: 3.0 lp/mR
Focus range: 20 m to infinity
Dioptre focus: -6 to + 2 dioptres
Eye relief: 27 mm
Weight: 1.1 kg with batteries

Manufacturer
Litton Systems Inc., Electron Devices Division, 1215 South 52nd Street, Tempe, AZ 85281-6987.
Status
Current. Production.

Litton M975/976 night vision binocular

M972/M973 night vision goggle

The M972/M973 night vision goggle system employs a single 18 mm proximity focussed image intensifier, either of 'Gen II Plus' type (M972 goggle) or third generation (M973 goggle).

Retrofit of the plug-in image intensifier is all that is required to take advantage of the improved low light performance of the third generation.

The goggle is lightweight, can be hand-held or head-mounted, and will accept either military style BA 1567 or BA 5567 or commercial 'AA' size alkaline or NiCd batteries. The unique open face mask design and long eye relief allows excellent peripheral vision and interface with protective masks.

The M972/M973 goggle is identical to that currently in production at Litton for the US Army, which was due for type-classification as AN/PVS-7 in late 1986.

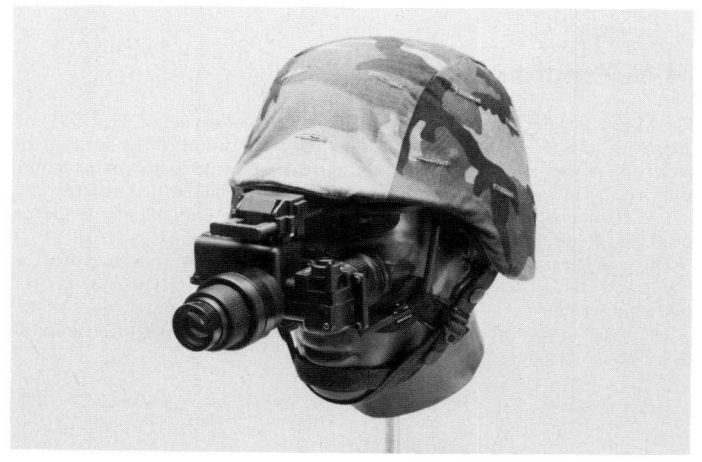

M972/M973 night vision goggle

DATA
Field-of-view: 40°
Magnification: × 1
Brightness gain: × 1850
Resolution: 0.68 lp/mR
Focus range: 250 mm to infinity
Dioptre focus: -6 to +2 dioptres
Eye relief: 25 mm
Weight: 680 g (mounted)

Manufacturer
Litton Systems Inc., Electron Devices Division, 1215 South 52nd Street, Tempe, AZ 85281-6987.
Status
Current. Production.

M909 high performance night vision goggles

These are based on the well-proven US military AN/PVS-5A design. The M909 incorporates a substantially improved object lens, higher gain and brighter second generation image intensifier tubes. These improvements give the M909 goggles twice the system gain of the AN/PVS-5A and a 33 per cent increase in useful range. All other applicable specifications of the AN/PVS-5A are met or exceeded in the M909.

DATA
Magnification: × 1
Field-of-view: 40°
Focus range: 300 mm to infinity
Objective: f/1.09 (T/1.15)
System gain: 1200
Dimensions: 119 × 173 × 165 mm
Weight: 880 g
Power supply: 2.7 V DC 1000 maH mercury cell
Battery life: 20 h

Manufacturer
Litton Systems Inc., Electron Devices Division, 1215 South 52nd Street, Tempe, AZ 85281-6987.
Status
Current. Production.

M909 night vision goggles

M-911A night vision modular system

The M-911A system is primarily designed as a hand-held medium range surveillance viewer. The modular concept, however, allows the user to tailor the optical configuration of the system to suit varied applications. The basic module incorporates the second generation image intensifier, the battery compartment, on-off switch, infra-red led, and basic lens interfaces. All optional objective and viewing lenses mount directly to this housing. The infra-red led provides covert light for night viewing to a distance of 6 m. There is a standard 1/4 × 20 tpi threaded tripod bush at the base of the housing, permitting attachment to a tripod or carrying strap.

A variety of object lenses are available, including two high-quality low light level night vision lenses especially designed to match the spectral response of the image intensifier tube. For less demanding applications, 'C' mount adapters allow the use of commercial CCTV, motion picture and still camera lenses.

The M911A accepts the military standard eyepiece common to the AN/PVS-5A and AN/PVS-5B goggles. It offers an adjustable dioptre range of +2 to -6. A camera relay lens can be fitted which allows connection of almost any commercial SLR camera by means of the T-2 mount system. For TV systems a T-2 to C mount adapter is available, and for the latest solid-state CCTV systems a special relay lens can be supplied which reduces the image by × 1.5 to match the CCTV format.

DATA

	with 75 mm lens	with 26.8 mm lens
Magnification	× 3	× 1
Gain	1200	1700
Field-of-view	13°	39°
Resolution	1.81 lp/mR	0.68 lp/mR
Focus range	2 m – infinity	25 cm – infinity
Weight	775 g	500 g
Battery life	40 h	40 h

Manufacturer
Litton Systems Inc., Electron Devices Division, 1215 South 52nd Street, Tempe, Arizona 85281-6987.
Status
Current. Production.

Litton M-911 body with camera relay lens and 75 mm f/1.3 objective

Litton M-911A viewer with 26.8 mm lens

M931 ranging night scope

This is a combined hand-held night viewer and laser rangefinder. Small size and light weight make the system ideal for mobile night operations involving forward observers, reconnaissance patrols or special forces. The completely self-contained system offers passive viewing capability and an active calibrated range readout which can function under moonlight or starlight conditions. The operator merely uses the night vision scope to detect his target, then triggers the laser unit to determine its range.

DATA
Magnification: × 3
System gain: 1000
Field-of-view: 13.6°
Resolution: 2 LP/mR
Focussing range: fixed infinity focus
Object lens: 75 mm f/1.3
Eyepiece magnification: × 9.4
Laser material: Gallenium arsenide
Laser wavelength: 850 mm
Average power: 5 mW
Range coverage: 20 – 1000 m
Range resolution: 10 m
Dimensions: 170 × 140 × 70 mm
Weight: 1.3 kg
Power supply: 4 × 3.5 V Lithium, 1 × 2.7 V mercury batteries
Battery life: 2 h continuous ranging, 20 – 30 h viewing
Image intensifier: 2nd generation

Manufacturer
Litton Systems Inc., Electron Devices Division, 1215 South 52nd Street, Tempe, Arizona 85281-6987.
Status
Current. Production.

Litton M-931 ranging night scope

NVS-500 night vision system

This is a second generation viewing or weapon-aiming system using a 25 mm second generation image intensifier tube and providing a choice of 95 mm f1.2 (with or without reticle) or 155 mm f2.0 catadioptric objective lenses. The combination of 95 mm focal length objective and a reticle projector offering an adjustment of 0.25 milliradian per click gives the device, when used as a weapon sight, great potential for accuracy.

The device is designed to meet military standards but also to be saleable at a price that would not bar it for civil law-enforcement uses. It can be used with cameras or video receiving equipment and available accessories include monocular and biocular viewers and a relay lens.

Power is supplied by two AA Penlight batteries which will give about 50 hours of operation. The power unit is interchangeable with that of the NVS-100 viewer (see previous entry).

DATA
Weight: 1.9 kg
Length: 279 mm
Diameter: 76 mm
Magnification: × 7
Resolution: 28 lp/mm
Luminance gain: 25 000 (ABC)
Objective lens: 95 mm T1.7 (155 mm available)
Relative aperture: 1.7
Battery life: 50 h continuous running

Manufacturer
Optic-Electronic Corp, 11545 Pagemill Road, Dallas, TX 75243.
Status
Current. In production since 1973.
Service
US and foreign police and law enforcement agencies.

NVS-500 device

NVS-900 night observation device

The NVS-900 is a long-range observation device using a second generation image intensifier. It can be used in conjunction with a laser rangefinder for artillery fire control or alone as a surveillance instrument. It can be easily mounted to the OEC GT-901 Goniometer tripod, permitting elevation from -400 to +400 mils and 6400 mils rotation in azimuth.

DATA
Magnification: × 9.7
Viewing range: (moonlight) 6000 m; (starlight) 4000 m
Field-of-view: 98 mils
Objective: 258 mm T/1.7
Dimensions: 483 × 244 mm
Weight: 9 kg
Power supply: 2 AA mercury cells
Battery life: 72 h

Manufacturer
Optic-Electronic Corp, 11545 Pagemill Road, Dallas, TX 75243.
Status
Current. Available.

NVS-900 night observation device

NV 38 night vision goggle

The NV 38 is a lightweight, head-mounted, self-contained night vision system. It employs a third-generation image intensification tube to amplify night-time illumination and is used by individuals for walking, driving, weapon firing, short-range surveillance, map reading and similar tasks. It consists of a single objective lens, an image intensifer tube assembly, and a beam splitter/dual eyepiece assembly. The unit is attached to a face mask which is held to the user's head by straps. The unit incorporates an infra-red light source which provides local illumination when required, for example for map-reading. It can be worn with all types of steel and Kevlar helmets and with the anti-gas respirator for use in NBC conditions.

DATA
Magnification: × 1
Weight, with batteries: 765 g
Field-of-view: 40°
Gain (minimum): 500
Angular resolution: 1.5 mrad
Focus range: 250 mm to infinity
Battery power: 2 × AA size alkaline or equivalent
Eyepiece dioptre range: +2 to -6
Objective: f/1.2

NV 38 cyclops night vision goggle

Manufacturer
Optic-Electronic Corp, 11545 Pagemill Road, Dallas, TX 75243.

Status
Current. Available.

Noctron V night vision viewer

The Noctron V is designed to fulfil the commerical needs for a second generation hand-held night-viewing device. The Noctron V will be used for direct viewing and low light level photography when attached to 35 mm SLR cameras or 16 mm movie/CCTV camera.

The standard unit features a 135 mm f1.8 telephoto lens, × 5 high performance biocular eyepiece and the latest technology 25 mm micro channel plate image tube. Designed to achieve superior imagery in low and high light levels with minimum bloom and streaking, using variable gain second generation tube technology.

A bayonet adaptor permits rapid interchange of all rear optic accessories. Other key features are: observer fatigue reduced through human engineering of balance and display characteristics; battery powered using two AA Penlight batteries, with a built-in battery power (LED) indicator; unique automatic circuit protects against accidental exposure of the tube to very high light levels; readily adapts to many commercially available telephoto and zoom lenses; special adaptor provides power from 12 V vehicle cigarette lighter outlet.

DATA
Basic System Characteristics
(135 mm f1.8 lens and × 5 biocular eyepiece)
Magnification: × 3
Field-of-view: 10.6°
Tube gain: 45–55 000
Dimensions: 320 × 80 × 210 mm
Weight: 1.98 kg
Detection range (standard unit): (man) 575 m; (vehicle) 792 m

Noctron V night vision viewer

Manufacturer
Varo Inc, Electron Devices, PO Box 469014, 2203 West Walnut Street, Garland, TX 75046. (An IMO Industries company.)

Status
Current. Available.

Service
Police and law enforcement agencies.

AN/PVS-5C passive night vision goggles

These improved standard infantry night vision goggles utilise two 18 mm micro-channel wafer image intensifiers to provide night vision capabilities for the user. New features include high-light cut-off for tube protection; operation on both 'AA' and military type batteries; a faster f/1.05 objective lens; self-contained IR illuminator; and a weight under 1 kg. They are intended for use by drivers of all vehicles, patrols, security surveillance, detection of IR sources, and, with the use of the light source, map reading and equipment maintenance. The focussing arrangements allow the user to focus down to 25 cm and up to infinity. The goggles have a soft face pad which fits over the front of the face from the nose to the forehead and allows it to be worn for long periods without fatigue.

DATA
Weight: 907 g
Magnification: × 1
Field-of-view: 40°
Resolution: 0.6 mRad
Focussing range: infinity to 250 mm
Objective: f/1.05
Image tube: two 18 mm micro-channel wafer tubes
Power source: BA 1567U or 'AA' batteries
Battery life: 15 – 100 h dependent upon battery type
Operational temperature: -45 to +45° C

Manufacturer
Varo Inc, Electron Devices, PO Box 469014, 2203 West Walnut Street, Garland, TX 75046. (An IMO Industries company.)

Varo AN/PVS-5C night vision goggles

Status
Current. Production.

Varo 'Nite-Eye' viewer

'Nite-Eye' is a lightweight second-generation passive night viewer using a high-performance 18 mm micro-channel wafer image intensfier and an f/1.05 object lens. The light weight and small size allow the user to have the benefits of passive night vision at his disposal under the most adverse conditions. It

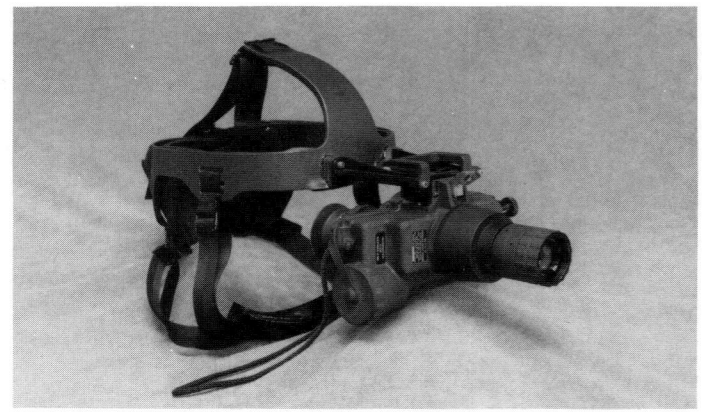
Varo 'Nite-Eye' viewer

also features a built-in intensified LED light source which projects a beam of light invisible to the naked eye so as to illuminate areas which are devoid of any ambient lighting.

Threaded lens mounts allow rapid interchange of object lenses to permit long-range viewing, 35 mm photography or TV camera surveillance. A standard ¼ in 20 tpi thread provides a means of attachment to a tripod or neck strap.

DATA
Weight, without objective: 425 g
Magnification: × 1 (26.8 mm lens); × 2.8 (72 mm lens)
Field-of-view: 40° (26.8 mm); 14° (72 mm)
Resolution: 0.8 lp/mR (26.8 mm); 2.16 lp/mR (72 mm)
Focus range: 250 mm to infinity (26.8mm); 10 m to infinity (72 mm)
Aperture: f/1.0 26.5 mm; f/1.5 72 mm
Dioptre adjustment: +2 to -6
Power supply: 2 × 'AA' batteries
Battery life: 50 h

Manufacturer
Varo Inc, Electron Devices, PO Box 469014, 2203 West Walnut Street, Garland, TX 75046. (An IMO Industries company.)
Status
Current. Production.

AN/PVS-7B single tube goggle

This improved single tube goggle is head-mounted, lightweight, features a high-light cut-off for tube protection and a quick-release lever to remove the goggle from its mask for hand-held observation. It is compatible with various helmets and gas masks, and can be teamed with a rifle and IR aiming spotlight. It also provides close focus capability for map reading and equipment maintenance, with the use of an IR light source. The AN/PVS-7B features a high performance second generation image tube with glass input, and can easily be upgraded to third generation.

DATA
Magnification: × 1
Field-of-view: 40°
Resolution: 32 lp/mm
Focussing range: 250 mm to infinity
Objective lens: 26 mm f/1.17, T/1.3
Dioptre adjustment: +2 to -6
Interpupillary adjustment: 55 to 71 mm
System gain: 1850
Weight: 680 g
Power supply: 2 × AA or 1 × BA-1567U

Varo AN/PVS-7B single tube goggle

Manufacturer
Varo Inc, Electron Devices, PO Box 469014, 2203 West Walnut Street, Garland, TX 75046. (An IMO Industries company.)

Status
Current. Production.

Night observation device (NOD II)

The NOD is a tripod-mounted night vision sight typically used for long-range surveillance when maximum performance and minimal operator fatigue is of primary importance. The exceptional performance of the NOD can be attributed to its high quality objective and the Varo-produced 25 mm micro-channel second generation plate image tube.

Automatic Brightness Control is a standard feature for the NOD. The ABC feature permits viewing during the twilight hours of dusk and dawn and protects the tube against bright flashes of light. The tube features muzzle-flash protection which prevents the tube from being damaged by high intensity short duration flashes of light. The flash protection circuit is designed to recover in time for the observer to see the round hit the target.

DATA
Weight: 11.6 kg
Dimensions: 410 × 260 × 260 mm
Magnification: × 9.4
Field-of-view: 5.6°
Objective focal length: 225 mm f/1.8
Battery: 2 × AA batteries

Manufacturer
Varo Inc, Electron Devices, PO Box 469014, 2203 West Walnut Street, Garland, TX 75046. (An IMO Industries company.)
Status
Current. Available.
Service
US and other armies.

NOD II night observation device

YUGOSLAVIA

PRD hand-held viewer

The PRD is a passive electro-optical binocular device for night observation and surveillance. Normally operating under moon or starlight, in the event of excess illumination the image intensifier is automatically switched off. As soon as operating light level conditions are regained, the device is automatically turned on.

DATA
Weight: 2.2 kg
Magnification: × 4
Field-of-view: 10°
Resolving power: 0.85 mRad at 1 × 10⁻³
Dioptre adjustment: ± 5 dioptres
Interpupillary adjustment: 58 to 72 mm
Focus range: 4 m to infinity
Power supply: 2.48 V NiCd cell
Battery life: 40 h

Manufacturer
Federal Directorate of Supply and Procurement, 9 Nemanjina Street, 11001 Beograd.
Status
Current. Production.

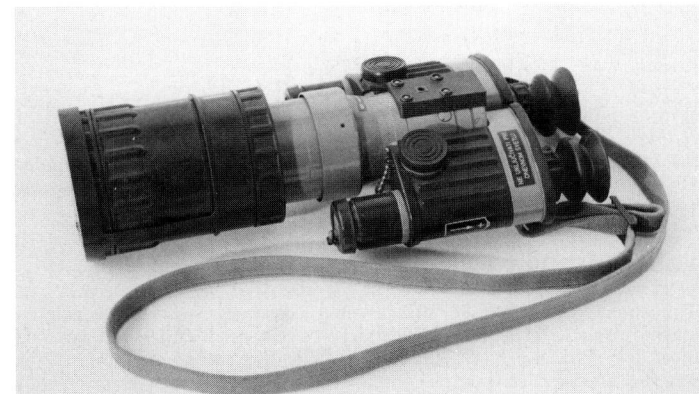
PRD hand-held binocular viewer

Service
Yugoslavian Army.

Sighting Equipment

BELGIUM

FN TR3 Tritium sighting system

The FN Tritium sighting system is FN's answer to low light level sighting conditions. The TR3 system considerably increases the soldier's combat effectiveness under poor light conditions, since he can easily distinguish the sights of his weapon and thus acquire the target more rapidly and with increased accuracy. It is a low-cost, easy to operate and absolutely maintenance-free sighting system.

FN weapons can be directly supplied with the TR3 sighting device, but the unit armourer can very easily retrofit existing weapons, replacing FN standard sights with the new tritium system.

Manufacturer
Fabrique Nationale Herstal SA, Branch Défense et Sécurité, B-4400 Herstal.
Status
Current. Production.

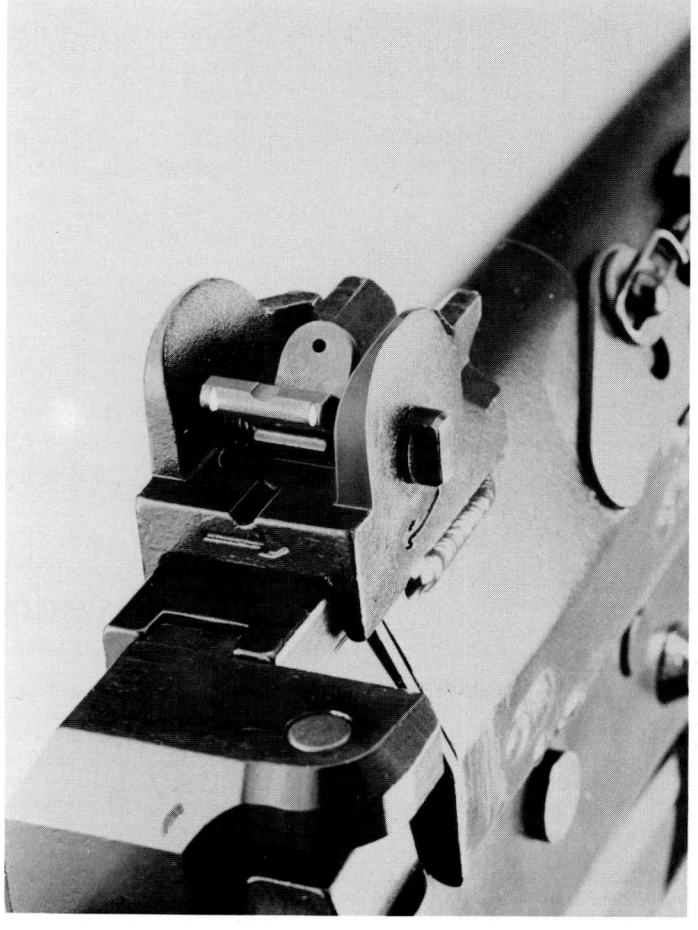

FN TR3 system rear sight, showing two tritium markers at each side of aiming notch; single marker is on foresight post

Laser Hit Marker

The Aims Optronics Laser Hit Marker is a laser spot projector which, by means of various adapters, can be attached to any small arm. When switched on it delivers a clearly visible red spot indicating the point of aim of the weapon. The firer need take no conscious aim, but with both eyes open merely lays the spot on the target and fires.

The Marker is a battery-powered Helium-Laser operating at 632.8 nm to deliver a visible spot; a compact model is available for distances up to 100 m and a newly developed laser beam control offers long distance aiming for snipers. For military applications a different laser, generating a spot only visible with night sights or night goggles, to a range of 300 m, can be provided. The marker can be mounted using any conventional type of telescope mounting rings to any type of small arm including pistols. Depending on the model of Marker, the laser beam can be adjusted either mechanically or optically in order to bore-sight the weapon at any selected range. Special X-Y collimators are available for beam control and long distance work.

Different power supplies accommodate the user's requirements in action or for training. Complete marksman training systems are available to improve accuracy, follow-through or to assess the shooter's overall performance.

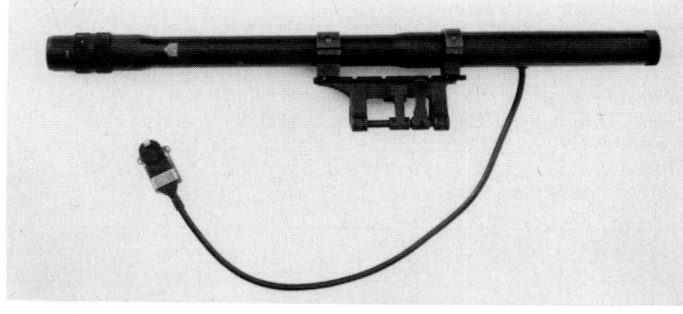

Aims Optronics Laser Hit Marker with mounting bracket

Manufacturer
Aims Optronics NV, SA, Rue F.Kinnenstraat 30, B-1950 Kraainem.
Status
Current. Production.

CANADA

CLASS Computerised Laser Sight System

CLASS is a new gunnery sight designed to be fitted to almost any direct or indirect fire weapon. It consists of a telescope sight, a rangefinding laser, a video sensor and a micro-computer coupled together to provide a full fire control solution. In use the gunner selects the ammunition type, sets the estimated wind speed, and fires the laser twice, ranging and ranging confirmation, at the target. The computer, using the information from the laser and the video sensor, proceeds to calculate the correct lead and super-elevation angles. The correct aim point is then injected into the optical path and a ready signal prompts the gunner to fire. The entire process takes less than three seconds to complete.

Although the point of aim is re-positioned, the graticule remains in the centre of the telescope field, a set of optical wedges being used to bend the field-of-view. This novel approach not only allows centralising of the graticule at all times but also permits lead and super-elevation angle corrections which are greater than the field-of-view. All target and firing data is presented in the sight eyepiece, above the field-of-view, so that the gunner need never take his eye from the sight once an engagement has begun.

CLASS computerised laser sight on Carl Gustav 84 mm RCL gun

DATA
LASER
Medium: NdYAG
Wavelength: 1.064 microns
Beam divergence: 0.7 mRad

RANGEFINDER
Range: 25 – 4000 m
Range increment: 5 m
Range accuracy: ± 3 m

SIGHT
Magnification: × 3.25
Field-of-view: 10°
Eye relief: > 30 mm

SYSTEM
Weight: 2.2 kg
Range super-elevation: 120 mRad
Pointing accuracy: 0.5 mRad

Manufacturer
Computing Devices Company, PO Box 8508, Ottawa, Ontario K1G 3M9.
Status
Advanced development.

Leitz Elcan small arms optical sight

Designed and developed by Ernest Leitz (Canada) Ltd, this sight was designed for use with a variety of rifles but is presently configured to be the prime sighting system for modified M16 rifles and Minimi light machine guns.

The Elcan sight has no moving parts. It is a robust lightweight optical sight that has been developed over the years, starting with the Canadian FNC1 rifle and progressing to its present use. Range corrections are accomplished by using externally mounted adjusting drums. The tip of the graticule pattern

vertical post is illuminated with a built-in tritium source. The sight is permanently sealed with dry nitrogen and is proofed against dust, moisture, insects, fungus and weather.

The optical sight, when used with the M16 rifle with modified upper receiver, replaces the carrying handle and iron sights. This permits the correct sight line and a proper cheek position for the firer.

Leitz Elcan small arms sight

DATA
Weight: 430 g ncluding mount
Dimensions: 160 × 72.5 × 43.5 mm
Magnification: × 3.5
Field-of-view: 7°
Exit pupil: 8 mm
Eye relief: 52 mm
Range adjustment: 0 – 1000 m

Manufacturer
Ernst Leitz (Canada) Ltd, 328 Ellen Street, Midland, Ontario L4R 2H2.
Status
Pre-production.
Service
Currently under trial in Canada and the USA.

CHINA, PEOPLE'S REPUBLIC

Type JWJ machine gun low light level sight

This is a second-generation image intensifying sight intended for use on machine guns and other crew-served weapons and also as an independent surveillance instrument. It is provided with bright light protection circuitry which prevents interference from flashes and bright illumination in the target area. An accessory handle (seen in the picture) is provided for when the sight is used for hand-held observation.

DATA
Magnification: × 4.5
Field-of-view: 10°
Focussing range: 20 m—infinity
Dioptric adjustment: ± 2.5 dioptres
Resolution: 0.5 mRad (E = 10^{-1} lux, C = 0.85); 1.2 mRad (E = 10^{-3} lux, C = 0.35)
Night vision range: person 500 m; vehicle 1000 m
Weight: 2.1 kg

Manufacturer
China North Industries Corp, 7A Yue Tan Nan Jie, Beijing.
Status
Current. Production.

Type JWJ night vision sight

FRANCE

TN2-1 SOPELEM night observation and driving sights

The TN2-1 sight is a nocturnal vision instrument providing for observation and movement by night, guiding the driver of an armoured vehicle by the commander (with his head outside the turret), and short-range night firing in combination with the PS I spotlight. It is a binocular sight fitted to a mask worn by the observer. It operates on the principle of electronic light intensification.

DATA
Binocular: monotube, 2nd generation, 18 mm wafer tube, double focus
Magnification: × 1
Field-of-view: 40°
Weight: 0.47 kg
Power supply: 2.7 V or 3.5 V military lithium batteries or 2 × 1.5 V standard commercial batteries

Manufacturer
SOPELEM, 19 Boulevard Ney, PO Box 264, 75866 Paris Cedex 18.
Status
Current. Production.
Service
French and Foreign Special Forces.

TN2-1 night observation and driving sight

SOPELEM PS 2 spotlight projector

The PS 2 spotlight projector is bore-sighted with the weapon, and emits a luminous beam in the close infra-red spectrum, indicating the pointing of the weapon on the target. It should be used with a night vision telescope, so that the firer can see the spot of light on the target. The SOPELEM TN2-1 sight (above) fulfils this function.

When the firer sees a target, using his night vision instrument as a surveillance device, he switches on the projector, the beam of which is invisible to the naked eye. The firer then lines up the spot on the target and fires. There is no need to take aim, merely to place the light spot on the target. The spotlight is adaptable to all weapons; NATO interface or specific interface can be used with all image intensifying night vision equipment.

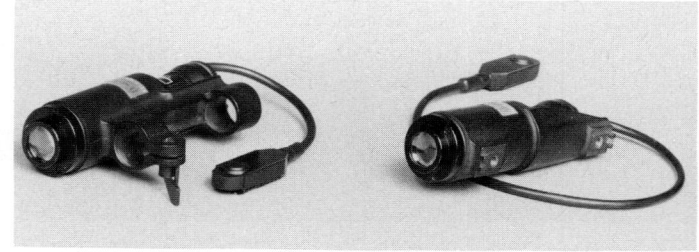

SOPELEM PS 2 spotlight projector

DATA
Emission wavelength: 820 nm
Power supply: 2 × AA batteries, 1.5 v
Weight: 250 g

Manufacturer
SOPELEM, 19 Boulevard Ney, PO Box 264, 75866 Paris Cedex 18.

Status
Current. Production.
Service
French Army and foreign Special Forces.

OB-50 SOPELEM night firing scope

The OB-50 SOPELEM night aiming telescope is designed for observation and aiming, without artificial lighting of the target, by intensification of residual light from the night sky. This technique ensures total discretion, and enables instantaneous detection of active infra-red emission. It incorporates a second-generation micro-channel 18 mm light intensifier tube fitted with double-proximity focus, and with built-in AGC. An illuminated micrometer and an eyepiece shade with shutter are also fitted.

The telescope is adaptable to all current types of infantry weapon and is particularly suitable for rifles, machine guns and light anti-tank rocket launchers.

DATA
Magnification: × 3.2
Field-of-view: 11°
Power supply: 2.7 V or 3.5 V military lithium batteries or 2 × 1.5 V commercial standard batteries
Length: 230 mm
Weight: (less support) 0.9 kg
Operating temperature: -40°C to + 55°C

Manufacturer
SOPELEM, 19 Boulevard Ney, PO Box 264, 75866 Paris Cedex 18.

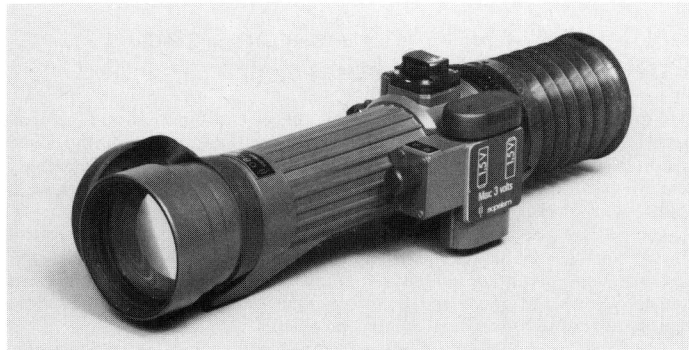
OB-50 SOPELEM night firing scope

Status
Production.
Service
French Army and foreign Special Forces.

MIRA thermal imaging night sight

MIRA (Milan Infra-Red Attachment) is an add-on sight unit which can be rapidly fitted to the existing Milan missile firing post sight to convert it to night operation. It is automatically harmonised with the optical path of the standard sight and reflects the IR picture into the optical sight so that the normal day sight eyepiece is still used for viewing. Using series-parallel scanning and a CMT detector, the sight picture is produced by LED visualisation. The sensor is cooled by a JT Cryogenic system.

DATA
Weight: 8 kg
Detection range: 4000 m
Recognition range: 2000 m
Duration: 2 h (cooling bottle and internal battery)

Manufacturers
Series production is carried out under the general management of: TRT, 88 rue Brillat Savarin, 75640 Paris Cedex.
Marconi Space and Defence Systems, The Grove, Warren Lane, Stanmore, Middlesex HA7 4LY, England.
Siemens AG, Hoffmannstrasse 51, D-8000 Munich 70, West Germany.
Status
Current. Production.
Service
French and British armies.

MIRA thermal imaging night sight on Milan firing post

GERMANY, WEST

INKAS infantry night combat system

INKAS is an integrated night-fighting system for the infantry soldier. It consists of an infra-red laser projector built into the weapon, and image-intensifying goggles worn by the man.

The laser spot projector is built into the cocking lever tube of the Heckler and Koch G3 rifle or G41 rifle, in front of the cocking handle and above the barrel. A small sliding switch is also fitted, behind the front sight on the left side. The fitting of this projector does not interfere with any of the rifle's normal functions, except that the bayonet cannot be fitted.

The goggles are of the usual second-generation pattern, attached to a face-mask and a retaining cap which fits beneath any normal headgear such as steel helmets. The goggles are specially manufactured for this system by Elektro-Spezial GmbH.

In use, the man wears the goggles and carries the rifle with the laser projector switched off. He uses the goggles to observe and seek targets. On detecting a target he identifies it and, if hostile, proceeds to engage. He

INKAS night fighting system

roughly aligns his rifle and then switches on the laser projector which sends an invisible beam, bore-sighted with the axis of the weapon. By merely laying the spot on to the target aim is taken and he can fire.

Although the makers provide the goggles with the integrated system, the laser target marker can be used with any image intensifying night goggles.

DATA
Weight of target marker unit: approx 200 g
Wavelength: 820 mm
Range: approx 200 m in starlight

Power supply: 3 V DC
Battery life: approx 3 h continuous operation; equivalent to perhaps 3000 engagements

Manufacturer
Heckler and Koch GmbH, D-7238 Oberndorf/Neckar.
Status
Project discontinued.

AN/PVS-4, AN/TVS-5 weapon sights, EURONOD-2 observation sight

All three sights are image intensifying devices using a 25 mm second-generation micro-channel plate tube, and are equipped with the same common standardised battery housing and eyepiece/eyeguard assembly which guarantees interchangeability. Only the objective lens assemblies are different and serve the required purpose.

The second-generation technology has built-in automatic gain control protecting the tube againt muzzle and detonation flash, which results in a uniform image without tendency to flare. Additional manual brightness control allows the operator to adjust screen luminance to his individual requirements and for contrast enhancement.

AN/PVS-4 and AN/TVS-5 are weapon sights with diffeerent magnifications, and use interchangeable projected graticules for every common weapon fully adjustable for adaptation to the scenery brightness. They can be easily adapted to the weapon by using appropriate backets.

EURONOD-2 is an observation sight with high magnification, offering a wide field of applications, particularly that of artillery fire control.

AN/PVS-4 rifle sight with G3 rifle mounting

DATA

	AN/PVS-4	AN/TVS-5	EURONOD-2
Magnification:	× 3.7	× 5.8	× 9.7
Field-of-view:	14°	9°	5.6°
Object lens:	95 mm f/1.6(T)	155 mm f/1.6(T)	258 mm f/1.7(T)
Focus range:	10 m – ∞	25 m – ∞	75 m – ∞
Length:	240 mm	310 mm	480 mm
Width:	120 mm	160 mm	240 mm
Height:	120 mm	170 mm	240 mm
Weight:	1.7 kg	3.6 kg	9.0 kg
Power supply:	2 × AA/LR6	2 × AA/LR6	2 × AA/LR6
Battery life:	50 h (20°C)	50 h (20°C)	50 h (20°C)
Operating range:	to 600 m	to 1200 m	to 6000 m

Manufacturer
Euroatlas GmbH, Zum Panrepel 2, D-2800 Bremen 45.
Status
Current. Available.

EUROVIS-4 weapon sight

This is a weapon sight of the latest generation, developed to meet military requirements. EUROVIS-4 weighs only 1 kg and is adaptable to any weapon, and due to its low weight even to very light weapons, by using an appropriate bracket.

It uses an 18 mm image intensifier tube of either 2nd or 3rd generation. The illuminated and controllable graticule, together with the focussing device with a positive grip usable with heavy gloves, ensures accurate aiming. Four times magnification gives a wide focussing range from 15 m to infinity, so that man-sized targets can be detected up to 600 m under normal weather conditions.

The image intensifier tubes use micro-channel plates and are provided with automatic gain control against muzzle flash and bright light sources. Excellent contrast and resolution capability, particularly at very low light levels, allow the operator to use the sight over long periods without fatigue. EUROVIS-4 fulfils all military environmental test requirements.

EUROVIS-4 weapon sight

DATA
Magnification: × 4
Field-of-view: 9°
Objective: 116 mm f/1.8(T)
Focussing range: 15 m to infinity
Weight: 1.0 kg
Power supply: 2 × AA/LR6
Battery life: > 50 h at 20°C
Operating range: to 1500 m

Manufacturer
Euroatlas GmbH, Zum Panrepel 2, D-2800 Bremen 45.
Status
Current. Available.

RT 5A laser illuminator

This illuminators is intended to be used with other Euroatlas night vision devices in the case of absolute darkness or in difficult observing conditions. It is also recommended where light barriers, for example bright light sources, are between the observer and his objective or where a dark objective is between bright areas. This laser torch is light and small and can easily be mounted on any vision device by using the adjusting and mounting bracket provided. The laser diode has a working range up to 1000 m at low current consumption. There is an adjusting wheel which permits the illumination field to be changed during operation within the ratio of 1:4.

DATA
Wavelength: 840 (–) 870 nm
Laser diode: double hetero-structure LB1
Radiant output: 6-10 mW
Pulse duty factor: 5%
Beamwidth: approx 1 mrad
Normal ocular hazard distance: 50 m
Working range: 1000 m
Length: 178 mm

Width: 75 mm
Height: 85 mm without mounting
Weight: 650 g
Power supply: 6 × AA/LR6 batteries

Manufacturer
Euroatlas GmbH, Zum Panrepel 2, D-2800 Bremen 45.
Status
Current. Available.

RT5A laser illuminator

LM-18 target marker

The LM-18 is a new generation covert infra-red target marker (aiming light). It is designed for weapon mountings and features an integral zeroing (bore-sighting) mechanism which dispenses with the need for costly zeroing mounts. It will enable rapid covert target acquisition by wearers of night vision goggles and other users of night vision systems.

The unit is powered by three readily available 'AA' size alkaline batteries. A quick release bracket interface enables rapid detachment and re-attachment to a weapon without the need for re-zeroing. Operation is by means of a remote wired 'press for on' switch which can be attached to any convenient point on the weapon. The unit is rugged and reliable.

DATA
Wavelength: 800-870 nm
Radiant output: 0.6 mW
Working range: up to 300 m
Spot size: 50 mm at 50 m range (others available)
Length: 120 mm
Diameter: 40 mm
Weight: 350 g
Power supply: 3 × AA/LR6 batteries
Waterproofing: 30 m maximum depth
Switch: remote, with strap to attach to weapon at any point

Manufacturer
Euroatlas GmbH, Zum Panrepel 2, D-2800 Bremen 45.
Status
Current. Available.

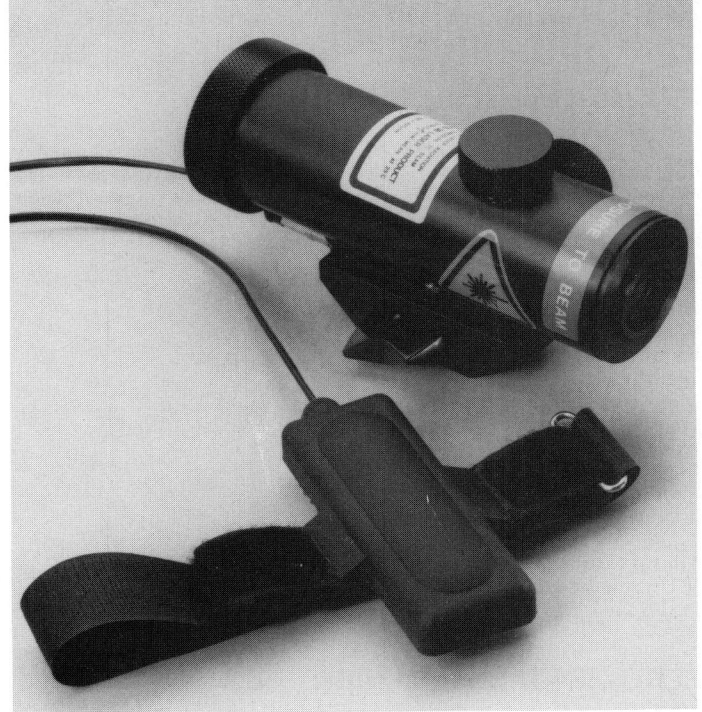

LM-18 target marker, with remote switch

Panzerfaust 3 target marker

This device is a dedicated laser spot projector intended to be used with the Panzerfaust 3 light anti-tank weapon. The laser spot projector attaches to the firing grip. The operator can wear night vision goggles and merely direct the spot on to the target, or a night vision telescope can be fitted above the spot projector through which the operator can take aim and distinguish the spot.

Manufacturer
Dynamit Nobel AG, Postfach 1261, D-5210 Troisdorf.
Status
Current. Production.

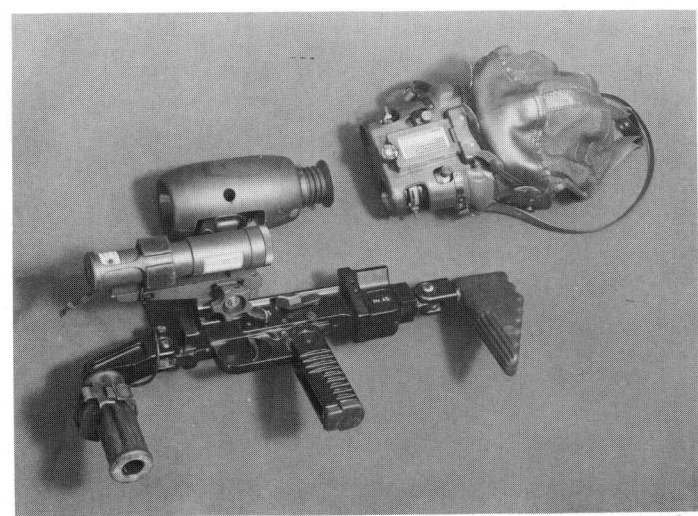

Panzerfaust 3 target marker

Telescope sight PSG-1

This is a conventional optical telescope sight intended for use with sniping rifles and particularly the Heckler and Koch PSG-1. In addition to serving to aim the weapon it can also be used as a long-range observation instrument. The maximum adjustable shooting distance is 600 m.

DATA
Length: 390 mm
Cap diameter: 56 mm
Height above rifle: 130 mm
Weight: 580 g
Magnification: × 6
Object lens: 42 mm
Exit pupil: 7 mm
Eye relief: 70 mm
Field-of-view: 4°
Elevation adjustment: 100 to 600 m
Graticule light: adjustable
Batteries: 3 × Varta 60DK

Hensoldt PSG-1 telescope sight

Manufacturer
M Hensoldt and Sohne Optische Werke AG, D-6330 Wetzlar 1.
Status
Current. Production.

ZF Telescope sight 4 x 24

The ZF is a monocular telescope with lens erecting system and four-power magnification. It is intended for sniping use and also as an observation instrument. All optical surfaces are coated with a wipe-proof reflection reducing coating, and the sight is so well sealed that even under sudden temperature changes and water-spray it remains usable.

DATA
Length: 228 mm
Width: 50 mm
Height above rifle: 95 mm
Weight: 350 g
Magnification: × 4
Object lens: 24 mm
Exit pupil: 6 mm
Eye relief: 60 mm
Field-of-view: 5° 40′ (100 mils)
Elevation adjustment: 100–600 m

Manufacturer
M Hensoldt and Sohne Optische Werke AG, D-6330 Wetzlar.

Hensoldt ZF telescope sight

Status
Current. Production.

FERO-Z24 Telescope sight

This is a monocular telescope sight with erecting lens system and is intended for use with sniping rifles and as an observation instrument. It has an illuminated graticule for use under poor light conditions. As with all Hensoldt military telescope sights the mount is to STANAG 2424 and will fit any NATO compatible weapon.

DATA
Length: 223 mm
Width: 46.5 mm
Height above rifle: 43 mm
Weight: 300 g
Magnification: × 4
Object lens: 24 mm
Exit pupil: 6 mm
Eye relief: 60 mm
Field-of-view: 6° (107 mils)
Elevation adjustment: 100–600 m at 100 m click-stops

Manufacturer
M Hensoldt and Sohne Optische Werke AG, D-6330 Wetzlar.

Hensoldt FERO-Z24 telescope sight

Status
Current. Production.

ZF 10 × 42 telescope sight

The ZF 10 × 42 telescope sight is a precision sight intended for shooting to ranges of 1000 m. It is a monocular/monobjective sight with a lens erecting system and 10-power magnification. The super-elevation angle is adjustable for shooting distances from 100 to 1000 m. All air/glass surfaces of the optical components are coated with a wipe-proof reflection reducing coating. The sight is so well sealed that even during sudden temperature changes and spray-water influences it remains usable. The graticule pattern can be changed to meet user requirements.

DATA
Length: 373 mm with caps and eyeguard
Diameter: 56 mm
Weight: 429 g
Magnification: x 10
Entrance pupil: 42 mm
Exit pupil: 4.2 mm
Eye relief: 70 mm
Field-of-view: 2.4°; 42 m at 1000 m
Dioptre setting: ± 2

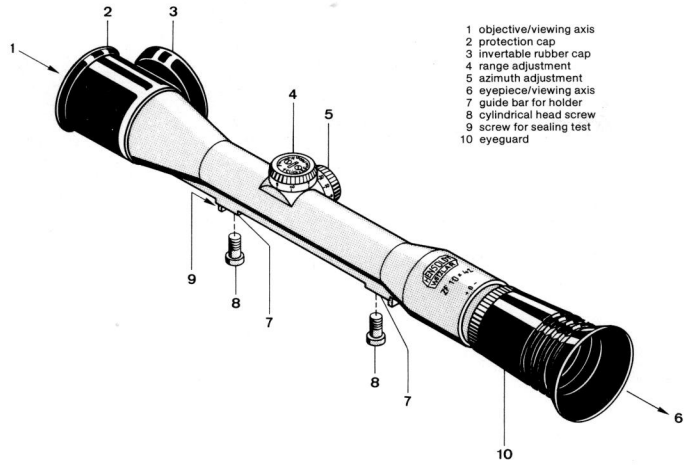

1 objective/viewing axis
2 protection cap
3 invertable rubber cap
4 range adjustment
5 azimuth adjustment
6 eyepiece/viewing axis
7 guide bar for holder
8 cylindrical head screw
9 screw for sealing test
10 eyeguard

Hensoldt ZF 10 × 42 telescope sight

Manufacturer
M Hensoldt and Sohne Optische Werke AG, D-6330 Wetzlar.
Status
Current. Production.

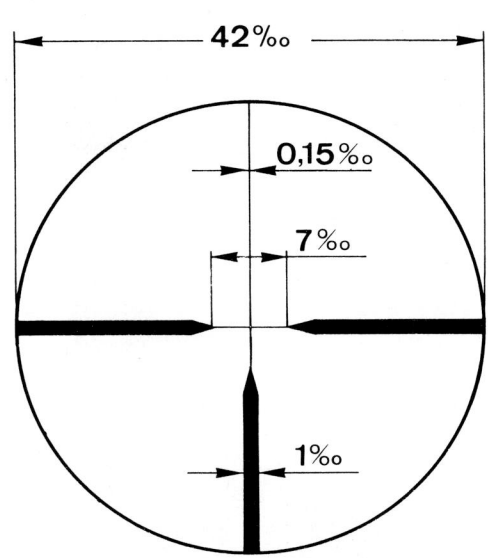

Standard graticule pattern of ZF 10 × 42 sight

ZF 6 × 36 telescope sight

This is a device for precise aiming with a rifle to a range of 1300 m. It serves the firer for target acquisition and aiming as well as general observation and the determination of target effects. The sight is a monocular/monobjective system and the super-elevation angle is adjustable for shooting at ranges from 100 to 600 m. For the 600 to 1300 m bracket there are sighting marks on the graticule. When shooting at these distances, the range adjustment is set to the 600 m mark and the additional super-elevation obtained by selection of the appropriate aiming mark. For use in intense sunlight conditions a grey (neutral density) filter can be fitted to the eyepiece.

DATA
Length: 255 mm with caps
Diameter: 55 mm
Weight: 390 g
Magnification: x 6
Entrance pupil: 36 mm
Exit pupil: 6 mm
Eye relief: 60 mm
Field-of-view: 4°; 70 m at 1000 m
Dioptre setting: fixed -0.5

Manufacturer
M Hensoldt and Sohne Optische Werke AG, D-6330 Wetzlar.
Status
Current. Production.

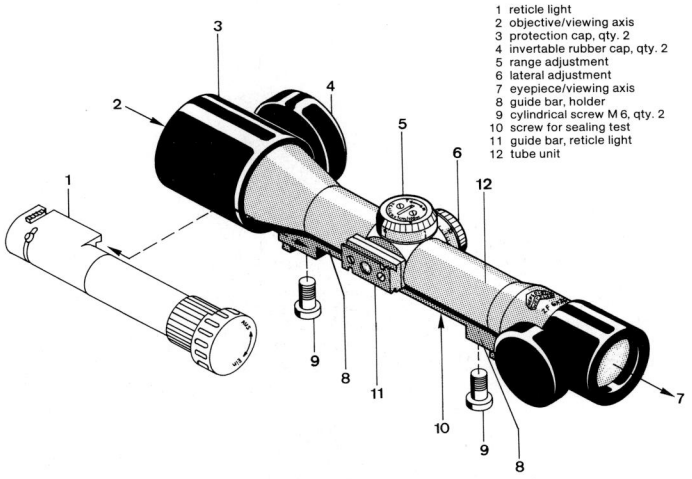

1 reticle light
2 objective/viewing axis
3 protection cap, qty. 2
4 invertable rubber cap, qty. 2
5 range adjustment
6 lateral adjustment
7 eyepiece/viewing axis
8 guide bar, holder
9 cylindrical screw M 6, qty. 2
10 screw for sealing test
11 guide bar, reticle light
12 tube unit

Hensoldt ZF 6 × 36 telescope sight

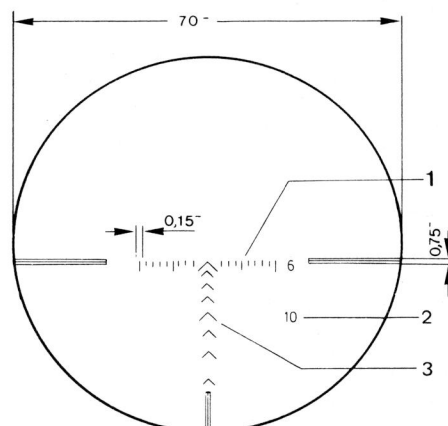

1 side scale
2 1000 m mark
3 sighting marks

Standard graticule of ZF 6 × 36 sight, showing long-range aiming marks

ZF 4 × 24 MG 3 telescope sight

The ZF 4 × 24 MG 3 serves for direct and indirect aiming of a mounted machine gun at ground targets up to a distance of 1600 m, for observation and for determination of distance when the target size is known. Due to the periscopic design of the sight, the gunner can aim and observe from behind cover.

DATA
Length: 210 mm
Height: 195 mm
Weight: 1650 g
Magnification: 4 ×
Eye relief: 45 mm
Field-of-view: 8°35': 150 m at 1000 m
Dioptre setting: ± 2.5 dioptres

Manufacturer
M Hensoldt and Sohne Optische Werke AG, D-6330 Wetzlar.
Status
Current. Production.

Hensoldt ZF 4 × 24 MG 3 telescope sight

ZF 4 × 24 MG 1 telescope sight

This is a similar sight to the MG 3 (above) but has a higher periscope tube, allowing for different mountings and heights of cover. It is used for aiming machine guns to ranges up to 1600 m and for rangefinding when the target size is known. Data is similar to that for the MG 3 sight except that the height is 285 mm, length 185 mm and the weight 1670 g.

Manufacturer
M Hensoldt and Sohne Optische Werke AG, D-6330 Wetzlar.
Status
Current. Production.

Hensoldt ZF 4 × 24 MG 1 telescope sight

Orion 80 passive night sight

This is a small-arm night sight using a three-stage image intensifier system. An adjustable reticle is projected into the image plane and its brightness can be increased or diminished to suit the target image.

DATA
Magnification: × 4
Field-of-view: 8°
Focus range: 20 m to infinity

Reticle adjustment: ± 5 mils in 0.5 mil steps
Eyepiece adjustment: ± 5 dioptres
Max diameter: approx 95 mm
Length: approx 290 mm
Weight: approx 1.8 kg
Power source: 2.5 V NiCd rechargeable battery giving 25 h operation between charges

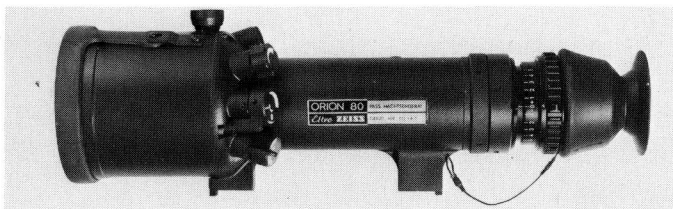

Orion 80 passive night sight

Manufacturer
Eltro GmbH, Gesellschaft für Strahlungstechnik, Kurpfalzring 106, Postfach 102120, 6900 Heidelberg 1, in co-operation with Carl Zeiss, Postfach 1380, 7082 Oberkochen.
Status
Current. Production.
Service
West German forces and other European forces and police units.

Orion 80 passive night sight on Heckler and Koch MP5A2 sub-machine gun

Orion 110 passive night sight

This weapon sight is similar in operation to the Orion 80 sight but larger and having a performance appropriate to larger weapons. Some of the basic electro-optical components, however, are common to both sights.

DATA
Magnification: × 6
Field-of-view: 6°
Focus range: 20 m to infinity
Reticle adjustment: ± 5 mils in 0.5 mil steps
Eyepiece adjustment: ± 5 dioptres
Max diameter: approx 120 mm
Length: approx 320 mm
Weight: approx 2.4 kg
Power source: 2.5 V NiCd rechargeable battery giving 25 h operation between charges

Orion 110 passive night sight on G3 rifle

Manufacturer
Eltro GmbH, Gesellschaft für Strahlungstechnik, Kurpfalzring 106, Postfach 102120, 6900 Heidelberg 1, in co-operation with Carl Zeiss, Postfach 1380, 7082 Oberkochen.

Status
Current. Production.
Service
West German forces.

Aiming point projector

This projector is primarily used with Heckler and Koch small arms and it is offered by Heckler and Koch as a fitment to its range of sub-machine guns. It is a powerful, narrow beam of white light, projected from a small tubular housing mounted above the weapon. The beam of light illuminates the target and is sufficiently large to allow the operator a limited ability to search with it. In the middle of the beam is a small black area which shows as a black dot. It is this dot which is the actual aiming mark. The operator swings his beam on to his target and refines the aim with the dot. Where the dot rests, the bullets will go.

It is an ideal arrangment for shooters who have to engage fleeting targets in difficult conditions. It enables the firer to shoot from the hip, keeping his head free from the sight pattern and so allowing him an unrestricted view of his target. The beam of light is quickly moved on to a chosen target, and the knowledge that it means that a weapon is pointing directly down the beam acts as a severe deterrent on taking retaliatory or evasive action.

The lamp is a 6 V, 10 W halogen type giving a range of up to 120 m in good weather conditions.

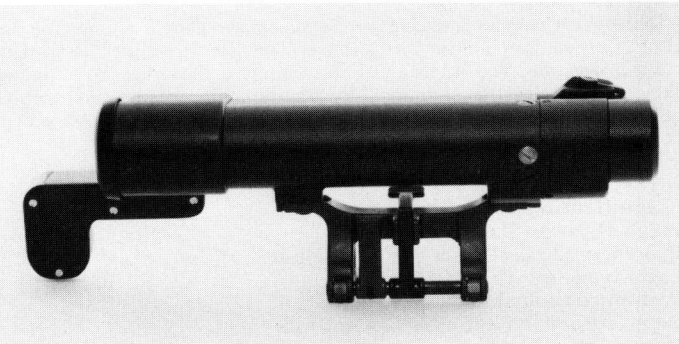

Hensoldt aiming point projector

DATA
Weight: (complete) 1.7 kg
Dimensions: 290 × 80 × 100 mm
Lamp: 6 V, 10 W halogen. Osram No 64225
Lamp life: approx 100 h

Illumination	White light	Black spot
25 m	1 m diameter	150 mm
50 m	2 m	300 mm
75 m	3 m	450 mm
100 m	4 m	600 mm

Range: (accuracy) 100 m; (search) 120 m

Batteries: rechargeable NiCd accumulators. Varta RSH 1.8, baby-size

Manufacturer
Hensoldt-Wetzlar. Marketed by Heckler and Koch GmbH, 7238 Oberndorf-Neckar.
Status
Current. Production.
Service
Several West German police forces. Military and police forces in many other countries.

ISRAEL

NVL-11 fire control night sight

The NVL-11 is a fire control sight unit for use on small anti-tank rocket launchers and similar short-range direct-fire weapons. It consists of a gated second-generation night vision telescope connected to a laser rangefinder and a ballistic calculator. The rangefinder has a maximum range of 970 m, thus limiting the type of weapon, while the ballistic calculator can be adjusted to the particular ballistics of the weapon within the range limitation. The sighting process is simple, involving acquiring the target, measuring the range, calculating the ballistic correction and applying it before opening fire. The use of this sight gives a night-time hit probability figure which is likely to be better than the daytime figure obtained with conventional sights.

DATA
Weight: 1.5 kg
Dimensions: 170 × 115 × 120 mm
Image intensifier: 18 mm 2nd generation wafer type
Magnification: x 3
Field-of-view: 13.6°
Wavelength: 850 nM
Average power: 5 mW
Beam pattern: vertical line, 3 mils high, 0.2 mil wide
Range: 20 to 970 m
Range resolution: 10 m

Manufacturer
International Technologies (Lasers) Ltd, 12 Hachoma Street, PO Box 4099, 75140 Rishon-Letzion.
Status
Current. Available.

NVL-11 fire control night sight

NVL-11 Mk 3 Fire Control night sight

The NV-11 Mark 3 is a computerised fire control night sight for small rocket launchers and recoilless guns, such as the RPG7, M72 LAW, Carl Gustav and 106mm M41. It is highly effective in all night firing situations, and a single system can support up to five different weapons or types of ammunition, with a maximum range of 1000 m.

The sight is based on a second-generation gated night vision unit. The fire control system is combined with a low power laser transmitter, optics and electronic circuits. The system measures the range to the target and computes the ballistic elevation required for the specific weapon/ammunition combination. It can be easily and quickly bore-sighted to the weapon in field conditions.

DATA
Magnification: 3 ×
Field-of-view: 13.6°
Eyepiece: adjustable from +2 to -6 dioptres
Laser wavelength: 850 nM
Average power: 3.5 mW
Beam spread: 3 mRad high, 0.2 mRad wide
Range: 20 to 990 m
Resolution: 10 m
Elevation angle: up to 70 mils
Accuracy of QE: 0.5 mRad
Power supply: 4 × 1.5 v alkaline batteries
Battery life: 3 hours ranging, or 20 hours viewing
Dimensions: 170 × 115 × 120 mm
Weight: 1.6 kg without batteries
Environmental: Qualified to MIL-STD-810D

NVL-11 Mark 3 fire control night sight

Manufacturer
International Technologies (Lasers) Ltd, 12 Hachoma Street, PO Box 4099, 75140 Rishon-Letzion.

Status
Current. Production.

AIM-1R visible laser aiming light

The AIM-1R is a small, lightweight, visible red laser aiming light that allows quick and accurate aiming of weapons by simply marking the target with a red dot. It is effective for indoor operations and for outdoor operations in twilight and night conditions, with an aiming range up to 150 m.

The unit will fit any weapon, from pistols to shoulder arms, and can be operated locally or by remote-control cable.

Applications include anti-terrorist units, infantry units, special forces, etc. It can be readily bore-sighted to any weapon.

DATA
Beam divergence: 1.0 mRad maximum
Average output power: 1 mW
Peak wavelength: 670 nM nominal
Power supply: 2 × 1.5 V AA alkaline
Battery life: approx. 5 hours
Dimensions: 92 × 49 × 57 mm excluding weapon adapter
Weight: 255 g excluding weapon adapter and batteries

Manufacturer
International Technologies (Lasers) Ltd, 12 Hachoma Street, PO Box 4099, 75140 Rishon-Letzion.
Status
Current. Production.

AIM-1R visible laser aiming light

S-8 target illuminator and designator

This is a high-intensity light projector which mounts beneath the fore-end handgrip of any weapon and is used to illuminate targets up to a range of 100-150 m. The light permits ready identification of the target and since the projector is bore-sighted to the weapon, placement of the light on the target guarantees the aim. Moreover the intensity of the light is such that the target will be disoriented and unable to fire back accurately. The unit, once mounted, becomes part of the weapon and does not materially affect the balance or handling. The operating switch has a safety catch and is spring-loaded for positive pressure only.

DATA
Bore-sight tolerance: ±2°
beam divergence: 0.7° at 50% peak intensity
Peak power intensity: 900 lux at 5 m
Lamp: 10 W quartz-halogen
Lamp life: 100 hours
Dimensions: 165 × 71 mm
Weight: <950 g including power pack

S-8 target illuminator and designator

Manufacturer
International Technologies (Lasers) Ltd, 12 Hachoma Street, PO Box 4099, 75140 Rishon-Letzion.
Status
Current. Production.

AIM-1DLR long-range IR laser aiming light

The AIM-1DLR is a laser aiming spotlight with unusual power; set at 'High' it will project a beam up to 3 km distance. It is intended for use on long-range light weapons such as heavy machine guns, light cannon, helicopter cannons and crew-served infantry weapons so that they can be fired in darkness by operators wearing night vision equipment. Customised mountings are available for a wide variety of weapons.

DATA
Wavelength: 820 – 850 nM
Beam divergence: 0.3 mRad
Output power: 20 mW (High mode, average)
Weight: 255 g less batteries and mount
Dimensions: 92 × 57 × 49 mm
Power source: 2 × 1.5 V AA cells or 28 V DC
Battery life: High 5 h; Low 50 h, using alkaline cells

Manufacturer
International Technologies (Lasers) Ltd, 12 Hachoma Street, PO Box 4099, 75140 Rishon-Letzion.
Status
Current. Available.

AIM-1DLR long-range laser aiming light mounted on light machine gun

SL-1 sniper's spotlight

The SL-1 is a laser illuminator for attachment to a rifle so as to provide the firer with a source of IR light capable of illuminating the scene viewed through a night vision sight when ambient light is insufficient to provide the necessary contrast for good viewing. The control gives four light levels to meet varying requirements, and can also be remote-controlled so as to allow the scene to be illuminated from an angle or to protect the firer against detection by IR-sensitive instruments in the target area. In normal use, the spotlight is fitted to the rifle and then aligned with the night sight so as to direct its light accurately into the field-of-view.

SL-1 spotlight mounted on sniping rifle

DATA
Source: GaAs laser diode
Wavelength: 850 ± 20 nM
Average power: 10 mW minimum
Beam divergence: 2 ± 0.5°
Weight: 550 g with batteries
Length: 155 mm
Diameter: 51 mm

Power source: 4 × 1.5 V AA alkaline batteries
Battery life: 5–15 h depending upon intensity mode

Manufacturer
International Technologies (Lasers) Ltd, 12 Hachoma Street, PO Box 4099, 75140 Rishon-Letzion.
Status
Current. Available.

BL-1 Borelight

The Borelight is a dry zeroing device for zeroing infra-red markers, designators and spotlights mounted on weapons. It can be used by day or night, is fast and accurate, and saves time and ammunition. The unit is supplied with a variety of spuds to fit various standard calibres. The selected spud is fitted to the Borelight and then entered into the muzzle of the weapon. The Borelight is adjusted to the same offset as the spotlight, and the light then adjusted until the beam is parallel with the bore. No night vision goggles or any other type of optical equipment is necessary.

DATA
Weight: approx 400 g
Dimensions: 190 × 43 × 96 mm
Spectral response: 300 – 1200 nM (visible to near IR)
Tube resolution: 60 lp/mm
Magnification: x 8
Field-of-view: 4°
Accuracy: ± 0.1 mrad

Manufacturer
International Technologies (Lasers) Ltd, 12 Hachoma Street, PO Box 4099, 75140 Rishon-Letzion.
Status
Current. Available.

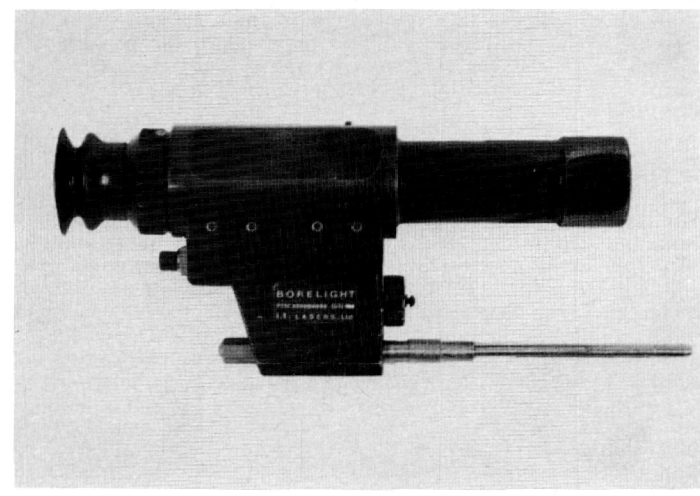

BL-1 Borelight

ORT-MS4 mini weapon night vision sight

The ORT-MS4 is a low weight second generation night sight. It is particularly suited for use on infantry weapons, or as a hand-held night observation device. The ORT-MS4's modular design incorporates an 18 mm second generation image intensifier. For professional night photography different objectives are available.

The instrument is self-contained and battery powered, and is supplied in a storage/carrying case with all necessary accessories.

ORT-MS4 weapon sight

DATA
Dimensions: 230 × 66 mm
Weight: 1190 g
Magnification: × 3.75
Field-of-view: 10°
Eyepiece focal length: 26.6 mm
Focus adjustment: + 2.5 to -6 dioptres
Objective: 100 mm fixed focus
Viewing range: moonlight 500 m; starlight 350 m

Manufacturer
ORTEK Ltd, PO Box 388, Sderot 80100.
Status
Current. Production.

Elbit Falcon optical sight

This sight uses a compact aircraft-type 'heads-up' display mounted on the front end of the rifle. The unique design does not incorporate a tube or telescope array, and therefore firing with both eyes open gives the user faster acquisition of a target as well as high precision in harsh battlefield conditions. Aiming is performed by placing a sharply-defined luminous red dot, which is projected in the display, on the target. Since parallax has been eliminated, the position of the dot on the screen is of no significance; provided the firer sees the dot anywhere on the screen and superimposes it on the target, he will hit.

The sight was specifically designed for the M16 and Galil assault rifles, but mounts are available for other types of rifle to permit fitting without the need for machining or other gunsmithing operations.

DATA
Weight: 300 g
Length: 182 mm
Lens diameter: 30 mm
Field-of-view: unrestricted
Lens magnification: 1 ×
Boresighting range: ± 14 mRad windage, ± 14 mRad elevation
Height: 51 mm
Dot diameter: 1.3 mrad
Power supply: 3.6 V Lithium AA cell
Battery life: 250 h day, 700 h night

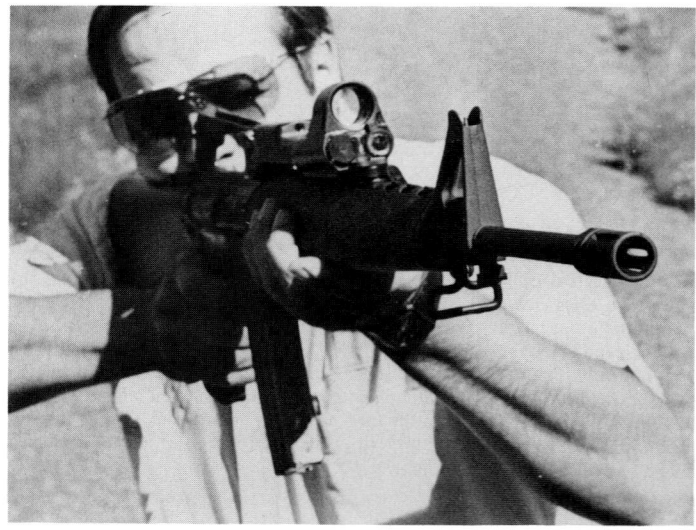

Elbit Falcon optical sight

Shelf life: 10 years
Environmental conditions: in accordance with MIL STD 810C

Manufacturer
Elbit Computers Ltd, PO Box 5390, Haifa 31053.
Status
Current. Production.

Hit-Eye 1500 optical sight

The Hit-Eye 1500 is an optical sight for assault rifles. It can be incorporated into the design of a new rifle or fitted, by means of mounts, to virtually any type of existing rifle. The graticule pattern is a circle, effective for aiming at targets at a few hundred metres range. Distance can be rapidly estimated with the aid of the graticule pattern, which is always centred in the field-of-view. Built-in illumination of the graticule permits firing at night and in poor light.

The sight is fitted with external knobs for zeroing adjustment in elevation and windage. External iron sights, with night aiming luminous points, may be fitted as an option.

DATA
Length: 135 mm
Diameter: 21 mm or 25.4 mm as required
Magnification: × 1.5
Field-of-view: 8°
Object lens diameter: 12 mm
Eye relief: 72 mm

Hit-Eye 1500 assault rifle sight

Manufacturer
ORTEK Ltd, PO Box 388, Sderot 80100.
Status
Current. Production.

Hit-Eye 3000 rifle sight

This is similar in principle to the Hit-Eye 1500, but of greater magnification and slightly larger. It also has a different graticule pattern, using the traditional cross wires, and the eyepiece is adjustable. The integrated mount fits into the carrying handle of the M16 rifle, but mounts for other rifles can be provided.

DATA
Length: 160 mm
Diameter: 35 mm
Magnification: × 3
Field-of-view: 6°
Object lens diameter: 21 mm
Eye relief: 90 mm
Dioptric adjustment: -5 to +5 dioptres
Elevation adjustment: (100-200) – 500 m

Manufacturer
ORTEK Ltd, PO Box 388, Sderot 80100.
Status
Current. Production.

Hit-Eye 3000 rifle sight

Hit-Eye 4000 launcher sight

Hit-Eye 4000 is a telescope sight for fitting to light anti-armour rocket launchers, though it can, of course, be mounted on virtually any direct-fire short-range weapon. Various mounts are available, and the sight can be zeroed by means of the mount. The sight is hermetically sealed, preventing internal fogging or fungal growth, is robust, has no moving parts, and can withstand the most extreme field conditions.

Hit-Eye 4000 provides an aiming capability up to 800 m range and can be provided with patterned graticules according to the ballistics of the weapon in use. The graticule is illuminated by a Betalight source, giving excellent aiming properties in night or poor light conditions.

DATA
Dimensions: 125 × 67 × 67 mm
Weight: 470 g
Magnification: × 4
Field-of-view: 10°
Eye relief: 35 mm

Manufacturer
ORTEK Ltd, PO Box 388, Sderot 80100.
Status
Current. Production.

Hit-Eye 4000 launcher sight

ITALY

M 166 mini weapon sight

The M 166 has been in production since 1977 and is in service with the Italian Army and government agencies. It is a small lightweight sight using the cascade tube technique for light intensification and it has an adjustable reticle for sighting.

DATA
Weight: 2 kg
Length: 410 mm
Diameter: 83 mm
Magnification: × 3
Field-of-view: 208 mil (11.7°)
Lens: 81.5 mm
Relative aperture: 1.4
Focus range: 15 m to infinity
Tube: 18 mm
Battery: 3 V lithium cell
Battery life: 60 h continuous operation
Temperature range: -40 to +50°C

Manufacturer
Aeritalia SpA, Avionic Systems and Equipment Group, Viale Europa 1, 20014 Nerviano (Milan).

M 166 and M 193 mini weapon sights

Status
Current. Production.
Service
Italian Government and military.

M 193 passive weapon sight

The M 193 is a second-generation edition of the M 166 and it fits to the same range of weapons. It has the usual advantages of a second generation over a first in that the gain is more uniform, there is less distortion and there is no, or very little, flare in bright lights.

DATA
Weight: 1.9 kg
Dimensions: 340 × 83 mm
Magnification: × 4.8
Field-of-view: 13°
Focus range: 15 m to infinity
Battery: 3 V lithium cell
Battery life: 60 h continuous operation

Manufacturer
Aeritalia SpA, Avionic Systems and Equipment Group, Viale Europa 1, 20014 Nerviano (Milan).
Status
Current. Available.

M 193 Aeritalia second-generation weapon sight

Service
Not known.

M176 night observation sight

This has been designed for artillery fire control and also as a sight for crew-served weapons such as the M40 recoilless rifle. A version without graticule is also available and is suitable for medium-range surveillance from helicopters, moving vehicles or static positions.

The M176 has a range of up to 1500 m in starlight conditions, and the weapon sight version has an adjustable illuminated graticule.

DATA
Magnification: × 5.3
Field-of-view: 10.5°
Resolution: 0.4 mil at 1 mlux
Focus range: 15 m to infinity
Objective: 136 mm f/1.2 catadioptric
Power supply: 3 V lithium cell
Battery life: 50 h
Weight: 2.2 kg
Operating temperature: -40 to +50°C

Manufacturer
Aeritalia, Avionic Systems and Equipment Division, Viale Europa 1, 20014 Nerviano (Milan).
Status
Current. Production.

M176 night sight

VTG 120 thermal imaging system

VTG 120 is a manportable apparatus for night surveillance and for the aiming of medium-range weapons. In particular, its application to the TOW missile system has been studied.

The equipment is completely independent of any external source; its battery and air bottle allow more than two hours of continuous operation. Empty bottles and battery can easily be replaced to extend operation.

It can be used to view directly, or it can be connected to an external CRT for remote monitoring of the observed scene. An electrically generated graticule can be positioned for aiming purposes.

A complete ancillary support equipment is available for use and maintenance, in particular:
    battery charger
    gas bottle charger
    bore-sight collimator

field test equipment
lens cleaning kit
test and adjustment equipment
cooling system purging kit
tripod mounting interface and
carrying case.

Standard performances on an operative target (tank) are detection at 3000 m and identification at 2000 m in standard meteorological conditions according to FINABEL 1.R.9.

The same set of modular subsystems has been arranged in different configurations in order to provide appropriate sights for the MILAN and MAF anti-tank missiles.

DATA
Field-of-view: (narrow) 20 × 40 mRad; (wide) 60 × 120 mRad
Resolution: (narrow) 0.16 mRad; (wide) 0.5 mRad
IR Band: 8 – 14 microns
Cooling system: Joule-Thomson minicooler
Detector: CMT 60-element array
Number of lines: 120
Power supply: 22 – 28 V DC NiCd battery

Manufacturer
Officine Galileo SpA, Via Albert Einstein 35, 50013 Campi Bisenzio (FI).
Status
Current. Production.

VTG 120 light thermal imager on TOW launcher

OGVN 7 miniaturised night sight

The OGVN 7 is a compact night sight for observation and aiming, suitable for installation on individual weapons. The optical system is equipped with second generation light intensifier tubes, with Automatic Brightness Control (ABC) and automatic switch-off when field-of-vision brightness is dangerously high. The telescope is equipped with a sighting graticule and bore-sighting devices.

DATA
Dimensions: 227 × 80 × 45 mm
Weight: 1.0 kg
Magnification: × 2.8
Field-of-view: 13°
Dioptre adjustment: ± 5 dioptres
Focussing: fixed, from 5 m to infinity
Resolution: 1 mrad at 0.001 lux
Power supply: 2 × 1.5 V alkaline-manganese batteries LR6
Battery life: 40 h at +25°C

Manufacturer
Officine Galileo SpA, Via Albert Einstein 35, 50013 Campi Bisenzio (FI).
Status
Current. Production.
Service
Under evaluation by Italian Army.

Officine Galileo OGVN 7 miniaturised night sight

OGVN 3 night vision sight

The OGVN 3 is a compact aiming telescope, suitable for installation on individual precision weapons. The high performance reflective telescope includes a second-generation micro-channel light intensifier tube with Automatic Brightness Control and automatic switch-off when field of vision brightness reaches dangerous levels.

The luminous protected aiming graticule may be adjusted in brightness and position as required for precision aiming.

DATA
Dimensions: 330 × 130 × 140 mm
Weight: 2.35 kg
Magnification: × 3.6
Field-of-view: 11°
Dioptre adjustment: ± 4 dioptres
Focussing: fixed, from 50 m to infinity
Resolution: 0.8 mrad at 10⁻³ lux
Power supply: 2 × 1.5 V alkaline-manganese batteries LR14
Battery life: 35 h at +25°C

Manufacturer
Officine Galileo SpA, Via Albert Einstein 35, 50013 Campi Bisenzio (FI).
Status
Current. Production.
Service
Italian Navy special forces.

Officine Galileo OGVN 3 night vision sight

NETHERLANDS

Type GK4MC passive mini weapon sight

The GK4MC small arms sight is a second-generation lightweight passive night vision device specially designed for accurate weapon aiming and surveillance at night.

It enables the user to engage the enemy and to observe targets without any artificial illumination, thus presenting full security from detection. The sight can also be used to detect hostile infra-red light sources.

The sight is primarily used as a night sight on basic infantry weapons such as rifles, machine guns and short-range anti-tank weapons. It can also be used as a hand-held or tripod-mounted observation device for direct viewing.

DATA
Dimensions: 295 × 144 × 98 mm
Weight: 1.5 kg
Magnification: × 4
Field-of-view: 10°
Battery power: 2 × 1.5 V DC, size AA, or 2 × NiCd cells, size AA, or one lithium cell size AA or size 2 × AA, with adapter

Manufacturer
Oldelft (B.V. Optische Industrie 'De Oude Delft'), PO Box 72, 2600 MD Delft.

GK4MC mini weapon sight

Status
Current. Production.
Service
Several armies.

Type MS4GT mini weapon sight

The Type MS4GT is a very low weight second generation sight particularly suited for use on infantry weapons. It incorporates a catadioptric object lens and an 18 mm micro-channel plate amplifier. The graticule may be illuminated, although this cuts the battery life by about 50 per cent. Mounting brackets are available to fit the sight to a variety of rifles, machine guns and platoon weapons. The Type MS4GT can also be used as a surveillance device or as an infra-red detection system.

DATA
Dimensions: 260 × 122 × 87 mm
Weight: approx 1 kg without mounting bracket
Magnification: × 4
Field-of-view: 11°
Battery: 2 × NiCd cells, size AA; 2 × dry cells, size AA; or, with adapter, one lithium cell size AA

Manufacturer
Oldelft (B.V. Optische Industrie 'De Oude Delft'), PO Box 72, 2600 MD Delft.
Status
Current. Production.

MS4GT mini weapon sight

Service
Several armies.

GK7MC aiming device

This passive night vision instrument is a lightweight aiming device suitable for medium-range weapons, such as machine guns, grenade launchers, recoilless guns and other crew-served weapons.

Its passive operation is based on a second generation image intensifier tube and does not require any infra-red or other artificial light source, thus presenting maximum security from detection. Explosion flashes, flares and other bright lights in the field-of-view will not bloom the image. Tube damage due to too great a light intensity is prevented by the attenuation characteristics of the micro-channel plates incorporated in the image intensifier tube.

It can also be employed as an observation device for long-range detection. The modular construction ensures simple maintenance and service.

GK7MC second generation passive aiming device

DATA
Dimensions: 363 × 174 × 134 mm
Weight: 2.2 kg
Magnification: × 6.7
Field-of-view: 6.5°
Power supply: 2 × AA NiCd or conventional cells, or one lithium cell with adapter

Manufacturer
Oldelft (B.V. Optische Industrie 'De Oude Delft'), PO Box 72, 2600 MD Delft.

Status
Current. Production.
Service
Several armies.

Type TM-007 laser target pointer

This small, lightweight, infra-red laser target pointer is specially designed for close-range combat with small arms at night, with the use of passive night vision goggles. It enables the user to aim at a target with great accuracy under low light conditions or in complete darkness, without visual recognition.

DATA
IR laser output power: 2 mW (optional 15 mW)
Wavelength: 820 nm nominal
Beam diameter: 5 mm nominal
Beam divergence: 0.35 mRad typical
Beam adjustment: + 10 and -10 mrad in steps of 0.75 mrad
Visibility with goggles: 300 m (spot 10 cm)
Operation temperature: -30° to 60°C
Dimensions: (length) 157 mm; (diameter) 34 mm
Weight, with batteries: 350 g
Batteries: 2 alkaline 1.5 V DC size AA
Operating time: approx 4 h continuous (2 mW type)

Manufacturer
Oldelft (B.V. Optische Industrie 'De Oude Delft'), PO Box 72, 2600 MD Delft.
Status
Current. Production.

TM-007 laser target pointer

UA 1119 second generation sight

The UA 1119 is a multi-purpose second generation binocular night sight designed for mounting on weapons and platforms. With a high-resolution image intensifier tube, × 4 magnification and shadow-type aiming graticule, the system provides high accuracy for aiming and observation in light conditions as low as 0.1 mlux.

It incorporates automatic brightness control and bright source protection to safeguard the user against dazzling from muzzle flashes and other sources of intense light. A daylight filter is incorporated in the lens cap to permit daylight operation. Lightweight and rugged, the sight is fitted with a standard NATO mounting and can be attached to a wide range of equipment by means of appropriate adaptors.

The UA 1119 can be operated from an external 24 V DC supply or from internal dry batteries. The mean time to repair and the maintenance costs are minimised since the electronic circuitry is integrated into the image intensifier tube.

This night sight is also available with a built-in CCD camera for use with standard CCIR video equipment, to enable images to be displayed on remote monitors and to be recorded for training or evaluation.

UA 1119 night sight

DATA
Magnification: × 4.2
Field-of-view: 9.5°
Resolution: 1 mrad at 1 mlux
Focus range: 25 m to infinity
Objective lens: f/1.7 refracting
Image intensifier: XX1380 series micro-channel plate tube
Power source: 24 V DC in external supply mode 2 × R14 (C-type) dry cells in battery mode
Operating voltage: 3 V DC nominal
Power consumption: 120 mW max in battery mode
Battery life: approx 40 h
Weight: 1.8 kg

Manufacturer
SIGNAAL-Usfa BV, P.O.Box 6034, 5600 HA, Eindhoven.
Status
Current. Production.

UA 1119 night sight with built-in video camera

UA 1134 compact individual weapon sight

For use on lightweight assault weapons, the UA 1134 employs a second generation image intensifier tube with computer-designed fixed focus objective and eyepiece lenses for maximum performance and ease of operation. The sight incorporates automatic brightness control and bright source protection to give a consistently bright image regardless of ambient light conditions while safeguarding the user against dazzle from intense point sources. It will also accept a third generation image intensifier without adaption.

The On/Off switch and the AZ/EL sighting adjustment knobs can be operated with gloved hands; no special tools are required. Once set for a specific weapon the sight remains accurate despite firing shocks or repeated removal and replacement.

Apart from changing the batteries and cleaning the lenses, the sight requires no field-level maintenance whatever. All components and sub-assemblies are interchangeable between units, thereby minimising downtime for repair.

DATA
Magnification: × 4
Field-of-view: 143 mils (8.0°)
Focus distance: fixed at 12 m (Other settings on request)
Sighting adjustment: ± 10 mrad adjustment range, click-stopped at 0.25 mrad intervals
Power source: 2 Penlite batteries (R6, AA) or 1 lithium battery
Operating temperature: standard batteries – -20 to +45°C; lithium batteries – -45 to +45°C
Weight: 1.1 kg

Manufacturer
SIGNAAL-Usfa BV, PO Box 6034, 5600 HA Eindhoven.

UA 1134 compact weapon night sight

Status
Current. Production.

UA 1116 second generation sight

Developed for use with small arms, the UA 1116 is an individual lightweight weapon sight. With minimised weight and size, the sight can be applied to the standard NATO sight bracket for mounting on all current types of infantry small arms and heavy machine guns.

The second generation micro-channel plate has its own integrated power supply and automatic brightness control together with protection against flare and dazzle. Objective and eyepiece lenses are fully adjustable, as is the illuminated graticule. Constant accuracy is maintained in spite of repeated firing shocks and frequent removal and replacement. The sight is virtually maintenance-free.

DATA
Weight: 1.6 kg
Magnification: × 4.2
Field-of-view: 9° 30′
Resolution: 1 mil at 1mlux (starlight) at target contrast of 30%
Focus range: 25 m to infinity
Power supply: 3 V DC nominal

Manufacturer
SIGNAAL-Usfa BV, P.O.Box 6034, 5600 HA, Eindhoven.
Status
Current. Production.
Service
NATO and Middle East armed forces.

UA 1116 in operation

Philips UA 1116 second generation sight

UA 1137/01 self-powered rifle sight

The UA 1137 is a revolutionary new image intensifier rifle sight which is powered by a novel internal energy course that can be manually re-energised during use. Containing no batteries, this maintenance-free rifle sight is for use on hand-held weapons in all ambient light conditions from full daylight to starlight.

Its wide horizontal field-of-view (22°) makes it suitable for both observation and aiming, while the vertical field-of-view of 10° ensures that the user is shielded mfrom sky illumination above and from own-weapon muzzle flash and barrel reflections below.

DATA
Magnification: × 1.5
Field-of-view: 22° horizontal, 10° vertical
Threshold recognition level: over 200 m
Power source: Internal manually regenerated source

Weapon interface: NATO STANAG 2324
Weight: less than 1.5 kg
Focal length: 50 mm

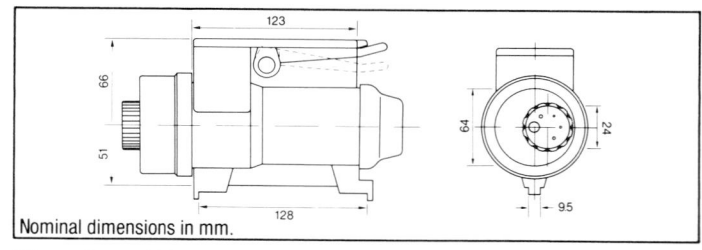
Nominal dimensions in mm.
Dimensions of Philips UA 1137 rifle sight

Aperture: f/1.2
Focusing: fixed (field 75 – 400 m)
Diopter adjustment: fixed at -1 diopter; other settings available to order
Eyepiece pupil: 6 mm diameter

Manufacturer
SIGNAAL-Usfa BV, P.O.Box 6034, 5600 HA, Eindhoven.
Status
Current. Production.

Philips UA 1137 self-powered rifle sight

NORWAY

Simrad KN200 image intensifier

The Simrad KN200 is a battery-powered image intensifier based on a new and unique design concept. The Simrad KN200 is intended as an add-on unit to optical sights on direct fire weapons, laser rangefinders and other daylight image devices such as TV cameras.

The unique design concept has overcome the problems usually connected with night sights, and compared to traditional night sights the KN200 offers several advantages. As the eyepiece of the existing day sight is also used at night, the graticule pattern remains the same and the position of the operator's eye remains unchanged day and night. It can be mounted and removed within a few seconds and no bore-sighting adjustments are necessary.

The use of a 100 mm objective aperture yields a very good range performance, enabling the operator to detect and engage targets within the practical range of the weapon during night operations. Sudden illumination does not have any effect on the sighting capabilities.

Change between second and third generation tubes requires no modification to the instrument.

DATA
Weight: 1.4 kg (1.46 kg with adjustable focus) including battery
Field-of-view: 10°
Focusing range: fixed or adjustable 25 m to infinity
Resolution: 0.5 mRad/lp (contrast 30%, illuminance 100 lux)
Objective lens: catadioptric, 100 mm f/1.0
Power supply: 2 × 1.5 v AA alkaline or 2 × C cells. Lithium optional
Battery life: >40 h at 20°C
Operational temperature range: -40 to +52°C

Manufacturer
Simrad Optronics A/S, PO Box 6114, N-0602 Oslo 6.
Status
Production.

Simrad KN200 image intensifer mounted on Simrad LP7 laser rangefinder

Simrad KN250 image intensifier

The Simrad KN250 Image Intensifier is based on a new and unique design which is patented in many countries. The KN250 is mounted as an add-on unit on telescope sights, laser rangefinders and other optical daylight imaging devices such as video cameras. As no boresighting adjustment is required, the mounting procedure only takes a few seconds.

The KN250 can use both 2nd and 3rd generation image intensifier tubes. Change between 2nd and 3rd generation tubes requires no modification of the instrument.

With this new concept the gunner can aim through the eyepiece of the day sight both day and night, an advantage not achieved with traditional types of night sight. Sudden illumination of the scene does not have any effect on the sighting capabilities.

DATA
Weight: 740 g (790 g with adjustable focus) including batteries
Magnification: × 1
Field-of-view: 12°
Resolution: 0.6 mRad/lp (contrast 30%, illuminance 100 lux)
Objective lens: catadioptric, 80 mm f/1.1 (T/1.4
Focusing: fixed or variable between 25 m and infinity

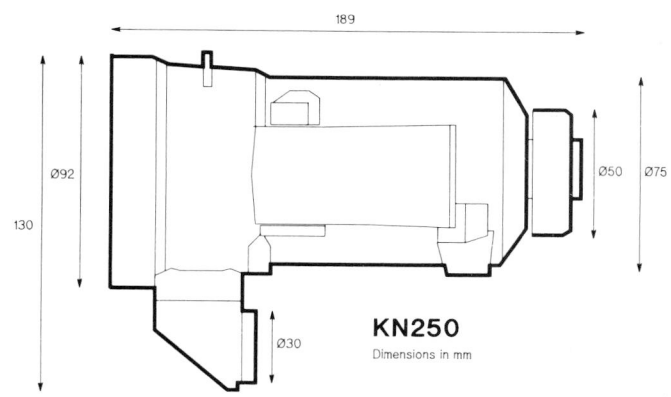

Simrad KN250 image intensifier; dimensions

Tube: 2nd or 3rd gen, 18 mm wafer tube
Power supply: 2 × 1.5 V alkaline AA cells.
Battery life: >40 h at 20°C
Operational temperature range: -40°C to +52°C

Manufacturer
Simrad Optronics A/S, PO Box 6114, N-0602 Oslo 6.
Status
Current. Production.

Simrad KN250 on Steyr AUG assault rifle

PAKISTAN

AN/PVS-4A individual weapon sight

The AN/PVS-4A is a weapon sight for use on rifles, machine guns and rocket launchers. It uses a 25 mm 2nd generation plate amplifier tube and has an internally adjustable, illuminated graticule projector; the graticule is interchangeable for use with different weapons. The sight has automatic gain control which allows it to cope with very large fluctuations in brightness at the target, and a bright source protection circuit prevents damage due to inadvertent exposure to daylight. A special cover permits daytime bore-sighting and training.

DATA
Length: 240 mm
Width: 120 mm
Height: 120 mm
Weight: 1.70 kg
Magnification: × 3.6
Field-of-view: 14.5°
Object lens: 95 mm T/1.6
Dioptre adjustment: +2-6
Power source: AA carbon zinc cell
Battery life: 15 h at 25°C

AN/PVS-4A weapon sight on MG3 machine gun

Manufacturer
Institute of Optronics Pakistan, PO Box 1596, Rawalpindi.

Status
Current. Production.

AN/TVS-5A crew-served weapon sight

This sight is intended for use on heavier weapons such as recoilless rifles, heavy machine guns and some types of wire guided anti-tank missile. It can also be tripod-mounted for use as a surveillance device.

The sight uses a 25 mm second generation plate amplifier tube, and has automatic gain control so that sudden bright lights in the target area will not affect the system. The internally mounted adjustable illuminated graticule can be interchanged to cater for different weapons or ammunition. There is a bright source protection circuit which prevents damage due to inadvertent exposure to daylight, and the sight is provided with a cover which allows daytime bore-sighting and training.

DATA
Length: 310 mm
Width: 160 mm
Height: 170 mm
Weight: 3.00 kg
Magnification: × 6
Field-of-view: 9°
Object lens: 155 mm T/1.7
Power supply: AA carbon zinc cells
battery life: 15 h at 20°C

AN/TVS-5A weapon sight on 106 mm recoilless rifle

Manufacturer
Institute of Optronics Pakistan, PO Box 1596, Rawalpindi.
Status
Current. Production.

SOUTH AFRICA

MNV miniature night sight

Equipped with an 18 mm second generation micro-channel intensifier, the MNV night sight is optimised for application in counter-insurgency operations.

Evolved from extensive combat experience, this instrument is of compact and rugged design and is easy to use. It features a fixed focus objective and is instantly ready for use as soon as switched on. The only adjustments required, prior to action, are bore-sighting and eyepiece dioptre setting.

Power supply is by means of two readily available Penlight batteries. Bore-sighting adjustment is incorporated in an integral quick-release mounting bracket. Mounts for a variety of rifles are available.

DATA
Weight: 950 g (without mounting bracket)
Length: 210 mm
Image intensifier: 18 mm micro-channel amplifier to MIL-I-49052B
Magnification: × 2.6
Field-of-view: 16°
Entrance pupil: 40 mm
Exit pupil: 7 mm
Eye relief: 24 mm
Eyepiece adjustment: -4 to +4 dioptres
Bore-sighting range: 20 mils continuous, azimuth and elevation
Recognition range: (man target) 200 m at 10^{-3} lux
Power supply: 2 × AA 1.5 V penlight batteries

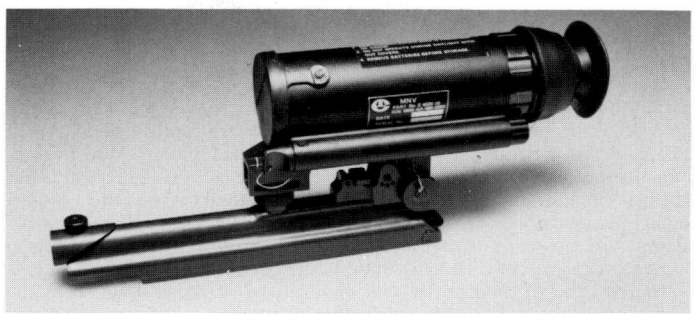

MNV sight mounted on receiver cover of R4 rifle

Manufacturer
Armscor, Private Bag X337, Pretoria.
Status
Current. Production.
Service
South African armed forces.

SPAIN

VNP-004 night sight

This has been developed in Spain specifically as a night sight for the MG42 machine gun, though there seems no reason why it should not be readily adaptable to other weapons. The VNP-004 is a second generation sight using a Philips XX1500 18 mm micro-channel amplifier, giving variable gain from × 30 000 to × 70 000 selectable by the operator. The sight is provided with a mounting of standard NATO dimensions.

DATA
Magnification: × 3.5
Field-of-view: 9° 30⁸ (169 mils)
Resolution: 0.6 mrad at 10^{-3} lux
Dioptre correction: +2 to -5
Focussing: fixed, parallax-free
Zero adjustment: graticule displacement
Dimensions: 340 × 150 × 120 mm
Weight: 3.20 kg
Power supply: 2 × 1.5 V batteries (R14 or equivalent)
Battery life: over 50 h

Manufacturer
ENOSA, Empresa Nacional de Optica SA, Avenida de San Luis 91, Madrid 33.
Status
Current. Available.

VNP-004 night sight

Service
Spanish Army.

AVN-01 laser aiming light

This is a laser spot projector capable of being fitted to virtually any weapon. When the user is wearing night vision goggles or using a night vision sight, the laser spot allows him to aim his weapon in conditions of darkness.

DATA
Laser: AsGa
Operating mode: pulse
Average power: 50 µW
Beam divergence: 2 mrad
Wavelength: 815 nm
Spectral bandwidth: 8 nm
Power supply: IEC-6F22 battery
Battery life: 100 hours

Manufacturer
ENOSA, Empresa Nacional de Optica SA, Avenida de San Luis 91, Madrid 33.
Status
Current. Available.

AVN-01 laser aiming light

SWEDEN

Aimpoint 2000 electronic sight unit

This electronic sight has been introduced by Aimpoint AB in Sweden and is claimed to enable the firer to aim more quickly and more accurately than he could with a conventional iron sight.

In the Aimpoint device the iron sights are replaced by a single battery-powered optical and electronic unit. The eyepiece of this unit is designed to be viewed by one eye while the other eye can be either open or closed – but is preferably open because the firer's field-of-view thus embraces more than can be seen through a telescope eyepiece and the sight cover is less obtrusive than it is if only one eye is used. The electronics of the system cause a small bright red dot to appear in the field-of-view coincident with the point of aim of the weapon and all that the firer has to do in order to aim correctly is to bring this dot into coincidence with the desired target.

The position of the dot depends only on the pointing direction of the weapon and is not altered by movement of the firer's head; when it is in coincidence with the target, therefore, the system is parallax-free. Other advantages of the system are that it can be used in all light and weather conditions, that it gives no signal to a potential enemy, that it is ruggedly made and, above all, that skill in using it is quickly acquired.

Originally this sight was developed for fast, precise shooting keeping both eyes open. Now a useful telescopic attachment has been produced which gives a × 3 magnified image and this substantially increases the range of accurate shooting. Two polarising filters are mounted and form a useful aid in brilliant light.

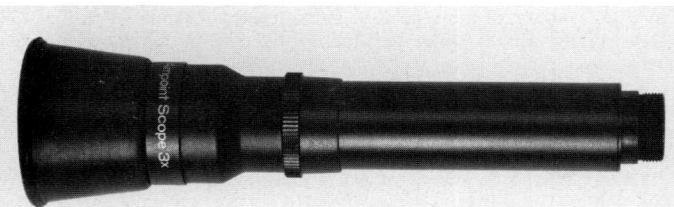

Aimpoint 2000 electronic sight unit and (above) telescope attachment

DATA
Length: 128 mm (185 mm with telescope)
Weight: 148 g (168 g with telescope)
Diameter: 25.4 mm
Material: anodised aluminium
Finish: black or stainless
Mounting: standard 1 in rings
Power supply: (long cap) one-piece lithium DL2/3A or two-piece mercury PX1/RM1N. (Short cap) one-piece lithium CR1/3N, 2L76BP or DL1/3N

Manufacturer
Aimpoint AB, Jägershillgatan 15, S-213 75 Malmö.
Status
Current. Production. More than 250 000 produced to date.
Service
Commercial sales world-wide.

Aimpoint 1000 electronic sight unit

The Aimpoint 1000 is to the same general specification as the Model 2000 described above, but differs in construction. The Model 1000 is cast in one piece and has built-in bases for the well-known Weaver system of sight mounts, whereas the Model 2000 requires the use of 1 in telescope mounting rings. In addition, the field-of-view of the Model 1000 is slightly less and there is no semi-automatic light adjustment circuit. Optically, the two sights are identical, both using the same type of red dot.

Manufacturer
Aimpoint AB, Jägershillgatan 15, S-213 75 Malmö.
Status
Current. Production.
Service
Commercial sales world-wide.

Aimpoint Model 1000 electronic sight unit

Weibull direct aiming sights

These sights, which arre adaptabe to all types of small-calibre direct-fire weapons from machine guns to light anti-aircraft guns, have been developed in close co-operation with military experts. The prevailing object has been to develop a unitary sight system which may be used for air, ground or sea targets or any combination of these. The graticule pattern, designed to fit specific weapons, is adapted to the ballistic performance and also to certain patterns of target behaviour which experience has shown can be expected for a large proportion of the time. The philosophy behind the system is that it is better to have a simple and effective system on a large number of weapons than a complicated and expensive system on a smaller number.

The Reflex Sight is equipped with four different graticules mounted on a revolving disc. Two of these are intended for air target combat – one for fast-moving targets, and the other for slow. One graticule is designed for seaborne targets, and is also suitable for ground targets, and the last graticule corresponds to a conventional ring sight. The sight embodies an optical collimating system, such that the aimer sees the graticule at an infinite distance, superimposed upon a view of the target area. It is not necessary for the aimer's eye to be in any specific place; provided he can see both the graticule and the target, the aim will be correct. For conditions where ambient light does not permit the graticule to be clearly visible, electric illumination is provided.

Custom-made graticules to correspond to any weapon are available, as are mounts and adapters to fit the sights to all types of weapon.

Manufacturer
Ingeniörsfirman JL Weibull AB, PO Box 43, S-232 20 Åkarp.

Weibull Reflex Sight for 7.62 mm FN-MAG machine gun

Status
Current. Production.
Service
Swedish Army.

Weibull Reflex Sight for 40 mm anti-aircraft gun

Type KS-2 anti-aircraft sight

This is a small anti-aircraft sight for use with light automatic weapons and features a new type of optical presentation of lead angle.

The sight consists of a combination prism and a slightly reducing telescopic system. The image from the telescopic system is combined with the directly observed target in the combination prism so that the eye sees two images of the target one of which is larger than the other. When the sight is aimed at a fixed target the images should coincide and form one image only. When the sight is aimed at a moving target the larger image should chase the small image with a certain multiple of target lengths between them. This multiple is easy to learn and is taken from a precalculated table for each type of weapon.

The angle between the two images is a fictitious lead angle and forms only one-fifth of the real lead angle. One attractive feature of the sight is therefore that the eye needs to receive information from only a small part of the field-of-view.

Silent zero device

This is the name given to a device for 'dry' zeroing the KS-2 without firing.

The instrument consists of a spud of appropriate calibre to which an arm with two small collimators is attached. Each collimator contains an aiming ring projected to infinity.

The spud is inserted into the muzzle of the weapon and turned until the collimators are in front of the KS-2 sight.

The weapon is then pointed towards the sky, a diffuse lamp or an illuminated wall. Two small rings are now visible in the sight; and when it is correctly zeroed these rings are concentric, otherwise the sight has to be aligned by means of its two zeroing screws.

The parallelism of the collimators with each other and with the spud is adjusted in the factory. The collimators are so small that there is no danger of misalignment. They are also protected by a frame.

DATA
Free eye movement: approx 32 × 32 mm
Field-of-view for 50 mm eye-to-sight separation: (direct) 480 mils; (through telescope) 350 mils
Engagement range: 100–1500 m depending on visibility
Weight: 1 kg including sight holder
Operating temperature: -40 to +40°C

Manufacturer
Bofors Aerotronics AB, S-181 84 Lidingö.
Status
Current. Production.
Service
Not known.

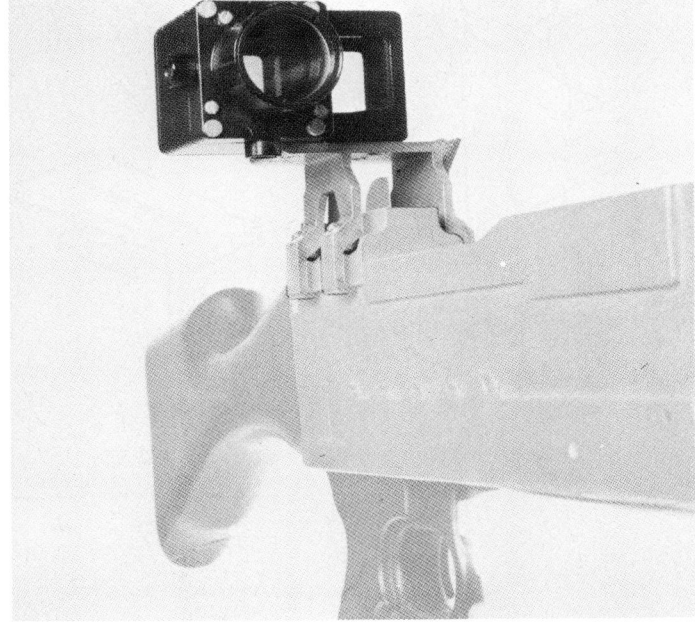

Type KS-2 anti-aircraft sight on machine gun

RS-420-5T Telescope gun sight

The RS-420 sight is a collimator sight with a built-in elevation device. It was developed primarily for use on 9 cm anti-tank guns and has been supplied in large quantities to the Swedish Army. The instrument is designed to withstand the shocks and strains encountered in field use.

For setting of super-elevation the objective is vertically adjustable by means of an elevation screw located at the bottom. It is provided with a circular scale graduated in distance.

Calibration, if necessary, is carried out by adjusting the scale which is locked in position by means of a locking-screw. In addition the scale can be exchanged for adapting the sight to different kinds of ammunition. The reticle is laterally adjustable with a knob. Turning the knob one dial unit moves the reticle one milliradian laterally. This knob is also adjustable for zero setting by bore-sighting.

Beneath the lateral adjustment screw a dovetail fitting for attaching the

Bofors RS-420 gun sight

illuminating device is supplied with the sight which comprises a housing with an incandescent lamp, a diaphragm to adjust illumination intensity and a built-in battery. The lamp illuminates the reticle as well as the elevation scale.

The eyepiece is provided with a removable, bellows-type rubber eye-cup.

DATA
Magnification: × 4
Objective diameter: 20 mm
Field angle: 10°
Exit pupil: 5 mm
Distance between rearmost lens surface and exit pupil 60 ± 5 mm
Resolving power, better than 15^8 within a central field with a radius of 1.5°
Parallax with lateral adjustment set at centre, max ± 0.5 mrad
Dioptre rating -0.25 to 0.5 dioptres

Manufacturer
Bofors Aerotronics AB, S-18184 Lidingö.
Status
Current. Available.
Service
Swedish Army.

NK-24 night sight

Night sight NK-24 is used together with the direct day sight RS-420-5T (above).

The NK-24 is a night sight of light intensifier type. The sight consists optically of three main parts: the catadioptric objective, the light intensifier tube, and the relay optics. The sight is also equipped with a breathing dessicator cartridge and a humidity indicator. There is also an adapter, of dovetail type, to fit the night sight to the day sight. The relay optics make it possible to bore-sight the direct sight RS-420-5T and the night vision sight NK-24. In front of the relay optics unit a cover is placed to protect a dichroite mirror. In twilight this cover can be opened and through the dichroite, semi-reflecting mirror, the image can be seen in the direct sight and reflected from the night sight. By superimposing the two images the two optical parts are aligned.

The night sight operates with built-in batteries. When not in use it is kept in a case.

DATA
Objective aperture: 100 mm diameter
Field-of-view: 10°
Magnification: × 1

Manufacturer
Bofors Aerotronics AB, S-18184 Lidingö.
Status
Current. Production.

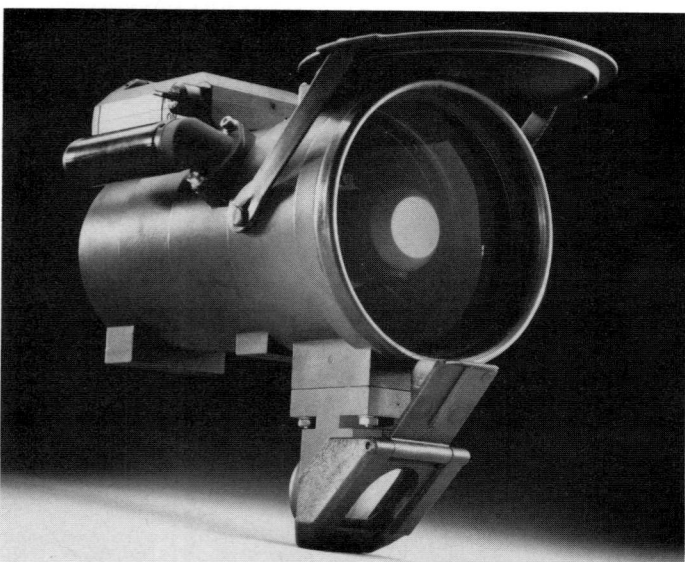

Bofors NK-24 night sight

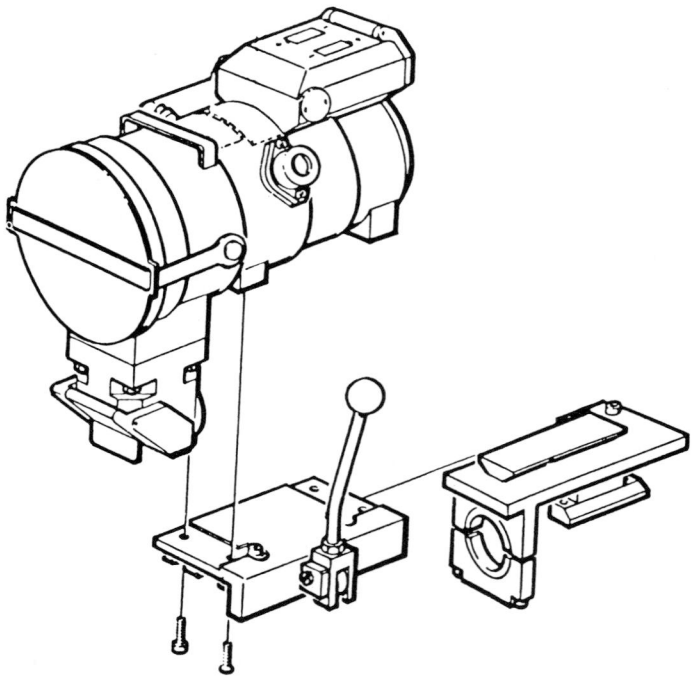

Bofors night sight NK-24 with adapter for fitting on RS-420-5T

SWITZERLAND

Kern-Mauser RV Reflex Sight

This sight has been developed by Kern in cooperation with Mauser, the well-known arms manufacturer. It makes shooting at moving targets faster and target acquisition easier. The sight is a reflex collimating sight which is used with both eyes open; the sighting eye sees a red dot and superimposes this on the picture of the target seen by the disengaged eye. The rifleman merely has to move his rifle until the red dot lies upon his desired point of aim. Due to the unique optical qualities of this type of sight, precise alignment of the rifleman's head is not necessary; provided he can see the red dot and place it on target, his aim is correct.

The RV sight requires no batteries. By day it uses reflected daylight to provide the red spot image, and at night the spot is illuminated by a Tritium light source with a service life of approximately 10 years. There is a polarising filter which permits continuous contrast adjustment.

The sight can be fitted to almost any type of weapon by means of a STANAG 2324 mount, a special low-mount bracket, or by proprietary mounts made by various manufacturers.

Manufacturer
Kern & Co., AG, Schachenallee, CH-5001 Aarau.
Status
Current. Production.

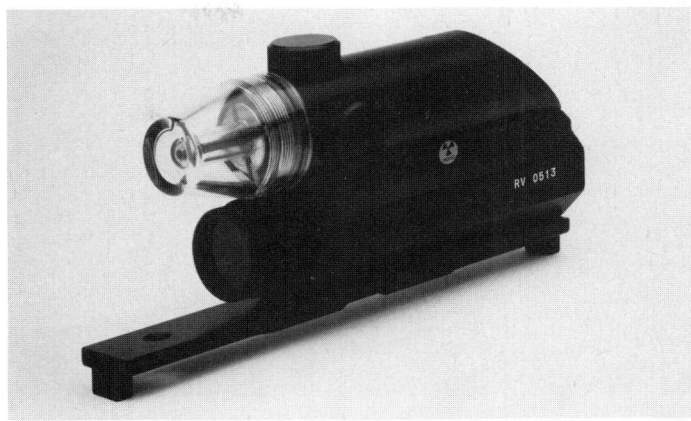

Kern-Mauser RV reflex sight

Wild SZR/SKS muzzle bore-sight

The Wild SZR/SKS is a universal muzzle bore-sight device, designed for efficient and precise calibration of weapons. It consists of an angled telescope and a wide range of interchangeable calibre bars for all standard bores from 7.62 to 155 mm.

The SZR2 bore-sight telescope can be used with a laser eyepiece, which projects the bore centreline as a visible red dot. This combination is particularly useful in industrial plants where large quantities of aiming devices must be aligned.

The SZR2-1 is the MIL-tested version of the bore-sight telescope. It has a fixed eyepiece and a graticule illumination window for use at night. Supplementary lenses are available for use at 10, 25 and 40 m distance.

DATA
Magnification: × 6
Field-of-view: 85 mils
Exit pupil diameter: 3 mm
Eye relief: 12 mm
Dioptric setting: -5 to + 5 dioptres
Graticule: cross-hairs with 1 mil graduations
Centring accuracy: ± 0.1 mil (telescope to calibre bar)
Dimensions of telescope: 111 × 94 × 42 mm
Weight of telescope: 300 g

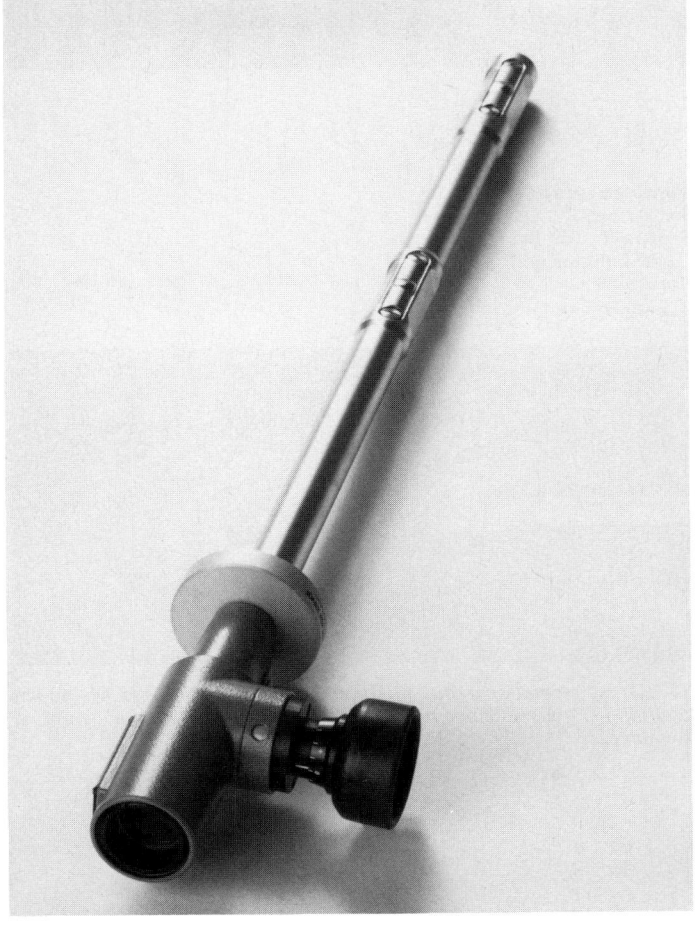

Wild SZR2-1 bore-sight telescope with SKS20 calibre bar

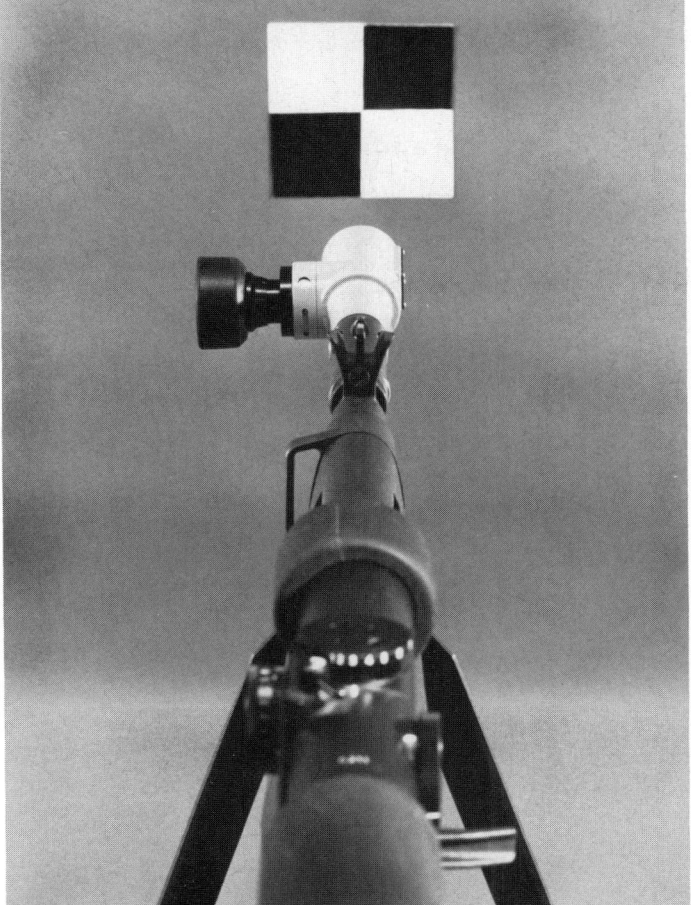

Wild SZR/SKS muzzle bore-sight

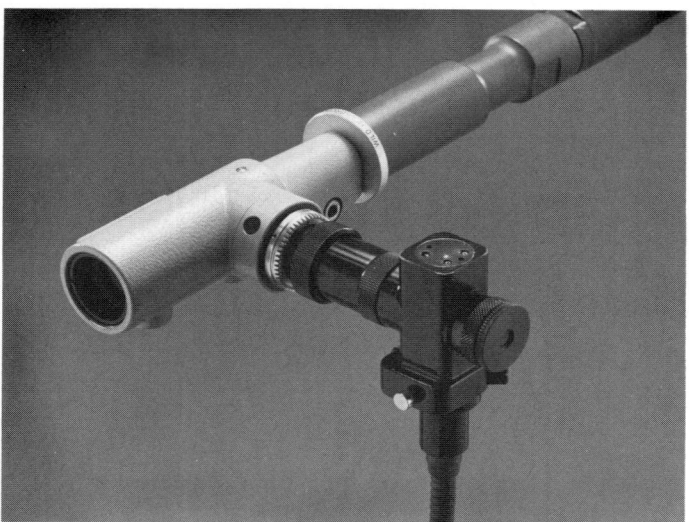

Manufacturer
Wild Leitz Ltd, Special Products Division, CH-9435 Heerbrugg.
Status
Current. Production.
Service
In service with several NATO armies.

Wild SZR2 bore-sight telescope with laser eyepiece

UNION OF SOVIET SOCIALIST REPUBLICS

NSP-2 IR sight

This sight is now obsolescent but is still seen occasionally. It is a conventional active infra-red (IR) type with a lamp and an image converter. It is heavy and suffers from the usual limitations of IR sights in that it is capable of being located and the user can only see directly down the beam.

NSP-2 IR sight showing size and bulk of this type of sight: lamp has its cover closed

NSP-3 first generation night sight

The NSP-3 should not be confused with the NSP-2 since it is an entirely different system and it uses a first generation II tube and is entirely passive in operation. It is issued on a scale of one per infantry section. There is an adjustable graticule though whether there is provision for automatic brightness control is not known, but it is thought to be unlikely.

DATA
Length: 490 mm
Weight: (with battery) 2.7 kg
Magnification: × 2.7
Field-of-view: 7°

Manufacturer
State arsenals.
Status
Current. Production.
Service
Not known.

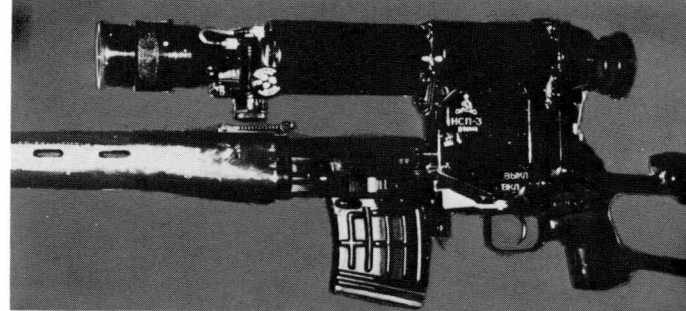

NSP-3 first generation night sight mounted on Dragunov rifle. Method of mounting is unusual since majority of length of sight body is overhung from bracket

NSP-3 first generation night sight, right-hand side

PGN-1 image intensifier sight

The PGN-1 bears a close resemblance to the NSP-3 but it has a slightly better performance and it is usually mounted on the RPG-7 rather than any of the small arms.

DATA
Length: 540 mm
Weight: (with battery) 3.5 kg
Magnification: × 3.4
Field-of-view: 5.7°
Range: 400–500 m

Manufacturer
State arsenals.
Status
Current. Production.
Service
Soviet and satellite armies.

PGN-1 image intensifier sight mounted on RPG-7

UNITED KINGDOM

SUSAT (Sight Unit Small Arms Trilux) L9A1

SUSAT (Sight Unit Small Arms Trilux) L9A1 has been developed by the Royal Armaments Research and Development Establishment (RARDE) to realise the full capabilities of the L85A1 rifle now entering service with the British Army.

SUSAT is smaller, lighter and more robust than its predecessor, SUIT (which will be found described in earlier editions of *Jane's Infantry Weapons*), and contains enhanced optics to improve twilight surveillance and weapon aiming tasks. Being of fixed focus, with the well-proven aiming needle for shooting, the only moving part on SUSAT is the adjuster for the variable intensity Trilux source.

SUSAT is fixed to the L85A1 rifle by means of a dedicated mount which allows for accurate weapon zeroing and range adjustment throughout the effective range of both Individual and Light Support weapons.

SUSAT can be mounted on to a wide range of rifles, machine guns and recoilless rifles by means of the combined zeroing/range/windage adjustable mount seen in the illustration.

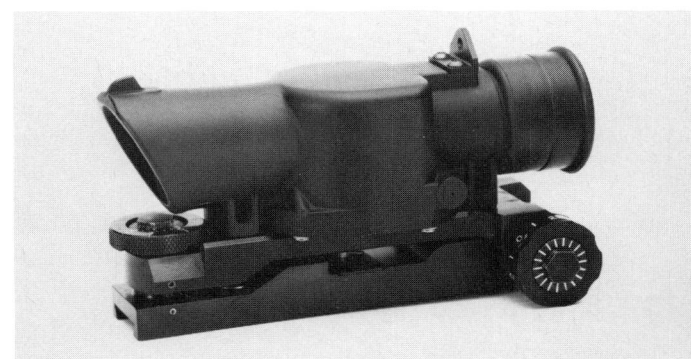

SUSAT sight complete with universal mount

DATA
Length: 145 mm
Width: 60 mm
Height: 60 mm
Weight: 470 g
Magnification: × 4
Field-of-view: 175 mils (10°)
Light transmission: more than 80%
Entrance pupil: 25.5 mm diameter
Exit pupil: 6.4 mm diameter
Eye relief: 25 mm
Eyepiece focus: -1 dioptre

Flatness of field: 0.25 dioptre
Veiling glare: 2° max

Manufacturer
United Scientific Instruments, 215 Vauxhall Bridge Road, London SW1V 1EN.
Status
Current. Production.
Service
British Army.

Swingfire combined sight

Initially for use in Striker in the direct fire role and with the FV438 in the separated fire role, the British Government funded the development of the Combined Sight which provides both visual and thermal imaging channels. Development was carried out by British Aerospace with Barr and Stroud.

In clear conditions, engagements can take place out to very nearly the full range of the optical mode, though in severe conditions of mist, rain and fog it may be reduced below this due to the attenuation effect of the water droplets. However, no matter how bad the weather, the performance of a TI sight is immeasurably superior to that of an optical system.

The sight can be used in both the mounted and dismounted role. In the

Crew carrying combined sight to separation fire position

Combined sight fitted in Striker

dismounted role it can be carried by one man, a second carries the base and reels out the separation cable. In action, the controller can switch from optical to TI mode without delay and at will. The method of engaging targets is exactly the same whichever viewing system is used and the only requirement is that the controller should be able to see his target.

The TI system is virtually immune to counter-measures and it greatly extends the versatility of the Swingfire missile system.

DATA
Height: 560 mm (tripod mount adds a further 250 mm)
Width: 340 mm
Depth: 280 mm
Weight: (as carried) 41 kg
Tripod and air bottle: 24 kg
Total emplaced weight: 65 kg
Cable: 80 m
Air bottle: 2 h on sight. In vehicle, 24 h continuous running

Manufacturer
British Aerospace plc, Army Weapons Division, Six Hills Way, Stevenage, Herts SG1 2DA.
Status
Current.
Service
In service with British Army.

Combined sight emplaced remote from firing vehicle. Large switch above eyepiece changes mode of viewing from thermal to optical and here is set on optical

IRS 218 night rifle sight

The Davin Optical IRS 218 is an image intensified rifle sight designed to allow the rifleman to engage targets at night. It can be used over all normal engagement ranges and offers superb performance based on the latest readily available image intensifier tubes.

Designed for the hostile battlefield environment, the IRS 218 is fully ruggedised, yet is light in weight ensuring minimal interference with normal weapon handling characteristics. It can be fitted on all common service rifles and features a non-reflecting lens and a special security eyecup to minimise the chance of detection.

An illuminated aiming mark injected into the eyepiece and imposed on the field-of-view is used for weapon aiming, while bore-sighting of the sight to the weapon is performed with a simple click-stop control for elevation and azimuth.

The IRS 218 design is simple but advanced and needs minimal maintenance support due to the high standards of manufacture.

DATA
Length (excluding eyecup): 204 mm
Diameter: 63 mm (max)
Weight (excluding batteries): 1–1.2 kg
Magnification: × 2.8
Field-of-view: 14.6 (26 m at 100 m)
Gain: 50 000
Tube: Mullard XX1500 or micro-channel plate
Battery: 2 × AA size alkaline cell
Battery life: 40 h (average)
Focus range: adjustable, 10 m to infinity
Mounting bracket: to fit most weapons
Objective lens aperture: F/1.4, 70 mm F/L

IRS 218 night rifle sight

Manufacturer
Davin Optical Limited, 13 Alston Works, Alston Road, Barnet, Hertfordshire EN5 4EL.
Status
Current. Available.
Service
Far East.

Kite individual weapon sight

Kite is a new lightweight high-performance night sight from Pilkington Optronics, designed to meet a demanding military specification. Made of a lightweight but highly robust material, Kite weighs only 1 kg. The advanced optical design affords a magnification of × 4 and a 9° field-of-view. The refractive lens system incoporates an injected graticule to assist with accurate weapon aiming, and the system is configured to accept second- or third-generation tubes.

The micro-channel plate image intensifier affords a good contrast image, particularly at low light levels, and will localise bright sources of light such as muzzle flash or flares. Kite is simple to operate, the only controls being a rotary on/off switch with graticule brightness adjustment and focussing controls. The modular design permits rapid sub-unit replacement and easy access to the battery compartment.

Kite has been selected for the British Army's SA80 weapon system.

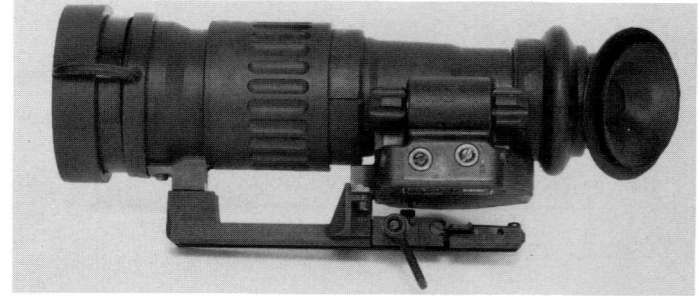

Pilkington Optronics Kite weapon sight

DATA
Field-of-view: 9°
Magnification: × 4
Weight: 1.0 kg
Range: 600 m (man-sized target under standard conditions)
Eye relief: 30 mm
Length: 255 mm
Diameter: 73 mm
Power supply: 2 × 1.5 V AA batteries. (UK MoD supplied with 2.7 V version)

Manufacturer
Pilkington Optronics Ltd, Glascoed Road, St Asaph, Clwyd LL17 OLL.
Status
Current. Production.
Service
British Ministry of Defence and overseas armed forces.

Kite weapon sight in use on FN FAL rifle

No 100 low level air defence sight

The No 100 low level air defence (LLAD) sight has been developed in co-operation with the Military Vehicles and Engineering Establishment as a solution to the problem of engagement by small arms of low speed, low flying aircraft. The system fits into the standard NATO Individual Weapon Sight category and can be readily attached to most light and general-purpose machine guns. The mounting attachment to the weapon enables the sight to be removed and refitted without readjustment.

The incorporation of photochromic glass in the field-of-view enables automatic adjustments of the contrast between the reticle and target dependent on the ambient light conditions, which is vital when engaging airborne targets against a bright background.

The reticle, which is optically injected into the field-of-view, provides a wide range of target speeds and approach angles. A special reticle for use against radio-ccontrolled model aircraft targets is available for training purposes.

The sight which is a fully sealed and desiccated instrument is environmentally tested. Adjustment is undertaken using external adjusting screws for both elevation and azimuth in two-mil divisions.

To use the reticle the gunner must first estimate the speed of the aircraft. He then positions the appropriate speed ring so that the nose of aircraft is just touching that ring, with the axis of the fuselage passing though the centre dot of the reticle. Provided the speed of the aircraft has been estimated correctly, the procedure will automatically apply the correct aim-off.

DATA
Magnification: × 1
Overall length: 120 mm
Overall height: 145 mm
Overall width: 45 mm
Weight: 475 g
Graticule illumination: ambient light
Environmental: completely sealed and desiccated. Operation and storage between -40 and +70°C

Manufacturer
Helio, 215 Vauxhall Bridge Road, London SW1V 1EN.
Status
Current. Production.

LLAD sight No 100, mounted on general-purpose machine gun

Watts RC35 air defence sight

Hall & Watts also manufacture the RC35 sight for weapon systems in the 20 – 40 mm calibre range. The RC35 is particularly suitable for retrofitting to towed anti-aircraft guns, heavy machine guns and automatic cannons.

DATA
Field-of-view: 18° × 12° (320 × 213 mils)
Aperture: 35 × 35 mm
Focal length: 157 mm
Length: 240 mm
Weight: 0.90 kg
Temperature range: -40° to +55°C

Manufacturer
Hall & Watts Ltd, 266 Hatfield Road, St Albans, Herts AL1 4UN.
Status
Current. Production.
Service
RC25 in service with British Army and others; RC35 with British and other navies.

Watts RC25 low level air defence sight

The RC25 is a reflex collimator sight designed for engaging low level aerial targets and static and crossing ground/surface/hovering targets. The sight is of unit magnification, self-illuminated, and designed to fit pintle-, tripod- and bipod-mounted machine guns in the 7.62 to 14.5 mm calibre range.

DATA
Field-of-view: 14.25° × 14.25° (250 × 250 mils)
Aperture: 25 mm
Focal length: 122 mm
Length: 190 mm
Weight: 625 g including zeroing mount
Temperature range: -32° to +60°C
Humidity: +40°C with rh 90-95%

Watts RC25 low level air defence sight

Watts C2A1 mortar and machine gun sight

The C2A1 is the latest sight in the C2 family of indirect fire sights. The sight was primarily designed for use with 81 mm mortars and machine guns deployed in the sustained fire or indirect fire role.

The C2A1 is a fixed focus optical instrument which uses tritium light sources to illuminate the elevation and azimuth scales, level vials and the telescope graticule. With the addition of a periscope extension the sight line can be raised by 355 mm permiting mortars to be used from weapon pits or armoured and soft-skinned vehicles.

Hall & Watts produce the C2 and C2A1 sights and spares under arrangements with Ernst Leitz of Canada.

DATA
C2A1 SIGHT
Magnification: × 1.8
Field-of-view: 180 mils
Weight of sight: 1.30 kg
Weight cased: 3.30 kg
Azimuth scale: 0 – 6400 mils
Elevation scale: 600 – 1600 mils
Machine gun scale: -200-+600 mils

PERISCOPE EXTENSION
Magnification: × 1
Field-of-view: 144 mils
Weight cased: 270 g
Length: 379.5 mm
Diameter: 19.1 mm

Watts C2A1 sight in use on 81 mm mortar

Watts C2A1 sight on general-purpose machine gun

Manufacturer
Hall & Watts Ltd, 266 Hatfield Road, St Albans, Herts AL1 4UN.
Status
Current. Production.

Service
In service with armies worldwide.

Watts Dry Zero device

The Watts Dry Zero device assists the maintenance of accurate individual weapon zeroing by providing bore-sighting, a reference of zero, and a zero-checking facility usable by day or night.

The Dry Zero device is a simple, robust and lightweight optical device, designed to be used by the soldier in field conditions to check and reset the zero of weapon optical sighting systems without the need to fire a shot after initial zero.

DATA
Graticule: grid
Focus: fixed at infinity
Illumination: Tritium
Sealing: dessicated and sealed
Calibre: 5.56 to 12.7 mm
Temperature range: -40° to +55°C
Operational length: 210 mm
Package length: 185 mm
Height: 130 mm
Diameter: 45 mm
Weight: < 500 g including carrying bag

Manufacturer
Hall & Watts Ltd, 266 Hatfield Road, St Albans, Herts AL1 4UN.
Status
Current. Production.

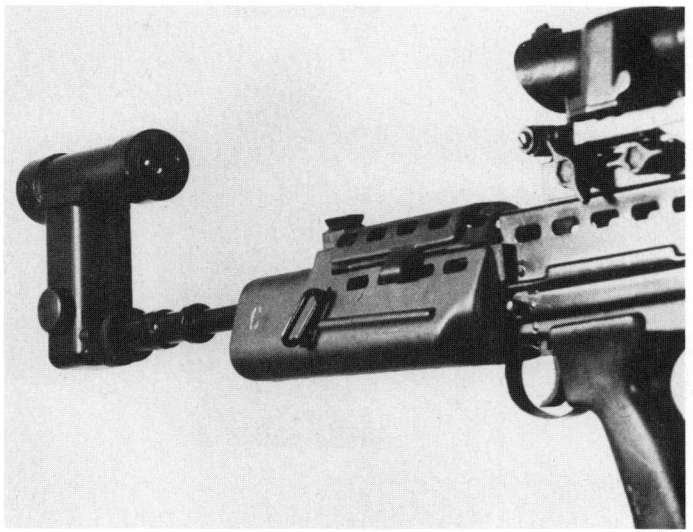

Watts Dry Zero device on L85 rifle

Armalon AM16ST sight unit

The AM16ST is a high precision, nitrogen sealed unit with coated glass lenses of the highest optical quality. The adjustments for windage and elevation are engineered to move the image so that the graticule pattern remains centred in the apparent field-of-view.

The sight has an integral mounting bracket, designed to fit AR15 and M16 pattern rifles. The trajectory compensating elevation drum markings indicate the type of ball ammunition for which the range settings are calibrated (for example M193/SS92) and whether the distances are in hundreds of metres or

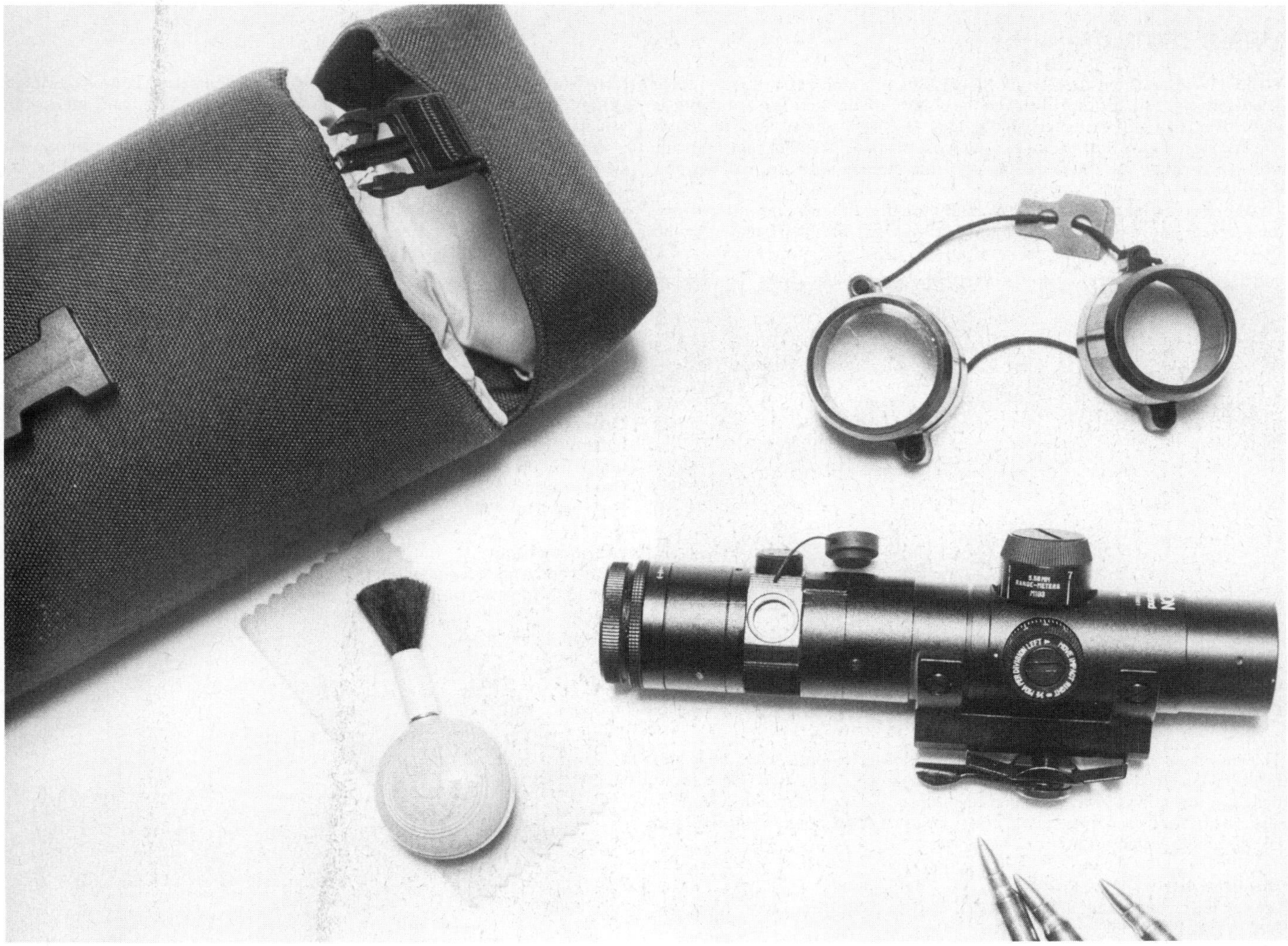

Armalon AM16ST sight with accessories

Armalon AM16ST sight on AR15 rifle

yards. If the drum is marked M193 or SS92 and not marked to indicate alternative barrel length or twist, the calibration assumes standard barrel length with rifling of one turn in 305 mm. If the drum is marked M855 or SS109, the assumption is a standard length barrel with rifling one turn in 178 mm.

DATA
Magnification: 4 ×
Oblective aperture: 21 mm
Field-of-view: 96.7 mils
Eye relief: approx 76 mm

Eyepiece focus: +2 to -3 dioptres, locking
Graticule system: multiple source black, red and green day and night illuminated
Weight: 450 g
Length: 190 mm
Diameter: 41 mm

Manufacturer
Armalon Ltd, 44 Harrowby Street, London W1H 5HX.
Status
Current; production.

WINS 1801 rifle sight

WINS (Weapon Intensified Night Sight) 1801 is a high performance image intensifier sight for rifles. It permits the soldier to aim and fire at targets at night over all normal engagement ranges. It is fully sealed, and its tough construction allows it to be used in the most hostile battlefield environment, while its light weight ensures minimal interference with normal weapon handling.

Aiming is carried out by the use of an illuminated aiming mark injected into the eyepiece and superimposed on the field-of-view. This mark may take the

form of a cross, an inverted V, a spot or such other mark as a purchaser might specify. Red, orange or yellow illumination of the aiming mark gives good visual contrast with the standard green night vision picture.

Zeroing of the sight to the weapon is easily performed, using simple and stable click-stop adjustments for elevation and azimuth. These adjust the internal position of the injected aiming mark.

The WINS 1801 may be fitted, by means of rigid adapters, to any standard military rifle such as the FN-FAL, M16, G3 or AK47. The sight may also be used as a hand-held viewer, and a pistol grip is available.

WINS 1801 rifle sight

DATA
Overall length: 204 mm
Max diameter: 63 mm
Weight (less batteries): 1.2 kg
Magnification: × 2.8
Field-of-view: 14.6° (260 mils)
Object lens: 70 mm f/1.4
Intensifier tube: Mullard XX1500 or enhanced micro-channel wafer
Gain: 30 000–50 000
Batteries: 2 × AA alkaline or NiCd, or 1 × TR132N mercury
Battery life: 40 h

Manufacturer
Solartron Defence Systems, Optronics Division, 580 Great Cambridge Road, Enfield, Middlesex EN1 3RX.
Status
Current. Production.

OE 8050 individual weapon sight

This new passive sight incorporates the latest in electro-optic technology and has been designed to meet the conflicting demands of the new generation of light support infantry weapons and medium-range surveillance and security tasks. The OE8050 accepts second and third generation image intensifiers and is compatible with most infantry weapons.

Manufacturer
Osprey Electronics Ltd, E27 Wellhead Industrial Centre, Dyce, Aberdeen AB2 OGD, Scotland.
Status
Current. Production.

Osprey OE8050 individual weapon sight

Type SS84 lightweight individual weapon sight

This lightweight enhanced second generation sight represents the latest advance in electro-optics technology from Pullin Controls. It has been designed for use with a wide variety of individual weapons from rifles to rocket launchers. Its modular construction provides ease of maintenance and the lens modules have been specifically designed to be compatible with third generation image intensifier tubes. The sight also has a colour-contrasted (red) graticule with brightness automatically controlled by scene luminance.

Type SS84 can be supplied without the graticule and with a pistol grip for general surveillance applications.

DATA
Length: 285 mm
Width: 90 mm
Height: 90 mm
Weight: 0.94 kg (including standard battery)
Magnification: × 3.8
Field-of-view: 176 mils (9.9°)
Graticule: illuminated, coloured red – automatically controlled by scene luminance
Graticule adjustment: ± 20 mils
Power supply: 2.7 V mercury battery as standard (options available include commercial AA size cells)
Image intensifier tube: 18 mm second generation micro-channel plate inverter

Manufacturer
GEC Sensors, Pullin Controls Division, Langston Road, Debden, Loughton, Essex IG10 3TW.

Type SS84 lightweight individual weapon sight

Status
Current. Production.
Service
Commercial and military sales to UK and overseas.

Type SS82 lightweight pocketscope/weapon sight

This compact, lightweight, enhanced second generation sight is based upon the modular SS80 design. It can be either hand-held for general surveillance and close-up tasks, or mounted on a wide variety of lightweight automatic weapons. With a weight of just 700 g it is the lightest as well as smallest sight in the Pullin Controls range of night vision devices.

DATA
Length: 230 mm
Width: 75 mm
Height: 70 mm
Weight: 700 g (including standard battery)
Magnification: × 2
Field-of-view: 354 mils (20.4°)
Graticule (if supplied): illuminated, colour red – automatically controlled by scene luminance
Graticule adjustment: ± 20 mils
Power supply: 2.7 V mercury battery as standard (options available include commercial AA size cells)
Image intensifier tube: 18 mm second generation micro-channel plate inverter

Manufacturer
GEC Sensors, Pullin Controls Division, Langston Road, Debden, Loughton, Essex IG10 3TW.
Status
Current.
Service
Commercial and military sales to UK and overseas.

Type SS82 lightweight pocketscope/weapon sight

Type SS20 Mark 2 individual passive weapon sight

This night sight has been in service with British forces since 1970. It was developed on a UK government contract starting in 1968.

The sight can be used for weapon aiming with any of a variety of individual weapons from rifles to rocket launchers and may also be used for surveillance. In its normal issue form it is contained in a light transit case complete with a universal adaptor, carrying bag, batteries, cleaning equipment and operating instructions.

DATA
Dimensions: 477.5 mm long × 110 mm wide × 185 mm high
Weight: 2.78 kg including battery
Magnification: × 3.75
Field-of-view: 180 mils
Graticule pattern: as required

Type SS20 Mark 2 individual passive weapon sight in use

Graticule adjustment: ± 24 mils in both elevation and azimuth in ¹/₂ mil steps
Operating range: up to 700 m in starlight for weapon aiming. 1 km for surveillance depending on light level and terrain
Power supply: 6.75 V mercury or rechargeable NiCd battery

Manufacturer
GEC Sensors, Pullin Controls Division, Langston Road, Debden, Loughton, Essex IG10 3TW.
Status
Current.
Service
British Army.

Type SS86 lightweight crew-served weapon sight

This lightweight, enhanced second-generation crew-served weapon sight is a further development based on the modular SS80 design. It can be fitted to anti-tank and infantry support weapons, and can also be mounted on a wide range of howitzers and field artillery for direct fire.

For convenience a right-angle eyepiece and graticules specific to customer requirements are available.

In a non-weapon role the sight can also be used as a medium/long-range night observation device.

DATA
Length: 365 mm
Width: 118 mm
Height: 118 mm
Weight: 2.30 kg including standard battery
Magnification: × 6
Field-of-view: 107 mils (6°)
Graticule: illuminated, colour red; automatically controlled by scene luminance
Graticule adjustment: ± 12.5 mils
Power supply: 2.7 V mercury battery standard. Options available including commercial AA cells
Image intensifier tube: 18 mm second generation micro-channel plate inverter

Manufacturer
GEC Sensors, Pullin Controls Division, Langston Road, Debden, Loughton, Essex IG10 3TW.

Type SS86 lightweight crew-served weapon sight

Status
Current. Production.

Ring Sight system

All Ring Sights are solid glass optical collimating sights. They are unit power (1 × magnification), allowing rapid target acquisition and an absence of scale effects.   The principle of operation is similar to that of some other collimating sights but the Ring Sight system differs from some in that the 'aiming' eye can see the target directly as well as the aiming mark which is projected at infinity into the field-of-view. For daylight firing the aiming mark is a circle which is generated, optically, using light reflected from the direction of the target and so adjusted that it always appears brighter than the target background. This ring marker is supplemented by an internal Tritium source. On some models this is a fixed 24-hour facility, while on others it merely requires the quick flip of a switch.

The fact that the firer, who can use one eye or both when aiming, can see the target as well as the marker with the aiming eye is claimed by the maker to have the advantage over other types of collimating sight, of avoiding the problems of eye wander and retinal rivalry. In addition the unit power enables the firer to improve his natural pointing ability.

The range of sights currently available is as follows:

Pistol sights
The LC-14-46 is a zeroable pistol sight that quickly slides on to any standard 20 mm dovetail. It has a cartwheel graticule giving quick acquisition, yet precise aiming at arm's length. Using simple mounts the sight can be fitted a

most makes of pistol, and special mounts can be made when requested. The sight has a flip-down microlight for low light and urban shooting.

Rifle sights
The LC-7-40 range of sights was originally designed for the AR15/M16 rifles. They weigh less than 4 g and fit snugly into the carrying handle with a single locknut. They are fully zeroable. There is a choice of graticules incorporating a ring for daytime use, with an integral microlight that lights up an 'Open T' at night time. There is no need to move any switch; depending upon the ambient light, one of the graticules will always be visible.

The LC-14-46 is a larger aperture sight for use on rifles unable to take rear-mounted sights because they incorporate sliding receiver covers, for example the AK series, FN-FAL etc. The sight is mounted on a mount fixed to the front gas barrel and, due to the larger aperture, the graticule can be seen very clearly by the firer.

The HC-10-62 was primarily designed for the new British SA80 (L85) rifle and is supplied in a detachable handle that slides on to the SA80 dovetail. The handle has an integral zeroing mechanism, and range setting and windage are optional. The optic is the first of a new range which allows both day and night graticules to be fully lit during the daytime, thereby allowing total flexibility under the most varied and adverse conditions. The graticules are illuminated not only by either the target light or the light from the sky, but by both at the

Ring Sight LC-7-40 on M16 rifle

Ring Sight LC-14-46-(05) on M60 machine gun

same time. The standard graticule has two daytime rings with four night-time radii. Alternative light sources can be supplied.

Riot gun and shotgun sights
The LC-14-46 is zeroable and is ideally suited for riot guns. Again, it is mounted to the weapon using a standard mount and has a flip-down light source for low light and urban conditions. The standard graticule is a cartwheel which allows the user to use the outer rings for both elevation and lead. For low velocity ammunition, a stadia reticle is recommended, giving a rapid rangefinding facility.

Machine pistol sights
The HC-18-80 was originally designed for the FN P90 machine pistol and has similar attributes to the HC-10-62 but with a larger aperture to allow for the increased eye relief. It is fully zeroable and has superb target acquisition.

Light and General Purpose machine gun sights
The LC-14-46 was designed for the M60 machine gun and is one of the few, if not the only, optical sight able to fit into the small space available between the interchangeable barrel and the feed cover. It is only 80 mm long and slides straight into the M60 dovetail. It can be supplied with a range of different interfaces for other weapons. It has a flip-down light source for low light capability and can be supplied with either a cartwheel graticule or an elliptical graticule for anti-helicopter firing. It is inexpensive and easy to install and use.

Heavy machine gun and Cannon sights
The LC-40-100 is the largest of the Ring Sight family and was designed as a backup sight for the BAe Air Defence Gunsight. It has subsequently been supplied in zeroable form for use on the .50 Browning M2HB machine gun

Ring Sight LC-14-46-(08) on Smith & Wesson revolver

where it has proved ideally suited for both ground and air defence roles. The large 40 mm aperture allows the firer great visibility and quicker acquisition of targets, and the specially designed graticule gives tangent elevation and lead marks. The flip-up light source instantly illuminates the centre two rings for low light, night-time, and flare situations.

The same sight is supplied with a zeroable or non-zeroable interface to fit the 20 mm and 30 mm cannon, both for Naval use and land vehicles and in the air defence role. Although the sight gives a phenomenal upgrade from the standard sight it is, nevertheless, inexpensive and easy to install.

DATA

	Sight	Aperture	Focal Length	Height	Width	Weight	Zeroable	Range Setting
LC-14-46 Pistol	14 mm	56 mm	80 mm	32 mm	27 mm	120 g	Yes	No
LC-14-46 Shotgun	14 mm	56 mm	80 mm	32 mm	27 mm	120 g	Yes	No
LC-14-46 M60 MG	14 mm	56 mm	80 mm	88 mm	62 mm	300 g	Yes	No
LC-7-40-M M16 M60 MG	7 mm	47 mm	63 mm	32 mm	14 mm	4 g	Yes	No
LC-7-40 Rifle	7 mm	47 mm	63 mm	32 mm	14 mm	4 g	Yes	No
HC-10-62 SA 80	10 mm	36 mm	65 mm	35 mm	16 mm	65 g	Yes	Optional
HC-18-80 FN P90	18 mm	46 mm	82 mm	50 mm	25 mm	190 g	Yes	No
LC-40-100 .50 M2HB	40 mm	132 mm	170 mm	120 mm	65 mm	1.7 kg	Yes	Reticle
LC-40-100 Cannon	40 mm	132 mm	130 mm	90 mm	55 mm	1.1 kg	Yes	Reticle

Manufacturer
Ring Sights Worldwide Ltd, PO Box 22, Bordon, Hants GU35 9PD.
Status
Current. Production.

Service
In use by military and police forces of several countries throughout the world.

Ring Sight RC35 for 20-40 mm guns

The unit is a Reflex Collimator Sight with no magnification, designed to provide surveillance, acquisition, target assessment and aiming. It is used with both eyes open. This enables the aimer to search for targets by moving his head – not the gun – then bringing the gun to bear without losing sight of the target. As the gun is aligned the sight graticule comes into view.

The graticule, bore-sighted to the gun, is projected to infinity and is always in focus. The standard graticule provided is for use by 20-30 mm guns in Low Level Air Defence and against surface targets. Alternative graticules for other calibres and applications are available. An emergency open sight is fitted on the left hand side.

A gaseous tritium light source, having an effective life of six years, is fitted for low light operation.

Manufacturer
BMARC, British Manufacture & Research Co Ltd, Springfield Road, Grantham, Lincolnshire NG31 7JB. (A subsidiary of Astra Holdings plc.).
Status
Current. Production.

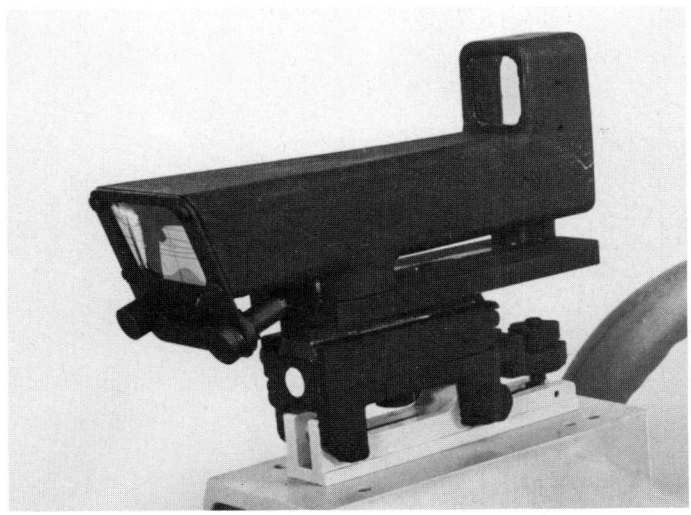

Ring Sight RC35 for 20 – 40 mm guns

Trilux night sights

Trilux night sights are simple modifications of rifle and machine gun iron sights and are designed to enable troops to fire accurately at dimly-distinguishable targets up to the maximum range of visibility. The principle involved is that of artificially enhancing the visibility of the foresight and enlarging the aperture of the backsight to allow for the expansion of the eye pupil in darkness.

The way in which the sighting principle is applied to a specific weapon depends on the design of the weapon's iron sights; and an indication of some of the different techniques is given by the accompanying illustrations. In all

cases, however, the enhancement of the visibility of the foresight is achieved by incorporating a Betalight self-powered light source in the night foresight. This source, which is visible only to the firer, shows him the position of his foresight and enables him to align it with a shadowy target. The light source has a maintenance-free useful life of 10 years.

Manufacturer
Saunders-Roe Developments Ltd, Millington Road, Hayes, Middlesex UB3 4NB.

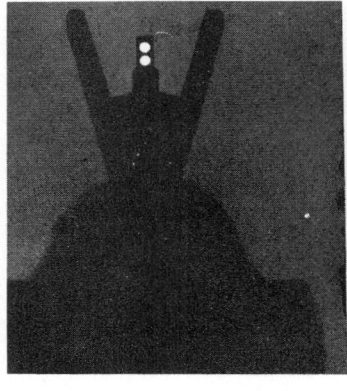

Appearance of Betalight foresight by night and day

Night sights for 7.62 mm FN FAL rifle

Night sights for 7.62 mm L1A1 rifle

Night sights for 5.56 mm AR-15 rifle

Status

Weapons for which sights are available include the FN FAL, the FN SLR L1A1, the G3 and the M16 rifles and the GPMG. The sights are currently in service with UK and NATO armies as well as commercial sales to armies throughout the world.

LS45 laser aiming system

The LS45 aiming system provides a lethal point-and-shoot capability to standard issue weapons. The target is acquired using an intense red dot of laser light and provides pinpoint accuracy, especially useful in security situations where the public are at risk. It is possible to aim from the hip, thus not alerting a target by assuming the conventional at-the-shoulder firing position.

A universal mounting clamp makes the LS45 equally suitable for rifles, light machine guns, pistols, shotguns and crossbows which are capable of being fitted with standard dovetail mounts. Elevation and windange adjustments allow zeroing to individual weapons, and a notch and blade are incorporated in the upper sight casing, for use in emergencies.

The system is capable of operating at ranges up to 500 m.

LS45 laser aiming system

DATA
Length: 180 mm
Width: 35 mm
Height: 67 mm
Weight: 340 g
Laser output: < 1 MW
Wavelength: 632.8 nm
Beam diameter: 0.34 mm
Dot size: 24 cm at 100 m
Divergence: 2.4 mrad
Power supply: 2×9 V alkaline cells
Battery life: in excess of 1500 3 s bursts

Manufacturer

Imatronic Ltd., Dorian House, 12A Rose Street, Wokingham, Berks RG11 1XU.
Distributors: United Scientific Instruments Ltd, United Scientific House, 215 Vauxhall Bridge Road, London SW1V 1EN: world-wide except US and Canada.
Status
Current. Production.

Model 1500 night sight

Model 1500 is a lightweight telescope weapon sight for use with small arms over short and intermediate ranges. The image intensifier used is a second generation pattern specially designed for this application, and has particularly good resolution, variable gain, good low-light performance and long operational life. It will give good vision in spite of highlights – headlights, street lights, pyrotechnic flares – in the field-of-view, and it is protected against ill effects from short term high illumination overload. The aiming graticule appears red against the green field of vision, making target acquisition and aiming fast and accurate. Bore-sighting is easily done by adjusting the position of the graticule by means of two adjuster screws.

DATA
Length: 265 mm
Max diameter: 76 mm
Main body diameter: 59 mm
Weight: 1.0 kg
Magnification: $\times 3$
Gain: $\times 45\,000$ nominal
Field-of-view: 10°
Object lens: 100 mm f/2.0
Power supply: $2 \times$ AA cells MN1500 or equivalent
Battery life: ca 80 h
Weapon mount: to requirement

Phychem Model 1500 night sight

P 840 infra-red pointer

The P 840 is a lightweight infra-red pointer for use in conjunction with night vision devices using image intensification.

The small diameter and light weight mean the pointer may be hand-held or fitted in place of a telescope sight. It is designed to suit imperial or metric small calibre weapon sight mountings. Pitch and azimuth adjustment knobs provide zeroing adjustment.

The P 840 operates from commercially available batteries which can be changed without removing the pointer from the weapon.

DATA
Dimensions: 248 × 32 × 46 mm
Weight: 290 g
Light output: Near-IR, wavelength 820 nm in the form of a collimated beam 5.4 m diameter, divergence L 0.3 mRad
Beam power: Approx 2 mW maximum

Manufacturer
Phychem Ltd, Turnpike Road Industrial Estate, Newbury, Berks RG13 2NS.
Status
Current. Available.

UNITED STATES OF AMERICA

Multi-purpose Universal Gunner's Sight (MUGS)

MUGS is a compact, lightweight ballistic fire control system incorporating an eye-safe laser and a sensor package which compensates for non-standard environmental effects upon munitions. Sensor data is matched with the ballistic data of the particular munition to produce an adjusted aimpoint. The system can store ballistic data for up to 10 weapons or munitions, which can be selectively recalled by the operator.

When operated, MUGS will determine the range to the selected target, ambient temperature, cross-wind at the gun bore line, barometric pressure, weapon cant and inclination. These factors are integrated with the ballistic data called from memory and the computer then calculates the quadrant elevation and lead angle to provide a new aimpoint. This enables the gunner to place accurate and effective fire on the target with the first shot.

The MUGS fire control system is particularly useful in extending the life of existing weapon systems. Testing conducted by McDonnell Douglas with

MUGS on the 106 mm recoilless rifle demonstrated that MUGS doubled the effective range of the weapon from 960 m to 2000 m. In addition to increasing effective range, MUGS also provides a cost saving by reducing the amount of ammunition necessary to complete a mission.

Six pre-production models were delivered to the US Marine Corps in October 1988 for evaluation on the 40 mm Mark 19 machine gun and on the .50 M2HB Browning machine gun. Several foreign governments have expressed interest for application on the 106 mm recoilless rifle.

DATA
Weight: 3.6 kg
Mounting weight: 450 g
Range of rangefinder: 90 – 4000 m
Range resolution: ± 5 m

MUGS sight mounted on 106 mm recoilless rifle

McDonnel Douglas MUGS sight unit

Beamwidth: 0.6 mrad
Wavelength: 1.5 microns
Magnification: × 8 — × 12
Field-of-view: 8° – 5°
Readout: 4 digit LCD
Power supply: lithium batteries
Battery life: 200 stationary, 100 moving target operations

Manufacturer
McDonnell Douglas Systems Co, Combat Systems Division, PO Box 516, St Louis, MO 63166.
Status
Advanced development.
Service
Evaluation with US Marine Corps.

Hughes Thermal Weapon Sight (TWS)

The TWS is a versatile manportable infra-red imaging sensor under development for the US Army Night Vision and Electro-Optics Laboratory, Fort Belvoir, Virginia. It is designed for the acquisition and sighting of targets by crew-served weapons during darkness or daylight and under adverse battlefield conditions.

In its various configurations the TWS is capable of recognising targets such

Hughes TWS weapon sight on M60 machine gun

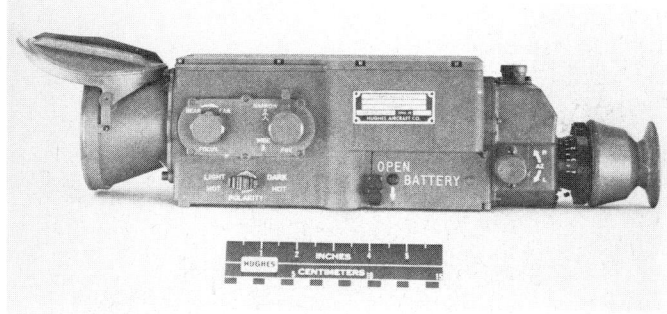

Hughes TWS system configured as a weapon sight

as groups of men, vehicles, tanks or aircraft at the necessary tactical ranges. In addition, the modular design allows for its application to other electro-optical sensor missions, such as perimeter surveillance, RPV payloads and light armoured vehicle control systems.

Since no single sensor configuration is capable of satisfying all the hardware's intended mission requirements, the TWS design uses modular telescope and eyepiece assemblies, each pair of which is tailored to a specific application, yet interfaces into a common sensor body.

No dimensions are given, since these depend upon the final development and the particular configuration.

Manufacturer
Hughes Aircraft Co, Electro-Optical and Data Systems Group, PO Box 902, Building E1/M.S. E104. El Segundo, CA 90245.
Status
Development.

Magnavox Short-Range Thermal Sight (SRTS)

The SRTS is a fully passive infra-red imaging rifle sight, designed for use on the M16A1/A2 rifles and the M203 grenade launcher. With suitable adapters, the SRTS is easily added to a variety of other weapons.

Operating passively in the infra-red spectrum, the sight provides the infantryman with the capability of target detection and firing in total darkness. User controls are minimised and are designed for ease of operation. Power is supplied from a standard Army throwaway lithium battery, which is the only consumable item required for system operation. Extensive use of modular components simplifies serviceability and maintainability.

Infra-red energy from the target scene being viewed is collected by a single element silicon aspheric lens. The emerging convergent beam is horizontally scanned by an oscillating mirror and then focussed on to a vertical linear array of 64 lead selenide detector elements which transduce the IR energy into electrical signals. Each detector's output is fed to a high gain pre-amplifier. The signals from the 64 pre-amplifiers are then multiplexed to a single composite video signal. The multiplexer is enabled by signals from the logic circuit which also generates horizontal and vertical retrace blanking and deflection signals for graticule generation. The composite video signal is then amplified and applied to a miniature cathode ray tube that is viewed through the monocular eyepiece.

DATA
Field-of-view: 6° horizontal, 4° vertical
Spectral band: 3.7 to 5 micrometers
Dimensions: 300 × 86 × 102 mm
Weight, with battery: < 1.8 kg
Power supply: disposable lithium cell
Battery life: 10 h

Manufacturer
Magnavox Electro-Optical Systems, 46 Industrial Avenue, Mahwah, NJ 07430-0615.
Status
Advanced development completed; US Army proceeding to Full-Scale Engineering Development.

Magnavox Short-Range Thermal Sight

Model 845 Mark II lightweight second generation sight

The Model 845 Mk II has been specifically designed to overcome the main objection to image intensifier sights, namely the size and weight which both combine to make it difficult for the firer to observe for more than a short period of time, or to shoot steadily without supporting his weapon.

The Model 845 Mk II is easily mounted on a wide variety of small arms and can double as a hand-held night viewer if necessary. It uses the same micro-channel plate as the Litton pocketscope and goggles (which see) so that it takes advantage of high volume manufacturing of the most expensive component. For higher performance under extremely low light level conditions, the M 845 Mk II can be factory-upgraded with a third generation intensifier tube.

Ease of use is enhanced by the fact that the eye relief is 51 mm which allows the firer to wear glasses, an unusual aspect in a night viewer. Unique features are the 'Redot' method of aiming, a LED battery condition indicator and a warning light on the edge of the field-of-view which warns when the battery has only one hour of life left. The actual battery is a commercial type that is available world-wide. The weapon sight incorporates a water-resistant flip-up lens cover with an internal filter for training or bore-sighting during daylight hours.

The designers have deliberately produced a low-cost system which is easy to use. To achieve this they have concentrated on a sight for relatively short-range night engagements, such as patrol actions and infantry defensive positions in average country, street fighting and general law enforcement operations.

DATA
Dimensions: (including eyeguard) 260 × 70 mm
Weight: (including battery) 1300 g
Height above mounting surface: 90 mm
Magnification: × 1.5
Field-of-view: 13.5°
Resolution: 1.5 lp/mR
Gain: 1000 × (nominal)
Tube: 18 mm channel plate, second generation
Battery: 2 × 1.5 V AA alkaline
Battery life: approx 40 h
Aiming mark: red LED spot
Environmental: Mil Spec

Manufacturer
Litton Systems Inc., Electron Devices Division, 1215 South 52nd Street, Tempe, AZ 85281-6987.

Litton M845 Mark II weapon sight

Status
Current. Available.
Service
Military and paramilitary use.

M921 Submersible night vision sight

The Litton M921 is a submersible second generation night sight for use on weapons from 5.56 mm rifles to light machine guns or for general surveillance. The submersible design provides watertight protection to a depth of 50 m for missions requiring underwater transport. It offers excellent resistance to salt-water corrosion, since there are no exposed threads and all external surfaces are hard Teflon coated. For increased sensitivity under extremely low light conditions the M921 can be factory upgraded with a third generation intensifier tube.

An open-cross graticule provides a highly effective sight picture, and there are click adjustments for azimuth and elevation.

DATA
Dimensions: (including eyeguard) 191 × 85 mm
Weight: (including battery) 2.10 kg
Height above mounting surface: 109 mm
Magnification: × 3
Field-of-view: 13°
Resolution: 1.8 lp/mR
Tube: 18 mm channel plate, second generation
Battery: BA 1567/U (mercury) or BA5567/U (lithium)
Battery life: approx 12 h

Manufacturer
Litton Systems Inc., Electron Devices Division, 1215 South 52nd Street, Tempe, AZ 85281-6987.
Status
Current. Available.
Service
Military and paramilitary use.

Litton M921 submersible weapon sight

M937/M938 individual weapon sights

The Litton Models 937 and 938 are compact, lightweight, battery-powered night vision sights for use on weapons from 5.56 mm assault rifles to light machine guns. The × 4 telescope provides a high resolution intensified image for effective aiming at medium- and long-range targets. The two models differ only in the type of image intensifying tube; the M937 uses a second generation tube, while the M938 uses a third generation tube for greater sensitivity and increased resolution under conditions of extremely low light.

A variable intensity amber coloured graticule, that can be superimposed on the green intensified image, provides the marksman with rapid and highly accurate aiming capability. There are precision click adjustments for both azimuth and elevation. Both models include a splash-proof lens cover which has a central pinhole for daylight training and bore-sighting.

DATA
Dimensions: 255 × 85 mm
Weight: (including battery) 1.10 kg
Height above mounting surface: 80 mm
Magnification: × 4
Field-of-view: 8.5°
Resolution: 2.8 lp/mR (2nd Gen); 3.2 lp/mR (3rd Gen)
Battery: 2 × 1.5 V AA alkaline
Battery life: 50 – 60 h

Manufacturer
Litton Systems Inc., Electron Devices Division, 1215 South 52nd Street, Tempe, AZ 85281-6987.
Status
Current. Available.
Service
Military and paramilitary use.

Litton M937 individual weapon sight

Litton AIM-1D/DLR laser aiming light

The AIM-1D/DLR laser aiming light allows a person wearing night vision goggles to aim a weapon. It will fit any weapon from handguns to crew-served weapons, rifles, shotguns and machine guns. It enables the use of night vision goggles to quickly aim and fire the weapon effectively. Moreover the added illumination provided by the laser beam will increase the effective viewing range of night vision goggles by about 20 per cent.

The AIM1D is constructed so that it can be operated at variable light intensities, either locally or by remote-control cable. Its electronics are a cast

Litton AIM-1D in use on M16 rifle

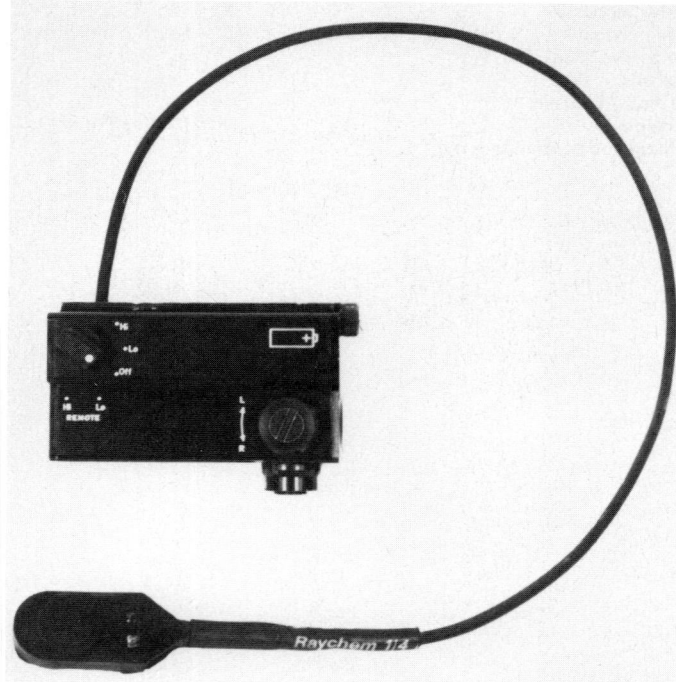

modular package, conforming to MILSPECS, with an option of providing underwater capability.

DATA
Beam divergence: 0.3 mrad
Average output power: 20 mW in high mode
Peak wavelength: 820-850 nm
Weight: 255 g
Dimensions: 92 × 57 × 49 mm
Power supply: 2 × 1.5 V DC AA batteries or 28 V DC
Battery life: 40 hours AIM-1D; 20 hours AIM-1D/DLR
Range: 500 m AIM-1D; 3 km AIM-1D/DLR

Manufacturer
Litton Systems Inc., Electron Devices Division, 1215 South 52nd Street, Tempe, AZ 85281-6987.
Status
Current. Available.
Service
Military and paramilitary use.

Litton AIM-1D/DLR laser aiming light

NVS-700 night vision system

This second generation individual weapon sight is now extensively used by military and police authorities in many countries. Both it and the NVS-800 crew-served weapon sight (which see) use the same 25 mm micro-channel plate image intensifier tube and have almost all parts in common except the objective lens and associated fittings. The smaller NVS-700 is suitable for mounting on the 5.56 mm M16 rifle and similar weapons.

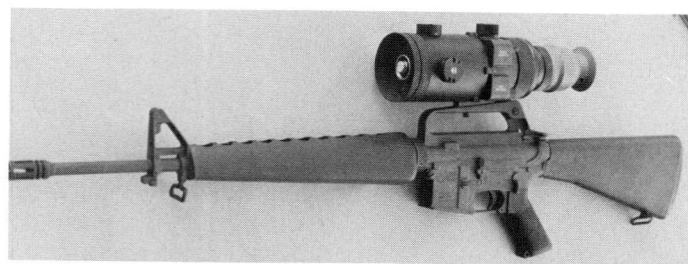

NVS-700 sight on M16 rifle

DATA
Magnification: × 3.5 (nominal)
Field-of-view: 253 mils
Viewing range: (moonlight) 25–700 m; (starlight) 25–450 m
Objective lens: (focal length) 95 mm T/1.7; (focus adjustment) 25 m to infinity
Eyepiece: (focal length) 26.5 mm; (focus adjustment) +3 to -6 dioptres
Length: 292 mm nominal
Diameter: 101.6 mm nominal
Weight: 1.814 kg
Operating temperature: (with arctic kit) -54 to +52°C
Power supply: 2 × AA mercury cells
Battery life: 60 h

Manufacturer
Optic-Electronic Corp, 11545 Pagemills Road, Dallas, TX 75243.
Status
Current. Production.
Service
Widespread military sales to foreign governments.

NVS-800 night vision system

Designed for use on heavy automatic weapons, recoilless guns and other armament of similar size, the NVS-800 sight closely resembles the NVS-700 (small starlight scope) individual weapon sight (see entry above) but has a greater range capability resulting from the use of a larger objective lens. The eyepiece, 25 mm second generation image intensifier and associated components are common to both sights and the data given below relates only to the differences between the two.

DATA
Magnification: × 6 (nominal)
Field-of-view: 156 mils
Viewing range: (moonlight) 25–2000 m; (starlight) 25–1200 m
Objective: 155 mm T/1.7
Length: 355.6 mm
Diameter: 165 mm
Weight: 3.856 kg
Power supply: 2 AA mercury cells
Battery life: 60 h

Manufacturer
Optic-Electronic Corp, 11545 Pagemills Road, Dallas, TX 75243.
Status
Current. Production.

NVS-800 night vision system

Service
US and foreign armies.

AN/PAQ-4 laser aiming light

The AN/PAQ-4 infra-red laser aiming light provides rapid night-time target acquisition and positive aiming for anti-terrorist squads, police, commando and jungle warfare units.

The light is robust and lightweight, battery powered, and can be mounted and bore-sighted to virtually all individual weapons including 5.56 mm and 7.62 mm rifles, light machine guns, recoilless guns and rocket launchers. It utilises hybrid electronics for controlled output and a Gallium Aluminium Arsenide laser diode tuned to an IR frequency above human eye level. The system is completely eye-safe, to allow for its use in training without the need for attenuation filters. The non-visible pulsed laser beam can only be seen through night vision goggles or other infra-red sensitive instruments.

AN/PAQ-4 laser aiming light

DATA
Dimensions: 160 × 42.5 × 53 mm
Weight: 300 g
Beam: circular, 2 mR diameter
Output wavelength: 850 mm
Spectral bandwidth: 8 nm
Power supply: 2 × BA 1567/U
Battery life: 100 h

Manufacturer
Optic-Electronic Corp, 11545 Pagemill Road, Dallas, TX 75243.

Status
Current. Available.

ITT F4961 day/night combat rifle sight

This is a compact, inexpensive, lightweight high-performance weapon sight which is capable of being used at any time of day or night. It incorporates proven third-generation technology and has been designed to make the maximum use of components common to other night vision devices and to permit low-profile attachment to the rifle.

During daylight an adjustable intensity red graticule is projected to the user as a reflex sight. At night the same graticule is projected on to the green image of the image-intensifier tube. During twilight or temporary brightness due to flares or explosions, both day and night scenes are visible to ensure the availability of one or other useful image.

The illuminated graticule is available either as a ballistic reticle pattern or as a simple aiming-point. Bore-sighting is carried out once, for both day and night alignment, and the necessary adjustment can be carried out in either day

or night mode by elevation and azimuth adjusters conveniently located on the top and left side of the sight. Operation of these adjustments is identical to that used by the widely-employed AN/PVS-4 weapon sight, ensuring troop familiarity.

The third-generation tube gives greatly improved resolution and system gain, and has a life almost four times longer than a second-generation device. It incorporates automatic brightness control in order to provide constant image brightness under varying light conditions, and bright-source protection to protect the tube during exposure to high light levels.

The sight has a built-in NATO STANAG mount base which can be attached directly to various weapons. Specific mounting brackets are also available for most current assault rifles.

ITT F4961 day/night rifle sight

DATA
Magnification: 1 ×
Field-of-view: 40°
Resolution: better than 30 lp/mm
Luminous gain: 20 000 to 35 000
Dimensions: 250 × 85 × 56 mm
Weight: 570 g
Power supply: 3.0 vDC; 2 × 1.5 v AA alkaline cells
battery life: 20 hours average

Manufacturer
ITT Defense, Electro-Optical Products Division, 7635 Plantation Road, Roanoke, VA 24019.
Status
Current; production.

AN/PVS-4 second generation weapon sight

Performing the same range of functions as the earlier first generation AN/PVS-2 Starlight scope but with superior characteristics, the AN/PVS-4 is a light, passive night vision sight using a 25 mm micro-channel plate inverter intensifier tube. The sight is easily attached to a number of different weapons or may be hand-held for night reconnaissance.

An adjustable internally projected reticle and interchangeable reticle pattern allows the sight to be bore-sighted to the various weapons without having to move the sight.

Image tube gain and reticle brightness are manually adjustable to compensate for different levels of ambient lighting.

Automatic gain control circuitry is employed to maintain the viewed scene illumination constant during periods of changing light level conditions, such

as the period from sunset to full darkness. This allows the operator of the sight to use the sight without having to readjust the tube gain control every few minutes during this period.

The tube features muzzle-flash protection which prevents the tube from being damaged by high intensity short duration flashes of light. The flash protection circuit is designed to recover in time for the observer to see the round hit the target.

DATA
Dimensions: 240 × 120 × 120 mm
Weight: 1.5 kg
Magnification: × 3.7
Field-of-view: 14° 30′
Focus range: 25 m to infinity
Objective focal length: 95 mm
Eyepiece focal length: 25 mm
Eye relief: 34 mm
Dioptre range: ±4
Power: 2 × 'AA' batteries
Battery life: 30 h

AN/PVS-4 second-generation weapon sight

Manufacturer
Varo Inc, Electron Devices, PO Box 469014, 2203 West Walnut Street, Garland, TX 75046. (An IMO Industries company.).
Status
Current. Production.
Service
US Army and Marine Corps. Numerous other armies all over the world.

AN/TVS-5 second generation crew-served weapon sight

This is a light passive night vision sight which uses 25 mm micro-channel plate inverter intensifier tube. It can be used with a range of different crew-served weapons and can also be tripod-mounted for night reconnaissance.

An adjustable internally projected reticle and interchangeable reticle patterns allow the sight to be bore-sighted to the various weapons without having to move the sight.

Image tube gain and reticle brightness are manually adjustable to compensate for different levels of ambient lighting.

AN/TVS-5 second generation weapon sight

Automatic gain control circuitry is employed to maintain the viewed scene illumination constant during periods of changing light level conditions, such as the period from sunset to full darkness. This allows the operator to use the sight without having to readjust the tube gain control every few minutes during this period.

The tube features muzzle-flash protection which prevents the tube from being damaged by high intensity short duration flashes of light. The flash protection circuit is designed to recover in time for the observer to see the round hit the target.

DATA
Dimensions: 310 × 160 × 170 mm
Weight: 3 kg
Magnification: × 6.2
Field-of-view: 9°
Focus range: 25 m to infinity
Dioptre range: ±4
Power supply: 2 × AA batteries. Battery life 30 h

Manufacturer
Varo Inc, Electron Devices, PO Box 469014, 2203 West Walnut Street, Garland, TX 75046. (An IMO Industries company.).
Status
Current. Production.
Service
US Army and Marine Corps. Also used by 20 other countries.

Aquila ™ Mini weapon sight

The new Aquila mini weapon sight is a 4 × passive individual weapon sight for infantry and special forces. The Model 2500 utilises the 18 mm MCP 2nd Generation Plus or Super Gen tube, while the Model 3000 features the 3rd Generation tube. The tubes are optically interchangeable.

This 1.1 kg sight operates on standard AA batteries and incorporates the basic on/off/graticule brightness, azimuth and elevation controls, in addition to the range and eyepiece dioptre adjustments.

DATA
Weight: 1.1 kg
Magnification: x4
Field-of-view: 8.3° (146 mils)
Objective focal length: 120 mm

Eyepiece focus: +2 to -5 dioptres
Graticule (red LED): 0.2 mil increments
Focus range: 25 m to infinity
Operating temperature: -54°C to +52°C

Manufacturer
Varo Inc, Electron Devices, PO Box 469014, 2203 West Walnut Street, Garland, TX 75046. (An IMO Industries company.).
Status
Current. Available.
Service
US Government agency.

Varo Aquila mini weapon sight

Varo Model 9886A infra-red aiming light

The Infra-red Aiming Light (IAL) is used in conjunction with night vision goggles to provide an aiming point that will allow the user to deliver accurate fire at night. The IAL consists of an aiming light assembly and a carrying bag. Mounting brackets are available for adapting to most weapons currently in use and can be supplied with the aiming light.

A borelight assembly with either a 5.56 mm or 7.62 mm mandrel can be supplied as an accessory to the IAL. The borelight assembly is a small lightweight infra-red source that is used to provide a fast and accurate means of boresighting the IAL to the bore of any weapon.

The IAL incorporates a mounting base that is compatible with the standard AN/PVS-4 type adapter brackets. Additional mounting brackets are available which will permit the IAL to be mounted to either the NATO STANAG base or to a Weaver type rail.

Controls and adjustments include a remote on/off switch and boresight adjustments. The boresight adjustment includes azimuth and elevation knobs indexed at 0.5 mil increments.

Varo Model 9886A infra-red aiming light

DATA
Optical:
Output power: 3.2 mW
Output peak wavelength: 820 nm
Output beam size: 0.2 mrad (max)
Spectral bandwidth: 12 nm
Range: > 400 m
Input DC current: 95 mA
Boresight adjustment: 0.5 mil clicks

Mechanical:
Weight: 345 g with batteries
Length: 208 mm
Width: 52 mm
Height: 75 mm

Electrical:
Power supply: 4 × AA batteries or 2 × BA5567/U lithium

Environmental:
Operating temperature: -54°C to +51°C
Storage temperature: -57°C to +65°C

Manufacturer
Varo Inc, Electron Devices, PO Box 469014, 2203 West Walnut Street, Garland, TX 75046. (An IMO Industries company.).
Status
Current. Available.

Star-Tron night vision equipment

The Star-Tron range of night vision equipment was originally marketed by the Smith & Wesson company (see *Jane's Infantry Weapons 1985/86*, pages 761 and 790). With the retirement of that company from this field, the Star-Tron division separated to become an independent firm and is once more offering the full range of night vision equipment as follows:

MK-101B Passive night vision system
MK-202B Passive night vision system
MK-303A Passive night vision system
MK-424 Series 1 Passive night vision system
MK-424 Series 3 and 4 Passive weapon sights
MK-426 Universal night vision system
MK-606A Series 3 Passive night vision system
MK-606A Series 4 Passive night vision system
MK-700 Passive Night Vision Riflescope
MK-850 Wide-Field Pocketscope.

Star-Tron has recently introduced its 'Star Glass' concept, a new pocketscope which enhances the night vision capability of the human eye by minimising the negative effect on the eye's dark adaption which is caused by the unnecessarily bright output screen of conventional night vision tubes. This has been achieved by developing an ultra-high resolution amplifier tube capable of resolving 100 line pairs/mm. This also permits the use of a lower system gain in order to effect the desired level of visual information.

Manufacturer
Star-Tron Technology Corp, PO Box 11526, 900 Freeport Road, Pittsburgh, PA 15238.
Status
Current. Production.
Service
US and foreign armies and police forces.

Personal Protection

BELGIUM

Browning bullet-proof vests

Browning supplies two types of bullet-proof garment. The first uses steel plate, and the second uses Kevlar woven armour protection.

Steel plate vests are protected by plates of hardened armour steel, tough enough to resist attack by armour-piercing bullets. Indeed, the company is so confident of its vest's ability to protect that it offers an insurance policy of BFr 1 million for the first year of purchase.

The vests are backed by a shock-absorbent Kevlar felt which reduces the trauma from bullet attack. The steel plates are contained in pockets, so arranged that their edges overlap but do not prevent movement of the body. The plates can be easily removed and when the jacket needs to be cleaned all the steel can be removed without difficulty. Browning has a patented system of jointing plates which are bent into a shallow 'vee' to cover each joint. This, it is claimed, almost entirely prevents the possibility of a bullet which is fired from the side forcing the plates apart and entering the body.

These vests are primarily intended for use by police and others exposed to short-range attack with little warning.

DATA
The protection provided is listed below:-

Types available	Resistance to bullets	Distance
16 layers Kevlar 29	.22 .25 auto/6.35 .32 auto/7.65 .38 special Cal 12	5 m
24 layers Kevlar 29	22 LR carbine .45 auto .38 special high velo 9 mm Para GP .357 Magnum	5 m
24 layers Kevlar 29 + 3 mm steel	9 mm Para GP PM 9 mm Superpenetrating GP .357 Magnum .44 Magnum Cal .12 Brenneke	5 m
24 layers Kevlar 29 + 5 mm steel	7.62 × 39 Sako 7.62 × 39 Kalashnikov	15 m
24 layers Kevlar 29 + 6.5 mm steel	7.62 × 51 SS77 FAL 5.56 × 45 SS109 FNC 7 in 5.56 × 45 SS92 FNC 12 in 5.56 × 45 P96 FNC 12 in	15 m
24 layers Kevlar 29 + ceramic + shock absorber	7.62 × 51 P80 FAL	15 m

VEST TYPES
P300: full protection to the waist and capable of being worn under a shirt
Weight: (3 mm) 5.2 kg; (5 mm) 6.8 kg; (6.5 mm) 10.9 kg

P200: larger than the previous vest, it protects the entire front of the trunk and the lower part of the back
Weight: (3 mm) 10 kg; (5 mm) 13.2 kg; (6.5 mm) 21.2 kg

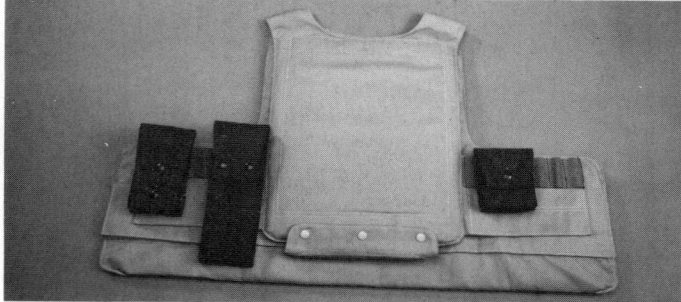

Ammunition pockets attached to P-304 jacket

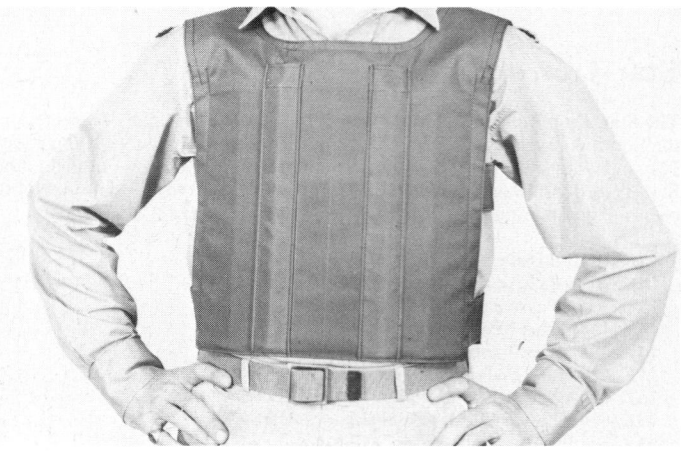

Type P300 vest. Two long pockets contain special deflecting joint plates

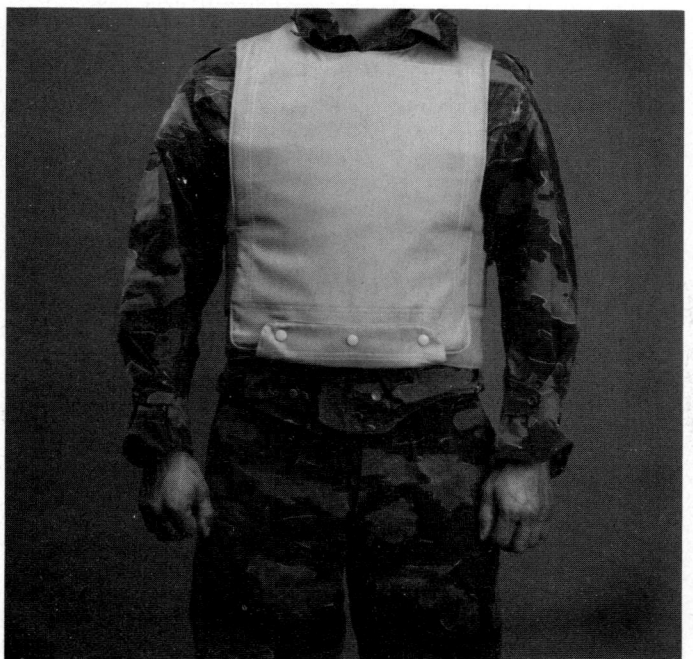

P-304 jacket

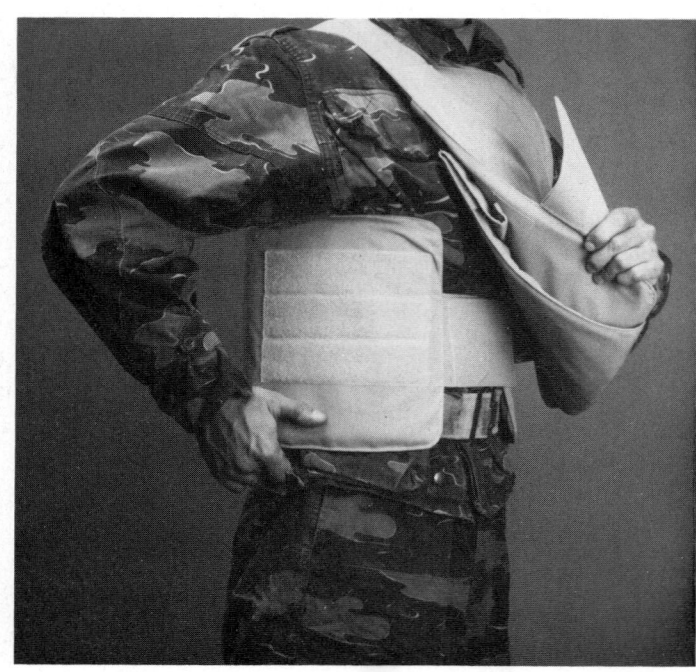

Showing Velcro fastening of P-304 jacket

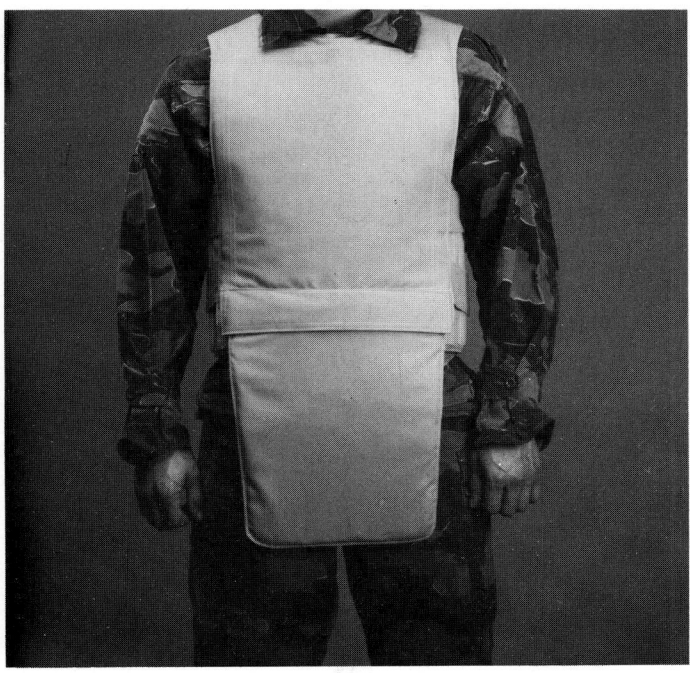

P-304 jacket with groin protection

KEVLAR LIGHT ARMOUR VEST

Browning also manufactures a 'Kevlar 29' light armour vest. The P304 jacket has been designed for maximum comfort and manoeuvrability. The Kevlar armour is sealed into a black polyethylene bag to avoid humidity and ultra-violet rays. At 5 m range the vest will stop handgun and sub-machine gun bullets up to 9 mm Parabellum, .44 Magnum and .45 ACP level.

The P304 is also available with steel plates of 3, 5 or 6.5 mm thickness. The degree of protection provided by this addition is as follows:

P304 + one 3 mm plate: will resist a burst of sub-machine gun 9 mm Parabellum

P304 + one 5 mm plate: will resist 7.62 × 39 mm ammunition

P304 + one 6.5 mm plate: will resist 7.62 × 51 mm ball

CERAMIC PLATE PROTECTION

The ballistic level of protection can also be increased by inserting ceramic plates in pockets in both the front and back of the jacket. If required, a trauma attenuation pack can be supplied to fit inside the jacket.

All fastenings are by Velcro for quick release. Removable or attached pockets for ammunition or other items are available.

The jacket is available in white, khaki, blue, sand or black colouring and in 24-ply Kevlar it weighs 4 kg.

All Browning vests are approved by the Belgian Proof House.

Manufacturer
Browning SA, B-4400 Herstal.
Status
Current. Available.
Service
Police departments in more than 40 countries.

FRANCE

SEMA personal armour

SEMA manufactures a complete range of protective armour devices for various applications, including bullet-proof glass for vehicles, bomb blankets, helmets and body armour.

Two types of body armour are made: the 'basic' model is a vest-type which may be worn over or under normal clothing; and the 'military' pattern which is an over-jacket fitted with ammunition pouches and pockets. There are varying degrees of protection available, based on the following ballistic classification:

Class A: 6.35 mm, 7.65 mm and .38 Special; Class 1: .38 Special High Velocity, .45 ACP, .357 Magnum semi-jacketed, 12-gauge buckshot; Class 2: as above, plus .357 Magnum Soft Point and 9 mm Parabellum British Mk 2Z; Class 3: 12-gauge Brenneke slug, .357 Magnum Speer HP, 9 mm Parabellum, Gévelot and FN, US M1 carbine soft point, Class 4: 9 mm Parabellum Geco 124 grains, .44 Magnum 240 grains; Class B: 5.56 mm, 7.62 mm × 39, 7.62 mm NATO AP and ball.

For Class A and Classes 1-4 reliance is entirely upon woven Kevlar fabric in various thicknesses; for Class B the armour is a combination of steel, Kevlar and ceramic plates. In general, the 'basic' models will provide protection to Class 2, 3 or 4 levels as desired; the 'military' jacket will provide protection to Class 4 as standard and may be uparmoured to Class B.

Manufacturer
Société d'Equipement Militaire et Administratif (SEMA), 9 rue de Lens, 92000 Nanterre.
Status
Current. Production.
Service
French police and Gendarmerie, and foreign armies and police forces.

SEMA military jacket

Herakles International body armour

Herakles International manufactures police, military and undercover body armour in various styles. The level of protection is basically against all hand gun and sub-machine gun bullets up to .44 Magnum power, with the added facility, in the military armour, of ceramic plates which will upgrade the protection to 7.62 and 5.56 mm NATO ball.

Military Jacket SM22: this is a sleeveless vest with high collar and groin protection. It can accept additional ceramic plates for high velocity protection and contains anti-trauma plates as standard. The cloth covering to the armour is resistant to fire, acid and seawater, and a special aeration system allows the body to breath and remain cool during long periods of wear.

Manufacturer
Herakles International, 20 rue Mercoeur, 44000 Nantes.
Status
Current. No longer in production.

Herakles military jacket SM22

SNPE body armour

SNPE manufactures a wide range of bullet-proof vests for use by military and police forces and also for civilian wear. Protection is provided by Kevlar cloth covered in special resin by a patented process. This Kevlar element is sealed into a polyester covering which protects against water and ultra-violet rays. The outer cloth of the garment is Kermel satin, coloured blue (other colours may be provided on demand), and the garment is closed and adjusted by a Velcro fastening. To meet higher threat levels reinforcing plates of ceramic material can be inserted into pockets in the front of the garment and secured by Velcro closures.

Manufacturer
SNPE, 12 Quai Henri IV, 75181 Paris-Cedex 04.
Status
Current. Production.

SNPE vest Type AF1

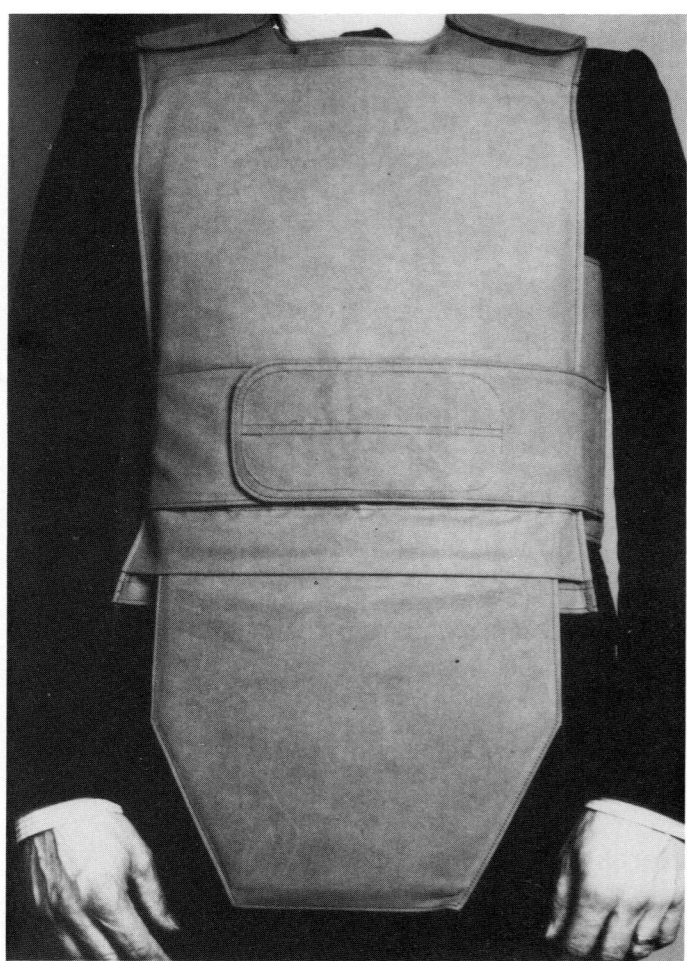

SNPE vest Type ST1

GERMANY, WEST

Mehler armoured jackets

Mehler Vario Systems GmbH, established in 1986, is a subsidiary of Val Mehler AG, one of the leading industrial textile companies. The annual turnover of Mehler Vario Systems is approximately DM 35 million. The company has been working with Kevlar for various applications since 1973 and has developed a range of armoured jackets for police, military and civilian applications. The stringent ballistic tests laid down by the West German Police Academy on behalf of all police forces in West Germany have been passed.

The following are the levels of protection commonly accepted:

Protection against fragments, (according to Mil Spec) soft core ammunition discharged from handguns, velocity up to 300 m/s;

West German Police Standard Class 1 – 9 mm full metal jacketed bullets fired from a sub-machine gun;

West German Police Standard Class 2 – 9 mm steel core and .357 Magnum KTW;

West German Police Standard Class 3 – 5.56 × 45 and 7.62 × 51 ball ammunition;

West German Police Standard Class 4 – 5.56 × 45 and 7.62 × 51 hard-core ammunition.

For protection classes 2, 3 and 4 special steel or ceramic plates are available, to be inserted into pockets in the garments. Also a trauma pack has been designed to minimise any shock to the human body.

The selection of the correct jacket for different possible applications depends on many variables, for example the type of ammunition to be defeated and the protected area in relation to weight. The company has prepared information detailing these criteria and advising on correct selection.

Military jacket type MIL 100

The MIL 100 is a fragmentation jacket with neck protector designed to NATO specification. It is in one piece, with a Velcro fastening in front and can be supplied with various pockets and attachments to meet customer requirements. The recommended protection level covers normal splinters and fragments, but it can be upgraded to cover standard pistol and sub-machine gun bullets if required.

DATA
Weight: (Medium size) 3.40 kg
Protected area: approx 5770 cm²
Resistance: fragmentation, and, on request, higher levels

Status
Current. Production.

MIL 100 fragmentation jacket

Fragmentation vest MIL-110

The MIL-110 is a light and flexible splinterproof vest, produced in accordance with NATO specifications. The preferred level of protection covers standard fragments at a V50 of 400 m/s but it can be upgraded to protect against 9 mm Parabellum, .44 Magnum and similar threat levels.

DATA
Weight: (Medium size, standard protection) 1.55 kg
Protected area: approx 3500 cm²

Status
Current. Production.

MIL-110 fragmentation vest

Fragmentation jacket MIL-120

This is a light military vest with neck and shoulder protection, produced according to NATO specifications. The preferred level of protection covers light bullets and standard fragments, though it can be supplied to give protection up to standard pistol and sub-machine gun bullets.

DATA
Weight: (medium size, standard) 2.60 kg
Protected area: approx 5300 cm^2

Status
Current. Production.

MIL-120 fragmentation vest

Bombsuit MIL-300

This is a light coverall garment with protective inserts affording all-round protection in the area of the trunk and in the front of the trouser legs. The normal level of protection covers fragments and pistol bullets up to 300 m/s.

DATA
Weight: 5.35 kg
Protected area: 7190 cm^2

Status
Current. Production.

MIL-300 Bombsuit

Slip-on vest S811

This type of vest offers complete all-round protection whilst allowing good freedom of movement. The neck and pelvic protectors are removable and do not interfere with mobility. The normal level of protection is to West German Police Standard I, and protection can be increased up to Class 4 level by inserts in the front and/or back pockets.

DATA
Weight: starting at approx 3 kg
Resistance: against fragmentation with inserts up to Class 4

Status
Current. Production.

S811 slip-on vest

Slip-on Vest Type T200

This vest was developed for special military assignments. It can be donned like a jacket and closes at the side. It is available in protection levels up to West German Police Standard I, and to increase the protection a Kevlar/ceramic plate can be inserted into the front of the vest; by adding a further plate, the level can be brought up to the maximum, affording protection against 7.62 × 54 API at 850 m/s striking velocity.

DATA
Weight: (Medium size, standard) 4.80 kg
Protected area: approx 5060 cm^2
Weight of Kevlar/ceramic plate: approx 3.11 kg

Manufacturer
Mehler Vario Systems GmbH, PO Box 760, Edelzellerstrasse 53, D-6400 Fulda.
Status
Current. Production.

Type T200 slip-on vest

ISRAEL

Rabintex bullet-proof vests

RAV-201

The RAV-201 has been designed to meet all Israeli, NATO and US standards. It has an outer covering of water-resistant nylon which is fire-retardant and has a reduced infra-red signature. This outer covering can be produced in a variety of shades and camouflage finishes. Beneath the outer covering, protection is provided by layers of Kevlar; the number of layers varies according to the threat level anticipated, ranging from levels 1 to 4. At the upper levels ballistic plates are inserted into pockets in the vest and a trauma pack is fitted to dissipate the striking energy. The Kevlar cloth in the vest is sealed in a water-repellent envelope, and the ballistic plates are constructed from Kevlar laminated with ceramics and composite materials.

Other models

RAV-100 is a concealable vest which can be worn underneath military or civil clothing. RAV-200 is intended for use by police and security personnel. RAV-300 is a special high-performance vest employing an increased proportion of composite material with a large ballistic plate insert.

Depending upon the threat level, a normal vest will weigh between 2.5 and 4.5 kg.

Manufacturer
Rabintex Industries Ltd, 3 Baruch Hirsch Street, Bnei Brak 51201.
Status
Current. Production.
Service
Israeli Defence Forces, military and police forces in NATO, Malaysia, Portugal, Singapore, USA and other countries.

Eagle bullet-proof clothing

Eagle bullet-proof vests are constructed of Kevlar 29 with Velcro closures and offer maximum protection at minimum weight. Both under- and over-clothing models are comfortable to wear, do not impede free movement, and require no special care. Every Eagle garment meets US Federal, Military and NATO standards for bullet-proof fabric.

Vests produced by Eagle include under-shirts and over-clothing vests in various degrees of ballistic protection, covering 11, 16 or 21 layers of Kevlar cloth. For greater levels of protection rigid plates of composite material can be fitted into the vests. The panels will increase protection and reduce trauma effect.

An armoured body suit (HH/KEV/24) is available for police and military personnel concerned in EOD duties. Easily worn and removed, this is a 24-layer suit weighing about 13 kg. This suit has been field-tested and has actually saved lives in operational situations.

Manufacturer
Eagle Military Gear Overseas Ltd, IBM House, 31 Shaul Hamelech Boulevard, PO Box 33737, Tel Aviv 64927.
Status
Current. Production.

Eagle AM-KEV/L-16 vest

Eagle armoured body suit

ITALY

Tecnocompositi Lasar lightweight armour systems

This company, part of the Montedison group, specialises in lightweight armour for protecting vehicles of all types and for uparmouring armoured vehicles and ships. In the body armour field it manufactures composite moulded shields and insert plates to be used in protective and bullet-proof garments.

Lasar Composite Bullet-proof Vests
These are anatomically moulded shields made of a Flexible Composite material which at equal weight is more efficient than a textile structure. Other advantages are that the composite material is non-ageing and impervious to environment.

Type Lasar 01. This is a three-piece (front, rear and pelvic shields) unit weighing 2.5 kg, covering 0.43 m² of body surface. The material is 3.5 mm thick and gives protection against handgun bullets up to .357 Magnum SP and JHP and 9 mm NATO bullets at close range.
Type G40 Super. This is also a three-piece unit but of 6.5 mm thickness and weighing 3.6 kg. It will give protection against handgun bullets up to .357 Magnum metal piercing and against bursts of 9 mm NATO sub-machine gun fire.
Lasar 12 Rigid under-shirt. This is a two-piece (front and rear shield) unit weighing 1.8 kg, covering 0.239 m. of body surface. The material is 4.5 mm thick and gives protection against handgun bullets to .357 Magnum JSP and SHP and 9 mm Parabellum Fiocchi M38 single shot.

Lasar insert shields for textile bullet-proof vests
These are lightweight shields which are to be inserted into pockets provided in existing armour garments, in order to improve the protective level. Standard types are:
Fibrosint LAS-710. This will defeat 7.62 × 51 mm NATO and 5.56 × 45 mm NATO ball and 7.62 × 39 mm steel-cored Soviet bullets at 10 m range. The shield is 10 mm thick and weighs 2.15 kg.
Fibrosint LAS-910. This defeats 7.62 × 51 mm NATO AP bullets at 25 m range. Thickness 11 mm, shield weight 2.4 kg.
Special shields can be developed to suit specific customer requirements.

Manufacturer
Tecnocompositi SpA, Lasar Division, Via Firenze 7/A, 22079 Villa Guardia (Como).

VEST LASAR 01   RIGID UNDER SHIRT LASAR 12

VEST G40 SUPER   INSERT PLATES LAS 710

Lasar body armour examples

Status
Current. Production.

Tecno Fibre Armour and Helmets

Tecno Fibre SpA is a member of the Advanced and Composite Materials Group, one of the operating companies of the Montedison Group. The company specialises in the production of military helmets in composite materials, and articles for personal protection.

Helmet TFH-A-1600. This is a 'Fritz' type helmet composed of a shell in composite material made up of several layers of Kevlar impregnated with thermosetting resin together with edge protection. The helmet interior has an adjustable nylon suspension with frontal anti-perspiration pad. The chinstrap is in two parts, provided with a buckle and slipping chin protection. There is a sponge-like shock absorber on the outer surface of the suspension ring. Weight 1.3 kg, protection V50 = 488 m/s for 1.1 g fragment according to MIL-P-46593A.

Helmet TFH-A-1800. This is similar to the TFH-A-1600 helmet described above but it designed to provide a higher level of protection. Weight 1.40 kg; protection V50 = 548 m/s.
Helmet TFH-A-2000. This is similar to the TFH-A-1600 helmet described above, but is designed to provide a higher level of protection. Weight 1.50 kg; protection level V50 = 610 m/s.
Helmet TFH-VN-1050. This is an anti-fragment helmet with a shell constructed of several layers of glass-fibre and ballistic nylon impregnated with thermosetting resin. There is a protective edging around the shell and an adjustable nylon suspension surrounded by a shock-absorbing strip which prevents impact on the helmet being transferred to the wearer's head. A chinstrap is in two parts, with buckle and slipping chin protection. The helmet weighs 850 g and offers protection of V50 = 320 m/s for a 1.1 g fragment according to MIL-P-46593A.
Helmet TFH-VN-1200. This is similar in all respects to the TFH-VN-1050 described above but offers a higher level of protection. Weight 1.05 kg, protection level V50 = 360 m/s.

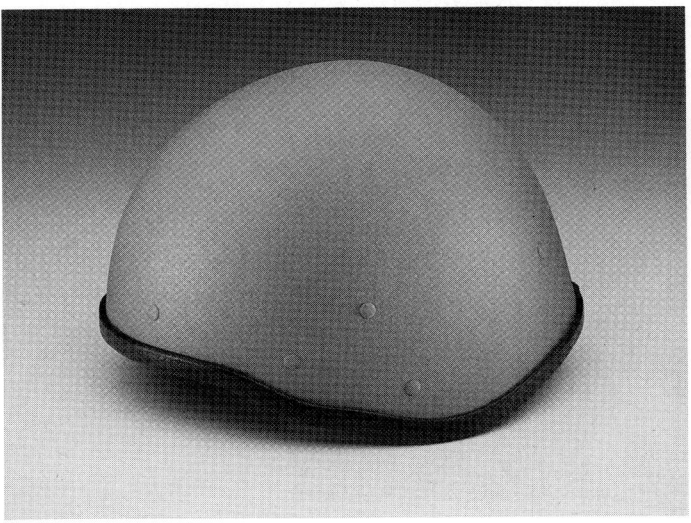

Tecno Fibre TFH-VN-1050 helmet; -1200 and -1600 models are of identical appearance

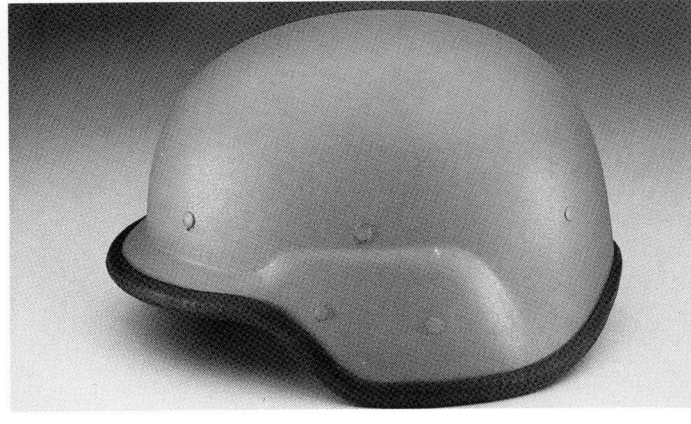

Tecno Fibre TFH-A-2000 helmet; -1600 and -1800 models are of identical appearance

Helmet TFH-VN-1600. This is similar in all respects to the TFH-VN-1200 described above but offers a higher level of protection. Weight 1.50 kg; protection level V50 = 470 m/s.

Helmet TFH-FM-2000. This is a steel-composite anti-fragment and anti-ballistic helmet, comprising a composite shell of several layers of Kevlar impregnated with thermosetting resin coupled to a steel shell. There is edge protection and an adjustable nylon suspension, with sponge-like shock absorbing material. The chinstrap is in two parts, with buckle and slipping chin protection. Weight 1.60 kg; protection level V50 = 610 m/s for 1.1 g fragment.

Tecno Fibre TFH-FM-2000 steel-composite helmet

Bullet Resistant Vest Type LASTEC 01

This is a bullet resistant vest for police and civil use. It is made in three pieces, front, rear and pelvic shield, with the possibility of superimposing the pelvic shield on the front. There are insert plates TEC 710 and TEC 910 which are placed in pockets provided on the front and rear of the vest so as to increase the level of protection.

DATA
Weight: 2.50 kg
Thickness: 3.5 mm
Protected area: 0.43 m²
Protection: handgun bullets up to .357 Magnum SP and JHP. and 9 mm NATO at close range.

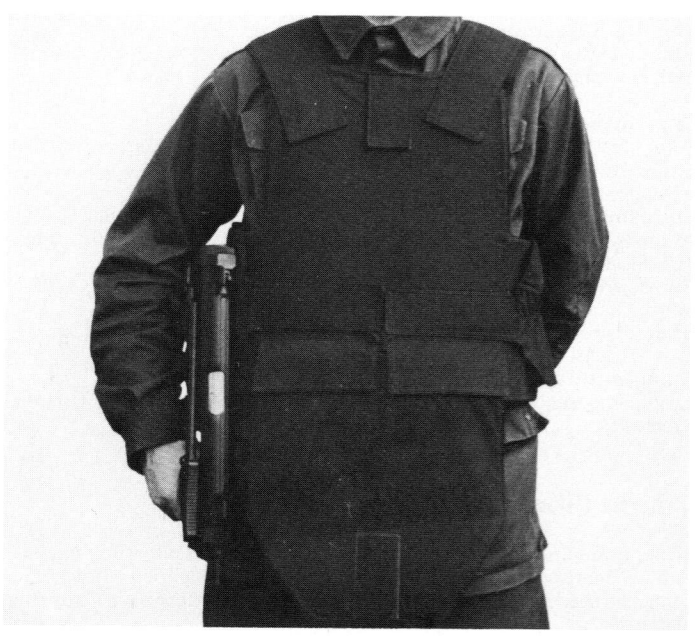

Lastec 01 bullet resistant vest

Bullet Resistant Vest Type G40S

This has been designed for use by Special Forces and is made in three pieces, front, rear and pelvic shield, with the possibility of superimposing the pelvic shield on the front. The insert plates TEC 710 and TEC 910 can be inserted into pockets in front and rear so as to improve the level of protection.

DATA
Weight: 3.60 kg
Thickness: 6.5 mm
Protected area: 0.43 m²
Protection: handgun bullets up to .357 Magnum MP, and bursts of 9 mm NATO sub-machine gun fire

Type G40S bullet resistant vest

Rigid Undershirt Type LASTEC 12

This is a rigid undershirt with a high level of efficiency combined with extremely low weight, very low thickness and anatomically moulded shape so as to fit beneath the shirt inconspicuously. It is 40 per cent lighter than equivalent soft protective garments. It is made in two pieces, front and rear.

DATA
Weight: 1.60 kg
Thickness: 4.5 mm
Protected area: 0.239 m^2
Protection: handgun bullets up to .357 Magnum JSP and SHP, and 9 mm Parabellum Fiocchi M38 single shot.

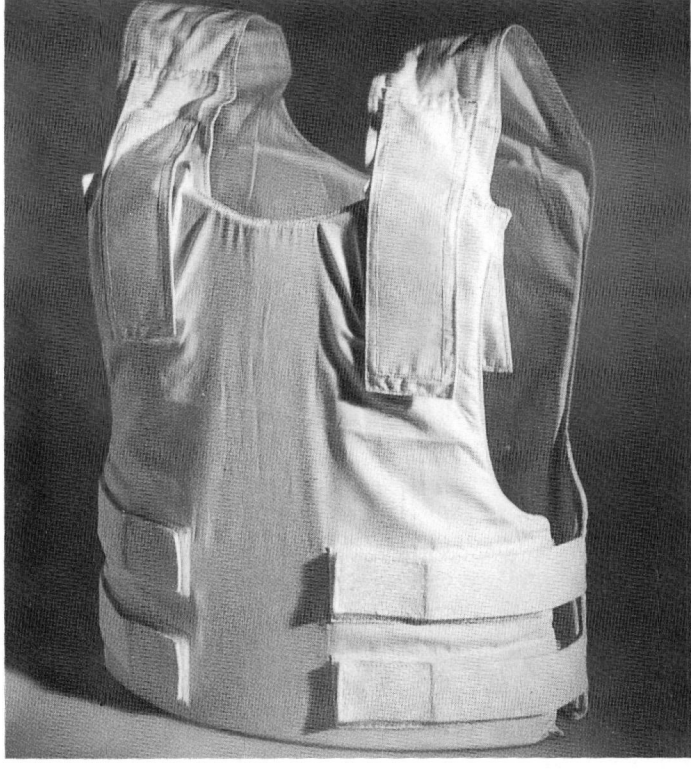

LASTEC 12 rigid undershirt

Armoured Skirts, Type LASTEC 105

Armoured skirts are lightweight systems capable of improving the ballistic protection of armoured vehicles of any class. They have the capability of destabilising and shattering kinetic energy penetrators such as AP, API, HVAP and APDS, thus defeating their ability to enter the vehicle. The skirts are light and easy to fit; on any structure having suitable pick-up points, skirt installation is quick and can be undertaken in the last phases of vehicle combat preparation, even by unskilled personnel.

DATA

	LASTEC 10	LASTEC 20ng	LASTEC 30
Weight	3 kg/m^2	6 kg/m^2	10 kg/m^2
Thickness	2.8 mm	5.6 mm	9 mm
Standard size		200 × 3000 mm	
Protection	Level 1	Level 2	Level 3

Manufacturer
TecnoFibre SpA, Via Guilio Terzi di S.Agata 8/10, 24030 Brembate Sopra (Bergamo).
Status
Current. Production.

UNITED KINGDOM

Bristol Armour

Bristol Armour products are designed and manufactured by Bristol Composite Materials Engineering Ltd (BCME) a wholly owned subsidiary of British Petroleum plc.

The Bristol Armour product range is extensive, including body armour, helicopter armour, bomb disposal suits, bomb search suits, helmets, ship armour and civil and military vehicle armour.

Various grades of flexible and structural armour are produced with protection levels varying from 17 grain (1.1 g) fragments through most calibres of small arms up to 23 mm cannon API. Armour is supplied to a wide variety of customers including military, naval, police and security forces as well as to civil users such as banks.

Type 12 body armour

The Type 12 jacket is designed to be worn over clothing by military and security forces. The jacket presents a slim, low profile appearance with built-in trauma attenuation for comfortable long term wear. The back has wraparound panels which attach to the front at each side. Two straps fit over the shoulder and attach to the front panel.

Double touch and close fasteners are provided. These allow for adequate adjustment within each of the three standard sizes, but hold the jacket firmly to the wearer. Double fastenings also prevent the jacket being accidentally removed in a scuffle or when moving through undergrowth. Rifle butt pads are attached to the front panel as standard, have a non-slip performance when wet. An optional extra is a detachable armoured pelvic panel.

Bristol Armour Type 12 jacket

The recently introduced TS range of armours used in the Type 12 jacket, and available in other styles, have only half the bulk of conventional armour/trauma attenuation systems and greatly improved flexibility for higher threat level protection.

Trauma attenuation

The trauma attenuation system used in the Type 12 jacket (and also available in the Type 10) is designed to meet current internationally recognised standards with a thickness of half that of conventional systems, yet giving flexibility.

Bristol helmets

The outer shell is a structure of bonded GRP material of similar construction to that used in Bristol GRP armour panels, but improved to give greater protection to both low and high velocity fragments and bullets, striking at all angles of incidence. There are three grades of helmet, each of different weight and offering different levels of protection. All are lightweight, in comparison with a similarly performing steel helmet, with specially shaped comfort and shock absorbing internal fittings which allow use for long periods without fatigue.

Trauma damage to the head is minimised by the use of a lightweight expanded polystyrene liner. Standard helmets are normally supplied with three size fittings. Conversion to smaller sizes is easily achieved by means of padding strips which also allow individual adjustment for maximum comfort. A three point terylene harness is used for maximum security. The outer shell can also be moulded from bonded Kevlar producing a lighter structure than GRP.

DATA
Grades: 9, 17 and 23
Weight: 1.0, 1.4 and 1.8 kg
Resistance: superior to steel of the same weight

Manufacturer
Bristol Composite Materials Engineering Ltd, Avonmouth Road, Avonmouth, Bristol BS11 9DU.

Bristol Armour lightweight helmet

Status
Current. Production.

Armourshield Ltd.

Armourshield designs, manufactures and supplies ballistic and fragmentation protective equipment to governments, police departments and public and private corporations throughout the world. The Armourshield product range includes body armour, screens and shields, structural protection (for temporary or permanent buildings) and protection for mobile units including land vehicles, marine craft, fixed-wing aircraft and helicopters. Protection against low velocity, fragmentation and high velocity rifle bullets can be achieved in all equipment with major weight savings over conventional units.

Advanced design and manufacturing techniques are the result of 14 years of extensive testing, the specific demands of users and the diverse climatic conditions in which the equipment has seen use. The UK MoD, US DoD and Netherlands MoD have all purchased direct from Armourshield during the past two years, and over 49 000 units of Armourshield body armour are in worldwide operational use.

Armourshield's considerable attention to research and development has lead to the introduction of the patented Blunt Trauma Shield, a unique energy absorber incorporated into all body armour as standard and at no extra cost to the purchaser. Ballistic performance and medical appraisal has been carried out by the UK MoD Establishment, Porton Down; this report is available on a direct Government-to-Government basis to approved enquirers.

Both body and structural armour units can incorporate new specialist materials which are fire-retardant, chemical, napalm and infra-red signature

protective. Adaptations are also available for Thermal Imaging countermeasures.

Series 25 Modular Armour

Model 777/25 This is a wrap-around vest protecting the upper body and with an optional removable groin protector. Fitted with Blunt Trauma Shields, it can be readily uparmoured by the insertion of a ceramic front plate, bringing it up to REV (below) standard. The vest alone weighs 2.5 kg. Two ceramic plates are available; the 'Midi' plate, is double-curved to fit closely to

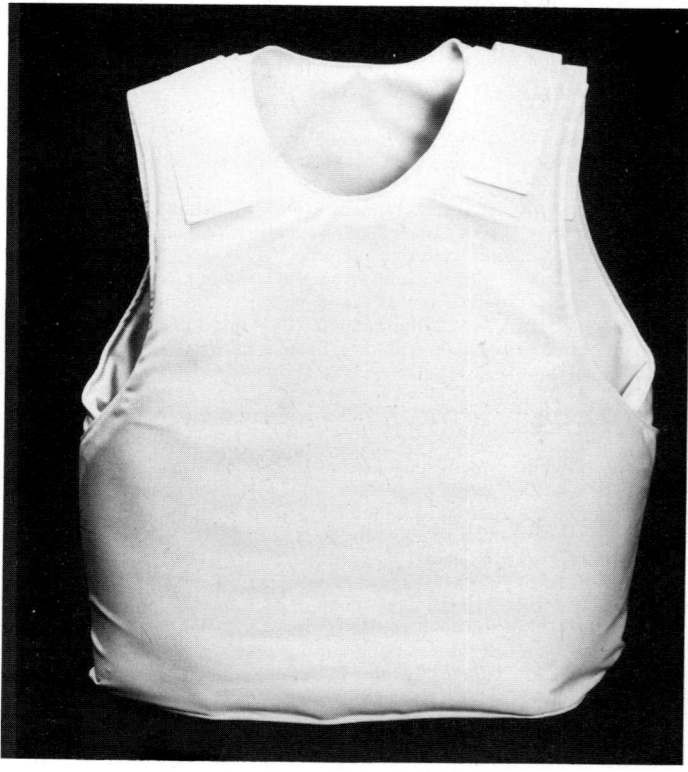
Model 777/25 vest

Example of Armour Shield armour after attack by 7.62 × 51 mm armour-piercing. Shots 1 and 2 are 30 mm apart. All bullets were stopped by armour

Model 777/25

Model REV/25 showing frontal ceramic plate

the body and can be equally well used in the front or back of the vest; it weighs 2.5 kg. The 'Maxi' plate is larger and thicker and has a single curvature, making it suitable for the front of the jacket only, and weighs 3.4 kg.

Model FW25 This vest gives full upper body protection with extra side protection. It weighs 2.95 kg and can be fitted with the 'Midi' or 'Maxi' plates described above.

Restricted Entry Vest (REV/25) This vest has been carefully designed and tailored to make it particularly suitable for applications where space is at a premium, such as exiting aircraft when parachute jumping, rappelling and abseiling, travelling in certain armoured vehicles and generally moving in confined spaces. In effect it is the Model 777/25 vest with the addition of the 'Midi' or 'Maxi' ceramic plates as standard.

General Purpose Vest GPV/25 This vest gives a greatly increased protected area, including the shoulders and under-arm area. The groin protector is of extra width so as to afford protection to the femoral arteries in the thighs, and it can be folded up in front of the vest if desired. A back collar can be fitted if required.

Both the REV and GPV vests carry the front ceramic plate in a preformed pouch, into which it drops and is retained by straps coming from the shoulders. In an emergency, should the plate be struck by a bullet and damaged, it is possible to release and replace it by taking the back plate out and placing it in the front plate pocket.

The Series 25 vests, without the addition of the ceramic plates, are capable of defeating attack up to and including steel-jacketed high velocity 9 mm Parabellum (for example Norma 19022), .357 Magnum, .44 Magnum and 7.62 mm Tokarev bullets when fired from 25 cm test barrels at above-standard velocities.

Optional Ceramic/Kevlar composite plates

There are three ceramic composite plates available:

the 'Mini' plate, 203 × 254 mm, weight approximately 1.3 kg. This fits into the Ultralite series of concealed vests as an option. It can be fitted into the front or back of the vests;

the 'Midi' plate, weighing 2.5 kg is double-curved and will fit in the front or back pockets of all garments and

the 'Maxi' plate, weighing 3.4 kg, is a large body-wrap, plate for use in the front pocket of the Series 25 vests.

All ceramic plates are capable of defeating, at a distance of 4.5 m from the weapon, 5.56 × 45 mm M193 and SS109 ball and P96 AP; 7.62 × 39 mm AP;

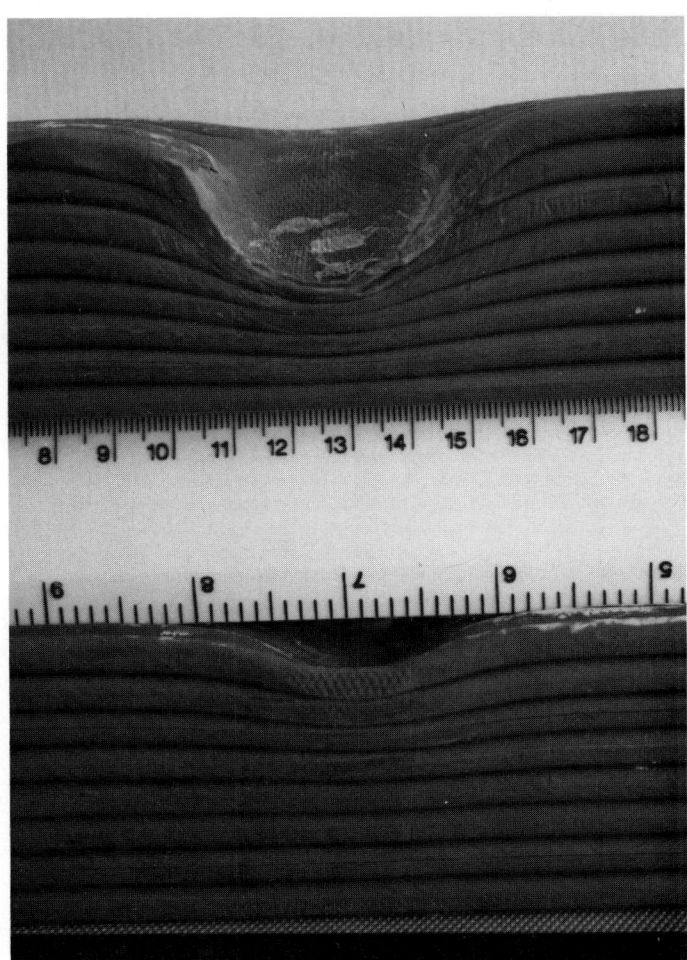

Plasticene blocks showing back-face signature through 18 layers of Kevlar, with and without Armourshield Blunt Trauma Pack. Attack by .357 Magnum from 3 m range.

Armourshield lightweight fragmentation jacket with front neck protection in place

GPV/25 general purpose vest

Armourshield lightweight fragmentation jacket in normal wear; front neck protector is attached to breast by Velcro fasteners when not in use

7.62 × 51 mm NATO ball and FN P80 AP; 7.62 × 54R Soviet ball; 7.62 × 63 mm (.30-06) ball and AP M2; .300 Weatherby Magnum soft-point; and 12 gauge rifled slugs and OO Buckshot.

Armourshield ultra-light flak jacket 777/FL17

This new jacket has been specially developed for military use by the forces of a European country. A principal feature of the design is the adherence to the client's requirement that the jacket should not have front opening and fastening, thus stopping the possibility of fragments striking front chest vital organs. There is, nevertheless, full through-ventilation, with the possibility of leaving the side fastening open when the wearer is not under combat conditions, but retaining full protection to both the front and back of the body. In addition there is patented blunt trauma protection for both the front vital organs and the spinal column, the trauma shield being positioned in accordance with medical advice given by the British Ministry of Defence. Neck protection is provided by a built-in rear protector which can be removed if necessary to permit full movement, and a detachable front protector.

Fully adjustable for height and width by means of Velcro fasteners, two standard sizes of jacket will fit almost any wearer. The jacket has been designed to be as light and thin as possible, consistent with a very high degree of fragmentation protection. It has been proved to British V50 17-grain (1.1 g) fragment at velocties over 500 m/s, and has also been successfully tested against NATO-standard 9 mm Parabellum bullets at 3 m range, velocity 400/410 m/s. Because of the patented blunt trauma shield the rear indentation was less than 10 mm in depth. By adding a ceramic composite plate the protection level can be raised to stop 7.62 × 51 mm NATO, 7.62 × 39 mm M43 and 5.56 × 45 mm M183 or SS109 bullets.

The jacket is in the form of a sleeveless vest, with internal openings to remove the ballistic inserts in order to permit the outer cover to be washed. The outer cover material is to standard UK camouflage pattern, is water-resistant and has IR signature protection.

Manufacturer
Armourshield Ltd, PO Box 456, Manchester M60 2LL.
Status
Current. Production.
Service
Various products are in use by UK Ministry of Defence, UK police forces, US Federal agencies and Department of Defense, US police and SWAT teams, and armies and security forces of several major countries.

Lightweight Body Armour

Lightweight Body Armour Ltd., a specialist manufacturer of advanced high-performance armour for military use, introduces three of the latest-generation armours.

NATO Troop Jacket

NTJ8470, the NATO troop jacket, was specifically designed for the West German armed forces and represents the most advanced personal armour system available. Lightweight, flexible, and offering the highest protection against fragments – which account for 70-80 per cent of all battlefield casualties – the NTJ8470 offers the most cost-effective answer to the protection of ground troops.

NATO Troop Jacket, showing side release and fastening

NATO Troop Jacket showing shoulder release and fastening

KQ90 showing groin protector in position

BCJ showing neck and shoulder protection

KQ90 Troop Jacket

KQ90, or Khamis al Qadaam, is a development of the NTJ8470 designed specifically for operations in the Middle East. The KQ90 is lightweight, rugged and requires the minimum of maintenance. Offering excellent all-round protection which can be worn over the uniform, the KQ90 can also be up-armoured rapidly to protect against high-velocity rifle attack, by the insertion of LBA ceramic armours in pockets provided.

Ballistic Combat Jacket

BCJ, the Ballistic Combat Jacket, is designed for military personnel who require a higher level of protection than that provided by an anti-fragmentation jacket. The BCJ can be provided with various strengths of Tetranike anti-ballistic armours to suit the specific operational requirements of each user.

Flexible armour panels

Lightweight Body Armour products are all supplied with the latest advanced high-technology Tetranike and Tetramid lightweight flexible armour panels. Products are supplied direct from the LBA factories in UK, Norway and West Germany, where they are constructed under strict quality control standards. The company has been assessed and registered by the British Ministry of Defence as operating quality control procedures in accordance with AQAP 4.

Research and Development

LBA Ltd have an advanced firing range available to customers, which allows live firing demonstrations, and provide a threat analysis service.

Manufacturer

Lightweight Body Armour Ltd, Hinton House, Daventry, Northants NN11 6QG.

Status

Current. Production.

RBR Armour

RBR Armour Ltd manufactures a wide range of bullet-resistant body armour which has been developed as a result of many years of development and testing on the company's own firing range, and after extensive consultation with medical experts on ballistic wounds, as well as with manufacturers of appropriate materials.

RBR 3001 jacket

The threats to be met vary continuously, which means that the armour designed to defeat them must be adapted accordingly. Often a compromise must be achieved between light weight and comfort on the one hand and level of protection on the other. The company has extensive experience to advise users on the optimum solution.

Body armour generally falls into two categories: flexible systems and hard systems. Flexible armour systems are based on the fabrics woven from special yarn such as Kevlar; they are designed to defeat threats from low velocity bullets and ballistic fragments. Hard armour systems are based on ceramic or steel plates faced with layers of Kevlar, glass fibre composites or similar materials; they are designed to counter high velocity rounds.

RBR manufactures a wide range of body armour, including many types of standard flexible garments. These include the 1000 Series of covert or concealed undervests; the 2000 Series of overvests for military and police forces; the 3000 Series of military fragmentation vests; the 4000 Series helicopter crew vest; the 5000 Series waistcoat and jackets for executives; the 6000 Series non-ballistic police garments; and the 7000 Series female vests.

Details of these garments, including details of their performance in resisting specific ammunition, types of weapon, velocities and impact energy, are available, based on certified tests.

In addition to the standard garments, the company can design, in consultation with the customer, personal armour to meet virtually any specified threat. Garments of agreed design can be manufactured quickly in large or small quantities, under strict quality control.

RBR 3001 military jacket

This gives protection against ballistic threats but remains light in weight so as to reduce fatigue in combat. It gives all-round protection against grenade fragments and sub-machine gun bursts, while double frontal protection stops handgun and close sub-machine gun attack. There is a pouch which accepts ceramic plate protection to protect vital organs against rifle threats. Groin protectors and all forms of pockets and attachment points for equipment can be provided. Depending on the level of protection selected, the jacket will withstand attack from all handgun and shotgun ammunition and rifle ball bullets.

Manufacturer

RBR Armour Ltd, 5 Townsend Street, London SE17 1HJ.

Status

Current. Production.

Armalite Body Armour

A-Series Special Forces armour

This jacket gives complete wrap-around upper torso protection, is multi-adjustable, and has quick-release straps. There are rifle pads on each shoulder face, one back pocket, two expandable box-pockets, epaulettes and grenade hooks. The cover is in fire-resistant or nylon fabric in a variety of colours or camouflage patterns. The basic vest is made either of Kevlar or ballistic nylon, as desired, is fitted with a complex impact reduction system, and may be given additional protection by means of ceramic/Kevlar composite plates at front and back. The weight varies from 2.0 to 7.17 kg depending upon the material and threat level selected.

F-Series General Purpose Lightweight armour

This consists of front and back panels connected by shoulder straps and side-closing straps with Velcro fastenings. It is available based on Kevlar or ballistic nylon and thus offers a wide range of protective levels with weights from 1.4 to 6.67 kg. Both front and back units are fitted with a pocket into which ceramic/Kevlar composite tiles can be inserted to give the maximum levels of protection. The garment is covered in proofed nylon or fire resistant material in various colours or in disruptive camouflage pattern as desired.

Manufacturer

Armalite Body Armour, PO Box 6, Heywood, Lancs OL10 4BW.

Status

Current. Production.

National Plastics GS Mark IV Helmet

The GS Mark IV helmet is a high-specification helmet designed to meet the demanding needs of today's defence forces.

The shell which provides the ballistic protection is manufactured from 23 layers of ballistic nylon fabric impregnated with a phenol/formaldehyde and polyvinyl/butyral resin system. The nylon-to-resin ratio is maintained at 80 to 20 parts by weight. The dense composite structure is formed by applying high pressure and heat to the nylon resin preform. On impact by high velocity fragments the shell is designed to absorb the kinetic energy.

The liner is formed from a high density polyethylene foam with a density of $100 \pm 10 \, kg/m^3$ which is designed to absorb the energy of blunt impact.

The front and rear comfort pads and chin support are manufactured from black gloving leather. All other components in contact with the skin are constructed from cotton webbing. The helmet shells are finished with one primer undercoat and two coats of matt polyurethane IRR finish NATO green paint.

The ballistic resistance is $V50 = 380 \, m/s$, using the standard 1.1 g fragment. The assembled shell will pass the shock absorption test equivalent to BS 5361, using a flat striker with an energy of 122 J and a drop height of 2.5 m. The finished shell conforms to infra-red reflectance requirements of STANAG 2338.

Manufacturer
National Plastics, Courtaulds Advanced Materials, 473 Foleshill Road, Coventry CV6 5AQ.
Status
Current. Production.

National Plastics GS Mark IV and AFV crewman's helmets

Service
British Army.

Highmark body armour

Highmark body armour is developed and produced in Northern Ireland to meet the personal protection requirements of civilians, military and police. The range of protective garments covers military and police jackets in both conventional and low profile styles and a number of protective undervests for wear beneath ordinary clothing.

Military body armour
This range includes an Infantry Jacket; Tactical Patrol Jacket; Tactical Assault Jacket; Fragmentation Jacket; and a Ballistic Hood capable of being worn with the various jackets. There are three levels of soft armour protection available: Mark IA protects against pistol and sub-machine gun bullets up to 9 mm calibre; Mark I covers the same group up to .44 Magnum calibre; Mark II covers the same as Mark I but to higher striking velocities. In addition ballistic plates can be inserted into the armour to upgrade protection: the Category III plate with Mark I or II armour protects against .30 Carbine and .357 KTW bullets; Category II plate against 7.62 and 5.56 mm NATO ball and 7.62 × 39 mm API; and Category I plate against 7.62 and 5.56 mm NATO AP bullets. The weight of the various garments varies according to the level of protection, the maximum weight is 3.5 kg.

Manufacturer
Highmark Enterprises, Ronoco House, 55 Adelaide Street, Belfast BT2 8FE, Northern Ireland.
Status
Current. Production.

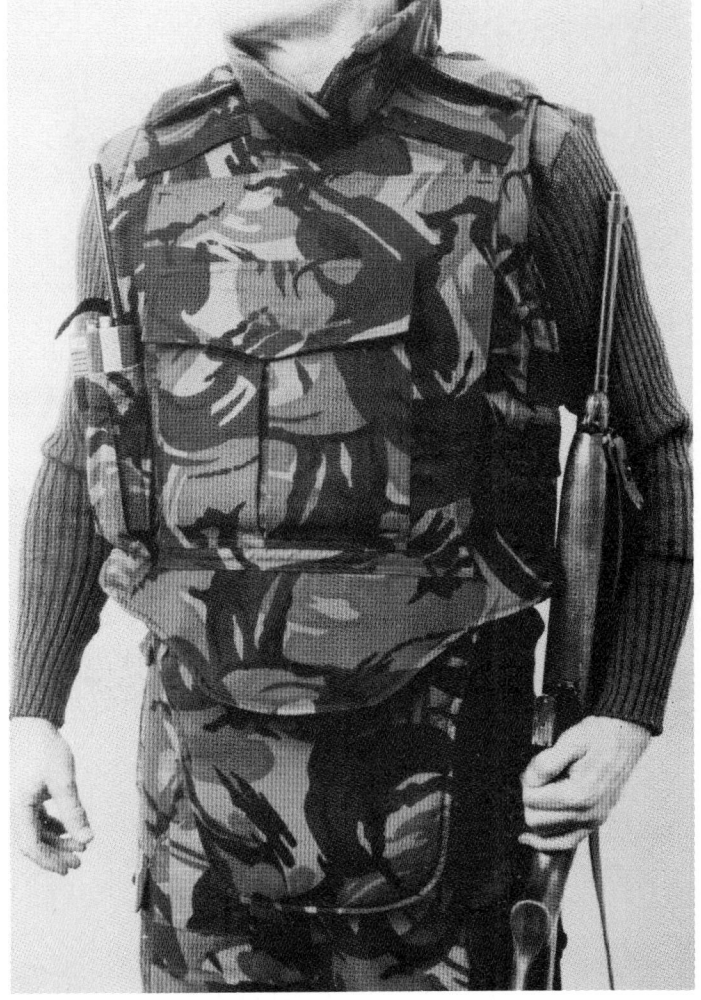
Highmark Tactical Assault Jacket

SA17 bullet-resistant jacket

This is marketed by Sas R and D Services Ltd and is a versatile, flexible and lightweight armoured jacket which gives protection against a wide range of threat levels but is sufficiently light to be worn in comfort even in hot climates. The jacket contains two protective layers; flexible Kevlar armour to stop and contain projectiles, backed by a trauma attenuation pack to cushion the shock due to high momentum impact. Uparmouring is achieved by using additional plates to break up high-energy ammunition from armour-piercing and high velocity rifle attack. These can be fitted rapidly into pockets in the jacket. The plates are available in three grades to meet different threat levels. The jacket cover is of waterproof nylon, reinforced to allow continuous use on active duty and the design gives maximum protection with minimum interference to normal body movements.

DATA
Weight: 2.7–4.4 kg depending on size. Uparmouring plates add between 1.2–2.9 kg, depending on selected level of protection

Resistance
Jacket only: up to 9 mm Parabellum, .357 and .44 Magnum and 12-bore shotgun
With addition of armour plates: up to and including .30-06 armour-piercing from 5 m range

Manufacturer
Sas R and D Services Ltd, 7 Melbourne Park, St Johns, Jersey, Channel Islands.

SA 14 Helmets

Sas R & D Services supplies a variety of types of riot control and military helmets. The riot control helmets are constructed from a one-piece thermoplastic shell with high resistance to impact and solvents. Inside this can be fitted shock-absorbing liners of rubber or polystyrene and an adjustable suspension system. Retention of the helmet is ensured by specially designed chinstraps of high strength polyester or nylon webbing, padded by simulated leather and with a chin cup. A polycarbonate visor can be fitted to some patterns, as can additional neck protection.

Military ballistic helmets can be had in either steel or composite materials, in conformity with the various MIL SPEC standards which have been laid down.

Manufacturer
Sas R and D Services Ltd, 7 Melbourne Park, St Johns, Jersey, Channel Islands.

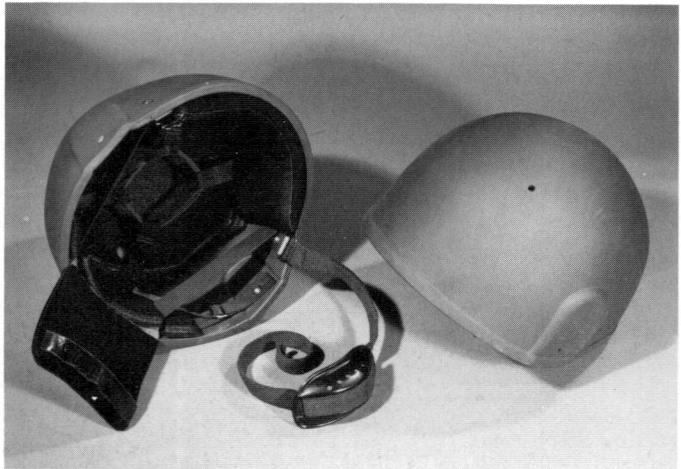

Sas Group Sa/14/07 composite ballistic helmet

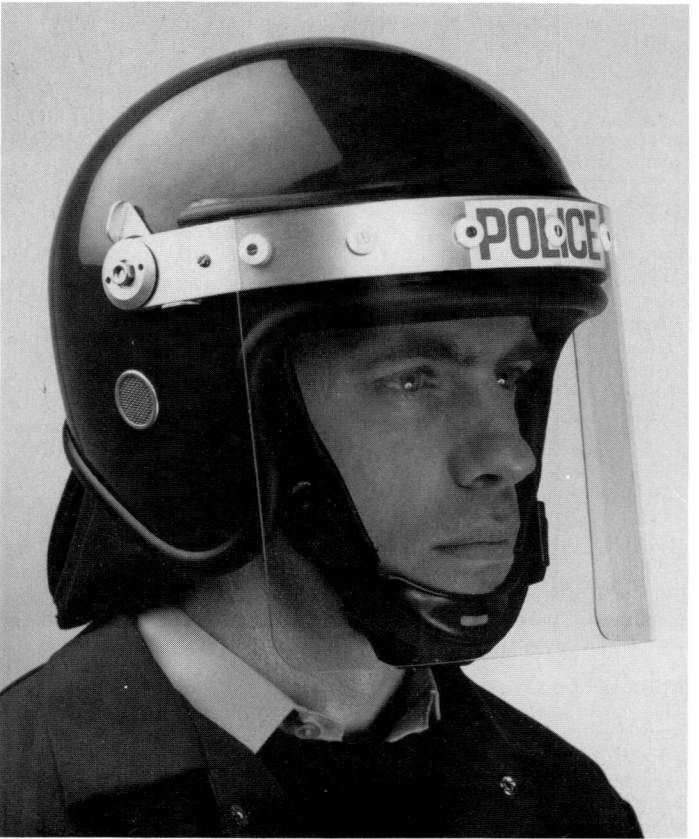

Sas Group SA14/004 riot helmet

UNITED STATES OF AMERICA

Second Chance personal armour

The Second Chance Body Armor manufactures personal armour for use by police, military and government agencies. It has concentrated for the past 16 years on persuading law enforcement agencies to equip their men and women with concealable vests of Kevlar Aramid material and the results of those who have worn them are impressive. By the end of 1988 over 450 lives had been saved by Second Chance protective garments: most of these having been shot at a range of 1 to 2 m with either handguns or shotguns. In every case these men and women were wearing their protection under their clothing so that it was concealed. The logic is that with visible protection the assailant will shoot for the head and kill.

Range of ammunition which has been fired at ballistic material incorporated in Second Chance soft body armour. Deformed projectiles are in foreground. Larger projectiles are single slugs from shotguns. Way in which material deforms bullet into mushroom shape is clearly shown

Military armour

Hardcorps

Hardcorps armour is quite specifically designed for military use and is intended to be worn over normal clothing. The Hardcorps product line is available in three distinct styles: Hardcorps 2, Hardcorps 3 and Hardcorps 4. Each style incorporates special laminated steel plating design to deter Soviet 7.62 mm armour piercing incendiary, .30-06 ball, 5.56 and 7.62 NATO ball. Hardcorps armour is available in two levels of ballistic protection as standard equipment with each Hardcorps unit. The K-47 armour plate stops the above ballistic rounds but when combined with the K-30 armour plate, Hardcorps armour will then be able to stop .30-06 armour piercing rounds in addition to Soviet and NATO AP rounds.

Hardcorps 2

The Hardcorps 2 military unit is an extremely cost-efficient body armour. This unit is 32 cm wide by 38 cm long, front and back. The ballistic panels between the steel plating and the torso offer Threat Level IIA protection (shell fragments and Magnum handgun firepower) and are carried in a heavy-duty jungle camouflage or black shell made of tough nylon pack-cloth. Included in the package is a 25.4 by 33 cm K-47/30 combination insert system for frontal high-power rifle protection. The armour plate is designed with a Kevlar 'wrap' to suppress fragment splash to throat and face. Specially designed shoulder straps offer airburst shrapnel protection and better vest weight distribution.

DATA
Weight: (without insert combination) 1.81 kg; (with K-47 plates) 4.9 kg; (with K-47/30 insert system) 6.79 kg

Hardcorps 3

The Hardcorps 3 unit is described as the 'all-purpose body armour system'. The unit consists of Threat Level IIA ballistic soft armour protection in the MiniRap style between the metal plating and the torso. The carrier is constructed of the same nylon pack-cloth material as the Hardcorps 2 but offers slightly more lower abdominal coverage. The carrier is available in jungle camouflage and black. There are three frontal pockets on the Hardcorps 3 designed to hold the K-47/30 armour plate system as part of the unit package. The overlapping plate design adds to the flexibility and comfort of the unit while protecting the wearer from the 'riding up' of projectiles when hit in the heat of combat. The top plate employs the 'splash-guard' lip design and measures 15.2 by 22.8 cm. The lower two plates measure 15.2 by 30.5 cm each. Optional back plate (25.4 by 33 cm) is available on request. The specially designed shoulder straps are standard equipment with this unit as are the deep side pockets for additional ammunition magazines or grenades.

DATA
Weight: (without insert combination) 2.74 kg; (with K-47 plates) 8.38 kg; (with K-47/30 insert system) 11.1 kg

Hardcorps 4

The Hardcorps 4 model is designed for use by attack forces. The unit can give critical 'kill-zone' coverage regardless of the size of the individual wearing it. The unit has front and back Threat Level IIA ballistic protective capability without the metal inserts. The outer shell material is constructed of nylon pack-cloth material available in jungle camouflage and black. The frontal plate system consists of four inserts, top and bottom plates each measure 15.2 by 22.8 cm with the top plate incorporating the special 'splash-guard' design. The two middle plates each measure 15.2 by 30.5 cm. All plates overlap each other for greater flexibility and ease of movement.

Standard features of the Hardcorps 4 are the ammunition pockets attached to the webbed side straps; the heavy duty back-pack; and the quick-release shoulder strap system.

DATA
Weight: (without insert combination) 3.63 kg; (with K-47 plates) 11.5 kg; (with K-47/30 insert system) 14.3 kg

Hardcorps 3

Hardcorps 2

Command Jac

The Command Jac carrier offers complete front, back, over-the-shoulder, neck and full wrap-around Model Y2A ballistic protection from .357 and .44 Magnum handgun firepower and 12ga rifled slugs. With the addition of the K30 frontal plate the wearer receives 396 cm² of added ballistic protection from .30 US Carbine and all known handgun armour-piercing ammunition. K-47 armour plate protection from armour-piercing rifle fire is available as an option to the wearer.

The Command Jac style is designed to accommodate the wearing of a pistol belt while still offering lower back and kidney protection. There are over a dozen frontal pockets for ammunition and accessories with back pockets for maps or optional additional armour plating.

DATA
Weight: (without insert combination) 3.97 kg; (with K-47 plates) 8.8 kg; (with K-47/30 insert system) 10.6 kg

Manufacturer
Second Chance Body Armor Inc, Box 578, Central Lake, Michigan 49622.
Status
Current. All styles available.

Command Jac

Protective Materials body armour

Protective Materials produces transparent and opaque armour of various types for small arms up through nuclear blast protection, and provides complete systems capability for initial design or upgrading of armour for tanks, APCs, light vehicles, equipment shelters, aircraft, naval vessels and security facilities. In the body armour field the company produces a full range of garments for military and police use.

Tactical Assault Armour
The tactical armour garments are primarily for military use and are available in various styles and ballistic protection levels. Garments in this range are the fragmentation jacket, which is to MIL SPEC B12370E and protects against fragments only; ballistic protection against small arms; military field jacket which is the standard US M65 field jacket with a removable Kevlar ballistic lining for protection against small arms fire; and the TAC jacket designed for the special weapons assault teams to protect against both small arms and rifle fire.

Manufacturer
Protective Materials Co, 7914 Queenair Drive, Gaithersburg, MD 20879.
Status
Current. Production.

Protective Materials Tactical Assault Armour

Point Blank armour

Point Blank manufactures a wide range of protective vests and jackets for different applications, including undercover operations and VIP defence. All vests are tested in accordance with Federal Standards for Ballistic Resistance of Police Body Armor (NIJ-STD 0101.03), not only for resistance to penetration but also for blunt trauma. All testing and certification is carried out by HP White Laboratories Inc, and stringent quality control is maintained throughout manufacture.

In late 1989 Point Blank adopted DuPont Kevlar 219, a new generation ballistic fibre which offers greater protection and comfort but with no additonal weight or bulk. This fibre has been incorporated into the Point

Blank High Performance series of garments, all of which have been tested to NIJ-STD 0101.03. All products are available in Kevlar 219 fibre, and the degree of protection offered at the various levels is somewhat higher than before.

Most garments are available in varying levels of ballistic protection: Model 10 level (Class I) protects against bullets from .22 long rifle at 320 m/s to .38 Spl at 259 m/s; Model 15 level (Class IIA) extends this to cover .45 Automatic Colt Pistol, .41 Magnum, .375 Magnum and 9 mm Parabellum at medium velocities; Model 20 level (Class II) takes in all previous levels but at higher velocities and also .44 Magnum; Model 30 level (Class IIIA) protects against

Hi-Risk Modular Tactical Armour

Kevlar helmet and fragmentation vest

Point Blank M-16 military/tactical armour

Point Blank NATO/SWAT vest

Point Blank Flame-resistant 'Nomex' utility suit

Point Blank Ballistic Flotation Vest

virtually all small arms bullets, shotgun charges, 9 mm sub-machine gun and fragment impact.

The newest and most up-to-date in tactical equipment is the Hi-Risk Modular tactical armour. This vest was developed by Point Blank in conjunction with the Los Angeles Police Department SWAT Team. The vest is available in two levels of protection and ceramic plates can be added to defeat armour-piercing ammunition.

For police and civil use there are a number of styles, including a waistcoat which can be worn openly since it resembles a conventional suit garment and a special female style vest for women police and security officers. An innovation is a ballistic raincoat which is well-suited to the protection of VIPs and which does not advertise its protective properties.

Recent introductions by Point Blank include a laminated Kevlar military helmet which meets full US military specifications; a ballistic T-shirt which can be worn beneath a normal shirt to afford concealed protection; and a military field jacket which is the standard M65 field jacket with a removable zip-out ballistic liner which is available in five levels of protection from Class I to Class IIIA. Of particular interest to counter-insurgency and rapid intervention forces is a 'Nomex' flame-resistant utility suit.

Manufacturer
Point Blank Body Armor, 185 Dixon Avenue, Amityville, NY 11701.
Status
Current. Available.

American Body Armor & Equipment

The American Body Armor & Equipment company manufactures a wide range of protective clothing ranging from undervests for male or female police officers to full-protection combat and EOD suits. The company developed a second-generation 'Armitron IIIA' system of construction which employs a hardened Kevlar fabric, shock plates, and careful contouring of the garment to fit the body.

These garments are available in varying grades to match the various accepted threat levels; in general terms the lowest grade of soft armour will protect against .38 Special bullets, the highest against high velocity 9 mm Parabellum. Beyond this point additional ceramic protection can be added in plate form, lifting the protective level to 7.62 × 39mm Soviet, 7.62 mm NATO and 5.56 mm NATO ball, 5,45 mm AK-74 ball and .30-06 ball.

SOV Tactical load-bearing vest
The SOV is a tactical load-bearing mid-vest constructed of a fishnet material known as Raschel knit. The detachable ammunition and other equipment pouches are constructed of 100 per cent nylon. The SOV is equipped with webs which accommodate the standard military pistol belt. This, in conjunction with the quick-release zip, provides for emergency removal of the vest. The back of the vest is laced with 550 lb (250 kg) parachute suspension line to assure size adjustment and form fit. The pouches are attached by a combination of Velcro and pull-the-dot snap fasteners. The SOV may be worn in conjunction with the American Body Armor tactical bullet resistant vests.

A1-TAC-FS tactical armour vest
The tactical armour vest ballistic package has been developed to defeat high velocity 9 mm sub-machine gun bullets. It is designed to withstand Threat Level IIIA and is reinforced with additional layers of Kevlar and is fitted with front and back pockets to hold steel and ceramic plates. It is available in four sizes and five colours.

AK-47 lightweight military body armour
The Model AK-47 vest is human engineered to be the lightest unit of tactical

A1-TAC-FS tactical armour vest

A1-TAC-ALT tactical assault vest

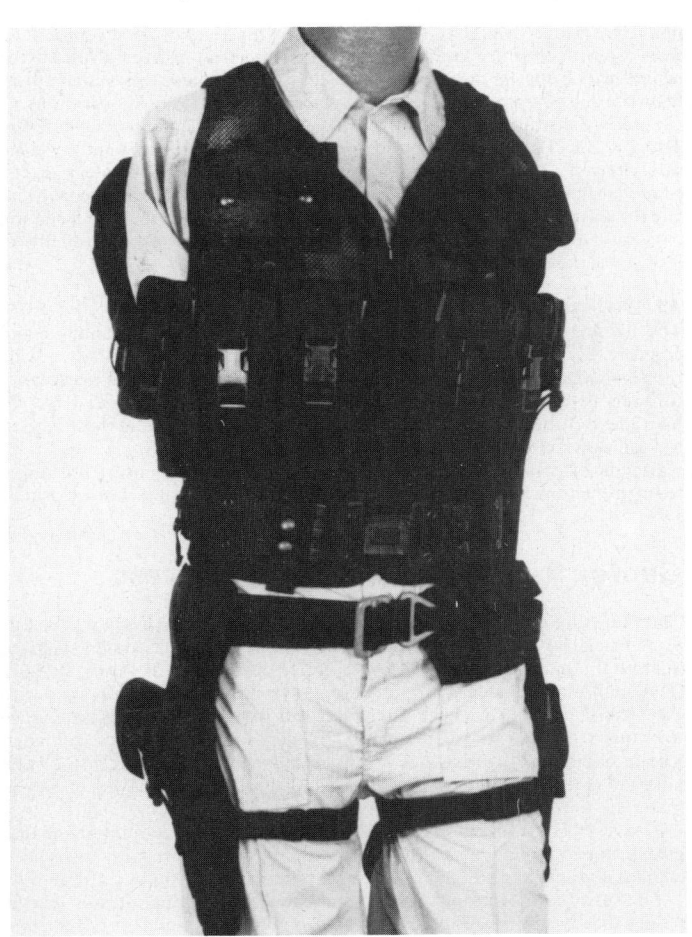

SOV Tactical load-bearing vest

A1-FLAK-USA fragmentation jacket

AK-47 lightweight military body armour

military armour available. Capable of defeating high velocity automatic weapons with ceramic plate inserts, this armour can be worn comfortably for extended periods of time in combat situations. Compatible with various load-bearing equipment, the AK-47 is perfectly suited for field operations.

Standard features include 'Black Magic' chemical toughening, offering Threat Levels II and IIA protection without plates installed; Spandex elastic side closures; front to back side overlap; front and rear plate pockets; over-the-shoulder ballistic protection; quick-release shoulder mechanism; and a ballistic collar. Optional equipment includes groin protector, carrying bag, and various sizes and weights of ceramic plate. The vest is available in four sizes and five colours.

A1-TAC-ALT tactical assault vest

The tactical assault vest has been designed for versatility, mobility and comfort. Because of its simplicity and capability of being worn with a wide range of load-bearing equipment, this vest has received wide acceptance throughout the world. It is offered in two ballistic levels, II and IIIA. To upgrade to rifle protection, Type III and Type IV ceramic plates must be used for additional efficiency.

Standard features include a Milspec 420-denier nylon cover; full upper torso protection; front and rear plate pockets; and Velcro and elastic closures.

Optional equipment includes groin protector, carrying bag, and a variety of ceramic plates of different sizes and Threat Levels. The vest is available in four sizes and five colours.

A1-FLAK-USA fragmentation jacket

This front-opening style jacket is identical in appearance to the US military fragmentation vest. However, the A1-FLAK-USA exceeds the performance of the US Military vest by also stopping most handgun, shotgun and sub-machine gun projectiles, while still being lighter in weight. The styling is designed to afford full mobility and comfort during prolonged periods of operation.

Ballistic Fragmentation V-50 at 1900 ft/s (580 m/s) for Threat Level IIIA.

Manufacturer
American Body Armor & Equipment Inc, One Kaplan Court, PO Box 1769, Fernandina Beach, FL 32034.
Status
Current. Production.
Service
Police and military forces world-wide.

Gentex fragmentation protective vest

This vest is manufactured to the requirements of MIL-B-44053 and consists of multiple layers of Kevlar cloth encased in inner and outer shells of lightweight ballistic nylon. The vest is a part of the US Army PASGT (Personal Armor System Ground Troops) system and is generally used with the PASGT protective helmet. The vest covers the upper torso and is designed to protect the wearer against typical fragmentation from mines, grenades, mortar bombs and artillery shell fire. Such fragments may penetrate the outer layers of cloth but will be slowed down and stopped by the layers of Kevlar material.

The vest has a front closure using Velcro strip, and a three-quarter collar to give protection to the neck and throat areas. It can be worn with any type of combat clothing and will not interfere with the wearer's combat efficiency.

The vest does not protect against small arms fire, though it may tend to reduce the severity of wounds, and may sometimes stop small arms bullets at angular impact or low velocity.

Manufacturer
Gentex Corp, LifeTex Business Group, Carbondale, Pennsylvania 18407.
Status
Current. Available.

Gentex protective vest

Gentex PASGT Helmet

The PASGT helmet was developed for US Army ground forces and paratroops in response to a military requirement to replace the M1 steel helmet. The helmet provides a 50 per cent higher level of protection against fragmentation from exploding munitions than does the M1. Its unique design gives protection to 12 per cent more vital head and neck area and its low contoured centre of gravity adds greater stability and makes the helmet more comfortable to wear. The cradle type suspension enables the helmet to be offset from the head for better ventilation and to allow space for deformation caused by impacting projectiles.

The helmet, although not intended to stop direct rifle fire, has

demonstrated a capability to defeat pistol-fired ammunition as well as other types of small arms fire which strike the helmet at lower velocity or at an angular impact.

The helmet shell is made of Kevlar ballistic cloth. In addition to the shell, the basic components include suspension assembly, headband and chinstrap. A foam impact pad and retention strap are added for parachute operations.

Gentex Corporation has produced variants of the PASGT helmet including lighter weight versions, snap-in-band suspension assemblies and special retention assemblies for parachute operations.

Gentex PASGT helmet

Manufacturer
Gentex Corporation, LifeTex Business Group, Carbondale, PA 18407.
Status
Current. Available.

PASGT helmet and PASGT vest

The PASGT helmet is made to US military standards and is presently being distributed to US and foreign military organisations worldwide. With construction utilising aramid ballistic material it is designed to defeat typical threats from common bursting munitions such as grenades, mines, mortars, bombs and artillery fire. This helmet offers 50 per cent greater ballistic protection than does the US M1 steel helmet and protects 12 per cent more head area.

The PASGT vest meets MIL SPEC MIL-H-4499A and MIL-B-44053A and is currently in production for the US military and various allies worldwide. It is designed to defeat typical threats from common bursting munitions, and although it is not rated against small arms fire, it is typically capable of defeating conventional Level IIA NIJ and PPAA STD-1989-05 Level A threats. This garment is also available in higher levels of ballistic protection up to and including armour-piercing rifle fire. In addition a flotation option can be ordered for marine applications.

Manufacturer
Silent Partner Body Armor Inc, Square One, Lafayette at the River, Gretna, LA 70053-5835.
Status
Current. Available.

Silent Partner PASGT vest and helmet

Covert Operations Protective System (COPS)

This covert operations vest is designed for stand-alone armour use. The cover is durable 400-denier waterproof nylon, and all straps are of MilSpec elastic and hook-and-loop. The ballistic protection is provided by Kevlar 29 1000-denier, 31 × 31 weave count, Zepel-D treated armour. The sculptured design guarantees maximum coverage from shoulder to waist; contoured side coverage is a standard feature, as is the eight point adjustment system. There is a 7 oz Sentinel hard armour insert for enhanced ballistic performance as well as knife-resistance, over the front centre chest area. Unlike any other vest currently available, the COPS is completely reversible; colour combinations include (but are not restricted to) black/white, navy blue/light brown, brown/tan and olive drab/jungle camouflage. Dual strap sets in the appropriate colours are included. Solid colour models in white, black, navy, brown and olive drab are also available.

All COPS armour meets or exceeds current government test standards. Protection levels offered include Levels I +, IIA, II, SMG (II +) and IIIA.

Manufacturer
Silent Partner Body Armor Inc, Square One, Lafayette at the River, Gretna, LA 70053-5835.
Status
Current. Available.

Silent Partner COPS vest

Ranger tactical vest

This garment incorporates state-of-the-art ballistic protection with an efficient tactical vest. The ballistic protection is entirely of Kevlar, and the free-standing collar offers virtually full neck protection without obstructing the use of either a ballistic helmet or Silent Partner's flexible armour Command Cap. Fully articulated armoured shoulder pads furnish additional coverage.

The tactical jacket exterior is of Cordura nylon with all interior construction of smooth and comfortable 400-denier waterproofed nylon. The front hook-and-loop closure ensures effective ballistic integrity with multiple overlapping plies of Kevlar.

Standard front design configuration includes ammunition pouches, grenade/shotgun cartridge loops and a built-in pistol holster. There is also a concealed front pocket for fitting a rifle-resistant hard armour insert.

The Ranger tactical vest is available in seven levels of ballistic protection; Levels I+, IIA, II, SMG (II+) and IIIA, with the possibility of adding two different rifle-resistant inserts, the Hex Hard Case (ceramic/laminate) and the 7 mm Zoned Armour Steel insert. The addition of these inserts lifts the protection to Levels III or IV.

Manufacturer
Silent Partner Body Armor Inc, Square One, Lafayette at the River, Gretna, LA 70053-5835.
Status
Current. Available.

Ranger tactical armoured jacket

Building Entry Team Armour (BETA)

The carrier of this garment is made from 1000-denier Cordura nylon. The front carrier component contains four ammunition/grenade/radio pouches. The back module provides attachment points for an optional back-pack. A 2 in (50 mm) nylon equipment belt is standard.

Front modules are offered in Enhanced Coverage (Small/Medium) or Doublewide (Large/X-Large) sizes. The most popular model, the SMG, uses both Level IIA and I+ front armour panels for superior integrity. Back coverage consists of a Level IIA Enhanced Coverage panel.

Each carrier module will accommodate as many as four flexible armour panels to handle high velocity SMG rounds and shotgun slugs. If even higher protection is desired, then a ceramic or steel insert plate can be added to provide protection up to 7.62 × 51 mm AP rifle fire.

Manufacturer
Silent Partner Body Armor Inc, Square One, Lafayette at the River, Gretna, LA 70053-5835.
Status
Current. Available.

BETA Building Entry Team Armour

Aircrewman body armour

This is for aircraft and helicopter crews and is designed to defeat high-powered rifle fire and threats from armour-piercing rifle ammunition. It is available in black, olive drab, or sage green, with a Nomex flame-resistant outer casing.

Manufacturer
Silent Partner Body Armor Inc, Square One, Lafayette at the River, Gretna, LA 70053-5835.
Status
Current. Available.

Aircrewman body armour

Boarding Party Vest (BPV)

This offers a unique combination of ballistic protection and safety in the marine environment, giving both small arms protection and approximately 11.5 kg of positive flotation, provided by puncture-proof foam. It is available in a wide range of colours in either standard nylon outer casing or flame-resistant Nomex. The vest is available in protection levels NIJ I to IV or PPA STD-198905 Levels A to E.

The vest can be configured with pockets, etc, according to the requirements of the purchaser.

Manufacturer
Silent Partner Body Armor Inc, Square One, Lafayette at the River, Gretna, LA 70053-5835.
Status
Current. Available.

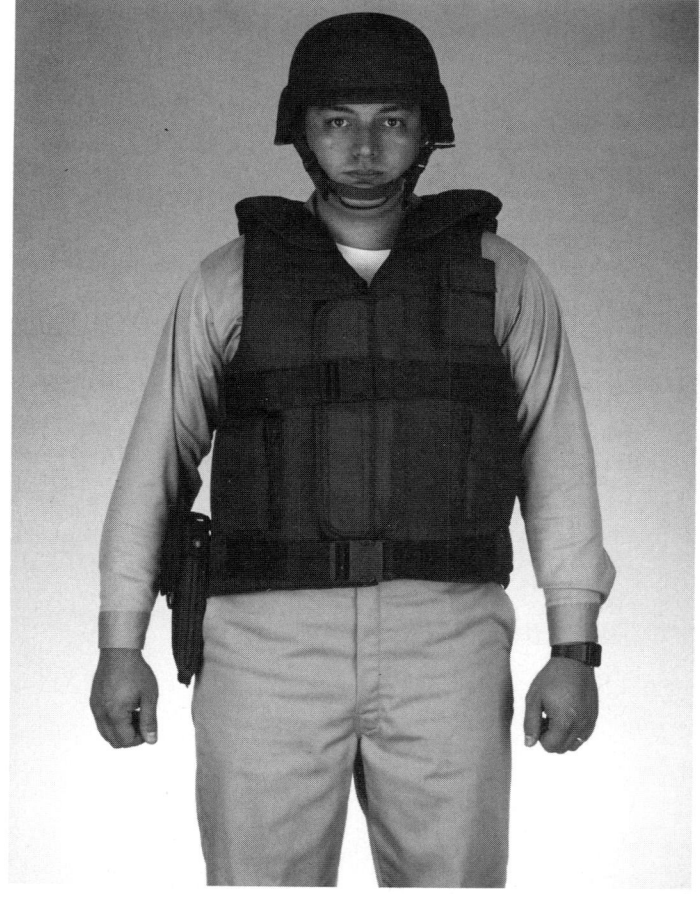

Boarding Party Vest

YUGOSLAVIA

Bullet resistant jacket Type 1

The Type 1 jacket is of the usual woven Kevlar construction and is designed to give protection against small arms such as pistols and sub-machine guns. It will protect against most 9 mm Parabellum and .357 Magnum class bullets at 5 m range. A Type 75C ceramic plate can be inserted into a pre-formed pocket, and this will improve the protection to a level which will withstand attack by 7.62 × 39 mm or 7.62 × 51 mm ball at 5 m range.

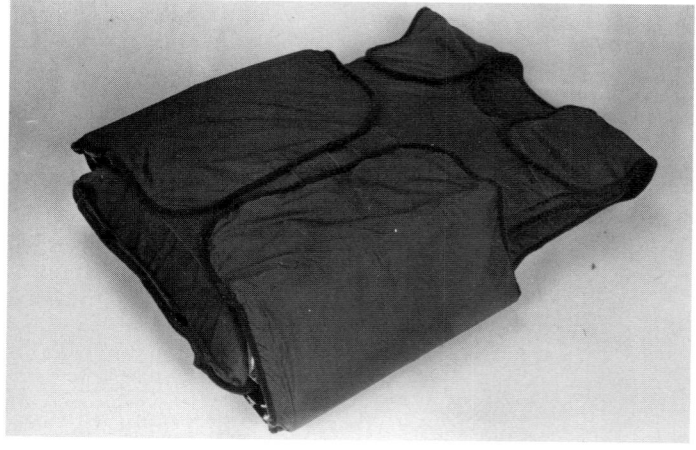

Bullet resistant jacket Type 1

Bullet resistant jacket Type 3

The Type 3 jacket is the military overjacket model, covered in camouflage pattern cloth and with built-in ammunition and utility pouches. The level of protection afforded is the same as that of Type 1; most sub-machine gun and pistol bullets at 5 m range, plus the ability to insert a ceramic plate to give protection against 7.62 mm ball bullets at 5 m range.

Bullet resistant jacket Type 3

MPC-1 helmet

The helmet shell is a laminated structure, made of light glass fibres, ballistic cloth and polyester resin. The shell offers high impact resistance and is very comfortable. The impact resistance is not impaired by treating the shell with paint, oil or other materials frequently harmful to helmets. A high rear contour enables the use of firearms from the prone position without the helmet being pushed forward over the wearer's eyes. The headband allows adaptation to different sizes and shapes of head.

The helmet, with headband, weighs 1.65 kg. An optional visor weighing 436 g, and a neck protector are available if required.

Manufacturer
Federal Directorate of Supply and Procurement, 9 Nemanjina Street, 11001 Beograd.
Status
Current. Production.

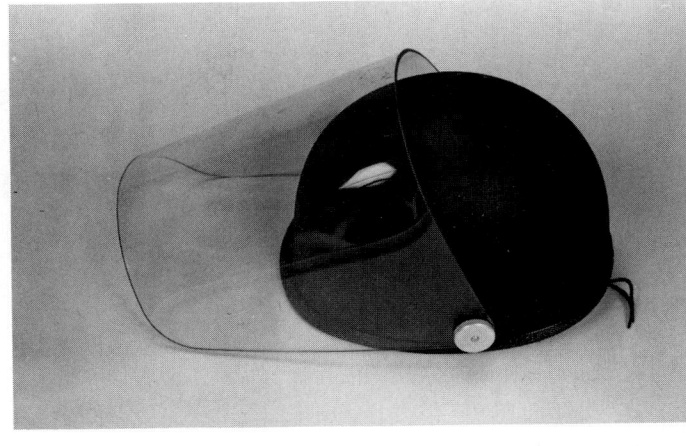

MPC-1 helmet with optional visor

Addenda

Pistols

UNITED KINGDOM

9 mm Spitfire

The Spitfire is a conventional recoil-operated self-loading pistol which uses the well-tried Browning dropping barrel method of locking the breech. A cam beneath the chamber acts as the controlling device to lower and raise the barrel so that lugs on the upper surface lock into recesses in the slide. The entire pistol, except for the grip surfaces and springs, is made from cast and machined stainless steel, and the machining is performed on computer-controlled tools to a tolerance of 5 microns. As a result the pistol is 'tight' throughout, with no slack in any of the components and with a very crisp trigger action.

The sights are adjustable for windage and zeroing and are extremely well-defined and clear. The action allows for double-action firing or for single-action operation, carrying the pistol cocked and locked. There is an ambidextrous safety catch on the frame where it can be operated easily by the thumb of either hand. The magazine holds 13 rounds and one further round can be loaded into the chamber, There is a filler piece on the bottom of the magazine which gives additional support to the little finger for firers with large hands.

The Spitfire has been carefully designed with the needs of police and special forces in mind and is robust and simple; at the same time, the quality of manufacture and finish is such that it can be used as a competition weapon in practical pistol and similar exercises.

DATA
Cartridge: 9 × 19 mm Parabellum
Operation: short recoil
Method of locking: Browning cam
Feed: 13-round magazine
Weight: 1000 g with empty magazine
Length: 180 mm
Barrel: 94 mm
Rifling: 6 groove, rh, one turn in 254 mm
Sights: fore; blade; rear: fully adjustable square notch
Muzzle velocity: 352 m/s

Manufacturer
John Slough of London, 35 Church Street, Hereford, Hereford and Worcs.
Status
Current. Production.
Service
Commercial sales; undergoing military evaluation.

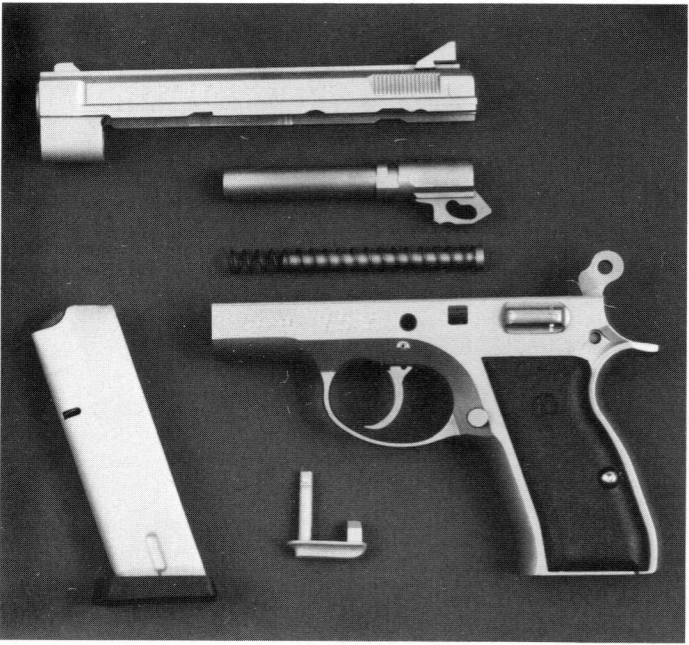

9 mm Slough Spitfire pistol

9 mm Slough Spitfire pistol field-stripped

UNITED STATES OF AMERICA

9 mm La France Nova

The Nova is the most compact and lightweight locked-breech pistol available in 9 mm Parabellum calibre. Produced by extensive modification of a Star Model BKM pistol, the Nova is ideal for concealed carrying. Electrofilm (dry-film lubricant) finish is standard, rendering the pistol virtually rustproof.

The pistol is recoil operated, using the usual Browning swinging link to separate the barrel lugs from the slide. Accuracy is exceptional for such a short weapon; typical groups measure 7-8 cm at 25 m range.

DATA
Cartridge: 9 × 19 mm Parabellum
Operation: recoil, semi-automatic
Feed: 6-round detachable box magazine
Weight empty: 600 g with empty magazine
Weight loaded: 680 g

La France Nova 9 mm pistol

Length: 157 mm
Barrel: 76 mm
Sights: (fore) Patridge blade, neon pink; (rear) fixed square notch
Muzzle velocity: 335 m/s

Manufacturer
La France Specialties, PO Box 178211, San Diego, CA 92117.
Status
Current. Limited production.

Sub-machine Guns

UNITED KINGDOM

9 mm Bushman IDW

The Bushman IDW (Individual Defence Weapon) is a very compact submachine gun which incorporates a patented rate regulator. This is factory adjusted to a standard rate of 450 rds/min, though this can be altered to suit the user's specification to any rate between 1 and 1400 rds/min, the later being the natural cadence of the weapon. The selected rate of 450 rds/min has been chosen as being the optimum rate which balances the various forces in the weapon to give absolutely steady operation; at this rate of fire there is no muzzle rise whatever, the gun merely oscillating gently about its centre of gravity and firing on the forward oscillation so that every shot is delivered from approximately the same muzzle location. As a result the weapon fires close groups whether held in one hand or two. Should the rate regulator fail for any reason, the gun reverts to its natural rate of 1400 rds/min, but even at this rate it is still quite controllable, there is minimal climb, and it can still be fired single-handed with quite acceptable accuracy. A further result of using the rate regulator is that the rate of fire, once set, is entirely unaffected by ammunition variations, and the gun will fire service ball, plastic practice or frangible ammunition with equal facility and at the same rate.

The Bushman IDW consists of an alloy frame and machined stainless steel receiver and barrel. The light bolt moves inside the receiver, and there is a cocking handle mounted at the rear of the receiver which can be gripped and pulled back, a similar action to cocking a common automatic pistol, after which the cocking handle is pushed back in to the receiver and remains stationary during firing. The box magazine is inserted into a housing in front of the pistol grip. The weapon fires from the open bolt, but the bolt is so light that there is no discernible shift of balance when the trigger is pressed, and single shots can be fired with accuracy comparable to a single-action pistol.

There are four independent safety devices. A thumb lever acts as a manual safety catch and fire selector, with positions for 'safe', 'fire' and 'auto'. With the elector on 'auto' the trigger becomes a double-pull type, the first 7 mm pull giving single shots, and a further 7 mm pull, against extra pressure, giving automatic fire. When set to 'safe' the lever directly locks the sear.

The grip safety positively locks the sear and totally inhibits the weapon from firing unless it is firmly pushed in by the act of gripping the pistol grip. It also acts as an indicator of the state of the weapon; when the weapon is cocked, the grip safety protrudes, when the weapon is uncocked the grip safety lies flush with the rear surface of the grip. The direction of action of the grip safety and the manual safety/selector switch are at 90° to each other, thus if the weapon is dropped, any impact which might tend to force one safety out will force the other into deeper engagement.

The magazine safety relies upon the position of the magazine in its housing; when inserted to a first stop the top round in the magazine is not in the feedway, and if the trigger is pressed and the bolt goes forward, it will not load a round. By giving the magazine a tap to seat it into the second, higher, stop the rounds are brought into the feedway and the gun will load and function normally. Operation of the magazine does not need a violent blow and can, with the short magazines, be conveniently performed with the little finger of the forward hand which is gripping the weapon. The weapon can thus be carried cocked, with a loaded magazine inserted to the first stop, in perfect safety and be ready for instant operation with the minimum of manipulation.

The bolt lock safety locks the bolt until the trigger is pulled and totally prevents the weapon from self-feeding and firing should it be dropped directly on to the end of the receiver. Even if the grip safety, manual safety and sear should all fail at once, the bolt lock will still prevent the weapon firing.

The sights are set into an 'Instafit' rib which can be inserted into and removed from the top of the receiver very easily. With the stock sight rib removed, alternative ribs carrying iron sights with Tritium night markers, laser spot markers, telescopes, or collimating sights can be fitted. Each alternate can be zeroed to the weapon and thereafter requires only about 20 seconds work, without special tools, to change.

Disassembly for cleaning and routine maintenance is simple and requires no tools. The mechanism can be exposed by removing cover plates, and the mechanism can be operated and the gun fired with these cover plates removed. The weapon is entirely made from milled steels and alloys, precision-machined; no plastics or stampings are used. The production weapons are currently in 9 mm Parabellum calibre, but by simply changing the barrel it becomes possible to fire .41 Action Express ammunition. A version in 10 mm calibre is under development.

9 mm Bushman Individual Defence Weapon

9 mm Bushman IDW showing size and method of holding

Length: 276 mm with standard barrel
Barrel: 82.5 mm standard; 152 mm and 254 mm optional
Rifling: 6 grooves rh, one turn in 406 mm
Sights: fore: blade; rear: V-notch, fully adjustable
Sight radius: 171.5 mm
Rate of fire: 450 rds/min regulated; 1400 rds/min unregulated
Muzzle velocity: 352 m/s, varying with ammunition
Effective range: 150 m

Manufacturer
Bushman Ltd., 10 Park Industrial Estate, Frogmore, St. Albans, Herts AL2 2DR.
Status
Current. Available.
Service
Undergoing military evaluation.

DATA
Cartridge: 9 × 19 mm Parabellum; .41 AE and 10 mm optional
Operation: blowback
Feed: 20-, 28- or 32-round box
Weight, empty: 2.92 kg

Rifles

GERMANY, EAST

7.62mm AMKS carbine

This is a shortened version of the MPiKMS-72 rifle (see page 160) and generally resembles it except for the shorter barrel and gas cylinder. The barrel is said to be 290 mm long, making the overall length, with butt folded, about 540 mm. A new design of plastic fore-end is used, but the magazine, pistol grip and folding butt appear to be the same as those used with the MPiKMS-72. An aluminium sound suppressor is said to be supplied as standard with each weapon, though the method of attaching the silencer is far from clear. It is notable that there is no form of muzzle compensator, as used on the AKSU-74, and thus one is inclined to suspect that the 290 mm barrel allows the propellant to be fully consumed before shot ejection.

East German 7.62 mm AKMS carbine (Jane's Soviet Intelligence Review/Ivo Sturzenegger)

National Inventories

In this section the infantry weapons believed to be in service with the military and paramilitary forces of the various countries are listed under the country headings.

Readers having information which will modify or complete any entry are invited to communicate with the Editor.

Pistols	Sub-machine guns	Rifles	Machine guns	Close support weapons	Mortars	Anti-tank
Afghanistan 7.62 mm Tokarev	7.62 mm PPSh41 7.65 mm CZ vz-61	7.62 mm Simonov SKS 7.62 mm AK47, AKM	7.62 mm RPD, RPK 12.7 mm DShK	30 mm AGS-17	82 mm M37 107 mm M38 120 mm M1943 160 mm M1943	73 mm SPG-9 82 mm RCL B-10 Snapper ATGW
Albania 7.62 mm Tokarev	7.62 mm PPD-40 7.62 mm PPS-41	7.62 mm M44 carbine 7.62 mm Simonov SKS 7.62 mm AK47, AKM	7.62 mm RP-46 7.62 mm RPD, RPK 12.7 mm DShK 14.5 mm KPV		82 mm M37 120 mm M43 160 mm M43	82 mm RCL T21 14.5 mm PTRS rifle
Algeria 9 mm MAC Mle 50 9 mm Beretta 34 9 mm Tokagypt	9 mm Vigneron 9 mm MAT 49 9 mm Uzi 9 mm Carl Gustav 45	7.62 mm AK47, AKM 7.62 mm Simonov SKS 7.62 mm Dragunov SVD 7.62 mm Beretta BM59	7.62 mm RP-46 7.62 mm RPD .30 BAR M1918A2 12.7 mm DShK		120 mm M43 160 mm M43	Sagger ATGW Snapper ATGW Spigot ATGW Spandrel ATGW Spiral ATGW
Angola 7.62 mm Tokarev 9 mm Stechkin 9 mm Makarov 9 mm CZ 27	7.65 mm Skorpion 9 mm Star Z-45 9 mm Uzi 9 mm FMBP-48	7.62 mm AK47, AKM 7.62 mm Simonov SKS 7.62 mm FN-FAL 7.63 mm G3	7.62 mm RP-46 7.62 mm RPD 7.62 mm vz/52 12.7 mm DShK	30 mm AGS-17	82 mm M43 120 mm M43	82 mm RCL B-10 107 mm RCL B-11 RPG-2 RPG-7 Sagger ATGW
Argentina 9 mm FN 35	9 mm FMK3 9 mm Sterling Mk 4 9 mm H&K MP5 .45 M3A1	5.56 mm FARA-83 5.56 mm Steyr AUG 7.62 mm FN-FAL 7.62 mm Beretta BM59 7.62 mm Steyr SSG69	7.62 mm FN-MAG .50 Browning M2HB	40 mm FMAP 40 mm HK69	81 mm Brandt 120 mm Brandt	89 mm RL M65 75 mm RCL M20 90 mm RCL M67 105 mm RCL M968 Cobra ATGW SS11, SS12 ATGW Bantam ATGW
Australia 9 mm FN 35	9 mm F1	5.56 mm F88 (AUG) 7.62 mm FN-FAL 7.62 mm Parker-Hale 82	5.56 mm F89 (Minimi) 7.62 mm Bren L4 7.62 mm M60 7.62 mm FN-MAG .50 Browning M2HB	40 mm M79 40 mm M203	81 mm F2 (L16)	66 mm LAW 84 mm Carl Gustav 106 mm RCL M40 MILAN ATGW
Austria 9 mm Glock 80 9 mm Walther P-38	9 mm Steyr MPi69	5.56 mm Steyr AUG 7.62 mm FN-FAL 7.62 mm Steyr SSG69	5.56 mm Steyr AUG/HBAR 7.62 mm MG 42/59 7.62 mm M74 .50 Browning M2HB	40 mm Steyr M203	60 mm Bohler 81 mm Bohler 120 mm Bohler 81 mm M1, M29 81 mm L16 120 mm M43	66 mm Law 74 mm Miniman 84 mm Carl Gustav 106 mm RCL M40 Panzerfaust 3
Bahrain 9 mm FN 35	9 mm Beretta M12 9 mm H&K MP5 9 mm Sterling Mk 4	7.62 mm Beretta BM59 7.62 mm FN-FAL 7.62 mm G3	7.62 mm FN-MAG		81 mm L16	120 mm RCL Mobat TOW ATGW
Bangladesh 9 mm FN 35	9 mm Sten 9 mm Sterling L2	7.62 mm FN-FAL 7.62 mm G3A2/3 7.62 mm AK47, AKM 7.62 mm Simonov SKS	7.62 mm Bren L4 7.62 mm RPD 7.62 mm HK11A1 7.62 mm HK21A1 12.7 mm DShK		82 mm Type 53 120 mm Type 53	106 mm RCL M40
Barbados 9 mm FN 35 .38 S&W	9 mm Sterling	7.62 mm FN-FAL	7.62 mm Bren L4 7.62 mm FN-MAG		3in UK	
Belgium 9 mm FN 35	9 mm Uzi 9 mm H&K MP5	5.56 mm FN-FNC 7.62 mm FN-FAL	5.56 mm Minimi 7.62 mm FN-MAG .50 Browning M2HB		60 mm NR493 81 mm NR475A1 107 mm M30	66 mm LAW Swingfire ATGW MILAN ATGW
Belize 9 mm FN 35	9 mm Sterling	5.56 mm M16A1 7.62 mm FN-FAL	7.62 mm Bren L4 7.62 mm FN-MAG			
Benin 7.62 mm Tokarev	9 mm MAT-49	7.5 mm MAS 49/56 7.62 mm AK47, AKM 7.62 mm Simonov SKS	7.62 mm RP-46 7.62 mm RPD 7.5 mm M24/29 7.5 mm AAT-52 .50 Browning M2HB 14.5 mm KPV		60 mm Brandt 81 mm MO-81-61	89 mm LRAC RPG-7
Bolivia .45 M1911A1 9 mm FN 35	.45 M3A1 9 mm Uzi 9 mm MAT-49	5.56 mm Steyr AUG 5.56 mm Galil 7.62 mm SIG 510-4 7.62 mm FN-FAL 7.62 mm G3	7.62 mm SIG 710-3 7.62 mm FN-MAG 7.62 mm M60 .50 Browning M2HB		60 mm Brandt 81 mm M29 107 mm M30	106 mm RCL M40

Pistols	Sub-machine guns	Rifles	Machine guns	Close support weapons	Mortars	Anti-tank
Botswana						
9 mm FN 35	9 mm Sten 9 mm Sterling	5.56 mm Galil 7.62 mm AK47, AKM 7.62 mm FN-FAL	7.62 mm Bren L4 7.62 mm FN-MAG 7.62 mm SGM		81 mm L16 120 mm M43	84 mm Carl Gustav RPG-7
Brazil						
9 mm Beretta M92	9 mm Beretta M12S 9 mm Walther MPK 9 mm Mekanika	5.56 mm M16A1 5.56 mm HK33E 7.62 mm FN-FAL	7.62 mm Madsen 7.62 mm FN-MAG .50 Browning M2HB	40 mm M79	60 mm Imbel 81 mm Imbel 81 mm M29 107 mm M30 120 mm Brandt(?)	3.5 in RL M20 57 mm RCL M18 75 mm RCL M20 106 mm RCL M40 Cobra ATGW
Brunei						
9 mm FN 35	5.56 mm Colt XM177 9 mm Sterling	5.56 mm M16A1 7.62 mm G3	5.56 mm Colt M16A1/HB 7.62 mm SIG 710-3 7.62 mm FN-MAG 7.62 mm HK21A1	40 mm M203	81 mm Tampella	SS-11 ATGW
Bulgaria						
7.62 mm Tokarev 9 mm Makarov	7.62 mm PPSh-41	7.62 mm AK47, AKM 7.62 mm Simonov SKS	7.62 mm RP-46 7.62 mm RPD, RPK 12.7 mm DShK 14.5 mm KPV		82 mm M43 120 mm M43 160 mm M43	82 mm RCL B-10 Snapper ATGW Sagger ATGW Spiral ATGW
Burkina Faso						
7.65 mm Walther PP 9 mm MAB PA-15 .38 Manurhin MR73	7.65 mm MAS-38 9 mm MAT-49 9 mm Beretta M12	5.56 mm SIG 540 5.56 mm AR70/.223 7.5 mm MAS-49/56 7.62 mm G3	7.5 mm M24/29 7.5 mm AAT-52 7.62 mm FN-MAG .50 Browning M2HB		60 mm Brandt 81 mm Brandt	89 mm LRAC 3.5 in RL M20 75 mm RCL Type 52 106 mm RCL M40
Burma						
9 mm FN 35	9 mm Sterling	5.56 mm M16A1 7.62 mm FN-FAL 7.62 mm G3	7.62 mm FN-MAG 7.62 mm MG3 .50 Browning M2HB	40 mm M79 40 mm M203	81 mm M29 82 mm M43 120 mm Tampella	84 mm Carl Gustav
Burundi						
7.65 mm Browning 1910 9 mm FN 35	9 mm Vigneron 9 mm MAT-49	7.5 mm MAS 49/56 7.62 mm FN-FAL 7.62 mm G3A3, A4	7.62 mm FN-MAG .30 Browning M1919A4 .50 Browning M2HB		82 mm M43	Blindicide 75 mm RCL Type 52
Cambodia						
9 mm FN 35 7.62 mm Tokarev	9 mm vz/23, 25	5.56 mm M16A1 7.62 mm FN-FAL 7.62 mm AK47, AKM	7.62 mm M60 7.62 mm RPD 7.62 mm Type 59 12.7 mm DShK 14.5 mm KPV		82 mm M43 120 mm Brandt	106 mm RCL M40 82 mm RCL B-10 107 mm RCL B-11
Cameroon						
7.65 mm MAS 1935S 9 mm FN 35 9 mm MAC Mle 50 .357 Manurhin MR73	9 mm MAT-49 9 mm H&K MP5	5.56 mm SIG 540 5.56 mm Steyr AUG 5.56 mm M16A1 7.5 mm MAS 49/56 7.62 mm FN-FAL	7.5 mm M24/29 7.62 mm HK21 .50 Browning M2HB 14.5 mm KPV	40 mm M203	60 mm Brandt 81 mm Brandt 81 mm L16 120 SB Brandt 120 mm rifled Brandt	57 mm RCL Type 52 89 mm LRAC 106 mm RCL M40 MILAN ATGW HOT ATGW TOW ATGW
Canada						
9 mm FN 35	9 mm C1 9 mm H&K MP5A2	5.56 mm C7 5.56 mm C8 7.62 mm FN-FAL 7.62 mm Parker-Hale	5.56 mm C9 7.62 mm FN-FAL/HB 7.62 mm FN-MAG 7.62 mm Browning C1 .50 Browning M2HB		60 mm M19 81 mm C3 107 mm M30 120 Brandt	66 mm LAW 84 mm Carl Gustav 106 mm RCL M40 TOW ATGW
Cape Verde						
	9 mm FMBP48	7.62 mm AK47, AKM 7.62 mm Simonov SKS	7.62 mm RPD 7.62 mm PK, RPK 7.62 mm MG3 7.62 mm SGM 12.7 mm DShK		82 mm M41 120 mm M43	89 mm LRAC RPG-7
Central African Republic						
.38 Manurhin MR73 9 mm Walther PP 9 mm MAC Mle 50	9 mm MAT-49 9 mm Uzi	5.56 mm SIG 541 5.56 mm M16A1 7.5 mm MAS 49/56 7.62 mm AK47, AKM	7.5 mm M24/29 7.5 mm AAT-52 7.62 mm RP-46 7.62 mm RPD, RPK 12.7 mm DShK		60 mm Brandt 81 mm Brandt 120 mm M43	89 mm LRAC RPG-7 106 mm RCL M40
Chad						
.38 Manurhin MR73 7.65 mm Walther PP 9 mm MAC Mle 50 9 mm Tokagypt 9 mm Walther P1 9 mm FN 35	9 mm MAT-49 9 mm Beretta M12 9 mm Beretta 38/49 9 mm Uzi	5.56 mm M16A1 7.62 vz/58 7.62 mm FN-FAL 7.62 mm CETME 58 7.62 mm SIG 542 7.62 mm G3 7.62 mm AK47, AKM	7.62 mm vz/52,53 7.62 mm AAT-52 7.62 mm PK, RPK 7.62 mm RPD 7.62 mm SGM 7.62 mm Yugo M70 12.7 mm DShK .50 Browning M2HB	30 mm AGS-17 40 mm M79	60 mm Brandt 81 mm Brandt 82 mm M43 120 mm Brandt	89 mm LRAC RPG-7 82 mm RCL B-10 106 mm RCL M40 Apilas TOW ATGW
Chile						
9 mm FN 35 9mm Beretta M92 9 mm SIG P220 9 mm Walther Pl .38 FAMAE	9 mm Madsen M53 9 mm Uzi 9 mm FAMAE 9 mm H&KM P5 9 mm Beretta M12S	5.56 mm Galil 5.56 mm HK33 5.56 mm M16A1 7.62 mm FN-FAL 7.62 mm G3 7.62 mm SIG 542 7.62 mm SIG 510	7.62 mm SIG 710-3 7.62 mm FN-FAL/HB 7.62 mm MG42/59 7.62 mm M60 .50 Browning M2HB		60 mm M19 81 mm M1 120 mm Brandt	3.5 in RL M20 106 mm RCL M40 MILAN ATGW Mamba ATGW

Pistols	Sub-machine guns	Rifles	Machine guns	Close support weapons	Mortars	Anti-tank
China, People's Republic						
7.62 mm Types 51, 54 7.62 mm Type 64 9 mm Type 59	7.62 mm Type 43 7.62 mm Type 50 7.62 mm Type 64	7.62 mm Carbine Type 53 7.62 mm Type 56 7.62 mm Type 68	7.62 mm Type 53 7.62 mm Type 56 7.62 mm Type 58 7.62 mm Type 67 12.7 mm Type 57 12.7 mm Type 77	30 mm AGS-17 35 mm W87	60 mm Type 31 82 mm Type 53 120 mm Type 53 160 mm M43	40 mm RL Type 56 40 mm RL Type 69 57 mm RCL Type 36 75 mm RCL Types 52, 56 82 mm RCL Type 65 90 mm RL Type 51 Red Arrow 8 ATGW
Colombia						
9 mm FN 35 .45 M1911A1	9 mm Madsen M46, 50, 53 9mm Walther MPK 9 mm Ingram 9 mm Uzi	5.56 mm Galil 7.62mm FN-FAL 7.62 mm G3 7.62 mm M14	7.62 mm M60 7.62 mm HK21E1 7.62 mm FN-MAG .30 BAR M1918 .30 Browning M1919A4 .50 Browning M2HB	40 mm M79	60 mm M19 81 mm M1 107 mm M30 120 mm Brandt	75 mm RCL M20 106 mm RCL M40 TOW ATGW
Comoros						
	9 mm MAT-49	7.5 mm MAS 49/56 7.62 mm AK47, AKM 7.62 mm SKS	7.5 mm AAT-52 7.62 mm RPD, RPK 7.62 mm Type 58 7.62 mm RP-46 7.62 mm SGM			
Congo						
7.65 mm Walther PP 9 mm MAC Mle 50 7.62 mm Tokarev	9 mm Vigneron 9 mm MAT-49 9 mm Franchi LF-57 9 mm Sola Super	7.5 mm MAS 49/56 7.62 mm FN-FAL 7.62 mm CETME 58 7.62 mm AK47, AKM	7.5 mm M24/29 7.62 mm RP-46 7.62 mm RPD, RPK 12.7 mm DShK 14.5 mm KPV		82 mm M41, 43 120 mm M43	57 mm RCL M18 RPG-7 Sagger ATGW
Costa Rica						
.45 M1911A1	9 mm Beretta 38/49	5.56 mm Galil 5.56 mm M16A1 5.56 mm Type 68 7.62 mm M14 7.62 mm FN-FAL .30 Garand M1	7.62 mm M60 .30 BAR M1918 .30 Browning M1919A4	40 mm M79		
Cuba						
9 mm Makarov 9 mm FN 35	9 mm vz/23,25 9 mm Star Z-45	7.62 mm vz/58 7.62 mm AK47, AKM	7.62 mm PK, RPK 7.62 mm RP-46 7.62 mm DPM 12.7 mm DShK		82 mm M41, 43 120 mm M43 160 mm M43	RPG-7 Sagger ATGW Snapper ATGW
Cyprus, Republic of						
9 mm FN 35	9 mm Sterling	7.62 mm G3 7.62 mm FN-FAL 7.62 mm vz/58	7.62 mm FN-MAG 7.62 mm SGM 12.7 mm DShK		81 mm M29 82 mm M41, 43 107 mm M2	57 mm RCL M18 106 mm RCL M40
Czechoslovakia						
7.62 mm CZ 52 7.65 mm CZ 83	7.62 mm Skorpion 7.62 mm vz/23,25	7.62 mm vz/58	7.62 mm vz/59 12.7 mm DShK 12.7 mm NSV		81 mm M48 81 mm M52 120 mm M43	RPG-7V RPG-75 82 mm RCL M59 Snapper ATGW Swatter ATGW Sagger ATGW Spiral ATGW
Denmark						
9 mm FN 35 9 mm SIG P210	9 mm Hovea M49 9 mm H&K MP5A3	5.56 mm M16A1 7.62 mm G3	7.62 mm MG42/59 .50 Browning M2HB		60 mm M51 81 mm M57 120 mm M50	84 mm Carl Gustav 106 mm RCL M40 Cobra ATGW TOW ATGW M72 LAW
Djibouti						
9 mm MAC Mle 50 9 mm MAB PA-15	9 mm MAT-49	5.56 mm SIG 540 5.56 mm FA-MAS 7.62 mm G3 7.62 mm CETME 58 7.62 mm FN-FAL	7.62 mm AAT-52 7.62 mm RPD, RPK 7.62 mm FN-MAG .50 Browning M2HB		60 mm Brandt 81 mm Brandt 120 mm Brandt	89 mm LRAC RPG-7 106 mm RCL M40 HOT ATGW
Dominican Republic						
.45 M1911A1 9 mm FB 35	9 mm Beretta M38/49 9 mm Uzi	7.62 mm G3 7.62 mm FN-FAL 7.62 mm CETME 58 7.62 mm M14	7.62 mm M60 7.62 mm FN-MAG .30 Browning M1919A4 .50 Browning M2HB	40 mm M79	81 mm M1 120 mm ECIA SL	106 mm RCL M40
Ecuador						
.45 M1911A1 9 mm FN 35	.45 M3A1 9 mm Uzi	5.56 mm Steyr AUG 5.56 mm SIG 540 5.56 mm M16A1 7.62 mm M14 7.62 mm FN-FAL	7.62 mm FN-MAG 7.62 mm Yugo MG42 .50 Browning M2HB	40 mm M203	81 mm M29 107 mm M30 160 mm M43	90 mm RCL M67 106 mm RCL M40
Egypt						
9 mm Helwan 9 mm Beretta M951	9 mm Akaba 9 mm Beretta 12S 9 mm Star Z-45 9 mm Port Said	7.62 mm SKS 7.62 mm AK47, AKM 7.62 mm SVD 7.92 mm Hakim	7.62 mm RPD 7.62 mm SGM 7.62 mm FN-MAG 7.62 mm M60 12.7 mm DShK .50 Browning M2HB		82 mm M37, 43 120 mm M43 160 mm M43	RPG-7V 107 mm RCL B-11 Snapper ATGW Sagger ATGW Swingfire ATGW MILAN ATGW HOT ATGW TOW ATGW

Pistols	Sub-machine guns	Rifles	Machine guns	Close support weapons	Mortars	Anti-tank
El Salvador 9 mm FN 35	9 mm H&K MP5 9 mm Uzi	5.56 mm HK33 5.56 mm M16A1 5.56 mm Galil 7.62 mm M14 7.62 mm G3	7.62 mm M60 7.62 mm Madsen .30 Browning M1919A4 .50 Browning M2HB	40 mm M79 40 mm M203	60 mm M19 81 mm M29 120 mm UBM-52	90 mm RCL M67 3.5 in RL M20 75 mm RCL M20
Equatorial Guinea 7.62 mm Tokarev	9 mm MAT-49	7.62 mm CETME 58 7.62 mm AK47, AKM 7.62 mm SKS	7.62 mm RPD, RPK 7.62 mm SGM 7.62 mm Type 67 7.62 mm RP-46 12.7 mm DShK		82 mm M37	
Ethiopia 9 mm Beretta M34 9 mm Makarov 7.62 mm vz/52	9 mm Beretta M38/49 9 mm Uzi	7.62 mm Type 56 7.62 mm Beretta BM59 7.62 mm vz/58 7.62 mm M14	.30 BAR M1918 7.62 mm RP-46 7.62 mm RPD, RPK 7.62 mm vz/52 12.7 mm DShK .50 Browning M2HB	40 mm M79	60 mm M19 81 mm M29 82 mm M43 107 mm M30 120 mm M43	Sagger ATGW Spigot ATGW Spandrel ATGW TOW ATGW
Fiji .45 M1911A1		5.56 mm M16A1 7.62 mm FN-FAL	7.62 mm M60	40 mm M79	81 mm M29	
Finland 9 mm Lahti M35 7.65 mm Parabellum 9 mm FN 35	9 mm Suomi M31 9 mm Suomi M44	7.62 mm M62, M76	7.62 mm M62 7.62 mm RPD 7.62 mm SGM 12.7 mm DShK, NSV		60 mm Tampella 81 mm Tampella 120 mm Tampella 120 mm M43 160 mm M43	M55 RL 66 LAW M72 SS-11 ATGW 95 mm RCL SM58-61 TOW ATGW Apilas
France 9 mm MAB PA-15 9 mm SIG P220 9 mm MAC Mle 50 9 mm Beretta 92	9 mm MAT-49 9 mm H&K MP5	5.56 mm FA-MAS 5.56 mm SG540 7.5 mm MAS 49/56 7.62 mm FR-F1, F2	7.62 mm AAT-52 .50 Browning M2HB		60 mm Brandt 81 mm Brandt 120 mm Brandt	Apilas 89 mm LRAC SS-11 ATGW MILAN ATGW HOT ATGW Eryx ATGW
Gabon .38 Manurhin MR73 9 mm MAC Mle 50 9 mm MAB PA-15	9 mm Beretta M12 9 mm MAT-49 9 mm Sterling 9 mm Uzi	5.56 mm FN-CAL 5.56 mm FA-MAS 5.56 mm M16A1 7.62 mm SIG 540 7.62 mm G3 7.62 mm AK47, AKM	7.62 mm AAT-52 7.62 mm FN-MAG .30 Browning M1919A4 .50 Browning M2HB	40 mm M203	81 mm Brandt 120 mm Brandt	89 mm LRAC 57 mm RCL M18 106 mm RCL M40 Armbrust
Gambia, The .38 Webley	9 mm Sterling	7.62 mm FN-FAL 7.62 mm FN-MAG .50 Browning M2HB	7.62 mm Bren L4			
Germany, East 9 mm Pistole M	7.62 mm PPSh-41	5.45 mm AK-74 7.62 mm Karabiner-S 7.62 mm MPiK, MPiKM 5.56 mm KKMPi69	5.45 mm RPK-74 7.62 mm LMGK (RPK) 7.62 mm PK, PKB, PKS 7.62 mm SGM 12.7 mm DShK, NSV		81 mm M37M 120 mm M43	RPG-7V, 7D 73 mm SPG-9 Snapper ATGW Sagger ATGW Spandrel ATGW Spigot ATGW
Germany, West 9 mm Pistole 1 9 mm Pistole 6 9 mm SIG P226	9 mm Uzi 9 mm H&K MP5 5.56 mm H&K MP53	7.62 mm G3 7.62 mm G3SG1 7.62 mm Mauser SP66	7.62 mm MG3	40 mm Granatpistole 40 mm HK79	120 mm Soltam 120 mm Brandt	Lanze 84 mm Carl Gustav Armbrust TOW ATGW MILAN ATGW HOT ATGW Panzerfaust 3
Ghana 9 mm FN 35	9 mm Sterling 9 mm H&K MP5	5.56 mm M16A1 5.56mm HK33 7.62 mm FN-FAL 7.62 mm G3	7.62 mm FN-MAG 7.62 mm Bren L4 12.7 mm DShK .50 Browning M2HB		81 mm Tampella 120 mm Tampella	84 mm Carl Gustav RPG-2
Greece 9 mm EP9S 9mm EP7	9 mm Steyr Mi69 9 mm H&K MP5	5.56 mm M16A2 5.56 mm HK33E 7.62 mm FN-FAL 7.62 mm G3	7.62 mm M60 7.62 mm HK11A1 7.62 mm FN-MAG .50 Browning M2HB	40 mm M79 40 mm M203	60 mm M2 81 mm M1/M29 107 mm M30	57 mm RCL M18 90 mm RCL M67 106 mm RCL M40 TOW ATGW MILAN ATGW Cobra ATGW
Guatemala .45 M1911A1 9 mm Star 9 mm FN 35	.45 M3A1 9 mm Uzi 9 mm Beretta 12	5.56 mm M16A1 7.62 mm Galil .30 Carbine M1	7.62 mm FN-MAG .30 Browning M1919A4 .50 Browning M2HB	40 mm M79	60 mm M2 81 mm M1 107 mm M30 120 mm ECIA SL	3.5 in RL M20 106 mm RCL M40

Pistols	Sub-machine guns	Rifles	Machine guns	Close support weapons	Mortars	Anti-tank
Guinea						
7.62 mm Tokarev	9 mm MAT-49 9 mm vz/23, 25 7.62 mm PPSh-41	7.62 mm vz/58 7.62 mm AK47, AKM 7.62 mm SKS	7.62 mm PK 7.62 mm SGM 12.7 mm DShK		82 mm M43 120 mm M43	RPG-2, -7 82 mm RCL B-10 Sagger ATGW Swatter ATGW
Guinea-Bissau						
7.62 mm Tokarev 7.62 mm CZ 52	9 mm vz/23, 25 9 mm FMBP M948	7.62 mm AK47, AKM 7.62 mm vz/52/57 7.62 mm SKS	7.62 mm PK, RPK 7.62 mm SGM 12.7 mm DShK 14.5 mm KPV		82 mm M43 120 mm M43	75 mm RCL Type 52 3.5 in RL M20 Sagger ATGW Spiral ATGW 82 mm RCL B-10
Guyana						
7.65 mm Walther PPK 9 mm Smith & Wesson M39	9 mm Sterling 9 mm Beretta M12	7.62 mm FN-FAL 7.62 mm G3 7.62 mm AK47, AKM 7.62 mm SKS	7.62 mm Bren L4 7.62 mm FN-MAG		81 mm L16 120 mm M43	
Haiti						
.45 M1911A1 9 mm Beretta 951	.45 Thompson 9 mm Uzi, Mini-Uzi	5.56 mm Galil 5.56 mm M16A1 7.62 mm G3 .30 Garand M1	7.62 mm M60 .30 Browning M1919A4		60 mm M2 81 mm M1	57 mm RCL M18 106 mm RCL M40
Honduras						
9 mm FN 35	9 mm Beretta 93R 9 mm Uzi, Mini-Uzi 9 mm H&K MP5	5.56 mm Mini-14 5.56 mm M16A1 7.62 mm FN-FAL 7.62 mm M14	7.62 mm M60 7.62 mm FN-MAG .50 Browning M2HB	40 mm M79 40 mm M203	60 mm M2 81 mm M1 120 mm Soltam	57 mm RCL 18 84 mm Carl Gustav 106 mm RCL M40
Hungary						
7.62 mm M48 7.65 mm M48	7.62 mm 48M	7.62 mm AMD 7.62 mm AK47, AKM 7.62 mm M48	7.62 mm RPK 7.62 mm PK, PKB, PKS 12.7 mm DShK		81 mm M37M 120 mm M43	RPG7V, 7D 73 mm SPG-9 107 mm RCL B-11 Snapper ATGW Sagger ATGW Spigot ATGW
India						
9 mm FN 35 9 mm Glock	9 mm Sterling 9 mm H&K MP5K	7.62 mm FN-FAL	7.62 mm Bren L4 7.62 mm FN-MAG .50 Browning M2HB		81 mm L16 82 mm M43 120 mm M43 160 mm M43	3.5 in RL M20 57 mm RCL M18 106 mm RCL M40 SS-11 ATGW MILAN ATGW Sagger ATGW
Indonesia						
9 mm Pindad 9 mm Beretta 92	9 mm Beretta M12	5.56 mm M16A1 5.56 mm SIG 541 5.56 mm FN-FNC 7.62 mm vz/52/57 7.62 mm FN-FAL 7.62 mm Beretta BM59	5.56 mm Minimi 7.62 mm FN-MAG 7.62 mm M60 12.7 mm DShK .50 Browning M2HB	40 mm M79 40 mm M203	60 mm Pindad 60 mm M2 81 mm Pindad 81 mm M29 120 mm UBM-52	3.5 in RL M20 75 mm RCL BO-10 105 mm RCL BO-11 106 mm RCL M40 84 mm Carl Gustav Entac ATGW
Iran						
.45 M1911A1 9 mm Beretta 92 9 mm SIG P220	9 mm Uzi 9 mm H&K MP5 9 mm Beretta M12	7.62 mm G3 7.62 mm AK47, AKM 7.62 mm Dragunov SVD	7.62 mm MG1A1 7.62 mm PK' RPK 12.7 mm DShK .50 Browning M2HB	30 mm AGS-17 40 mm M79	60 mm M19 81 mm M29 107 mm M30 120 mm Soltam M65	3.5 in RL M20 RPG-7V 57 mm RCL M18 75 mm RCL M20 106 mm RCL M40 Dragon ATGW Entac ATGW TOW ATGW
Iraq						
7.62 mm Tokarev 9 mm FN 35 9 mm CZ 75	9 mm Sterling	7.62 mm AK47, AKM 7.62 mm SKS 7.62 mm SVD	7.62 mm FN-MAG 7.62 mm RPD 7.62 mm SGM 12.7 mm DShK		81 mm M37 120 mm M43 160 mm M43	RPG-7 Sagger ATGW HOT ATGW MILAN ATGW
Ireland						
9 mm FN 35	9 mm FFV M45 9 mm Uzi 9 mm HK53	5.56 mm Steyr AUG 7.62 mm FN-FAL	7.62 mm FN-MAG .50 Browning M2HB		60 mm Brandt 81 mm Swedish 120 mm Swedish	84 mm Carl Gustav 90 mm FFV RCL gun MILAN ATGW
Israel						
9 mm Beretta M951	9 mm Uzi, Mini-Uzi	5.56 mm Galil 5.56 mm M16A1 7.62 mm FN-FAL 7.62 mm M14 7.62 mm AK47, AKM	7.62 mm FN-FAL/HB 7.62 mm FN-MAG .30 Browning M1919A4 12.7 mm DShK .50 Browning M2HB	40 mm M79 40 mm M203 40 mm Mk19	52 mm IMI 81 mm Soltam 120 mm Soltam M65 160 mm M66	84 mm Carl Gustav 106 mm RCL M40 Dragon ATGW TOW ATGW Sagger ATGW MAPATS ATGW
Italy						
9 mm Beretta M34 9 mm Beretta M951 9 mm Beretta M92	9 mm Beretta M38/49 9 mm Beretta M12S 9 mm Franchi LF-57	5.56 mm AR70/90 7.62 mm BM59	7.62 mm MG42/59 7.62 mm M73 .50 Browning M2HB		81 mm OTO M62 120 mm Brandt	3.5 in RL M20 75 mm RCL M20 106 mm RCL M40 Cobra ATGW TOW ATGW MILAN ATGW 80 mm Folgore Apilas

Pistols	Sub-machine guns	Rifles	Machine guns	Close support weapons	Mortars	Anti-tank
Ivory Coast 9 mm MAB PA-15 9 mm MAS Mle 50 .357 Manurhin MR73	9 mm MAT-49	5.56 mm SIG 540 7.5 mm MAS 49/56 7.62 mm G3 7.62 mm FN 30-11	7.5 mm M24/29 7.5 mm AAT-52 .50 Browning M2HB		81 mm Brandt 120 mm Brandt	89 mm LRAC 106 mm RCL M40
Jamaica 9 mm FN 35	9 mm Sterling	5.56 mm M16A1 7.62 mm FN-FAL	7.62 mm FN-FAL/HB 7.62 mm FN-MAG .50 Browning M2HB		81 mm L16A1	
Japan 9 mm SIG P220 .38 New Nambu	9 mm H&K MP5 9 mm SCK M66	7.62 mm Type 64	7.62 mm Type 62 .50 Browning M2HB		60 mm M1 81 mm M1 81 mm Type 64 107 mm M30 120 mm Brandt	84 mm Carl Gustav 106 mm RCL M40 Type 64 ATGW TOW ATGW KAM-3D ATGW
Jordan 9 mm FN 35 9 mm Glock	9 mm Sterling 9 mm H&K MP5K	5.45 mm AK47 5.56 mm AR70/90 5.56 mm M16A1, A2 7.62 mm G3	7.62 mm FN-MAG 7.62 mm HK21E 7.62 mm M60 .50 Browning M2HB	40 mm M79 40 mm M203	81 mm M29 107 mm M30 120 mm Brandt	106 mm RCL M40 TOW ATGW Dragon ATGW HOT ATGW Apilas
Kenya 9 mm FN 35	9 mm Sterling 9 mm Uzi 9 mm H&K MP5	7.62 mm FN-FAL 7.62 mm G3	7.62 mm HK21A1 7.62 mm Bren L4 7.62 mm FN-MAG 7.62 mm AAT-52	40 mm M79	81 mm L16 120 mm Brandt	84 mm Carl Gustav 120 mm RCL Wombat MILAN ATGW Swingfire ATGW
Korea, North 7.62 mm Type 68 7.65 mm Type 64	7.62 mm Type 49	7.62 mm Type 63 (SKS) 7.62 mm Type 58 (AK47) 7.62 mm Type 68 (AKM)	7.62 mm Type 64 (RPK) 7.62 mm RPD 7.62 mm PK, PKB, PKS 12.7 mm DShK		82 mm M37 120 mm M43 160 mm M43	RPG-2 RPG-7V 82 mm SPG-82 82 mm RCL B-10 Snapper ATGW Sagger ATGW
Korea, South .45 M1911A1	.45 M3A1	5.56 mm K2 5.56 mm M16A1 .30 Carbine M1 .30 Garand M1	5.56 mm K3 7.62 mm M60 7.62 mm FN-MAG .30 Browning M1919A4 .50 Browning M2HB	40 mm M79 40 mm M203	60 mm M2/KM19 81 mm KM29 107 mm M30	66 mm LAW M72 3.5 in RL M20 106 mm RCL M40 TOW ATGW
Kuwait 9 mm FN 35	9 mm Sterling	7.62 mm FN-FAL	7.62 mm FN-MAG .50 Browning M2HB		81 mm L16	TOW ATGW HOT ATGW
Laos 9 mm Makarov 9 mm CZ 70 7.62 mm Tokarev	9 mm MAT-49 7.62 mm PPSh-41	7.62 mm AK47, AKM 7.62 mm Type 56 7.62 mm SKS	7.62 mm RP-46, RPD 12.7 mm DShK		81 mm M1 82 mm M43 107 mm M2A1 107 mm M1938 120 mm M43	57 mm RCL M18 106 mm RCL M40 107 mm RCL B-11
Lebanon 9 mm FN 35 9 mm Walther P-38 9 mm Colt Commander	9 mm MAT-49 9 mm Sterling	5.56 mm FA-MAS 5.56 mm FN-CAL 5.56 mm M16A1 5.56 mm SIG 540 7.62 mm FN-FAL 7.62 mm G3	7.5 mm M24/29 7.62 mm AAT-52 7.62 mm FN-MAG .50 Browning M2HB	40 mm M79 40 mm M203	60 mm Brandt 81 mm Brandt 81 mm M29 120 mm Brandt	89 mm RL M65 RPG-7 106 mm RCL M40 TOW ATGW MILAN ATGW
Lesotho .38 Enfield	9 mm Sterling	5.56 mm Beretta AR70/90 5.56 mm M16A1 5.56 mm Galil 7.62 mm AK47, AKM	7.62 mm FN-MAG 7.62 mm Bren L4 7.62 mm SGM			
Liberia .45 M1911A1 9 mm FN 35	9 mm Uzi	5.56 mm M16A1, A2 7.62 mm FN-FAL .30 Carbine M1	5.56 mm M16/HB 7.62 mm FN-FAL/HB 7.62 mm SIG 710-3 7.62 mm M60 .30 BAR M1918 .30 Browning M1919A4 .50 Browning M2HB	40 mm M203	60 mm M31 81 mm M29 107 mm M30	3.5 in RL M20 106 mm RCL M40
Libya 9 mm Beretta M951 7.62 mm Tokarev 9 mm Stechkin 9 mm Makarov	9 mm Beretta M12 9 mm Sterling L34 7.65 mm Skorpion	7.62 mm FN-FAL 7.62 mm Beretta BM59 7.62 mm AK47, AKM 7.62 mm SKS 7.62 mm G3 7.62 mm vz/58	7.62 mm RP-46 7.62 mm RPD, RPK 7.62 mm FN-MAG 12.7 mm DShK		81 mm M1 82 mm M37 107 mm M30 120 mm M43 160 mm M43	RPG-7 84 mm Carl Gustav 106 mm RCL M40 107 mm RCL B-11 Swingfire ATGW Sagger ATGW TOW ATGW
Luxembourg 9 mm FN 35	9 mm Uzi 9 mm H&K MP5	7.62 mm FN-FAL	7.62 mm FN-MAG .50 Browning M2HB		81 mm M29	106 mm RCL M40 TOW ATGW

Pistols	Sub-machine guns	Rifles	Machine guns	Close support weapons	Mortars	Anti-tank
Madagascar 7.65 mm Walther PP 7.62 mm Tokarev 7.62 mm vz/52 9 mm MAB PA-15	9 mm MAT-49	7.62 mm SKS 7.62 mm AK47, AKM 7.62 mm Type 68	7.5 mm AAT-52 12.7 mm DShK .50 Browning M2HB		60 mm M1 82 mm M37 120 mm M43	89 mm LRAC RPG-7 106 mm RCL M40 SS-11 ATGW
Malawi 9 mm FN 35	9 mm Sterling	7.62 mm FN-FA1 7.62 mm G3 7.62 mm CETME L	7.62 mm FN-MAG 14.5 mm KPV		81 mm L16	3.5 in RL M20
Malaysia 9 mm FN 35 9 mm H&K P9S	9 mm Sterling	5.56 mm M16A1 5.56 mm AR70/90 5.56 mm HK33E 7.62 mm FN-FAL 7.62 mm G3SG1	5.56 mm AS70/90 7.62 mm Bren L4 7.62 mm FN-MAG 7.62 mm HK11A1 7.62 mm HK21E .50 Browning M2HB	40 mm M79 40 mm M203	81 mm L16	84 mm Carl Gustav 3.5 in RL M20 120 mm RCL Mobat 106 mm RCL M40 SS-11 ATGW
Mali 7.65 mm Walther PP	9 mm MAT-49	7.5 mm MAS 49/56 7.62 mm AK47, AKM	7.5 mm AAT-52 7.62 mm PK, RPK 7.62 mm SGM 12.7 mm DShK		82 mm M43 120 mm M43	RPG-2 RPG-7 Sagger ATGW Swatter ATGW
Malta 7.62 mm Tokarev 9 mm Beretta M34 9 mm Makarov	9 mm H&K MP5K 9 mm Sterling 9 mm Uzi	7.62 mm FN-FAL 7.62 mm AKM	7.62 mm RPD, RPK 12.7 mm DShK			RPG-7
Mauritania 9 mm MAC Mle 50 9 mm MAB PA-15 7.62 mm Tokarev	9 mm MAT-49 9 mm Star Z-45	7.5 mm MAS 49/56 7.62 mm FR-F1 .30 Carbine M1	7.92 mm MG42 7.5 mm AAT-52 7.62 mm FN-MAG .30 Browning M1919A4 .50 Browning M2HB		60 mm Brandt 81 mm Brandt 120 mm Brandt 120 mm Ecia	RPG-7 57 mm RCL M18 75 mm RCL M20 106 mm RCL M40 MILAN ATGW
Mauritius .38 Manurhin MR73 7.65 mm Walther PP	9 mm H&K MP5	5.56 mm SIG 540	7.5 mm AA-52 7.62 mm Bren L4		81 mm Brandt	
Mexico .45 M1911A1 9 mm H&K P7M13	9 mm Mendoza 9 mm HK53 9 mm H&K MP5	5.56 mm M16A1 5.56 mm HK33E 7.62 mm FN-FAL 7.62 mm G3	5.56 mm Ameli 7.62 mm FN-MAG 7.62 mm HK21A1 .50 Browning M2HB		60 mm M2 81 mm M1 107 mm M30 120 mm Brandt	106 mm RCL M40 MILAN ATGW
Morocco 7.65 mm HK4 9 mm HK VP70 9 mm MAC Mle 50 9 mm MAB PA-15	9 mm Beretta 38/49 9 mm MAT-49 9 mm H&K MP5	5.56 mm Beretta 70/223 5.56 mm M16A1 5.56 mm Steyr AUG 7.62 mm AK47, AKM 7.62 mm FN-FAL 7.62 mm G3 7.62 mm Valmet M76 7.62 mm Beretta BM59	7.62 mm AAT-52 7.62 mm RPD 7.62 mm M60 .50 Browning M2HB		60 mm M2 81 mm Ecia LN 82 mm M37 120 mm M43 120 mm Brandt 120 mm Ecia	66 mm M72 LAW RPG-7V 89 mm LRAC 3.5 in RL M20 106 mm RCL M40 Dragon ATGW MILAN ATGW TOW ATGW HOT ATGW
Mozambique 7.62 mm Tokarev 9 mm Stechkin 9 mm Makarov 9 mm FN 35 9 mm Walther P-38	9 mm Star Z-75 9 mm FMBP M48 9 mm FMBP M63 9 mm vz/23, 25 9 mm Franchi LF-57	7.62 mm AK47, AKM 7.62 mm vz/58 7.62 mm FN-FAL 7.62 mm SKS	7.62 mm PK, RPK 7.62 mm SGM 7.92 mm vz/53 12.7 mm DShK	30 mm AGS-17	82 mm M43 120 mm M43	82 mm RCL B-10 107 mm RCL B-12 Sagger ATGW Spigot ATGW
Nepal 9 mm FN 35	9 mm Sterling	7.62 mm FN-FAL	7.62 mm Bren L4 7.62 mm FN-MAG		81 mm M29 107 mm M30 120 mm M43	
Netherlands 9 mm FN 35 9 mm Walther P5 9 mm H&K P9S	9 mm Uzi 9 mm H&K MP5	7.62 mm FN-FAL 7.62 mm Steyr SSG69	7.62 mm FN-MAG 7.62 mm FN-FAL/HB .50 Browning M2HB		81 mm M1 107 mm M30 120 mm Brandt	66 mm M72 LAW 84 mm Carl Gustav 106 mm RCL M40 Dragon ATGW TOW ATGW
New Zealand 9 mm FN 35	9 mm Sterling 9 mm H&K MP5	5.56 mm Steyr AUG 5.56 mm M16A1 7.62 mm FN-FAL 7.62 mm Parker-Hale 82	7.62 mm FN-FAL/HB 7.62 mm Bren L4 7.62 mm FN-MAG .50 Browning M2HB	40 mm M79 40 mm M203	81 mm L16 81 mm Brandt 120 mm Soltam	66 mm M72 LAW 84 mm Carl Gustav 106 mm RCL M40
Nicaragua .45 M1911A1 9 mm Makarov	9 mm Uzi 9 mm vz/23, 25 9 mm Madsen M50	5.56 mm M16A1 5.56 mm Galil 5.56 mm SIG 540	7.62 mm M60 7.62 mm RPD, RPK 12.7 mm DShK .50 Browning M2HB	30 mm AGS-17 40 mm M79	60 mm M29 82 mm M43 120 mm M43	
Niger 7.65 mm Walther PP 9 mm MAB PA-15	9 mm MAT-49 9 mm H&K MP5 9 mm Uzi	7.5 mm AS 49/56 7.62 mm G3 7.62 mm M14	7.62 mm HK11 7.62 mm HK21 7.62 mm AAT-52 .50 Browning M2HB		60 mm Brandt 81 mm Brandt 120 mm SB Brandt 120 mm rifled Brandt	89 mm LRAC 57 mm RCL M18 75 mm RCL M20

Pistols	Sub-machine guns	Rifles	Machine guns	Close support weapons	Mortars	Anti-tank
Nigeria 9 mm FN 35 9 mm SIG P220 9 mm Beretta M951	9 mm Sterling 9 mm Beretta M12 9 mm Franchi LF-57 9 mm vz/23, 25 9 mm HK MP5 9 mm Uzi	5.56 mm M16A1 5.56 mm AR70/223 5.56 mm FN-FNC 5.56 mm SIG 540 7.62 mm FN-FAL 7.62 mm G3 7.62 mm Steyr SSG69 7.62 mm vz/52 7.62 mm Beretta BM59	5.56 mm AUG/HBAR 7.62 mm HK21 7.62 mm FN-MAG 7.62 mm RP-46 7.62 mm RPD, RPK 12.7 mm DShK .50 Browning M2HB		60 mm M29 81 mm L16 81 mm M43	89 mm LRAC 3.5 in RL M20 RPG-7 84 mm Carl Gustav Swingfire ATGW
Norway 9 mm Glock 9 mm Walther P1 9 mm HK P7M8 .45 M1912	9 mm MP40 9 mm H&K MP5	7.62 mm G3	7.62 mm MG3 .50 Browning M2HB	40 mm HK79	81 mm L16 107 mm M30	66 mm M72LAW 3.5 in RL M20 57 mm RCL M18 75 mm RCL M20 84 mm Carl Gustav 106 mm RCL M40 TOW ATGW
Oman 9 mm FN 35	9 mm Sterling	5.56 mm Steyr AUG 5.56 mm M16A1 5.56 mm SIG 540 7.62 mm FN-FAL	7.62 mm FN-MAG .50 Browning M2HB	40 mm M79 40 mm M203	60 mm Brandt 81 mm L16 107 mm M30 120 mm Brandt	TOW ATGW MILAN ATGW
Pakistan 9 mm Walther P-38	9 mm Sterling 9 mm H&K MP5	7.62 mm G3 7.62 mm Type 56 (AKM)	7.62 mm RPD 7.62 mm MG1A3 .50 Browning M2HB 12.7 mm Type 54		60 mm PMT 81 mm PMT 120 mm PMT	75 mm RCL Type 52 RPG-7 106 mm RCL M40 Cobra ATGW TOW ATGW
Panama 9 mm FN 35	9 mm Uzi	5.56 mm M16A1 5.56 mm Type 65 .30 Garand M1	5.56 mm M16A1/HB 7.62 mm FN-MAG 7.62 mm M60 .30 Browning M1919A4 .50 Browning M2HB	40 mm M203	60 mm M19 81 mm M29	
Papua New Guinea 9 mm FN 35	9 mm Sterling	5.56 mm Steyr AUG (F88) 7.62 mm FN-FAL	7.62 mm FN-MAG			
Paraguay 9 mm FN 35 9 mm HK P9S 9 mm HK VP70Z	9 mm Madsen 9 mm Carl Gustav 9 mm Uzi	5.56 mm AR70/90 5.56 mm SIG 540 7.62 mm FN-FAL	7.62 mm FN-FAL/HB 7.62 mm Madsen .50 Browning M2HB	40 mm M79	81 mm Brandt 107 mm M2 107 mm M30	75 mm RCL M20
Peru 9 mm FN 35 9 mm Star 30M	9 mm Uzi 9 mm SIMA 79 9 mm Star Z-45, Z-62	5.56 mm M16A1 7.62 mm FN-FAL 7.62 mm G3 7.62 mm AK47, AKM 7.62 mm Steyr SSG69	7.62 mm FN-MAG 7.62 mm M60 .50 Browning M2HB 12.7 mm DShK		81 mm Brandt 120 mm Brandt 120 mm Ecia	106 mm RCL M40 Cobra ATGW SS-11 ATGW
Philippines .45 M1911A1 9 mm FN 35 9 mm Glock	.45 M3A1 9 mm Uzi	5.56 mm M16A1 5.56 mm Galil 7.62 mm M14 7.62 mm G3	5.56 mm Ultimax 100 7.62 mm M60 7.62 mm FN-MAG .50 Browning M2HB	40 mm M79 40 mm M203	81 mm M29 107 mm M30	75 mm RCL M20 90 mm RCL M67 106 mm RCL M40
Poland 9 mm P-64 9 mm Makarov	9 mm PM-63	7.62 mm PMK 7.62 mm PMK-DGN 7.62 mm Dragunov SVD	7.62 mm RPD 7.62 mm RPK 7.62 mm PK, PKS		82 mm M37M 120 mm M43 160 mm M160	RPG-7V 82 mm RCL B-10 73 mm RCL SPG-9 Sagger ATGW Snapper ATGW
Portugal 9 mm Walther P1 9 mm FN 35 9 mm HK VP70M	9 mm FMBP 63, 76 9 mm Uzi 9 mm Sterling 9 mm Star Z-45	5.56 mm HK33E 7.62 mm G3	7.62 mm FN-MAG 7.62 mm HK21 7.62 mm MG42/59 .50 Browning M2HB	40 mm M79	60 mm FBP 81 mm FBP 107 mm M30 120 mm Brandt	3.5 in RL M20 75 mm RCL M20 106 mm RCL M40 TOW ATGW SS-11 ATGW
Qatar 9 mm HK4	9 mm Sterling 9 mm H&K MP5	5.56 mm M16A1 5.56 mm Valmet M76 7.62 mm G3 7.62 mm AK47, AKM	5.56 mm M16A1/HB 7.62 mm HK21 7.62 mm Valmet 62 7.62 mm FN-MAG .50 Browning M2HB	40 mm M203	81 mm L16	HOT ATGW
Romania 7.62 mm Tokarev 9 mm Makarov	9 mm Orita 9 mm vz/24, 26	7.62 mm AK47, AKM	7.62 mm RPD, RPK 7.62 mm SGM 7.62 mm PK, PKS		82 mm M37M 120 mm M43	RPG-7V 73 mm RCL SPG-9 Snapper ATGW
Rwanda 7.65 mm Browning M1910 9 mm FN 35 9 mm MAB PA-15	9 mm Vigneron 9 mm Uzi	7.5 mm MAS 49/56 7.62 mm FN-FAL	7.62 mm FN-MAG 7.62 mm FN-FAL/HB .50 Browning M2HB		81 mm Brandt	Blindicide

Pistols	Sub-machine guns	Rifles	Machine guns	Close support weapons	Mortars	Anti-tank
St Vincent & Grenadines						
		5.56 mm M16A1 7.62 mm FN-FAL	7.62 mm M60	40 mm M79		
Sao Tome & Principe						
	7.62 mm PPS-43	7.62 mm AK47, AKM 7.62 mm SKS 7.62 mm SGM 12.7 mm DShK 14.5 mm KPV	7.62 mm MG3 7.62 mm PK			RPG-7
Saudi Arabia 9 mm FN 35 9 mm HK P9S	9 mm MPi69 9 mm Beretta M12 9 mm H&K MP5	5.56 mm Steyr AUG 5.56 mm HK33E 7.62 mm G3 7.62 mm Steyr SSG69	5.56 mm Steyr AUG/HBAR 7.62 mm MG3 7.62 mm FN-MAG .50 Browning M2HB	40 mm M79 40 mm HK69	81 mm L16 81 mm M29 107 mm M30	75 mm RCL M20 90 mm RCL M67 TOW ATGW HOT ATGW
Senegal 7.65 mm Walther PP 9 mm MAB PA-15 .38 Manurhin MR73	9 mm MAT-49 7.65 mm MAS-38 5.56 mm HK53	5.56 mm SIG 540 5.56 mm KH33E 5.56 mm FA-MAS 7.5 mm MAS 49/56 7.62 mm G3	7.5 mm AAT-52 7.62 mm HK21 .50 Browning M2HB		81 mm Brandt 120 mm Brandt 120 mm Rifled Brandt	89 mm LRAC RPG-7 MILAN ATGW
Seychelles .38 Manurhin MR73 7.65 mm Walther PP	9 mm MAT-49	5.56 mm SIG 540 7.62 mm AK47, AKM 7.62 mm SKS 7.5 mm MAS 49/56	7.62 mm Bren L4 7.62 mm RPD, RPK 7.62 mm RP-46 7.62 mm FN-MAG 7.62 mm AAT-52 12.7 mm DShK		82 mm M43	RPG-7
Sierra Leone 9 mm FN 35 7.62 mm Tokarev	9 mm Sterling 7.62 mm PPSh-41 7.62 mm PPS-43	7.62 mm AK47, AKM 7.62 mm SKS 7.62 mm FN-FAL	7.62 mm FN-MAG 7.62 mm RPD .303 Vickers 12.7 mm DShK		60 mm Brandt 81 mm Brandt	84 mm Carl Gustav 3.5 in RL M20
Singapore 9 mm FN 35 9 mm SIG P226 9 mm HK P7M8	9 mm Sterling 9 mm H&K MP5	5.56 mm M16A1 5.56 mm SAR80 5.56 mm SR88	5.56 mm M16A1/HB 5.56 Ultimax 100 7.62 mm FN-MAG .50 Browning M2HB	40 mm M203 40 mm CIS-40GL	60 mm Soltam/ODE 81 mm Soltam 120 mm Soltam 120 mm Tampella	84 mm Carl Gustav 3.5 in RL M20 106 mm RCL M40 TOW ATG
Somalia 7.62 mm vz/52 7.62 mm Tokarev 9 mm Makarov	9 mm Sterling 9 mm Uzi 9 mm Beretta M12 9 mm vz/23, 25	5.56 mm SAR80 5.56 mm M16A1 7.62 mm AK47, AKM 7.62 mm vz/58 7.62 mm M14 7.62 mm G3	7.62 mm AAT-52 7.62 mm RPD, RPK 7.62 mm SGM 7.62 mm RP-46 12.7 mm DShK .50 Browning M2HB	40 mm M79	81 mm M1941 120 mm M43	RPG-2 RPG-7 89 mm LRAC MILAN ATGW Sagger ATGW
South Africa 9 mm Z88 9 mm FN 35 9 mm Star	9 mm Uzi 9 mm Sterling 9 mm BXP	5.56 mm R4, R5 7.62 mm FN-FAL 7.62 mm G3	7.62 mm SS-77 7.62 mm FN-MAG 7.62 mm Browning M1919 .50 Browning M2HB	30 mm AGS-17 40 mm Armscor 40 mm MGL-6	60 mm M1 81 mm M3 120 mm Brandt	RPG-7 Entac ATGW
Spain 9 mm Llama 82 9 mm Astra A80 9 mm Star 30M 9 mm H&K P9S	9 mm Star Z-45, 62 9 mm Star Z-70B, 84 9 mm H&K MP5	5.56 mm CETME L, LC 5.56 mm HK33E 7.62 mm CETME C	5.56 mm Ameli 7.62 mm FN-MAG 7.62 mm MG1A3 7.62 mm MG 42/59 .50 Browning M2HB	40 mm M79	60 mm Ecia L 81 mm Ecia LL 105 mm Ecia L 120 mm Ecia SL	89 mm RL M69 106 mm RCL M40 Dragon ATGW MILAN ATGW TOW ATGW
Sri Lanka 7.62 mm Type 50 9 mm FN 35	9 mm Sterling 9 mm H&K MP5A3	5.56 mm FN-FNC 5.56 mm SAR80 5.56 mm M16A1 7.62 mm SKS 7.62 mm Type 56	5.56 mm Minimi 7.62 mm Type 58 7.62 mm FN-MAG 7.62 mm HK11 7.62 mm HK21	40 mm M203 40 mm HK69	82 mm M43 107 mm M43 120 mm M43	82 mm RCL M60 106 mm RCL M40
Sudan 9 mm Helwan 9 mm H&K P9S 9 mm FN 35	9 mm Sterling 9 mm H&K MP5 9 mm Beretta M12 9 mm Uzi	7.62 mm SKS 7.62 mm G3 7.62 mm AK47, AKM	7.62 mm HK21 7.62 mm RPD, RPK 7.62 mm RP-46 7.62 mm M60 7.62 mm SGM 7.62 mm MG3		81 mm M37M 82 mm M43 120 mm SB Brandt 120 mm Rifled Brandt	RPG-7 Swingfire ATGW
Surinam 9 mm FN 35	9 mm Uzi	7.62 mm FN-FAL 7.62 mm AKM	7.62 mm FN-MAG 7.62 mm Bren L4		81 mm M29	106 mm RCL M40
Swaziland						
	9 mm Sterling 9 mm Uzi	5.56 mm Galil 5.56 mm SIG 540 5.56 mm AR-18 7.62 mm FN-FAL	7.62 mm FN-MAG 7.62 mm Bren L4			

Pistols	Sub-machine guns	Rifles	Machine guns	Close support weapons	Mortars	Anti-tank
Sweden 9 mm m/07 9 mm FN 35 9 mm SIG P210	9 mm FFV M45(D)	5.56 mm AK-5 7.62 mm AK-4	7.62 mm M36 7.62 mm M39 7.62 mm FN-MAG		81 mm M29 120 mm M41D	74 mm Miniman 84 mm Carl Gustav 90 mm RCL Pvpjäs 1110 TOW ATGW BILL ATGW
Switzerland 9 mm SIG P210	9 mm SIG 310	5.56 mm Stgw 90 7.5 mm Kar 55 7.5 mm Stgw 57	7.5 mm M25 7.5 mm MG51		60 mm Brandt 81 mm Mw33 81 mm Mw72	83 mm RL RR50 83 mm RL RR58 106 mm RCL M40
Syria 7.62 mm Tokarev 9 mm Makarov	9 mm vz/23, 25	7.62 mm AK47, AKM 7.62 mm FN-FAL 7.62 mm Steyr SSG69	7.62 mm RPD, RPK 7.62 mm SGM 12.7 mm DShK, NSV		82 mm M41, 43 120 mm M43 160 mm M43	82 mm RCL SPG-9 RPG-7V Sagger ATGW Snapper ATGW HOT ATGW
Taiwan 9 mm FN 35 .45 M1911A1	.45 M3A1 9 mm M3A1	5.56 mm M16A1 5.56 mm Type 65 7.62 mm M14	5.56 mm Type 75 7.62 mm Type 74 7.62 mm M60 .30 Browning M1919 .50 Browning M2HB	60 mm M2	3.5 in RL M20 81 mm M29 81 mm Type 75	75 mm RCL Type 52 90 mm RCL M67 TOW ATGW
Tanzania 7.62 mm vz/52 9 mm Stechkin 9 mm FN 35	9 mm Sterling 9 mm vz/23,25	7.62 mm AK47, AKM 7.62 mm SKS 7.62 mm FN-FAL 7.62 mm G3 7.62 mm vz/58	7.62 mm RP-46 7.62 mm RPD, RPK 7.62 mm FN-MAG 7.62 mm SGM 12.7 mm DShK		82 mm M43 120 mm M43	RPG-7 75 mm RCL Type 52 TOW ATGW
Thailand 9 mm SIG P226 9 mm Glock .45 M1911A1	9 mm Uzi 9 mm H&K MP5	5.56 mm HK33 5.56 mm M16A1, A2	5.56 mm Minimi 7.62 mm M60 .50 Browning M2HB	40 mm M79 40 mm M203	60 mm M19 81 mm M29 107 mm M30 120 mm Brandt	66 mm M72 LAW 106 mm RCL M40 TOW ATGW Dragon ATGW Armbrust M72 LAW
Togo .38 Manurhin MR73 7.65 mm Walther PP 9 mm FN 35 9 mm MAB PA-15 9 mm SIG P240	9 mm MAT-49 9 mm Uzi	5.56 mm SIG 540 7.62 mm AK47, AKM 7.62 mm G3 7.62 mm Hakim 7.62 mm Steyr SSG69	7.5 mm AAT-52 7.62 mm MG3 7.62 mm RPD, RP-46 12.7 mm DShK .50 Browning M2HB 14.5 mm KPV		81 mm M37M 82 mm M43	89 mm LRAC 57 mm RCL M18 75 mm RCL Type 56 82 mm RCL Type 65
Tonga .38 Smith & Wesson	9 mm Sten	5.56 mm FN-FNC	.303 Bren .303 Vickers .50 Browning M2HB			
Trinidad & Tobago 9 mm FN 35	9 mm Sterling	5.56 mm M16A1 5.56 mm Galil 7.62 mm FN-FAL	7.62 mm Bren L4 7.62 mm FN-MAG 7.62 mm M60		60 mm M2	82mm B-300
Tunisia 9 mm Beretta M34 9 mm Beretta M951 9 mm FN 35 9 mm MAC Mle 50 9 mm MAB PA-15	9 mm MAT-49 9 mm Sterling 9 mm Uzi 9 mm Beretta 38/49 9 mm Beretta 12	5.56 mm Steyr AUG 7.62 mm FN-FAL 7.62 mm M14 .30 Carbine M1	5.56 mm Steyr AUG/HBAR 7.62 mm M60 7.62 mm AAT-52 .30 Browning M1919A4 .50 Browning M2HB		60 mm Brandt 81 mm M29 82 mm M43 107 mm M30 120 mm Brandt	89 mm LRAC MILAN ATGW TOW ATGW
Turkey 9 mm MKE	.45 M3A1 9 mm Rexim 9 mm H&K MP5	5.56 mm M16A2 7.62 mm G3 7.62 mm FN-FAL	7.62 mm MG3 7.62 mm FN-MAG .50 Browning M2HB	40 mm M79 40 mm M203	60 mm M2 81 mm UTI 107 mm M2 120 mm HY12D1	3.5 in RL M20 106 mm RCL M40 Cobra ATGW MILAN ATGW TOW ATGW
Uganda 7.62 mm Tokarev 9 mm FN 35	7.65 mm Skorpion 9 mm Sterling 9 mm Uzi	5.56 mm M16A1 7.62 mm SKS 7.62 mm FN-FAL 7.62 mm G3	7.62 mm M60 7.62 mm RPD, RPK 7.62 mm PK 7.62 mm Bren L4 7.62 mm FN-MAG 12.7 mm DShK		60 mm Brandt 81 mm L16 82 mm M43	Sagger ATGW
Union of Soviet Socialist Republics 5.45 mm PSM 7.62 mm Tokarev 9 mm Makarov	5.45 mm AKSU-74	5.45 mm AK-74 7.62 mm AKM 7.62 mm Dragunov 7.62 mm SKS	5.45 mm RPK-74 7.62 mm RPK, RPKS 7.62 mm PK, PKB, PKS 12.7 mm DShK, NSV 14.5 mm KPV	30 mm AGS-17 40 mm BG-15	82 mm M37M 107 mm M107 120 mm M43 160 mm M160	RPG-7V, 7D 73 mm RPG-18 73 mm RCL SPG-9 Swatter ATGW Sagger ATGW Spiral ATGW Spigot ATGW Spandrel ATGW

Pistols	Sub-machine guns	Rifles	Machine guns	Close support weapons	Mortars	Anti-tank
United Arab Emirates						
9 mm FN 35	5.56 mm HK53	5.56 mm M16A1	5.56 mm HK23E	40 mm M203	81 mm L16	84 mm Carl Gustav
9 mm H&K P7M13	9 mm H&K MP5	5.56 mm FA-MAS	5.56 mm Minimi		120 mm Brandt	SS-11 ATGW
	9 mm Sterling	7.62 mm G3	7.62 mm FN-MAG			TOW ATGW
		7.62 mm FN-FAL	.50 Browning M2HB			
		7.62 mm AKM				
United Kingdom						
9 mm FN 35	9 mm Sterling L2	5.56 mm L85A1	5.56 mm L86A1	40 mm M79	51 mm	66 mm M72 LAW
	9 mm Sterling L34	7.62 mm FN-FAL	7.62 mm FN-MAG (GPMG)	40 mm M203	81 mm L16A1	84 mm Carl Gustav
		7.62 mm L39A1	7.62 mm Bren L4			LAW-80
		7.62 mm L42A1	7.62 mm Browning L3			Swingfire ATGW
		7.62 mm L96A1	.50 Browning M2HB			MILAN ATGW
						TOW ATGW
United States of America						
9 mm Beretta 92F (M9)	9 mm Colt	5.56 mm M16A1, A2	5.56 mm M249 (Minimi)	40 mm M79	60 mm M19	66 mm M72 LAW
.45 M1911A1	9 mm H&K MP5	7.62 mm M14	7.62 mm M60	40 mm M182	81 mm M29A1	90 mm RCL M67
	.45 M3A1	7.62 mm M21	.50 Browning M2HB	40 mm M203	81 mm M252 (L16)	106 mm RCL M40
				40 mm Mk19 Mod3	107 mm M30	Dragon ATGW
					120 mm Tampella	TOW ATGW
						FFV AT-4
Uruguay						
9 mm H&K P7M8	9 mm Uzi	5.56 mm M16A1	7.62 mm FN-MAG		81 mm M1	106 mm RCL M40
9 mm HK4	9 mm H&K MP5	7.62 mm FN-FAL	.30 Browning M1919A4, A6		107 mm M30	MILAN ATGW
9 mm FN 35	9 mm Star Z-45	.30 Garand M1	.30 BAR M1918A2			
9 mm SIG P220			.50 Browning M2HB			
Venezuela						
9 mm FN 35	9 mm Uzi	5.56 mm FN-FNC	7.62 mm FN-MAG		60 mm Brandt	106 mm RCL M40
9 mm HK4	9 mm Beretta M12	7.62 mm M14	7.62 mm M60		81 mm Brandt	SS-11 ATGW
	9 mm Walther MPK	7.62 mm FN-FAL	.50 Browning M2HB		120 mm Brandt	
Vietnam						
7.62 mm Tokarev	7.62 mm K-50M	7.62 mm AK47, AKM	7.62 mm Type 53, 67		60 mm M19	RPG-2, -7
7.62 mm Type 68	7.62 mm MAT-49	7.62 mm SKS	7.62 mm DPM, RPD		82 mm M43	82 mm RCL B-10
	7.62 mm PPSh-41	7.62 mm Type 56	7.62 mm SGM		81 mm M29	75 mm RCL Type 52, 57
			12.7 mm DShK		107 mm M30	107 mm RCL B-11
					120 mm M43	Sagger ATGW
					160 mm M43	
Yemen, North						
9 mm Beretta M951	9 mm Beretta 38/49	5.56 mm M16A2	7.62 mm RPD	40 mm M79	81 mm M29	75 mm RCL M20
		7.62 mm vz/52	12.7 mm DShK	82 mm M43	Dragon ATGW	
		7.62 mm FN-FAL	.50 Browning M2HB		120 mm M43	Vigilant ATGW
		7.62 mm SKS				TOW ATGW
Yemen, South						
7.62 mm Tokarev	9 mm Sterling	7.62 mm AK47, AKM	7.62 mm RPD, RPK		82 mm M43	RPG-7
	9 mm Port Said	7.62 mm SKS	7.62 mm SGM		120 mm M43	
			12.7 mm DShK		160 mm M43	
Yugoslavia						
7.62 mm M57	7.62 mm M49/57	7.62 mm M59/66	7.62 mm M53	128 mm RL M71	50 mm M8	44 mm RL M57
7.62 mm M70	7.62 mm M56	7.62 mm M70, 70A	7.62 mm M65A, B		60 mm M57	64 mm RL RBR-M80
9 mm M65			7.92 mm M72		81 mm M31	75 mm RCL M20
			.50 Browning M2HB		81 mm M68	82 mm RCL M60
			12.7 mm NSV		120 mm M74, M75	105 mm RCL M65
						Sagger ATGW
						Snapper ATGW
						Dragon ATGW
						TOW ATGW
Zaire						
9 mm FN 35	9 mm Uzi	5.56 mm FN-CAL	5.56 mm M16A1/HB		60 mm Brandt	RPG-7
9 mm S&W M39	9 mm Franchi LF-57	5.56 mm FN-FNC	7.62 mm Bren L4		81 mm PRB	57 mm RCL M18
9 mm H&K P7M13	9 mm H&K MP5	5.56 mm SIG 540	7.62 mm FN-MAG		81 mm M37M	75 mm RCL M20
		5.56 mm M16A1	7.62 mm M60		107 mm M30	89 mm LRAC
		7.62 mm CETME 58	7.62 mm FN-FAL/HB		120 mm SB Brandt	106 mm RCL M40
		7.62 mm AK47, AKM	12.7 mm DShK		120 mm Rifled Brandt	SS-11 ATGW
		7.62 mm FN-FAL	.50 Browning M2HB			Entac ATGW
Zambia						
7.62 mm Tokarev	9 mm Sterling	7.62 mm FN-FAL	7.62 mm PK			RPG-7
7.62 mm vz/52	9 mm H&K MP5	7.62 mm G3	7.62 mm SGM			44 mm RM M57
9 mm Stechkin		7.62 mm AK47, AKM	12.7 mm DShK			57 mm RCL M18
						75 mm RCL M20
						84 mm Carl Gustav
						Sagger ATGW
Zimbabwe						
9 mm FN 35	9 mm Sterling	5.56 mm AR70/223	5.56 mm Ultimax 100		81 mm L16	RPG-7
7.62 mm Tokarev	9 mm Uzi	5.56 mm R4	7.62 mm FN-MAG		82 mm M43	40 mm Type 69
.45 M1911A1	9 mm Walther MPK	7.62 mm FN-FAL	7.62 mm Bren L4		120 mm M43	106 mm RCL M40
		7.62 mm M14	7.62 mm RPD, RPK			107 mm RCL B-12
		7.62 mm M14	12.7 mm DShK			
		7.62 mm AK47, AKM	.50 Browning M2HB			

Index of Manufacturers

NOTE: Telephone, Telex and Facsimile (Teletext) numbers are given in the international convention; they must be prefixed by the appropriate national dialling-out code.

Where a discrepancy exists between data on the text pages and in this table, the table should be accepted as the most up-to-date and correct version. This table is composed after the text has been completed, and there are often last-minute changes in address and other details which cannot be corrected in the text.

Dayton Metals Corporation
PO Box 435, Araneta Center Post Office, Fiesta
Carnibal Building, Cubao, Quezon City,
Philippine Republic
Mortars   501

Degen & Co AG
Grittweg 9, CH-4435 Niederdorf, Switzerland
Tel: +41-61-978-282 TX: 966093 DEGN CH
Fax: +41-61-978490
Mortar ammunition   680

Detonics Firearms Industries
13456 SE 27th Place, Bellevue, WA 98005, USA
Tel: +1-206-624-9090
Pistol   68

Diehl GmbH & Co
Ammunition Division, Fischbachstrasse 16,
D-8505 Röthenbach, West Germany
Tel: +49-911-509-1 TX: 622591 MD D
Fax: +49-911-509-2510
Anti-tank weapons   422
Grenades   598
Mortar ammunition   652

Diemaco Inc
1036 Wilson Avenue, Kitchener, Ontario N2C 1J3
Canada
Tel: +1-519-893-6840 TX: 069-55164
Fax: +1-519-893-3144
Machine guns   304
Rifles   144

DISA Systems Group
9 Stationsvej, DK-3550 Slangerup, Denmark
Tel: +45-233-4565 TX: 42559 DISAS DK
Fax: +45-233-4545
Machine gun mountings   312

Dynamit-Nobel AG
Postfach 1261, Kaiserstrasse 1, D-5210 Troisdorf,
West Germany
Tel: +49-2241-890 TX: 885666 DN D
Fax: +49-2241-89-1540
Anti-tank weapons   420
Sighting equipment   791

Dynamit-Nobel Wien GmbH
Postfach 74, Opernring 3-4, A-1015, Vienna,
Austria
Tel: +43-222-565646 TX: 114177 NOBEL A
Grenades   574
Pyrotechnics   695

E

Eagle Military Gear Overseas Ltd
IBM House, 31 Shaul Hamelech Boulevard, PO
Box 33737, Tel Aviv, 64927, Israel
Tel: +972-3-210281 TX: 342286 EAGLE IL
Body armour   838

El-Op Electro-Optics Industries Ltd
Advanced Technology Park, Kiryat Weizmann,
PO Box 1165, Rehovot 76110, Israel
Tel: +972-8-486211 TX: 361944 RIL IL
Fax: +972-8-486214
Surveillance equipment   754

Elbit Computers Ltd
PO Box 5390, Haifa 31053, Israel
Tel: +972-4-517111 TX: 46774
Fax: +972-4-520002
Sighting equipment   799

Electronique Serge Dassault
55 Quai Marcel Dassault, 92214 Saint Cloud,
France
Tel: +331-4911-8000 TX: 250787 F ESD SCLOU
Mortar fire control   533

Eltro GmbH
Postfach 102120, D-6900 Heidelberg 1, West
Germany
Tel: +49-6221-7051 TX: 04-61811
Sighting equipment   795

Elviemek SA
Atrina Centre, 32 Kifissias Avenue, Athens,
Greece
Tel: +30-1-682-8601 TX: 214-4258 ELV GR
Fax: +30-1-684-1524
Grenades   601
Pyrotechnics   717

ENARM
Rua Homen de Mello 66, Apt 201, Tijuca, Rio de
Janiero, Brazil
Sub-machine guns   88

ENOSA (Empresa Nacional de Optica SA)
Avenida de San Luis 91, Madrid 33, Spain
Tel: +34-1-202-6040 TX: 42719 OPTI E
Sighting equipment   807

Esperanza y Cia SA
PO Box 2, Avenida de Xemien 12, Marquina,
48270 Vizcaya, Spain
Tel: +34-4-686-6025 TX: 31170 ECMA E
Fax: +34-4-686-6026
Mortars   507
Mortar ammunition   676

Euroatlas GmbH
Zum Panrepel 2, Postfach 450241, D-2800,
Bremen 45, West Germany
Tel: +49-421-486930 TX: 244504 EURAT D
Fax: +49-421-486-9341
Sighting equipment   790
Surveillance equipment   750

Eurometaal NV
Postbus 419, 1500 EX, Zaandam, Netherlands
Tel: +31-75-504-911 TX: 19303 EMZ NL
Fax: +31-75-169-396
Grenades   612

Euromissile
12 rue de la Redoute, 92260 Fontenay-aux-Roses,
France
Tel: +331-4661-7311 TX: 204691 EUROM F
Fax: +331-4661-6467
Anti-tank weapons   418

Explosivos Alaveses SA
Apartado 198, 01080 Vitoria, Alava, Spain
Tel: +34-222-350154 TX: 35508 EXPAL E
Grenades   616
Pyrotechnics   726

F

**Fabrica Militar de Armas Portatiles
'Domingo Matheu'**
Avenida Ovidio Lagos 5250, 2000-Rosario,
Argentina
Grenades   570
Sub-machine guns   80
Machine guns   293
Pistols   3
Rifles   128

FAMAE
Pedro Montt 1606, Santiago, Chile
Pistols   10
Mortars   472
Pyrotechnics   703
Sub-machine guns   89

**Federal Directorate of Supply &
Procurement**
9 Nemanjina Street, 11005 Belgrade 9, Yugoslavia
Tel: +38-11-621-522 TX: 11360 YU SDPR
Fax: +38-11-324-981
Anti-tank weapons   461
Body armour   858
Grenades   637
Light support weapon   289
Mortars   531
Mortar ammunition   690
Sub-machine guns   127
Surveillance equipment   785

Federal Laboratories Inc
PO Box 305, Saltsburg, PA 15681, USA
Tel: +1-412-639-3511 TX: 86-6294
Fax: +1-412-639-3888
Pyrotechnics   742

FEG Arms and Gas Appliances Factory
Soroksari ut 158, H-1095 Budapest, Hungary
Tel: +36-1-477-920 TX: 22-4213 H
Pistols   30

**Ferranti International, Weapons
Equipment Division**
St Mary's Road, Moston, Manchester M10 0BE,
England
Tel: +44-61-681-2071 TX: 667857
Fax: +44-61-682-2500
Grenade Fuzes   631

**Ferranti International, Weapons
Equipment Division**
10 Spring Lakes, Deadbrook Lane, Aldershot,
Hants GU12 4HA England
Tel: +44-4868-25467 TX: 859570 G
Fax: +44-4868-7039
Air launcher   274

FFV Ordnance
S-63187 Eskilstuna, Sweden
Tel: +46-16-155340 TX: 46075 FFVHK S
Fax: +46-16-124354
Anti-tank weapons   431
Grenades   619
Mortar ammunition   679
Rifles   197
Sub-machine guns   115

FHE-Brandt Switzerland
PO Box 23, CH-1000 Lausanne 6, Switzerland
Close support weapon   270

FN Herstal SA
B-4400 Herstal, Belgium
Tel: +32-41-640-800 TX: 41223 FABNA B
Fax: +32-41-645452
Grenades   576
Machine guns   295
Machine gun ammunition   551
Pistols   5
Sighting equipment   786
Sub-machine guns   84
Rifles   135

**Ford Aerospace and Communications
Corp**
Aeronutronic Division, PO Box A, Ford Road,
Newport Beach, CA 92658-9983, USA
Tel: +1-714-641-2500 TX: 678470
Anti-tank weapons   459

Franchi, Luigi, SpA
Via del Serpente 12, Zona Industriale, Fornaci,
(Brescia), Italy
Tel: +39-30-341161 TX: 300208 FRARM I
Fax: +39-30-347415
Machine guns   329,330
Pistols   47
Rifles   184
Shotguns   264
Sub-machine guns   106

Fuji Optical Co
324 I-Chome, Uetake-machi, Omiya City, Japan
Tel: +81-486-63111 TX: 22885 FUJINON J
Surveillance equipment   755

G

GEC Sensors, Pullin Controls Division
Langston Road, Debden, Loughton, Essex IG10
3TW, England
Tel: +44-81-508-5522 TX: 23855
Fax: +44-81-502-5522
Sighting equipment   819
Surveillance equipment   774

General Electric Aerospace

Lakeside Avenue, Burlington, VT 05401-4985, USA
Tel: +1-802-657-6000 TX: 510-299-0028

Cannon	399
Machine guns	375,377

Gentex Corp

PO Box 315, Carbondale, PA 18407, USA
Tel: +1-717-282-8226 TX: 831-883 GENTEX CRBL
Fax: +1-717-282-8555

Body armour	854

GIAT: Groupement Industriel des Armements Terrestres

10 Place Georges Clemenceau, 92211 Saint Cloud, France
Tel: +331-4602-5200 TX: 260010 F

Cannon	383
Machine guns	316
Rifles	156
Sub-machine guns	95

Glock GmbH

PO Box 50, A-2232 Deutsche-Wagram, Austria
Tel: +43-2247-2460 TX: 133307 GLOCK A
Fax: +43-2247-246012

Grenades	575
Pistols	4

Göncz Co

11526 Burbank Boulevard, North Hollywood, CA 91601, USA
Tel: +1-818-505-0408

Carbines	252
Pistol	76

Greek Powder & Cartridge Co (Pyrkal)

1 Ilioupoleos Avenue, Hymettus, GR-17236 Athens, Greece
Tel: +30-1-975-1857 TX: 221986 EEPK GR
Fax: +30-1-970-5009

Mortar ammunition	655

Grendel Inc

PO Box 560908, Rockledge, FL 32956-0908, USA
Tel: +1-407-636-1211 Fax: +1-407-633-6710

Rifles	236

H

Haley & Weller Ltd

Wilne, Draycott, Derbyshire DE7 3QJ, England
Tel: +44-3317-2475 TX: 378215 HALWEL G
Fax: +44-3317-3046

Grenades	628
Pyrotechnics	736

Hall & Watts Defence Optics Ltd

266 Hatfield Road, St Albans, Herts AL1 4UN, England
Tel: +44-727-59288 TX: 267001 WATTS G
Fax: +44-727-35683

Sighting equipment	815
Surveillance equipment	776

Hawk Engineering Inc

42 Sherwood Terrace, Suite 101, Lake Bluff, IL 60044, USA
Tel: +1-312-295-2340 Fax: +1-312-295-3319

Light support weapons	284

Heckler & Koch GmbH

Postfach 1329, D-7238 Oberndorf/Neckar, West Germany
Tel: +49-7423-791 TX: 760313 HUKO D
Fax: +49-7423-79406

Grenade launchers	262
Machine guns	316,327
Pistols	25
Pyrotechnics	709
Rifles	162,225
Sighting equipment	790
Sub-machine guns	97

Heckler & Koch Inc

21480 Pacific Boulevard, Sterling, VA 22170-8903, USA
Tel: +1-703-450-1900 TX: 710-955-0846
Fax: +1-703-450-8160

Close Assault Weapon System	284

Helio

215 Vauxhall Bridge Road, London SW1V 1EN, England
Tel: +44-71-821-8080 TX: 262748 G

Sighting equipment	815

Hellenic Arms Industry (E.B.O.) SA

160 Kifissias Avenue, 11525 Athens, Greece
Tel: +30-647-2611 TX: 21-8562 EBO GR
Fax: +30-647-2715

Cannon	387
Machine guns	327
Mortars	488
Pistols	29
Rifles	172
Sub-machine guns	101

Helwan Machine Tools

23 Talat Harb Street, PO Box 1582, Cairo, Egypt

Mortars	475

Hensoldt Wetzlar Optische Werke AG

Postfach 1760, D-6330 Wetzlar 1, West Germany
Tel: +49-6441-4041 TX: 483884
Fax: +49-6441-404203

Sighting equipment	792,795

Highmark Manufacturing Co Ltd

Ronoco House, 55 Adelaide Street, Belfast BT2 8FE, Northern Ireland
Tel: +44-232-233476 TX: 747421 HYMARK G
Fax: +44-232-233979

Body Armour	847

Hilton Gun Co

Station Road, Hatton, Derbys DE6 5EL England
Tel: +44-283-814463 TX: 341711

Grenade launcher	275
Multi-purpose gun	275

Hirtenberg AG

A-2552 Hirtenberg, Austria
Tel: +43-2256-8184 TX: 14447 PATRON A
Fax: +43-222-8114-342

Mortar ammunition	639

Honeywell Inc, Defense Systems Group

Armament Systems Division, 7225 Northland Drive, Brooklyn Park, MN 55428, USA
Tel: +1-612-536-4547 TX: 291134 HON MTKA
Fax: +1-612-536-4545

Anti-tank weapons	455
Light support weapon	280

Howa Machinery Ltd

Shinkawa-cho, Nishikasugai-Gun, 452 Aichi, Japan
Tel: +81-52-502-1111 TX: 0443-9931 HOWA J
Fax: +81-52-409-3777

Rifles	186

Hsing Hua Co

PO Box 8746, Taipei, Taiwan
Tel: +886-2761-5367 TX: 11171 MSCSF

Grenades	620

Hughes Aircraft Co, E-O & Data Systems Group

PO Box 902, Bldg E1, MS E-104, El Segundo, CA 90245, USA
Tel: +1-213-616-1023

Sighting equipment	824

Hughes Aircraft Co, Missile Systems Group

8433 Fallbrook Avenue, Bldg 261, MS N-27, Canoga Park, CA 91304-0445, USA
Tel: +1-818-702-3816

Anti-tank weapons	459

Hunting Engineering Ltd

Reddings Wood, Ampthill, Beds MK45 2HD, England
Tel: +44-525-403431 TX: 82105

Anti-tank weapons	450

Hydroar SA

Rue de Rocio 196, Vila Olimpia, Sao Paulo SP, CEP 04552, Brazil
Tel: +55-11-815-6922 TX: 011-30841 HYIM BR

Anti-tank weapons	405
Flame thrower	259

I

Imatronic Ltd

Dorian House, 12A Rose Street, Wokingham, Berks RG11 1XY, England
Tel: +44-734-784903 TX: 848584 GEM G
Fax: +44-734-771233

Sighting equipment	822

IMBEL: Industria de Materiel Bélico de Brasil

Avenida das Nações Unidas 13.797, Bloco III, 1° Andar, CEP 04794 São Paulo SP, Brazil
Tel: +55-11-531-5055 TX: (011)-37481 IMBL BR

Pistols	10
Sub-machine guns	86
Rifles	139,142

INDEP: Industrias Nacionais de Defesa EP

Rua Fernando Palha, Aptdo 8106, 1802 Lisboa-Codex, Portugal
Tel: +351-1-384-370 TX: 12514 INDFBP P
Fax: +351-1-382-830

Machine guns	333
Mortars	502
Mortar ammunition	673
Rifles	191
Sub-machine guns	111

Indian Ordnance Factory Board

10A Auckland Road, Calcutta 700 001, India

Mortars	489
Mortar ammunition	657

Indian State Arms Factory

Kalpi Road, Kanpur 208009, India

Rifles	174

Industria Nacional de Armas SA,

São Paulo, Brazil

Sub-machine guns	87

Instalaza SA

Monreal 27, 50002 Zaragoza, Spain
Tel: +34-76-293422 TX: 58952 INTZ E
Fax: +34-76-299331

Anti-tank weapons	429
Grenades	618

International Technologies (Lasers) Ltd

12 Hachoma Street, PO Box 4099, 75140 Rishon-Letzion, Israel
Tel: +972-3961-6567 TX: 381095 ITL IL
Fax: +972-3961-6563

Sighting equipment	796
Surveillance equipment	752

Intertechnik GmbH

Industriezelle 56, Postfach 100, A-4040 Linz, Austria
Tel: +43-732-2892 TX: 02-1522
Fax: +43-732-2892-123

Anti-tank weapons	403

Intratec USA Inc

12405 SW 130th Street, Miami, FL 33186, USA
Tel: +1-305-232-8121

Pistols	77

Israel Military Industries
PO Box 1044, Ramat Hasharon, 47100 Israel
Tel: +972-3542-5222 TX: 03-3719 MISBIT IL
Fax: +972-3540-6908

Anti-tank weapons	423,424
Grenades	604
Machine guns	328
Mortars	490
Mortar ammunition	659,662
Pistols	32
Rifles	175
Sub-machine guns	102

ITM: Industrial Technology & Machines AG
Postfach 260, CH-4503 Solothurn, Switzerland
Tel: +41-65-228618 TX: 934631 ITM CH
Fax: +41-65-228317

Pistols	61
Rifles	200

ITT Defense
Electro-Optic Products Division
7635 Plantation Road, Roanoke, VA 24019, USA
Tel: +1-703-563-0371 TX: 829458 ITT EOPD
Fax: +1-703-362-7370

Sighting equipment	829

J

JAI: Jorgen Andersen Ingeniorsfirma AS
1 Produktionsvej, DK-2660 Glostrup, Denmark
Tel: +45-291-8888 TX: 35378 JAI DK
Fax: +45-291-3252

Surveillance equipment	762

Johnson Firearms Specialties Inc
PO Box 12204, Salem, OR 97309, USA
Tel: +1-503-581-3244

Rifle accessories	245

Junghans Feinwerktechnik GmbH
PO Box 110, D-7230 Schramberg, West Germany
Tel: +49-7422-181 TX: 762811 JU D
Fax: +49-7422-18400

Fuzes	423,652

K

Kawasaki Heavy Industries Ltd
World Trade Centre Building, 4-1
  Hamamatsu-cho, 2-Chome, Minato-ku, Tokyo
  105, Japan
Tel: +81-3435-2479 TX: 242-4371 KAWAJU J
Fax: +81-3432-3977

Anti-tank weapons	427

Kern & Co AG
Schachenalee, CH-5001 Aarau, Switzerland
Tel: +41-6425-1111 TX: 981106 CH

Sighting equipment	811

Kia Machine Tool
4th Floor, KIA Building, 15 Yoido-dong,
  Yongdeungpo-gu, Seoul, South Korea
Tel: +82-2783-9418 TX: 25754 KIAMT K
Fax: +82-2782-4864

Mortars	500

Knight's Armament Co
1306 29th Street, Vero Beach, FL 32960, USA
Tel: +1-407-562-5697 Fax: +1-407-569-2955

Pistol	76

KOHEMA: Korea Heavy Machinery Industries Ltd
446 Shindorim-Dong, Kuro-ku, Seoul, South
  Korea
Tel: +82-262-3851 TX: 28453

Anti-tank weapons	454

Korea Explosives Co Ltd
34 Seosomoon-Dong, Chung-ku, Seoul, South
  Korea
Tel: +82-2753-0381 TX: K-23684 KOMITE
  SEOUL Fax: +82-2752-3475

Grenades	609
Mortar ammunition	666
Pyrotechnics	722

L

La France Specialties
PO Box 178211, San Diego, CA 92117, USA
Tel: +1-619-293-3373

Pistol	68,860
Rifles	241

Laboratório de Pesquisa de Armamento Automatico Lda
Rua Homen de Melo 66, Grupo 201, Tijuca, 20510
Rio de Janiero, Brazil

Rifles	141

Laboratório de Projetos de Armamento Automático Lda
Rua Homen de Melo 66, Grupo 201, Tijuca, 20510
Rio de Janiero, Brazil

Sub-machine guns	87

Lacroix, Etienne, Tous Artifices SA
6 Boulevard de Joffery, BP 213, 31601 Muret
  Cedex, France
Tel: +33-6151-0337 TX: 531478 LACART F
Fax: +33-6151-4277

Grapnel launcher	487
Mortar ammunition	648
Pyrotechnics	706

La Precisa; Stabilimenti di Teano SpA
81057 Teano (Caserta), Italy

Grenades	609

Leitz, Ernst, Canada Ltd
328 Ellen Street, Midland, Ontario L4R 2H2,
  Canada
Tel: +1-705-526-5401 TX: 06-875-561 ELCAN
  MID Fax: +1-705-526-5831

Mortar sights	470
Sighting equipment	787

Lightweight Body Armour Ltd
Hinton House, Daventry, Northants NN11 6QG,
  England
Tel: +44-327-61282 TX: 312112 ARMOUR G
Fax: +44-327-60656

Body armour	845

Litton Electron Devices, Electro-Optics Dept
1215 S. 52nd Street, Tempe, AZ 85281-6987, USA
Tel: +1-602-968-4471 TX: 910-9500-149
  Fax: +1-602-968-4471 Ext 223

Sighting equipment	826
Surveillance equipment	780

Llama-Gabilondo y Cia SA
PO Box 290, Portal de Gamarra 50, 01080 Vitoria,
  Spain
Tel: +34-945-262400 TX: 35517 LLAMA E
  Fax: +34-945-262444

Pistols	50

Losfeld-Industries
15 rue Thiebault, 94220 Charenton, France
Tel: +331-4368-1031 TX: 262053 LOSFIND F

Grenades	594

Luchaire Défense
180 Boulevard Haussmann, 75382 Paris Cedex 08,
  France
Tel: +331-4562-4022 TX: 650312
  Fax: +331-4563-2851

Anti-tank weapons	415
Grenades	591

Lyttleton Engineering Works (Pty) Ltd
Private Bag X5, Lyttleton, Verwoerdburg 0140,
  South Africa
Tel: +27-12-620-2290 TX: 322561

Machine guns	335
Rifles	194

M

McDonnell Douglas Helicopter Co
5000 East McDowell Road, Mesa, AZ
  85205-9797, USA
Tel: +1-602-891-9021

Cannon	397
Machine guns	377

McDonnell Douglas Systems Co, Combat Systems Division
PO Box 516, St Louis, MO 63166, USA
Tel: +1-314-232-0232 TX: 44857

Anti-tank weapons	456
Light support weapons	278
Sighting equipment	823

Magnavox Electronic Systems Co
1313 Production Road, Fort Wayne, IN 46808,
  USA
Tel: +1-219-429-6000 TX: 228472 MAGNAVOX
  FWA D

Mortar fire control	537
Sighting equipment	825

Makina ve Kimya Endustrisi Kurumu
Tandogan Medyani, 06330 Ankara, Turkey
Tel: +90-4223-2011 TX: 42223 MKGAS TR
  Fax: +90-4222-2241

Machine guns	341
Mortars	514
Pistols	63
Rifles	205
Sub-machine guns	116

Manroy Engineering
Hobbs Lane, Beckley, E Sussex TN31 6TS,
  England
Tel: +44-79726-555 TX: 95249 MANROY G
  Fax: +44-79726-374

Machine guns	362
Machine gun mounts	363
Pyrotechnics	735

Manufacture d'Armes Automatiques de Bayonne (MAB)
Lotissment Industrielle des Pontots, 64100
  Bayonne, France

Pistols	18

Manufacture Nationale d'Armes de Chatellerault (MAC)
Chatellerault, France

Pistols	17

Manufacture Nationale d'Armes de St Etienne (MAS)
St Etienne, France

Pistols	17
Rifles	157
Sub-machine guns	96

Manufacture Nationale d'Armes de Tulle
Tulle, France

Sub-machine guns	96

Marconi Command & Control Systems
Chobham Road, Frimley, Camberley, Surrey
  GU16 5PE, England
Tel: +44-276-63311 TX: 858289 MCCS G
  Fax: +44-276-29784

Mortar fire control	536
Surveillance equipment	768

Marquardt Company, The
16555 Saticoy Street, Van Nuys, CA 91409-9104,
  USA
Tel: +1-818-989-6400 TX: 651-420

Light support weapon	279

Martin Marietta Inc
Electronics & Missiles Group, PO Box 555837
  MP-325, Orlando, FL 32855-5837, USA
Tel: +1-407-356-2136

Anti-tank weapons	459

J L Weibull AB
PO Box 43, S-23220 Akarp, Sweden
Tel: +46-404-65080 TX: 33159 EXPOFRA S
    Sighting equipment   808

Wild Leitz Ltd
CH-9435 Heerbrugg, Switzerland
Tel: +41-71-703-131 TX: 881-222-31
  Fax: +41-71-703-145
    Bore-sighting equipment   811
    Surveillance equipment   763

Wormald Vigilant Ltd
PO Box 19545, 211 Maces Road, Christchurch,
  New Zealand
Tel: +64-3-895-897 TX: 49343 ALARMS NZ
    Mortar fire control   534

Y

Yangji Metal Industrial Co
Dae Chang Building 605, 42 Jan Gyo Dong,
  Chung-gu, Seoul, South Korea
    Mortars   500

Z

Zavodi Crvena Zastava
29 Novembra 12, 11000 Belgrade, Yugoslavia
Tel: +38-11-323-981 TX: 12118 YU
    Cannon   401
    Machine guns   378
    Pistols   78
    Rifles   253
    Sub-machine guns   125

Zeiss, Carl
Postfach 1380, D-7082 Oberkochen, West
  Germany
Tel: +49-7364-201 TX: 713-7510
    Sighting equipment   795

Zengrange Ltd
Greenfield Road, Leeds, Yorks LS9 8DB,
  England
Tel: +44-532-489048 TX: 557621 ZEN G
  Fax: +44-532-492349
    Mortar fire control   536

Alphabetical Index

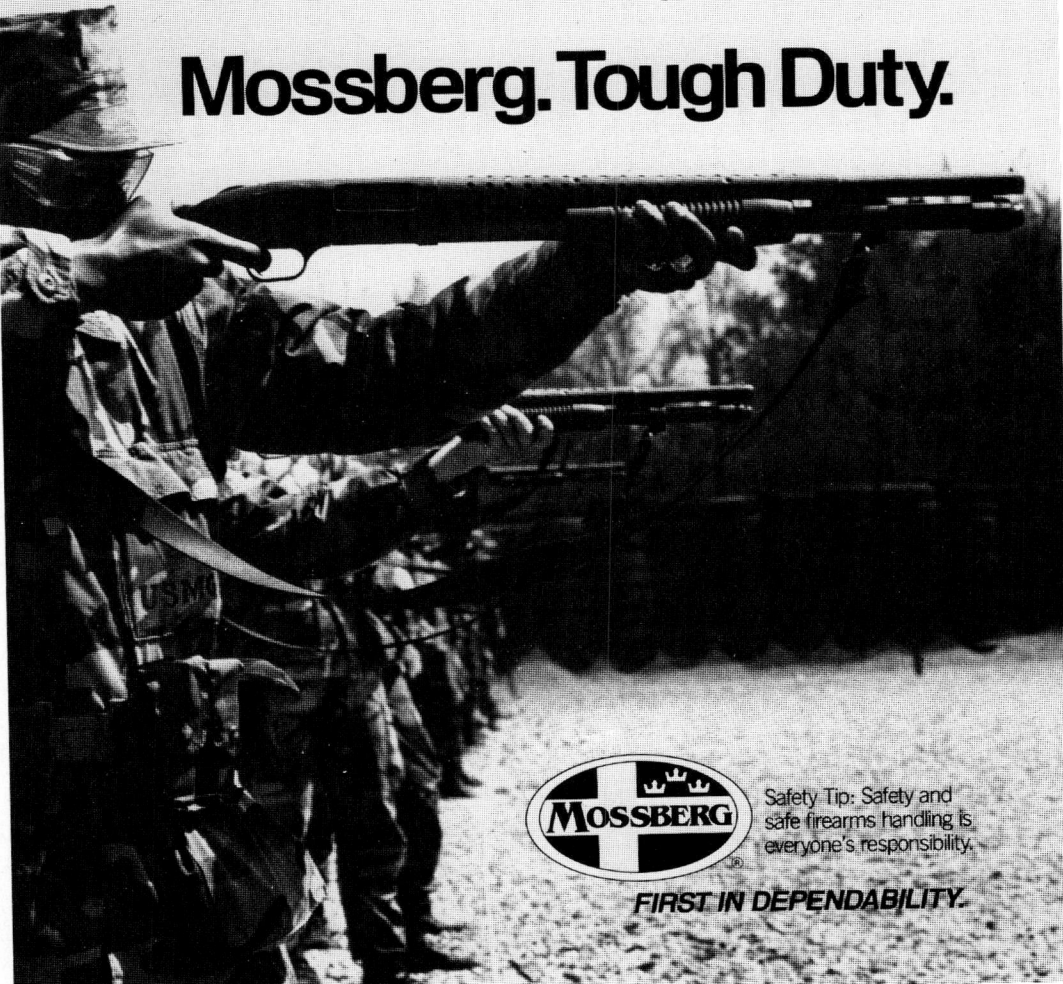

Oldelft
allows you a shot
in the dark

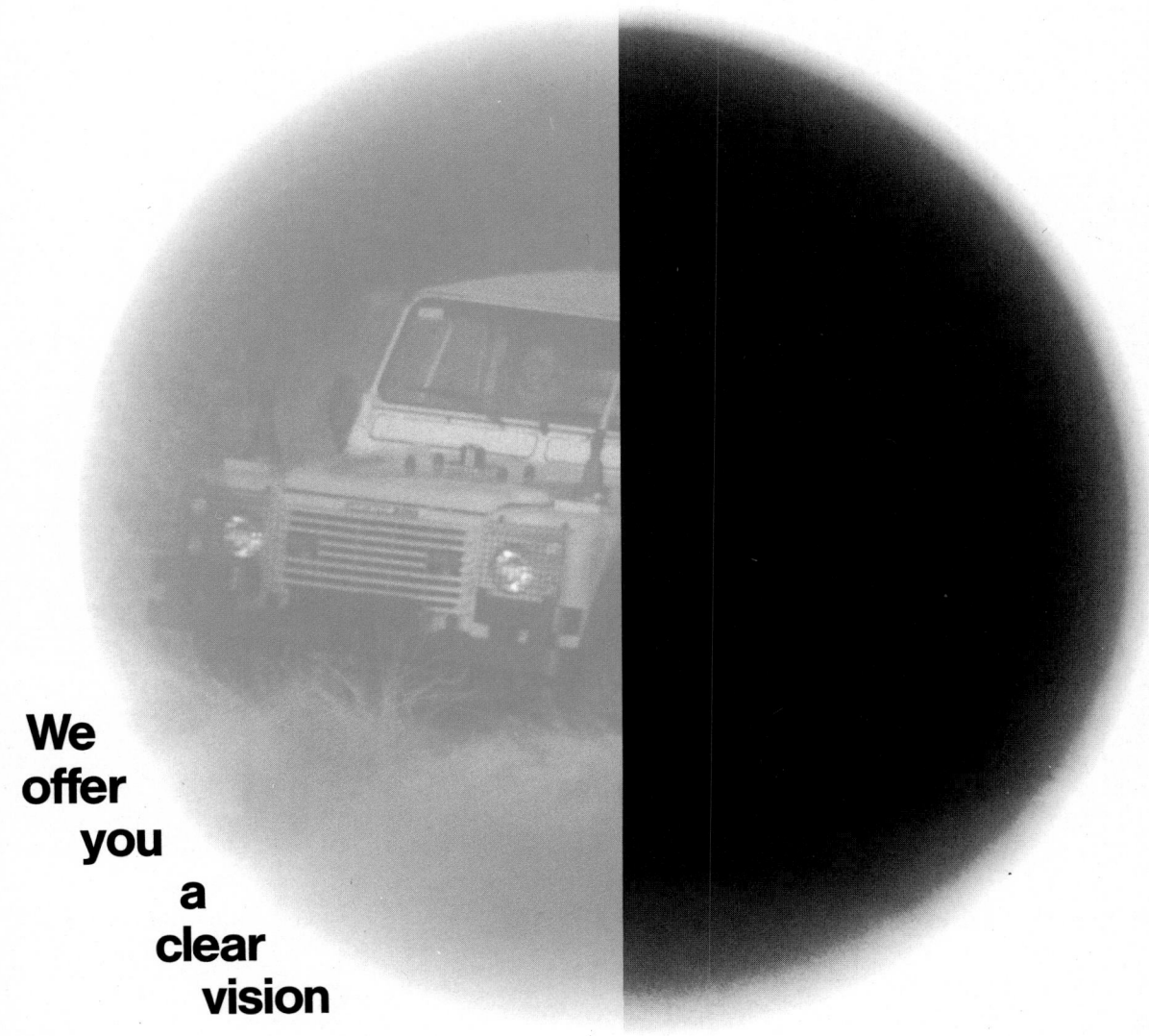

**We
offer
you
a
clear
vision**

For more information about our electro-optical systems please contact

Oldelft

B.V. Optische Industrie "De Oude Delft," P.O. Box 72, 2600 MD Delft, The Netherlands, Tel. (015)601901, Tlx 38011, Fax (015)145762

Printed and bound in Great Britain by Biddles Ltd, Guildford and King's Lynn